IMAGING ANATOMY

Ultrasound

SECOND EDITION

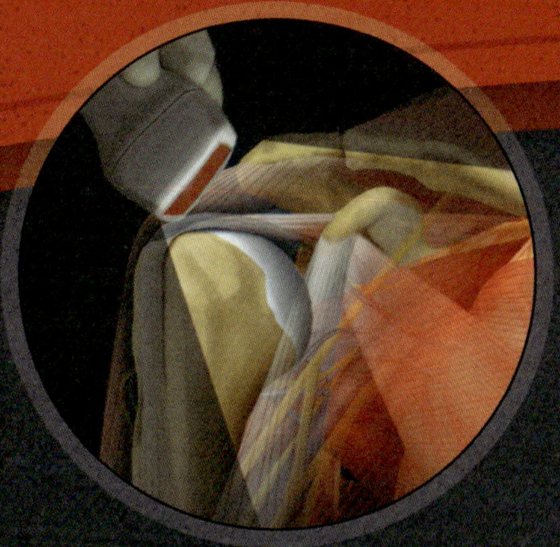
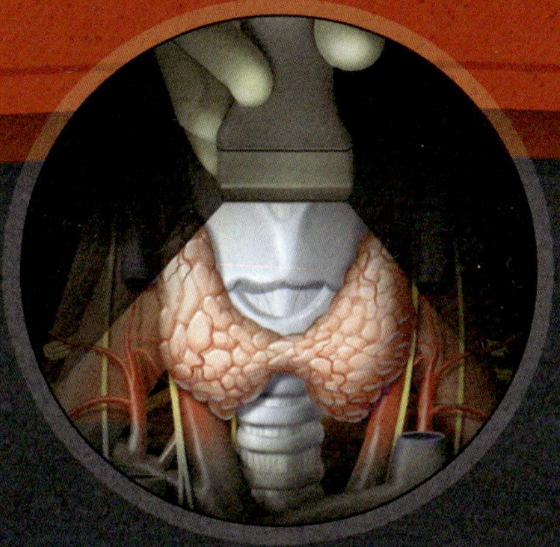
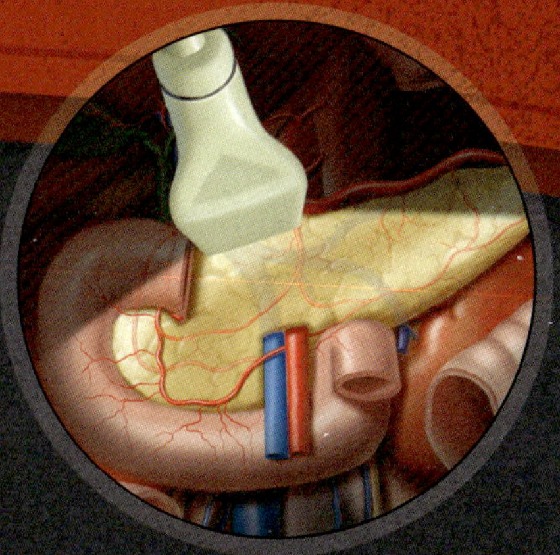
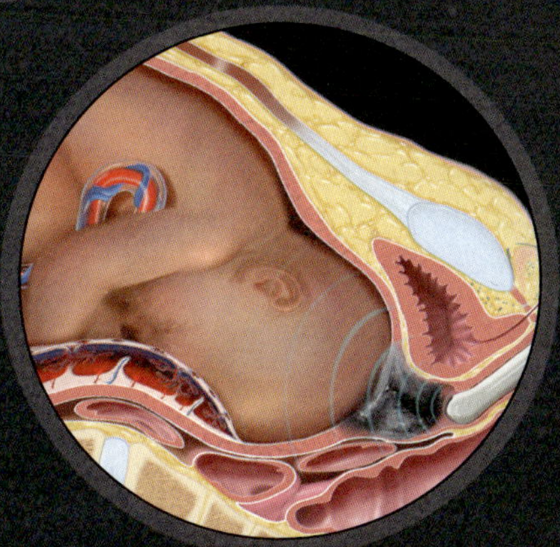
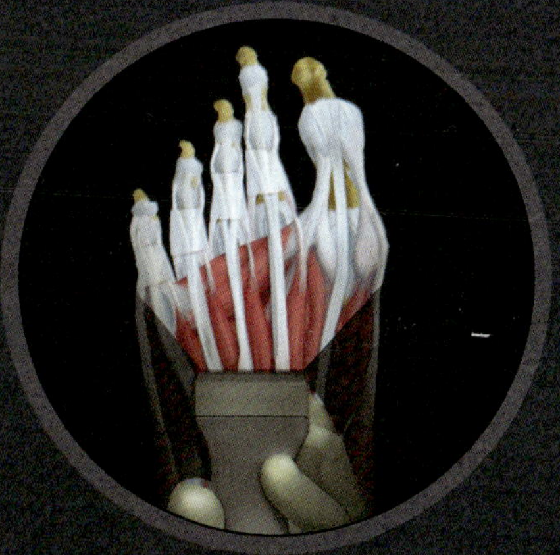

WOODWARD

GRIFFITH | ANTONIO | AHUJA

WONG | KAMAYA | WONG-YOU-CHEONG

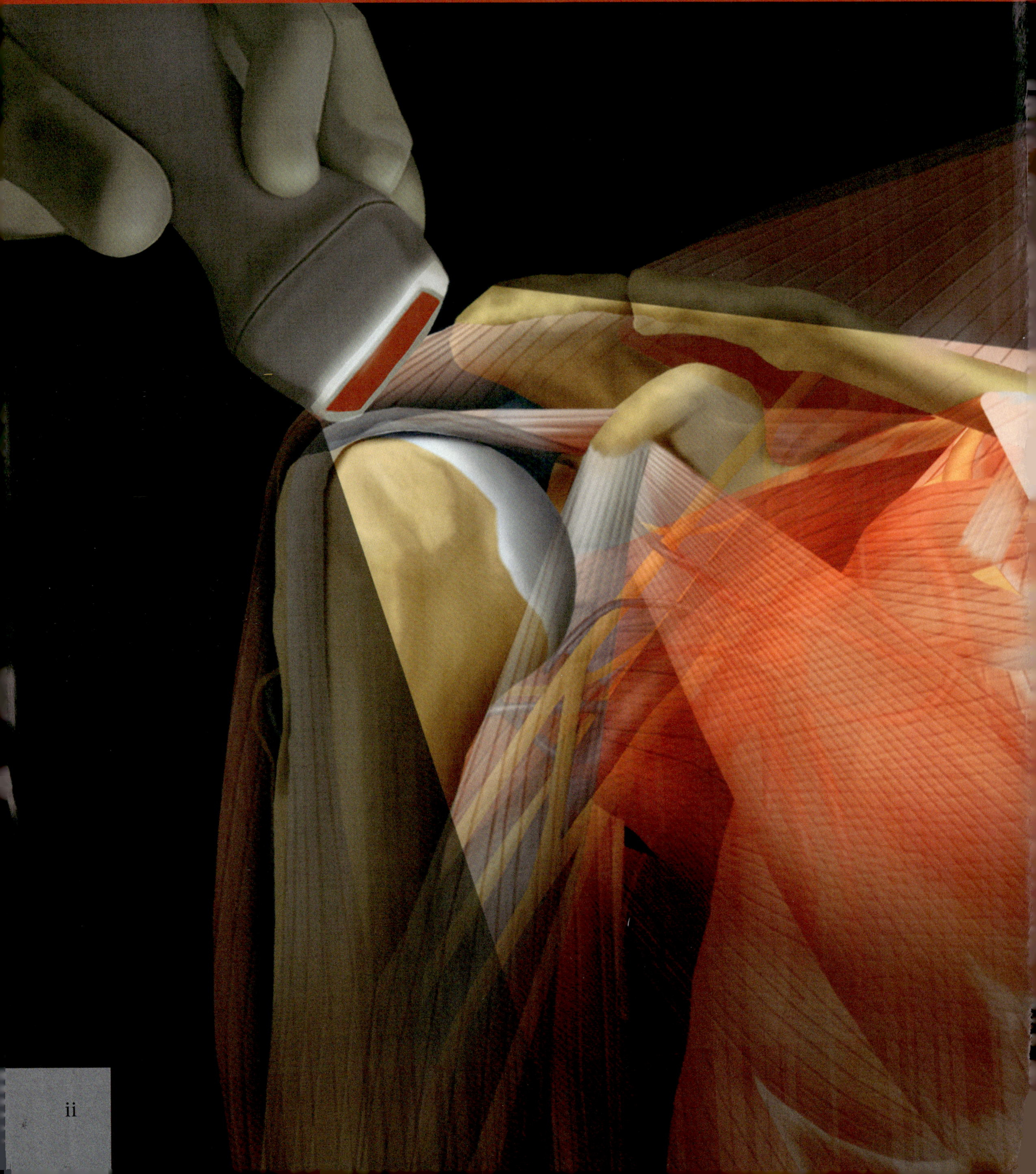

Ultrasound

SECOND EDITION

Paula J. Woodward, MD
Professor of Radiology
David G. Bragg, MD and Marcia R. Bragg Presidential Endowed Chair in Oncologic Imaging
Adjunct Professor of Obstetrics and Gynecology
University of Utah School of Medicine
Salt Lake City, Utah

James F. Griffith, MD, MRCP, FRCR
Professor
Department of Imaging and Interventional Radiology
The Chinese University of Hong Kong
Hong Kong (SAR), China

Gregory E. Antonio, MD, DRANZCR, FHKCR
Honorary Professor
Department of Imaging and Interventional Radiology
The Chinese University of Hong Kong
Consultant Radiologist
Scanning Department
St. Teresa's Hospital
Hong Kong (SAR), China

Anil T. Ahuja, MBBS (Bom), MD (Bom), FRCR, FHKCR, FHKAM (Radiology)
Professor of Diagnostic Radiology & Organ Imaging
Faculty of Medicine
The Chinese University of Hong Kong
Prince of Wales Hospital
Hong Kong (SAR), China

K. T. Wong, MBChB, FRCR, FHKCR, FHKAM (Radiology)
Consultant & Clinical Associate Professor (Honorary)
Department of Imaging and Interventional Radiology
Prince of Wales Hospital
Faculty of Medicine
The Chinese University of Hong Kong
Hong Kong (SAR), China

Aya Kamaya, MD, FSRU, FSAR
Associate Professor of Radiology
Director, Stanford Body Imaging Fellowship
Stanford University School of Medicine
Stanford, California

Jade Wong-You-Cheong, MBChB, MRCP, FRCR
Professor
Department of Diagnostic Radiology
and Nuclear Medicine
University of Maryland School of Medicine
Director of Ultrasound
University of Maryland Medical Center
Baltimore, Maryland

ELSEVIER

1600 John F. Kennedy Blvd.
Ste 1800
Philadelphia, PA 19103-2899

IMAGING ANATOMY: ULTRASOUND, SECOND EDITION ISBN: **978-0-323-54800-7**

Notices

Knowledge and best practice in this field are constantly changing. As new research and experience broaden our understanding, changes in research methods, professional practices, or medical treatment may become necessary.

Practitioners and researchers must always rely on their own experience and knowledge in evaluating and using any information, methods, compounds, or experiments described herein. In using such information or methods they should be mindful of their own safety and the safety of others, including parties for whom they have a professional responsibility.

With respect to any drug or pharmaceutical products identified, readers are advised to check the most current information provided (i) on procedures featured or (ii) by the manufacturer of each product to be administered, to verify the recommended dose or formula, the method and duration of administration, and contraindications. It is the responsibility of practitioners, relying on their own experience and knowledge of their patients, to make diagnoses, to determine dosages and the best treatment for each individual patient, and to take all appropriate safety precautions.

To the fullest extent of the law, neither the Publisher nor the authors, contributors, or editors, assume any liability for any injury and/or damage to persons or property as a matter of products liability, negligence or otherwise, or from any use or operation of any methods, products, instructions, or ideas contained in the material herein.

Publisher Cataloging-in-Publication Data

Names: Woodward, Paula J.
Title: Imaging anatomy. Ultrasound / [edited by] Paula J. Woodward.
Other titles: Ultrasound.
Description: Second edition. | Salt Lake City, UT : Elsevier, Inc., [2017] | Includes bibliographical references and index.
Identifiers: ISBN 978-0-323-54800-7
Subjects: LCSH: Human anatomy--Handbooks, manuals, etc. | Ultrasonic imaging--Handbooks, manuals, etc. | MESH: Ultrasonography--methods--Atlases. | Anatomy, Cross-Sectional--Atlases.
Classification: LCC QM25.I43 2017 | NLM WN 17 | DDC 616.07'543--dc23

International Standard Book Number: 978-0-323-54800-7

Cover Designer: Tom M. Olson, BA
Cover Art: Richard Coombs, MS

Printed in Canada by Friesens, Altona, Manitoba, Canada

Last digit is the print number: 9 8 7 6 5 4 3 2 1

Preface

Anatomy is the fundamental infrastructure upon which all comprehension of the human body builds, in both health and disease. It is essential to everyone who practices medicine but is critical to those of us who perform and interpret ultrasound. You cannot understand what is abnormal without a thorough understanding of what is normal. That is why we wrote this book. This second edition of *Imaging Anatomy: Ultrasound* is the single most detailed and inclusive ultrasound anatomy text available on the market.

I have always found studying anatomy a bit like reading the dictionary—there is a lot of fantastic information, but there isn't much of a plot. Such a necessary topic is often difficult to approach. We have taken it as our mission, however, to break down those barriers and create an accessible anatomy text. Here is our story:

- The Characters: Each anatomic area (Brain & Spine, Head & Neck, Thorax, Abdomen, Pelvis, Extremities, and Developmental Anatomy) has its own complete cast of fascinating characters (organs). There is no hero in this book though; each is as important as the next, from the Parotid Gland to the Pelvic Floor to the Metatarsals and Toes. They all have their vital role to play.

- The Story Line: Every chapter begins with Gross Anatomy, followed by Imaging Anatomy, which includes best imaging techniques, helpful tips, and potential pitfalls. The tale is presented in an engaging, reader-friendly style. Convoluted descriptions are abandoned as key anatomic principles are outlined in a succinct, bulleted format for quick reference.

- The Illustrations: Never before has there been such a beautifully illustrated ultrasound anatomy text. The graphics, created by our own very talented group of medical illustrators, are of extraordinary quality. Those alone would make this book worth the read. But then following the graphics are extended galleries of detailed, extensively labeled, high-quality ultrasound images. A page turner for certain.

- The Authors: Given the expansive scope of this book, it required experts in all the various anatomic regions. I am quite fortunate to have some brilliant sonologists leading and editing their areas: Drs. James Griffith (Musculoskeletal), Anil Ahuja (Head & Neck), and Aya Kamaya & Jade Wong-You-Cheong (Abdomen & Pelvis). In addition to the physicians, I must acknowledge the talented sonographers whose fine work is highlighted throughout this book.

- The Editorial Staff: To publish any book (especially one of this complexity) takes an incredible group of individuals working behind the scenes to make it happen. I would like to thank the wonderful Elsevier Salt Lake City editorial and production staff, medical illustrators, and image editors—with a special shout out to Matt Hoecherl, who helped me immeasurably. I'm extremely lucky to work with you guys.

It is with a great deal of pride that we present to you the second edition of *Imaging Anatomy: Ultrasound*. While it might not be an epic thriller, it does have a compelling narrative to keep the reader engaged and informed throughout.

Paula J. Woodward, MD
Professor of Radiology
David G. Bragg, MD and Marcia R. Bragg Presidential Endowed Chair in Oncologic Imaging
Adjunct Professor of Obstetrics and Gynecology
University of Utah School of Medicine
Salt Lake City, Utah

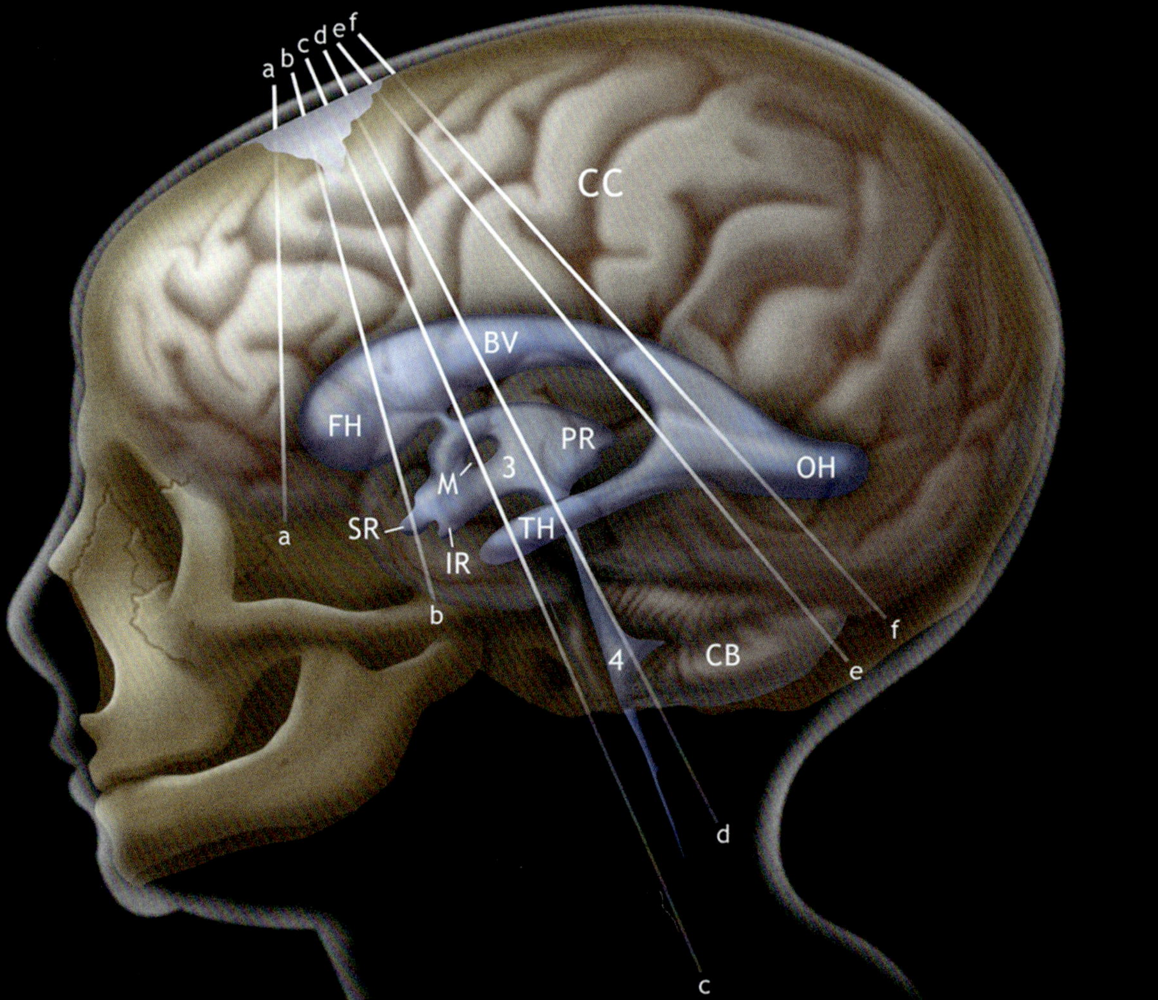

a b c d e f
CC
BV
FH
PR
3
M
OH
SR
IR
TH
a
b
4
CB
e
f
d
c

Acknowledgments

Lead Editor

Matt W. Hoecherl, BS

Text Editors

Arthur G. Gelsinger, MA
Nina I. Bennett, BA
Terry W. Ferrell, MS
Lisa A. Gervais, BS
Karen E. Concannon, MA, PhD
Megg Morin, BA

Image Editors

Jeffrey J. Marmorstone, BS
Lisa A. M. Steadman, BS

Illustrations

Richard Coombs, MS
Lane R. Bennion, MS
Laura C. Wissler, MA

Art Direction and Design

Tom M. Olson, BA
Laura C. Wissler, MA

Production Coordinators

Rebecca L. Bluth, BA
Angela M. G. Terry, BA
Emily C. Fassett, BA

ELSEVIER

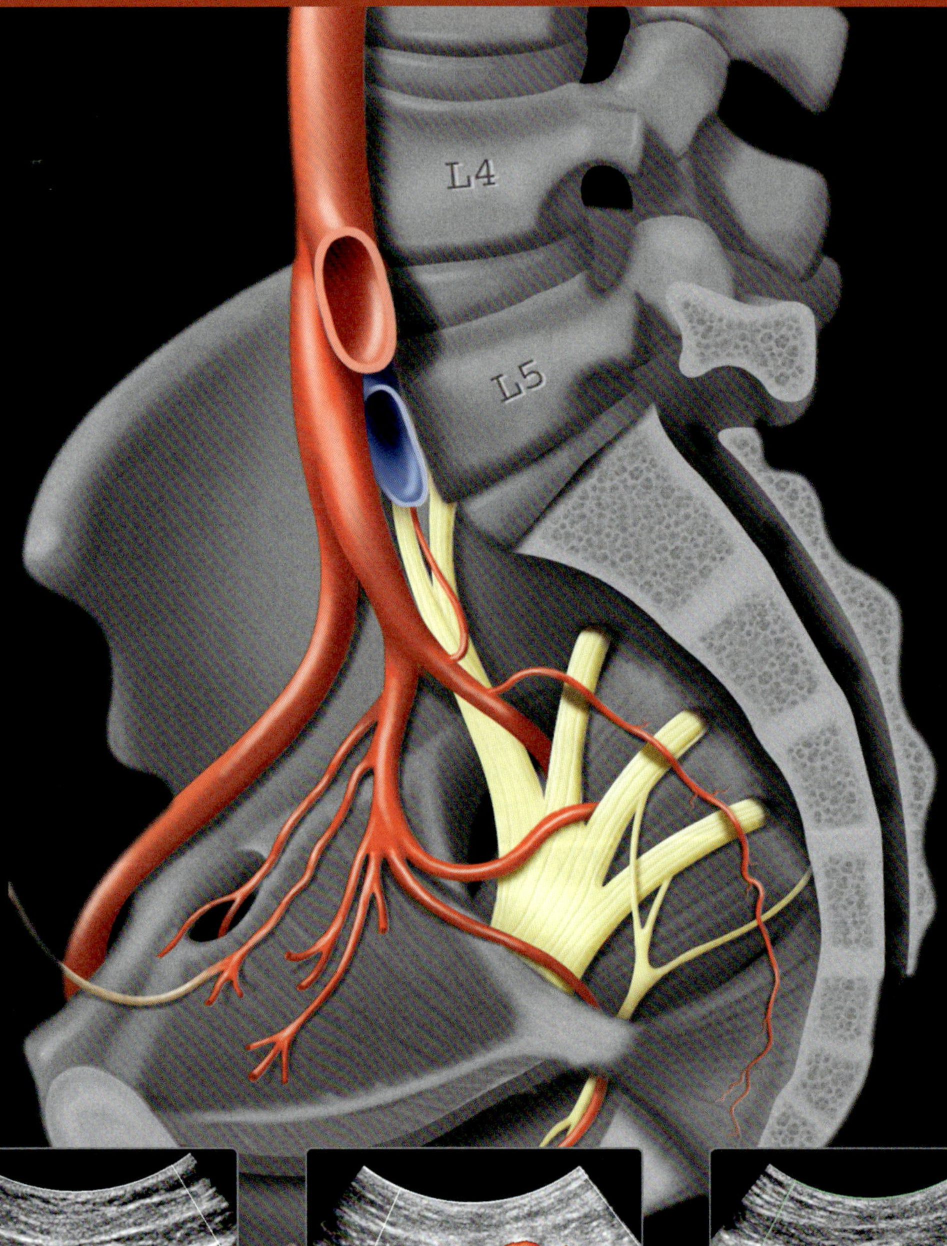
L4
L5

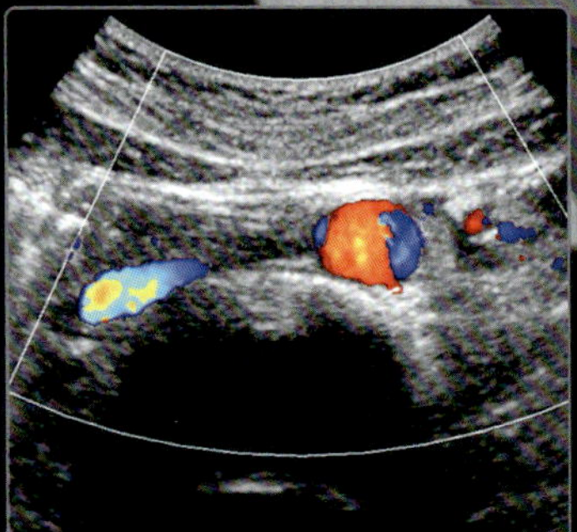

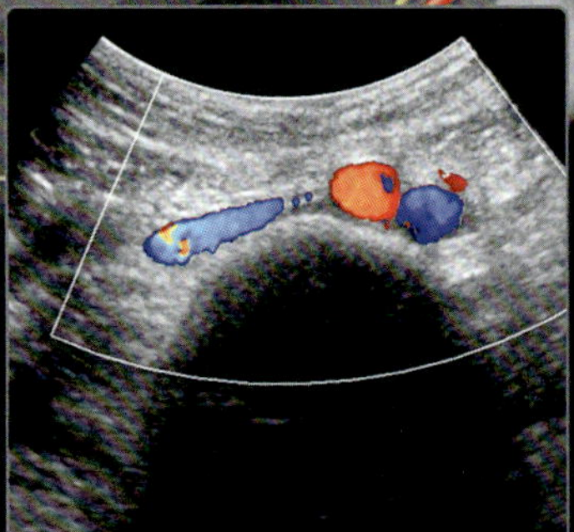

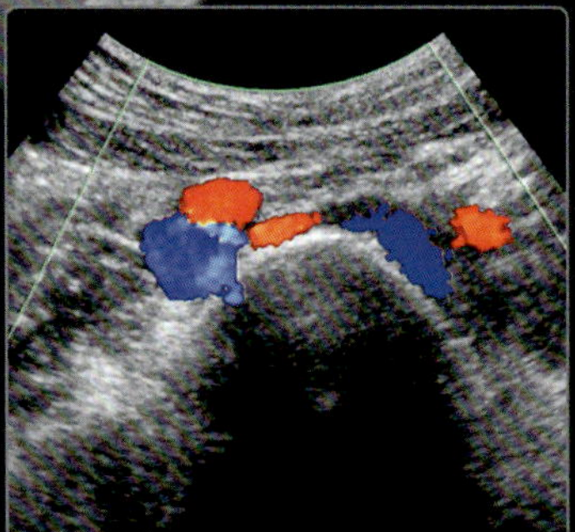

Sections

TABLE OF CONTENTS

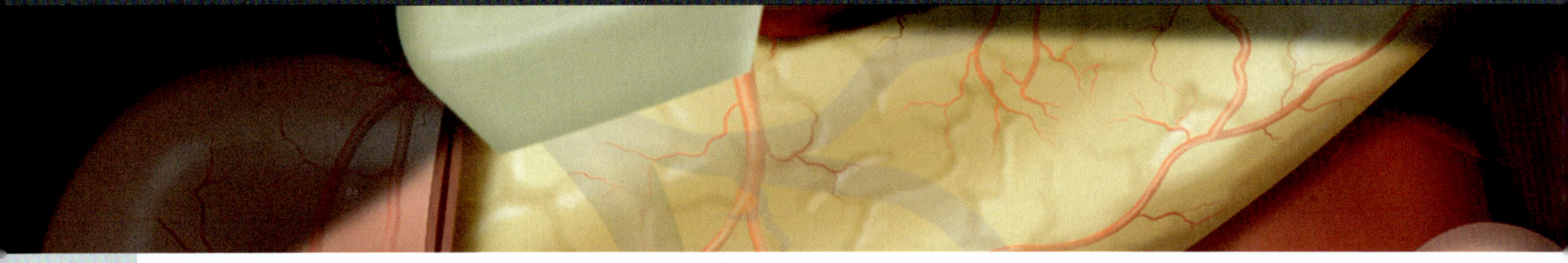

TABLE OF CONTENTS

TABLE OF CONTENTS

IMAGING ANATOMY
Ultrasound

SECOND EDITION

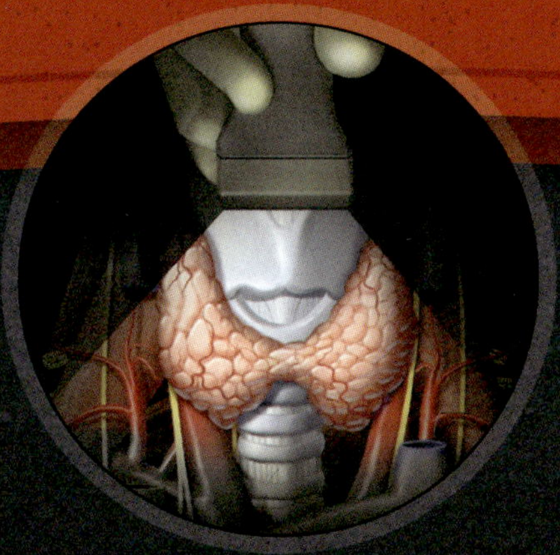
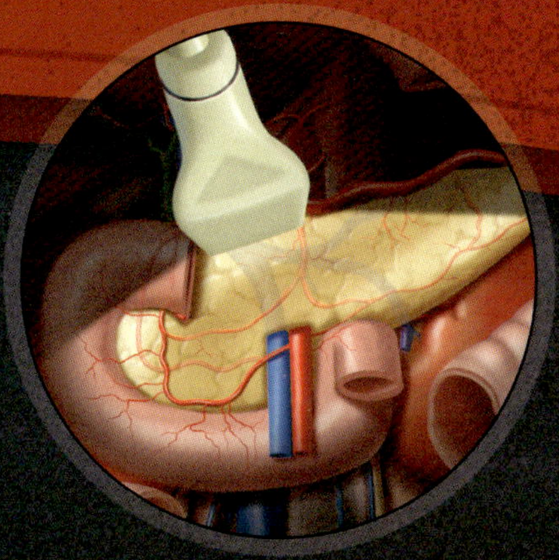
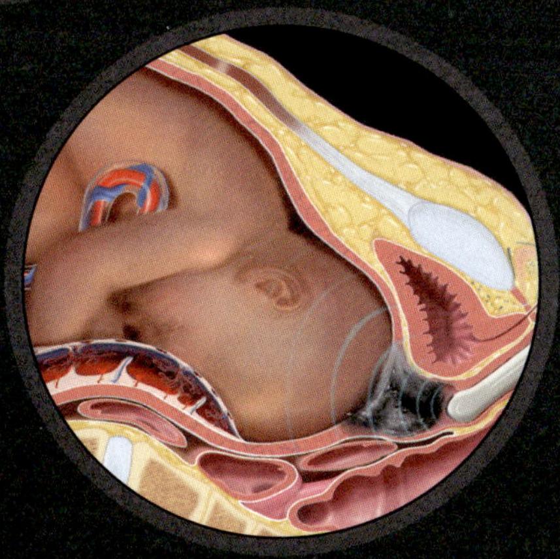
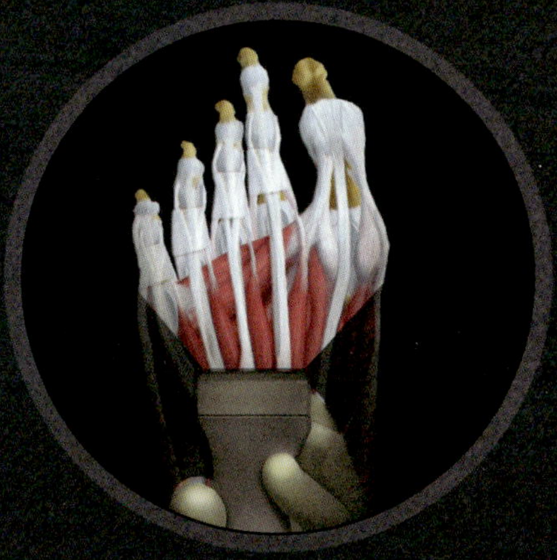

WOODWARD

GRIFFITH | ANTONIO | AHUJA
WONG | KAMAYA | WONG-YOU-CHEONG

SECTION 1

Brain and Spine

TERMINOLOGY

Definitions

- Fontanelle: Broad areas of connective tissue at junction of major sutures

GROSS ANATOMY

Overview

- **Scalp**
  - Scalp has 5 layers
    - Skin (epidermis, dermis, hair, sebaceous glands)
    - Subcutaneous tissue (very vascular fibroadipose tissue)
    - Epicranial tissue (scalp muscles, galea aponeurotica)
    - Subaponeurotic tissue (loose areolar connective tissue)
    - Pericranium (periosteum of skull)
- Skull (28 separate bones, mostly connected by fibrous sutures)
  - Cranium has several parts
    - Calvarial vault
    - Cranial base
    - Facial skeleton
  - Calvarial vault composed of several bones
    - 2 frontal bones separated by metopic suture
    - Paired parietal bones
    - Squamous occipital bone
    - Paired squamous temporal bones
  - 3 major serrated fibrous joints (sutures) connect bones of vault
    - Coronal suture
    - Sagittal suture
    - Lambdoid suture
  - Outer, inner tables
    - 2 thin plates of compact cortical bone
    - Separated by diploic space (cancellous bone containing marrow)
  - Endocranial surface
    - Lined by outer (periosteal) layer of dura
    - Grooved by vascular furrows
    - May have areas of focal thinning (arachnoid granulations), foramina (emissary veins)

IMAGING ANATOMY

Overview

- Scalp: Hypoechoic, 5 layers cannot be further separately resolved
- Calvarium echogenic outer/inner tables; diploic space filled with fatty marrow and appears hypoechoic, suture appears as gap between echogenic calvarium
- Frontal bones
  - Frontal sinuses show wide variation in aeration
  - Frontal bones often appear thickened, hyperostotic (especially in older females)
- Parietal bones
  - Areas of parietal thinning, granular foveolae (for arachnoid granulations) common adjacent to sagittal suture
  - Inner tables often slightly irregular (convolutional markings caused by gyri), grooved by paired middle meningeal arteries + vein
- Occipital bone
  - Deeply grooved by superior sagittal, transverse sinuses
  - Internal occipital protuberance marks sinus confluence (torcular Herophili)
- Temporal bones
  - Thin, inner surface grooved by middle meningeal vessels
  - Outer surface grooved by superficial temporal artery
- Fontanelle: Provide acoustic window for US examination of underlying brain parenchyma
  - **Anterior fontanelle**
  - Between 2 frontal and 2 parietal bones, usually disappears by age 2
  - When fused, corresponds to bregma: Meeting of sagittal, coronal sutures
  - **Posterior fontanelle**
  - Small, usually closes between 3-6 months of age
  - When fused, corresponds to lambda: Meeting of sagittal, lambdoid sutures
  - Pterion
    - Anterolateral fontanelle; closes between 3-6 months of age
    - H-shaped junction between frontal, parietal bones + greater sphenoid wing, squamous temporal bone
  - Asterion
    - Posterolateral fontanelle, persists until 2 years of age
  - **Mastoid fontanelle**
  - Located at junction of temporosquamous and lambdoid sutures
  - Persists until 2 years

ANATOMY IMAGING ISSUES

Imaging Recommendations

- High-frequency linear array transducers provide superb resolution of near-field structures
- Good skin-to-transducer coupling achieved by copious use of acoustic coupling gel
- Superficial standoff pad can be used to increase depth of focal zone
- US can be used to evaluate cranial sutures and assists diagnosis of craniosynostosis (premature fusion of sutures)

EMBRYOLOGY

Embryologic Events

- Skull base formed from endochondral ossification
- Calvarial vault forms via membranous ossification
  - Curved mesenchymal plates appear at day 30
  - Extend toward each other, skull base
  - As paired bones meet in midline, metopic and sagittal sutures are induced (coronal suture is present from onset of ossification)
  - Unossified centers at edges of parietal bone form fontanelles
  - Vault grows rapidly in 1st postnatal year

SCALP AND CALVARIAL VAULT

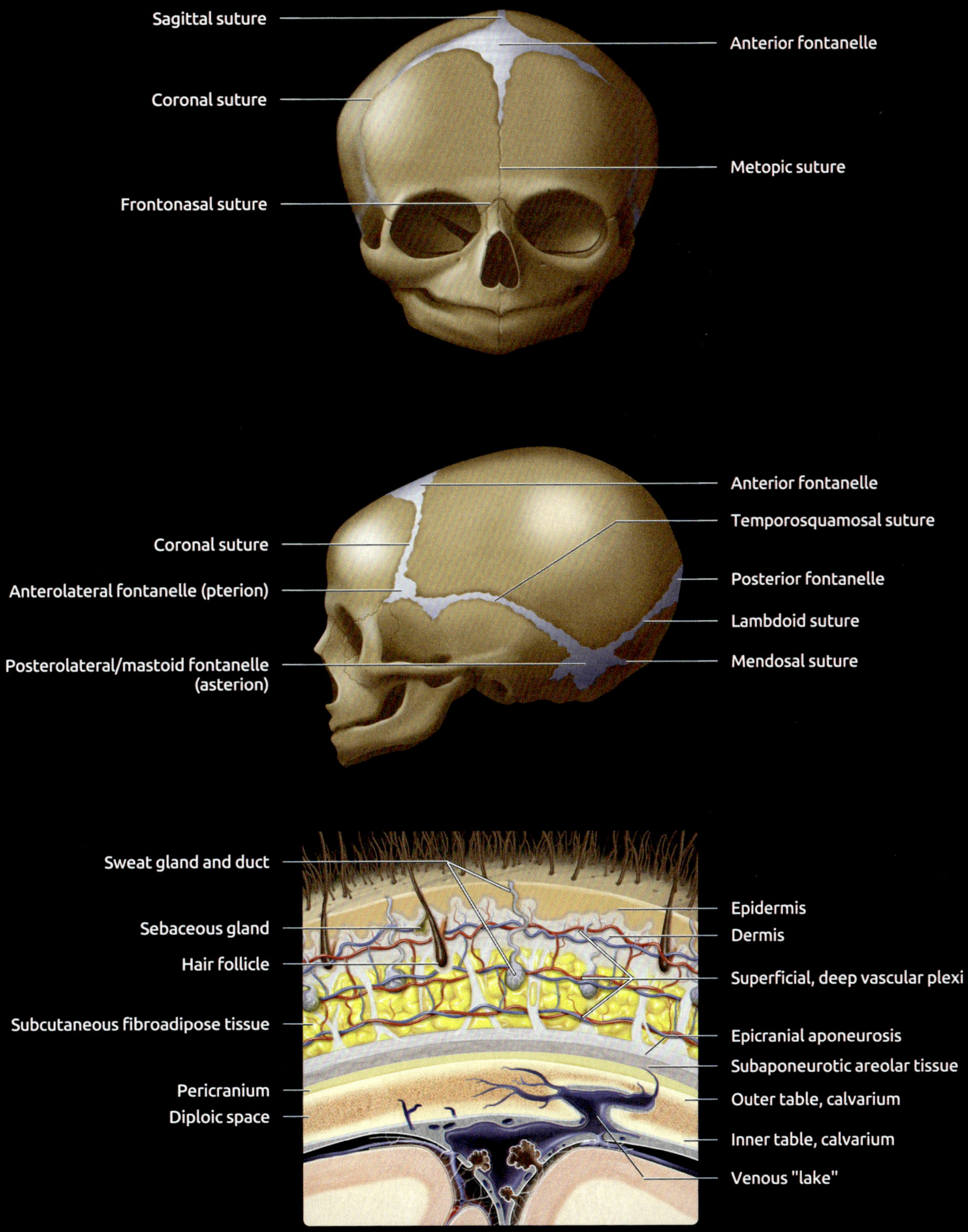

(Top) *Graphic depiction of an infant cranium, frontal view, is shown. The anterior fontanelle is present between 2 frontal and 2 parietal bones, which usually close by 2 years of age. When fused, this site corresponds to bregma: The meeting point of sagittal and coronal sutures.* **(Middle)** *Lateral view of an infant calvarial vault is shown. The posterior fontanelle is small and usually closes by 3-6 months of age. When fused, this corresponds to lambda: The meeting of sagittal and lambdoid sutures. The anterolateral fontanelle (pterion) closes at ~ 3 months of age. The posterolateral fontanelle (asterion) often persists until 2 years of age.* **(Bottom)** *Scalp and calvarium are depicted in cross section. The 5 scalp layers are depicted. Skin consists of epidermis and dermis. Hair follicles and a sebaceous gland, the subcutaneous fibroadipose tissue, and sweat glands and ducts as well as superficial and deep cutaneous vascular plexi are shown.*

US AND T2WI MR, SCALP

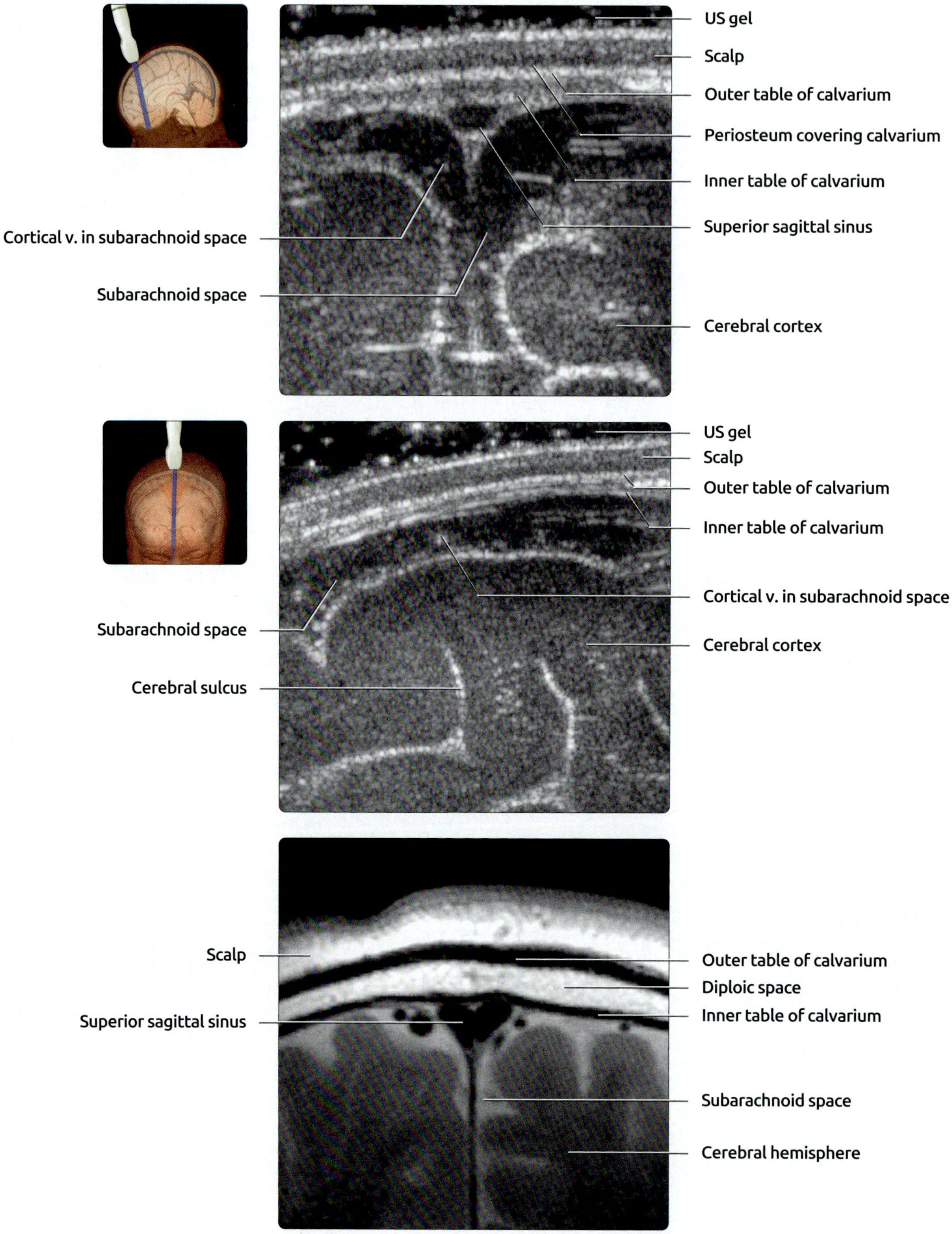

(Top) *Anterior coronal US scan through the anterior fontanelle shows the scalp covering the frontal bone of the calvarial vault. The 5 layers of the scalp are as follows: Skin, connective tissue consisting of lobules of fat, artery and emissary vein, aponeurosis, loose connective tissue, which accounts for mobility of the scalp on the underlying bone and periosteum adhering to the outer table of skull. These 5 layers, however, cannot be resolved by US.* **(Middle)** *Midsagittal scan through the anterior fontanelle shows the scalp covering the frontal bone of the calvarial vault. Beneath the inner table of the calvarial vault is the anechoic subarachnoid space.* **(Bottom)** *Coronal T2 MR shows the hypointense outer and inner table of calvarium. The scalp and diploic space are hyperintense. The superior sagittal sinus appears as signal void structure below the inner table of calvarium.*

US AND T2WI MR CALVARIUM

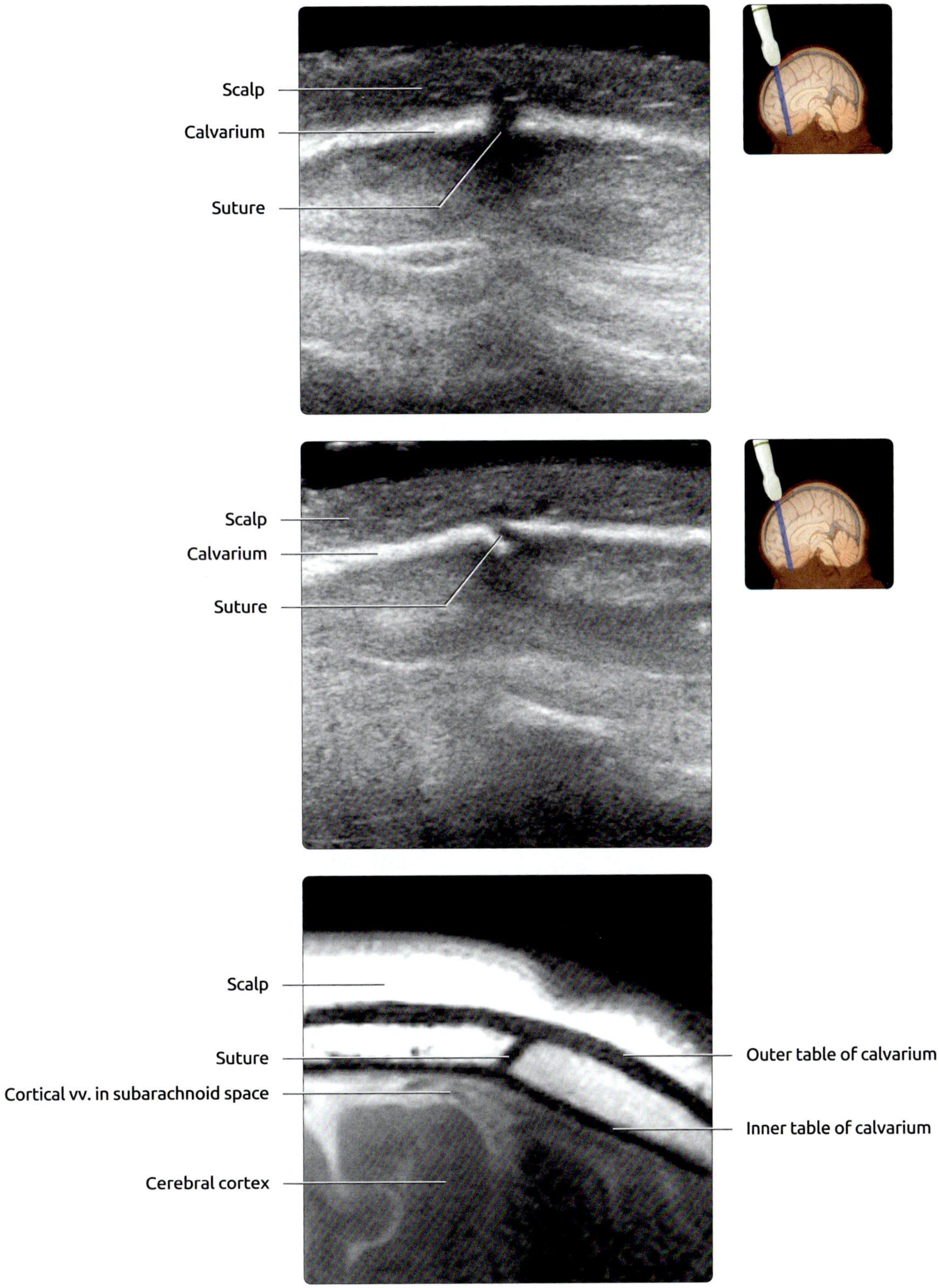

(Top) *Coronal US through the anterior fontanelle shows the scalp and calvarium. There is discontinuity in a hypoechoic band extending from the outer to inner table of the calvarial vault. This represents a normal suture. The scalp appears hypoechoic compared to the echogenic outer and inner tables of the calvarium. The 5 layers of the scalp cannot be resolved by US.* **(Middle)** *Coronal US through the anterior fontanelle shows another suture of the calvarial vault. The width and curvature of sutures is variable and should not be mistaken for a bony fracture.* **(Bottom)** *Sagittal T2 MR of the scalp and calvarium is shown. The suture line appears with the same signal intensity as the outer and inner table of calvarium. Cortical veins can be seen within the hyperintense subarachnoid space, which is immediately under the inner table.*

GROSS ANATOMY

Supratentorial Structures

- **Gyri**: Complex convolutions of brain cortex; hypoechoic on ultrasound (US)
- **Sulci** (fissure): CSF-filled grooves or clefts that separate gyri; echogenic on US
  - Sulci separate gyri, **fissures** separate hemispheres/lobes
- **Frontal lobe**
  - **Central sulcus separates frontal, parietal lobes**
  - Precentral gyrus contains primary motor cortex
  - Premotor cortex: Within gyrus just anterior to precentral gyrus (motor cortex)
  - 3 additional major gyri: Superior frontal gyrus, middle frontal gyrus, & inferior frontal gyrus
    - Superior sulcus separates superior & middle gyri
    - Inferior sulcus separates middle & inferior gyri
  - Orbital gyri cover base of frontal lobe; gyrus rectus medially
- **Parietal lobe**
  - **Posterior to central sulcus**
  - Separated from occipital lobe by parietooccipital sulcus (medial surface)
  - Postcentral gyrus: Primary somatosensory cortex
  - Superior & inferior parietal lobules lie posterior to postcentral gyrus
  - Supramarginal gyrus lies at end of sylvian fissure
  - Angular gyrus lies ventral to supramarginal gyrus
  - Medial surface of parietal lobe is precuneus, in front of parietooccipital sulcus
- **Occipital lobe**
  - **Posterior to parietooccipital sulcus**
  - Primary visual cortex on medial occipital lobe
  - Cuneus on medial surface
- **Temporal lobe**
  - **Inferior to sylvian fissure**
  - Superior temporal gyrus: Primary auditory cortex
  - Middle temporal gyrus: Connects with auditory, somatosensory, visual association pathways
  - Inferior temporal gyrus: Higher visual association area
  - Includes major subdivisions of limbic system
- **Insula**
  - Lies deep in floor of sylvian fissure, overlapped by frontal, temporal, parietal opercula
- **Limbic system**
  - Includes amygdala, hippocampus, thalamus, hypothalamus, basal ganglia, & cingulate gyrus
    - Cingulate gyrus extends around corpus callosum
  - Important role in emotion, behavior, & long-term memory
- **White matter tracts**: 3 major types of fibers
  - Association fibers: Interconnect different cortical regions in same hemisphere
  - Commissural fibers: Interconnect similar cortical regions of opposite hemispheres
    - **Corpus callosum** is largest commissural fiber, links cerebral hemispheres
  - Projection fibers: Connect cerebral cortex with deep nuclei, brainstem, cerebellum, spinal cord
    - Internal capsule is major projection fiber
- **Basal ganglia**
  - Paired deep gray matter nuclei
  - Caudate nucleus, lentiform nucleus (including putamen, globus pallidus)
- **Thalamus**
  - Paired nuclear complexes, serve as relay station for most sensory pathways

Posterior Fossa (Infratentorial) Structures

- Protected space surrounded by calvarium & bounded by tentorium cerebelli superiorly & foramen magnum inferiorly
- Posterior fossa contents
  - Brainstem (midbrain, pons, & medulla oblongata) anteriorly, cerebellum posteriorly
  - Cerebral aqueduct & 4th ventricle
  - CSF cisterns containing cranial nerves, vertebrobasilar arterial system & veins
- **Cerebellum**
  - Integrates coordination & fine-tuning of movement, & regulation of muscle tone
  - 3 surfaces: Superior (tentorial), inferior (suboccipital), anterior (petrosal)
  - **2 hemispheres & midline vermis**
    - Divided into lobes & lobules by transverse fissures
    - Major fissures: Primary (tentorial), horizontal (petrosal), prebiventral/prepyramidal (suboccipital) cerebellar fissures
  - Connected to brainstem by **3 paired peduncles**
    - Superior cerebellar peduncle (brachium conjunctivum) connects cerebellum to cerebrum via midbrain
    - Middle cerebellar peduncle (brachium pontis) connects to pons
    - Inferior cerebellar peduncle (restiform body) connects to medulla
- **Brainstem**
  - **3 anatomic divisions**
    - **Midbrain (mesencephalon)**: Upper brainstem, connects pons & cerebellum with forebrain
    - **Pons**: Bulbous midportion of brainstem, relays information from brain to cerebellum
    - **Medulla**: Caudal (inferior) brainstem, relays information from spinal cord to brain
  - Functional divisions
    - Ventral part: Large descending white matter tracts; contains midbrain cerebral peduncles, pontine bulb, medullary pyramids
    - Dorsal part: Tegmentum, common to midbrain, pons & medulla; contains cranial nerve nuclei & reticular formation

Ventricular System & Subarachnoid Space

- Cerebral ventricles consist of paired lateral, midline 3rd, & 4th ventricles
- Communicate with each other as well as central canal of spinal cord & subarachnoid space
- Direction of CSF flow
  - Lateral ventricles → foramen of Monro → 3rd ventricle → cerebral aqueduct → 4th ventricle → foramina of Luschka & Magendie → subarachnoid space
  - Bulk of CSF resorption through arachnoid granulations in superior sagittal sinus

- **Lateral ventricles**
  - Paired, C-shaped, curve posteriorly from temporal horns, arch around/above thalami
  - Each has body, atrium, **3 horns (frontal, temporal, & occipital)**
    - Occipital horn typically largest
    - Asymmetry is common, often L > R
    - Sizes change with maturity, more prominent in preterm infants
  - **Atrium/trigone**: Confluence of horns
    - Contains glomus of choroid plexus
  - Lateral ventricles communicate with each other & 3rd ventricle via Y-shaped foramen of Monro
- **3rd ventricle**
  - Thin, usually slit-like, between thalami
    - May not see fluid, just bright echogenic line on US
  - 80% have central adhesion between thalami (massa intermedia)
  - Communicates with 4th ventricle via cerebral aqueduct (of Sylvius), passing through dorsal midbrain
- **4th ventricle**
  - Infratentorial, diamond-shaped cavity (rhomboid fossa) along dorsal pons & upper medulla
  - Fastigium: Blind ending, dorsally pointed midline outpouching from body of 4th ventricle
    - Important marker for true midline vermian plane on US
  - Communicates with subarachnoid space via foramina of Magendie & Luschka
  - Terminates inferiorly at obex, which communicates with central canal of spinal cord
- **Choroid plexus**
  - Produces CSF
  - Glomus (enlargement of choroid plexus in atrium) thickest area
  - Tapers & extends anteriorly to foramen of Monro & roof of 3rd ventricle
  - Tapers laterally into roof of temporal horns
  - Present in roof of 4th ventricle but **never extends into frontal or occipital horns**
- **Subarachnoid space/cisterns**
  - CSF spaces between pia & arachnoid
  - Numerous trabeculae, septa, membranes cross subarachnoid space & create smaller compartments termed cisterns
    - **Supratentorial/peritentorial cisterns**: Suprasellar, interpeduncular, ambient (perimesencephalic), quadrigeminal cistern, & cistern of velum interpositum
    - **Infratentorial (posterior fossa) cisterns**: Prepontine, premedullary, superior cerebellar, cisterna magna, & cerebellopontine
  - All cisterns communicate with each other & with ventricular system
- Midline cystic structures (normal variants)
  - **Cavum septi pellucidi**: Anterior to foramen of Monro, between anterior horns of lateral ventricles
    - 85% closed by 3-6 months after birth, some remain open into adulthood
      - Once closed called septum pellucidum
  - **Cavum vergae**: Posterior to foramen of Monro, interposed between bodies of lateral ventricles
    - Posterior extension of cavum septi pellucidi
    - Begins to close from posterior to anterior from 6-month gestation; 97% closed by full term
  - **Cavum velum interpositum**: Potential space above choroid in roof of 3rd ventricle & below fornices
    - Typically seen in premature infants

ANATOMY IMAGING ISSUES

Imaging Approaches

- **Anterior fontanelle** most commonly used approach
  - **Sagittal scans**
    - Midline scan: Best view for corpus callosum, cerebellar vermis
    - Sweep side-to-side from this position documenting key areas
      - **Caudothalamic groove**: Most common site of germinal matrix hemorrhage
      - Size of lateral ventricle
      - Far lateral to assess degree of sulcal development
  - **Coronal scans**
    - Important to maintain symmetrical imaging of each 1/2 of brain
    - Symmetrical structures (from anterior to posterior) include: Frontal horns, bodies & trigones of lateral ventricles; caudate nuclei, putamen, internal capsule, & thalami
    - Midline structures (from anterior to posterior) include: Interhemispheric fissure, genu & anterior body of corpus callosum, cavum septi pellucidi, 3rd ventricle, brainstem
- **Posterior fontanelle**
  - Best view to evaluate occipital horns for intraventricular hemorrhage
    - Can misinterpret clot adherent to choroid plexus from anterior fontanelle approach alone
- **Mastoid fontanelle**
  - Located at junction of squamosal, lambdoidal, occipital sutures
  - Transducer placed about 1 cm behind helix of ear & 1 cm above tragus
  - Allows assessment of brainstem & posterior fossa
  - Best view for 4th ventricle, posterior cerebellar vermis, cerebellar hemispheres, & cisterna magna
- **Transtemporal**
  - Temporal bone anterior to ear is thin, allowing imaging of brainstem even after sutural closure
  - Best view for cerebral peduncles & 3rd ventricle

Imaging Pitfalls

- Need to know changing appearance with gestational age at birth; normal gyral pattern in 26-week preterm infant would be abnormal in term infant
- Slit-like lateral ventricles common in infants, not to be mistaken for cerebral edema
- Glomus of choroid plexus can be bulbous & irregular, not to be mistaken for blood clot
  - Evaluate with color Doppler & posterior fontanelle view
- Echogenic material in frontal or occipital horns is clot; choroid does not extend into these horns

GYRI AND SULCI

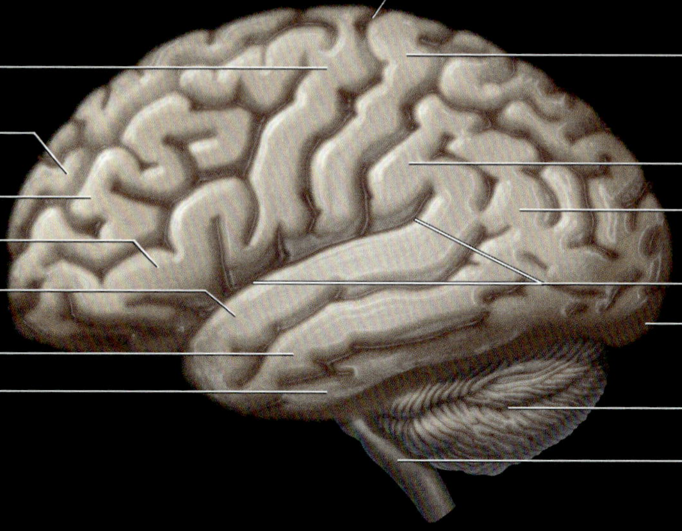

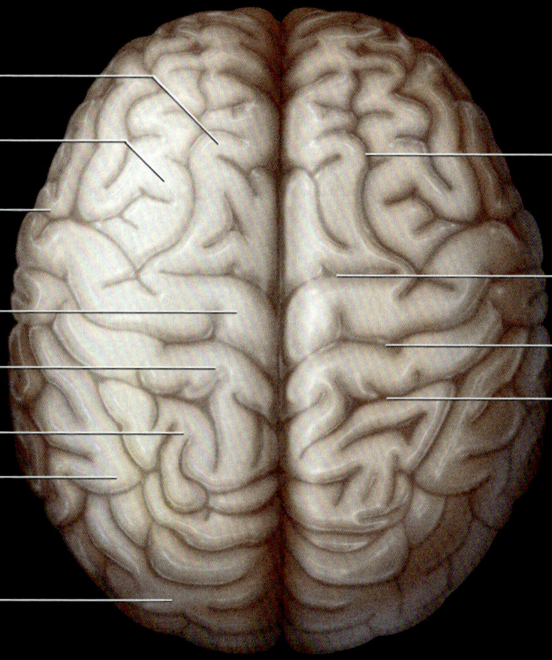

(Top) *Lateral surface of the brain depicts the major gyri and sulci. The frontal lobe extends from the frontal pole to the central sulcus. The supramarginal and angular gyri are part of the parietal lobe. The superior temporal gyrus contains the primary auditory cortex, and also forms the temporal operculum. The insular cortex lies within the sylvian fissure beneath the frontal, temporal, and parietal opercula.* **(Bottom)** *Surface anatomy of the cerebral hemisphere, seen from above, shows the gyri and lobules on the left, and the sulci on the right. The central (Rolandic) sulcus separates the anterior frontal lobe from the posterior parietal lobe. The precentral gyrus of the frontal lobe is the primary motor cortex while the postcentral gyrus of the parietal lobe is the primary sensory cortex. On ultrasound, the sulci appear echogenic while the adjacent gyri are hypoechoic.*

MIDLINE, SUBARACHNOID SPACE

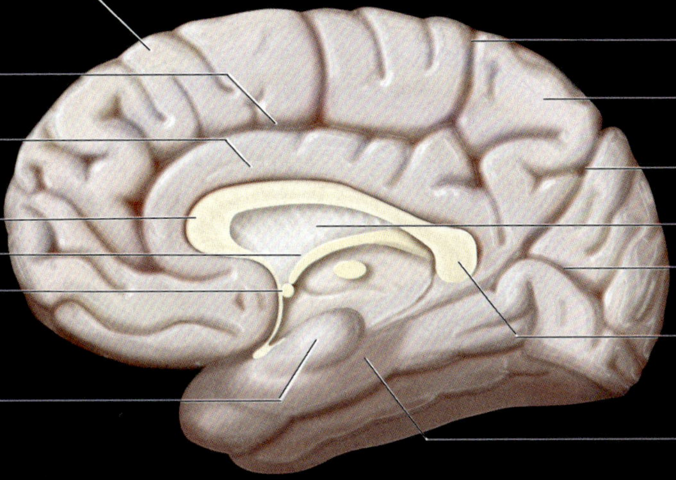

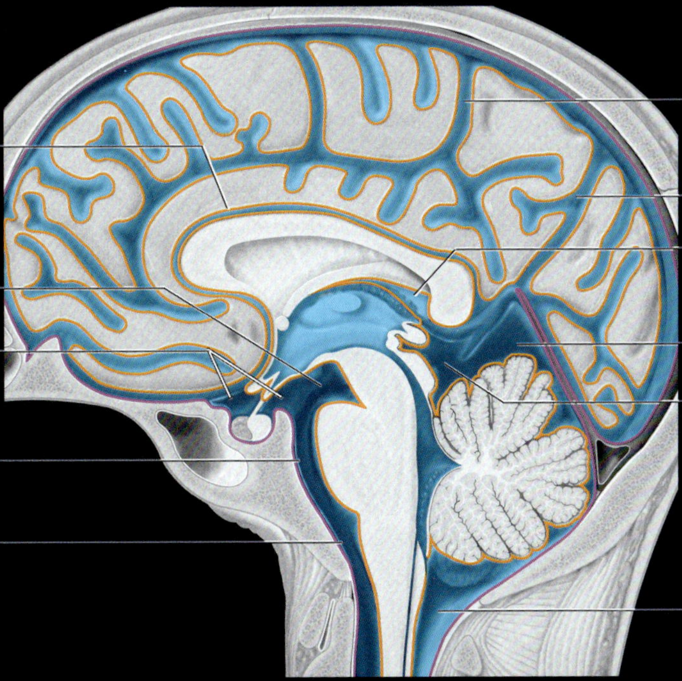

(Top) *This midline sagittal graphic shows a medial view of the cerebral hemisphere. The corpus callosum represents the major commissural fiber. The fornix and cingulate gyrus are important in the limbic system. The cingulate gyrus is involved with emotion formation and processing, learning, and memory.* **(Bottom)** *Sagittal midline graphic through the interhemispheric fissure depicts subarachnoid spaces with CSF (blue) between the arachnoid (purple) & pia (orange). The central sulcus separates the frontal lobe (anterior) from the parietal lobe (posterior). The pia mater is closely applied to the brain surface, whereas the arachnoid is adherent to the dura. The ventricles communicate with the cisterns and subarachnoid space via the foramina of Luschka and Magendie. The cisterns normally communicate freely with each other.*

VENTRICULAR SYSTEM

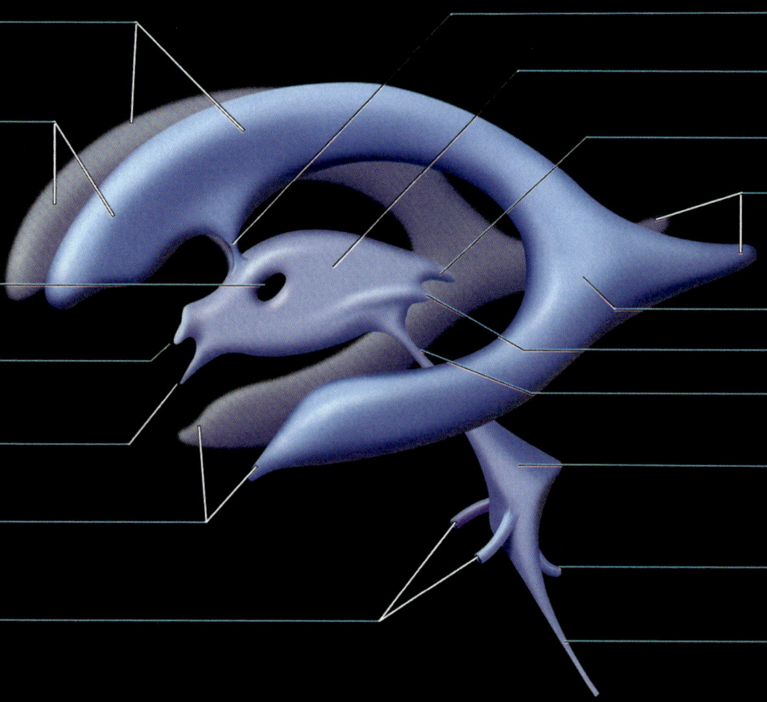

Schematic 3D representation of the ventricular system, viewed in the sagittal plane, demonstrates the normal appearance and communicating pathways of the cerebral ventricles. CSF flows from the lateral ventricles through the foramen of Monro into the 3rd ventricle, and from there through the cerebral aqueduct into the 4th ventricle. CSF exits the 4th ventricle through the foramina of Luschka and Magendie to the subarachnoid space.

STANDARD US PLANES VIA ANTERIOR FONTANELLE

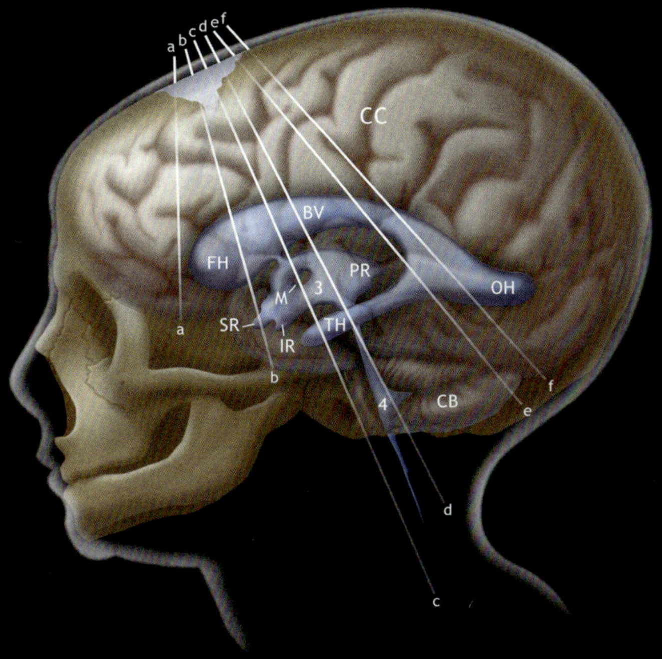

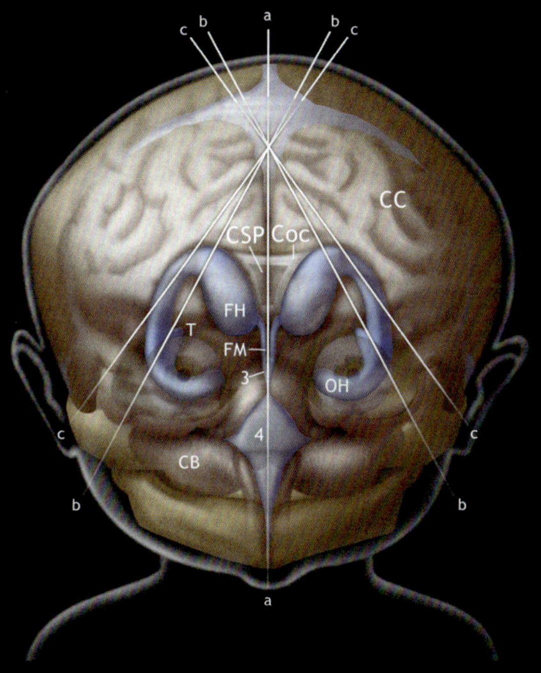

(Top) *Graphic shows the common coronal planes used in ultrasound brain scanning: Plane A to F from front to back. Cerebral cortex (CC); body of lateral ventricle (BV); frontal horn (FH); occipital horn (OH); massa intermedia (M); pineal recess (PR); 3rd ventricle (3); temporal horn (TH); supraoptic recess (SR), infundibular recess (IR); 4th ventricle (4); cerebellum (CB).* **(Bottom)** *Graphic shows the common sagittal planes used in ultrasound brain scanning: Plane A to C from midline to lateral. Cerebellum (CB); cerebral cortex (CC); corpus callosum (Coc); cavum septi pellucidi (CSP); frontal horn (FH); foramen of Monro (FM); occipital horn (OH); temporal horn (T); 3rd ventricle (3); 4th ventricle (4).*

CORONAL US VIA ANTERIOR FONTANELLE

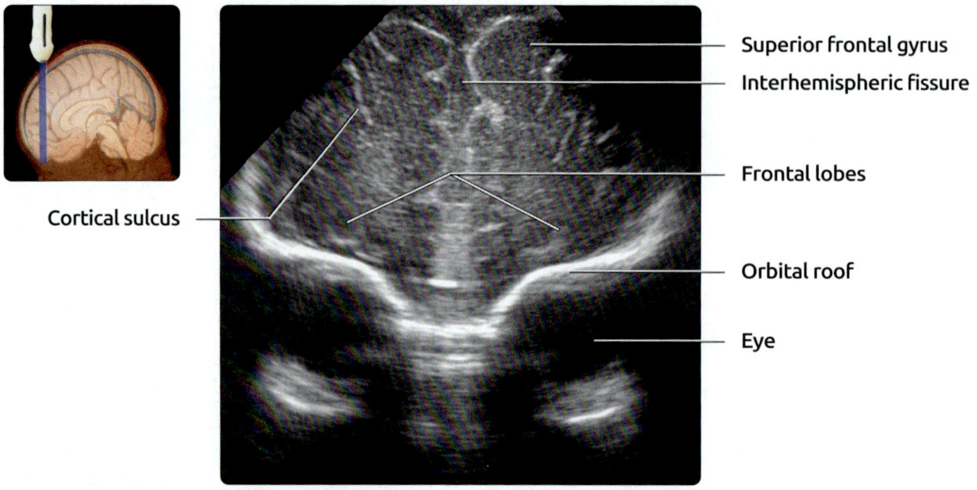

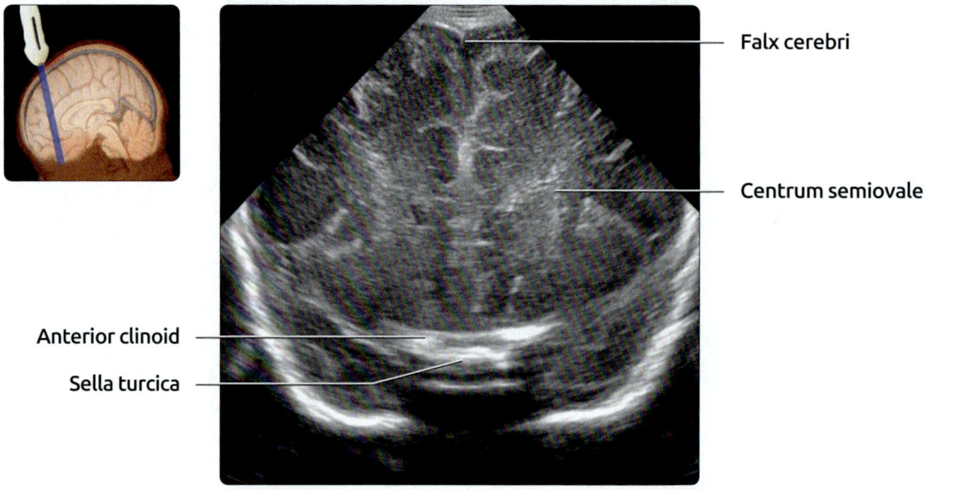

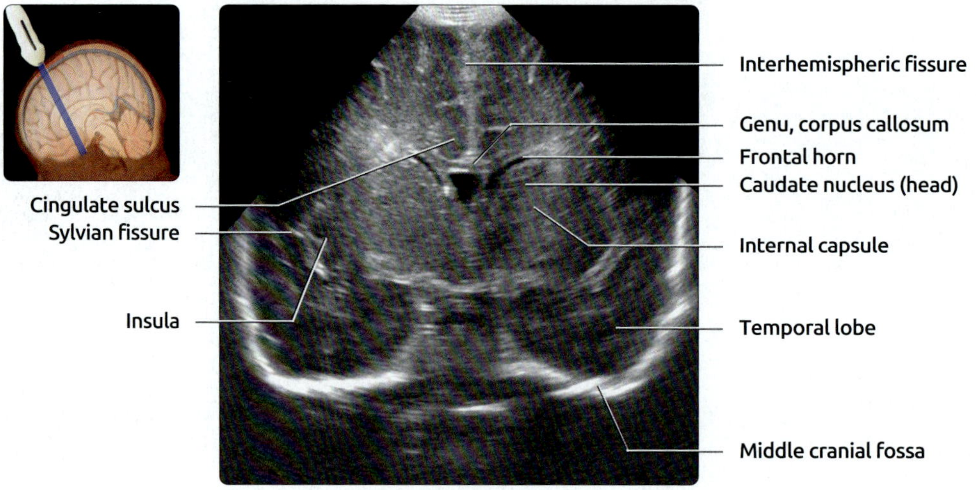

(Top) *The 1st of 9 coronal ultrasounds of the brain through the anterior fontanelle in a term infant shows the frontal lobes lie in the anterior cranial fossa with orbital cavities deep to the floor of the skull base.* **(Middle)** *An image centered more posteriorly demonstrates a slightly more echogenic white matter region of the brain parenchyma known as the centrum semiovale. Parts of the skull base, including the sella turcica and anterior clinoid, can be seen.* **(Bottom)** *Image acquired just anterior to the foramen of Monro. The frontal horns of lateral ventricles are now seen. No choroid plexus should be present in the frontal horns. Any intraventricular echogenic material seen at this level should raise the suspicion of blood clot. The head of the caudate nucleus is inferior and lateral to the frontal horn and is separated from the lentiform nucleus by the internal capsule.*

CORONAL T1 MR

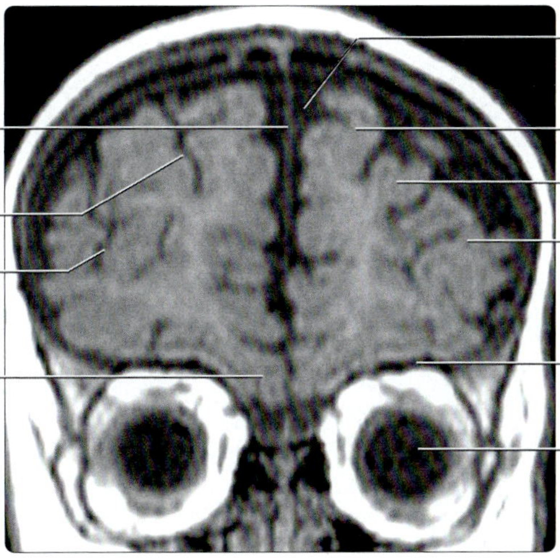

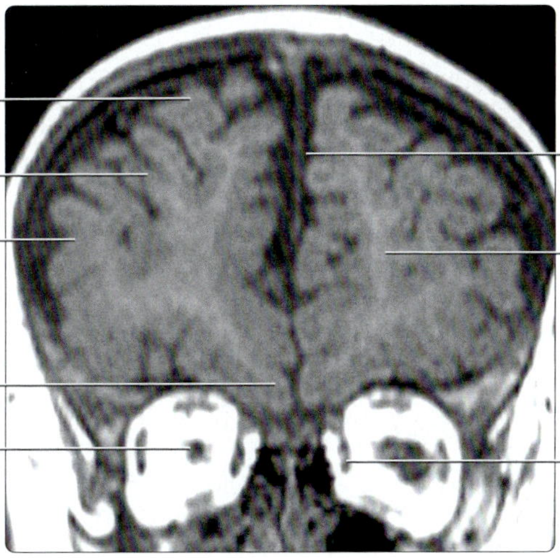

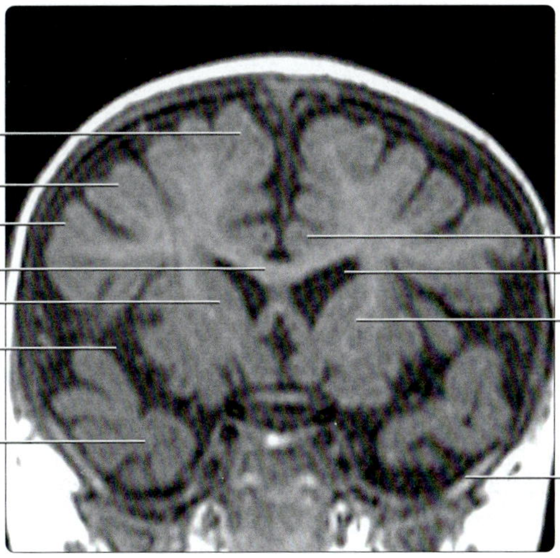

(Top) *The 1st of 9 coronal T1 MR images through the cerebral hemispheres from anterior to posterior is shown. The images are taken through planes/levels corresponding to those commonly used for ultrasound scans through the anterior fontanelle. The 3 major frontal gyri are shown: Superior frontal gyrus, middle frontal gyrus, and inferior frontal gyrus, separated by the superior and inferior frontal sulci. The straight gyrus (gyrus rectus) is the most medial, covering the base of the frontal lobe.* **(Middle)** *Slightly more posteriorly, the major white matter tracts, the centrum semiovale, are seen.* **(Bottom)** *This image shows the frontal horns. Immediately below each frontal horn is the caudate head, separated from the lentiform nucleus by the internal capsule.*

CORONAL US VIA ANTERIOR FONTANELLE

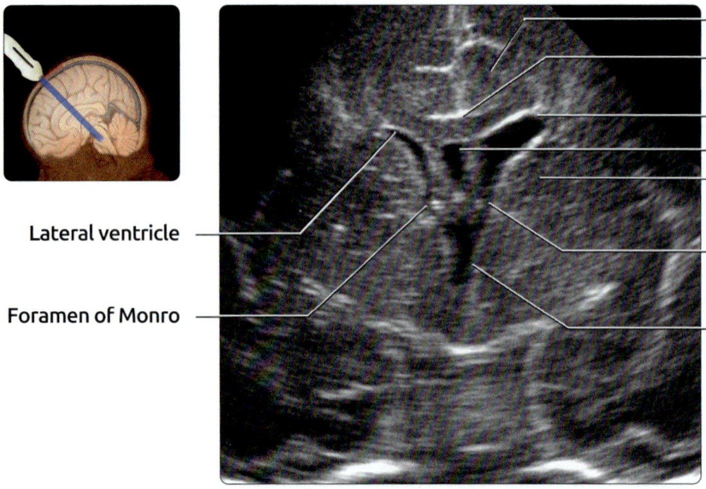

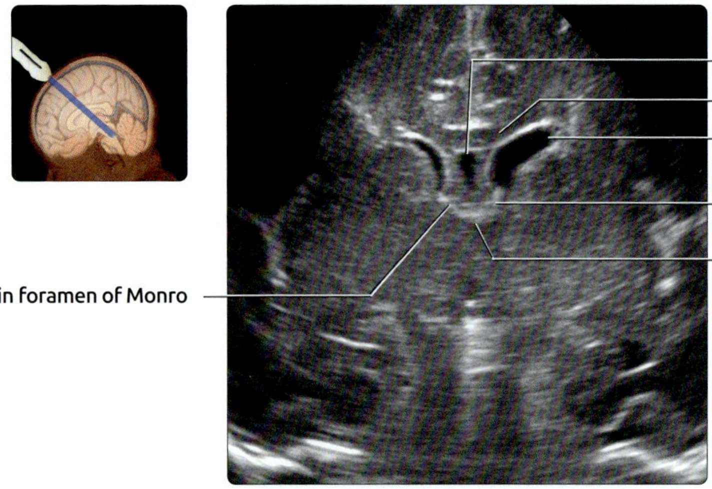

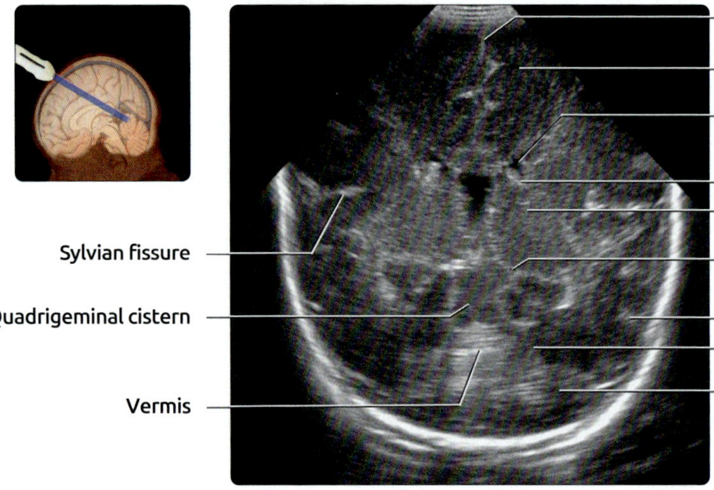

(Top) *The 4th of 9 coronal ultrasounds through the anterior fontanelle in a term infant is shown. This image is taken at the level of the foramen of Monro. The lateral ventricles are seen with the body of the caudate nucleus and anterior portions of the thalami below. It is not uncommon that the ventricles are asymmetric.* **(Middle)** *Just slightly more posterior, the choroid plexus is present on the floor of the lateral ventricles and roof of the 3rd ventricle. The 3 echogenic foci of the choroid plexus, 1 on the roof of the 3rd ventricle and 2 located bilaterally on the floor of the lateral ventricles, are known as the 3-dot sign.* **(Bottom)** *A more posterior coronal image at the level of the quadrigeminal cistern is shown. Another ultrasound landmark, known as the echogenic star, is seen, which comprises the choroidal fissures as the upper limbs and tentorium cerebelli as the lower limbs. Inferiorly, the vermis appears echogenic, while the cerebellar hemispheres on both sides are hypoechoic.*

CORONAL T1 MR

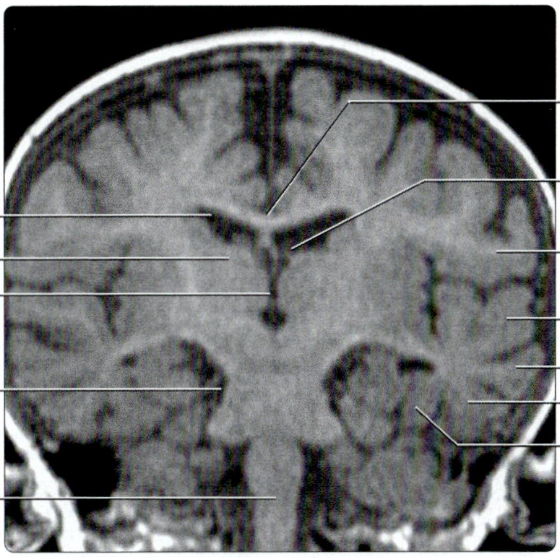

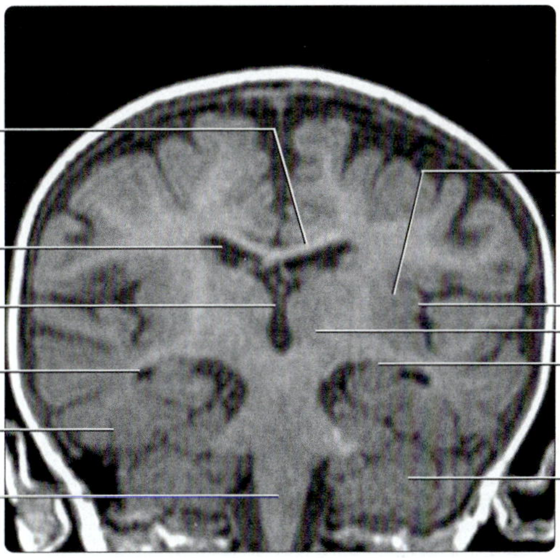

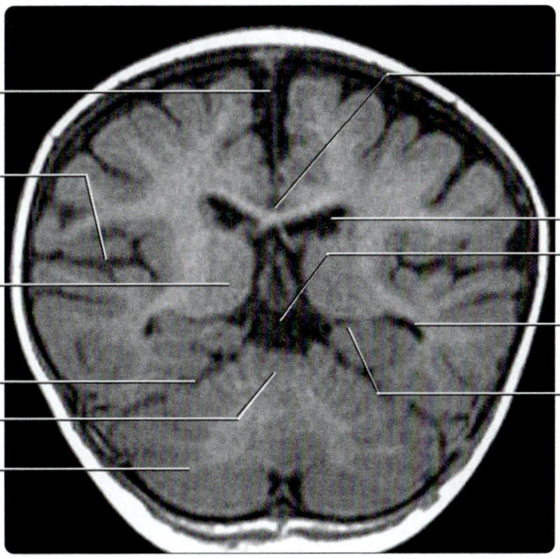

(Top) *The 4th of 9 coronal T1 MR images through the cerebral hemispheres from anterior to posterior is shown. The images are taken through planes/levels corresponding to those commonly used for ultrasound through the anterior fontanelle. This image is taken at the level of the foramen of Monro where both lateral ventricles unite, becoming the 3rd ventricle in the midline.* **(Middle)** *This image shows the thalami on either side of the 3rd ventricle.* **(Bottom)** *This image slightly more posterior shows the quadrigeminal cistern in the midline. Together with choroidal fissures and tentorium cerebelli on both sides, it gives rise to the characteristic echogenic star appearance on coronal ultrasound scanning.*

CORONAL US VIA ANTERIOR FONTANELLE

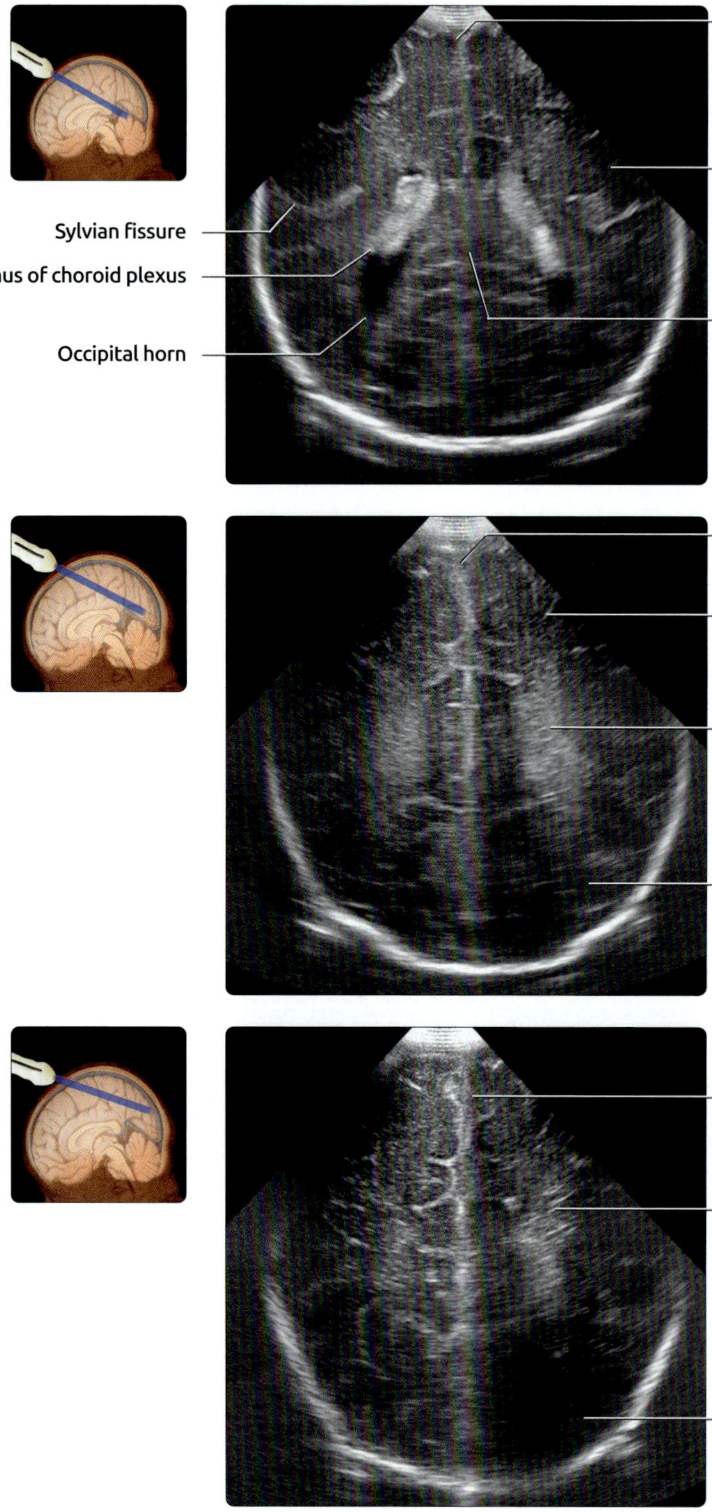

(Top) *The 7th of 9 coronal ultrasounds obtained through the anterior fontanelle in a term infant is shown. This image is taken at the trigone of the lateral ventricles. The glomus of the choroid plexus appears highly echogenic, nearly occupying the whole trigone.* **(Middle)** *This image, slightly posterior to the trigone, shows mildly echogenic white matter regions within the corona radiata, lateral and parallel to both trigones of the lateral ventricles. These regions are known as the periventricular halo, a normal finding, present in almost all normal mature and premature neonates. The echogenicity of the halo should be less than that of the choroid plexus and symmetrical in appearance.* **(Bottom)** *The most posterior coronal image shows the cortex of the occipital lobe with multiple echogenic sulci extending medially from the lateral margin of the brain. The falx is in midline.*

CORONAL T1 MR

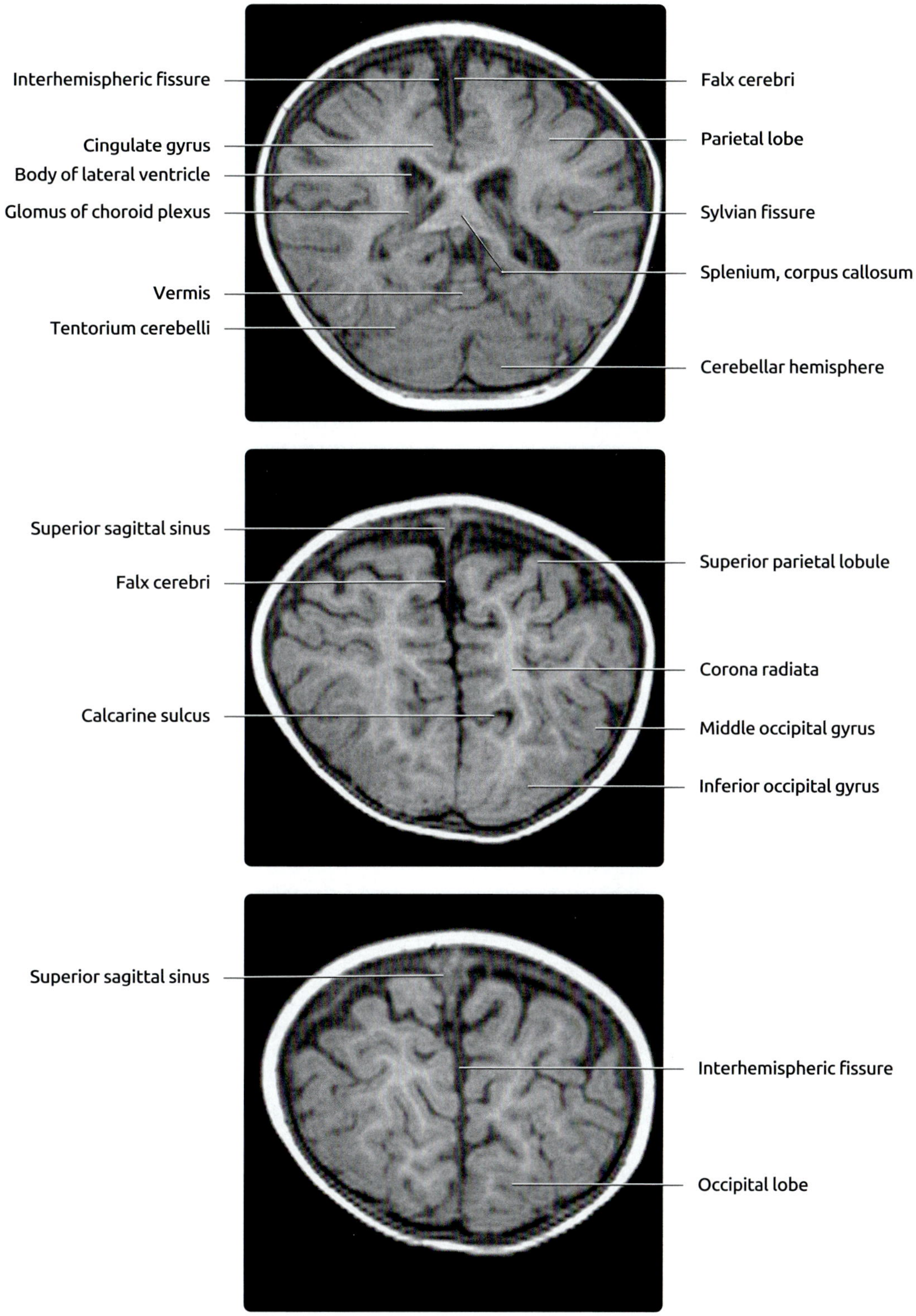

(Top) *The 7th of 9 coronal T1 MR images through the cerebral hemispheres from anterior to posterior are shown. The images are taken through planes/levels corresponding to those commonly used for ultrasound scans through the anterior fontanelle. The glomus of the choroid plexus is prominent within the trigones of the lateral ventricles.* **(Middle)** *More posterior image shows the posterior parietal lobes and occipital lobes. Cerebral hemispheres are separated by the interhemispheric fissure, which contains the falx cerebri. The ventricular system and cerebellum are no longer seen at this level. The primary visual cortex is on the medial aspect of the occipital lobe.* **(Bottom)** *This most posterior image shows gyri of the occipital lobe and a portion of the superior sagittal sinus, which arches posteriorly to the torcular Herophili.*

SAGITTAL US VIA ANTERIOR FONTANELLE

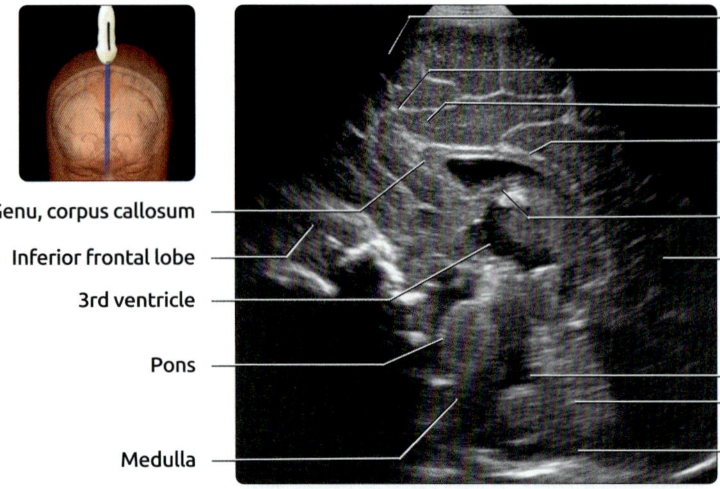

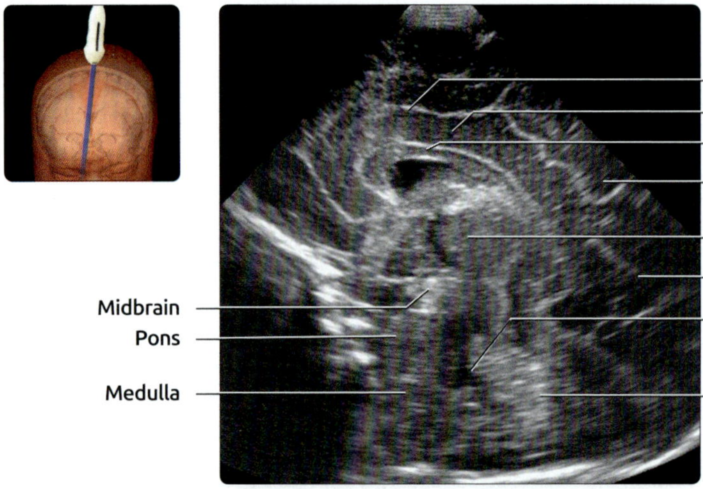

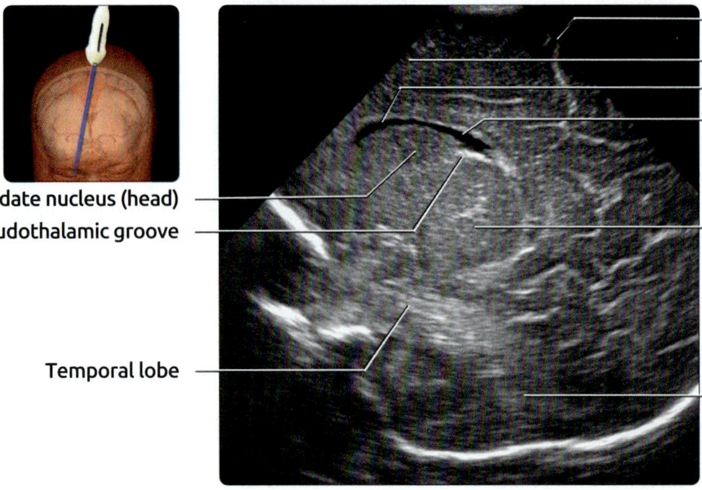

(Top) *The 1st of 6 sagittal ultrasounds of the brain through the anterior fontanelle in term infant is shown. This image, obtained in the midline, shows the corpus callosum as a hypoechoic curving line. Callosal and cingulate sulci are parallel to and above the corpus callosum. The midline also allows evaluation of the posterior fossa structures, including the brainstem anteriorly and the vermis posteriorly. The 4th ventricle is well seen in this plane and appears as a triangular fluid-filled structure at the level of the mid vermis.* **(Middle)** *Parasagittal image obtained angling slightly lateral to midline: The thalamus and echogenic sulci can now be seen more clearly.* **(Bottom)** *Parasagittal image obtained by angling more laterally shows the caudothalamic groove, the junction between the caudate nucleus and the thalamus. This is the area of the vascular germinal matrix, which is vulnerable to hemorrhage in preterm infants.*

SAGITTAL T1 MR

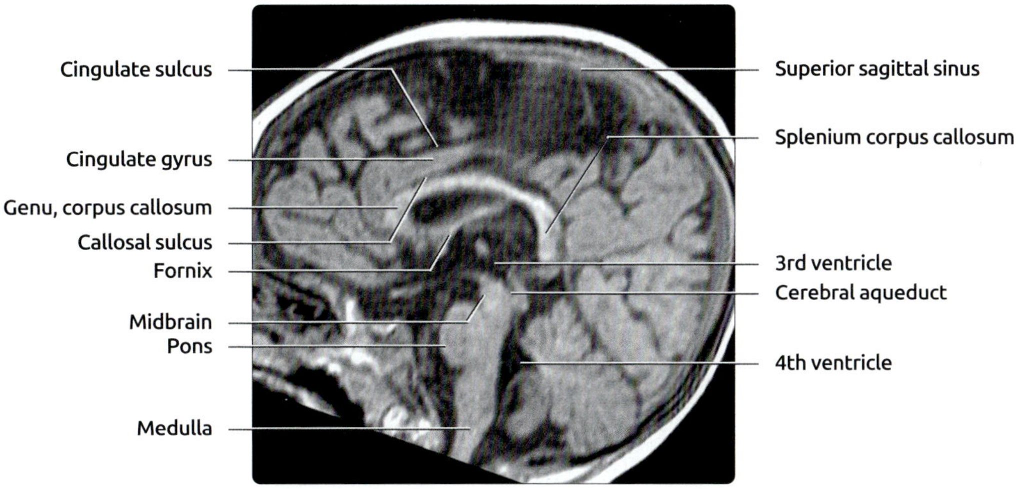

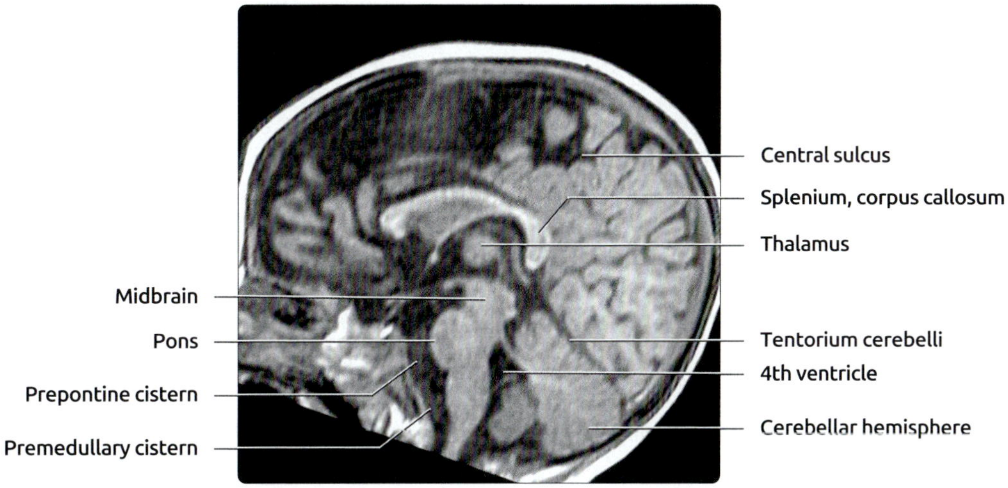

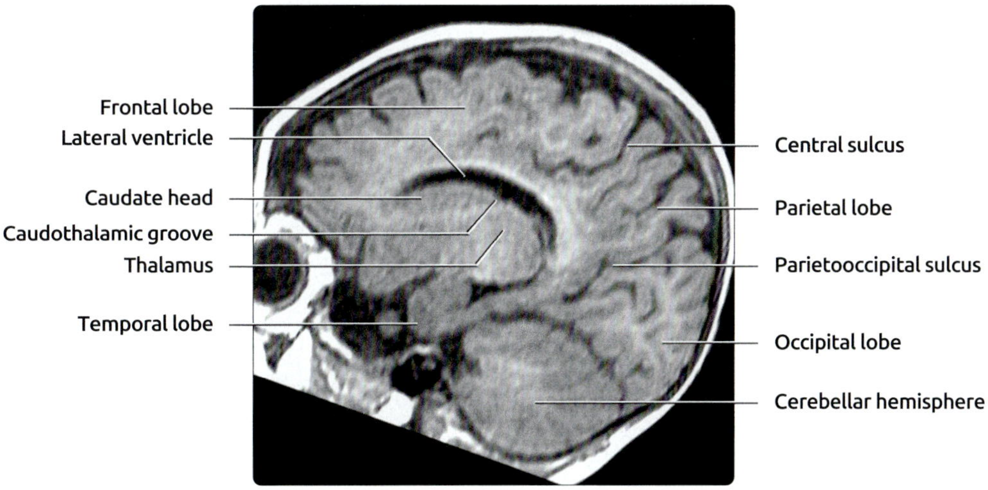

(Top) *The 1st of 6 sagittal T1 MR images through the cerebral hemispheres from midline to lateral is shown. The images are taken through planes/levels corresponding to those commonly used for ultrasound scans through the anterior fontanelle. The midline sagittal image shows the corpus callosum, the largest commissural fiber connecting both cerebral hemispheres.* **(Middle)** *Parasagittal image just off the midline is shown. The tentorium cerebelli is a dural fold separating the brain into supratentorial and infratentorial compartments.* **(Bottom)** *More lateral image shows the caudothalamic groove between the caudate head and thalamus. The parietooccipital sulcus is an important landmark, differentiating the parietal from the occipital lobes.*

SAGITTAL US VIA ANTERIOR FONTANELLE

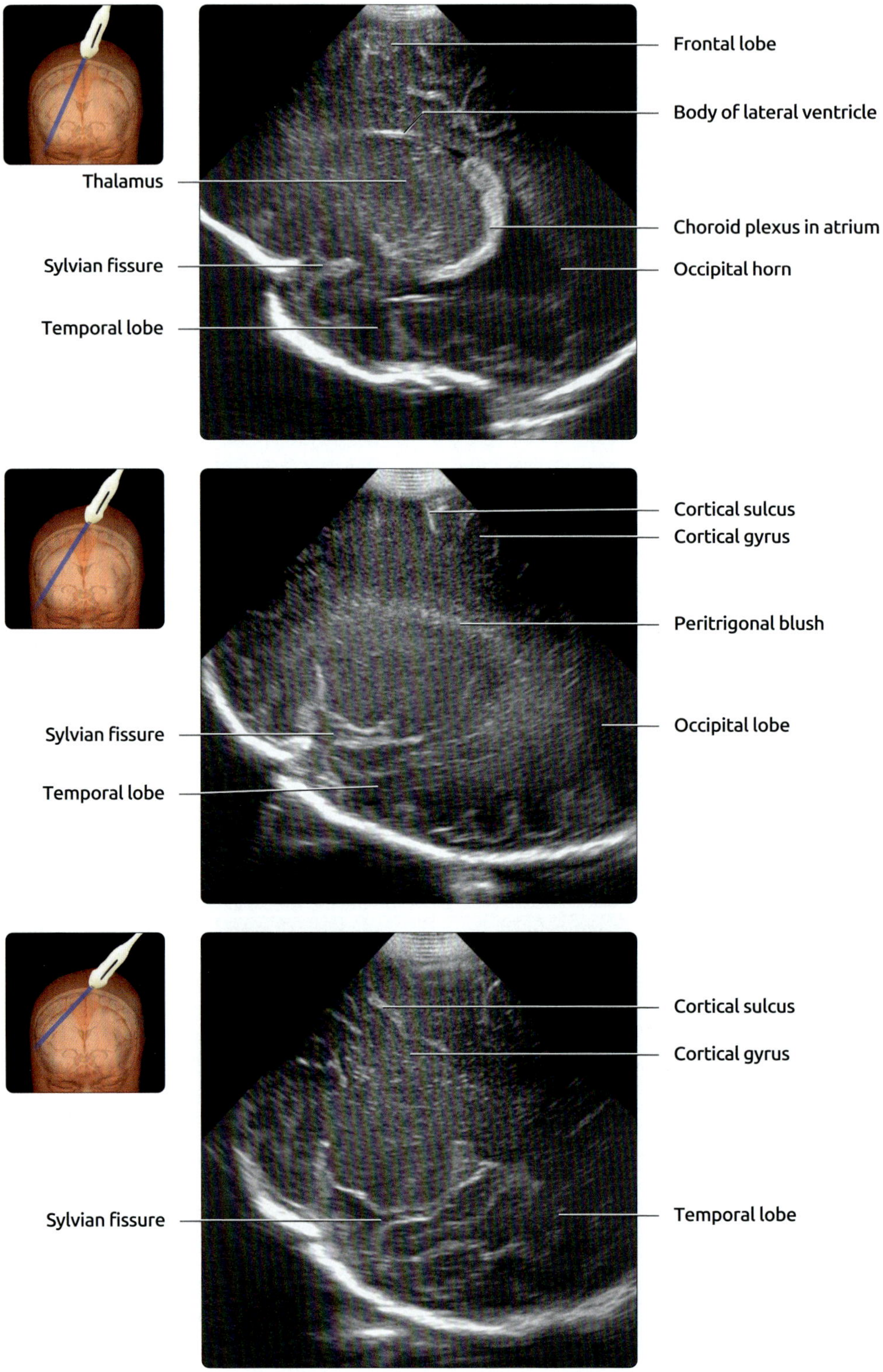

(Top) *This parasagittal image shows the glomus of the choroid plexus in the trigone. The glomus tapers anteriorly as it courses along the floor of the lateral ventricle to the foramen of Monro and continues along the roof of the 3rd ventricle. It also tapers posteriorly from the trigone into the temporal horn of each lateral ventricle. Glomus may appear bulbous and irregular at the trigone and should not be mistaken as a blood clot.* **(Middle)** *This parasagittal image is obtained just lateral to the lateral ventricle. The echogenic white matter of the brain just posterior and superior to the ventricular trigone is known as the peritrigonal blush or halo, representing radiating white fiber tracts (corona radiata). The peritrigonal blush is more prominent in premature than in term neonates.* **(Bottom)** *This is the last and most lateral sagittal image obtained, showing the mature sulcal pattern with hyperechoic sulci and hypoechoic gyri.*

SAGITTAL T1 MR

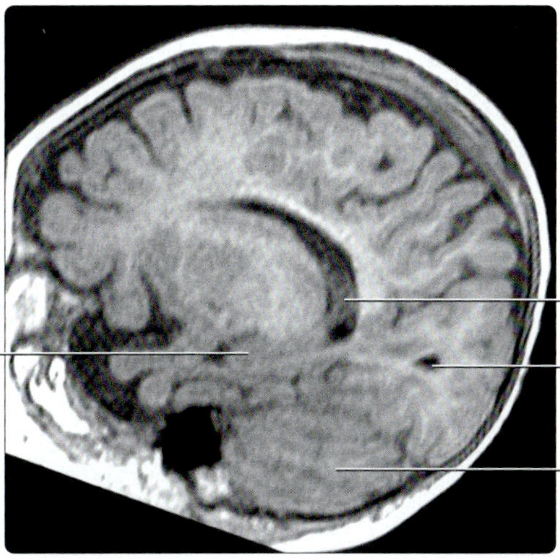

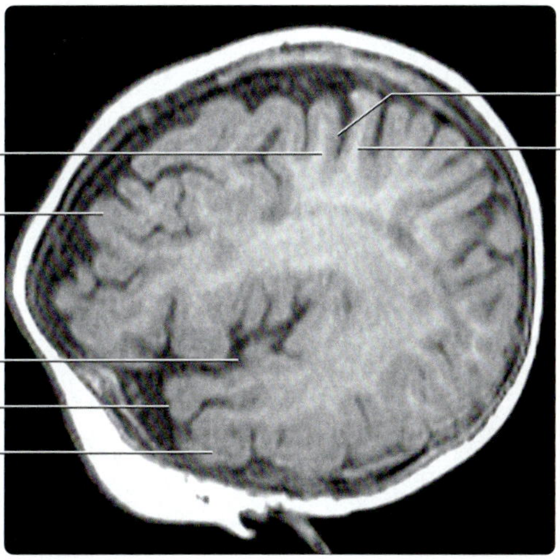

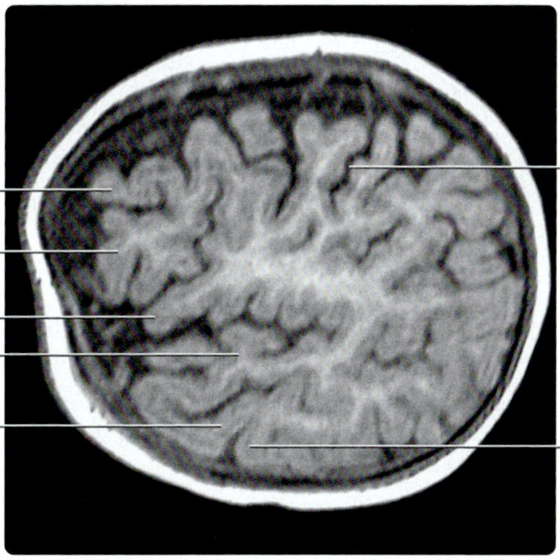

(Top) *This image shows a prominent choroid plexus within the atrium of the lateral ventricle, which tapers posteriorly and extends into the temporal horn.* **(Middle)** *This parasagittal image shows the sylvian fissure bound superiorly by the frontal operculum and inferiorly by the temporal operculum. The central sulcus separates the frontal lobe anteriorly from the parietal lobe posteriorly.* **(Bottom)** *This image shows the most lateral portion of the sylvian fissure. The temporal lobe is inferior to the sylvian fissure. The superior temporal gyrus contains the primary auditory cortex. The middle temporal gyrus connects auditory, somatosensory, and visual association pathways. The inferior temporal gyrus is the higher visual association area.*

PREMATURE INFANT (23 WEEKS 6 DAYS)

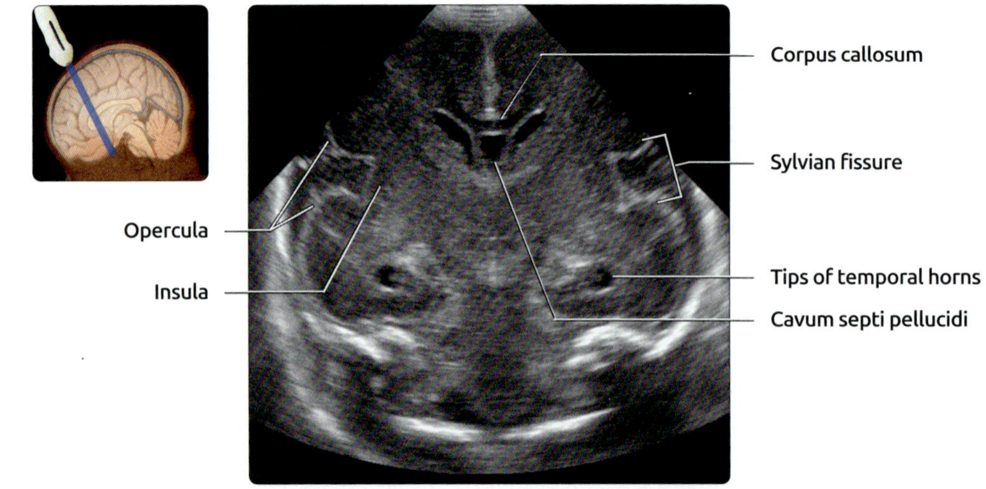

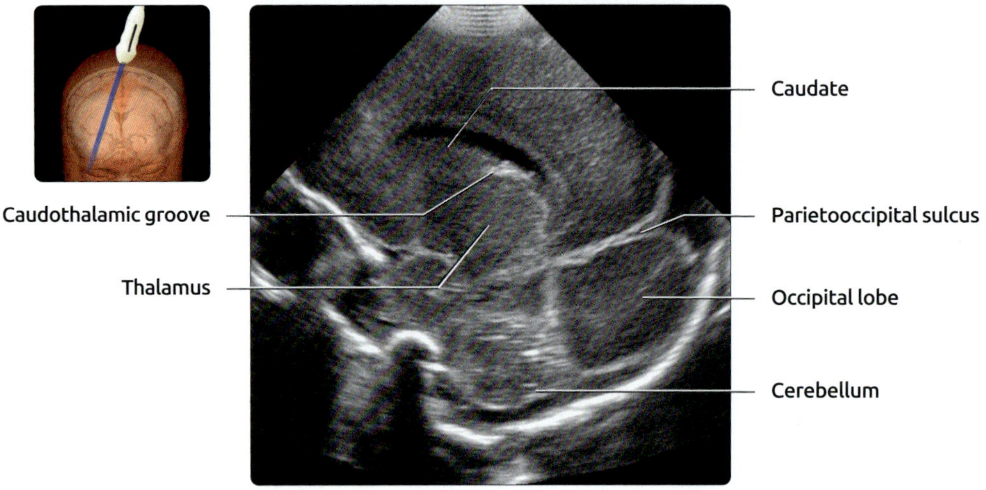

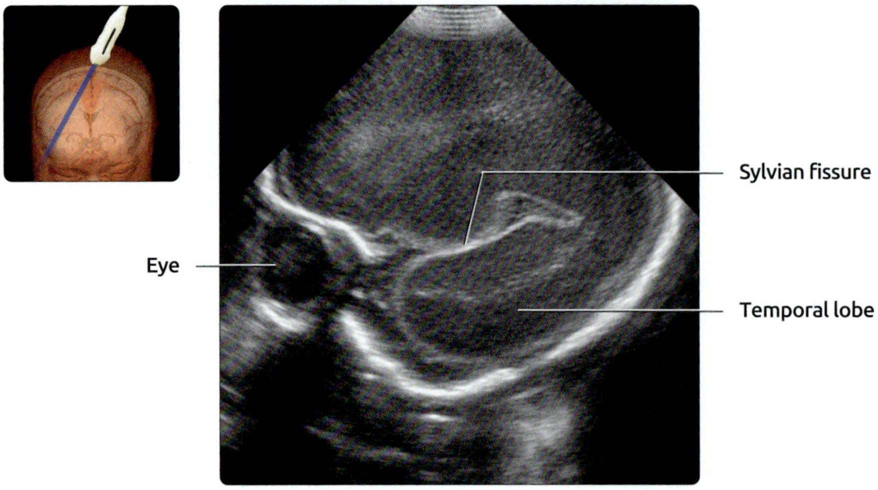

(Top) *This coronal image of a very premature infant, born at 23-weeks 6-days gestational age, shows a very large, square, open sylvian fissure. The opercula have not yet grown to cover the insula.* **(Middle)** *Sagittal image through the caudothalamic groove in the same case shows the parietooccipital sulcus. The cortex otherwise appears "flat" without gyri/sulcal formation.* **(Bottom)** *Another sagittal image further lateral shows similar findings with no cortical gyri/sulci seen.*

SYLVIAN FISSURE AT DIFFERENT AGES

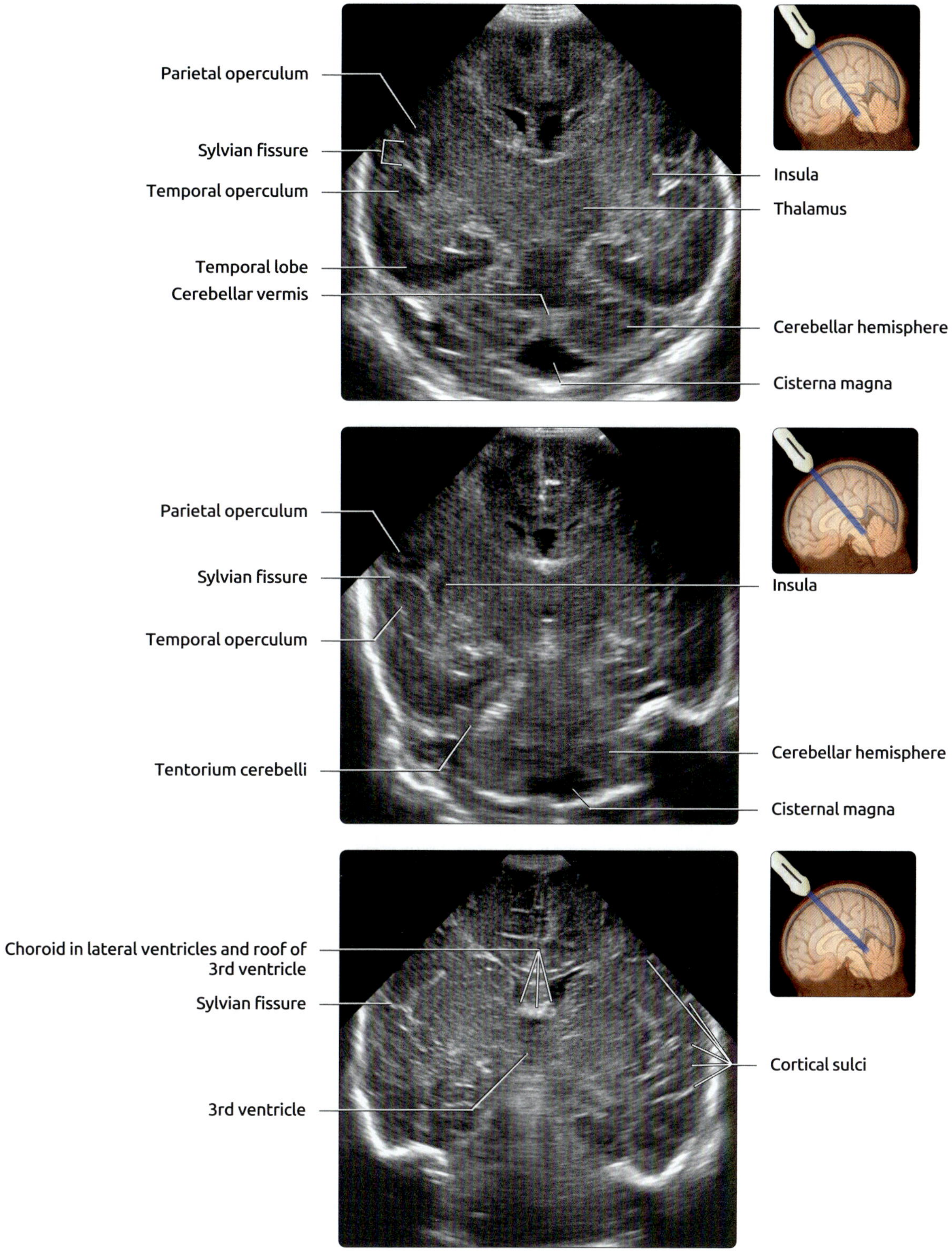

(Top) *A different infant born at 29 weeks 1 day shows more advanced development of the sylvian fissures. The frontal, temporal, and parietal lobes all have opercula, which have grown to cover the insula.* **(Middle)** *At 31 weeks 6 days the opercula have grown to cover the insula.* **(Bottom)** *Another coronal image through the level of the sylvian fissure in a full-term infant shows multiple gyri and sulci over the convexities of the brain. It is important to understand the developmental anatomic changes; lack of cortical sulci may be normal for preterm infants, depending on the gestational age at delivery, but is very abnormal at term.*

SAGITTAL US VIA POSTERIOR FONTANELLE

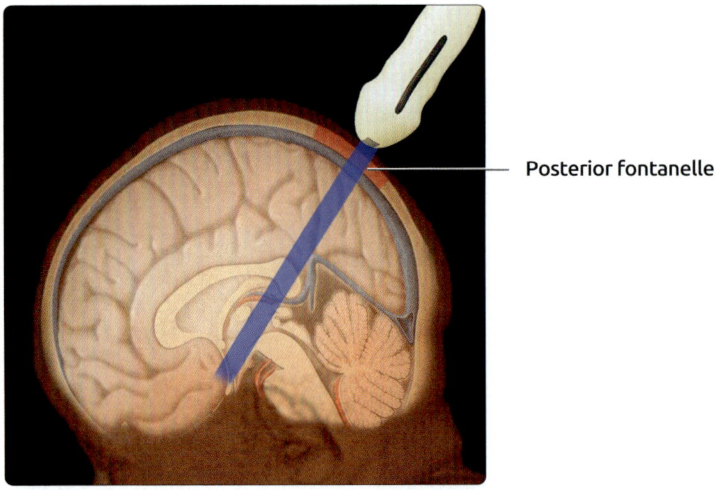

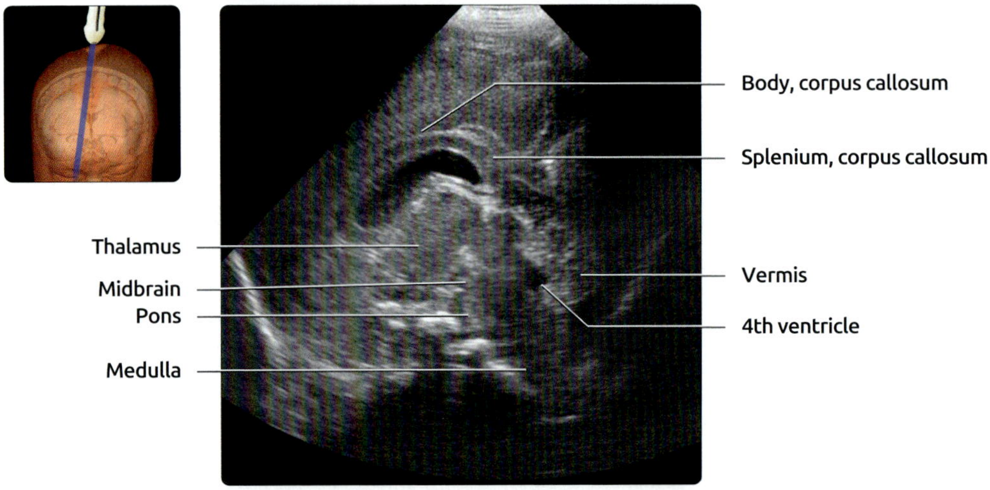

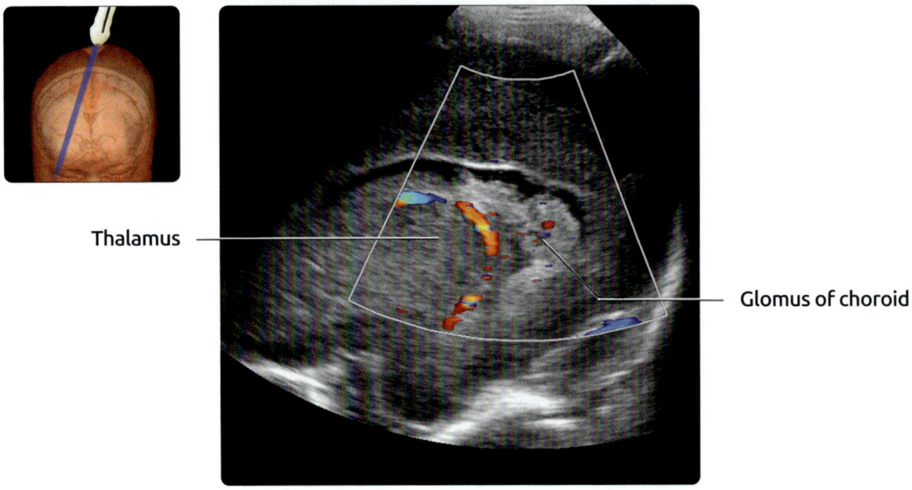

(Top) *Although routine scanning is performed via the anterior fontanelle, the posterior fontanelle is another alternative, particularly when it is difficult seeing more posterior structures in the brain.* **(Middle)** *This scan through the posterior fontanelle in a 26-week premature infant was performed to better evaluate the corpus callosum. The splenium is particularly well seen in this view.* **(Bottom)** *Color Doppler image shows flow within the choroid plexus of the glomus. The posterior fontanelle view can be helpful to differentiate bulky choroid from clot. The occipital horn does not contain choroid plexus, and any echogenic material in the occipital horn should raise the suspicion of intraventricular hemorrhage.*

AXIAL US THROUGH TEMPORAL BONE

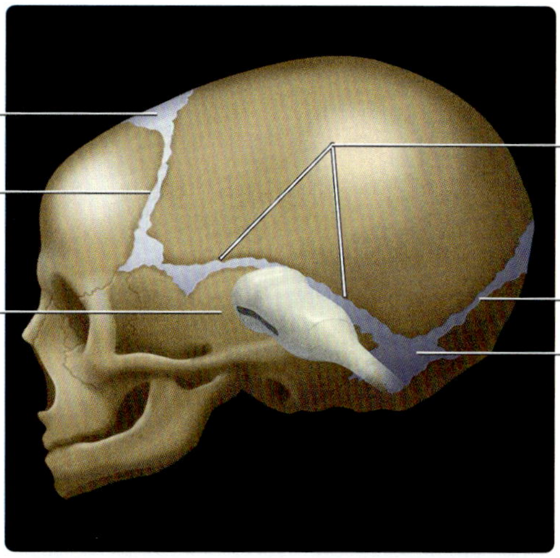

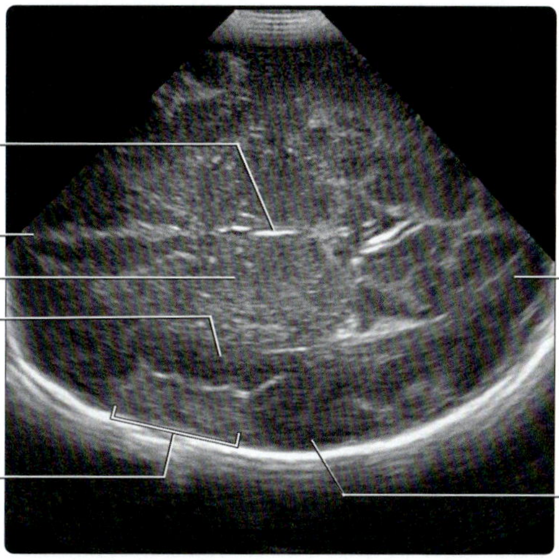

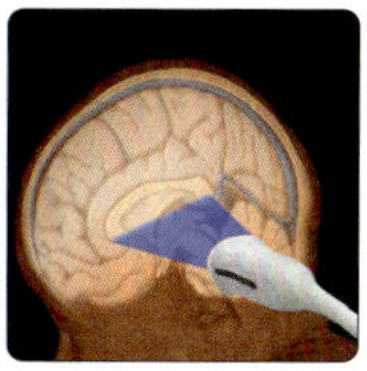

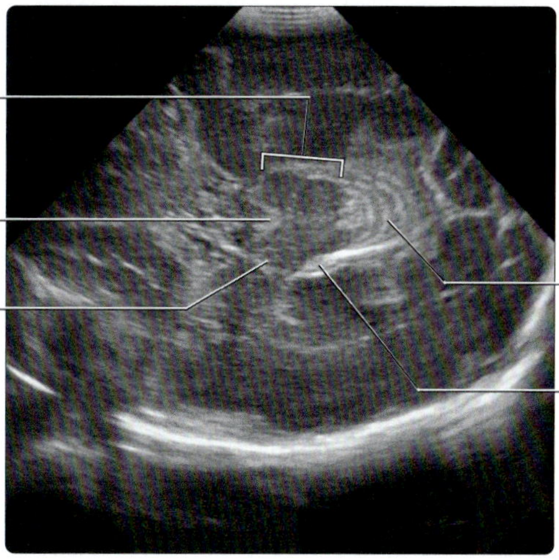

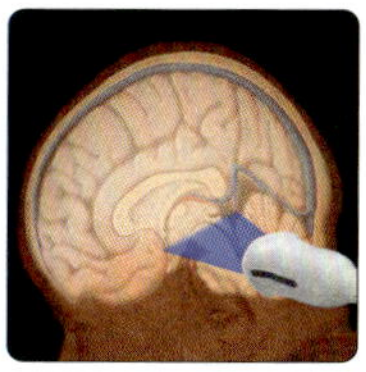

(Top) *Graphic of the transtemporal acoustic window is shown. The transducer is placed more anterior and superior than the mastoid fontanelle approach. The temporal bone anterior to the ear is thin enough to allow imaging of the brainstem even after closure of the temporosquamosal suture. This acoustic window allows the best assessment of cerebral peduncles and the 3rd ventricle.* **(Middle)** *Transtemporal axial scan in a 29-week premature infant shows intracranial anatomy in a plane similar to CT or MR.* **(Bottom)** *The temporal bone anterior to the ear is thin, allowing imaging of the brainstem even after sutural closing. This is the best view for the cerebral peduncles and midbrain.*

CEREBELLUM AND POSTERIOR FOSSA

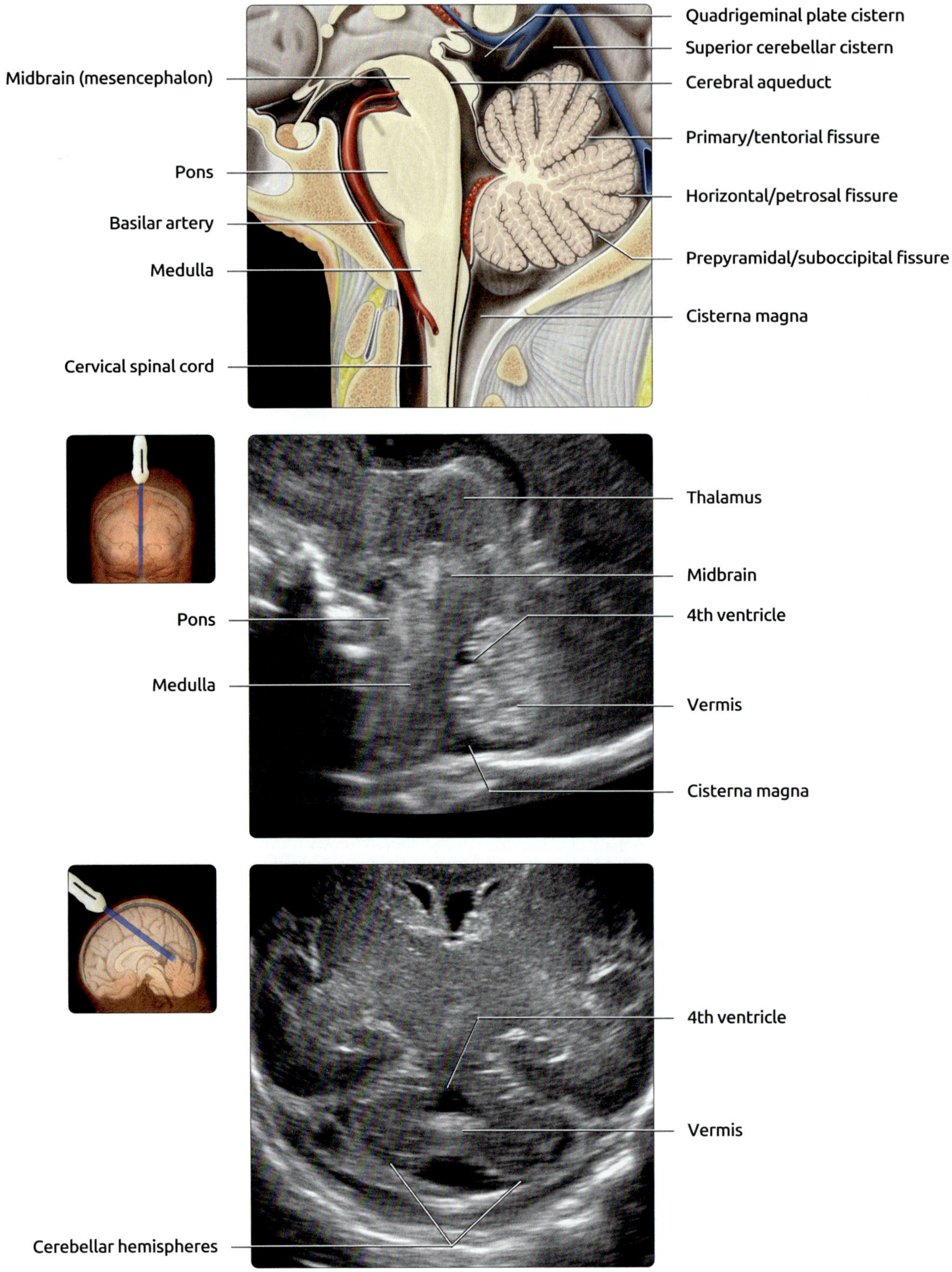

(Top) *Sagittal midline graphic of the posterior fossa shows the anterior brainstem and posterior cerebellum separated by the 4th ventricle. The brainstem consists of midbrain (mesencephalon), pons, and medulla. The cerebellum has superior (tentorial), inferior (suboccipital), and anterior (petrosal) surfaces. The primary (tentorial) fissure and horizontal (petrosal) fissures divide the vermis and cerebellar hemispheres into lobules.* **(Middle)** *This midline sagittal view shows the brainstem. The pons is easily identified by its anterior bulge.* **(Bottom)** *A coronal image through the anterior fontanelle of a premature infant, born at 29 weeks 1 day, shows symmetric cerebellar hemispheres. The vermis is midline, covers the 4th ventricle, and is more echogenic than the rest of the cerebellum.*

POSTERIOR FOSSA VIA MASTOID APPROACH

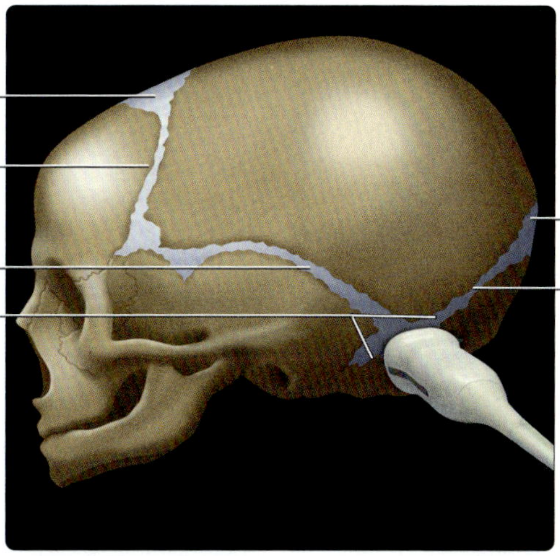

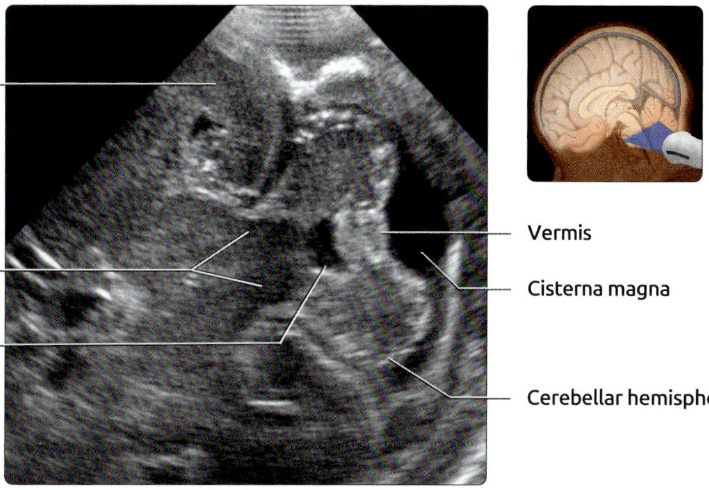

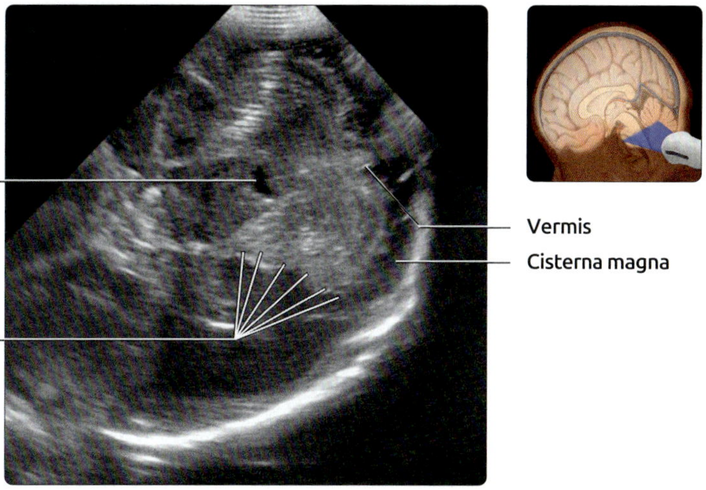

(Top) *The mastoid/posterolateral fontanelle is located at the junction of temporosquamosal, lambdoidal, and occipital sutures. It allows assessment of brainstem and posterior fossa structures, which are not well demonstrated in the standard planes through the anterior fontanelle. The transducer is placed ~ 1 cm behind the helix of ear and 1 cm above the tragus. This acoustic window allows the best visualization of 4th ventricle, posterior cerebellar vermis, cerebellar hemispheres, and cisterna magna.* **(Middle)** *The mastoid approach allows for detailed evaluation of the posterior fossa structures. This is a premature infant (27 weeks 5 days), which is evident by the lack of cortical gyri and cerebellar folia.* **(Bottom)** *Another mastoid view in an infant born at 37 weeks shows maturation with extensive folia on the surface of the cerebellum.*

CAVUM SEPTI PELLUCIDI ET VERGAE

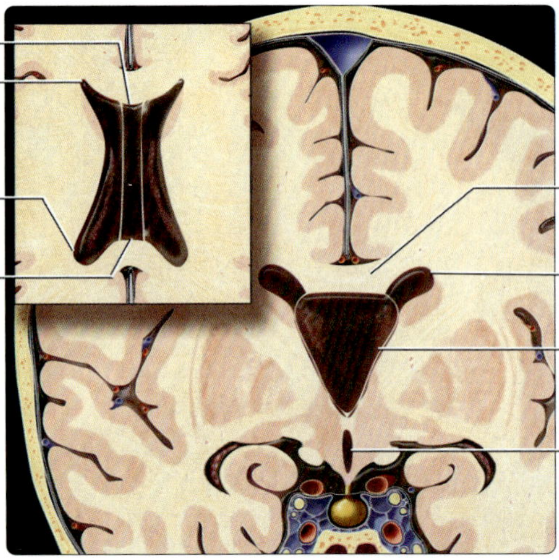

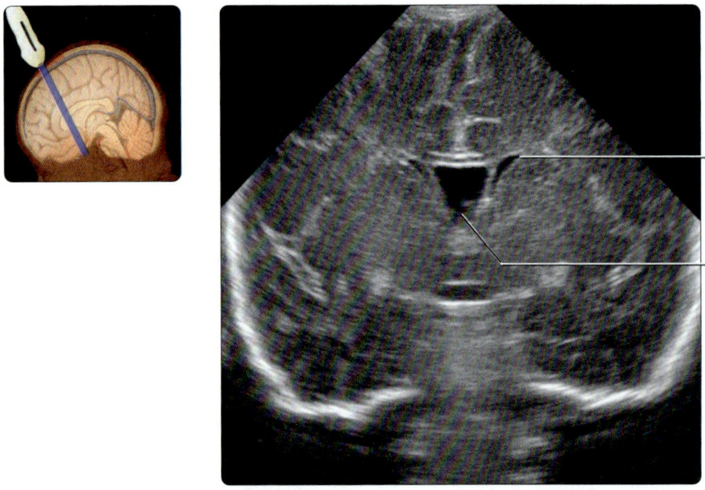

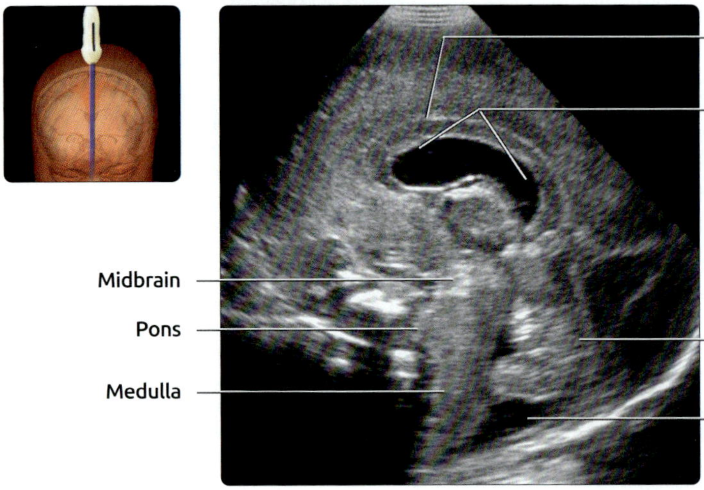

(Top) *Coronal graphic with an axial insert shows a classic cavum septi pellucidi with a posterior extension, the cavum vergae. It creates a finger-like CSF collection between the lateral ventricles.* **(Middle)** *The cavum septi pellucidi can be quite large, especially in premature infants and should not be confused with an elevated 3rd ventricle or intracranial cyst.* **(Bottom)** *Midline sagittal image in this 27-week premature infant shows a cavum septi pellucidi continuing posteriorly into the cavum vergae. This is a common finding in premature infants. The cavum vergae is closed in 97% of full-term infants and the cavum septi pellucidi is closed in 85% of infants by 3-6 months of age; however, it can remain open until adulthood.*

CAVUM VELUM INTERPOSITUM

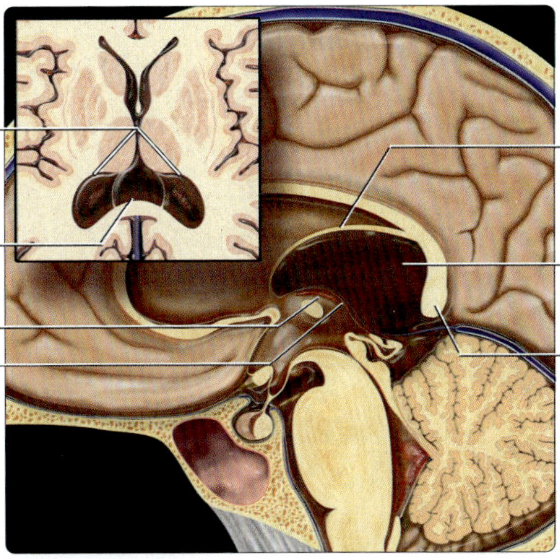

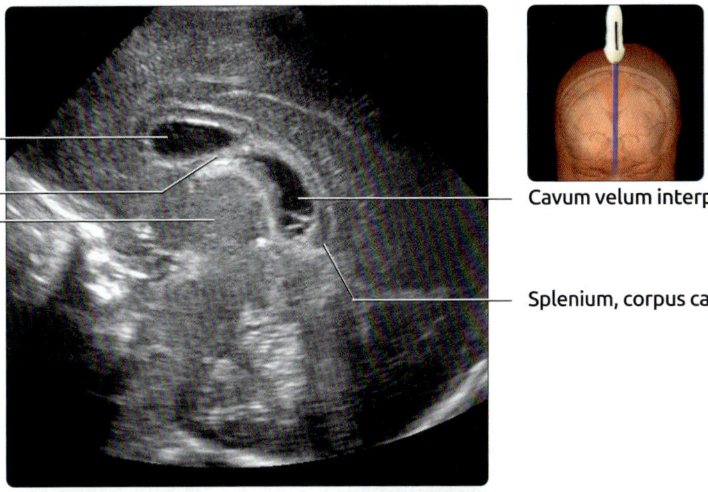

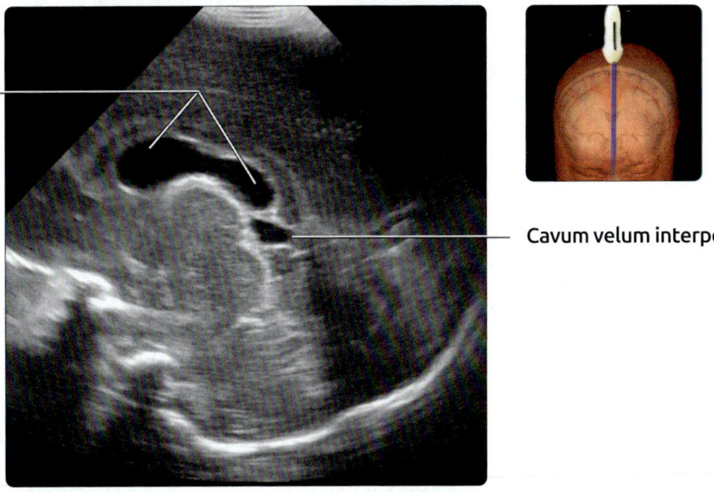

(Top) *Sagittal graphic with an axial insert shows a cavum velum interpositum. Note the elevation and splaying of the fornices. Also noted is the inferior displacement of the internal cerebral veins and 3rd ventricle.* **(Middle)** *Midline sagittal ultrasound shown a mildly complex cavum velum interpositum.* **(Bottom)** *This premature infant has a cavum septi pellucidi, vergae, and interpositum. Like a cavum septi pellucidi and vergae, a cavum velum interpositum is more common in premature infants.*

VASCULAR ANATOMY

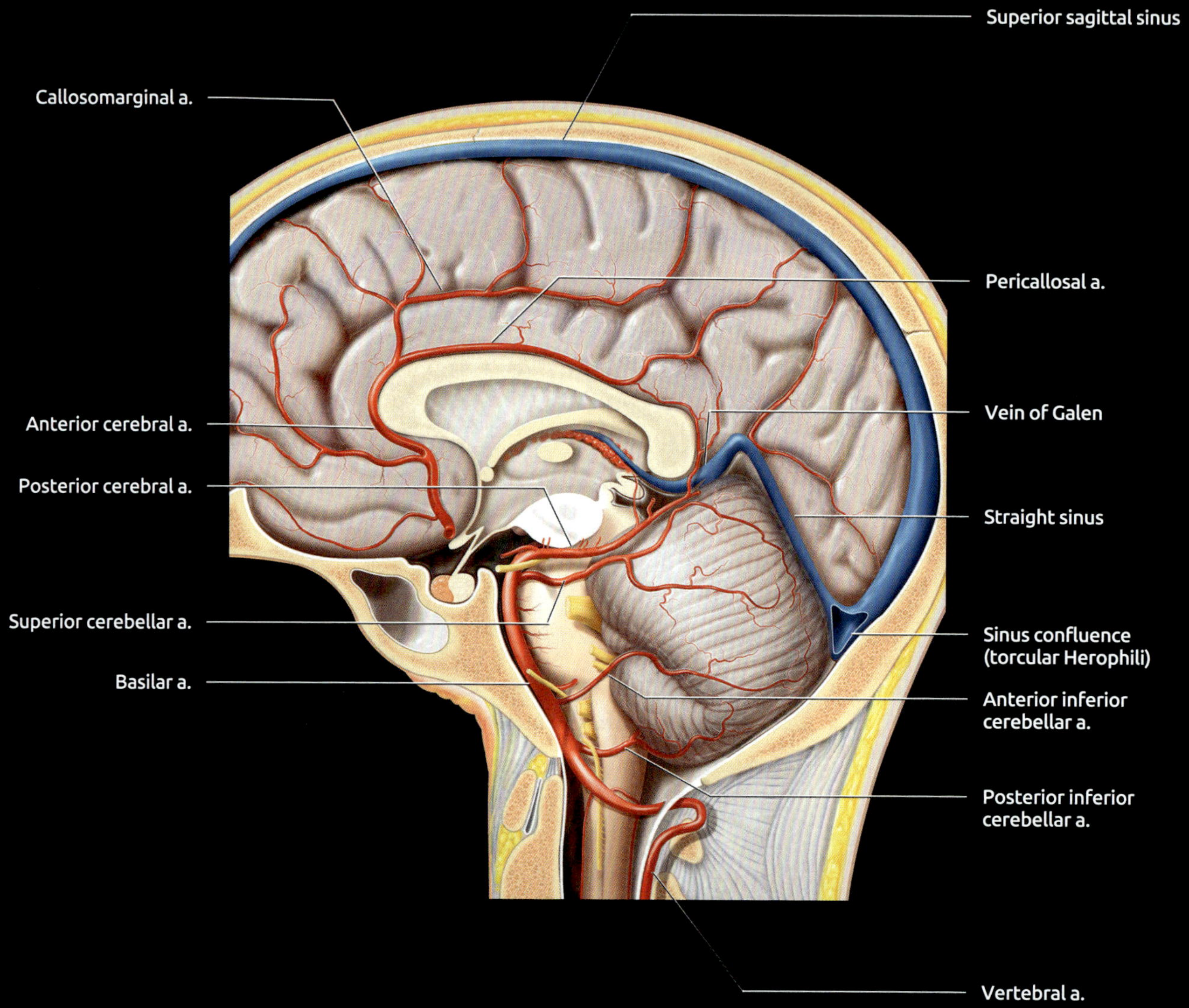

This graphic shows the arteries, sinuses, and veins, which can be seen seen on a routine midline sagittal view. The anterior cerebral artery and its 2 main branches, the pericallosal and callosomarginal arteries, are easily seen on midline color Doppler ultrasound. The basilar artery is also easily identified, running anterior to the brainstem. The middle cerebral and posterior cerebral arteries are better evaluated in an axial plane using a transtemporal or mastoid approach.

MIDLINE COLOR DOPPLER

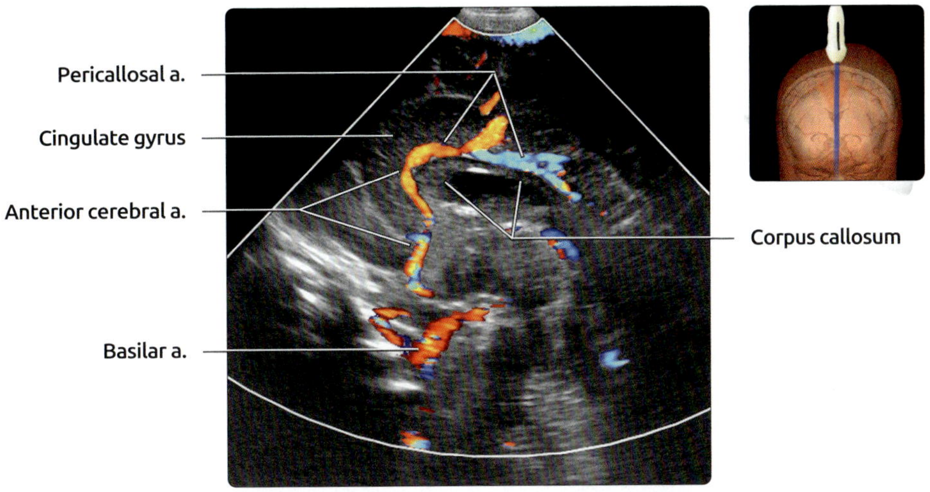

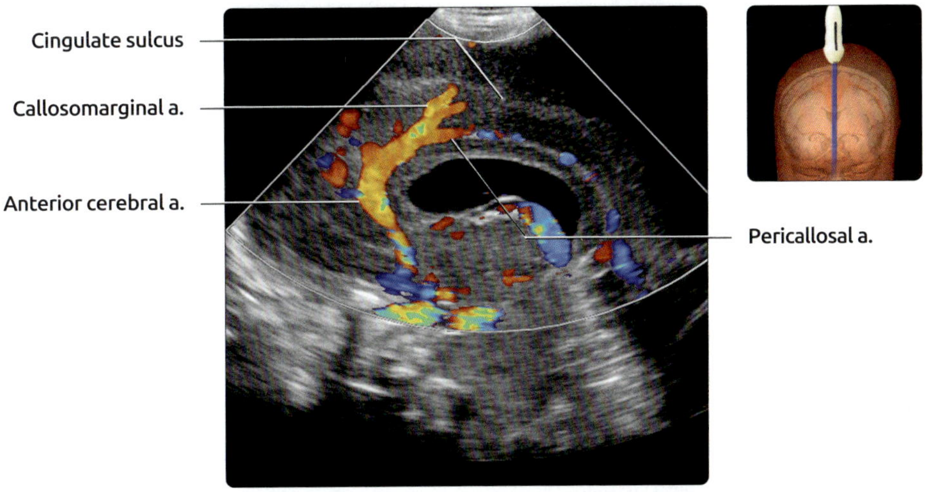

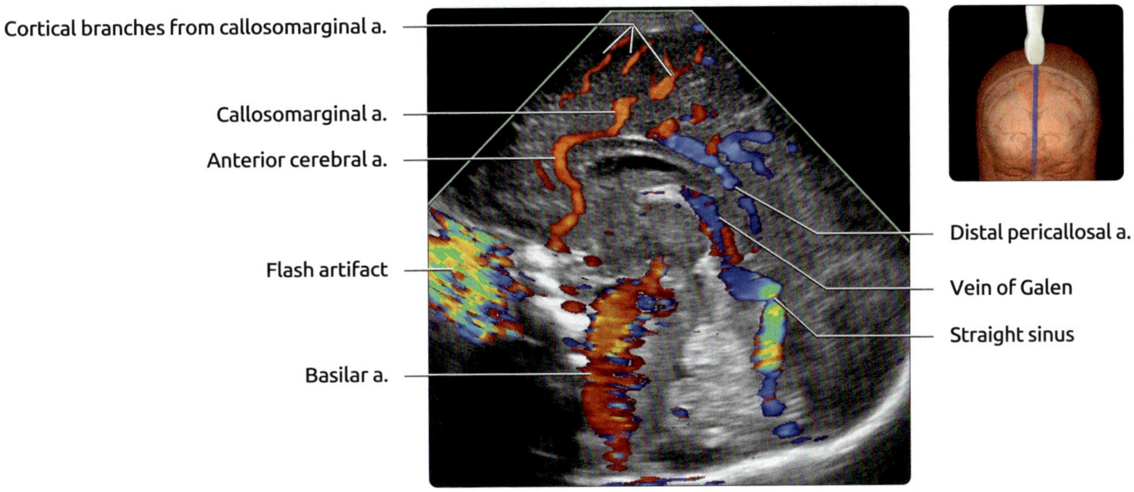

(Top) *Midline sagittal color Doppler image, obtained via the anterior fontanelle, shows the pericallosal artery running in the callosal sulcus, just above the corpus callosum. In a normal newborn, the pericallosal artery should be close to the surface of the corpus callosum. While in callosal agenesis, this artery remains far from the 3rd ventricle and takes an upward oblique direction.* **(Middle)** *The anterior cerebral artery divides into the pericallosal artery, which continues along the corpus callosum, and the callosomarginal artery, which courses superiorly to travel above the cingulate gyrus within the cingulate sulcus.* **(Bottom)** *Multiple cortical branches of the callosomarginal artery are seen traveling within cortical sulci.*

CIRCLE OF WILLIS

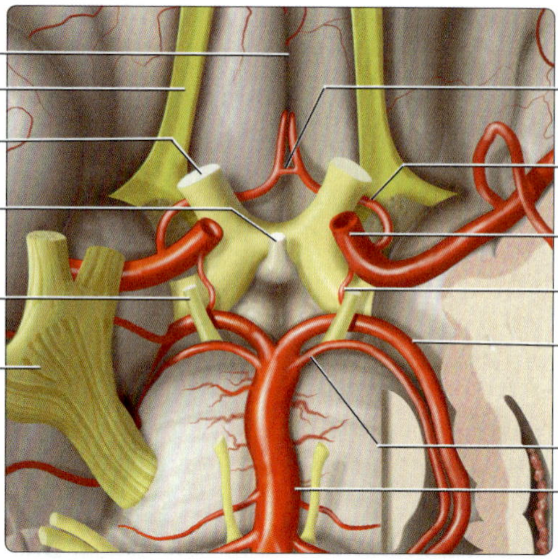

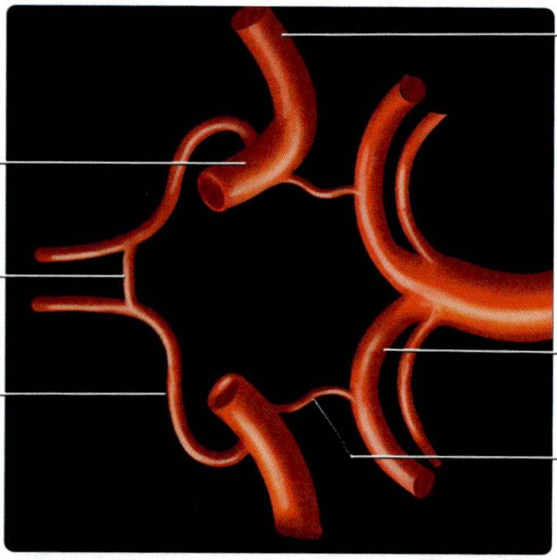

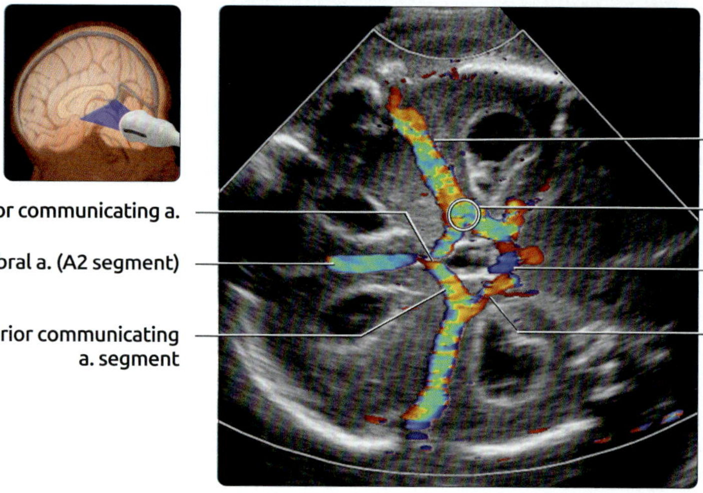

(Top) *The circle of Willis illustrated in situ shows its intricate relationship to adjacent structures. It located in the suprasellar cistern just below the diencephalon. The hypothalamus, infundibular stalk, and optic chiasm lie in the middle of the circle. The horizontal (A1) anterior cerebral artery segment passes above the optic nerves (CNII); the posterior communicating artery passes above the oculomotor nerves (CNIII). The anterior communicating artery is near the midline, below the interhemispheric fissure.* **(Middle)** *The circle of Willis is shown in isolation and turned 90° counterclockwise to match the plane of the ultrasound.* **(Bottom)** *Transtemporal axial color Doppler in a premature infant with ventriculomegaly from an intracranial hemorrhage shows the circle of Willis. The anterior and middle cerebral arteries are the terminal branches of the internal carotid artery. The posterior cerebral artery is the terminal branch of the basilar artery. These 3 crucial arteries communicate via the anterior and posterior communicating arteries in a complete circle of Willis. The transtemporal approach provides the best window for evaluating the circle of Willis.*

CEREBRAL ARTERIES

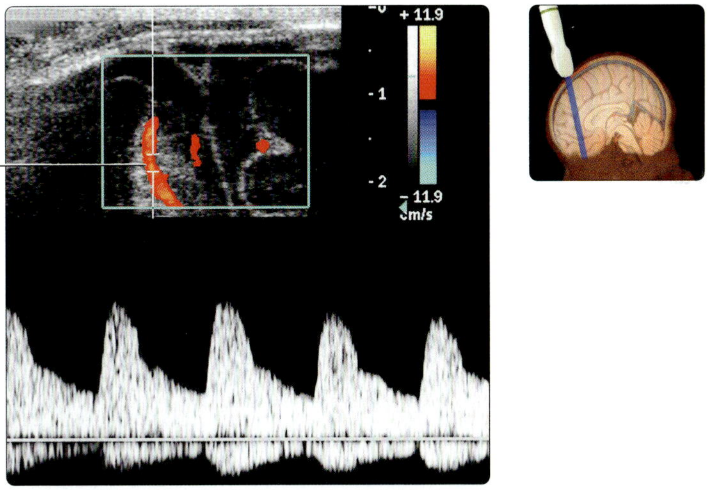

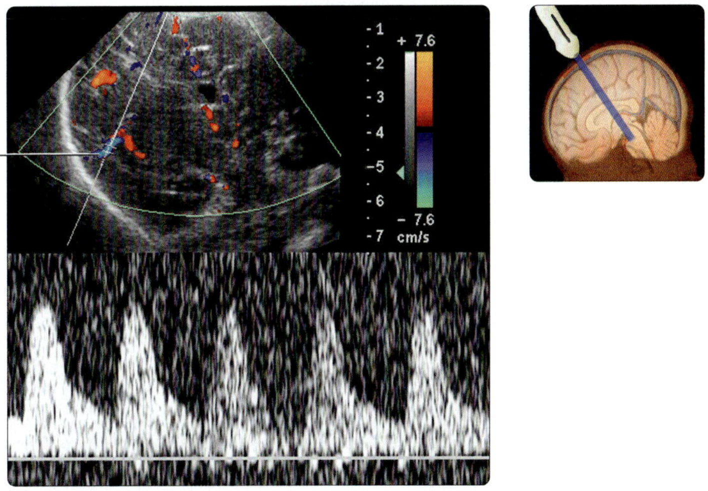

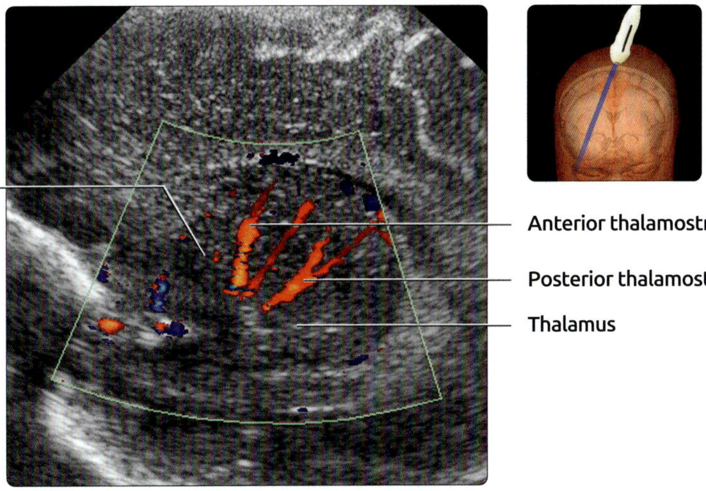

(Top) *Doppler waveform of a cortical branch of the anterior cerebral artery shows a low-resistance waveform with abundant diastolic flow. Velocity of the cerebral artery is a reliable reflection of intracranial pressure.* **(Middle)** *Doppler waveform of the middle cerebral artery in the coronal plane is obtained by angling the beam laterally toward the sylvian fissure. A low-resistance arterial waveform is again noted.* **(Bottom)** *Color Doppler ultrasound obtained by a parasagittal scan through the anterior fontanelle shows the thalamostriate arteries. Anteriorly, the caudate nucleus is supplied by the anterior thalamostriate artery while the thalamus posteriorly is supplied by the posterior thalamostriate artery. The thalamostriate arteries arise from the middle cerebral artery.*

CEREBRAL VEINS AND SINUSES

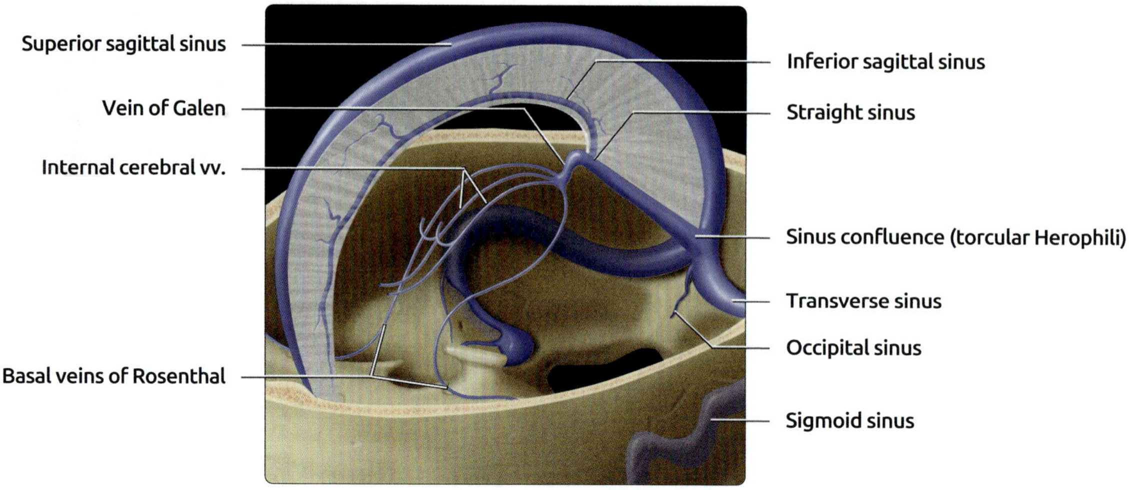

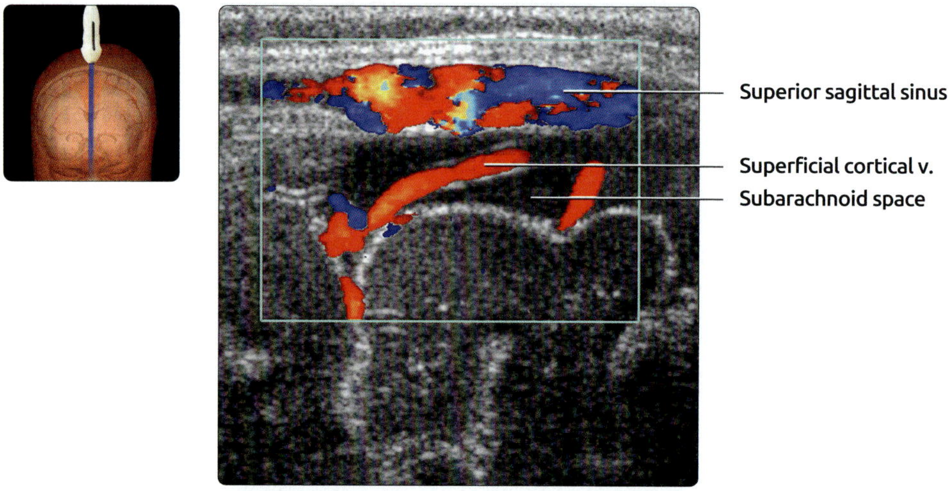

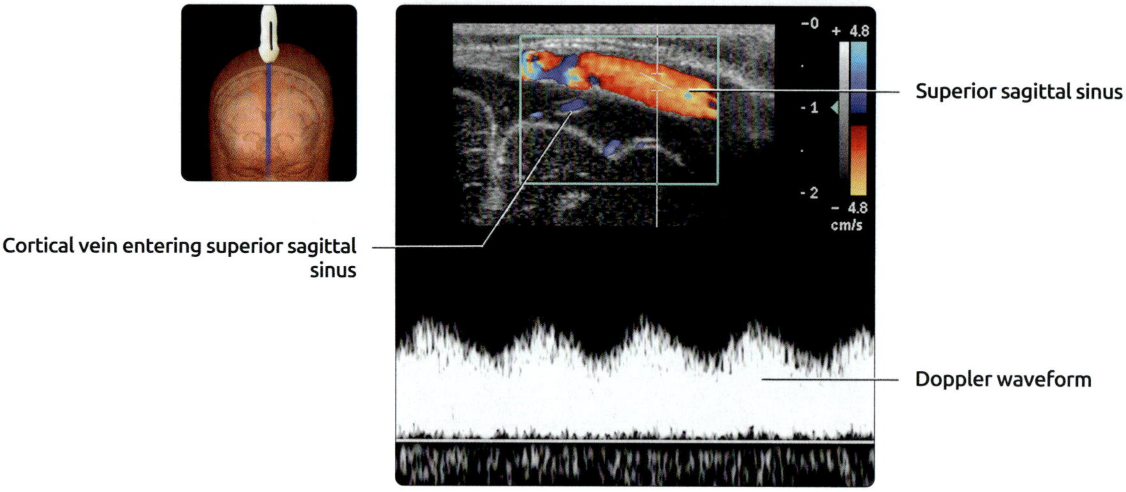

(Top) *This graphic shows the the connections between the major dural sinuses and deep cerebral veins. The internal cerebral veins and basal veins of Rosenthal drain into the vein of Galen, which in turn drains into the straight sinus.* **(Middle)** *Sagittal color Doppler scan through the anterior fontanelle shows a superficial cortical vein traversing the subarachnoid space and draining into the superior sagittal sinus.* **(Bottom)** *Spectral Doppler waveform of the superior sagittal sinus shows the waveform of the sinus to be pulsatile, under the effect of transmitted cardiac pulsations.*

CEREBRAL VEINS AND SINUSES

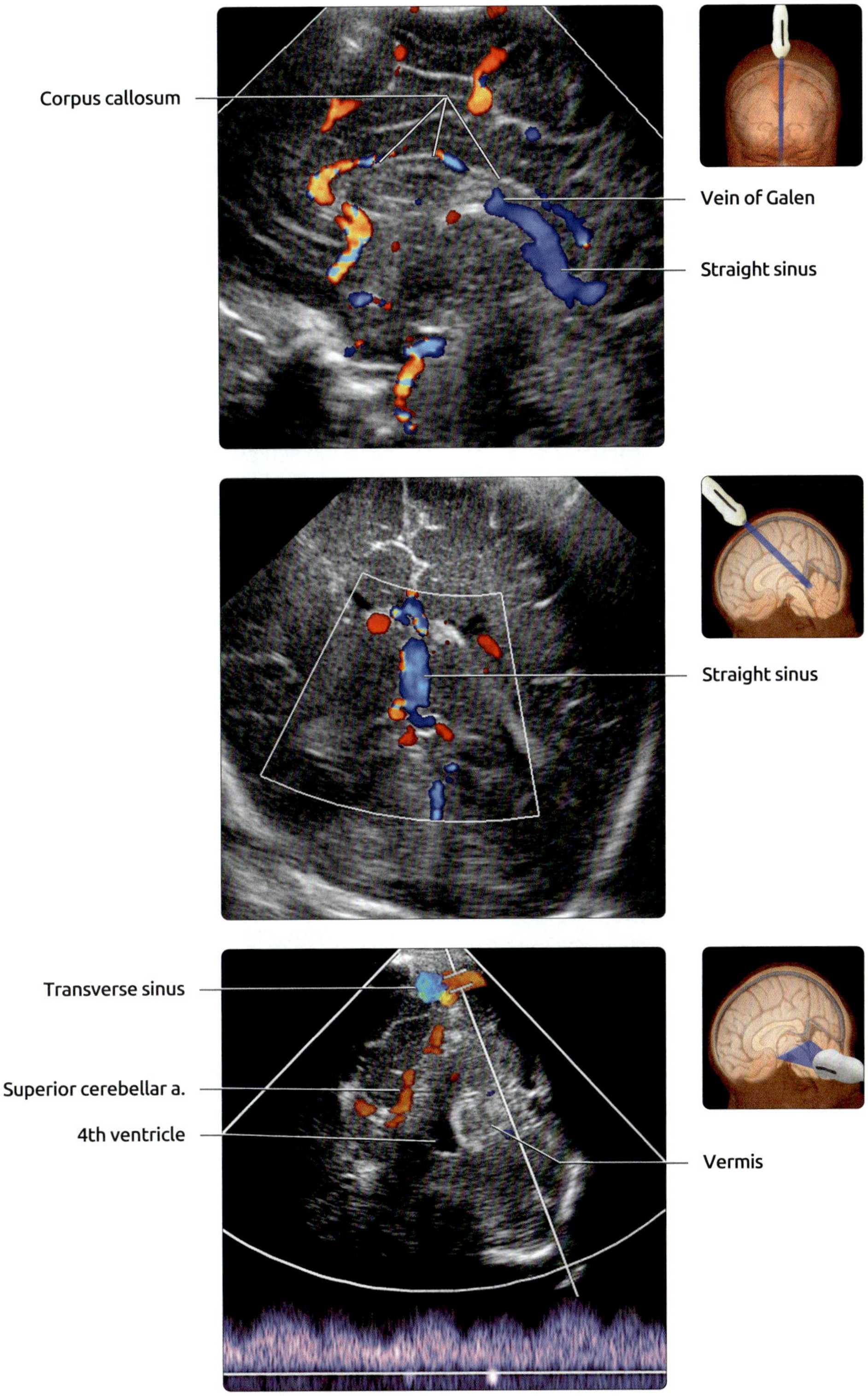

(Top) *Color Doppler ultrasound obtained by a midline sagittal scan through the anterior fontanelle shows the relationship of the vein of Galen and straight sinus. The vein of Galen is seen under the splenium of the corpus callosum. It receives drainage from the paired internal cerebral veins and basal veins of Rosenthal. The vein of Galen continues inferiorly into the straight sinus. The straight sinus, superior sagittal sinus, and transverse sinuses converge posteriorly, forming the torcular Herophili.* **(Middle)** *In the angled coronal plane, the straight sinus is seen in the midline between the lateral ventricles.* **(Bottom)** *Other sinuses can be evaluated using different acoustic windows. The transverse sinus can be accessed via a mastoid fontanelle approach, as shown here. The cerebral venous system is valveless, and pulsed Doppler waveforms typically show cardiac pulsations.*

GROSS ANATOMY

Overview

- Eye is embedded in ocular fat, moved by ocular muscles, and surrounded by bony orbit
- **Anterior segment of globe**
  - Cornea: Transparent; covers iris and pupil
  - Anterior chamber: Directly behind cornea; filled with aqueous fluid for maintaining ocular pressure
  - Iris: Colored diaphragm in anterior chamber
  - Pupil: Opening in center of iris
  - Lens: Behind pupil; capable of image focusing
- **Posterior segment of globe**
  - Vitreous body (humor): Clear, jelly-like substance in posterior chamber
  - Retina: Inner wall membrane of globe; receives image and converts it into nerve impulses
  - Macula: Center of retina; daylight and color vision
  - Fovea: Macular center; highest spatial resolution
  - Sclera: Dense, fibrous outer coating of globe
  - Choroid: Vascular layer between retina and sclera
- **Optic nerve**: Transmits nerve impulses from retina to brain
  - Surrounded by dural sheath
  - Extends posteriorly to optic chiasm and tract

IMAGING ANATOMY

Internal Contents

- **Ocular structures**
  - Lens: Well-defined, oval, echolucent structure behind iris
  - Vitreous cavity: Echolucent posterior compartment
  - Retina: Layer subjacent to vitreous
  - Macula: Appears as short, bright reflector
  - Sclera: Highly reflective; thickness ↑ with ↓ eye size (anterior: 0.6 mm and posterior: 1.0 mm)
  - Choroid: Normally indistinguishable from overlying retina and underlying sclera; vascular with color Doppler
- **Extraocular muscles**
  - **Rectus muscles**: Periglobal, hypoechoic structures with origin from annulus of Zinn
    - Medial rectus: Thickest and easiest to examine
    - Lateral rectus: Patient's nose may interfere with transducer movement during examination
    - Inferior rectus: Difficult to examine
    - Superior rectus and levator palpebrae muscles: Usually depicted as 1 complex
  - Superior and inferior oblique muscles difficult to visualize
- **Nerves**
  - **Optic nerve: Cranial nerve II (CNII)**
    - Hypoechoic linear structure exiting globe posteromedially
    - Easily identified at equator of eye
  - Remainder of intraocular nerves not usually identify by ultrasound unless enlarged
    - Oculomotor nerve (CNIII): Motor to medial, superior, and inferior recti and levator palpebrae; parasympathetic motor to iris
    - Trochlear nerve (CNIV): Motor to superior oblique
    - Trigeminal nerve (CNV): Sensory from orbit and eyelids (V1)
    - Abducens nerve (CNVI): Motor to lateral rectus
- **Arteries and veins**
  - **Ophthalmic artery**
    - Major blood supply to orbit
      - Largest ocular artery; fastest Doppler velocity
    - Enters orbit through optic foramen, below and lateral to optic nerve
    - Crosses over nerve to nasal side of orbit
      - Straight portion easily seen on nasal side orbit
      - Often tortuous in posterior orbit
    - Blood velocities vary with age, systemic blood pressure, smoking, and posture
      - High-resistance waveform; resembles that of external carotid artery
      - Dicrotic notch is usually present
  - **Lacrimal artery**
    - Large branch of ophthalmic artery
    - Runs along upper border of lateral rectus
    - Often originates before entering orbit
    - Sometimes derived from anterior branches of middle meningeal artery
  - **Central retinal artery and vein**
    - Most readily detectable orbital vessels
    - Run parallel to each other within optic nerve
    - Artery on nasal side and vein on temporal side
    - Flow unaffected by blood pressure and postural changes but affected by ↑ intraocular pressure
  - **Superior ophthalmic vein**
    - Largest vein detected in center of retrobulbar orbit
    - Runs obliquely and superior to optic nerve
    - Ends in cavernous sinus via superior orbital fissure
    - Enhance visualization by Valsalva maneuver
  - **Inferior ophthalmic vein**
    - Runs in inferior orbit and has 2 branches
    - One joins pterygoid venous plexus via inferior orbital fissure
    - Another frequently drains into superior ophthalmic vein
  - Posterior ciliary arteries
    - Seen on the nasal and temporal side of optic nerve

ANATOMY IMAGING ISSUES

Imaging Recommendations

- Insonation power maintained at level of low mechanical index (< 0.23)
- Avoid prolonged scanning to ↓ chance of lens damage
- Minimal acoustic jelly and gentle transducer pressure applied on closed eyelid during scanning to minimize eye movement due to patient discomfort

CLINICAL IMPLICATIONS

Clinical Importance

- Assessment of ophthalmic artery: Potential for collateral flow in proximal carotid occlusion
- Evaluation of extraocular muscle thickening
- Investigation of central retinal artery and vein occlusion; cranial arteritis; nonarteritic ischemic optic neuropathy; diabetic retinopathy; retinal detachment; ocular metastasis

OPTIC NERVE AND EXTRAOCULAR MUSCLES

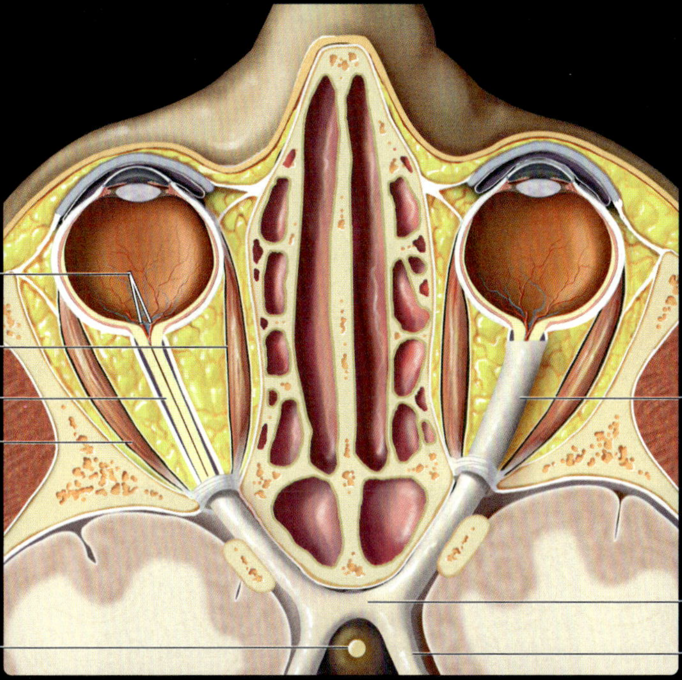

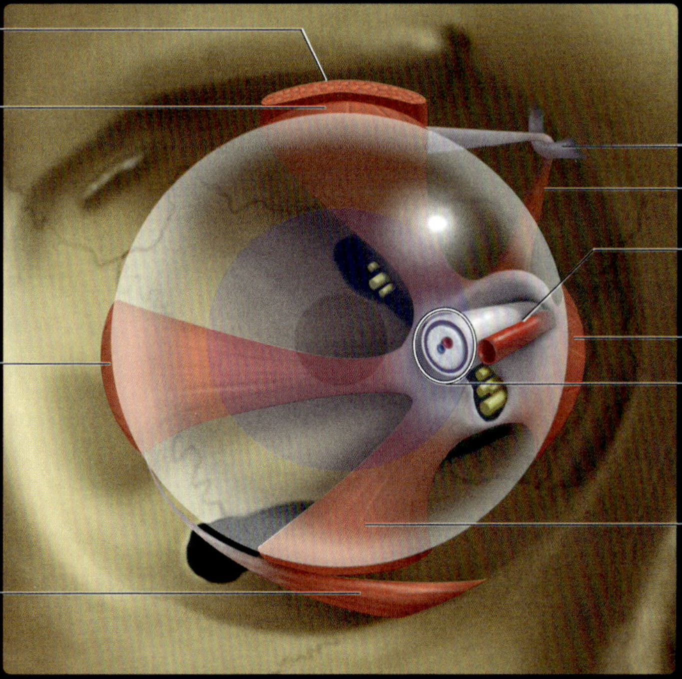

(Top) *The eyes are embedded in ocular fat and protected by the surrounding bony orbit. The optic nerve, chiasm, and tract form the afferent visual pathway. The medial and lateral rectus muscles are seen in this plane. The medial rectus muscle is the thickest and most easily evaluated by ultrasound.* **(Bottom)** *Frontal graphic of the right orbit is shown. The rectus muscles originate at the annulus of Zinn at the orbital apex and insert at the corneoscleral junction of the eye, forming a muscle cone. The superior oblique muscle courses through the trochlea, providing for the angled pulley motion of this muscle. The inferior oblique inserts at the inferolateral aspect of the eye. The oblique muscles are more difficult to evaluate by ultrasound.*

EXTRAOCULAR MUSCLES

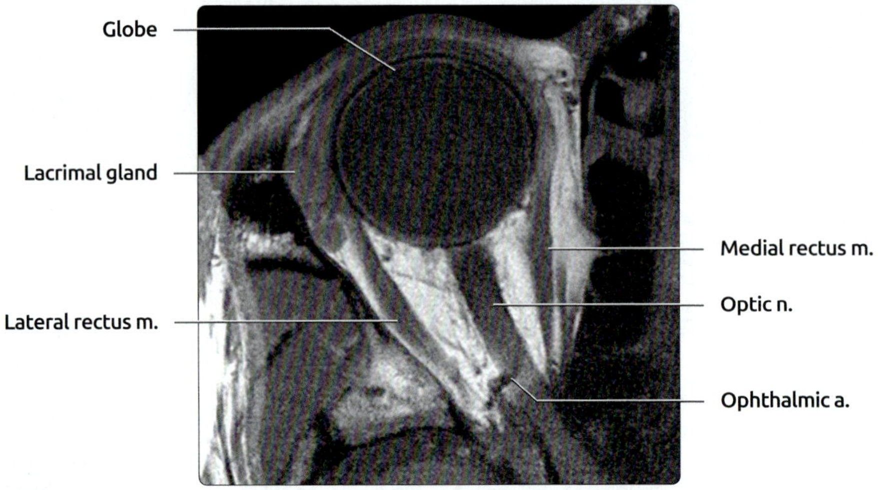

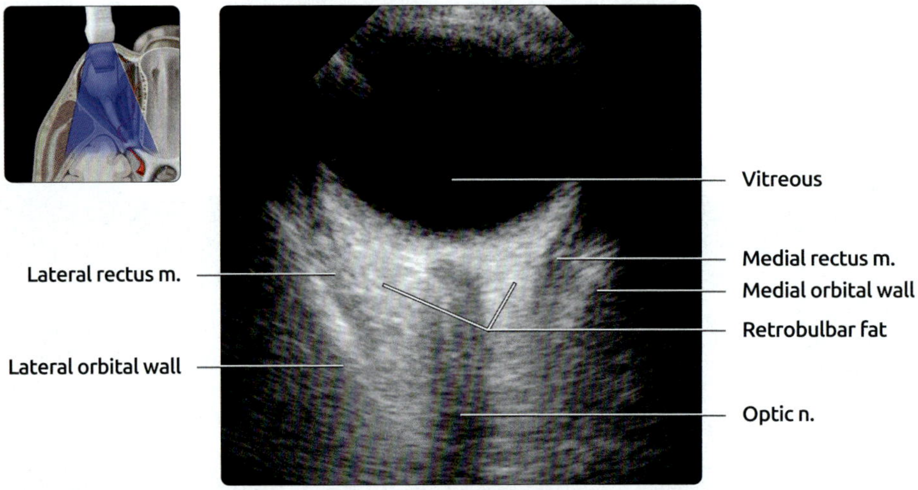

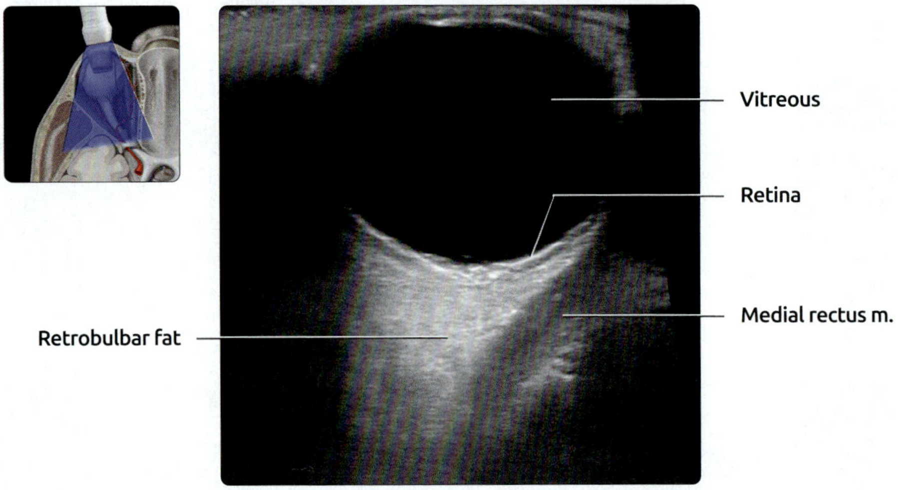

(Top) *Axial high-resolution T1 MR at the level of the midorbit shows the medial and lateral rectus muscles on either side of the optic nerve. The ophthalmic artery is seen as it is crossing over the optic nerve from the temporal to the nasal side of orbit.* **(Middle)** *Axial grayscale ultrasound of the right orbit shows the hypoechoic medial and lateral recti running obliquely within the bony orbit. The optic nerve is surrounded by echogenic retrobulbar fat.* **(Bottom)** *This medially angled transverse ultrasound just above the optic nerve better shows the medial rectus muscle. It is the largest and most easily imaged of the recti muscles. Medial transducer angulation to evaluate the lateral rectus is often limited by the patient's nose.*

EXTRAOCULAR MUSCLES

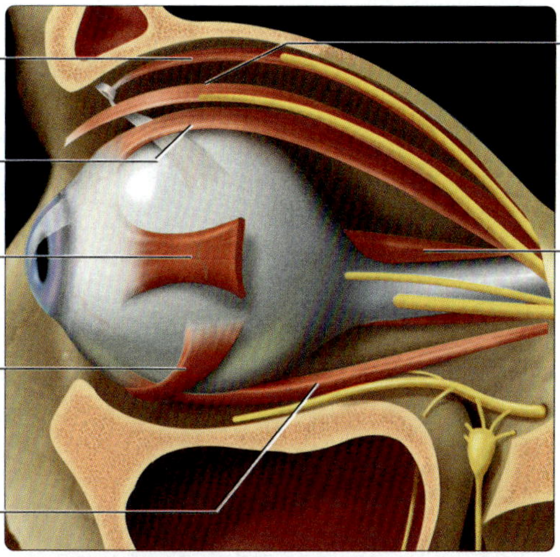

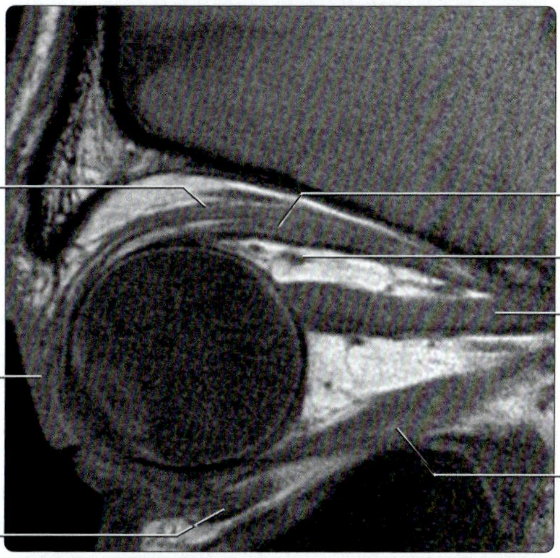

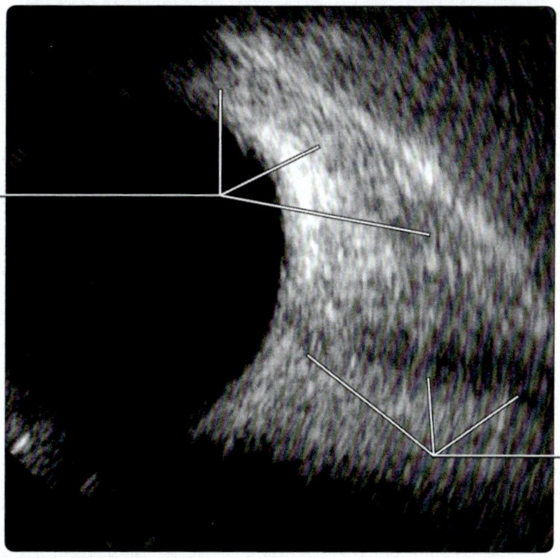

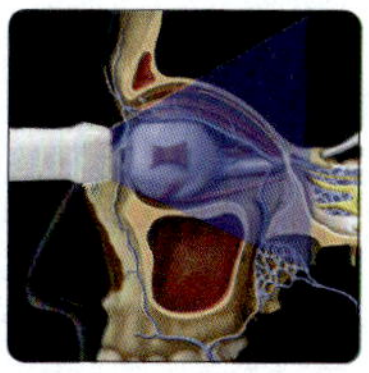

(Top) *This graphic of the left orbit demonstrates the extraocular muscles in a sagittal plane. The superior and inferior oblique muscles attach to the sclera posterolaterally and are more difficult to evaluate with ultrasound. The levator palpebrae superioris muscle originates from the annulus of Zinn with the recti muscles. It courses above the superior rectus to insert on the upper eyelid.* **(Middle)** *High-resolution oblique sagittal MR of the eye shows the intimate relationship between the superior rectus and levator palpebrae superioris muscles.* **(Bottom)** *Sagittal grayscale ultrasound of the right orbit, oriented to match the MR, shows the superior rectus and levator palpebrae muscle complex. Although the superior rectus and levator muscles sometimes can be imaged as distinct structures, these 2 muscles are usually displayed as 1 complex with conventional technique.*

ANTERIOR GLOBE

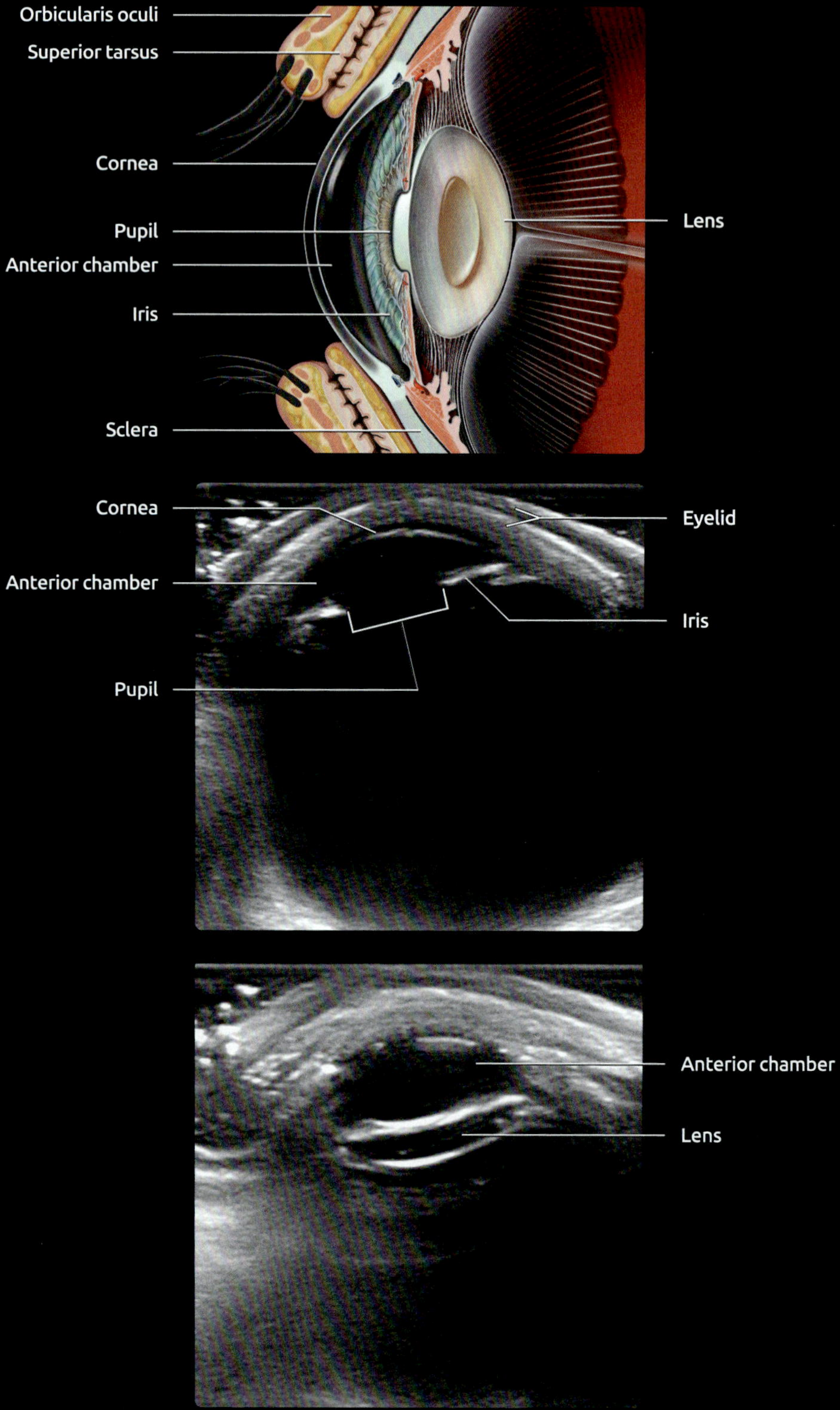

(Top) *This graphic shows the anterior eye, including (from front to back) the transparent avascular cornea, the anterior chamber filled with aqueous fluid, beneath which is the iris, pupil, and lens.* **(Middle)** *A midline axial image shows the cornea, anterior chamber, iris, and pupil.* **(Bottom)** *Angling the transducer slightly cephalad in the same patient shows the lens. It has a characteristic well-defined, oval shape*

VASCULAR SUPPLY

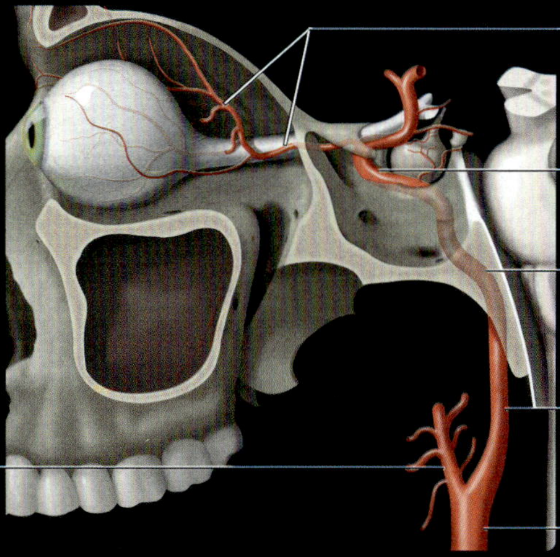

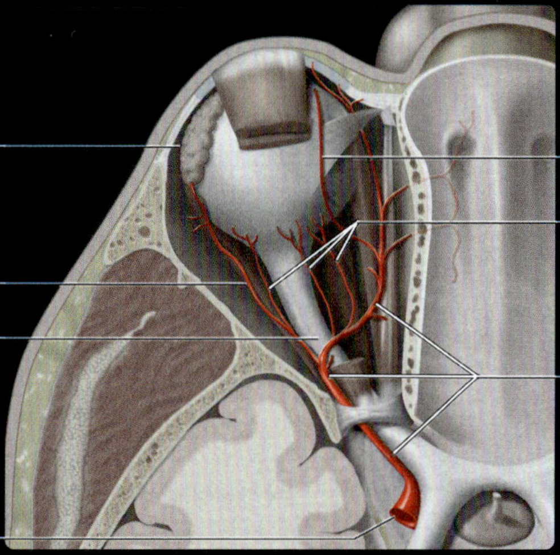

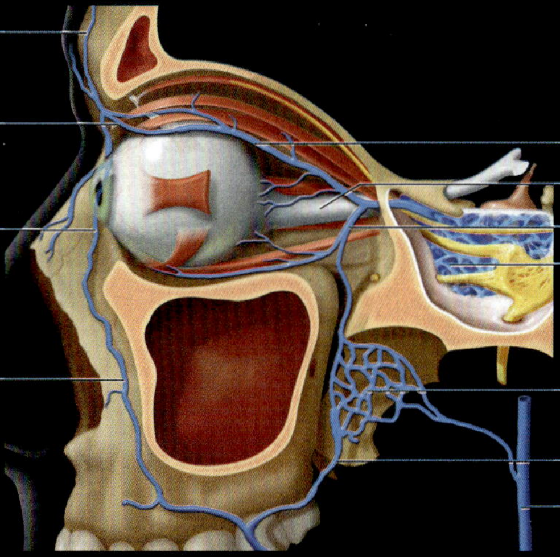

(Top) *The ophthalmic artery is the 1st intradural branch of the internal carotid artery and is the major blood supply of the orbit.* **(Middle)** *Axial graphic (superior view) shows the relationship of the ophthalmic artery and its branches. The ophthalmic artery crosses over the optic nerve from the temporal to nasal side in the posterior orbit. The lacrimal artery is one of the largest branches, which runs along the upper border of the lateral rectus muscle to supply the lacrimal gland.* **(Bottom)** *The superior ophthalmic vein is the largest orbital vein. It begins at the nasofrontal vein, passes through the superior orbital fissure, and ends in the cavernous sinus. The inferior orbital vein runs in the inferior orbit with one branch joining the superior ophthalmic vein and another exiting via the inferior orbital fissure to the pterygoid plexus.*

OPTIC NERVE AND VESSELS

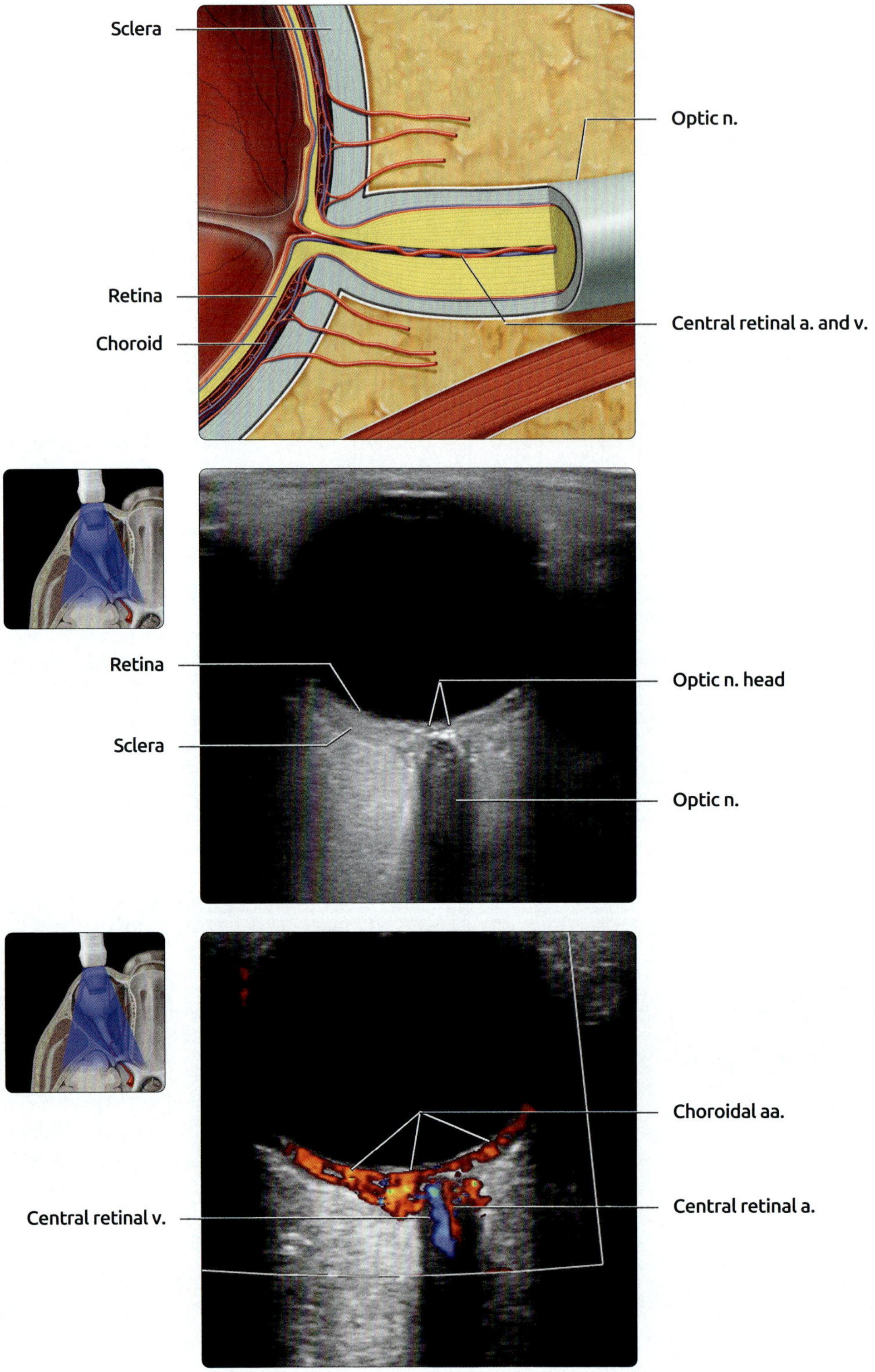

(Top) *Sagittal graphic of the posterior right eye at the level of the optic nerve head shows the central retinal artery and vein. These are the most easily identifiable orbital vessels. They run parallel to each other in the optic nerve with the artery on the nasal side and vein on the temporal side.* **(Middle)** *Axial grayscale ultrasound through the midglobe of the right eye shows the hypoechoic optic nerve.* **(Bottom)** *Color Doppler ultrasound in the same patient easily shows the parallel central retinal artery on the medial (nasal) side and the vein on the lateral (temporal) side.*

OPHTHALMIC ARTERY

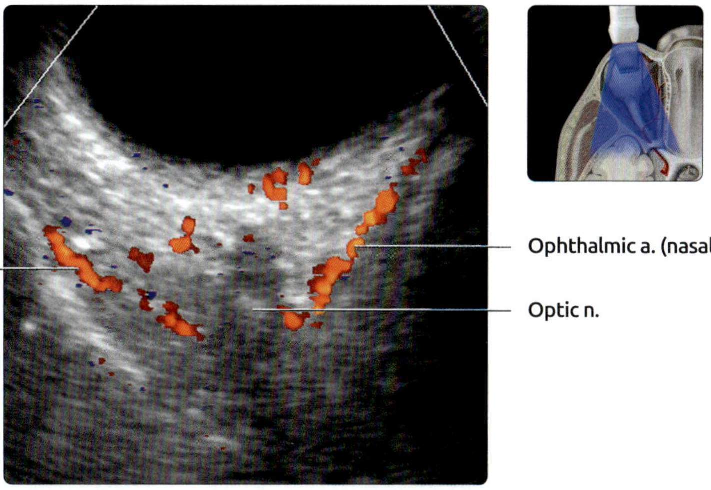

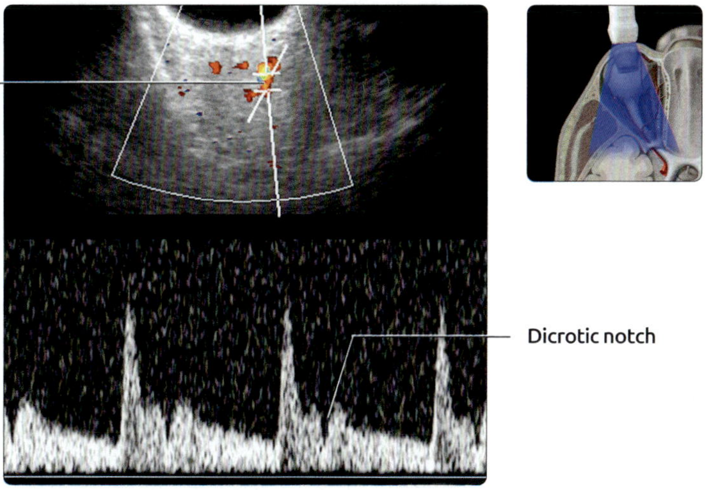

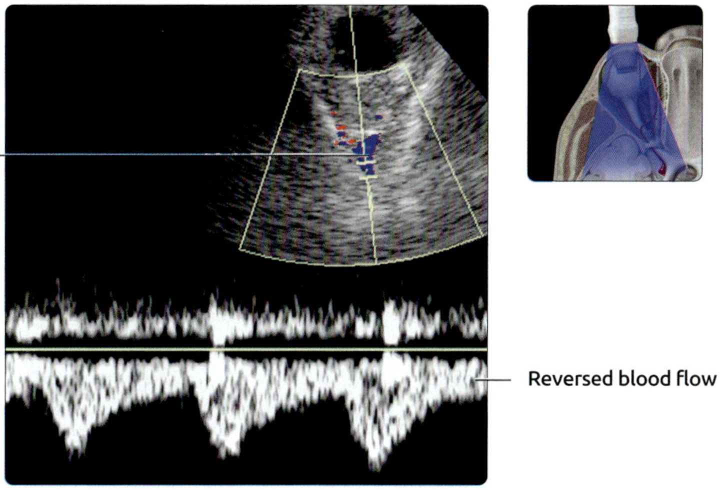

(Top) *Axial color Doppler ultrasound of the right orbit demonstrates the ophthalmic artery continuing along the nasal side of the orbit. The lacrimal artery is an early, large branch of the ophthalmic artery, which courses along the temporal side of the orbit. The ophthalmic artery is the largest artery in the orbit with the fastest blood flow and is readily detectable on color Doppler examination. It is more tortuous in the posterior orbit but is relatively straight in the nasal orbit. Normal flow direction is directed toward the transducer.* **(Middle)** *Axial spectral Doppler ultrasound of normal right ophthalmic artery is shown. The characteristic waveform is antegrade and high resistance with a dicrotic notch. Note the waveform resembles that of the external carotid artery.* **(Bottom)** *The ophthalmic artery can provide collateral circulation as shown in this patient with proximal internal carotid artery occlusion. Flow is reversed, flowing away from the transducer (blue) with retrograde flow confirmed on the Doppler tracing. In addition, the flow resistance is lowered with absent dicrotic notch. The resulting waveform is similar to that of the normal internal carotid artery.*

OPHTHALMIC VEINS

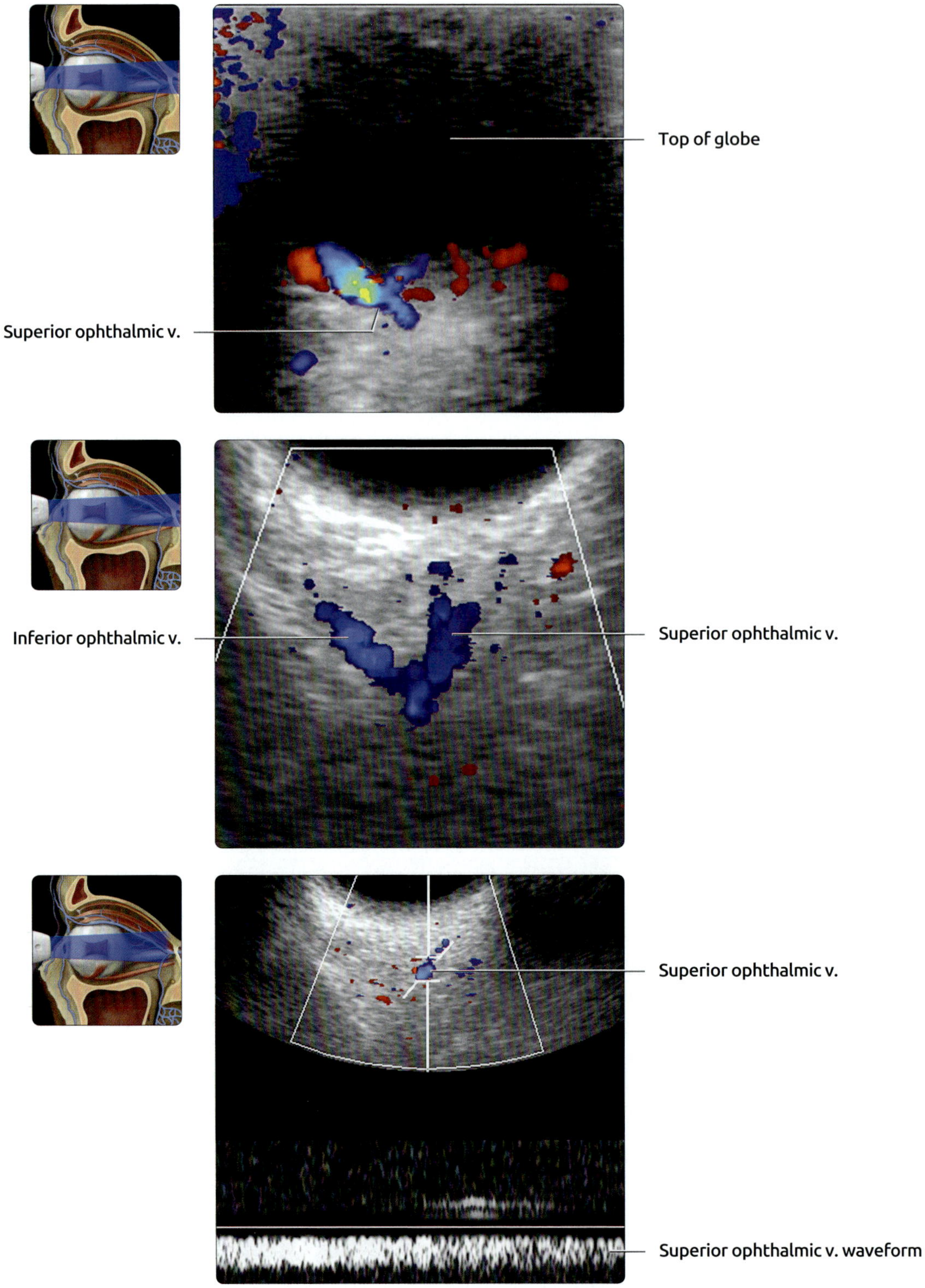

(Top) *Axial color Doppler ultrasound of the most superior aspect of the right eye shows the ophthalmic vein as it courses over the globe.* **(Middle)** *A more posterior image shows the junction of the superior and inferior ophthalmic veins above the optic nerve. The superior ophthalmic vein runs obliquely in the center of the retrobulbar orbit receiving a branch of the inferior ophthalmic vein before it ends in the cavernous sinus via the superior orbital fissure. Note the flow direction of both veins is away from the transducer.* **(Bottom)** *Axial spectral Doppler ultrasound shows a right superior ophthalmic vein waveform, which is continuous and constant on the negative side of the trace. The flow pattern in this vessel is variable, it may be constant or phasic. In addition, flow direction may be reversed during a Valsalva maneuver.*

3D ULTRASOUND, RIGHT ORBIT

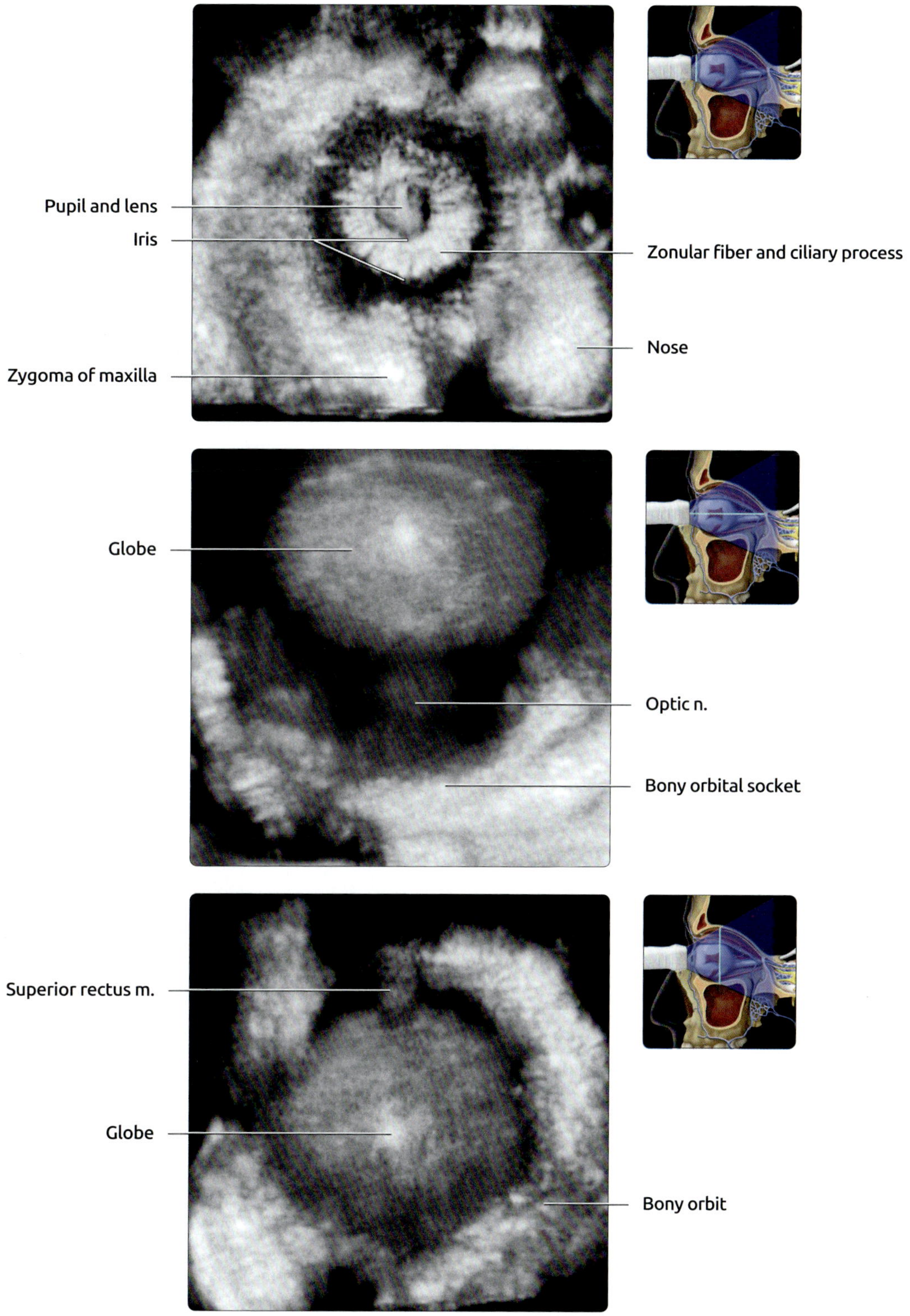

(Top) *3D ultrasound of the right orbit (frontal view) using a high-resolution 12-MHz linear 3D transducer demonstrates the anterior chamber of the eye. The pupil and lens are seen surrounded by the highly reflective iris, zonular fibers, and ciliary processes.* **(Middle)** *3D ultrasound of the right orbit (axial view) shows the optic nerve exiting from the posterior aspect of the eye. This view is useful in the investigation of retrobulbar abnormality of the orbit.* **(Bottom)** *3D ultrasound of the right orbit (coronal view) shows the attachment of the superior rectus muscle to the eye. With proper manipulation of 3D-rendered images in 3 orthogonal planes, different extraocular muscle attachments to the eye can be demonstrated.*

OPTIC NERVE MENINGIOMA

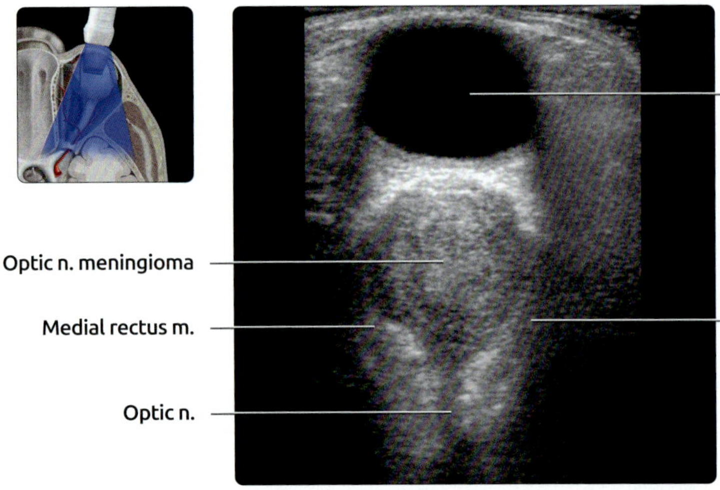

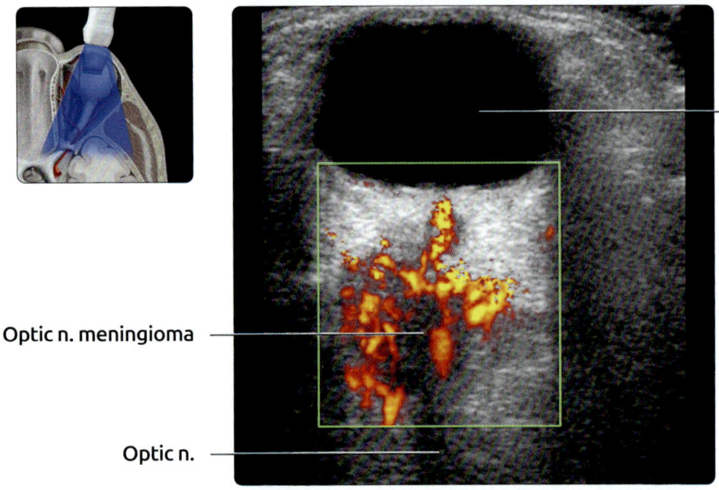

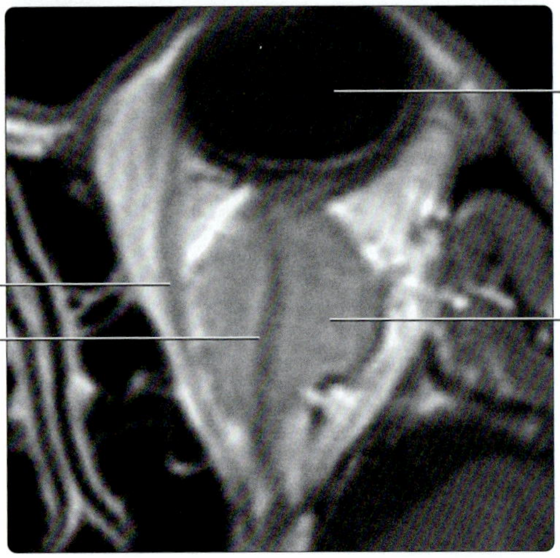

(Top) *Axial grayscale ultrasound of the left eye shows a histologically proven optic nerve sheath meningioma. The mass is large and moderately echogenic encasing the optic nerve and abutting the medial and lateral rectus muscles. Although the mass is identified by ultrasound, further evaluation with MR is required for complete evaluation.* **(Middle)** *Corresponding axial color Doppler ultrasound shows increased vascularity within the mass.* **(Bottom)** *Corresponding axial T1 C+ MR of the left eye shows a soft tissue mass with homogeneous contrast enhancement encasing a normal-looking optic nerve. Features are consistent with an optic nerve meningioma.*

OCULAR MELANOMA

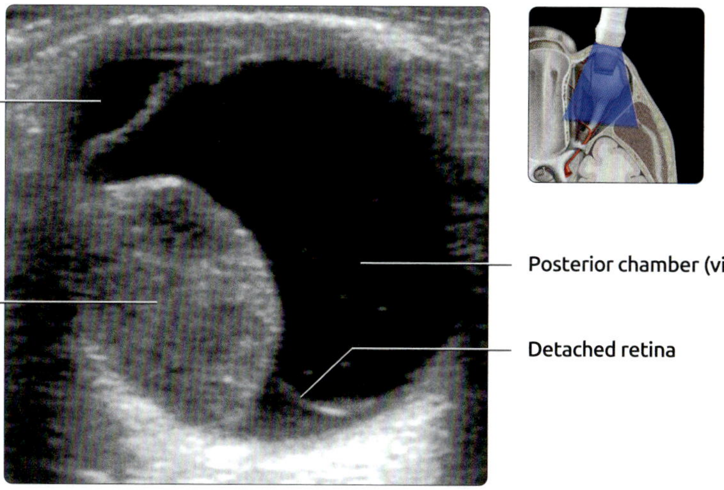

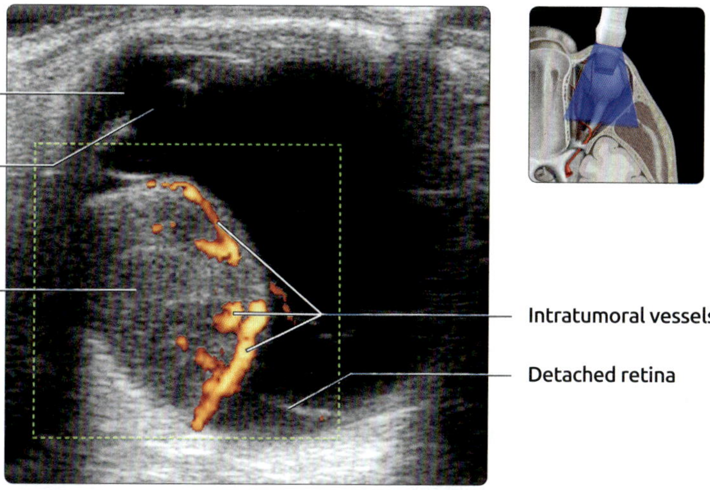

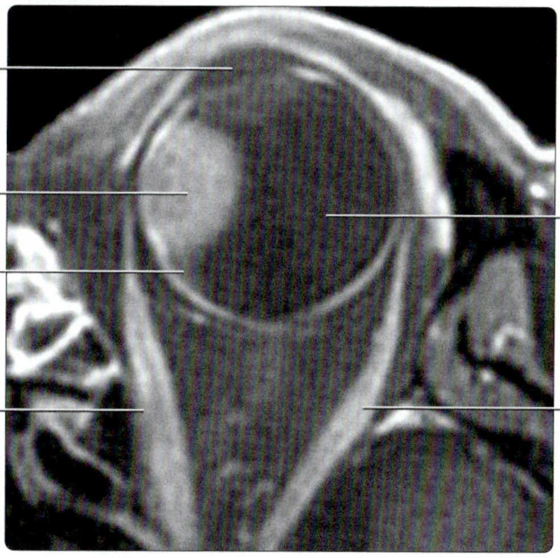

(Top) *Axial grayscale ultrasound of the left eye shows a histologically proven ocular melanoma. It is arising from the medial wall of the posterior chamber behind the lens. The mass is large, heterogeneous, and occupies nearly 1/3 of the posterior chamber with part of the retina "peeled" off from the underlying sclera.* **(Middle)** *Corresponding axial color Doppler ultrasound shows multiple small vessels within the mass.* **(Bottom)** *Corresponding axial T1 C+ MR of the left eye with fat saturation shows the ocular melanoma arising from the sclera of the medial wall of the posterior chamber. The mass shows homogeneous enhancement and lifts off part of the retina from the underlying sclera.*

TERMINOLOGY

Abbreviations

- **Carotid arteries**
  - Common carotid (CCA); internal carotid (ICA)
- **Cerebral arteries**
  - Anterior (ACA); middle (MCA); posterior (PCA)
- **Communicating arteries**
  - Anterior (ACoA); posterior (PCoA)
- **Posterior fossa vertebrobasilar arteries**
  - Vertebral (VA); basilar (BA); posterior inferior cerebellar artery (PICA); anterior inferior cerebellar artery (AICA); superior cerebellar artery (SCA)
- **Sinuses**
  - Superior sagittal (SSS); sphenoparietal (SPS); straight (SS); transverse (TS); cavernous (CS)
- **Cerebral veins**
  - Basal vein of Rosenthal (BV); internal cerebral (ICV); great vein of Galen (GV); deep middle cerebral (dMCV); anterior cerebral (ACV)

GROSS ANATOMY

Intracranial Internal Carotid Artery

- Complex course with several vertical/horizontal segments, 3 genus (1 petrous, 2 cavernous)
- 6 intracranial segments (cervical ICA = C1)
  - Petrous (C2), lacerum (C3), cavernous (C4), clinoid (C5), ophthalmic (C6), communicating (C7)

Anterior Cerebral Artery

- Smaller, more medial terminal branch of ICA with 3 segments
- **A1**: Horizontal or precommunicating segment
  - Extends anteromedially above optic nerve & chiasm, below medial olfactory stria
- **A2**: Vertical or postcommunicating segment
  - From ACoA junction, ascends in interhemispheric fissure, anterior to corpus callosum rostrum
- **A3**: Distal segment
  - Pericallosal artery; begins distal to origin of callosomarginal artery
- **ACoA**: Communicates between A1 segments; completes anterior portion of circle of Willis
- **Cortical branches** supply anterior 2/3 of medial hemispheres
  - Medial orbitofrontal artery (1st cortical branch)
  - Frontopolar artery arises from mid A2
  - Pericallosal artery begins distal to origin of callosomarginal artery
  - Callosomarginal artery is smaller of 2 distal ACA branches
- **Perforating branches** supply medial basal ganglia, corpus callosum genu, internal capsule anterior limb

Middle Cerebral Artery

- Larger, lateral terminal branch of supraclinoid ICA with 4 main segments
- **M1**: Horizontal segment
  - From terminal ICA bifurcation to MCA bifurcation
- **M2**: Extends from MCA bifurcation to periinsular sulci
  - Superior trunk: Prefrontal, precentral, central sulcus (rolandic), anterior parietal, orbitofrontal arteries
  - Inferior trunk: Posterior parietal, middle temporal, posterior temporal, occipital temporal. angular (terminal) arteries
- **M3**: Courses inferolaterally through sylvian fissure
- **M4**: Exits sylvian fissure & ramifies over cerebral convexity
- **Cortical branches** supply most of lateral surface of cerebral hemispheres; except convexity & inferior temporal gyrus
- **Penetrating lenticulostriate branches** arise from M1 & supply basal ganglia, internal & external capsules

Posterior Cerebral Artery

- BA terminates into 2 PCAs, with 4 segments
- **P1**: Precommunicating segment; extends laterally from BA to junction with PCoA
- **P2**: Ambient segment; swings posterolaterally around midbrain
- **P3**: Quadrigeminal segment; short segment within lateral aspect of quadrigeminal cistern behind midbrain
- **P4**: Calcarine segment; PCA terminates above tentorium in calcarine fissure
- **Cortical branches** supply posterior 1/3 of medial hemisphere surface, inferior temporal lobe, & occipital lobe (including visual cortex)
  - Anterior temporal artery arises from P2; courses anterolaterally & anastomoses with MCA branches
  - Posterior temporal artery arises from P2; courses posterolaterally along hippocampal gyrus
  - Medial & lateral terminal trunks; major branches include parietooccipital & calcarine arteries
- **Penetrating branches** supply mesencephalon, thalami, posterior limb of internal capsule, optic tract
  - Thalamoperforating arteries arise from P1
  - Thalamogeniculate arteries arise from P2
- Ventricular/choroidal branches arise from P2 & supply choroid plexus of 3rd/lateral ventricles, thalami, posterior commissure, & cerebral peduncles

Vertebrobasilar System

- **V3** segment (atlas loop) is extracranial/extraspinal; exits at transverse foramen (atlas) to enter into foramen magnum
- **V4** segment is intradural/intracranial
  - Perforates dura, enters skull through foramen magnum, courses superomedially behind clivus
  - Branches include anterior & posterior spinal arteries, meningeal branches, perforating branches to medulla & PICA
    - **PICA**: Arises from distal VA, curves around/over cerebellar tonsil, gives off perforating medullary, choroid, tonsillar, cerebellar branches
- **BA** large median artery formed by union of VAs
  - It courses superiorly in prepontine cistern & has multiple branches
    - Pontine, midbrain perforating branches
    - **AICA**: Courses inferolaterally, lies ventromedial to CNVII & VIII; often loops into IAC; its size is inversely proportional to size of PICA
    - **SCA**: Arise from distal BA, course posterolaterally around mesencephalon below CNIII, tentorium; lie above CNV, often in contact with it
    - **PCA** (terminal BA branches)

Intracranial Veins

- **Dural venous sinuses** lie between 2 layers of dura mater & drain intracranial veins
  - **Posterosuperior sinuses**: SSS, inferior sagittal, 2 TS, SS & occipital
  - **Anteroinferior sinuses**: 2 CS, 2 superior petrosal, 2 inter-CS, 2 inferior petrosal & basilar plexus
- Intracranial veins: **Cerebral & cerebellar veins**
  - **Superficial ("external") cerebral veins**: Divided into superior, middle, & inferior divisions
  - **Deep ("internal") cerebral veins**
    - Drain deep parts of hemisphere & are in pairs
    - Run parallel with each other posteriorly beneath splenium of corpus callosum
    - **BV**: Receives ACV, dMCV, inferior striate veins
    - **GV**: Receives ICVs, BV, pericallosal veins, & veins draining superior aspect of posterior fossa
      - Curves backward & upward around splenium of corpus callosum & into SS
    - **Cerebellar veins**: Superior & inferior
      - Superior cerebellar veins run anteromedially across superior vermis to end in SS & ICVs
      - Inferior cerebellar veins (larger) end in TS, superior petrosal, & occipital sinuses

ANATOMY IMAGING ISSUES

Imaging Recommendations

- Can assess intracranial stenosis & occlusive disease, monitor recanalization after thrombolysis & collateral formation
  - Preoperative compression test to evaluate collateralizing capacity of circle of Willis
- Performed using **low-frequency transducer (1.8-3.6 MHz)**
- Grayscale ultrasound: Identify main anatomical structures in correct plane, followed by color Doppler ultrasound
- Unless specified, conventional color coding used
  - Red: Flow toward transducer
  - Blue: Flow away from transducer
- Vessel identification depends on window used, beam angulation, insonation depth, & flow direction
- Basal skull arteries are extremely variable in size, development, & course; supplementary CTA/MRA is useful to enhance its diagnostic accuracy
- For difficult cases, compression test is helpful for arterial identification & assessment of collaterals
  - Must be performed by experienced investigator
  - Exclude risk of embolism in extracranial arteries prior to compression
  - For anterior circulation, compress CCA in lower neck by 2 fingers
  - For vertebrobasilar system, compress VA at mastoid slope
- Normal adults: Highest velocities in MCA or ACA
- Flow velocities in basal cerebral arteries show consistent decrease with increasing age

Transtemporal Approach (Most Common Approach)

- Transducer on temporal bone superior to zygomatic arch
- **Axial mesencephalic plane**
  - Identify hypoechoic, butterfly-shaped mesencephalon as landmark
  - Assess C5-C7 segments, A1 segment, M1 & M2 segments, P1 & P2 segments, PCoA
  - Also assess dMCV, BV, GV, SS, & contralateral TS
- **Axial ventricular plane**
  - Tilting of transducer 10° upward from mesencephalic plane
  - Identify hypoechoic 3rd ventricle, echogenic pineal gland, & choroid plexus of trigone
  - Assess A2 segment, M2 & M3 segments, & P3
  - Assess midline shift due to MCA infarction
- **Anterior/posterior coronal planes**
  - Assess C4-C7 segments, A1 segment, M1-M3 segments, PCA, & distal BA
- **Normal mean velocities & depth of insonation**
  - Terminal ICA: 39 ± 9 cm/sec (60-67 mm)
  - MCA: 62 ± 12 cm/sec (30-67 mm)
  - ACA: 50 ± 11 cm/sec (60-80 mm)
  - PCA: 39 ± 10 cm/sec (55-80 mm)

Transfrontal Approach

- Insonation depth of 10-16 cm
- **Paramedian frontal bone window**
  - Slightly lateral t midline of forehead
  - Identify echogenic orbital roof, hypoechoic 3rd ventricle, & corpus callosum
- **Lateral frontal bone window**
  - Just above lateral aspect of eyebrow
  - Identify echogenic falx cerebri, sylvian fissure, & hypoechoic mesencephalon
- Assess A1 & A2 segments, M1 segment, PCA, PCoA, pericallosal artery
- Also assess ICV, GV, SS

Transforaminal/Suboccipital Approach

- Transducer placed between squama occipitalis & spinous process of 1st cervical vertebra
- Ultrasound beam aimed at nasal bridge
- Identify echogenic processus transversus & clivus
- Assess VAs & BA, which are visualized as Y
- Normal mean velocities & depth of insonation
  - VA: 38 ± 10 cm/sec (40-85 mm)
  - BA: 41 ± 10 cm/sec (> 80 mm)

Transorbital Approach

- Problem of insonation of orbital lens
- Low mechanical index < 0.23 highly recommended
- Assess intraocular vessels, C4-C6 segments
- Normal mean velocities & depth of insonation
  - Ophthalmic artery: 21 ± 5 cm/sec (40-60 mm)
  - Carotid siphon: 47 ± 10 cm/sec (60-80 mm)

Submandibular Approach

- Assess distal C1 & C2 segments
- Normal mean velocities & depth of insonation
  - Cervical ICA: 37 ± 9 cm/sec (35-70 mm)

Imaging Pitfalls

- Extensive variations; incomplete circle of Willis
- Variable probe-to-vessel angle
- Misdiagnosis: Hyperdynamic collaterals for stenosis; vasospasm for stenosis; displacement of basal vessels by mass lesion for occlusion

DISTAL INTERNAL CAROTID ARTERY

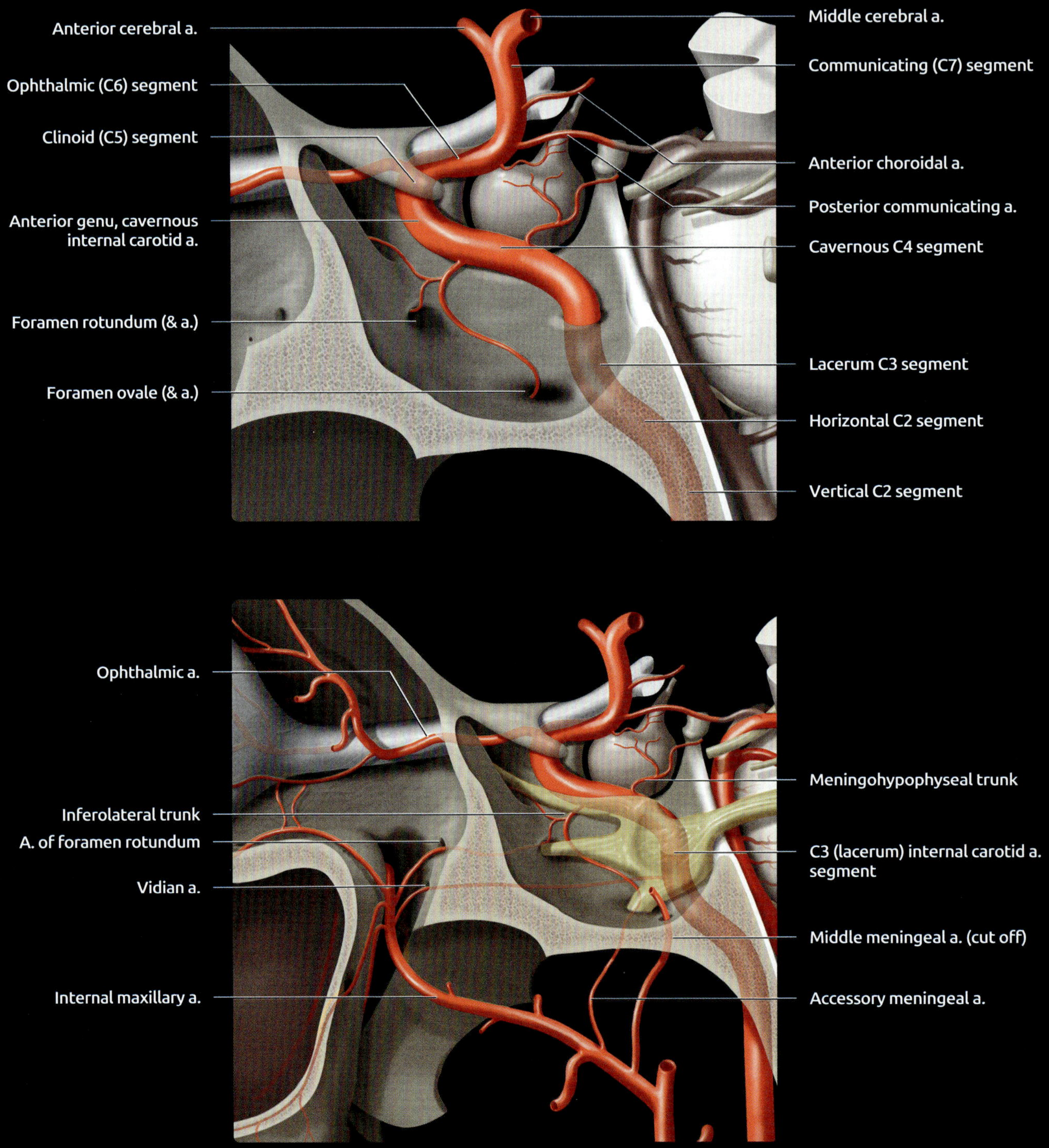

(Top) *Graphic shows intracranial internal carotid artery (ICA) segments. The C2 segment runs in the carotid canal and continues as the C3 segment after leaving the canal. The C4 segment, an extension of C3, has branches anastomosing extensively with external carotid artery (ECA) branches. The C5 segment ends near the anterior clinoid process and the C6 segment extends to just below the posterior communicating artery (PCoA). The C7 segment, after giving rise to the PCoA, branches into anterior cerebral artery (ACA) and middle cerebral artery (MCA).* **(Bottom)** *Graphic shows numerous ICA to ECA anastomoses through cavernous and deep facial branches of the 2 arteries, respectively. These include numerous anastomoses in and around the orbit; the small vidian artery anastomosing between the internal maxillary artery and petrous C2 segment and the accessory meningeal artery, an important branch that may supply part of the trigeminal ganglion, anastomosing with the inferolateral trunk of the cavernous ICA.*

DISTAL INTERNAL CAROTID ARTERY

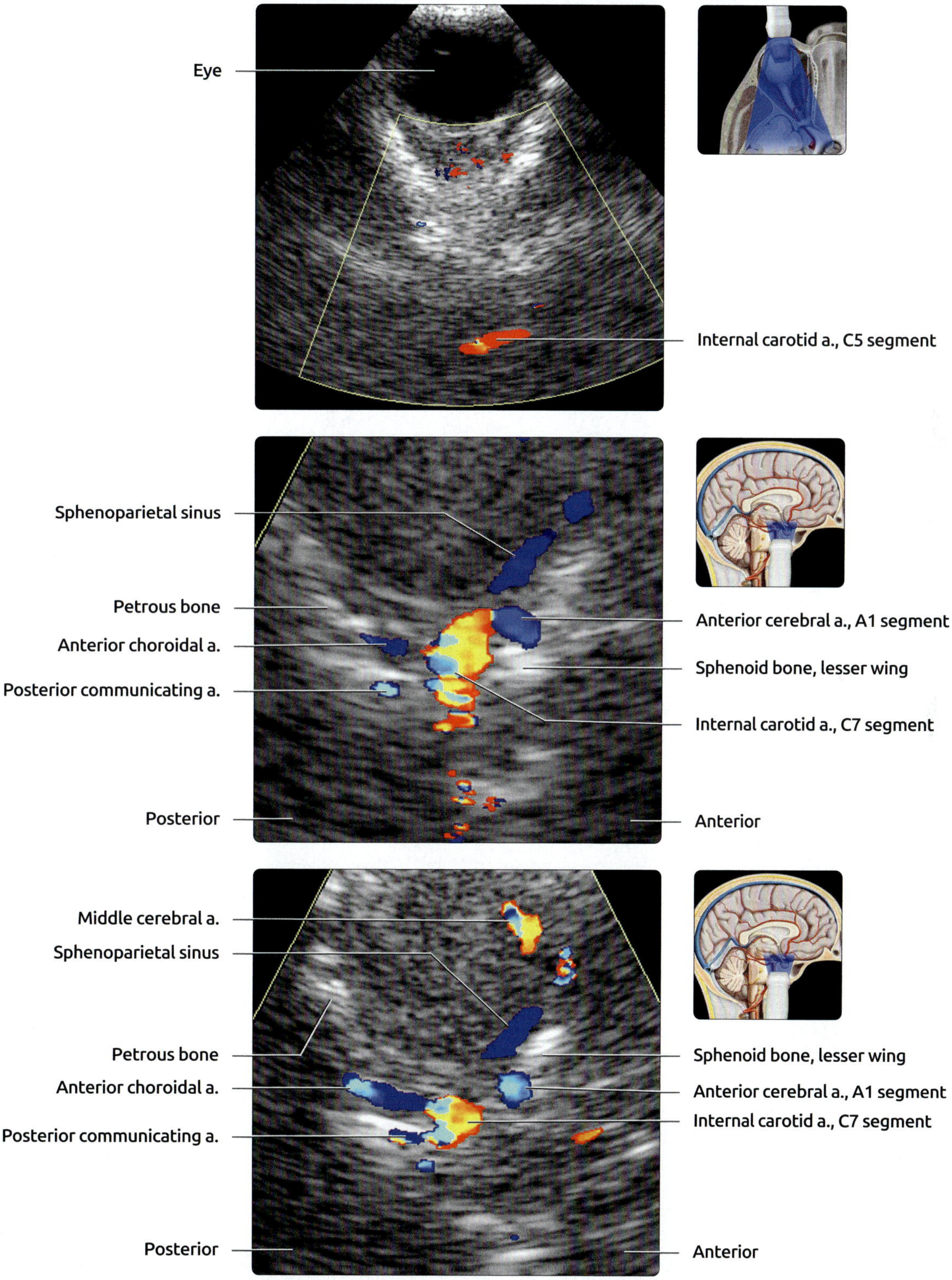

(Top) *Axial color Doppler ultrasound (transorbital approach) shows the C5 segment of the ICA with flow toward the transducer. C4-C6 segments can be assessed by this approach.* **(Middle)** *Axial color Doppler ultrasound (transtemporal approach) at a plane slightly caudad to the standard mesencephalic plane shows the C7 segment of the ICA, ACA, and sphenoparietal sinus.* **(Bottom)** *Axial oblique color Doppler ultrasound (transtemporal approach) with slight caudad tilting from the standard axial mesencephalic plane shows the anterior choroidal artery and PCoA branching out from the communicating segment of the ICA.*

ANTERIOR CEREBRAL ARTERY

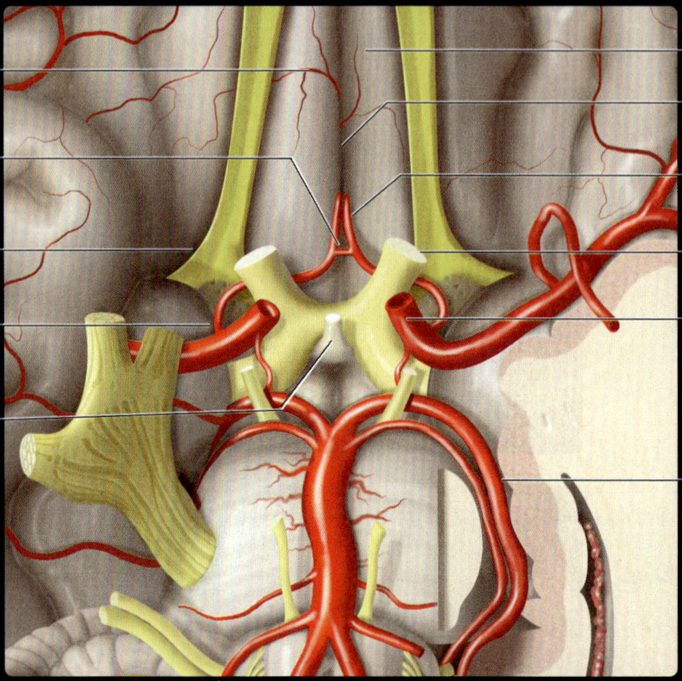

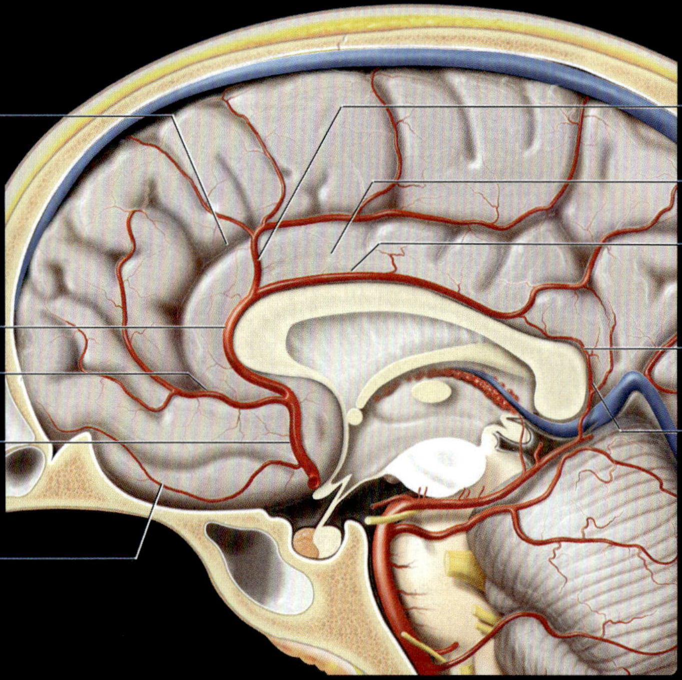

(Top) *Submentovertex graphic shows the relationship of the circle of Willis and its components to the cranial nerves. Note that the normal course of the horizontal (A1) segment is over the optic nerves.* **(Bottom)** *Sagittal (midline) graphic through the interhemispheric fissure shows the relationship of the ACA and its branches to the underlying brain parenchyma. The A2 segment ascends in front of the 3rd ventricle within the cistern of the lamina terminalis. The A3 segment curves around the corpus callosum genu. The branch point of the distal ACA into the pericallosal and callosomarginal arteries varies. Almost the entire anterior 2/3 of the medial hemisphere surface is supplied by the ACA and its branches. Branches of the posterior and anterior cerebral arteries anastomose around the corpus callosum genu.*

ANTERIOR CEREBRAL ARTERY

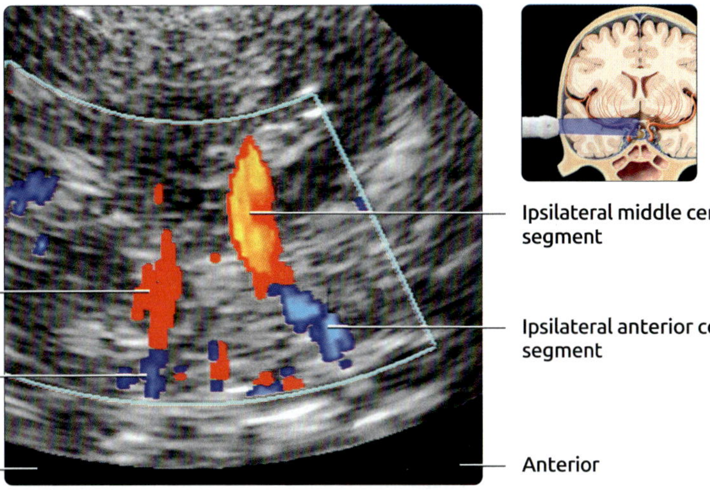

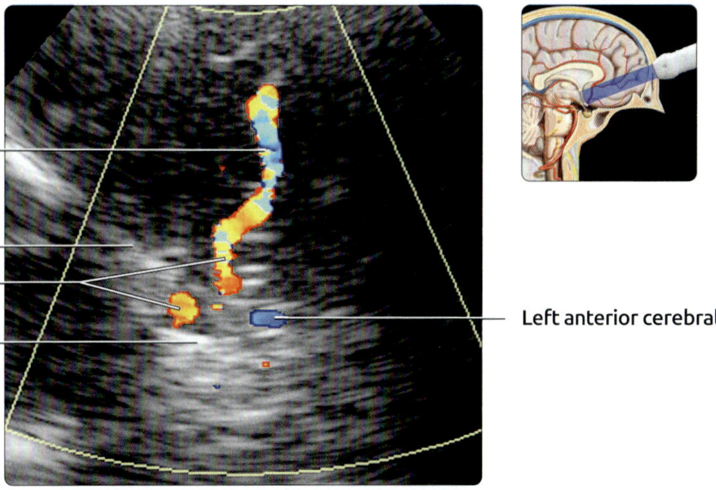

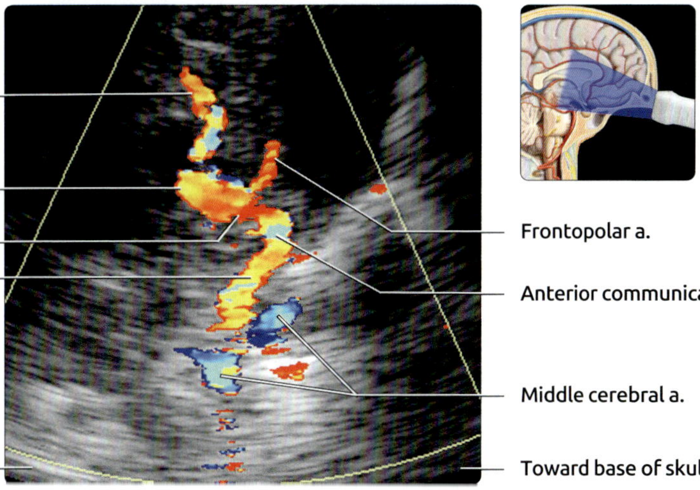

(Top) *Axial color Doppler ultrasound (transtemporal approach) shows the MCA/ACA junction on the mesencephalic plane, which is the most useful plane for showing the circle of Willis. However, it is limited in depicting more distal segments of the ACA, which can be visualized using a transfrontal approach.* **(Middle)** *Axial color Doppler ultrasound (paramedian frontal approach) shows the right A2 segment running along the interhemispheric fissure. Echogenic sphenoid bone and sella turcica provide important landmarks for identification of the A2 segment, which starts close to the sella turcica and tip of the sphenoid bone.* **(Bottom)** *Sagittal color Doppler ultrasound (paramedian frontal approach) shows the frontopolar artery arising anteriorly from the mid A2 segment and the callosomarginal artery branching out from the distal A2 segment. Note the pericallosal artery (A3 segment) can be seen by this approach.*

ANTERIOR CEREBRAL ARTERY

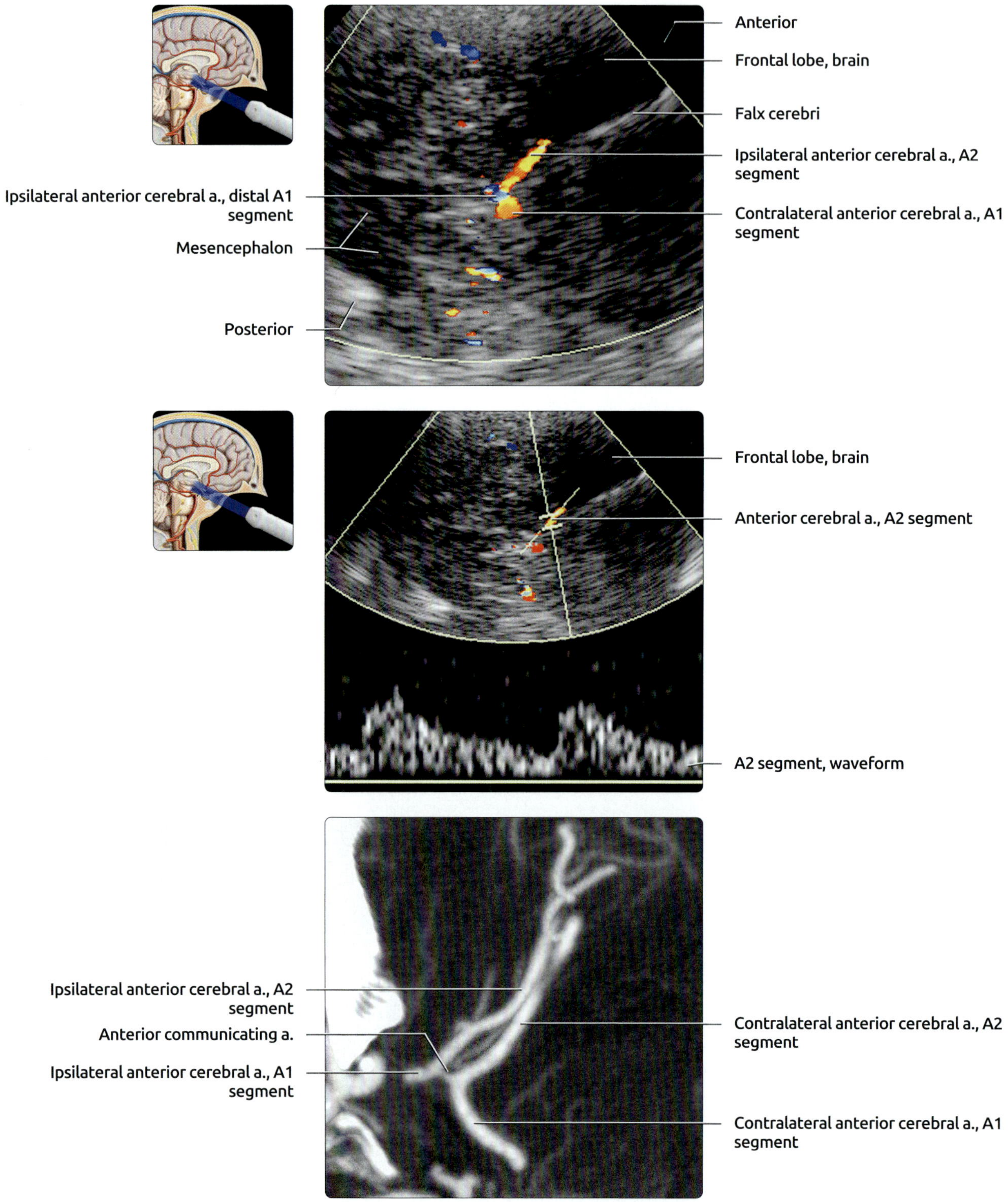

(Top) *Axial color Doppler ultrasound (lateral frontal approach) shows the A2 segment as seen through the frontal bone at the lateral eyebrow. The ipsilateral A2 segment is often seen as a prominent vessel running along the hyperechoic falx cerebri with flow toward the transducer. Sometimes the contralateral distal A1 or proximal A2 segment can be depicted coursing obliquely toward the ipsilateral A2 segment, mimicking the anterior communicating artery (ACoA), which is infrequently discernible in normal subjects.* **(Middle)** *Axial spectral Doppler ultrasound (lateral frontal approach) of the A2 segment shows the typical low-resistance waveform with forward flow.* **(Bottom)** *Corresponding oblique axial reformatted CT arteriogram shows the A1 segments becoming A2 segments after the ACoA junction. They then ascend parallel in the interhemispheric fissure.*

ANTERIOR CEREBRAL ARTERY

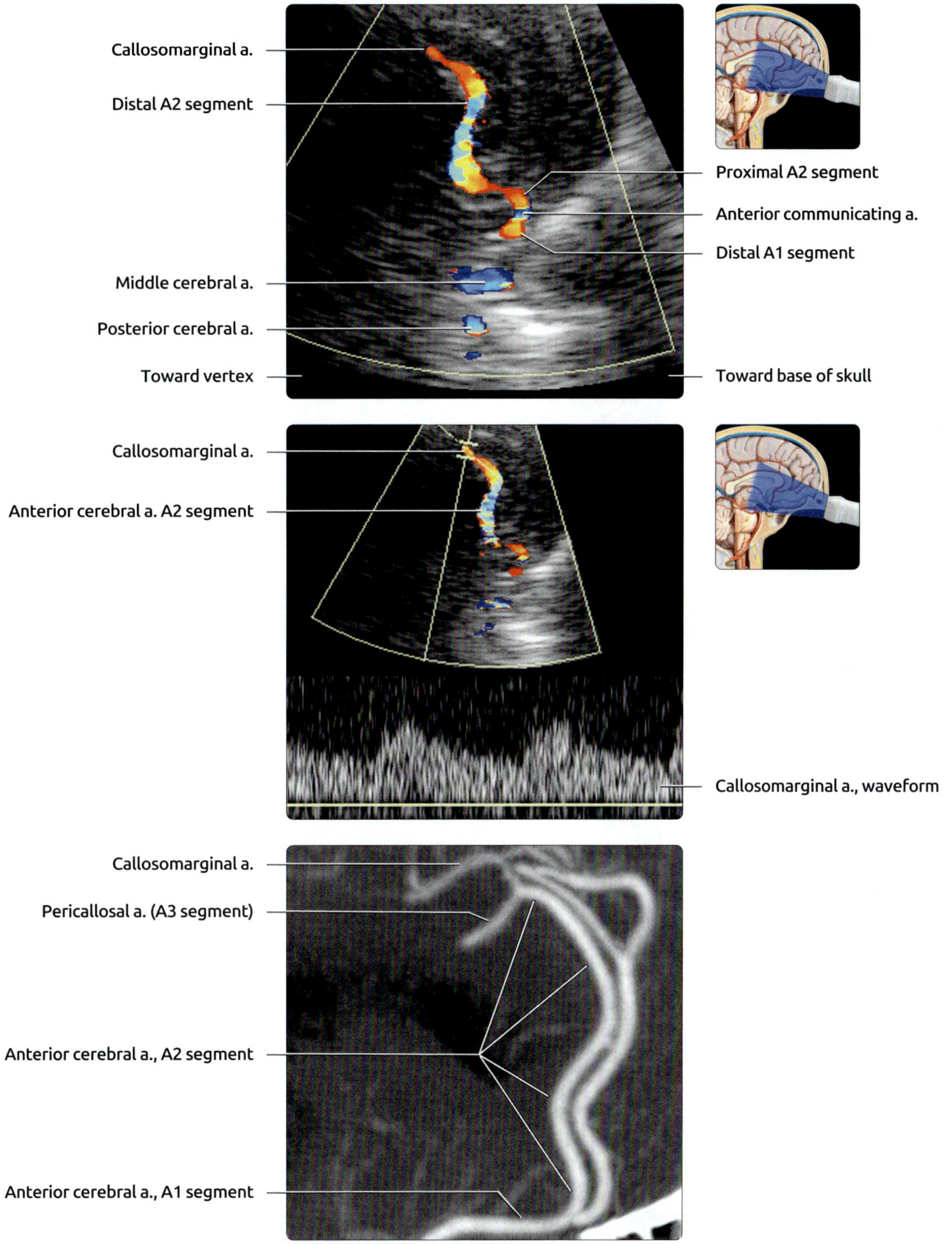

(Top) *Sagittal color Doppler ultrasound (paramedian frontal approach) illustrates the A2 segment giving off the callosomarginal artery, which is the smaller branch of the distal ACA. It runs posterosuperiorly in cingulate sulcus and on the superior surface of the cingulate gyrus. Its flow is directed toward the transducer.* **(Middle)** *Sagittal spectral Doppler ultrasound (paramedian frontal approach) shows the antegrade, low-resistance waveform of callosomarginal artery of the A2 segment.* **(Bottom)** *Corresponding oblique sagittal reformatted CT arteriogram shows the A2 segment giving off the callosomarginal artery and becoming the A3 segment (pericallosal artery). A similar branching pattern is seen on the opposite side.*

MIDDLE CEREBRAL ARTERY

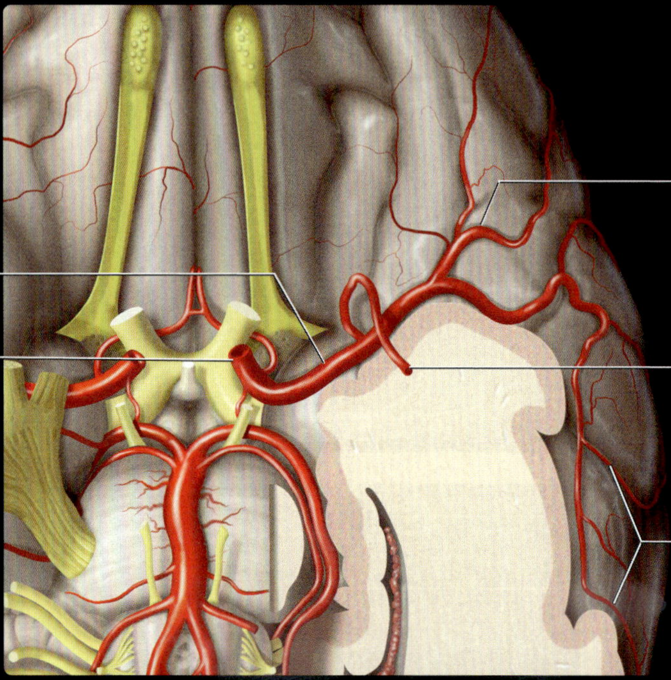

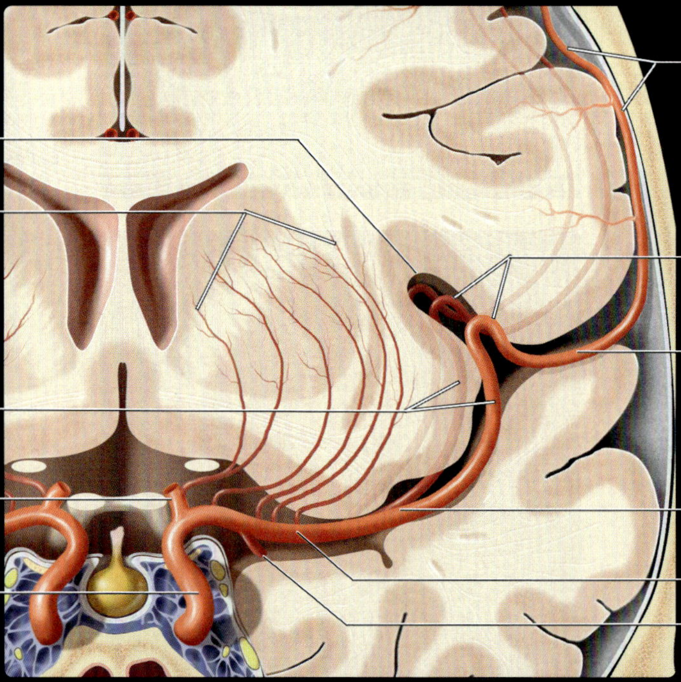

(Top) *The MCA and its relationship to adjacent structures is depicted on these graphics. The submentovertex graphic shows the left temporal lobe sectioned through the temporal horn of the lateral ventricle. The MCA supplies much of the lateral surface of the brain and is the larger of the 2 terminal branches of the ICA.* **(Bottom)** *AP graphic shows the MCA and its relationship to the adjacent brain. The MCA courses through the sylvian fissure, and the M1-M4 segments are well delineated. A few medial and numerous lateral lenticulostriate arteries arise from the top of the horizontal (M1) MCA segment, course superiorly through the anterior perforated substance, and supply the lateral basal ganglia and external capsule.*

MIDDLE CEREBRAL ARTERY

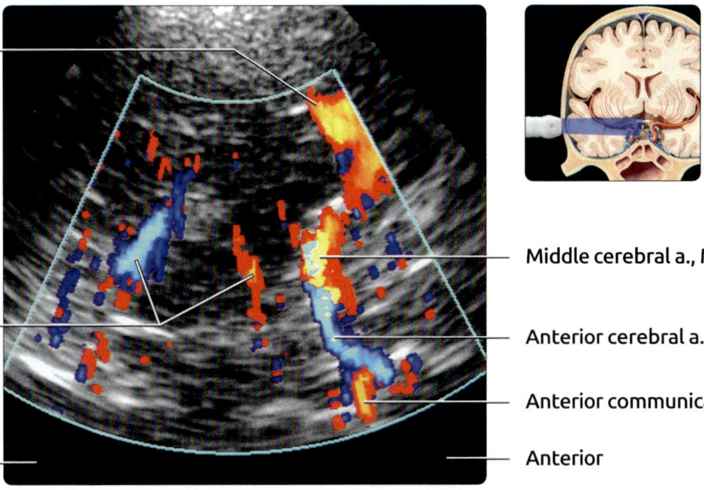

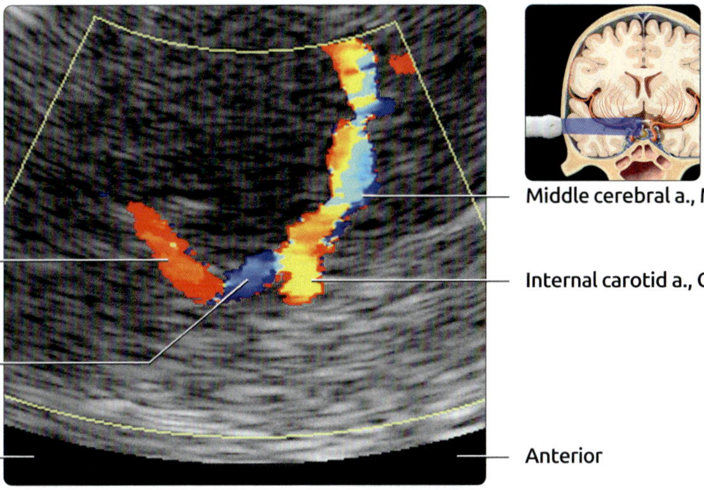

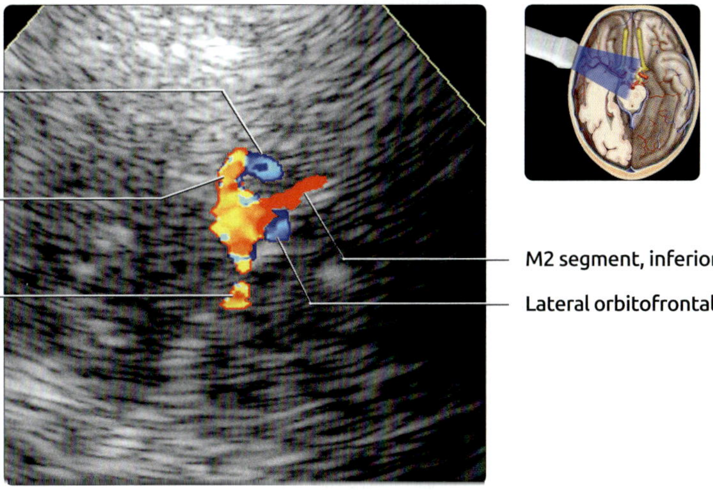

(Top) *Axial color Doppler ultrasound (transtemporal approach) on the mesencephalic plane illustrates the MCA appearing in the anterior aspect of the circle of Willis. The origin of the M1 segment can be determined accurately with identification of the MCA/ACA junction.* **(Middle)** *Axial color Doppler ultrasound (transtemporal approach) on the standard mesencephalic plane depicts the M1 segment arising from the communicating segment of the ICA with flow direction toward the transducer.* **(Bottom)** *Axial oblique color Doppler ultrasound of the MCA bifurcation (transtemporal approach) shows the superior and inferior trunks of the M2 segments. Note the MCA bifurcates in ~ 75% of normal subjects, and the rest trifurcate with an individual anterior temporal artery.*

MIDDLE CEREBRAL ARTERY

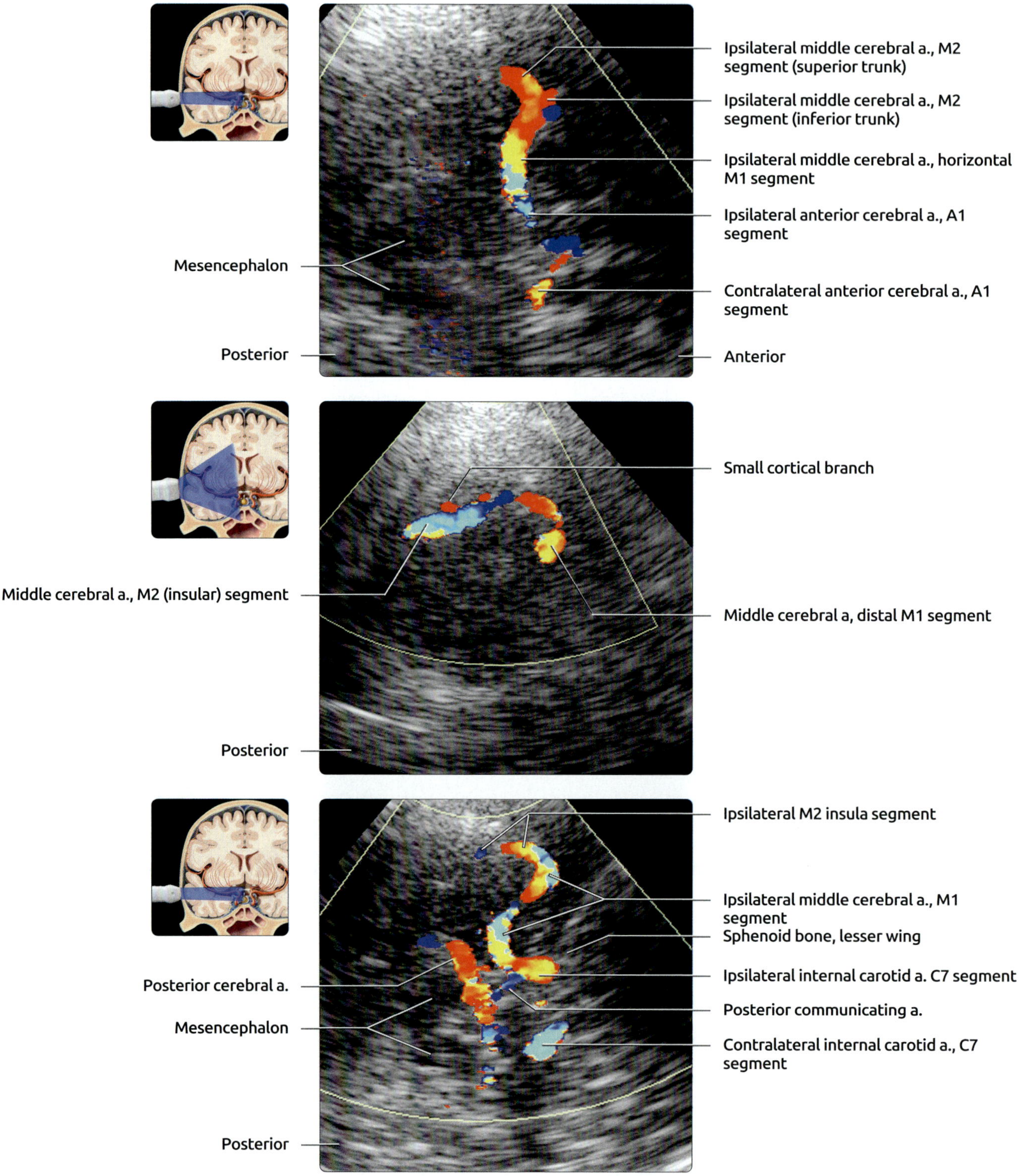

(Top) *Axial color Doppler ultrasound (transtemporal approach) of a young adult shows the M1 segment of the MCA on the standard mesencephalic plane where it starts from the ICA termination to MCA bi-/trifurcations. In young adults, the M1 usually bows dorsally.* **(Middle)** *Oblique coronal color Doppler ultrasound (transtemporal approach) of the same adult shows the M2 segment in the insula. Note that the flow direction in the M2 segment is directed toward the transducer in the proximal segment and away from the transducer as it ascends within the sylvian fissure.* **(Bottom)** *Axial color Doppler ultrasound (transtemporal approach) of the MCA in an elderly patient shows a tortuous M1 segment. Its distal portion bows ventrally, bringing the entire segment closer to sphenoid bone. The M1 segment originates near the medial aspect of the lesser wing of the sphenoid and traverses laterally and horizontally to end in the limen insulae.*

MIDDLE CEREBRAL ARTERY

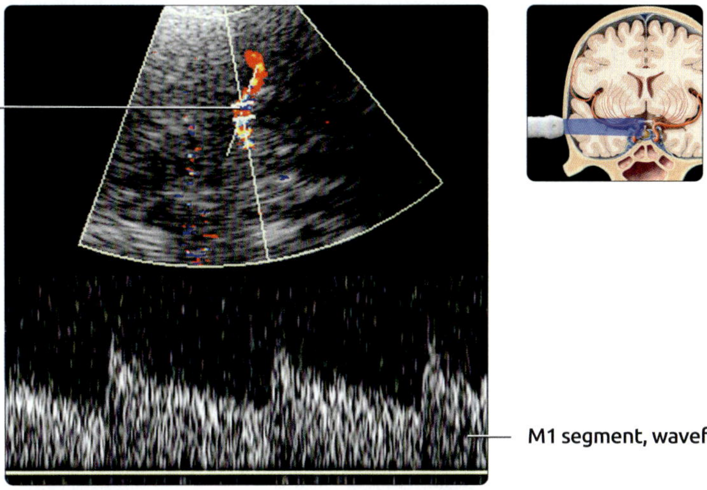

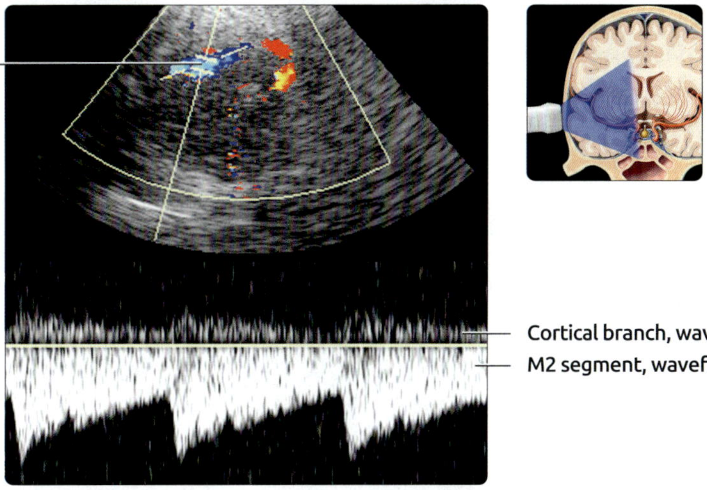

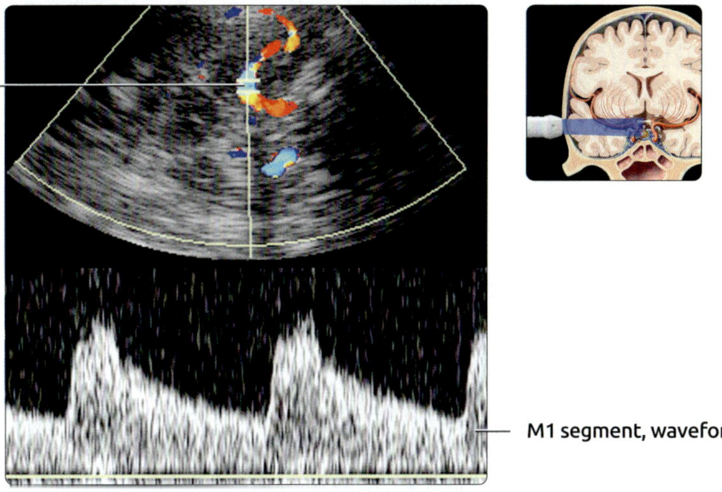

(Top) *Axial spectral Doppler ultrasound (transtemporal approach) of the MCA in a young adult shows that the waveform pattern is of low resistance, and the mean velocity is within normal range of 62 ± 12 cm/sec. The clinical importance of assessing the MCA is its high association with stroke in diseased arteries.* **(Middle)** *Oblique, coronal, spectral Doppler ultrasound (transtemporal approach) of a young adult shows a waveform of the M2 segment, which is away from the transducer. The waveform is inseparable from that of a small neighboring cortical branch with flow toward the transducer.* **(Bottom)** *Axial spectral Doppler ultrasound (transtemporal approach) of an elderly patient shows Doppler waveform of the M1 segment. Note that accurate flow velocity measurement in an elderly patient may be difficult due to uncertainty in angle correction in a tortuous vessel.*

POSTERIOR CEREBRAL ARTERY

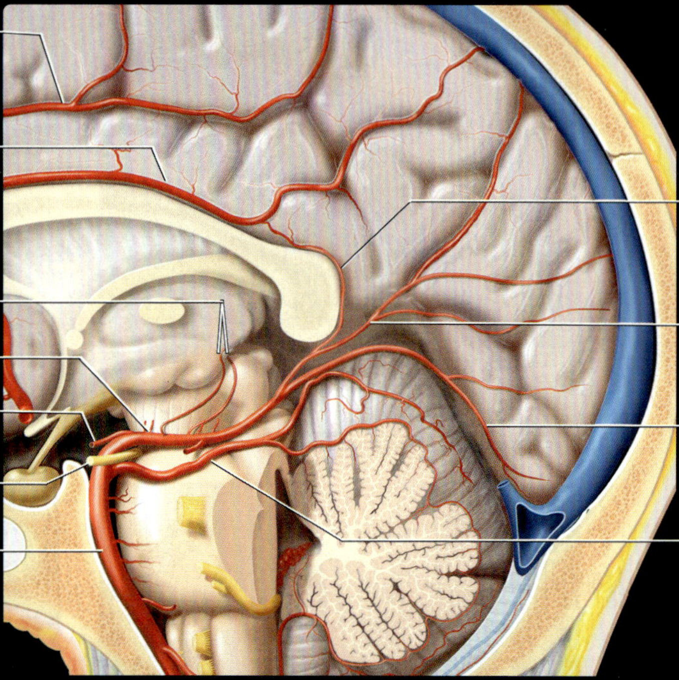

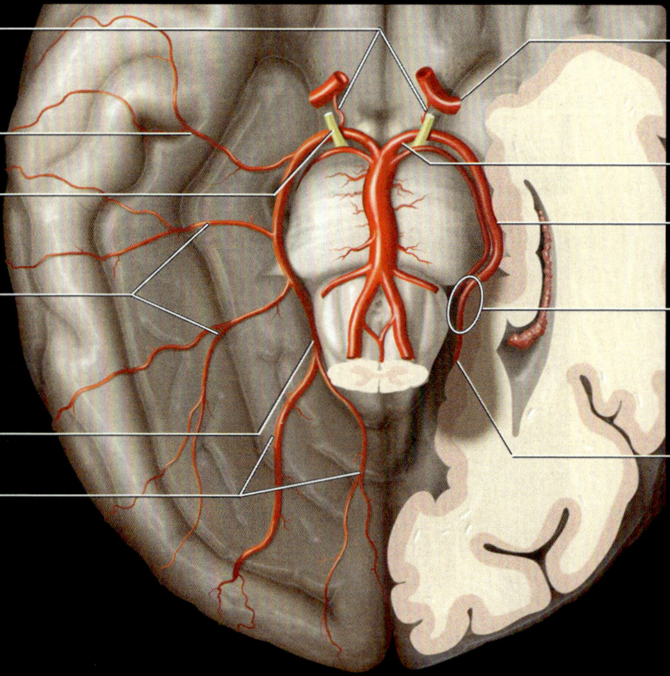

(Top) *Lateral graphic shows the posterior cerebral artery (PCA) and its branches. The PCA has central (perforating), choroidal, and cortical branches as well as a small branch to the corpus callosum splenium. The tentorium and CNIII lie between the PCA above and the superior cerebellar artery below.* **(Bottom)** *Submentovertex graphic shows the PCA and the relationship of its segments to the midbrain. The PCA supplies the occipital lobe and almost all of the inferior surface of the temporal lobe (except for its tip). The precommunicating (P1) PCA segment extends from the basilar bifurcation to the PCoA junction. The ambient (P2) segment swings posterolaterally around the midbrain. The quadrigeminal segment (P3) lies behind the midbrain. The PCA terminal segment is the calcarine (P4) segment.*

POSTERIOR CEREBRAL ARTERY

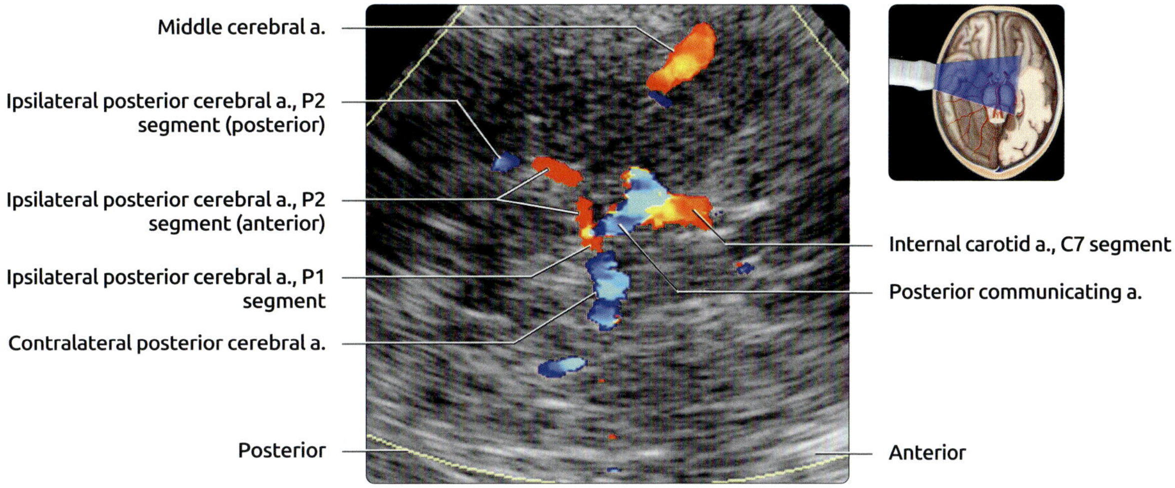

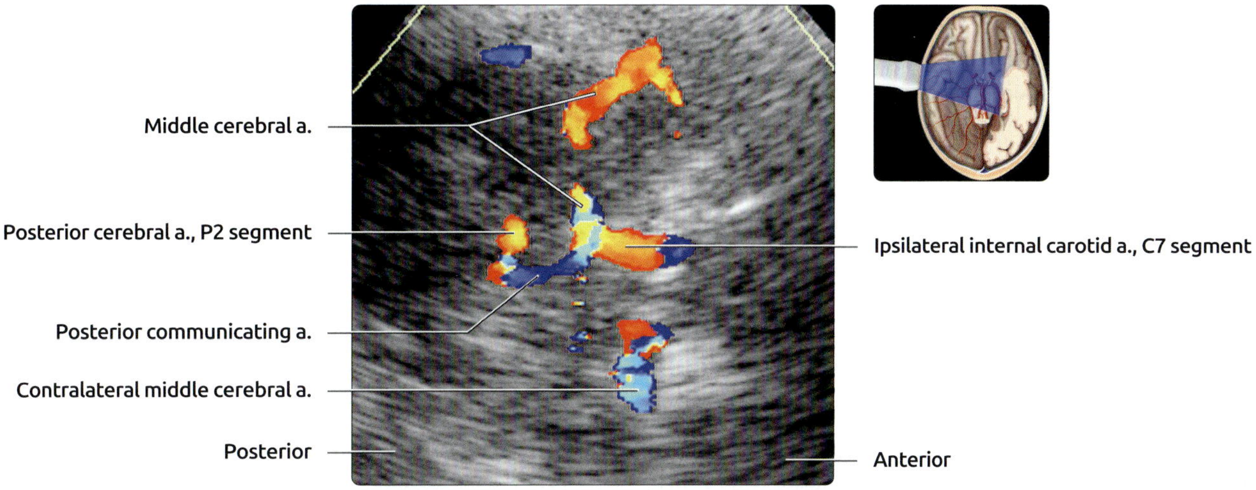

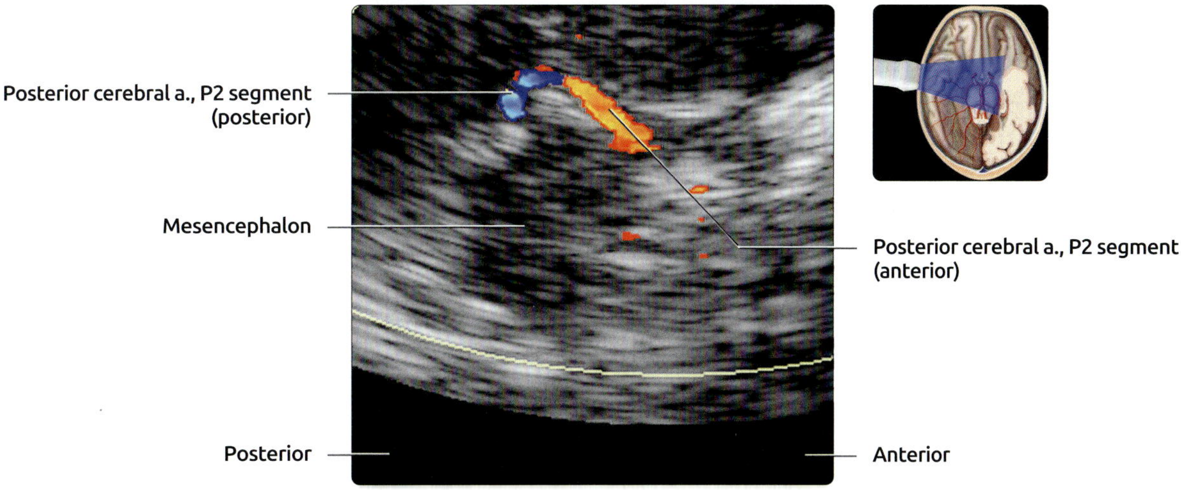

(Top) *Axial color Doppler ultrasound (transtemporal approach) on the mesencephalic plane shows the P1 segment extending laterally from the basilar artery to its junction with the PCoA and continuing as the P2 segment around the mesencephalon. The flow direction of P1 and anterior P2 segments is toward the transducer, whereas the posterior P2 segment is away from the transducer.* **(Middle)** *Axial color Doppler ultrasound (transtemporal approach) shows the PCoA is identified as a vascular structure connecting the ICA C7 segment to the PCA with flow away from the transducer. The clinical importance of investigation of PCoA is due to the prevalence of aneurysm formation and effectiveness of collateralization.* **(Bottom)** *Axial color Doppler ultrasound (transtemporal approach) shows the P2 segment curving posteriorly around mesencephalon with change in flow direction around the curvature.*

POSTERIOR CEREBRAL ARTERY

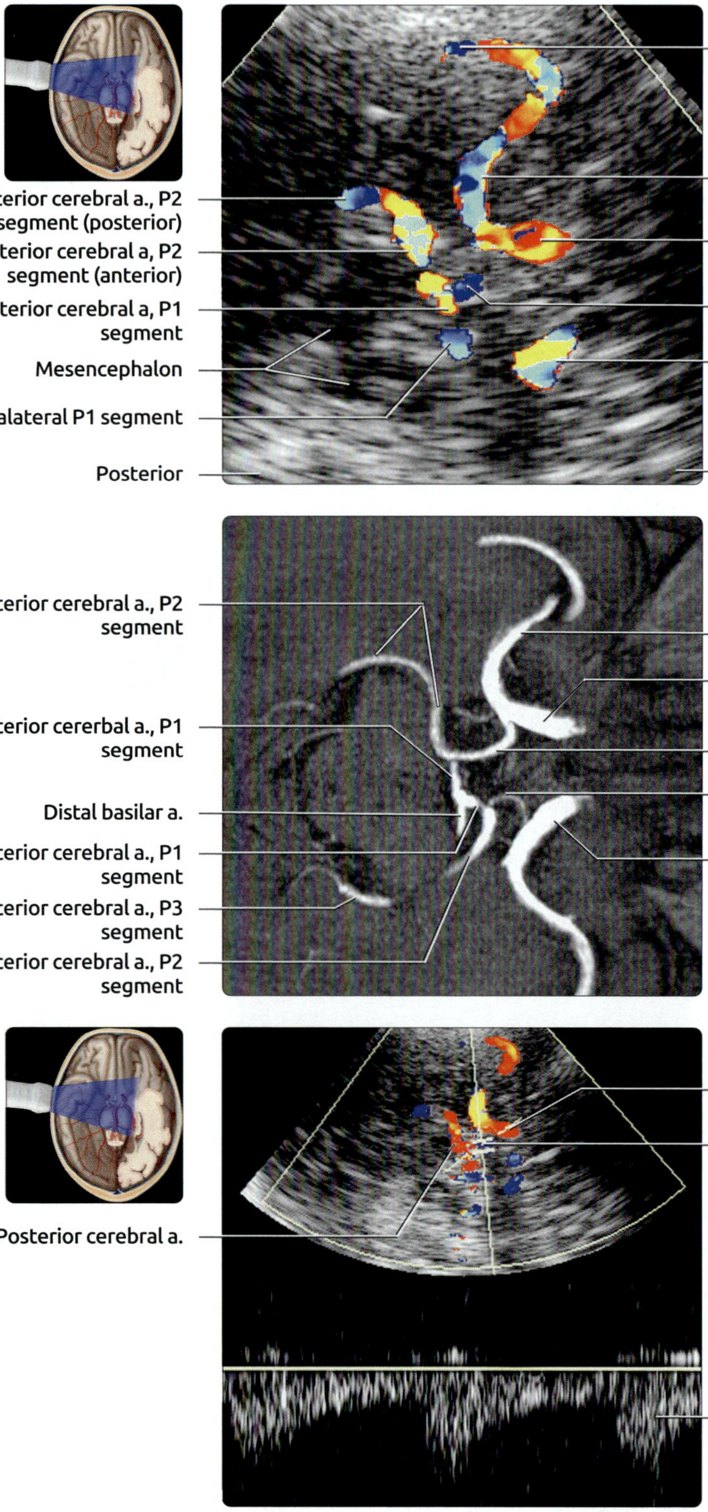

(Top) *Axial color Doppler ultrasound (transtemporal approach) on the mesencephalic plane shows the P1 segment emerging in the interpeduncular cistern with the P2 segment encircling the mesencephalon with flow toward transducer in P1 and P2 segments. A short P3 segment is also noted with flow away from the transducer. With adequate acoustic penetration, the contralateral arterial segments can be visualized with flow direction opposite to the ipsilateral arteries.* **(Middle)** *Corresponding MR arteriogram in a similar projection plane shows the bilateral PCAs and PCoAs in the circle of Willis.* **(Bottom)** *Axial spectral Doppler ultrasound (transtemporal approach) of the PCoA on standard the mesencephalic plane shows normal backward flow from the ICA to the PCA in this artery. Note only 75% of the PCoA is discernible transcranially.*

POSTERIOR CEREBRAL ARTERY

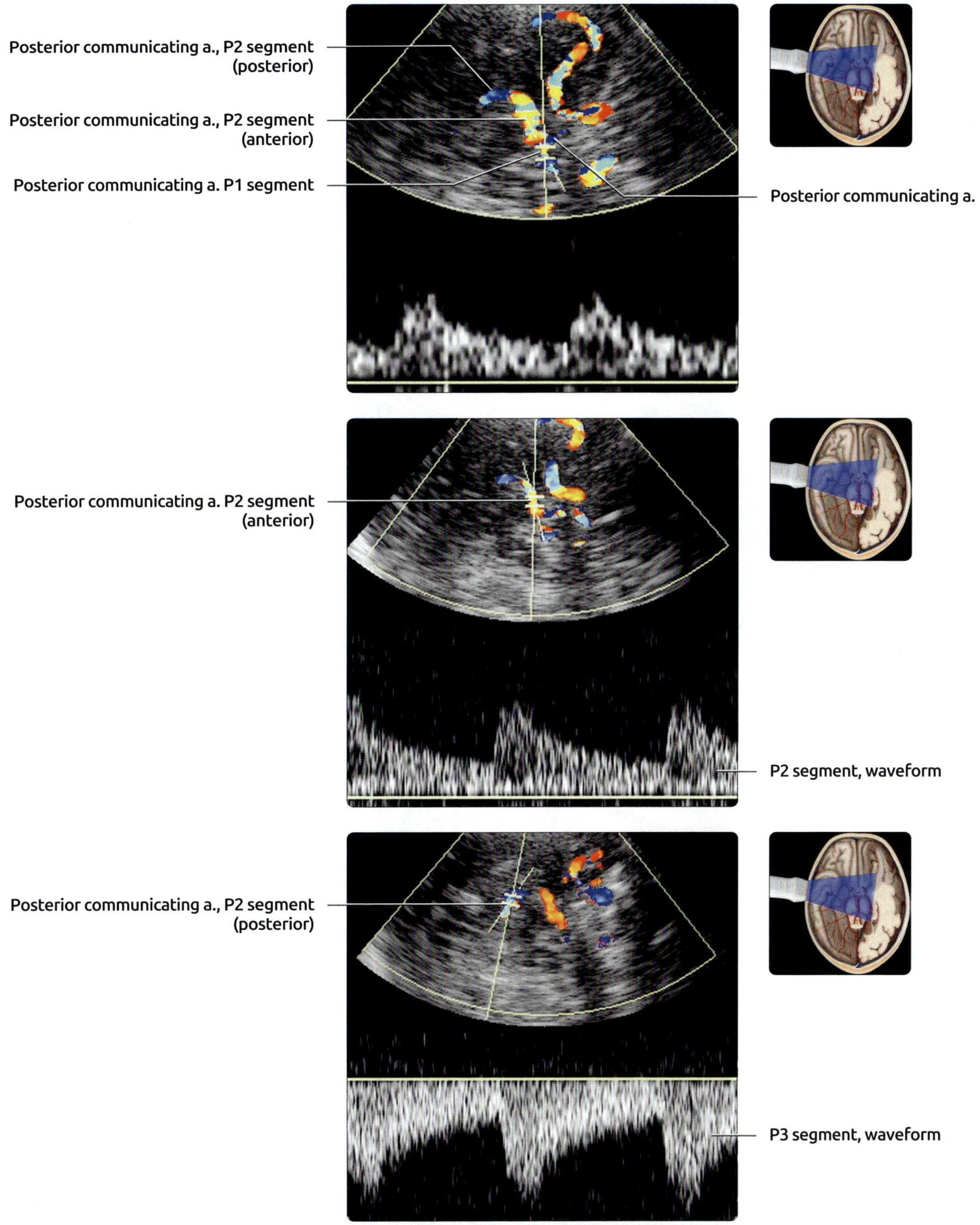

(Top) *Axial spectral Doppler ultrasound (transtemporal approach) of the P1 segment shows the waveform of P1 flow is of low resistance and on the positive side of the trace. Note that the P1 segment is short and often hypoplastic or absent, making its visualization difficult.* **(Middle)** *Axial spectral Doppler ultrasound (transtemporal approach) of the anterior P2 segment on the axial mesencephalic plane shows normal antegrade low-resistance flow with a waveform pattern similar to the P1 segment.* **(Bottom)** *Axial spectral Doppler ultrasound (transtemporal approach) of the posterior P2 segment on the axial ventricular plane shows the waveform of this segment is of low resistance and on the negative side of the trace, denoting flow away from transducer.*

VERTEBROBASILAR SYSTEM

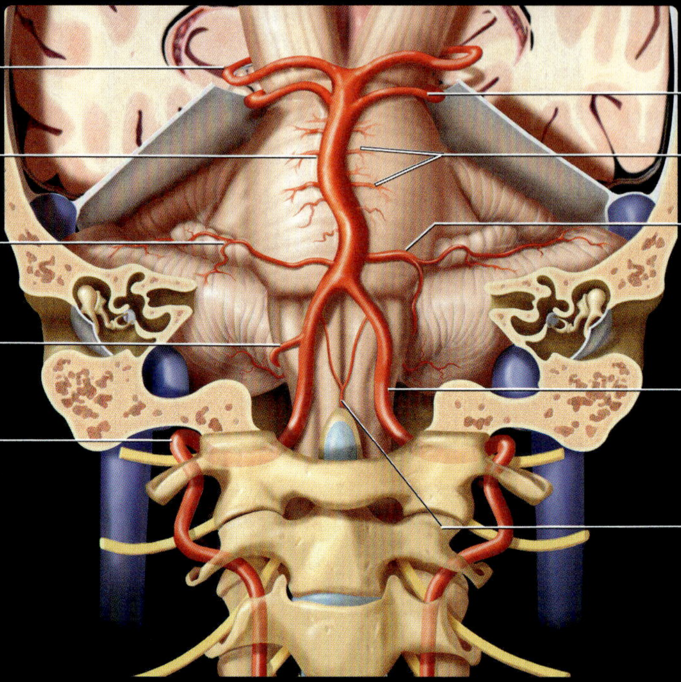

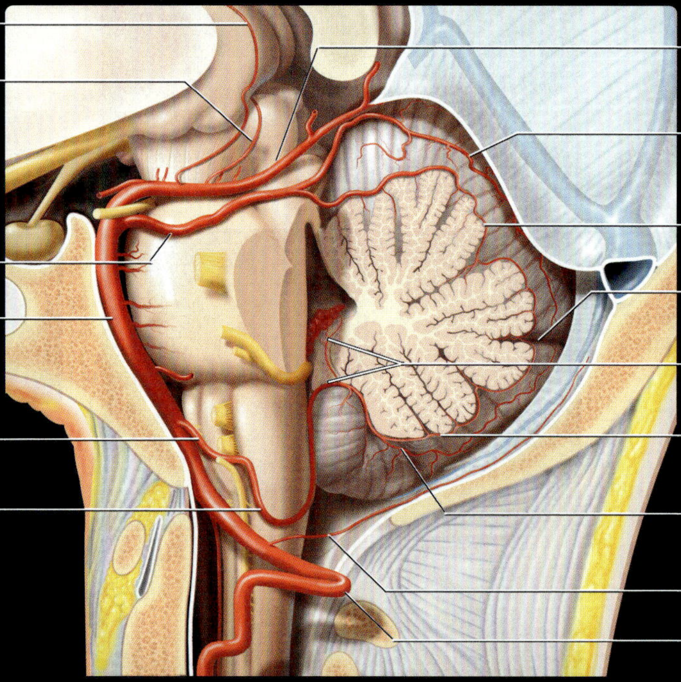

(Top) *Frontal graphic depicts the vertebrobasilar system. The V3 is the short extraspinal vertebral artery (VA) segment that extends from the top of the C1 to the foramen magnum. The V4 is the intradural (intracranial) segment. A right posterior inferior cerebellar artery (PICA) is shown originating from the VA, while the right anterior inferior cerebellar artery (AICA) is from the basilar artery. A combined AICA-PICA trunk is a common normal variant and is shown on the left side.* **(Bottom)** *Lateral graphic depicts the vertebrobasilar system. Note the relationship of the PICA loops to the medulla, cerebellar tonsil. Watershed between the SCA, PICA is often near the great horizontal fissure of the cerebellum.*

VERTEBROBASILAR SYSTEM

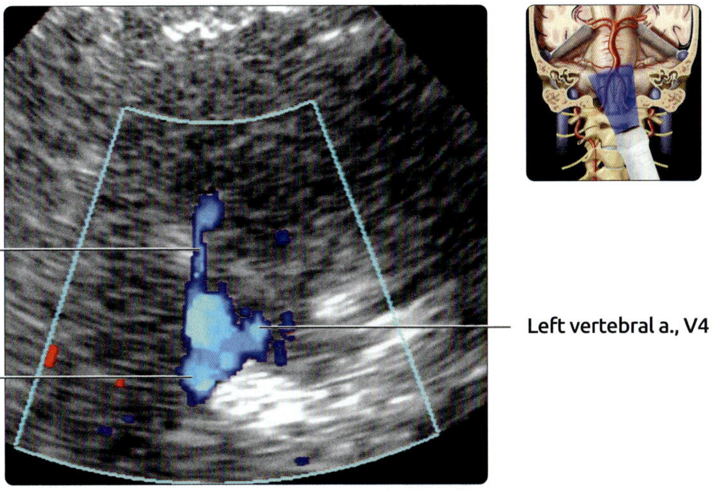

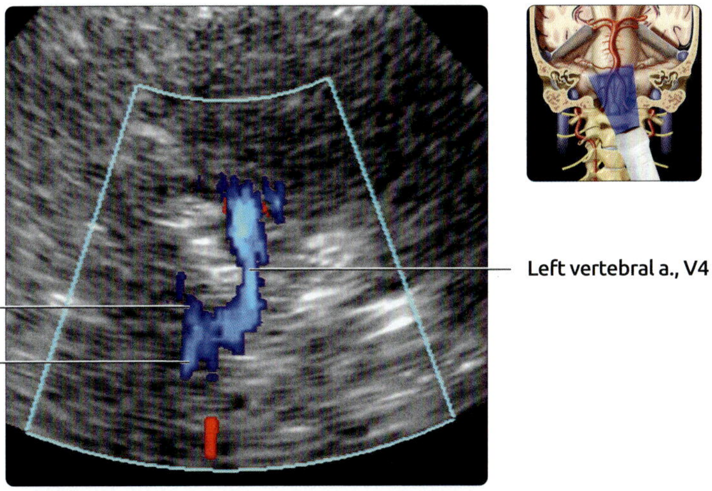

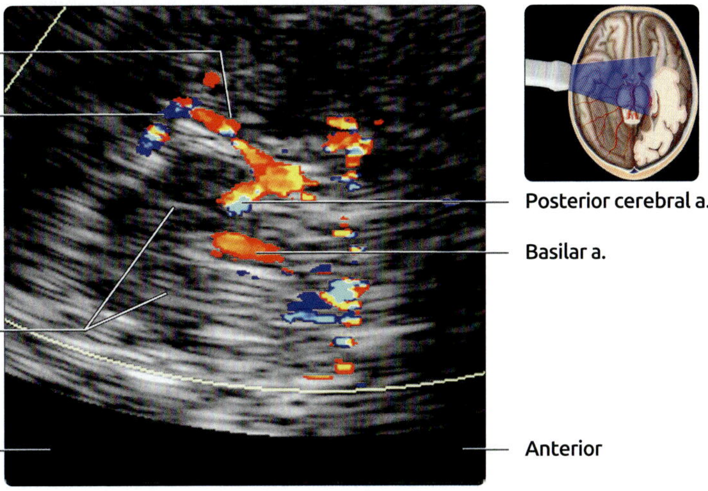

(Top) *Axial color Doppler ultrasound (suboccipital approach) through the foramen magnum shows the right V4 segment of the VA and proximal basilar artery with flow directed away from the transducer. Note that reversed basilar artery flow toward the transducer would signify severe intracranial subclavian steal.* **(Middle)** *Axial color Doppler ultrasound (suboccipital approach) through the foramen magnum shows the left V4 segment of the VA with flow directed away from the transducer. The convergence of 2 V4 segments forming the basilar artery should give rise to a Y configuration. However, the Y configuration may not be always obtainable because the 3 vessels are not always on the same plane.* **(Bottom)** *Axial color Doppler ultrasound (transtemporal approach) on the standard mesencephalic plane shows the distal portion of basilar artery, which terminates into the PCAs in the interpeduncular/suprasellar cistern.*

VERTEBROBASILAR SYSTEM

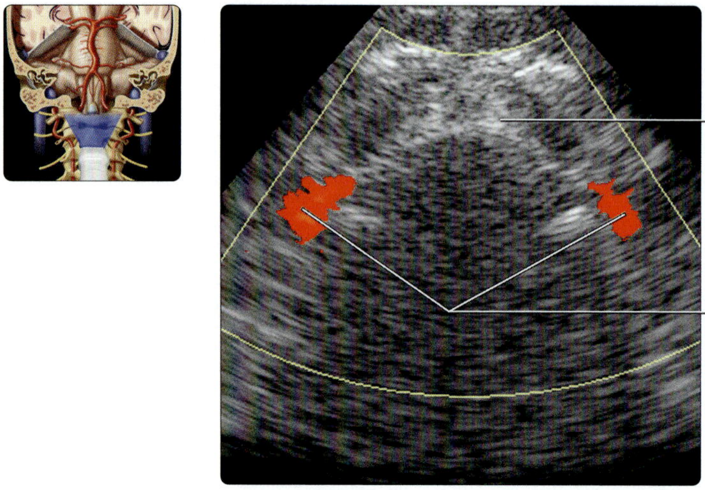

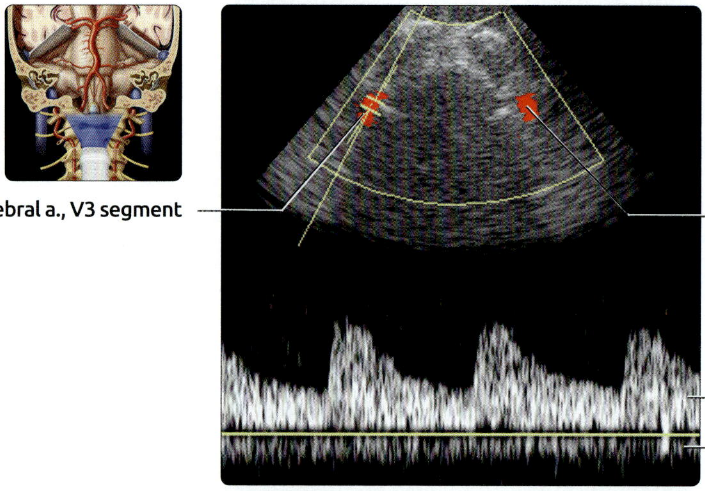

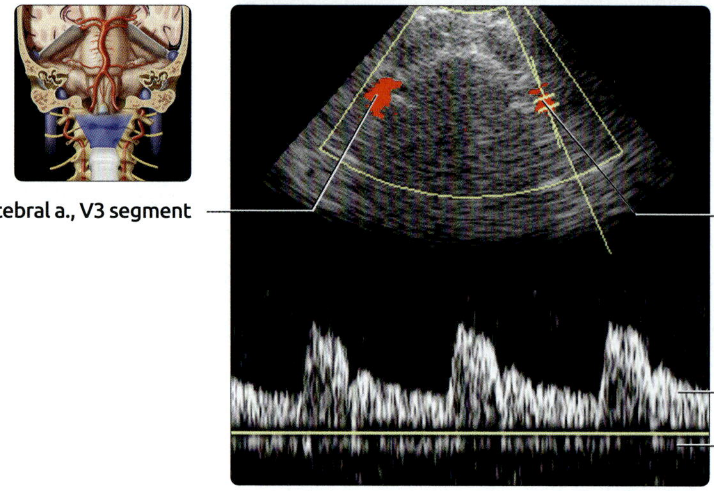

(Top) *Axial color Doppler ultrasound (suboccipital approach) at a level just above the atlas shows the 2 short extraspinal V3 segments exiting from the transverse foramina of the atlas and running posteromedially at the horizontal groove on the posterior arch (atlas). The flow direction of these vessels is toward the transducer.* **(Middle)** *Corresponding spectral Doppler ultrasound of the right V3 segment demonstrates the characteristic antegrade low-resistance flow pattern of this vessel. Note the Doppler signal from the ipsilateral vertebral vein is frequently inseparable from the arterial signal and is detected on the opposite side of the trace.* **(Bottom)** *Corresponding spectral Doppler ultrasound of the left V3 segment shows similar waveform as the contralateral artery.*

VERTEBRAL ARTERY V4 SEGMENT

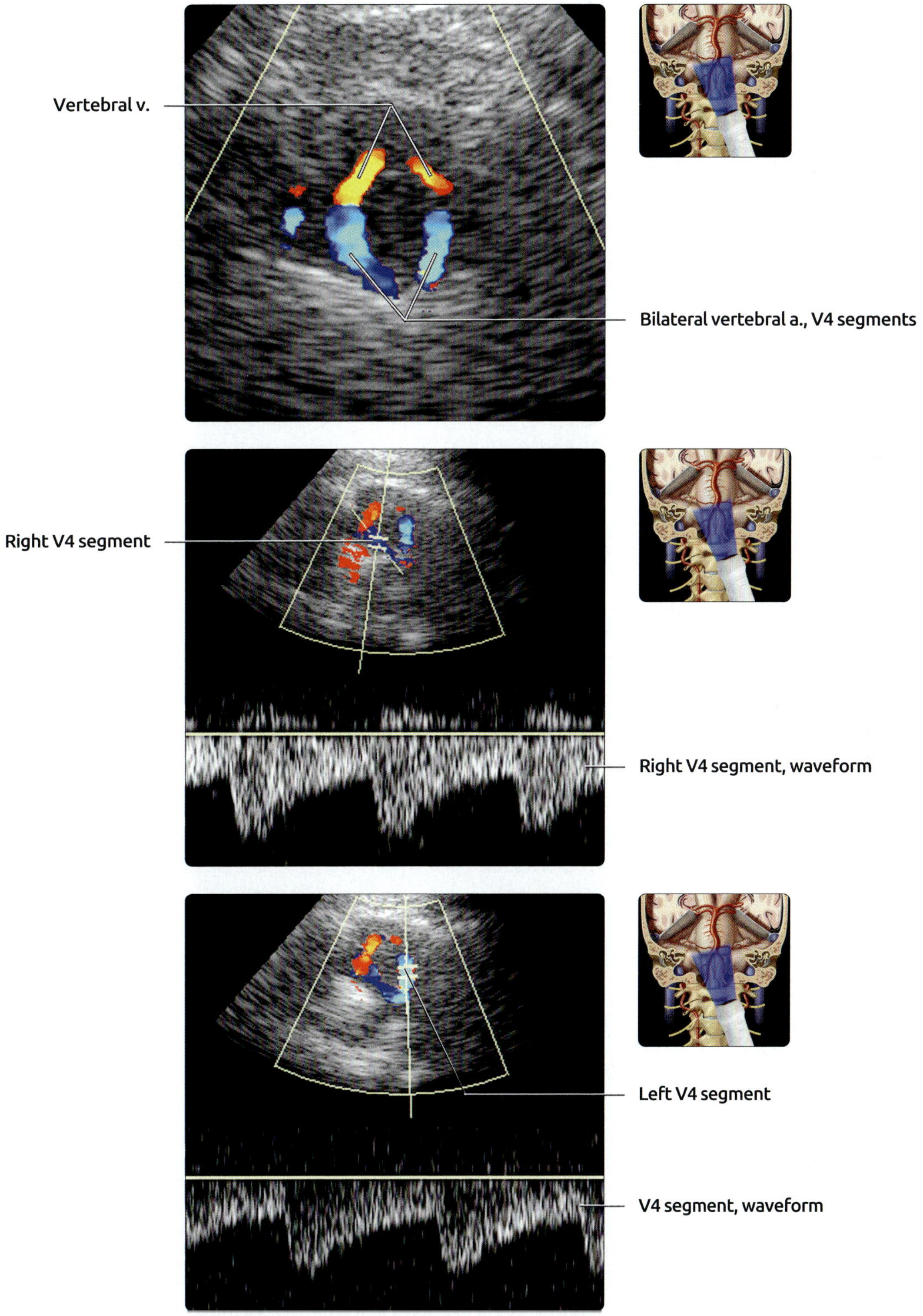

(Top) *Axial color Doppler ultrasound (suboccipital approach) through the foramen magnum shows the 2 intradural V4 segments of the VA with flow directed away from the transducer. The 2 vertebral veins formed in the suboccipital triangle from numerous small tributaries are also visualized with flow toward the transducer.* **(Middle)** *Corresponding spectral Doppler waveform of the right V4 segment shows the normal antegrade low-resistance flow pattern similar to the V3 segment, but with opposite flow direction, as evident by the negative tracing.* **(Bottom)** *A spectral Doppler waveform of the left V4 segment shows a similar tracing.*

INTRACRANIAL VEINS AND SINUSES

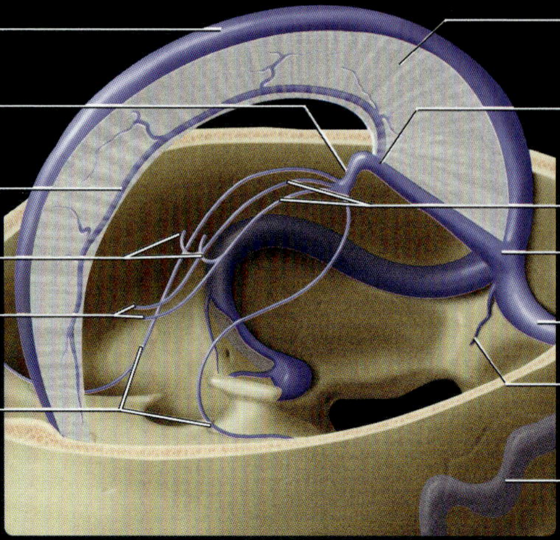

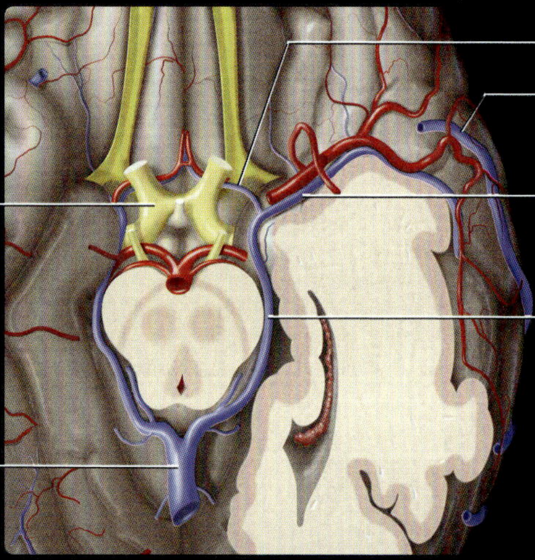

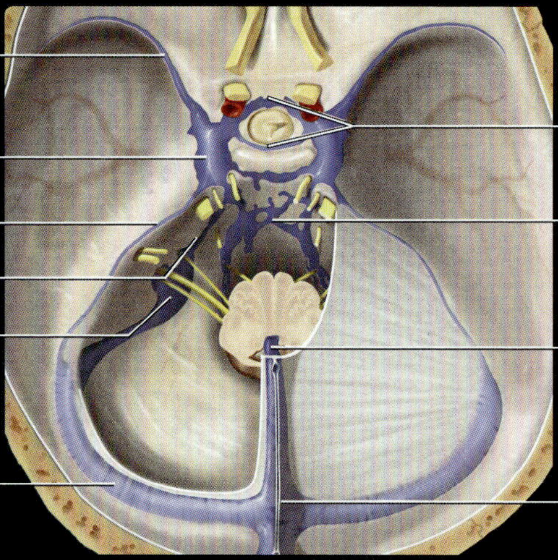

(Top) *This 3D rendering of the falx cerebri with major dural sinuses and deep veins shows the interconnections between these 2 venous systems.* **(Middle)** *Graphic shows an inferior intracranial view of the deep venous structures. Intracranial veins are valveless veins, which pierce the arachnoid membrane and meningeal layer of dura mater and drain into the cranial venous sinuses.* **(Bottom)** *Graphic shows the superior intracranial view of the dural venous sinuses. The cerebral hemispheres, midbrain, and pons as well as the left 1/2 of the tentorium cerebelli have been removed. Note the numerous interconnections between both halves of the cavernous sinus, the basilar venous plexus, and the petrosal sinuses.*

INTRACRANIAL VEINS

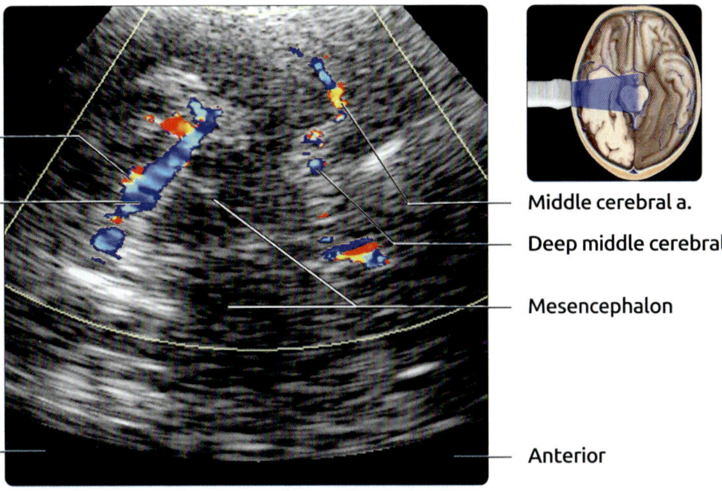

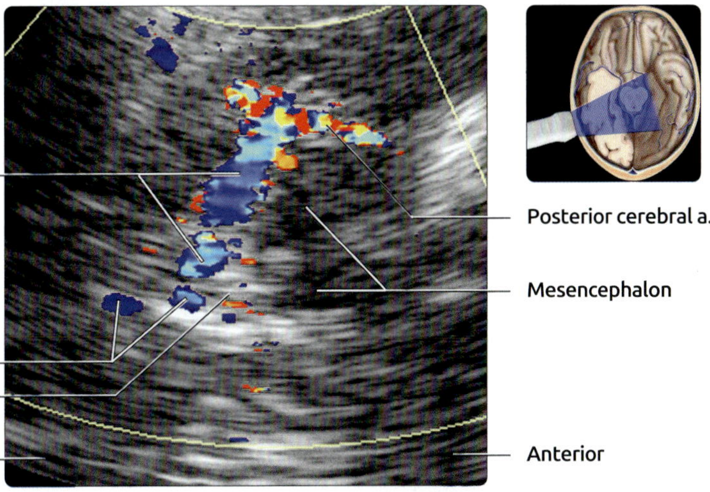

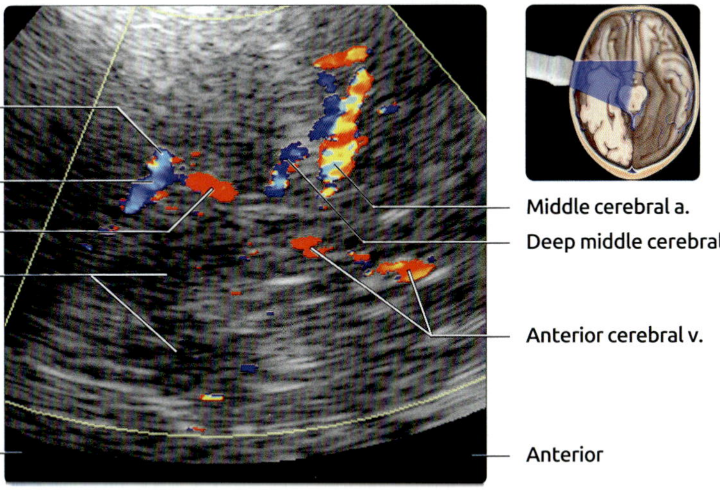

(Top) *Axial color Doppler ultrasound (transtemporal approach) illustrates the basal vein of Rosenthal (BV) running posteriorly along the mesencephalon with flow signal away from the transducer. The ipsilateral PCA is often seen closely related to it with opposite flow direction.* **(Middle)** *Axial color Doppler ultrasound (transtemporal approach) demonstrates the great vein of Galen (GV). The vein is identified in the midline posterior to mesencephalon in the hyperechoic triangular region of the pineal body and quadrigeminal cistern.* **(Bottom)** *Axial color Doppler ultrasound (transtemporal approach) shows the 2 tributaries of the BV: The anterior cerebral vein (ACV) and deep middle cerebral vein (dMCV). The ACV courses posteriorly along the base of the brain and joins the dMCV, which runs in the sylvian fissure to form the BV just anterior to the mesencephalon. Note that both the ACV and MCV are accompanied by their respective arteries.*

INTRACRANIAL VEINS

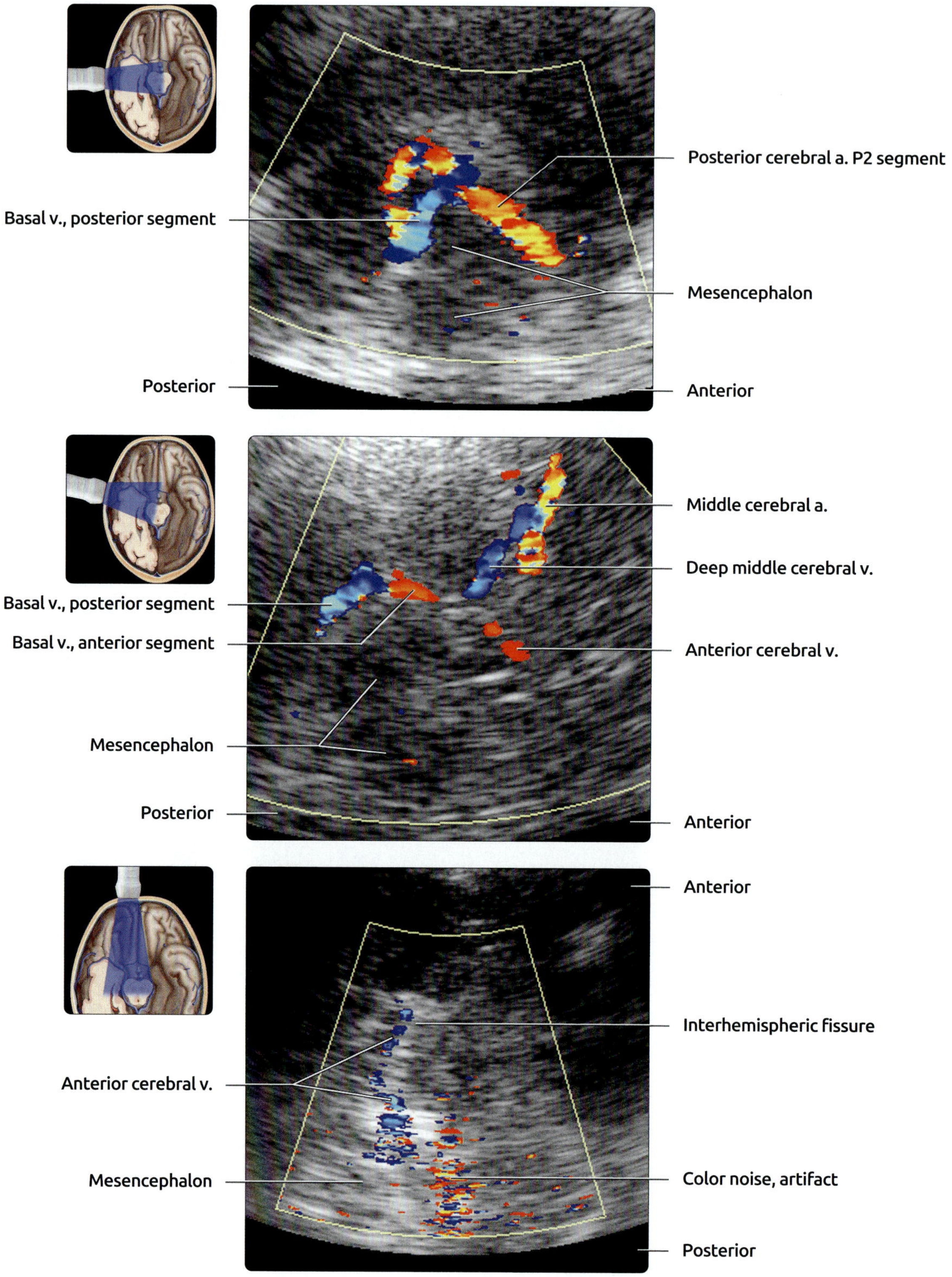

(Top) *Axial color Doppler ultrasound (transtemporal approach) shows the posterior segment of the BV. The vein is seen surrounding the mesencephalon and runs almost parallel to the PCA. The blue color coding of the vessel indicates flow directed away from the transducer.* **(Middle)** *Axial color Doppler ultrasound (transtemporal approach) with a more anterior tilting approach is shown. The ACV is seen running posteriorly to join the dMCV before draining into the anterior segment of BV. The flow in the ACV and anterior segment of BV is directed toward the transducer, as denoted by the red color coding.* **(Bottom)** *Axial color Doppler ultrasound (paramedian frontal approach) shows the ACV, which courses posteriorly within the interhemispheric fissure; this small vein, after joining the dMCV, forms the BV. Identification of this vessel is difficult due to its slow flow.*

INTRACRANIAL VEINS

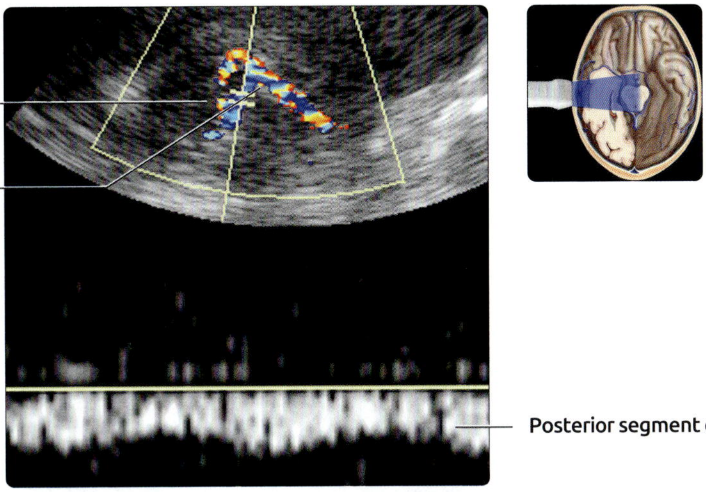

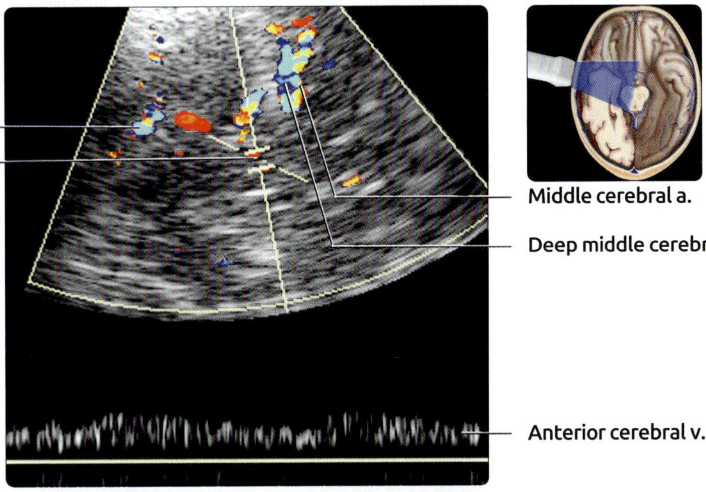

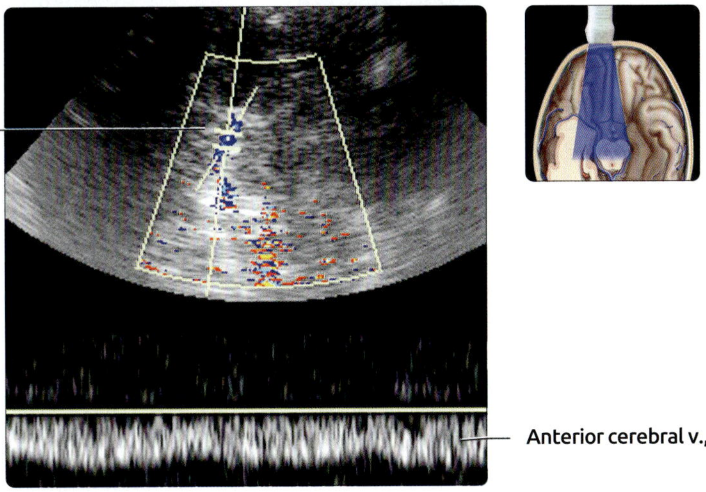

(Top) *Axial spectral Doppler ultrasound (transtemporal approach) shows the BV waveform (posterior segment) with typical respiratory phasicity.* **(Middle)** *Axial spectral Doppler ultrasound (transtemporal approach) shows the Doppler waveform of the ACV. Note the waveform is on the positive scale, representing flow toward the transducer.* **(Bottom)** *Axial spectral Doppler ultrasound shows the ACV waveform obtained from the paramedian frontal approach with the flow away from the transducer. Note the flow direction of a vessel depends on the scanning approach used, and the ACV flow is toward the transducer using transtemporal approach but away from the transducer using the transfrontal approach.*

GROSS ANATOMY

Vertebral Bodies

- Vary in size, shape depending on region
- **Cervical**: Upper 7 vertebrae
  - **C1 (atlas)**: No body, spinous process; circular shape
  - **C2 (axis)**: Body with bony peg (dens/odontoid process)
  - C3-C6 similar in size, shape; C7 marked by longest spinous process
- **Thoracic**: Bodies heart-shaped, central canal round, short pedicles, broad laminae
  - Costal articular facets on body/transverse processes
- **Lumbar**: Large body, thick pedicles, broad laminae
- **Sacrum**: Fusion of 5 segments
- **Coccyx**: Fusion of 3-5 segments

Spinal Cord

- Suspended within thecal sac
- Anchored to dura by **denticulate ligaments**
- **2 widened segments**: Cervical enlargement (C3-T2) and lumbar enlargement (T9-T12)
- Cord tapers to diamond-shaped point (**conus medullaris**), normally ends at T12 to L2-L3, most common at T12-L1
- In contrast to brain, gray matter is on inside with white matter on periphery of cord
  - Central gray matter, formed by columns ("horns") of neuronal cell bodies, is roughly H-shaped
- **Filum terminale**: Connective tissue extension of pia mater extending inferiorly from conus
  - Fuses distally into dura, attaches to dorsal coccyx
- **Cauda equina**: "Horse's tail" of lumbar, sacral, coccygeal nerve roots below conus
- **Nerve roots**
  - Cervical spine: 8 nerves, which exit above vertebral body
    - C8 nerve root exits above T1
  - Remainder of nerve roots exit below vertebral body
  - Paired dorsal, ventral roots exit from their respective hemicords
    - Unite in/near intervertebral foramen to form ganglion

IMAGING ANATOMY

Vertebral Bodies

- Ossified vertebral body appears echogenic
  - Cartilaginous tip at spinous process appears hypoechoic
- Cartilaginous coccyx is hypoechoic
  - Ossified coccygeal vertebral bodies have rounded central nucleus rather than square contour as in sacrum

Spinal Cord

- Cord is predominately hypoechoic
  - **Central echogenic complex**: Interface between myelinated ventral white matter commissure and CSF within ventral median fissure
  - **Central spinal canal (ependymal canal)**: CSF-containing space throughout length of cord; contiguous with ventricular system
    - Typically do not see fluid in canal unless dilated
  - **Conus medullaris**: Terminal part of cord, should taper gradually
- **Cauda equina**: Multiple, linear, diverging nerve roots
  - Drape dependently within thecal sac, undulate with each CSF pulsation
- **Filum terminale** should be < 2 mm in diameter
  - Hypoechoic center with more echogenic outer margin
  - Dorsal extension toward coccyx

ANATOMY IMAGING ISSUES

Imaging Recommendations

- Nonossified posterior-median intraneural synchondrosis provides ample acoustic window in newborns
- Spinal cord is best visualized by US within 1st month after birth for term infants
- Transverse scan of cord possible in older infant as cartilaginous gap in vertebral ring allows penetration of US beam
- Determination of position of tip of conus medullaris should always be included in neonatal spine US

Imaging Approaches

- Infants are preferably scanned in prone position
- Decubitus position is adopted to calm struggling baby by bottle or breast feeding
- Scan both longitudinally and transversely using high-frequency linear US transducer
- Ways to define vertebral levels where conus ends
  - Count downward from 12th rib
  - Count upward after defining lumbosacral junction
    - For more clear delineation of L5-S1, accentuate lumbar lordosis by elevation of shoulders
- Craniocervical junction can be assessed by scanning base of skull through foramen magnum

Imaging Pitfalls

- Counting of vertebral level upward from last ossified vertebral body can be misleading due to variability in ossification of coccygeal vertebral bodies
- In older infants, ossification may preclude an adequate examination

CLINICAL IMPLICATIONS

Clinical Importance

- **Tethered cord**
  - Conus terminates at or above inferior L2 vertebra in ≥ 98% of normal population
    - Conus at normal position by 0-2 months of age
  - Conus terminating below L2-L3 disc is abnormal at any postnatal age but significance questionable in absence of signs/symptoms
    - Important to evaluate appearance of nerve roots and filum terminale as well as conus level
    - Follow-up scan for borderline cases
  - Conus located over mid L3 or lower, or with lack of normal nerve-root pulsations, requires further evaluation with MR
  - Filum terminale thickened (> 2 mm at L5-S1 on axial/transverse images)
  - Cord may appear taut or directly apposed to dorsal thecal sac
    - Lack of conus motion with CSF pulsations
    - Lack of dependent ventral shift of conus when prone

CERVICAL SPINE

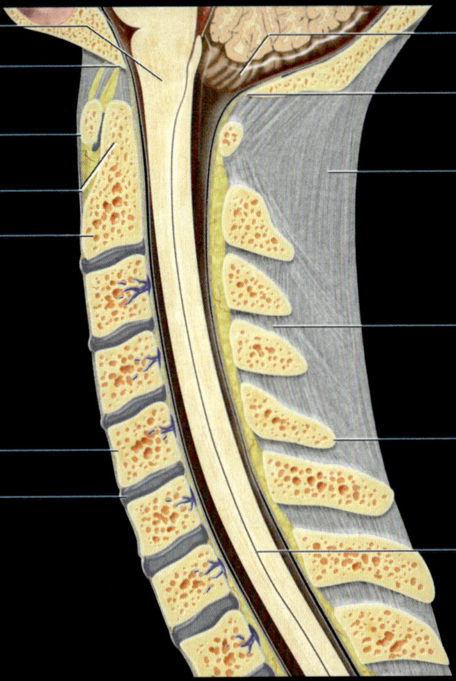

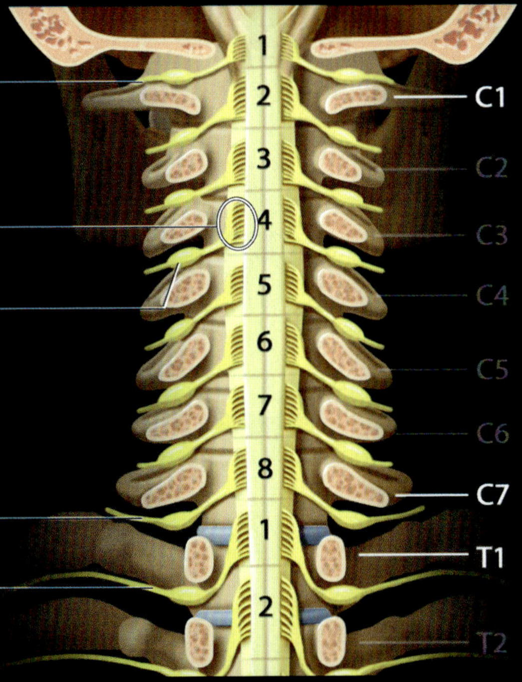

(Top) *Sagittal midline graphic of the cervical spine and cord shows a gentle lordotic curve and smooth alignment of the adjacent vertebrae. The cerebellar tonsils are normally above the line between the basion and opisthion. There is smooth transit of the cervicomedullary junction without kinking.* **(Bottom)** *Graphic of the cervical spine, with key vertebral bodies highlighted, shows the nerve rootlets coalescing; after forming ventral and dorsal nerve roots, they will unite in the intervertebral foramen to form a spinal ganglion. The cervical nerves are unique as they exit above their respective vertebral body. C8 nerve root exits above T1; below this, the nerve root exits below its vertebral body.*

SPINAL CORD

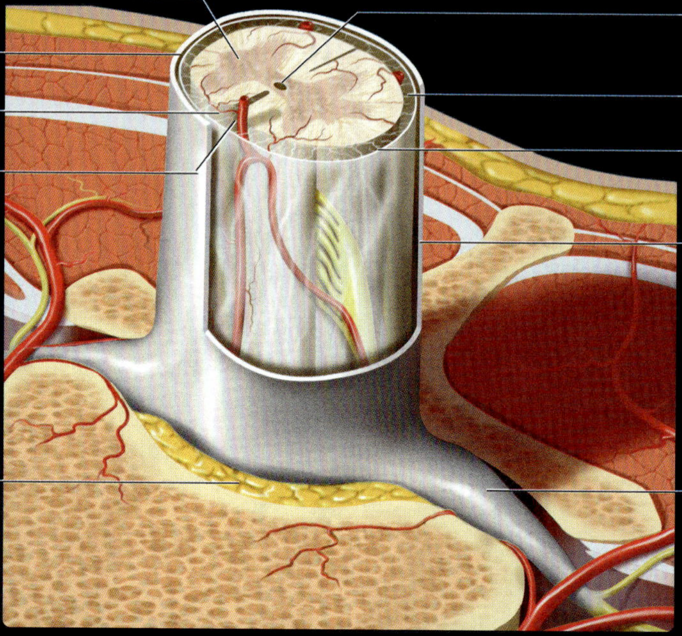

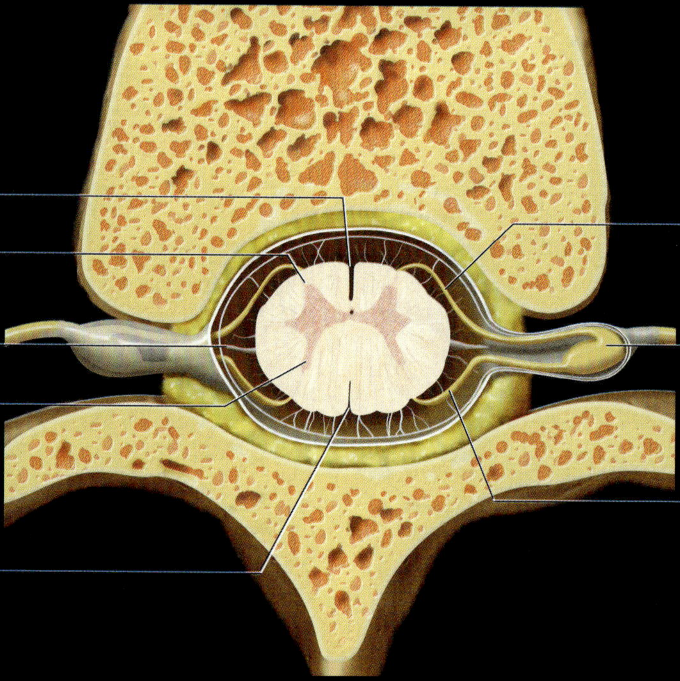

(Top) *Cutaway graphic of the spinal cord and its coverings demonstrates the meningeal layers and their relationship to adjacent regional structures.* **(Bottom)** *Axial graphic shows the internal anatomy of the distal thoracic spinal cord. In contrast to the brain, the gray mater is on the inside, while the white matter is along the periphery. The gray matter forms columns that are roughly H-shaped in cross section. The deep ventral median fissure divides the ventral hemicords, while the smaller dorsal median sulcus divides the dorsal hemicords. The dorsal and ventral nerve roots arise from the dorsolateral and ventrolateral sulci, respectively, and unite in the intervertebral foramen to form the spinal ganglion.*

CAUDA EQUINA

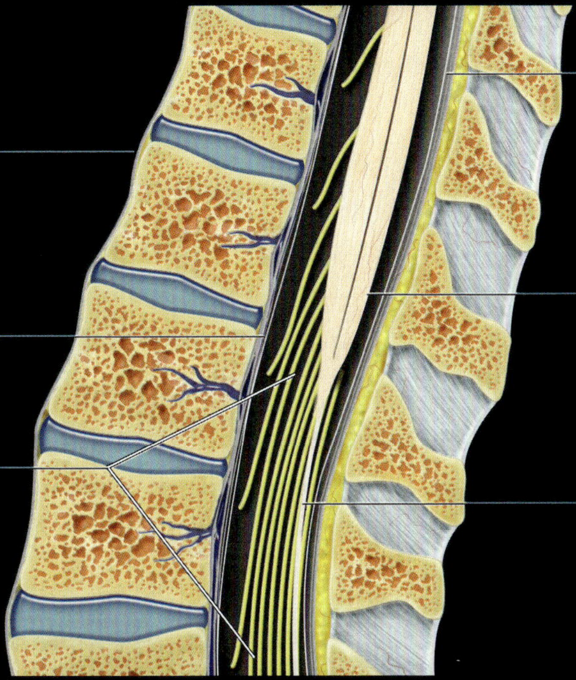

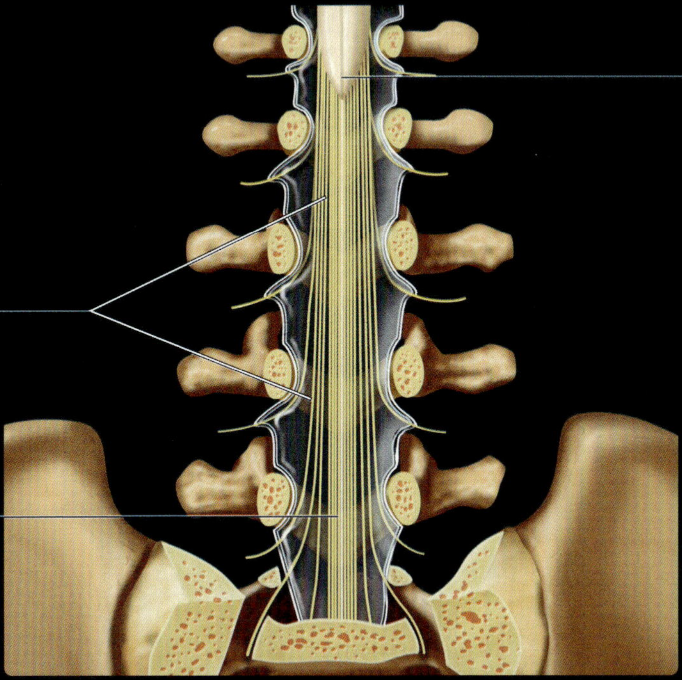

(Top) *Sagittal graphic of the thoracolumbar junction demonstrates a normal conus and cauda equina anatomy. The filum terminale lies among the cauda equina roots and affixes the conus dorsally to the terminal thecal sac.* **(Bottom)** *Coronal graphic through the middle of the spinal canal shows the distal spinal cord and nerve roots of the cauda equina. Note the cord ends in a diamond-shaped point, the conus medullaris. Lumbar nerve roots exit the thecal sac just under the pedicles of their same-numbered vertebral segments. The filum terminale is a strand of connective tissue that extends inferiorly from the conus to the dorsal coccyx. It typically contains no functional neural tissue and no fat. A normal conus should end above the inferior margin of L2. Extension below the the L2-L3 disc space is concerning for possible tethered cord.*

CERVICAL SPINE

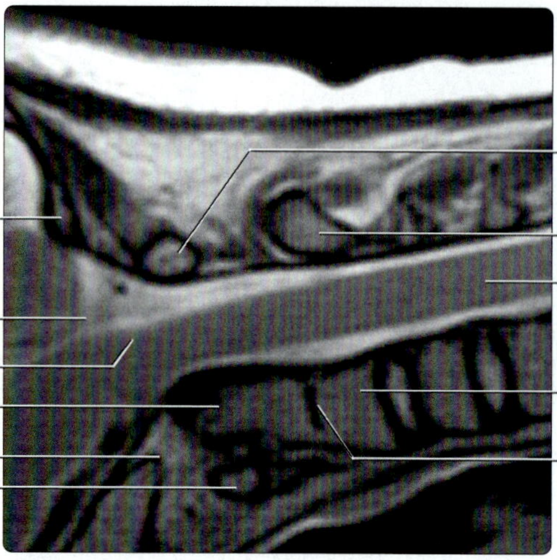

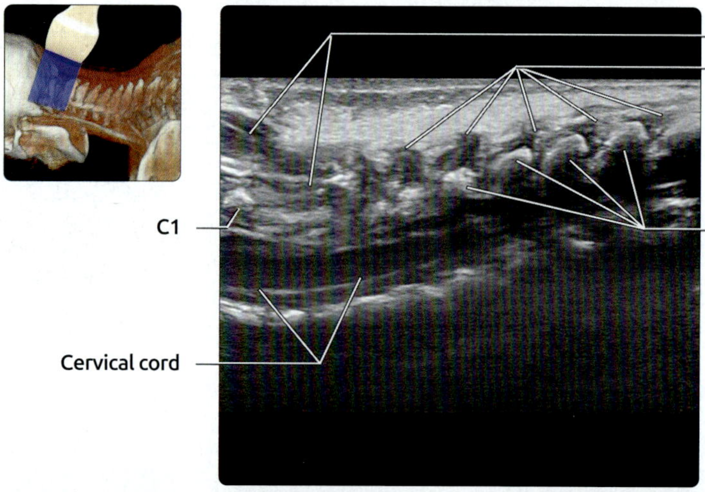

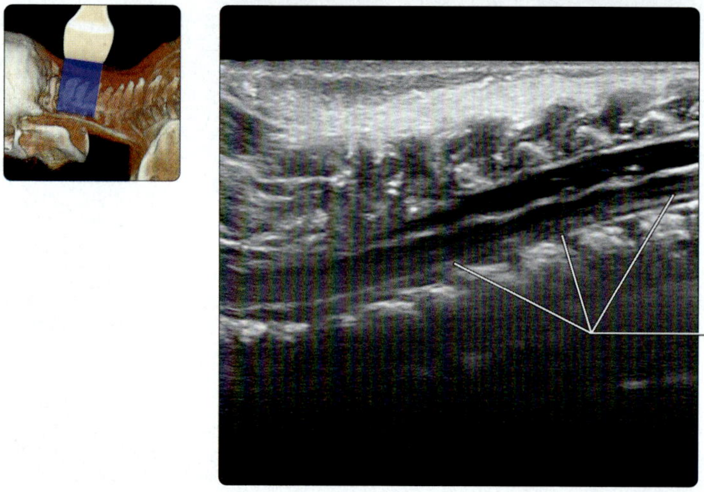

(Top) *Sagittal T2 MR of the cervical spine is displayed in the prone position to match the spinal ultrasound. High-signal CSF is seen within the foramen magnum and surrounding the spinal cord. There is a smooth transition of the cervicomedullary junction. The cerebellar tonsils should be normally above the line between basion and opisthion.* **(Middle)** *The spinous processes of the cervical spine are longer and overlap more than in the lower spine making visualization of the cord more difficult, particularly in the transverse plane. Note the cartilaginous portion of the spinous processes appears hypoechoic, while the ossified portion of the vertebral body is echogenic and shadowing. The upper cervical cord is seen by scanning laterally through the neck and angling toward the spinal canal.* **(Bottom)** *Slightly lower and angled medially, the lower portion of the cervical cord can be seen.*

THORACIC SPINE

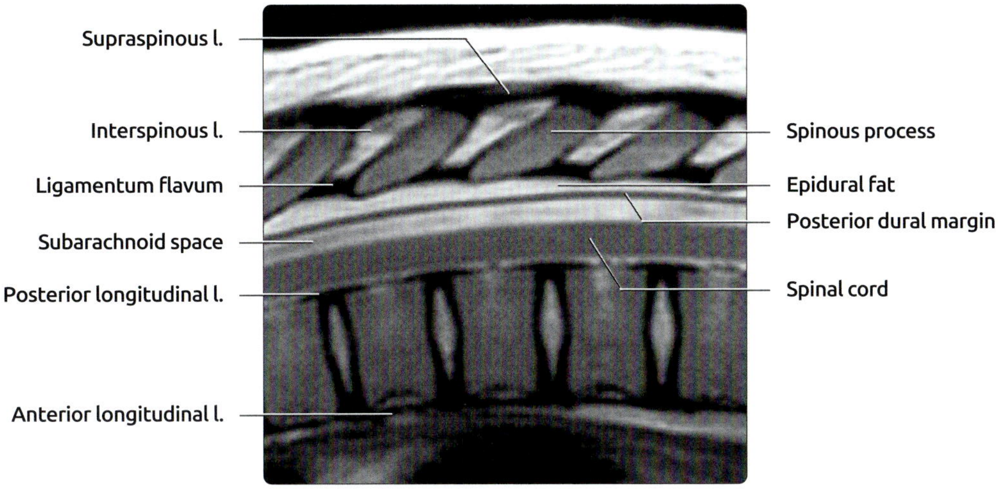

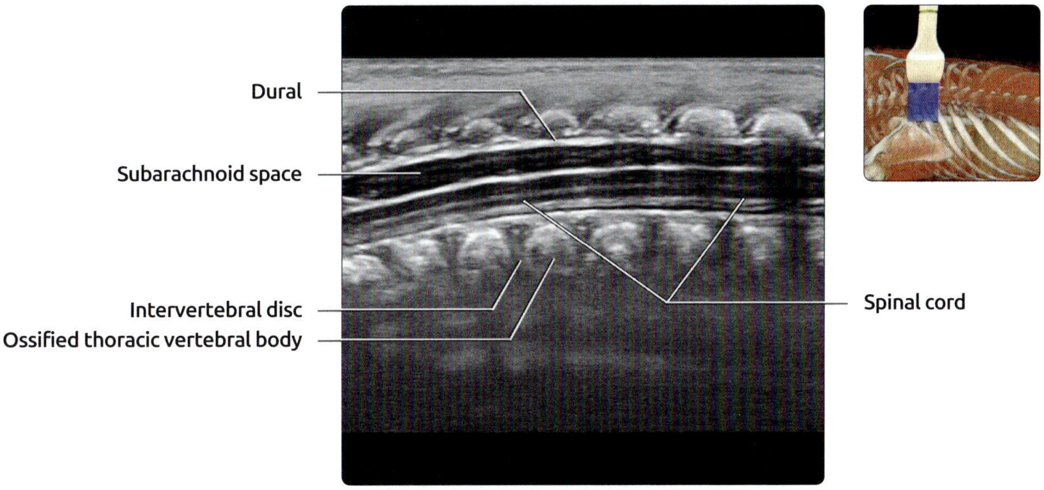

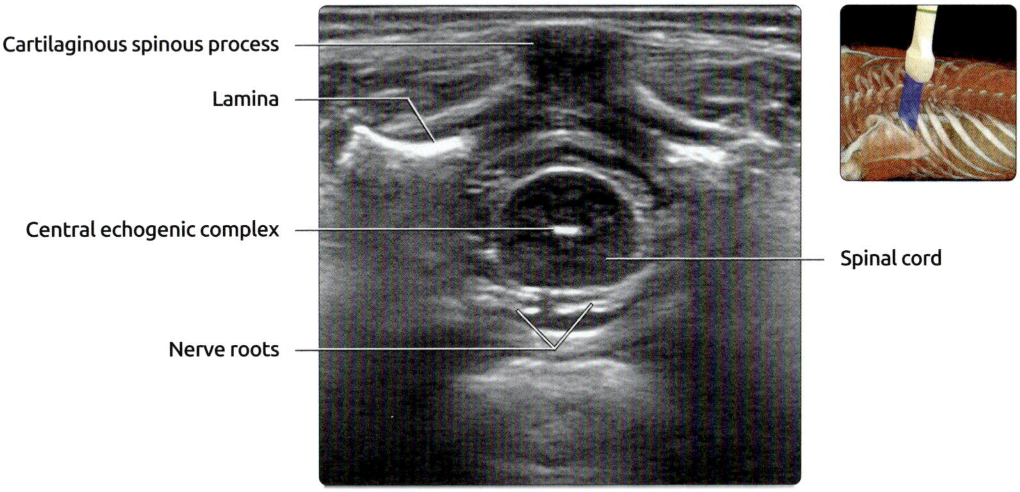

(Top) *This image at mid thoracic level shows this segment of the spinal cord is more narrow when compared with the higher level (C3 to T2) and lower level (T9 to T12).* **(Middle)** *Longitudinal ultrasound at the mid thoracic level of the spinal cord shows the narrowest portion of the cord. The cord should lie dependently within the thecal sac.* **(Bottom)** *This transverse scan is taken at T12. At this point, the cord has started to widen again. The spinal cord is predominately hypoechoic with a central echogenic complex. Nerve roots are seen around the cord as it is approaching the conus medullaris.*

LONGITUDINAL US, CONUS

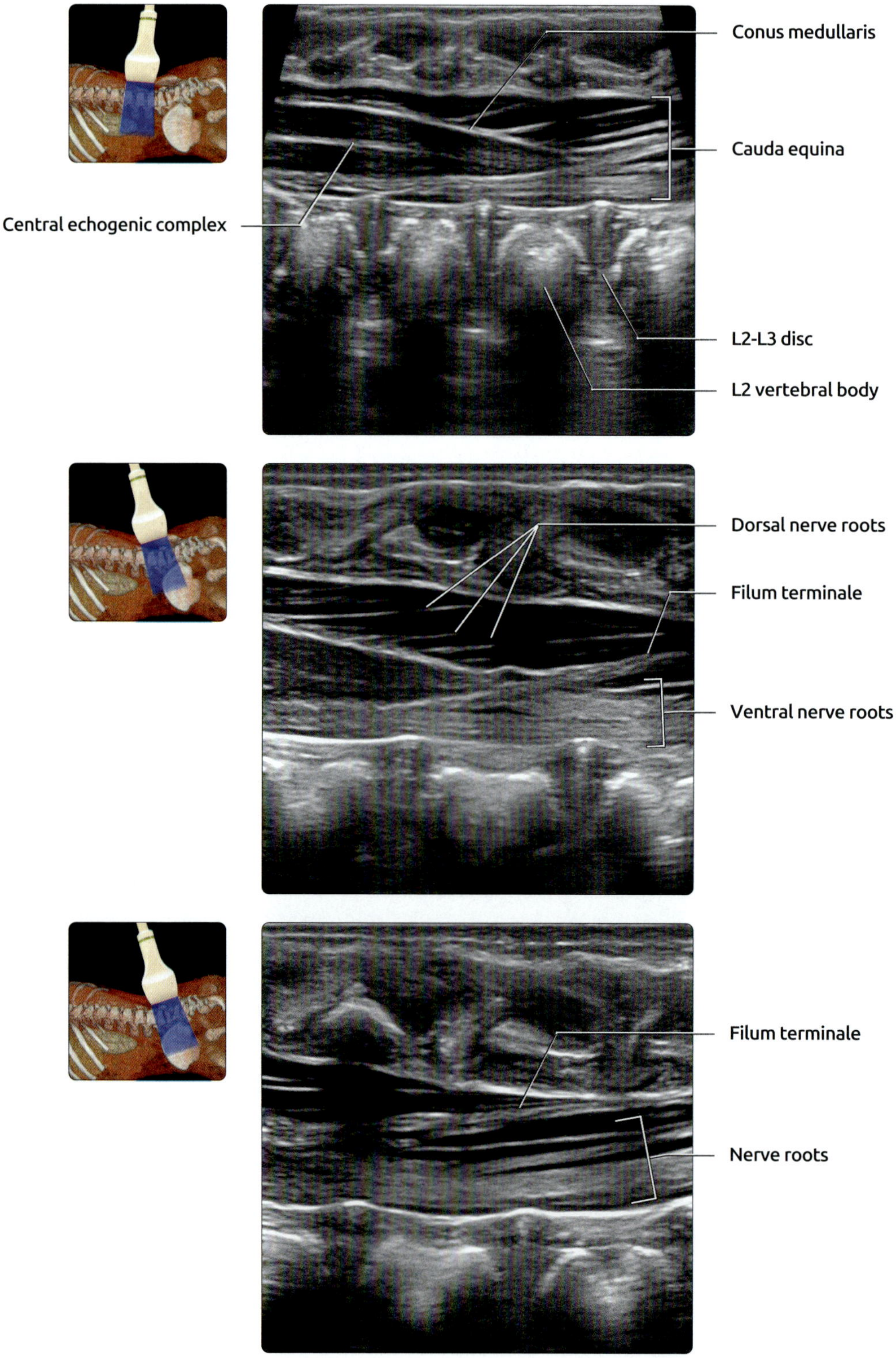

(Top) *The caudal portion of the cord tapers gradually forming the conus medullaris and should terminate above the L2-L3 disc space as in this case. Contrary to popular misunderstanding, this central echogenic complex is predominately a reflection of echoes from the interface between the ventral white commissure and CSF within the ventral median fissure rather than from the central canal. The nerve roots surrounding the conus tail form the cauda equina (horse's tail).* **(Middle)** *The filum terminale is connective tissue continuation of the pia mater extending inferiorly from the conus. It extends dorsally through the thecal sac to insert on the dura of the dorsal coccyx. The nerve roots and filum should should move freely and undulate with CSF pulsations. The dorsal nerve roots are seen as they move to exit the canal, while the ventral nerve roots lie dependently within the thecal sac.* **(Bottom)** *More distally, the filum terminale is seen moving dorsally to insert on the dura at the level of the coccyx.*

AXIAL US, CONUS

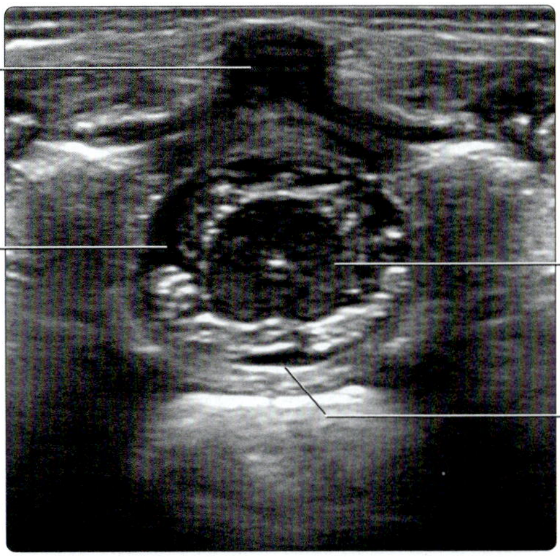

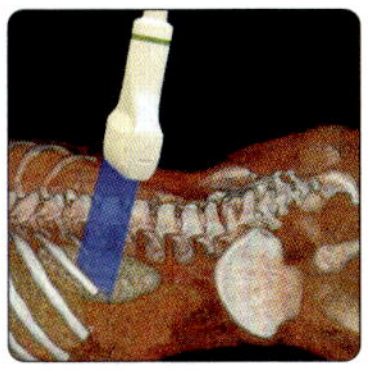

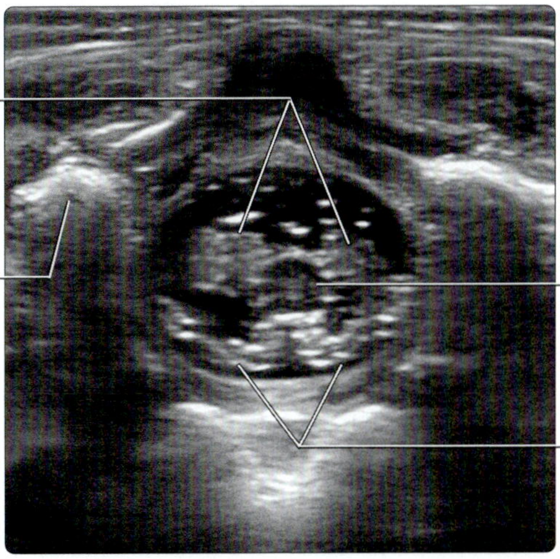

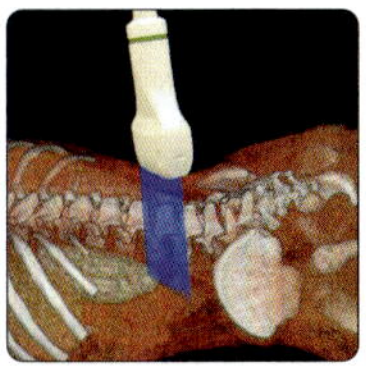

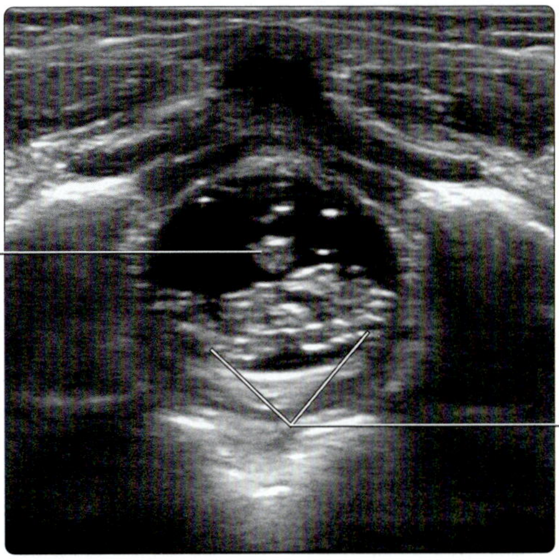

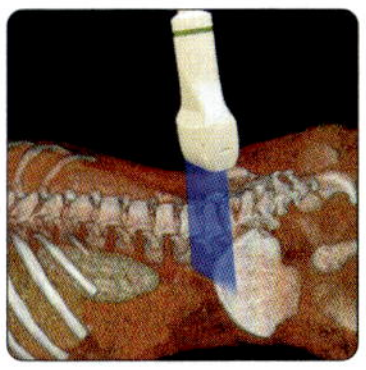

(Top) *Axial ultrasound in the upper lumbar spine at the level of the conus medullaris shows the normal hypoechoic cord with central echogenic complex. The spinal cord lies within the subarachnoid space, which is bound by the dura mater and filled with CSF. Echogenic nerve rootlets are seen draped around the cord.* **(Middle)** *Axial ultrasound scanning between L2 and L3 shows the tip of the conus. It is surrounded by free-floating nerve roots, collectively known as cauda equina. The conus normally should be above the L2-L3 disc space.* **(Bottom)** *Axial ultrasound at L4 level shows the filum terminale and nerve rootlets of cauda equina, all of which should float freely within the thecal sac. The filum terminale is the prolongation of the pia mater and attaches dorsally at the level of the coccyx. The filum terminale can be clearly distinguished from the nerve roots of the cauda equina and should be < 2 mm in diameter.*

SACRUM AND COCCYX

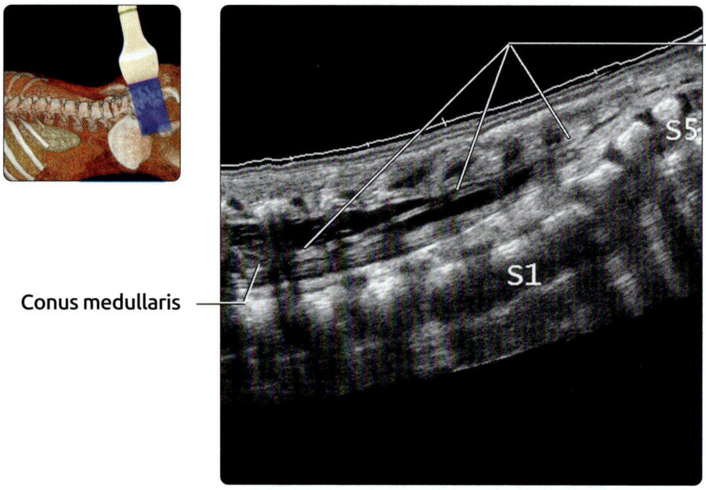

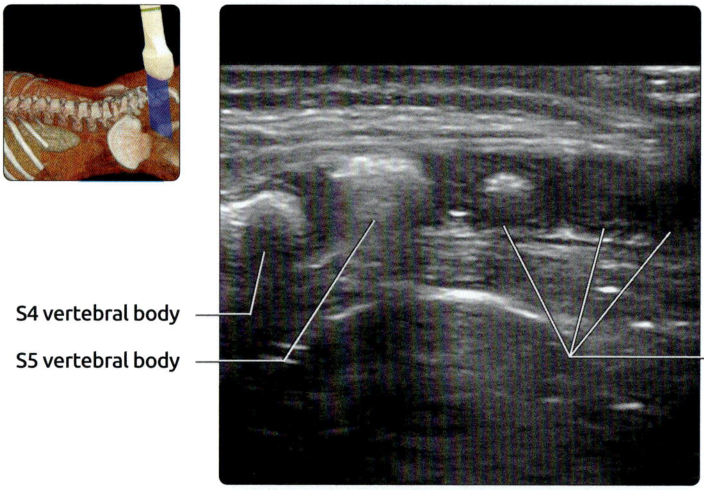

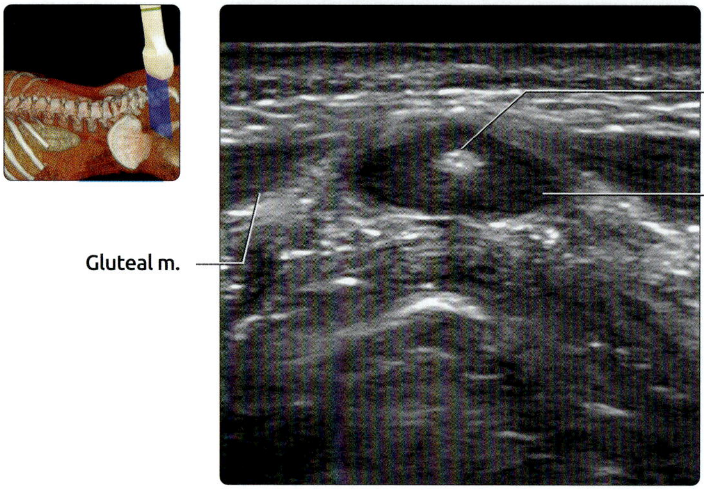

(Top) *This extended field of view shows the filum terminale extending dorsally within the thecal sac to insert on the dura at the level of the coccyx. One of the most difficult and crucial things in spinal sonography is accurately numbering the vertebral bodies. The angle at L5-S1 can be accentuated by raising up the shoulders; confirm your impression by counting down from the last rib. Only after confirming in both directions can you be confident in your determination of the conus level.* **(Middle)** *The coccygeal segments are the last vertebral bodies to ossify so they often appear partially or completely hypoechoic by ultrasound.* **(Bottom)** *The ossification center of a coccygeal vertebral body is rounded, helping to distinguish it from a sacral vertebral body, which has a square ossification center.*

ABNORMAL CORD

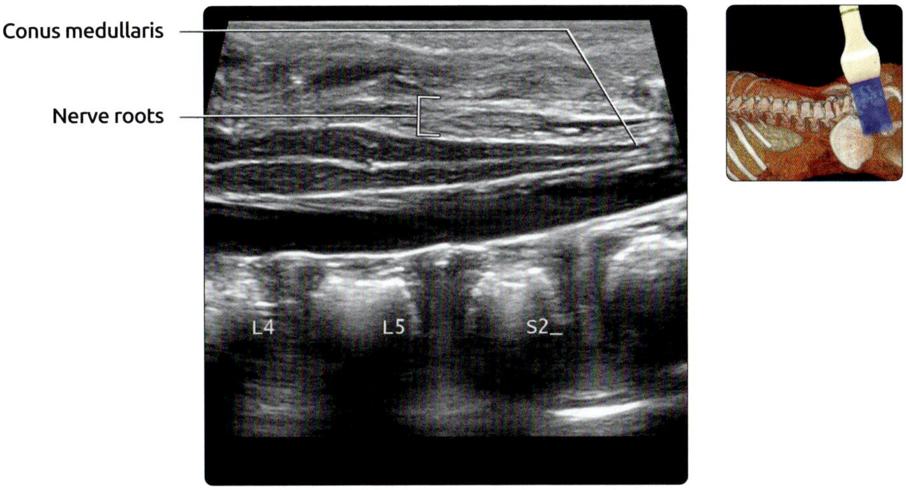

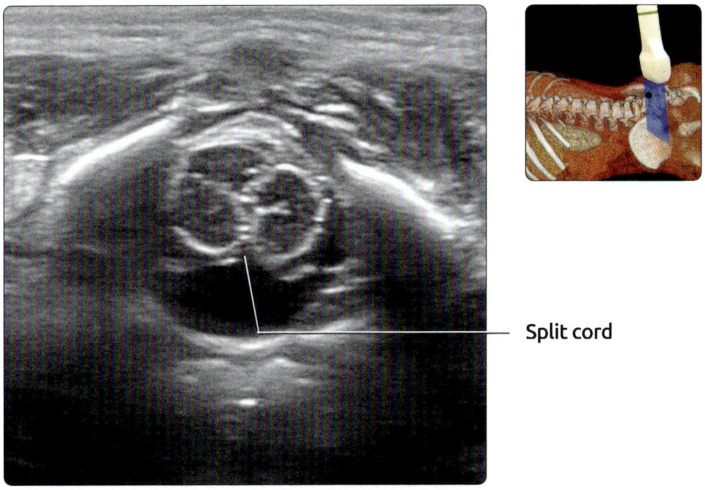

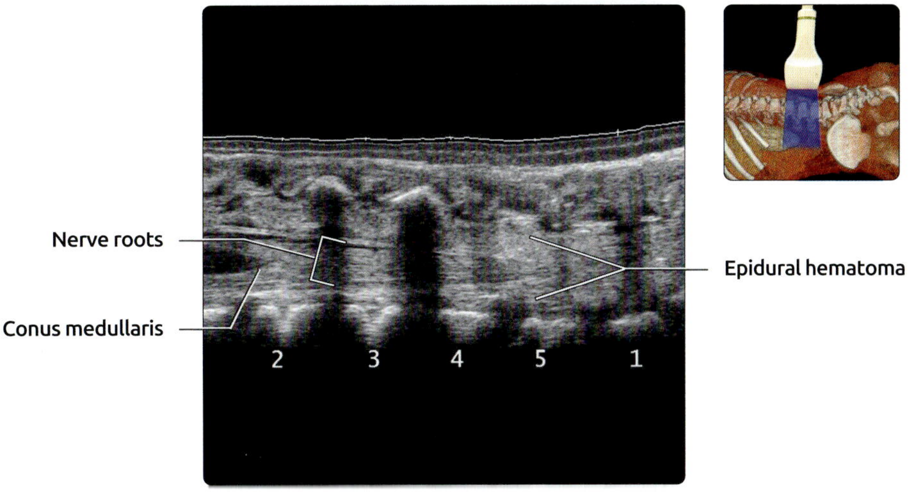

(Top) *In this case of tethered cord, the conus is ending very low at the level of S3. Also note the cord appears stuck to the dorsal surface of the sac and not lying dependently as would normally be seen. No free-floating nerve roots were seen during the real-time examination and are shown here in a conglomerate mass.* **(Middle)** *In the axial plane, not only is the position abnormal, but there is a cleft seen within the cord. A split cord was confirmed by MR.* **(Bottom)** *This scan was performed after a failed attempt at a lumbar puncture. The scan shows the thecal sac is surrounded by a hyperechoic epidural hematoma. The nerve roots are compressed to the middle of the sac.*

SECTION 2

Head and Neck

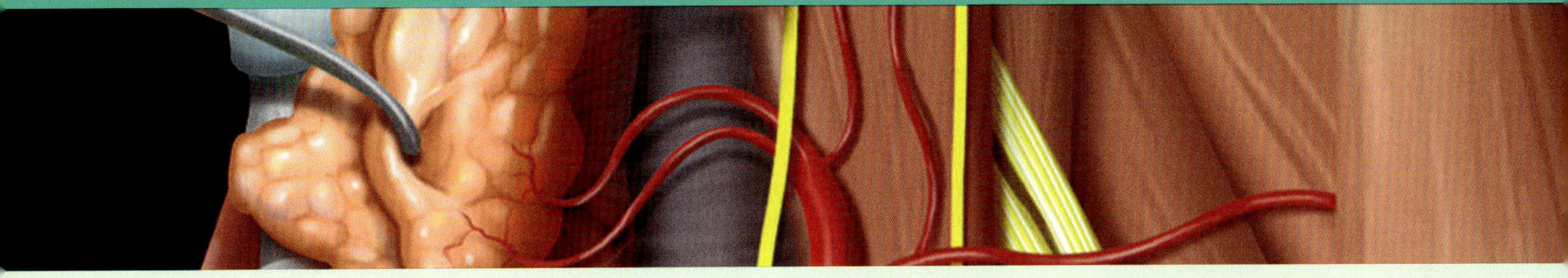

TERMINOLOGY

Abbreviations

- Infrahyoid neck (IHN)

Definitions

- Suprahyoid neck (SHN): Spaces from skull base to hyoid bone (excluding orbits, paranasal sinuses, and oral cavity) including parapharyngeal, pharyngeal mucosal, masticator, parotid, carotid (CS), buccal, retropharyngeal (RPS), and perivertebral (PVS) spaces
- Infrahyoid neck (IHN): Spaces below hyoid bone to thoracic inlet, including visceral space (VS), posterior cervical space (PCS), anterior cervical space (ACS), CS, RPS, and PVS

IMAGING ANATOMY

Overview

- Fascial spaces of SHN and IHN are key for cross-sectional imaging
  - Concept is difficult to apply with ultrasound
- Ultrasound anatomy is based on division of neck into anterior and posterior triangles
  - Anterior triangle: Bounded anteriorly by midline and posteriorly by posterior margin of sternomastoid muscle
    - Further divided into suprahyoid and infrahyoid portions
    - Suprahyoid portion: Divided by anterior belly of digastric muscle into submental and submandibular triangles
    - Infrahyoid portion: Divided by superior belly of omohyoid muscle into muscular and carotid triangles
  - Posterior triangle: Bound anteriorly by posterior margin of sternomastoid muscle and posteriorly by anterior border of trapezius muscle
    - Apex formed by mastoid process, base of triangle formed by clavicle
    - Subdivided by posterior belly of omohyoid muscle into occipital triangle (superior) and supraclavicular triangle (inferior)
- **Submental region**
  - Key structures include anterior belly of digastric muscle, mylohyoid, genioglossus and geniohyoid muscles, sublingual glands, and lingual artery
- **Submandibular region**
  - Key structures include submandibular gland, mylohyoid muscle, hyoglossus muscle, anterior and posterior bellies of digastric muscle, facial vein, and anterior division of retromandibular vein (RMV)
- **Parotid region**
  - Key structures include parotid gland, masseter and buccinator muscles, RMV, and external carotid artery (ECA)
- **Cervical region**
  - Upper cervical region: Skull base to hyoid bone/carotid bifurcation
    - Key structures include internal jugular vein (IJV), carotid bifurcation, jugulodigastric node, and posterior belly of digastric muscle
  - Midcervical region: Hyoid bone to cricoid cartilage
    - Key structures include IJV, common carotid artery (CCA), vagus nerve, and lymph nodes
  - Lower cervical region: Cricoid cartilage to clavicle
    - Key structures include IJV, CCA, superior belly of omohyoid, and lymph nodes
- **Supraclavicular fossa**
  - Key structures include trapezius, sternomastoid, omohyoid muscles, brachial plexus elements, and transverse cervical nodes
- **Posterior triangle**
  - Bordered anteriorly by sternomastoid muscle and posteriorly by trapezius muscle
  - Floor formed by scalene muscles, levator scapulae, and splenius capitis muscles
- **Midline**
  - Key structures include hyoid bone, strap muscles, thyroid, larynx, and tracheal rings

ANATOMY IMAGING ISSUES

Imaging Recommendations

- Use of high-resolution transducers is essential
- Color/power Doppler examination provides useful supplementary information to grayscale ultrasound
- US is very sensitive in identifying abnormalities (and in characterizing many head and neck soft tissue lesions)
  - Combination with fine-needle aspiration cytology (FNAC) provides specificity and increased diagnostic accuracy
- US + FNAC usually provides adequate information for patient management
- Cross-sectional imaging (CT, MR) may be required for
  - Large mass, when detailed anatomical extent is not fully examined by US
  - Deep-seated lesion with suboptimal US visualization and evaluation
  - Preoperative assessment of relevant adjacent structures (e.g., bone involvement)

Imaging Approaches

- Ultrasound imaging protocol
  - Start in submental region by scanning in transverse plane
  - Next, scan submandibular region in transverse and longitudinal/oblique planes
  - Then scan parotid region in transverse and longitudinal planes
  - Then examine upper cervical, midcervical, and lower cervical regions in transverse plane
  - Then examine supraclavicular fossa with transducer held transversely
  - Then scan posterior triangle transversely along a line drawn from mastoid process to ipsilateral acromion
  - Finally, scan midline and thyroid gland in both transverse and longitudinal planes
- This protocol is robust and can be tailored to suit individual clinical conditions
- Transverse scans quickly identify normal anatomy and detect abnormalities
- Any abnormality identified is further examined in longitudinal/oblique planes (grayscale and Doppler)
- In restless children, it may not be possible to follow above protocol
  - It would therefore be best to evaluate primary area of interest first, before child becomes uncooperative

SCANNING AREAS AND LYMPH NODE GROUPS

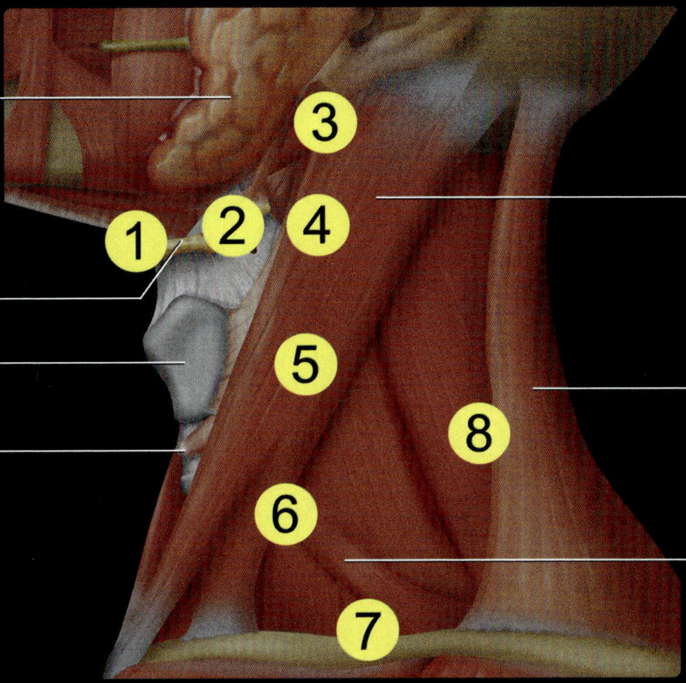

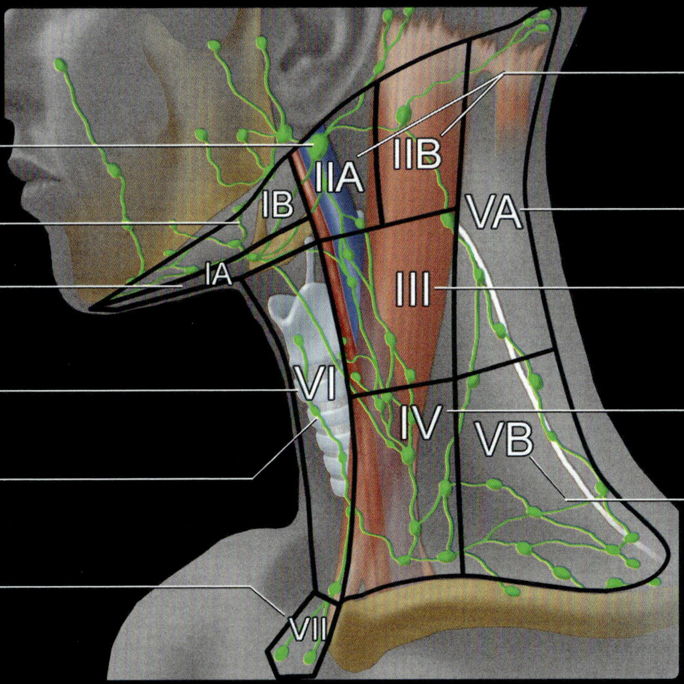

(Top) *Schematic diagram shows the protocol for ultrasound examination of the neck with 8 regions scanned in order: (1) Submental region, (2) submandibular region, (3) parotid region, (4) upper cervical region, (5) midcervical region, (6) lower cervical region, (7) supraclavicular fossa, and (8) posterior triangle. The above protocol is robust and helps to adequately evaluate the neck for common clinical conditions. Note that deep structures cannot be adequately assessed by ultrasound.* **(Bottom)** *Lateral oblique graphic of the neck shows the anatomic locations of the major nodal groups of the neck. Division of the internal jugular nodal chain into high, middle, and low regions is defined by the level of the hyoid bone and cricoid cartilage. Similarly, the spinal accessory nodal chain is divided into high & low regions by the level of the cricoid cartilage.*

NECK SPACES

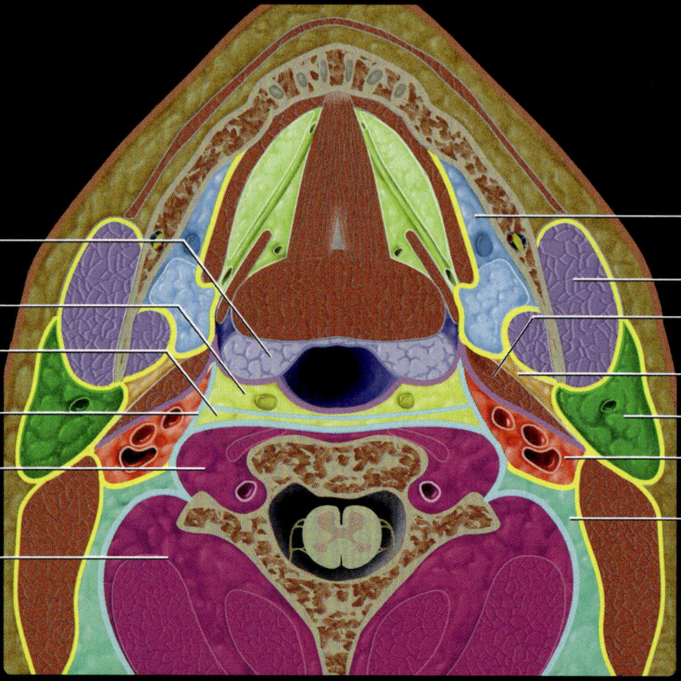

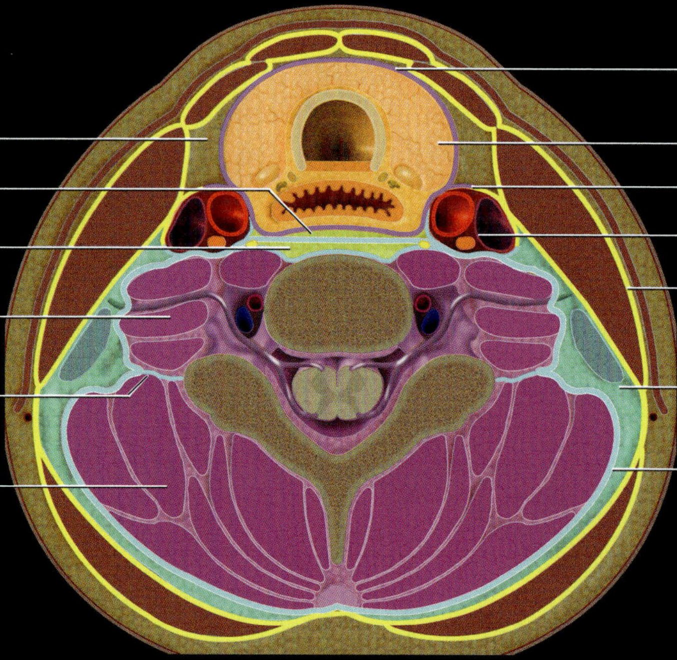

(Top) *Axial graphic shows the suprahyoid neck spaces at the level of the oropharynx. The superficial (yellow line), middle (pink line), and deep (turquoise line) layers of deep cervical fascia (DCF) outline the suprahyoid neck spaces. Notice that the lateral borders of the retropharyngeal and danger spaces are called the alar fascia and represent a slip of the deep layer of DCF.* **(Bottom)** *Axial graphic depicts the fascia and spaces of the infrahyoid neck. The 3 layers of DCF are present in the suprahyoid and infrahyoid neck. The carotid sheath is made up of all 3 layers of DCF (tricolor line around carotid space). Notice the deep layer completely circles the perivertebral space, diving in laterally to divide it into prevertebral and paraspinal components. Although the spaces are not adequately demonstrated by US, it is important to be familiar with the concept in order to understand neck anatomy.*

TRANSVERSE ULTRASOUND

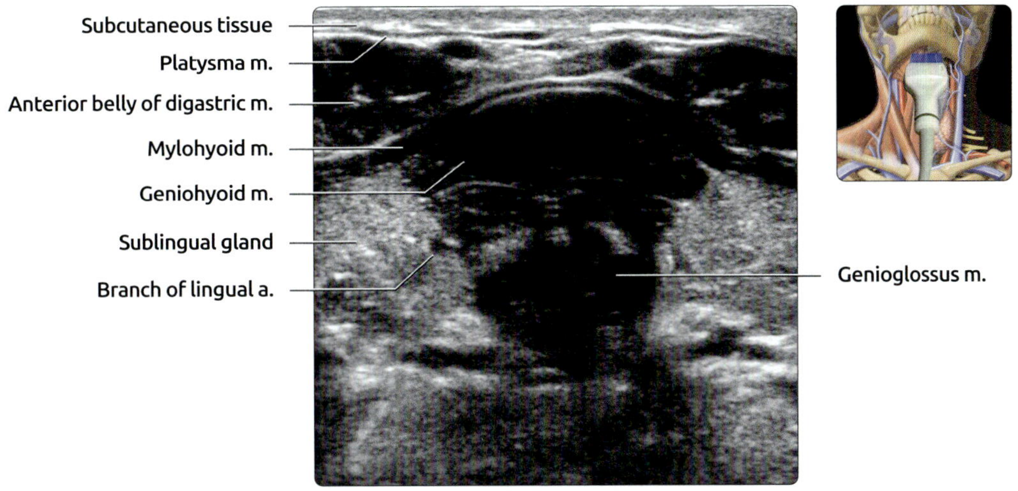

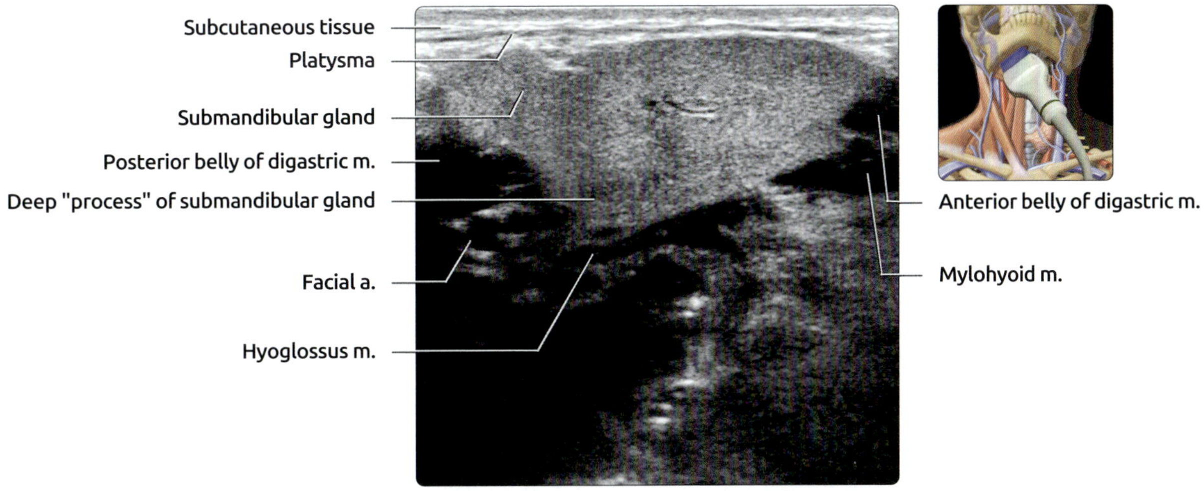

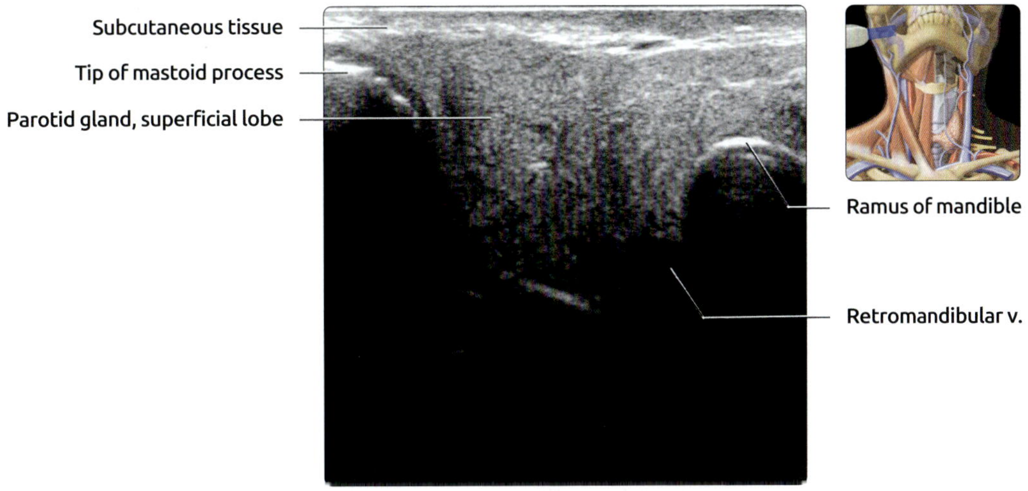

(Top) *Standard transverse grayscale ultrasound shows the submental region. The mylohyoid muscle is an important landmark for the division of the sublingual (deep to mylohyoid muscle) and submandibular (superficial to mylohyoid muscle) spaces. Part of the extrinsic muscles of the tongue, including the geniohyoid and genioglossus, are visualized.* **(Middle)** *Standard transverse grayscale ultrasound shows the submandibular region. The submandibular gland is the key structure with its homogeneous echotexture. The gland sits astride the mylohyoid and posterior belly of the digastric muscles.* **(Bottom)** *Standard transverse grayscale ultrasound shows the parotid region. Note that the deep lobe is obscured by shadowing from the mandible and cannot be evaluated. The retromandibular vein serves as a landmark for the intraparotid facial nerve.*

TRANSVERSE ULTRASOUND

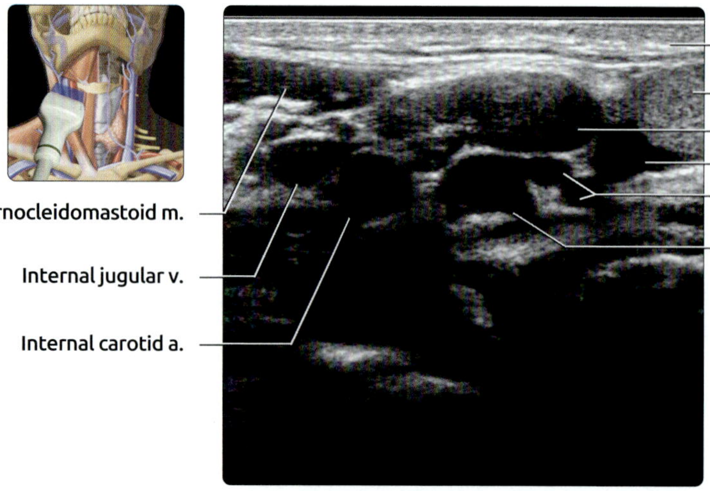

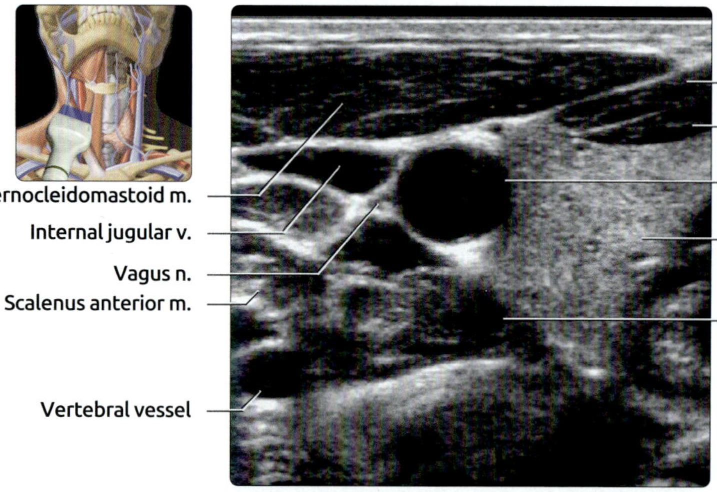

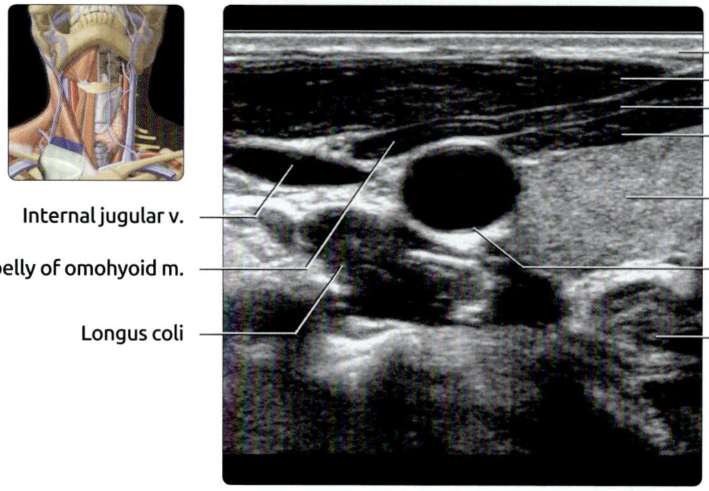

(Top) *Standard transverse grayscale ultrasound shows the upper cervical level. Key structures include the internal jugular vein, the proximal internal and external carotid arteries, and the jugular chain lymph nodes. The jugulodigastric node is the most prominent and consistently seen on ultrasound.* **(Middle)** *Standard grayscale ultrasound shows the midcervical level. Note the vagus nerve is clearly seen on ultrasound.* **(Bottom)** *Standard grayscale ultrasound shows the lower cervical level. The thyroid gland is related to the common carotid and internal jugular vein laterally. The anterior strap muscles (including the sternohyoid and sternothyroid muscles) and the superior belly of the omohyoid are clearly visualized.*

TRANSVERSE ULTRASOUND

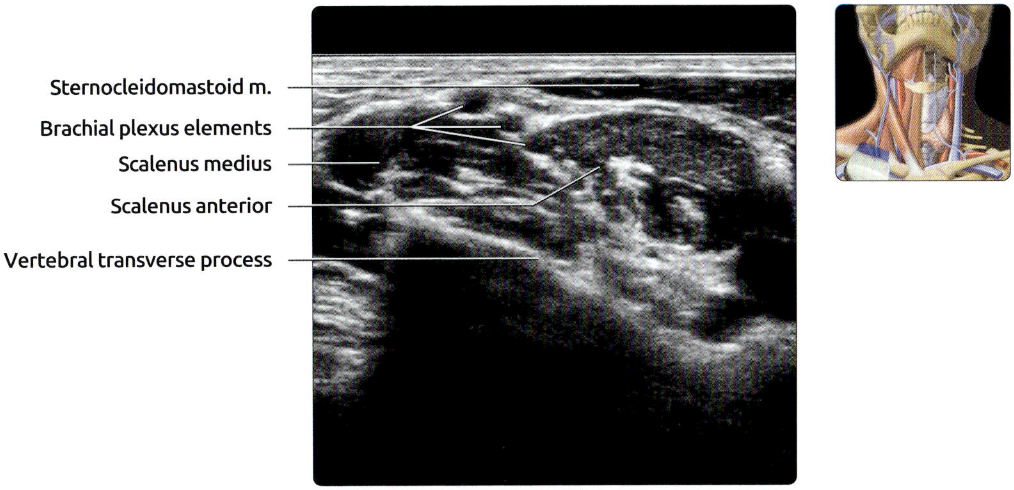

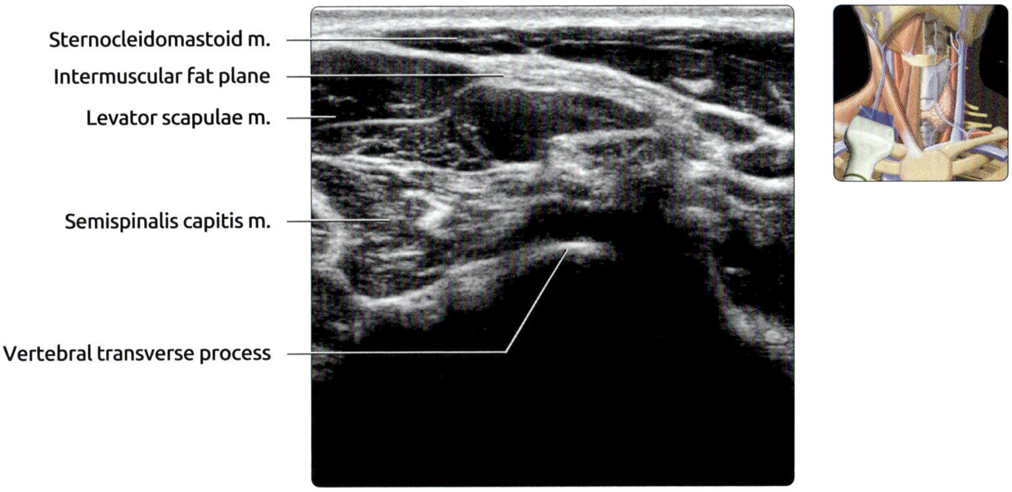

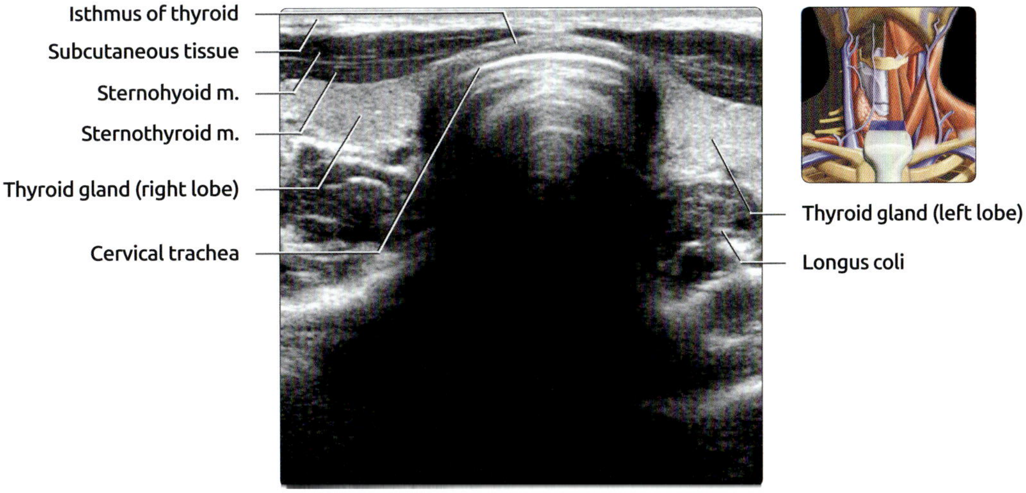

(Top) *Standard grayscale ultrasound shows the supraclavicular fossa. Note that the trunks of the brachial plexus are consistently seen on high-resolution ultrasound at this site.* **(Middle)** *Standard transverse grayscale ultrasound shows the posterior triangle. Note that the intermuscular fat plane is visible. The spinal accessory nerve and lymph nodes are important contents of the posterior triangle.* **(Bottom)** *Standard transverse grayscale ultrasound shows the midline of the lower anterior neck. The isthmus of the thyroid gland, the trachea, and the longus coli are key structures to be identified.*

TERMINOLOGY

Synonyms

- Sublingual space (SLS), submental triangle

Definitions

- Sublingual region: Paired nonfascial-lined spaces of oral cavity in deep oral tongue, above floor of mouth, superomedial to mylohyoid muscle

IMAGING ANATOMY

Overview

- Borders of submental triangle are readily defined on ultrasound
  - Floor is formed by mylohyoid muscle
  - Apex is limited anteriorly by symphysis menti
  - Base is bounded posteriorly by hyoid bone
  - Anterior belly of digastric muscle represents sides of triangle
- SLS is deep space of oral cavity superomedial to mylohyoid muscles
  - Contains key neurovascular structures of oral cavity
  - Includes glossopharyngeal nerve (CNIX), hypoglossal nerve (CNXII), lingual nerve (branch of V3), lingual artery and vein

Anatomy Relationships

- SLS relationships
  - SLS in deep oral tongue superomedial to mylohyoid muscle and lateral to genioglossus-geniohyoid muscles
  - Communication between SLSs occurs in midline anteriorly as narrow isthmus beneath frenulum
  - SLS communicates with submandibular space (SMS) and inferior parapharyngeal space (PPS) at posterior margin of mylohyoid muscle
    - There is no fascia dividing posterior SLS from adjacent SMS
    - Therefore, there is direct communication with SMS and PPS in this location

Internal Contents

- Major muscles forming borders of submental triangle
  - Anterior belly of digastric muscle
    - Marks lateral border of submental triangle
  - Mylohyoid muscle
    - Muscle of floor of mouth
    - Muscular sling between medial aspect of mandibular bodies
    - Anterior attachment to mandible inferior to origins of genial muscles
    - Separates SLS (deep to mylohyoid muscle plane) from SMS (superficial to mylohyoid muscle)
  - Genioglossus and geniohyoid muscles
    - Form root of tongue
    - Together with hyoglossus muscle, they make up major extrinsic muscles of tongue
- Posterior aspect of SLS is divided into medial and lateral compartments by hyoglossus muscle
- Lateral compartment contents
  - Hypoglossal nerve
    - Motor to intrinsic and extrinsic muscles of tongue
    - Intrinsic muscles of tongue include inferior lingual, vertical, and transverse muscles
  - Lingual nerve: Branch of mandibular division of trigeminal nerve (CNV3) combined with chorda tympani branch of facial nerve
    - Lingual nerve branch of CNV3: Sensation to anterior 2/3 of oral tongue
    - Chorda tympani branch of facial nerve: Anterior 2/3 of tongue taste and parasympathetic secretomotor fibers to submandibular ganglion/gland
  - Sublingual glands and ducts
    - Lie in anterior SLS bilaterally
    - ~ 5 small ducts open under oral tongue into oral cavity
    - With age, sublingual glands atrophy, becoming difficult to see on imaging
  - Submandibular glands and submandibular ducts
    - Submandibular gland deep margin extends into posterior opening of SLS
    - Submandibular duct runs anteriorly to papillae in anteromedial subfrenular mucosa
- Medial compartment contents
  - Glossopharyngeal nerve (CNIX)
    - Provides sensation to posterior 1/3 of tongue
    - Carries taste input from posterior 1/3 of tongue
    - Located more cephalad in medial compartment compared to lingual artery and vein
  - Lingual artery and vein
    - Vascular supply to oral tongue
    - Seen running just lateral to genioglossus muscle

ANATOMY IMAGING ISSUES

Imaging Recommendations

- High-resolution ultrasound is ideal imaging tool for evaluating submental masses
- Major structures are best seen on transverse scans with patient's neck in slight hyperextension
- For more deep-seated lesions (e.g., deep to root of tongue), MR is necessary for better anatomical assessment
  - Ultrasound may help in directing needle for guided biopsy of such lesions

Key Concepts

- What defines mass as primary to SLS
  - Center of lesion is superomedial to mylohyoid muscle and lateral to genioglossus muscle
- Common lesions in submental region include
  - Congenital lesions: Epidermoid/dermoid cyst
  - Enlarged lymph node: Reactive, inflammatory or neoplastic (metastatic/lymphomatous nodes)
  - Inflammatory conditions: Ranula, abscess
  - Sublingual gland lesions: Sialadenitis, calculus, benign/malignant salivary gland tumor

SUBLINGUAL SPACE

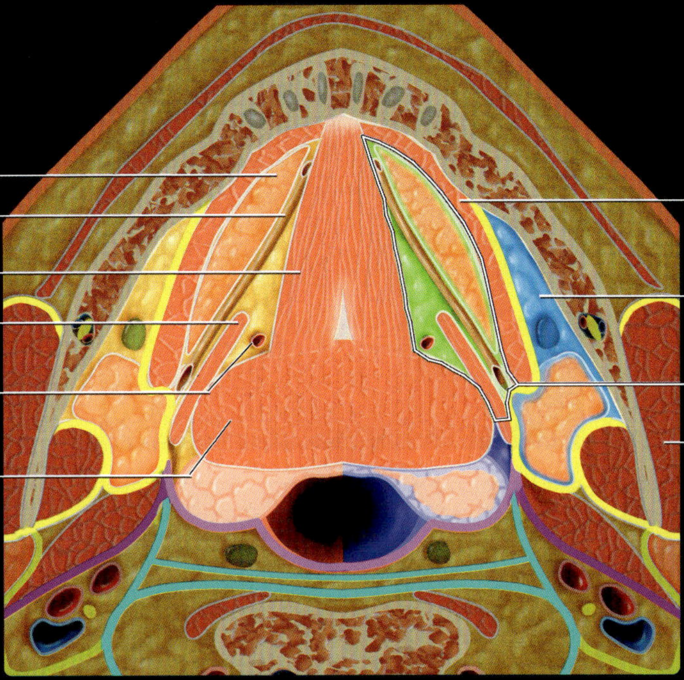

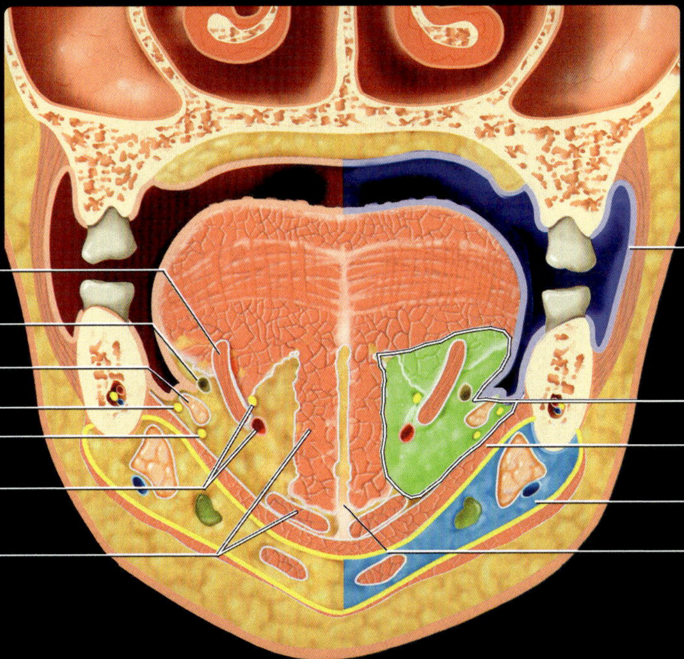

(Top) *Axial graphic through the body of the mandible shows the sublingual space (on patient's left, shaded in green) situated superomedial to the mylohyoid muscle and lateral to the genioglossus muscle. Notice the absence of fascia surrounding the sublingual space. The yellow line represents the superficial layer of deep cervical fascia.* **(Bottom)** *Coronal graphic through the oral cavity shows position of the mylohyoid muscle, which is the landmark in this area. The sublingual space is shaded in green. The medial sublingual space compartment contains the glossopharyngeal nerve (CNIX) and lingual artery/vein, and the lateral sublingual space compartment contains the submandibular duct, sublingual gland, lingual nerve, and hypoglossal nerve (CNXII). The fascia-lined (yellow line) submandibular space is inferolateral to the mylohyoid muscle.*

TRANSVERSE ULTRASOUND

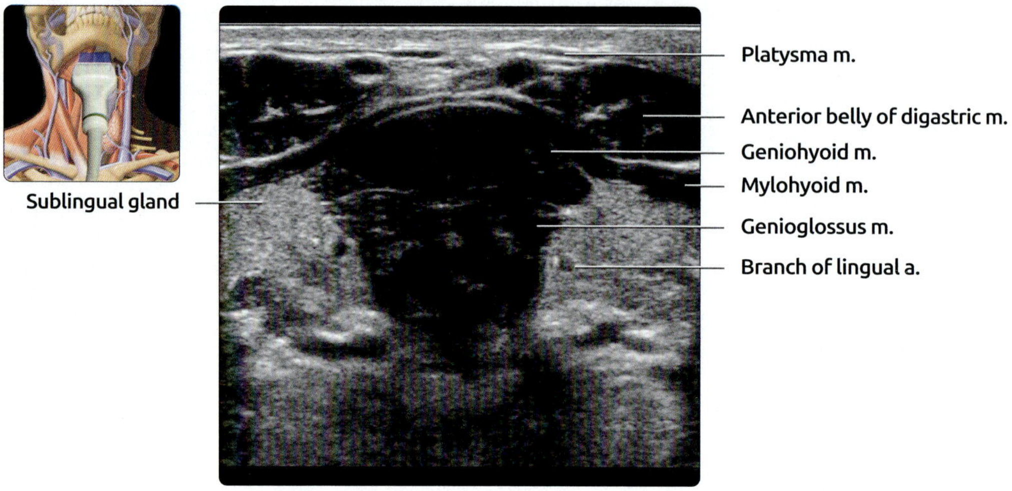

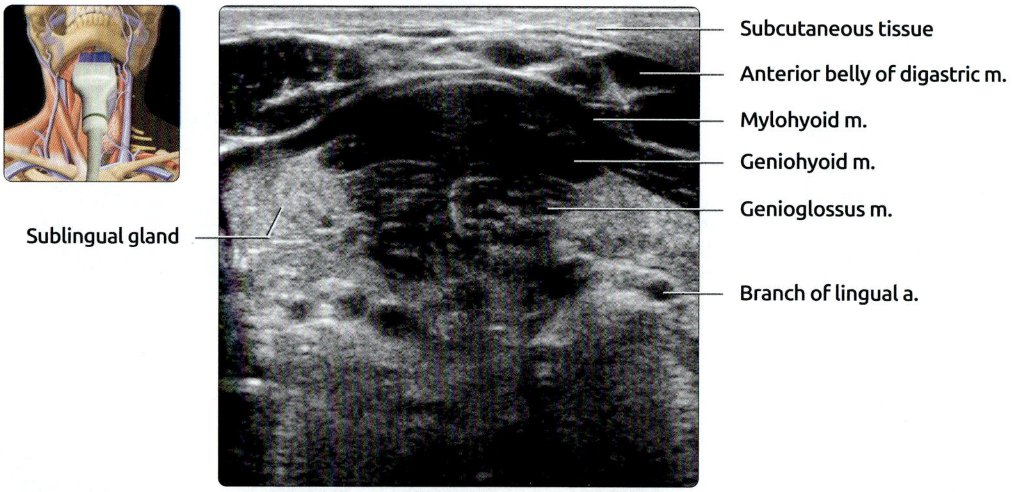

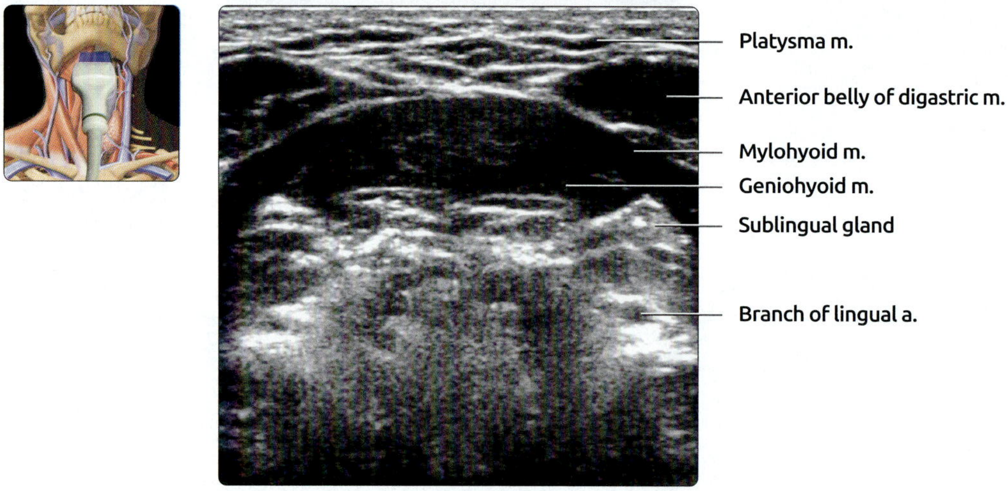

(Top) *Anterior transverse grayscale ultrasound of the submental and sublingual region is shown. The mylohyoid muscle is the landmark for division of the sublingual space (deep to the mylohyoid plane) and submandibular space (superficial to the muscle plane). The sublingual gland appears as homogeneous, hyperechoic structures lateral to the geniohyoid/genioglossus muscle. Branches of lingual artery can be easily picked up on transverse plane. The submandibular duct sits alongside the lingual vessels, and a submandibular calculus may impact at this site.* **(Middle)** *More posterior transverse grayscale ultrasound allows the clear depiction of extrinsic muscles of the tongue at the root.* **(Bottom)** *Transverse grayscale ultrasound shows the submental region in a more posterior location.*

POWER DOPPLER ULTRASOUND AND CORONAL MR

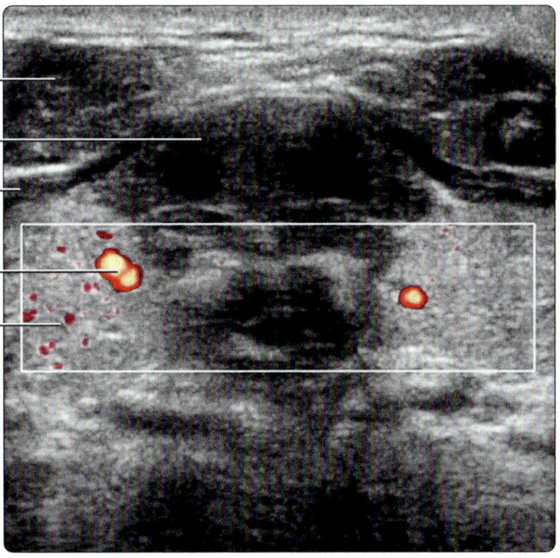

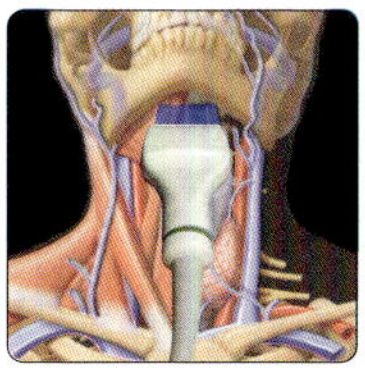

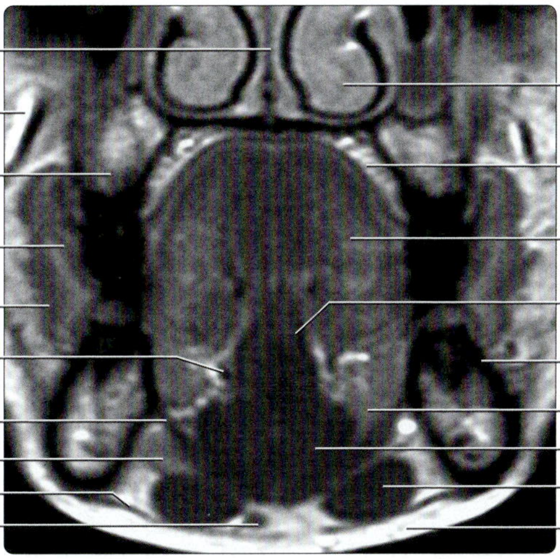

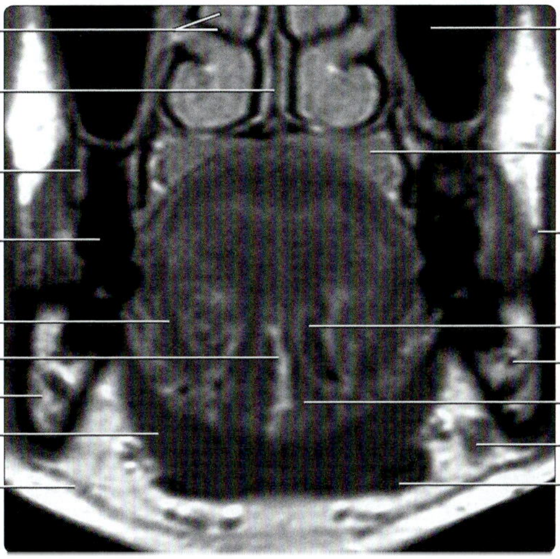

(Top) *Power Doppler ultrasound of the submental region shows the presence of color flow within the branches of the lingual artery. The use of Doppler examination aids in differentiation from the dilated submandibular duct.* **(Middle)** *Correlative coronal T1WI MR shows the floor of the mouth and tongue. The mylohyoid muscle is the landmark separating the sublingual and submandibular spaces.* **(Bottom)** *Correlative coronal T1WI MR shows the floor of the mouth and tongue in a location more posterior. For optimal use of ultrasound, the operator must also be familiar with the correlative anatomy on other imaging modalities.*

LONGITUDINAL AND TRANSVERSE ULTRASOUND

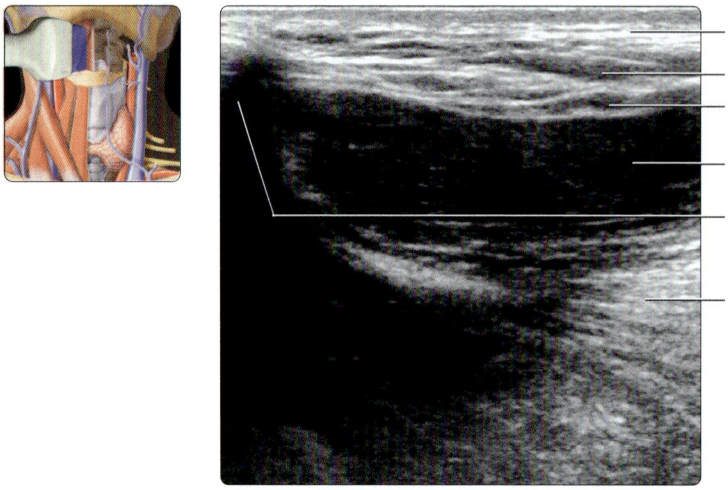

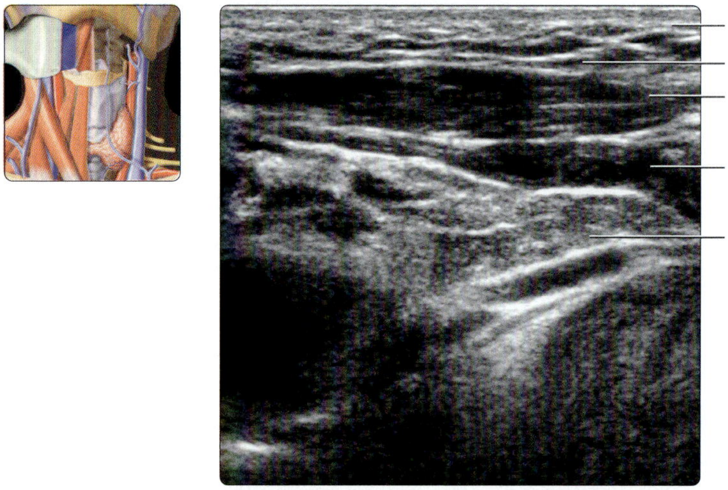

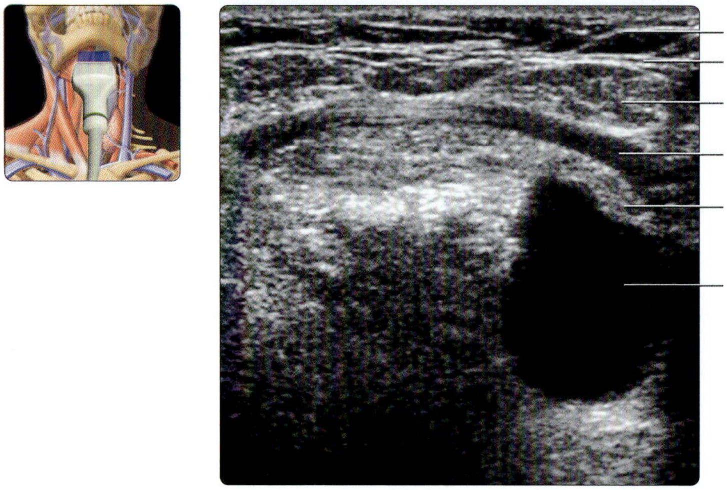

(Top) *Longitudinal grayscale ultrasound of the submental region shows the relationship of the mylohyoid, geniohyoid, and genioglossus muscles. Note that scanning just off the midline will show more of the anterior belly of the digastric muscle rather than the mylohyoid muscle anteriorly.* **(Middle)** *Parasagittal longitudinal grayscale ultrasound shows the submental region. The sublingual gland is visualized within the sublingual space (deep to the mylohyoid muscle) underneath the anterior belly of the digastric and mylohyoid muscles.* **(Bottom)** *Transverse grayscale ultrasound shows a well-circumscribed, anechoic, cystic lesion in the left sublingual space (i.e., deep to the mylohyoid muscle plane). The appearance is suggestive of a ranula; relationship to the mylohyoid determines whether it is a simple or diving ranula.*

SAGITTAL MR AND TRANSVERSE ULTRASOUND

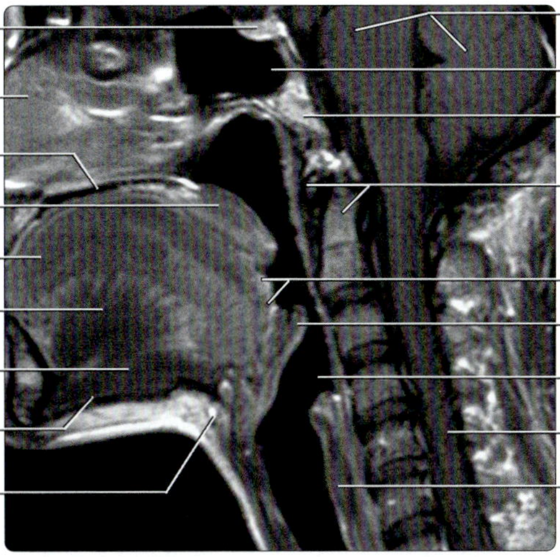

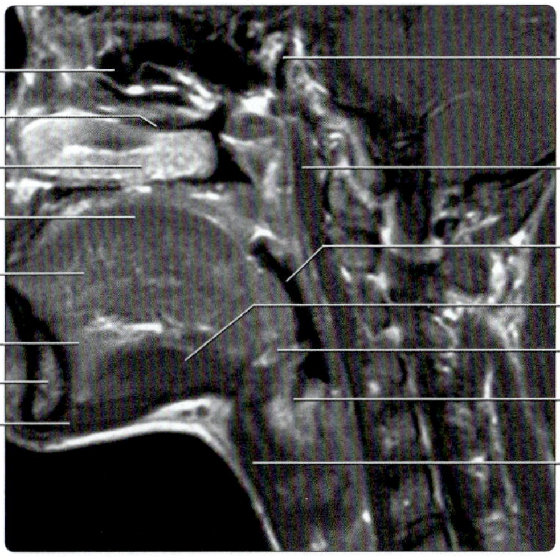

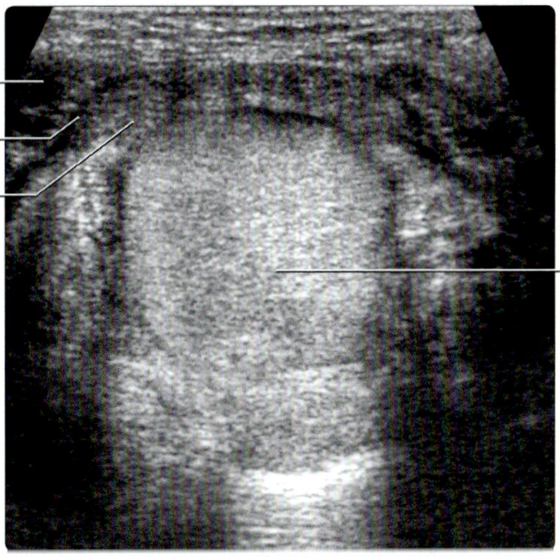

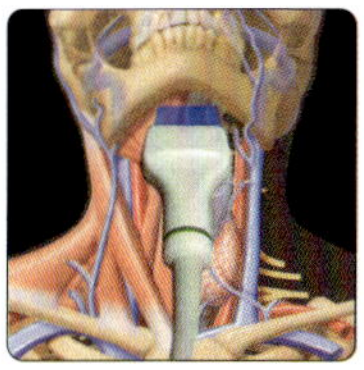

(Top) *Sagittal T1WI MR shows the floor of the mouth close to the midline. Note the positions of the mylohyoid and geniohyoid muscles between the mandible anteriorly and hyoid bone posteriorly.* **(Middle)** *Sagittal T1WI MR shows the floor of the mouth in the paramedian plane. Note that the anterior belly of the digastric muscle is now seen as it extends anteromedially to insert on the inner cortex of the mandible.* **(Bottom)** *Transverse grayscale ultrasound of the submental region shows a well-circumscribed, homogeneous, hyperechoic, midline mass deep to the mylohyoid, geniohyoid, and genioglossus muscles. The appearances and anatomical location of the lesion are suggestive of an epidermoid cyst. Congenital lesions in the neck are site specific, and familiarity with the correlative anatomy is often the best clue to their diagnosis.*

TERMINOLOGY

Abbreviations

- Submandibular space (SMS)

Definitions

- Fascial-lined space inferolateral to mylohyoid muscle
  - Contains submandibular gland, lymph nodes, and anterior belly of digastric muscles

IMAGING ANATOMY

Overview

- One of the distinct locations within oral cavity that may be used to develop location-specific differential diagnoses
  - Other locations include oral mucosal space/surface, sublingual space, and root of tongue

Anatomy Relationships

- Inferolateral to mylohyoid muscle
- Deep to platysma muscle
- Cephalad to hyoid bone
- Communicates posteriorly with sublingual space and inferior parapharyngeal space at posterior margin of mylohyoid muscle
- Continues inferiorly into infrahyoid neck as anterior cervical space

Internal Contents

- Submandibular gland
  - 1 of 3 major salivary glands
  - Divided anatomically into superficial and deep lobes by mylohyoid muscle
  - Superficial lobe is larger and in SMS itself
    - Superficial layer, deep cervical fascia forms submandibular gland capsule
    - Crossed by facial vein and cervical branches of facial nerve (marginal mandibular branch)
  - Smaller deep lobe, often called deep "process"
    - Tongue-like extension of gland that wraps around posterior aspect of mylohyoid muscle
    - Projects into posterior aspect of sublingual space
    - Submandibular duct projects off deep lobe into sublingual space
  - Submandibular gland innervation
    - Parasympathetic secretomotor supply from chorda tympani branch of facial nerve
    - Comes via lingual branch of CNV3
- Submental (level IA) and submandibular (level IB) nodal groups
  - Receive lymphatic drainage from anterior facial region
    - Including oral cavity, anterior sinonasal, and orbital areas
  - Few elliptical lymph nodes with preserved internal architecture is constant normal finding
- Anterior belly of digastric muscle
  - Divides suprahyoid portion into submental and submandibular triangles
- Hyoglossus muscle
  - Deep to mylohyoid muscle
  - Marks anterior margin of submandibular gland
  - Submandibular duct runs between hyoglossus muscle and mylohyoid muscle
- Facial vein and artery pass through SMS
  - Facial vein courses anteriorly and superiorly to submandibular gland
- Anterior division of retromandibular vein
  - Outlines posterior border of submandibular gland
- Caudal loop of CNXII
  - Passes through SMS before looping anteriorly and cephalad into tongue muscle
- Tail of parotid gland may "hang down" into posterior SMS

ANATOMY IMAGING ISSUES

Questions

- Major clinical-radiological question when mass is present in SMS: Is lesion nodal or submandibular gland in origin
  - If "beaking" of submandibular gland tissue around lesion margin is present, and lesion is completely surrounded by glandular parenchyma, lesion origin is in submandibular gland
  - Fatty cleavage plane between mass and submandibular gland identifies lesion as nodal in origin
  - Internal architecture (e.g., presence of echogenic hilum) helps to identify lymph node
- Consider major differential diagnoses for mass in submandibular region
  - Congenital lesion: Epidermoid cyst, cystic hygroma
  - Inflammatory condition: Submandibular gland sialadenitis/abscess, diving ranula, chronic sclerosing sialadenitis (Kuttner tumor), Sjögren syndrome
  - Lymph node enlargement: Reactive, inflammatory, or neoplastic (secondary or lymphomatous)
  - Benign salivary gland tumor, lipoma
  - Malignant salivary gland tumor

Imaging Recommendations

- Scan submandibular region in transverse and longitudinal/oblique planes, as these best demonstrate floor of submandibular region, hyoglossus, and mylohyoid muscles
- Always establish origin of mass (i.e., submandibular glandular or extraglandular mass), as this will help to narrow differential diagnosis
- Remember to evaluate glandular/extraglandular ductal dilatation and lymph nodes at this location

Imaging Pitfalls

- Distinction between submandibular glandular mass and enlarged lymph node can be difficult, especially if mass is large
- Lesions of parotid tail may appear in posterior submandibular region clinically
- Coronal MR helps to evaluate and localize large masses at this site

SUBMANDIBULAR SPACE

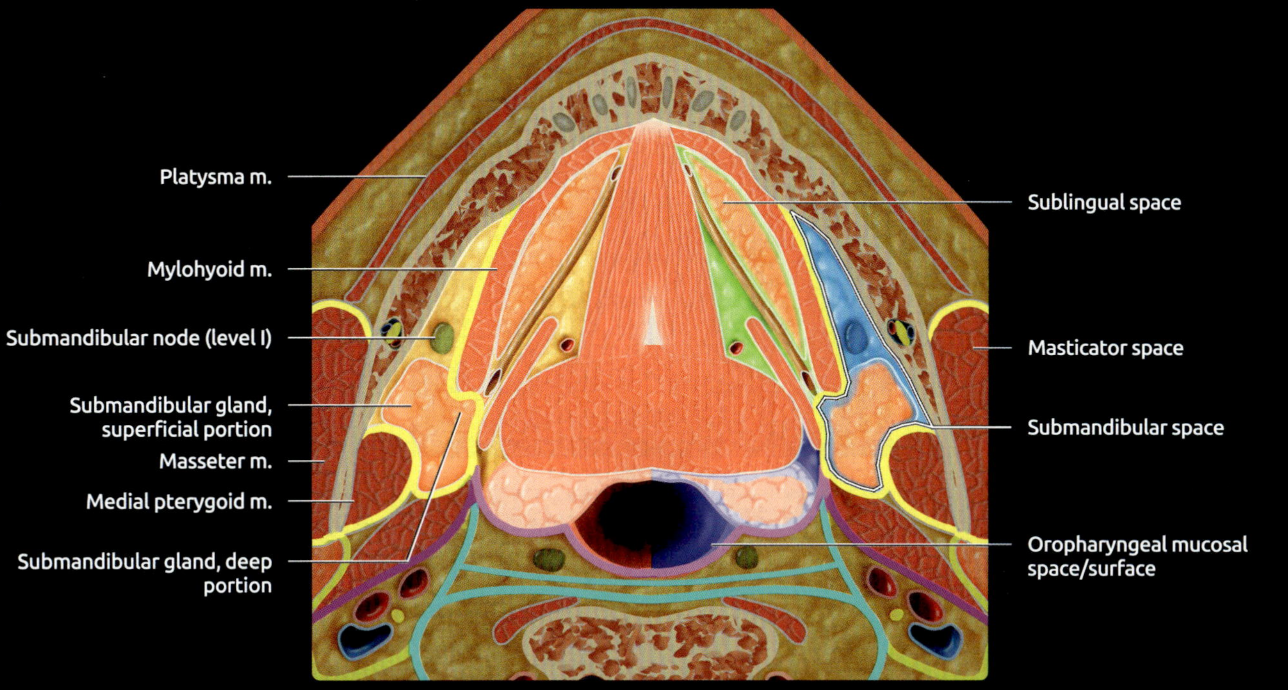

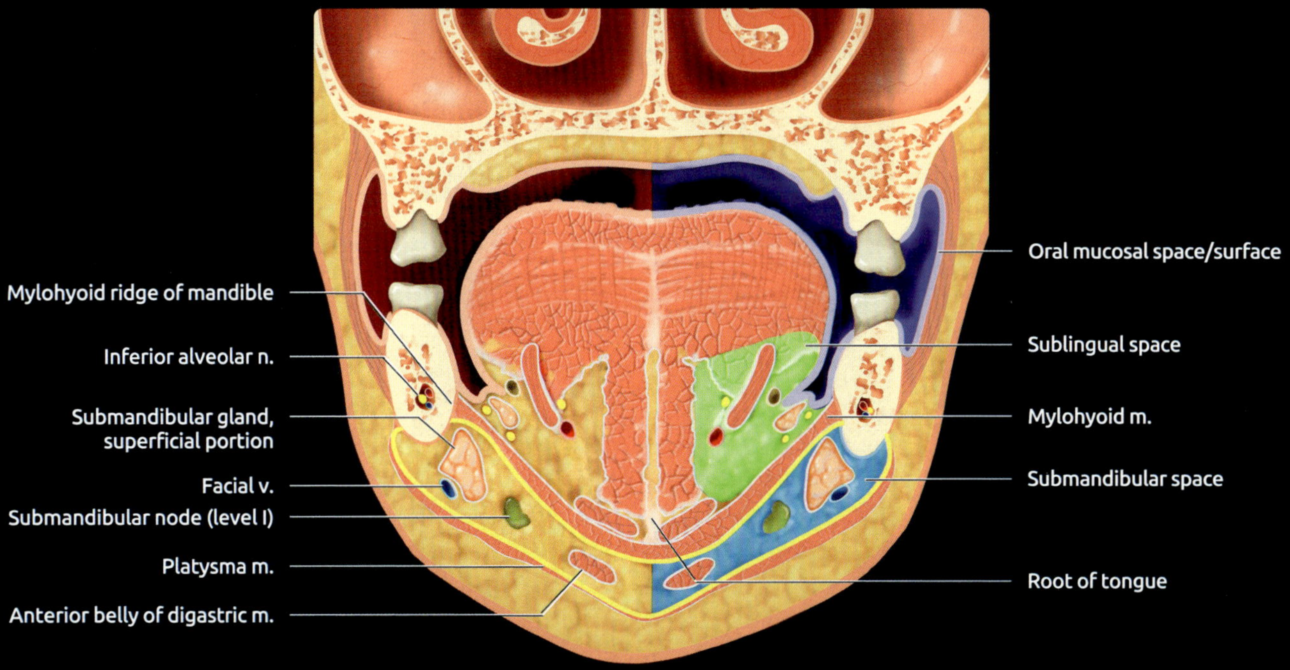

(Top) *Axial graphic shows the oral cavity with emphasis on the submandibular space (SMS), shaded in light blue on the patient's left. The SMS is inferolateral to the mylohyoid muscle. Note that the principal structures of the SMS are the submandibular gland and lymph nodes.* **(Bottom)** *In this coronal graphic through the oral cavity, the SMS is shaded in light blue. The superficial layer of the deep cervical fascia (yellow line) is seen lining the vertical horseshoe-shaped SMS inferolateral to the mylohyoid muscle. The contents of the SMS are the anterior belly of the digastric muscle, submandibular nodes, submandibular gland, and facial vein. Note that the platysma muscle forms the superficial margin of the SMS.*

TRANSVERSE ULTRASOUND

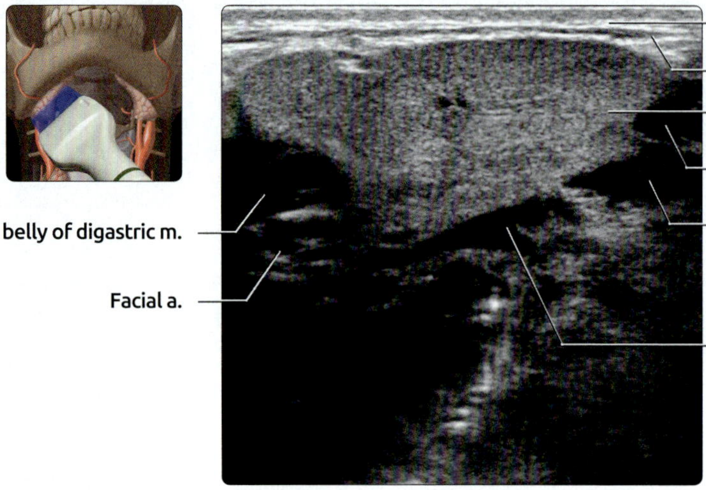

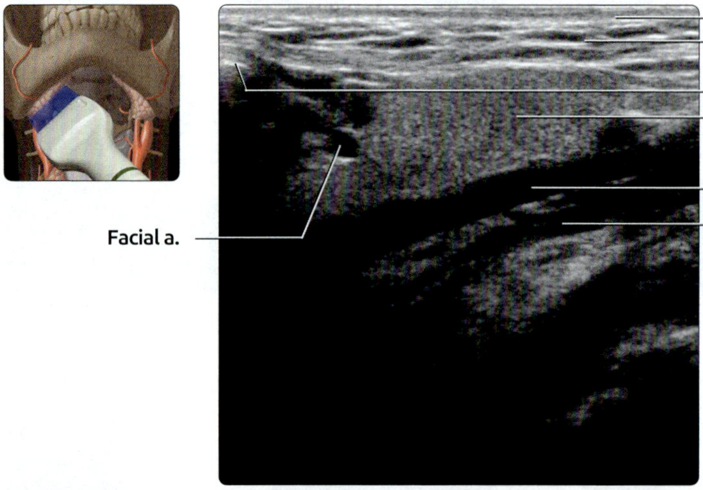

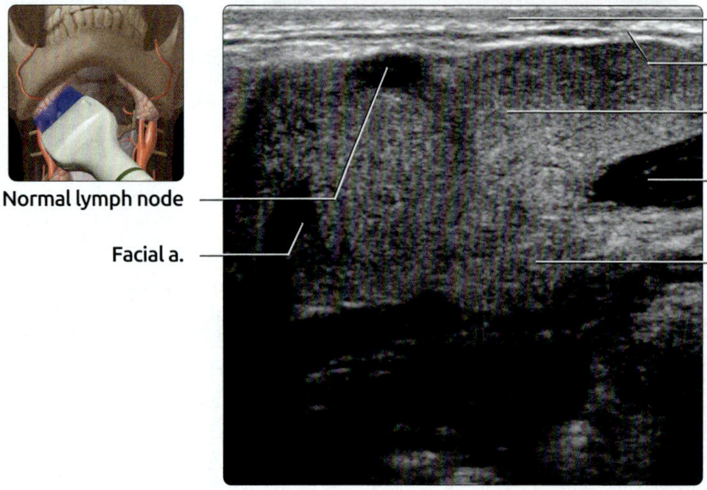

(Top) *Transverse grayscale ultrasound shows the submandibular region. The submandibular gland sits astride the posterior belly of the digastric and mylohyoid muscles. The hyoglossus muscle is seen deep to the submandibular gland.* **(Middle)** *Transverse grayscale ultrasound of the submandibular region (slightly more posterior scan) shows the consistent relationship of the submandibular gland superficial to the mylohyoid and hyoglossus muscles. The submandibular duct runs between these 2 muscles.* **(Bottom)** *Transverse grayscale ultrasound shows the submandibular gland. The gland is divided into superficial and deep lobes, demarcated by the free posterior edge of the mylohyoid muscle. Normal lymph nodes are a constant finding in this region.*

AXIAL MR AND POWER DOPPLER ULTRASOUND

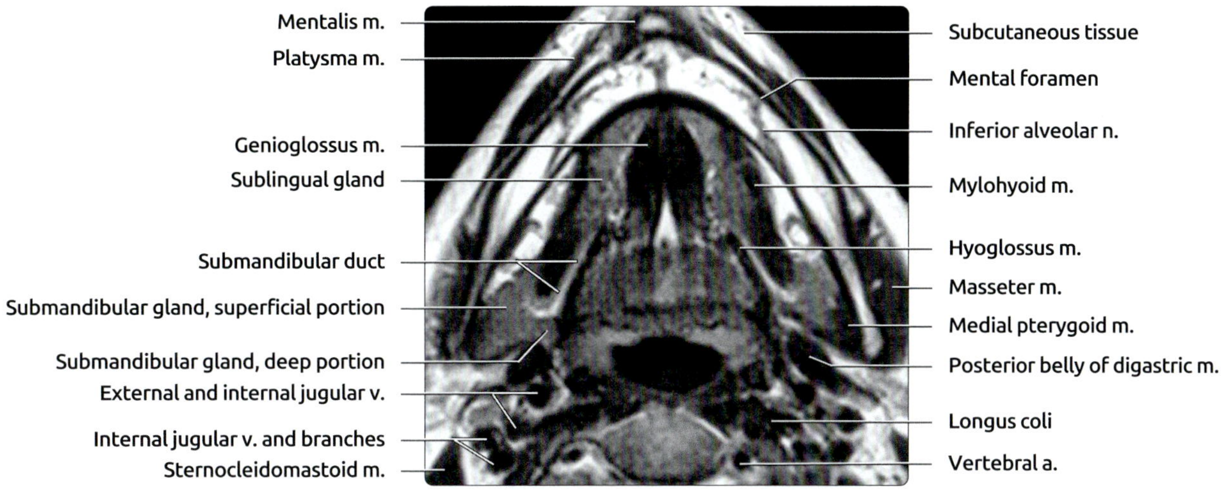

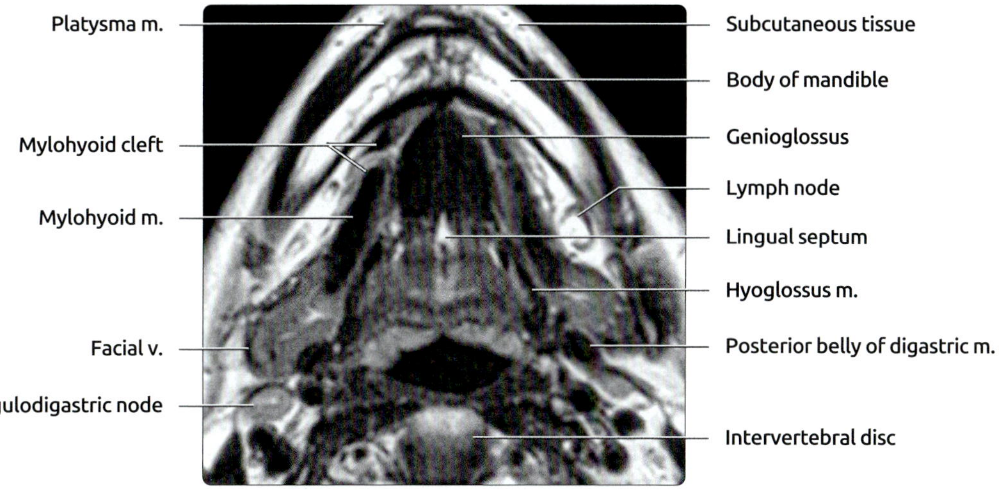

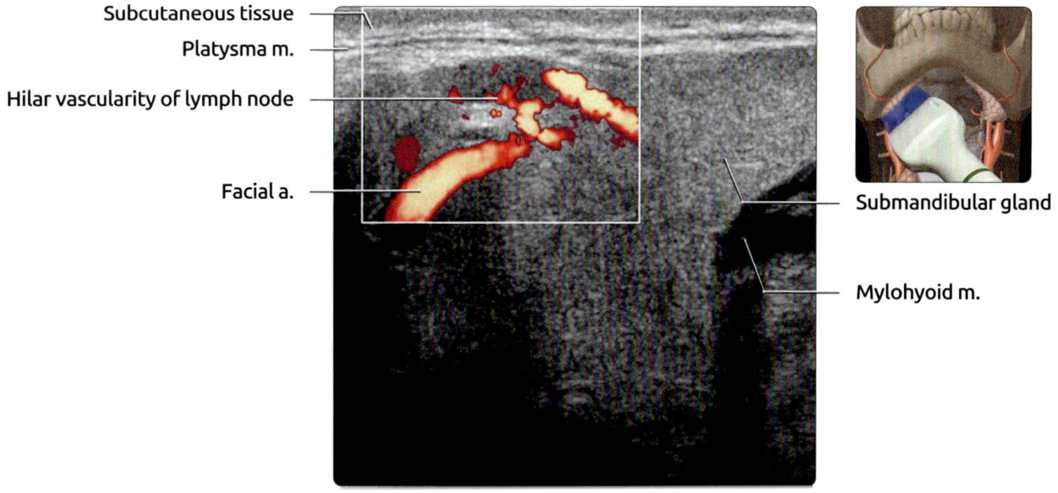

(Top) *Axial T2WI MR shows the floor of the mouth. The SMS contains submandibular glands, fat, and lymph nodes. Note the high-signal submandibular ducts entering the posterior aspect of sublingual spaces bilaterally.* **(Middle)** *In a more inferior image, both submandibular glands are seen wrapping around the posterior margins of the mylohyoid muscles. The neurovascular pedicle to each side of the tongue is closely related to the hyoglossus muscle.* **(Bottom)** *Transverse power Doppler ultrasound of submandibular gland shows vascular flow within the facial artery. Note the presence of normal hilar vascularity within the lymph node.*

TRANSVERSE ULTRASOUND

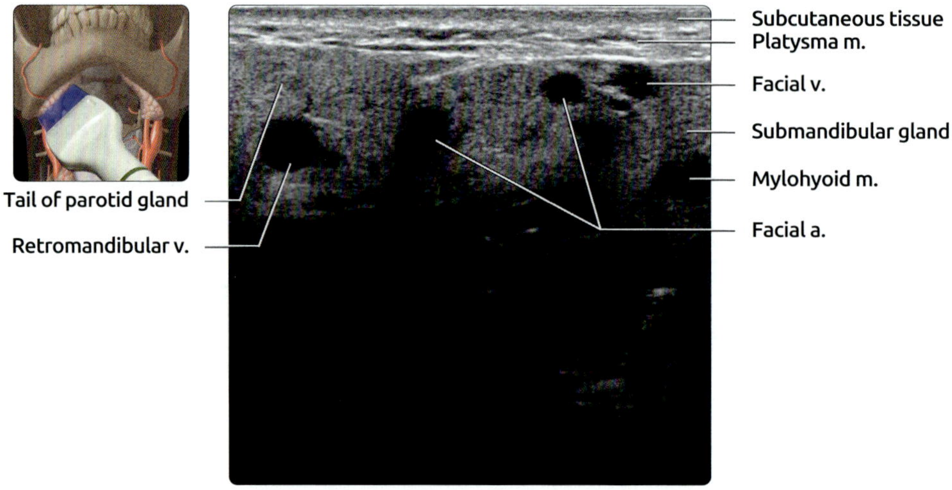

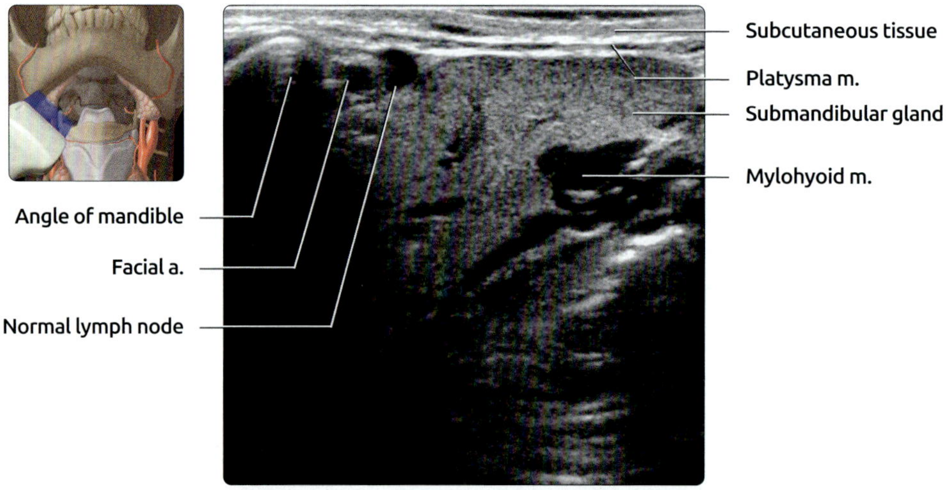

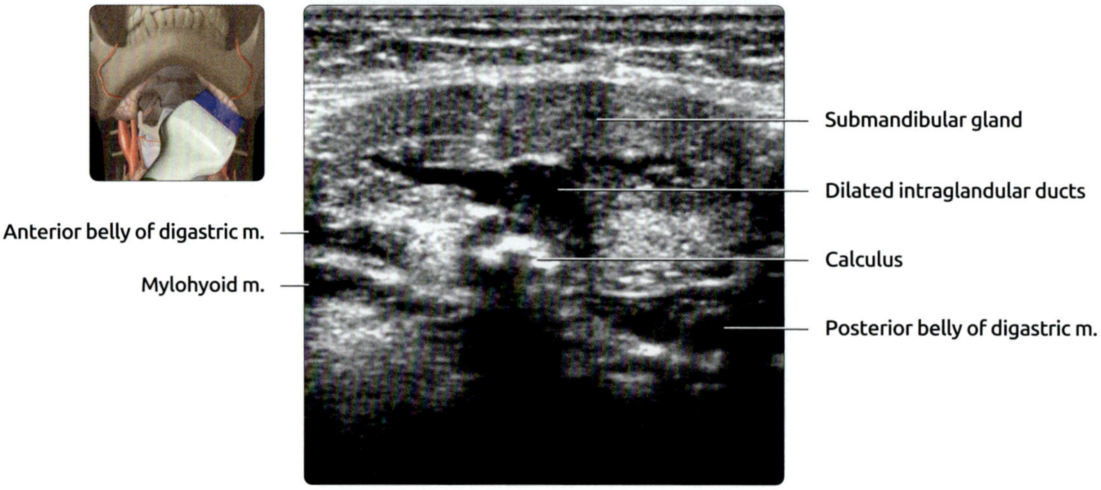

(Top) *Transverse grayscale ultrasound shows the posterior submandibular region. Note the close proximity of the submandibular gland to the tail of the parotid gland. On ultrasound, it may be difficult to localize the origin of large lesions at this site. Displacement of vessels often provides the clue.* **(Middle)** *Longitudinal grayscale ultrasound shows the submandibular region. The submandibular gland is located inferior and posterior to the mandible and superficial to the mylohyoid muscle.* **(Bottom)** *Transverse grayscale ultrasound of the left submandibular gland shows a large obstructing calculus with intraglandular ductal dilatation. Note the glandular parenchyma appears heterogeneous and hypoechoic, compatible with sialadenitis secondary to obstruction.*

POWER DOPPLER ULTRASOUND AND PATHOLOGY

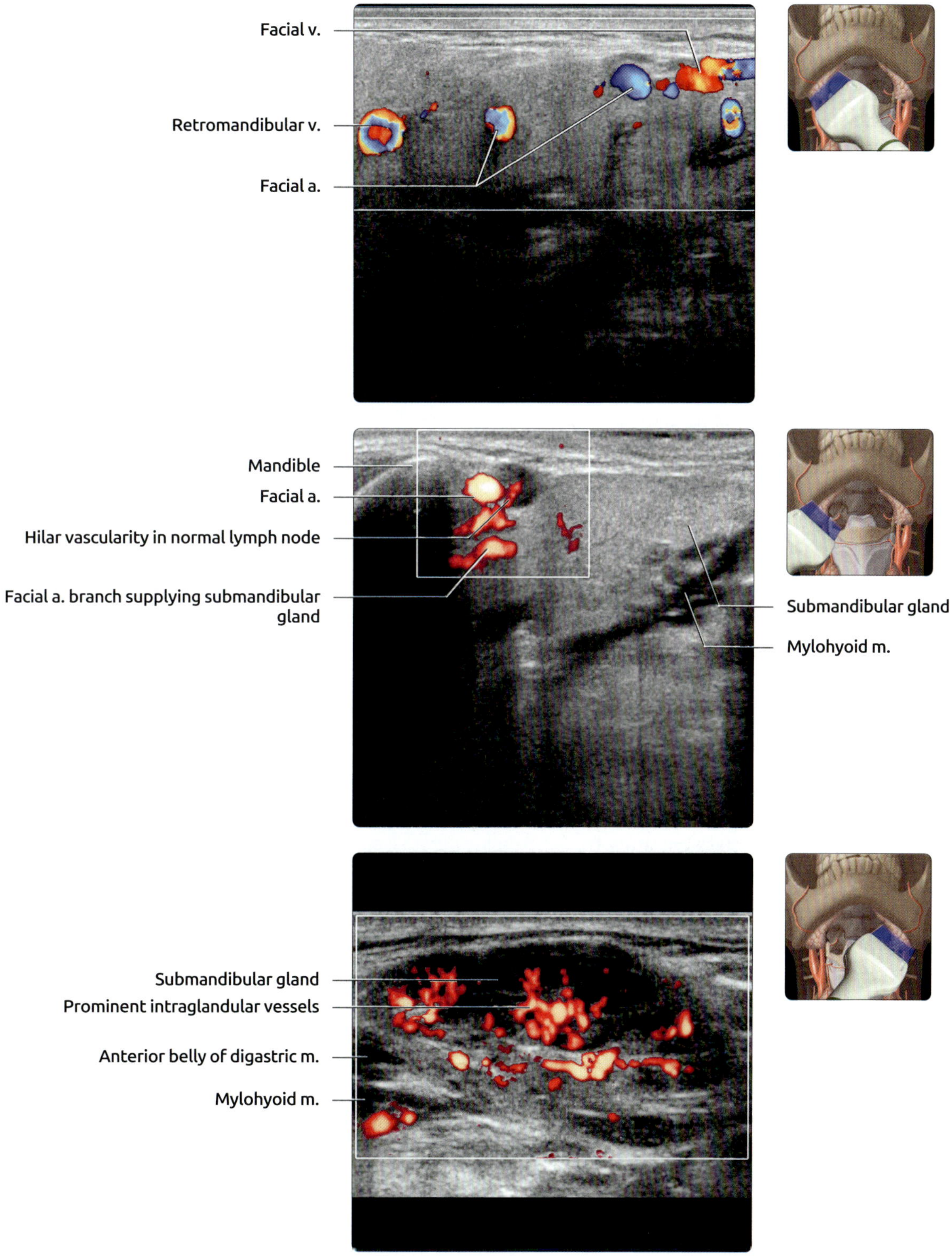

(Top) *Transverse color Doppler ultrasound helps to identify and confirm important vascular landmarks in the posterior submandibular region, including the retromandibular vein and facial artery.* **(Middle)** *Longitudinal power Doppler ultrasound of the submandibular region shows the relationship of the facial artery to the superficial portion of the submandibular gland. The hilar vascularity of a normal lymph node and the vessels supplying the submandibular gland are seen.* **(Bottom)** *Transverse power Doppler ultrasound of the left submandibular gland shows an enlarged, heterogeneous submandibular gland with patchy hypoechoic "nodular" areas. Note prominent intraglandular vessels running through "nodules" with no mass effect/displacement. No ductal dilatation or calculus is seen. These changes are also present in the contralateral gland (not shown), suggesting chronic sclerosing sialadenitis (Kuttner tumor).*

TERMINOLOGY

Abbreviations

- Parotid space (PS)

Definitions

- Paired lateral suprahyoid neck spaces enclosed by superficial layer of deep cervical fascia containing parotid glands, nodes, and extracranial facial nerve branches

IMAGING ANATOMY

Extent

- Extends from external auditory canal (EAC) and mastoid tip superiorly to below angle of mandible (parotid tail)

Internal Contents

- Parotid gland
  - Divided anatomically into superficial lobe and deep lobe by extracranial facial nerve
    - Superficial lobe: Constitutes ~ 2/3 of parotid glandular parenchyma
    - Deep lobe: Smaller component, projects into lateral parapharyngeal space
- Extracranial facial nerve (CNVII)
  - Exits stylomastoid foramen as single trunk; ramifies within PS lateral to retromandibular vein (RMV)
  - Ramifying intraparotid facial nerve creates surgical plane between superficial and deep lobes
- External carotid artery (ECA)
  - Medial and smaller of 2 vessels seen just behind mandibular ramus in PS
- RMV
  - Lateral and larger of 2 vessels seen just behind mandibular ramus in parotid
  - Formed by union of superficial temporal vein and maxillary vein
  - Intraparotid facial nerve branches course just lateral to RMV
- Intraparotid lymph nodes
  - ~ 20 lymph nodes found in each parotid gland
  - Parotid nodes are 1st-order drainage for EAC, pinna, and surrounding scalp
- Parotid duct
  - Emerges from anterior PS, runs along surface of masseter muscle
  - Duct then arches through buccal space to pierce buccinator muscle at level of upper 2nd molar
- Accessory parotid glands
  - Project over surface of masseter muscle
  - Present in ~ 20% of normal anatomic dissections
- Masseter muscle
  - Muscle of mastication related to outer surface of mandibular ramus
  - Parotid duct runs anteriorly on its surface
- Buccinator muscle
  - Deep muscle of buccal space, extends anteriorly and just medially to anterior margin of masseter muscle
    - Parotid duct pierces to enter buccal mucosa at upper 2nd molar level

ANATOMY IMAGING ISSUES

Questions

- Is deep lobe of parotid gland involved
  - For parotid mass, it is important to determine location and extent of involvement in relation to extracranial facial nerve (i.e., superficial/deep lobe involvement)
    - Difference in surgical approach and risk of perioperative facial nerve injury
  - Intraparotid facial nerve is not visible with USG, CT, or MR, except proximally with high-resolution MR
  - On US, RMV is used as marker for division of parotid gland into superficial and deep lobes (due to close proximity to CNVII)

Imaging Approaches

- Scan in both transverse and longitudinal planes
  - Transverse scans define anatomic location of salivary gland masses in relation to ECA and RMV
  - Longitudinal scans help to better evaluate lesions in parotid tail and for Doppler examination
- USG cannot evaluate deep lobe mass or deep extension of superficial masses
  - Lower frequency transducer (e.g., 5 MHz) with gel block/standoff pad helps to evaluate large parotid mass with suspicious deep lobe extension
  - MR/CT is required for full anatomical delineation
  - US helps to direct needle for guided biopsy
- Always evaluate masseter muscle as its lesions clinically mimic parotid pathology
- Normal intraglandular ducts are seen as echogenic streaks within parotid parenchyma
  - When dilated, seen as 2 bright lines separated by fluid within
  - Extraglandular portion of duct is seen on US only if it is dilated

CLINICAL IMPLICATIONS

Clinical Importance

- Although US cannot visualize parotid deep lobe, it is still ideal initial imaging modality to evaluate parotid masses, as most are located in superficial lobe
  - US characterizes common salivary masses and safely guides fine-needle aspiration cytology/biopsy for confirmation

EMBRYOLOGY

Embryologic Events

- PS undergoes late encapsulation in embryogenesis

Practical Implications

- Late encapsulation results in intraparotid lymph nodes
- Warthin tumor arises within this lymphoid tissue (intraparotid > > periparotid > upper cervical)
- Parotid nodes are 1st-order drainage for malignancies of adjacent scalp, EAC, pinna, and deep face
- No such nodes in submandibular gland due to early encapsulation; therefore, no Warthin tumor or nodal metastases in submandibular gland

PAROTID GLAND AND SPACE

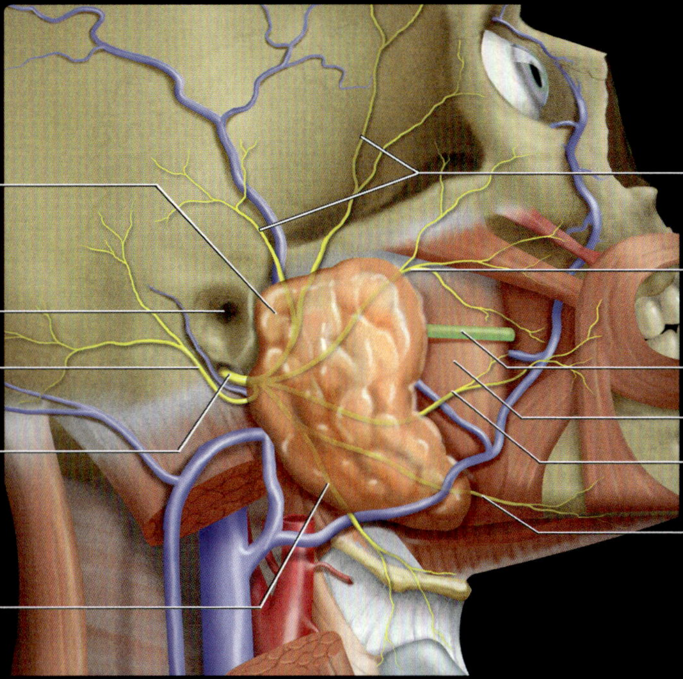

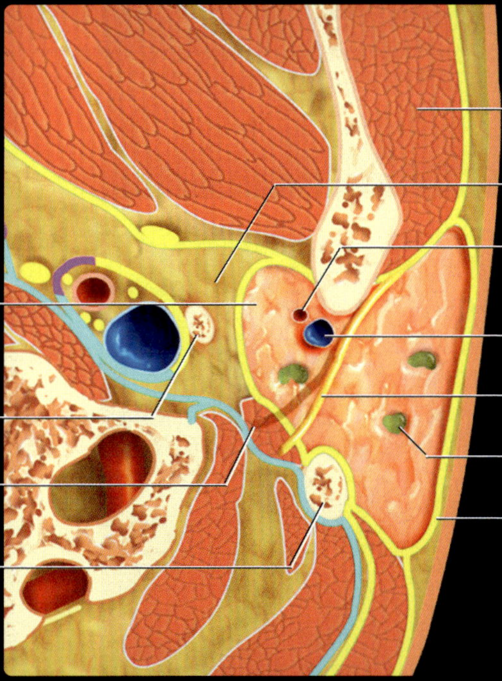

(Top) *Lateral schematic diagram shows the parotid region. The parotid gland is situated in front of the external auditory meatus and below the zygomatic arch. The parotid duct emerges from the anterior margin and passes superficial to the masseter muscle. The facial nerve, after emerging from the stylomastoid foramen, enters the parotid gland and divides into terminal branches to supply muscles of facial expression.* **(Bottom)** *Axial graphic shows the parotid space (PS) at the level of C1 vertebral body. The intraparotid course of the facial nerve extends from just medial to the mastoid tip to a position just lateral to the retromandibular vein, dividing the parotid gland into superficial and deep lobes.*

TRANSVERSE US

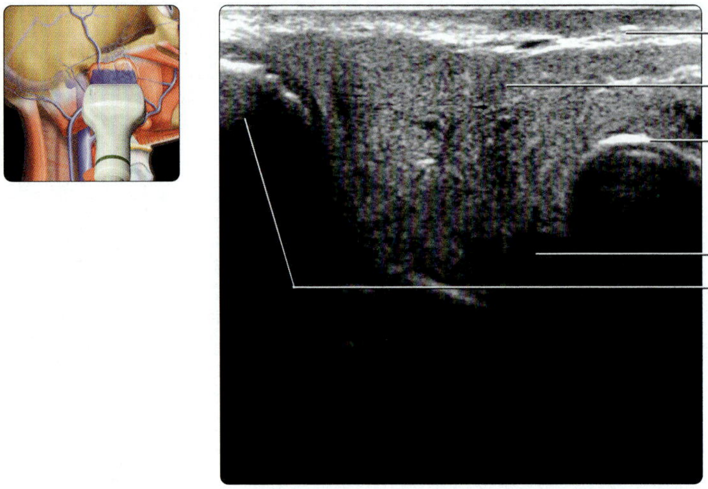

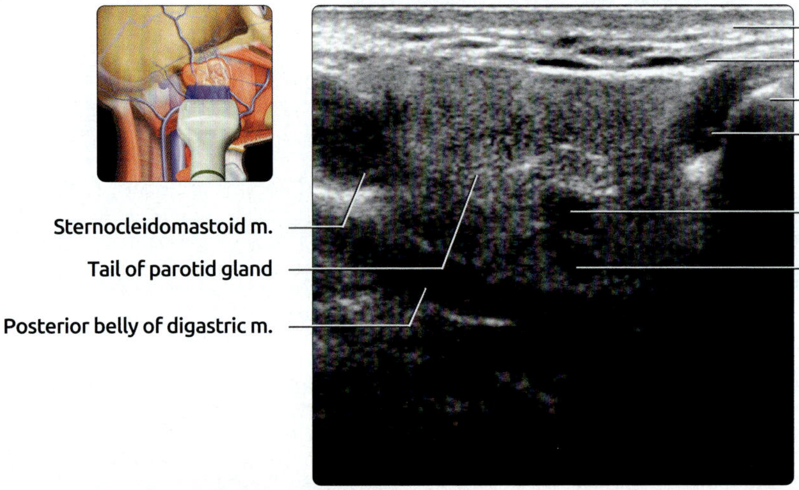

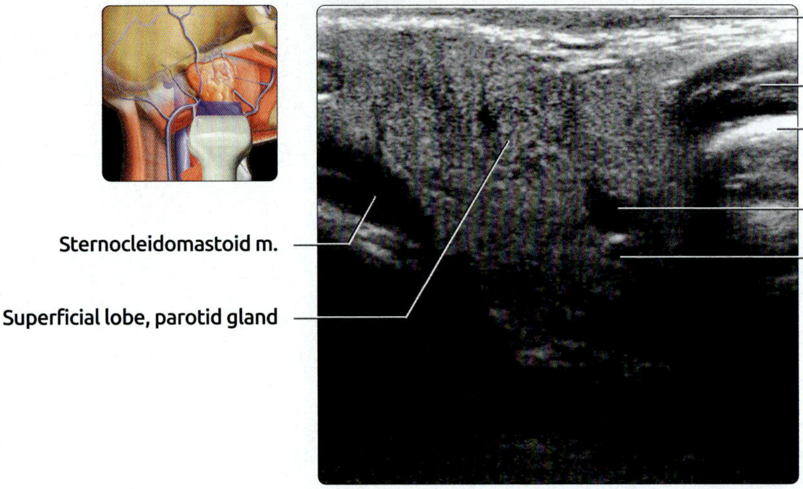

(Top) *Transverse grayscale US shows the parotid region. Note its relationship to the mastoid process and the mandibular ramus. The glandular parenchyma shows a homogeneous, hyperechoic pattern. The retromandibular vein is visualized as a round, anechoic structure within the parotid gland.* **(Middle)** *Transverse grayscale US shows the parotid tail region. The sternocleidomastoid muscle and the posterior belly of the digastric muscle are related to the posterior margin of the parotid tail. The retromandibular vein and external carotid artery serve as markers to infer the location of CNVII.* **(Bottom)** *Transverse grayscale US shows the parotid gland. The retromandibular vein is usually larger and lateral to the external carotid artery within the parotid gland. Note that the deep lobe is obscured by shadowing from the mandibular ramus.*

AXIAL T1 MR AND POWER DOPPLER US

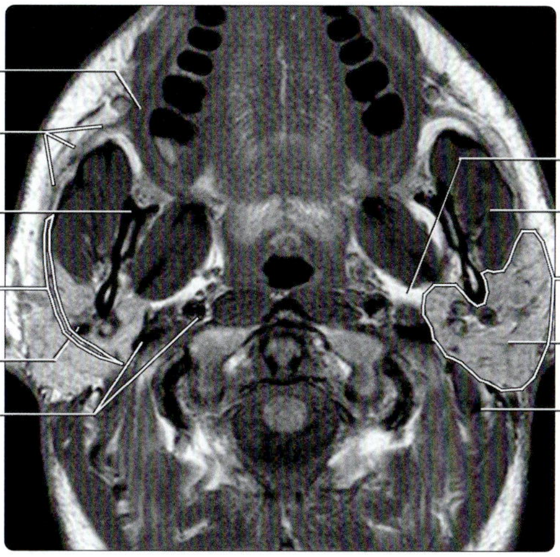

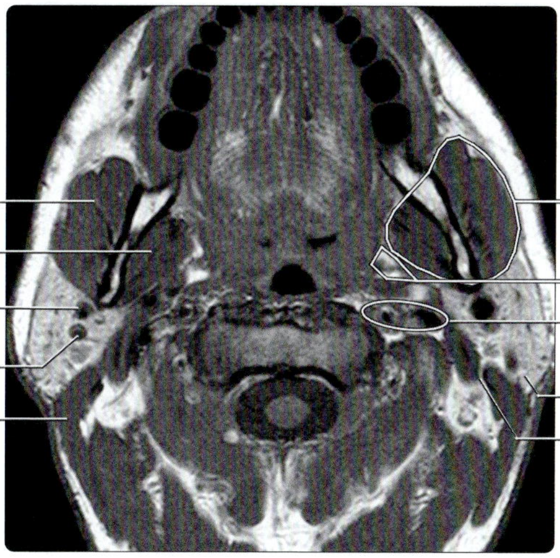

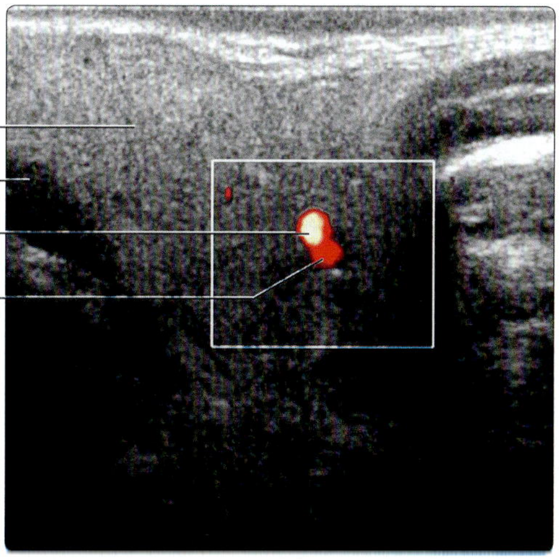

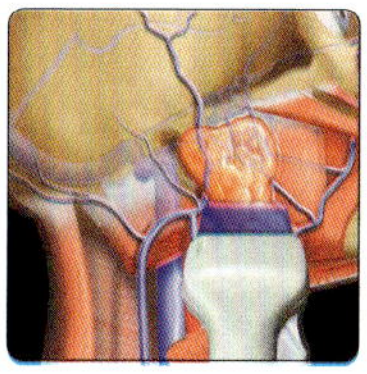

(Top) *Axial T1-weighted MR at the oral pharyngeal level shows the parotid gland as a homogeneous T1-hyperintense structure posterolateral to the mandibular ramus and masseter muscle. Note the projected intraparotid facial nerve course drawn on the right.* **(Middle)** *Axial T1-weighted MR shows a more inferior level of the parotid glands. The PS relates anteriorly to the masticator space, medially to the parapharyngeal space, and is separated medially by the posterior belly of digastric muscle from the carotid space.* **(Bottom)** *Power Doppler US helps to depict the retromandibular vein and external carotid artery, which are sometimes difficult to see in patients with bright fatty parotid glands. The retromandibular vein and external carotid artery help to infer location of CNVII.*

LONGITUDINAL US

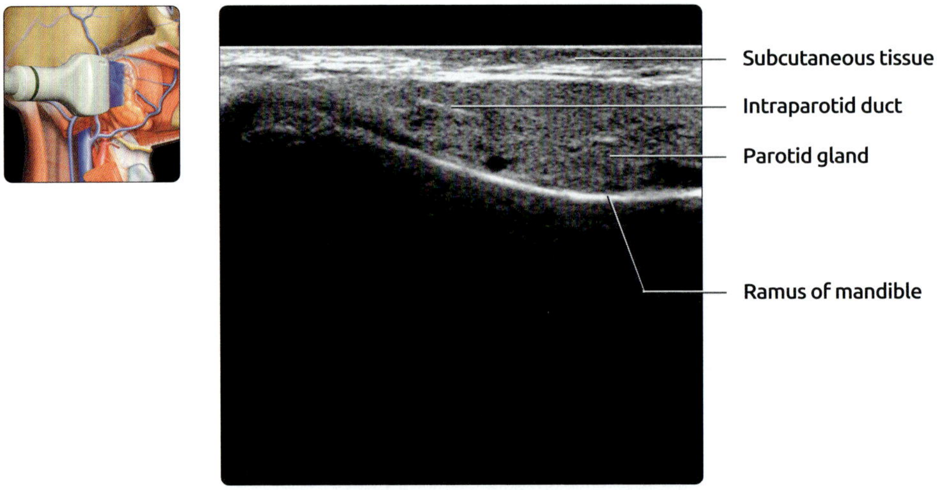

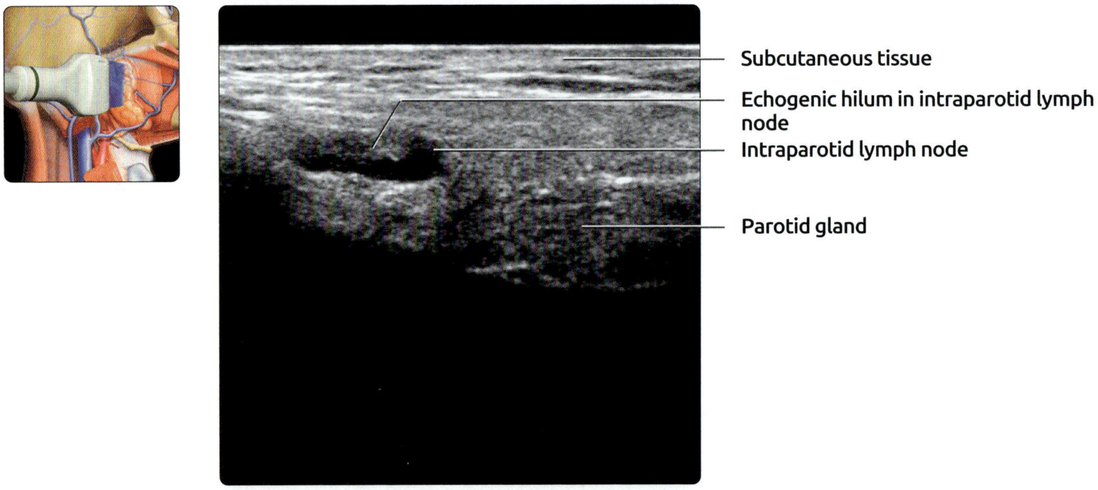

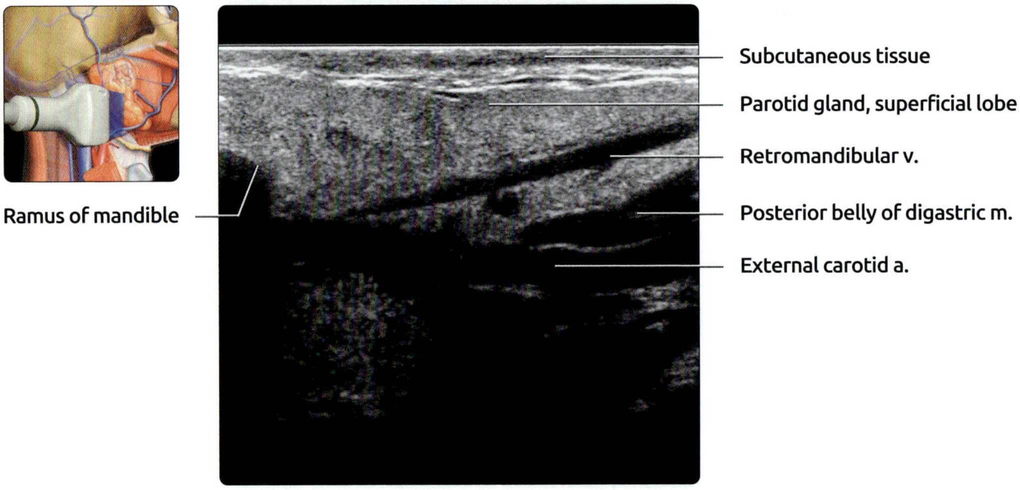

(Top) *This is the 1st image in a series of longitudinal grayscale US scans of the parotid gland. The parotid gland is superficial to the ramus of the mandible.* **(Middle)** *Second image shows a normal intraparotid lymph node in the superficial lobe of the parotid gland. On high-resolution US, normal nodes are invariably seen in the parotid tail and in the pretragal parotid gland. The elliptical shape and normal internal architecture with echogenic hilum suggest its benign nature.* **(Bottom)** *Third image at the plane of the retromandibular vein is shown. Such anatomy is best seen in children and young adults where there is not much fat deposition in the gland.*

CORONAL T1 MR AND POWER DOPPLER US

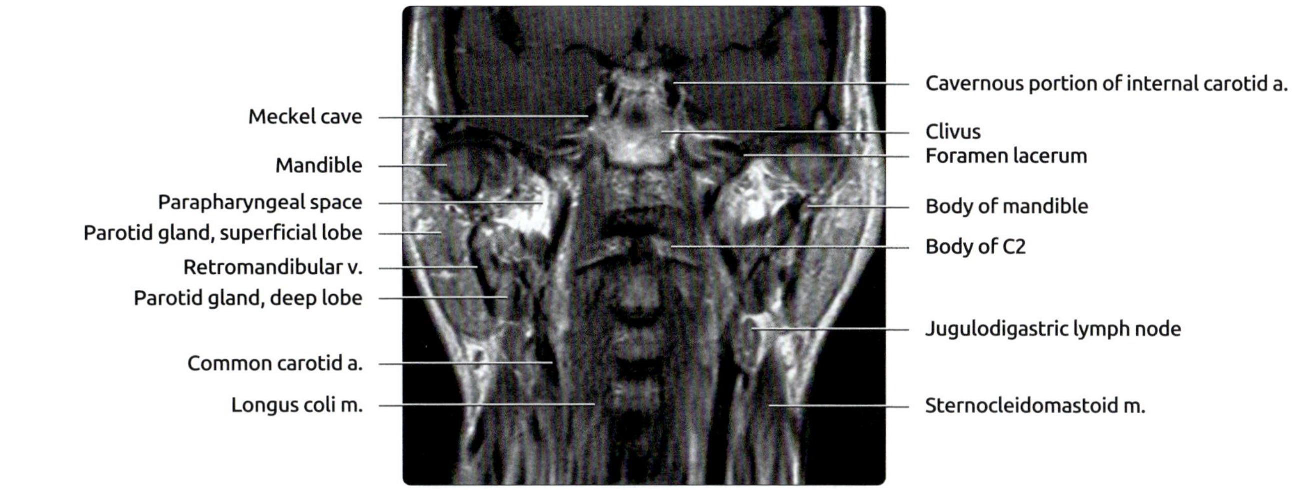

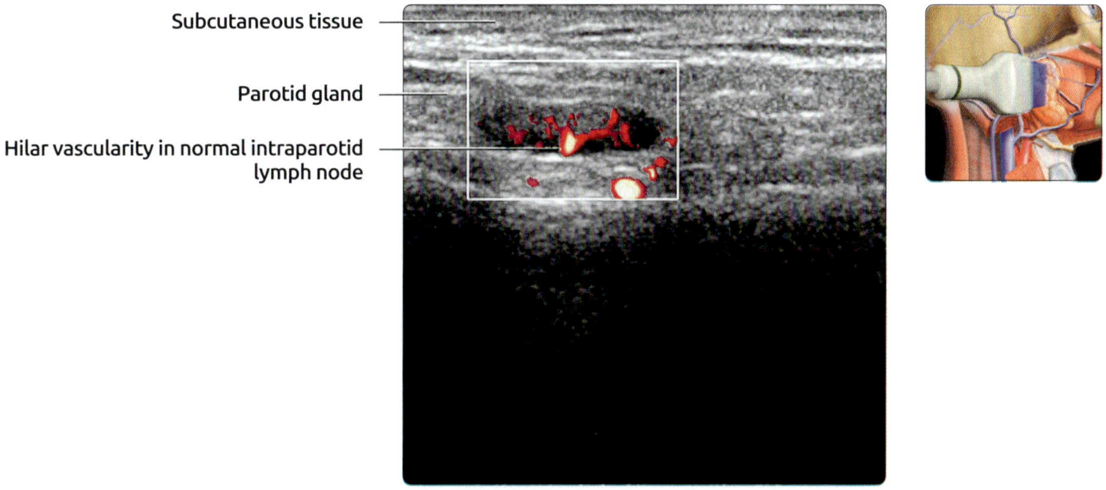

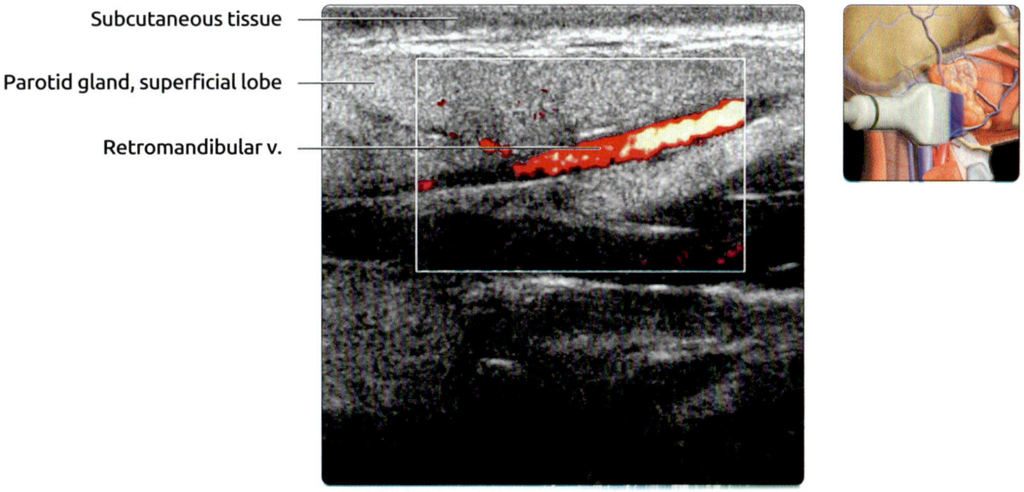

(Top) *Coronal T1-weighted MR shows the parotid glands. The retromandibular vein is seen as a signal-void tubular structure traversing the parotid gland vertically, helping to suggest the location of CNVII.* **(Middle)** *Longitudinal power Doppler US of the parotid gland shows the presence of hilar vascularity within a normal intraparotid lymph node. On high-resolution US, more nodes are seen in children compared to adults.* **(Bottom)** *Longitudinal power Doppler US of the parotid gland clearly delineates the retromandibular vein.*

TRANSVERSE US

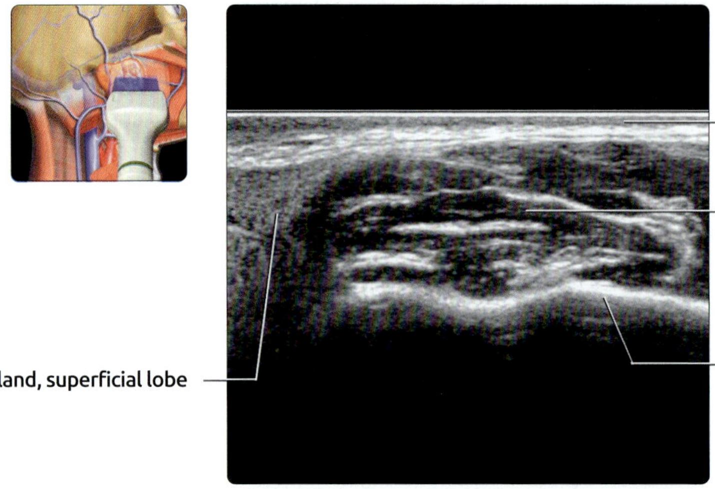

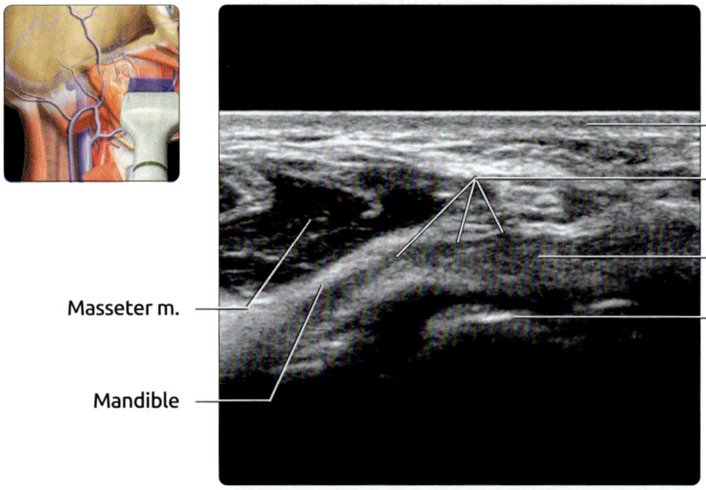

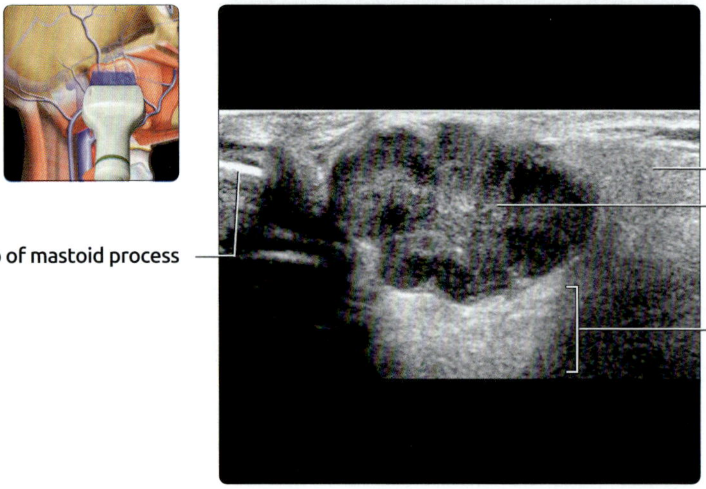

(Top) *Transverse grayscale US of the anterior parotid region shows the masseter muscle superficial to the ramus of the mandible and closely related to the parotid gland. Note that masseter muscle lesions can mimic parotid pathology clinically.* **(Middle)** *Transverse grayscale US of the anterior parotid region/facial region shows the buccinator muscle as a thin, hypoechoic structure extending anteriorly and just medial to the anterior margin of the masseter muscle. The buccal space, which lies lateral to the buccinator muscle, is fat filled and contains the facial nerve, vein, artery, and the parotid duct.* **(Bottom)** *Transverse grayscale US of the right parotid gland shows a well-defined, solid, heterogeneous, hypoechoic mass with a lobulated margin in the superficial lobe. Despite the solid nature of the tumor, there is intense posterior acoustic enhancement, often seen in benign mixed tumors. Pathology confirmed benign mixed tumor.*

AXIAL T1 MR AND TRANSVERSE US

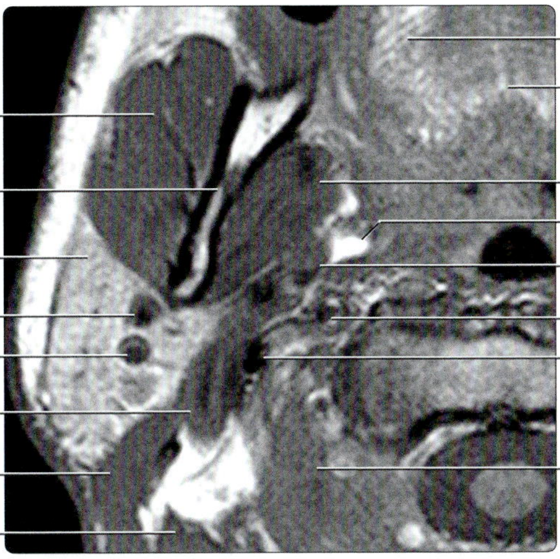

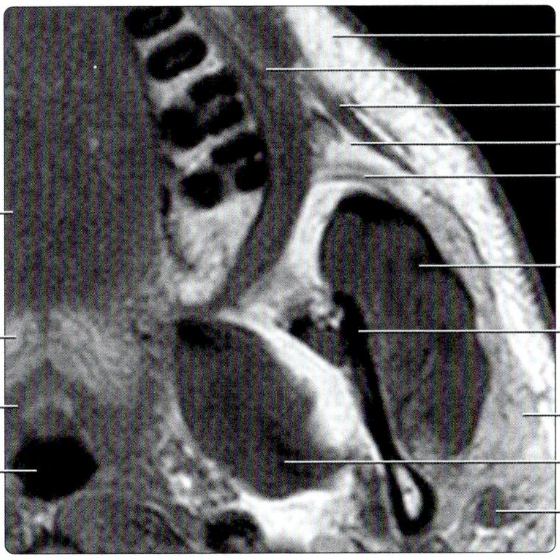

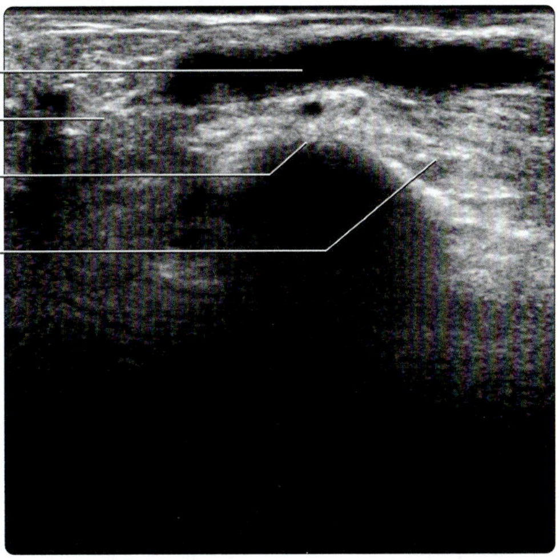

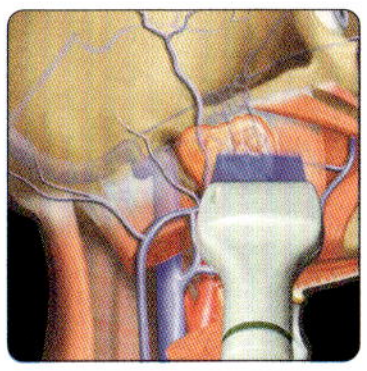

(Top) *Axial T1-weighted MR of the right parotid region shows the relationship of the parotid gland to the masseter muscle and the ramus of the mandible, the parapharyngeal space and the posterior belly of the digastric muscle, and the upper sternocleidomastoid muscle.* **(Middle)** *Axial T1-weighted MR shows the anterior parotid region. The parotid duct courses anteromedially within the buccal fat space and pierces the buccinator muscle at the level of the upper 2nd molar into the buccal mucosa.* **(Bottom)** *Transverse grayscale US of the right parotid gland shows a grossly dilated parotid duct superficial to the ramus of the mandible and masseter muscles. The ductal dilatation is due to distal parotid ductal stricture close to the orificial opening. Note that the parotid gland itself is atrophic and heterogeneous, secondary to chronic infection.*

TERMINOLOGY

Abbreviations

- Internal carotid artery (ICA)
- External carotid artery (ECA)
- Internal jugular vein (IJV)
- Jugulodigastric (JD) lymph node

Synonyms

- Carotid triangle
- Suprahyoid anterior triangle

Definitions

- Portion of anterior triangle adjacent to major vessels of carotid sheath
- Extends from skull base superiorly to hyoid bone inferiorly

IMAGING ANATOMY

Overview

- Cervical region is divided into upper, mid, and lower cervical levels in order to identify relevant groups of jugular cervical lymph nodes
- On ultrasound, upper cervical region is best scanned transversely from submandibular/parotid tail region to carotid bifurcation
- Major structures of upper cervical level: Cervical portion of ICA, ECA with origins of major branches, carotid bifurcation, IJV, posterior belly of digastric muscle, and JD lymph node

Internal Contents

- Cervical portion of ICA
  - 1 of 2 branches from common carotid artery
    - Runs lateral or posterolateral to ECA
    - Usually of larger caliber than ECA
    - Low-resistance arterial flow waveform on Doppler ultrasound
  - Supplies anterior part of brain, eye, and its appendages
  - Divided into bulbous, cervical, petrous, cavernous, and cerebral portions
  - Only first 2 portions lie extracranially and are accessible by ultrasound
  - No branch in extracranial portions
- ECA
  - Runs medial to cervical ICA
  - Smaller than ICA
  - High-resistance arterial flow waveform on Doppler ultrasound
  - Plays pivotal role in collateral circulation if arterial occlusion of ICA/vertebral artery
  - 1st branch: Superior thyroid artery
    - Readily detected by grayscale and Doppler examination
- Carotid bifurcation
  - ~ at level of hyoid bone (i.e., division between upper and midcervical levels)
  - Bulbous dilatation of proximal ICA beyond carotid bifurcation: Carotid bulb
  - Carotid body is located at carotid bifurcation
- IJV
  - Inferior continuation of sigmoid sinus from level of jugular foramen at skull base
  - Right usually of larger caliber than left
  - Lateral/posterolateral to ICA
  - Acts as landmark for jugular cervical lymph nodes
- Posterior belly of digastric muscle
  - Key structure in separating parotid region superiorly from upper cervical level inferiorly
  - Runs anteroinferiorly from mastoid process to hyoid bone
  - Emerges deep to sternocleidomastoid muscle to abut tail of parotid gland
  - Major vessels run deep to muscle
    - From posterior to anterior: IJV, ICA, and ECA
- JD lymph node
  - Largest and most superior lymph node of deep jugular chain, a.k.a. "sentinel" node of internal jugular chain
  - Resides close to carotid bifurcation/IJV
  - Orientated along line of digastric muscle
  - Commonly involved in head and neck cancer
- Vagus nerve
  - Descends from skull base within carotid sheath
  - Sandwiched between ICA medially and IJV
  - More difficult to see than in mid and lower cervical levels

ANATOMY IMAGING ISSUES

Questions

- What are common differential diagnoses for mass in upper cervical level
  - Enlarged lymph nodes: Reactive, inflammatory, and neoplasm (metastases/lymphoma)
  - Congenital lesion: 2nd branchial cleft cyst
  - Vascular lesion: IJV varix, IJV thrombosis
  - Neoplasm: Vagal schwannoma, carotid body tumor
- What are common sites
  - JD lymph node is common site of nodal metastases from head and neck cancer
  - Common sites of primary include oral cavity (including tonsils and tongue), nasopharyngeal carcinoma
  - JD node also commonly involved in lymphoma
    - Multiple, often bilateral nodal involvement, pseudocystic/reticulated internal architecture on ultrasound

Imaging Recommendations

- Scan in transverse plane from submandibular region/parotid tail down to carotid bifurcation
  - Longitudinal scanning for more detailed assessment of internal architecture of lesion and evaluation of intralesional vascularity on Doppler
- Color flow imaging helps to identify major vessels and their anatomic relation to node/mass
  - Also provides flow information of cervical masses to characterize their nature
- US provides safe real-time guidance for fine-needle aspiration cytology (FNAC)/biopsy to further enhance diagnostic yield
  - FNAC/biopsy is not recommended for suspected carotid body tumor due to risk of uncontrolled bleeding

UPPER CERVICAL ANATOMY

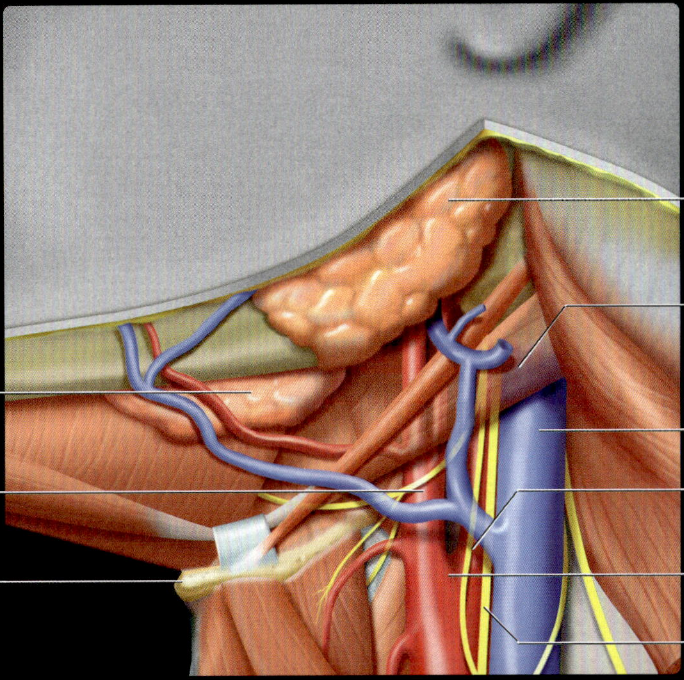

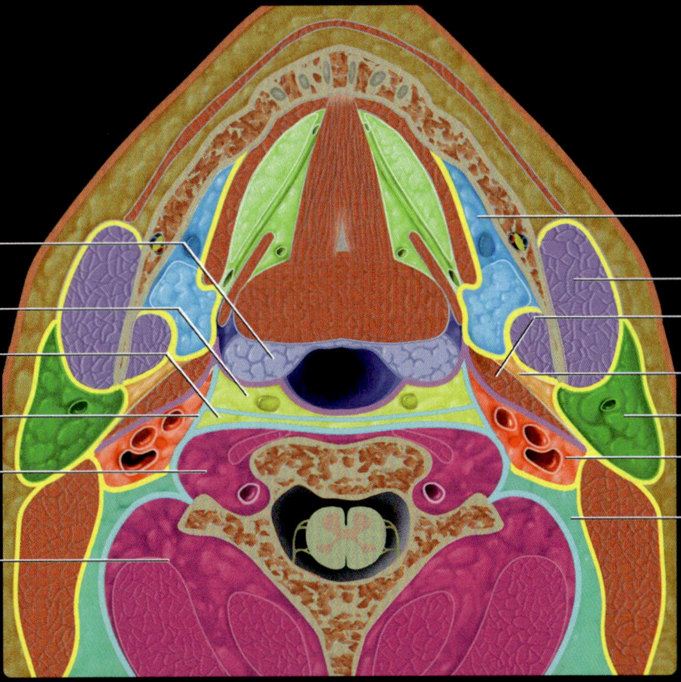

(Top) *Schematic diagram shows the key structures in the upper cervical level. The hyoid bone and the carotid bifurcation are 2 anatomical landmarks for the inferior margin of the upper cervical level. With high-resolution ultrasound, the jugulodigastric lymph node will be consistently seen at this site and should not be mistaken for pathology. Visualization of a similar node on the opposite side helps.* **(Bottom)** *Axial graphic shows the suprahyoid neck spaces at the level of the oropharynx. The superficial (yellow line), middle (pink line), and deep (turquoise line) layers of the deep cervical fascia outline the suprahyoid neck spaces. Ultrasound assessment of the upper cervical level involves scanning transversely along the major vessels of the carotid sheath (i.e., the internal carotid artery and internal jugular vein).*

TRANSVERSE ULTRASOUND

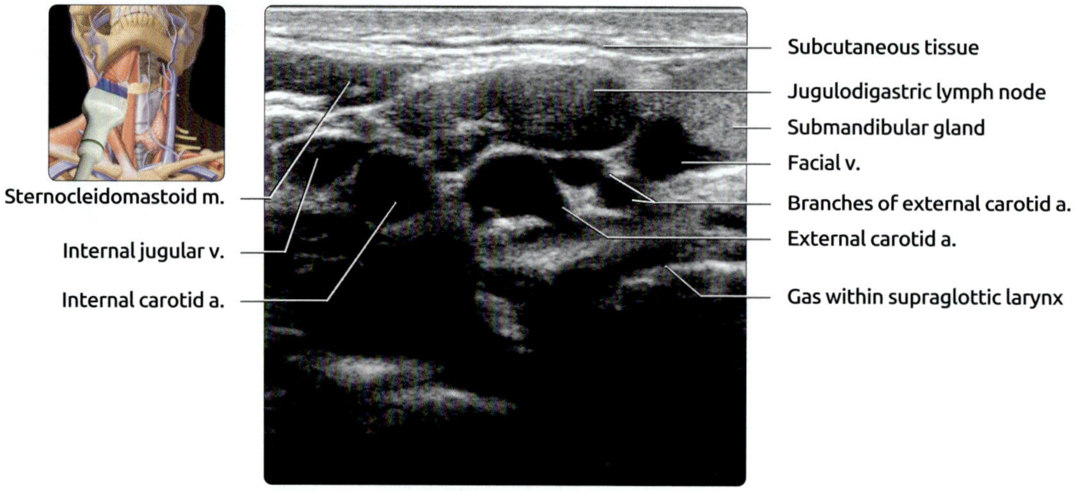

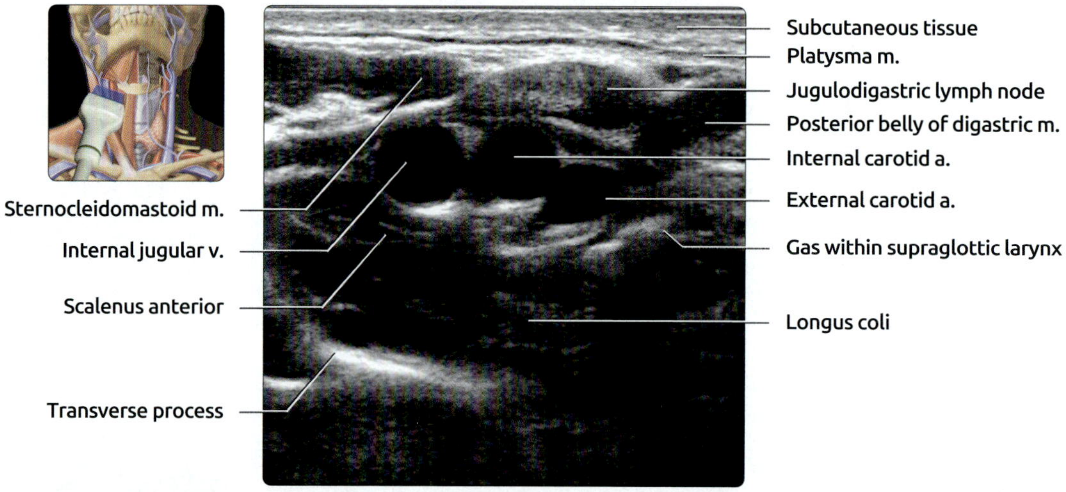

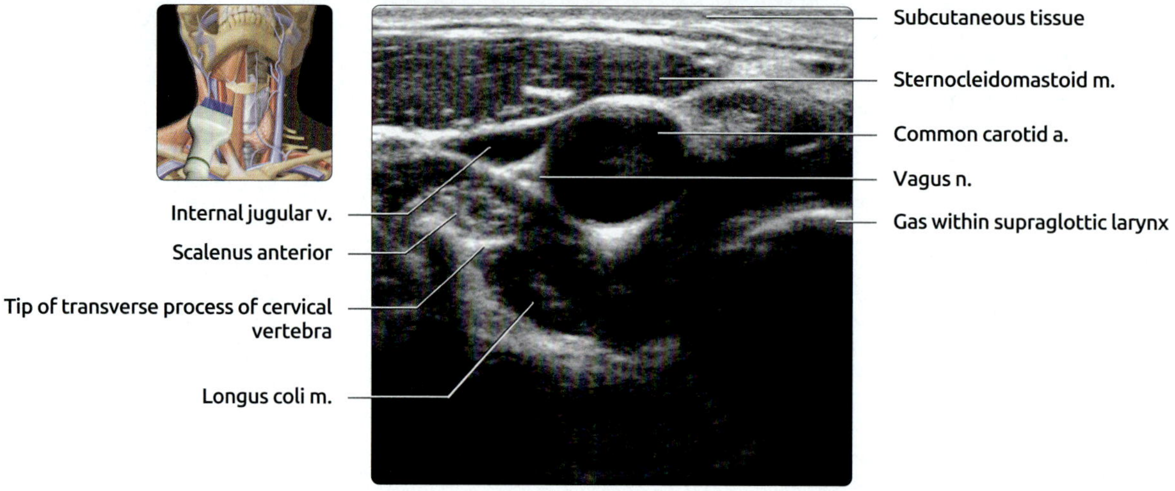

(Top) *First image in a series of consecutive transverse grayscale ultrasound images of the upper cervical level clearly identifies key vascular landmarks, including the internal and external carotid arteries and the internal jugular vein. The uppermost and largest deep cervical lymph node (i.e., the jugulodigastric lymph node) is consistently seen on ultrasounds in the upper cervical level anterior to the carotid arteries. It is usually elliptical, hypoechoic, and with echogenic hilum.* **(Middle)** *Second image shows the carotid bifurcation in the transverse plane. The external carotid artery is usually more medial and smaller than the internal carotid artery.* **(Bottom)** *Third image below the level of carotid bifurcation clearly shows the common carotid artery, internal jugular vein, and vagus nerve to be major structures within the carotid sheath.*

AXIAL CECT

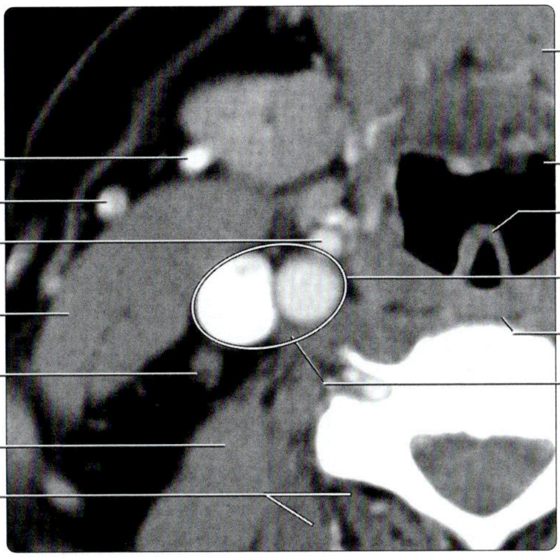

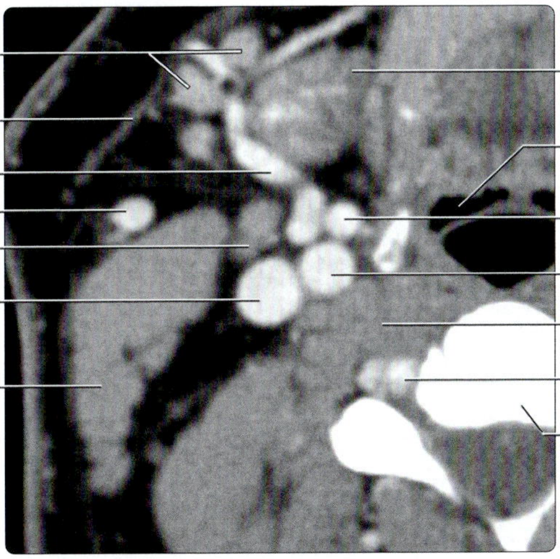

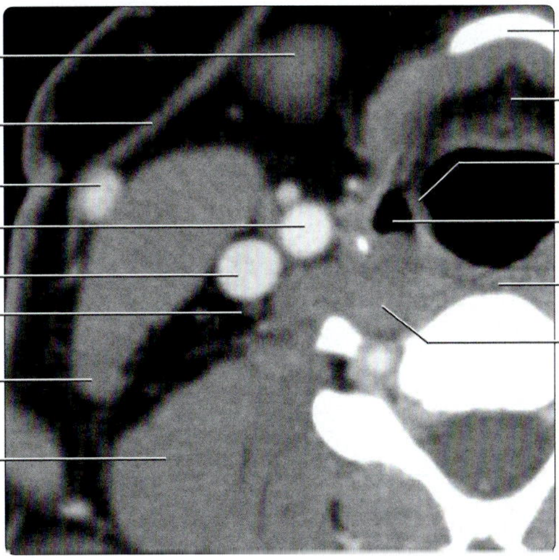

(Top) *At the level just above the hyoid bone, the carotid bifurcation can be seen. The vagus nerve is seen within the carotid sheath.* **(Middle)** *CECT image shows the suprahyoid neck at the level of the free margin of epiglottis. The jugulodigastric lymph node is commonly seen; however, the internal architecture is better assessed on ultrasound than CECT.* **(Bottom)** *At the level of the hyoid bone, the carotid space now contains the common carotid artery, internal jugular vein, and vagus nerve only. The submandibular space is seen anteriorly and is predominantly fat filled at this level. Familiarity with cross-sectional anatomy is key to optimal ultrasound examination of the neck, as lesions at this level are site specific.*

TRANSVERSE ULTRASOUND

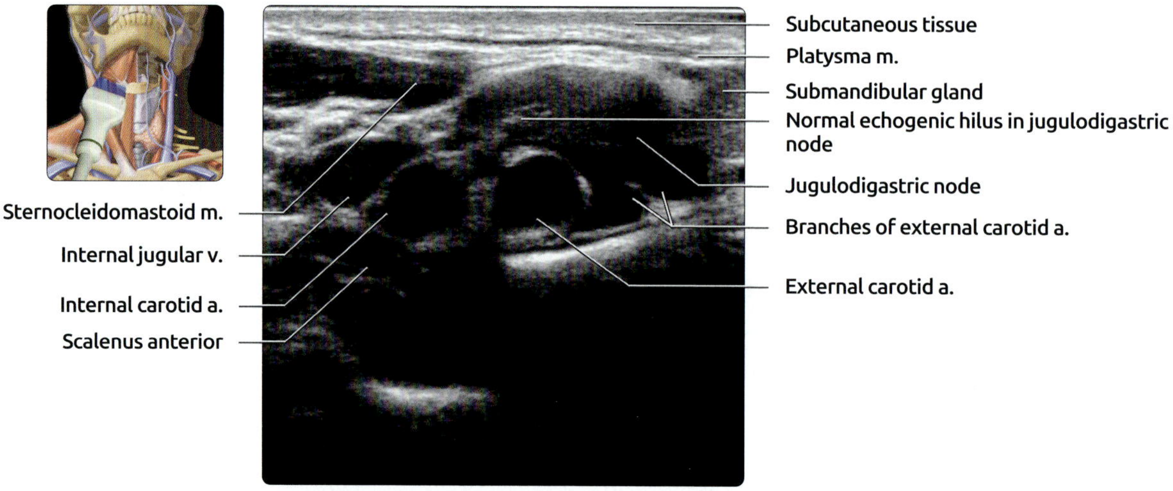

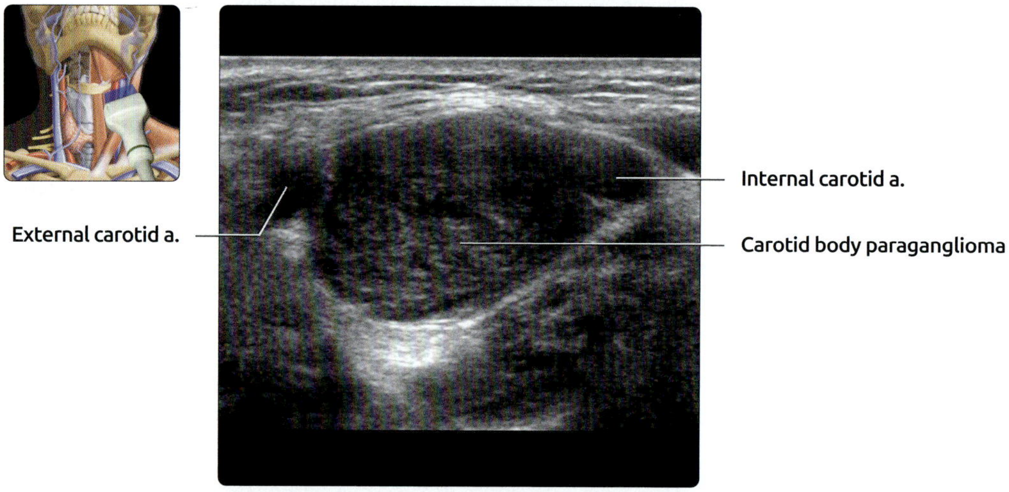

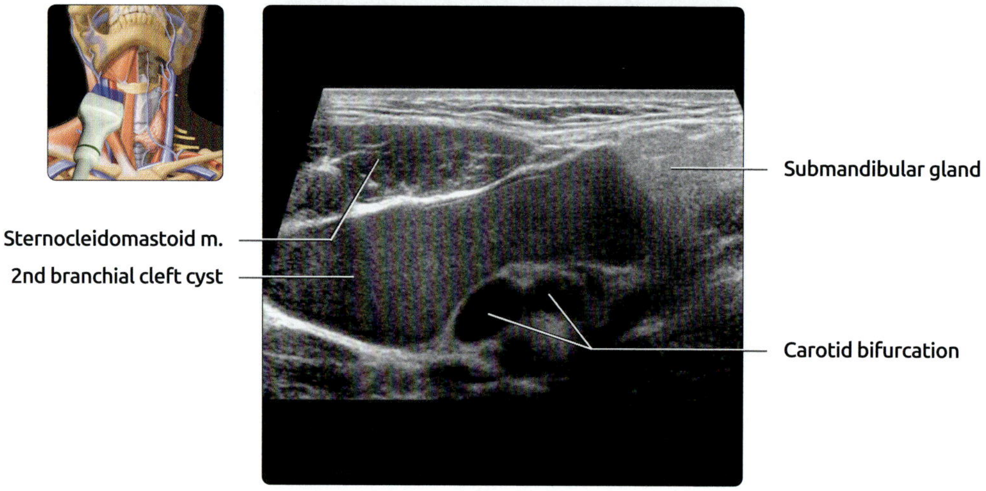

(Top) *Transverse grayscale ultrasound shows the upper cervical level. The normal jugulodigastric lymph node is elliptical in shape with preserved echogenic hilum. Its anatomical relationship with the major neck vessels is better appreciated on power Doppler ultrasound.* **(Middle)** *Transverse grayscale ultrasound of the left upper cervical level shows a large, heterogeneous, hypoechoic mass centered at the carotid bifurcation, splaying the internal and external carotid arteries. Appearances and location are strongly indicative of carotid body paraganglioma.* **(Bottom)** *Transverse grayscale ultrasound of the right upper cervical level shows a well-circumscribed cystic mass with a pseudosolid appearance along the medial edge of the sternomastoid muscle, posterior to the submandibular gland and superficial to major vessels within the carotid sheath. Location and appearances are suggestive of 2nd branchial cleft cyst. This pseudosolid appearance is commonly seen in congenital neck cysts, including thyroglossal duct cysts.*

POWER DOPPLER ULTRASOUND

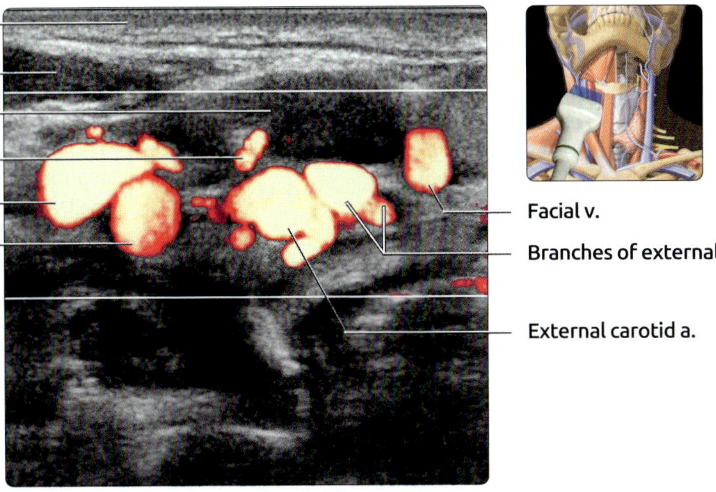

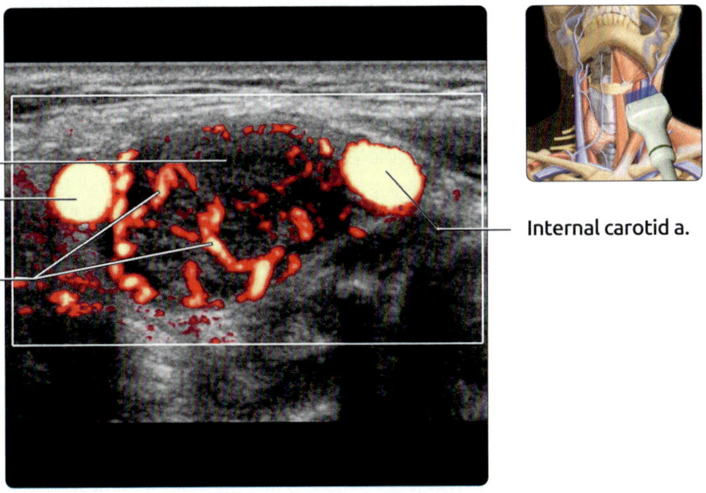

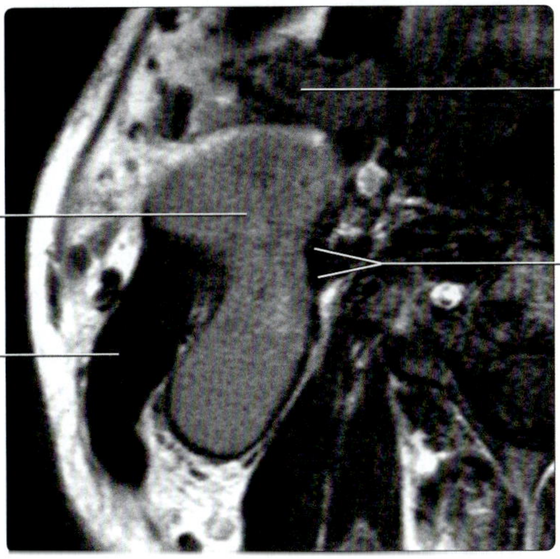

(Top) *Transverse power Doppler ultrasound shows the upper cervical level. The major vessels, including the internal and external carotid arteries and the internal jugular vein, show color flow. Note the presence of normal hilar vascularity within the echogenic hilum of the jugulodigastric lymph node.* **(Middle)** *Transverse power Doppler ultrasound shows marked intratumoral vessels in carotid body paraganglioma. Note the characteristic splaying of internal and external carotid arteries.* **(Bottom)** *Axial T1WI MR shows a lobulated right upper neck lesion with well-circumscribed margins located posterior to the submandibular gland, along the medial edge of the sternocleidomastoid muscle and superficial to major arteries of the carotid sheath. Location of lesion is typical for 2nd branchial cleft cyst. Note T1-hyperintense signal within the 2nd branchial cleft cyst due to proteinaceous intracystic content.*

TERMINOLOGY

Definitions

- Portion of anterior triangle adjacent to major vessels of carotid sheath
- Extends from hyoid bone superiorly to cricoid cartilage inferiorly

IMAGING ANATOMY

Overview

- Major lymphatic drainage of head and neck region is via deep jugular cervical lymph nodes, which are distributed along upper, mid, and lower cervical levels
- Key structures at midcervical level
  - Common carotid artery (CCA), internal jugular vein (IJV), vagus nerve, lymph nodes, omohyoid muscle, and esophagus

Internal Contents

- CCA
  - Arises from aortic arch directly on left from brachiocephalic trunk on right
  - Ascends neck within carotid sheath
    - Medial to IJV
    - Vagus nerve sandwiched between CCA and IJV
  - No named branches in midcervical level
  - Bifurcates into internal carotid artery and external carotid artery at level of hyoid bone
  - Low-resistance arterial flow pattern
- IJV
  - Inferior continuation of sigmoid sinus from jugular foramen at skull base
  - Major deep venous drainage from brain and neck
  - Descends within carotid sheath
    - Lateral to CCA at midcervical level
  - Joins subclavian vein to form brachiocephalic vein
  - Ultrasound appearance
    - Occasionally presence of slow venous flow may mimic IJV thrombus
      - Real-time visualization of layering and sharp linear interface help to distinguish it from thrombus
      - Make sure IJV is compressible and has respiratory phasicity
- Vagus nerve (CNX)
  - Extracranial segment starts superiorly from jugular foramen at skull base
  - Descends along posterolateral aspect of carotid artery within carotid sheath
    - Passes anterior to aortic arch on left and subclavian artery on right
  - Major autonomic nerve supply to visceral organs in thorax and abdomen
  - Ultrasound appearance
    - On transverse scan, appears as round, hypoechoic structure with central echogenicity
      - Close relation to CCA, IJV is clue to its identification
    - On longitudinal scan, seen as long tubular hypoechoic structure with fibrillar pattern
- Lymph nodes
  - Numerous lymph nodes are seen along deep jugular chain in midcervical level
  - Receive lymphatic drainage from upper jugular chain and directly from adjacent structures, including larynx, hypopharynx, and thyroid gland
    - Commonly involved in nodal metastases from common head and neck cancers
  - Usually located anterior to vessels of carotid sheath
  - Ultrasound appearance
    - Small, elliptical in shape, presence of echogenic hilus, preserved hilar vascularity on Doppler imaging
- Omohyoid muscle
  - Arises from anterior portion of body of hyoid bone
  - Runs obliquely to cross anterior to CCA and deep to sternocleidomastoid muscle
  - Intermediate tendon overlies IJV
    - Occasionally mistaken for lymph node
  - Then runs obliquely across inferior posterior triangle to attach to posterior aspect of lateral clavicle
- Cervical esophagus
  - Junction of cricopharyngeus muscle and proximal cervical esophagus is located in midcervical level
  - Lies posterior/posterolateral to trachea and medial to CCA
    - More commonly slightly off midline to left
    - Asking patient to swallow during real-time sonography helps to accurately identify esophagus
  - Ultrasound appearance
    - May be mistaken for parathyroid adenoma or paratracheal lymph node
    - Have patient swallow to confirm if any question

ANATOMY IMAGING ISSUES

Imaging Approaches

- Scan in transverse plane along major vessels of carotid sheath (keep CCA and IJV in center of image)
  - From carotid bifurcation down to level of cricoid cartilage
- Longitudinal/oblique plane for better assessment of internal architecture and vascularity of any lesion detected on scanning in transverse plane
- Color flow imaging helps to identify major vessels and characterize lesions in this region
- Ultrasound also provides safe real-time guidance for fine-needle aspiration cytology or biopsy

CLINICAL IMPLICATIONS

Clinical Importance

- Common differential diagnoses for mass at midcervical level
  - Enlarged lymph nodes: Reactive, inflammatory, or neoplastic (metastases/lymphoma) disease
  - Inflammatory: Abscess
  - Congenital lesion: Lymphatic malformation, off-midline thyroglossal duct cyst
  - Neoplasm: Vagal schwannoma, esophageal lesion

MIDCERVICAL ANATOMY

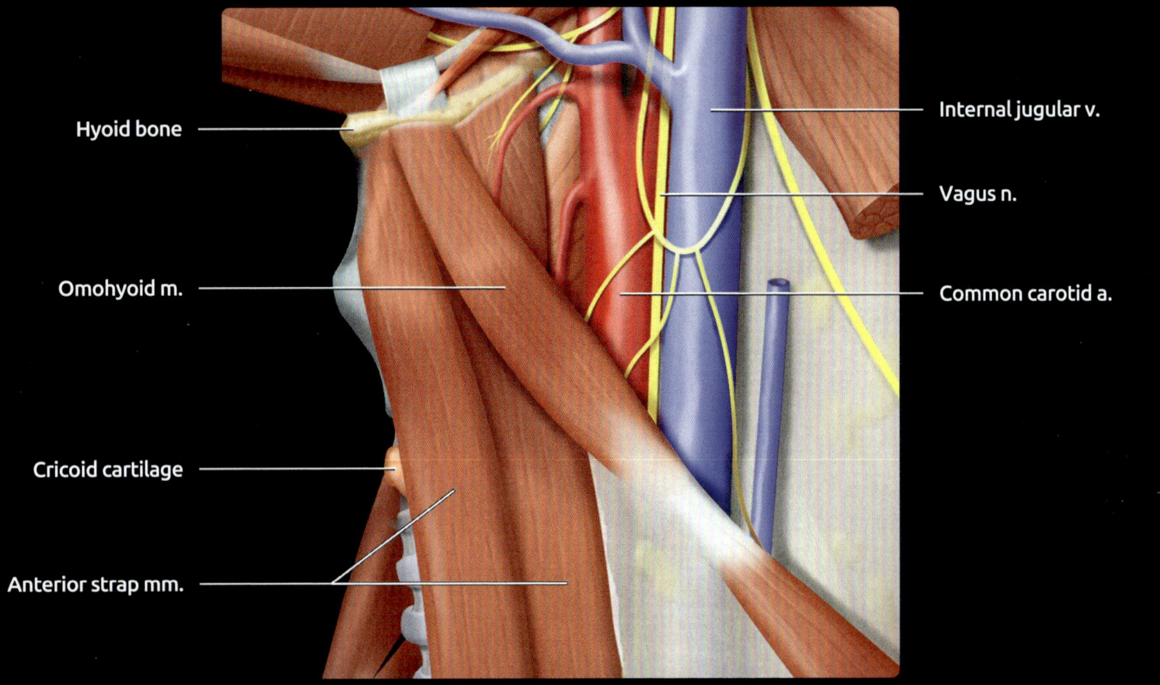

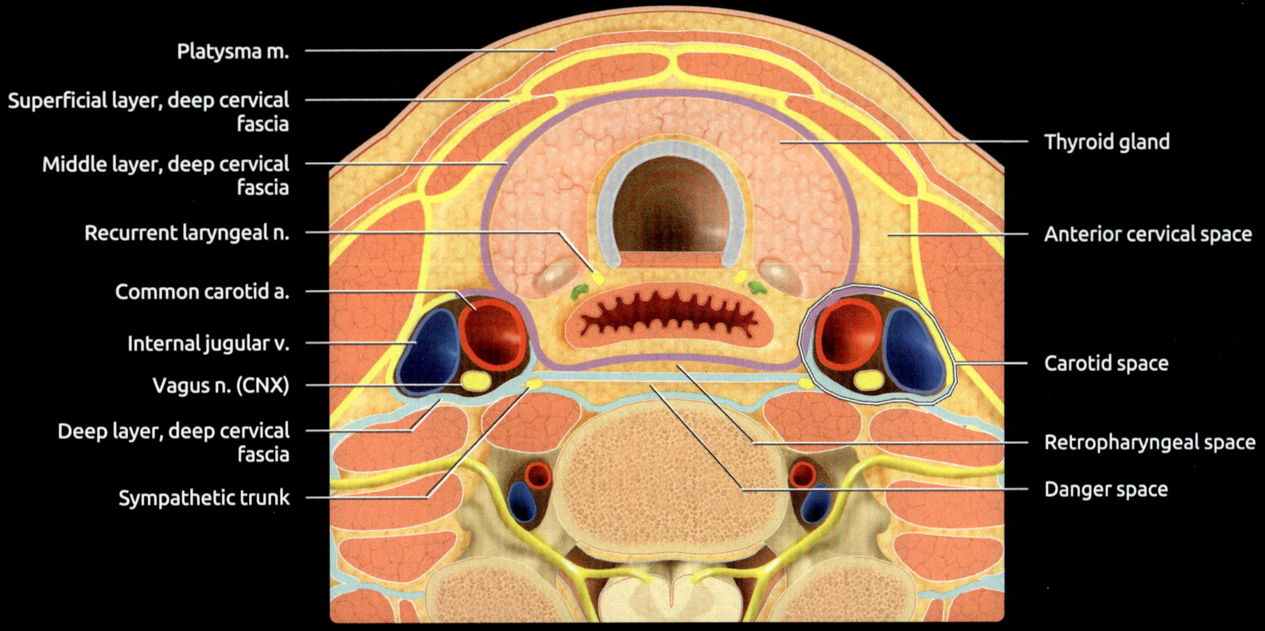

(Top) *Schematic diagram shows the midcervical level, which contains key structures, including the common carotid artery, internal jugular vein, and vagus nerve within the carotid sheath. The hyoid bone and the cricoid cartilage are the anatomical landmarks for the superior and inferior borders of the midcervical level, respectively. With high-resolution ultrasound, normal small jugular chain lymph nodes may be seen at this site and should not be mistaken for pathology.* **(Bottom)** *Axial graphic shows the midcervical level in the infrahyoid neck. Note that the carotid sheath contains all 3 layers of the deep cervical fascia (tricolor line). In the infrahyoid neck, the carotid sheath is tenacious throughout its length. The infrahyoid carotid space contains the common carotid artery, internal jugular vein, and vagus cranial nerve.*

TRANSVERSE ULTRASOUND

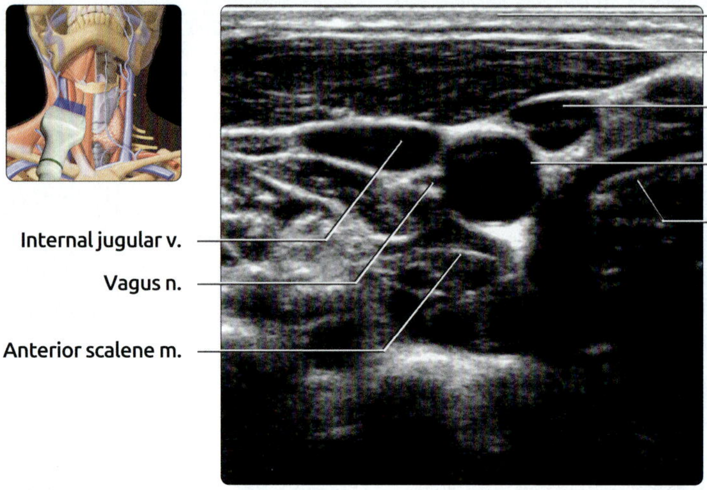

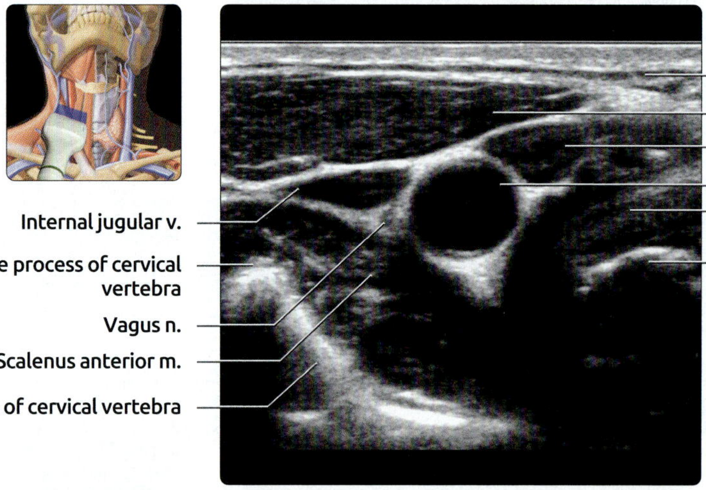

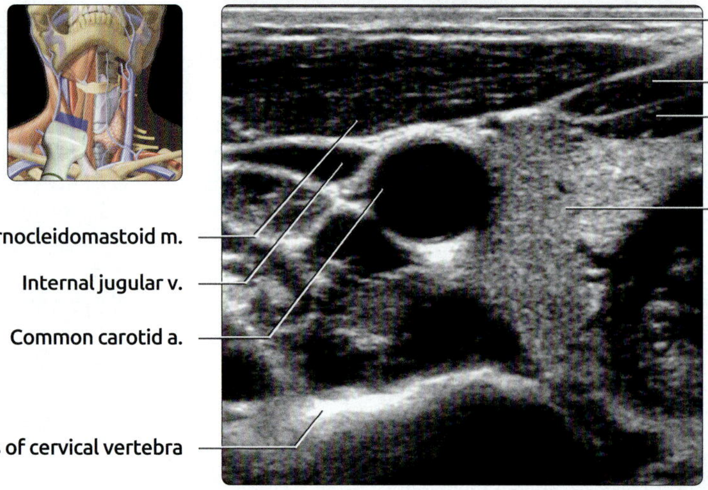

(Top) *First image shows consecutive transverse grayscale ultrasound of midcervical level. Deep cervical lymph nodes are commonly found along and anterior to the major vessels of the carotid sheath. These are commonly hypoechoic and elliptical with a normal echogenic hilum and hilar vascularity.* **(Middle)** *Second image shows midcervical level in transverse plane. At this level, the common carotid artery, internal jugular vein, and vagus nerve are the main structures within the carotid sheath. The vagus nerve is usually located between the common carotid artery and the internal jugular vein and appears as a small, round, hypoechoic nodule with a central echogenic dot.* **(Bottom)** *Third image shows the midcervical level. The cricoid cartilage may not be routinely seen on ultrasound, and visualization of the superior pole of the thyroid gland ~ coincides with this level.*

AXIAL CECT

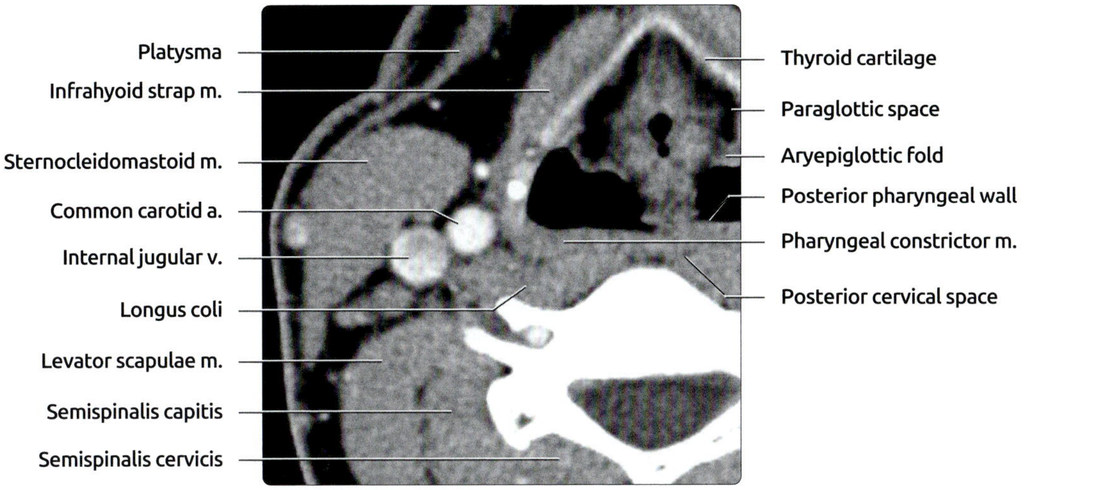

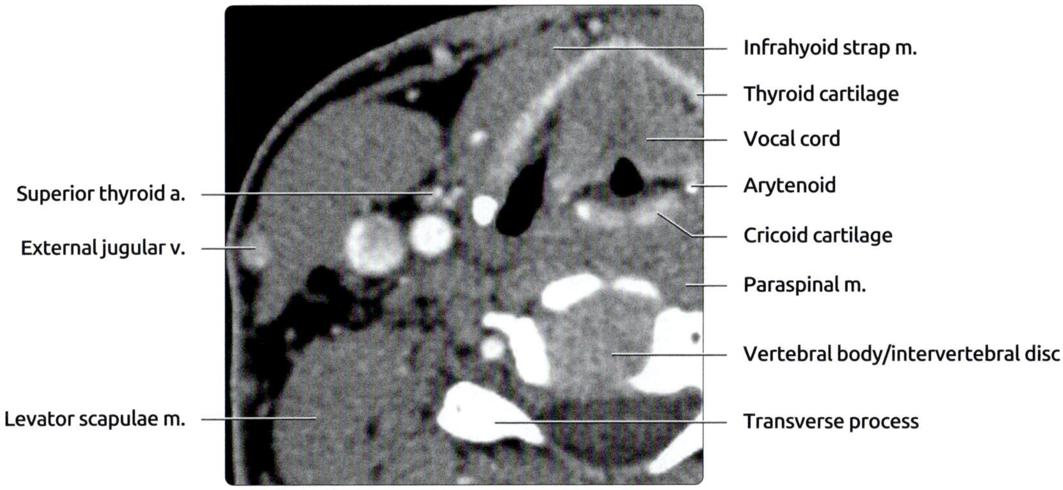

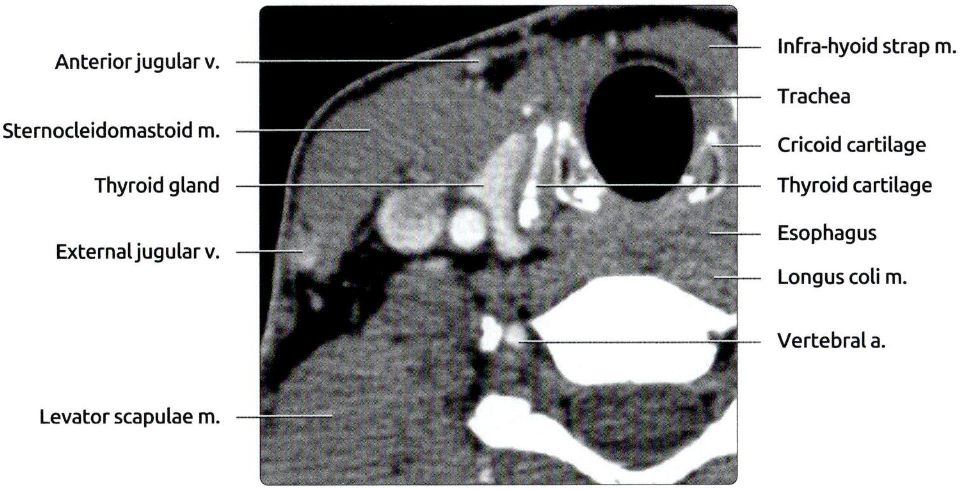

(Top) *CECT shows the upper thyroid cartilage level. Major structures of the midcervical level, including the common carotid artery, internal jugular vein, and sternocleidomastoid muscle, are well demonstrated. Small lymph nodes are commonly seen adjacent to major vessels of the carotid sheath.* **(Middle)** *CECT shows the level of the lower thyroid lamina. Apart from the common carotid artery and internal jugular vein, branches of the external carotid artery, such as the superior thyroid artery, can be demonstrated.* **(Bottom)** *CECT shows the level of the cricoid cartilage. The upper pole of the thyroid gland begins to be included in cross section. Note the presence of calcification/ossification in the thyroid lamina and cricoid cartilages, which is a normal age-related change.*

TRANSVERSE GRAYSCALE ULTRASOUND

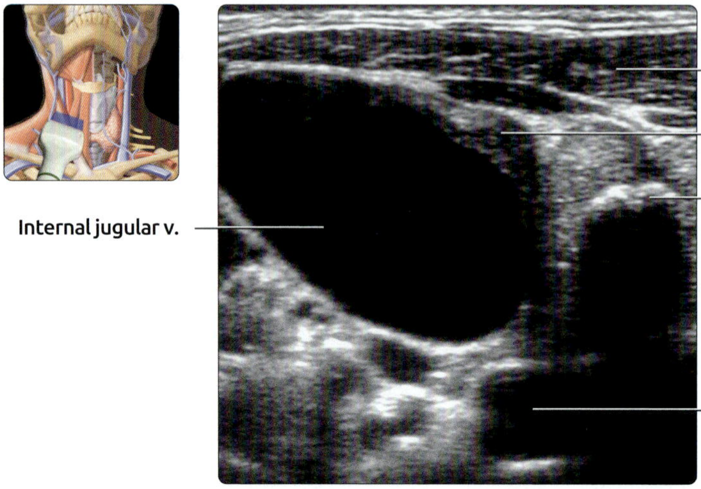

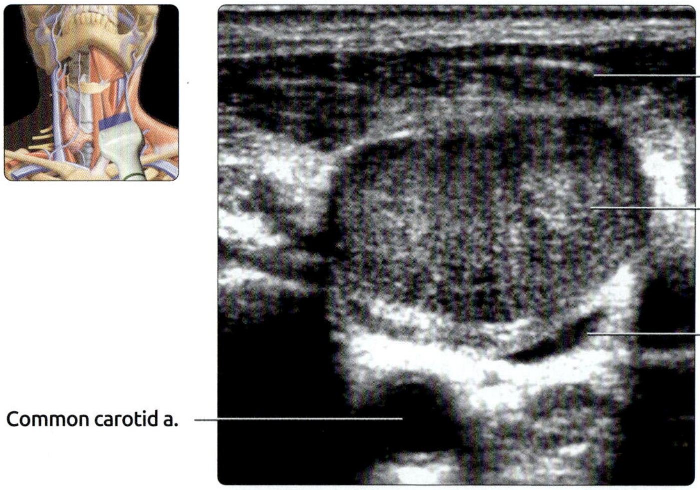

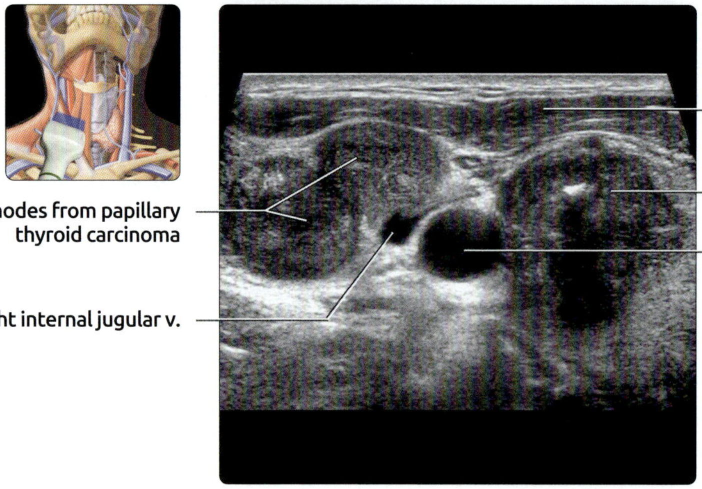

(Top) *Transverse grayscale ultrasound of the right midcervical level shows an eccentric, hypoechoic thrombus in the anteromedial wall of the right internal jugular vein. Note the presence of an adjacent thyroid carcinoma in the right lobe.* **(Middle)** *Transverse grayscale ultrasound of the left midcervical level shows an enlarged, round, solid, hypoechoic, deep cervical lymph node with loss of echogenic hilum. The adjacent left internal jugular vein is compressed. Pathology showed metastatic squamous cell carcinoma.* **(Bottom)** *Transverse grayscale ultrasound of the right midcervical level shows multiple enlarged, round, solid, slightly hyperechoic lymph nodes with punctate calcification suggesting metastases from primary papillary thyroid carcinoma. Note thyroid papillary carcinoma seen as an ill-defined, solid, hypoechoic nodule with punctate and coarse calcification in the right lobe of the thyroid gland. The right internal jugular vein is compressed by enlarged lymph nodes but remains patent.*

TRANSVERSE COLOR, DOPPLER, AND GRAYSCALE ULTRASOUND

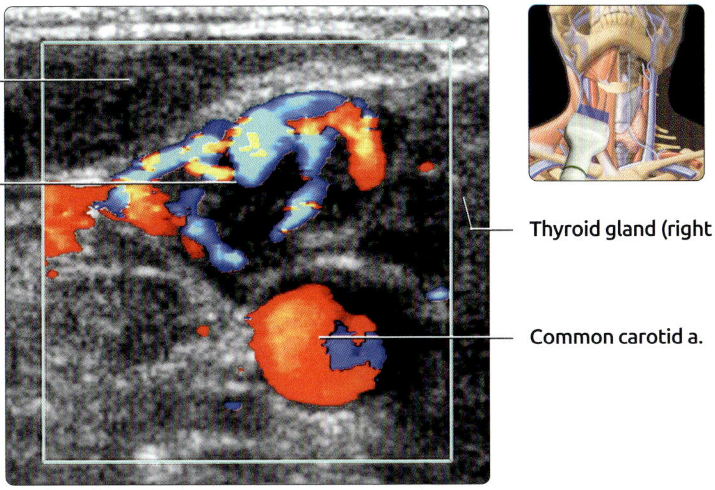

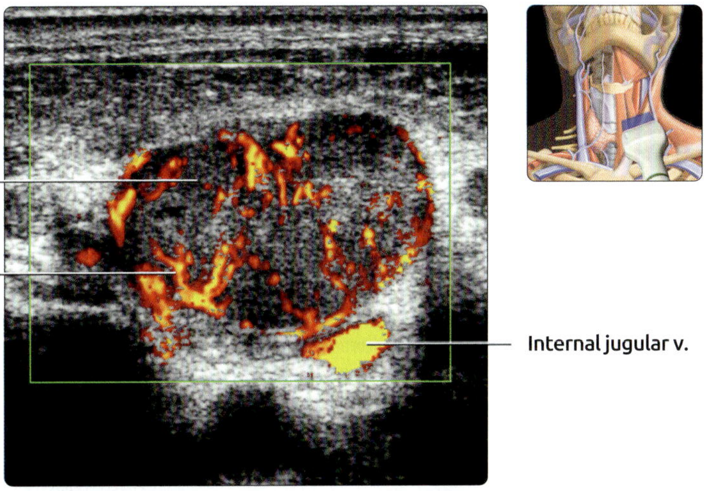

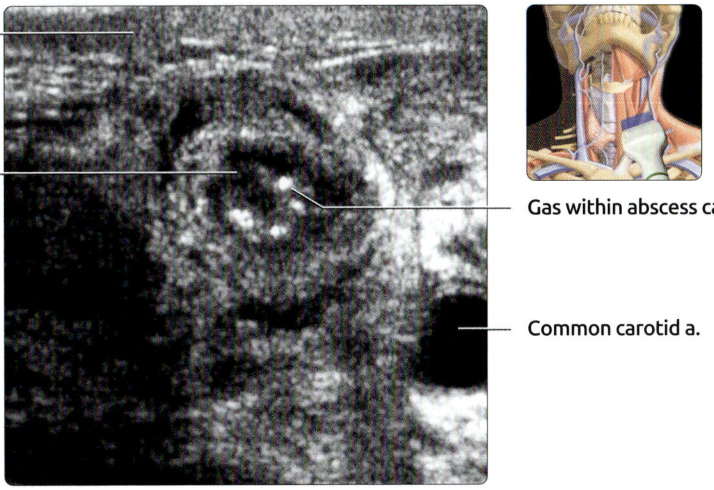

(Top) *Transverse color Doppler ultrasound in a patient with occlusive internal jugular vein thrombus shows the presence of vascularity within the internal jugular vein thrombus, suggesting it is a tumor thrombus rather than a bland venous thrombus.* **(Middle)** *Transverse power Doppler ultrasound shows the presence of peripheral/subcapsular intranodal vessels in a metastatic lymph node from head and neck squamous cell carcinoma. The vascularity is completely different from the hilar pattern of a normal cervical lymph node.* **(Bottom)** *Transverse grayscale ultrasound shows a neck abscess with fluid and gas within. The adjacent soft tissue is edematous and thickened. Note its close proximity to the common carotid artery, putting it at a risk of rupture.*

TERMINOLOGY

Definitions

- Portion of lower anterior neck adjacent to carotid sheath, below level of cricoid cartilage and above level of clavicle

IMAGING ANATOMY

Overview

- Key structures in lower cervical level
  - Vessels: Common carotid artery (CCA), internal jugular vein (IJV), subclavian artery
  - Vagus nerve
  - Scalenus anterior muscle
  - Lymph nodes
- Key structures in supraclavicular fossa
  - Trapezius, sternocleidomastoid, omohyoid muscles
  - Brachial plexus (BP) elements
  - Transverse cervical chain lymph nodes

Internal Contents

- Subclavian artery
  - Arises from brachiocephalic trunk on right and aortic arch on left
  - Major arterial supply to upper limb
  - Contributes to arterial supply to neck structures and brain via vertebral artery
  - Junction of subclavian artery and CCA is readily identified on scanning in transverse plane in lower cervical level
    - Location marks root of neck
  - Origin of subclavian artery can be seen by angling transducer inferiorly behind medial head of clavicle
- Scalenus anterior muscle
  - Runs inferiorly from transverse processes of cervical spine
  - Passes posterior to IJV to dip behind clavicle
  - Lies between 2nd part of subclavian artery posteriorly and subclavian vein anteriorly
  - Related posteriorly to scalenus medius muscle
    - BP roots/rami lie between scalenus anterior muscle and scalenus medius muscle in supraclavicular fossa
    - Scanning inferiorly in transverse plane, BP roots/rami appear as small, round, hypoechoic structures emerging from behind lateral border of scalenus anterior muscle
- BP
  - Formed from ventral rami of C5-T1 ± minor branches from C4, T2
  - Divided into roots/rami, trunks, division, cords, and branches
    - Roots/rami: Originate from spinal cord levels C5-T1 enter posterior triangle by emerging between scalenus anterior and medius muscle
    - Trunks: Upper (C5-C6), middle (C7), lower (C8-T1)
    - Divisions: Formed by each trunk, dividing into anterior and posterior branches in supraclavicular fossa
    - Cords: Lateral, medial, posterior cords descend behind clavicle to leave posterior triangle and enter axilla
    - Branches: In axilla
- Trapezius muscle
  - Anterior border marks posterior margin of posterior triangle and supraclavicular fossa and is easily recognized
  - Distal portion is attached to lateral clavicle
- Inferior belly of omohyoid muscle
  - Runs obliquely from intermediate tendon to traverse inferior portion of supraclavicular fossa
  - Divides occipital triangle superiorly from supraclavicular triangle inferiorly
- Transverse cervical chain lymph nodes
  - Seen adjacent to transverse cervical artery and vein, which arise from thyrocervical trunk and IJV
  - Related to and just superior to inferior belly of omohyoid muscle

ANATOMY IMAGING ISSUES

Imaging Approaches

- From midcervical level, proceed to scan in transverse plane along carotid sheath until medial head of clavicle (keep CCA and IJV in center of image)
- Then sweep laterally in transverse plane above mid-/lateral portion of clavicle to assess supraclavicular fossa

Consider

- **Enlarged lymph node is most common cause of mass in lower cervical level and supraclavicular fossa**
  - Terminology
    - Omohyoid node: Deep cervical chain node just superior to omohyoid (where omohyoid crosses IJV)
    - Virchow node: Signal node; lowest node of deep cervical chain nodes
    - Troisier node: Most medial node of transverse cervical chain nodes
  - Reactive nodes
    - Enlarged, cortical hypertrophy, preserved echogenic hilum and hilar vascularity
  - Metastatic nodes
    - Round, hypoechoic, peripheral/subcapsular vascularity, may have central necrosis
  - Lymphomatous nodes
    - Large, heterogeneous, reticulated/pseudocystic appearance, increased peripheral and central vascularity, bilateral involvement
  - Tuberculous lymphadenitis
    - Matted, necrotic, enlarged nodes with soft tissue edema and hypovascular/displaced hilar vascularity on power Doppler ultrasound
- Isolated metastatic lymph node in supraclavicular fossa is unusual from primary in head and neck region
  - Careful search for infraclavicular primary is indicated
  - Common primary from lung, breast, esophagus, and colorectal cancers
- Differential diagnoses
  - BP schwannoma
  - Lipoma
  - Venous malformation
  - Lymphatic malformation

LOWER CERVICAL AND SUPRACLAVICULAR ANATOMY

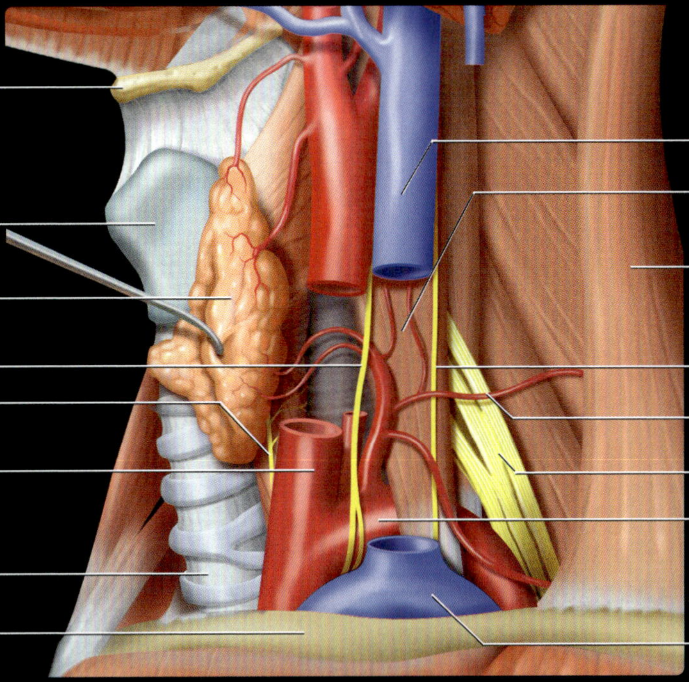

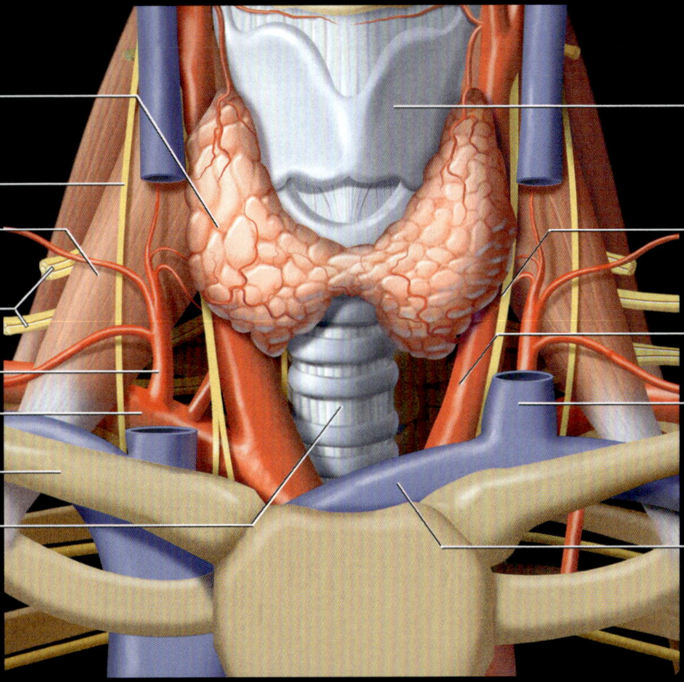

(Top) *Graphic in lateral projection shows the anatomical relationship of major structures in the lower cervical level and supraclavicular fossa, including the common carotid artery, internal jugular vein, subclavian vessels, and brachial plexus. In assessing the supraclavicular fossa on ultrasound, adequate visualization is achieved by sweeping the transducer laterally in the transverse plane from the medial head of the clavicle.* **(Bottom)** *Graphic in frontal projection shows the lower cervical level and medial portion of the supraclavicular fossa. The major lymph nodes in these regions are mainly located close to the major vessels of the carotid sheath, including the Virchow node and the Troisier node. Presence of isolated malignant nodes at this site usually points to an infraclavicular primary. Proximity of these nodes to adjacent pulsatile vessels makes Doppler examination at this site suboptimal/difficult.*

TRANSVERSE ULTRASOUND

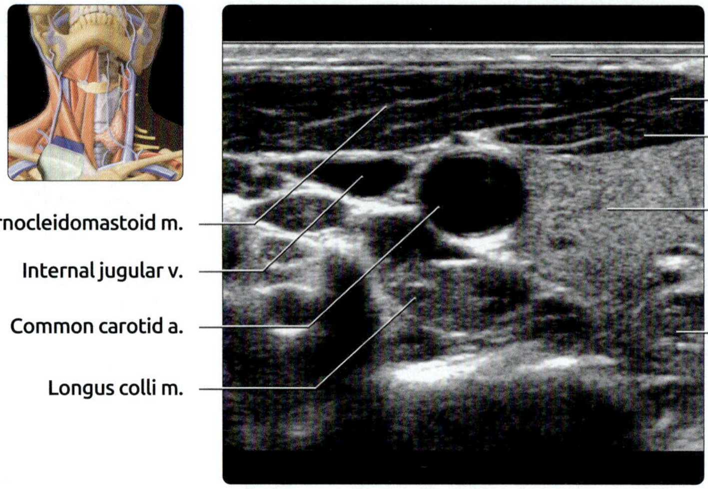

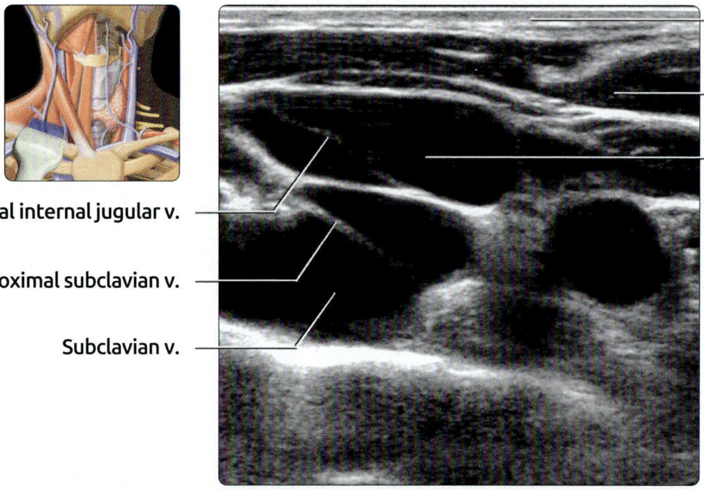

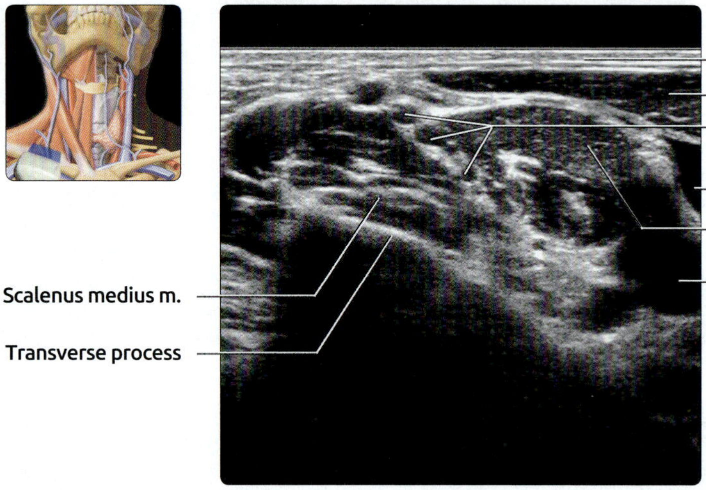

(Top) *First image in a series of transverse grayscale ultrasound images shows the lower cervical level. Major structures at this level include the common carotid artery, internal jugular vein, thyroid gland, and overlying muscles of the anterior neck.* **(Middle)** *Second image shows the level of the medial supraclavicular fossa. Note the proximity of the proximal internal jugular vein to the subclavian vein at this site, which join to form the brachiocephalic vein. Supraclavicular lymph nodes are commonly found adjacent to these major vessels.* **(Bottom)** *Third image shows the lateral supraclavicular fossa. The scalenus anterior and medius muscles are clearly visualized with the trunks of the brachial plexus between them. Ultrasound is often used to guide brachial plexus blocks. Ultrasound also helps to exclude brachial plexus involvement by metastases at this site.*

AXIAL CECT AND POWER DOPPLER ULTRASOUND

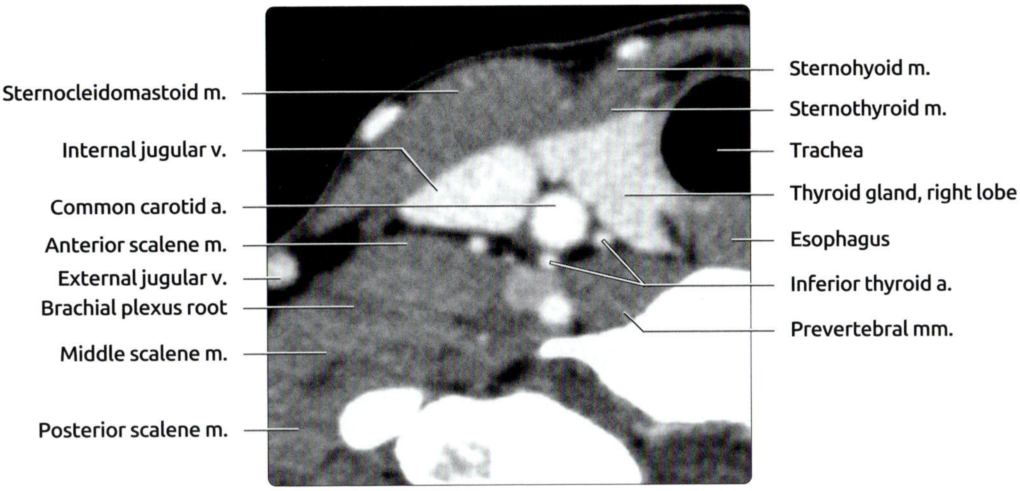

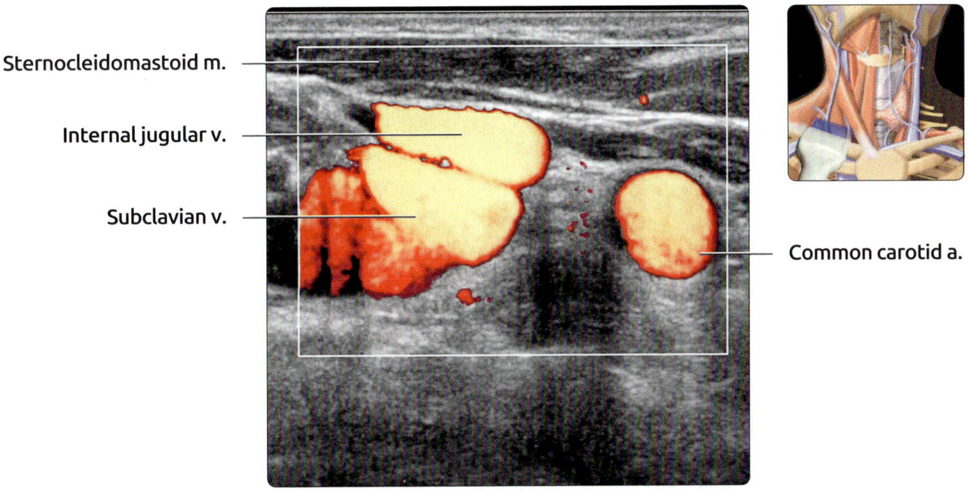

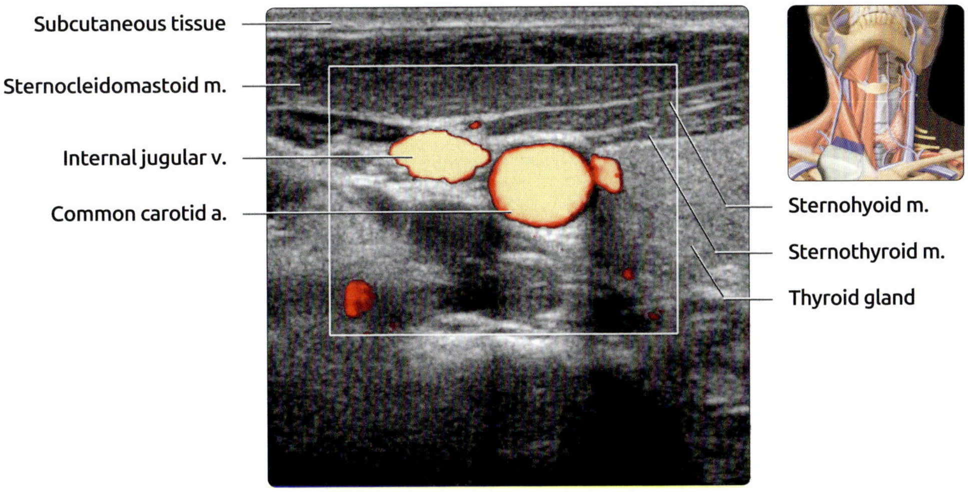

(Top) *Axial CECT shows the lower cervical level. The relationship of the thyroid gland with the adjacent structures, including the carotid sheath, strap muscles, trachea, and esophagus is demonstrated.* **(Middle)** *Transverse power Doppler ultrasound in the supraclavicular fossa helps to delineate vascular structures in this region, including the confluence of the internal jugular vein and the subclavian vein. The proximal portion of the common carotid artery from the brachiocephalic trunk is also seen at this level.* **(Bottom)** *Transverse power Doppler ultrasound of the lower cervical level allows the clear depiction of the color-filled common carotid artery and internal jugular vein. Nodes are often seen at this site and are readily evaluated by ultrasound and, if needed, biopsied using ultrasound guidance.*

LONGITUDINAL ULTRASOUND AND PATHOLOGY

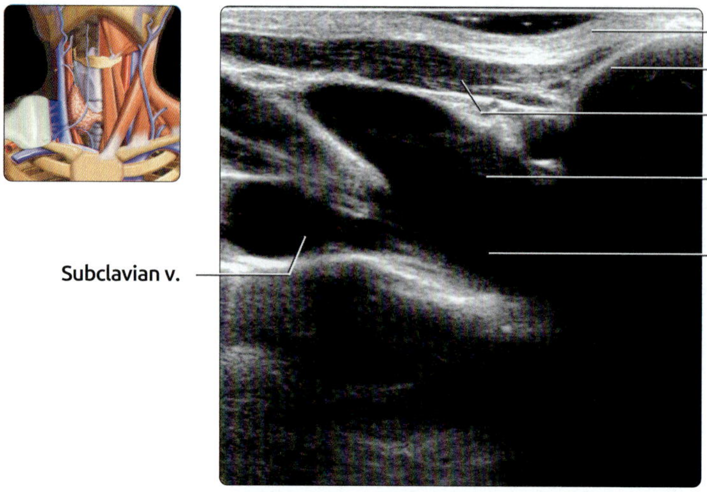

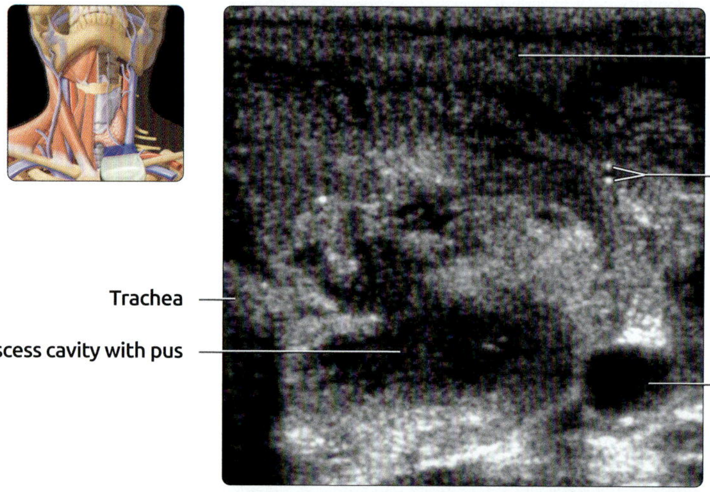

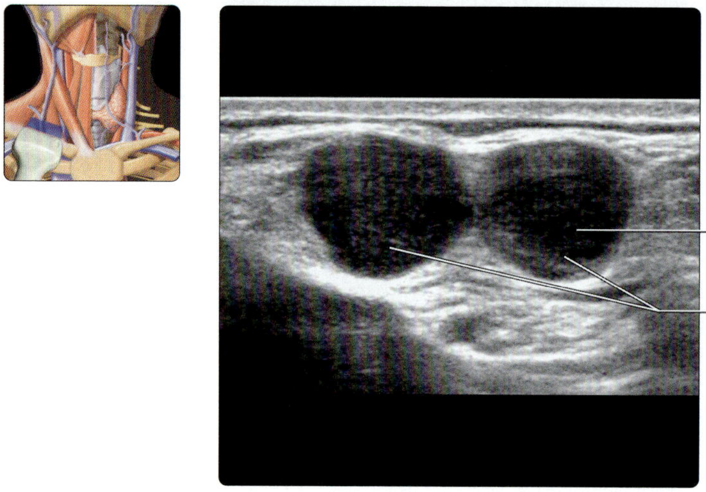

(Top) *Longitudinal grayscale ultrasound of the supraclavicular fossa shows the confluence of the internal jugular vein and the subclavian vein to form the brachiocephalic vein just above the medial portion of the clavicle. Nodes often nestle under these vessels, making biopsy access difficult. Pulsation from the vessels also makes Doppler evaluation of nodes at this level suboptimal.* **(Middle)** *Transverse grayscale ultrasound of the left lower cervical level shows abscess formation in the left lobe and perithyroidal soft tissue due to acute suppurative thyroiditis. Note the presence of pus and echogenic foci due to gas bubbles.* **(Bottom)** *Transverse grayscale ultrasound of the right supraclavicular fossa shows multiple enlarged, round, predominantly solid, hypoechoic lymph nodes with intranodal cystic necrosis and no adjacent soft tissue edema. Appearances are highly suspicious of metastatic lymph nodes. Isolated metastatic nodes at this site point to an infraclavicular primary, commonly from lung, breast, or esophagus.*

CECT AND PATHOLOGY

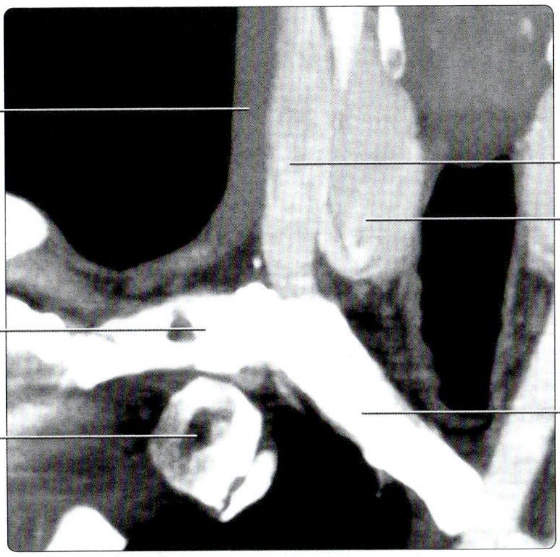

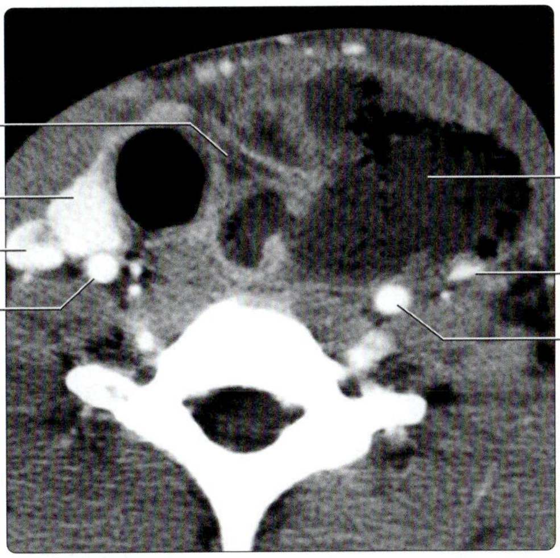

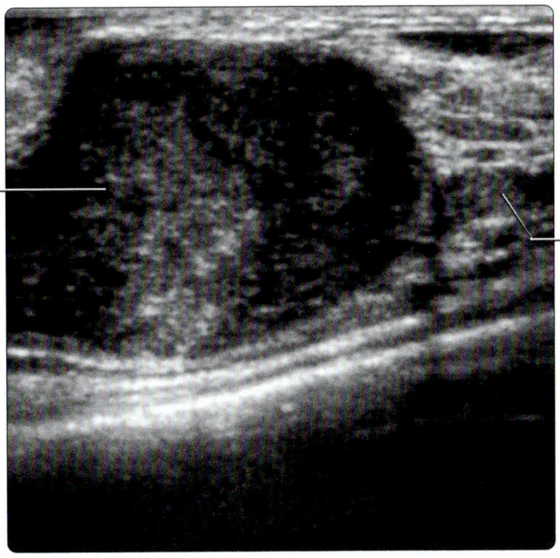

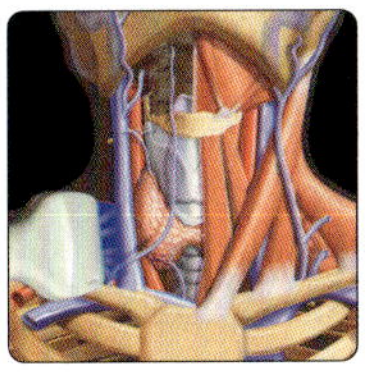

(Top) *Coronal reformatted CECT of the lower neck and the supraclavicular fossa illustrates the venous anatomy.* **(Middle)** *Axial CECT of acute suppurative thyroiditis shows a large heterogeneous abscess with thick peripheral enhancement involving the left lobe of the thyroid gland and perithyroidal soft tissue. Note the presence of fluid and gas within the abscess cavity and marked subcutaneous thickening in left lower neck.* **(Bottom)** *Longitudinal grayscale ultrasound of the right supraclavicular fossa shows a solid, hypoechoic, lobulated mass contiguous with a thickened nerve. The appearances suggest a brachial plexus schwannoma. The continuation with a nerve is the clue to its diagnosis.*

TERMINOLOGY

Abbreviations

- Posterior cervical space (PCS)

Definitions

- Posterolateral fat-containing space in neck with complex fascial boundaries that extends from posterior mastoid tip to clavicle behind sternocleidomastoid muscle

IMAGING ANATOMY

Overview

- Posterolateral fat-filled space just deep and posterior to sternomastoid muscle
- Posterior border bound by anterior edge of trapezius muscle

Extent

- PCS extends from small superior component near mastoid tip to broader base at level of clavicle
- When viewed from side, appears as tilting tent

Anatomy Relationships

- Superficial space lies superficial to PCS
- Deep to PCS is perivertebral space
  - Anterior PCS is superficial to prevertebral component of perivertebral space
  - Posterior PCS is superficial to paraspinal component of perivertebral space

Internal Contents

- Fat is primary component of PCS
- Floor is formed by muscles running obliquely: Scalene muscles, levator scapulae, and splenius capitis muscles (from anterior to posterior)
  - Subdivided by inferior belly of omohyoid muscle into occipital and subclavian triangles
- Muscular floor is covered by superficial and deep layers of deep cervical fascia
- Spinal accessory nerve (CNXI)
  - Arises from nerve cells in anterior gray column of upper 5 segments of spinal cord
  - Ascends alongside spinal cord and enters skull through foramen magnum
  - Unites with cranial root to exit through jugular foramen
  - Spinal portion then separates from cranial root
  - Motor supply to soft palate, pharynx, larynx, sternocleidomastoid, and trapezius muscles
- Spinal accessory lymph node chain
  - Level 5 spinal accessory nodes (SAN) further subdivide into A and B levels at hyoid bone
    - Level 5A: SAN above cricoid cartilage level
    - Level 5B: SAN below cricoid cartilage level
- Preaxillary brachial plexus
  - Segment of brachial plexus emerging from anterior and middle scalene gap passes through PCS
  - Leaves PCS with axillary artery into axillary fat
- Dorsal scapular nerve
  - Arises from brachial plexus (spinal nerves C4 and C5)
  - Motor innervation to rhomboid and levator scapulae muscles
- Transverse cervical artery and vein
  - Arises from thyrocervical trunk of subclavian artery and internal jugular vein respectively
  - Course in inferior posterior triangle and parallel to clavicle

ANATOMY IMAGING ISSUES

Imaging Approaches

- Scanning is usually undertaken in transverse plane
- From mastoid tip superiorly to acromion process inferiorly
  - Spinal accessory chain lymph nodes run from point midway between mastoid process and angle of mandible to outer 1/3 of clavicle

Imaging Pitfalls

- On transverse scan, tips of transverse process of cervical vertebrae may be seen as echogenic structures with posterior acoustic shadowing
  - Do not mistake these for calcified lymph nodes
  - Longitudinal scan helps to clarify

Consider

- On ultrasound, normal-looking nodes are routinely found in posterior triangle of healthy individuals
- Most lesions in posterior triangle arise from spinal accessory chain lymph nodes
  - Reactive lymphadenopathy
    - Elliptical, preserved internal architecture and hilar vascularity
  - Infective lymphadenitis, such as tuberculous lymphadenitis
    - Enlarged hypoechoic necrotic nodes with matting and soft tissue edema, avascular or displaced hilar vascularity
  - Metastatic lymph nodes
    - Primary from nasopharyngeal carcinoma and squamous cell carcinoma (SCC) from other H&N sites
    - Enlarged, round, hypoechoic nodes with intranodal necrosis and peripheral vascularity
  - Lymphomatous nodes
    - Usually bilateral neck involvement
    - Enlarged, heterogeneous, reticulated/pseudocystic appearance, increased hilar and peripheral vascularity, hilar >> peripheral
- Other diseases that may occur in posterior triangle include
  - Congenital lesion: Lymphangioma, usually transspatial
  - Benign tumor: Lipoma, nerve sheath tumor

CLINICAL IMPLICATIONS

Clinical Importance

- CNXI runs in floor of posterior triangle
  - Accessory cranial neuropathy results when CNXI is injured
  - Most commonly injured during neck dissection for malignant SCC nodes
  - Less commonly injured by extranodal spread of SCC
- SAN are main normal occupants of PCS

POSTERIOR CERVICAL SPACE

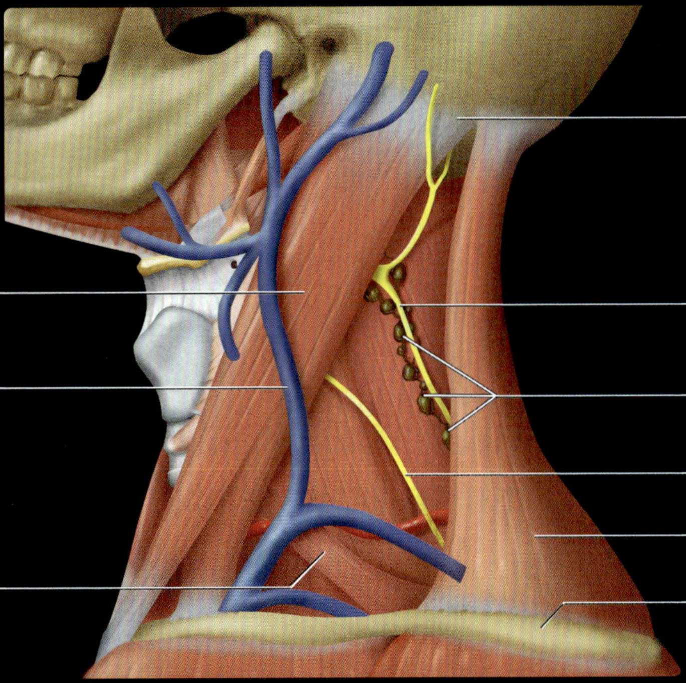

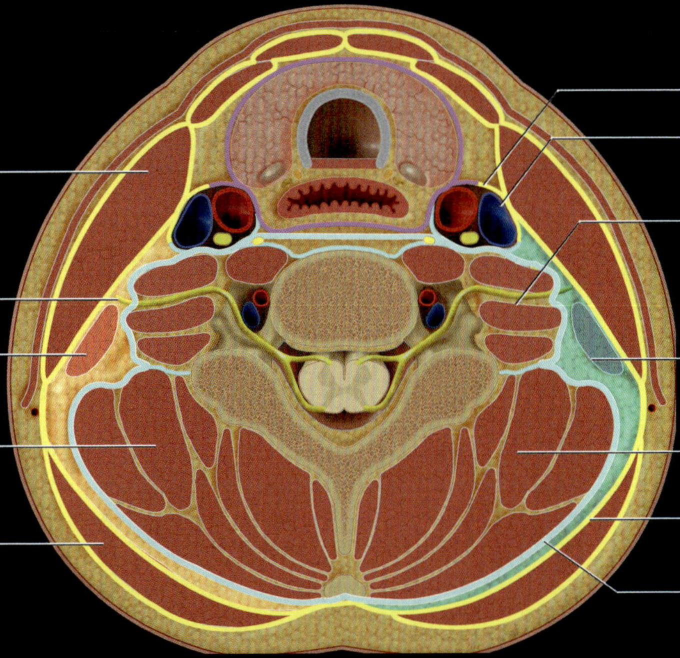

(Top) *Lateral graphic of the neck shows the posterior triangle as a "tilting tent" with its superior margin at the level of the mastoid tip and its inferior border at the clavicle. Note that it has 2 main nerves in its floor, the accessory nerve (CNXI) and the dorsal scapular nerve. The spinal accessory nodal chain is its key occupant with regards to the kind of lesions found in the posterior triangle.* **(Bottom)** *Axial graphic through the thyroid bed of the infrahyoid neck depicts the posterior cervical space (PCS) with its complex fascial borders. The superficial layer of the deep cervical fascia is its superficial border, while the deep layer of the deep cervical fascia is its deep border. Note the tricolor carotid sheath is its anteromedial border. The brachial plexus roots travel through the PCS on their way to the axillary apex.*

TRANSVERSE ULTRASOUND

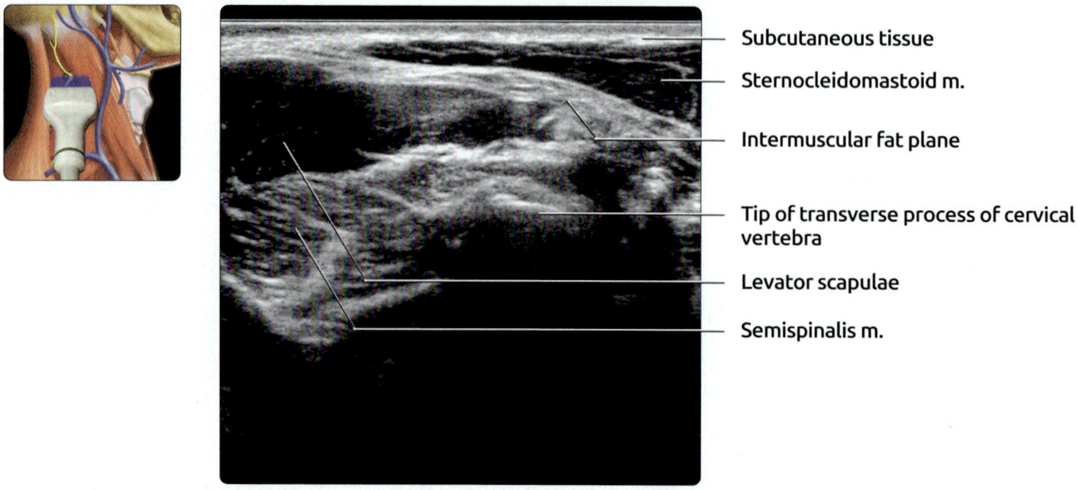

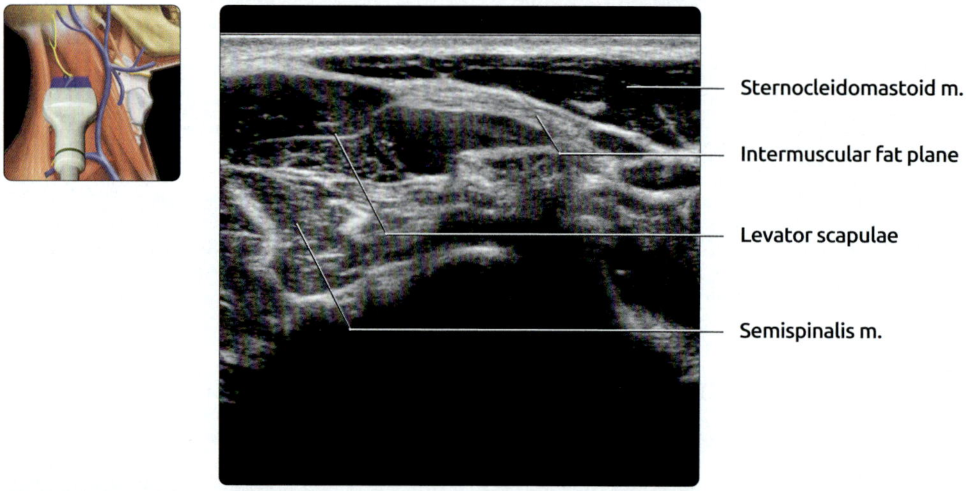

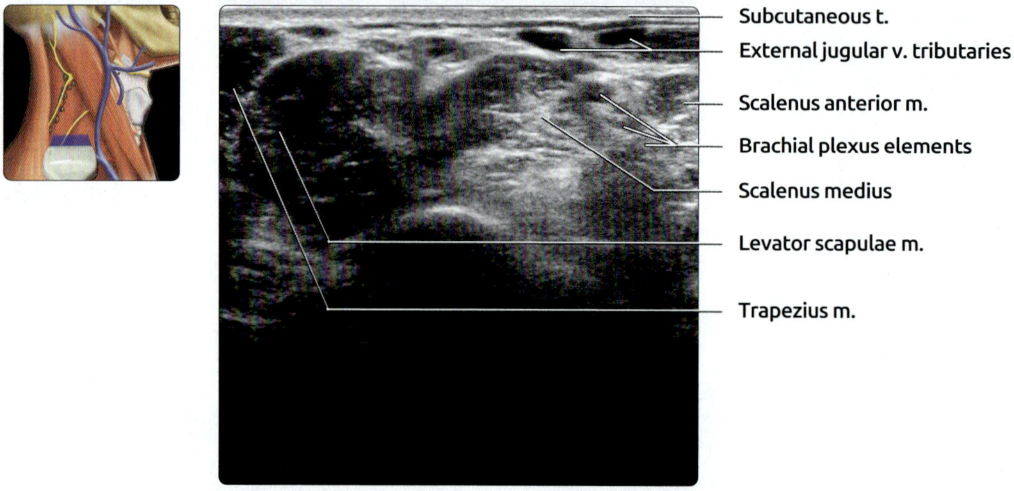

(Top) *First of 3 transverse grayscale ultrasound images shows the posterior triangle. The sternocleidomastoid muscle marks the anterior border of the posterior triangle. Muscles form the floor of the posterior triangle. Note the accessory nerve and nodes lie in the intermuscular fat plane.* **(Middle)** *Standard transverse grayscale ultrasound of the posterior triangle shows the intermuscular fat plane, which is best screened in this plane. Once pathology is detected, further examination, particularly Doppler, is best done longitudinally.* **(Bottom)** *Lower level of the posterior triangle is shown. The trapezius muscle marks the posterior margin of the posterior triangle. The main bulk of the levator scapulae muscle forms the muscular floor.*

AXIAL MR

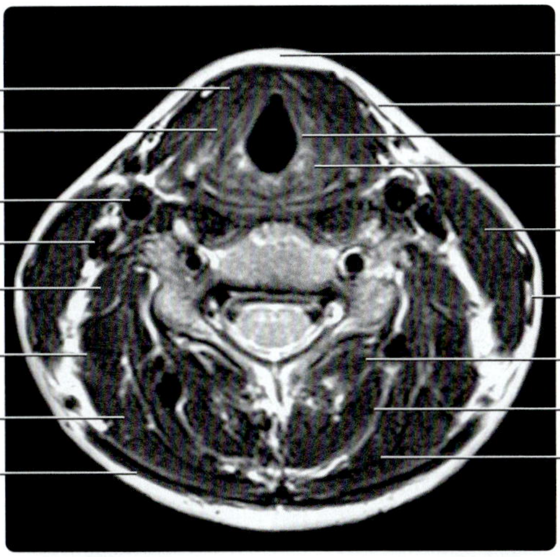

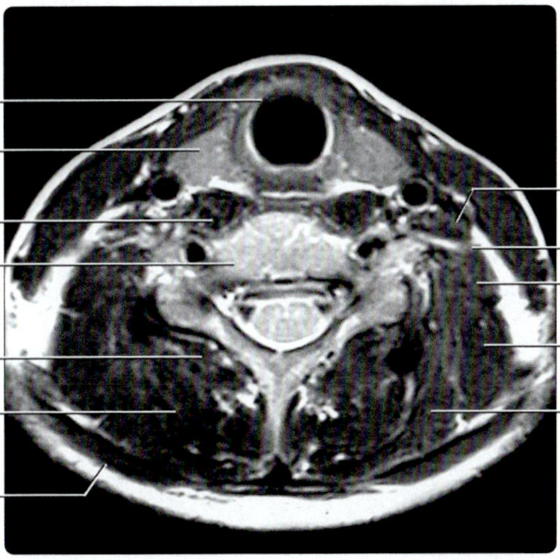

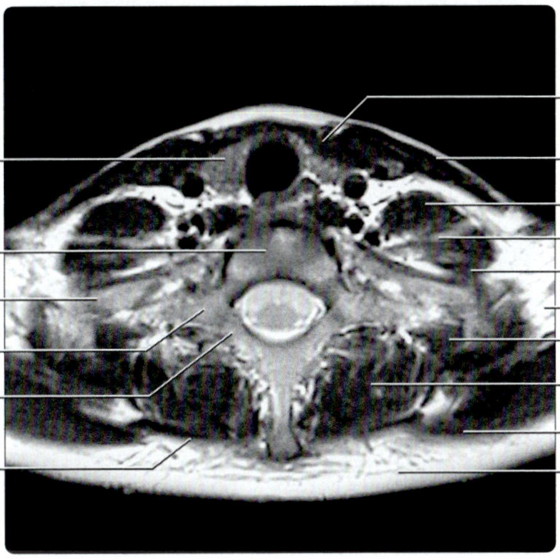

(Top) *Axial PD MR of the neck at the level of the vocal cord shows the largely fat-filled posterior triangle with muscular floor.* **(Middle)** *Axial PD MR of the neck at the level of thyroid gland shows the scalenus medius muscle begins at the midcervical level. The nerve roots and the trunk of the brachial plexus is seen emerging between the scalenus anterior and medius muscles.* **(Bottom)** *Axial PD MR of the lower posterior triangle shows the trunks of the brachial plexus between the scalenus anterior and the medius muscles diverging posterior to the clavicle to the axillary region. Note the content of the posterior triangle is predominantly fat with large muscles forming its boundaries.*

POSTERIOR TRIANGLE PATHOLOGY

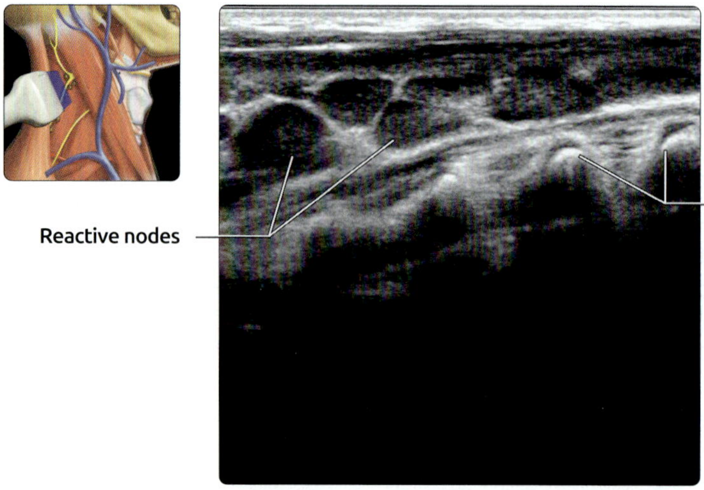

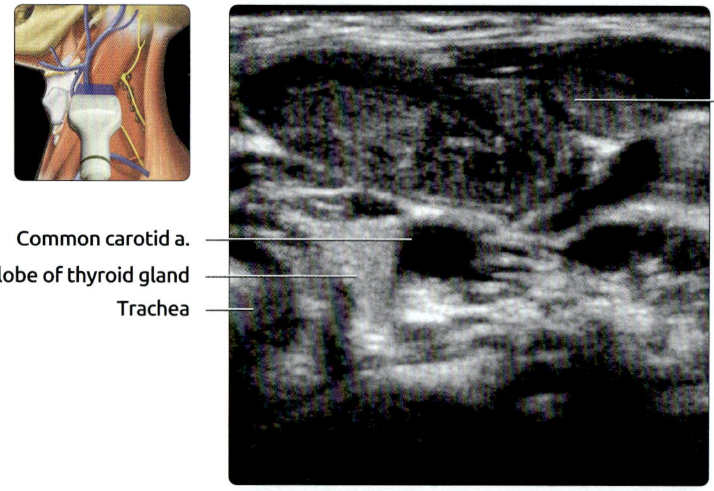

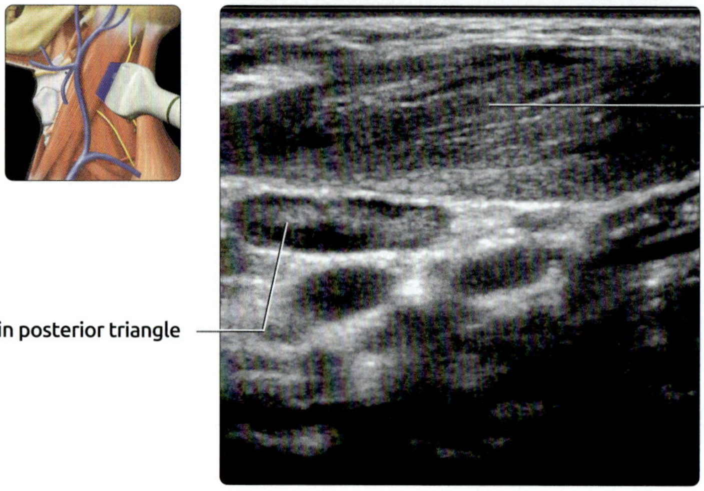

(Top) *Longitudinal grayscale ultrasound of the posterior triangle shows a chain of reactive accessory nodes. Note their location in the intermuscular fat plane. Do not mistake the tips of the transverse processes with calcified nodes.* **(Middle)** *Transverse grayscale ultrasound in an infant with torticollis shows a heterogeneously enlarged left sternocleidomastoid muscle. The appearance is compatible with a sternocleidomastoid pseudotumor.* **(Bottom)** *Longitudinal grayscale ultrasound of the left posterior triangle confirms sternocleidomastoid muscle hypertrophy. Part of the normal muscular striations are preserved.*

POSTERIOR TRIANGLE PATHOLOGY

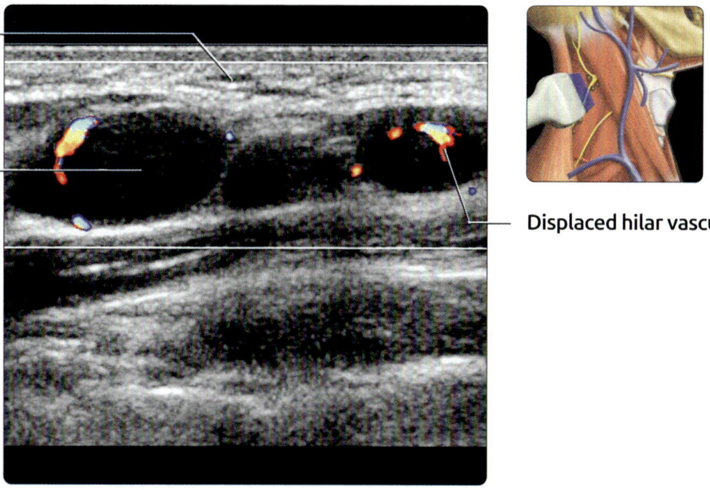

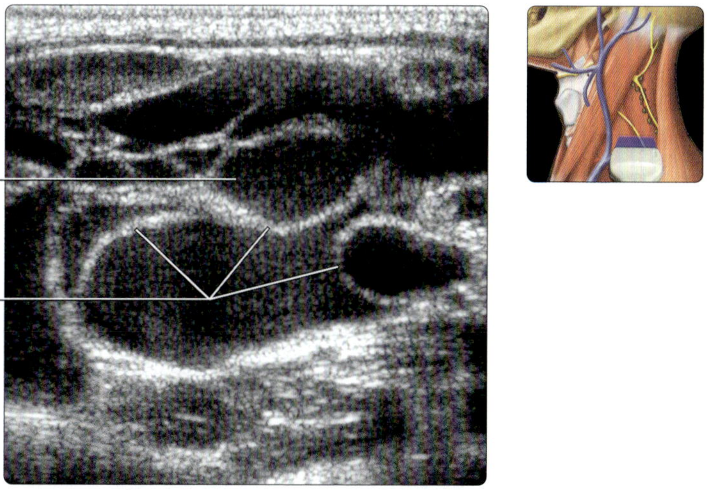

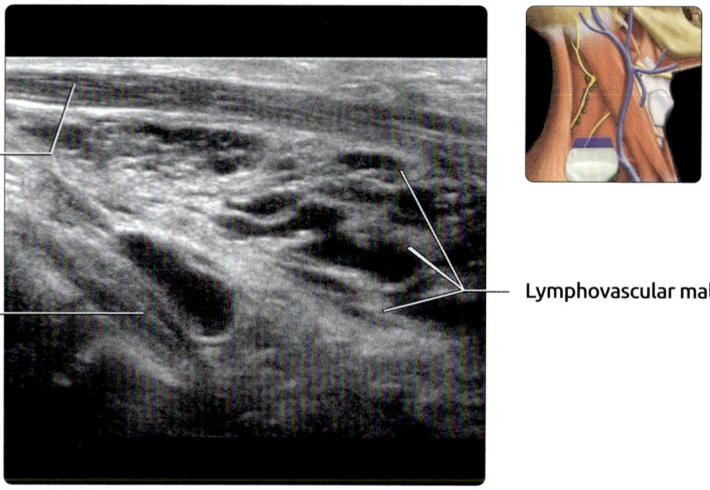

(Top) *Longitudinal color Doppler ultrasound in the posterior triangle shows multiple enlarged, hypoechoic lymph nodes with displaced hilar vascularity. Pathology showed tuberculous lymphadenitis. Once pathology is detected in transverse scans, it is better evaluated on a longitudinal scan, particularly if performing a Doppler examination.* **(Middle)** *Transverse grayscale ultrasound shows a large, multiseptated, cystic mass occupying the left posterior triangle in a child. The appearance is suggestive of a lymphatic malformation. These lesions are transspatial and frequently extend into other neck spaces.* **(Bottom)** *Oblique transverse grayscale ultrasound shows a child who underwent 2 treatments of ultrasound-guided sclerotherapy for a large, multiloculated lymphatic malformation. Note that small cystic locules persist; however, much of the lymphatic malformation has been replaced by fatty tissue. Ultrasound safely guides intralesional sclerosant and readily monitors posttreatment change in size and appearance.*

IMAGING ANATOMY

Overview

- H- or U-shaped gland in anterior cervical neck formed from 2 elongated lateral lobes with superior and inferior poles connected by median isthmus

Anatomy Relationships

- Thyroid gland lies anterior and lateral to trachea in visceral space of infrahyoid neck
- Extends from level of 5th cervical vertebra to 1st thoracic vertebra
- Posteromedially are tracheoesophageal grooves
  - Contains paratracheal lymph nodes, recurrent laryngeal nerve, parathyroid glands
- Posterolaterally are carotid spaces
  - Contains common carotid artery, internal jugular vein, vagus nerve
- Anteriorly are infrahyoid strap muscles
- Anterolaterally are sternocleidomastoid muscles

Internal Contents

- Thyroid gland
  - 2 lateral lobes (i.e., right and left lobes)
    - Measure ~ 4 cm in height
    - Each lobe has upper and lower poles
    - Lateral lobes are commonly asymmetric in size
  - Lateral lobes are joined by midline isthmus
  - **Pyramidal lobe** present in 30-50% of cases
    - Extends superiorly from isthmus toward hyoid bone
    - More common on left
- **Arterial supply**
  - **Superior thyroid arteries**
    - 1st anterior branch of external carotid artery
      - Runs superficially on anterior border of lateral lobe
      - Sends branch deep into gland before curving toward isthmus where it anastomoses with contralateral artery
    - Proximal course closely associated with superior laryngeal nerve
  - **Inferior thyroid arteries**
    - Arise from thyrocervical trunk, branch of subclavian artery
    - Ascend vertically, then curve medially to enter tracheoesophageal groove
    - Most branches penetrate posterior aspect of lateral thyroid lobe
    - Closely associated with recurrent laryngeal nerve
  - Thyroidea ima occasionally present (3%)
    - Single vessel originating from aortic arch or brachiocephalic artery
    - Enters thyroid gland at inferior border of isthmus
- **Venous drainage**
  - 3 pairs of veins arise from venous plexus on surface of thyroid gland
  - Superior and middle thyroid veins drain into internal jugular veins
  - Inferior thyroid veins drain into brachiocephalic veins
- **Lymphatic drainage**
  - Lymphatic drainage is extensive and multidirectional
  - Initial lymphatic drainage courses to periglandular nodes
    - Prelaryngeal, pretracheal (Delphian), and paratracheal nodes along recurrent laryngeal nerve
    - Paratracheal nodes drain along recurrent laryngeal nerve into mediastinum
    - Regional drainage occurs laterally into internal jugular chain (levels 2-4) and spinal accessory chain (level 5), higher in neck along internal jugular vein

ANATOMY IMAGING ISSUES

Imaging Approaches

- Ultrasound appearance
  - Normal thyroid parenchyma has fine, uniform echoes and is hyperechoic compared to adjacent muscles
  - Echogenic thyroid capsule is clearly visualized and helps to differentiate thyroid lesions from extrathyroidal masses
- Both longitudinal and transverse scans are required for comprehensive ultrasound assessment of thyroid gland
  - Transverse scan helps to locate thyroid nodules, their relationship to trachea, major vessels in carotid sheath, and to evaluate internal architecture and extrathyroid extension
  - Longitudinal scan helps to evaluate internal architecture, vascularity on Doppler, and extrathyroidal extension
- Examination of thyroid nodules includes
  - Ultrasound features of thyroid nodule
  - Assessment of adjacent structures (including trachea, esophagus, strap muscles, carotid artery, and internal jugular vein) and cervical lymph nodes

EMBRYOLOGY

Embryologic Events

- Thyroid gland originates from 1st and 2nd pharyngeal pouches (medial anlage)
- Originates as proliferation of endodermal epithelial cells on median surface of developing pharyngeal floor termed foramen cecum
- Bilobed thyroid gland descends anterior to pharyngeal gut along thyroglossal duct
- Inferior descent of thyroid gland anterior to hyoid bone and laryngeal cartilages

Practical Implications

- **Thyroglossal duct cyst**: Results from failure of involution of portion of thyroglossal duct
- **Thyroid tissue remnants**: From sequestration of thyroid tissue along thyroglossal duct
  - Seen anywhere along course of thyroglossal duct from foramen cecum to superior mediastinum
- **Ectopic thyroid gland**: From incomplete descent of thyroid into low neck
  - Seen anywhere along course from foramen cecum in tongue base to superior mediastinum
  - Most common location is just deep to foramen cecum in tongue base (i.e., lingual thyroid)

THYROID GLAND, VASCULAR SUPPLY

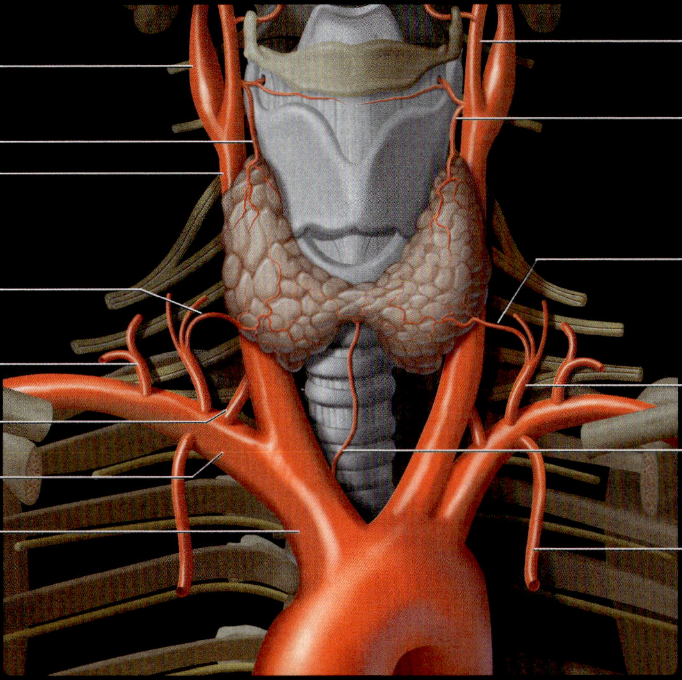

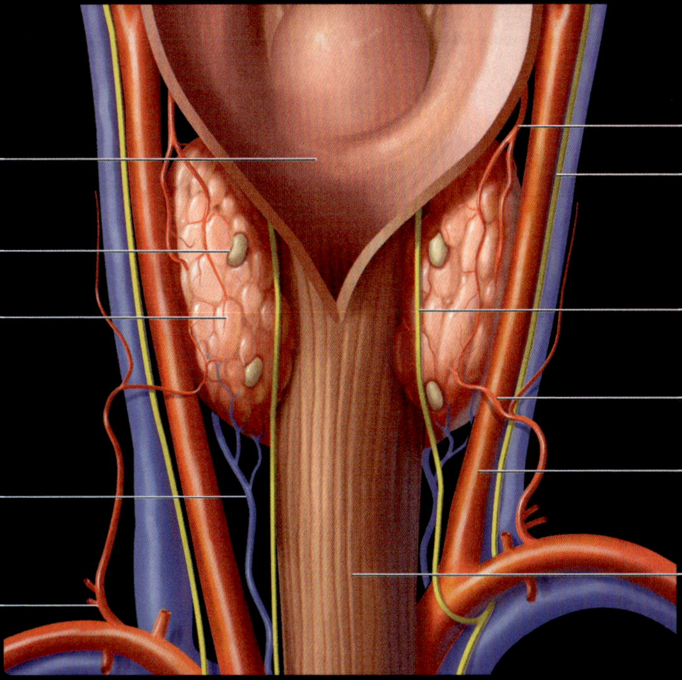

(Top) *Anterior graphic shows the arterial supply to the thyroid. The superior thyroid artery is the 1st branch of the external carotid artery. The inferior thyroid artery arises from the thyrocervical trunk, a branch of the subclavian artery. The thyroidea ima is a variant branch, which arises from the aortic arch or brachiocephalic artery, and enters the inferior border of the isthmus.* **(Bottom)** *Coronal graphic from behind shows both the arterial and venous vascular supply. Note the close proximity of the recurrent laryngeal artery, which runs in the tracheoesophageal groove along the posterior border of the thyroid gland.*

TRANSVERSE ULTRASOUND

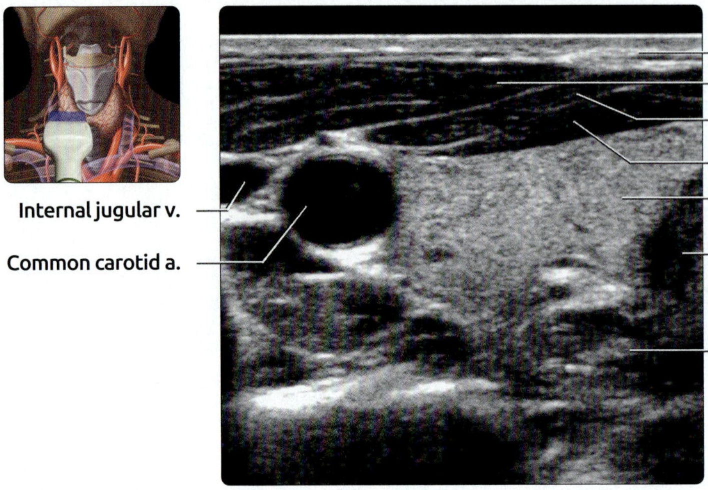

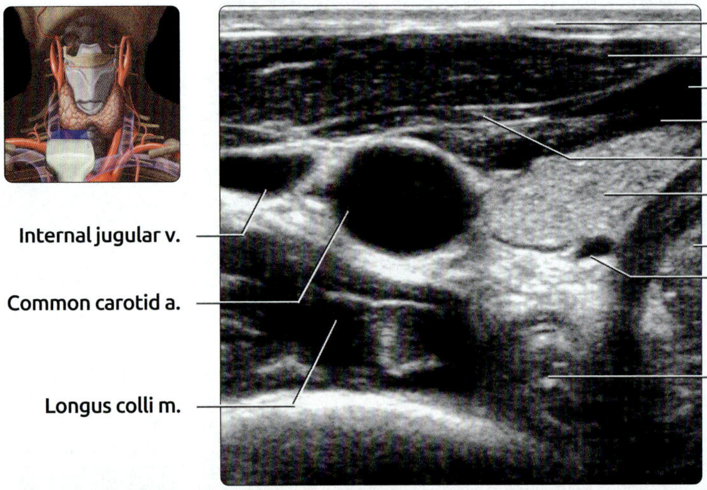

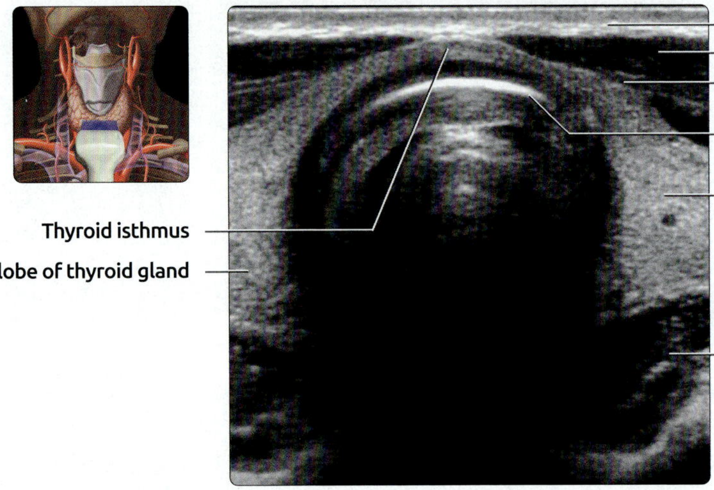

(Top) *Transverse grayscale ultrasound of the right lobe of the thyroid gland shows the homogeneous, hyperechoic echo pattern of the glandular parenchyma. Note its close anatomical relationship with the major vessels of the carotid sheath (internal jugular vein and common carotid artery) laterally, the trachea medially, and the cervical esophagus posteromedially.* **(Middle)** *Transverse grayscale ultrasound shows the level of the inferior pole of the thyroid gland. The inferior thyroid artery is a consistent finding related to and supplying the inferior pole.* **(Bottom)** *Midline transverse grayscale ultrasound shows the thyroid isthmus connecting the 2 lobes. The isthmus lies on the anterior surface of the trachea. In view of the intimate anatomical relationship between the thyroid gland and the trachea, a local tumor invasion to the trachea from malignant thyroid carcinoma is commonly seen, rendering surgical excision more extensive than total thyroidectomy.*

AXIAL CECT

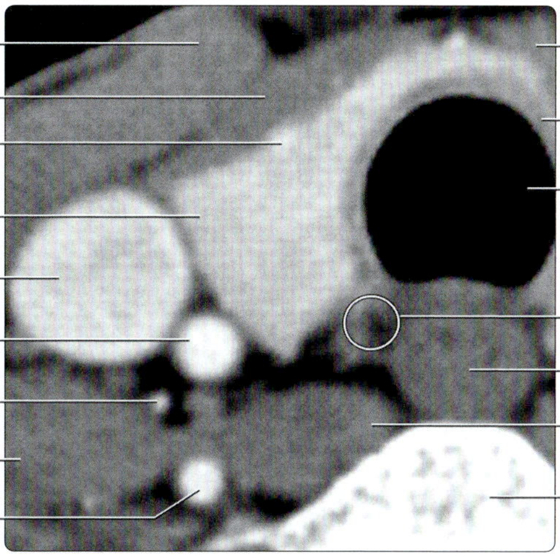

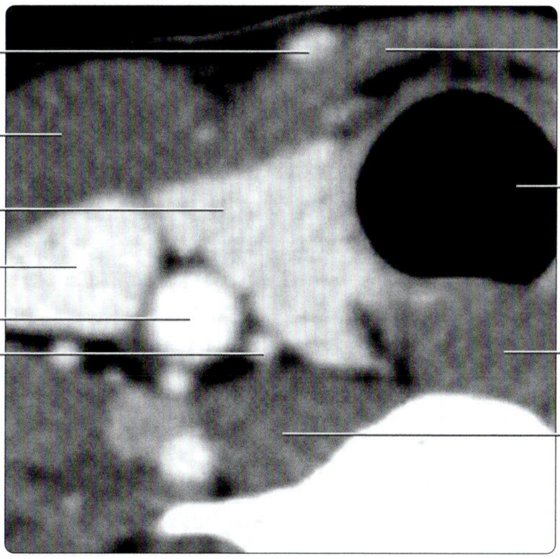

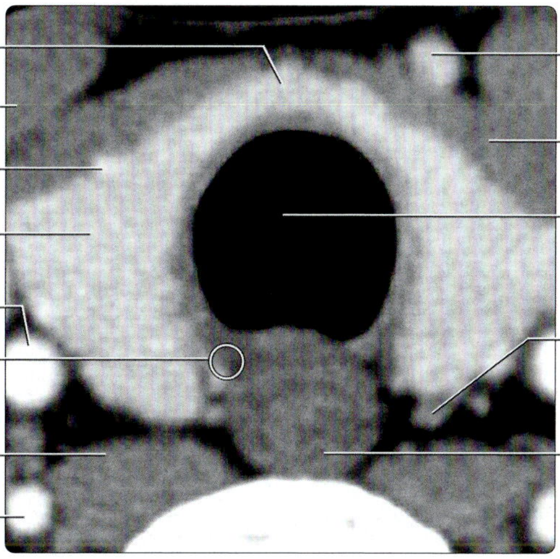

(Top) *Correlative CECT shows the right thyroid gland. The thyroid gland is seen as a triangular, homogeneously enhancing structure embracing the anterior aspect of the cervical trachea. The inferior thyroid artery seen on this CT is the main trunk, while that seen on the ultrasound is a branch.* **(Middle)** *Axial CECT shows the lower pole of the thyroid gland.* **(Bottom)** *Axial CECT shows the neck at midline, indicating the tracheoesophageal groove. Remember that the recurrent laryngeal nerve, paratracheal nodes, and parathyroid glands may be seen on ultrasound in this location, but they normally cannot be well demonstrated on routine CT images. Always evaluate extensions of the tumor into the trachea, tracheoesophageal groove, nodes, strap muscles, common carotid artery, and internal jugular vein.*

LONGITUDINAL ULTRASOUND

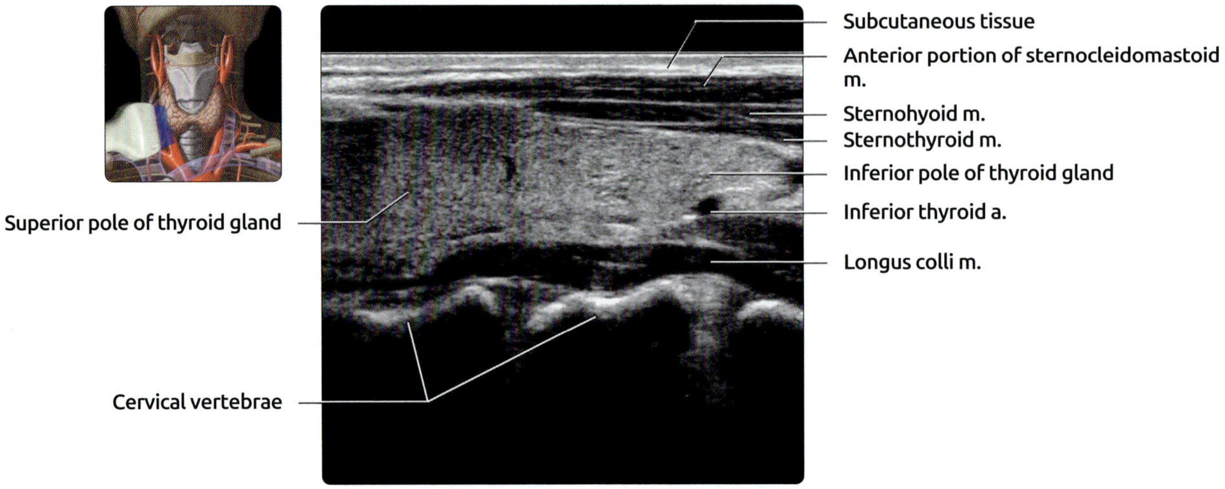

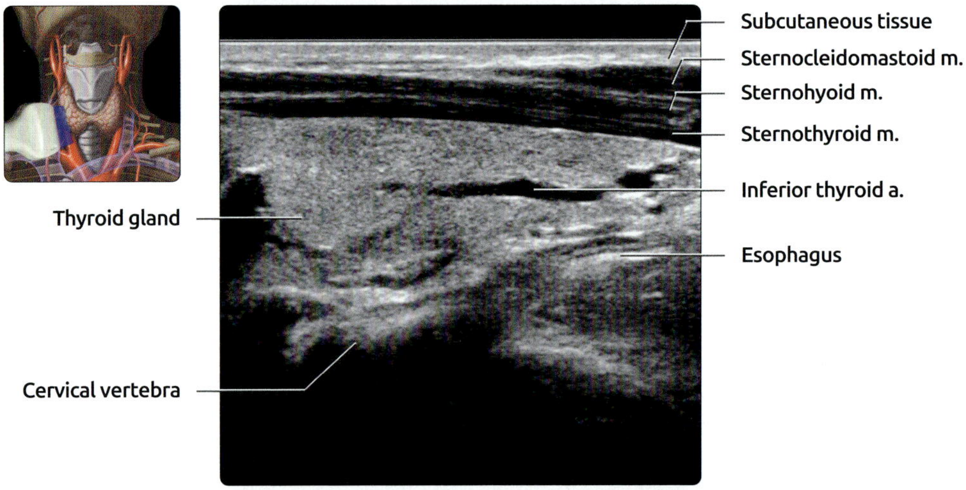

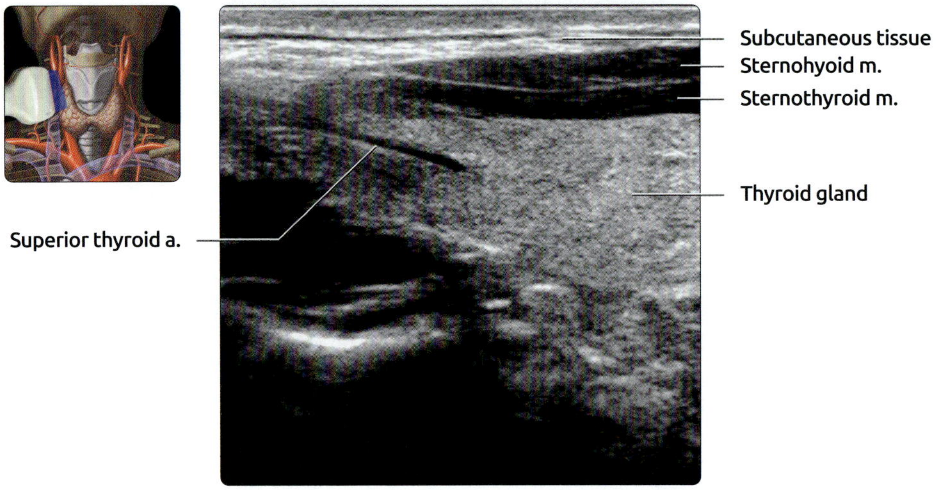

(Top) *Parasagittal longitudinal grayscale ultrasound shows the thyroid gland. The homogeneous, hyperechoic echo pattern of the glandular parenchyma is better assessed on longitudinal scans. Part of the tortuous course of the inferior thyroid artery is seen in relation to the lower pole.* **(Middle)** *Parasagittal longitudinal grayscale ultrasound shows the inferior thyroid artery coursing superiorly from the inferior pole within the glandular parenchyma.* **(Bottom)** *Parasagittal longitudinal grayscale ultrasound shows the superior thyroid artery, the 1st anterior branch of the external carotid artery, running inferiorly within and supplying the upper pole of the thyroid gland. Longitudinal scans best evaluate the glandular parenchyma and vascularity.*

LONGITUDINAL ULTRASOUND

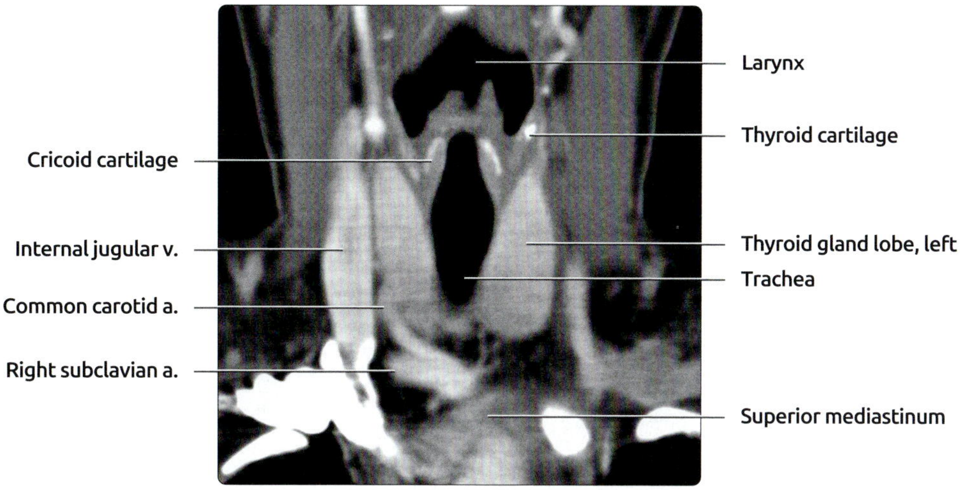

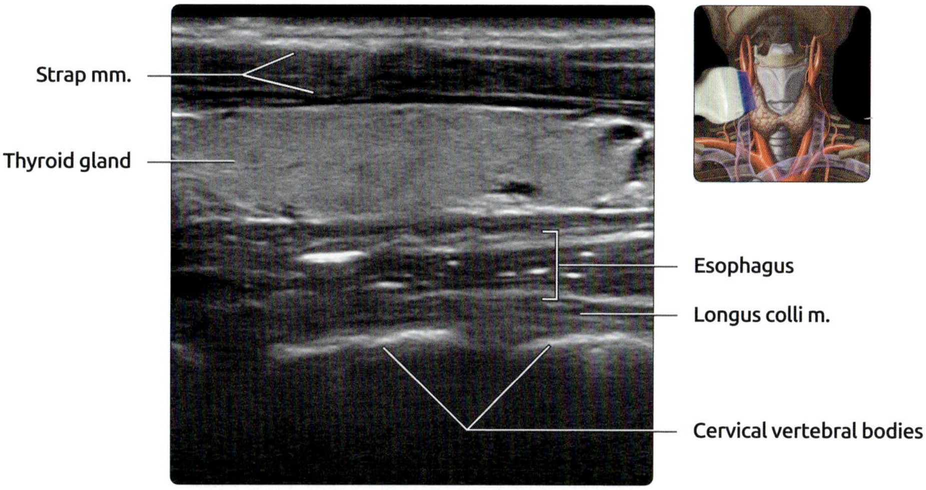

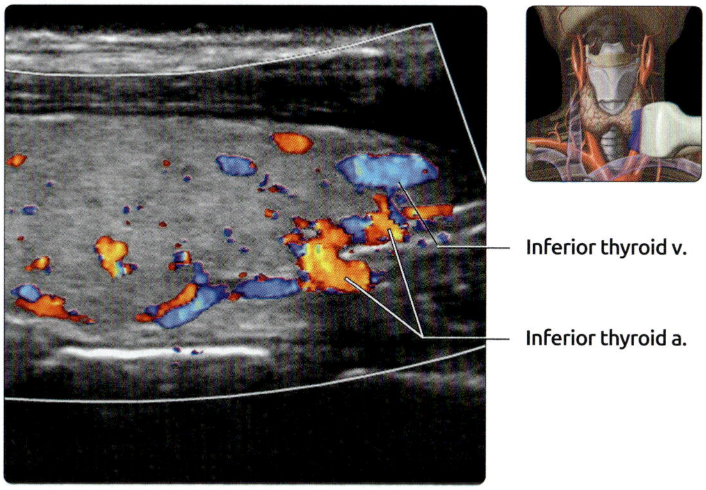

(Top) *In this image, the H- or U-shaped lobes of the thyroid gland are particularly well seen. Note the intimate relationship between the superomedial thyroid gland and the larynx. Remember that for thyroid malignancies the 1st-order nodes are the paratracheal nodes, which drain inferiorly into the superior mediastinum.* **(Middle)** *The thyroid gland is also immediately related to the esophagus as shown in this longitudinal scan taken from the lateral neck.* **(Bottom)** *This longitudinal color Doppler ultrasound shows both the inferior thyroid artery and vein. The thyroid gland is a very vascular organ. Three pairs of veins arise from a venous plexus on the surface of the gland. The superior and middle veins drain to the internal jugular vein and the inferior thyroid vein drains into the brachiocephalic vein.*

PYRAMIDAL LOBE

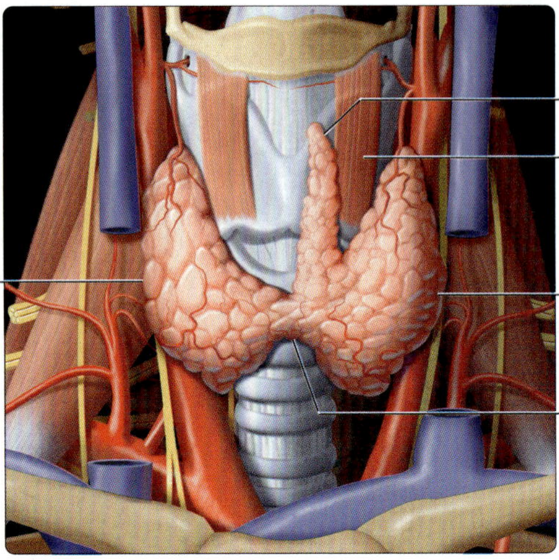

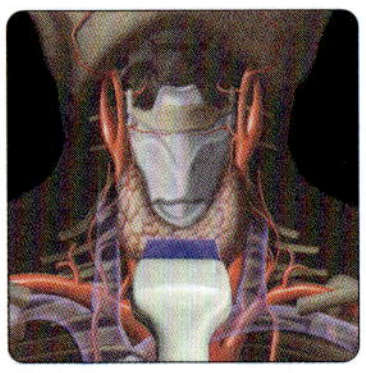

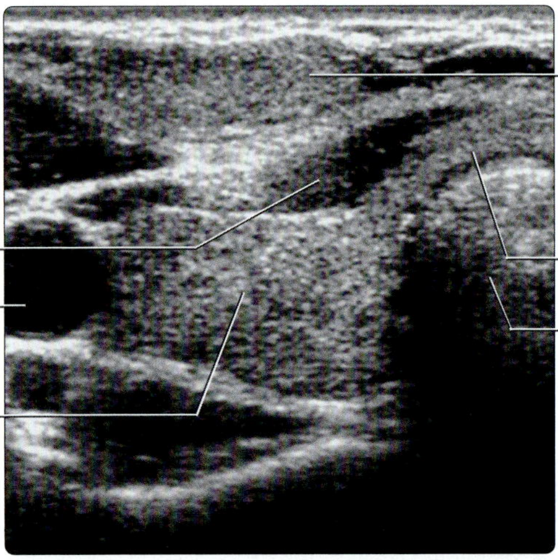

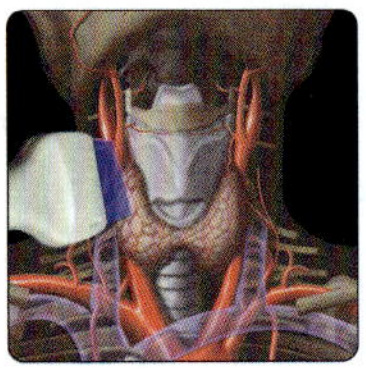

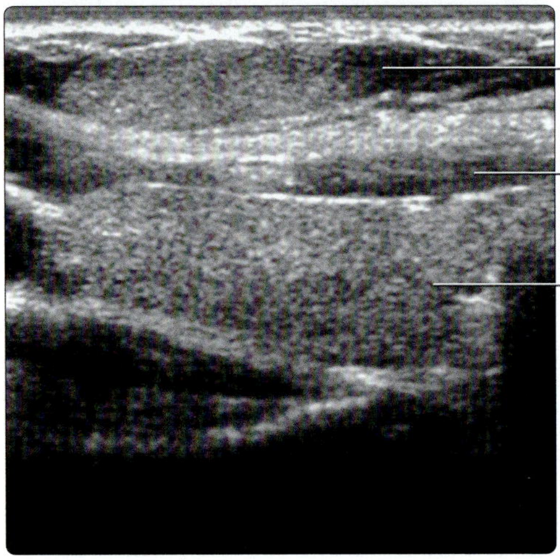

(Top) *There are many variants in the size and shape of the thyroid gland. This graphic shows the most common, a pyramidal lobe, which formed along the thyroglossal duct, and is present to some degree in 30-50% of the population. They more commonly project to the left but can be in the midline or to the right. Note that it is anterior to the strap muscles, while the main thyroid lobe is posterior.* **(Middle)** *Transverse grayscale ultrasound shows the pyramidal lobe of the thyroid gland. The echo pattern is identical to the normal right thyroid lobe (i.e., homogeneous, hyperechoic). It is located anterior to the anterior strap muscle. This should not be mistaken for a mass. Careful scanning will show it connects to the isthmus of the thyroid.* **(Bottom)** *A longitudinal ultrasound in the same patient shows the pyramidal lobe anterior to the strap muscle.*

THYROGLOSSAL DUCT CYST

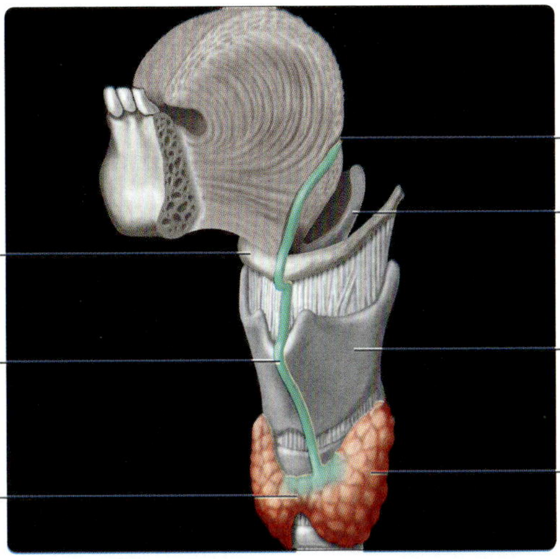

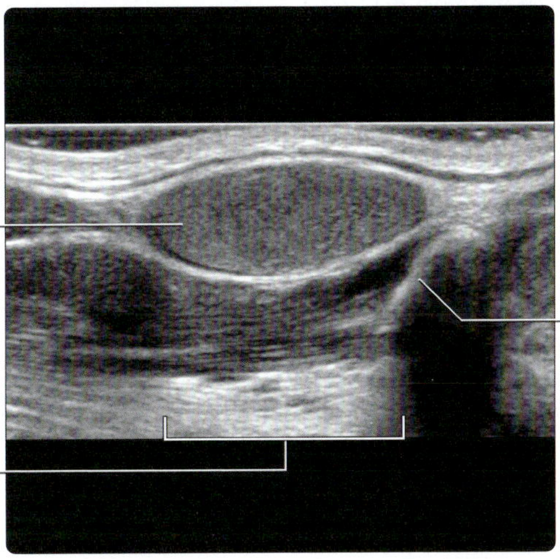

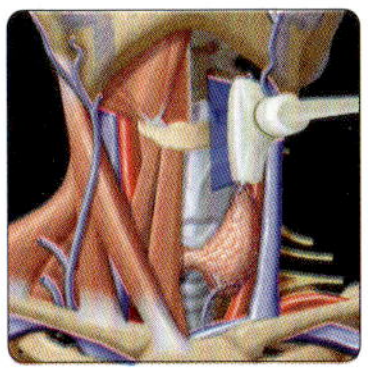

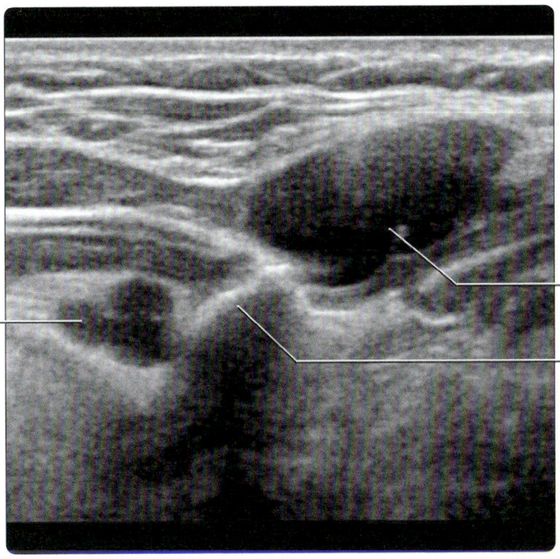

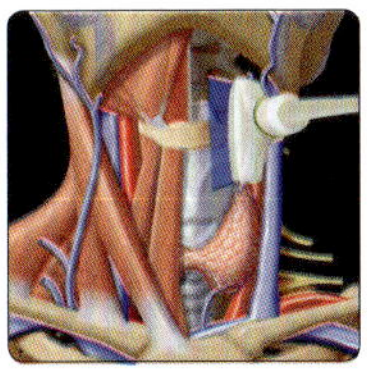

(Top) *Sagittal oblique graphic displays the thyroglossal duct tract as it traverses the cervical neck from its origin at the foramen cecum to its termination in the anterior and lateral visceral space of the infrahyoid neck. Failure of complete involution results in persistent secretion of the epithelial cells, which line the duct and the formation of a cyst.* **(Middle)** *Longitudinal grayscale ultrasound shows a well-defined, suprahyoid thyroglossal duct cyst with uniform internal echoes and posterior enhancement (pseudosolid appearance). This homogeneous appearance is due to intracystic proteinaceous content.* **(Bottom)** *Longitudinal grayscale ultrasound shows multiple thyroglossal duct cysts both above (suprahyoid) and below (infrahyoid) the hyoid bone. These may be found midline/paramidline between foramen cecum at the tongue base and the thyroid bed.*

TERMINOLOGY

Abbreviations

- Parathyroid gland (PTG)

Definitions

- Endocrine glands that control calcium metabolism by producing parathormone

IMAGING ANATOMY

Anatomy Relationships

- Small lentiform glands posterior to thyroid gland in visceral space
  - Extracapsular in most cases but can be located within thyroid gland
- **Located in region of tracheoesophageal groove**
- Normal measurements
  - ~ 6 mm in length, 3- to 4-mm transverse, and 1-2 mm in anteroposterior diameter
- Normal glands are often not seen on imaging (US/CT/MR)
  - If specifically targeted, normal PTG may be seen by using high-frequency transducer
  - Appear as small, well-circumscribed, hypoechoic nodules posterior to thyroid gland separated by echogenic thyroid capsule
- Variable number, but typically 4
  - 2 superior and 2 inferior PTGs
  - May be as many as 12
- **Superior PTGs**
  - More constant in position as compared with lower PTGs
  - Lie on posterior border of middle 1/3 of thyroid in 75%
  - 25% found behind upper or lower 1/3 of thyroid
  - 7% found below inferior thyroidal artery
  - Rarely found behind pharynx or esophagus
- **Inferior PTGs**
  - More variable in location
  - 50% of inferior glands lie lateral to lower pole of thyroid gland
  - 15% lie within 1 cm of inferior thyroid poles
  - 35% position is variable, residing anywhere from angle of mandible to lower anterior mediastinum
  - Intrathyroidal PTG are rare
- Arterial supply
  - Superior PTG supplied by superior thyroid artery
  - Inferior PTG supplied by inferior thyroid artery

ANATOMY IMAGING ISSUES

Imaging Approaches

- Usually targeted exam looking for parathyroid adenoma (PTA)
- US technique
  - Best 1st examination for localizing most PTA
  - Use high-resolution linear array transducer (7.5-10 MHz)
  - Identifies 95% of PTA weighing > 1 gram
  - Easier to start scanning in transverse plane with patient's neck hyperextended
  - Start above thyroid at angle of mandible and move downward, scanning through thyroid to level of clavicle
  - Angulate transducer at clavicle to see any obvious lesion in mediastinum
  - PTAs demonstrate hypervascularity on color flow imaging
  - Color flow imaging best done in longitudinal plane
- Nuclear scintigraphy
  - Tc-99m sestamibi concentrates in PTA
  - Useful for detection of ectopic PTA (most common site below inferior thyroid pole)

Imaging Pitfalls

- Parathyroid lesion may be confused with other lesions or normal anatomical structures
  - Paratracheal lymph nodes
    - Look for echogenic hilum and typical color Doppler pattern with central vessel
  - Thyroid nodules in subcapsular location
  - Longus colli muscle, esophagus, blood vessels also potential mimics
    - Scan in multiple planes and use color Doppler
- Detection of parathyroid lesion is limited in
  - Obese patients with short neck
  - Ectopic PTG (e.g., in mediastinum)
  - Postoperative neck

Consider

- Main indication for imaging of PTGs is localization of PTA causing hyperparathyroidism with hypercalcemia

CLINICAL IMPLICATIONS

Clinical Importance

- PTA is most common cause of primary hyperparathyroidism
- US localization facilitates minimally invasive parathyroidectomy
- US safely guides percutaneous injection of absolute alcohol in PTA

EMBRYOLOGY

Embryologic Events

- **Superior PTGs** develop from 4th branchial pouch along with primordium of thyroid gland
- **Inferior PTGs** develop from 3rd branchial pouch along with anlage of thymus
  - Descend variable distance with thymic anlage in thymopharyngeal duct tract
  - May descend into anterior mediastinum as far as pericardium

Practical Implications

- Abnormal PTG descent may cause inferior PTG to occupy "ectopic" sites
  - May be of critical importance when searching for PTA
    - In cases where surgical exploration for PTA is done without imaging, no PTA may be found if PTG is ectopic

PARATHYROID GLANDS

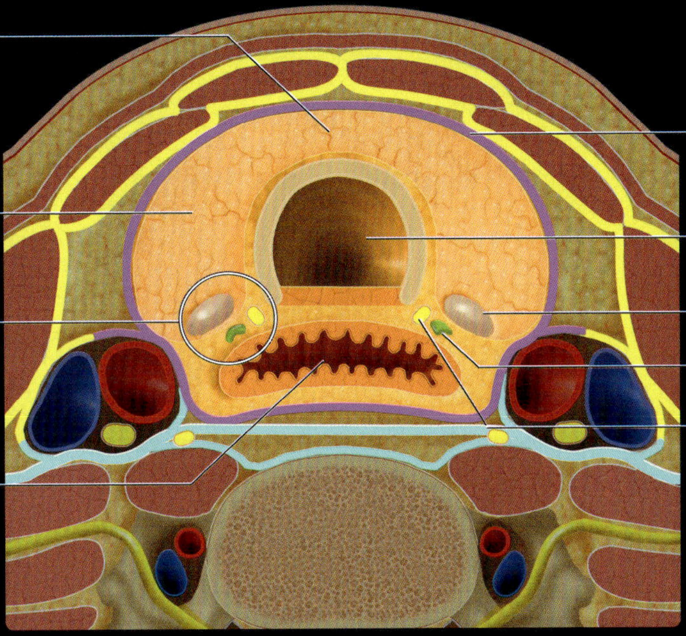

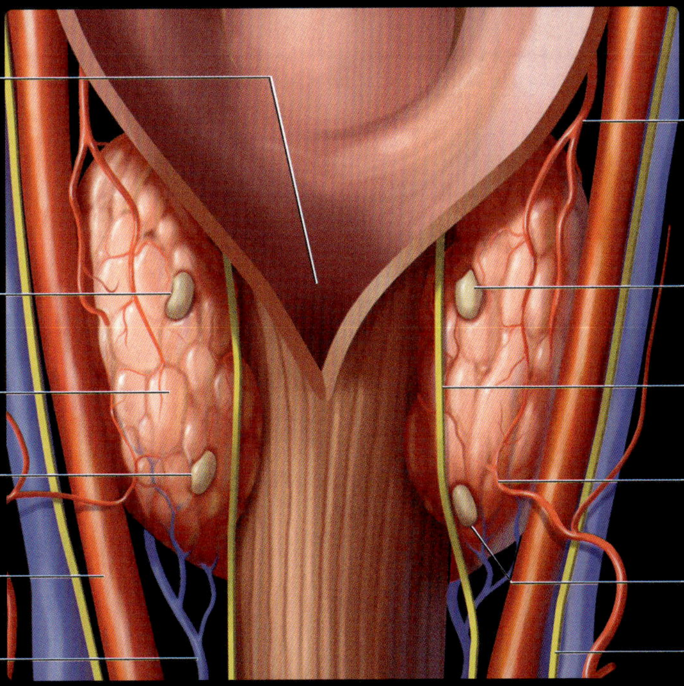

(Top) *Axial graphic at the thyroid level depicts the superior parathyroid glands in the visceral space just posterior to the thyroid gland. Note that there are 3 key structures found in the area of the tracheoesophageal groove: The recurrent laryngeal nerve, paratracheal lymph node chain, and parathyroid glands.* **(Bottom)** *Coronal graphic illustrates the esophagus, parathyroid glands, and thyroid gland from behind. The drawing depicts the typical anatomic relationships of the paired superior and inferior parathyroid glands in the visceral space. Note the arterial supply to superior and inferior parathyroid glands is from the superior and inferior thyroid arteries, respectively.*

NORMAL PARATHYROID

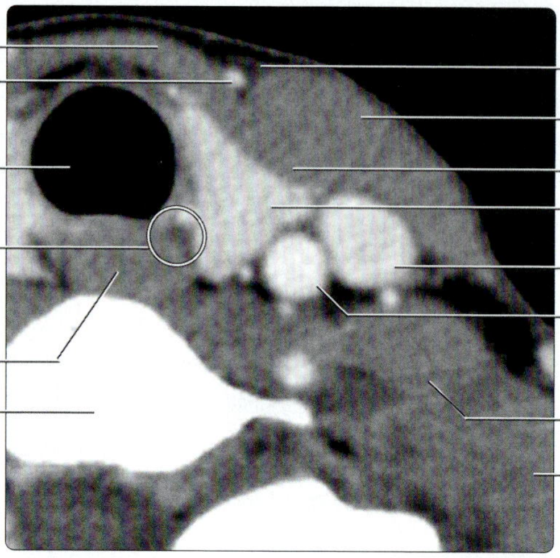

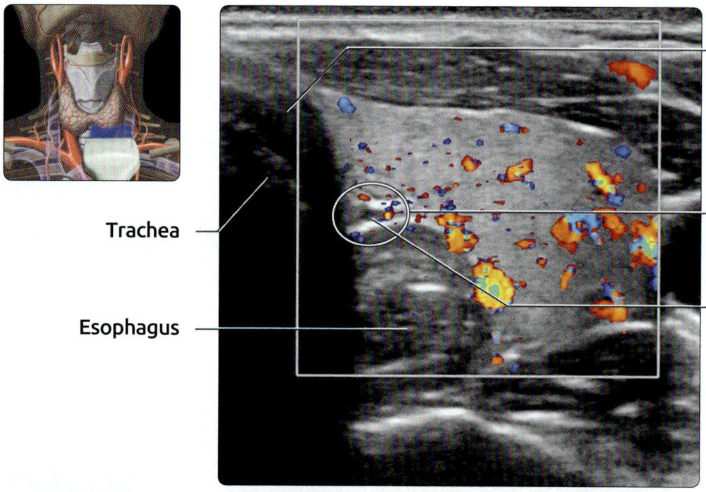

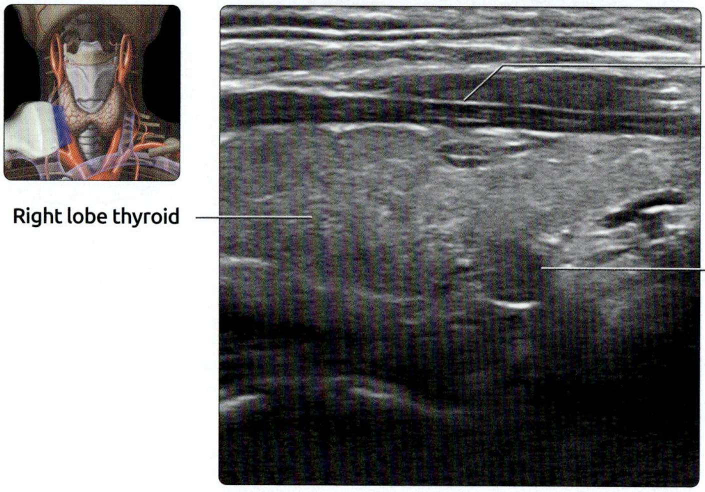

(Top) *Axial CECT of the neck shows the region of the lower pole of the thyroid's left lobe. A normal parathyroid gland is normally not seen. The anatomical location is commonly at the tracheoesophageal groove, posterior to the thyroid gland.* **(Middle)** *Transverse color Doppler US at a similar level shows a small, ovoid, hypoechoic area of soft tissue in the tracheoesophageal groove. Although it is difficult to tell with certainty, this is likely a normal parathyroid gland.* **(Bottom)** *Longitudinal scan of the right thyroid shows a larger, hypoechoic nodule in the expected location of the parathyroid gland, posterior the middle 1/3 of the thyroid gland. This is the typical location of the superior parathyroid glands. Careful analysis during a real-time scan should be done to prove they are outside the thyroid capsule and not a thyroid nodule. A parathyroid gland may also sometimes be confused with a small lymph node, but a normal lymph node should have an echogenic hilum.*

PARATHYROID ADENOMA

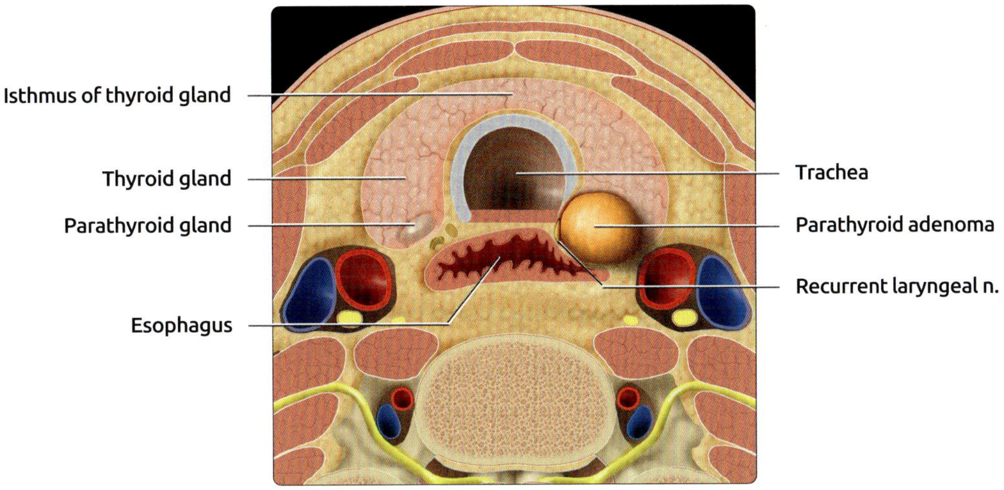

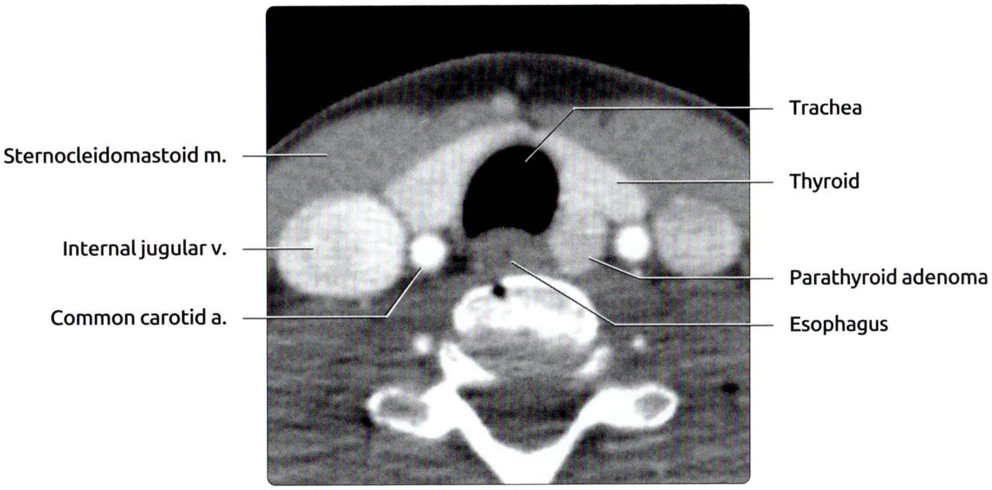

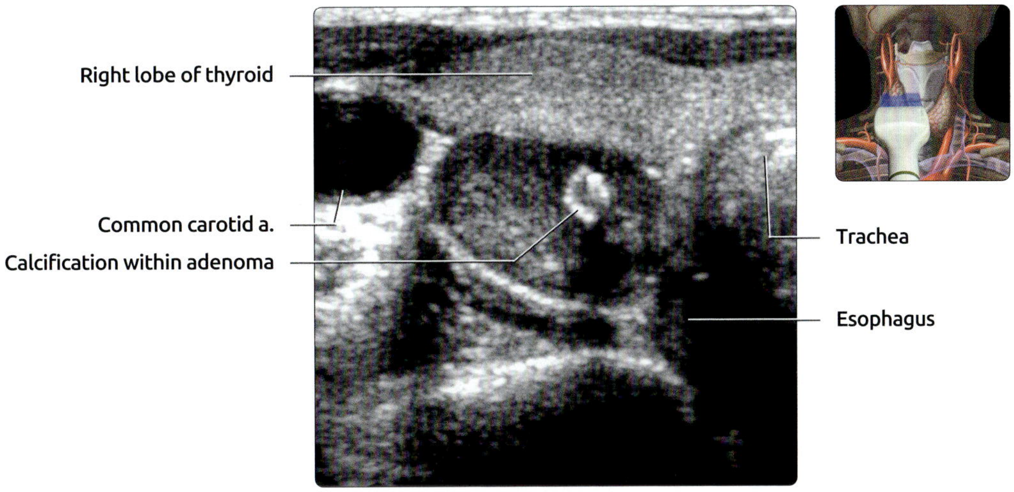

(Top) *Axial graphic shows a well-circumscribed mass in the left tracheoesophageal groove causing mass effect on the recurrent laryngeal nerve, esophagus, trachea, and left thyroid lobe. This is typical of a parathyroid adenoma, but a recurrent laryngeal nerve schwannoma or nodal disease in the paratracheal nodal chain could also cause such an appearance.* **(Middle)** *Axial CECT in a patient with acute hoarseness shows a well-defined, round mass in left tracheoesophageal groove. The mass is distinct from the posterior aspect of the left thyroid lobe and appears slightly less enhancing than the thyroid, typical of a parathyroid adenoma. The hoarseness was from impingement of the recurrent laryngeal nerve, which improved after surgery.* **(Bottom)** *Transverse grayscale US shows a well-circumscribed, solid, round, hypoechoic parathyroid adenoma posterior to the right thyroid lobe in the region of the tracheoesophageal groove. This patient had biochemical evidence of primary hyperparathyroidism. Note that the presence of calcification within an adenoma is uncommon; it is more commonly seen with carcinoma.*

PARATHYROID ADENOMA

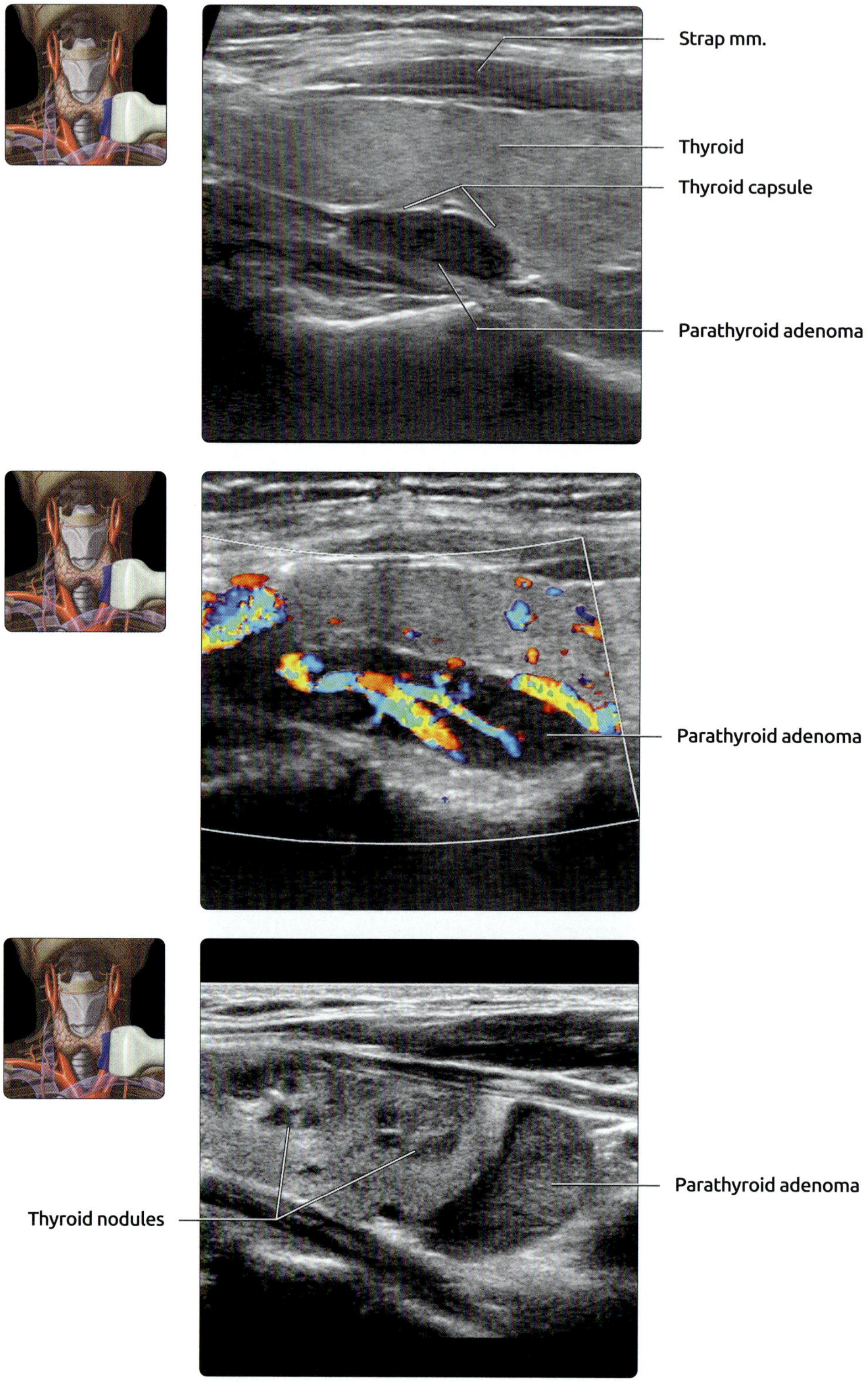

(Top) *Longitudinal US of the left lobe of the thyroid in a patient with hypercalcemia shows a well-defined, ovoid, hypoechoic mass. Note the distinct thyroid capsule, confirming this is posterior to the thyroid and therefore likely a parathyroid gland rather than a thyroid nodule.* **(Middle)** *Color Doppler US in the same patient shows increased vascularity, typical of a parathyroid adenoma.* **(Bottom)** *Longitudinal grayscale US shows a hypoechoic mass below the lower pole of the thyroid gland in a patient with hypercalcemia. Multiple thyroid nodules are also seen.*

PARATHYROID ADENOMA

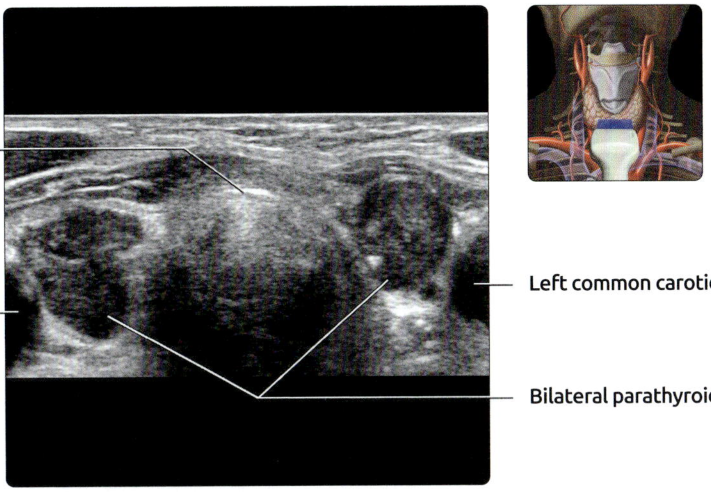

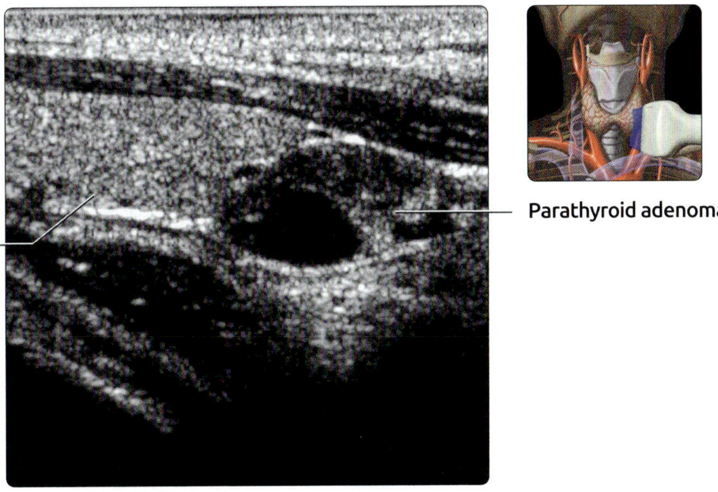

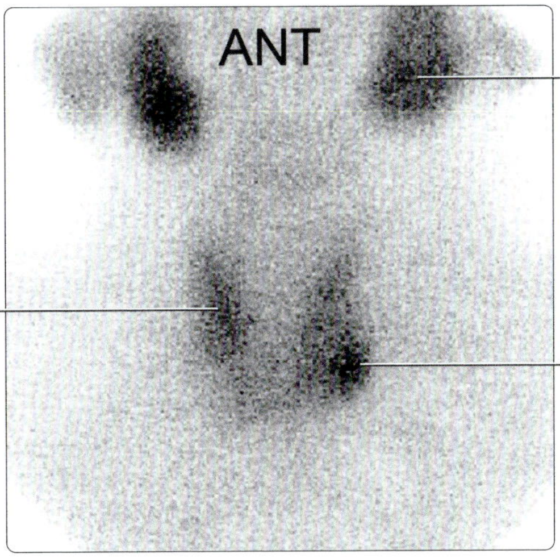

(Top) *Transverse grayscale US shows multiple parathyroid adenomas on either side of the trachea. Of patients with primary hyperparathyroidism, 2-3% may have multiple parathyroid adenomas.* **(Middle)** *Longitudinal grayscale US shows a well-circumscribed parathyroid adenoma inferoposterior to the lower pole of left lobe of the thyroid gland. The echotexture of the adenoma is heterogeneous.* **(Bottom)** *Planar sestamibi scintigraphy in the same patient shows a solitary focus of increased tracer uptake superimposed on the lower pole of the left thyroid gland. The scintigraphic features are suggestive of solitary hyperfunctioning parathyroid adenoma.*

IMAGING ANATOMY

Overview

- **Larynx**: Cartilaginous skeleton bounded by ligaments and muscles at junction of upper and lower airway
  - Cranial margin at level of glossoepiglottic and pharyngoepiglottic folds with caudal margin defined by lower edge of cricoid
    - Superior connection: Oropharynx
    - Inferior connection: Trachea
- **Hypopharynx**: Caudal continuation of pharyngeal mucosal space, located between oropharynx and esophagus
  - Extends from level of glossoepiglottic and pharyngoepiglottic folds superiorly to inferior border of cricoid cartilage (cricopharyngeus muscle)
    - Superior connection: Oropharynx
    - Inferior connection: Cervical esophagus

Internal Contents

- Laryngeal cartilages
  - Thyroid cartilage: Largest laryngeal cartilage
    - "Shields" larynx
    - 2 anterior laminae meet anteriorly at acute angle
    - Superior thyroid notch at anterosuperior aspect
    - Superior cornua are elongated and narrow, and they attach to thyrohyoid ligament
    - Inferior cornua are short and thick, articulating medially with sides of cricoid cartilage
  - Cricoid cartilage: Only complete ring in endolarynx
    - Provides structural integrity
    - 2 portions: Posterior lamina and anterior arch
    - Lower border of cricoid cartilage is junction between larynx above and trachea below
  - Arytenoid cartilage: Paired pyramidal cartilages that sit on top posterior aspect of cricoid cartilage
    - Vocal and muscular processes are at level of true vocal cord (TVC)
    - Vocal processes: Anterior projections of arytenoids where posterior margins of TVC attach
  - Corniculate cartilage: Rests on top of superior process of arytenoid cartilage within aryepiglottic (AE) fold
- Supraglottic larynx
  - Extends from tip of epiglottis above to laryngeal ventricle below
  - Contains vestibule, epiglottis, preepiglottic fat, AE folds, false vocal cord (FVC), paraglottic space, and arytenoid cartilages
  - Epiglottis: Leaf-shaped cartilage, larynx lid with free margin (suprahyoid), fixed portion (infrahyoid)
    - Petiole is "stem" of leaf that attaches epiglottis to thyroid lamina via thyroepiglottic ligament
    - Hyoepiglottic ligament attaches epiglottis to hyoid
    - Glossoepiglottic fold is midline mucous membrane covering hyoepiglottic ligament
  - Preepiglottic space: Fat-filled space between hyoid bone anteriorly and epiglottis posteriorly
  - AE folds: Projects from cephalad tip of arytenoid cartilages to inferolateral margin of epiglottis
    - Represents superolateral margin of supraglottis, dividing it from pyriform sinus (hypopharynx)
  - FVC: Mucosal surfaces of laryngeal vestibule of supraglottis
  - Paraglottic spaces: Paired fatty regions beneath FVC and TVC
    - Superiorly, they merge into preepiglottic space, which terminates inferiorly at under surface of TVC
- Glottic larynx
  - TVC and anterior and posterior commissures
    - Composed of thyroarytenoid muscle (medial fibers are "vocalis muscle") covered by mucosa
    - Anterior commissure: Midline, anterior meeting point of TVC
- Subglottic larynx
  - Subglottis extends from under surface of TVC to inferior surface of cricoid cartilage
  - Mucosal surface of subglottic area is closely applied to cricoid cartilage
- Hypopharynx: Consists of 3 regions
  - Piriform sinus: Anterolateral recess of hypopharynx
    - Between inner surface of thyrohyoid membrane (above), thyroid cartilage (below), and AE folds (laterally)
    - Pyriform sinus apex (inferior tip) at level of TVC
    - Anteromedial margin of pyriform sinus is posterolateral wall of AE fold (marginal supraglottis)
  - Posterior hypopharyngeal wall: Inferior continuation of posterior oropharynx wall
  - Post cricoid region: Anterior wall of lower hypopharynx
    - Interface between hypopharynx and larynx
    - Extends from cricoarytenoid joints to lower edge of cricoid cartilage

ANATOMY IMAGING ISSUES

Imaging Recommendations

- Role of US in laryngeal cancer is limited, particularly in era of MDCT and MR
- US may serve supplementary role to clinical examination and CT/MR in assessing superficial extent of laryngeal tumor
- US combined with fine-needle aspiration cytology is useful for nodal staging of laryngeal tumor
- Although role of US in imaging of larynx is very limited, sonologist doing neck US should be familiar with its anatomy in order not to mistake its appearances for abnormalities

Imaging Sweet Spots

- Real-time US is well suited to quickly evaluate vocal cord mobility in children presenting with hoarseness and stridor
- US also guides safe percutaneous vocal cord injection for patients with cord palsy, provided degree of cartilaginous calcification does not obscure US visualization of vocal cord

Imaging Pitfalls

- Presence of motion and calcification/ossification within laryngeal cartilage (common in adult) restricts detailed sonographic assessment of internal laryngeal structures

SUPRAGLOTTIC STRUCTURES

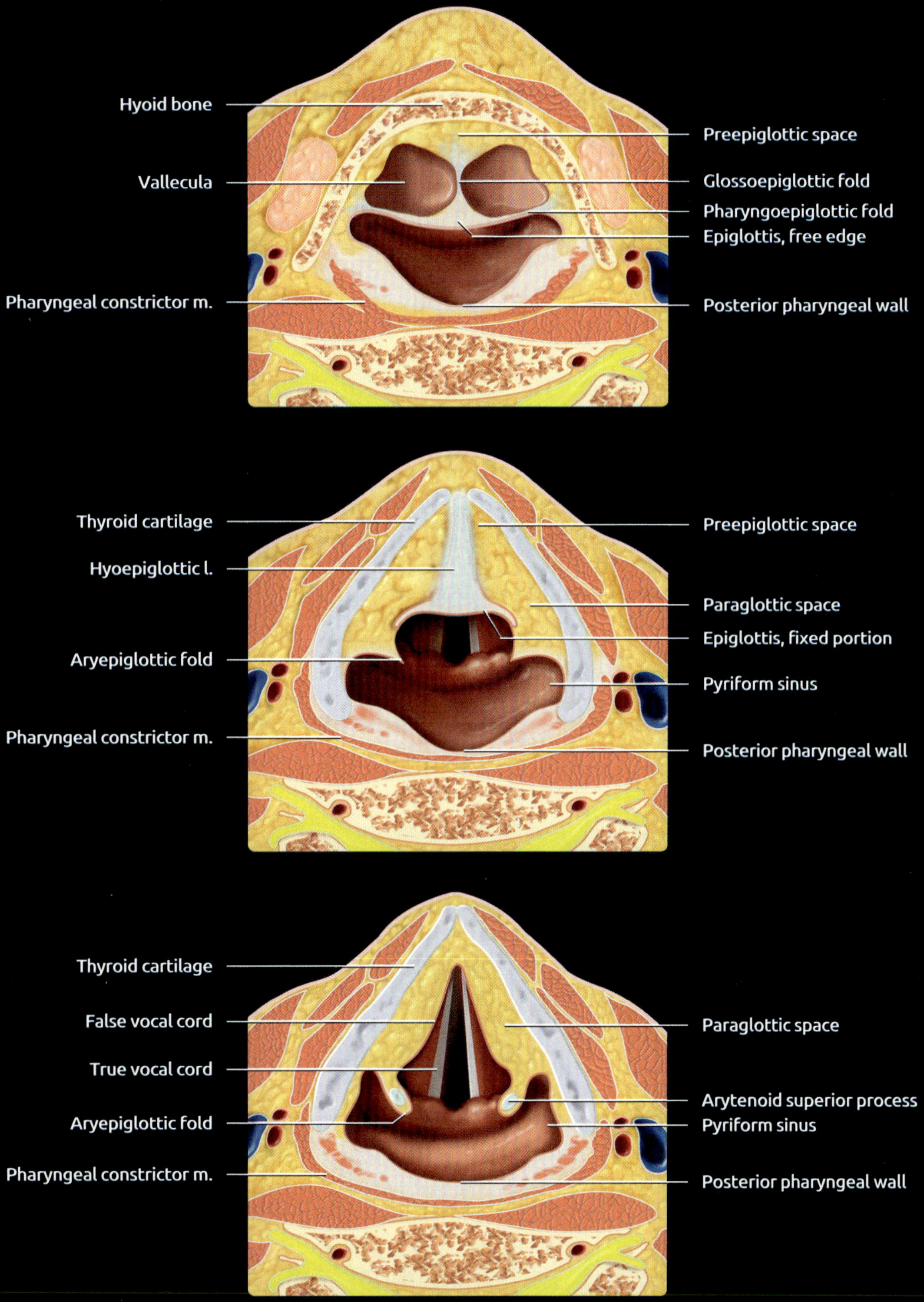

(Top) *Axial graphic of the larynx and hypopharynx shows the roof of hypopharynx at the hyoid bone level and the high supraglottic structures. The free edge of the epiglottis is attached to the hyoid bone via the hyoepiglottic ligament, which is covered by the glossoepiglottic fold.* **(Middle)** *Graphic at the midsupraglottic level shows the hyoepiglottic ligament dividing the lower preepiglottic space. No fascia separates the preepiglottic space from the paraglottic space. These 2 endolaryngeal spaces are submucosal in locations where tumors can hide from clinical detection. The aryepiglottic fold represents the junction between the larynx and hypopharynx.* **(Bottom)** *Graphic at the low supraglottic level shows false vocal cords formed by mucosal surfaces of laryngeal vestibule. The paraglottic space is beneath the false vocal cords, a common location for submucosal tumor spread.*

GLOTTIC AND SUBGLOTTIC STRUCTURES

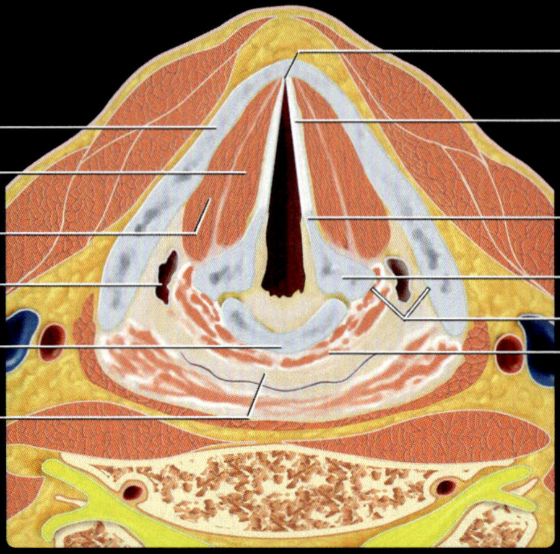

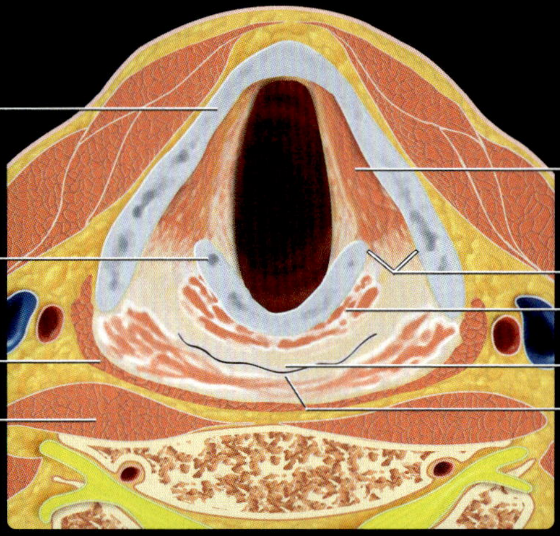

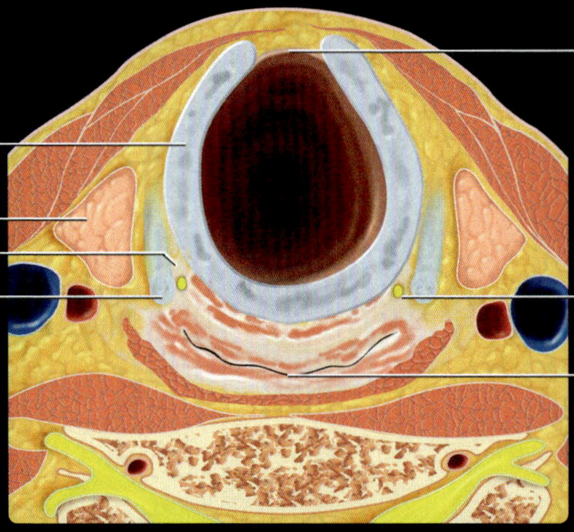

(Top) *Graphic at the glottic, true vocal cord level shows the thyroarytenoid muscle, which makes up the bulk of the true vocal cord. The medial fibers of the thyroarytenoid muscle are known as the vocalis muscle. Pyriform sinus apex is seen at the glottic level.* **(Middle)** *Graphic at the level of the undersurface of the true vocal cord shows the posterior lamina of the cricoid cartilage. Postcricoid hypopharynx represents the anterior wall of the lower hypopharynx and extends from the cricoarytenoid joints to the lower edge of the cricoid cartilage at the cricopharyngeus muscle. The posterior wall of the hypopharynx represents the inferior continuation of the posterior oropharyngeal wall and extends to the cervical esophagus.* **(Bottom)** *Graphic at the subglottic level shows the cricothyroid joint immediately adjacent to the recurrent laryngeal nerve, which is located in the tracheoesophageal groove.*

LARYNX

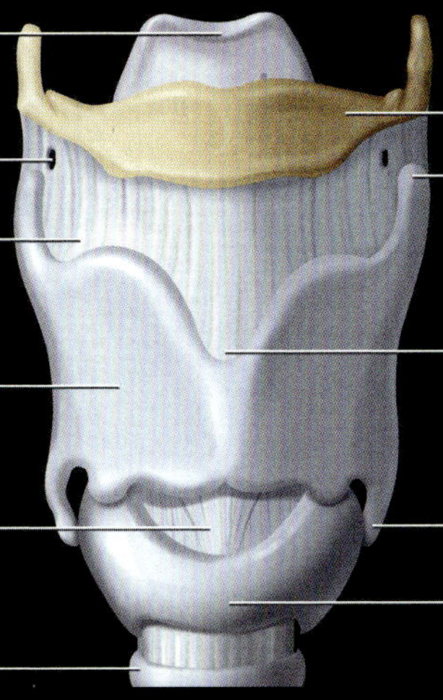

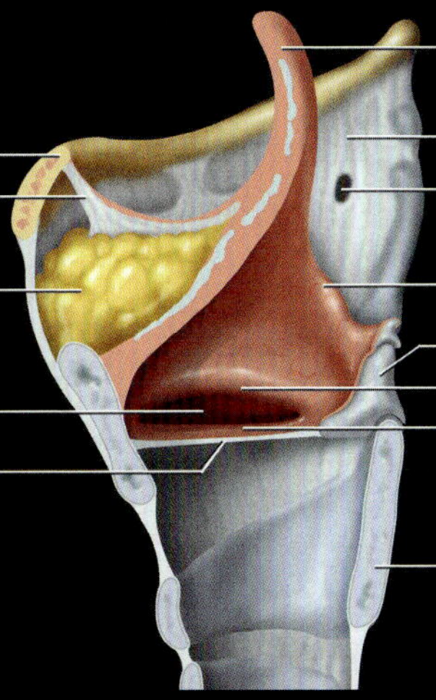

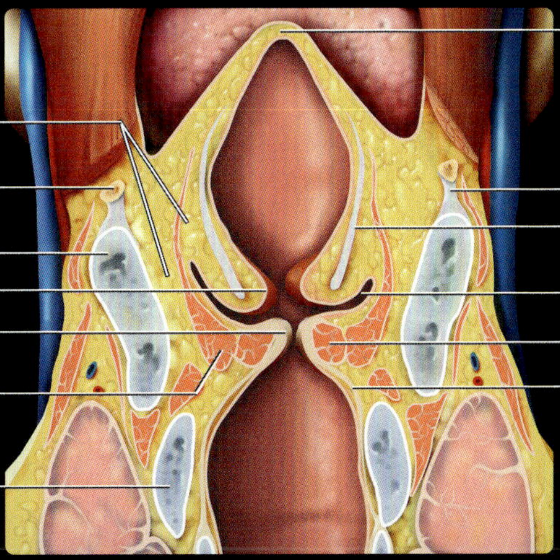

(Top) *Anterior graphic depicts the laryngeal cartilage, which provides the structural framework for the soft tissues of the larynx. Note that 2 large anterior laminae "shield" the larynx. The thyrohyoid membrane contains an aperture through which the internal branch of the superior laryngeal nerve and associated vessels course.* **(Middle)** *Sagittal graphic of the midline larynx shows the laryngeal ventricle air space, which separates false vocal cords above from true vocal cords below.* **(Bottom)** *Coronal graphic (posterior view) shows the false and true vocal cords separated by the laryngeal ventricle. The quadrangular membrane is a fibrous membrane that extends from the upper arytenoid and corniculate cartilages to the lateral epiglottis. The conus elasticus is a fibroelastic membrane that extends from the vocal ligament of the true vocal cord to the cricoid.*

TRANSVERSE ULTRASOUND

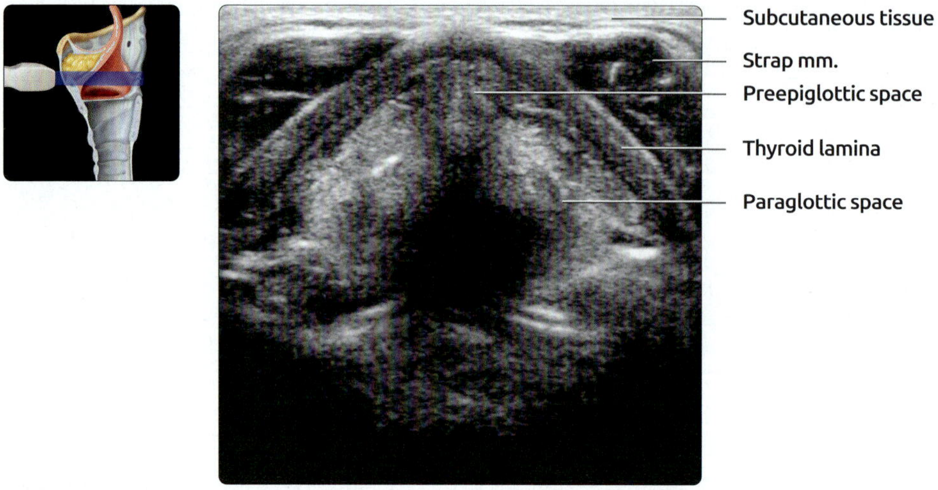

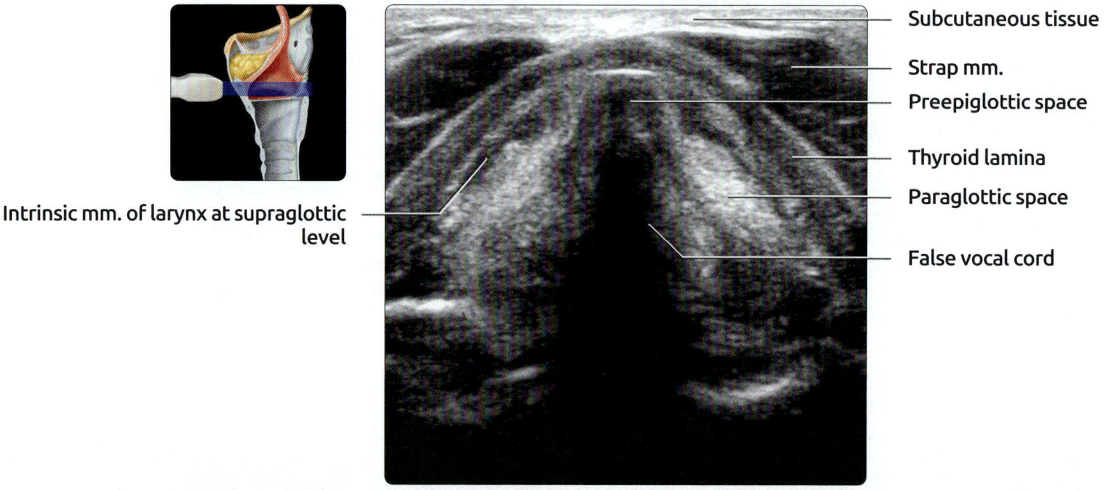

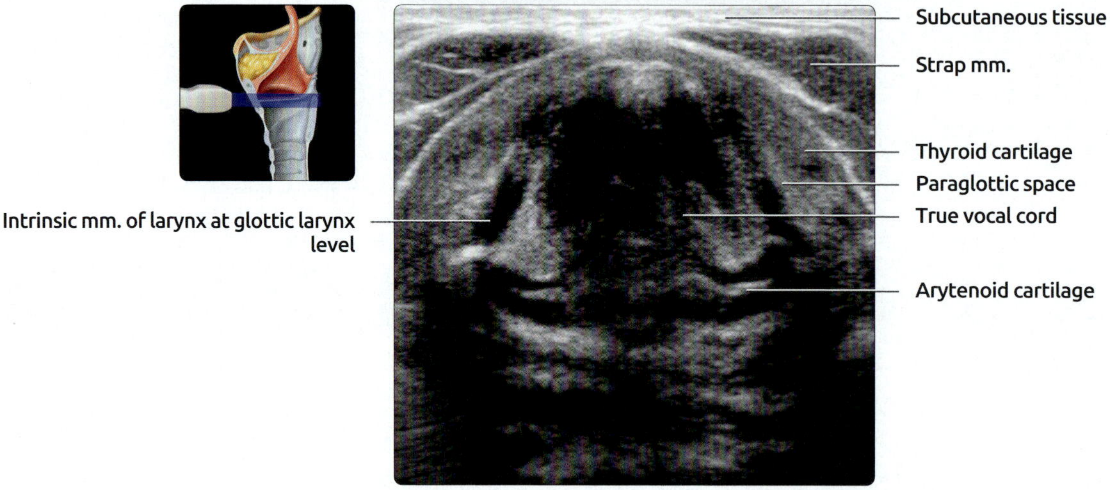

(Top) *Axial grayscale ultrasound of the larynx at the supraglottic larynx level shows that the thyroid laminae are the largest cartilaginous structures of the larynx and appear as thin, hypoechoic bands that join at the midline anteriorly. The hyperechoic, fat-filled paraglottic and preepiglottic spaces are important surgical landmarks for the staging of laryngeal carcinoma.* **(Middle)** *Transverse grayscale ultrasound of the larynx at the level of the false vocal cords shows abundant fat in the paraglottic spaces. The echo-poor intrinsic muscles of the larynx are embedded within the echogenic paraglottic fat.* **(Bottom)** *Transverse grayscale ultrasound of the larynx at the level of the true vocal cords shows that the arytenoid cartilage appears as echogenic foci posteriorly with attachments to the true vocal cords, which have distinct echo-poor appearances.*

AXIAL CECT

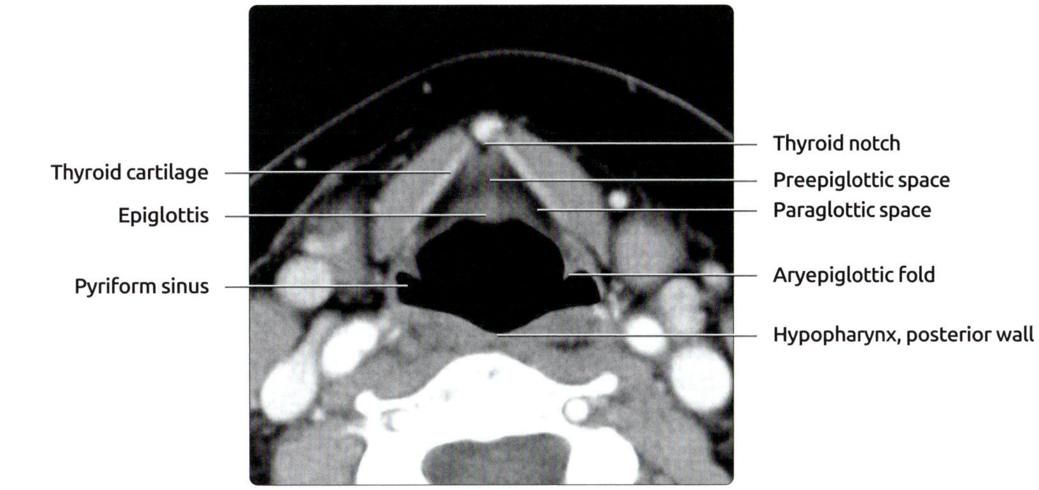

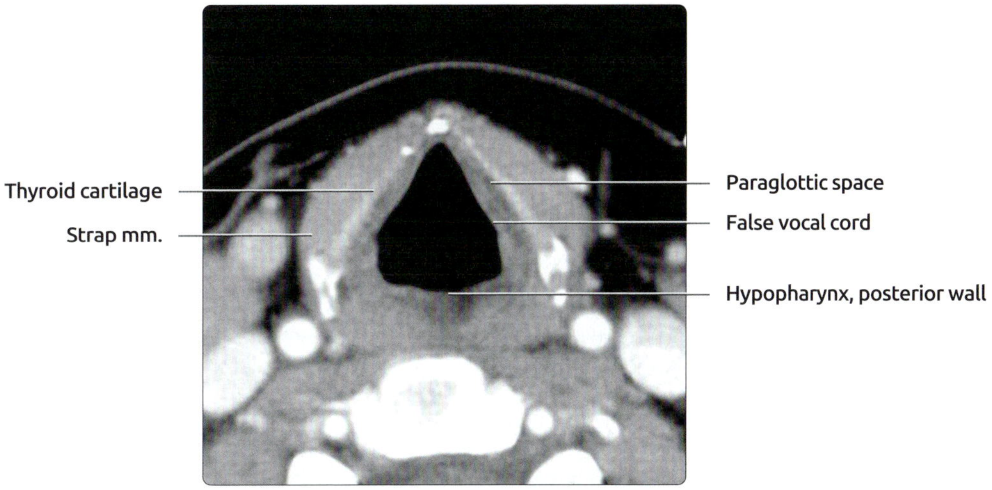

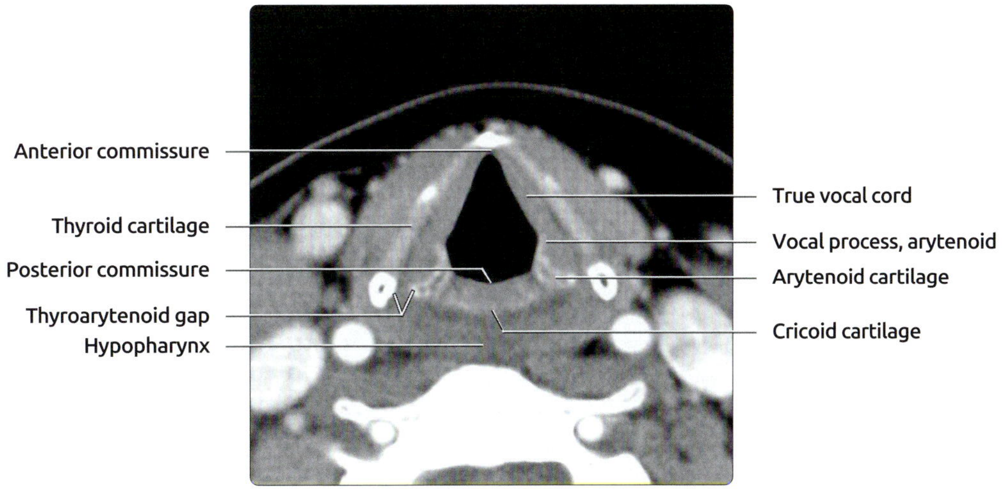

(Top) *Axial CECT of the high supraglottic level shows that the preepiglottic and paraglottic spaces are continuous. This allows tumors to spread submucosally in these locations. The aryepiglottic fold, part of the larynx, represents a transition between the larynx and hypopharynx.* **(Middle)** *Axial CECT shows the low supraglottic level at the false vocal cord level. The paraglottic space represents the deep fatty space beneath the false vocal cords. Tumors that cross the laryngeal ventricle and involve false and true vocal cords are considered transglottic.* **(Bottom)** *Axial CECT at the glottic level shows the true vocal cords in abduction during quiet respiration. True vocal cord level is identified on CT when the arytenoid and cricoid cartilages are seen and muscle fills the inferior paraglottic space. The anterior and posterior commissures of the true vocal cords should be < 1 mm in normal patients. The postcricoid hypopharynx is typically collapsed.*

LONGITUDINAL ULTRASOUND

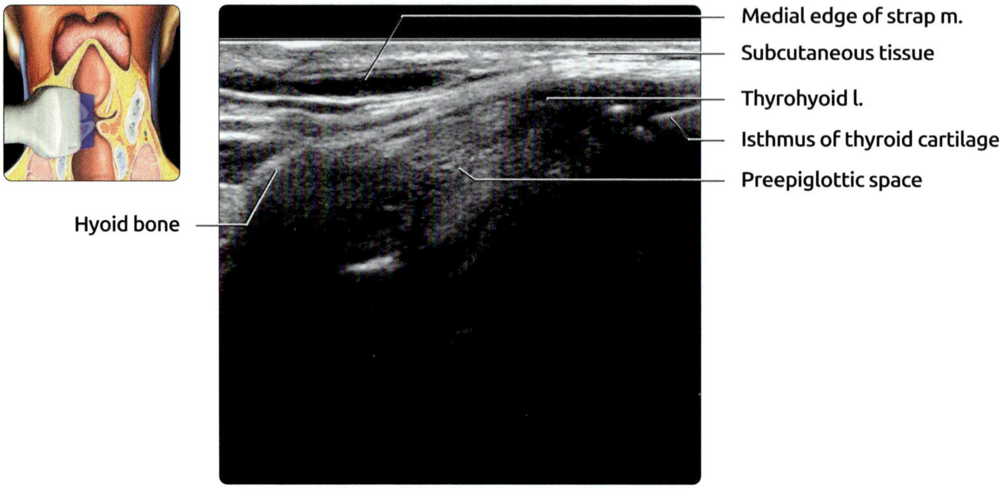

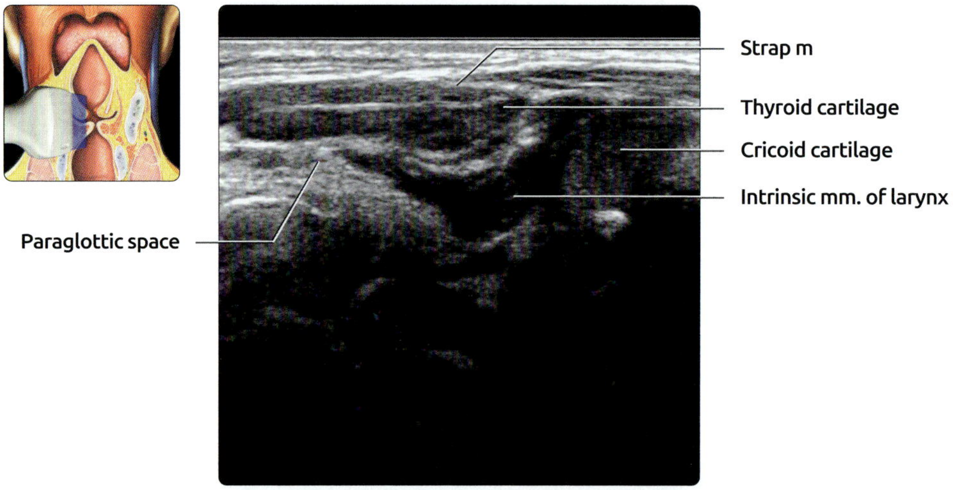

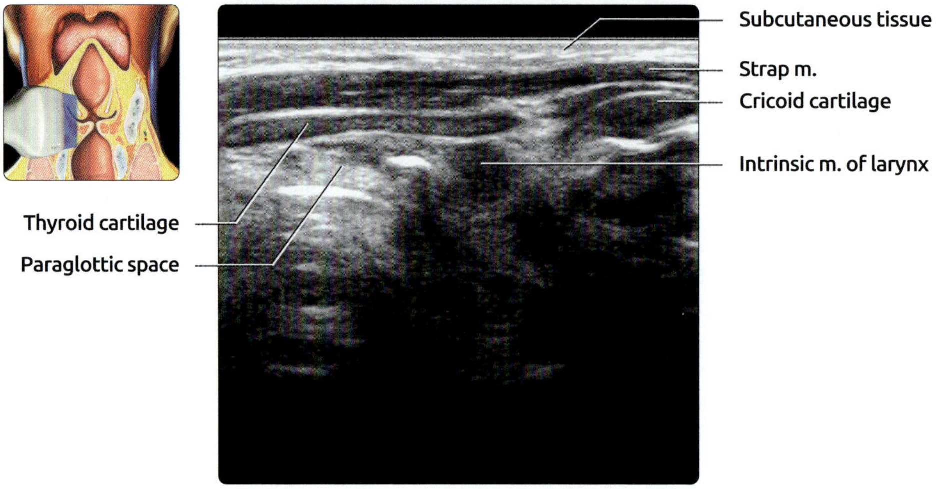

(Top) *Midline sagittal longitudinal grayscale ultrasound of the supraglottic larynx shows the fat-filled, echogenic, preepiglottic space underneath the thyrohyoid membrane. Tumor spread at this location is readily assessed by ultrasound.* **(Middle)** *Parasagittal longitudinal grayscale ultrasound of the larynx shows sonolucent thyroid and cricoid cartilages with no laryngeal calcification/ossification in a young adult. The paraglottic space is fat filled and appears echogenic. The intrinsic muscles of the larynx are embedded within the paraglottic space and are hypoechoic on ultrasound.* **(Bottom)** *Parasagittal longitudinal grayscale ultrasound shows the larynx further lateral. Gas within the laryngeal lumen appears highly echogenic, casting posterior acoustic shadowing.*

SAGITTAL AND CORONAL REFORMATTED NECT

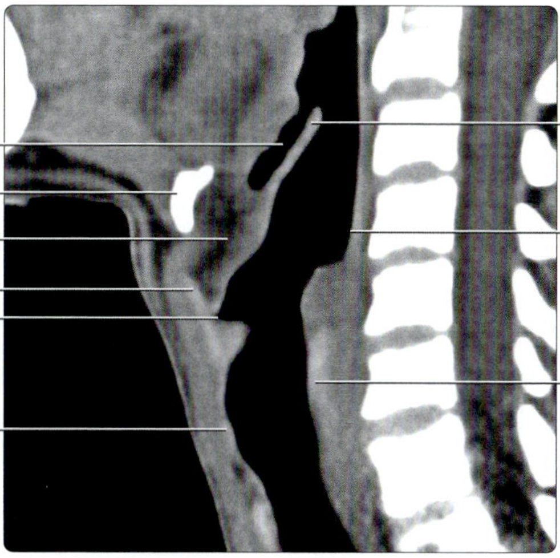

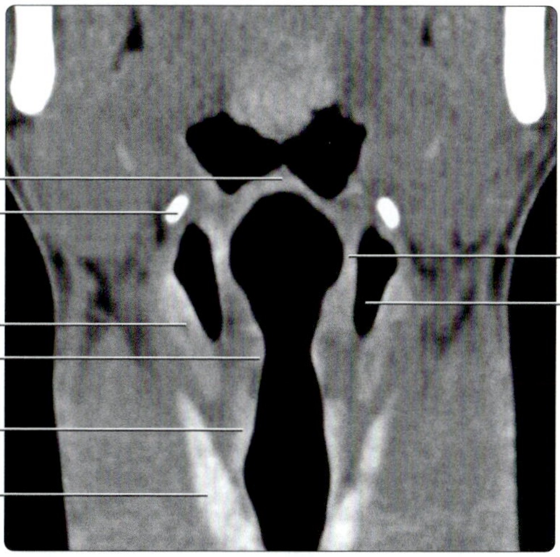

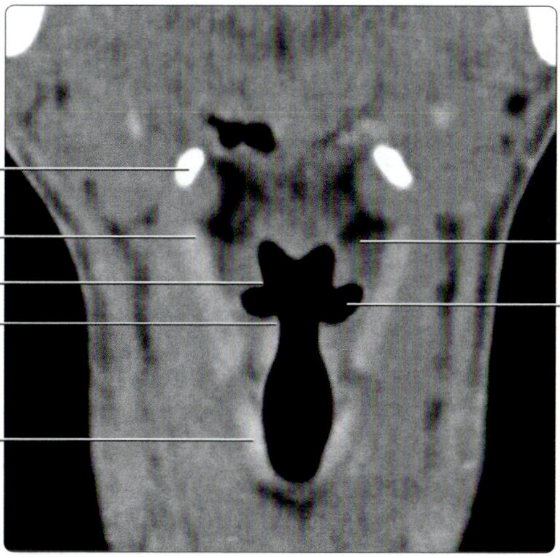

(Top) *Parasagittal reformatted NECT shows the laryngeal ventricle air space that separates the false vocal cords above from the true vocal cords below.* **(Middle)** *In this coronal reformatted NECT, aryepiglottic folds are well seen as they extend from the lateral epiglottis to the arytenoid cartilage. The pyriform sinus is the most common location for tumors of the hypopharynx.* **(Bottom)** *In this coronal reformatted NECT, the laryngeal ventricle is visible as an air space between false vocal cords above and true vocal cords below. When a tumor crosses the laryngeal ventricle to involve the true and false cords, it is transglottic, which has important treatment implications. Coronal imaging is particularly useful for evaluation of transglottic disease. Note that ultrasound is unable to demonstrate such detailed anatomy, particularly of deeper structures.*

IMAGING ANATOMY

Overview

- Trachea and esophagus pass through visceral space of neck
- Trachea
  - 10- to 13-cm flexible tube made of cartilage and fibromuscular membrane
    - Extends in midline from inferior larynx at ~ 6th cervical vertebral body to carina at upper margin of 5th thoracic vertebral body
- Esophagus
  - ~ 25-cm muscular tube of longitudinal and circular smooth muscle
    - Extends in midline from inferior hypopharynx at ~ 6th cervical vertebral body to 11th thoracic vertebral body
  - Descends behind trachea and thyroid, lying in front of lower cervical vertebrae
    - Inclines slightly to left in lower cervical neck and upper mediastinum, returning to midline at T5 vertebral body level

Cervical Trachea

- Boundaries
  - Anterior
    - Infrahyoid strap muscles, isthmus of thyroid gland
  - Lateral
    - Lobes of thyroid gland
    - **Tracheoesophageal groove**: Contains recurrent laryngeal nerve, paratracheal nodes, parathyroid glands
  - Posterior
    - Cervical esophagus
- **Tracheal cartilages**
  - Each cartilage is imperfect ring of cartilage surrounding anterior 2/3 of trachea
  - Flat deficient posterior portion is completed with fibromuscular membrane
  - Cross-sectional shape of trachea is that of letter D with flat posterior side
  - Smooth muscle fibers in posterior membrane (trachealis muscle) attach to free ends of tracheal cartilages and provide alteration in tracheal cross-sectional area
  - Hyaline cartilage calcifies with age
  - Minor salivary glands are sporadically distributed in tracheal mucosa
- Blood supply
  - Inferior thyroid arteries and veins
- Lymphatic drainage
  - Level VI pretracheal and paratracheal nodes
- US appearance
  - Hypoechoic tracheal ring composed of hyaline cartilage that is incomplete posteriorly
  - Ring-down artifacts from air in tracheal lumen
  - Midline behind thyroid isthmus

Cervical Esophagus

- Begins at lower border of cricoid cartilage as continuation of hypopharynx
  - Upper limit is defined by cricopharyngeus muscle, which encircles it from front to back
- Boundaries
  - Anterior
    - Cervical trachea
  - Anterolateral
    - Tracheoesophageal groove structures
  - Lateral
    - Common carotid artery, internal jugular vein, vagus nerve
  - Posterior
    - Retropharyngeal fascia/muscle
- Usually in slightly off-midline position to left in cervical portion
- Active peristalsis in antegrade (i.e., downward) direction
- Blood supply
  - Inferior thyroid arteries and veins
- Lymphatic drainage
  - Level VI paratracheal nodes
- US appearance
  - Thick-walled tubular structure with alternating concentric echogenic/hypoechoic rings (**gut signature**)
    - Alternating echogenic/hypoechoic layers representing mucosal, submucosal, muscular, and serosal layers on longitudinal scan
  - Presence of air in lumen; moves on swallowing

ANATOMY IMAGING ISSUES

Imaging Recommendations

- US is best done with patient's head slightly hyperextended
- Ask patient not to swallow during US
- Transverse and longitudinal scans are necessary for comprehensive US examination
- Assess adjacent structures and regional neck nodes

Imaging Pitfalls

- Tracheal ring calcification and intraluminal air render complete assessment of trachea difficult
- Intraluminal gas may obscure posterior esophageal wall
- Esophagus is mobile tube and slips side to side in neck depending on direction head is turned
  - On US, esophagus may be on right when head is turned to left

CLINICAL IMPLICATIONS

Clinical Importance

- **Recurrent laryngeal nerve** is located in tracheoesophageal groove
  - Though nerve itself is not usually visible on US, its expected course should be carefully examined in patients with vocal cord palsy
- Often invaded/compressed/displaced by extrinsic neck mass, particularly thyroid
  - Carefully evaluate both trachea and esophagus when neck mass is seen
- Primary tracheal tumors are rare

TRACHEA AND ESOPHAGUS

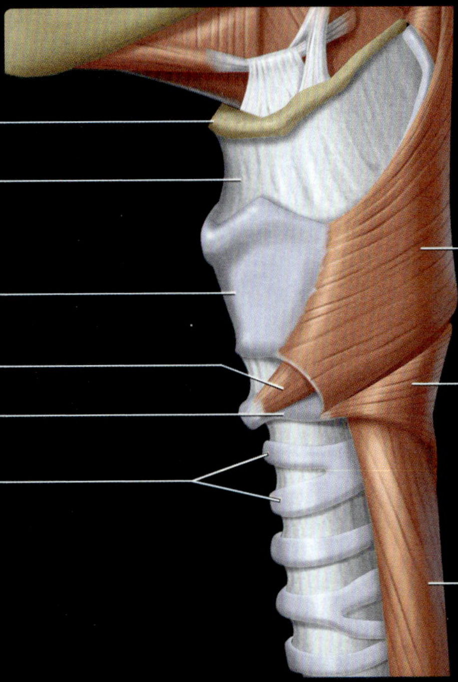

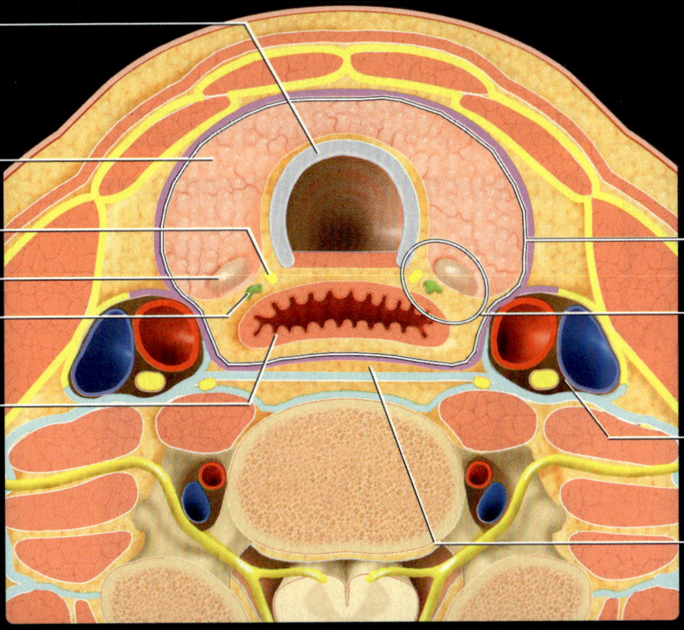

(Top) *Lateral graphic shows junction of larynx and hypopharynx with the trachea and esophagus. The cricopharyngeus muscle is a constrictor muscle and separates the hypopharynx from the cervical esophagus. The esophagus is composed of outer longitudinal muscles and an inner circular muscle layer. The 1st tracheal ring is broadest of all tracheal cartilages and is often merged to cricoid cartilage or 2nd tracheal ring. The rings are incomplete, and the posterior fibromuscular membrane is directly juxtaposed to the esophagus.* **(Bottom)** *Axial graphic shows the anteroposterior relationship of the trachea and esophagus in the lower neck. Important components of the tracheoesophageal groove include the recurrent laryngeal nerve, paratracheal nodes, and parathyroid glands.*

TRANSVERSE ULTRASOUND

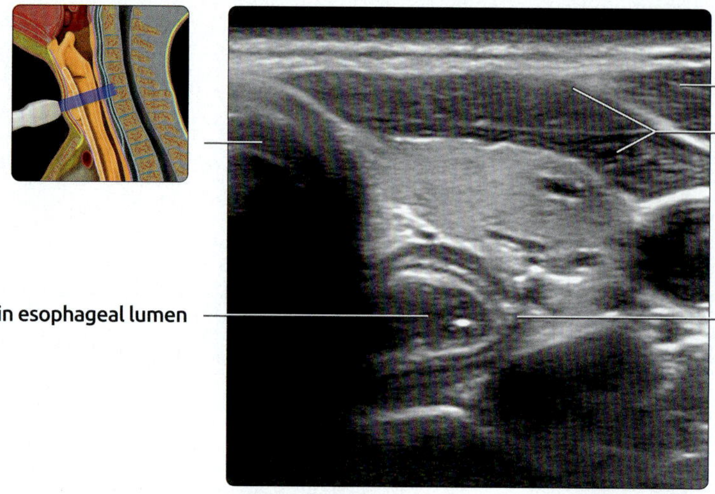

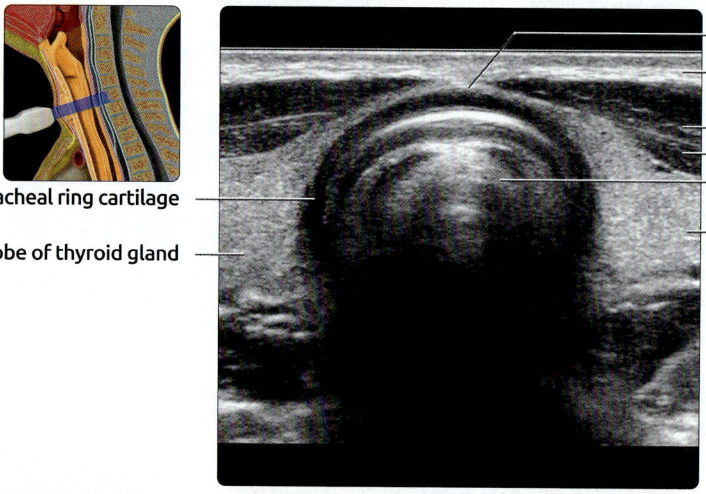

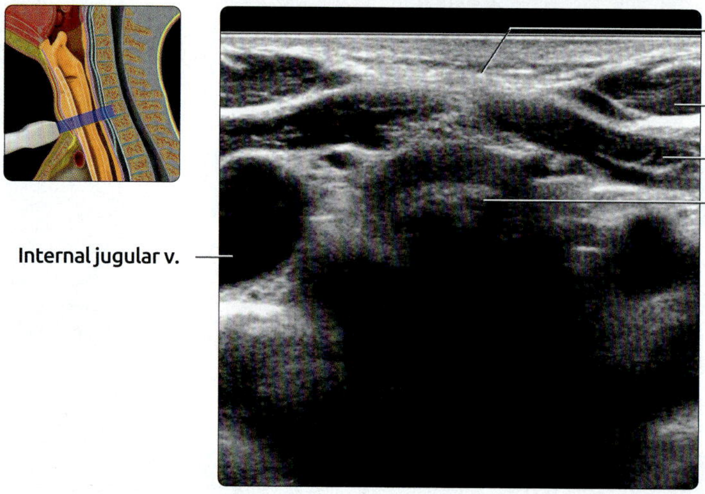

(Top) *Transverse grayscale US of the left lower cervical level shows the location of the cervical esophagus posterior to the left lobe of the thyroid gland and posterolateral to the trachea. It is easily recognized by the alternating hypo-/hyperechoic rings (gut signature). If there is any question, have the patient swallow. The recurrent laryngeal nerve is located in the tracheoesophageal groove. The nerve is not visualized on US.* **(Middle)** *Transverse grayscale US of the midline anterior neck at the level of thyroid gland shows the trachea as a midline structure underneath the isthmus of the thyroid gland and related laterally to the thyroid lobes. Note the hypoechoic tracheal ring composed of hyaline cartilage that is incomplete posteriorly.* **(Bottom)** *Transverse grayscale US of the suprasternal region shows the lower cervical trachea underneath the insertion sites of the strap muscles.*

AXIAL CECT

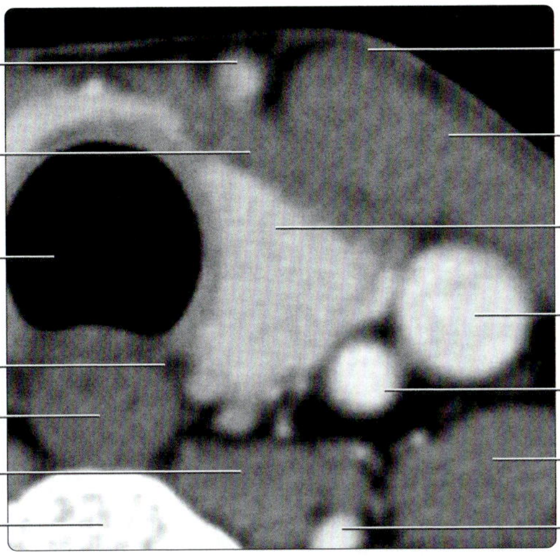

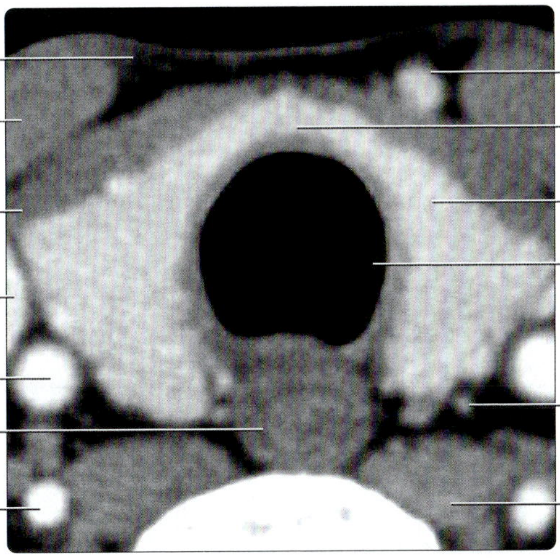

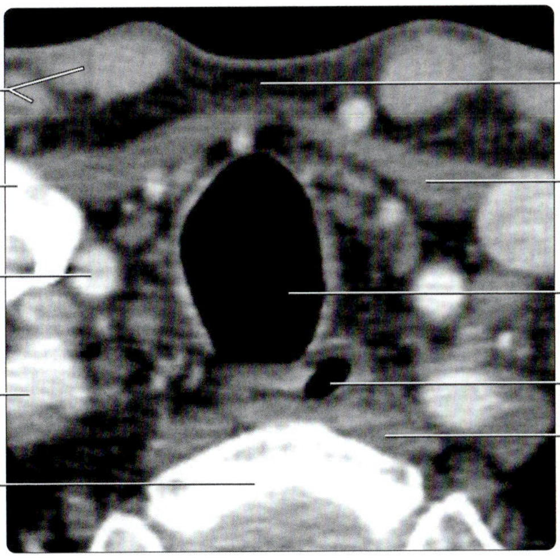

(Top) *Axial CECT of the lower cervical level shows the close anatomical relationship of the trachea and the esophagus with adjacent structures, such as the thyroid gland.* **(Middle)** *Axial CECT of the lower cervical level shows the trachea surrounded by thyroid lobes and the isthmus and esophagus posterior to the trachea.* **(Bottom)** *Axial CECT at the level of the suprasternal region shows that the esophagus at this level is usually slightly off midline to the left in relation to the trachea. The esophagus and trachea are surrounded by mediastinal fat and related to the major vessels in the superior mediastinum. Although US is able to detect the spread of a thyroid tumor to the trachea and esophagus (and vice versa), CT and MR delineate the involvement much better.*

LONGITUDINAL ULTRASOUND AND CT

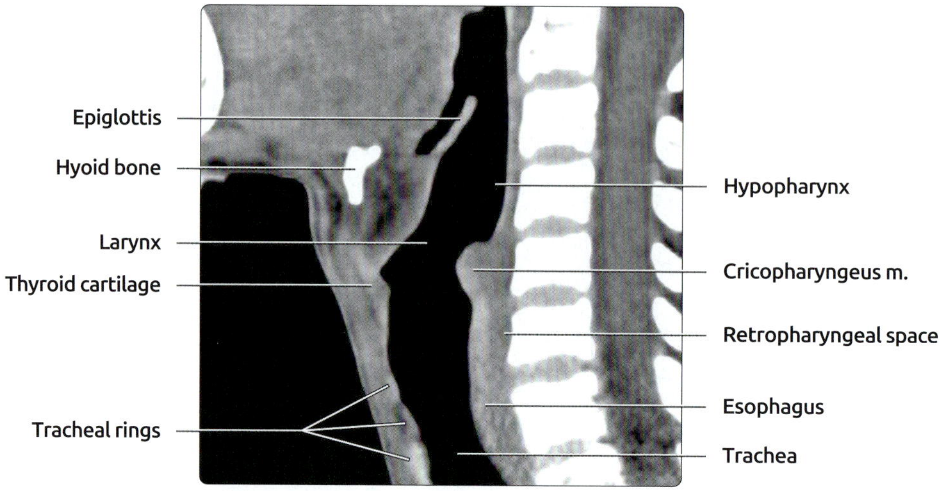

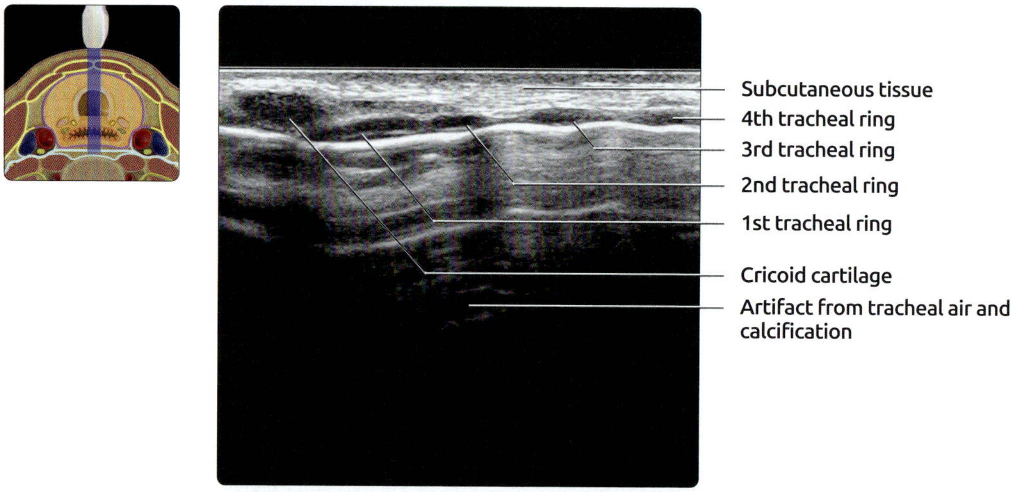

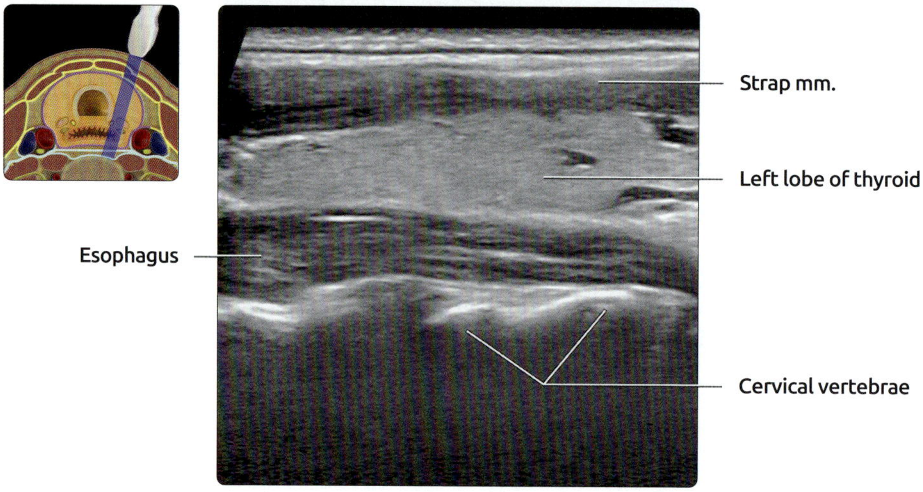

(Top) *Sagittal NECT reformation of the midline infrahyoid neck shows noncalcified tracheal rings anteriorly. These rings form an arch around the trachea and are incomplete posteriorly. The posterior wall is composed of a thick fibromuscular membrane and is immediately adjacent to the esophagus.* **(Middle)** *Longitudinal grayscale US of the midline anterior neck shows the presence of hypoechoic tracheal rings along the cervical portion of the trachea. Note the hypoechoic, noncalcified, cricoid cartilage above the tracheal rings.* **(Bottom)** *Longitudinal grayscale US of the lower left neck at the thyroid gland level shows the cervical esophagus posterior to the left lobe of the thyroid gland. It is a long tubular structure with alternating echogenic/hypoechoic layers representing the mucosal, submucosal, muscular, and serosal layers.*

PATHOLOGY

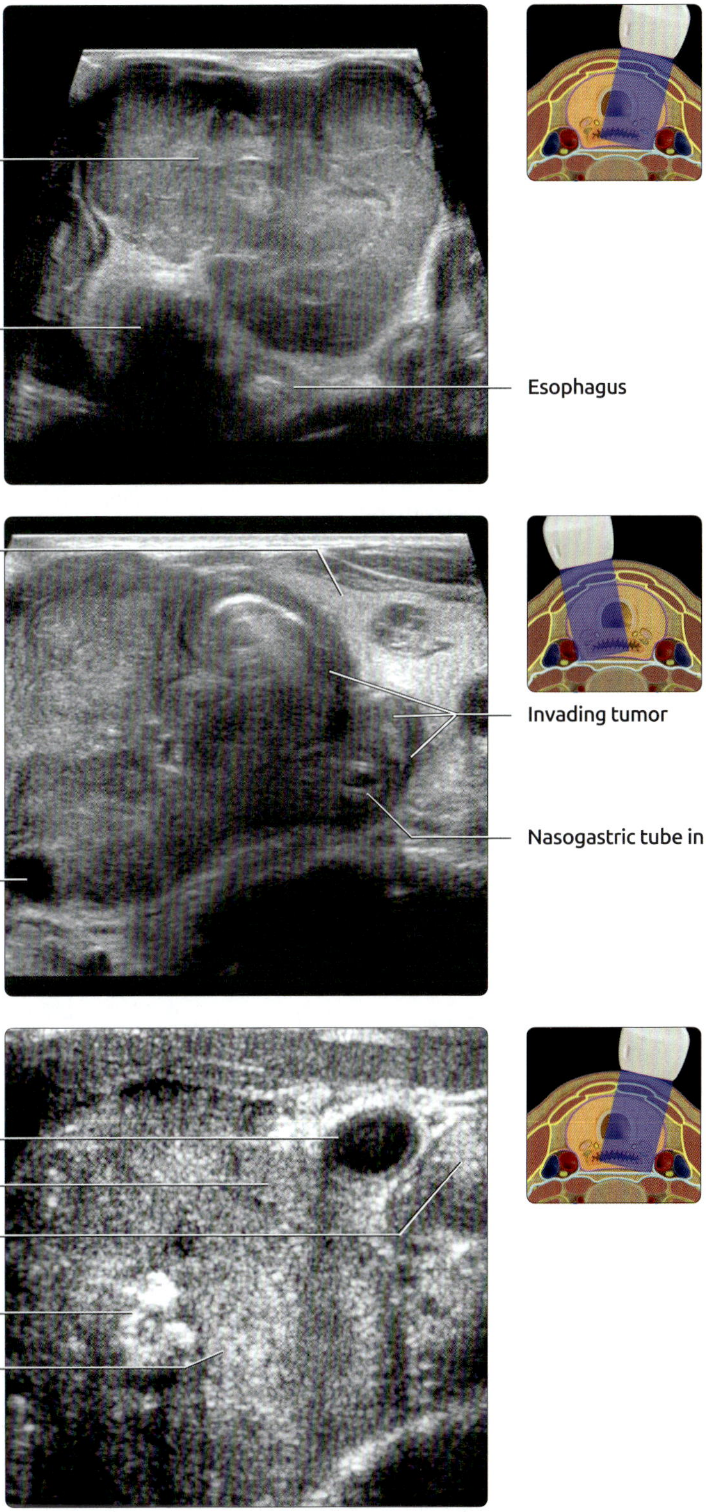

(Top) *Transverse grayscale US in a patient with uterine leiomyosarcoma shows large metastasis to the thyroid. The trachea and esophagus are displaced but not invaded. Always evaluate both of these structures when a neck mass, particularly one involving the thyroid, is identified.* **(Middle)** *Transverse grayscale US shows a diffusely enlarged, hypoechoic right thyroid lobe due to infiltration by a tumor arising from the esophagus. Note the encasement of the right common carotid artery.* **(Bottom)** *Transverse grayscale US of the left lower cervical level shows a large heterogeneous mass posterior to and invading the lower pole of the left thyroid gland, compatible with known locally extensive esophageal carcinoma. Note the presence of left lower jugular lymph node metastases.*

IMAGING ANATOMY

Overview

- 10th cranial nerve (CNX) is mixed (sensory, taste, motor, parasympathetic)
  - Parasympathetic nerve supplying regions of head and neck and thoracic and abdominal viscera
  - Additional vagus nerve (CNX) components
    - Motor to soft palate (except tensor veli palatini muscle), pharyngeal constrictor muscles, larynx, and palatoglossus muscle of tongue
    - Visceral sensation from larynx, esophagus, trachea, and thoracic and abdominal viscera
    - Sensory nerve to external tympanic membrane, external auditory canal (EAC), and external ear
- 4 major segments: Intraaxial, cisternal, skull base, and extracranial
- Intraaxial segment
  - Vagal nuclei are in upper and middle medulla
    - Contain motor, sensory (including taste from epiglottis), and parasympathetic fibers
  - Fibers to and from these nuclei exit lateral medulla in postolivary sulcus inferior to glossopharyngeal nerve (CNIX) and superior to bulbar portion of accessory nerve (CNXI)
- Cisternal segment
  - Exits lateral medulla in postolivary sulcus between CNIX and bulbar portion of CNXI
  - Travels anterolaterally through basal cistern together with CNIX and bulbar portion of CNXI
- Skull base segment
  - Passes through posterior pars vascularis portion of jugular foramen (JF)
    - Accompanied by CNXI and jugular bulb
    - Superior vagal ganglion is found within JF
- Extracranial segment
  - Exits JF into nasopharyngeal carotid space
  - Inferior vagal ganglion lies just below skull base
  - Descends along posterolateral aspect of internal carotid artery (ICA) into thorax
    - Passes anterior to aortic arch on left and subclavian artery on right
  - Forms plexus around esophagus and major blood vessels to heart and lungs
  - Gastric nerves emerge from esophageal plexus and provide parasympathetic innervation to stomach
  - Innervation to intestines and visceral organs follows arterial blood supply to that organ
- Extracranial branches in head and neck
  - Auricular branch (Arnold nerve)
    - Sensation from external surface of tympanic membrane, EAC, and external ear
    - Passes through mastoid canaliculus extending from posterolateral JF to mastoid segment of facial nerve (CNVII) canal
  - Pharyngeal branches
    - Pharyngeal plexus exits just below skull base
    - Sensory to epiglottis, trachea, and esophagus
    - Motor to soft palate and pharyngeal constrictor muscles
  - Superior laryngeal nerve
    - Motor to cricothyroid muscle
    - Sensory to mucosa of supraglottis
- Recurrent laryngeal nerve
  - On right, recurs at cervicothoracic junction, passing posteriorly around subclavian artery
  - On left, recurs in mediastinum, passing posteriorly under aorta at aortopulmonary window
  - Nerves recur in tracheoesophageal grooves
  - Motor to all laryngeal muscles except cricothyroid muscle
  - Sensory to mucosa of infraglottis

ANATOMY IMAGING ISSUES

Imaging Recommendations

- Extracranial segment is only portion accessible for USG evaluation (upper, mid, lower cervical regions)
  - Lies between ICA/common carotid artery (CCA) and internal jugular vein (IJV) on transverse scans
  - Linear hypoechoic structure with central echogenic fibrillar pattern on longitudinal scan
  - On axial scans, seen as round, hypoechoic structure with central echogenic focus
  - Best seen from level of carotid bifurcation down to lower cervical region
  - Color/power Doppler helps to distinguish it from small vessels in vicinity of major vessels of carotid sheath
- More easily visualized on USG in patients with previous radiotherapy
  - Appears diffusely thickened with smooth contour
- USG readily identifies CNX schwannoma in upper, mid, or lower cervical region
  - Appears as round/ovoid solid hypoechoic mass
  - Related to ICA/CCA and IJV
    - No splaying of carotid bifurcation, which occurs in carotid body tumor
  - CNX is contiguous with mass
  - Increased intranodular vascularity on power Doppler
  - USG features obviate need for fine-needle aspiration/biopsy
- Recurrent laryngeal nerve cannot be confidently visualized on USG
  - In patient with vocal cord palsy, USG may help to detect abnormality in tracheoesophageal groove

Imaging Pitfalls

- USG cannot assess intrathoracic portion of CNX
  - CT is imaging modality of choice if CNX lesion in mediastinum is suspected

CLINICAL IMPLICATIONS

Clinical Importance

- CNX dysfunction
  - Proximal symptom complex (lesion between medulla and hyoid bone)
    - Multiple cranial nerves involved (CNIX-CNXII) with oropharyngeal and laryngeal dysfunction
  - Distal symptom complex (lesion below hyoid bone)
    - Isolated CNX involvement with laryngeal dysfunction only

VAGUS NERVE

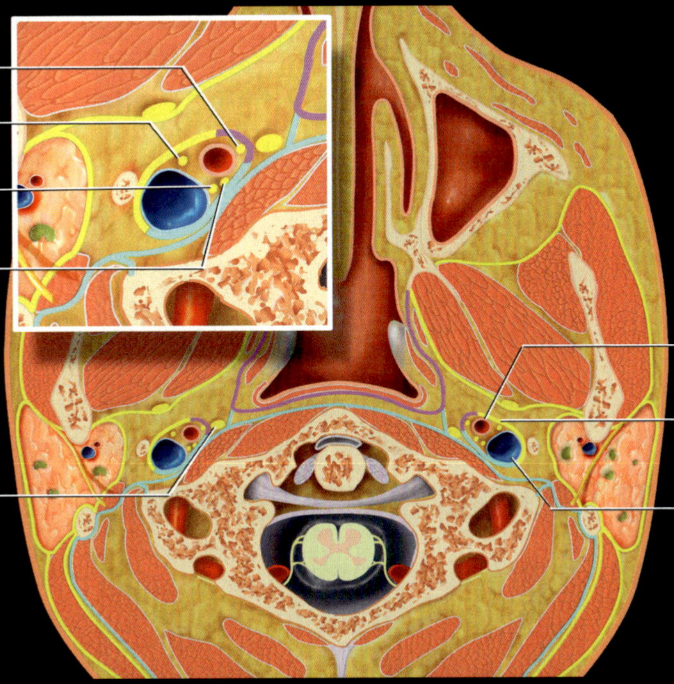

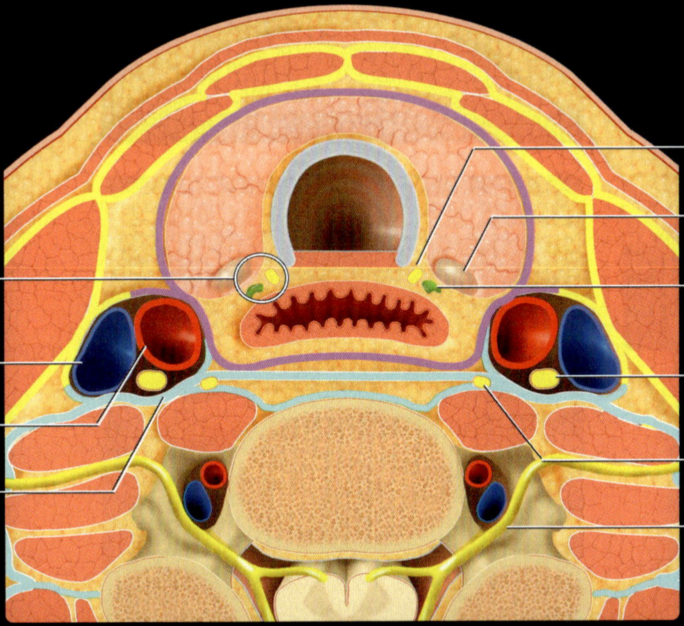

(Top) *Axial graphic of nasopharyngeal carotid spaces shows the extracranial vagus nerve situated posteriorly in the gap between the internal carotid artery and the internal jugular vein. Note that at this level the glossopharyngeal nerve (CNIX), accessory nerve (CNXI), and hypoglossal nerve (CNXII) are all still within the carotid space. This site is not accessible on ultrasound.* **(Bottom)** *Axial graphic through the infrahyoid carotid spaces at the level of the thyroid gland demonstrates that the vagus trunk is the only remaining cranial nerve within the carotid space. It remains in the posterior gap between the common carotid artery and the internal jugular vein. Note the recurrent laryngeal nerve in the tracheoesophageal groove within the visceral space. Remember that the left recurrent laryngeal nerve turns cephalad in the aortopulmonary window in the mediastinum, whereas the right recurrent nerve turns at the cervicothoracic junction around the subclavian artery.*

TRANSVERSE, POWER DOPPLER, AND LONGITUDINAL ULTRASOUNDS

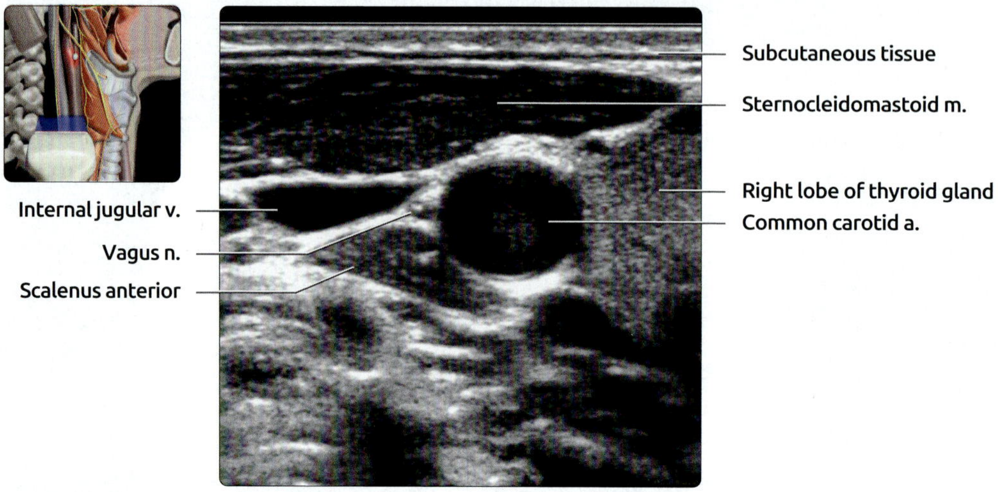

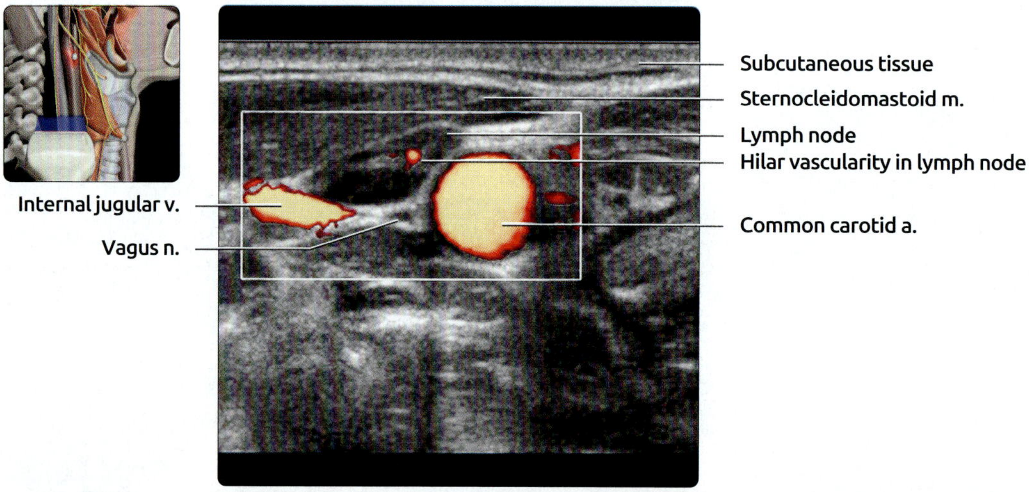

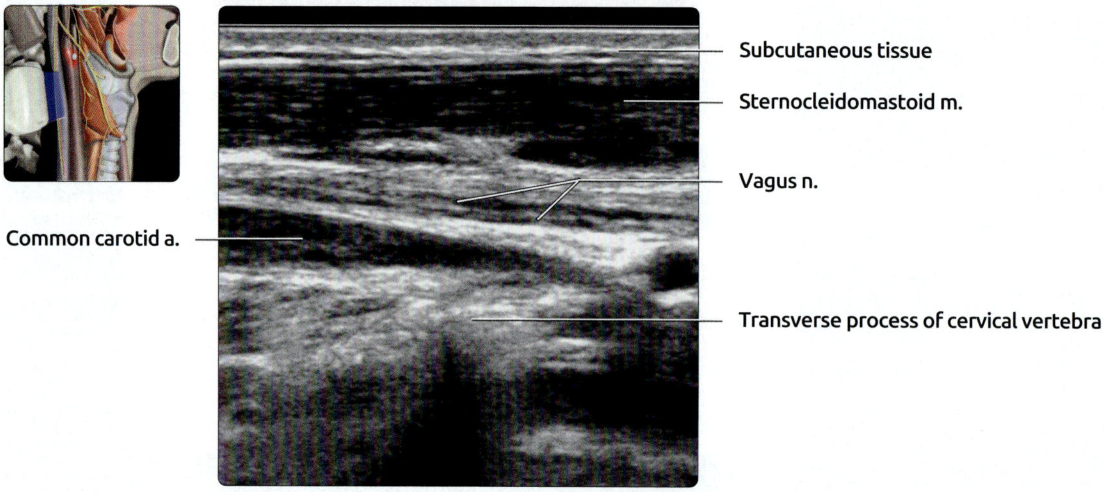

(Top) *Transverse grayscale ultrasound of the lower cervical level at the thyroid gland level shows the vagus nerve as a small, round, hypoechoic structure that exhibits central echogenicity within the carotid sheath and is located between the common carotid artery and the internal jugular vein.* **(Middle)** *Power Doppler ultrasound of the midcervical level in the transverse plane demonstrates the avascular nature of the vagus nerve adjacent to the common carotid artery and internal jugular vein. Note the presence of hilar vascularity in the adjacent normal deep cervical lymph node.* **(Bottom)** *Longitudinal grayscale ultrasound shows the vagus nerve, which appears as a long, thin, tubular, hypoechoic structure with a central echogenic fibrillary pattern within. On ultrasound, the vagus nerve is readily seen from the carotid bifurcation to the lower cervical region.*

AXIAL AND CORONAL CECT

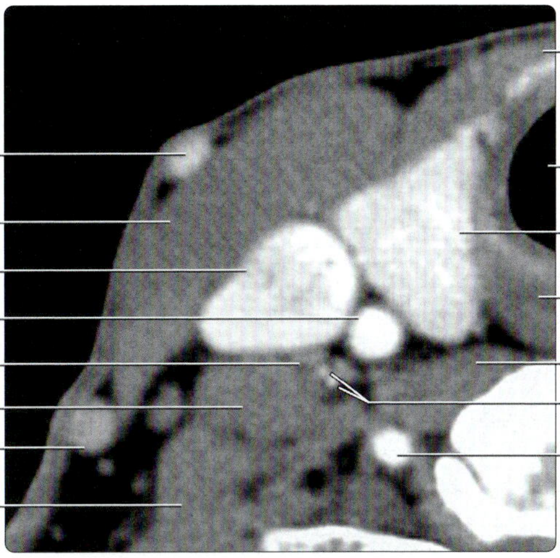

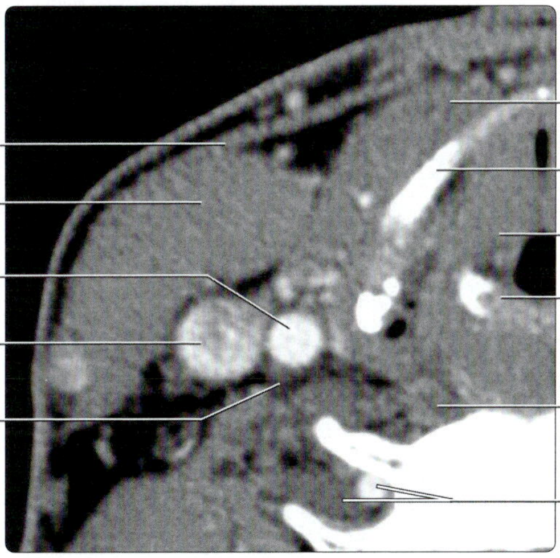

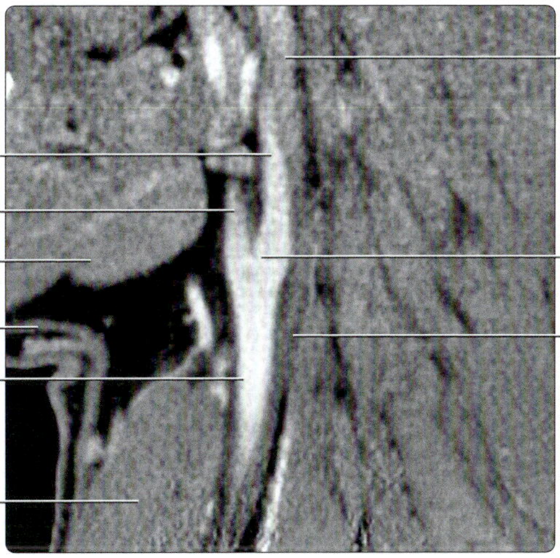

(Top) *Axial CECT of the neck at the midcervical level shows the vagus nerve as an isodense dot in the posterior aspect of the carotid sheath. The inferior thyroid arteries are seen as contrast-enhanced dots in its proximity.* **(Middle)** *Axial CECT of the neck in a different patient also shows the vagus nerve as an isodense dot in the posterior aspect of the carotid space.* **(Bottom)** *Oblique sagittal reformatted CECT of the neck shows the course of the vagus nerve. It is closely related to the posterior aspect of the common carotid artery. Although CT demonstrates the vagus nerve in the neck, high-resolution ultrasound clearly evaluates its internal architecture.*

PATHOLOGY

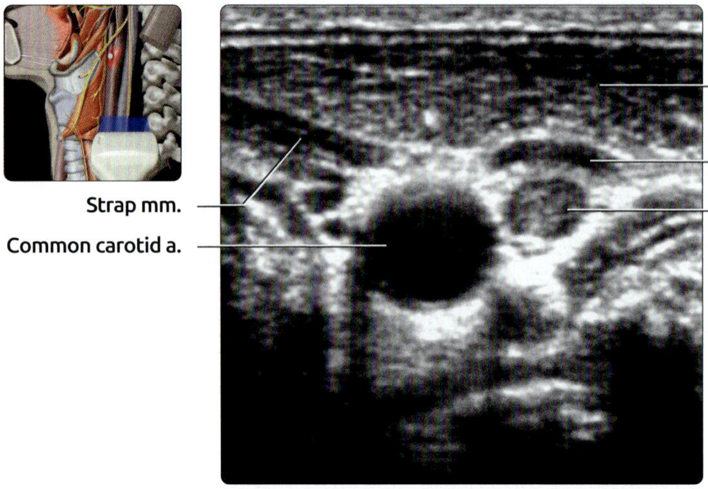

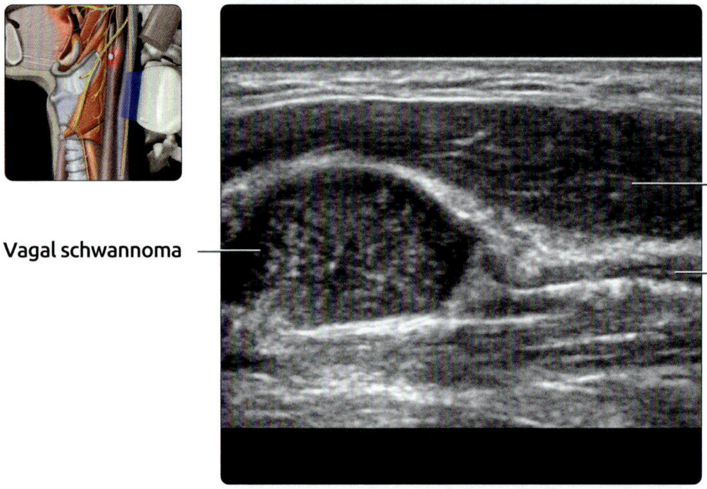

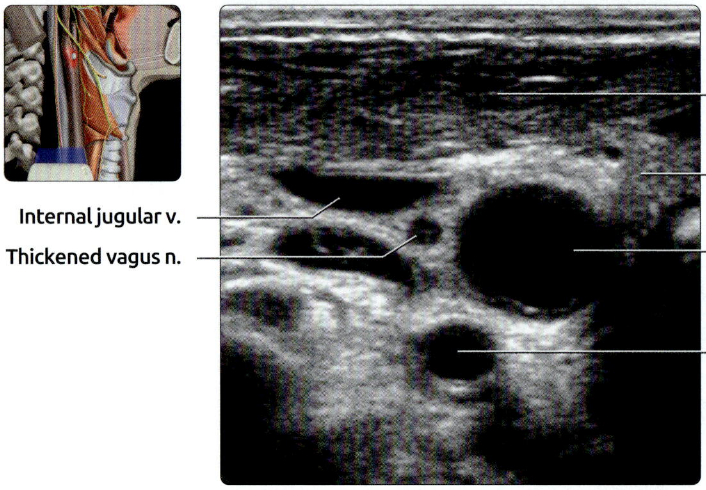

(Top) *Transverse grayscale ultrasound of left midcervical level shows a well-circumscribed, hypoechoic mass closely related to the left internal jugular vein and common carotid artery. The anatomical location helps to identify the mass as originating from the vagus nerve.* **(Middle)** *Longitudinal grayscale ultrasound shows a oblong, solid, hypoechoic masses contiguous with the vagus nerve inferiorly. Scanning in a longitudinal plane is best for proving a mass is arising from the nerve.* **(Bottom)** *Transverse grayscale ultrasound of the lower cervical level in a patient with previous neck irradiation for head and neck cancer shows that the vagus nerve is diffusely thickened with smooth contour as a result of postirradiation change.*

PATHOLOGY

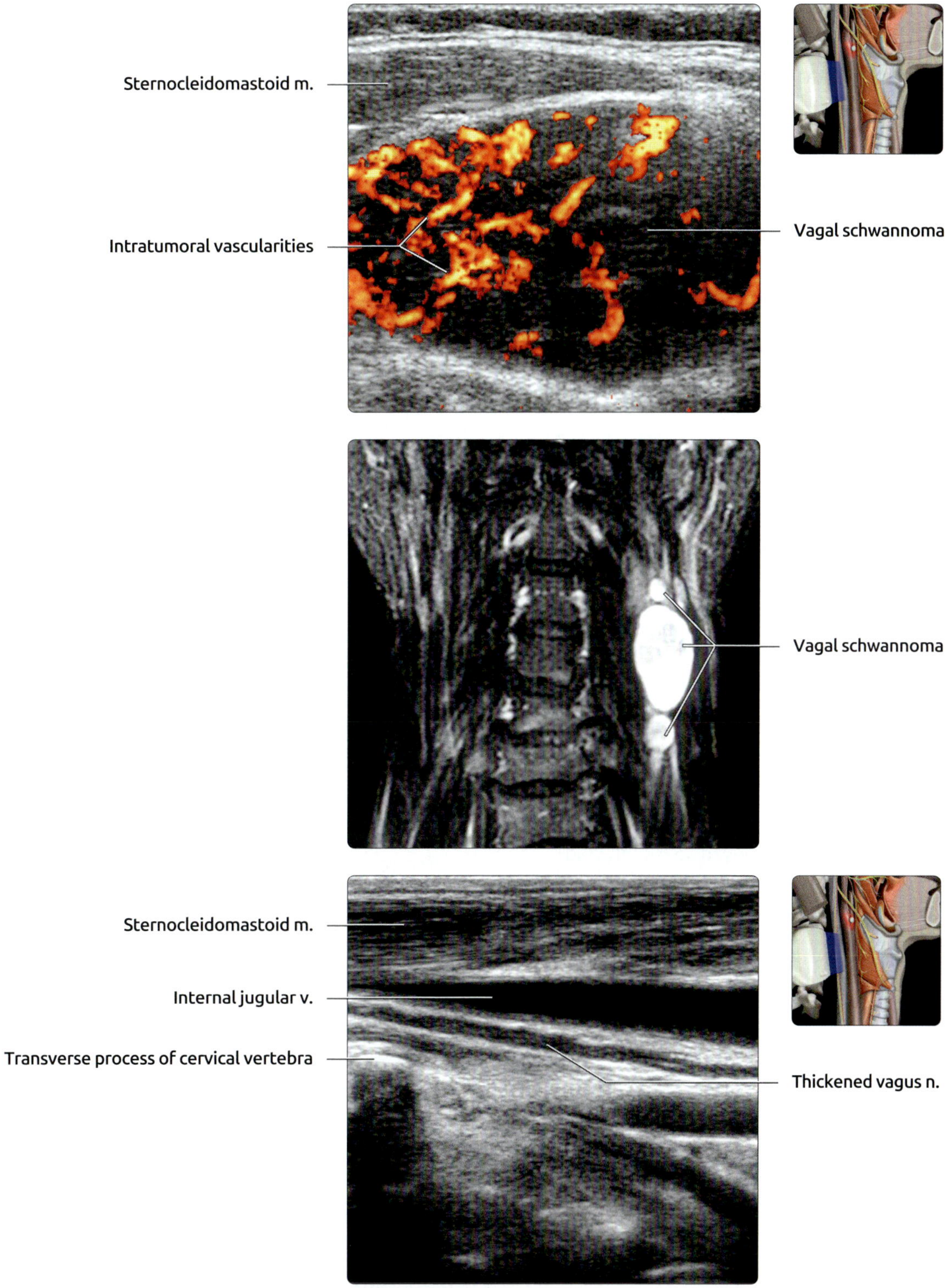

(Top) *Longitudinal power Doppler ultrasound of a vagal schwannoma reveals marked increased intratumoral vascularity. Ultrasound readily identifies a vagal nerve schwannoma and obviates the need for fine-needle aspiration cytology or biopsy.* **(Middle)** *Coronal fat-suppressed T2WI MR shows marked T2 hyperintensity of vagal schwannoma.* **(Bottom)** *Longitudinal grayscale ultrasound shows the diffusely thickened vagus nerve in relation to the internal jugular vein.*

GROSS ANATOMY

Overview

- Common carotid artery (CCA) divides into external carotid artery (ECA) and internal carotid artery (ICA)
- ECA is smaller of 2 terminal branches
  - Supplies most of head and neck (except eye, brain)
  - Numerous anastomoses with ICA and vertebral artery; important source of collateral blood flow
- ICA has no normal extracranial branches

IMAGING ANATOMY

Common Carotid Artery

- Right CCA originates from brachiocephalic trunk; left CCA originates from aortic arch
- Courses superiorly in carotid space, anteromedial to internal jugular vein
- Divides into ECA and ICA at ~ C3-C4 level

Cervical Internal Carotid Artery

- 90% posterolateral to ECA
- Carotid bulb
  - Focal dilatation of ICA at its origin from CCA
  - Flow reversal occurs in carotid bulb
- Ascending cervical segment
  - Courses superiorly within carotid space
  - Enters carotid canal of skull base (petrous temporal bone)
  - No named branch in neck

External Carotid Artery

- Smaller and medial compared with ICA
- Has 8 major branches in neck
- **Superior thyroid artery**
  - 1st ECA branch (may arise from CCA bifurcation)
  - Arises anteriorly and courses inferiorly to apex of thyroid
  - Supplies superior thyroid and larynx
  - Anastomoses with inferior thyroid artery (branch of thyrocervical trunk)
- **Ascending pharyngeal artery**
  - Arises from posterior ECA (or CCA bifurcation)
  - Courses superiorly between ECA and ICA
  - Visceral branches, muscular branches, and neuromeningeal branches
- **Lingual artery**
  - 2nd anterior ECA branch
  - Loops anteroinferiorly, then superiorly to tongue
  - Major vascular supply to tongue, oral cavity, and submandibular gland
- **Facial artery**
  - Originates just above lingual artery
  - Curves around mandible, then passes anterosuperiorly across cheek, closely related to submandibular gland
  - Supplies face, palate, lip, and cheek
- **Occipital artery**
  - Originates from posterior aspect of ECA
  - Courses posterosuperiorly between occiput and C1
  - Supplies scalp, upper cervical musculature, and posterior fossa meninges
- **Posterior auricular artery**
  - Arises from posterior ECA above occipital artery
  - Courses superiorly to supply pinna, scalp, external auditory canal, and chorda tympani
- **Superficial temporal artery**
  - Smaller of 2 terminal ECA branches
  - Runs superiorly behind mandibular condyle, across zygoma
  - Supplies scalp and gives off transverse facial artery
- **Maxillary artery**
  - Larger of 2 terminal ECA branches
  - Arises within parotid gland, behind mandibular neck
  - Gives off middle meningeal artery (supplies cranial meninges)

ANATOMY IMAGING ISSUES

Imaging Recommendations

- Normal US appearances of carotid arteries
  - CCA diameter: 6.3 ± 0.9 mm, smooth and thin intima, antegrade low-resistance arterial flow
  - ICA diameter: 4.8 ± 0.7 mm, smooth and thin intima, antegrade low-resistance flow
  - ECA diameter: 4.1 ± 0.6 mm, smooth and thin intima, antegrade high-resistance flow
- In assessing carotid arteries on US, following parameters should be examined
  - Intimal-medial thickness
    - Distance between leading edges of lumen-intima interface and media-adventitia interface at far edge
    - 0.5 - 1.0 mm in healthy adults
  - Presence of atherosclerotic plaques
    - Eccentric/concentric, noncircumferential/circumferential
    - Calcified plaque/soft plaque
  - Luminal diameter/area reduction
    - Should be measured on true cross-sectional view of affected artery
    - Color flow imaging helps to detect residual lumen in tight stenosis or in assessing indeterminate total occlusion
  - Spectral Doppler analysis
    - Arterial flow pattern: Low-resistance/high-resistance flow, antegrade/retrograde flow, special waveform (e.g., damped waveform, preocclusive "thump")
    - Peak systolic velocity measurement
    - Systolic velocity ratio measurement

Imaging Pitfalls

- Scanning technique must be meticulous to produce reliable Doppler US results
- Obliquity of imaging plane in relation to cross section of artery may wrongly estimate degree of stenosis

CLINICAL IMPLICATIONS

Clinical Importance

- Doppler parameters give physiologic assessment of degree of stenosis
- Consider acute idiopathic carotidynia for tender mass around distal carotid, near bifurcation
  - Vessel wall thickening, no luminal narrowing or velocity elevation

NORMAL ARTERIAL ANATOMY

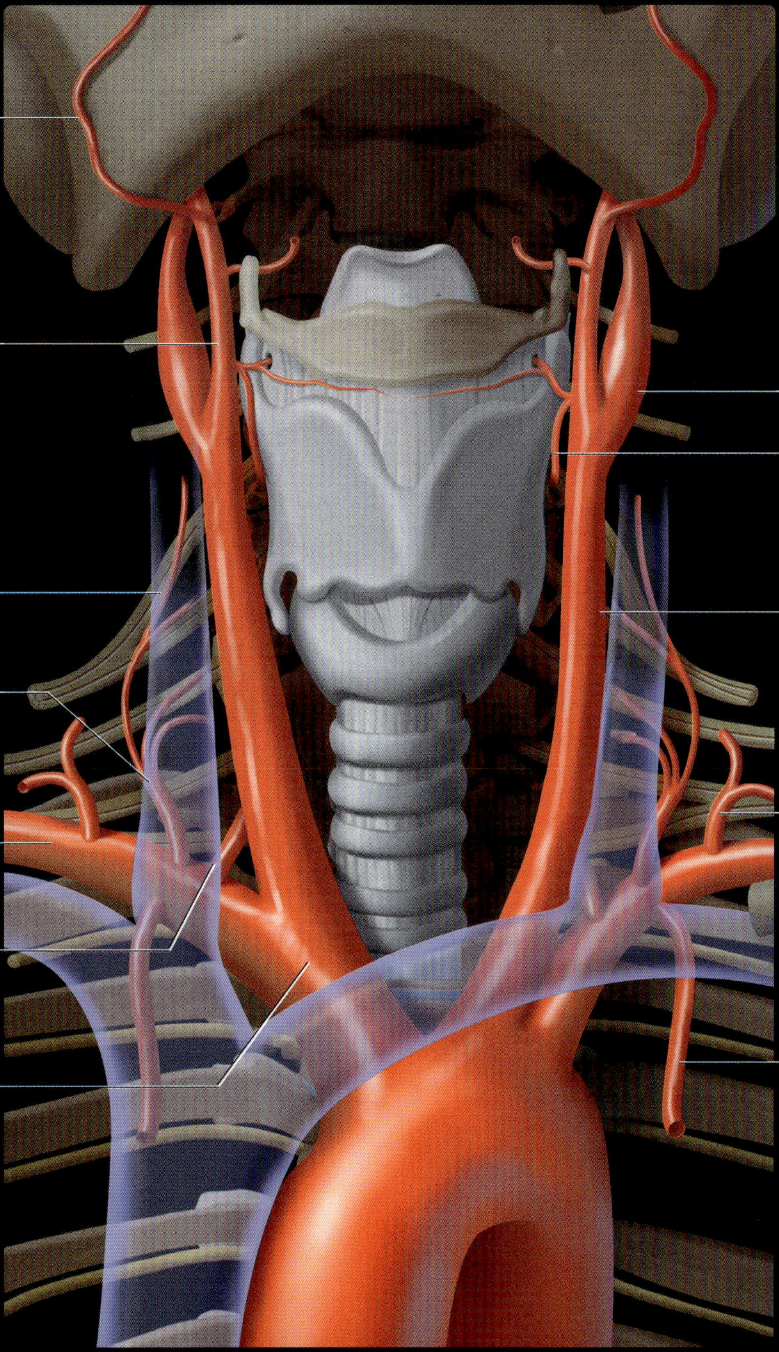

Anterior graphic of the neck, with the veins ghosted, shows the normal arterial anatomy. The right common carotid artery (CCA) arises from the brachiocephalic artery, while the left common carotid artery arises from the aorta. They ascend medial to the jugular veins and bifurcate into the internal (ICA) and external (ECA) carotid arteries at the level of the upper thyroid cartilage. The ECAs supply most of the head and neck, while the ICAs supply the brain and eye. The ICAs have no extracranial branches.

AORTIC ARCH

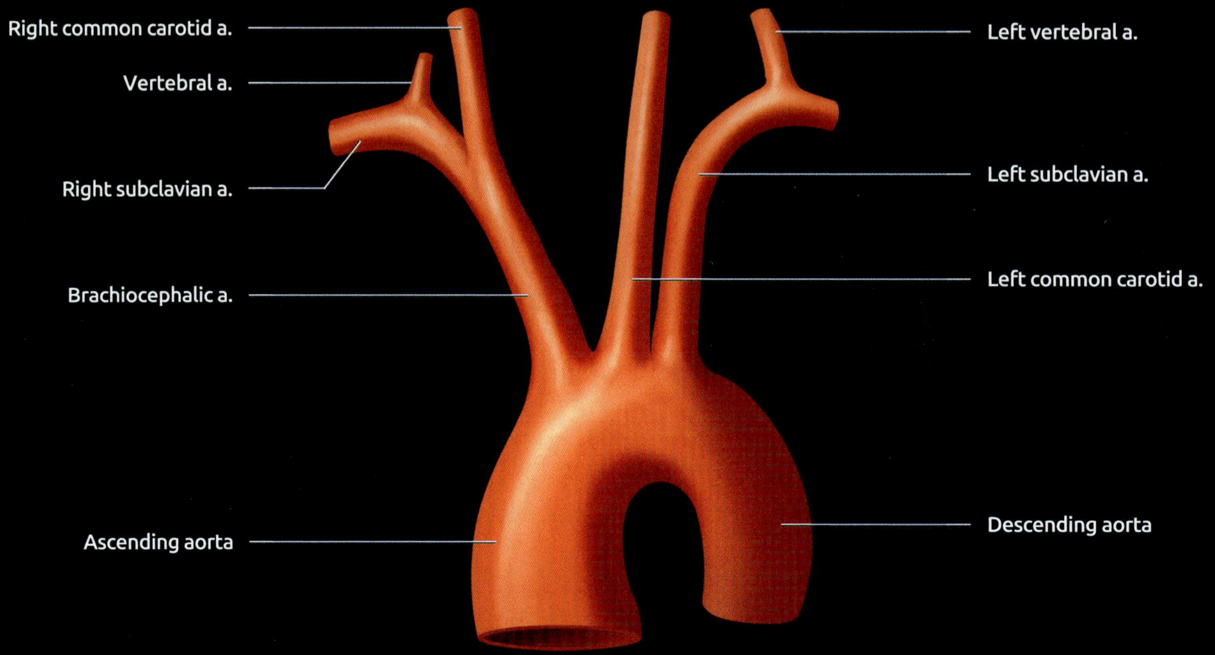

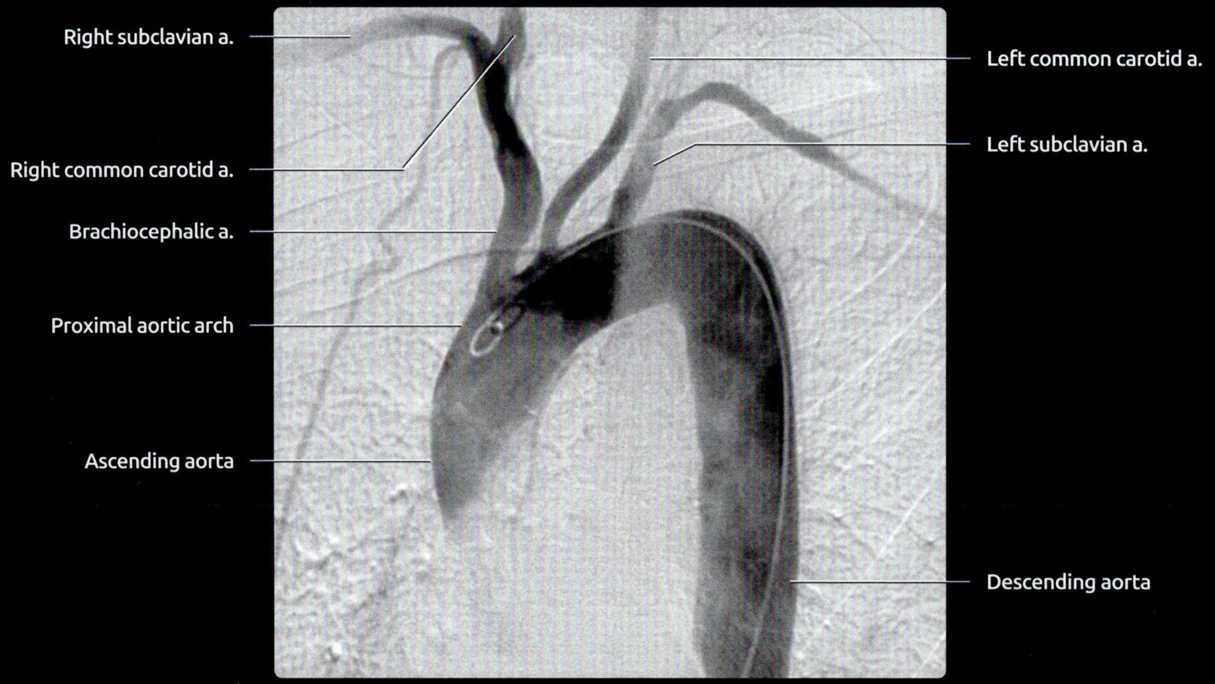

(Top) *Graphic shows the classic branching pattern of the 3 great arteries from the aortic arch: The brachiocephalic, left common carotid, and left subclavian arteries. This is seen in ~ 80% of cases. The brachiocephalic and the left CCA may have a common origin (10-20% of cases).* **(Bottom)** *Left anterior oblique thoracic aortogram demonstrates the anatomy of the aortic arch and its branches. Contrast injected into the distal ascending aorta opacifies the aortic arch and its branches. The 1st branch is the brachiocephalic artery followed by the left common carotid and left subclavian arteries. The irregular contour of the descending aorta and the luminal narrowing of the left subclavian artery are secondary to atherosclerosis.*

COMMON CAROTID ARTERY

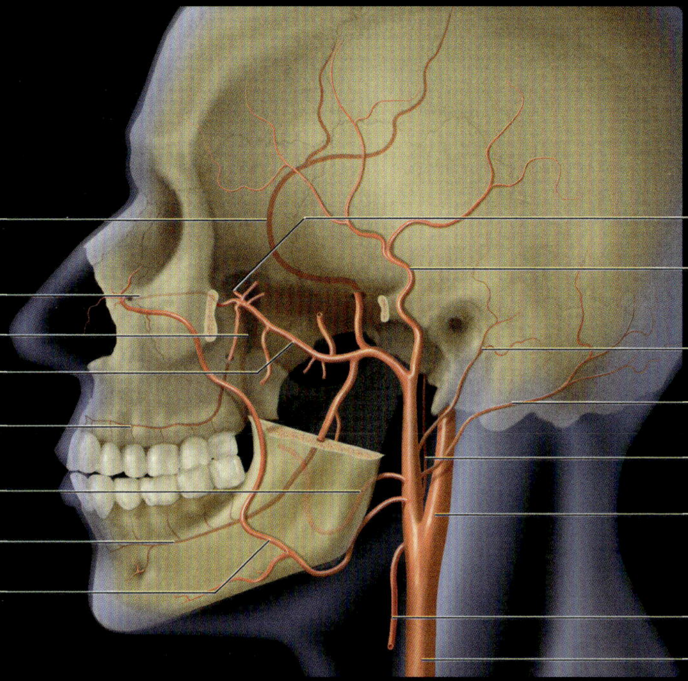

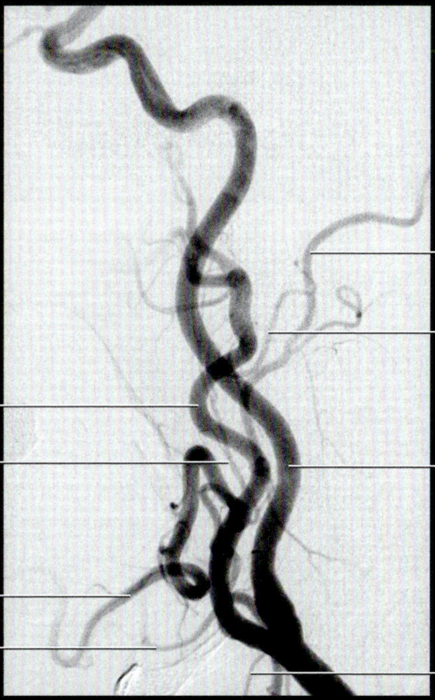

(Top) *Lateral graphic depicts the CCA and its 2 terminal branches, the ECA and internal ICA. The scalp and the superficial facial structures are removed to show the deep ECA branches. The ECA terminates by dividing into the superficial temporal and internal maxillary arteries (IMA). Within the pterygopalatine fossa, the IMA divides into numerous deep branches. Its distal termination is the sphenopalatine artery, which passes medially into the nasal cavity. Numerous anastomoses between the ECA branches (e.g., between the facial and maxillary arteries) and between the ECA and the orbital and cavernous branches of the ICA provide potential sources for collateral blood flow.* **(Bottom)** *The early arterial phase of CCA angiogram is shown with bony structures subtracted. The major ECA branches are opacified.*

CAROTID BIFURCATION

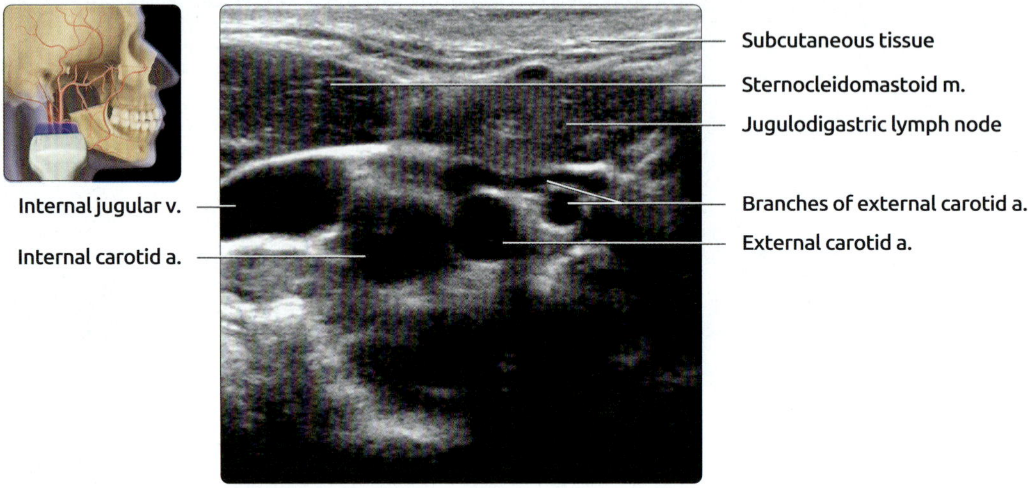

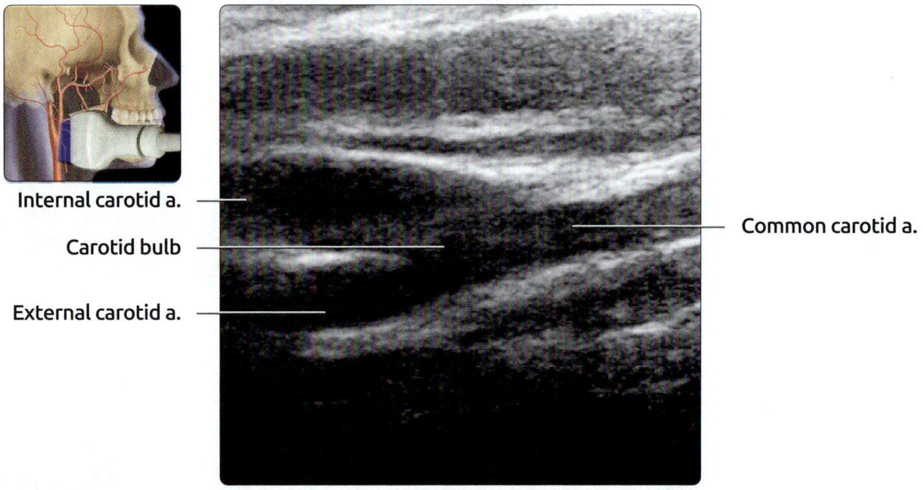

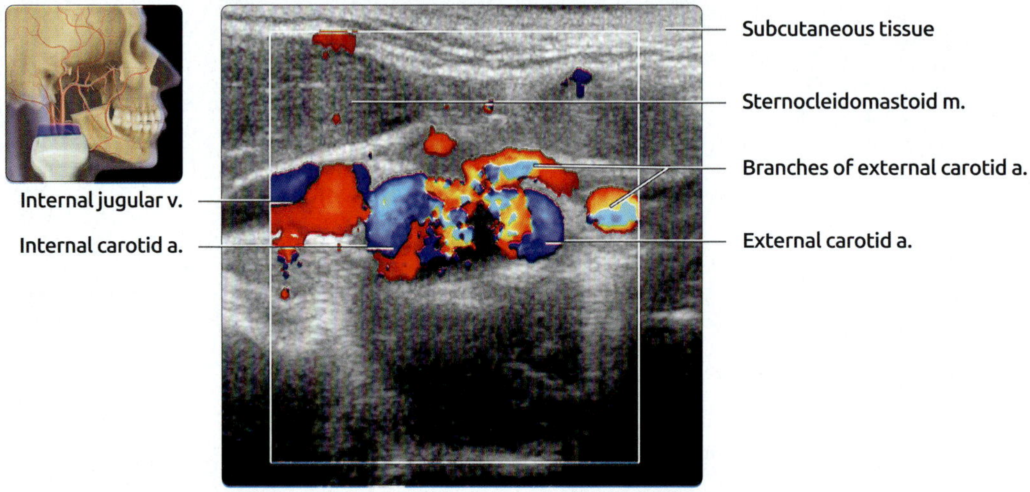

(Top) *Transverse grayscale ultrasound shows the upper cervical level at the carotid bifurcation. The CCA bifurcates into the ICA and ECA. The former is usually of larger caliber, laterally located, and has no branches in the neck.* **(Middle)** *Longitudinal grayscale ultrasound in coronal orientation demonstrates the carotid bifurcation. The proximal portion of the ICA is usually mildly dilated and is termed carotid bulb. At this site, the color/spectral Doppler study is more complex due to a disturbance of laminar flow and should not be misinterpreted as an abnormality.* **(Bottom)** *Color Doppler ultrasound of carotid bifurcation in a transverse plane demonstrates turbulent flow in the carotid bulb. Branches of ECA are easier to depict than on grayscale examination.*

CAROTID BIFURCATION

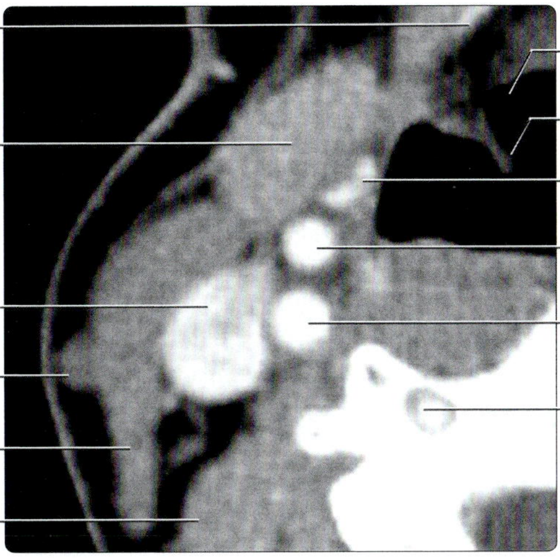

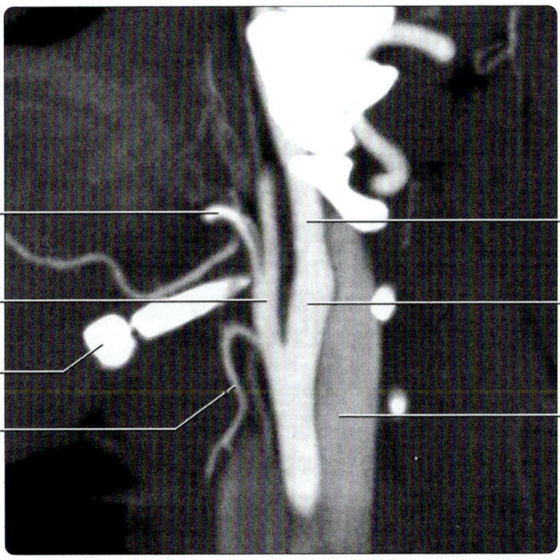

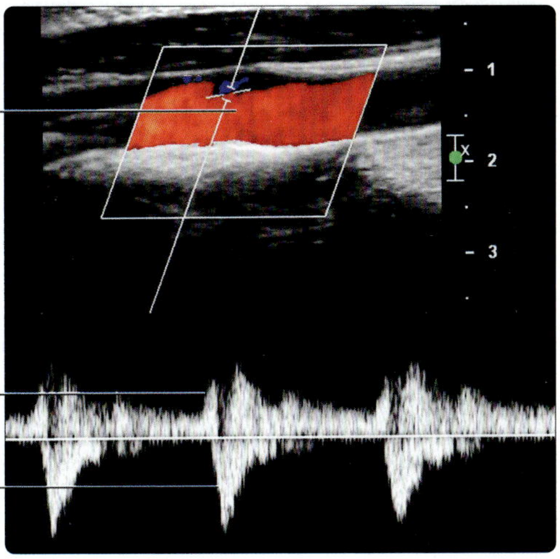

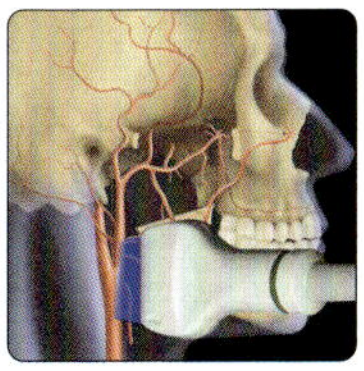

(Top) *Axial CECT shows the upper neck just beyond the carotid bifurcation. The ICA is larger and more posterolateral in position than the ECA.* **(Middle)** *Maximum-intensity projection in the sagittal plane demonstrates carotid bifurcation into the ECA and ICA at the level of the hyoid bone. Note the branching nature of the ECA in contrast to the ICA.* **(Bottom)** *Spectral Doppler ultrasound shows the carotid bulb, which has a different flow pattern than the rest of the ICA. In early systole, blood flow is accelerated in a forward direction. As the peak systole is approached, a large separation zone with flow reversal develops. Flow separation should be seen in normal individuals, and its absence should raise the suspicion of plaque formation.*

INTERNAL CAROTID ARTERY

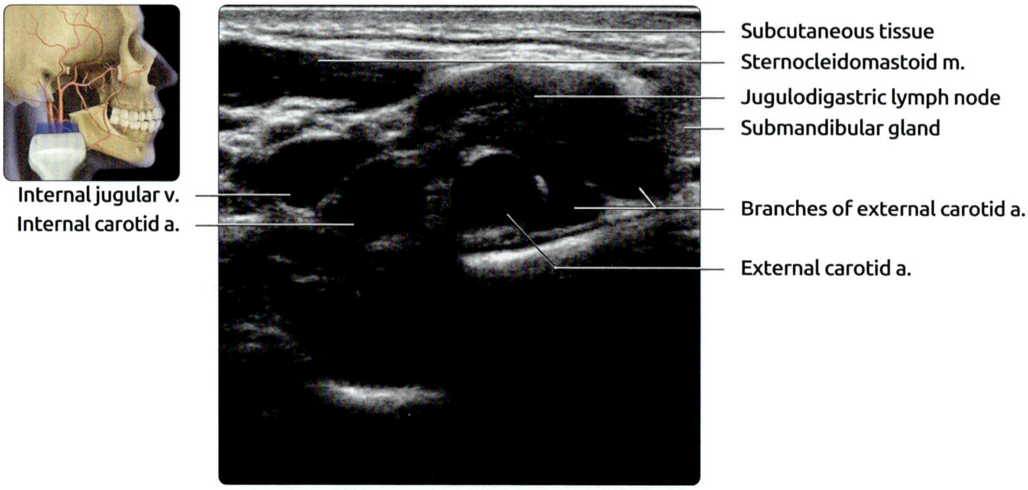

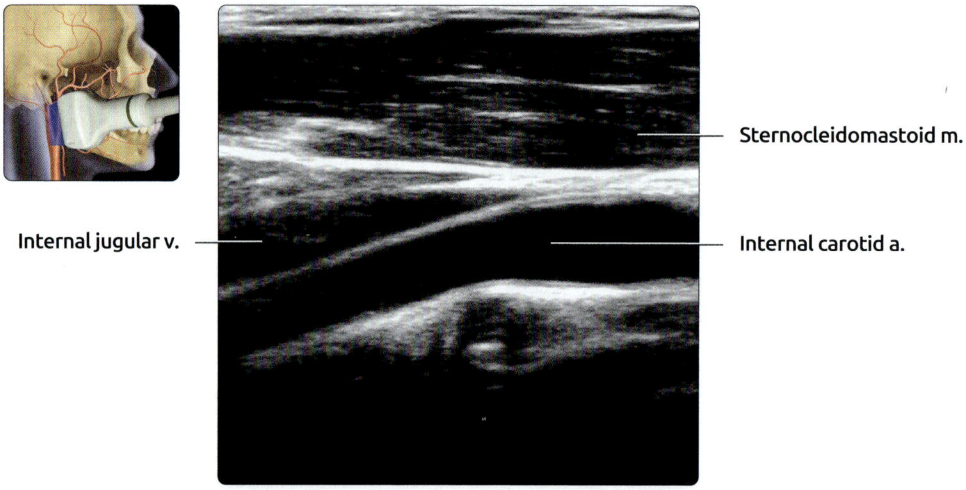

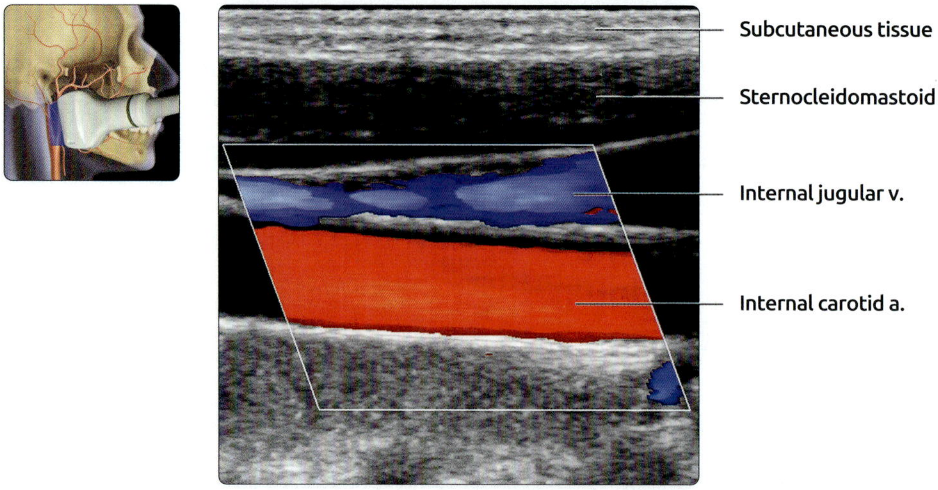

(Top) *Transverse grayscale ultrasound of the upper neck, just beyond the carotid bifurcation, shows the close anatomical relationship of the upper neck with the internal jugular vein, ECA, and jugulodigastric lymph node.* **(Middle)** *Longitudinal grayscale ultrasound shows the ICA. Note its smooth wall with no intimal thickening; it is free of atherosclerotic plaque in a normal individual. No branch is seen in the cervical region.* **(Bottom)** *Color Doppler ultrasound shows the ICA and the internal jugular vein in the longitudinal plane. Note that the normal antegrade flow is toward the cranial direction of the ICA and opposite the caudal direction of flow in the adjacent internal jugular vein.*

INTERNAL CAROTID ARTERY

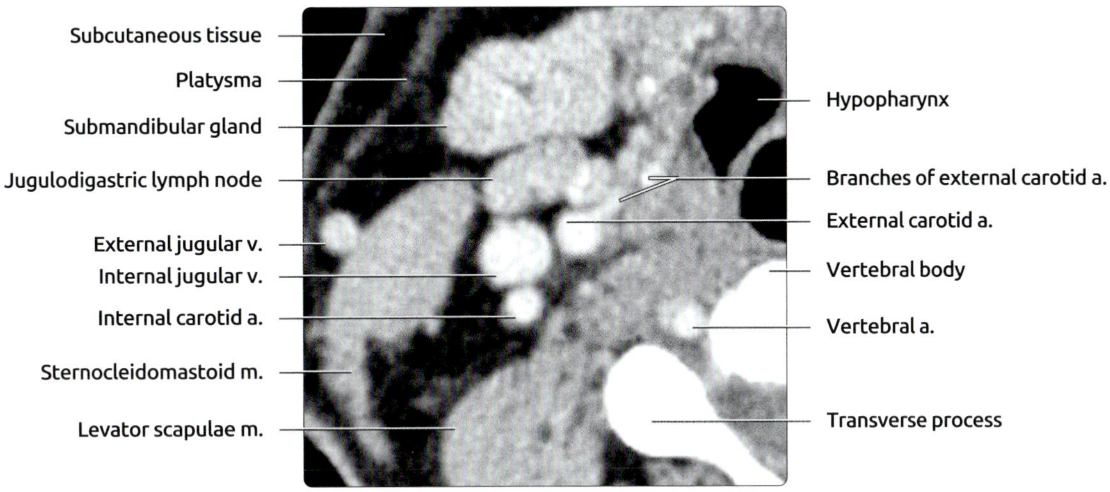

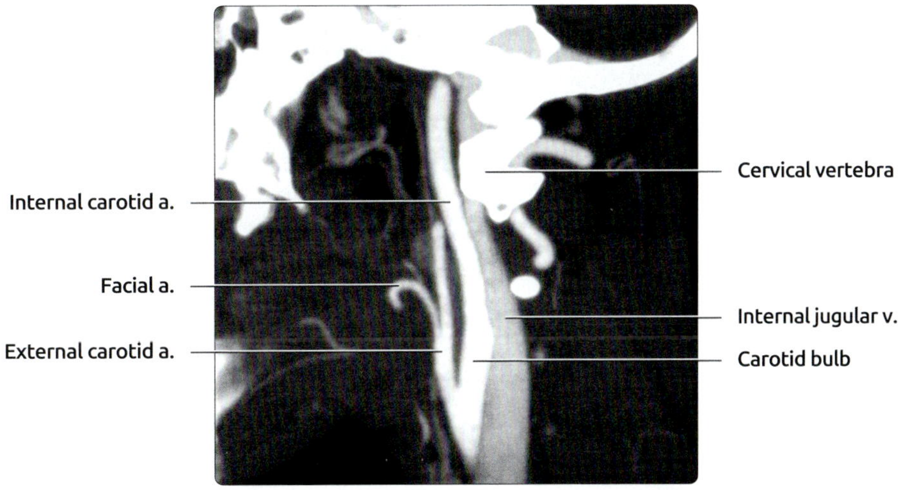

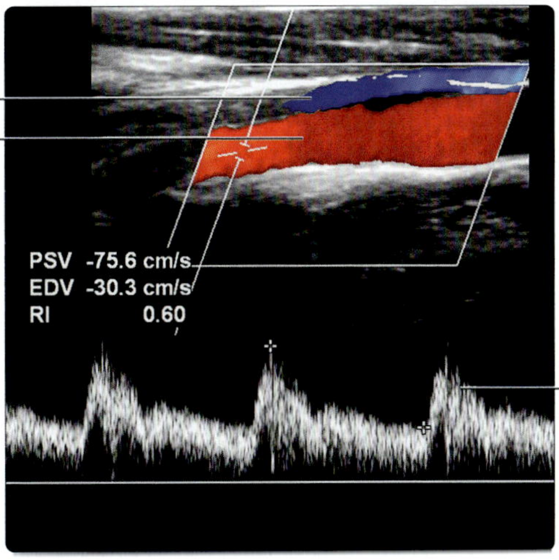

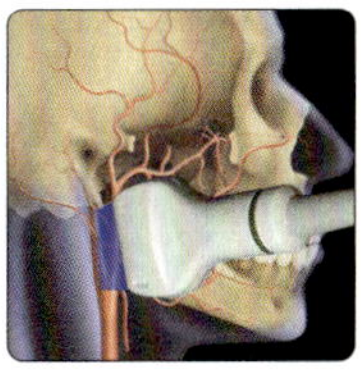

(Top) *Axial CECT of the upper neck shows the anatomical relation of the ICA with the internal jugular vein and branches of the ECA.* **(Middle)** *Maximum-intensity projection CECT in the sagittal plane shows the normal configuration and contour of the cervical portion of the ICA. Note the lack of an arterial branch from the cervical ICA in the neck in contrast to the ECA. Note the mild dilatation of the ICA at its origin (carotid bulb).* **(Bottom)** *Spectral Doppler ultrasound in the longitudinal plane shows the cervical portion of the ICA, which has a low-resistance flow pattern with antegrade flow in the diastolic phase. The waveform is different from that of the carotid bulb.*

EXTERNAL CAROTID ARTERY

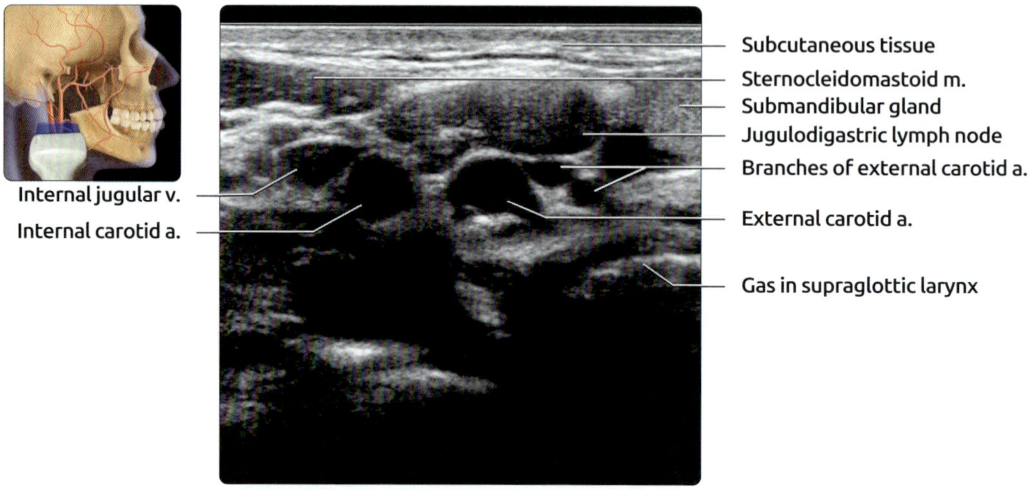

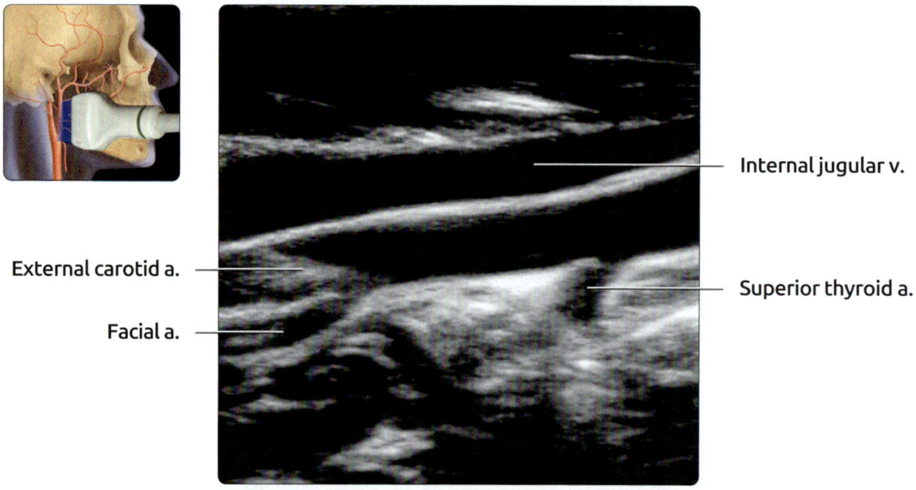

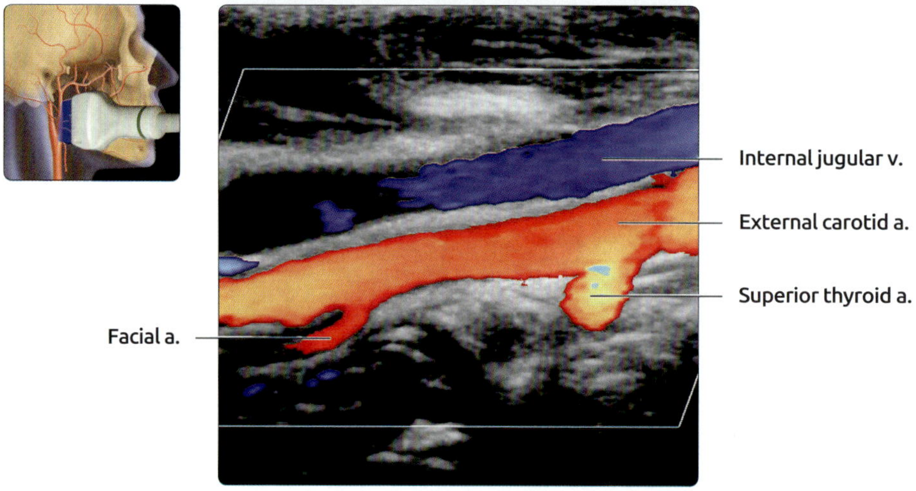

(Top) *Transverse grayscale ultrasound shows the upper neck above the carotid bifurcation. The position of the ECA medial to the ICA and internal jugular vein, posterior to the jugulodigastric lymph node, is well demonstrated.* **(Middle)** *Longitudinal grayscale ultrasound shows the ECA. Two anterior branches, the superior thyroid artery and the facial artery, are seen arising from the proximal portion of the ECA. They course inferiorly to the upper pole of the thyroid and superiorly to the facial region.* **(Bottom)** *Color Doppler ultrasound shows the ECA in the longitudinal plane. The antegrade flow in the cranial direction of the ECA is demonstrated. Note the opposite flow direction of the adjacent internal jugular vein.*

EXTERNAL CAROTID ARTERY

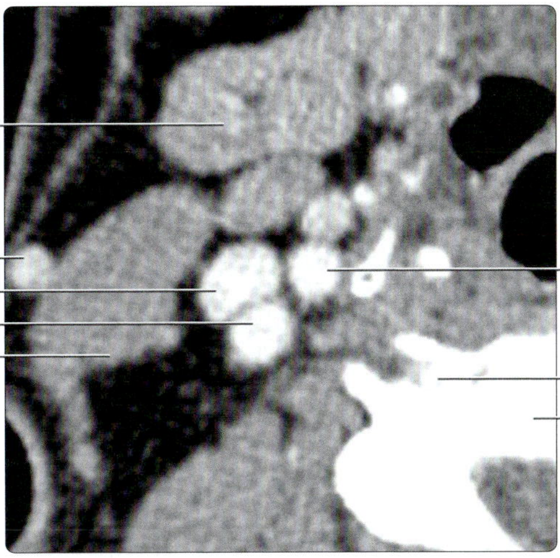

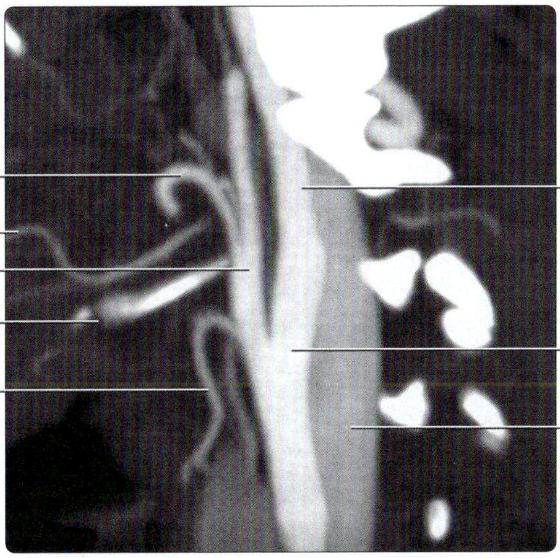

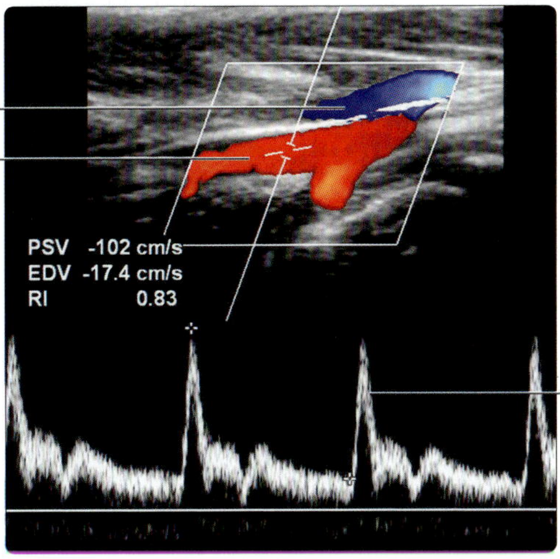

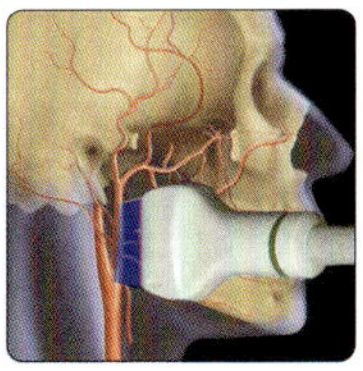

(Top) *Axial CECT of the upper neck at the hyoid bone level shows the relationship of the ECA with the adjacent ICA and the internal jugular vein.* **(Middle)** *Maximum-intensity projection CECT in the sagittal plane shows the normal contour and configuration of the ECA. Note some of its major branches, including the superior thyroid artery, lingual artery, and facial artery, in its proximal portion.* **(Bottom)** *Spectral Doppler ultrasound of the ECA in longitudinal plane shows a high-resistance flow pattern with a low diastolic component. Contrarily, the CCA and ICA are of a low-resistance pattern with a high diastolic component.*

GRAYSCALE AND DOPPLER ULTRASOUND

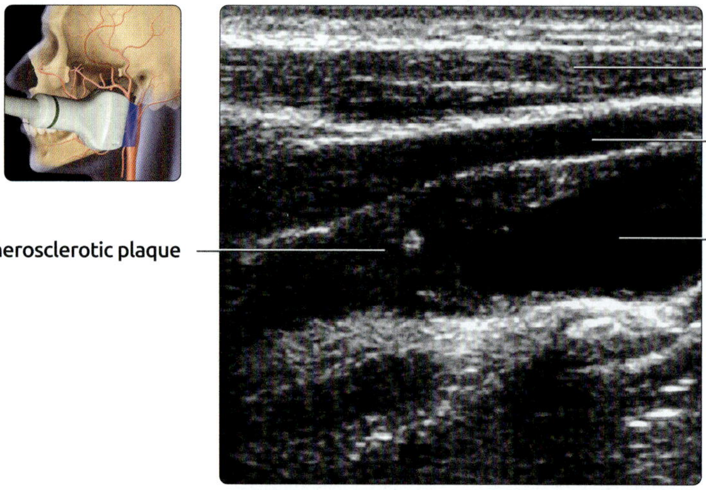

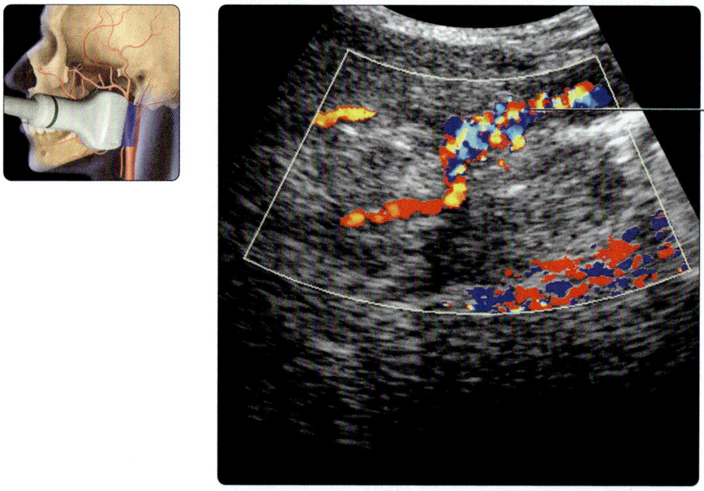

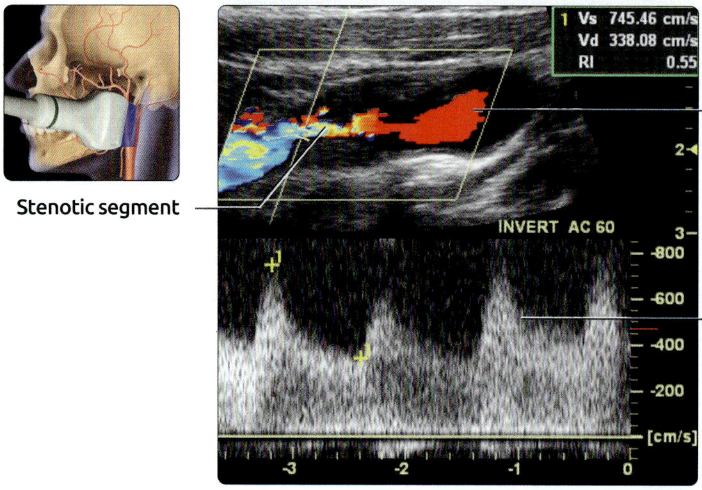

(Top) *Longitudinal grayscale ultrasound of the cervical portion of the ICA shows hypoechoic atherosclerotic plaque with marked luminal narrowing.* **(Middle)** *Longitudinal color Doppler ultrasound in the same patient helps to demonstrate the turbulent arterial flow through the severe stenotic segment in the proximal ICA. Color flow imaging is a useful tool to distinguish severe stenosis from complete occlusion.* **(Bottom)** *Spectral Doppler ultrasound of the proximal ICA demonstrates markedly elevated peak systolic and peak diastolic velocity, indicating severe stenosis.*

GRAYSCALE AND DOPPLER ULTRASOUND

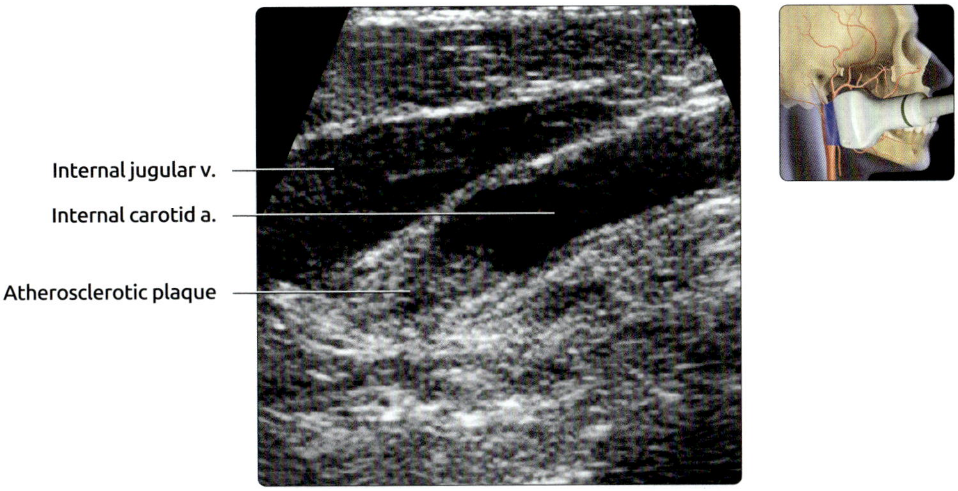

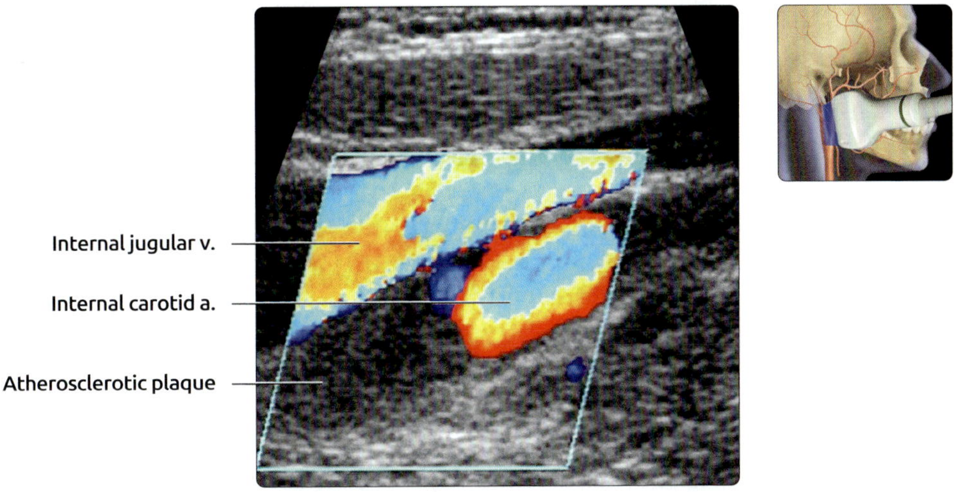

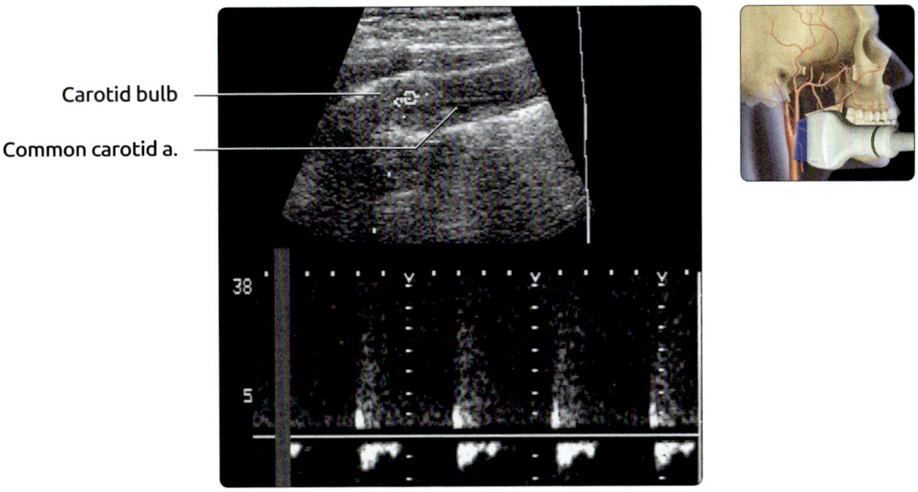

(Top) *Longitudinal grayscale ultrasound of the cervical portion of the ICA shows a slightly hyperechoic atherosclerotic plaque causing complete arterial occlusion.* **(Middle)** *Color Doppler ultrasound in the same patient reveals an absence of arterial flow in the occluded segment of the ICA.* **(Bottom)** *Spectral Doppler ultrasound at the carotid bifurcation shows no detectable signal within the occluded segment and preocclusive "thump" proximal to the occluded segment. The carotid bulb was occluded on grayscale imaging (not shown).*

COMMON CAROTID ARTERY

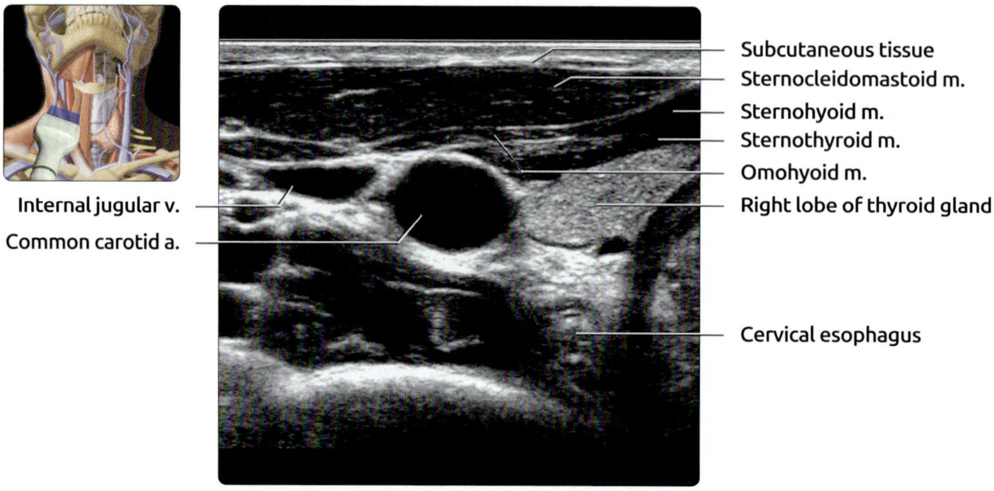

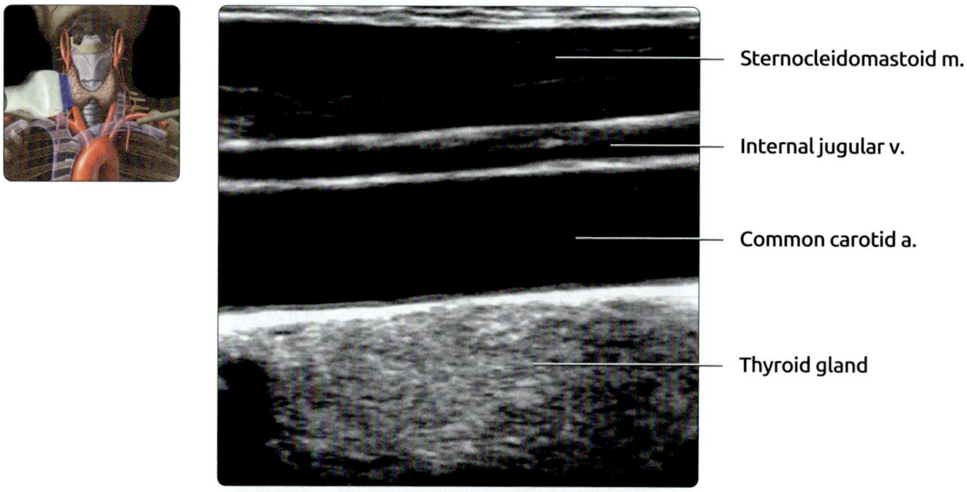

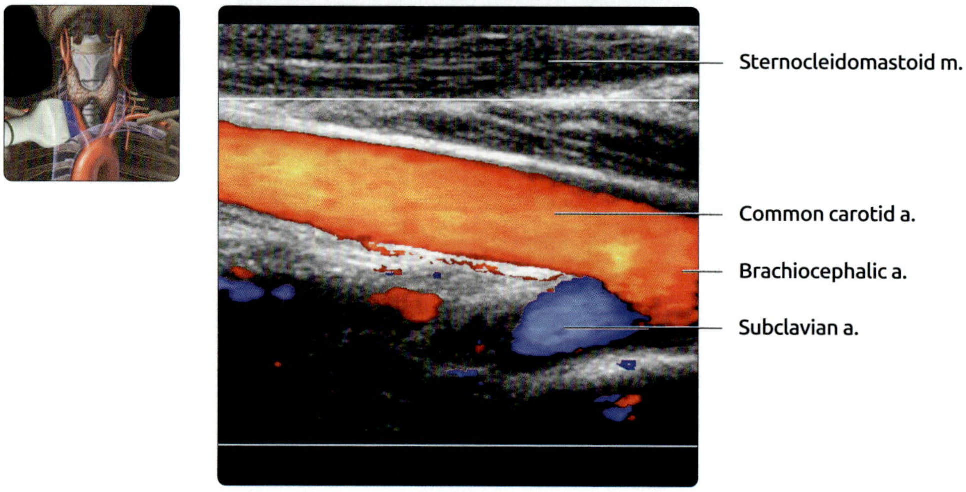

(Top) *Transverse grayscale ultrasound shows the distal CCA at the level of the upper pole of the thyroid gland. Note that the wall in a normal individual is smooth with no intimal thickening or atherosclerotic plaque. The lumen is circular in cross section. There is no major named branch in the neck apart from the termination into ECA and ICA at the level of the hyoid bone.* **(Middle)** *Longitudinal grayscale ultrasound of the CCA shows the smooth outline of the intimal layer.* **(Bottom)** *Color Doppler ultrasound of the proximal CCA at the root of the neck in the longitudinal plane demonstrates the normal antegrade arterial flow in the cranial direction. Its origin along with the subclavian artery from the right brachiocephalic artery is also well demonstrated.*

COMMON CAROTID ARTERY

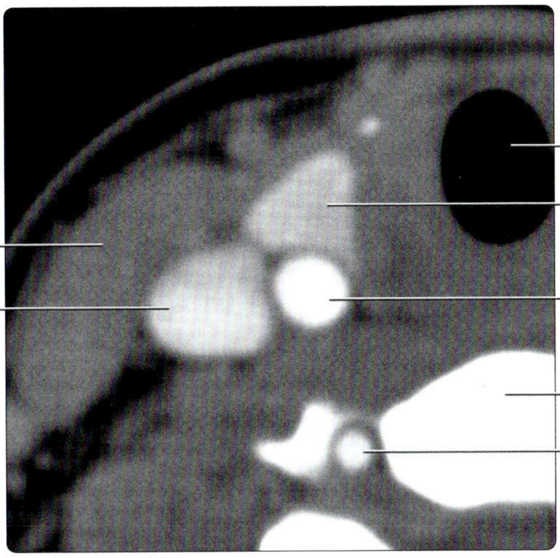

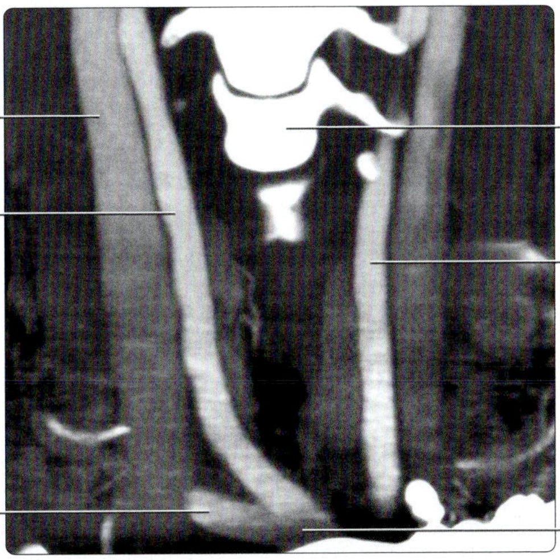

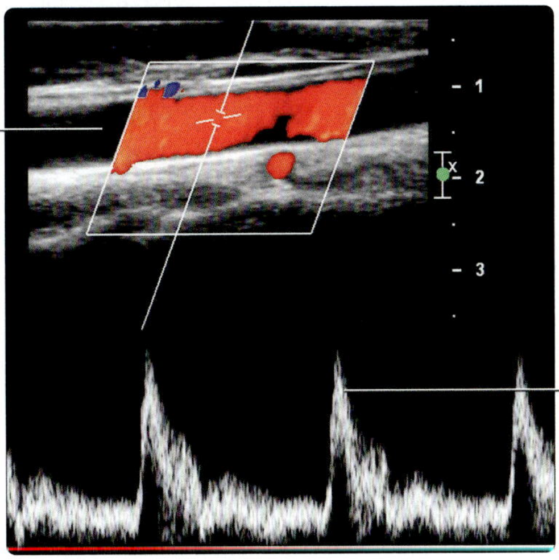

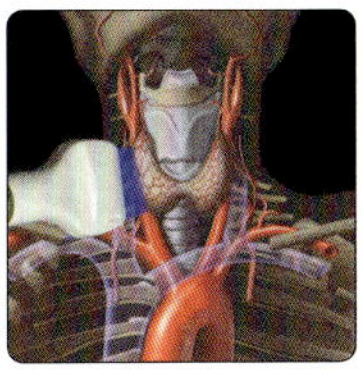

(Top) *Axial CECT of the lower neck shows contrast-filled CCA related anteriorly and medially to the right lobe of the thyroid gland and laterally to the internal jugular vein.* **(Middle)** *Coronal reformatted CECT shows the normal contour and vertical course of the CCA. It originates from the brachiocephalic artery on the right at the root of the neck with the right subclavian artery.* **(Bottom)** *Spectral Doppler ultrasound of the CCA shows low-resistance arterial flow with a forward diastolic component. The scanning technique must be meticulous to produce a reliable Doppler assessment.*

IMAGING ANATOMY

Overview

- **Vertebral artery** (VA): 4 segments
  - V1 segment (extraosseous segment)
    - Arises from 1st part of subclavian artery
    - Courses posterosuperiorly to enter C6 transverse foramen
    - Branches: Segmental cervical muscular, spinal branches
  - V2 segment (foraminal segment)
    - Ascends through C6-C3 transverse foramina
    - Turns superolaterally through inverted L-shaped transverse foramen of axis (C2)
    - Courses short distance superiorly through C1 transverse foramen
    - Branches: Anterior meningeal artery, unnamed muscular/spinal branches
  - V3 segment (extraspinal segment)
    - Exits top of atlas (C1) transverse foramen
    - Lies on top of C1 ring, curving posteromedially around atlantooccipital joint
    - As it passes around back of atlantooccipital joint, turns sharply anterosuperiorly to pierce dura at foramen magnum
    - Branches: Posterior meningeal artery
  - V4 segment (intradural/intracranial segment)
    - After VA enters skull through foramen magnum, courses superomedially behind clivus
    - Unites with contralateral VA at or near pontomedullary junction to form basilar artery (BA)
    - Branches: Anterior, posterior spinal arteries, perforating branches to medulla, posterior inferior cerebellar artery (PICA)
    - Arises from distal VA, curves around/over tonsil, gives off perforating medullary, choroid, tonsillar, cerebellar branches
- **BA**
  - Courses superiorly in prepontine cistern (in front of pons, behind clivus)
  - Bifurcates into its terminal branches, posterior cerebral arteries (PCAs), in interpeduncular or suprasellar cistern at or slightly above dorsum sellae
  - Branches: Pontine, midbrain perforating branches (numerous), anterior inferior cerebellar artery (AICA), superior cerebellar arteries (SCAs), PCAs (terminal branches)

Vascular Territory

- VA
  - Anterior spinal arteries: Upper cervical spinal cord, inferior medulla
  - Posterior spinal arteries: Dorsal spinal cord to conus medullaris
  - Penetrating branches: Olives, inferior cerebellar peduncle, part of medulla
  - PICA: Lateral medulla, choroid plexus of 4th ventricle, tonsil, inferior vermis/cerebellum
- BA
  - Pontine perforating branches: Central medulla, pons, midbrain
  - AICA: Internal auditory canal, CNVII and VIII, anterolateral cerebellum
  - SCA: Superior vermis, superior cerebellar peduncle, dentate nucleus, brachium pontis, superomedial surface of cerebellum, upper vermis

Normal Variants, Anomalies

- Normal variants
  - VA: Variation in size from right to left, dominance common; origin from aortic arch in 5%
- Anomalies
  - VA/BA may be fenestrated or duplicated (may have increased prevalence of aneurysms)
  - Embryonic carotid-basilar anastomoses (e.g., persistent trigeminal artery)

ANATOMY IMAGING ISSUES

Imaging Recommendations

- V1 and V2 segments are amenable to USG examination
- Examination usually starts in V2 segment and proceeds downward to V1 segment, then to its origin
- Examination of V2 segment
  - Transducer oriented longitudinally in midcervical region between trachea and sternocleidomastoid muscle
  - Angle transducer laterally from common carotid artery (CCA) and locate V2 segment posterior to acoustic shadowing of transverse processes
- Examination of V1 segment
  - Trace caudally from V2 to its origin
  - Left VA more difficult to visualize than right VA
  - Do not confuse with vertebral vein lying adjacent to VA, which can appear pulsatile
    - Color flow imaging helps to differentiate
- Normal waveform of VA on spectral Doppler analysis
  - Low-resistance flow
  - Similar to that of CCA but with lower amplitude
  - Peak systolic velocity: 59 ± 17 cm/sec; end-diastolic velocity: 19 ± 8 cm/sec
  - Flow velocity asymmetry is common and related to caliber of VA

Imaging Pitfalls

- VA distal to V2 cannot be properly assessed by USG
  - Abnormalities in spectral Doppler waveform of VA at V1/V2 segment provide clue for disease beyond V2

EMBRYOLOGY

Embryologic Events

- Plexiform longitudinal anastomoses between cervical intersegmental arteries → VA precursors
- Paired plexiform dorsal longitudinal neural arteries (LNAs) develop, form precursors of BA
- Transient anastomoses between dorsal LNAs develop and internal carotid arteries (ICAs) appear (primitive trigeminal/hypoglossal arteries, etc.)
- Definitive VAs arise from 7th cervical intersegmental arteries, anastomose with LNAs
- LNAs fuse as temporary connections with ICAs regress → definitive BA, vertebrobasilar circulation formed

GRAPHIC AND VOLUME-RENDERED CTA

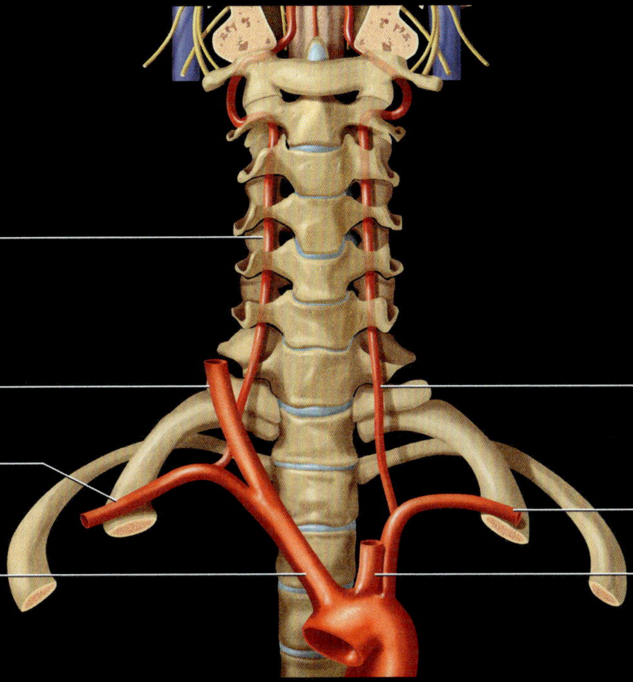

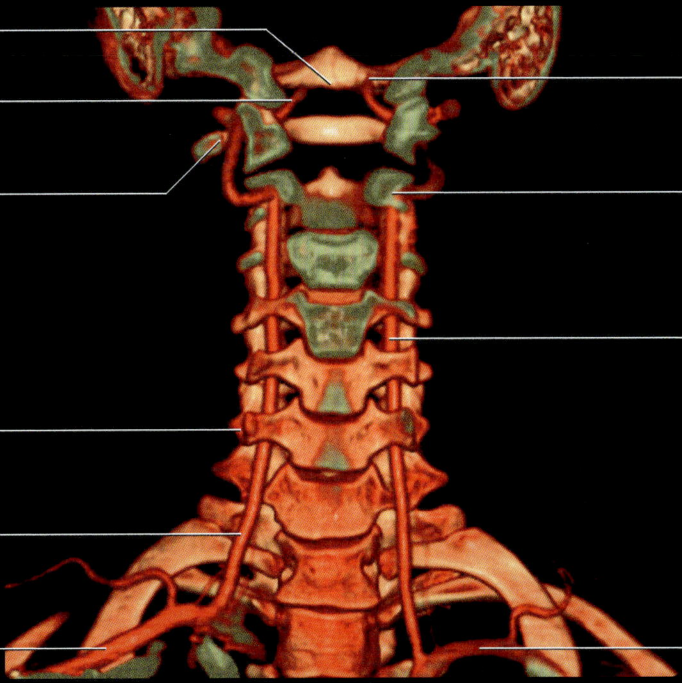

(Top) *AP graphic shows 2 of the 3 extracranial segments of the vertebral arteries (VAs) and their relationship to the cervical spine. The extraosseous (V1) VA segments extend from the superior aspect of the subclavian arteries to the C6 transverse foramina. The V2 (foraminal) segment extends from C6 to the VA exit from the C1 transverse foramina.* **(Bottom)** *3D-VRT CTA shows the extracranial VAs, which originate from the superior aspect of the subclavian arteries. The VAs typically enter the transverse foramina of C6 and ascend almost vertically to C2, where they make a 90° turn laterally in the L-shaped C2 transverse foramina before ascending vertically again to C1.*

TRANSVERSE, LONGITUDINAL GRAYSCALE AND COLOR DOPPLER

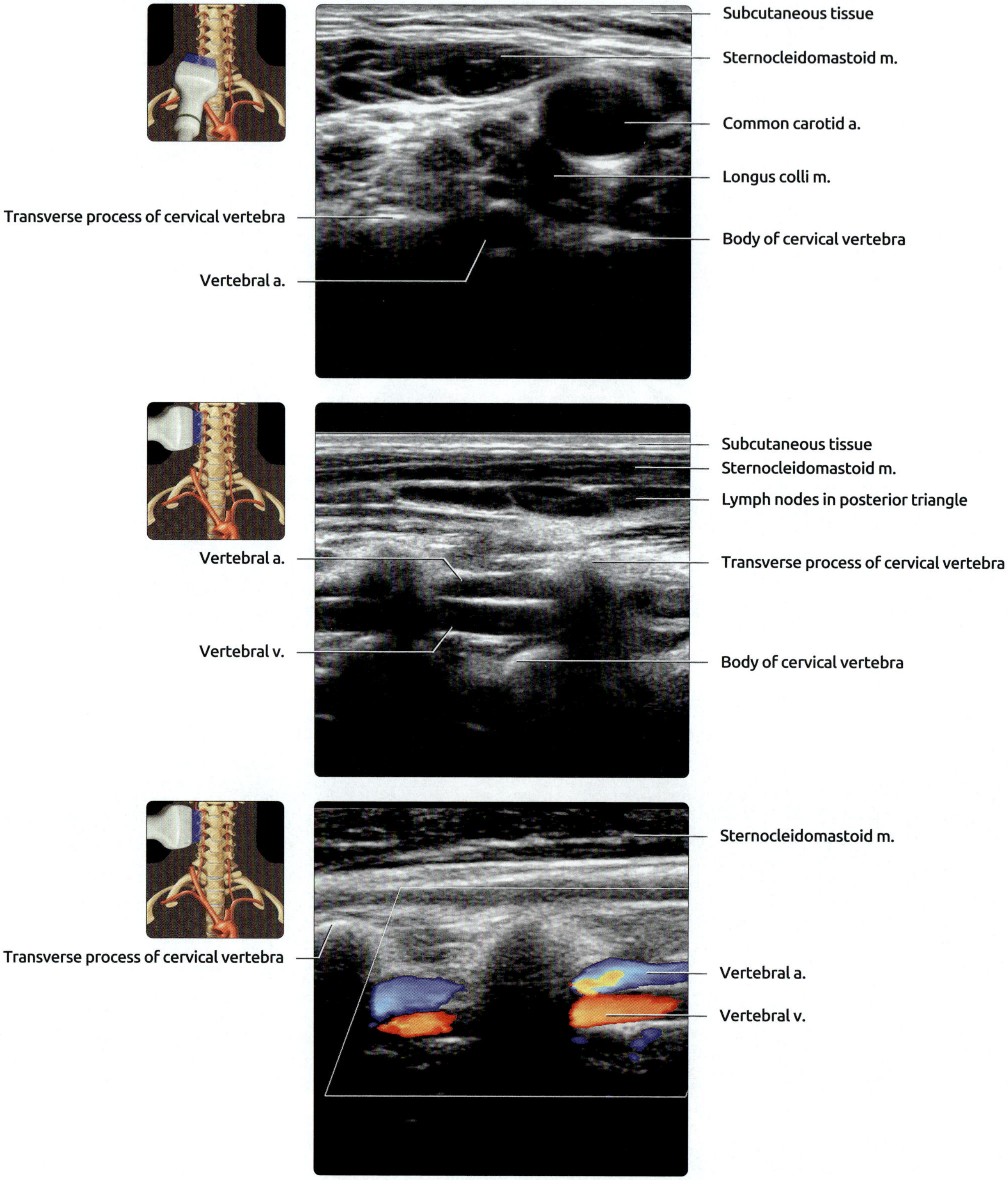

(Top) *Transverse grayscale ultrasound of the lower neck shows the proximal V1 segment of the VA, which arises from the 1st part of the subclavian artery and courses superiorly to enter the transverse foramina of the lower cervical vertebra. Note its posterior relationship to the longus colli at this level.* **(Middle)** *Longitudinal grayscale ultrasound of the posterior neck demonstrates the V2 segment of the VA within the transverse foramina of cervical vertebrae. Note the presence of dense posterior acoustic shadowing from the transverse processes, obscuring a clear view of the underlying vertebral vessels.* **(Bottom)** *Color Doppler ultrasound shows the V2 segment of the VA in the longitudinal plane. Note the opposite flow direction of the vertebral vein (i.e., craniocaudal direction) as compared with that of the VA (caudocranial direction).*

AXIAL AND CORONAL CECT, SPECTRAL DOPPLER ULTRASOUND

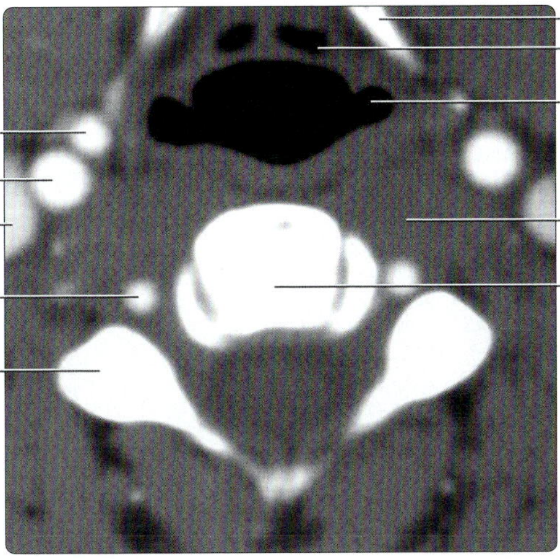

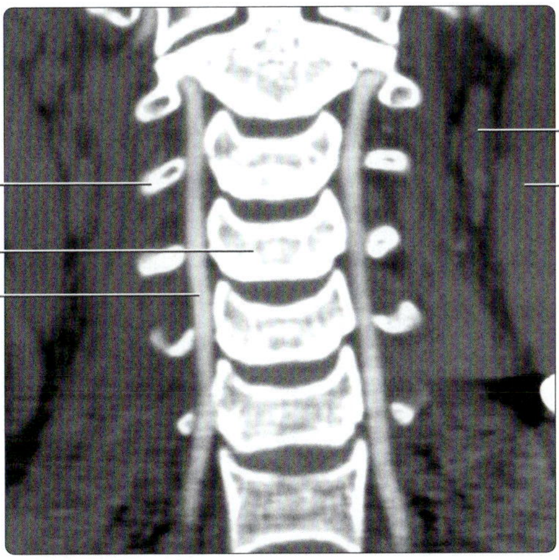

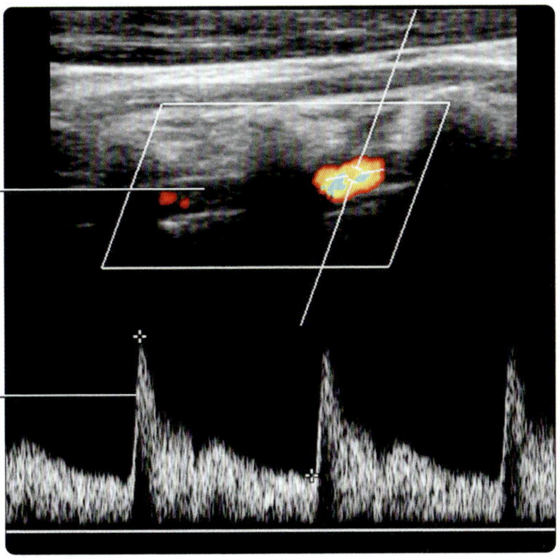

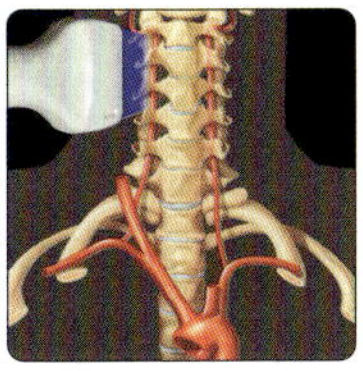

(Top) *Axial CECT of the neck at the level of the hyoid bone shows the VA running in a caudocranial direction within the foramen transversarium of the cervical vertebrae. This portion is amenable for ultrasound examination.* **(Middle)** *Coronal reformatted CECT of the neck shows the vertical course of the VAs through the transverse foramina of C6-C2 vertebrae. Note its close anatomical relationship with the transverse processes and bodies of the cervical vertebrae.* **(Bottom)** *Spectral Doppler ultrasound of the V2 segment of the VA is of low resistance, similar to that of the common carotid artery but with lower amplitude. Spectral analysis of the V2 segment provides a clue to stenosis/occlusion proximally and distally. For example, a high-resistance flow pattern without a diastolic flow component is often associated with a distal flow obstruction.*

GRAYSCALE AND DOPPLER ULTRASOUND

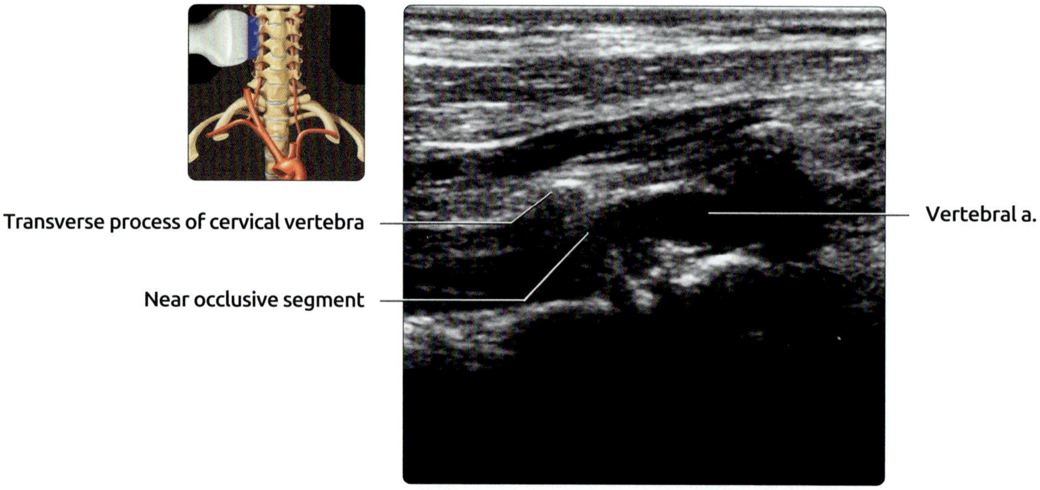

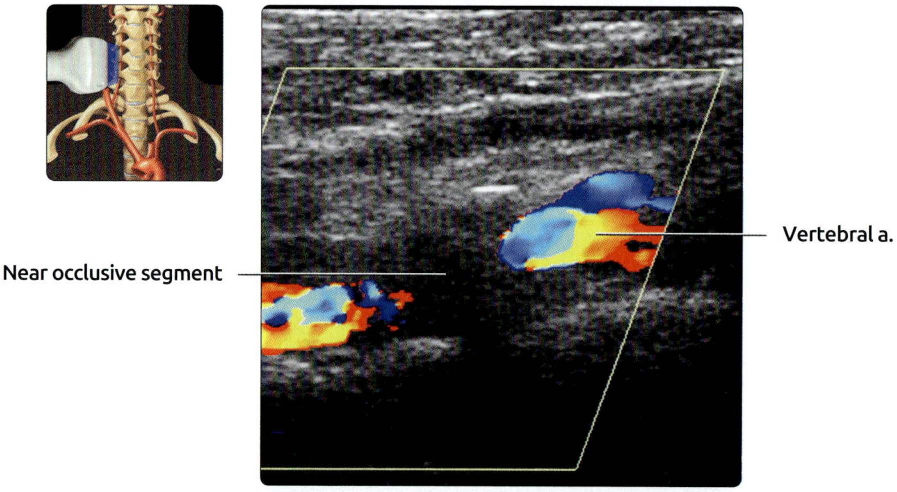

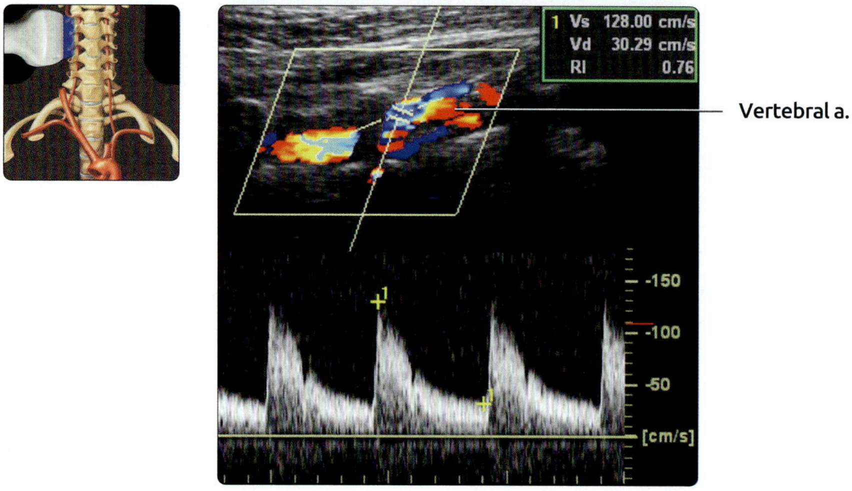

(Top) *Longitudinal grayscale ultrasound of the V2 segment of the VA shows the presence of hypoechoic atherosclerotic plaque, causing near complete occlusion.* **(Middle)** *Color Doppler ultrasound of the VA in the same patient shows a lack of arterial color flow within the nearly occluded segment.* **(Bottom)** *Spectral Doppler ultrasound in the same patient shows a high-resistance flow pattern with elevated peak systolic and diastolic velocities.*

GRAYSCALE AND DOPPLER ULTRASOUND

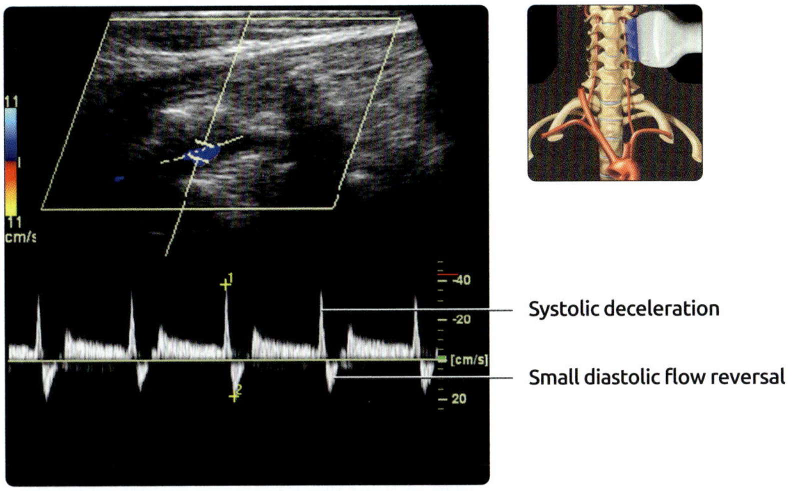

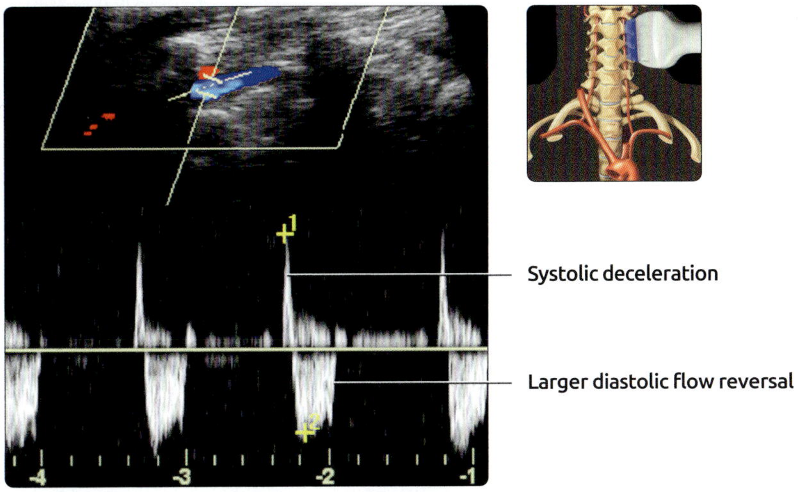

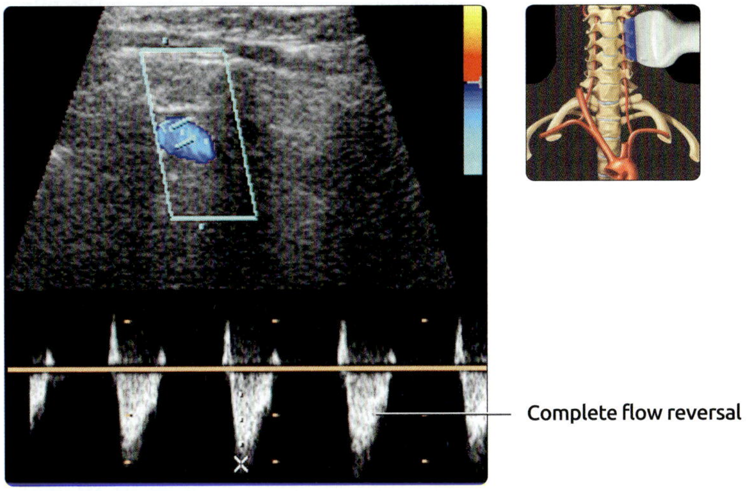

(Top) *Spectral Doppler ultrasound shows a mild degree of subclavian steal syndrome. There is systolic deceleration of the vertebral flow in an antegrade direction with small diastolic flow reversal.* **(Middle)** *Spectral Doppler ultrasound shows a moderate degree of subclavian steal syndrome. The degree of systolic deceleration and diastolic flow reversal is more pronounced with alternating vertebral flow demonstrated.* **(Bottom)** *Spectral Doppler ultrasound shows severe subclavian steal syndrome. There is near complete reversal of flow in the VA with relative absent antegrade systolic flow. This pattern is commonly associated with the occurrence of vertebrobasilar symptoms.*

GROSS ANATOMY

Overview

- Major extracranial venous system composed of facial veins, neck veins, scalp, skull (diploic), and orbital veins
- **Facial veins**
    - Facial vein
        - Begins at angle between eye, nose
        - Descends across masseter, curves around mandible
        - Joins internal jugular vein (IJV) at hyoid level
        - Tributaries from orbit (supraorbital, superior ophthalmic veins), lips, jaw, facial muscles
    - Deep facial vein
        - Receives tributaries from deep face, connects facial vein with pterygoid plexus
    - Pterygoid plexus
        - Network of vascular channels in masticator space between temporalis/lateral pterygoid muscles
        - Connects cavernous sinuses and clival venous plexus to face/orbit tributaries
        - Drains into maxillary vein
    - Retromandibular vein (RMV)
        - Formed from union of maxillary and superficial temporal veins
        - Lies within parotid space
        - Passes between external carotid artery and CNVII to empty into IJV
- **Neck veins**
    - **External jugular vein (EJV)**
        - From union of retromandibular and posterior auricular veins
        - Courses inferiorly on surface of sternocleidomastoid muscle
        - Drains into subclavian vein in supraclavicular fossa
        - Receives tributaries from scalp, ear, and face
        - Size, extent highly variable
    - **IJV**
        - Caudal continuation of sigmoid sinus from jugular foramen at skull base
        - Jugular bulb = dilatation at origin
        - Courses inferiorly in carotid space posterolateral to internal/common carotid arteries underneath sternocleidomastoid muscle
        - Unites with subclavian vein to form brachiocephalic vein
        - Size highly variable; significant side-to-side asymmetry common; right usually larger than left
    - Subclavian vein
        - Proximal continuation of axillary vein in thoracic inlet
        - EJV drains into subclavian vein
        - Subclavian vein joins IJV to form brachiocephalic vein
    - Vertebral venous plexus
        - Suboccipital venous plexus
        - Tributaries from basilar (clival) plexus, cervical musculature
        - Interconnects with sigmoid sinuses, cervical epidural venous plexus
        - Terminates in brachiocephalic vein

IMAGING ANATOMY

Overview

- Low pressure inside; easily compressible
    - Light probe pressure with good surface contact between transducer and skin to ensure optimal visualization
    - Use of Valsalva maneuver helps to distend major neck veins
- IJV
    - Largest vein of neck
    - Deep cervical chain lymph nodes commonly found along its course
    - Beware of thrombosis in patients with previous central venous catheterization or adjacent tumors
    - Always check for compressibility and phasicity on respiration
    - Presence of vascularity in IJV thrombosis is usually seen with tumor thrombus rather than bland venous thrombus
- Subclavian vein
    - Accessible on USG by inferior tilting of transducer in supraclavicular fossa
    - Venous valves are present in most patients
    - Thrombosis/stenosis commonly seen in patients on chronic hemodialysis or with previous subclavian venous catheterization
- RMV
    - Serves as landmark on USG to infer position of intraparotid portion of facial nerve
    - Anterior division of RMV sandwiched between submandibular gland anteriorly and parotid tail posteriorly
        - Its displacement helps to determine origin of mass in posterior submandibular region

ANATOMY IMAGING ISSUES

Imaging Pitfalls

- Neck veins are often overlooked, as most sonologists pay more attention to arteries than veins in neck
- Not all neck veins are readily assessed by ultrasound
    - Only large and superficial veins are clearly seen
- Asymmetric IJVs are common; 1 IJV may be many times size of contralateral IJV
    - IJV venous varix: Extreme dilatation of IJV upon Valsalva maneuver with clinically palpable neck lump
- Slow flow within IJV may appear as low-level hyperechoic intraluminal "mass"
    - May mimic IJV thrombus
    - Moving nature of echoes on real-time ultrasound and sharp linear near-field interface help to distinguish artifacts from slow flow and IJV thrombus

CLINICAL IMPLICATIONS

Clinical Importance

- Ultrasound safely guides needle for venous access
- Absence of respiratory phasicity is strong indicator of abnormality

FACE AND NECK VEINS

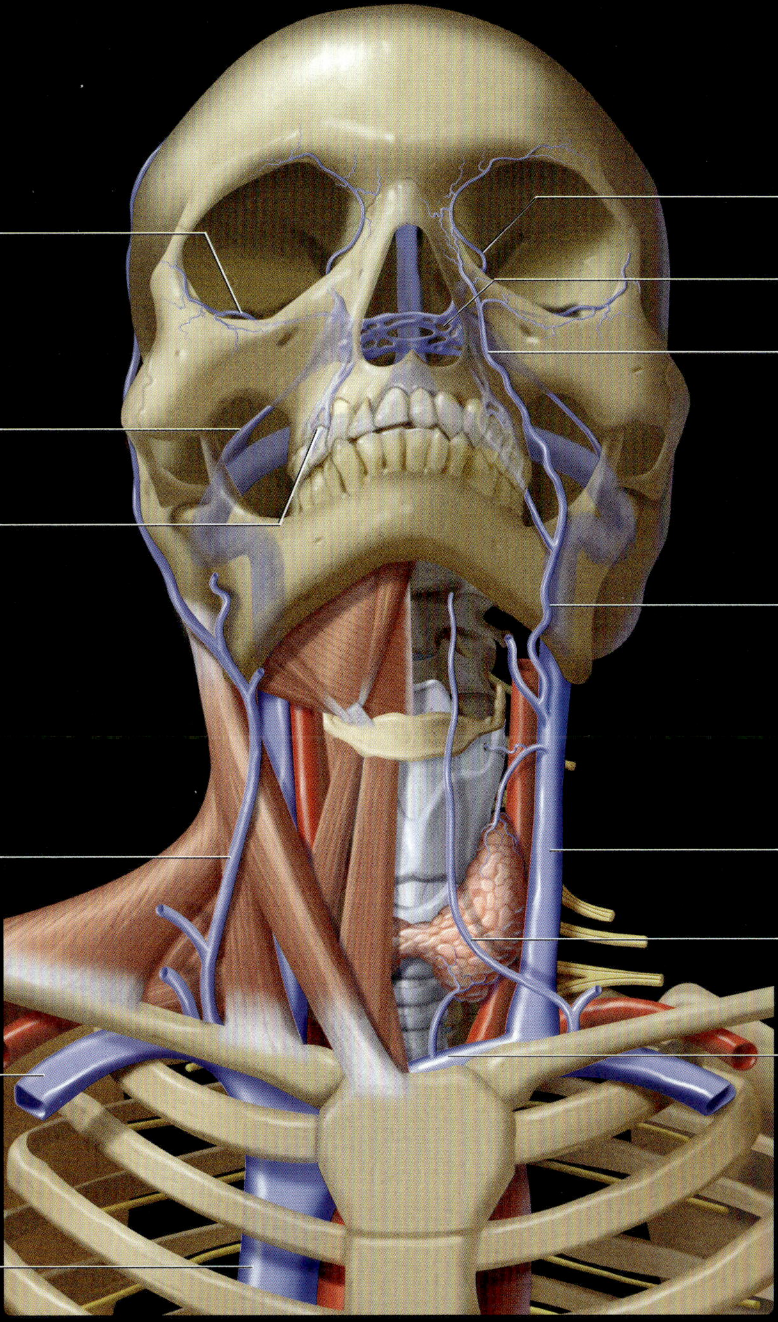

Anteroposterior view of the extracranial venous system depicts the major neck veins, their drainage into the mediastinum, and their numerous interconnections with the intracranial venous system. The pterygoid venous plexus receives tributaries from the cavernous sinus and provides an important potential source of collateral venous drainage if the transverse or sigmoid sinuses become occluded.

GRAYSCALE AND COLOR DOPPLER ULTRASOUND (IJV)

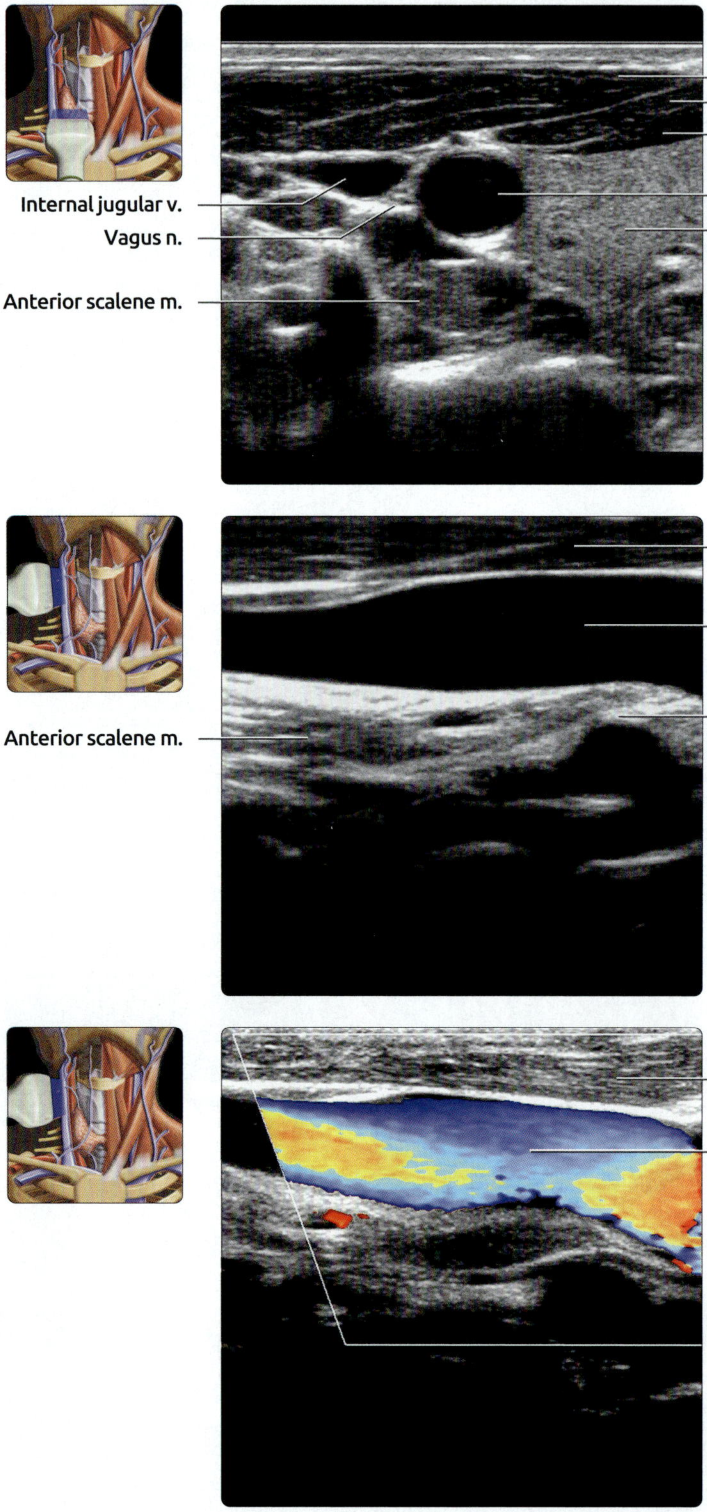

(Top) *Transverse grayscale ultrasound of the lower cervical level shows the normal anatomical relationship between the internal jugular vein and the adjacent structures. It is underneath the sternocleidomastoid muscle and lateral to the common carotid artery and vagus nerve within the carotid sheath.* **(Middle)** *Longitudinal grayscale ultrasound shows the internal jugular vein in the midcervical level. The internal jugular vein appears as a tubular anechoic structure coursing in a vertical direction. It should be examined with light probe pressure and with compression to exclude a venous thrombosis.* **(Bottom)** *Corresponding color Doppler ultrasound in the longitudinal plane shows color flow filling the entire lumen of the internal jugular vein. The use of color Doppler helps to identify and evaluate the presence and nature of an internal jugular vein thrombus.*

CECT AND SPECTRAL DOPPLER ULTRASOUND (IJV)

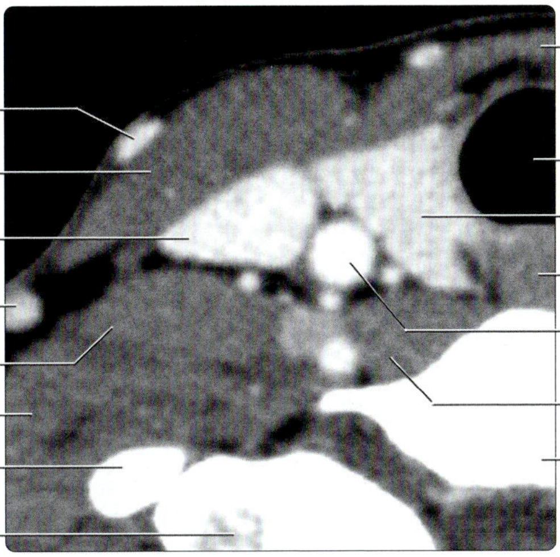

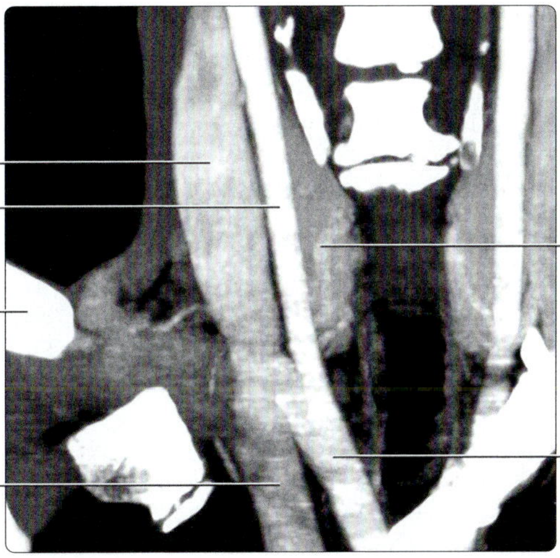

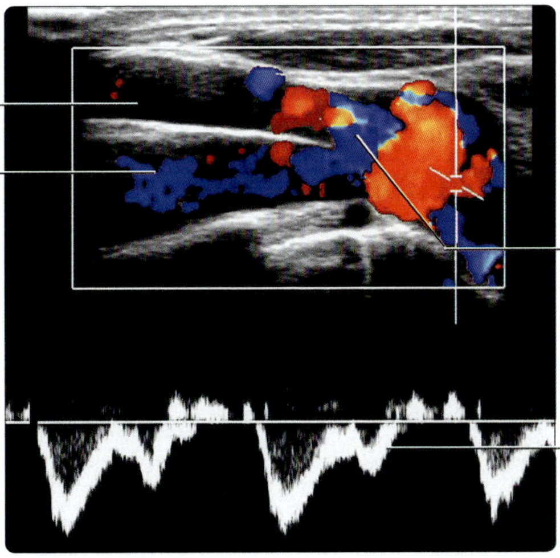

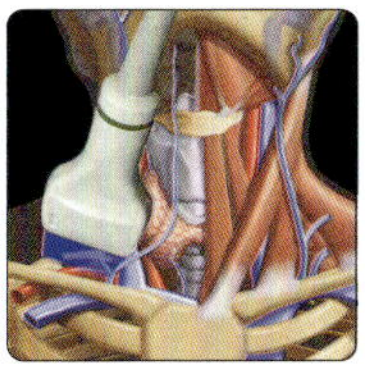

(Top) *Axial CECT of the lower neck shows the internal jugular vein, which is usually larger than and lateral to the common carotid artery. The external jugular vein is in a subcutaneous location.* **(Middle)** *Coronal reformatted CECT of the lower neck shows the close anatomical relationship of the internal jugular vein and common carotid artery within the carotid sheath. The internal jugular vein continues inferiorly below the clavicle to join the subclavian vein to form the brachiocephalic vein.* **(Bottom)** *Transverse spectral Doppler ultrasound shows the internal jugular vein at the level of the supraclavicular fossa at the junction with the subclavian vein. The normal biphasic venous waveform, which varies with respiratory motion, can be easily demonstrated and helps to exclude the presence of obstructing venous thrombus.*

GRAYSCALE AND COLOR DOPPLER ULTRASOUND (EJV)

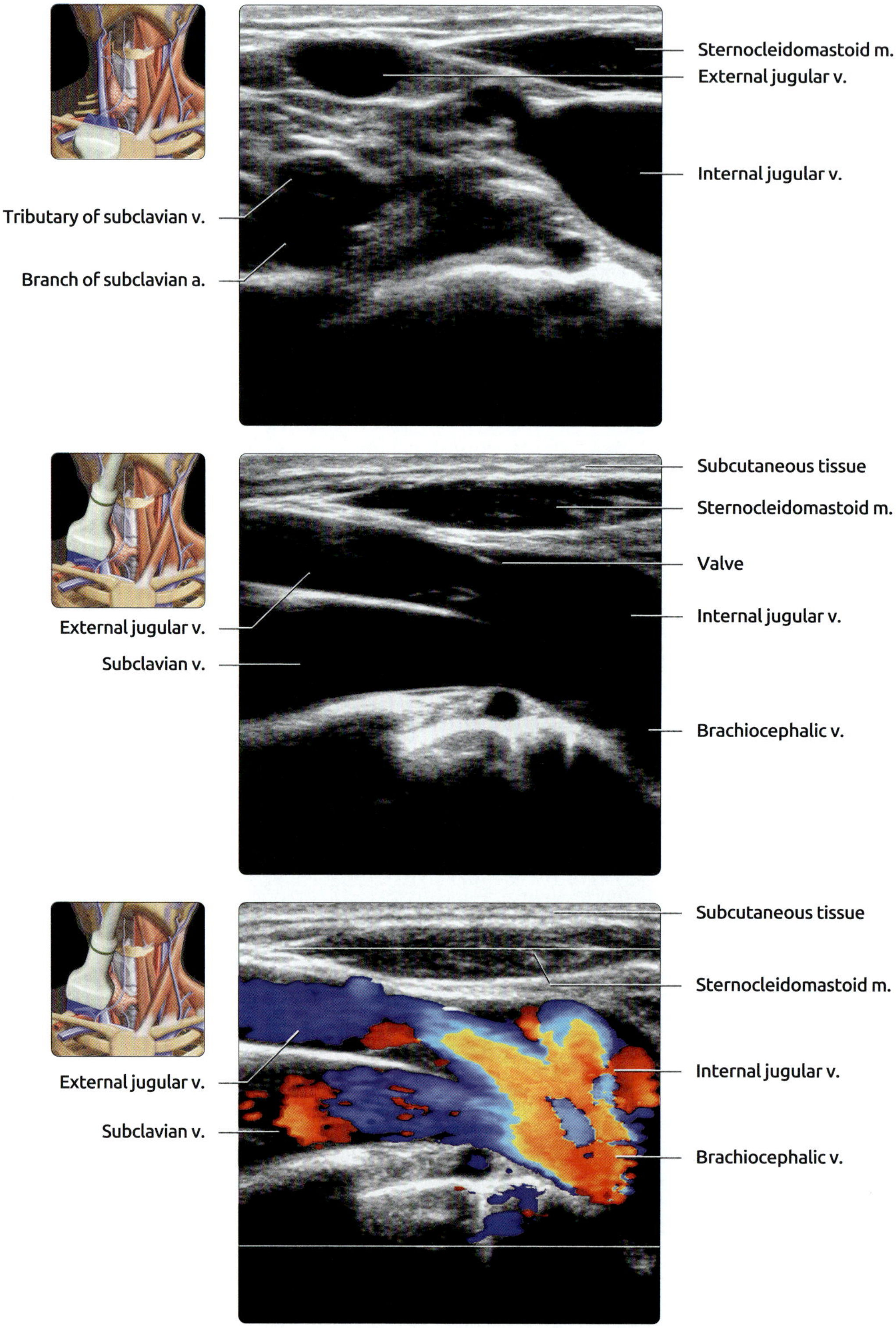

(Top) *Transverse grayscale ultrasound of right lower neck shows the location of the external jugular vein in relation to the sternocleidomastoid muscle. It appears as a distended, round, anechoic structure on Valsalva maneuver using light transducer pressure.* **(Middle)** *Transverse grayscale ultrasound shows the external jugular vein at the supraclavicular level, at the site of union with the subclavian vein, close to the terminal portion of the internal jugular vein. Valve leaflets are commonly seen within the major veins at the thoracic inlet level.* **(Bottom)** *Corresponding transverse color Doppler ultrasound at the supraclavicular level helps to depict the venous drainage of the external jugular vein to the subclavian vein. Note that the subclavian vein joins the internal jugular vein to form the brachiocephalic vein.*

CECT AND SPECTRAL DOPPLER ULTRASOUND (EJV)

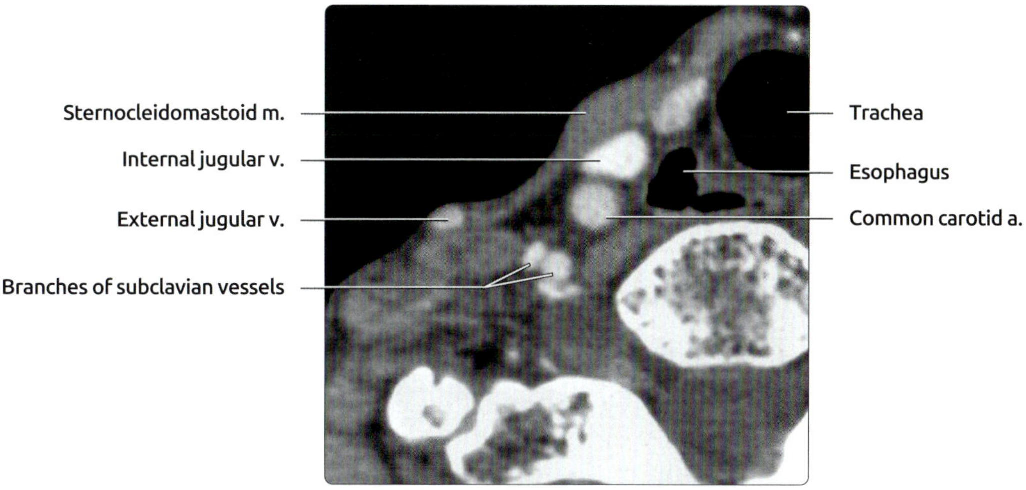

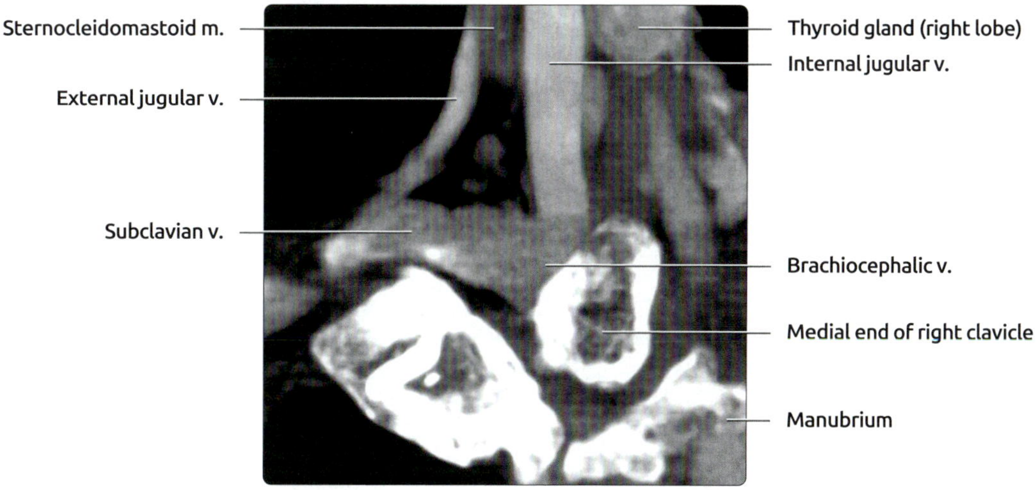

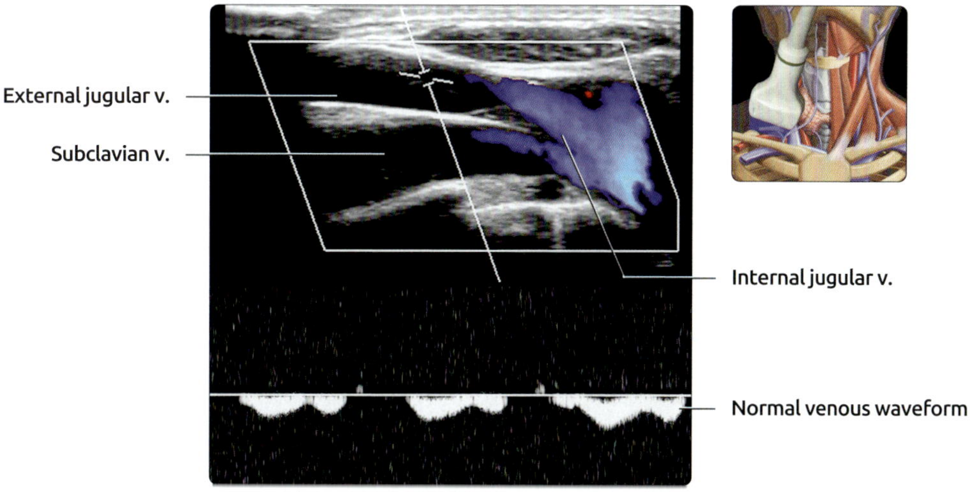

(Top) *Axial CECT shows the right lower neck. Note the superficial anatomical location of the external jugular vein. Thus, light probe pressure is necessary for the assessment of the external jugular vein on ultrasound because with increasing pressure, the vein will be compressed.* **(Middle)** *Coronal reformatted CECT shows the right lower neck. Note the drainage of the external jugular vein to the subclavian vein, which joins the internal jugular vein to form the brachiocephalic vein at the thoracic inlet level.* **(Bottom)** *Spectral Doppler ultrasound interrogating the terminal portion of the external jugular vein shows a normal, phasic, low-pressure venous waveform, which helps to confirm its patency.*

LONGITUDINAL AND TRANSVERSE ULTRASOUND

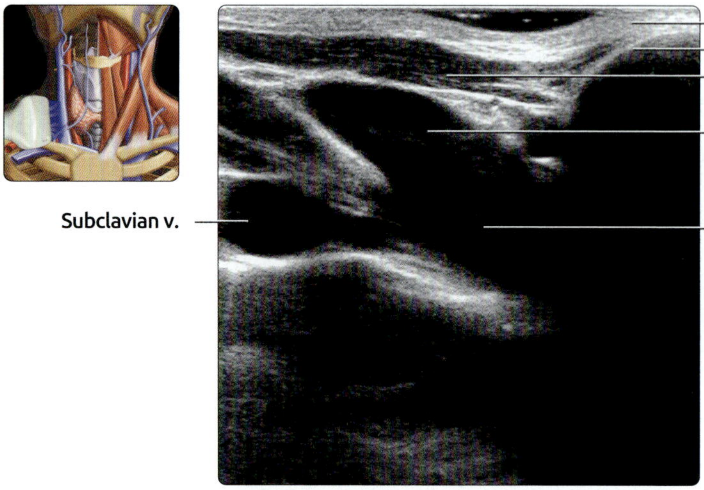

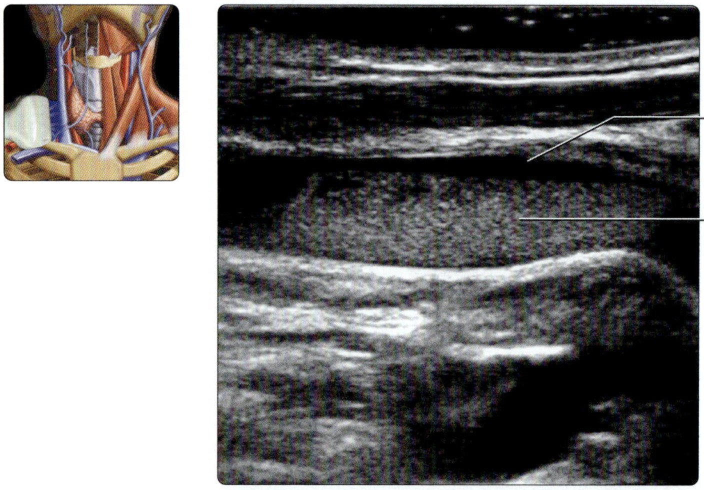

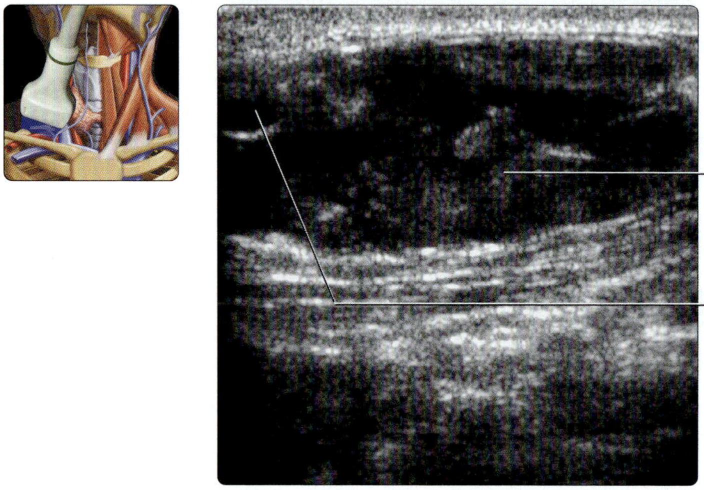

(Top) *Longitudinal grayscale ultrasound at the supraclavicular level shows the union of internal jugular vein and subclavian vein to form the brachiocephalic vein. The more distal portion of the brachiocephalic vein is obscured by the overlying clavicle and is therefore not assessed by ultrasound.* **(Middle)** *Longitudinal grayscale ultrasound of the right internal jugular vein shows a pseudothrombus phenomenon due to slow venous flow within the internal jugular vein. Note the layering with a sharp linear border in the internal jugular vein lumen in the near field, which is the clue to distinguish it from venous thrombus.* **(Bottom)** *Transverse grayscale ultrasound of the right posterior triangle shows a well-defined hypoechoic mass with multiple internal sinusoidal spaces. The lesion is inseparable from the external jugular vein. Surgery confirmed a venous vascular malformation (VVM) arising from the external jugular vein.*

CECT, COLOR AND POWER DOPPLER ULTRASOUND

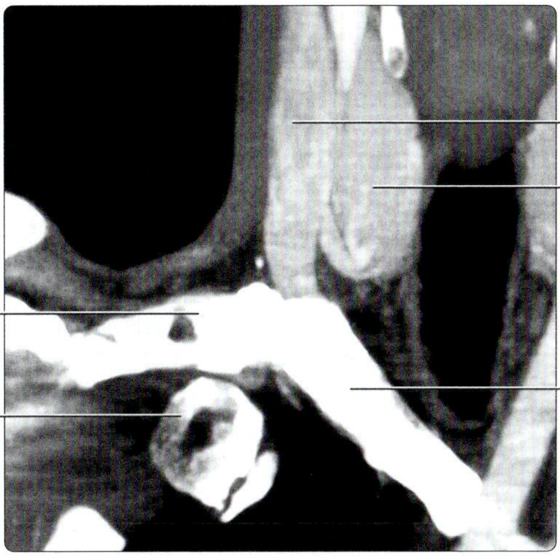

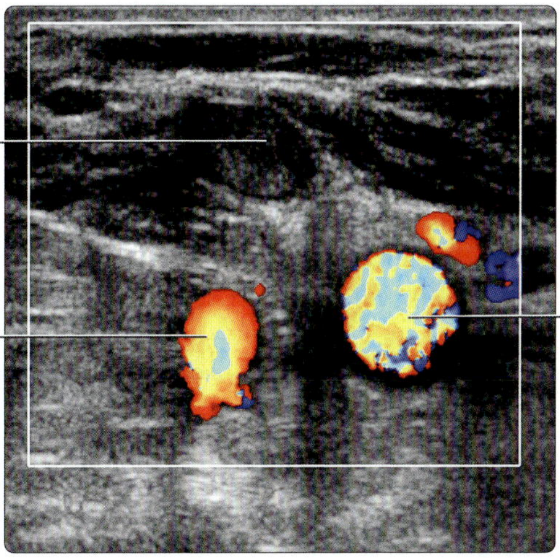

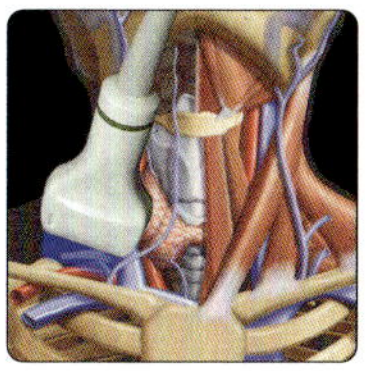

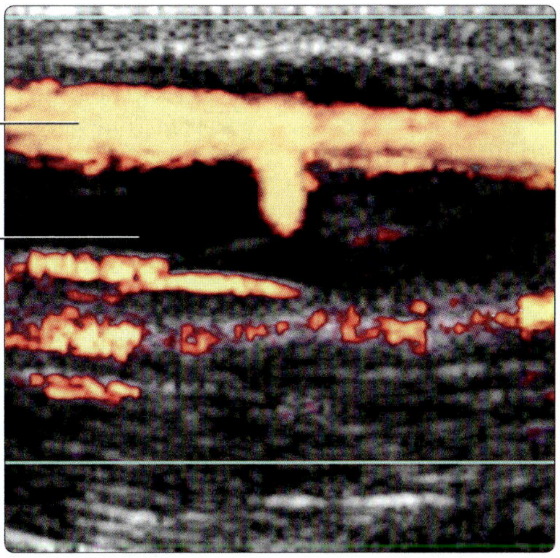

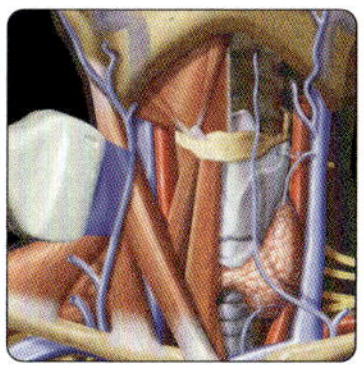

(Top) *Coronal reformatted CECT of the right supraclavicular fossa shows the formation of the brachiocephalic vein by union of the subclavian vein (dense contrast filling due to injection in ipsilateral antecubital fossa) and internal jugular vein. The brachiocephalic vein can be fully assessed on CECT as compared with ultrasound.* **(Middle)** *Transverse color Doppler ultrasound of the right supraclavicular level reveals intraluminal, hypoechoic, avascular echoes causing occlusion of the internal jugular vein. The appearances are of a bland venous thrombus due to prolonged central venous catheterization.* **(Bottom)** *Longitudinal power Doppler ultrasound shows the external jugular vein VVM. Note the intimate relationship of the VVM and the external jugular vein. Sinusoidal spaces are usually not color filled due to very slow flow within.*

TERMINOLOGY

Synonyms

- Internal jugular chain (IJC): Deep cervical chain
- Spinal accessory chain (SAC): Posterior triangle chain
- Transverse cervical chain: Supraclavicular chain
- Anterior cervical chain: Prelaryngeal, pretracheal, paratracheal nodes
- Paratracheal node: Recurrent laryngeal node

Definitions

- Jugulodigastric node: "Sentinel" (highest) node, found at apex of IJC at angle of mandible
- Virchow node: "Signal" node, lowest node of deep cervical chain
- Troisier node: Most medial node of transverse cervical chain
- Omohyoid node: Deep cervical chain node superior to omohyoid as it crosses jugular vein
- Delphian node: Pretracheal node

IMAGING ANATOMY

Overview

- In normal adult neck, may be up to 300 lymph nodes
  - Internal structures: Capsule, cortex, medulla, hilum
- US appearances of normal cervical lymph node
  - Small, oval/reniform shape with well-defined margin
  - Homogeneous, hypoechoic cortex with echogenic hilum
  - Hilar vascularity on color/power Doppler examination
- Imaging-based nodal classification
  - **Level I: Submental and submandibular nodes**
    - Level IA: Submental nodes: Found between anterior bellies of digastric muscles
    - Level IB: Submandibular nodes: Found around submandibular glands in submandibular space
  - **Level II: Upper IJC nodes: From posterior belly of digastric muscle to hyoid bone**
    - Level IIA: Level II node anterior, medial, lateral or posterior to IJV; if posterior to IJV, node must be inseparable from IJV; contains jugulodigastric nodal group
    - Level IIB: Level II node posterior to IJV with fat plane visible between node and IJV
  - **Level III: Mid IJC nodes**
    - From hyoid bone to inferior margin of cricoid cartilage
  - **Level IV: Lower IJC nodes**
    - From inferior cricoid margin to clavicle
  - **Level V: Nodes of posterior cervical space/spinal accessory chain**
    - SAC nodes lie posterior to back margin of sternocleidomastoid muscle
    - Level VA: Upper SAC nodes from skull base to bottom of cricoid cartilage
    - Level VB: Lower SAC nodes from cricoid to clavicle
  - **Level VI: Nodes of visceral space**
    - Found from hyoid bone above to top of manubrium below
    - Midline group of cervical lymph nodes
    - Includes prelaryngeal, pretracheal, and paratracheal subgroups
  - **Level VII: Superior mediastinal nodes**
    - Between carotid arteries from top of manubrium above to innominate vein below
- Other nodal groups not included in standard imaging-based nodal classification
  - Parotid nodal group: Intraglandular or extraglandular
  - Retropharyngeal (RPS) nodal group: Medial RPS nodes and lateral RPS nodes (Rouvière node)
  - Facial nodal group

ANATOMY IMAGING ISSUES

Imaging Approaches

- Nodal metastases from primary tumors are site specific; therefore, it is critical to understand usual patterns of lymphatic spread
- Equivocal nodes outside usual pattern less suspicious
- Likely location of primary tumor can be suspected in patients presenting with nodal mass
- Nodal disease outside usual pattern may suggest aggressive tumor or prompt search for 2nd primary

Imaging Pitfalls

- Retropharyngeal (RPS) nodes and superior mediastinal nodes cannot be assessed by US

Key Concepts

- Useful US features suspicious of malignancy
  - Shape: Round, long:short axis ratio < 2
  - Loss of echogenic hilum
  - Presence of intranodal necrosis (cystic/coagulation)
  - Presence of extracapsular spread: Ill-defined margin
  - Peripheral/subcapsular flow on color/power Doppler ultrasound
  - Increased intranodal intravascular resistance: Resistive index (RI) > 0.8, pulsatility index (PI) > 1.6
  - Internal architecture: Punctate calcifications in metastatic node from papillary thyroid carcinoma, reticulated/pseudocystic appearance of lymphomatous node
- No single finding sensitive or specific enough; these signs should be used in combination
- FNAC helps to improve diagnostic accuracy
- Tuberculous nodes mimic metastatic nodes
  - Differentiating features: Intranodal necrosis, nodal matting, soft tissue edema and displaced hilar vascularity/avascularity, calcification (post treatment)

CLINICAL IMPLICATIONS

Clinical Importance

- Presence of malignant SCCa nodes on staging associated with 50% ↓ in long-term survival
  - If extranodal spread present, further 50% ↓
- Location of metastatic nodes in neck may help predict site of primary tumor
  - RPS and posterior triangle nodes seen in nasopharyngeal carcinoma, and lower cervical nodes in lung cancer
  - When Virchow node found on imaging without upper neck nodes, primary not in neck, and whole-body imaging warranted

LYMPH NODE GROUPS

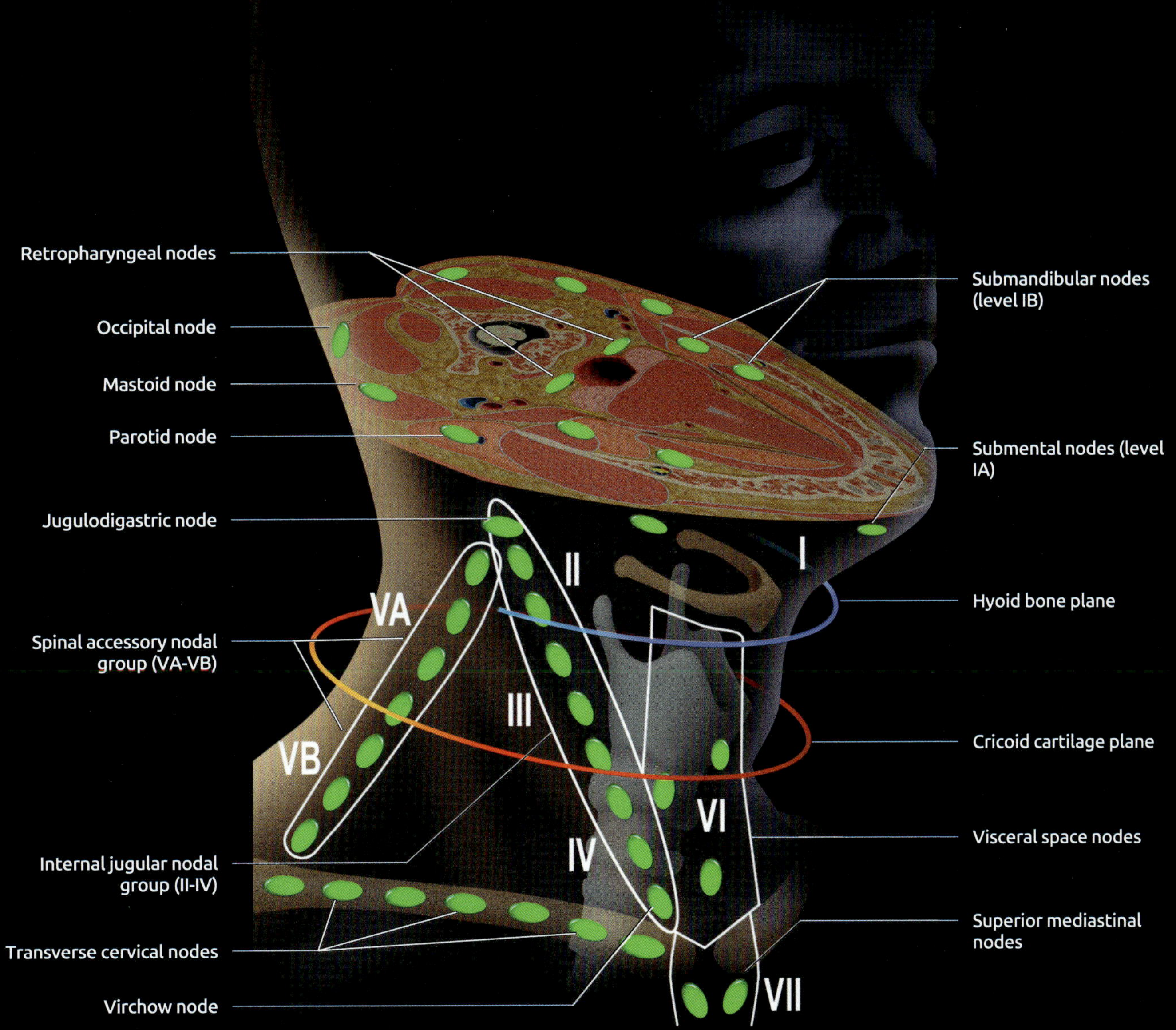

Lateral oblique graphic of the cervical neck depicts an axial slice through the suprahyoid neck. The retropharyngeal nodes behind the pharynx are often clinically occult. The hyoid bone (blue arc) and cricoid cartilage (orange circle) planes are highlighted, as they serve to subdivide the internal jugular and spinal accessory nodal group levels. Regional lymph nodes are staged the same way for almost all head and neck tumors, largely based on size, bilaterality, and number of nodes involved. N stages for most pharyngeal and oral cavity tumors are determined by this generic head and neck nodal classification. Tumors of the nasopharynx, however, have a unique N classification scheme.

FACE AND NECK LYMPH NODES

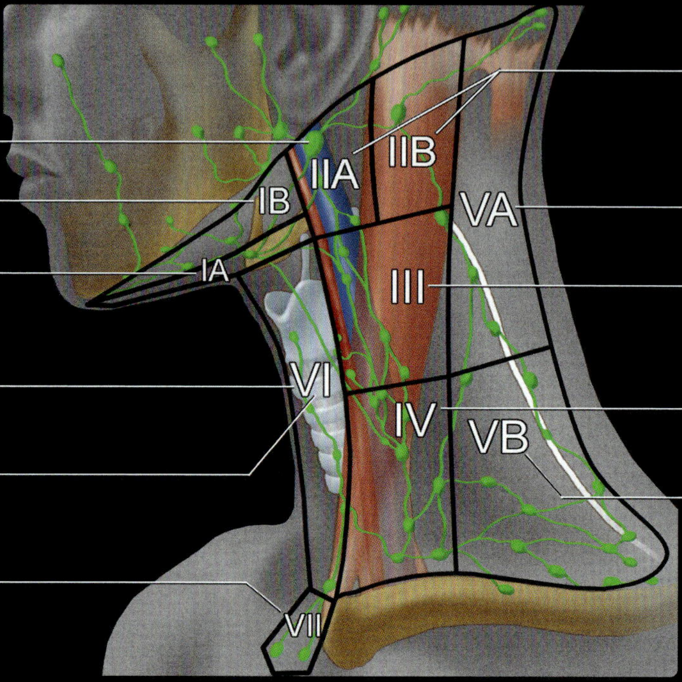

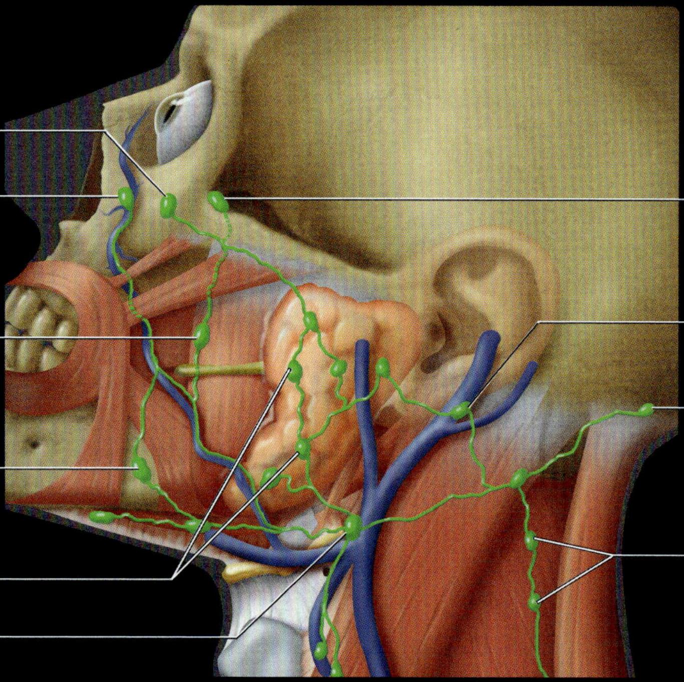

(Top) *Lateral oblique graphic of the neck shows the anatomic locations for the major nodal groups of the neck. Division of the internal jugular nodal chain into high, middle, and low regions is defined by the level of the hyoid bone and cricoid cartilage. Similarly, the spinal accessory nodal chain is divided into high and low regions by the level of the cricoid cartilage.* **(Bottom)** *Lateral view shows facial nodes plus parotid nodes. None of these nodes bear level numbers but instead must be described by their anatomic location. Note that the internal jugular chain is the final common pathway for all lymphatics of the upper aerodigestive tract and neck.*

AXIAL CECT

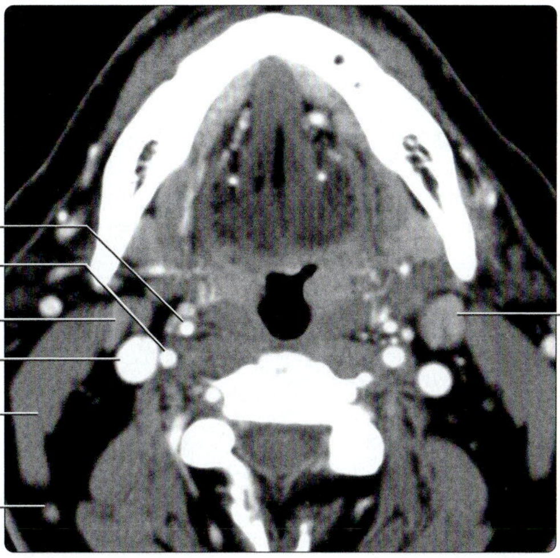

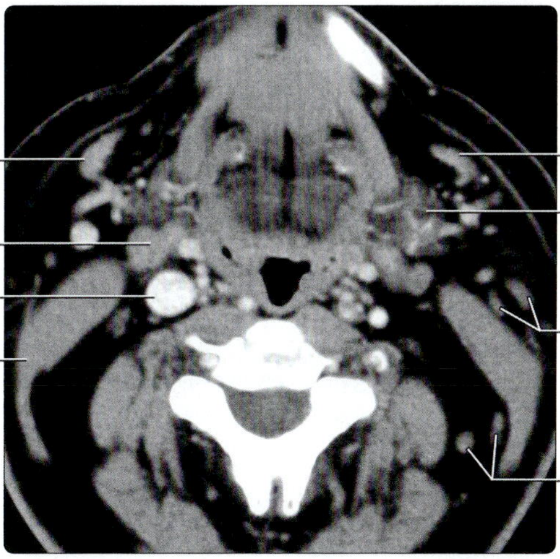

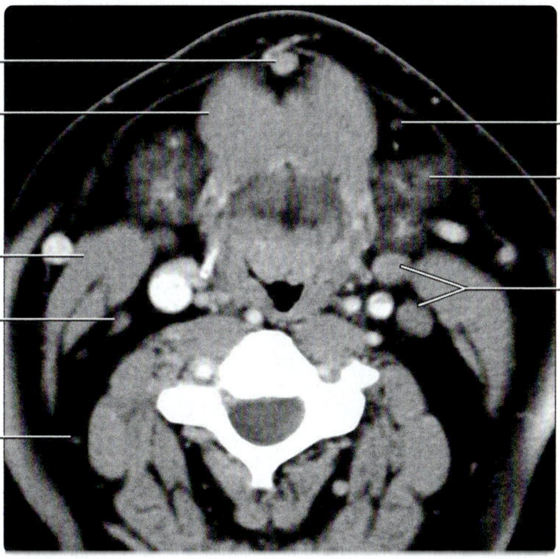

(Top) *First of 3 axial CECT images of the suprahyoid neck, presented from superior to inferior, demonstrates lymph nodes in the internal jugular (level II) and spinal accessory chains (level V). The jugulodigastric node is the highest or "sentinel" node of the internal jugular chain.* **(Middle)** *In this image, the internal jugular & spinal accessory lymph nodes are seen along with submandibular nodes (level IB) anterolateral to the submandibular glands in the submandibular space. Note that the internal jugular nodes are immediately adjacent to the carotid space, whereas the spinal accessory nodes are in the posterior cervical space.* **(Bottom)** *In this image just above the hyoid bone, a submental (level IA) node is seen between the anterior bellies of the digastric muscles. Note also the submandibular (level IB), high internal jugular (level IIA & IIB), & spinal accessory (level VA) nodes.*

TRANSVERSE AND LONGITUDINAL US AND PATHOLOGY

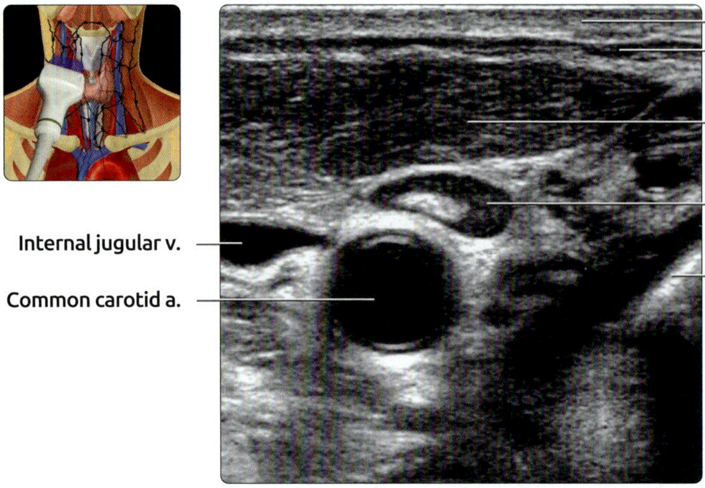

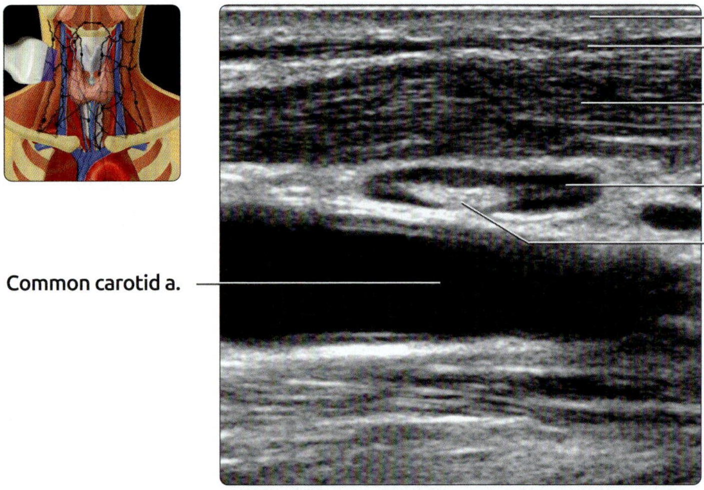

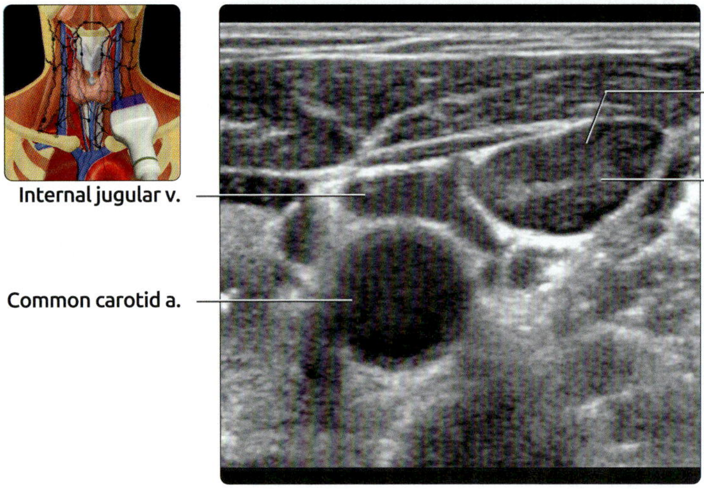

(Top) *Transverse grayscale ultrasound of the midcervical level shows the normal appearance of a cervical lymph node (i.e., ovoid shape with echogenic hilum). It is commonly found anterior to the carotid artery/internal jugular vein.* **(Middle)** *Longitudinal grayscale ultrasound of the midcervical level shows a normal elliptical hypoechoic lymph node with echogenic hilum anterior to the common carotid artery.* **(Bottom)** *Transverse grayscale ultrasound shows a hypoechoic node with cortical hypertrophy and echogenic hilum along the deep cervical/jugular chain. This is the classic appearance of a reactive node. Note its relation to internal jugular vein & common carotid artery. This is a common site of reactive nodes, which are often bilateral & symmetric.*

POWER DOPPLER US AND PATHOLOGY

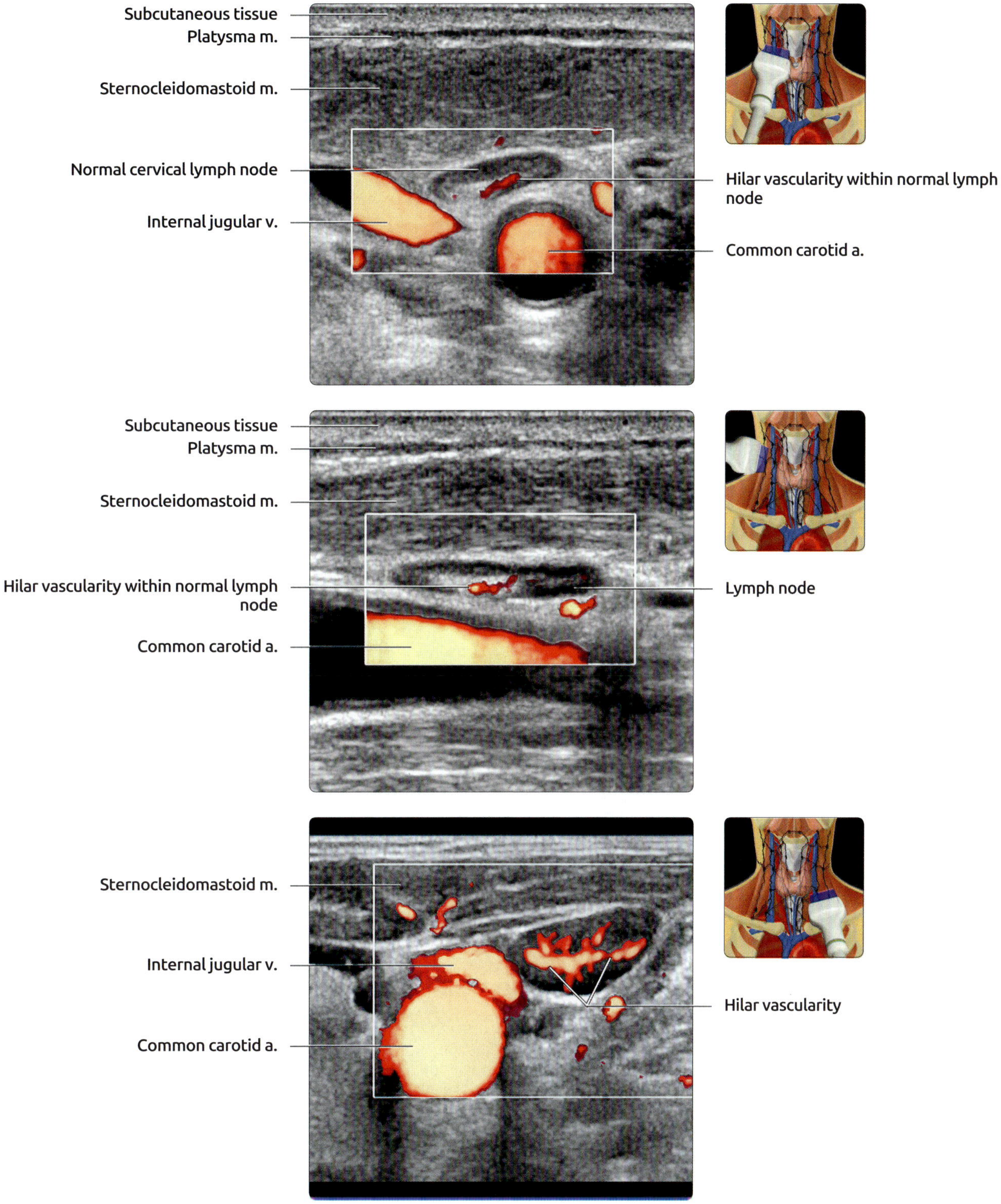

(Top) *Transverse power Doppler ultrasound shows the presence of hilar vascularity within the echogenic hilum of a normal cervical lymph node.* **(Middle)** *Longitudinal power Doppler ultrasound shows hilar vascularity within the echogenic hilum of a normal cervical lymph node. The presence of echogenic hilum and hilar vascularity are good signs of cervical lymph node benignity.* **(Bottom)** *Power Doppler ultrasound of a reactive node clearly defines radiating hilar vascularity and relation to IJV and CCA.*

PATHOLOGY

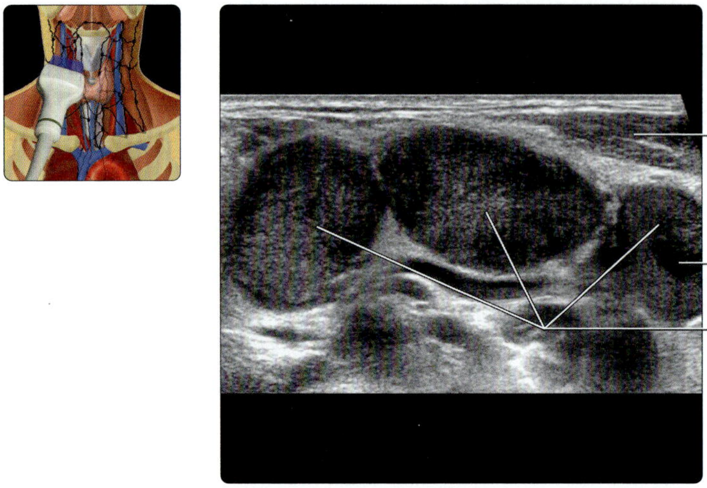

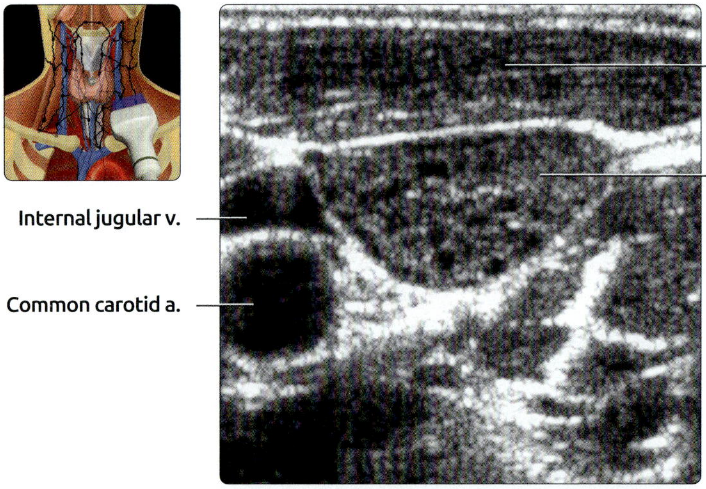

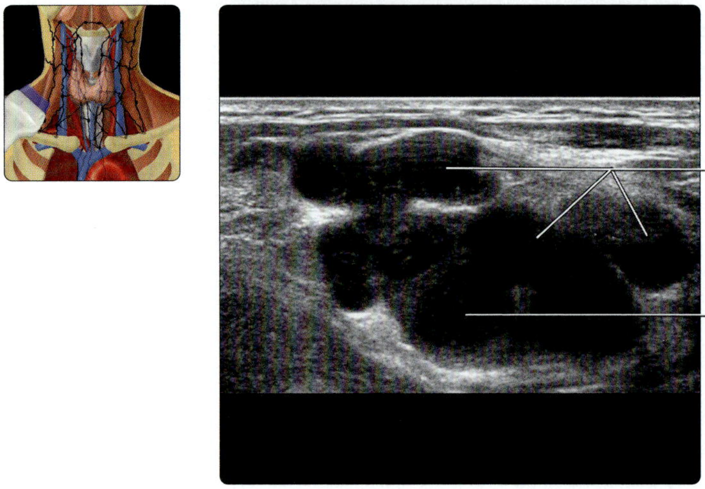

(Top) *Transverse grayscale ultrasound of the upper neck shows multiple enlarged, round, predominantly solid, hypoechoic lymph nodes. Patient has known history of H&N cancer. Overall features are consistent with metastatic nodes. Presence of intranodal cystic necrosis in a patient with known primary malignancy is indicative of the metastatic nature of lymph nodes.* **(Middle)** *Transverse grayscale ultrasound of a lymphomatous lymph node in mid deep jugular chain demonstrates the typical reticulated echo pattern.* **(Bottom)** *Transverse grayscale ultrasound shows multiple matted, enlarged, heterogeneous, hypoechoic lymph nodes in the posterior triangle. Some of them demonstrate intranodal necrosis. A mild degree of edema is noted in the adjacent soft tissue. Features are compatible with tuberculous lymphadenitis.*

PATHOLOGY

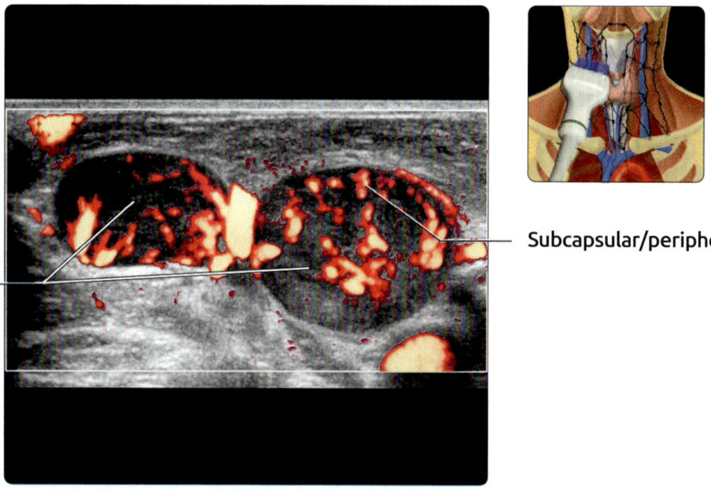

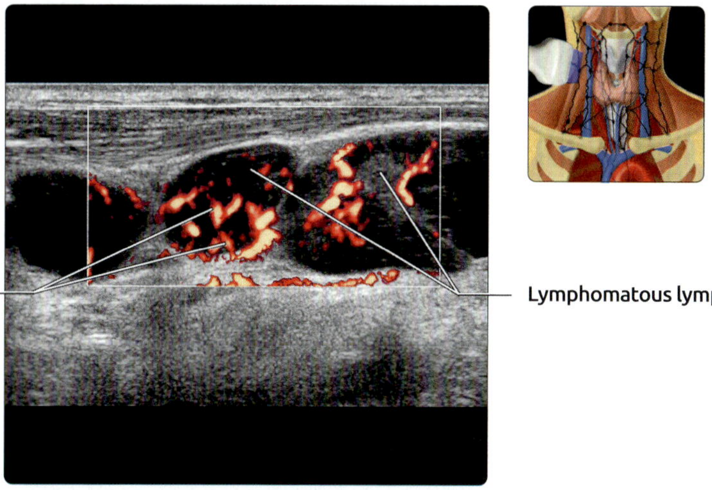

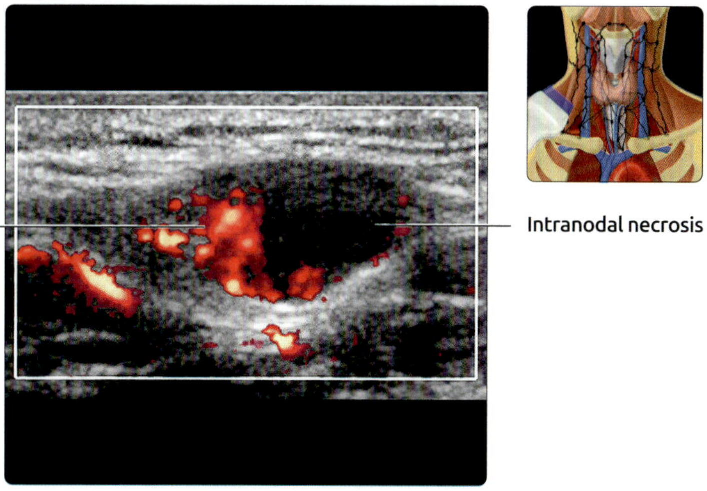

(Top) *Transverse power Doppler ultrasound shows multiple subcapsular/peripheral intranodal vessels in multiple round, hypoechoic, solid lymph nodes at the upper cervical level. Pathology confirmed metastatic squamous cell carcinoma.* **(Middle)** *Longitudinal power Doppler ultrasound of multiple lymphomatous lymph nodes shows chaotic peripheral and central intranodal vessels. Note that hilar vascularity is more prominent than peripheral vascularity.* **(Bottom)** *Transverse power Doppler ultrasound shows a tuberculous lymph node in posterior triangle that is predominantly hypovascular with displaced hilar vascularity. The hypovascular portion corresponds to intranodal caseating necrosis.*

SECTION 3

Thorax

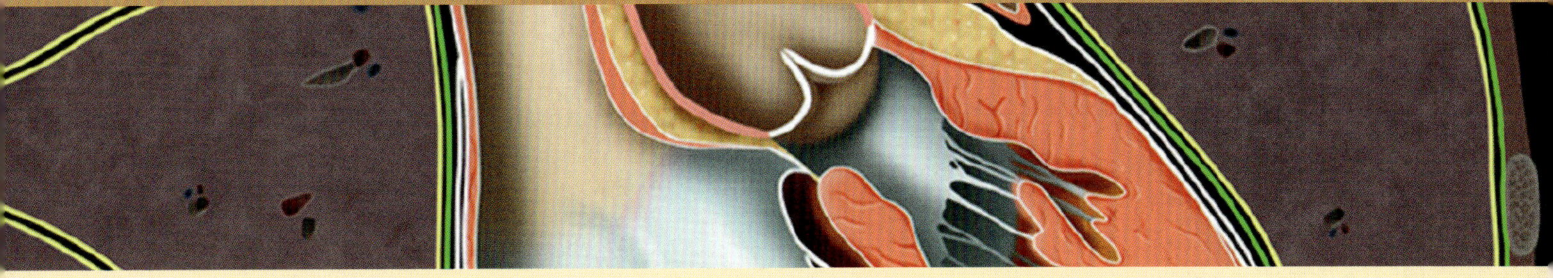

TERMINOLOGY

Synonyms

- Terminology is confusing as anatomic and clinical definitions vary
- **Thoracic outlet is clinical term** for junction of head and neck
- Anatomically this area is actually referred to as **thoracic inlet**, with thoracic outlet being opening into abdominal cavity
- Superior thoracic aperture also synonym but technically is only bony borders

Definitions

- Transition area between cervical/brachial spaces and thoracic cavity
- Opening at superior end of rib cage bordered by bony ring
  - Posteriorly: T1 vertebral body
  - Laterally: Right and left 1st ribs and their costal cartilages
  - Anteriorly: Superior border of manubrium

IMAGING ANATOMY

Overview

- Contents passing through outlet include
  - Trachea
  - Esophagus
  - Great vessels for head and neck and arms
  - Nerves
- Thoracic outlet is clinically divided into **3 compartments**
  - From medial to lateral: Interscalene triangle, costoclavicular space, retropectoralis minor space

Interscalene Triangle

- Boundaries
  - Anterior scalene (anteriorly)
  - Middle and posterior scalene (posteriorly)
  - 1st rib (inferiorly)
- Contents
  - Subclavian artery (inferiorly)
  - 3 trunks of brachial plexus (upper and middle trunks in superior part of triangle, inferior trunk behind subclavian artery)
- Subclavian vein is anterior to anterior scalene muscle and is thus not within interscalene triangle

Costoclavicular Space

- Boundaries
  - Clavicle (superiorly)
  - Subclavius muscle (anterior)
  - 1st rib and middle scalene muscle (posterior)
- Contents
  - Subclavian artery and vein
  - 1st part of axillary vein
    - Axillary artery begins at lateral border of 1st rib as direct continuation of subclavian artery
    - Axillary artery ends laterally at inferior border of teres major muscle
    - Posterior to axillary vein
  - 1st part of axillary vein
    - Anterior to axillary artery
  - 2 divisions and 2 cords of brachial plexus form within this space
    - Posterior cord of brachial plexus is formed by posterior division of upper, middle, and lower trunks
    - Lateral cord is formed by anterior division of upper and middle trunks
    - Medial cord is formed by anterior division of lower trunk

Retropectoralis Minor Space

- Boundaries
  - Posterior border of pectoralis minor muscle (anteriorly)
  - Subscapularis muscle (posterosuperiorly)
  - Anterior chest wall (posteroinferiorly)
- Contents
  - 2nd part of axillary vein and artery
  - 3 cords of brachial plexus: Posterior, lateral, and medial
    - Names of 3 cords of brachial plexus indicate their relationship with 2nd part of axillary artery
    - Lateral to pectoralis major muscle, cords divide into 5 terminal branches

Clinical Significance

- **Thoracic outlet syndrome**
  - Compression syndrome of neurovascular structure(s)
    - Causes arm pain when arm is adducted
  - Typically cervical rib or elongated C7 transverse process present
  - Neurological compression is common in both costoclavicular space and interscalene triangle
  - Arterial compression is most common in costoclavicular space followed by interscalene triangle
    - Causes narrowing with poststenotic aneurysm
  - Venous compression may cause subclavian vein thrombosis

ANATOMY IMAGING ISSUES

Imaging Recommendations

- Ultrasound provides distinct advantages in evaluation
  - Allows simultaneous, real-time grayscale (anatomical) and color Doppler (functional) evaluation of vessels to assess compression and flow restriction
  - Can be performed with dynamic maneuvers, which exacerbates thoracic outlet syndrome clinically, allowing simultaneous imaging assessment
  - Allows patients to be scanned in upright position, which frequently exacerbates compression (CT and MR requires patients to be scanned supine)

Imaging Pitfalls

- Ultrasound cannot evaluate lungs
- Shadowing from lung apices, ribs, and clavicle may make complete evaluation difficult
- 8th cervical and 1st thoracic nerve roots are not adequately visualized, especially in patients with short necks

GRAPHICS, THORACIC OUTLET

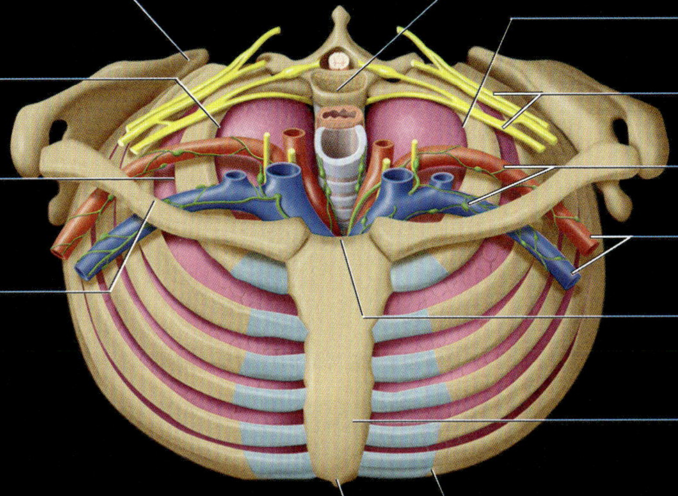

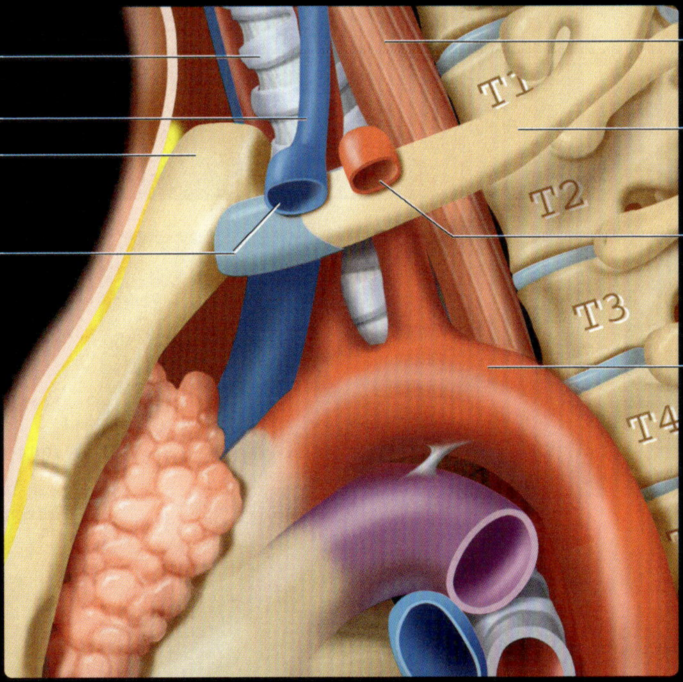

(Top) *Graphic depicts the thoracic outlet that is bounded by the T1 vertebral body, right and left 1st ribs and their costal cartilages, and the manubrium of sternum. Vascular structures allow blood flow to enter and exit the thorax through this ring. The apex of the adjacent axillary region is also shown, which is bounded by the clavicle, scapula, and outer border of the 1st rib.* **(Bottom)** *Lateral graphic of the thoracic outlet shows the subclavian artery and vein as they pass out of the thorax over the 1st rib. At the lateral margin of the 1st rib, they are renamed the axillary artery and vein, and continue as the vascular supply to the arm. The trachea and esophagus are 2 important structures that pass through the thoracic outlet.*

SUPERIOR THORACIC APERTURE

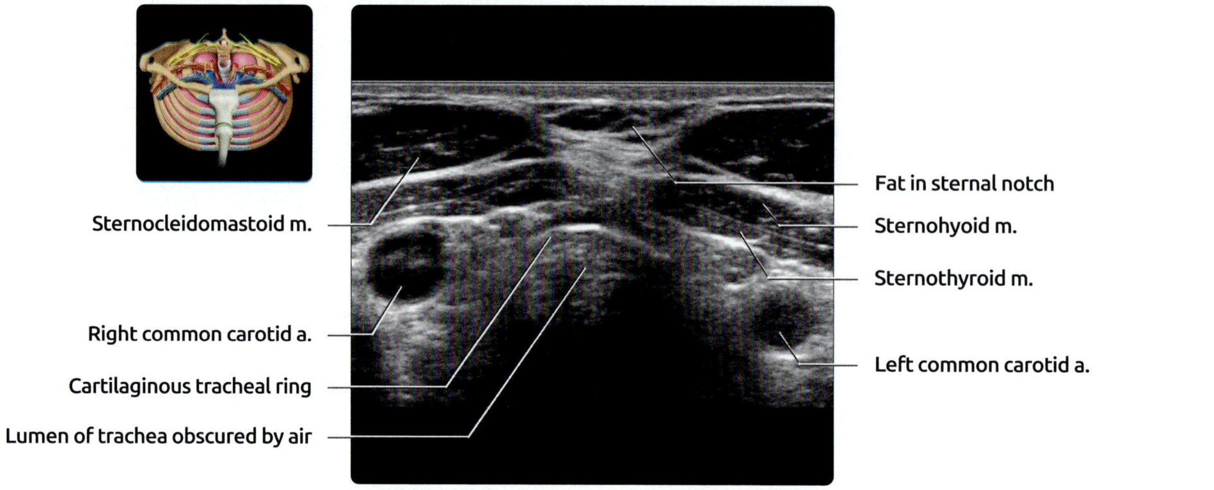

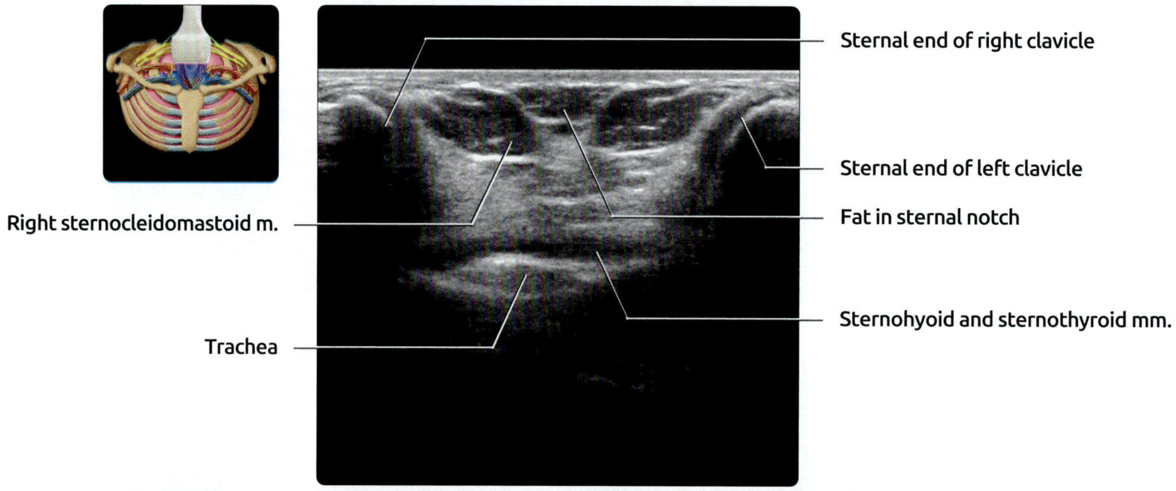

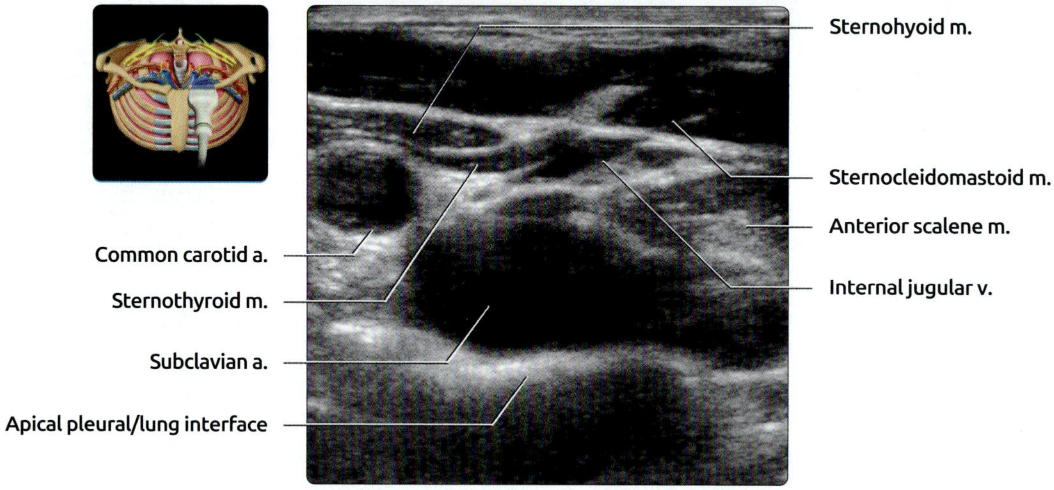

(Top) *Transverse scan of superior thoracic aperture at the sternal notch is shown. The trachea is located in the center of the image, where its cartilaginous ring is well-visualized, but the lumen is obscured by gas. Strap muscles of the neck, which are anterior to the trachea, are well-delineated.* **(Middle)** *Transverse scan at sternal notch level with caudal tilting of transducer to better demonstrate the thoracic outlet is shown. The strap muscles appear thinner using this scanning plane. A retrosternal extension of goiter can be demonstrated in this view of the superior thoracic aperture.* **(Bottom)** *Transverse scan of left supraclavicular fossa is shown. The lung apex is located deep to the vessels and immediately behind the subclavian vein. This view is good for searching for supraclavicular lymphadenopathy.*

SUPERIOR THORACIC APERTURE

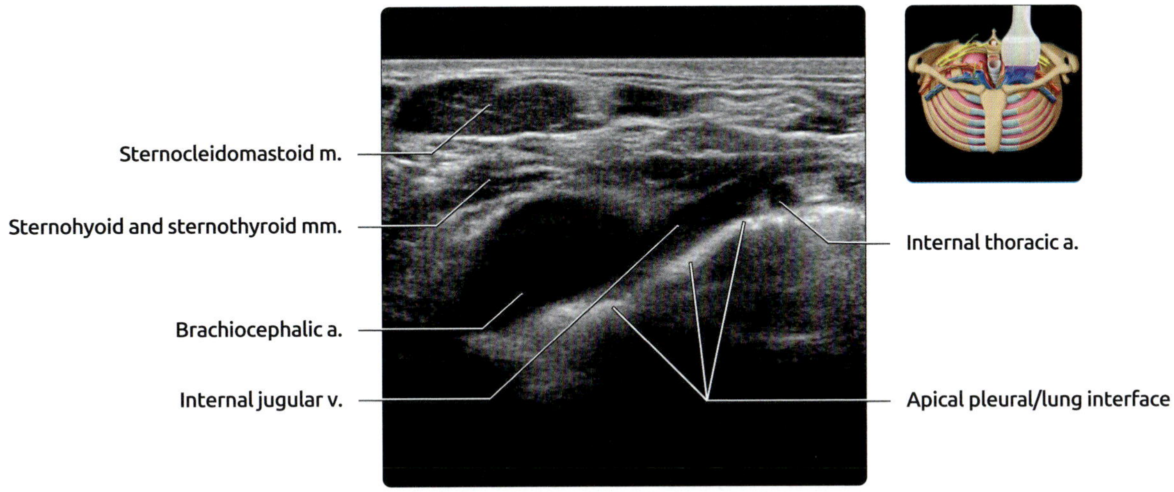

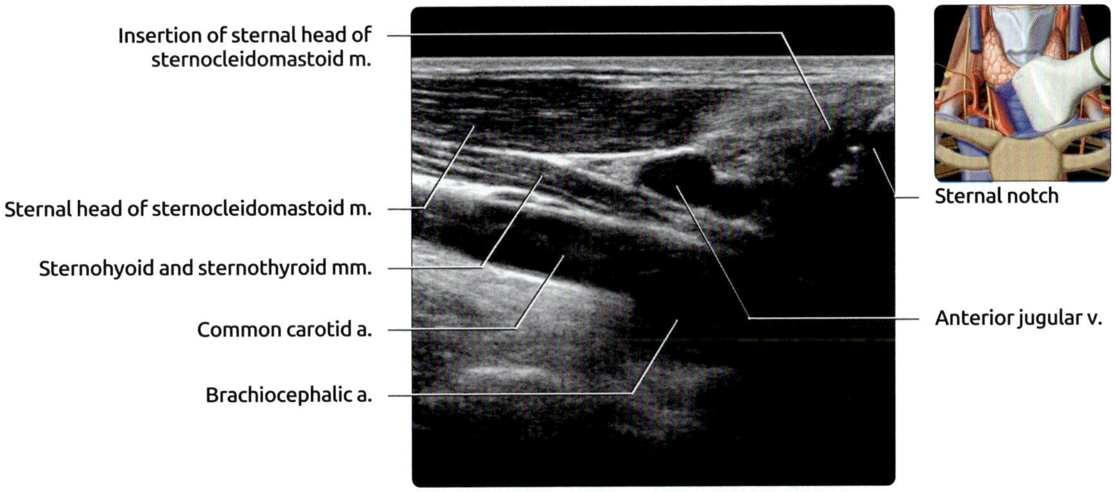

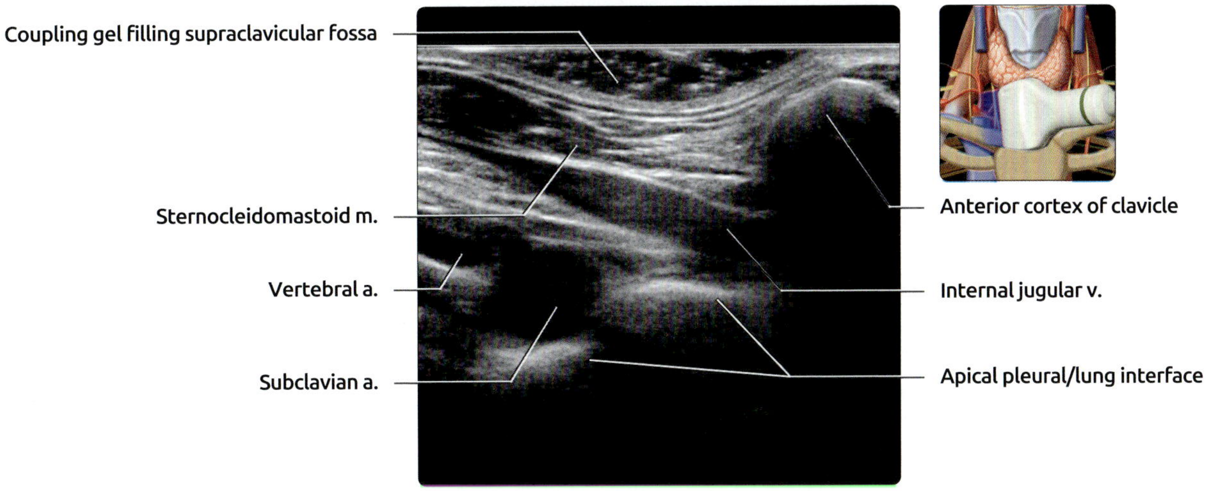

(Top) *Transverse scan of left supraclavicular fossa with caudal tilting of the transducer is shown. The apical pleural/lung interface is visualized immediately behind the brachiocephalic artery, internal jugular vein, and internal thoracic artery.* **(Middle)** *Longitudinal scan of superior thoracic aperture at the sternal notch is shown. Insertion of sternocleidomastoid at the sternal notch appears echogenic and tapered compared to the muscle belly. The anterior jugular vein crosses anterior to strap muscles and common carotid artery.* **(Bottom)** *Longitudinal scan of supraclavicular fossa along the internal jugular vein is shown. The supraclavicular fossa is filled with coupling gel to provide good transducer contact. The apical pleural/lung interface is visualized immediately behind internal jugular vein and subclavian artery.*

INTERSCALENE TRIANGLE

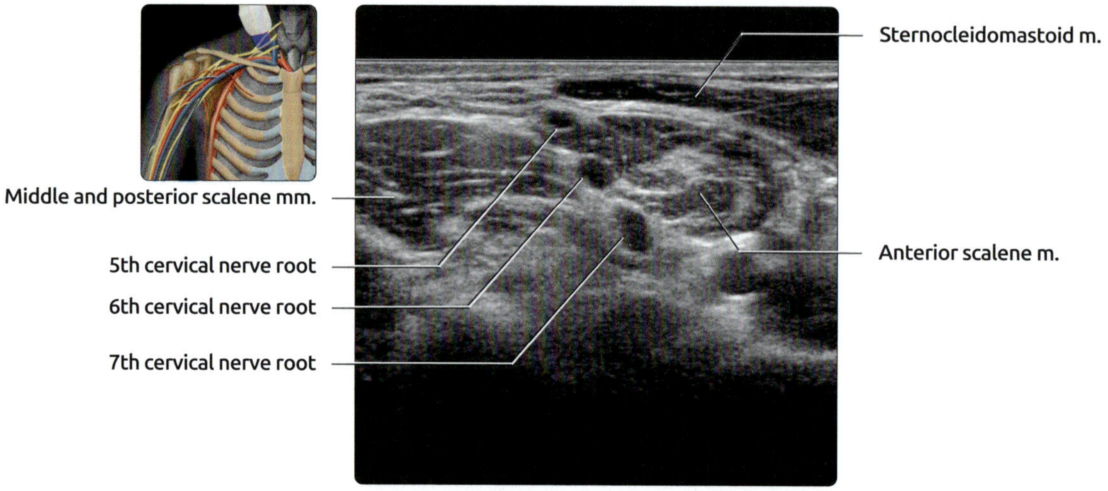

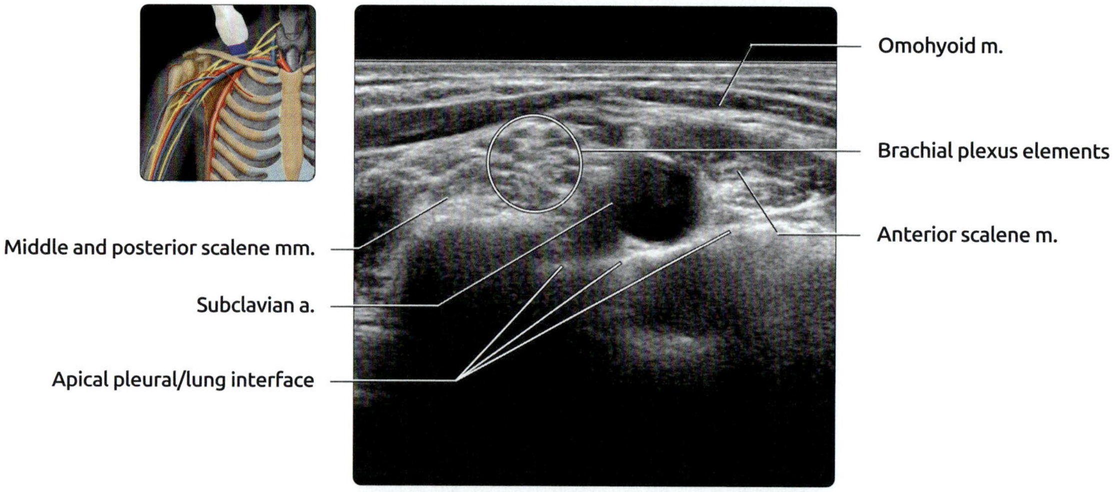

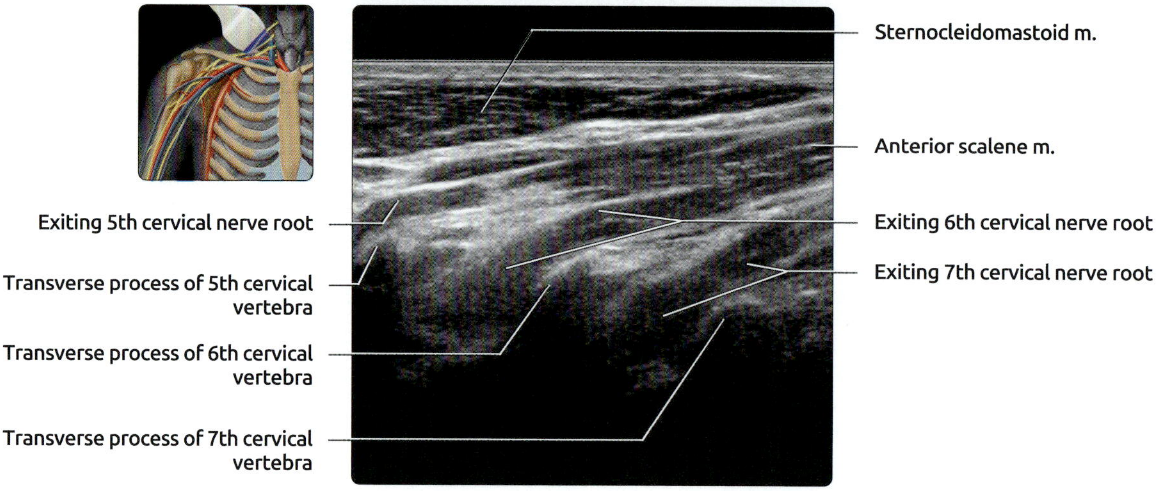

(Top) *Oblique coronal scan of upper interscalene triangle is shown. The 5th, 6th, and 7th cervical nerve roots are well-demonstrated in cross section, lying between anterior and middle scalene muscles. The 8th cervical and 1st thoracic nerve roots are usually more difficult to visualize on ultrasound because of their deeper location behind subclavian artery.* **(Middle)** *An oblique coronal scan of lower interscalene triangle shows the brachial plexus elements. They appear as a cluster of well-delineated, round, hypoechoic structures that result from the divisions of the nerve trunks. The subclavian artery is seen posterior and medial to the brachial plexus.* **(Bottom)** *This oblique coronal scan of interscalene triangle is focused on the nerve roots. The extraforaminal parts of the 5th, 6th, and 7th cervical nerve roots are well-visualized as they exit from the intervertebral foramina in the downward and outward direction.*

COSTOCLAVICULAR SPACE

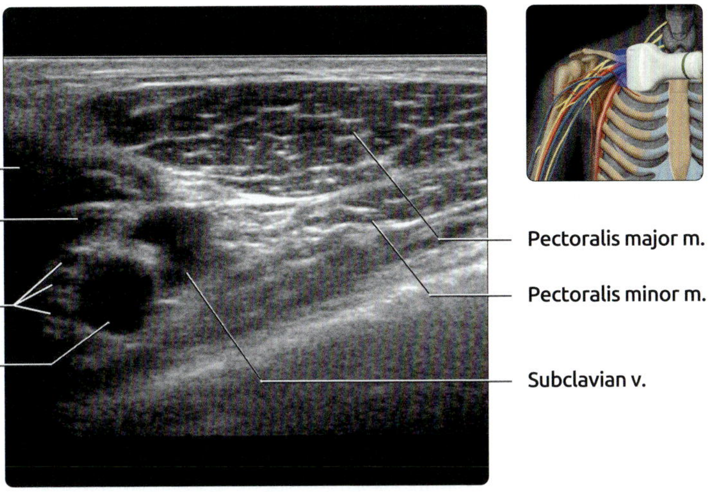

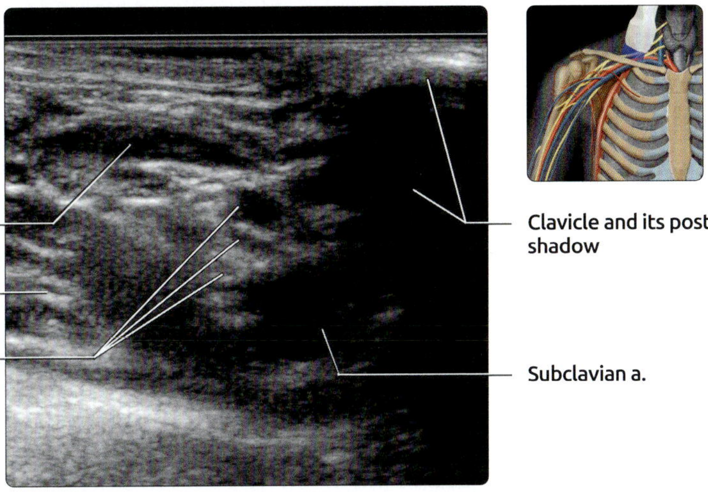

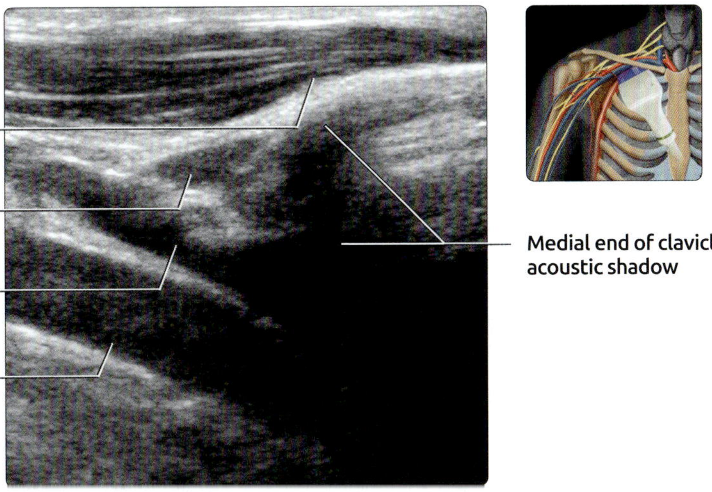

(Top) *Sagittal scan of costoclavicular space is shown using an infraclavicular approach. This space is limited superiorly by the clavicle and anteriorly by the subclavius muscle and contains the subclavian vein anteriorly, the subclavian artery immediately posterior to it, and 3 cords of brachial plexus.* **(Middle)** *Sagittal scan of costoclavicular space from the supraclavicular approach shows the brachial plexus and subclavian artery. The subclavian vein is obscured by the shadowing posterior to the clavicle.* **(Bottom)** *The costoclavicular space is shown in a transverse plane. The long axis of subclavian artery and vein are visualized in this view, but the brachial plexus is difficult to visualize because of its proximity to the clavicle.*

RETROPECTORALIS MINOR SPACE

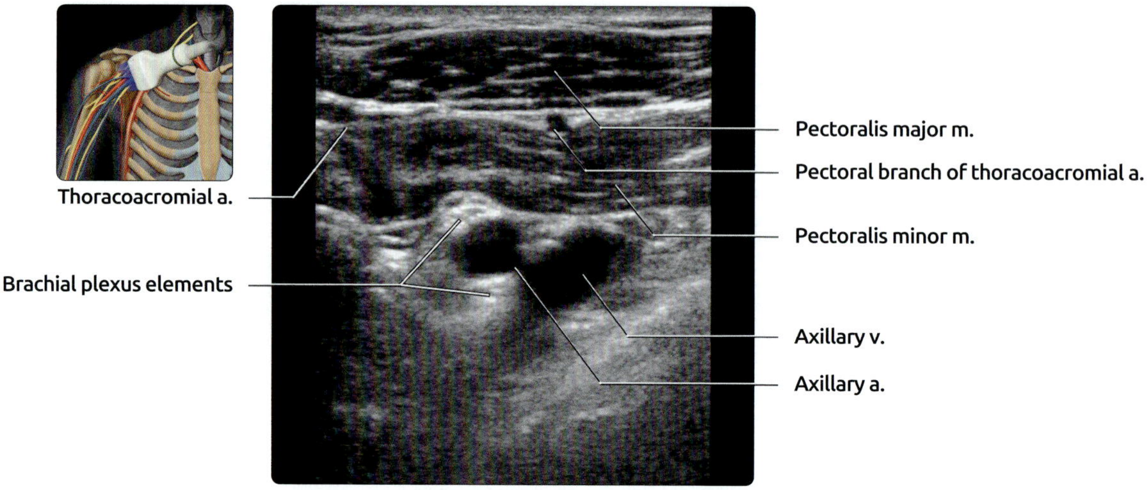

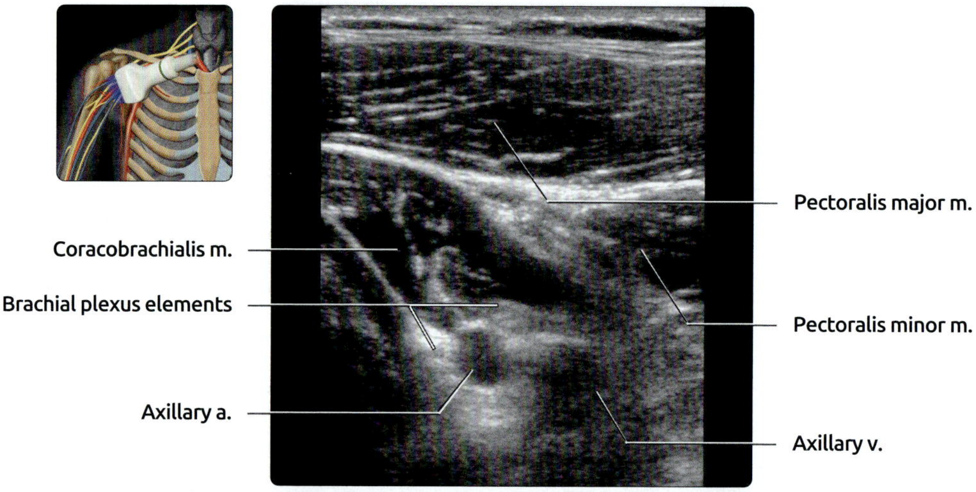

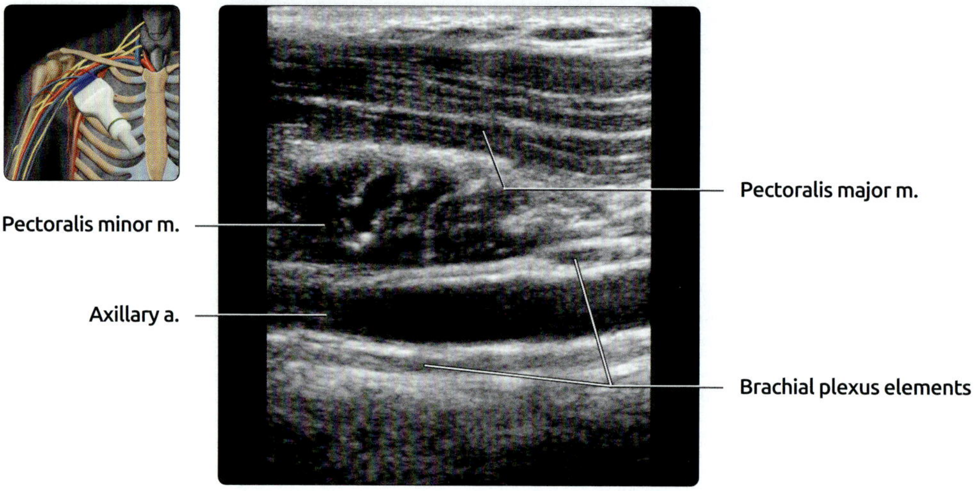

(Top) *A sagittal scan of retropectoralis minor space is shown. This space is bound anteriorly by the pectoralis minor muscle and posteriorly by the anterior chest wall. The axillary artery and vein are well-depicted, and the cords of brachial plexus are visualized above and posterior to the axillary artery.* **(Middle)** *Just lateral to the retropectoralis minor space the brachial plexus and axillary artery and vein are moving away from the pectoralis minor muscle, but with a similar configuration as seen in the retropectoralis minor space.* **(Bottom)** *A transverse scan of the retropectoralis minor space shows the brachial plexus running parallel to, above, and posterior to the axillary artery. On a longitudinal scan, the axillary vein cannot be visualized together with the axillary artery and brachial plexus in the same image because of their triangular configuration.*

DYNAMIC INTERROGATION

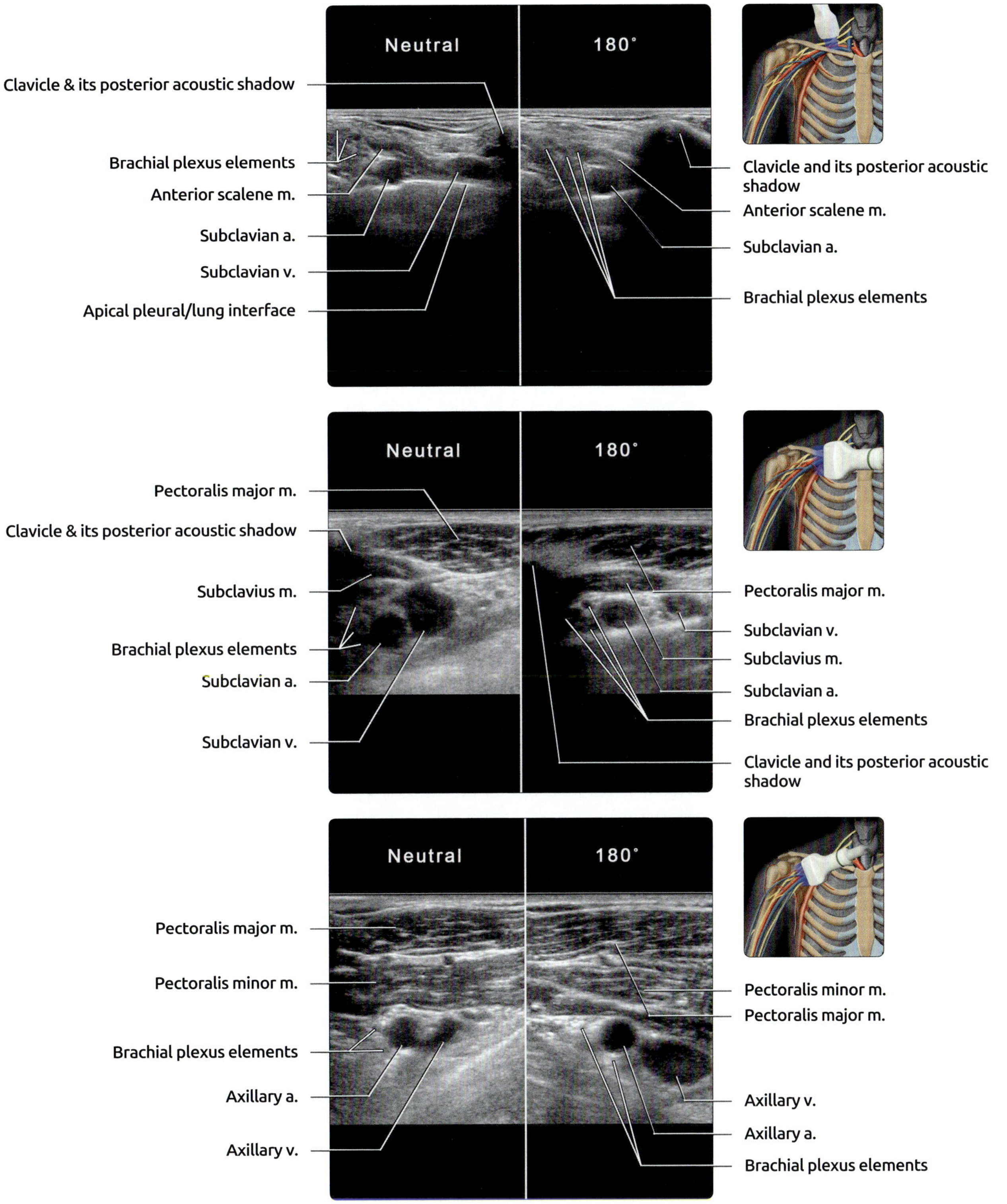

(Top) *Dual sagittal image of lower interscalene triangle with the arm in neutral position (left image) and 180 degree abduction (right image) is shown. The space in the interscalene triangle decreases with the arm abducted. The brachial plexus and subclavian arterial caliber show no significant change.* **(Middle)** *Dual sagittal image of costoclavicular space (from infraclavicular approach) with the arm in a neutral position (left image) and the arm at 180 degrees (right image) is shown. The subclavian vein appears to be compressed by ~ 50%, which is due to the compressive effect of arm abduction in costoclavicular space.* **(Bottom)** *Dual sagittal image of retropectoralis minor space with arm in neutral position (left image) and 180 degree abduction (right image) is shown. The axillary artery and brachial plexus show no significant change. The axillary vein is more distended with arm abducted.*

SPECTRAL DOPPLER INTERROGATION

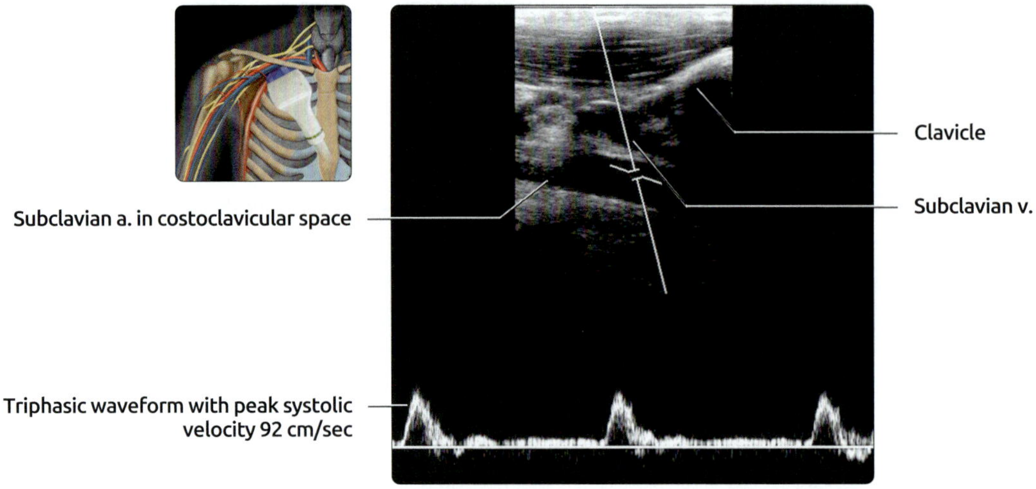

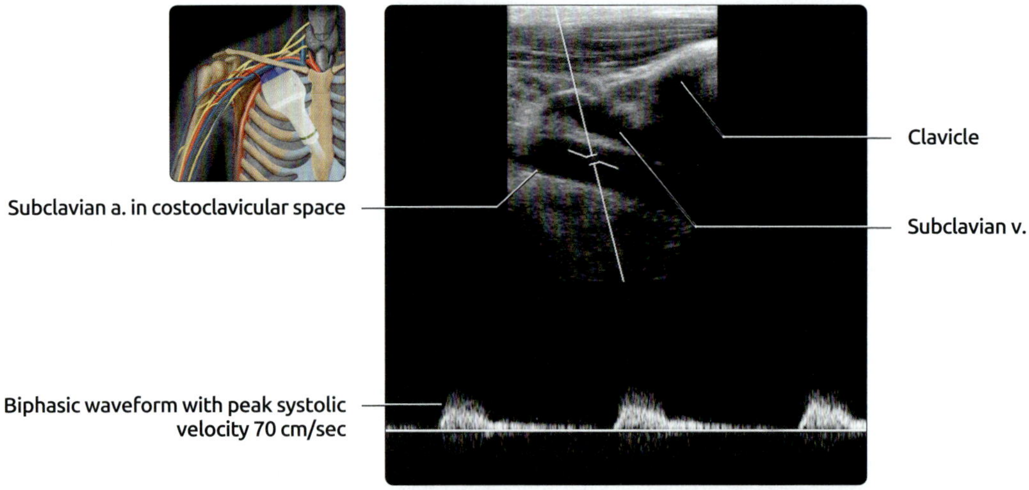

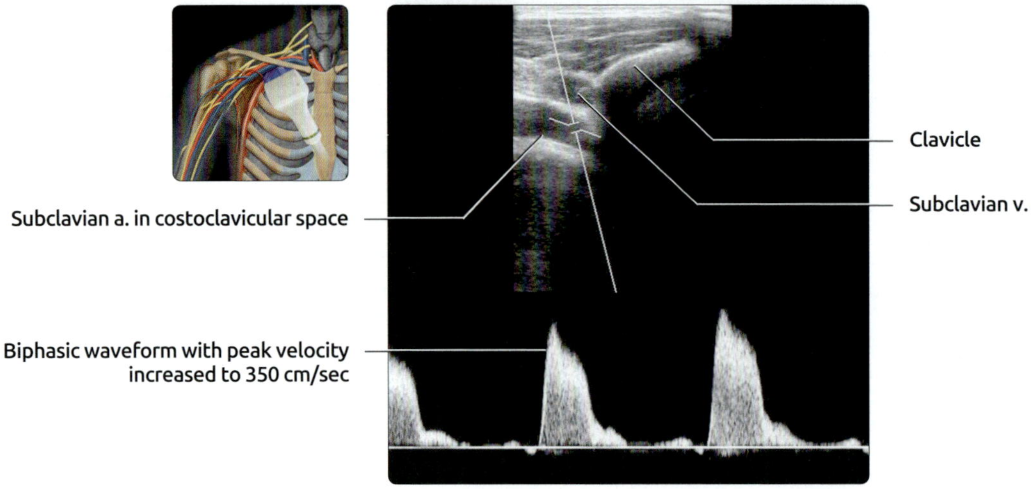

(Top) *Doppler spectrum of the subclavian artery in costoclavicular space is shown with the arm in a neutral position. A normal triphasic waveform is recorded with peak systolic velocity 92 cm/sec.* **(Middle)** *Doppler spectrum of subclavian artery in costoclavicular space is shown with the arm in 90 degrees abduction (same subject as previous image). The waveform changes from triphasic to biphasic, with no significant change in peak systolic velocity.* **(Bottom)** *Doppler spectrum of subclavian artery is shown with the arm in 180 degrees abduction (same subject as previous image). Biphasic waveform is still present, with the peak systolic velocity markedly increased to 350 cm/sec. Features are indicative of arterial compression in costoclavicular space. The costoclavicular space is the most frequent site of arterial compression.*

SPECTRAL DOPPLER INTERROGATION

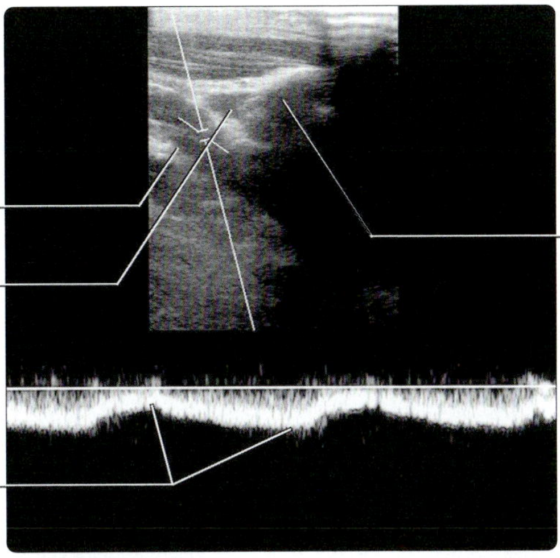

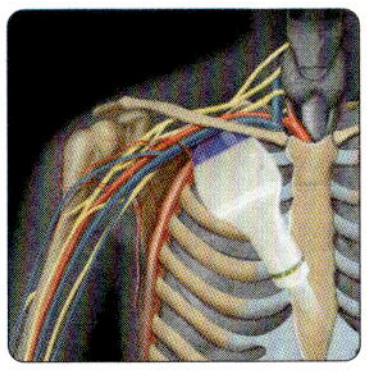

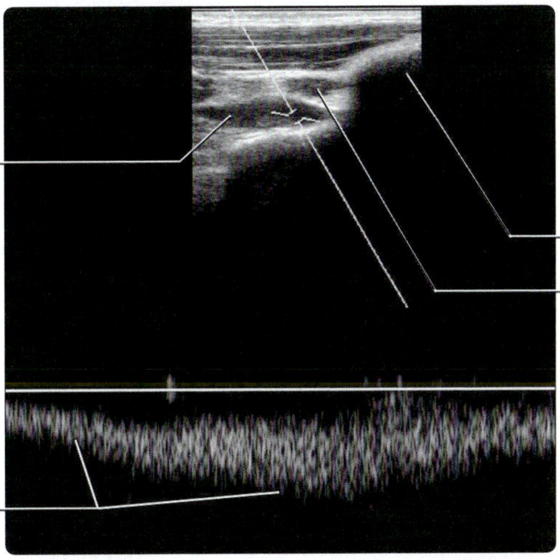

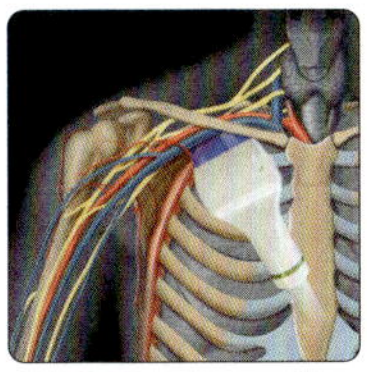

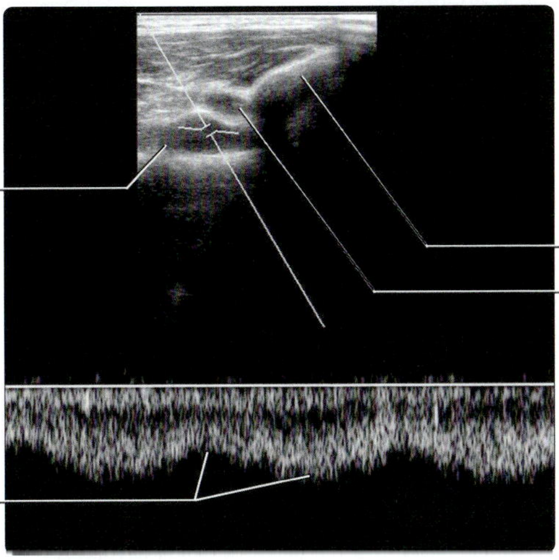

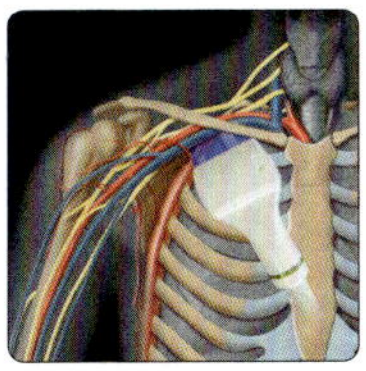

(Top) *Doppler spectrum of the subclavian vein in costoclavicular space is shown with the arm in a neutral position. Normal venous waveform with phasic changes are recorded.* **(Middle)** *Doppler spectrum of the subclavian vein in costoclavicular space is shown with the arm in 90 degrees abduction (same subject as previous image). Normal venous waveform with phasic changes is again recorded. The velocity increases from 15-30 cm/sec.* **(Bottom)** *Doppler spectrum of subclavian waveform in costoclavicular space is shown with arm 180 degrees abduction (same patient as previous image). The velocity increases to 40 cm/sec while normal venous waveform with phasicity is preserved. This indicates a mild compressive effect of arm abduction on subclavian vein.*

GROSS ANATOMY

Overview

- Pleura: Continuous surface epithelium and underlying connective tissue
- Visceral pleura adheres to pulmonary surfaces
- Parietal pleura is continuation of visceral pleura
  - Lines corresponding 1/2 of thoracic wall
  - Covers ipsilateral diaphragm and ipsilateral mediastinal surface
- Visceral and parietal pleurae form right and left pleural cavities
  - Potential spaces containing small amount of serous pleural fluid
- Combined thickness of visceral and parietal pleurae and fluid-containing pleural space is < 0.5 mm
- Visceral pleura directly apposes and slides freely over parietal pleura during respiration

Pleural Space

- Potential space; normally contains 2-10 mL of fluid
- Fluid production capacity: 100 mL/h; fluid absorption capacity: 300 mL/h
- Fluid flux normally from parietal pleura capillaries to pleural space; absorbed by microscopic stomata in parietal pleura

Costodiaphragmatic Recesses

- Pleura extends caudally beyond inferior lung border
- Costal and diaphragmatic pleura separated by narrow slit, costodiaphragmatic recess
- Extends ~ 5 cm below inferior border of lung during quiet inspiration
- Caudal extent at 12th rib posterolaterally

Visceral Pleura

- Covers lung parenchyma surfaces
- Blood supply and drainage
  - Supply by systemic bronchial vessels, drainage by pulmonary and bronchial veins
  - Lymphatic drainage to deep pulmonary plexus within interlobar and peribronchial spaces toward hilum
- Histology
  - Mesothelial layer, thin connective tissue layer, chief layer of connective tissue, vascular layer, limiting lung membrane (connected to chief layer by collagen and elastic fibers)
  - Single layer of flat mesothelial cells separated by basal lamina from underlying lamina propria of loose connective tissue

Parietal Pleura

- Covers nonparenchymal surfaces
- Forms lining of thoracic cavities
- Blood supply and drainage
  - Supply from adjacent chest wall (intercostal, internal mammary, diaphragmatic arteries)
  - Drainage to bronchial veins (diaphragmatic pleural drainage to inferior vena cava and brachiocephalic trunk)
- Histology
  - Single layer of parietal mesothelial cells over loose, fat-containing areolar connective tissue; bounded externally by endothoracic fascia

ANATOMY-BASED IMAGING ISSUES

Ultrasound Technique

- Provides detailed imaging of costal surfaces of pleura
  - Use high-frequency linear transducer
- Ribs cause posterior acoustic shadowing
  - Scanning in both inspiration and expiration helps examine pleura, which may have been obscured by ribs

Clinical Uses

- Differentiation of solid pleural masses from fluid
- Assessment of echogenicity and morphology of fluid collections
- Guide pleural drainage or biopsy

CLINICAL IMPLICATIONS

Pleural Effusion

- Categorized as transudate- or exudate-based on composition of fluid obtained by thoracentesis
  - **Transudate** not associated with pleural disease
    - Systemic abnormalities (cardiac failure, pericardial disease, cirrhosis, pregnancy, hypoalbuminemia, overhydration, renal failure)
  - **Exudate** indicates presence of pleural disease
    - Pneumonia, empyema, tuberculosis, neoplasm, pulmonary embolism, collagen vascular disease
  - Split pleura sign: Thickened visceral and parietal pleura encasing abnormal fluid collection
    - Concerning for empyema in febrile patient
    - May also occur in noninfected collections, malignant effusion, and hemothorax
- Ultrasound
  - Anechoic effusions may be either transudative or exudative
  - Septations suggestive of exudative effusion
    - Often loculated

Pleural Thickening

- **Focal thickening**
  - Asbestos exposure
    - Pleural plaques occur 15-20 years after exposure
    - Focal collections of acellular collagen on parietal pleura (costal, diaphragmatic, and mediastinal pleura)
      - Discontinuous areas of pleural thickening, predominantly along 6th-8th ribs and on parietal pleura at domes of diaphragms
    - May be noncalcified or calcified
    - There is usually sparing apices and costophrenic sulci
  - Localized fibrous tumor
    - Solitary lenticular, round, or lobulated neoplasm
    - Benign (80%) or malignant
  - Bronchogenic carcinoma may focally invade pleura or produce diffuse pleural thickening
- **Diffuse thickening**
  - May be benign (fibrothorax) or malignant (metastases, mesothelioma, lymphoma, invasive thymoma)

PLEURAL SPACES

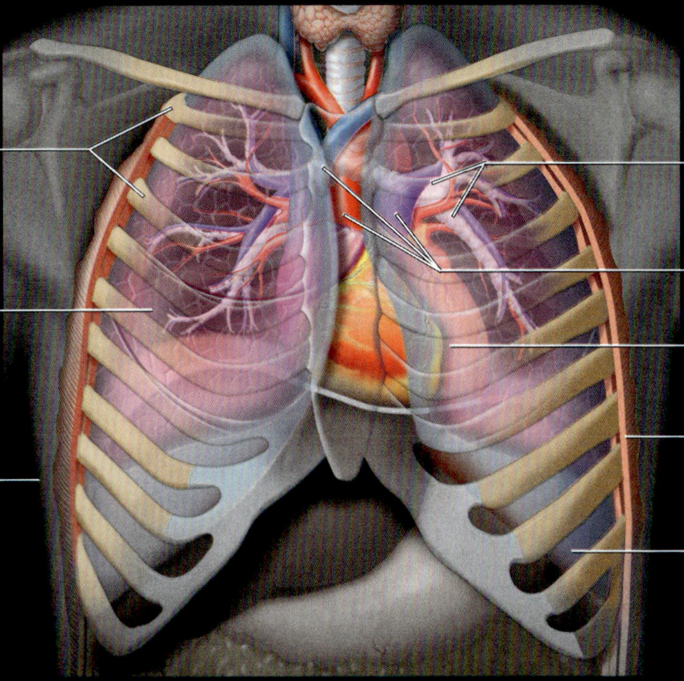

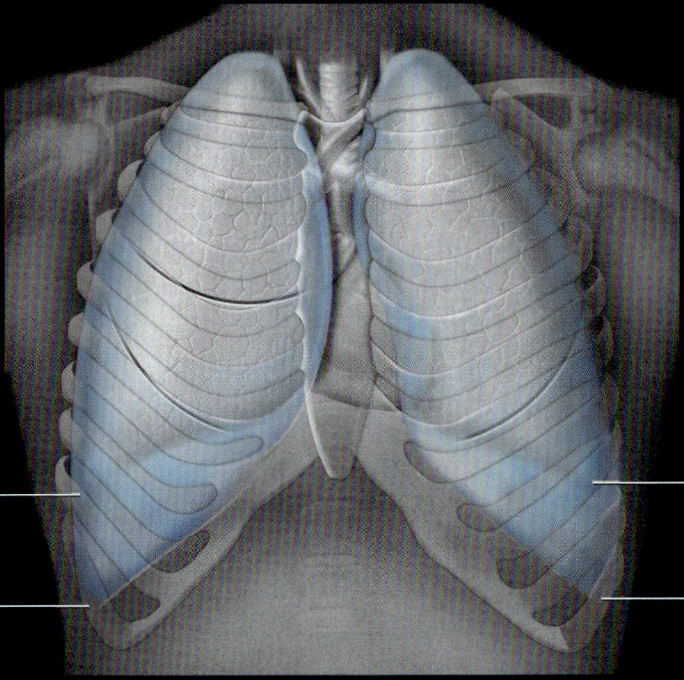

(Top) *Graphic shows the complex and diverse structures and organs of the thorax, including the primary organs of respiration, thoracic cardiovascular system, and proximal gastrointestinal tract.* **(Bottom)** *Graphic shows the anatomy of the pleura with all the other structures removed. The apposed pleural surfaces create a potential space that normally contains a small amount of fluid, which lubricates the pleural surfaces and reduces friction during respiratory motion. The visceral pleura covers the pulmonary surfaces. The parietal pleura covers the nonpulmonary surfaces within the thoracic cavity and extends more caudally than the lungs. The inferior reflection of parietal pleura extends within the costophrenic sulci to the level of the upper kidneys.*

PLEURAL SPACES

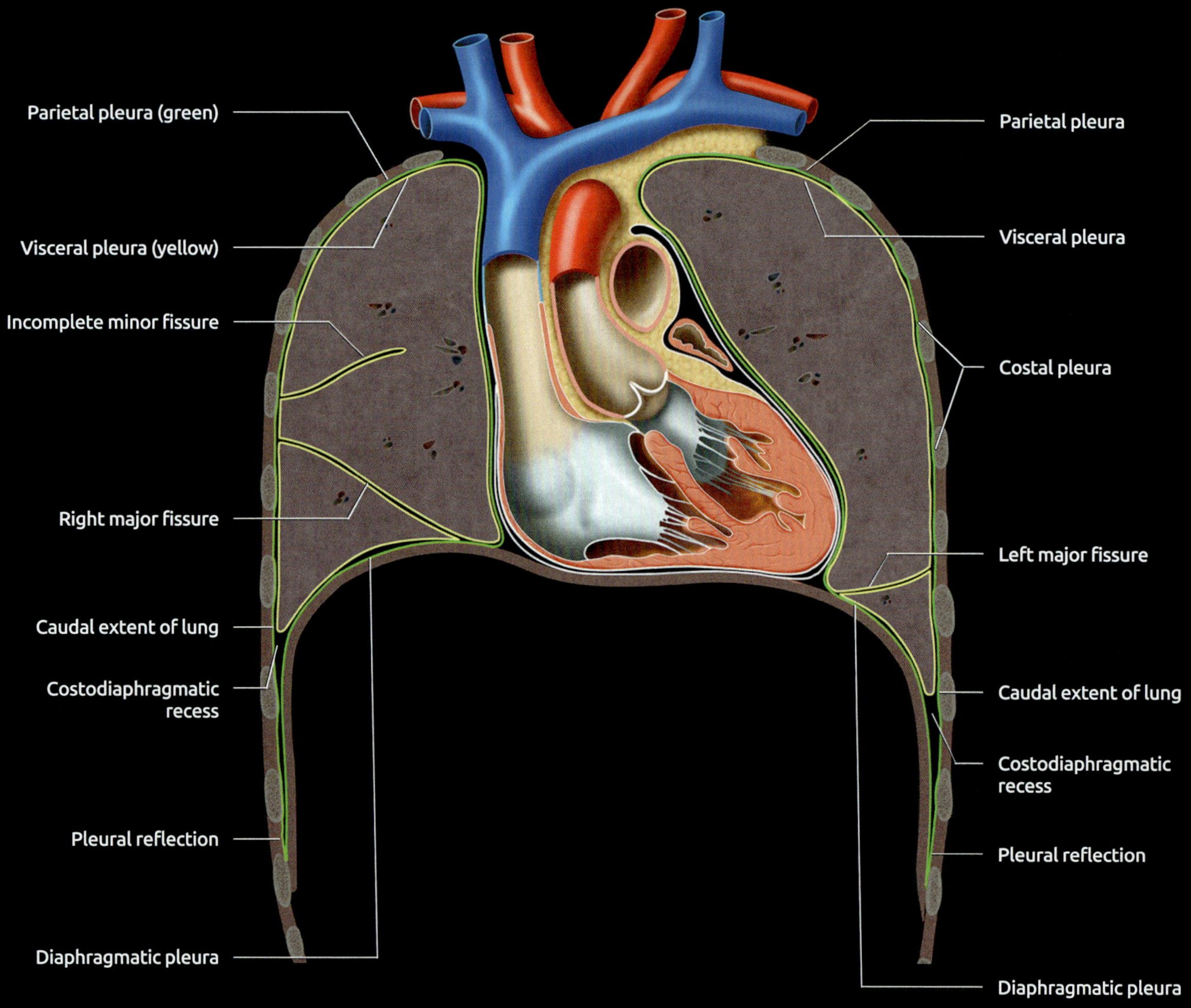

Graphic shows the extent and distribution of the pleura as visualized in the coronal plane. Visceral pleura (yellow) covers the surfaces of both lungs and forms interlobar fissures that may be complete or incomplete in their extension to the hila. Parietal pleura (green) lines both thoracic cavities and may be designated by its location as costal, diaphragmatic, or mediastinal pleurae. Inferiorly, the parietal pleura extends deeply into the costodiaphragmatic recesses where costal and diaphragmatic pleura are in apposition.

GRAPHIC, PLEURAL SPACES

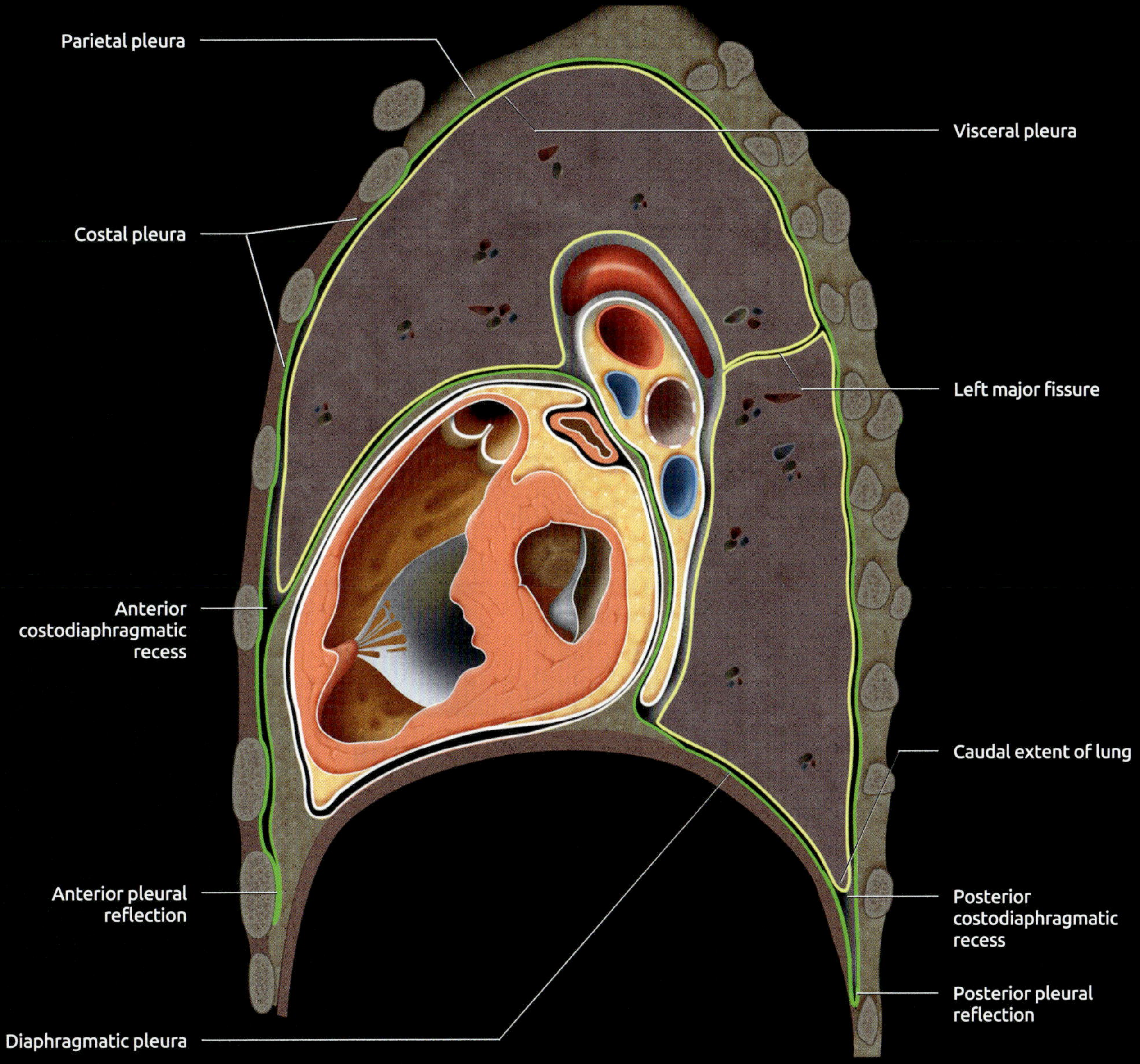

Graphic shows the extent of parietal (green) and visceral (yellow) pleura as visualized in the sagittal plane in the left midclavicular zone. Notice the depth of the pleural reflections. When evaluating for a pleural effusion by ultrasound, it is best to have the patient in the upright position and scan along the posterior costodiaphragmatic recess.

PLEURA, INTERCOSTAL WINDOW

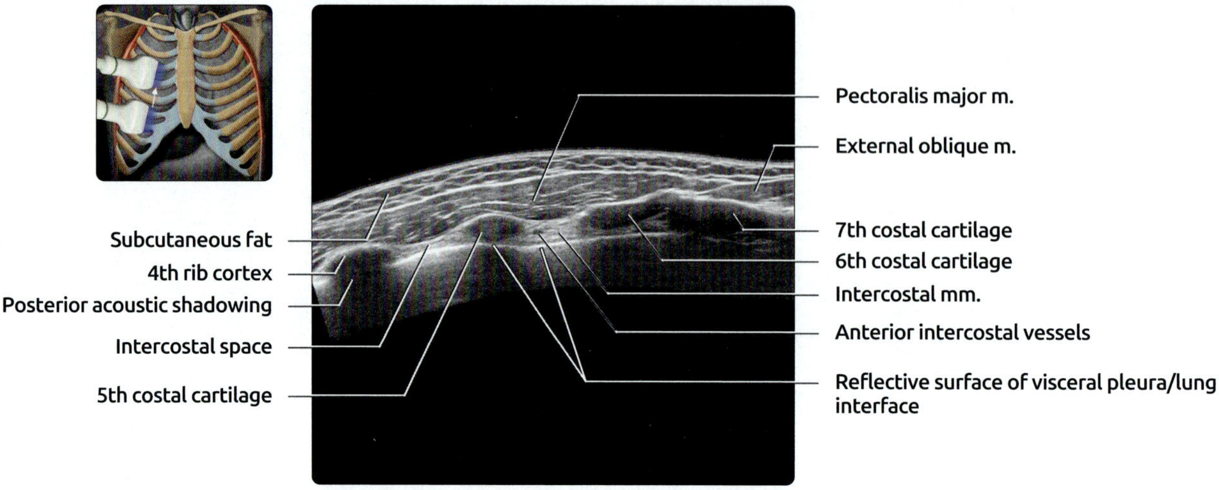

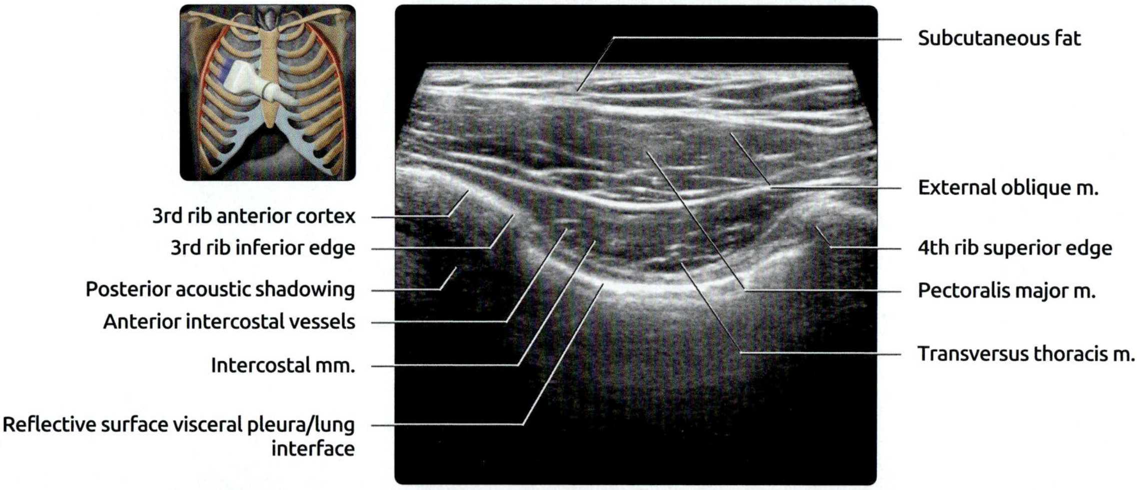

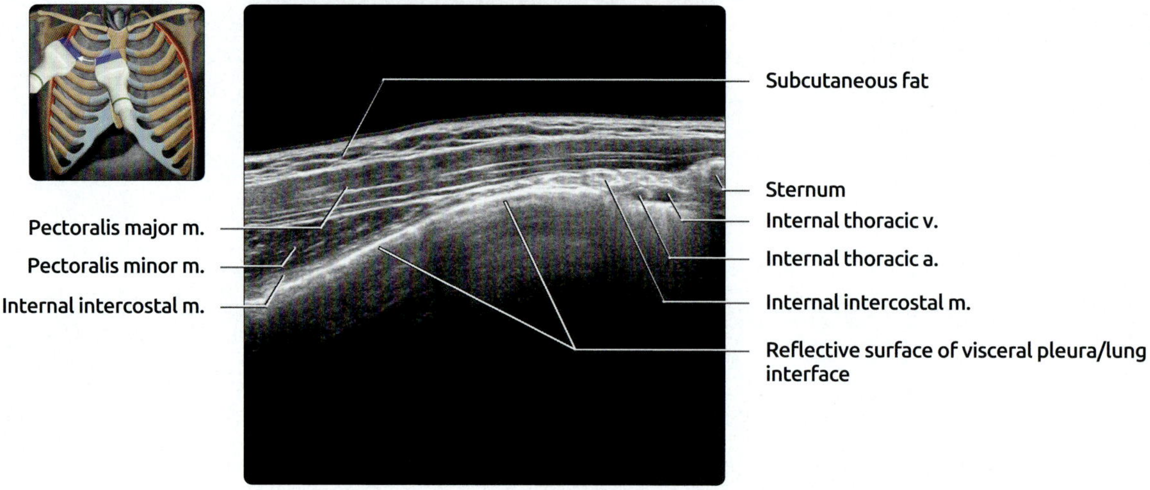

(Top) *Panoramic sagittal scan of the lower anterior chest wall and pleura is shown. The cortex of the ribs produces significant posterior acoustic shadowing, which obscures the underlying pleura; this is less of an issue with costal cartilage (which is relatively sonolucent). By scanning during respiration (and thus pleural movement), all of the pleura can be evaluated with ultrasound.* **(Middle)** *Sagittal scan of the anterior pleura is shown. The intercostal muscles run in between the edges of the ribs. The innermost intercostal muscle is separated from the external and internal intercostal muscles by the intercostal vessels and nerve.* **(Bottom)** *Oblique transverse scan of the anterior pleura just lateral to the sternum using an intercostal window is shown. Uninterrupted strips of pleura can be demonstrated by placing the transducer obliquely along the intercostal space.*

PLEURA

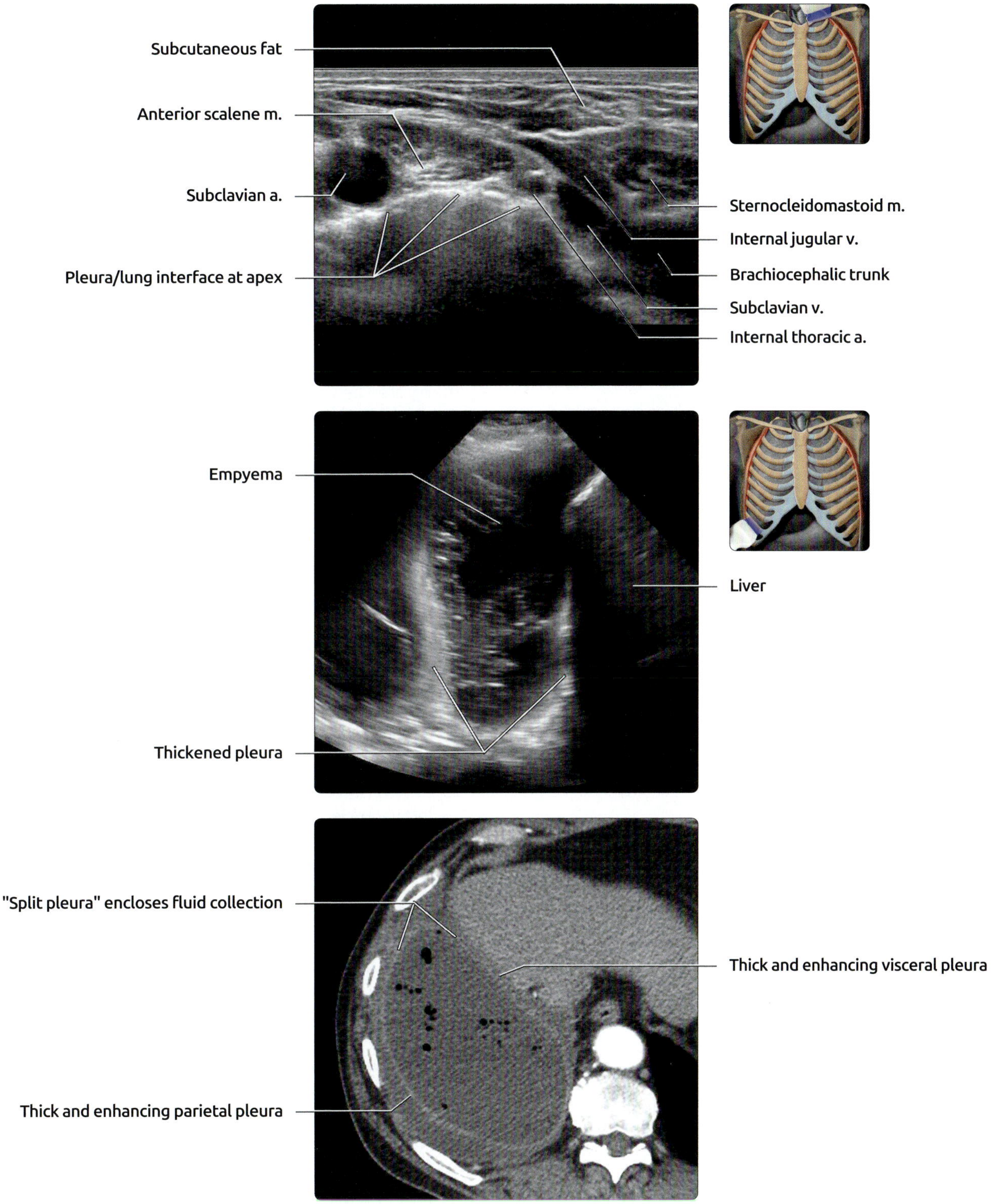

(Top) *Oblique transverse scan of the apical pleura is shown. The apical pleura is adjacent to important cervical structures. Pathology in the lung apex, such as a Pancoast tumor, can therefore extend readily into the neck.* **(Middle)** *Longitudinal ultrasound of the right hemithorax in a patient with pneumonia, persistent fevers, and elevated white count shows a loculated, complex fluid collection. Note the thickened surrounding pleura. Aspiration, using ultrasound guidance, was performed confirming an empyema.* **(Bottom)** *Axial CECT (soft tissue window) of a patient with empyema demonstrates the split pleura sign associated with a loculated fluid collection in the posterolateral aspect of the right inferior pleural space. Enhancing, smoothly thickened pleurae "split" to enclose the abnormal fluid collection.*

IMAGING ANATOMY

Overview

- Sheet of skeletal muscle that merges centrally to form central aponeurotic tendon
- Separates thoracic cavity from abdominal cavity
- Descends to increase thoracic volume and decrease intrathoracic pressure
  - Chief muscle of respiration

Muscular Components

- Costal/rib contribution
  - Muscle fibers from internal surface of ribs 7-12
  - Forms left and right hemidiaphragms
- Lumbar contribution
  - Originates as diaphragmatic crura and from 3 arcuate ligaments
- Sternal contribution
  - Slips of muscle from xiphoid process running posteriorly to insert on central tendon
  - Sternocostal hiatus on each side of these central slips of muscle
    - Internal thoracic vessels run through hiatus to enter abdomen

Central Tendon

- Aponeurosis with interlacing fibers
- Peripheral muscular component inserts centrally onto this aponeurosis
- Trefoil appearance due to 3 parts
  - Heart rests upon central leaf
    - Right side of central leaf contains foramen for inferior vena cava
  - Left and right leaves represent domes of hemidiaphragms

Diaphragmatic Crura

- **Left crus**
  - Narrower and shorter than right crus
  - Attached to left side of aorta, left anterolateral surfaces of L1 and L2 vertebral bodies and discs
- **Right crus**
  - Broader and longer than left crus
  - Attached to right side of aorta, right anterolateral surfaces of L1-L3 vertebral bodies and discs
  - Surrounds esophageal hiatus

Arcuate Ligaments

- Median arcuate ligament
  - Joins medial aspects of both crura
  - Runs in arch over anterior surface of aorta
  - Contributes fibers to right diaphragmatic crus
- Medial arcuate ligament
  - Fibrous thickening of thoracolumbar fascia over proximal aspect of psoas major muscle
- Lateral arcuate ligament
  - Fibrous thickening of thoracolumbar fascia over quadratus lumborum muscle
  - Runs in arch from L1 transverse process to ipsilateral 12th rib

Diaphragmatic Hiatus

- **Vena caval foramen**
  - Located in right posterior margin of central leaf of central tendon
  - At T8/9 intervertebral disc level
  - Inferior vena cava wall adheres to this foramen
    - Diaphragmatic movement on inspiration dilates inferior vena cava lumen and helps increase flow of blood to right atrium
- **Esophageal hiatus**
  - Located in posterocentral (left of midline) aspect of muscular component of diaphragm
  - At T10 vertebral level
  - Encircled by right crus fibers, which constricts esophagus on inspiration and thus prevents reflux
  - Contains left gastric artery branches, anterior and posterior vagal trunks, and esophagus
- **Aortic hiatus**
  - Located behind median arcuate ligament (thus outside diaphragm) and thus not constricted by respiration
  - At T12 vertebral level
  - Contains aorta, azygos vein, and thoracic duct

Phrenic Nerve (C3-C5 Ventral Rami)

- Right phrenic nerve
  - Runs posterolateral to right brachiocephalic vein and superior vena cava
  - Continues between mediastinal pleura and right parietal pericardium, anterior to right lung hilum
  - Finally runs on right side of inferior vena cava to enter right hemidiaphragm lateral to vena caval foramen
- Left phrenic nerve
  - Runs between left common carotid artery and left subclavian artery
  - Passes on left side of aortic arch
  - Runs on left parietal pericardium (over left auricle and left ventricle)
  - Enters left hemidiaphragm lateral to left pericardial margin

ANATOMY IMAGING ISSUES

Imaging Recommendations

- Sonographic imaging of diaphragm is best done using abdominal contents (liver, spleen) as window
  - Gas in lungs precludes using thoracic approach to examine diaphragm (except when there is sizable pleural effusion displacing intervening lung)
- US allows real-time evaluation of diaphragmatic movement
  - M-mode can track diaphragmatic excursion
  - Essential for investigating neurological or muscular diaphragmatic abnormalities

Imaging Pitfalls

- Diaphragmatic insertions on thoracoabdominal wall may appear as slips of muscle and tendon rather than sheet
  - Can be mistaken for peritoneal nodules

DIAPHRAGM

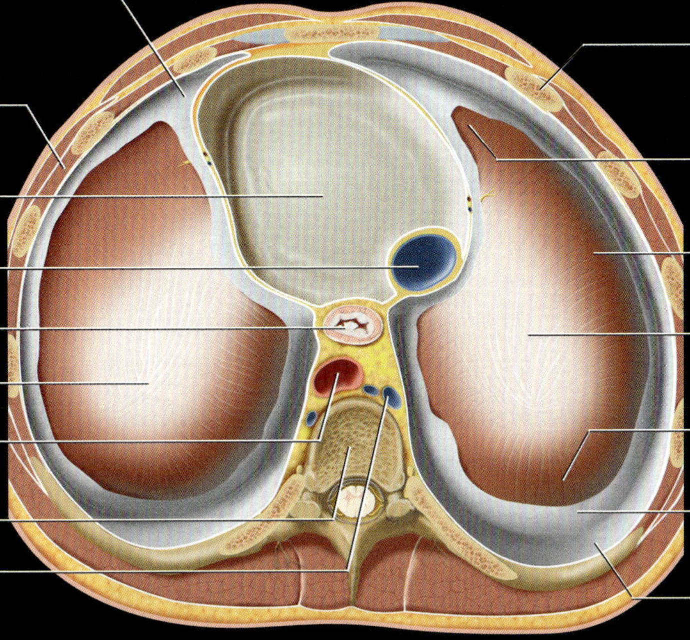

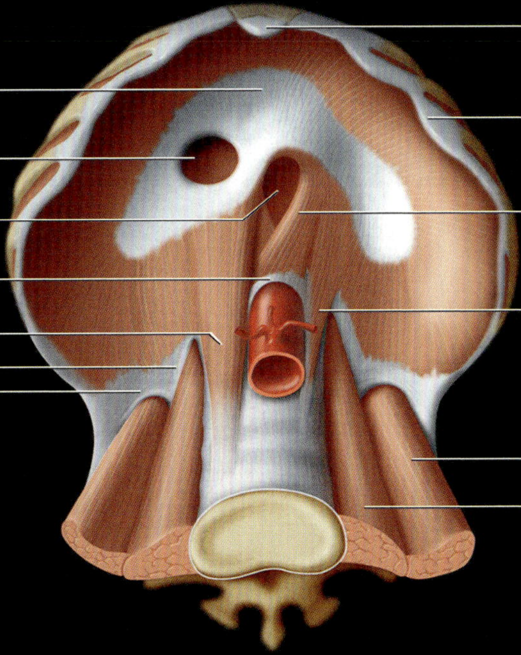

(Top) *Graphic of the diaphragm looking from above is shown. The diaphragm is composed of a central tendon and a peripheral muscular component. The central tendon is made up of 3 leaves: Central leaf (on which rests the pericardium) and left and right leaves form the dome of each hemidiaphragm.* **(Bottom)** *Graphic shows the abdominal surface of the diaphragm looking from below. Note the origins of the diaphragm from the sternum, costal cartilages, and lumbar vertebrae and insertion into the trefoil-shaped central tendon, the fibrous aponeurosis of the diaphragmatic muscle fibers. The inferior vena cava hiatus is through the central tendon. The esophageal hiatus is surrounded by the right crus. The median arcuate ligament unites the crura and passes over the aorta just above the celiac axis. The right crus is longer and thicker than the left, and both insert into the anterior longitudinal ligament of the lumbar spine. The psoas passes behind the medial arcuate ligament, and the quadratus lumborum passes behind the lateral arcuate ligament.*

DIAPHRAGM

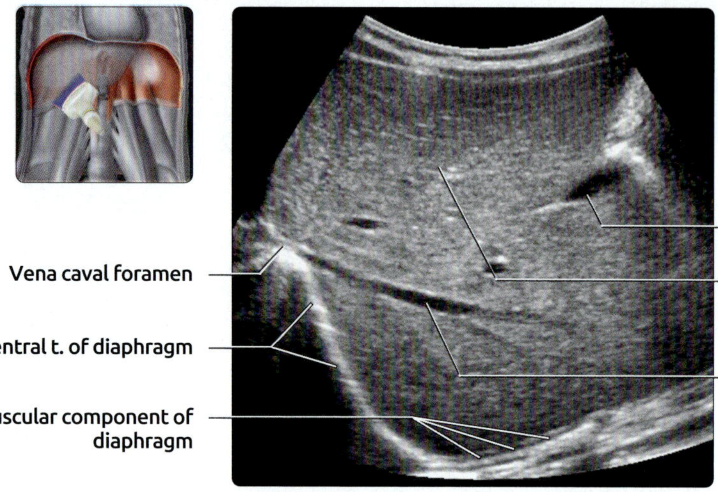

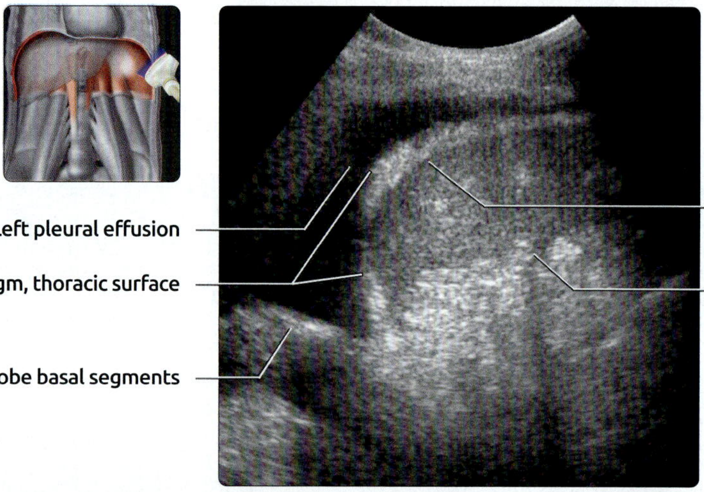

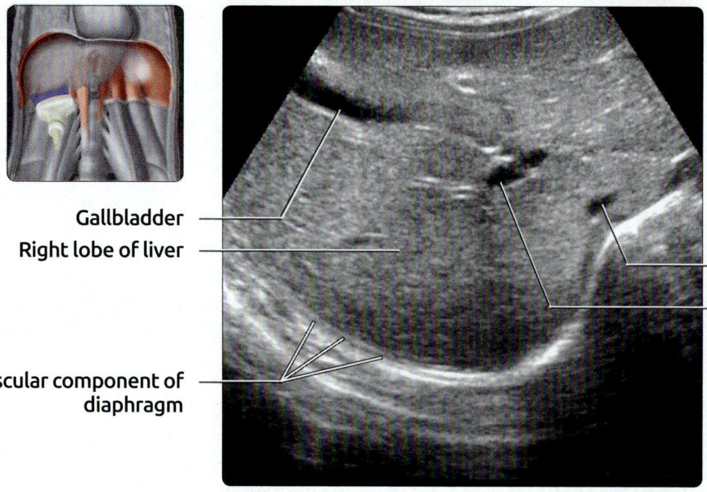

(Top) *Oblique sagittal scan of the upper abdomen using a subcostal window is shown. The dome of the diaphragm is well demonstrated with US due to its strong reflection (a result of the gas-containing lung on the other side of the diaphragm). The vena caval foramen is at the right posterior edge of the middle leaf of the central tendon of diaphragm.* **(Middle)** *Oblique coronal scan using a left lower intercostal window in a patient with left pleural effusion is shown. The thoracic surface of the diaphragm is outlined by the pleural effusion. Without the effusion, this diaphragmatic surface is difficult to visualize.* **(Bottom)** *Oblique transverse scan of the upper abdomen is shown. The costal part of the muscular portion of the diaphragm arises from the internal surface of the inferior 6 ribs. The contraction of this muscular component causes depression of the diaphragm on inspiration.*

DIAPHRAGMATIC MOVEMENT

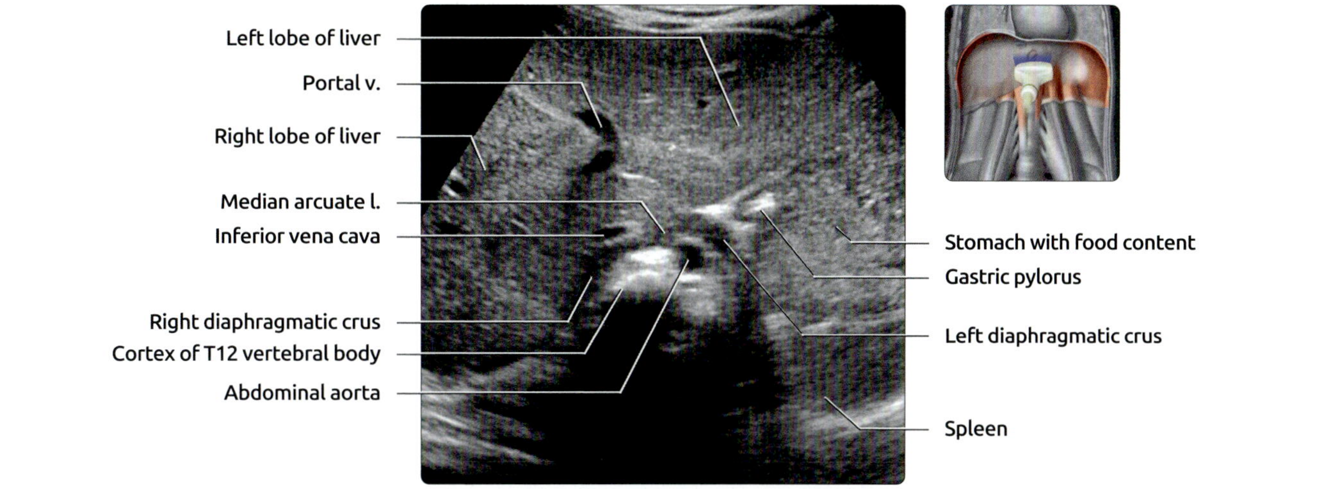

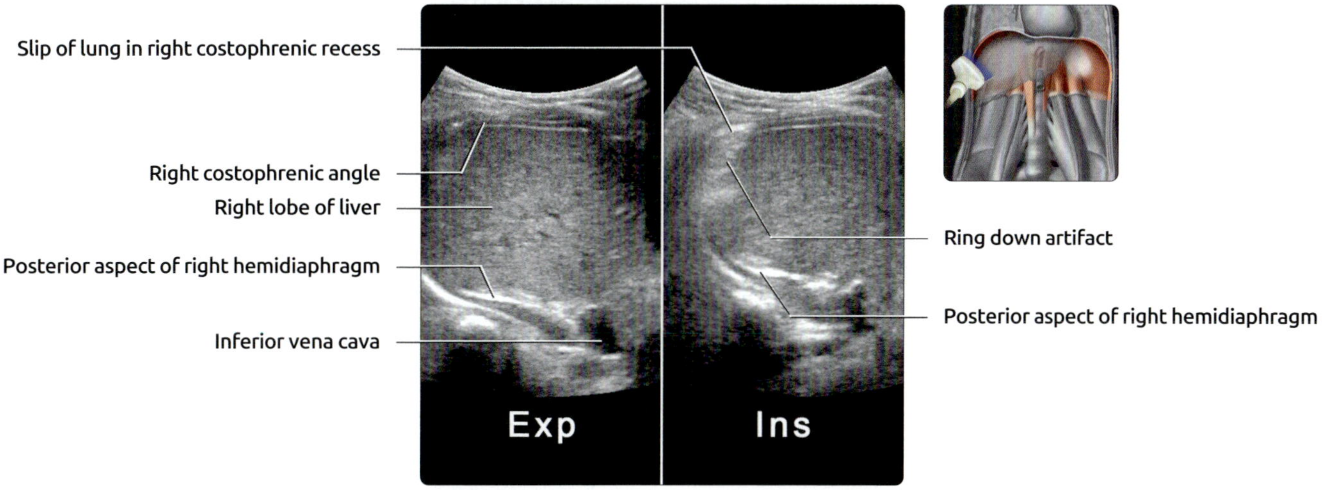

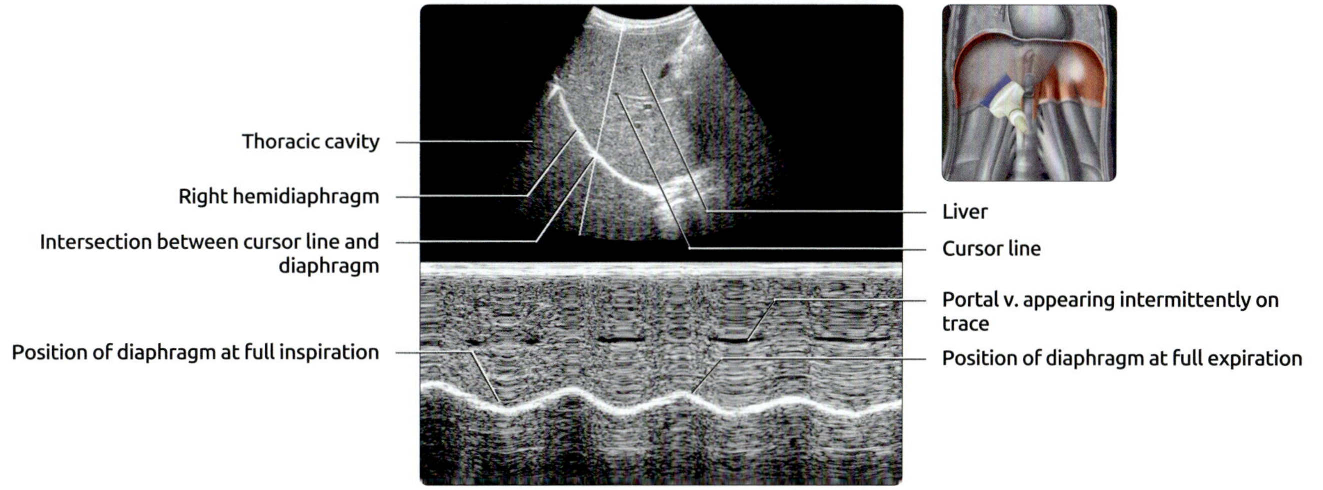

(Top) *Transverse scan at the T12 vertebral body level is shown. The aorta emerges under the median arcuate ligament to enter the abdomen. Thus, the aorta is outside the diaphragm and therefore not constricted by it during inspiration. The diaphragmatic crura lie on both sides of the aorta, taking origin from the cortical surface of the lumbar vertebrae. The crura are hypoechoic and should not be mistaken for lymph nodes.* **(Middle)** *Oblique transverse scans of the right hemidiaphragm using the same intercostal window during full expiration (left image) and full inspiration (right image) are shown. With inspiration, a slip of lung may enter the costophrenic recess and cause posterior acoustic shadowing (gas ring down).* **(Bottom)** *Diaphragmatic tracing obtained in M-mode is shown. The cursor line was placed to cut through the liver and right hemidiaphragm. The respiratory motion of the diaphragm (bright line) can thus be traced.*

GENERAL ANATOMY AND FUNCTION

Chest Wall Anatomy

- Skin, subcutaneous fat
- Blood vessels, lymphatics, nerves
- Bone, cartilage
- Muscles
- Endothoracic fascia, fibroelastic connective tissue between inner aspect of chest wall and costal pleura

Function

- Musculoskeletal cage: Surrounds cardiorespiratory system
  - Effects respiration by expanding and contracting during ventilation

Surface Landmarks

- **Suprasternal (jugular) notch**: At superior manubrium of sternum
  - Between sternal ends of clavicles
- **Sternal angle**: Landmark for internal thoracic anatomy
  - Anterior projection at level of costal cartilage of 2nd rib
- **Costal margin**: Inferior margins of lowest ribs and costal cartilages

SKELETAL STRUCTURES

Sternum

- Flat, broad bone forms anterior thoracic wall composed of 3 parts (manubrium, body, xiphoid process)
- **Manubrium** forms superior part of sternum
- **Body** articulates with manubrium superiorly, xiphoid process inferiorly, bilateral costal cartilages of 2nd-7th ribs
- **Xiphoid** process variable size, shape, ossification; articulates with body of sternum superiorly

Ribs

- 12 pairs, symmetrically arrayed; numbered in accordance with attached vertebral body
- True ribs (1-7) attach to sternum by costal cartilages (synovial joints)
- False ribs (8-10) articulate by costal cartilages with costal cartilage of 7th rib
- Floating ribs (11-12) do not articulate with sternum or rib costal cartilages
  - Short costal cartilages terminate in abdominal wall muscle
- Vertebral articulation
  - Head articulates with demifacets of 2 adjacent vertebral bodies
  - Neck located between head and tubercle of each rib
  - Tubercle articulates with vertebral transverse process
- Body: Longest part of each rib
- Angle: Most posterior part
- Costal groove on inner surface of inferior border; accommodates intercostal neurovascular bundle

MUSCLES

Pectoral

- Pectoralis major: Largest muscle in breast and pectoral region
  - Originates from anterior chest wall, sternum, and clavicle
  - Adducts, flexes, and medially rotates arm
- Pectoralis minor: Deep to pectoralis major
  - Originates from chest wall, inserts onto coracoid process of scapula
  - Stabilizes scapula

Intercostal

- External: Contained within 11 intercostal spaces; extend from tubercle of ribs to costochondral junction
- Internal: Middle layer; occupy 11 intercostal spaces; extend from border of sternum to angle of ribs
- Innermost: Form inner layer of chest wall muscles with subcostales and transversus thoracis muscles

Serratus Anterior

- Thin muscular sheet which overlies lateral thoracic cage and intercostal muscles
  - Arises from upper 8 ribs; wraps around rib cage; inserts along medial border of anterior surface of scapula

VESSELS AND NERVES

Arteries

- **Internal thoracic (internal mammary)**: Branch of subclavian artery
  - Descends posterior to first 6 costal cartilages supplying intercostal arteries
  - Blood supply for anterior chest wall

Veins

- Azygos vein receives drainage from posterior intercostal veins, hemiazygos and accessory hemiazygos veins

Nerves

- Anterior rami of thoracic spinal nerves (T1-T11) supply skin, tissues of chest wall; form intercostal nerves
- Intercostal nerves run in costal groove, between internal and innermost intercostal muscles
- Brachial plexus: Branching network of nerve roots, trunks, divisions, cords, and branches
  - Spinal roots form 3 trunks; behind clavicle, each dividing into anterior and posterior divisions

IMAGING

Ultrasound

- Allows detailed examination of intercostal spaces and contents
- Intercostal muscles are seen as hypoechoic structures
- Intercostal membrane is seen as hyperechoic layer
  - However, intercostal structures are often thin, making it difficult to separate them into distinct layers
- Only anterior cortex of ribs evaluated (medulla obscured)

Computed Tomography

- Helical CT and multiplanar reformations optimal for visualization of osseous lesions

RIBS AND INTERCOSTAL SPACES

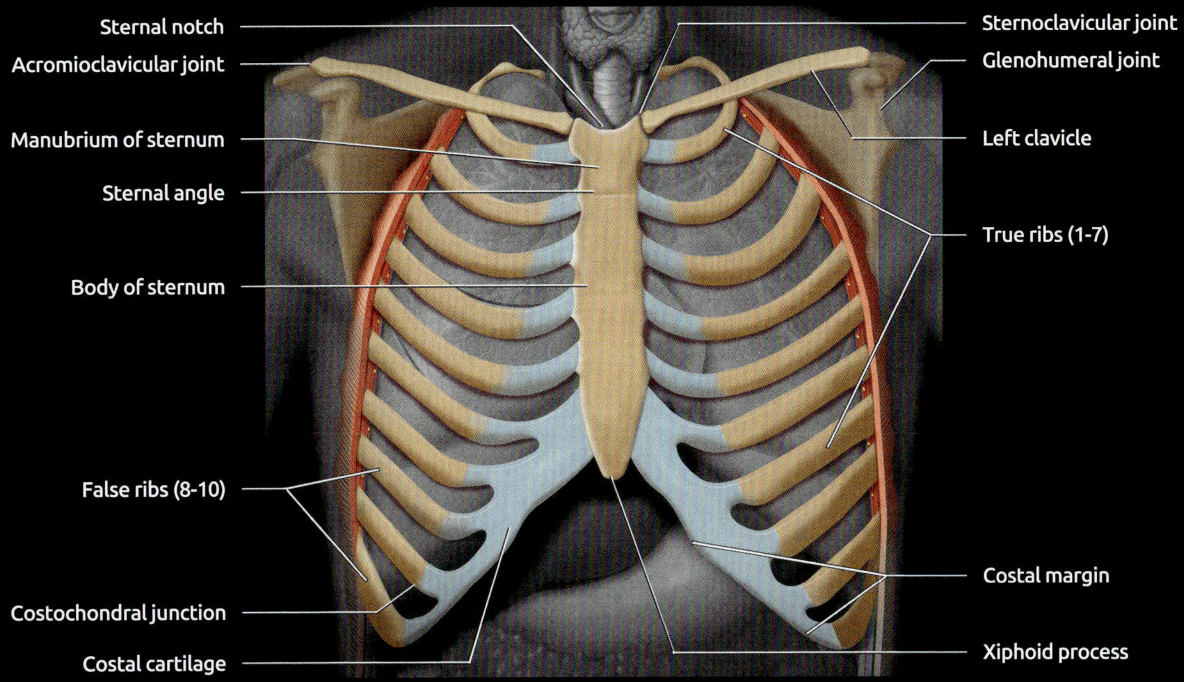

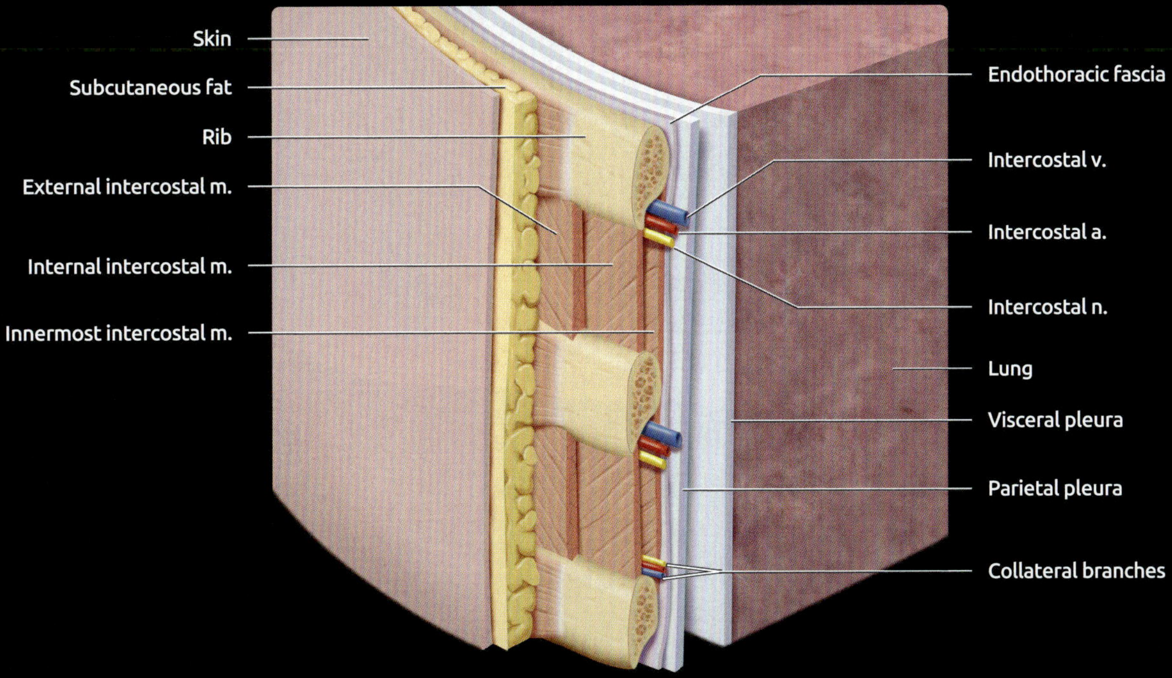

(Top) *Graphic depicts the chest wall structures forming a musculoskeletal thoracic cage that surrounds the cardiorespiratory organs and effects respiration by expanding and contracting during ventilation.* **(Bottom)** *Graphic demonstrates details of the intercostal region, showing 3 layers of intercostal muscles (external, internal, and innermost) between the ribs. The costal groove along the inferomedial aspect of each rib accommodates the intercostal neurovascular bundle (vein, artery, and nerve). Small collateral branches of the major intercostal vessels and nerves may be present above the body of the subjacent rib. The endothoracic fascia forms a connective tissue layer between the inner aspect of the chest wall and the costal parietal pleura.*

CHEST WALL MUSCULATURE

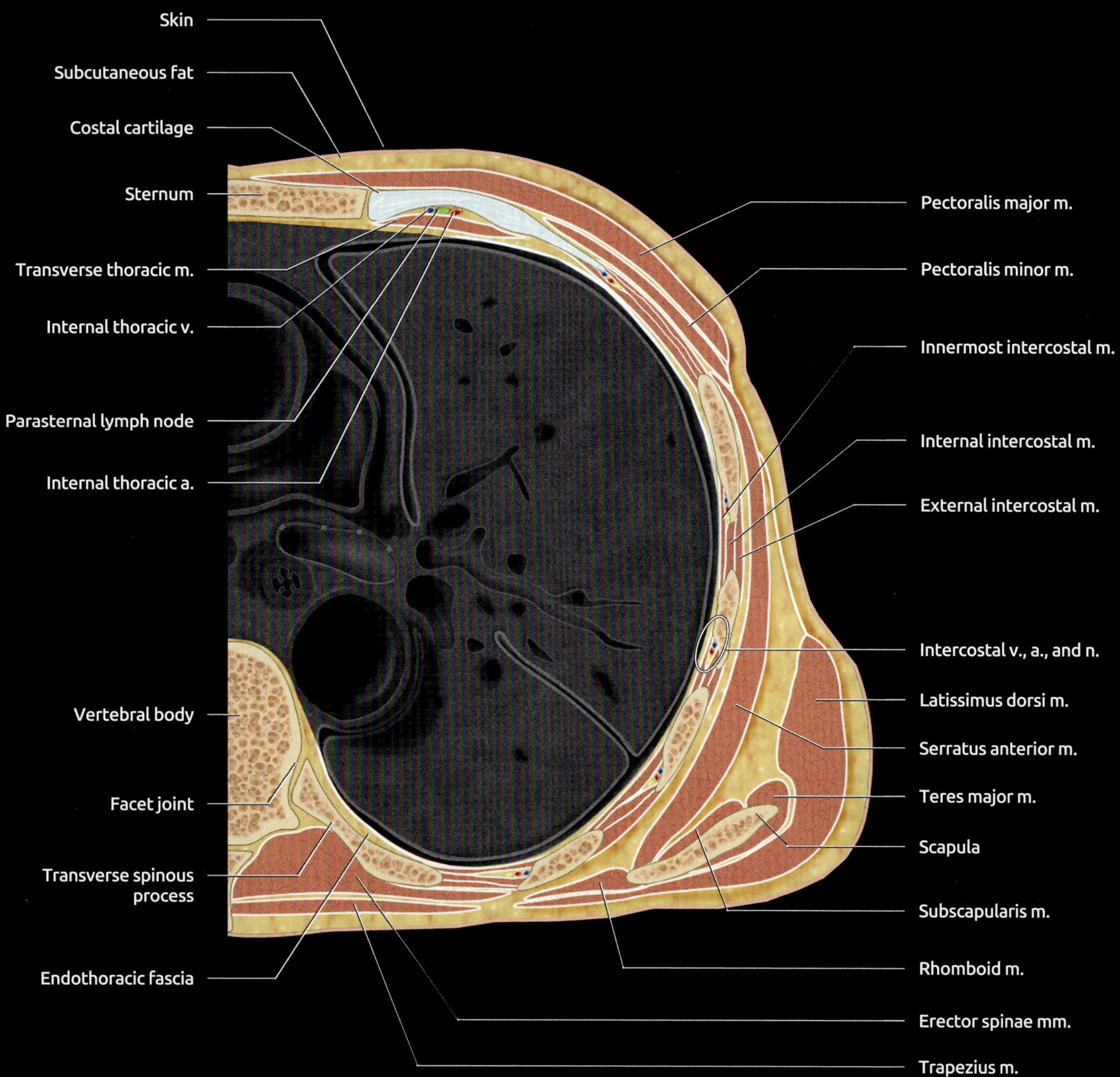

Graphic depicts the chest wall layers as visualized in the axial plane, including skin, subcutaneous fat, blood vessels, lymphatics, and musculoskeletal structures. The innermost layer, the endothoracic fascia, is a fibroelastic connective tissue layer between the inner aspect of the chest wall and the pleura.

MR, CHEST WALL

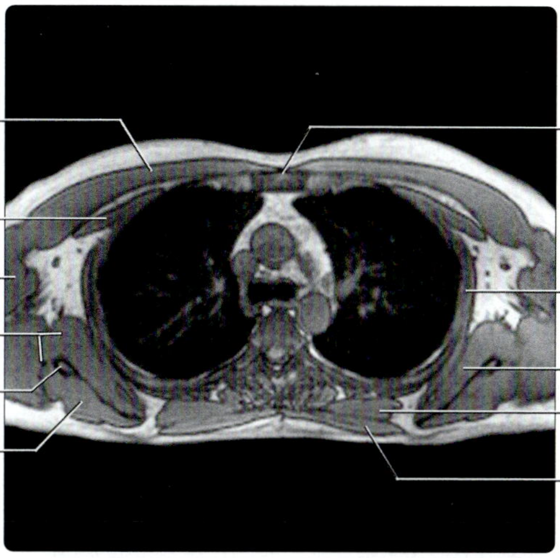

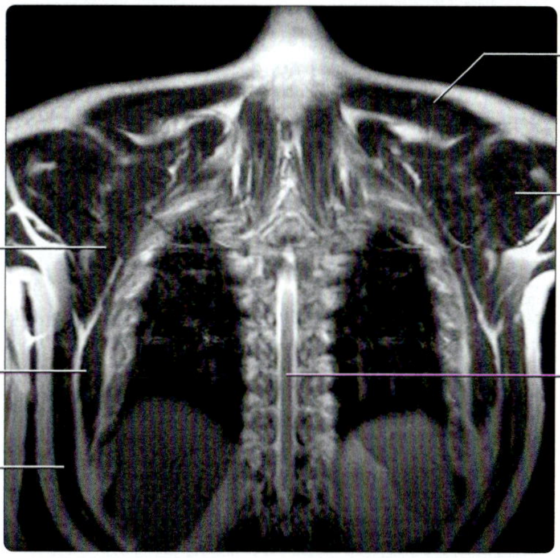

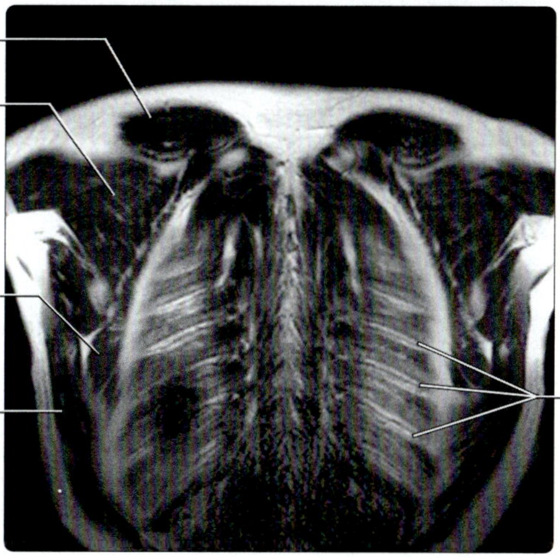

(Top) *Axial T1 MR at the level of the aorticopulmonary window shows the major muscles of the chest wall.* **(Middle)** *Coronal T1 MR of the chest through the level of the thoracic spinal canal is shown. The serratus anterior is a thin muscular sheet that overlies the lateral thoracic cage and intercostal muscles. It wraps around the rib cage and inserts along the medial border of the scapula.* **(Bottom)** *Coronal T1 MR through the level of the posterior ribs is shown. The neurovascular bundles run along the inferior aspect of the ribs in the costal groove.*

RIBS AND INTERCOSTAL SPACES

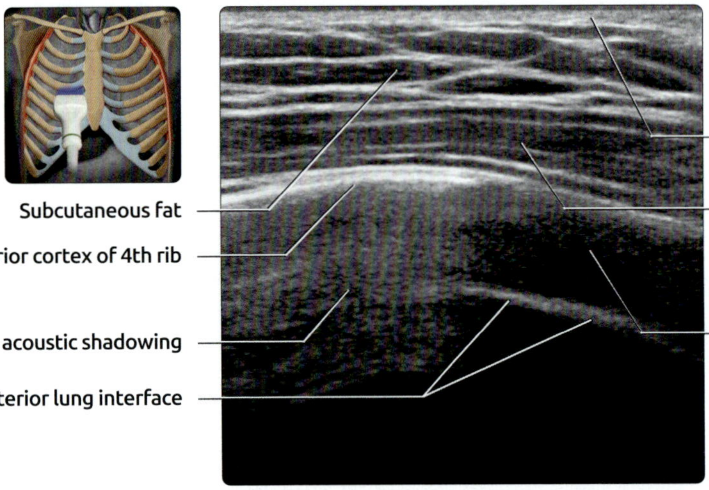

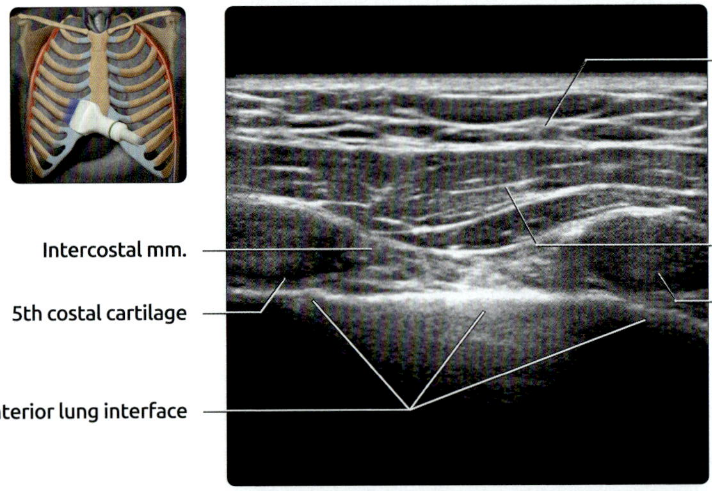

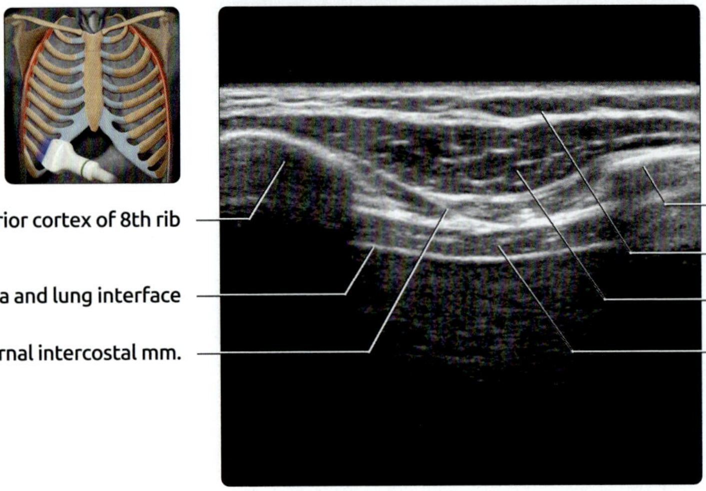

(Top) *Transverse view of the costochondral junction of the 4th rib is shown. The costal cartilage appears homogeneously hypoechoic and does not produce significant posterior acoustic shadowing, thus allowing the underlying pleural surface to be seen. The cortex of the 4th rib casts a strong posterior shadow, obscuring the underlying pleura.* **(Middle)** *Oblique sagittal view of the intercostal space between the costal cartilages in the anterior chest wall is shown. The costal cartilages are well defined and oval-shaped on cross section. Intercostal muscles between costal cartilages and overlying pectoralis muscle can be clearly visualized.* **(Bottom)** *More laterally, oblique sagittal view shows intercostal spaces between the ribs. Areas behind the ribs are obscured by acoustic shadowing. The intercostal muscles in here are better differentiated compared to more medially.*

RIBS AND INTERCOSTAL SPACES

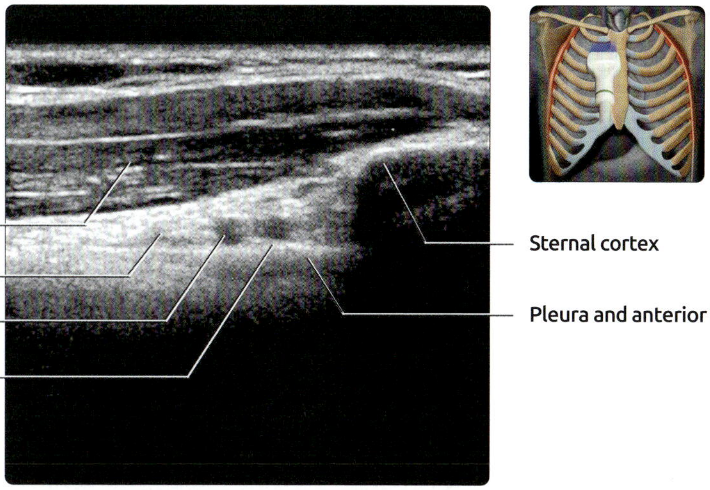

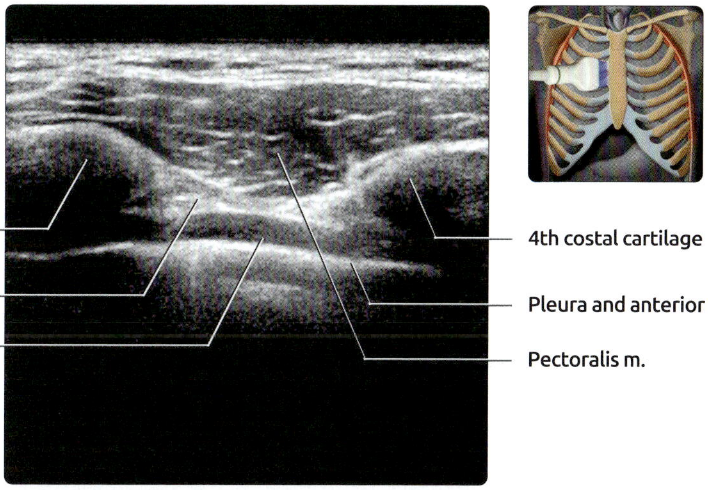

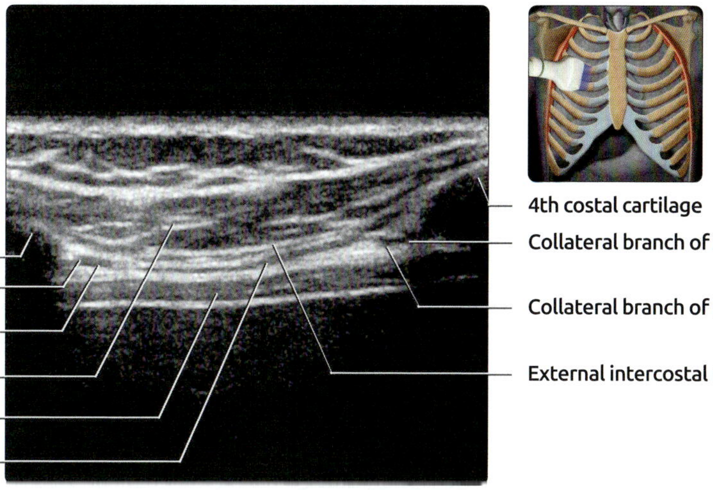

(Top) *Transverse view of the anterior intercostal space just lateral to the sternum is shown. The internal thoracic artery and vein next to the sternum are well visualized with this approach, as are the overlying pectoralis muscle and underlying pleura.* **(Middle)** *Sagittal view of the anterior intercostal space shows the internal thoracic artery. Long axis of the internal thoracic artery can be well delineated by scanning along the lateral margin of the sternum.* **(Bottom)** *Sagittal view of the anterior intercostal space and the anterior intercostal vessels is shown. The anterior intercostal artery and anterior intercostal vein are identified just inferior to rib above. Smaller collateral branches of the intercostal artery and vein can be identified just superior to the rib above. At the anterior intercostal space, the external intercostal muscle is seen as a thin membrane.*

TERMINOLOGY

Definitions

- Lobe (segment): Drainage territory defined by each major duct
- Terminal duct: Main duct for segment extending to nipple
  - 2 parts: Extralobular terminal duct (ELTD) and intralobular terminal duct (ILTD)
- Lobule: Composed of ILTD and complex system of tiny ducts terminating in blind-ending acini (alveoli)
- Terminal ductal lobular unit (TDLU): Lobule + ELTD
- Cooper ligaments: Suspensory ligaments extending from anterior mammary fascia into dermis to support breast tissue

LOBE/SEGMENT

Composition

- Major ducts and branches → ELTDs → ILTDs → acini

Clinical Considerations

- Average of 15-20 lobes per breast
- Lobar volume and anatomy variable
- Most, but not all, lobes drain to corresponding nipple duct orifice (some share common ducts)

DUCTAL SYSTEM

Duct Orifices

- Usually 8-12 per nipple
- Arranged radially in nipple crevices
- Some major ducts merge deep to nipple surface
- Quadrant of duct orifice on nipple surface does not always correspond to quadrant of lobe

Lactiferous Sinus (Ampullary Segment)

- Widened duct segment just deep to nipple orifice
- Average diameter: 4-5 mm

Major Ducts

- Average diameter: 1 mm
- Arborize into segmental and subsegmental branches of variable length and number giving rise to terminal ducts
- Branches may extend into multiple breast quadrants
- Drain 20-40 lobules, each containing 10-100 acini

TERMINAL DUCT LOBULAR UNIT

Overview

- Functional, glandular unit of breast
- Composition
  - ELTD + ILTD + acini surrounded by loose stromal matrix of collagen and reticular fibers
    - 10-100 acini drain into each ILTD
- May arise directly from major ducts or lactiferous sinuses
- Multiple rows of TDLUs arise from distal segmental/subsegmental ducts

TDLU Proliferation

- Late adolescence
- Pregnancy and lactation
- Exogenous hormones: Birth control pills and hormone replacement therapy (HRT)
- Postovulatory (secretory) phase of menstrual cycle

TDLU Regression

- Postpartum/postlactation and menopause
- Hormone-blocking medications: Estrogen receptor antagonists and aromatase inhibitors

ZONAL ANATOMY

Premammary Zone

- Between skin and anterior mammary fascia
- Contains subcutaneous fat, nerves, lymphatics, blood vessels, and Cooper ligaments

Mammary Zone

- Between anterior and posterior mammary fasciae
- Contains TDLUs and ducts, stromal fat, stromal connective tissue, nerves, lymphatics, and blood vessels

Retromammary Zone

- Between posterior mammary fascia and chest wall
- Contains fat and posterior suspensory ligaments

INNERVATION

Overview

- Breast innervated by 3 nerve groups arising from intercostal and cervical nerves: Medial, lateral, and superior mammary branches
- Sympathetic fibers control vasomotor response
- Glandular secretory functions controlled hormonally

Medial Breast Innervation

- Anterior cutaneous branches of 2nd-6th intercostal nerves passing through intercostal spaces near sternum
- Medial branches innervate skin over parasternal region

Lateral Breast Innervation

- Lateral cutaneous branches of 2nd-6th intercostal nerves
- Perforate chest wall in mid-axillary line
- Deep branch of 4th intercostal nerve courses along anterior pectoralis major fascia and has perforators to nipple-areolar complex
- 2nd intercostal (intercostobrachial) nerve innervates axillary tail and may be injured during axillary nodal dissection

Superior Breast Innervation

- Cutaneous branches of 2nd and 3rd intercostal nerves
- Medial, intermedius, and lateral branches of supraclavicular nerve via cervical plexus

Nipple-Areolar Complex Innervation

- Nerve plexus arising from lateral cutaneous branches of 3rd-5th intercostal nerves and anterior cutaneous branches of 2nd-5th intercostal nerves
- Greatest contribution by deep branch of lateral cutaneous 4th intercostal nerve
- Nerves course along superficial fascia, becoming subdermal at areola

VASCULAR SUPPLY

Arteries

- Vessels enter via superolateral, superomedial, and deep aspects of breast

- Medial and lateral mammary branches and intramammary anastomoses
- **Medial mammary arteries**
  - Originate from internal thoracic artery (a.k.a. internal mammary artery), branch of subclavian artery
  - Emerge from 2nd-4th intercostal spaces
  - Supply medial and central breast
- **Lateral mammary arteries**
  - Multiple origins; supply lateral breast
    - Lateral thoracic artery (branch of 2nd portion of axillary artery) provides dominant supply
    - Pectoral branches of thoracoacromial artery (branch of 2nd portion axillary artery)
    - Perforators from 2nd-4th posterior intercostal arteries
    - Superior thoracic artery (branch of 1st portion of axillary artery)

Venous Drainage

- Superficial and deep system drainage with intramammary anastomoses
- **Superficial system** usually do not parallel arteries
  - Subareolar plexus of nipple-areolar complex creates circumferential radiating draining veins
    - Superomedial breast drainage to 2nd and 3rd intercostal spaces into internal thoracic vein
    - Inferior breast drainage to inframammary fold into 4th and 5th intercostal spaces via perforators into lateral or medial drainage routes
- Deep system accompany arteries supplying region of breast
  - Superolateral veins drain into subclavian vein
  - Intercostal veins to internal mammary vein
  - Inferolateral veins drain into lateral thoracic vein (tributary to axillary vein)

LYMPHATICS AND LYMPH NODES

Breast Lymphatic Drainage

- Deep breast tissues to superficial lymphatics to periareolar lymphatic (Sappey) plexus
- 75% drainage to axilla via lateral and medial trunks extending from areola to axilla
  - Axilla drains into subclavian lymphatic trunk
- 25% drainage to internal mammary nodes
- Anastomotic lymphatic channels may communicate with contralateral skin and breast; aberrant drainage common after mastectomy or reduction

Axillary Lymph Nodes

- **Surgical lymph node levels**
  - Level I nodes: Lateral/inferior to pectoralis minor muscle
  - Level II nodes: Deep/posterior to pectoralis minor muscle
  - Level III nodes: Medial and superior to pectoralis minor muscle

Internal Mammary Lymph Nodes

- Located in parasternal intercostal spaces, < 5 mm in diameter
- Predominantly drain far medial and deep medial breast

Intramammary Lymph Nodes

- Most common in far lateral, axillary, and posteromedial aspects of breast
- 25-28% of normal women have intramammary nodes visible on mammography
- May be difficult to distinguish intramammary node in axillary tail from axillary lymph node

IMAGING ISSUES

Ultrasound

- Image capture and reporting
  - Standardized reporting according to American College of Radiology BI-RADS Atlas most commonly used to report both normal and abnormal findings and guide management
  - Include location of probe on every image and cine clip including breast laterality (right breast vs. left breast), clockface, centimeters from nipple and probe position (transverse or longitudinal)
  - Include symptom (if applicable) in addition to probe location on captured image (i.e., "palpable" or "pain")
  - Report echogenicity of finding relative to normal breast subcutaneous fat
- Technical considerations
  - Use high-frequency linear-array transducer with center frequency of at least 12 MHz (preferably higher)
  - Subcutaneous fat is reference echogenicity for other structures and gain should be set so that fat is mid gray
- Normal findings
  - Overall breast tissue composition variable
  - Fat: Hypoechoic echotexture
  - Ducts frequently visible under nipple as linear, branching anechoic or hypo- to isoechoic channels depending on content
    - Physiologic duct dilatation occurs in menopause, pregnancy, and lactation
  - Cooper ligaments visible as hyperechoic curved lines within subcutaneous fat
  - Interlobular stromal fibrous tissue usually hyperechoic
  - Glandular elements usually iso- or slightly hypoechoic
  - Cysts anechoic, round or oval masses that are common and considered normal finding when patient asymptomatic
  - Normal lymph nodes always visible within axilla if no prior surgery and often visible in breast
    - Morphology
      - Elliptical shape with circumscribed margins
      - Thin, C-shaped, hypoechoic cortex (≤ 3 mm thick) and hyperechoic fatty hilum
      - Morphology more important than size in determining normal vs. abnormal
    - Variable size: May be > 2 cm in length, especially if fatty
    - Doppler often shows normal hilar vascularity
  - Normal internal mammary nodes may be visible in intercostal spaces
    - Smaller than axillary nodes: Average size 4-6 mm
    - Size criteria more important than morphology; different from axillary nodes
  - Vessels (arteries and veins) commonly visible
    - Distinguish arteries from veins by direction of flow, pulsatility, waveform; normal veins compressible

TANNER STAGES OF BREAST DEVELOPMENT

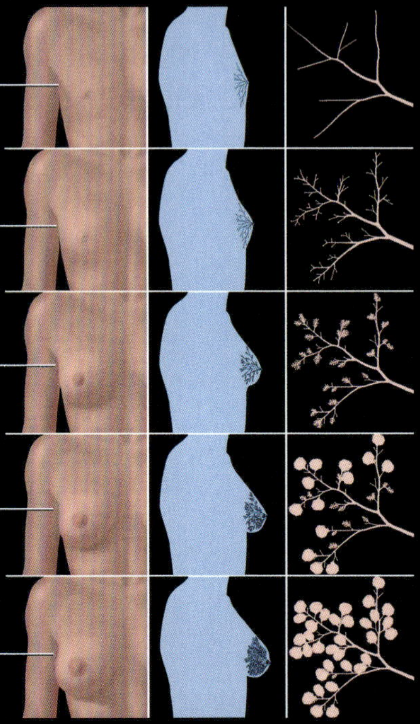

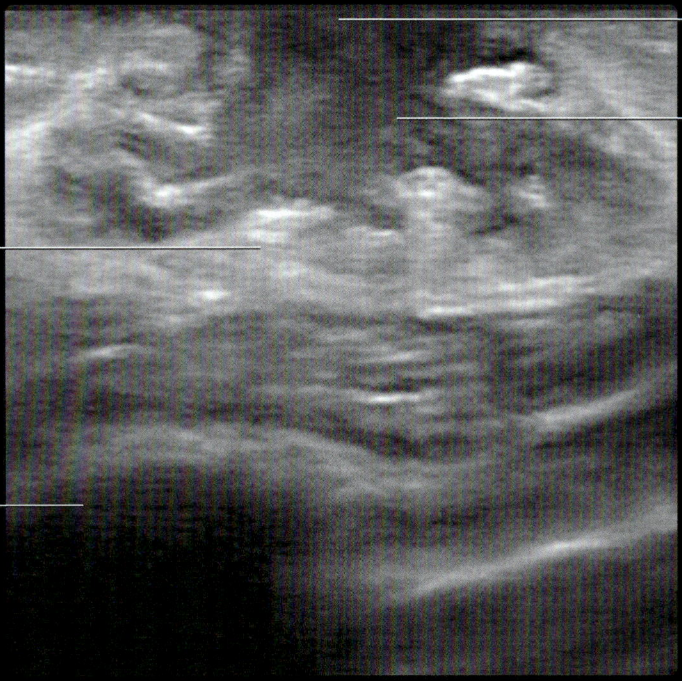

(Top) *During childhood, main ducts branch and give rise to terminal buds, the precursors of terminal ductal lobular units (TDLU). This diagram demonstrates the 5 Tanner phases of breast development with the left column showing progressive changes in breast appearance as viewed from the front, the middle column representing glandular changes as viewed from the side, and the right column depicting corresponding changes in the ducts and TDLU. (I) Nipple elevation but no palpable glandular elements and minimal duct branching; (II) projection of nipple and breast as mound with small terminal buds projecting from ductal branches; (III) increased glandular and areolar tissue with primitive TDLU; (IV) separate nipple-areolar complex as secondary mound and complete TDLU; (V) final development with smooth breast and areolar contour and proliferation of TDLU.* **(Bottom)** *Transverse ultrasound shows the subareolar region of a 10-year-old girl. The developing breast tissue consists of proliferating ducts and periductal stromal fibrous tissue, resulting in a hypoechoic mass with finger-like projections deep to the nipple. Clinically, this is a palpable disc of tissue deep to the nipple.*

LOBE, SEGMENTAL ANATOMY

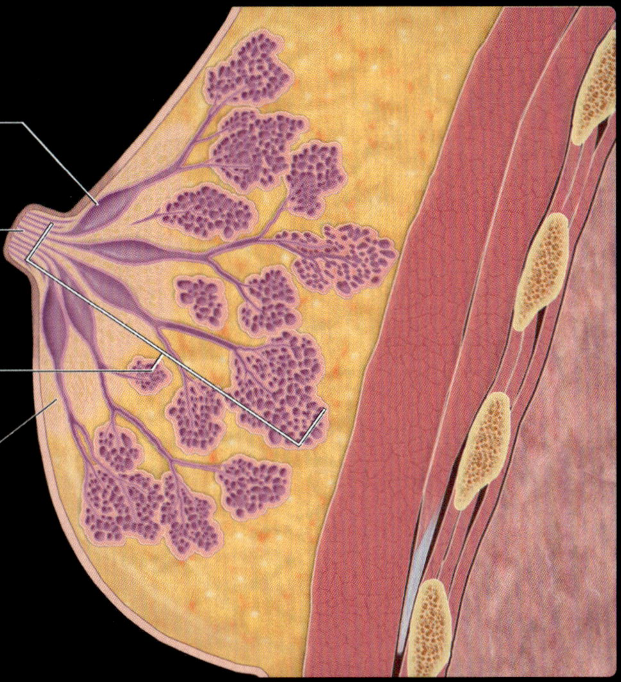

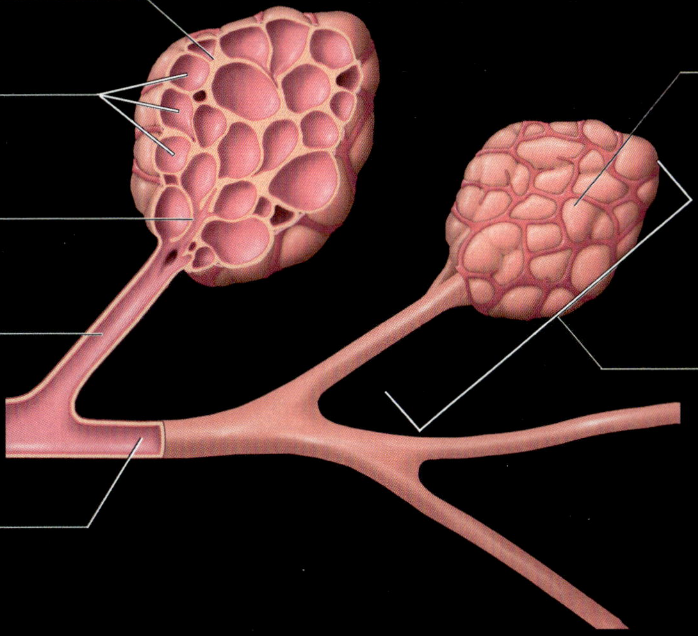

(Top) *Sagittal graphic shows components of a lobe/segment from the duct orifice on the nipple surface to the TDLUs. Breasts average 15-20 lobes, draining into 8-12 duct orifices. Variability the in amount of glandular elements (TDLUs) and interlobular fibrous stromal tissue accounts for differences in ultrasound echotexture and mammographic density.* **(Bottom)** *The TDLU arises from branches of distal subsegmental ducts and is composed of the extralobular terminal duct (ELTD), the intralobular terminal duct (ILTD), and multiple acini arranged around the ILTD. The TDLU on the left has been sectioned to demonstrate the relationship between acini and the ILTD. Note that 10-100 acini drain into each ILTD. The intralobular stroma is composed of collagen and reticular fibers.*

LYMPHATIC DRAINAGE AND LYMPH NODES

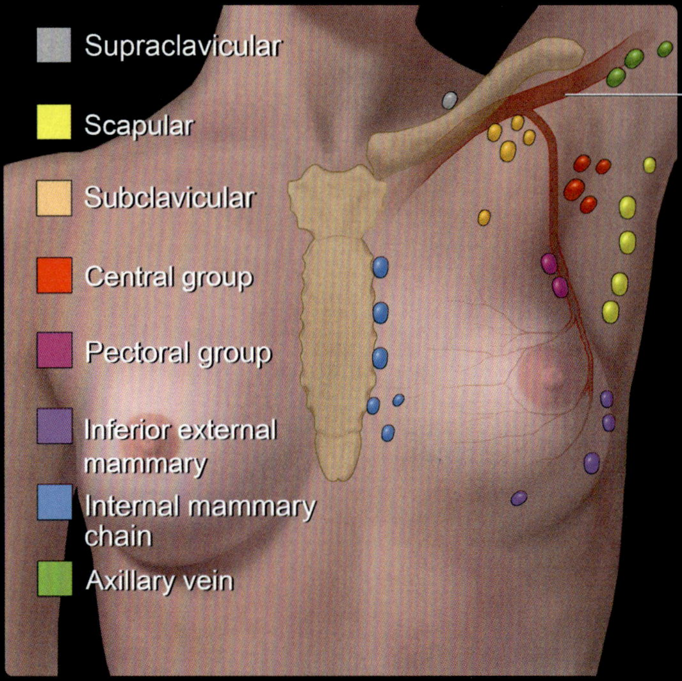

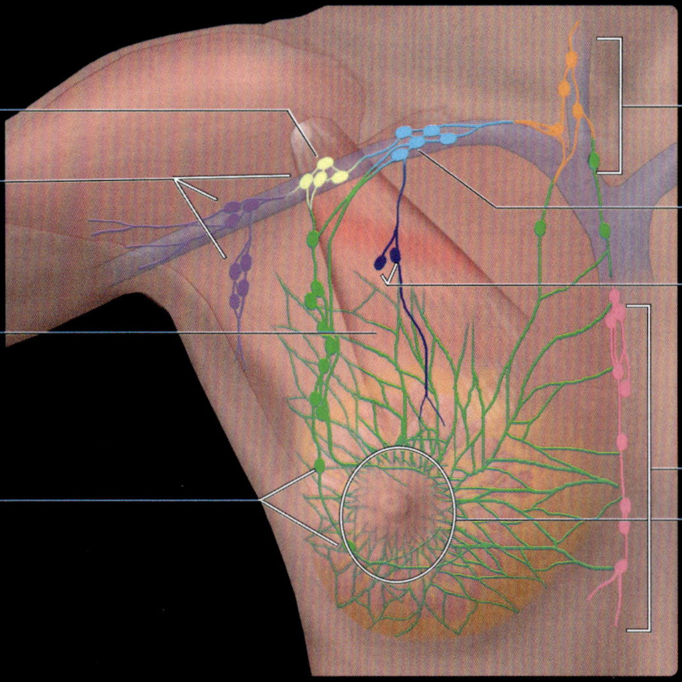

(Top) *Each major lymph node group has well-defined anatomy. The axillary vein group is medial and posterior to the axillary vein, the pectoral group projects at the lower margin of the pectoral muscle, the scapular group projects at the intersection of posterior axilla and scapula, the central group projects posterior to pectoralis minor muscle into the axillary fat, the interpectoral group (not labeled) projects between pectoralis major and pectoralis minor, the subclavicular group projects at the apex of the axilla, and the inferior external mammary group is lateral and inferior to the breast and the internal mammary chain in the parasternal intercostal spaces.* **(Bottom)** *Surgical lymph node levels are most commonly used to describe the location of axillary nodes using the pectoralis minor muscle as a point of reference. Nodes lateral/inferior to the pectoralis minor are level I. Nodes deep/posterior, including interpectoral (Rotter) nodes, are level II. Nodes medial and superior are level III.*

NEUROVASCULAR SUPPLY

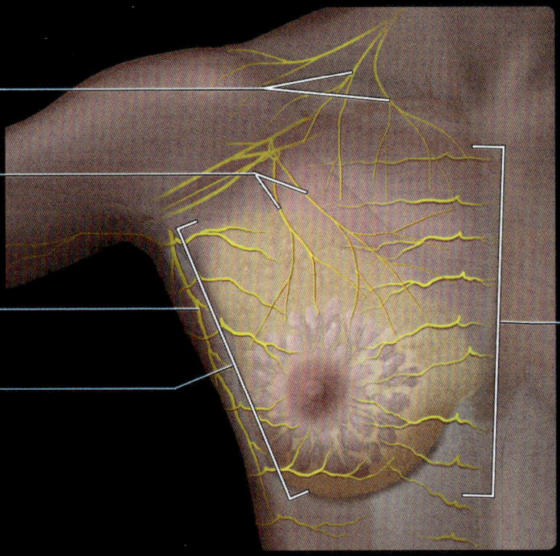

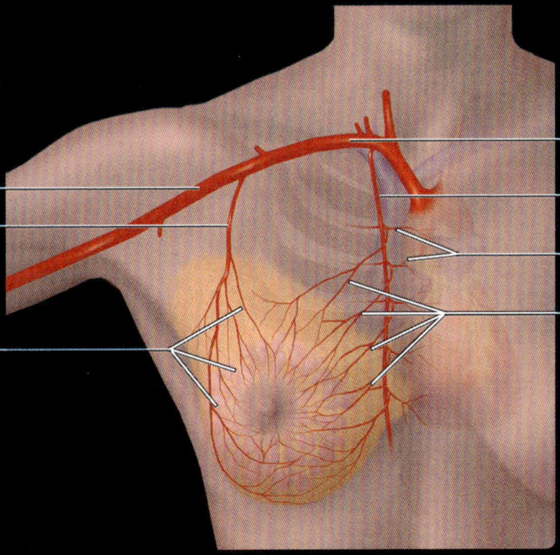

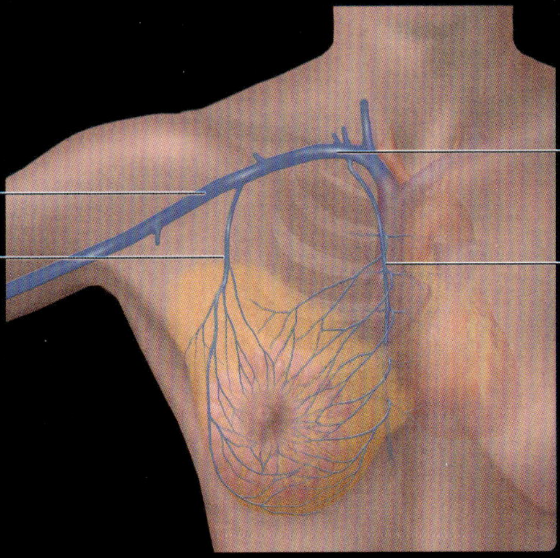

(Top) *The superior breast is innervated by supraclavicular branches from the cervical plexus and brachial plexus. The remaining breast is innervated by anterior and lateral branches of the 2nd-6th intercostal nerves. The long thoracic nerve does not innervate breast tissue but is important to isolate during axillary surgery to avoid denervation of the serratus anterior muscle.* **(Middle)** *The majority of the arterial supply to the medial and central breast is provided by the internal thoracic (a.k.a. internal mammary) artery. The arterial supply to the lateral breast arises from the axillary artery via lateral thoracic, superior thoracic, and thoracoacromialis arteries and intercostal perforators.* **(Bottom)** *Superficial draining veins join the deep draining venous system. Superficial superomedial draining veins drain into the internal thoracic (a.k.a. internal mammary) vein. The deep venous system is composed of the lateral thoracic vein, which is a tributary to the axillary vein; the superolateral vein, which is a tributary to the subclavian vein; and the posterior intercostals, which are tributaries to the vertebral plexus.*

BREAST PARENCHYMAL VARIATION

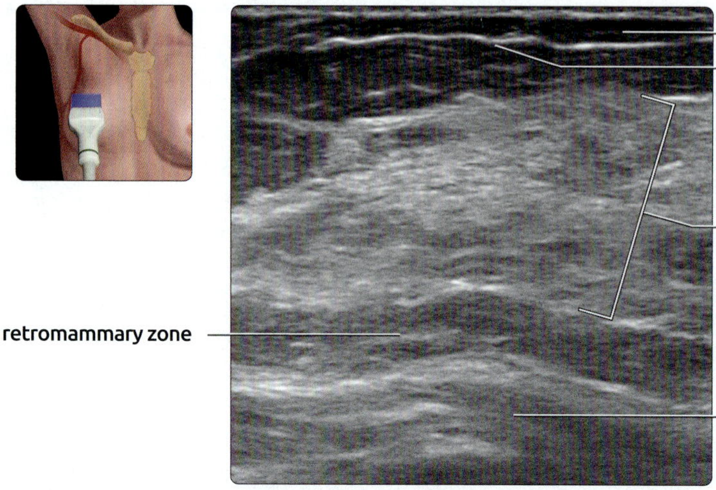

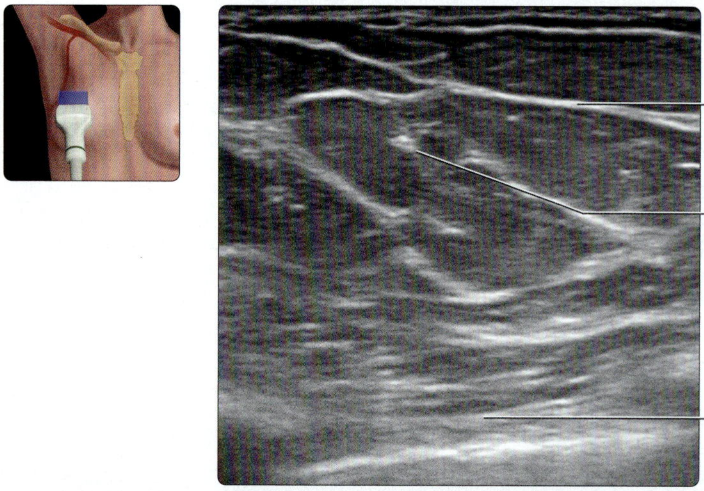

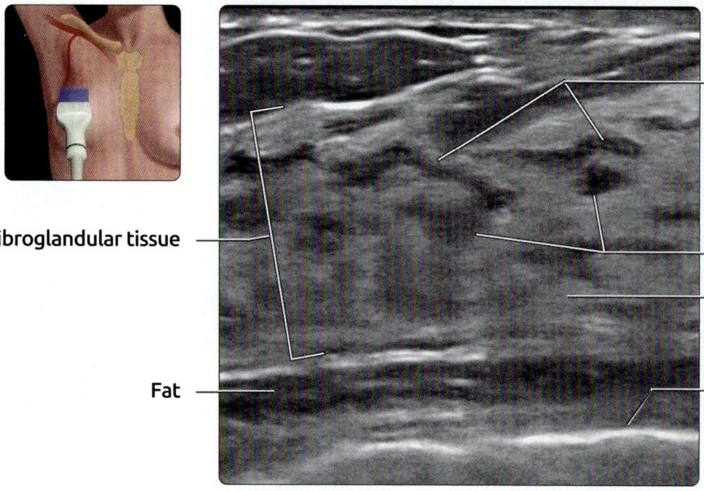

(Top) *Transverse view of the breast shows homogeneous background echotexture (fibroglandular), which fills the mammary zone. Normal fat is seen in the premammary and retromammary zones. Isoechoic and hypoechoic masses are more visible within the echogenic fibrous tissue.* **(Middle)** *Transverse view of the breast shows homogeneous background echotexture (fat). The echogenic ligaments and bands of fibrous tissue are easily seen coursing through the fat. Delineating the mammary zones is difficult in predominantly fatty breasts. Isoechoic masses can be difficult to detect in this type of predominantly fatty breast tissue.* **(Bottom)** *Transverse view of the breast shows heterogeneous background echotexture in which there is dense fibroglandular tissue composed of a mix of echogenic fibrous tissue, hypoechoic and anechoic ducts, and multiple tiny, hypoechoic masses, which typically represent cysts &/or benign proliferative change of TDLUs. A thin band of fat represents the retromammary zone.*

BREAST PARENCHYMAL VARIATION

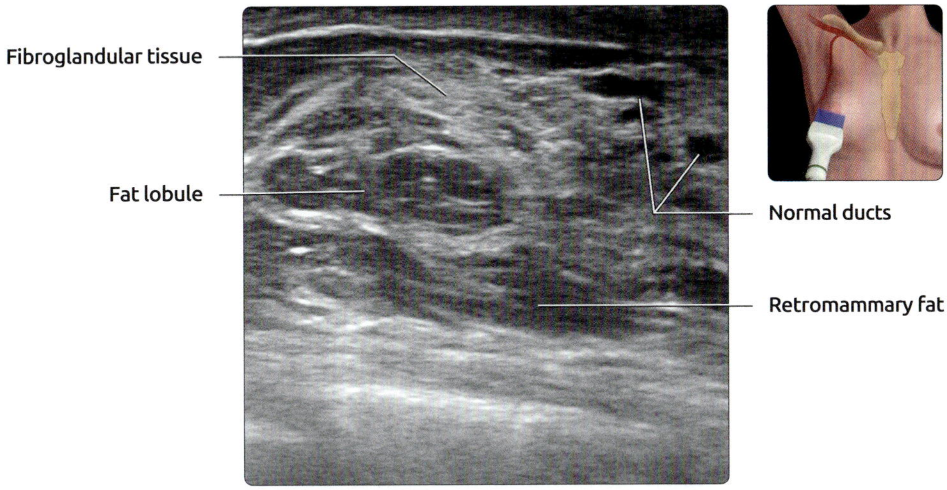

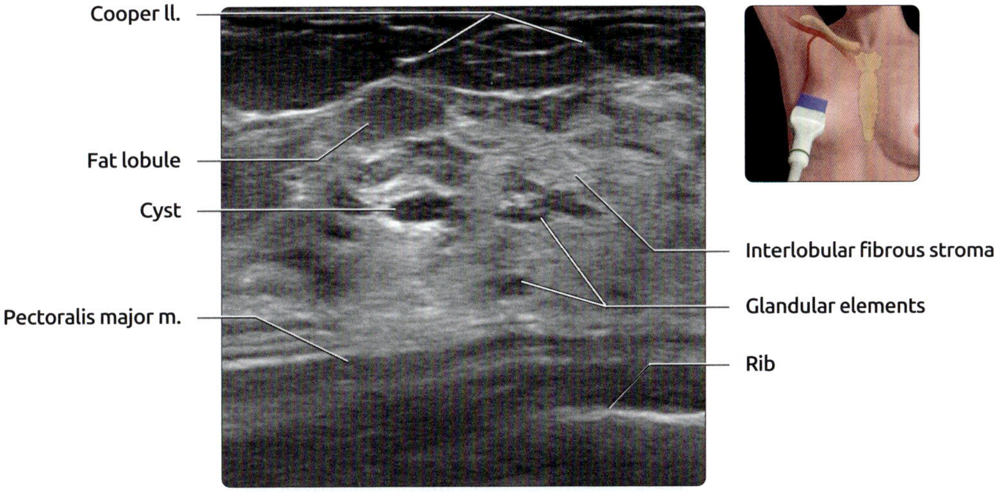

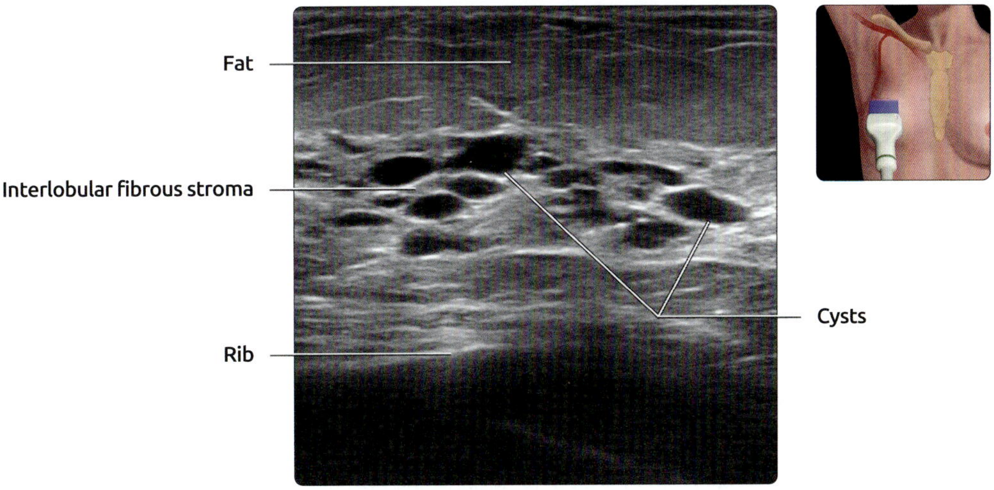

(Top) *Transverse ultrasound shows heterogeneous fibroglandular tissue with intermixed lobules of fat adjacent to ducts in the subareolar region. The areolar skin thickens adjacent to nipple.* **(Middle)** *Transverse ultrasound of the breast shows heterogeneous fibroglandular tissue containing hypoechoic areas representing glandular elements as well as a more discrete small cyst. Small cysts are common within fibroglandular tissue and considered a normal finding.* **(Bottom)** *Transverse ultrasound of the breast shows multiple anechoic cysts clustered together within fibroglandular tissue. This represents fibrocystic change, which is considered a normal variant. Fibrocystic change often waxes and wanes and may cause chronic or cyclical breast pain.*

NIPPLE, INTRAMAMMARY NODE

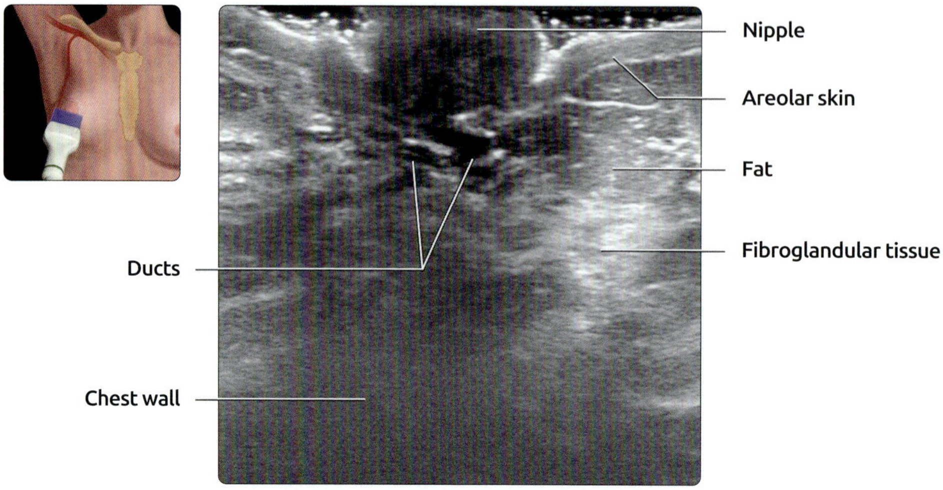

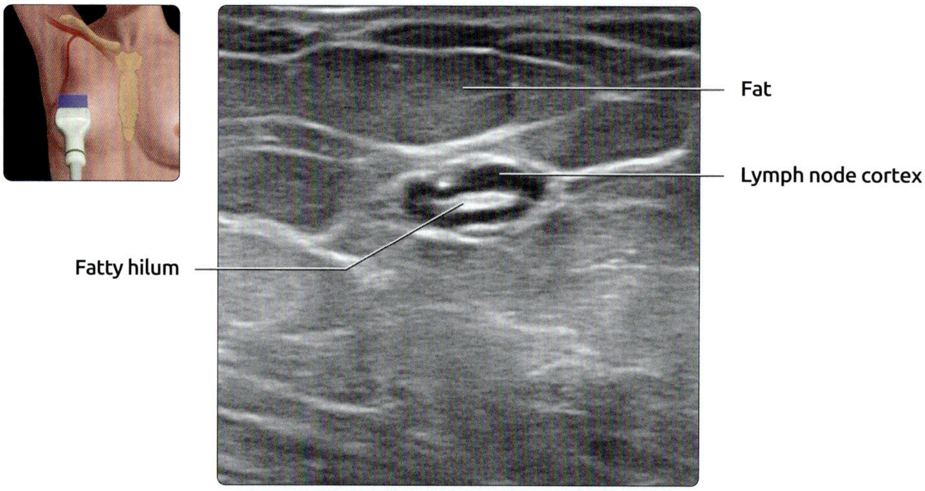

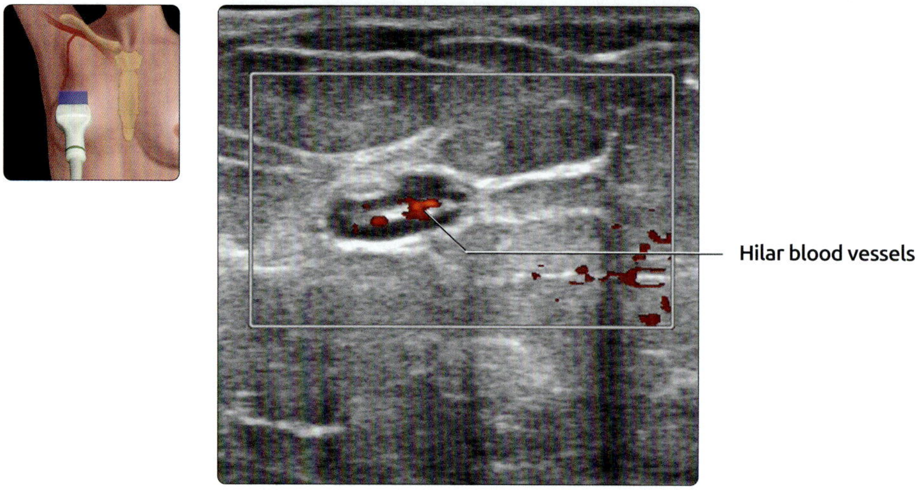

(Top) *Transverse ultrasound of the nipple areolar complex obtained with gel standoff and minimal transducer pressure shows normal nipple, which is indented by the probe and small, anechoic subareolar ducts. The nipple often shadows intensely, limiting evaluation of the subareolar region. Positional changes of the nipple and probe help to circumvent shadowing to allow visualization of the underlying tissue.* **(Middle)** *Longitudinal ultrasound of the upper outer quadrant breast shows a normal-appearing intramammary lymph node characterized by its reniform shape, thin anechoic cortex, and hyperechoic fatty hilum.* **(Bottom)** *Longitudinal ultrasound of the same lymph node with power Doppler shows vascular flow in the hilum, which should be visible in superficial lymph nodes. An artery is often seen coursing near the lymph node (not shown).*

BREAST DUCTS

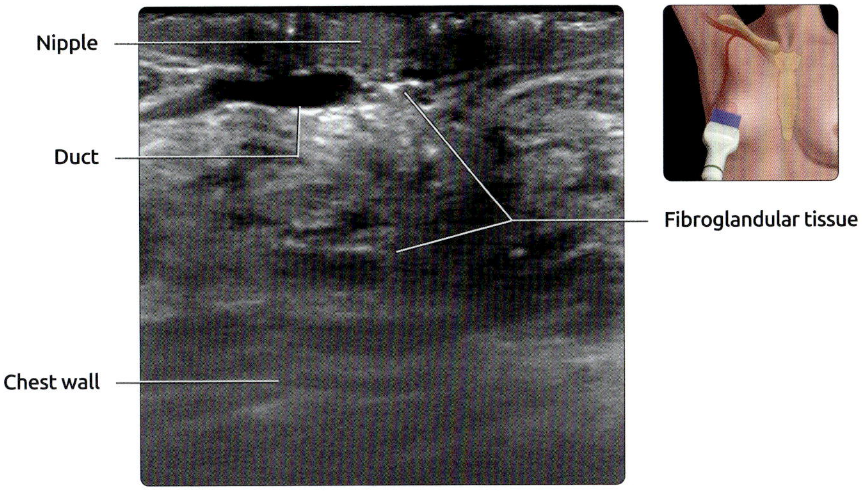

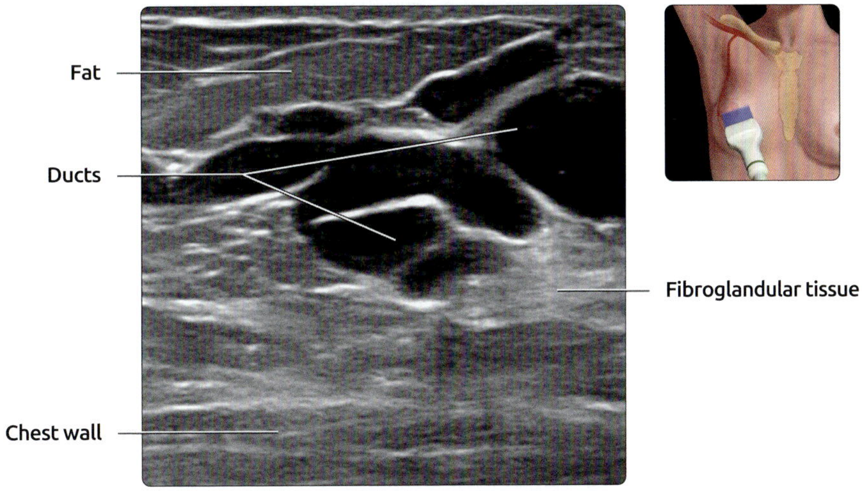

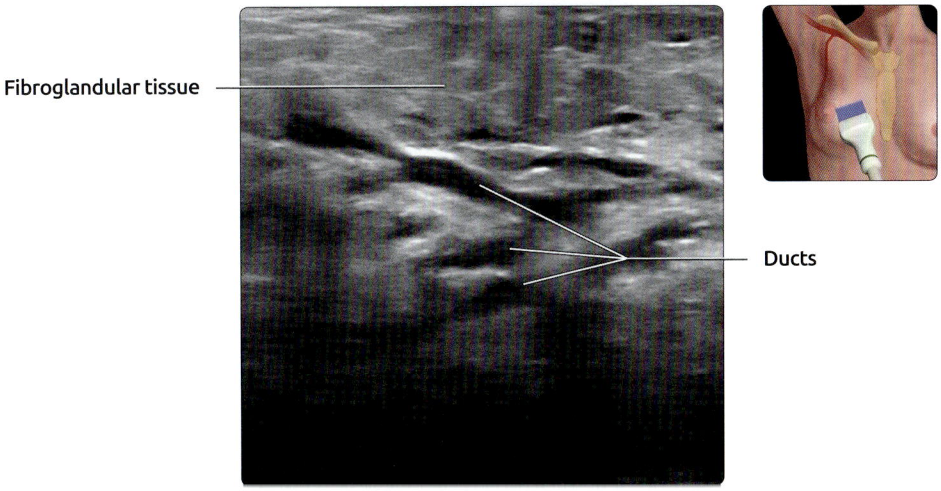

(Top) *Radial ultrasound of the nipple areolar complex shows a more sessile (rather than rounded) nipple with a mildly ectatic subareolar duct. Multiple ectatic ducts containing anechoic or mildly echogenic fluid is a common finding; however, a solitary dilated duct should raise concern for an intraductal mass or duct obstruction.* **(Middle)** *Radial ultrasound of the periareolar breast tissue shows markedly dilated anechoic ducts. This was due to prior subareolar duct excision with associated chronic fluid within residual ducts. The duct walls are smooth as expected with benign duct ectasia.* **(Bottom)** *Transverse ultrasound of upper outer quadrant breast tissue with lactational change shows multiple fluid-filled ducts. Ducts dilated with fluid are not commonly seen so diffusely outside of the subareolar region except in the setting of lactation. Fibroglandular tissue with lactational change is isoechoic rather than hyperechoic in the nonlactating breast.*

LACTATION-RELATED CHANGES

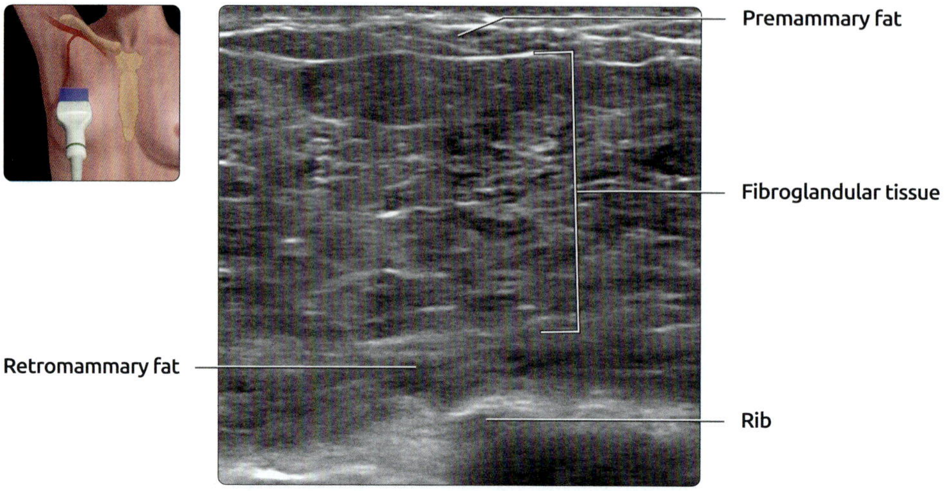

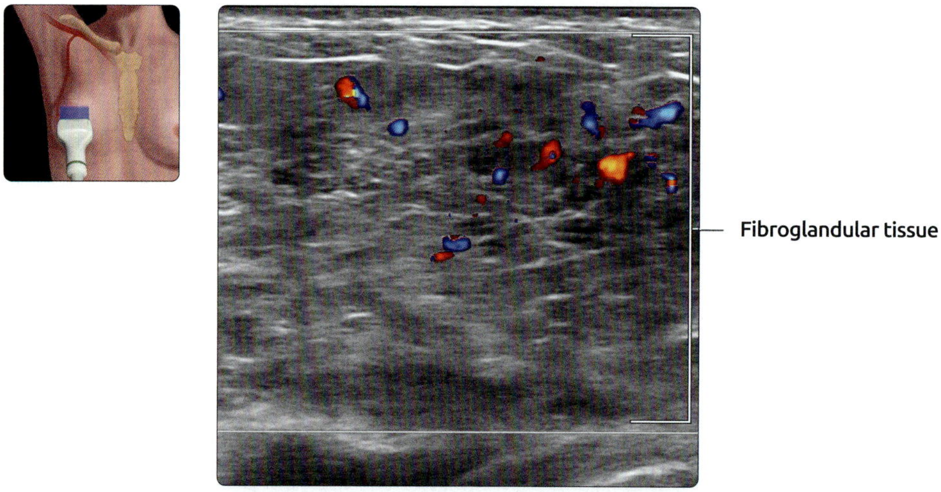

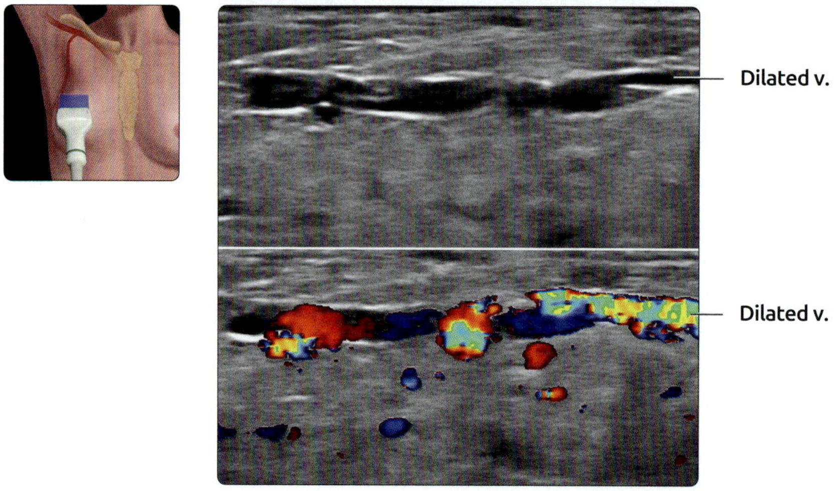

(Top) *Transverse ultrasound of a breast with lactational change shows isoechoic fibroglandular tissue, which expands the mammary zone. The premammary and retromammary fat are compressed.* **(Middle)** *Transverse ultrasound of the same tissue with color Doppler shows hypervascularity of the fibroglandular tissue, which is expected in lactational tissue due to the marked increase in blood flow related to milk production.* **(Bottom)** *Longitudinal ultrasound without and with color Doppler of breast tissue with lactation change shows an anechoic tubular structure that could represent a duct or blood vessel. Color Doppler proves the finding is a blood vessel rather than a duct. This is a vein that is more dilated than typical for normal breast tissue, but it is normal in a patient who is breast feeding.*

AXILLARY LYMPH NODES

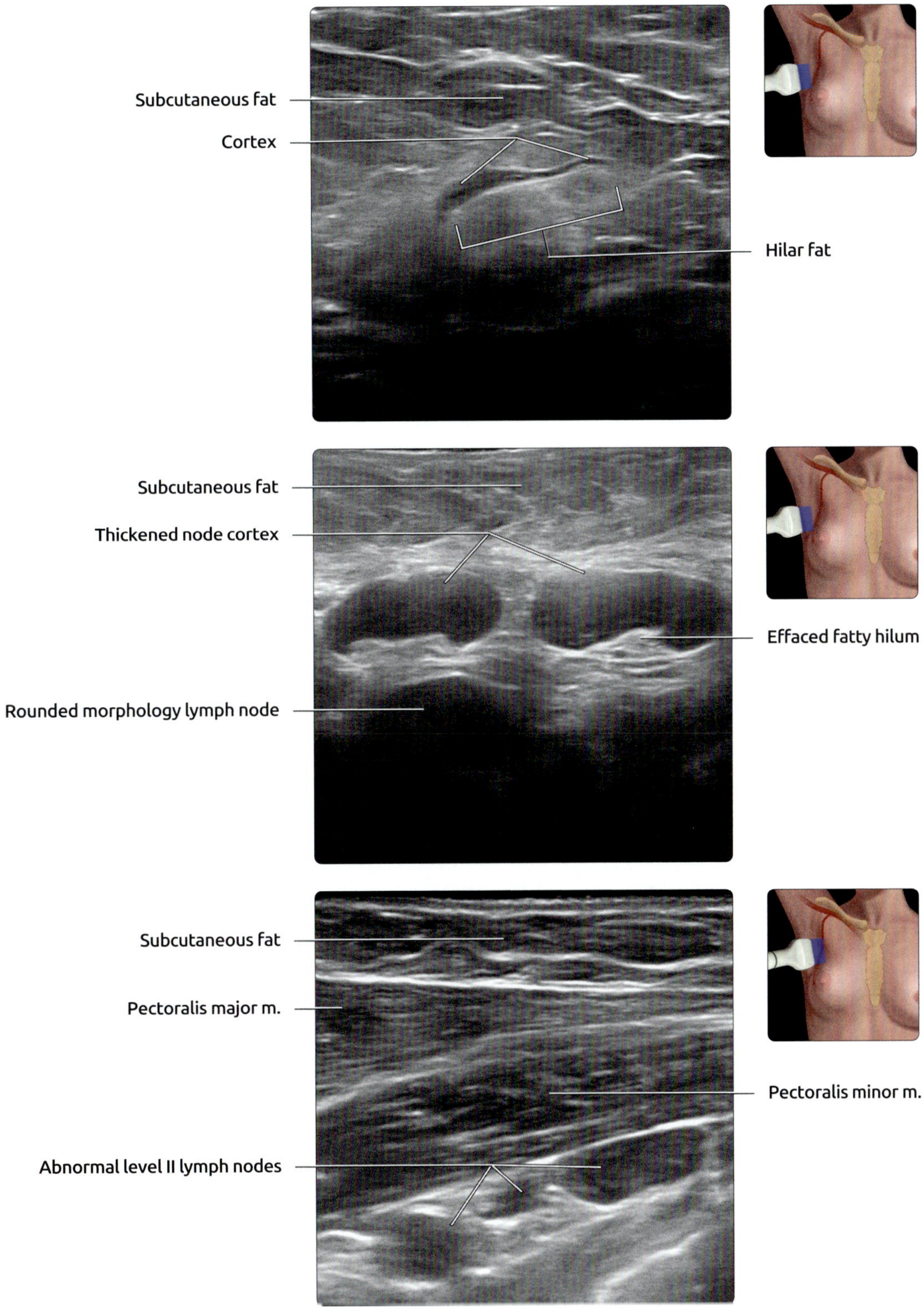

(Top) *Longitudinal ultrasound of the axilla shows a normal morphology lymph node with uniformly thin, hypoechoic cortex (which should measure < 3-4 mm), circumscribed margins, and hyperechoic fatty hilum.* **(Middle)** *Longitudinal ultrasound of the axilla shows multiple abnormal morphology lymph nodes. The adjacent superficial nodes maintain a reniform shape but have markedly thickened cortex, which effaces the hilar fat. The deeper node shows a more rounded morphology in addition to cortical thickening, which are both abnormal findings.* **(Bottom)** *Longitudinal ultrasound of the axillary region shows visible abnormal level II axillary lymph nodes, which are characterized by their location deep to the pectoralis minor muscle. Though lymph nodes exist in this location, they are not usually visible on ultrasound unless abnormal.*

SECTION 4

Abdomen

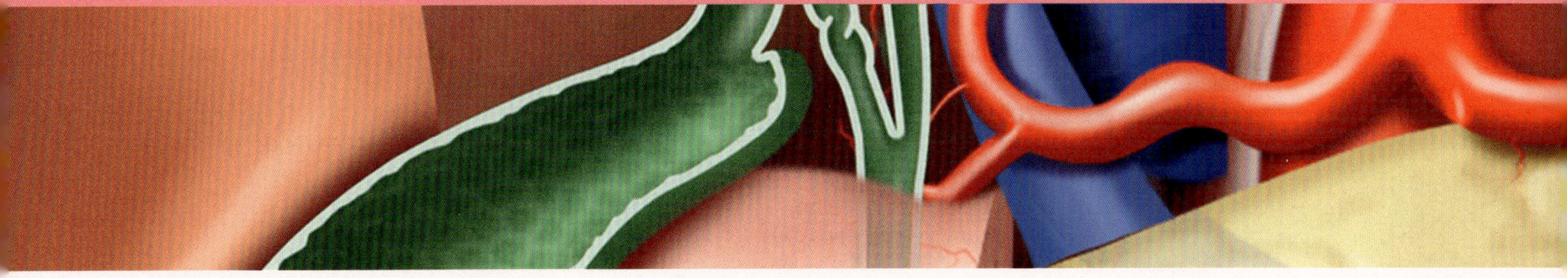

GROSS ANATOMY

Overview

- Liver: Largest gland and largest internal organ (average weight: 1,500 g)
  - Functions
    - Processes all nutrients (except fats) absorbed from gastrointestinal (GI) tract; conveyed via portal vein
    - Stores glycogen, secretes bile
  - Relations
    - Anterior and superior surfaces smooth and convex
    - Posterior and inferior surfaces indented by colon, stomach, right kidney, duodenum, inferior vena cava (IVC), and gallbladder
  - Covered by peritoneum except along gallbladder fossa, porta hepatis, and bare area
    - **Bare area**: Nonperitoneal posterior superior surface where liver abuts diaphragm
    - **Porta hepatis**: Portal vein, hepatic artery, and bile duct located within hepatoduodenal ligament
  - **Falciform ligament**
    - Extends from liver to anterior abdominal wall
    - Separates right and left subphrenic peritoneal recesses (between liver and diaphragm)
    - Marks plane separating medial and lateral segments of left hepatic lobe
    - Carries round ligament (ligamentum teres), fibrous remnant of umbilical vein
  - **Ligamentum venosum**
    - Remnant of ductus venosus
    - Separates caudate from left hepatic lobe
- **Vascular anatomy (unique dual afferent blood supply)**
  - **Portal vein**
    - Carries nutrients from gut and hepatotrophic hormones from pancreas to liver along with oxygen
      - Contains 40% more oxygen than systemic venous blood
    - 75-80% of blood supply to liver
  - **Hepatic artery**
    - Supplies 20-25% of blood
    - Liver less dependent than biliary tree on hepatic arterial blood supply
    - Usually arises from celiac artery
    - Variations common, including arteries arising from superior mesenteric artery
  - **Hepatic veins**
    - Usually 3 (right, middle, and left)
    - Many variations and accessory veins
    - Collect blood from liver and return it to IVC
    - Confluence of hepatic veins just below diaphragm and entrance of IVC into right atrium
  - **Portal triad**
    - At all levels of size and subdivision, branches of hepatic artery, portal vein, and bile ducts travel together
    - Blood flows into hepatic sinusoids from interlobular branches of hepatic artery and portal vein → hepatocytes, which detoxify blood and produce bile
      - Blood collects into central veins → hepatic veins
      - Bile collects into ducts → stored in gallbladder and excreted into duodenum
- **Segmental anatomy**
  - 8 hepatic segments
    - Each receives secondary or tertiary branch of hepatic artery and portal vein
    - Each drained by its own bile duct (intrahepatic) and hepatic vein branch
  - Caudate lobe = segment 1
    - Has independent portal triads and hepatic venous drainage to IVC
  - Left lobe
    - Lateral superior = segment 2
    - Lateral inferior = segment 3
    - Medial superior = segment 4a
    - Medial inferior = segment 4b
  - Right lobe
    - Anterior inferior = segment 5
    - Posterior inferior = segment 6
    - Posterior superior = segment 7
    - Anterior superior = segment 8

IMAGING ANATOMY

Internal Contents

- **Capsule**
  - Reflective Glisson capsule making borders of liver well defined
- **Left lobe**
  - Contains segments 2, 3, 4a, and 4b
  - Longitudinal scan
    - Triangular in shape
    - Rounded upper surface
    - Sharp inferior border
  - Transverse scan
    - Wedge-shaped tapering to left
  - Liver parenchyma echoes are mid gray with uniform, sponge-like pattern interrupted by vessels
- **Right lobe**
  - Contains segments 5, 6, 7, and 8
  - Liver parenchymal echoes similar to left lobe
  - Sections of right lobe show same basic shape, though right lobe usually larger than left
- **Caudate lobe**
  - Longitudinal scan
    - Almond-shaped structure posterior to left lobe
  - Transverse scan
    - Seen as extension of right lobe
- **Portal veins**
  - Have thicker reflective walls than hepatic veins; portal veins have fibromuscular walls
  - Wall reflectivity also depends on angle of interrogation; portal veins cut at more oblique angle may have less apparent wall
  - Can be traced back toward porta hepatis
  - Normal portal flow is hepatopetal on color Doppler; absent or reversal of flow may be seen in portal hypertension
  - Normal velocity 13-55 cm/sec

- Portal waveform has undulating appearance due to variations with cardiac activity and respiration
- Branches run in transverse plane
- Hepatic portal vein anatomy is variable

- **Hepatic veins**
  - Appear as echolucent defects within liver parenchyma with no reflective wall: Large sinusoids with thin or absent wall
  - Branches enlarge and can be traced toward IVC
  - Flow pattern has triphasic waveform
    - Resulting from transmission of right atrial pulsations into veins
      - A wave: Atrial contraction
      - S wave: Systole (tricuspid valve moves toward apex)
      - D wave: Diastole
  - Right hepatic vein
    - Runs in coronal plane between anterior and posterior segments of right hepatic lobe
  - Middle hepatic vein
    - Lies in sagittal or parasagittal plane between right and left hepatic lobe
  - Left hepatic vein
    - Runs between medial and lateral segments of left hepatic lobe
    - Frequently duplicated
  - 1 of 3 major branches of hepatic veins may be absent
    - Absent right hepatic vein ~ 6%
    - Less commonly middle and left hepatic vein
- **Hepatic artery**
  - Flow pattern has low-resistance characteristics with large amount of continuous forward flow throughout diastole
    - Normal velocity: 30-70 cm/sec
    - Resistive index ranges 0.5-0.8, increases after meal
  - Common hepatic artery usually arises from celiac axis
  - Classic configuration: 72%
    - Celiac axis → common hepatic artery → gastroduodenal artery and proper hepatic artery → latter gives rise to right and left hepatic artery
  - Variations from classic configuration
    - Common hepatic artery arising from superior mesenteric artery (replaced hepatic artery): 4%
    - Right hepatic artery arising from superior mesenteric artery (replaced right hepatic artery): 11%
    - Left hepatic artery arising from left gastric artery (replaced left hepatic artery): 10%
- **Bile ducts**
  - Normal peripheral intrahepatic bile ducts too small to be demonstrated
  - Normal right and left hepatic ducts measuring few millimeters usually visible
  - Normal common duct
    - Most visible in its proximal portion just caudal to porta hepatis: < 5 mm
    - Distal common duct should typically measure < 6-7 mm
    - In elderly, generalized loss of tissue elasticity with advancing age leads to increase in bile duct diameter: < 8 mm (somewhat controversial)

ANATOMY IMAGING ISSUES

Imaging Recommendations

- Transducer
  - 2.5- to 5.0-MHz curvilinear or vector transducer generally most suitable
  - Higher frequency linear transducer (i.e., 7-9 MHz) useful for evaluation of liver capsule and superficial portions of liver
- Left lobe
  - Subcostal window with full inspiration generally most suitable
- Right lobe
  - Subcostal window
    - Cranial and rightward angulation useful for visualization of right lobe below dome of hemidiaphragm
    - Can sometimes be obscured by bowel gas
  - Intercostal window
    - Usually gives better resolution for parenchyma without influence from bowel gas
    - Right lobe just below hemidiaphragm may not be visible due to obscuration from lung bases
    - Important to tilt transducer parallel to intercostal space to minimize shadowing from ribs

Imaging Pitfalls

- Because of variations of vascular and biliary branching within liver (common), frequently impossible to designate precise boundaries between hepatic segments on imaging studies

CLINICAL IMPLICATIONS

Clinical Importance

- Liver US often 1st-line imaging modality in evaluation for elevated liver enzymes
  - Diffuse liver disease, such as hepatic steatosis, cirrhosis, hepatomegaly, hepatitis, and biliary ductal dilatation well visualized on US
  - Documentation of patency of portal vein, hepatic vein waveforms, and hepatic arterial velocities helpful in evaluation for etiologies of elevated liver function tests
- Liver metastases common
  - Primary carcinomas of colon, pancreas, and stomach commonly metastasize to liver
    - Portal venous drainage usually results in liver being initial site of metastatic spread from these tumors
  - Metastases from other non-GI primaries (breast, lung, etc.) commonly spread to liver hematogenously
- Primary hepatocellular carcinoma
  - Common worldwide
    - Risk factors include cirrhosis of any etiology and chronic viral hepatitis B in certain populations
    - Chronic hepatitis C with stage 3 fibrosis and nonalcoholic steatohepatitis may also have increased risk of HCC
    - US commonly used for screening and surveillance in patients at risk for development of hepatocellular carcinoma typically at 6-month intervals

HEPATIC VISCERAL SURFACE

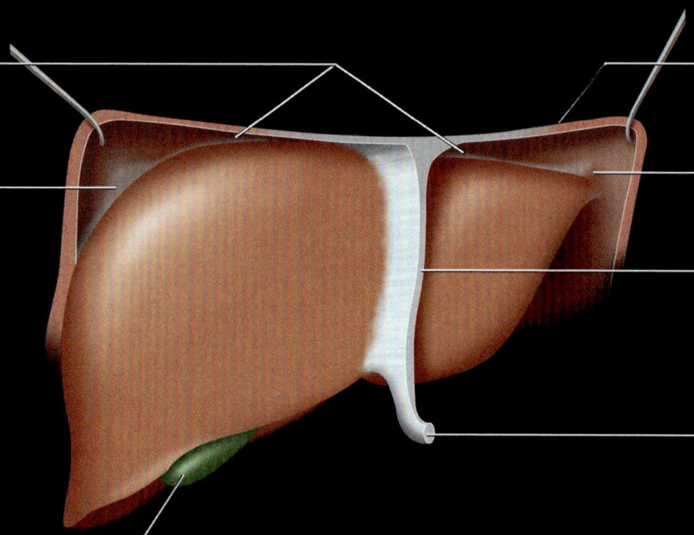

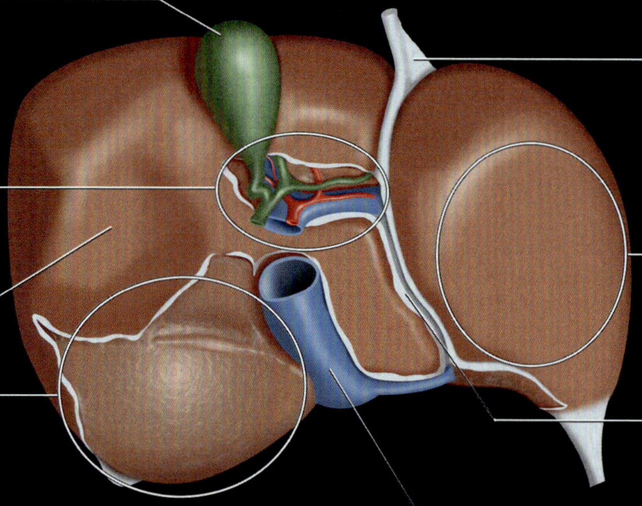

(Top) *The anterior surface of the liver is smooth and molds to the diaphragm and anterior abdominal wall. Generally, only the anterior/inferior edge of the liver is palpable on a physical exam. The liver is covered with peritoneum, except for the gallbladder bed, porta hepatis, and the bare area. Peritoneal reflections form various ligaments that connect the liver to the diaphragm and abdominal wall, including the falciform ligament, the inferior edge that contains the ligamentum teres, and the obliterated remnant of the umbilical vein.* **(Bottom)** *Graphic shows the liver inverted, which is somewhat similar to the surgeon's view of the upwardly retracted liver. The structures in the porta hepatis include the portal vein (blue), hepatic artery (red), and the bile ducts (green). The visceral surface of the liver is indented by adjacent viscera. The bare area is not easily accessible.*

HEPATIC ATTACHMENTS AND RELATIONS

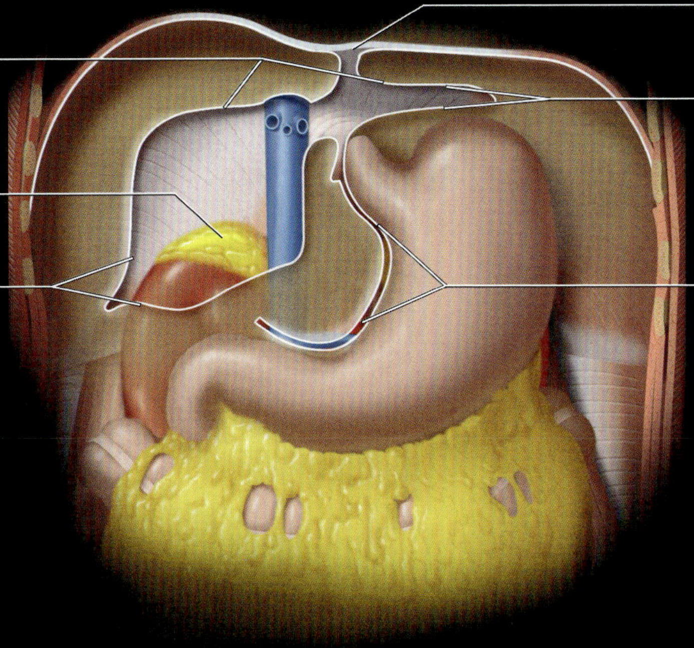

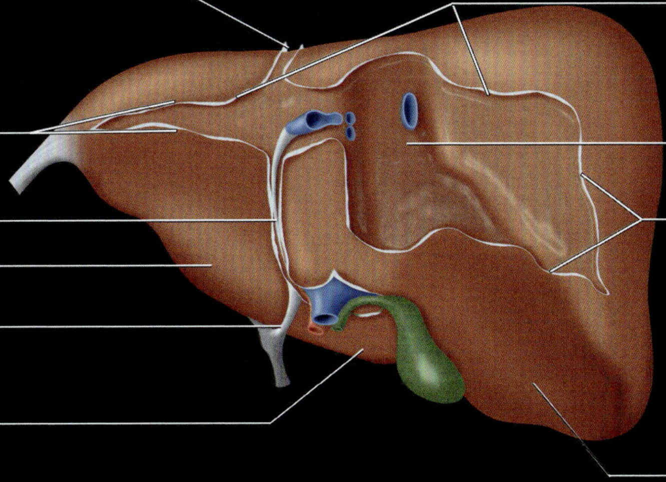

(Top) *The liver is attached to the posterior abdominal wall and diaphragm by the left and right triangular and coronary ligaments. The falciform ligament attaches the liver to the anterior abdominal wall. The bare area is in direct contact with the right adrenal gland, kidney, and inferior vena cava (IVC).* **(Bottom)** *Posterior view of the liver shows the ligamentous attachments. While these may help to fix the liver in position, abdominal pressure alone is sufficient, as evidenced by orthotopic liver transplantation, after which the ligamentous attachments are lost without the liver shifting position. The diaphragmatic peritoneal reflection is the coronary ligament whose lateral extensions are the right and left triangular ligaments. The falciform ligament separates the medial and lateral segments of the left lobe.*

HEPATIC VESSELS AND BILE DUCTS

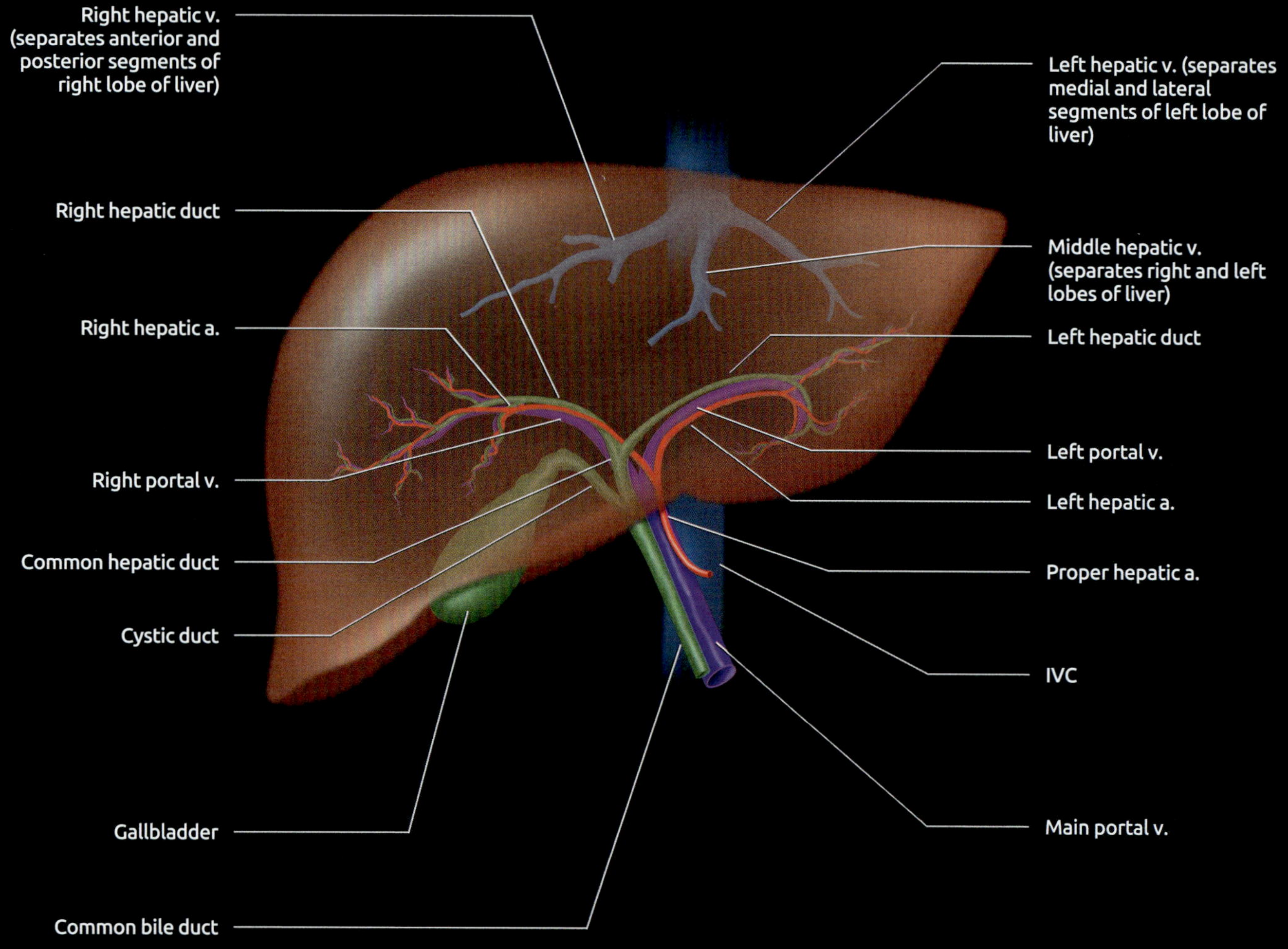

Graphic emphasizes that at every level of branching and subdivision, the portal veins, hepatic arteries, and bile ducts course together, constituting the portal triad. Each segment of the liver is supplied by branches of these vessels. Conversely, hepatic venous branches lie between hepatic segments and interdigitate with the portal triads but never run parallel to them.

HEPATIC ARTERIAL ANATOMY

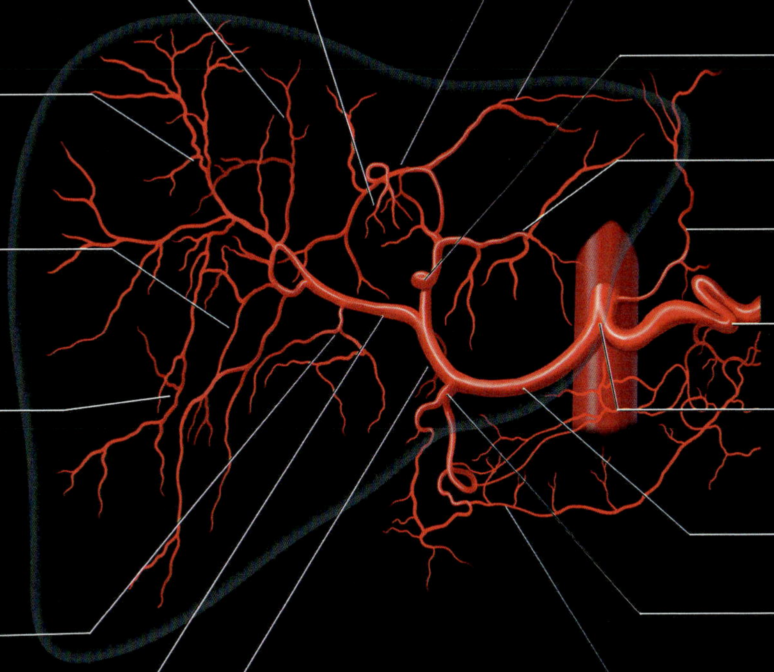

Graphic demonstrates the conventional hepatic arterial supply to the liver. The celiac artery arises at roughly the T12 level before dividing into the common hepatic artery, left gastric artery, and splenic artery. The common hepatic artery gives off the gastroduodenal artery inferiorly and becomes the proper hepatic artery, which then divides into the right and left hepatic arteries at the liver hilum. The left hepatic artery courses superiorly and slightly to the left before giving off branches to segments 2-4. In some instances, the segment 4 artery may arise directly from the proper hepatic artery and is then termed the middle hepatic artery. The right hepatic artery divides into anterior and posterior branches, which take a upward vertical course and horizontal course, respectively. The anterior branch gives off arteries supplying segments 5 and 8, while the posterior branches supplies segments 6 and 7. Segment 1 (caudate) is typically supplied by small branches of either the right or left hepatic arteries (or both).

HEPATIC ARTERIAL VARIANTS

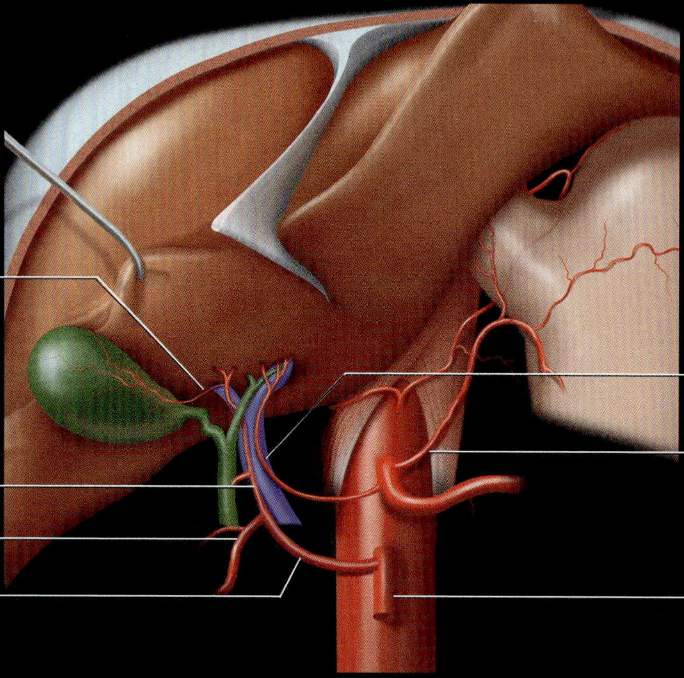

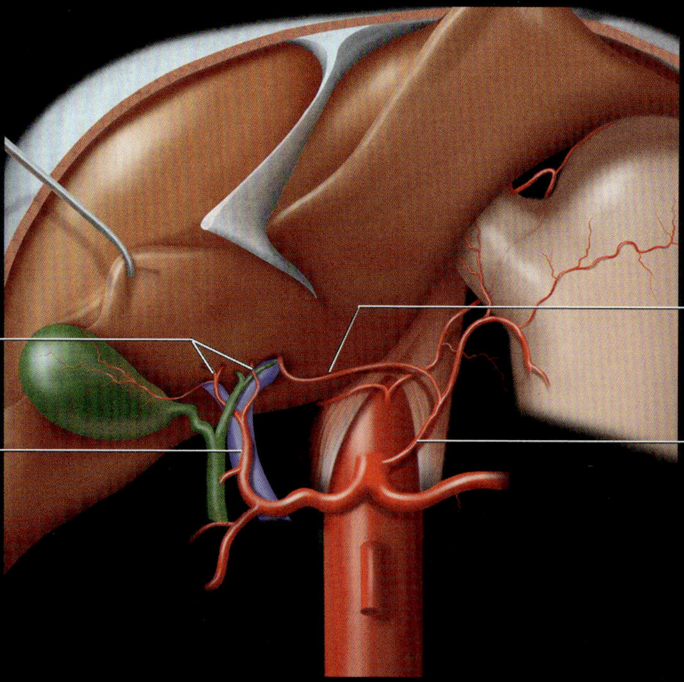

(Top) *Graphic depicts a separate origin of the left hepatic artery from the celiac trunk. In addition, the right hepatic artery is "replaced," arising from the superior mesenteric artery. The gastroduodenal and cystic arteries arise from the replaced right hepatic, as is common with this variation.* **(Bottom)** *Graphic depicts an "accessory" left hepatic artery arising from the left gastric artery. An accessory artery is a vessel in addition to those originating from the conventional depiction. In this case, there is a left hepatic artery arising from the proper hepatic artery as well. All of these variations are common and have major implications for patients undergoing any sort of upper abdominal surgery, especially partial hepatic resection or liver transplantation.*

HEPATIC ARTERIAL VARIANTS

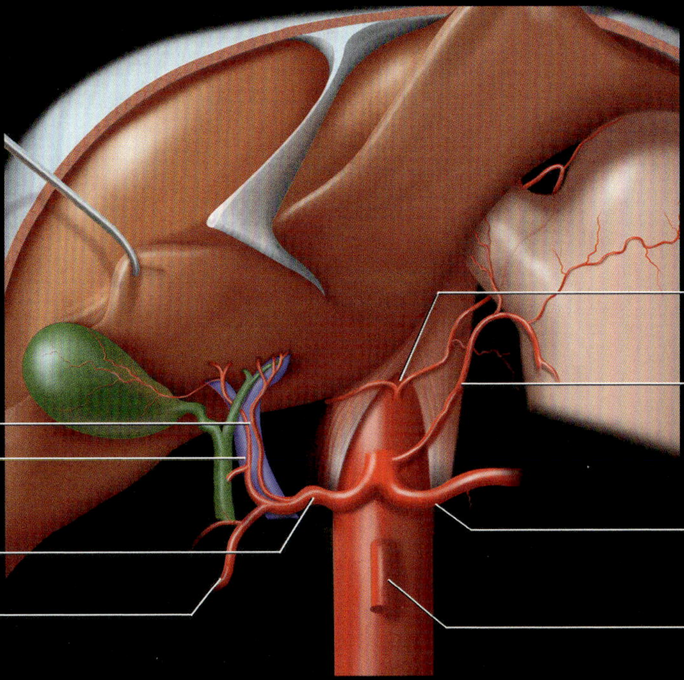

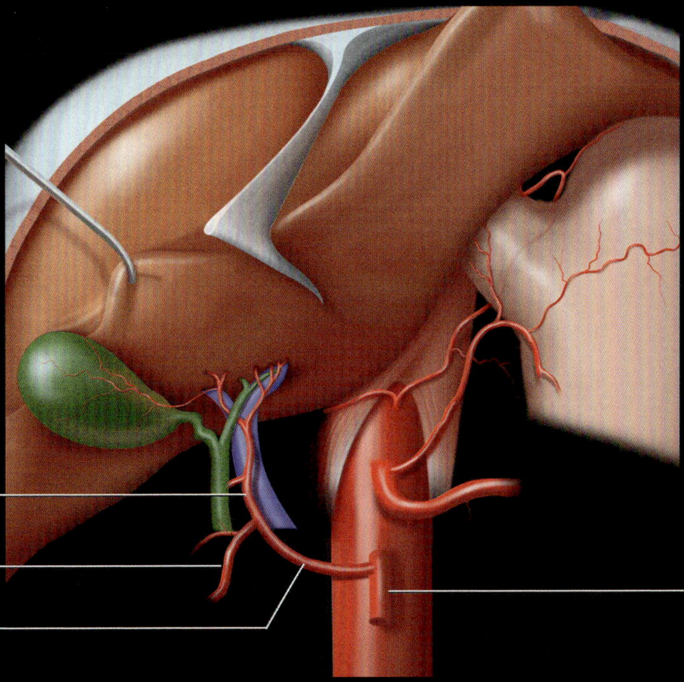

(Top) *In over 40% of individuals, there are variations in the origin and course of the hepatic arteries that differ from the "conventional" depiction. In this graphic, the left hepatic artery arises from the common hepatic artery, proximal to the origin of the gastroduodenal artery. The gallbladder and extrahepatic common bile duct are supplied by the right hepatic artery, as usual. The hepatic artery courses parallel to the portal vein and lies between the vein and the bile duct.* **(Bottom)** *Graphic shows a completely replaced hepatic artery arising from the superior mesenteric artery. In this setting, the hepatic artery passes through or behind the head of the pancreas and the portal vein and may be inadvertently ligated during pancreatic surgery if this variant is not recognized.*

TRANSVERSE LEFT LOBE OF LIVER

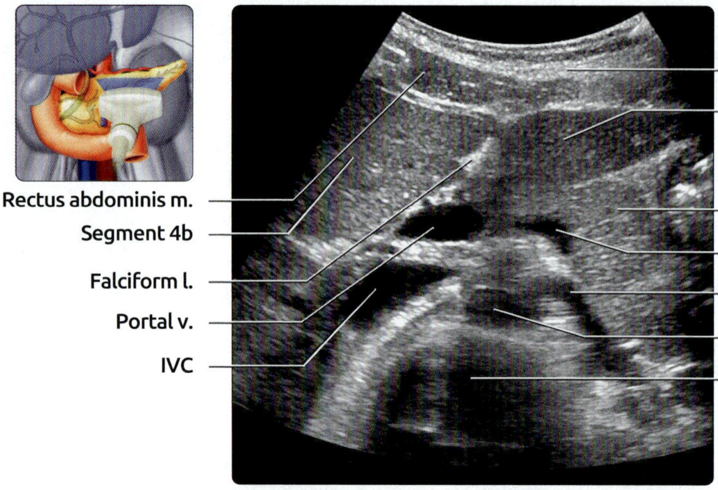

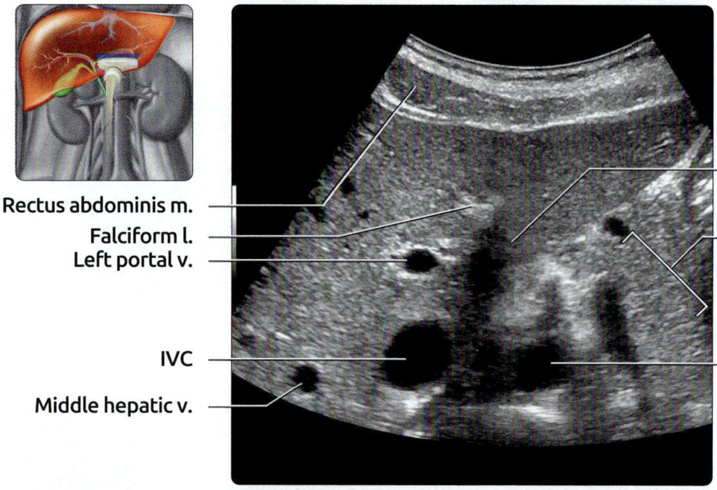

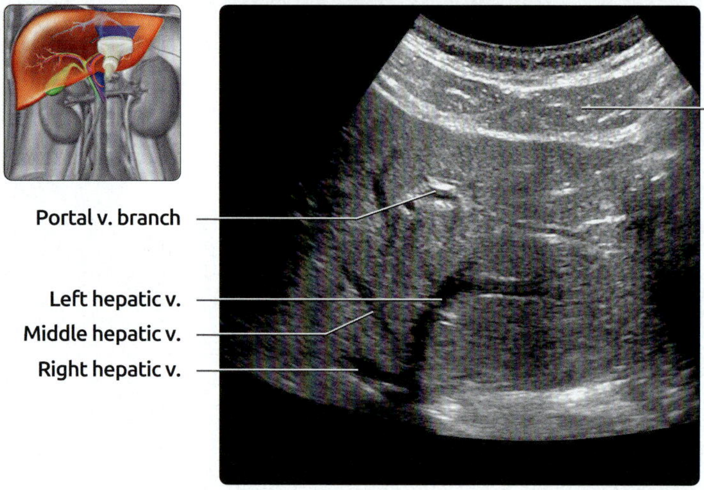

(Top) *Transverse grayscale US of the left lobe of the liver is shown centered at the level of the falciform ligament and pancreas.* **(Middle)** *Transverse grayscale US of the left lobe of the liver is shown.* **(Bottom)** *Transverse grayscale US of the left lobe of the liver is shown centered at the level of the left hepatic vein.*

LEFT LOBE OF LIVER: LEFT PORTAL VEIN

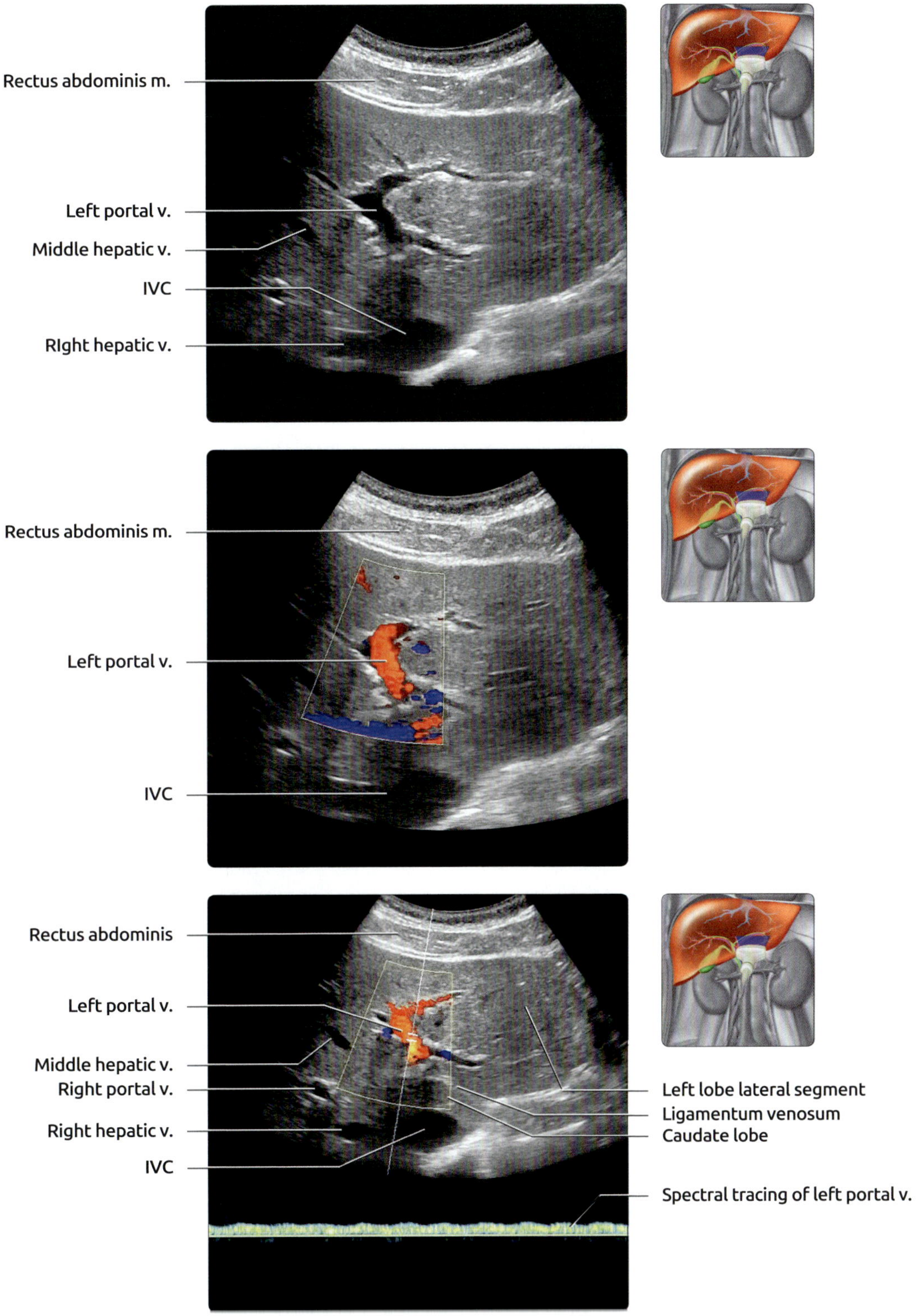

(Top) *Transverse grayscale US of the left lobe of the liver is shown centered at the left portal vein.* **(Middle)** *Transverse color Doppler US of the left lobe of the liver is shown centered at the level of the left portal vein. Flow in the left portal vein is directed toward the transducer, indicating that the flow is hepatopetal and therefore normal.* **(Bottom)** *Spectral tracing of the left portal vein on this transverse pulsed Doppler US shows that the flow is monophasic, directed toward the transducer, with a mildly undulating waveform related to slight transmission of the cardiac cycle, which is a normal appearance for the portal vein.*

LEFT LOBE OF LIVER: LEFT HEPATIC VEIN

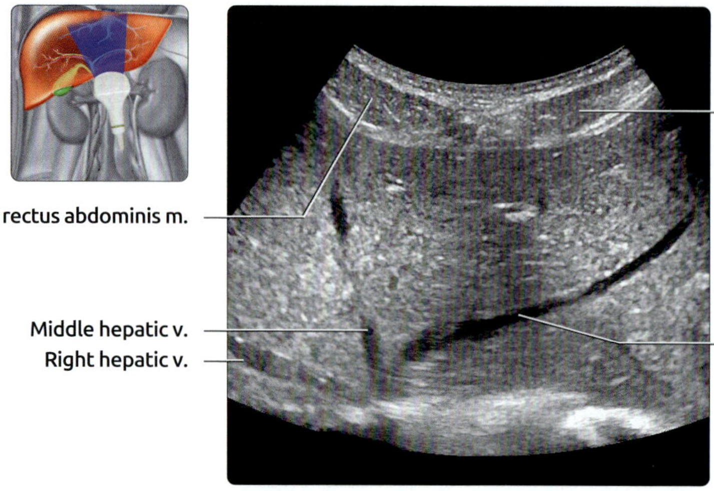

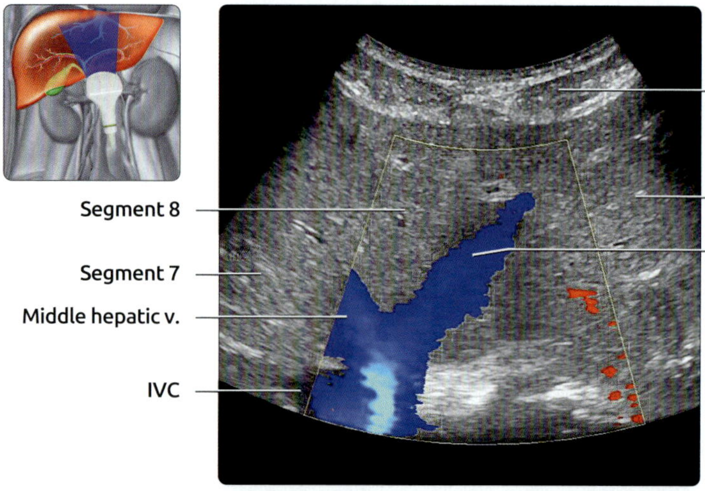

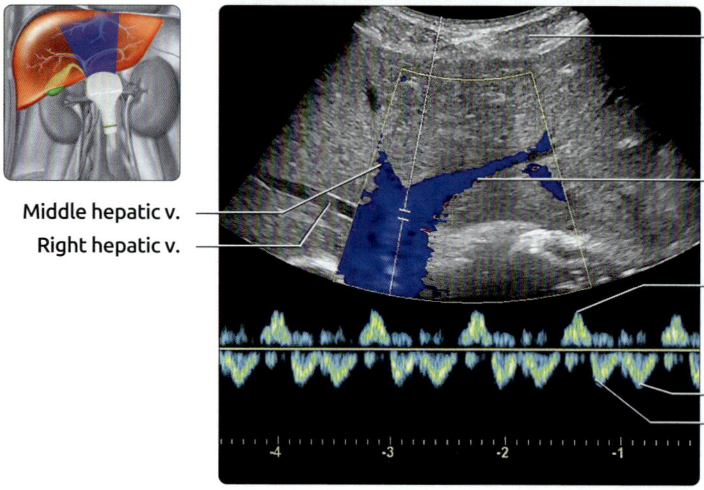

(Top) *Transverse grayscale US of the liver centered at the left hepatic lobe shows the right, middle, and left hepatic veins as they join into the intrahepatic IVC.* **(Middle)** *Transverse color Doppler US of the liver, centered at the confluence of the hepatic veins, shows that the flow direction is away from the transducer, directed toward the IVC.* **(Bottom)** *Spectral tracing of the left hepatic vein near the confluence with the IVC shows a characteristic triphasic waveform pattern, which represents reflection of cardiac motion.*

LONGITUDINAL LEFT LOBE OF LIVER

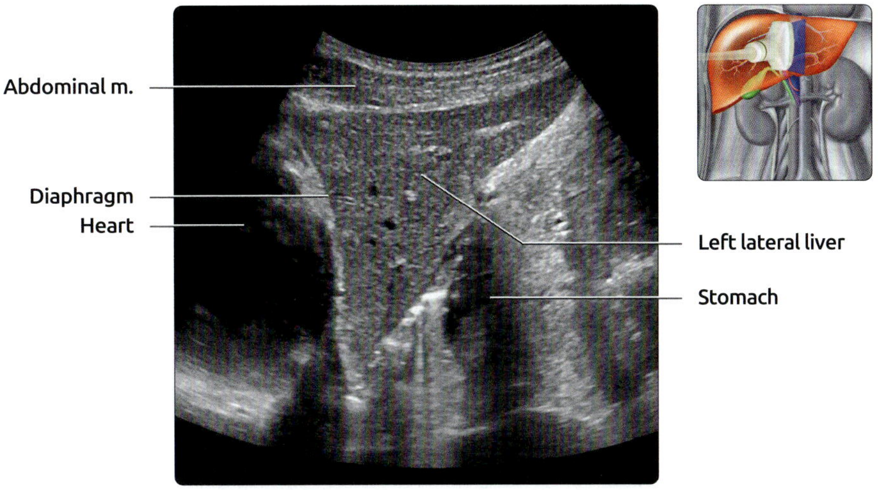

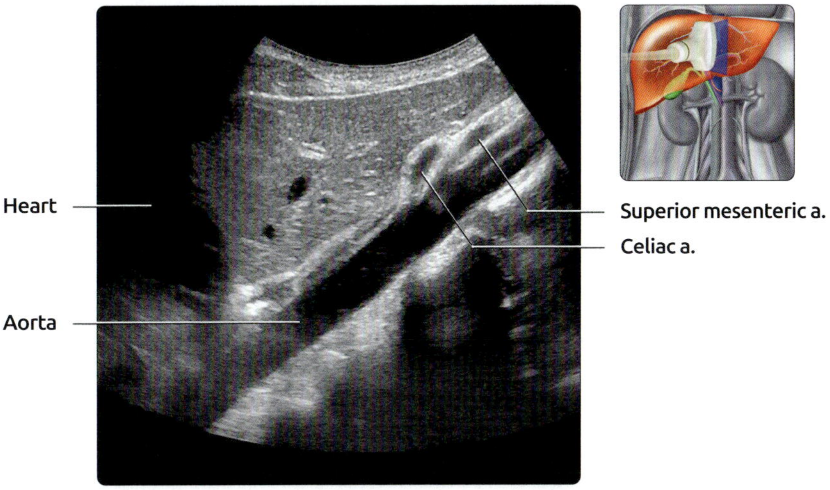

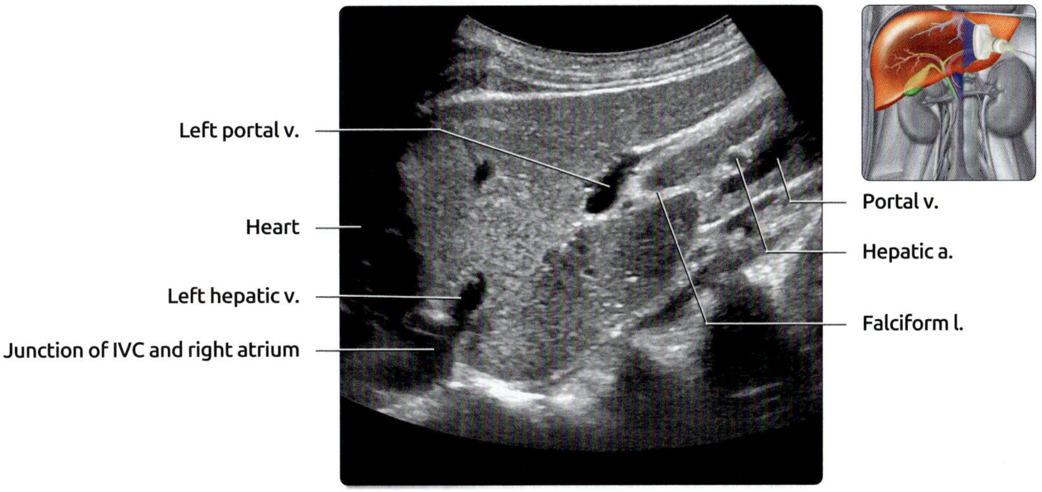

(Top) *Longitudinal grayscale US of the left lobe of the liver shows a triangular-shaped cross section. The heart is partially visualized above the diaphragm.* **(Middle)** *Longitudinal grayscale US view of the left lobe of the liver at the level of the aorta shows the aorta posterior to the liver, the celiac artery, and superior mesenteric artery arising from the aorta.* **(Bottom)** *Longitudinal grayscale US of the left lobe of the liver shows the left hepatic vein and left portal vein in cross section.*

TRANSVERSE RIGHT LOBE OF LIVER

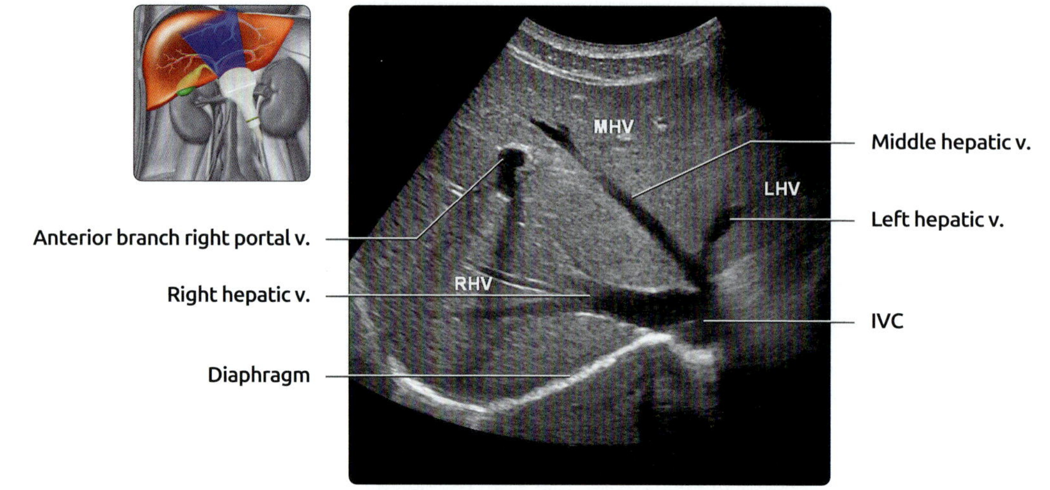

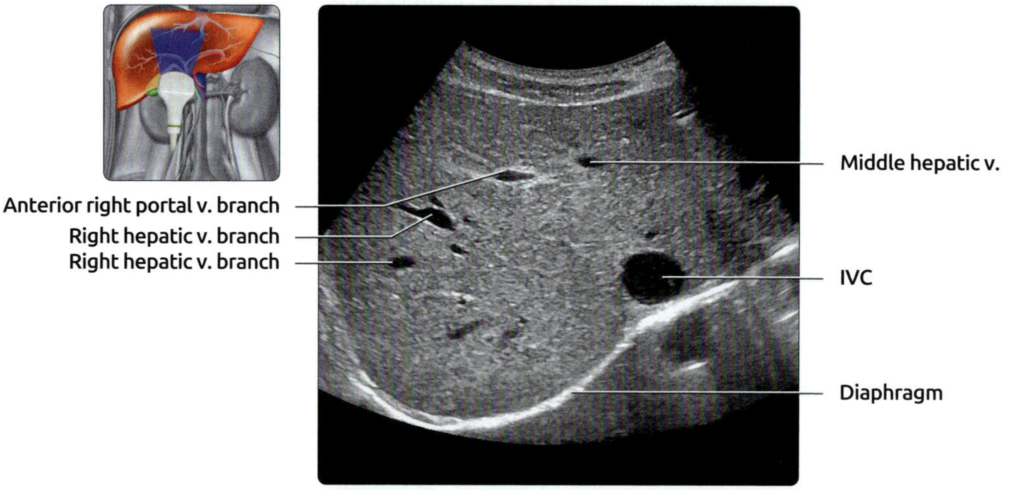

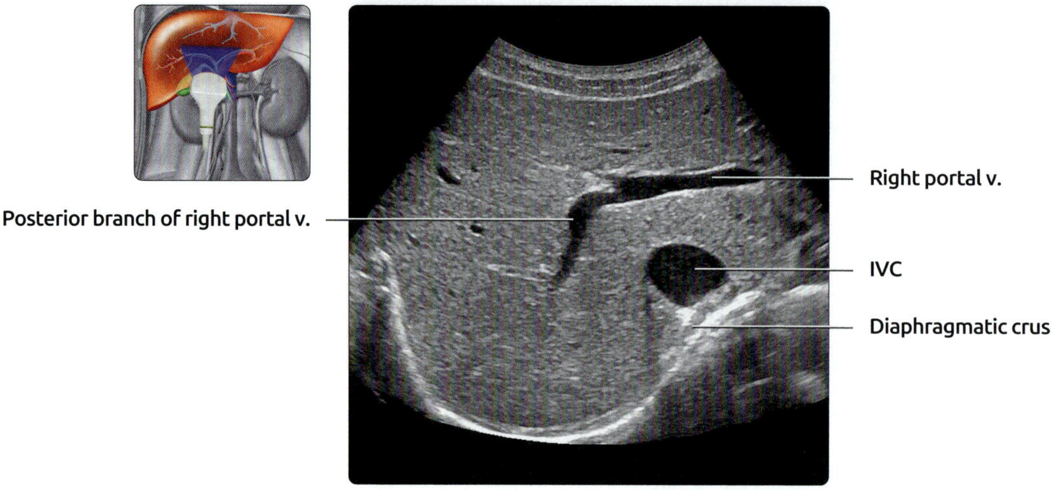

(Top) *Transverse grayscale US at the level of the hepatic vein confluence shows the right, middle, and left hepatic veins as they join with the IVC posteriorly.* **(Middle)** *Transverse grayscale US of the liver just below the confluence of the hepatic veins shows the IVC and more peripheral portions of the right and left hepatic veins.* **(Bottom)** *Transverse grayscale US of the right lobe of the liver, centered at the right portal vein, shows the posterior branch of the right portal vein, which is typically directed away from the transducer.*

RIGHT LOBE OF LIVER: RIGHT HEPATIC VEIN

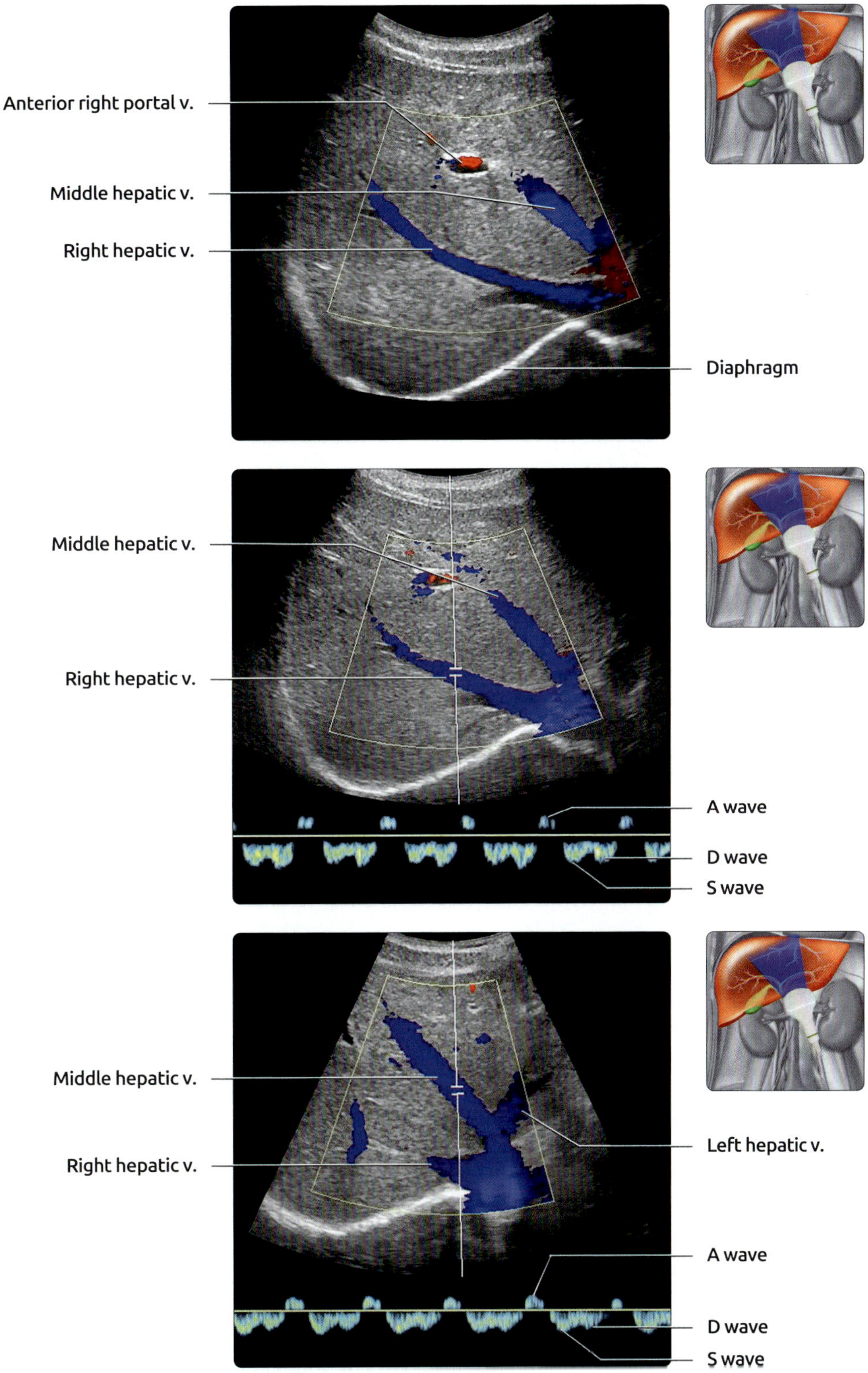

(Top) *Transverse color Doppler US of the right lobe of the liver shows that the right and middle hepatic veins are directed away from the transducer and flowing toward the IVC.* **(Middle)** *Spectral tracing of the right hepatic vein shows a typical triphasic waveform with A, S, and D waves representing reflection of cardiac motion in the hepatic veins.* **(Bottom)** *Spectral tracing of the middle hepatic vein shows a typical triphasic waveform with A, S, and D waves representing reflection of cardiac motion in the hepatic veins.*

MAIN PORTAL VEIN

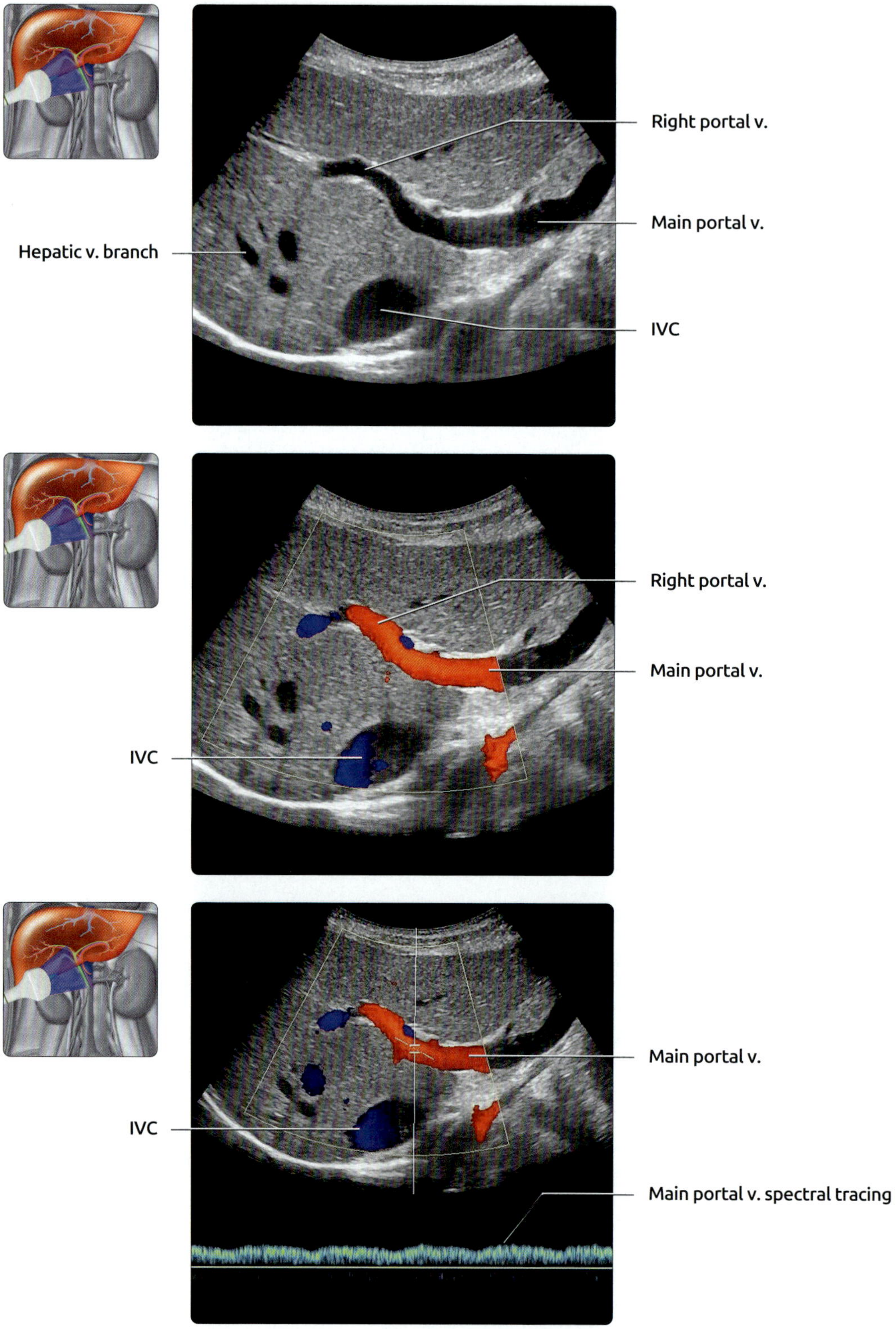

(Top) *Longitudinal oblique grayscale US is shown centered at the level of the main and right portal veins.* **(Middle)** *Longitudinal oblique color Doppler US, centered at the level of the main and right portal veins, shows that flow in the portal vein is directed toward the liver (hepatopetal).* **(Bottom)** *Longitudinal oblique spectral Doppler US of the main portal vein shows that the flow is hepatopetal, with gentle undulation reflecting the cardiac and respiratory cycle.*

PORTA HEPATIS

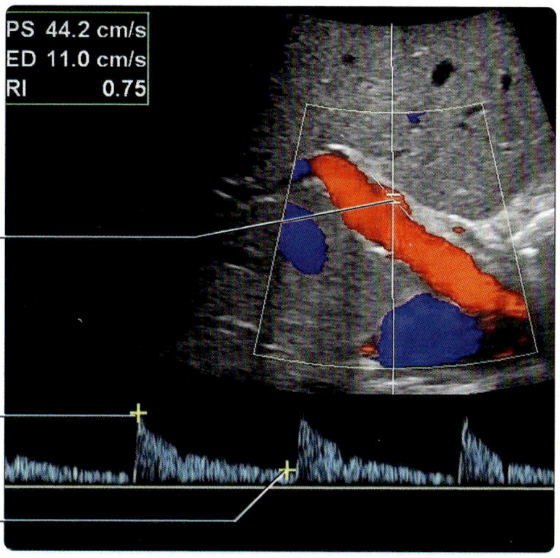

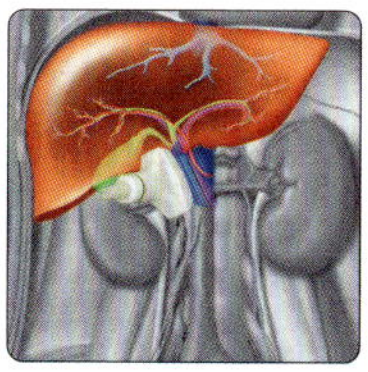

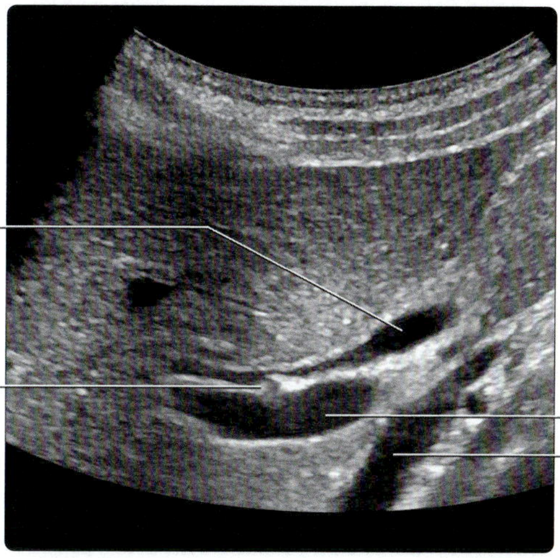

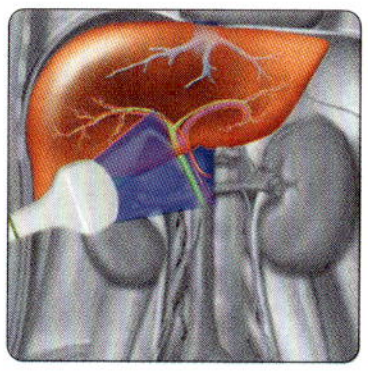

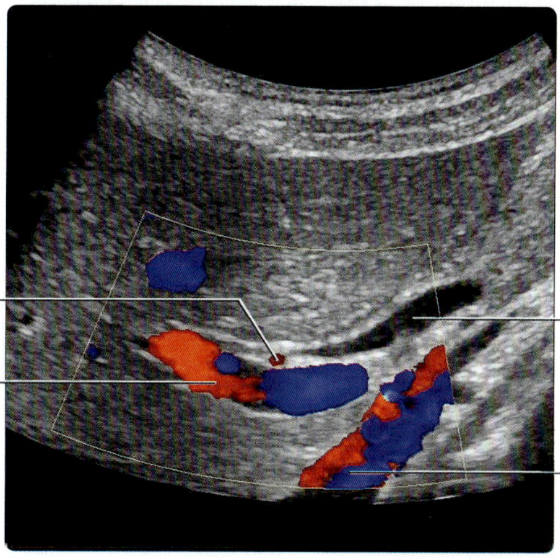

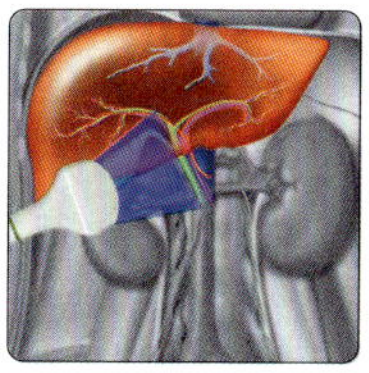

(Top) *Longitudinal oblique spectral tracing of the main hepatic artery shows a typical low-resistance waveform with brisk upstroke and forward diastolic flow. In this case, the hepatic artery velocity is 44 cm/sec, which is normal. When measuring velocity, proper angle correction is the key to obtaining accurate velocities.* **(Middle)** *Oblique grayscale US of the liver, centered at the porta hepatis, shows the common bile duct anterior to the right hepatic artery and portal vein. The IVC is seen posterior to the portal vein.* **(Bottom)** *Oblique color Doppler US of the liver, centered at the porta hepatis, shows the common bile duct is anterior to the portal vein, and the right hepatic artery is between these 2 structures. This is the typical anatomy in this location, although anatomic variants of the right hepatic artery may occur in which the hepatic artery may be located anterior to the common bile duct.*

LONGITUDINAL LIVER

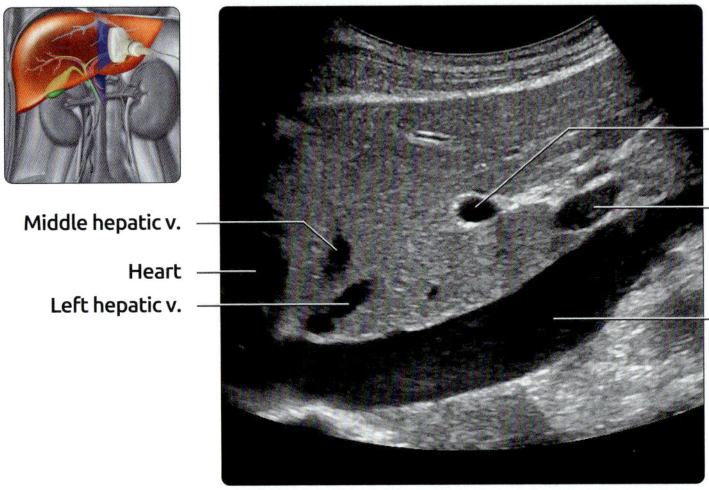

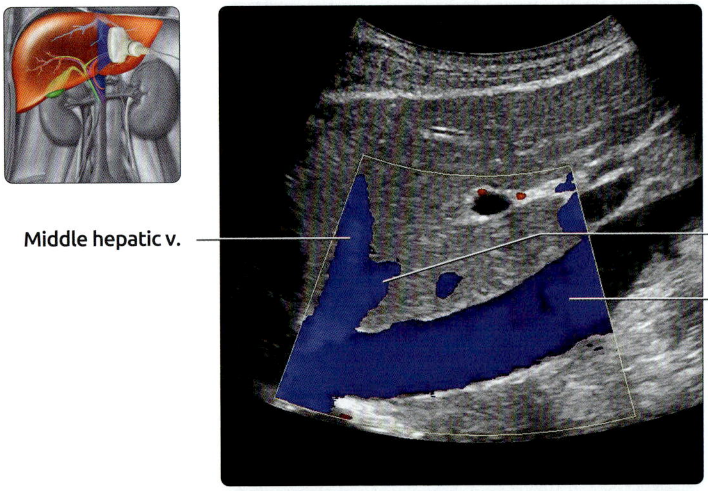

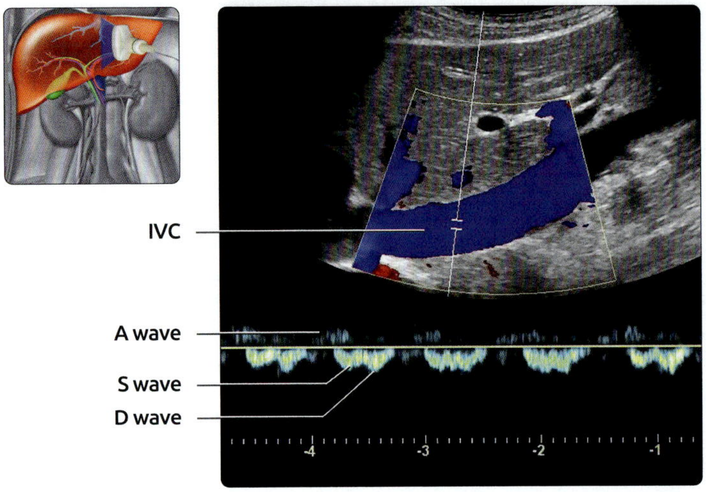

(Top) *Longitudinal grayscale US of the right lobe of the liver is shown centered at the level of the IVC.* **(Middle)** *Longitudinal color Doppler US of the liver is shown at the level of the IVC.* **(Bottom)** *Spectral tracing of the IVC shows a typical triphasic waveform with A, S, and D waves representing reflection of cardiac motion in the IVC.*

OTHER VIEWS OF LIVER

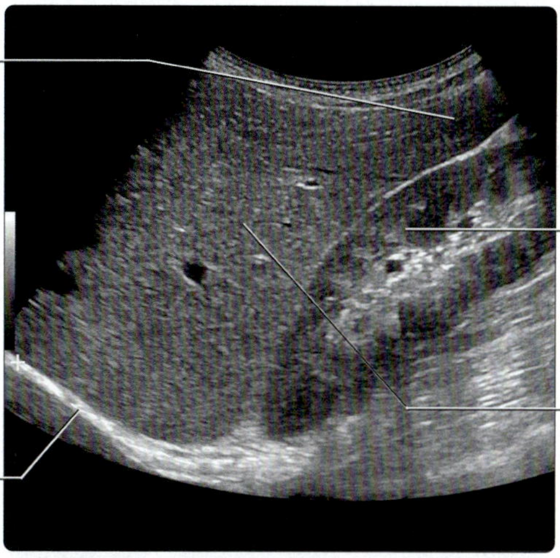

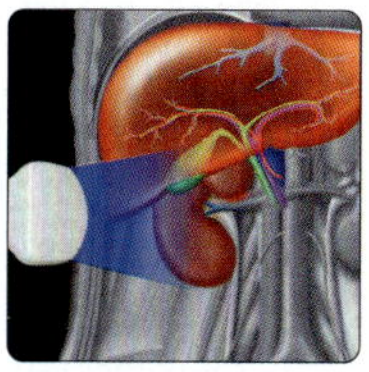

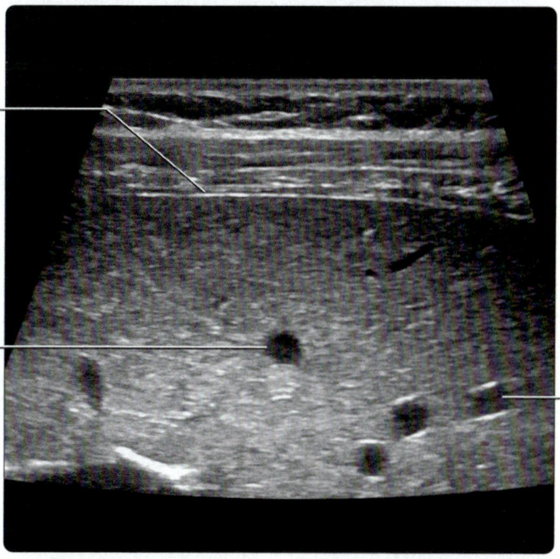

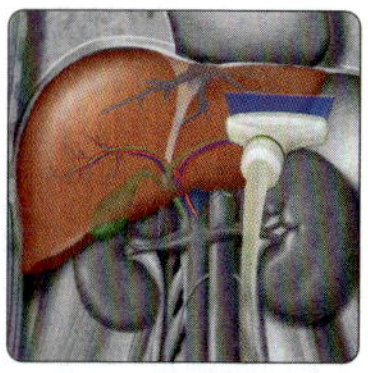

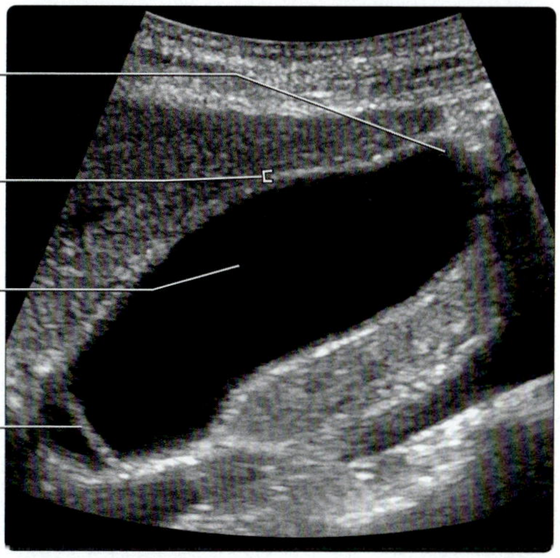

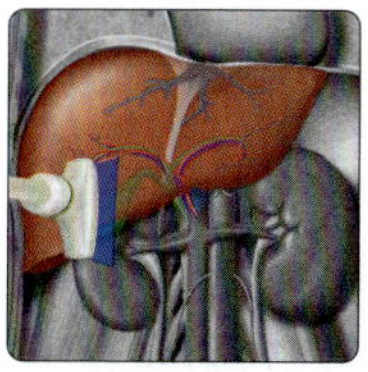

(Top) *Longitudinal grayscale US of the right lobe of the liver shows the liver ends just above the inferior margin of the right kidney. Normal hepatic length should be < 15.0-15.5 cm. Notice that the normal hepatic parenchyma is slightly hyperechoic compared to the normal kidney.* **(Middle)** *Transverse high-resolution US of the liver capsule, as seen here, is typically obtained with higher frequencies (7-9 MHz). Subtle nodularity of the capsule and small subcapsular liver lesions that may not be as well visualized with standard (3-5 MHz) frequencies are best visualized with this view. Note the hepatic veins have no discernible wall, whereas the portal veins have slightly echogenic walls.* **(Bottom)** *Longitudinal oblique US shows a normal gallbladder with anechoic fluid within the lumen and normal appearance of the gallbladder wall. Normal gallbladder wall thickness (< 3 mm) should be measured at the interface with the liver. A fold in the gallbladder neck is incidentally seen in this patient.*

LIVER SEGMENTAL ANATOMY

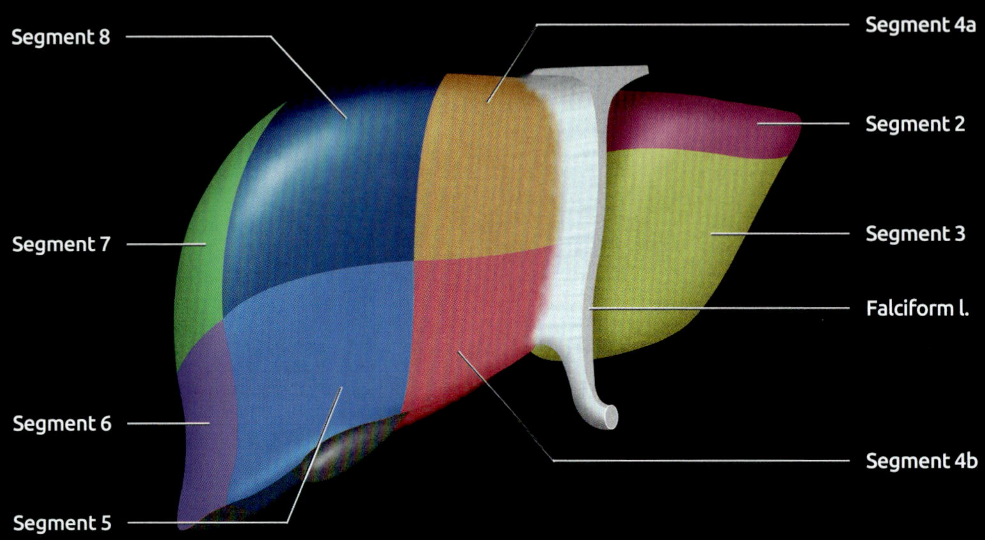

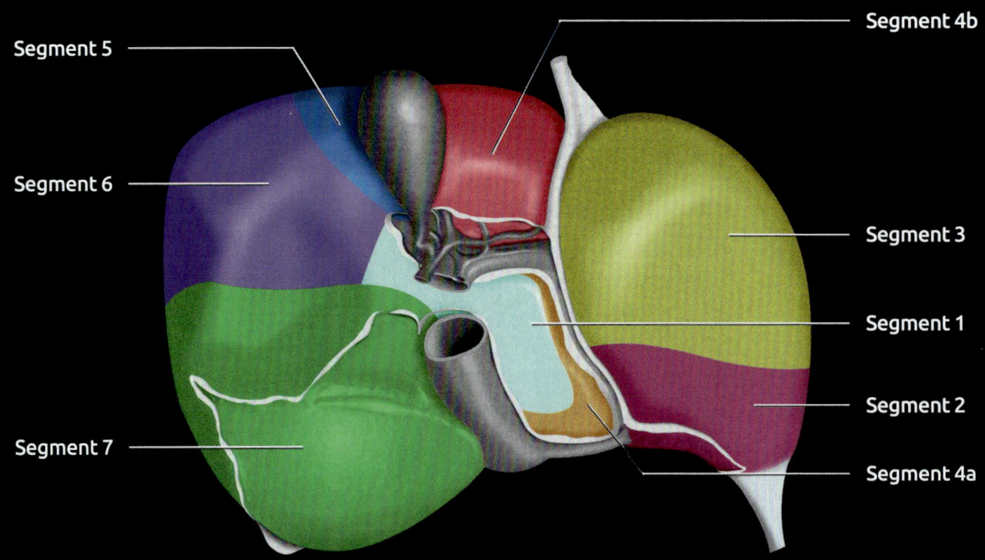

(Top) *1st of 2 graphics demonstrating the segmental anatomy of the liver in a somewhat idealized fashion is shown. Segments are numbered in a clockwise direction, starting with the caudate lobe (segment 1), which cannot be seen on this frontal view. The falciform ligament divides the lateral (segments 2 and 3) from the medial (segments 4a and 4b) left lobe. The horizontal planes separating the superior from the inferior segments follow the course of the right and left portal veins. An oblique vertical plane through the middle hepatic vein, gallbladder fossa, and IVC divides the right and left lobes.* **(Bottom)** *Inferior view of the liver shows that the caudate is entirely posterior, abutting the IVC, ligamentum venosum, and porta hepatis. In this view, a plane through the IVC and gallbladder divides the left and right lobes.*

LIVER SEGMENTAL ANATOMY

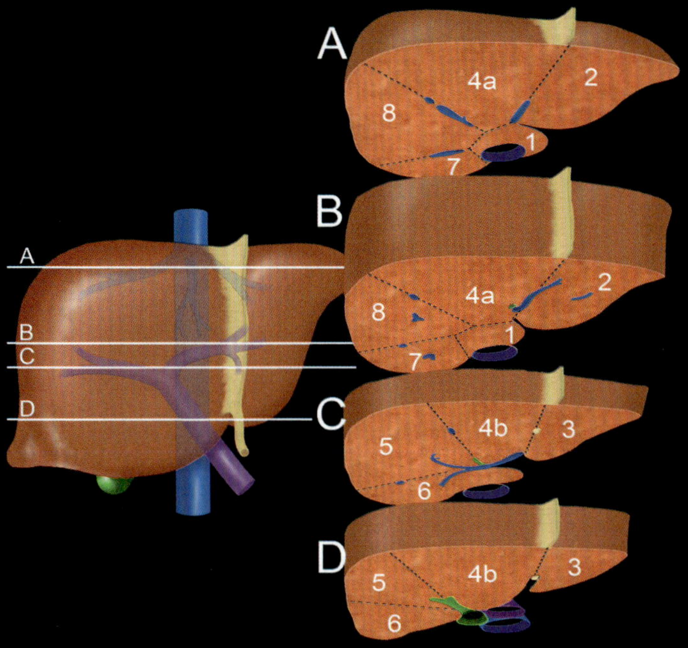

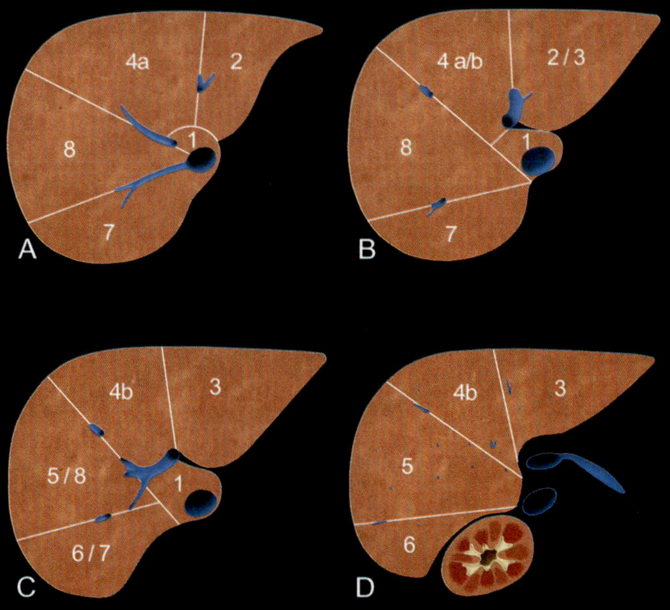

(Top) *Illustration demonstrates the division of the Couinaud segments of the liver at 4 different levels of the liver. The Couinaud segments are defined by the hepatic veins (hepatic vein plane) and the portal veins (portal vein plane).* **(Bottom)** *Another illustration focuses on the segments at each of the 4 levels shown in the previous graphic, with graphic A the most superior (at the level of the IVC-hepatic vein confluence) and graphic D the most inferior (at the level of the portal-splenic confluence). The left hepatic vein divides segments 2 and 3 from 4a and 4b, which the plane of the left portal vein divides the superior segments of the left lobe (2 and 4a) from the inferior segments (3 and 4b). The middle hepatic vein divides segments 4a/b from 5 and 8, while the right hepatic vein divides segments 5 and 8 from 6 and 7. The right portal vein divides the superior segments of the right lobe (7 and 8) from the inferior segments (5 and 6).*

LIVER SEGMENTAL ANATOMY (CT)

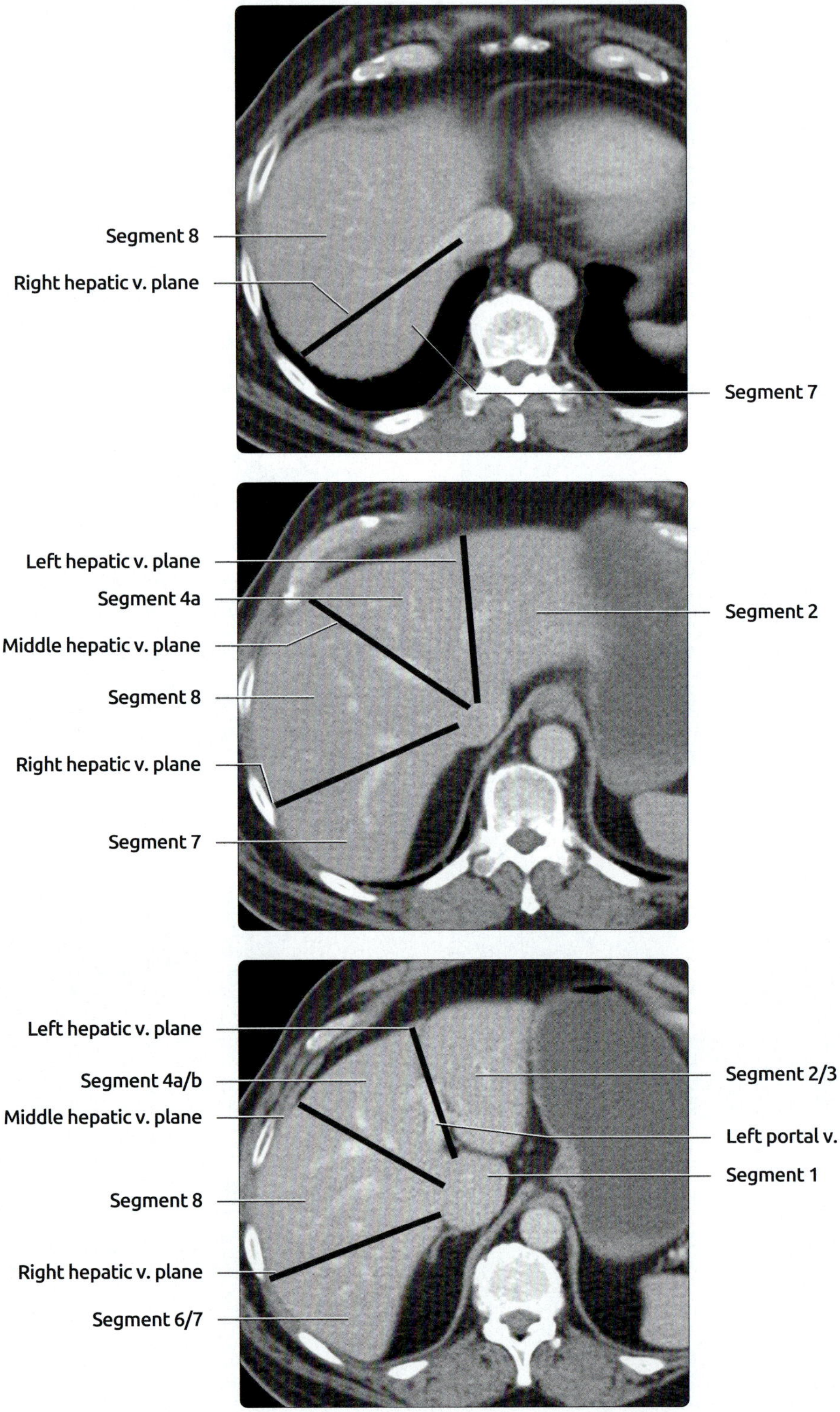

(Top) *Axial CECT through the superior liver demonstrates the right hepatic vein coursing through the right lobe, dividing segments 7 and 8. The right hepatic vein plane also divides segments 5 and 6 below the level of the portal veins.* **(Middle)** *A slightly more caudal image (still above the portal vein plane) demonstrates the left hepatic vein plane (line extrapolated through the plane of the left hepatic vein) dividing segments 2 and 4a of the liver, the middle hepatic vein plane (line extrapolated through the middle hepatic vein) dividing segments 4a and 8, and the right hepatic vein plane (line extrapolated through the middle hepatic vein) dividing segments 7 and 8.* **(Bottom)** *More caudal level now demonstrates the portal veins coming into view, which divide the liver into superior and inferior segments. The portal vein plane demarcates the division between segments 2 and 3, 4a and 4b, 5 and 8, and 6 and 7.*

LIVER SEGMENTAL ANATOMY (CT)

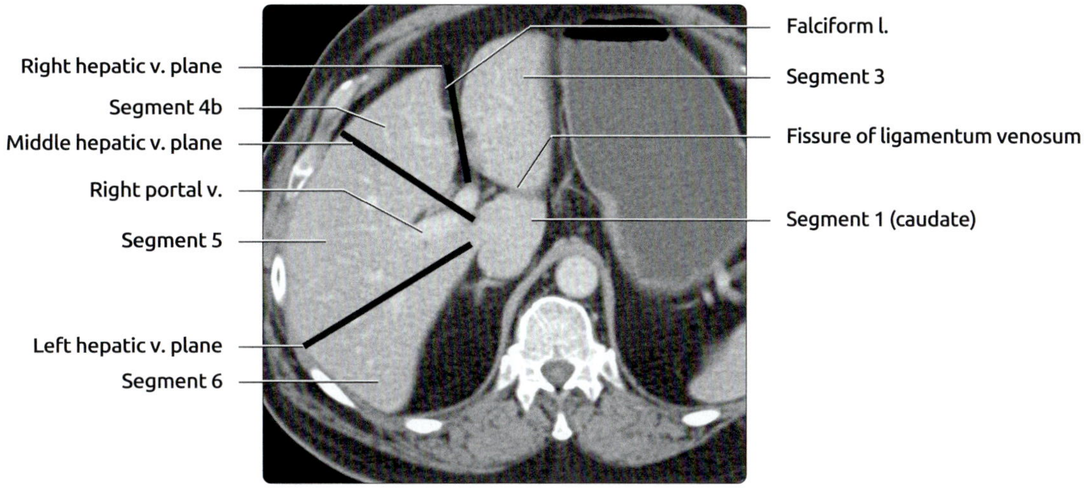

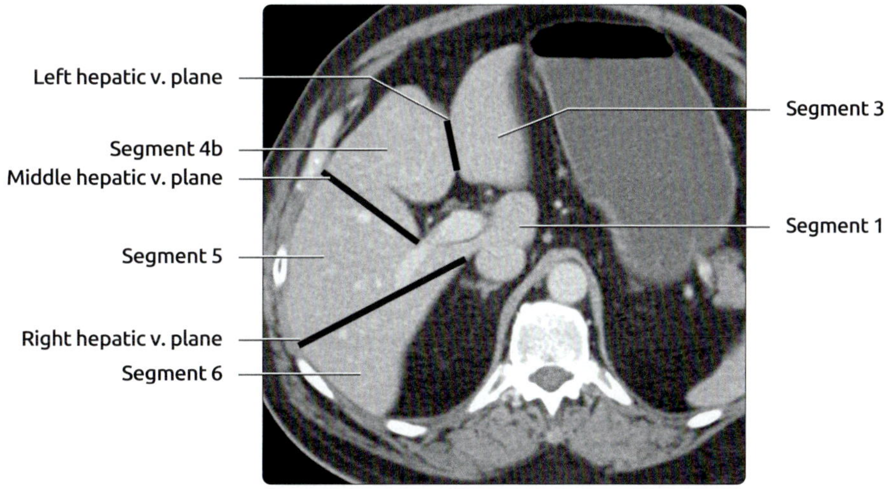

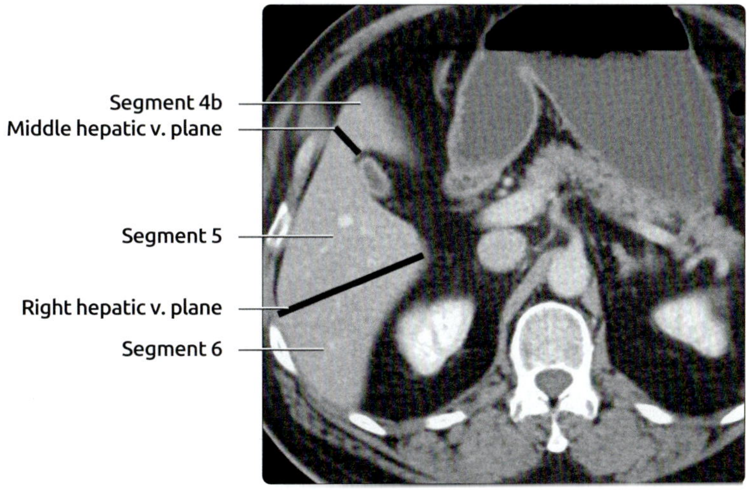

(Top) *Axial CECT below the portal vein plane now demonstrates segments 3, 4b, 6, and 7. Note the location of the caudate lobe surrounding the IVC and posterior to the fissure of the ligamentum venosum.* **(Middle)** *Axial CECT more inferiorly continues to demonstrates the inferior segments of the liver, including 3, 4b, 5, and 6.* **(Bottom)** *Final CECT through the inferior liver now shows that most of the left hepatic lobe has gone out of view in this patient, leaving only the inferior segments of the right hepatic lobe (5 and 6) defined by the middle and right hepatic veins.*

LIVER SEGMENTAL ANATOMY: TRANSVERSE RIGHT LIVER (US)

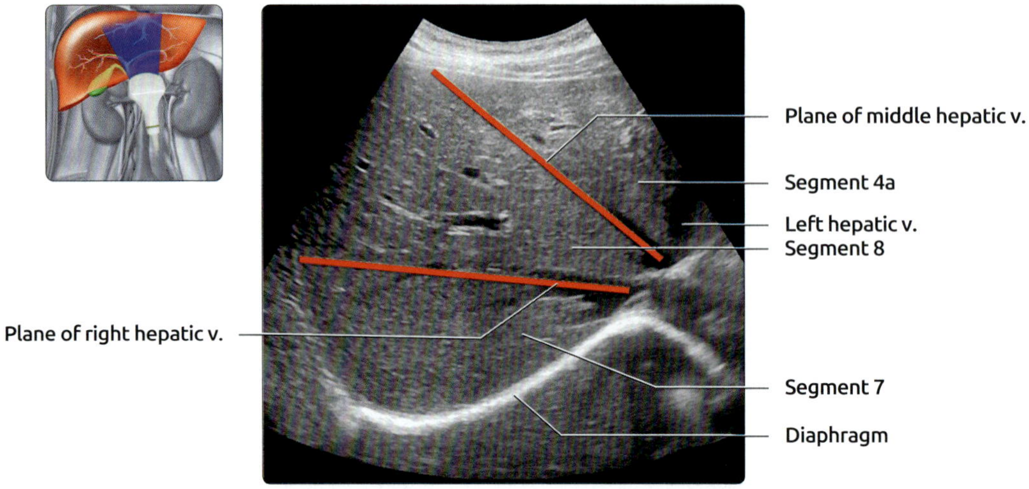

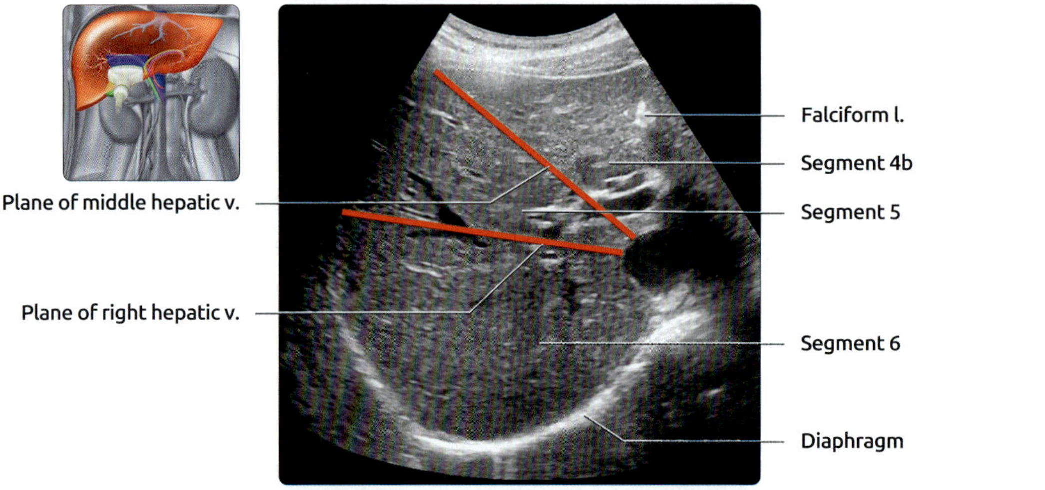

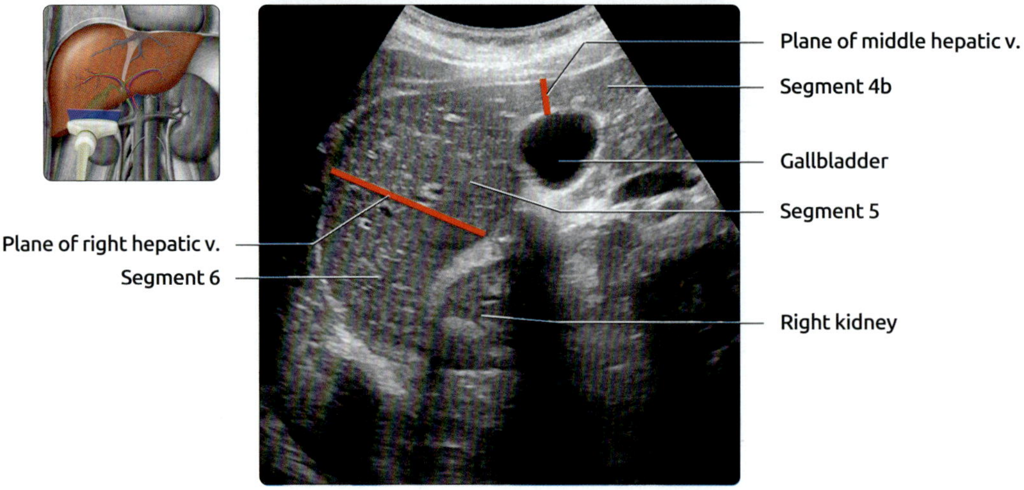

(Top) *Transverse US of the upper right lobe of the at the level of the confluence of the hepatic veins shows the right hepatic vein separates segment 7 (superior posterior segment of the right lobe of the liver) from segment 8 (superior anterior segment of the right lobe of the liver), and the middle hepatic vein separates segment 8 from segment 4a (superior medial segment of the left lobe of the liver).* **(Middle)** *Transverse US of the right lobe of the liver just below the level of the portal vein shows the right hepatic vein, which demarcates the anterior from posterior segments of the right lobe of the liver. The plane of the middle hepatic vein separates the left lobe from the right lobe of the liver. A horizontal plane in line with the main portal vein demarcates the upper from lower liver segments.* **(Bottom)** *Transverse US of the right lobe of the liver inferiorly at the level of the gallbladder shows the inferior anterior segment of the right lobe of the liver (segment 5) and inferior posterior segment of the right lobe of the liver (segment 6). The demarcation between the 2 segments is created by drawing a plane vertically from the right hepatic vein. A vertically oriented plane at the level of the gallbladder and middle hepatic vein separates the right and left lobe of the liver.*

LIVER SEGMENTAL ANATOMY: TRANSVERSE LEFT LIVER (US)

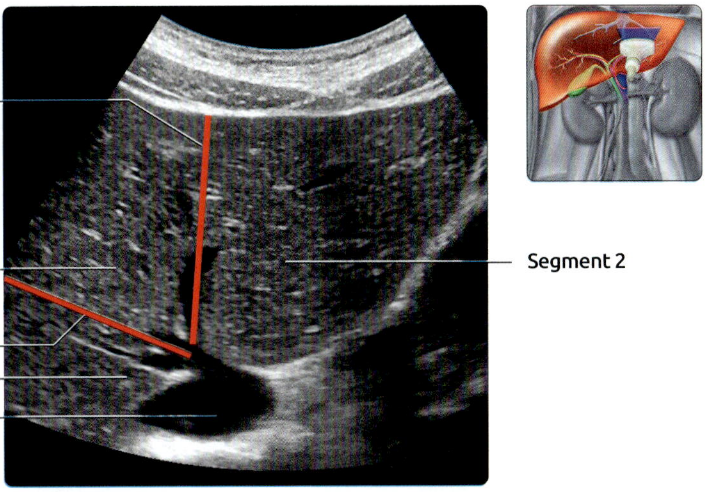

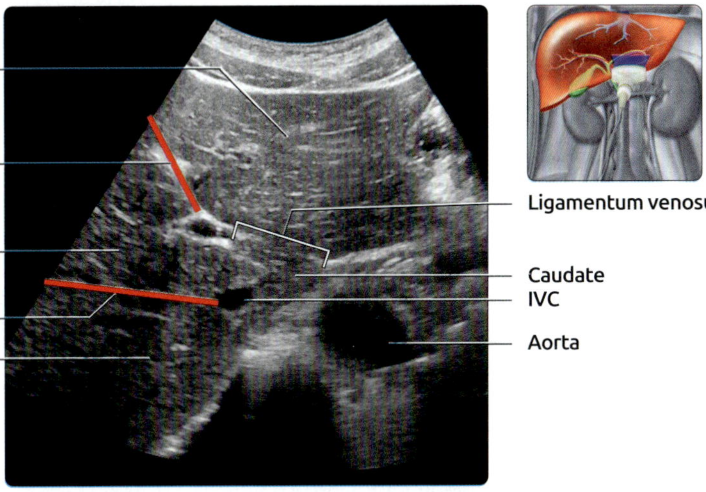

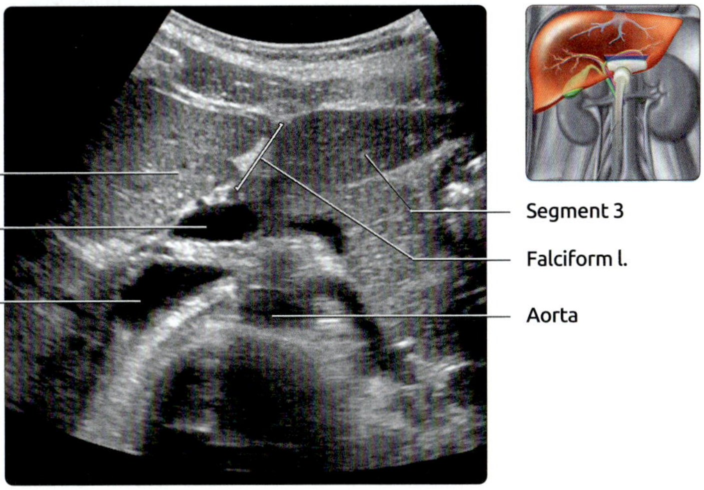

(Top) *Transverse grayscale US of the left lobe of the liver at the level of the confluence of the hepatic veins and IVC shows the left hepatic vein separates the superior lateral segment of the left lobe of the liver (segment 2) from the superior medial segment of the left lobe of the liver (segment 4a).* **(Middle)** *Transverse grayscale US shows the left lobe of the liver at a level just inferior to the left portal vein. The caudate lobe of the liver (segment 1) abuts the ligamentum venosum, IVC, and left portal vein. The falciform ligament separates the inferior lateral segment of the left lobe of the liver (segment 3) from the inferior medial segment of the left lobe of the liver (segment 4b).* **(Bottom)** *Transverse grayscale US shows the inferior aspect of the left lobe of the liver at the level of the pancreas. The falciform ligament separates the inferior lateral segment of the left lobe of the liver (segment 3) from the medial inferior segment of the left lobe (segment 4b).*

TERMINOLOGY

Abbreviations

- Extrahepatic biliary structures
  - Gallbladder (GB)
  - Cystic duct (CD)
  - Right hepatic (RH) and left hepatic (LH) ducts
  - Common hepatic duct (CHD)
  - Common bile duct (CBD)

Definitions

- Proximal/distal biliary tree
  - Proximal refers to portion of biliary tree that is closer in proximity to liver and hepatocytes
  - Distal refers to caudal end closer to ampulla and bowel
- Central/peripheral
  - Central refers to biliary ducts close to porta hepatis
  - Peripheral refers to higher-order branches of intrahepatic biliary tree extending into hepatic parenchyma

IMAGING ANATOMY

Overview

- Biliary ducts carry bile from liver to duodenum
  - Bile is produced continuously by liver, stored and concentrated by GB, and released intermittently by GB contraction in response to presence of fat in duodenum
  - Hepatocytes form bile → bile canaliculi → interlobular biliary ducts → collecting bile ducts → right and left hepatic ducts → CHD → CBD → intestines
- **Common bile duct**
  - Forms in free edge of lesser omentum by **union of CD and CHD**
  - Length of duct: 5-15 cm, depending on point of junction of cystic and CHD
  - Descends posterior and medial to duodenum, lying on dorsal surface of pancreatic head
  - Joins with pancreatic duct to form hepaticopancreatic **ampulla of Vater**
  - Ampulla opens into duodenum through major duodenal (hepaticopancreatic) papilla
  - Distal CBDt is thickened into sphincter of Boyden and hepaticopancreatic segment is thickened into a sphincter of Oddi
    - Contraction of these sphincters prevents bile from entering duodenum; forces it to collect in GB
    - Relaxation of sphincters in response to parasympathetic stimulation and cholecystokinin (released by duodenum in response to fatty meal)
- Vessels, nerves, and lymphatics
  - **Arteries**
    - Hepatic arteries supply intrahepatic ducts
    - Cystic artery supplies proximal common duct
    - RH artery supplies middle part of common duct
    - Gastroduodenal and pancreaticoduodenal arcade supply distal common duct
    - Cystic artery supplies GB (usually from RH artery; variable)
  - **Veins**
    - From intrahepatic ducts → hepatic veins
    - From common duct → portal vein (in tributaries)
    - From GB directly into liver sinusoids, bypassing portal vein
  - Nerves
    - Sensory: Right phrenic nerve
    - Parasympathetic and sympathetic: Celiac ganglion and plexus; contraction of GB and relaxation of biliary sphincters is caused by parasympathetic stimulation, but more important stimulus is from hormone cholecystokinin
  - Lymphatics
    - Same course and name as arterial branches
    - Collect at celiac lymph nodes and node of omental foramen
    - Nodes draining GB are prominent in porta hepatis and around pancreatic head
- **Gallbladder**
  - ~ 7-10 cm long, holds up to 50 mL of bile
  - Lies in shallow fossa on visceral surface of liver
  - Vertical plane through GB fossa and middle hepatic vein divides LH and RH lobes
  - May touch and indent duodenum
  - Fundus is covered with peritoneum and relatively mobile; body and neck attached to liver and covered by hepatic capsule
  - Fundus: Wide tip of GB, projects below liver edge (usually)
  - Body: Contacts liver, duodenum, and transverse colon
  - Neck: Narrowed, tapered, and tortuous; joins CD
  - CD: 3-4 cm long, connects GB to CHD; marked by spiral folds of Heister; helps to regulate bile flow to and from GB
- **Normal measurements**
  - CBD/CHD
    - < 6-7 mm in patients without history of biliary disease in most studies
    - Controversy about dilatation related to previous cholecystectomy and old age
  - Intrahepatic ducts
    - Normal diameter of 1st and higher-order branches < 2 mm or < 40% of diameter of adjacent portal vein
    - 1st- (i.e., LH duct and RH duct) and 2nd-order branches are normally visualized
    - Visualization of 3rd and higher-order branches is often abnormal and indicates dilatation

ANATOMY IMAGING ISSUES

Imaging Recommendations

- Patient should fast for at least 4-6 hours prior to US examination to ensure GB is not contracted after meal, ideally fasting for 8-12 hours (overnight)
- Complete assessment includes scanning liver, porta hepatis region, and pancreas in sagittal, transverse, and oblique views
- Subcostal and right intercostal transverse views help align bile ducts and GB along imaging plane for optimal visualization
- Usually structures are better assessed and imaged with patient in full-suspended inspiration and in left lateral oblique position

- Harmonic imaging provides improved contrast between bile ducts and adjacent tissues, leading to improved visualization of bile ducts, luminal content, and wall
- For imaging of gallstone disease, special maneuvers are recommended
  - Move patient from supine to left lateral decubitus position
    - Demonstrates mobility of gallstones
    - Gravitates small gallstones together to appreciate posterior acoustic shadowing
  - Set focal zone at level of posterior acoustic shadowing
    - Maximizes effect of posterior acoustic shadowing to confirm gallstone(s)

Imaging Approaches

- Transabdominal USis ideal initial investigation for suspected biliary tree or GB pathology
  - Cystic nature of bile ducts and GB (especially if these are dilated) provides inherently high-contrast resolution
  - Acoustic window provided by liver and modern state-of-the-art US technology provides good spatial resolution
  - Common indications of US for biliary and GB disease include
    - Right upper quadrant/epigastric pain
    - Abnormal liver function test or jaundice
    - Suspected gallstone disease
    - Pancreatitis
  - US plays key role in multimodality evaluation of complex biliary problems
- Supplemented by various imaging modalities, including MR/MRCP and CT

Imaging Pitfalls

- **Common pitfalls in evaluation of GB**
  - Posterior shadowing may arise from GB neck, Heister valves of CD, or adjacent gas-filled bowel loops
    - May mimic cholelithiasis
    - Scan after repositioning patient in prone or left lateral decubitus positions
  - Food material within gastric antrum/duodenum
    - Mimics GB filled with gallstones or GB containing milk of calcium
    - During real-time scanning, carefully evaluate peristaltic activity of involved bowel with oral administration of water
- **Common pitfalls in US evaluation of biliary tree**
  - Redundancy, elongation, or folding of GB neck on itself
    - Mimics dilatation of CHD or proximal CBD
    - Avoided by scanning patient in full-suspended inspiration
    - Careful real-time scanning allows separate visualization of CHD/CBD medial to GB neck
  - Presence of gas-filled bowel loops adjacent to distal extrahepatic bile ducts
    - Obscure distal biliary tree and render detection of choledocholithiasis difficult
    - Scan with patient in decubitus positions or after oral intake of water
  - Gas/particulate material in adjacent duodenum and pancreatic calcification
    - Mimic choledocholithiasis within CBD
  - Presence of gas within biliary tree
    - May mimic choledocholithiasis, differentiated by presence of reverberation artifacts
    - Limits US detection of biliary calculus

Key Concepts

- Direct venous drainage of GB into liver bypasses portal venous system, often results in sparing of adjacent liver from generalized steatosis (fatty liver)
- Nodal metastasis from GB carcinoma to peripancreatic nodes may simulate primary pancreatic tumor
- Sonography: Optimal means of evaluating GB for stones and inflammation (acute cholecystitis); best done in fasting state (distends GB)
- Intrahepatic bile ducts follow branching pattern of portal veins
  - Usually lie immediately anterior to portal vein branch; confluence of hepatic ducts just anterior to bifurcation of right and main portal veins

CLINICAL IMPLICATIONS

Clinical Importance

- In patients with obstructive jaundice, US plays key role
  - Differentiates biliary obstruction from liver parenchymal disease
  - Determines presence, level, and cause of biliary obstruction
- Common variations of biliary arterial and ductal anatomy result in challenges to avoid injury at surgery
  - CD may run in common sheath with bile duct
  - Anomalous RH ducts may be severed at cholecystectomy
- Close apposition of GB to duodenum can result in fistulous connection with chronic cholecystitis and erosion of gallstone into duodenum

Function & Dysfunction

- Obstruction of CBD is common
  - Gallstones in distal bile duct
  - Carcinoma arising in pancreatic head or bile duct
  - Result is jaundice due to back up of bile salts into bloodstream

Embryologic Events

- Abnormal embryological development of fetal ductal plate can lead to spectrum of liver and biliary abnormalities, including
  - Polycystic liver disease
  - Congenital hepatic fibrosis
  - Biliary hamartomas
  - Caroli disease
  - Choledochal cysts

GALLBLADDER IN SITU

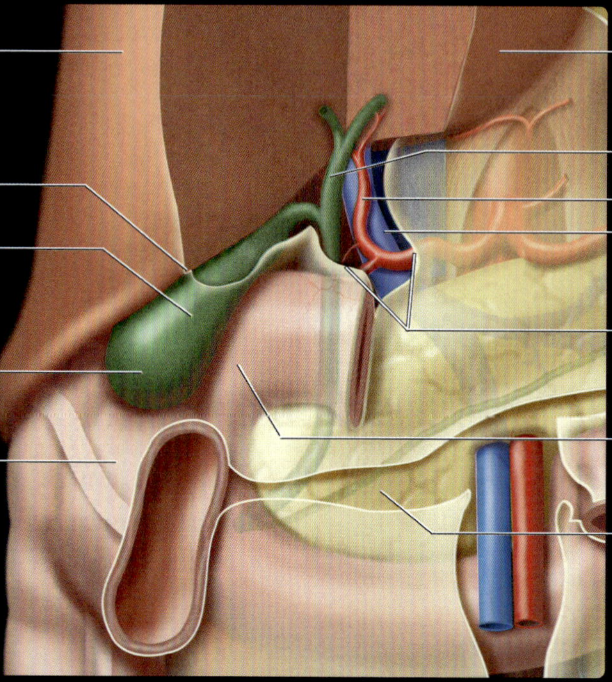

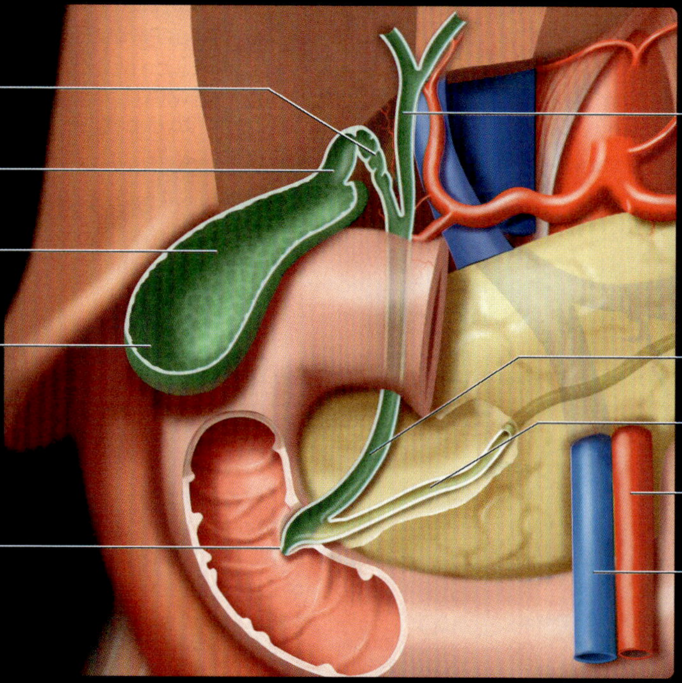

(Top) *Graphic shows that the gallbladder is covered with peritoneum, except where it is attached to the liver. The extrahepatic bile duct, hepatic artery, and portal vein run in the lesser omentum. The fundus of the gallbladder extends beyond the anterior-inferior edge of the liver and can be in contact with the hepatic flexure of the colon. The body (main portion of the gallbladder) is in contact with the duodenum.* **(Bottom)** *The neck of the gallbladder narrows before entering the cystic duct, which is distinguished by its tortuous course and irregular lumen. The duct lumen is irregular due to redundant folds of mucosa, called the spiral folds of Heister, that are believed to regulate the rate of filling and emptying of the gallbladder. The cystic duct joins the hepatic duct to form the common bile duct, which passes behind the duodenum and through the pancreas to enter the duodenum.*

ANATOMIC VARIATIONS OF BILIARY TREE

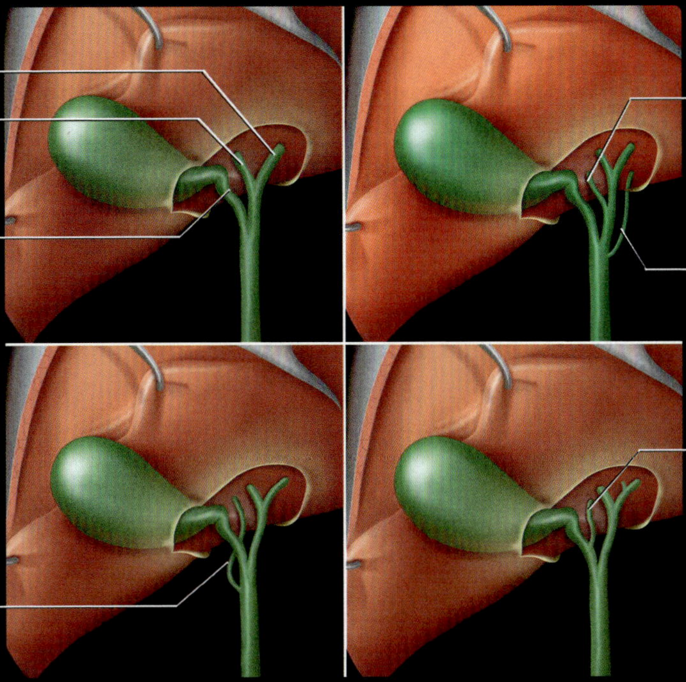

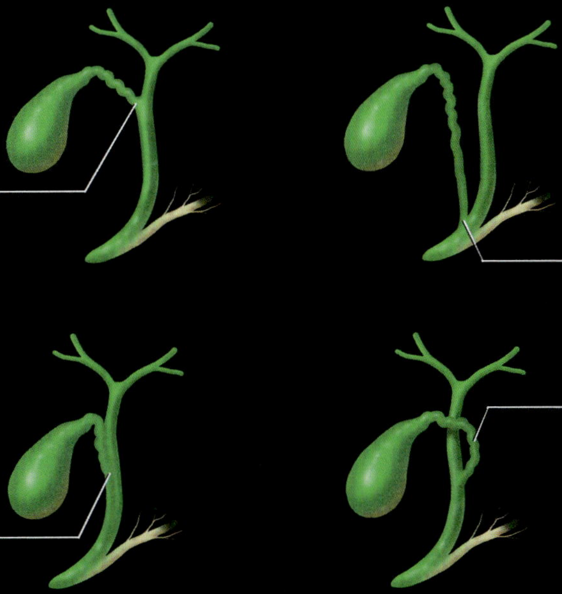

(Top) *Graphic shows the conventional arrangement of the extrahepatic bile ducts, but variations are common (20% of population) and may lead to inadvertent ligation or injury at surgery (such as during cholecystectomy), where the cystic duct is clamped and transected. Most accessory ducts are on the right side and usually enter the common hepatic duct, but they may enter the cystic or common bile duct. Accessory left ducts enter the common bile duct. While referred to as accessory, these ducts are the sole drainage of bile from at least 1 hepatic segment. Ligation or laceration can lead to significant hepatic injury or bile peritonitis.* **(Bottom)** *The course and insertion of the cystic duct are highly variable, leading to difficulty in isolation and ligation at cholecystectomy. The cystic duct may be mistaken for the common hepatic or common bile duct.*

BILIARY TREE

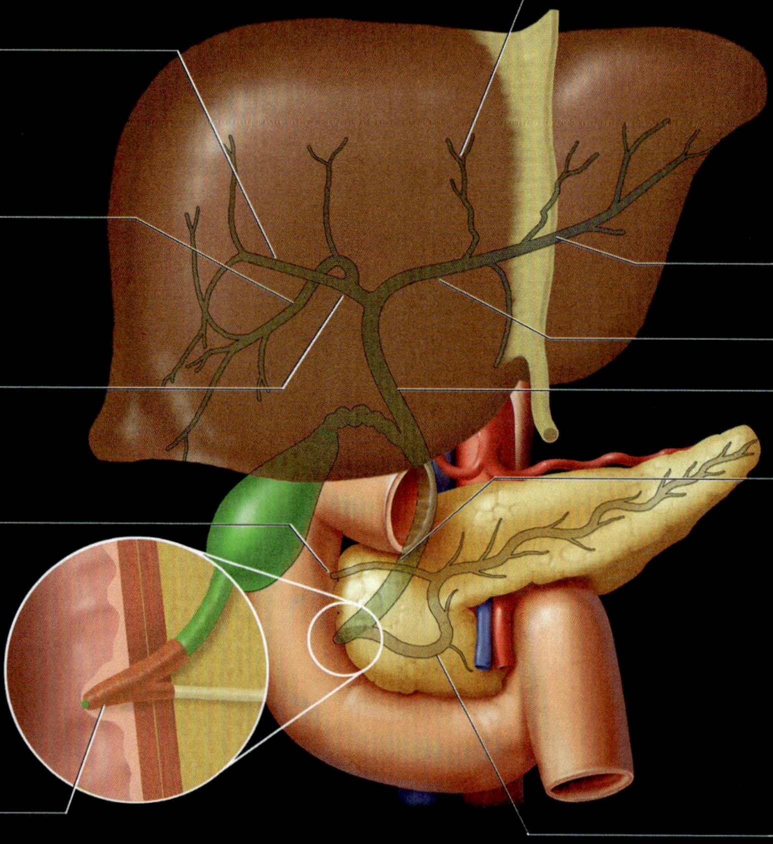

Note the distribution of the larger intrahepatic bile ducts. The common bile duct usually joins with the pancreatic duct in a common channel or ampulla (of Vater) but may enter the major duodenal papilla separately. The distal bile duct has a sphincteric coat of smooth muscle, the choledochal sphincter (of Boyden), which regulates bile emptying into the duodenum. When contracted, this sphincter causes bile to flow retrograde into the gallbladder for storage. The common hepaticopancreatic ampulla may be surrounded by a smooth muscle sphincter (of Oddi).

BILIARY TREE

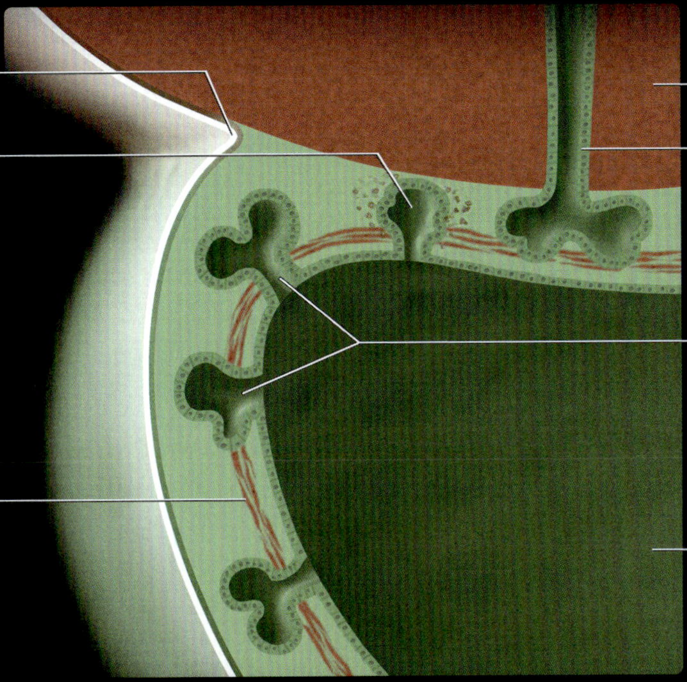

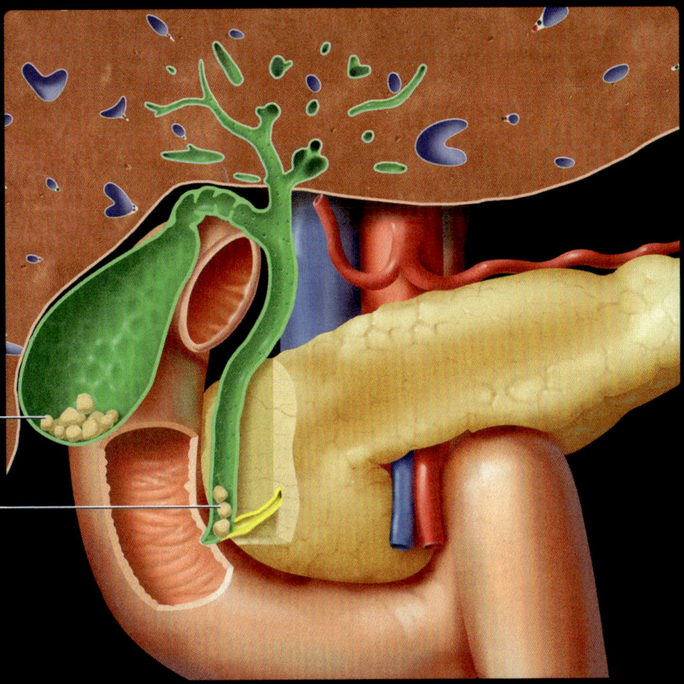

(Top) *The gallbladder body and neck are adherent to the liver and may be bridged by aberrant bile ducts (of Luschka). Mucous glands are found in the gallbladder neck. Rokitansky-Aschoff sinuses are pseudodiverticula that extend into the wall and may collect debris, becoming inflamed.* **(Bottom)** *Graphic demonstrates cholelithiasis (stones in the gallbladder) and choledocholithiasis (stones in the bile ducts). Gallstones are extremely common and may remain asymptomatic. Stones that become impacted, even temporarily, in the gallbladder neck may cause inflammation and distention of the gallbladder, clinically referred to as acute cholecystitis. Stones that pass through the cystic duct often cause biliary colic (spasms of right upper quadrant pain), as they often become trapped within the common bile duct, causing obstruction.*

RIGHT HEPATIC LOBE

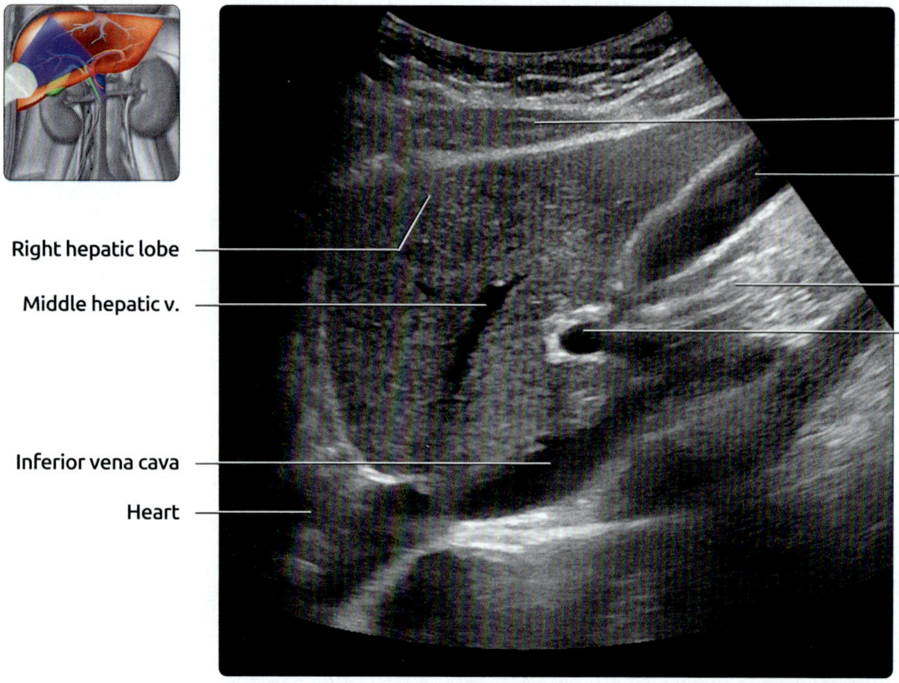

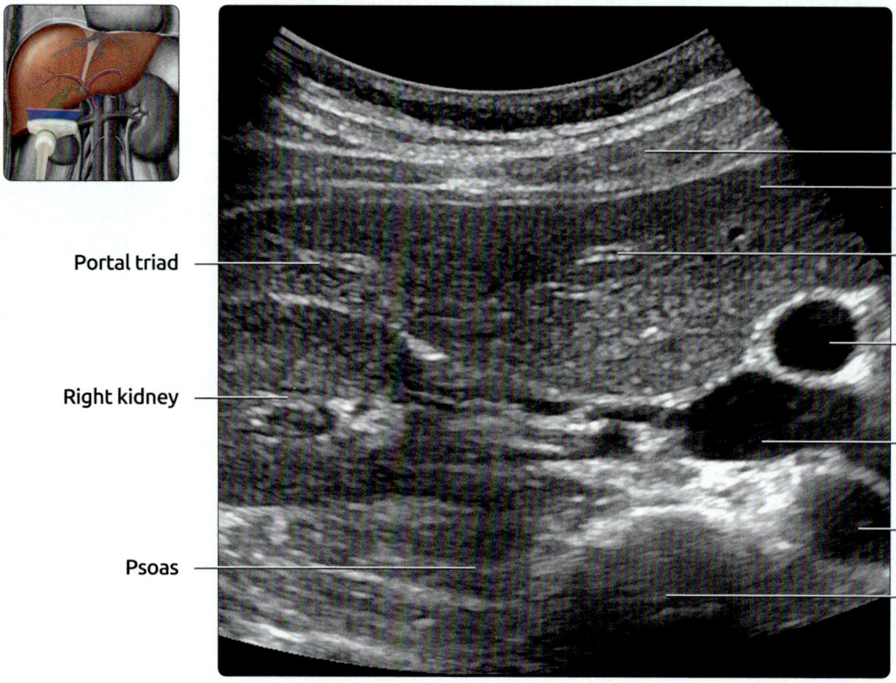

(Top) *Subcostal, longitudinal ultrasound shows the right hepatic lobe and gallbladder with patient in the left lateral decubitus position. Ideally, a patient must fast for at least 4-6 hours to allow for adequate gallbladder distension.* **(Bottom)** *Subcostal, transverse ultrasound of the right hepatic lobe demonstrates its anatomical relationships with major vessels and the right kidney. The intrahepatic bile ducts are localized within the portal triads, which are visible by the prominent echogenic walls of the portal veins in these triads. The portal triad contains the portal vein, bile duct, and hepatic artery. Normally, the intrahepatic bile ducts and hepatic arteries are not readily visible unless they are dilated.*

LEFT HEPATIC LOBE

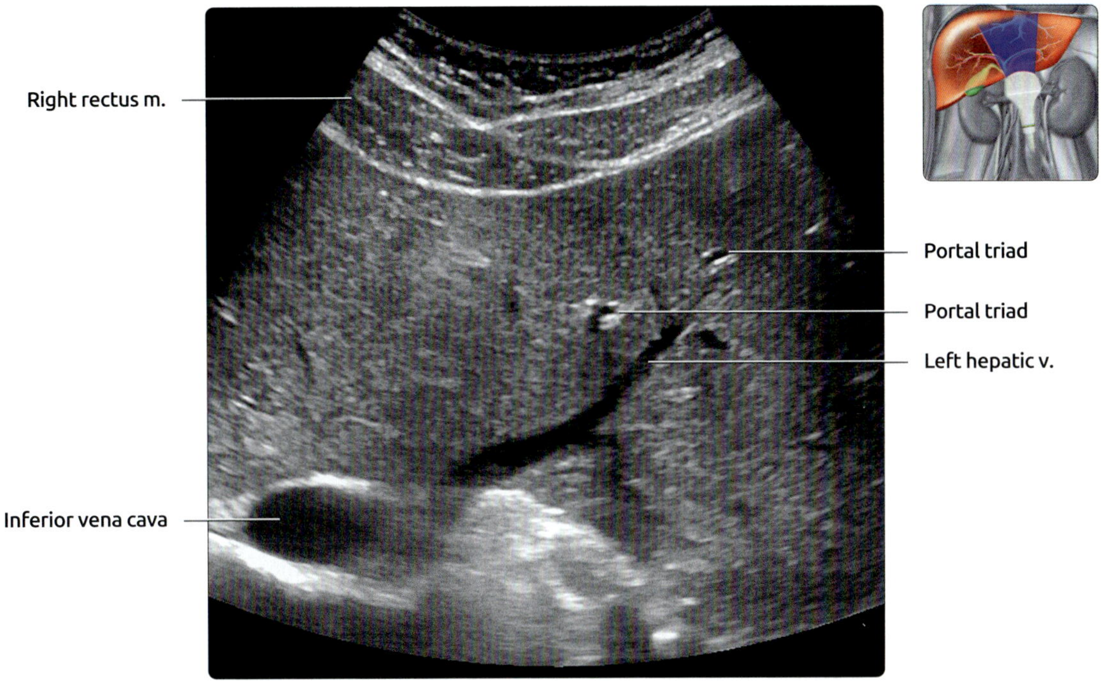

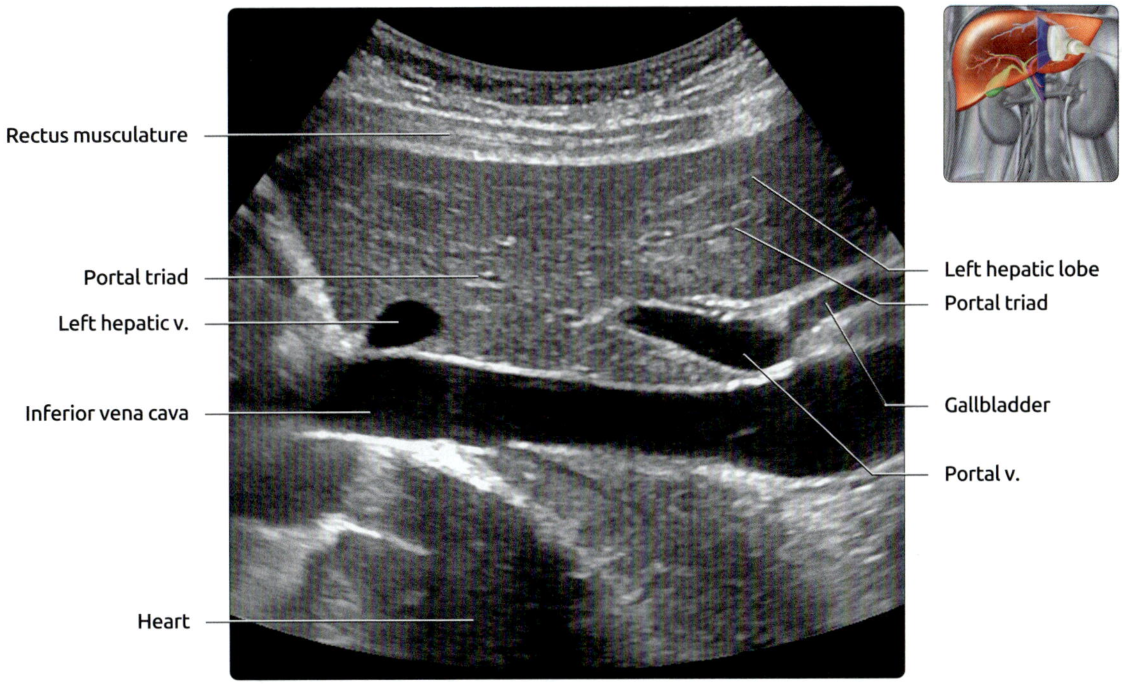

(Top) *Subxiphoid, transverse grayscale ultrasound of the left hepatic lobe shows the left hepatic vein and several portal triads. The portal triads are identified by the prominent echogenic walls of the portal veins. The portal triad contains a portal vein, a bile duct, and a hepatic artery. The hepatic artery and bile duct are not readily visible unless they are dilated.* **(Bottom)** *Longitudinal ultrasound of the left hepatic lobe near the confluence of the left hepatic vein with the inferior vena cava is shown. Portal triads are recognizable by the prominent walls of the portal veins.*

GALLBLADDER

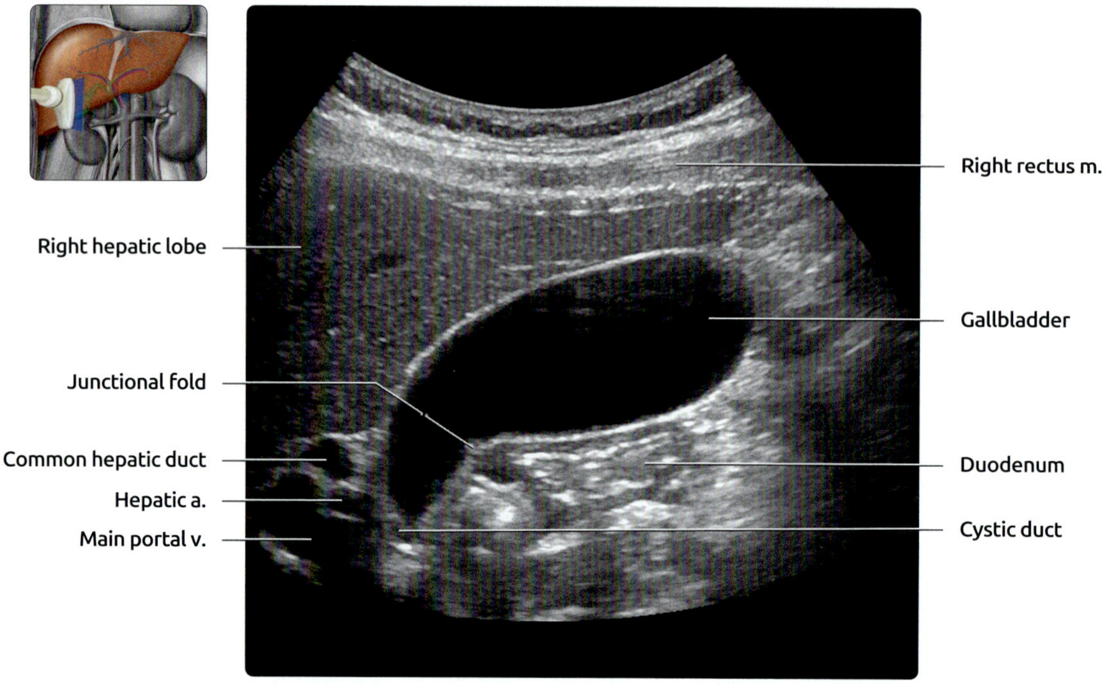

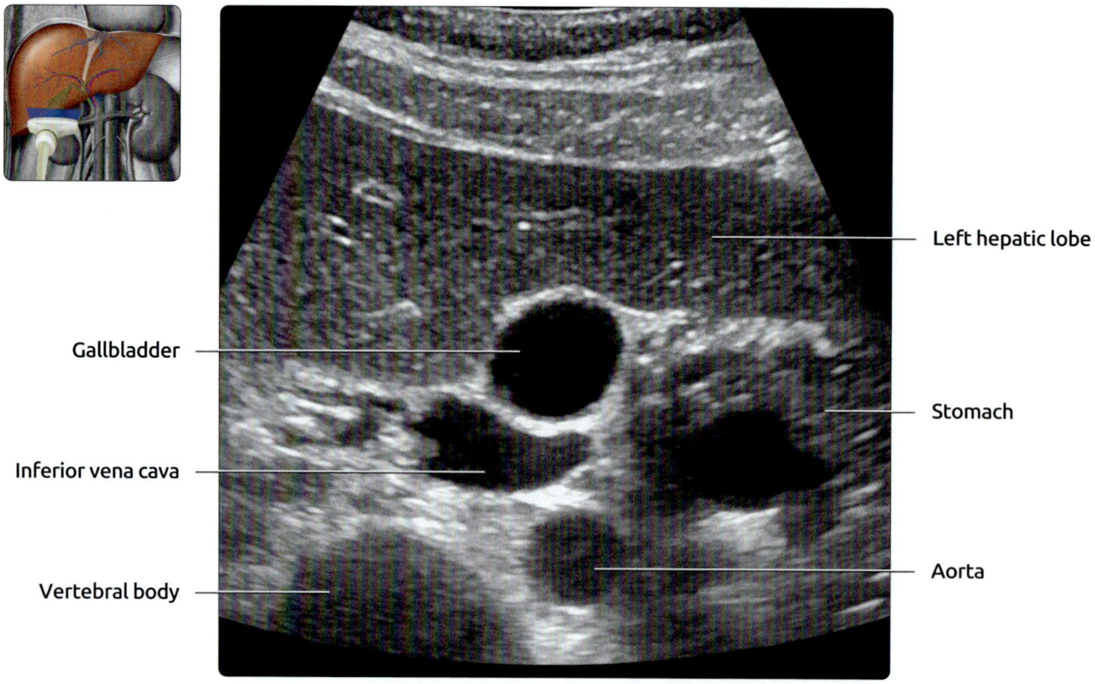

(Top) *Subcostal, longitudinal grayscale ultrasound shows the gallbladder with patient in the left lateral decubitus position. The gallbladder is distended with bile, causing increased transmission of echoes. A patient must fast for at least 4-6 hours to allow for adequate gallbladder distension.* **(Bottom)** *Transverse ultrasound of a distended gallbladder in the left lateral decubitus position is shown. The gallbladder is best evaluated if the patient has fasted at least 4-6 hours, ideally 8-12 hours.*

COMMON BILE DUCT

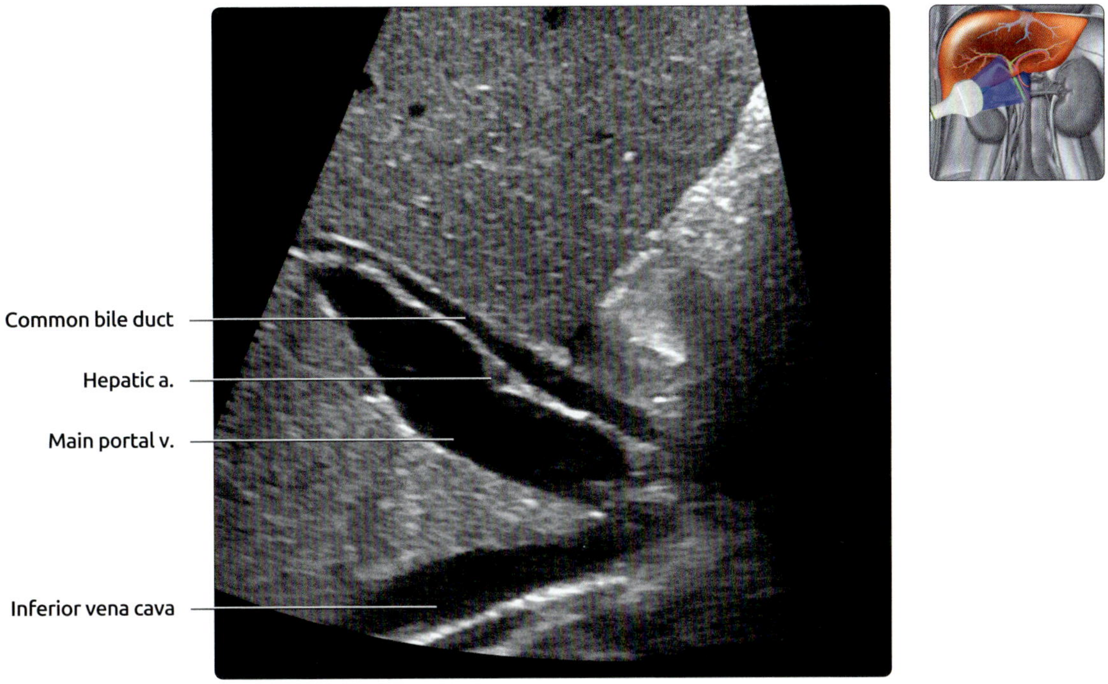

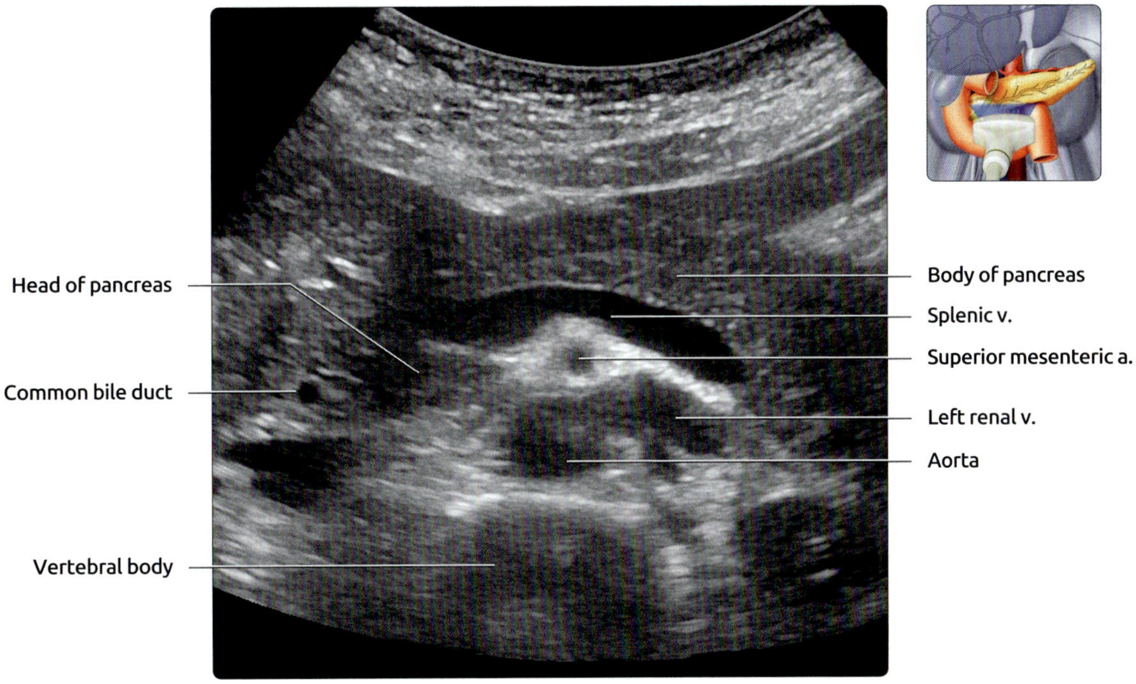

(Top) *Subcostal, longitudinal oblique ultrasound shows the common bile duct and the portal vein in the hepatoduodenal ligament. The common bile duct is normally identified anterior to the portal vein. The distal common bile duct pierces the pancreatic head. Also demonstrated is the hepatic artery, which normally traverses between the portal vein and common bile duct.* **(Bottom)** *Transverse ultrasound at the level of the pancreas demonstrates the distal common bile duct as it pierces the pancreatic head. Overlying bowel gas may obscure this area; positioning the patient upright or in a left lateral decubitus position may help visualization.*

RIGHT INTRAHEPATIC DUCTS

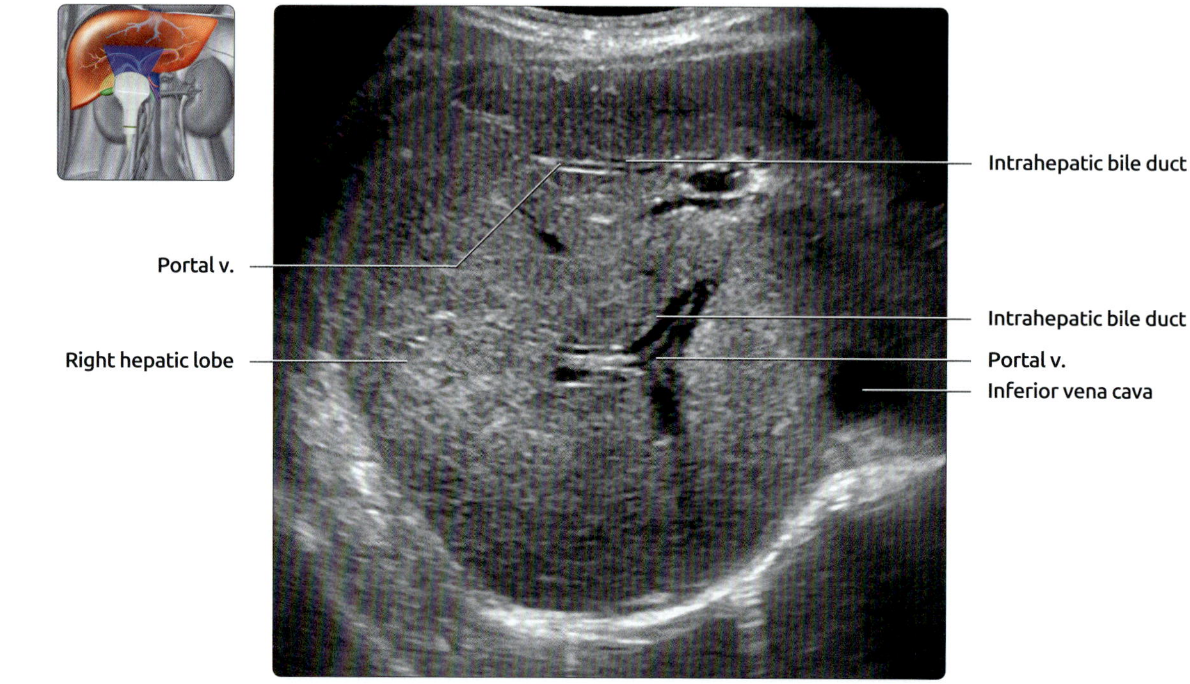

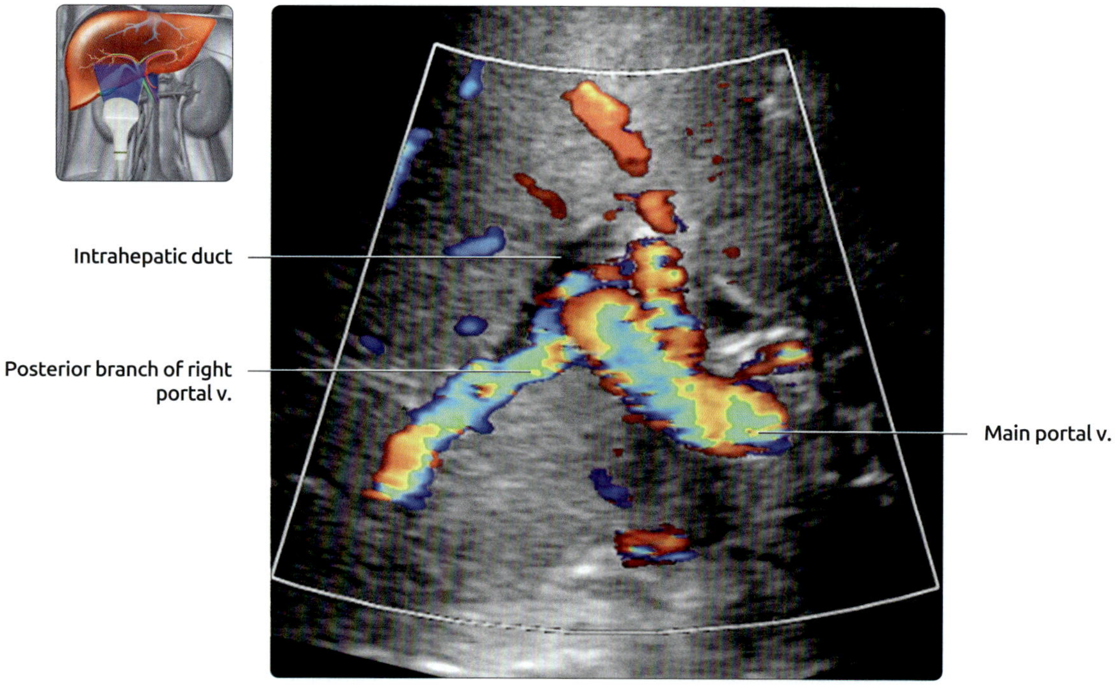

(Top) *Transverse ultrasound through the right hepatic lobe in a patient with biliary obstruction demonstrates dilated right intrahepatic bile ducts. Normally, intrahepatic ducts are not readily visualized unless they are dilated. Parallel linear hypoechoic structures are not normally seen in the liver, and this may represent a dilated bile duct or dilated hepatic artery adjacent to the normally visualized portal venous structure in the portal triad.* **(Bottom)** *Transverse color Doppler ultrasound through the right hepatic lobe in a patient with biliary obstruction demonstrates dilated right intrahepatic bile ducts. Normally, intrahepatic ducts are not readily visualized unless they are dilated. Color Doppler ultrasound helps to determine whether the dilated structure is a dilated hepatic artery or dilated bile duct.*

LEFT INTRAHEPATIC DUCTS

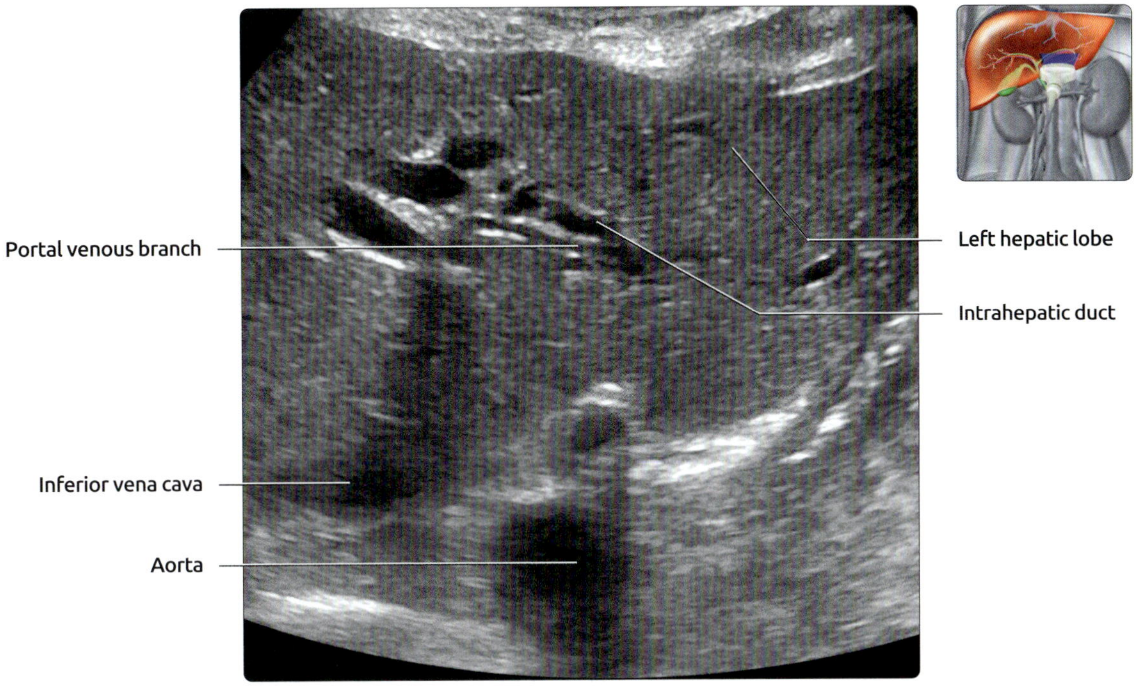

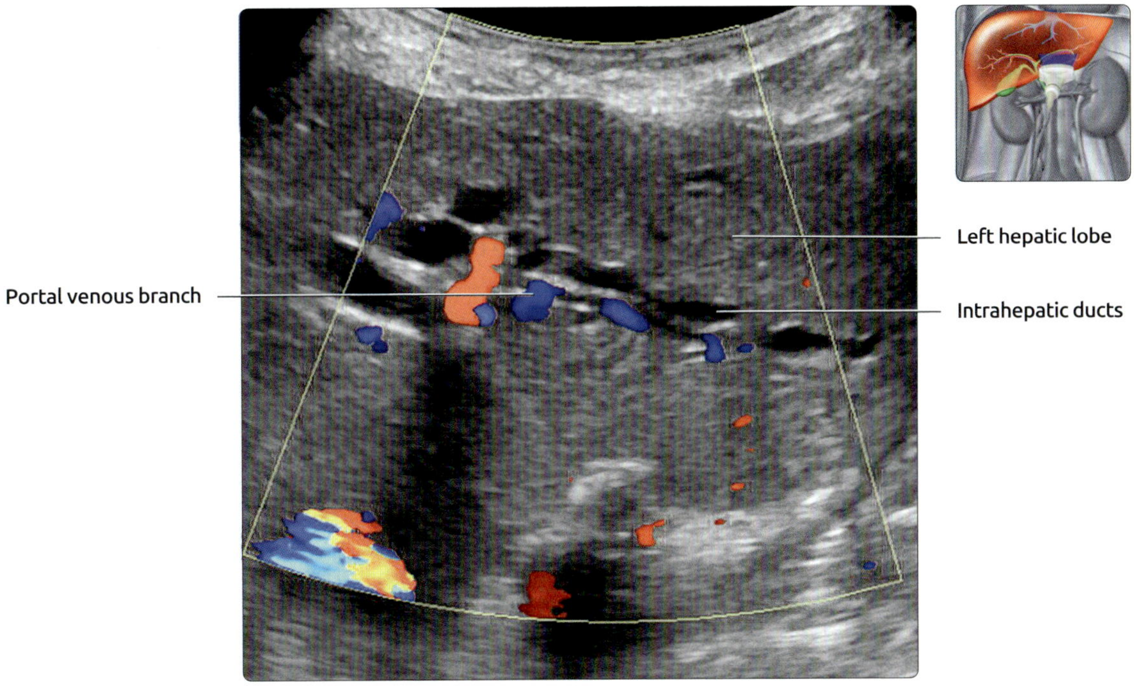

(Top) *Transverse ultrasound through the left hepatic lobe in a patient with biliary obstruction demonstrates dilated left intrahepatic bile ducts. Normally, intrahepatic ducts are not readily visualized unless they are dilated.* **(Bottom)** *Transverse color Doppler ultrasound through the left hepatic lobe in a patient with biliary obstruction demonstrates dilated left intrahepatic bile ducts. Normally, intrahepatic ducts are not readily visualized unless they are dilated. Color Doppler ultrasound is helpful to determine whether the dilated structure is vascular or biliary, as dilated hepatic arteries can be mistaken for dilated bile ducts. The absence of flow allows one to determine that these are biliary in origin.*

GROSS ANATOMY

Overview

- Intraperitoneal lymphatic organ located posterior to stomach and intimately associated with retroperitoneum (pancreatic tail and left kidney)
- Surrounded by peritoneum (except at hilum) and suspended by several ligaments
  - **Gastrosplenic ligament**
    - Left anterior margin of lesser sac
    - Connects spleen to greater curvature of stomach
    - Carries short gastrics and left gastroepiploic arteries and venous branches to spleen
  - **Splenorenal ligament**
    - Left posterior margin of lesser sac
    - Connects spleen to left kidney and pancreatic tail
    - Carries splenic artery and vein to splenic hilum
  - Splenocolic ligament: Between spleen and splenic flexure of colon
  - Splenophrenic ligament: Between spleen and inferior surface of diaphragm
- **Normal size is variable**; no universal consensus
  - Generally, normal adult spleen considered 12 cm length x 7 cm width x 4 cm thickness
    - Length = longest diameter in longitudinal plane; width = longest transverse (anterior-posterior) diameter; thickness = maximal thickness in transverse plane at hilum
  - Splenic index (product of length, thickness, and width): Normally 120-480 cm^3
- Functions
  - Manufactures lymphocytes, filters blood (removes damaged red blood cells and platelets)
  - Acts as blood reservoir: Can expand or contract in response to changes in blood volume
- Histology
  - Soft organ with fibroelastic capsule and comprised of pulp
    - White pulp: Lymphoid nodules/tissue primarily surrounding vasculature
    - Red pulp: Sinusoidal spaces containing blood
  - Trabeculae: Extensions of capsule into parenchyma; carry arterial and venous branches
  - Splenic cords (plates of cells) lie between sinusoids; red pulp veins drain sinusoids
- **Vasculature**
  - **Splenic artery** arises from celiac axis in > 90%; 8% directly from aorta
    - Often very tortuous
  - **Splenic vein** runs in groove along dorsal surface of pancreatic body and tail
    - Receives inferior mesenteric vein (IMV)
    - Combined splenic vein and IMV join superior mesenteric vein to form portal vein

IMAGING ANATOMY

Overview

- Homogeneous echogenicity
  - Echogenicity: Pancreas > spleen > liver > kidney
  - Radiating pattern of segmental arteries and veins
- **Splenic artery**
  - Low-resistance waveform; tortuosity of vessel results in turbulence and spectral broadening
  - Normal diameter: 4-8 mm; peak systolic velocity (PSV): 25-45 cm/sec
- **Splenic vein**
  - Normal diameter: 5-10 mm; PSV: 9-18 cm/sec
  - Splenic vein at midline is useful landmark for locating pancreas
    - Pancreas lies anterior to splenic vein
  - Diameter increases between 50-100% from quiet respiration to deep inspiration; increase of < 20% suggests portal hypertension
  - Spectral Doppler waveform typically shows band-like flow profile with minimal respiratory fluctuations

ANATOMY IMAGING ISSUES

Imaging Recommendations

- Patient positioned supine or right decubitus position (left side up) with left arm raised
- Place transducer parallel to ribs in 10th or 11th intercostal space at left midaxillary line, searching for best window
  - Due to rib angle, this results in oblique view, which by convention is called longitudinal or transverse (depending on transducer orientation)
  - Transverse US view of spleen does not correlate directly to axial CT view
- End expiration may be helpful; lung base may obscure spleen in full inspiration
- Spleen poorly accessed from posterior (obscured by left lung base), anterior, or subcostal approach (obscured by stomach and colon)
- Assess splenic vein at hilum and midline for patency and flow direction
- Can use spleen as acoustic window to visualize tail of pancreas

Key Concepts

- Spleen has highly variable size and shape
  - Easily indented and displaced by masses and even loculated fluid collections

EMBRYOLOGY

Practical Implications

- **Accessory spleen** (splenunculus, splenule)
  - Found in 10-30% of population, and may be multiple
  - Usually small, near splenic hilum
  - Can enlarge and simulate mass, especially after splenectomy
  - Ectopic intrapancreatic splenule can mimic pancreatic tail mass; should not be > 3 cm from tail tip
- **Wandering spleen**: Spleen may be on long mesentery
  - Found in any abdominopelvic location; risk of torsion
- **Asplenia and polysplenia** (heterotaxy syndromes)
  - Rare congenital conditions of altered left/right orientation of organs
  - Associated with cardiovascular anomalies, intestinal malrotation, etc.
- **Splenosis**: Peritoneal implantation of splenic tissue after traumatic splenic injury, can mimic polysplenia

LIGAMENTS AND VESSELS

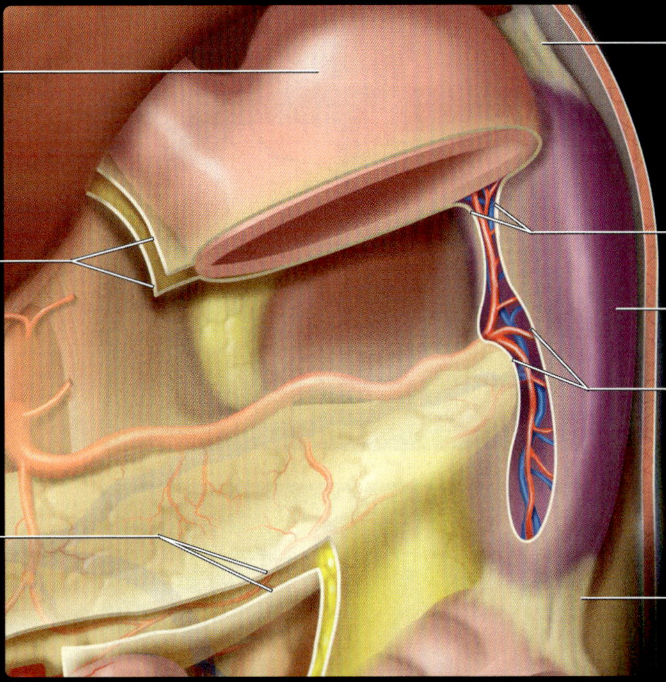

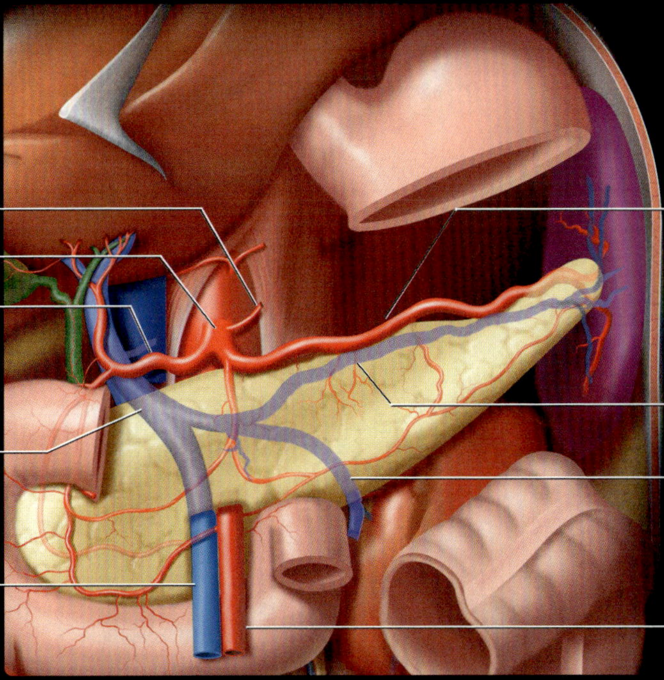

(Top) *Graphic shows the liver is retracted upward and the stomach transected to reveal the pancreas and spleen. The spleen is affixed to surrounding organs by several ligamentous attachments. The gastrosplenic ligament connects the spleen to the greater curve of the stomach and transports the short gastrics and left gastroepiploic vessels. The splenorenal ligament connects the spleen to the left kidney and pancreatic tail and carries the splenic artery and vein into the spleen at the hilum. The splenocolic ligament extends to the splenic flexure of the colon and the splenophrenic ligament to the diaphragm.* **(Bottom)** *Coronal graphic with the mesenteric reflections removed to show the splenic vascular supply. The splenic artery arises from the celiac axis and is often quite tortuous. The splenic vein runs posterior to the body of the pancreas and receives the inferior mesenteric vein. It joins the superior mesenteric vein behind the neck of the pancreas to form the portal vein.*

SPLENIC SHAPE

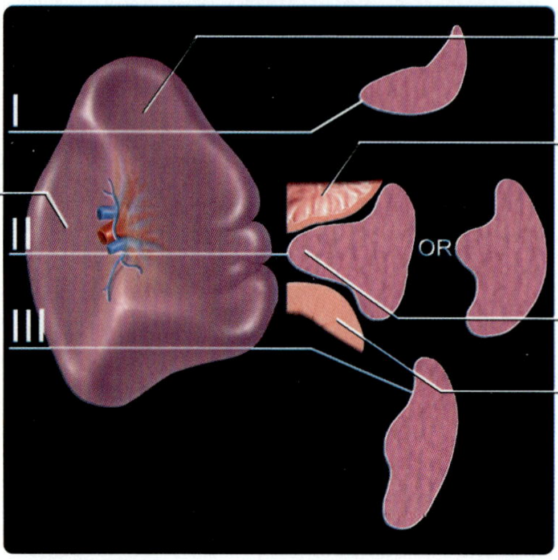

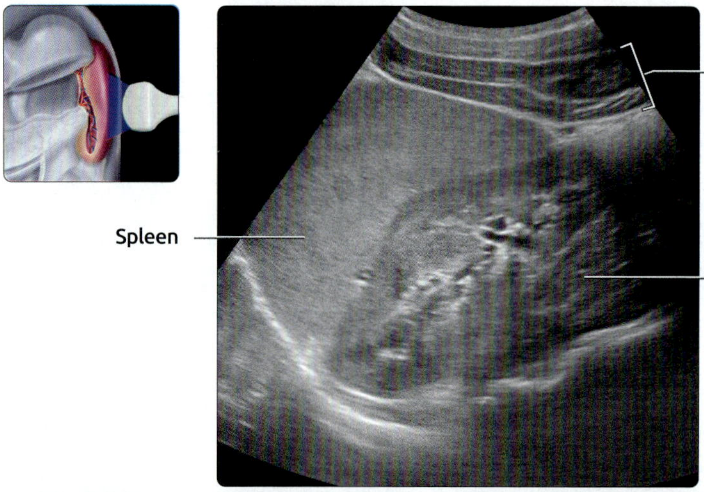

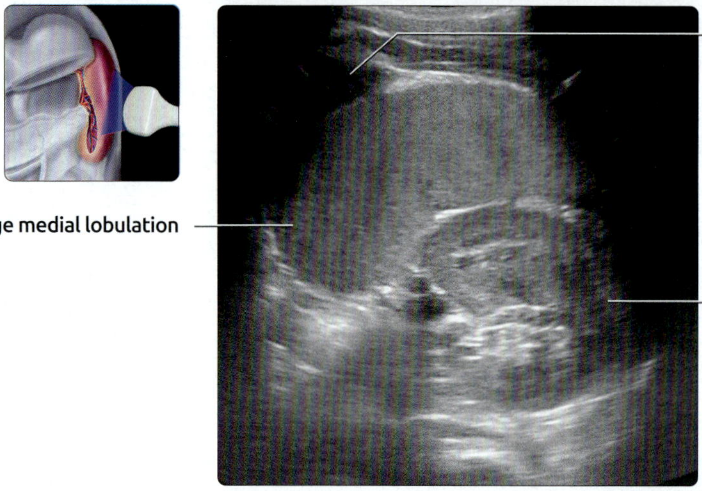

(Top) *Graphic shows the medial surface of the spleen and representative axial sections at 3 levels through the parenchyma. The spleen is of variable shape and size, even within the same individual, varying with states of nutrition and hydration. It is a soft organ that is easily indented by adjacent organs. The medial surface is often quite lobulated as it is interposed between the stomach and the kidney.* **(Middle)** *Longitudinal ultrasound shows the spleen conforming to the shape of the left kidney. In this plane, the spleen appears quite small.* **(Bottom)** *The oblique transverse plane in the same patient shows a large medial lobulation. Volume measurements are more accurate in determining the size of the spleen than any individual measurements, but even that has large individual variability.*

SPLENIC SHAPE

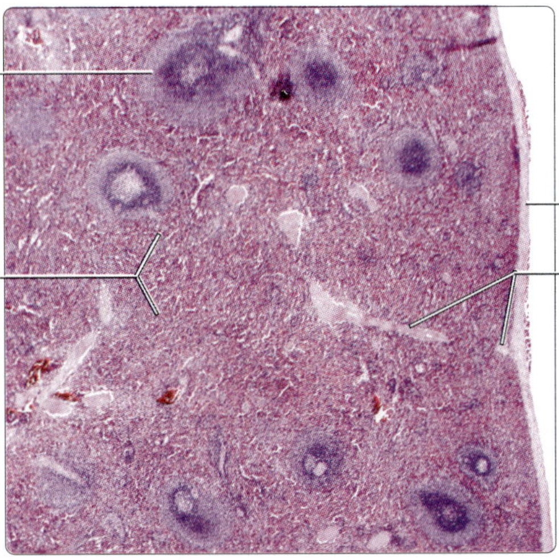

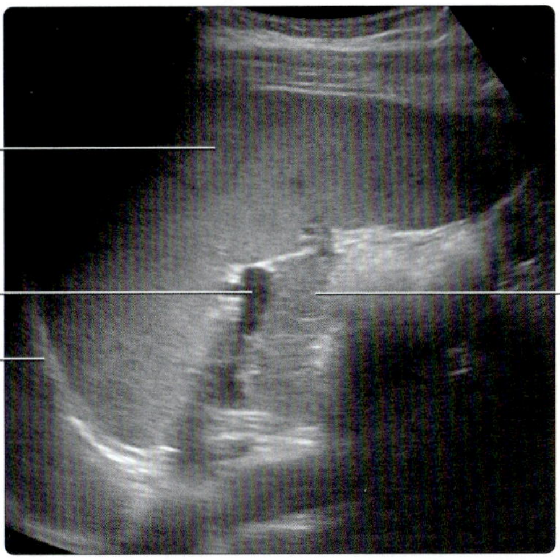

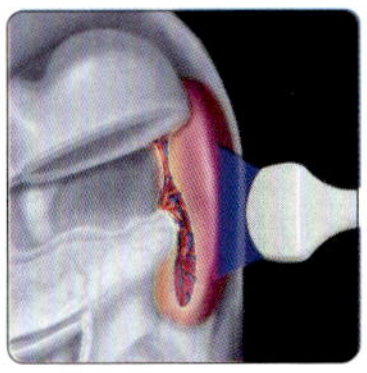

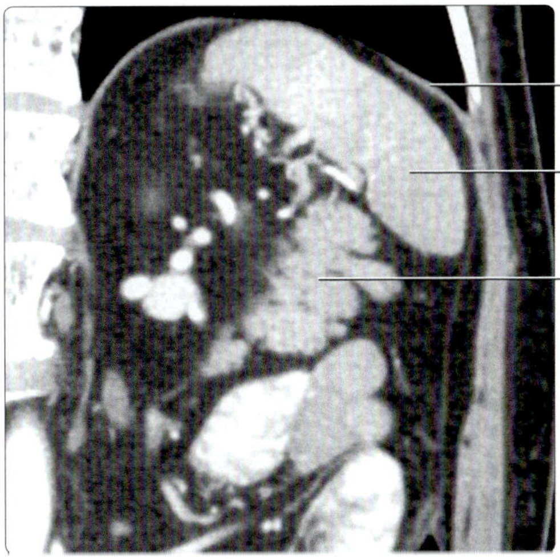

(Top) *Histologic section of a normal spleen viewed at low power shows white pulp and red pulp. A thin splenic capsule with slivers of branching trabeculae is also noted. The red pulp is "spongy" and compressible.* **(Middle)** *This longitudinal view of the spleen through the left flank shows the intimate relationship with the tail of the pancreas. In this case, the spleen was long (13.5 cm), but the splenic index and volume were normal.* **(Bottom)** *Oblique correlative multiplanar reconstruction CT of the spleen is shown. Note the extension of the pancreatic tail toward the splenic hilum. This allows the pancreatic tail to be visualized by ultrasound using the spleen as an acoustic window.*

SPLENIC VESSELS

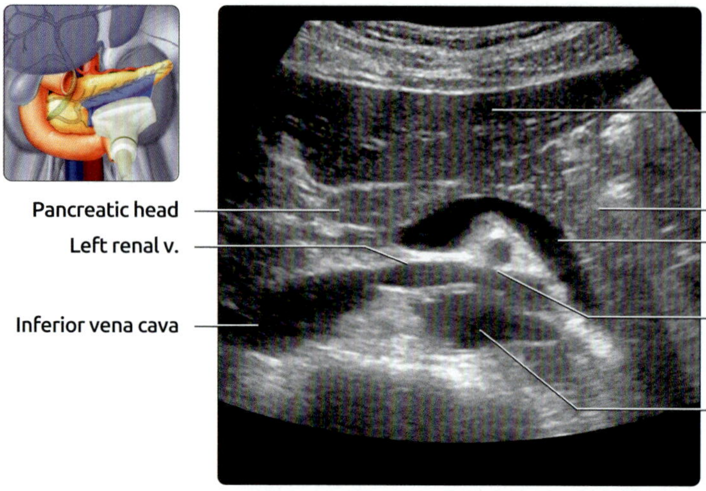

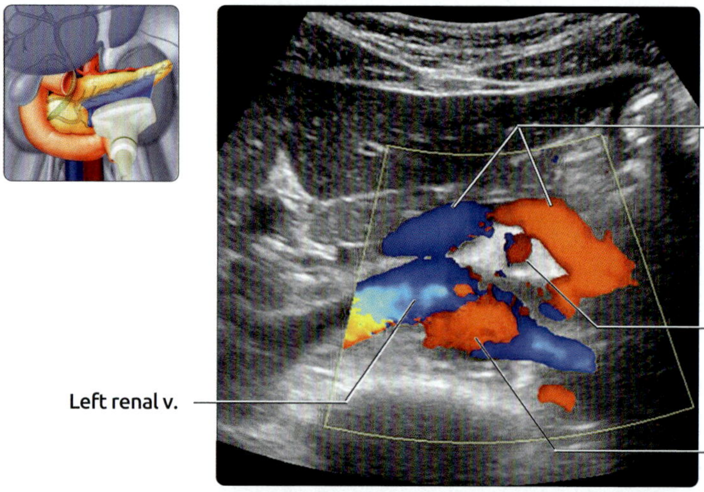

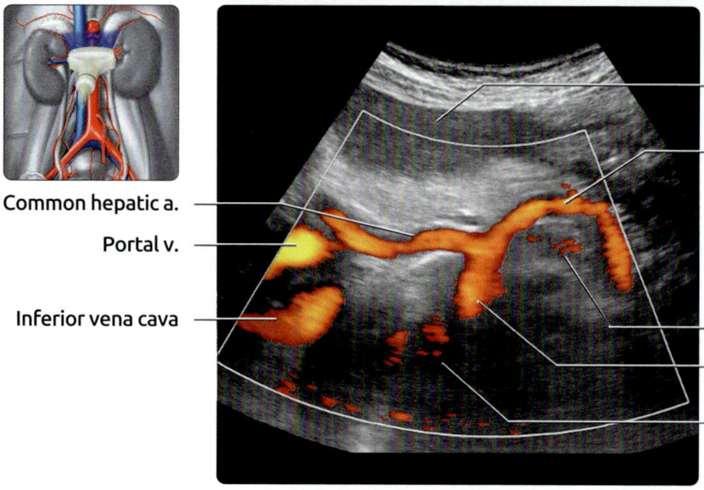

(Top) *Midline transverse anterior grayscale ultrasound of the splenic vein is shown. The splenic vein is located deep to the pancreatic body. Note the left renal vein course between the superior mesenteric artery and aorta.* **(Middle)** *Color Doppler ultrasound of the same area shows the normal direction of flow in the splenic vein toward the liver (hepatopetal). Note the change in color from red to blue, which is due to the position of the transducer, aligned at the midpoint of the vein. Using the information provided by the color bar, the red portion of the splenic vein is blood flowing toward the transducer (away from the spleen), and the blue is blood flowing away from the transducer (toward the the liver).* **(Bottom)** *Power Doppler ultrasound of the upper abdomen at midline shows the origin of the splenic artery from the celiac axis. The celiac artery branches into the splenic artery, common hepatic artery, and left gastric artery (not shown). After its takeoff, the splenic artery typically has a tortuous course. The branching of the celiac axis, as shown in this image, has been referred to as the seagull sign.*

SPLENIC VESSELS

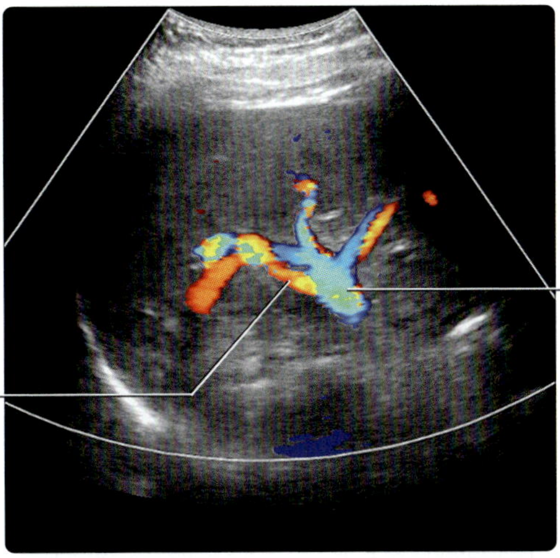

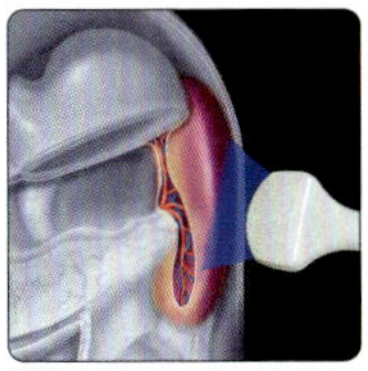

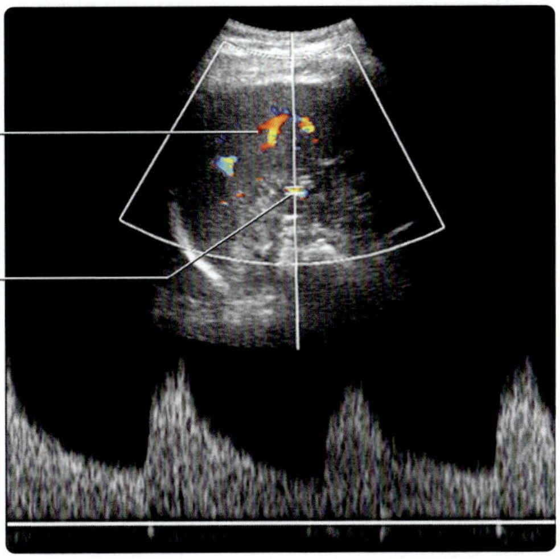

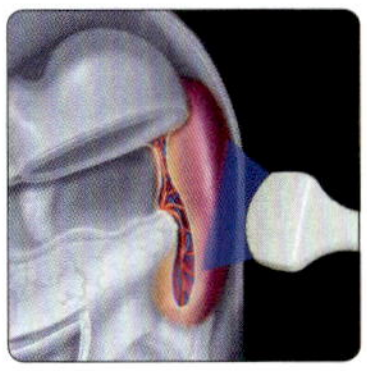

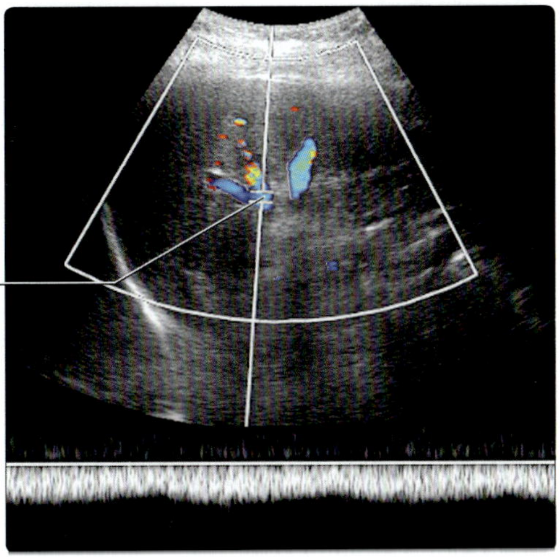

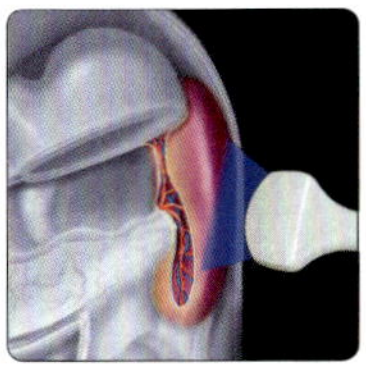

(Top) *Longitudinal oblique color Doppler ultrasound demonstrates the branching of the splenic arteries and veins in the splenic hilum.* **(Middle)** *Spectral Doppler waveform ultrasound of the distal splenic artery at the splenic hilum is shown. Because of a tortuous course, flow in this vessel is typically turbulent. The splenic artery has a low-resistance waveform (ample flow throughout diastole). Normal peak systolic velocity for the splenic artery is 25-45 cm/sec.* **(Bottom)** *Spectral Doppler waveform ultrasound of the splenic vein at the hilum shows a typical band-like flow profile with minimal respiratory fluctuations; flow is directed away from the transducer (away from the spleen). Normal peak systolic velocity of the splenic vein is 9-18 cm/sec.*

SPLENULE

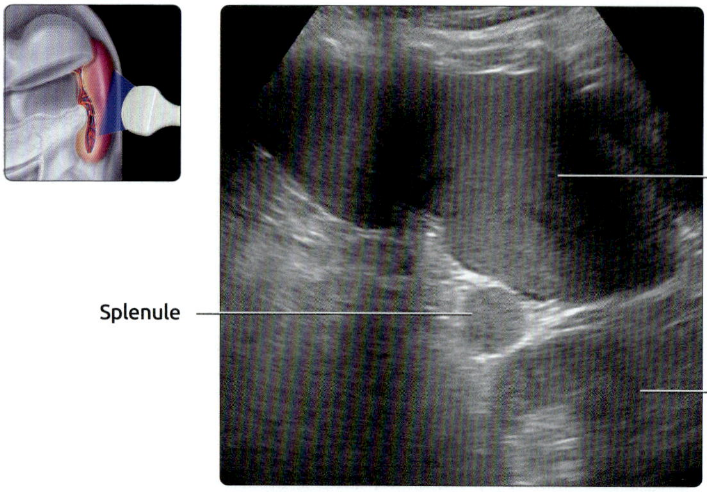

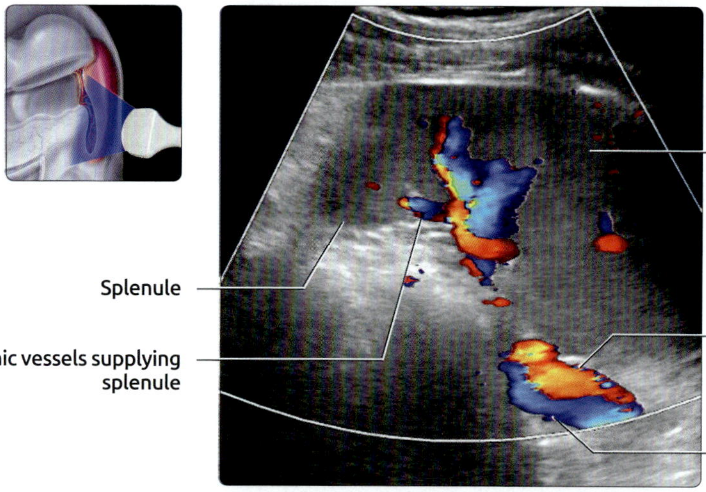

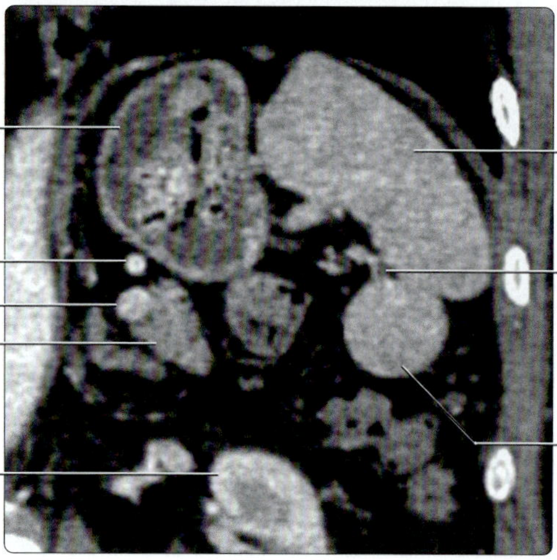

(Top) *Transverse oblique intercostal ultrasound shows a splenule between the spleen and left kidney. Splenules should have the same echogenicity and echotexture as the spleen, though this may depend on the sonographic window.* **(Middle)** *Longitudinal oblique intercostal color Doppler ultrasound shows a splenule adjacent to the spleen tip. The identification of vascular supply from the splenic vessels may also aid in identification of a splenule.* **(Bottom)** *Correlative coronal CECT of the left upper quadrant demonstrates a splenule along the inferior tip of the spleen. Branch vessels from the splenic artery and vein supplying the splenule are visualized.*

SPLENOSIS AND HETEROTAXY

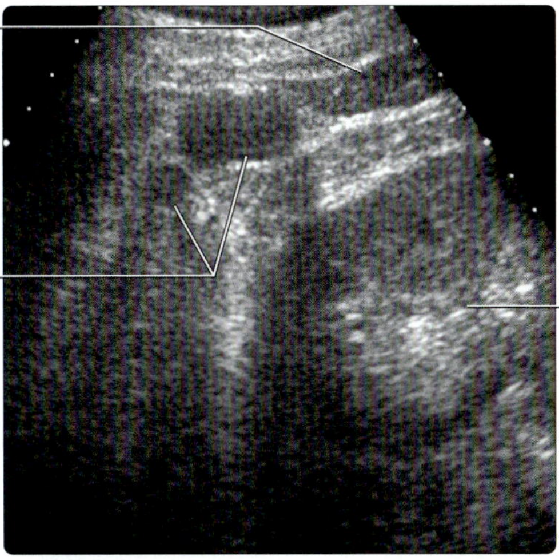

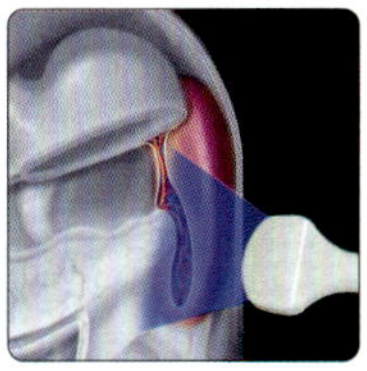

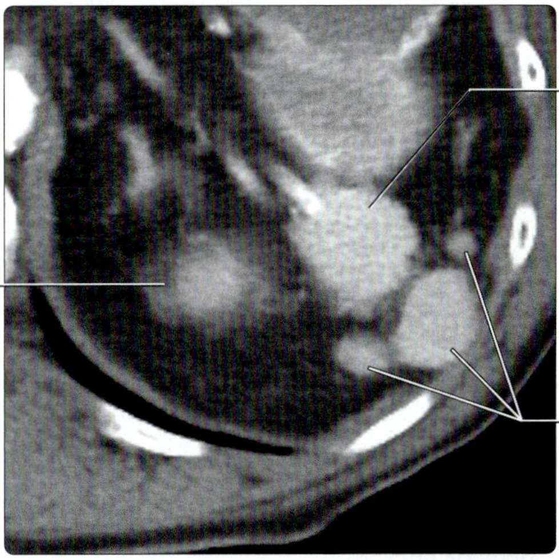

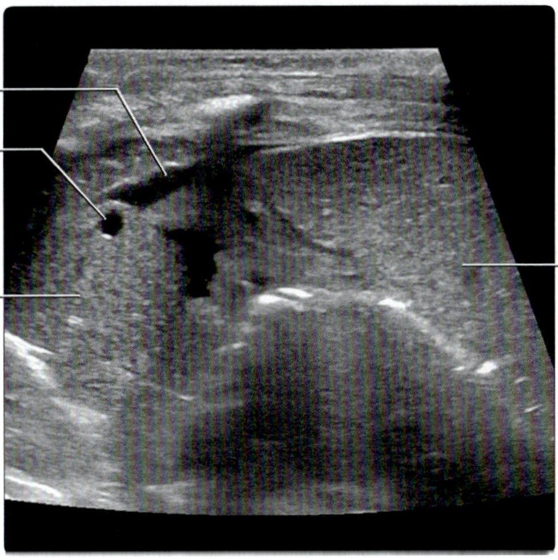

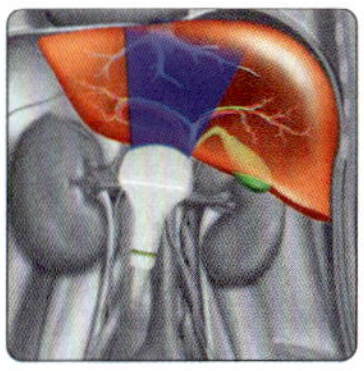

(Top) *Longitudinal oblique intercostal grayscale ultrasound through the left upper quadrant in a patient with prior splenectomy shows rounded, hypoechoic structures adjacent to the upper pole of the left kidney; this represents splenosis.* **(Middle)** *Correlative axial CECT through the left upper quadrant in a patient with prior splenic trauma and splenectomy shows residual splenosis.* **(Bottom)** *Transverse anterior ultrasound in the setting of heterotaxy with polysplenia demonstrates the relationship between the right-sided spleen and the left-sided liver. The classification of heterotaxy syndromes is complex; there is a spectrum ranging from classic asplenia to classic polysplenia. Heterotaxy with polysplenia (i.e., left double sidedness or left isomerism) may present with multiple spleens (resembling splenules) or a single spleen, as in this case.*

GROSS ANATOMY

Overview

- Pancreas: Accessory digestive gland lying behind stomach in anterior pararenal space (APS) of retroperitoneum
  - Exocrine function: Pancreatic acinar cells secrete pancreatic juice → pancreatic duct → duodenum
  - Endocrine: Pancreatic islet cells (of Langerhans) secrete insulin, glucagon, and other polypeptides → portal venous system

Divisions

- Head: Thickest part; lies to right of superior mesenteric artery and vein (SMA, SMV)
  - Attached to "C" loop of duodenum (2nd and 3rd parts)
  - Uncinate process: Head extension, posterior to SMV
  - Bile duct lies along posterior surface of head, joins with pancreatic duct (of Wirsung) to form hepatopancreatic ampulla (of Vater)
  - Main pancreatic and bile ducts empty into major papilla in 2nd portion of duodenum
- Neck: Thinnest part; lies anterior to SMA, SMV
  - SMV joins splenic vein behind pancreatic neck to form portal vein
- Body: Main part; lies to left of SMA, SMV
  - Splenic vein lies in groove on posterior surface of body
  - Anterior surface is covered with peritoneum forming back surface of omental bursa (lesser sac)
- Tail: Lies between layers of splenorenal ligament in splenic hilum

Internal Structures

- Pancreatic duct (of Wirsung) runs length of pancreas, turning inferiorly through head to join bile duct
- Accessory pancreatic duct (of Santorini) opens into duodenum at minor duodenal papilla
  - Usually communicates with main pancreatic duct
  - Variations are common, including dominant accessory duct draining most pancreatic juice
- Vessels, nerves, and lymphatics
  - Arteries to head mainly from gastroduodenal artery
    - Pancreaticoduodenal arcade of vessels around head also supplied by SMA branches
  - Arteries to body and tail from splenic artery
  - Veins are tributaries of SMV and splenic vein → portal vein
  - Autonomic nerves from celiac and superior mesenteric plexus
    - Parasympathetic stimulation of pancreatic secretion, but pancreatic juice secretion is mostly under hormonal control (secretin, from duodenum)
  - Lymphatics follow blood vessels
    - Collect in splenic, celiac, superior mesenteric and hepatic nodes

IMAGING ANATOMY

Overview

- Pancreas can be localized on ultrasound by
  - Typical parenchymal architecture: Homogeneously isoechoic/hyperechoic echo pattern when compared with overlying liver
  - Surrounding anatomical landmarks: Body anterior to splenic vein; neck anterior to SMA/SMV
- Variations in reflectivity related to degree of fatty infiltration; uncinate process and posterior pancreatic head are relatively echo-poor in 25% of subjects (lack of intraparenchymal fat)

ANATOMY IMAGING ISSUES

Imaging Recommendations

- Use 2- to 5-MHz transducers, or up to 9 MHz for smaller patients
- Techniques to combat overlying stomach and bowel gas include
  - Displacement of intervening bowel gas by gentle graded compression with transducer
  - Overnight fasting or fasting > 6-8 hours
  - Noneffervescent fluid can be given orally to fill gastric fundus
    - Scanning delayed for few minutes to allow fluid to settle
    - Patient can lie on left side to allow imaging of body and tail of pancreas
    - Patient can then be turned right to allow gastric fluid to flow to stomach antrum and duodenum, allowing imaging of head and uncinate process
- CT is preferred imaging modality for imaging of pancreas
- MRCP (± secretin) or ERCP useful for defining pancreatic duct

Imaging Pitfalls

- Ultrasound examination of pancreas is often limited by overlying bowel gas

Key Concepts

- Shape, size, and texture of pancreas are quite variable
  - Largest in young adults
  - Atrophy and fatty infiltration with age (> 70), obesity, diabetes, corticosteroids, Cushing disease
  - Pancreatic duct also becomes more prominent with age (normal < 3 mm diameter)
  - Focal bulge or mass effect is abnormal
- Location behind lesser sac
  - Acute pancreatitis often results in lesser sac fluid (may mimic pseudocyst)
- Pancreas lies in APS
  - Inflammation (from pancreatitis) easily spreads to duodenum and descending colon; also lie in APS
  - Inflammation easily spreads into mesentery and mesocolon; roots of these lie just ventral to pancreas
- Obstruction of pancreatic duct
  - Relatively common result of chronic pancreatitis (fibrosis &/or stone occluding pancreatic duct), or pancreatic ductal carcinoma
- Acute pancreatitis
  - Relatively common result of gallstone (lodged in hepatopancreatic ampulla causing bile to reflux into pancreas) or damage from alcohol abuse

PANCREAS IN SITU

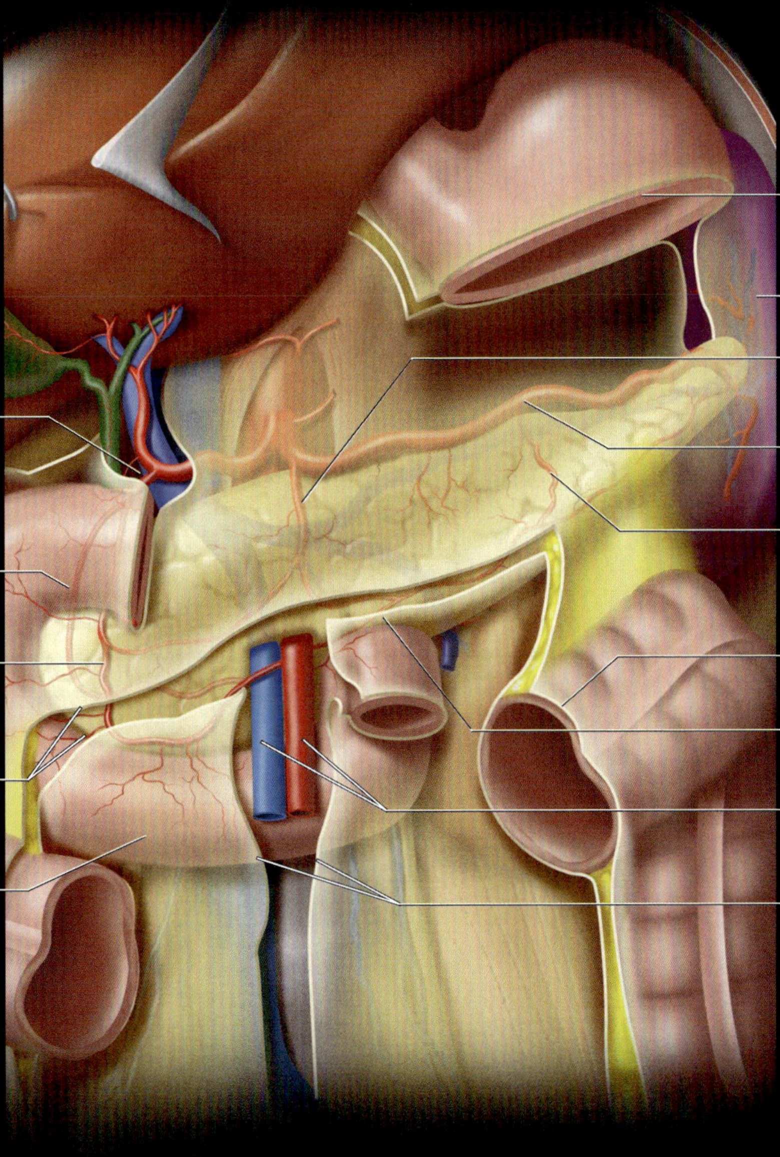

Graphic shows the arterial supply to the body & tail of the pancreas through terminal branches of the splenic artery, which are variable in number and size. The 2 largest are usually the dorsal (superior) and great pancreatic arteries, which arise from the proximal and distal splenic artery, respectively. The arteries to the pancreatic head and duodenum come from the pancreaticoduodenal arcades that receive flow from the celiac and superior mesenteric arteries. The superior mesenteric vessels pass behind the neck of the pancreas and in front of the 3rd portion of the duodenum. The root of the transverse mesocolon and small bowel mesentery arise from the surface of the pancreas and transmit the blood vessels to the small bowel and transverse colon. The splenic vein runs along the dorsal surface of the pancreas. The splenic vessels and pancreatic tail insert into the splenic hilum.

PANCREAS, TRANSVERSE VIEW

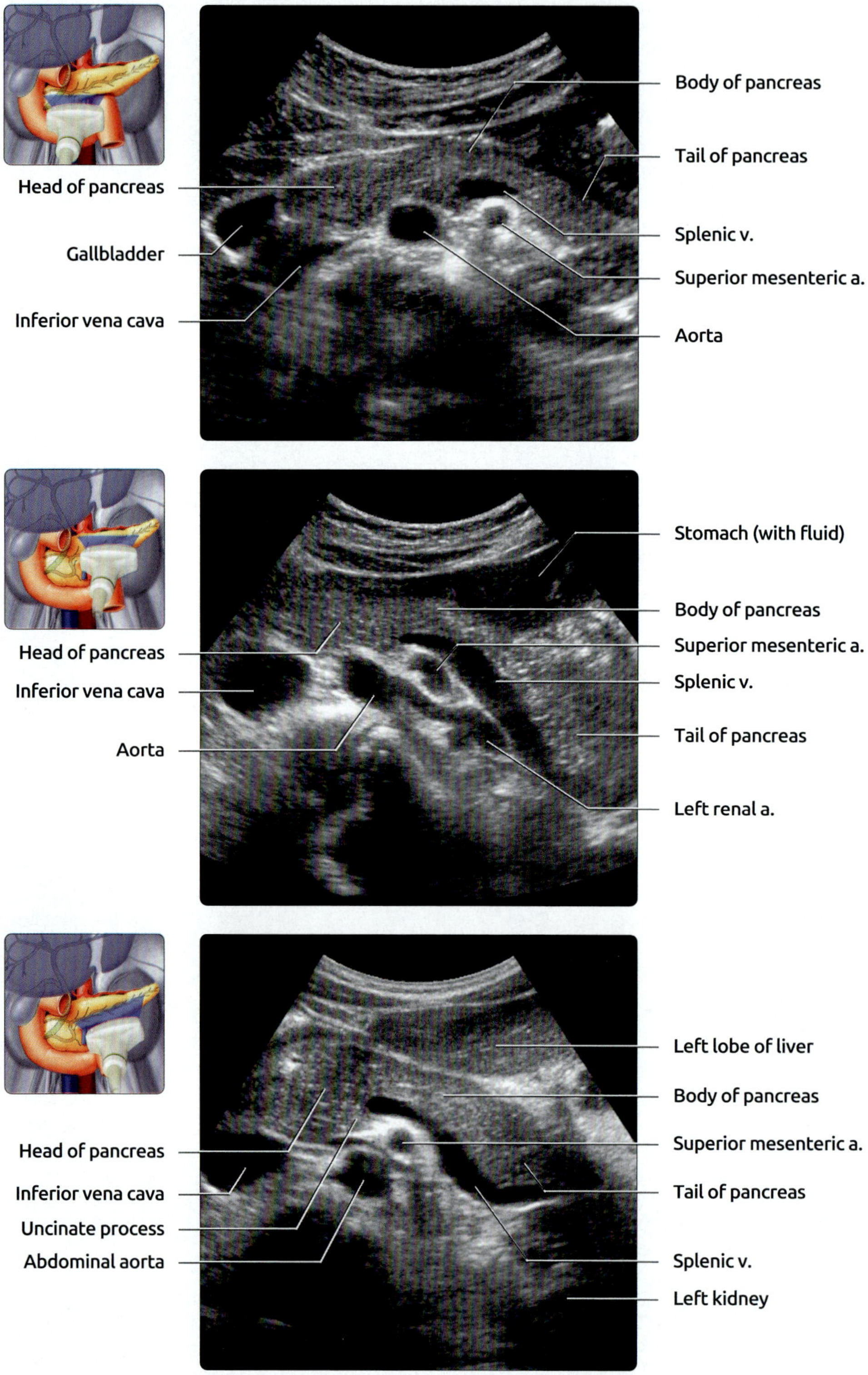

(Top) *Transverse transabdominal grayscale ultrasound at the epigastrium is shown. Anatomically, the pancreatic axis from head to tail is directed superiorly and to the left. This lower transverse section demonstrates the bulk of the pancreatic head.* **(Middle)** *Transverse transabdominal grayscale ultrasound at the epigastrium is shown, slightly higher than the previous image. Note that the pancreatic body and tail have now come into view.* **(Bottom)** *Oblique transabdominal grayscale ultrasound at the epigastrium is shown. The transducer is tilted slightly cranially and laterally to the left to follow the pancreatic axis, thus imaging the pancreas in its entirety. The splenic vein courses along the posterior pancreas and provides an excellent landmark in locating the pancreas. The superior mesenteric artery is more posteriorly located and has a characteristic dot shape as it is imaged end-on.*

PANCREAS, TRANSVERSE VIEW, CT

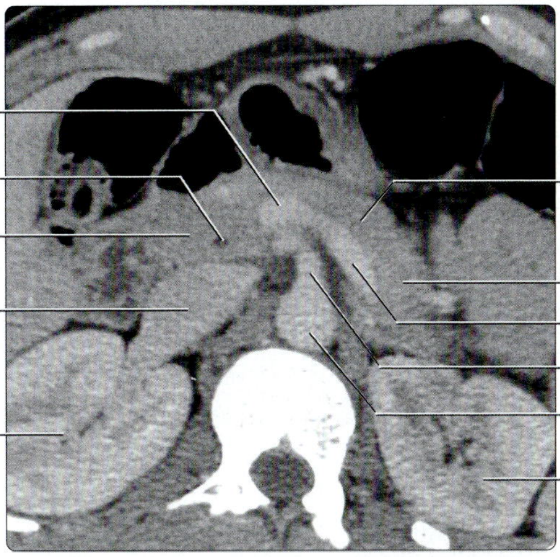

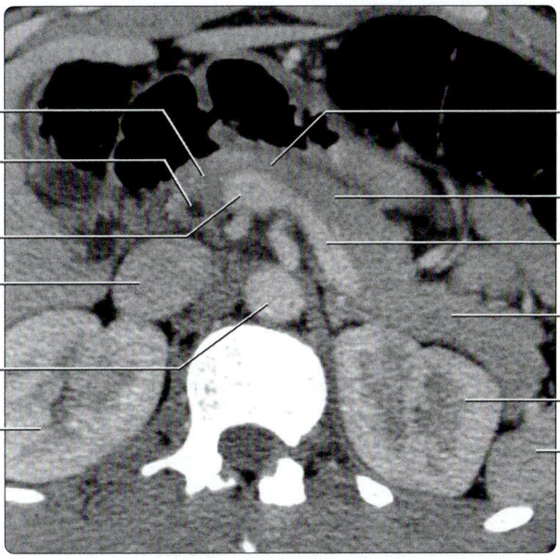

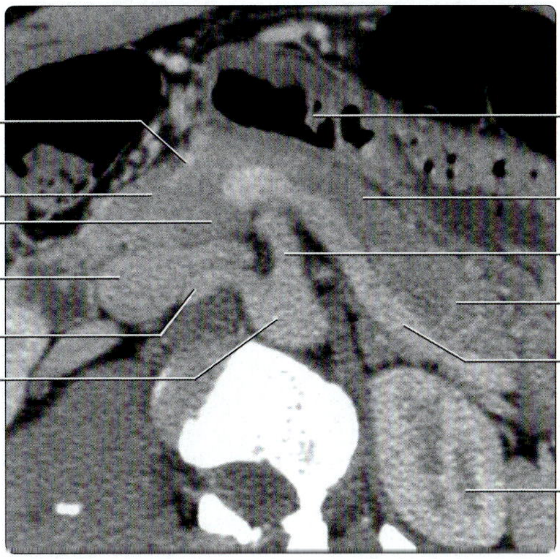

(Top) *Correlative transverse CECT of the pancreas is shown at the level of the origin of the superior mesenteric artery. Note the common bile duct within the pancreatic head before it exits into the duodenum.* **(Middle)** *Correlative transverse CECT of the pancreas shows the course of the pancreatic tail, which goes posteriorly and forms close relations with the left kidney and spleen.* **(Bottom)** *Correlative oblique CECT follows the pancreatic axis and demonstrates its head, body, and tail. The splenic vein courses along the posterior pancreas, following its contour. The anteriorly located stomach may be distended with fluid and used as an acoustic window during ultrasound.*

PANCREATIC ARTERIAL ANATOMY

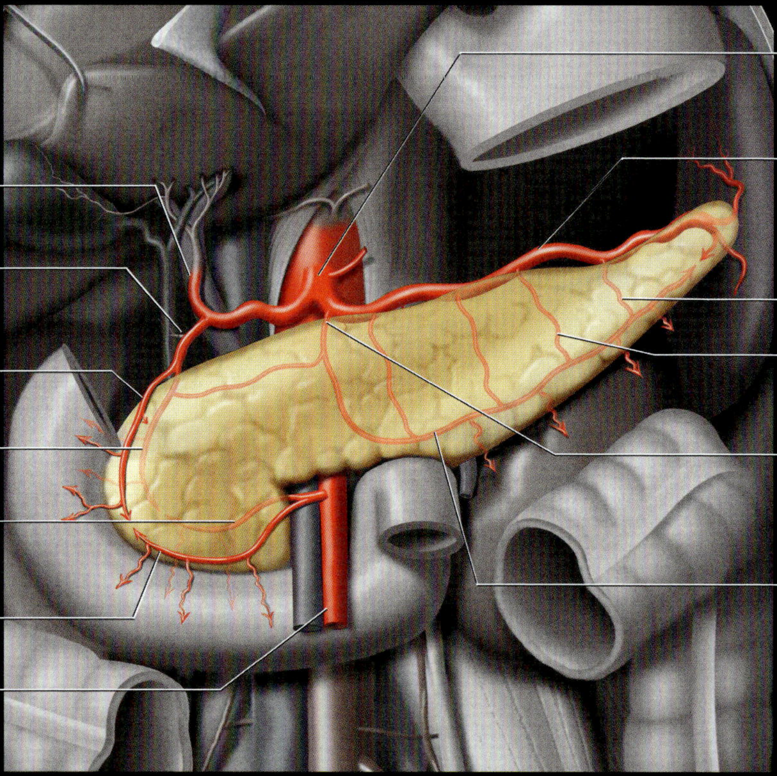

Graphic demonstrates the arterial supply to the pancreas. The pancreatic head is primarily supplied by the anterior and posterior pancreaticoduodenal arcades, including anterior and posterior superior pancreaticoduodenal arteries arising from the gastroduodenal artery and anterior and posterior inferior pancreaticoduodenal arteries arising from the superior mesenteric artery (SMA). The blood supply to the body and tail segments is primarily via the splenic artery with the 2 biggest branches, including the dorsal pancreatic artery and the pancreatic great (magna) artery, which arise from proximal and midportions of the splenic artery, respectively.

PANCREAS VASCULATURE, TRANSVERSE VIEW

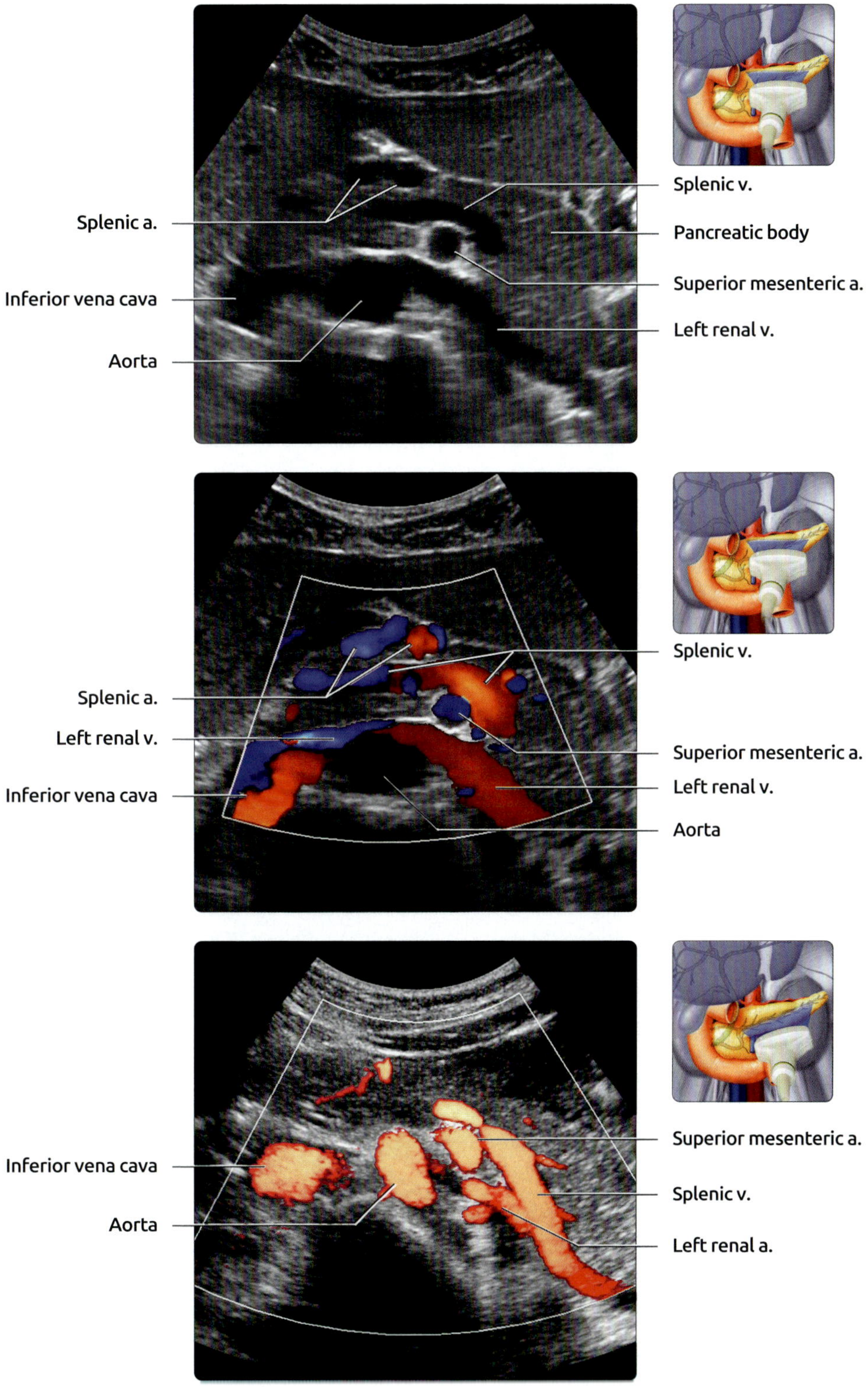

(Top) *Transverse subxiphoid grayscale ultrasound shows the left renal vein coursing posterior to the pancreatic body, superior mesenteric artery, and splenic vein. Two segments of the tortuous splenic artery are visible in cross section anteriorly.* **(Middle)** *Transverse color Doppler ultrasound is shown at the same level. There is a small amount of caudal tilt so that the SMA is blue, away from the transducer, and the IVC is mostly red, toward the transducer. The splenic and left renal veins on the right side of the image are toward the transducer (red) then switch to blue as they travel away from the transducer toward the SMV and IVC, respectively. Note the mixing in the IVC secondary to the left renal vein inflow. The splenic artery is tortuous and the 2 visible segments flow opposite directions.* **(Bottom)** *Transverse subxiphoid ultrasound in a different patient is shown using color power Doppler. This is more sensitive for detecting vascular flow but fails to provide information on flow direction, although some newer ultrasound machines support directional color power Doppler.*

PANCREAS, SAGITTAL VIEW

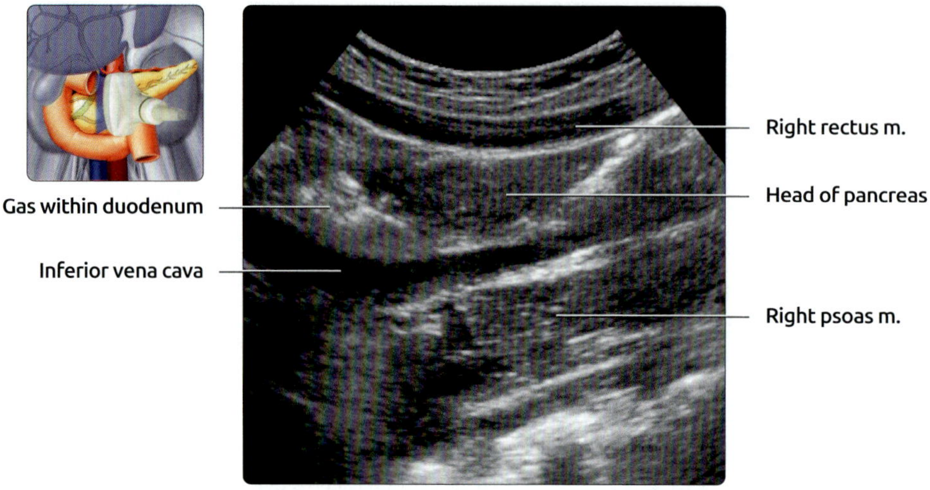

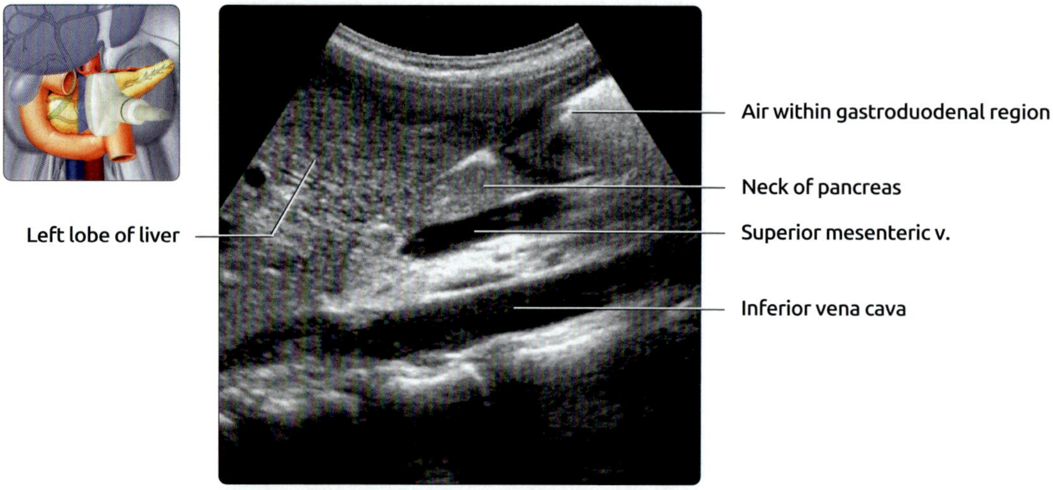

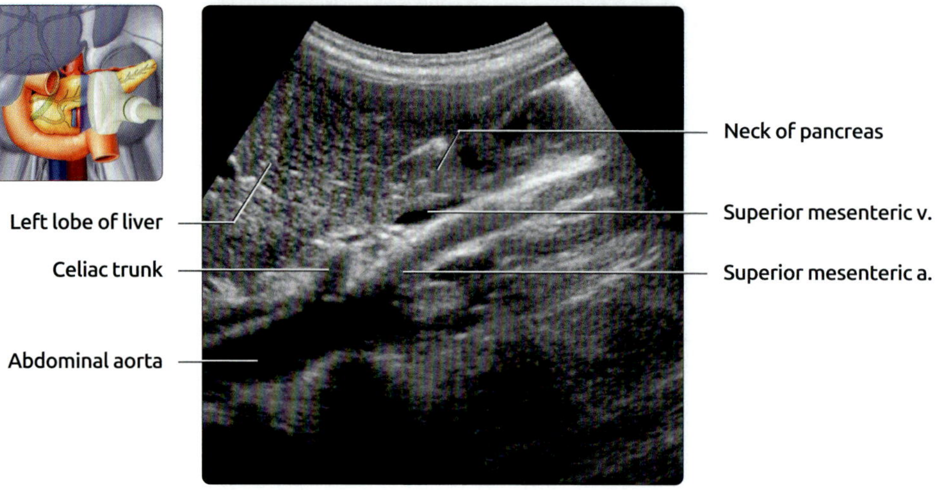

(Top) *Longitudinal transabdominal grayscale ultrasound at the epigastrium, right paramedian region, is shown. Note the relationship of the pancreatic head with the posteriorly located inferior vena cava.* **(Middle)** *Longitudinal transabdominal grayscale ultrasound at the epigastrium, right paramedian region, is shown continuing medially from the previous image. Note the superior mesenteric vein coming into view; this is a good landmark for locating the neck of the pancreas on sagittal ultrasound.* **(Bottom)** *Longitudinal transabdominal grayscale ultrasound at the epigastrium, right paramedian region, is shown slightly more medial to the previous image. The origin of the superior mesenteric artery arising from the abdominal aorta is brought into view. The SMA is also a useful marker for identifying the neck of the pancreas on sagittal ultrasound.*

PANCREAS, SAGITTAL VIEW

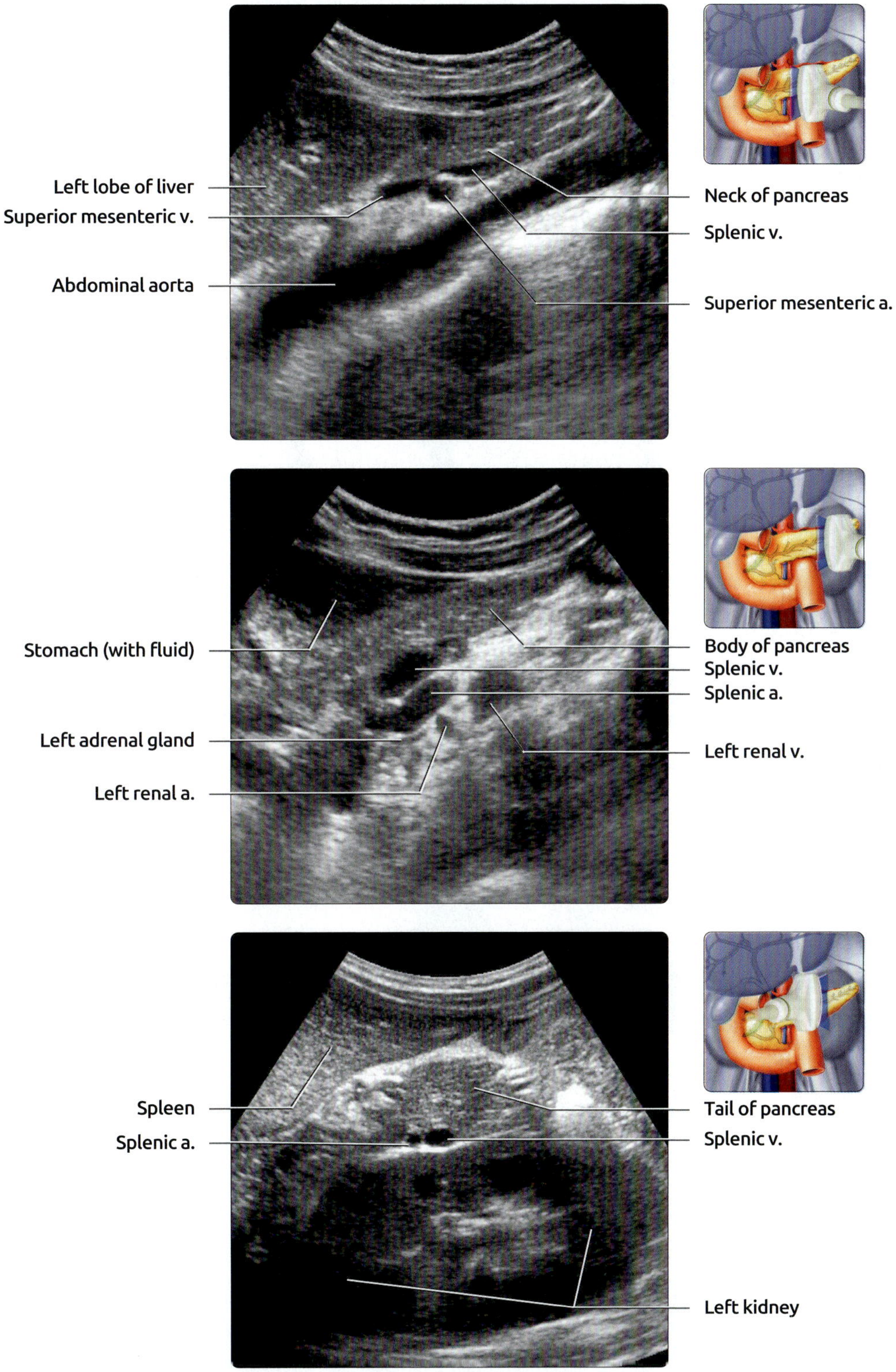

(Top) *Longitudinal oblique transabdominal grayscale ultrasound at the midline epigastrium is shown, revealing the SMA that has taken off from the abdominal aorta, its relationship with the splenic vein, and the anteriorly located pancreatic neck. The inferior mesenteric vein joins the splenic vein here.* **(Middle)** *Longitudinal transabdominal grayscale ultrasound at the epigastrium, left paramedian region, is shown by sweeping the transducer laterally from the top image. The body of the pancreas lies to the left of the SMA (not shown). The stomach lies superiorly and may be filled with fluid for use as an acoustic window. The splenic vein maintains its course behind the pancreas.* **(Bottom)** *Longitudinal transabdominal grayscale ultrasound at the epigastrium is shown, continuing the scan laterally from the middle image. The pancreatic tail is identified between the spleen and the left kidney.*

PANCREATIC DUCT VARIATIONS

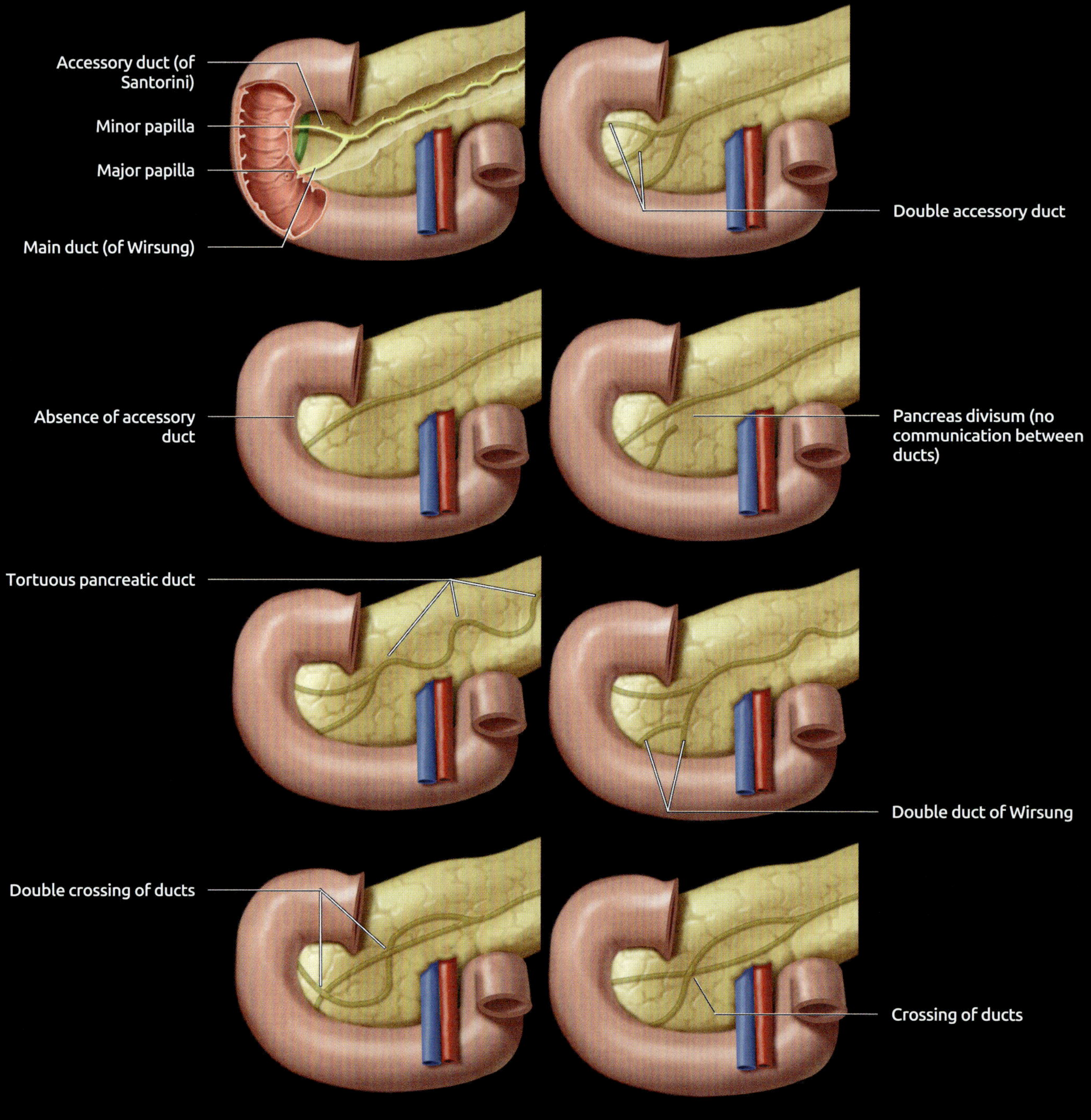

The accessory duct (of Santorini) originates with the dorsal pancreatic anlage, which is the larger bud from the embryologic foregut, comprising the pancreatic body and tail. The main duct (of Wirsung) originates with the ventral, smaller, anlage that develops into the pancreatic head and uncinate process. Usually, the main and accessory pancreatic ducts fuse, and the main duct becomes the primary conduit for drainage of secretions into the duodenum. The pancreatic duct courses through the center of the gland and is joined by tributaries that enter it at right angles. In the head, the duct turns caudally and dorsally and runs parallel to the common bile duct before joining it at the ampulla of Vater and entering the major papilla. The accessory duct usually enters the duodenum more proximally through the minor papilla.

ANNULAR PANCREAS

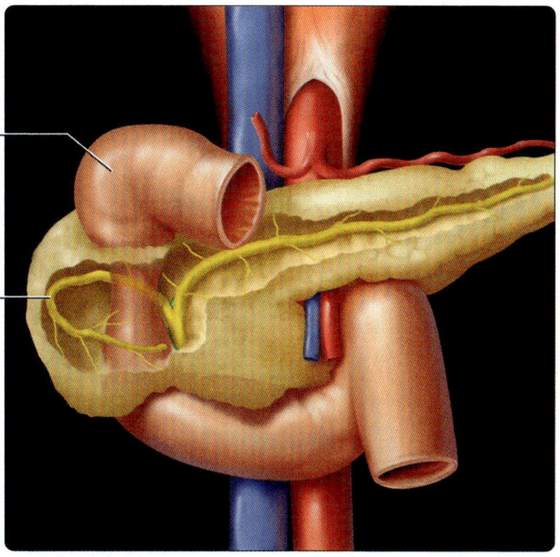

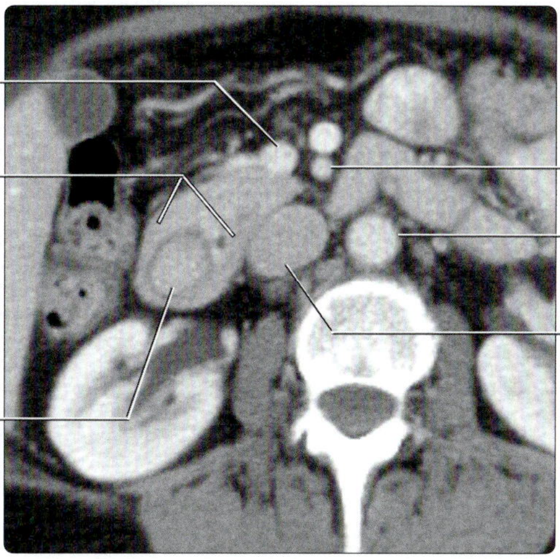

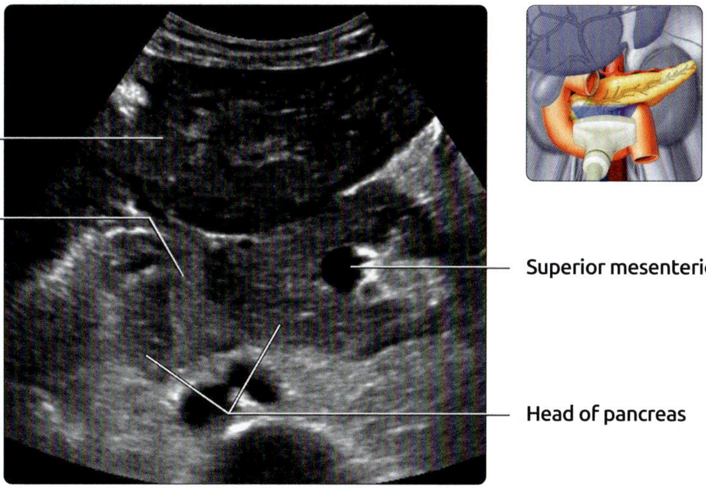

(Top) *Graphic shows annular pancreas in which there is abnormal rotation and fusion of the ventral and dorsal pancreatic anlage resulting in a circumferential mass of pancreatic tissue that encircles and narrows the duodenum. The duct draining the pancreatic head encircles the 2nd portion of the duodenum. This condition may remain asymptomatic or may result in obstruction of the duodenum, often in neonates.* **(Middle)** *Axial CECT shows the 2nd portion of duodenum completely encircled by pancreatic tissue, consistent with an annular pancreas.* **(Bottom)** *Transverse transabdominal grayscale ultrasound at the midline epigastrium, angled slightly caudally through the liver, shows the descending segment of the duodenum coursing through the pancreatic head in this patient with annular pancreas.*

GROSS ANATOMY

Overview

- Kidneys are paired, bean-shaped, retroperitoneal organs
  - Function:
    - Removal of excess water, salts, and wastes of protein metabolism from blood
    - Regulation of water and electrolyte balance
    - Secretion of hormones which control blood pressure, bone and blood production

Anatomic Relationships

- Located in retroperitoneum, within perirenal space, surrounded by renal fascia (of Gerota)
- Each adult kidney is ~ 9-14 cm in length, 5 cm in width, 3 cm in thickness
- Both kidneys lie on quadratus lumborum muscles, lateral to psoas muscles, between T12-L3

Internal Structures

- Kidneys are hollow centrally with renal sinus occupied by fat, renal pelvis, calyces, vessels, and nerves
- Renal hilum: concavity where artery enters and vein and ureter leave renal sinus
- Renal pelvis: Funnel-shaped expansion of upper end of ureter
  - Receives major calyces (infundibula) (2 or 3), each of which receives minor calyces (2-4)
- Renal papilla: Pointed apex of renal pyramid of collecting tubules that excrete urine
  - Each papilla indents a minor calyx
  - 7-10 papilla per kidney
- Renal cortex: Outer part, contains renal corpuscles (glomeruli, vessels), proximal portions of collecting tubules and loop of Henle
- Renal medulla: Inner part, contains renal pyramids, distal parts of collecting tubules, and loops of Henle
- Vessels, nerves, and lymphatics
  - Artery
    - Usually 1 for each kidney
    - Arise from aorta at about L1-L2 vertebral level
  - Vein
    - Usually 1 for each kidney
    - Lies in front of renal artery and renal pelvis
  - Nerves
    - Autonomic from renal and aorticorenal ganglia and plexus
  - Lymphatics
    - To lumbar (aortic and caval) nodes

IMAGING ANATOMY

Overview

- Well-defined retroperitoneal bean-shaped structures, which move with respiration

Internal Contents

- **Renal capsule**
  - Normal kidneys are well-defined due to presence of renal capsule and are less reflective than surrounding fat
- **Renal cortex**
  - Renal cortex has reflectivity that is less than adjacent liver or spleen
  - If renal cortex brighter than normal liver (hyperechoic), high suspicion of renal parenchymal disease
- **Medullary pyramids**
  - Medullary pyramids are less reflective than renal cortex
- **Corticomedullary differentiation**
  - Margin between cortex and pyramids is usually well-defined in normal kidneys
  - Margin between cortex and pyramids may be lost in presence of generalized parenchymal inflammation or edema
- **Renal sinus**
  - Echogenic due to the fat that surrounds blood vessels and collecting systems
  - Outline of renal sinus is variable, from smooth to irregular
  - Renal sinus fat may increase in obesity, steroid use, and sinus lipomatosis
  - Renal sinus fat may decrease in cachectic patients and neonates
  - If sinus echoes are indistinct in noncachectic patient, tumor infiltration or edema should be considered
- **Collecting system (renal pelvis and calyces)**
  - Not usually visible in dehydrated patient
  - May be seen as physiological "splitting" of renal sinus echoes in patients with a full bladder undergoing diuresis
  - Physiological "splitting" of renal sinus echoes is common in pregnancy
    - Causes of dilatation of pelvicalyceal system include mechanical obstruction by enlarging uterus, hormonal factors, increased blood flow, and parenchymal hypertrophy
    - May occur as early as 12 weeks into pregnancy
    - Seen in up to 75% of right kidneys at 20 weeks into pregnancy, less common on left side, thought to be due to cushioning of ureter from gravid uterus by sigmoid colon
    - Obvious dilatation of pelvicalyceal system can be seen in 2/3 of patients at 36 weeks
    - Changes usually resolve within 48 hours after delivery
  - Possible obstruction can be excluded by performing post micturition images of collecting system and looking for ureteral jets in the bladder with color Doppler
    - AP diameter of renal pelvis in adults should be < 10 mm
- **Renal arteries**
  - Normal caliber 5-8 mm
  - 2/3 of kidneys are supplied by single renal artery arising from aorta
  - 1/3 of kidneys are supplied by 2 or more renal arteries arising from aorta
    - Main renal artery may be duplicated
    - Accessory renal arteries may arise from aorta superior or inferior to main renal artery
    - Accessory renal arteries may enter kidney either in hilum or at poles
    - Extrahilar accessory renal arteries may arise from ipsilateral renal artery, ipsilateral iliac artery, aorta, or retroperitoneal arteries
  - Spectral Doppler

  - Open systolic window, rapid systolic upstroke occasionally followed by secondary slower rise to peak systole with subsequent diastolic delay but persistent forward flow in diastole
  - Continuous diastolic flow is present due to low resistance in renal vascular bed
  - Low resistance flow pattern is also present in intrarenal branches
  - Normal peak systolic velocity (PSV) 75-125 cm/s, not more than 180 cm/s
    - > 200 cm/s is abnormal
  - Resistive index (RI) is (peak systolic velocity - end diastolic velocity)/peak systolic velocity; normal < 0.7
  - Pulsatility index (PI) is (peak systolic velocity - end diastole velocity)/mean velocity, normal < 1.8
- **Renal veins**
  - Normal caliber 4-9 mm
  - Formed from tributaries that coalesce at renal hilum
  - Right renal vein is relatively short and drains directly into IVC
  - Left renal vein receives left adrenal vein from above and left gonadal vein from below
  - Left renal vein crosses midline between aorta and superior mesenteric artery
  - Spectral Doppler
    - Normal PSV 18-33 cm/s
    - Spectral Doppler in right renal vein mirrors pulsatility in IVC
    - Spectral Doppler in left renal vein may show only slight variability of velocities consequent upon cardiac and respiratory activity

Size

- Bipolar length is found by rotating transducer around its vertical axis such that the longest craniocaudal length can be identified
- Normal size between 10-15 cm
- Volume measurements
  - May be more accurate, but is time consuming
  - 3D ellipsoidal formula can be used for volume estimation
    - Length x AP diameter x transverse diameter x 0.5
  - Consistency and changes in volume over time more important

ANATOMY IMAGING ISSUES

Imaging Recommendations

- Right kidney
  - Liver used as acoustic window
  - Transducer placed in subcostal or intercostal position
  - Varying degree of respiration is useful
  - Raising patient's right side and scanning laterally/posterolaterally may be useful
- Left kidney
  - More difficult to visualize due to bowel gas from small bowel and splenic flexure
  - Usually easier to search for left kidney using posterolateral approach with left side raised
  - Full right lateral decubitus with pillow under right flank and left arm extended above head may be useful in difficult cases
    - Spleen can be used as acoustic window for imaging upper pole of left kidney
- Posterior approach for both kidneys
  - Useful for interventional procedures such as renal biopsy, nephrostomy
  - Use bolster or pillow under the patient's abdomen to decrease lordosis
  - Image quality may be impaired by thick paraspinal muscles and ribs shadowing
- Renal arteries
  - Origins best seen from midline anterior approach
  - Right renal artery can usually be followed from origin to kidney
  - Left renal artery often requires posterolateral coronal transducer scanning position for visualization
- Renal veins
  - Best seen on transverse scan from anterior approach
  - May also be seen on coronal scan from posterolateral coronal

Key Concepts

- Accessory renal vessels
  - Accurate diagnosis necessary when planning surgery (e.g., resection, transplantation)
  - Due to limitations of ultrasound, CT Arteriography, magnetic resonance angiography or digital subtraction angiography are more sensitive and accurate
- Normal variants may mimic disease
  - Dromedary hump and hypertrophied column of Bertin may be mistaken for renal tumors
- Congenital anomalies very common
  - Leading cause of renal failure in children
  - Early diagnosis important

EMBRYOLOGY

Embryologic Events

- Congenital structural anomalies include abnormal renal number, position, structure and vessels
  - Abnormal number: absence of one or both kidneys; supernumerary kidney
  - Abnormal position: pelvic kidney, crossed fused renal ectopia, malrotation, ptosis
  - Abnormal structure:
    - Duplication: results from lack of fusion and commonly produces an enlarged kidney with 2 separate hila and pelvicalyceal systems, these may join or continue as 2 ureters
      - Ureters may be completely separate until they join the bladder or join proximal to the bladder
      - "Duplex kidney": Bifid renal pelvis with single ureter
    - Hypertrophied column of Bertin (lobar dysmorphism; Fetal lobulation; Hilar lip
    - Pelviureteric junction obstruction
  - Often accompanied by anomalies of other systems
  - VATER acronym: Vertebral, anorectal, tracheoesophageal, radial ray, renal anomalies

RENAL FASCIA AND PERIRENAL SPACE

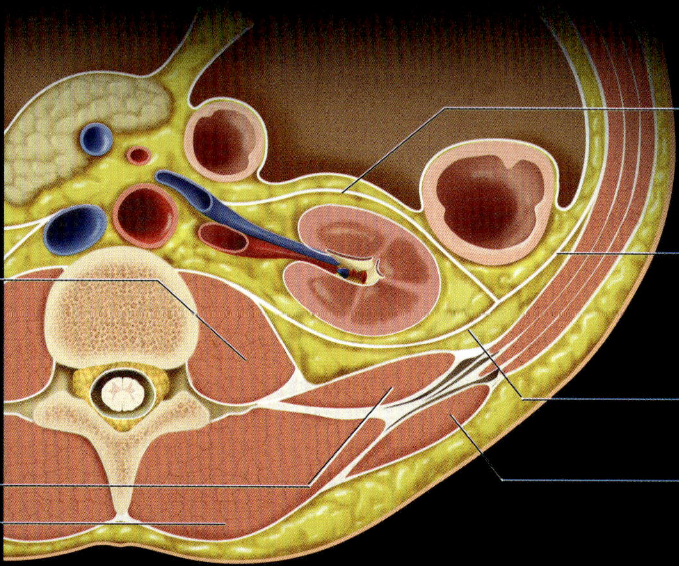

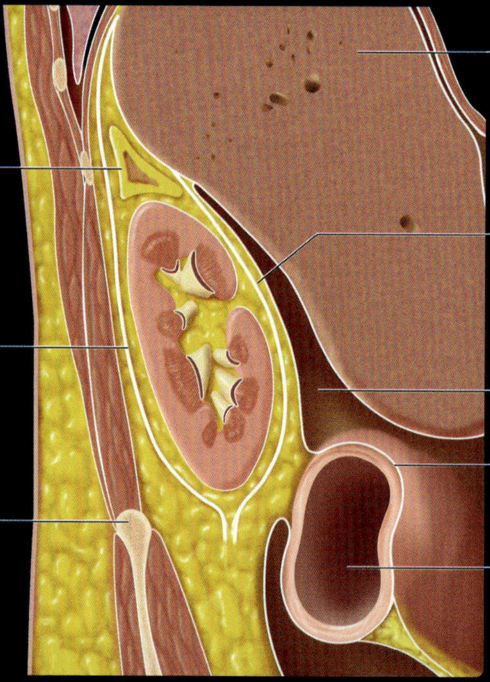

(Top) *The anterior and posterior layers of the renal fascia envelop the kidneys and adrenals along with the perirenal fat. Medial to the kidneys, the course of the renal fascia is variable (and controversial). The posterior layer usually fuses with the psoas or quadratus lumborum fascia. The perirenal spaces do not communicate across the abdominal midline. However, the renal and lateroconal fasciae are laminated structures that may be distended with fluid collections to form interfascial planes that do communicate across the midline and also inferiorly to the extraperitoneal pelvis.* **(Bottom)** *Sagittal section through the right kidney shows the renal fascia enveloping the kidney and adrenal gland. Inferiorly, the anterior and posterior renal fasciae come close together at about the level of the iliac crest. Note the adjacent peritoneal recesses.*

KIDNEYS IN SITU

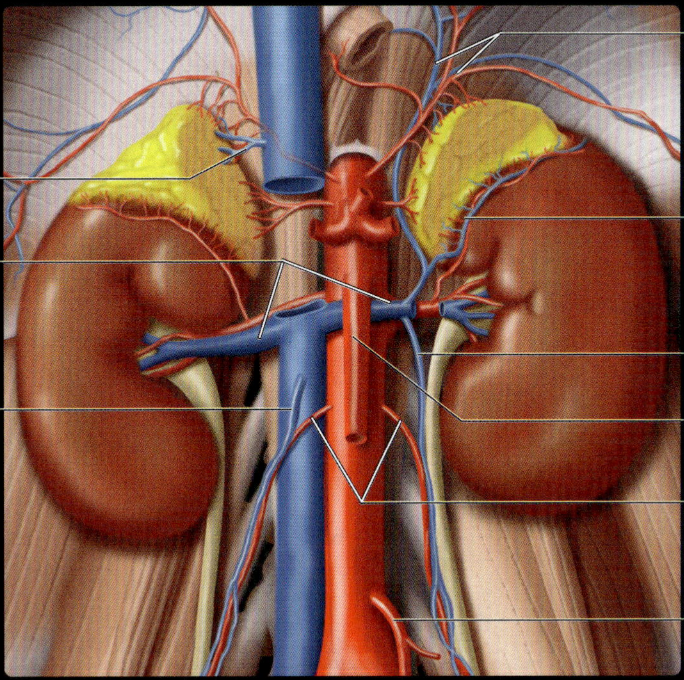

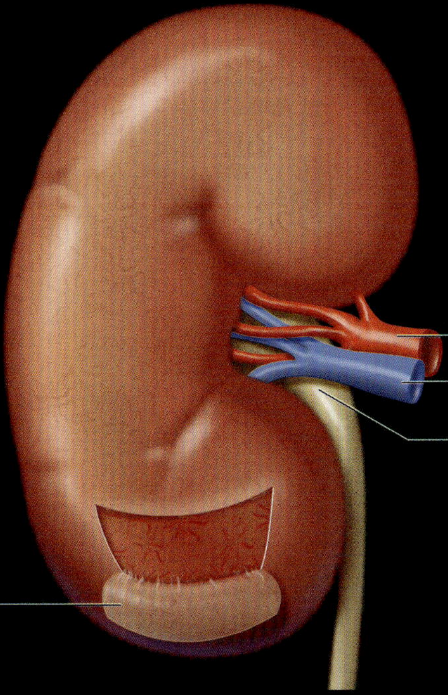

(Top) *The kidneys are retroperitoneal organs that lie lateral to the psoas, on the quadratus lumborum muscles. The oblique course of the psoas muscles results in the lower pole of the kidney lying lateral to the upper pole. The right kidney usually lies 1-2 cm lower than the left, due to inferior displacement by the liver. The adrenal glands lie above and medial to the kidneys, separated by a layer of fat and connective tissue. The peritoneum covers much of the anterior surface of the kidneys. The right kidney abuts the liver and the hepatic flexure of the colon and duodenum, while the left kidney is in close contact with the pancreas (tail), spleen, and splenic flexure.* **(Bottom)** *The fibrous capsule is stripped off with difficulty. Subcapsular hematomas do not spread far along the surface of the kidney, but compress the renal parenchyma, unlike most perirenal collections.*

RENAL ARTERY

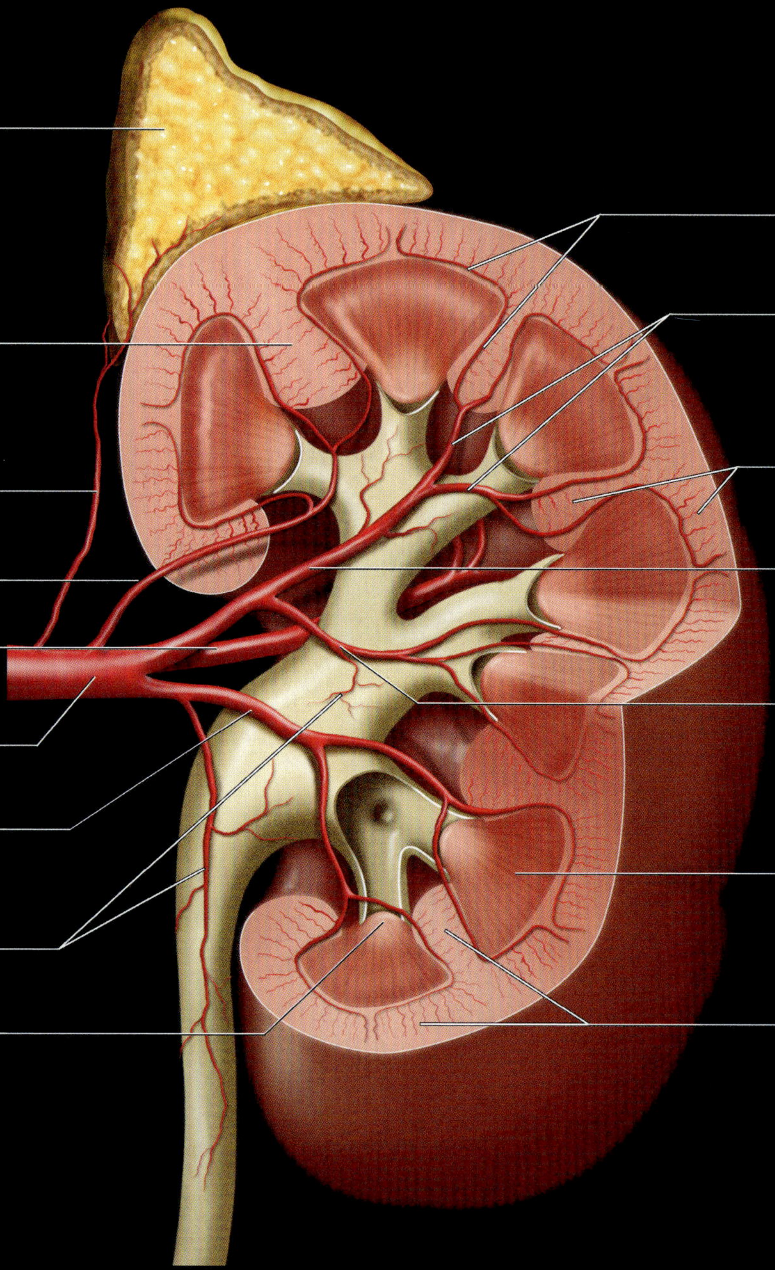

The kidney is usually supplied by a single renal artery, the 1st branch of which is the inferior adrenal artery. It then divides into 5 segmental arteries, only 1 of which (the posterior segmental artery) passes dorsal to the renal pelvis. The segmental arteries divide into the interlobar arteries that lie in the renal sinus fat. Each interlobar artery branches into 4-6 arcuate arteries that follow the convex outer margin of each renal pyramid. The arcuate arteries give rise to the interlobular arteries that lie within the renal cortex, including the cortical columns (of Bertin) that invaginate between the renal pyramids. The interlobular arteries supply the afferent arterioles to the glomeruli. The arterial supply to the kidney is vulnerable, as there are no effective anastomoses between the segmental branches, each of which supplies a wedge-shaped segment of parenchyma.

RENAL ARTERY VARIANTS

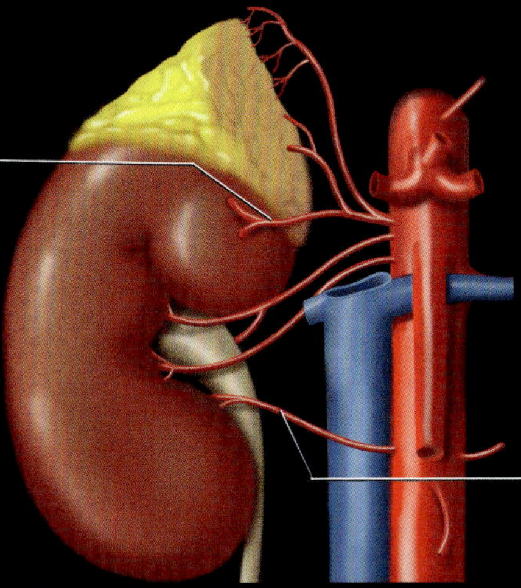

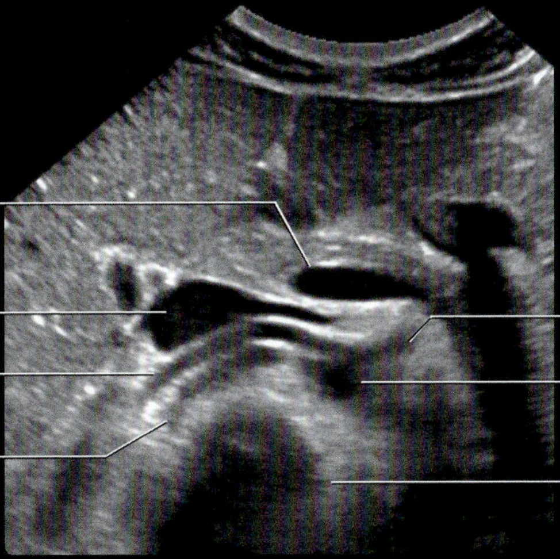

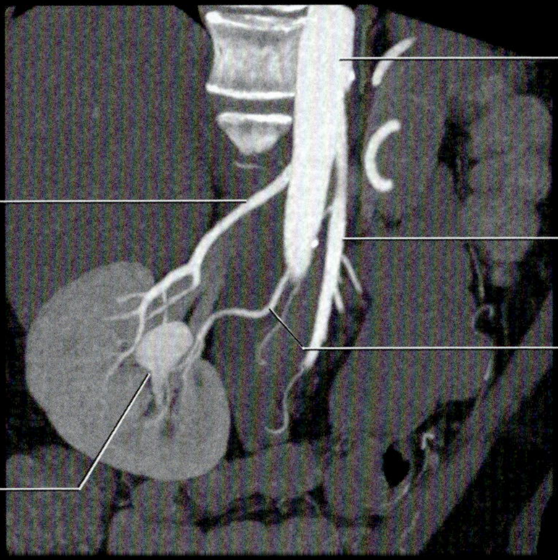

(Top) *Graphic depicts supernumerary renal arteries arising directly from the aorta. Some of these enter the kidney at locations other than the renal hilum, close to the renal poles. These "polar" or "extrahilar" arteries may be ligated or transected unintentionally during renal or other surgeries. These are sometimes referred to as "accessory" renal arteries, but each is an end artery and the sole arterial supply to a substantial portion of the renal parenchyma.* **(Middle)** *Transverse ultrasound shows 2 right renal arteries running posterior to the Inferior vena cava.* **(Bottom)** *Combined arterial/excretory phase CT in a potential renal donor shows 2 right renal arteries. Their origins are relatively far apart, which is a factor limiting the sensitivity of ultrasound for multiple renal arteries.*

RENAL VEIN VARIANTS

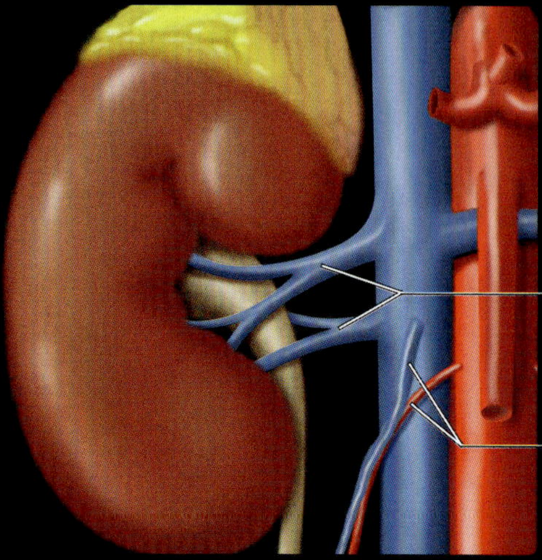

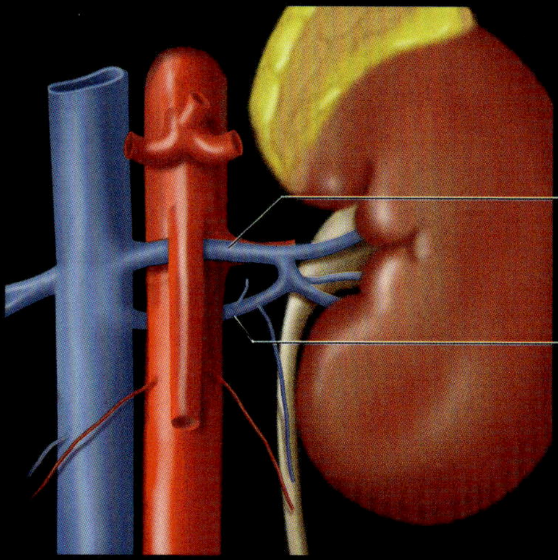

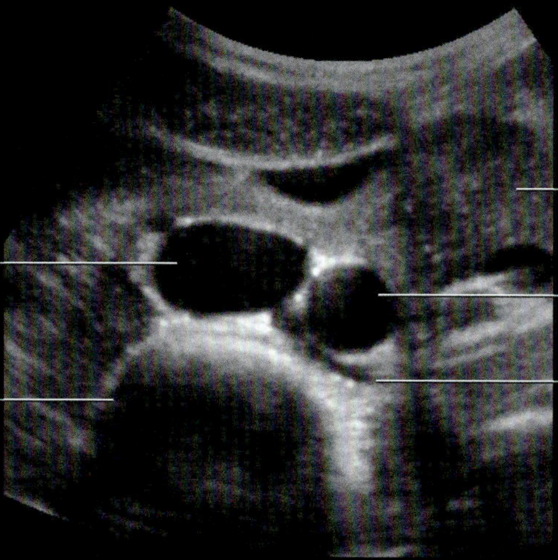

(Top) *Persistence of the collar of veins on the right results in supernumerary right renal veins that encircle the renal pelvis.* **(Middle)** *Anomalies of the renal veins are less common than those of the arteries but are encountered in clinical practice and may have important implications. All anomalies are variations of the embryologic development and persistence of portions of the paired longitudinal channels, the subcardinal and supracardinal veins, which form a ladder-like collar around the aorta. Normally, only the anterior components persist, becoming the renal veins, which course anterior to the aorta. Persistence of the whole collar results in a circumaortic renal vein, which is depicted in this graphic. This anomaly is more common than an isolated retroaortic renal vein.* **(Bottom)** *Transverse oblique ultrasound shows an incidental retroaortic left renal vein.*

RENAL VEIN VARIANTS

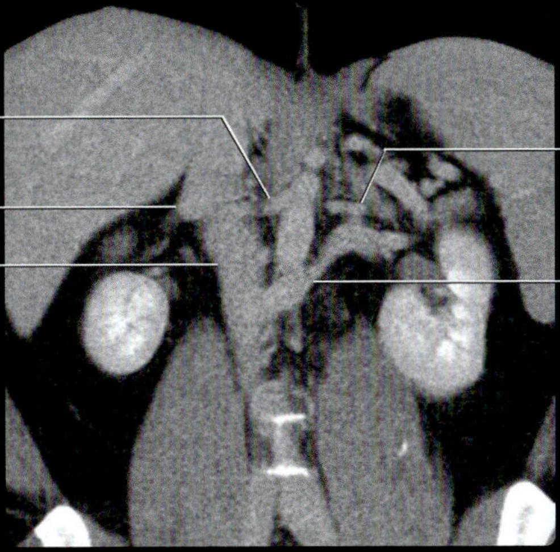

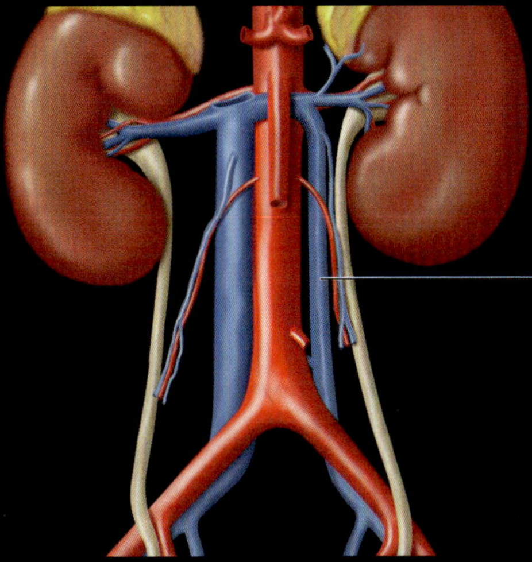

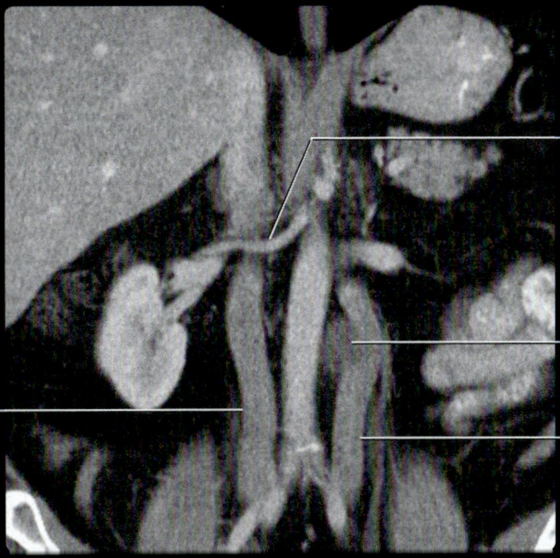

(Top) *The left renal vein is retrocaval.* **(Middle)** *Persistence of the left supracardinal vein below the kidney results in a "duplicated" inferior vena cava.* **(Bottom)** *Incidental finding of a duplicated vena cava is shown in a patient with mixed germ cell tumor and retroperitoneal adenopathy.*

CT UROGRAM

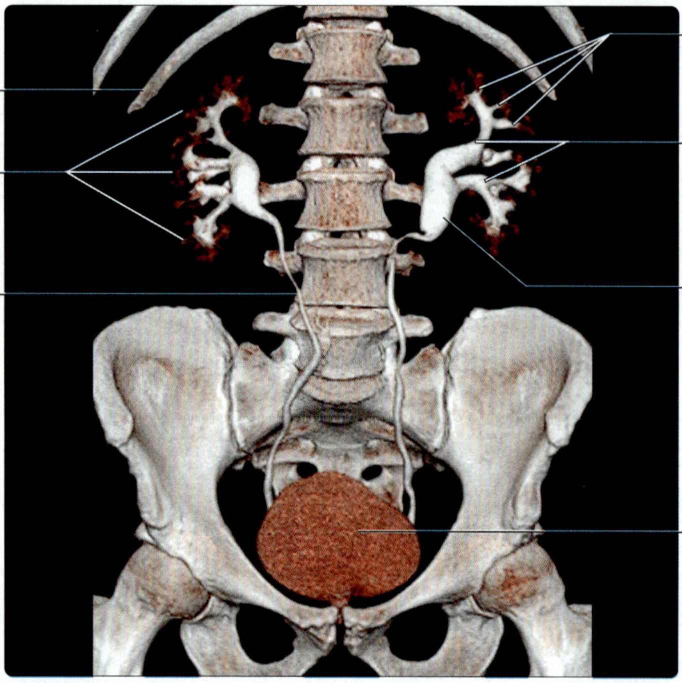

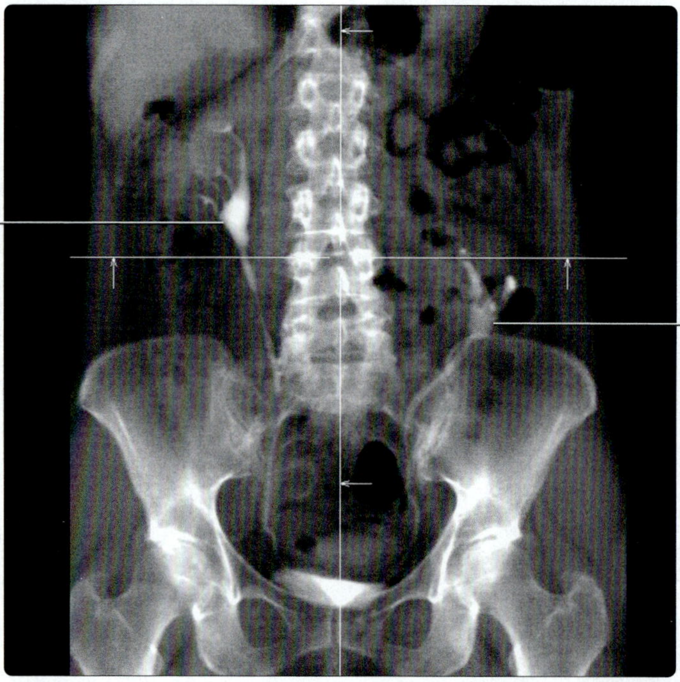

(Top) *Volumetric multidetector CT can be viewed as a surface-rendered 3D image to simulate an excretory urogram. Postprocessing includes multiplanar reformation with window leveling, bone subtraction, and arbitrary color maps. Here, opacified urine is displayed as white. Less dense urine within the renal tubules in the pyramids and the diluted urine within the bladder are displayed as red. The CT scan was obtained in suspended inspiration, resulting in caudal displacement of the kidneys. In the supine position at quiet breathing, the upper poles of the kidneys usually lie in front of the 12th ribs.* **(Bottom)** *Coronal maximum-intensity projection obtained during a CT urogram shows a congenital low-lying left kidney. Both kidneys are malrotated with the renal pelves directed anteriorly.*

RIGHT KIDNEY, ANTERIOR ABDOMEN SCAN

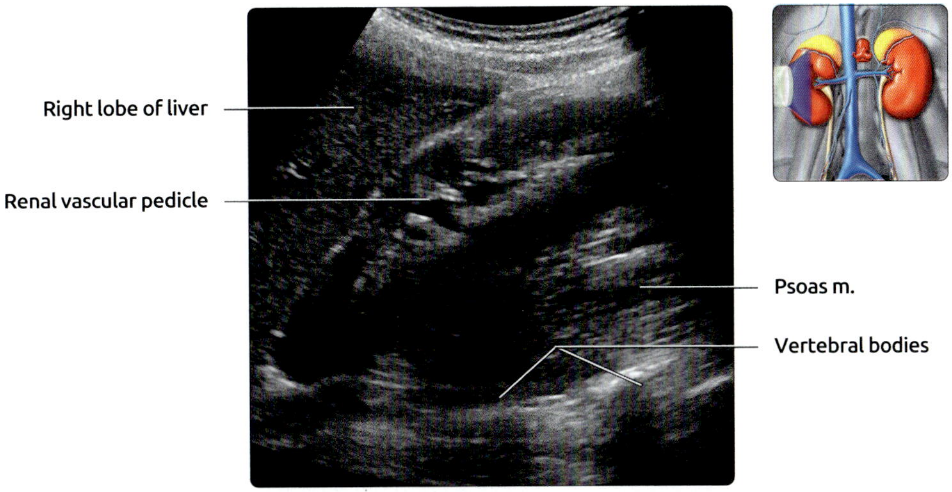

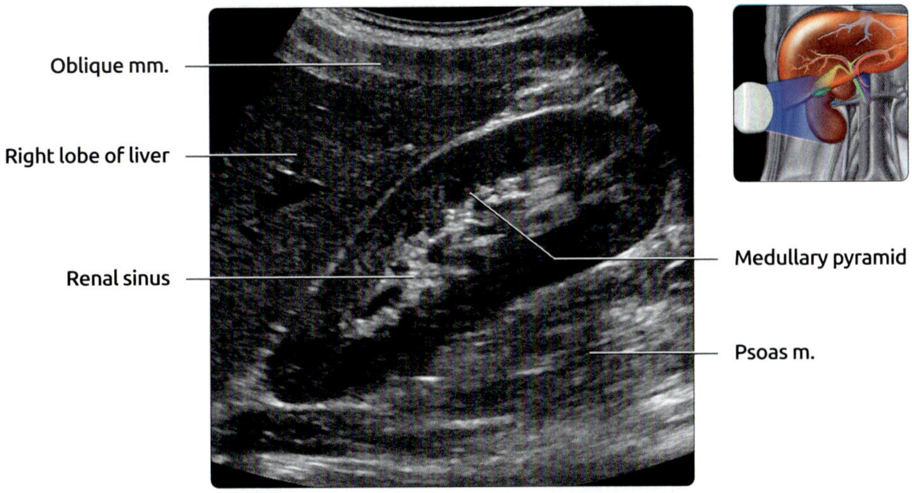

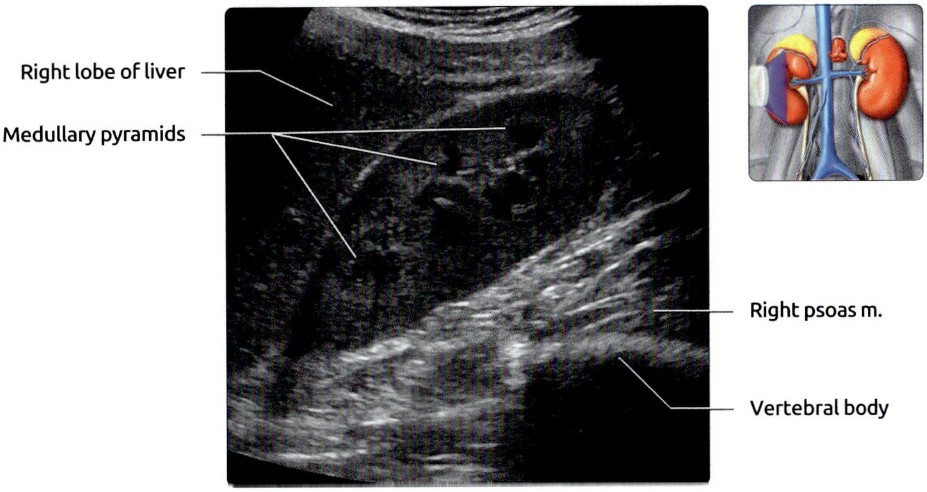

(Top) *Longitudinal oblique grayscale ultrasound shows the right kidney using the liver as an acoustic window with the probe angled medially toward the renal hilum and vascular pedicle.* **(Middle)** *Longitudinal grayscale ultrasound shows the right kidney using the liver as an acoustic window. This approach usually provides excellent visualization of the right kidney and is useful for measuring bipolar renal length.* **(Bottom)** *Longitudinal oblique grayscale ultrasound of the right kidney using the liver as an acoustic window, obtained with more lateral angulation (when compared with the previous 2 images), cuts through the renal parenchyma on the lateral aspect of the right kidney. Note that the echogenic sinus is not demonstrated.*

RIGHT KIDNEY, CT CORRELATION

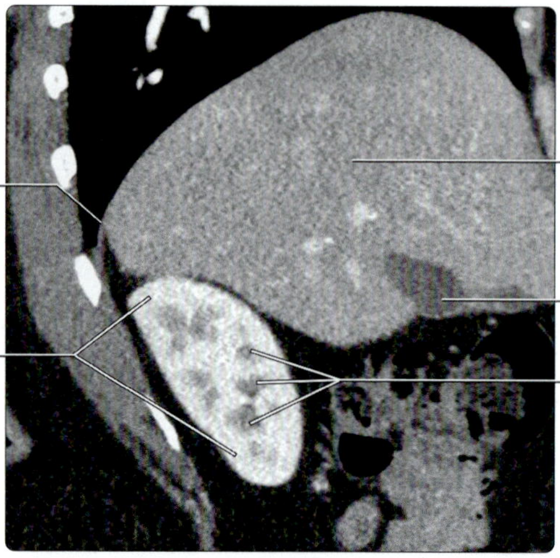

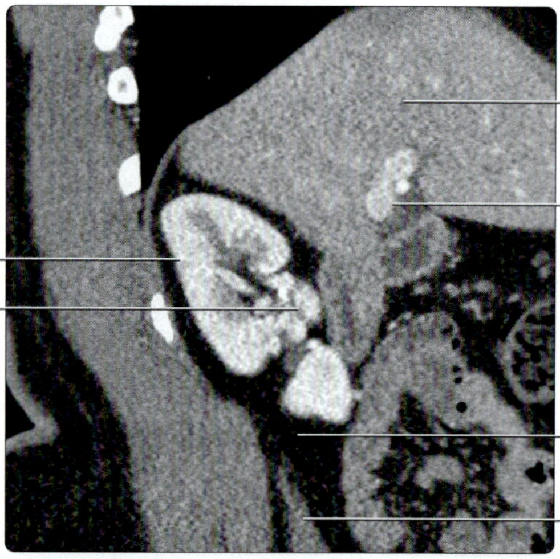

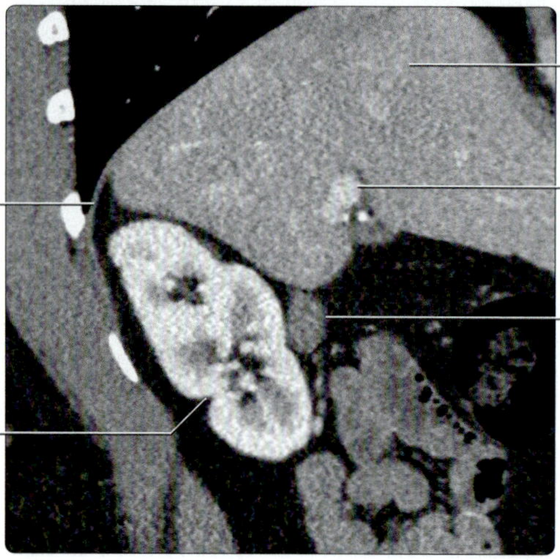

(Top) *Correlative longitudinal CT multiplanar reconstruction of the right kidney is shown. Real-time ultrasound can be performed in any plane; thin-slice CT is acquired in the axial plane and then reconstructed in other planes. These images have been reconstructed to mirror the planes used in ultrasound. CT is typically performed with intravenous contrast and acquired at different time points to image the renal vessels, cortex, medulla, and collecting system.* **(Middle)** *Correlative longitudinal oblique CT multiplanar reconstruction of the right kidney cutting through the right renal vein is shown. The plane of this image is angulated more medially when compared with the previous image.* **(Bottom)** *Correlative longitudinal oblique CT multiplanar reconstruction of the right kidney through the right portal vein is shown. This plane is angled more laterally when compared with the previous 2 images.*

RIGHT KIDNEY, ANTERIOR ABDOMEN SCAN

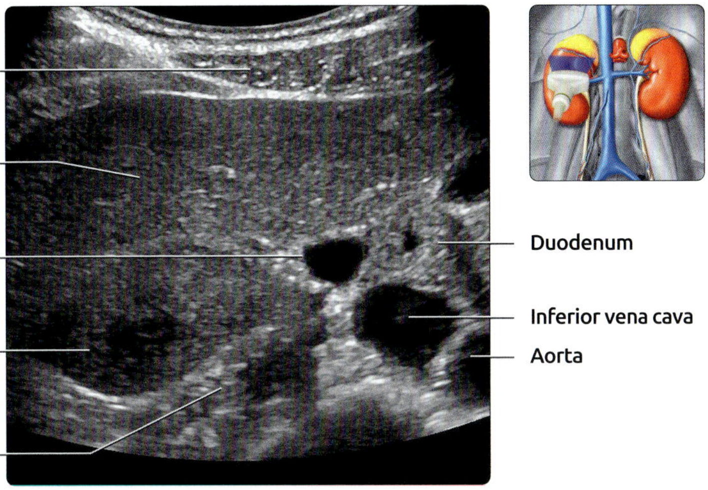

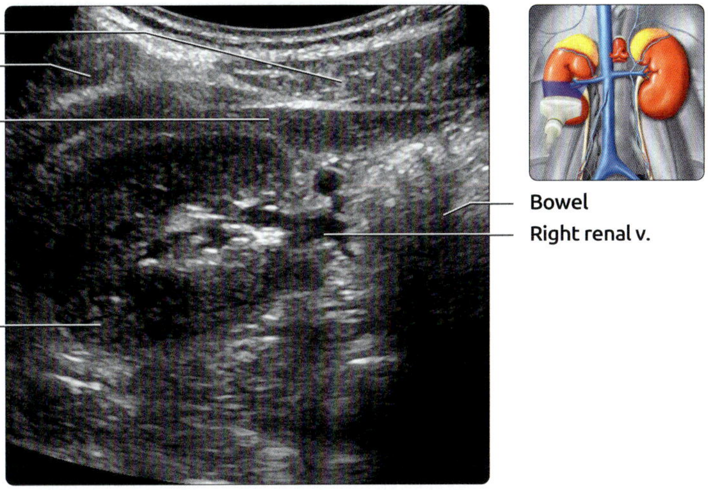

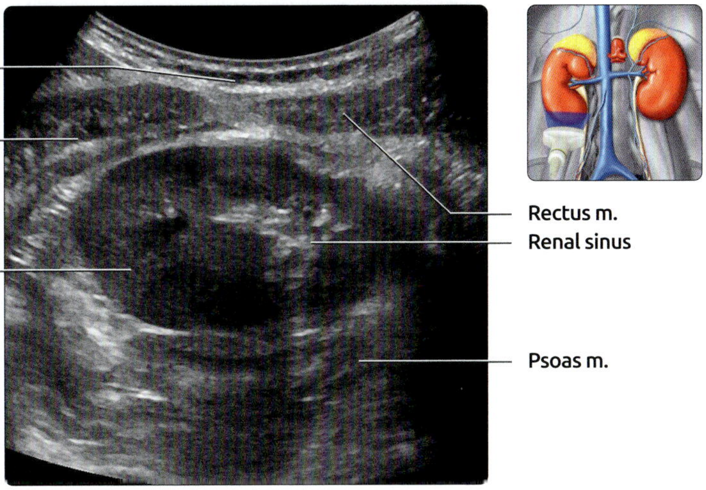

(Top) *Transverse grayscale ultrasound of the upper pole of the right kidney is shown.* **(Middle)** *Transverse grayscale ultrasound of the mid pole of the right kidney shows the renal hilum with the renal vein. Note that the pelvicalyceal system within the renal sinus echoes is not usually visible unless dilated.* **(Bottom)** *Transverse grayscale ultrasound of the lower pole of the right kidney is shown. The renal parenchymal echogenicity is less than the adjacent liver or spleen. If the renal parenchyma is brighter than normal liver, renal parenchymal disease should be suspected.*

RIGHT KIDNEY, CT CORRELATION

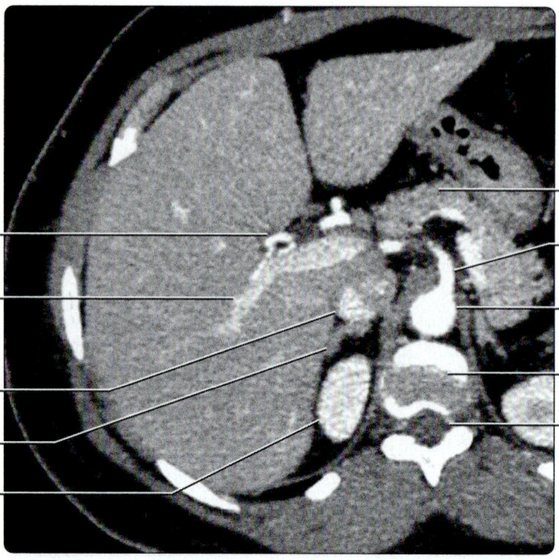

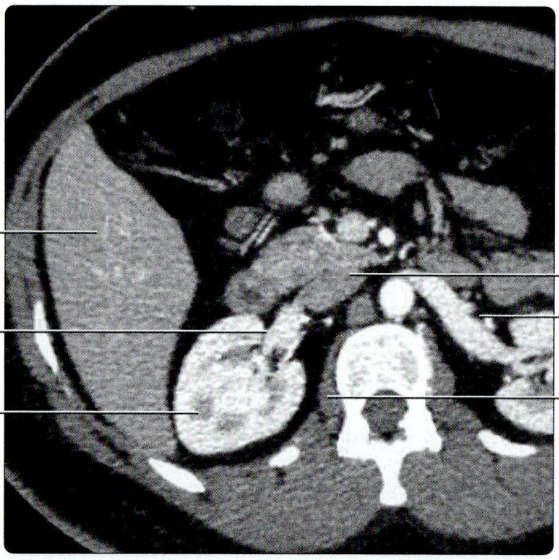

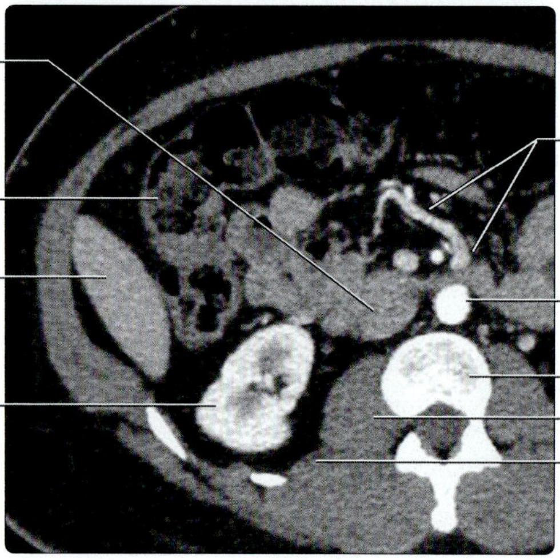

(Top) *1st in a series of 3 correlative transverse CT images shows the right kidney from the upper pole to the lower pole. These are planes commonly used when performing ultrasound. This image shows the upper pole of the right kidney.* **(Middle)** *Transverse CT at the level of the mid pole of the right kidney shows the renal hilum with the renal vein. The right and left renal veins are opacified, but the vena cava is not yet opacified.* **(Bottom)** *Correlative transverse CT shows the lower pole of the right kidney. The psoas and quadratus lumborum are variable in thickness depending on patient gender and level of activity.*

RIGHT KIDNEY, POSTERIOR ABDOMEN SCAN

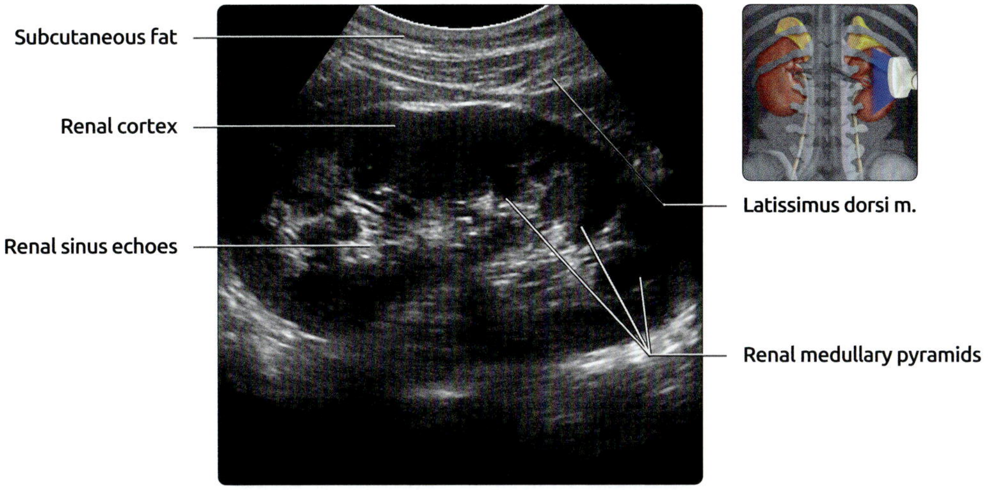

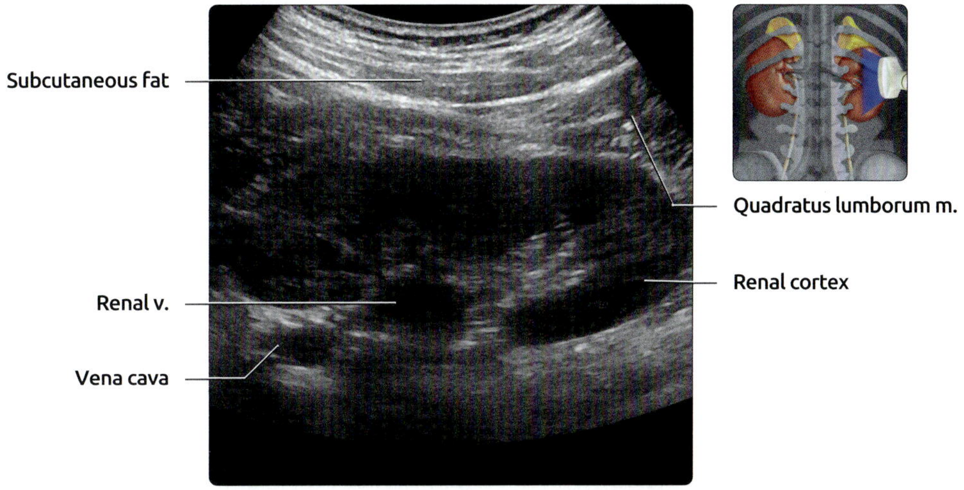

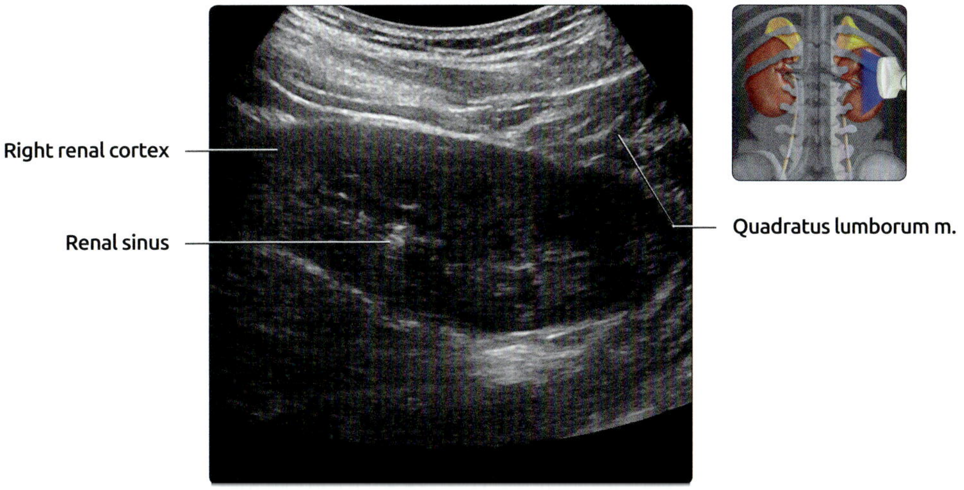

(Top) *Longitudinal grayscale ultrasound of the right kidney, scanning from the posterior approach, shows normal medullary pyramids. This approach is useful for intervention, such as renal biopsy or percutaneous nephrostomy, as the kidney is just deep to skin and muscle. This approach is also a good way for standardizing renal length measurements in children.* **(Middle)** *Oblique scan through the longitudinal axis shows the right kidney with the patient prone. This approach may be problematic due to the presence of ribs. Deep inspiration may help to bring the kidney down from below the ribs.* **(Bottom)** *Longitudinal view shows the lateral right kidney from a posterior approach.*

RIGHT KIDNEY, POSTERIOR ABDOMEN SCAN

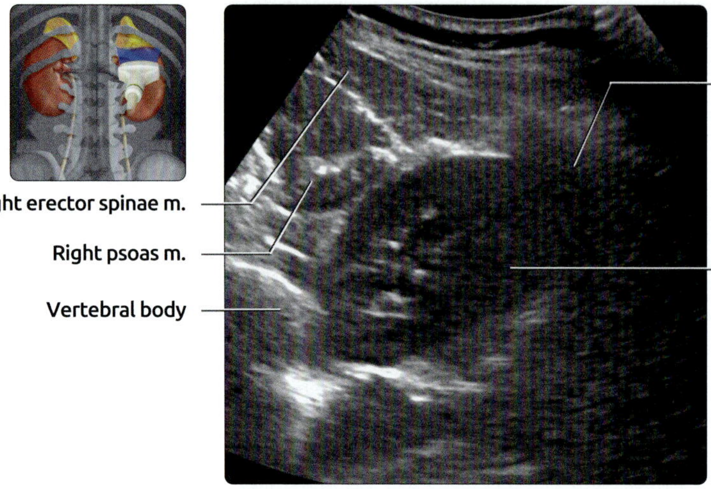

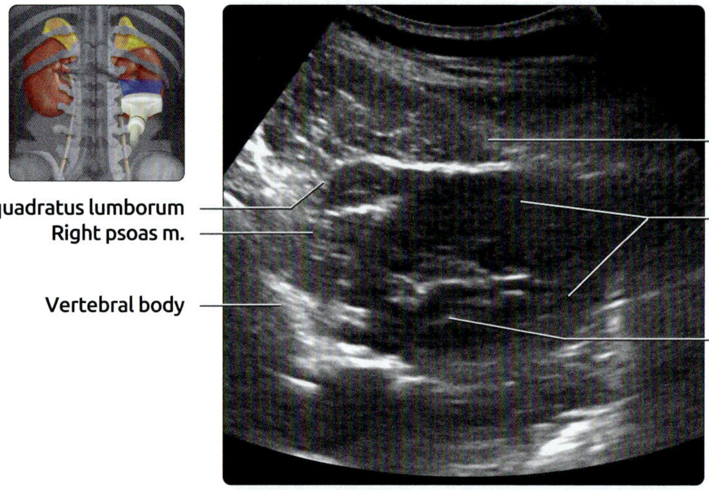

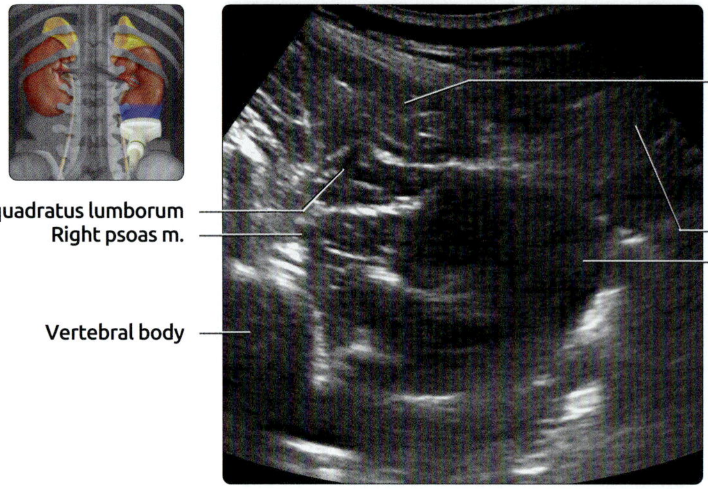

(Top) *Transverse grayscale ultrasound of the right kidney, scanning from the posterior approach, is shown. Scanning through the posterior approach is useful while performing interventional procedures, such as nephrostomy or renal biopsy. However, visualization/image quality may be impaired by thick paraspinal muscles and rib shadowing. This image shows the upper pole of the right kidney.* **(Middle)** *Transverse grayscale ultrasound from the posterior approach shows the mid pole of the right kidney.* **(Bottom)** *Transverse grayscale ultrasound from the posterior approach shows the lower pole of the right kidney.*

RIGHT MAIN RENAL ARTERY AND VEIN

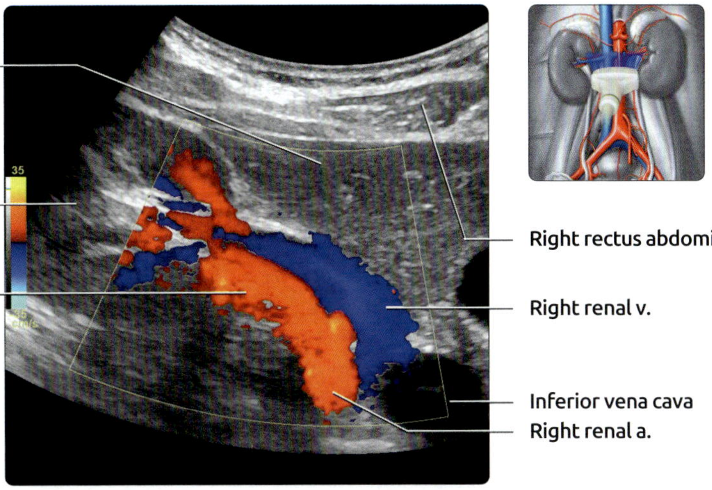

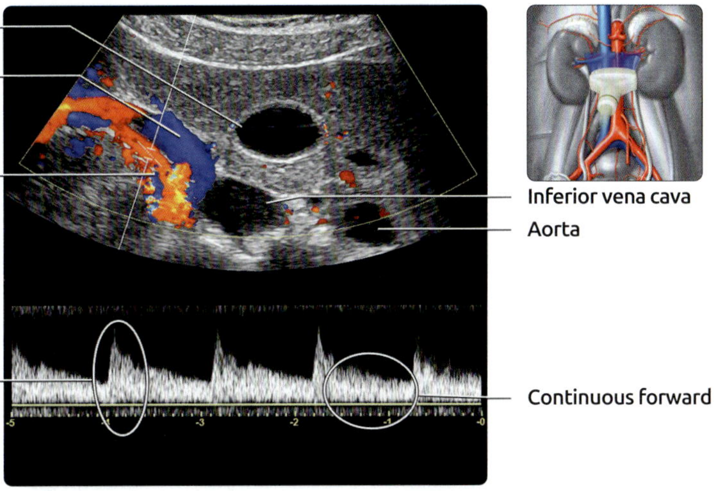

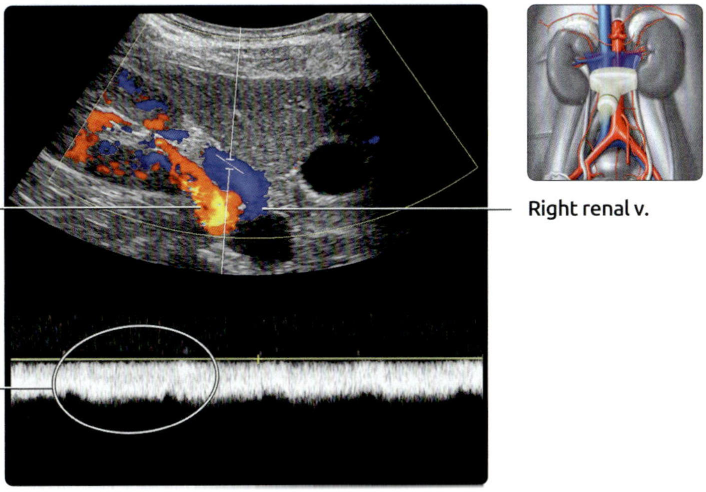

(Top) *Transverse color Doppler ultrasound shows the right renal hilum. Note that the right renal artery lies posterior to the renal vein and inferior vena cava. The renal artery normally measures 5-8 mm in caliber.* **(Middle)** *In this spectral Doppler waveform of the right renal artery, note the low-resistance renal waveform with continuous forward systolic and diastolic flow. Normal PSV ranges from 60-140 cm/sec; not > 180 cm/sec. Normal resistivity index is < 0.7 and normal pulsatility index is < 1.8. Note the artifactual absence of color signal in the aorta and vena cava secondary to Doppler angle.* **(Bottom)** *Spectral Doppler waveform of the right renal vein is shown, which usually mirrors the pulsatility in the inferior vena cava. The renal vein normally measures 4-9 mm in caliber. Normal PSV ranges from 18-33 cm/sec.*

RIGHT INTRARENAL ARTERY AND VEIN

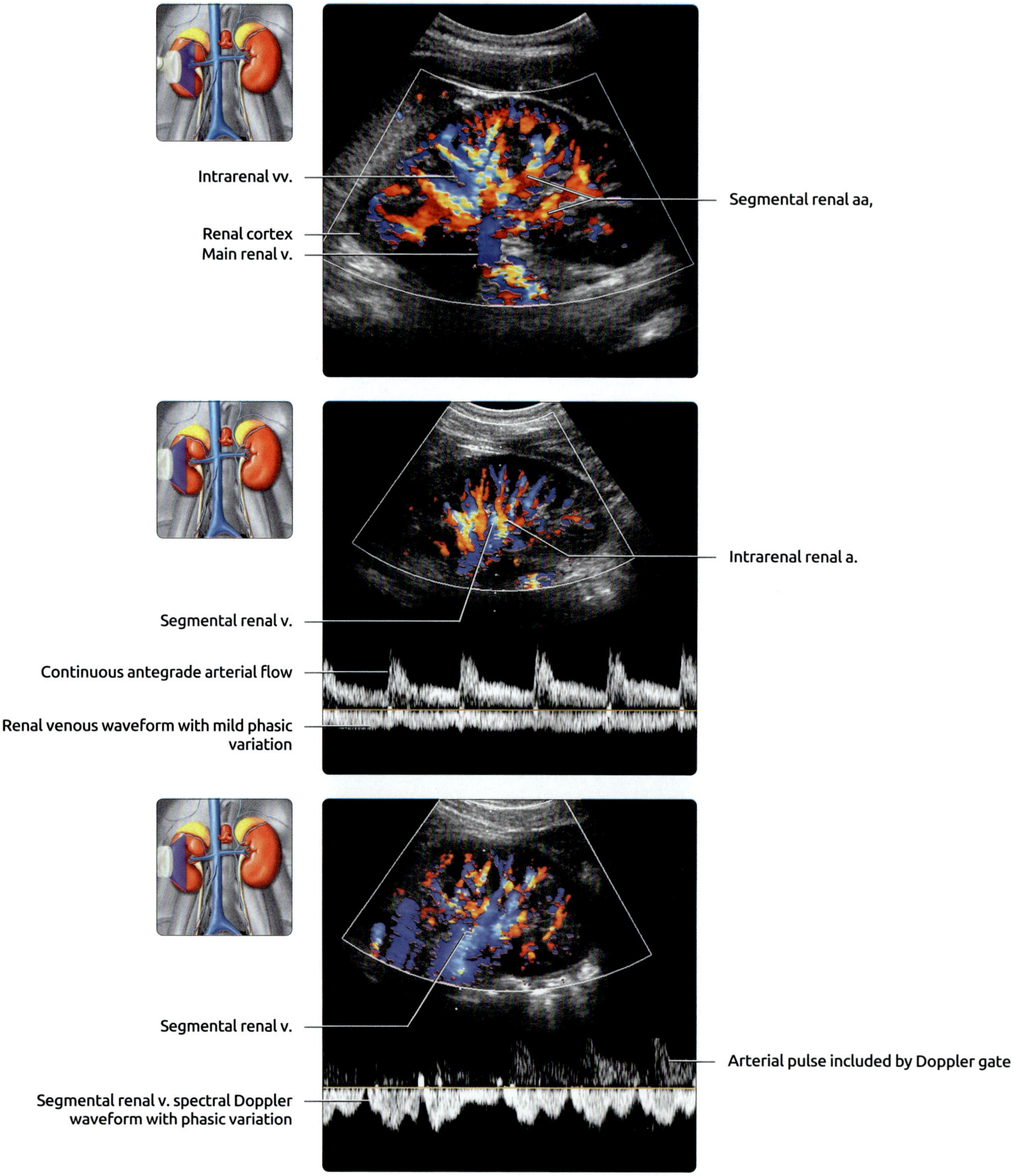

(Top) *Longitudinal color Doppler ultrasound of the right kidney shows renal artery branches as red and renal vein branches as blue. However, as color Doppler colors can be changed by the operator, it is important to look at the color scale bar (not included).* **(Middle)** *In this spectral Doppler waveform of a right segmental renal artery branch, note the low-resistance arterial Doppler waveform with continuous diastolic flow, similar to that seen in the more proximal renal artery. The venous waveform shows minimal phasicity.* **(Bottom)** *In this spectral Doppler waveform of a right segmental renal vein branch, note the phasic variation in the renal vein, which can vary depending on systemic venous pressure and cardiac and fluid status.*

LEFT KIDNEY, POSTERIOR SCAN

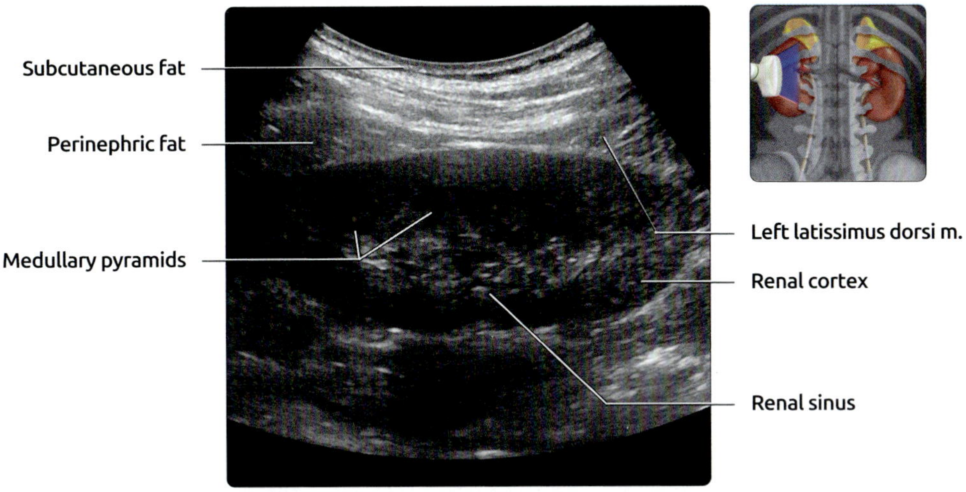

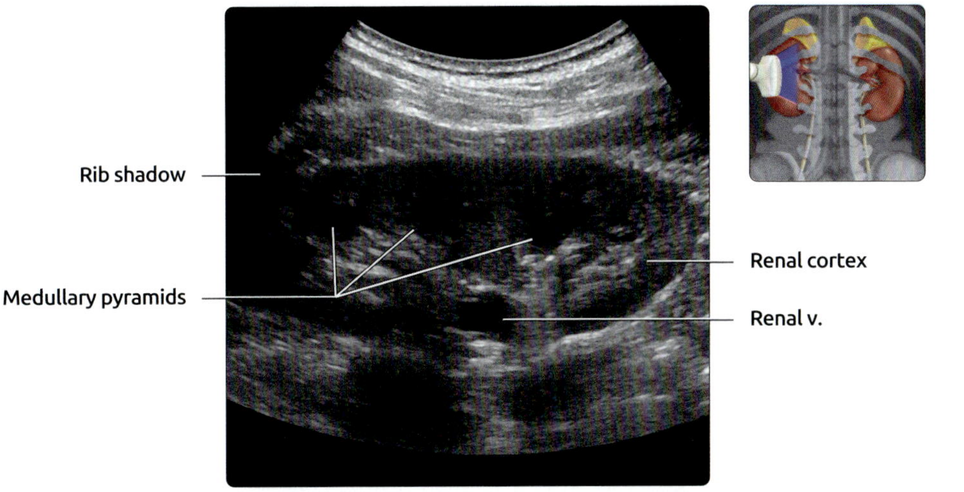

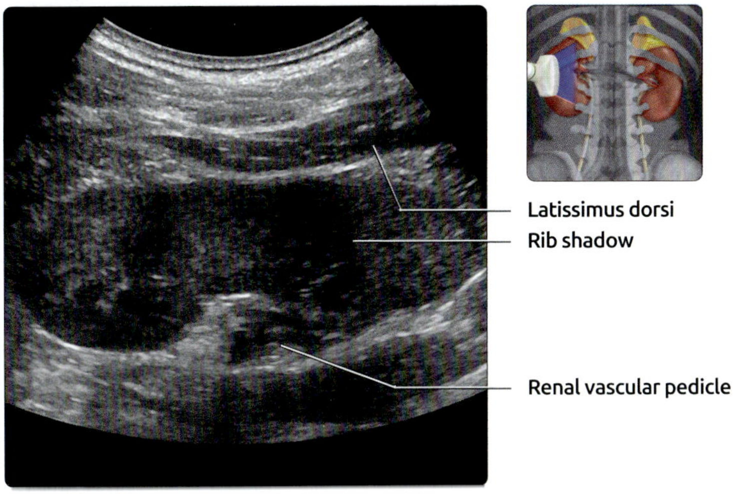

(Top) *Longitudinal grayscale ultrasound of the left kidney, scanning from the posterolateral approach, is shown. This approach avoids interference from bowel gas shadowing. However, the ribs and lung base may degrade the image. Deep inspiration can decrease artifact from ribs and lung base. Note the regular distribution of the renal medullary pyramids.* **(Middle)** *Longitudinal grayscale ultrasound shows the left kidney, scanning from the posterolateral approach, with the transducer angling more posteriorly when compared with the previous image. Note shadowing from rib degrading imaging of the upper pole.* **(Bottom)** *Longitudinal grayscale ultrasound shows the left kidney, scanning from the posterolateral approach, with the transducer angling more anteriorly when compared with the previous 2 images. The renal sinus is no longer in the field of view.*

LEFT KIDNEY, CT CORRELATION

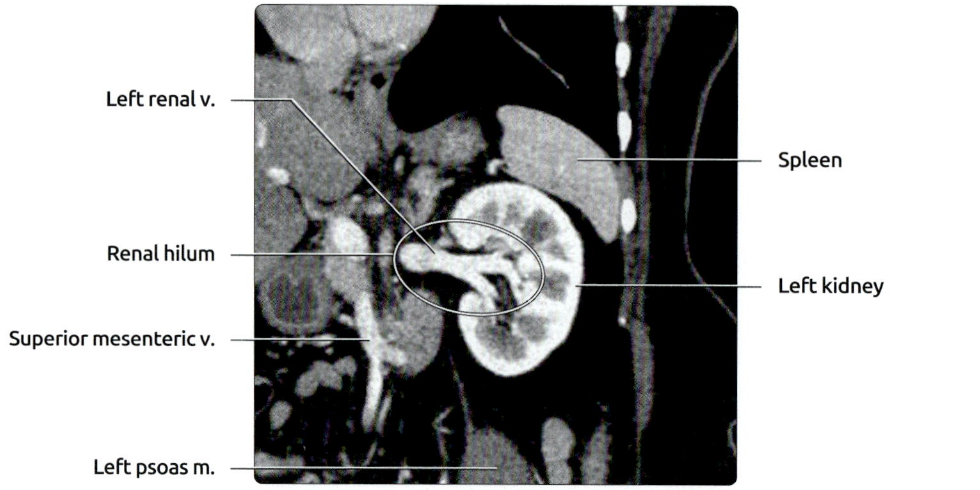

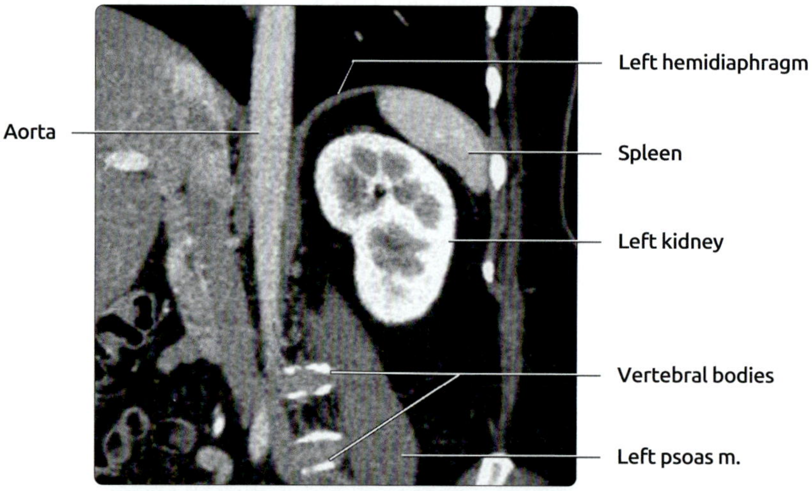

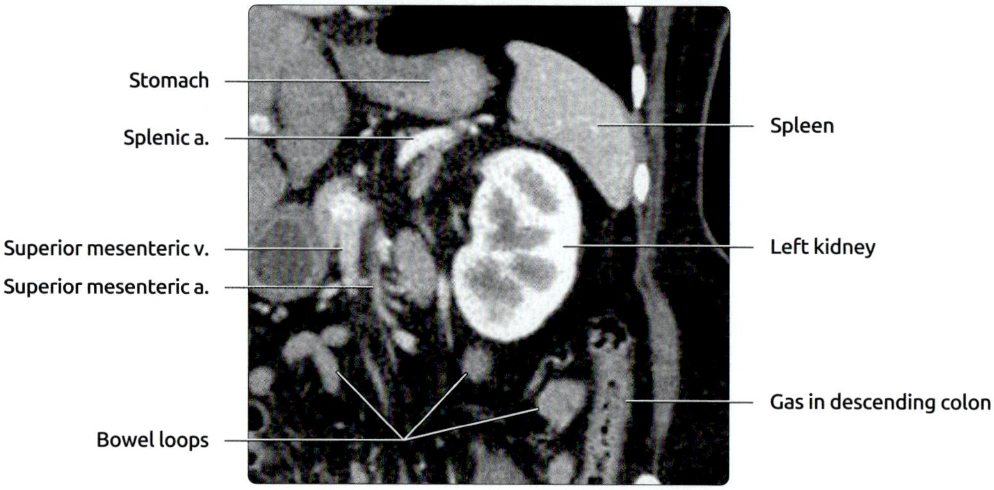

(Top) *The 1st in a series of 3 correlative longitudinal oblique multiplanar reconstruction CT images of the left kidney, through planes commonly used when examining the patient with ultrasound, shows clear visualization of the kidney, renal pedicle, and surrounding structures. This makes multidetector row CT the imaging modality of choice (over ultrasound) for evaluation of renal vascular anatomy and pathology.* **(Middle)** *Correlative longitudinal oblique multiplanar reconstruction CT shows the left kidney in a plane more posterior to the previous image.* **(Bottom)** *Correlative longitudinal oblique multiplanar reconstruction CT shows the left kidney in a plane more anterior than the previous 2 images. Note that ultrasound has a much more limited field of view compared to CT using this approach.*

LEFT KIDNEY, POSTERIOR TO POSTEROLATERAL ABDOMEN SCAN

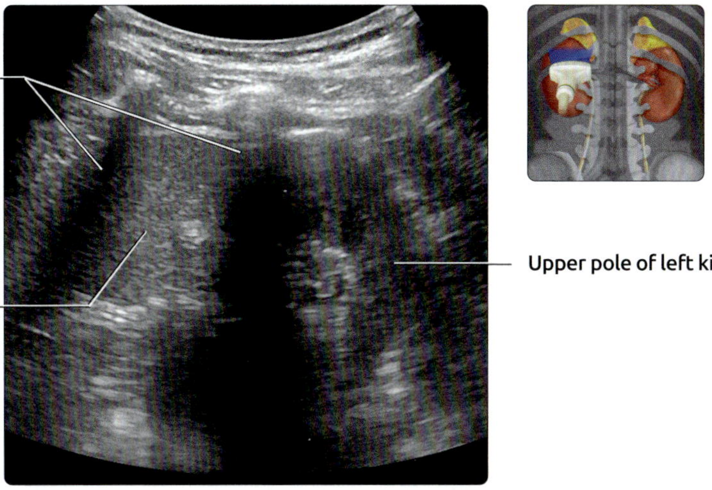

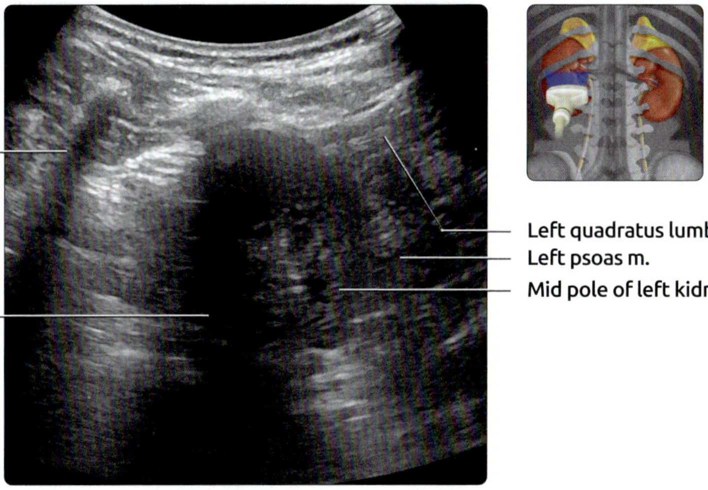

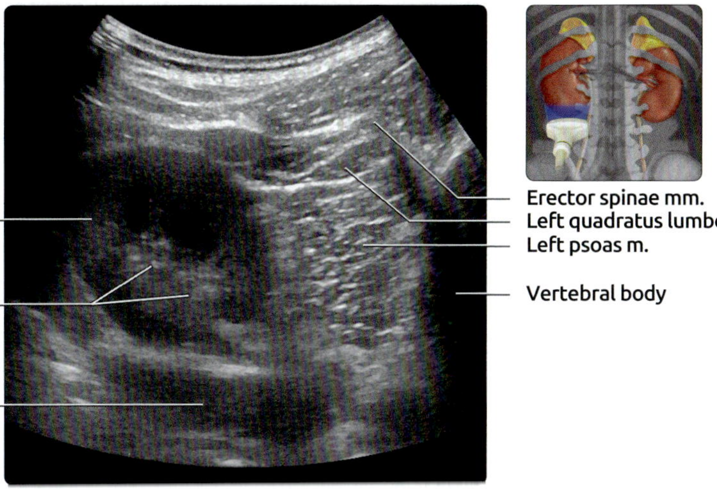

(Top) *Transverse ultrasound of the left upper kidney shows that ribs often interfere with good visualization of the upper kidney regardless of the approach.* **(Middle)** *Transverse grayscale ultrasound of the mid pole of the left kidney, using the posterior lateral approach, shows persistent degradation by rib shadows. Deep inspiration will move the kidney away from the lower ribs and improve image quality.* **(Bottom)** *Transverse ultrasound of the lower pole of the left kidney, using the posterior approach with the patient prone, decreases interference from ribs. There is a suitable window for nontargeted renal biopsy.*

LEFT KIDNEY, ANTERIOR TO ANTEROLATERAL TRANSVERSE SCAN

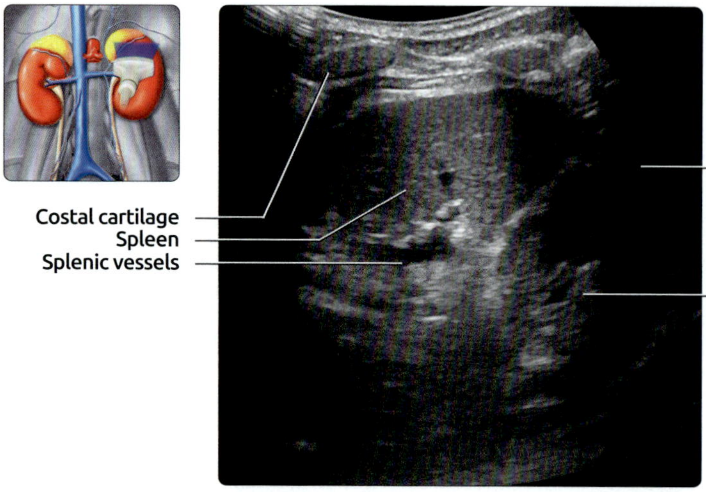

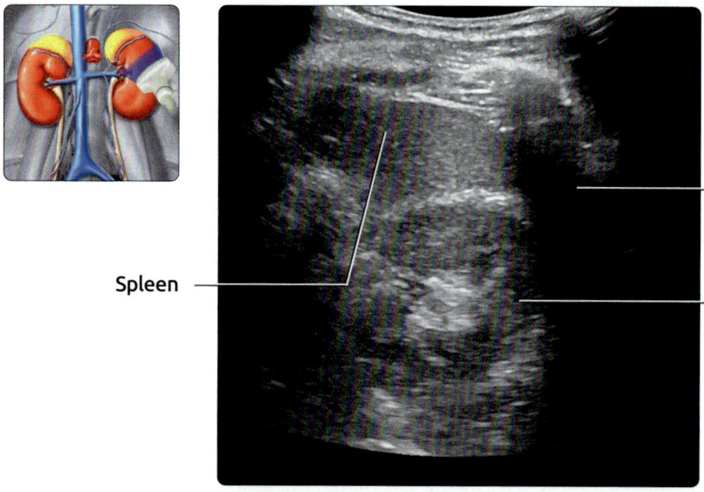

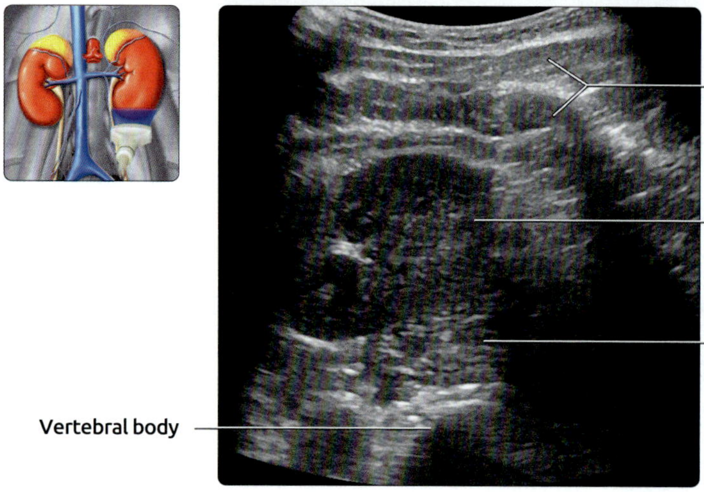

(Top) *Transverse grayscale ultrasound shows the upper pole of the left kidney using the anterolateral approach. Note the proximity of the kidney to the spleen and the interfering rib shadows.* **(Middle)** *Transverse grayscale ultrasound shows the mid pole of the left kidney using the posterolateral approach.* **(Bottom)** *Transverse grayscale ultrasound shows the lower pole of the left kidney using the posterolateral approach.*

LEFT KIDNEY, ANTEROLATERAL APPROACH

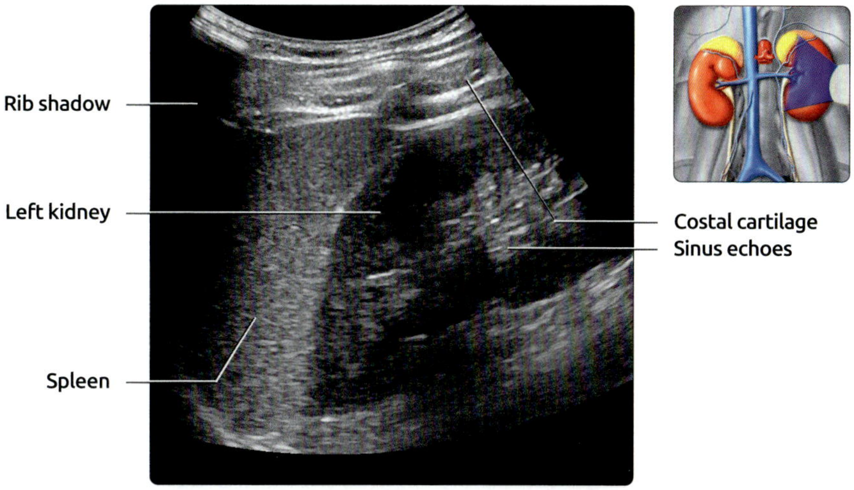

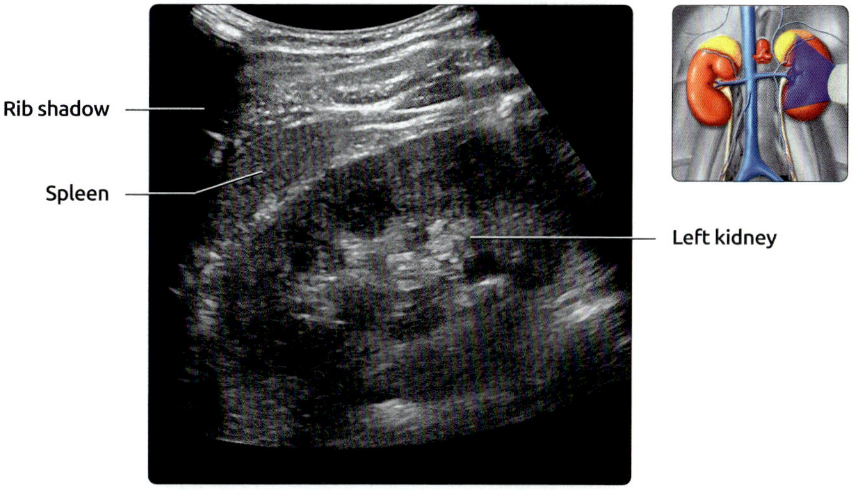

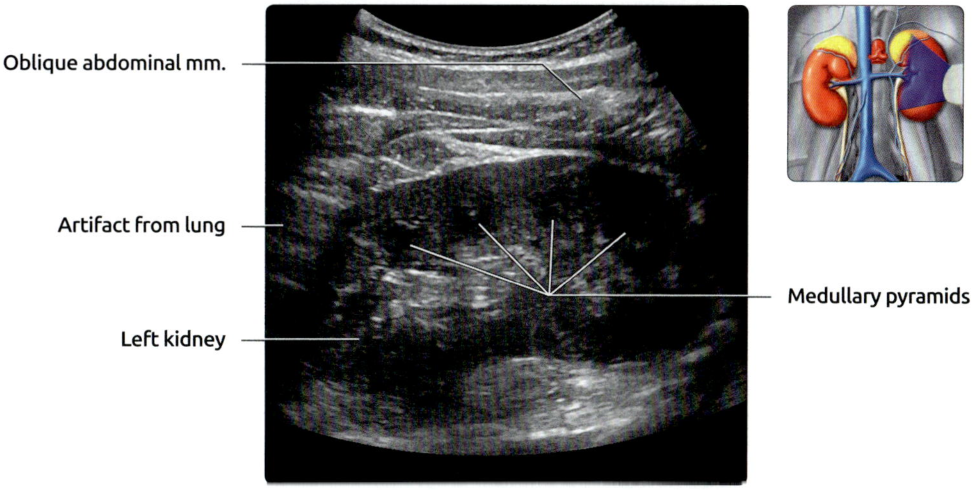

(Top) *Longitudinal grayscale ultrasound shows the upper pole of the left kidney using the anterolateral approach. Note that the spleen may occasionally be used as an acoustic window for the left kidney.* **(Middle)** *Longitudinal ultrasound shows the left kidney using the anterolateral approach.* **(Bottom)** *Longitudinal grayscale ultrasound shows the left kidney using the anterolateral approach and angling posteriorly. The lung base causes reverberation artifact, which partially obscures the upper pole.*

LEFT KIDNEY, CT CORRELATION

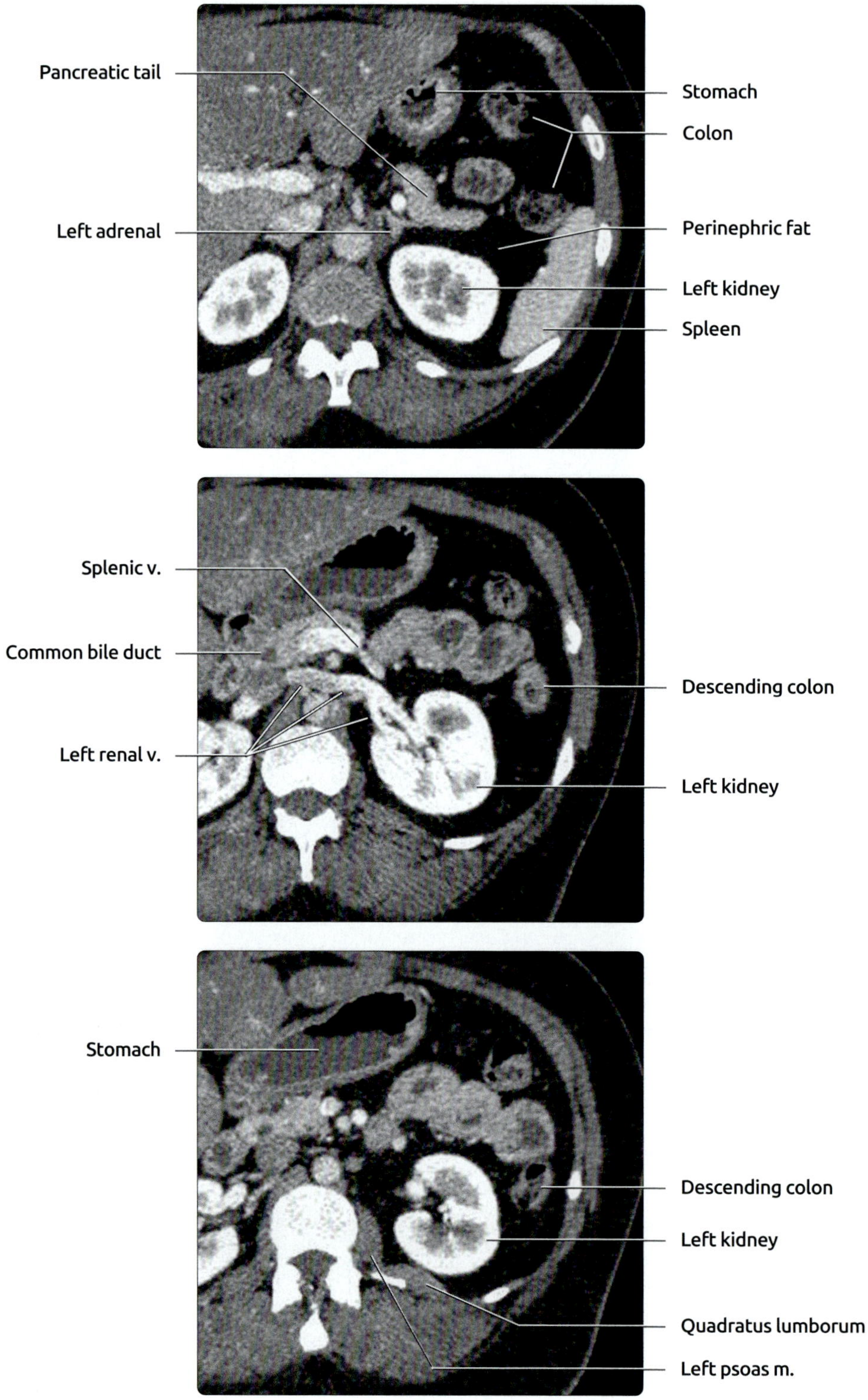

(Top) *The 1st in a series of 3 correlative transverse CT images of the left kidney, through planes commonly used when examining the kidney, is shown using the anterior approach. Note the relationship of the spleen to the upper pole of the left kidney, allowing it to be used as an acoustic window, particularly in patients with splenomegaly. This transverse CT shows the upper pole of the left kidney.* **(Middle)** *Correlative transverse CT shows the mid pole of the left kidney at the level of the left renal vein.* **(Bottom)** *Correlative transverse CT shows the lower pole of the left kidney.*

LEFT MAIN RENAL ARTERY AND VEIN

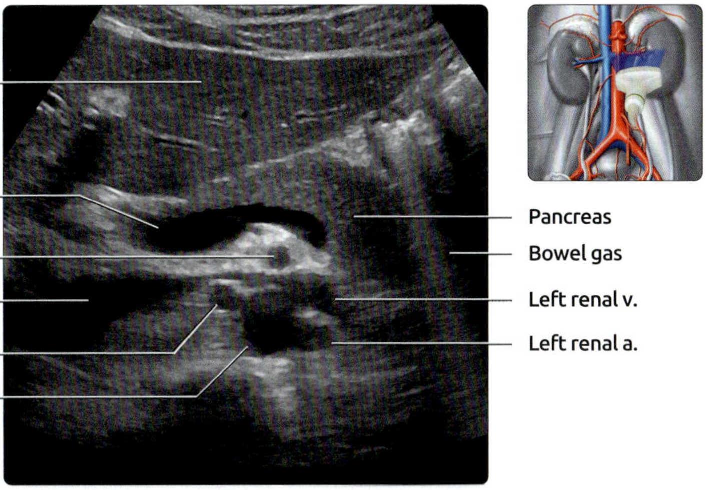

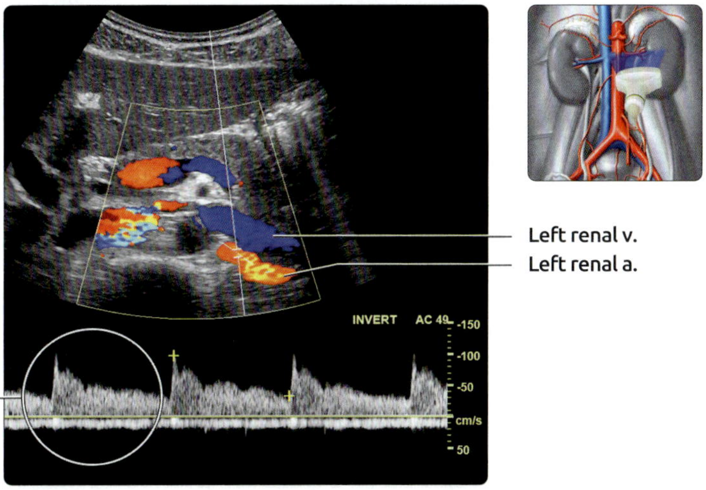

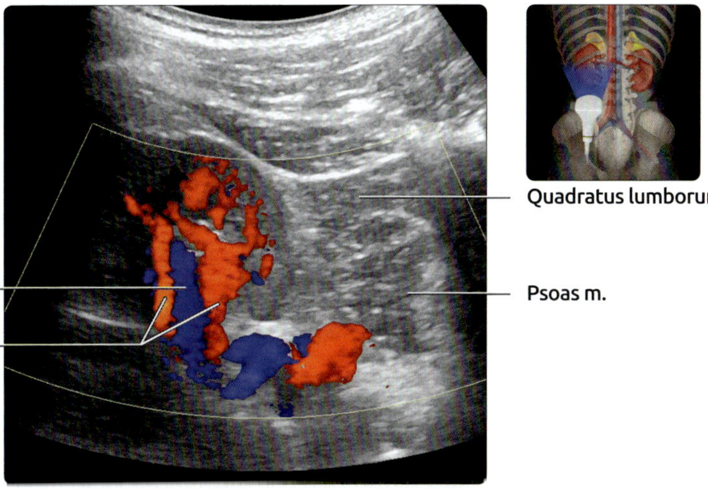

(Top) *Transverse ultrasound in the anterior midline shows that the left renal artery arises from the anterolateral aorta just around or below the level of the superior mesenteric artery. The normal caliber of the renal artery ranges from 5-8 mm. The left renal vein courses between the aorta and superior mesenteric artery.* **(Middle)** *Spectral Doppler waveform of the left renal artery is shown. The normal PSV in the artery ranges from 60-140 cm/sec with a normal resistive index < 0.7 and pulsatility index < 1.8. There is variability in venous velocity consequent upon cardiac and respiratory activity. The renal vein normally measures 4-9 mm in caliber with a PSV of 18-33 cm/sec.* **(Bottom)** *Transverse color Doppler waveform of the left renal hilum obtained from a posterior approach is shown. An anterior approach may not be feasible secondary to bowel gas. There are 2 left renal arteries anterior and posterior to the left renal vein.*

LEFT INTRARENAL ARTERY AND VEIN

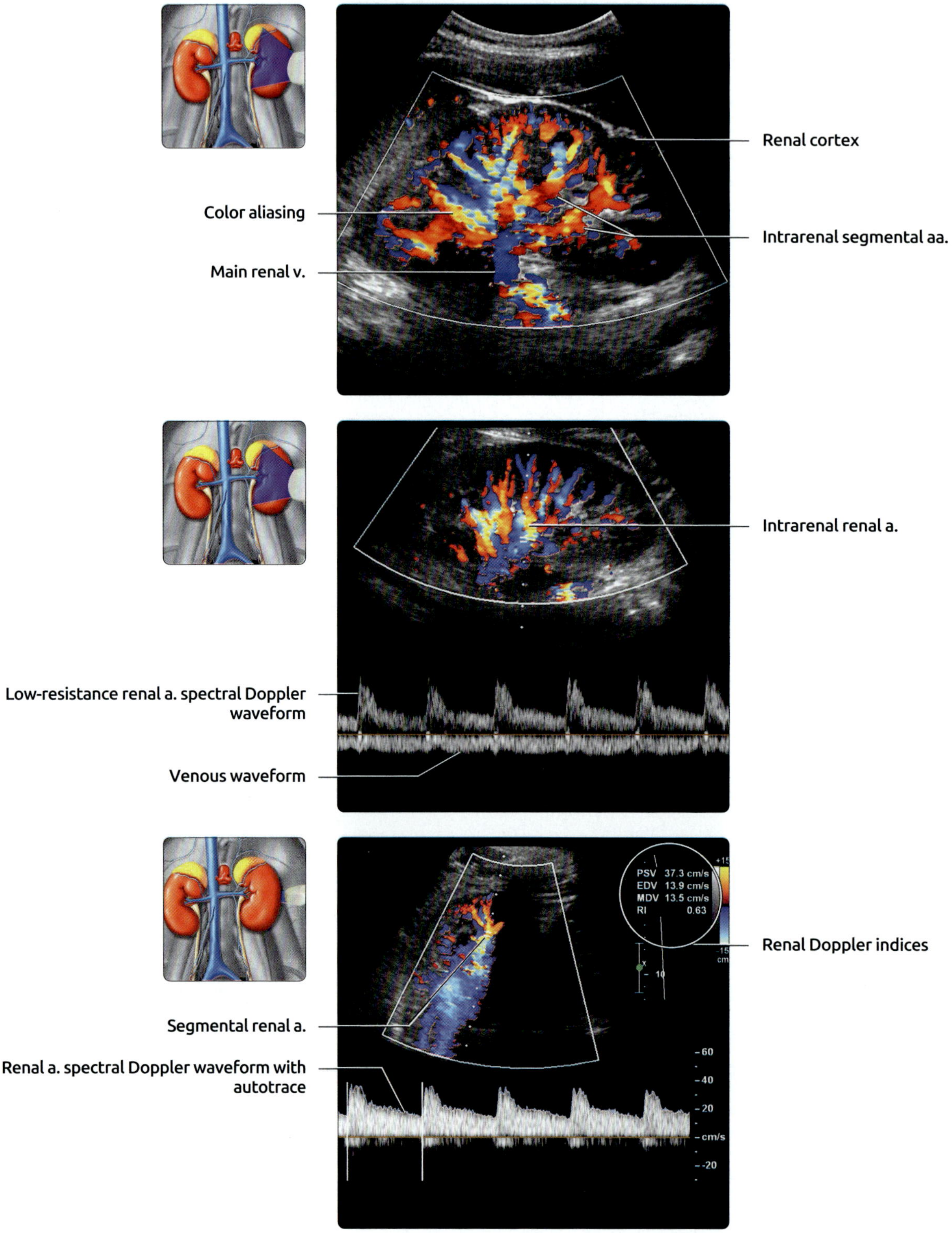

(Top) *Longitudinal color Doppler ultrasound of the left kidney shows renal artery branches as red and renal vein branches as blue. There is some color aliasing in the veins, shown as yellow.* **(Middle)** *In this spectral Doppler waveform of a left segmental renal artery, there is continuous flow throughout the cardiac cycle with a low- resistance pattern. The segmental renal vein shows mild phasicity.* **(Bottom)** *Spectral Doppler waveform of a left segmental renal artery in the transverse plane is shown. Doppler indices are normal: PSV of 37.3 cm/s, EDV of 13.9 cm/sec, and resistive index (RI) of 0.63.*

HYPERTROPHIED COLUMN OF BERTIN

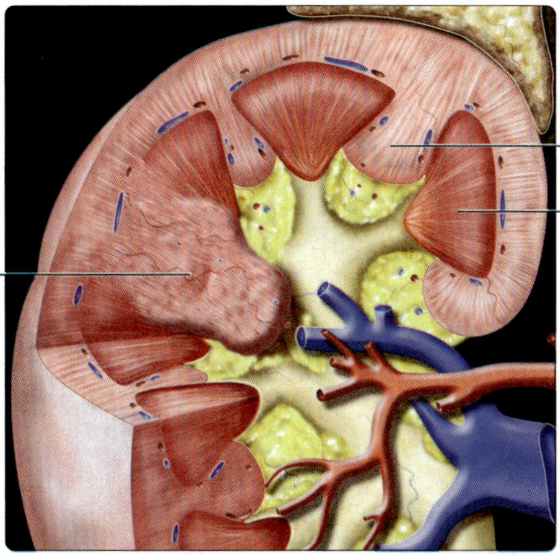

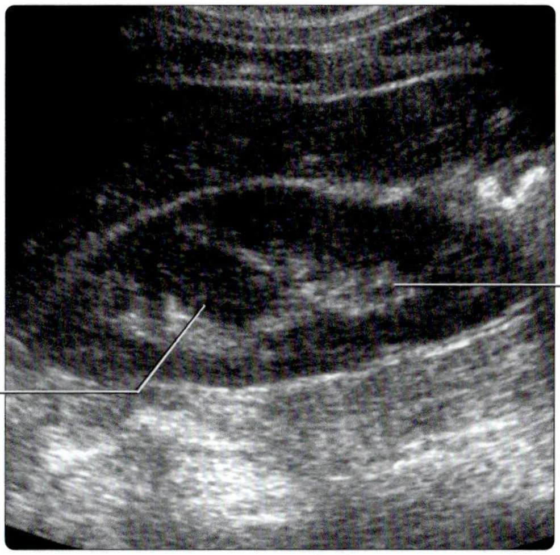

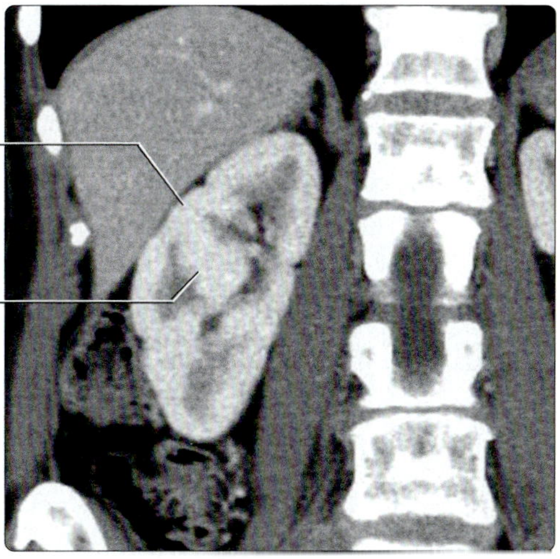

(Top) *Graphic depicts a hypertrophied column of Bertin, which is a rounded enlargement of the septal cortical tissue that separates the renal pyramids. This is normal tissue with the same imaging features as other parts of the renal cortex, but it may protrude into the renal sinus fat and may be mistaken for a renal mass. This generally tends to be more of a diagnostic dilemma on ultrasound relative to either CT or MR.* **(Middle)** *Longitudinal ultrasound of the right kidney shows an ovoid lesion, isoechoic to cortex, indenting the renal sinus. The patient was asymptomatic.* **(Bottom)** *Coronal CECT in the same patient shows that the column of Bertin enhances just like renal cortex. On the delayed phase, the column of Bertin remained isodense to cortex.*

RENAL DUPLICATION

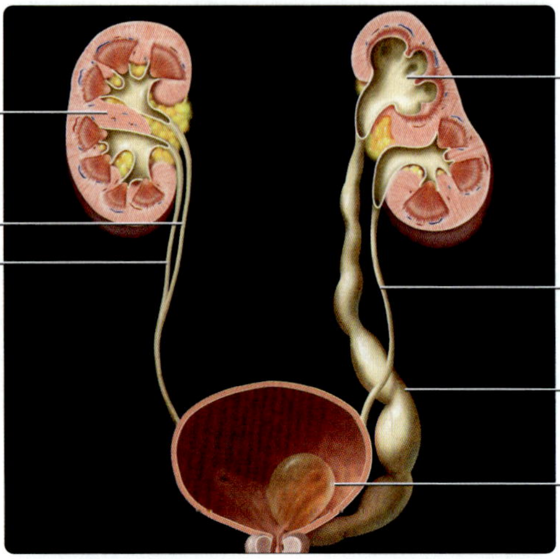

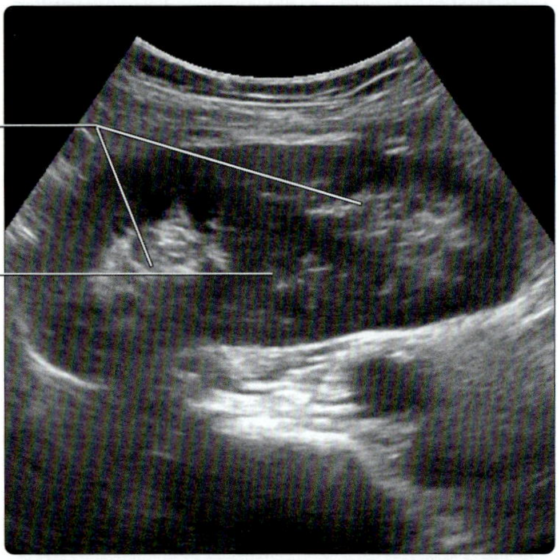

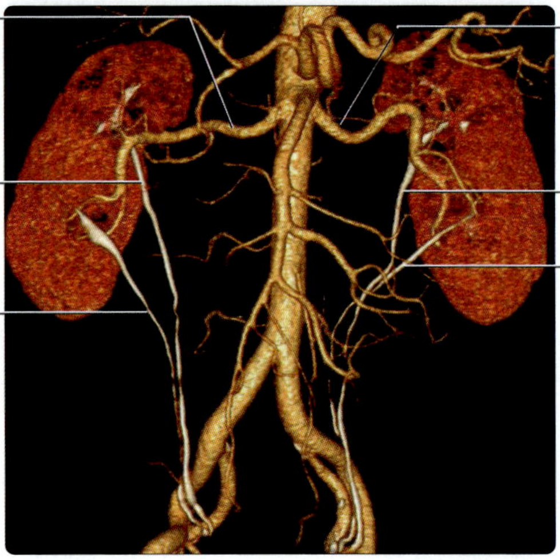

(Top) *Graphic shows bilateral renal duplication. On the right, there are 2 ureters exiting the renal pelvis with fusion in the mid third. On the left, the duplication is complete with the ureter of the superior moiety inserting lower than normal and terminating in a ureterocele. The ureter of the lower pole moiety inserts superior to the upper pole moiety (Weigert-Meyer rule) and is prone to reflux.* **(Middle)** *Longitudinal ultrasound through the left kidney shows separation of the renal sinus echo by cortical tissue indicating duplication.* **(Bottom)** *Anterior volume-rendered CT angiogram of a potential renal donor is shown. There are duplicated ureters bilaterally, joining near the bladder.*

HORSESHOE KIDNEY

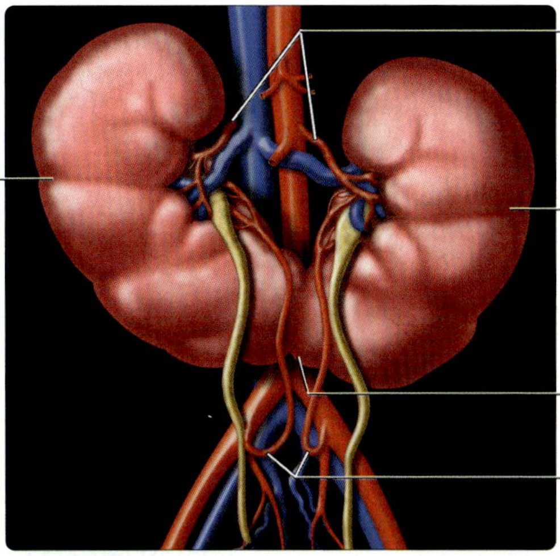

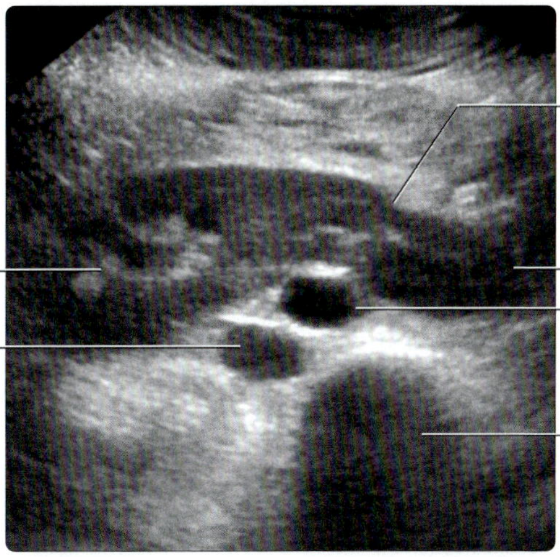

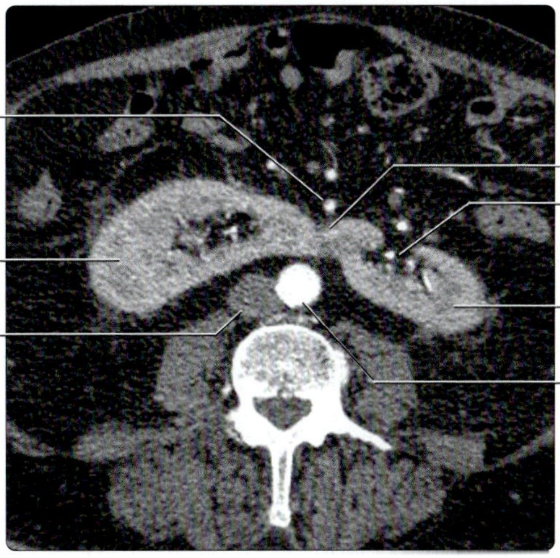

(Top) *Graphic depicts a horseshoe kidney with fusion of the lower poles in the midline. The midline bridge may consist of fibrous or renal tissue. Note the multiple renal arteries bilaterally with inferior arteries arising from the distal aorta or common iliac arteries. The inferior mesenteric artery (not shown) prevents further ascent of the midline bridge.* **(Middle)** *Transverse midline ultrasound shows a horseshoe kidney using an anterior approach. There is a bridge of cortical tissue between the 2 moieties. Because horseshoe kidneys are malrotated, pelvic dilation is not uncommon.* **(Bottom)** *Arterial phase CECT of a horseshoe kidney shows malrotation with anteriorly directed renal pelves. The inferior mesenteric artery prevents ascent of the kidneys.*

GROSS ANATOMY

Overview

- Adrenal (suprarenal) glands are part of both endocrine and nervous systems
- Lie within perirenal space, bounded by perirenal (Gerota) fascia
- **Right adrenal** is more apical in location
  - Lies anterolateral to right crus of diaphragm, medial to liver, posterior to inferior vena cava (IVC)
  - Often pyramidal in shape, inverted V shape on transverse section
- **Left adrenal** is more caudal
  - Lies medial to upper pole of left kidney, lateral to left crus of diaphragm, posterior to splenic vein and pancreas
  - Often crescentic in shape, lambda, tricorn hat, or triangular on transverse section
- **Adrenal cortex**
  - Derived from mesoderm
  - Has important endocrine functions
  - Secretes corticosteroids (cortisol, aldosterone) and androgens
- **Adrenal medulla**
  - Derived from neural crest
  - Part of sympathetic nervous system
  - Chromaffin cells secrete catecholamines (mostly epinephrine) into bloodstream

Vessels, Nerves, and Lymphatics

- Adrenal gland has very rich vascular and nervous connections
- **Arteries**
  - **Superior adrenal arteries**: 6-8 from inferior phrenic arteries
  - **Middle adrenal artery**: 1 from abdominal aorta
  - **Inferior adrenal artery**: 1 from renal arteries
- **Veins**
  - Right adrenal vein drains into IVC
  - Left adrenal vein drains into left renal vein (usually after joining left inferior phrenic vein)
- **Nerves**
  - Extensive sympathetic connection to adrenal medulla
  - Presynaptic sympathetic fibers from paravertebral ganglia end directly on secretory cells of medulla
- Lymphatics drain to lumbar (aortic and caval) nodes

ANATOMY IMAGING ISSUES

Imaging Recommendations

- Transducer: 2-5 MHz for adults
  - High-frequency (7.5- to 10.0-MHz) linear transducer may be used in neonates
- Complex shape requires multiplanar evaluation
- US may be used for initial screening and detection of adrenal lesions, followed by CT/MR for further characterization
- Right adrenal gland
  - Intercostal transverse approach, using liver as acoustic window
  - Direct anterior abdomen scanning is limited by overlying bowel loops and depth to gland
- Left adrenal gland
  - Intercostal at midaxillary line, using spleen or left kidney as acoustic window
  - In pediatric subjects and thin adults, direct transabdominal US at epigastrium
    - Stomach may be distended with fluid to serve as acoustic window
- **US appearance**
  - Relative to body size, adrenal glands are much larger and more easily identified in neonatal population
    - Also makes them more vulnerable to hemorrhage
  - Cortex is hypoechoic with hyperechoic medulla
    - Creates ice cream sandwich appearance
  - Adrenal glands often difficult to see and easily overlooked in adult patients (especially if obese), unless specifically targeted

CLINICAL IMPLICATIONS

Clinical Importance

- Rich blood supply reflects important endocrine function
- Adrenal glands are designed to respond to stress (trauma, sepsis, surgery, etc.) by secreting cortisol and epinephrine
  - Overwhelming stress may result in adrenal hemorrhage, acute adrenal insufficiency (addisonian crisis)
- Adrenal glands are common site for hematologic metastases (lung, breast, melanoma, etc.)
- **Adrenal cortex lesions**
  - **Adrenal adenomas**
    - Very common (at least 2% of general population) but usually cause no symptoms
    - US appearance is nonspecific; best evaluated by CT and MR sequences that show lipid-rich mass
  - **Cushing syndrome (excess cortisol)**
    - Signs: Truncal obesity, hirsutism, hypertension, abdominal striae
    - Causes: Pituitary tumors (↑ ACTH), exogenous (steroids) > adrenal adenoma > carcinoma
  - **Conn syndrome (excess aldosterone)**
    - Signs: Hypertension, hypokalemic alkalosis
    - Causes: Adrenal adenomas > hyperplasia > carcinoma
  - **Addison syndrome (adrenal insufficiency)**
    - Signs: Hypotension, weight loss, altered pigmentation
    - Causes: Autoimmune > adrenal metastases > adrenal hemorrhage > adrenal infection
  - **Myelolipoma**
    - Benign tumor composed of mature adipose tissue and variable amount of hematopoietic elements
    - Echogenic on US
- **Adrenal medulla lesions**
  - **Neuroblastoma**
    - Malignant tumor of sympathetic chain
    - Most common extracranial solid malignancy in children (median age at diagnosis: 15-17 months)
    - Location: Adrenal gland > retroperitoneum > posterior mediastinum
  - **Pheochromocytoma**
    - Tumor arising from chromaffin cells of adrenal medulla or extraadrenal paraganglia
    - Location: Adrenal gland (90%), sympathetic chain from neck to bladder (10%)

ADRENAL GLANDS

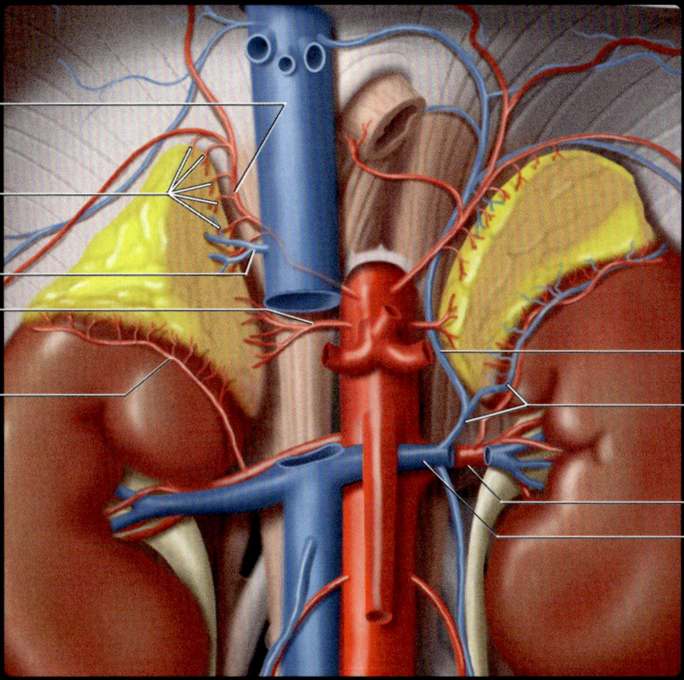

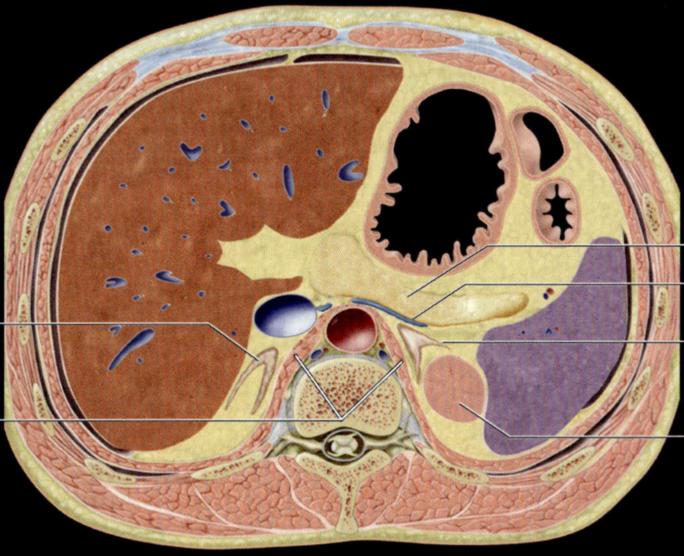

(Top) *The size of the adrenal glands is exaggerated in this illustration to facilitate demonstration of the vascular anatomy. The adrenal glands have a very rich vascular supply, reflecting their critical role in maintaining homeostasis and responding to stress. The superior adrenal arteries are short branches of the inferior phrenic arteries bilaterally. The middle adrenal arteries are short vessels arising from the aorta. The inferior adrenal arteries are branches of the renal arteries. The left adrenal vein drains into the left renal vein (usually after joining with the inferior phrenic vein), while the right adrenal vein drains directly into the inferior vena cava (IVC).* **(Bottom)** *Graphic shows the right adrenal gland lying above the right kidney, while the left adrenal gland lies partly in front of the upper pole of the left kidney. The left adrenal gland lies directly posterior to the splenic vein and body of pancreas and lateral to the left crus of the diaphragm. The right adrenal gland lies lateral to the crus, medial to the liver, and directly behind the IVC.*

ADRENAL GLANDS, NEONATE

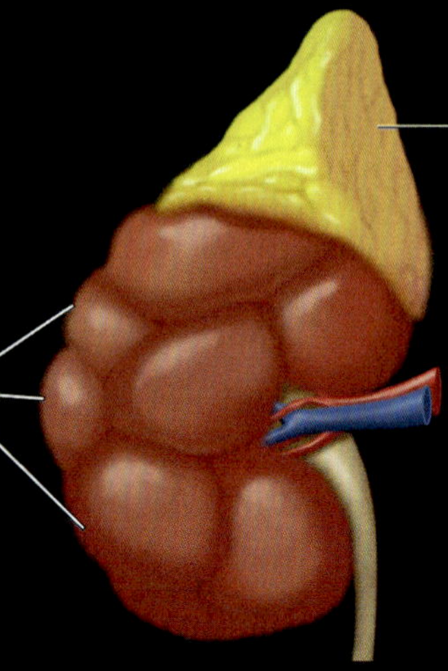

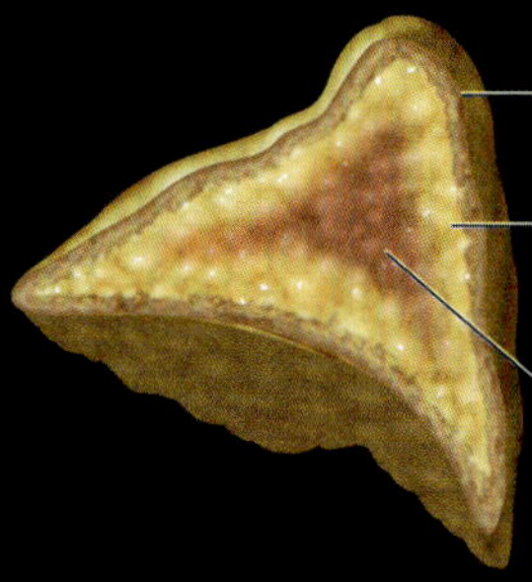

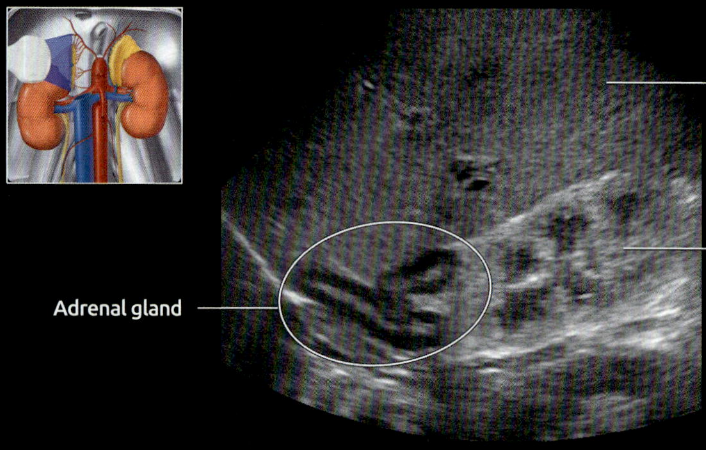

(Top) *Graphic shows the appearance of the adrenal gland and kidney in a fetus or neonate. The adrenal gland is much larger relative to the body size than in the adult, making it easier to see on US. The kidney has a lobulated appearance, reflecting the ongoing fusion of the individual renal lobes, each comprised of 1 renal pyramid and its associated renal cortex. This appearance can sometimes persist into adulthood.* **(Middle)** *The adrenal gland is essentially 2 organs in a single structure. The cortex is an endocrine gland, secreting primarily cortisol, aldosterone, and androgenic steroids. The adrenal medulla is part of the autonomic nervous system and secretes epinephrine and norepinephrine.* **(Bottom)** *A longitudinal US of the right adrenal gland in a neonate shows its folded, triangular morphology and relative large size when compared to the kidney.*

ADRENAL GLANDS, NEONATE

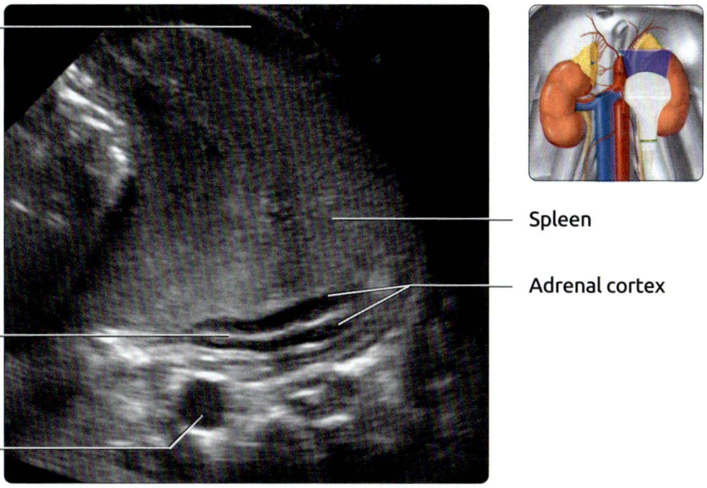

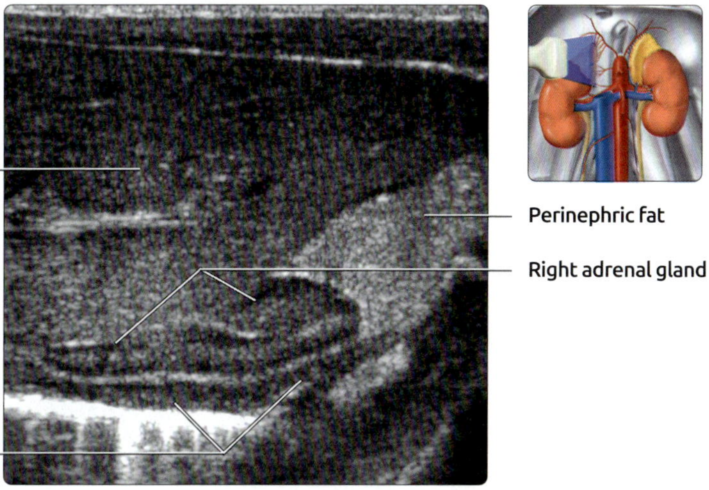

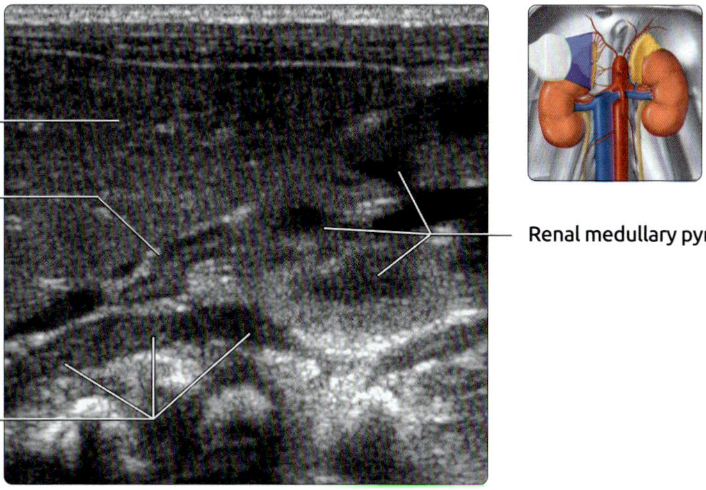

(Top) *Transverse US of the left adrenal gland in a newborn shows the typical trilaminar, ice cream sandwich appearance with a hypoechoic cortex and hyperechoic medulla.* **(Middle)** *Longitudinal US of the right adrenal gland is shown using the liver as acoustic window. Using a high-frequency linear transducer better demonstrates the adrenal gland in neonates and infants; note the typical trilaminar appearance.* **(Bottom)** *Another image in the same patient with the transducer angulated slightly more laterally shows the lateral limb of the right adrenal gland. The crus of the right hemidiaphragm is seen posteriorly; it is more hypoechoic and lacks the adrenal glands trilaminar appearance.*

ADRENAL GLANDS, ADULT

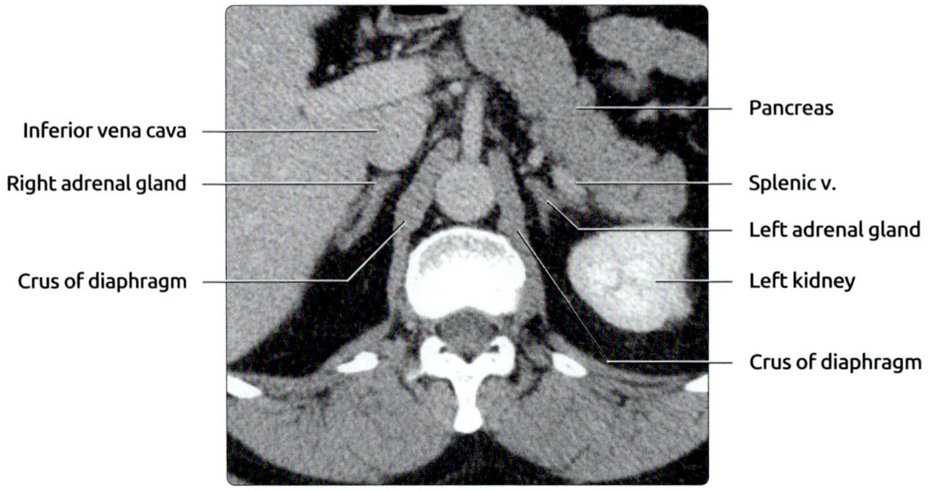

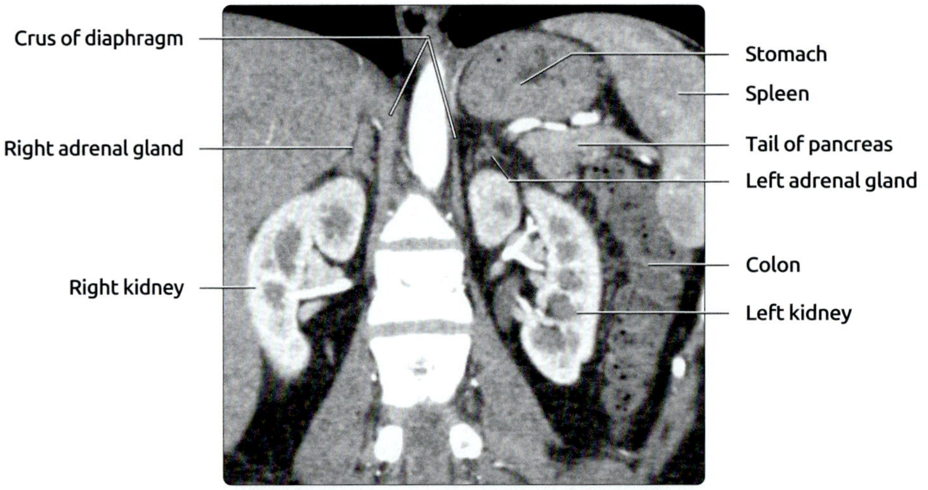

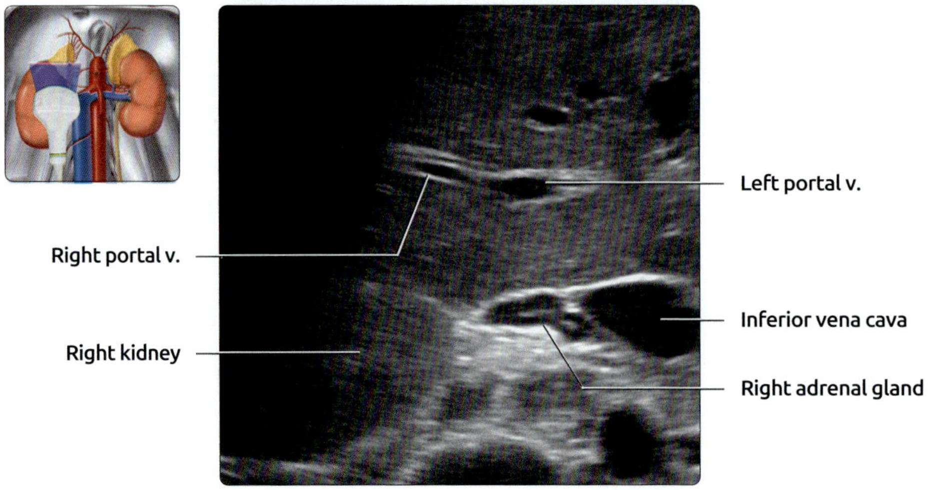

(Top) *The right adrenal gland is usually suprarenal, touches the back of the IVC and lies lateral to the right crus and medial to the liver. The left adrenal gland lies medial to upper pole of left kidney, lateral to left crus of diaphragm, and posterior to splenic vein and pancreas.* **(Middle)** *This coronal view demonstrates the relation between the adrenals and adjacent organs.* **(Bottom)** *Axial US in a 19-year-old woman shows the adrenal gland adjacent to the IVC. Note the trilaminar appearance is still seen in this young adult.*

ADRENAL GLANDS, ADULT

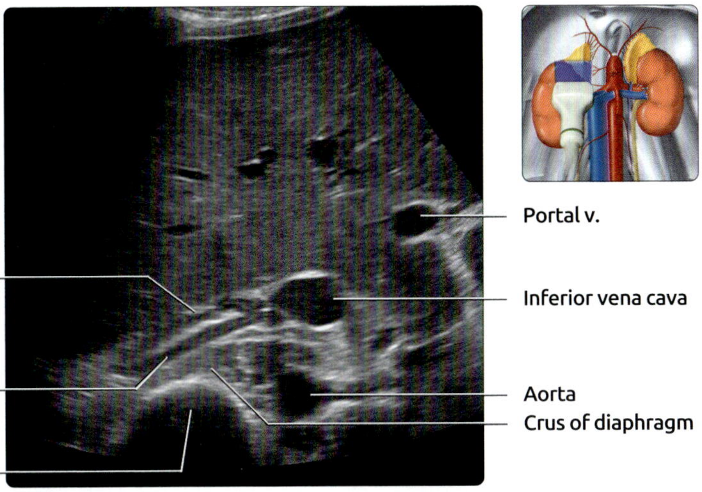

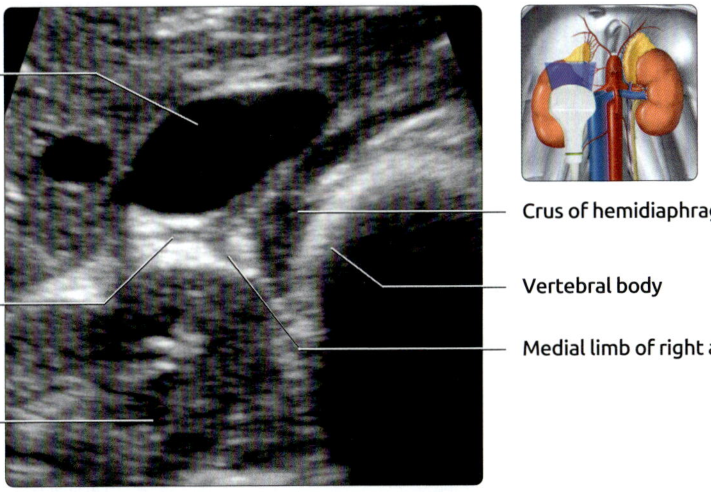

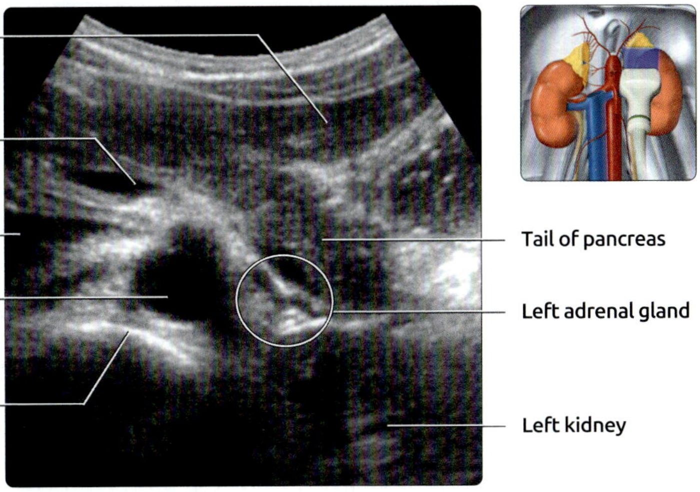

(Top) *Using the liver as an acoustic window, the right adrenal gland is particularly well seen in this thin adult. Two limbs are easily identified.* **(Middle)** *Zoomed transverse grayscale US shows the right adrenal gland. Note the medially located crus of the right hemidiaphragm, which may be mistaken for the gland. The crus is more hypoechoic, and its contour hugs the vertebral body.* **(Bottom)** *Subxiphoid transverse grayscale US performed at the epigastrium, level of the pancreatic body and tail, is shown. The left adrenal gland is seen as a lambda structure and surrounded by hyperechoic perirenal fat. It is anterior to the left kidney and to the left of the abdominal aorta. The adrenal gland can be easily missed if not specifically targeted.*

GROSS ANATOMY

Divisions

- **Esophagus**
  - Cervical and thoracic segments
- **Stomach**
  - Hollow muscular organ between esophagus and small intestine
  - Location: Intraperitoneal, in left upper quadrant, bordered superiorly by left hemidiaphragm, posterolaterally by spleen, posteroinferiorly by pancreas
    - Greater omentum attached from greater curvature and drapes over small and large intestines
    - Lesser omentum attached from lesser curvature to porta hepatis, covers lesser sac
  - Function
    - Gastric acid production for breakdown of large molecules of food into smaller molecules in preparation for small intestinal absorption
    - Storage of food
  - Sections
    - Gastroesophageal junction/cardia, lower esophageal sphincter
    - Fundus and body: Delineated by horizontal plane passing through cardia
    - Antrum/pylorus: Lower section facilitating entry of gastric contents into duodenum
  - Curvatures
    - Greater curvature: Lateral wall of stomach
    - Lesser curvature: Medial wall of stomach
  - Rugae/internal ridges increase surface area for digestion
  - Arterial supply
    - Right and left gastric arteries supply lesser curvature
    - Right and left gastroepiploic arteries supply greater curvature
    - Short gastric artery supplies fundus
  - Venous drainage
    - Follow arteries and drain into portal vein and its tributaries
- **Small bowel**
  - Between stomach and large intestine
  - ~ 4-7 meters in length
  - Centrally located in abdomen
  - Intraperitoneal, except for 2nd-4th portions of duodenum
  - Function: Further breakdown of food molecules from stomach with eventual absorption
  - Intraluminal extensions/folds valvulae conniventes increase surface area for absorption
    - Abundant in proximal small bowel, decrease in number in distal small bowel loops
  - **Duodenum**
    - C-shaped hollow tube connecting stomach with jejunum
    - Begins with duodenal bulb, ends in ligament of Treitz (duodenojejunal junction)
    - Arterial supply and venous drainage: Superior and inferior pancreaticoduodenal artery, pancreaticoduodenal veins
  - **Jejunum**
    - Connects duodenum with ileum
    - ~ 2.5 meters in length
    - Begins at ligament of Treitz
    - Along with ileum, suspended by mesentery
    - Arterial supply and venous drainage: Superior mesenteric artery and vein
  - **Ileum**
    - Connects jejunum with ascending colon
    - ~ 3.5 meters in length
    - Along with jejunum, suspended by mesentery
    - Arterial supply and venous drainage: Superior mesenteric artery and vein
- **Large bowel**
  - Between small bowel and anus
  - ~ 1.5 meters in length
  - Peripherally located in abdomen
    - Cecum and appendix, transverse colon, and rectosigmoid intraperitoneal
    - Ascending colon, descending colon, and middle rectum retroperitoneal
    - Distal rectum extraperitoneal
  - Function: Absorption of remaining water, storage, and elimination of waste
  - Sections
    - Ascending colon: Located in right side of abdomen, includes cecum where appendix arises
    - Hepatic flexure: Turn of colon at liver
    - Transverse colon: Traverses upper abdomen
    - Splenic flexure: Turn of colon at spleen
    - Descending colon: Left side of abdomen
    - Sigmoid/rectum: At posterior pelvis
  - With taenia coli: 3 bands of smooth muscle just under serosa
    - Haustration: Sacculations in colon resulting from contraction of taenia coli
    - Epiploic appendages: Small fat accumulations on viscera
  - Arterial supply
    - Superior mesenteric artery supplies colon from appendix through splenic flexure
    - Ileocolic branch supplies cecum
    - Right colic branch supplies ascending colon
    - Middle colic branch supplies transverse colon
    - Inferior mesenteric artery supplies descending colon through rectum
    - Left colic branch supplies descending colon
    - Sigmoid branches supply sigmoid
    - Superior rectal artery supplies superior rectum
    - Middle and inferior rectum supplied by arteries of same name originating from internal iliac artery
  - Venous drainage
    - Superior and inferior mesenteric veins
- **Anus**
  - External opening of rectum
    - Termination of gastrointestinal tract
  - With sphincters for controlling defecation
  - Internal anal sphincter
    - Thin ring of smooth muscle surrounding anal canal, deep to submucosa
    - Under involuntary control

- Continuous with muscularis propria of rectum
- Forms incomplete ring in females
- External anal sphincter
  - Thick ring of skeletal muscle around internal anal sphincter
  - Under voluntary control
  - 3 parts from superior to inferior: Deep, superficial, and subcutaneous
- Longitudinal muscle
  - Thin muscle between internal and external anal sphincters
  - Conjoined muscle from muscularis propria of rectum and levator ani

Histology

- Bowel wall throughout GI tract has uniform general histology, comprised of 4 layers
  - Mucosa
    - Functions for absorption and secretion
    - Composed of epithelium and loose connective tissue
    - Lamina propria
    - Muscularis mucosa (deep layer of mucosa)
  - Submucosa
    - Consists of fibrous connective tissue
    - Contains Meissner plexus
  - Muscularis externa
    - Muscular layer responsible for peristalsis or propulsion of food through gut
    - Contains Auerbach plexus
  - Serosa
    - Epithelial lining continuous with peritoneum

IMAGING ANATOMY

Overview

- GI tract extends from mouth to anus
- Esophagus, which is intrathoracic, is difficult to visualize with external ultrasound due to rib cage and air-containing lungs
  - Endoluminal ultrasound performed to assess mural pathology
- Stomach to rectum lie within abdomen and pelvis
- Stomach, 1st part of duodenum, jejunum, ileum, transverse colon, and sigmoid colon suspended within peritoneal cavity by peritoneal folds and are mobile
- 2nd-4th part of duodenum, ascending colon, descending colon, and rectum typically extraperitoneal/retroperitoneal
  - Retroperitoneal structures have more fixed position and are easy to locate
- Stomach located in left upper quadrant
  - Identified by presence of rugae/mural folds
  - Prominent muscular layer facilitates identification of pylorus
- Small bowel loops are located centrally within abdomen
  - Abundant valvulae conniventes helps identify jejunal loops
  - Jejunalization of ileum seen in celiac disease to compensate for atrophy of folds in proximal small bowel
  - Contents of jejunal loops usually liquid and appear hypoechoic/anechoic
- Cecum and colon identified by haustral pattern
  - Located peripherally in abdomen
  - Contain feces and gas
  - Haustra seen as prominent curvilinear echogenic arcs with posterior reverberation
  - Cecum identified by curvilinear arc of hyperechogenicity (representing feces and gas) in right lower quadrant blind ending caudally
  - Not uncommonly, cecum high lying and may be horizontally placed
  - Sigmoid colon variable length and mobile
  - Junction of left colon with sigmoid colon identified in left iliac fossa by tracing descending colon
  - Rectosigmoid junction has fixed position and is identified with full bladder, which acts as acoustic window
- Appendicular base normally located in right lower quadrant
  - Length and direction of tip vary
  - Retrocecal appendix and pelvic appendix can be difficult to locate transabdominally
    - Transvaginal ultrasound examination useful to identify pelvic appendix
- Normal measurements of bowel caliber
  - Small bowel < 3 cm
  - Large bowel
    - Cecum < 9 cm
    - Transverse colon < 6 cm
- Stratified appearance of bowel wall on histology is depicted by 5 distinct layers on ultrasound as **alternating echogenic/sonolucent (hypoechoic) appearance (gut signature)**
  - Interface of lumen and mucosa: Echogenic
  - Muscularis mucosa: Hypoechoic
  - Submucosa: Echogenic
  - Muscularis propria/externa: Hypoechoic
  - Serosa: Echogenic
- Normal bowel wall thickness < 3 mm

Bowel Motility

- Bowel is hollow viscus and is constantly mobile due to peristalsis
  - Assessing direction of flow of contents often challenging
  - When visibility permits, direction of flow can be determined by following long segments of bowel in continuous fashion
- Fixed points of bowel easy to assess with transabdominal ultrasound
  - Pylorus, "C loop" of duodenum, and ileocecal junction useful landmarks to assess direction of content flow
- Different bowel pathologies have potential to alter normal gut motility
- Real-time dynamic ultrasound provides useful information regarding bowel mobility, which can aid in diagnosis of underlying condition
  - Cine function useful to store dynamic images for review
- Abnormal bowel identified as thickened or dilated segments
  - Thickened bowel demonstrates reduced peristalsis
    - Stands out among normally peristalsing loops of normal bowel

GASTROINTESTINAL TRACT IN SITU

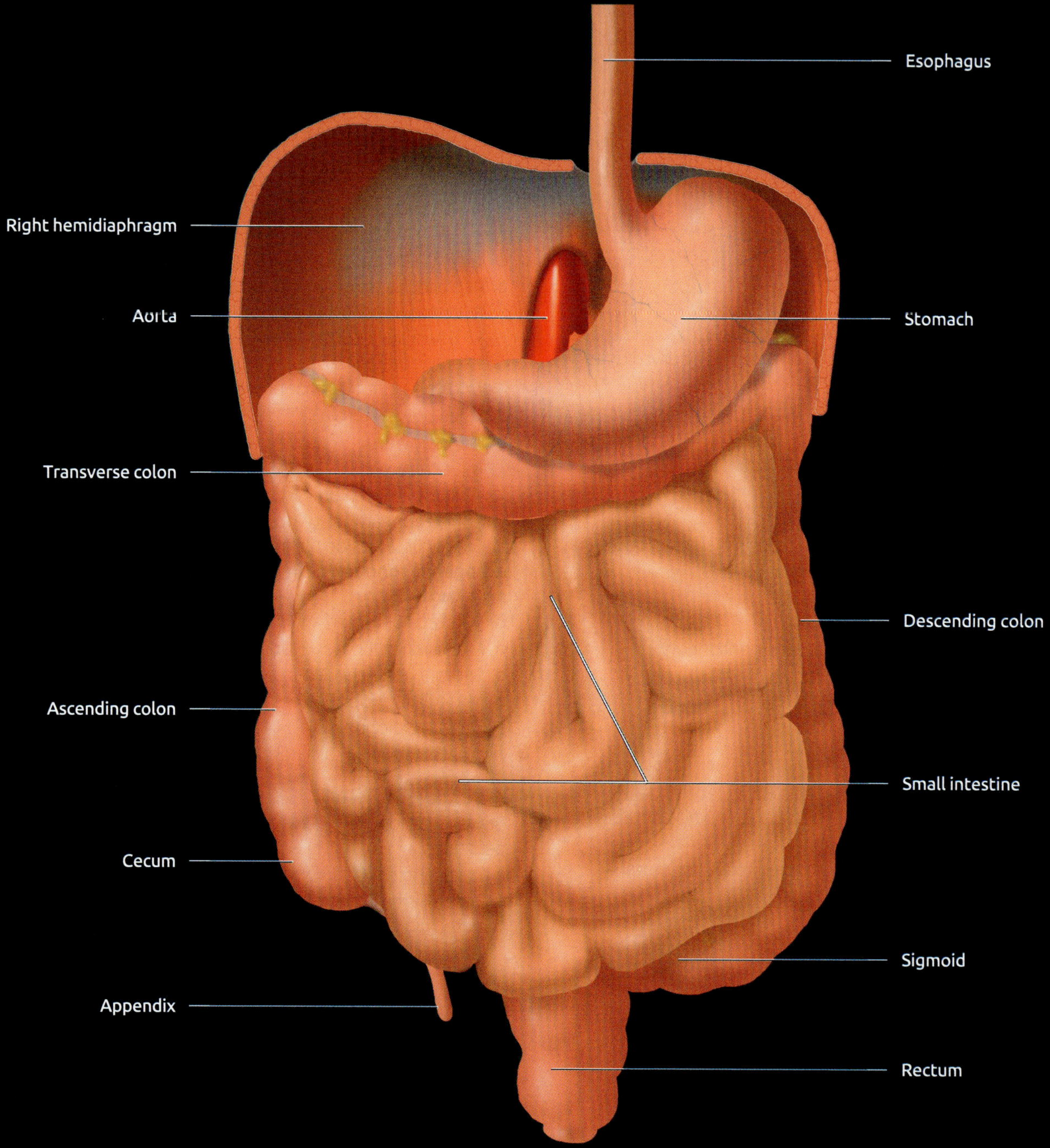

Graphic shows the gastrointestinal tract in situ. The liver and the greater omentum have been removed. Note the relatively central location of the small intestine compared with the peripherally located large intestine. Most of the bowel segments are intraperitoneal, apart from the 2nd-4th parts of the duodenum, the ascending and descending colon, and the middle 1/3 of the rectum, which are retroperitoneal. The distal 1/3 of the rectum is extraperitoneal.

STOMACH AND DUODENUM IN SITU

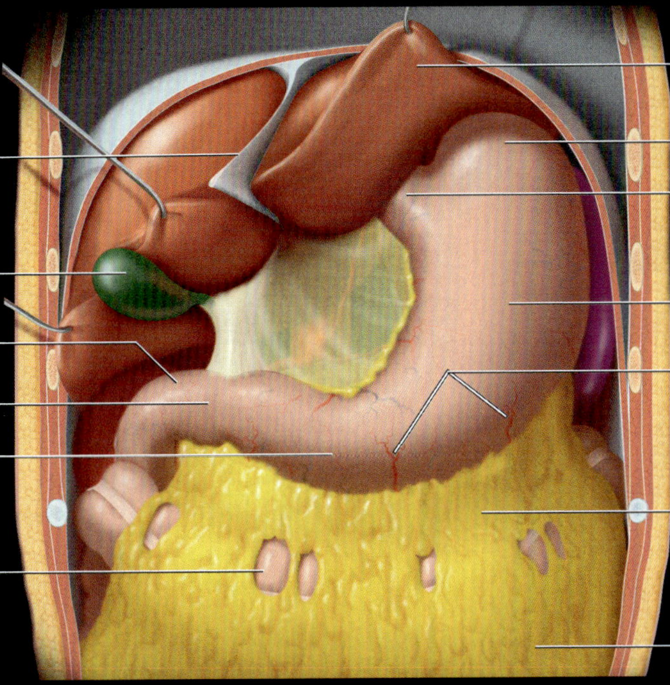

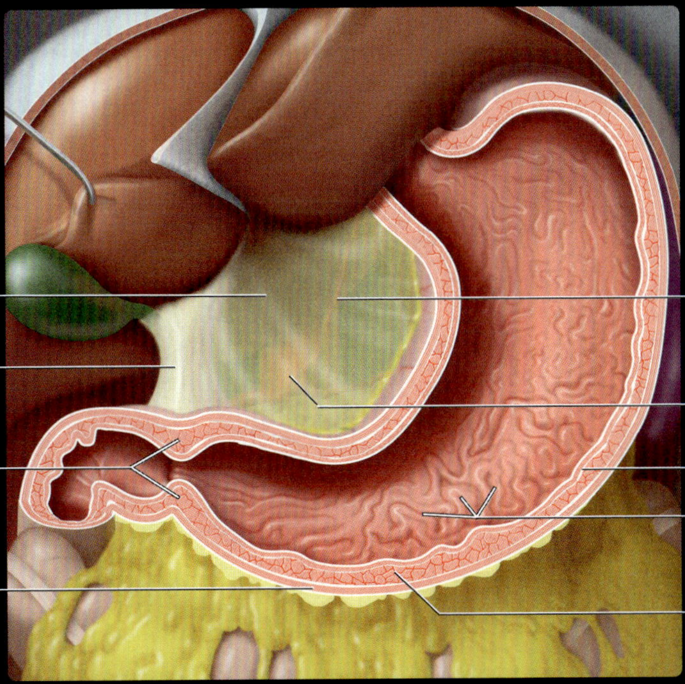

(Top) *Graphic shows the stomach and proximal duodenum in situ. The liver and gallbladder have been retracted upward. Note that the lesser curvature and anterior wall of the stomach touch the underside of the liver, and the gallbladder abuts the duodenal bulb. The greater curvature is attached to the transverse colon by the gastrocolic ligament, which continues inferiorly as the greater omentum, covering most of the colon and small bowel.* **(Bottom)** *Graphic shows the lesser omentum extending from the stomach to the porta hepatis, divided into the broader and thinner hepatogastric ligament and the thicker hepatoduodenal ligament. The lesser omentum carries the portal vein, hepatic artery, common bile duct, and lymph nodes. The free edge of the lesser omentum forms the ventral margin of the epiploic foramen. Note the layers of gastric muscle; the middle circular layer is thickest.*

DUODENUM

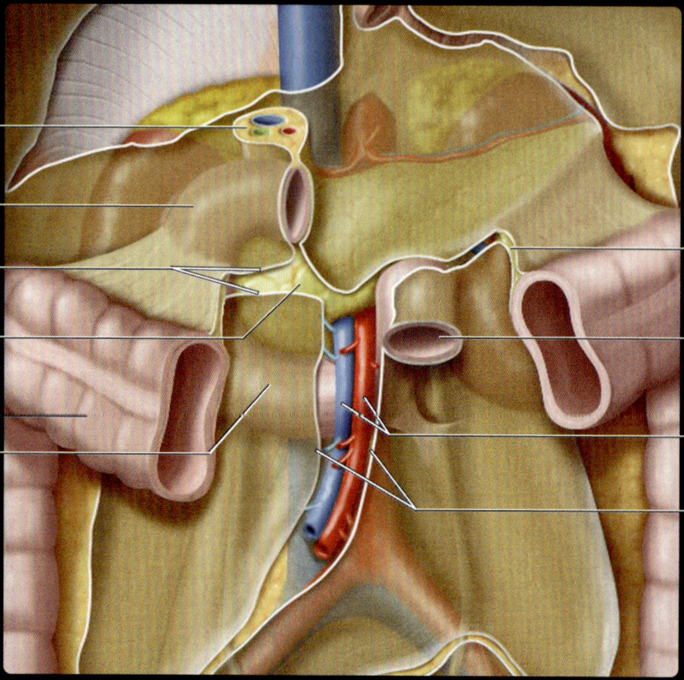

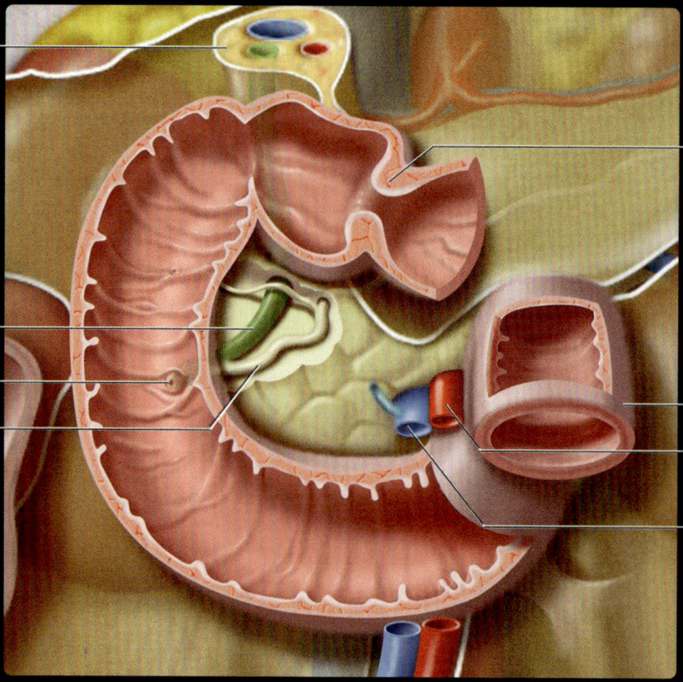

(Top) *The duodenum is retroperitoneal, except for the bulb (1st part). The proximal jejunum is intraperitoneal. The hepatoduodenal ligament attaches the duodenum to the porta hepatis and contains the portal triad (bile duct, hepatic artery, portal vein). The root of the transverse mesocolon and mesentery both cross the duodenum. The 3rd portion of the duodenum crosses in front of the aorta and inferior vena cava (IVC) and behind the superior mesenteric vessels. The 2nd portion of the duodenum is attached to the pancreatic head and lies close to the hilum of the right kidney.* **(Bottom)** *Graphic shows the duodenal bulb suspended by the hepatoduodenal ligament. The duodenal-jejunal flexure is suspended by the ligament of Treitz, an extension of the right crus. The major pancreaticobiliary papilla enters the medial wall of the 2nd portion of the duodenum.*

SMALL INTESTINE

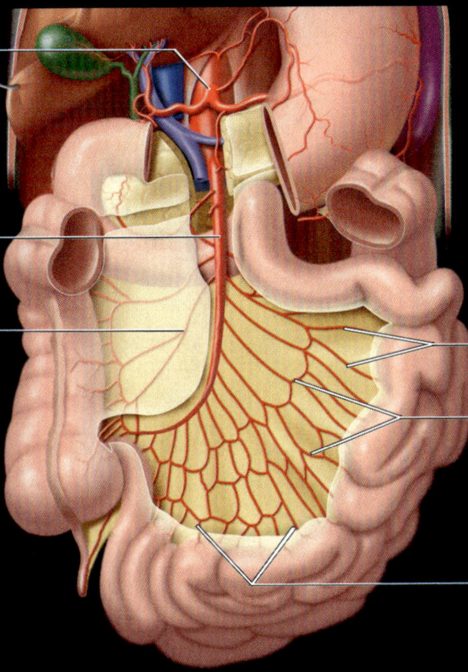

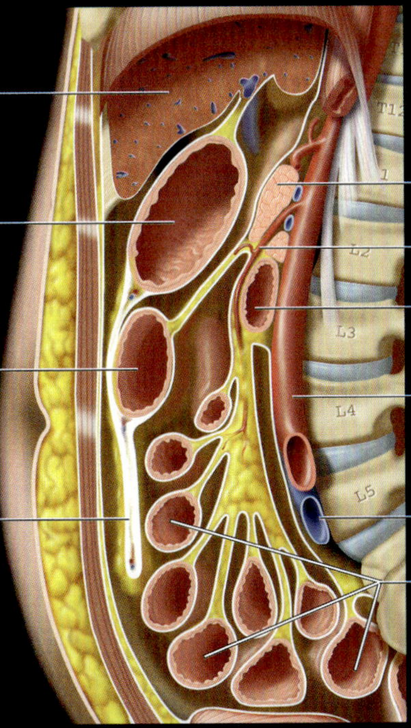

(Top) *Graphic shows the vascular supply of the entire small intestine from the superior mesenteric artery (SMA). The small bowel segments are displaced inferiorly. The SMA arises from the anterior abdominal aorta and gives off the inferior pancreaticoduodenal branch that supplies the duodenum and pancreas. Arising from the left side of the SMA are numerous branches to the jejunum and ileum. Jejunal arteries are generally larger and longer than those of the ileum. After a straight course, the arteries form multiple intercommunicating, curvilinear arcades.* **(Bottom)** *Graphic shows the sagittal section of the central abdomen, revealing the jejunum and ileum suspended in a radial pattern by the mesentery. Note the overlying greater omentum attached from the inferior portion of the stomach to drape the small bowel segments and transverse colon.*

COLON, RECTUM, AND ANUS

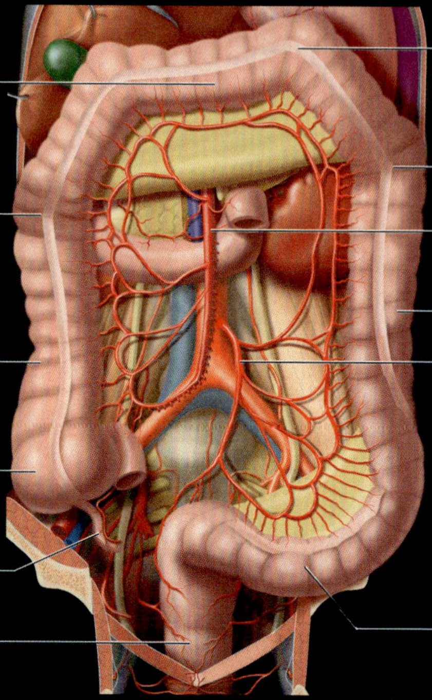

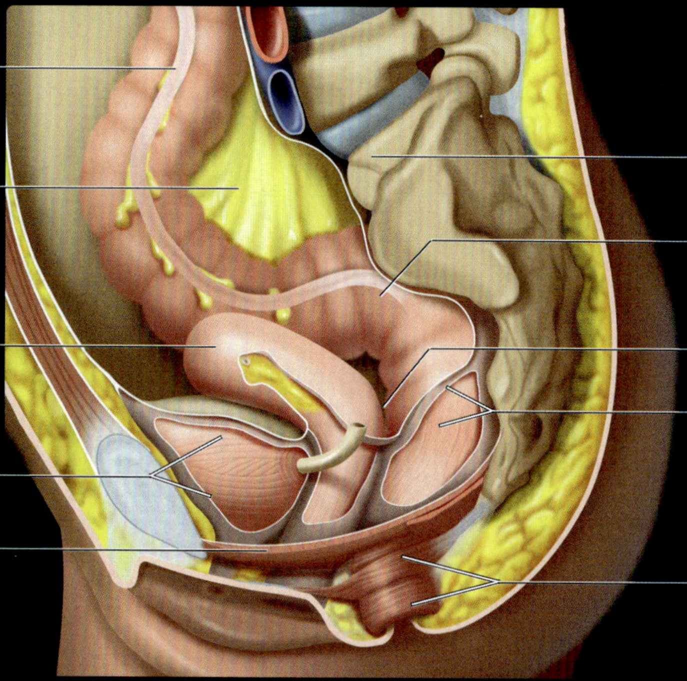

(Top) *Graphic shows the colon in situ. The transverse colon has been retracted upward to demonstrate the arterial supply of the colon from the superior and inferior mesenteric arteries. The SMA supplies the colon from the appendix through the splenic flexure, and the inferior mesenteric artery (IMA) supplies the descending colon through the rectum. Note the band of smooth muscle (taenia coli) running along the length of the intestine, which terminates in the vermiform appendix; these result in sacculations/haustrations along the colon, giving it a segmented appearance.* **(Bottom)** *The sigmoid colon is on a mesentery, while the rectum is retroperitoneal. The anterior surface of the rectum has a peritoneal covering, which extends deep in the pelvis in women, forming the rectouterine pouch (of Douglas) as it is reflected along the posterior surface of the uterus. The rectum is narrowed as it passes through the pelvic diaphragm and then enters the anal canal with 3 levels of the anal sphincter (deep, superficial, and subcutaneous). The rectum has a continuous external longitudinal coat of muscle, unlike the colon with its discontinuous rows of taeniae.*

STOMACH

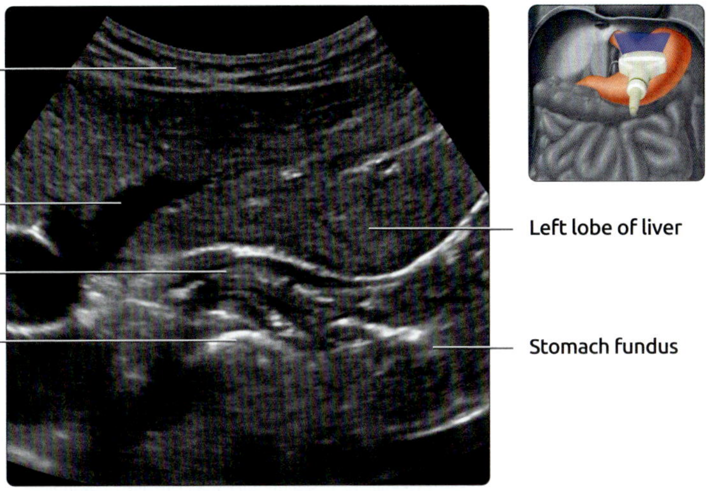

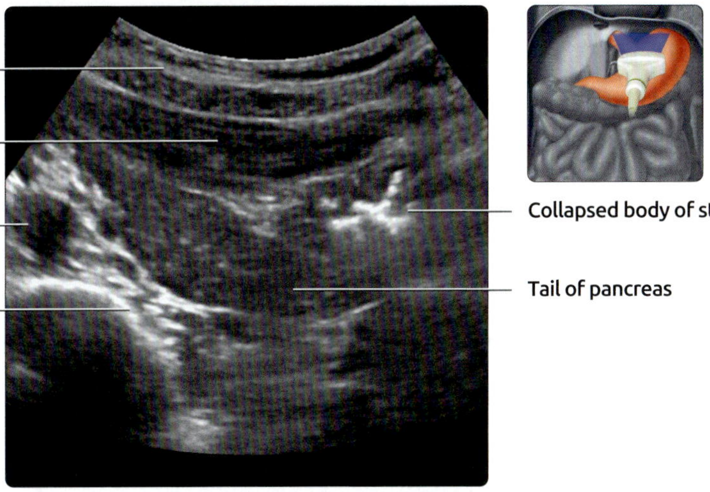

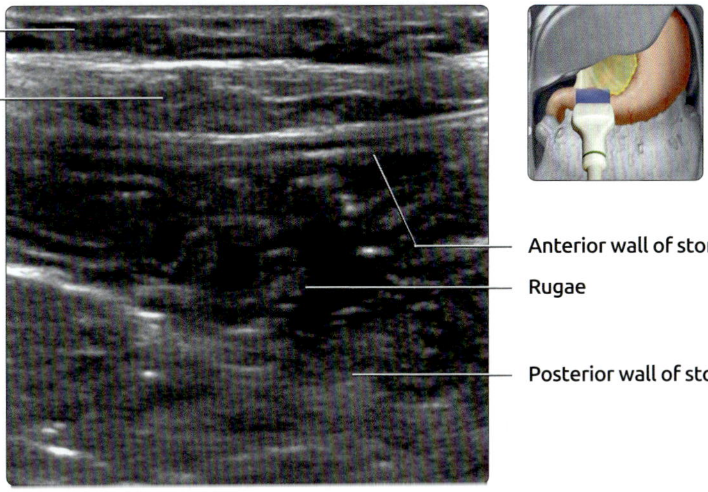

(Top) *Transverse oblique ultrasound at the epigastric region shows the gastroesophageal junction, which can be traced to the fundus of the stomach. Note the relationship to the adjacent structures.* **(Middle)** *Transverse oblique ultrasound through the left upper quadrant shows the collapsed stomach body with the rugal folds. Echogenic gas is seen between the rugae. The tail of the pancreas is seen posterior to the stomach, and the left lobe of the liver is anterior to the stomach.* **(Bottom)** *High-resolution transverse ultrasound through the epigastric region shows the gastric body tapering to the gastric antrum. Note the gastric folds (rugae). The stomach wall shows the gut signature.*

GASTRODUODENAL REGION/DUODENUM

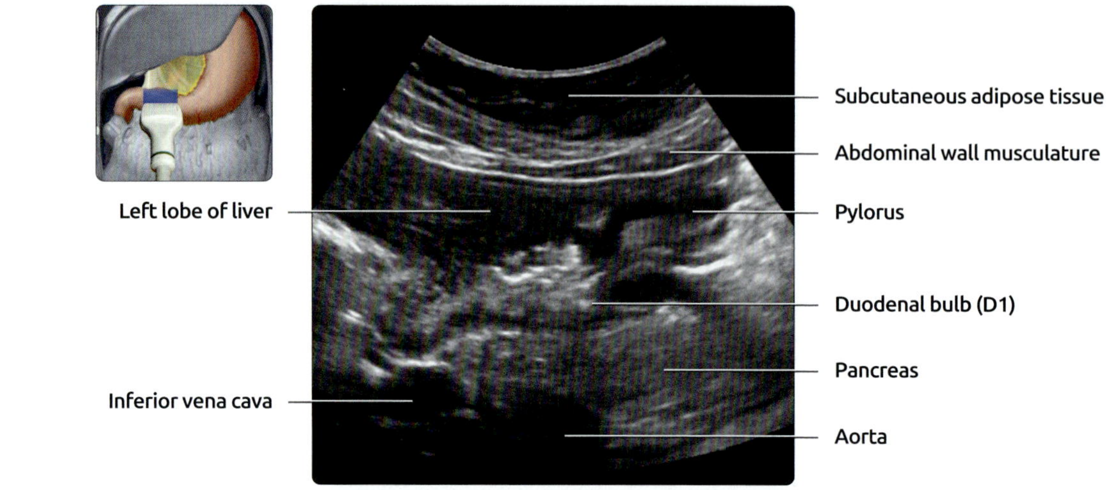

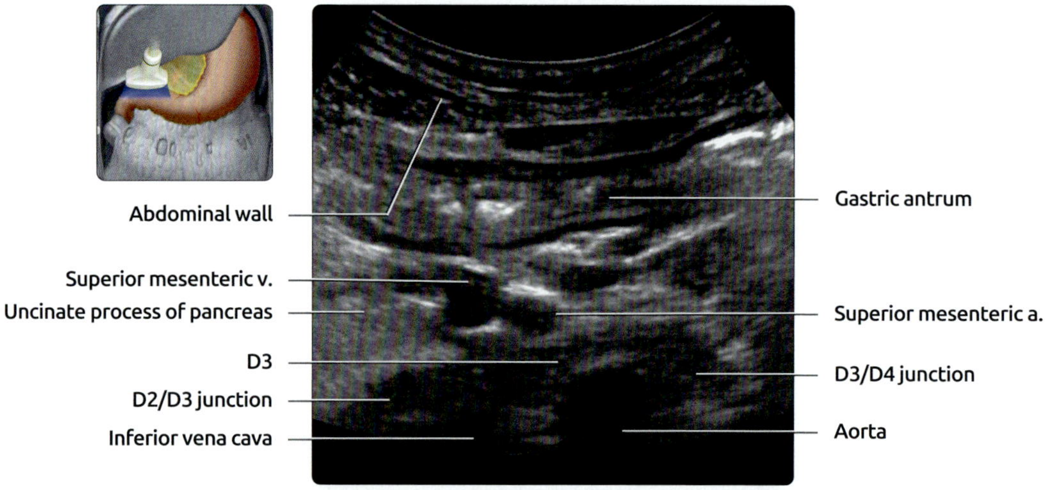

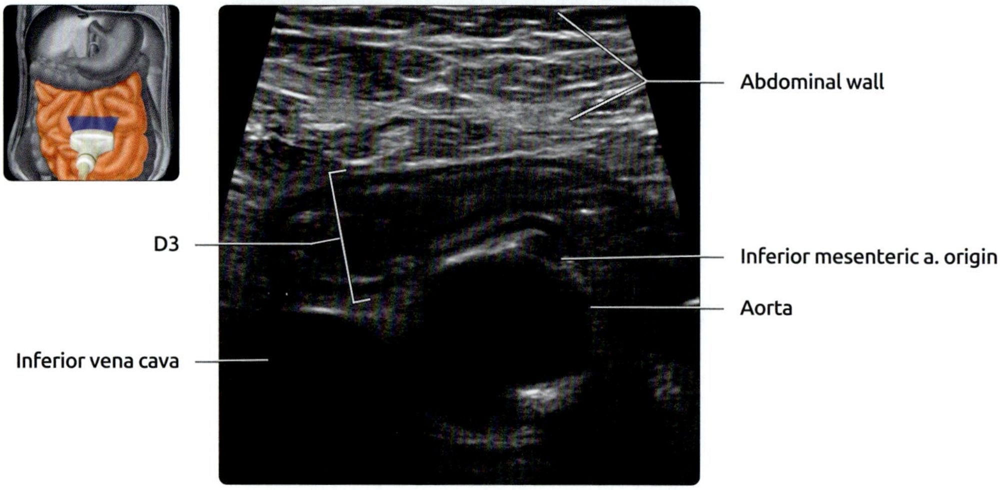

(Top) *Transverse oblique ultrasound through the epigastric region shows the pylorus with the hypoechoic prominent muscular wall leading to the duodenal bulb.* **(Middle)** *Transverse ultrasound through the upper abdomen shows the gastric antrum anteriorly compressed by the curvilinear probe and collapsed 3rd part of the duodenum (D3) posteriorly. The SMA and the superior mesenteric vein (SMV) are seen in the plane in between. The IVC and aorta are seen posterior to the D3.* **(Bottom)** *Axial midline high-resolution ultrasound through the upper abdomen shows fluid distended in the D3 located across, in the upper retroperitoneum. Note the aorta and IVC posterior to the D3. Gut signature is seen in the wall of the D3.*

SMALL BOWEL

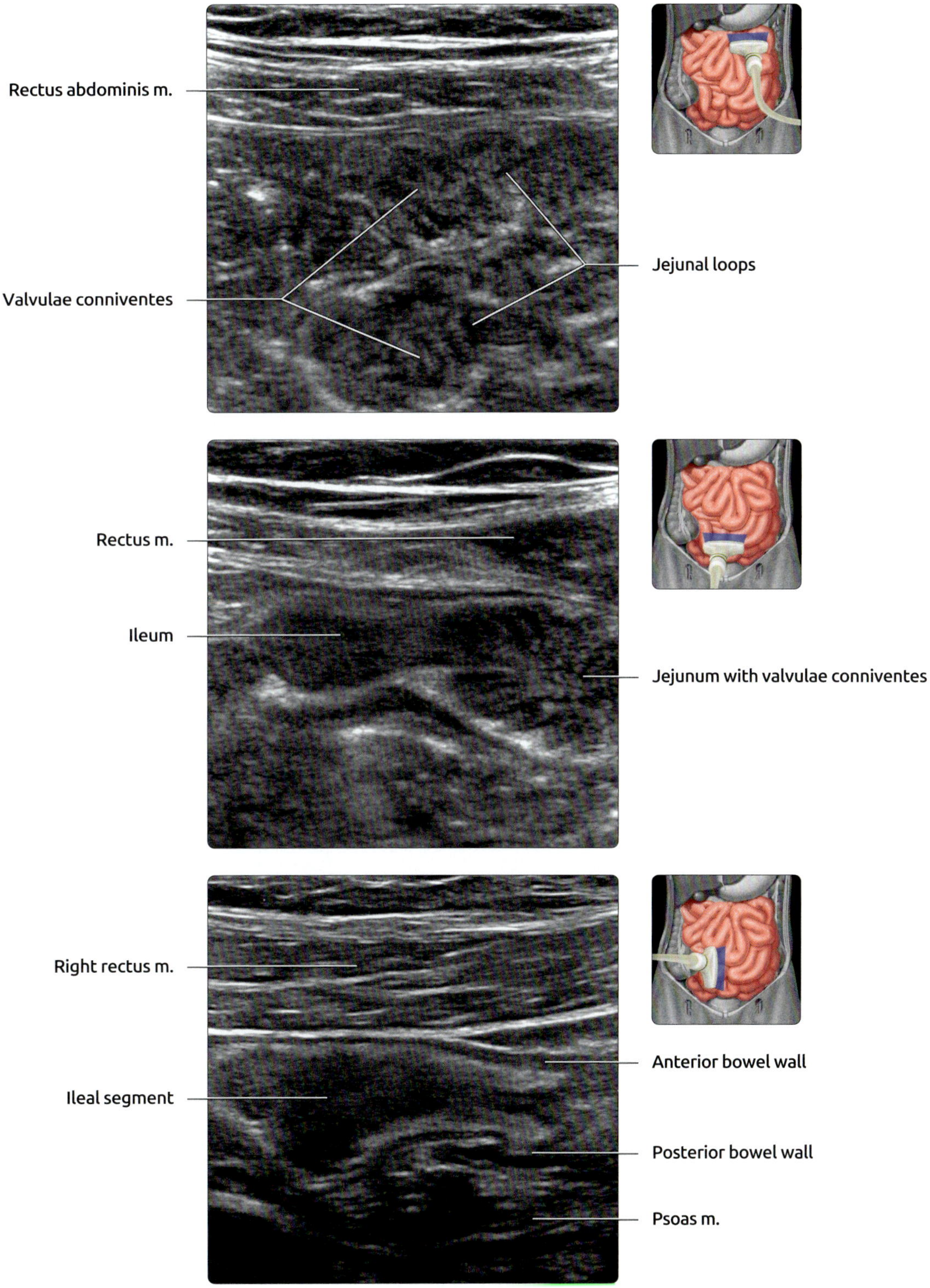

(Top) *Transverse oblique ultrasound through the left flank shows jejunal loops with mucosal folds, valvulae conniventes.* **(Middle)** *Transverse ultrasound through the lower abdomen close to midline shows jejunal ileal transition from a segment with folds to a segment with no folds.* **(Bottom)** *Sagittal oblique ultrasound through the right lower quadrant shows a normal ileal segment. Note the lack of folds and a normal gut signature in the wall.*

SMALL BOWEL AND LARGE BOWEL

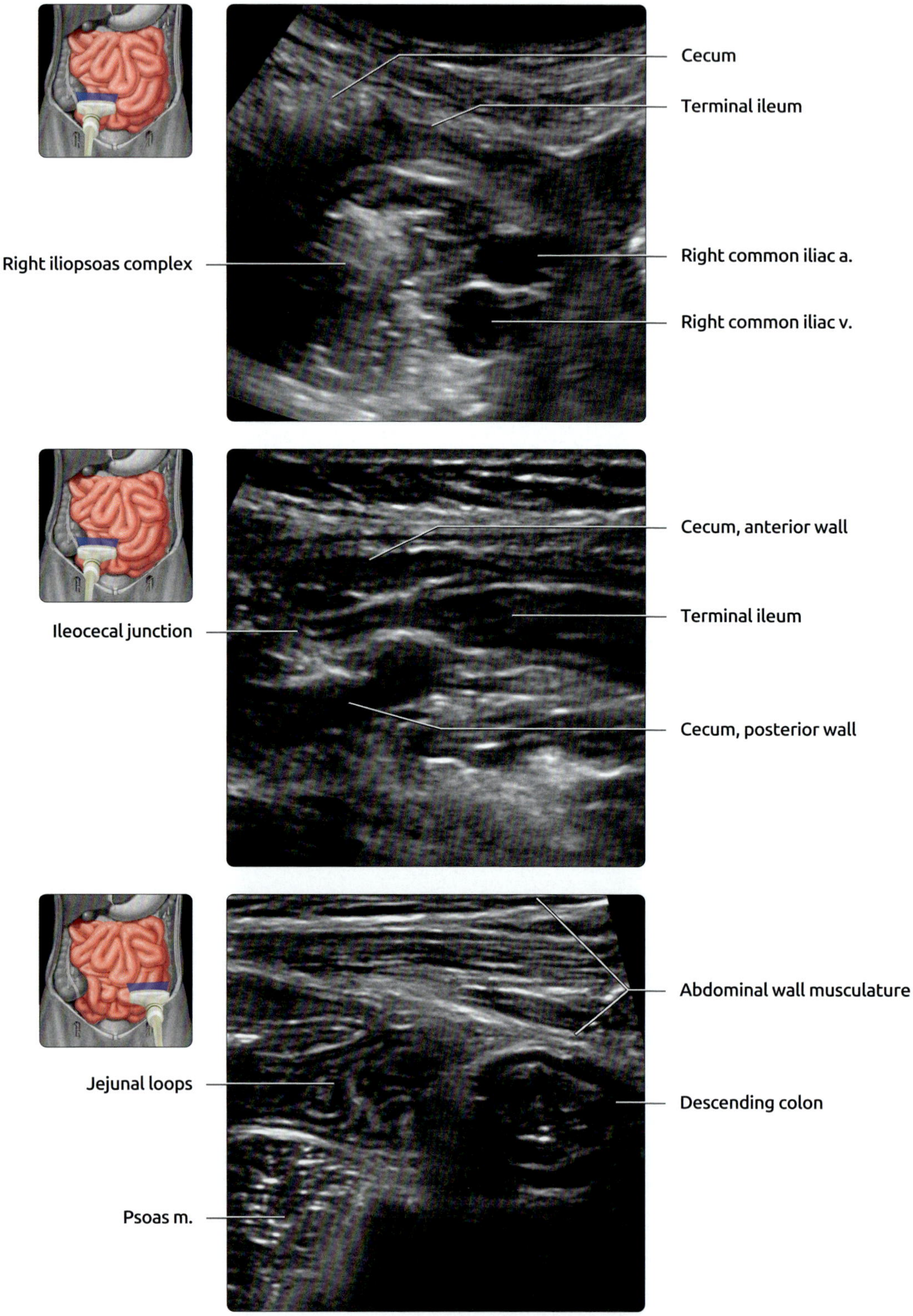

(Top) *Transverse oblique ultrasound through the right iliac fossa (RIF) with graded compression shows a compressed, normal terminal ileum (between the abdominal wall musculature anteriorly and psoas muscle posteriorly) leading to the cecum.* **(Middle)** *Transverse high-resolution ultrasound through the RIF in the same patient shows the ileocecal junction and the ileocecal valve.* **(Bottom)** *Transverse ultrasound through the left iliac fossa shows a short-axis view of the descending colon with gut signature. Medial to the descending colon, normal jejunal loops can be seen with valvulae conniventes.*

ILEOCECAL JUNCTION

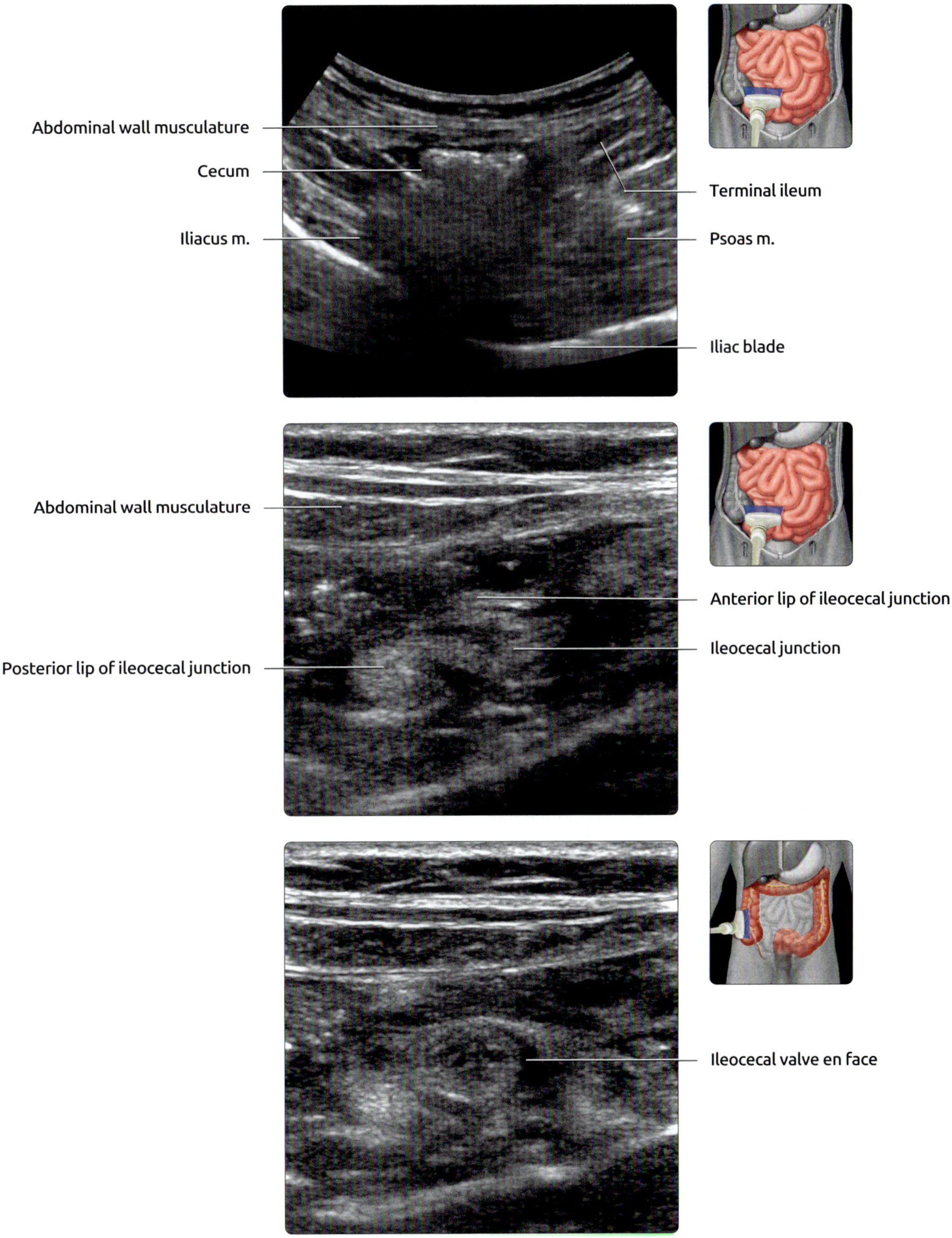

(Top) *Transverse ultrasound through the RIF shows the cecum, which is represented by curvilinear echogenicity with posterior reverberation. Note the terminal ileum compressed by the probe between the abdominal wall musculature and the iliopsoas complex posteriorly.* **(Middle)** *High-resolution transverse ultrasound through the RIF shows echogenic (fatty) lips of a normal ileocecal valve.* **(Bottom)** *Oblique right sagittal ultrasound through the RIF in the same patient shows the ileocecal junction end on.*

APPENDIX

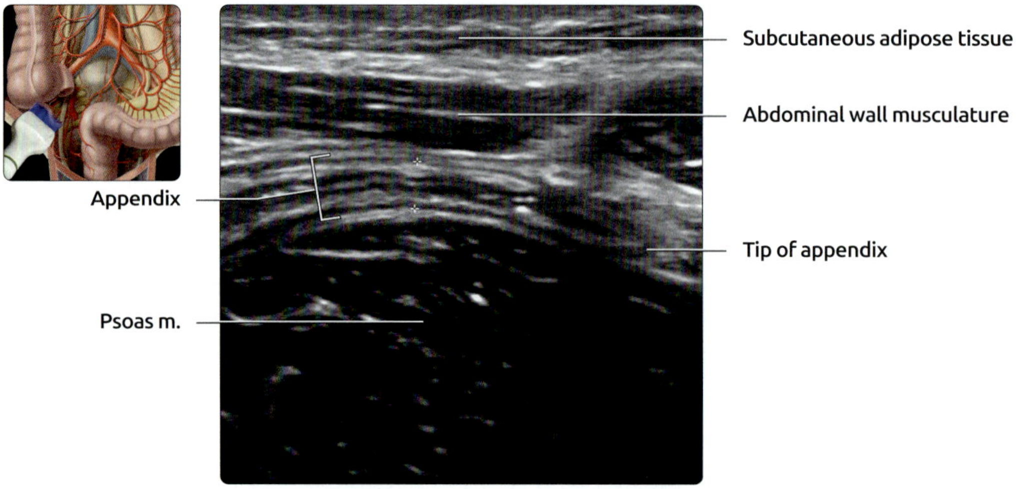

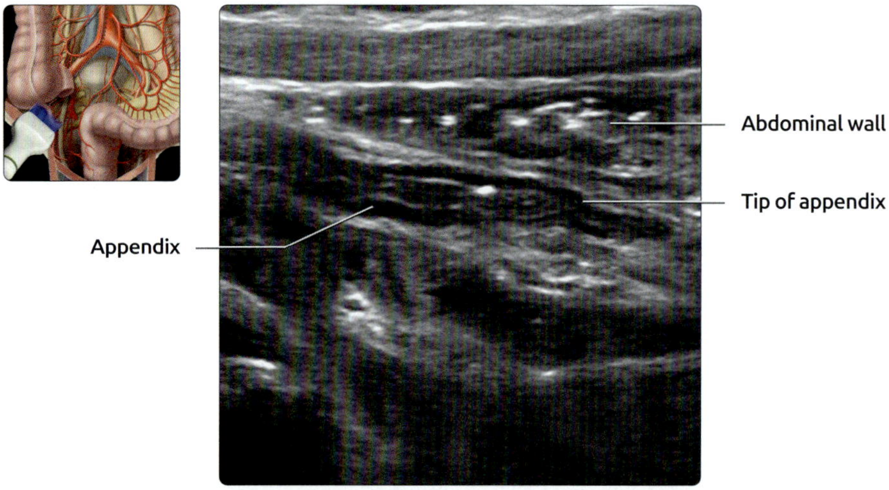

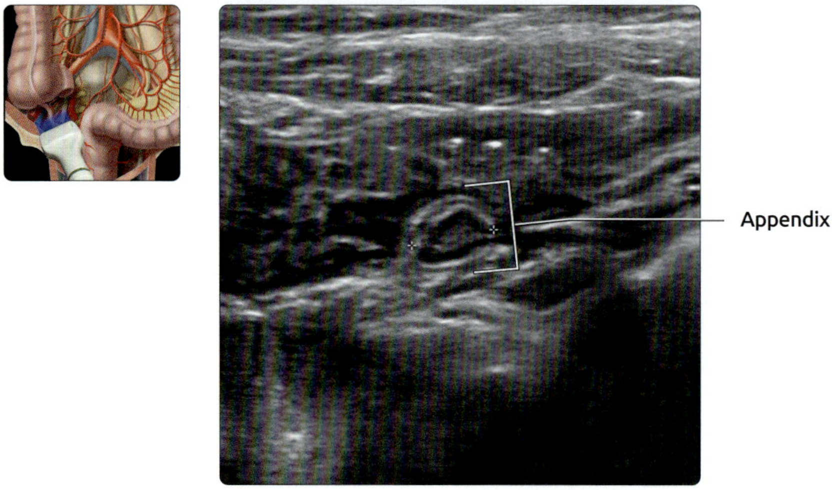

(Top) *Coronal oblique ultrasound through the RIF shows a long-axis normal appendix with a stratified mural appearance. Note the absence of periappendiceal inflammatory changes.* **(Middle)** *Transverse oblique ultrasound through the RIF shows a blind-ending tubular structure with a gut signature representing the long-axis view of a normal appendix. (Courtesy A. Law, MD.)* **(Bottom)** *Transverse oblique ultrasound through the RIF shows the short-axis view of a normal appendix with preservation of gut signature. Note the normal appearance of periappendicular fat. (Courtesy A. Law, MD.)*

LARGE BOWEL

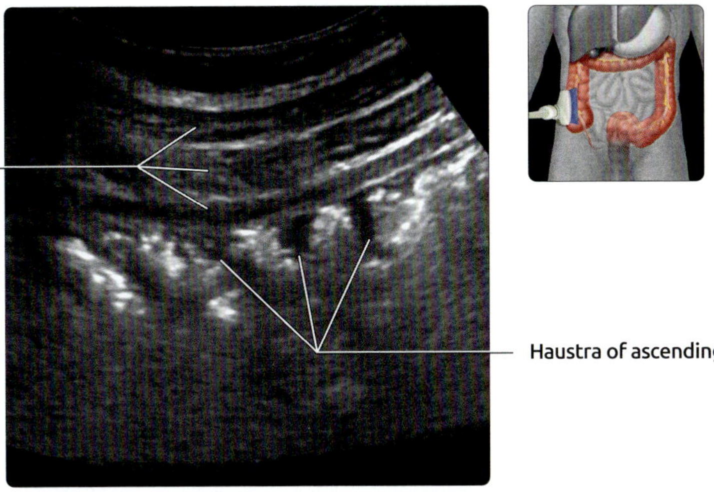

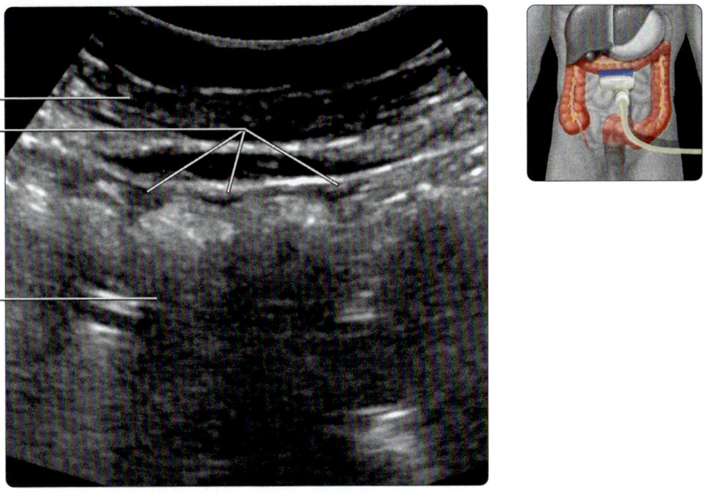

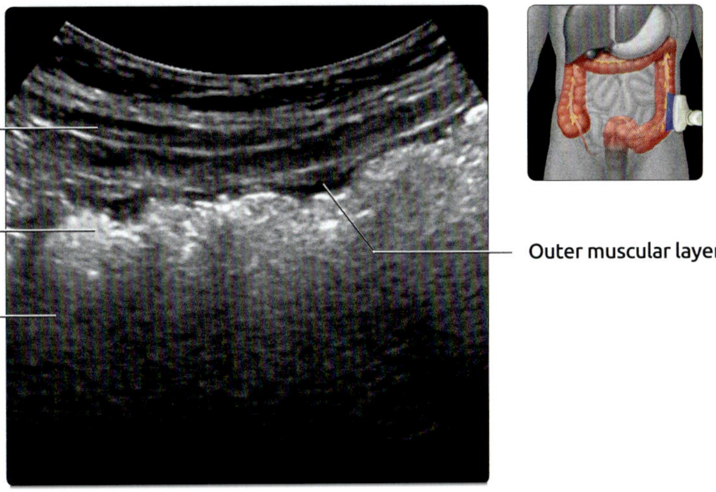

(Top) *Right sagittal ultrasound shows a normal ascending colon with curvilinear arcs of echogenicity from luminal gas/feces reflecting the normal haustra pattern.* **(Middle)** *Transverse ultrasound through the epigastric region close to the midline shows the normal haustral pattern of intraluminal gas/feces within the horizontally lying transverse colon. Note the posterior reverberation artifact from a gas-filled colon.* **(Bottom)** *Left sagittal ultrasound through the left side of the abdomen shows the descending colon represented by arcs of echogenicity with posterior reverberation artifacts due to gaseous contents of the colon. Note the normal haustral pattern. Images obtained with curvilinear probe or linear probe with virtual convex are better for orientation of anatomy within the peritoneal cavity.*

LARGE BOWEL

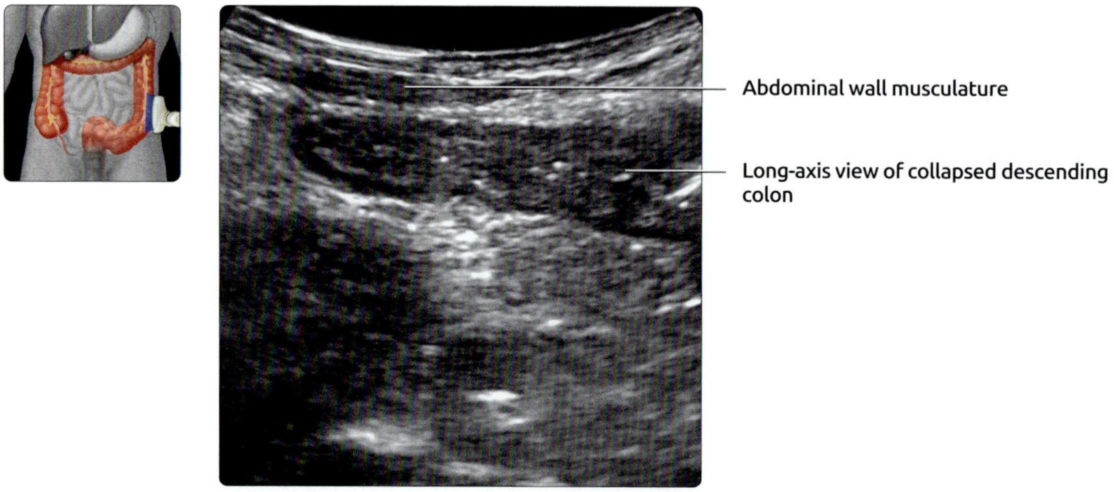

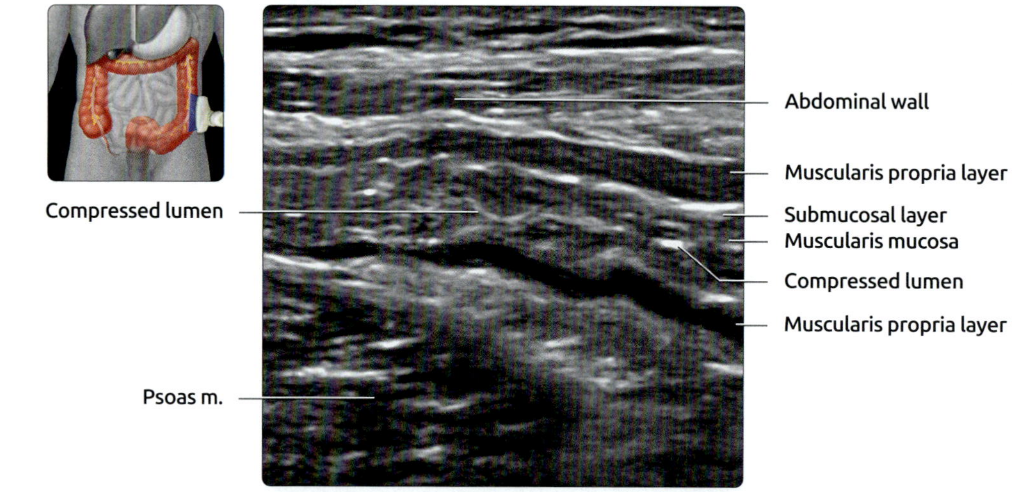

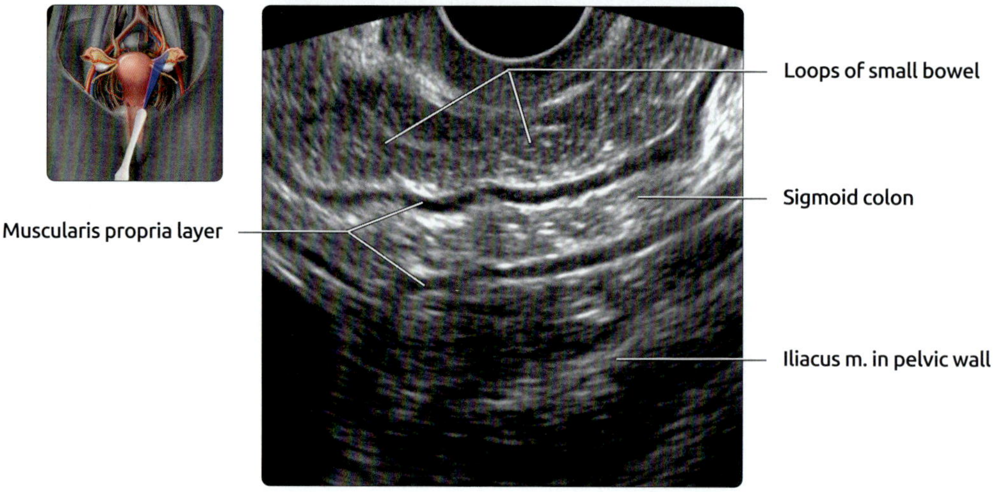

(Top) *Left sagittal ultrasound shows the normal descending colon in a collapsed state and compressed by the ultrasound probe. Note the gut signature. This is an alternative appearance to the descending colon when it is collapsed and empty.* **(Middle)** *High-resolution left sagittal ultrasound obtained with a higher frequency linear probe from the same patient shows the collapsed descending colon with gut signature. The hypoechoic outer layer represents the muscularis propria layer.* **(Bottom)** *Transvaginal ultrasound shows the sigmoid colon and pelvic loops of small bowel. Note the hypertrophied outer muscularis propria layer of the sigmoid colon, seen in patients with irritable bowel syndrome and early diverticular disease.*

RECTOSIGMOID REGION

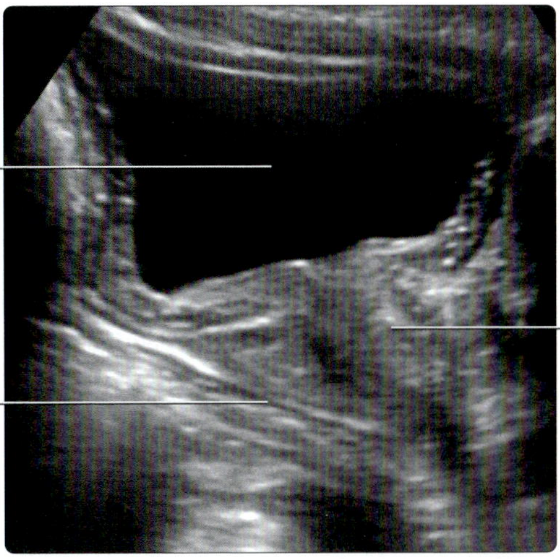

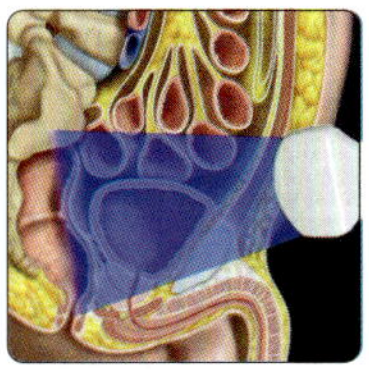

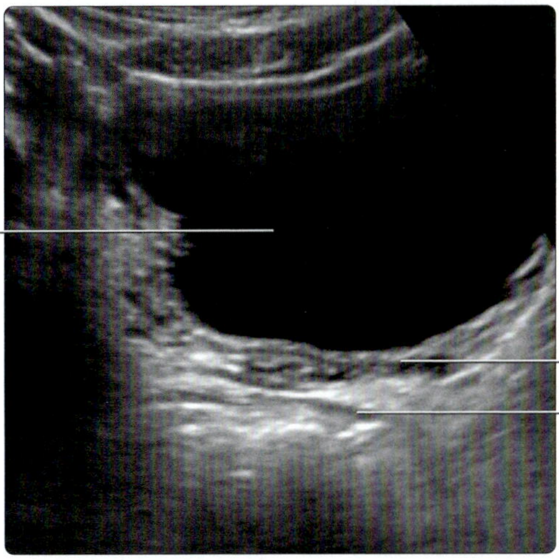

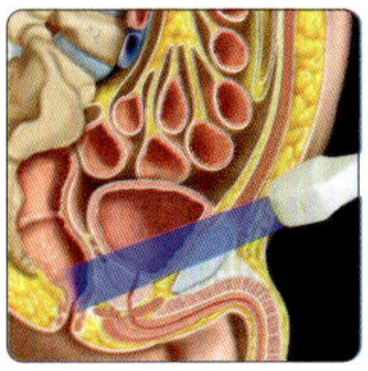

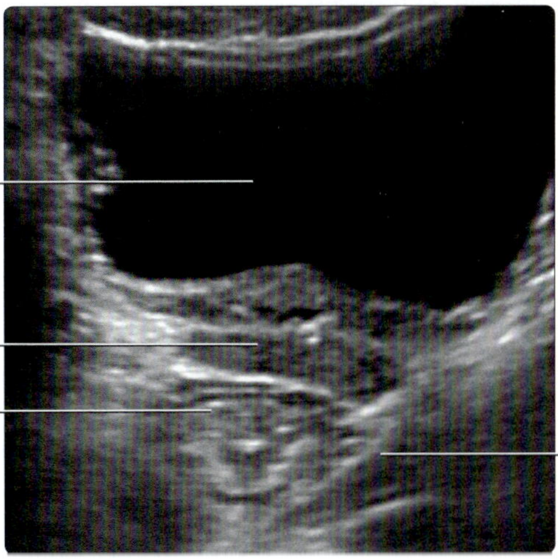

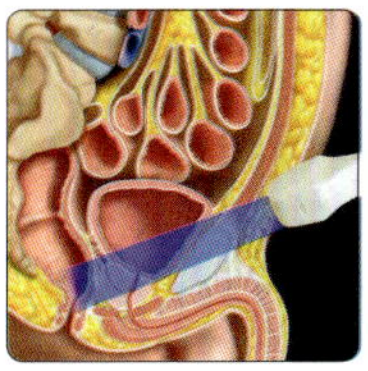

(Top) *Midline sagittal ultrasound in a male patient shows the anterior wall of the rectosigmoid region and its relationship to the prostate gland anteriorly.* **(Middle)** *Transverse ultrasound through the pelvis in the same patient (with cystic fibrosis) shows the anterior wall of the rectosigmoid region and its relationship to the seminal vesicles anteriorly (note the small seminal vesicles seen here).* **(Bottom)** *Transverse ultrasound at a lower level in the same patient shows the lower rectum, pelvic floor, and anterior relationship to the prostate gland.*

GROSS ANATOMY

Overview

- Major lymphatic vessels and nodal chains lie along major blood vessels (aorta, inferior vena cava, iliac)
- Lymph nodes carry same name as vessel they accompany
- Lymph from alimentary tract, liver, spleen, and pancreas passes along celiac, superior mesenteric chains to nodes
  - Efferent vessels from alimentary nodes form intestinal lymphatic trunks
  - Cisterna chyli (chyle cistern)
    - Formed by confluence of intestinal lymphatic trunks and right and left lumbar lymphatic trunks, which receive lymph from nonalimentary viscera, abdominal wall, and lower extremities
    - May be discrete sac or plexiform convergence
- Thoracic duct: Inferior extent is chyle cistern at L1-L2 level
  - Formed by convergence of main lymphatic ducts of abdomen
  - Ascends through aortic hiatus in diaphragm to enter posterior mediastinum
  - Ends by entering junction of left subclavian and internal jugular veins
- Lymphatic system drains surplus fluid from extracellular spaces and returns it to bloodstream
  - Important function in defense against infection, inflammation, and tumor via lymphoid tissue present in lymph nodes, gut wall, spleen, and thymus
  - Absorbs and transports dietary lipids from intestine to thoracic duct and bloodstream
- Lymph nodes
  - Composed of cortex and medulla
  - Invested in fibrous capsule, which extends into nodal parenchyma to form trabeculae
  - Internal honeycomb structure filled with lymphocytes that collect and destroy pathogens
  - Hilum: In concave side, with artery and vein, surrounded by fat

Abdominopelvic Nodes

- Preaortic nodes
  - Celiac nodes: Drainage from gastric nodes, hepatic nodes, and pancreaticosplenic nodes
  - Superior and inferior mesenteric nodes: Drainage from mesenteric nodes
- Lateral aortic nodes
  - Drainage from kidneys, adrenal glands, ureter, posterior abdominal wall, testes and ovary, uterus, and fallopian tubes
- Retroaortic nodes
  - Drainage from posterior abdominal wall
- External iliac nodes
  - Primary drainage from inguinal nodes
  - Flow into common iliac nodes
- Internal iliac nodes
  - Drainage from inferior pelvic viscera, deep perineum, and gluteal region
  - Flow into common iliac nodes
- Common iliac nodes
  - Drainage from external iliac, internal iliac, and sacral nodes
  - Flow into lumbar (lateral aortic) chain of nodes
- Superficial inguinal nodes
  - In superficial fascia parallel to inguinal ligament, along cephalad portion of greater saphenous vein
  - Receive lymphatic drainage from superficial lower extremity, superficial abdominal wall, and perineum
  - Flow into deep inguinal and external iliac nodes
- Deep inguinal nodes
  - Along medial side of femoral vein, deep to fascia lata and inguinal ligament
  - Receive lymphatic drainage from superficial inguinal and popliteal nodes
  - Flow into external iliac nodes

IMAGING ANATOMY

Overview

- CT is test of choice for cancer staging
- May be supplemented by PET/CT in select cancers
- US may be useful in children or thin adults
  - Normal nodes are elliptical with echogenic fatty hilum and uniform hypoechoic cortex
  - Normal lymph nodes rarely detected on abdominal US
- Normal diameter of lymph node varies depending on location
  - Short-axis diameter
    - Abdominopelvic < 10 mm
    - Hepatogastric ligament < 8 mm
    - Retrocrural < 6 mm

ANATOMY IMAGING ISSUES

Imaging Recommendations

- Transducer: 2-5 MHz or 5-9 MHz for thinner patients
- Patient examined in supine position with < 4 hours of fasting to decrease bowel gas
- Graded compression technique to clear overlying bowel loops

CLINICAL IMPLICATIONS

Clinical Importance

- Nodal enlargement is nonspecific, may be neoplastic, inflammatory, or reactive
- Normal-sized lymph nodes may harbor metastatic malignancy
- Node morphology is more specific for pathology
  - Abnormal nodes have replacement or loss of fatty hilum
  - Look for central necrosis, cystic change, or calcification
- Lymphoma
  - Multiple enlarged hypoechoic or anechoic nodes
- Metastatic lymphadenopathy
  - More echogenic and heterogeneous nodes compared to lymphomatous nodes
- Infectious/reactive lymphadenopathy
  - Nonspecific sonographic features
  - May contain necrotic centers in mycobacterial infection

RETROPERITONEAL LYMPH NODES

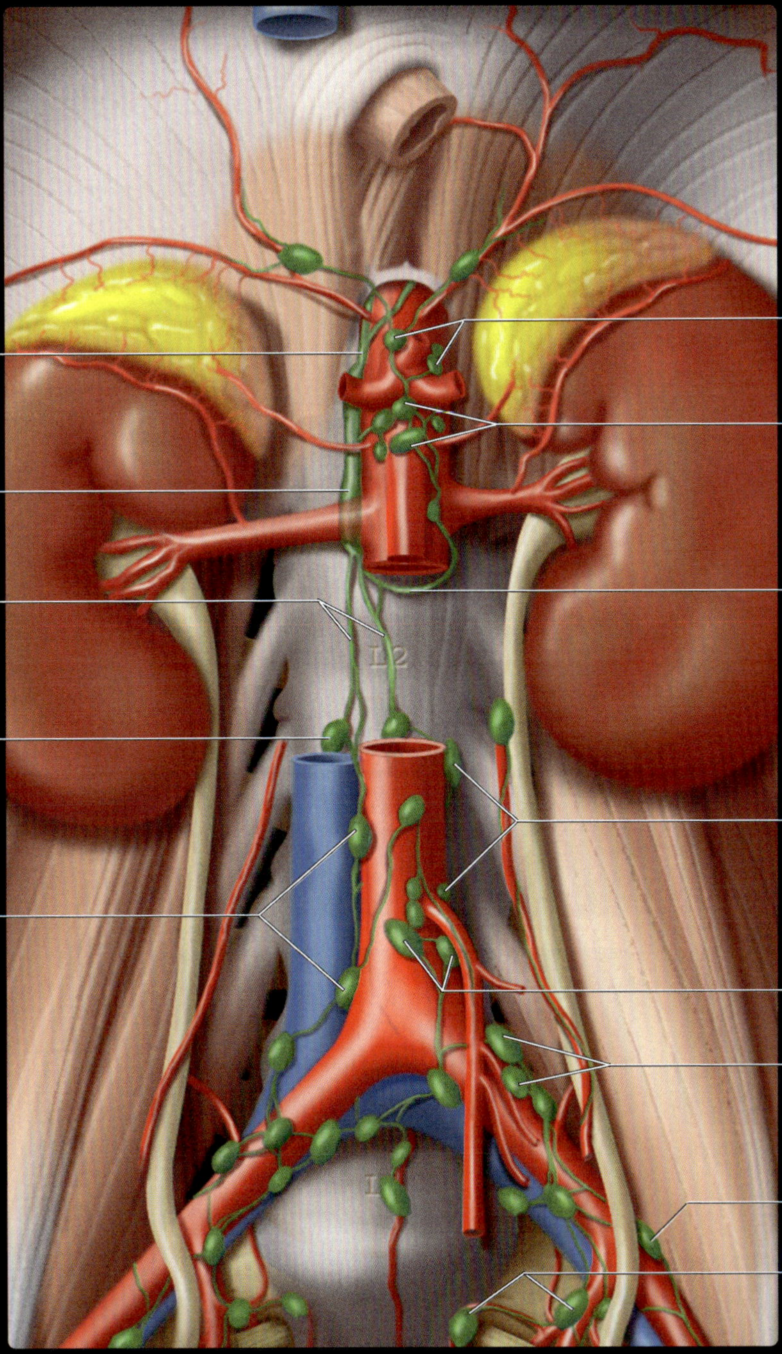

Graphic shows that the major lymphatics and lymph nodes of the abdomen are located along and share the same name as the major blood vessels, such as the external iliac nodes, celiac, and superior mesenteric nodes. The paraaortic and paracaval nodes are also referred to as the lumbar nodes and receive afferents from the lower abdominal viscera, abdominal wall, and lower extremities; they are frequently involved in inflammatory and neoplastic processes. The lumbar trunks join with an intestinal trunk (at ~ the L1 level) to form the cisterna chyli, which may be a discrete sac or a plexiform convergence. The cisterna chyli and other major lymphatic trunks join to form the thoracic duct, which passes through the aortic hiatus to enter the mediastinum. After picking up additional lymphatic trunks within the thorax, the thoracic duct empties into the left subclavian or innominate vein.

NONENLARGED NODES AND PATHOLOGIC NODES

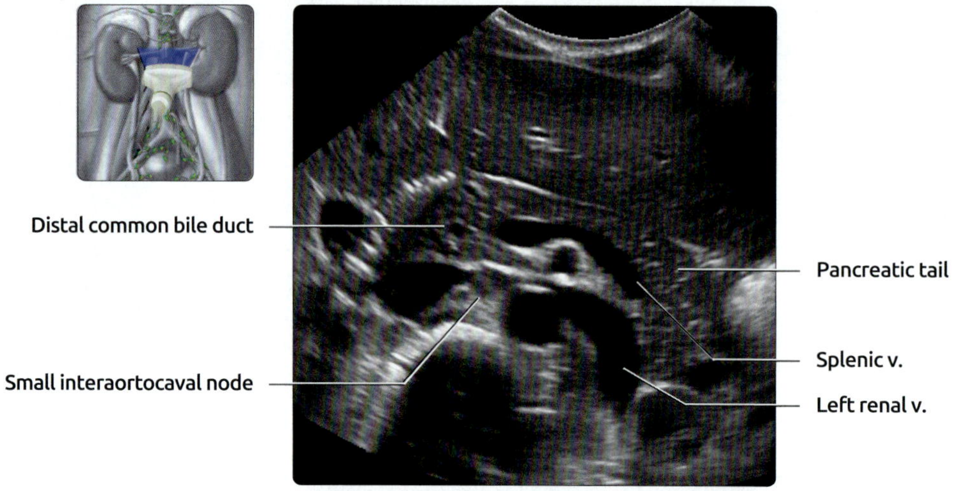

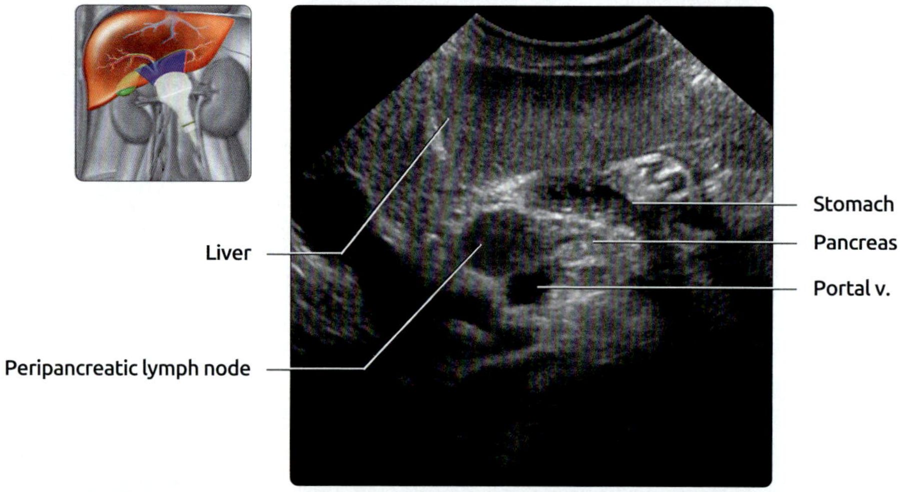

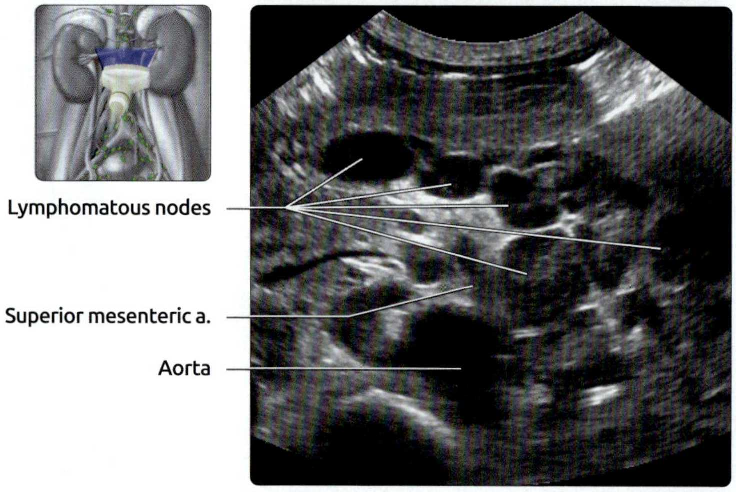

(Top) *Transverse US at the level of the pancreas and splenic vein shows a small interaortocaval node.* **(Middle)** *Transverse US of the epigastric region shows an enlarged hypoechoic peripancreatic lymph node anterior to the portal vein in a patient with hepatitis C. Normal-sized lymph nodes are rarely seen in adult abdominal US.* **(Bottom)** *Transverse US of the upper midline abdomen in a patient with lymphoma shows multiple abnormal enlarged hypoechoic lymph nodes around the superior mesenteric artery.*

LYMPHANGIOGRAM

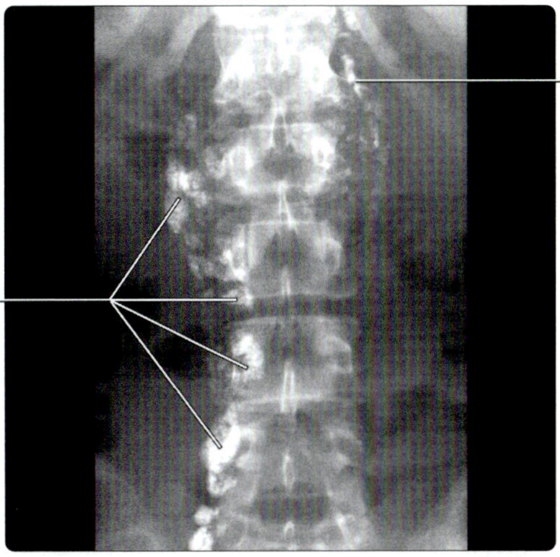

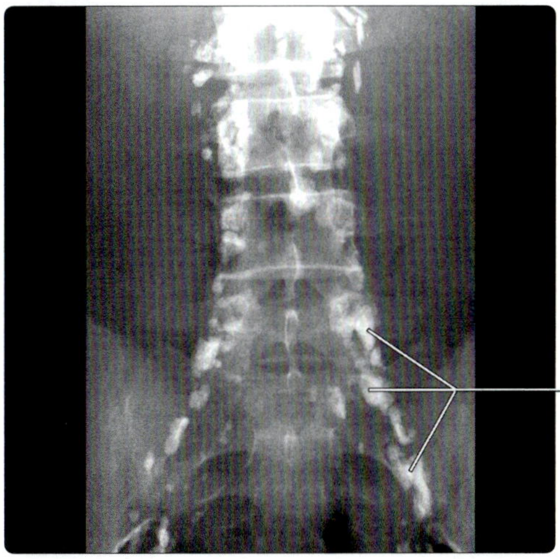

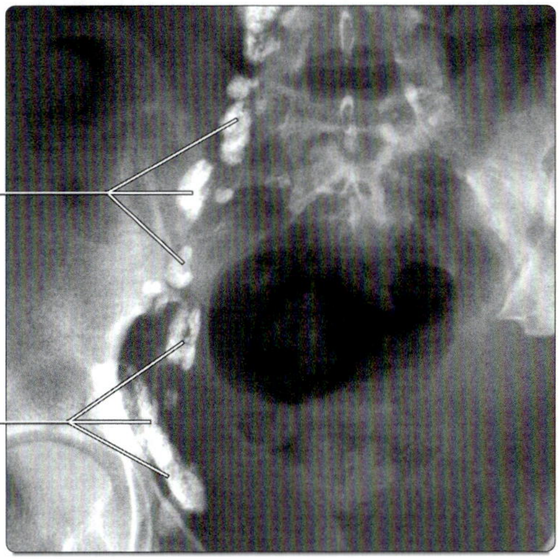

(Top) *This is the 1st of 3 images from a lymphangiogram. Iodinated oil is slowly infused into the lymphatics of the foot to produce opacification of the lymph channels and nodes. Note the subcentimeter (short-axis) diameter of these normal retroperitoneal lymph nodes.* **(Middle)** *Lymphatic channels and lymph nodes parallel the course of major blood vessels and share similar names, such as these common iliac nodes.* **(Bottom)** *With the availability of CT, MR, and PET/CT, lymphangiograms are rarely performed; however, they provide a unique depiction of the lymphatic system.*

TERMINOLOGY

Definitions

- "Proximal" and "distal" in arterial and venous systems apply to position of arterial and venous segment in relation to heart (rather than direction of flow)
- Aneurysm is focal increase in caliber of artery with diameter of dilated segment measuring at least 1.5x > adjacent unaffected segments

GROSS ANATOMY

Overview

- **Abdominal aorta**
  - Enters abdomen at T12 level, bifurcates at L4
  - Level of origins of major branches: Celiac axis (T12), superior mesenteric artery (SMA) (L1), renal arteries (L1/2), inferior mesenteric artery (IMA) (L3), common iliac arteries (L4)
- **Inferior vena cava (ICV)**
  - Blood from alimentary tract passes through portal venous system before entering IVC through hepatic veins
  - Begins at L5 level with union of common iliac veins
  - Leaves abdomen via IVC hiatus in diaphragm at T8 level
  - IVC tributaries correspond to paired visceral and parietal branches of aorta
  - IVC development has complex embryology
    - Various anomalies are common (up to 10% of population), especially at and below level of renal veins
    - All are variations of persistence/regression of embryologic sub- and supracardinal veins

IMAGING ANATOMY

Overview

- Not all branches of abdominal aorta and tributaries of IVC can be well seen on ultrasound examination
- Major arterial branches of abdominal aorta seen on ultrasound
  - Celiac artery, common hepatic artery, splenic artery, SMA, IMA, renal arteries, common iliac arteries
- Major venous tributaries draining into IVC
  - Common iliac veins, renal veins, hepatic veins

Internal Contents

- **Abdominal aorta**
  - Normal peak systolic velocity (PSV): 60-110 cm/sec
  - Spectral Doppler waveform
    - Upper: Narrow, well-defined systolic complex with forward flow during diastole
    - Mid: Reduced diastolic flow
    - Distal: Absent diastolic flow, similar to lower limb arteries
  - Normal caliber: 15-25 mm
    - Upper: 22 mm above renal arteries
    - Middle: 18 mm below renal arteries
    - Lower: 15 mm above bifurcation
  - Best ultrasound imaging plane: Both transverse and longitudinal
- **Celiac axis**
  - Normal PSV: 98-105 cm/sec
  - Spectral Doppler demonstrates low-resistance flow with high end-diastolic velocities
  - Flow velocity not dependent on food intake
  - Normal caliber: 6-10 mm
  - Best ultrasound imaging plane: Transverse plane: To show typical T-shaped bifurcation
- **Common hepatic artery**
  - Normal PSV: 70-120 cm/sec
  - Spectral Doppler shows low-resistance flow characteristics with large amount of continuous flow in diastole
  - Normal caliber: 4-10 mm
  - Best ultrasound imaging plane
    - Start transverse midline to follow common hepatic artery to right of T-shaped bifurcation from celiac axis
    - Gastroduodenal artery may be seen to arise from common hepatic artery along anterosuperior aspect of pancreas; thereafter, common hepatic artery becomes proper hepatic artery
- **Splenic artery**
  - Normal PSV: 70-110 cm/sec
  - Spectral Doppler shows typically turbulent flow due to tortuosity of vessel
  - Normal caliber: 4-8 mm
  - Best ultrasound imaging plane
    - Transverse midline approach: Shows proximal portion of artery well
    - Intercostal through spleen, using it as acoustic window: Useful for showing distal splenic artery around hilum
- **SMA**
  - Normal PSV: 97-142 cm/sec
  - Spectral Doppler demonstrates high-impedance flow with low diastolic velocities during fasting due to relative vasoconstriction
  - End-diastolic velocity increases after meal due to vasodilation of mesenteric branches
    - Typically 30-90 min after meals
  - Normal caliber: 5-8 mm
  - Best ultrasound imaging plane
    - Longitudinal midline approach: Best for evaluation of SMA blood flow
    - Transverse plane: Useful for identifying short anteriorly directed stump, shows dot-like appearance, surrounded by distinctive triangular mantle of fat
- **IMA**
  - Normal PSV: 93-189 cm/sec
  - Spectral Doppler demonstrates high-impedance flow with low diastolic velocities during fasting due to relative vasoconstriction
  - End-diastolic velocity increases after meal due to vasodilation of mesenteric branches
  - Normal caliber: 1-5 mm
  - Best imaging plane: Transverse plane following line of aorta
    - Origin of IMA arises from below origins of renal arteries and may be anterior or slightly to left of midline
- **Renal arteries**
  - Normal PSV: 60-140 cm/sec

  - Not > 180 cm/sec
- Spectral Doppler demonstrates open systolic window, rapid systolic upstroke occasionally followed by secondary slower rise to peak systole with subsequent gradual diastolic decline but with persistent forward flow in diastole
- Normal caliber: 5-8 mm
- Best imaging plane
  - Transverse anterior midline approach: Best for identification of origins of renal arteries
  - Posterolateral approach using kidneys as acoustic window: Useful for visualization of distal portions of renal arteries

- **Common iliac arteries**
  - Normal spectral Doppler shows characteristic triphasic waveform
    - Initial high-velocity peak forward-flow phase resulting from cardiac systole
    - Brief phase of reverse flow in early diastole
    - Low velocity forward flow in diastole
  - Normal caliber: 8-12 mm
  - Best imaging plane: Transverse anterior approach and oblique anterior approach along long axis of iliac arteries
  - Stenosis causing 1-19% diameter reduction
    - Triphasic waveform with minimal spectral waveform broadening
    - PSV increase < 30% relative to adjacent proximal segment; proximal and distal waveform remain normal
  - Stenosis causing 20-49% diameter reduction
    - Triphasic waveform usually maintained, though reverse flow component may be diminished
    - Spectral broadening is prominent with filling in of area under systolic peak
    - PSV 30-100% increase relative to adjacent proximal segment, proximal and distal segment remain normal
  - Stenosis causing 50-99% diameter reduction
    - Monophasic waveform with loss of reverse flow component, forward flow throughout cardiac cycle
    - Extensive spectral broadening
    - PSV > 100% increase relative to adjacent proximal segment
    - Distal waveform is monophasic with reduced systolic velocity
  - Occlusion
    - No flow on color or spectral Doppler
    - Preocclusive thump may be heard proximal to site of obstruction
    - Distal waveforms are monophasic with reduced systolic velocities
- **IVC**
  - Normal PSV: 44-118 cm/sec
  - Spectral Doppler shows slow flow that varies with respiration and cardiac pulsation
  - Normal caliber: 5-29 mm during quiet inspiration
  - Best imaging plane: Both transverse and longitudinal
- **Common iliac veins**
  - Spectral Doppler shows 5 normal characteristics
    - Spontaneous flow, phasic flow, flow ceases with Valsalva maneuver, flow augmentation with distal compression, unidirectional flow towards heart
  - Best imaging plane: Transverse anterior approach and oblique anterior approach along long axis of iliac veins
- **Renal veins**
  - Normal PSV: 18-33 cm/sec
  - Right renal vein is relatively short and drains directly into IVC
  - Left renal vein runs slightly longer course
    - Receives left gonadal vein, usually courses anterior to aorta before joining IVC
  - Spectral Doppler of right renal vein mirrors pulsatility of IVC
  - Spectral Doppler of left renal vein may show only slight variability of flow velocities consequent upon cardiac and respiratory activity
  - Normal caliber: 4-9 mm
  - Best imaging plane: Transverse anterior approach
- **Hepatic veins**
  - Normal PSV: 16-40 cm/sec
  - Spectral Doppler shows triphasic waveform due to transmitted cardiac activity
  - Best imaging plane: Transverse/oblique subcostal approach with cranial angulation

ANATOMY IMAGING ISSUES

Imaging Recommendations

- Use 2- to 5-MHz transducer
- Fasting for 12 hours is recommended to reduce interference by bowel gas
  - Imaging patient in morning after overnight fast is most convenient protocol
  - Satisfactory Doppler signal for aorta and IVC can be obtained in up to 90% of patients
- Lateral decubitus position and graded compression to displace intervening bowel gas may be useful
- Angle correction is crucial in spectral Doppler assessment
- Detailed delineation of branches of aorta and IVC better assessed on CTA or MRA
- Digital subtractive angiography usually reserved for when intervention may be required (e.g., embolization of mesenteric artery branches in GI bleed, renal artery stenting, etc.)

Imaging Pitfalls

- Bowel gas, patient body habitus, and operator dependence are main factors contributing to suboptimal ultrasound examination of aorta and IVC

AORTA

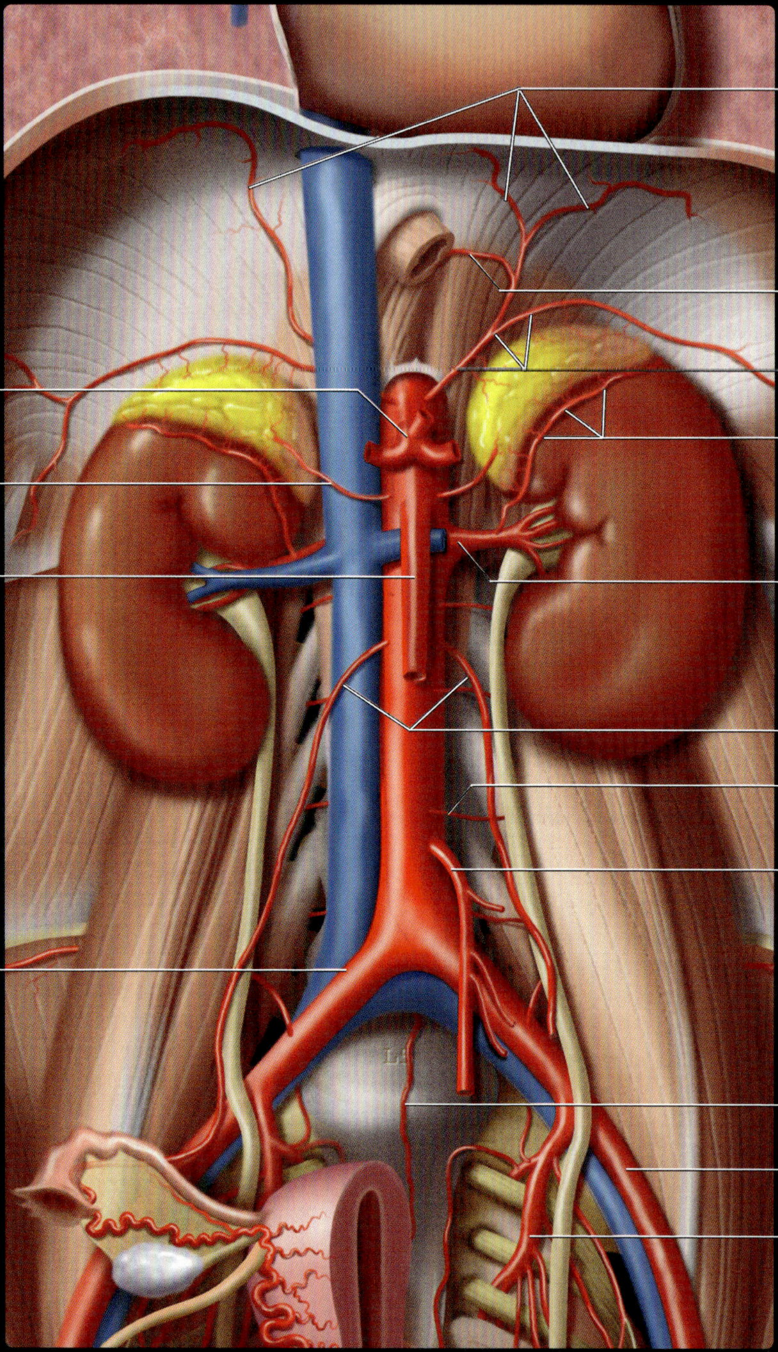

Graphic shows the major arteries to the GI tract arise as unpaired vessels from the aortic midline plane and include the celiac, superior, and inferior mesenteric arteries. Branches to the urogenital and endocrine organs arise as paired vessels in the lateral plane and include the renal, adrenal, and gonadal (testicular or ovarian) arteries. The diaphragm and posterior abdominal wall are supplied by paired branches in the posterolateral plane, including the inferior phrenic and lumbar arteries (4 pairs, only 1 of which is labeled in this graphic). The anterior abdominal wall is supplied by the inferior epigastric and deep circumflex iliac arteries, both branches of the external iliac artery. The inferior epigastric artery turns superiorly to run in the rectus sheath where it anastomoses with the superior epigastric artery, a terminal branch of the internal mammary (thoracic) artery.

CT ANGIOGRAM

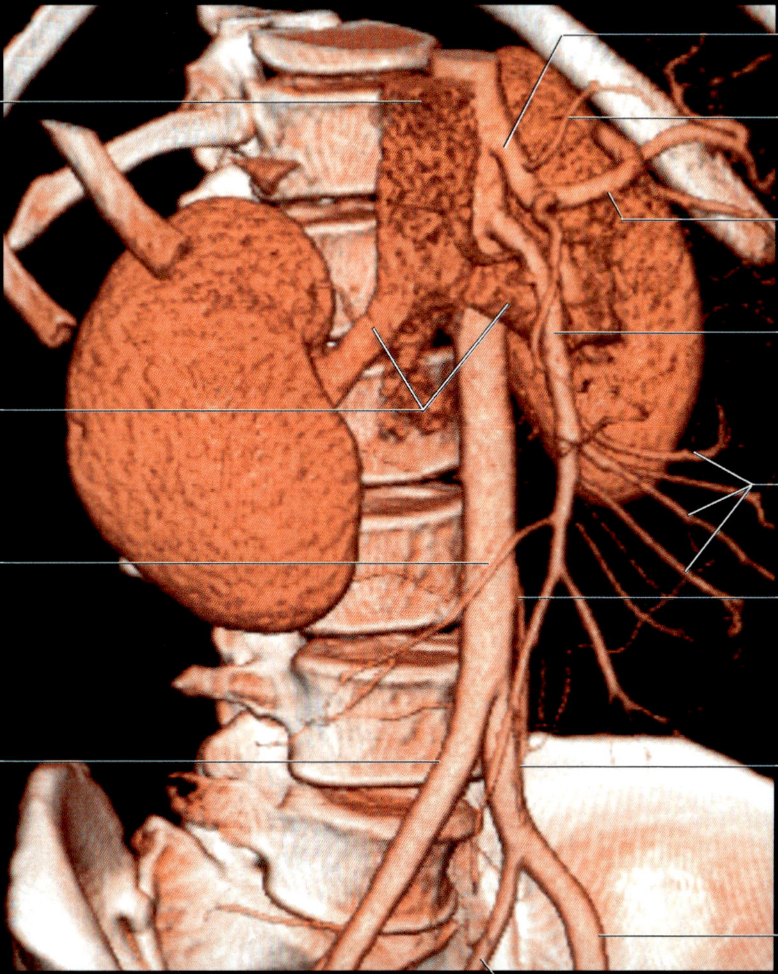

Volume-rendered CT angiogram shows the aorta and some of its major abdominal branches. This image was obtained in the late arterial phase of imaging, resulting in some opacification of the renal veins and the suprarenal inferior vena cava (IVC). The infrarenal IVC is not yet opacified because the circulation to the lower abdominal organs and legs is neither as abundant nor rapid as it is to the kidneys.

INFERIOR VENA CAVA

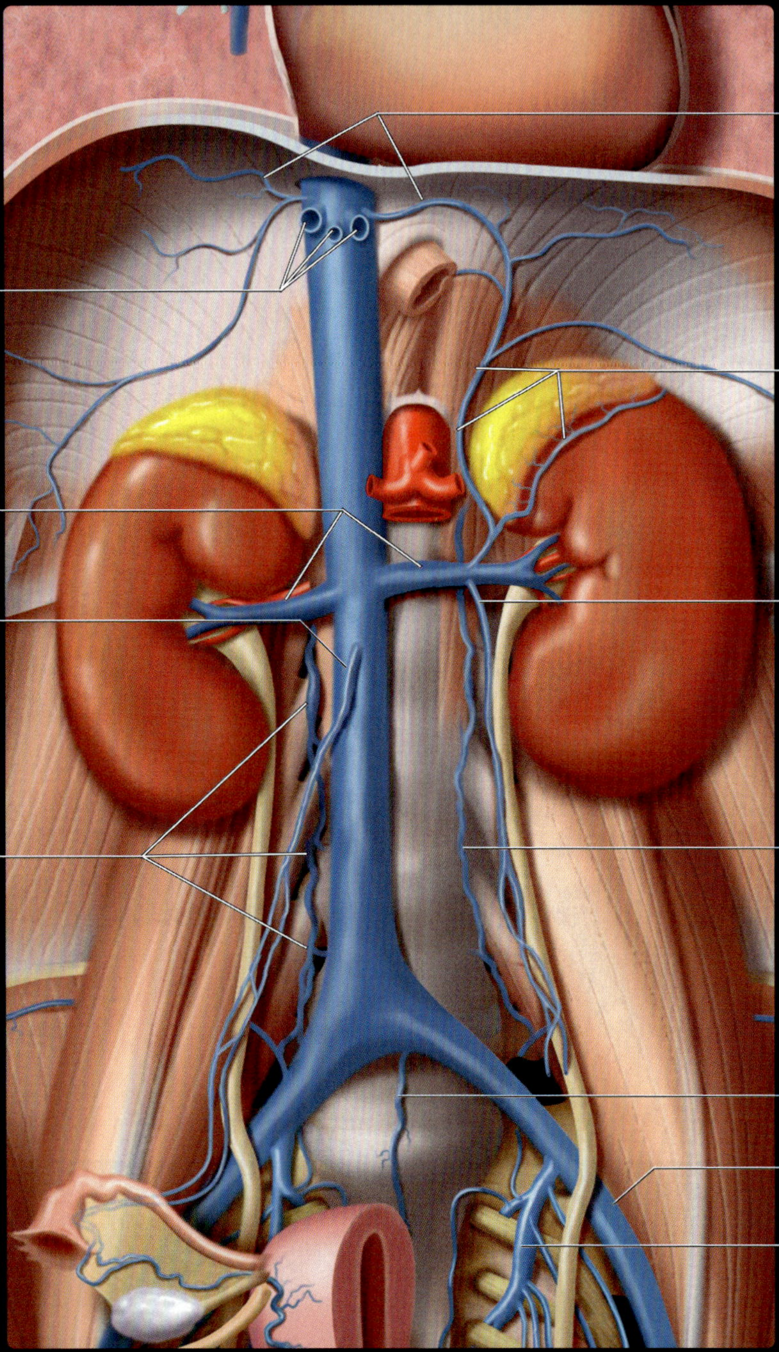

The IVC begins at the L5 level with the confluence of the common iliac veins, which are themselves the result of the confluence of the internal and external iliac veins. Note the ascending lumbar veins, which anastomose freely between the IVC, azygous and hemiazygos veins, and renal veins. These are an important pathway for collateral flow in the event of obstruction of the IVC or one of its major tributaries, and these veins play an important role in spread of tumor and infection from the pelvis and spine to the thorax, upper spine, and brain. The right renal vein rarely receives tributaries, while the left receives the gonadal, adrenal, and lumbar veins. The left adrenal vein also anastomoses with the inferior phrenic vein. The hepatic veins return blood from the liver as they join the IVC just below its hiatus in the diaphragm at ~ the T8 level.

COMMON VARIATIONS OF INFERIOR VENA CAVA

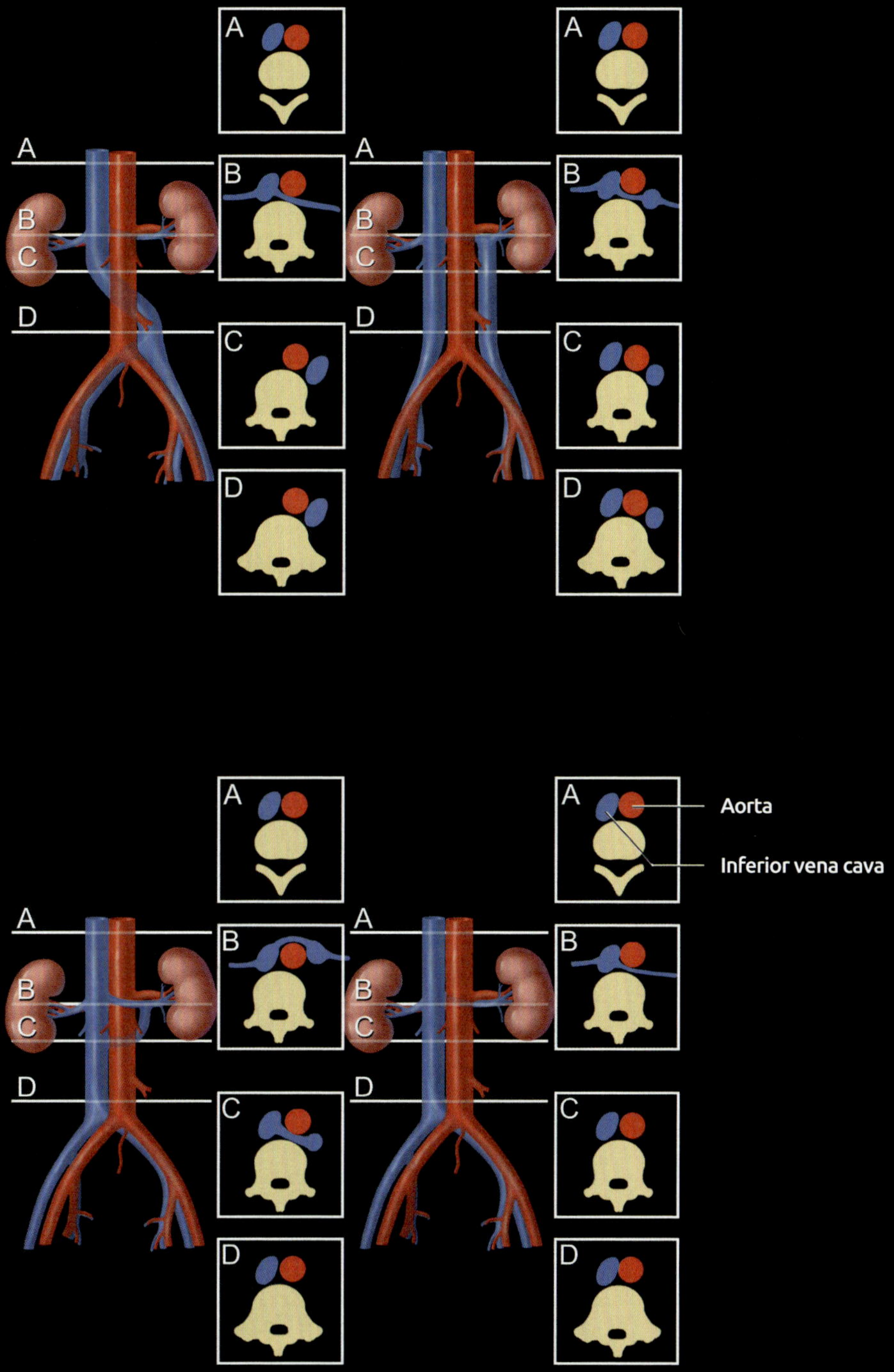

(Top) *The first 2 of 4 graphics illustrating common variations of the IVC is shown. The labeled lines on the frontal graphics correspond to the levels of the axial sections. The left graphic shows transposition of the IVC in which the infrarenal portion of the IVC lies predominantly to the left side of the aorta. A more common anomaly is shown on the right graphic, a "duplication" of the IVC in which the left common iliac vein continues in a cephalad direction without crossing over to join the right iliac vein. Instead, it joins the left renal vein and then crosses over to the right. The suprarenal IVC has a conventional course and appearance.* **(Bottom)** *The left graphic shows a circumaortic left renal vein with the smaller, more cephalic vein passing in front of the aorta and the larger vein passing behind and caudal. The right graphic shows a completely retroaortic renal vein.*

PROXIMAL AORTA, SAGITTAL VIEW

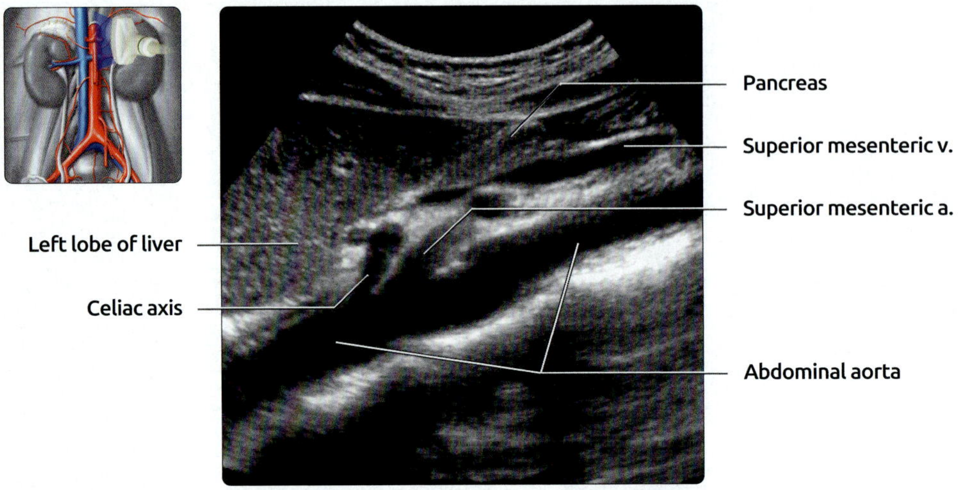

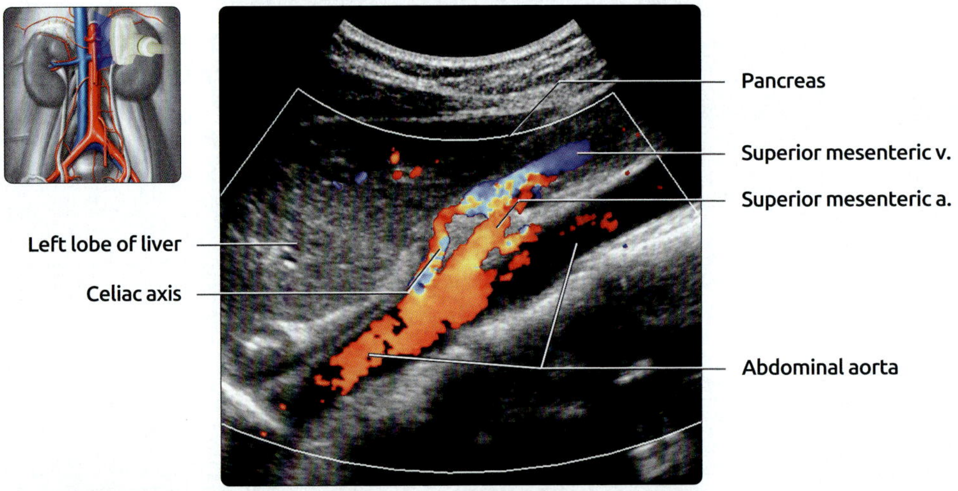

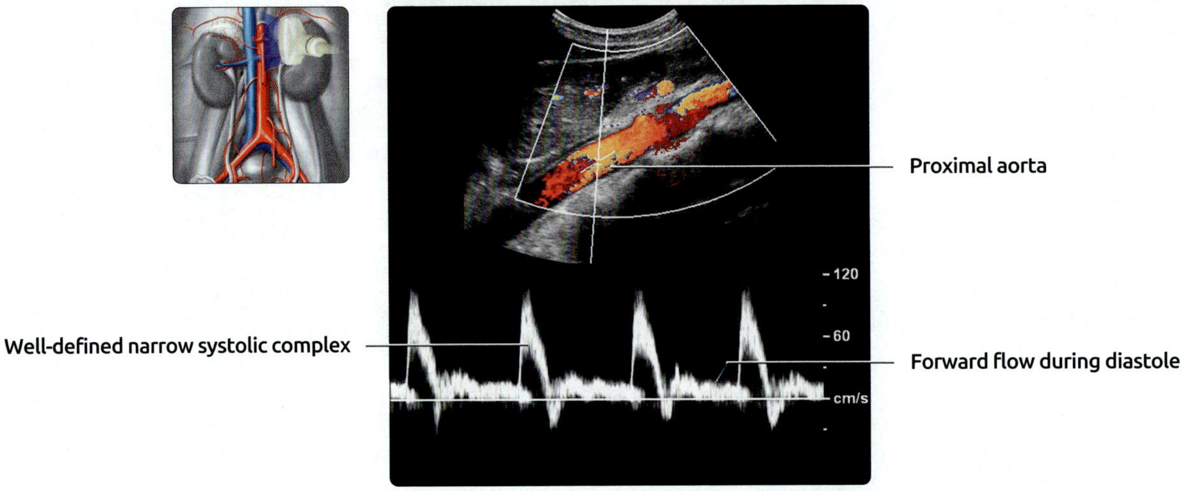

(Top) *Longitudinal grayscale ultrasound of the proximal abdominal aorta shows the origins of the celiac axis and superior mesenteric artery. The origin of the celiac axis is usually seen at T12, while the origin of the superior mesenteric artery is usually seen at L1 immediately below the origin of the celiac axis on this view. The normal diameter of the abdominal aorta is 15-25 mm, and the average diameter of the proximal abdominal aorta above the renal arteries is around 22 mm.* **(Middle)** *Longitudinal color Doppler ultrasound of the proximal abdominal aorta shows the origins of the celiac axis and superior mesenteric artery. Apart from assessment of flow through the aorta, this is also a useful view for assessment of the flow in the superior mesenteric artery.* **(Bottom)** *Spectral Doppler ultrasound of the proximal aorta shows narrow, well-defined systolic complex with forward flow during diastole.*

MIDDLE AND DISTAL AORTA, SAGITTAL VIEW

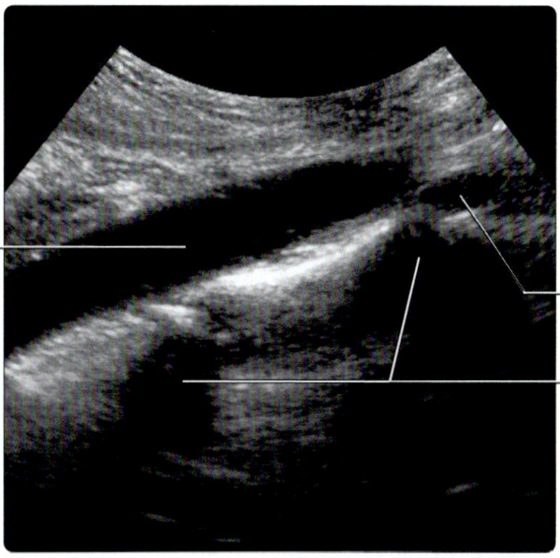

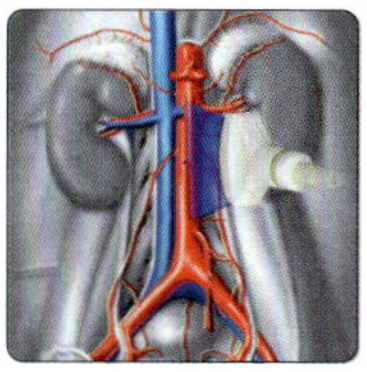

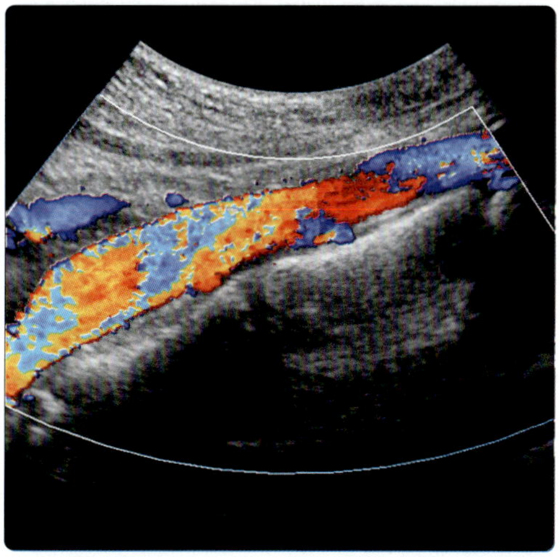

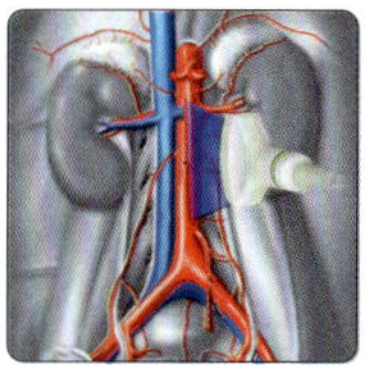

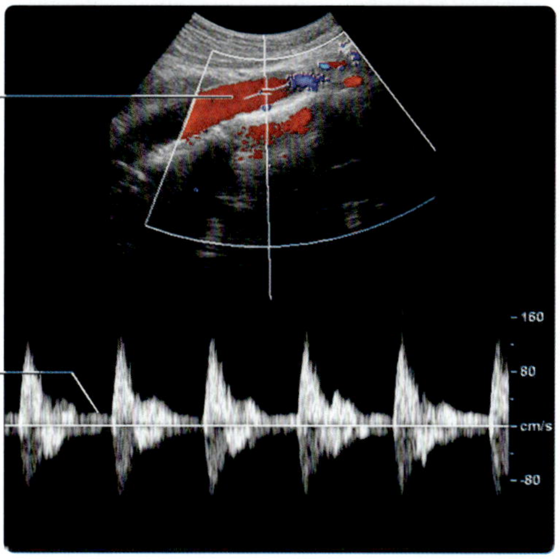

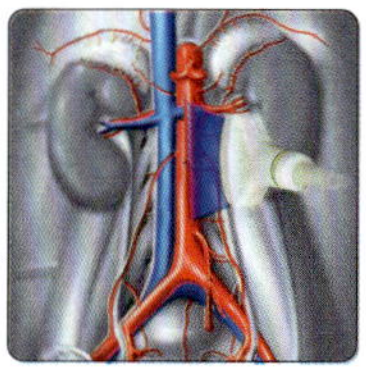

(Top) *Longitudinal grayscale ultrasound of the mid/distal abdominal aorta shows one of the common iliac arteries coming off the aortic bifurcation. The aortic bifurcation is usually seen at the L4 level. The average diameter of the mid portion of the abdominal aorta is around 18 mm below the renal arteries and 15 mm above the aortic bifurcation.* **(Middle)** *Color Doppler ultrasound of the mid/distal abdominal aorta shows one of the common iliac arteries coming off the aortic bifurcation.* **(Bottom)** *Spectral Doppler ultrasound of the mid/distal abdominal aorta shows reduced diastolic flow compared with the proximal aorta. The spectral Doppler waveform that is to be expected in the distal abdominal aorta usually shows absent diastolic flow, similar to that seen in the lower limb arteries.*

CELIAC AXIS

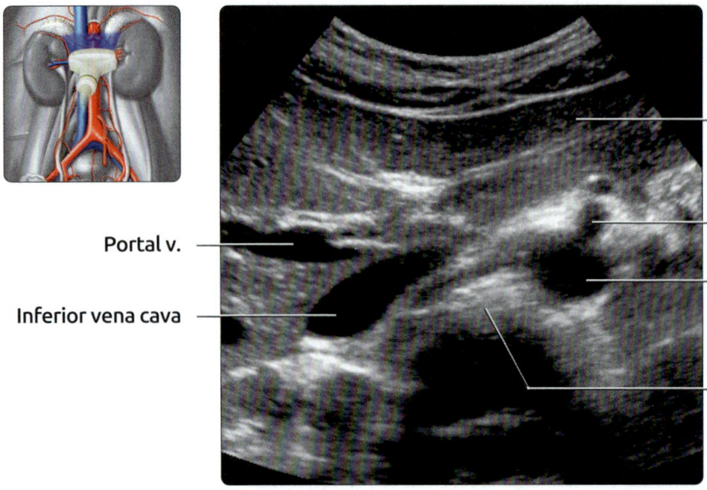

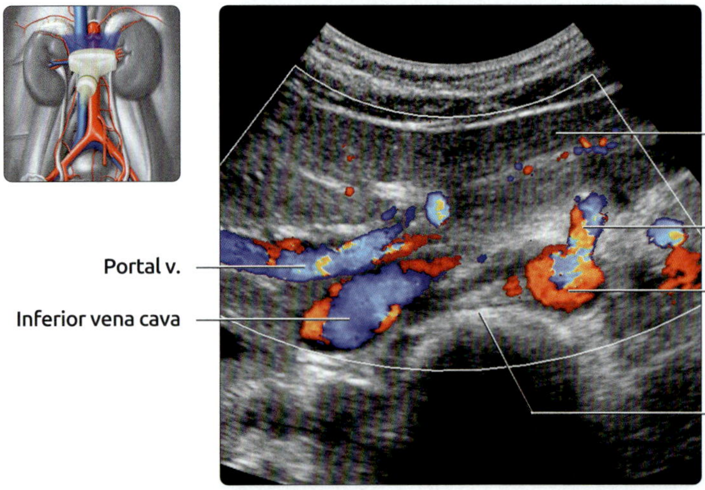

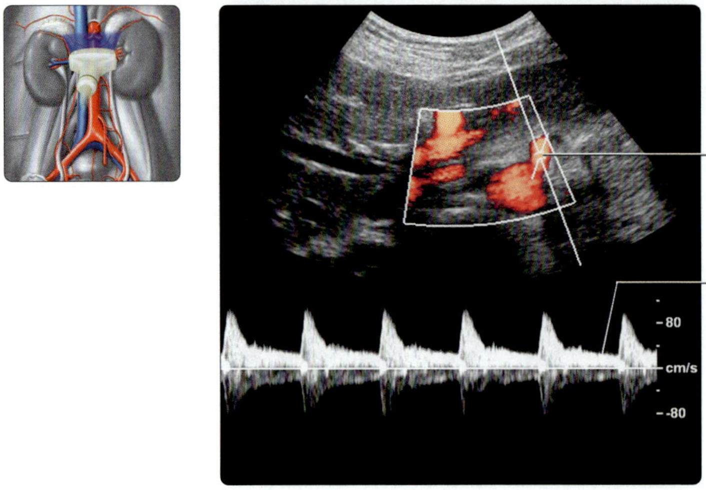

(Top) *Transverse grayscale ultrasound of the abdominal aorta at the level of the celiac axis via a midline approach is shown. The celiac axis is the 1st branch of the abdominal aorta seen arising anteriorly, usually at the level of T12. The normal caliber of the celiac axis is between 6-10 mm. Note the normal position of the IVC to the right of the abdominal aorta.* **(Middle)** *Transverse color Doppler ultrasound of the abdominal aorta at the level of the celiac axis via a midline approach is shown. As the celiac axis is perpendicular to the transducer, a small amount of cranial angulation was used to improve the color Doppler signal obtained.* **(Bottom)** *Spectral Doppler ultrasound of the celiac axis shows low-resistance flow with high end-diastolic velocities. Flow velocity in the celiac axis is not dependent on food intake, and the normal peak systolic velocity (PSV) ranges from 98-105 cm/sec.*

SPLENIC ARTERY

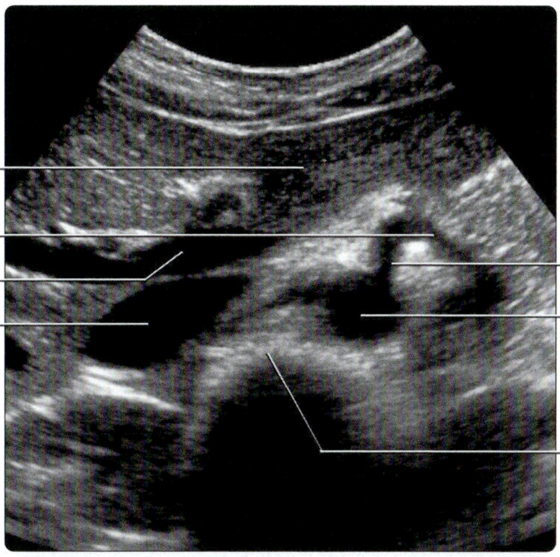

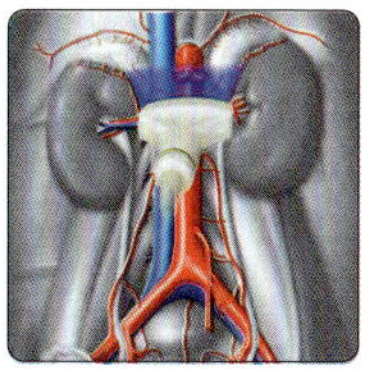

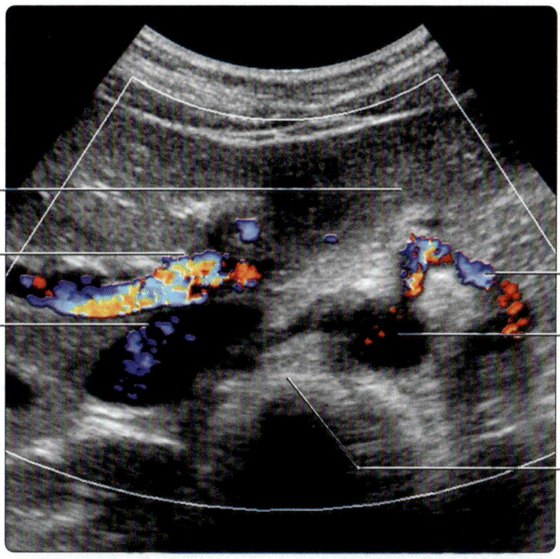

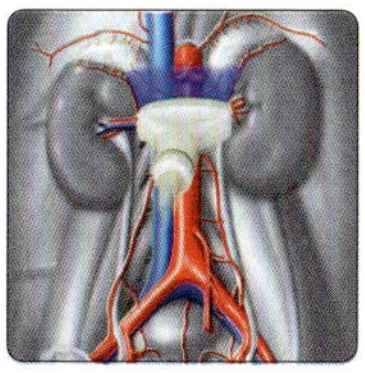

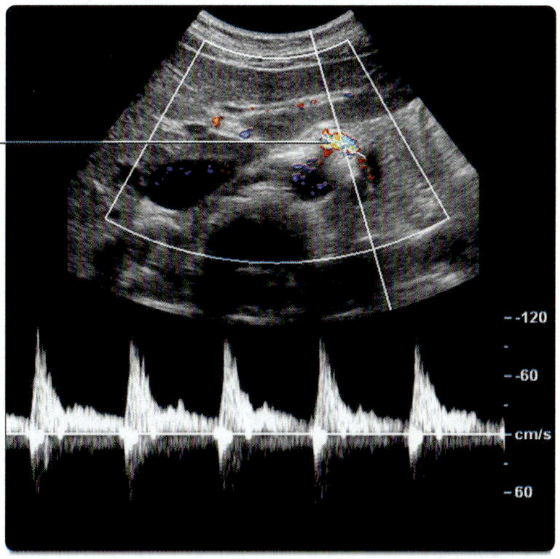

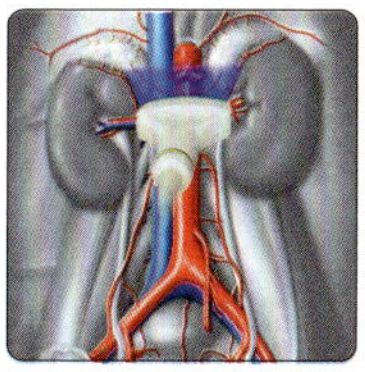

(Top) *Transverse grayscale ultrasound via a midline approach shows the origin of the splenic artery as it branches from the celiac axis and hooks to the left. This is usually the best view for visualizing the proximal portion of the splenic artery. The more distal splenic artery is tortuous in its course. An intercostal approach using the spleen as an acoustic window may be useful for showing the distal splenic artery around the hilum. The normal caliber of the splenic artery ranges from 4-8 mm.* **(Middle)** *Transverse color Doppler ultrasound via a midline approach shows flow within the proximal splenic artery.* **(Bottom)** *Spectral Doppler ultrasound of the proximal splenic artery shows typically turbulent flow due to tortuosity of the vessel. The normal PSV in the splenic artery ranges between 70-110 cm/sec.*

COMMON HEPATIC ARTERY

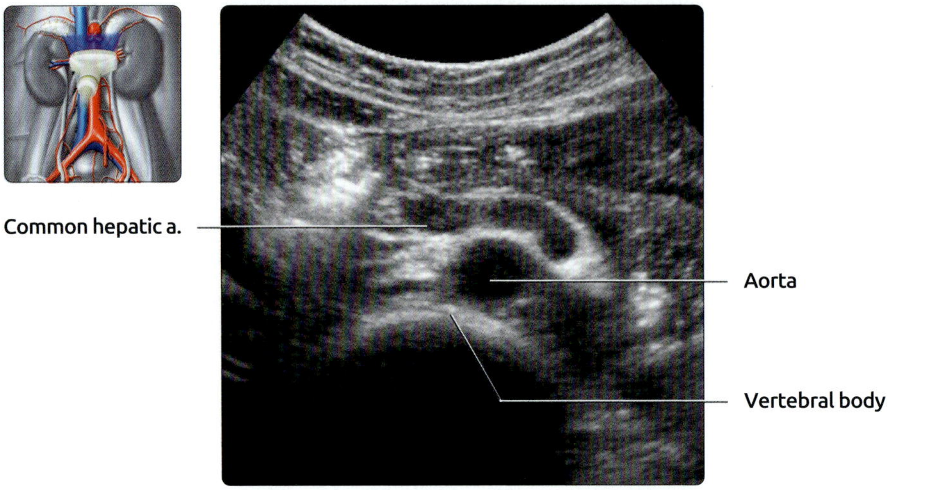

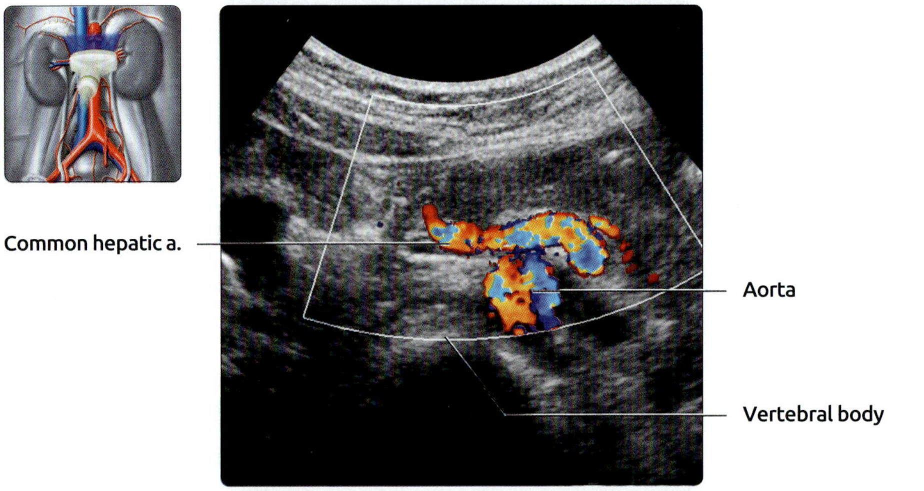

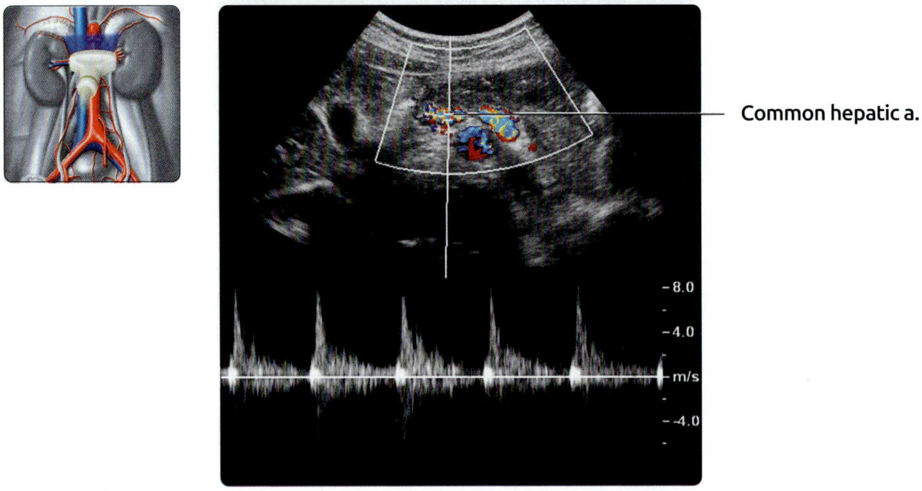

(Top) *Transverse grayscale ultrasound via a midline approach shows the proximal portion of the common hepatic artery as it branches to the right off the T-shaped bifurcation of the celiac axis (not imaged). The normal diameter of the common hepatic artery ranges from 4-10 mm.* **(Middle)** *Color Doppler ultrasound shows flow within the proximal portion of the common hepatic artery. The gastroduodenal artery may be seen to arise from the common hepatic artery along the anterosuperior aspect of the pancreas; thereafter, the common hepatic artery becomes the proper hepatic artery.* **(Bottom)** *Spectral Doppler ultrasound of the common hepatic artery shows low-resistance flow characteristics with large amount of continuous flow in diastole. The normal PSV for the common hepatic artery ranges from 70-120 cm/sec.*

SUPERIOR MESENTERIC ARTERY

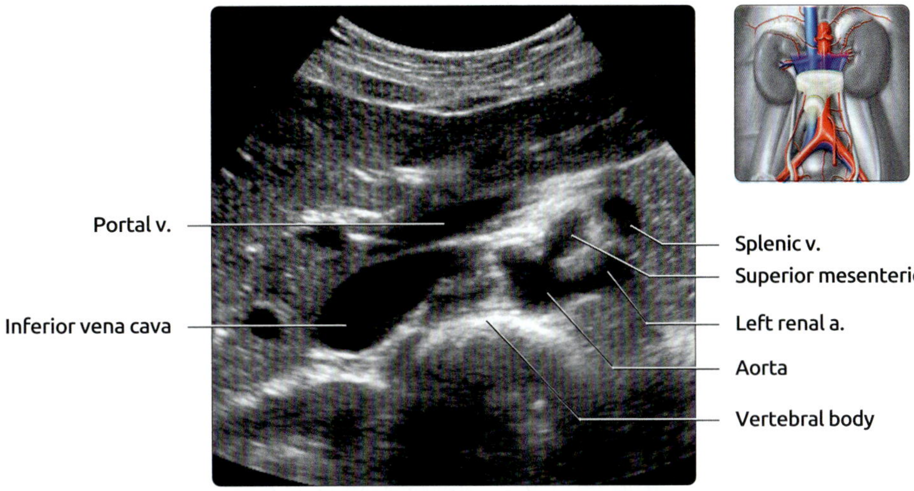

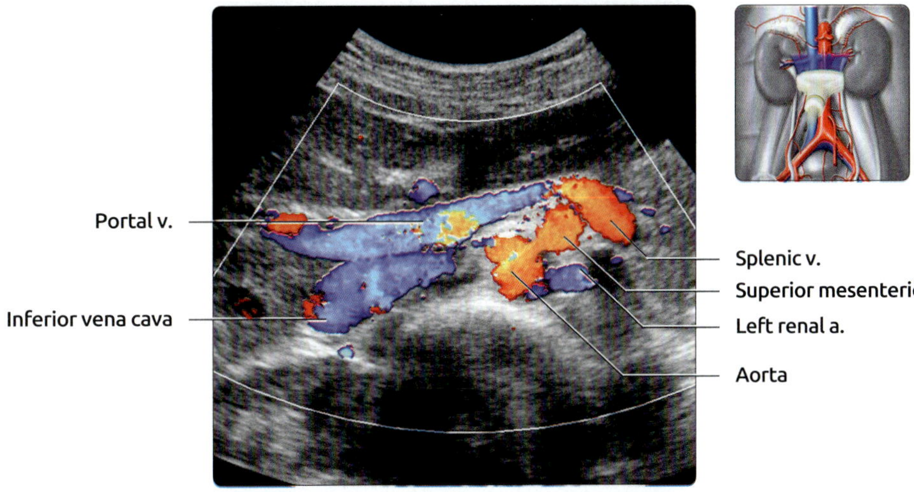

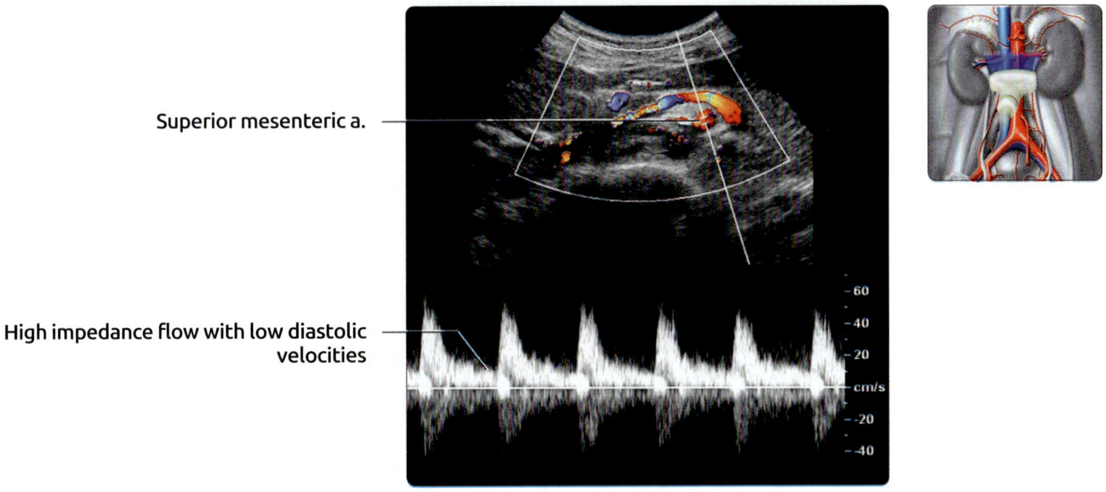

(Top) *Transverse grayscale ultrasound of the origin of the superior mesenteric artery seen arising anteriorly from the aorta is shown. The origin of the superior mesenteric artery is usually seen at L1 between the celiac axis (T12) and the renal arteries (L1/2). The normal caliber of the superior mesenteric artery ranges between 5-8 mm.* **(Middle)** *Transverse color Doppler ultrasound of the proximal superior mesenteric artery is shown. The transducer has been angled slightly cranially with the arterial blood coming toward the transducer shown in red and the venous blood going away from the transducer shown in blue.* **(Bottom)** *Spectral Doppler ultrasound of the proximal portion of the superior mesenteric artery is shown. High-impedance flow with low diastolic velocities is observed during fasting due to relative vasoconstriction. End-diastolic velocity increases after meals due to vasodilation of the mesenteric branches, typically within 30-90 min.*

RIGHT RENAL ARTERY

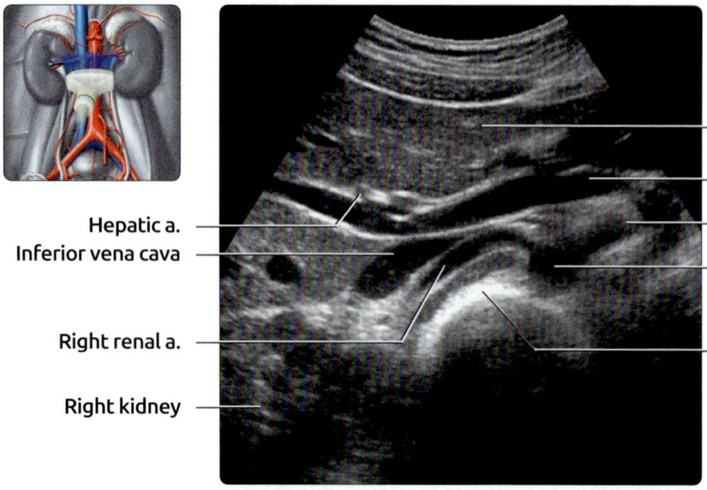

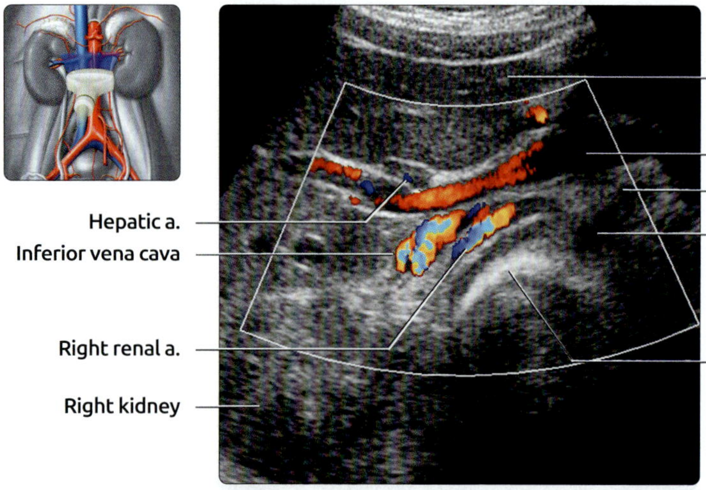

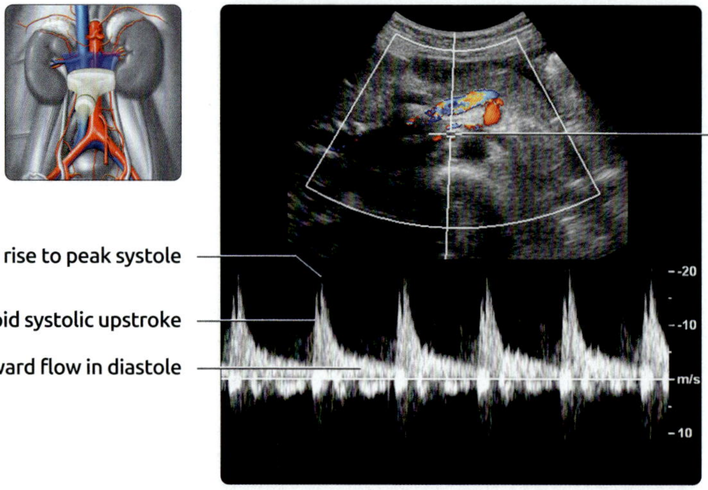

(Top) *Transverse grayscale ultrasound shows the proximal right renal artery as it branches from the aorta and courses behind the IVC. This midline anterior approach is usually the best for evaluating the origin of the renal arteries. The normal diameter of the renal arteries ranges from 5-8 mm. The renal arteries arise around the L1/2 level at or below the level of the superior mesenteric artery.* **(Middle)** *Transverse color Doppler ultrasound of the abdominal aorta shows the proximal right renal artery as it branches from the aorta and courses behind the IVC.* **(Bottom)** *Spectral Doppler ultrasound of the right renal artery shows an open systolic window, rapid systolic upstroke followed by a secondary slower rise to peak systole. There is a subsequent gradual decrease in flow but it remains persistently forward throughout diastole. The normal PSV ranges from 60-140 cm/sec but not more than 180 cm/sec.*

LEFT RENAL ARTERY

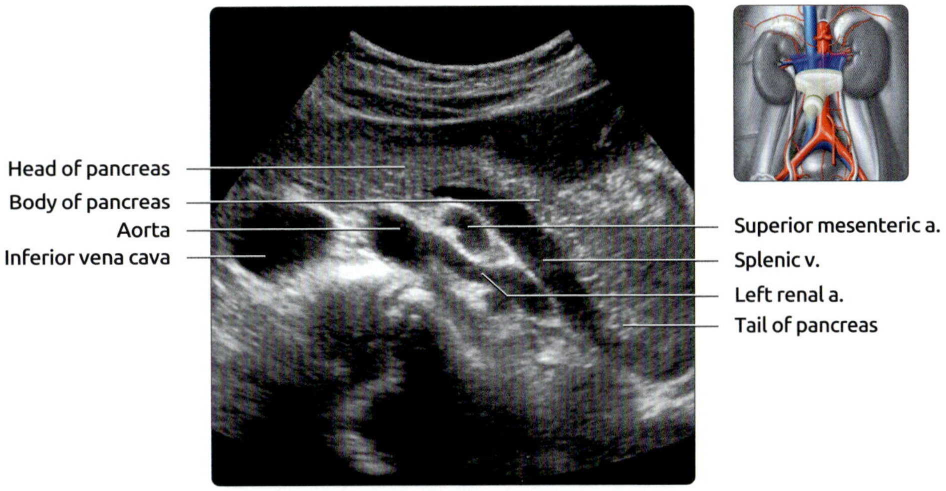

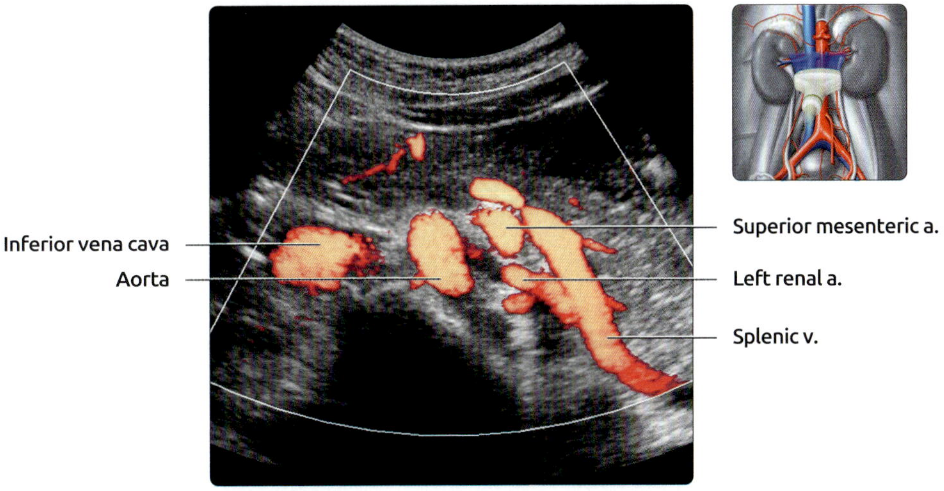

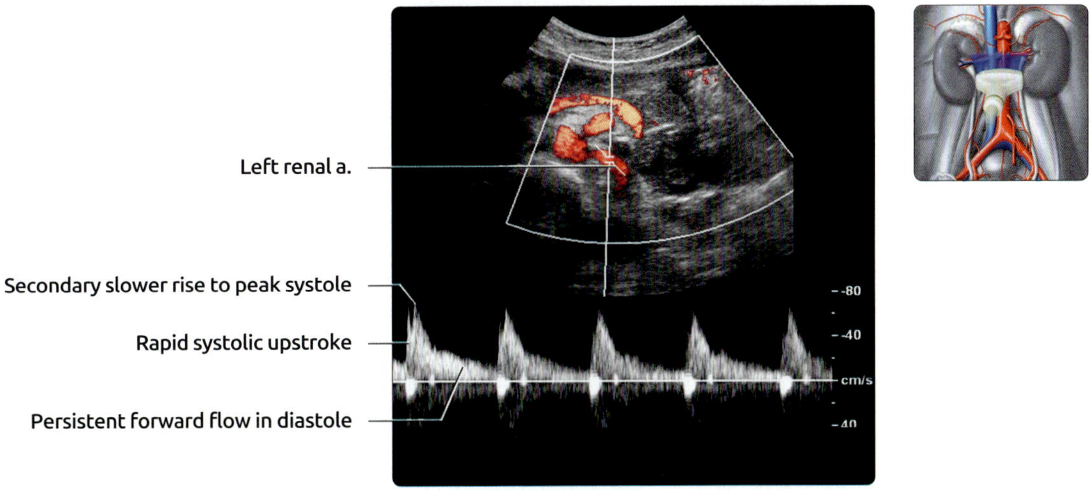

(Top) *Transverse grayscale ultrasound shows the proximal left renal artery as it branches from the aorta and courses posterior to the superior mesenteric artery and splenic vein. This midline anterior approach is usually the best for evaluating the origin of the renal arteries. The normal diameter of the renal arteries ranges from 5-8 mm. The renal arteries arise around the L1/2 level.* **(Middle)** *Power Doppler ultrasound shows flow in the proximal left renal artery as it branches from the aorta and courses posterior to the superior mesenteric artery and splenic vein.* **(Bottom)** *Spectral Doppler ultrasound of the left renal artery shows an open systolic window, rapid systolic upstroke followed by a secondary slower rise to peak systole with subsequent gradual decrease in velocity but persistent forward flow throughout diastole. The normal PSV ranges from 60-140 cm/sec but not more than 180 cm/sec.*

INFERIOR MESENTERIC ARTERY

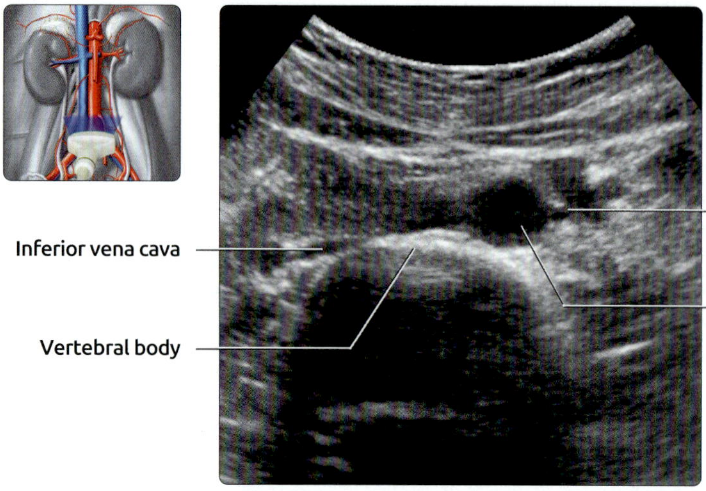

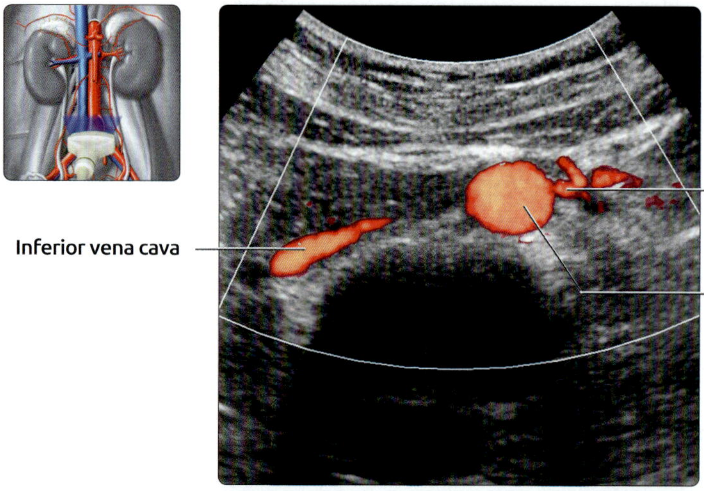

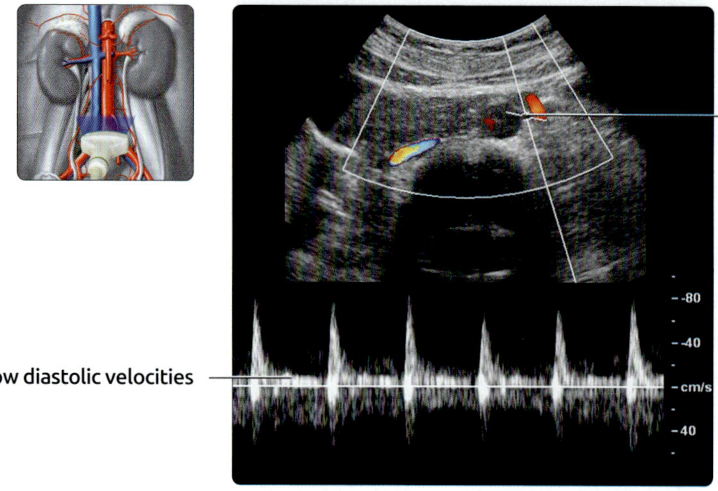

(Top) *Transverse grayscale ultrasound shows the distal abdominal aorta at the level of the origin of the inferior mesenteric artery. The inferior mesenteric artery arises from the anterior or left anterolateral aspect of the abdominal aorta at the L3 level. The transverse plane following the line of the aorta is the best imaging plane for identification of the origin of the inferior mesenteric artery. The normal caliber of the inferior mesenteric artery ranges from 1-4 mm.* **(Middle)** *Transverse power Doppler ultrasound shows flow in the proximal inferior mesenteric artery.* **(Bottom)** *Spectral Doppler ultrasound of the inferior mesenteric artery shows high-impedance flow with low diastolic velocities during fasting due to relative vasoconstriction. End-diastolic velocity increases after meal due to vasodilation of the mesenteric branches. Normal PSV ranges from 93-189 cm/sec.*

AORTIC BIFURCATION

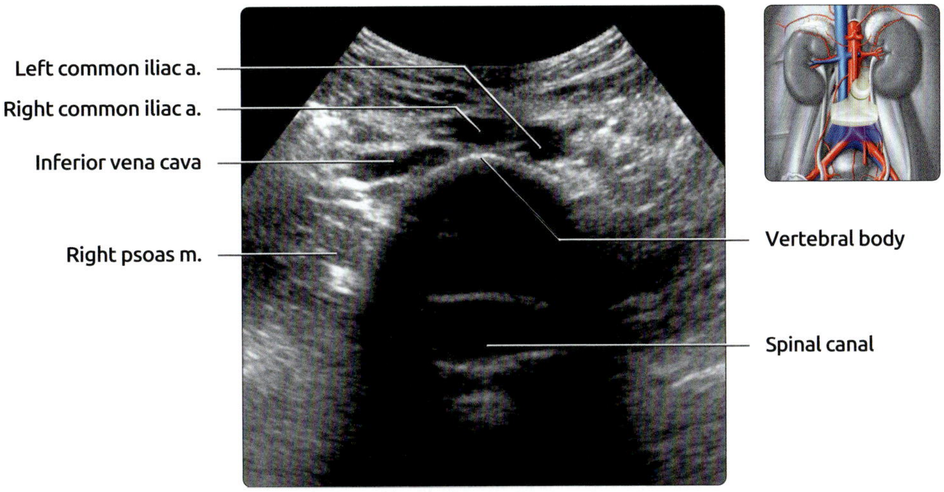

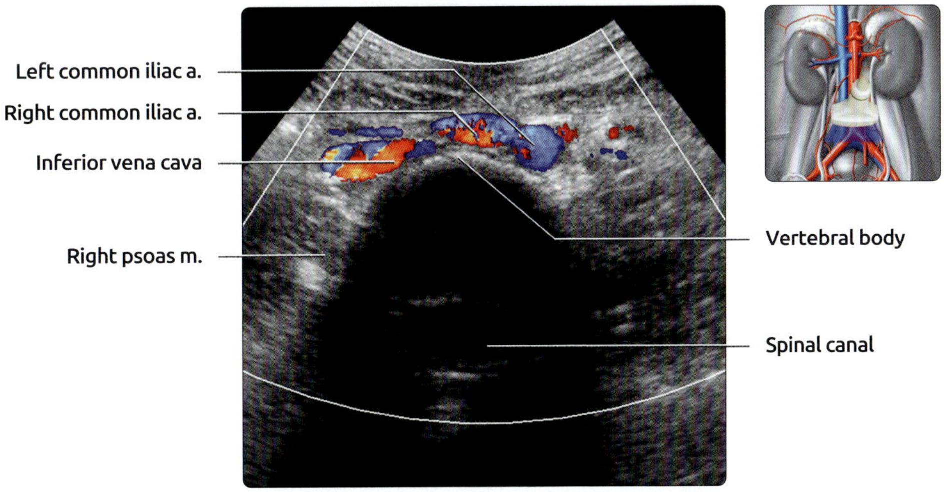

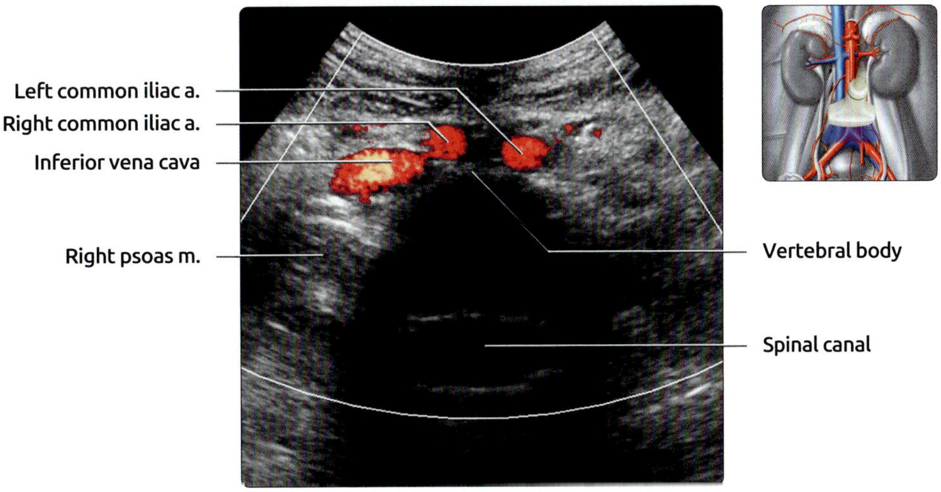

(Top) *Transverse grayscale ultrasound of the aortic bifurcation shows the origins of the right and left common iliac arteries. The aortic bifurcation is usually seen at the L4 level.* **(Middle)** *Transverse color Doppler ultrasound shows flow in the origins of the right and left common iliac arteries.* **(Bottom)** *Transverse power Doppler ultrasound shows flow in the origins of the right and left common iliac arteries. Power Doppler is less angle dependent and demonstrates flow more readily, particularly in vascular structures that are close to a right angle with the transducer.*

RIGHT COMMON ILIAC ARTERY

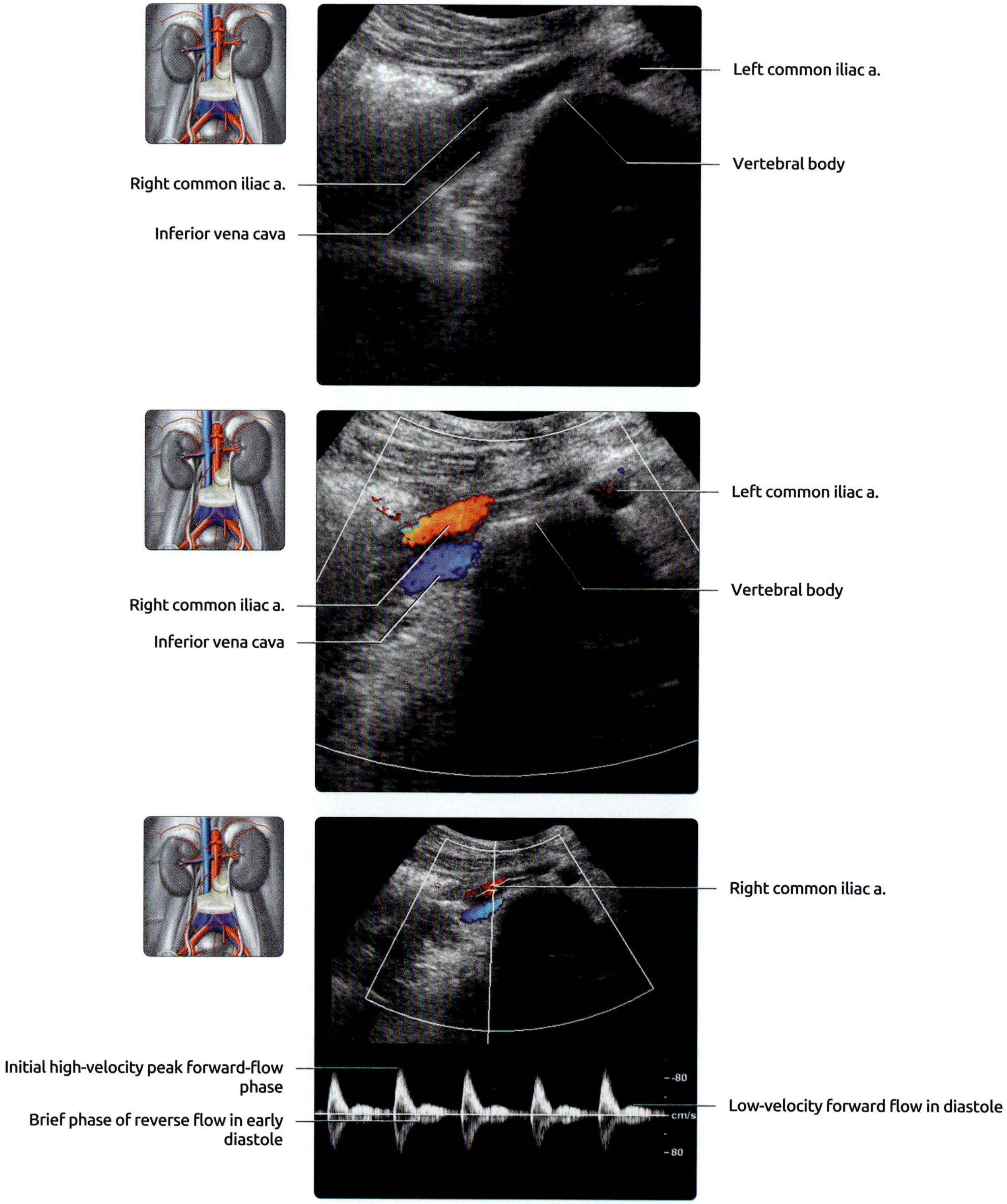

(Top) *Oblique grayscale ultrasound shows the course of the right common iliac artery as it branches off the aortic bifurcation. The proximal common iliac artery is 1st identified on the transverse plane, and the transducer is then angulated along the long axis of the right common iliac artery. The normal diameter of the common iliac arteries ranges from 8-12 mm.* **(Middle)** *Oblique color Doppler ultrasound shows flow in the proximal right common iliac artery.* **(Bottom)** *Spectral Doppler ultrasound of the right common iliac artery shows a triphasic waveform. An initial high-velocity peak forward flow phase, resulting from cardiac systole, is followed by a brief phase of reversed flow in early diastole, and low-velocity forward flow in the remainder of diastole.*

LEFT COMMON ILIAC ARTERY

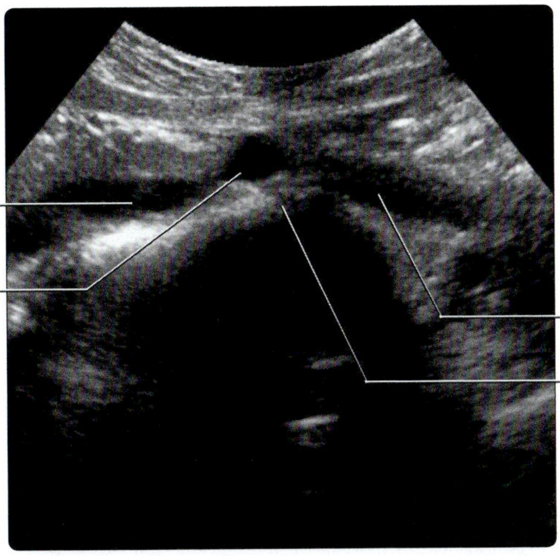

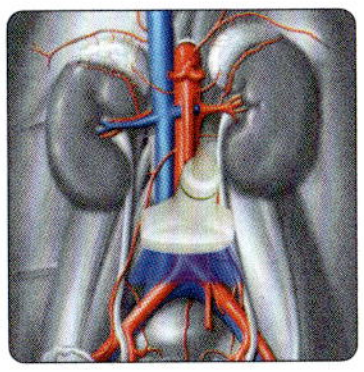

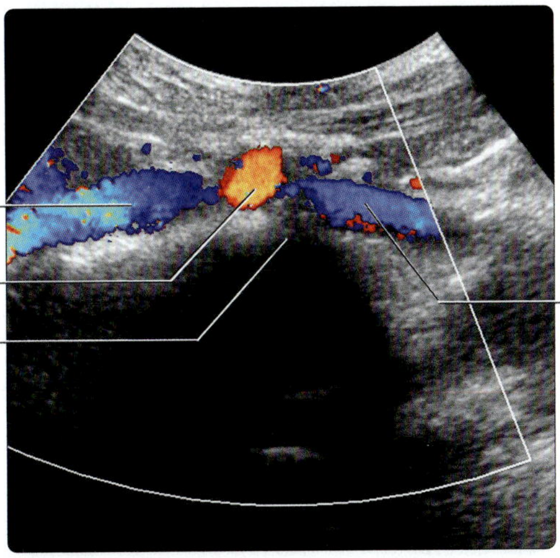

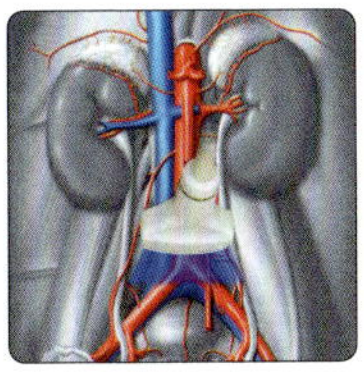

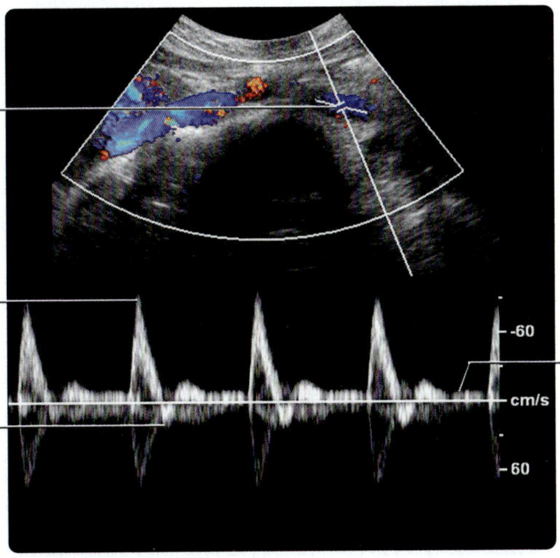

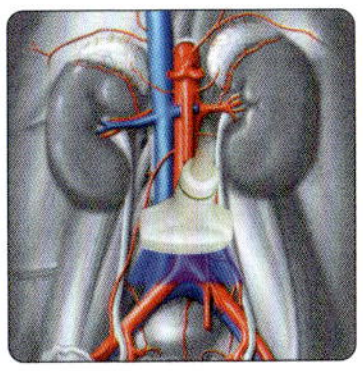

(Top) *Oblique grayscale ultrasound shows the course of the left common iliac artery as it branches off the aortic bifurcation. The proximal common iliac artery is 1st identified on the transverse plane, and the transducer is then angulated along the long axis of the left common iliac artery. The normal diameter of the common iliac arteries ranges from 8-12 mm.* **(Middle)** *Oblique color Doppler ultrasound shows flow in the proximal left common iliac artery.* **(Bottom)** *Spectral Doppler ultrasound of the left common iliac artery shows the characteristic triphasic waveform. An initial high-velocity peak forward flow phase, resulting from cardiac systole, is followed by a brief phase of reversed flow in early diastole, and low-velocity forward flow in the remainder of diastole.*

PROXIMAL INFERIOR VENA CAVA, SAGITTAL VIEW

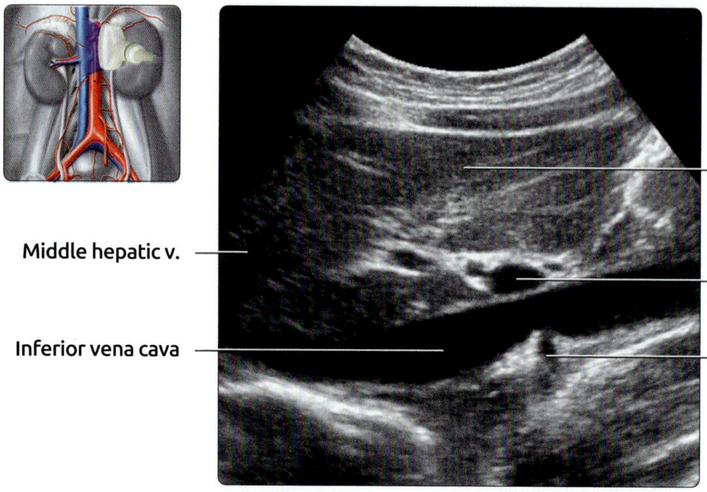

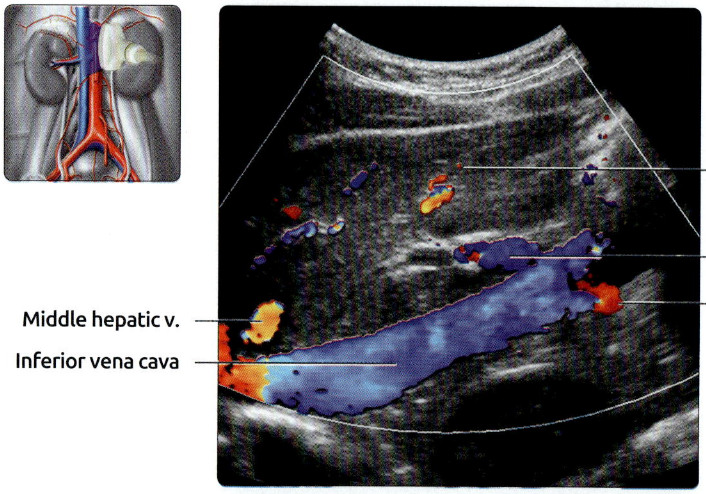

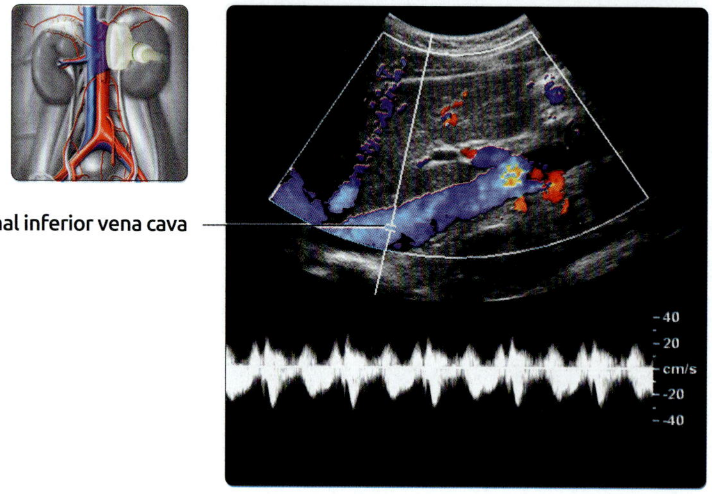

(Top) *Longitudinal grayscale ultrasound of the epigastric region shows the proximal IVC. The normal caliber of the IVC ranges from 5-29 mm during quiet respiration with the diameter being larger in the proximal portion. The diameter increases by ~ 10% during deep inspiration.* **(Middle)** *Longitudinal color Doppler ultrasound of the epigastric region shows flow within the proximal IVC (shown in blue).* **(Bottom)** *Spectral Doppler ultrasound of the proximal IVC shows slow flow that varies with respiration and cardiac pulsation. The pulsatile Doppler flow pattern contributed by atrial pressure changes is more apparent in the proximal portion of the IVC. The normal PSV ranges from 48-115 cm/sec.*

MID INFERIOR VENA CAVA, SAGITTAL VIEW

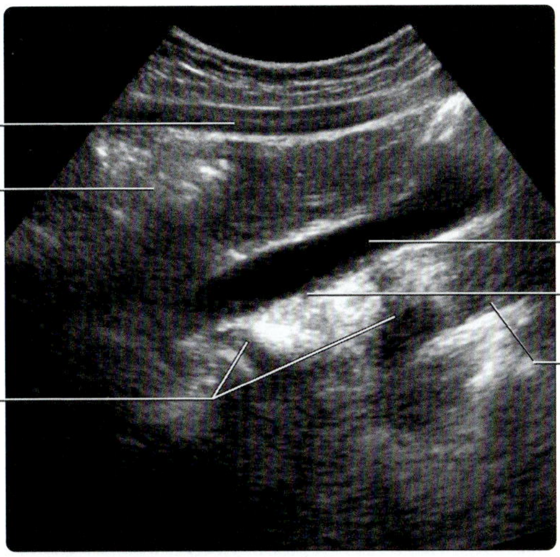

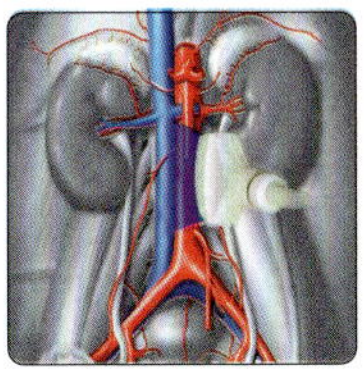

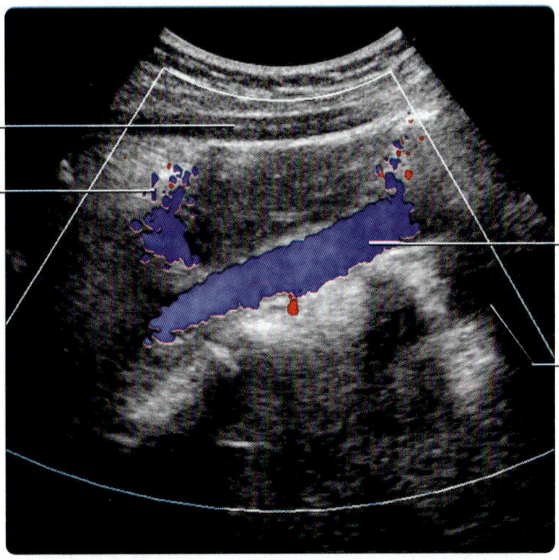

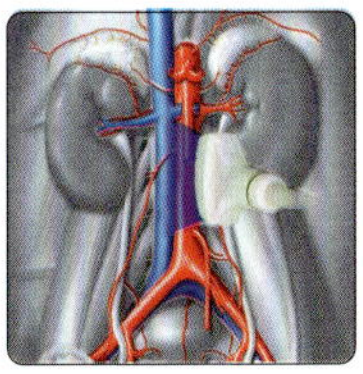

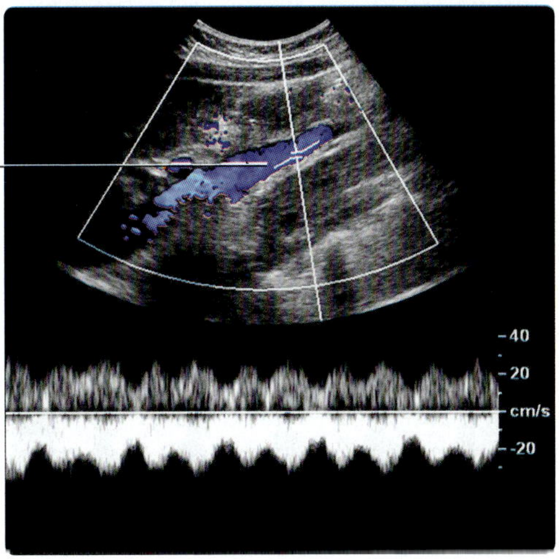

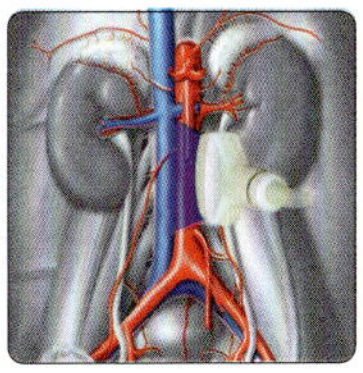

(Top) *Longitudinal grayscale ultrasound of the epigastric region shows the middle portion of the IVC lying to the right of the lumbar vertebrae. Therefore, the right psoas may be included in the image. The normal caliber of the IVC ranges from 5-29 mm during quiet respiration with the diameter being larger in the proximal portion. The diameter increases by ~ 10% during deep inspiration.* **(Middle)** *Longitudinal color Doppler ultrasound shows flow in the mid portion of the IVC.* **(Bottom)** *Spectral Doppler ultrasound of the mid portion of the IVC shows less pulsatility than that seen in the proximal IVC. This is due to the reduction in influence of the right atrium.*

DISTAL INFERIOR VENA CAVA, SAGITTAL VIEW

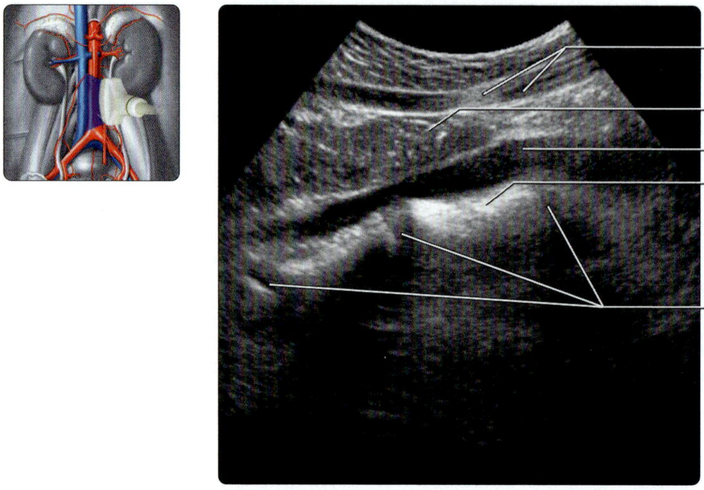

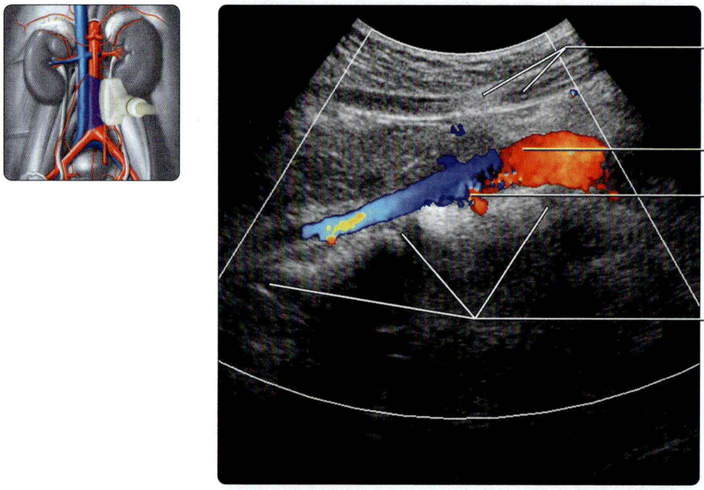

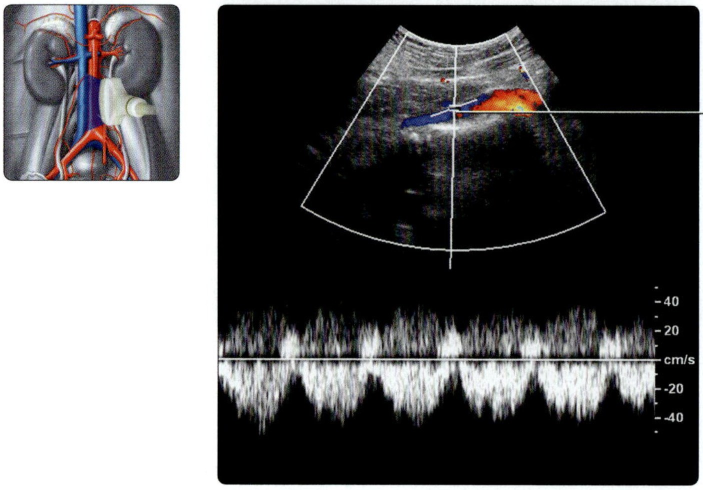

(Top) *Longitudinal grayscale ultrasound of the distal IVC below the level of the renal veins is shown. Left-sided IVC and other congenital anomalies are encountered in the IVC below the renal veins in ~ 10% of the population. The diameter of the distal IVC is normally smaller than that seen in the proximal IVC. Note that in this subject, the anterior border of the vertebral bodies and the intervertebral discs can be visualized.* **(Middle)** *Longitudinal color Doppler ultrasound shows flow in the distal IVC above the common iliac vein confluence.* **(Bottom)** *Spectral Doppler ultrasound of the distal IVC shows only variations from respiration with no influence from the right atrium, unlike that seen in the proximal IVC.*

PROXIMAL INFERIOR VENA CAVA, TRANSVERSE VIEW

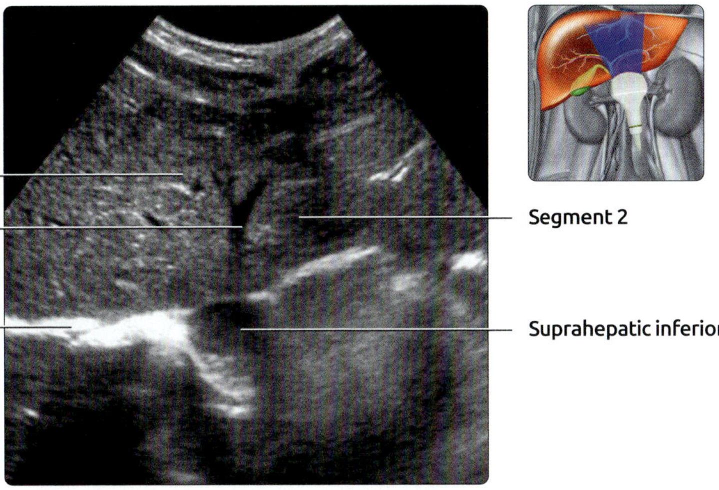

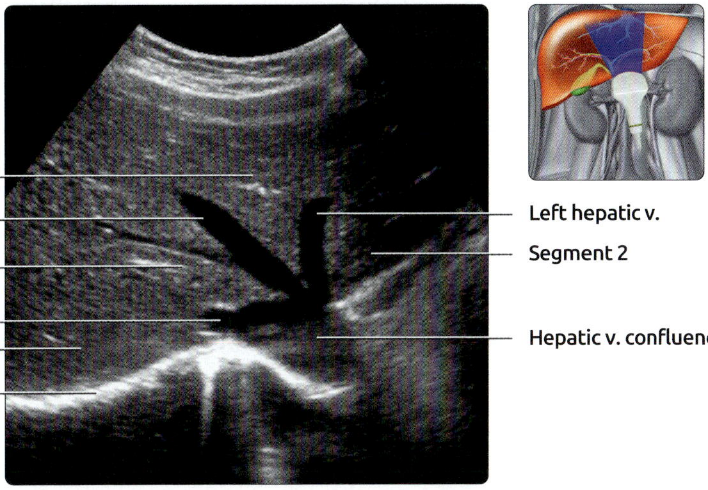

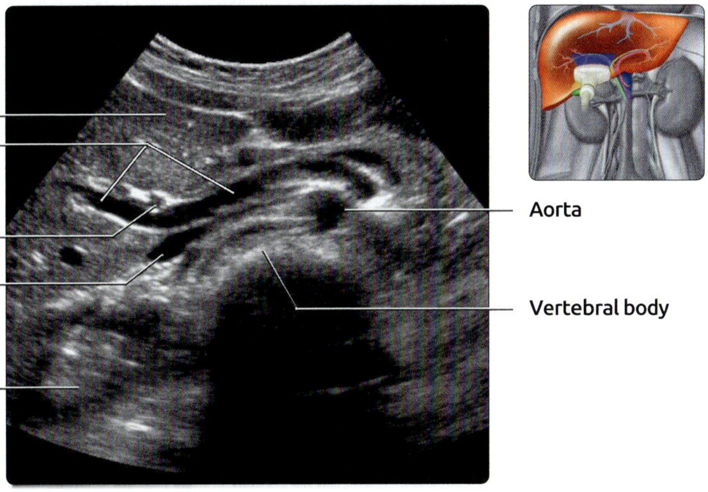

(Top) *Transverse grayscale ultrasound with cranial angulation shows the suprahepatic IVC proximal to the hepatic confluence. The IVC leaves the abdomen via the IVC hiatus in the diaphragm at the T8 level.* **(Middle)** *Transverse grayscale ultrasound shows the hepatic vein confluence. Cranial angulation is often required to produce this image and is a very useful view for designation of the segments of the liver. The middle hepatic vein defines the plane that separates the right and left lobe of the liver. The right hepatic vein defines the plane that separates the anterior and posterior segments of the right lobe. The left hepatic vein defines the plane that separates the medial and lateral segments of the left lobe.* **(Bottom)** *Transverse grayscale ultrasound shows the distal intrahepatic IVC at the level of the extrahepatic portal vein.*

LEFT HEPATIC VEIN

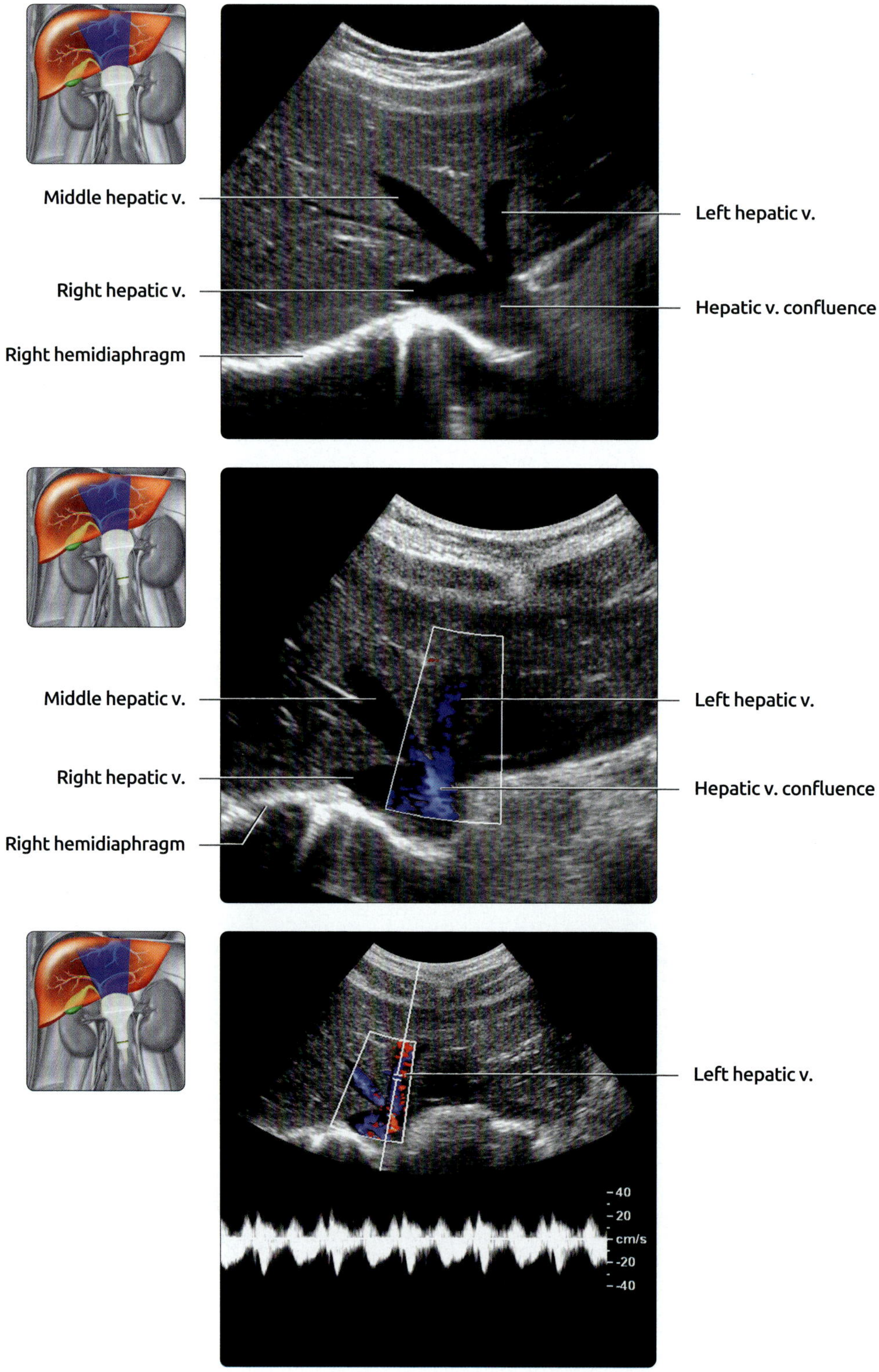

(Top) *Transverse grayscale ultrasound shows the left hepatic vein as it drains into the hepatic vein confluence. It may join the middle hepatic vein before joining the proximal IVC. The left hepatic vein runs between the medial and lateral segments of the left lobe and is frequently duplicated.* **(Middle)** *Transverse color Doppler ultrasound shows flow within the left hepatic vein.* **(Bottom)** *Spectral Doppler ultrasound of the left hepatic vein shows the normal chaotic pulsatile waveform, which results from a combination of phasic variation and transmission of right atrial pulsations to the vein. The normal PSV in hepatic veins ranges from 16-40 cm/sec.*

MIDDLE HEPATIC VEIN

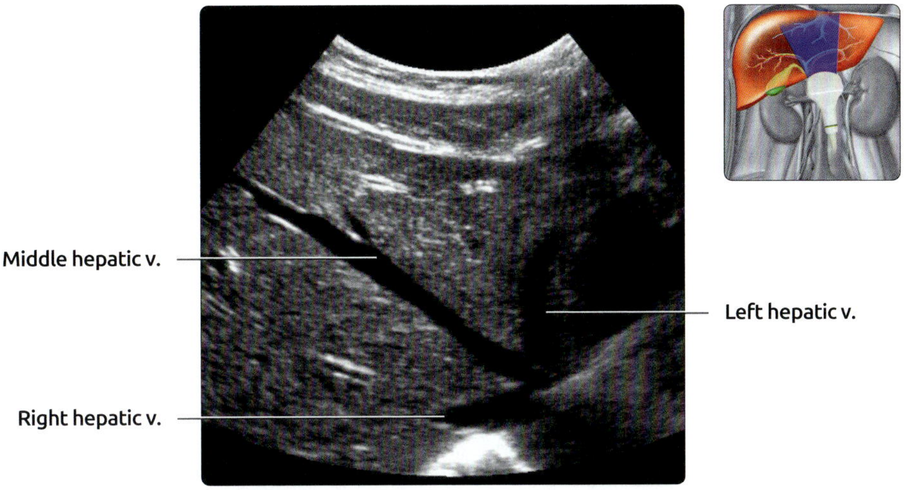

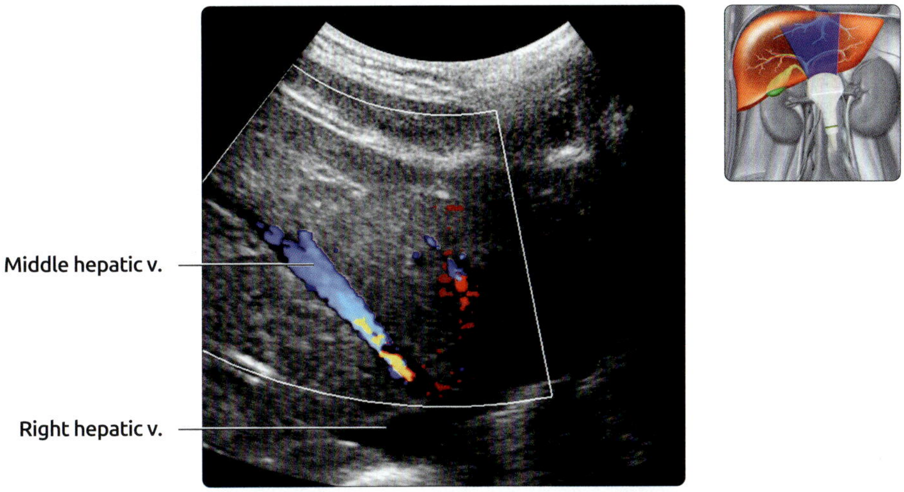

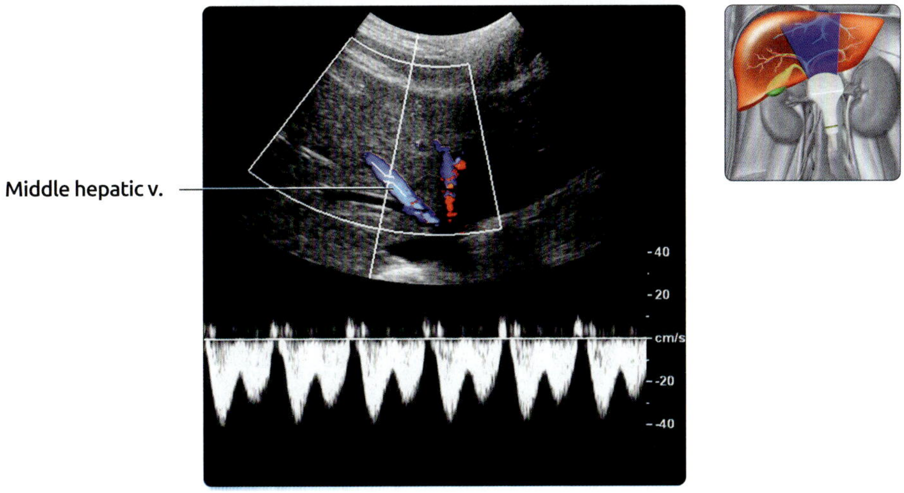

(Top) *Transverse grayscale ultrasound of the middle hepatic vein as it drains directly into the hepatic vein confluence is shown. The middle hepatic vein lies in a sagittal or parasagittal plane between the right and left lobes of the liver.* **(Middle)** *Color Doppler ultrasound shows flow within the middle hepatic vein.* **(Bottom)** *Spectral Doppler ultrasound of the middle hepatic vein shows the normal chaotic pulsatile waveform, which results from a combination of phasic variation and transmission of right atrial pulsations to the vein. The normal PSV in hepatic veins ranges from 16-40 cm/sec.*

RIGHT HEPATIC VEIN

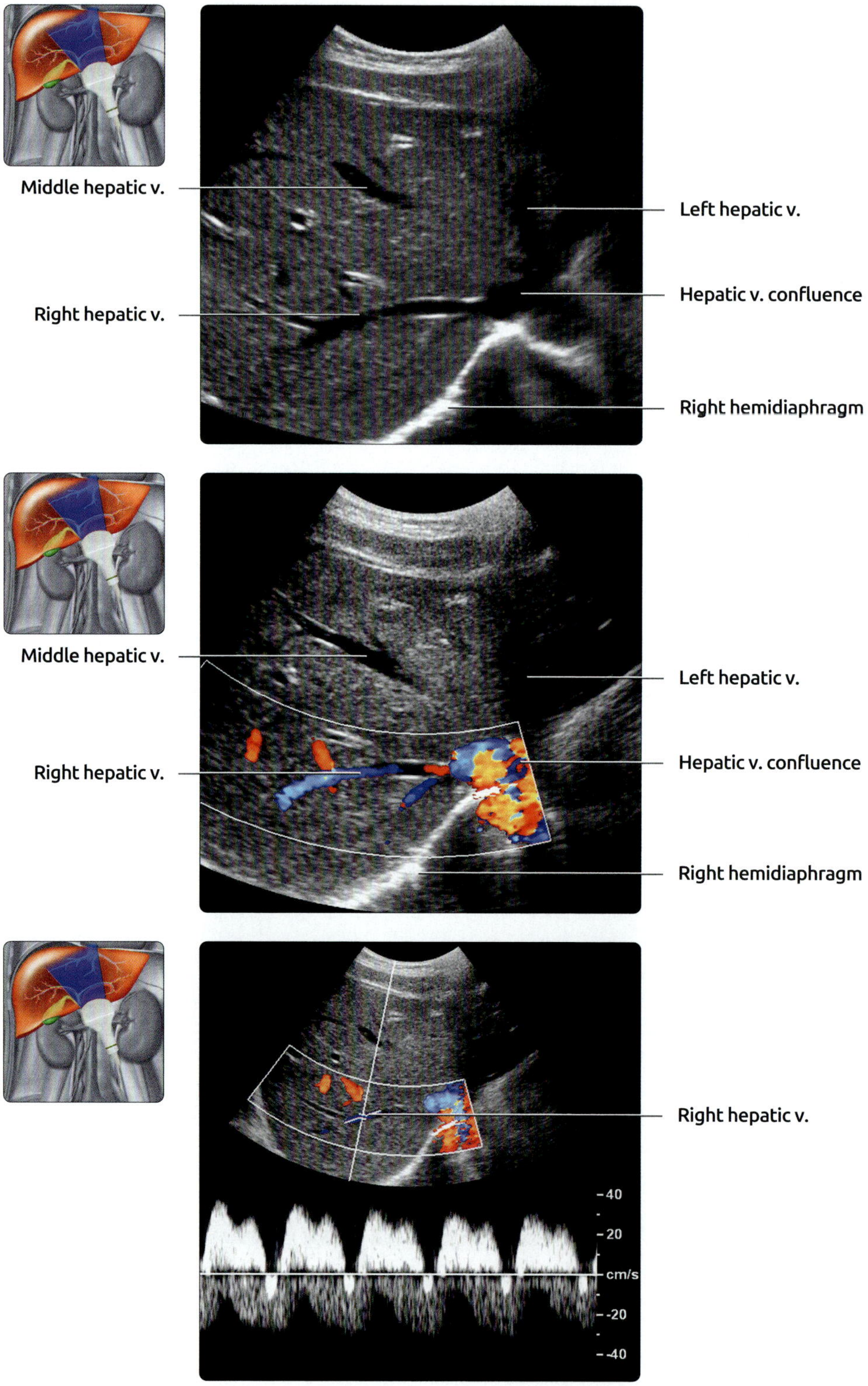

(Top) *Transverse grayscale ultrasound of the right hepatic vein as it enters the hepatic vein confluence is shown. The right hepatic vein runs in a coronal plane between the anterior and posterior segments of the right lobe of the liver and may be absent in 6% of the population.* **(Middle)** *Transverse color Doppler ultrasound shows flow in the right hepatic vein. Note that the mid portion of the vein is perpendicular to the transducer thereby creating an area of lack of signal in the mid portion of the vein.* **(Bottom)** *Spectral Doppler ultrasound of the right hepatic vein shows the normal chaotic pulsatile waveform, which results from a combination of phasic variation and transmission of right atrial pulsations to the vein. The normal PSV in hepatic veins ranges from 16-40 cm/sec.*

RIGHT RENAL VEIN

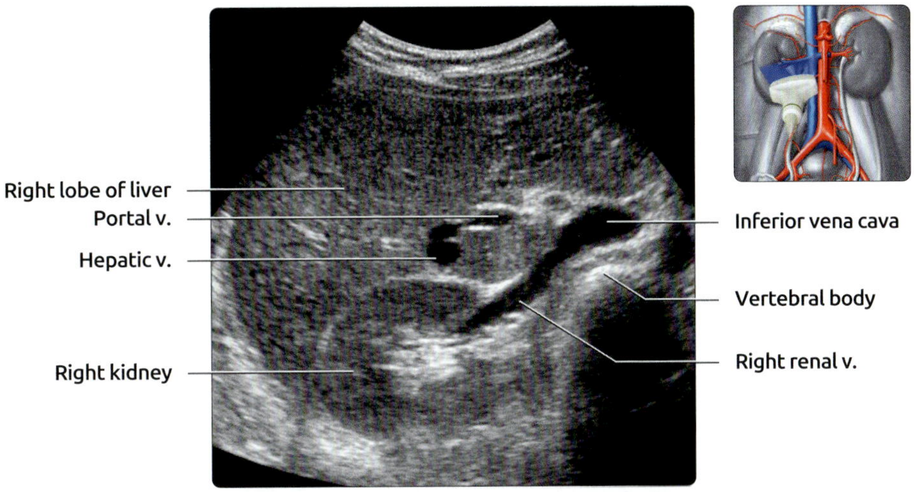

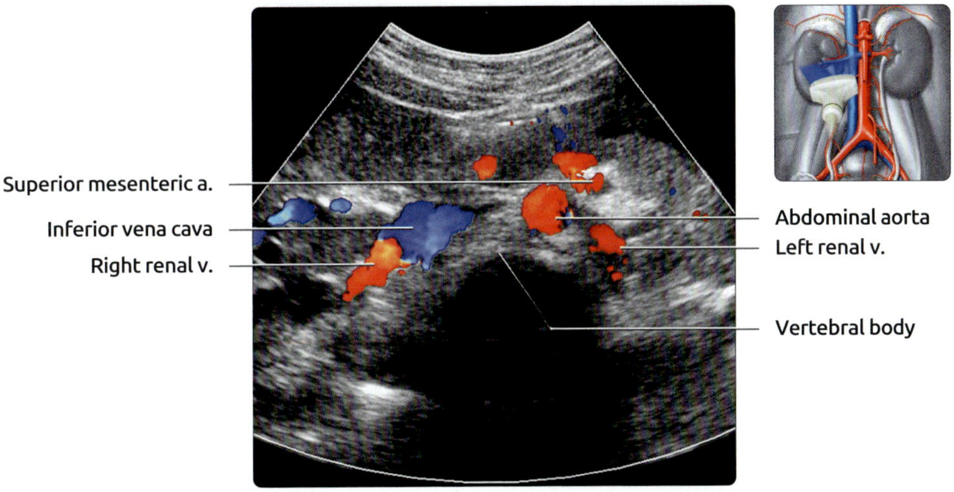

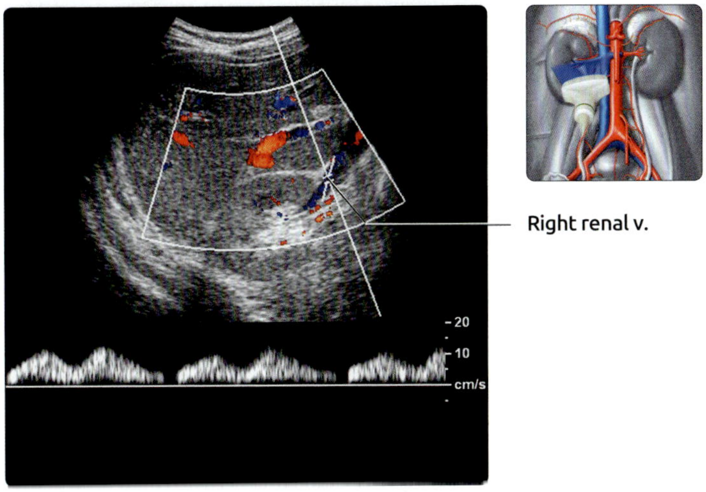

(Top) *Transverse grayscale ultrasound shows the right renal vein utilizing the liver as an acoustic window. This is usually the best approach for imaging the proximal right renal vein. The right renal vein is relatively short and drains directly into the IVC. The normal caliber of the right renal vein is between 4-9 mm.* **(Middle)** *Transverse color Doppler ultrasound shows flow in the right renal vein. The image is obtained with slight cranial angulation so that flow toward the transducer is registered as red (right renal vein, aorta, superior mesenteric artery, left renal vein) and flow away from the transducer in the IVC is registered as blue.* **(Bottom)** *Spectral Doppler ultrasound of the right renal vein mirroring the pulsatility in the IVC is shown. The normal PSV ranges from 18-33 cm/sec.*

LEFT RENAL VEIN

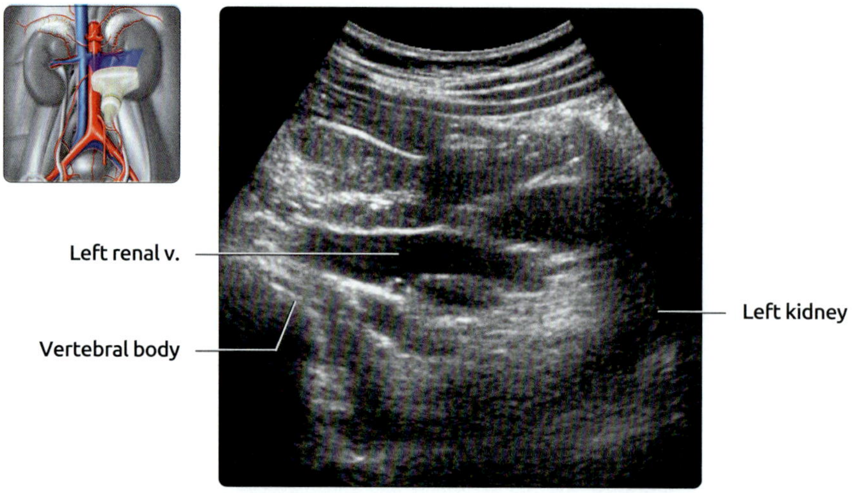

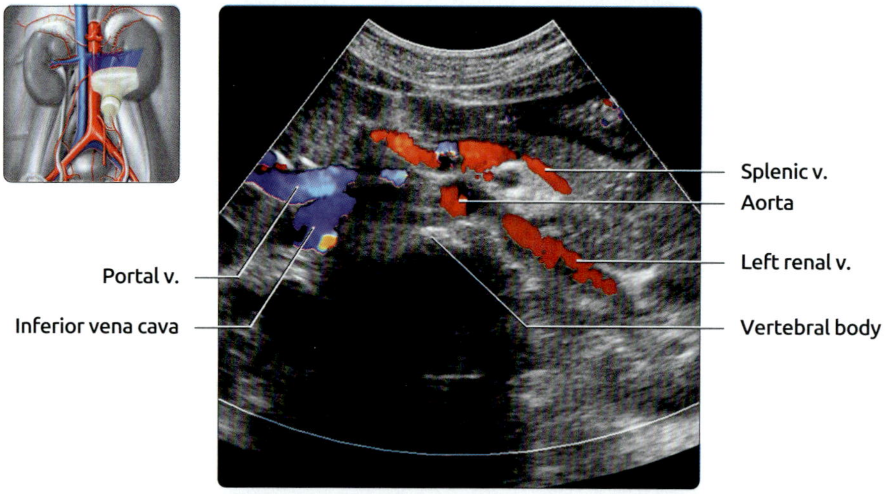

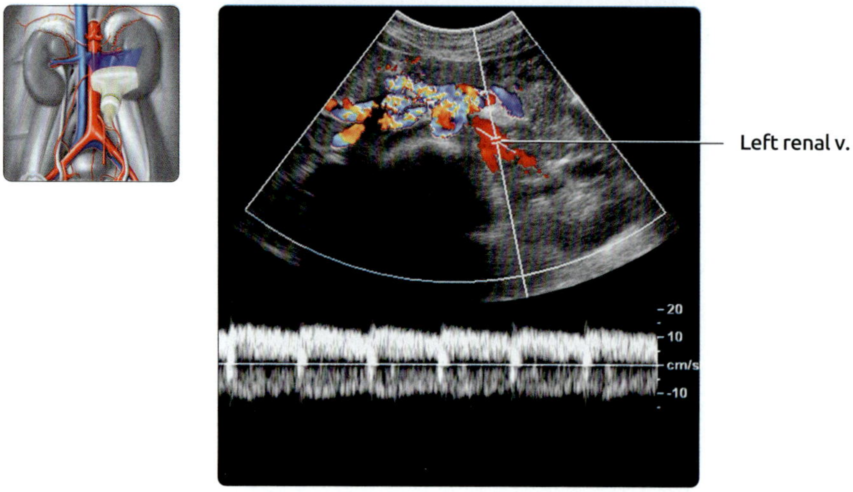

(Top) *Transverse grayscale ultrasound of the left renal vein via the anterior approach is shown. This approach may be affected by intervening bowel loops. The left renal vein has a longer course to the IVC compared with the right renal vein. The normal caliber of the left renal vein is between 4-9 mm.* **(Middle)** *Transverse color Doppler ultrasound with slight cranial angulation shows flow in the left renal vein in red. Flow toward the transducer is shown in red (aorta, splenic vein) and flow away from the transducer is shown in blue (IVC, portal vein). Note that the left renal vein usually courses anterior to the aorta before entering the IVC.* **(Bottom)** *Spectral Doppler of the left renal vein shows slight variability of flow velocities consequent upon cardiac and respiratory activity.*

COMMON ILIAC VEINS

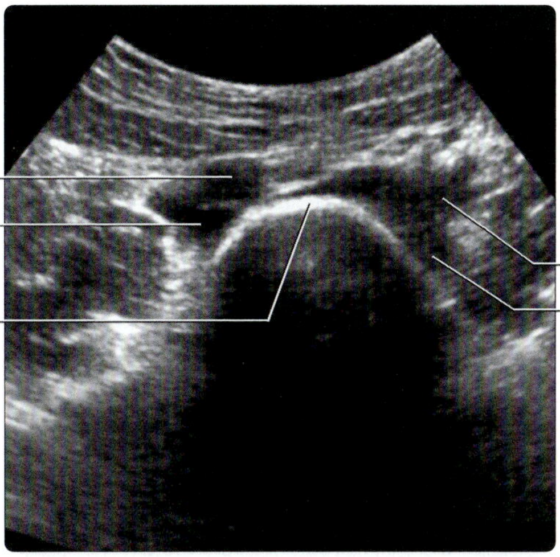

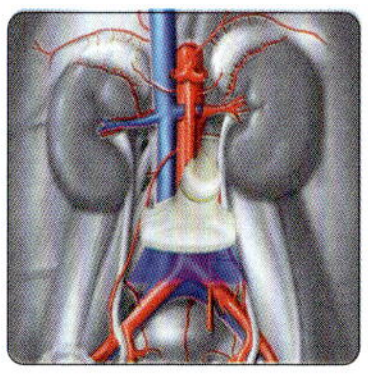

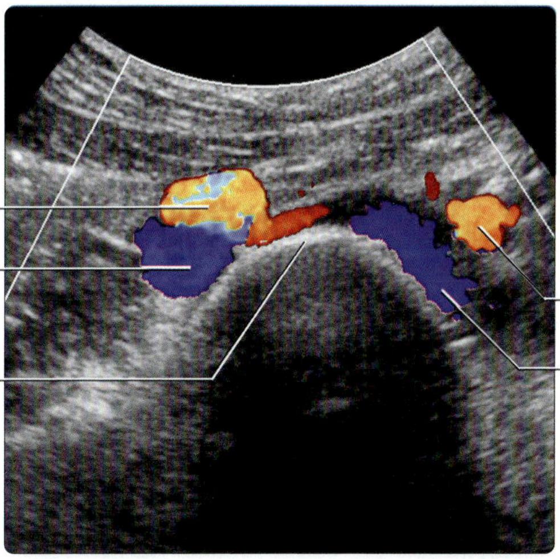

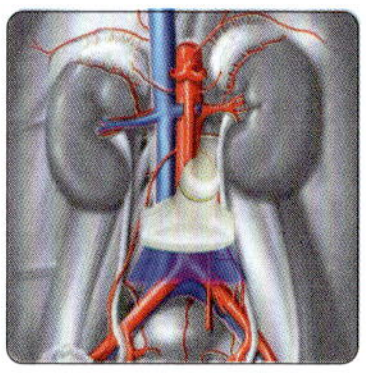

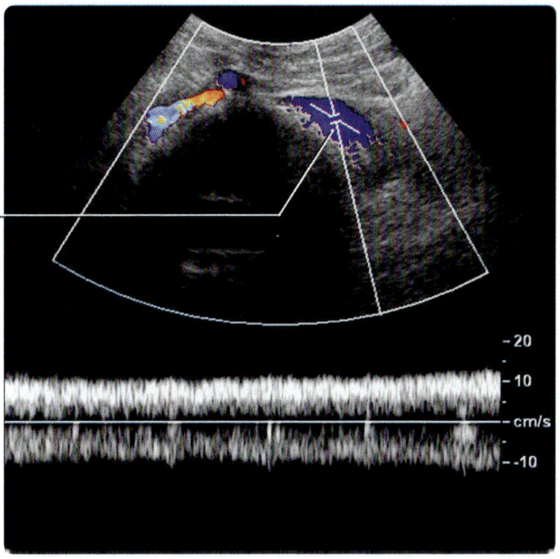

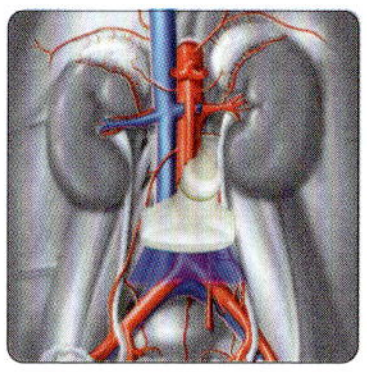

(Top) *Transverse grayscale ultrasound shows the common iliac veins just below where they converge to form the IVC. Note at this level the common iliac arteries are located anterior to the common iliac veins. Examination of the common iliac veins with ultrasound is often limited by intervening bowel loops and the patient's body habitus.* **(Middle)** *Transverse color Doppler ultrasound of the proximal common iliac veins below the origin of the IVC shows flow in the common iliac veins in blue and flow in the common iliac arteries in red.* **(Bottom)** *Spectral Doppler ultrasound of the left common iliac vein is shown. The normal spectral Doppler of the common iliac veins has 5 characteristics: Spontaneous flow, phasic flow, flow ceases with Valsalva maneuver, flow augmentation with distal compression, and unidirectional flow toward the heart.*

AORTIC ANEURYSM AND DISSECTION

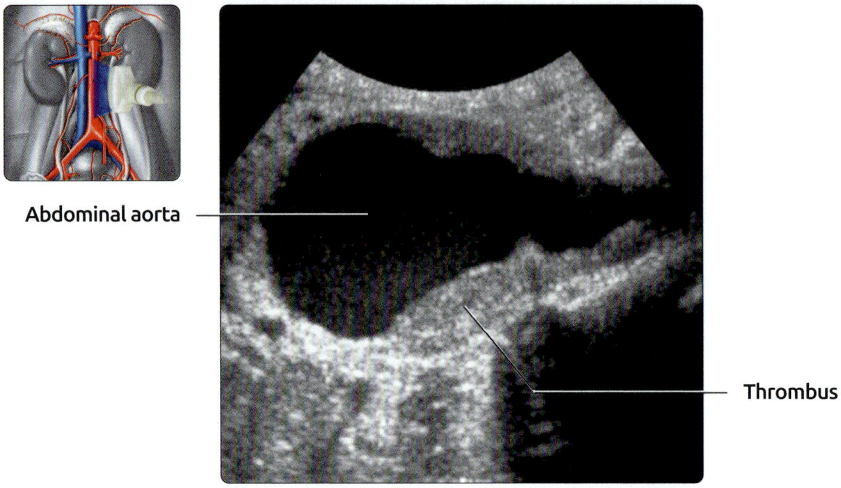

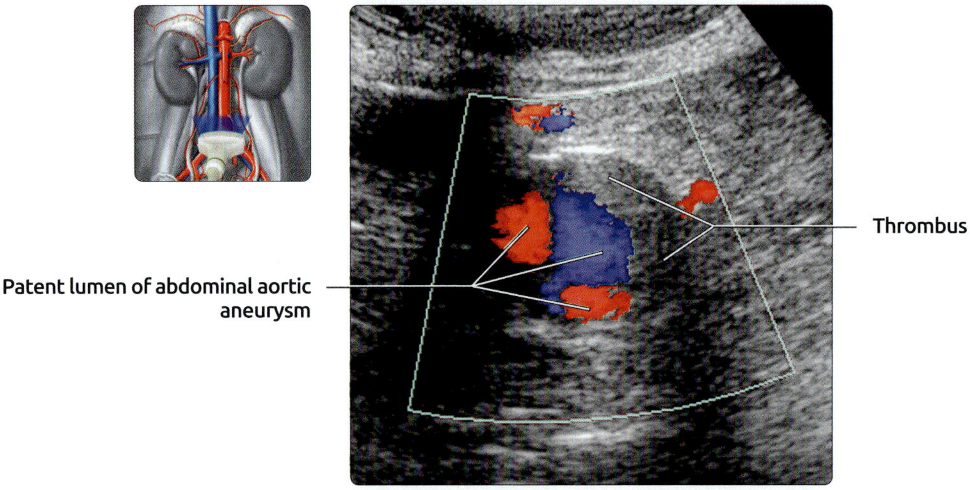

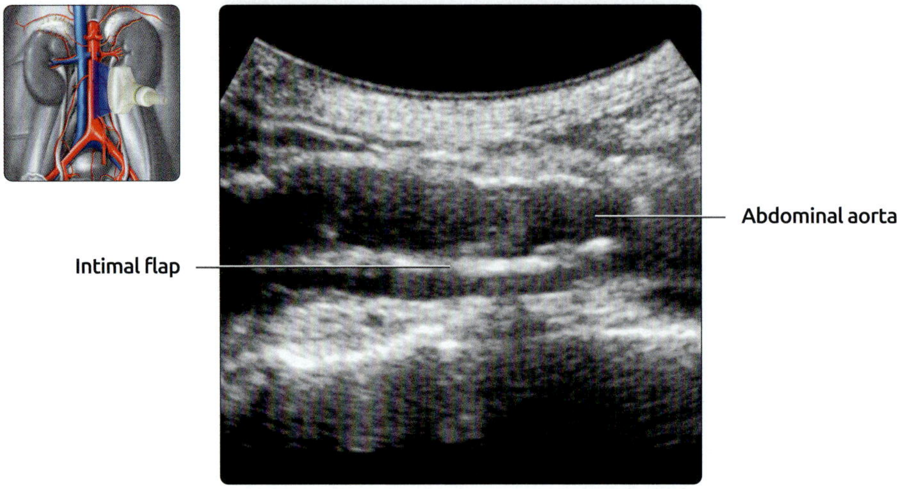

(Top) *Oblique transabdominal grayscale ultrasound shows a distal abdominal aortic aneurysm with no involvement of the bifurcation. A vessel is considered aneurysmal when it is dilated 1.5x or more compared to normal adjacent segments. The abdominal aorta is considered to be dilated when its caliber exceeds 3 cm.* **(Middle)** *Transverse color Doppler ultrasound shows an abdominal aortic aneurysm with circumferential mural thrombus in the lumen of an aortic aneurysm.* **(Bottom)** *Longitudinal grayscale ultrasound shows dissection of the abdominal aorta with an echogenic intimal flap. Dissection in the abdominal aorta is more commonly seen as an extension of a dissection of the thoracic aorta rather than a localized event in the abdominal aorta.*

INFERIOR VENA CAVA OBSTRUCTION

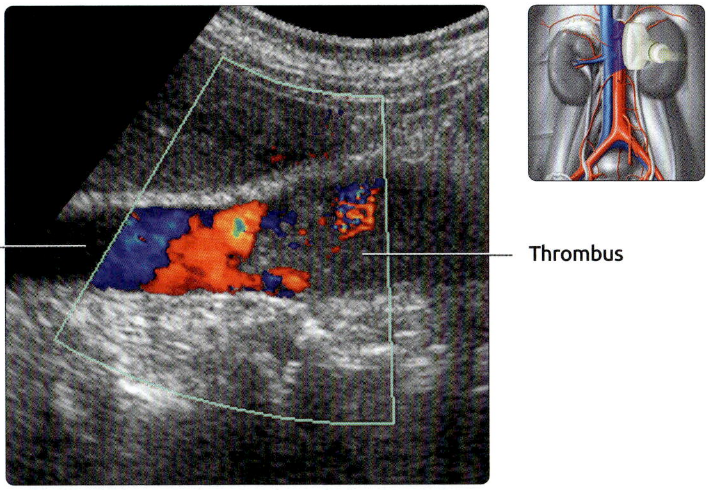

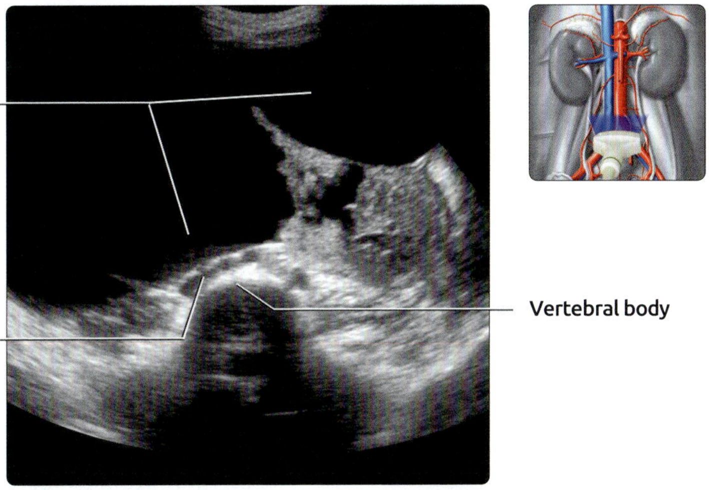

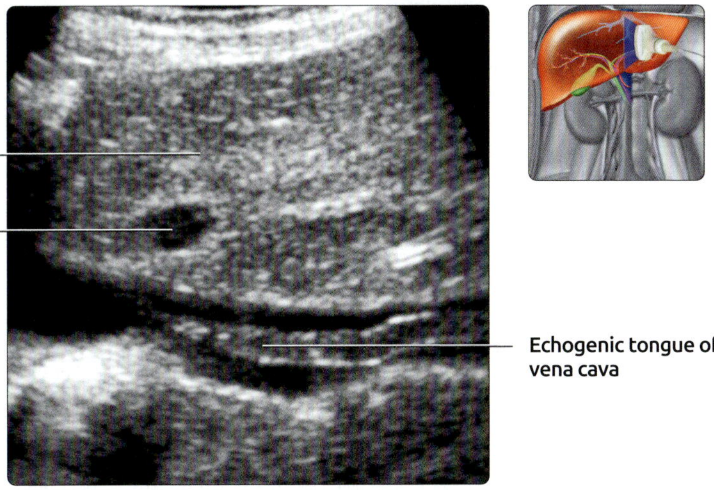

(Top) *Longitudinal color Doppler ultrasound shows thrombosis within the IVC. Tumors such as renal cell carcinoma and hepatocellular carcinoma have a propensity to invade venous structures and are both causes of tumor thrombus extension into the IVC. Note the color Doppler flow within the thrombus, suggesting it is a tumor thrombus rather than a bland stasis/occlusive thrombus.* **(Middle)** *Transverse transabdominal grayscale ultrasound shows a large complex cystic tumor compressing the IVC.* **(Bottom)** *Longitudinal grayscale ultrasound shows an echogenic "tongue" of thrombus extending from the iliac veins into the partially patent IVC. This patient is at high risk of pulmonary embolism.*

TERMINOLOGY

Definitions

- Peritoneal cavity: Potential space in abdomen between visceral and parietal peritoneum, usually containing only small amount of peritoneal fluid (for lubrication)
- Abdominal cavity: Not synonymous with peritoneal cavity
  - Contains all of abdominal viscera (intra- and retroperitoneal)
  - Limited by abdominal wall muscles, diaphragm, and (arbitrarily) pelvic brim

GROSS ANATOMY

Divisions

- Greater sac of peritoneal cavity
- Lesser sac (omental bursa)
  - Communicates with greater sac via epiploic foramen (of Winslow)
  - Bounded anteriorly by caudate lobe, stomach, and greater omentum
    - Posteriorly by pancreas, left adrenal, and kidney
    - On left by splenorenal and gastrosplenic ligaments
    - On right by epiploic foramen and lesser omentum

Compartments

- Supramesocolic space
  - Divided into right and left supramesocolic spaces, which are separated by falciform ligament
    - Right supramesocolic space: Composed of right subphrenic space, right subhepatic space, and lesser sac
    - Left supramesocolic space: Divided into left perihepatic spaces (anterior and posterior) and left subphrenic (anterior perigastric and posterior perisplenic)
- Inframesocolic compartment
  - Divided into right inframesocolic space, left inframesocolic space, paracolic gutters, and pelvic cavity
  - Pelvic cavity is most dependent part of peritoneal cavity in erect and supine positions

Peritoneum

- Thin serous membrane consisting of single layer of squamous epithelium (mesothelium)
  - Parietal peritoneum lines abdominal wall
  - Visceral peritoneum (serosa) lines abdominal organs

Mesentery

- Double layer of peritoneum that encloses organ and connects it to abdominal wall
- Covered on both sides by mesothelium and has core of loose connective tissue containing fat, lymph nodes, blood vessels, and nerves passing to and from viscera
- Most mobile parts of intestine have mesentery, while ascending and descending colon are considered retroperitoneal (covered only by peritoneum on anterior surface)
- Root of mesentery is its attachment to posterior abdominal wall
- Root of small bowel mesentery is ~ 15 cm and passes from left side of L2 vertebra downward and to right
  - Contains superior mesenteric vessels, nerves, and lymphatics
- Transverse mesocolon crosses almost horizontally in front of pancreas, duodenum, and right kidney

Omentum

- Multilayered fold of peritoneum that extends from stomach to adjacent organs
- Lesser omentum joins lesser curve of stomach and proximal duodenum to liver
  - Hepatogastric and hepatoduodenal ligament components contain common bile duct, hepatic and gastric vessels, and portal vein
- Greater omentum
  - 4-layered fold of peritoneum hanging from greater curve of stomach like apron, covering transverse colon and much of small intestine
    - Contains variable amounts of fat and abundant lymph nodes
    - Mobile and can fill gaps between viscera
    - Acts as barrier to generalized spread of intraperitoneal infection or tumor

Ligaments

- All double-layered folds of peritoneum, other than mesentery and omentum, are peritoneal ligaments
- Connect 1 viscus to another (e.g., splenorenal ligament) or viscus to abdominal wall (e.g., falciform ligament)
- Contain blood vessels or remnants of fetal vessels

Folds

- Reflections of peritoneum with defined borders, often lifting peritoneum off abdominal wall (e.g., median umbilical fold covers urachus and extends from dome of urinary bladder to umbilicus)

Peritoneal Recesses

- Dependent pouches formed by peritoneal reflections
- Many have eponyms [e.g., Morison pouch for posterior subhepatic (hepatorenal) recess; pouch of Douglas for rectouterine recess]

ANATOMY IMAGING ISSUES

Imaging Recommendations

- Transducer: Typically 2-5 MHz for abdominal survey and deep recesses, up to 9 MHz for thinner patients
- High-frequency linear transducer 8-15 MHz may be used to evaluate anterior abdominal wall and parietal peritoneum
- Patient examined supine with additional decubitus positions to determine if fluid collection is free or loculated
- Peritoneal cavity and its various mesenteries and recesses are usually not apparent on imaging studies unless distended or outlined by intraperitoneal fluid or air

PERITONEAL CAVITY

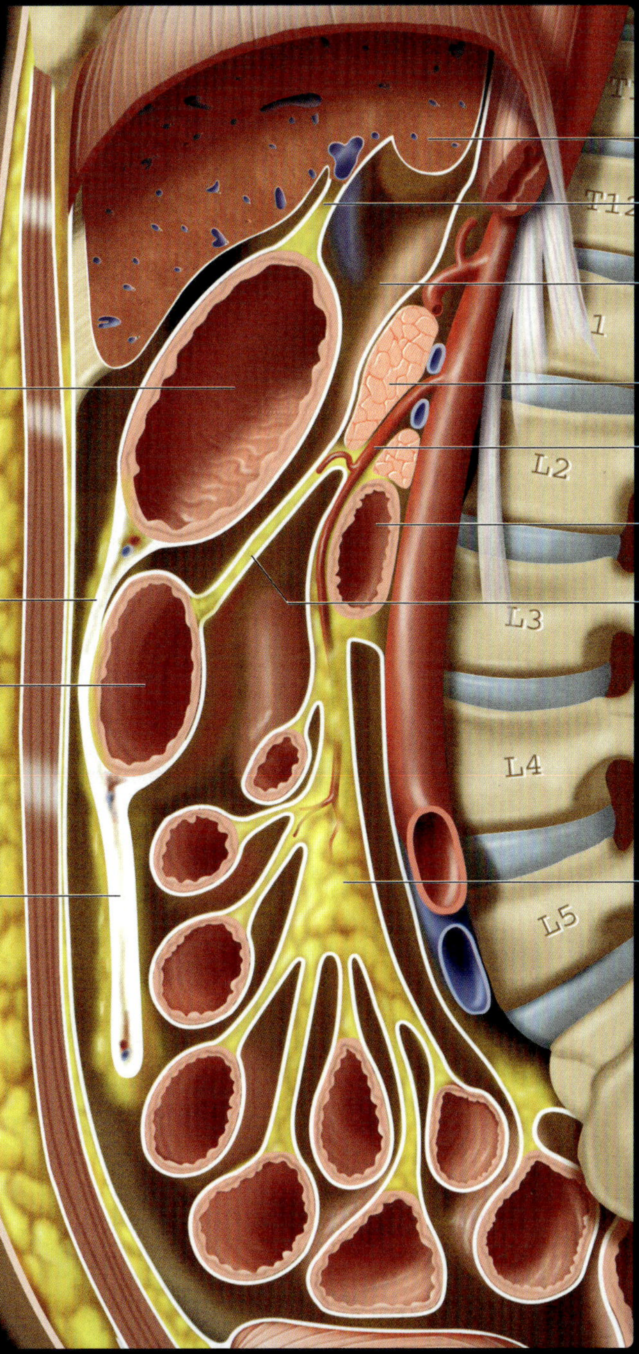

Graphic of a sagittal section of the abdomen shows the peritoneal cavity artificially distended to show anatomy. Note the margins of the lesser sac in this plane, including the caudate lobe of the liver, stomach, gastrocolic ligament anteriorly and pancreas posteriorly. The hepatogastric ligament is part of the lesser omentum and carries the hepatic artery and portal vein to the liver. The mesenteries are multilayered folds of the peritoneum that enclose a layer of fat and convey blood vessels, nerves, and lymphatics to the intraperitoneal abdominal viscera. The greater omentum is a 4-layered fold of the peritoneum that extends down from the stomach, covering much of the colon and small intestine. The layers are generally fused together caudal to the transverse colon. The gastrocolic ligament is part of the greater omentum.

PERITONEAL DIVISIONS AND COMPARTMENTS

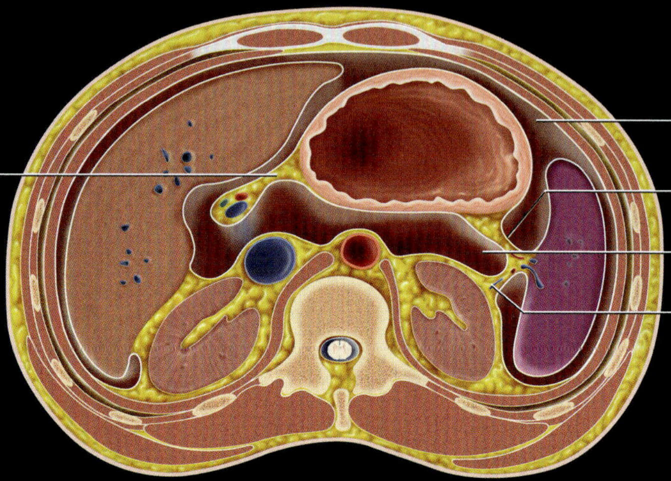

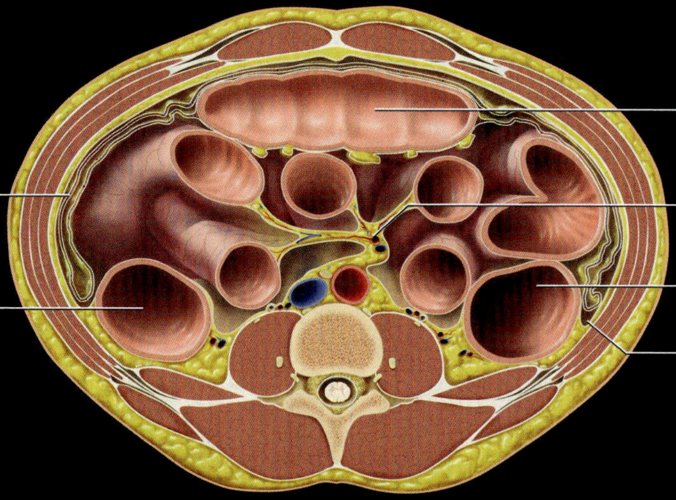

(Top) *The borders of the lesser sac (omental bursa) include the lesser omentum, which contains the common bile duct and hepatic and gastric vessels. The left border includes the gastrosplenic ligament (with short gastric vessels) and the splenorenal ligament (with splenic vessels).* **(Bottom)** *The paracolic gutters are formed by reflections of the peritoneum covering the ascending and descending colon and the lateral abdominal wall. Note the innumerable potential peritoneal recesses lying between the bowel loops and their mesenteric leaves. The greater omentum covers much of the bowel like an apron.*

PERITONEAL DIVISIONS AND COMPARTMENTS

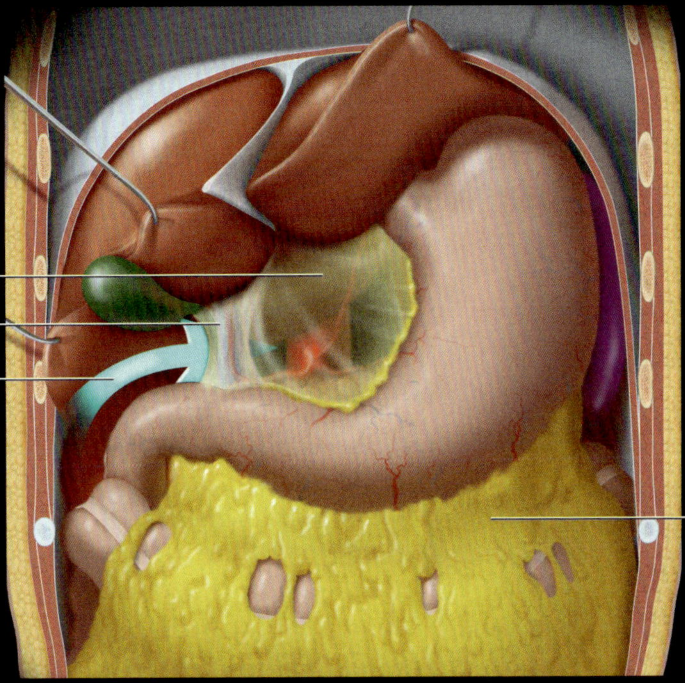

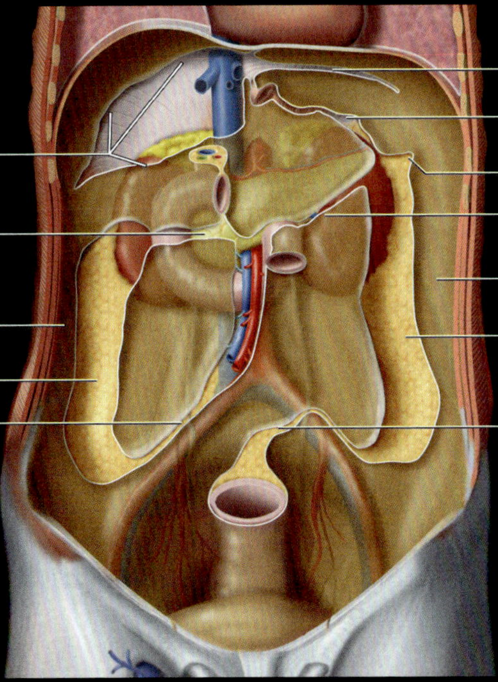

(Top) *In this graphic, the liver has been retracted upward. The lesser omentum is comprised of the hepatoduodenal and hepatogastric ligaments. It forms part of the anterior wall of the lesser sac and contains the common bile duct, hepatic and gastric vessels, and portal vein. The aorta and celiac artery can be seen through the lesser omentum, as they lie just posterior to the lesser sac.* **(Bottom)** *Frontal view of the abdomen, with all of the intraperitoneal organs removed, shows that the root of the transverse mesocolon divides the peritoneal cavity into supramesocolic and inframesocolic spaces that communicate only along the paracolic gutters. The coronary and triangular ligaments suspend the liver from the diaphragm. The superior mesenteric vessels traverse the small bowel mesentery, whose root crosses obliquely from the upper left to the lower right posterior abdominal wall.*

RIGHT SUPRAMESOCOLIC SPACE

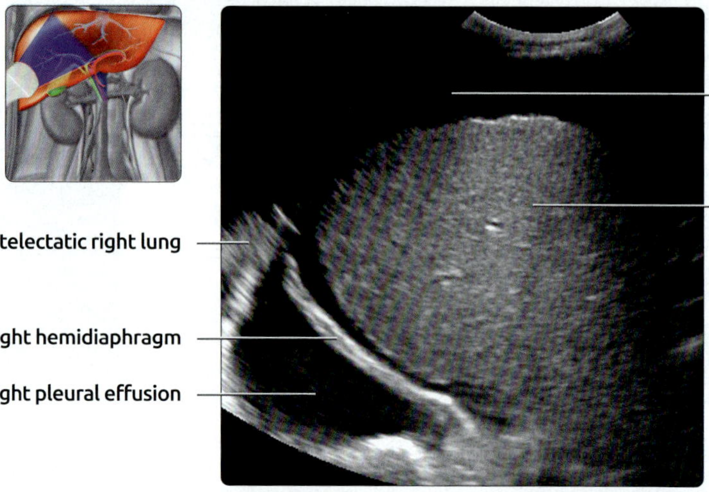

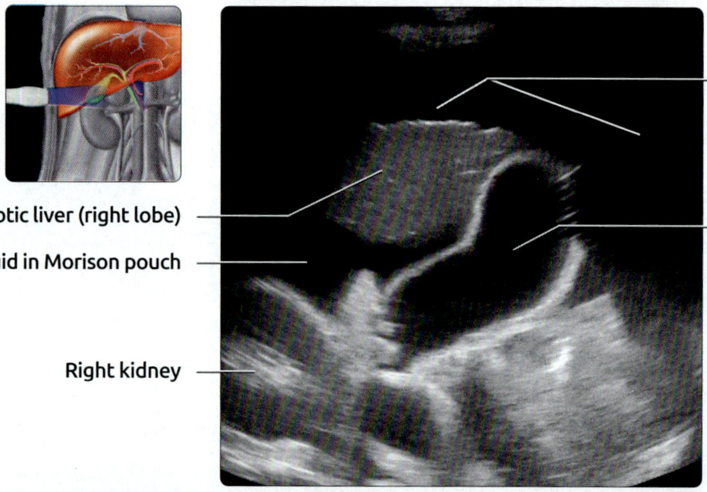

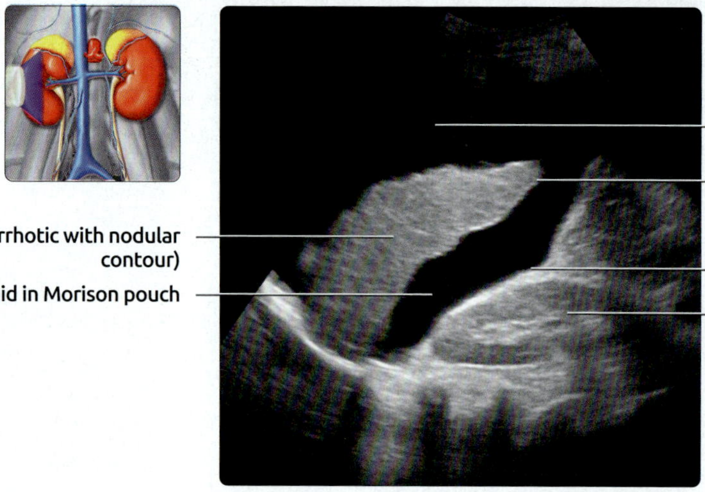

(Top) *Intercostal oblique grayscale ultrasound (in a patient with cirrhosis) shows the dome of the right lobe of the liver and moderate fluid in the right subphrenic region extending anterior to the liver. The fluid is separated from the right-sided pleural effusion by the right diaphragmatic leaf.* **(Middle)** *Subcostal oblique transverse ultrasound of the right upper quadrant shows fluid in the right anterior subhepatic space and in the hepatorenal space. The ascites are secondary to hepatic cirrhosis, and the gallbladder is physiologically distended.* **(Bottom)** *Longitudinal transabdominal grayscale ultrasound shows fluid in the right posterior subhepatic space, also known as the Morison pouch, and hepatorenal fossa. This space is continuous with the right anterior subhepatic space and right paracolic gutter.*

RIGHT SUPRAMESOCOLIC SPACE: LESSER SAC

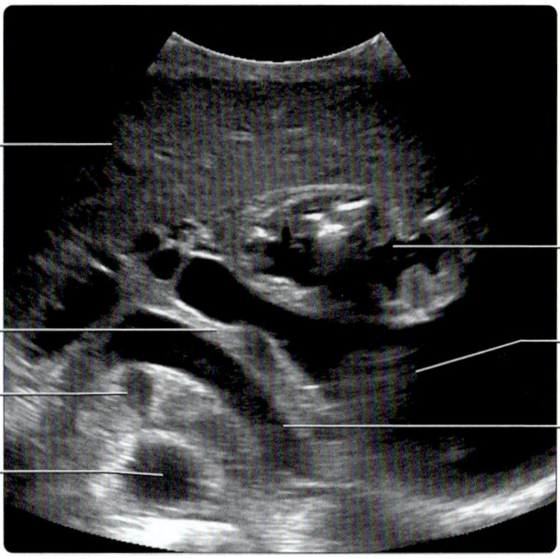

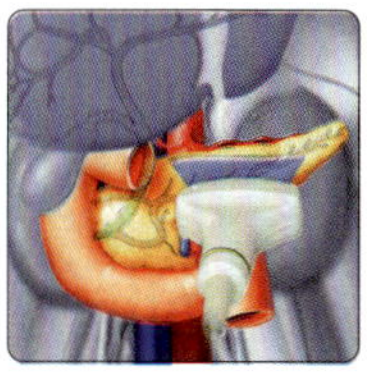

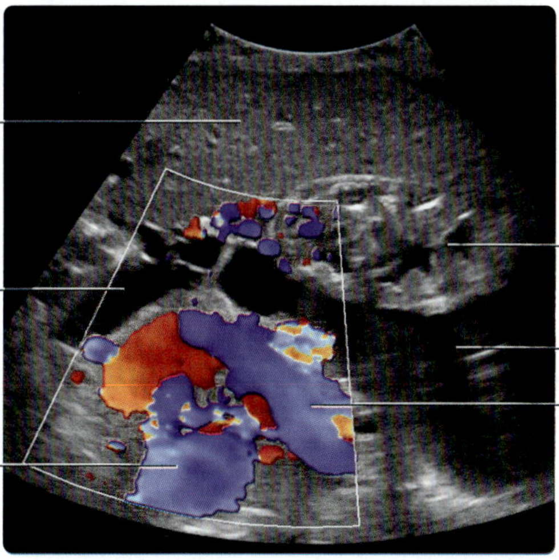

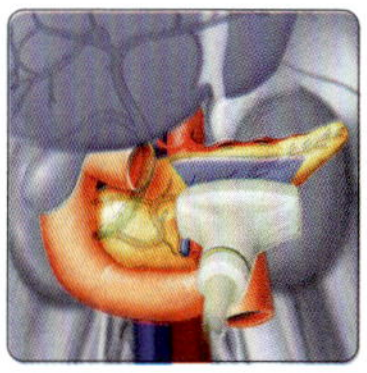

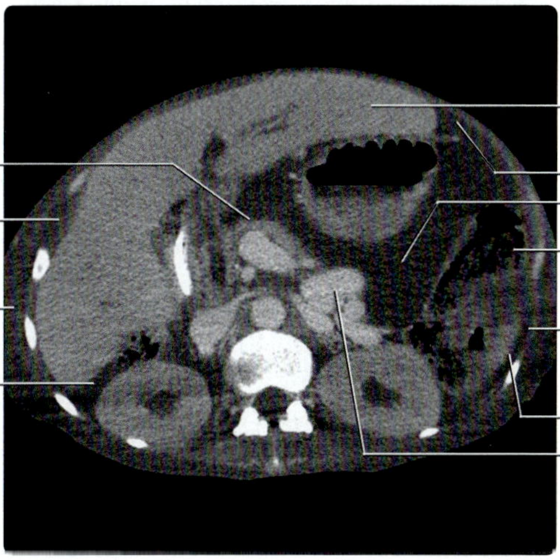

(Top) *Subxiphoid transverse grayscale ultrasound shows a fluid collection in the lesser sac, which extends to the left, behind the stomach and anterior to the pancreas. The lesser sac is part of the right supramesocolic space and communicates with the rest of the peritoneal cavity through the epiploic foramen (of Winslow).* **(Middle)** *Subxiphoid transverse color Doppler ultrasound of the same patient shows moderate fluid in the lesser sac posterior to the stomach. The splenic vein was dilated in this patient with portal hypertension status post liver transplant.* **(Bottom)** *Axial CECT of the same patient shows fluid in the lesser sac and peritoneal cavity as well as diffuse anasarca.*

LEFT SUPRAMESOCOLIC SPACE

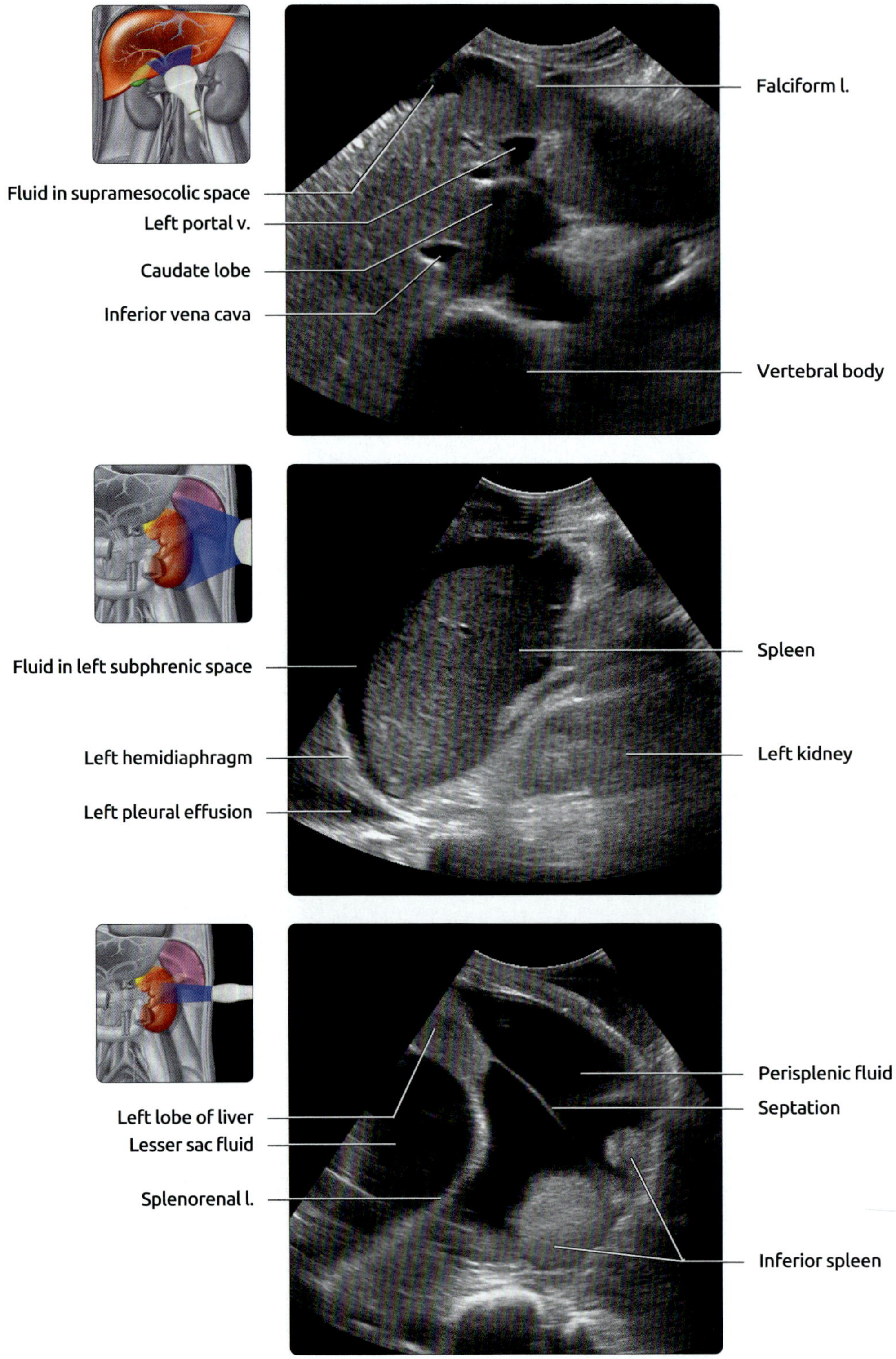

(Top) *Subxiphoid transverse grayscale ultrasound shows fluid anterior to the left lobe of the liver that is localized to the left posterior subhepatic space. Incidental calculi are seen within a dilated intrahepatic biliary duct.* **(Middle)** *Longitudinal grayscale ultrasound of the left upper quadrant shows a small amount of perisplenic fluid extending under the left hemidiaphragm. The left subphrenic space is separated from the right subphrenic space by the falciform ligament.* **(Bottom)** *Transverse grayscale ultrasound of the left upper quadrant reveals fluid in the perisplenic space and lesser sac.*

INFRAMESOCOLIC SPACE

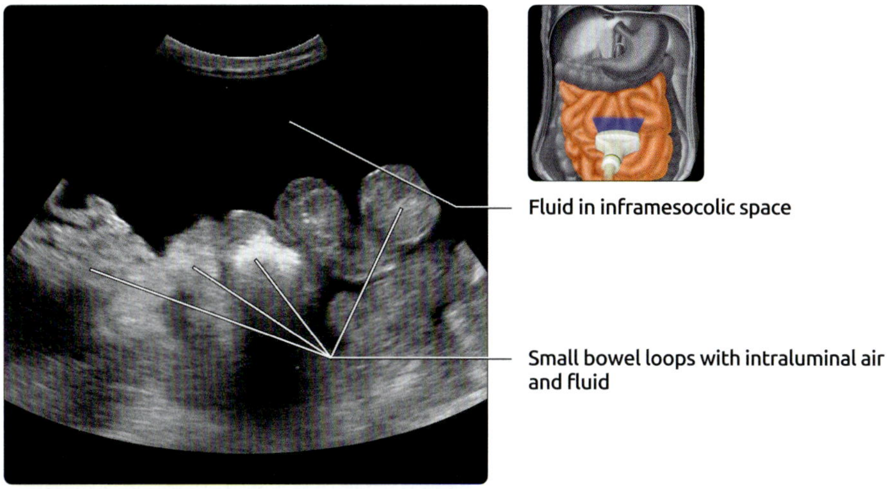

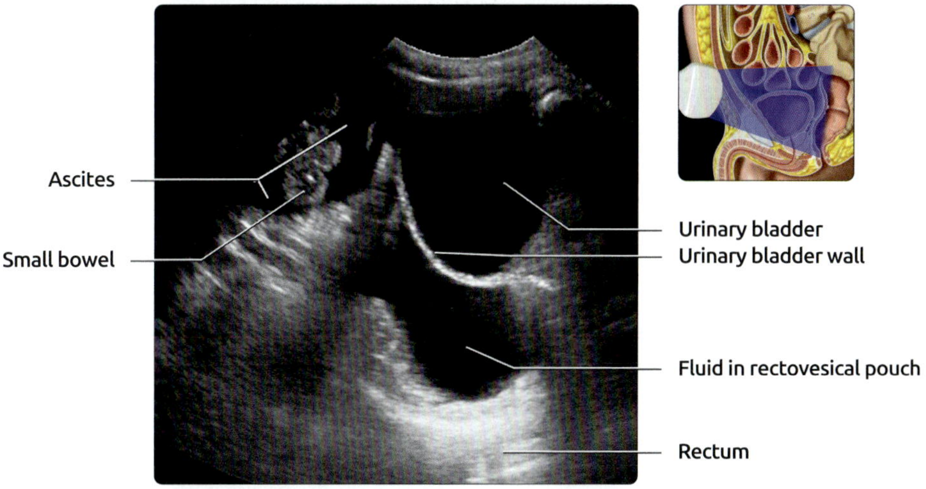

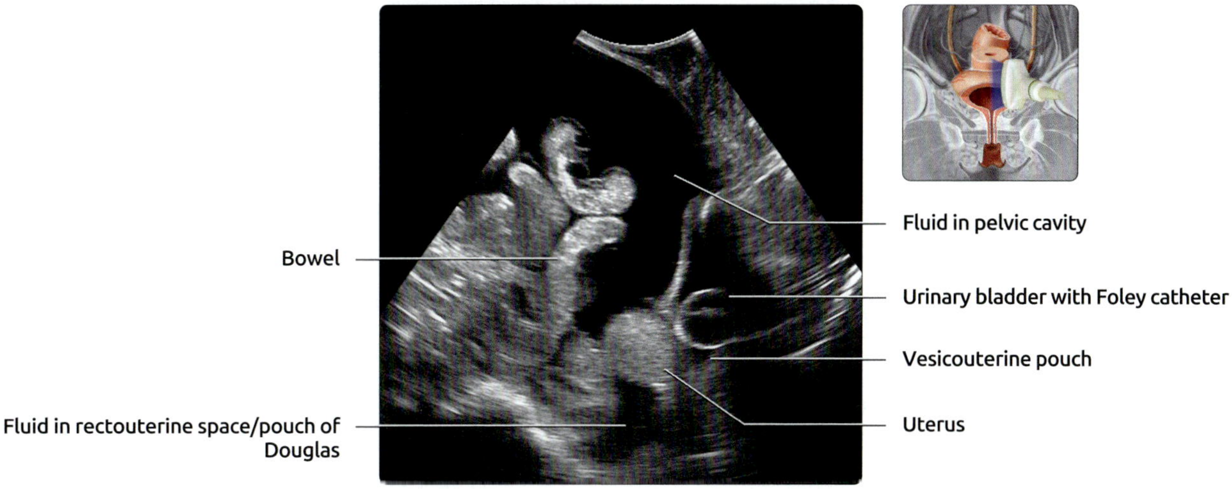

(Top) *Transverse transabdominal ultrasound of the central abdomen reveals moderate to large ascites with floating small bowel loops. The left inframesocolic space is larger compared to the right and communicates directly with the pelvic cavity.* **(Middle)** *Longitudinal ultrasound of the midline suprapubic region in a male patient demonstrates intraperitoneal fluid between bowel loops and extending into the dependent rectovesical pouch. There is a distended urinary bladder.* **(Bottom)** *Longitudinal grayscale ultrasound of the female pelvis shows free fluid. The uterus divides the pelvic cavity into the vesicouterine and rectouterine (pouch of Douglas) spaces. In this case, the vesicouterine space contains minimal fluid.*

TERMINOLOGY

Definitions

- Abdomen: Region between diaphragm and pelvis

GROSS ANATOMY

Anatomic Boundaries of Anterior Abdominal Wall

- Superiorly: Xiphoid process and costal cartilages of 7th-10th ribs
- Inferiorly: Iliac crest, iliac spine, inguinal ligament, and pubis
- Inguinal ligament is inferior edge of aponeurosis of external oblique muscle

Muscles of Anterior Abdominal Wall

- Consist of 3 flat muscles (external oblique, internal oblique, and transverse abdominal), and 1 strap-like muscle (rectus)
- Combination of muscles and aponeuroses (sheet-like tendons) act as corset to confine and protect abdominal viscera
- Linea alba is fibrous raphe stretching from xiphoid to pubis
  - Forms central anterior attachment for abdominal wall muscles
  - Formed by interlacing fibers of aponeuroses of oblique and transverse abdominal muscles
  - Rectus sheath also formed by these aponeuroses as they surround rectus muscle
- Linea semilunaris is vertical fibrous band at lateral edge of rectus sheath bilaterally
  - Aponeuroses of internal and transversus abdominis join in linea semilunaris before forming rectus sheath
- External oblique muscle
  - Largest and most superficial of 3 flat abdominal muscles
  - Origin: External surfaces of ribs 5-12
  - Insertion: Linea alba, iliac crest, pubis via broad aponeurosis
- Internal oblique muscle
  - Middle of 3 flat abdominal muscles
  - Runs at right angles to external oblique
  - Origin: Posterior layer of thoracolumbar fascia, iliac crest, and inguinal ligament
  - Insertion: Ribs 10-12 posteriorly, linea alba via broad aponeurosis, pubis
- Transversus abdominis (transversalis) muscle
  - Innermost of 3 flat abdominal muscles
  - Origin: Lowest 6 costal cartilages, thoracolumbar fascia, iliac crest, inguinal ligament
  - Insertion: Linea alba via broad aponeurosis, pubis
- Rectus abdominis muscle
  - Origin: Pubic symphysis and pubic crest
  - Insertion: Xiphoid process and costal cartilages 5-7
  - Rectus sheath: Strong, fibrous compartment that envelops each rectus muscle
    - Contains superior and inferior epigastric vessels
- Actions of anterior abdominal wall muscles
  - Support and protect abdominal viscera
  - Help flex and twist trunk, maintain posture
  - Increase intraabdominal pressure for defecation, micturition, and childbirth
  - Stabilize pelvis during walking, sitting up
- Transversalis fascia
  - Lies deep to abdominal wall muscles and lines entire abdominal wall
  - Separated from parietal peritoneum by layer of extraperitoneal fat

Muscles of Posterior Abdominal Wall

- Consist of psoas (major and minor), iliacus, and quadratus lumborum
- Psoas: Long, thick, fusiform muscle lying lateral to vertebral column
  - Origin: Transverse processes and bodies of vertebrae T12-L5
  - Insertion: Lesser trochanter of femur (passing behind inguinal ligament)
  - Action: Flexes thigh at hip joint; bends vertebral column laterally
- Iliacus: Large triangular sheet of muscle lying along lateral side of psoas
  - Origin: Superior part of iliac fossa
  - Insertion: Lesser trochanter of femur (after joining with psoas tendon)
  - Action: "Iliopsoas muscle" flexes thigh
- Quadratus lumborum: Thick sheet of muscle lying adjacent to transverse processes of lumbar vertebrae
  - Invested by lumbodorsal fascia
  - Origin: Iliac crest and transverse processes of lumbar vertebrae
  - Insertion: 12th rib
  - Actions: Stabilizes position of thorax and pelvis during respiration, walking; bends trunk to side

Paraspinal Muscles

- Also called erector spinae muscles
  - Invested by lumbodorsal fascia
- Composed of 3 columns
  - Iliocistalis: Lateral
  - Longissimus: Intermediate
  - Spinalis: Medial
- Origins: Sacrum, ilium, and spines of lumbar and 11th-12th thoracic vertebrae
- Insertions: Ribs and vertebrae with additional muscle slips joining columns at successively higher levels
- Action: Extends vertebral column

ANATOMY IMAGING ISSUES

Imaging Recommendations

- High-frequency (5-12 MHz) linear transducer for anterior abdominal wall and paraspinal muscles
- 3-5 MHz for posterior abdominal wall muscles
- Supine position for examination of anterior and lateral abdominal wall
  - Image during Valsalva maneuver and in standing position to increase abdominal pressure and elicit hernias
  - Prone position for ultrasound of paraspinal muscles
- Compare with contralateral side to check for symmetry

ANTERIOR ABDOMINAL WALL

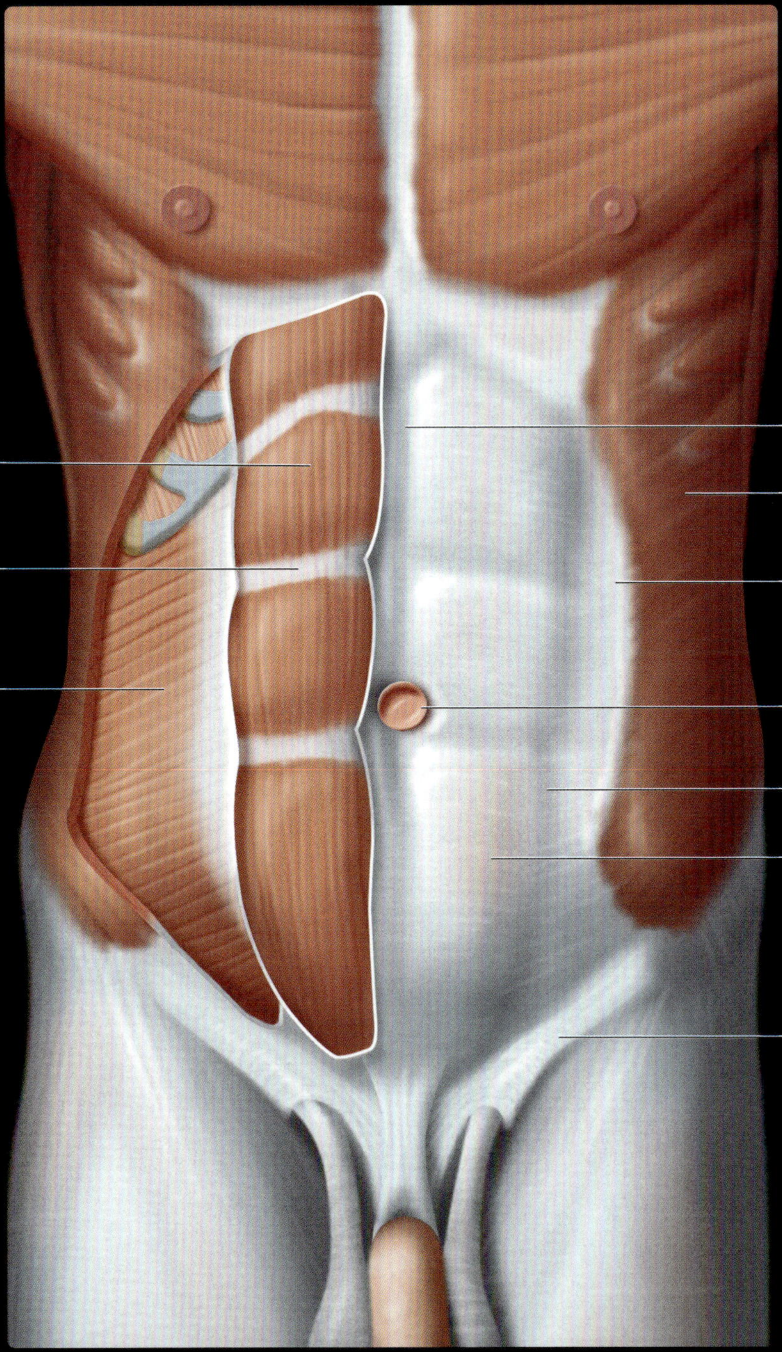

Graphic shows the aponeuroses of the internal and external oblique and transverse abdominal muscles are 2-layered and interweave with each other, covering the rectus muscle, constituting the rectus sheath and linea alba. About midway between the umbilicus and symphysis, at the arcuate line, the posterior rectus sheath ends (arcuate line), and the transversalis fascia is the only structure between the rectus muscle and parietal peritoneum.

POSTERIOR ABDOMINAL WALL

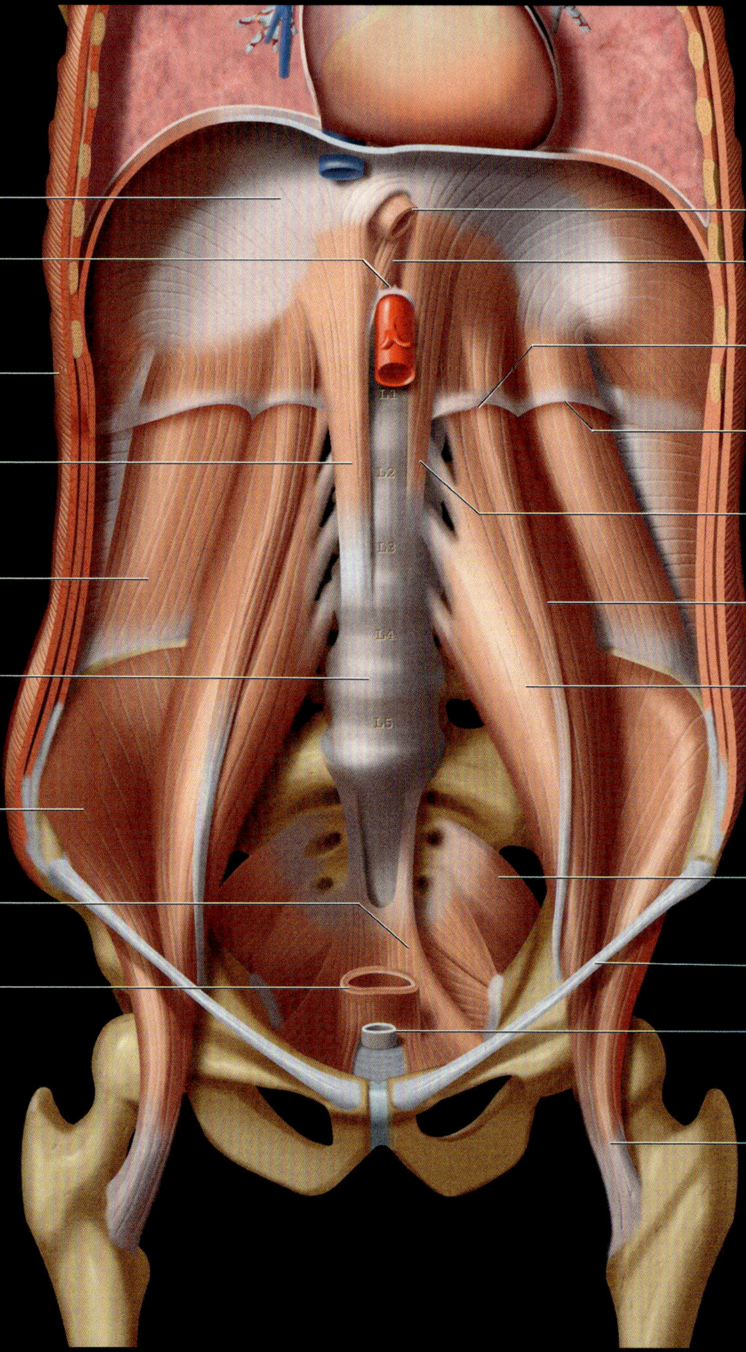

Graphic shows the lumbar vertebrae are covered and attached by the anterior longitudinal ligament and the diaphragmatic crura are closely attached to it, as are the origins of the psoas muscles, which also arise from the transverse processes. The iliacus muscle arises from the iliac fossa of the pelvis and inserts into the tendon of the psoas major, constituting the iliopsoas muscle, which inserts onto the lesser trochanter. The quadratus lumborum arises from the iliac crest and inserts onto the 12th rib and transverse processes of the lumbar vertebrae. The diaphragmatic and transverse abdominal fibers interlace. The psoas and quadratus lumborum pass behind the diaphragm under medial and lateral arcuate ligaments.

MUSCLES OF BACK IN SITU

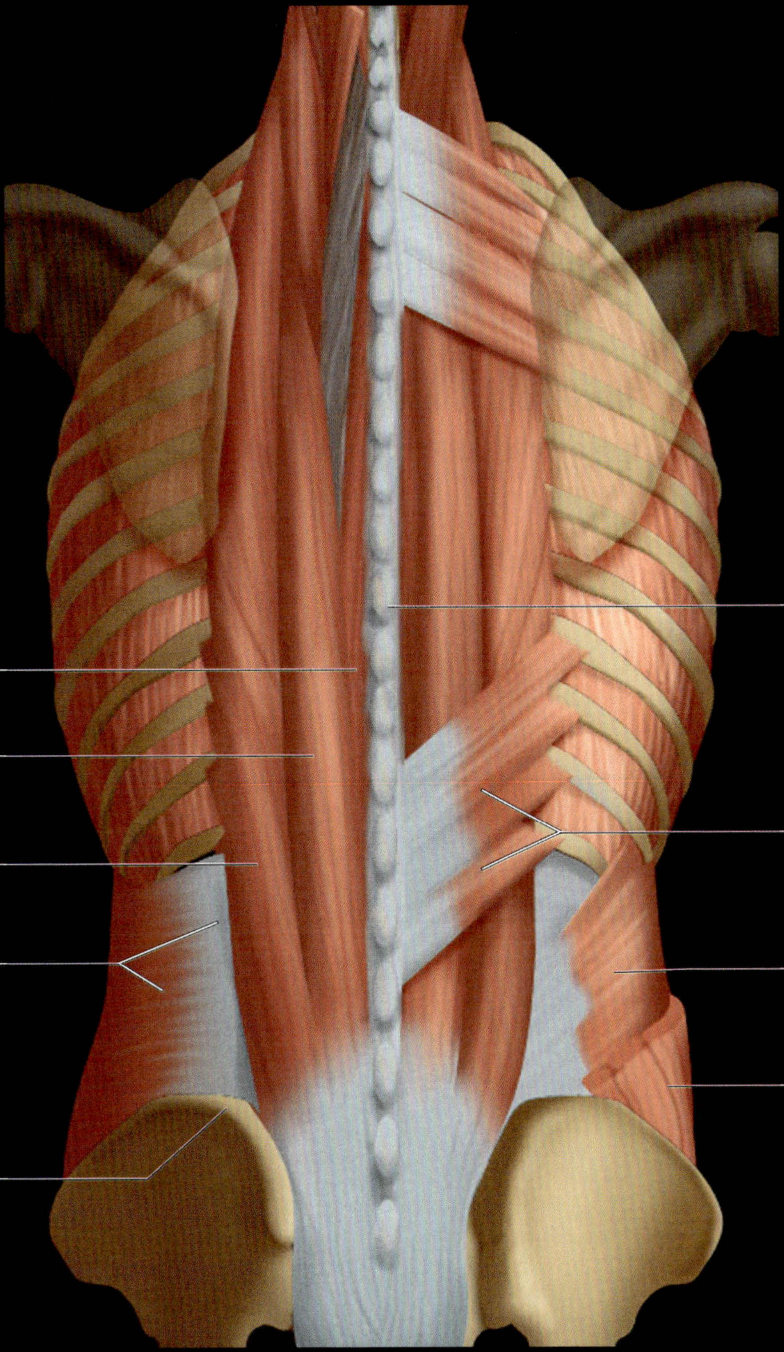

Graphic shows the paraspinal muscles and muscles of the back. The latissimus dorsi muscles are not included. The erector spinae have thick, tendinous origins from the sacral, iliac crests, the lumbar, and 11th to 12th thoracic spinous processes. Superiorly, the muscle becomes fleshy, and in the upper lumbar region subdivides to become the iliocostalis, longissimus, and spinalis muscles (from lateral to medial), tapering as they insert into the vertebrae and ribs. The erector muscles flank the spinous processes and span the length of the posterior thorax and abdomen. They are responsible for extension of the vertebral column.

ANTERIOR ABDOMINAL WALL

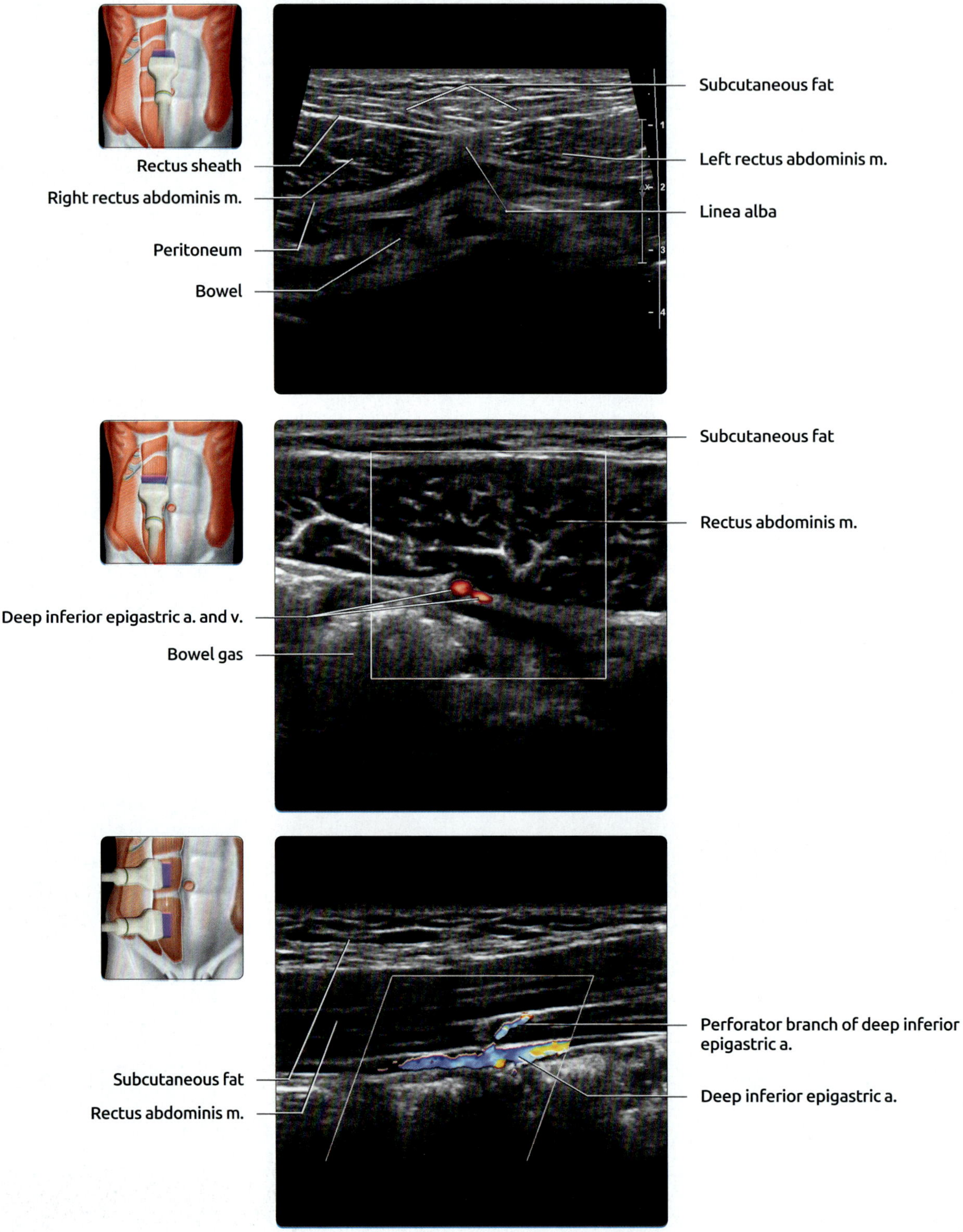

(Top) *Transverse grayscale ultrasound of the midline anterior abdominal wall shows the paired rectus abdominis muscles separated by the linea alba. The rectus abdominis muscles are comparable in echogenicity and thickness. The surrounding rectus sheath is seen as a fine, thin, echogenic structure around the muscles.* **(Middle)** *Transverse power Doppler ultrasound of a rectus abdominis muscle in the lower abdomen shows the deep inferior epigastric artery and vein. Branches of the superior epigastric artery that arise in the upper abdomen anastomose with branches of the inferior epigastric artery at the umbilicus.* **(Bottom)** *Longitudinal color Doppler ultrasound shows a perforating branch of the deep inferior epigastric artery extending into the rectus muscle. These perforators are important for breast reconstruction with abdominal wall flaps.*

ANTEROLATERAL ABDOMINAL WALL

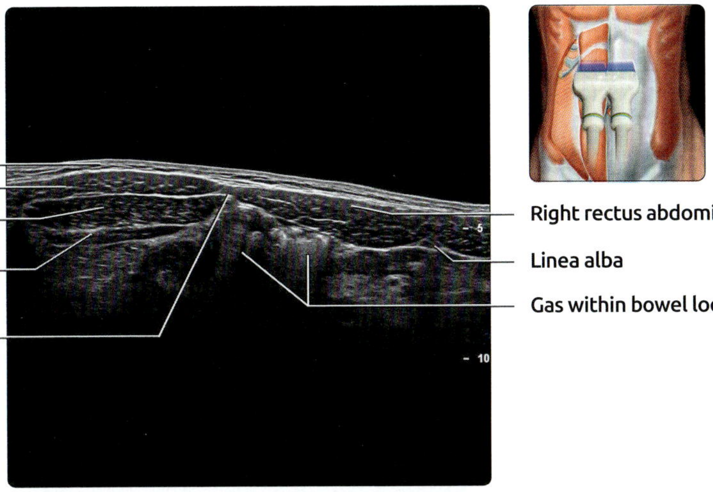

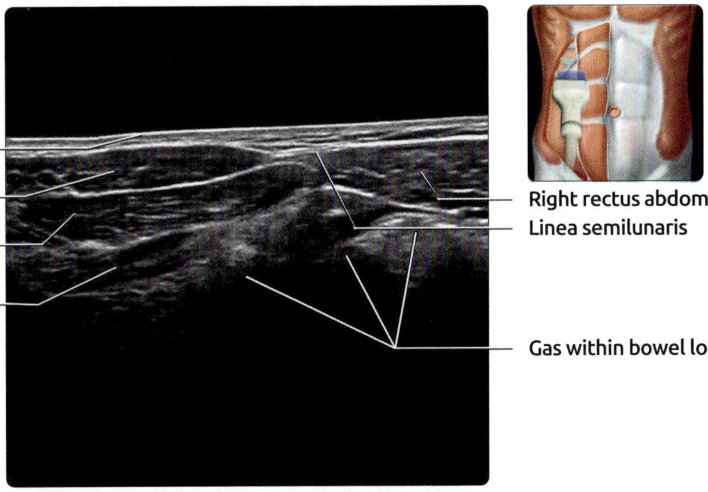

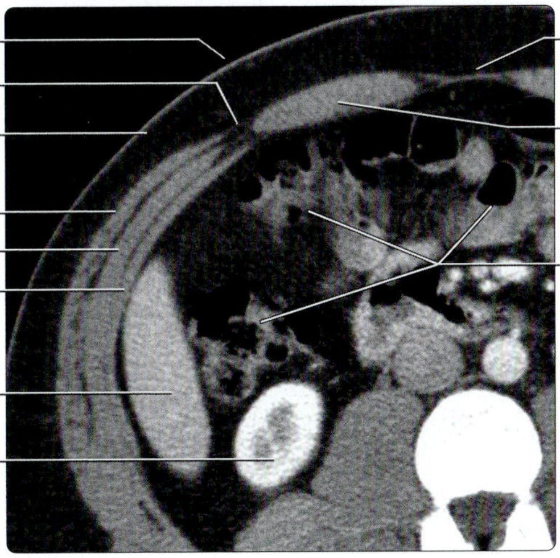

(Top) *Transverse extended FOV grayscale ultrasound shows the relationship of the medially located rectus abdominis and the laterally located oblique and transverse abdominal muscles. Medially, the external and internal oblique and the transversus abdominal muscles form aponeuroses that comprise the rectus sheath, with the muscles thinning at the linea semilunaris. The linea alba is thin in the lower abdomen.* **(Middle)** *Transverse grayscale ultrasound at the right anterolateral abdominal wall shows the relationship of the lateral abdominal wall muscles in better detail. Note the oblique and transverse abdominal muscles taper medially as they become aponeuroses.* **(Bottom)** *Correlative axial CECT illustrates the muscles of the abdominal wall. The rectus abdominis muscle in the anterior abdominal wall is shown, along with the oblique and transverse abdominal muscles in the anterolateral abdominal wall and their aponeuroses.*

POSTERIOR ABDOMINAL WALL

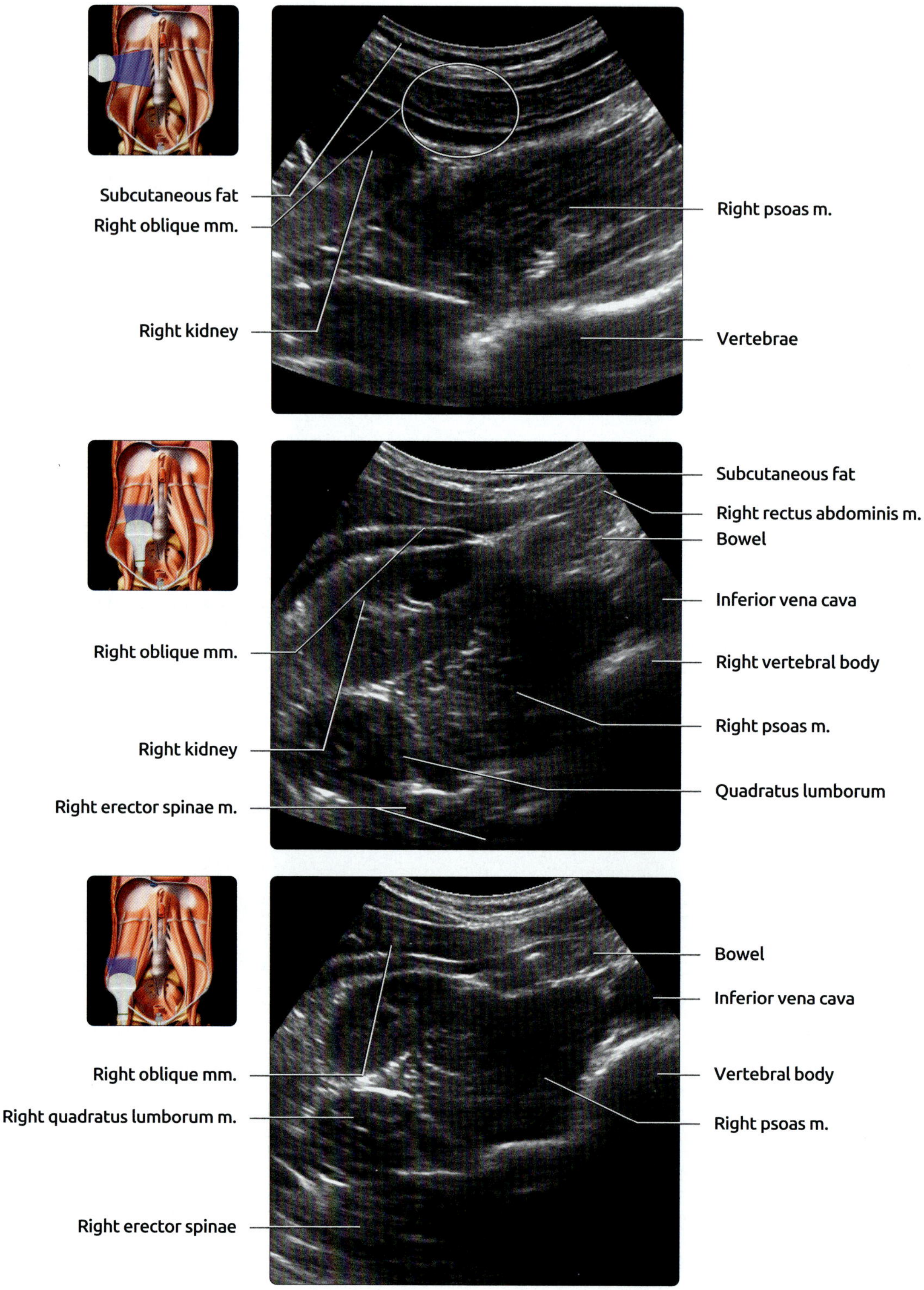

(Top) *Longitudinal oblique grayscale ultrasound through the lower right abdomen shows the right psoas muscle, which originates from the lumbar spine and inserts into the proximal femur.* **(Middle)** *Transverse grayscale ultrasound of right mid abdomen using the kidney as an acoustic window is shown. The kidney is anterior and lateral to the psoas and anterior to the quadratus lumborum. The psoas runs along the paravertebral region in its entire abdominal course. The quadratus lumborum originates from the iliolumbar ligament and iliac crest to insert into the last rib and lumbar transverse processes. It is easily identified as the muscle on which the kidney rests.* **(Bottom)** *Transverse grayscale ultrasound of the right upper abdomen, continuing the scan inferiorly, shows the relationship of the posterior abdominal wall muscles are maintained.*

POSTERIOR ABDOMINAL WALL, CT CORRELATION

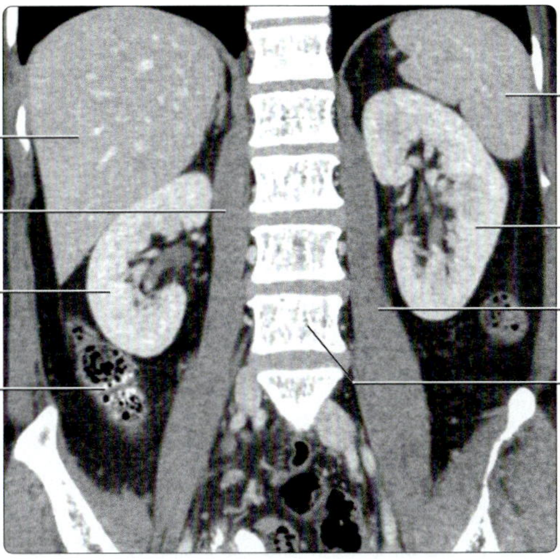

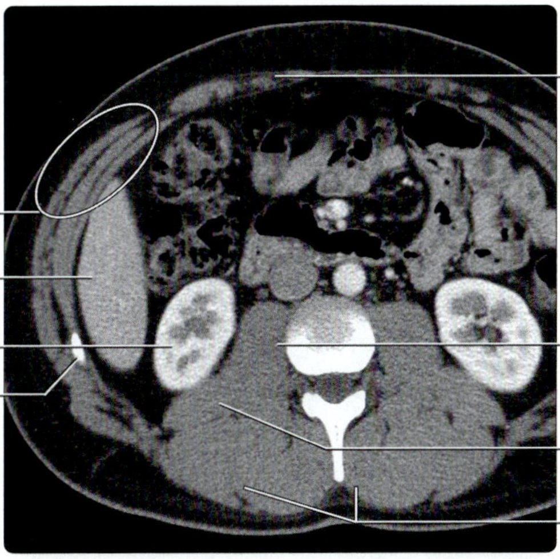

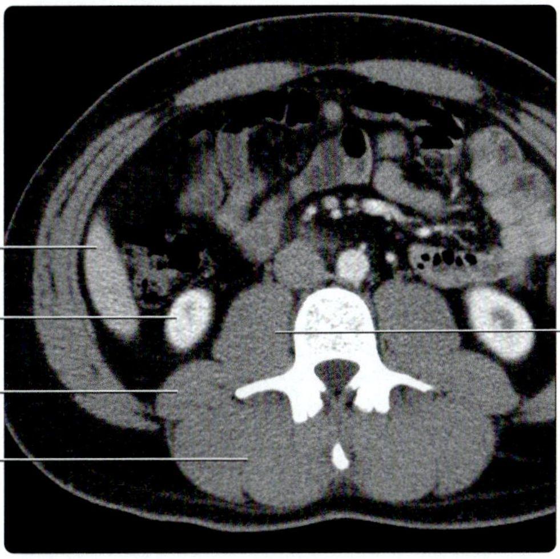

(Top) *Correlative coronal CECT shows the paralumbar location of the psoas muscles and their medial location relative to the kidneys. The psoas muscles originate from the lumbar and 12th thoracic vertebral bodies and their transverse processes and run past the pelvic brim, where they course inferolaterally to be joined by the iliacus muscle.* **(Middle)** *Correlative axial CECT better illustrates the anatomic relationships of the kidney with the posterior abdominal wall muscles. The kidney is lateral to the psoas muscle and rests upon the quadratus lumborum muscle. The erector spinae muscles are immediately posterior to the quadratus lumborum, and the 2 muscles are invested by the lumbodorsal fascia.* **(Bottom)** *Correlative axial CECT at the level of the inferior pole of the right kidney is shown. The psoas muscle and quadratus lumborum muscles, seen in their midsections, are now thicker.*

POSTERIOR ABDOMINAL WALL

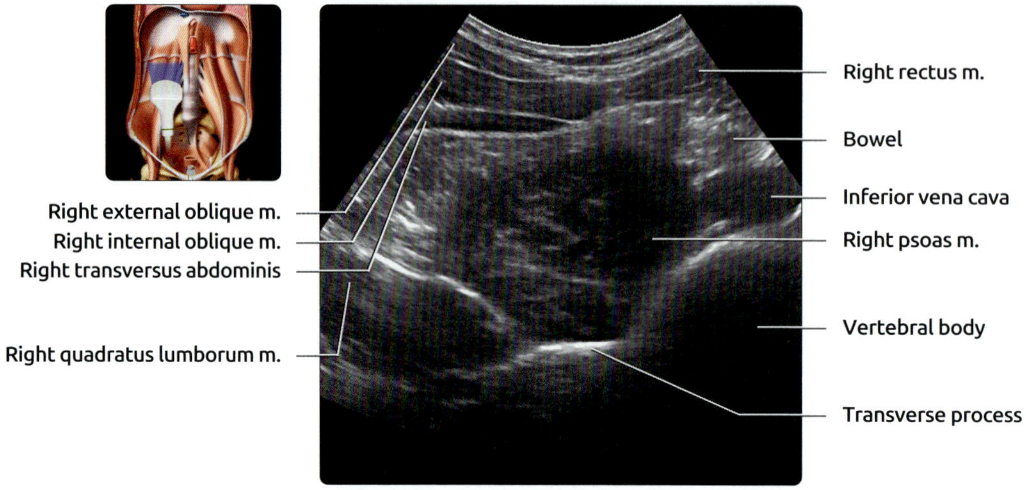

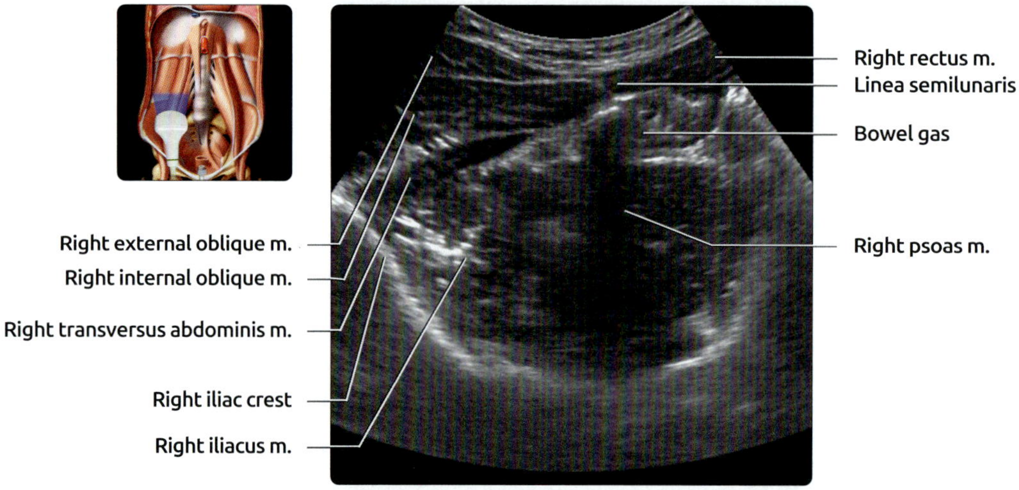

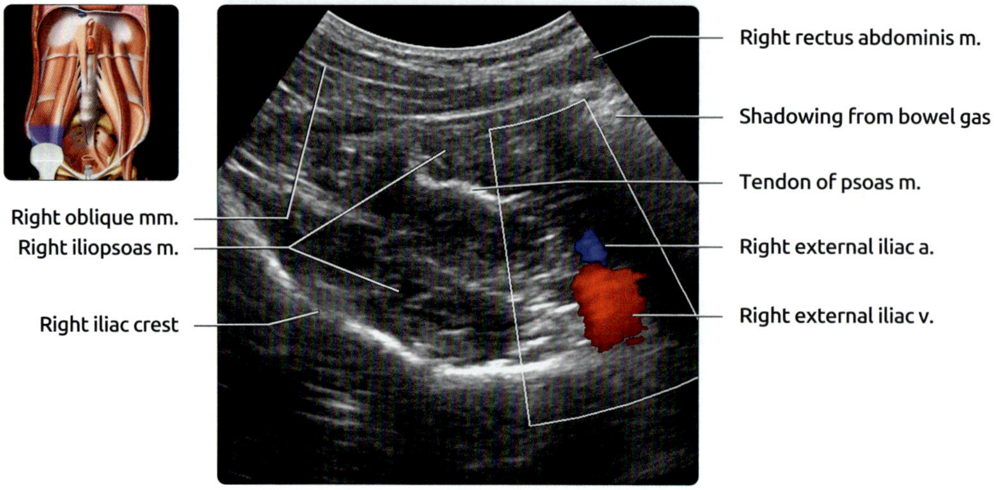

(Top) *Transverse grayscale ultrasound in the lower abdominal region shows the right psoas muscle, composed of the psoas minor, which rests upon the psoas major. The 2 muscles cannot be separated clearly on ultrasound. Because of their depth, the paraspinal muscles cannot be demonstrated in detail.* **(Middle)** *Transverse grayscale ultrasound of the right lower abdomen, continued from the previous image, shows that the distal psoas muscle has diminished in size. It rests on the medial portion of the iliacus muscle; the latter is a flat muscle that fills the iliac fossa. Both continue inferiorly together.* **(Bottom)** *Distally, the fibers from the iliacus muscle converge and insert into the lateral side of the psoas muscle to form the iliopsoas muscle. Common iliac vessels can be seen medially.*

POSTERIOR ABDOMINAL WALL, CT CORRELATION

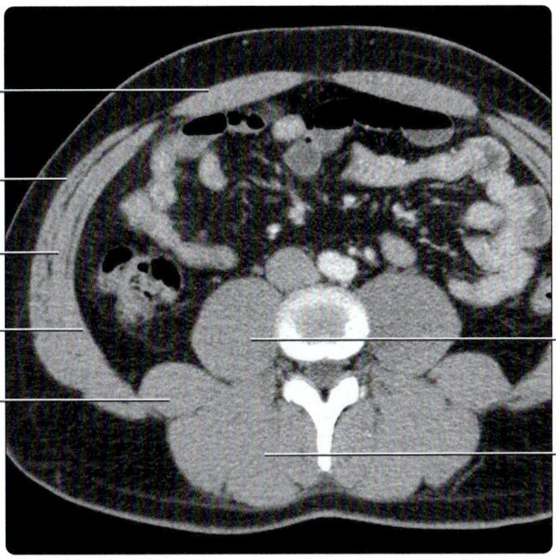

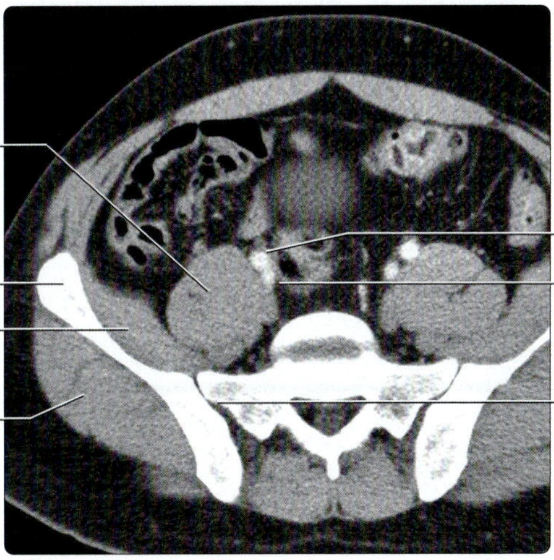

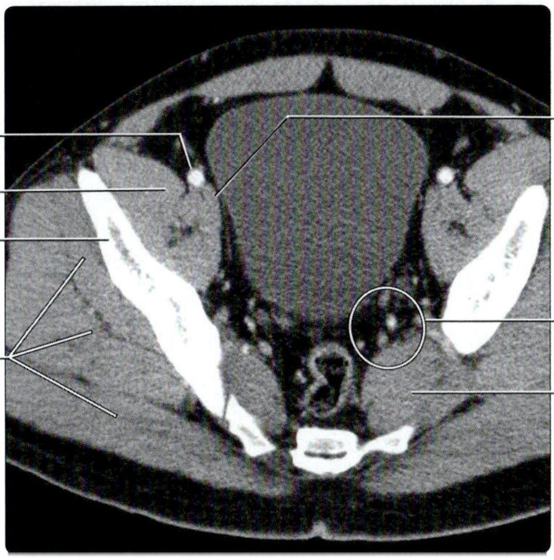

(Top) *Correlative axial CECT below the kidneys shows the quadratus lumborum muscle is more laterally located and the psoas muscle is directly anterior to the erector spinae muscle.* **(Middle)** *Axial correlative CECT shows the psoas muscle has begun its dorsolateral course and is now anterior to the iliacus muscle. The iliacus muscle is easily identified as a flat muscle filling the iliac fossa, arising from the upper 2/3 of the iliac fossa, inner lip of the iliac crest, anterior sacroiliac and the iliolumbar ligaments, and base of the sacrum.* **(Bottom)** *Correlative axial CECT shows the psoas and iliacus muscles have converged and are now indistinguishable from one another. The resultant iliopsoas muscle passes beneath the inguinal ligament and becomes tendinous as it inserts into the lesser trochanter of the femur.*

PARASPINAL MUSCLES

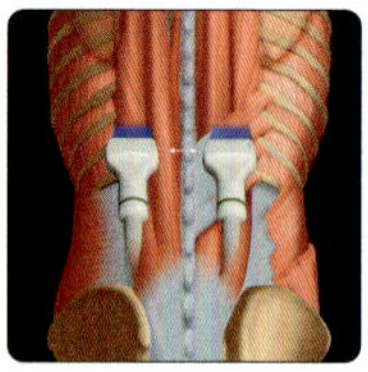

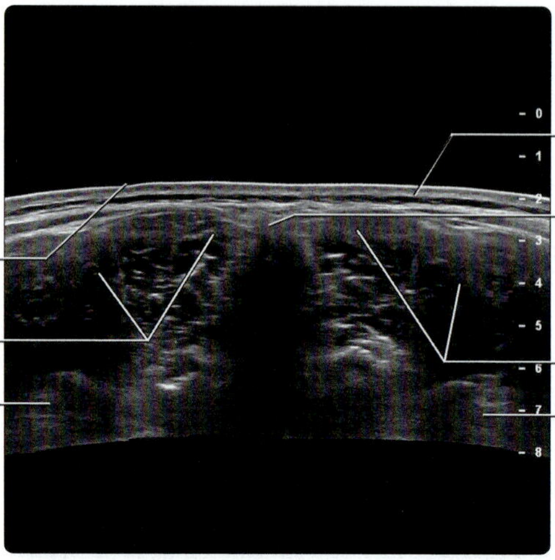

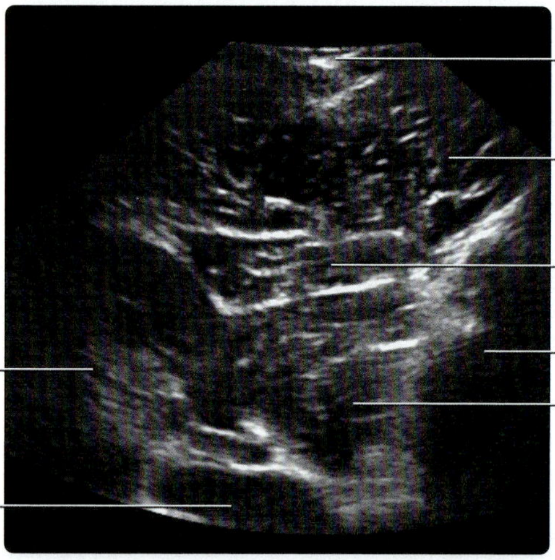

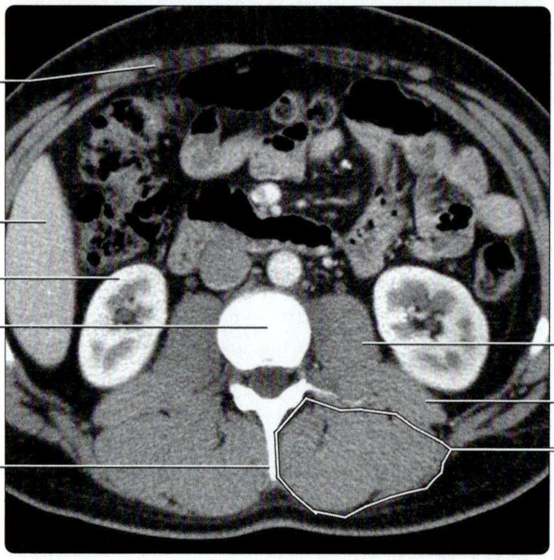

(Top) *Transverse extended FOV grayscale ultrasound of the back (with the patient prone) shows the erector spinae muscles flanking the spinous process. They are invested by the lumbodorsal fascia, which also invests the anteriorly located quadratus lumborum muscle. The kidneys are partially demonstrated.* **(Middle)** *Transverse oblique grayscale ultrasound shows the left erector spinae muscle (with the patient prone). The 3 columns (iliocostalis, longissimus, and spinalis muscles, from lateral to medial) comprising the erector spinae are not clearly separated from one another on ultrasound. They are identified collectively as a thick, fleshy muscle lateral to the spinous process.* **(Bottom)** *Correlative axial CECT of the paraspinal muscles at the level of the kidneys shows the erector spinae muscles originate from a broad and thick tendon, which originates from the sacrum and iliac crest, lumbar, and 11th and 12th thoracic spinous processes.*

MR

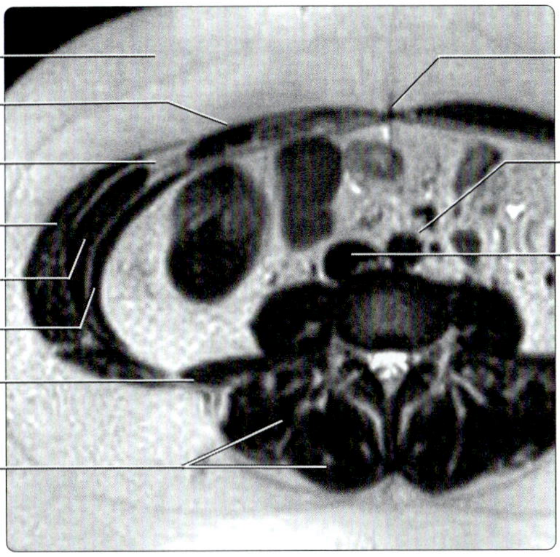

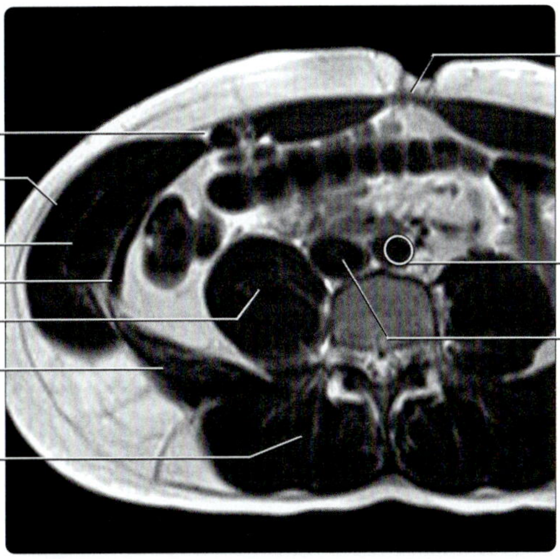

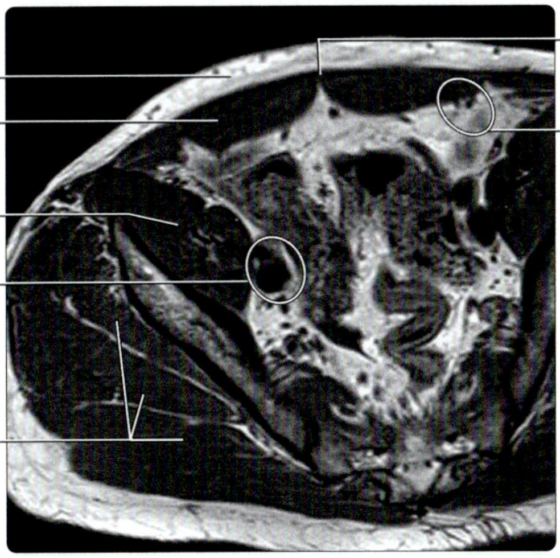

(Top) *Axial T2 HASTE MR in an older patient with muscle atrophy shows fat in between the individual muscles of the anterior and posterior abdominal wall.* **(Middle)** *Axial T2 HASTE MR in a younger male patient shows more bulky abdominal wall musculature with little intermuscular fat.* **(Bottom)** *Axial T1 MR at a lower level shows the iliopsoas as 1 muscle bundle.*

SECTION 5
Pelvis

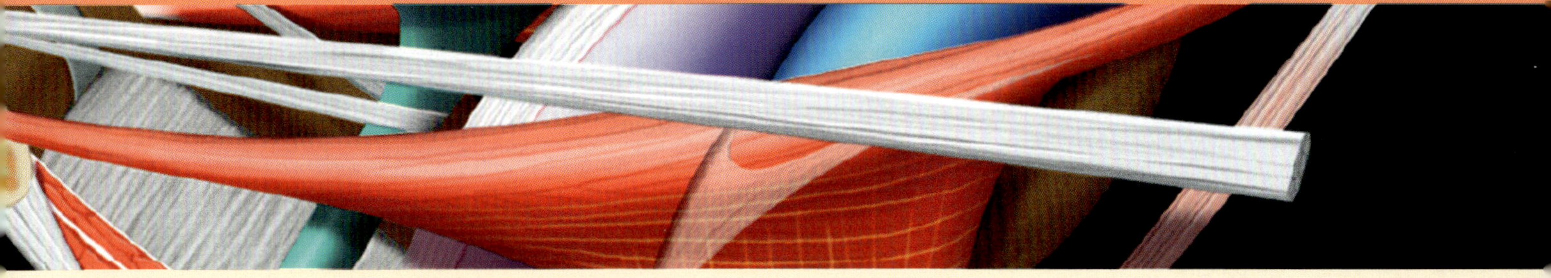

GROSS ANATOMY

Arteries

- Abdominal aorta
  - Testicular and ovarian arteries originate below renal arteries
  - Median (middle) sacral artery is small, unpaired branch from posterior aspect of distal aorta
  - Divides into common iliac arteries at L4-5
- Common iliac arteries
  - Run anterior to iliac veins and inferior vena cava
  - Usually no major branches
    - Rarely, gives off aberrant iliolumbar or accessory renal arteries
  - ~ 4 cm long
- External iliac artery
  - No major branches
  - Exits pelvis beneath inguinal ligament
  - Larger than internal iliac artery
  - Inferior epigastric (medial) and deep iliac circumflex (lateral) arteries demarcate junction between external iliac and common femoral arteries
- Internal iliac (hypogastric) artery
  - Principal vascular supply of pelvic organs
  - Divides into anterior and posterior trunk
    - Anterior trunk to pelvic viscera
    - Posterior trunk to pelvic musculature
- Anterior trunk of internal iliac artery
  - Branching pattern quite variable
  - Umbilical artery
    - Only pelvic segment remains patent after birth
    - Remainder becomes fibrous medial umbilical ligament
  - Obturator artery
    - Exits pelvis through obturator canal to supply medial thigh muscles
  - Superior vesicle artery
    - Supplies bladder and distal ureter
    - Gives off branch to ductus deferens in males
  - Inferior vesicle artery (male)
    - May arise from middle rectal artery
    - Supplies prostate, seminal vesicles, and lower ureters
  - Uterine artery (female)
    - Passes over ureter at level of cervix ("water under the bridge")
    - Anastomoses with vaginal and ovarian arteries
  - Vaginal artery (female)
  - Middle rectal artery runs above pelvic floor and anastomoses with superior and inferior rectal arteries to supply rectum
    - Also anastomoses with inferior vesicle artery
  - Internal pudendal artery
    - Supplies external genitalia (penis, clitoris) and rectum
  - Inferior gluteal (sciatic) artery
    - Largest and terminal branch of anterior division of hypogastric artery
    - Supplies muscles of pelvic floor, thigh, buttocks and sciatic nerve
- Posterior division of internal iliac artery
  - Iliolumbar artery
    - Ascends laterally to supply iliacus, psoas, and quadratus lumborum muscles
  - Lateral sacral artery
    - Runs medially toward sacral foramina to anastomose with middle sacral artery
  - Superior gluteal artery
    - Largest and terminal branch of posterior division
    - Supplies piriformis and gluteal muscles

Veins

- External iliac vein
  - Upward continuation of femoral vein at level of inguinal ligament
  - Receives inferior epigastric, deep iliac circumflex, and pubic veins
- Internal iliac vein begins near upper part of greater sciatic foramen
  - Gluteal, internal pudendal and obturator veins have origins outside pelvis
  - Pelvic viscera drain into multiple, deep pelvic venous plexuses
    - These drain into veins, which roughly parallel pelvic arteries
- Right gonadal vein drains into IVC, left gonadal vein drains into left renal vein
- Common iliac vein is formed by union of external and internal iliac veins
  - Unites with contralateral side to form IVC

IMAGING ANATOMY

Overview

- CT angiography (CTA) and MR angiography (MRA) are imaging modalities of choice to evaluate pelvic vessels
  - Ultrasound is limited to demonstrating common iliac, external iliac, and proximal internal iliac vessels

ANATOMY IMAGING ISSUES

Imaging Recommendations

- Transducer: 2-5 MHz
- Patient examined in supine position
  - Place transducer lateral to rectus muscles, angulating medially
- Fasting for 4 hours may help decrease overlying bowel gas

Imaging Pitfalls

- Pelvic vessels are usually obscured by overlying bowel gas

CLINICAL IMPLICATIONS

Clinical Importance

- Abdominal aortic aneurysms may extend to involve iliac arteries
- Rich, complex collateral circulation helps ensure delivery of blood to pelvic organs and lower limbs in event of proximal obstruction
- Patients with deep venous thrombosis of lower limbs may have involvement of iliac veins

ILIAC ARTERIES AND VEINS IN SITU

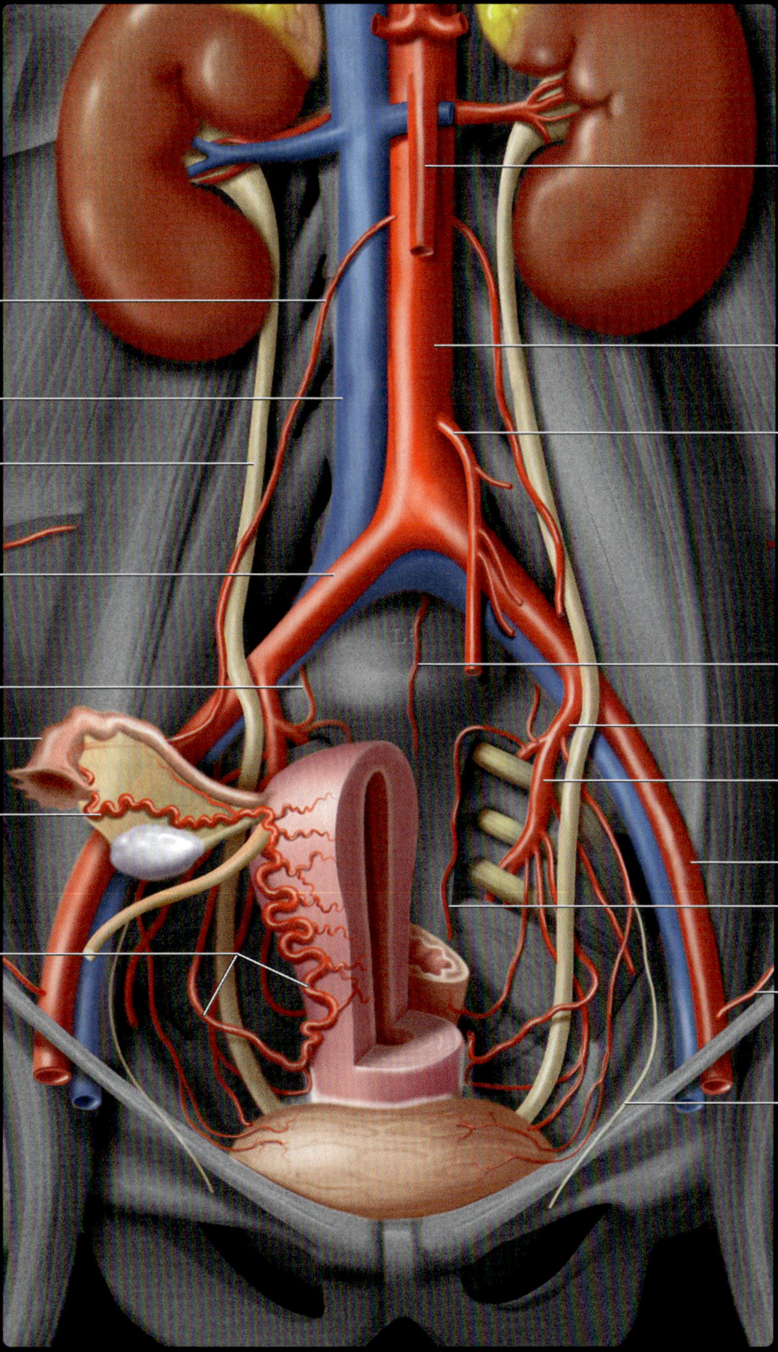

Frontal graphic shows the abdominal aorta, inferior vena cava, and the iliac vessels in a female. The inferior mesenteric artery is the smallest of the anterior mesenteric branches of the aorta and continues in the pelvis as the superior rectal artery. The paired ovarian arteries arise from the aorta below the renal arteries and pass inferiorly on the posterior abdominal wall to enter the pelvis. The ureters cross anterior to the bifurcation of the common iliac arteries on their way to the urinary bladder. The common iliac artery divides into the external iliac artery, which supplies the lower extremity and the internal iliac (hypogastric) artery, which supplies the pelvis. The internal iliac artery divides into an anterior trunk for the pelvic viscera and a posterior trunk for the muscles of the pelvis.

COMMON ILIAC ARTERY AND BRANCHES

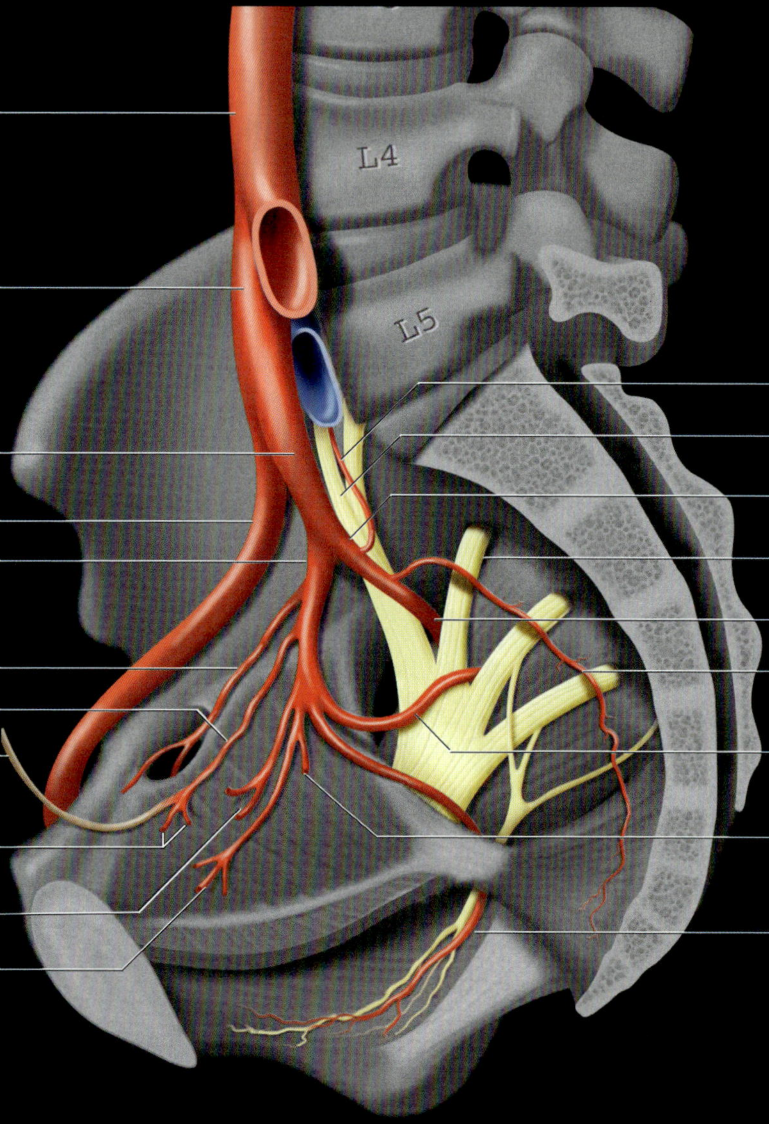

Graphic shows the pelvic arteries and their relation to the sacral nerves. The superior gluteal artery passes posteriorly and runs between the lumbosacral trunk and the anterior ramus of the S1 nerve, whereas the inferior gluteal artery usually runs between the S1-2 or S2-3 nerve roots to leave the pelvis through the inferior part of the greater sciatic foramen. Only the proximal portion of the umbilical arteries remains patent after birth, while the distal portion obliterates forming the medial umbilical ligaments. Arteries to the deep pelvic viscera include the superior and inferior vesicle, uterine, middle rectal, and internal pudendal. The individual branching pattern is quite variable.

ILIAC VEINS IN SITU

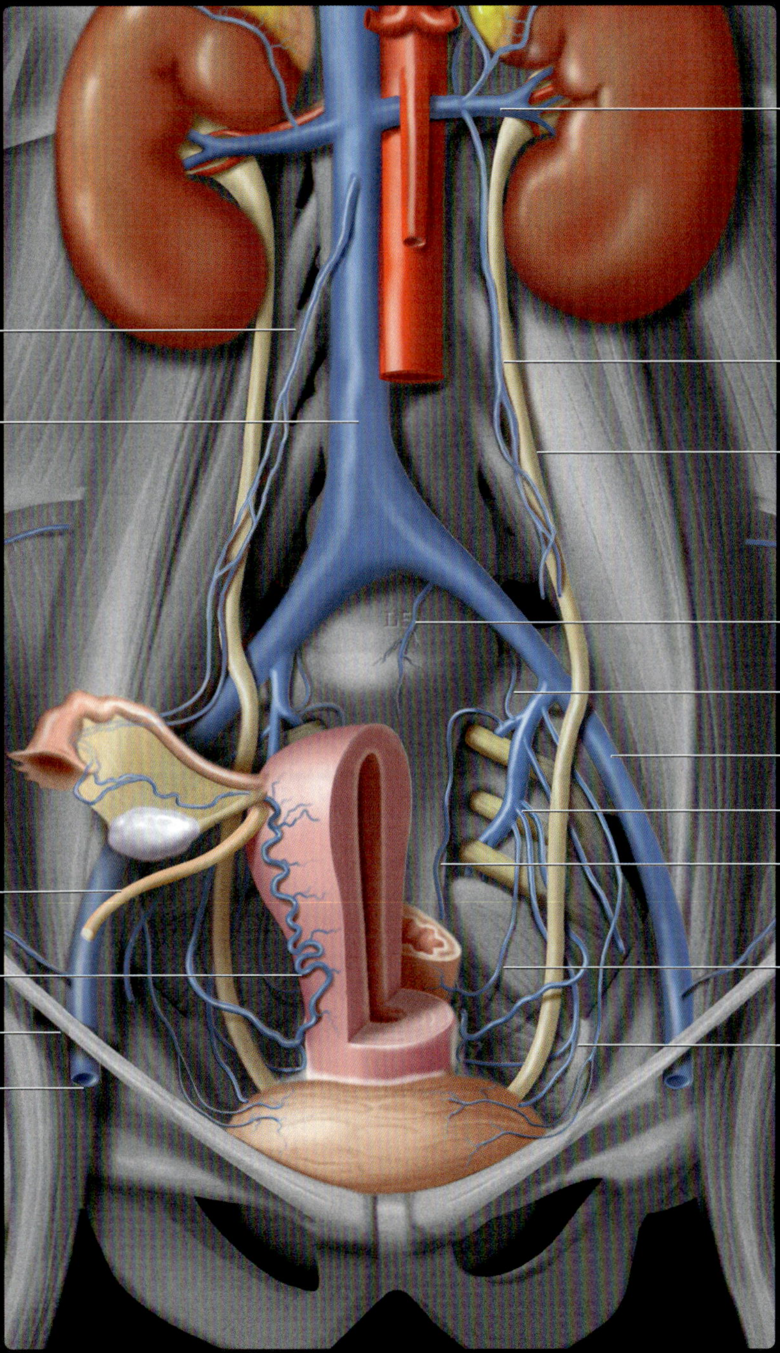

Graphic of the veins of the pelvis is shown. The left ovarian vein drains into the left renal vein, whereas the right ovarian vein drains directly into the inferior vena cava. Multiple intercommunicating pelvic venous plexuses (rectal, vesicle, prostatic, uterine, and vaginal) drain mainly to the internal iliac veins. There is a communication between the pelvic veins and the intraspinal epidural plexus of veins through the sacral venous plexus.

AORTIC BIFURCATION

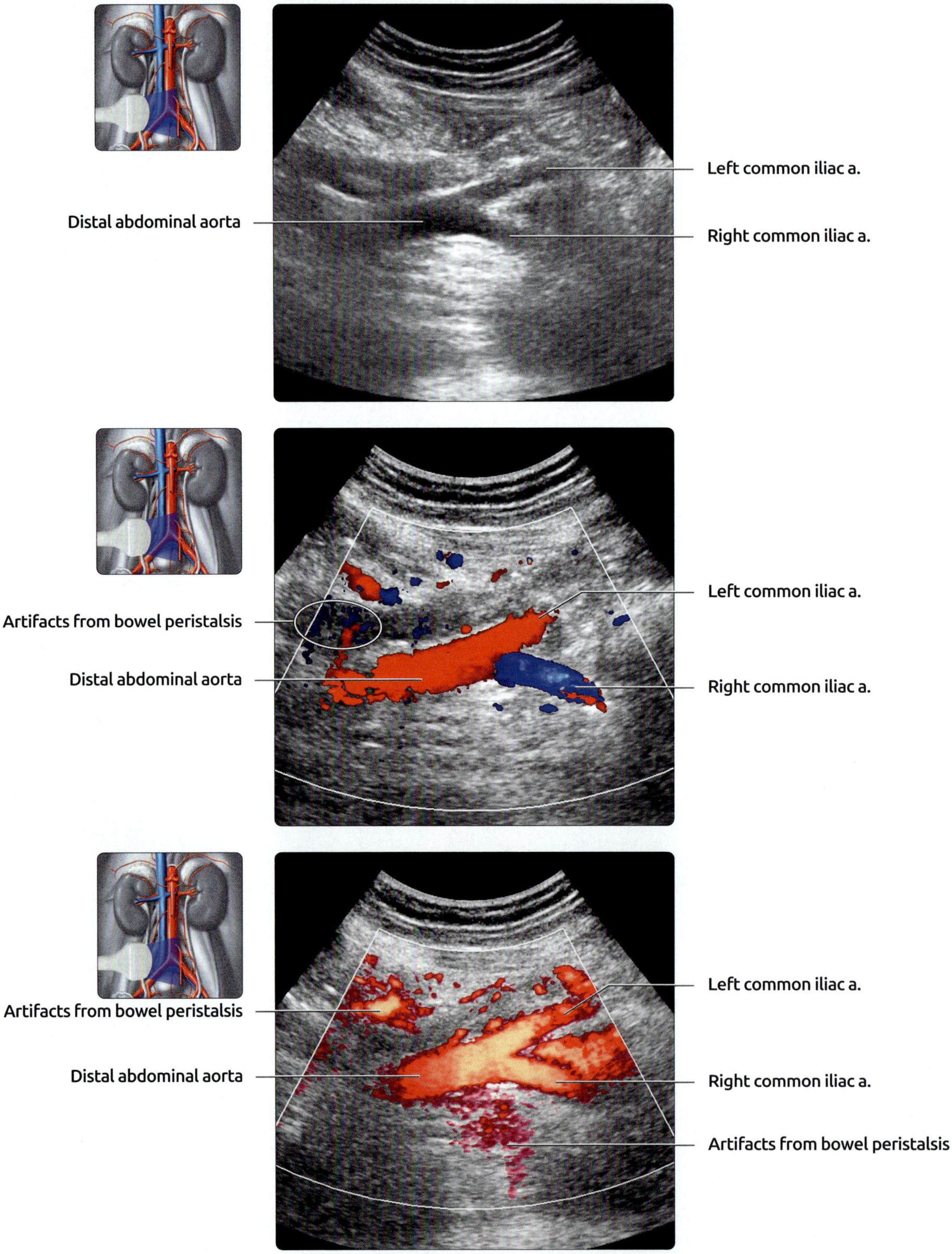

(Top) *Coronal grayscale ultrasound shows the bifurcation of the distal aorta into the common iliac arteries. This occurs at the level of the L4 vertebra and corresponds to the umbilicus, serving as a useful landmark for transducer placement for common iliac artery insonation.* **(Middle)** *Coronal color Doppler ultrasound demonstrates color flow in the distal aorta and bifurcation. The right common iliac artery assumes a blue color (as opposed to the red color of the distal aorta and right common iliac artery) owing to its flow direction. Peristalsis of adjacent bowel segments also demonstrates color on color Doppler ultrasound, rendering artifacts.* **(Bottom)** *Coronal power Doppler is more sensitive than color Doppler in demonstrating blood flow without providing information on flow direction. There is also significant increase in image artifacts from peristalsis.*

COMMON ILIAC ARTERY

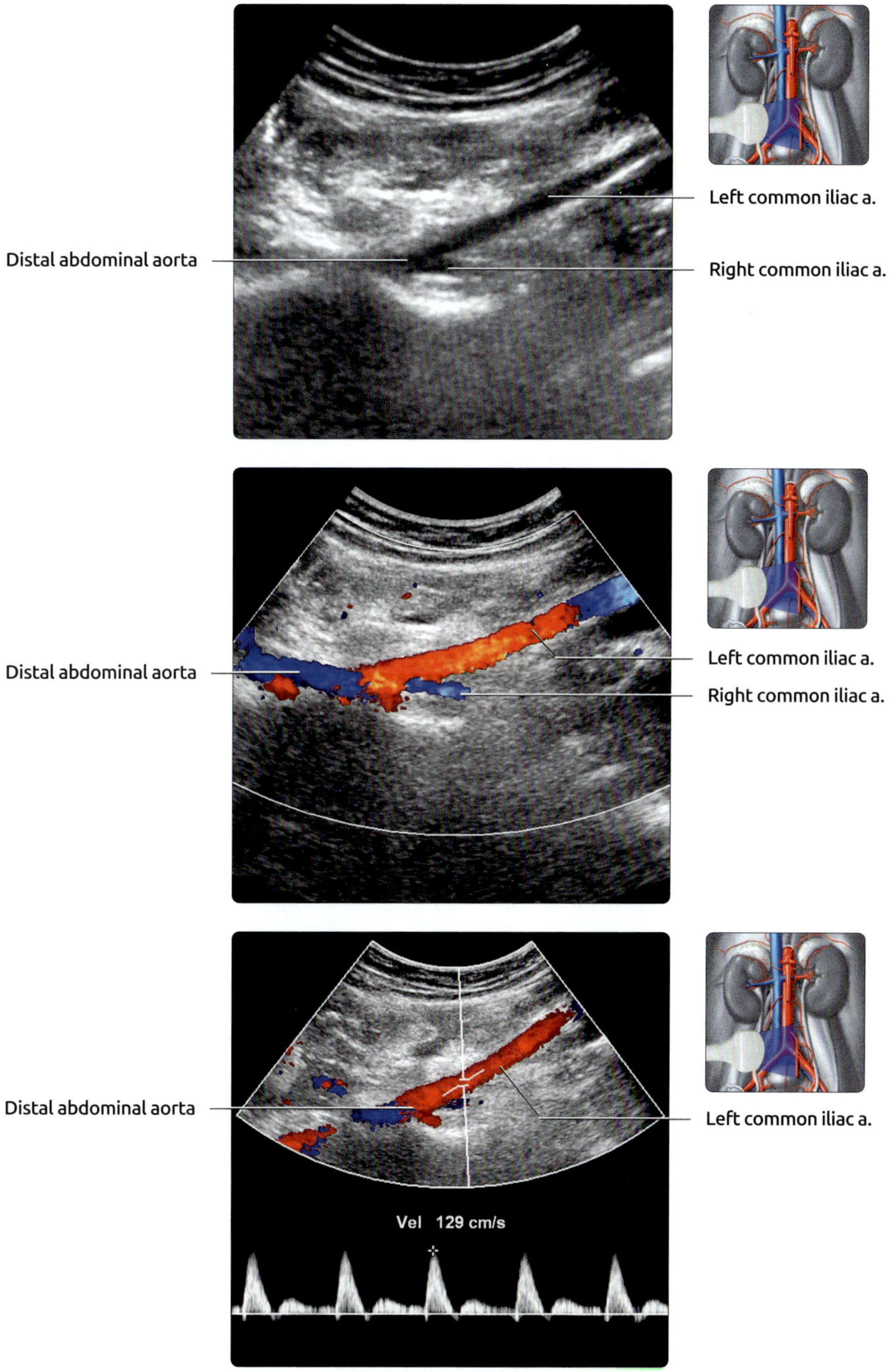

(Top) *Coronal transabdominal grayscale ultrasound is shown, angulating the transducer to demonstrate the course of the left common iliac artery. The common iliac artery is about 5 cm long with diameters of 1.3 cm (females) and 1.5 cm (males).* **(Middle)** *Coronal transabdominal color Doppler of the left common iliac artery (shown in red) with consistent intense color indicating uniform mean flow velocity is shown. This is a useful plane for examining abdominal aortic aneurysms when there is extension into the common iliac arteries.* **(Bottom)** *Transabdominal color pulsed Doppler ultrasound of the left common iliac artery shows peak systolic velocity of 129 cm/sec, within the normal range of 80-187 cm/sec. Note normal triphasic spectral waveform.*

INTERNAL ILIAC ARTERY

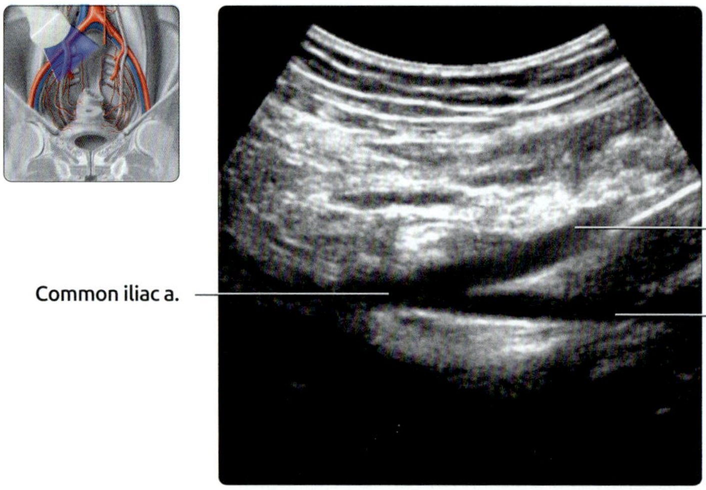

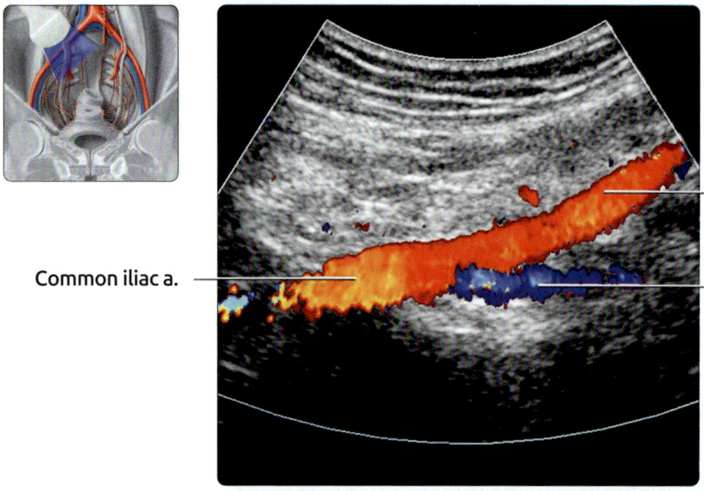

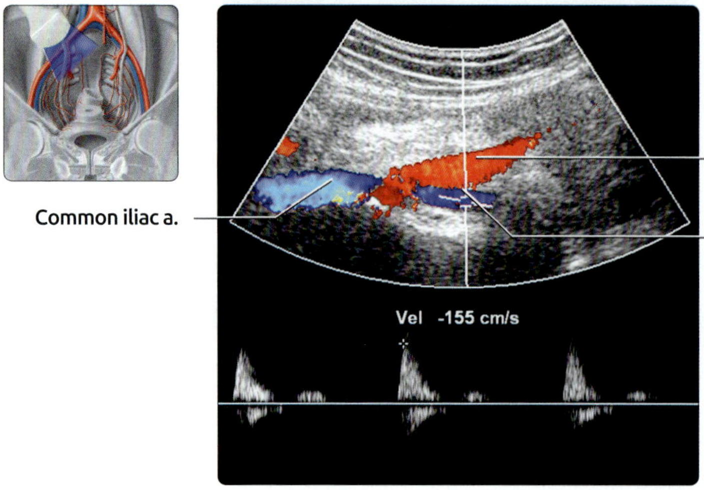

(Top) *Oblique transabdominal grayscale ultrasound of the distal common iliac artery demonstrates its bifurcation into the external iliac and internal iliac arteries. The internal iliac artery has a smaller caliber compared to the external iliac artery and courses more posteriorly. The internal iliac artery divides into 2 trunks, which are usually too deep to be demonstrated on ultrasound.* **(Middle)** *Longitudinal transabdominal color Doppler ultrasound shows intense red color in the distal common iliac and external iliac arteries, suggesting uniform mean velocity in the arterial segments. The branches of the internal iliac artery (shown in blue) supply the wall and viscera of the pelvis, including the reproductive organs.* **(Bottom)** *Pulsed Doppler ultrasound shows the internal iliac artery, which is usually investigated in graft kidneys and some cases of erectile dysfunction.*

EXTERNAL ILIAC ARTERY

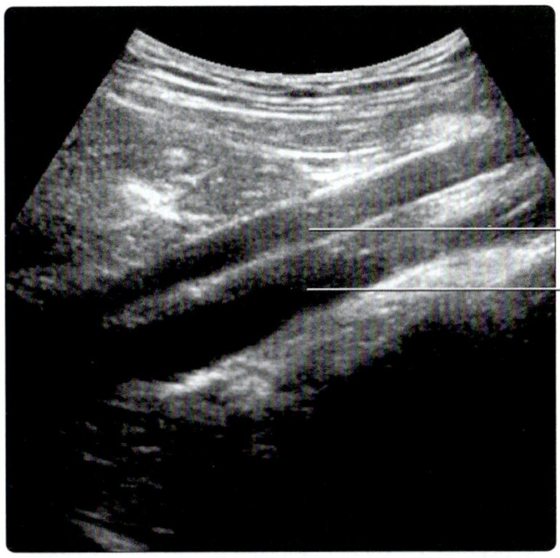

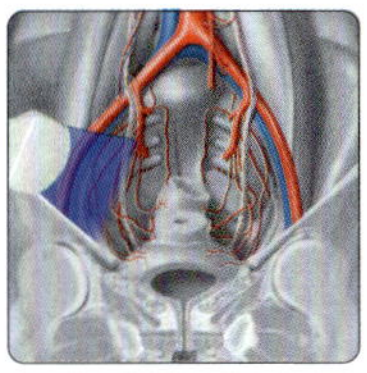

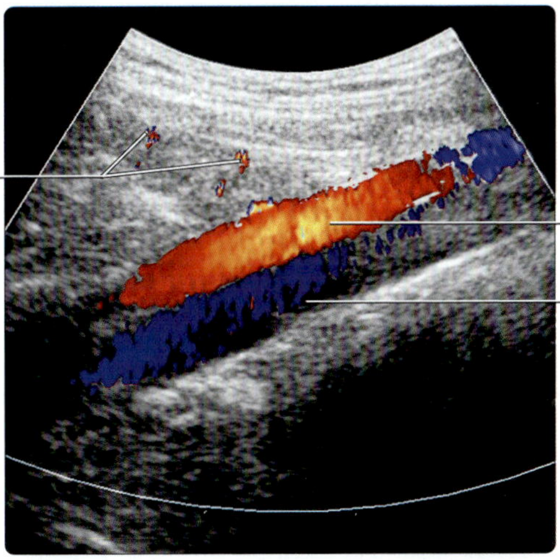

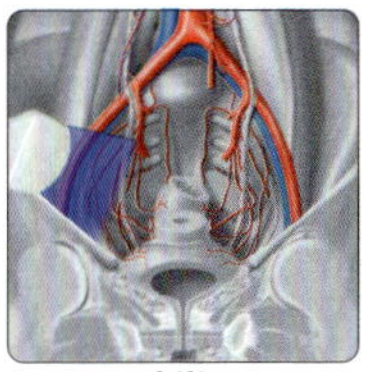

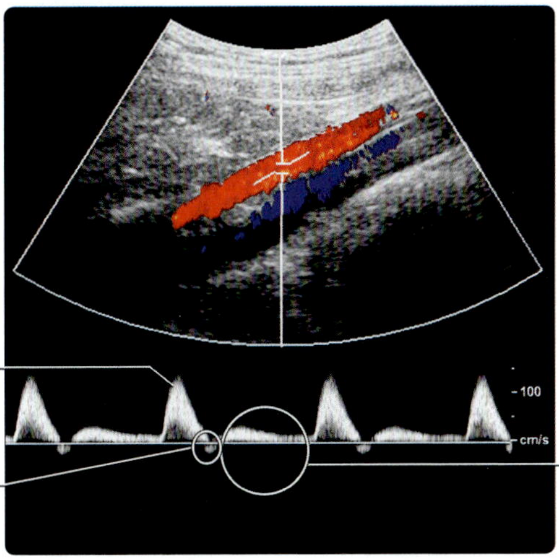

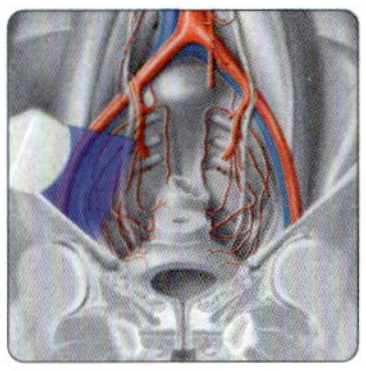

(Top) *Longitudinal transabdominal grayscale ultrasound shows the external iliac artery, usually easily demonstrated owing to its superficial location and absence of overlying bowel gas. Normal diameters are up to 11 mm in females and 12 mm in males.* **(Middle)** *Longitudinal transabdominal color Doppler ultrasound shows the relationship of the external iliac artery (shown in red) with the external iliac vein (shown in blue), which is located posteriorly. The 2 vessels run along the same course as they enter the thigh.* **(Bottom)** *Pulsed Doppler ultrasound of the external iliac artery is shown; the waveform resembles those from lower extremity arteries. Note high-velocity forward flow during systole and low-velocity forward flow during diastole. There is an intervening short reversal of flow in early diastole due to peripheral resistance. Peak systolic velocity of 129 cm/sec is within the normal range (< 140 cm/sec).*

ILIAC VESSELS, TRANSVERSE

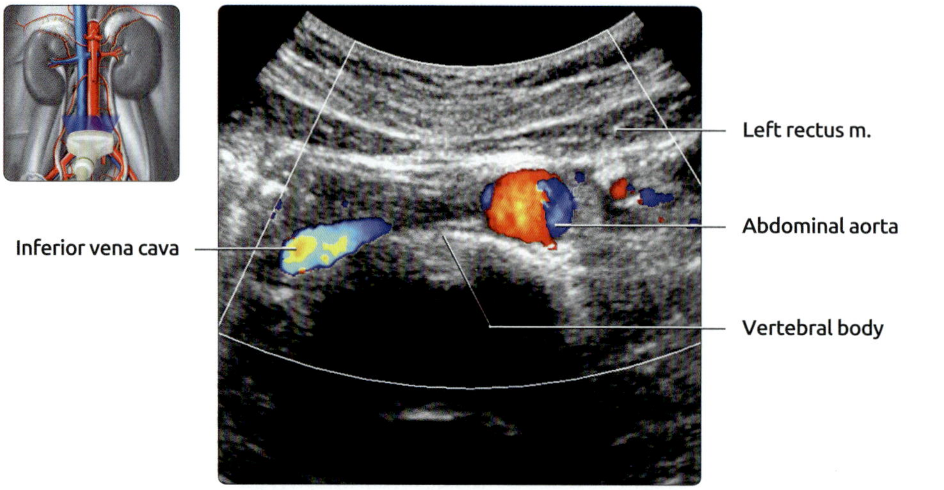

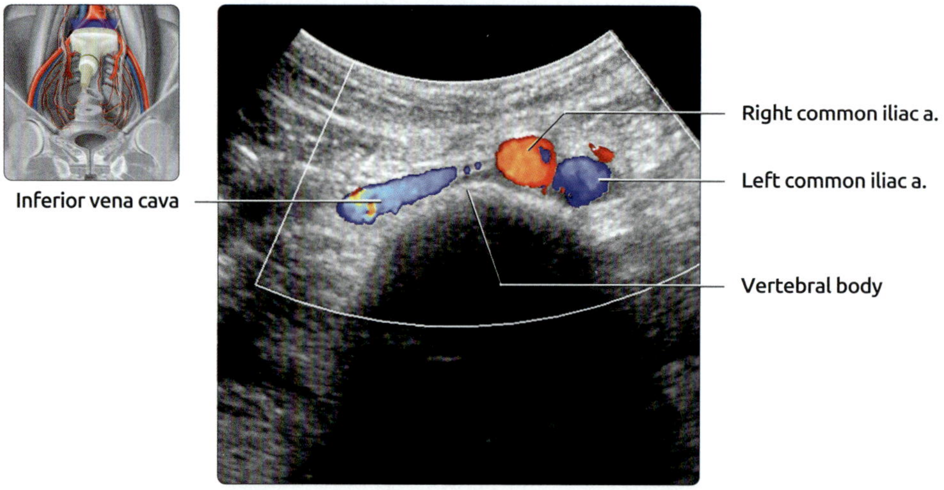

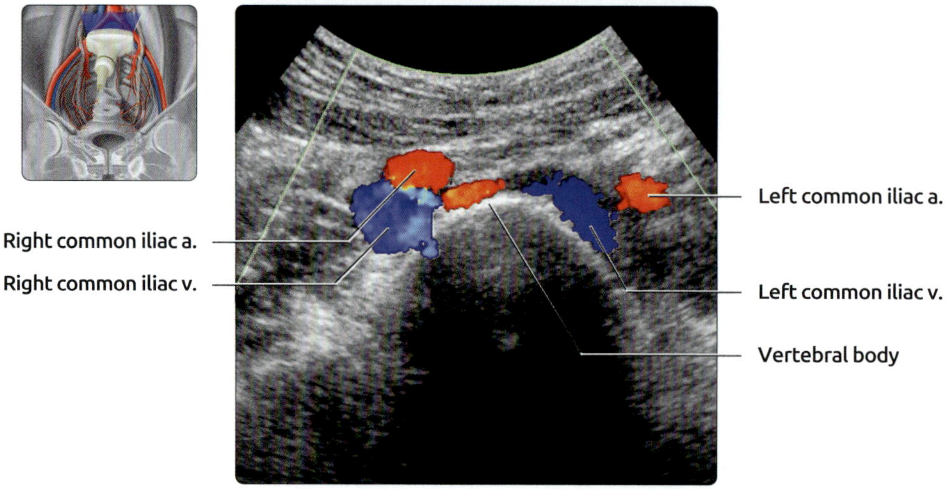

(Top) *Transverse transabdominal color Doppler ultrasound at the supraumbilical level shows the distal aorta (shown in red) and inferior vena cava (shown in blue), both of which have not yet bifurcated. The inferior vena cava is to the right of the abdominal aorta; a left-sided inferior vena cava is rarely encountered (0.2-0.5%) and may be associated with other vascular anomalies such as circumaortic or retroaortic renal vein.* **(Middle)** *Transverse transabdominal color Doppler ultrasound at the infraumbilical level is shown. The distal aorta has bifurcated into the paired common iliac arteries at the level of L4, occurring more proximally than the formation of the inferior vena cava.* **(Bottom)** *Transverse transabdominal color Doppler ultrasound continued from the above image. The paired common iliac veins (in blue) are now identified, which run posterior to their arterial counterparts (in red).*

ILIAC VESSELS, CT

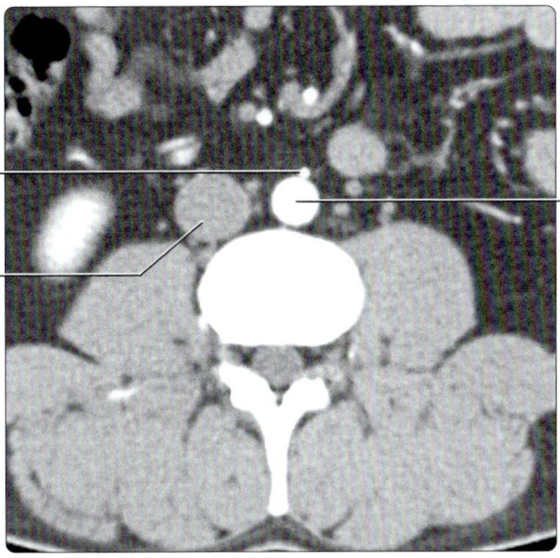

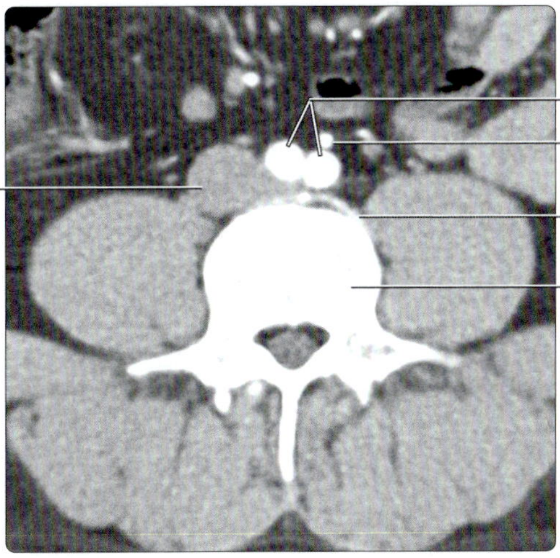

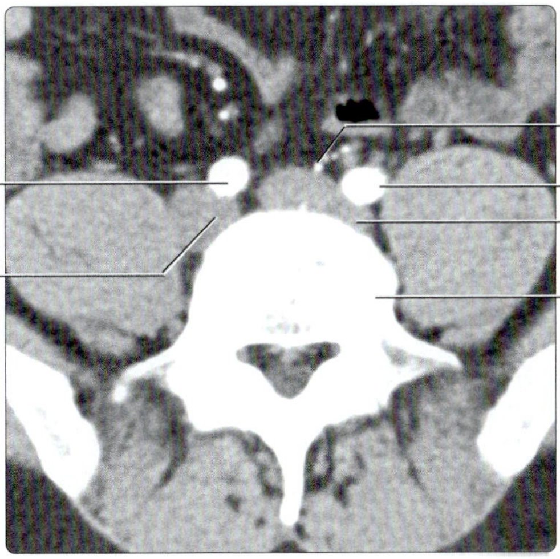

(Top) *First of 3 axial CECT images of the pelvic vessels is shown. The abdominal aorta rests on the vertebral body; its distal portion gives off the inferior mesenteric artery, the smallest of the mesenteric arteries. The inferior vena cava is identified to the right of the abdominal aorta and spine.* **(Middle)** *The aorta bifurcates at the level of L4 into the 2 common iliac arteries. Despite its diminutive caliber, the lumbar artery is identified on CT.* **(Bottom)** *The common iliac arteries usually give off no visceral branches. They may, however, give origin to accessory renal arteries. At this lower level, the termination of the common iliac veins are identified just before forming into the inferior vena cava. The left common iliac vein is longer than the right as it traverses the spine to form the IVC, which is located in the right paraspinal region.*

ILIAC VESSELS, TRANSVERSE

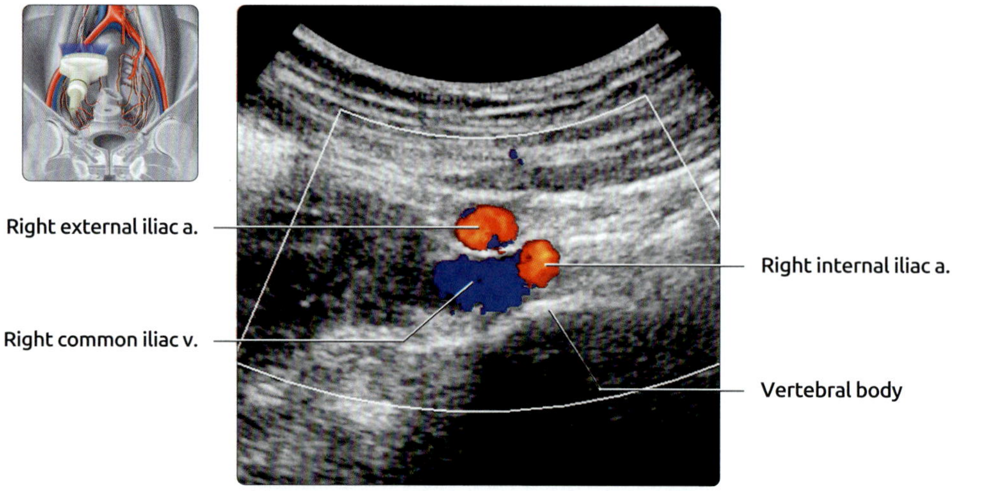

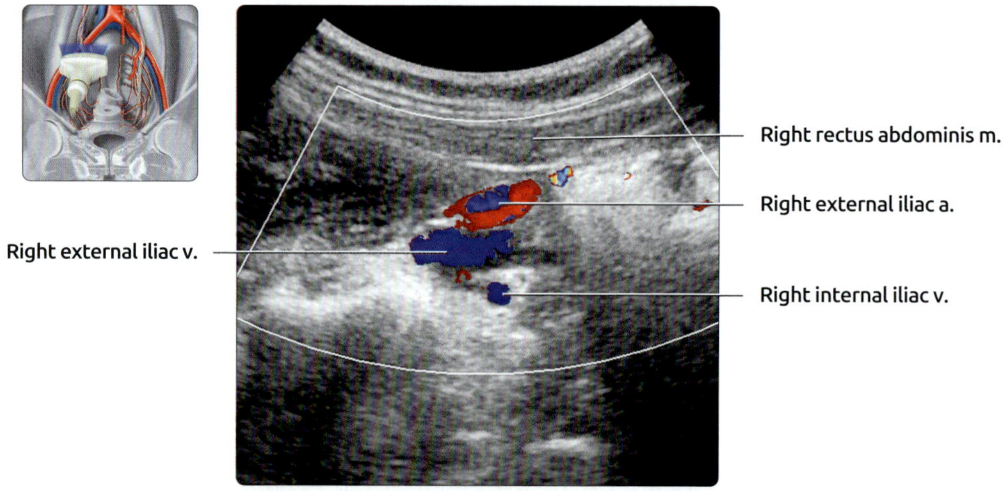

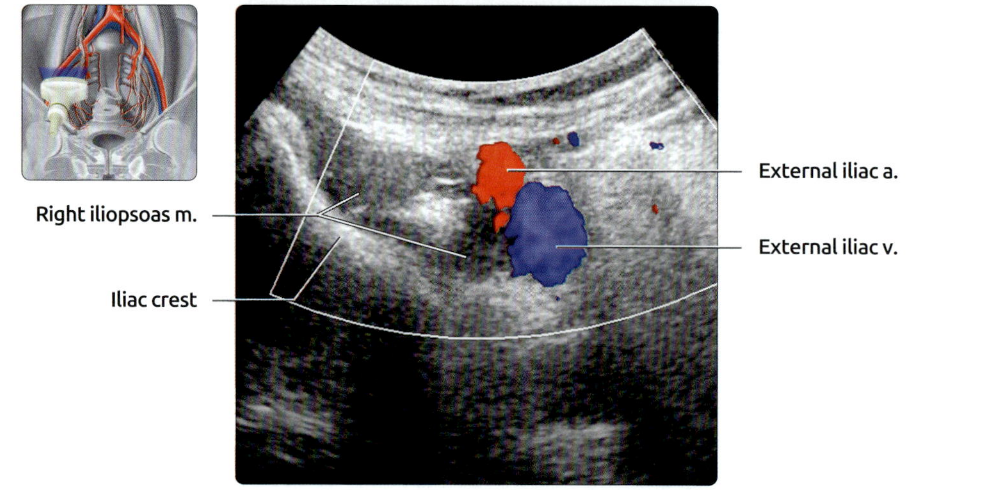

(Top) *Transverse transabdominal color Doppler ultrasound shows dichotomous branching of the right common iliac artery into the larger caliber external iliac artery (red) and smaller caliber internal iliac artery (red). The common iliac vein (blue) maintains its posterior location in relation to the 2 arteries.* **(Middle)** *Transverse transabdominal color Doppler ultrasound, continued more inferiorly in the same patient demonstrates the right external iliac vein (blue) and smaller internal iliac vein (blue). The right external iliac artery (red) is identified anterior to its venous counterpart.* **(Bottom)** *Transverse color Doppler ultrasound at the right iliac fossa is shown. The right external iliac artery (red) and vein (blue) course medial to the psoas major/iliopsoas muscle and maintain their relationship until they exit the pelvis beneath the inguinal ligament. Note the larger diameter of the vein compared to the artery.*

ILIAC VESSELS, CT

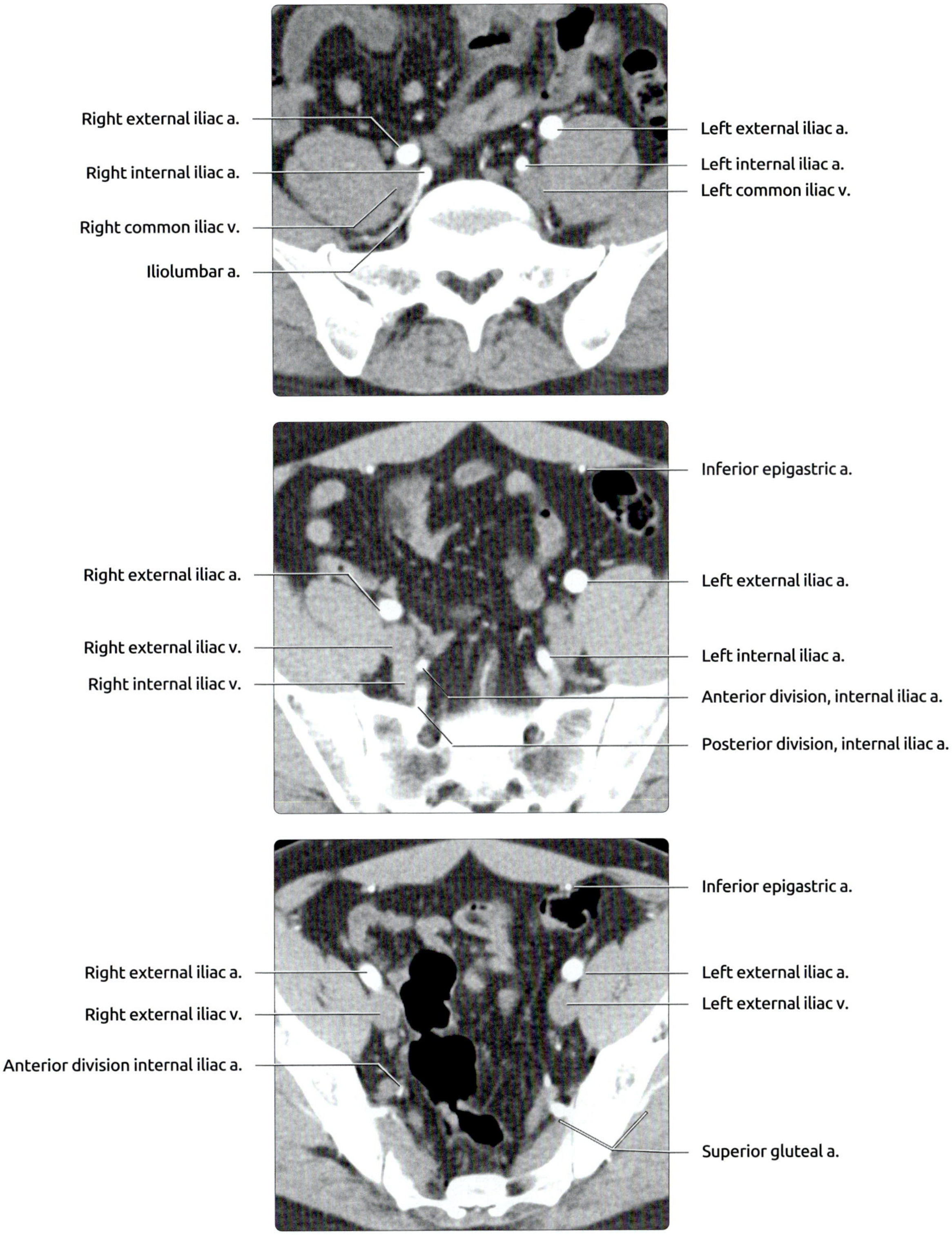

(Top) *First of 3 axial CECTs of the pelvic vessels is shown. The common iliac arteries bifurcate more proximal to the formation of the common iliac veins. The iliolumbar artery, usually a branch of the posterior trunk of the internal iliac artery, arises in this subject from the main internal iliac artery. CT is the imaging modality of choice in the examination of smaller pelvic vessels.* **(Middle)** *The internal iliac artery divides into an anterior and posterior trunk. The anterior trunk mainly supplies the pelvic viscera, whereas the posterior trunk supplies the pelvic musculature. Note their deep location relative to the external iliac vessels, which limits ultrasound examination of internal iliac vessel branches and tributaries.* **(Bottom)** *The external iliac artery and vein continue in an anterolateral direction as they course out of the pelvis and into the thigh.*

COMMON ILIAC VEIN

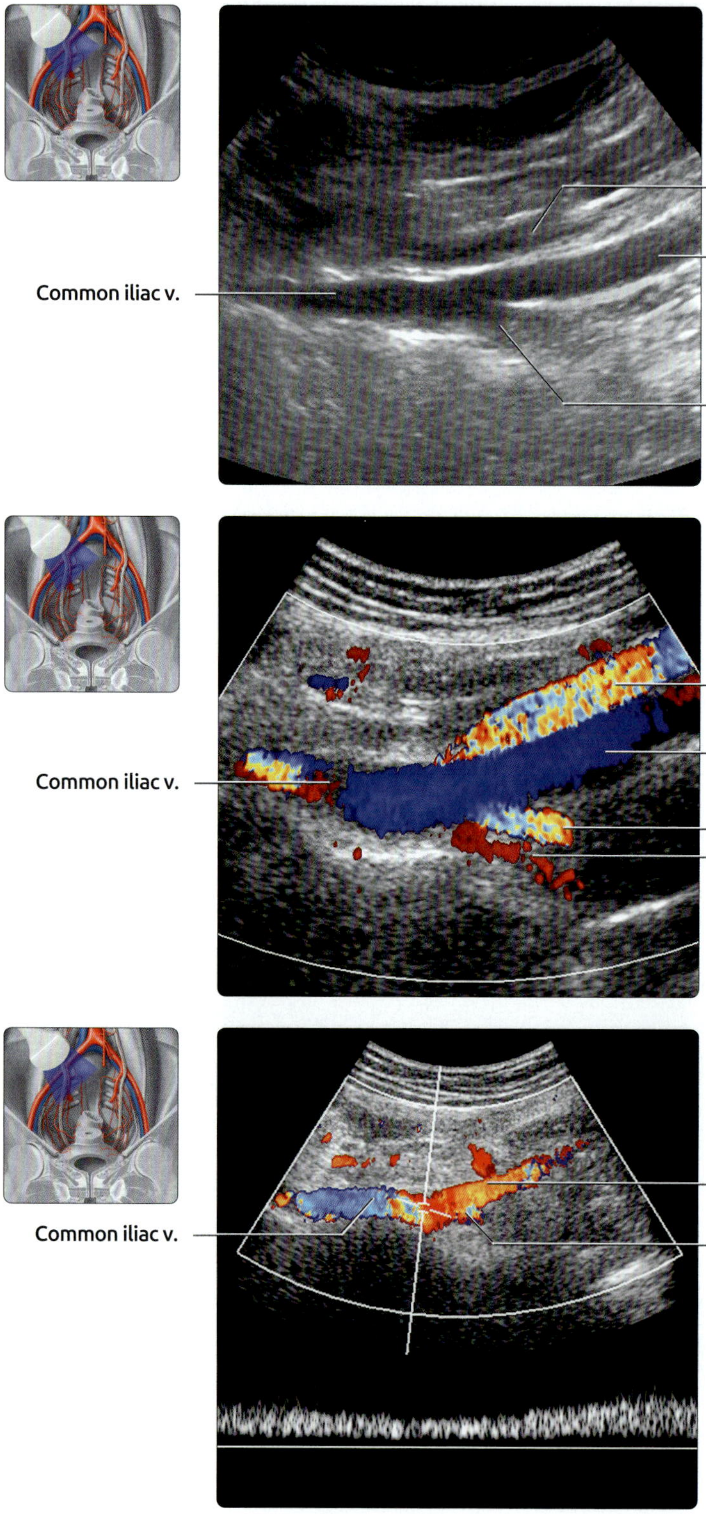

(Top) *Longitudinal transabdominal grayscale ultrasound shows the external and internal iliac veins converging to form the common iliac vein.* **(Middle)** *Longitudinal transabdominal color Doppler ultrasound provides information on the direction of blood flow in the pelvic vessels, facilitating identification and differentiation of the veins from arteries. The external iliac vein (shown in blue) is a large-caliber vessel seen posterior to its arterial counterpart (shown in multicolor). Note its uniform intensity in color flow. The internal iliac vein (shown in red) is smaller in diameter and directed inferomedially.* **(Bottom)** *Pulsed Doppler ultrasound of the common iliac vein shows uniform velocity with mild phasic changes associated with respiration.*

INTERNAL ILIAC VEIN

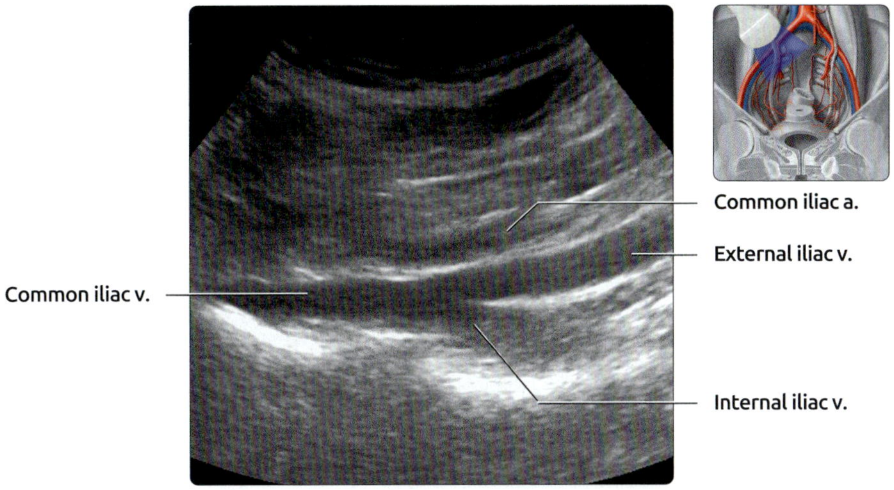

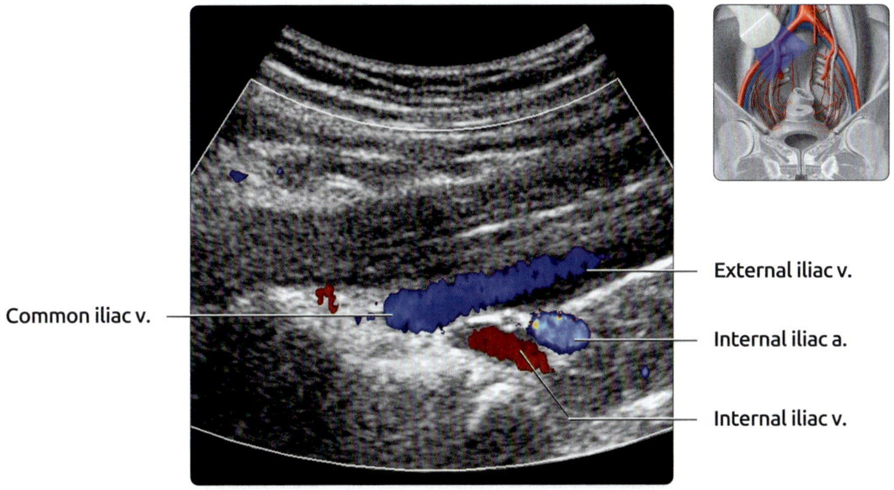

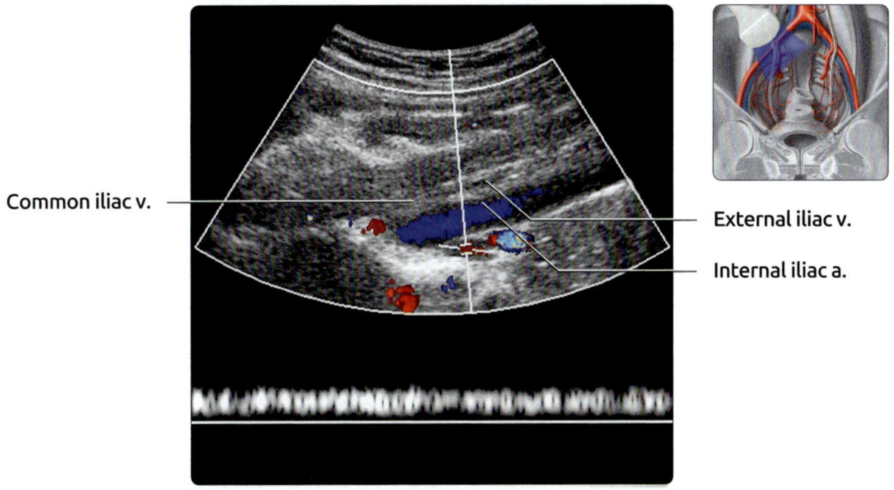

(Top) *Longitudinal transabdominal grayscale ultrasound shows the external and internal iliac veins converging to become the common iliac vein. The internal iliac vein drains various veins from outside the pelvis, the sacrum, and from the venous plexuses connected to pelvic viscera.* **(Middle)** *Longitudinal transabdominal color Doppler ultrasound shows a short segment of the internal iliac vein (shown in red) draining into the common iliac vein (shown in blue). Owing to its depth, the internal iliac vein cannot be identified on ultrasound in its entirety.* **(Bottom)** *Pulsed Doppler ultrasound of the internal iliac vein shows continuous venous flow and uniform velocity devoid of phasic changes and unaffected by respiration.*

EXTERNAL ILIAC VEIN

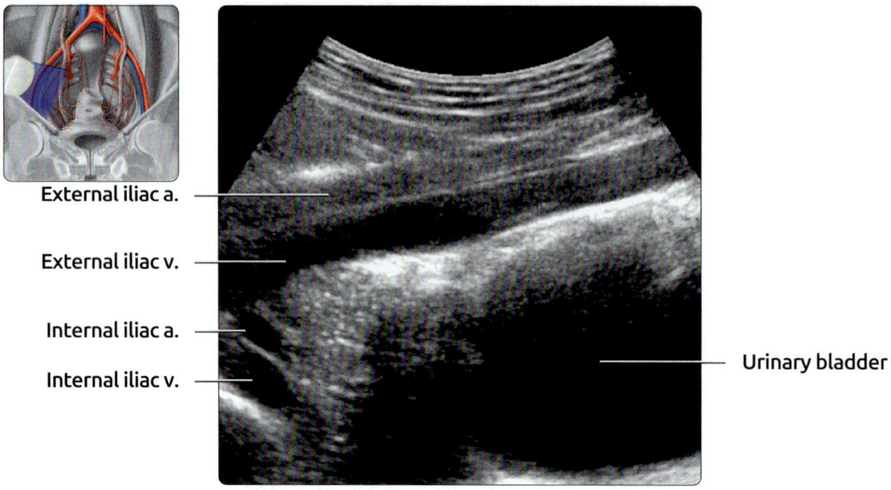

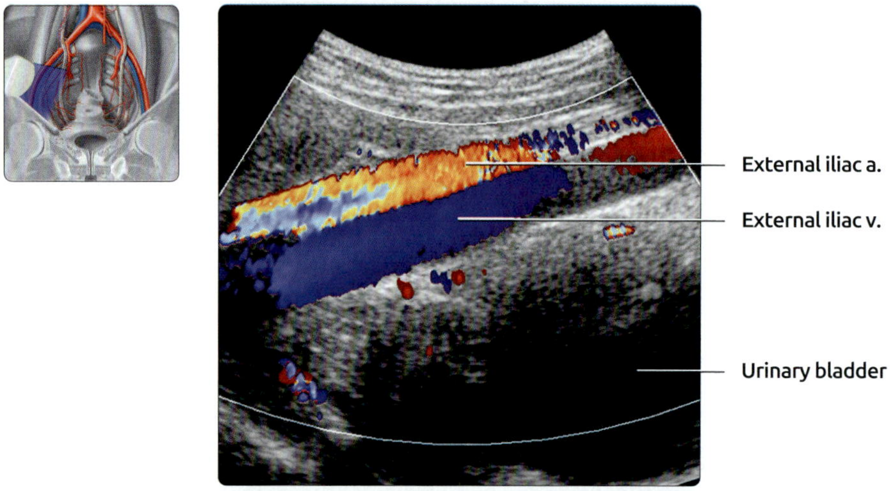

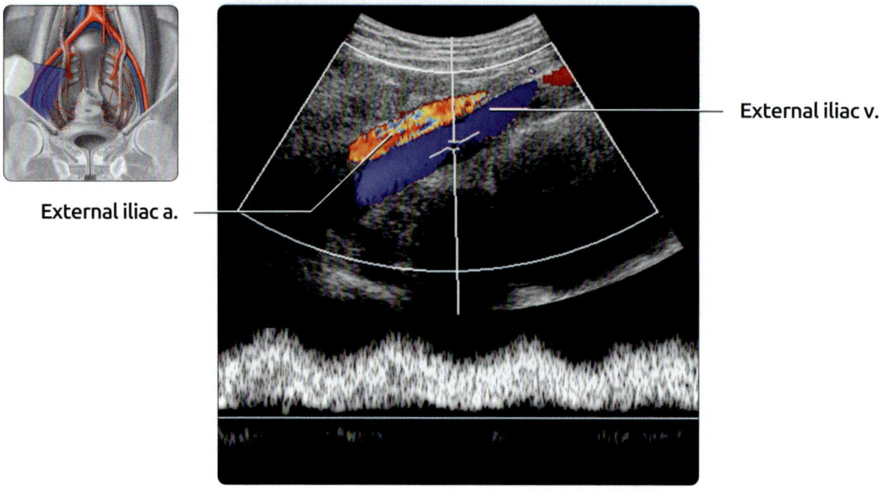

(Top) *Longitudinal transabdominal grayscale ultrasound at the lower abdomen is shown. The external iliac vein runs parallel and posterior to the external iliac artery. The 2 vessels are readily identified owing to their superficial location. The transducer is angulated medially, and the urinary bladder is included in the image.* **(Middle)** *Longitudinal transabdominal color Doppler ultrasound shows different flow directions in the external iliac artery (shown in red) and vein (shown in blue). The artery is anterior to the vein, and the 2 vessels run beneath the inguinal ligament to enter the thigh. Owing to its superficial location, the external iliac vein is easily examined during ultrasound investigation for deep venous thrombosis in the lower limbs.* **(Bottom)** *Pulsed Doppler ultrasound of the external iliac vein is shown. Normal spectral waveform shows continuous flow and exaggerated phasic changes from deep inspiration/expiration.*

ATHEROSCLEROTIC DISEASE

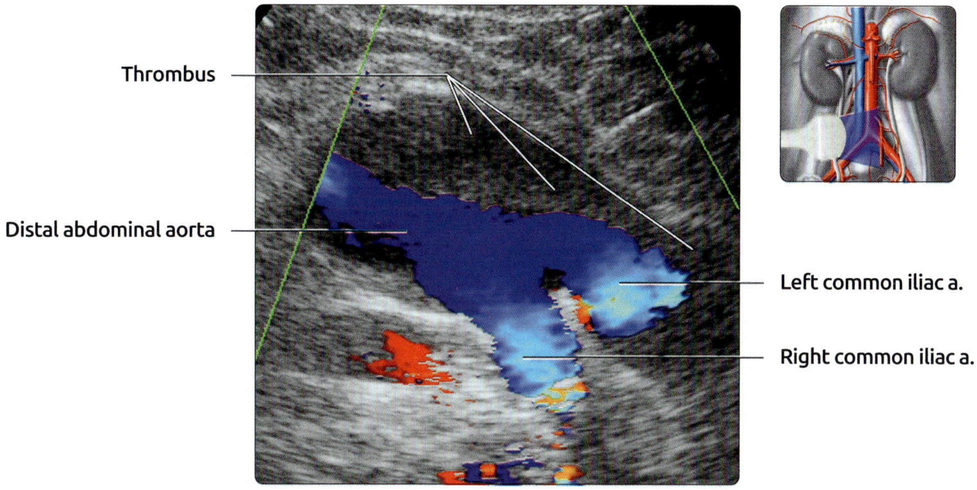

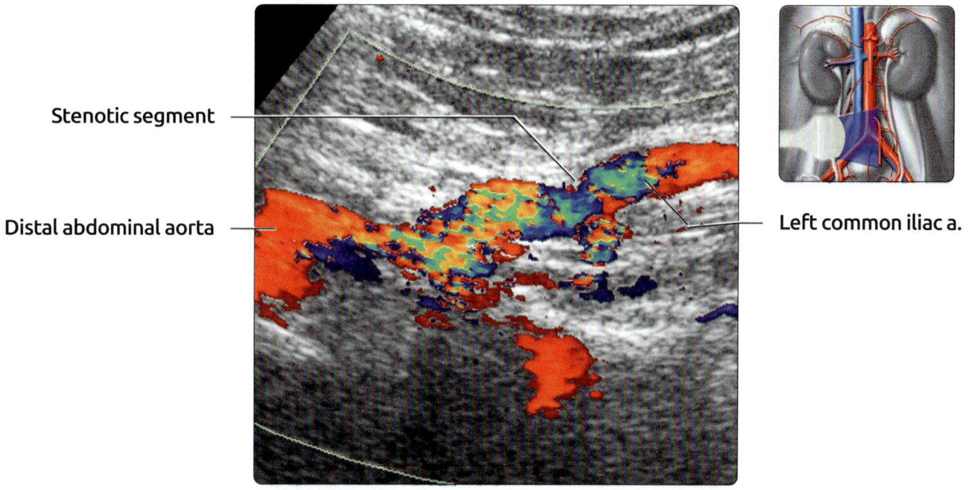

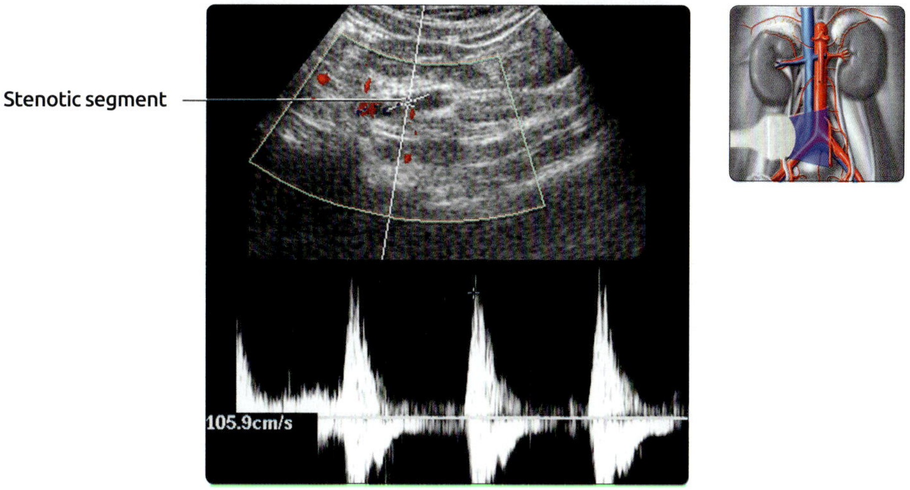

(Top) *Longitudinal color Doppler ultrasound shows a moderate amount of mural thrombus in the infrarenal aorta (shown in blue) just above the bifurcation, causing narrowing of the lumen. The thrombus extends into the left common iliac artery.* **(Middle)** *Oblique color Doppler ultrasound shows the common iliac artery with a > 50% diameter reduction. High-velocity turbulent flow is seen as "aliasing" artifact on color Doppler imaging at the stenotic segment.* **(Bottom)** *Corresponding oblique pulsed color Doppler ultrasound shows spectral Doppler trace at the common iliac artery stenosis with characteristic high-velocity bidirectional waveform.*

GROSS ANATOMY

Ureters

- Muscular tubes (25-30 cm long) that carry urine from kidneys to bladder
- **Course**
  - In abdomen, **retroperitoneal location**
    - Proximal ureters lie in perirenal space
    - Mid ureters lie over psoas muscles slightly medial to tips of L2-L5 transverse process
  - In pelvis, lie anterior to sacroiliac joints crossing common iliac artery bifurcation near pelvic brim
    - **Lie anterior to internal iliac vessels** and course along pelvic sidewall
    - At level of ischial spines, ureters curve anteromedially to enter bladder at level of seminal vesicles (men) or cervix (women)
  - **Ureterovesical junction (UVJ)**: Ureters pass obliquely through muscular wall of bladder for ~ 2 cm
    - Creates valve effect with bladder distension, preventing vesicoureteral reflux (VUR)
- **3 points of physiological narrowing**
  - Ureteropelvic junction
  - Pelvic brim (crossing over common iliac artery)
  - UVJ
- **Vessels, nerves, and lymphatics**
  - Arterial branches are numerous and variable, arising from aorta and renal, gonadal, internal iliac, vesicle, and rectal arteries
  - Venous branches and lymphatics follow arteries with similar names
  - Innervation
    - Autonomic from adjacent sympathetic and parasympathetic plexuses
      - Responsible for ureteral peristalsis
    - Also carry pain (stretch) receptors
      - Stone in abdominal ureter perceived as back and flank pain
      - Stone in pelvic ureter may project to scrotum or labia
  - Lymphatics to external and internal iliac nodes (pelvic ureter), aortocaval nodes (abdomen)

Bladder

- Hollow, distensible viscus with strong, muscular wall and normal adult capacity of 300-600 mL of urine
- Lies in **extraperitoneal** (retroperitoneal) pelvis
- **Peritoneum covers dome** of bladder
  - Reflections of peritoneum form deep recesses in pelvic peritoneal cavity
  - **Rectovesical pouch** (between rectum and bladder) is most dependent recess in men (and in women following hysterectomy)
  - **Vesicouterine pouch** (between bladder and uterus) and **rectouterine pouch of Douglas** (between rectum and uterus)
    - Rectouterine pouch most dependent in women
- Bladder is surrounded by extraperitoneal fat and loose connective tissue
  - **Perivesical space** (contains bladder and urachus)
  - **Prevesical or retropubic space (of Retzius)** between bladder and symphysis pubis
    - Communicates superiorly with infrarenal retroperitoneal compartment
    - Communicates posteriorly with presacral space
  - Spaces can expand to contain large amounts of fluid (as in extraperitoneal rupture of bladder and hemorrhage from pelvic fractures)
- Bladder wall composed mostly of **detrusor muscle**
  - **Trigone** of bladder: Triangular structure at base of bladder with apices marked by 2 ureteral orifices and internal urethral orifice
- **Vessels, nerves, and lymphatics**
  - Arteries from internal iliac
    - Superior vesicle arteries and other branches of internal iliac arteries in both sexes
  - Venous drainage
    - Men: Vesicle and prostatic venous plexuses → internal iliac and internal vertebral veins
    - Women: Vesicle and uterovaginal plexuses → internal iliac vein
  - Autonomic innervation
    - Parasympathetic from pelvic splanchnic and inferior hypogastric nerves (causes contraction of detrusor muscle and relaxation of internal urethral sphincter to permit emptying of bladder)
    - Sensory fibers follow parasympathetic nerves

IMAGING ANATOMY

Overview

- Normal ureters are small in caliber (2-8 mm) and are difficult to appreciate on ultrasound
- Fluid-distended urinary bladder is anechoic with posterior acoustic enhancement
- Urinary bladder changes in shape and position depending on intraluminal volume of urine
  - In its nondistended state, urinary bladder is retropubic in location, lying anterior to uterus in females and rectum in males
  - In markedly distended state, urinary bladder may occupy abdominopelvic area
  - Urinary bladder wall changes in thickness depending on state of distension of urinary bladder and is normally 3-5 mm in thickness

ANATOMY IMAGING ISSUES

Imaging Recommendations

- Transducer: Curvilinear 2-5 MHz
- **Ureters**
  - Ureters are normally not seen on ultrasound unless they are dilated; when dilated, overlying bowel gas may still limit ureteral evaluation in transabdominal approach
    - Proximal dilated ureters may be well seen using kidney as window in coronal oblique plane
    - Middle portion of dilated ureter may be identified in pediatric patients or thin adults using transabdominal approach
    - Dilated terminal ureter/UVJ are seen best along posterolateral aspect of urinary bladder on transverse view

- Can also be evaluated by endovaginal sonography in women
- Ureteral caliber may slightly increase as result of overfilled urinary bladder
  - Distended bladder may cause ureteral and pelvicalyceal dilation and rescanning post void is beneficial to exclude obstruction
- Color Doppler
  - Assess normal ureteral jets at and helps exclude complete ureteral obstruction
  - Look for **twinkle artifact** from obstructing stone
- **Bladder**
  - Recommend fluid intake prior to examination to ensure optimal distension of urinary bladder
    - In fully distended state, urinary bladder is easily visualized using transabdominal approach
  - Examine patient in supine position with transabdominal suprapubic approach
    - Perform scanning in sagittal and transverse planes
    - Patient may be placed in decubitus position to determine mobility and differentiated intravesical masses from debris or stones
    - With poor distention, caudal transducer angulation is needed to visualize urinary bladder in its retropubic location
  - Nature of cystic structure in pelvis may be ascertained by asking patient to void or by inserting Foley catheter
  - Transvaginal ultrasound may be used in women for evaluation of suspect bladder neck lesions, UVJ stone, or ureterocele
  - Advantages of ultrasound
    - Radiation-free, real-time assessment with high spatial resolution of bladder and bladder wall
    - Real-time assessment of intraluminal masses in bladder for mobility and vascularity
    - Real-time imaging guidance for bladder intervention, e.g., placement of percutaneous suprapubic catheters
    - Real-time assessment of ureteral jets using color Doppler imaging; particularly useful in pregnant patients with dilated collecting system

Imaging Pitfalls

- **Reverberation artifacts** are commonly encountered behind anterior wall of urinary bladder
  - Appear as regularly spaced lines at increasing depth as result of repeated reflection of ultrasound signals between highly reflective interfaces close to transducer
  - May be reduced or avoided by changing scanning angle or by moving transducer or using spacer
- Underdistended bladder may give false impression of wall thickening and limits intraluminal assessment
  - Have patient drink water and rescan with better distention
- Large midline ovarian or pelvic cystic mass may simulate bladder on transabdominal ultrasound
  - Attention to normal bladder shape, rescanning after voiding to confirm empty bladder, or transvaginal imaging is helpful to differentiate

CLINICAL IMPLICATIONS

Clinical Importance

- Ureters are at high risk of inadvertent injury during abdominal or gynecological surgery due to close proximity to uterine (in uterosacral ligament) and gonadal arteries (at pelvic brim)
- **Ectopic ureter**
  - Usually (80%) associated with complete ureteral duplication; more common in females
  - In complete duplication, upper moiety inserts ectopically inferiorly and distally to lower moiety (**Weigert-Meyer rule**) and can be associated with ureterocele
    - Ureterocele may cause obstruction of upper pole moiety; also distorts UVJ of normally inserting lower pole moiety causing predisposition to VUR
  - Ectopic ureteral insertion in females can occur in urethra or vagina, leading to urinary incontinence
- **Ureterocele**
  - Cystic dilation of intramural portion of ureter bulging into bladder
    - Orthotopic: Normal insertion of single ureter
    - Ectopic: Inserts below trigone, mostly in duplicated system
- **Ureteral duplication**
  - Bifid ureter drains duplex kidney, but ureters unite before entering bladder
- **Urachal anomalies**
  - Patent fetal urachus forms conduit between umbilicus and bladder
  - Urachus is normally obliterated to form median umbilical ligament
  - May persist as cyst, diverticulum, or rarely, fistula
  - Risk of infection or carcinoma (adenocarcinoma)
- **Bladder diverticula** are common
  - Congenital: Hutch diverticulum (near UVJ)
  - Acquired (usually due to chronic bladder outlet obstruction), associated with trabeculated bladder wall
  - Can lead to infection, stones, tumor
- **Trauma**
  - **Extraperitoneal bladder rupture**
    - Urine and blood distend prevesical space (Retzius)
    - Urine often tracks posteriorly into presacral space, superiorly into retroperitoneal abdomen
    - High association with pelvic fractures
  - **Intraperitoneal bladder rupture**
    - Urine flows up paracolic gutters into peritoneal recesses and surrounds bowel
    - Bladder ruptures along dome, which is in contact with intraperitoneal space
    - Usually caused by blunt trauma to overdistended bladder

URETERS AND URINARY BLADDER IN SITU

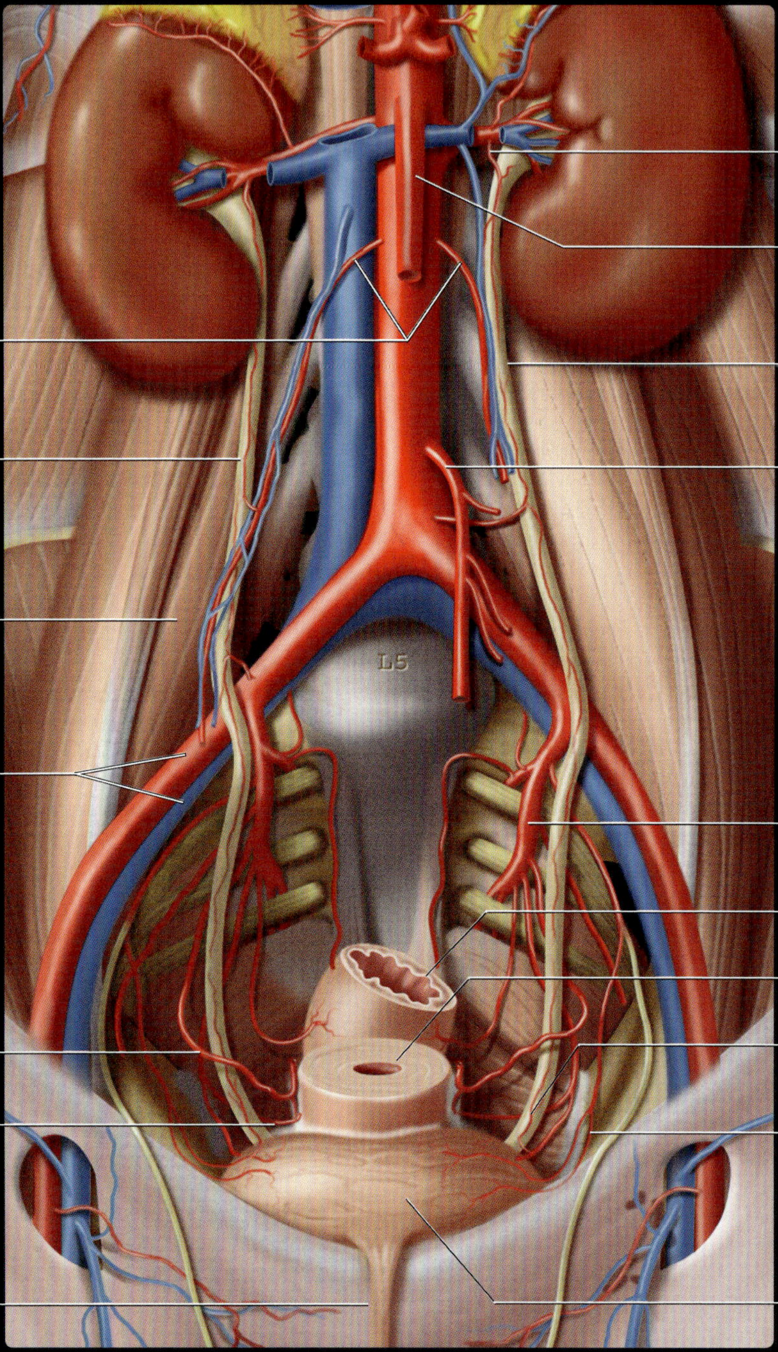

The ureters receive numerous and highly variable arterial branches from the aorta and the renal, gonadal, and internal iliac arteries. These vessels are short and can be easily ruptured by retraction of the ureter during surgical procedures. The arterial supply to the bladder is also quite variable. Both genders receive supply from the superior vesicle arteries and from various branches of the internal iliac arteries. Branches to the prostate and seminal vesicles (men) also send branches to the inferior bladder wall. In women, branches to the vagina send arteries to the base of the bladder. Note how the ureters deviate anteriorly as they cross the external (or common) iliac vessels and pelvic brim. This may constitute a point of relative narrowing where the passage of ureteral calculi (stones) may be impeded. In the abdomen, the ureters course along the psoas muscles.

URINARY BLADDER IN SITU, MALE

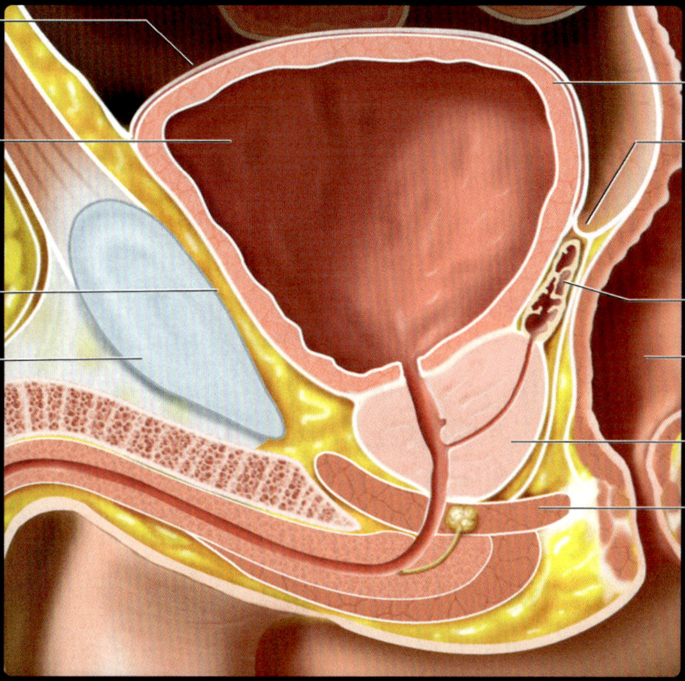

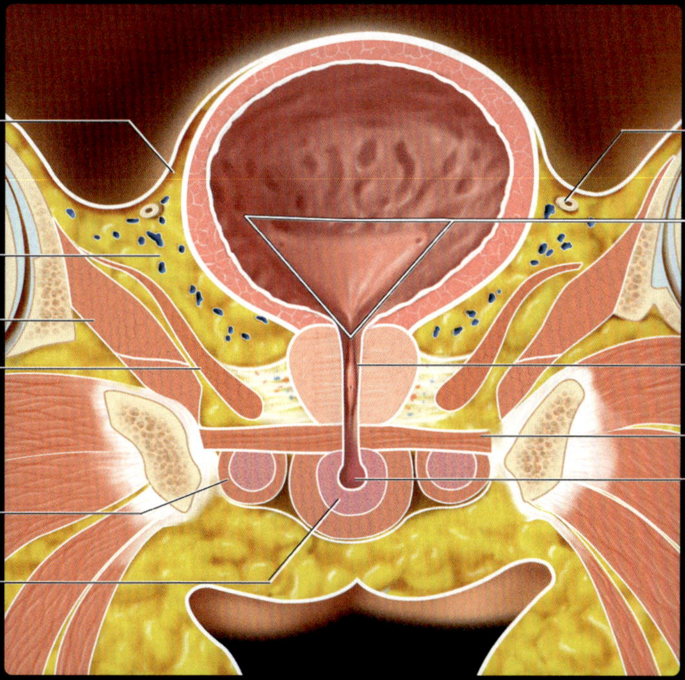

(Top) *Graphic of a sagittal section of the male bladder shows that it rests on the prostate, which separates it from the muscular pelvic floor. The bladder wall is muscular, strong, and very distensible. In males, the urinary bladder is directly anterior to the rectum, and the rectovesical pouch is the deepest point in the pelvis.* **(Bottom)** *A coronal section of the male bladder shows the anatomic relationships of the bladder with its surrounding structures. The trigone is a triangle formed by the ureteral orifices and urethral outlet. The ureters enter the bladder through an oblique anteromedial course that helps to prevent urinary reflux into the ureters.*

URINARY BLADDER IN SITU, FEMALE

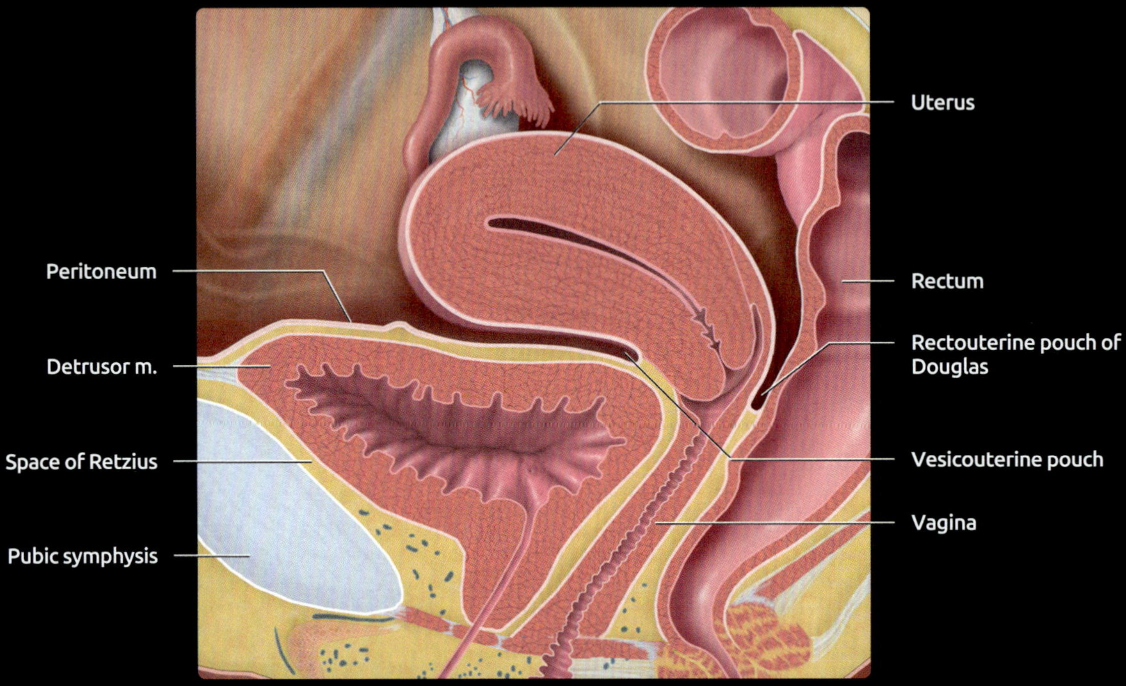

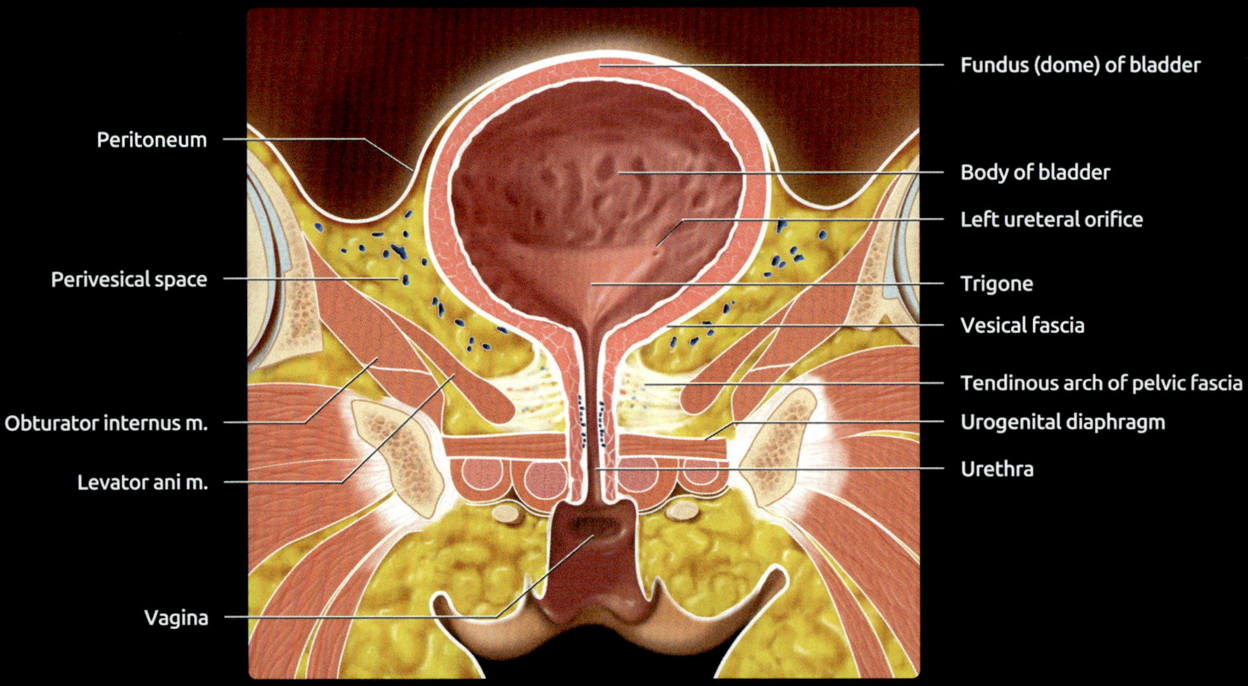

(Top) *Sagittal graphic shows a nondistended bladder in a female. When decompressed, the bladder wall can appear quite thick and can erroneously be interpreted as abnormal. The dome of the bladder is covered with peritoneum. The bladder is surrounded by a layer of loose fat and connective tissue (the prevesical space of Retzius and perivesical spaces) that communicate superiorly with the retroperitoneum. Note the vagina/uterus in the female pelvis, which intervenes between the urinary bladder and rectum.* **(Bottom)** *A frontal (coronal) section of the female bladder shows that it rests almost directly on the muscular floor of the pelvis. The dome of the bladder is covered with peritoneum. The trigone is the distinct triangular base of the bladder whose apices are formed by the ureteral and urethral orifices.*

URETERS

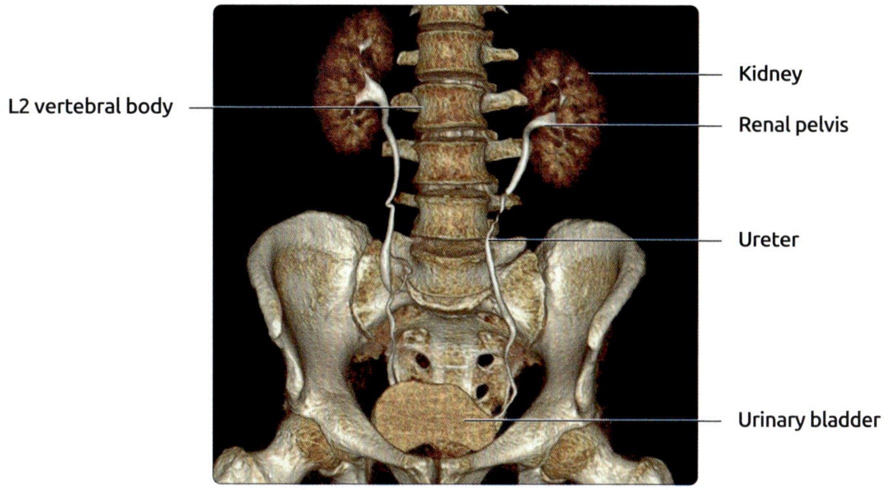

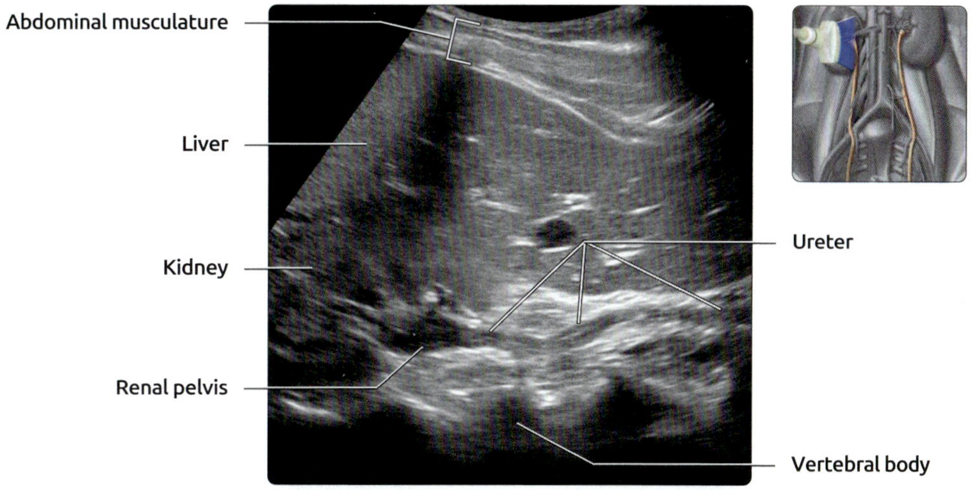

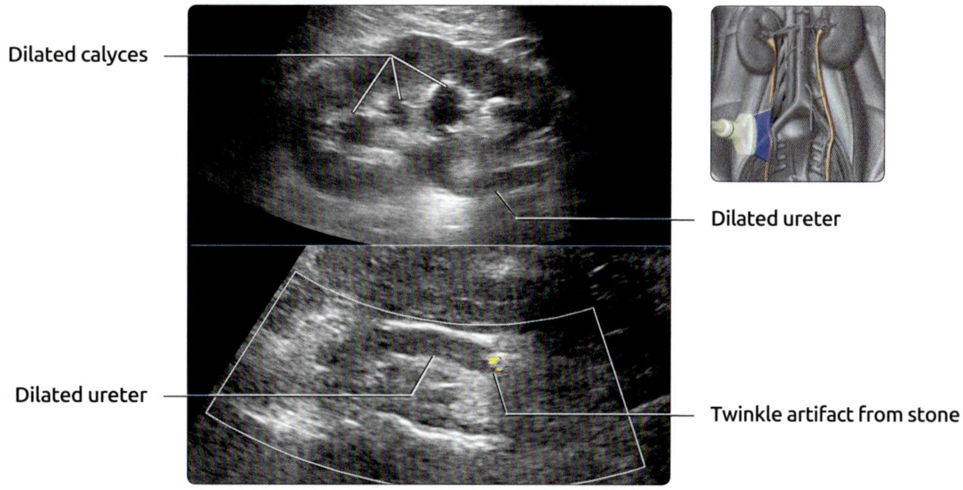

(Top) *Volume-rendered 3D reformatted image from a CT urogram shows the normal course of the ureters overlying L3-L5 transverse process.* **(Middle)** *The ureter is identified in this very thin teenage girl by scanning through the right flank and using the liver as an acoustic window. Even in thin patients, the ureter is usually not seen secondary to overlying bowel gas.* **(Bottom)** *This composite image shows an obstructing midureteral stone just above the pelvic rim. The pelvic rim, along with the ureteropelvic and ureterovesical junctions, are areas of narrowing and are the most likely locations for a stone to lodge. Even if you can't seen the entire ureter, specifically target these areas on your scan when evaluating for an obstructing stone.*

URETERS

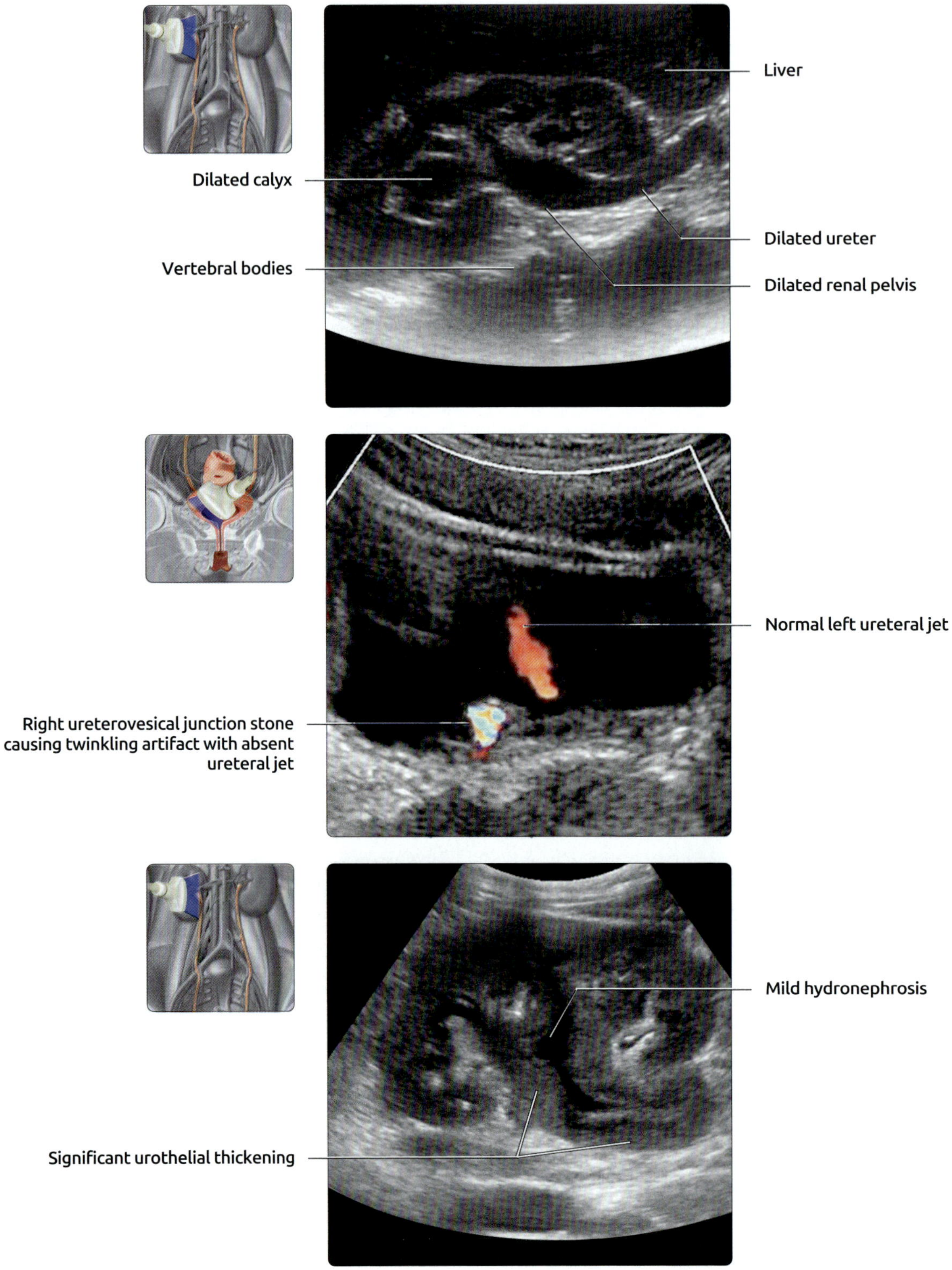

(Top) *Longitudinal transabdominal grayscale ultrasound of the right kidney shows a dilated pelvicalyceal system and proximal ureter. The ureter is normally not visible on ultrasound unless it is dilated, as seen here.* **(Middle)** *Oblique color Doppler ultrasound of the suprapubic region in the same patient shows a calculus at the right ureterovesical junction causing twinkling artifact with an absent ureteral jet. A normal left ureteric jet is seen.* **(Bottom)** *Longitudinal transabdominal ultrasound of the left kidney shows significant urothelial thickening of the pelvicalyceal system and proximal ureter in this patient with known extramedullary hematopoiesis.*

BLADDER

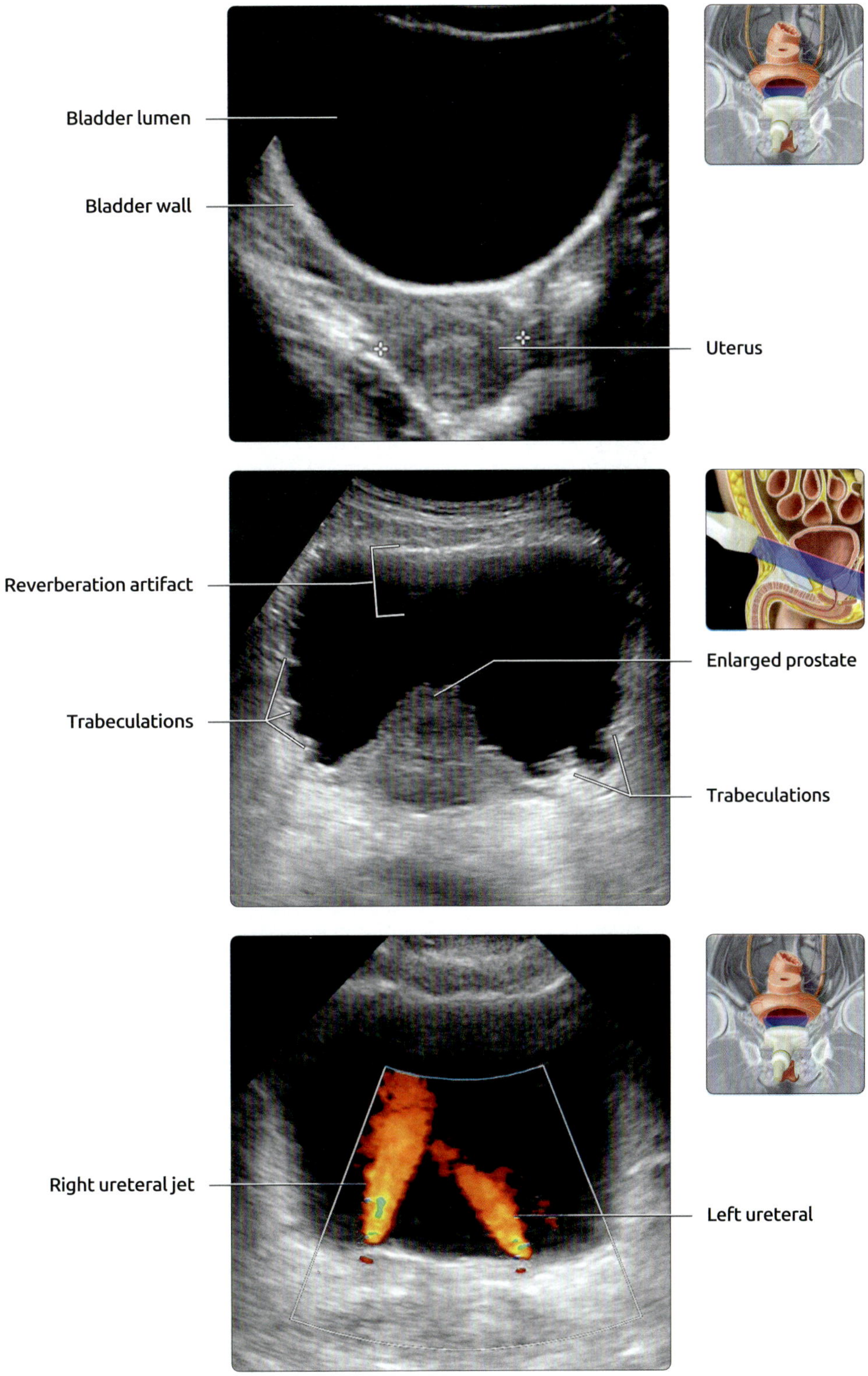

(Top) *Transverse transabdominal grayscale ultrasound shows a distended bladder with a smooth wall. The transducer must be angled caudally to image the urinary bladder, especially when it is not well distended and assumes a retropubic location.* **(Middle)** *Transverse ultrasound of the bladder shows wall thickening and trabeculations in this patient with benign prostatic hypertrophy and chronic bladder outlet obstruction. Note the reverberation artifact, which compromises evaluation of the anterior bladder wall.* **(Bottom)** *Transverse color Doppler image at the level of the ureteral orifices shows bilateral, symmetric, ureteral jets. Always evaluate for the presence of ureteral jets when evaluating for obstruction.*

URINARY BLADDER

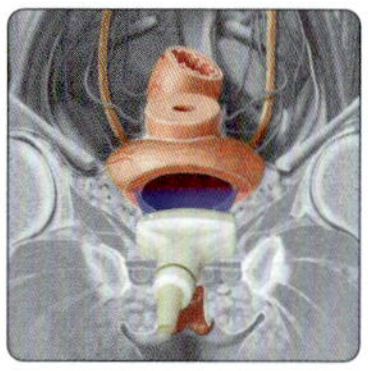

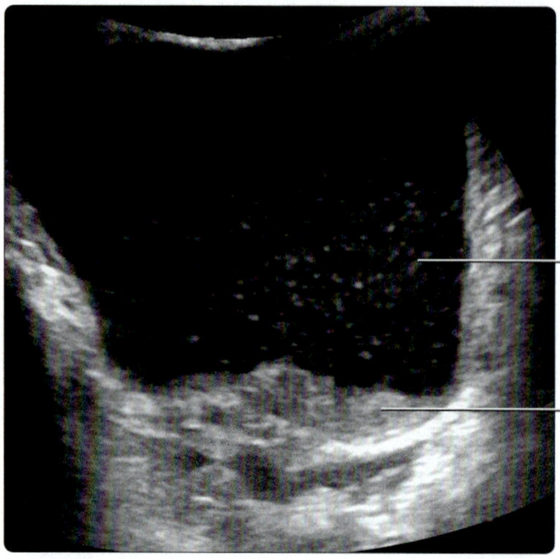

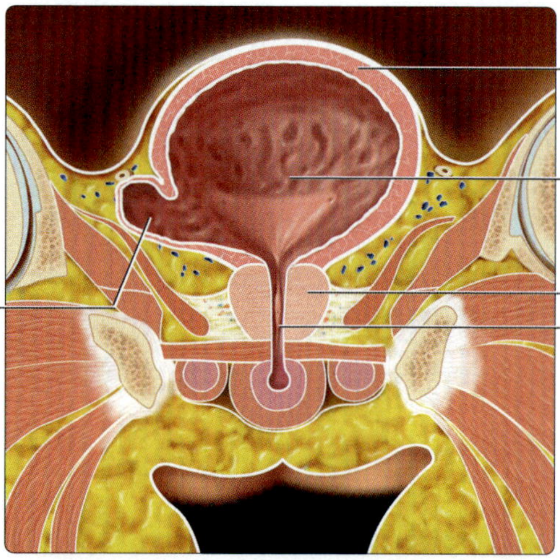

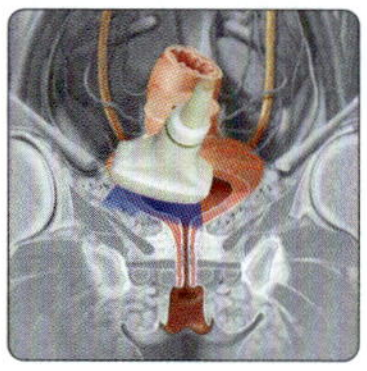

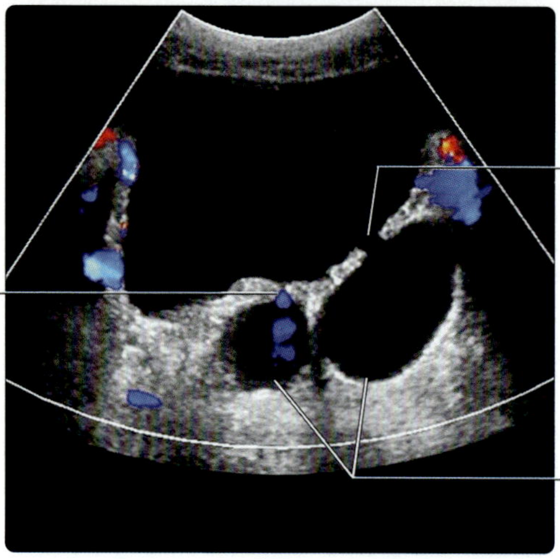

(Top) *Transverse transabdominal ultrasound of the bladder shows floating internal echoes with layering echogenic debris in this patient with cystitis. This could be confused with wall thickening, and the patient should be put in a decubitus position to document debris movement.* **(Middle)** *Graphic shows a diverticulum arising from the lateral urinary bladder wall with herniation of the mucosa and submucosa through the muscular wall.* **(Bottom)** *Transverse oblique transabdominal color Doppler ultrasound through the bladder shows 2 well-distended bladder diverticula along the left posterolateral bladder. A urine jet is identified in one of the diverticular necks.*

WEIGERT-MEYER RULE

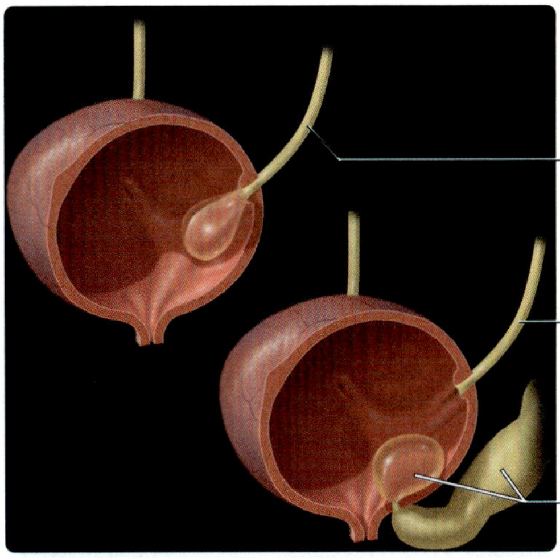

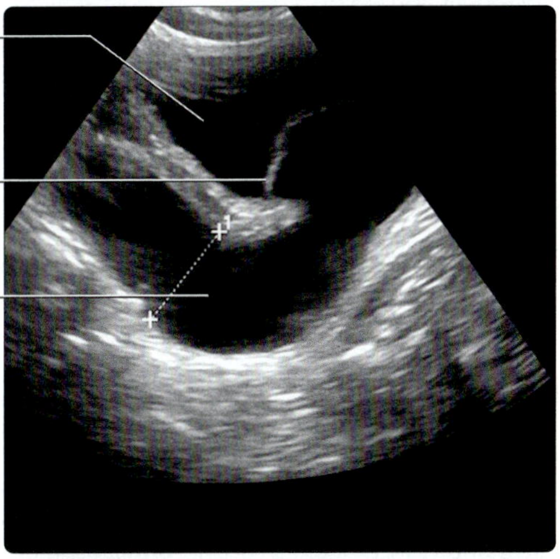

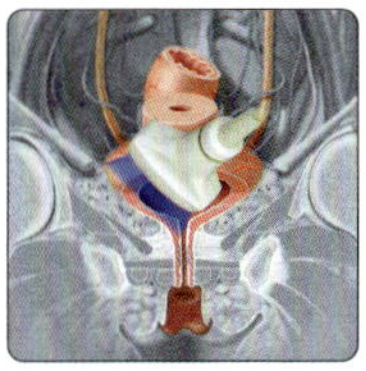

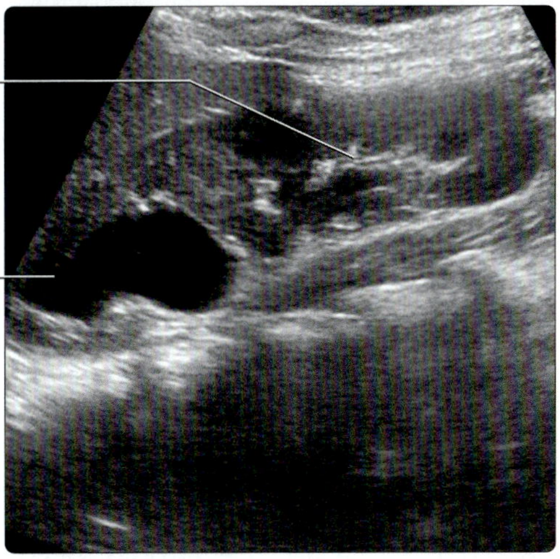

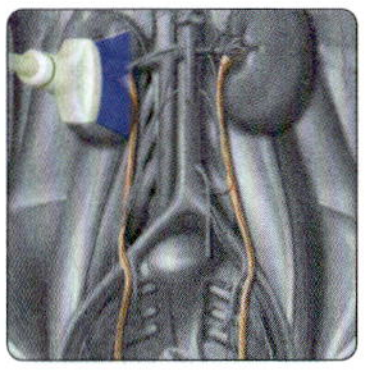

(Top) *Graphic illustrates an orthotopic ureterocele in a single ureter system (left, upper) and an ectopic ureterocele in a duplicated ureter system (right, lower). Note the hydroureter accompanying the ectopic ureterocele. The ectopic ureterocele inserts inferior and medial to the normally inserting ureter (Weigert-Meyer rule).* **(Middle)** *Longitudinal oblique transabdominal grayscale ultrasound at the suprapubic region shows a dilated ureter seen terminating in a ureterocele; the patient had complete duplication of the collecting system.* **(Bottom)** *Longitudinal transabdominal grayscale ultrasound through the right kidney in the same patient shows a dilated, obstructed upper pole moiety with a decompressed inferior moiety. The lower pole moiety is at risk for reflux and may also be dilated.*

GROSS ANATOMY

Prostate

- Walnut-sized gland beneath bladder and in front of rectum
  - Normal prostate in young male ~ 3 cm length x 4 cm width x 2 cm depth
  - Normal weight ~ 20-30 grams
- Inverted conical shape
  - **Base**: Superior portion, continuous with bladder neck
  - **Apex**: Inferior portion, continuous with striated sphincter
- Capsule: Composed of condensed fibromuscular band, not true capsule
  - Does not completely envelop prostate: Absent at base and not clearly defined at apex
  - Capsular components are inseparable from prostatic stroma and periprostatic connective tissue
- Posteriorly, **Denonvilliers fascia** (thin layer of connective tissue) separates prostate and seminal vesicles from rectum
- Laterally, prostate is cradled by pubococcygeal portion of levator ani
- Toward apex, puboprostatic ligaments extend anteriorly to affix prostate to pubic bone
  - Apex is continuous with striated external urethral sphincter
- **Ejaculatory ducts** form at junction of vas deferens and seminal vesicles and enter prostate base
- Prostatic urethra
  - **Verumontanum (a.k.a. colliculus seminalis)**
    - Midway between base and apex where urethra makes ~ 35° bend anteriorly
      - Openings of prostatic utricle and ejaculatory ducts
    - Divides prostatic urethra into proximal (preprostatic) and distal (prostatic) segments
  - **Preprostatic sphincter**: Thickened circular smooth muscle in proximal segment (a.k.a. involuntary internal urethral sphincter, periurethral zone)
    - Thought to function during ejaculation to prevent retrograde flow of seminal fluid
    - May also have resting tone, which maintains closure of preprostatic urethra, thereby aiding urinary continence
    - Contains small periurethral glands completely enclosed in sphincter
      - Although these glands constitute < 1% of glandular prostate, they can contribute significantly to prostatic volume as one site of origin of **benign prostatic hyperplasia (BPH)**
  - **Urethral crest**: Narrow longitudinal ridge on midline posterior wall
    - Multiple small openings draining prostatic ducts
  - **Prostatic utricle**: Small, superoposteriorly directed vestigial blind pouch with opening in verumontanum, ~ 6 mm long
    - Müllerian remnant (homologous with uterus and vagina)
- **Neurovascular bundles (NVB)**
  - Lie posterolaterally to prostate
  - Carry nerves and vascular supply to corpora cavernosa
- Vascular supply
  - Most commonly, arterial supply from inferior vesical artery
    - Often divides into 2 main branches: Urethral arteries and capsular artery
      - Urethral arteries supply periurethral glands and transition zone (TZ) → main supply for BPH
      - Bulk of capsular artery runs posterolaterally with cavernous nerves in NVB and ends at pelvic diaphragm
  - Venous drainage via periprostatic plexus; receives blood from dorsal vein of penis; drains into internal iliac veins
- Nerve supply: Pelvic plexuses arising from S2-4 (parasympathetic) and L1-2 (sympathetic) fibers
- Lymphatic drainage chiefly to obturator and internal iliac nodes; small portion may initially pass through presacral group or, less commonly, external iliac nodes

Lobar Anatomy (Lowsley)

- Based on studies on human fetal prostate; distinct lobes do not exist in prepubertal and normal adult prostate
- Lobes: Anterior, median, posterior, and 2 lateral
  - Currently used in context of BPH
    - Lateral lobes: Hyperplasia of glands in TZ
    - Median lobe: Hyperplasia of periurethral glands in preprostatic sphincter or TZ and may project into bladder
- Largely replaced by zonal anatomy

Zonal Anatomy (McNeal)

- Prostate is histologically composed of ~ 70% glandular and 30% nonglandular elements
- **2 nonglandular elements: Prostatic urethra and anterior fibromuscular stroma (AFMS)**
  - AFMS is contiguous with bladder muscle and external urethral sphincter, up to 1/3 of prostatic mass
  - AFMS runs anteriorly from bladder neck to striated urinary sphincter
- **Peripheral zone (PZ)**: ~ 70% glandular tissue, covers posterolateral aspects of gland
  - Surrounds central zone (CZ) and prostatic (distal) urethra
  - Ducts drain into prostatic sinuses along urethra
  - ~ 70-75% prostatic adenocarcinomas arise in this zone
- **CZ**: ~ 25% glandular tissue; cone-shaped zone around ejaculatory ducts with widest portion making majority of prostatic base
  - Only 1-5% prostate adenocarcinoma originate in this zone; mainly involved by secondary invasion
- **TZ:** ~ 5-10% glandular tissue, 2 separate lobules surround preprostatic urethra (urethra proximal to verumontanum)
  - 1 site of origin of BPH, along with periurethral glands
  - ~ 20-25% of prostate adenocarcinoma arise in this zone
- Periurethral glands in preprostatic sphincter: < 1% glandular tissue, a site of origin of BPH
- **Prostate pseudocapsule ("surgical capsule")**
  - Visible boundary between TZ and PZ representing compressed tissue
  - Frequently, calcified corpora amylacea (laminated bodies formed of secretions and degenerate cells) highlight plane between PZ and TZ

Seminal Vesicles and Ejaculatory Ducts

- **Seminal vesicles**

- Sac-like structures superolateral to prostate, lateral outpouchings of vas deferens
- Secrete fructose-rich fluid (energy source for sperm)
- Arterial supply: Vesiculodeferential artery (branch of superior vesical artery)
  - May have additional supply from inferior vesical artery
- Venous drainage into pelvic venous plexus
- Lymphatic drainage into external and internal iliac nodes

- **Ejaculatory ducts**
  - Located on either side of midline
  - Formed by union of seminal vesicle duct and vas deferens
  - Start at base of prostate and run forward and downward through gland in CZ

IMAGING ANATOMY

Ultrasound

- **Prostate**
  - Transabdominal ultrasound can assess size but **transrectal ultrasound (TRUS)** required for detailed assessment
  - Normal TZ is typically uniformly more echogenic than inner gland
  - Inner gland (TZ and CZ) is often distinguishable from PZ
    - Heterogeneous TZ in BPH
  - TRUS is mainstay of many image-guided prostate interventions
    - Prostate biopsy: Either abnormal digital rectal exam or elevated prostate-specific antigen (PSA)
    - Brachytherapy, cryotherapy, high-intensity focused ultrasonography (HIFU), and BPH evaluation
- Prostate volume measurement
  - Prolate ellipse volume for 3 unequal axes: Width x height x length x 0.523
  - 1 cc of prostate tissue ~ 1 g; prostate weighs ~ 20 g in young men
  - Prostatic enlargement when gland is > 40 g
- **Seminal vesicles and vasa deferentia**
  - Cystic appearance on TRUS, should be symmetric

MR

- Best modality for evaluating extracapsular spread and staging of prostatic carcinoma
- T1WI: Homogeneous intermediate signal intensity
  - Often best sequence for NVB, which are at 5- and 7-o'clock positions
- T2WI: Depicts zonal anatomy
  - AFMS is low signal intensity
  - PZ is high signal intensity, ≥ periprostatic fat
  - PZ is surrounded by thin, low signal true capsule
  - CZ and TZ are similar T2 signal intensity, < PZ
  - Seminal vesicles are bright, similar to other fluid containing structures
- Tumors are low signal on T2WI, bright on diffusion weighted imaging, and show faster, stronger enhancement and rapid washout on dynamic contrast enhancement

ANATOMY IMAGING ISSUES

Imaging Recommendations

- Transducer
  - 7-10 MHz rectal transducer (end-firing or transverse panoramic)
  - 3.5-6 MHz curved linear transducer for transabdominal ultrasound
  - Perform in at least 2 orthogonal planes (axial and sagittal)
- Patient position
  - TRUS: Left lateral decubitus with flexed hips and knees or in lithotomy position
  - Transabdominal ultrasound: Supine, using urinary bladder as acoustic window (transvesical)
    - Fluid intake to ensure bladder distension

Imaging Pitfalls

- Abnormal vascularity on power Doppler ultrasound may be seen in hypertrophy, inflammation, and cancer
  - Useful for directing biopsy
- Transabdominal ultrasound of prostate is limited to evaluation of prostate size

Transrectal Biopsy of Prostate

- Most transrectal transducers have needle guidance system
- Periprostatic block with local anesthesia injected along NVB; may also use anesthetic gel and intraprostatic injection of local anesthetic
- Complications
  - Common: Hematuria, hematochezia, and hematospermia
  - Other: Acute prostatitis, UTI, sepsis

CLINICAL IMPLICATIONS

Function

- Main function is to add nutritional secretions to sperm to form semen during ejaculation
- Also plays role in controlling flow of urine; prostate muscle fibers are under control of involuntary nervous system and contract to slow and stop urine

Zonal Distribution of Prostatic Disease

- **Prostate adenocarcinomas**
  - 75% in PZ
    - Up to 80% of prostatic cancers in PZ are hypoechoic
  - 20% in TZ
  - 5% in CZ
- **BPH**: Nodular stromal and epithelial hyperplasia in periurethral (preprostatic) glands and TZ
  - Compresses CZ and PZ
  - Can cause bladder outlet obstruction from urethral compression &/or increased smooth muscle tone along bladder neck, prostate, and urethra

Spread of Prostate Carcinoma

- Signs of extraprostatic extension of prostatic carcinoma
  - Asymmetry of NVB
  - Obliteration of rectoprostatic angle
  - Irregular bulge in prostatic contour

PROSTATE

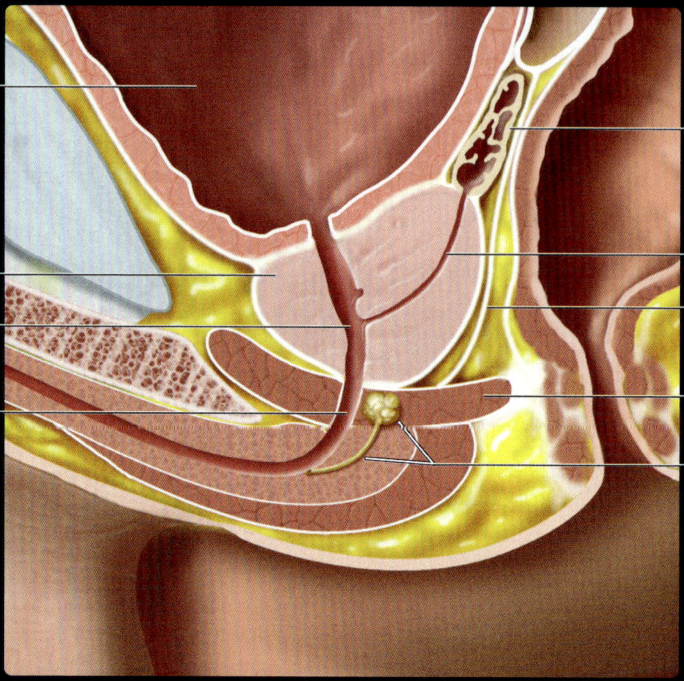

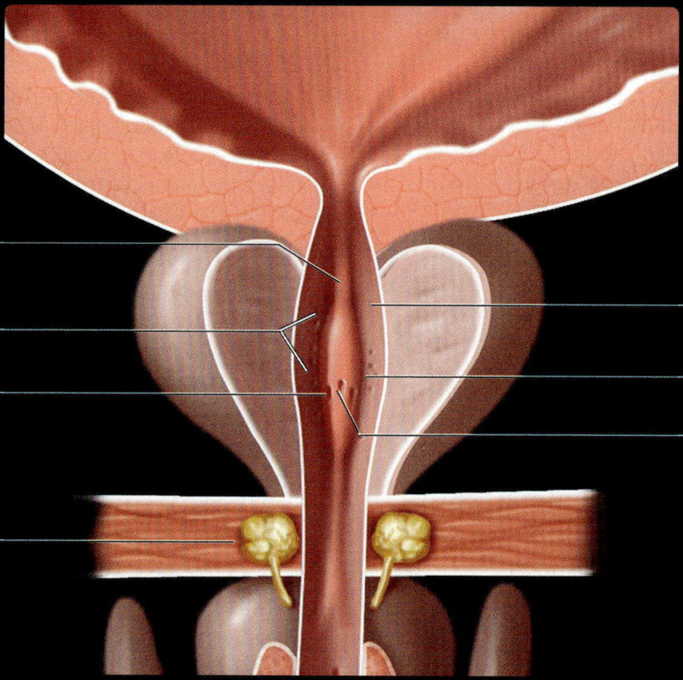

(Top) *Graphic illustrates the relationship between the prostate and the male pelvic organs. The prostate surrounds the upper part of the urethra (prostatic urethra). The base of the prostate is continuous with the bladder neck, and its apex is continuous with external sphincter. The posterior surface is separated from the rectum by the rectovesical septum (Denonvilliers fascia).* **(Bottom)** *Graphic shows the topography of the posterior wall of the prostatic urethra. The urethral crest is a mucosal elevation along the posterior wall with the verumontanum being a mound-like elevation in the midportion of the crest. The utricle opens midline onto the verumontanum with the ejaculatory ducts opening on either side. The prostatic ducts are clustered around the verumontanum and open into the prostatic sinuses, which are depressions along the sides of the urethral crest.*

VAS DEFERENS AND SEMINAL VESICLES

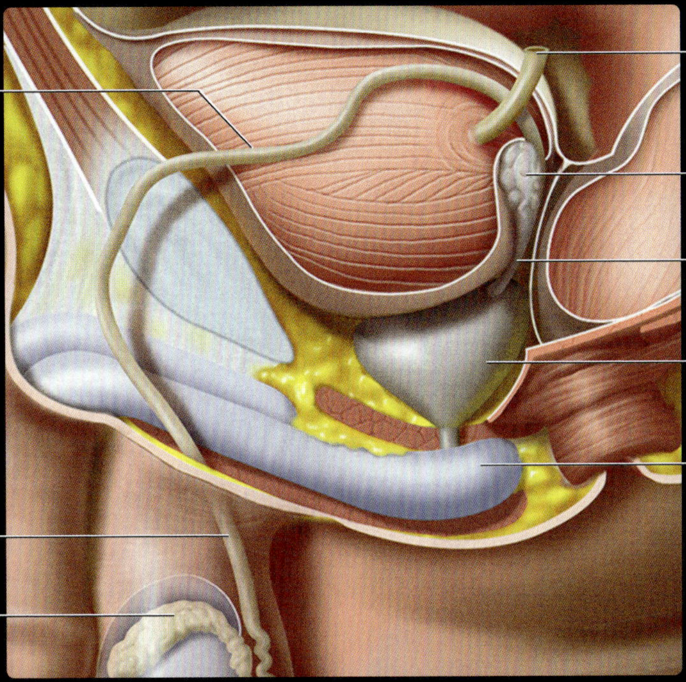

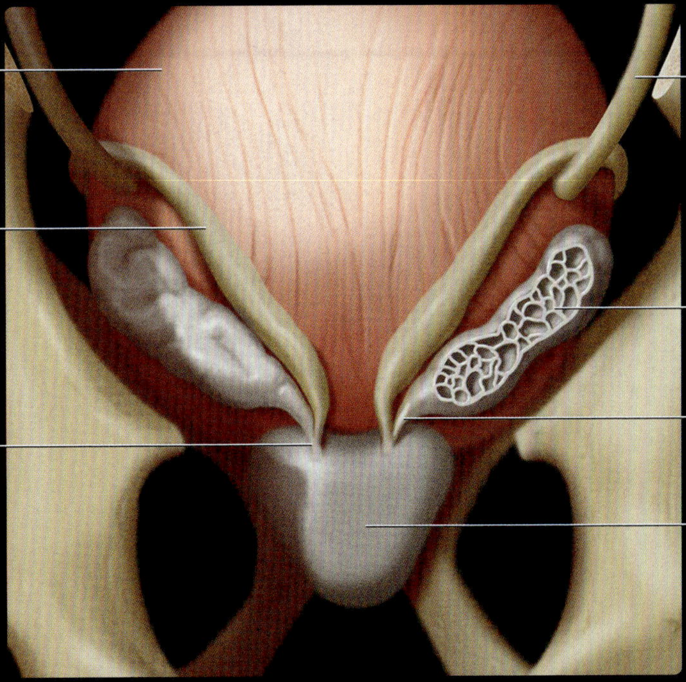

(Top) *This lateral view shows the position of the prostate deep in the pelvis. The vas deferens leaves the scrotum as a component of the spermatic cord, which courses through the inguinal canal into the pelvis.* **(Bottom)** *Posterior view of the prostate gland and seminal vesicles is shown. The cut surface of the seminal vesicle shows its highly convoluted fold pattern. The vas deferens crosses superior to the ureterovesical junction and continues along the posterior surface of the urinary bladder medial to the seminal vesicle. In the base of the prostate, it is directed forward and joined at an acute angle by the duct of the seminal vesicle to form the ejaculatory duct. The ejaculatory ducts course anteriorly and downward through the prostate to slit-like openings on either side of the orifice of the prostatic utricle.*

ZONAL ANATOMY OF THE PROSTATE

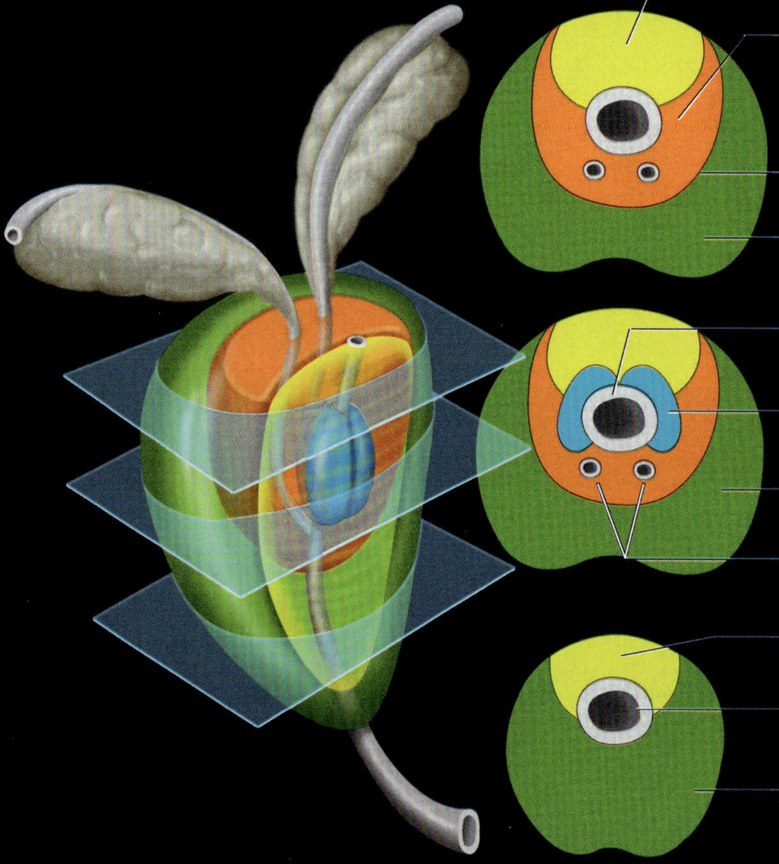

Graphic depiction of the prostate with axial drawings of the zonal anatomy at 3 different levels is shown. The transition zone (TZ) (in blue) is anterolateral to the verumontanum. The central zone (CZ) (in orange) surrounds the ejaculatory ducts and encloses the periurethral glands and the TZ. It is conical in shape and extends downward to about the level of the verumontanum. The peripheral zone (PZ) (in green) surrounds the posterior aspect of the CZ in the upper 1/2 of the gland, and it surrounds the urethra in the lower half, below the verumontanum. The prostatic pseudocapsule is a visible boundary between the CZ and PZ. The anterior fibromuscular stroma (in yellow) covers the anterior part of the gland and is thicker superiorly and thins inferiorly in the prostatic apex.

ZONAL ANATOMY OF THE PROSTATE

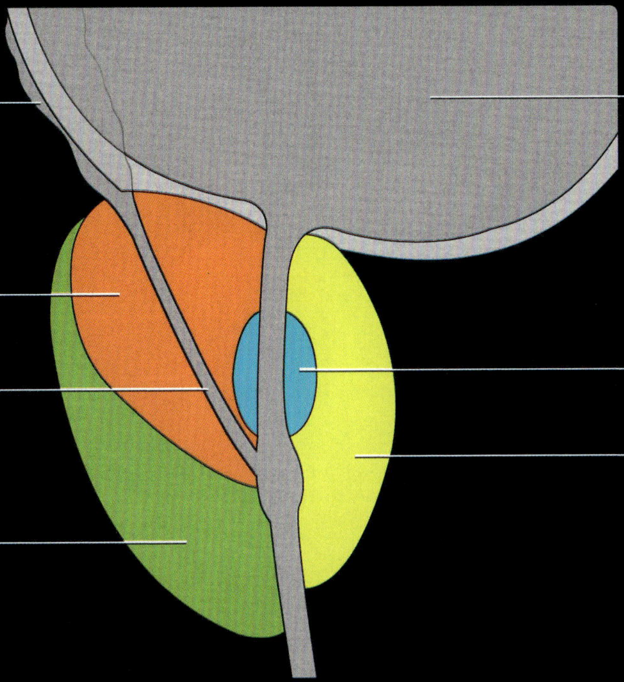

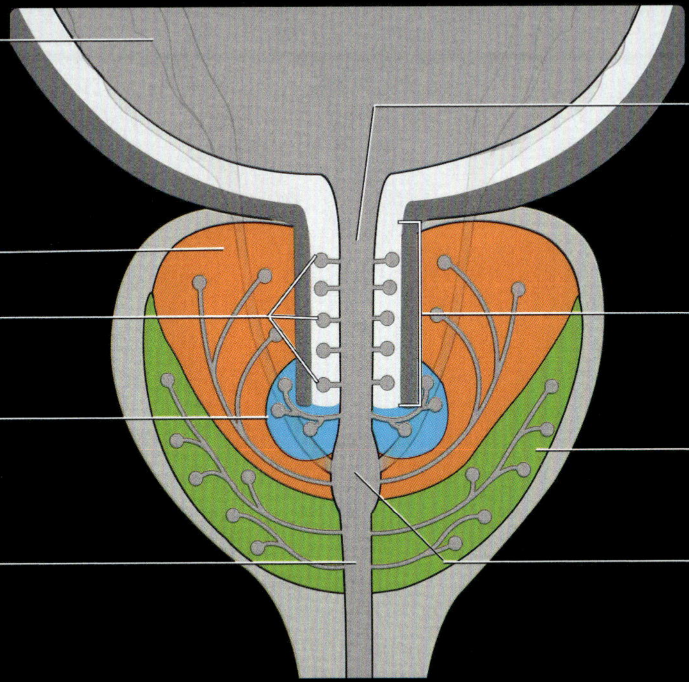

(Top) *Graphic illustrates the zonal anatomy of the prostate in the sagittal plane. The CZ surrounds the proximal urethra posterosuperiorly, enclosing both the periurethral glands and the TZ. It forms most of the prostatic base. The PZ surrounds both the CZ and the distal prostatic urethra.* **(Bottom)** *The proximal 1/2 of the prostatic urethra is surrounded by preprostatic sphincter, which extends inferiorly to the level of the verumontanum and encloses the periurethral glands. The preprostatic sphincter is thought to function during ejaculation to prevent retrograde flow and may also contribute to resting tone. The TZ is a downward extension of the periurethral glands around the verumontanum. Periurethral glands are < 1% of the normal prostate but are one of the sites of origin for benign prostatic hyperplasia (BPH) and can enlarge significantly.*

NORMAL PROSTATE AND BENIGN PROSTATIC HYPERPLASIA

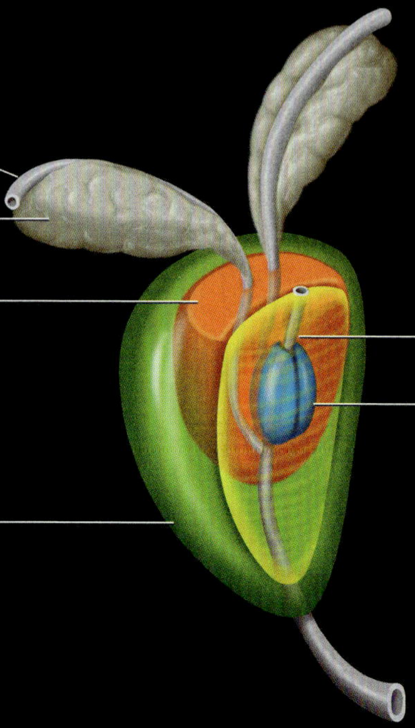

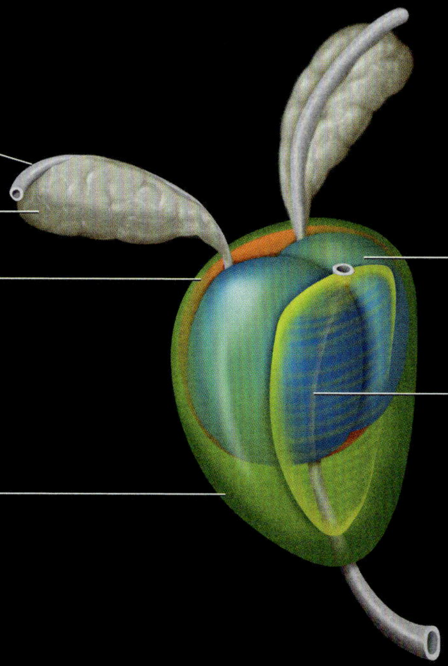

(Top) *First of 2 graphics comparing zonal anatomy in a young man to an older man with BPH is shown. The TZ in young men is small in size, comprising 5% of the volume of the glandular tissue of the prostate. It surrounds the anterolateral aspect of the urethra at the level of verumontanum in a horseshoe fashion.* **(Bottom)** *With the development of BPH, there is enlargement of the TZ. This causes enlargement of the prostate and compression of the CZ and PZ. BPH mainly involves the TZ, though other zones may also be involved. The enlarged TZ causes compression of the prostatic urethra, the primary reason for development of urinary obstructive symptoms in patients with BPH.*

GROSS IMAGES, PROSTATE

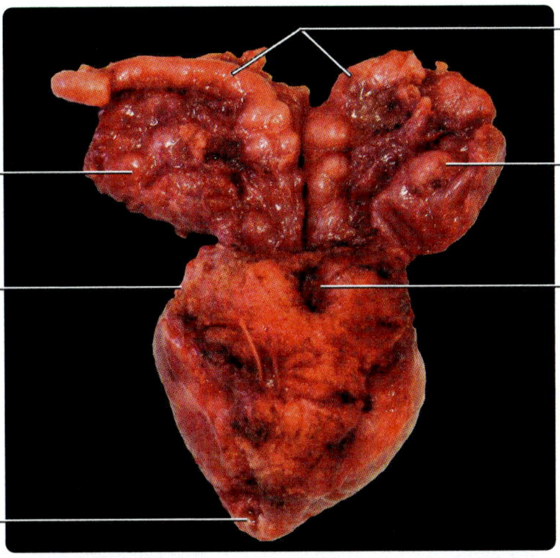

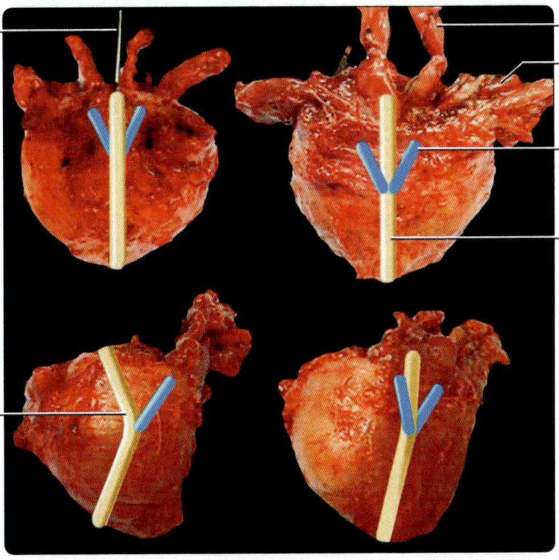

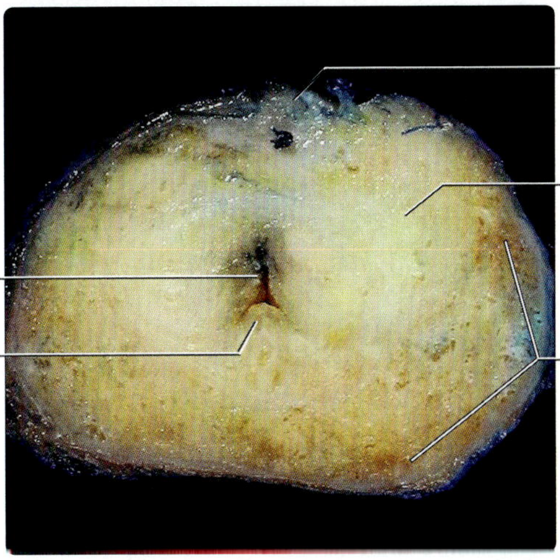

(Top) *Anterior-superior view of a resected prostate shows the urethra at the base. Paired bilateral seminal vesicles are attached to the posterior aspect of the base, lateral to the tubular vas deferens. (From DP: Genitourinary.)* **(Middle)** *These different views of the prostate depict the course of the prostatic urethra (yellow), which has a 35° anterior angulation halfway between base and apex, at the level of the verumontanum; this divides the prostatic urethra into proximal and distal segments. The ejaculatory ducts (blue) have a straighter course from the base of the seminal vesicles to the verumontanum. (From DP: Genitourinary.)* **(Bottom)** *This cross section of the prostate at the level of the verumontanum shows the PZ, which has a more spongy appearance, surrounding the hypertrophied, nodular TZ. (From DP: Genitourinary.)*

VARYING DEGREES OF BENIGN PROSTATIC HYPERPLASIA

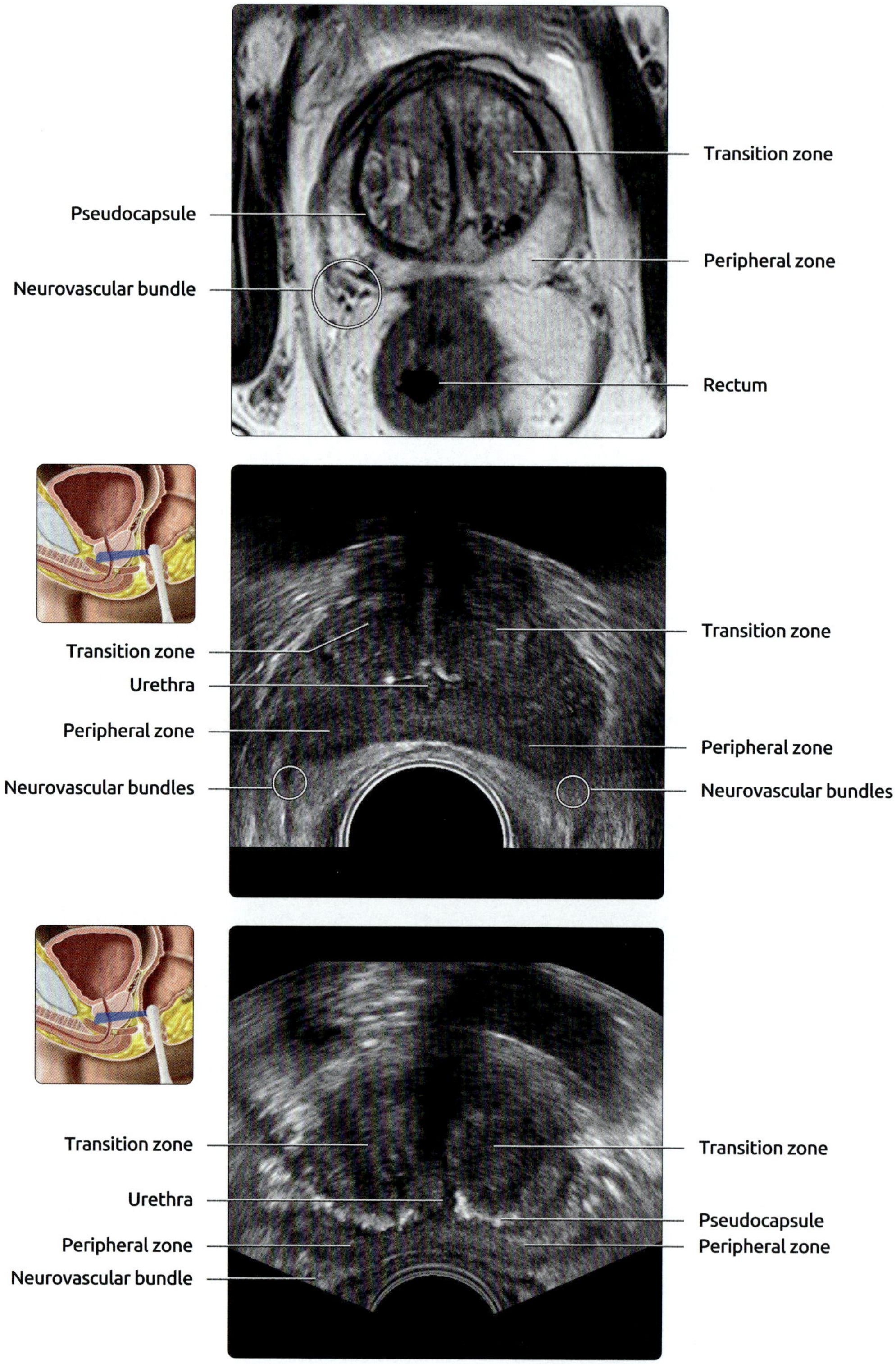

(Top) *Axial T2 MR shows nodular enlargement of the TZ, a typical appearance of BPH. The PZ is high signal intensity. There is a prominent low-signal pseudocapsule between the TZ and PZ.* **(Middle)** *Transverse TRUS at the level of the midprostate shows the 2 lobes of the TZ on either side of the urethra. The more homogeneous PZ is along the posterolateral aspects of the prostate. The neurovascular bundles course through the retroprostatic fat at the 5- and 7-o'clock positions but are usually not identified as discrete structures. Note the periurethral calcifications.* **(Bottom)** *Transverse TRUS of the midprostate gland demonstrates an enlarged TZ in a patient with BPH. The pseudocapsule separates the TZ from the PZ. Frequently, the pseudocapsule will be outlined by calcifications, which represent calcified corpora amylacea (laminated bodies formed by secretions and degenerate cells). The pseudocapsule is also referred to as the surgical capsule, as it is a landmark for transurethral resection of the prostate (TURP). Modern medical treatments for symptomatic BPH has made TURP procedures much less common than in the past.*

VARYING DEGREES OF BENIGN PROSTATIC HYPERPLASIA

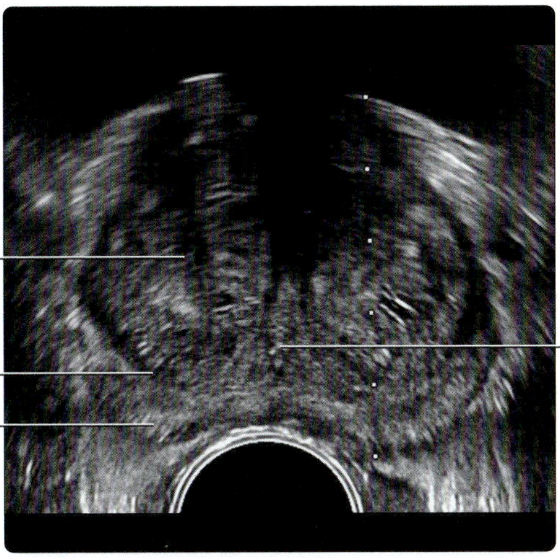

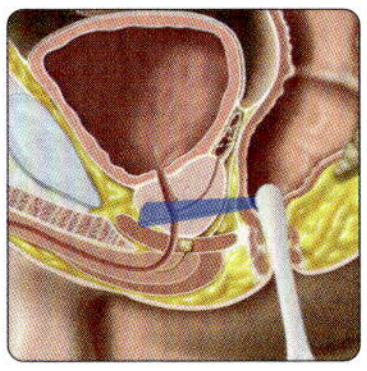

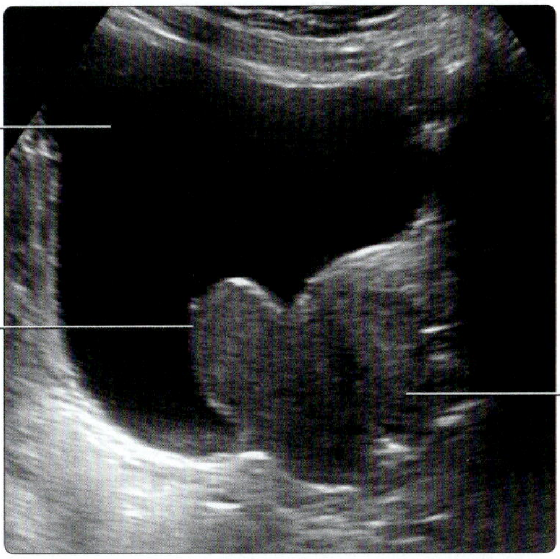

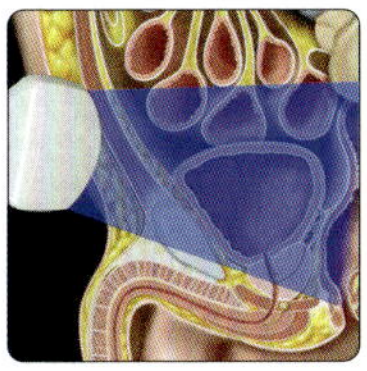

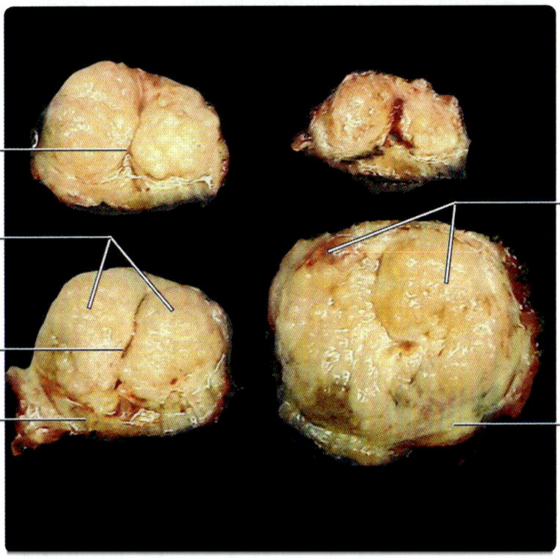

(Top) *Transverse TRUS of the midprostate in a different patient with BPH shows heterogeneous enlargement of the 2 lobes of the TZ, which flank the urethra. Tiny cystic spaces within the TZ represent cystic BPH nodules vs. retention cysts, which are often indistinguishable by imaging. The more hyperechoic PZ is being compressed along the posterolateral aspects of the prostate.* **(Middle)** *This longitudinal transabdominal postvoid image shows a markedly enlarged prostate with a hypertrophied median lobe projecting into the bladder. The median represents hyperplasia of the periurethral glands and is common with BPH. It typically projects cephalad into the bladder and can be associated with significant urinary retention. There is a large postvoid residual in this case.* **(Bottom)** *Consecutive cross sections of a prostate gland with BPH show marked nodular enlargement of the TZ. It is compressing the prostatic urethra, which was the cause of the patient's bladder-obstructive symptoms. The PZ is compressed and attenuated by the hyperplastic TZ. (From DP: Genitourinary.)*

SAGITTAL PROSTATE

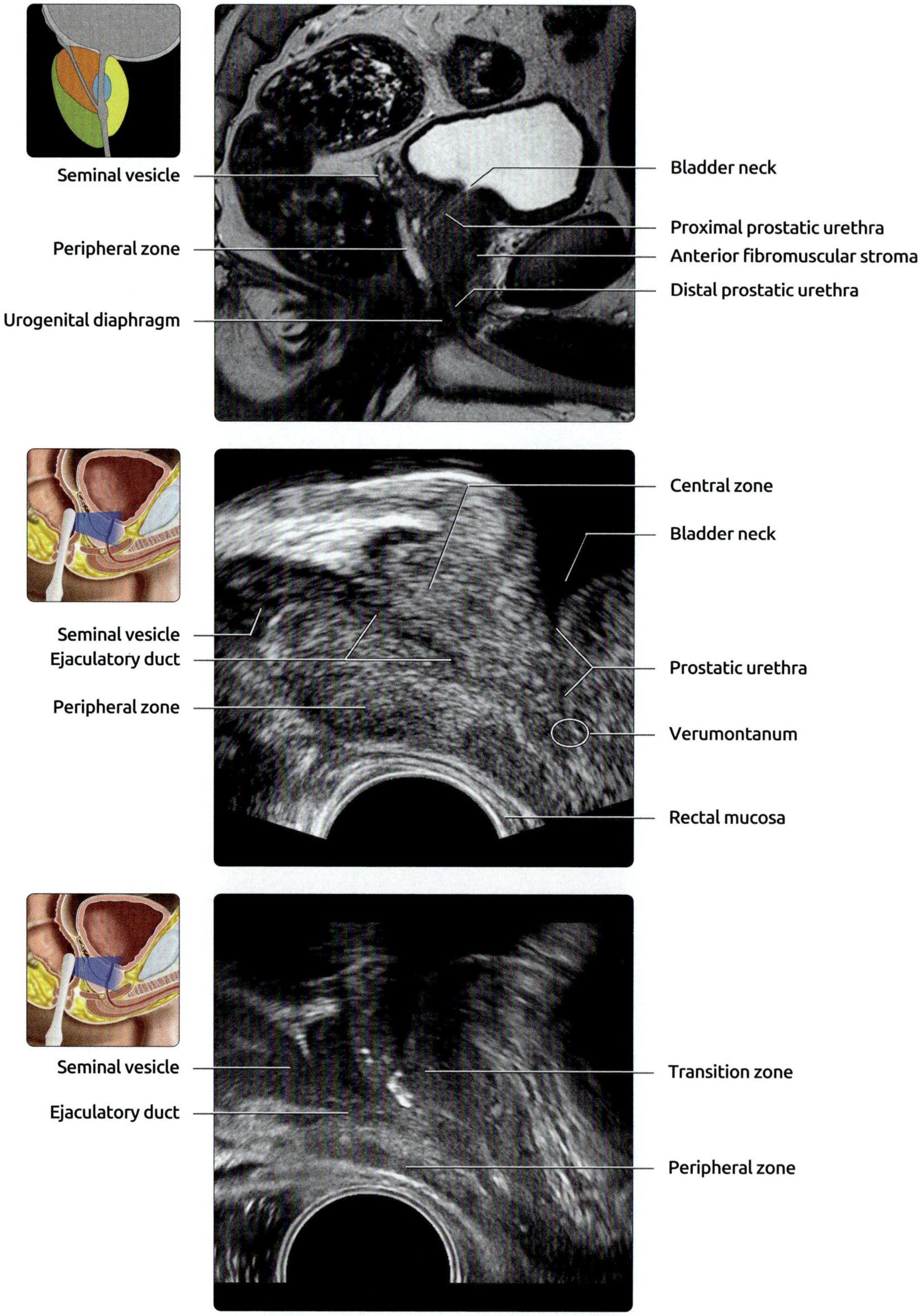

(Top) *Sagittal T2 MR shows the bladder neck and proximal prostatic urethra, which has a posterior course. It makes a 35° anterior angulation at the level of the verumontanum.* **(Middle)** *Sagittal TRUS in a patient without significant BPH shows the course of the proximal prostatic urethra anterior to the CZ. The ejaculatory ducts start at the base of the prostate and run forward and downward through the CZ. They terminate at the verumontanum on either side of the utricle orifice. The verumontanum is the boundary between the proximal and distal segments of the prostatic urethra.* **(Bottom)** *Parasagittal TRUS shows the ejaculatory duct emerging after fusion of the vas deferens and seminal vesicle, entering the prostate base. The ejaculatory ducts are surrounded by the CZ, which is not readily distinguishable on TRUS especially in the setting of BPH. Calcifications are seen within the more anterior TZ. The PZ runs along the posterolateral aspect of the prostate.*

SEMINAL VESICLES AND VAS DEFERENS

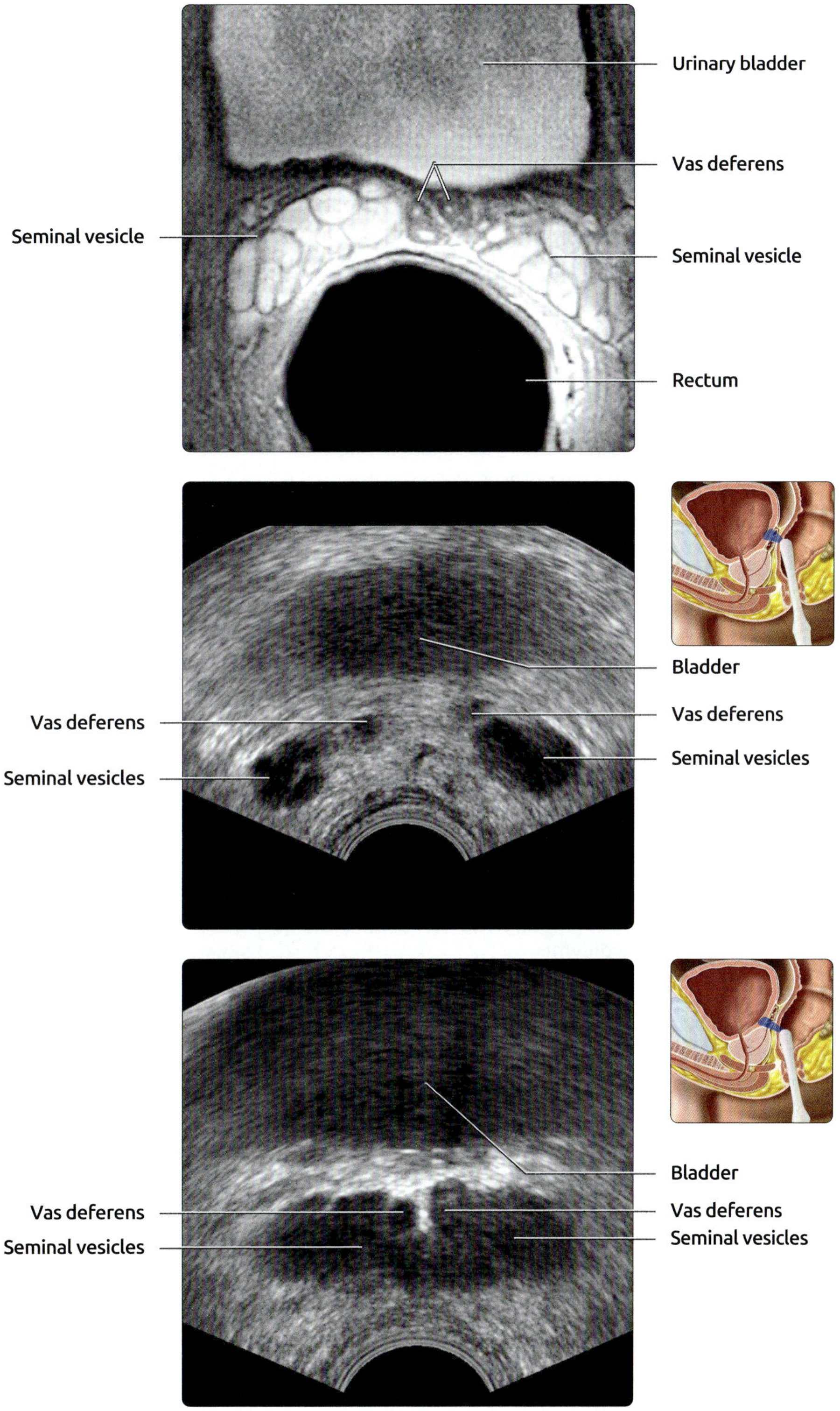

(Top) *Axial T2 MR shows the seminal vesicles between the urinary bladder and the rectum. The low-signal walls of the vas deferens should not be confused with tumor extension from prostate carcinoma.* **(Middle)** *Transverse TRUS at the level of the seminal vesicles shows bilateral seminal vesicles and the vas deferens.* **(Bottom)** *Transverse TRUS shows the vas deferens converging with the seminal vesicles. Their union will form the ejaculatory ducts, which enter the prostate base and course within the prostate enclosed within the CZ. The ejaculatory ducts empty into the urethra at the verumontanum.*

GROSS ANATOMY

Testis

- Densely packed seminiferous tubules separated by thin fibrous septa
  - 200-300 lobules in adult testis
  - Each has 400-600 seminiferous tubules
  - Total length of seminiferous tubules 300-980 meters
- **Seminiferous tubules** converge posteriorly to form larger ducts (tubuli recti)
  - Drain into rete testis at testicular hilum
- **Rete testis** converges posteriorly to form 15-20 efferent ductules
  - Penetrate posterior tunica albuginea at mediastinum to form head of epididymis
- **Tunica albuginea** forms thick fibrous capsule around testis
- **Mediastinum testis** is thickened area of tunica albuginea where ducts, nerves, and vessels enter and exit testis
- **Testicular appendage (appendix testis)**
  - Small, nodular protuberance from surface of testis
  - Remnant of müllerian system

Epididymis

- Crescent-shaped structure running along posterior border of testis
- Efferent ductules form head (**globus major**)
  - Unite to form single, long, highly convoluted tubule in body of epididymis
- Tubule continues inferiorly to form epididymal tail (**globus minor**)
  - Attached to lower pole of testis by loose areolar tissue
- Tubule emerges at acute angle from tail as **vas deferens** (a.k.a. ductus deferens)
  - Continues cephalad within spermatic cord
  - Eventually merges with duct of seminal vesicle to form ejaculatory duct
- **Epididymal appendage (appendix epididymis)**
  - Small nodular protuberance from surface of epididymis
  - Remnant of wolffian system

Spermatic Cord

- Contains vas deferens, nerves, lymphatics, and connective tissue
- Begins at internal (deep) inguinal ring and exits through external (superficial) inguinal ring into scrotum
- Arteries
  - **Testicular artery**
    - Branch of aorta
    - Primary blood supply to testis
  - **Deferential artery**
    - Branch of inferior or superior vesicle artery
    - Arterial supply to vas deferens
  - **Cremasteric artery**
    - Branch of inferior epigastric artery
    - Supplies muscular components of cord and skin
- Venous drainage
  - **Pampiniform plexus**
    - Interconnected network of small veins
    - Merges to form testicular vein
    - Left testicular vein drains to left renal vein
    - Right testicular vein drains to inferior vena cava
- **Lymphatic drainage**
  - Testis follows venous drainage
    - Right side drains to interaortocaval chain
    - Left side drains to left paraaortic nodes near renal hilum
  - Epididymis may also drain to external iliac nodes
  - Scrotal skin drains to inguinal nodes

EMBRYOLOGY

Testis

- Testis develop from genital ridges, which extend from T6-S2 in embryo
- **Composed of 3 cell lines**
  - **Germ cells**
    - Form in wall of yolk sac and migrate along hindgut to genital ridges
    - Form spermatogenic cells in mature testes
  - **Sertoli cells**
    - Supporting network for developing spermatozoa
    - Form tight junctions (**blood-testis barrier**)
    - Secrete müllerian-inhibiting factor
      - Causes paramesonephric (müllerian) ducts to regress
      - Embryologic remnant may remain as appendix testis
  - **Leydig cells**
    - Principal source of testosterone production
    - Lies within interstitium
    - Causes differentiation of mesonephric duct (wolffian) ducts
      - Each duct forms epididymis, vas deferens, seminal vesicle, ejaculatory duct
      - Embryologic remnant may remain as appendix epididymis
- **Scrotum** derived from labioscrotal folds
  - Folds swell under influence of testosterone to form twin scrotal sacs
    - Point of fusion is median raphe, which extends from anus, along perineum, to ventral surface of penis
  - **Processus vaginalis**, sock-like evagination of peritoneum, elongates through abdominal wall into twin sacs
    - Aids in descent of testes, along with **gubernaculum** (ligamentous cord extending from testis to labioscrotal fold)
    - Results in component layers of adult scrotum
- **Testicular descent**
  - Between 7-12th week of gestation, testes descend into pelvis
    - Remain near internal inguinal ring until 7th month, when they begin descent through inguinal canal into twin scrotal sacs
    - Testes remain retroperitoneal throughout descent
    - Testes intimately associated with posterior wall of processus vaginalis
  - Component layers of spermatic cord and scrotum form during descent through abdominal wall
    - Transversalis fascia → **internal spermatic fascia**

- □ Transversus abdominis muscle is discontinuous inferiorly and does not contribute to formation of scrotum
  - – Internal oblique muscle → **cremasteric muscle and fascia**
  - – External oblique muscle → **external spermatic fascia**
  - – **Dartos muscle and fascia** embedded in loose areolar tissue below skin
- ○ Processus vaginalis closes and forms **tunica vaginalis**
  - – Mesothelial-lined sac around anterior and lateral sides of testis
  - – Visceral layer of tunica vaginalis blends imperceptibly with tunica albuginea

ANATOMY-BASED IMAGING ISSUES

Imaging Recommendations

- Palpate scrotal contents and take history prior to US examination
- High-frequency (10- to 15-MHz) linear transducer
- Patient in supine position
  - ○ Penis positioned on anterior abdominal wall
  - ○ Towel draped over thighs to elevate scrotum
  - ○ Additional positions with patient upright or with patient performing Valsalva maneuver
    - – Important for evaluation of inguinal hernia or varicocele

IMAGING ANATOMY

Sonographic Anatomy

- Testes
  - ○ Ovoid, homogeneous, medium-level, granular echotexture
  - ○ **Mediastinum testis** may appear as prominent echogenic line emanating from posterior testis
  - ○ **Blood flow**
    - – Testicular artery pierces tunica albuginea and arborizes over periphery of testis
    - – Multiple, radially arranged vessels travel along septa
    - – May have prominent **transmediastinal artery**
    - – **Tunica vasculosa**: Vascular plexus in periphery of testis, beneath tunica albuginea
    - – Low-velocity, low-resistance waveform on Doppler imaging with continuous forward flow in diastole
- **Epididymis**
  - ○ Isoechoic to slightly hyperechoic compared with testis
  - ○ Best seen in longitudinal plane
  - ○ Head has rounded or triangular configuration
  - ○ Head 10-12 mm, body and tail often difficult to visualize
    - – May be helpful to follow course of epididymis in transverse plane if difficult to visualize in longitudinal plane
- **Spermatic cord**
  - ○ Scan along course of inguinal canal
    - – May be difficult to differentiate from surrounding soft tissues
  - ○ Slow flow in pampiniform plexus may make identification on color Doppler difficult
    - – Use provocative maneuvers (Valsalva, standing), especially when looking for varicocele

CLINICAL IMPLICATIONS

Hydrocele

- Fluid between visceral and parietal layers of tunica vaginalis
- Small amount of fluid is normal
- Larger hydroceles may be either congenital (patent processus vaginalis) or acquired

Cryptorchidism

- Failure of testes to descend completely into scrotum
- Most lie near external inguinal ring
- Associated with decreased fertility and testicular carcinoma
  - ○ Risk of carcinoma is increased for both testes, even if other side is normally descended

Varicocele

- Idiopathic or secondary to abdominal mass
  - ○ Idiopathic more common on left
- Vessel diameter > 3 mm abnormal

Dilated Rete Testes

- Clusters of dilated tubules in mediastinum testis
  - ○ Do not confuse with mass
- Empty into epididymis
- Often associated with epididymal cysts

Scrotal Calculi

- Free-floating calcifications within tunica vaginalis
- May result from torsion of appendix testis or epididymis

Torsion

- Occurs most commonly when tunica vaginalis completely surrounds testis and epididymis
  - ○ Testis is suspended from spermatic cord (like **bell clapper**) rather than being anchored posteriorly
- Normal grayscale appearance with early torsion
  - ○ Becomes heterogeneous and enlarged with infarction
- Color and spectral Doppler required for diagnosis
  - ○ Some flow may be seen even if torsed but will be decreased compared to normal side
  - ○ Venous flow compromised 1st, then diastolic flow, and finally systolic flow
- Look in inguinal canal for twist in spermatic cord (**whirlpool sign**)

Testicular Carcinoma

- Most common malignancy in young men
  - ○ 95% are germ cell tumors
    - – Seminoma (most common pure tumor), embryonal, yolk sac tumor, choriocarcinoma, teratoma
    - – Mixed germ cell tumor (components of 2 or more cell lines) most common overall
  - ○ Remainder of primary tumors are sex cord (Sertoli cells) or stromal (Leydig cells)
  - ○ Lymphoma, leukemia, and metastases more common in older men
- Most metastasize via lymphatics in predictable fashion
  - ○ Right-sided 1st echelon nodes: Interaortocaval chain at 2nd vertebral body
  - ○ Left-sided 1st echelon nodes: Left paraaortic nodes in area bounded by renal vein, aorta, ureter, and inferior mesenteric artery

TESTIS AND EPIDIDYMIS

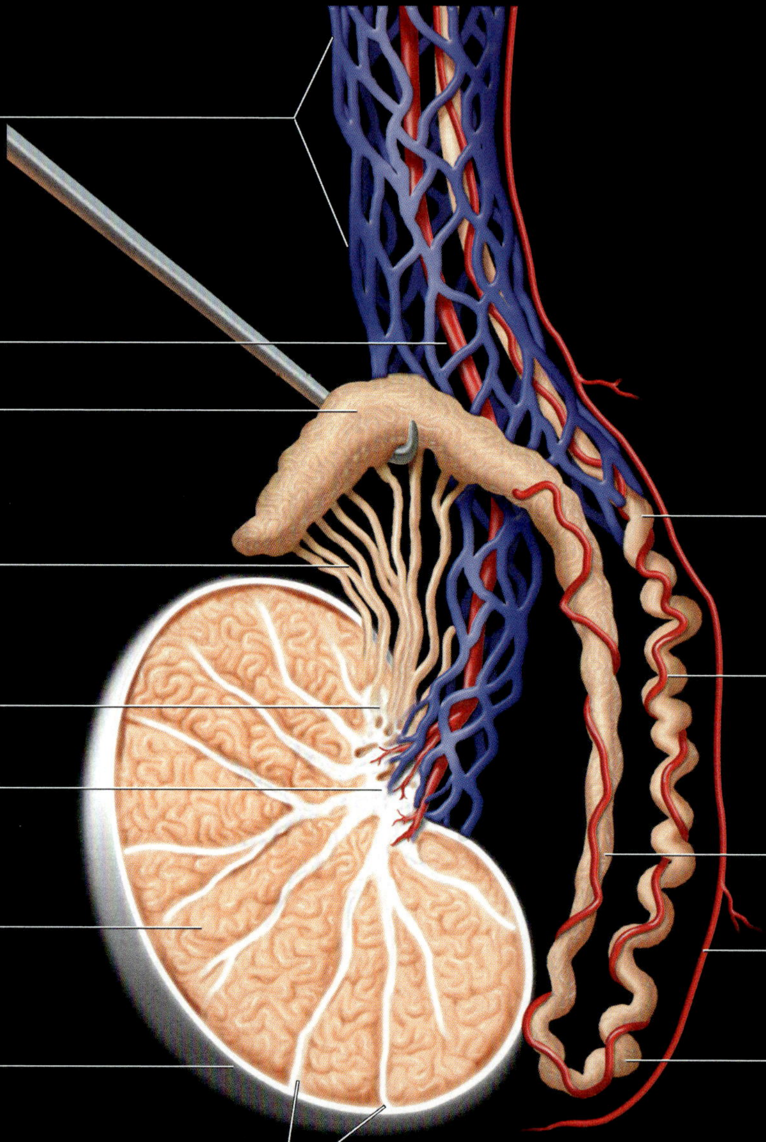

Graphic shows the testis is composed of densely packed seminiferous tubules, which are separated by thin fibrous septa. These tubules converge posteriorly, eventually draining into the rete testis. The rete testis continues to converge to form the efferent ductules, which pierce through the tunica albuginea at the mediastinum testis and form the head of the epididymis. Within the epididymis these tubules unite to form a single, highly convoluted tubule in the body, which finally emerges from the tail as the vas deferens. In addition to the vas deferens, other components of the spermatic cord include the testicular artery, deferential artery, cremasteric artery, pampiniform plexus, lymphatics, and nerves.

EPIDIDYMIS AND SCROTAL WALL LAYERS IN SITU

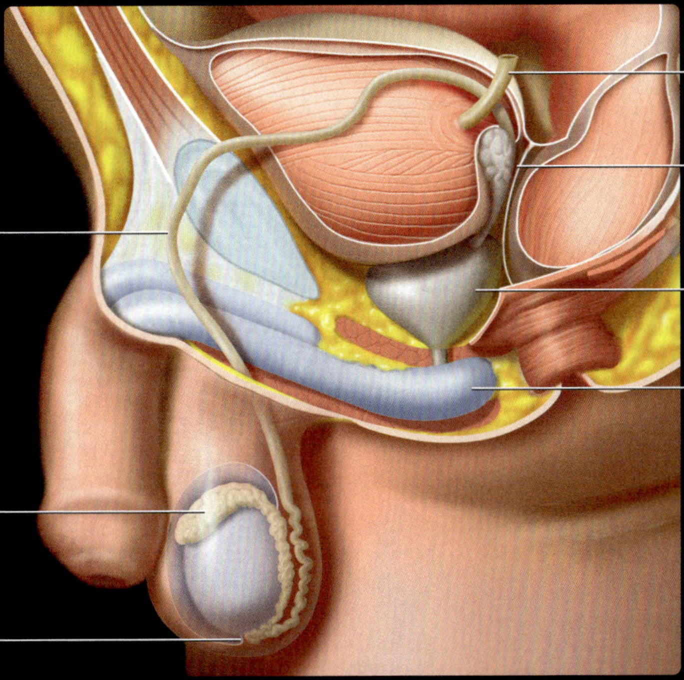

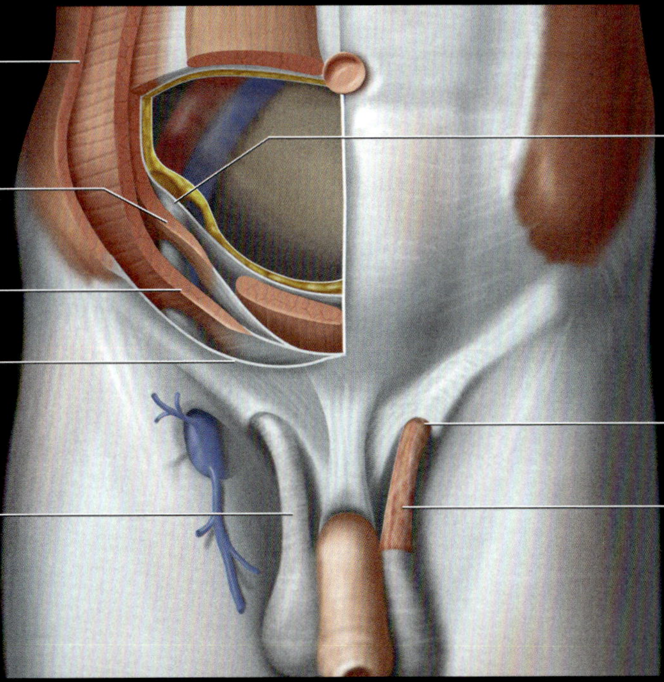

(Top) *Graphic shows that the tail of the epididymis is loosely attached to the lower pole of the testis by areolar tissue. The vas deferens (also referred to as the ductus deferens) emerges from the tail at an acute angle and continues cephalad as part of the spermatic cord. After passing through the inguinal canal, the vas deferens courses posteriorly to unite with the duct of the seminal vesicle to form the ejaculatory duct. These narrow ducts have thick, muscular walls composed of smooth muscle, which reflexly contract during ejaculation and propel sperm forward.* **(Bottom)** *The muscle layers of the pelvic wall have been separated to show the spermatic cord as it passes through the inguinal canal. The cremasteric muscle is derived from the internal oblique muscle, while the external spermatic fascia is formed by the fascia of the external oblique muscle.*

SCROTAL DEVELOPMENT

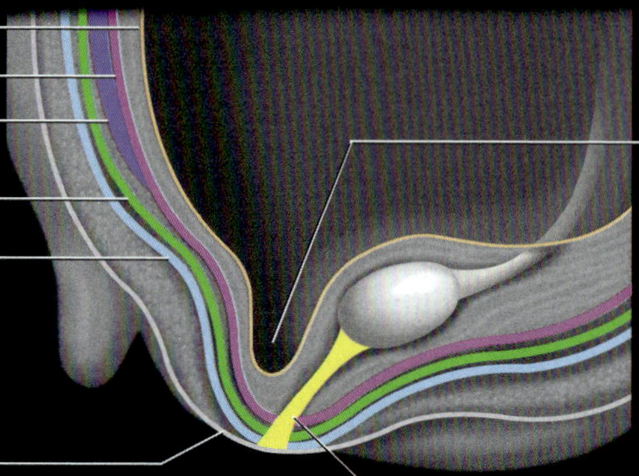

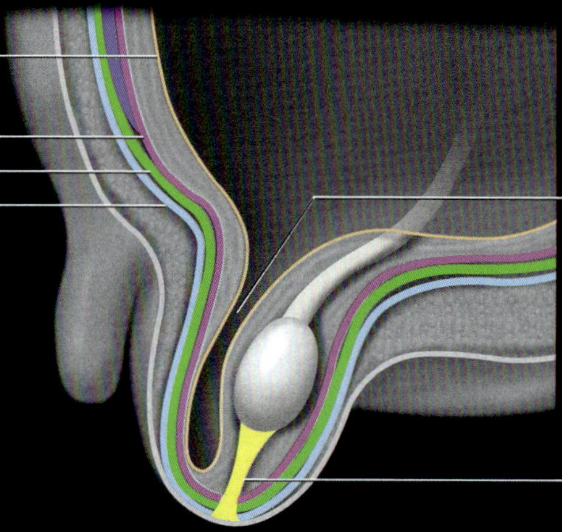

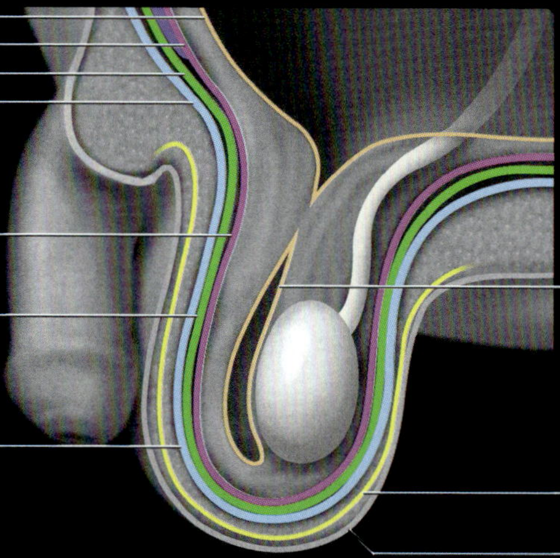

(Top) *The processus vaginalis is a sock-like evagination of the peritoneum, which elongates caudally through the abdominal wall. It forms just anterior to the developing testes and, along with the gubernaculum (a ligamentous cord extending from the testis to the labioscrotal fold), aids in their descent.* **(Middle)** *As the processus vaginalis evaginates, it becomes ensheathed by fascial extensions of the abdominal wall, which ultimately form the layers of the scrotum and spermatic cord. The transversus abdominis muscle is discontinuous inferiorly and does not contribute to the formation of the scrotum.* **(Bottom)** *The abdominal wall derivative layers of the scrotum are as follows: Transversalis fascia → internal spermatic fascia, internal oblique muscle → cremasteric muscle and fascia, external oblique muscle → external spermatic fascia. The dartos muscle is embedded in the loose areolar tissue and is closely associated with the skin. The various layers of the scrotum cannot usually be discerned with imaging. The superior portion of the processus vaginalis closes and forms an isolated mesothelial-lined sac, the tunica vaginalis.*

SCROTUM AND TESTES

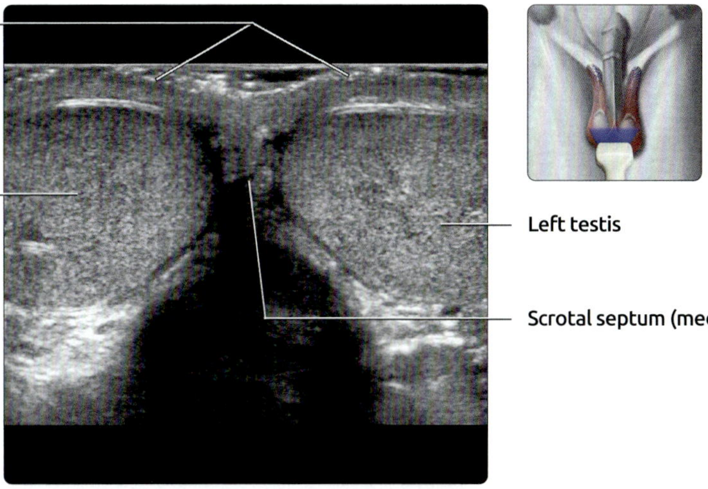

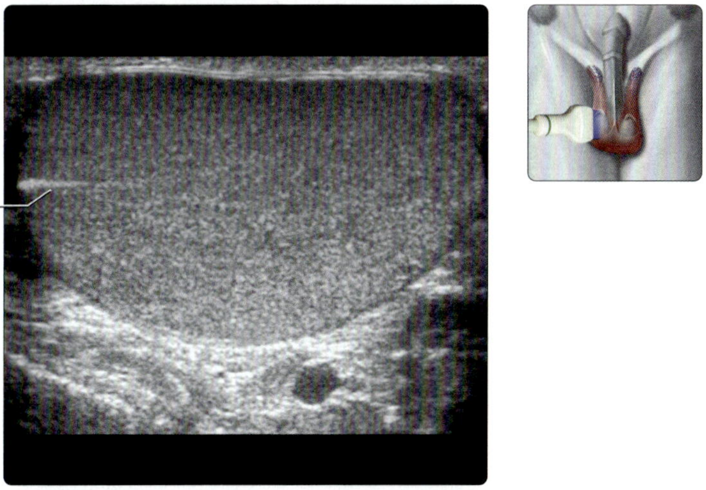

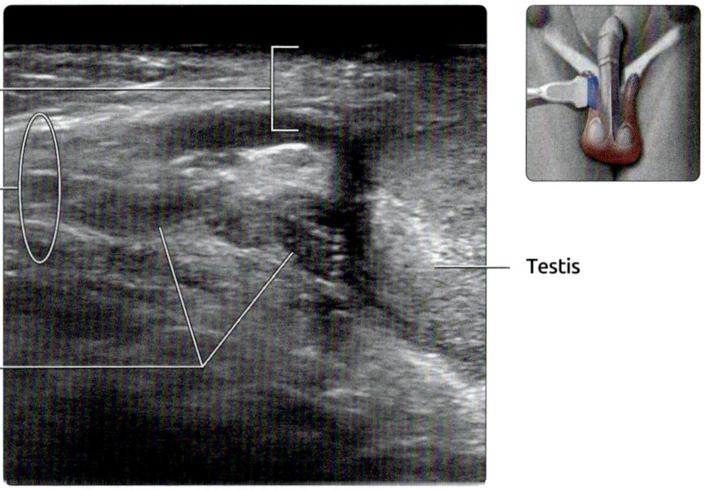

(Top) *Transverse grayscale ultrasound shows both testes. This is a useful approach for comparing the appearance of the testes, which should have similar, homogeneous, medium-level, granular echotexture.* **(Middle)** *Longitudinal ultrasound shows the ovoid shape. The tunica albuginea may form an echogenic linear band where it invaginates at the mediastinum testis. The mediastinum testis has a craniocaudal linear course and is where the efferent ductules, vessels, and lymphatics pierce through the capsule.* **(Bottom)** *Longitudinal scan at the superior pole of the testis shows the spermatic cord as it is entering the inguinal canal. It is important to look at the spermatic cord for a twist (whirlpool sign) in cases of suspected torsion.*

TUNICA VAGINALIS, ALBUGINEA, AND VASCULOSA

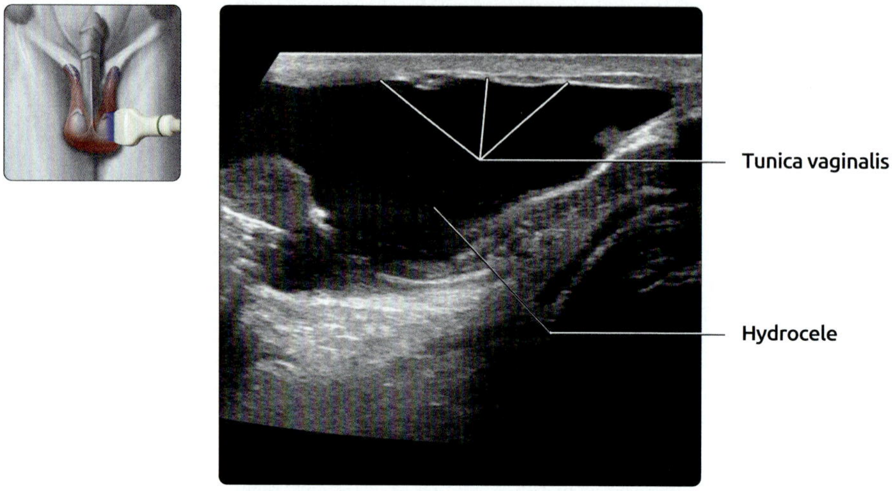

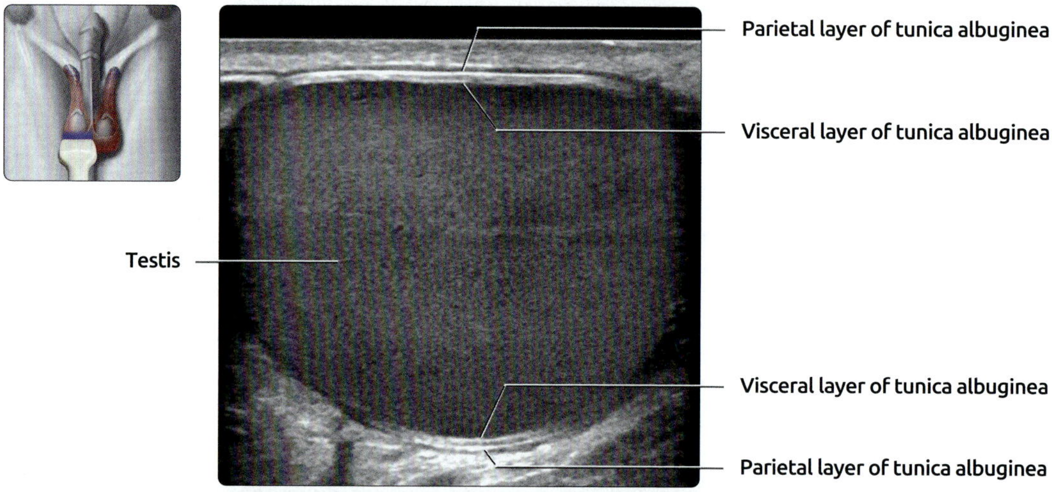

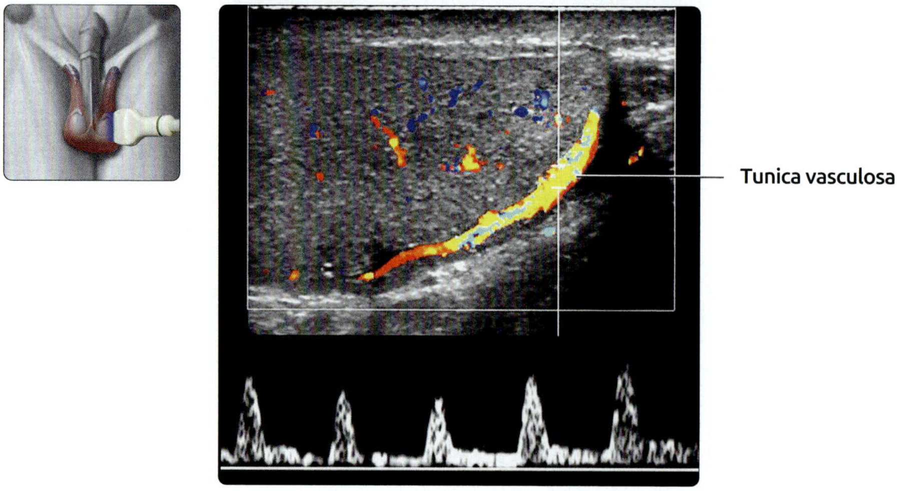

(Top) *Sagittal grayscale ultrasound of the left scrotum shows the outermost serous membrane covering the testis and epididymis, the tunica vaginalis. There is small fluid within the tunica vaginalis (hydrocele).* **(Middle)** *Transverse grayscale ultrasound of the right testis shows 2 thin echogenic layers covering the testis, representing the parietal and visceral layers of tunica albuginea, which is the fibrous covering of the testis.* **(Bottom)** *Sagittal color Doppler ultrasound shows the vascular plexus (tunica vasculosa) along the outermost layer of the testis, just beneath the tunica albuginea.*

SCROTAL DOPPLER

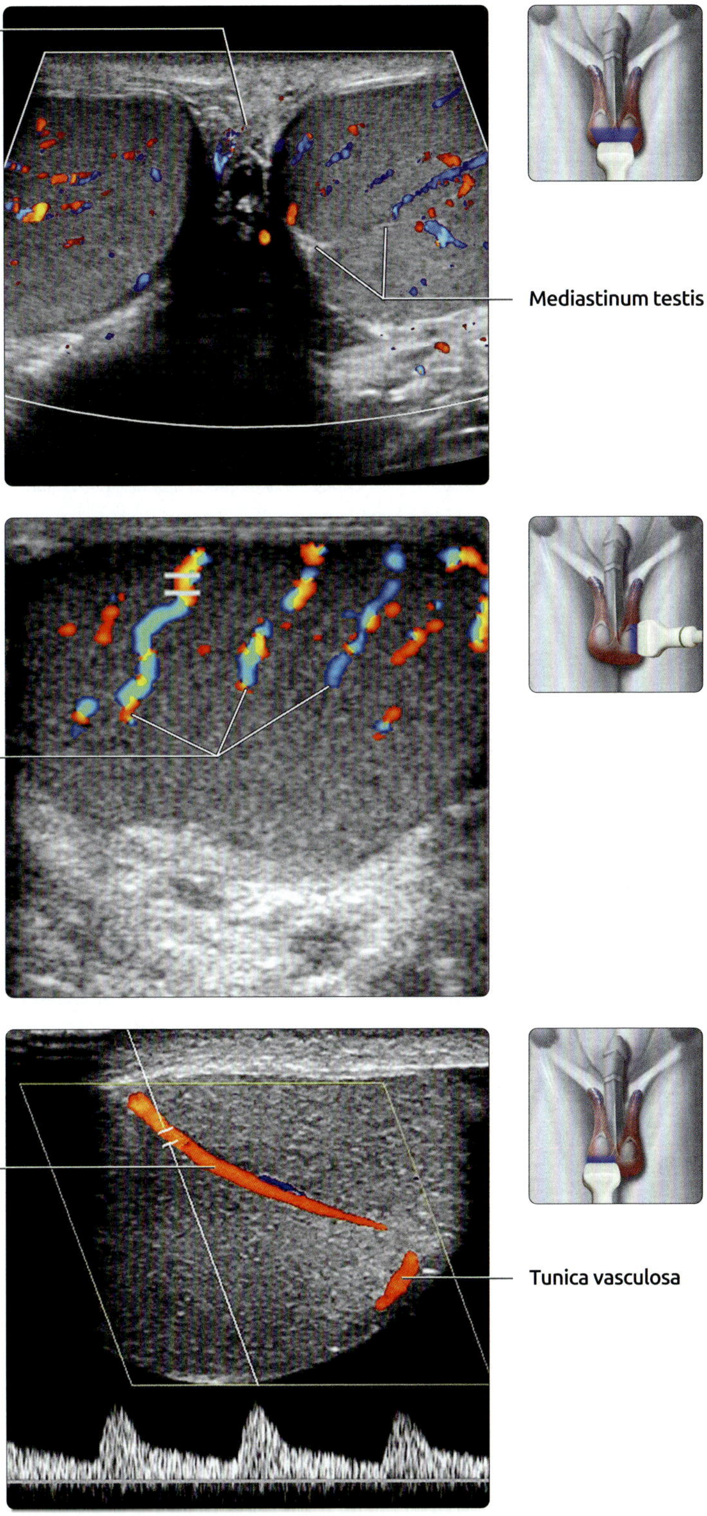

(Top) *Transverse color Doppler ultrasound shows normal symmetric blood flow in the testes. It is important to compare flow between testes to determine if the symptomatic side has increased or decreased flow when compared to the asymptomatic side.* **(Middle)** *Longitudinal color Doppler ultrasound shows prominent, radially arranged vessels within the testis.* **(Bottom)** *Color Doppler ultrasound shows a prominent transmediastinal artery with pulsed Doppler showing normal low-resistance arterial flow. Also shown is the tunica vasculosa, a vascular plexus that runs beneath the tunica albuginea.*

SCROTAL DOPPLER

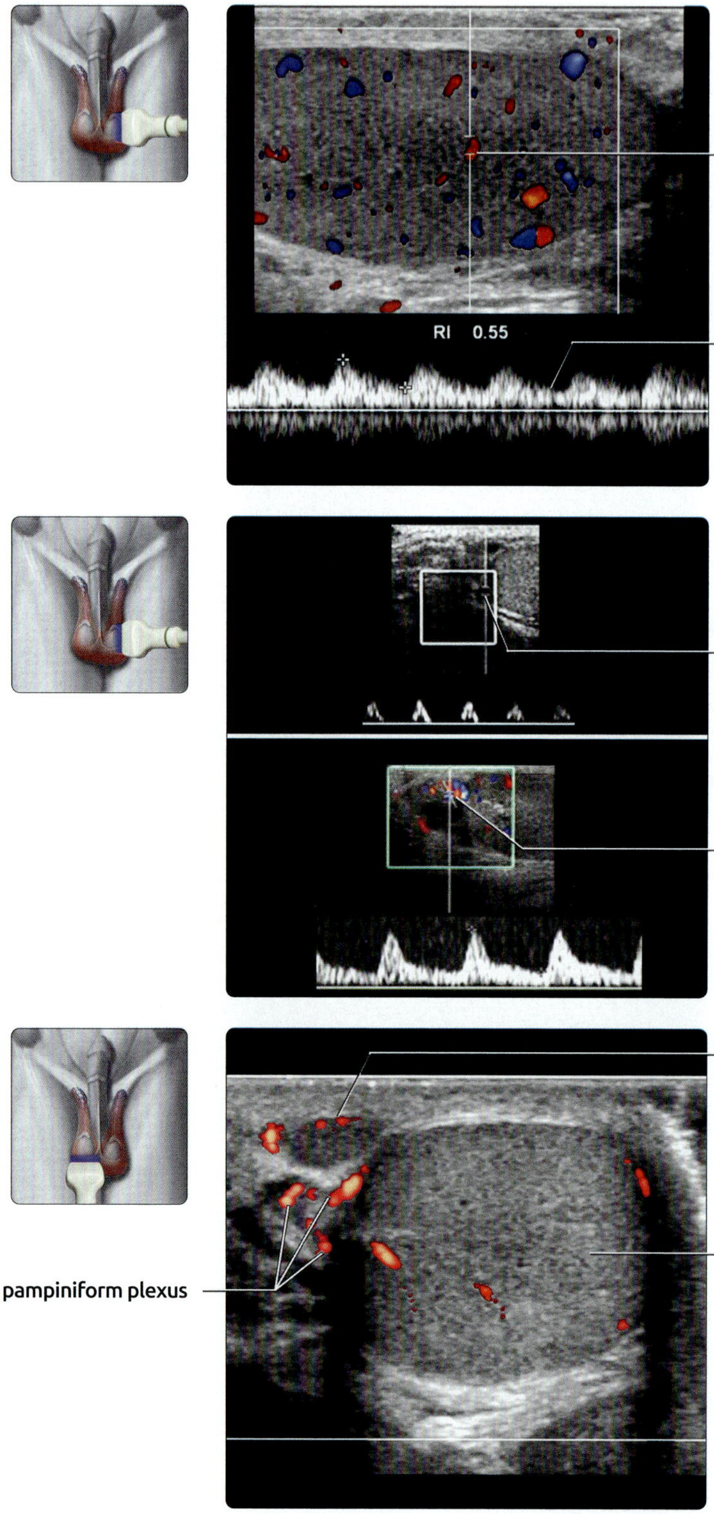

(Top) *Sagittal color Doppler ultrasound of a normal left testis shows normal blood flow with a normal spectral waveform of an intratesticular artery. The artery should have a low-resistance waveform and the resistive index (RI) should be between 0.48-0.75 (mean RI: 0.62).* **(Middle)** *Two color Doppler ultrasounds show the epididymal arterial supply. The upper image demonstrates the normal cremasteric artery with a low-flow, high-resistance pattern. The lower image shows the normal epididymal artery, a branch of the testicular artery with a low-resistance waveform.* **(Bottom)** *Transverse power Doppler ultrasound shows flow within the pampiniform plexus of the spermatic cord. If looking for a varicocele, it is important to use provocative maneuvers, including Valsalva and scanning in the standing position, to distend the veins.*

EPIDIDYMIS

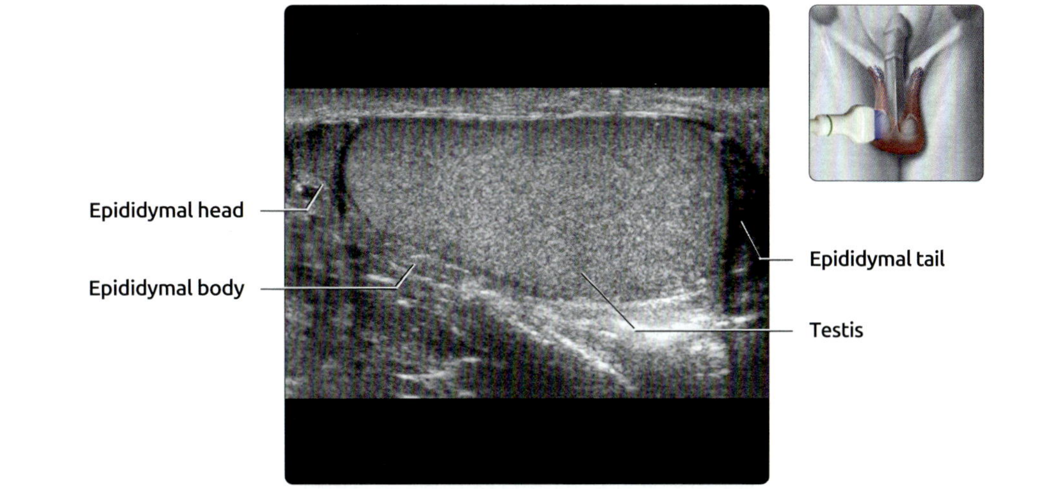

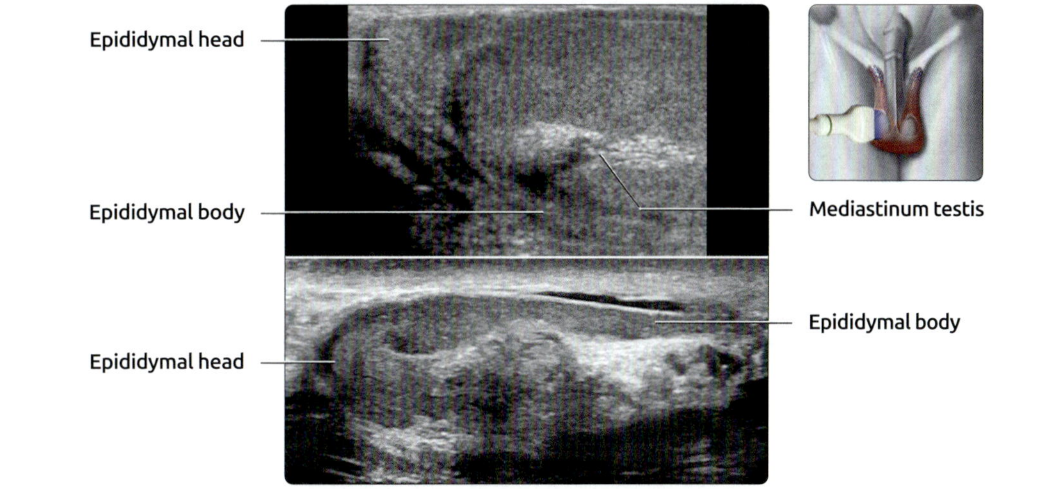

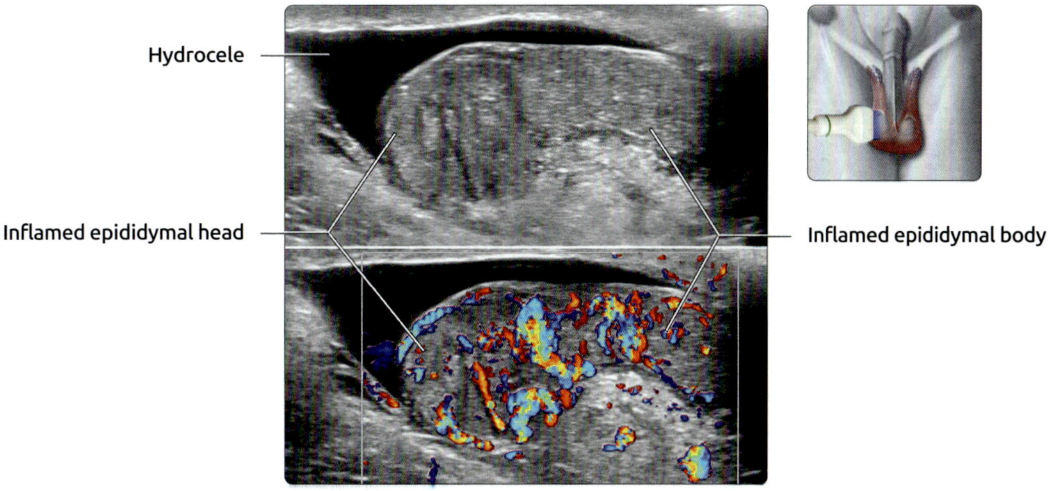

(Top) *Longitudinal ultrasound shows both the epididymal head (globus major) and tail (globus minor). The epididymal head measures ~ 10-12 mm and is iso- to slightly hyperechoic compared to the testis. The body and tail are often more difficult to visualize and may be slightly less echogenic than the head.* **(Middle)** *This composite image shows the normal epididymal head and body. The body may be difficult to visualize from an anterior scanning plane (above), but by moving the transducer more posteriorly (below), it is often better visualized.* **(Bottom)** *This composite image of a longitudinal grayscale ultrasound (above) and a color Doppler ultrasound (below) in a patient with acute epididymitis shows enlargement and hyperemia of both the epididymal head and body.*

RETE TESTIS AND TUBULAR ECTASIA

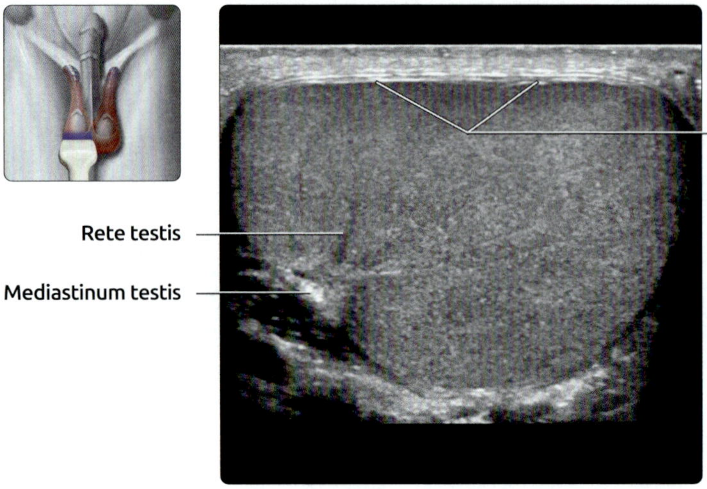

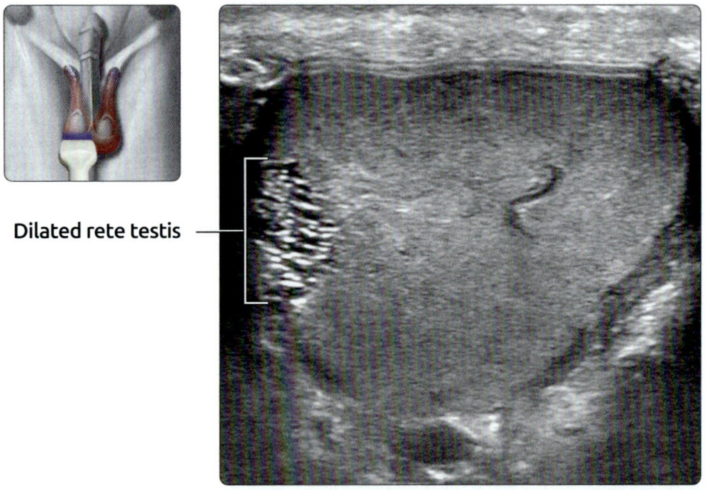

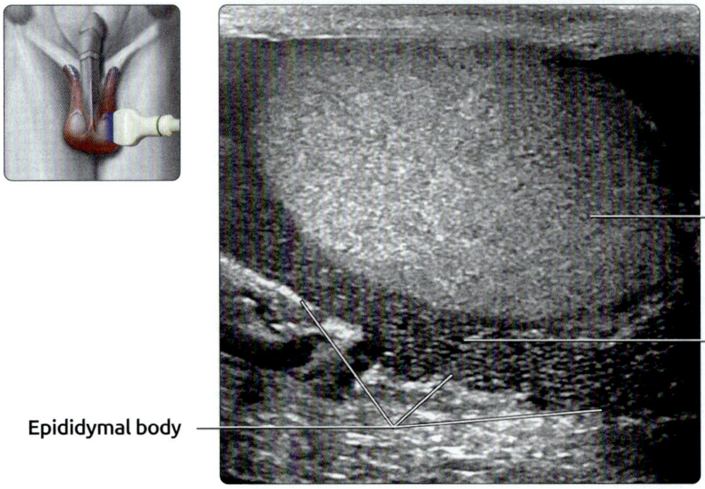

(Top) *Transverse grayscale ultrasound of the normal testis shows diffuse low-level internal echoes in the parenchyma, an echogenic mediastinum testis, and the striated pattern of the rete testis as it converges to the mediastinum. The testis is covered with 2 thin, echogenic layers of tunica albuginea.* **(Middle)** *Transverse ultrasound of the the right testis shows tubular ectasia of the rete testis. Tubular ectasia is located posteriorly by the mediastinum and is frequently bilateral. It may give an impression of a mass, but careful scanning shows the "mass" is actually a series of dilated tubules.* **(Bottom)** *Sagittal grayscale ultrasound of the epididymis demonstrates tubular ectasia within the epididymis.*

TESTICULAR AND EPIDIDYMAL APPENDAGES

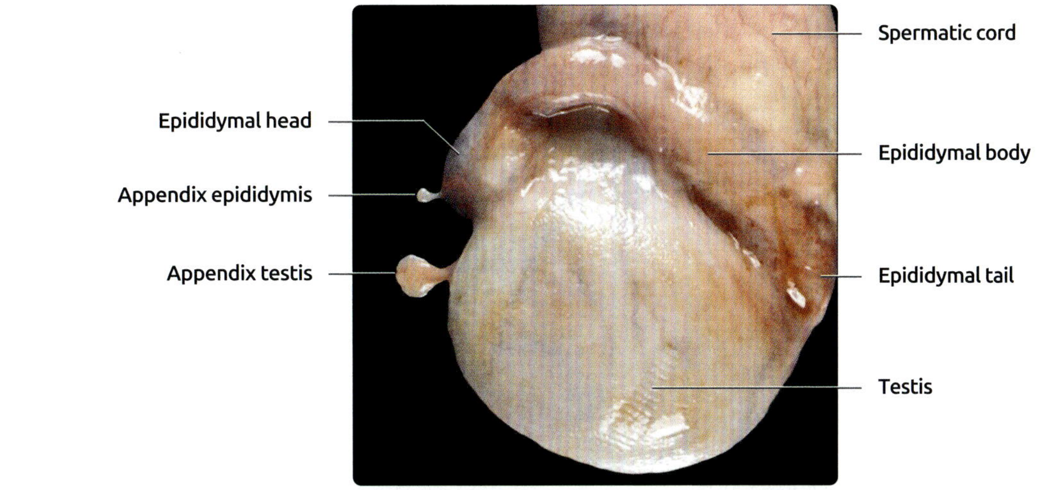

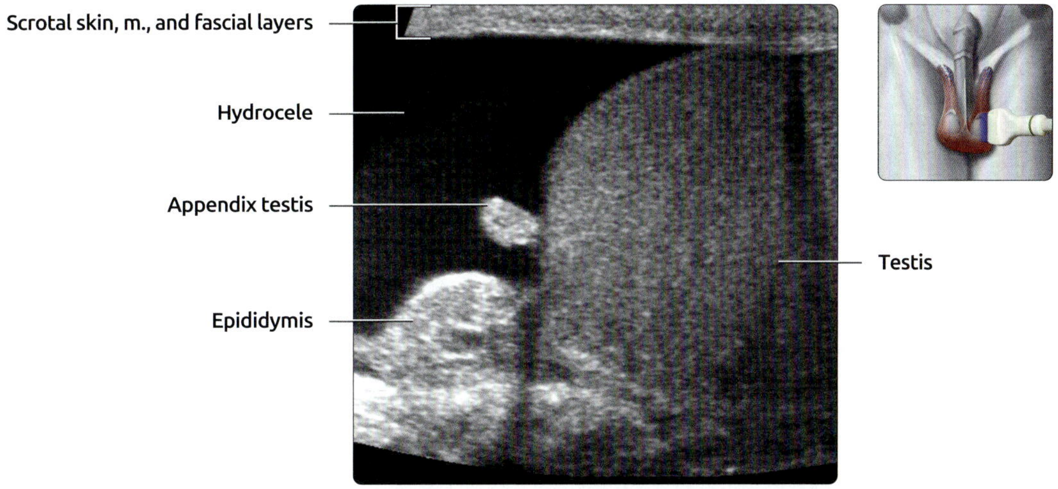

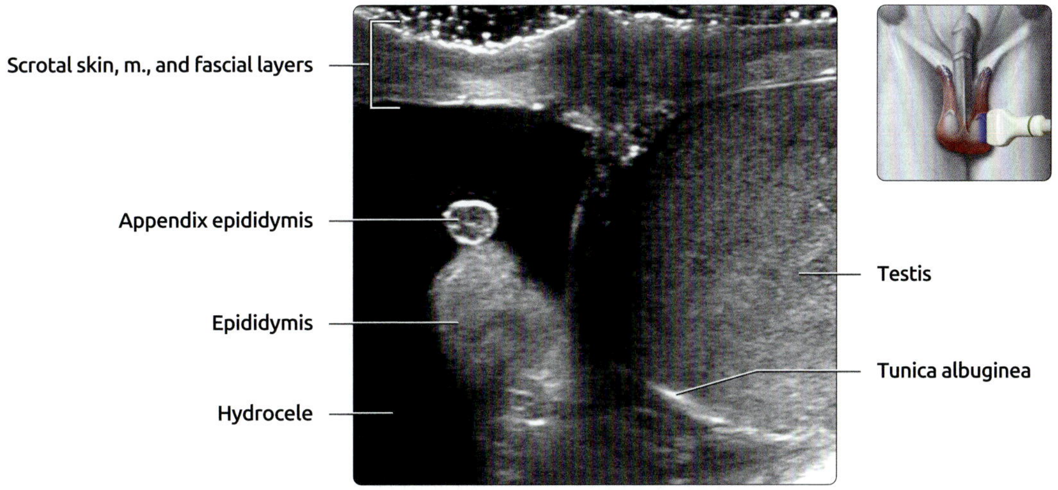

(Top) *Gross specimen shows the the testis, epididymis, and spermatic cord. Two embryologic remnants form small appendages from the testis and epididymis.* **(Middle)** *Ultrasound of the testis in a patient with a hydrocele shows a small, nodular protuberance from the surface of the testis. This is the appendix testis, which is a remnant of the müllerian system.* **(Bottom)** *Longitudinal ultrasound of the upper testis and epididymis shows a small "tag" of tissue projecting from the epididymis. This is an appendix epididymis, which is a remnant of the wolffian system. Both the appendix testis and appendix epididymis are usually not visible sonographically unless there is a hydrocele. They are usually of no clinical significance; however, they can torse and be a cause of scrotal pain.*

IMAGING ANATOMY

Penis

- **Composed of 3 cylindrical shafts**
  - **2 corpora cavernosa**: Main erectile bodies
    - On dorsal surface of penis
    - Diverge at root of penis (**crura**) and are invested by **ischiocavernosus muscles**
    - Chambers traversed by numerous trabeculae, creating sinusoidal spaces
    - Multiple fenestrations between corpora, creating multiple anastomotic channels
  - **1 corpus spongiosum**: Contains urethra
    - On ventral surface, in groove created by corpora cavernosa
    - Becomes **penile bulb** (urethral bulb) at root and is invested by **bulbospongiosus muscle**
    - Forms **glans penis** distally
    - Also erectile tissue but of far less importance
- **Tunica albuginea** forms capsule around each corpora
  - Thinner around spongiosum than cavernosa
- All 3 corpora surrounded by deep fascia (**Buck fascia**) and superficial fascia (**Colles fascia**)
- **Suspensory ligament** of penis (part of fundiform ligament) is inferior extension of abdominal rectus sheath
- Main arterial supply from internal pudendal artery
  - **Cavernosal artery** runs within center of each corpus cavernosum
    - Gives off **helicine arteries**, which fill trabecular spaces
    - Primary source of blood for erectile tissue
  - **Paired dorsal penile arteries** run between tunica albuginea of corpora cavernosa and Buck fascia
    - Supplies glans penis and skin
  - Multiple anastomoses between cavernosal and dorsal penile arteries
- **Venous drainage** of corpora cavernosa
  - Emissary veins in corpora pierce through tunica albuginea → circumflex veins → deep dorsal vein of penis → retropubic venous plexus
  - Superficial dorsal vein drains skin and glans penis
- Primary innervation from terminal branches of internal pudendal nerve

Normal Erectile Function

- Neurologically mediated response eliciting smooth muscle relaxation of cavernosal arteries, helicine arteries, and cavernosal sinusoids
- Blood flows from helicine arteries into sinusoidal spaces
- Sinusoids distend, eventually compressing emissary veins against rigid tunica albuginea
  - Venous compression prevents egress of blood from corpora, which maintains erection

Urethra

- Posterior urethra (prostatic, membranous) and anterior urethra (bulbous and penile)
- **Prostatic urethra**: Traverses prostate
  - **Verumontanum** is 1-cm ovoid mound along ureteral crest (smooth muscle ridge on posterior wall)
  - Prostatic utricle, prostatic ducts, and ejaculatory ducts enter in this segment
- **Membranous urethra**: Short course through urogenital diaphragm (level of external urethral sphincter)
  - Contains bulbourethral glands (**Cowper glands**)
- **Bulbous urethra**: Below urogenital diaphragm to suspensory ligament of penis at penoscrotal junction
- **Penile urethra**: Pendulous portion, distal to suspensory ligament
- Both penile and bulbous urethra lined by mucosal urethral glands (**glands of Littre**)
- **2 points of fixations**: Urogenital diaphragm (membranous urethra) and penoscrotal junction

ANATOMY IMAGING ISSUES

Imaging Recommendations

- Transducer: High-frequency (7.5- to 10-MHz) linear transducer
- Patient is supine with penis lying on anterior abdominal wall
- Transducer placed on ventral side of penis
  - Corpus spongiosum easily compressed so use ample gel and gentle pressure
- For erectile dysfunction studies, vasodilating agent is injected into dorsal 2/3 of shaft
- **Cavernosal artery evaluation**
  - In flaccid state, there is little diastolic flow
  - At onset of erection, there is dilatation with increase in both systolic and diastolic flow
  - At maximum erection, venous drainage is blocked
    - Waveform changes to high resistance with reversal of diastolic flow
    - Peak systolic velocity > 30 cm/sec
    - Cavernosal artery diameter increase > 75%
- Urethra
  - Imaging of anterior urethra is optimal with distension
    - Gel may be injected retrograde or patient may be asked to void during scan
  - Posterior urethra is best imaged by transrectal ultrasound of prostate

CLINICAL IMPLICATIONS

Erectile Dysfunction

- Complex and often multifactorial including vascular, neurogenic, and psychologic factors
- **Arteriogenic impotence** effects inflow
  - Usually internal pudendal and penile arteries
  - Blockage may be as high as distal aorta (**Leriche syndrome**)
- **Venogenic impotence** effects outflow
  - Ineffective venoocclusion with continuous outflow of blood from sinusoids

Trauma

- Need to carefully evaluate tunica albuginea in suspected penile fracture
- Membranous urethra most commonly injured with pelvic trauma and fractures
- Bulbous urethra most often from straddle injury

Peyronie Disease

- Plaque formation on tunica albuginea
- Painful erections with shortening and curvature of penis

PENIS AND PERINEUM

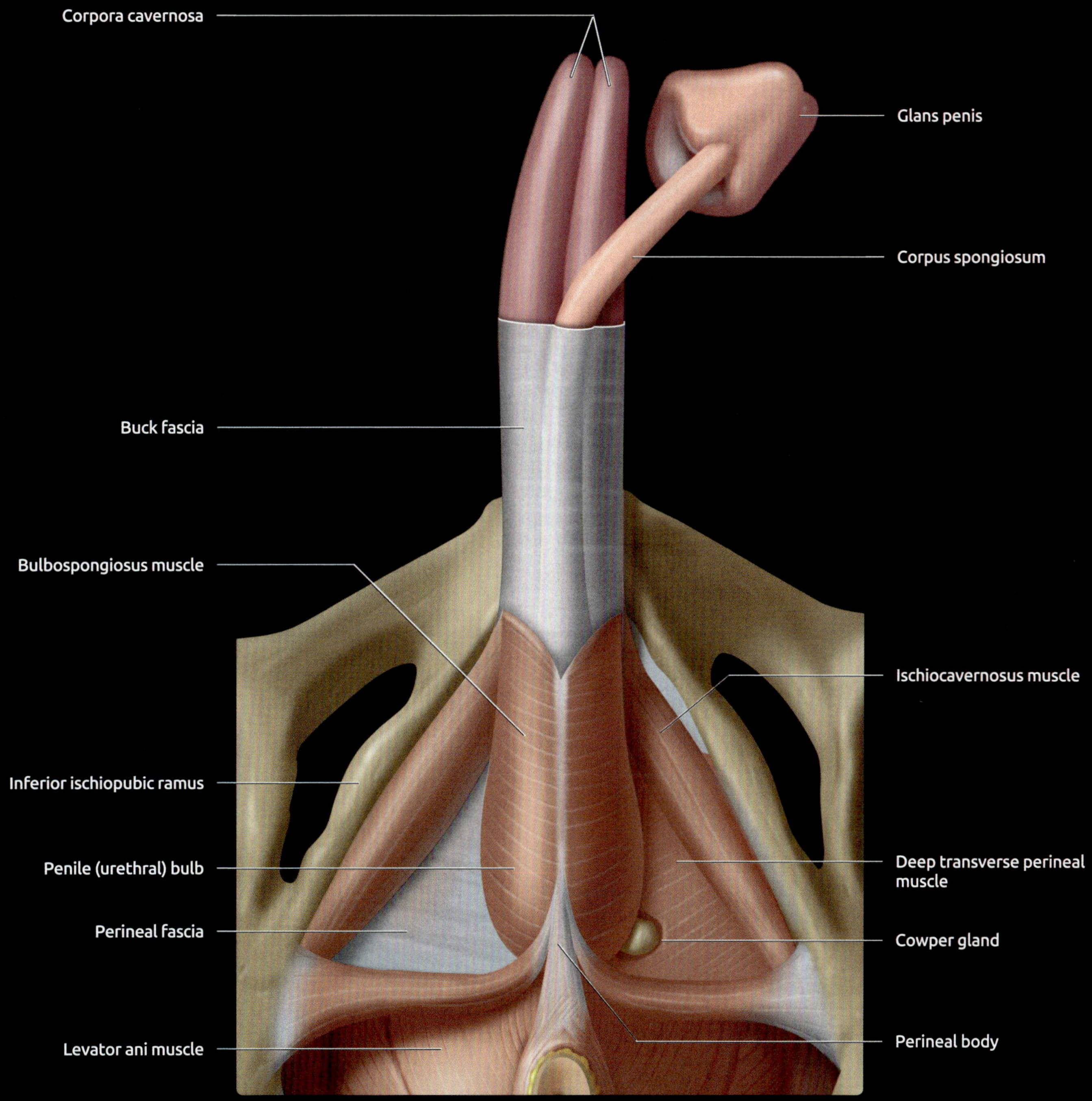

At the root of the penis, the corpora split into a triradiate form. The crura of the corpora cavernosa diverge and each crus is invested by an ischiocavernosus muscle. The corpus spongiosum remains midline and is invested by the bulbospongiosus muscle. It widens at its attachment to the urogenital diaphragm and is called the penile or urethral bulb. The root of the penis is firmly anchored between the urogenital diaphragm above and Colles fascia below (cut away to show perineal muscles). Additional supports include fascial attachments to the medial surface of the pubic rami and and pubic symphysis.

PENIS AND URETHRA

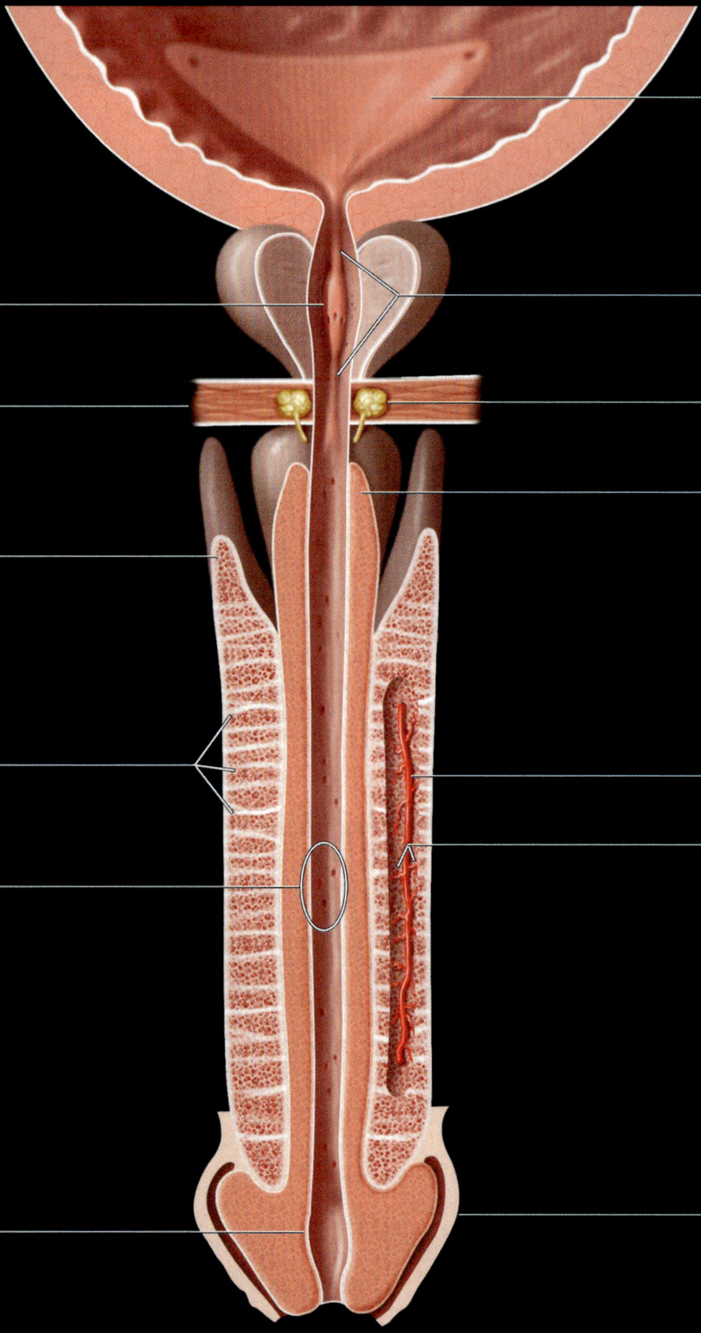

Graphic shows the posterior wall of the urethra from the bladder base to the fossa navicularis. The verumontanum is the most prominent portion of the urethral crest. The prostatic utricle (an embryologic remnant of the müllerian system) enters in the center of the verumontanum, along with the ejaculatory ducts, which are just distal on either side. Cowper glands are within the urogenital diaphragm, but their ducts course distally ~ 2 cm to enter the bulbous urethra. Multiple, small, mucosal glands (glands of Littre) line the mucosa of the anterior urethra. The corpora cavernosa are the primary erectile bodies. They are traversed by numerous trabeculae, creating sinusoidal spaces. The cavernosal artery runs within the center of each corpus cavernosum and gives rise to the helicine arteries, which flow into the sinusoids.

PENIS AND URETHRA

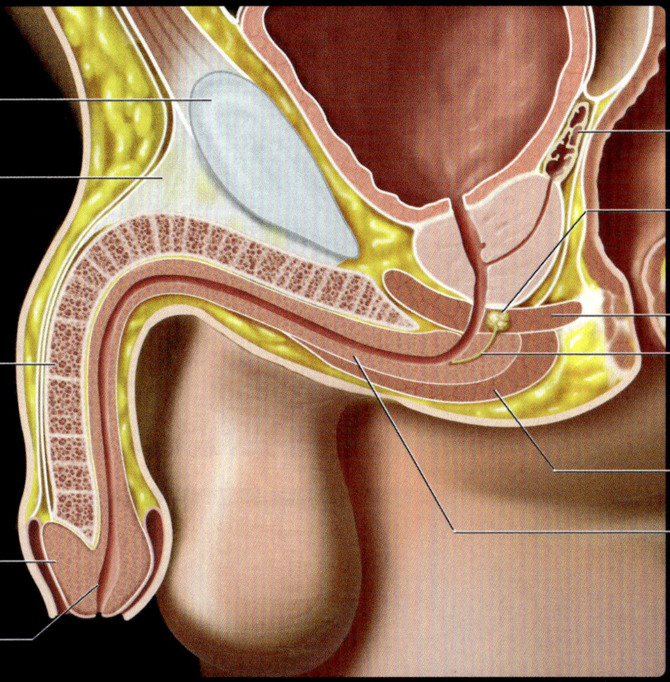

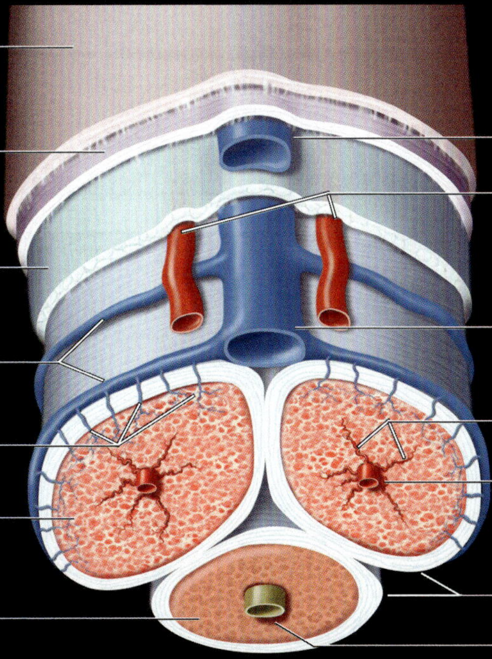

(Top) *Sagittal midline graphic shows the course of the urethra. The anterior urethra runs within the corpus spongiosum. The external (pendulous) portion of the penis begins below the pubic symphysis at the penoscrotal junction. It is fixed at this position by the suspensory ligament of the penis, which is part of the fundiform ligament, an inferior extension of the rectus sheath.* **(Bottom)** *Cross section of the penis viewed from above shows the 3 corpora, each of which is invested by a tunica albuginea. The 3 corpora are surrounded by Buck fascia. The cavernosal arteries run within the corpora cavernosa and give rise to the helicine arteries, which fill the sinusoids. Normal drainage is via the emissary veins → circumflex veins → deep dorsal vein of penis. With an erection, the emissary veins are compressed against the strong tunica albuginea, occluding venous drainage. The dorsal penile arteries and deep dorsal vein run beneath the Buck fascia. The penile arteries supply the glans and skin, but there are significant anastomoses between both the arterial and venous systems.*

NORMAL ERECTILE FUNCTION

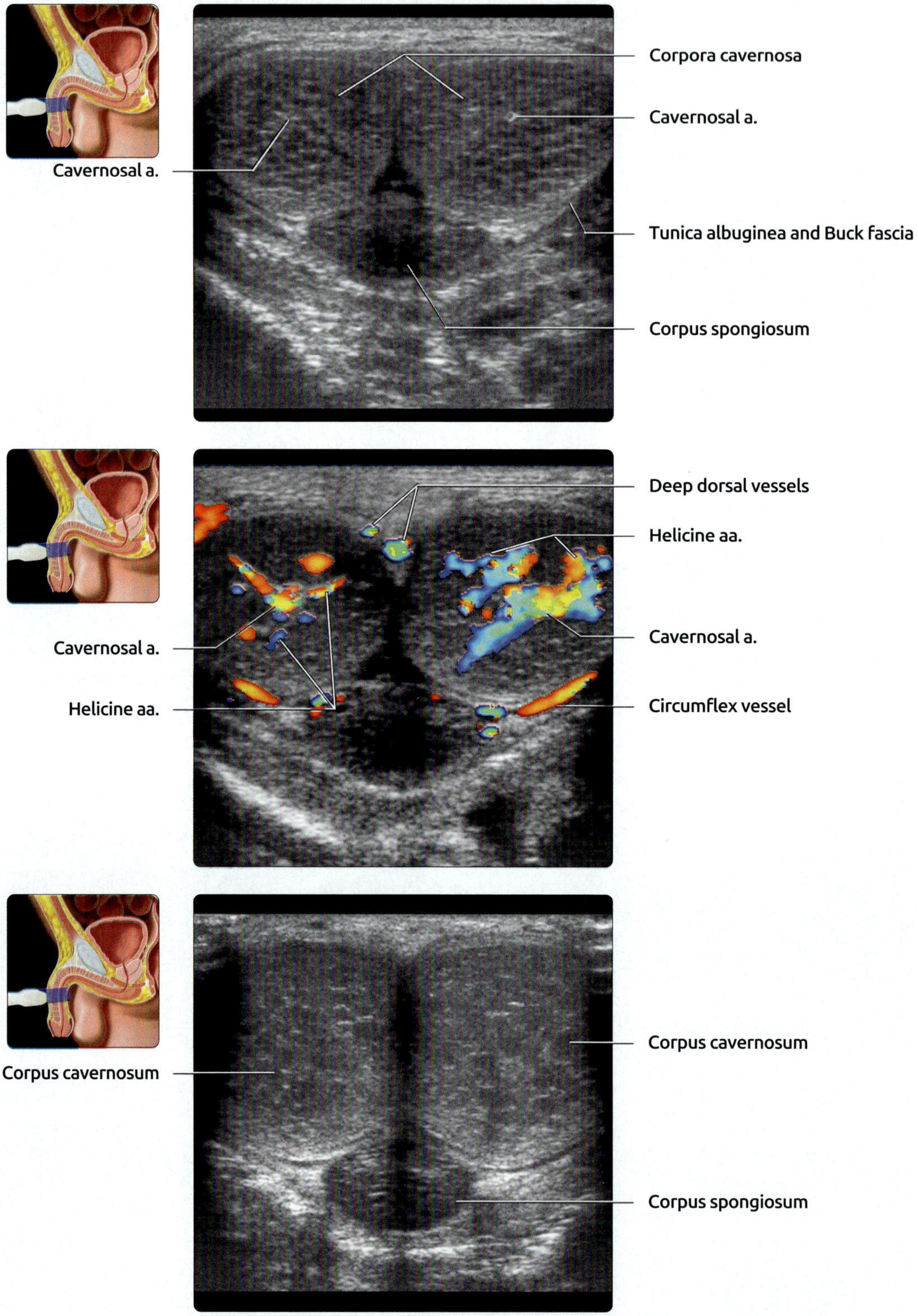

(Top) *The 1st of 3 cross-sectional ultrasounds of the penis showing the changing appearance during a normal erection is shown. The paired corpora cavernosa are the primary erectile bodies. They are composed of a complex network of trabeculae, which create the sinusoids and give a sponge-like appearance on ultrasound. Each corpus is surrounded by a tough tunica albuginea, which appears as a thin echogenic line. Buck fascia, which surrounds all 3 corpora, is intimately associated with the tunica and cannot be distinguished as a separate structure.* **(Middle)** *Color Doppler ultrasound at the onset of an erection shows arterial inflow. There is dilation of the cavernosal and helicine arteries as they begin to fill the sinusoidal spaces.* **(Bottom)** *With continued arterial inflow and compressed venous outflow, the corpora become maximally distended and rigid.*

NORMAL ERECTILE FUNCTION

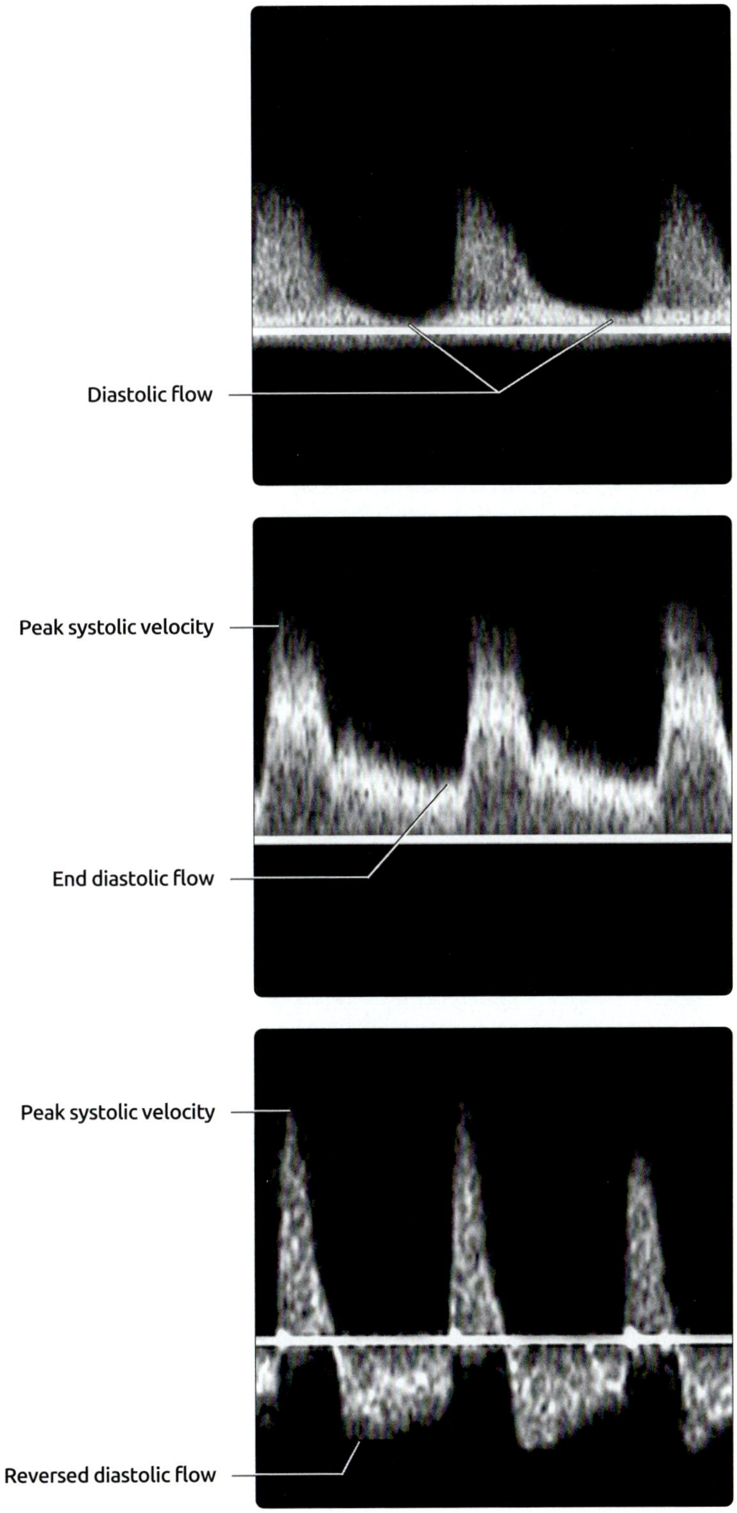

(Top) *The 1st of 3 Doppler tracings showing the changing flow patterns during a normal erection is shown. In the flaccid state, there is little to no diastolic flow.* **(Middle)** *At onset of erection, there is vasodilation of the cavernosal and helicine arteries (increased inflow) and smooth muscle relaxation within the sinusoids (decreased resistance). This causes a marked increase in both the peak systolic velocity (PSV) and end diastolic flow.* **(Bottom)** *In the fully erect state, the emissary veins are compressed against the rigid tunica albuginea, preventing outflow of blood. This dramatically increases the arterial resistance resulting in absent or reversed diastolic flow. A PSV > 30 cm/sec is considered normal, while < 25 cm/sec is strong evidence of arteriogenic impotence (25-30 borderline). Venogenic impotence is due to failure of draining vein occlusion. The Doppler tracing would show continuous forward flow in diastole.*

NORMAL ERECTILE FUNCTION

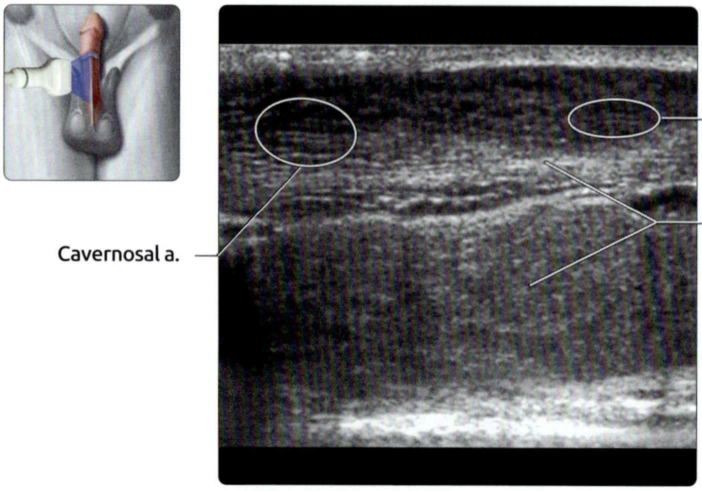

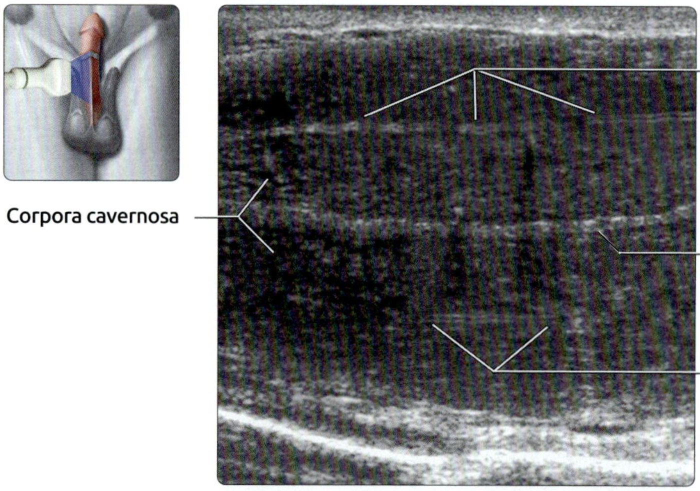

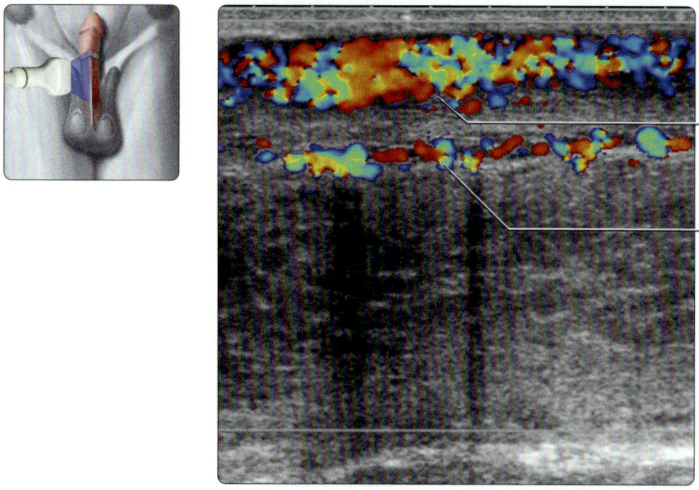

(Top) *The 1st of 3 longitudinal ultrasounds show the corpora cavernosa. In the flaccid state, the cavernosal arteries are tortuous and therefore can only be intermittently visualized in the longitudinal plane.* **(Middle)** *In the erect state, the cavernosal arteries assume a straighter course and can be easily imaged. Magnified views of the cavernosal artery should be obtained to measure vessel diameter. Most measure < 1 mm in the flaccid state. The diameter should increase by 75% with an erection. Failure to increase in size, coupled with an abnormal Doppler waveform, is strong evidence of arteriogenic impotence. Despite the fact that each corpora is sheathed within its own tunica albuginea, there is significant communication across the septum between the 2 corpora cavernosa. Therefore, when performing a study, only 1 corpus cavernosum needs to be injected with the vasodilating agent.* **(Bottom)** *Color Doppler shows flow in both the cavernosal artery and deep dorsal vein of the penis. As an erection starts to wane, or in venogenic impotence, significant flow is seen in the draining vein.*

MR

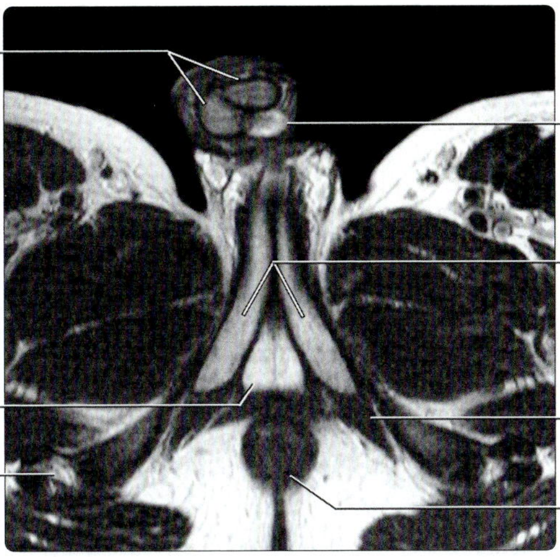

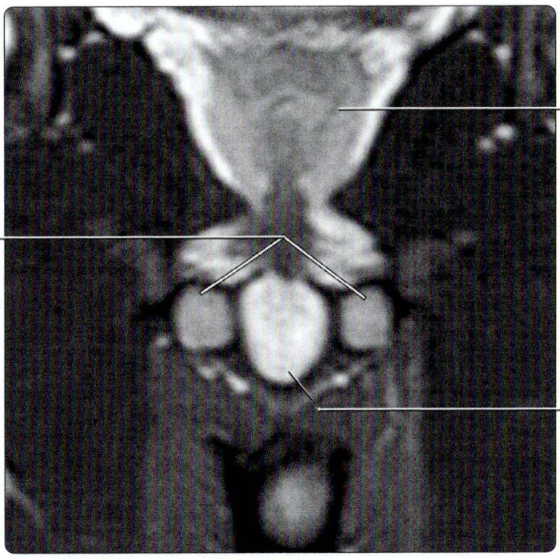

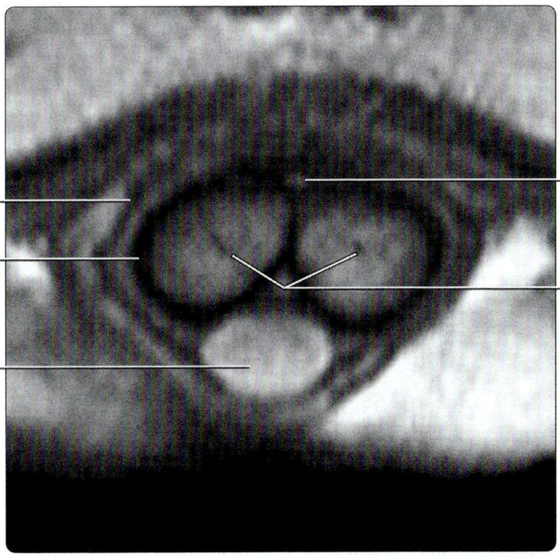

(Top) *Axial T2 MR of the pelvis shows the shaft and root of the penis. Upon entering the pelvis, the 3 corpora are each invested within a muscle layer. The ischiocavernosus muscles surround the crura of the corpora cavernosa, and the bulbospongiosus muscle surrounds the root of the corpus spongiosum. These muscles are innervated by the perineal nerves and assist in erection, ejaculation, and the sensation of orgasm.* **(Middle)** *Coronal STIR MR at the level of the prostate shows all 3 corpora at the same level. The corpora cavernosa crura diminish in size as they continue to diverge and attach to the inferior ischiopubic rami. The corpus spongiosum remains midline and enlarges to form the penile bulb (a.k.a. the urethral bulb).* **(Bottom)** *T2 MR cross section of the penis shows the low-signal tunica albuginea surrounding each of the corpora (more prominent around the cavernosa). Buck fascia cannot be separated from the tunica albuginea.*

URETHRA

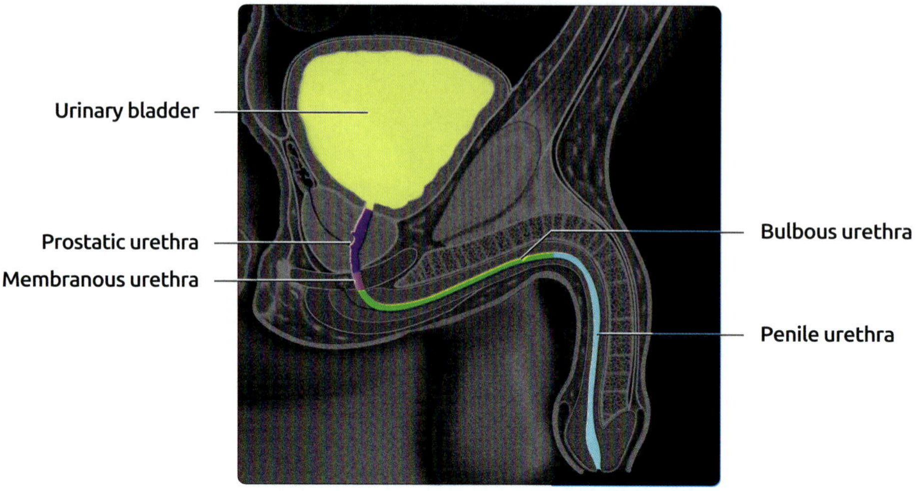

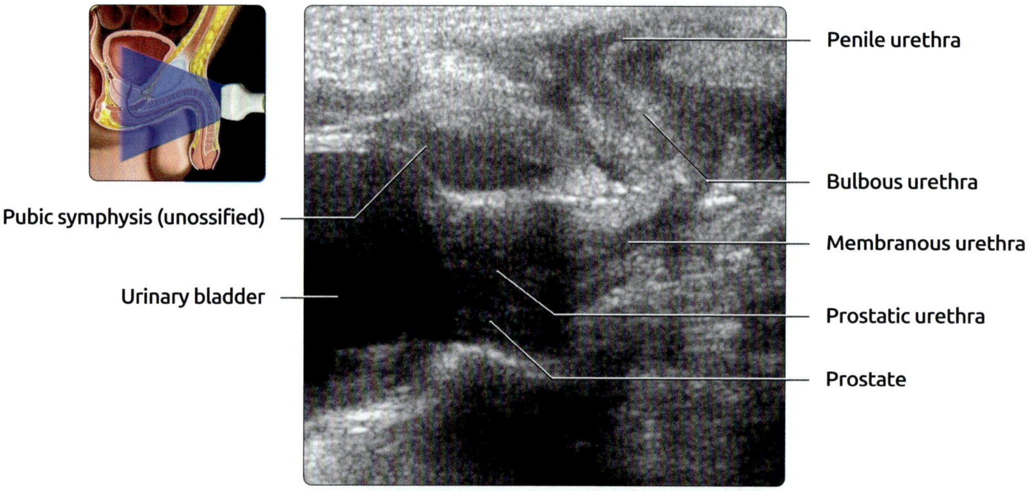

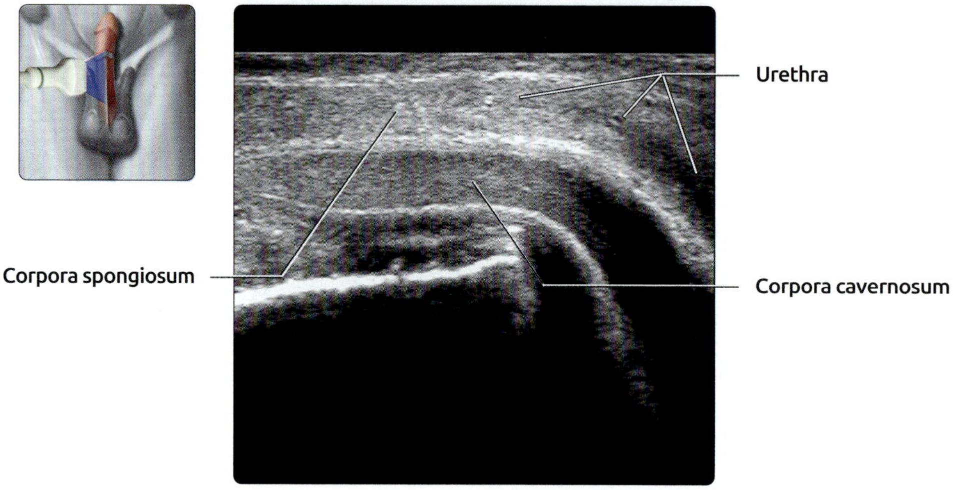

(Top) *Graphic shows the urethra's 2 major divisions: Anterior and posterior urethra, each with two parts. The posterior urethra is composed of the prostatic and membranous portions, and the anterior urethra is composed of the bulbous and penile portions. The prostatic urethra begins at the bladder base and extends to the apex of the prostatic gland. The membranous urethra traverses the urogenital diaphragm. It is the shortest portion of the urethra but the area most vulnerable to injury. The bulbous urethra extends from the bottom of the urogenital diaphragm to the suspensory ligament of the penis. The penile urethra is distal to the suspensory ligament and travels through the pendulous portion of the penis. It widens into the fossa navicularis at the distal glans.* **(Middle)** *Longitudinal transpelvic/perineal ultrasound of a male neonate shows the urethral segments.* **(Bottom)** *Longitudinal scan shows the penis positioned on the anterior abdominal wall and scanned on the ventral side. The urethra is identified within the substance of the corpora spongiosum. One of the paired corpora cavernosa, which is more hypoechoic relative to the spongiosum, is shown.*

PEYRONIE DISEASE

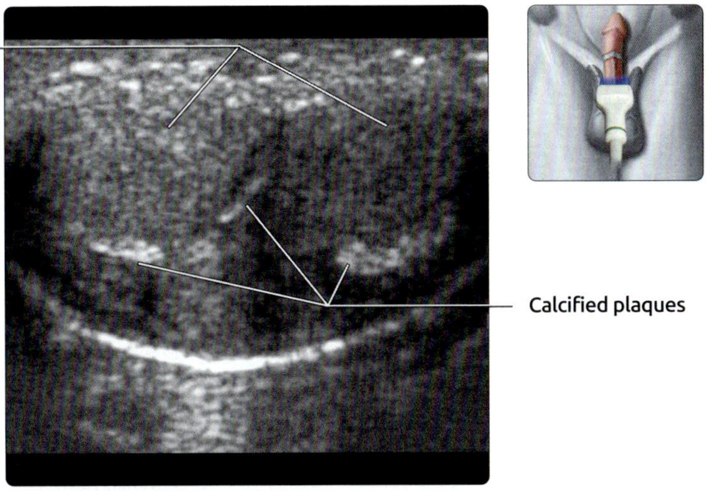

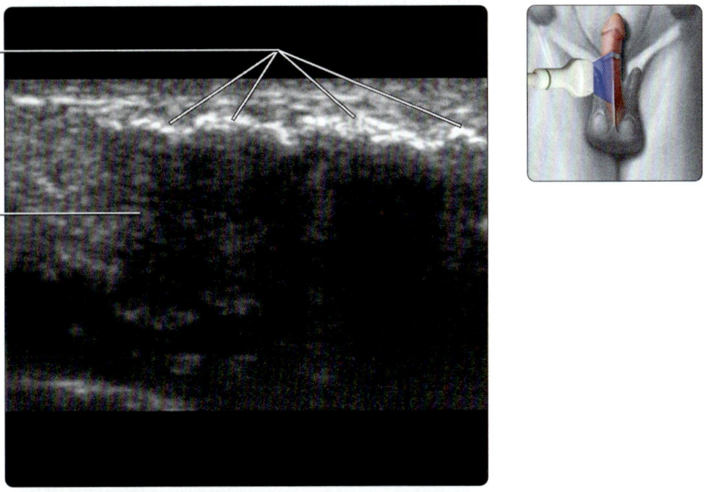

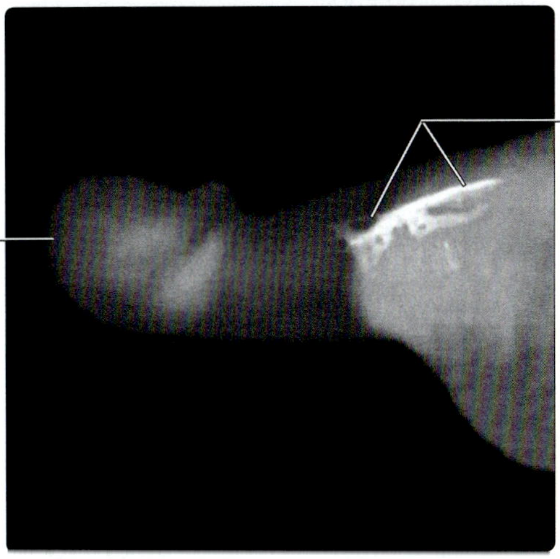

(Top) *The 1st of 3 images in a patient with Peyronie disease shows a cross-sectional ultrasound of the calcified plaques within the tunica albuginea of the corpora cavernosa.* **(Middle)** *Longitudinal ultrasound of 1 corpus cavernosum shows densely calcified plaques with significant posterior shadowing.* **(Bottom)** *Lateral radiograph of the penis shows not only the calcified plaques, but also shortening and upward curvature of the penis. Peyronie disease is an incompletely understood inflammatory process with fibroblastic proliferation and calcification involving the tunica albuginea. It manifests as pain and deformity with erection. The penis may shrink in size with disease progression.*

GROSS ANATOMY

Overview

- Anatomical divisions
  - **Body (corpus)**: Upper 2/3 of uterus
    - Fundus: Superior to ostia of fallopian tubes
  - **Cervix**: Lower 1/3 of uterus
    - Isthmus: Junction of body and cervix
- **Parametrium**: Outer layer, part of visceral peritoneum
- **Myometrium**: Smooth muscle that forms main bulk of uterus
- **Endometrium**: Inner layer
  - Stratum functionalis (inner): Thicker, varies with cyclical changes
  - Stratum basalis (outer): Thin, does not change

Anatomic Relationships

- Extraperitoneal location in midline of true pelvis
- Uterine position
  - **Flexion** is axis of uterine body relative to cervix
  - **Version** is axis of cervix relative to vagina
  - Anteversion with anteflexion is most common
- Peritoneum extends over bladder dome anteriorly and rectum posteriorly
  - **Vesicouterine pouch (anterior cul-de-sac)**: Anterior recess between uterus and bladder
  - **Rectouterine pouch of Douglas (posterior cul-de-sac)**: Posterior recess between vaginal fornix and rectum; most dependent portion of peritoneum in female pelvis
- **Supporting ligaments**
  - Broad ligaments: Double layer of peritoneum
    - Extends laterally to pelvic wall and forms supporting mesentery for uterus
  - Round ligaments: Arise from uterine cornu
    - Course anteriorly, through inguinal canal to insert on labia majora
  - Uterosacral ligaments (posteriorly), cardinal ligaments (laterally), and vesicouterine ligaments (anteriorly) form from connective tissue thickening of broad ligament by cervix
- **Fallopian tubes** connect uterus to peritoneal cavity
  - 4 segments: Interstitial, isthmus, ampulla, infundibulum
- **Arteries**: Dual blood supply
  - Uterine artery (UA) arises from internal iliac artery, anastomoses with ovarian artery
  - Arcuate arteries arise from UAs; seen in outer 1/3 of myometrium
  - Radial arteries arise from arcuate arteries and penetrate vertically into myometrium
  - Basal and spiral arteries arise from radial arteries to supply stratum basalis and stratum functionalis
- Venous drainage mirrors arteries
  - Parametrial venous network prior to drainage into uterine or ovarian veins

Endometrial Variations With Menstrual Cycle

- **Proliferative phase (follicular phase of ovary)**
  - End of menstrual phase to ovulation (~ 14 days)
  - Estrogen induces proliferation of functionalis layer
- **Secretory phase (luteal phase of ovary)**
  - Ovulation to beginning of menstrual phase
  - Progesterone induces secretion of glycogen, mucus, and other substances
- Menstrual phase
  - Sloughing of functionalis layer

Uterine Variations With Age

- Neonatal: Prominent size secondary to effects of residual maternal hormone stimulation
- Infantile: Corpus < cervix (1:2)
- Prepubertal: Corpus = cervix (1:1)
- Reproductive: Corpus > cervix (2:1)
  - 7.5-9.0 cm (length)
  - 4.5-6.0 cm (breadth)
  - 2.5-4.0 cm (thickness)
- Postmenopausal: Overall reduction in size, similar to prepubertal uterus

IMAGING ANATOMY

Myometrium

- Inner layer (junctional zone): Thin and hypoechoic, < 12 mm
- Middle layer: Thick, homogeneously echogenic
- Outer layer: Thin, hypoechoic layer peripheral to arcuate vessels

Endometrium

- **Proliferative phase**
  - Early: Thin, single echogenic line
  - Progressive hypoechoic thickening (4-8 mm), classic trilaminar appearance
- **Secretory phase**
  - Increased echogenicity and thickening up to 16 mm
- **Menstrual phase**
  - Early: Cystic areas within echogenic endometrium indicating endometrial breakdown
  - Progressive heterogeneity with mixed cystic (blood) and hyperechoic (clot or sloughed endometrium) regions

ANATOMY IMAGING ISSUES

Imaging Recommendations

- Sonohysterography to evaluate endometrial pathology
- 3D ultrasound additive in many cases, particularly müllerian duct anomalies and intrauterine device evaluation

EMBRYOLOGY

Embryologic Events

- Organogenesis phase: Uterus formed from paired paramesonephric (müllerian) ducts
- Fusion phase: Paired ducts fuse in midline to form uterus and upper vagina
  - Unfused portions remain as fallopian tubes
- Resorption phase: Resorption of uterine septum

Practical Implications

- Müllerian duct anomalies occur during 1 of 3 phases of formation
  - Organogenesis: Uterine agenesis, hypoplasia, unicornuate
  - Fusion: Didelphys, bicornuate
  - Resorption: Septate, arcuate

UTERUS

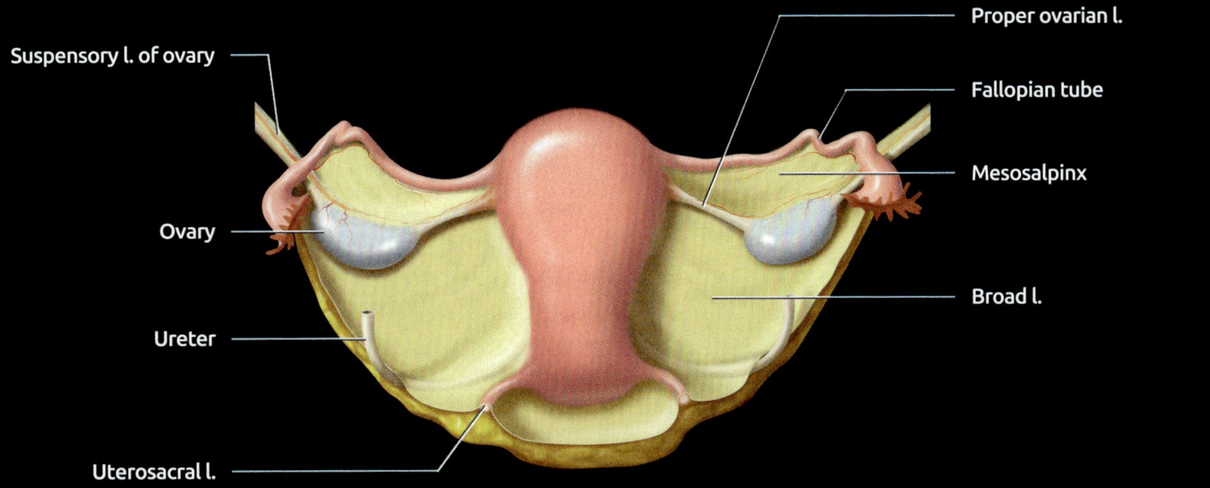

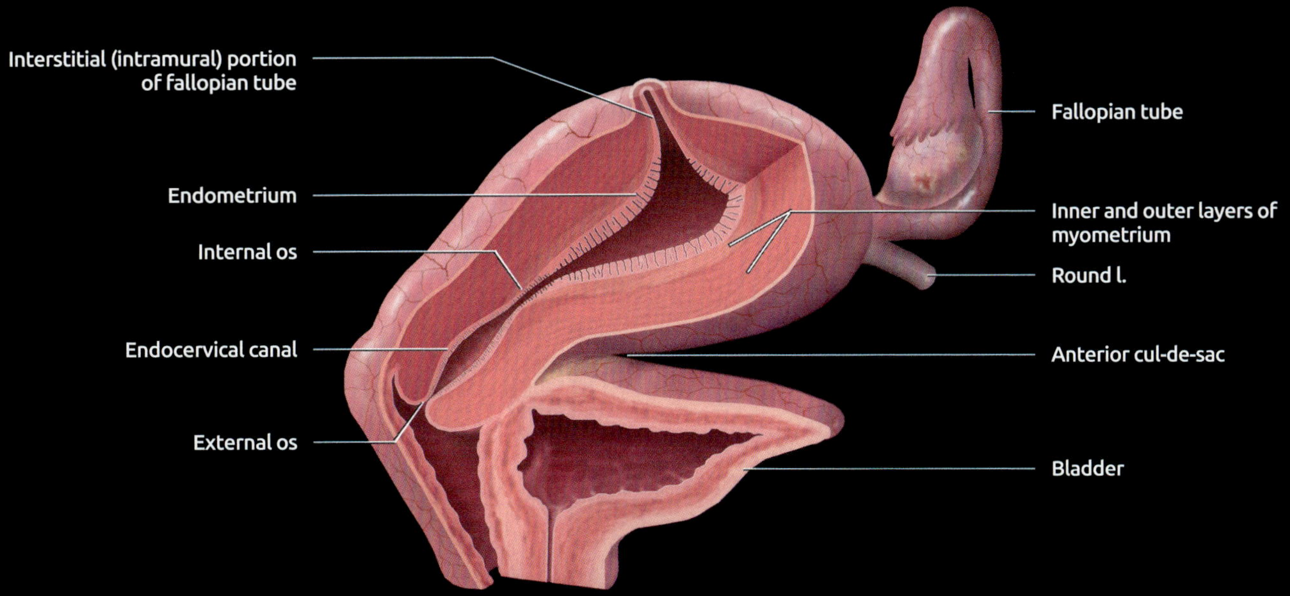

(Top) *Illustration shows the posterior aspect of the uterus and ovaries with the other tissues removed. The ovary is suspended from the pelvic side wall by the suspensory ligament of the ovary and from the uterus by the proper ovarian ligament. These ligaments separate the mesosalpinx above from the broad ligament below.* **(Bottom)** *The uterus is composed of a glandular endometrium and muscular myometrium. The smooth muscle within the inner portion of the myometrium is more compacted and relatively hypovascular.*

VASCULAR SUPPLY

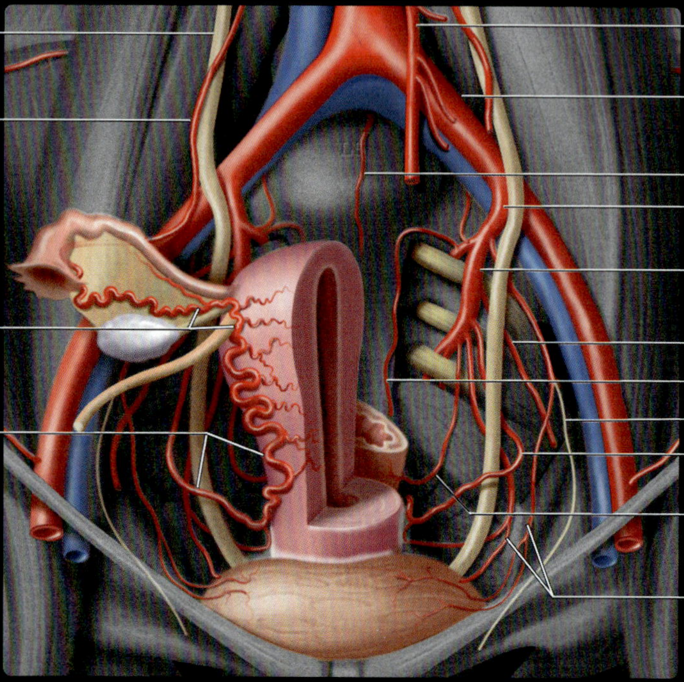

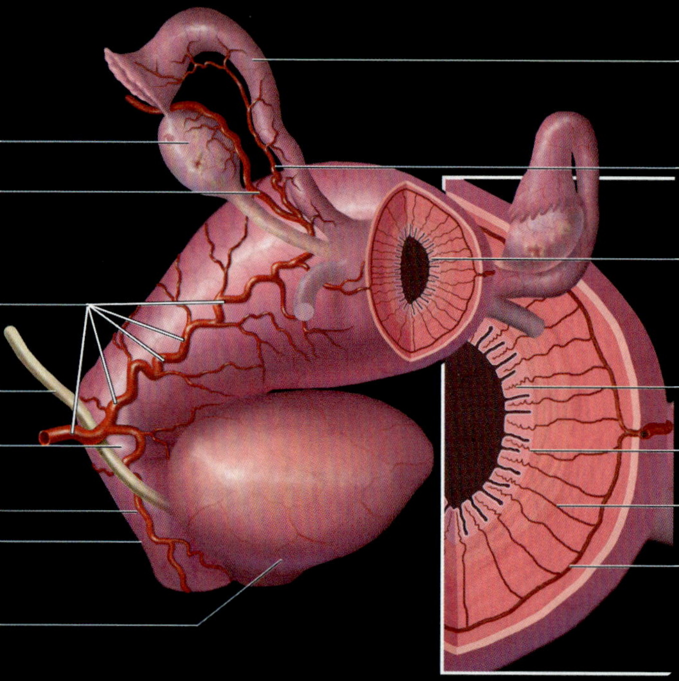

(Top) *Frontal graphic shows the arterial supply to the pelvis. The uterus has a dual blood supply with contributions from both the uterine and ovarian arteries. The ovarian artery originates from the aorta, and the uterine artery is a branch off the anterior trunk of the internal iliac artery.* **(Bottom)** *Graphic shows the uterine vasculature. The descending segment of the uterine artery, after branching off from the internal iliac artery, runs medially toward the cervix. The ascending segment ascends laterally along the uterine wall to meet the ovarian and fallopian arteries. The transverse segments cross the cardinal ligament and anastomose extensively with each other to form the arcuate arteries, which give rise to the radial arteries penetrating the myometrium vertically. The arteries then branch into spiral and basal arteries in the functional and basal layers of the endometrium respectively.*

VASCULAR SUPPLY

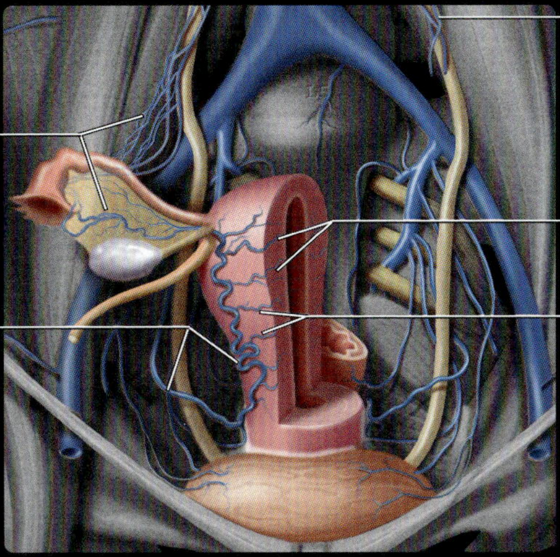

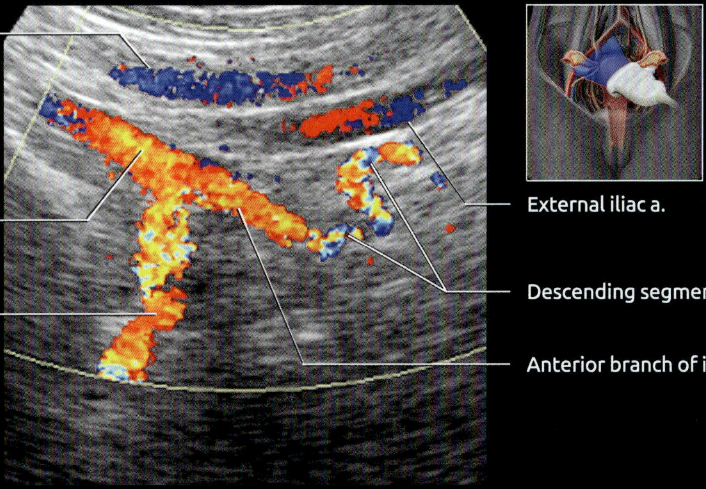

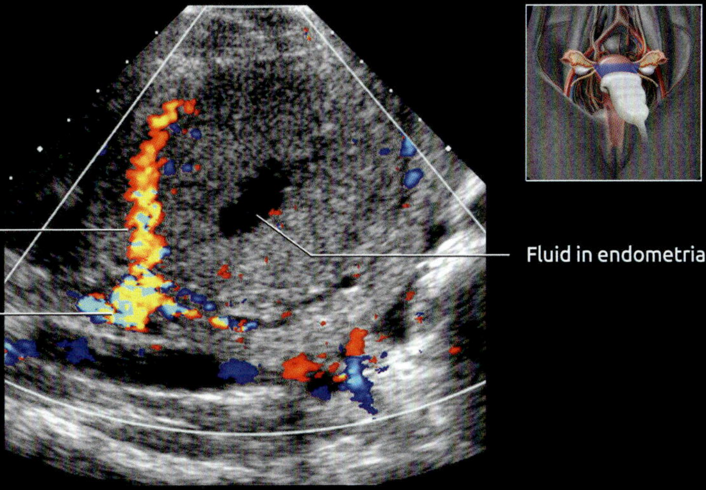

(Top) *This graphic shows the venous drainage from of the uterus, which follows the arterial system. There is venous drainage via both the uterine and ovarian veins.* **(Middle)** *Longitudinal transabdominal color Doppler ultrasound shows the uterine artery arising from the anterior branch of the internal iliac artery, which is near the level of the uterine cervix. The uterine artery can also be assessed with transvaginal ultrasound by locating it lateral to the cervix.* **(Bottom)** *Color Doppler ultrasound in a patient who has just undergone a dilatation and curettage for a failed 1st trimester pregnancy is shown. There is very prominent flow in the arcuate arteries, which run in the outer 1/3 of the myometrium.*

CUL-DE-SACS AND LIGAMENTS

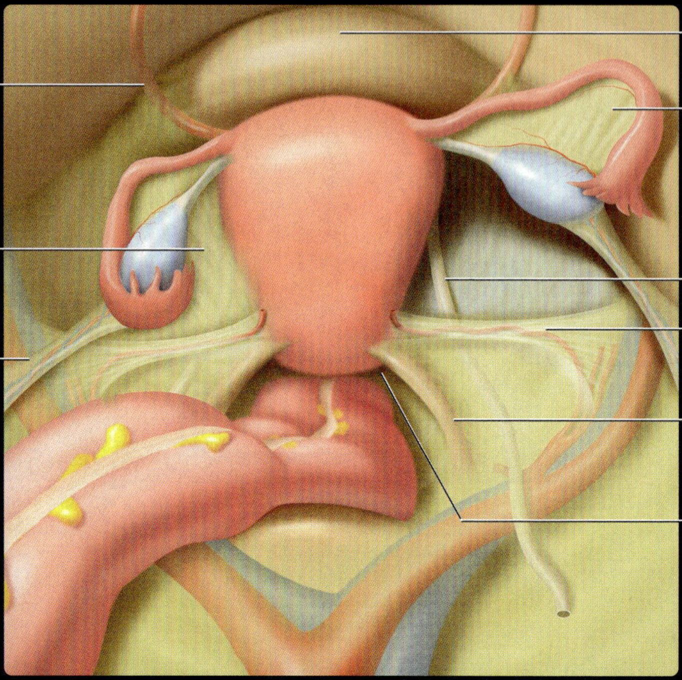

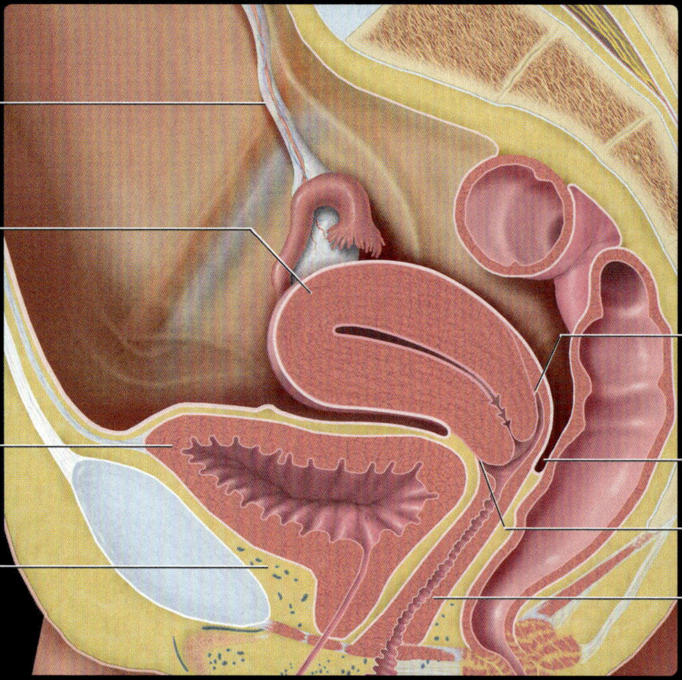

(Top) *Uterus viewed in situ from above and behind shows its positioning and major ligaments. The uterus is covered by a sheet of peritoneum, creating a double layer (the broad ligament), which sweeps laterally to attach to the pelvic wall. Areas of thickening at its base are the cardinal ligaments, which attach to the lateral pelvic wall, and the uterosacral ligaments, which attach to the sacrum. The uterosacral ligaments form the lateral borders of the posterior cul-de-sac (rectouterine pouch or pouch of Douglas). The round ligaments arise from the cornu of the uterus and course anteriorly to pass through the inguinal canal and insert on the labia majora. They offer little support for the uterus.* **(Bottom)** *Sagittal graphic of the female pelvis shows the bladder, uterus, and rectum, all of which are extraperitoneal. Note the posterior vaginal fornix extends more cephalad than the anterior vaginal fornix, and the posterior cul-de-sac is the most dependent portion of the peritoneal cavity.*

3D UTERUS

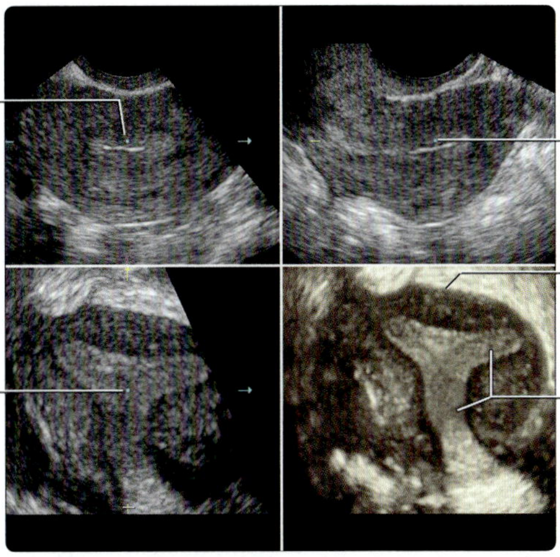

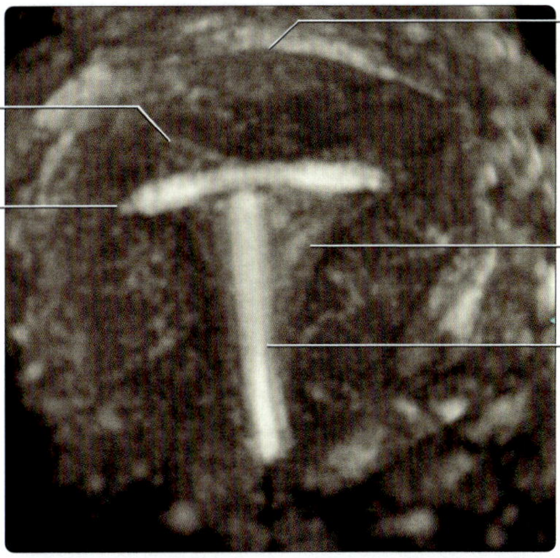

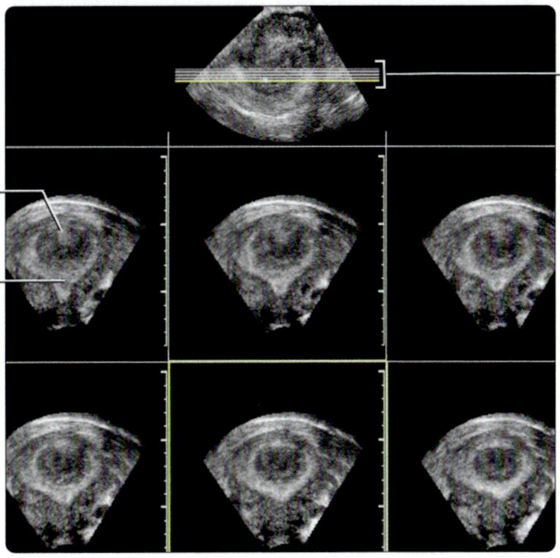

(Top) *This set of images shows a 3D data set of a normal uterus. A dedicated 3D ultrasound probe provides automated acquisition of a volume data set. This data is displayed in 3 simultaneous orthogonal planes. A central localizer point on each image allows the operator to know the precise location in all 3 planes and manipulate the image to the optimal projection. In the coronal image, the entire endometrial and fundal contours are displayed. This has particular utility in evaluating müllerian duct anomalies.* **(Middle)** *3D images are also helpful for showing intrauterine uterine device (IUD) position. In this case, the right arm of the IUD is projecting into the myometrium.* **(Bottom)** *3D data sets can be manipulated and sliced through like a CT or MR. This shows a set of 6 slices through a submucosal fibroid, which is projecting into the endometrial cavity. The lines on the localizer image on top correspond to slices below (the yellow line indicates the level of the image in the yellow box).*

NORMAL SAGITTAL IMAGES OF UTERUS

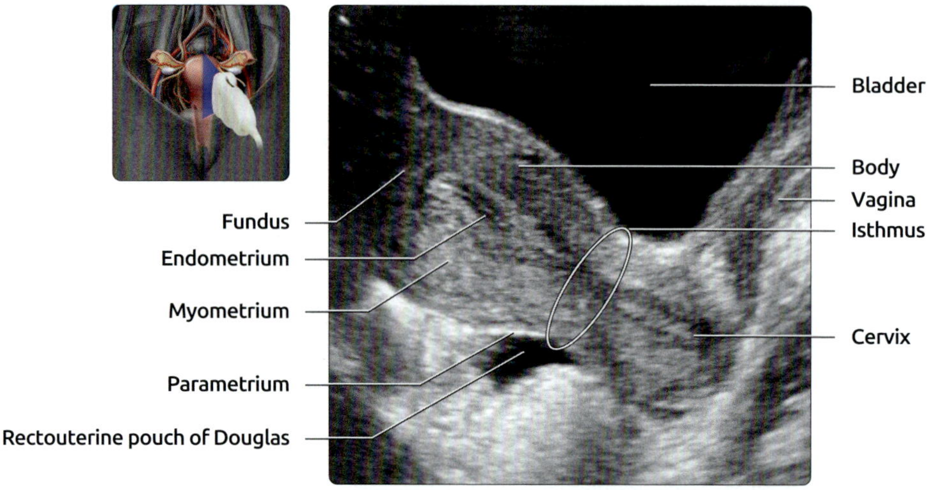

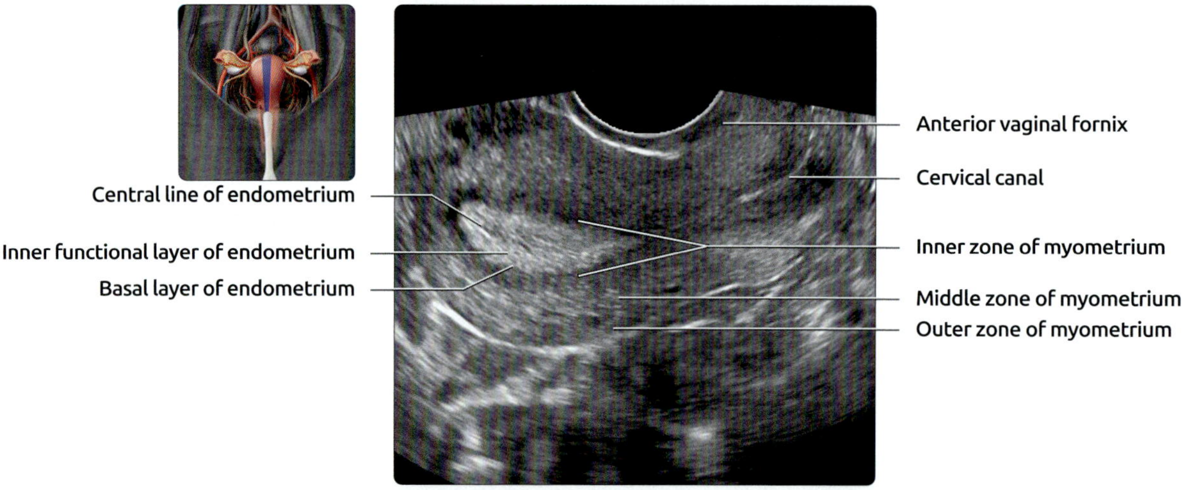

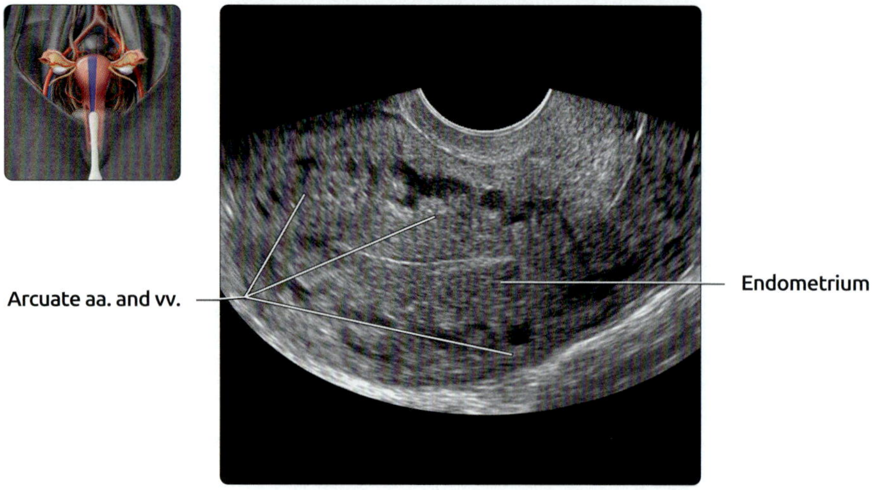

(Top) *Longitudinal TA ultrasound shows a normal anteverted uterus. Version refers to the angle the cervix makes with the vagina. In this case, the cervix is angled anteriorly, and the uterus continues in a straight line with the cervix. This is the most common position found in the female pelvis.* **(Middle)** *Longitudinal TV ultrasound obtained in the secretory phase demonstrates different zones. The smooth muscle within the inner zone of the myometrium is more compact, making it more hypoechoic (subendometrial halo). The majority of myometrium is homogeneously echogenic with the outer zone being less echogenic.* **(Bottom)** *Longitudinal transvaginal scan obtained in the early proliferative phase shows prominent arcuate arteries and veins, which run in the outer 1/3 of the myometrium. These may become calcified following menopause.*

NORMAL VARIATIONS, UTERINE POSITION

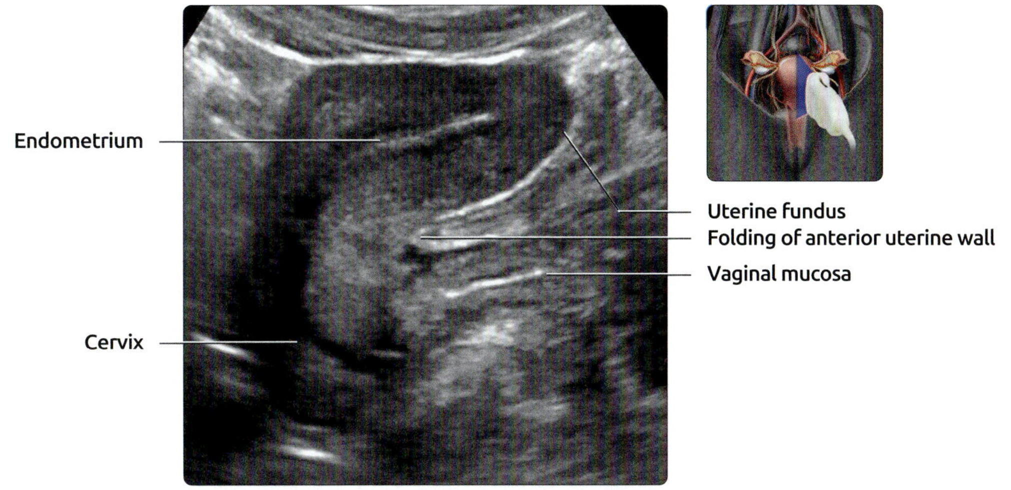

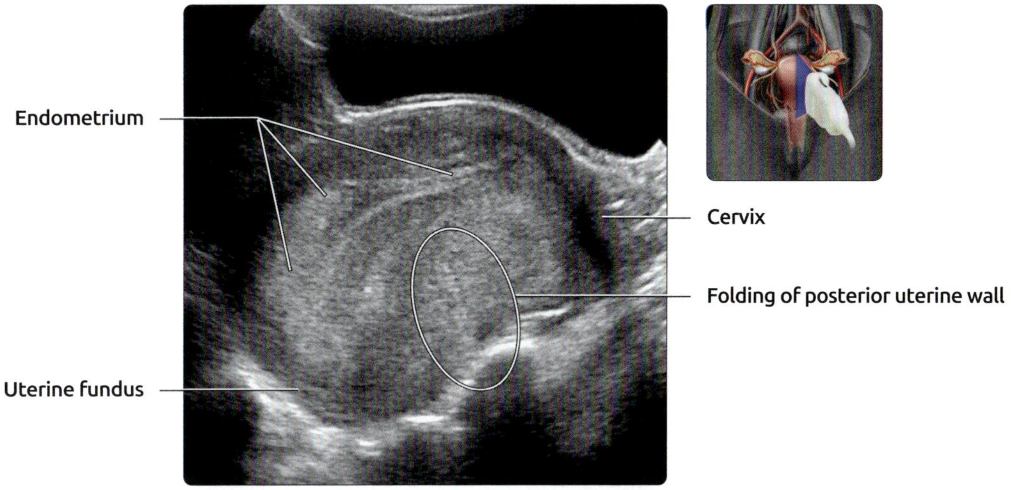

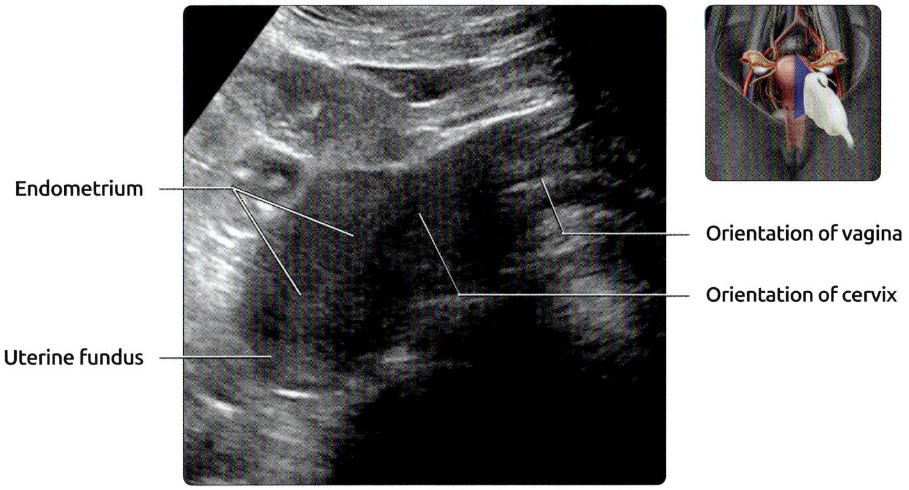

(Top) *Longitudinal TA ultrasound of an anteverted, anteflexed uterus shows the uterine cervix is angled forward with respect to the vagina, and the uterine body is angled forward with respect to the cervix. Version refers to the angle of the cervix relative to the vagina. Flexion refers to the angle of the uterine body relative to the cervix, i.e., the uterus and the cervix are not in a straight line.* **(Middle)** *Uterine retroflexion is shown. This is an anteverted uterus with exaggerated retroflexion in which the uterus resembles a boxing glove. Folding of the posterior uterine wall may be confused with an intramural fibroid.* **(Bottom)** *Uterine retroversion is shown. The orientation of the uterus and cervix is posterior with respect to the vagina. Retroversion frequently limits transabdominal evaluation of the uterus, as seen here.*

UTERINE VARIATIONS WITH AGE

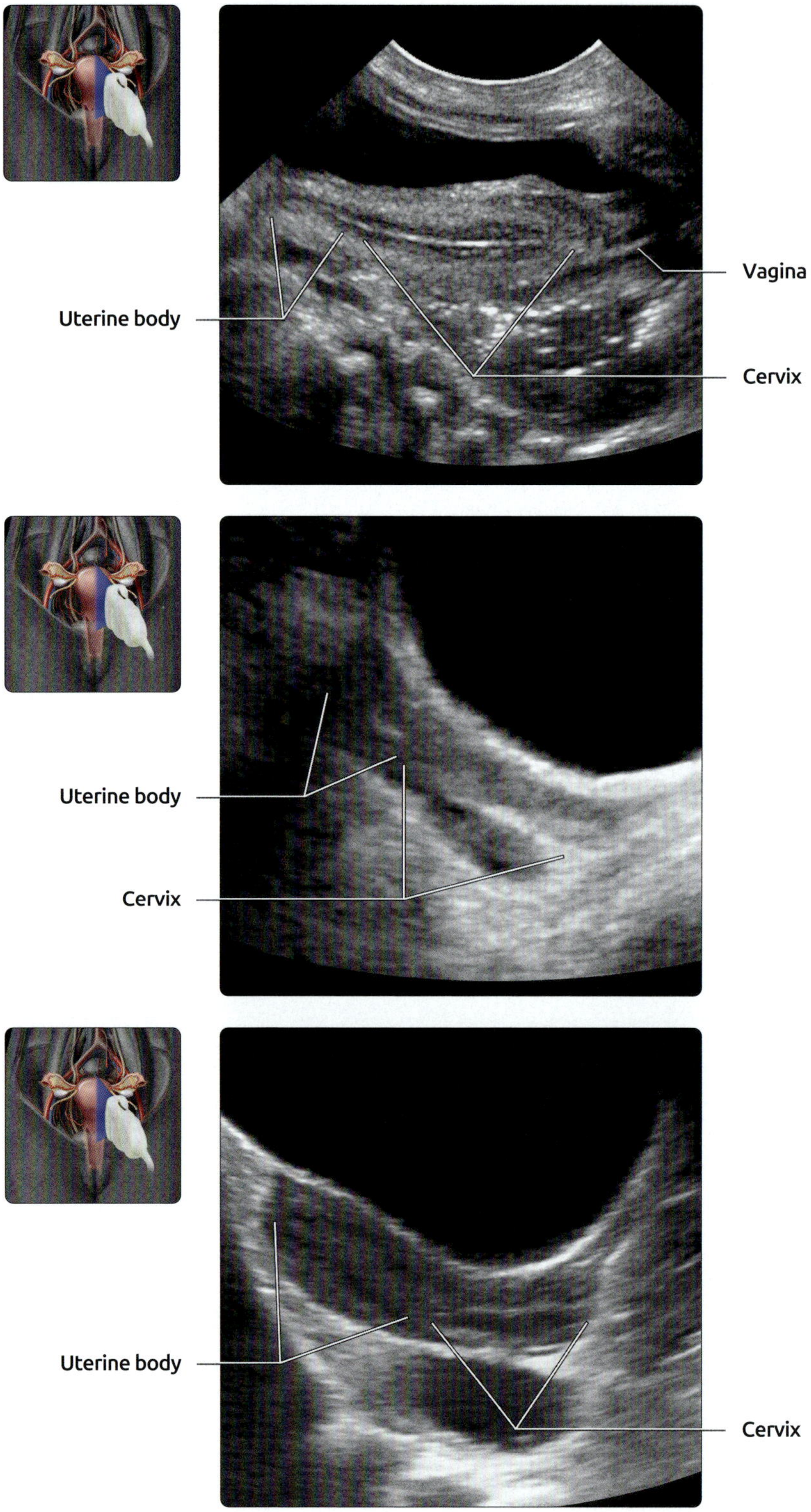

(Top) *Longitudinal TA ultrasound shows an immediate neonatal uterus (day 2). The uterus is prominent with a bulbous cervix and a rudimentary body. The endometrium is seen as a thin, echogenic line, which may be due to stimulation by the residual maternal hormones.* **(Middle)** *Longitudinal TA ultrasound shows a prepubertal uterus in a 8-year-old patient. The uterus demonstrates a tubular appearance with the length of the cervix nearly double that of the uterine body.* **(Bottom)** *Longitudinal TA ultrasound shows an early pubertal uterus in a 12-years-old patient. The body length of the uterus approximates the cervical length with the endometrium, changing in appearance and thickness during the menstrual cycle. At this time, the uterine body grows dramatically until it reaches the adult size.*

UTERINE VARIATIONS WITH AGE

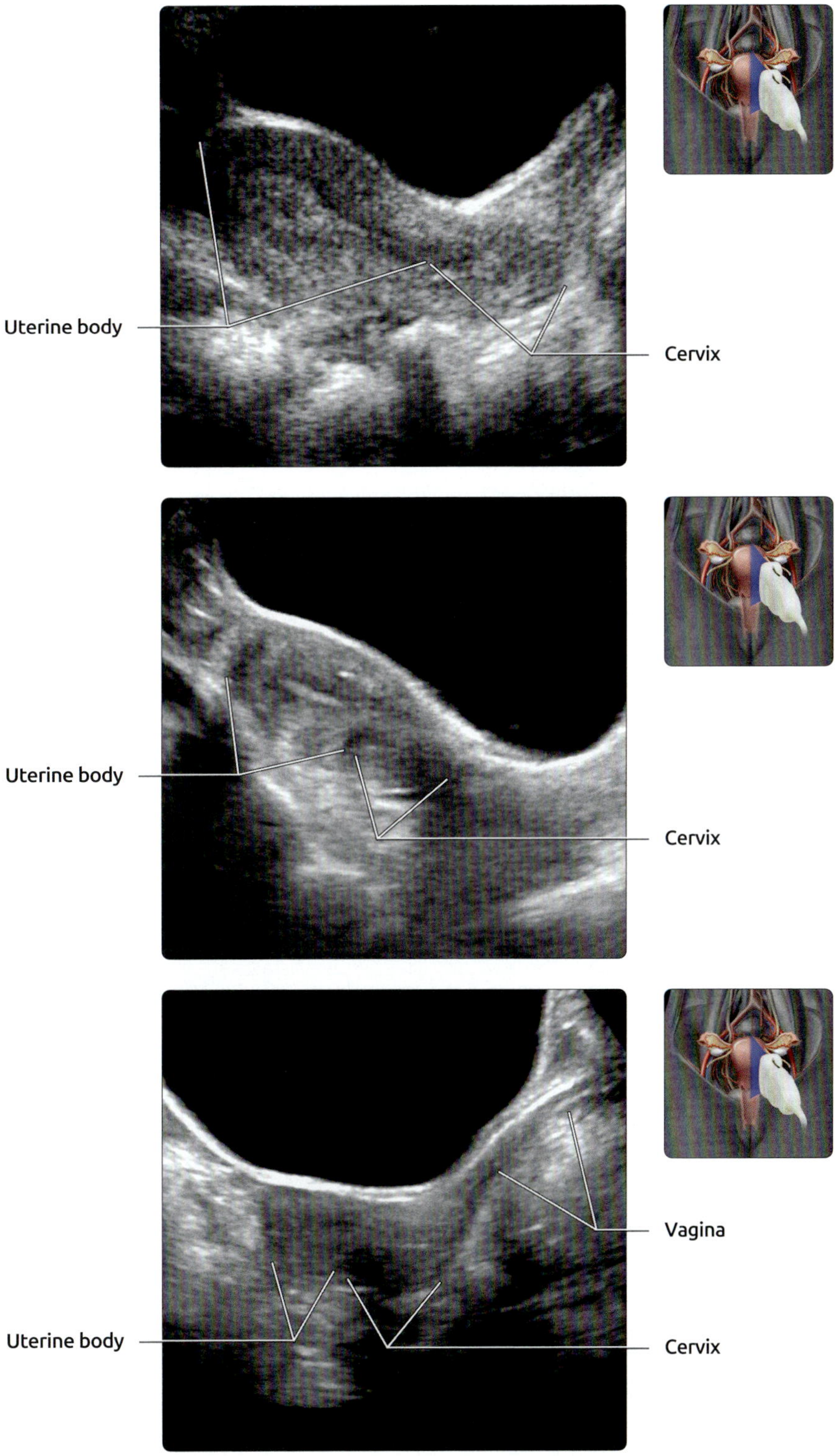

(Top) *Longitudinal TA ultrasound shows a nulliparous uterus. The normal adult uterus should attain a pear-shaped or hourglass appearance with the length of the uterine body double that of the cervix. The size of a nulliparous uterus is usually smaller than that of a parous uterus.* **(Middle)** *Longitudinal TA ultrasound shows an early postmenopausal uterus, which is atrophic with prominent reduction in body size relative to the cervix.* **(Bottom)** *Longitudinal TA ultrasound shows a later postmenopausal uterus. Note that the cervix: body ratio is similar to that of a prepubertal uterus.*

CYCLIC CHANGES OF ENDOMETRIUM

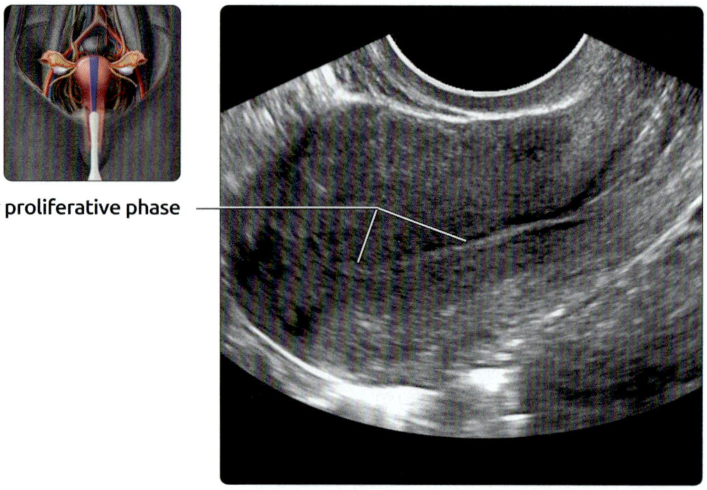

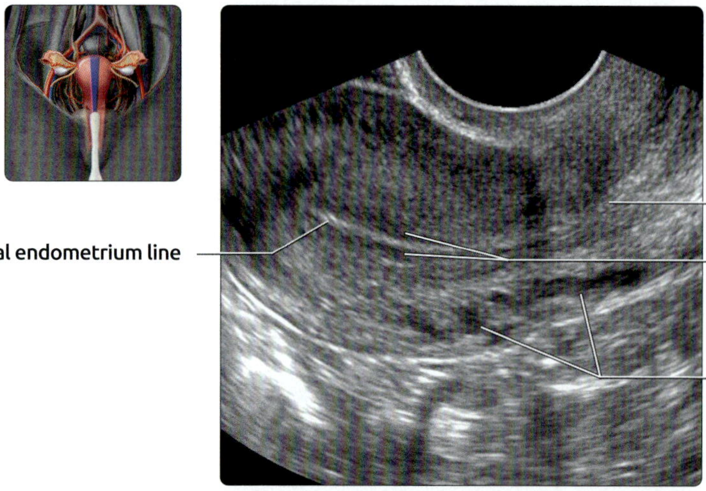

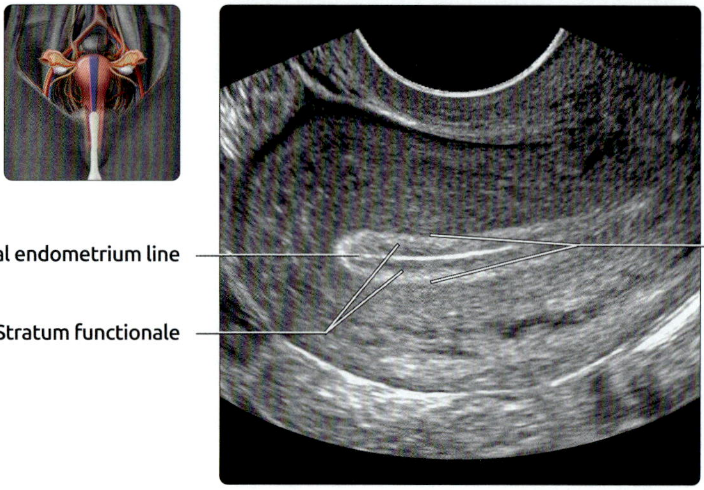

(Top) *Longitudinal TA ultrasound shows endometrium in the postmenstrual or early proliferative phase. Note the endometrium is thin and echogenic.* **(Middle)** *Longitudinal TA ultrasound of the endometrium during mid proliferative phase shows the endometrium progressively thickened and slightly more echogenic.* **(Bottom)** *Longitudinal TA ultrasound of endometrium in the periovulatory phase shows thickening of the stratum functionalis with an echogenic central line and a layered, trilaminar appearance.*

CYCLIC CHANGES OF ENDOMETRIUM

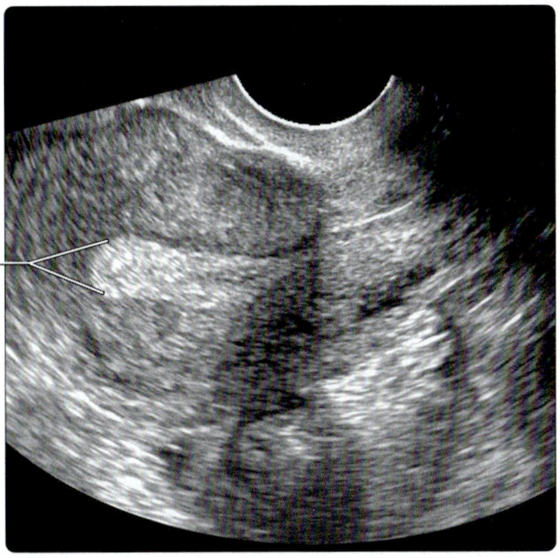

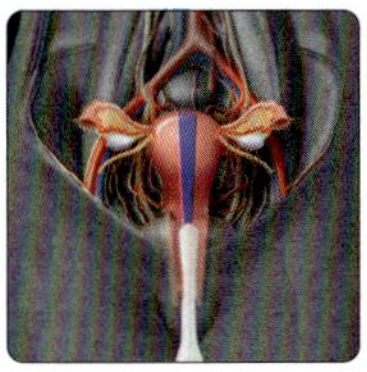

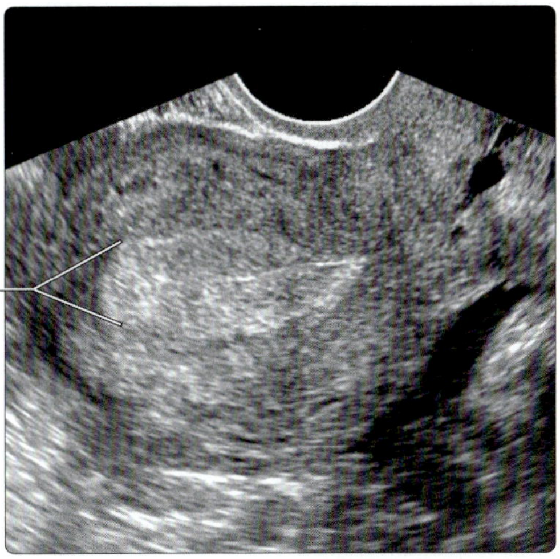

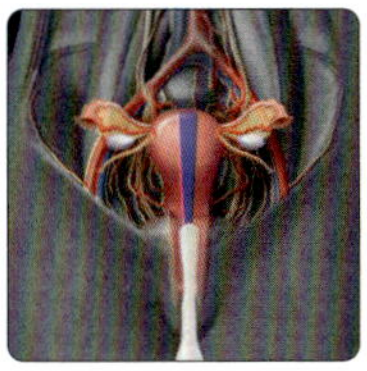

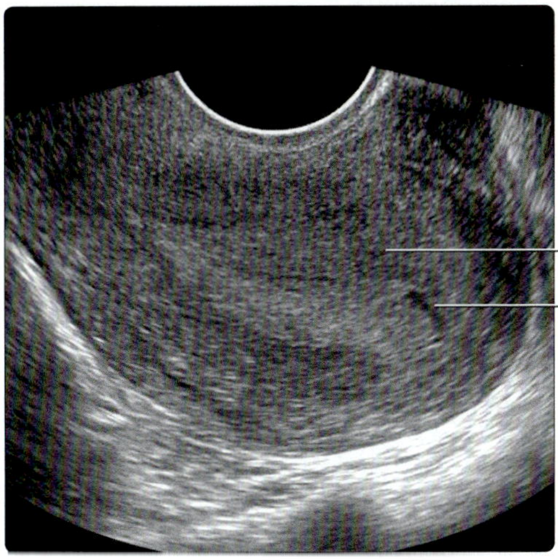

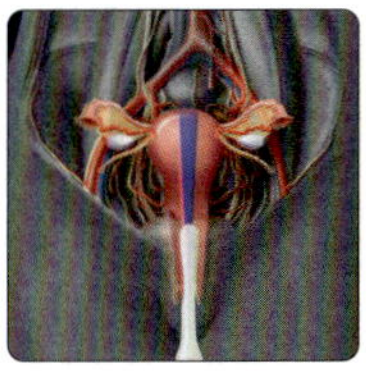

(Top) *Longitudinal TV scan shows the endometrium during the early secretory phase. The endometrium becomes progressively thickened and more echogenic with loss of the trilaminar appearance. These are cyclic endometrial changes in a neutral uterus.* **(Middle)** *Longitudinal TV ultrasound in the late secretory phase shows the endometrium is uniformly thickened and echogenic. The normal maximal endometrial thickness should not exceed 1.6 cm, through transmission can sometimes be visualized secondary to the mucus-filled glands.* **(Bottom)** *Longitudinal TV ultrasound shows a thickened endometrium just prior to menstruation. Echogenicity has decreased and is more heterogeneous than in the secretory phase. A small amount of fluid can be seen within the endometrial cavity.*

INTRAUTERINE ARTERIES

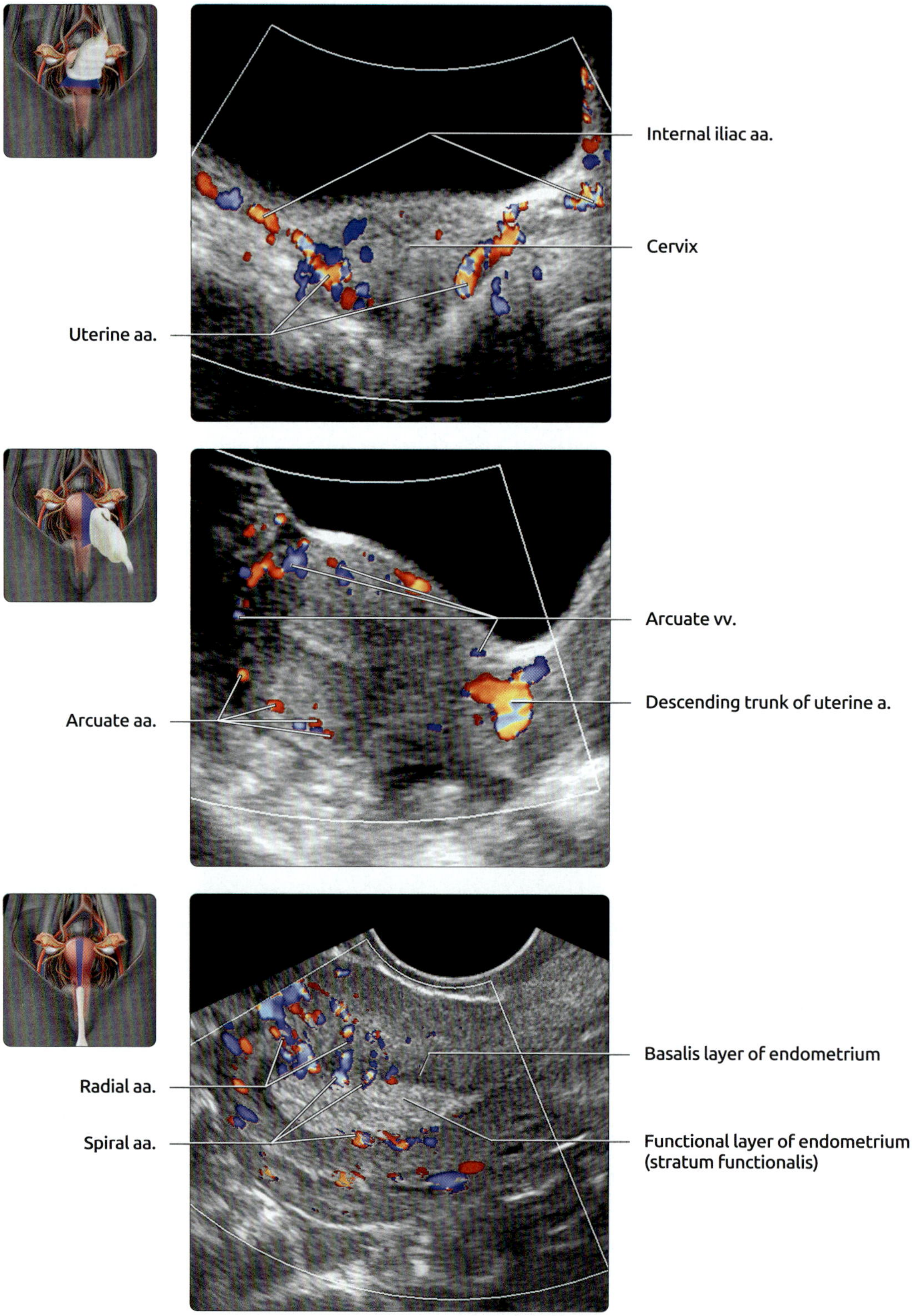

(Top) *Transverse TA color Doppler ultrasound shows descending branches of both uterine arteries running medially at the level of the cervix. Care must be taken not to confuse these with the iliac arteries, which lie more laterally.* **(Middle)** *Longitudinal TA color Doppler ultrasound shows the arcuate arteries and veins located at the periphery of the uterus. The arcuate arteries commonly calcify with advancing age.* **(Bottom)** *Longitudinal TV color Doppler ultrasound shows arcuate arteries branching into radial arteries, which run vertically in the myometrium. These in turn give rise to the basal and spiral arteries, which supply the basal and functional layers of the endometrium, respectively. The spiral arteries penetrate deep into the stratum functionalis of the endometrium, which sheds during menstruation.*

FALLOPIAN TUBES

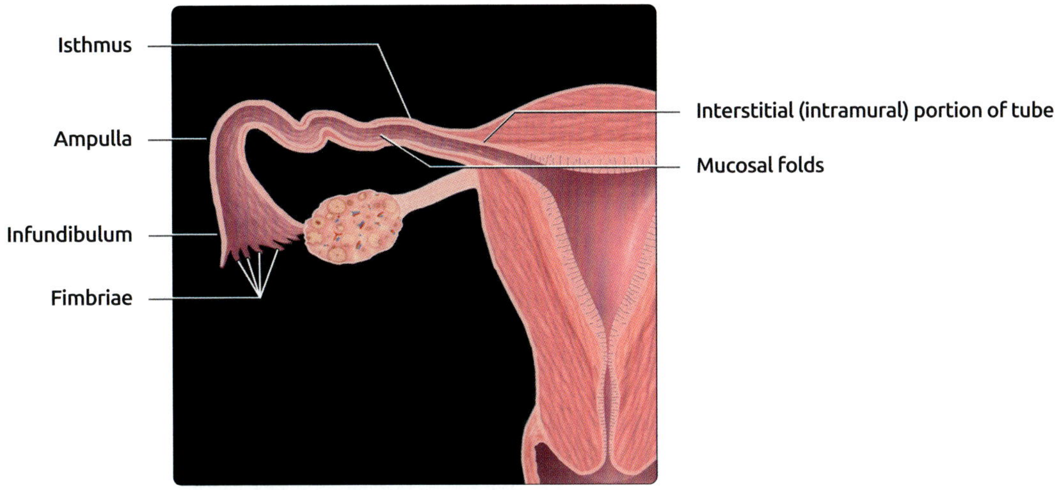

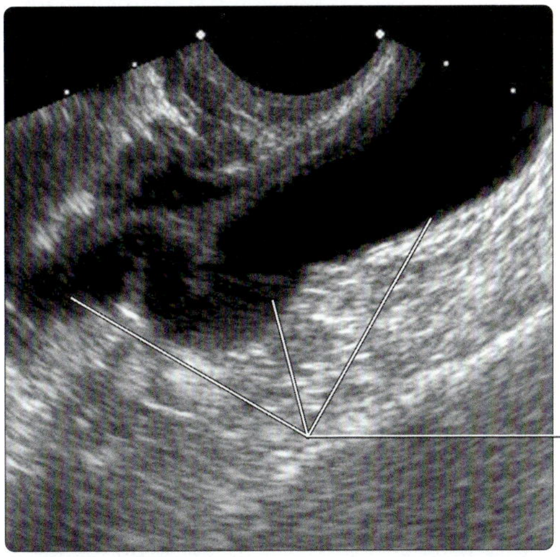

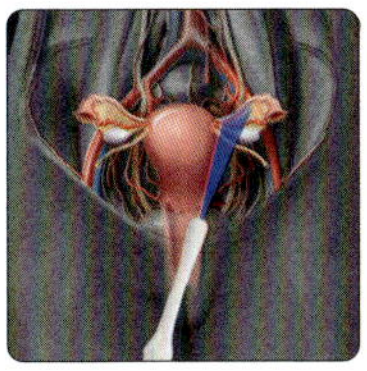

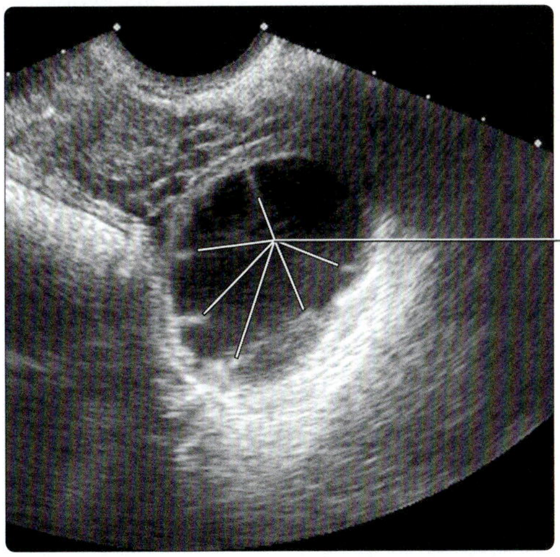

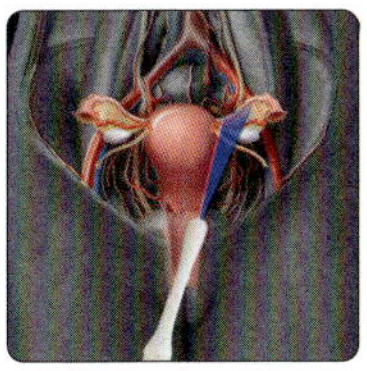

(Top) *A graphic of the right fallopian tube shows the 4 segments, including the interstitial (intramural) portion, isthmus, ampulla, and infundibulum, which is ringed by the fimbriae.* **(Middle)** *First of 2 transvaginal ultrasound images of a left-sided hydrosalpinx is shown. During real-time scanning, it is important to try to elongate the tube, as in this image. This allows differentiation from a cystic ovarian mass.* **(Bottom)** *In the cross-sectional plane, a hydrosalpinx will often display a cogwheel appearance. The mucosal folds of the fallopian tube project into the lumen. As the hydrosalpinx becomes more chronic, these folds will become thick and nodular.*

GROSS ANATOMY

Overview

- Begins at inferior narrowing of uterus (isthmus)
  - Supravaginal portion: Endocervix
  - Vaginal portion: Ectocervix
- Endocervical canal: Spindle-shaped cavity communicates with uterine body and vagina
- Internal os: Opening into uterine cavity
- External os: Opening into vagina
- Largely fibrous stroma with high proportion of elastic fibers interwoven with smooth muscle
- Endocervical canal lined by mucus-secreting columnar epithelium
  - Epithelium in series of small, V-shaped folds (plicae palmatae)
- Ectocervix lined by stratified squamous epithelium
- Squamocolumnar junction near external os but exact position variable
- Nabothian cysts are commonly seen
  - Represent obstructed mucus-secreting glands
- Entire cervix is extraperitoneal
  - Anterior: Peritoneum reflects over dome of bladder above level of internal os
  - Posterior: Peritoneum extends along posterior vaginal fornix, creating rectouterine pouch of Douglas (cul-de-sac)
- Arteries, veins, nerves, and lymphatics
  - Arterial supply
    - Descending branch of uterine artery from internal iliac artery
  - Venous drainage
    - To uterine vein and drains into internal iliac vein
  - Lymphatics
    - Drain into internal and external iliac lymph nodes
  - Innervation
    - Sympathetic and parasympathetic nerves from branches of inferior hypogastric plexuses
- Variations with pregnancy
  - Nulliparous: Circular external os, arterial waveform shows high resistivity index (RI)
  - During pregnancy: Changes become apparent by ~ 6 weeks of gestation
    - Softened and enlarged cervix due to engorgement with blood with decreased RI of uterine artery
    - Hypertrophy of mucosa of cervical canal: Increased echogenicity of mucosal layer
    - Increased secretion of mucous glands: Increased volume of mucus ± mucus plug in cervical canal
  - Parous: Larger vaginal part of cervix, external os opens out transversely with anterior and posterior lips
- Variations with age: Cervix grows less with age than uterus
  - Neonatal: Adult configuration due to residual maternal hormonal stimulation
  - Infantile: Cervix predominant with cervix:corpus length ratio ~ 2:1
  - Prepubertal: Cervix:corpus length ratio ~ 1:1
  - Reproductive: Uterus predominant, cervix:corpus length ratio ≥ 1:2
  - Postmenopausal: Overall reduction in size

Anatomy Relationships

- Anterior
  - Supravaginal cervix: Superior aspect of posterior bladder wall
  - Vaginal cervix: Anterior fornix of vagina
- Posterior
  - Supravaginal cervix: Rectouterine pouch of Douglas
  - Vaginal cervix: Posterior fornix of vagina
- Lateral
  - Supravaginal cervix: Bilateral ureters
  - Vaginal cervix: Lateral fornices of vagina
- Ligamentous support: Condensations of pelvic fascia attached to cervix and vaginal vault
  - Transverse cervical (cardinal) ligaments
    - Fibromuscular condensations of pelvic fascia
    - Pass to cervix and upper vagina from lateral walls of pelvis
  - Pubocervical ligaments
    - 2 firm bands of connective tissue
    - Extend from posterior surface of pubis, position on either side of neck of bladder and then attach to anterior aspect of cervix
  - Sacrocervical ligaments
    - Fibromuscular condensations
    - Attach posterior aspect of cervix and upper vagina from lower end of sacrum
    - Form 2 ridges, 1 on either side of rectouterine pouch of Douglas

IMAGING ANATOMY

Ultrasound

- Transabdominal scan
  - Mucus within endocervical canal usually creates echogenic interface
  - In periovulatory phase, cervical mucus becomes hypoechoic due to high fluid content
  - Mucosal layer: Echogenic
    - Thickness and echogenicity shows cyclical changes similar to endometrium
  - Submucosal layer: Hypoechoic
  - Cervical stroma: Intermediate to echogenic
- Transvaginal scan
  - Angle of insonation should be optimized for best visualization
  - Imaging may be improved with withdrawal of probe into mid vagina

MR

- Important in local staging of cervical cancer
- Uniform intermediate signal on T1WI
- Zonal anatomy on T2WI
  - Endocervical canal: High signal
  - Cervical stroma: Predominately low signal, contiguous with junctional zone
  - Outer layer of smooth muscle (variably present): Intermediate signal
  - Parametrium: Variable signal intensity
    - Cardinal ligament and associated venous plexuses high signal
    - Sacrocervical ligament low signal

GRAPHICS OF CERVIX ANATOMY

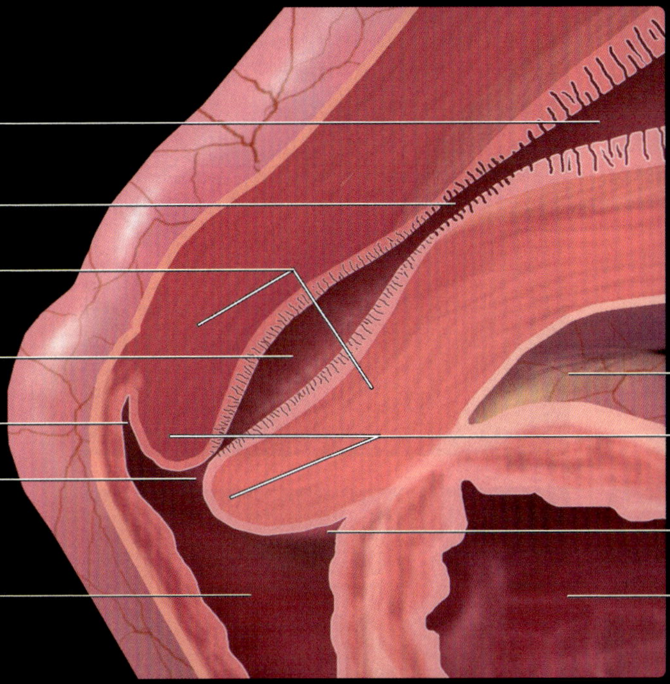

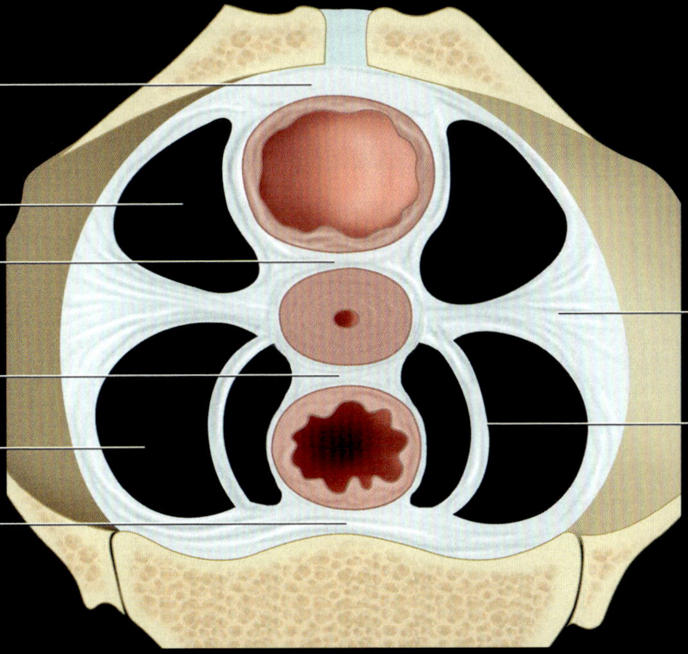

(Top) *Median sagittal graphic shows the cervix, which begins at the isthmus, the inferior narrowing portion of the uterus. It has a supravaginal portion (endocervix) and a vaginal portion (ectocervix), which divides the vagina into shallow anterior fornix, deep posterior, and lateral fornices.* **(Bottom)** *Graphic shows the female pelvic ligaments and spaces at the cervical/vaginal junction. The ligaments are visceral ligaments, which are composed of specialized endopelvic fascia and contain vessels, nerves, and lymphatics. Some of the main supporting ligaments for the uterus are attached to the cervix, which are cardinal and uterosacral ligaments. The spaces are largely filled with loose connective tissue and are used as dissection planes during surgery.*

TRANSVAGINAL ULTRASOUND OF CERVIX

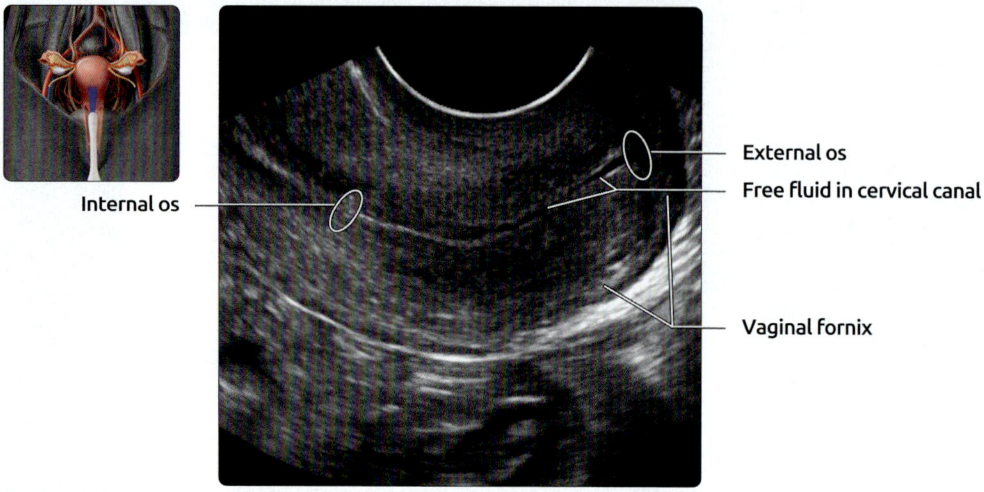

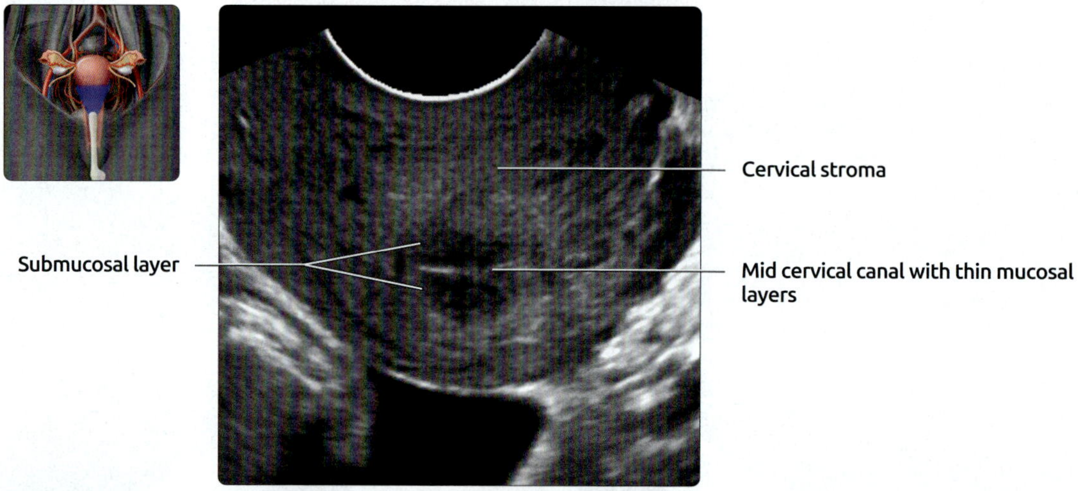

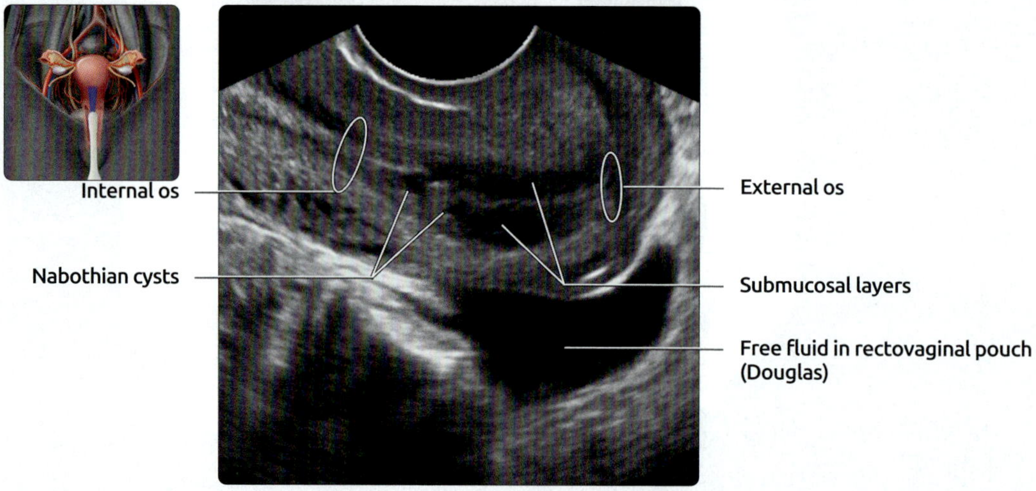

(Top) *Sagittal transvaginal ultrasound of the cervix shows hypoechoic fluid present in the endocervical canal. The endocervical canal is rich in mucus-secreting glands. The mucus secreted is usually slightly echogenic but becomes hypoechoic during periovulatory phase. As the transducer abuts the anterior lip of the cervix, the posterior wall of the vaginal fornix can be seen to cover the external os and extend along the posterior lip of the cervix.* **(Middle)** *Transverse transvaginal ultrasound of the cervix at the lower endocervical canal shows a typical appearance with an echogenic mucosal layer, a hypoechoic band of the submucosal layer, and intermediate echogenic stroma. The submucosal layer is filled with mucus-secreting glands leading to its hypoechoic appearance.* **(Bottom)** *Longitudinal transvaginal ultrasound shows the transducer abutting the anterior lip of the external os. The submucosal layer is thickened with typical low echogenicity.*

TRANSVAGINAL ULTRASOUND OF CERVIX

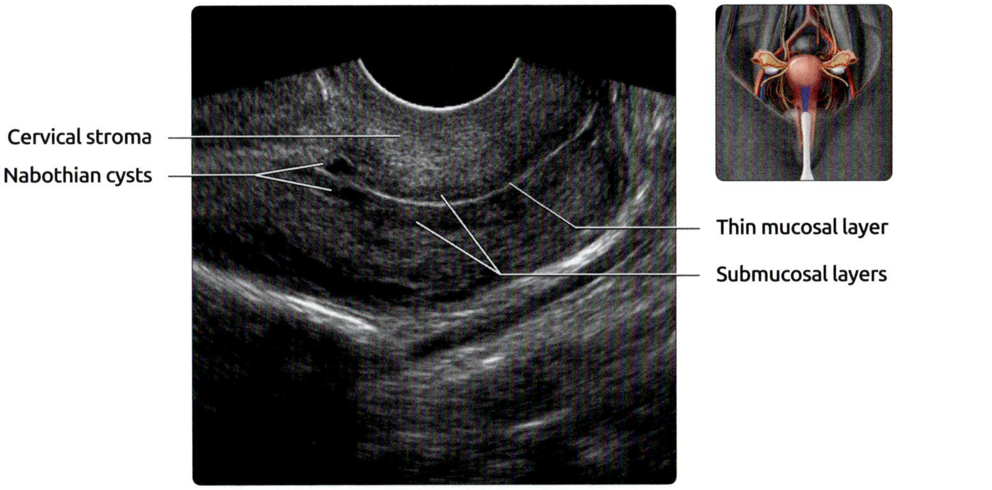

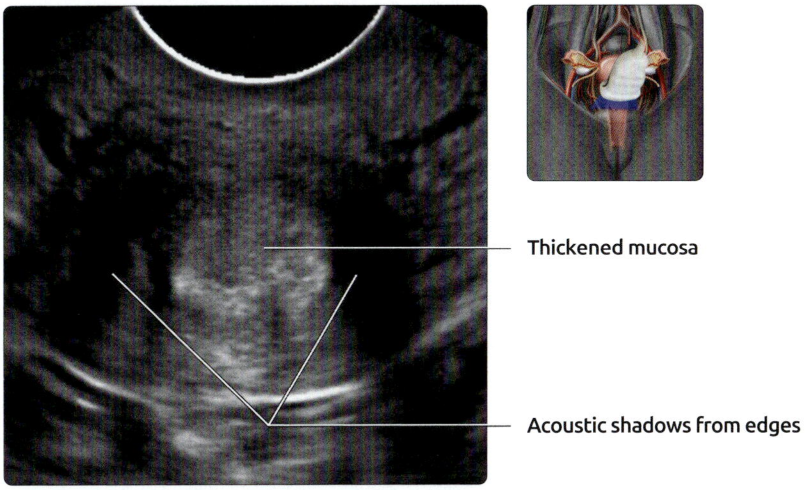

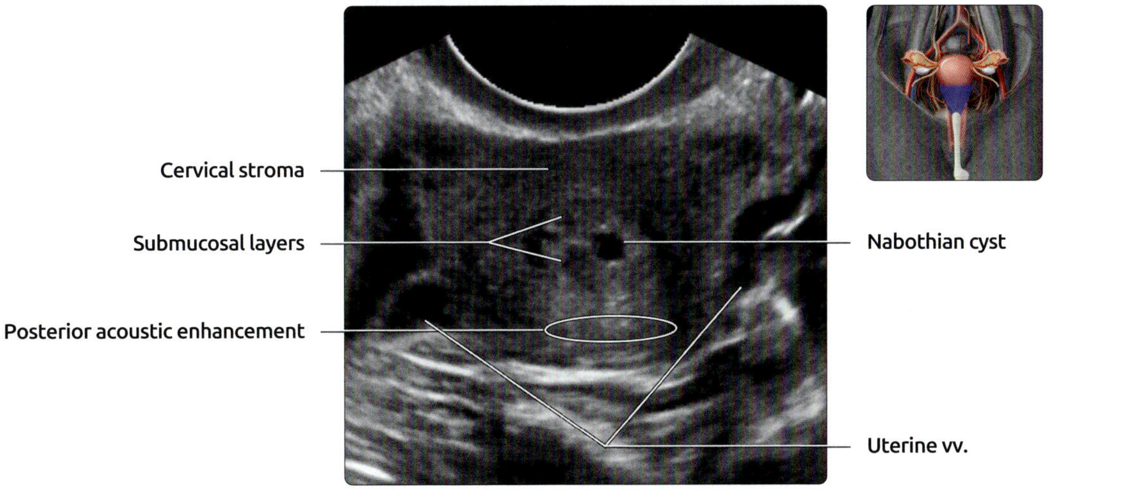

(Top) *Longitudinal transvaginal ultrasound of cervix shows 2 small nabothian cysts adjacent to the internal os. Nabothian cyst is a common sonographic finding in the cervix and is usually anechoic but sometimes can contain internal debris. It is generally of no clinical significance.* **(Middle)** *Transverse transabdominal ultrasound at the level of the cervix of a nonpregnant uterus commonly shows thickened mucosal layers. Note that during the menstrual cycle, the thickness and echogenicity of the mucosal layer undergoes changes as the endometrium does. When thickened, it typically casts shadowing from its edges.* **(Bottom)** *Transverse transvaginal ultrasound of the cervix at the mid endocervical canal shows the echogenic mucosal layers and thickened submucosal layers. A simple nabothian cyst is present with minimal posterior acoustic enhancement. The submucosal layer is filled with mucus-secreting glands leading to its hypoechoic appearance and posterior acoustic enhancement.*

TRANSABDOMINAL ULTRASOUND OF CERVIX

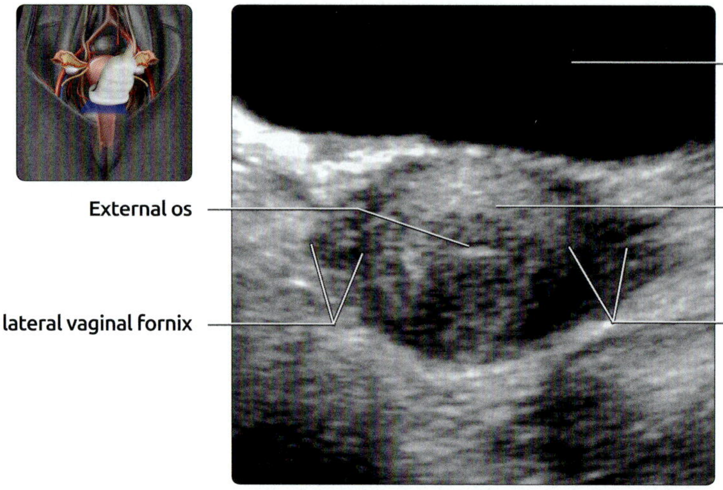

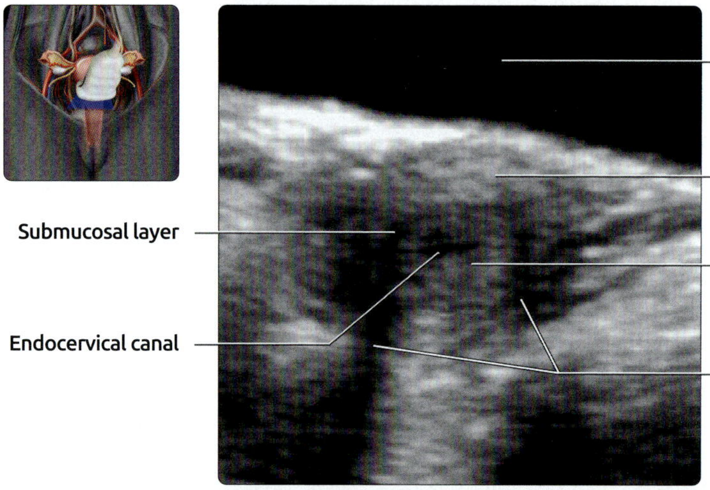

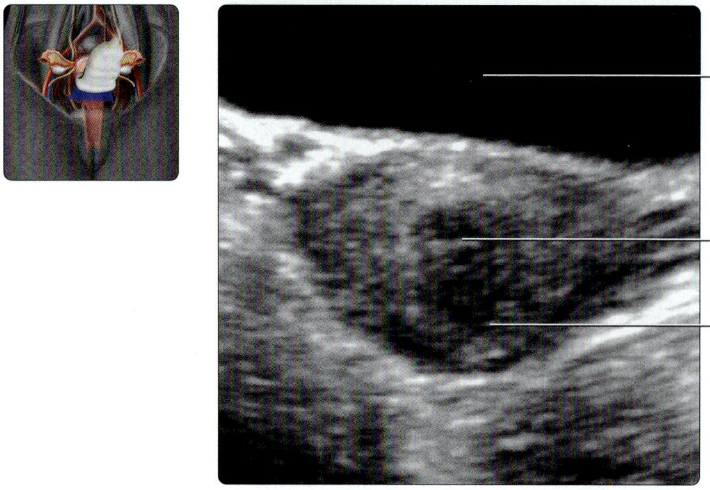

(Top) *Transverse transabdominal ultrasound shows the ectocervix at the level of external os. The lateral vaginal fornices are seen as relatively hypoechoic areas on each side of the ectocervix.* **(Middle)** *Transverse transabdominal ultrasound of the mid endocervix shows the mildly thickened and echogenic mucosal layer. Note that during the menstrual cycle, the thickness and echogenicity of the mucosal layer undergoes changes as the endometrium does. When thickened, it typically casts shadowing from its edges.* **(Bottom)** *Transverse transabdominal ultrasound shows the upper cervix at the level of the internal os, which opens into the uterine cavity. Identification of the internal os is clinically significant in pregnancy for placental site localization.*

CHANGES OF CERVIX DURING PREGNANCY

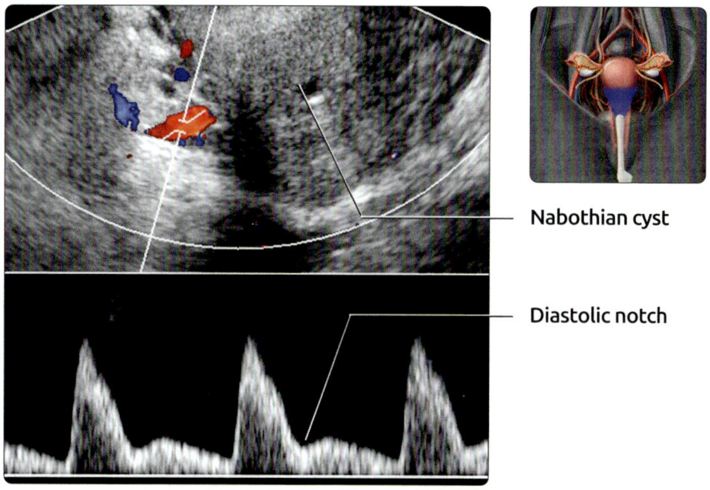

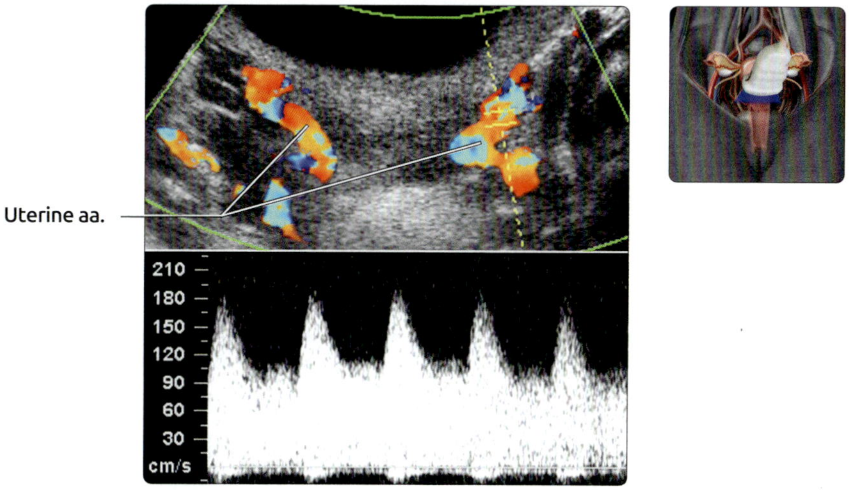

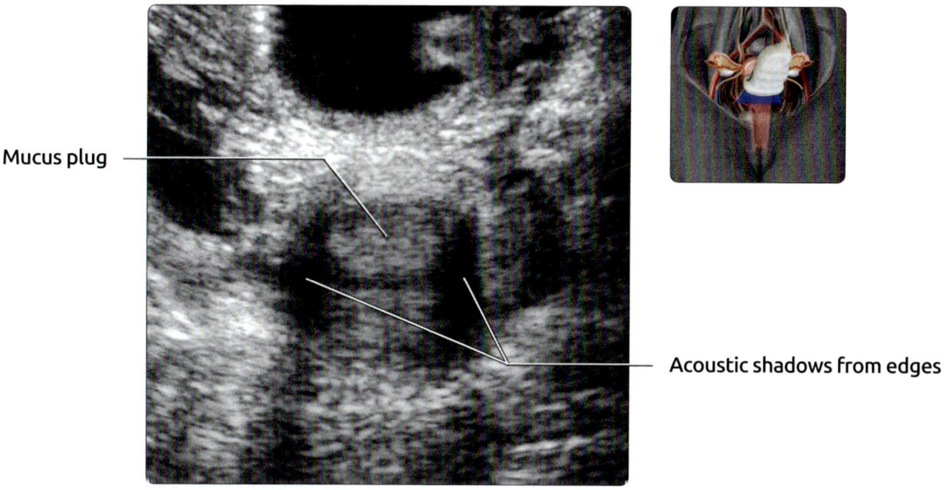

(Top) *Transverse transvaginal ultrasound shows spectral waveform of the uterine artery at the lateral margin of a nonpregnant cervix. There is typical high-resistance flow with a diastolic notch. In normal women, the Doppler waveform usually demonstrates a high-resistance pattern except in late secretory phase.* **(Middle)** *As softening of the cervix due to engorgement with blood becomes apparent by 6 weeks after conception, the changes can be reflected in the uterine artery with high-velocity, low-resistance flow seen on this transabdominal spectral Doppler examination.* **(Bottom)** *Transverse transabdominal ultrasound shows the typical thick and echogenic mucus plug in a pregnant cervix casting dense shadows from its edges.*

TERMINOLOGY

Abbreviations

- Vaginal artery (VA), uterine artery (UA)

GROSS ANATOMY

Overview

- Muscular tube formed by smooth muscle and elastic connective fibers
- Serves as excretory duct for uterus, female organ for copulation, and part of birth canal
- Extends up and back from vestibule of external genitalia to surround cervix of uterus
- Has anterior and posterior walls, normally in apposition, with longer posterior wall
- Superiorly, cervix projects downward and backward into vagina and divides vagina into shallow anterior, deep posterior, and lateral fornices
- Upper 1/2 of vagina lies above pelvic floor, lower 1/2 lies within perineum
- Lined with stratified squamous epithelium
- Inner mucosal surface of wall form rugae when collapsed
- Thin mucosal fold called hymen surrounds entrance to vaginal orifice
- Outer surface (adventitial coat) is thin fibrous layer continuous with surrounding endopelvic fascia
- Vasculature
    - Arterial supply
        - VA: Can branch directly from internal iliac artery (anterior trunk) or sometimes from inferior vesical artery or UA
        - Vaginal branches of UA
        - Branches of VA and UA anastomose to form 2 median longitudinal vessels: Azygos arteries, 1 in front and 1 behind vagina
    - Venous drainage
        - Form venous plexus around vagina
        - Eventually drains to internal iliac veins
- Variations with age
    - Menarche: 7-10 cm long
    - Postmenopausal: Shrinks in length and diameter; fornices virtually disappear

Anatomic Relationships

- Anterior
    - Superior: Bladder base
    - Inferior: Urethra
- Posterior
    - Upper 1/3: Rectouterine pouch of Douglas
    - Middle 1/3: Ampulla of rectum
    - Lower 1/3: Perineal body
- Lateral
    - Upper 1/3: Ureters
    - Middle 1/3: Levator ani and pelvic fascia
    - Lower 1/3: Bulb of vestibule, urogenital diaphragm, and bulbospongiosus muscles
- Ligamentous supports
    - Upper 1/3: Levator ani muscles, transverse cervical (cardinal), pubocervical, and sacrocervical ligaments
    - Middle 1/3: Urogenital diaphragm
    - Lower 1/3: Perineal body

IMAGING ANATOMY

Ultrasound

- Transabdominal US with distended bladder is standard imaging technique
    - Caudal angulation on both longitudinal and transverse scans
    - Commonly found at/near sagittal midline of pelvis
    - Length and wall thickness vary in response to bladder and rectal filling
    - Combined thickness of anterior and posterior vaginal walls should not exceed 1 cm for transabdominal scan with distended bladder
    - Characteristic appearance of 3 parallel lines
        - Highly echogenic mucosa centrally, may be difficult to visualize if stretched by distended bladder
        - Moderately hypoechoic muscular walls
- Transperineal US with nondistended bladder for assessment of uterine prolapse or for difficult cases
    - Vagina, especially vaginal canal, is less well-defined

EMBRYOLOGY

Embryologic Events

- Uterus and upper vagina are formed from paired müllerian (paramesonephric) ducts
- Paired ducts meet in midline and fuse, forming uterovaginal canal
- Lower vagina is formed from urogenital sinus

CLINICAL IMPLICATIONS

Uterine Prolapse

- Ligamentous support of pelvic organs may be damaged or become lax, leading to uterine prolapse or prolapse of vaginal walls
- Cystocele: Sagging of bladder with bulging of anterior vaginal wall
- Rectocele: Sagging of ampulla of rectum with bulging of posterior vaginal wall
- Best to be investigated by transperineal US supplemented with 3D

Müllerian Duct Anomalies

- Failure of müllerian duct development ± fusion
- Vagina most commonly affected in uterus didelphys (class III anomaly); vaginal septum seen in ~ 75% of cases

Pelvic Abscess

- Common site: Rectouterine pouch of Douglas
- Feasible for transvaginal US-guided drainage of pelvic abscess without doing major operation

Persistent Sexual Arousal Syndrome

- Persistent sexual arousal during sleep in postmenopausal women
- VA blood flow as 1 diagnostic aid
- VA normally shows high-resistance flow
- During sexual arousal, increased blood flow to VA with low-resistance spectral waveform

VAGINA IN SITU AND ARTERIAL SUPPLY

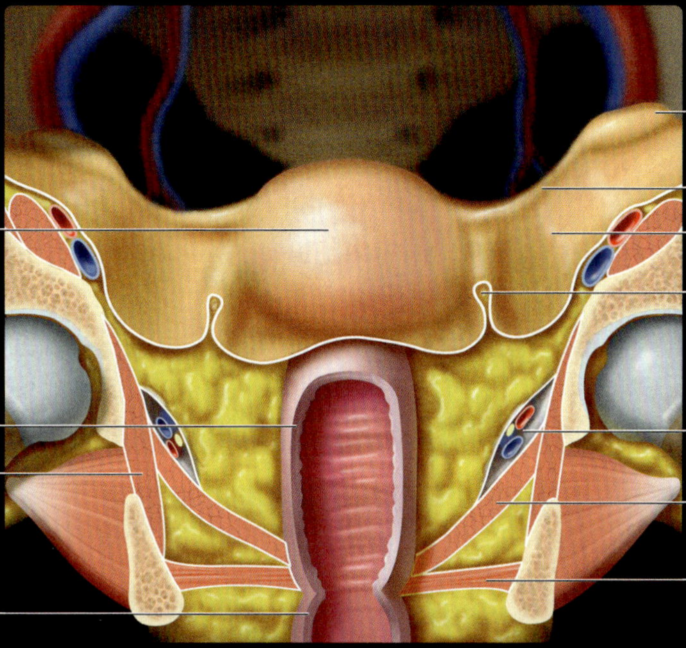

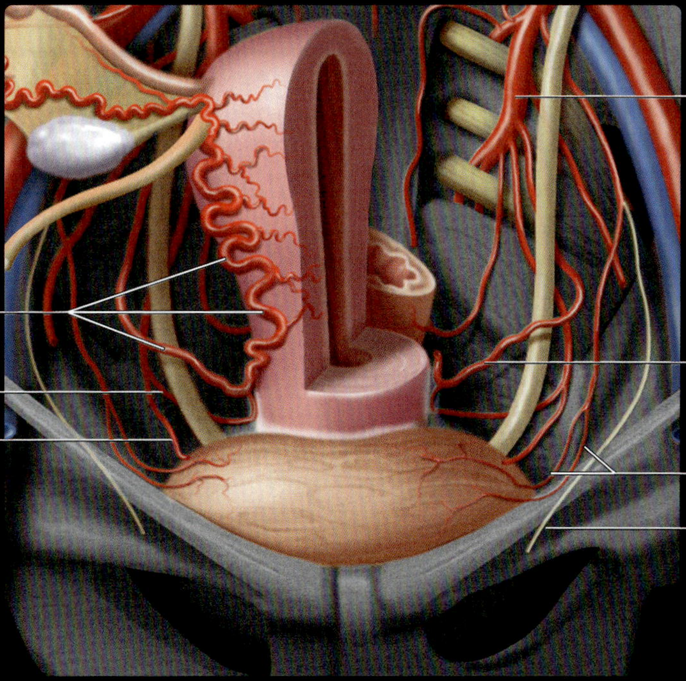

(Top) *Coronal view shows the pelvic floor at the level of the vagina. The levator ani muscles form the pelvic floor through which the urethra, vagina, and rectum pass and are the main support for the pelvic organs. The deep transverse perineal muscle and fascia, along with the urethral sphincter, form the urogenital diaphragm, which is the main support of the lower vagina.* **(Bottom)** *Frontal graphic shows the iliac vessels. The internal iliac artery divides into an anterior trunk and posterior trunk. The vaginal artery (VA) can branch off directly from the anterior trunk of the internal iliac artery or sometimes from the inferior vesical artery or uterine artery (UA). The arterial supply of the vagina includes the VA and vaginal branch of the descending trunk of UA.*

VAGINAL IMAGING BY VARIOUS TECHNIQUES

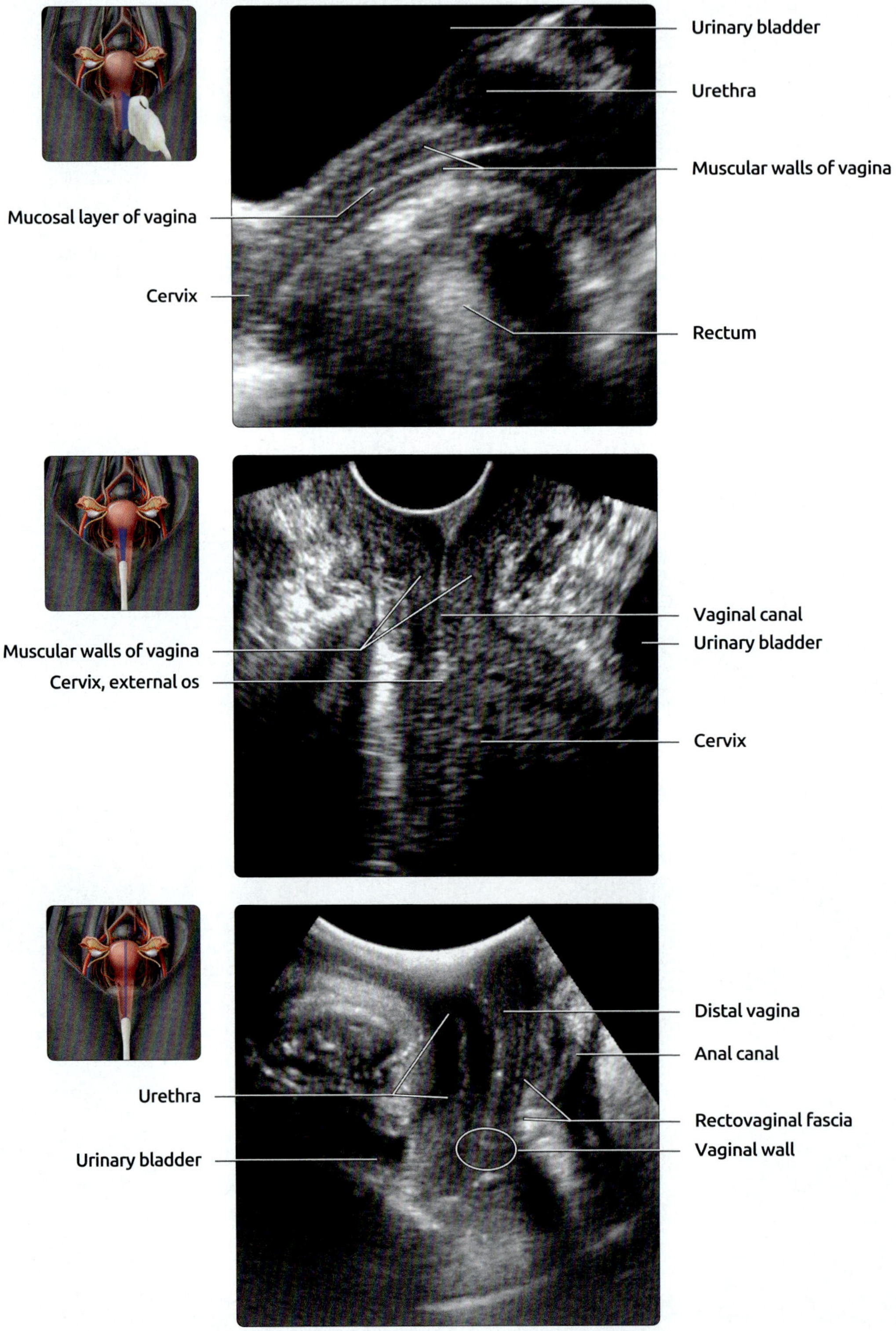

(Top) *Transabdominal midline sagittal US of the vagina shows characteristic triple-line echoes, i.e., hypoechoic muscular walls interfaced by echogenic mucosa. When looking for the vagina using transabdominal US, it is best to view with a distended bladder, starting at the midline near the cervical level and tilting the transducer further caudally.* **(Middle)** *Longitudinal transvaginal US of the vagina again shows the characteristic triple-line echo pattern. Using transvaginal US, gradually withdraw the high-frequency vaginal transducer so as to outline the vaginal canal.* **(Bottom)** *Transperineal sagittal US shows the vagina sandwiched between the urethra anteriorly and the rectum posteriorly. Note that the vaginal canal is barely visible in the absence of intraluminal acoustic jelly or fluid.*

TRANSVERSE US OF VAGINA

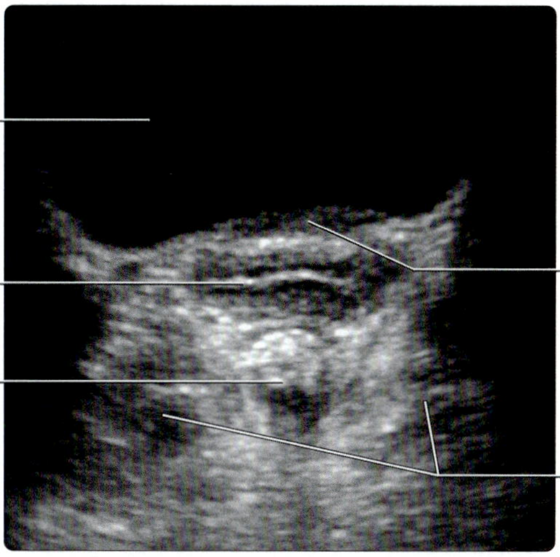

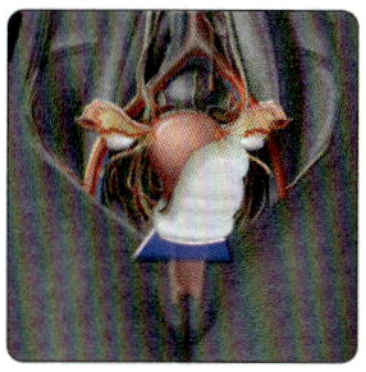

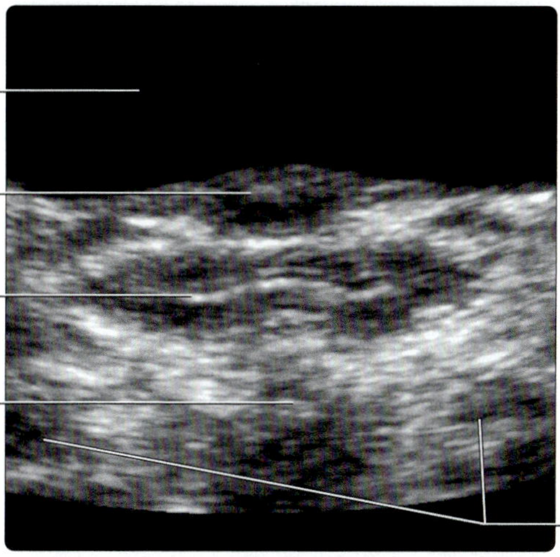

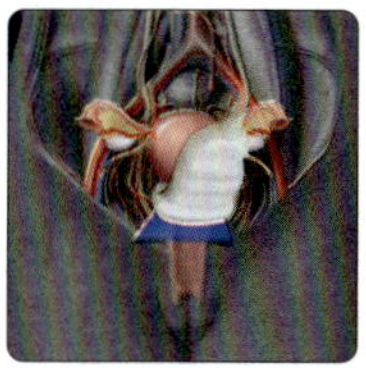

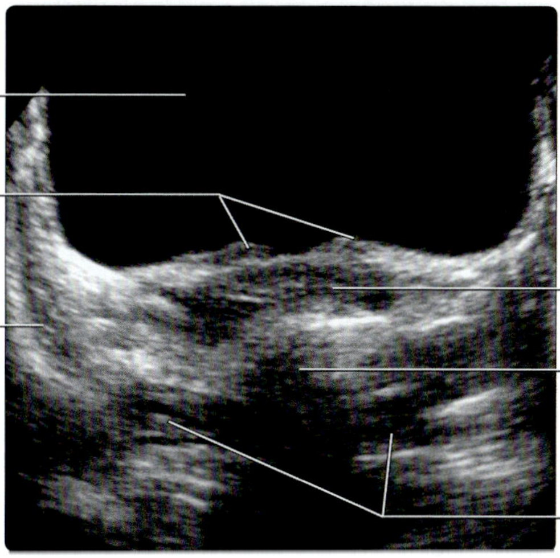

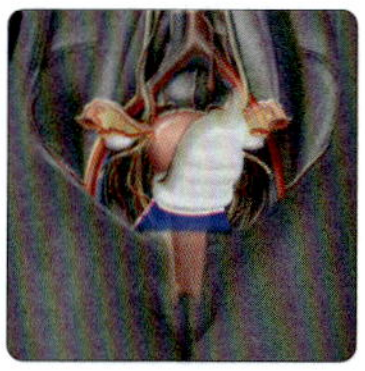

(Top) *Transverse transabdominal US shows the mid to lower vagina at the level of anal canal. For transabdominal US of the vagina, caudal angulation of the US probe is needed on both longitudinal and transverse scans. Note that the vaginal canal is better demonstrated on transabdominal US because the angle of insonation is more favorable, approaching a right angle.* **(Middle)** *Transverse transabdominal US of the mid vagina shows the levator ani muscles adjacent to the posterolateral aspect of the vagina.* **(Bottom)** *The upper vagina is shown at the level of the ureteric orifices. The ureters run lateral to the lateral fornices of the vagina, cross anteriorly, and then enter into the posterior wall of the bladder. This is a useful plane for investigation of the ureteric jets.*

COLOR DOPPLER US OF VAGINAL ARTERY

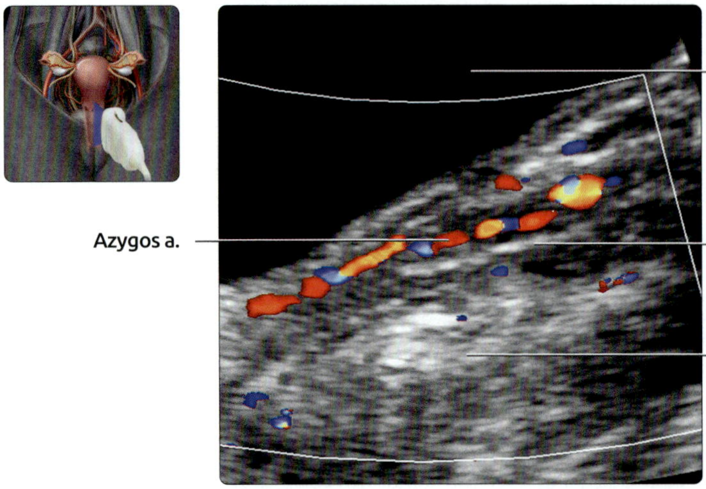

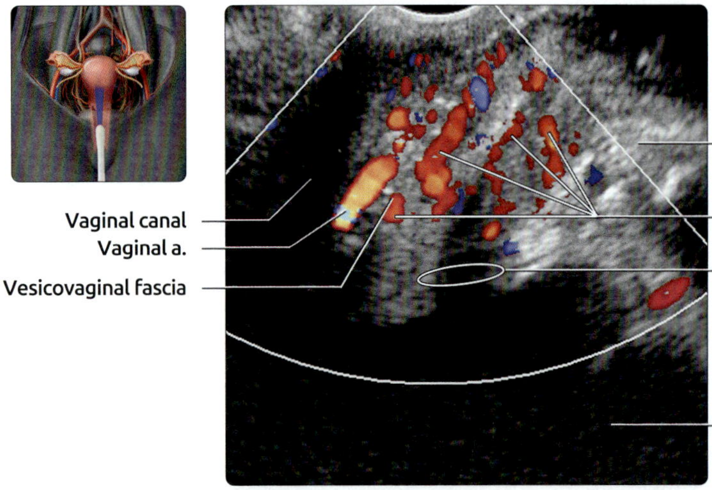

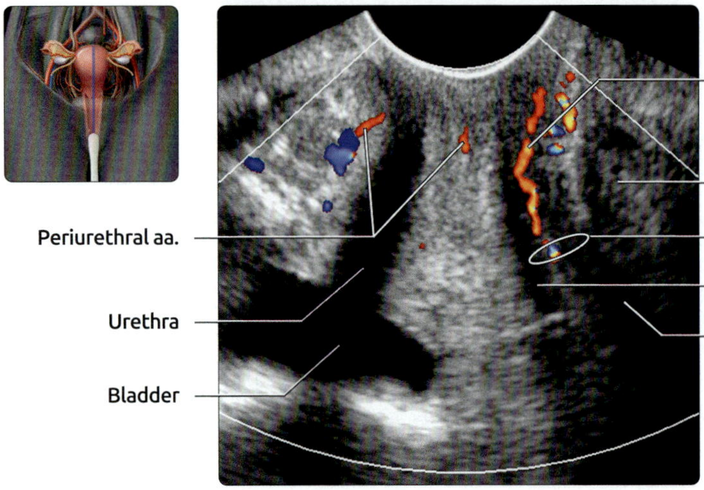

(Top) *Longitudinal transabdominal color Doppler US of the vagina shows the longitudinally running azygos artery, which arises from the anastomosis of vaginal branches of the UA and branches of the VA.* **(Middle)** *Longitudinal transvaginal color Doppler US shows the highly vascularized vagina with multiple small branches of the VA running along the vesicovaginal fascia.* **(Bottom)** *Transperineal sagittal color Doppler US shows a tortuous branch of the VA running along the vesicovaginal fascia in the vagina.*

SPECTRAL WAVEFORM OF VAGINAL ARTERY

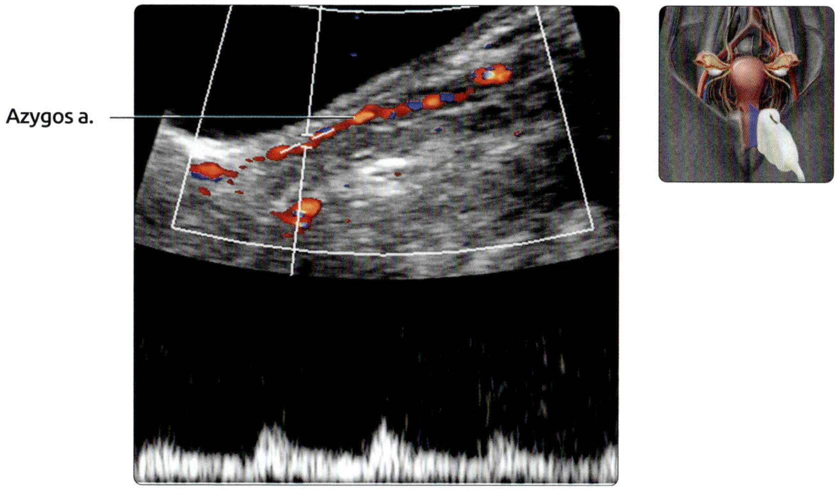

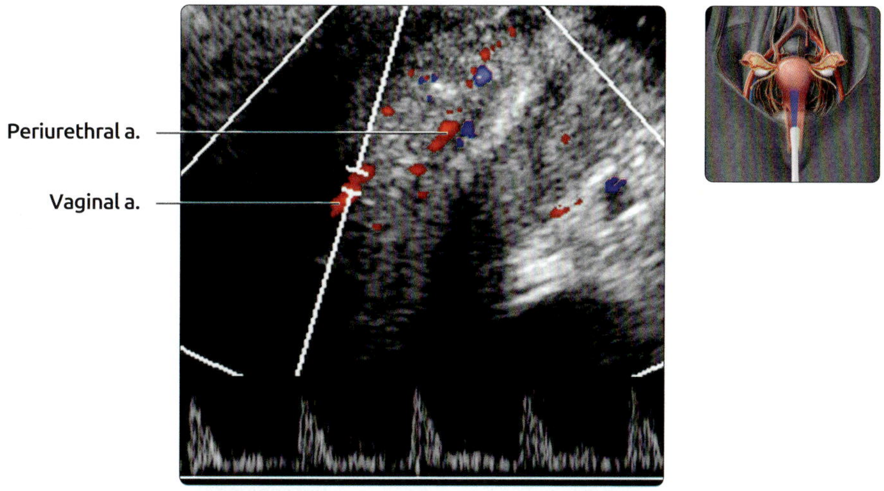

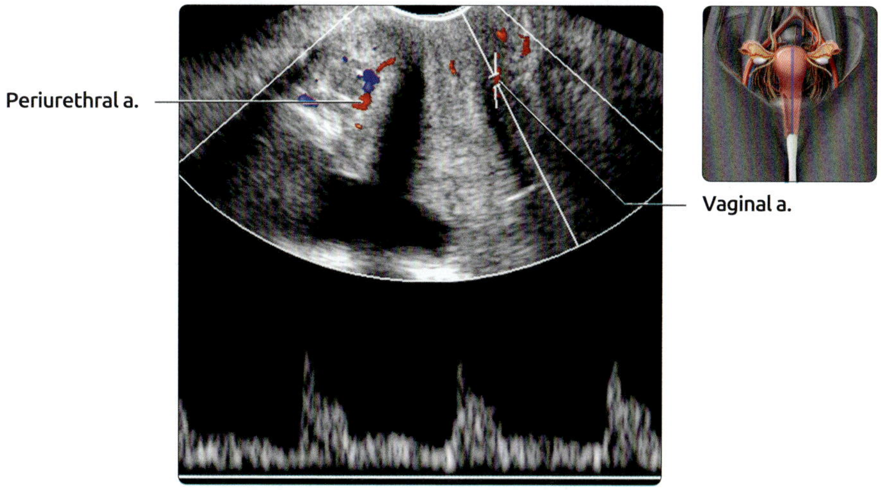

(Top) *Transabdominal spectral Doppler US of the azygos artery shows a low-resistance flow during mid cycle. The findings are most probably due to the influence of cyclical/hormonal change.* **(Middle)** *Spectral waveform of VA by transvaginal scan shows high-resistance flow, which is the most common pattern in normal females.* **(Bottom)** *Spectral Doppler US shows the VA by transperineal scan. The typical high-flow resistance in the VA may decrease during sexual arousal, cyclically or related to hormonal changes. This phenomenon is useful for investigation and management of sexual dysfunction in postmenopausal women.*

GROSS ANATOMY

Overview

- Ovaries located in true pelvis, although exact position variable
  - Only pelvic organ entirely inside peritoneal sac
  - Laxity in ligaments allows some mobility
  - Location affected by parity, bladder filling, ovarian size, and uterine size/position
  - Located within ovarian fossa in nulliparous women
    - Lateral pelvic sidewall below bifurcation of common iliac vessels
    - Anterior to ureter
    - Posterior to broad ligament
  - Position more variable in parous women
    - Pregnancy displaces ovaries, seldom return to same spot
- Fallopian tube drapes over much of surface
  - Partially covered by fimbriated end
- Composed of medulla and cortex
  - Vessels enter and exit ovary through medulla
  - Cortex contains follicles in varying stages of development
  - Surface covered by specialized peritoneum called germinal epithelium
- Ligamentous supports
  - Suspensory ligament of ovary (infundibulopelvic ligament)
    - Attaches ovary to lateral pelvic wall
    - Contains ovarian vessels and lymphatics
    - Positions ovary in craniocaudal orientation
  - Mesovarium
    - Attaches ovary to broad ligament (posterior)
    - Transmits nerves and vessels to ovary
  - Proper ovarian ligament (utero-ovarian ligament)
    - Continuation of round ligament
    - Fibromuscular band extending from ovary to uterine cornu
  - Mesosalpinx
    - Extends between fallopian tube and proper ovarian ligament
  - Broad ligament
    - Below proper ovarian ligament
- Arterial supply: Dual blood supply
  - Ovarian artery is branch of aorta, arises at L1/L2 level
    - Descends to pelvis and enters suspensory ligament
    - Continues through mesovarium to ovarian hilum
    - Anastomoses with uterine artery
- Drainage via pampiniform plexus into ovarian veins
  - Right ovarian vein drains to inferior vena cava
  - Left ovarian vein drains to left renal vein
- Lymphatic drainage follows venous drainage to preaortic lymph nodes at L1 and L2 levels

Physiology

- ~ 400,000 follicles present at birth, but only 0.1% (400) mature to ovulation
- Variations in menstrual cycle
  - Follicular phase (days 0-14)
    - Several follicles begin to develop
    - By days 8-12, dominant follicle develops, while remainder start to regress
  - Ovulation (day 14)
    - Dominant follicle, typically 2.0-2.5 cm, ruptures and releases ovum
  - Luteal phase (days 14-28)
    - Luteinizing hormone induces formation of corpus luteum from ruptured follicle
    - If fertilization occurs, corpus luteum maintains and enlarges to corpus luteum cyst of pregnancy

Variations With Age

- At birth: Large ovaries ± follicles due to influence of maternal hormones
- Childhood: Volume < 1 cm^3, follicles < 2-mm diameter
- Above 8 year old: ≥ 6 follicles of > 4-mm diameter
- Adult, reproductive age: Mean volume ~ 10 ± 6 cm^3, max 22 cm^3
- Postmenopausal: Mean ~ 2-6 cm^3, max 8 cm^3 and may contain few follicle-like structures

IMAGING ANATOMY

Ultrasound

- Scan between uterus and pelvic sidewall
  - Ovaries often seen adjacent to internal iliac vessels
- Medulla mildly hyperechoic compared to hypoechoic cortex
- Dominant follicle around time of ovulation
  - Cumulus oophorus: Nodule or cyst along margin of dominant follicle represents mature ovum
- Corpus luteum may have thick, echogenic ring
  - Doppler: Vascular wall or "ring"
  - Hemorrhage common
- Echogenic foci common
  - Nonshadowing, 1-3 mm
  - Represent specular reflectors from walls of tiny unresolved cysts or small vessels in medulla
- Doppler: Low-velocity, low-resistance arterial waveform
- Volume (0.523 x length x width x height) more accurate than individual measurements

ANATOMY IMAGING ISSUES

Imaging Recommendations

- Transabdominal (TA) US with full bladder is good for overview of pelvic organs
  - Detects ovaries and masses superior to uterus that may be missed by transvaginal (TV) US
- TV US is excellent in assessing detail of ovaries and characterizing lesions compared to TA US
  - Lesions higher in pelvis can be missed because of limited field of view
- Postmenopausal ovaries can be difficult to detect because of atrophy, paucity of follicles and surrounding bowel

LIGAMENTOUS SUPPORT AND ANATOMY OF OVARY

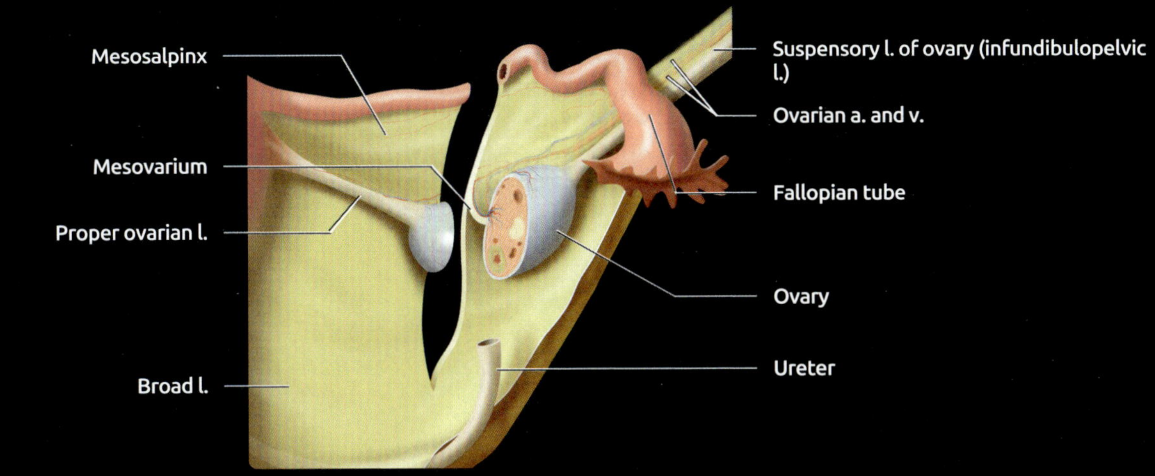

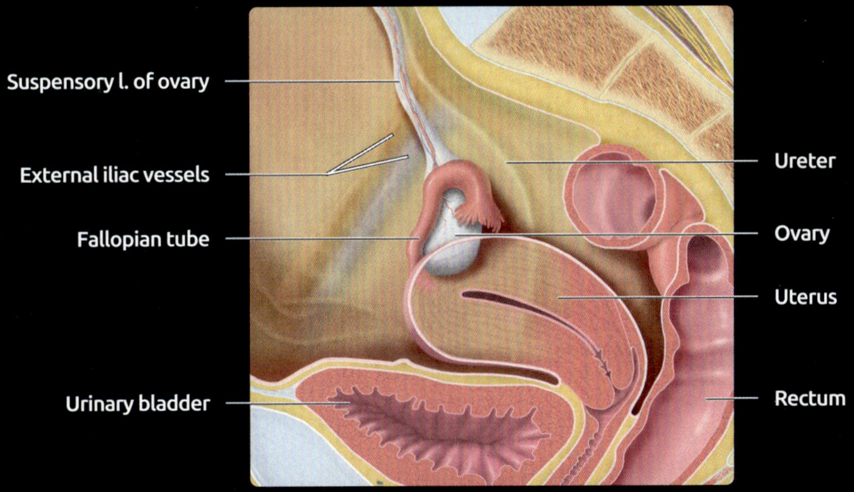

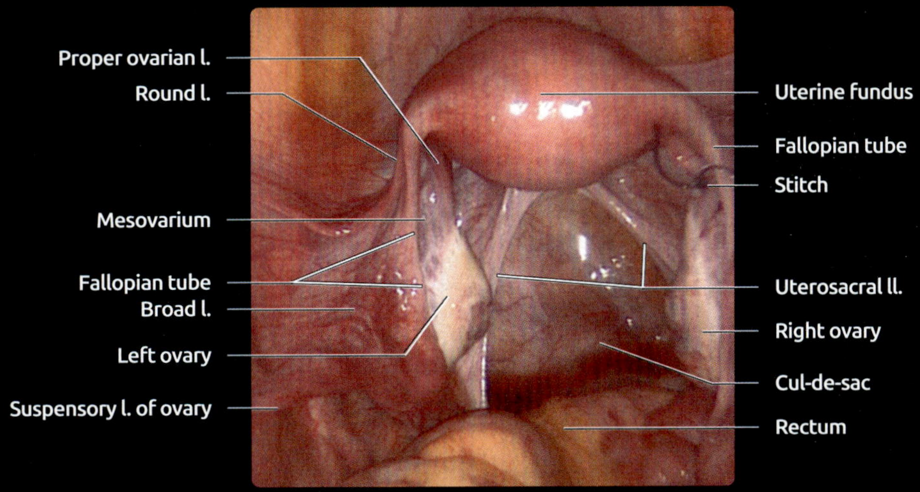

(Top) *Posterior view of the ligamentous attachment of the ovary is shown. The ovary is attached to the pelvic sidewall by the suspensory ligament (infundibulopelvic ligament) of the ovary, which transmits the ovarian artery and vein. These vessels enter the ovary through the mesovarium, a specialized ligamentous attachment between the ovary and broad ligament. The ovary is attached to the uterus by the proper ovarian ligament, which divides the mesosalpinx above from the broad ligament below.* **(Middle)** *Sagittal graphic of the female pelvis shows the location of the ovary, which lies in the ovarian fossa, the area below the iliac bifurcation, posterior to the external iliac vessels, and anterior to the ureter.* **(Bottom)** *Photograph during laparoscopy viewing the uterine fundus from above demonstrates the ligamentous structures of the ovary and uterus.*

NORMAL OVARY, VARIATIONS WITH AGE

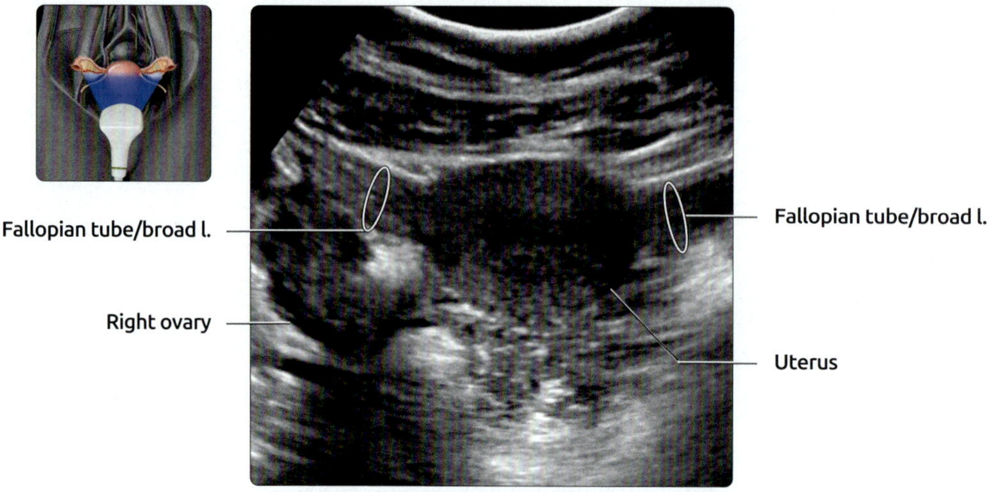

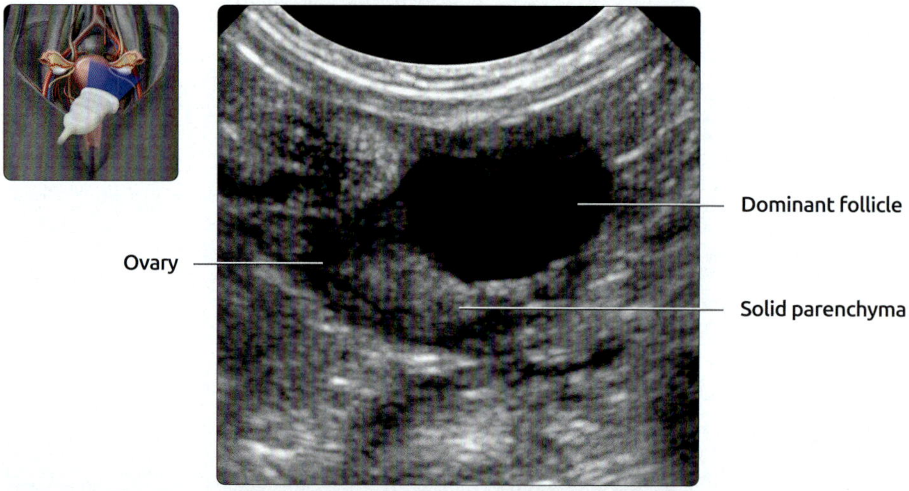

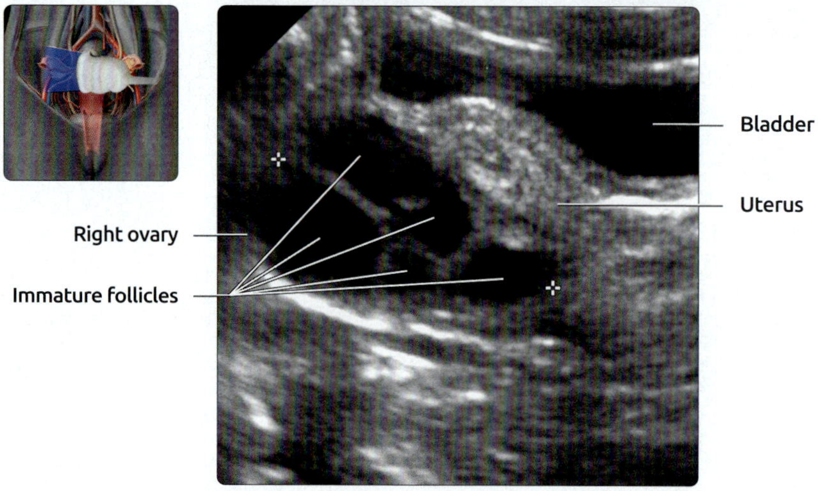

(Top) *Transverse TA ultrasound at the level of the uterine fundus in a 23-year-old woman shows the right ovary in the typical position of the ovarian fossa. The fallopian tube and broad ligament can sometimes be seen as a band of tissue connecting the ovary to the uterine horn. Ovarian ligaments can be lax making ovarian position quite variable from above the fundus to the posterior rectouterine pouch of Douglas.* **(Middle)** *Transverse TA ultrasound of the ovary in a neonate is shown. The size of the ovary is enlarged with a dominant follicle related to stimulation from residual maternal gonadotrophins. Visible follicles may persist until 9 months of age or longer.* **(Bottom)** *Longitudinal TA ultrasound of the ovary of a 5-month-old girl is shown. The ovary is slightly prominent due to stimulation from maternal hormones. The ovary is small (total volume of 1.7 cc) with immature follicles of variable size (usually < 0.9 cm). The size of the ovaries change very little in the first 6 years of life.*

NORMAL OVARY, VARIATIONS WITH AGE

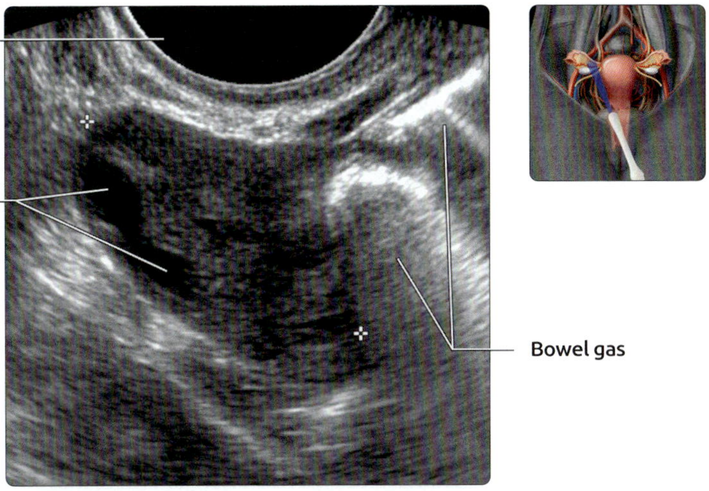

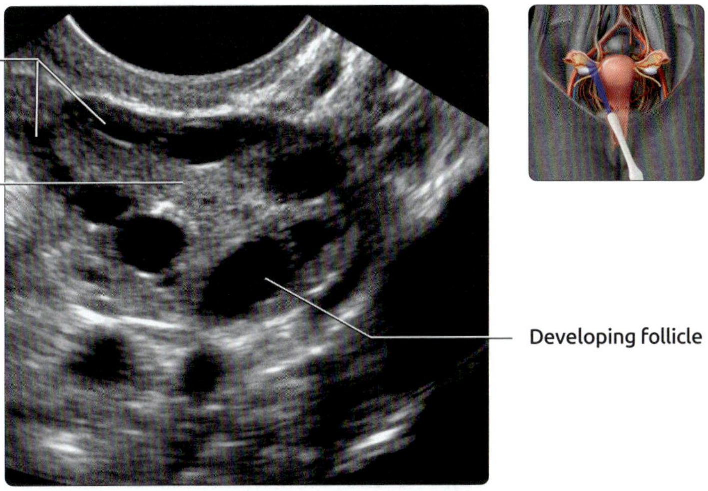

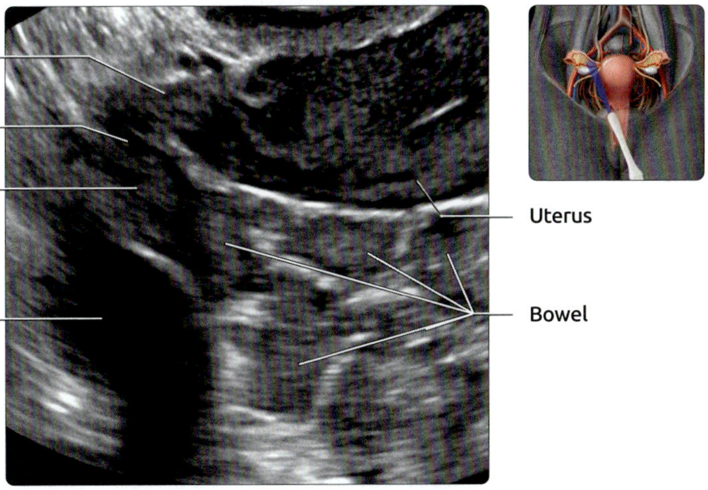

(Top) *Transverse TV ultrasound of the right ovary in a 18-year-old woman shows an oval ovary with immature follicles. This image also demonstrates the superior resolution of TV ultrasound, from the ability to bring the ovary to the near field, as well as the higher frequencies used.* **(Middle)** *Transverse TV ultrasound of an adult ovary is shown. There are multiple developing follicles of variable size around the echogenic ovarian stroma (medulla) where the ovarian vessels and lymphatics enter and exit. Ovulation usually occurs when the follicle enlarges to be between 2.0 and 2.5 cm.* **(Bottom)** *TV ultrasound in a 74-year-old postmenopausal woman demonstrates atrophic right ovary containing tiny cysts. Normal ovaries are variably detected in the postmenopausal woman because of their small size, lack of follicles, and surrounding bowel loops.*

COLOR DOPPLER IMAGING OF OVARIAN ARTERY

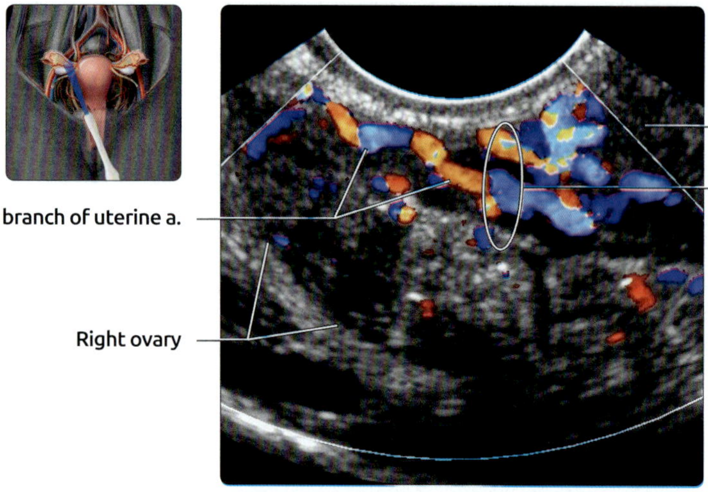

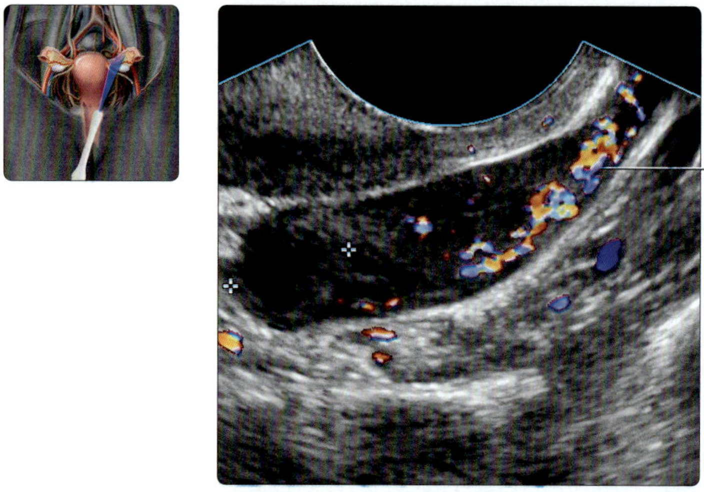

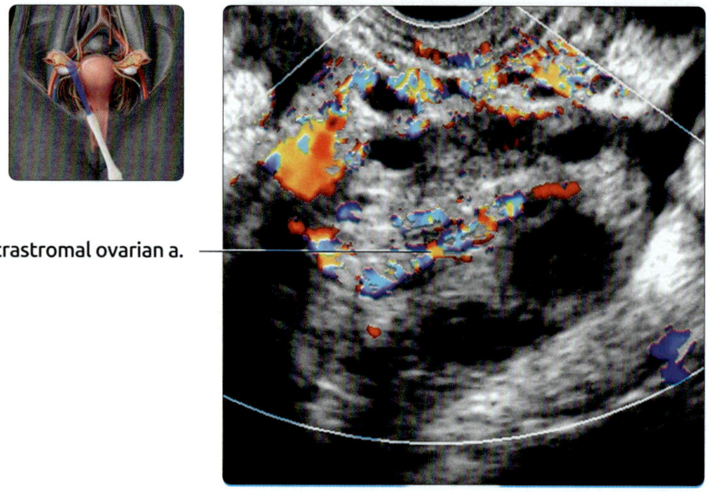

(Top) *Transverse TV ultrasound shows the ovarian branch of the uterine artery running in the broad ligament. It begins at the uterine horn and goes to the ovary through the proper ovarian ligament/mesovarium anastomosing with the ovarian artery.* **(Middle)** *Transverse TV ultrasound of the left ovary demonstrates involuting corpus luteum (calipers). The ovarian vessels are seen entering the ovary from the suspensory ligament of the ovary.* **(Bottom)** *Longitudinal color Doppler TV ultrasound shows an intrastromal ovarian artery running in the ovarian medulla. Note the ovarian vascularity will progressively increase after menstruation and approach a maximum in the luteal phase.*

SPECTRAL WAVEFORM OF OVARIAN ARTERY

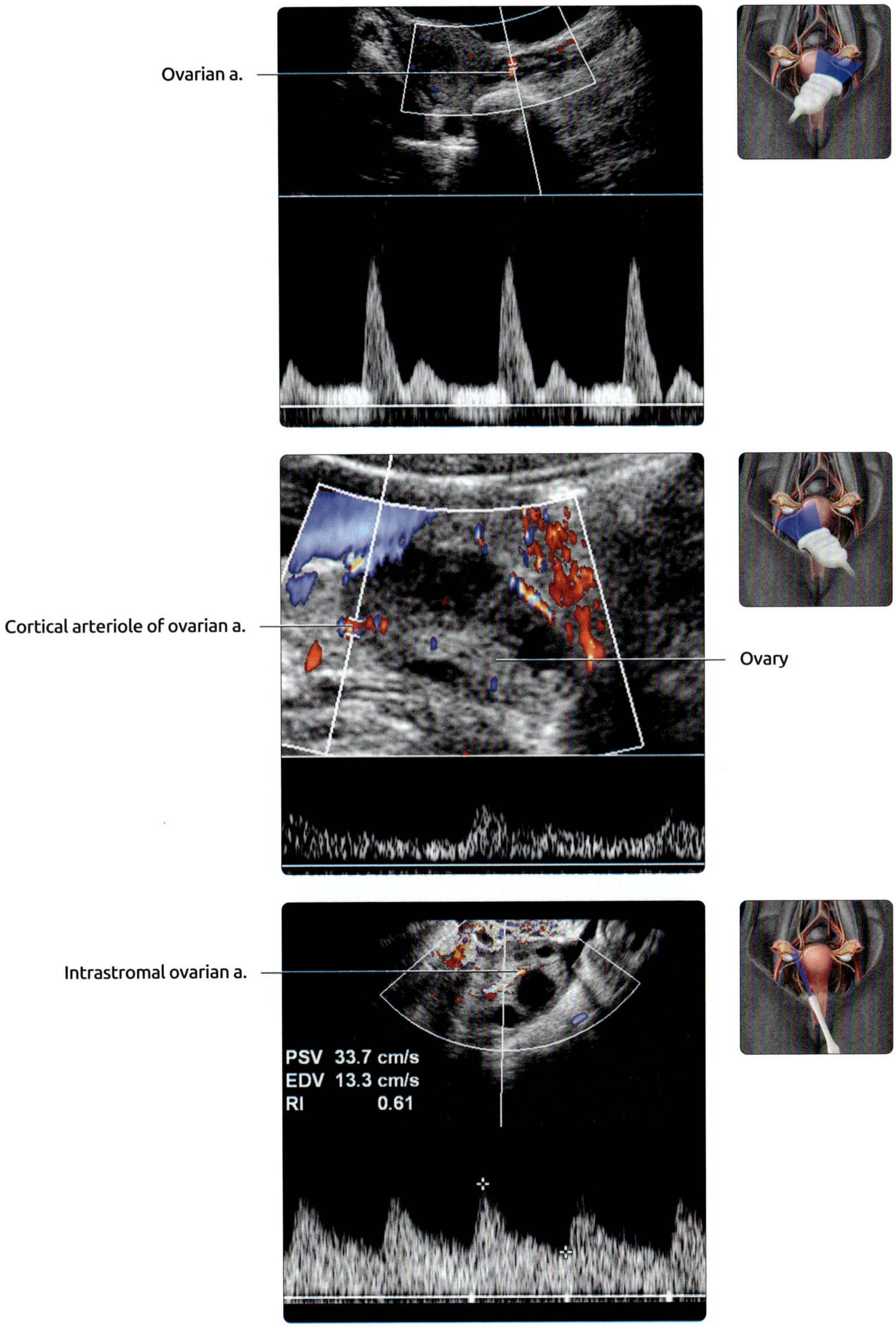

(Top) *Transverse spectral Doppler TA ultrasound shows a normal ovarian artery with a high-resistance flow pattern suggestive of an inactive state of the ovary.* **(Middle)** *Transverse spectral Doppler TA ultrasound shows the waveform of the cortical arteriole of the ovarian artery.* **(Bottom)** *Transverse spectral Doppler TV ultrasound of the intrastromal ovarian artery as a continuation of the straight cortical arteriole shows a typical low-resistance, low-velocity waveform during the luteal phase.*

CYCLIC CHANGES OF OVARY

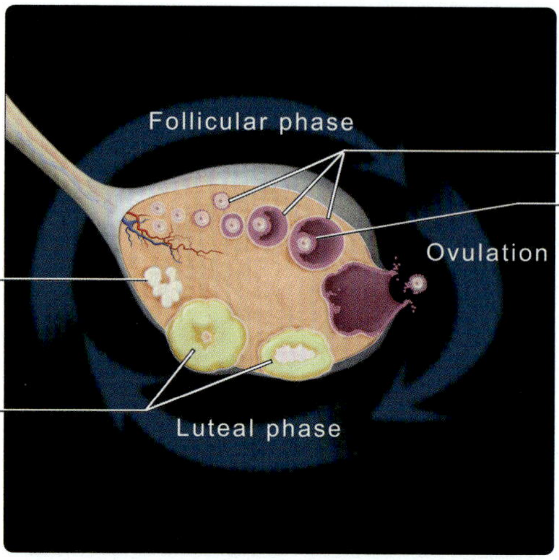

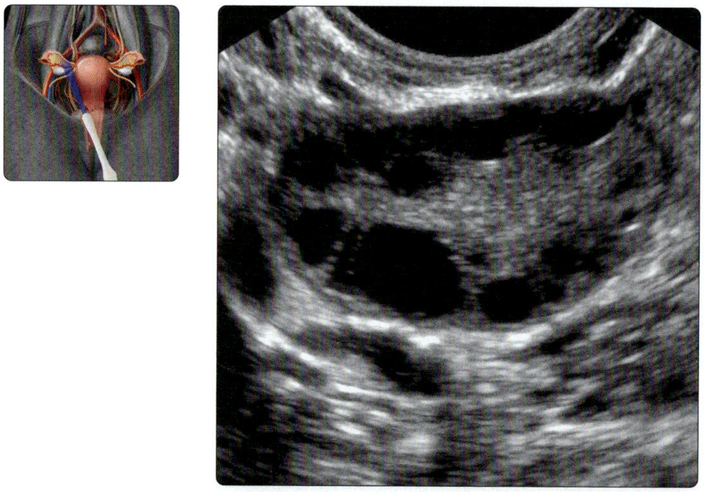

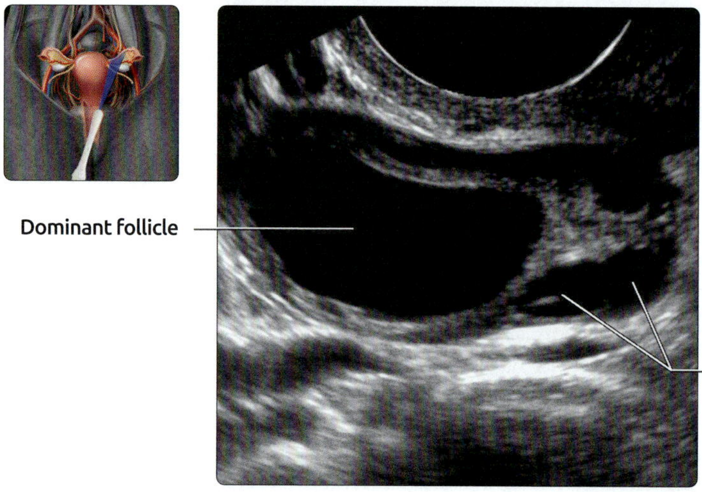

(Top) *During the follicular phase of the menstrual cycle, several follicles begin to develop, but by days 8-12, a dominant follicle has formed, and the remainder begin to regress. On day 14, the follicle ruptures, and the egg is released. After ovulation, a corpus luteum forms, and if fertilization does not occur, the corpus luteum degenerates into a corpus albicans.* **(Middle)** *Longitudinal TV ultrasound of the ovary at the early follicular phase is shown. Note the developing follicles of variable size at the periphery of the ovary.* **(Bottom)** *TV ultrasound shows a dominant follicle developed in the late follicular phase days before ovulation. This should not be confused for a pathologic cyst.*

CYCLIC CHANGES OF OVARY

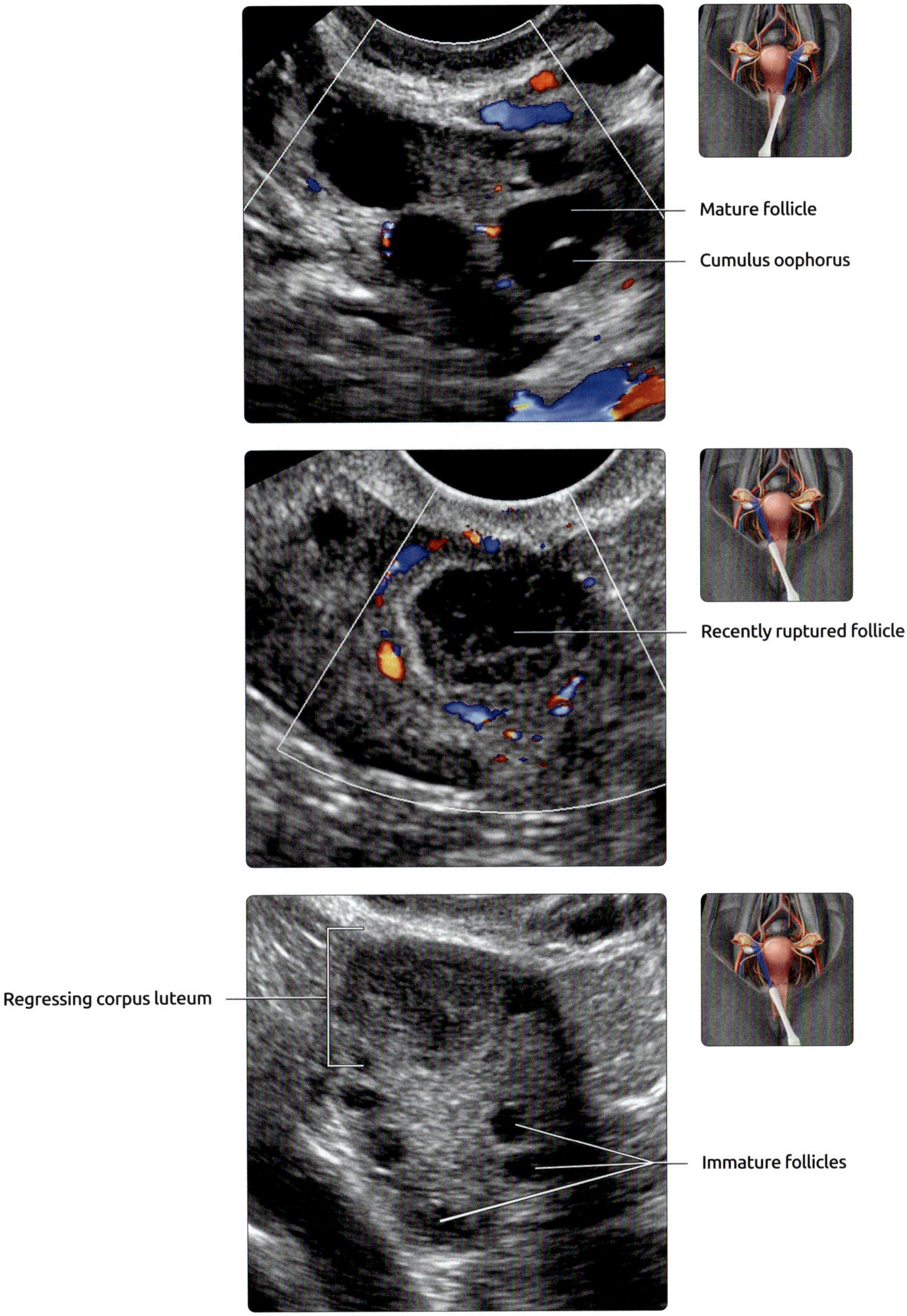

(Top) *Transverse TV color Doppler ultrasound of the ovary demonstrates a large mature follicle with a small cyst on its wall representing a cumulus oophorus. The size of a mature follicle can reach up to 25 mm before ovulation.* **(Middle)** *Longitudinal TV color Doppler ultrasound of the ovary shows a dominant follicle immediately after its rupture at ovulation. Note the partially collapsed wall resulting from loss of part of the liquor folliculi and the hypoechoic internal contents representing blood.* **(Bottom)** *Transverse TV ultrasound of the ovary shows a regressing corpus luteum with typical hypoechoic, thick, crenulated wall, and echogenic internal contents representing blood.*

CYCLIC CHANGES OF INTRAOVARIAN ARTERY

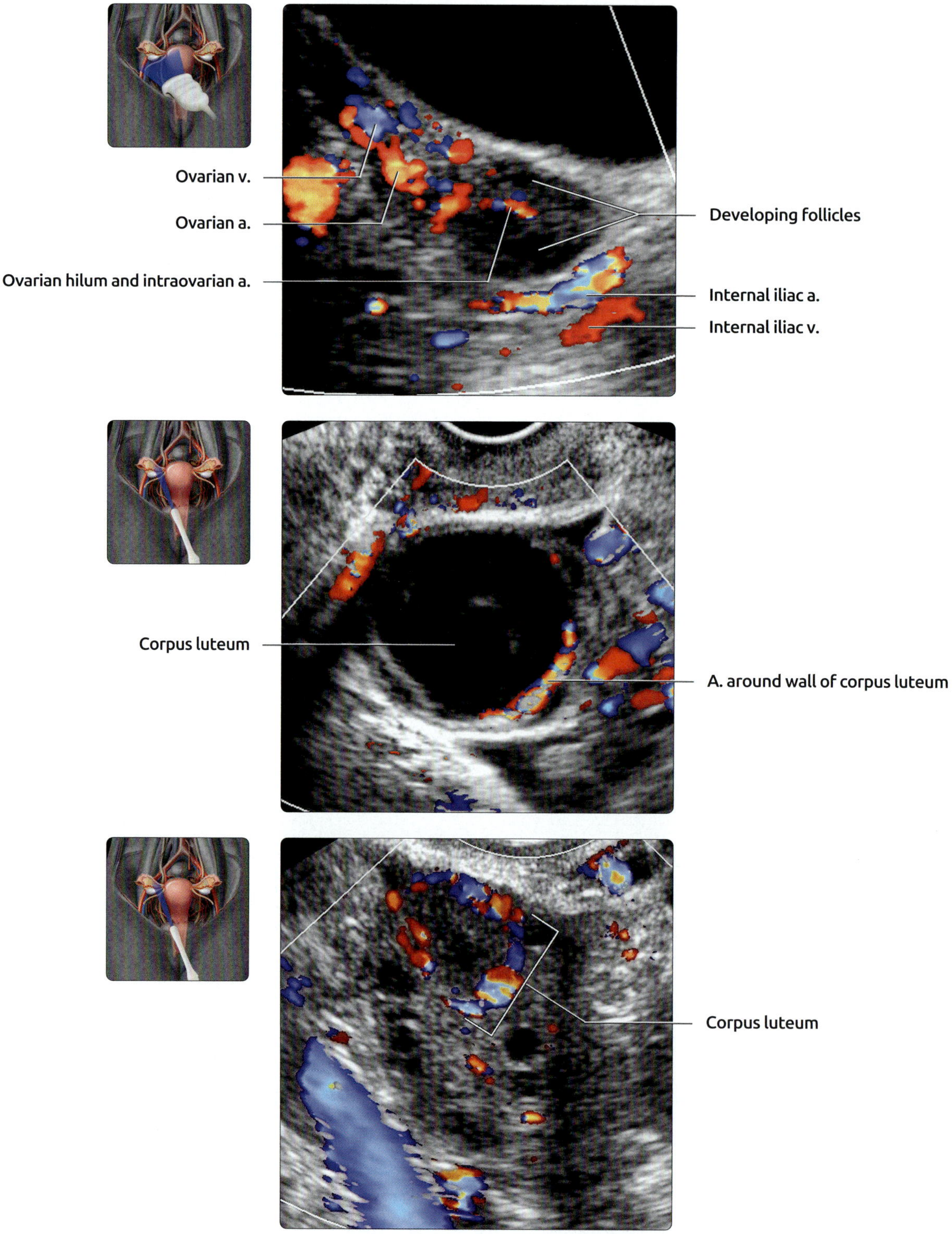

(Top) *Color Doppler TA ultrasound shows an inactive ovary. The ovary demonstrates a hilar artery surrounded by small developing follicles in the early follicular phase. Note the nondominant ovary may show a similar appearance as an inactive ovary.* **(Middle)** *Longitudinal TV ultrasound of the ovary in the early luteal phase is shown. A corpus luteum with low-level internal echoes is seen following ovulation. The wall of the corpus luteum usually displays the most intense color pattern.* **(Bottom)** *Transverse TV ultrasound in midluteal phase is shown. The ovary shows a regressing corpus luteum with typical peripheral color Doppler vascularity.*

CYCLIC CHANGES OF INTRAOVARIAN ARTERY

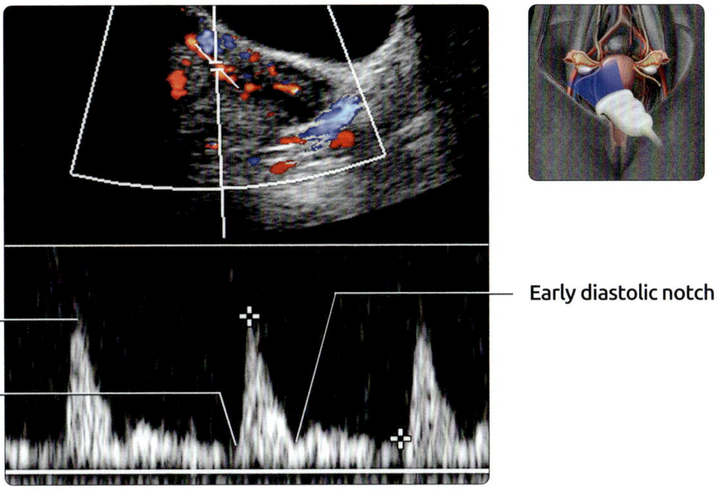

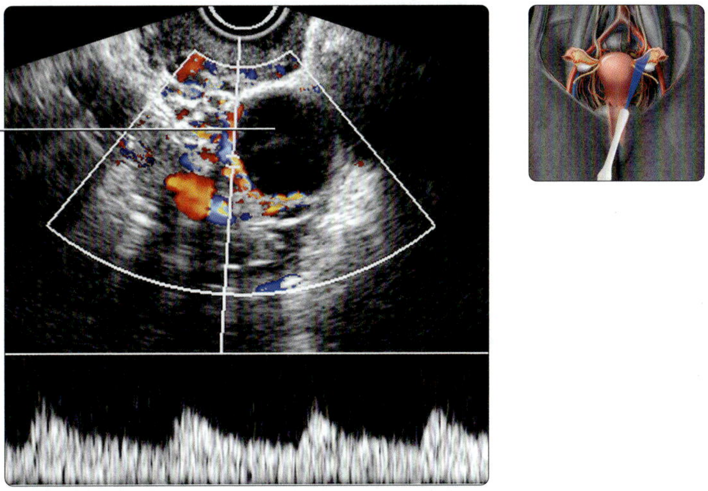

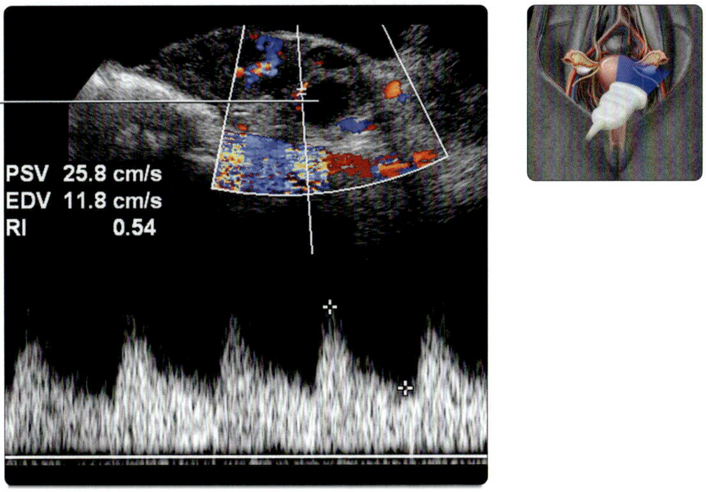

(Top) *Transverse spectral Doppler TA ultrasound of the ovarian artery is shown. The ovarian artery blood flow shows a high-resistance flow pattern with low end-diastolic velocity and an early diastolic notch. This notch indicates initial resistance to forward flow through the ovarian parenchyma. The flow resistance is at its maximum during the first 8 days of the cycle.* **(Middle)** *Spectral Doppler TV ultrasound of the intraovarian artery in early luteal phase is shown. The ovarian artery has a low-resistance flow, which reaches the lowest level in early luteal phase. At this time, the intraovarian vascularity is easily detectable.* **(Bottom)** *Spectral Doppler of TA ultrasound of the intraovarian artery in midluteal phase is shown. The ovarian arterial flow is of medium resistance, and the flow resistance will gradually increase through to the regenerative phase.*

GROSS ANATOMY

Functional Anatomy

- **Classic 3-compartment approach**
  - **Anterior**: Includes urinary bladder, urethra, and urethral support system
  - **Middle**: Includes vagina (anterior and posterior wall) and uterocervical support
  - **Posterior**: Contains rectum and supporting structures
- **Active** (pelvic floor muscles) and **passive** (pelvic bones, supportive connective tissue) **conceptual approach**
- **Multilayered system approach**: Considers passive and active components of pelvic floor as integrated multilayer system, organized from cranial to caudal
  - **1st layer: Endopelvic fascia**
  - **2nd layer: Pelvic diaphragm**
  - **3rd layer: Urogenital diaphragm**
  - **4th layer: Superficial external genital muscles**
- **Functional supporting systems approach**: New, more function-based classification of pelvic floor support system
  - **Urethral support system**: Structures that maintain urinary continence
  - **Vaginal support system**: Supporting elements that prevent prolapse
  - **Maintenance of anal continence**: Supporting elements and anal sphincter complex

Bony Pelvis & Walls

- **Osseous structures**: 2 iliac bones, sacrum and coccyx
  - Pubic bones meet in midline at pubic symphysis
  - Pelvis is divided into 2 parts by pelvic brim
    - False pelvis above forms part of abdominal cavity
    - True pelvis below pelvic brim
- Walls
  - **Anterior wall** formed by posterior surfaces of bodies of pubic bone, symphysis pubis (SP), and pubic rami
  - **Posterior wall** formed by coccyx and sacrum, piriformis muscles, and their covering parietal pelvic fascia
  - **Lateral wall** formed by Ilium, ischium, and obturator internus muscle and fascia
- Supporting ligaments: Sacrotuberous and sacrospinous

Pelvic Diaphragm

- **Formed by coccygeus and levator ani muscles**
  - Acts as shelf to support pelvic organs
- **Coccygeus muscle**
  - Broad musculotendinous structure forming posterior part of pelvic diaphragm
- **Levator ani muscle**: Divided anatomically into 3 components (differentiated according to origin and direction of fiber bundles)
  - **Puborectalis muscle**
    - Arises from superior and inferior pubic rami
    - Unites with contralateral puborectalis muscle posterior to rectum, forming sling
    - Does not insert onto any skeletal structure
  - **Pubococcygeus muscle**
    - Arises from back of pubic bone and anterior part of obturator fascia
    - Inserts onto lateral aspect of coccyx
  - **Iliococcygeus muscle**
    - Arises from fascia overlying obturator internus
    - Inserts onto lateral aspect of coccyx, overlapping with fibers of pubococcygeus muscle in staggered arrangement
  - Levator ani muscle group **works at rest and during stress to counteract intraabdominal pressure and support pelvic organs**
- **Urogenital hiatus**: Opening within levator ani muscle through which urethra, vagina, and rectum pass (and through which prolapse occurs)
  - Bounded ventrally by pubic bones and laterally by puborectalis muscle

Ligaments & Fascia

- Complex network of connective tissue
- **Ligaments**: Forms well-defined layer composed of specialized aggregation of connective tissue
  - Arcus tendineus levator ani (ATLA): Condensation of obturator fascia
    - Major attachment site of levator ani muscles
  - Arcus tendineus fascia pelvis (ATFP)
    - Provides lateral anchoring sites for anterior vaginal wall that underlies and supports urethra
- **Endopelvic fascia**: Forms diffuse layer of less well-defined connective tissue beneath parietal peritoneum
  - Important for passive support of visceral organs and pelvic floor
- **Vaginal support**: Can be divided into 3 levels
  - Level I (suspension): Upper portion of vagina adjacent to cervix
    - Suspended from above by relatively long connective tissue fibers of upper paracolpium
  - Level II (attachment): Midportion of vagina
    - Attaches vaginal wall more directly to ATFP
  - Level III (fusion): From introitus to 2-3 cm above hymenal ring
    - Near introitus, vagina is fused laterally to levator ani
    - Posteriorly, attached to perineal body
    - Anteriorly, blends with urethra
- **Urethral support**: Urethral ligaments, level III endopelvic fascia, and puborectalis muscle

Urogenital Diaphragm & Perineum

- Cavity of pelvis is divided by pelvic diaphragm into main pelvic cavity above and perineum below
- **Urogenital diaphragm**: Fibromuscular layer directly below pelvic diaphragm
  - Spans anterior pelvic outlet and is attached to pubic bones
  - Crossed by urethra and vagina (continuous sheet in males)
- **Perineal body**: Fascial condensation posterior to vagina
  - Insertion site of perineal muscle and external anal sphincter
- Perineum is divided into anterior and posterior parts by line drawn across ischial tuberosities
  - Anterior: **Urogenital triangle**
  - Posterior: **Anal triangle**
- **Superficial external genital muscles** include superficial transverse perineal, bulbospongiosus, and ischiocavernosus

IMAGING ANATOMY

Overview

- **SP**: Appears as echogenic structure with highly reflective surface anterior to urethra
  - On Valsalva maneuver, inferior margin of SP serves as line of reference for maximal bladder descent, uterovaginal prolapse, and rectocele
- **Lower urinary tract**
  - Bladder neck and urethra form funnel
  - **Urethra**: Multilayer; cylindrical; posterior to SP
    - Central: Mucosa/submucosa; hypoechoic
    - Concentric: Muscular lissosphincter; echogenic
    - Outer: Striated rhabdosphincter
  - **Periurethral artery**
    - Identified in urethral mucosa parallel to lumen
    - Resistivity index (RI): Postmenopausal > premenopausal
- **Anterior vaginal septum**: Hypoechoic vesicovaginal fascia between urethra and vagina
- **Vagina**: Posterior to urethra
  - Longitudinal: Hypoechoic structure; poorly defined lumen unless outlined by air, fluid, or jelly
  - Transverse: Visualized as H-shaped hypoechoic structure
  - Vaginal artery: High RI; ↓ during sexual arousal
- **Posterior vaginal septum**: Perineal body and rectovesical fascia
  - Perineal body: Fibrous and muscular tissue between anus and lower vagina
  - Rectovesical fascia: Hypoechoic fascia between rectum and mid vagina
- **Anus**
  - Mucosa: Thin hypoechoic innermost layer
  - Submucosa: Echogenic layer
  - Internal anal sphincter: Concentric hypoechoic ring surrounding central mucosa
  - External anal sphincter: Hyperechoic ring encircling internal anal sphincter
- **Levator ani muscle**: 3 components
  - Puborectalis: Hyperechoic sling around anorectal junction
  - Pubococcygeus: Runs from pubic bone posteriorly towards coccyx
  - Iliococcygeus: Lies posterolaterally between ischial spine and coccyx; usually small

ANATOMY IMAGING ISSUES

Imaging Recommendations

- Bladder neck position and motility can reliably be assessed by transperineal US
- Parting of labia necessary to improve image quality
- Bladder filling required depending on applications
- Stool and gas may interfere with exam; consider prior enema for better diagnostic accuracy

Imaging Approaches

- Performed with patient in dorsal lithotomy with hips flexed or slightly abducted or in standing position
- 3.5- to 7-MHz curvilinear transducer is commonly used; higher frequency for superficial structures
- 3D US advantages: Produces 3 orthogonal planes and allows sequential assessment
- Urethra
  - Midsagittal scan most useful projection
  - Different pelvic floor maneuvers key for detecting functional abnormalities
    - Resting, straining by Valsalva maneuver, contracting by withholding maneuver used to demonstrate bladder neck mobility
  - Retrovesical angle between proximal urethra and trigone is useful to assess bladder neck descent
- Anorectal canal
  - Axial scan: Most useful projection
  - Anal sphincter complex is best evaluated with 3D transperineal US; transrectal scan for mucosa and wall

CLINICAL APPLICATIONS

Normal Function

- Pelvic organs lie between high abdominal pressure and low atmospheric pressure
- Muscles give active support of pelvic floor; ligaments give passive support to hold pelvic organs in place
- When pelvic muscles function properly, pelvic floor is closed
  - Ligaments and fasciae are under no tension
  - Fasciae simply act to stabilize pelvic organs in their position above levator ani muscle

Abnormal Function

- When pelvic muscles relax or are damaged, ligaments are put under strain
  - If damaged pelvic floor muscles cannot close levator hiatus, connective tissues must support pelvic organs for extended periods
  - Connective tissue will eventually fail to hold vagina and other pelvic organs in place
- **Stress urinary incontinence**
  - Funneling of internal urethral meatus on Valsalva or at rest is suggestive of incontinence
  - Retrovesical angle > 160° on Valsalva is often associated with funneling
  - Bladder neck descent > 2.5 cm on Valsalva is strongly associated with urodynamic stress incontinence
  - Color Doppler US may demonstrate urine leakage through urethra on Valsalva maneuver
- **Fecal incontinence**
  - Largely due to anal sphincter musculature damage during vaginal delivery
  - Anal sphincter defects are best diagnosed with 3D transperineal US
- **Prolapse**
  - Anterior compartments (cystocele) better demonstrated than posterior compartment (rectocele)
  - Lateral defects of endopelvic fascia well-demonstrated by 3D transperineal US
- **Postoperative assessment**
  - Status of bladder neck after colposuspension
  - Demonstration of fascial and synthetic slings; bulking agents

BONES AND LIGAMENTS

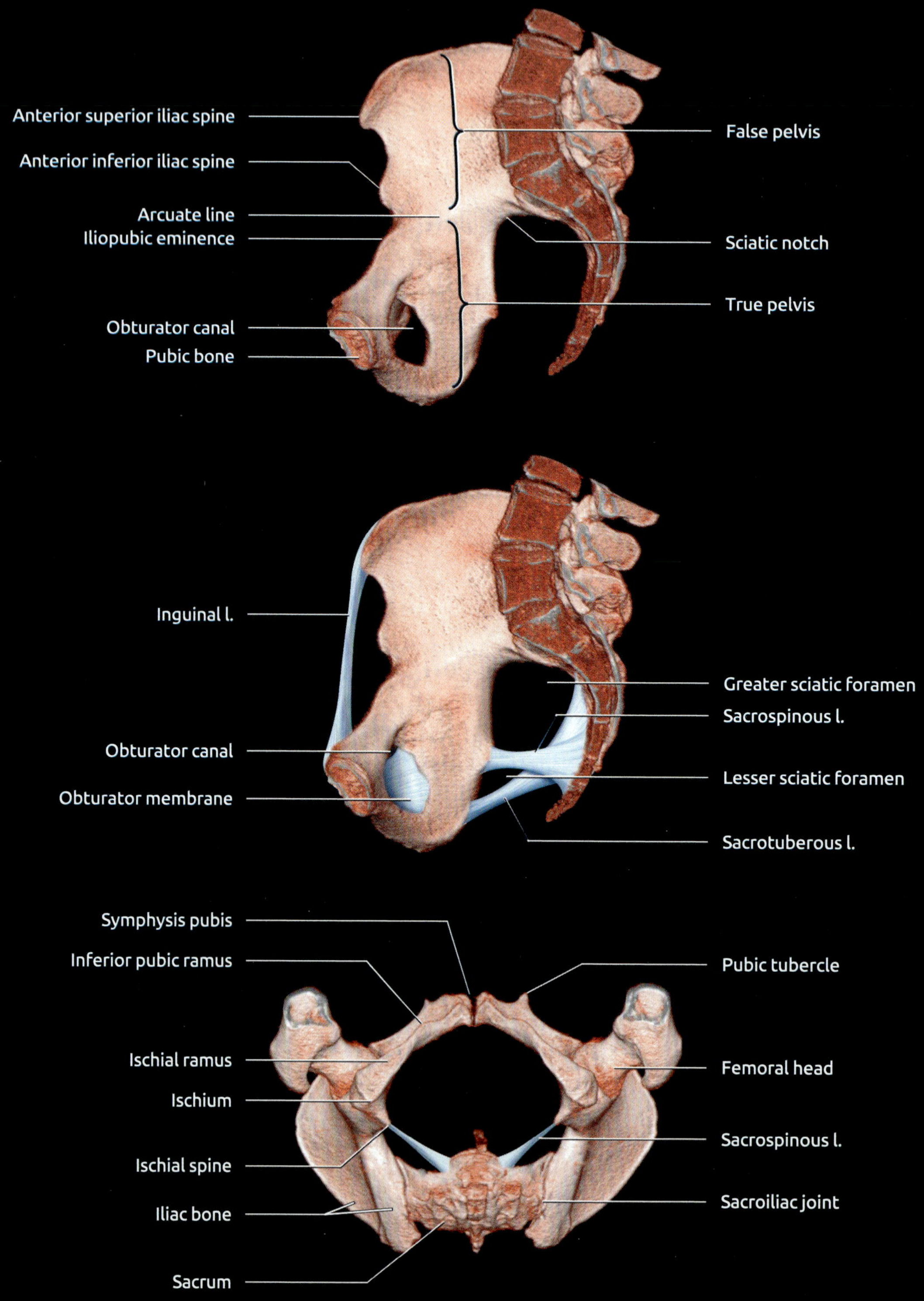

(Top) *3D CT reconstruction of the female pelvis viewed from the medial surface. The arcuate line is a bony prominence, which courses from the sacral promontory anteriorly toward the iliopubic eminence. The false pelvis is above the arcuate line, while the true pelvis is below it.* **(Middle)** *3D CT reconstruction with graphic enhancement of the ligaments shows a medial view of the pelvic sidewall. The sacrospinous ligament extends between the sacrum and ischial spine. The sacrotuberous ligament extends from the lateral part of sacrum, coccyx, and posterior inferior iliac spine to insert on the ischial tuberosity. The greater sciatic foramen is above the sacrospinous ligament and the lesser sciatic foramen is below it.* **(Bottom)** *3D CT reconstruction shows the pelvic outlet viewed from below and graphically enhanced to show the sacrospinous ligament. The pelvic outlet is formed by the ischiopubic rami, ischial spines, inferior symphysis pubis, sacrospinous ligaments, and coccyx.*

PELVIC DIAPHRAGM

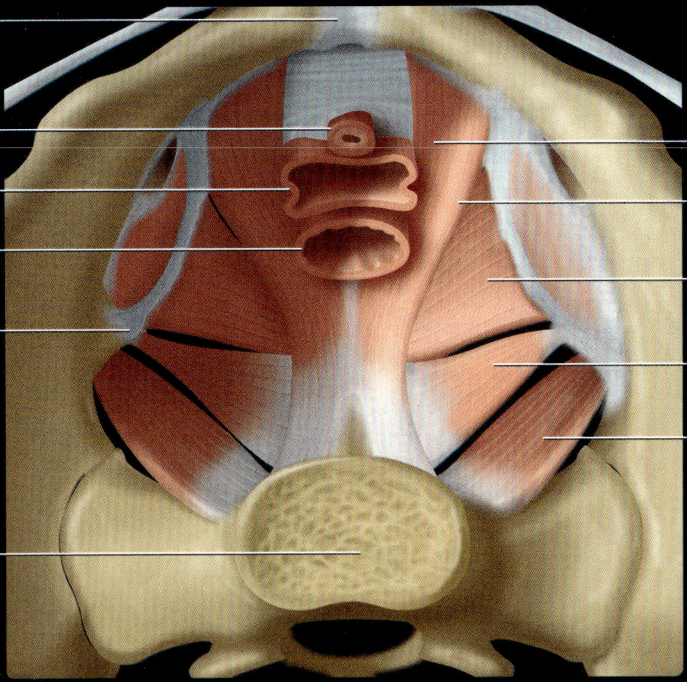

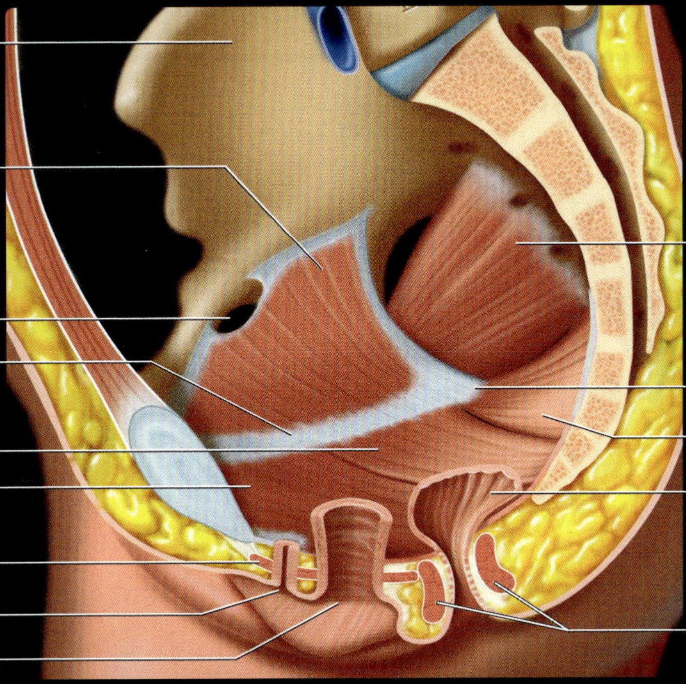

(Top) *Female pelvic diaphragm, superior view, shows the passage of urethra, vagina, and rectum through the pubococcygeus muscle.* **(Bottom)** *The true pelvis is bowl-shaped; therefore, the designation of walls is somewhat arbitrary. The lateral wall of the true pelvis is formed by part of the ilium and ischium below the pelvic inlet, the obturator internus muscle and its covering membrane, and the sacrotuberous and sacrospinous ligaments. The pelvic floor is formed by the pelvic diaphragm (coccygeus and levator ani muscles and fascia). The levator ani is composed of 3 separate muscles: Pubococcygeus, iliococcygeus, and puborectalis. The levator ani is attached to the pubic bones anteriorly, the ischial spines laterally, and to the arcus tendineus levator ani (thickening in the obturator fascia) between the bony attachments. The pelvic diaphragm separates the pelvic cavity from the perineum.*

UROGENITAL DIAPHRAGM

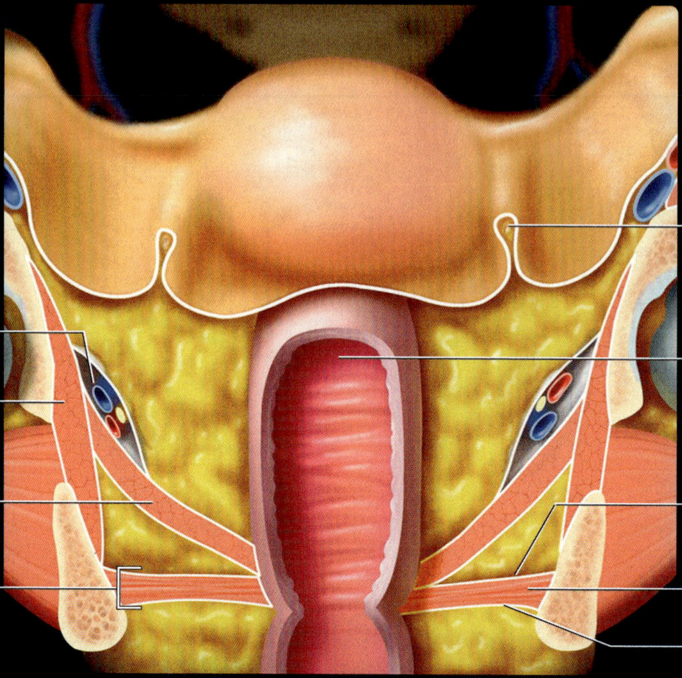

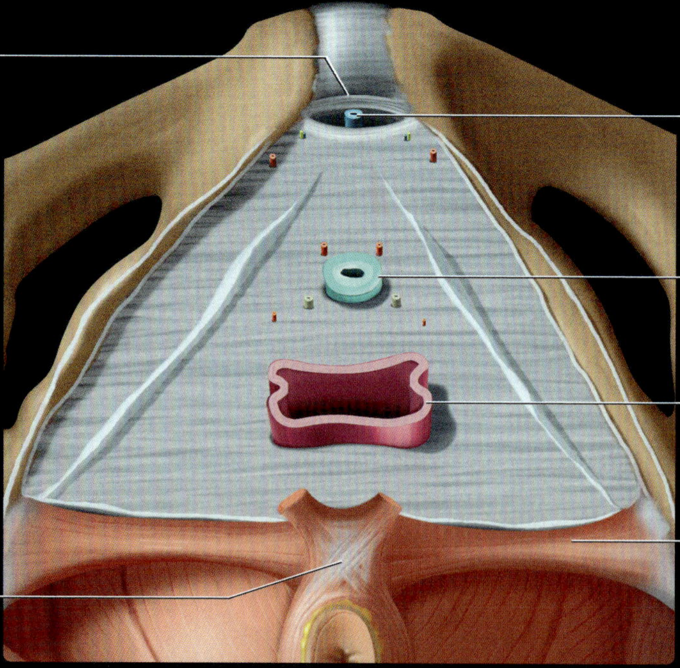

(Top) *Coronal graphic of the pelvic floor shows the urogenital (UG) diaphragm. The UG diaphragm is the fibromuscular layer directly below the pelvic diaphragm (levator ani muscles). It is a trilaminar structure with the deep transverse perineal muscle sandwiched between superior and inferior fascial layers. It is part of the perineum, which is located below the levator ani and includes the external genitalia.* **(Bottom)** *Graphic shows the inferior view of the UG diaphragm. It is triangular in shape and attaches laterally to the pubic bones. At the most anterior (ventral) aspect of the perineal membrane (covering fascia), the base of the arcuate ligament is separated from the anterior border of the UG diaphragm by an opening for the deep dorsal vein of the clitoris. Both the urethra and vagina pass through the UG diaphragm.*

PERINEUM

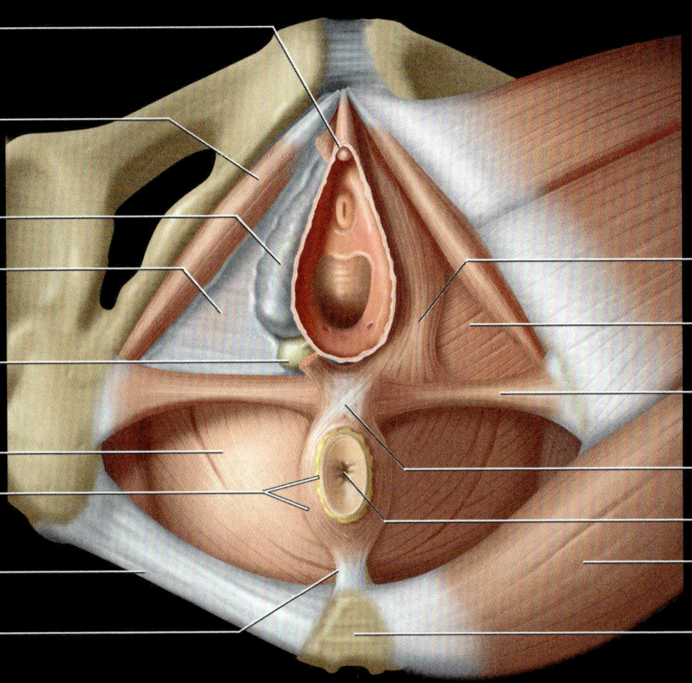

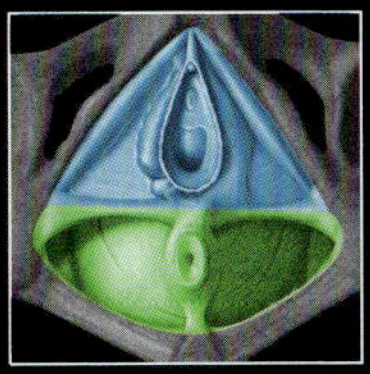

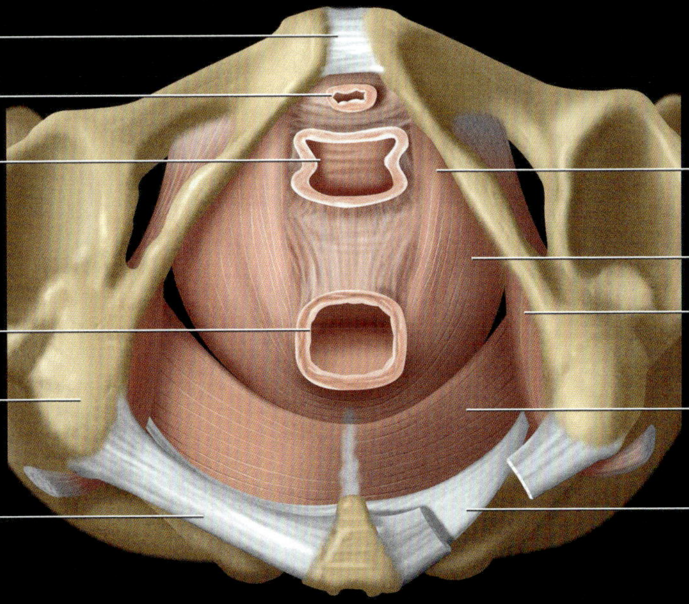

(Top) *The perineum is a diamond-shaped region bordered by the 2 ischiopubic rami and the 2 sacrotuberous ligaments. A horizontal line, along the superficial transverse perineal muscle, connecting the 2 ischial tuberosities divides the perineum into the anterior UG triangle (blue in inset) and the posterior anal triangle (green in inset). The perineal body lies at the midpoint of this line, just anterior to the anal canal and provides attachment for muscles and ligaments that support the perineum.* **(Bottom)** *A deeper dissection, with the superficial external genital muscles removed, shows the muscles of the levator ani (puborectalis, pubococcygeus, and iliococcygeus).*

VAGINA AND SUPPORTING STRUCTURES

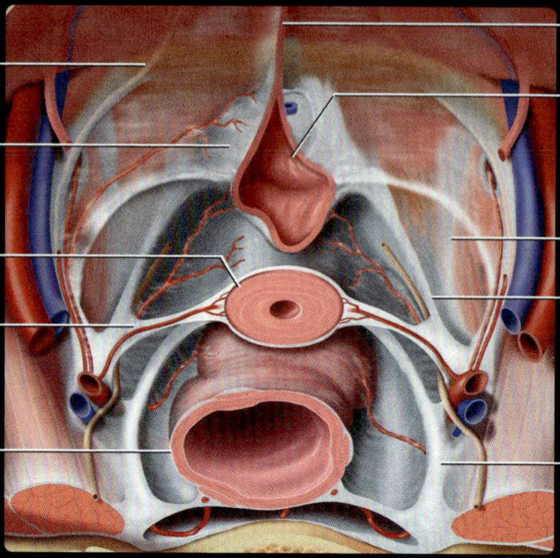

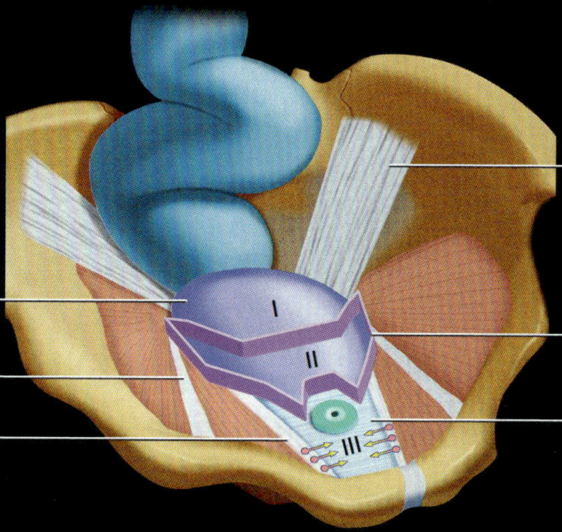

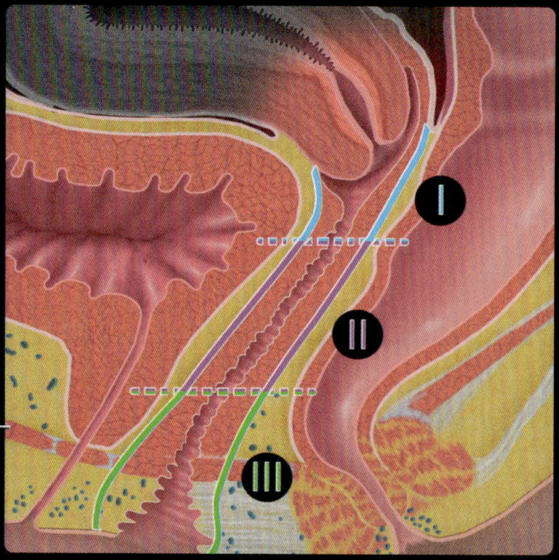

(Top) *Endopelvic fascia, a complex network of connective tissue, forms a continuous adventitial layer covering the pelvic diaphragm and viscera. Ligaments are a more well-defined aggregate of connective tissue.* **(Middle)** *This schematic diagram (uterus removed) shows the type of support the vagina (purple) receives at various levels. In level I (suspension), the paracolpium suspends the vagina from the lateral pelvic walls. In level II (attachment), the vagina is attached to the arcus tendineus fasciae pelvis and the superior fascia of levator ani. In level III (fusion), the vagina, near the introitus, is fused laterally to the levator ani.* **(Bottom)** *Lateral schematic of the 3 levels of endopelvic fascia support. Level I, the upper 2-3 cm, gives support to the upper vagina and uterus. Level II supports the midportion of the vagina and bladder. Level III is the distal 2-3 cm. At this level, there is no intervening paracolpium, and the anterior vaginal wall is fused with the lower posterior urethra. Level III provides urethral and vesical neck support.*

URETHRA AND SUPPORTING STRUCTURES

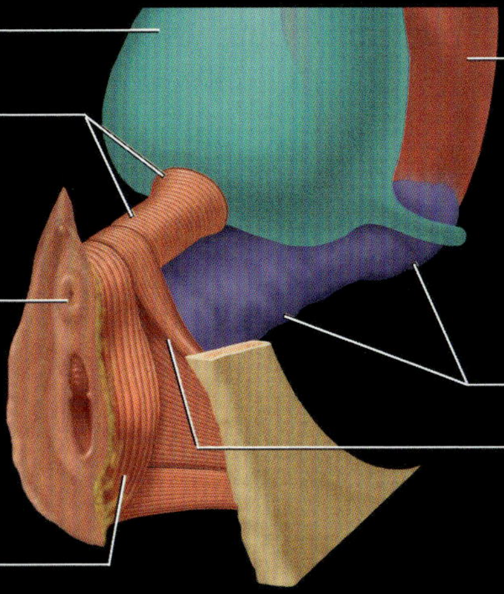

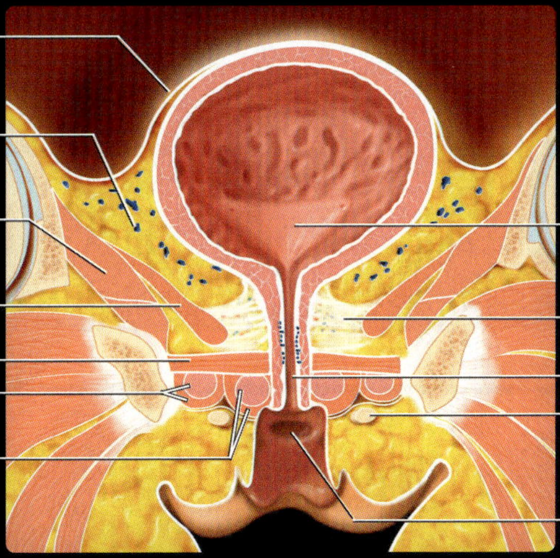

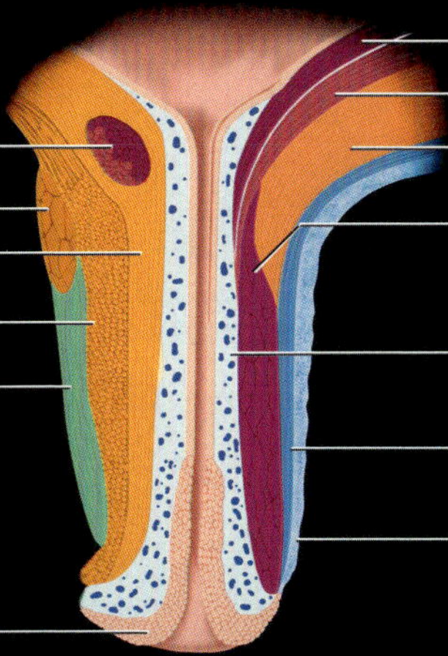

(Top) *Graphic shows the external urethral sphincter. It has 2 different components, an upper circumferential sphincteric portion, and 2 lower, arch-like muscular bands (compressor urethrae and urethrovaginal sphincter).* **(Middle)** *Coronal graphic shows the bladder rests on the muscular floor of the pelvis and is supported by the endopelvic fascia suspended between the arcus tendineus fascia pelvis. The urethra passes through the UG (opening in the levator ani) and then through the UG diaphragm.* **(Bottom)** *Midsagittal section of the urethra shows its histologic complexity. The lamina propria is rich in collagen and elastic components, and has an extensive vascular plexus that functions in maintaining urinary continence by coapting and creating a seal. This is surrounded by 2 layers of smooth muscle, an inner longitudinal and outer circular layer. The urethra undergoes marked histologic and morphologic changes with aging. Striated muscle decreases and is replaced by connective tissue. The vascular plexus is also affected by decreased estrogen levels. All of these changes can adversely affect continence.*

PASSIVE AND ACTIVE COMPONENTS OF PELVIC SUPPORT

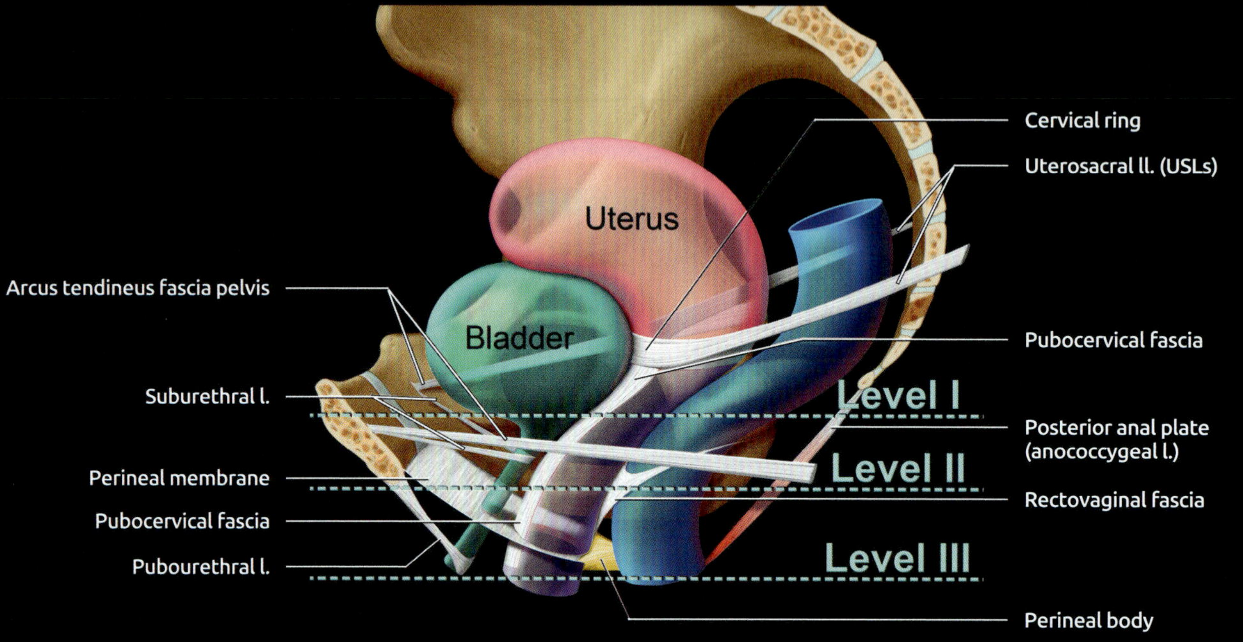

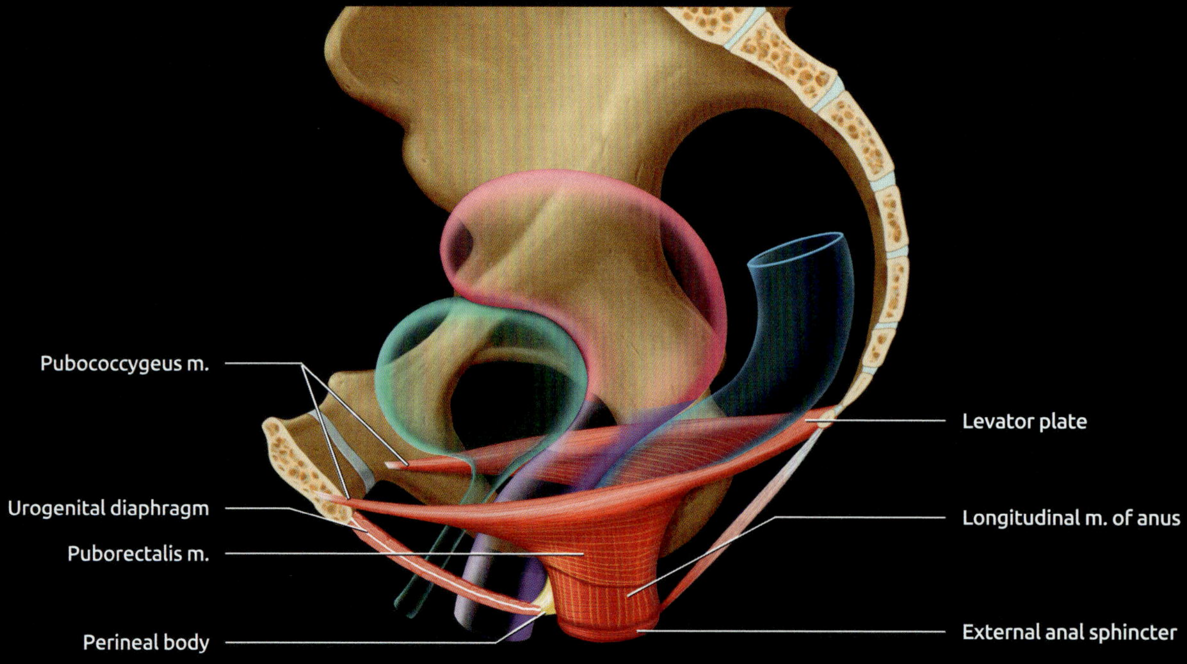

(Top) *This series of 2 graphics illustrates the passive and active conceptual approach to the pelvic floor. Passive components include the bony pelvis and supportive connective tissue. The supportive connective tissue is either in the form of a diffuse ill-defined layer (the endopelvic fascia), or as well-defined specialized aggregations of connective tissue (ligaments). The 3 endopelvic fascial levels include level I (upper vagina adjacent to the cervix), level II (midportion of the vagina), and level III (from the introitus to 2-3 cm above the hymenal ring).* **(Bottom)** *Graphic shows the main active component of pelvic support system, the levator ani muscle. The levator ani muscle is a wide sheet of muscle whose main components are the puborectalis muscle, which forms a sling around the junction of the rectum and anal canal, the pubococcygeus muscle, which passes posteriorly to insert into the anococcygeal body, and the iliococcygeus muscle, whose fibers fuse to form the levator plate and insert on coccyx.*

PASSIVE AND ACTIVE COMPONENTS OF PELVIC SUPPORT

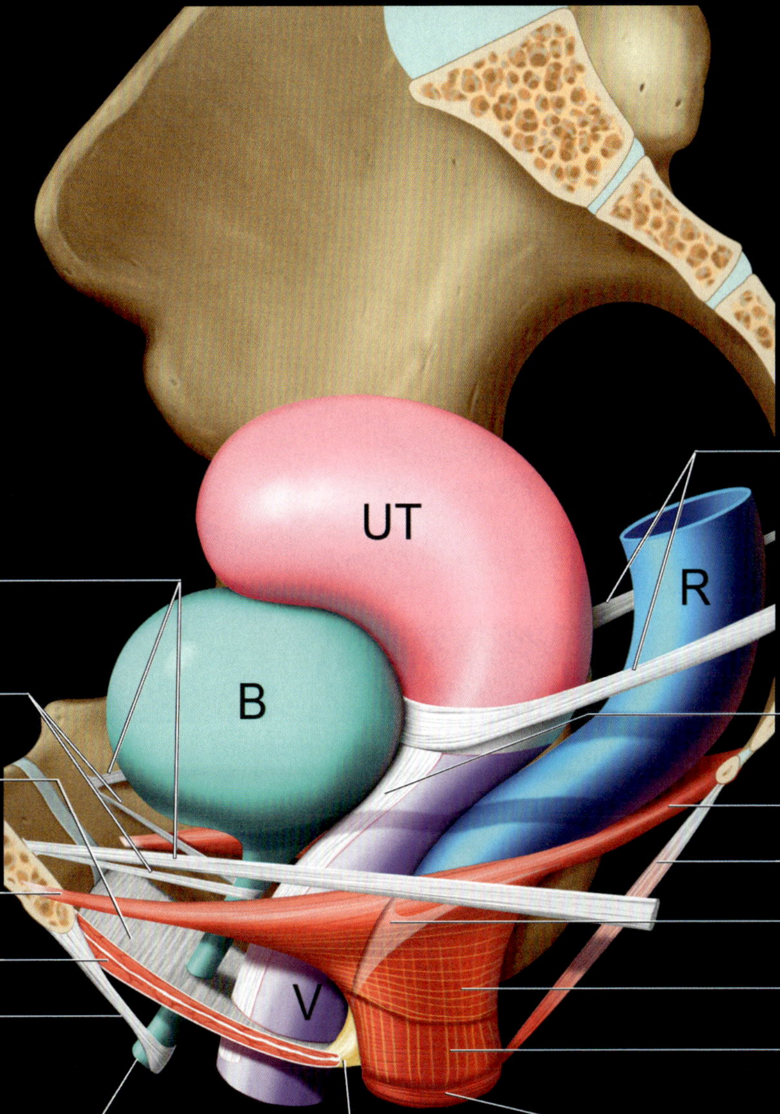

Graphic of the pelvis illustrates the multilayered system approach that considers the passive and active components of pelvic floor as an integrated multilayer system. From cranial to caudal, the pelvic support system consists of endopelvic fascia, pelvic diaphragm, perineum, and the external genital muscles. The muscles (levator ani) give active support to the pelvic floor, whereas the ligaments give passive support holding organs in place. When the levator ani is functioning properly, the pelvic floor is closed and the ligaments and fasciae are under no tension. When the musculature is damaged and cannot close the levator hiatus, ligaments are put under strain and will eventually fail resulting in pelvic organ prolapse.

RECTUM, ANAL CANAL, AND SPHINCTER COMPLEX

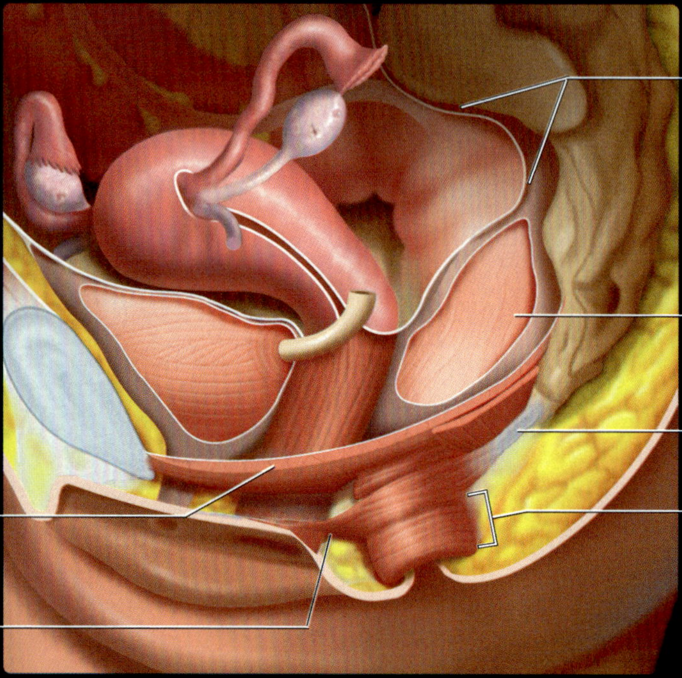

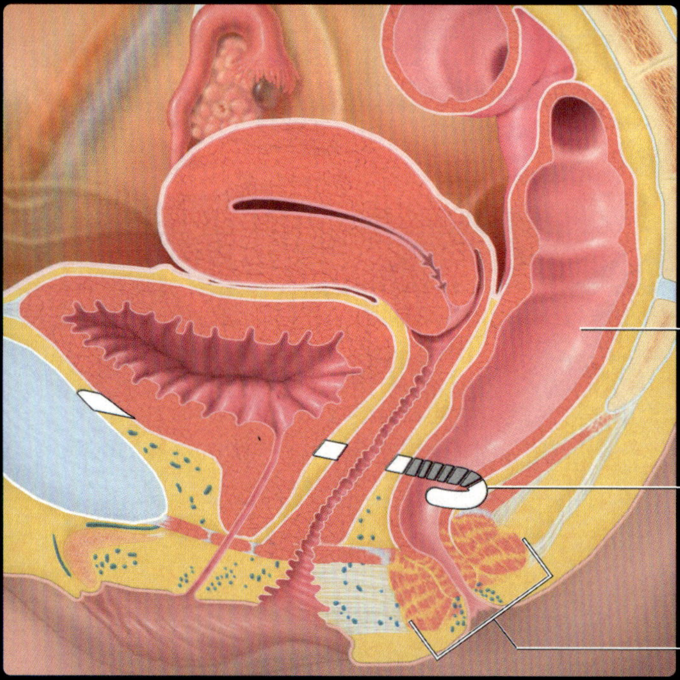

(Top) *The rectum is formed from the terminal portion of the colon. It begins at the level of the 3rd sacral segment and ends at the anus. Anatomically, it is divided into 2 sections: The rectum proper (10-12 cm in length), and the anal canal (3-4 cm in length). The intraperitoneal portion is related anteriorly to the upper vagina and uterus. The extraperitoneal rectum is related anteriorly to the posterior vaginal wall and rectovaginal septum. The inferior rectum has no mesentery but is enveloped in fat and is bordered by the mesorectal fascia (mesorectum). The anal sphincter envelops the anal canal and is composed of several cylindrical layers.* **(Bottom)** *The ampullary portion of the rectum rests on the pelvic diaphragm; at this level, it turns ~ 90° posteriorly. The anal sphincter is tilted anteriorly in the sagittal plane. The puborectalis muscle is schematically shown forming a sling around the rectum; it is responsible for creating the anal rectal angle. The anal canal is fixed posteriorly to the sacrum by the presacral fascia (fascia of Waldeyer).*

RECTUM, ANAL CANAL, AND SPHINCTER COMPLEX

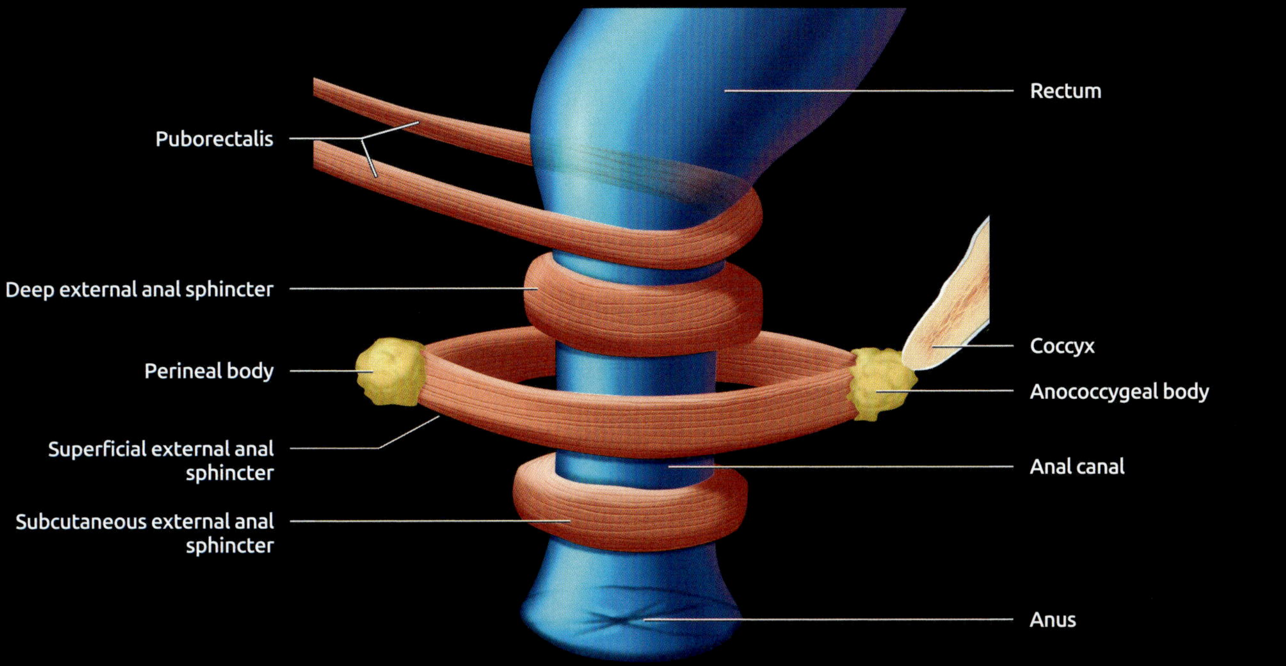

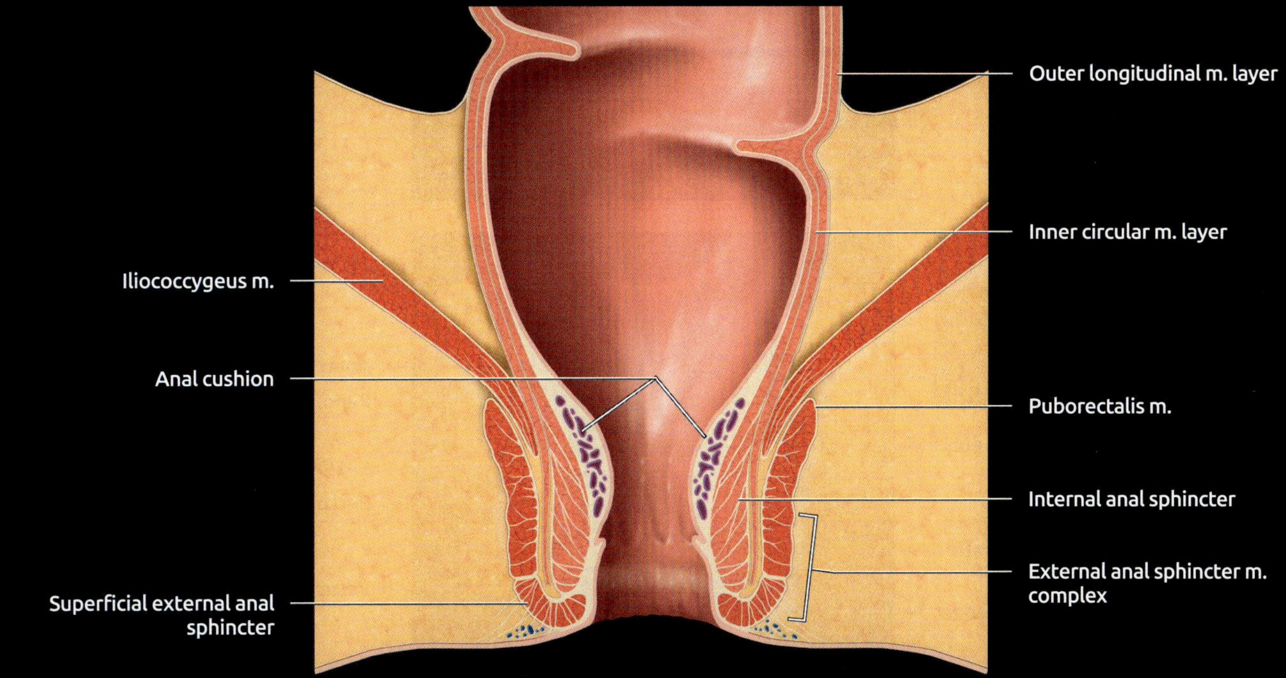

(Top) *This schematic diagram shows the arrangement of the puborectalis muscle and the external anal sphincter (EAS) complex. The superficial EAS attaches to the perineal and anococcygeal bodies.* **(Bottom)** *The internal anal sphincter (IAS) is a continuation of the circular muscle layer of the muscularis propria of the rectum. The EAS is composed of multiple components and constitutes the outer and inferior part of the anal sphincter complex. The lowermost part of the anal canal is surrounded by the superficial external anal sphincter.*

TRANSPERINEAL ULTRASOUND ANATOMY

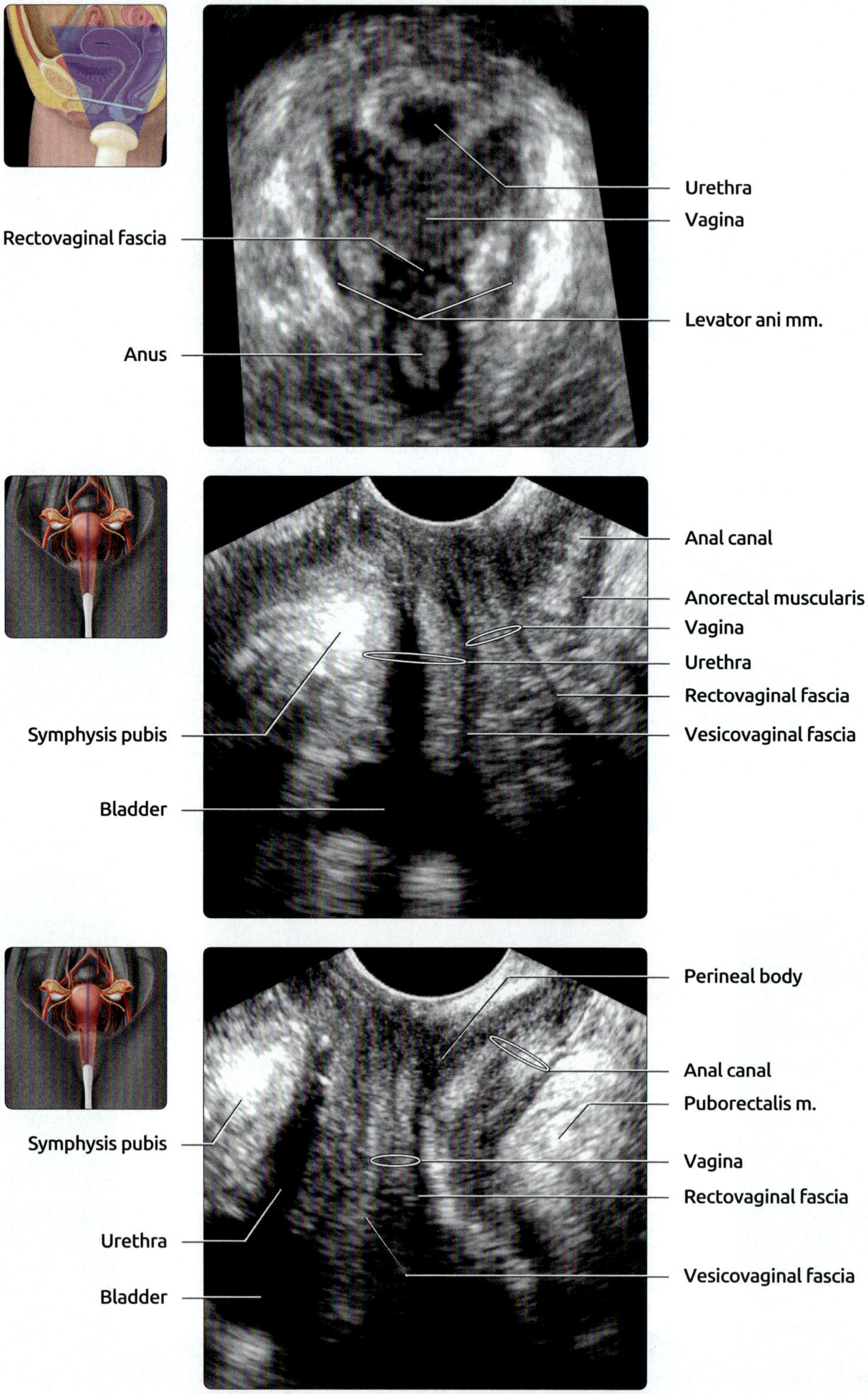

(Top) *3D volume US (axial plane) shows the relationship of the urethra, vagina, and anus. The urethra is seen imbedding in the anterior wall of the vagina while the anus is separated from the vagina by the rectovaginal fascia.* **(Middle)** *Midsagittal scan of the anterior triangle shows the urethra and vagina. The urethra is seen as a hypoechoic tubular structure immediately posterior to the symphysis pubis and is separated from the vagina by the vesicovaginal fascia.* **(Bottom)** *Midsagittal scan in the same woman as previous image, but tilted more posteriorly to show the vaginal septum, is shown. The perineal body is seen as the distal attachment between the vagina and the anus, whereas the rectovaginal fascia is depicted as a hypoechoic lining separating the mid vagina from the rectum. Note the hyperechogenicity of the puborectalis muscle is similar to that of the symphysis pubis.*

TRANSPERINEAL ULTRASOUND ANATOMY

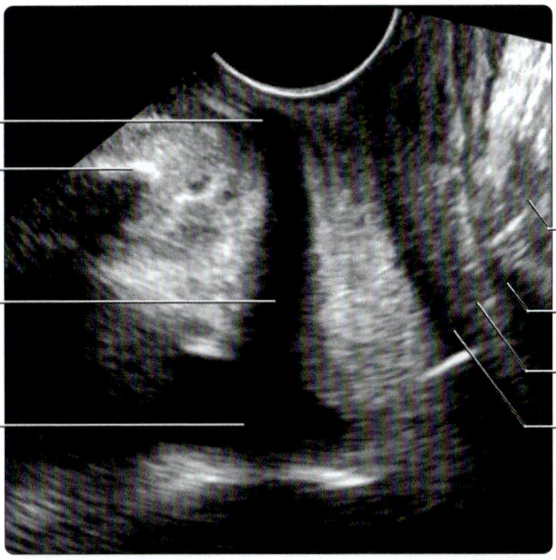

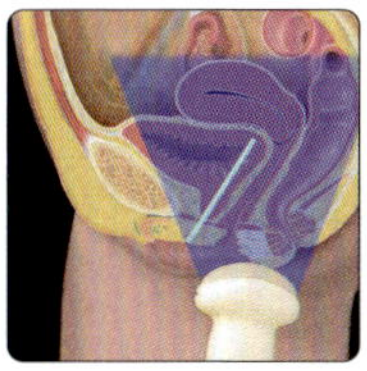

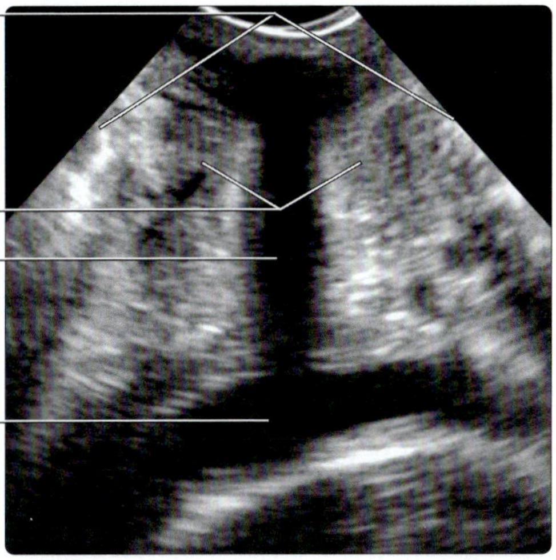

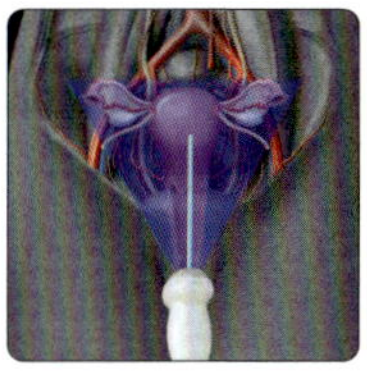

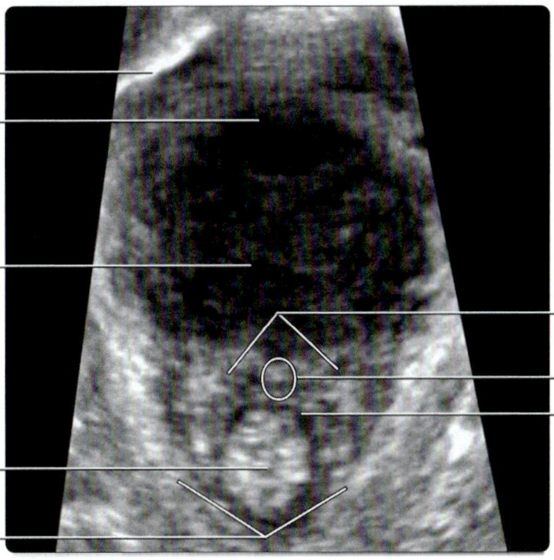

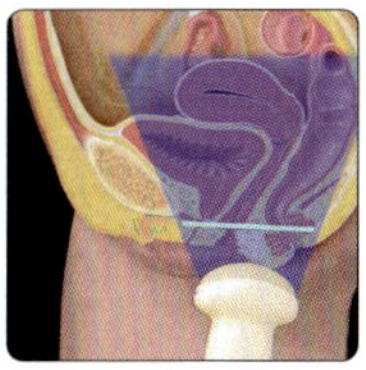

(Top) *Midsagittal plane of 3D transperineal US shows the relationship between the urethra, vagina, and anorectal canal. Note the vaginal canal is barely visible unless it is outlined by fluid or acoustic jelly.* **(Middle)** *Corresponding coronal plane shows the urethra and bladder. Note the lateral walls of the vagina appear as slightly hypoechoic areas on either side of the urethra.* **(Bottom)** *Corresponding axial plane at the level just caudal to the symphysis pubis shows the distal urethra "resting" on the vagina. The anus is supported by a muscle complex formed by the superficial transverse perineal muscles, which insert into the perineal body and external anal sphincter. Note the hyperechoic puborectalis forms a "sling" around the anus.*

HIGH-RESOLUTION 3D US OF URETHRA

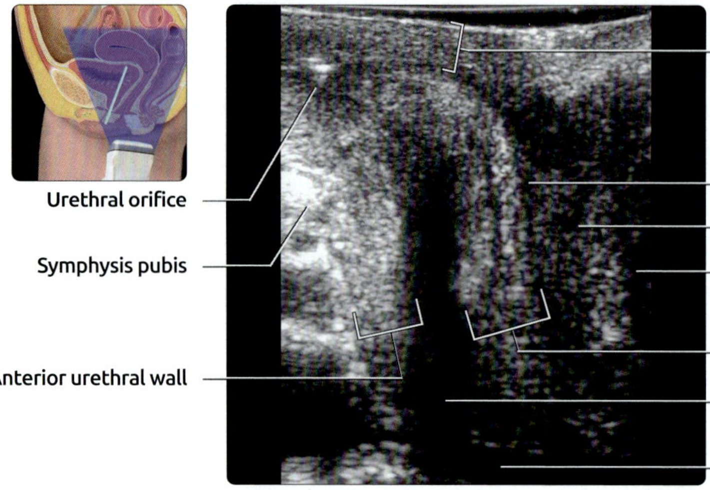

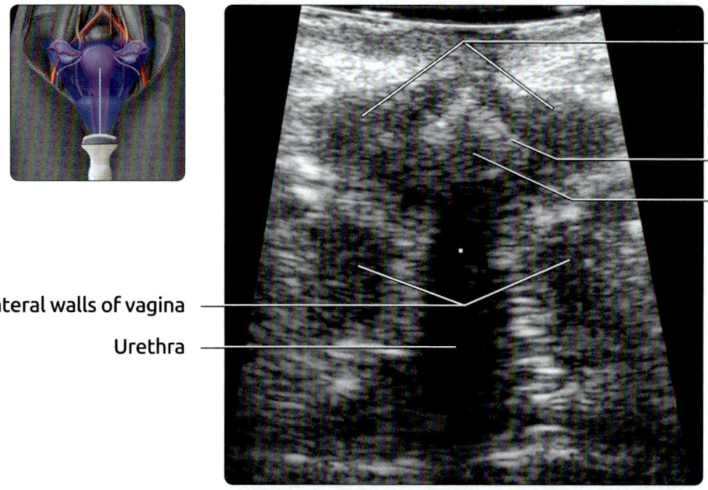

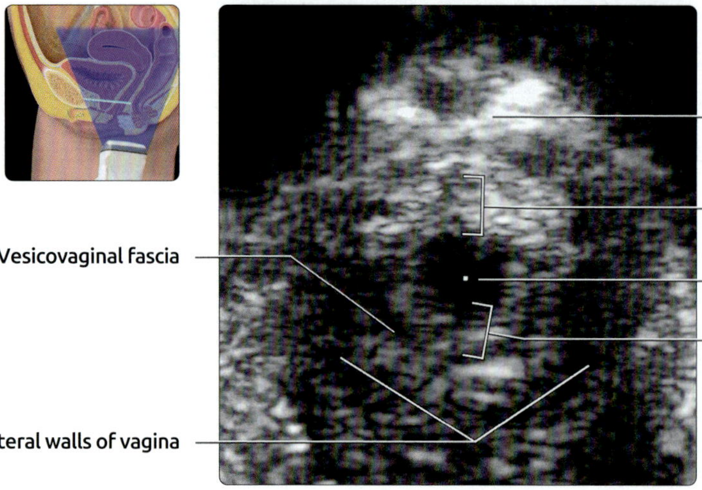

(Top) *Midsagittal scan of the urethra using a high-resolution 12 MHz linear 3D transducer is shown. With this technology, the muscular wall of the urethra can be clearly depicted, and measurement of the urethral wall thickness is feasible.* **(Middle)** *Corresponding coronal plane shows the urethral orifice in cross section. The urethral orifice is encircled by the echogenic urethral sphincter muscle and arched over by the urethral compressor muscles.* **(Bottom)** *Corresponding axial plane shows the urethra in cross section. Although the wall thickness of urethral wall is discernible, differentiation of different layers of the urethra is generally not feasible with the standard transperineal technique.*

MEASUREMENT OF BLADDER NECK DESCENT

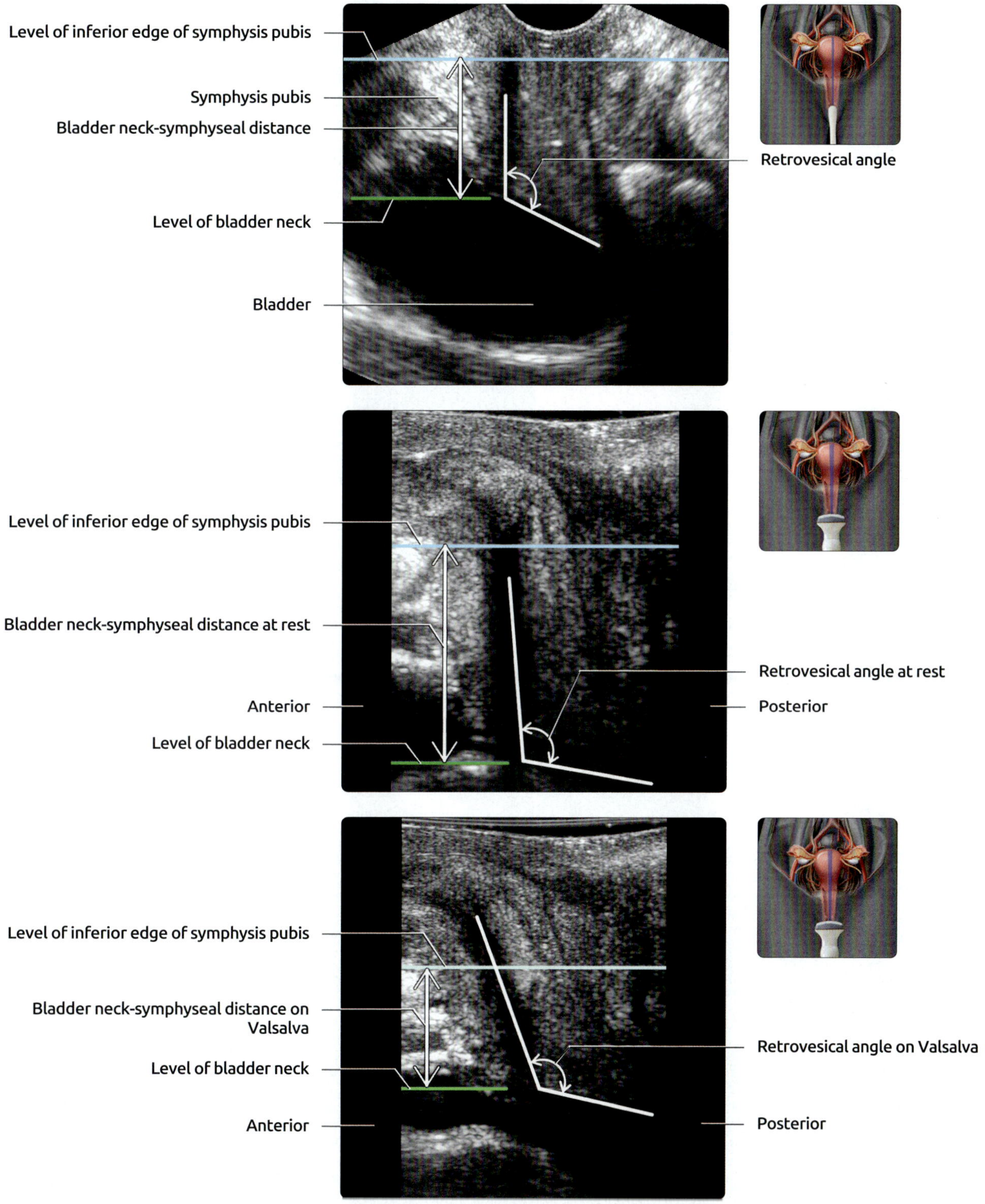

(Top) *Midsagittal plane of the lower urinary tract demonstrates measurements of bladder neck-symphyseal distance (BSD) and retrovesical angle (RVA) used in the evaluation of bladder neck descent as a cause of urinary stress incontinence. BSD is the distance between the inferior edge of the SP and bladder neck. RVA is the angle between proximal urethra and trigone. A significant change of BSD and RVA between resting and Valsalva maneuver reflects severe bladder neck descent.* **(Middle)** *Bladder neck position at rest is shown. The RVA measured at rest normally ranges between 90-120°.* **(Bottom)** *Change of bladder neck position on Valsalva maneuver. The proximal urethra demonstrates a posteroinferior rotational descent that shortens the BSD and widens the RVA. A shortening of BSD > 2.5 cm or a widening of RVA > 160° is indicative of significant bladder descent.*

STRESS URINARY INCONTINENCE: INTRINSIC SPHINCTERIC DEFICIENCY

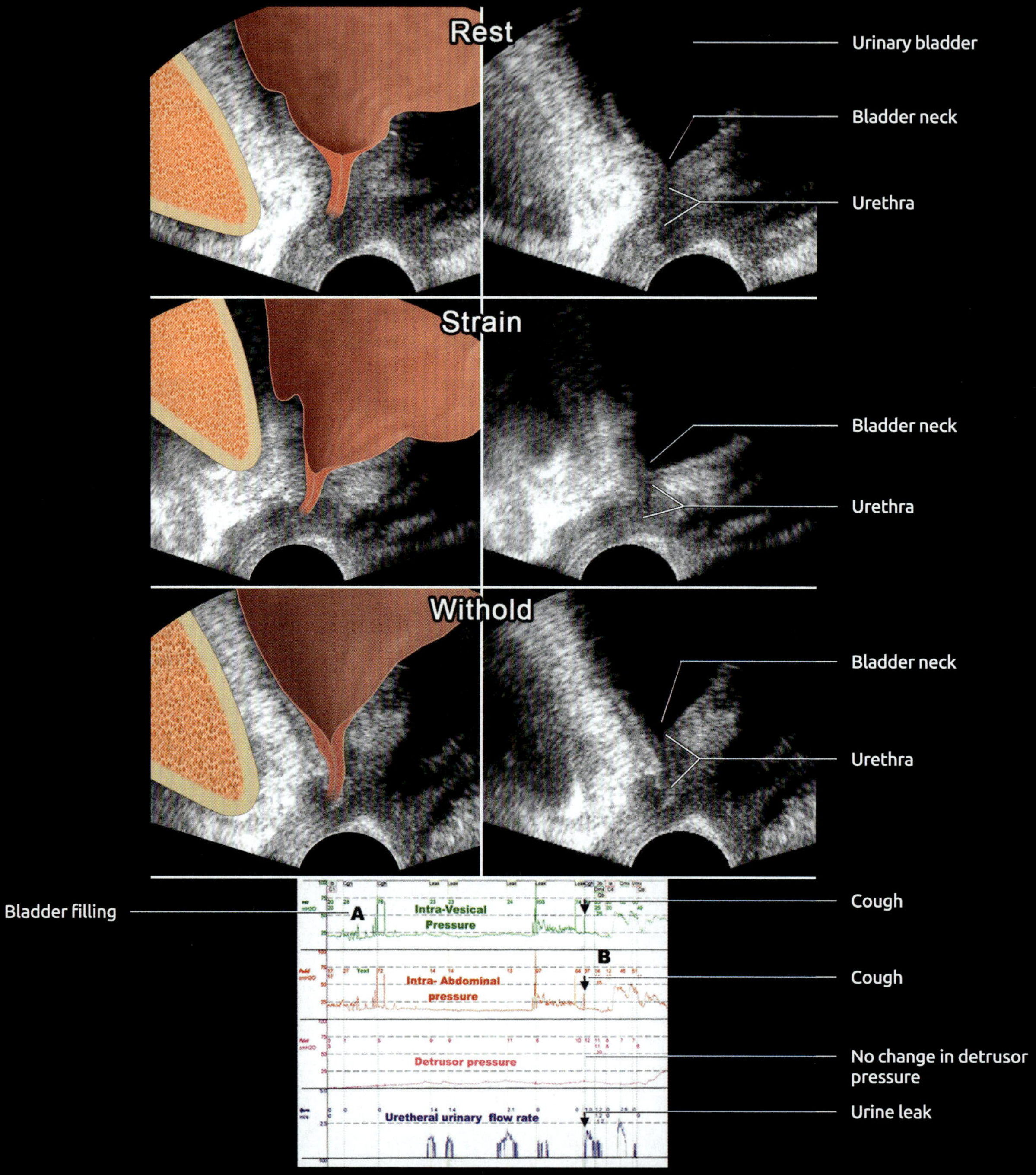

Transperineal ultrasound evaluation of a patient with urinary incontinence shows a sagittal view of the urine-filled bladder, bladder neck, and symphysis pubis. Images at rest revealed intrinsic malfunction of the urethral sphincter characterized by an open vesical neck. During straining and withholding (active contraction of the pelvic floor), the bladder neck was persistently open with funneling, without the expected narrowing during withholding. The patient had urine leakage throughout the exam. Subtracted cystometrogram shows a stable bladder with no rise in detrusor pressure during filling. When a cough is elicited, there is a sharp, isolated pressure spike on the intravesical and intraabdominal tracings, but there are no spikes on the subtracted detrusor tracing. The presence of leakage occurring with coughing confirms that this is stress urinary incontinence due to ineffective urethral closure rather than detrusor overactivity.

URGE URINARY INCONTINENCE: DETRUSOR INSTABILITY

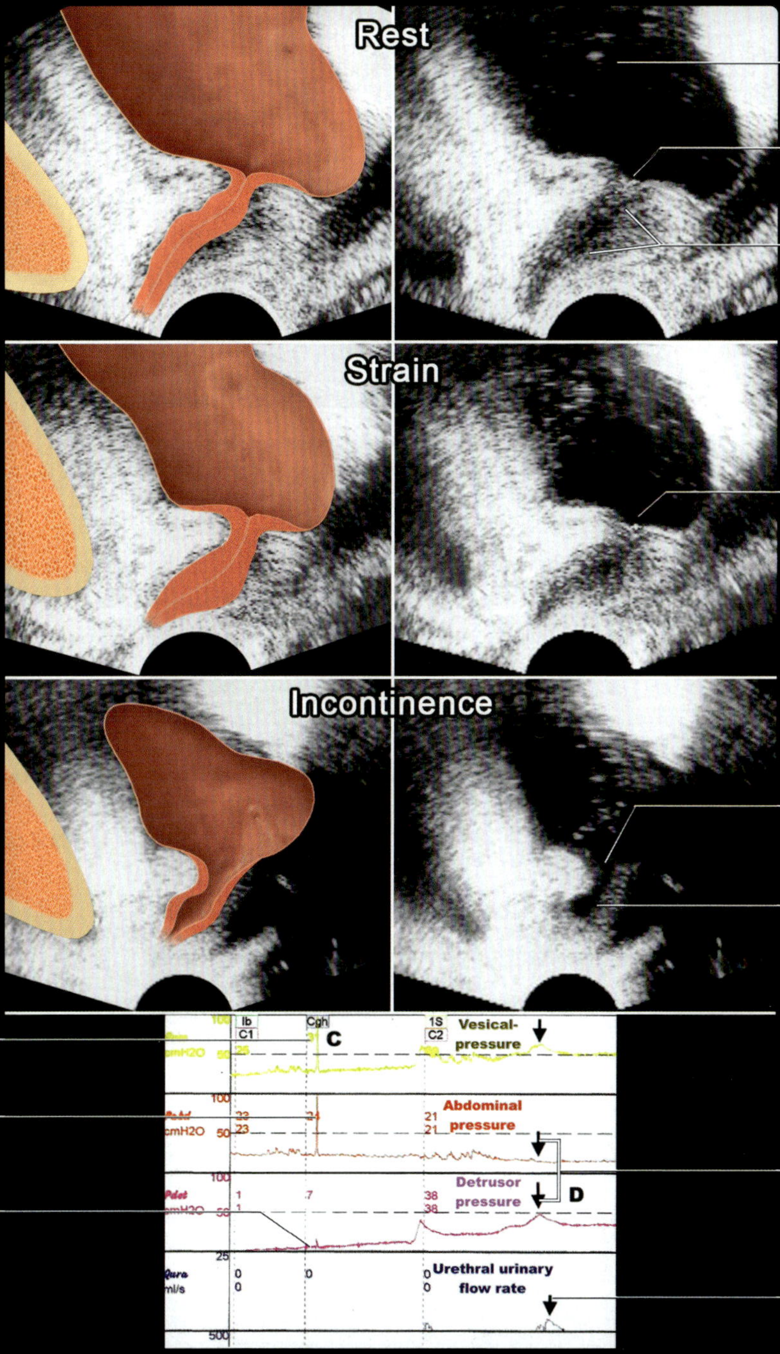

Transperineal evaluation of a patient with urinary incontinence shows a sagittal view of the urine-filled bladder, bladder neck, and symphysis pubis. Note the relatively high position of the bladder neck. The bladder neck does not descend and remains above the inferior margin of the pubis symphysis during maximum straining. During the examination, there was a sudden descent and opening of the bladder neck accompanied by passage of urine, which the patient was unable to stop. Subtracted cystometrogram shows detrusor instability. There is a normal sharp spike in vesical and abdominal pressures during coughing with the subtracted detrusor pressure remaining stable. However, there was a spontaneous increase in detrusor pressure without an increase in the abdominal pressure, indicating that the pressure originated from the detrusor muscle. These findings are consistent with urge urinary incontinence related to detrusor muscle instability.

COLOR DOPPLER US, URETHRA AND VAGINA

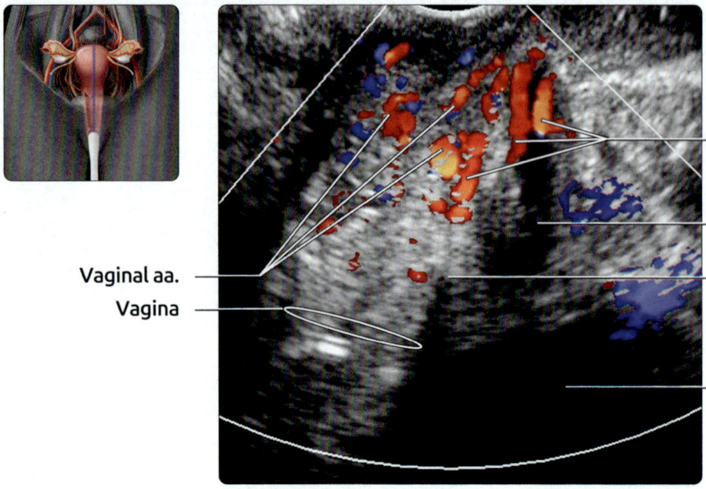

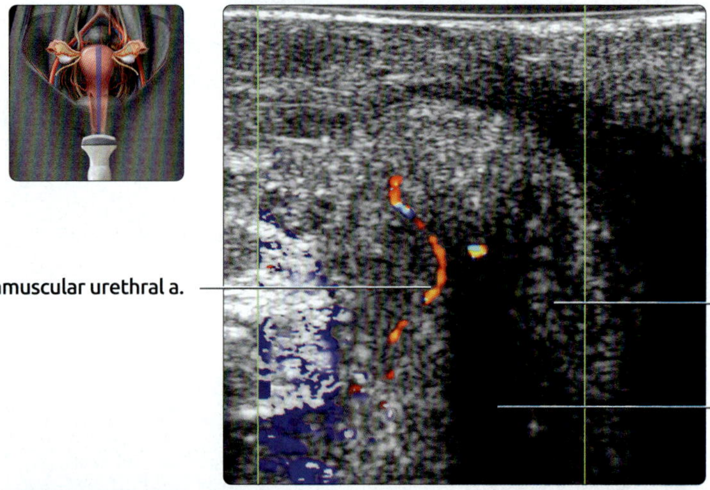

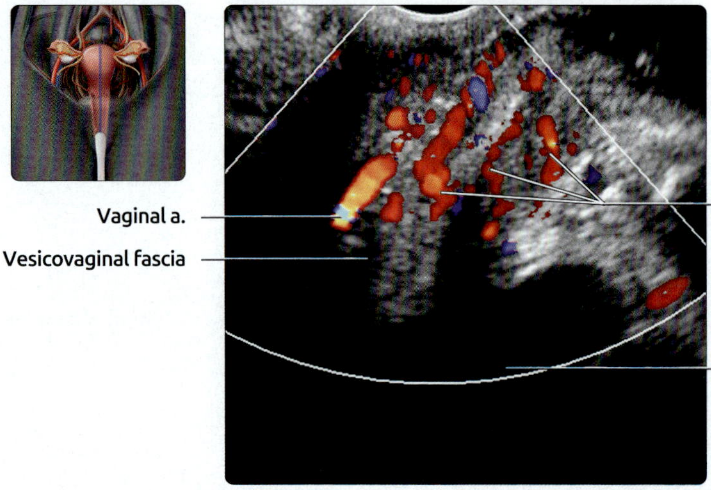

(Top) *Midsagittal color Doppler US of the urethra shows 2 periurethral arteries near the mucosa of the urethra and numerous vessels in the vagina. Care must be taken not to confuse vaginal arteries with periurethral arteries.* **(Middle)** *High-resolution color Doppler US (midsagittal scan) of the urethra shows a small artery running in the muscular layer of the urethra. Note urethral vascularity is influenced by hormonal changes during the menstrual cycle, pregnancy, and post menopause. It is higher in premenopausal women than those in menopause.* **(Bottom)** *Midsagittal color Doppler US of the vaginal artery, which is in the vicinity of the periurethral arteries and is easily confused with them. Special attention should be paid to differentiate vaginal and periurethral arteries. These are separated by the vesicovaginal fascia.*

SPECTRAL DOPPLER US, URETHRA AND VAGINA

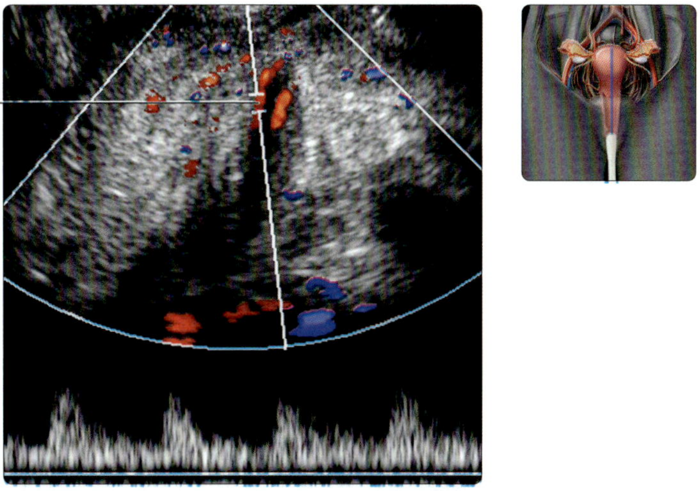

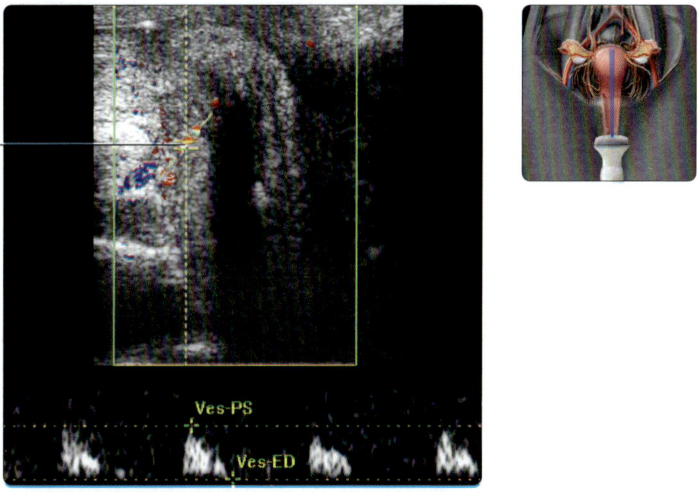

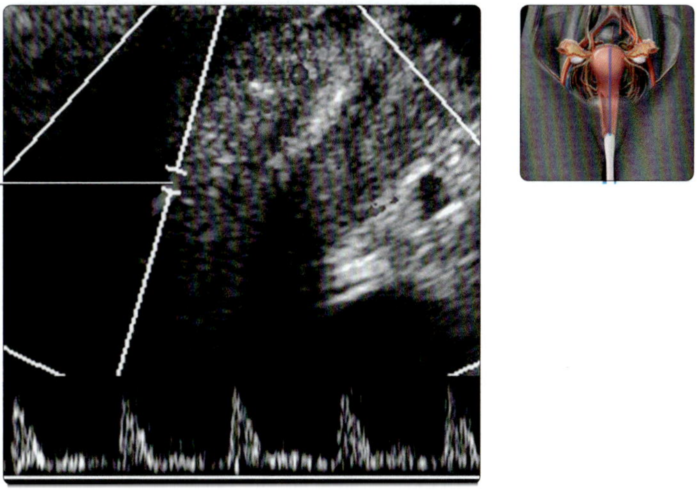

(Top) *Midsagittal spectral Doppler scan of the periurethral artery shows normal, low-resistance blood flow. It has been suggested that decrease in periurethral vascularity is related to stress incontinence in postmenopausal women.* **(Middle)** *Doppler waveform of intramuscular urethral artery demonstrates high-flow resistance with absent diastolic velocity in this artery.* **(Bottom)** *Doppler waveform of the vaginal artery demonstrates high-resistance blood flow with a low diastolic velocity component. Flow resistance may be lowered during the state of sexual arousal.*

AXIAL TRANSPERINEAL US, ANAL CANAL

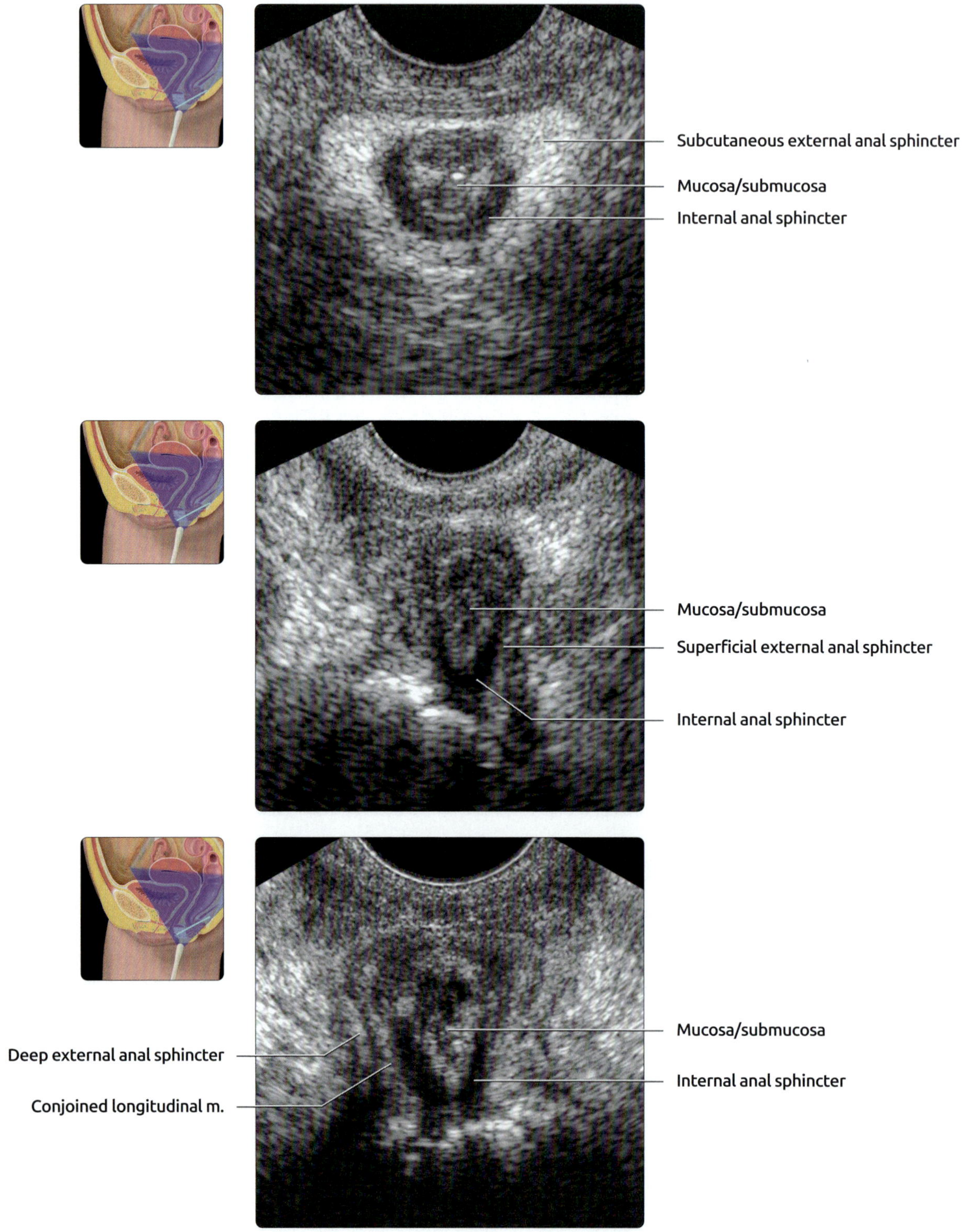

(Top) *3D ultrasound evaluation of the anal orifice is shown. The echogenic mucosa/submucosa of the anus is surrounded by 2 concentric rings. The inner hypoechoic ring represents the IAS and the outer hyperechoic ring denotes the subcutaneous EAS.* **(Middle)** *In the midanal anal canal, the IAS appears as a hypoechoic ring that is asymmetrical in thickness. With advancing age, IAS may lose its uniform thickness and echogenicity.* **(Bottom)** *In the high anal canal, the conjoined longitudinal muscle is identified as a moderately echogenic band between the IAS and EAS. However, it is not always distinguishable from the EAS along the entire anal canal.*

3D VOLUME RENDERED US, VAGINA

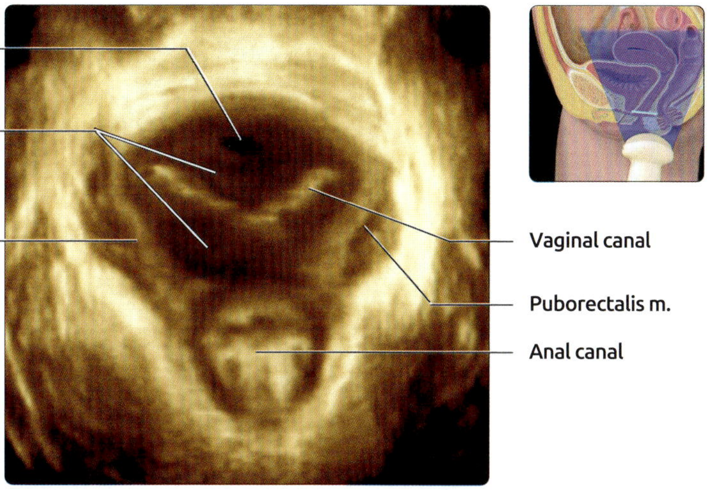

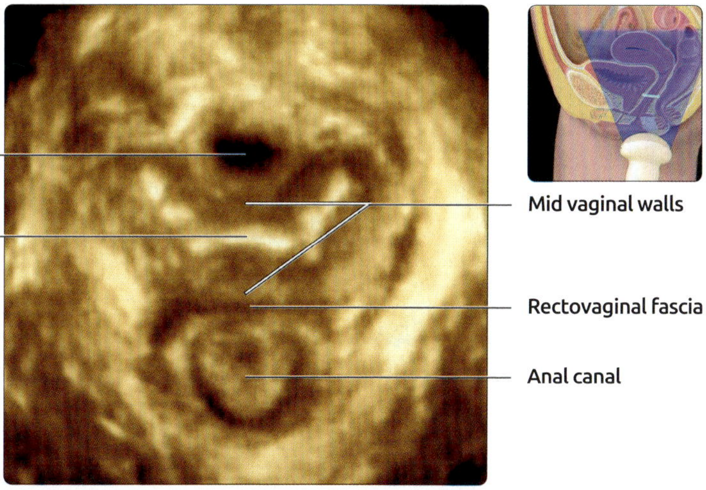

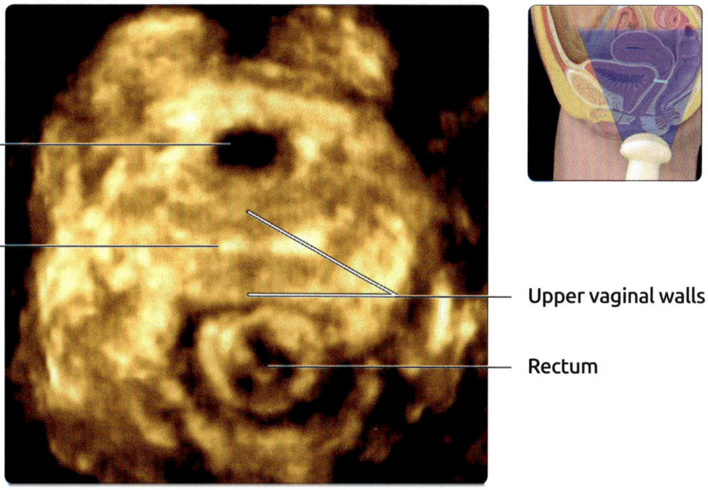

(Top) *3D volume rendered axial scan of the lower vagina at the level of the symphysis pubis shows the vagina "sandwiched" between the urethra and the anus. The vaginal canal is enhanced by acoustic jelly and air bubbles trapped within it following transvaginal scan. It resembles a widened letter "V."* **(Middle)** *At a higher lever in the mid vagina, it appears as a "cradle" partially invaginating the urethra in its anterior wall. The vaginal canal changes its shape from "V" to "U."* **(Bottom)** *In its upper portion, the vagina appears flattened and loses its anterolateral extensions. The vaginal canal becomes an echogenic straight line.*

3D VOLUME RENDERED US, PELVIC FLOOR

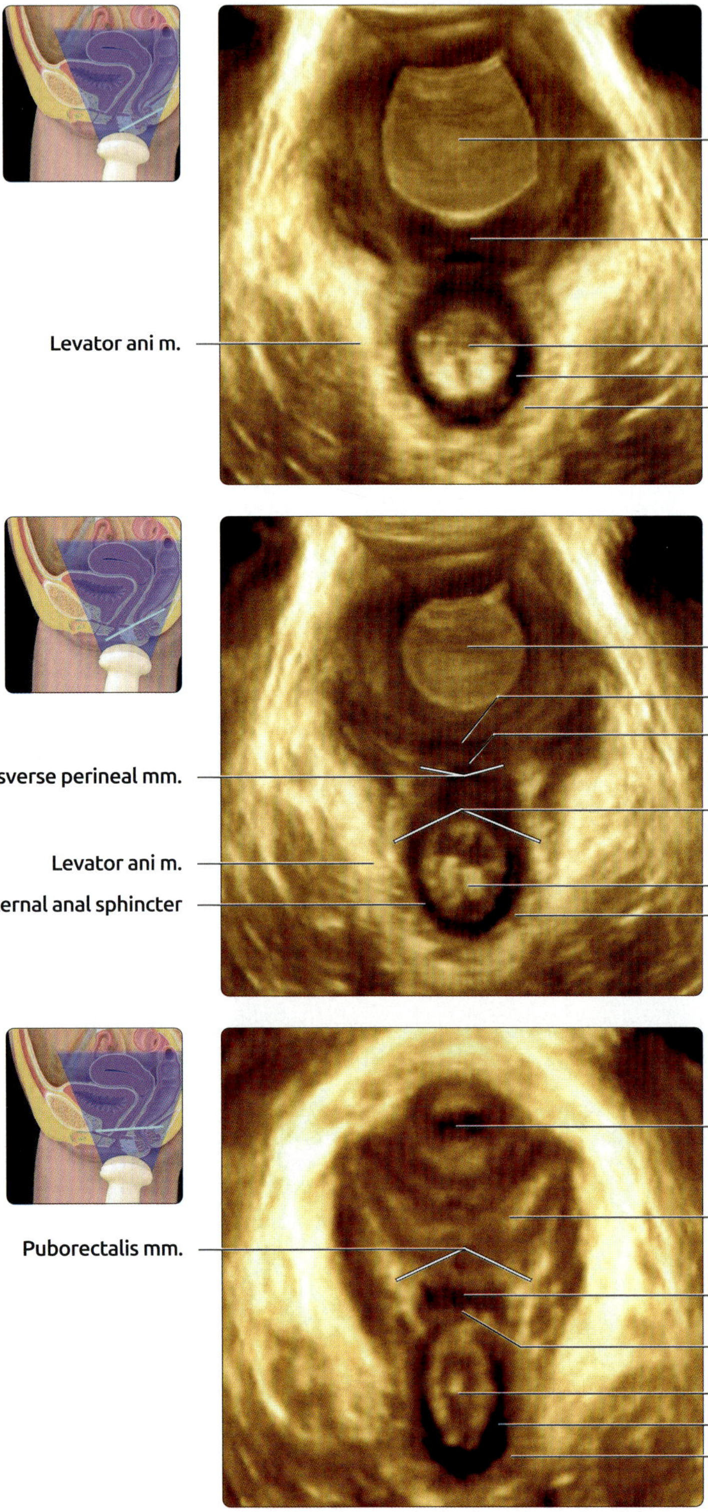

(Top) *Axial scan of the mid anal canal is shown. The superficial EAS is seen as an incomplete ring with absent anterior portion. It is a frequent finding either due to a natural gap in normal females or sphincter rupture in multiparous women.* **(Middle)** *Axial scan of the anorectal junction and the perineal body is shown. The deep EAS is seen to be integral with the puborectalis, which forms a "sling" around the anorectal junction. The EAS and transverse perineal muscles meet in the perineal body, which provides fundamental support to all musculoligamentous components of the pelvis.* **(Bottom)** *Axial scan of the puborectalis muscle in the anterior triangle is shown. The muscles are seen as 2 echogenic linear structures running alongside the lower vagina and urethra that are projected on an end-on view.*

3D VOLUME RENDERED US, PELVIC FLOOR

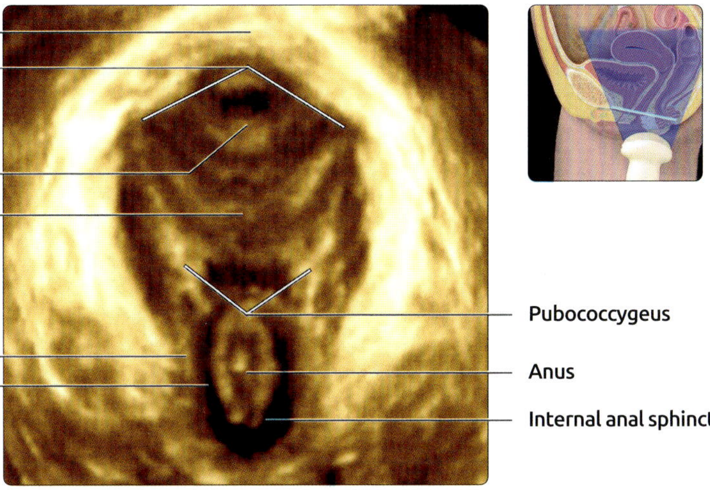

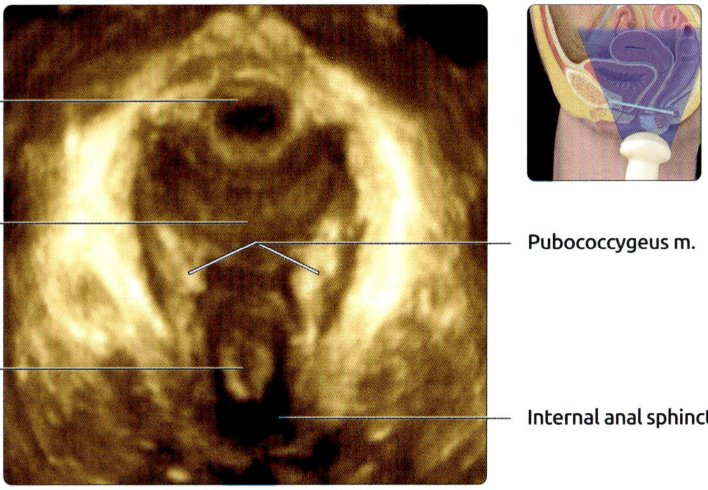

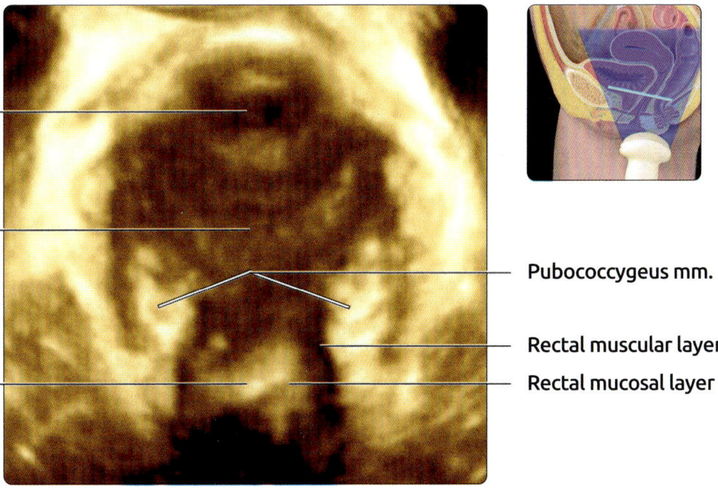

(Top) *Axial plane of the paravaginal support is shown. It is seen as the pubovaginal attachment lateral to the urethra. Identification of the paravaginal support is important because disruption of this structure during vaginal delivery may be related to anterior vaginal prolapse and stress urinary incontinence.* **(Middle)** *Axial scan of the pubococcygeus in the anterior triangle is shown. The echogenic muscle arises from the pubic bone and runs posteriorly towards the ischial spine. It also crosses the midline to form rectal and vaginal hiatus.* **(Bottom)** *Axial scan of the distal rectum is shown. The rectal wall shows 2 distinct layers: The innermost echogenic mucosa and outermost hypoechoic muscular layer, which is a continuation of the IAS and conjoined longitudinal muscle.*

SECTION 6
Upper Extremity

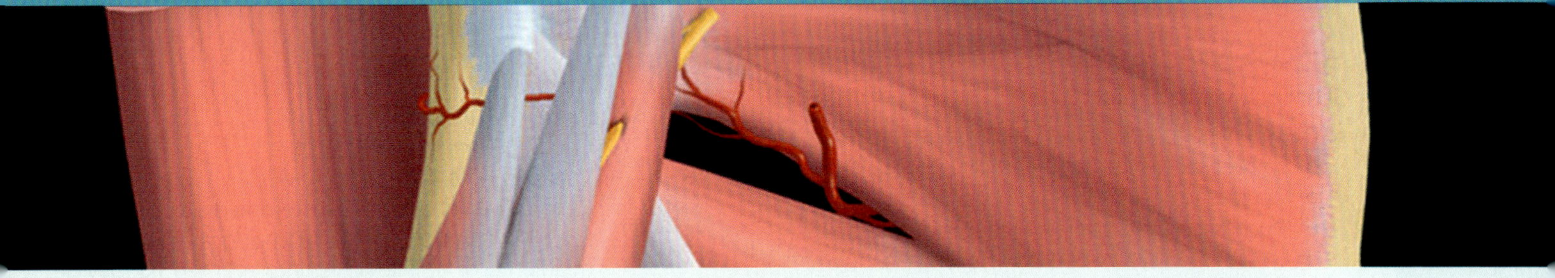

TERMINOLOGY

Abbreviations

- Sternoclavicular joint (SCJ)
- Acromioclavicular joint (ACJ)

GROSS ANATOMY

Sternoclavicular Joint

- Between medial end of clavicle & manubrium
  - Synovial sellar-type (saddle) joint
  - Medial end of clavicle = large & bulbous
  - Much larger than manubrial concavity
  - < 1/2 of medial clavicle articulates with manubrium
    - Stability through capsuloligamentous structures
- Intraarticular disc
  - Attached to joint capsule anteriorly & posteriorly
  - Complete or incomplete ± perforations
  - Thickest posterosuperiorly (3 mm)
- Ligaments of SCJ
  - Capsular ligaments
    - Cover anterosuperior & posterior aspects of SCJ
    - Prevent upward displacement of medial clavicle, which may be caused by downward force on shoulder
    - Anterior stronger than posterior portion
  - Interclavicular ligament
    - Connects superomedial aspect of clavicle to capsular ligaments & upper manubrium
    - Covers anterosuperior & posterior aspects of joint
    - Prevents excessive upward motion of clavicle
  - Costoclavicular ligaments
    - Unite inferior surface medial end clavicle to upper surface of 1st rib
    - Anterior fibers arise from anteromedial surface of 1st rib & resist upward motion
    - Posterior fibers arise lateral to anterior fibers & resist downward motion
- Muscle attachments to medial clavicle & sternum
  - Pectoralis major from anterior aspect medial 2/3 of clavicle (clavicular head)
  - Sternocleidomastoid from posterior surface medial 1/3 of clavicle (clavicular head)
  - Sternohyoid & sternothyroid muscles separate great vessels from SCJ

Acromioclavicular Joint

- Synovial joint between lateral end of clavicle & medial end of acromion
  - Articular surface of clavicle oriented posterolaterally whereas articular surface of acromion oriented anteromedially
    - Angle of inclination between opposing articular surfaces varies with clavicle overriding acromion (50%), vertical orientation between acromion & clavicle (25%), clavicle underriding acromion (5%), & mixed pattern (20%)
    - Maximum width of normal joint on ultrasound = 5 mm if < 35 years & < 4.4 mm if > 35 years
    - Maximum thickness of capsule from bony surface = 2.7 mm if < 35 years & < 3.6 mm if > 35 years
- Intraarticular disc
  - Undergoes rapid degeneration beginning in 2nd decade → marked degeneration of disc by 4th decade
- Ligaments of ACJ
  - Superior AC ligament
    - Stronger & thicker (2.0-5.5 mm) than thin or absent inferior AC ligament
    - Inserts along lateral clavicle (8 mm) & medial acromion (10 mm)
  - Coracoclavicular ligaments
    - Conoid & trapezoid ligaments
    - Vary significantly in length & width
    - Conoid ligament located posteromedially
    - Inserts to conoid tubercle, which is located where middle 1/3 of clavicle curves into lateral 1/3 of clavicle
    - Mainly prevents upward movement of clavicle
    - Trapezoid ligament located anterolaterally
    - Inserts to trapezoid ridge, which runs along inferior surface, of lateral 1/3 of clavicle
    - Mainly prevents lateral compression of clavicle against acromion
  - Muscle attachments to lateral clavicle
    - Deltoid attached to anterior surface lateral 1/3 of clavicle
    - Trapezius attached to posterior surface lateral 1/3 of clavicle

ANATOMY IMAGING ISSUES

Imaging Recommendations

- High-resolution linear transducer
- Align transducer transversely along SCJ or ACJ
- ACJ laxity can be assessed by pulling down on arm while observing change in joint width on ultrasound
  - Compare with contralateral side
- Main clinical presentation of SCJ is painless lump
  - Mild degrees of capsular thickening readily apparent clinically since joint is just beneath skin surface
    - Clinical swelling often due to relative forward positioning of apparently swollen SCJ due to axial rotation of upper trunk
    - Occasionally due to mild capsular swelling ± mild subluxation secondary to SC osteoarthritis
  - Main clinical presentation of ACJ is pain due to osteoarthritis, ACJ impingement, inflammatory arthropathy, & subluxation/dislocation

Imaging Pitfalls

- SCJ or ACJ
  - Normally, step-off between medial clavicle & manubrium &, to lesser degree, between lateral clavicle & acromion
  - Should not be interpreted as subluxation
  - Acromion normally elevates from rest position during arm adduction
  - ACJ index = ACJ width of uninjured side/ACJ width of injured side = 1.0 normally
  - Determine whether
    - ACJ not subluxed (similar to opposite side): Grade 1
    - ACJ partially subluxed (clavicle subluxed < 50% depth of ACJ): Grade 2
    - ACJ severely subluxed or dislocated (clavicle subluxed > 50% depth of ACJ): Grade 3

TRANSVERSE US, STERNOCLAVICULAR JOINT

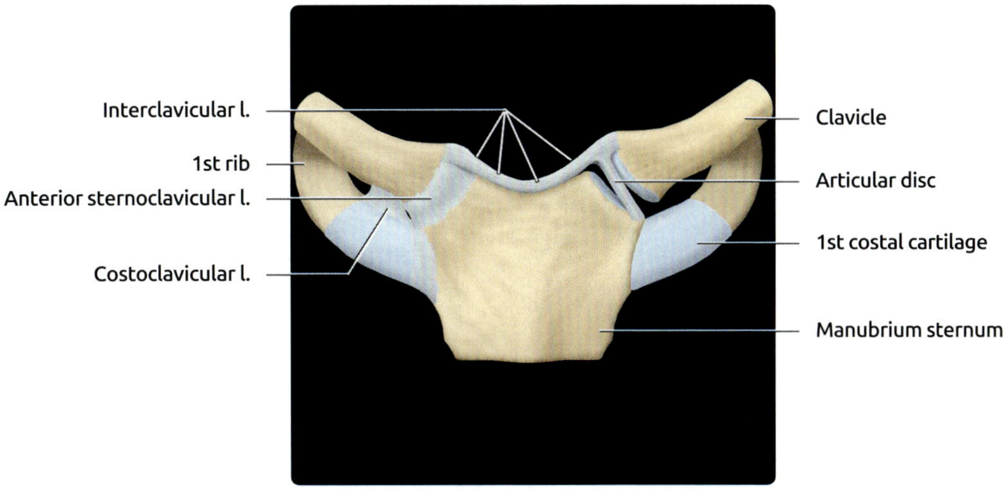

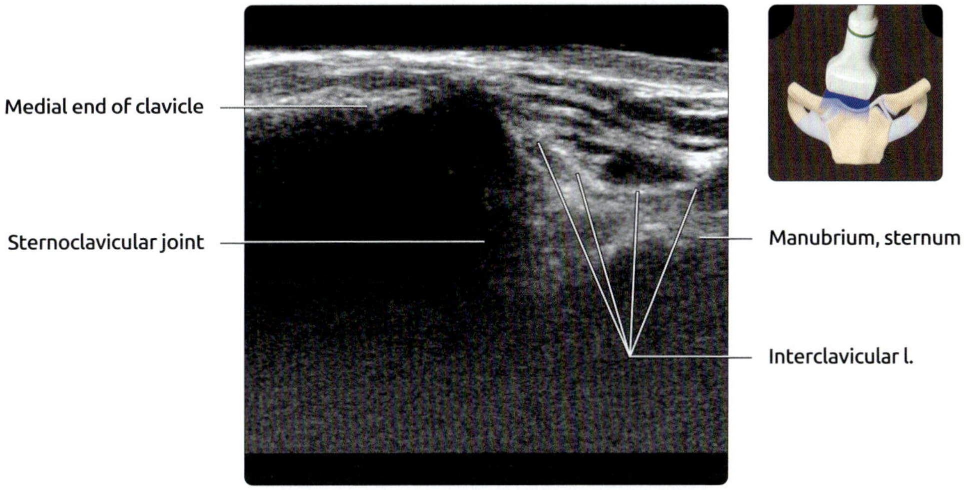

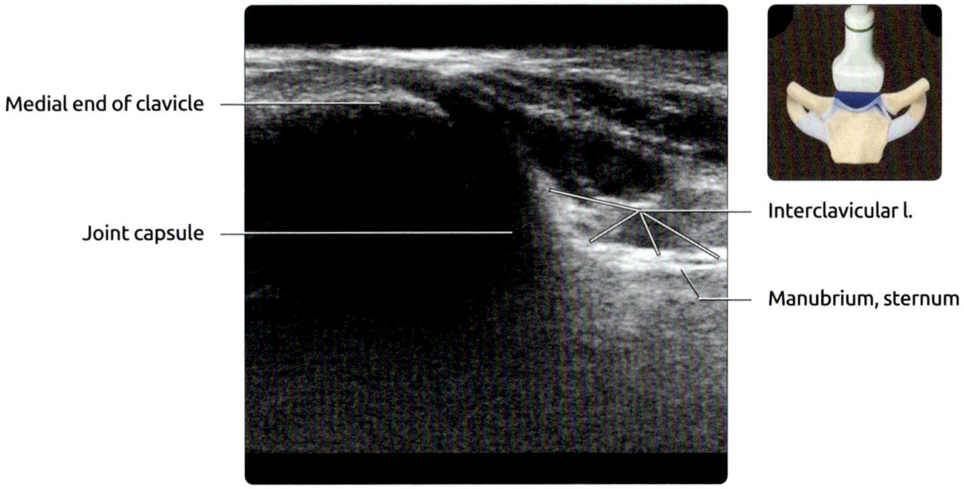

(Top) *Graphic shows the anterior aspect of the sternoclavicular joint. Note the joint capsule, articular disc, and interclavicular ligament.* **(Middle)** *Transverse grayscale ultrasound shows the anterosuperior aspect of the sternoclavicular joint. The medial clavicle is much larger than the articulating surface of the manubrium. The thin interclavicular ligament is closely applied to the superior aspect of manubrium, and its connection with the medial ends of both clavicles is depicted.* **(Bottom)** *Transverse grayscale ultrasound shows the superior aspect of the sternoclavicular joint. The costoclavicular ligament prevents upward movement of the medial clavicle when the lateral clavicle or shoulder is depressed.*

LONGITUDINAL US, STERNOCLAVICULAR JOINT

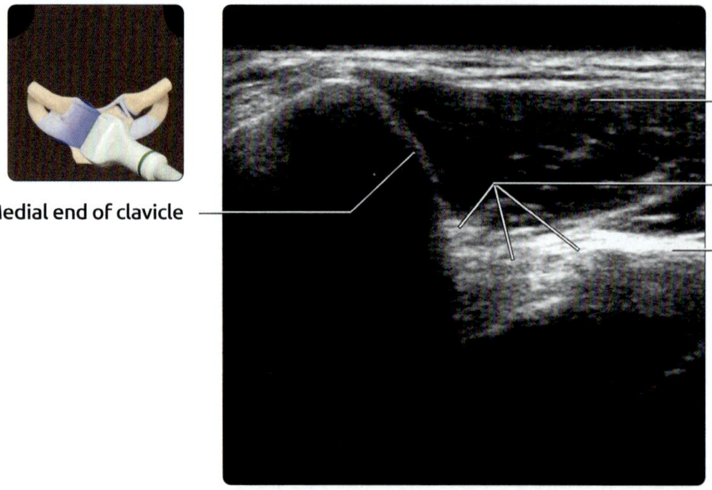

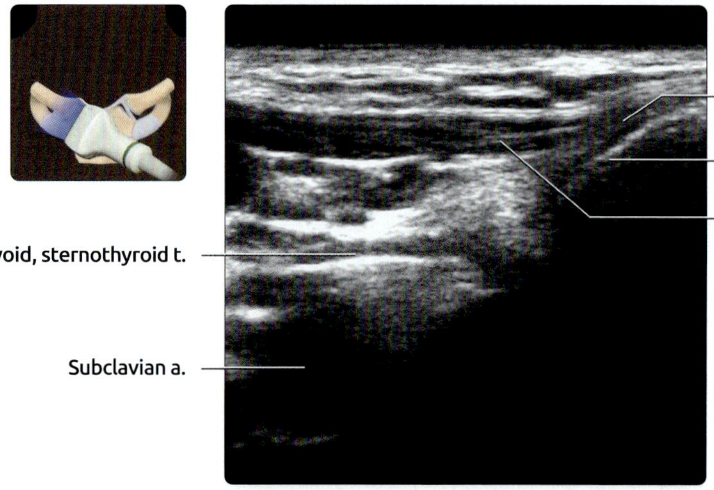

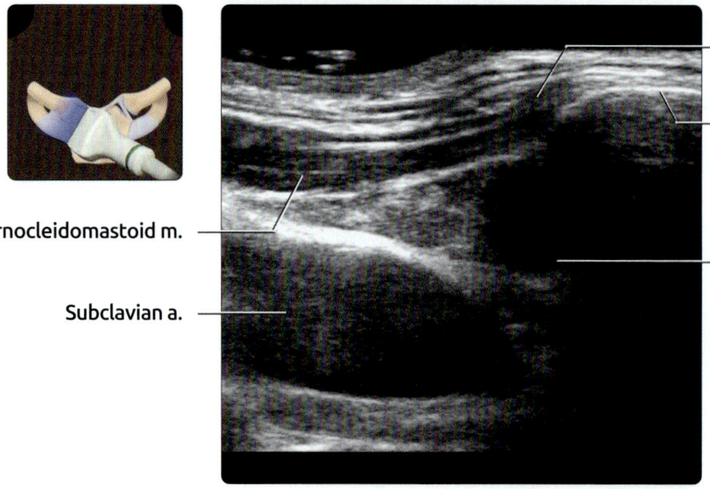

(Top) *Longitudinal grayscale US shows sternoclavicular joint. Costoclavicular ligament prevents upward movement of the medial clavicle when the shoulder is depressed. Pectoralis major muscle arises from the medial 1/2 of the anterior surface of the clavicle, as well as from the sternum, upper costal cartilages, and upper part of external oblique aponeurosis.* **(Middle)** *Longitudinal grayscale US shows the sternoclavicular joint region. Sternocleidomastoid is attached to the upper surface of the medial end of the clavicle, as well as the upper anterior surface of the manubrium. Sternohyoid and sternothyroid are attached to the posterior aspect of the sternum, as well as the clavicle and 1st costal cartilage.* **(Bottom)** *Longitudinal grayscale US shows the sternoclavicular joint. Great vessels lie posterior to the sternoclavicular joint and may get injured in posterior dislocation. All tendinous attachments should be assessed if dislocation is present as they may also be injured.*

US, ACROMIOCLAVICULAR JOINT

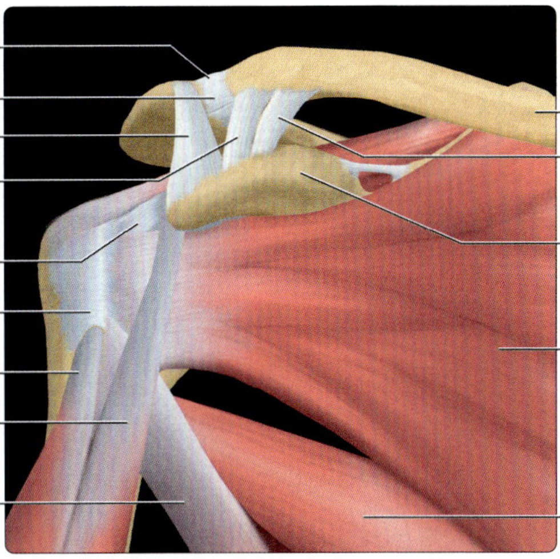

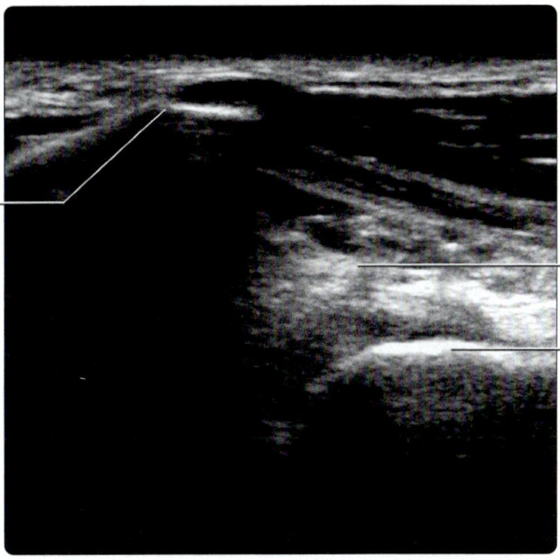

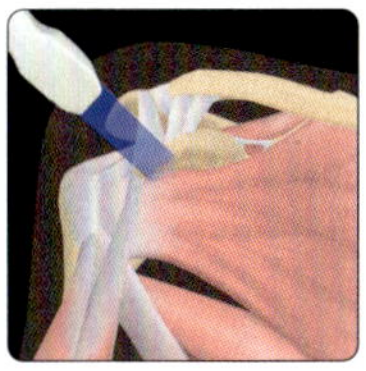

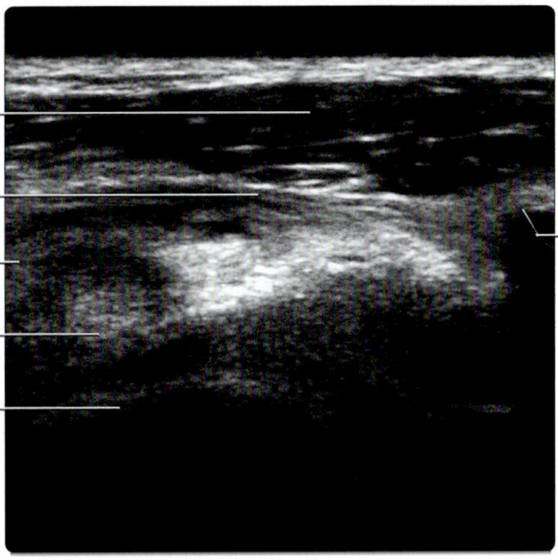

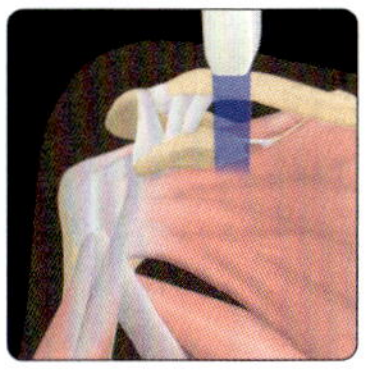

(Top) *Anterior graphic shows the shoulder in superficial dissection.* **(Middle)** *Longitudinal grayscale ultrasound shows the acromioclavicular joint region. The coracoclavicular ligament is demonstrated but is not as clearly depicted on US as it is on MR exam. These ligaments prevent upward and lateral movement of the clavicle.* **(Bottom)** *Transverse grayscale ultrasound of the acromioclavicular joint region shows the coracoacromial ligament. The supraspinatus tendon and intervening bursa can impinge against the coracoacromial ligament during arm abduction.*

TRANSVERSE US, ACROMIOCLAVICULAR JOINT

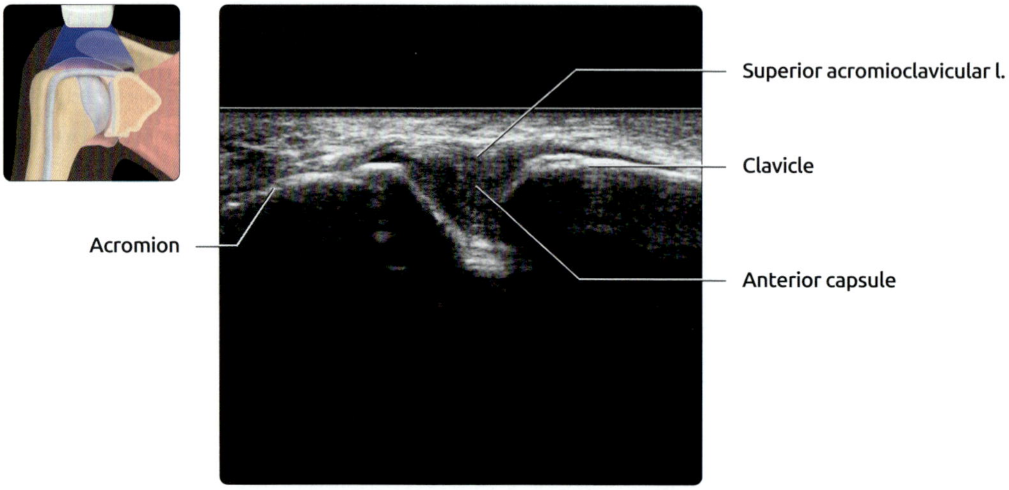

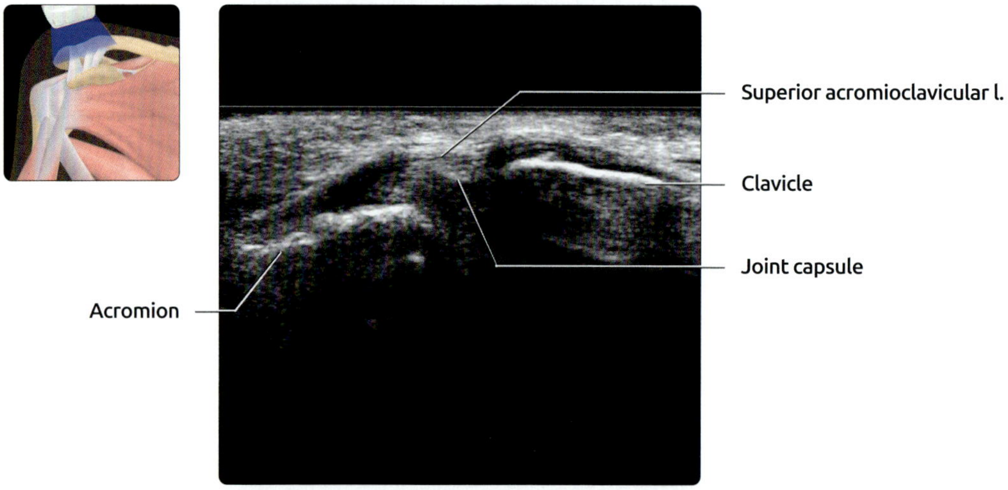

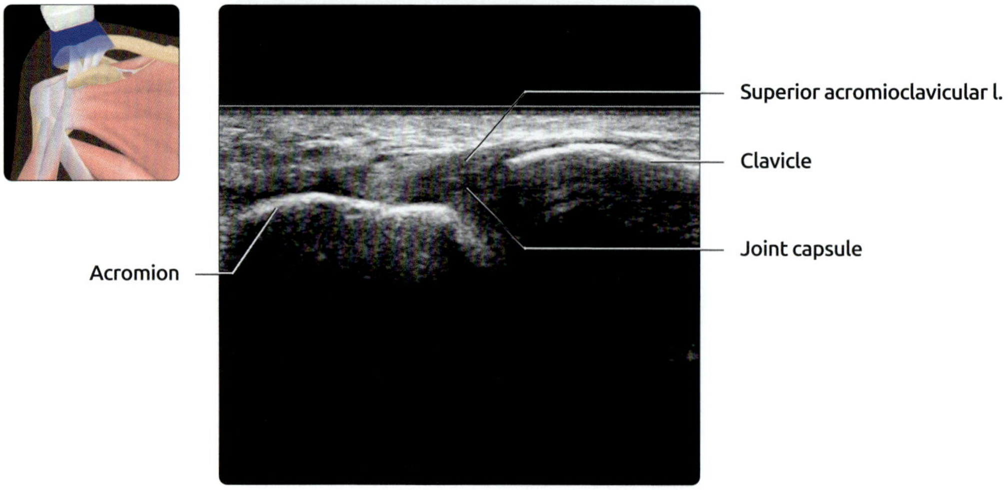

(Top) *Transverse grayscale ultrasound shows the anterior aspect of the acromioclavicular joint. The joint capsule of the acromioclavicular joint is thin with a strong supporting superior acromioclavicular ligament.* **(Middle)** *Transverse grayscale ultrasound shows the anterosuperior acromioclavicular joint. Separation of the clavicle and acromion can be readily appreciated. Note how opposing bone margins are not vertically aligned.* **(Bottom)** *Transverse grayscale ultrasound shows the superior aspect of the acromioclavicular joint. In this image, the clavicle slightly overrides the acromion. This is a normal configuration.*

TRANSVERSE US, ACROMIOCLAVICULAR JOINT

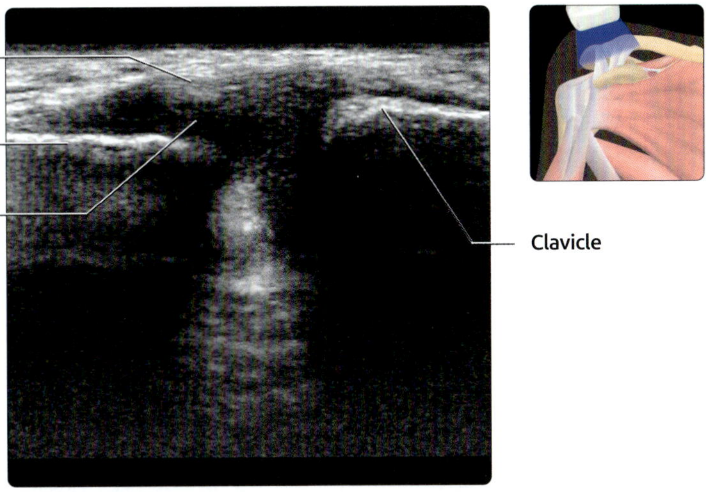

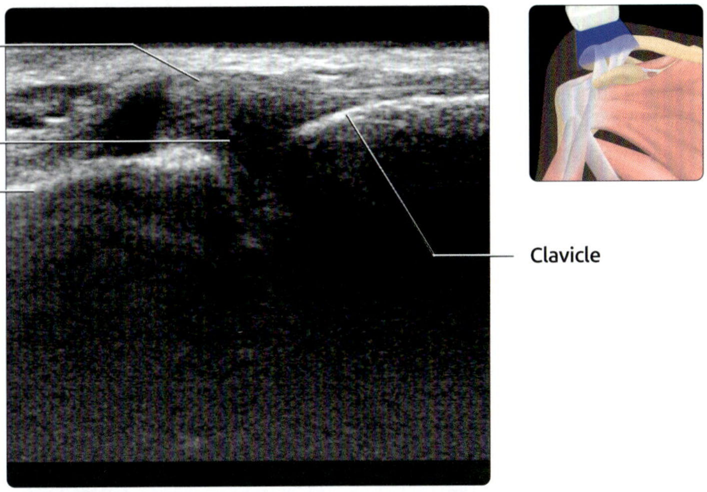

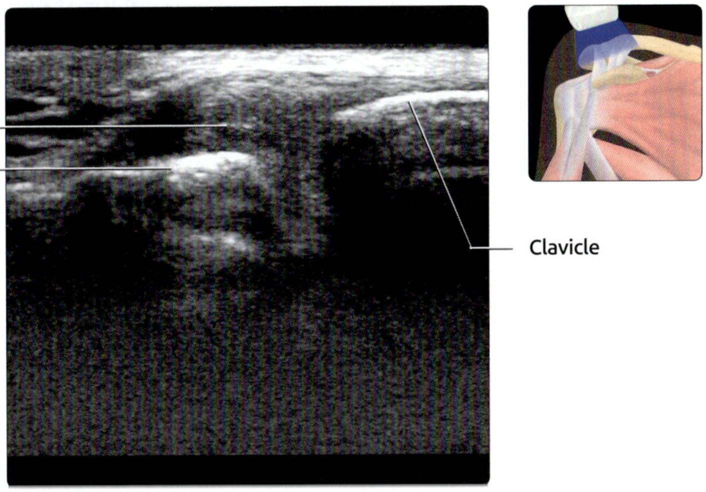

(Top) *Transverse grayscale ultrasound shows the anterosuperior aspect of the acromioclavicular joint with the arm positioned by the side of the body. Note that the clavicle slightly overrides the acromion.* **(Middle)** *Transverse grayscale ultrasound shows the acromioclavicular joint with the arm in abducted position. The acromion is now level with the lateral aspect of the clavicle. Note how the joint capsule bulges superiorly and the opposing bones are approximated with the arm abducted.* **(Bottom)** *Transverse grayscale ultrasound shows the acromioclavicular joint with the arm in an adducted position. The acromion is now depressed relative to this lateral end of the clavicle.*

IMAGING ANATOMY

Overview

- **Rotator cuff**
  - Consists of supraspinatus, infraspinatus, teres minor, subscapularis muscles, and tendons
  - Cuff tendons blend with shoulder joint capsule
  - Supraspinatus and infraspinatus tendons are inseparable at insertion
  - Anterior 2.25 cm of tendon comprises supraspinatus tendon insertional area
- **Supraspinatus muscle**
  - Origin: Supraspinatus fossa of scapula
  - Insertion: Superior facet (horizontal orientation) and anterior portion of middle facet of greater tuberosity
    - Broad insertional area
  - Nerve supply: Suprascapular nerve
  - Blood supply: Suprascapular artery and circumflex scapular branches of subscapular artery
  - Action: Abduction of humerus
  - Muscle consists of 2 distinct portions
    - Anterior portion is larger, fusiform in shape, has dominant tendon, and is more likely to tear
    - Posterior portion is flat and has terminal tendon
  - Most commonly injured rotator cuff tendon
- **Infraspinatus muscle**
  - Origin: Infraspinatus fossa of scapula
  - Insertion: Mid to posterior aspects of middle facet of greater tuberosity; centrally positioned within tendon
  - Nerve supply: Suprascapular nerve, distal fibers
  - Blood supply: Suprascapular artery and circumflex scapular branches of subscapular artery
  - Action: External rotation of humerus and resists posterior subluxation
- **Teres minor muscle**
  - Origin: Lateral scapular border, middle 1/2
  - Insertion: Inferior facet (vertical orientation) of greater tuberosity
  - Nerve supply: Axillary nerve
  - Blood supply: Posterior circumflex humeral artery and circumflex scapular branches of subscapular artery
  - Action: External rotation of humerus
  - Least commonly injured rotator cuff tendon
- **Subscapularis muscle**
  - Origin: Subscapular fossa of scapula
  - Insertion: Lesser tuberosity and up to 40% may insert at surgical neck
  - Some fibers cross over to lateral lip of bicipital groove, reinforcing and blending with transverse ligament
  - Nerve supply: Subscapular nerve, upper and lower
  - Blood supply: Subscapularis artery
  - Action: Internal rotation of humerus, also adduction, extension, depression, and flexion
  - 4-6 tendon slips converge into main tendon; multipennate morphology increases strength
- **Rotator cuff tendon blood supply**
  - Derived from adjacent muscle, bone, and bursae
  - Normal hypovascular regions in tendons
    - Termed critical zone: ~ 1 cm proximal to insertion
    - Vulnerable to degeneration and calcific deposition
    - However, insertional area is more prone to tearing than critical zone
- **Biceps tendon, long head**
  - Origin: Superior glenoid labrum (biceps anchor)
    - Portions may attach to supraglenoid tubercle, anterosuperior labrum, posterosuperior labrum, and coracoid base
  - Runs along superior aspect of shoulder to bicipital groove
  - Action: Stabilizes and depresses humeral head
  - Anatomic variants: Anomalous intra- and extraarticular origins from rotator cuff and joint capsule
  - Tendon sheath communicates with glenohumeral joint and normally contains small amount of fluid
- **Subacromial-subdeltoid fat plane**
  - Subacromial and subdeltoid portions
    - ± subcoracoid extension in some patients
  - Fat plane is superficial to bursa
  - May be interrupted or absent in normal patients
  - Attached along free border of coracoacromial ligament and deep surface of deltoid muscle, and humeral neck
- **Rotator cuff interval**
  - Space between supraspinatus and subscapularis tendon through which biceps tendon passes
  - **Borders of rotator cuff interval**
    - Triangular-shaped space
    - Reflections of glenohumeral ligament and coracohumeral ligament form biceps reflection pulley
    - Biceps reflection pulley stabilizes biceps tendon within rotator cuff interval
    - Superior border: Leading edge of supraspinatus
    - Inferior border: Superior aspect of subscapularis tendon
    - Lateral border: Long head of biceps tendon and bicipital groove
    - Medial border: Base of coracoid process
  - **Contents of rotator interval**
    - Long head of biceps tendon; biceps reflection pulley
- **Coracoacromial ligament**
  - Forms coracoacromial arch along with acromion and coracoid process
    - Reinforces inferior aspect of acromioclavicular joint
  - Extends from distal coracoid to subacromial area
  - Broad insertion to undersurface acromion
    - Ligament is thicker at acromion (normal thickness < 2.5 mm) and may be associated with spurs
- **Glenoid labrum**
  - Triangular-shaped rim of fibrocartilage, which extends around periphery of glenoid

ANATOMY IMAGING ISSUES

Imaging Approaches

- Tendons best seen when on stretch
  - High-resolution linear transducer
  - Long-axis (longitudinal) & transverse view of each tendon
  - Each part of tendon needs to be examined; anisotropy prevents all parts of curved rotator cuff tendons from being seen at same time
  - Need to realign ("toggle") probe to see different parts of tendons

- Supraspinatus tendon
  - Arm extended and internally rotated behind lumbar region (Crass position)
    - If too painful, hand in behind hip ("back pocket") with elbow close to body (modified Crass position)
- Infraspinatus and teres minor tendons
  - Arm flexed and internally rotated with hand placed on contralateral shoulder
  - Teres minor tendon located posteroinferior to infraspinatus tendon
- Subscapularis tendon: Arm neutral and externally rotated
- Long head of biceps tendon
  - Arm neutral and externally rotated
  - Vary degree of external rotation for optimal view of biceps tendon
  - Check for tendon subluxation
- Subacromial-subdeltoid bursa
  - Stretching tendons may squeeze fluid from area of bursa under inspection
  - Examine in all positions and also in neutral position
  - Fluid collects preferentially just lateral to acromion and proximal humerus and near coracoacromial ligament
- Coracoid process and coracoacromial ligament
  - Neutral position
- Acromioclavicular joint
  - Neutral position
  - Can pull down on arm to assess joint laxity
- Glenohumeral joint
  - Neutral position
  - Best seen from posterior aspect of joint
  - Passive movement of arm during scanning can help in identifying posterior glenoid labrum
- Spinoglenoid notch
  - Neutral position just medial to glenohumeral joint
- Supraspinatus and infraspinatus muscles
  - Neutral position with hands resting on thigh
  - Examine thickest part of muscles from behind (in coronal and sagittal planes)
  - ↓ muscle bulk, ↑ echogenicity, and ↓ visibility of central tendon are signs of atrophy with fatty replacement
    - Compare muscle echogenicity to that of trapezius or deltoid muscle

Imaging Sweet Spots

- Look for tears particularly at anterior leading edge of supraspinatus tendon
  - Unexplained bursal fluid is good secondary sign of rotator cuff tear
- Bursal fluid is often best seen with arm in neutral position or ↓ internal rotation (hand in back pocket)

Imaging Pitfalls

- **Anisotropy**
  - Echoes are optimally reflected when transducer is parallel to tendon fibers
  - Rotator cuff tendons are prone to anisotropy due to curved course
  - If transducer is not at right angles to tendon, it will appear either isoechoic or hypoechoic to muscle
    - May simulate tendinosis or partial tear
- **Tendon edges**
  - Interfaces of tendons with adjacent structures may simulate tears
  - All pathology should be confirmed in 2 planes
- **Rotator cuff cable**
  - Thick band of fibers running perpendicular to supraspinatus tendon
  - Located on deeper aspect of tendon just proximal to insertional area
  - May reinforce critical zone supraspinatus fibers
  - Cable thicker in young subjects but more easily seen in elderly subjects due to supraspinatus tendinosis
  - Can simulate tendinosis or partial-thickness tear
- **Tendinous interspace at rotator cuff interval**
  - Interspace between leading (anterior) edge of supraspinatus and long head of biceps tendon may simulate tear
  - Overcome by recognizing ovoid or rounded shape of biceps tendon
  - Rotator cuff interval best seen with external rotation
- **Focal thinning at supraspinatus-infraspinatus junction**
  - Mild diffuse thinning of supraspinatus and infraspinatus tendon junction is normal finding
  - Should not be mistaken for tendon attenuation or partial-thickness tear
- **Musculotendinous junction**
  - Supraspinatus tendon
    - Hypoechoic muscle extending along superficial aspect of tendon may simulate subacromial-subdeltoid bursal distension
    - Interdigitating tendons of anterior and posterior portions may simulate tendinosis or tear
  - Infraspinatus tendon
    - Muscle fibers surrounding centrally positioned tendon may be confused with tear
  - Subscapularis tendon
    - 4-6 tendon slips converging into main tendon may simulate tendinosis
- **Fibrocartilaginous insertion**
  - Thin layer of fibrocartilage exists between tendon and bone at insertional area
  - Steeper tendon insertion = thicker fibrocartilaginous layer
  - This thin hypoechoic layer of fibrocartilage may simulate avulsive tear
- **Subacromial-subdeltoid fat plane**
  - Fat plane lies mainly superficial to bursa and deep to deltoid muscle
  - Normal bursa is very thin
  - Thickness of echogenic fat plane is variable among patients though usually similar from side to side
  - May be wrongly interpreted as bursal fluid
    - Look for intrabursal fluid ± hyperemia (latter is feature of inflammatory arthropathy)
- **Fluid in biceps tendon sheath**
  - Communicates with glenohumeral joint
  - Small amount of fluid is normal
    - Do not misinterpret as long head of biceps tenosynovitis
  - ↑ fluid in biceps tendon sheath usually reflects ↑ fluid in glenohumeral joint

MUSCLES AND LIGAMENTS

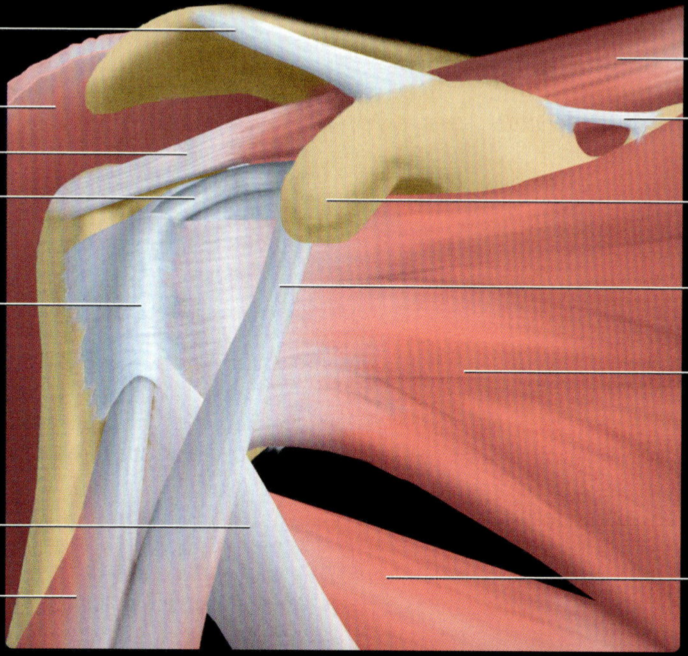

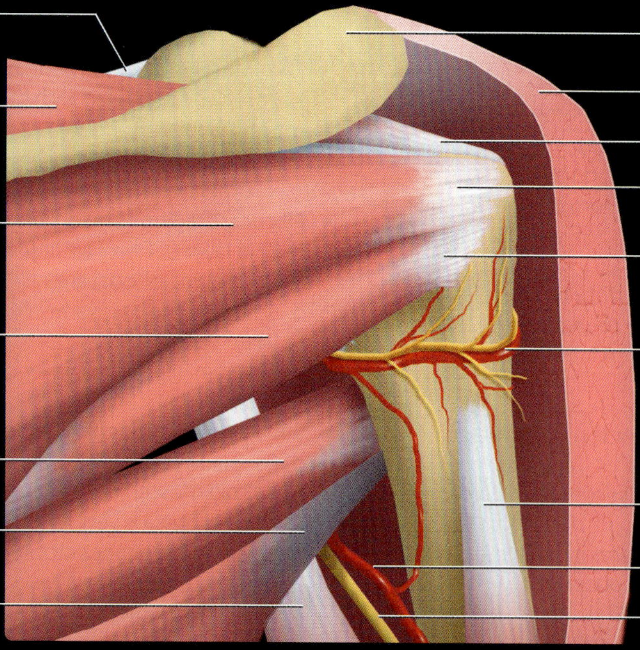

(Top) *Anterior graphic of the shoulder illustrates the rotator cuff and adjacent structures. The rotator cuff consists of supraspinatus, infraspinatus, teres minor, and subscapularis muscles and tendons. The biceps tendon courses the rotator cuff interval between the supraspinatus and subscapularis tendons, then descends along the bicipital groove, which is covered by the transverse ligament.* **(Bottom)** *Posterior graphic of the shoulder illustrates the rotator cuff and adjacent structures. The infraspinatus and teres muscles and tendons form the posterior wall of the rotator cuff. Inferior to the teres minor muscle and superior to the teres major is the axillary nerve and posterior circumflex humeral vessels running through the quadrilateral space.*

DEEP STRUCTURES

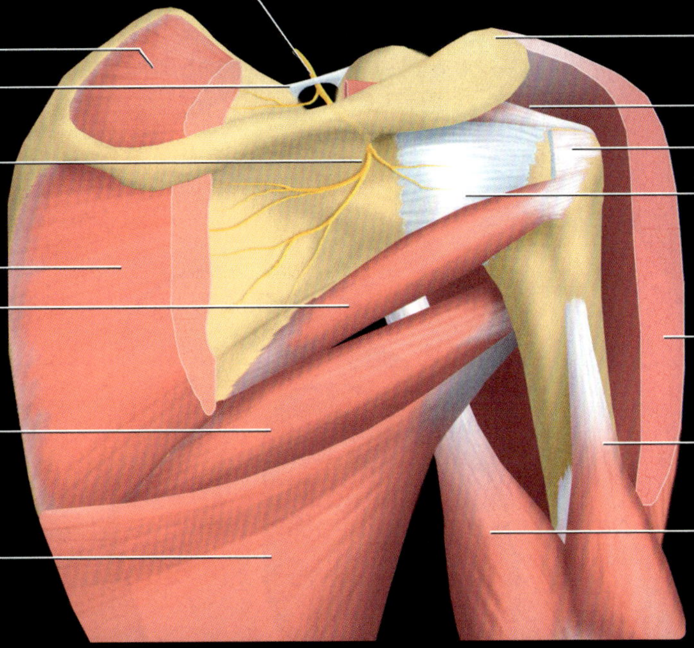

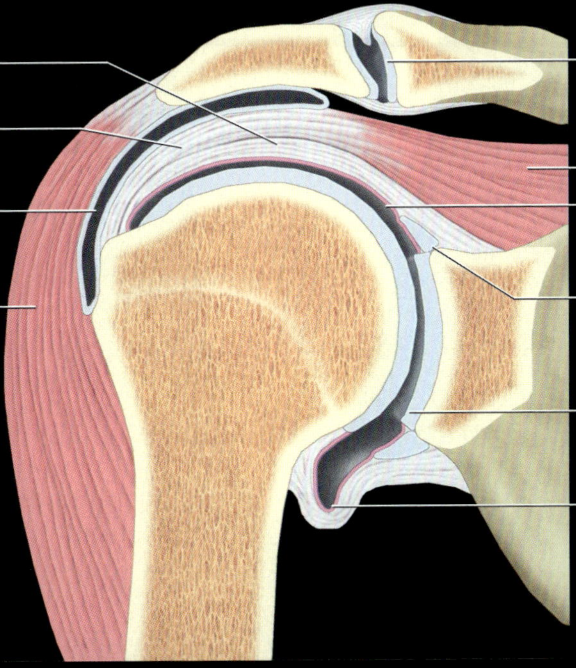

(Top) *Deep scapulohumeral dissection shows the course of the suprascapular nerve. The nerve enters the supraspinous fossa through the suprascapular notch, below the superior transverse scapular ligament. The nerve then passes beneath the supraspinatus and curves around the lateral border of the spine of the scapula to enter the infraspinous fossa.* **(Bottom)** *Graphic shows the coronal section through the midportion of the shoulder joint. Note the subacromial-subdeltoid bursa is situated in the space between the deltoid, acromion, acromioclavicular joint, distal clavicle superiorly, and the supraspinatus tendon and muscle inferiorly. There is no direct communication between the bursa and glenohumeral joint unless there is full-thickness tear of the supraspinatus tendon. The bursa is the main pain-producing structure around the shoulder. The wide expanse of the subacromial-subdeltoid bursa enables one to understand why shoulder symptoms are poorly localized. Patients often complain of generalized shoulder pain.*

SAGITTAL NORMAL LABRUM

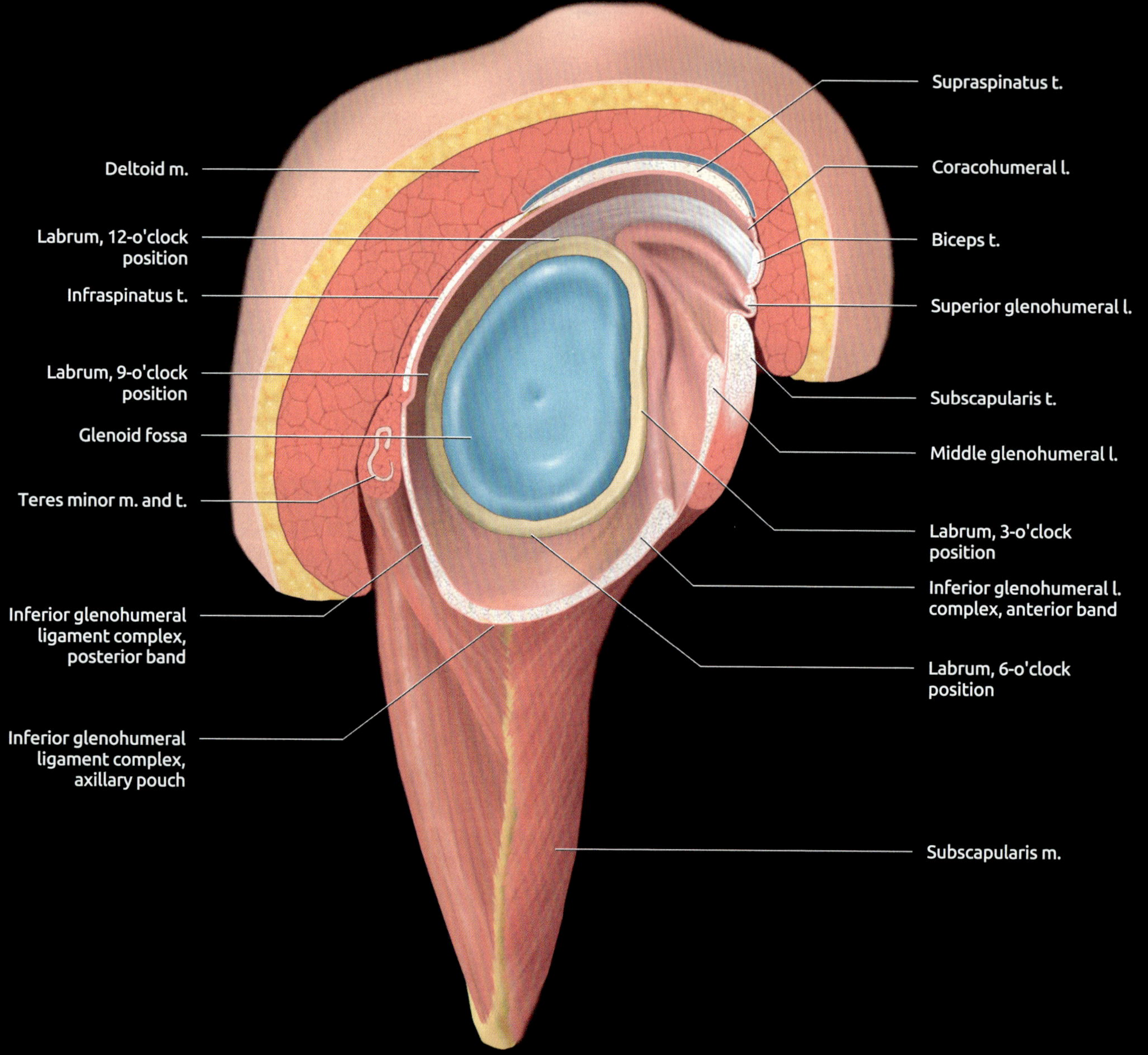

Sagittal graphic shows the glenoid fossa. The labrum lines the edge of the glenoid, increasing the circumference and depth of the shoulder joint. The labrum plays an important role in stabilization of the shoulder joint. At the inferior aspect of the glenoid, the labrum is barely discernible. The superior, middle, and inferior glenohumeral ligaments are distinctive thickenings of the anterior shoulder capsule and cannot be seen as separate structures on ultrasound examination.

LONGITUDINAL US, SUPRASPINATUS INSERTIONAL AREA

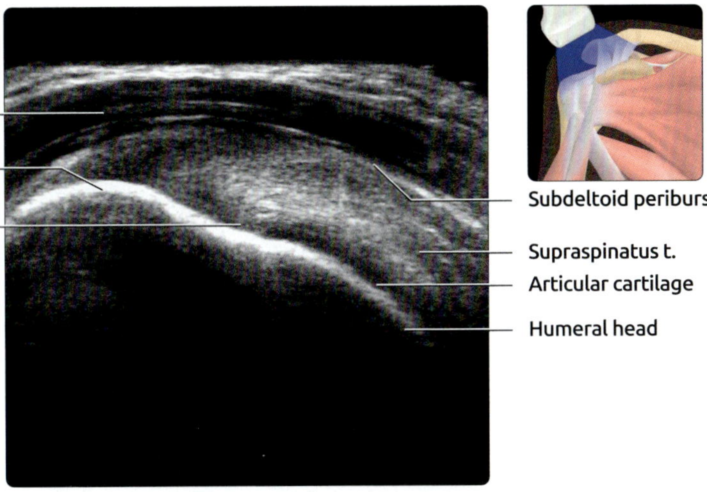

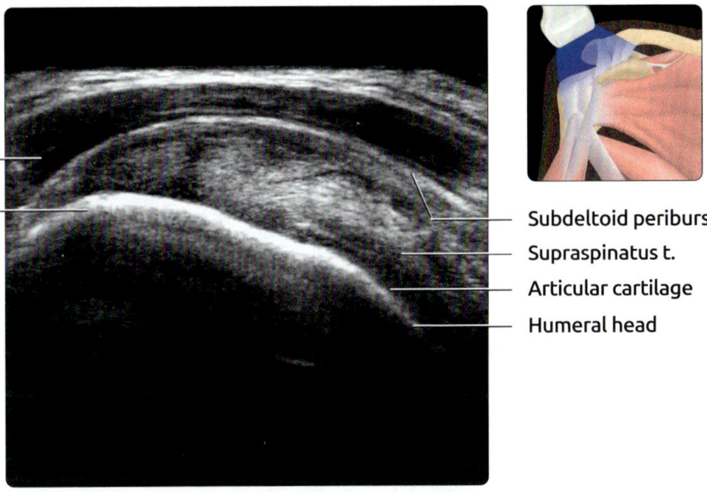

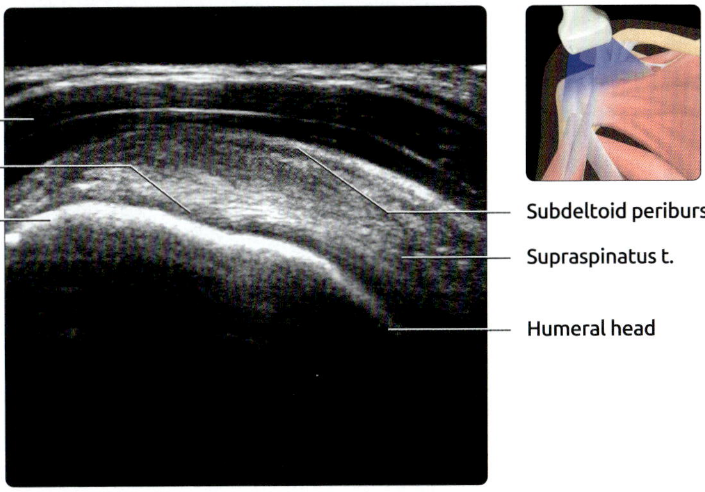

(Top) *Longitudinal grayscale US shows the anterior fibers of the supraspinatus tendon insertional area. The supraspinatus tendon inserts over a wide area (footprint) on the anterior aspect of the greater tuberosity. Many tears of the supraspinatus tendon involve avulsion of the tendon from its insertional site.* **(Middle)** *Longitudinal grayscale US shows the supraspinatus tendon midfibers. There is often a thin hypoechoic line at the insertional area. This represents Sharpey fibers and fibrocartilage.* **(Bottom)** *Longitudinal grayscale US shows the supraspinatus tendon posterior fibers. The anterior, middle, and posterior fibers should be evaluated in turn.*

TRANSVERSE US, SUPRASPINATUS INSERTIONAL AREA

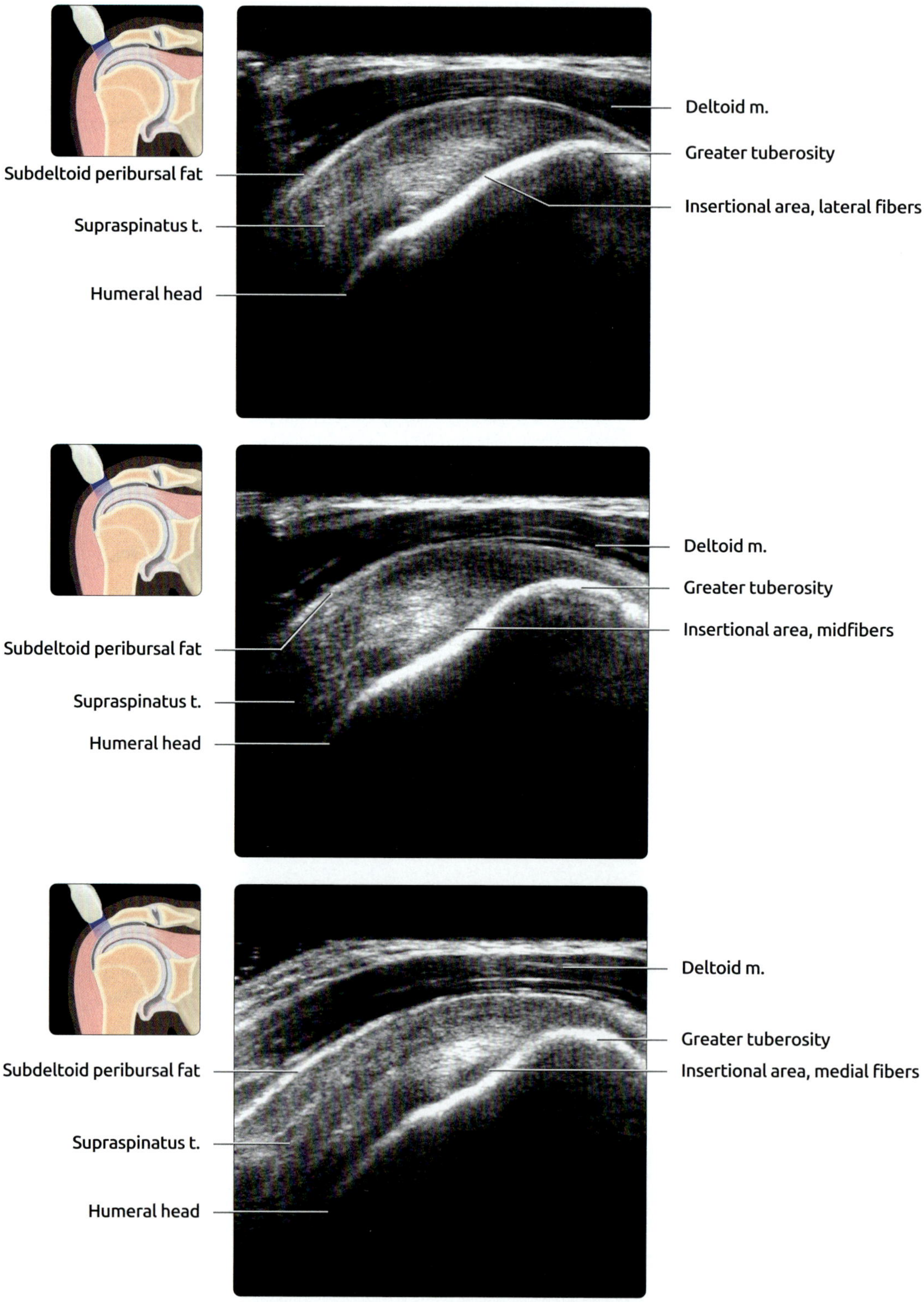

(Top) *Transverse grayscale US shows the supraspinatus tendon insertional area. Angulation and slight movement of the transducer will allow visualization of the lateral, middle, and medial fibers, respectively. The leading anterior edge of the supraspinatus is a common site of tear (rim rent tear).* **(Middle)** *Transverse grayscale US shows the supraspinatus tendon at the insertional area of midfibers. The fibrillar echotexture of the supraspinatus tendon can be seen but is prone to anisotropy.* **(Bottom)** *Transverse grayscale US shows the supraspinatus midfibers insertional area. Subdeltoid peribursal fat is variable in depth. The normal bursa cannot be depicted. It is seen only when distended with fluid or thickened due to synovitis.*

TRANSVERSE AND LONGITUDINAL US, SUPRASPINATUS

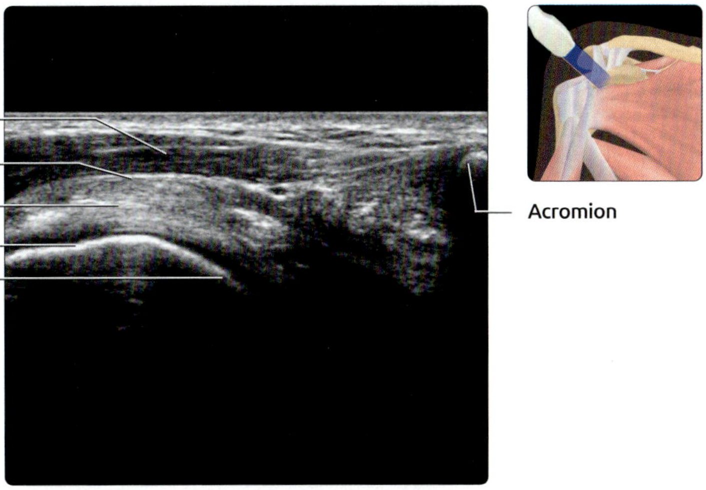

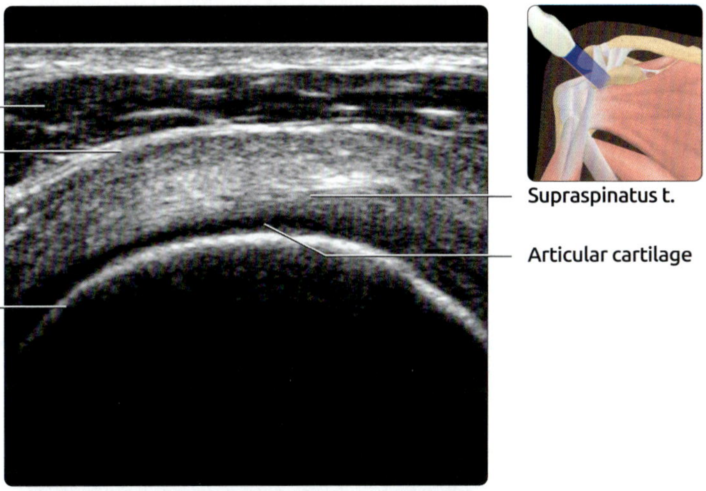

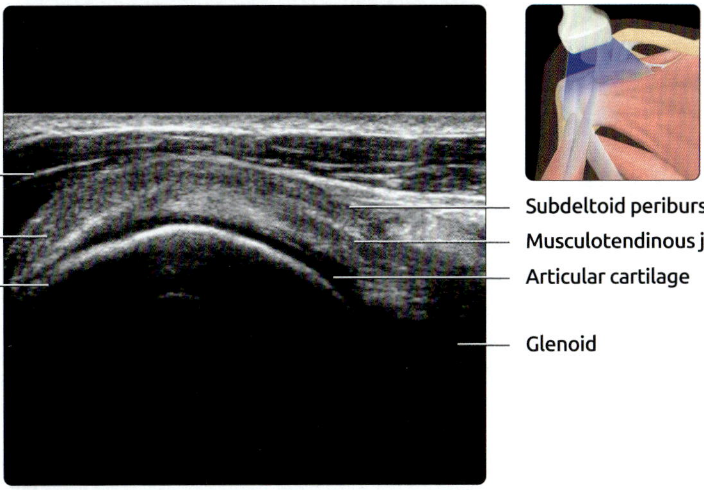

(Top) *Transverse grayscale US shows the supraspinatus tendon just medial to the insertional area. The critical zone is located just medial to the insertional area. This is a relatively hypovascular area. It is prone to calcific tendinosis and tears, though most tears tend to occur at the insertional site.* **(Middle)** *Transverse grayscale US shows the supraspinatus medial to the insertional area. The fibrillar pattern of the supraspinatus tendon is prone to anisotropy. With tendinosis, the fibrillar pattern is disrupted. The tendon becomes more hypoechoic and thickened.* **(Bottom)** *Longitudinal grayscale US shows the supraspinatus tendon at a slightly more medial aspect. Tears of the musculotendinous junction are relatively uncommon.*

GRAPHIC AND TRANSVERSE US, ROTATOR CUFF INTERVAL

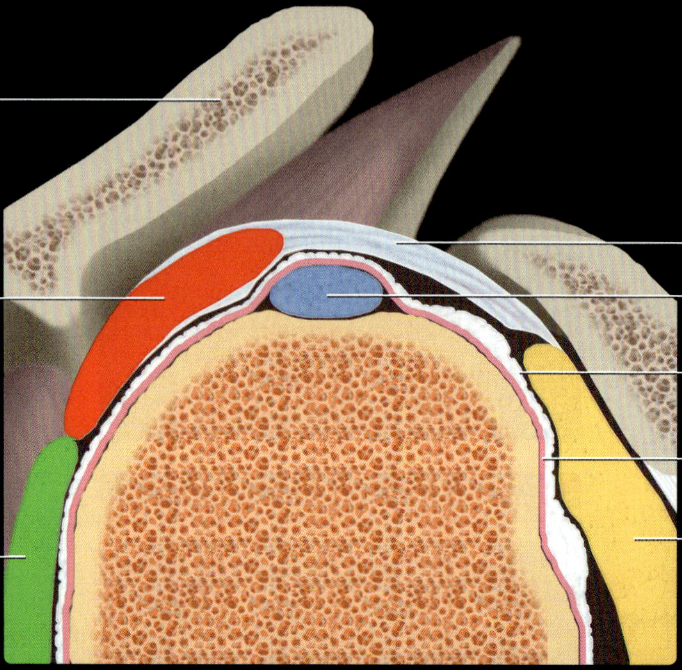

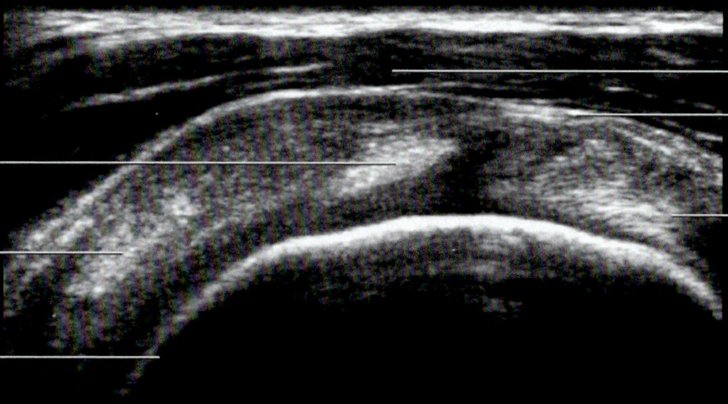

(Top) *Graphic shows the relationship of the coracohumeral ligament to the rotator cuff tendons. The coracohumeral ligament is not a true ligament but a folded portion of the glenohumeral capsule that extends from the coracoid process to the humerus. The undersurface is lined by synovium. Portions of the coracohumeral ligament pass superficial and deep to the supraspinatus tendon. The coracohumeral ligament attaches to the superior border of the subscapularis tendon as well as to the greater tuberosity.* **(Bottom)** *Transverse grayscale US shows the shoulder at the rotator cuff interval. The rotator cuff interval represents the space between the leading anterior edge of the supraspinatus tendon and the adjacent superior edge of the subscapularis. It contains the biceps tendon. This should not be mistaken for a tear of the anterior aspect of the supraspinatus tendon. Patients with adhesive capsulitis may show increased hypoechogenicity and hyperemia of the rotator cuff interval as a result of inflammatory fibrovascular soft tissue overgrowth.*

TRANSVERSE US, BICEPS TENDON

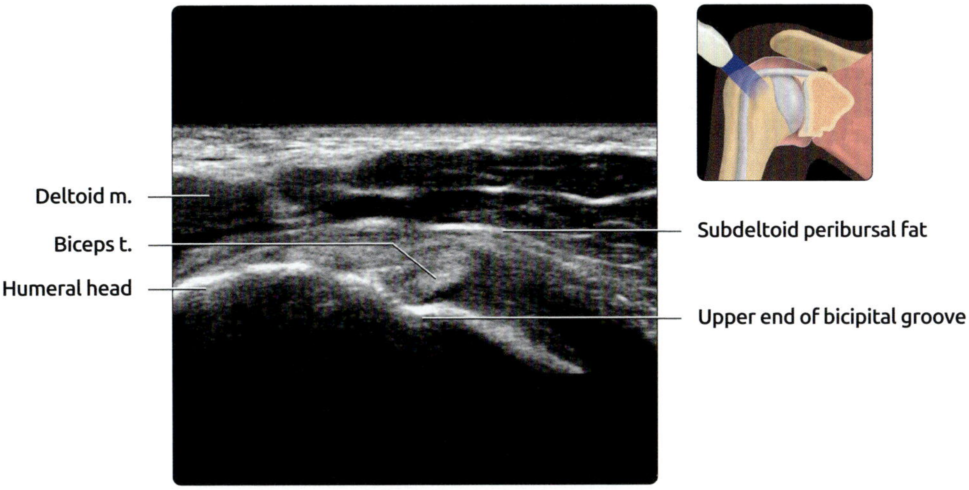

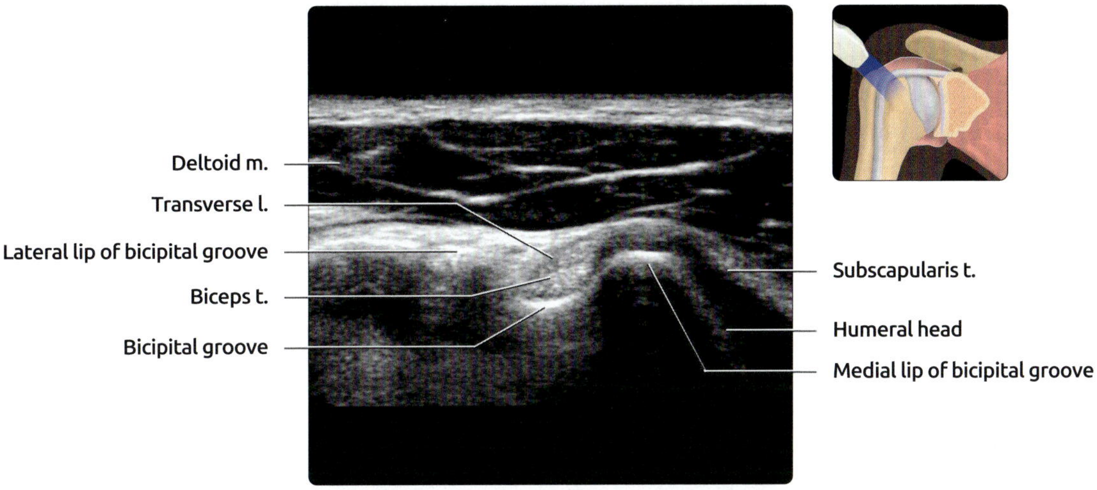

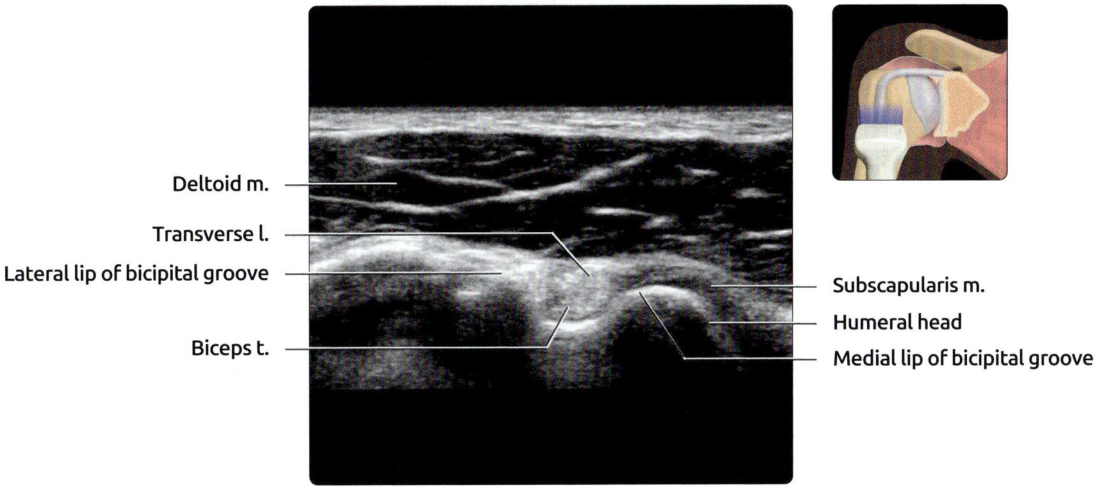

(Top) *Transverse grayscale US at the proximal aspect of the biceps tendon is shown. The bicipital groove forms a fibroosseous tunnel for the biceps tendon. The normal biceps tendon has an ovoid configuration. With tendinosis, the tendon becomes larger and more rounded in appearance.* **(Middle)** *Transverse grayscale US shows the biceps tendon. The biceps tendon is held in position by the transverse ligament. The subscapularis tendon inserts into the lesser tuberosity, which forms the medial lip of the bicipital groove. The biceps tendon may sublux medially from the groove and may be associated with subscapularis tendon injury.* **(Bottom)** *Transverse grayscale US shows relations of the biceps tendon. The transverse ligament is not a distinct entity but consists of a fibrous expansion of both the pectoralis major tendon and subscapularis tendon inserting into the lateral lip of the bicipital groove.*

LONGITUDINAL US, BICEPS TENDON

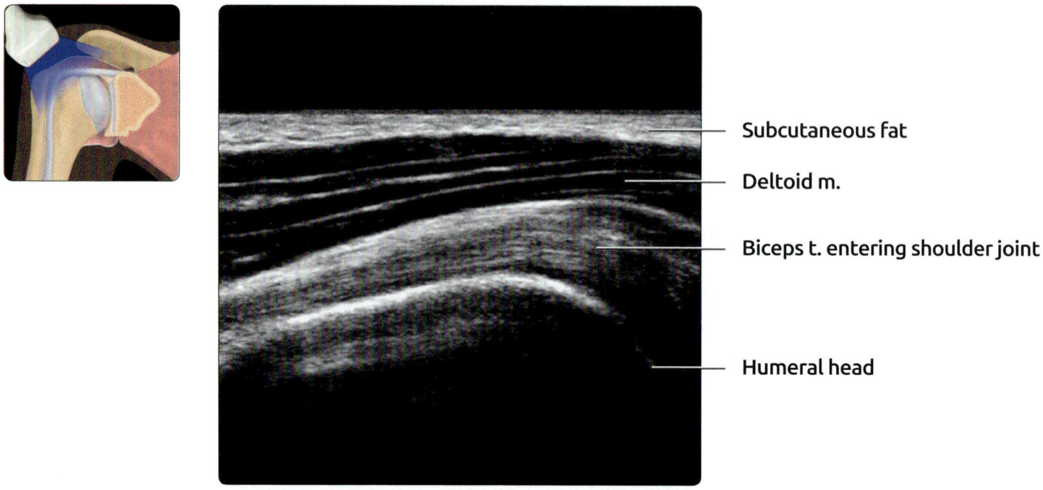

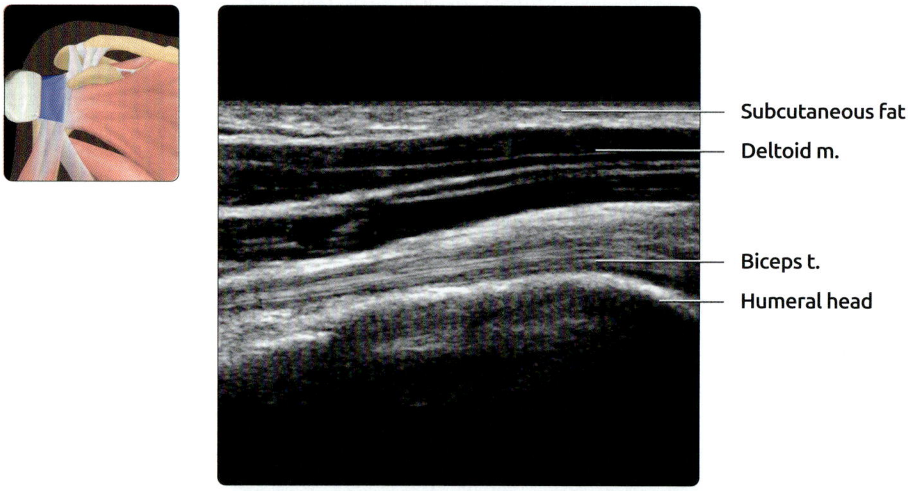

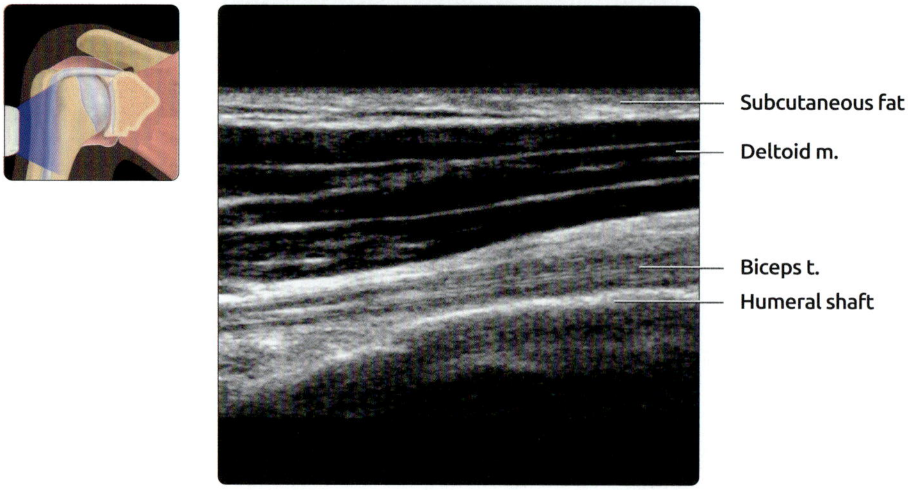

(Top) *Longitudinal grayscale US shows the uppermost section of the biceps tendon. The intraarticular portion of the biceps long head tendon cannot be fully depicted on longitudinal imaging. On longitudinal section, the biceps tendon seems to expand at its upper end as it becomes more ovoid in contour. This location is also the most common site of biceps tendinosis, which also manifests as enlargement of the biceps tendon.* **(Middle)** *Longitudinal grayscale US shows the upper section of the biceps tendon. A small amount of fluid in the biceps tendon sheath, which is continuous with the glenohumeral joint, is normal and should not be mistaken for tenosynovitis. No fluid is depicted in this image.* **(Bottom)** *Longitudinal grayscale US shows the lower aspect of the biceps tendon. Most incomplete tears of the biceps tendons are longitudinal in orientation and are best depicted on transverse imaging.*

LONGITUDINAL US, SUBSCAPULARIS

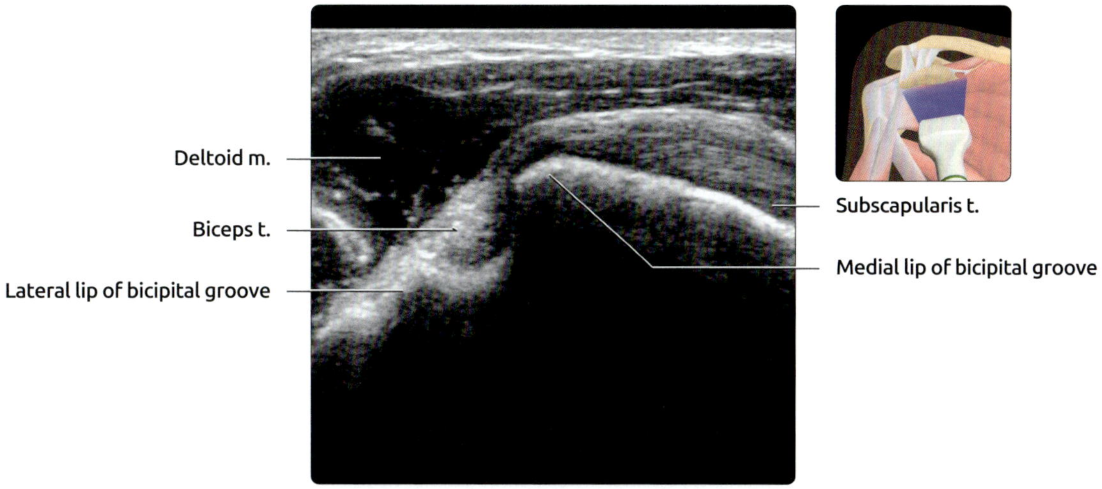

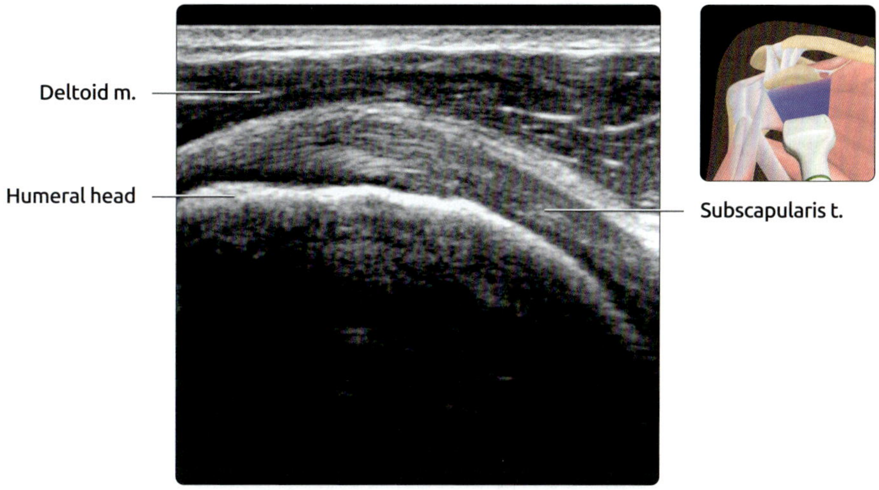

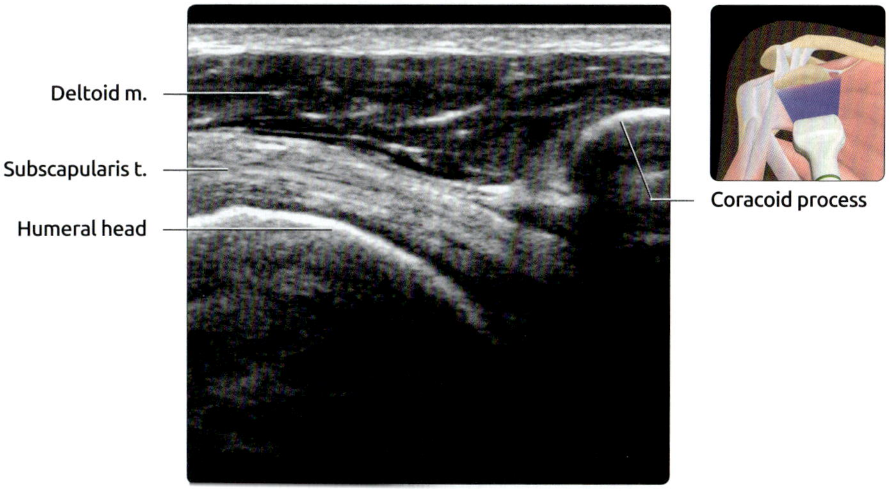

(Top) *Longitudinal grayscale US shows the subscapularis insertion. The subscapularis inserts into the medial lip of the bicipital groove but has a fibrous expansion traversing the biceps tendon through which it also gains attachment to the lateral lip of the bicipital groove.* **(Middle)** *Longitudinal grayscale US shows the subscapularis insertional area. Complete tears of the subscapularis are uncommon and usually follow a severe traumatic event. Partial tears are more common and usually involve the superior edge of the tendon.* **(Bottom)** *Longitudinal grayscale US shows the subscapularis tendon. The subscapularis moves beneath the coracoid process during internal-external rotation. Impingement may potentially occur at this location (subcoracoid impingement).*

TRANSVERSE US, SUBSCAPULARIS TENDON

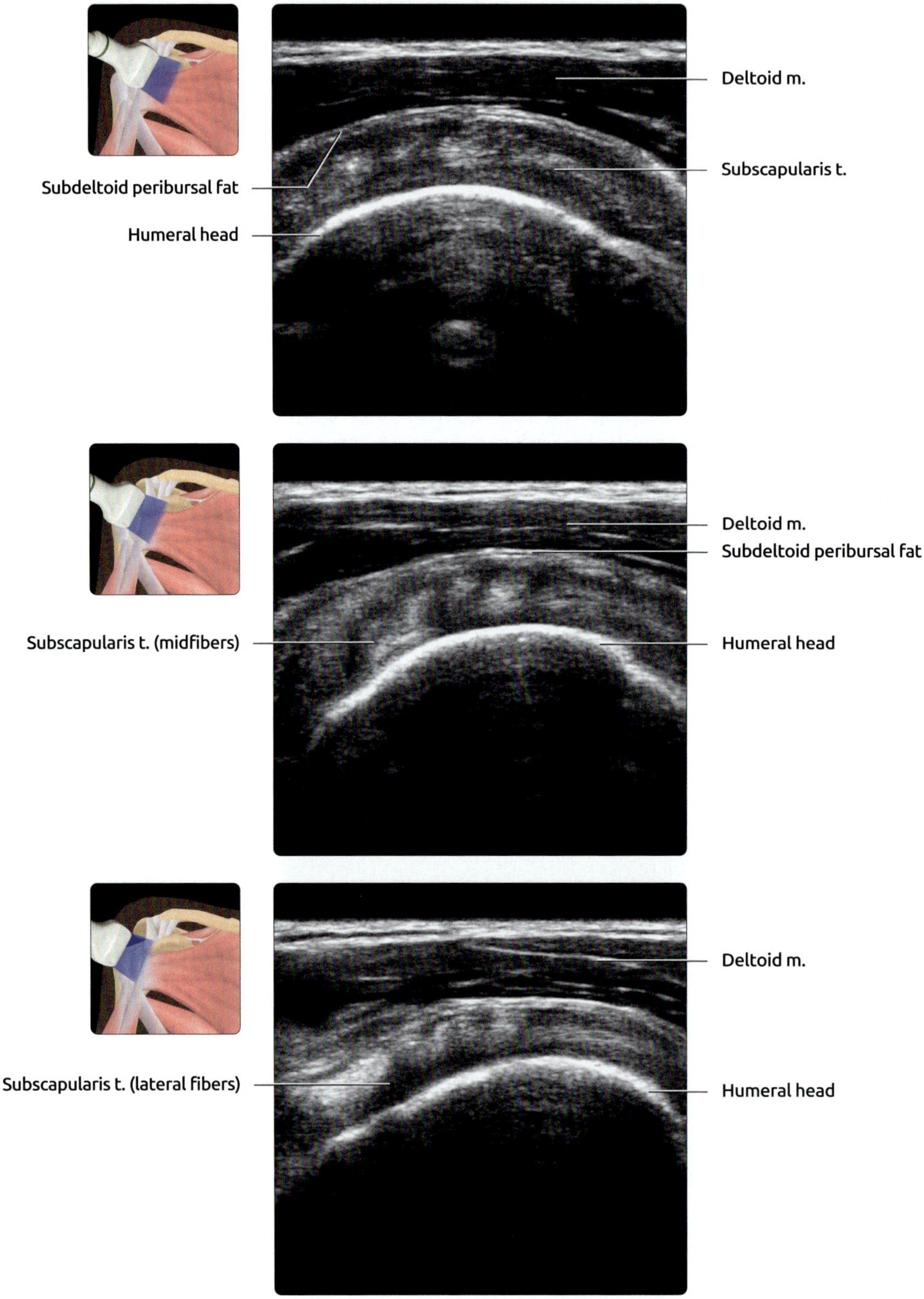

(Top) *Transverse grayscale US shows the subscapularis tendon midfibers. As they converge toward the insertion, fiber bundles of the multipennate subscapularis tendon give the tendon a mixed echogenic appearance. This is normal and should not be mistaken for tendinosis.* **(Middle)** *Transverse grayscale US shows the subscapularis at the level of midfibers.* **(Bottom)** *Transverse grayscale US shows the subscapularis tendon. Tears of the subscapularis tendon usually occur just proximal to the insertional area. These tears may involve the fascial covering of the biceps tendon, facilitating biceps tendon dislocation.*

LONGITUDINAL AND TRANSVERSE US, INFRASPINATUS

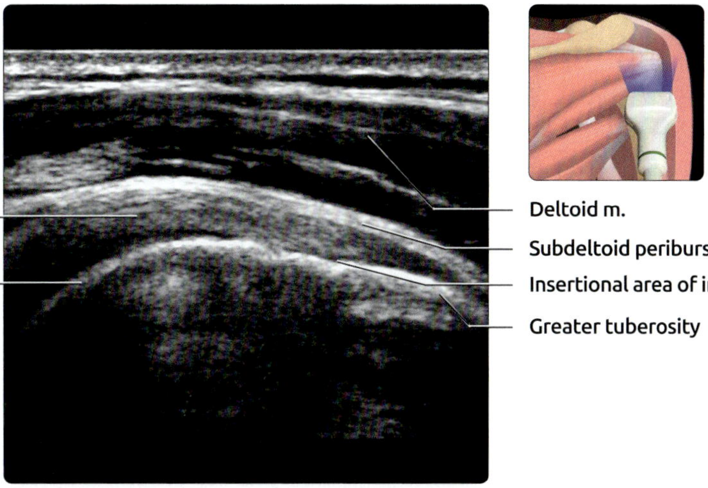

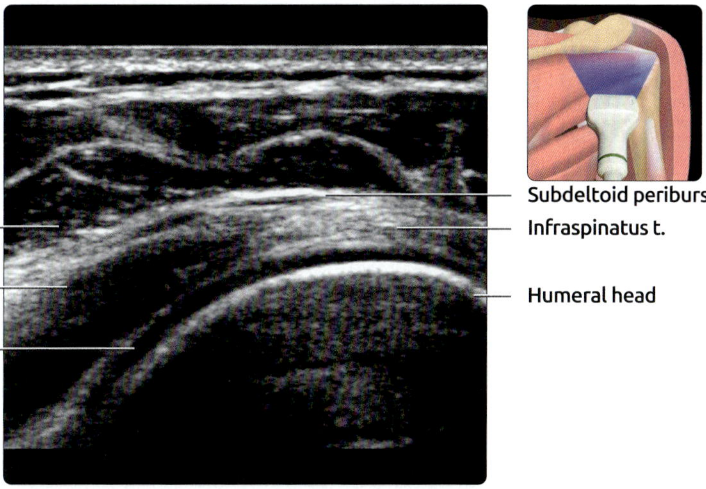

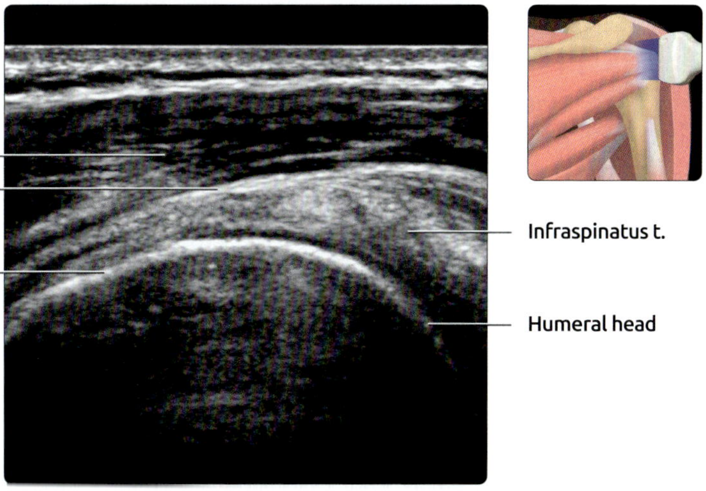

(Top) *Longitudinal grayscale US shows the infraspinatus tendon insertional area. The infraspinatus tendon is less commonly torn than the supraspinatus tendon. Most tears are avulsive-type tears involving the insertional area.* **(Middle)** *Longitudinal grayscale US shows the infraspinatus musculotendinous junction. The muscle fibers interdigitate with the tendon at the musculotendinous junction and should not be mistaken for tears/tendinosis.* **(Bottom)** *Transverse grayscale US shows the infraspinatus insertional area. All tears of the rotator cuff tendons should be confirmed in both planes (transverse and longitudinal).*

LONGITUDINAL AND TRANSVERSE US, TERES MINOR

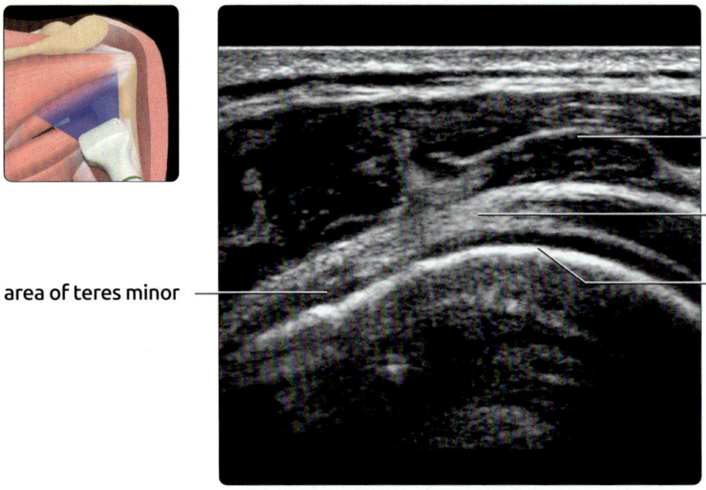

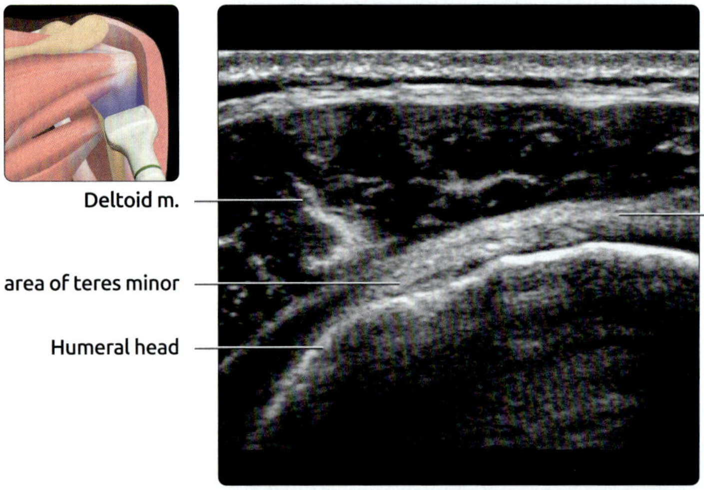

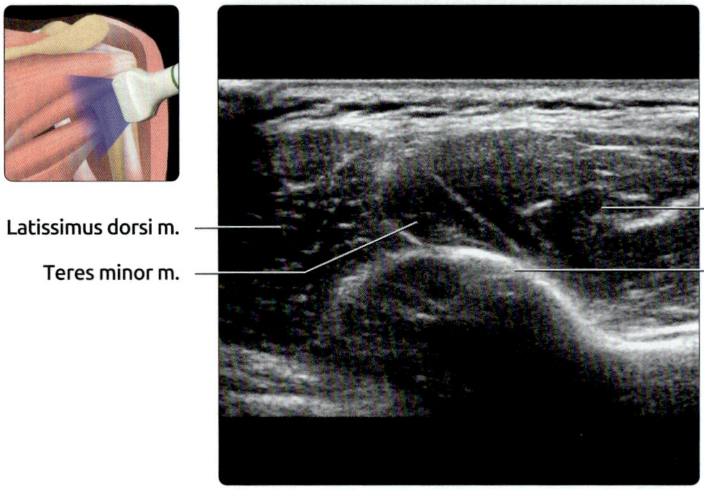

(Top) *Longitudinal grayscale US shows the teres minor muscle insertional area. The teres minor muscle is usually not torn in isolation, but it may be torn in massive rotator cuff tears. It is a small muscle and tendon seen at the posteroinferior edge of the infraspinatus tendon.* **(Middle)** *Longitudinal grayscale US shows the teres minor insertional area.* **(Bottom)** *Transverse grayscale US shows the teres minor. The teres minor muscle is best depicted on longitudinal imaging at the inferior aspect of the infraspinatus tendon. Isolated atrophy of the teres minor muscle can occur in quadrilateral space syndrome due to compression of the axillary nerve (part of the deltoid muscle may also be affected).*

SHOULDER JOINT AND POSTERIOR ASPECT

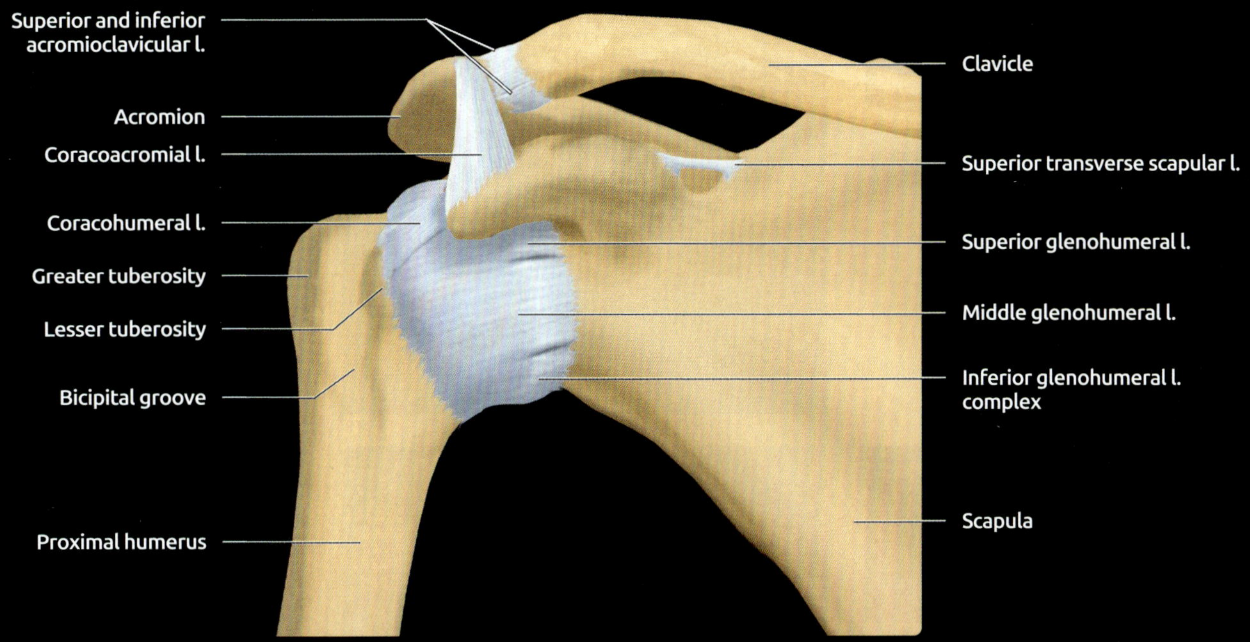

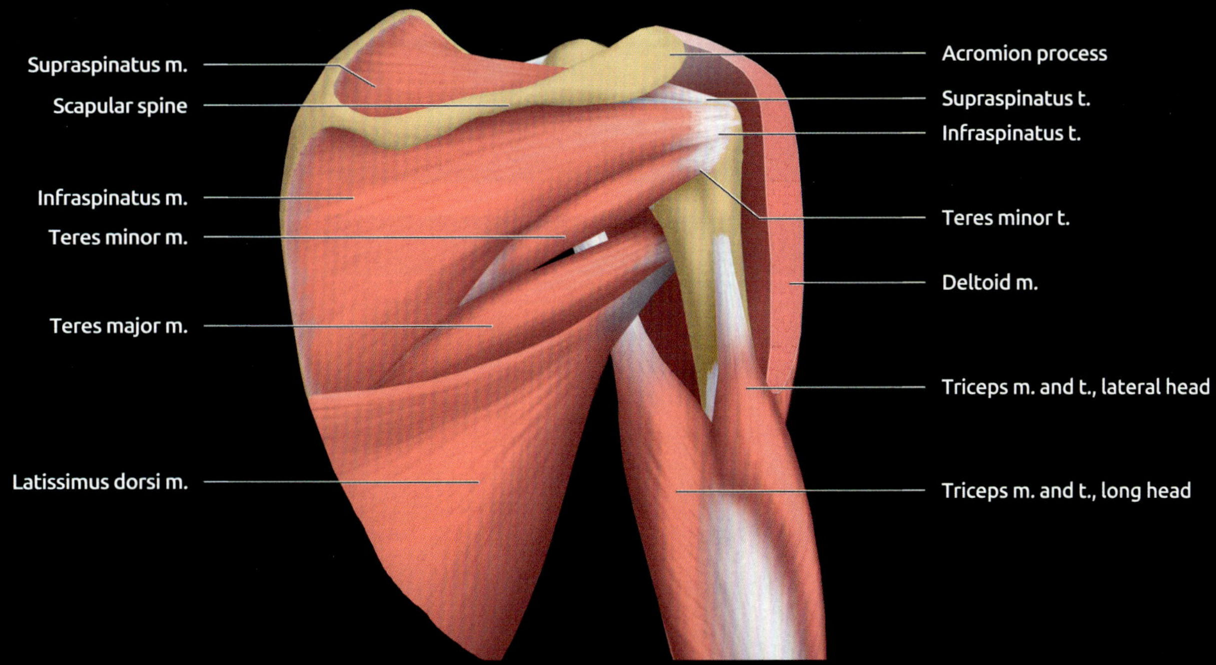

(Top) *Anterior graphic illustrates the right shoulder in deep dissection. The muscles have been removed.* **(Bottom)** *Posterior graphic illustrates the shoulder. Superficial scapulohumeral dissection demonstrates the musculature.*

LONGITUDINAL US, POSTERIOR ASPECT

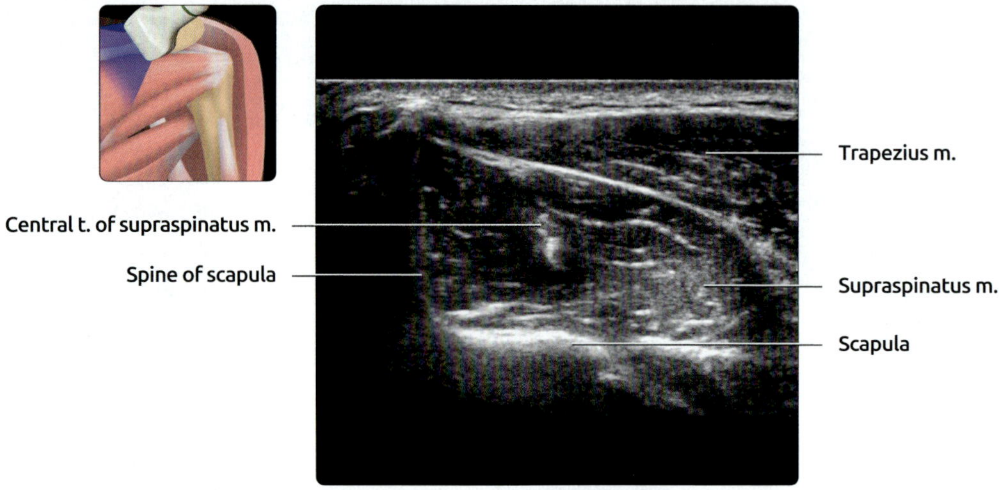

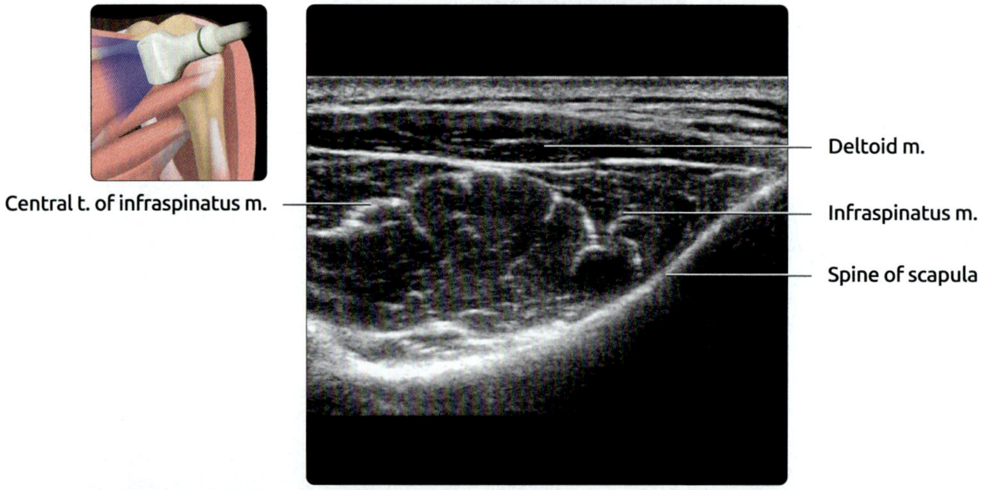

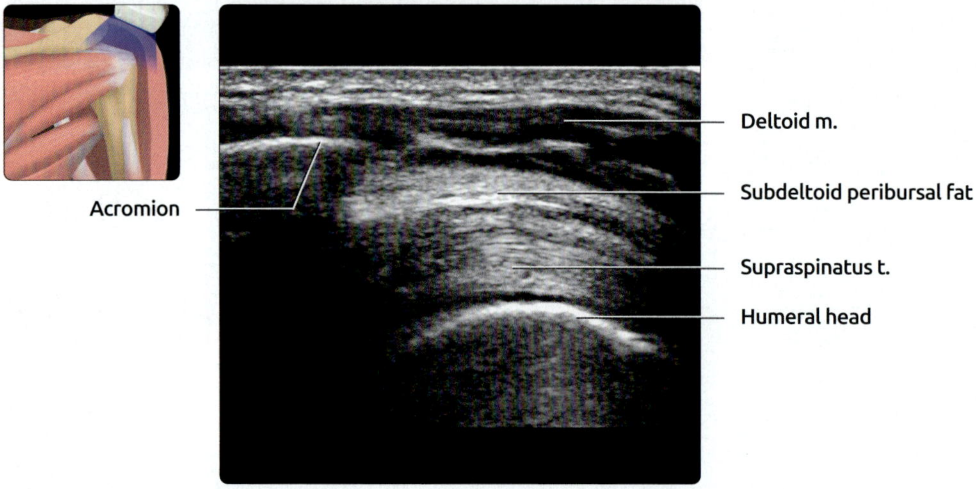

(Top) *Longitudinal grayscale US shows the posterior shoulder at the supraspinatus muscle. Muscle atrophy is common in rotator cuff pathology. Ultrasound is almost as sensitive as MR in depiction of muscle atrophy. Atrophy is seen as a reduction in muscle bulk with increase in muscle echogenicity. As a result, the central tendon is less readily seen.* **(Middle)** *Longitudinal grayscale US shows the shoulder on posterior view. Infraspinatus muscle atrophy usually accompanies supraspinatus muscle atrophy to a lesser or greater degree. Similar to the supraspinatus muscle, muscle bulk is reduced and becomes more echogenic with the central tendon becoming less easy to see.* **(Bottom)** *Longitudinal grayscale US shows the supraspinatus tendon. The supraspinatus tendon slides below the acromion during abduction. This relationship may be seen from either the anterior or posterior aspect of the shoulder. Dynamic imaging may reveal impingement of the supraspinatus against the acromion during abduction.*

LONGITUDINAL AND TRANSVERSE US, POSTERIOR GLENOID LABRUM AND GLENOHUMERAL JOINT

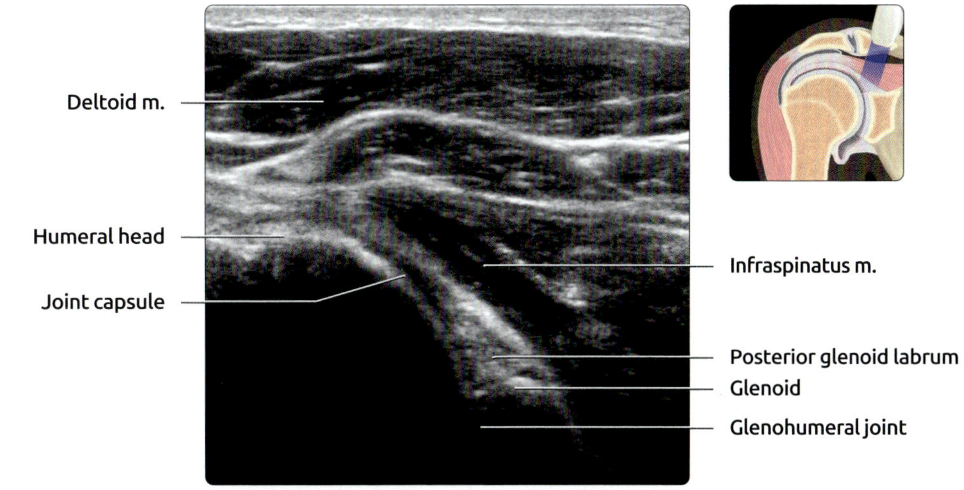

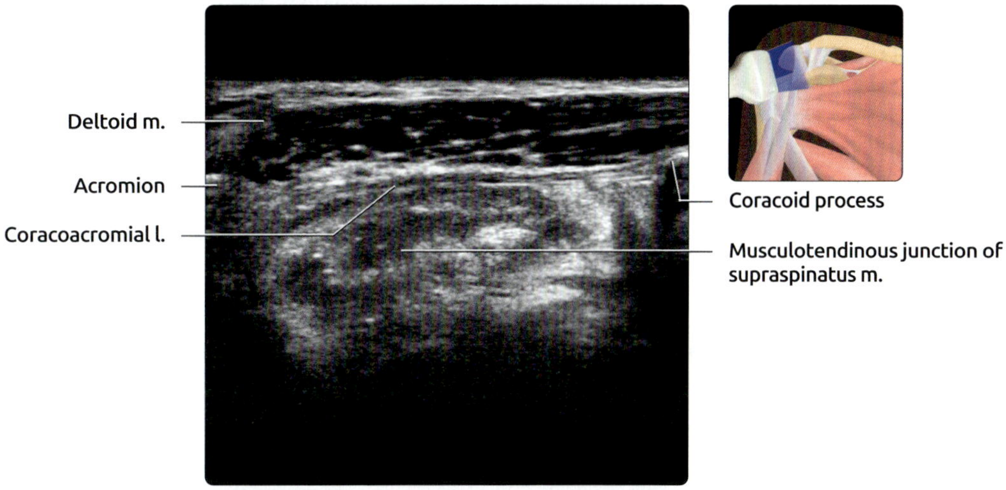

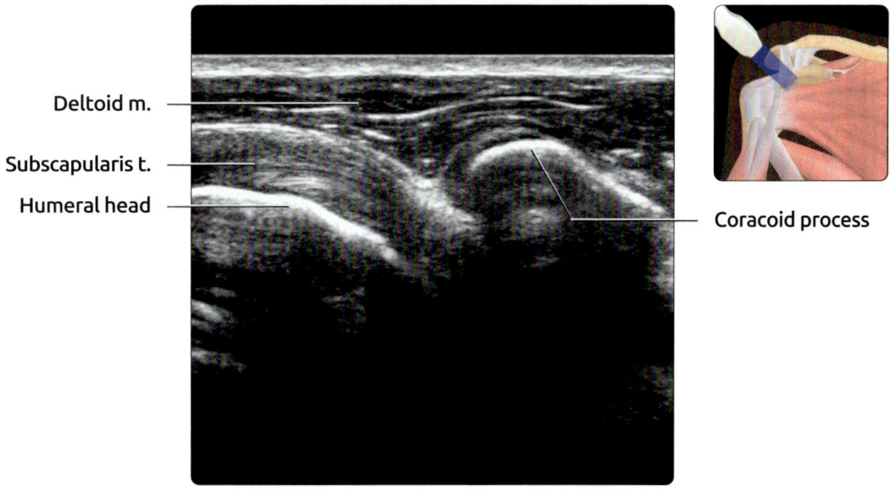

(Top) *Longitudinal grayscale US shows the posterior shoulder. The glenohumeral joint is most easily seen from the posterior aspect. This is a good site for US-guided joint injection. Joint effusions are also best seen in this area.* **(Middle)** *Longitudinal grayscale US shows the coracoacromial ligament. The coracoacromial ligament extends from the coracoid process to the anteroinferior margin of the acromion. Impingement of the supraspinatus tendon and the overlying subacromial-subdeltoid bursa may occur during shoulder abduction.* **(Bottom)** *Transverse grayscale US shows the coracoid process. The coracoid process is close to the humeral head and intervening subscapularis. Occasionally, the coracohumeral distance is reduced and subcoracoid impingement may potentially occur.*

TERMINOLOGY

Definitions

- Fat-filled space between upper limb and thoracic wall

IMAGING ANATOMY

Extent

- Axilla is shaped like pyramid with top layers shaved off
- Consists of apex, base, and 4 walls
  - Apex
    - Bounded by scapula, 1st rib, and mid 1/3 of clavicle
    - Arm communicates with posterior triangle of neck via apex of axilla
  - Anterior wall
    - Composed of pectoralis major and minor muscles
  - Posterior wall
    - Composed of teres major, latissimus dorsi, and subscapularis muscles
  - Medial wall
    - Composed of serratus anterior, upper ribs, and intercostal spaces
  - Lateral wall
    - Medial and lateral lips of bicipital groove into which anterior and posterior walls are inserted
  - Base
    - Composed of axillary fascia, subcutaneous fat, and skin
- Contents of axilla
  - Axillary artery and vein
  - Cords and branches of brachial plexus
  - Coracobrachialis and biceps muscles
  - Lymph nodes and vessels
  - Fat
- Axillary artery
  - Continuation of subclavian artery
  - Lies on posterior wall of axilla
  - Surrounded by cords and branches of brachial plexus
  - Vein lies medial to artery
  - Arterial branches
    - Superior thoracic artery
    - Acromiothoracic artery
    - Lateral thoracic artery
    - Subscapular artery
    - Anterior and posterior circumflex humeral arteries
- Axillary vein
  - Continuation of brachial vein
    - Tributaries correspond to branches of axillary artery and also cephalic vein
- Brachial plexus
  - Cord and terminal branches of brachial plexus pass through axilla
  - 3 cords: Lateral, medial, and posterior
  - Lateral cord gives rise to
    - Lateral pectoral nerve
    - Musculocutaneous nerve: Pierces coracobrachialis to descend down arm between biceps and brachialis muscles
    - Contributes to median nerve
  - Medial cord gives rise to
    - Medial pectoral nerve
    - Ulnar nerve
    - Contributes to median nerve
    - Medial cutaneous nerves of arm and forearm
  - Posterior cord gives rise to
    - Subscapular nerve
    - Thoracodorsal nerve
    - Axillary nerve, which passes through quadrilateral space posteriorly around surgical neck of humerus, accompanied by posterior circumflex humeral artery
    - Radial nerve

Anatomy Relationships

- **Quadrilateral space**
  - Superior border: Teres minor muscle
  - Inferior border: Teres major muscle
  - Lateral border: Surgical neck of humerus
  - Medial border: Long head of triceps muscle
  - Contents: Axillary nerve and posterior circumflex humeral artery
    - Axillary nerve supplies teres minor muscle, deltoid muscle, posterolateral cutaneous region of shoulder and upper arm
- **Triangular space**
  - Located medial to quadrilateral space
  - Superior border: Teres minor muscle
  - Inferior border: Teres major muscle
  - Lateral border: Long head of triceps muscle
  - Contents: Circumflex scapular artery
    - Branch of subscapular artery supplying infraspinatus fossa

ANATOMY IMAGING ISSUES

Key Concepts

- Quadrilateral space syndrome = neurovascular compression syndrome due to compression of axillary nerve or posterior humeral circumflex artery in quadrilateral space
  - Can have complications that are purely neurologic, purely vascular, or both
  - Effects
    - Point tenderness of quadrilateral space
    - Poorly localized shoulder pain ± paresthesia radiating top lateral arm
    - Symptoms aggravated by abduction and external rotation of arm
    - Teres minor ± deltoid atrophy
    - Intermittent, ischemic-type pain
  - Due to
    - Humeral fracture
    - Fibrous bands secondary to trauma
    - Muscle hypertrophy ± fibrotic bands seen in throwing athletes, tennis players, or volleyball players
    - Paralabral cysts: Most common cause of mass in this region, high associations with labral tears
    - Glenohumeral joint dislocation
    - Extreme or prolonged abduction of arm during sleep

MUSCLES

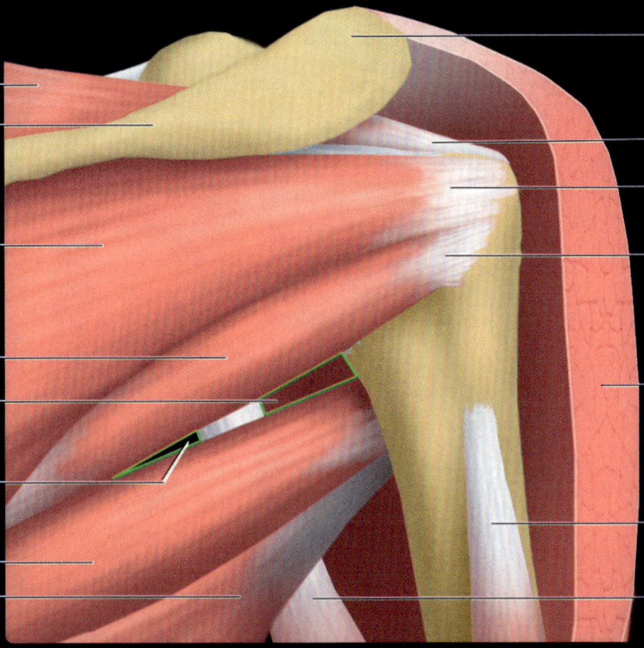

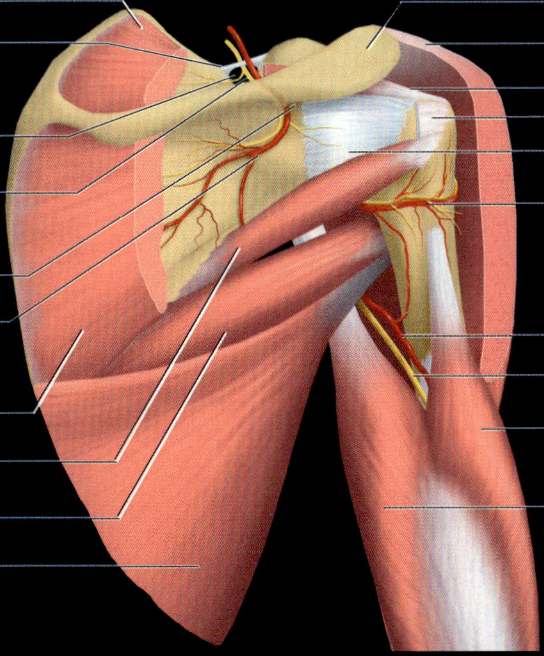

(Top) *Posterior graphic shows the shoulder. Superficial scapulohumeral dissection shows the location of the quadrilateral space and triangular space (each outlined in green). The quadrilateral space transmits the axillary nerve and the posterior circumflex humeral artery, while the much less important triangular space transmits the scapular circumflex vessels.* **(Bottom)** *Graphic of the deep scapulohumeral dissection shows the major neurovascular structures, including those in the quadrilateral space.*

DEEPER STRUCTURES

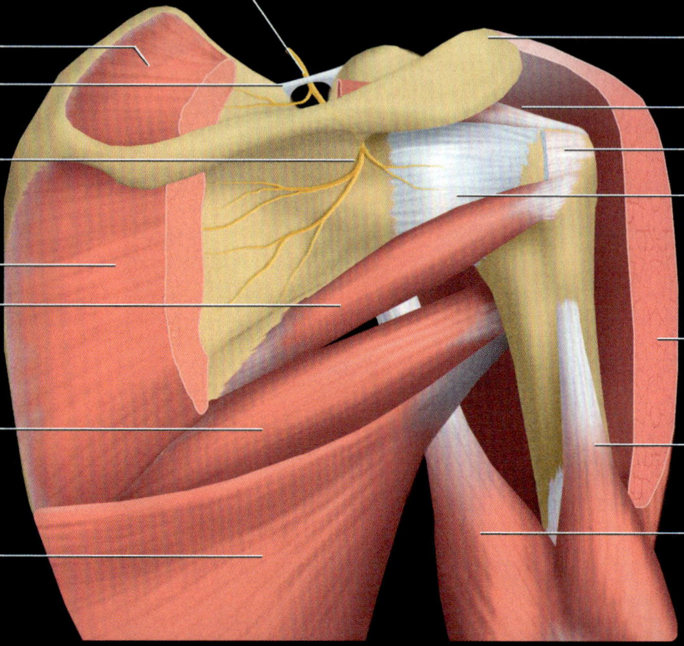

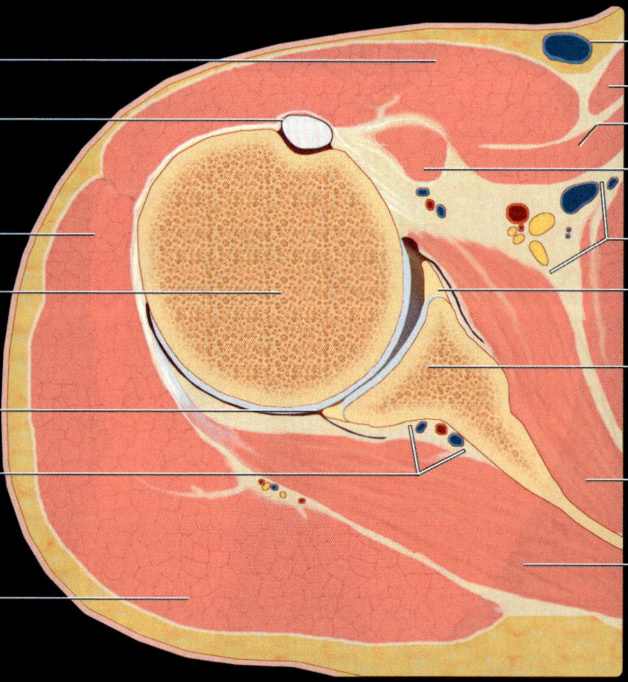

(Top) *Graphic of deep scapulohumeral dissection shows the course of the suprascapular nerve.* **(Bottom)** *Axial graphic shows the location of the suprascapular artery, nerve, and vein branches just below the level of the spinoglenoid notch. The suprascapular nerve is a mixed motor and sensory nerve. It arises from the anterior rami of the C5 and C6 roots. It passes deep to the superior transverse scapular ligament of the suprascapular notch. It then passes deep to the inferior transverse scapular ligament of the spinoglenoid notch. It supplies the supraspinatus and the infraspinatus muscles with sensory fibers to the acromioclavicular joint and the glenohumeral joint capsules.*

RELATIONSHIP OF NEUROVASCULAR STRUCTURES

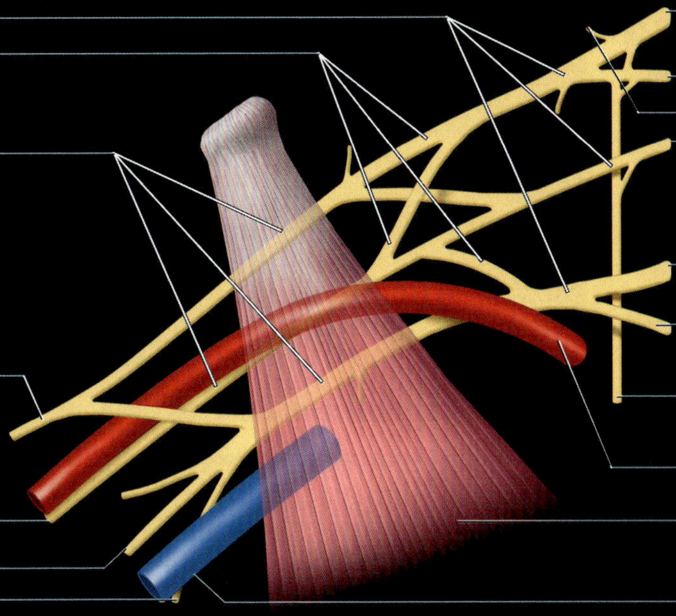

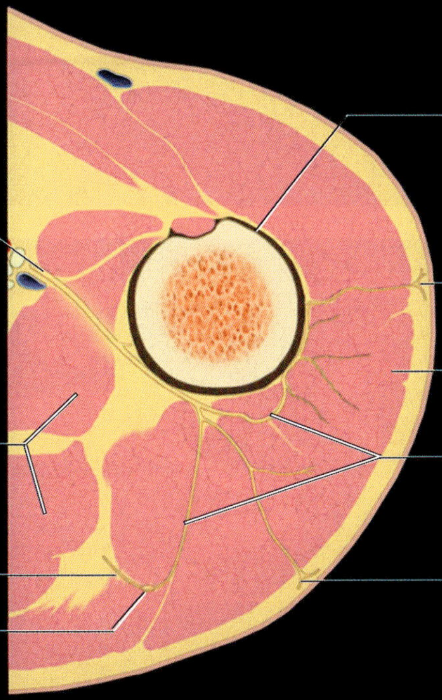

(Top) *Graphic shows relations of the brachial plexus in the axilla. The brachial plexus arises from the anterior rami of the C5, C6, C7, C8, and T1 nerves. They 1st unite to form the superior, middle, and inferior trunks. The trunks then divide and reunite to form the lateral, posterior, and middle cords. Beyond the lateral margin of the pectoralis minor muscle, they continue as the terminal branches of the plexus (axillary, musculocutaneous, radial, medial, and ulnar nerves).* **(Bottom)** *Graphic shows a section through the upper arm with the axillary nerve and its branches.*

TRANSVERSE US, AXILLA

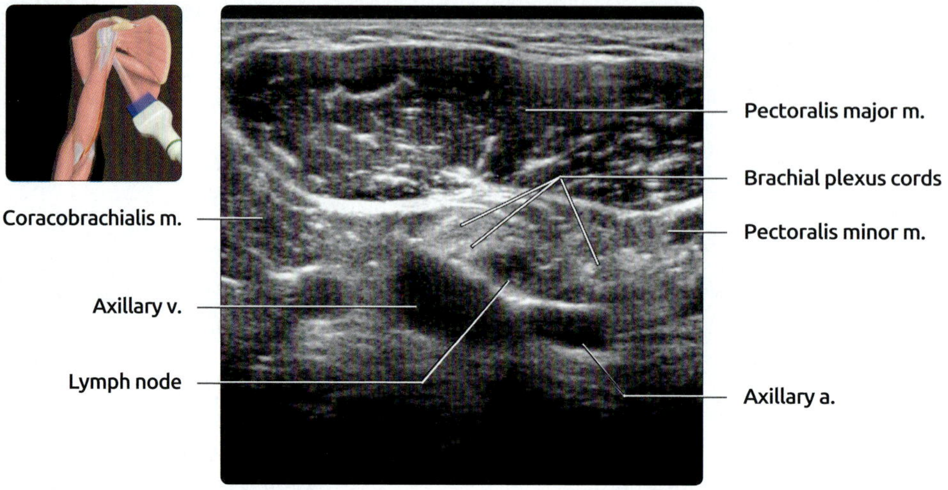

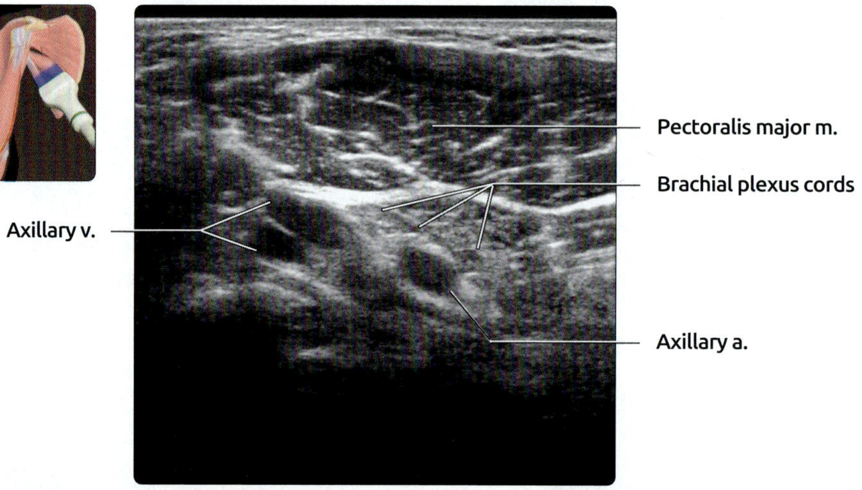

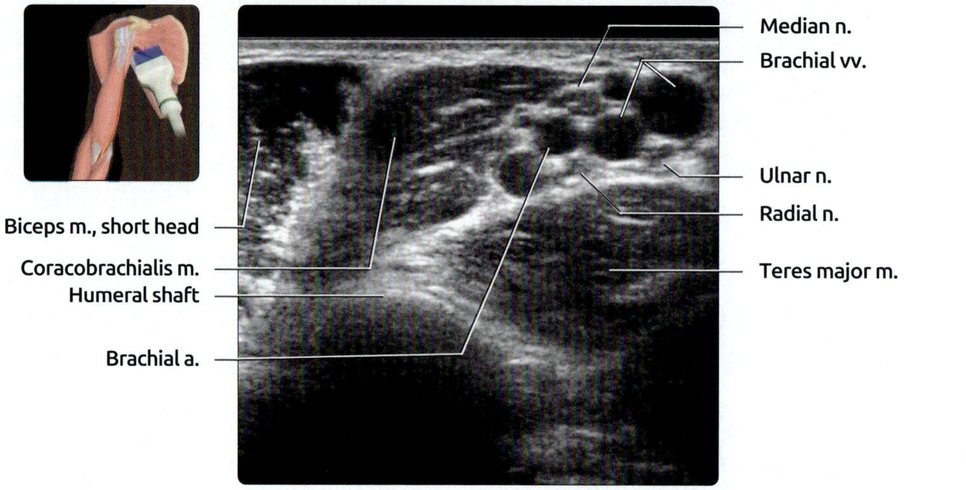

(Top) *Transverse grayscale ultrasound shows the proximal axilla with the arm abducted. The contents of the axilla are inspected with the arm fully abducted.* **(Middle)** *Transverse grayscale ultrasound shows the midportion of the axilla. The axillary artery and vein pass through the axilla surrounded by the cords of the brachial plexus. The cords (medial, lateral, and posterior) are named relative to their position alongside the axillary artery. Lymph nodes with variably fatty hila are commonly seen in the axilla around the neurovascular bundle.* **(Bottom)** *Transverse grayscale ultrasound shows the distal aspect of the axilla.*

TRANSVERSE US, ANTERIOR AXILLARY WALL

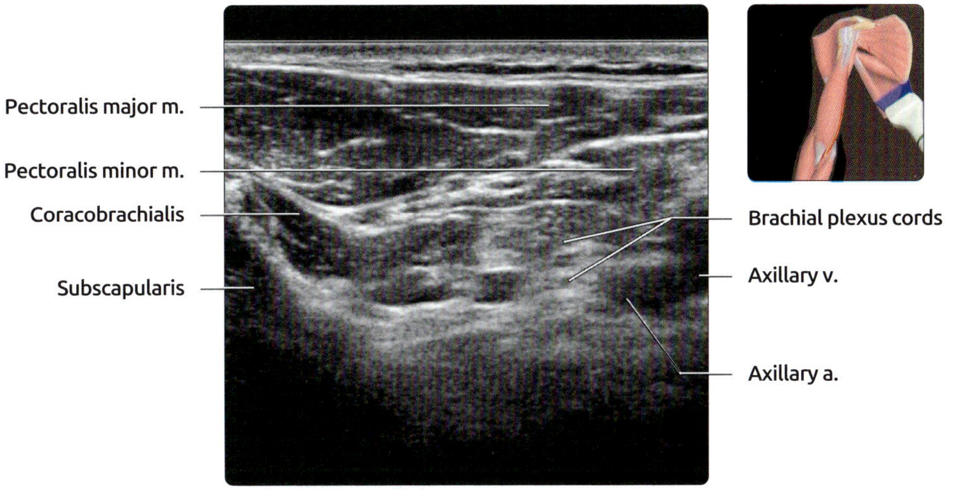

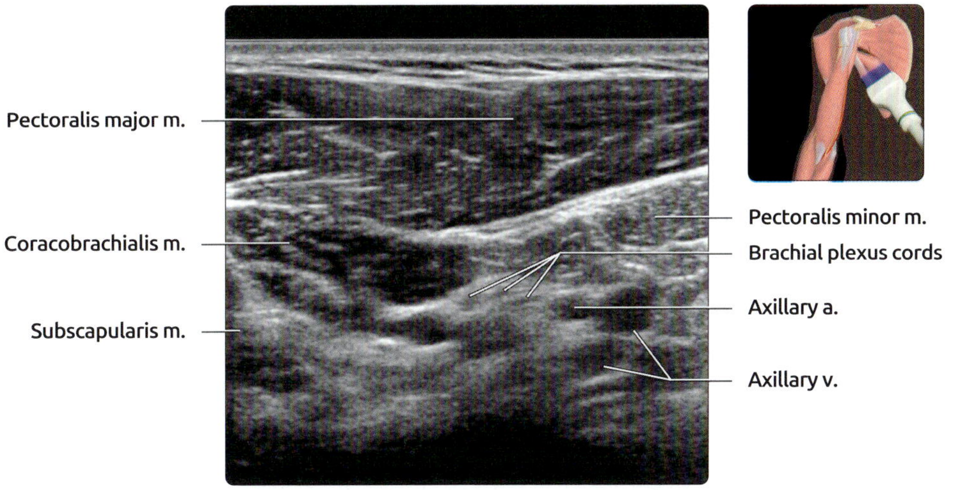

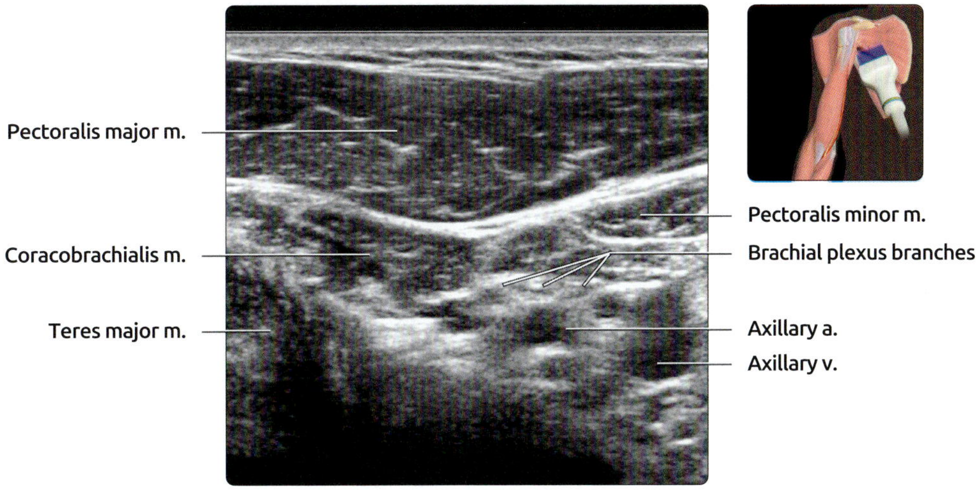

(Top) *Transverse grayscale ultrasound shows the proximal anterior wall of the axilla. The anterior wall of the axilla can be examined with the arm by the side.* **(Middle)** *Transverse grayscale ultrasound shows the midsection of the anterior wall of the axilla. The anterior wall of the axilla is formed by the pectoralis major and pectoralis minor muscles.* **(Bottom)** *Transverse grayscale ultrasound shows the distal anterior wall of axilla.*

TRANSVERSE US, QUADRILATERAL SPACE

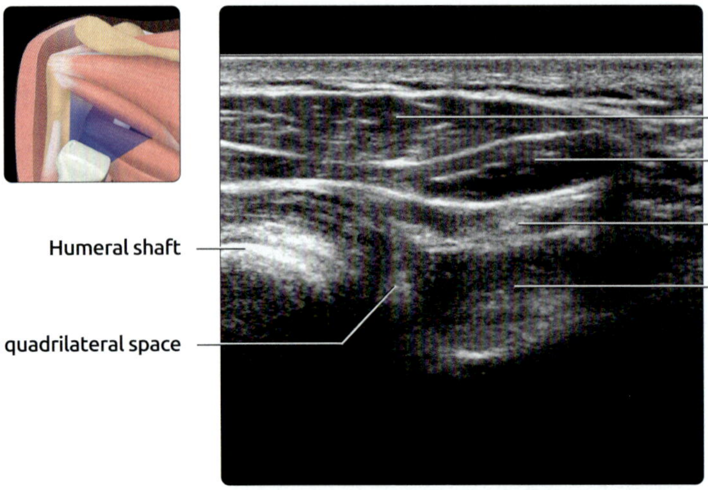

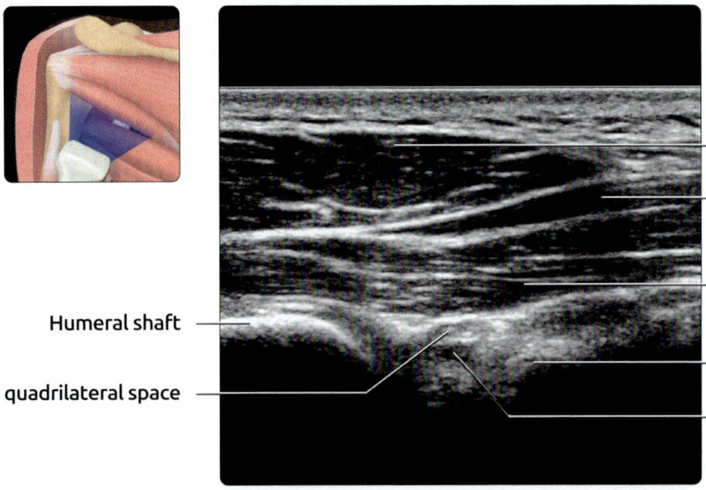

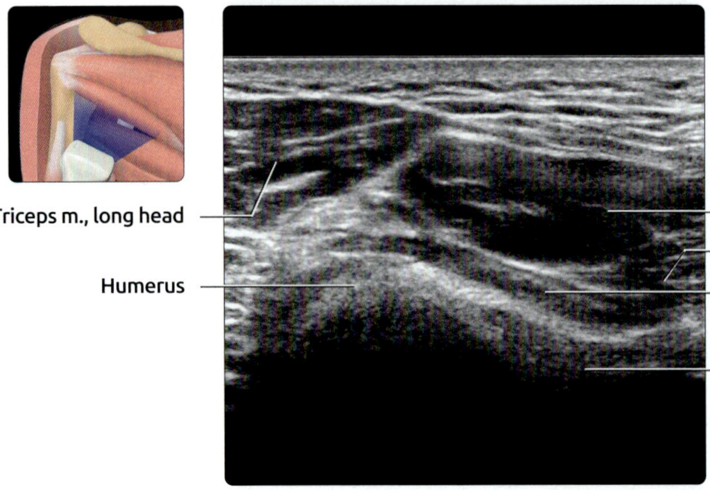

(Top) *Transverse grayscale ultrasound of the arm shows the anterior aspect of the quadrilateral space (quadrangular space) with the arm abducted. This space contains the axillary nerve and posterior circumflex humeral vessels.* **(Middle)** *Transverse grayscale ultrasound shows the quadrilateral space with the arm adducted. The contents of the space are more easily seen with the arm adducted.* **(Bottom)** *Transverse grayscale ultrasound in the more posterior position of the arm shows the posterior circumflex humeral artery posterior to the quadrilateral space.*

TRANSVERSE US, POSTERIOR AXILLARY WALL

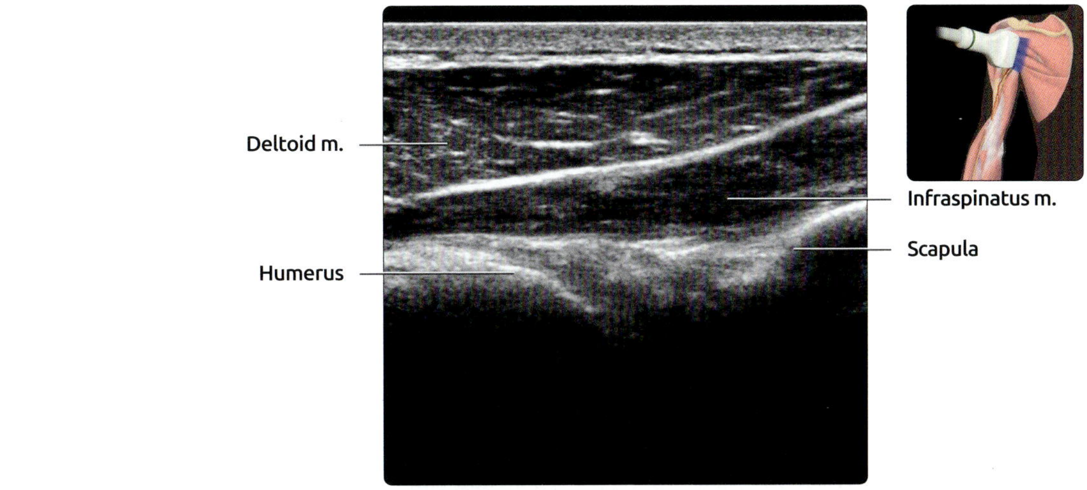

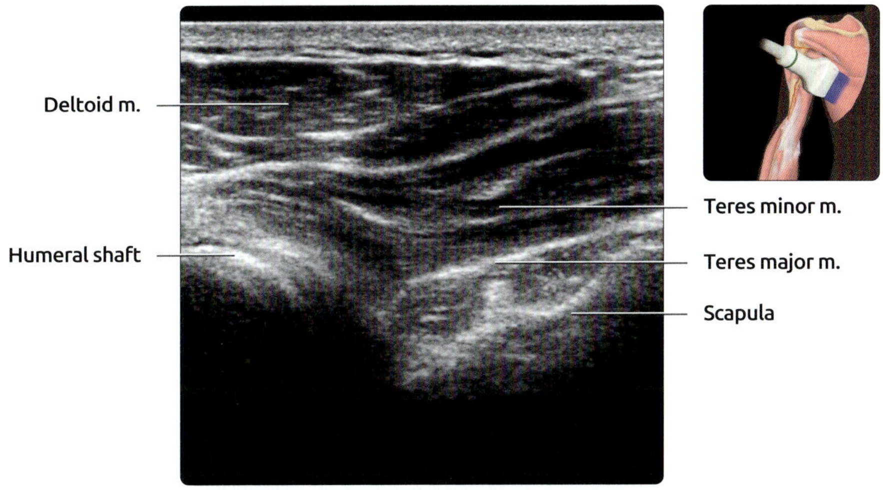

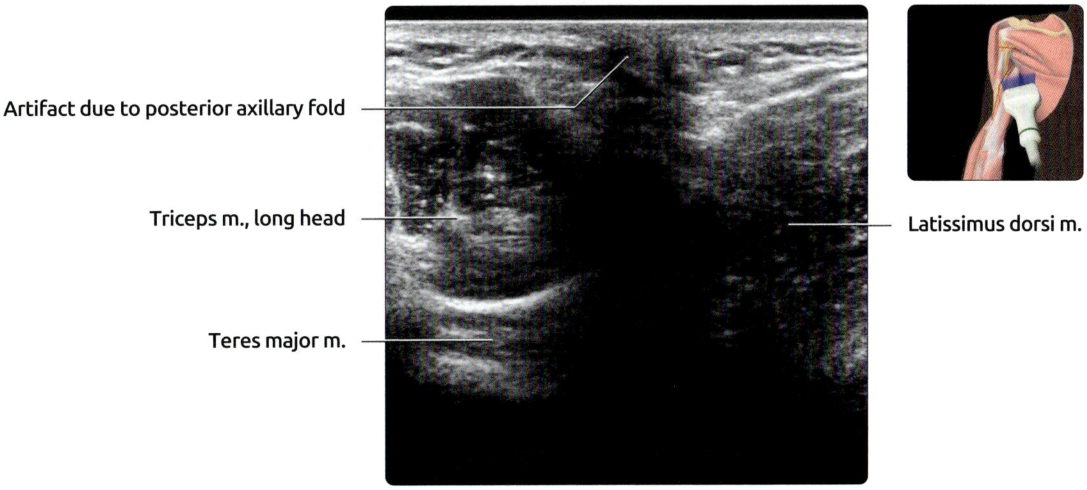

(Top) *Transverse grayscale ultrasound shows the upper aspect of the posterior axillary wall.* **(Middle)** *Transverse grayscale ultrasound shows the middle aspect of the posterior wall of the axilla. The posterior wall of the axilla is composed of the teres major, latissimus dorsi, and subscapularis muscles.* **(Bottom)** *Transverse grayscale ultrasound shows the lowermost aspect of the axilla.*

IMAGING ANATOMY

Overview

- Muscles of upper arm are divided into anterior and posterior compartments

Anatomy Relationships

- **Anterior compartment of arm**
  - Coracobrachialis muscle
    - Origin: Coracoid process tip, in common with and medial to short head biceps tendon
    - Insertion: Medial surface of humeral midshaft, between brachialis and triceps muscle origins
    - Nerve supply: Musculocutaneous nerve, perforates muscle
    - Blood supply: Brachial artery, muscular branches
    - Action: Flexes and adducts shoulder, supports humeral head in glenoid
  - Biceps muscle, short head
    - Origin: Coracoid process tip, in common with and lateral to coracobrachialis tendon
    - Insertion: Radial tuberosity after joining long head
    - Nerve supply: Musculocutaneous nerve
    - Blood supply: Brachial artery, muscular branches
    - Action: Flexes elbow and shoulder, supinates forearm
  - Biceps muscle, long head
    - Origin: Predominantly supraglenoid tubercle; also superior glenoid labrum and coracoid base
    - Insertion: Radial tuberosity after joining with short head
    - Tendon runs along bicipital groove, which is covered by transverse ligament
    - Transverse ligament spans proximal end of bicipital groove and comprises deeper layer formed from subscapularis tendon fibers and superficial fibrous layer in continuity with supraspinatus tendon and coracohumeral ligament
    - Nerve supply: Musculocutaneous nerve
    - Blood supply: Brachial artery, muscular branches
    - Action: Flexes elbow, supinates forearm
    - Variants, biceps muscle: 3rd head in 10% arising at upper medial aspect of brachialis muscle; 4th head can arise from lateral humerus, bicipital groove, or greater tuberosity
  - Brachialis muscle
    - Origin: Distal 1/2 of anterior humeral shaft and 2 intermuscular septa
    - Insertion: Tuberosity of ulna and anterior surface of coronoid process
    - Nerve supply: Musculocutaneous nerve plus branch of radial nerve
    - Blood supply: Brachial artery, muscular branches, and recurrent radial artery
    - Action: Flexes forearm
    - Covers anterior aspect of elbow joint
- **Posterior compartment of arm**
  - Triceps muscle, long head
    - Origin: Infraglenoid tubercle of scapula
    - Insertion: Proximal olecranon and deep fascia of arm after joining with lateral and medial heads
    - Nerve supply: Radial nerve
    - Blood supply: Deep brachial artery branches
    - Action: Elbow extension, adducts humerus when arm is extended
  - Triceps muscle, lateral head
    - Origin: Posterior and lateral humeral shaft, lateral intermuscular septum
    - Insertion: Proximal olecranon and deep fascia of arm after joining with long and medial heads
    - Nerve supply: Radial nerve
    - Blood supply: Deep brachial artery branches
    - Action: Elbow extension
  - Triceps muscle, medial head
    - Origin: Posterior humeral shaft from teres major insertion to near trochlea, medial intermuscular septum
    - Insertion: Proximal olecranon and deep fascia of arm after joining with lateral and long heads
    - Nerve supply: Radial and branches of ulnar nerve
    - Blood supply: Deep brachial artery branches
    - Action: Elbow extension
  - Anconeus muscle
    - Origin: Lateral epicondyle of humerus
    - Insertion: Lateral cortex of olecranon and posterior 1/4 of ulna
    - Nerve supply: Radial nerve
    - Blood supply: Deep brachial artery branch
    - Action: Assists elbow extension, abducts ulna
- **Fascia**
  - Brachial fascia
    - Continuous with fascia covering deltoid and pectoralis major
    - Varies in thickness; thin over biceps and thick over triceps muscles
    - Lateral intermuscular septum from lower aspect of greater tuberosity to lateral epicondyle
    - Medial intermuscular septum from lower aspect of lesser tuberosity to medial epicondyle
    - Perforated by ulnar nerve, superior ulnar collateral artery, and posterior branch of inferior ulnar collateral artery
    - Provides traction on deep fascia of forearm
  - Bicipital fascia
    - a.k.a. lacertus fibrosus
    - Arises from medial side of distal biceps tendon at level of elbow joint
    - Passes superficial to brachial artery
    - Continuous with deep fascia of forearm

ANATOMY IMAGING ISSUES

Imaging Recommendations

- Examine with arm extended in supine and prone positions
- Transverse plane most helpful to delineate borders of anterior and posterior compartments and relationship to neurovascular structures

Imaging Pitfalls

- Muscle may appear echogenic, simulating fatty replacement or edema when beam not aligned parallel to muscle fibers

ARM

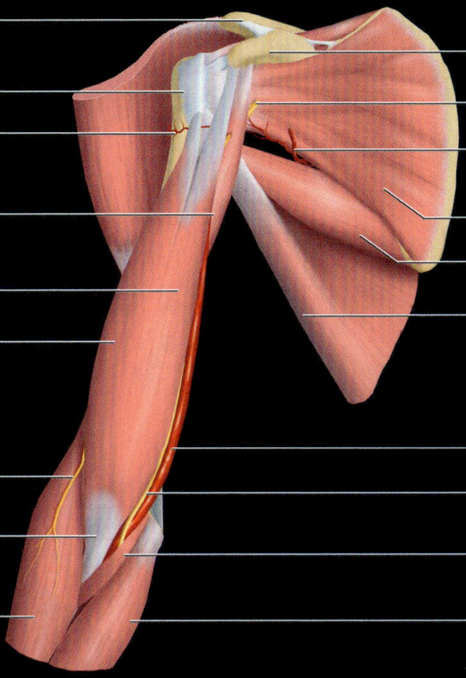

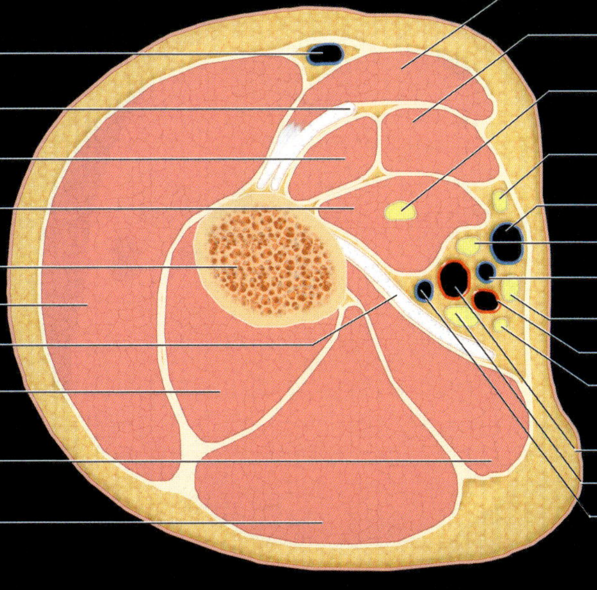

(Top) *Anterior graphic of the arm is shown. The short head of the biceps originates from the coracoid process tip, in common with the coracobrachialis tendon. The long head of the biceps tendon originates from the biceps labral complex at the supraglenoid tubercle and extends along the rotator cuff interval and bicipital groove of humerus to join with the short head and form the biceps muscle. The brachialis muscle originates from the distal 1/2 of the anterior humeral shaft. The transverse ligament spans the proximal bicipital groove from medial to lateral lip. Anatomical studies suggest that the transverse ligament is formed from fibers of the subscapularis tendon as well as a more superficial discrete fibrous band.* **(Bottom)** *Graphic shows the upper humeral level. Note that the coracobrachialis muscle is lying deep to the biceps muscle. In the posterior compartment, the lateral head and long heads of the triceps muscles can be seen at this level. All the neurovascular bundles are lying on the medial side of the upper arm.*

AXIAL, ARM

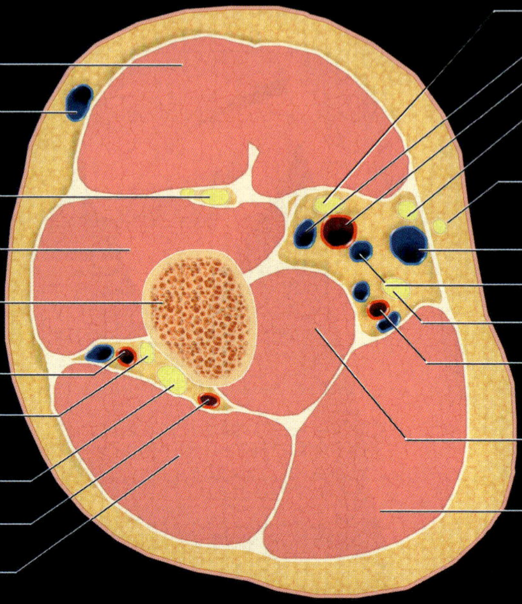

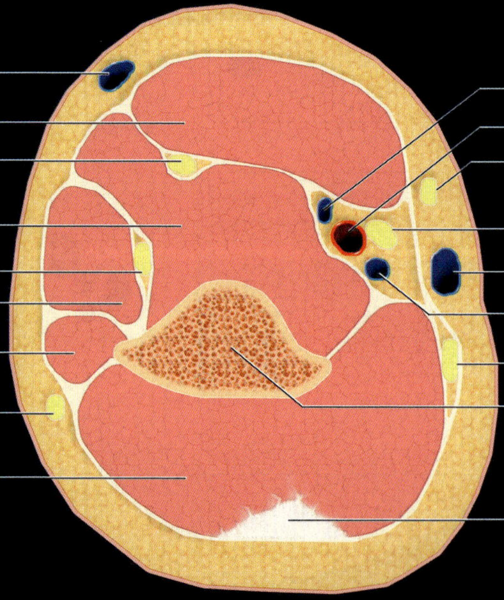

(Top) *Axial graphic shows the right arm at the midhumeral level. The posterior compartment of the arm consists of 3 heads of triceps muscle, and the medial head of the triceps can be delineated in this level. In the anterior compartment, the brachialis can be delineated at this midarm level.* **(Bottom)** *Axial section shows the distal humeral level. The deep brachial artery and radial nerve course along the posterolateral humerus reaching the anterolateral aspect of the humerus at this level. The median nerve is lying within the intermuscular septum. The biceps muscle is thinning anteriorly, and the triceps tendon appears.*

TRANSVERSE US, ANTEROMEDIAL ASPECT

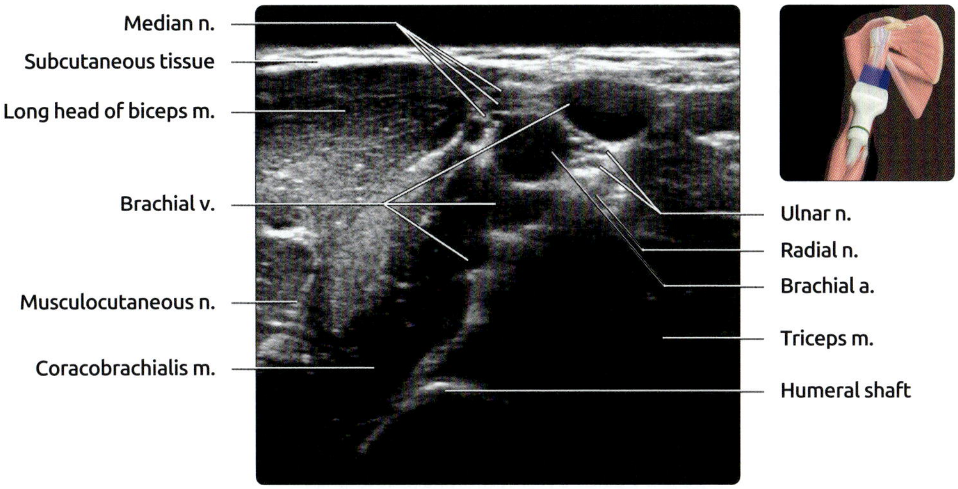

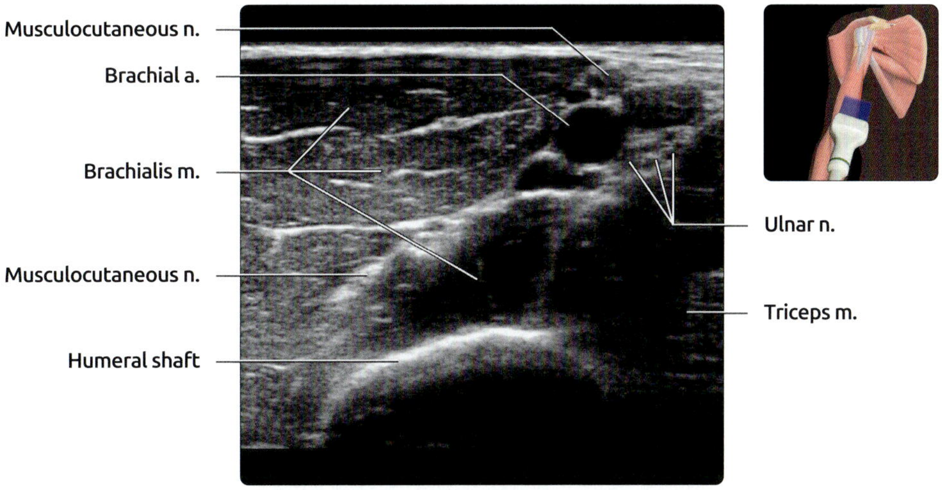

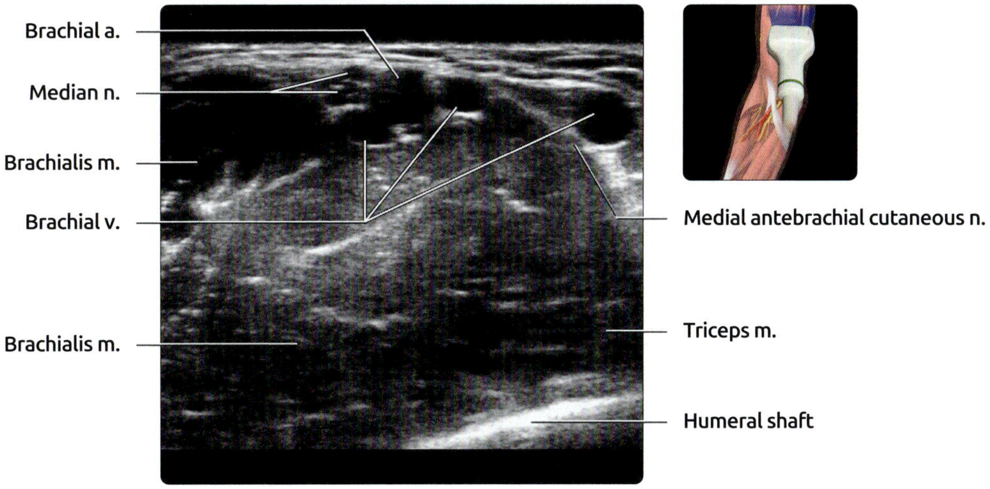

(Top) *Transverse grayscale ultrasound shows the anteromedial aspect of the upper arm. The terminal branches of the brachial plexus (namely, the median nerve, ulnar nerve, and radial nerve) all lie around the brachial vein and artery. They can be recognized by their position relative to the brachial artery, the radial nerve lying deep, median nerve laterally, and ulnar nerve medially.* **(Middle)** *Transverse grayscale ultrasound shows the anteromedial aspect of the midarm. Compression of the median or ulnar nerve is uncommon in the midarm. The neurovascular bundle helps define the separation of the anterior and posterior compartments of the arm medially.* **(Bottom)** *Transverse grayscale ultrasound shows the anteromedial aspect of the distal arm. In the distal arm, the median nerve alone is closely associated with the brachial neurovascular bundle.*

TRANSVERSE US, ANTERIOR ASPECT

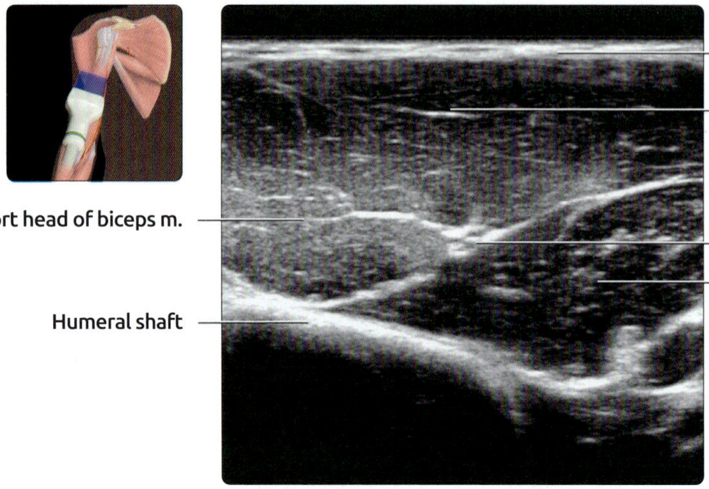

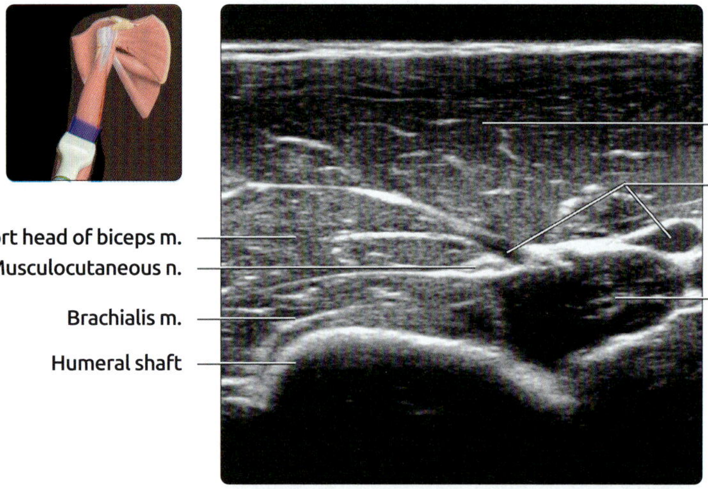

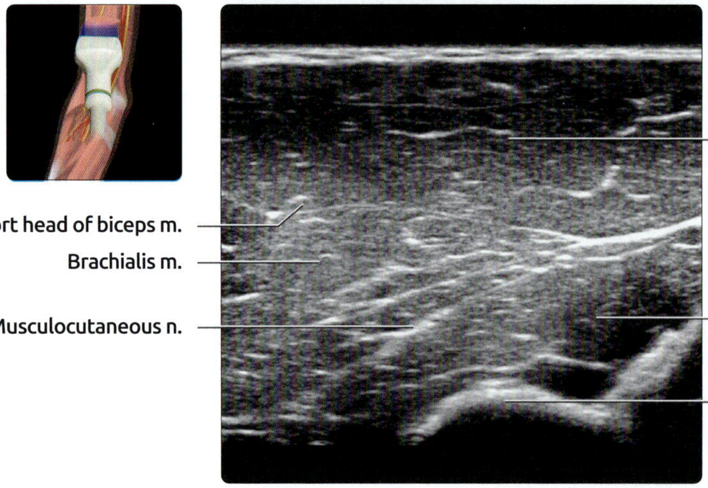

(Top) *Transverse grayscale ultrasound shows the proximal 1/3 of the arm. The anatomy of the upper arm is quite straightforward. The arm muscles are divided into 2 compartments. The anterior compartment is composed of the biceps, brachialis, and coracobrachialis muscles. The biceps muscle is midline, the brachialis muscle is midline and lateral, and the coracobrachialis muscle is midline and medial.* **(Middle)** *Transverse grayscale ultrasound shows the mid 1/3 of the anterior arm. Most of the neurovascular bundles in the arm lie anteromedially.* **(Bottom)** *Transverse grayscale ultrasound shows the distal 1/3 of the anterior arm. The musculocutaneous nerve is the main critical neurovascular structure of the anterior arm lying between the biceps, coracobrachialis, and brachialis muscles.*

TRANSVERSE US, ANTEROLATERAL ASPECT

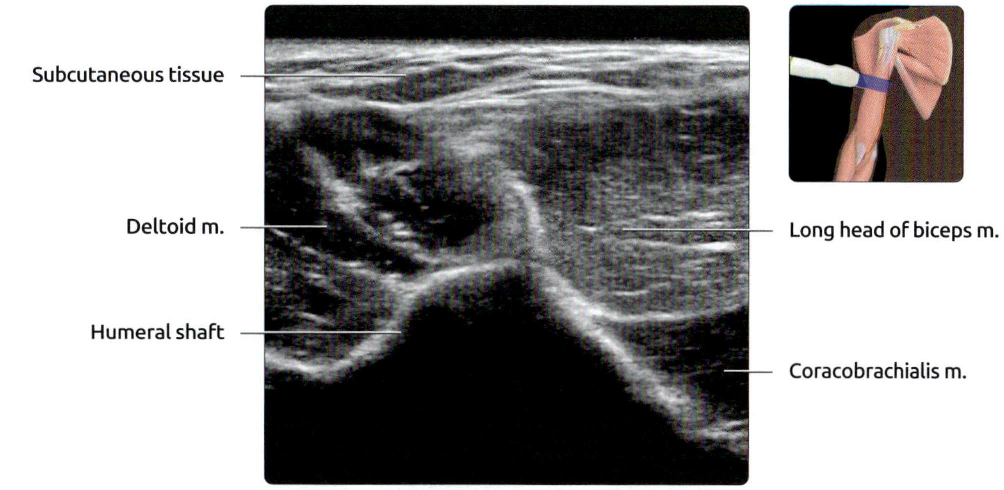

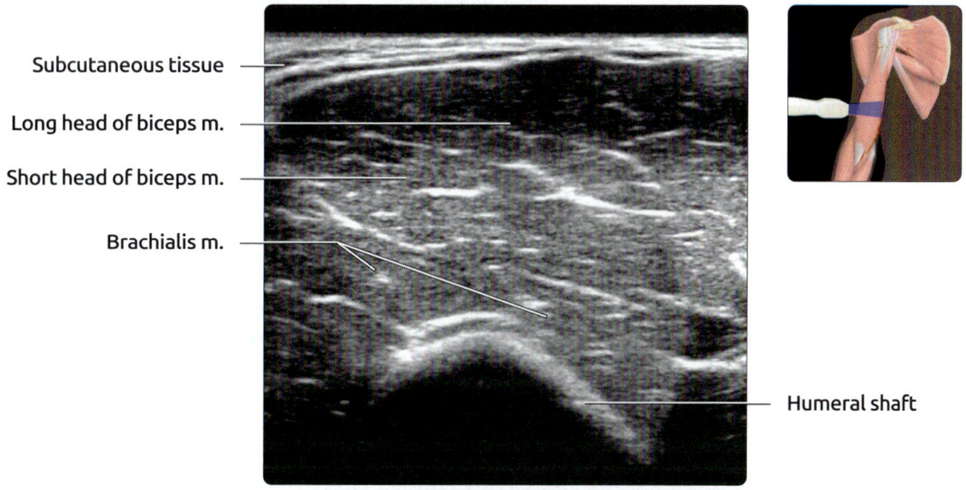

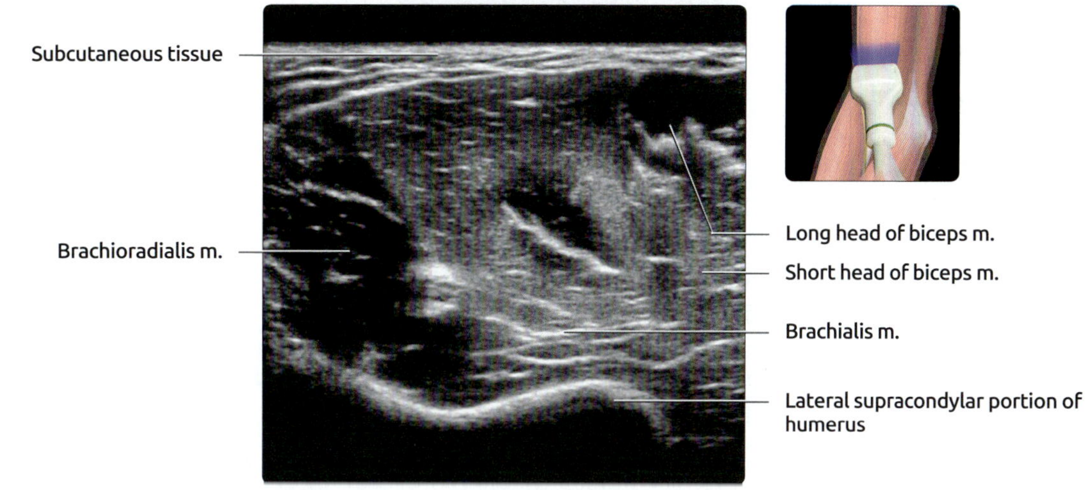

(Top) *Transverse grayscale ultrasound shows the anterolateral aspect of the proximal arm. This is an area prone to injury during contact sports. Fortunately, there is no neurovascular bundle in the area that would be exposed to injury.* **(Middle)** *Transverse grayscale ultrasound shows the anterolateral section of the midarm. Both the biceps muscle and brachialis muscle may be prone to overuse muscle injury if too vigorous exercise is undertaken. The biceps muscle may appear echogenic if the transducer is aligned at an angle to the transverse plane.* **(Bottom)** *Transverse grayscale ultrasound shows the anterolateral aspect of the distal arm. Both heads of the biceps converge to form the distal biceps tendon just proximal to the antecubital fossa.*

POSTERIOR ARM

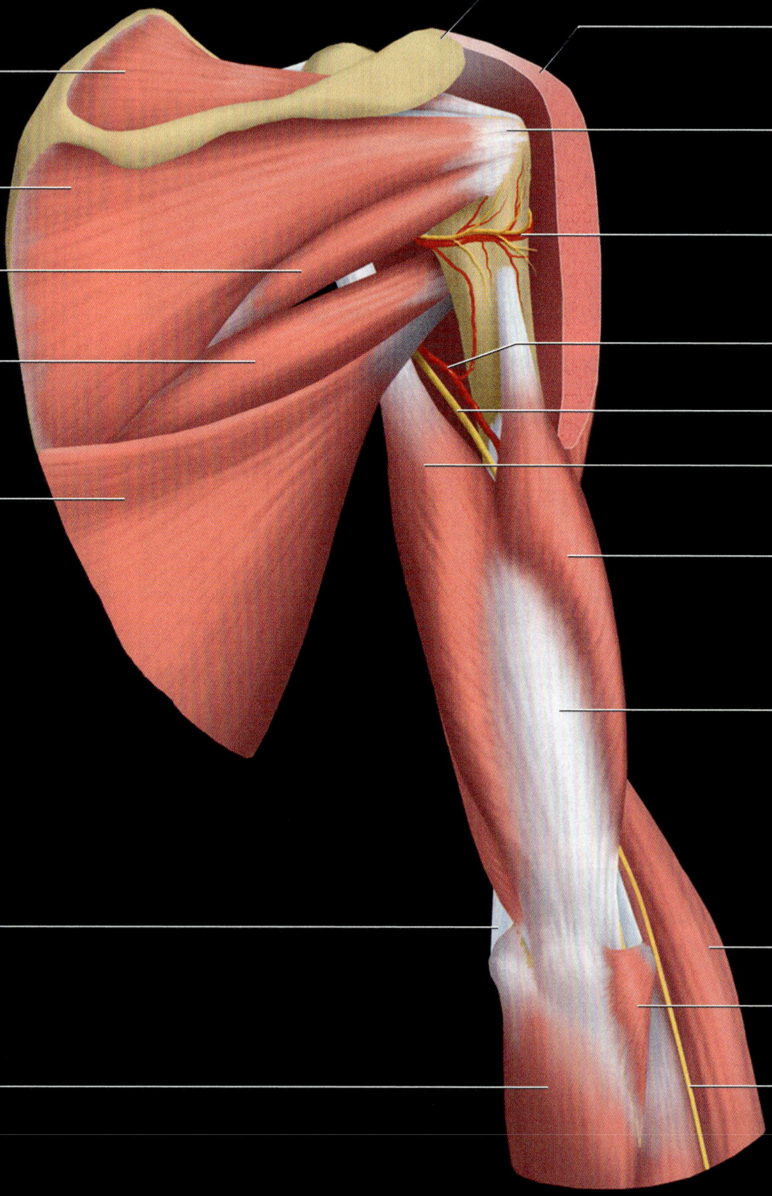

Graphic shows the superficial dissection of the posterior arm. Note that the quadrilateral space is an anatomic space bounded by the long head of the triceps, teres minor and teres major muscles, and cortex of the humerus. Axillary nerve and posterior circumflex humeral artery are within the space. Any narrowing of this space will cause quadrilateral space syndrome.

TRANSVERSE US, POSTERIOR ASPECT

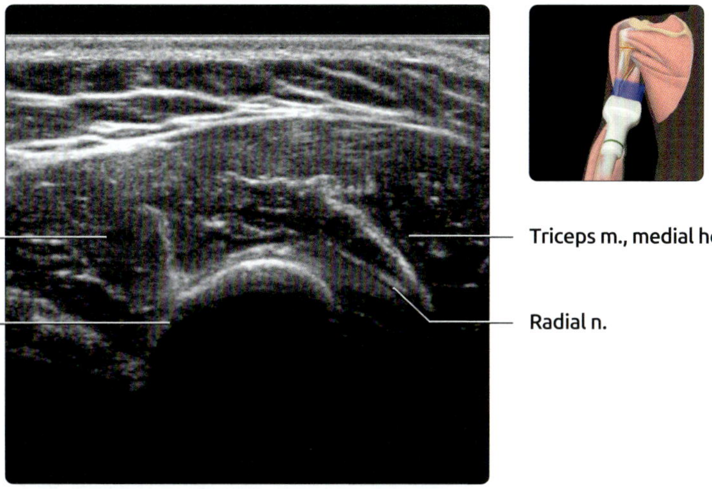

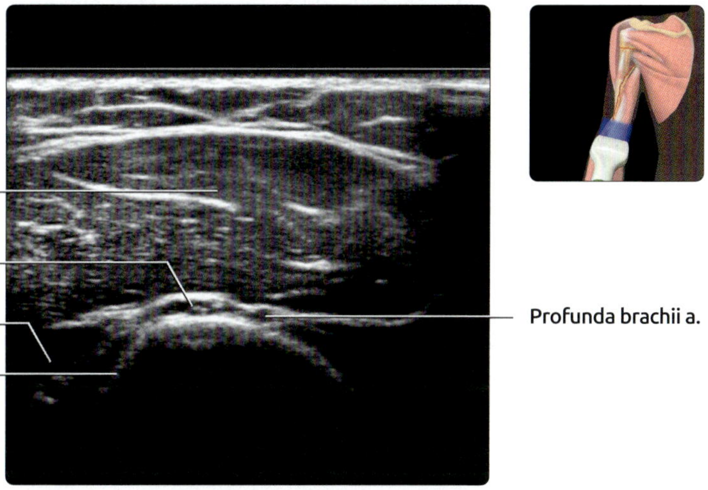

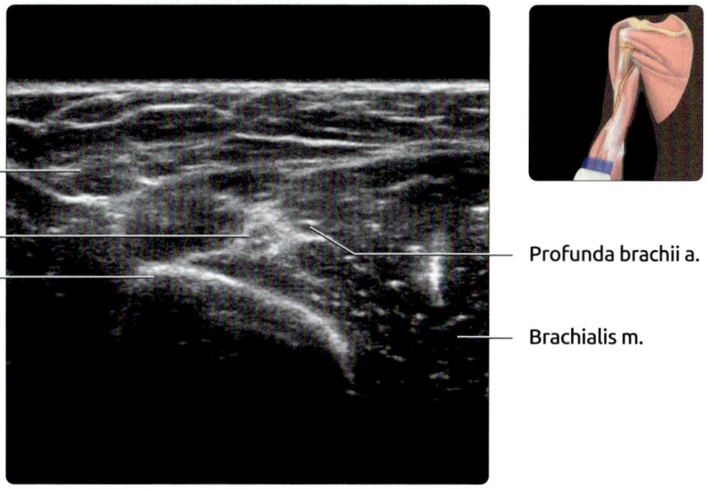

(Top) *Transverse grayscale ultrasound shows the posterior upper arm. The triceps muscle occupies nearly the whole of the posterior upper arm. The radial nerve supplies the triceps and supinator muscles.* **(Middle)** *Transverse grayscale ultrasound shows the posterior midarm. The radial nerve runs in the spiral groove on the posterior aspect of the midarm. At this level, it is prone to injury from humeral shaft fractures, fibrosis, or fibrotic bands following trauma or extrinsic compression.* **(Bottom)** *Transverse grayscale ultrasound shows the posterior aspect of the distal arm. The radial nerve pierces the lateral intermuscular septum just proximal to the lateral epicondyle. It then runs between the brachialis and brachioradialis muscles and passes anterior to the lateral epicondyle. The nerve may be compressed by the lateral intermuscular septum.*

GROSS ANATOMY

Arteries

- Brachial artery
- Continuation of axillary artery
- Extends from lower border of teres major to neck of radius
- Accompanies median nerve
- Branches of brachial artery in arm
  - Superior ulnar collateral arises medially, descends with ulnar nerve posterior to medial epicondyle, and then forms anastomosis with branches of ulnar artery
  - Inferior ulnar collateral arises distal to superior ulnar collateral artery, descends anterior to medial epicondyle, and then forms anastomosis with branches of ulnar artery
  - Profunda brachii: Main branch of brachial artery
    - Arises just distal to teres major, accompanies radial nerve, and passes backwards in spiral groove on posterior surface of humerus
    - Gives off another branch, which passes upward and anastomoses with arteries around shoulder joint
    - Gives 2 lower medial branches, which descend to elbow joint on its medial side and take part in anterior and posterior anastomosis around elbow joint
- Brachial artery lies superficial in its course in arm
- Covered by bicipital aponeurosis at elbow where it may be compressed against medial surface of humerus
- Brachial artery divides at neck of radius into radial artery and ulnar artery
  - High takeoff of radial artery from brachial artery is common anatomical variant
  - Radial artery passes inferiorly and laterally, lying on tendon of biceps brachii
  - Ulnar artery passes down and medially, deep to pronator teres
  - Within antecubital fossa, radial and ulnar arteries give off recurrent branches to elbow joint

Veins

- Categorized as superficial and deep venous system
  - Veins of upper limb are variable in number and position
- Brachial vein
  - Forms deep venous system
  - Singular or paired, continue into axillary vein
  - Accompanied by brachial artery and nerves of upper limb
- Cephalic vein
  - Lies anterolaterally in superficial fascia of antecubital fossa and gives off medial cubital vein, which joins basilic vein
  - Runs in deltopectoral groove before piercing clavipectoral fascia to drain into axillary vein
- Basilic vein
  - Passes upward on medial side of forearm, receiving anterior and posterior tributaries, and then passes to anteromedial aspect of elbow where it receives medial cubital vein
  - Continues on medial aspect of biceps brachii until it pierces deep fascia to run cranially, medial to brachial artery
  - Joins brachial vein and becomes axillary vein at lower border of teres major

ANATOMY IMAGING ISSUES

Imaging Recommendations

- Venous examination
  - Examination of upper limb veins is normally performed with patient supine and arm abducted to about 90°
  - Use of high-frequency linear transducer (5-10 MHz) is recommended
    - Determination of compressibility
    - Flow pattern: Normal venous waveform
  - Augmentation of flow is obtained by manual compression of forearm or upper arm
    - Alternatively, patient clenches fist to increase venous flow
    - Flow can also be augmented by deep inspiration
- Arterial examination
  - In patients with possible arterial compression syndromes, arteries are examined in various positions of shoulder abduction so that any compression may be accentuated

Imaging Pitfalls

- Clavicle can produce mirror artifact of subclavian vessels
- Brachial vein duplication is common
- Due to proximity, arm venous flow is affected by respiration
  - Deep inspiration will augment flow
- Angle of insonation should not be > 60° as this will reduce sensitivity of Doppler and color flow signals
- Color signal scale should match flow velocity
- Doppler gate should be located at center of artery or vein as this is site of peak flow velocity
- Arteriography: Gold standard for assessment of accuracy of arterial Doppler ultrasound with respect to stenoses or occlusion
  - Other methods of visualizing arterial tree are computed tomography angiography and magnetic resonance angiography

CLINICAL IMPLICATIONS

Clinical Importance

- Brachial artery may be used for access in angiography, either by high brachial approach or in antecubital fossa, where it lies medial to biceps tendon and lateral to median nerve
- Spontaneous thrombosis of axillary vein occasionally occurs following excessive movements of arm at shoulder joint
- Diagnosis of exclusion of deep vein thrombosis in upper or lower limb; spontaneous or related to indwelling catheters
- Vein mapping prior to bypass grafts
- Localization of veins for venous access and cannulation

ARTERIES

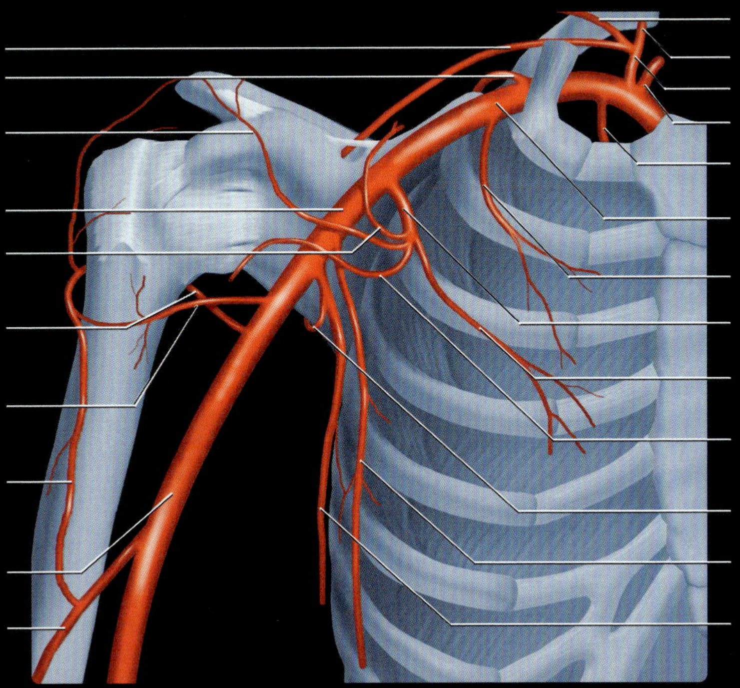

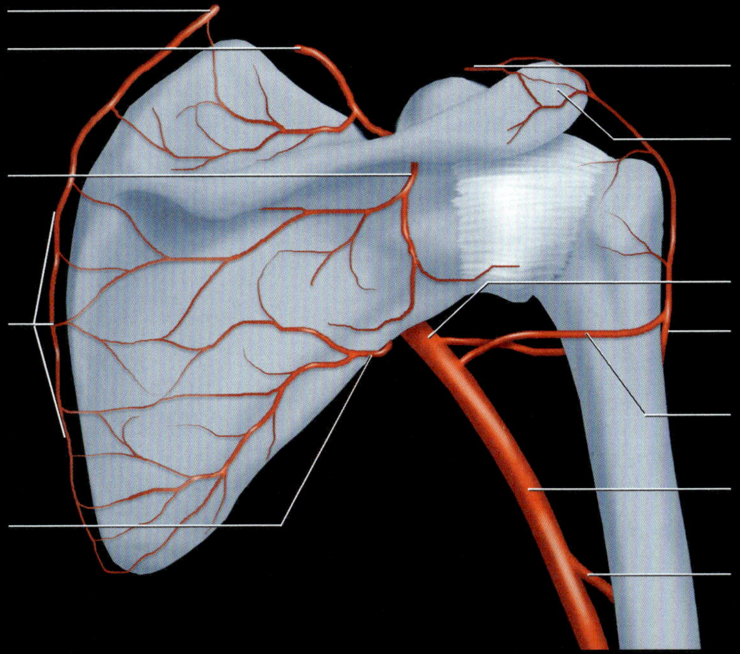

(Top) *Anterior graphic shows arterial supply to the shoulder. The shoulder is predominantly supplied by the anterior and posterior circumflex humeral, suprascapular, and circumflex scapular arteries.* **(Bottom)** *Posterior graphic shows arterial supply to the shoulder. Extensive collateral blood vessels include anastomoses with intercostal arteries.*

AXIAL, RIGHT ARM ARTERIES AND VEINS

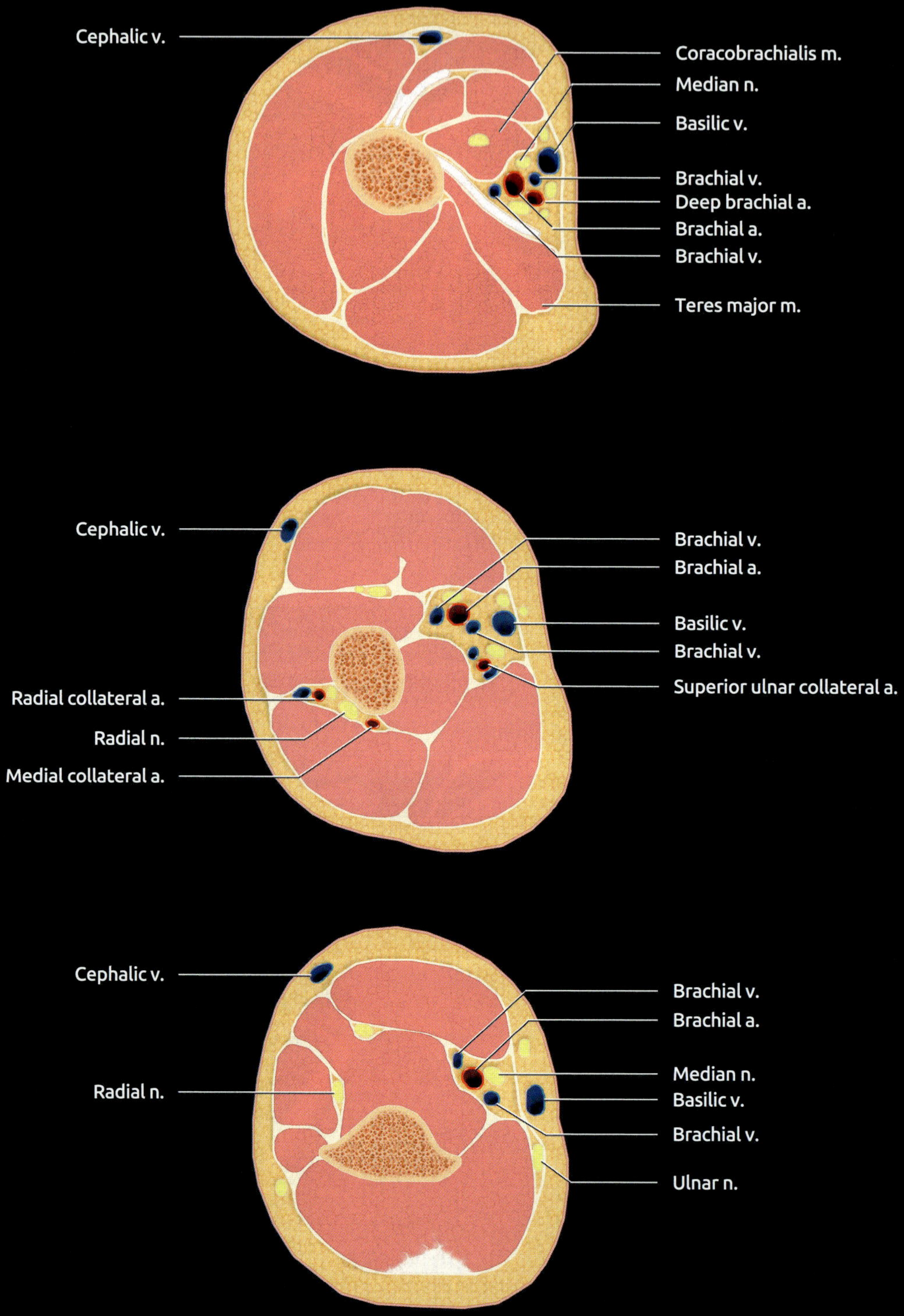

(Top) *Axial graphic shows the upper right arm. The brachial artery is a continuation of the axillary artery and accompanies the median nerve. The deep brachial artery (profunda brachii artery) is the main branch of brachial artery and accompanies the radial nerve. The basilic vein lies medially and the cephalic vein anterolaterally.* **(Middle)** *Axial graphic shows the right arm at the midhumeral level. Note that the profunda brachii artery accompanies the radial nerve and passes backward in spiral groove of humeral shaft before dividing into medial and radial collateral arteries. Superior ulnar collateral artery arises from the brachial artery and descends with the ulnar nerve.* **(Bottom)** *Axial graphic shows the right arm at the distal humeral level. Note how the median nerve is on the medial side of the brachial artery at this level.*

TRANSVERSE US, PROXIMAL ARM

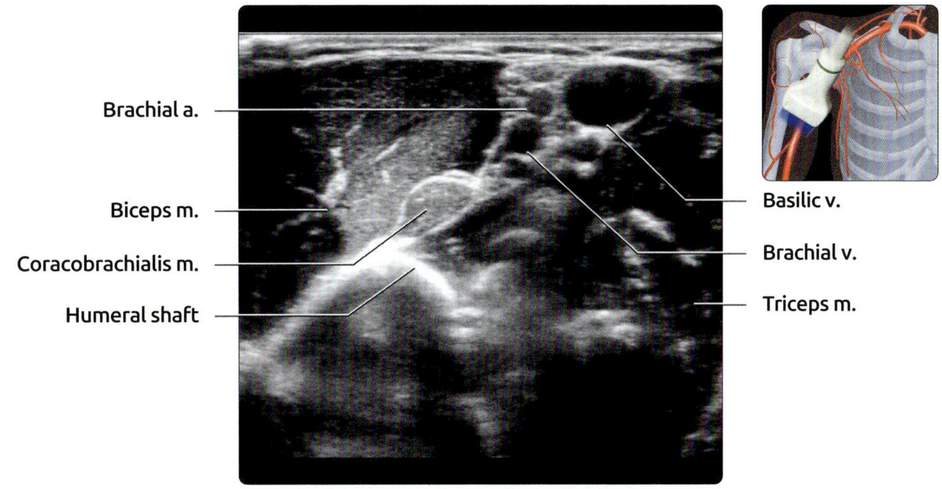

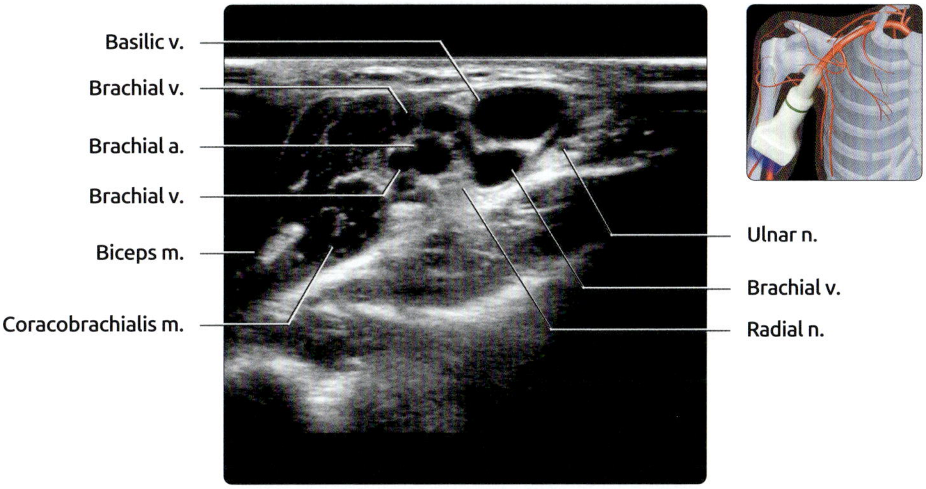

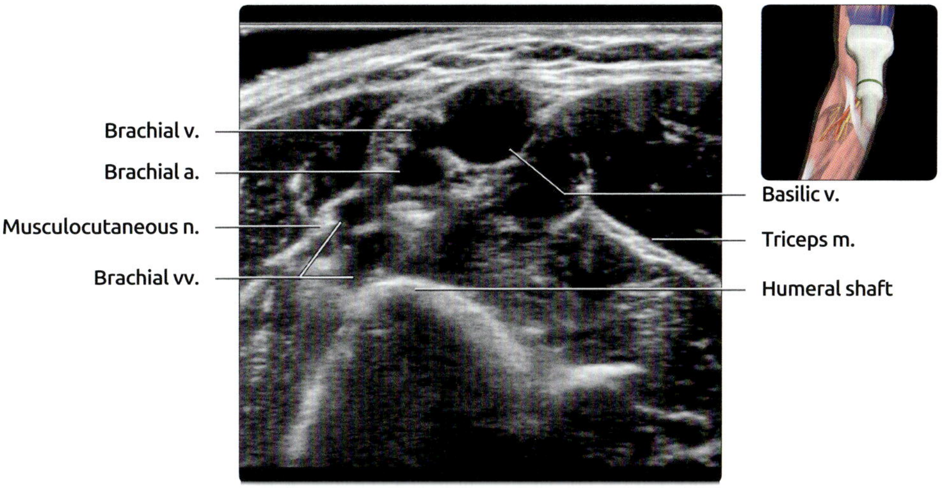

(Top) *Transverse grayscale ultrasound shows the vessels of the proximal arm. The basilic vein lies deep to the investing fascia in the arm. The veins are highly variable in location and are surrounded by nerves; hence, ultrasound guidance is helpful when cannulating the veins of the arm.* **(Middle)** *Transverse grayscale ultrasound shows the vessels of the midarm. The basilic vein is formed on the ulnar side of the wrist and is joined by the median cubital vein (from cephalic vein at the elbow). In some cases, the brachial artery may bifurcate in the midarm or higher.* **(Bottom)** *Transverse grayscale ultrasound shows the vessels of the distal arm. The brachial artery can be seen clinically as a pulsatile vessel in "locomotor brachialis," i.e., in arteriosclerosis.*

COLOR DOPPLER, BRACHIAL ARTERY

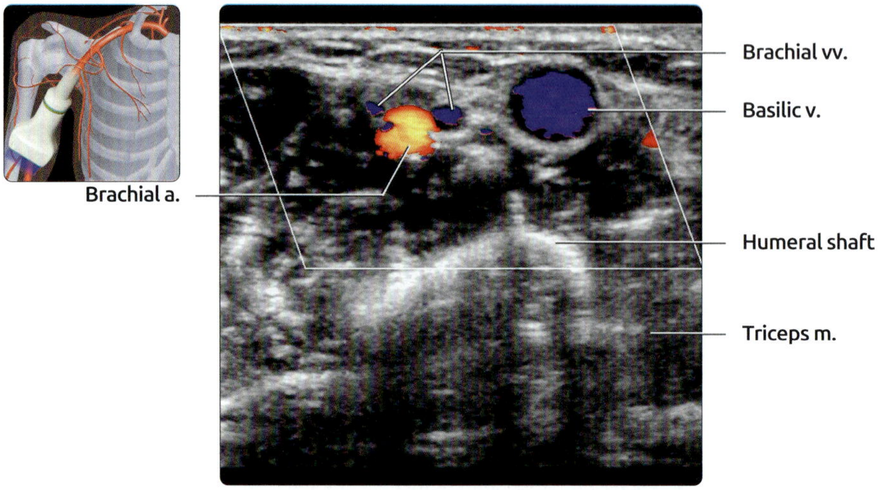

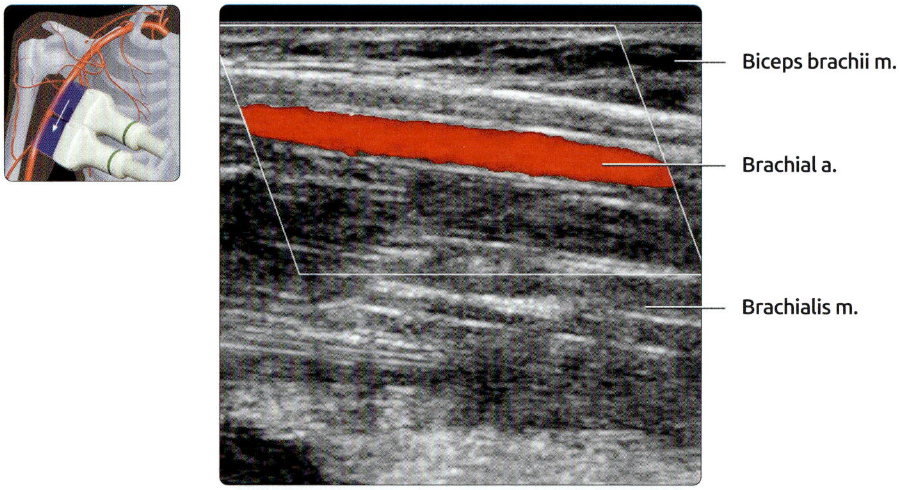

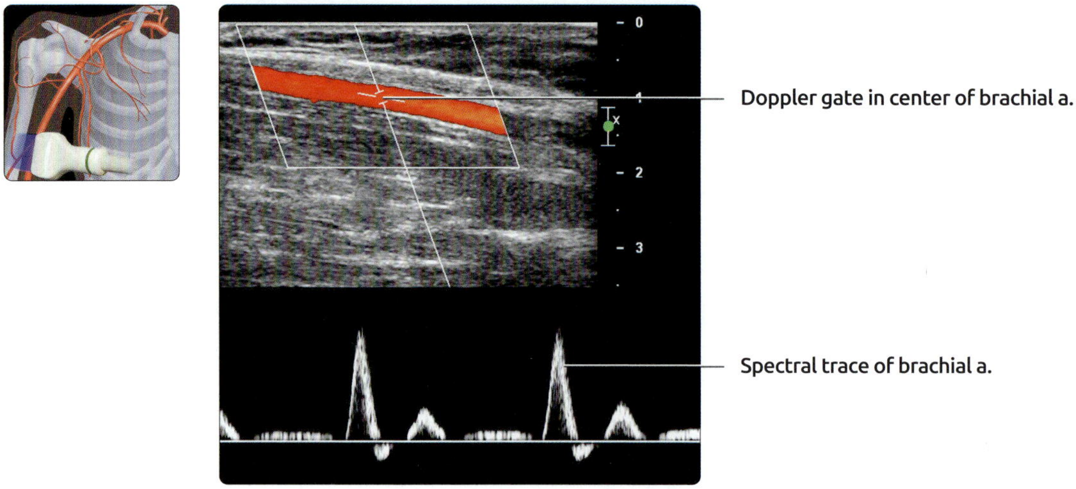

(Top) *Transverse color Doppler ultrasound shows the brachial artery. Note the accompanying paired brachial veins.* **(Middle)** *Longitudinal color Doppler ultrasound shows the brachial artery. The normal diameter of the brachial artery at the antecubital fossa ranges from 5.5-6.0 mm.* **(Bottom)** *Longitudinal color Doppler spectral analysis ultrasound shows the brachial artery. Normal triphasic flow pattern of the large arteries is seen, compared to the biphasic flow of more distal arteries, but changes in temperature or even maneuvers, such as fist clenching, may dramatically change the characteristics of upper extremity waveforms, especially in and near the hands. Peak systolic velocity of the brachial artery ranges from 50-100 cm/sec (average 57 cm/sec); end diastolic velocity 7-12 cm/sec and a pulsatility index around 5 is considered normal.*

LONGITUDINAL COLOR DOPPLER, BRACHIAL VEIN

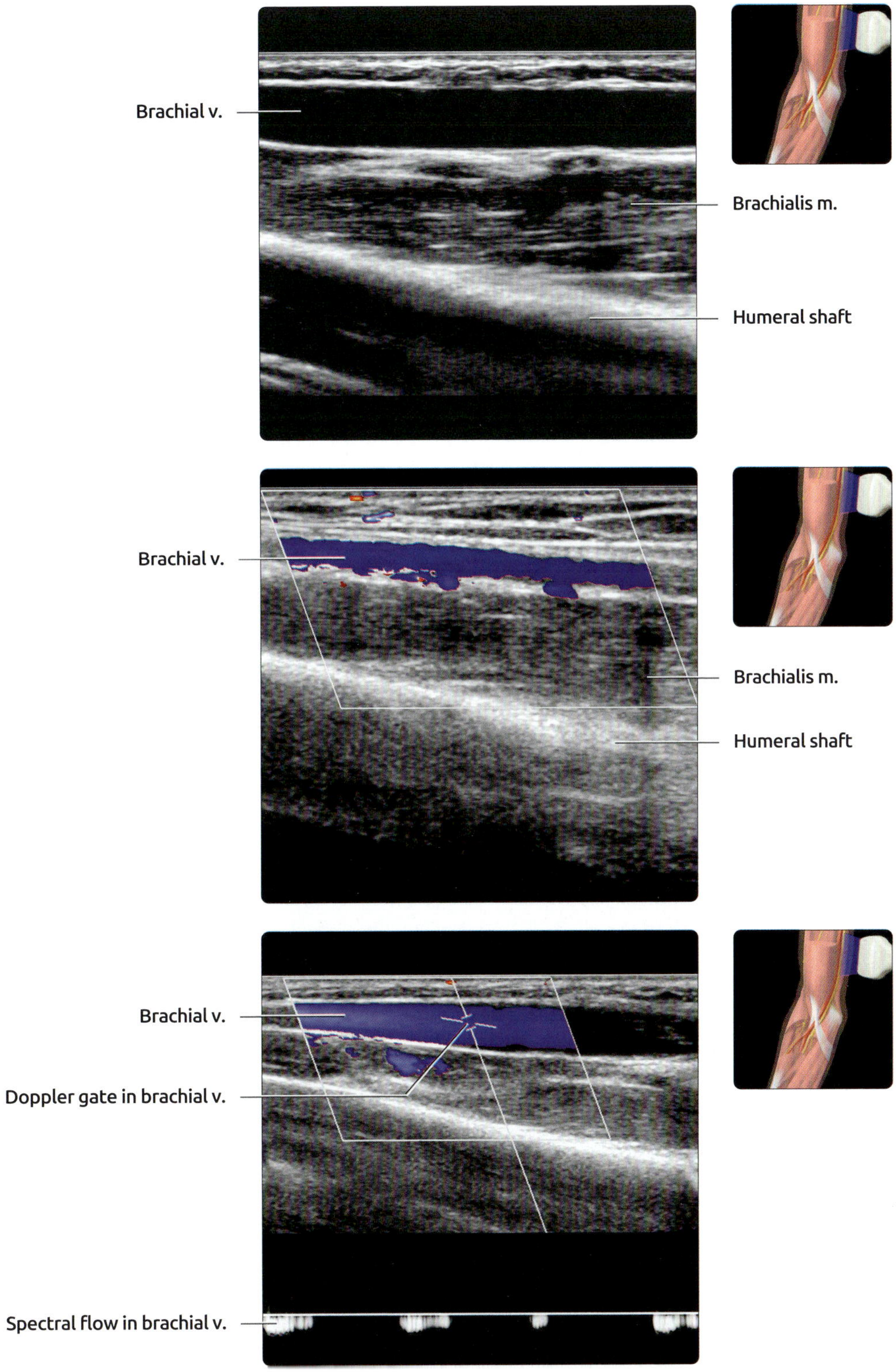

(Top) *Longitudinal grayscale ultrasound shows the brachial vein. Normal brachial vein in the upper arm measures around 2.5-2.7 mm. The venous system of the upper limb is categorized as a superficial and deep system. Brachial vein constitutes the deep venous system. Brachial veins can be paired or singular.* **(Middle)** *Longitudinal color Doppler ultrasound shows the brachial vein. The ulnar and radial veins collect blood from the deep palmar arch and run in pairs alongside the corresponding arteries. They unite slightly above the elbow joint and flow into the brachial veins, which may be either paired or singular.* **(Bottom)** *Spectral Doppler ultrasound shows the flow pattern in the brachial vein. The flow is intermittent due to respiration. Examinations using proximal compression are not clinically relevant. Distal compression, which produces augmentation sounds, helps locate the veins.*

LONGITUDINAL US, DYNAMIC

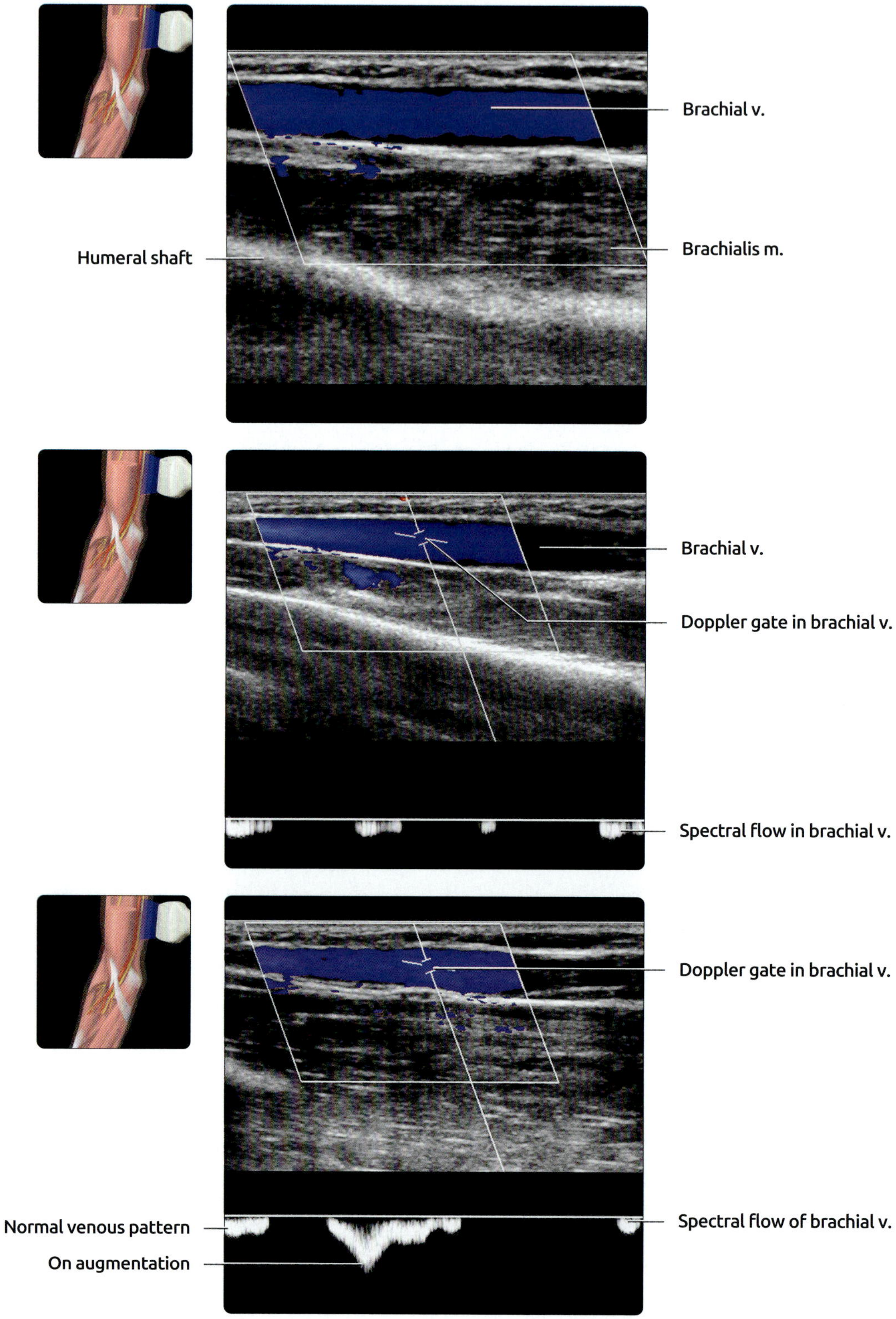

(Top) *Longitudinal color Doppler ultrasound shows the brachial vein. Using Doppler mode, phasic flow during normal respiration is observed.* **(Middle)** *Spectral Doppler ultrasound shows the flow pattern in the brachial vein. In the upper extremities, including the subclavian vein, the behavior of venous flow during respiration is the opposite of what it is in the lower body region. During inspiration, pressure in the thorax decreases, and the venous blood flows toward the heart. During expiration, the thoracic pressure increases and the flow decreases, or even ceases, during deep respiration.* **(Bottom)** *Spectral Doppler ultrasound shows the phenomenon of augmentation, demonstrating the patency of the veins. It involves squeezing of the distal arm muscles (causing motion artifacts from movement), which will increase venous flow. Deep inspiration will have the same effect.*

TRANSVERSE AND LONGITUDINAL US, CEPHALIC VEIN

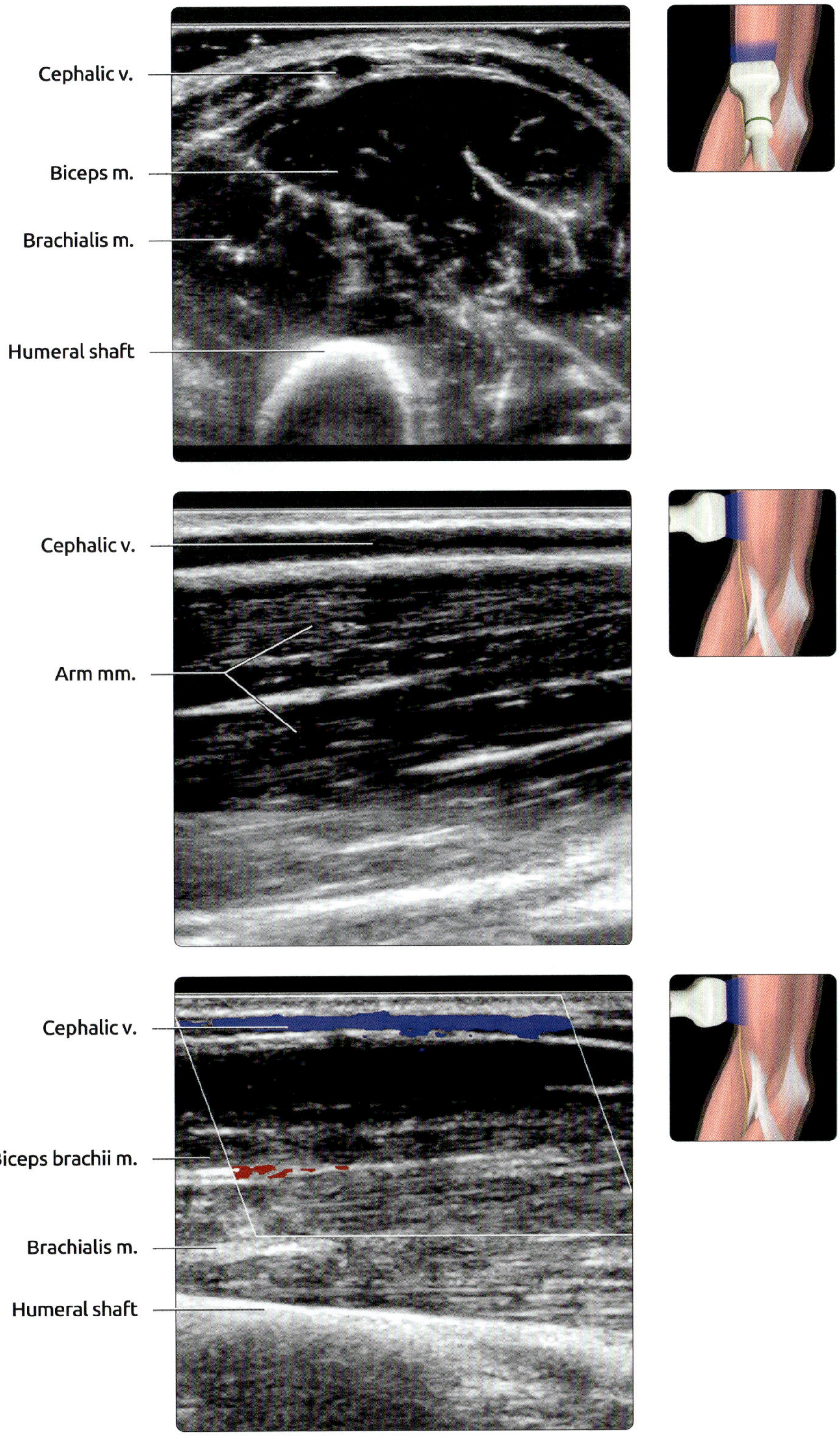

(Top) *Transverse grayscale ultrasound at the proximal arm shows the cephalic vein. The normal diameter of the cephalic vein just above the elbow measures 2.5-2.9 mm.* **(Middle)** *Longitudinal grayscale ultrasound at the proximal arm shows the cephalic vein. The most important veins of the arm are the cephalic vein and basilic veins. The superficial cephalic vein unites the deep-seated brachial and axillary veins. Assessment of valvular insufficiency has no clinical role in the upper limb vessels.* **(Bottom)** *Longitudinal color Doppler ultrasound in the distal arm shows the cephalic vein. Color flow duplex sonography allows ready assessment of the important upper arm veins (the paired brachial veins, cephalic veins, and basilic veins). Continuous color intensity makes it possible to exclude thromboses.*

GROSS ANATOMY

Joint Capsule Attachments

- Posterior
  - Proximally attaches to humerus proximal to olecranon fossa and capitellum; distally attaches to proximal ulna
- Anterior
  - Proximally attaches to humerus proximal to coronoid and radial fossae; distally attaches to coronoid process and annular ligament
- Anterior and posterior elbow fat pads are intracapsular but extrasynovial
- Capsule of elbow is weakest anteriorly; it is strengthened on both sides by ligaments
- Synovial recesses
  - Olecranon recesses
    - Largest; superior, medial, and lateral around olecranon process
  - Anterior humeral recess
    - Proximal to coronoid fossa
  - Annular recess
    - Surrounds radial neck
  - Ulnar collateral ligament recess
    - Deep to ligament
  - Radial collateral ligament recess
    - Deep to ligament
  - Capsule of elbow is weakest anteriorly
    - Capsule is strengthened on both sides by collateral ligaments

Ligaments

- **Ulnar collateral ligament**
  - Extends from distal aspect of medial epicondyle to coronoid and olecranon portions of ulna
  - Separated from overlying common flexor tendon by thin layer of fat
  - 3 bands: Anterior, posterior, and transverse
    - Anterior band: Most important; cord-like, 4-6 mm thick, taut in extension, extends from medial epicondyle to coronoid process (sublime tubercle); divided into superficial and deep layers
    - Posterior band: Less important; fan-like and thinner, taut in flexion, extends from medial epicondyle to olecranon
    - Transverse band: Functionally unimportant; forms base of triangle between anterior and posterior bands
- **Radial collateral ligament**
  - Triangular in shape
  - Apex on lateral epicondyle → blends distally with annular ligament
  - Lies deep to common extensor tendon
  - Provides origin for superficial head of supinator muscle
- **Annular ligament**
  - Attached to anterior and posterior aspects of radial notch of ulna forming collar around radial head
    - Anterior attachment becomes taut in supination
    - Posterior attachment becomes taut in extreme pronation
    - Provides origin for superficial head of supinator muscle

Tendons

- **Triceps tendon**
  - 3 heads combine to insert onto olecranon process
  - Medial fibers insert exactly on medial margin of olecranon
  - Lateral fibers may fan out as lateral cubital retinaculum to connect with ulna and antebrachial fascia
- **Biceps tendon**
  - Flat tendon forms ~ 7 cm proximal to elbow joint
  - Twists 90° as it passes deep so that anterior surface faces laterally
    - Expands close to attachment into radial tuberosity; attachment area is ~ 3 cm^2
  - Short and long heads have separate adjoined attachments
  - Also attaches to bicipital aponeurosis
    - Bicipital aponeurosis completely encircles ulnar forearm flexor muscles
    - Merges with investing fascia of forearm
    - May be important in stabilizing distal aspect of biceps tendon
- **Common extensor tendon**
  - Arises from lateral epicondyle
  - Superficial to radial collateral ligament
  - Composed of extensor-supinator group
    - Extensor carpi radialis brevis, extensor carpi radialis longus, extensor digiti minimi, extensor digitorum communis
- **Common flexor tendon**
  - Arises from medial epicondyle
  - Superficial to ulnar collateral ligament
  - Composed of flexor-pronator group
    - Flexor carpi radialis, flexor carpi ulnaris, flexor digitorum superficialis, pronator teres, palmaris longus

Bursae

- **Subtendinous olecranon bursa**
  - Between triceps tendon and olecranon
- **Subcutaneous olecranon bursa**
  - Between skin and olecranon process
- **Bicipitoradial bursa**
  - Between biceps tendon and radial tuberosity
- **Radioulnar bursa**
  - Between extensor digitorum and radiohumeral joint

ANATOMY IMAGING ISSUES

Imaging Recommendations

- High-frequency linear transducer
- Large amount of gel for superficial structures
- Examination is usually tailored to site of symptoms
- Lateral elbow is best seen with elbow fully or slightly flexed
- Medial elbow is best seen with elbow extended
- Distal biceps tendon is best seen with forearm in full supination
- Distal biceps tendon insertion is best seen from posterior part of elbow with forearm fully pronated and elbow flexed with transducer aligned transversely over radial tuberosity
- Effusion will displace fat pads away from bone

CORONAL ELBOW

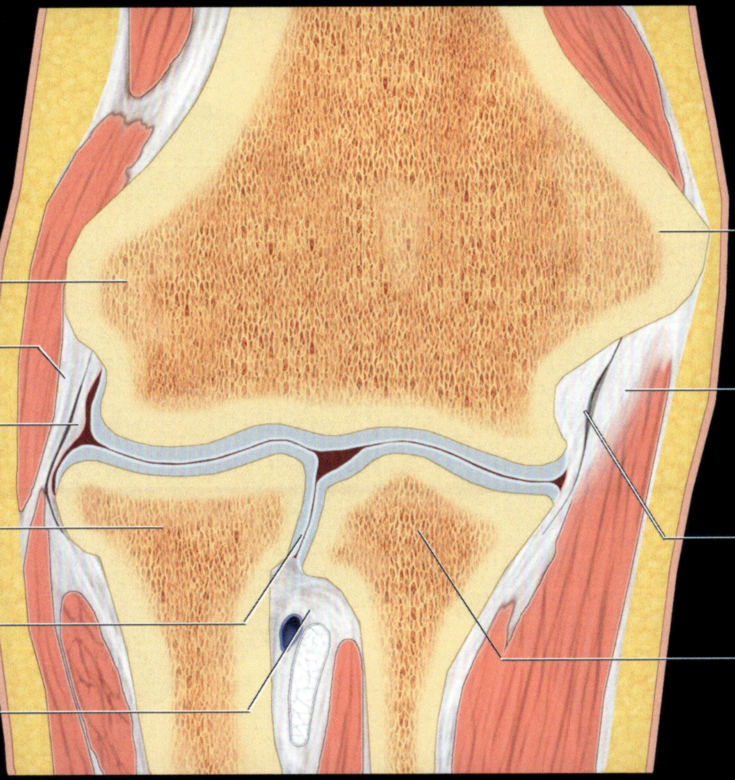

Coronal section through the level of the epicondyles shows the collateral ligaments deep to the common tendon groups. Although the radial collateral ligament is shown, the section is too anterior to show most of the ulnar collateral ligament, which originates just posterior to the radial collateral ligament.

PROXIMAL ELBOW CROSS SECTION

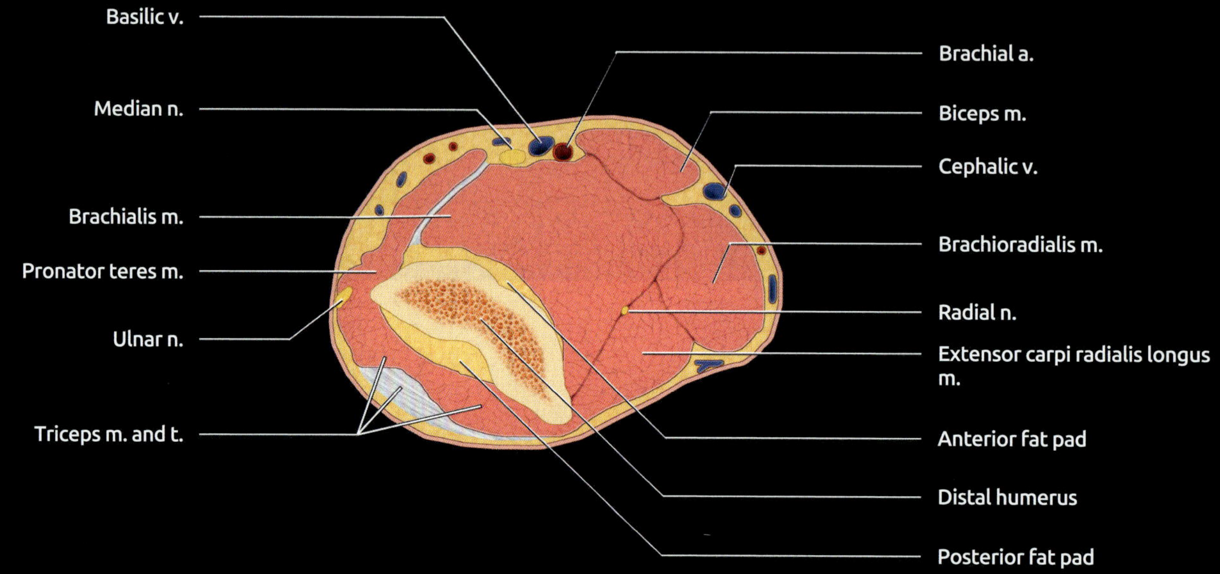

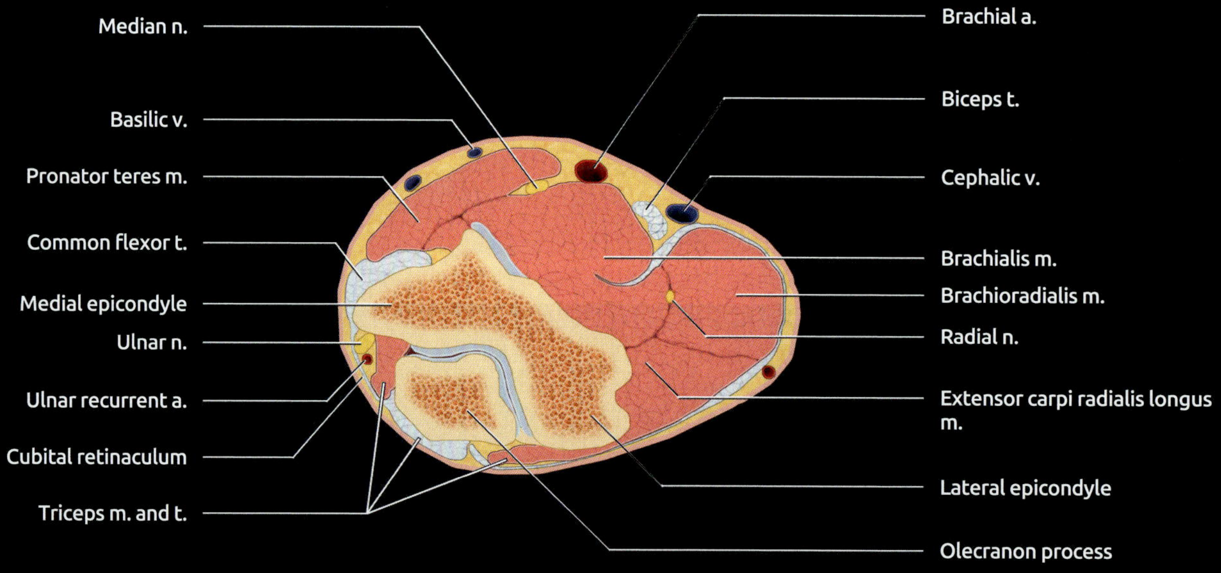

(Top) *Axial graphic shows the supracondylar region of the humerus. The anterior and posterior fat pads are seen in the coronoid and olecranon fossae, respectively. The brachialis muscle accounts for the bulk of the anterior compartment of the distal arm.* **(Bottom)** *Axial graphic shows the epicondylar region of the distal humerus. The triceps muscle thins as its tendon attaches to the olecranon. The ulnar nerve and posterior ulnar recurrent artery are held in the cubital tunnel by the cubital retinaculum (formed from the ligament of Osborne and the fascia of flexor carpi ulnaris tendon).*

MID ELBOW CROSS SECTION

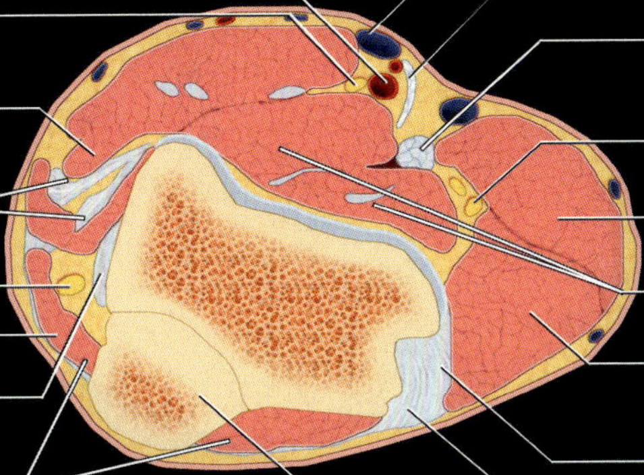

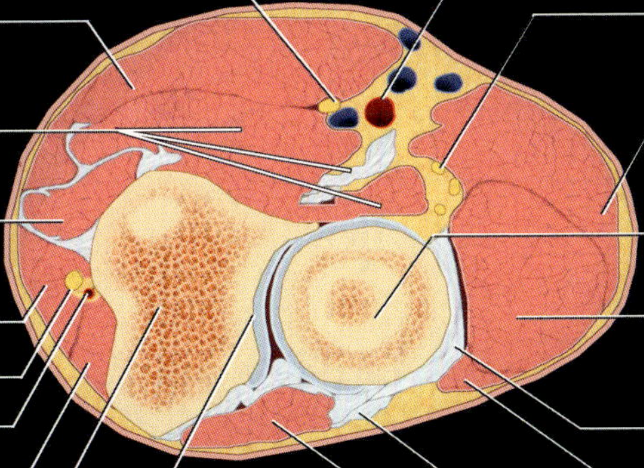

(Top) *Axial graphic immediately proximal to the elbow joint is shown. The common extensor tendon overlies the radial collateral ligament and may be difficult to distinguish at this level. The ulnar nerve has exited the cubital tunnel and is entering the flexor carpi ulnaris.* **(Bottom)** *Axial graphic at the level of the proximal radioulnar joint is shown. The articulating surfaces of the proximal radioulnar joint are well seen with the radial head being held in the radial notch of the ulna by the annular ligament. The lateral ulnar collateral ligament blends with the posterior aspect of the annular ligament.*

DISTAL ELBOW CROSS SECTION

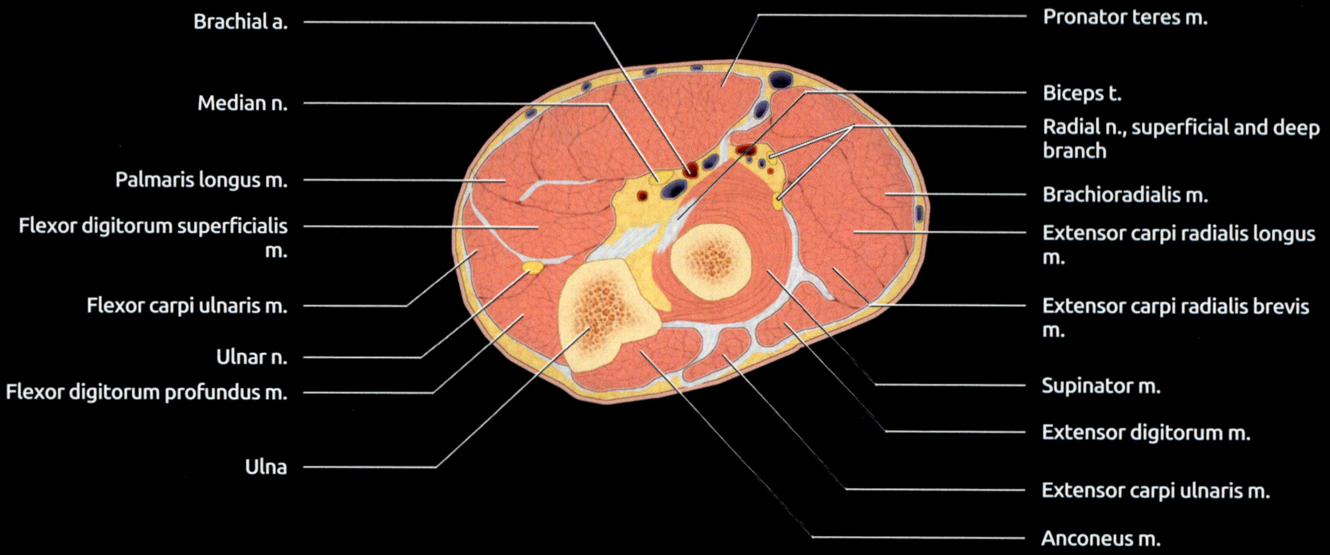

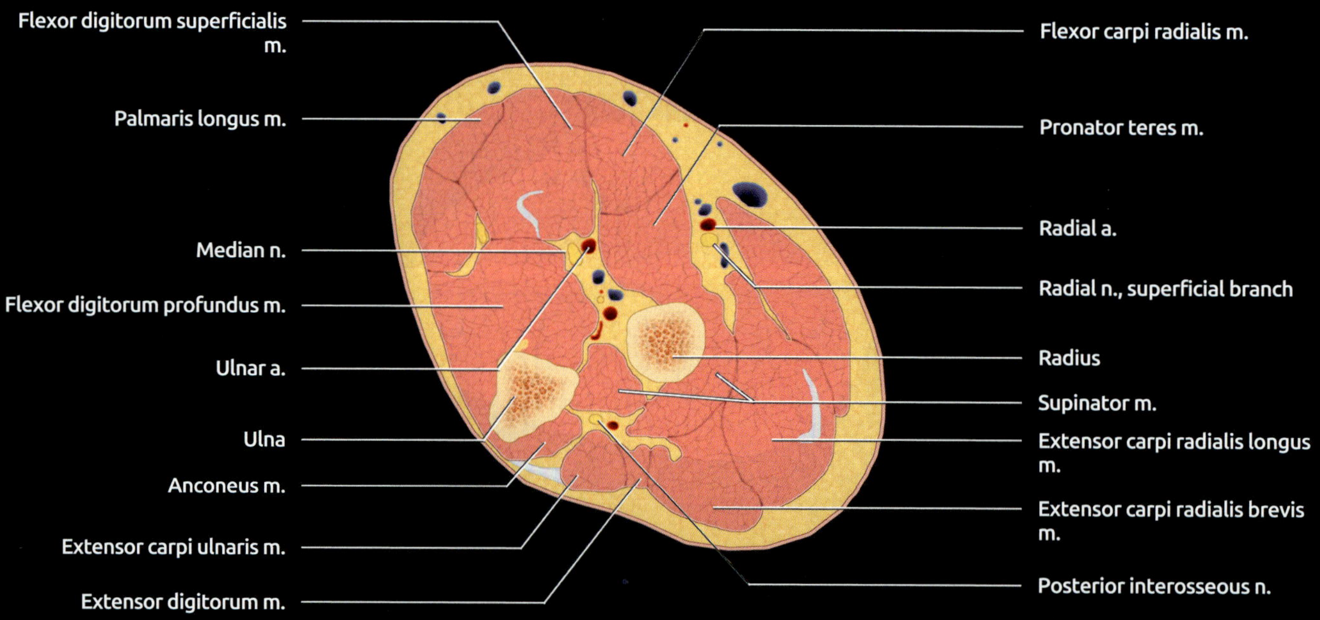

(Top) *Graphic shows the axial elbow at a level immediately above the radial tuberosity. The brachialis tendon inserts on the ulnar tuberosity, and the biceps tendon approaches its insertion on the radial tuberosity, which is more distal than the brachialis insertion.* **(Bottom)** *At the level of the proximal forearm, the muscles are starting to align themselves into the anterior (flexor) compartment and the posterior (extensor) compartment.*

LONGITUDINAL US, ANTERIOR ELBOW

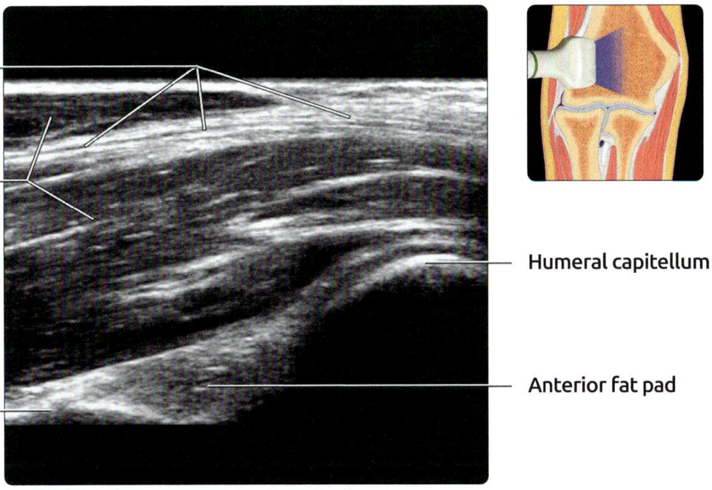

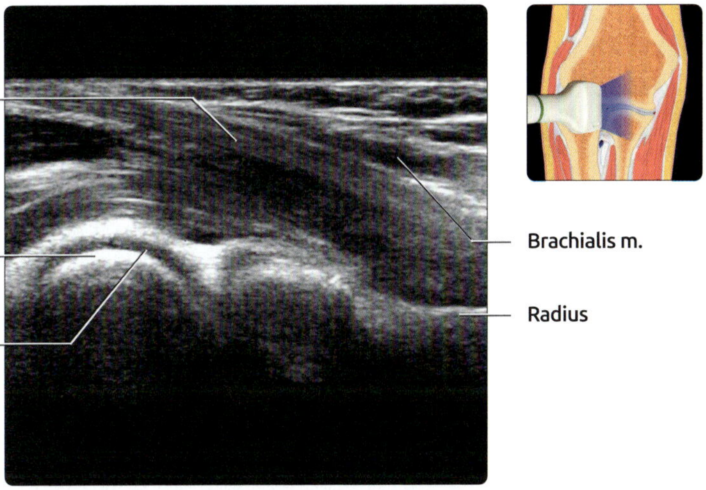

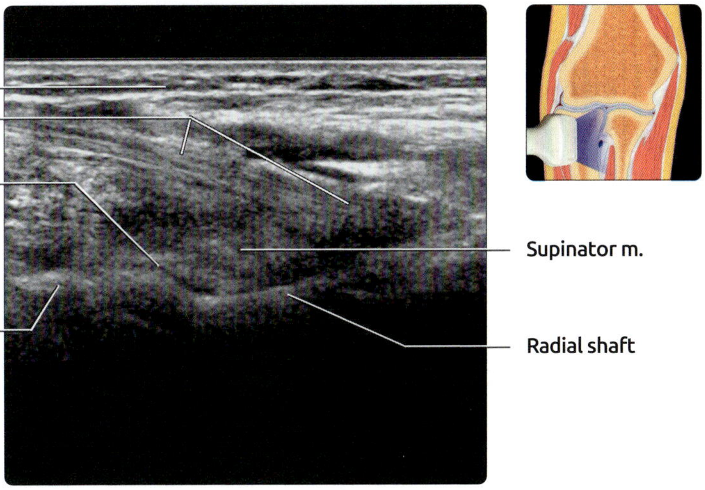

(Top) *Longitudinal grayscale ultrasound shows the biceps tendon proximal to the elbow. Most biceps tendon tears occur near the shoulder. Tears of the distal end of the biceps tendon are uncommon and mostly occur when unexpected force is applied to the bent arm.* **(Middle)** *Longitudinal grayscale ultrasound shows the biceps tendon at the elbow. With a torn biceps tendon, a bulge in the arm ("Popeye" muscle) may appear.* **(Bottom)** *Longitudinal grayscale ultrasound shows the biceps tendon just distal to the elbow. The biceps tendon flattens and rotates 90° as it approaches the insertion into the posterior aspect of the radial tuberosity. The distal biceps tendon is best visualized on longitudinal section with the forearm fully supinated. To see the insertional area of distal biceps tendon, fully pronate the forearm and flex the elbow with transducer placed posteriorly over the radial tuberosity.*

TRANSVERSE US, ANTERIOR ELBOW

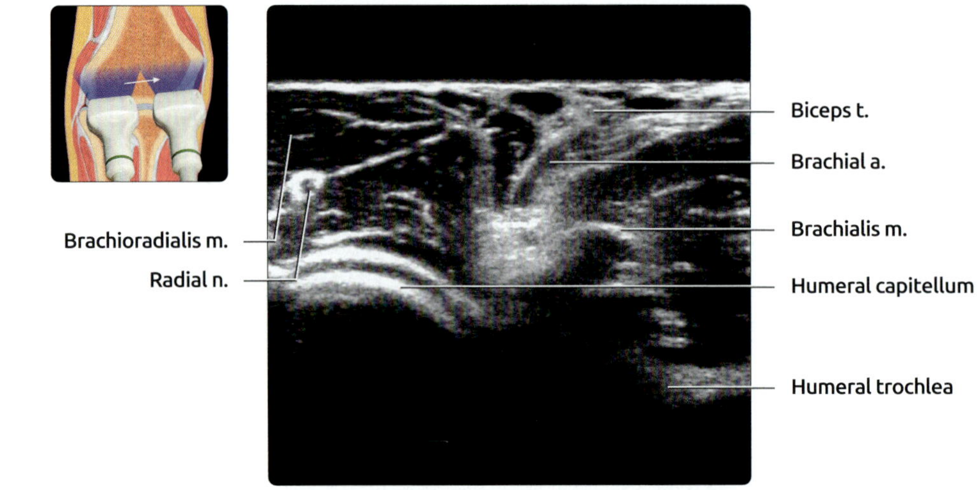

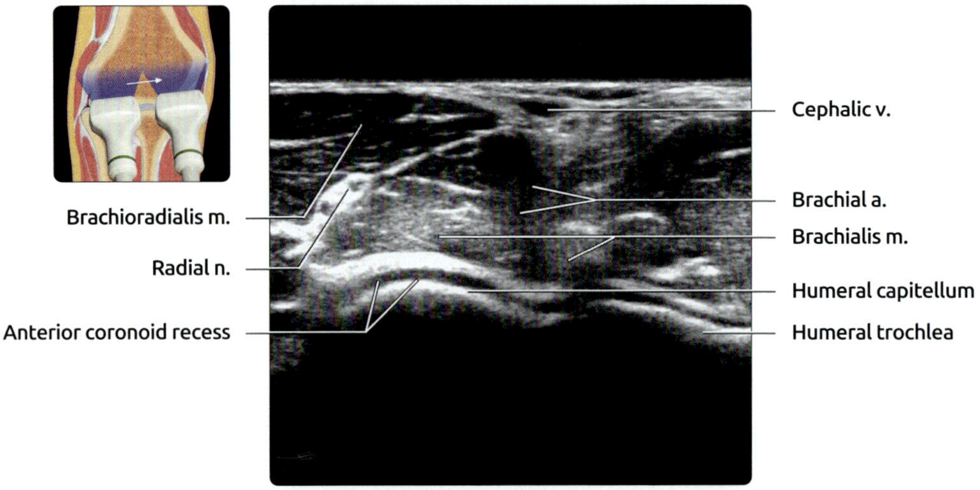

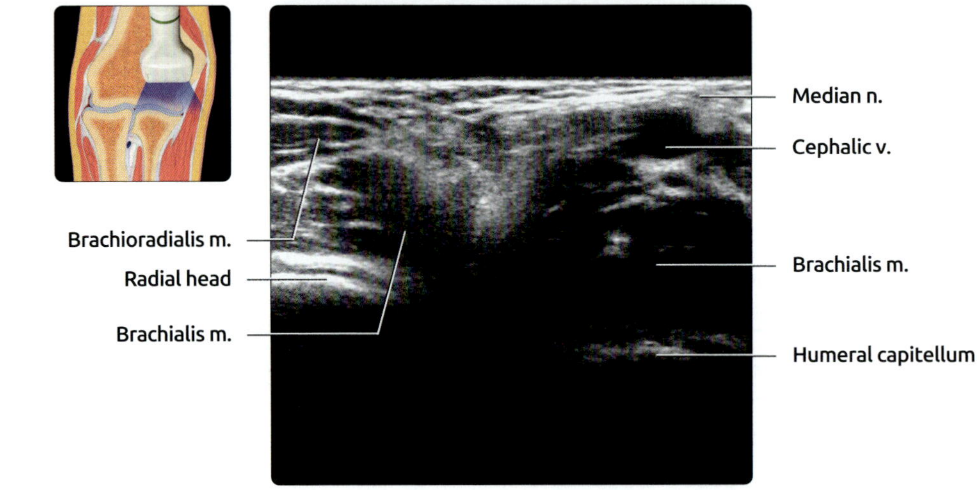

(Top) *Transverse grayscale ultrasound shows the proximal end of the antecubital fossa. Flexion and extension occur at the humeroulnar articulation, while pronation and supination involve pronation of the radius around the ulna.* **(Middle)** *Transverse grayscale ultrasound shows the midportion of the antecubital fossa. The median cubital vein unites the cephalic vein with the basilic vein. It is often used for venipuncture and venous access. It lies superficial to the bicipital aponeurosis.* **(Bottom)** *Transverse grayscale ultrasound shows the distal aspect of the antecubital fossa. The elbow joint is supplied by arteries from the profunda brachii, brachial arteries, and recurrent branches of the radial and ulnar arteries.*

TRANSVERSE US, ANTECUBITAL FOSSA

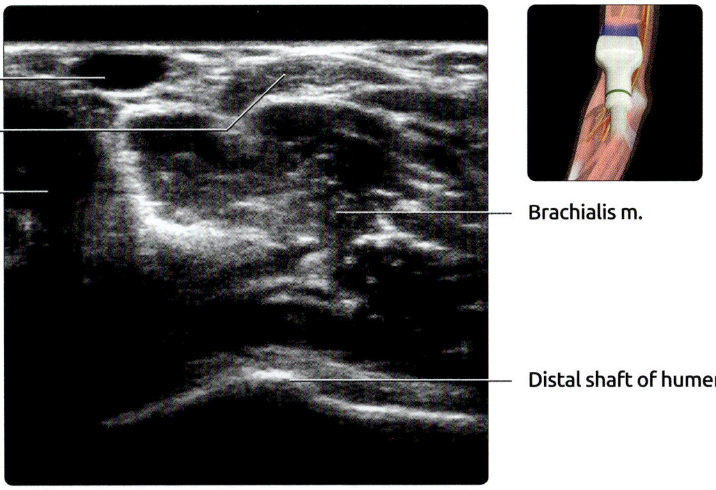

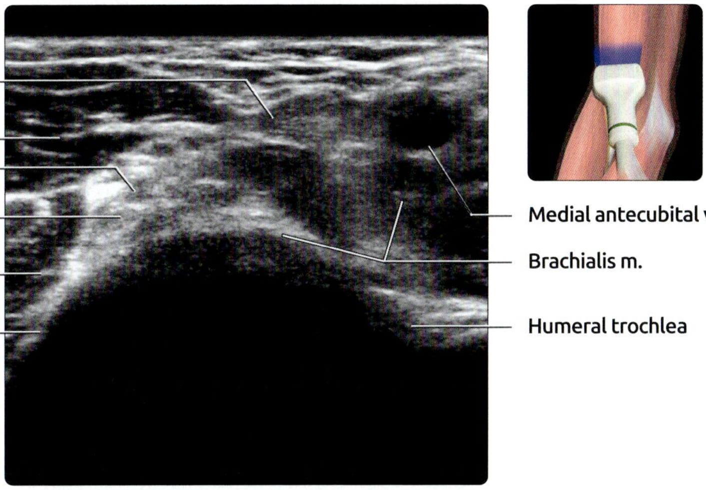

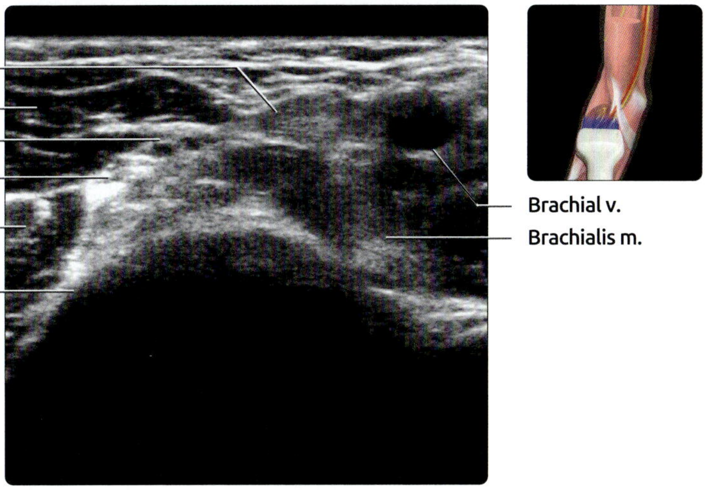

(Top) *Transverse grayscale ultrasound shows the anterior elbow and proximal antecubital fossa. The antecubital fossa or cubital fossa is the area anterior to the elbow joint.* **(Middle)** *Transverse grayscale ultrasound shows the antecubital fossa at the midportion. The antecubital fossa contains the radial nerve, biceps brachii tendon, brachial artery, and median nerve from lateral to medial orientation.* **(Bottom)** *Transverse grayscale ultrasound shows the distal antecubital fossa. The bicipital aponeurosis forms the roof of the antecubital fossa. The superficial veins of the elbow lie above the bicipital aponeurosis.*

TRANSVERSE US, BICEPS TENDON

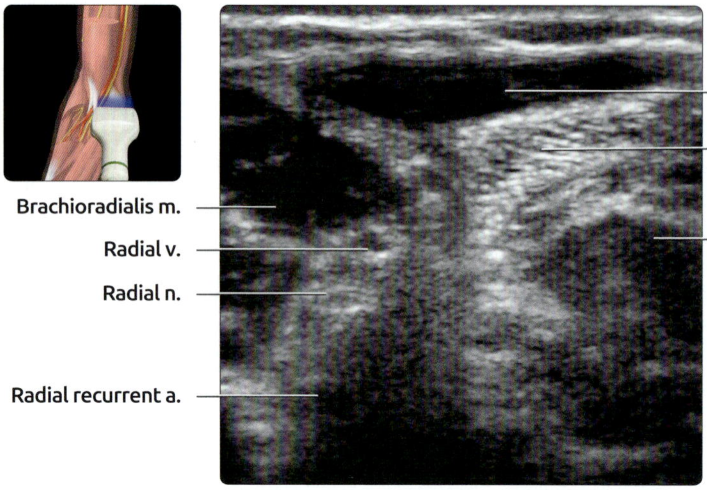

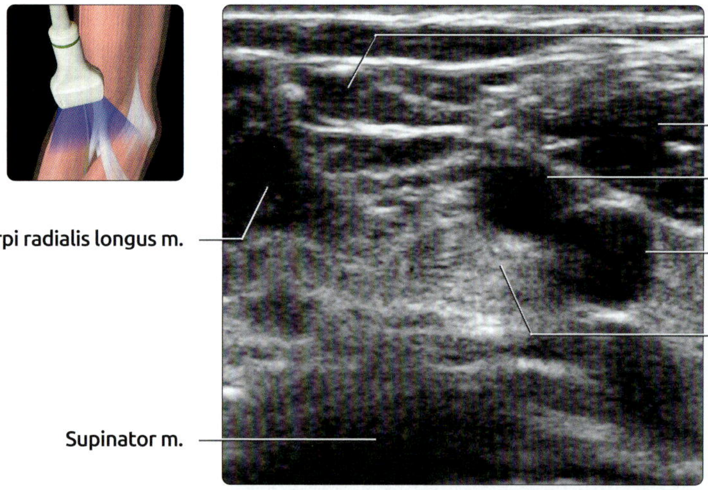

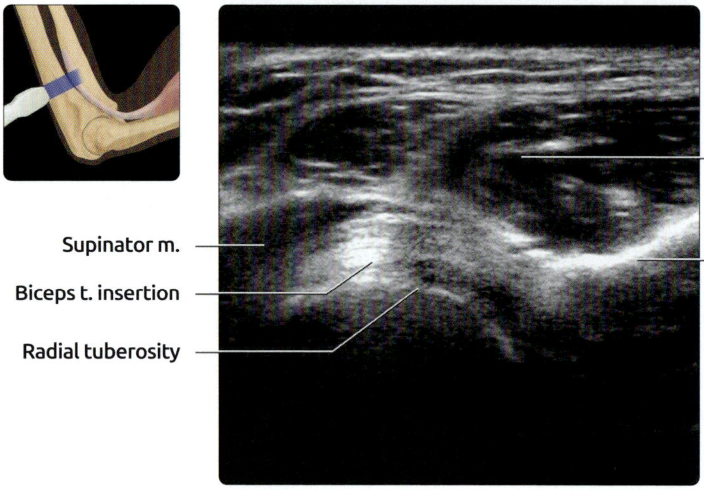

(Top) *Transverse grayscale ultrasound shows the biceps tendon at the upper antecubital fossa. The biceps tendon forms a few centimeters above the elbow joint.* **(Middle)** *Transverse grayscale ultrasound shows the biceps tendon at the midcubital fossa. The biceps tendon dips deeply and rotates as it passes to its insertion. The distal aspect of the biceps tendon and the insertion is difficult to see on ultrasound of the anterior surface of the elbow. Maximum supination will increase its visibility.* **(Bottom)** *Transverse grayscale ultrasound shows the biceps tendon insertional area. The insertional area of the biceps tendon is best seen with the elbow flexed and the wrist fully pronated. This positions the radial tuberosity posteriorly, allowing the insertional area to be viewed from the posterior aspect as shown here.*

TRANSVERSE AND LONGITUDINAL US, ANTEROLATERAL ELBOW

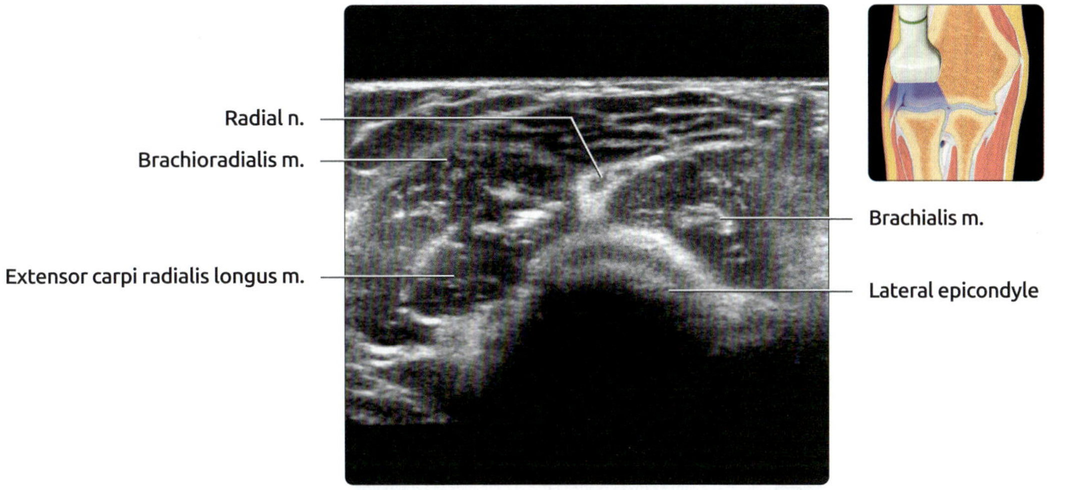

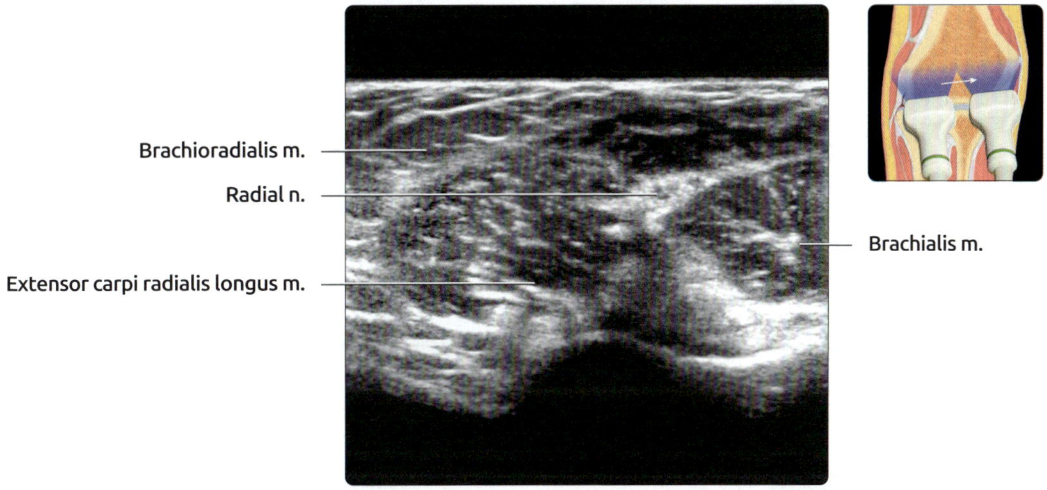

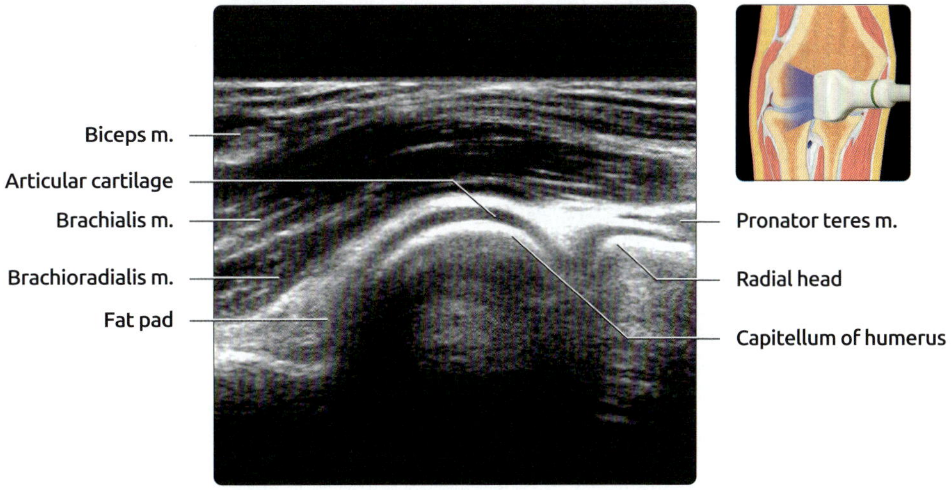

(Top) *Transverse grayscale ultrasound shows the proximal anterolateral aspect of the elbow. The radial nerve is located on the anterolateral aspect of the elbow. It can be readily identified deep to the brachioradialis muscle.* **(Middle)** *Transverse grayscale ultrasound shows the midsection of the anterolateral elbow. The radial nerve passes anterior to the epicondyle deep to the brachioradialis muscle.* **(Bottom)** *Longitudinal grayscale ultrasound shows the anterolateral aspect of the elbow. The fat pads are intracapsular but extrasynovial. The synovial space lies deep to the fat pad on the bone surface. Distension of the fat pads allows these to become visible radiographically in elbow joint effusions.*

LATERAL ELBOW

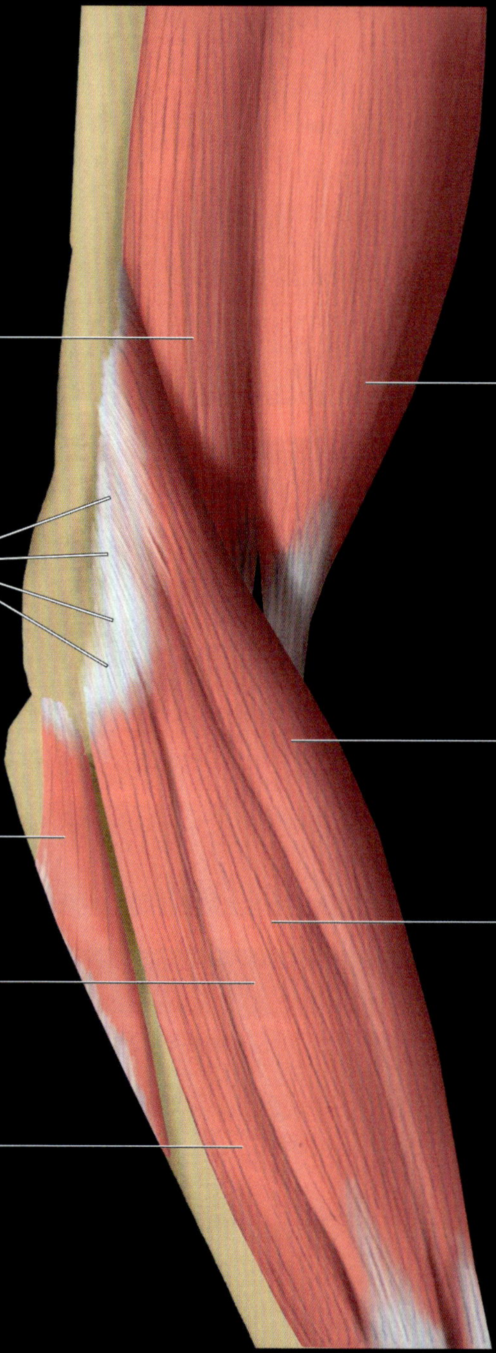

Side view of the lateral aspect of the elbow shows the extensor-supinator group and its common extensor tendon origin attaching to the lateral epicondyle and supracondylar aspect of the humerus. The common extensor tendon origin is composed of the extensor carpi radialis brevis, extensor digitorum, extensor digiti minimi, and extensor carpi ulnaris. The extensor digitorum origin and extensor carpi radialis brevis are very closely aligned in this section.

LONGITUDINAL US, COMMON EXTENSOR ORIGIN

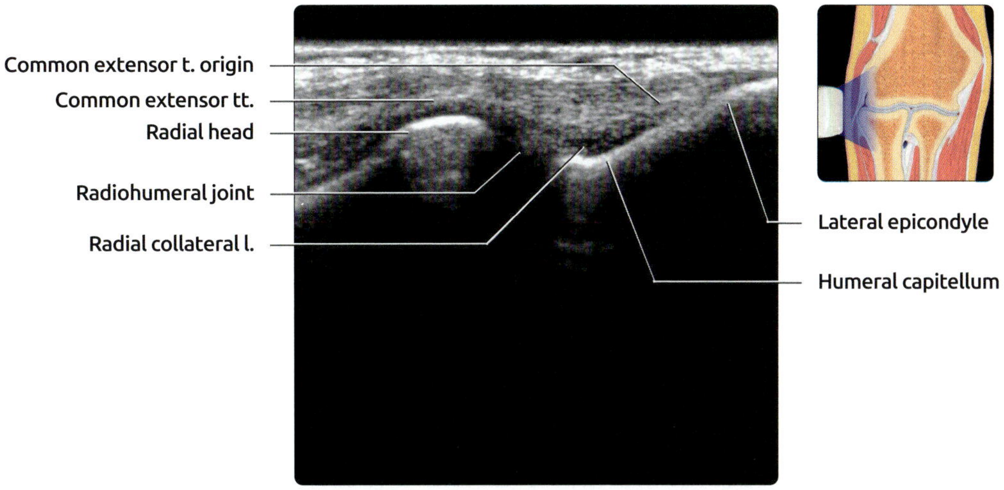

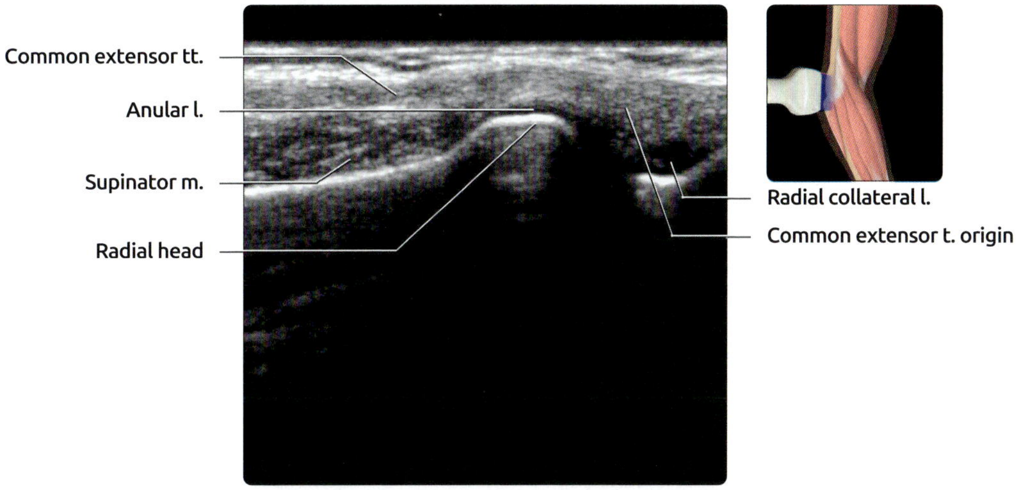

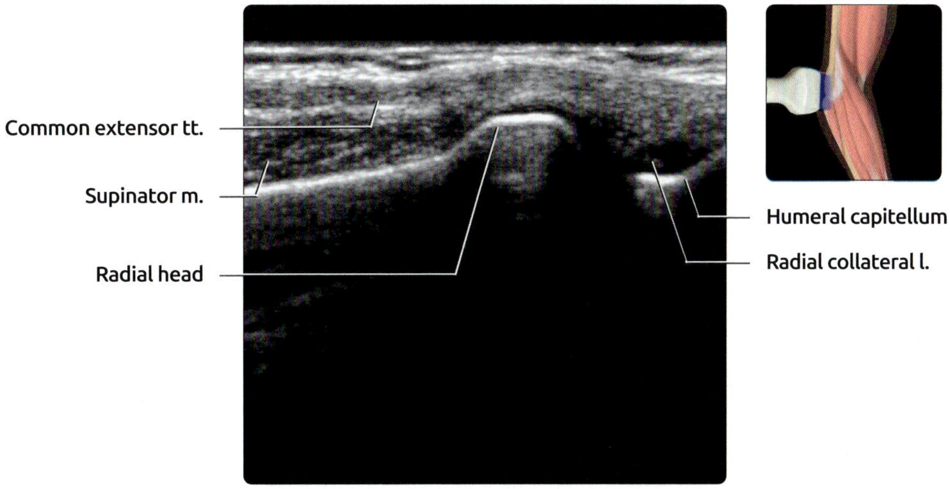

(Top) *Longitudinal grayscale US at the common extensor tendon origin is shown. Common extensor tendon origin has a broad attachment on the posteroinferior aspect of the lateral epicondyle. Tendon fibers cannot be separated into individual components at or near the insertion. Fibers from extensor digitorum make up most of the superficial portion, while fibers from extensor carpi radialis brevis make up most of the deep fibers. Contribution from extensor carpi ulnaris and extensor digiti minimi is only small.* **(Middle)** *Longitudinal grayscale US at common extensor tendon origin with the wrist in neutral position is shown. The radial collateral ligament is less distinct than the ulnar collateral ligament. It runs from the inferior aspect of the lateral epicondyle to gain attachment to the annular ligament.* **(Bottom)** *Longitudinal grayscale US at common extensor tendon origin with the wrist in extension is shown. Wrist extension slightly straightens extensor tendons, and dynamic movement during US can help separate movable tendons from fixed radial collateral ligament.*

TRANSVERSE US, COMMON EXTENSOR TENDONS

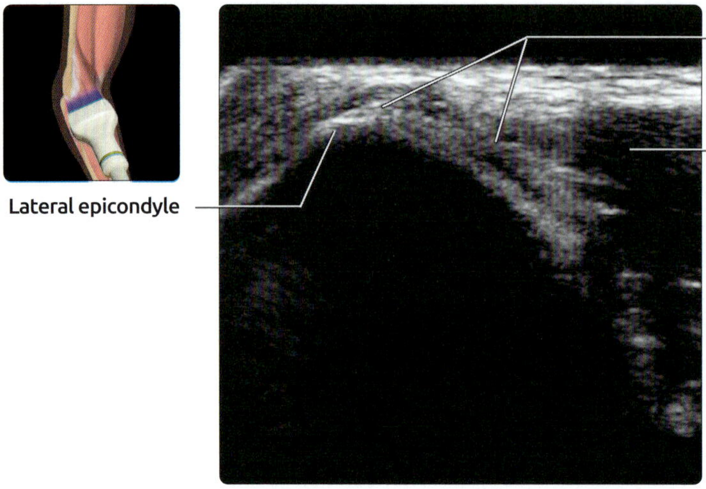

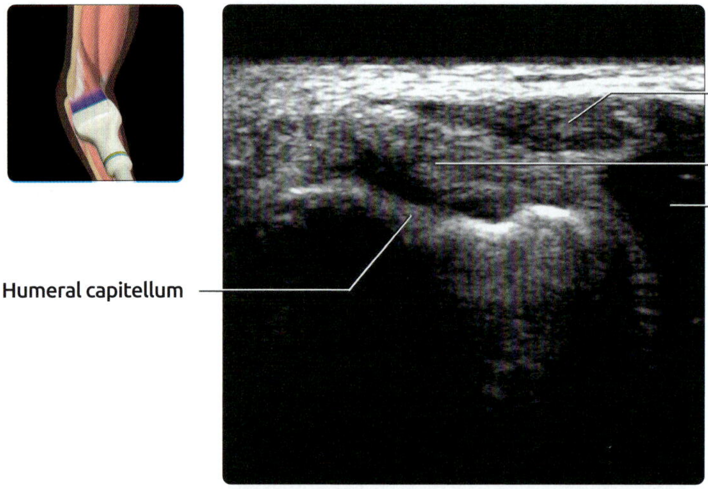

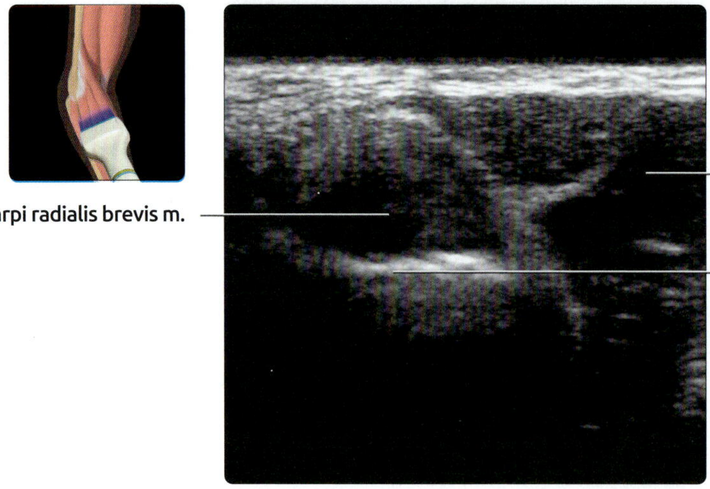

(Top) *Transverse grayscale ultrasound shows the common extensor origin. The common extensor tendon gains broad insertion onto the posterior aspect of the lateral epicondyle.* **(Middle)** *Transverse grayscale ultrasound shows the common extensor origin just distal to its attachment. The tendons begin to diverge just distal to the attachment though they are not yet separate.* **(Bottom)** *Transverse grayscale ultrasound shows the common extensor tendons even more distal to the above section. The individual components of the common tendon origin begin to be seen near the elbow joint.*

TRANSVERSE AND LONGITUDINAL US, LATERAL ELBOW

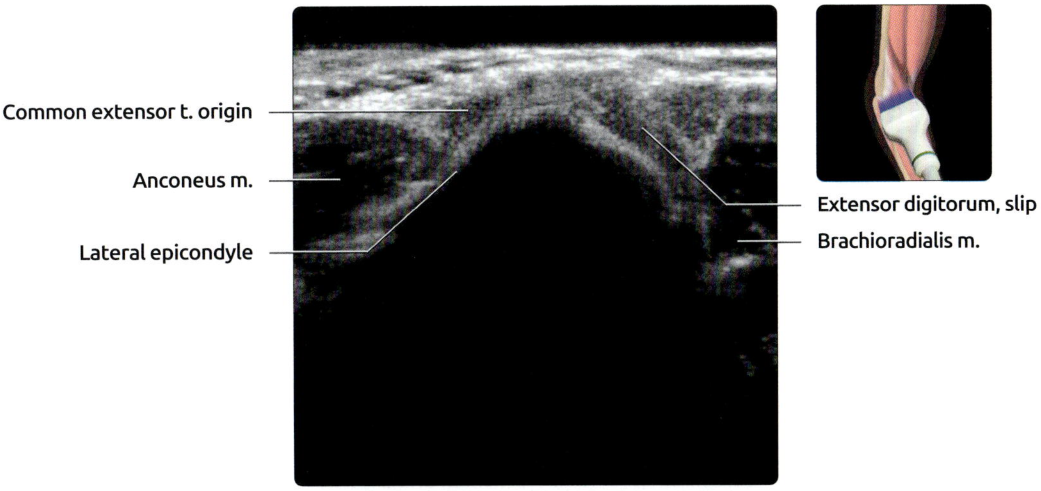

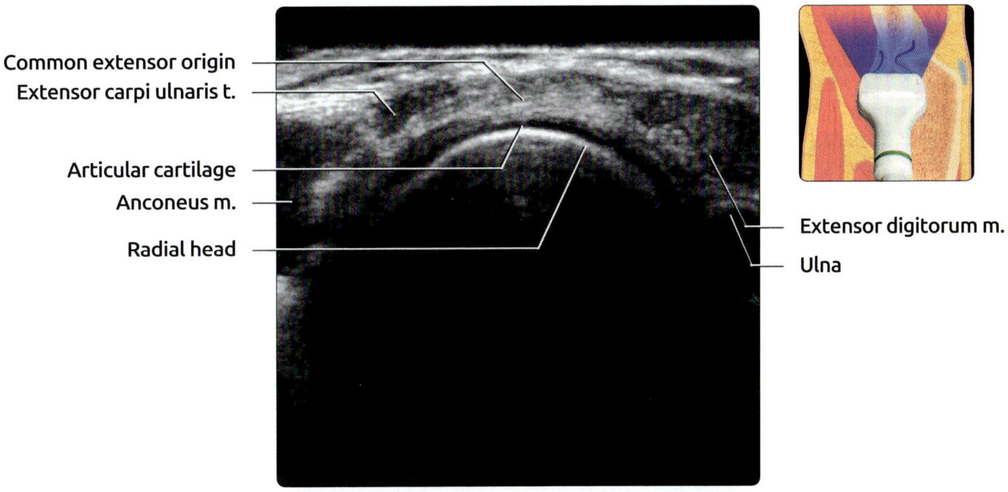

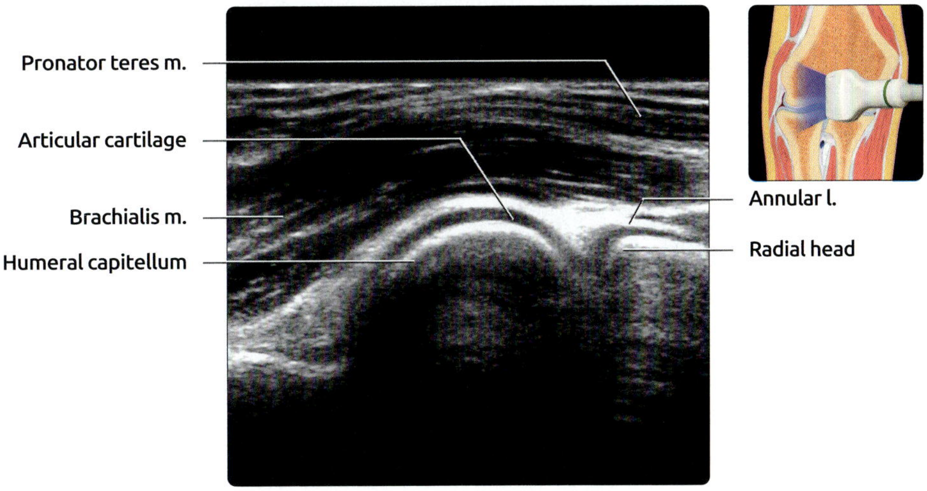

(Top) *Transverse grayscale ultrasound shows the common extensor tendon origin. The tendon caliber is often better appreciated on transverse, rather than on longitudinal, imaging.* **(Middle)** *Transverse grayscale ultrasound shows the common extensor tendons at the level of the radial head. The extensor tendons that originate from the common tendon origin can be traced distally, allowing them to be seen as separate tendons (extensor digitorum, extensor digiti minimi, extensor carpi ulnaris, and extensor carpi radialis brevis).* **(Bottom)** *Longitudinal grayscale ultrasound shows the common extensor tendon origin. The caliber of the common extensor tendon origin can be best appreciated on transverse imaging.*

TRANSVERSE AND LONGITUDINAL US, ANTEROMEDIAL ELBOW

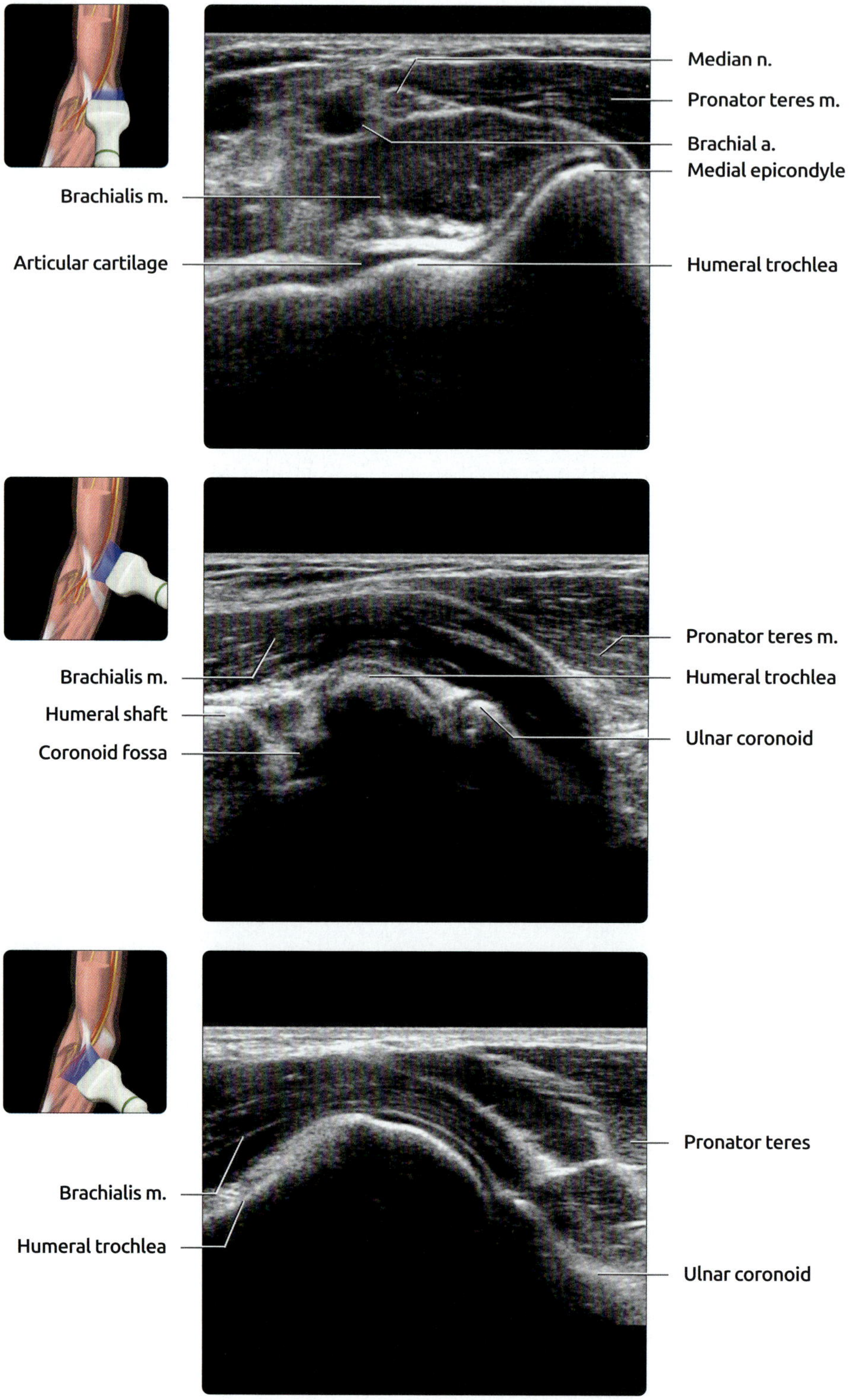

(Top) *Transverse grayscale ultrasound shows the anteromedial aspect of elbow. The pronator teres has 2 heads: 1 from the common flexor origin (medial epicondyle) and supracondylar region and 1 from the medial side of the coronoid process. The median nerve lies deep to the pronator teres and may be trapped between these 2 heads as it enters the forearm.* **(Middle)** *Longitudinal grayscale ultrasound shows the anteromedial aspect of the elbow. Centrally at the junction of the distal humerus between the coronoid fossa anteriorly and the olecranon fossa posteriorly, the humerus is very thin and potentially simulating a bone defect.* **(Bottom)** *Longitudinal grayscale ultrasound shows the anteromedial aspect of the elbow. The brachialis muscle is closely applied to the anterior aspect of the elbow joint as it descends to its insertion on the ulnar tuberosity and the coronoid process of the ulna.*

FLEXOR ASPECT

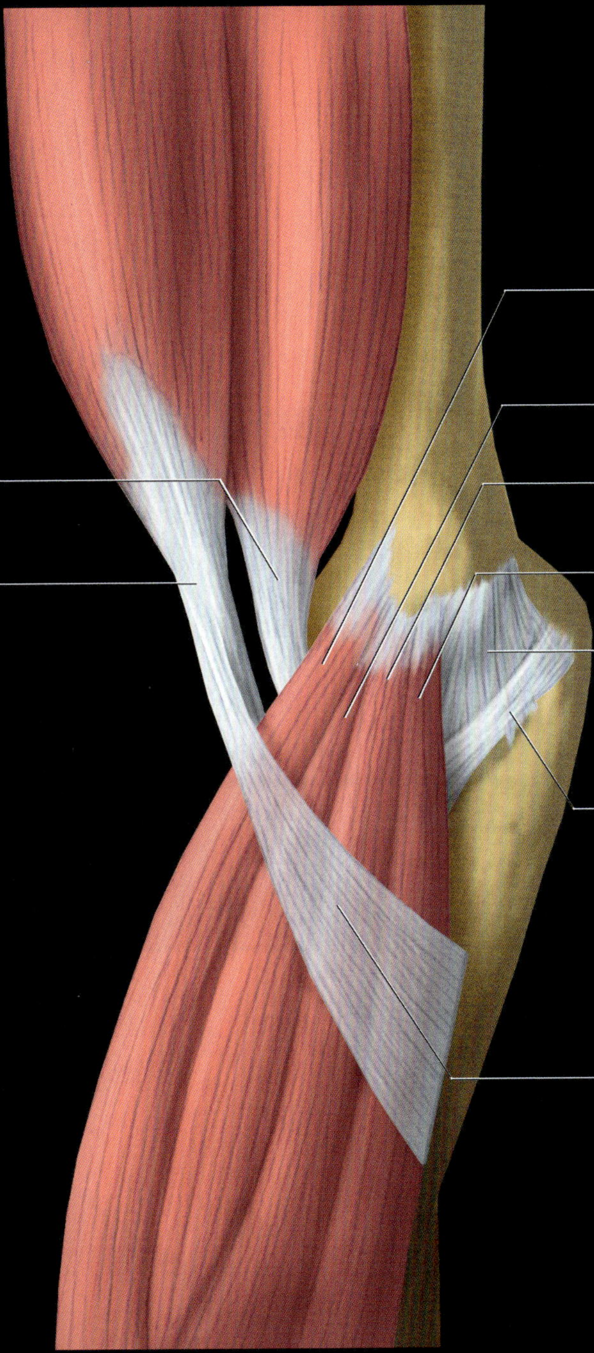

Side view of the medial aspect of the elbow shows the flexor-pronator group and its common flexor tendons attaching to the medial epicondyle. The anterior band of the ulnar collateral ligament is deep to the common flexor tendons. The biceps aponeurosis blends with the anterior aspect of the common flexor muscle mass. The flexor digitorum profundus (not shown) arises from the proximal and midulna, posterior and deep to the flexor carpi ulnaris. It does not act on the elbow and is not part of the flexor-pronator group.

LIGAMENTS AROUND ELBOW

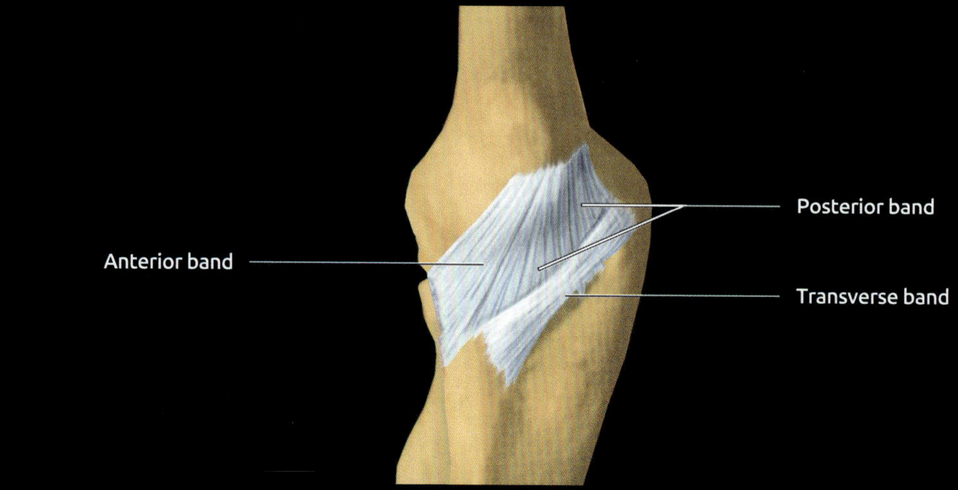

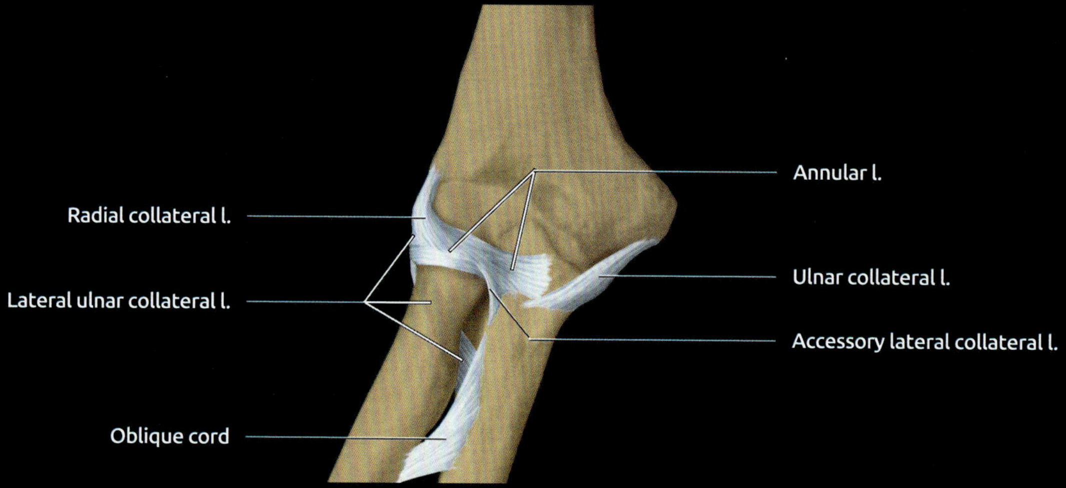

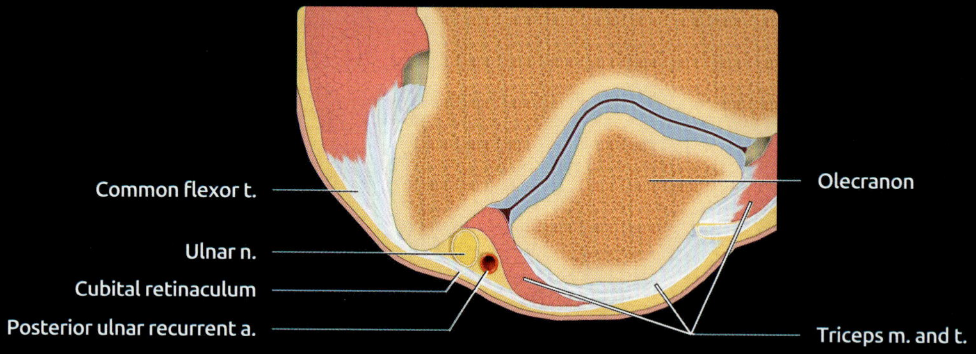

(Top) *Side view of the medial aspect of the elbow shows the 3 components of the ulnar collateral ligament: Anterior, posterior, and transverse band.* **(Middle)** *Anterior view of the elbow shows the radial collateral ligament complex, which is composed of the radial collateral ligament (provides varus stability), lateral ulnar collateral ligament (provides posterolateral stability), annular ligament (holds radial head against radial notch of the ulna), and accessory lateral collateral ligament (reinforces annular ligament). The oblique cord is part of the proximal radioulnar joint.* **(Bottom)** *Axial graphic shows the cubital tunnel. The ulnar nerve may be compressed within the cubital tunnel (cubital tunnel syndrome) by a mass, posttraumatic osseous deformity, or aneurysm of the recurrent ulnar artery. The ulnar nerve may also sublux out of the cubital tunnel leading to a low-grade traumatic neuritis and symptoms of carpal tunnel syndrome.*

LONGITUDINAL US, COMMON FLEXOR ORIGIN

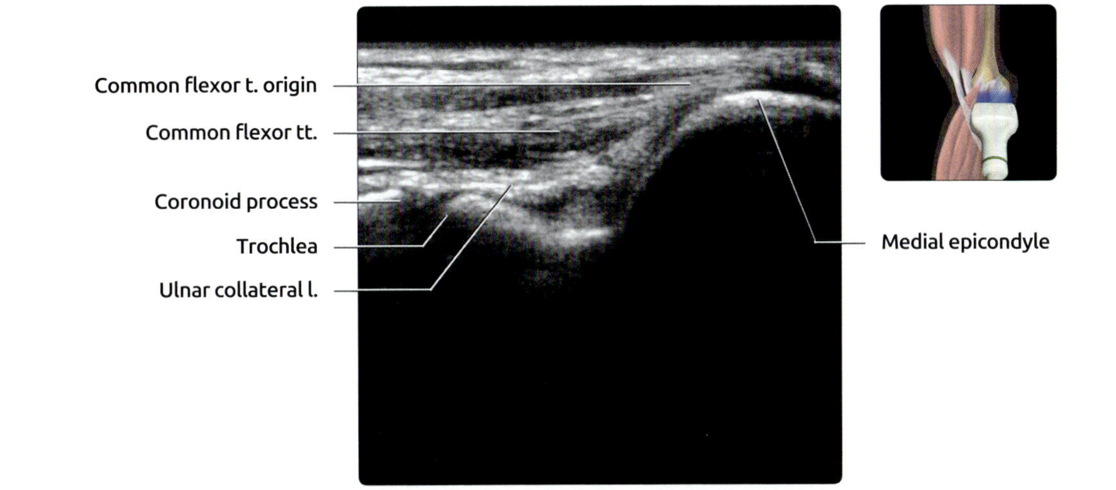

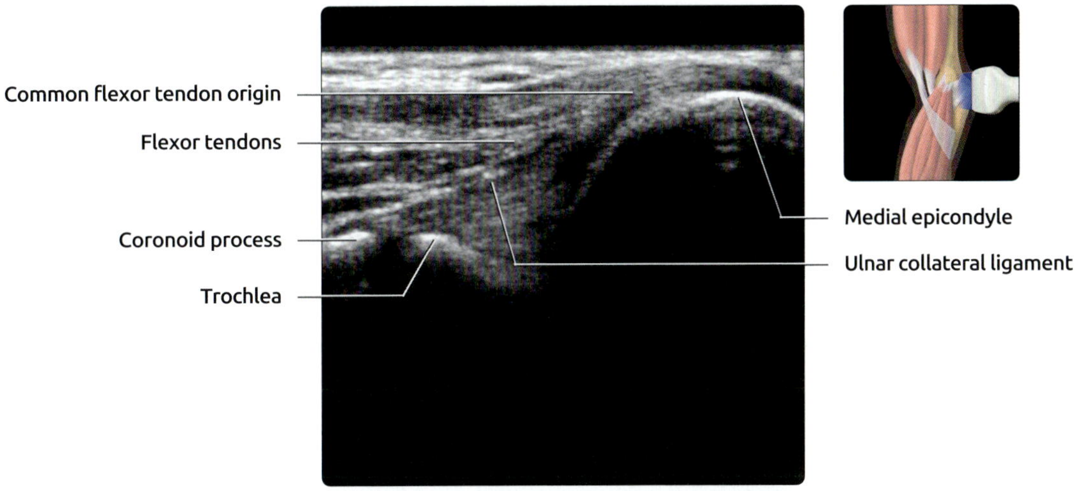

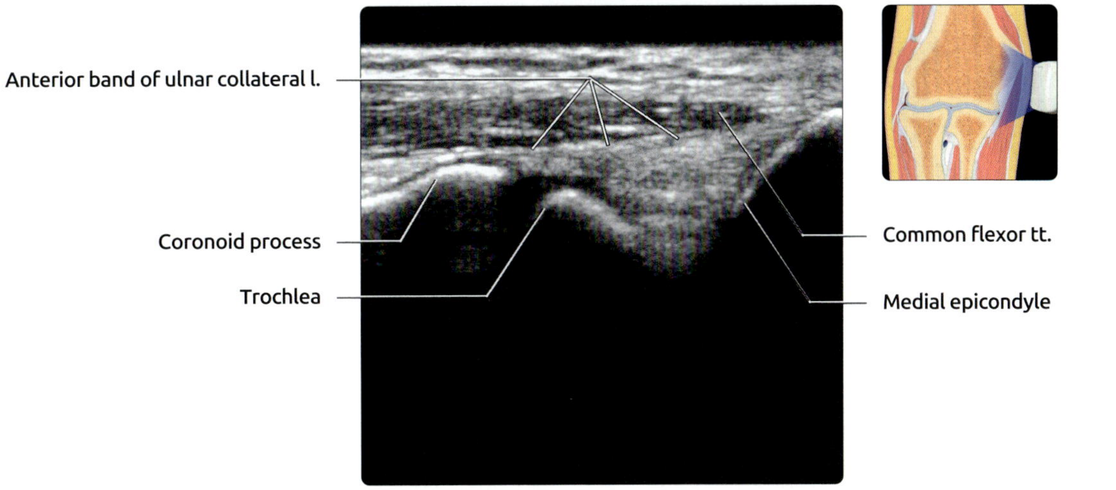

(Top) *Longitudinal grayscale ultrasound shows the common flexor tendon origin. These tendons (pronator teres, flexor carpi radialis, palmaris longus, flexor digitorum superficialis, flexor carpi ulnaris) are inseparable at or close to the origin.* **(Middle)** *Longitudinal grayscale ultrasound shows the ulnar collateral ligament. The ulnar collateral ligament runs from the medial epicondyle to the coronoid process. The ulnar collateral ligament lies just deep to the flexor tendons and may be injured along with these tendons.* **(Bottom)** *Longitudinal grayscale ultrasound shows the ulnar collateral ligament anterior band. The anterior band (medial epicondyle-sublime tubercle) is the strongest component of the ulnar collateral ligament.*

TRANSVERSE US, COMMON FLEXOR ORIGIN

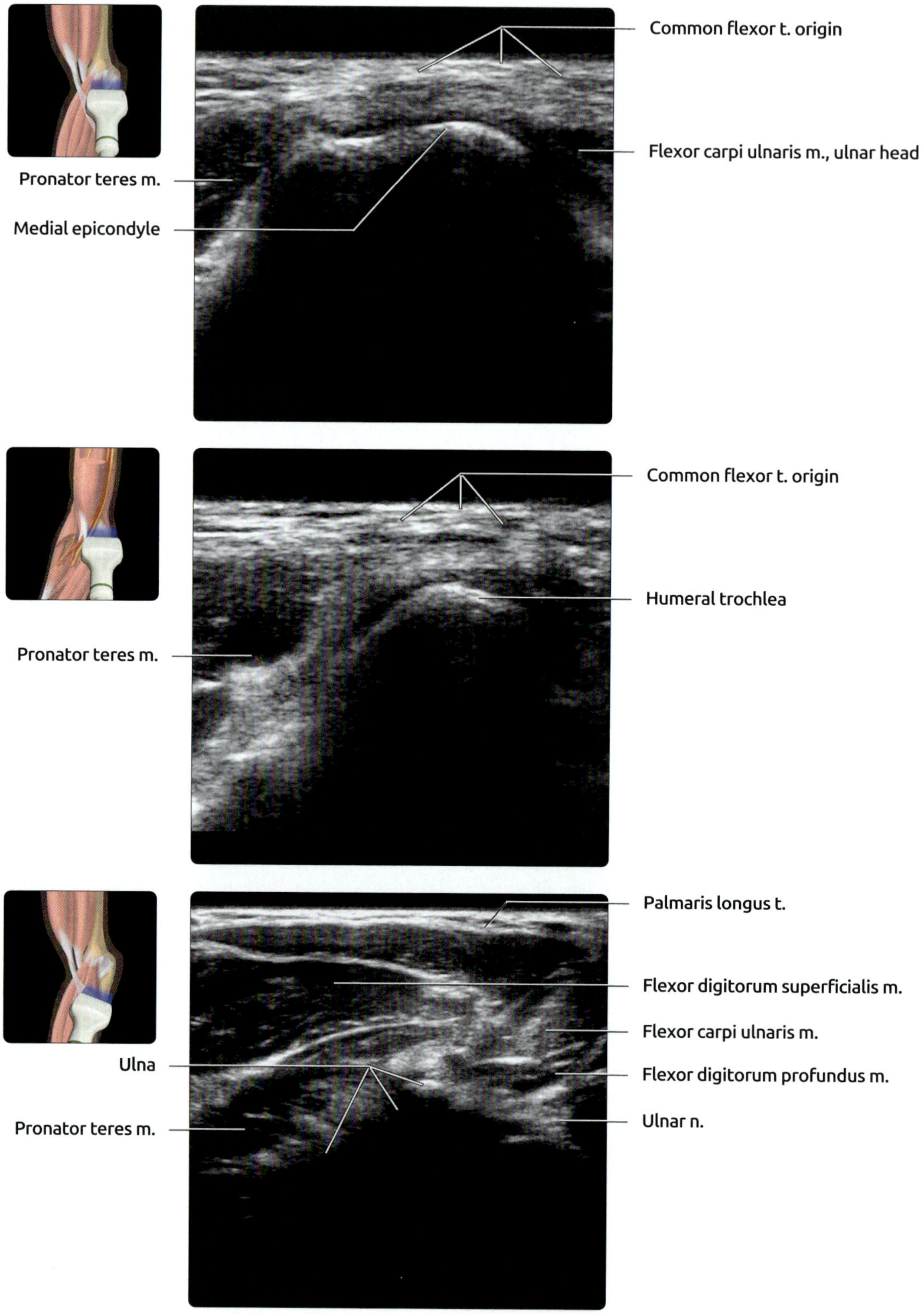

(Top) *Transverse grayscale ultrasound shows the common flexor tendon origin. Tendinosis of the common flexor tendon is termed golfer's elbow or medial epicondylitis. Yet most people who get this injury are involved in other sports, particularly tennis.* **(Middle)** *Transverse grayscale ultrasound shows the common flexor tendons just distal to the elbow. The different components of the common flexor tendons can be seen by following the tendon distally and proximally.* **(Bottom)** *Transverse grayscale ultrasound shows the forearm more distal to the common flexor origin.*

TRANSVERSE US, POSTERIOR ELBOW

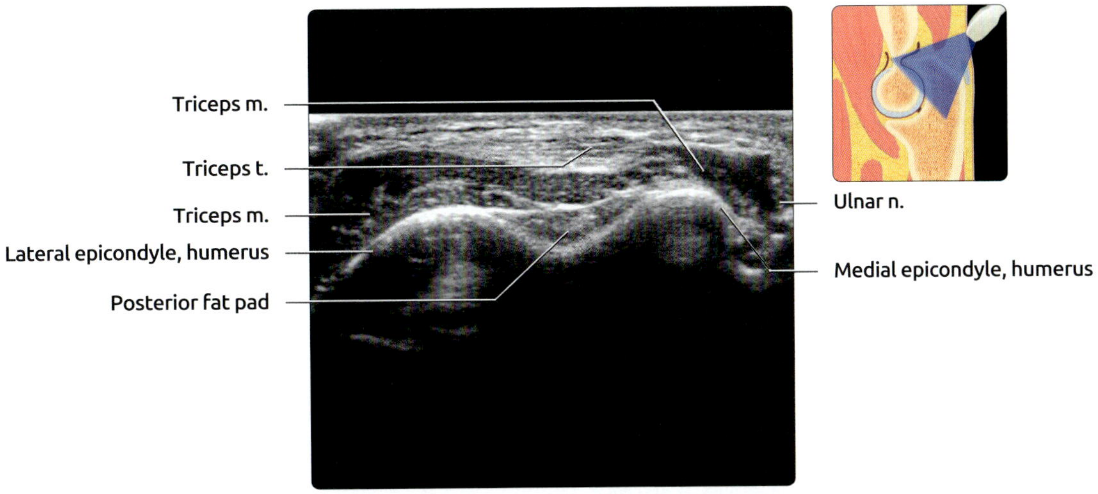

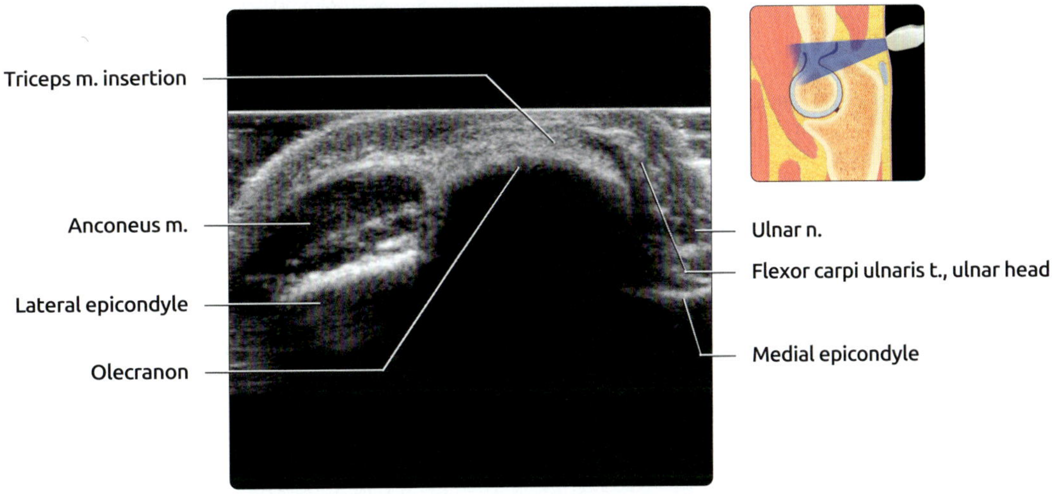

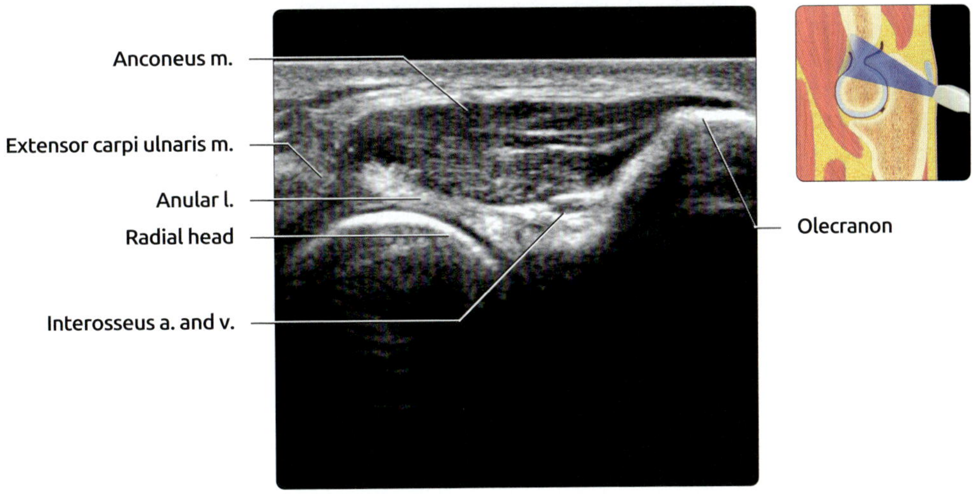

(Top) *Transverse grayscale ultrasound shows the posterior elbow. There are 3 fat pads in the elbow: 2 anterior fat pads (radial and coronoid fossa) and 1 posterior fat pad (olecranon fossa). The synovium extends from the articular surface of the humerus and lines these fossae, i.e., it lies deep to the fat pads, which are enclosed by the elbow capsule. The fat pads therefore are intraarticular but extrasynovial. Fluid distension of the synovial cavity elevates the fat pads.* **(Middle)** *Transverse grayscale ultrasound shows the elbow at the level of the olecranon. The anconeus is a small muscle on the posterior aspect of the elbow that extends from the lateral epicondyle to the olecranon.* **(Bottom)** *Transverse grayscale ultrasound shows the proximal radioulnar joint. The annular ligament is attached to the anterior and posterior margins of the radial notch of the ulna.*

IMAGING ANATOMY

Flexors

- Deep flexor group
  - Flexor digitorum profundus
    - Origin: Proximal ulna
    - Insertion: Index, middle, ring, and little finger distal phalangeal bases
  - Flexor pollicis longus
    - Origin: Radius, interosseous membrane, coronoid process
    - Insertion: Thumb distal phalangeal base
  - Pronator quadratus
    - Origin: Medial distal volar ulna
    - Insertion: Lateral distal dorsal radius
- Superficial
  - Flexor carpi radialis (FCR)
    - Origin: Medial epicondyle/common flexor tendon
    - Insertion: 2nd metacarpal (MC) base; slip to 3rd MC base
  - Palmaris longus (PL)
    - Origin: Medial epicondyle/common flexor tendon
    - Insertion: Volar distal flexor retinaculum and palmar aponeurosis
  - Flexor carpi ulnaris (FCU)
    - Origin: Medial epicondyle (humeral head) and medial proximal ulna (ulnar head)
    - Insertion: Pisiform
  - Flexor digitorum superficialis
    - Origin: Medial epicondyle and ulnar coronoid process (humeroulnar head); volar proximal radius (radial head)
    - Insertion: Middle phalanges index, middle, ring, and little fingers

Extensors

- Deep
  - Abductor pollicis longus (APL)
    - Origin: Dorsal lateral ulna, dorsal midradius
    - Insertion: 1st MC radial base
  - Extensor pollicis brevis (EPB)
    - Origin: Dorsal midradius
    - Insertion: Thumb proximal phalangeal base
  - Extensor pollicis longus (EPL)
    - Origin: Dorsal midulna
    - Insertion: Thumb distal phalanx base
    - Variant: Fused with EPB
  - Extensor indicis
    - Origin: Dorsal midulna and interosseous membrane
    - Insertion: Extensor hood of index finger
- Superficial
  - Extensor carpi radialis longus (ECRL)
    - Origin: Lateral epicondyle/common extensor tendon
    - Insertion: Dorsal radial 2nd MC
  - Extensor carpi radialis brevis (ECRB)
    - Origin: Lateral epicondyle/common extensor tendon
    - Insertion: Dorsal radial 3rd MC
  - Extensor digitorum (ED)
    - Origin: Lateral epicondyle/common extensor tendon
    - Insertion: Middle and distal phalanges of index, middle, ring, little fingers
  - Extensor digiti minimi (EDM)
    - Origin: Lateral epicondyle/common extensor tendon
    - Insertion: Extensor hood of little finger proximal phalanx with slip to ring finger
  - Extensor carpi ulnaris (ECU)
    - Origin: Common extensor tendon and dorsal ulna
    - Insertion: Dorsal ulnar 5th MC base

Anomalous Muscles

- May present as soft tissue mass
  - May create neural compression
- Accessory PL
  - Superficial to FD tendons, medial to FCR
- ED manus brevis
  - Arises from distal radius or dorsal radiocarpal ligament
  - Inserts on 2nd MC
  - May be tender or present as mass
- Extensor carpi radialis intermedius
  - Arises from humerus or as accessory slip from ECRB or ECRL
  - Inserts on 2nd &/or 3rd MC
- Extensor carpi radialis accessory
  - Arises from humerus or ECRL
  - Inserts on 1st MC bone, abductor pollicis brevis (APB), or 1st dorsal interosseous muscle

ANATOMY IMAGING ISSUES

Imaging Pitfalls

- Many variations of flexor and extensor muscles and tendons
- EPB may be absent or fused with EPL
- PL absent in 10%; short tendon with extended muscle belly may compress median nerve
- Multiple tendon slips can mimic longitudinal tendon tears (e.g., APL)
- Small amount of fluid common in tendon sheath (e.g., ECRB, ECRL, ECU)

Imaging Issues

- APL and EPB intersect ECRB and ECRL just proximal to extensor retinaculum; may impinge at musculotendinous intersection ("intersection syndrome")
- Sites of radial nerve entrapment in forearm
  - Posterior interosseous nerve within radial tunnel under arcade of Frohse
  - Posterior interosseous nerve between deep and superficial heads of supinator muscle
  - Superficial radial nerve between brachioradialis and ECRL in mid- to distal forearm
- Sites of ulnar nerve entrapment in forearm
  - Cubital tunnel
  - Beyond cubital tunnel, it passes between humeral and ulnar origins of FCU muscle
- Sites of medial nerve entrapment in forearm
  - Between heads of pronator teres

COMMON EXTENSOR TENDON

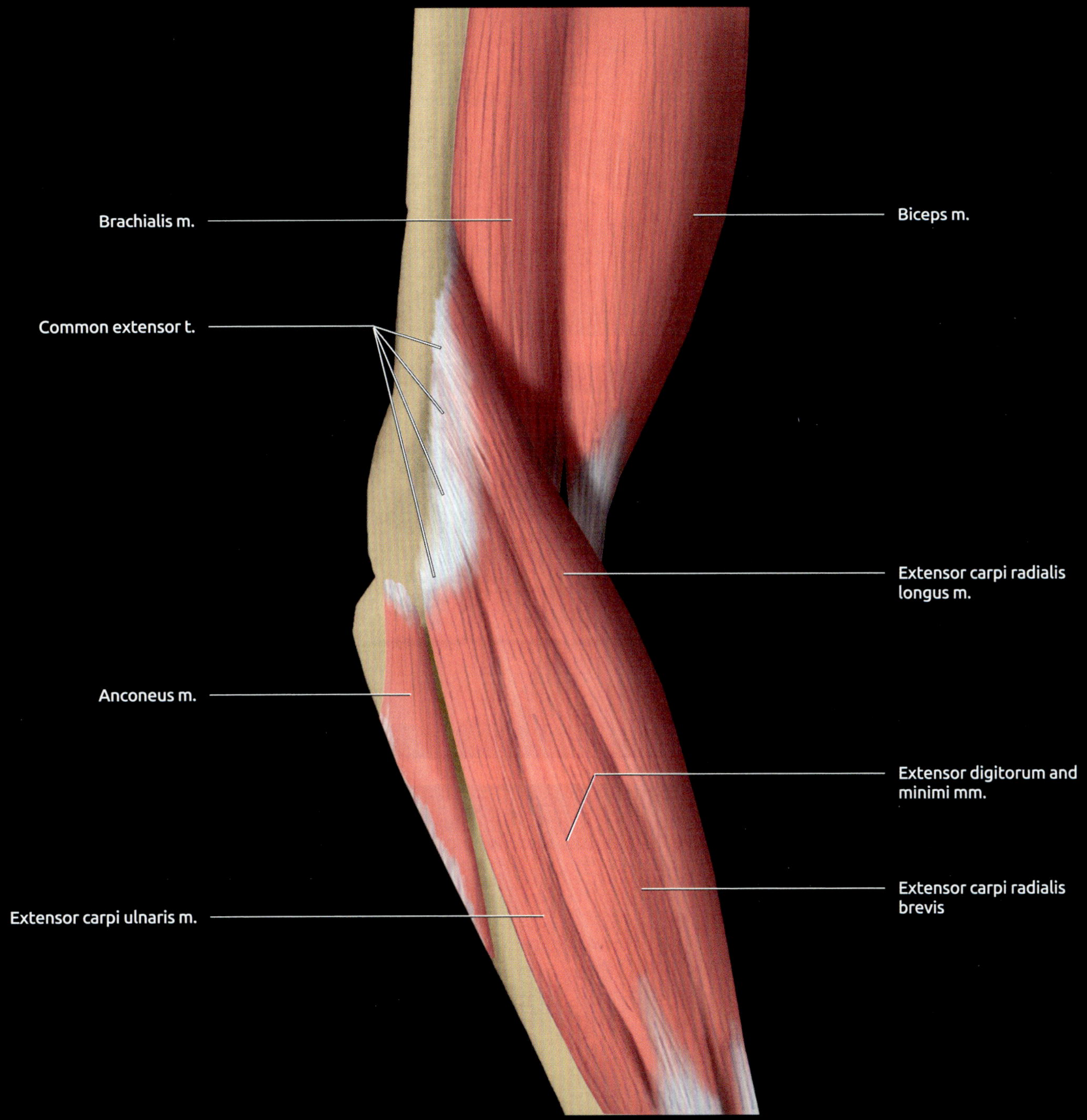

Graphic of a side view of the lateral aspect of the elbow shows the extensor-supinator group and its common extensor tendon origin attached to the lateral epicondyle and supracondylar aspect of the humerus. The common extensor tendon origin is composed of the extensor carpi radialis brevis, extensor digitorum, extensor digiti minimi, and extensor carpi ulnaris.

TRANSVERSE US, ANTERIOR FOREARM

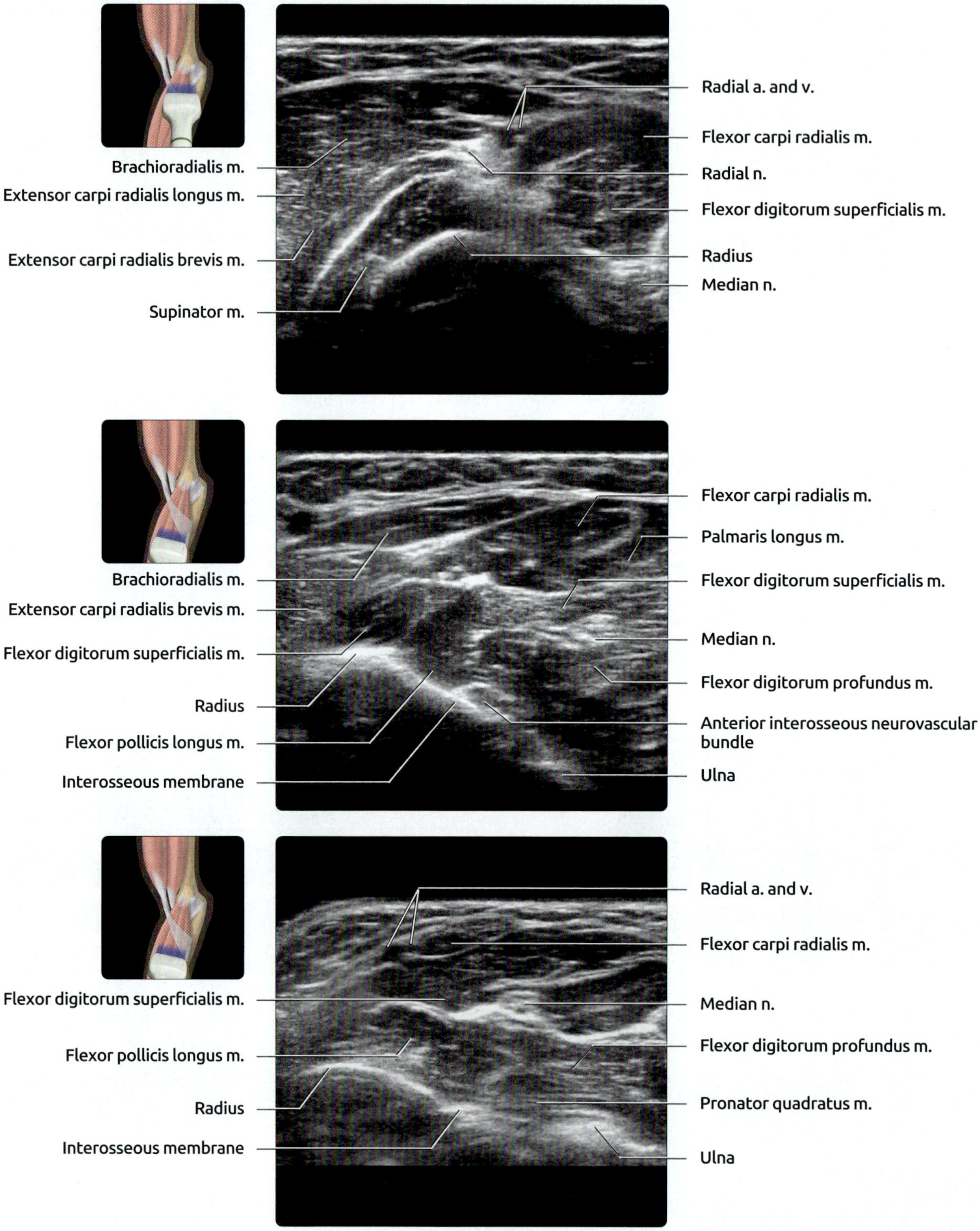

(Top) *Transverse grayscale ultrasound shows the anterior aspect of the proximal 1/3 of the forearm. The anterior aspect of the forearm consists of 5 flexor muscles: 3 superficial and 2 deep.* **(Middle)** *Transverse grayscale ultrasound shows the anterior aspect of the midforearm. The 3 superficial flexor muscles are, from ulnar to radial aspect, the flexor carpi ulnaris, flexor digitorum superficialis, and flexor carpi radialis. The 2 deep flexor muscles are flexor digitorum profundus and flexor pollicis longus.* **(Bottom)** *Transverse grayscale ultrasound shows the anterior aspect distal 1/3 of the forearm. The pronator quadratus is an easily recognized muscle that helps to pronate the forearm.*

TRANSVERSE US, ANTEROMEDIAL FOREARM

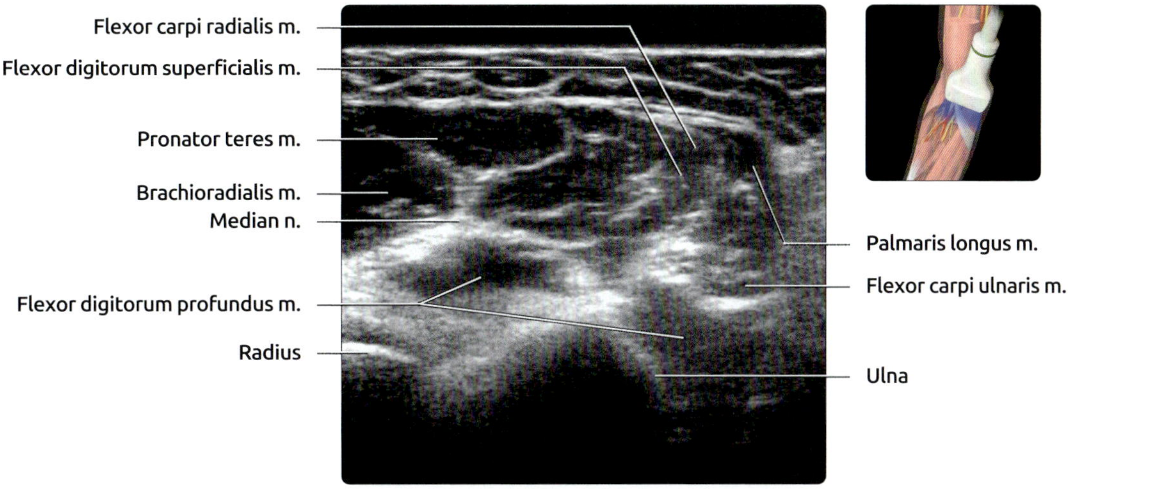

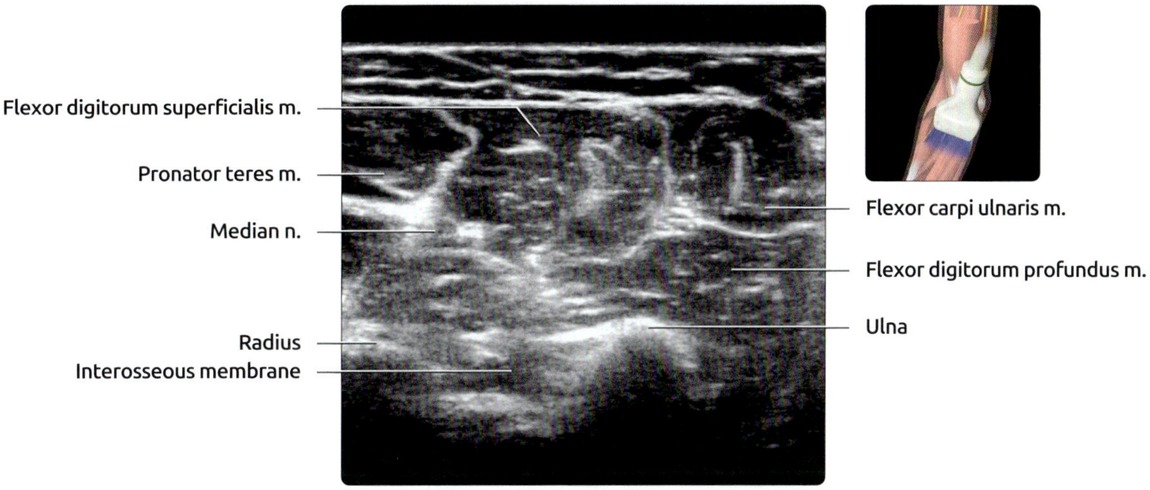

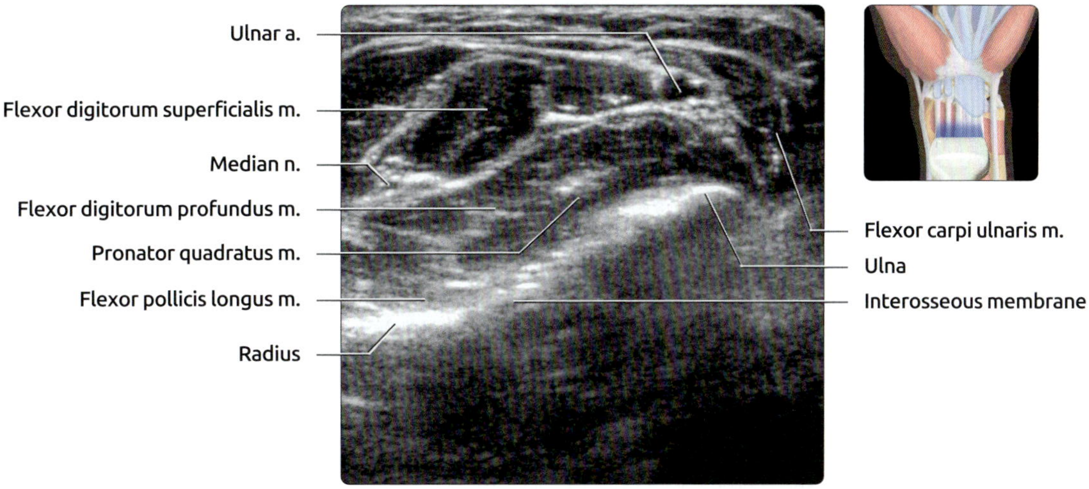

(Top) *Transverse grayscale ultrasound shows the anteromedial aspect of the proximal 1/3 of the forearm. The main flexors of the wrist are contained along the anterior and anteromedial aspect of the forearm. These are divided into 2 groups: Superficial and deep.* **(Middle)** *Transverse grayscale ultrasound shows the anteromedial aspect of the middle 1/3 of the forearm. The superficial group consists of the flexor carpi ulnaris, the flexor digitorum superficialis, and the flexor carpi radialis. The deep group consists of the flexor digitorum profundus and the flexor pollicis longus.* **(Bottom)** *Transverse grayscale ultrasound shows the anteromedial aspect of the distal 1/3 of the forearm. The ulnar nerve lies deep to the flexor carpi ulnaris, while the median nerve lies between the flexor digitorum superficialis and flexor digitorum profundus muscles.*

TRANSVERSE US, ANTEROLATERAL FOREARM

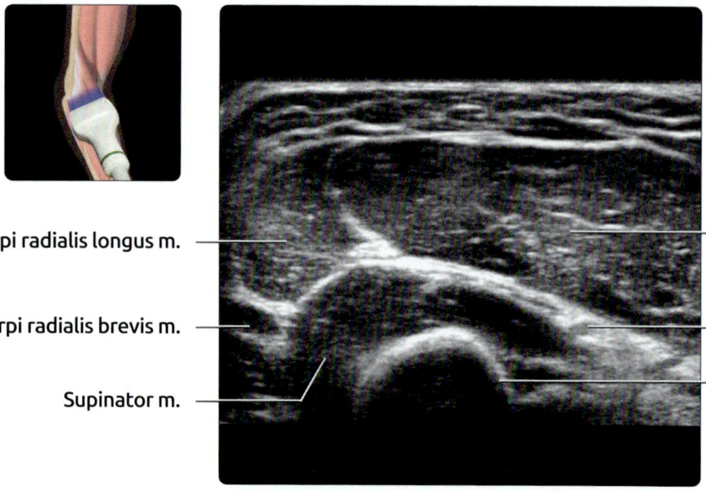

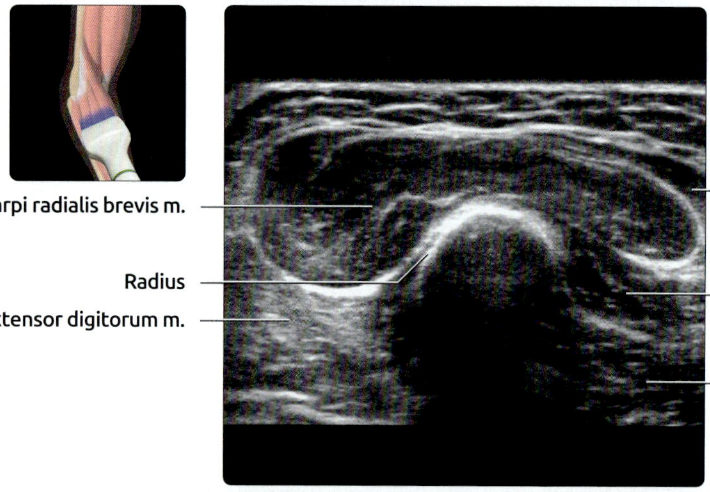

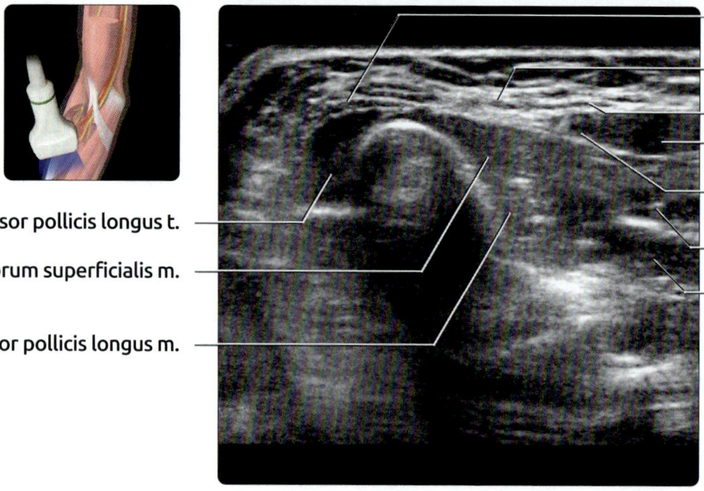

(Top) *Transverse grayscale ultrasound shows the anterolateral proximal 1/3 of the forearm. The anterolateral aspect of the forearm comprises 4 extensor muscles running along the course of the radius.* **(Middle)** *Transverse grayscale ultrasound shows the anterolateral aspect of the middle 1/3 of the forearm. The anterolateral group comprises the extensor carpi radialis brevis, extensor carpi radialis longus, brachioradialis, and the pronator teres.* **(Bottom)** *Transverse grayscale ultrasound shows the anterolateral aspect of the distal 1/3 of the forearm.*

ANTERIOR FOREARM, ARTERIES AND NERVES

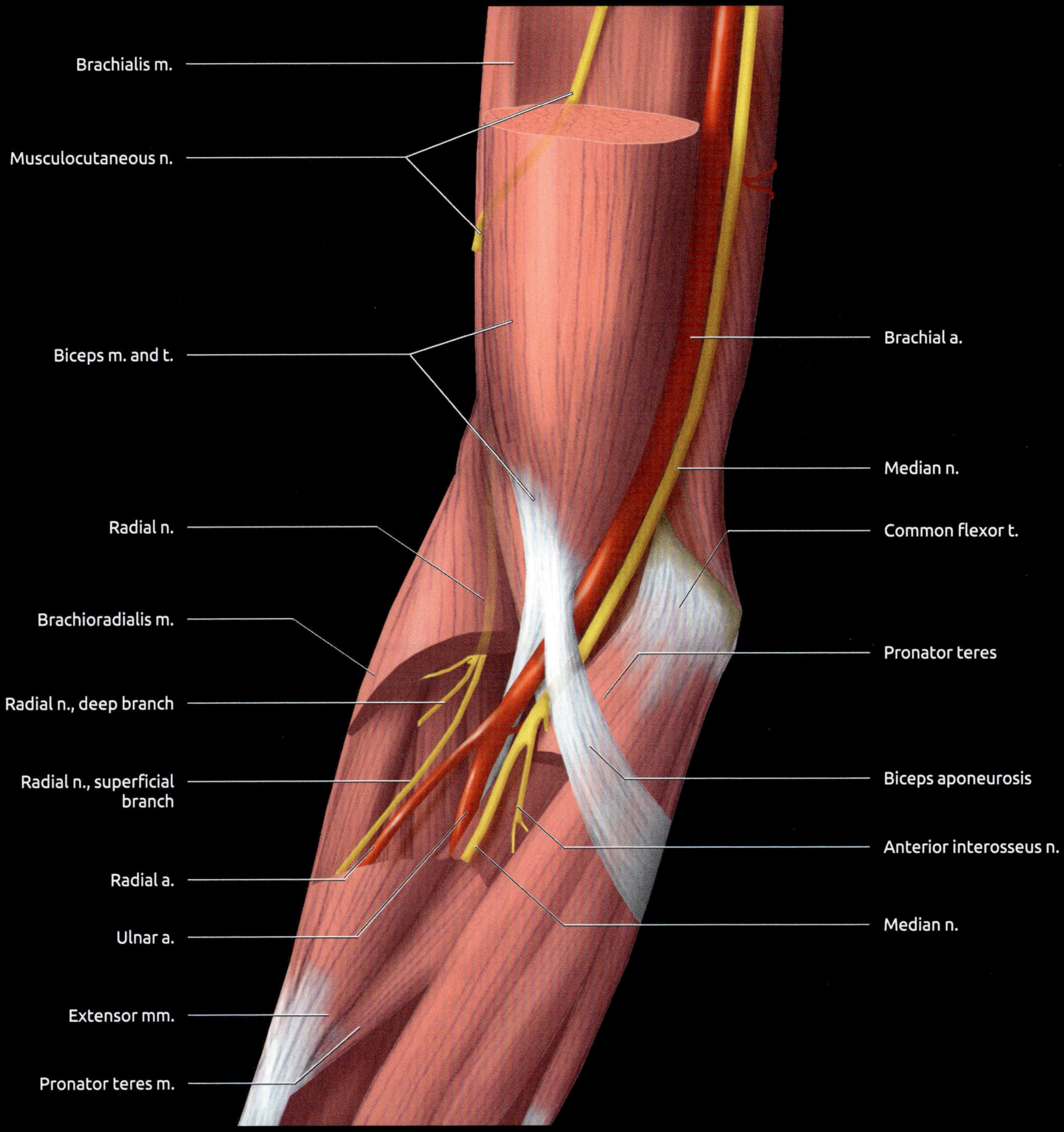

Graphic of the anterior view of the cubital fossa shows the median nerve and brachial artery passing underneath the biceps aponeurosis. The anterior interosseus nerve arises from the median nerve as the median nerve passes between the 2 heads of the pronator teres muscle. The radial nerve, deep to the brachioradialis muscle, is seen dividing into the superficial and deep branches.

COMMON FLEXOR TENDON

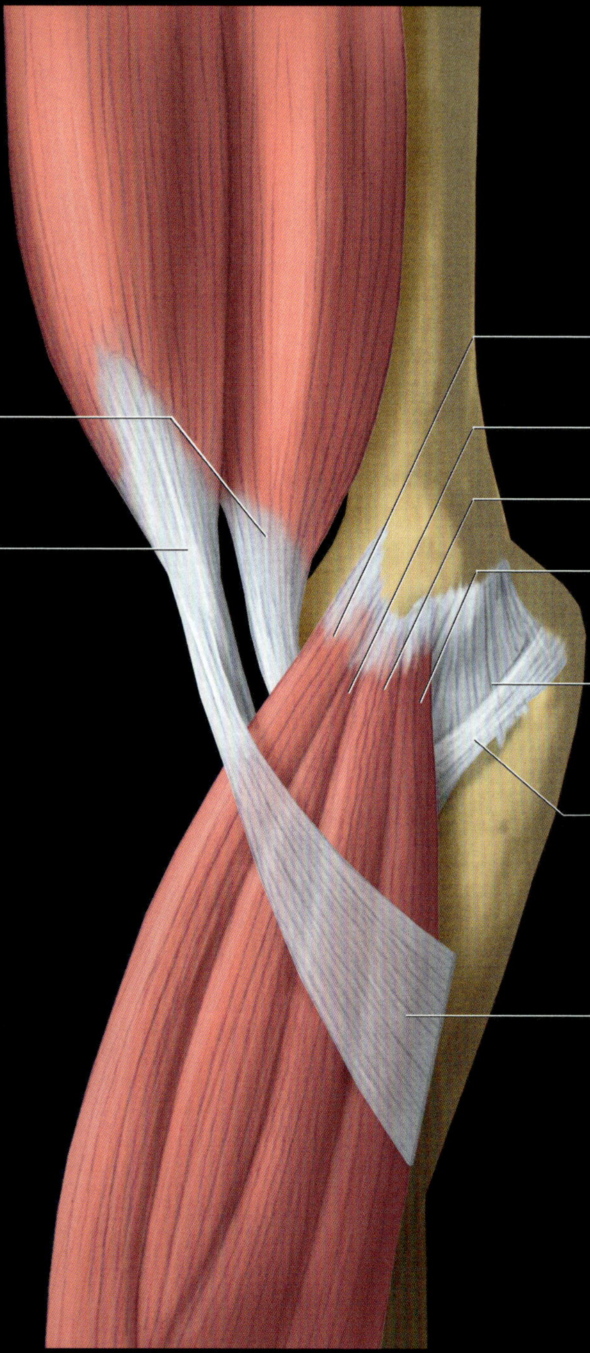

Graphic of the side view of the medial aspect of the elbow shows the flexor-pronator group and the common flexor tendon origin attaching to the medial epicondyle. The anterior band of the ulnar collateral ligament is deep to the common flexor tendon. The biceps aponeurosis blends with the anterior aspect of the common flexor muscle mass. The flexor digitorum profundus arises from the proximal ulna, posterior and deep to the flexor carpi ulnaris. It does not act on the elbow and is not part of the flexor-pronator group.

TRANSVERSE US, POSTERIOR ASPECT

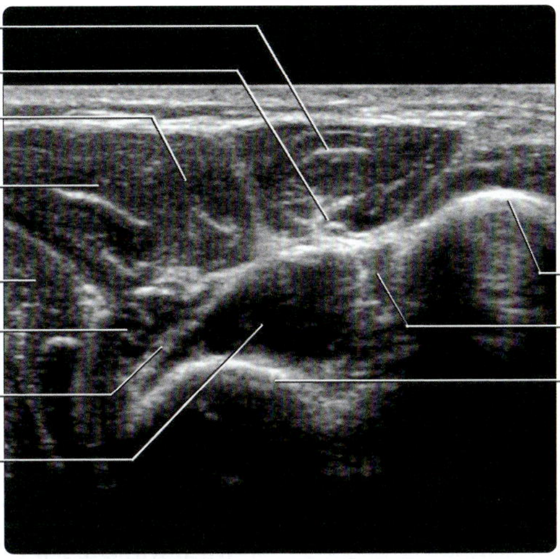

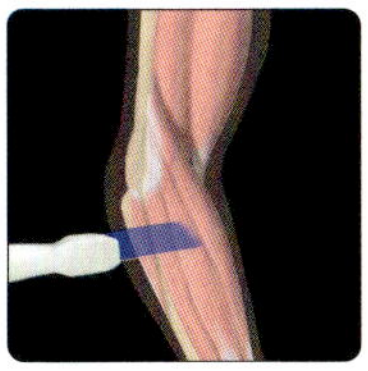

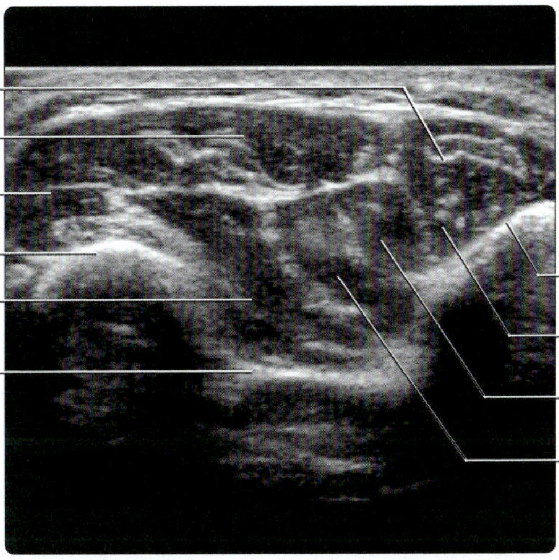

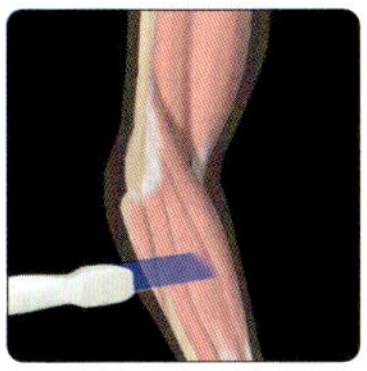

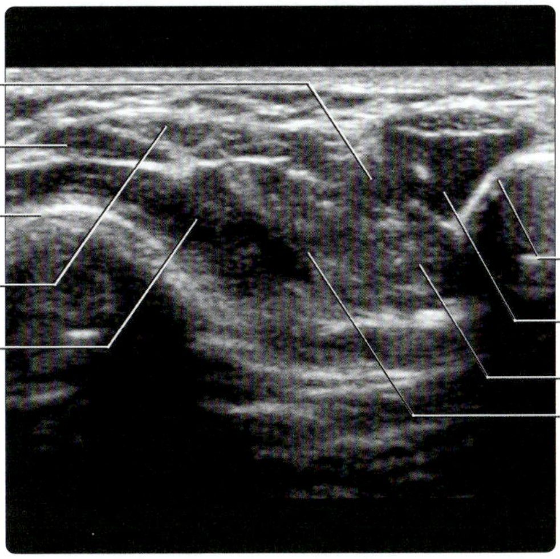

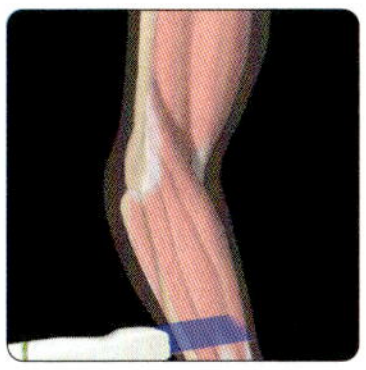

(Top) *Transverse grayscale ultrasound shows the posterior aspect of the proximal 1/3 of the forearm. The posterior aspect of the forearm contains the extensor muscles. They are contained in 2 groups: Superficial and deep.* **(Middle)** *Transverse grayscale ultrasound shows the posterior aspect of the middle 1/3 of the forearm. The superficial group comprises (ulnar to radial) the extensor carpi ulnaris, extensor digiti minimi, and extensor digitorum. The deep group comprises the extensor pollicis longus, extensor pollicis brevis, and abductor pollicis longus.* **(Bottom)** *Transverse grayscale ultrasound shows the posterior aspect of the distal 1/3 of the forearm.*

GROSS ANATOMY

Arteries

- Brachial artery
  - Continuation of axillary artery
  - Located in cubital fossa, medial to biceps tendon and deep to biceps aponeurosis
  - Accompanies median nerve
  - Gives off 2 medial branches, which descend toward elbow joint on medial side of brachial artery and contribute to anterior and posterior anastomosis around elbow
  - Brachial artery divides into radial artery and ulnar artery in cubital fossa
- Radial artery
  - Medial to distal biceps tendon
  - Covered by brachioradialis muscle
  - Distally, leaves forearm and moves laterally, crossing floor of anatomical snuffbox
  - Terminates in deep palmar arch of hand
  - Branches
    - Radial recurrent artery: Runs proximally along lateral side of elbow to form anastomosis with branches of deep brachial artery
    - Muscular branches to lateral side of forearm
    - Distal anastomotic branches: Palmar carpal arch, superficial palmar arch, and dorsal carpal arch
- Ulnar artery
  - Proximally deep to pronator teres
  - Distally lies on flexor digitorum profundus and lateral to ulnar nerve
  - Anterior and posterior ulnar recurrent arteries form anastomosis around medial side of elbow with branches of brachial artery
  - Branches
    - Common interosseus artery: Arises in distal aspect of cubital fossa, 2 cm below origin of ulnar artery from brachial artery
    - Anterior interosseus artery: Runs distally on interosseus membrane and ends in dorsal carpal arch; reaches posterior compartment by means of perforating branches; terminates at wrist joint after piercing interosseus membrane; supplies deep muscles of forearm
    - Posterior interosseus artery: Enters posterior compartment proximal to interosseus membrane, between supinator and abductor pollicis longus, and supplies posterior muscles
    - Muscular branches to medial side of forearm
    - Distal anastomotic branches: Palmar carpal arch, dorsal carpal arch

Veins

- Veins of upper limb are variable in number and position
  - In hand, there are 2 sets of veins: Deep and superficial
    - Deep veins follow arteries traveling as paired venae comitantes; radial, ulnar, and brachial veins
  - Superficial veins of hand and forearm are drained by cephalic and basilic veins
- Cephalic vein
  - Begins on lateral aspect of wrist by draining blood from venous plexus on dorsum of hand
    - Ascends along lateral side of forearm and upper arm in superficial fascia
    - As vein drains cranially, it receives tributaries from anterior and posterior aspects of forearm
    - Cephalic vein lies laterally in superficial fascia of antecubital fossa and gives off median cubital vein
    - Median vein of forearm flows into medial cubital vein
- Basilic vein
  - Begins on medial side of wrist and drains medial part of venous plexus on dorsum of hand
  - Basilic vein passes upward on medial side of forearm, receiving anterior and posterior tributaries
  - Passes to anteromedial aspect of elbow where it receives median cubital vein

ANATOMY IMAGING ISSUES

Imaging Recommendations

- Venous examination
  - Examination of upper limb veins is normally performed with patient supine and arm abducted to ~ 90°
  - Use of high-frequency linear transducer (5-10 MHz) is recommended
    - Determination of compressibility and flow pattern is normally performed
    - Phasicity with respiration noted
  - Augmentation of flow is obtained by manual compression of forearm
  - Alternatively, augmentation can be achieved by asking patient to clench fist
- Arterial examination
  - In patients with possible arterial compression syndromes, arteries are examined in various degrees of abduction so that any compression may be accentuated
  - If angle of insonation is > 60°, Doppler and color sensitivity ↓
  - Keep Doppler gate in center of lumen for peak flow velocity
  - Match color signal scale with flow velocity

Imaging Pitfalls

- Gold standard for assessment of accuracy of arterial Doppler ultrasound is usually arteriography

CLINICAL IMPLICATIONS

Clinical Importance

- Brachial artery in antecubital fossa may be used for arterial access for angiography
- High takeoff of radial artery from brachial artery is common anatomical variant
  - On arterial Doppler studies, if either radial or ulnar artery is difficult to trace distally from elbow, then artery should be sought at wrist and followed proximally
- Hemodialysis arteriovenous fistulae are generally constructed by joining cephalic veins to radial artery at wrist

FOREARM ARTERIES

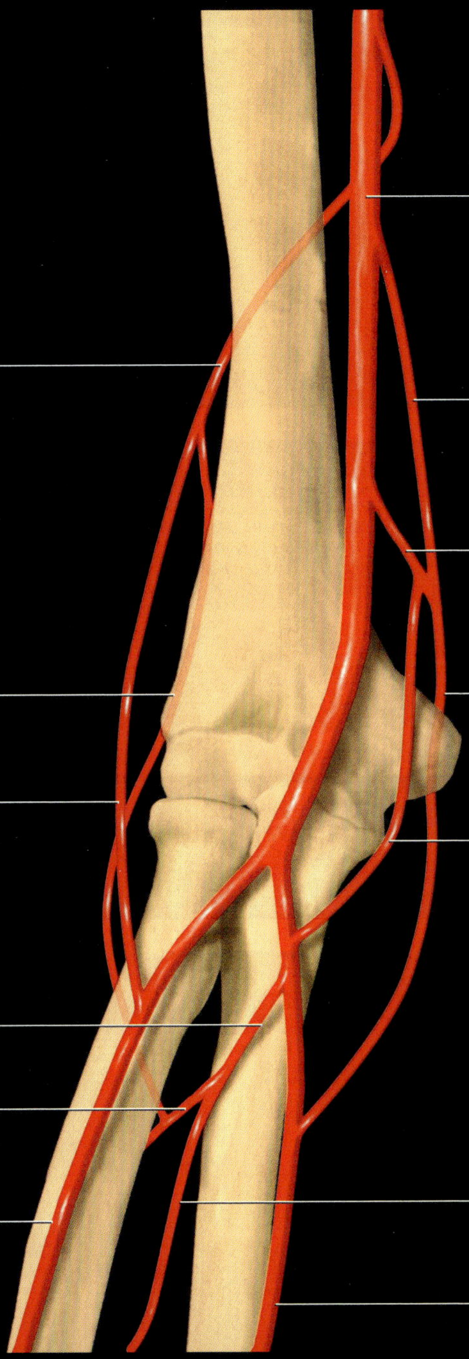

Graphic shows the brachial artery, the major artery of the arm. It divides into the radial and ulnar arteries in the proximal forearm.

TRANSVERSE US, PROXIMAL FOREARM

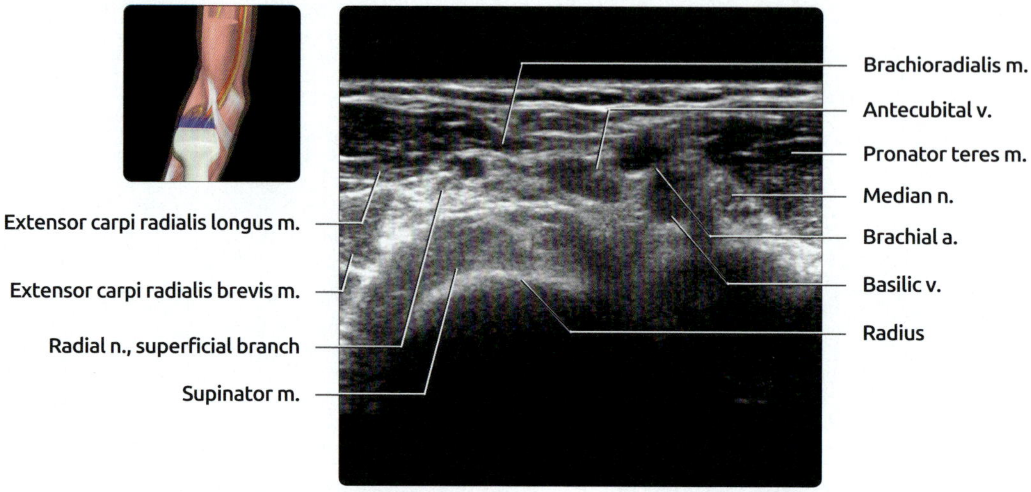

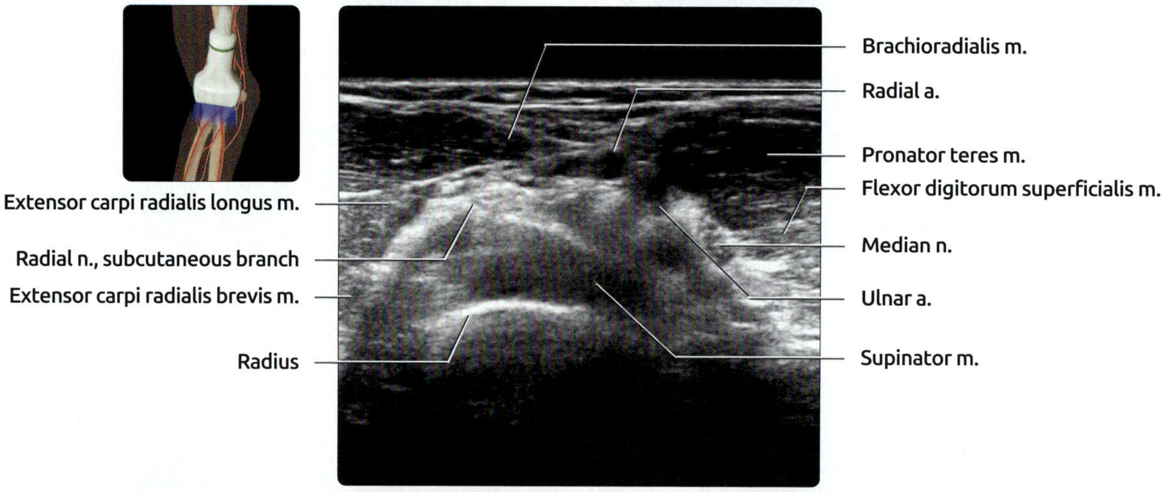

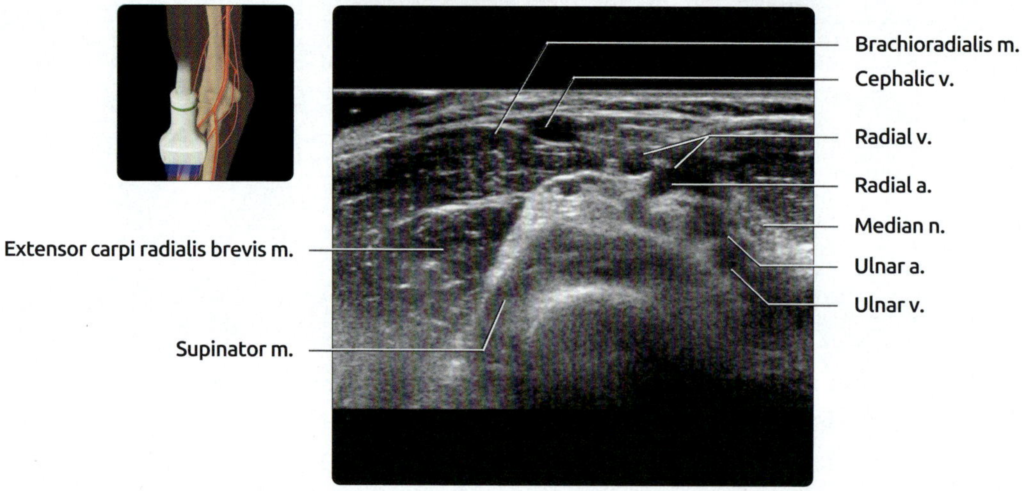

(Top) *Transverse grayscale ultrasound shows forearm vessels at the volar aspect of the proximal forearm. The brachial artery divides into the radial and ulnar arteries in the proximal arm. The forearm veins are very variable in number and appearance. Both main veins should be assessed when investigating deep vein thrombosis.* **(Middle)** *Transverse grayscale ultrasound shows radial vessels on the volar aspect of proximal forearm just distal to bifurcation of the brachial artery.* **(Bottom)** *Transverse grayscale ultrasound shows radial vessels on the volar aspect of midforearm. The superficial branch of the radial nerve roughly follows the radial artery, while the ulnar nerve closely follows the ulnar artery.*

TRANSVERSE US, RADIAL ARTERY

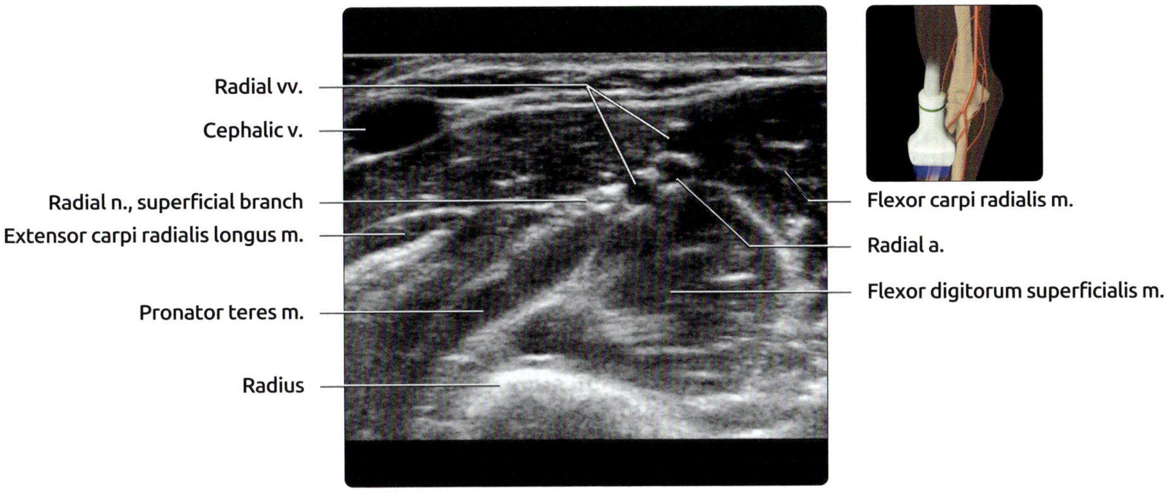

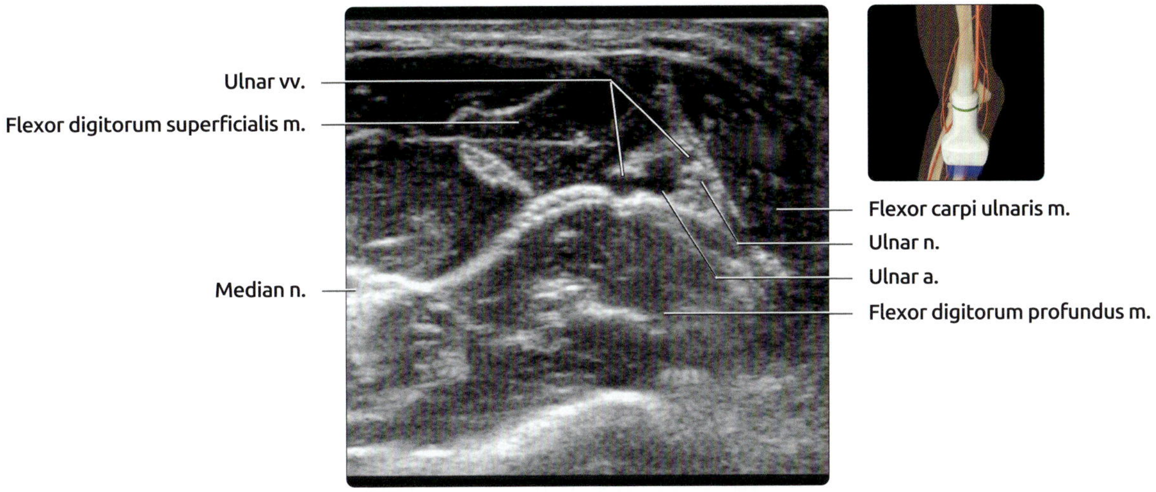

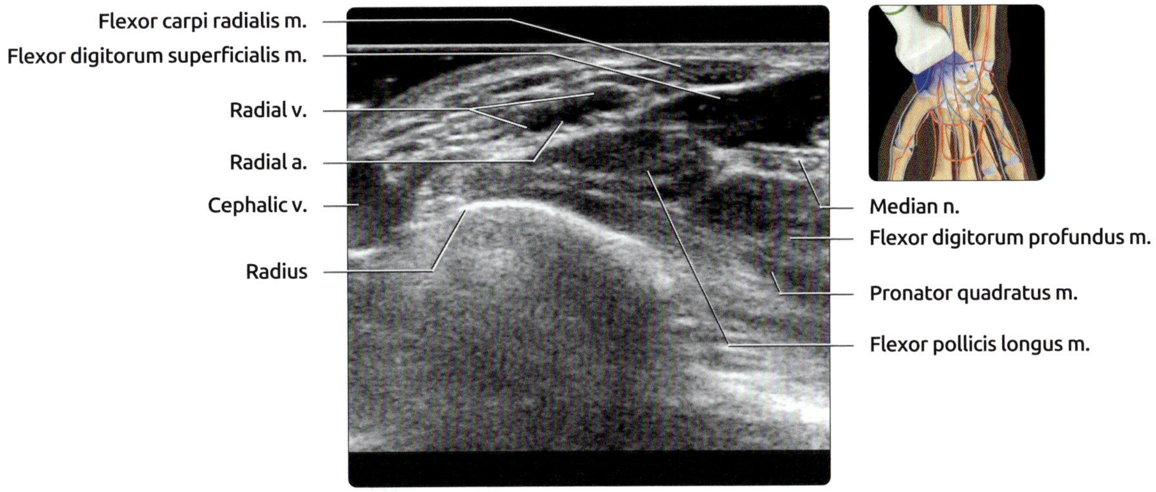

(Top) *Transverse grayscale ultrasound shows the radial artery on the volar aspect of the midforearm. The radial artery serves as a landmark to divide the anterior (flexor) and posterior (extensor) compartments of the forearm.* **(Middle)** *Transverse grayscale ultrasound shows the ulnar artery on the volar aspect of the midforearm. The ulnar artery is more variable in position than the radial artery. It may arise more proximal in the arm region. When the ulnar artery arises proximally, it often has an aberrant course and may lie superficial to the flexor muscles in the forearm.* **(Bottom)** *Transverse grayscale ultrasound shows the radial artery on the volar aspect of distal forearm. The radial artery is sometimes used instead of the long saphenous vein in coronary artery surgery. The caliber of the artery can be checked prior to surgery by ultrasound.*

TRANSVERSE AND LONGITUDINAL US, ULNAR ARTERY

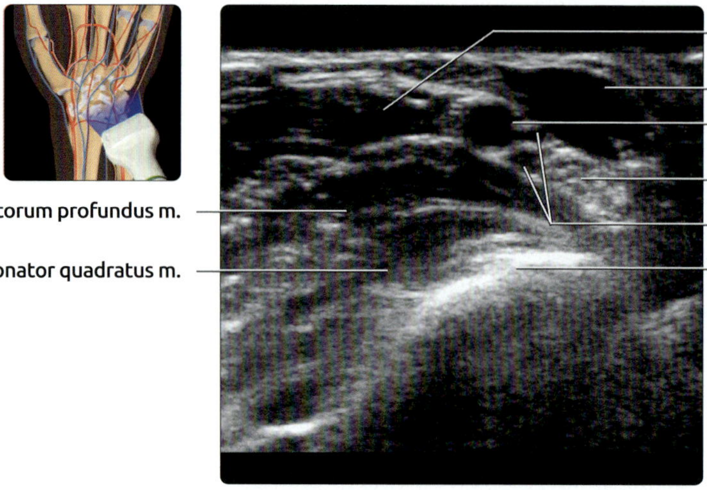

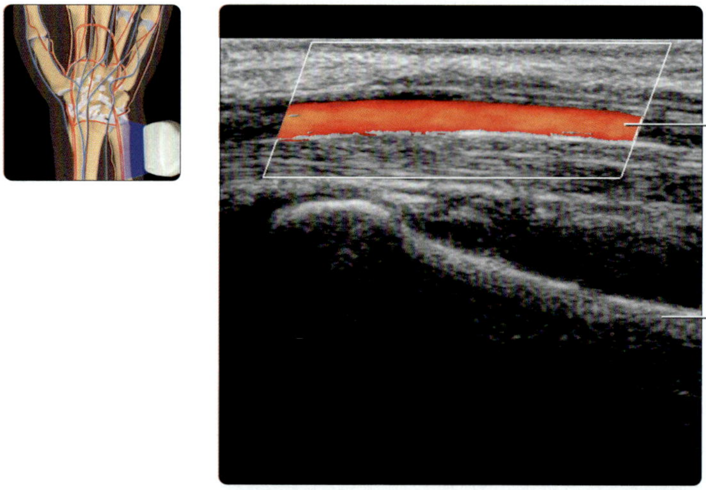

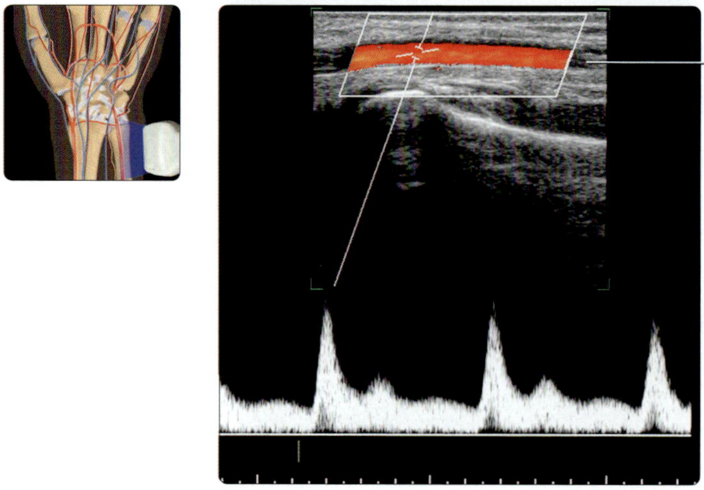

(Top) *Transverse grayscale ultrasound shows the ulnar artery on the volar aspect of the distal forearm. The Allen test is performed clinically to check dual arterial supply to the hand prior to radial artery cannulation or excision.* **(Middle)** *Longitudinal color Doppler ultrasound shows the ulnar artery on the volar aspect of distal forearm. Flow velocity and direction is denoted by the color signal. The normal diameter of the ulnar artery (1.9-2.2 mm) is measured near the wrist.* **(Bottom)** *Longitudinal spectral Doppler analysis of the ulnar artery on the volar aspect of the distal forearm is shown. The spectral gate is placed in the middle of the arterial lumen. Normal unidirectional flow is present with a systolic peak and good diastolic flow. The peak systolic velocity of the ulnar artery ranges between 40-90 cm/sec, while end-diastolic velocity ranges between 5-12 cm/sec. Pulsatility index is variable, ranging 3.9-5.0.*

ARTERIES AT DISTAL FOREARM AND HAND

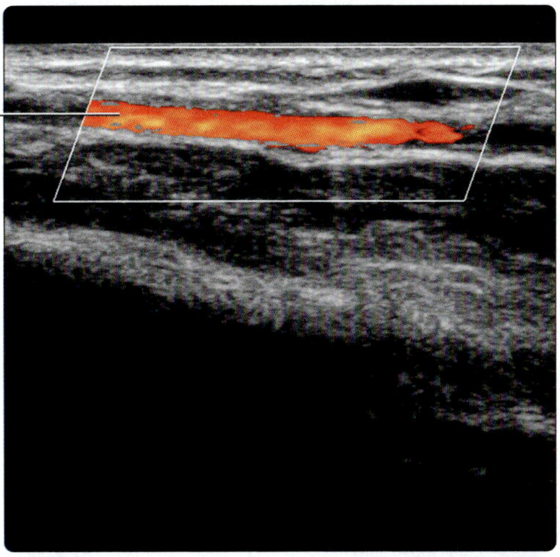

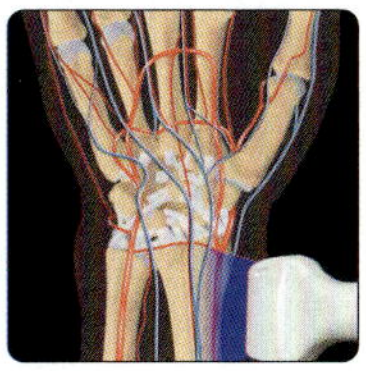

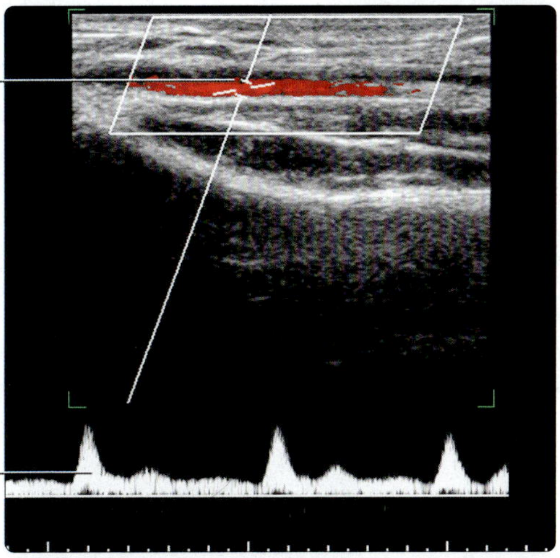

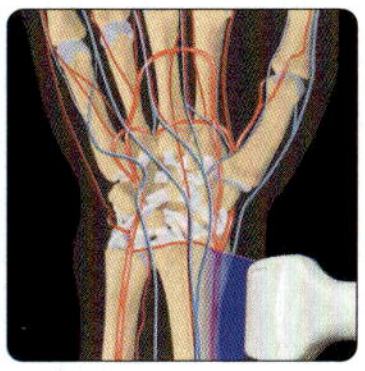

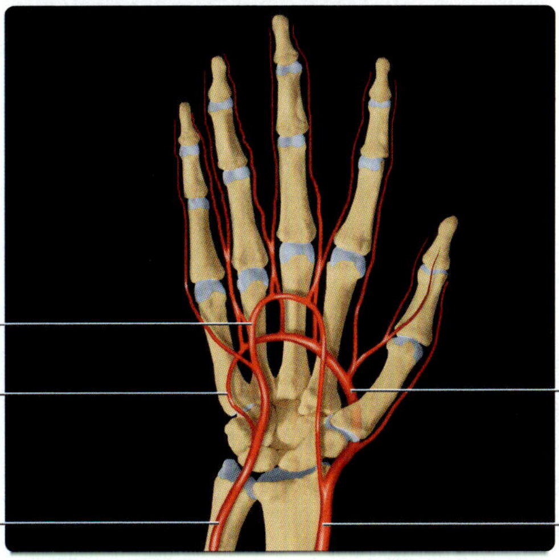

(Top) *Longitudinal color Doppler ultrasound shows the radial artery on the volar aspect of the distal forearm just proximal to the wrist. The normal diameter of the radial artery just proximal to the wrist ranges 2.2-2.8 mm.* **(Middle)** *Spectral analysis shows the radial artery just proximal to the wrist. Spectral analysis is best performed with the artery visualized longitudinally rather then transversely. The normal angle corrected peak systolic velocity of the radial artery ranges 40-90 cm/sec, the end diastolic velocity ranges 5-10 cm/sec, and pulsatility index is variable, ranging 3.9-5.0.* **(Bottom)** *Graphic shows the distal aspect of the radial and ulnar arteries in the forearm and their terminal course in the wrist and hand to form the superficial and deep palmar arterial arches.*

VENOUS SYSTEM, ANTERIOR

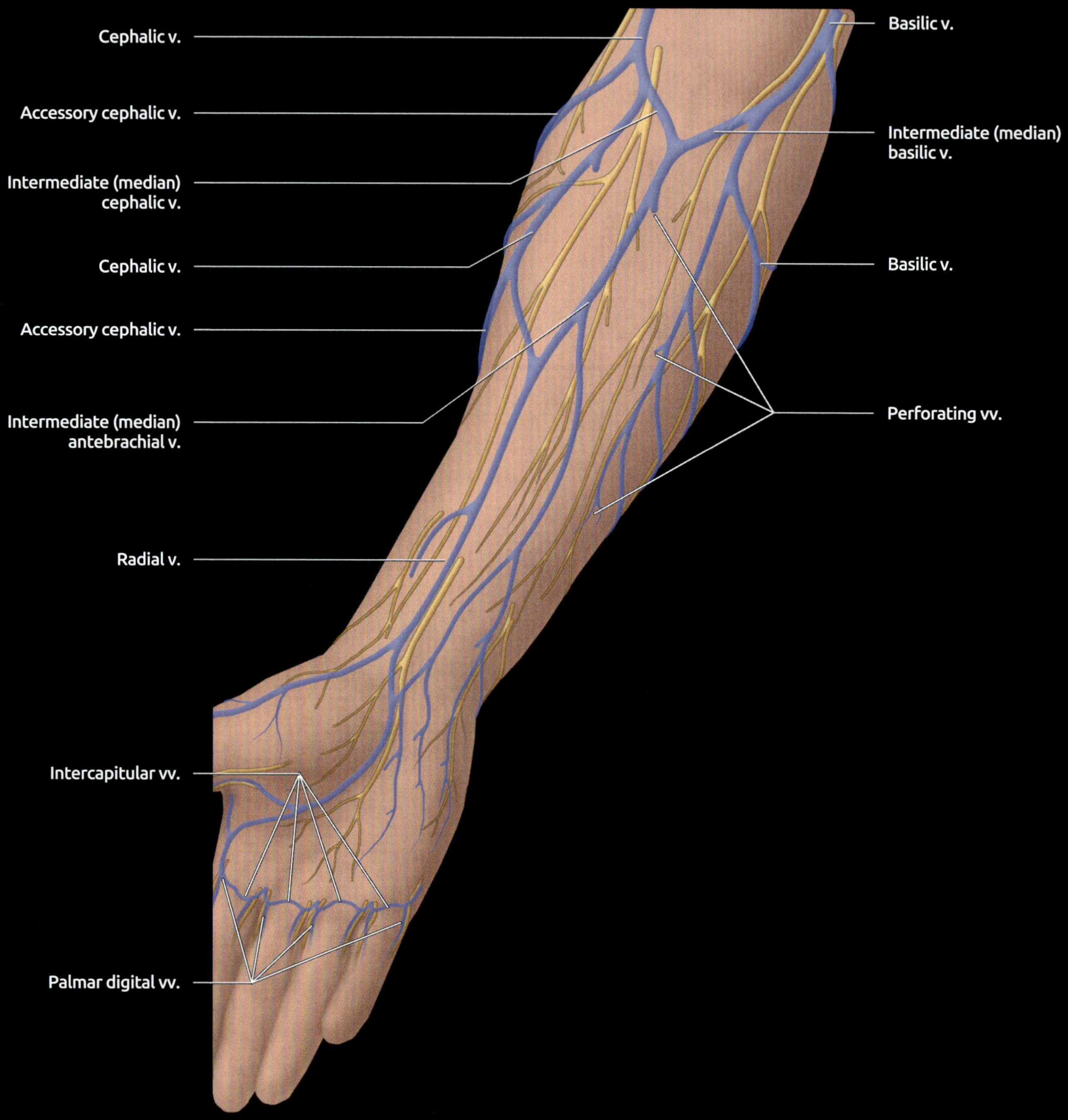

Graphic shows the superficial venous system of the forearm. The most important collecting vein of the forearm and arm is the cephalic vein. The cephalic vein proceeds from the radial margin of the dorsal venous network of the hand and follows a course diagonally above the radial margin of the forearm to the cubital fossa. From there, it continues laterally in the distal arm, moving anteromedially in the upper arm to enter the axillary vein just below the clavicle. It contains 6-10 valves. The ulnar and radial veins collect blood from the deep palmar arch and course in pairs alongside the corresponding arteries. Blood is continuously shunted from these superficial veins in the subcutaneous tissue to deep veins via the perforating veins. Although the upper extremities have many communicating veins, they have no clinical significance and are therefore not specifically named and located.

VENOUS SYSTEM, POSTERIOR

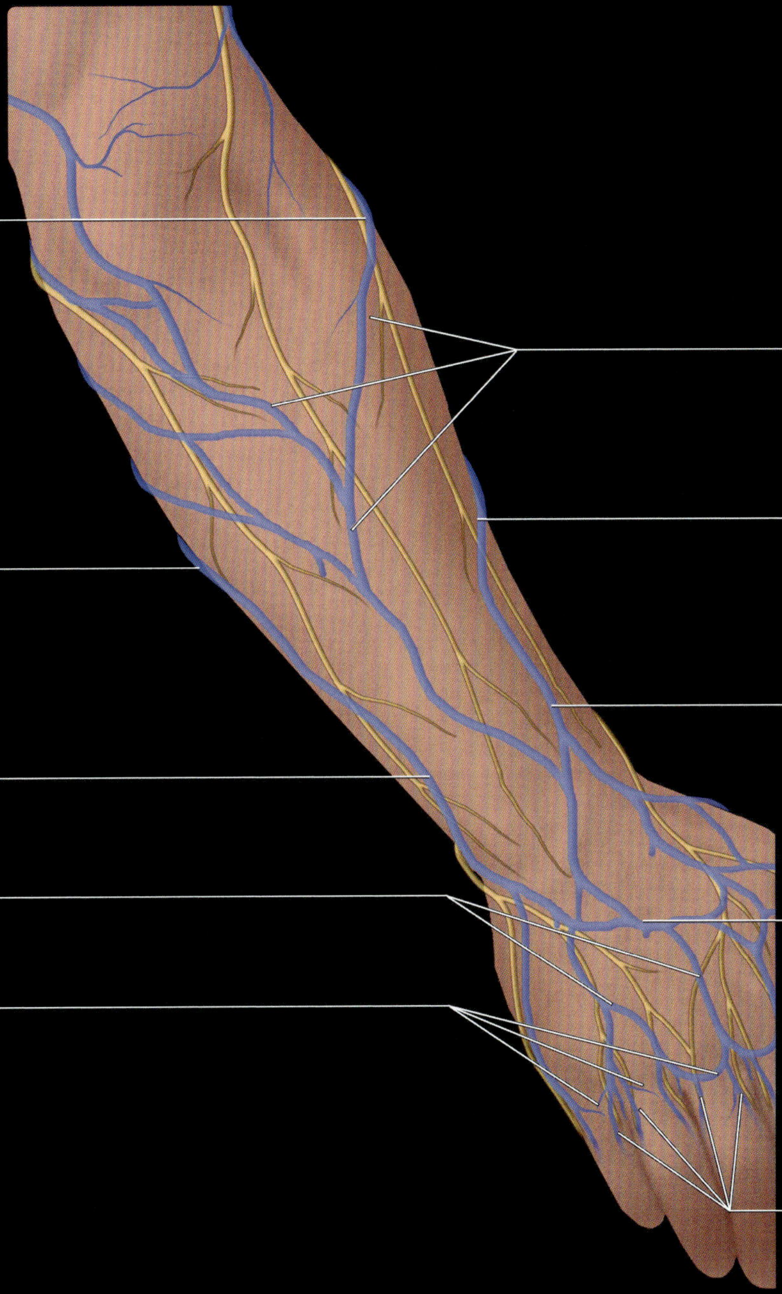

Graphic shows the posterior view of the superficial venous system of the forearm. The basilic vein starts from the ulnar margin of the dorsal venous plexus and travels onward to the bend of the elbow, where it is joined by the cephalic vein. It contains 4-8 valves. There are 2 additional small branches that pass superficially on the volar side of the forearm, the median antebrachial vein and the median cubital vein, which flow into the cephalic vein and the basilic vein somewhat above the bend of the elbow. Although the upper extremities have many communicating veins, they have no clinical significance and are therefore not specifically named and located.

TERMINOLOGY

Definitions

- Articulation complex comprising distal radioulnar, radiocarpal, ulnocarpal, pisotriquetral, midcarpal, and carpometacarpal joints

IMAGING ANATOMY

Osseous Structures

- Distal radius: Lister tubercle on dorsal surface
- Distal ulna: Ulnar variance refers to length of ulnar head relative to distal radius: Ulnar minus or ulnar plus
- Proximal carpal row: Scaphoid, lunate, triquetrum, pisiform
  - Scaphoid: Proximal and distal pole separated by waist
    - Tuberosity = volar prominence of distal pole
  - Lunate: 1/2 moon-shaped
  - Triquetrum: Triangular-shaped
  - Pisiform: Pea-shaped, sesamoid-type bone to which flexor carpi ulnaris attaches and continues distally as pisohamate and pisometacarpal ligaments
- Distal carpal row: Trapezium, trapezoid, capitate, hamate
  - Trapezium: Saddle-shaped bone linking carpus and thumb
  - Trapezoid: Wedge-shaped bone
  - Capitate: Head (proximal), neck (midportion), and body (bulky distal portion)
  - Hamate: Hook (hamulus) arises from palmar surface

Ligaments

- Extrinsic (on palmar or volar aspects of wrist) or intrinsic (between carpal bones)
- Major wrist stabilizers: Volar ligaments
- Extrinsic ligaments
  - Palmar: Radioscaphocapitate, radiolunotriquetral, radioscapholunate, ulnotriquetral, ulnolunate, scaphotriquetral
  - Dorsal: Scaphotriquetral, radiotriquetral, ulnotriquetral, radial collateral
- Intrinsic ligaments
  - Proximal interosseous: Scapholunate, lunotriquetral
  - Distal interosseous: Trapeziotrapezoid, trapeziocapitate, capitohamate

Muscles and Tendons

- **Flexors, deep**
  - Flexor digitorum profundus: Originated from ulna; inserted to index, middle, ring, and little finger distal phalangeal bases
  - Flexor pollicis longus: Originated from radius, interosseous membrane, and coronoid process of ulna; inserted to distal phalangeal base of thumb
- **Flexors, superficial**
  - Flexor carpi radialis: Originated from medial epicondyle; inserted to 2nd metacarpal base
  - Palmaris longus: Originated from medial epicondyle; inserted to palmar aponeurosis
  - Flexor carpi ulnaris: Originated from medial epicondyle and medial olecranon/proximal ulna; inserted to pisiform
  - Flexor digitorum superficialis: Originated from medial epicondyle and coronoid process of ulna and anterior radius; inserted to middle phalangeal bases of digits 2-5
- **Extensors, deep**
  - Abductor pollicis longus: Originated from ulna; inserted to radial aspect 1st metacarpal base
  - Extensor pollicis brevis: Originated from radius; inserted to proximal phalangeal base of thumb
  - Extensor pollicis longus: Originated from midulna; inserted to distal phalangeal base of thumb
  - Extensor indicis: Originated from midulna; joins ulnar side of extensor digitorum tendon inserting into 2nd digit extensor hood
- **Extensors, superficial**
  - Extensor carpi radialis longus: Originated from lateral supracondylar ridge of humerus; inserted to dorsal radial 2nd metacarpal base
  - Extensor carpi radialis brevis: Originated from lateral humeral epicondyle; inserted to dorsal radial 3rd metacarpal base
  - Extensor digitorum: Originated from lateral humeral epicondyle; inserted to distal phalangeal bases of digits of 2-5
  - Extensor digiti minimi: Originated from lateral humeral epicondyle; inserted to extensor hood of little finger
  - Extensor carpi ulnaris: Originated from lateral humeral epicondyle; inserted to 5th metacarpal base

Retinacula

- **Flexor retinaculum**
  - Also called transverse carpal ligament: Attached to pisiform and hook of hamate, scaphoid, and trapezium
- **Extensor retinaculum**
  - Attaches to ulnar styloid process, triquetrum, and pisiform medially; crosses obliquely to attach to Lister tubercle and radial styloid process laterally
    - Sends septa to radius, creating 6 compartments for extensor tendons
  - Compartment contents
    - 1st: Abductor pollicis longus and extensor pollicis brevis
    - 2nd: Extensor carpi radialis longus and brevis
    - 3rd: Extensor pollicis longus
    - 4th: Extensor digitorum and extensor indicis
    - 5th: Extensor digiti minimi
    - 6th: Extensor carpi ulnaris

Anatomic Spaces

- **Carpal tunnel**
  - Margins: Carpal bones (dorsal margin); flexor retinaculum (volar margin); pisiform and hook of hamate (ulnar margin); scaphoid and trapezium (radial margin); radiocarpal joint (proximal margin); and metacarpal base (distal margin)
    - Contents: Flexor digitorum superficialis, flexor digitorum profundus, flexor pollicis longus, median nerve
- **Guyon canal**
  - Margins: Fascial extension from flexor retinaculum, volar carpal ligament (volar margin), pisiform and flexor carpi ulnaris (ulnar margin), flexor retinaculum (radial and dorsal margins)
    - Contents: Ulnar artery and vein, ulnar nerve

CARPAL BONES AND WRIST COMPARTMENTS

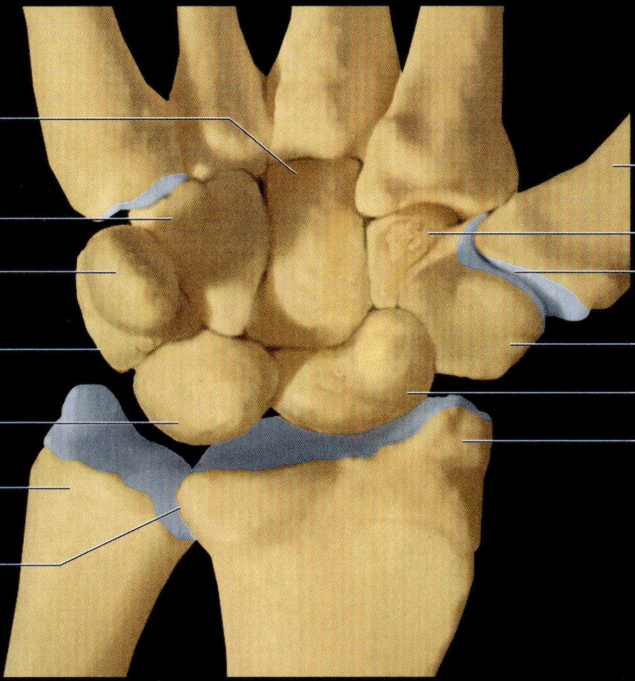

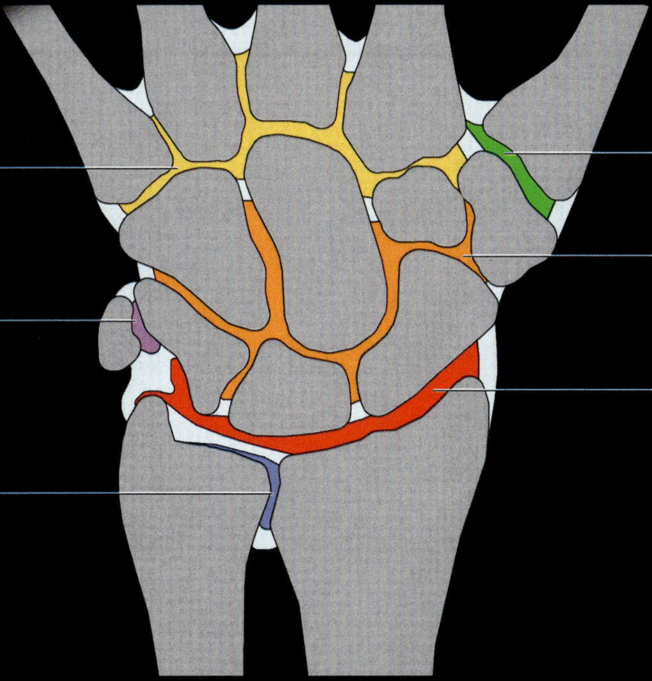

(Top) *Graphic shows the bones of the wrist joint.* **(Bottom)** *Graphic shows the 5 different wrist compartments. Distal radioulnar compartment is separated from the radiocarpal compartment by the triangular fibrocartilage complex. Pisotriquetral compartment is separated from the radiocarpal compartment in 20%. Midcarpal compartment is separated from the radiocarpal compartment by the scapholunate and lunotriquetral ligaments and typically communicates with the carpometacarpal joints. First carpometacarpal compartment is separated from the common carpometacarpal compartment by the trapeziometacarpal ligament.*

DORSAL TENDONS AND TENDON SHEATHS

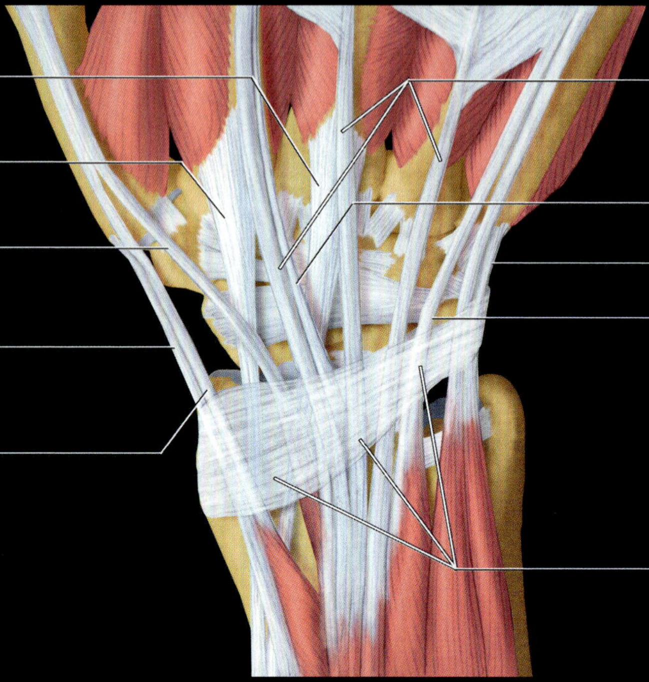

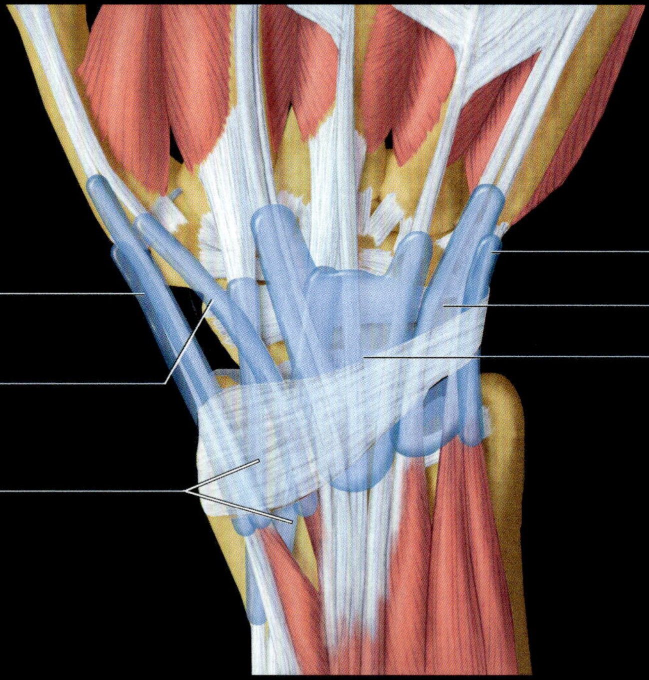

(Top) *Dorsal extensor tendons pass deep to the extensor retinaculum, separated into 6 compartments by fibrous attachments of the retinaculum to underlying bone. Compartment contents include abductor pollicis longus and extensor pollicis brevis, extensor carpi radialis longus and brevis, extensor pollicis longus, extensor digitorum and extensor indicis, extensor digiti minimi, and extensor carpi ulnaris.* **(Bottom)** *Separate tendon sheaths enclose the dorsal extensor tendons in compartments 1-6 individually.*

TENDONS: RELATIONS TO DORSAL AND VOLAR WRIST

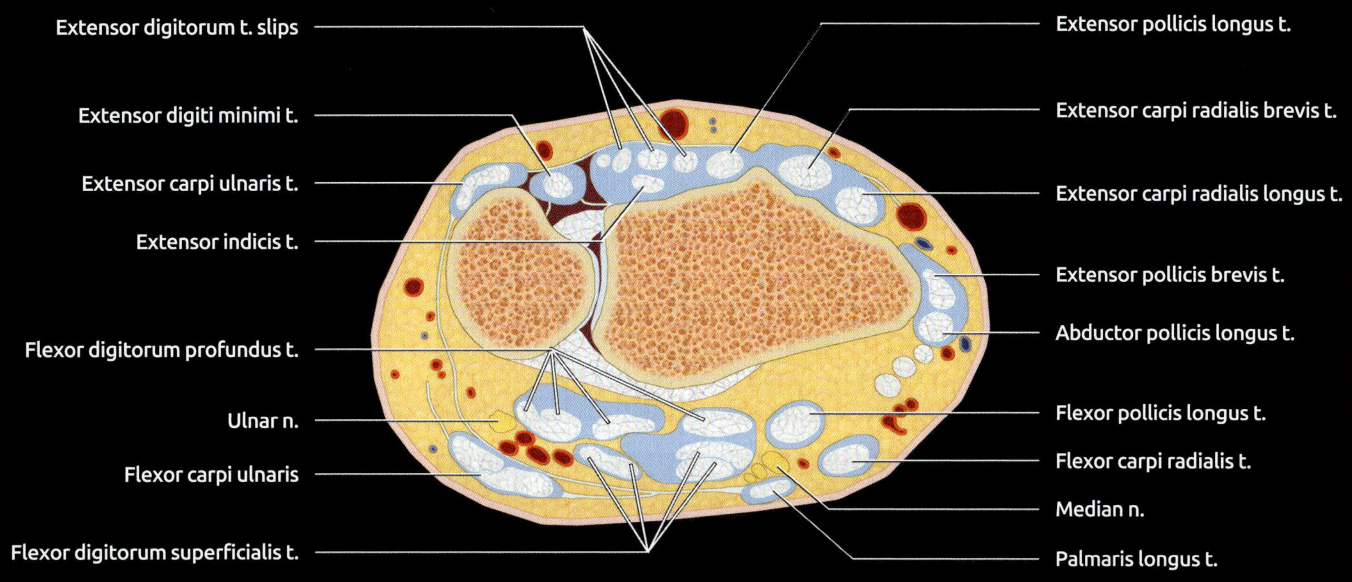

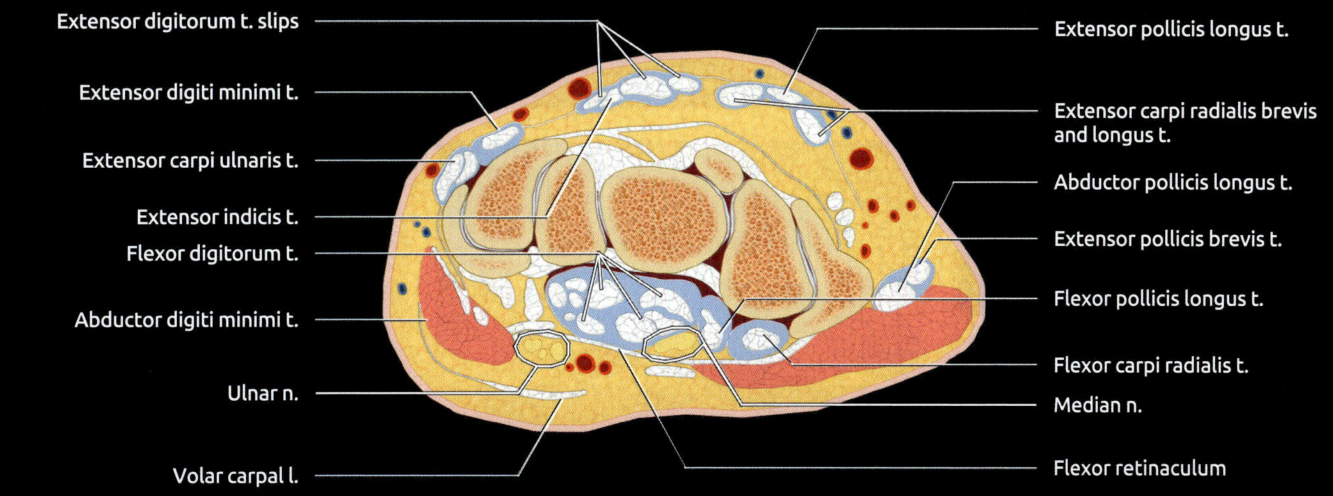

(Top) *Graphic representation shows tendons in the proximal wrist. Extensor tendons are deep to the extensor retinaculum, whereas flexor tendons are proximal to the flexor retinaculum at this level in the wrist.* **(Bottom)** *Midcarpal tunnel is shown. Median nerve is slightly flattened as it passes deep to the flexor retinaculum and remains superficial to the flexor pollicis longus. Ulnar nerve, artery, and veins lie lateral to the pisiform and may divide near the pisiform into deep and superficial branches.*

TRANSVERSE US, RADIAL ASPECT

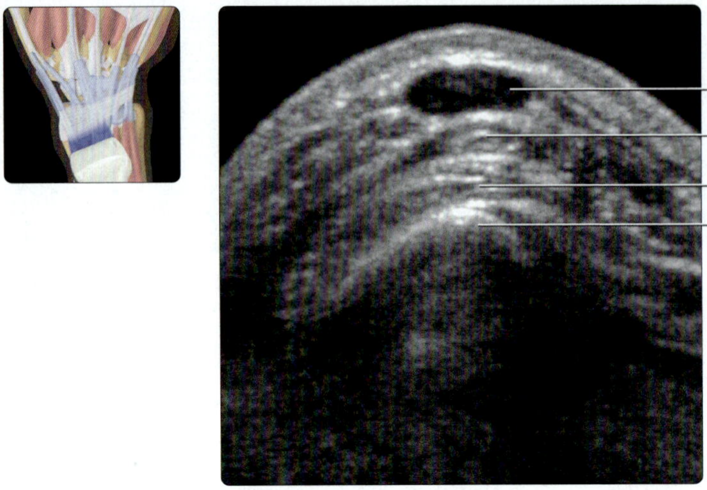

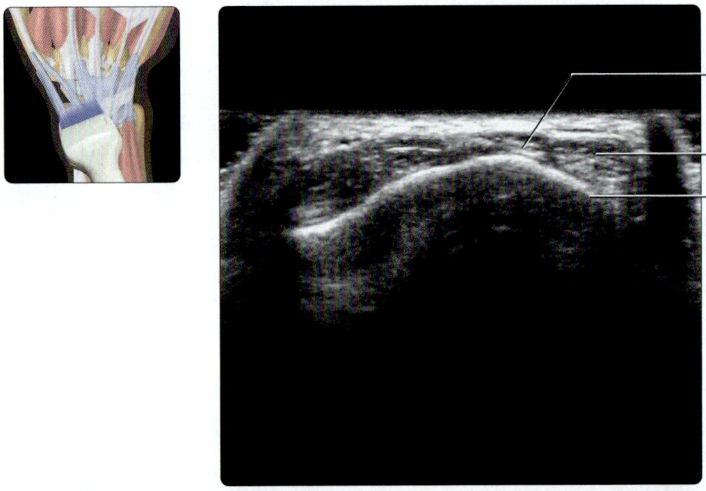

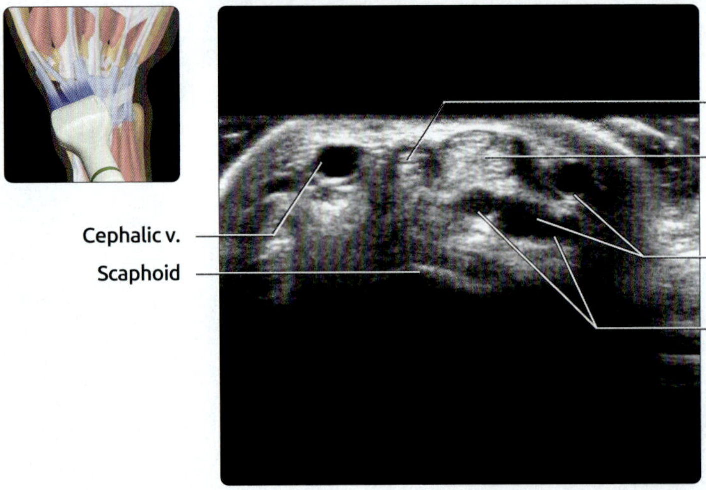

(Top) *Transverse grayscale US shows the 1st extensor compartment proximal to the distal radius. The 1st compartment contains the extensor pollicis brevis and the larger abductor pollicis longus tendons. The 1st compartment tendons run obliquely over 2nd compartment tendons a few centimeters proximal to the wrist. Pain may occur at this intersection (intersection syndrome).* **(Middle)** *Transverse grayscale US shows the 1st extensor compartment at the distal radius. Extensor pollicis brevis extends to the base of the proximal phalanx of the thumb, whereas abductor pollicis longus inserts onto the base of the 1st metacarpal bone. They form the radial aspect of the anatomical snuff box. These tendons are swollen in patients with de Quervain disease.* **(Bottom)** *Transverse grayscale US shows the 1st extensor compartment at the level of the scaphoid. Dorsal branch of the radial artery passes deep to the tendons of the 1st extensor compartment to enter the dorsum of the hand.*

TRANSVERSE US, DORSAL WRIST

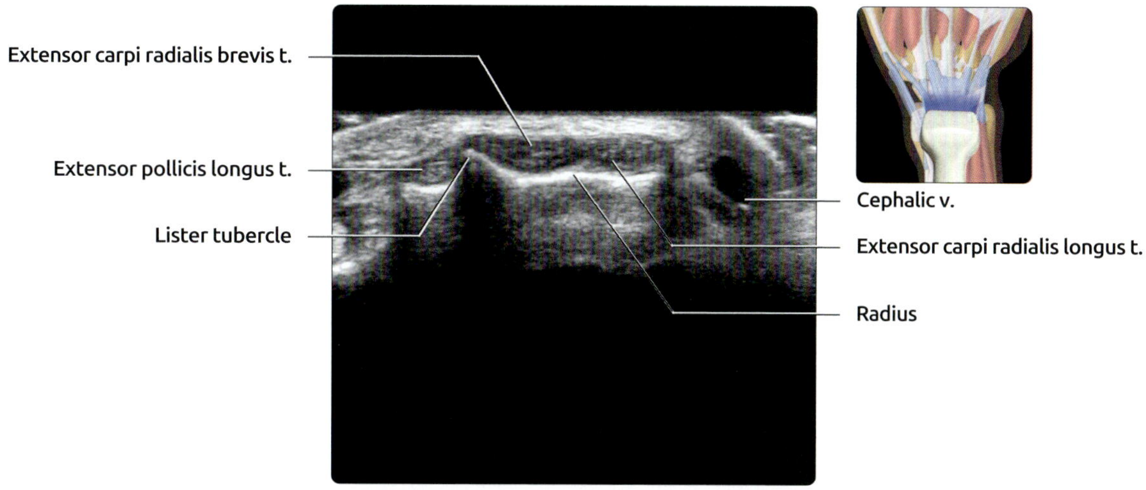

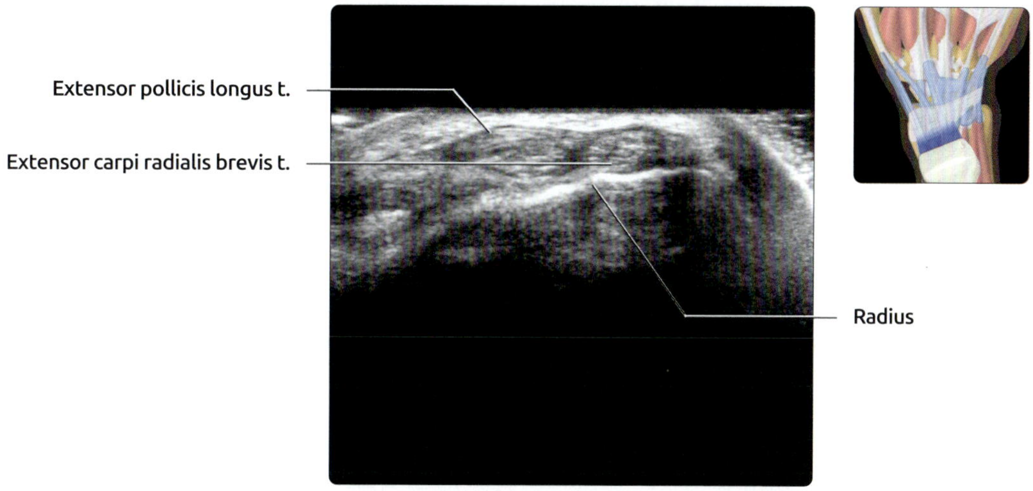

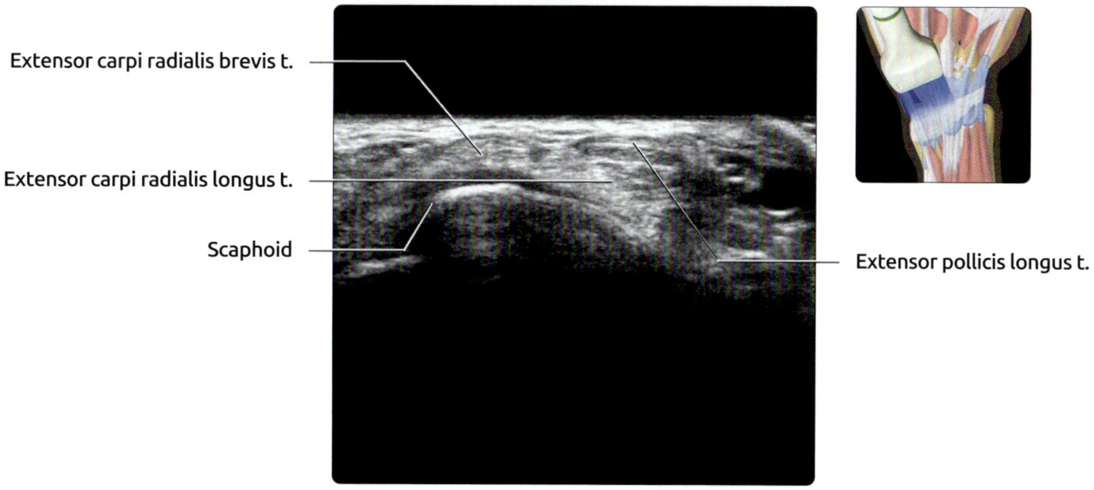

(Top) *Transverse grayscale US shows the 2nd and 3rd extensor compartments at the distal radius level. The 2nd extensor compartment comprises the extensor carpi radialis longus and extensor carpi radialis brevis. Extensor carpi radialis longus inserts into the base of the index finger metacarpal, whereas the extensor carpi radialis brevis inserts into the middle finger metacarpal. The 2nd compartment is separated from the 3rd compartment, containing the extensor pollicis longus, by the Lister tubercle.* **(Middle)** *Transverse grayscale US shows the 2nd and 3rd extensor compartments at the distal radius level. The 2nd compartment forms the dorsal aspect of the anatomical snuff box.* **(Bottom)** *Transverse grayscale US shows the 2nd extensor compartment at the scaphoid level. The extensor pollicis longus tendon hooks around the Lister tubercle and crosses superficial to the 1st compartment tendons as it runs toward its insertion on the base of the distal phalanx of the thumb. It is prone to rupture in inflammatory arthropathies and distal radial fractures.*

TRANSVERSE US, EXTENSOR DIGITORUM

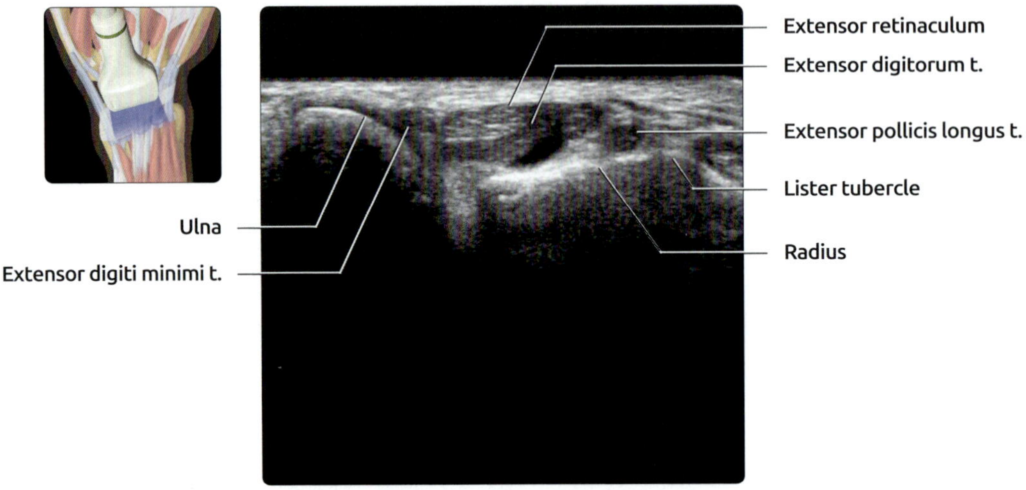

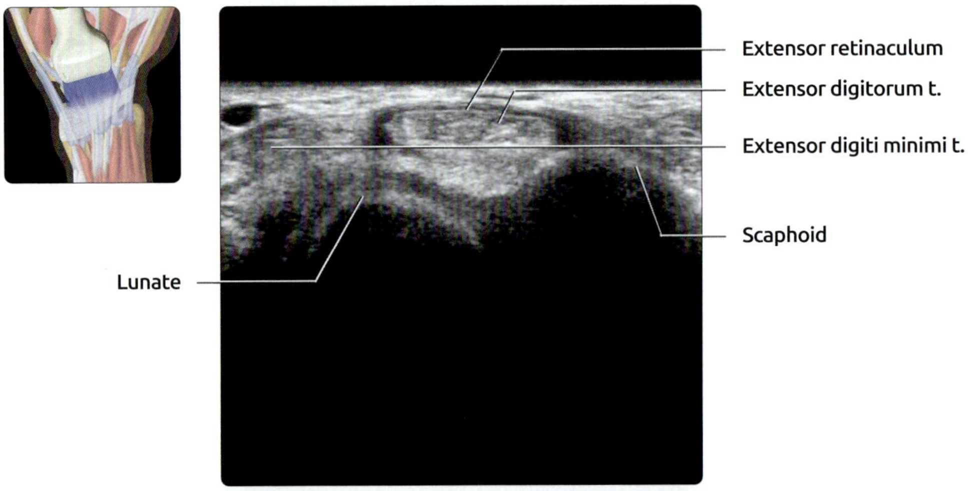

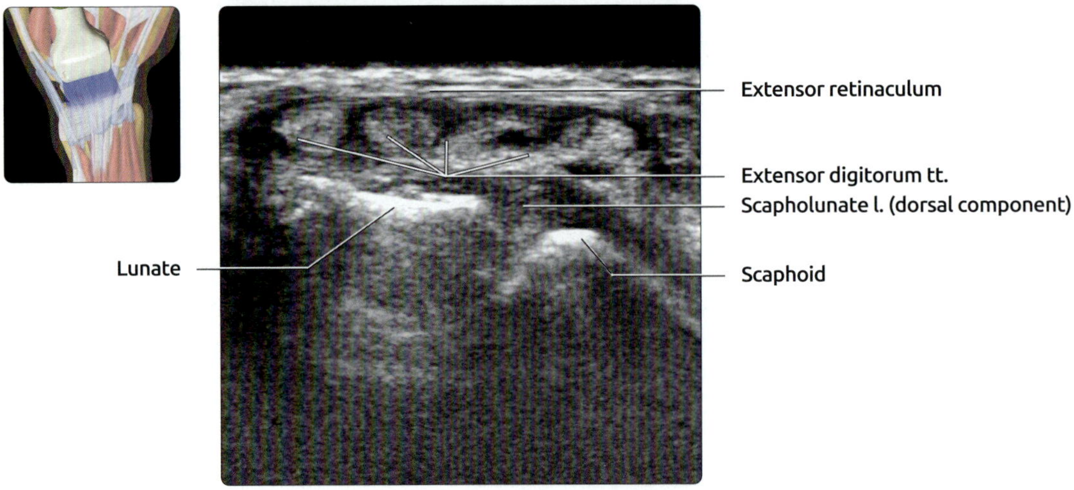

(Top) *Transverse grayscale US shows the 4th extensor compartment at the distal radius. The 4th extensor compartment contains the 4 extensor digitorum tendons and the extensor indicis tendon, which lies radial to the extensor digitorum tendons. The extensor digiti minimi has a separate compartment.* **(Middle)** *Transverse grayscale US shows the 4th extensor compartment at the level of the proximal carpal row. The extensor retinaculum is a thickened continuation of the antebrachial fascia. It is attached to the anterior aspect of the distal radius, styloid process of the ulna, and the triquetral and pisiform bones.* **(Bottom)** *Transverse grayscale US shows the 4th extensor compartment at the level of the proximal carpal row. The extensor retinaculum holds the extensor tendons in place. Because it is closely applied to the extensor tendons, with anisotropy, it may appear hypoechoic and be confused with tenosynovitis.*

TRANSVERSE US, EXTENSOR CARPI ULNARIS

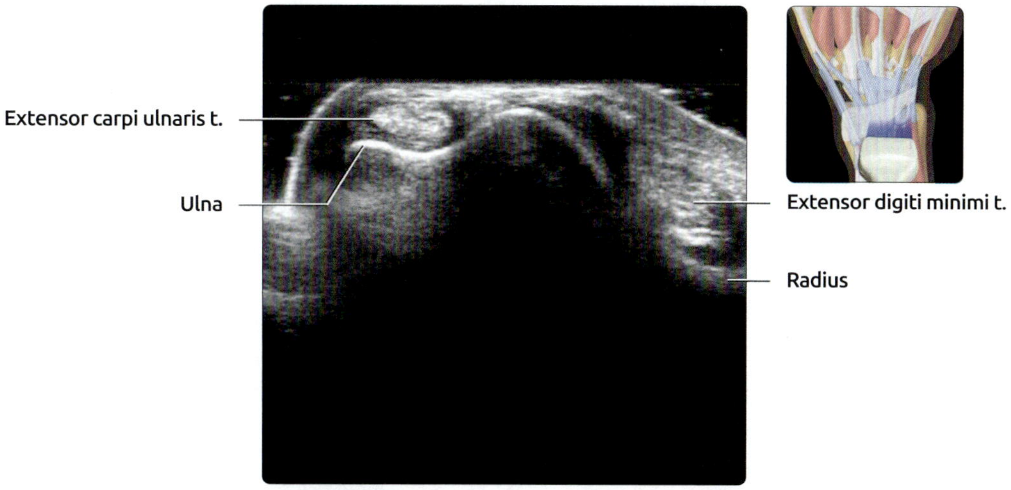

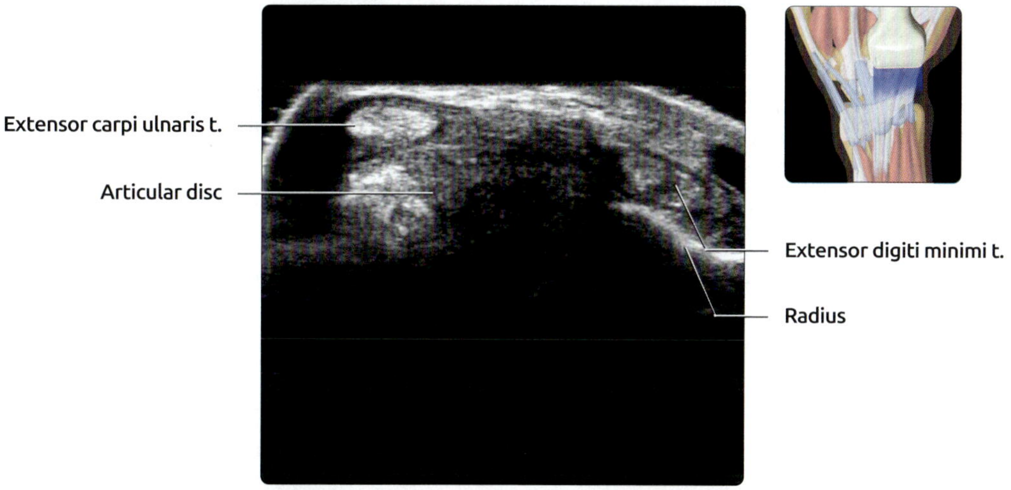

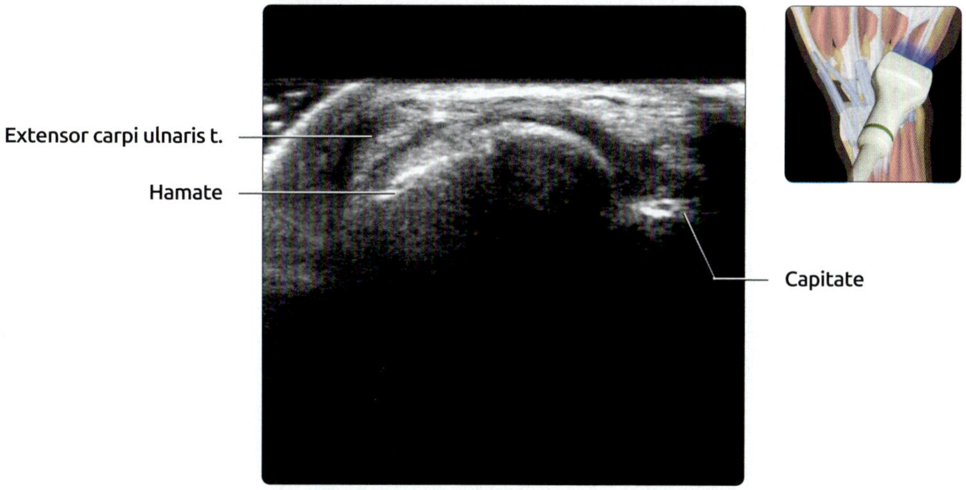

(Top) *Transverse grayscale US shows the 5th and 6th extensor compartments. The 5th compartment contains the extensor digiti minimi tendon. This tendon is joined by the extensor digitorum tendon to the little finger just proximal to the metacarpophalangeal joint. The 6th compartment contains the extensor carpi ulnaris tendon, which runs in a groove in the distal ulna. Its position in the groove will change with pronation and supination.* **(Middle)** *Transverse grayscale US shows the 5th and 6th extensor compartments. The extensor carpi ulnaris tendon often has a midline irregular hypoechoic line within its substance close to the insertion. This should not be mistaken for a longitudinal tear.* **(Bottom)** *Transverse grayscale US at insertion of the extensor carpi ulnaris tendon is shown. The extensor carpi ulnaris tendon widens as it passes the lunate bone on its way to insert into the base of the 5th metacarpal.*

VOLAR WRIST

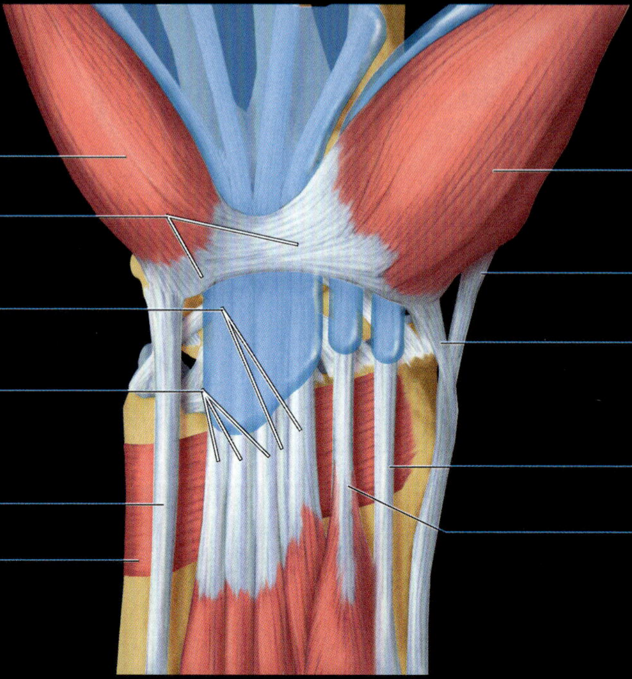

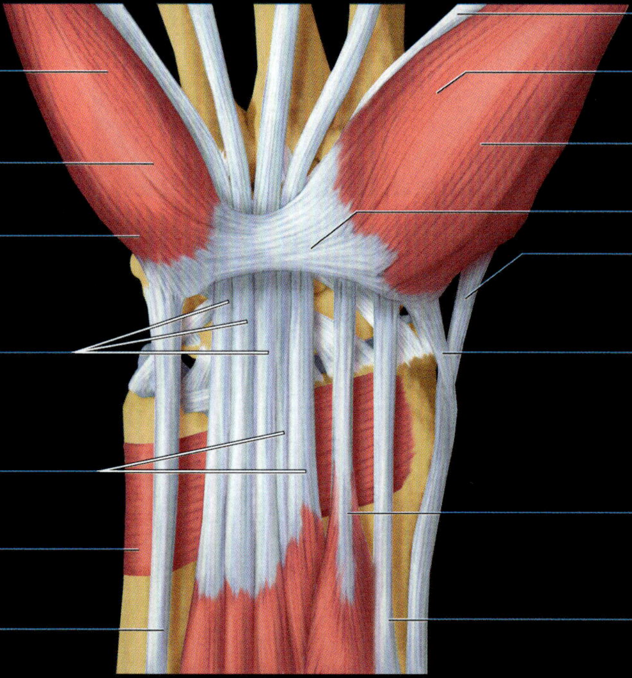

(Top) *Graphic shows tendons and retinaculum of the volar wrist. The flexor retinaculum spans the palmar arch, attaching to the tubercle of the scaphoid, pisiform, hook of hamate, and ridge of the trapezium. The thenar eminence musculature includes abductor pollicis brevis, opponens pollicis, flexor pollicis brevis, and adductor pollicis. The hypothenar eminence musculature includes palmaris brevis, adductor digiti minimi, flexor digiti minimi brevis, and opponens digiti minimi.* **(Bottom)** *Volar muscles and tendons are displayed with their relation to the flexor retinaculum. Note the muscles of thenar and hypothenar eminences arise from the retinaculum itself. The flexor digitorum and flexor pollicis longus tendons pass deep to the retinaculum. On the radial side, the retinaculum splits to accommodate the flexor carpi radialis tendon.*

TRANSVERSE US, VOLAR WRIST

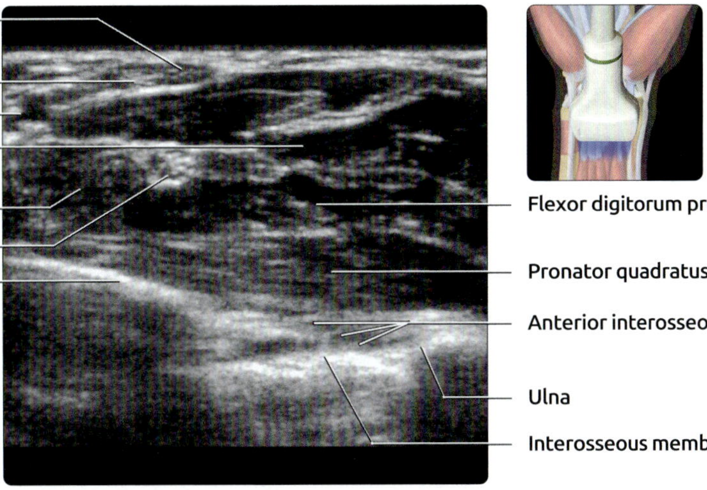

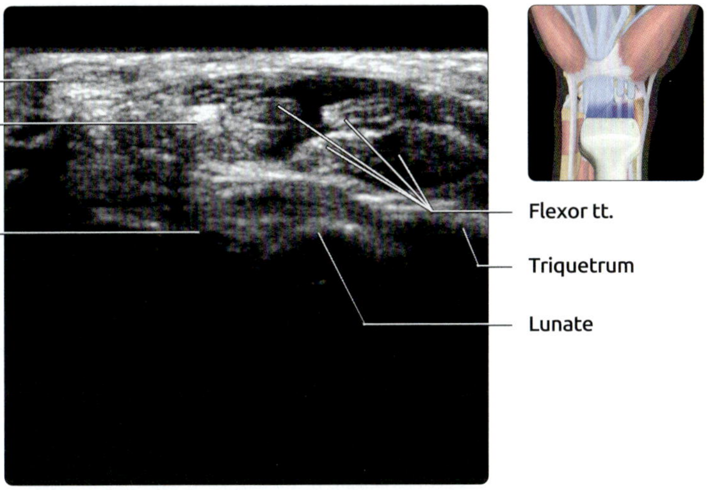

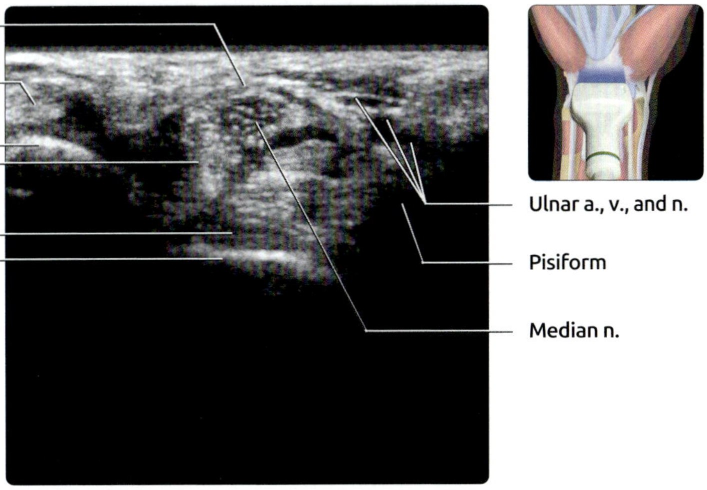

(Top) *Transverse grayscale US shows the volar aspect of the distal forearm just proximal to the wrist. In addition to the tendons that pass through the carpal tunnel, the flexor carpi ulnaris, flexor carpi radialis, and palmaris longus tendons also traverse the wrist joint.* **(Middle)** *Transverse grayscale US shows the volar aspect of the wrist just proximal to the carpal tunnel. The 4 tendons of flexor digitorum superficialis, 4 tendons of flexor digitorum profundus, and flexor pollicis longus tendons pass through the carpal tunnel. The median nerve dips deeply as it enters the carpal tunnel.* **(Bottom)** *Transverse grayscale US shows the volar aspect of the wrist at the tunnel inlet. The inlet (and outlet) of the carpal tunnel can be best recognized by identifying the proximal and distal margins of the flexor retinaculum. In evaluating carpal tunnel syndrome, the caliber of the nerve should be measured proximal to the tunnel, at the tunnel inlet, and at the tunnel outlet.*

US, VOLAR WRIST

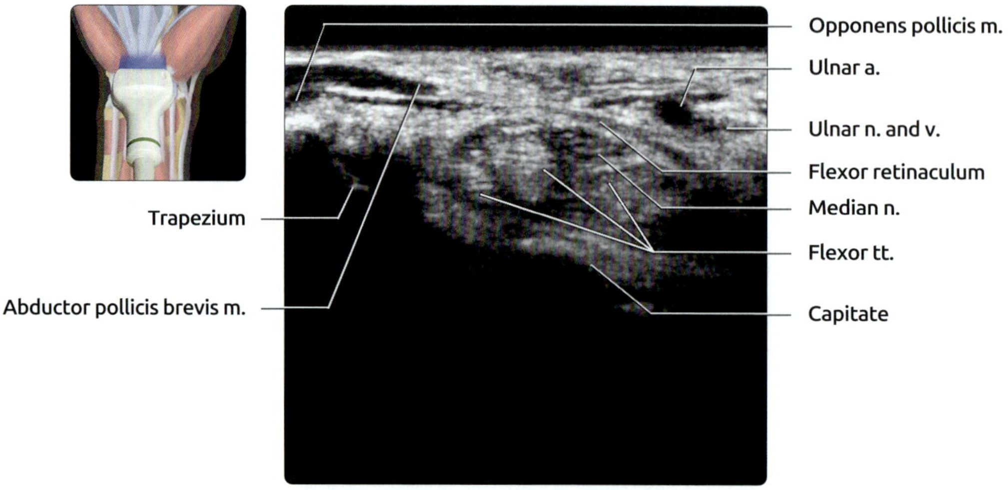

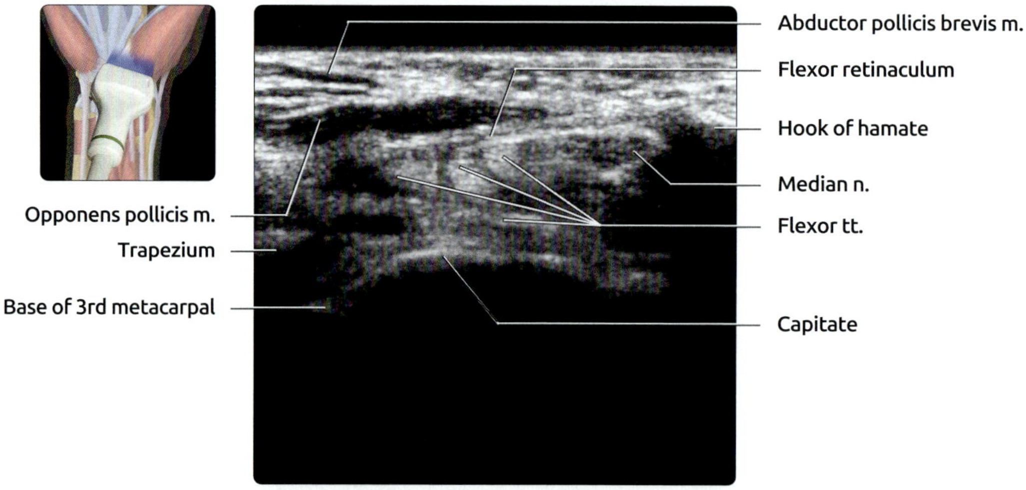

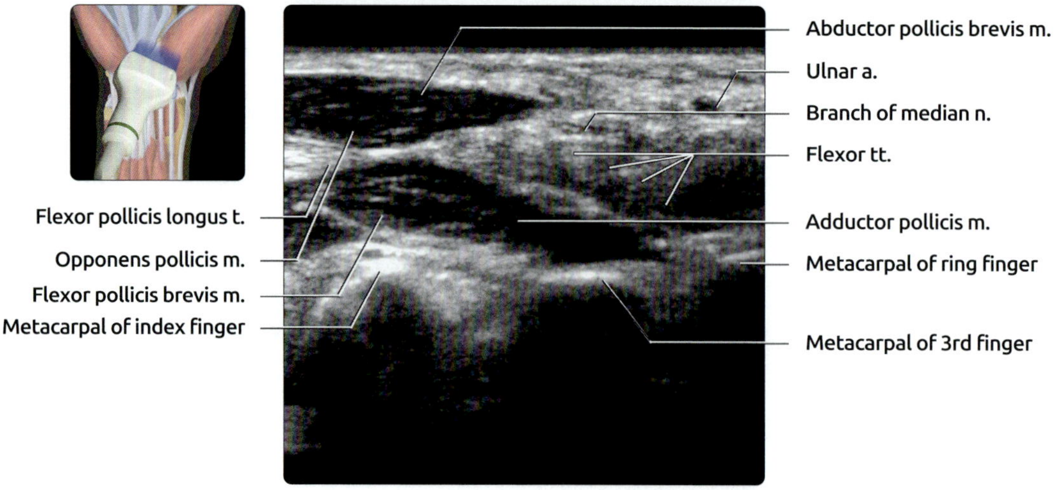

(Top) *Transverse grayscale US shows the volar aspect at midcarpal tunnel. The median nerve lies in the carpal tunnel just deep to the retinaculum. You may need to use anisotropy to clearly identify the margins of the median nerve separate from the adjacent flexor tendons.* **(Middle)** *Transverse grayscale US shows the volar aspect of the wrist at the tunnel outlet. The tunnel outlet is considered to be the narrowest part of the carpal tunnel.* **(Bottom)** *Transverse grayscale ultrasound shows the volar aspect of the wrist just beyond the tunnel outlet. The median nerve divides into its terminal branches just beyond the tunnel outlet.*

US, SCAPHOID

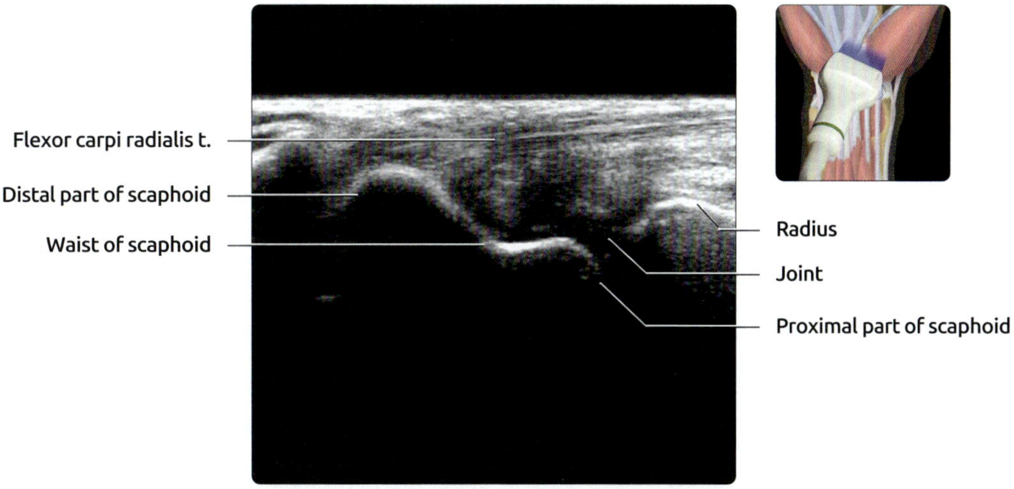

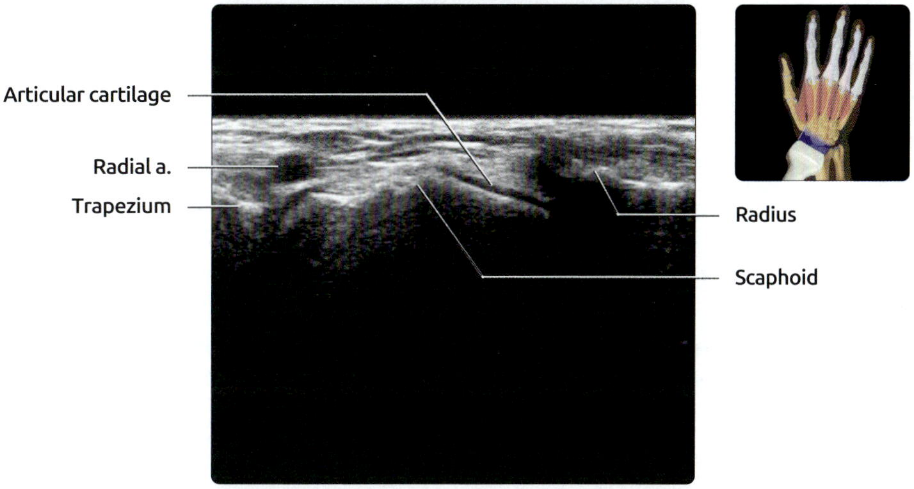

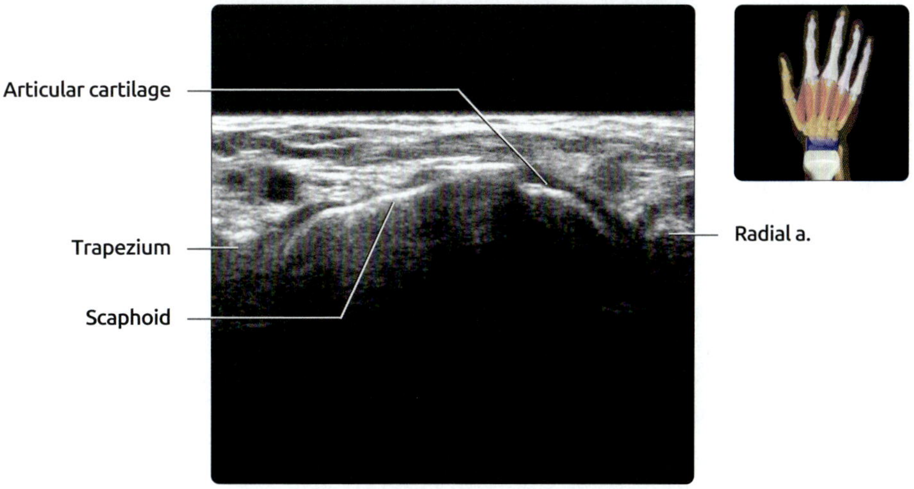

(Top) *Grayscale US longitudinal to the palmar aspect of the scaphoid bone is shown. US is a useful means of diagnosing a scaphoid fracture. Angulation of the transducer along the long axis of the scaphoid allows the palmar cortical outline to be appreciated.* **(Middle)** *Grayscale US longitudinal to dorsal aspect of the scaphoid bone is shown. There is often mild cortical irregularity of the scaphoid surface, particularly on the dorsal side. The absence of surrounding edema, hematoma, periosteal thickening, and cortical discontinuity allows one to differentiate this normal appearance from a fracture.* **(Bottom)** *Grayscale US transverse to dorsal aspect of the scaphoid bone is shown.*

GRAPHICS AND US, TRIANGULAR FIBROCARTILAGE

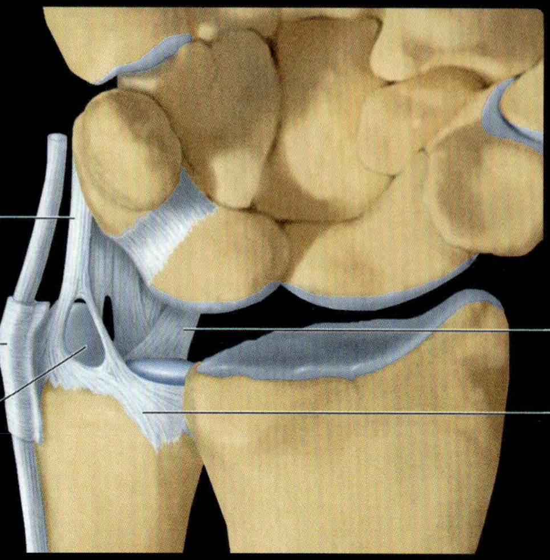

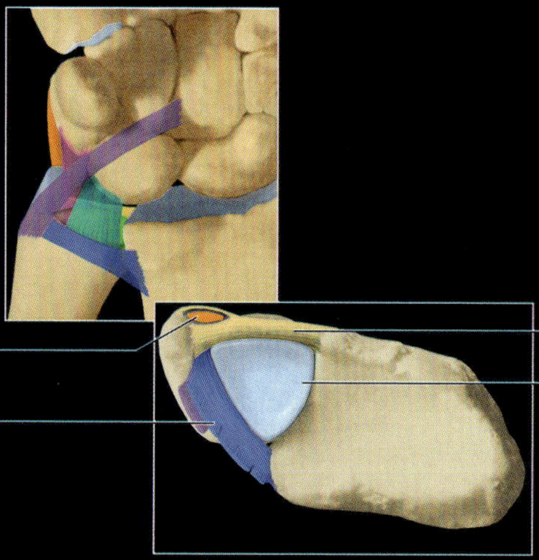

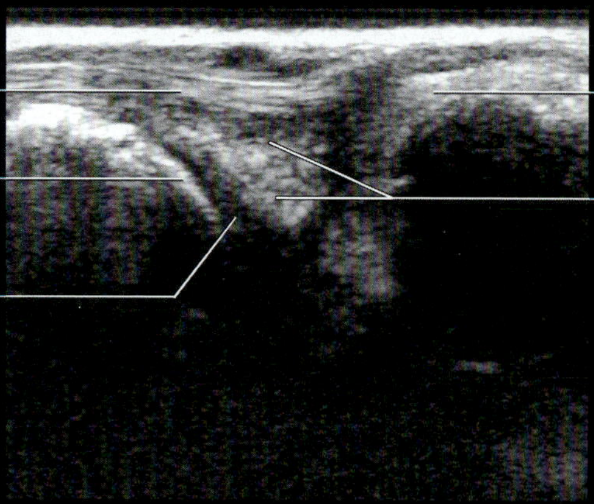

(Top) *Graphic shows supporting structures of the triangular fibrocartilage complex. The ulnocarpal ligaments and the volar radioulnar ligament are on the volar side. At the ulnar border, there is the ulnar collateral ligament. On the dorsal surface, there is the extensor carpi ulnaris tendon and its subsheath as well as the dorsal radioulnar ligament.* **(Middle)** *Graphic depicts the axial view of the articular disc of triangular fibrocartilage complex. The articular disc is inseparable from supporting dorsal and volar radioulnar ligaments. The disc is widest at its radial attachment. Central tears are more common, while peripheral tears, being better vascularized, have the capacity to heal.* **(Bottom)** *Longitudinal grayscale US shows the ulnar aspect of triangular fibrocartilage complex. The fibrocartilaginous articular disc is of different echotexture to hypoechoic hyaline cartilage. US is not as sensitive at depicting triangular fibrocartilage complex tears as MR. The extensor carpi ulnaris tendon provides an acoustic window though which to see the articular disc of the triangular fibrocartilage complex.*

VOLAR AND DORSAL LIGAMENTS

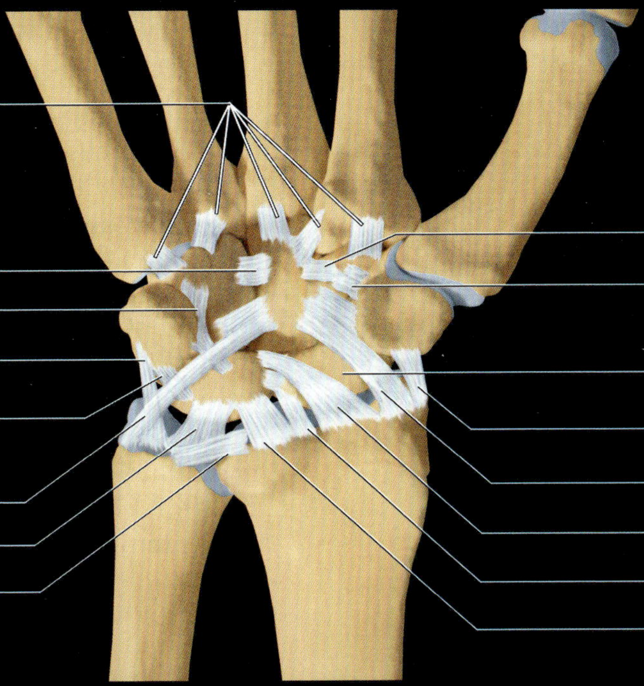

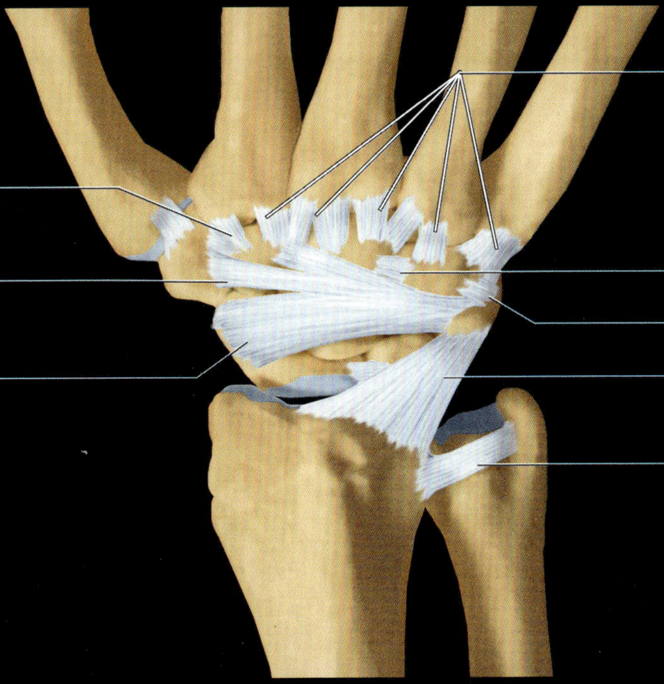

(Top) *Graphic shows volar intrinsic and extrinsic ligaments. The extrinsic ligaments connect the bones of the forearm (radius and ulna) and those of the carpus. The intrinsic ligaments connect carpal bones to carpal bones.* **(Bottom)** *Dorsal ligaments stabilize and restrict motion but are less critical to the stability of the wrist structures than the volar ligaments. On the volar aspect of the wrist, there is a triangular area of weakness (called the space of Poirier) between the lunate and capitate, which is not covered by any ligaments.*

EXTRINSIC FLEXOR MUSCLES: DIGITS 2-5

Flexor Digitorum Superficialis

- Origin: Common flexor tendon origin (medial humeral epicondyle) & midradius
- Insertion: Volar plate proximal interphalangeal (IP) joints & bases of middle phalanges digits 2-5
- Innervation: Median nerve
- Superficialis tendons split at bases of proximal phalanges into 2 tendon slips
  - Slips pass around flexor digitorum profundus (FDP) tendon before inserting into middle phalangeal bases
  - Flexor digitorum superficialis (FDS) forms "tunnel" for FDP
- Flexes metacarpophalangeal joints (aided by lumbricals & interossei) & proximal IP joints

Flexor Digitorum Profundus

- Origin: Proximal 1/2 of radius & interosseus membrane
- Insertion: Volar plates, distal IP joints, & bases of distal phalanges of digits 2-5
- Innervation: Ulnar & median nerves
- Flexes distal IP joints & proximal IP, metacarpophalangeal (aided by lumbricals & interossei) joints of digits 2-5

INTRINSIC MUSCLES: HYPOTHENAR AREA

Hypothenar Eminence

- Mirror image of thenar eminence
- Muscles (superficial to deep): Abductor digiti minimi (ADM), flexor digiti minimi (FDM), opponens digiti minimi (ODM)
  - Origin: ADM from pisiform; FDM & ODM from flexor retinaculum & hook of hamate
  - Insertion: ADM & FDM → conjoint insertion ulnar aspect base of proximal phalanx little finger
    - ODM to proximal 2/3 metacarpal shaft little finger
  - Innervation: All by ulnar nerve

INTRINSIC MUSCLES: THENAR AREA

Thenar Eminence

- Muscles: Abductor pollicis brevis (APB), flexor pollicis brevis (FPB), opponens pollicis (OP)
  - Origin: Flexor retinaculum & tubercle of trapezium
  - Insertion: APB, FPB conjoint insertion at base of proximal phalanx radial aspect
    - OP to volar aspect 1st metacarpal shaft
- Innervation: All by median nerve except FPB, which has also partially supplied by ulnar nerve

Adductor Pollicis Muscle

- Oblique (proximal) & transverse (distal) heads
- Origin: Capitate, trapezoid, 2nd & 3rd metacarpals
- Insertion: Ulnar aspect base of proximal phalanx of thumb, IP joint, volar plate of thumb
  - Also contributes fibers to extensor hood of thumb, which forms adductor aponeurosis
- Innervation: Ulnar nerve
- Actions: Adducts thumb toward 3rd digit & aids extension of IP joint
- Separated from thenar muscles by fascial plane & innervated by ulnar rather than median nerve

INTRINSIC MUSCLES: PALMAR AREA

Lumbricals

- Numbered 1-4 from radial to ulnar
- Origin: Radial aspect FDP tendon index to little fingers
- Insertion: Radial aspect extensor tendon expansion index to little fingers
- Innervation: Most radial 2 lumbricals by median nerve, most ulnar 2 lumbricals by ulnar nerve
- Actions: Flex metacarpophalangeal joints & extend IP joints

Palmar Interossei

- Numbered 1-3 from radial to ulnar
- Origin: Mesial palmar aspect diaphyses of 2nd, 4th, & 5th metacarpals
- Insertion: Mesial lateral bands & mesial aspect bases of proximal phalanges of same digit as origin
- Innervation: Ulnar nerve
- Actions: Adduct digits & assist lumbricals with flexion of metacarpophalangeal joints & extension of IP joints of digits 2, 4, & 5

Dorsal Interossei

- Numbered 1-4 from radial to ulnar
- Origin: FDP tendons just distal to carpal tunnel
- Insertion: Radial lateral bands of digits 2-5
- Innervation: Ulnar nerve
- Actions: Extends IP joints 2-5 & abducts digits 2-5

TENDONS

Tendon Sheaths

- Common flexor sheath (a.k.a. ulnar bursa)
  - Contains FDS & FDP tendons
  - Begins just proximal to carpal tunnel
  - Ends just beyond carpal tunnel for digits 2-4
  - Encloses 5th digit flexor tendons over entire course to distal IP joint
- Flexor pollicis longus tendon sheath (a.k.a. radial bursa)
  - Encloses flexor pollicis longus tendon from just proximal to carpal tunnel to tendon insertion at base of distal phalanx thumb
  - ± communication with ulnar bursa at level of carpal tunnel in ~ 50% of normal population
    - Potential route for spread of infection
- Individual flexor digital sheaths (digits 2-4)
  - Encloses flexor tendons from level of metacarpal necks to bases of distal phalanges

Extensor Expansion (or Extensor Hood)

- Expansions of extensor digitorum, extensor indicis, & extensor digiti minimi tendons on dorsum of fingers
  - Help prevent lateral translation of tendon
- 3 bands of extensor expansion
  - Single, wide middle band passes over dorsum of proximal phalanx & proximal 1/2 of middle phalanx
  - 2 cord-like lateral bands extend from proximal phalanx to distal phalanx
    - Lumbricals, palmar & dorsal interossei insert into these lateral bands

PALMAR ASPECT

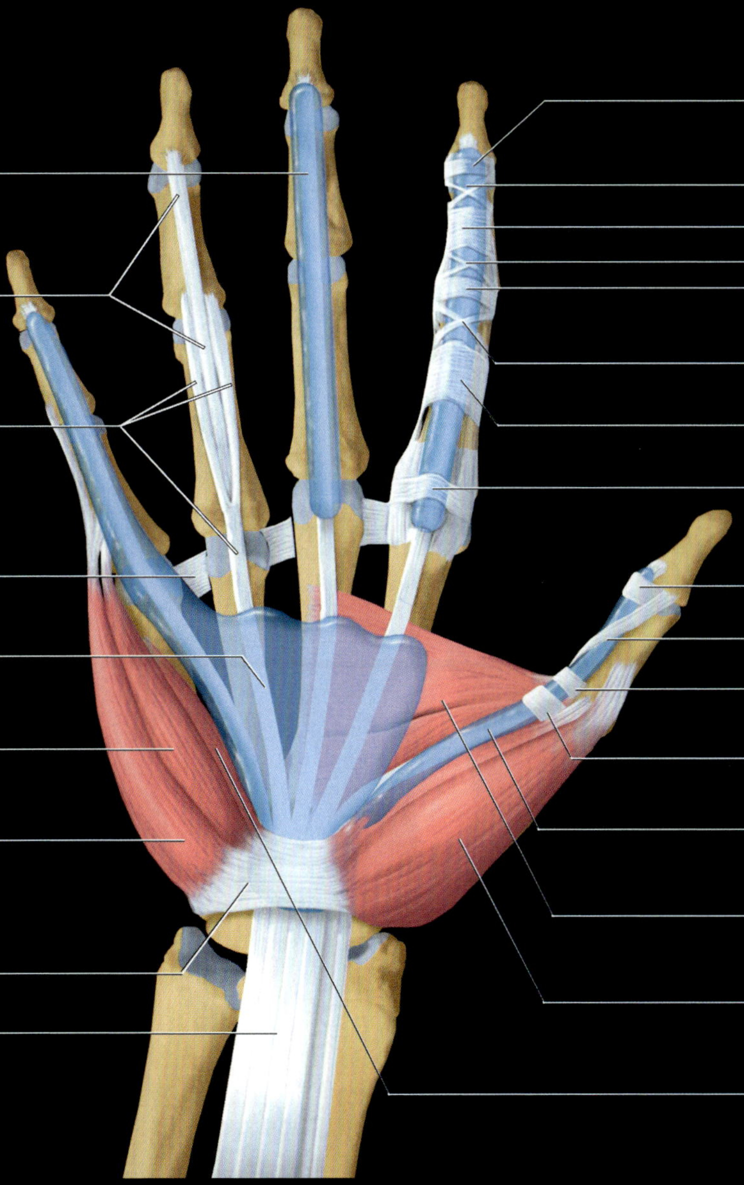

Pulley system for the 3rd-5th digits is identical to that demonstrated for the 2nd digit. Common digital sheath of the 4th digit is removed to show the relationship of flexor digitorum superficialis (FDS) and flexor digitorum profundus (FDP) tendons. Deep transverse metacarpal ligaments connect the volar plates (not shown) of digits 2-5. Although there is overlap of the radial and ulnar bursae in this image, these structures normally do not invariably communicate. The radial and ulnar bursae may communicate as a normal variation in ~ 50% of the population. Similarly, any 1 or more of the common digital sheaths may communicate with the ulnar bursa in up to 10% of the normal population. These different bursal communications are important as they provide routes for more extensive spread of infection.

LUMBRICAL AND INTEROSSEI MUSCLES

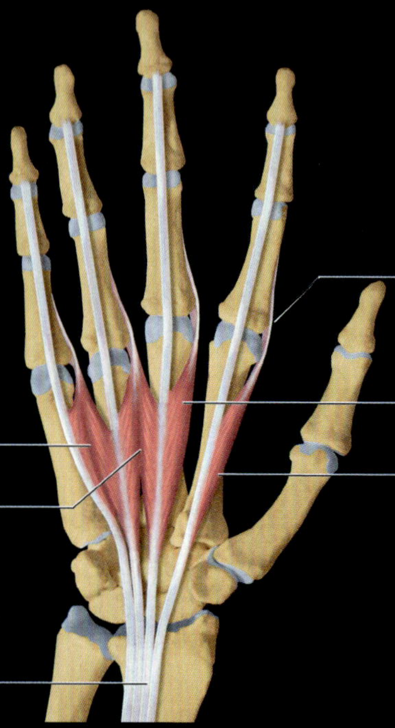

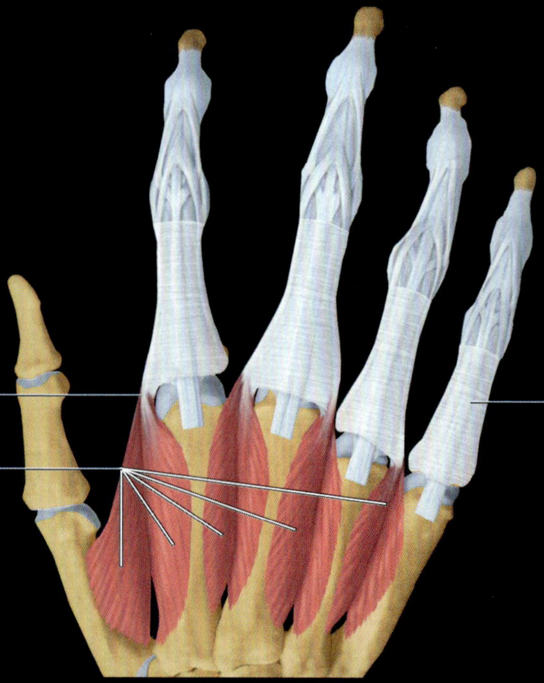

(Top) *Lumbricals originate from the FDP tendons as shown. Note how the 1st and 2nd lumbricals arise only from tendons to the 2nd and 3rd digits, respectively (unipennate), whereas the 3rd and 4th lumbrical arise from the tendons to both the 3rd and 4th and the 4th and 5th digits, respectively (bipennate).* **(Bottom)** *Graphic of the dorsal surface of the extensor mechanism of the hand and digits 2-5 demonstrates the complex relationship of the various fiber bands. As in imaging, distinction of individual fiber bands is often difficult and must be inferred by knowledge of where the structure "should" be with respect to more easily identifiable structures (such as bones and joints).*

PALMAR INTEROSSEI

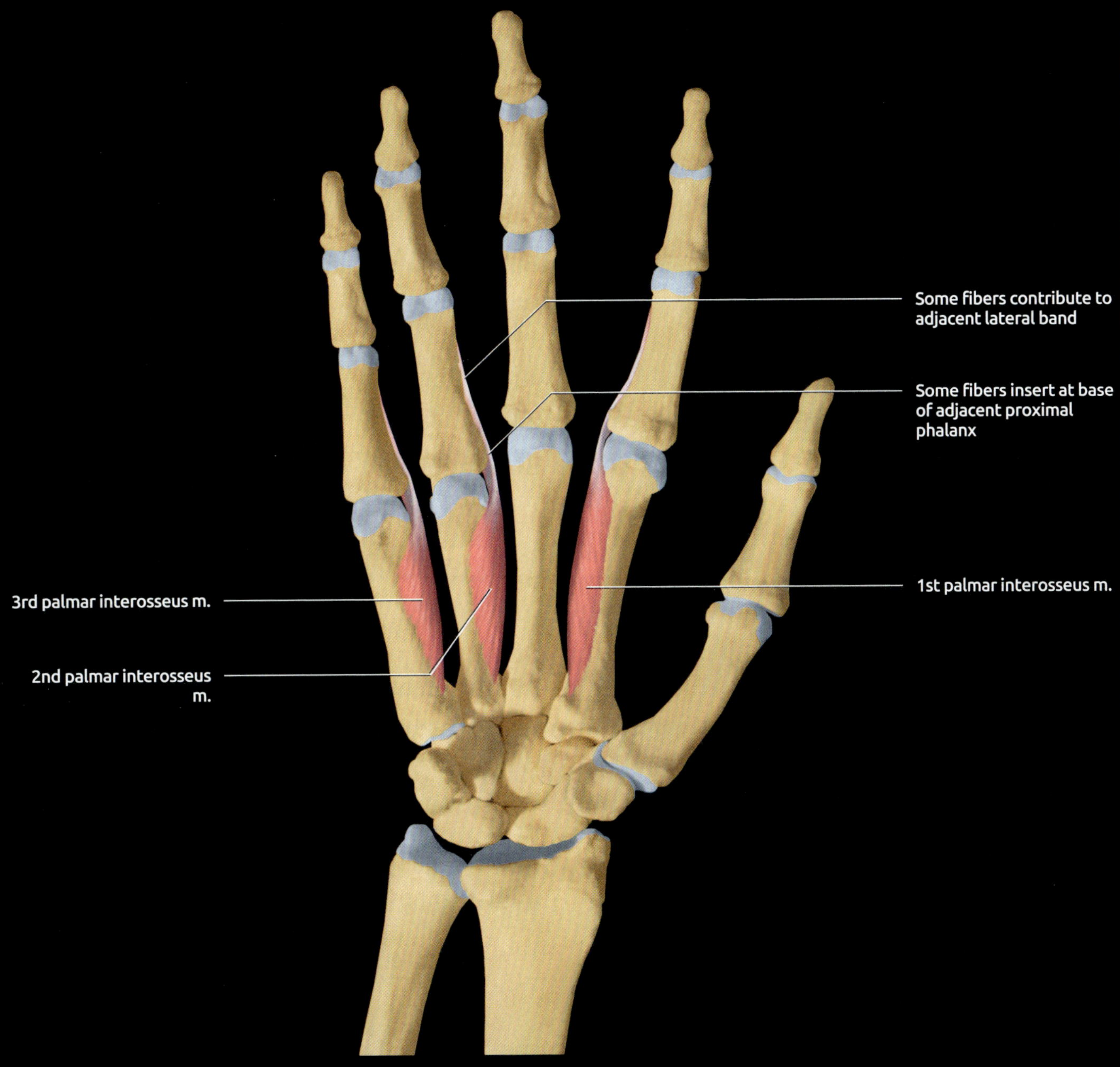

Palmar interossei insert at both bases of adjacent proximal phalanges as well as adjacent lateral bands. The 2nd and 3rd palmar interossei contribute to the radial lateral bands of the 4th and 5th digits, respectively, whereas the 1st palmar interosseus contributes to the ulnar lateral band of the 2nd digit.

TRANSVERSE US, PALM

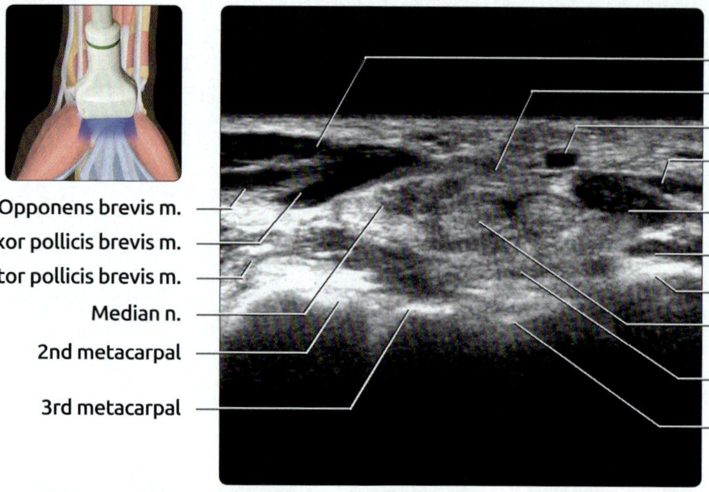

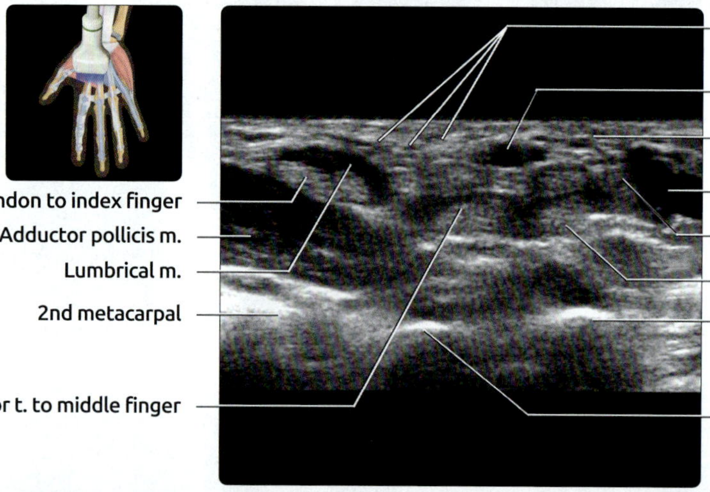

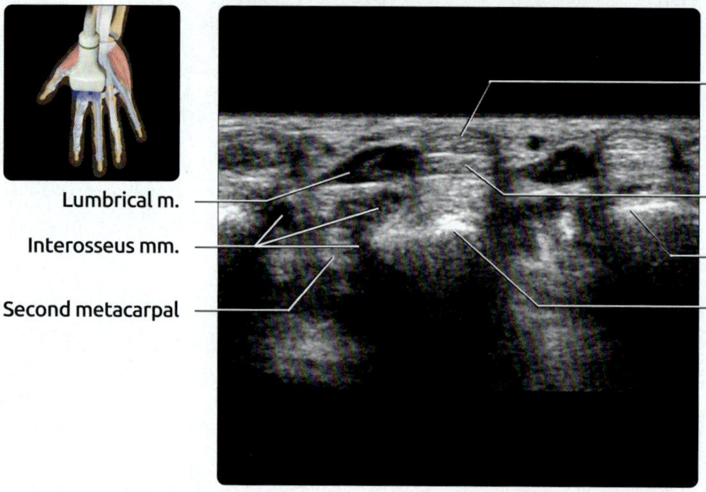

(Top) *Transverse grayscale ultrasound shows the proximal aspect of the palm. The palmar aponeurosis is thin and difficult to demonstrate in the mid- to distal palm especially.* **(Middle)** *Transverse grayscale ultrasound shows the midportion of the palm. Beneath the aponeurosis lies the flexor tendons, the lumbrical muscles, the muscles of the thenar and hypothenar eminences, and, most deeply, the interossei. There are 3 palmar interossei and 4 dorsal interossei muscles.* **(Bottom)** *Transverse grayscale ultrasound shows the distal portion of the palm.*

THENAR-HYPOTHENAR EMINENCE

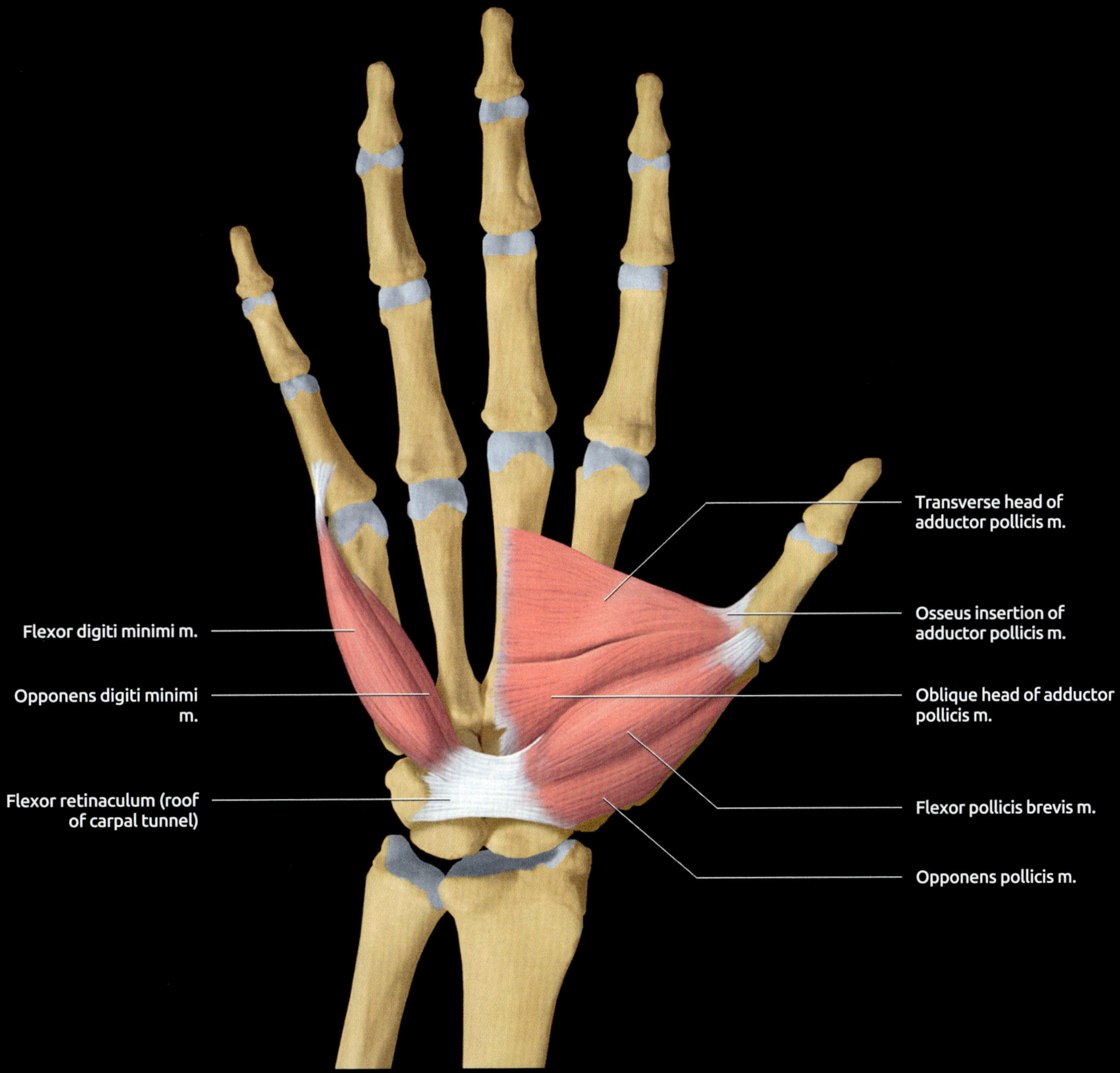

Graphic shows the thenar and hypothenar muscles of the hand. Strictly speaking, the adductor pollicis is not one of the thenar group of muscles since it has a separate fascia and separate nerve supply.

TRANSVERSE US, THENAR EMINENCE

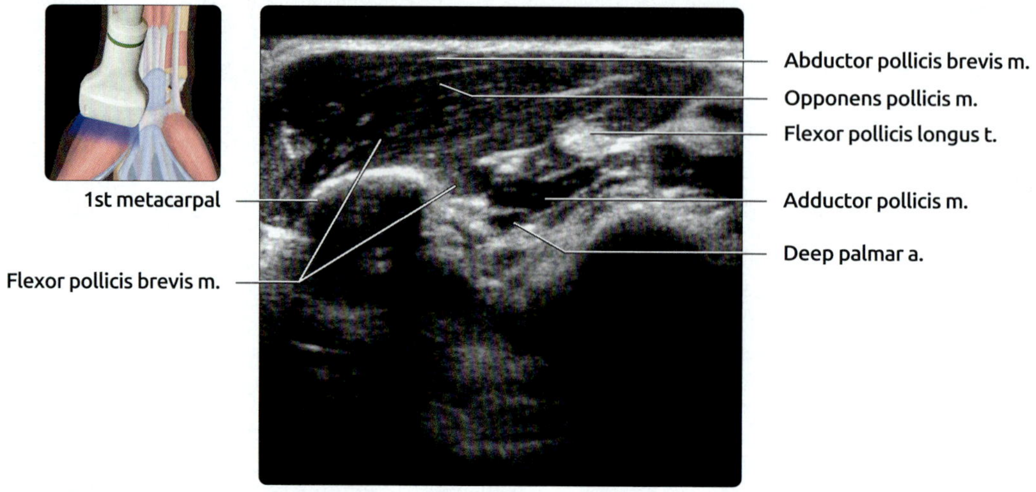

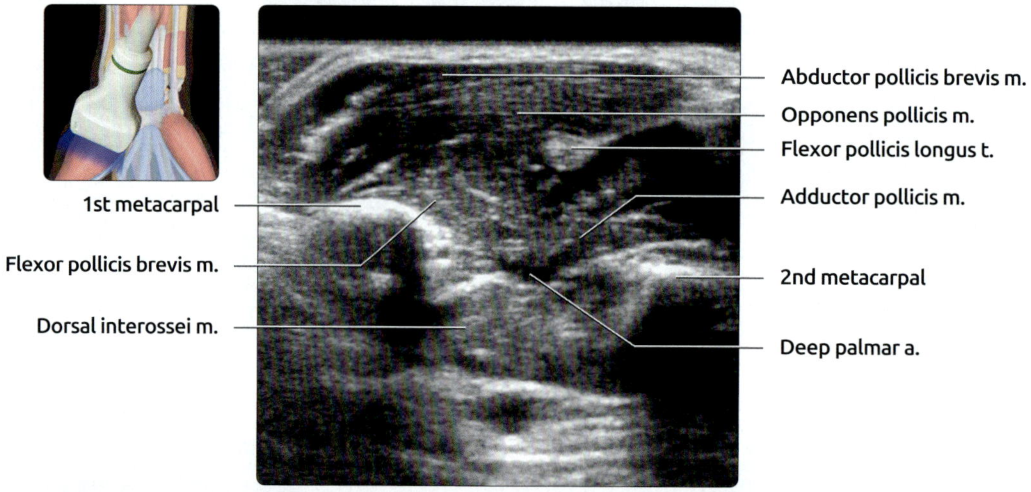

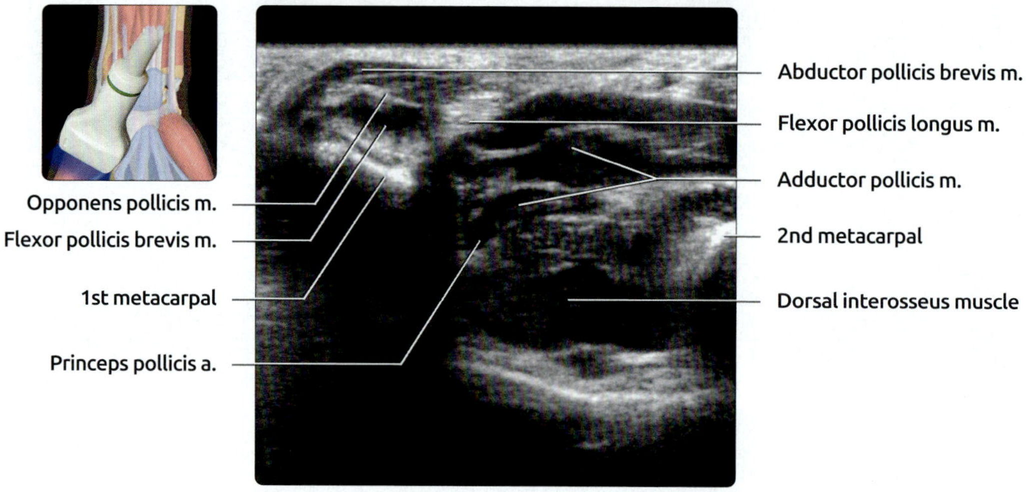

(Top) *Transverse grayscale ultrasound shows the proximal aspect of the thenar eminence. The thenar eminence comprises 4 muscles: Abductor, opponens, flexor, and adductor pollicis from above downward. The adductor muscle is, strictly speaking, not part of the thenar eminence since it is separated by a fascial layer and has a separate innervation.* **(Middle)** *Transverse grayscale ultrasound shows the midportion of the thenar eminence. The flexor pollicis longus tendon is easily recognized as an echogenic tendon alongside the flexor pollicis brevis, deep to the opponens pollicis and superficial to the adductor pollicis muscle.* **(Bottom)** *Transverse grayscale ultrasound shows the distal portion of the thenar eminence.*

LONGITUDINAL US, THENAR EMINENCE

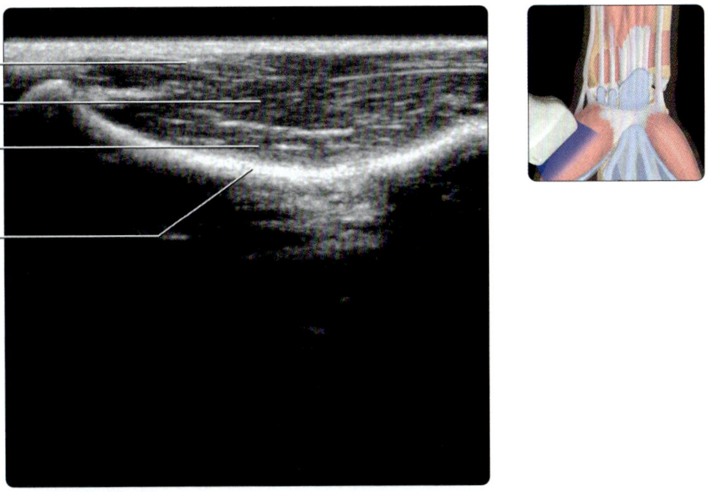

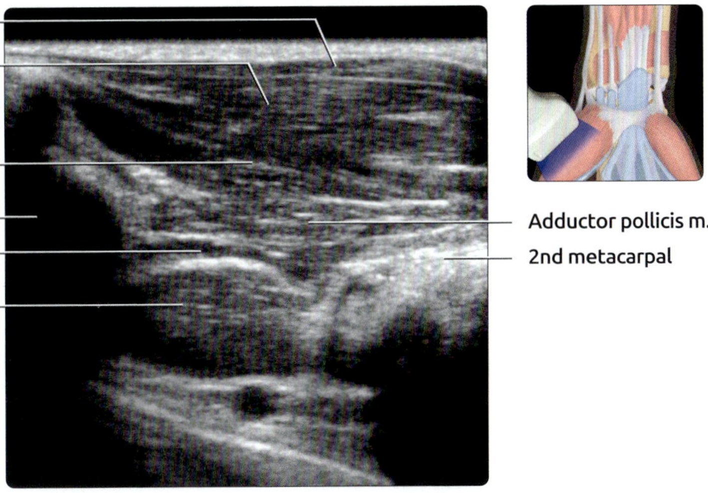

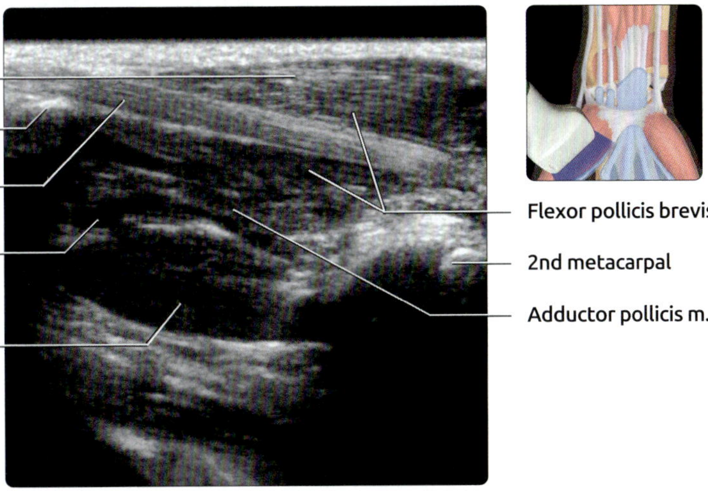

(Top) *Longitudinal grayscale ultrasound shows the proximal aspect of the thenar eminence. Use the mnemonic ABOF to remember the muscles of the thenar eminence.* **(Middle)** *Longitudinal grayscale ultrasound shows midthenar eminence. The abductor pollicis brevis, opponens pollicis, and flexor pollicis brevis (ABOF) lie superficial to deep in that order.* **(Bottom)** *Longitudinal grayscale ultrasound shows the distal aspect of the thenar eminence. The flexor pollicis longus tendon passes through the more distal aspect of the thenar eminence.*

TRANSVERSE US, HYPOTHENAR EMINENCE

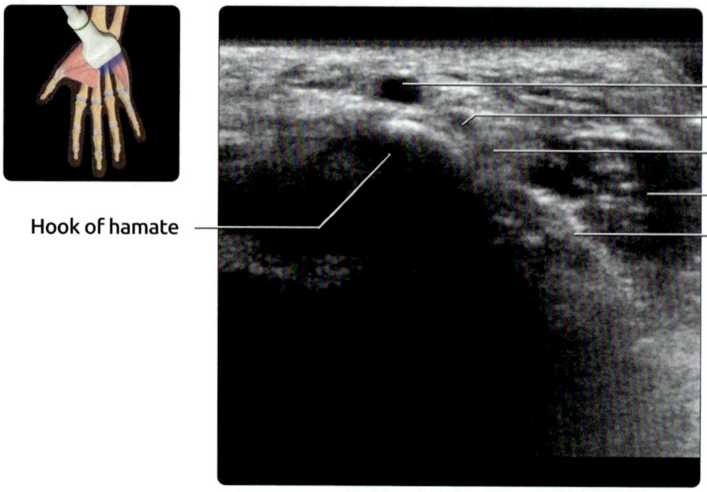

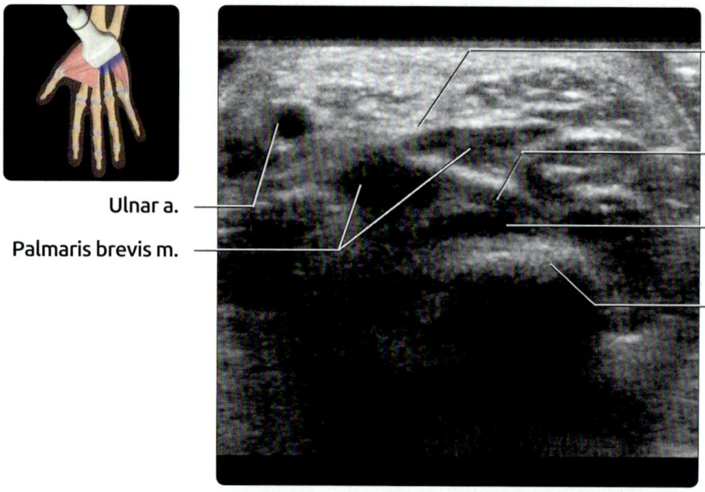

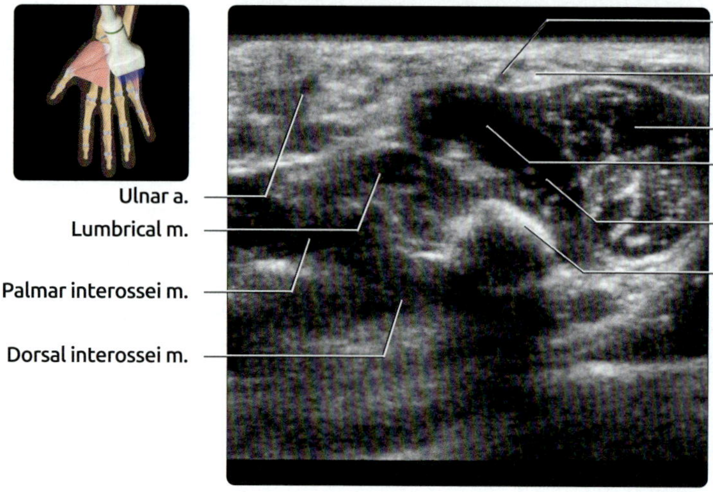

(Top) *Transverse grayscale ultrasound shows the proximal 1/3 of the hypothenar eminence. The hypothenar eminence along with the other muscles of the hand, except for the thenar, are supplied by the ulnar nerve.* **(Middle)** *Transverse grayscale ultrasound shows the middle 1/3 of the hypothenar eminence. Hypothenar muscle atrophy can occur with injury to the deep branch of the ulnar nerve just distal to Guyon canal.* **(Bottom)** *Transverse grayscale ultrasound shows the distal 1/3 of the hypothenar eminence. The order of muscles in the hypothenar eminence is different from the thenar eminence with abductor digiti minimi, flexor digiti minimi, and opponens digiti minimi from superficial to deep.*

DORSAL ASPECT OF HAND

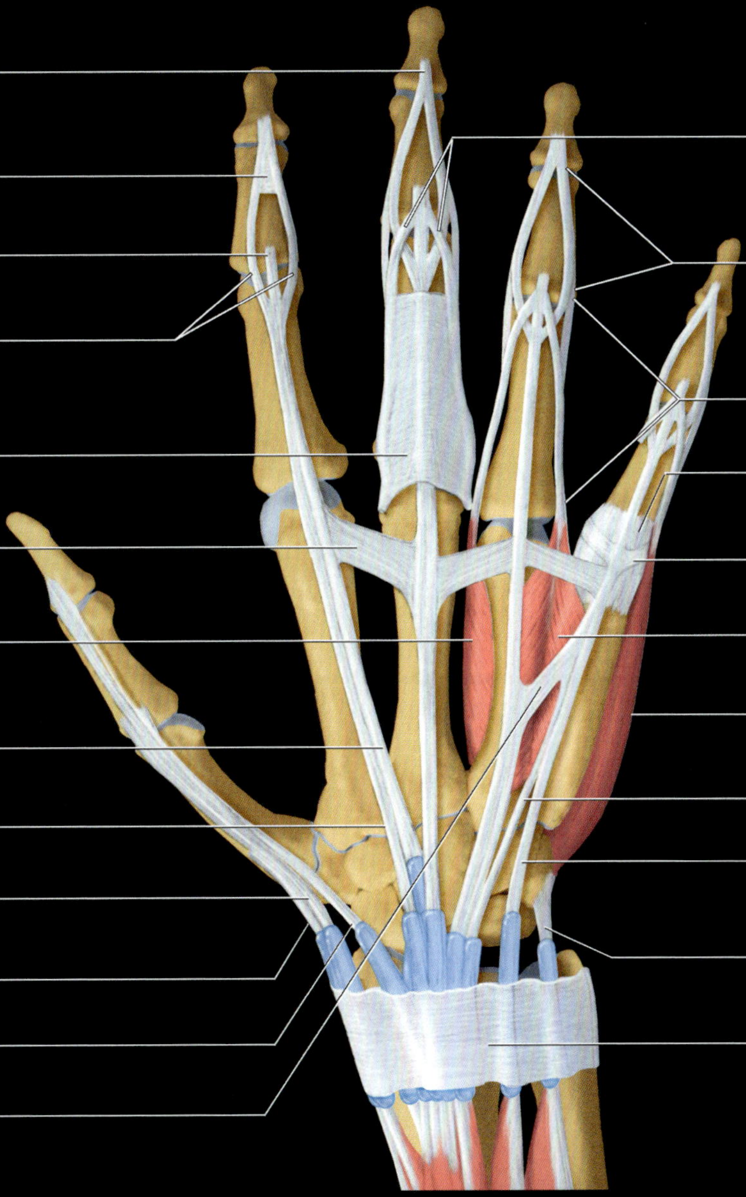

Extensor mechanism composite shows different components of the extensor mechanism on different digits. With the exception of the extensor indicis, extensor digiti minimi, abductor digiti minimi, and 4th extensor digitorum communis (EDC) tendon contribution to the 5th EDC tendon, any structure on any of the 2nd-5th digits can be extrapolated to any and all of the other 2nd-5th digits. Note that the 2nd extensor retinaculum compartment and its contents (extensor carpi radialis longus and brevis tendons) are not included in this graphic. Although the abductor pollicis longus tendon travels in the 1st compartment of the extensor retinaculum, its insertion (not shown) is actually on the radial aspect of the volar surface of the base of the 1st metacarpal.

DORSAL INTEROSSEI

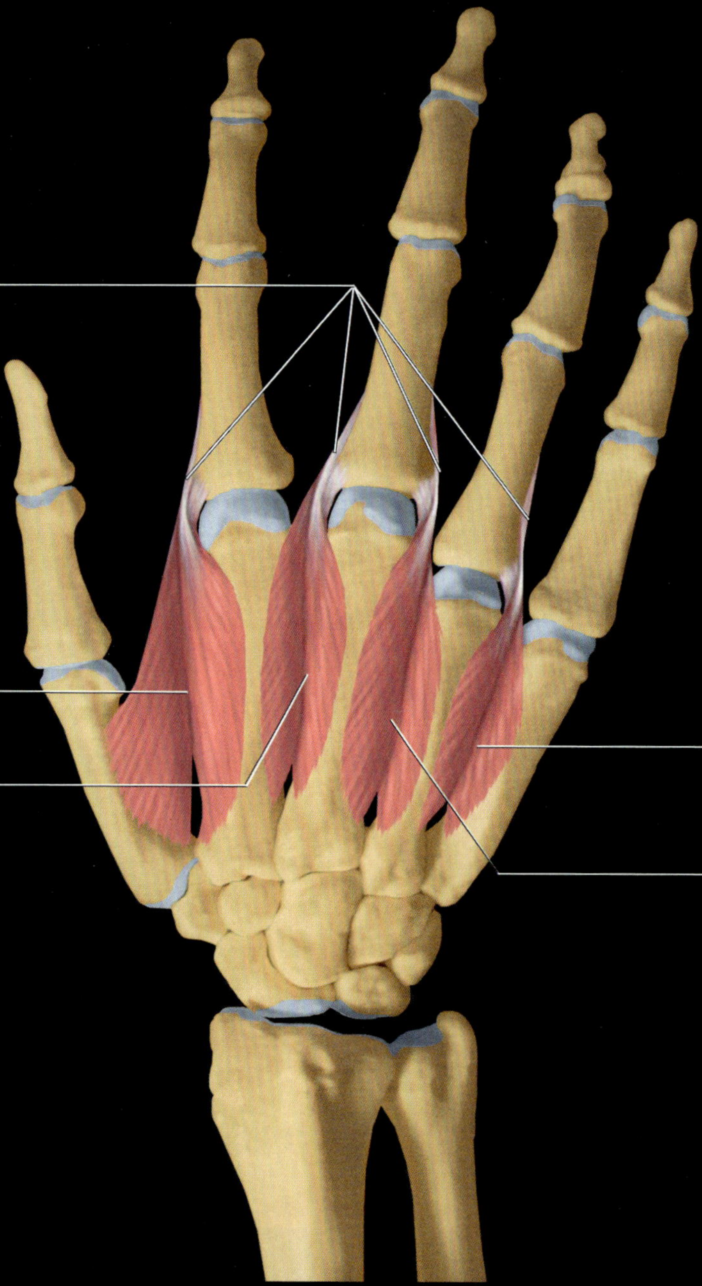

Tendons of the dorsal interossei muscle (along with tendons of the palmar interossei and tendons of the lumbricals) help form lateral bands. Note how each interosseus muscle only contributes fibers to the mesial-most adjacent lateral band. The dorsal interossei do not contribute any lateral band fibers to the 1st or 5th digits.

TRANSVERSE US, DORSUM HAND

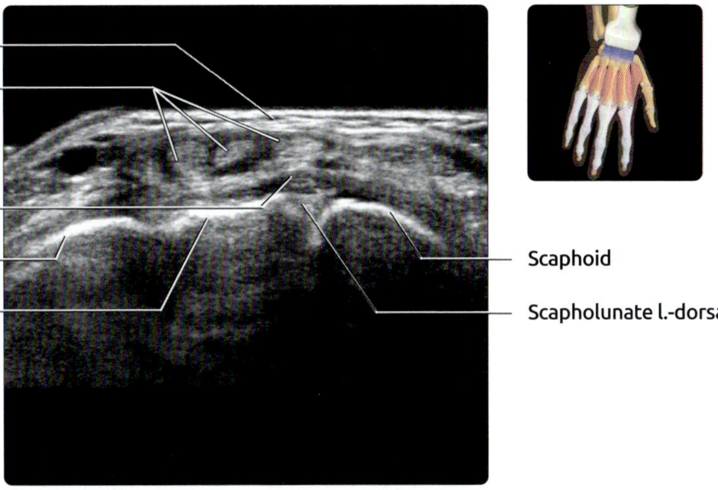

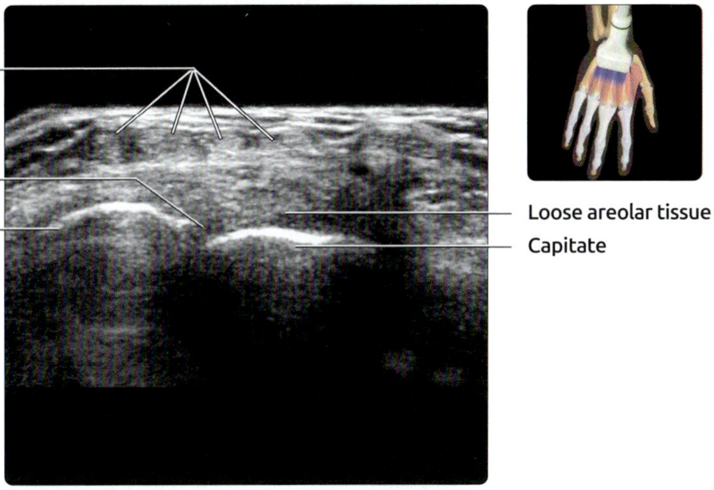

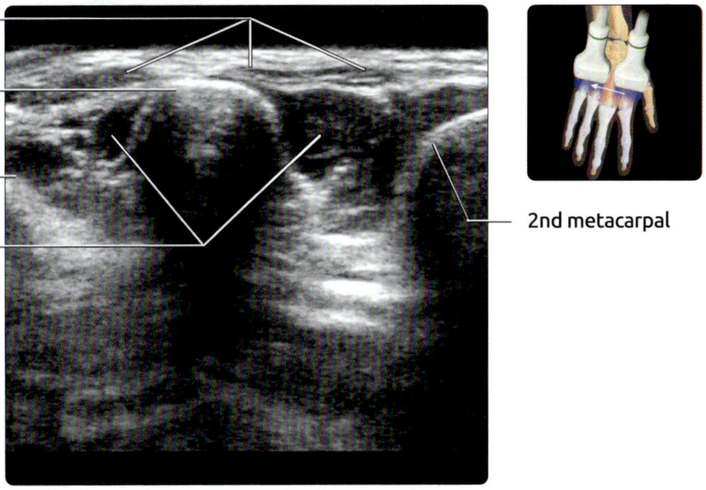

(Top) *Transverse grayscale ultrasound shows the proximal aspect of the dorsum hand. The tissues on the dorsum hand are less tightly bound than the tissues on the palmar aspect of the hand, and this part of the hand tends to swell much more with inflammatory or edematous processes.* **(Middle)** *Transverse grayscale ultrasound shows the mid aspect of the dorsum of hand. Loose areolar tissue separates the extensor tendons from the metacarpal bones. The extensor tendons do not have a complete tendon sheath from the level of the extensor retinaculum to the insertion.* **(Bottom)** *Transverse grayscale ultrasound shows the distal aspect of the dorsal hand. There are 4 dorsal interossei, 1 for each metacarpal interspace.*

TERMINOLOGY

Abbreviations

- Abductor pollicis longus (APL)
- Extensor pollicis longus (EPL)
- Extensor pollicis brevis (EPB)
- Flexor retinaculum (FR)
- Pronator quadratus (PQ)

GROSS ANATOMY

Arteries

- Radial artery
  - Terminal branch of brachial artery
  - Located superficial to PQ, then continues dorsally around radial styloid process, then passes deep to APL and EPB, across anatomic snuffbox and deep to EPL
  - Branches of radial artery
    - Superficial palmar branch continues distally superficial to FR through thenar eminence to anastomose with ulnar artery as superficial palmar arch
  - Deep palmar arch is continuation of radial artery in hand
    - Main radial artery continues dorsally at radial styloid across anatomic snuffbox deep to EPB and APL
    - Passes over dorsal aspect of scaphoid bone
    - Passes between heads of 1st dorsal interosseus muscle to enter palm, joins ulnar artery to form deep palmar arch
    - Deep palmar arch is located at level of metacarpal bases 1 cm proximal to superficial palmar arch
    - Lies deep to long flexor tendons and superficial to interossei
    - Deep branch of ulnar nerve lies on convexity of deep palmar arch
    - Princeps pollicis artery, radial artery of index finger, and palmar metacarpal arteries arise from deep palmar arch
- Ulnar artery
  - Terminal branch of brachial artery
  - Located superficial to PQ, then runs between flexor carpi ulnaris and flexor digitorum superficialis tendons
  - Branches of ulnar artery
  - Common interosseus artery branches in forearm divide into anterior and posterior interosseus arteries
    - Anterior interosseus artery travels distally; at proximal PQ, gives small branch that pierces membrane, enters dorsal forearm, and anastomoses with posterior interosseus artery
    - Median artery (branch of anterior interosseus artery) arises in forearm and may persist accompanying median nerve into carpal tunnel, contributing to superficial palmar arch
    - Posterior interosseus artery travels distally between deep and superficial extensors to anastomose with anterior interosseus artery
  - Superficial palmar arch is continuation of ulnar artery in hand
    - Located along line extended in arch across palm from thumb web space
    - Similar to metacarpal shaft level ~ 1 cm distal to deep palmar arch
    - Located deep to palmar aponeurosis but superficial to long flexor tendons and digital nerves
    - Common palmar digital arteries arise from superficial palmar arch
    - Run distally on surface of 2nd, 3rd, and 4th lumbrical muscles
    - Near metacarpophalangeal joints, each common palmar digital artery divides into 2 proper digital arteries
    - Run in subcutaneous fat of digits (lateral aspects)

Veins

- Dorsal venous network of hand and network formed by dorsal metacarpal veins
  - Found on dorsum of hand
  - Unite to form cephalic vein (radial side of wrist) and basilic vein (ulnar side of wrist)
    - Basilic and cephalic veins travel proximally in superficial tissues
    - Both veins unite at antecubital fossa via median cubital vein
    - Continues proximally as cephalic vein until it joins brachial vein in axilla to form axillary vein
  - Layout of superficial veins in hand and forearm is highly variable
- Causes of deep venous thrombosis in upper limb extremity are different from that of lower limb
  - Related to intravenous cannulations, infusions, radiotherapy, effort-induced thrombosis, and malignant obstruction

IMAGING ANATOMY

Overview

- Extensive volar and dorsal anastomoses typically exist

ANATOMY IMAGING ISSUES

Imaging Recommendations

- Assessment of caliber of small arteries of hand best performed by digital subtraction angiography as they are close to limit of resolution by ultrasound
  - For example, when investigating suspected Raynaud disease or Buerger disease

Imaging Pitfalls

- Vessel visualization on standard imaging is limited
  - Radial and ulnar artery: 2-3 mm, typically visualized
  - Vascular arches: 1 mm, inconsistently visualized
  - Multiple branches and anastomoses, < 1 mm, rarely visualized
- Entry points of nutrient vessels into bone should not be mistaken for erosions
- If angle of insonation of ultrasound beam is at or near 90° to vessel, Doppler and color sensitivity will ↓
- Optimize gate size for small-caliber arteries and position at center of arterial lumen to detect peak flow velocity

HAND VASCULATURE

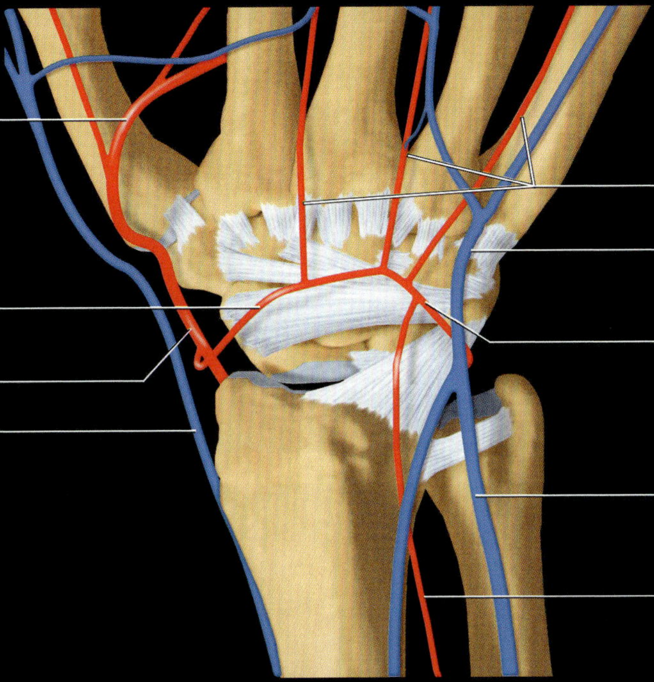

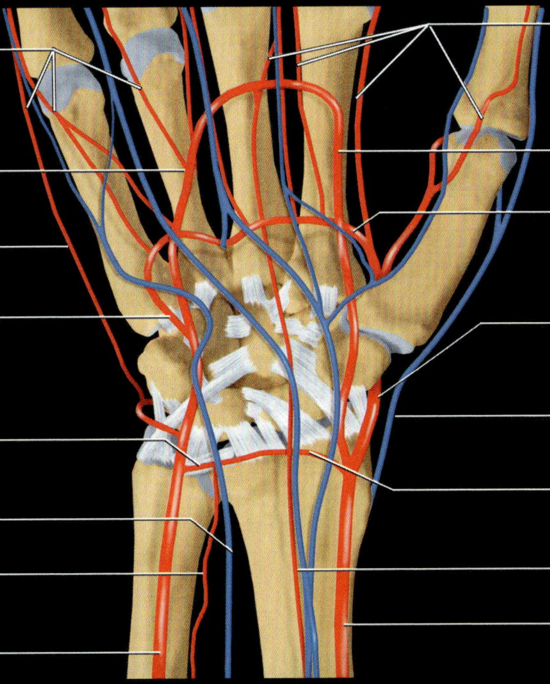

(Top) *Vasculature of the dorsal wrist is shown. The dorsal carpal arch supplies the distal radius, distal carpal row, and lateral proximal carpal row. The venous plexus drains into 2 main venous systems, the cephalic and basilic, with multiple anastomotic communications.* **(Bottom)** *Vasculature of the volar wrist is shown. Three major arterial arches are contributed to by the radial, ulnar, and interosseous arteries: Palmar (volar) carpal, deep, and superficial palmar arches. A dorsal venous plexus drains into the cephalic and basilic veins.*

TRANSVERSE US, WRIST AND PALM

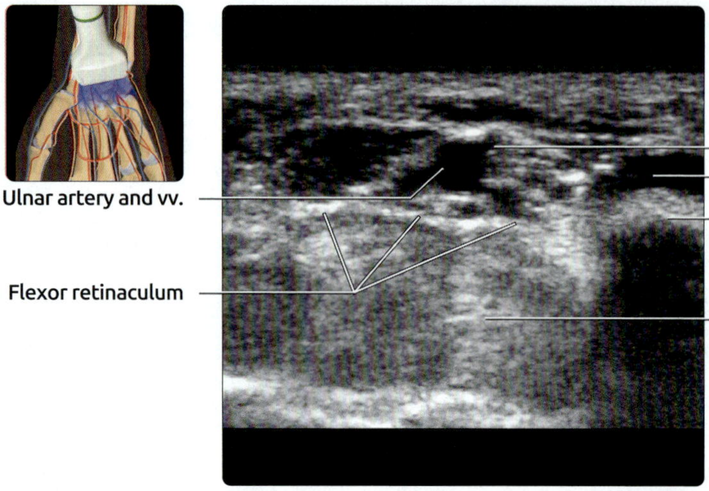

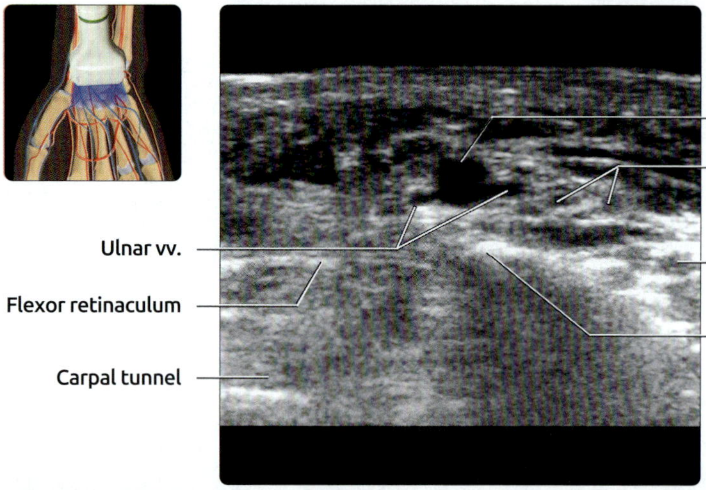

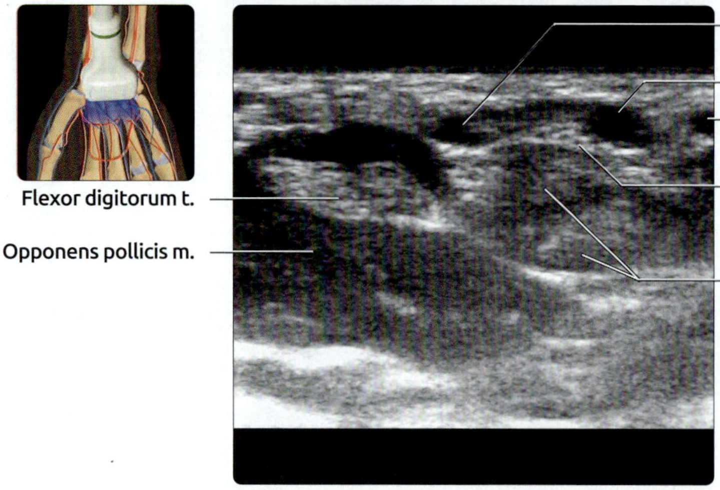

(Top) *Transverse grayscale ultrasound shows the ulnar artery at the level of the pisiform bone within Guyon canal. The ulnar artery at the wrist lies on the radial side of the pisiform. Between the ulnar artery and the pisiform lies the ulnar nerve. The ulnar artery divides either within or just beyond the Guyon canal.* **(Middle)** *Transverse grayscale ultrasound shows the ulnar artery at the level of the hook of hamate. Here, it is prone to compression injury (hypothenar hammer syndrome), which may damage its walls leading to either thrombosis, occlusion, or aneurysms. It is usually the superficial branch of the ulnar artery that is most prone to hypothenar hammer syndrome.* **(Bottom)** *Transverse grayscale ultrasound shows the ulnar artery at the level of the midpalm. The superficial branch of the ulnar artery contributes to the superficial palmar arch.*

TRANSVERSE US, DORSAL ASPECT

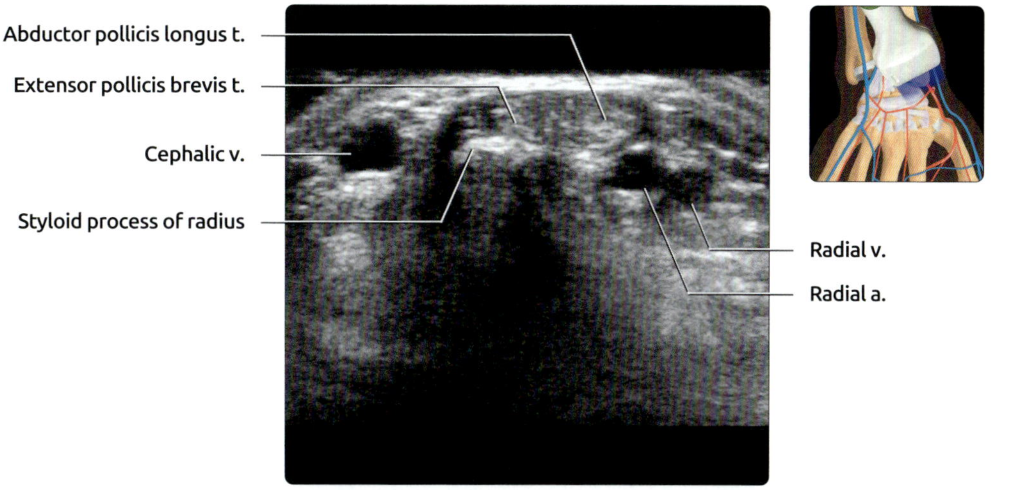

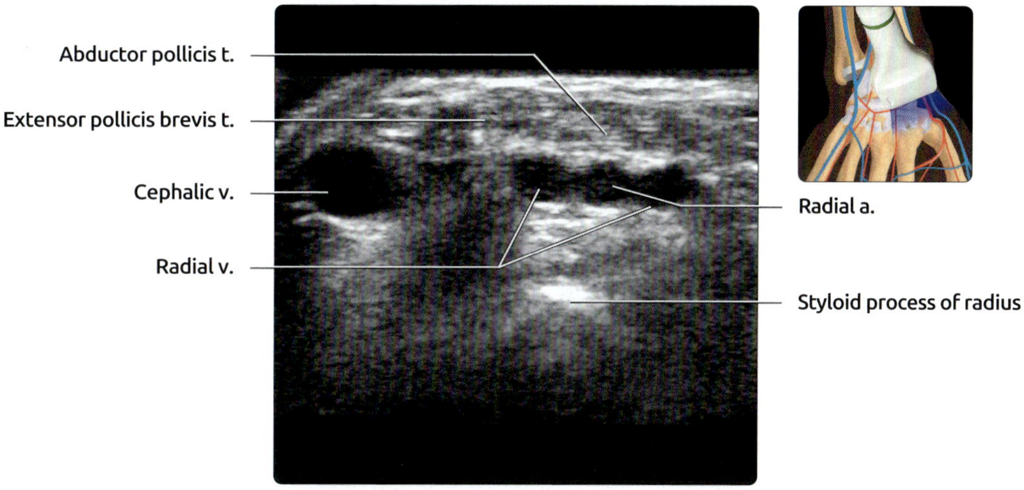

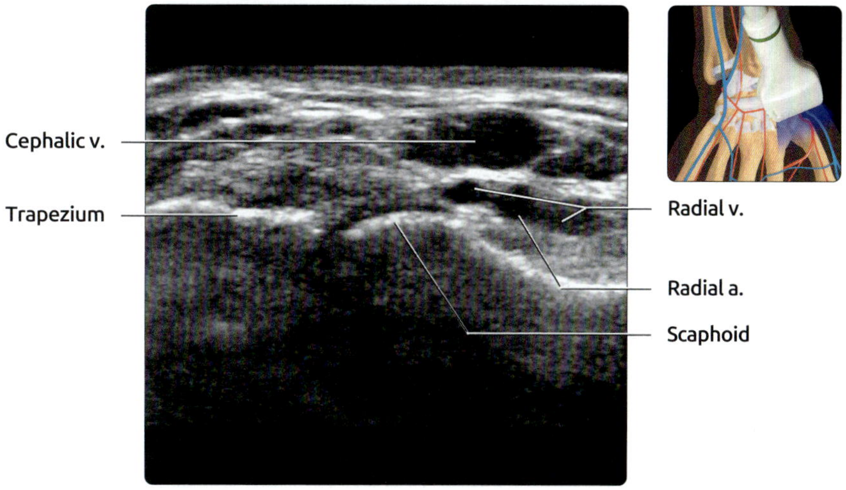

(Top) *Transverse grayscale ultrasound shows the radial artery just proximal to the wrist. The radial artery runs a superficial course parallel with the shaft of the radius in the distal forearm. At the wrist, it lies alongside the 1st extensor compartment, comprising abductor pollicis and extensor pollicis brevis tendons.* **(Middle)** *Transverse grayscale ultrasound shows the radial artery at the wrist joint level. The radial artery runs deep to the 1st extensor compartment tendons. Just before it winds around the wrist, it gives off the superficial palmar branch, which helps form the superficial palmar arch.* **(Bottom)** *Transverse grayscale ultrasound shows the radial artery at the dorsum of the hand. The deep branch of the radial artery runs on the dorsal surface of the scaphoid and trapezium bones and then passes deep between the oblique and transverse heads of the adductor pollicis muscle to enter the palm to form the deep palmar arch along with the deep branch of the ulnar artery.*

TRANSVERSE US, PALMAR ASPECT

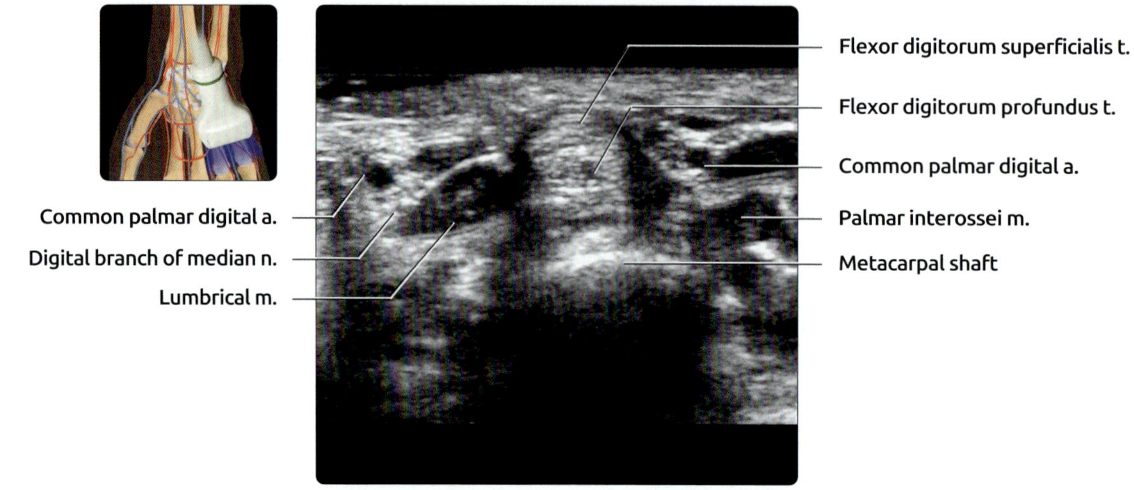

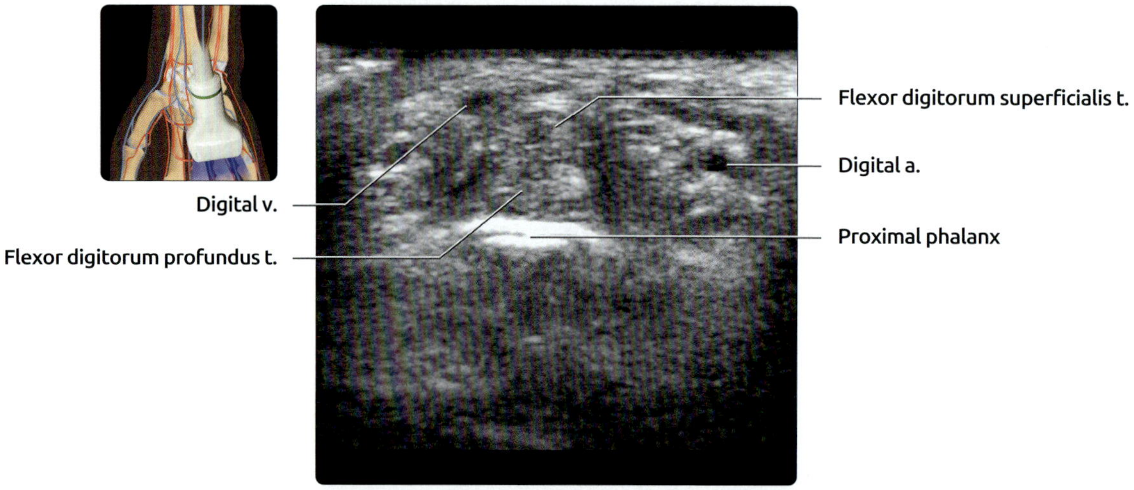

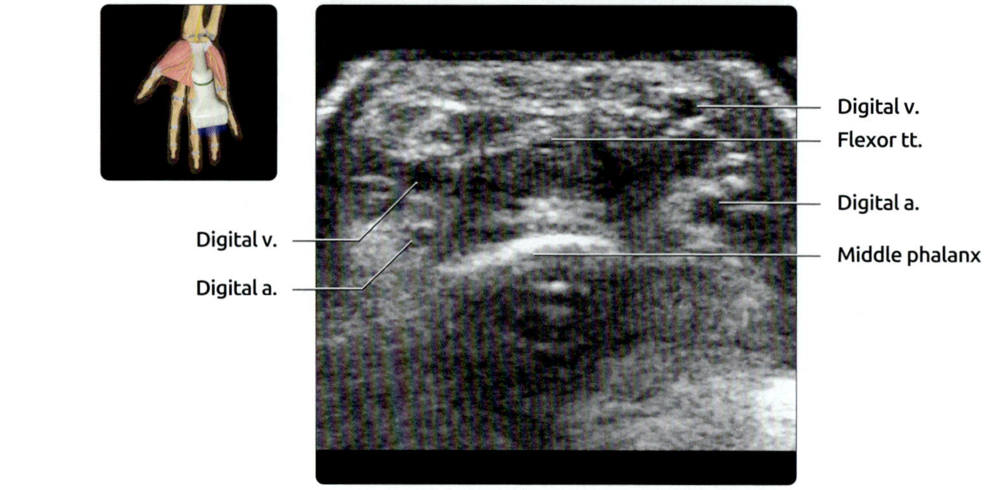

(Top) *Transverse grayscale ultrasound shows the vessels of the midpalm at the midmetacarpal shaft level. The common palmar digital arteries arise from the superficial palmar arch and divide at the finger bases into proper digital arteries. Each finger has 4 (2 volar, 2 dorsal) proper digital arteries.* **(Middle)** *Transverse grayscale ultrasound shows the digital vessels at the proximal phalanx level. The palmar digital arteries lie dorsal to the digital nerves. The palmar digital arteries are usually bigger than the dorsal digital arteries. Ulnopalmar arteries are twice as big as the radiopalmar arteries (1.8 mm vs. 1.1 mm). Peak systolic velocity in hand arteries is variable, ranging 30-70 cm/sec.* **(Bottom)** *Transverse grayscale ultrasound shows vessels at the middle phalanx level. The veins of the fingers and the hand are variable in number and location. Dorsal veins are of variable size, ranging 1.0-1.5 mm.*

HAND ARTERIES

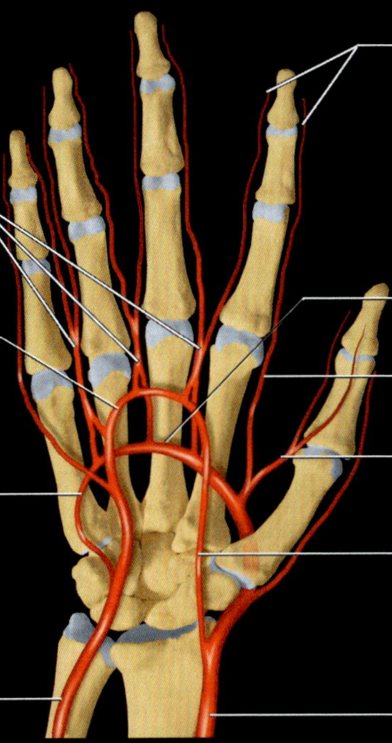

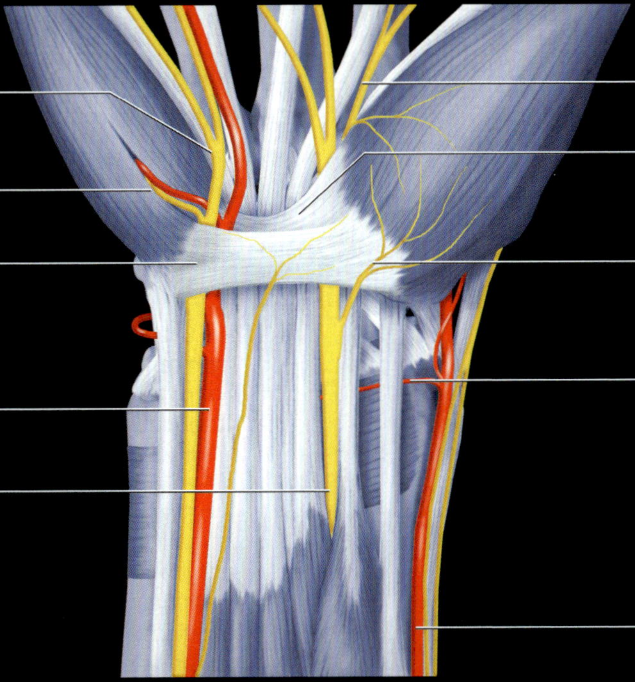

(Top) *Graphic shows the hand arteries. After giving off its contribution to the superficial palmar arch, the radial artery travels around the radial aspect of the wrist to the dorsum of the hand where it travels deep in the anatomic snuffbox. It then dives into the 1st interspace between the heads of the 1st dorsal interosseus muscle as well as between the transverse and oblique heads of the adductor pollicis muscle before forming the deep palmar arch.* **(Bottom)** *Graphic shows the relationship of the nerves and arteries to the volar carpal ligament (superficial fibers of flexor retinaculum that form the roof of Guyon canal) and flexor retinaculum (which forms the roof of the carpal tunnel). The ulnar nerve (and accompanying artery) pass deep to the volar ligament before branching into the deep and superficial branches.*

TERMINOLOGY

Abbreviations

- Abductor pollicis brevis (APB)
- Extensor pollicis longus (EPL)
- Extensor pollicis brevis (EPB)
- Flexor pollicis longus (FPL)
- Flexor pollicis brevis (FPB)
- Metacarpophalangeal (MCP)
- Opponens pollicis (OP)
- Adductor pollicis (AP)

GROSS ANATOMY

Muscles

- Thenar muscle
  - Composed of 4 muscles
    - APB
    - FPB
    - OP
    - AP
  - Origin: All originate from flexor retinaculum and tubercle of trapezium
    - APB also originates from tubercle of scaphoid
  - Insertion: APB and FPB share combined insertion at base of proximal phalanx radial aspect
    - OP inserts onto volar aspect 1st metacarpal
  - AP is often grouped with thenar muscles but is anatomically distinct from them
    - AP is separated from thenar muscles by fascial plane and innervated by ulnar nerve rather than median nerve
- Extensor muscles
  - EPL
    - Origin: Midulna and interosseus membrane
    - Insertion: Dorsal aspect base of distal phalanx of thumb
    - Extends both MCP and interphalangeal joints of thumb
  - EPB
    - Origin: Distal radius and interosseus membrane
    - Insertion: Dorsal aspect base of proximal phalanx of thumb
    - Extends MCP joint of thumb
- FPL
  - Origin: Midradius and interosseus membrane
  - Insertion: Base of distal phalanx of thumb
  - Flexes MCP and interphalangeal joints of thumb
  - FPL tendon sheath (a.k.a. radial bursa)
    - Encompasses FPL tendon from just proximal to carpal tunnel to its insertion (distal phalanx of thumb)
    - Radial bursa occasionally communicates with ulnar bursa (common flexor sheath) at level of carpal tunnel
- Extensor expansion
  - Formed by fibers from AP on ulnar side and APB on radial side
  - Aids extension of interphalangeal joint
  - AP contribution to extensor expansion is a.k.a. adductor aponeurosis

Tendon Pulley System

- More variable than finger tendon pulley system
- In addition to standard A1 and A2 pulleys, have oblique and variable annular pulleys (Av)
- A1 pulley: At MCP joint level
- Av pulley: Along proximal 1/2 of proximal phalanx, variable in position, runs either obliquely or transversely across proximal phalanx
- Oblique pulley: Runs obliquely just proximal to A2 pulley
- A2 pulley: At interphalangeal joint level

IMAGING ANATOMY

Overview

- On dorsal transverse imaging, extensor expansions are seen as thin, regular, hypoechoic bands arising from edges of extensor tendon
- On palmar transverse imaging, annular pulleys are seen as thin, ring-shaped structures covering flexor tendons and inserting in palmar plate

ANATOMY IMAGING ISSUES

Imaging Recommendations

- Thenar muscles can be identified reliably on ultrasound
  - All thenar muscles can be identified by placing transducer in orthograde position, parallel to direction of muscle fibers
- Extensor and flexor tendons of thumb bear similar ultrasound features to other flexor and extensor tendons of hand

CLINICAL IMPLICATIONS

Clinical Importance

- Radial bursa may communicate with ulnar bursa at level of carpal tunnel, allowing spread of infection
- Ulnar collateral ligament of MCP joint of thumb is especially prone to tearing in forced abduction and extension
  - Referred to as gamekeeper's thumb or skier's thumb
  - If ulnar collateral ligament is torn, position of its proximal free end with respect to adductor aponeurosis is of importance
    - If proximal portion is displaced superficial to adductor aponeurosis, it is referred to as Stener lesion, and surgery is usually required
    - Stener lesion: Interposition of adductor aponeurosis between proximally retracted torn ulnar collateral ligament and distal portion
    - If proximal portion remains deep to adductor aponeurosis, conservative management may suffice
- A1 pulley implicated in pathogenesis of trigger thumb
  - Normal thickness of A1 pulley < 0.5 mm
  - In trigger thumb, A1 pulley is thickened to ~ 1 mm or greater
  - ± restricted movement of FPL tendon under A1 pulley during passive flexion on real-time US imaging

DEEP DISSECTION OF HAND

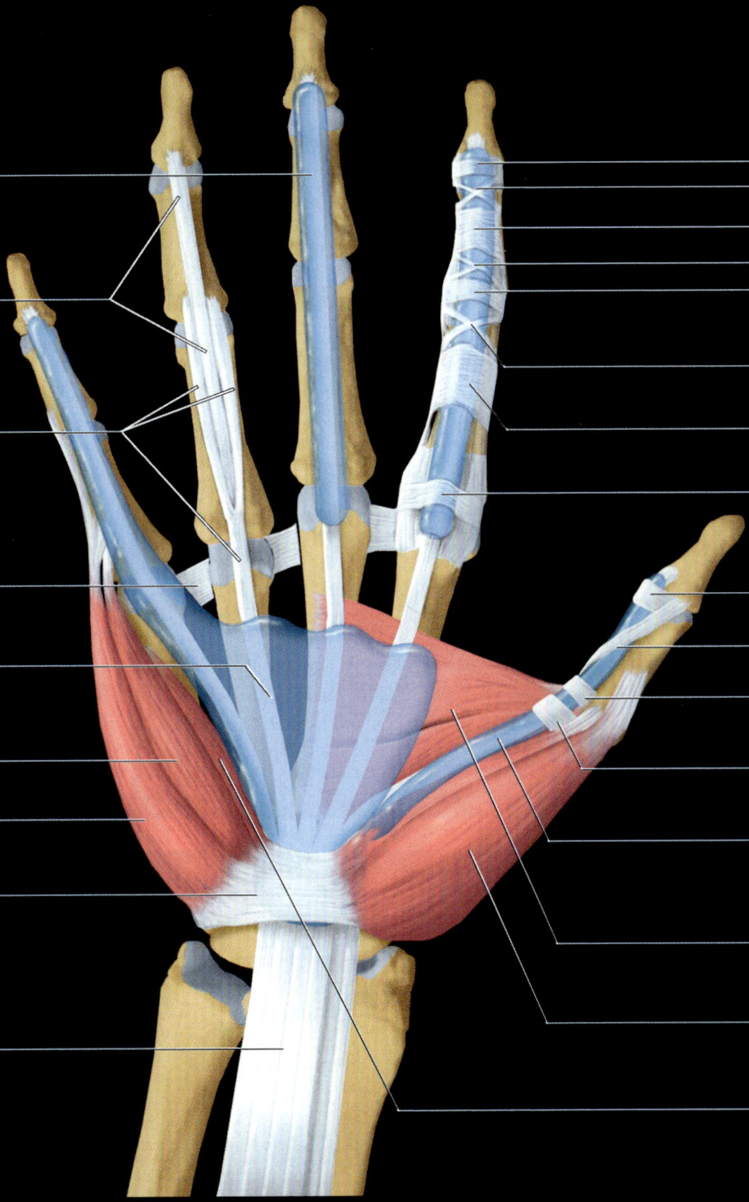

Graphic shows the pulley system of the 2nd digit. The pulley system for 3rd-5th digits, which are identical to that demonstrated for 2nd digit, are removed in this graphic. The common digital sheath of the 4th digit is removed to show the relationship of the flexor digitorum superficialis and flexor digitorum profundus tendons. Deep transverse metacarpal ligaments connect volar plates (not shown) of digits 2-5. Although there is overlap of radial and ulnar bursae in this image, these structures normally do not communicate. It is, however, important to know that radial and ulnar bursae may communicate as a normal variation in a small percentage of population. Similarly, 1 or more of common digital sheaths may communicate with radial bursa in up to 10% of the normal population. These normal variant bursal communications are important as they can provide routes for more extensive spread of infection.

US, CARPOMETACARPAL JOINT

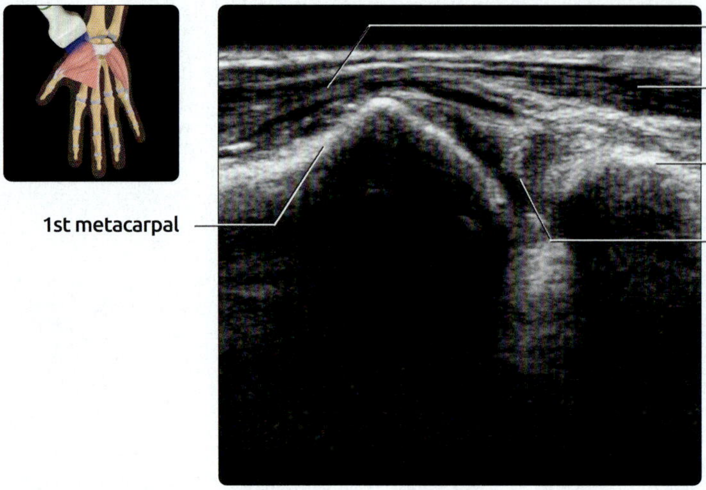

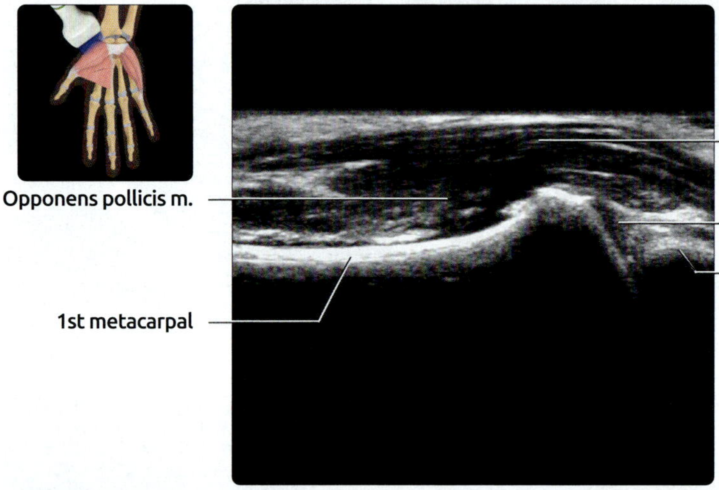

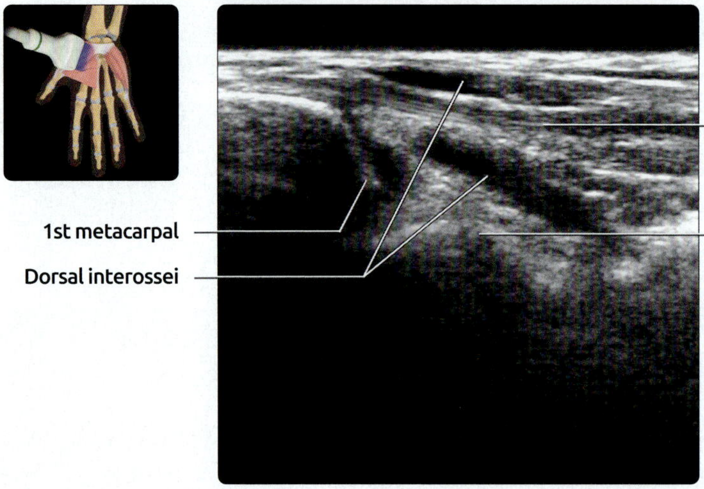

(Top) *Longitudinal grayscale ultrasound shows the palmar aspect of the carpometacarpal joint with radial deviation of the thumb. The carpometacarpal joint of the thumb is by far the most mobile of the carpometacarpal joints. It is best viewed by radially deviating the thumb.* **(Middle)** *Longitudinal grayscale ultrasound shows the palmar aspect of the carpometacarpal joint of the thumb. Because of its mobility, the carpometacarpal joint of the thumb is the only carpometacarpal joint prone to osteoarthritis. This gives rise to deep pain in the thenar eminence.* **(Bottom)** *Longitudinal grayscale ultrasound shows the dorsal aspect of the carpometacarpal joint of the thumb. The large dorsal component of the 1st metacarpal base relative to the articulating surface on the trapezium leads to a normal step-off on the dorsal aspect of this joint.*

US, METACARPOPHALANGEAL JOINT

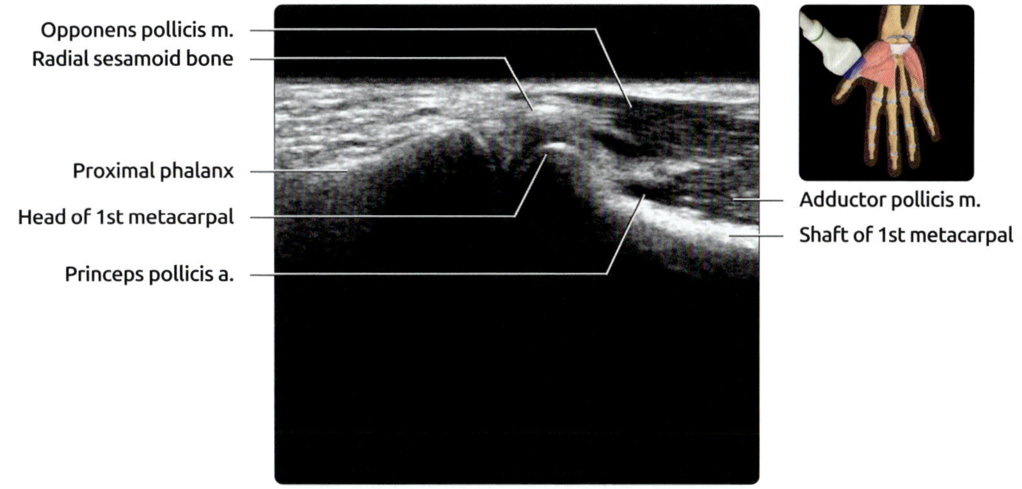

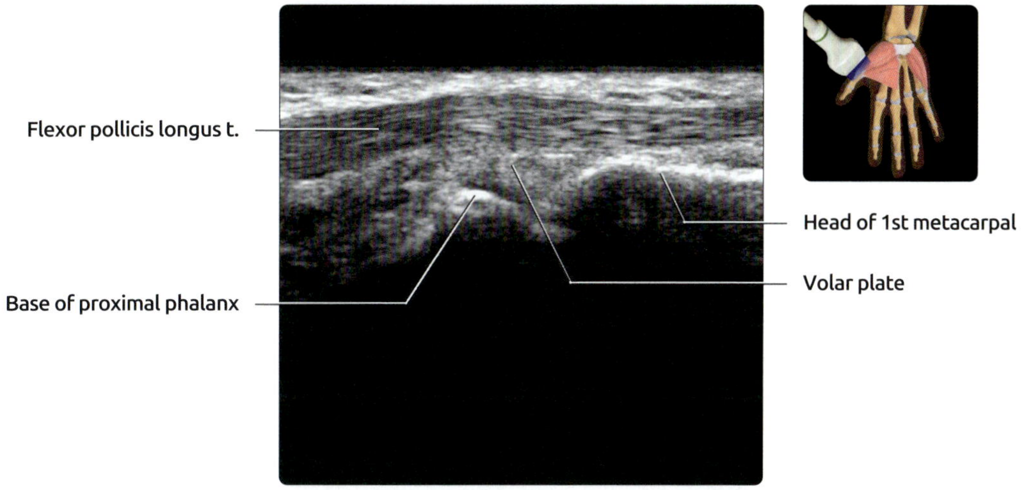

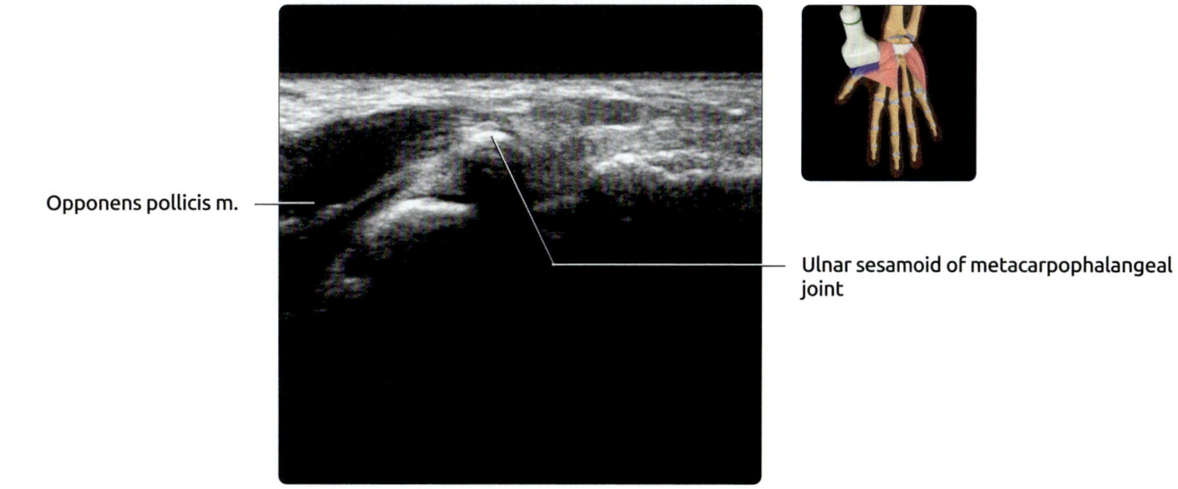

(Top) *Longitudinal grayscale ultrasound shows the metacarpophalangeal joint of the thumb. The collateral ligament complex makes up most of the lateral aspects of the metacarpophalangeal and interphalangeal joint capsules. The ulnar collateral ligament of the 1st metacarpophalangeal joint is especially prone to tearing during forced abduction and extension. Such an injury is commonly referred to as gamekeeper's thumb or skier's thumb.* **(Middle)** *Longitudinal grayscale ultrasound shows the metacarpophalangeal joint of the thumb. The volar plate constitutes the majority of the volar aspect of the joint capsule of the metacarpophalangeal joint. The volar plate may be avulsed with forced extension.* **(Bottom)** *Transverse grayscale ultrasound shows the 1st metacarpophalangeal joint. The metacarpal head is usually involved in erosions before the proximal phalanx. The metacarpophalangeal joint is an area prone to erosions in inflammatory arthritis.*

ULNAR COLLATERAL LIGAMENT, THUMB

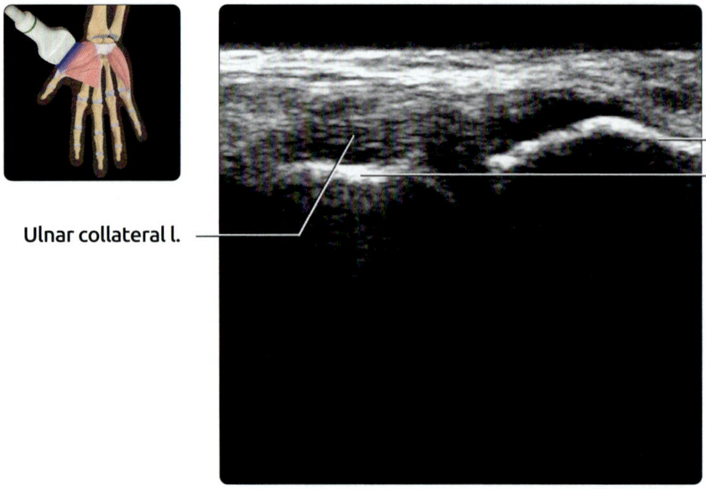

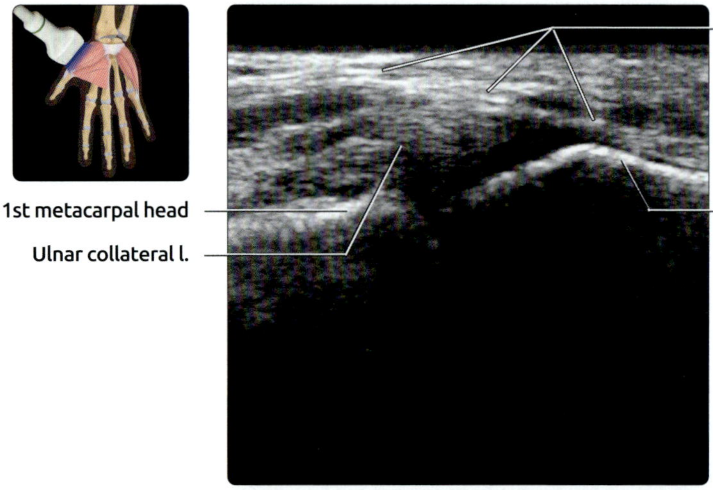

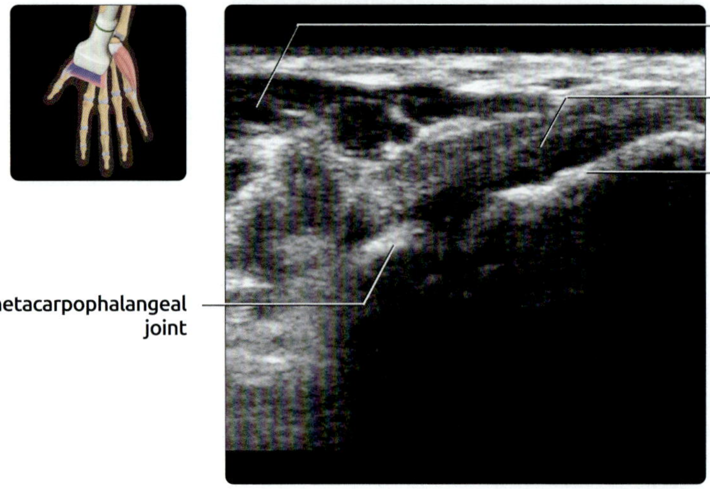

(Top) *Longitudinal grayscale ultrasound along the palmar aspect of the ulnar collateral ligament of the thumb is shown. The ulnar collateral ligament of the metacarpophalangeal joint is commonly injured by forced abduction of the thumb. Avulsion normally occurs at the distal (i.e., proximal phalangeal) attachment.* **(Middle)** *Longitudinal grayscale ultrasound shows the ulnar collateral ligament at the midportion. The adductor pollicis aponeurosis overlies the ulnar collateral ligament. The thin adductor aponeurosis is attached to the medial side of the proximal phalanx base and the ulnar sesamoid bone. If the torn ulnar collateral ligament displaces above the adductor aponeurosis (Stener lesion), this is an indication for operative repair.* **(Bottom)** *Transverse grayscale ultrasound shows the metacarpophalangeal joint of the thumb. This shows the normal thick ulnar collateral ligament in transverse section.*

TRANSVERSE US PALMAR ASPECT, THUMB

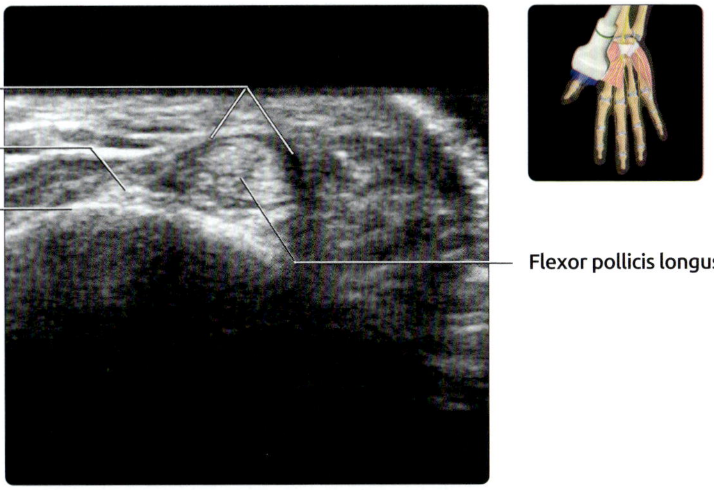

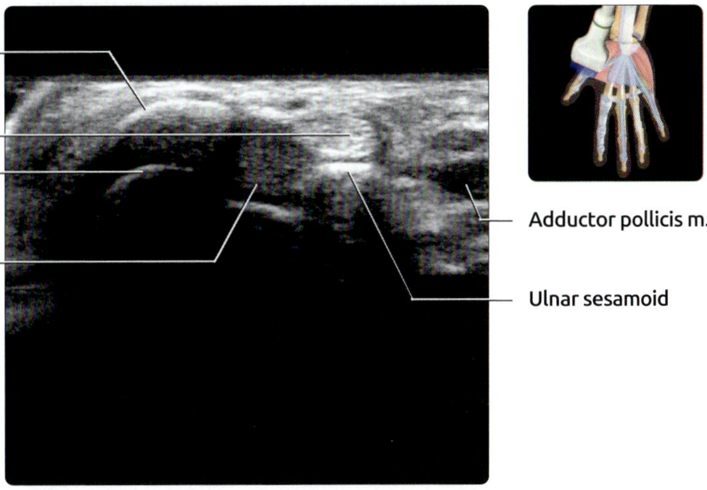

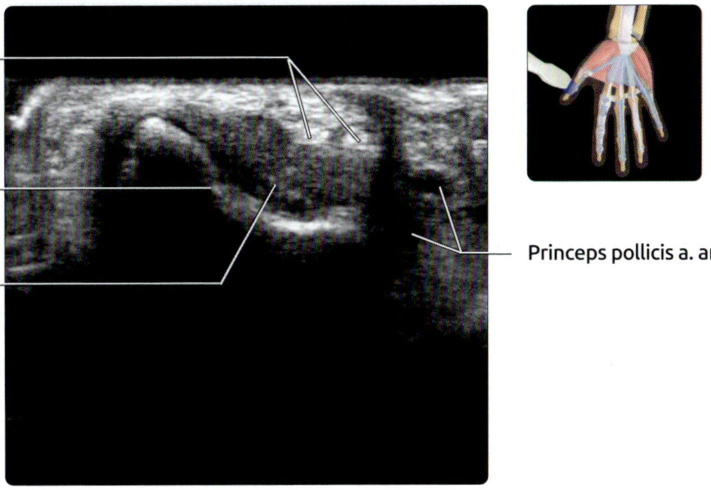

(Top) *Transverse grayscale ultrasound along the flexor aspect of the thumb at the metacarpal head level is shown. The flexor pollicis brevis inserts into the radial sesamoid and the base of the proximal phalanx. The flexor pollicis longus inserts into the base of the distal phalanx.* **(Middle)** *Transverse grayscale ultrasound along the flexor aspect of the thumb at the metacarpophalangeal joint level is shown. The flexor pollicis longus tendon passes over the ulnar sesamoid bone of the metacarpophalangeal joint.* **(Bottom)** *Transverse grayscale ultrasound along the flexor aspect of the thumb at the interphalangeal joint level is shown. The flexor pollicis longus tendon is held in position primarily by the A1 pulley at the metacarpophalangeal joint and the A2 pulley at the interphalangeal joint. The flexor pollicis longus tendon may divide into 2 slips at its insertion.*

EXTENSOR ASPECT, THUMB

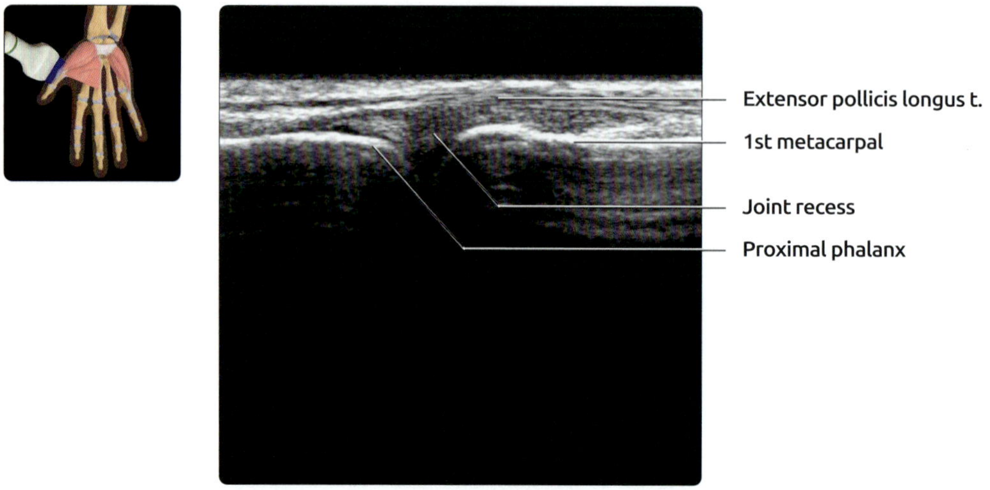

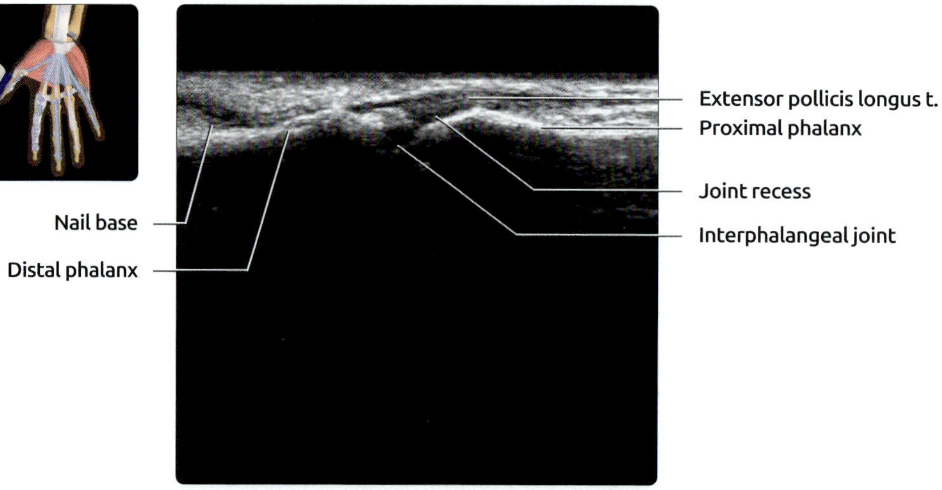

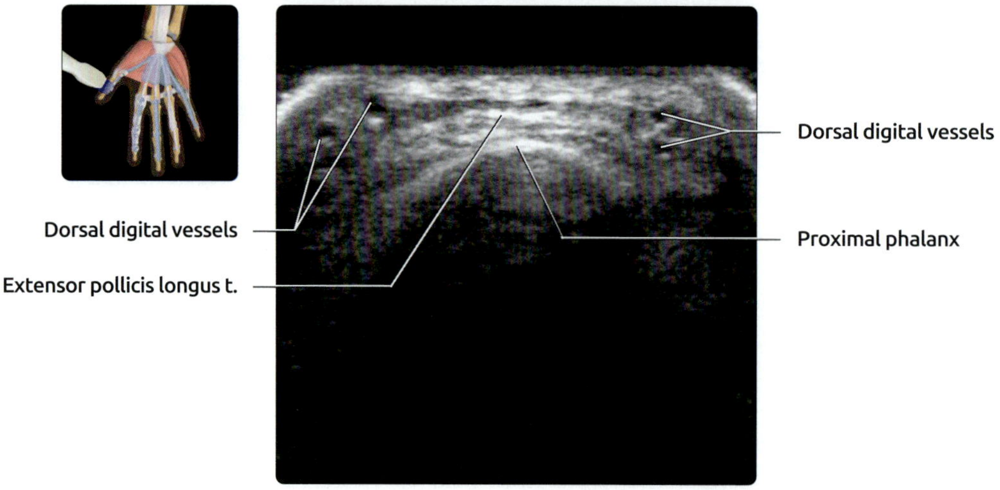

(Top) *Longitudinal grayscale ultrasound along the extensor aspect of the thumb at the metacarpophalangeal joint level is shown. The extensor pollicis longus tendon thins as it merges with the extensor expansion on the dorsum of the thumb.* **(Middle)** *Longitudinal grayscale ultrasound along the extensor aspect of the thumb at the interphalangeal joint level is shown. The extensor pollicis longus tendon inserts into the distal phalanx.* **(Bottom)** *Transverse grayscale ultrasound along the extensor aspect of the thumb at the proximal phalanx level is shown. The thin extensor expansion of the extensor pollicis longus is visible.*

TRANSVERSE US, THENAR EMINENCE

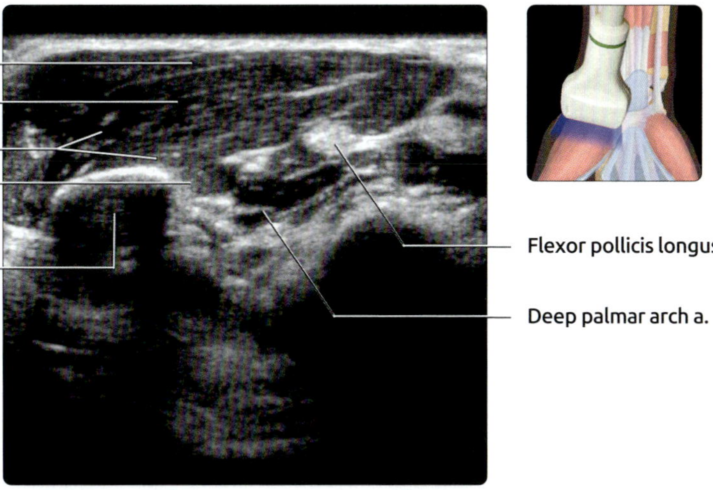

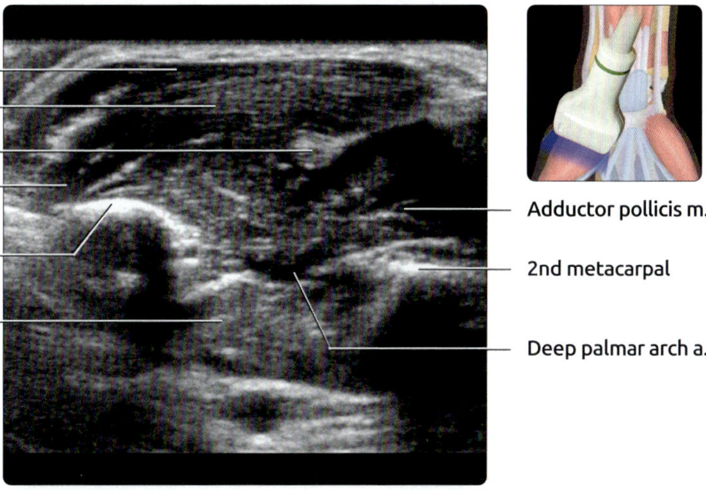

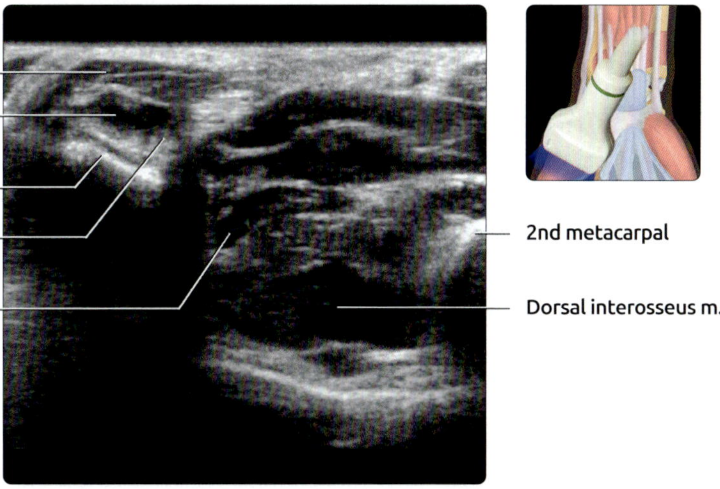

(Top) *Transverse grayscale ultrasound along the proximal aspect of the thenar eminence is shown. The thenar eminence is composed of 4 muscles: The abductor, opponens, flexor, and adductor pollicis from superficial to deep. The adductor muscle is, strictly speaking, not part of the thenar eminence since it is separated by a fascial layer and has a separate innervation.* **(Middle)** *Transverse grayscale ultrasound shows the midportion of the thenar eminence. The flexor pollicis longus tendon is easily recognized as an echogenic tendon alongside the flexor pollicis brevis, deep to the opponens pollicis and superficial to the adductor pollicis muscle.* **(Bottom)** *Transverse grayscale ultrasound shows the distal portion of the thenar eminence.*

LONGITUDINAL US, THENAR EMINENCE

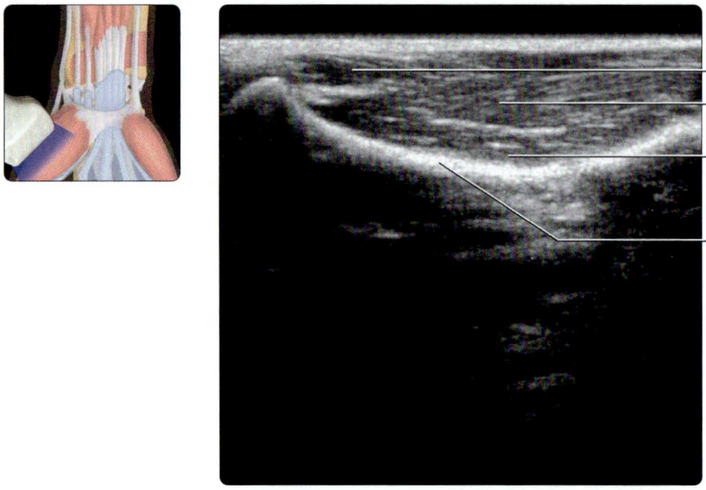

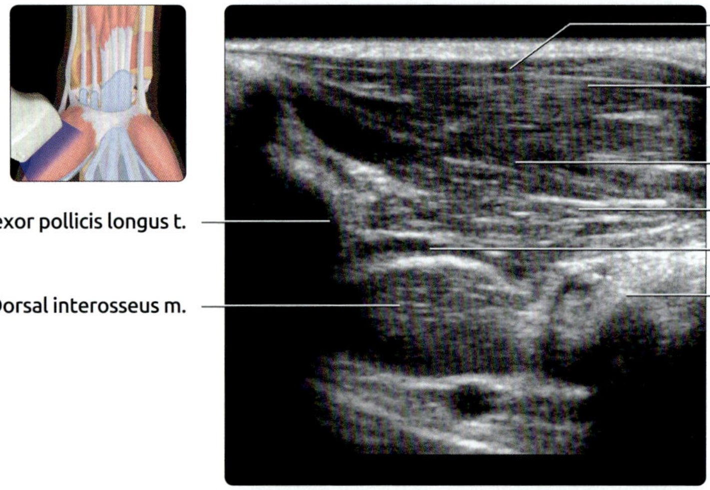

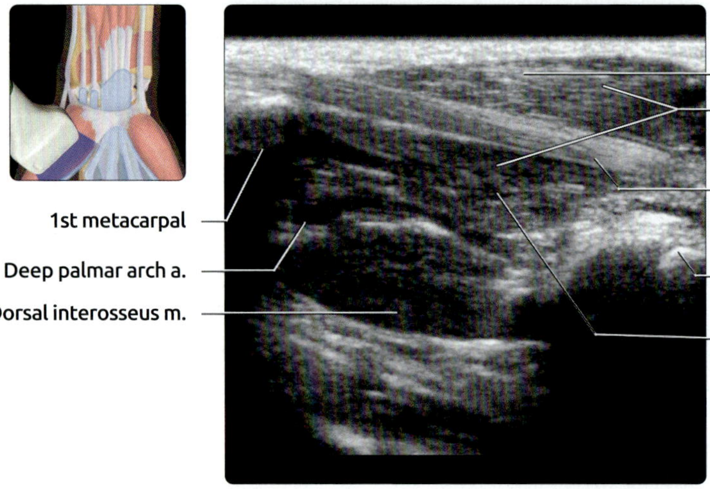

(Top) *Longitudinal grayscale ultrasound along the proximal aspect of thenar eminence is shown. ABOF, a mnemonic to remember the muscles of the thenar eminence, stands for abductor pollicis brevis, opponens pollicis, and flexor pollicis brevis, which lie superficial to deep in that order.* **(Middle)** *Longitudinal grayscale ultrasound shows the midthenar eminence. The abductor pollicis brevis and flexor pollicis brevis share a combined insertion at the lateral base of the proximal phalanx of the 1st digit.* **(Bottom)** *Longitudinal grayscale ultrasound along the ulnar aspect of the thenar eminence is shown. Although often grouped with the thenar muscles due to proximity, the adductor pollicis is distinct from them, as it is separated from the thenar muscles by a fascial plane and innervated by the ulnar nerve as opposed to the median nerve.*

GRAPHICS, STENER LESION

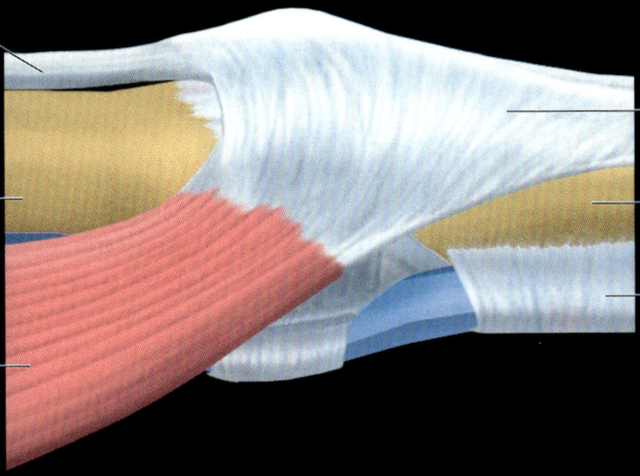

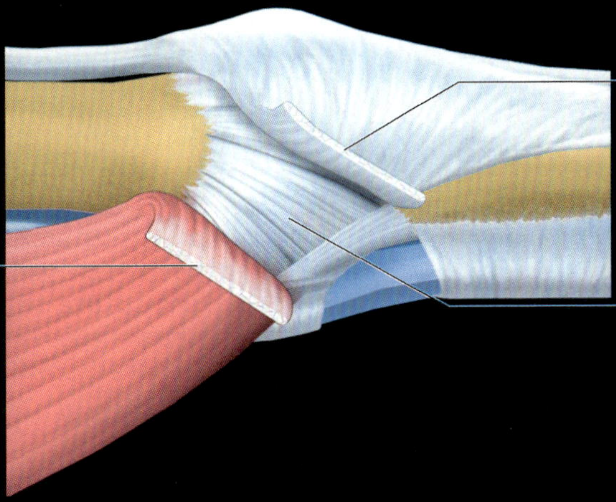

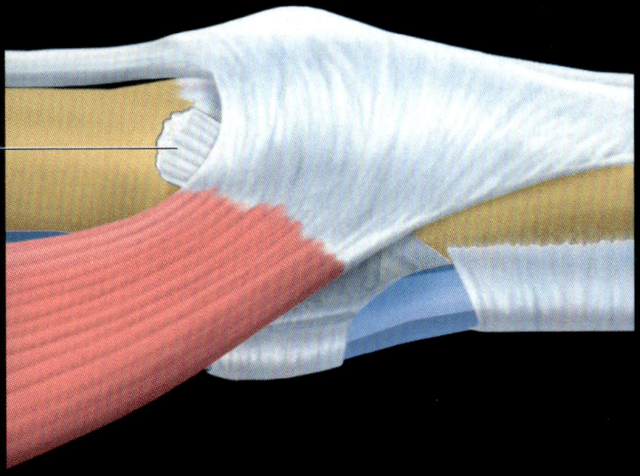

Graphic shows the ulnar aspect of the metacarpophalangeal joint of the thumb. On the top image, the adductor tendon aponeurosis attaches to the extensor expansion on the dorsum of the thumb. The middle image shows a reflected adductor aponeurosis with the ulnar collateral ligament beneath. When a complete tear of the ulnar collateral ligament occurs, shown in the bottom image, the proximal end of the ligament may be retracted back over the adductor aponeurosis. This is referred to as a Stener lesion and does require surgical intervention.

TERMINOLOGY

Abbreviations

- Extensor digitorum communis (EDC)
- Extensor digiti minimi (EDM)
- Interphalangeal (IP) joint
- Metacarpophalangeal joint (MCP) joint

GROSS ANATOMY

Muscles

- Flexor aspect
  - Flexor digitorum superficialis
    - Origin: Common flexor tendon origin and midradius
    - Insertion: Volar plates of proximal IP joints and bases of middle phalanges digits 2-5
    - Superficialis tendon is superficial to profundus tendon in palm until it divides at level of proximal 1/3 of proximal phalanx
    - 2 slips of superficialis tendon pass around profundus tendon and reunite deep to profundus tendon prior to insertion
    - Flexes both MCP joints, aided by lumbricals and interossei and proximal IP joints of digits 2-5
  - Flexor digitorum profundus (FDP)
    - Origin: Proximal radius, midradius, and interosseus membrane
    - Profundus tendon passes through divided superficialis tendon to insert at base of distal phalanx
    - Insertion: Volar plates of distal IP joints and bases of distal phalanges digits 2-5
    - Flexes distal and proximal IP joints as well as MCP joints aided by lumbricals and interossei
- Palmar interossei
  - Denoted 1-3 from radial to ulnar
  - Origin: Mesial palmar diaphyses of 2nd, 4th, and 5th metacarpals
  - Insertion: Mesial lateral bands and proximal phalanges have same base as digit of origin
  - Adducts digits and assists lumbricals with flexion MCP and extension IP joints in digits 2, 4, and 5
- Extensor aspect
  - EDC
    - Origin: Common extensor tendon origin (lateral humeral epicondyle)
    - Insertion: As central slip at bases of middle phalanges and proximal IP joint capsule
    - Separate EDC tendon to 5th digit tendon is present in 50% of general population
  - Extensor indicis
    - Origin: Posterior ulna, distal ulna, and interosseus membrane
    - Insertion: Blends with 2nd digit EDC tendon and extensor hood
  - EDM
    - Origin: Common extensor tendon origin (lateral humeral epicondyle)
    - Insertion: 2 tendons of EDM fuse with one another and with 5th digit EDC tendon prior to insertion at base of proximal phalanx 5th digit
- Lumbricals
  - Numbered 1-4 from radial to ulnar
  - Origin: FDP tendons, just distal to carpal tunnel
  - Insertion: Radial lateral bands of digits 2-5
    - Extend IP and flex MCP joints of digits 2-5
- Dorsal interossei
  - Numbered 1-4 from radial to ulnar
  - Origin: Dorsolateral metacarpal diaphyses
  - Insertion: Adjacent lateral bands of digits 2-4
    - Extend 2nd-5th IP joints and abduct digits 2-5

Tendon Pulley System

- Annular pulley
  - Retinaculum that retain flexor tendons at 5 points = annular pulleys
  - Prevent tendon bowstringing during finger flexion
  - **Odd**-numbered pulleys at **oval** parts (i.e., joints)
  - **Even**-numbered pulleys at **elegant** waists (i.e., shafts)
  - A2 and A4 are functionally most important pulleys to prevent bowstringing
    - A1: MCP joint and base of proximal phalanx
    - A2: Middle and distal 3rds of proximal phalanx
    - A3: Proximal IP joint
    - A4: Midshaft of middle phalanx
    - A5: Distal IP joint
  - Also cruciform pulleys, C1-C3
- Extensor expansion
  - Begins just proximal to MCP joints and terminates just proximal to proximal IP joint
  - Dorsal expansion of fibers oriented perpendicular to long axis of extensor tendons
  - Fibers of extensor expansion interdigitate with EDC tendons to prevent lateral translation

Tendon Sheaths

- 2nd-4th flexor synovial sheaths extend from metacarpal neck to bases of distal phalanges
- Different for 5th digit flexor tendons, which are covered by synovial sheath over their entire course (to level of distal IP joint)

ANATOMY IMAGING ISSUES

Imaging Approaches

- Extensor tendons in digits often seen best on longitudinal imaging
- Pulleys are seen as focal hypoechoic thickening overlying tendon sheath
  - As pulleys are widened at base, thickness of pulleys should be measured at 10 or 2 o'clock positions
  - Do not misdiagnose as tendon sheath thickening
  - A1 and A2 pulley can be depicted in most cases
  - Injury inferred by bowstringing of flexor tendons
  - Compare with normal fingers

CLINICAL IMPLICATIONS

Clinical Importance

- Digital sheaths may connect to ulnar bursa in up to 10% of population, providing route for spread of infection from digits 2-4 to common flexor sheath and vice versa
- Pulley injury inferred by bowstringing of flexor tendons

TENDON SHEATHS

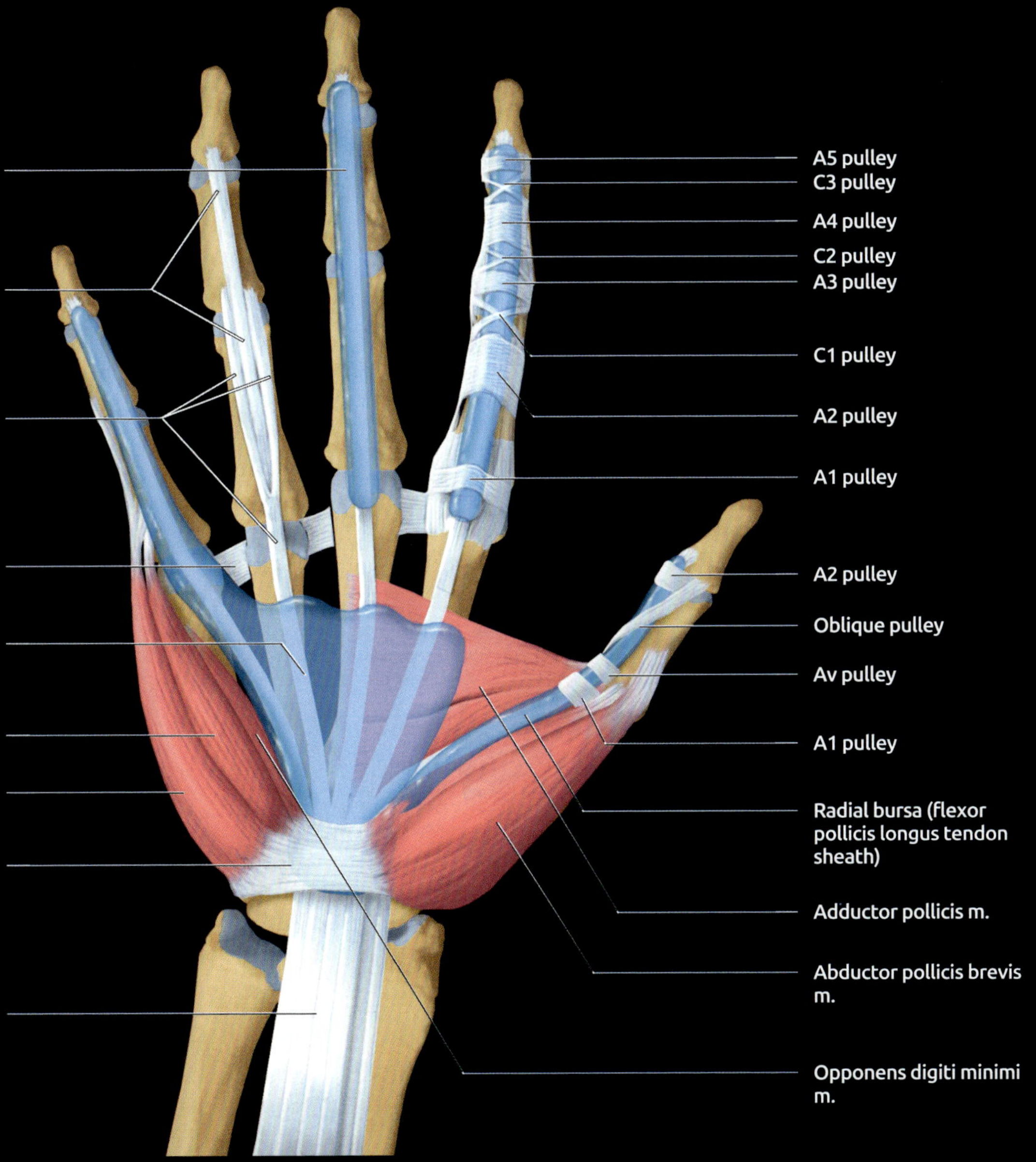

Graphic of pulley system on the palmar side of the 3rd-5th digits is identical to that demonstrated for the 2nd digit. The common digital sheath of the 4th digit is removed to show the relationship of the flexor digitorum superficialis and flexor digitorum profundus tendons. Deep transverse metacarpal ligaments connect the volar plates (not shown) of digits 2-5. Although there is overlap of the radial and ulnar bursae in this image, these structures normally do not communicate. It is, however, important to know that the radial and ulnar bursae may communicate as a normal variation in a small percentage of the population. Similarly, any one or more of the digital sheaths may communicate with the radial bursa in up to 10% of the normal population. These normal variant bursal communications are important as they can provide routes for more extensive spread of infection.

LIGAMENTS AND PULLEYS

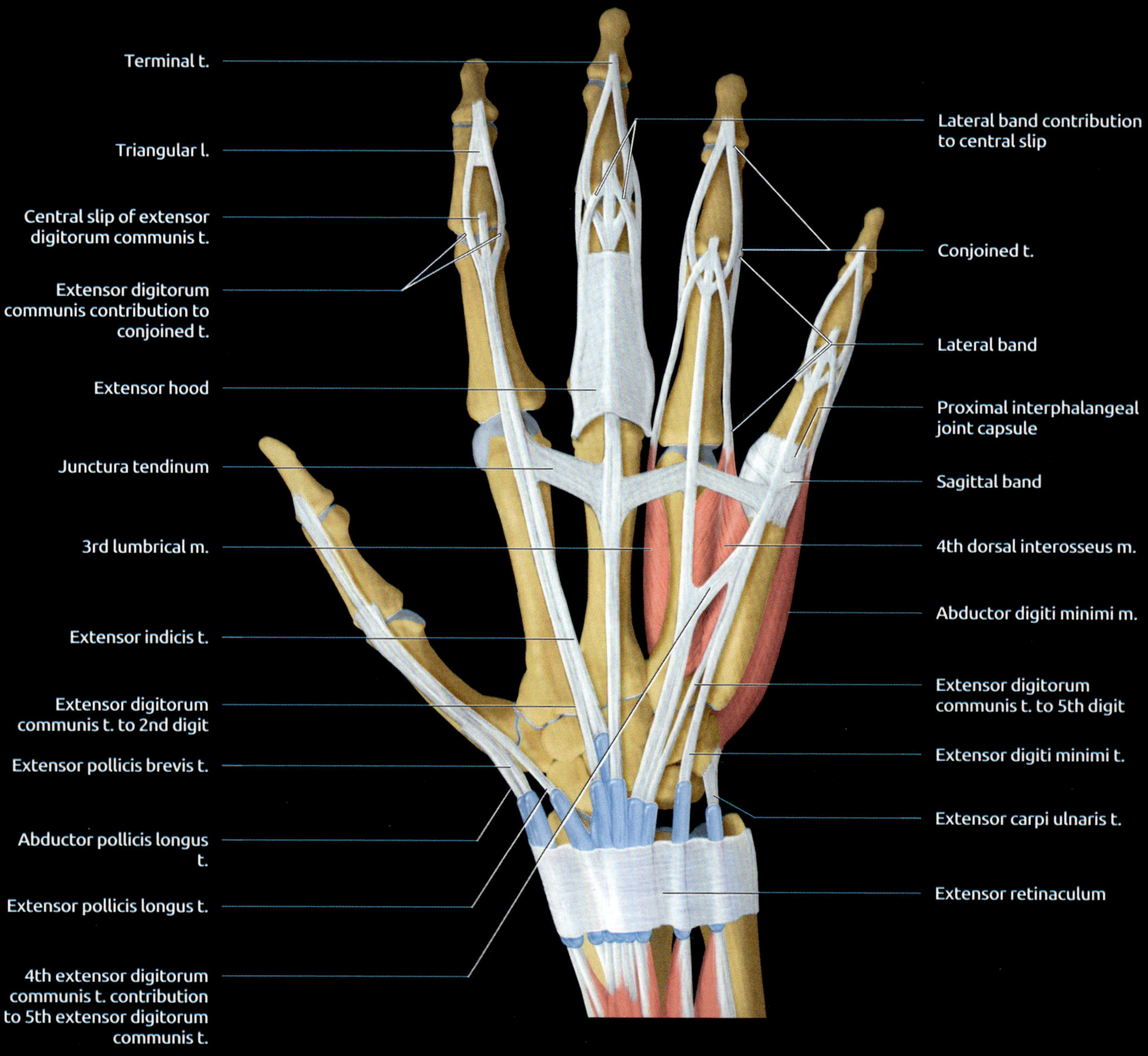

Graphic of the dorsum hand and finger extensor mechanism shows different components of the extensor mechanism on different digits. With the exception of the extensor indicis, extensor digiti minimi, abductor digiti minimi, and 4th extensor digitorum communis tendon contribution to the 5th extensor digitorum communis tendon, any structure on any of the 2nd-5th digits can be extrapolated to any and all of the other 2nd-5th digits. Note that the 2nd extensor retinaculum compartment and its contents (extensor carpi radialis longus and brevis tendons) are not included in this graphic. Although the abductor pollicis longus tendon travels in the 1st compartment of the extensor retinaculum, its insertion (not shown) is actually on the radial aspect of the volar surface of the base of the 1st metacarpal.

GRAPHIC AND LONGITUDINAL US, METACARPOPHALANGEAL JOINT

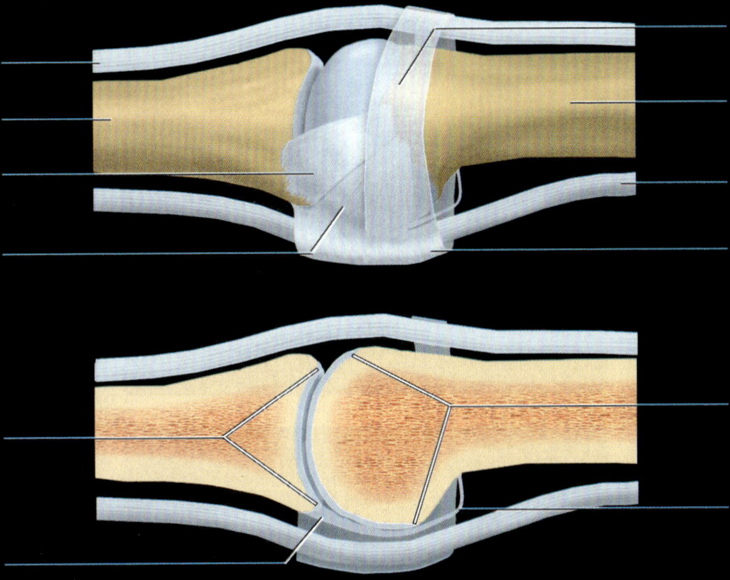

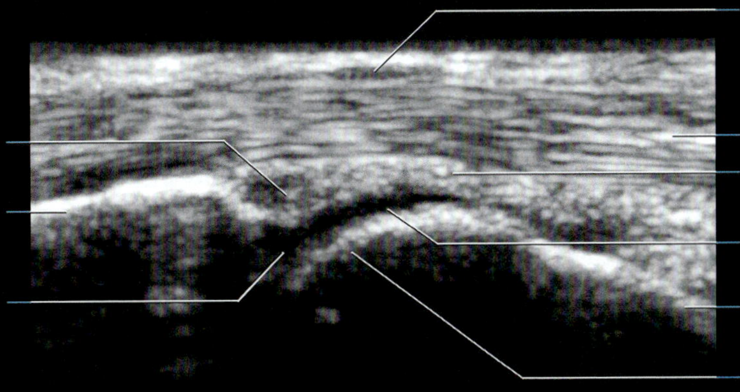

(Top) *Surface and cut-section graphics show the metacarpophalangeal (MCP) joint. The collateral ligament complex is composed of 2 separate fiber bands. The main collateral ligament inserts on the base of the adjacent phalanx. The accessory collateral ligament inserts on the volar plate. The volar plate is thick distally and thin and redundant proximally. With the exception of the sagittal band (present only at MCP joints), anatomy in this graphic is duplicated in all MCP and interphalangeal (IP) joints in the hand.* **(Bottom)** *Longitudinal grayscale US shows the middle finger at the MCP joint.*

LONGITUDINAL US

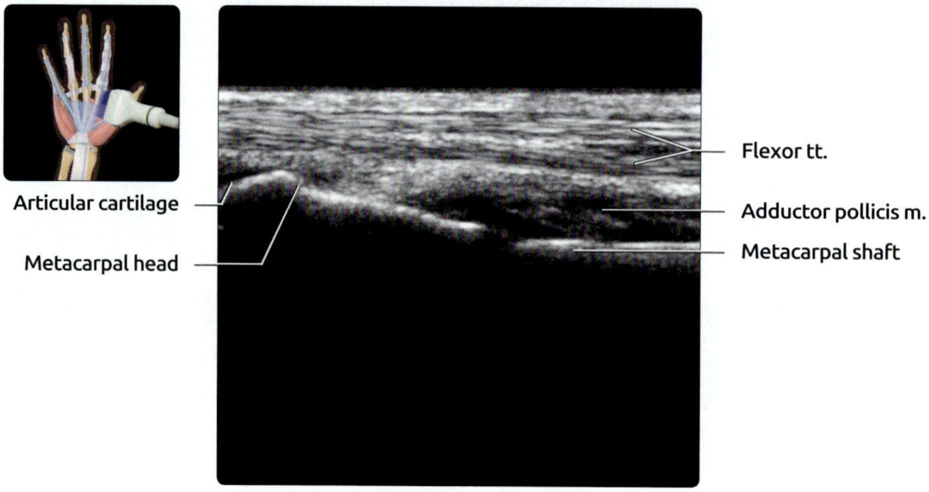

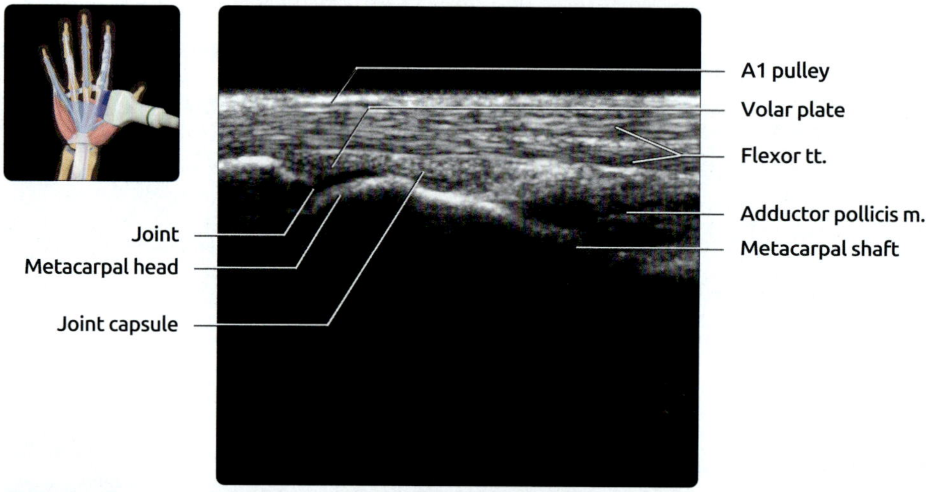

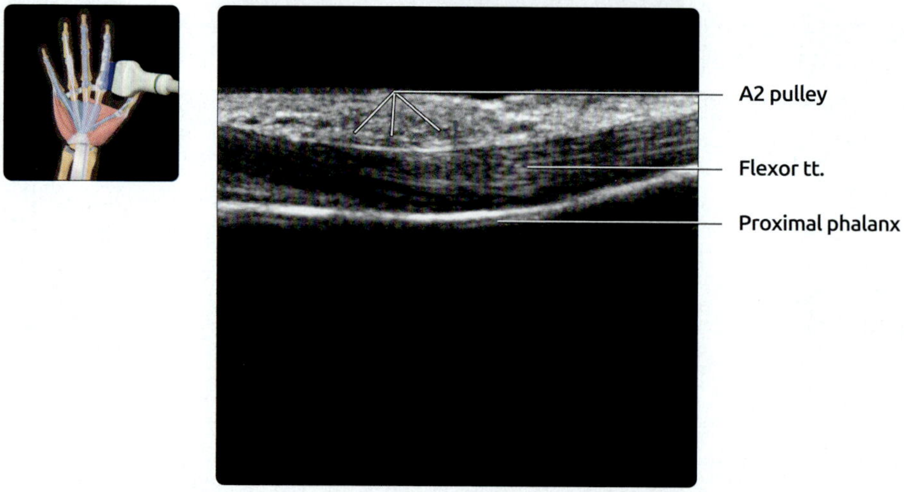

(Top) *Longitudinal grayscale US shows the finger at the metacarpal shaft level. From the base of the metacarpals (carpal tunnel outlet) to the neck of the metacarpals, the flexor tendons of the 2nd-4th fingers (i.e., index, middle, and ring fingers) are devoid of flexor tendon sheaths. A normal flexor tendon sheath cannot be seen on ultrasound and can only be seen when it is distended. Do not confuse the flexor tendon pulleys with tendon sheath thickening.* **(Middle)** *Longitudinal grayscale US shows fingers at the MCP joint level. MCP joints are narrowed dorsally and flare in the volar direction, increasing contact with the proximal phalanx as the joint is flexed. The capsule of the MCP joint extends from the metacarpal neck level to the base of proximal phalangeal joint.* **(Bottom)** *Longitudinal grayscale US shows the A2 pulley of the middle finger at the proximal phalanx.*

LONGITUDINAL US

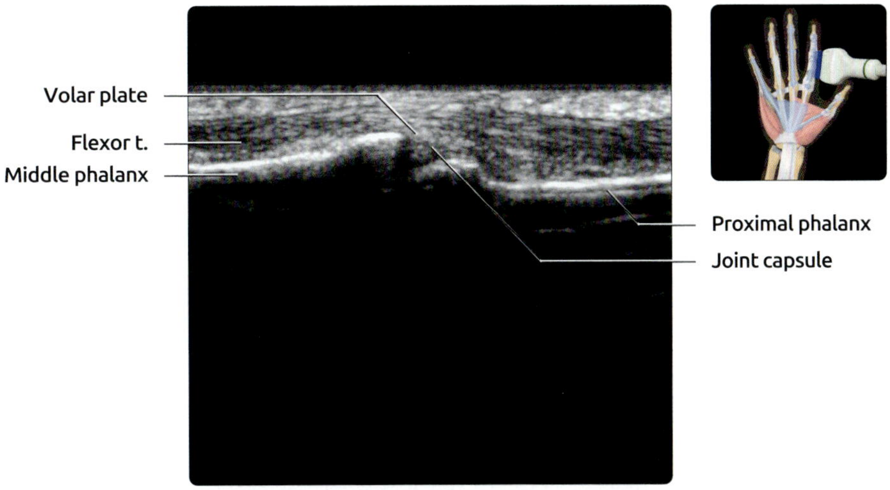

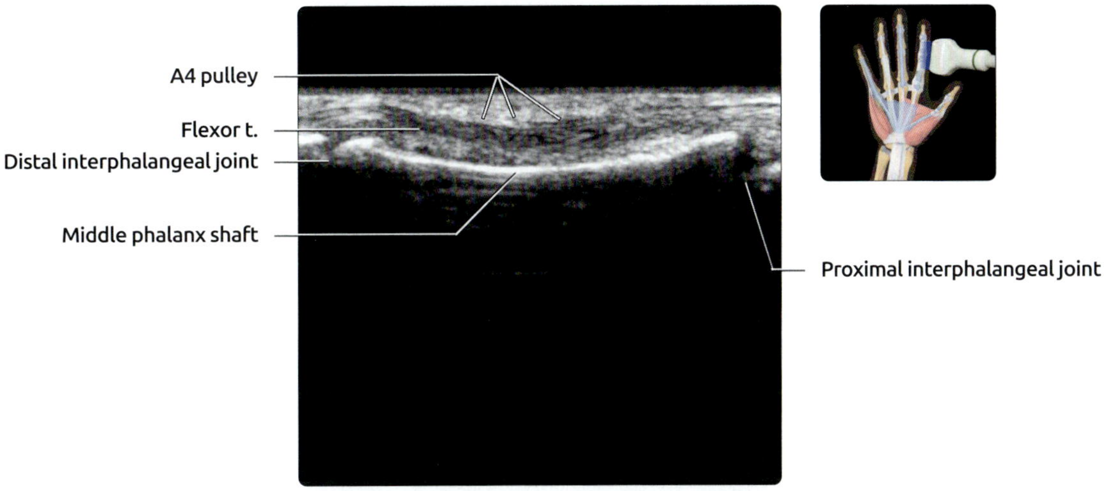

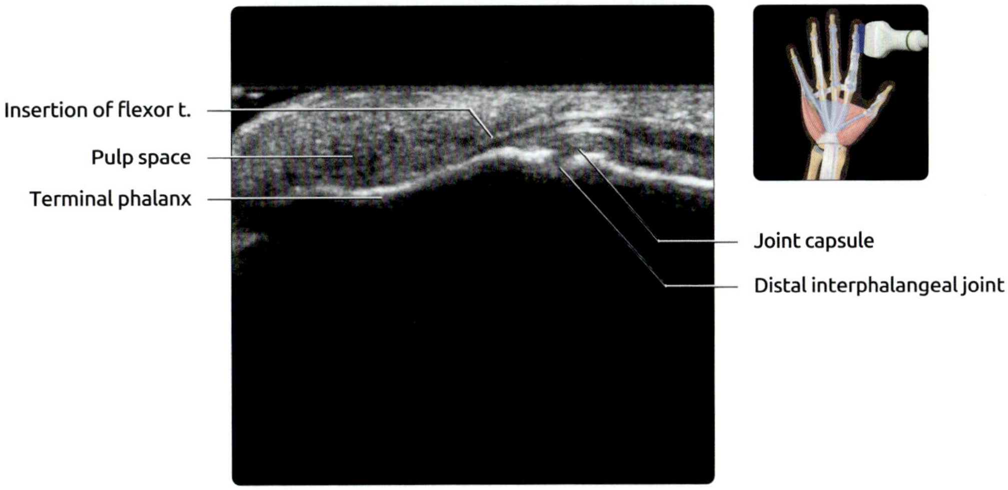

(Top) *Longitudinal grayscale US shows the proximal IP joint of the middle finger. The volar plate is a fibrocartilaginous thickening on the volar aspect of the MCP joint and proximal IP joints. At the proximal IP joint, it is firmly attached to the base of the middle phalanx and has a thinner insertion onto the distal end of the proximal phalanx. It may become avulsed during hyperextension injuries often associated with a small avulsion fracture. There is an inability to fully extend and flex the joint, a condition that usually improves with time unless the joint is locked.* **(Middle)** *Longitudinal grayscale US shows the middle finger at the middle phalanx. The A4 pulley prevents bowstringing of flexor tendons at the middle phalanx.* **(Bottom)** *Longitudinal grayscale US shows the distal IP joint of middle finger. An avulsion of the flexor tendon from the distal phalangeal base is usually accompanied by a small avulsion fracture.*

TRANSVERSE US

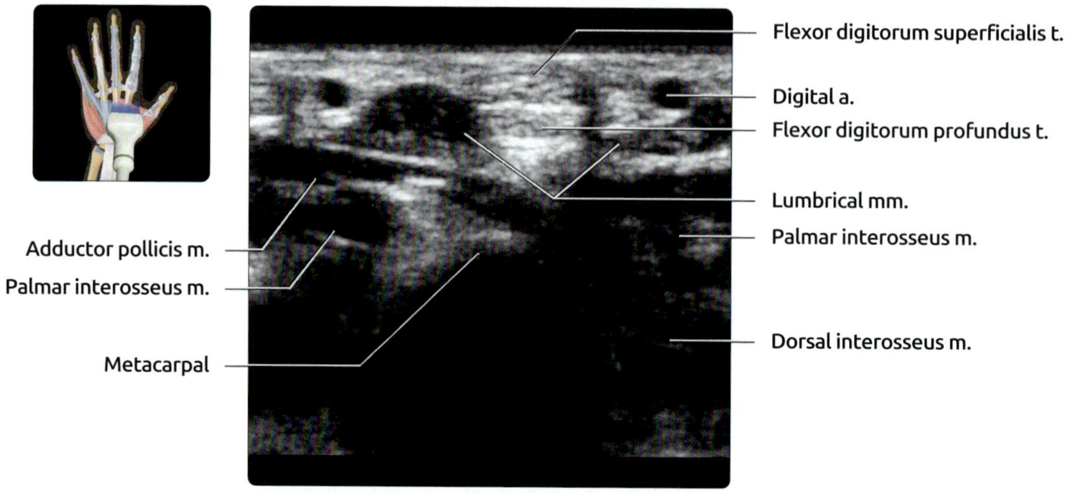

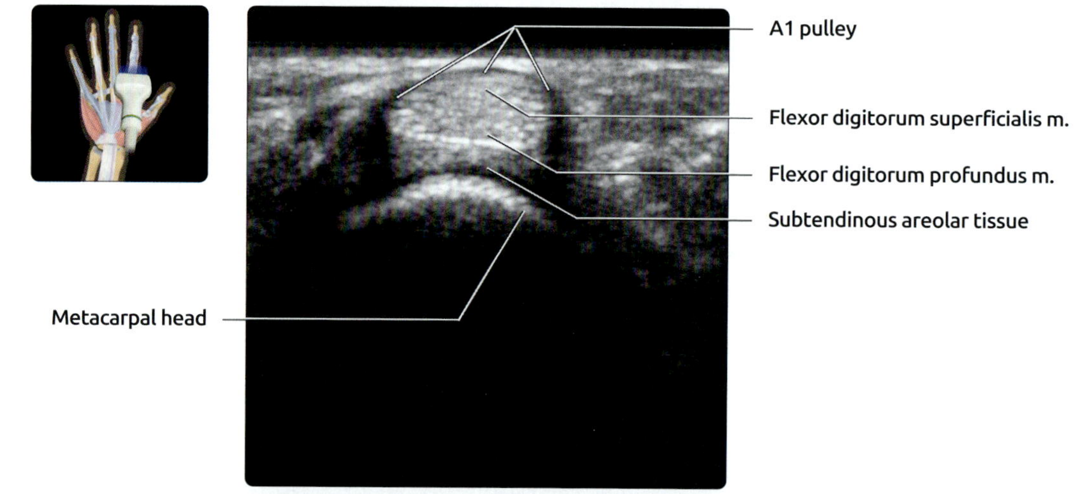

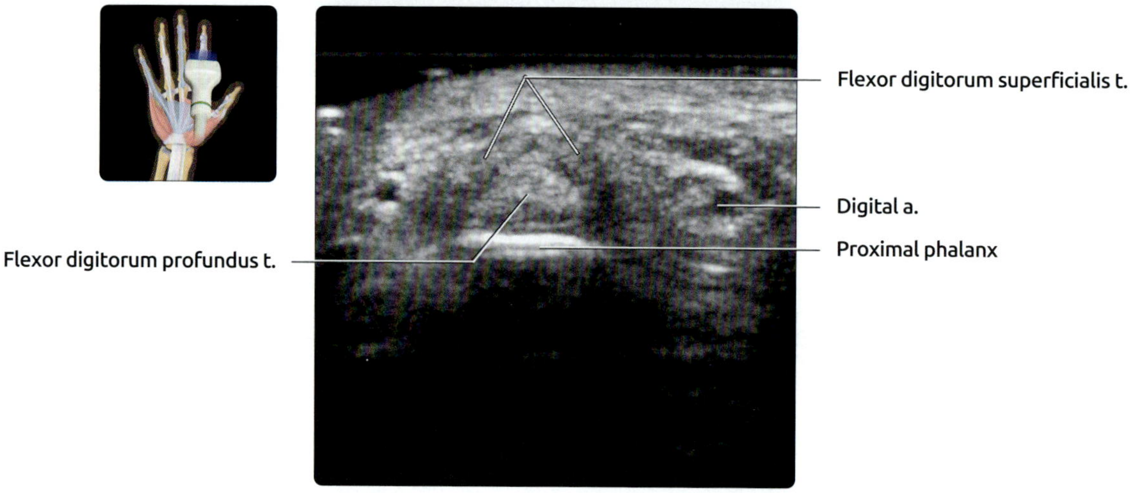

(Top) *Transverse grayscale US shows the middle finger at the metacarpal shaft. There are 4 lumbrical muscles in the hand. These are unusual muscles in that they do not originate from the skeleton but instead originate from the flexor digitorum tendons and insert into the extensor expansions.* **(Middle)** *Transverse grayscale US shows the middle finger at the metacarpal head. This image is of the A1 pulley level. The A1 pulley extends from the metacarpal head to the proximal phalanx base. It is seen as a thin hypoechoic to hyperechoic rim (depending on the beam angulation) around the flexor tendons. It broadens considerably at the base, which cannot be clearly depicted.* **(Bottom)** *Transverse grayscale US shows the middle finger at the proximal phalanx. The flexor digitorum superficialis tendon splits just proximal to the insertion, enabling the flexor digitorum profundus tendon to push through the gap.*

TRANSVERSE US

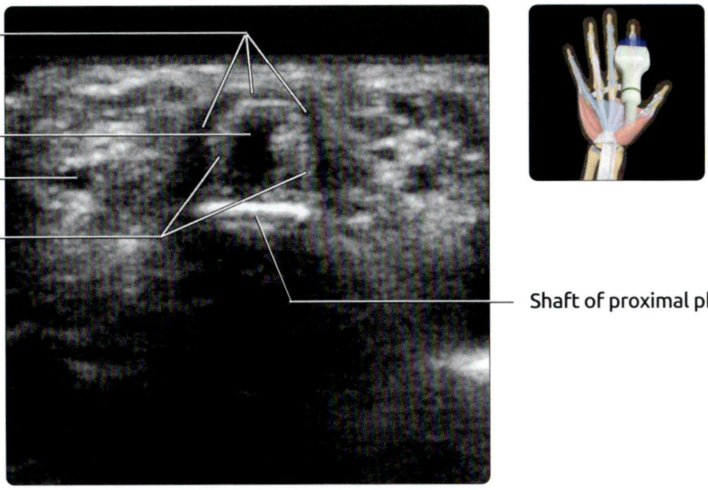

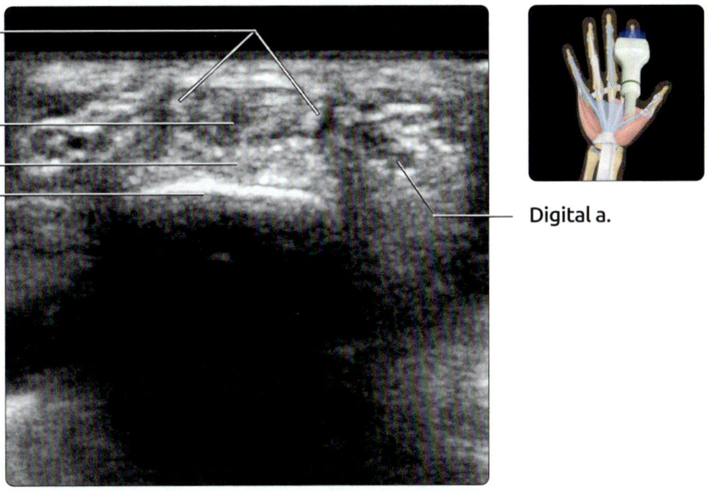

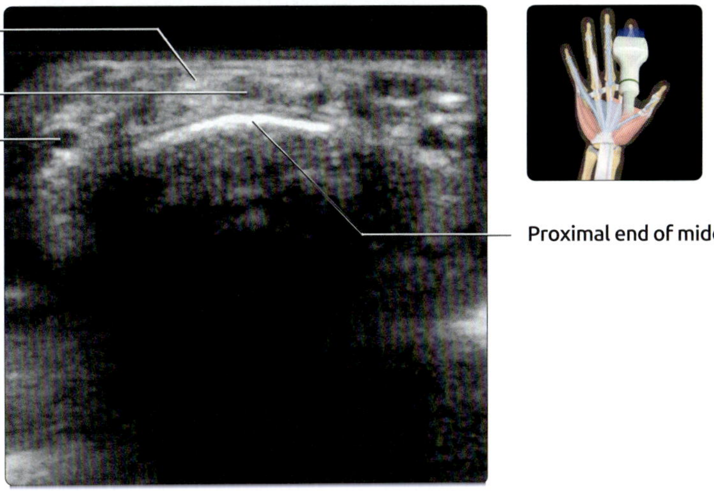

(Top) *Transverse grayscale US shows the middle finger shaft of the proximal phalanx. Do not confuse thickening of the A2 pulley with tendon sheath thickening.* **(Middle)** *Transverse grayscale US shows the middle finger proximal phalangeal head region just distal to the A2 pulley. The small slips of the flexor digitorum superficialis tendon descend on either side of the flexor digitorum profundus tendon to insert onto the volar aspect of the base of the middle phalanx.* **(Bottom)** *Transverse grayscale US shows the middle finger at the proximal end of the middle phalanx. The insertions of the slips of the flexor digitorum superficialis are small. As such, particularly in the presence of edema, it may be difficult to confirm isolated avulsion of 1 slip on US. In this situation, the flexor digitorum superficialis tendon will not be retracted.*

TRANSVERSE US AND GRAPHIC

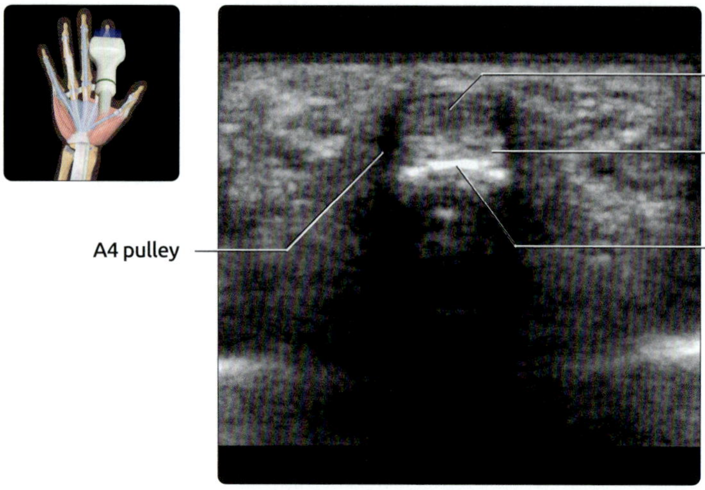

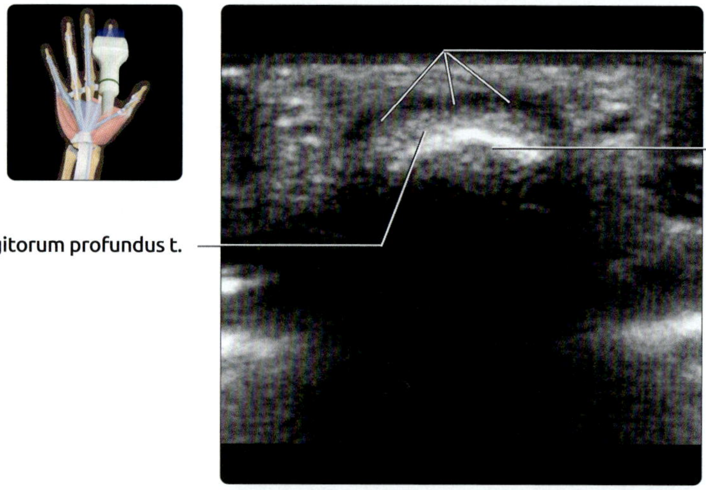

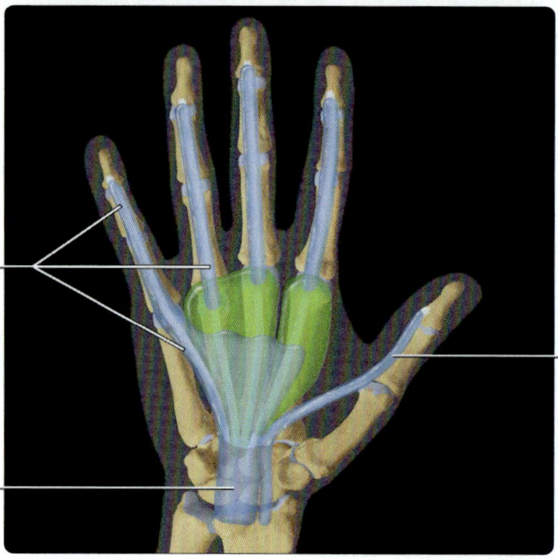

(Top) *Transverse grayscale US shows the middle finger at the shaft of the middle phalanx. Do not confuse the subtendinous connective tissue as part of the tendon.* **(Middle)** *Transverse grayscale US shows the middle finger at the base of the distal phalanx. The flexor digitorum profundus tendon gains broad attachment to the base of the distal phalanx and is supported by the A5 pulley just proximal to the insertion.* **(Bottom)** *Graphic shows the tendon sheaths of the fingers (blue). Digital sheaths may connect to the ulnar bursa (green) in up to 10% of the population, providing a route for the spread of infection from digits 2-4 to the common flexor sheath and vice versa.*

LONGITUDINAL US, EXTENSOR ASPECT

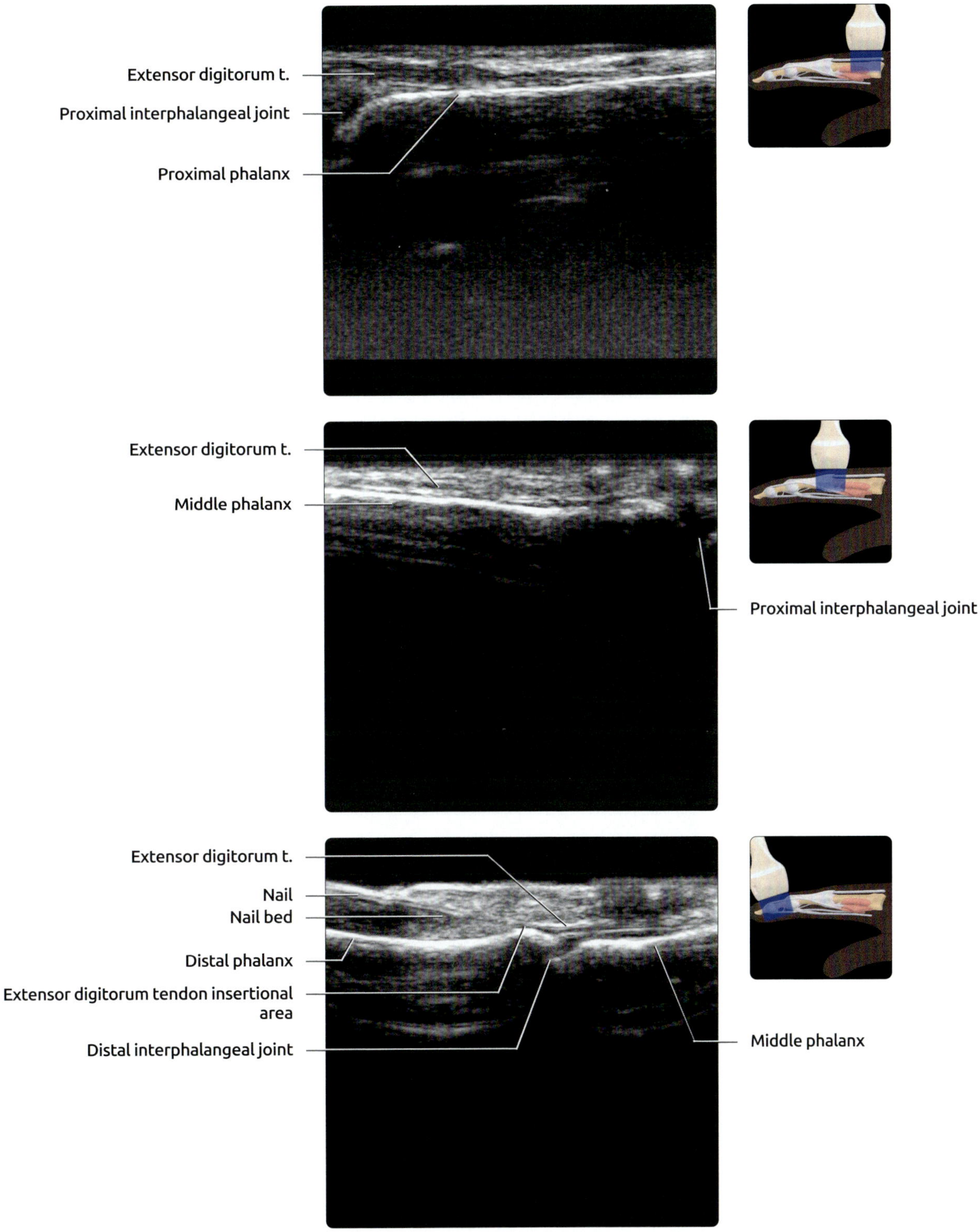

(Top) *Longitudinal grayscale US shows the dorsal aspect of the proximal phalanx. The extensor expansion on the dorsum of the fingers is thin and flat and does not have a synovial sheath.* **(Middle)** *Longitudinal grayscale US shows the dorsal aspect of the middle phalanx. No pulleys are present on the extensor aspect of the fingers. The extensor expansion (or hood) are attached to the phalanges and the lateral bands.* **(Bottom)** *Longitudinal grayscale US shows the dorsal aspect of the distal phalanx. The extensor digitorum is attached to the base of the distal phalanx. This is a common site of avulsion, often with a small avulsion fracture.*

TRANSVERSE US, EXTENSOR ASPECT

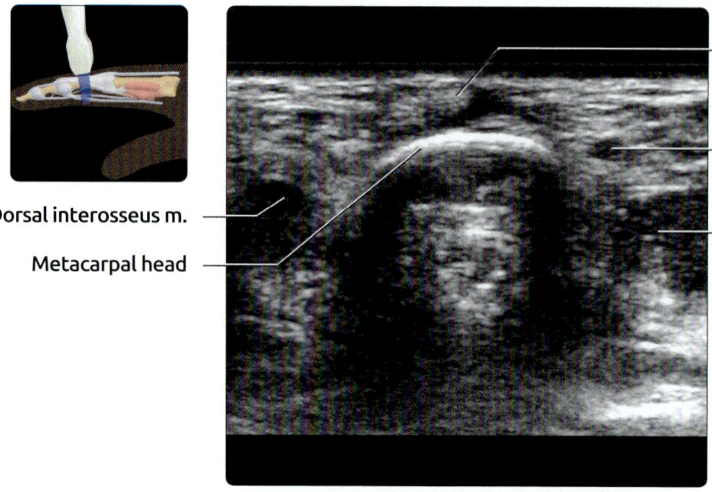

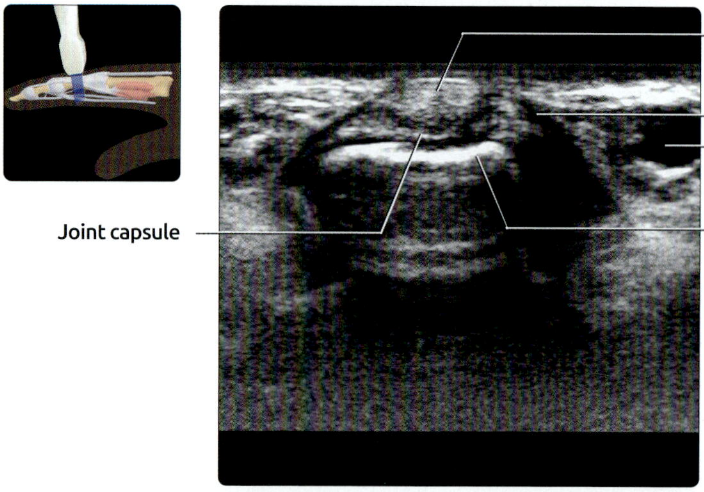

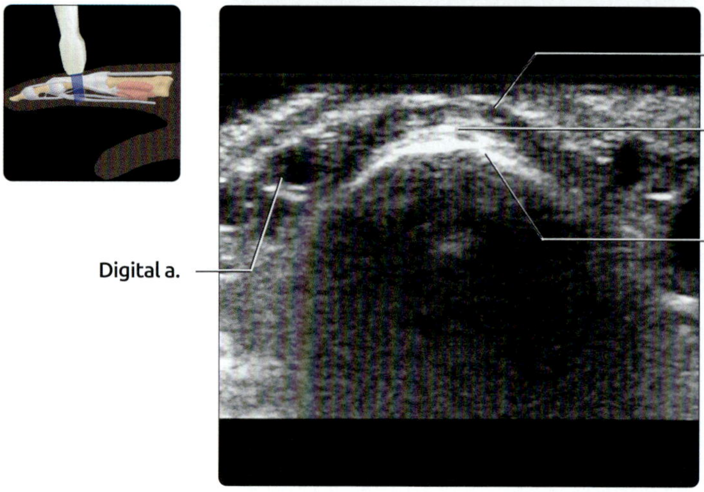

(Top) *Transverse grayscale US shows the metacarpal head/neck region of the extensor aspect of the fingers. The extensor tendons of the hand do not have a complete synovial sheath. They are covered by thin paratenon. The tendons have an ovoid configuration.* **(Middle)** *Transverse grayscale US shows the extensor aspect of the fingers at the metacarpal head region. At the proximal phalanx extending over the middle phalanx, the extensor tendons spread out and connect to the extensor expansion or hood.* **(Bottom)** *Transverse grayscale US shows the extensor aspect of the fingers at the proximal phalanx level. With expansion of the extensor tendon, it becomes very thin such that the normal tendon is difficult to illustrate on the dorsum of the fingers. If injured, the tendon swells and is usually more readily seen.*

AXIAL FINGER AND HAND DERMATOMES

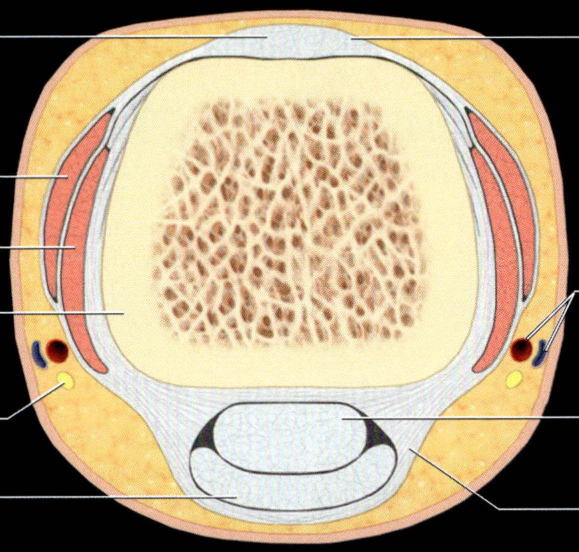

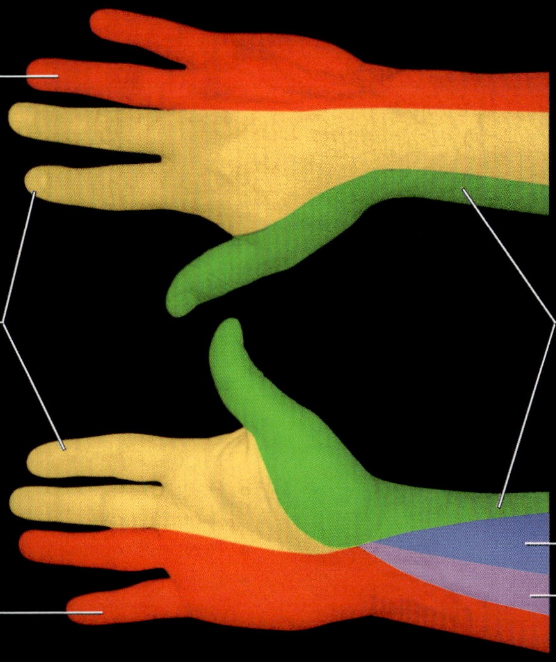

(Top) *Graphic shows the transverse section of a finger depicting the relationships of the digital nerve, vessels, and tendons to the phalanx. On the dorsal aspect, the extensor expansion interdigitate with the extensor digitorum communis tendons to prevent lateral translation. On the palmar side, the collateral ligaments extend from the palmar aponeurosis in the palm to fuse with the fibrous digital sheaths. Other thickenings of the fibrous digital sheaths form annular pulleys and cruciform bands. The common digital artery and the accompanying nerve intervene between the lumbrical muscle and the tendons.* **(Bottom)** *Dermatomes of the hand and wrist correlated to corresponding cervical or thoracic nerves. It is necessary for a physician to test the integrity of spinal cord segment C5 to T1 in the upper limb. Usually, physicians will test the C6 to C8 dermatomes in the hand, i.e., the thumb represents C6 dermatome, the middle finger tip represents C7 dermatome, and the little finger represents C8 dermatome.*

GROSS ANATOMY

Overview

- Brachial plexus is major neural supply to upper limbs
  - Formed from C5-T1 roots combining to form 3 trunks, 6 divisions, 3 cords, and 5 motor/sensory terminal branches (classic description)
    - 5 motor/sensory terminal branches are ulnar, median, musculocutaneous, axillary, and radial nerves
  - In reality, up to 30 variations on this classic pattern exist with no predominant pattern
    - Cannot rely on having definite number of trunks, divisions, or cords at any particular location

Ultrasound Anatomy

- 5 specific areas should be assessed to ensure complete evaluation
  - Extraforaminal, intrascalene supraclavicular, infraclavicular, and axillary areas
- **Extraforaminal area (5 roots)**
  - Single large fascicle arising from spinal cord levels C5 to T1
  - Divides and recombines along length of brachial plexus
  - Epineurium is thin proximally and becomes thicker more distally along brachial plexus
- **Intrascalene area (5 roots → 3 trunks)**
  - Posterior to clavicular head of sternocleidomastoid muscle between scalenus anterior muscle anteromedially and scalenus medius muscle posterolaterally
  - Roots are present in superior interscalene area proximally
  - C5 and T1 roots are more difficult to see due to their deep location
  - **Roots unite to form trunks** in inferior interscalene area distally
    - **Upper (C5-6), middle (C7), lower (C8, T1) trunks**
    - Average cross-sectional area of these trunks is 0.07 cm^2 (upper), 0.09 cm^2 (middle), and 0.09 cm^2 (lower)
    - Upper and middle trunks are cephalad and posterior to subclavian artery
    - Upper (C5-6) trunk or part of it may run through or anterior to anterior scalene muscle
    - Lower trunk is posterior to subclavian artery and can be more difficult to see because of its depth
    - Deep cervical artery runs posteriorly between middle and lower trunks
    - Minor nerves arising directly from trunks: Suprascapular nerve and nerve to subclavius muscle
  - Brachial plexus enters supraclavicular fossa by emerging between scalenus anterior and medius muscles
- **Supraclavicular area (6 divisions)**
  - Trunk divides into anterior and posterior divisions in supraclavicular fossa
  - Lie superoposterior and lateral to subclavian artery ("bunch of grapes")
  - Anterior divisions innervate anterior (flexor) muscles
  - Posterior divisions innervate posterior (extensor) muscles
  - Transverse cervical artery runs laterally in this region, usually below divisions of brachial plexus
- **Infraclavicular area (3 cords)**
  - Passageway between supraclavicular fossa and apex of axilla
  - About 3 cm inferomedial to coracoid process, cords position themselves lateral, medial, and posterior to axillary artery
  - Lateral cord innervates anterior (flexor) muscles
  - Medial cord innervates anterior (flexor) muscles
  - Posterior cord innervates posterior (extensor) muscles
    - Posterior cord is most difficult to see as it lies posterior to brachial artery
  - Cords tend to be more hyperechoic than hypoechoic trunks or divisions as brachial plexus contains increasing number of fascicles and more connective tissue distally
  - Overlying cords is pectoralis major muscle medially, pectoralis minor muscle laterally, while serratus anterior covers chest wall medially
- **Axillary area (5 terminal branches)**
  - Pyramidal space beneath glenohumeral joint
  - **Lateral cord** → median nerve and musculocutaneous nerve (C5-6)
  - **Medial cord** → median nerve and ulnar nerve (C8-T1)
  - **Posterior cord** → axillary nerve (C5-6), radial nerve (C5-T1), thoracodorsal nerve (C6-8) as well as upper (C6-7) and lower (C5-6) subscapular nerves
  - Both lateral and medial cords contribute to median nerve
  - **Nerves are arranged around axillary artery** with ulnar nerve at 2-3 o'clock, radial nerve at 5-6 o'clock, and median nerve at 10-11 o'clock
  - Median, ulnar, and radial nerves are enclosed within sheath separate from musculocutaneous nerve
  - In most cases, musculocutaneous nerve branches early from lateral cord
    - Musculocutaneous nerve is easily seen by moving probe over medial bicipital groove proximally where nerve is seen between coracobrachialis and biceps muscles
    - Musculocutaneous nerve then descends along lateral side of arm between biceps and brachialis muscles

ANATOMY IMAGING ISSUES

Imaging Approaches

- Ultrasound is most efficient way of seeing brachial plexus as perineural fat allows good visualization of nerves
- Best to evaluate brachial plexus on transverse scanning though roots; also well seen on longitudinal scanning
- Routinely evaluate each of 5 designated areas
- Color Doppler imaging helps distinguish brachial plexus from vascular structures
- Infraclavicular portion of brachial plexus is least well seen on ultrasound
  - It is covered by pectoralis muscle and lies relatively deep at ~ 4 cm from skin surface

BRACHIAL PLEXUS

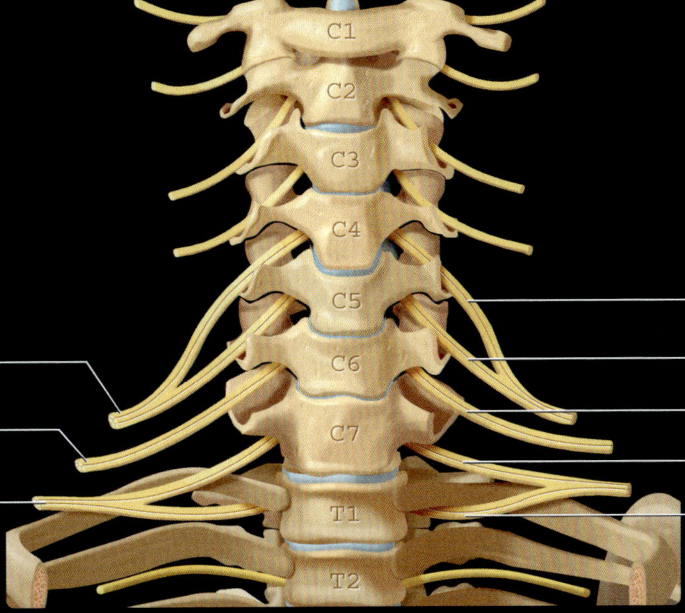

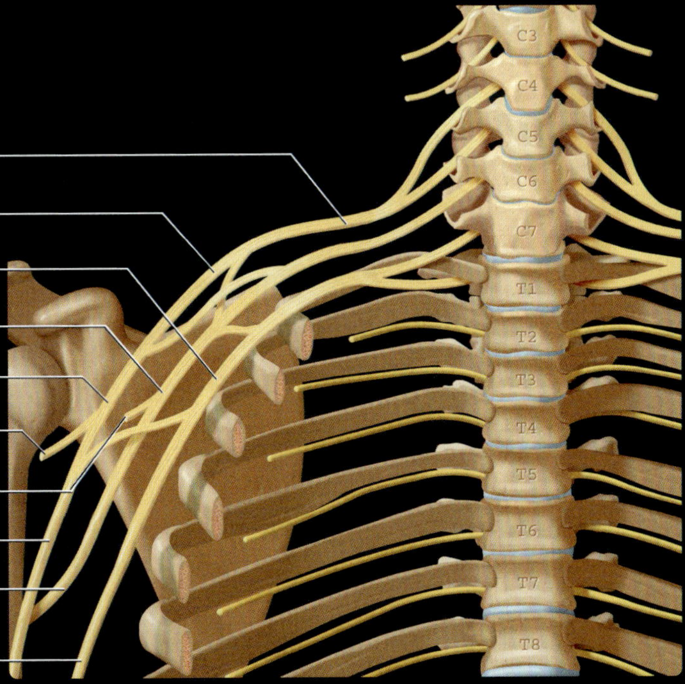

(Top) *Coronal graphic of the cervical spine and supraclavicular brachial plexus demonstrates the cervical ventral primary rami combining to form the brachial plexus. The C1-C7 roots exit above the same numbered pedicle, C8 exits above the T1 pedicle, and more caudal roots exit below their numbered pedicle. The trunks are visualized on ultrasound mainly in the intrascalene area.* **(Bottom)** *Coronal graphic of the brachial plexus demonstrates the more distal plexus elements extending into the axilla. The trunks recombine into posterior and anterior divisions that form the cords. The posterior cord forms the radial and axillary nerves. The medial cord forms the ulnar nerve, while the lateral cord forms the musculocutaneous nerve. The median nerve is formed from branches of both the lateral and medial cords.*

BRACHIAL PLEXUS AND PERIPHERAL NERVES

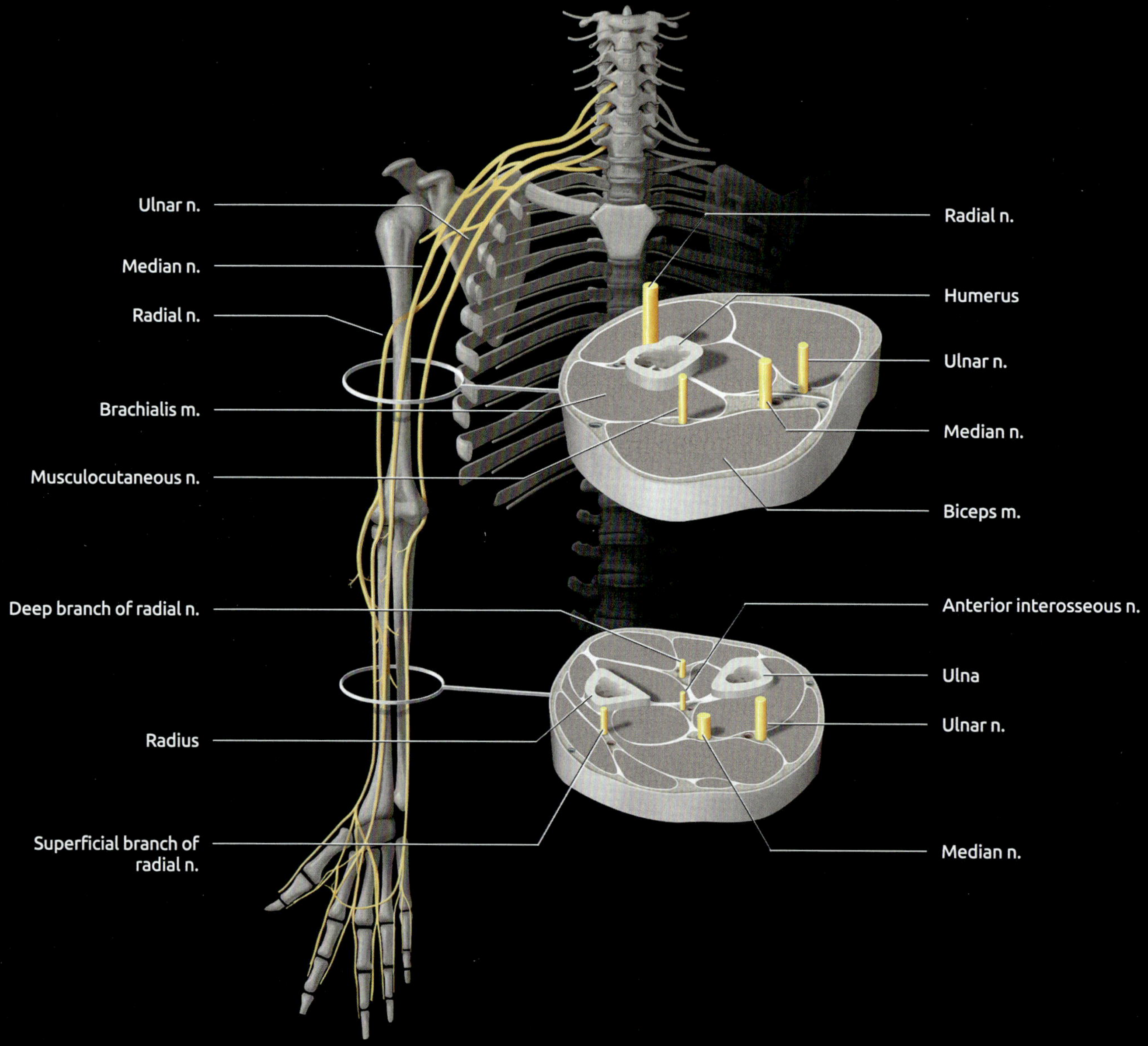

This graphic shows 3 of the 5 terminal branches of the brachial plexus (median, radial, and ulnar nerves), as they course down the arm. The other 2 terminal branches are the musculocutaneous and axillary nerves. The axillary nerve innervates the teres minor and deltoid muscles, while the musculocutaneous nerve passes through the coracobrachialis muscle and descends between the biceps and brachialis muscles.

CORONAL AND SAGITTAL T1 MR

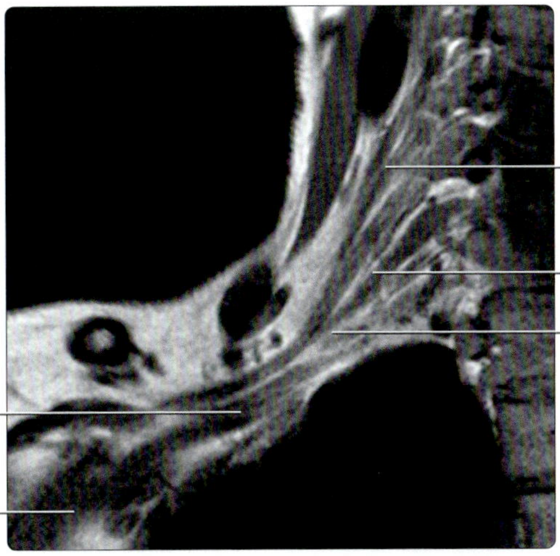

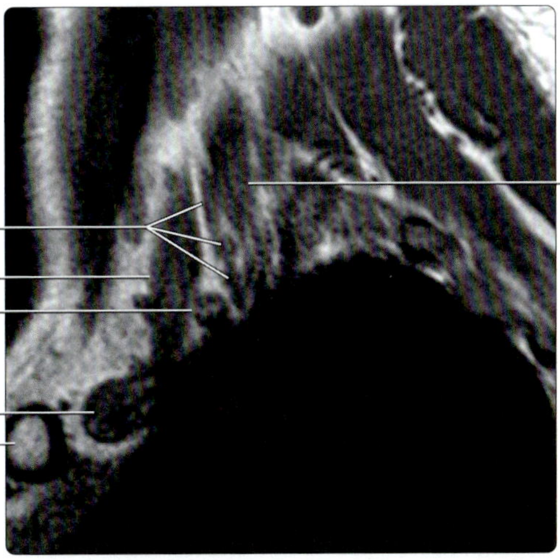

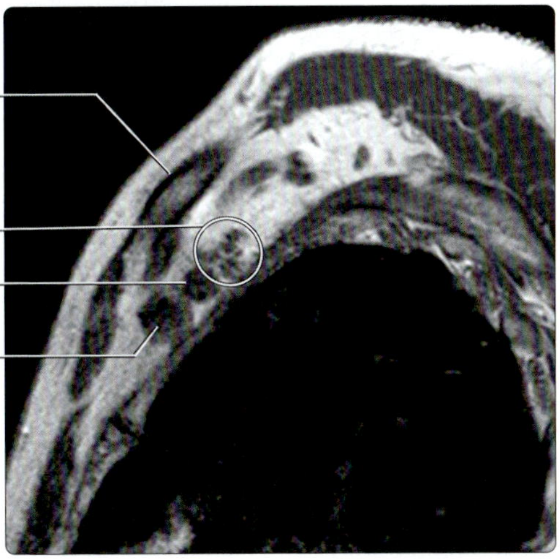

(Top) *Oblique coronal T1WI MR with large field of view shows the normal brachial plexus. The divisions and cords run alongside the subclavian artery. The approximate locations of the roots, trunks, divisions, cords, and peripheral nerves are shown.* **(Middle)** *Sagittal T1WI MR shows the hypoechoic roots/trunks of the brachial plexus in the intrascalene area between the scalenus anterior and medius muscles. The scalenus anterior muscle separates the subclavian artery from the subclavian vein.* **(Bottom)** *A more lateral sagittal T1WI MR shows the infraclavicular region (costocervical space) with the hypointense divisions/cords of the brachial plexus located dorsocephalad to the subclavian artery.*

TRANSVERSE AND LONGITUDINAL ULTRASOUND

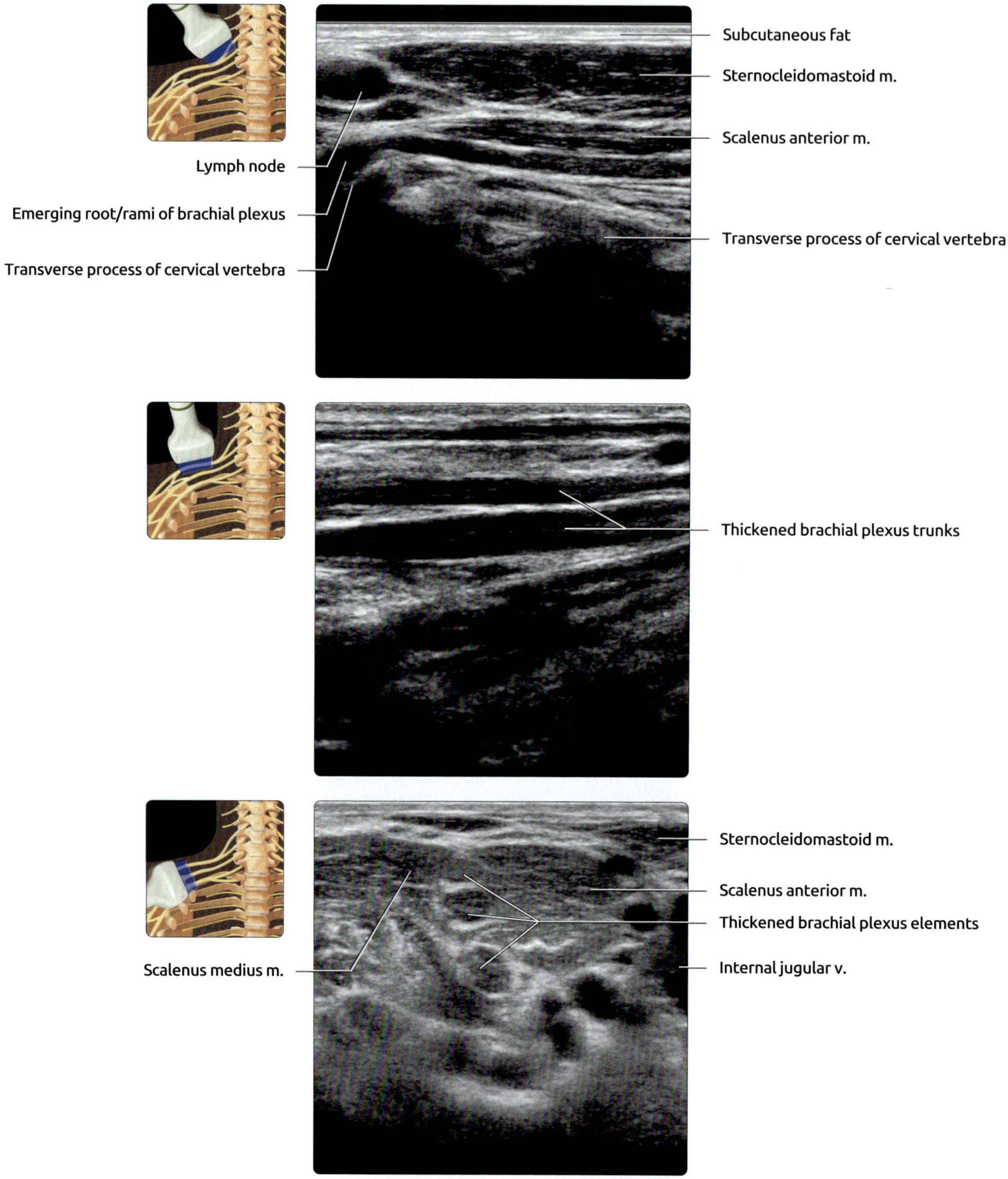

(Top) *Longitudinal grayscale ultrasound of the posterior triangle of the neck shows the root and trunk of the brachial plexus, which appears as a thin, tubular, hypoechoic structure related superficially to the scalenus anterior muscle and deeply to the cervical vertebrae.* **(Middle)** *Longitudinal grayscale ultrasound of right posterior triangle/supraclavicular fossa confirms the elongated linear, hypoechoic, thickened elements of the brachial plexus. The patient had past history of neck irradiation for metastatic neck nodes, and the nerve thickening is likely secondary to postradiation change.* **(Bottom)** *Transverse grayscale ultrasound of right lower posterior triangle/supraclavicular fossa shows round smooth hypoechoic nodules between the scalenus anterior and medius muscles. The anatomical location suggests the possibility of thickened brachial plexus elements.*

TRANSVERSE AND LONGITUDINAL ULTRASOUND

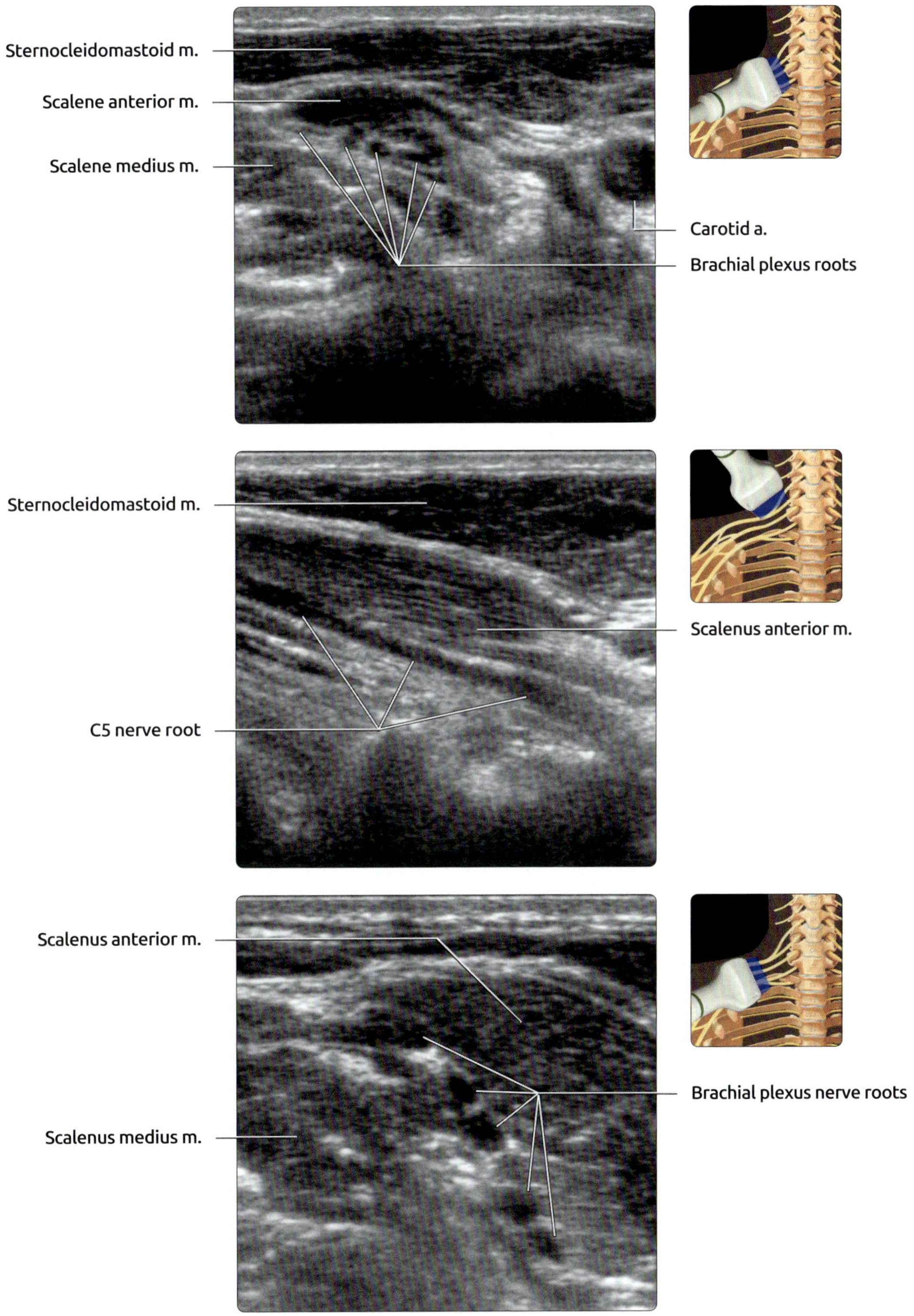

(Top) *Transverse ultrasound of the anterolateral lower neck shows the hypoechoic roots of the brachial plexus as they exit the neural foramina and move to the intrascalene area.* **(Middle)** *Longitudinal ultrasound shows the longitudinal section of the hypoechoic C5 nerve root as it emerges from the exit of the foraminal area and descends to the intrascalene area.* **(Bottom)** *Transverse ultrasound at the entry into the intrascalene area shows 5 hypoechoic roots of the brachial plexus between the scalenus anterior and medius muscles. The hypoechoic roots are clearly seen against the adjacent hyperechoic intermuscular fat.*

MR AND ULTRASOUND

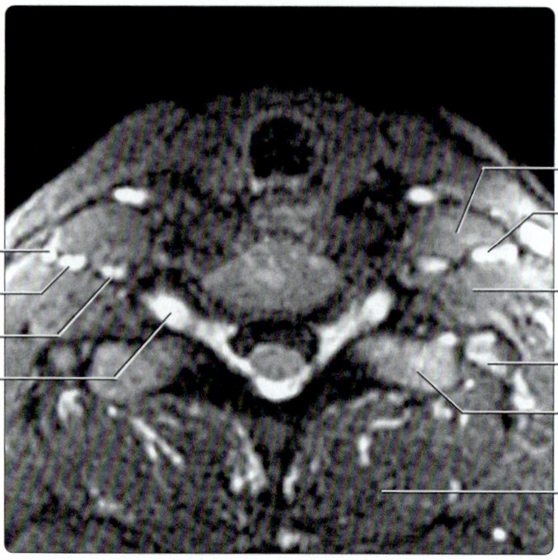

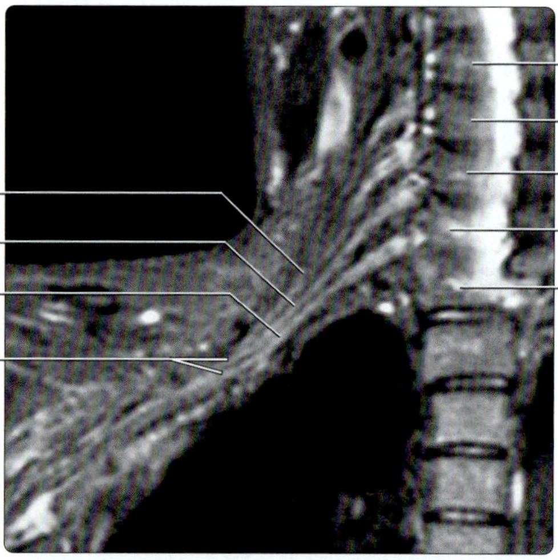

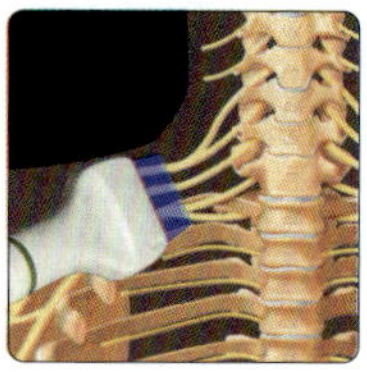

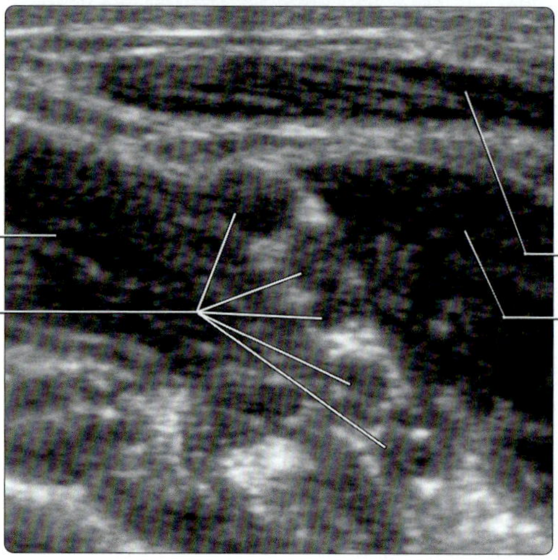

(Top) *Axial STIR MR at the C7/T1 level depicts the linear alignment of the C5 through C8 ventral primary rami. C5 and C6 are closely approximated and form the left upper trunk.* **(Middle)** *Reformatted oblique coronal T2WI MR of the brachial plexus shows that the trunks and divisions of the brachial plexus continue from cervical nerve roots to the axilla.* **(Bottom)** *Transverse ultrasound at the exit from the intrascalene area shows 5 hypoechoic roots and trunks of the brachial plexus between the scalenus anterior and medius muscles. The roots and trunks of the brachial plexus are readily seen between the scalene muscles, but there is considerable variability in their appearance due to the inherent variability in the brachial plexus regarding the formation of the individual trunks, divisions, and cords.*

TRANSVERSE ULTRASOUND

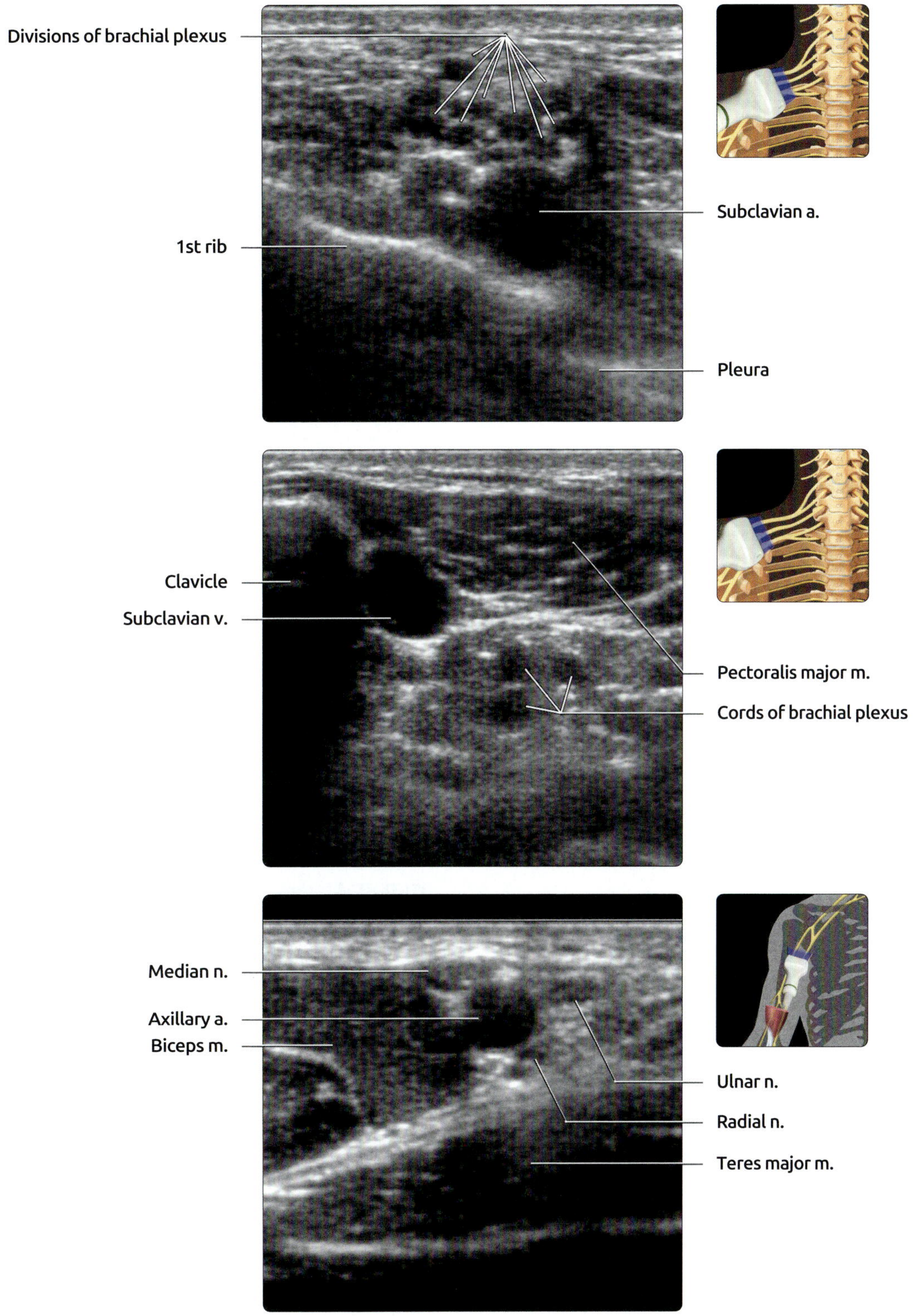

(Top) *Transverse ultrasound at supraclavicular area shows hypoechoic divisions of the brachial plexus located dorsocranial to the subclavian artery akin to a "cluster of grapes." In a small percentage of subjects, some of the divisions may lie caudal to the subclavian artery.* **(Middle)** *Transverse ultrasound at the infraclavicular area shows hypoechoic cords of the brachial plexus located dorsocaudal to the subclavian artery. More medially, the cords are covered by the pectoralis major muscle, while more laterally, they are covered also by the pectoralis minor muscle.* **(Bottom)** *Transverse ultrasound at the upper axillary region shows the axillary artery surrounded by the median, ulnar, and radial nerves. The position of these nerves can vary with regard to the axillary artery, so each nerve should be traced distally to confirm its identity.*

TERMINOLOGY

Abbreviations

- Extensor digitorum (ED)
- Extensor carpi radialis brevis (ECRB)
- Extensor digiti minimi (EDM)
- Extensor carpi ulnaris (ECU)
- Extensor pollicis longus (EPL)
- Abductor pollicis longus (APL)
- Extensor indicis (EI)

GROSS ANATOMY

Arm

- Radial nerve is largest branch of posterior cord of brachial plexus
  - Receives contributions from cervical roots C5-C8
  - Contains motor and sensory components that supply muscles of extensor compartments of arm, forearm, and hand
- Lies between coracobrachialis and teres major muscles and then between bellies of medial and lateral head of triceps
- Runs deep to triceps in spiral groove and then deep to brachioradialis
- ~ 10 cm proximal to lateral epicondyle of humerus, radial nerve penetrates lateral intermuscular septum to enter anterior space of upper arm
- Just anterior to lateral epicondyle, nerve bifurcates into sensory (superficial branch of radial nerve) and motor (deep branch of radial nerve and posterior interosseous nerve) components

Forearm

- Branches of radial nerve
  - Divides at level of lateral epicondyle
  - Superficial
    - Purely sensory
    - Direct continuation of radial nerve
    - Courses distally, deep to brachioradialis, to posterior compartment distal forearm and then to dorsal wrist
    - Divides into lateral branch (supplies radial wrist and thumb skin) and medial branch (supplies mid and ulnar wrist skin)
    - Dorsal digital nerves supply ulnar thumb, index, middle, and radial ring fingers
  - Deep
    - Purely motor
    - Enters supinator, then exits distally and runs posteriorly as posterior interosseous nerve
    - Supplies ECRB, supinator, ED, EDM, ECU, EPL, APL, and EL

Hand: Branches of Radial Nerve

- Not usually seen on routine imaging
- No motor innervation in hand
- Sensory: Dorsal surface from radiocarpal joint to just distal to proximal interphalangeal joints for digits 1-3 and radial 1/2 of 4th digit
  - Gives terminal branches to supply skin of lateral 2/3 dorsum of wrist and hand and lateral 2.5 digits

ANATOMY IMAGING ISSUES

Imaging Recommendations

- Use high-frequency (at least ≥ 7.5-MHz) transducer
- Radial nerve is best identified at lateral aspect of distal arm where it runs deep to brachialis
  - Identified deep to triceps in spiral groove midhumeral level
  - Or in proximal aspect arm medially
    - Lies just deep to brachial artery
- Assess shape, size, echotexture, and integrity of nerve
  - Trace medium-sized branches by following their course as they branch from parent nerve
- Examination of nerves along transverse plane is preferable to longitudinal scanning as it allows easier, continuous following of nerves throughout limb

Imaging Approaches

- Ultrasound
  - Nerve echotexture
    - Transverse: Uniformly dispersed hypoechoic dots (nerve fascicles) separated by hyperechoic epineurium
    - Longitudinal: Parallel hypoechoic tracts of uniform caliber
    - Overall gain control and focus settings optimized for adequate visualization of nerve

Imaging Pitfalls

- Smaller nerves and nerve branches (1-2 mm) are difficult to identify, and their location may only be inferred by adjacent vessels
- Anisotropy artifacts may affect scanning, especially in short axis; tilting transducer during scanning helps to achieve best view
- Structural compression is easier to appreciate than functional compression

CLINICAL IMPLICATIONS

Clinical Importance

- Sites of potential entrapment of radial nerve
  - Lateral to long head of triceps due to fibrous arch
  - Spiral groove between medial and lateral heads of triceps muscle due to fibrous arch, fractures, tourniquet
  - Between brachialis and brachioradialis muscles at antecubital fossa due to muscle (body builders)
  - Posterior interosseous nerve by arcade of Frohse (superficial proximal margin of supinator) or recurrent radial artery (leash of Henry) crossing just proximal to arcade
  - Posterior interosseous nerve between deep and superficial heads of supinator muscle (supinator syndrome)
  - Superficial radial nerve between brachioradialis and extensor carpi radialis longus in mid to distal forearm (Wartenberg syndrome)

RADIAL NERVE

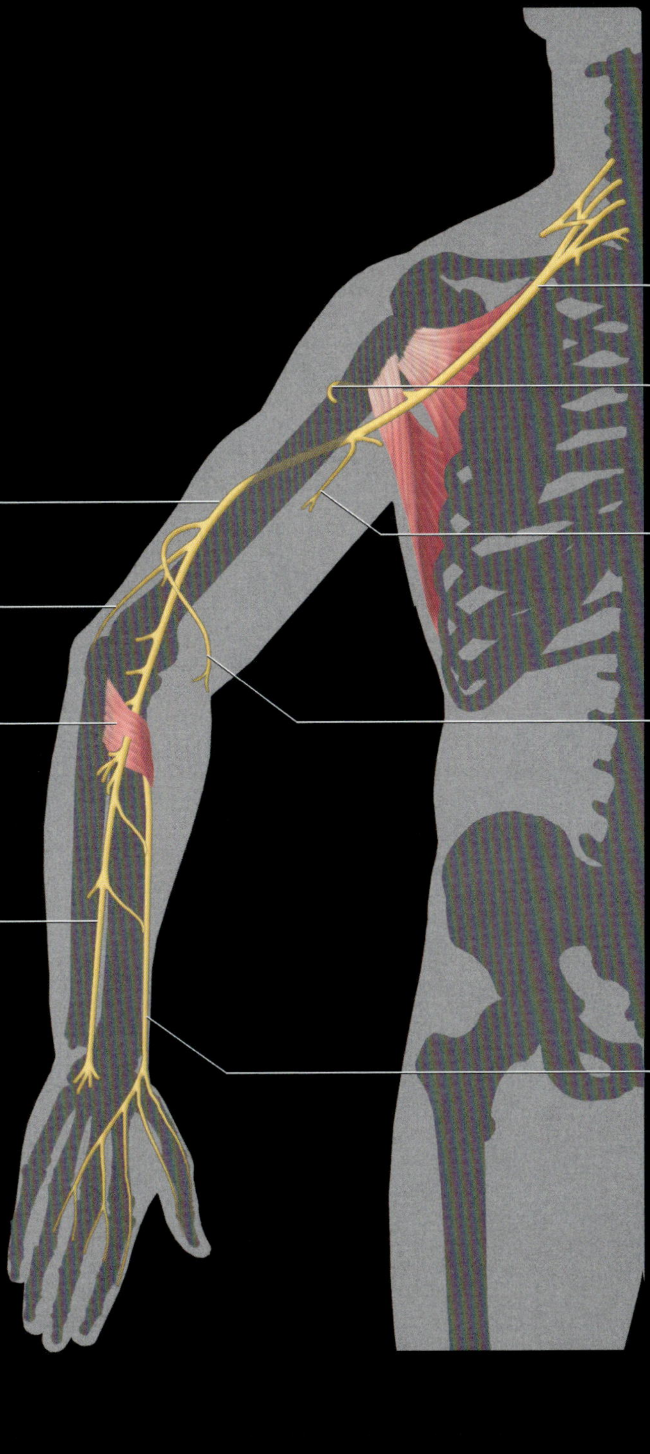

Graphic shows the course and distribution of the radial nerve. The radial nerve arises from the posterior cord of the brachial plexus (C5-C8, T1). It spirals posterolaterally around the humerus with the deep brachial artery located anterolateral between the brachialis and the brachioradialis. It gives off the posterior cutaneous nerve of the forearm, which passes posterior to the lateral condyle and supplies the posterior forearm. It divides into deep and superficial branches at the lateral epicondyle. The deep branch is purely motor and pierces the supinator muscle and winds around the lateral aspect of the radial neck. The superficial branch is purely sensory and is located in the anterolateral aspect of the forearm, superficial to the supinator and pronator teres muscles.

POSTERIOR ARM

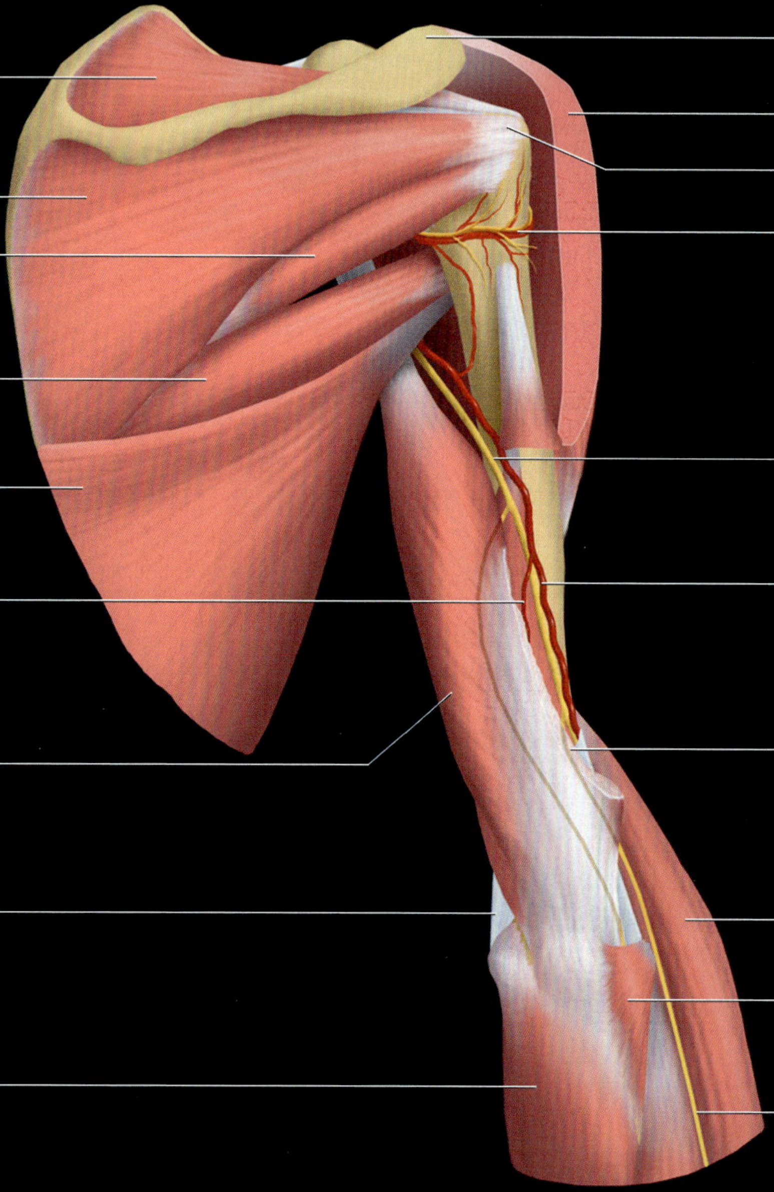

Graphic shows the course of the radial nerve in the posterior aspect of the arm. The radial nerve spirals posterolaterally around the humerus with the deep brachial artery. It gives off the posterior cutaneous nerve of the forearm, which passes posterior to the lateral condyle and supplies the posterior forearm. It is located anterolaterally, between the brachialis and brachioradialis, where it supplies the triceps, anconeus, brachioradialis, and the lateral portion of the brachialis.

TRANSVERSE US, ARM

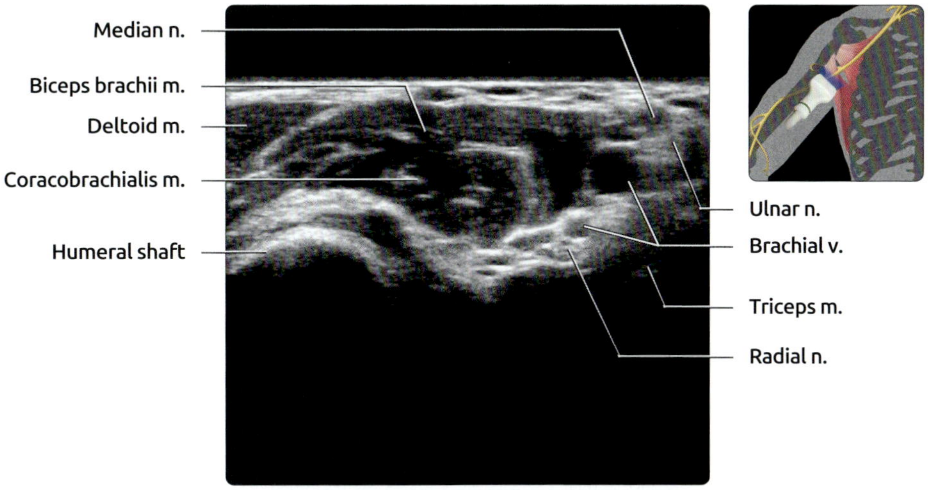

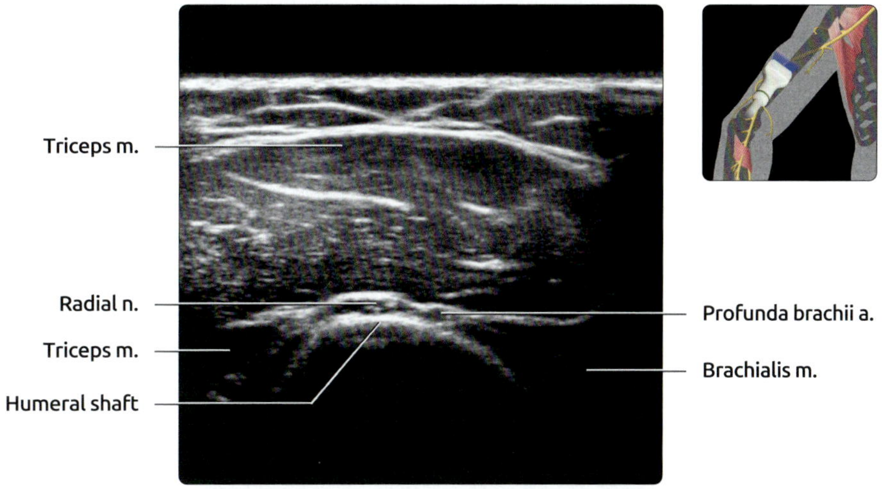

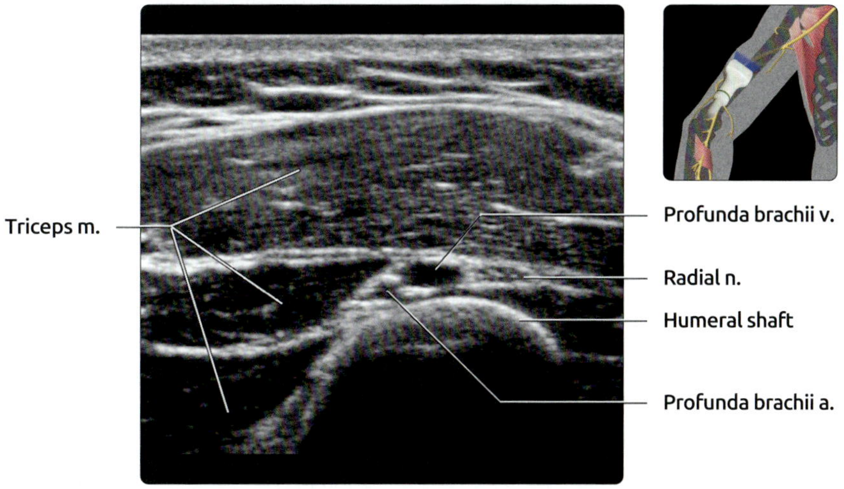

(Top) *Transverse grayscale ultrasound shows the radial nerve in the proximal 1/3 of the arm. The radial nerve is a continuation of the posterior cord of the brachial plexus. It supplies the triceps and the supinator. In the proximal arm, it lies deep to the brachial artery and vein.* **(Middle)** *Transverse grayscale ultrasound shows the radial nerve at the mid 1/3 of the arm. The radial nerve may be injured as it passes around the posterior aspect of the mid to distal 1/3 of the humeral shaft. This injury may occur at the time of fracture, during fracture manipulation or fixation, or as a result of postoperative fibrosis.* **(Bottom)** *Transverse grayscale ultrasound shows the radial nerve at the mid 1/3 of the arm (coned view). The radial nerve is accompanied in the spiral groove by the profunda brachii artery and vein.*

TRANSVERSE US, SPIRAL GROOVE

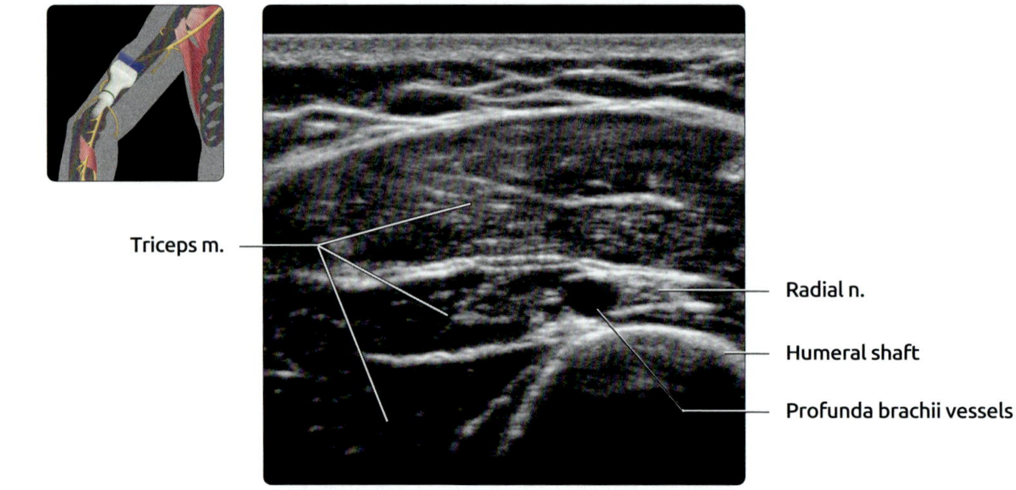

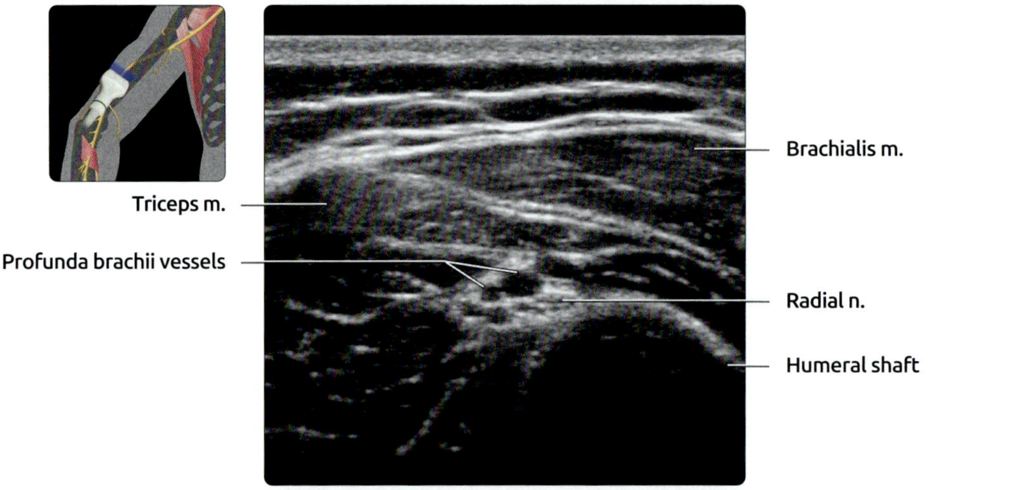

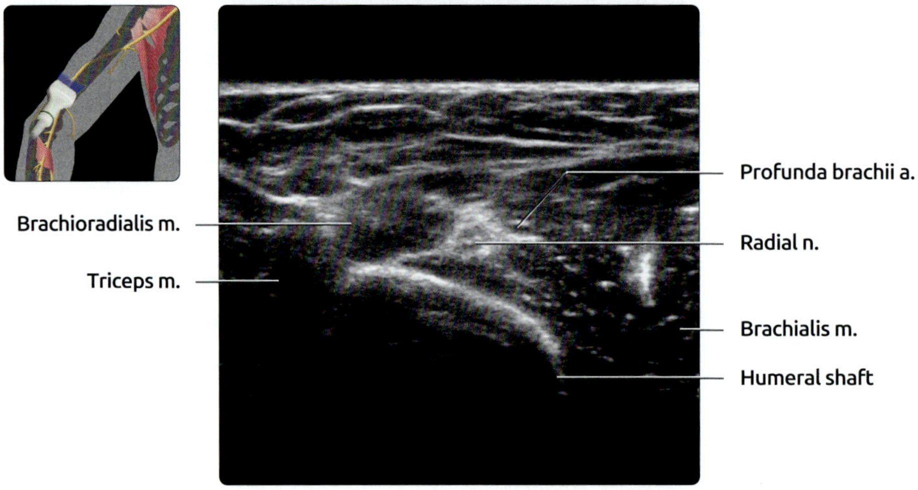

(Top) *Transverse grayscale ultrasound shows the radial nerve in the midpart of the spiral groove (coned view). The surgeon must be aware of the site of the radial nerve when performing fixation of the humerus, as the fixator screws or plate may potentially damage the radial nerve.* **(Middle)** *Transverse grayscale ultrasound shows the radial nerve in the distal aspect of the spiral groove in the midhumerus region.* **(Bottom)** *Transverse grayscale ultrasound shows the radial nerve on the volar aspect of the distal 1/3 of the arm. The radial nerve passes from the posterior to the anterior compartment by piercing the lateral intermuscular septum. Compression of the radial nerve at the lateral intermuscular septum may occur. The radial nerve may be compressed between the brachioradialis and brachialis muscles in very muscular individuals.*

TRANSVERSE US, POSTERIOR FOREARM

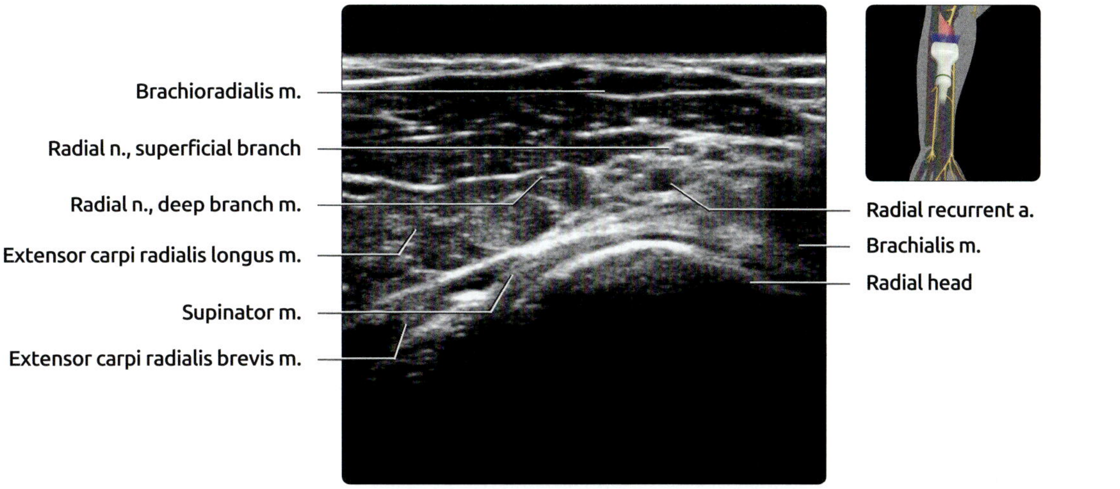

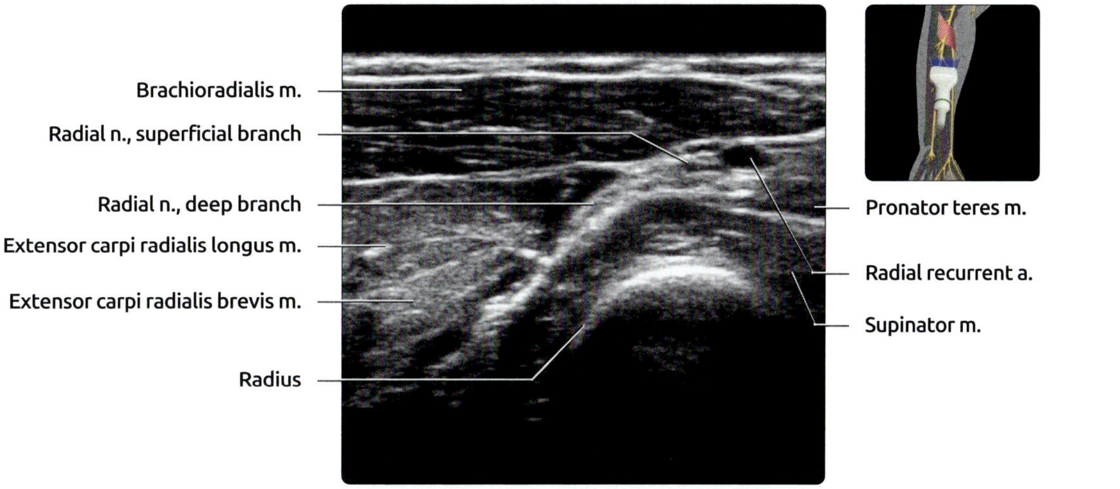

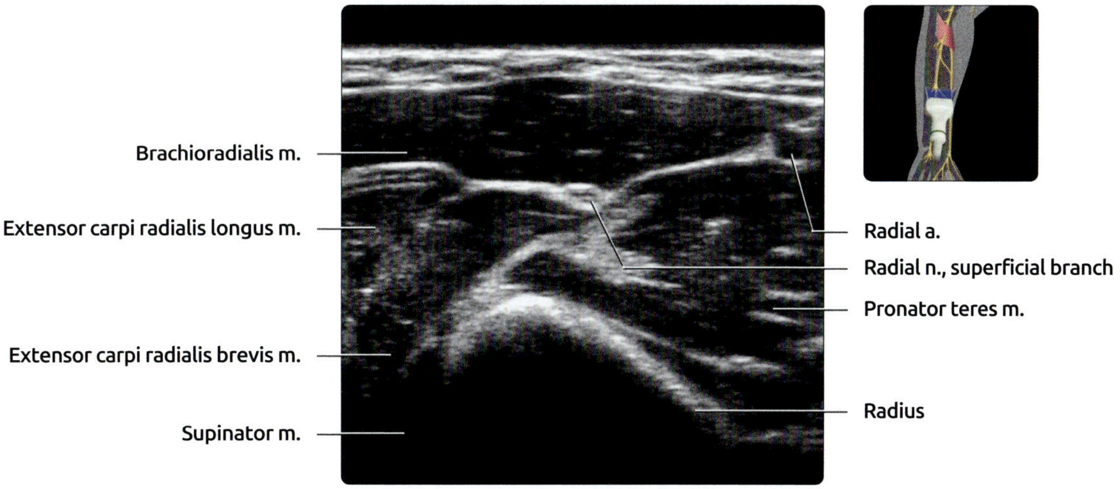

(Top) *Transverse grayscale ultrasound shows the radial nerve course in the proximal forearm. The radial nerve passes in front of the lateral epicondyle. It can be readily identified deep to the brachioradialis muscle. It divides into superficial and deep branches in the region of the radial head. The radial nerve is prone to compression between the radial head and the supinator muscle (radial tunnel syndrome).* **(Middle)** *Transverse grayscale ultrasound shows the radial nerve at the proximal forearm. The superficial branch of the radial nerve continues to descend in the forearm deep to the brachioradialis muscle. The deep branch pierces the supinator muscle, after which it is known as the posterior interosseus nerve.* **(Bottom)** *Transverse grayscale ultrasound shows the radial nerve in the midforearm. The deep branch of the radial nerve may be compressed as it passes through the supinator muscle (arcade of Frohse).*

TRANSVERSE US, FOREARM

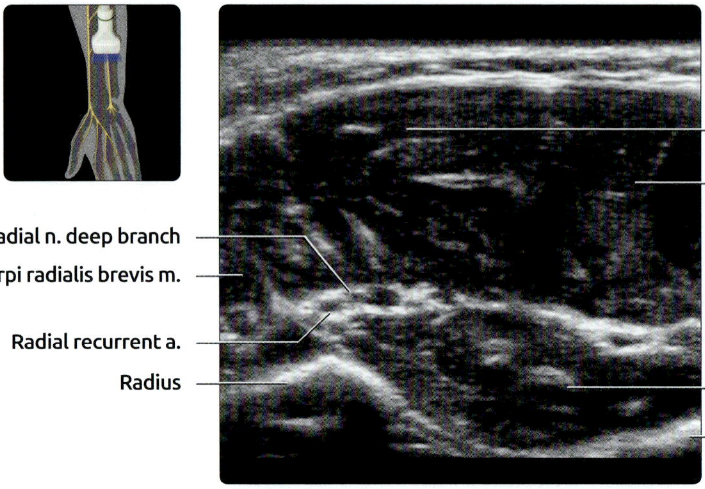

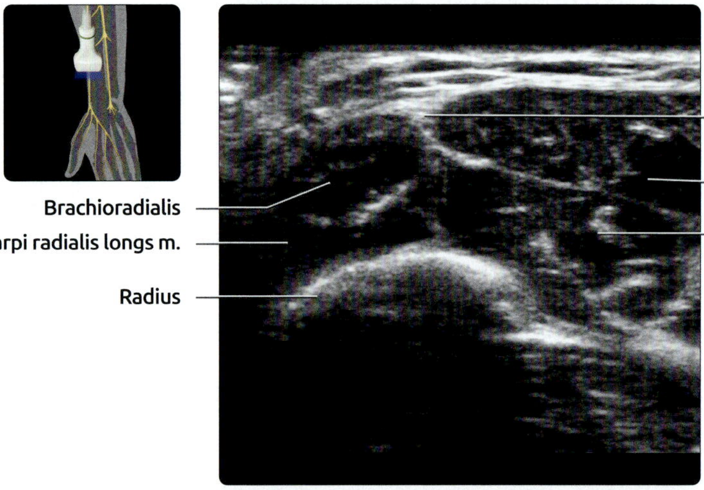

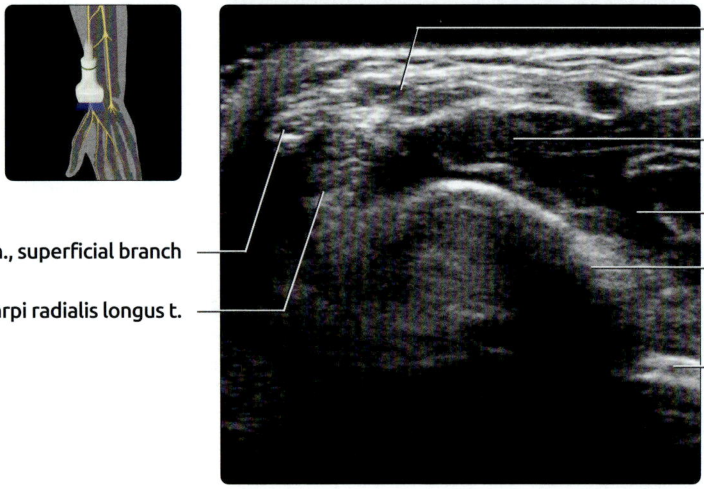

(Top) *Transverse grayscale ultrasound shows the deep branch of the radial nerve on the volar aspect of the distal forearm.* **(Middle)** *Transverse grayscale ultrasound shows the superficial branch of the radial nerve on the volar aspect of the distal forearm. About 8 cm proximal to the radial styloid, the superficial branch of the radial nerve emerges from underneath the brachioradialis muscle to lie between the brachioradialis and the extensor carpi radialis longus tendons. It may be compressed in this location (Wartenberg syndrome).* **(Bottom)** *Transverse grayscale ultrasound shows the superficial branch of the radial nerve on the radial side of distal forearm close to the wrist. It now lies close to the skin.*

ELBOW AND WRIST

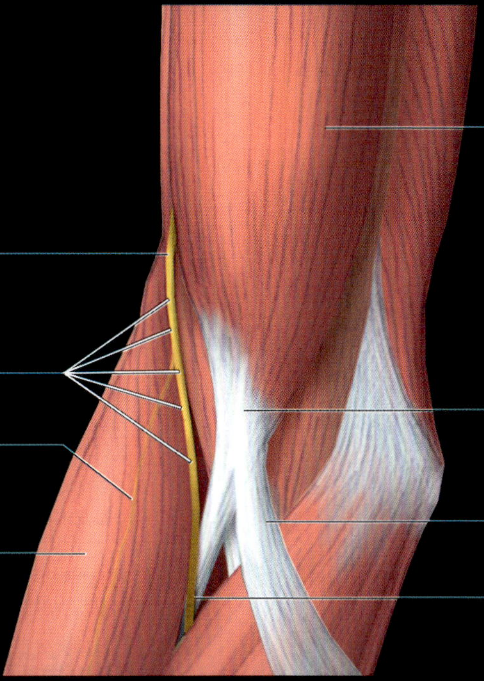

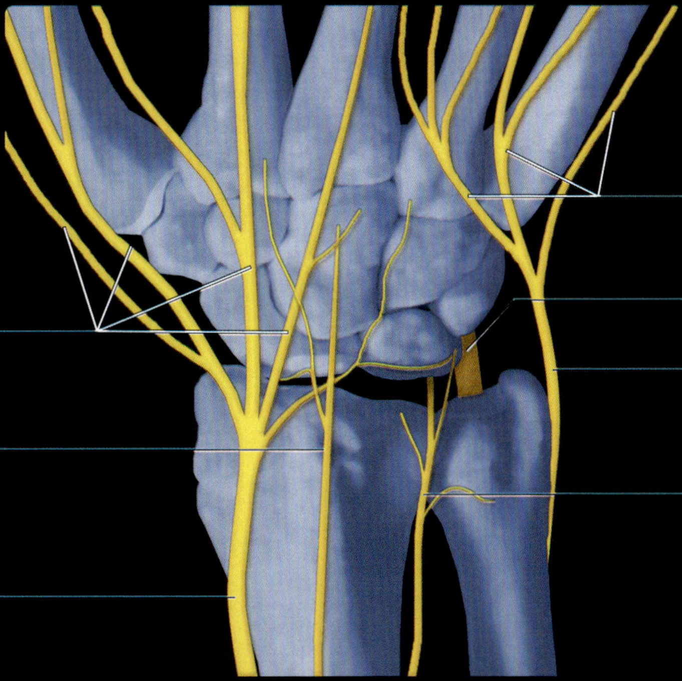

(Top) *Graphic shows the radial nerve course at the anterolateral aspect of the elbow. The deep branch of the radial nerve is potentially prone to impingement at the superficial proximal margin of the supinator muscle (arcade of Frohse) resulting in posterior interosseus nerve syndrome.* **(Bottom)** *Graphic shows the nerves of the dorsal wrist. The radial nerve branches in the forearm with the superficial, posterior cutaneous, and posterior interosseus branches supplying the wrist and hand. The ulnar nerve provides branches to both the dorsal and volar wrist.*

TERMINOLOGY

Abbreviations

- Flexor digitorum profundus (FDP)
- Flexor digitorum superficialis (FDS)
- Flexor carpi ulnaris (FCU)
- Flexor pollicis longus (FPL)
- Pronator quadratus (PQ)
- Palmaris longus (PL)

GROSS ANATOMY

Arm: Median Nerve

- Arises from both medial and lateral cords of brachial plexus (C6-C8, T1)
- Lies medial to brachial artery and vein
- Located deep to biceps aponeurosis in cubital fossa
- Gives articular branches to elbow joint
- Supplies pronator teres, PQ, and flexors of anterior compartment forearm (except FCU and medial 1/2 of FDP, which are supplied by ulnar nerve)

Forearm: Median Nerve and Branches

- Enters forearm by passing between heads of pronator teres; located in forearm between FDP and FDS
    - Courses distally and attached to deep surface of FDS by fascial sheath
    - Anterior interosseus nerve
        - Arises from median nerve at level of pronator teres
        - Located in forearm anterior to interosseus membrane, between FPL and FDP
- At wrist, median nerve emerges from lateral side of FDS, becomes more superficial, and runs deep to PL tendon toward carpal tunnel
    - Supplies FPL, PQ, and lateral 1/2 of FDP

Wrist and Hand: Median Nerve and Branches

- Median nerve lies along axis of ring finger, i.e., to ulnar side of midline
- Lies deep to antebrachial fascia and PL before dipping deeply to enter carpal tunnel
- At distal radioulnar joint: Nerve is rounded, deep to PL, medial and superficial to FCR and FPL, and lateral and superficial to FDS
- Carpal tunnel is fibroosseous tunnel located immediately distal to volar wrist crease
    - Formed by carpal bones and flexor retinaculum (transverse carpal ligament)
    - Flexor retinaculum attaches to pisiform, hook of hamate, scaphoid, and trapezium
    - 9 structures pass through carpal tunnel
        - 4 FDS tendons, 4 FDP tendons, FPL tendon, and median nerve
        - Median nerve is located superficially in carpal tunnel just deep to retinaculum
    - Tunnel outlet is narrowest part of carpal tunnel
- Branches
    - Motor to thenar and 1st and 2nd lumbrical muscles
    - Sensory to radial 1/2 of palm, radial 3.5 digits
        - Palmar cutaneous branch typically arises proximal to carpal tunnel
    - Median nerve divides just distal to carpal tunnel

ANATOMY IMAGING ISSUES

Imaging Recommendations

- Examination of nerve in transverse plane is preferable to longitudinal scanning since it allows nerve to be followed continuously throughout limb
- Use proximal and distal borders of retinaculum as markers of carpal tunnel

Imaging Approaches

- Ultrasound
    - Use high-frequency 12- to 17-MHz transducer

Imaging Pitfalls

- Median nerve may be bifid
    - If quantifying size of nerve, both components should be measured and summated
- Angle transducer at right angles to nerve as angulation will ↑ nerve caliber
    - May need to make use of anisotropy to best differentiate nerve from tendons in carpal tunnel
- Avoid too much transducer pressure as this will ↓ nerve caliber and ↑ flattening
- Median nerve may be difficult to see at tunnel outlet in subjects with thick palmar skin

CLINICAL IMPLICATIONS

Clinical Importance

- Normal cross-sectional area of median nerve ranges from 3.9-9.0 mm^2
    - Mean fascicular diameter ranges from 0.3-0.5 mm
    - Nerve thickening is considered if size changes beyond above-mentioned limits (e.g., > 10 mm^2 cross-sectional area)
    - Larger median nerve = greater the likelihood of carpal tunnel syndrome
- Potential sites of median nerve compression
    - Deep to ligament of Struthers
        - Ligament between medial epicondyle and anomalous bony spur arising from distal humeral diaphysis (present in 1% population)
        - May compress median nerve or brachial artery or both
    - Between brachialis and bicipital aponeurosis
        - Bicipital aponeurosis = medial fascial expansion of biceps tendon that extends superficial to brachial artery and median nerve and merges with investing fascia forearm
    - Between heads of pronator teres
        - Either median nerve or anterior interosseus nerve may be compressed
    - Carpal tunnel syndrome
        - Pain, paresthesia, numbness due to compression of nerve in carpal tunnel
        - Most common compressive neuropathy
        - Nerve swollen proximal to inlet ± at inlet ± at outlet ± distal to outlet
        - Do not use opposite side as comparative standard as subclinical compression common

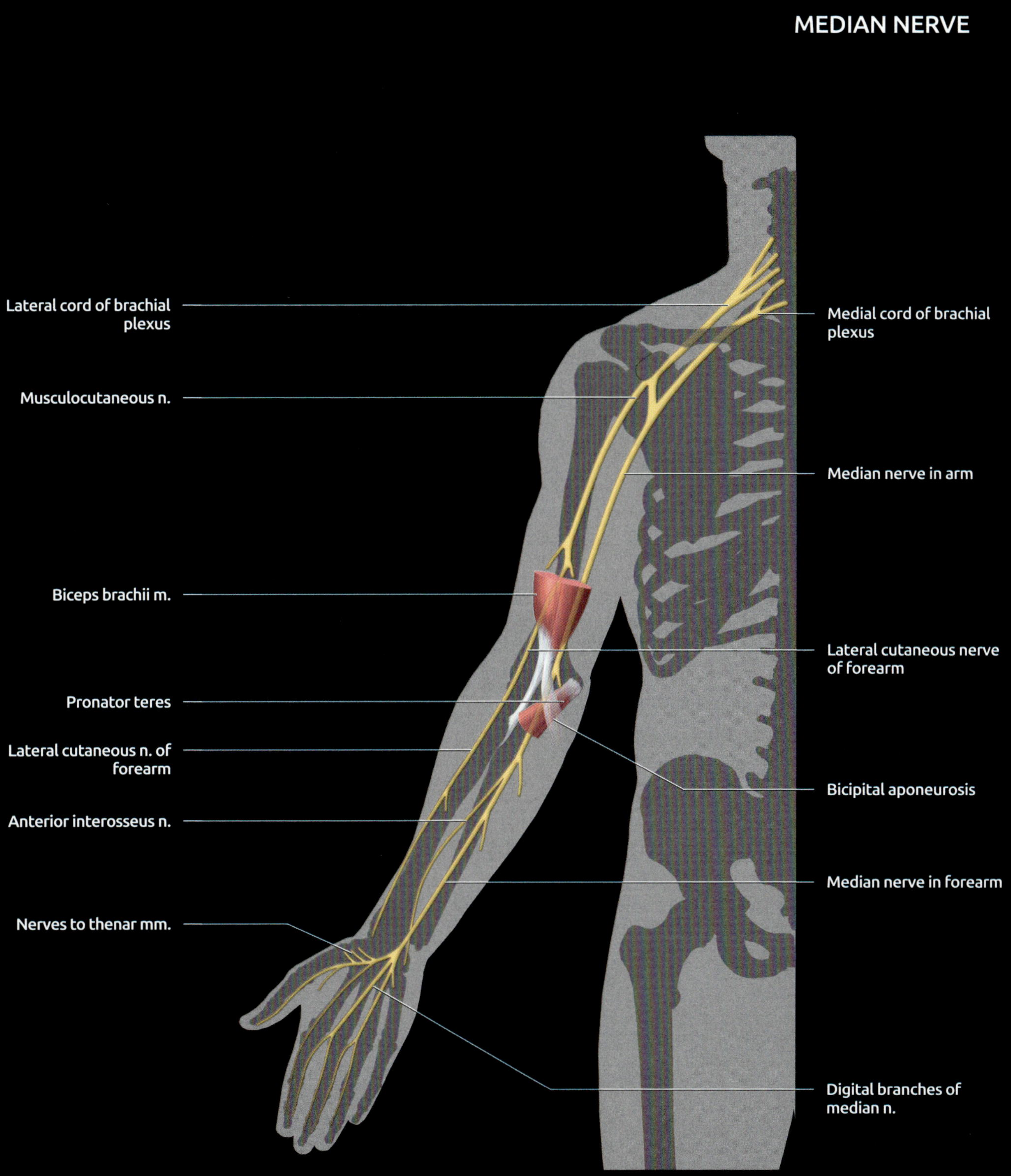

Graphic shows the course and distribution of the median and musculocutaneous nerves. The median nerve arises from the medial and lateral cords of the brachial plexus (C6-C8, T1). It travels together with other nerves of the arm. In the elbow, it is located deep to the biceps aponeurosis, enters the forearm by passing between the heads of the pronator teres, and is located in the forearm between the flexor digitorum superficialis and profundus muscles. The musculocutaneous nerve arises from the lateral cord of brachial plexus (C5, C6, C7). It lies between the brachialis and biceps brachii muscles and supplies both of them. It then becomes superficial at the elbow and continues laterally as the lateral cutaneous nerve of the forearm, which innervates the skin of the lateral side of the forearm.

CUBITAL FOSSA

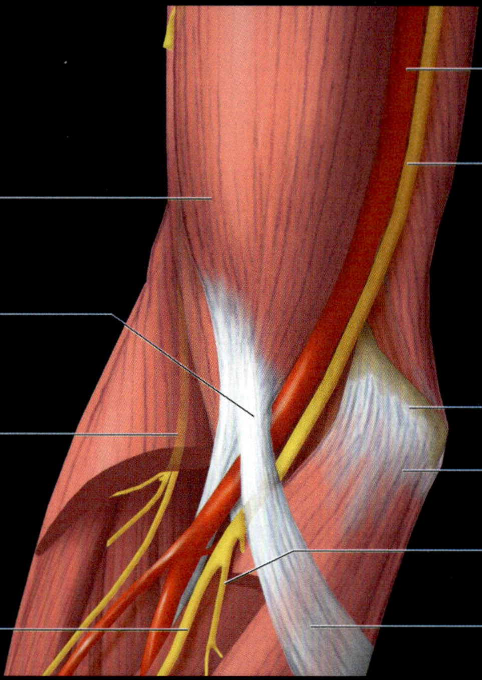

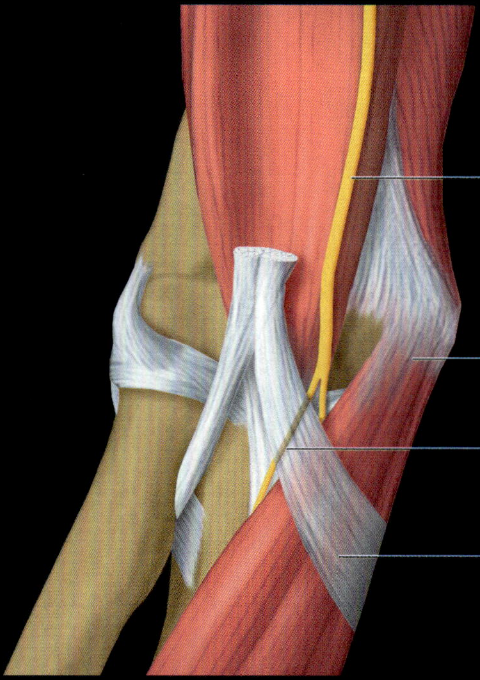

(Top) *Graphic of the anterior view of the cubital fossa shows the median nerve and brachial artery passing underneath the biceps aponeurosis. The anterior interosseus nerve arises from the median nerve as the median nerve passes between 2 heads of the pronator teres muscle.* **(Bottom)** *The median nerve may be entrapped between 2 heads of the pronator teres muscle or by the overlying biceps aponeurosis. The median nerve gives rise to the articular branches to the elbow joint. It supplies the pronator teres, pronator quadratus, and flexors of the anterior compartment of the forearm (except the flexor carpi ulnaris and medial 1/2 of the flexor digitorum profundus, which are supplied by the ulnar nerve).*

TRANSVERSE US

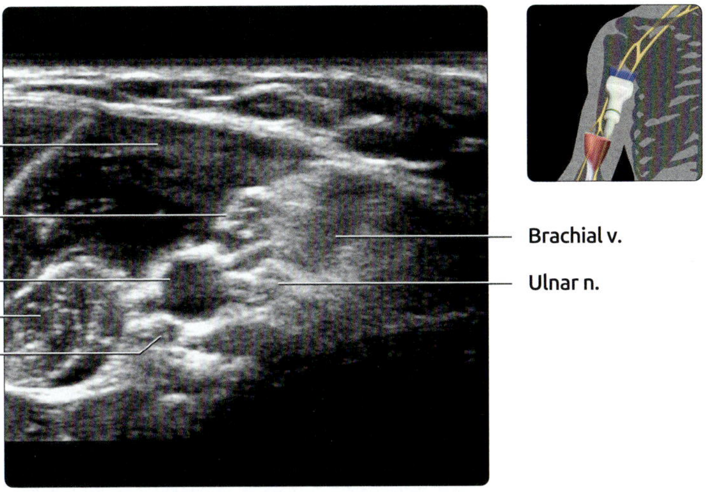

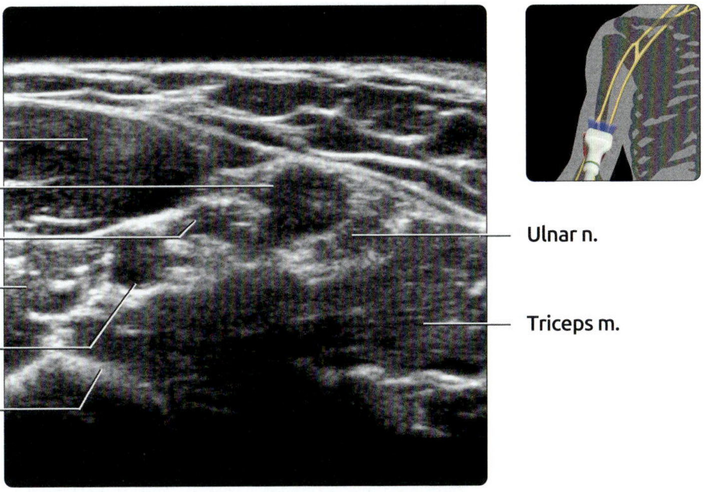

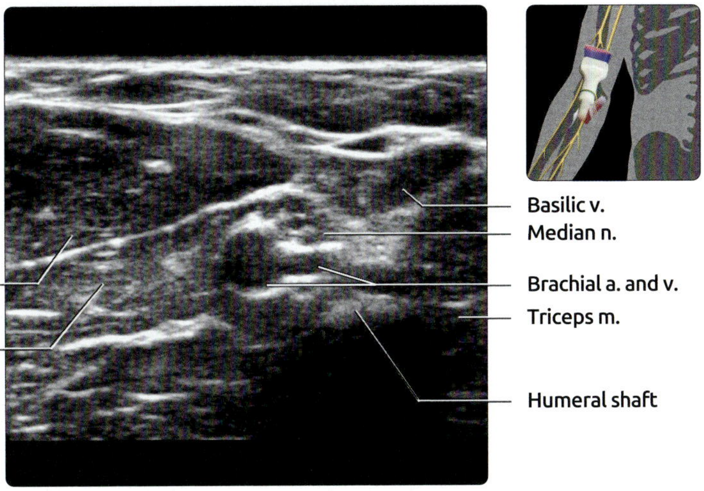

(Top) *Transverse grayscale ultrasound shows the median nerve at the proximal arm. The median nerve arises from the lateral and medial cords of the brachial plexus. In the proximal arm, it lies just lateral to the brachial vein and artery.* **(Middle)** *Transverse grayscale ultrasound shows the median nerve at the midarm. Little happens to the median nerve in the upper to midarm. It gives off no branches and is not prone to compression by normal structures.* **(Bottom)** *Transverse grayscale ultrasound shows the median nerve of the distal arm. The median nerve passes over the brachial artery and vein from lateral to medial. There is occasionally a ligament (of Struthers) in the distal arm with a bony spur attached to the humerus ~ 5 cm proximal to the medial epicondyle. The pronator teres gains additional attachment from this ligament. The median nerve may be compressed deep to this ligament.*

TRANSVERSE US

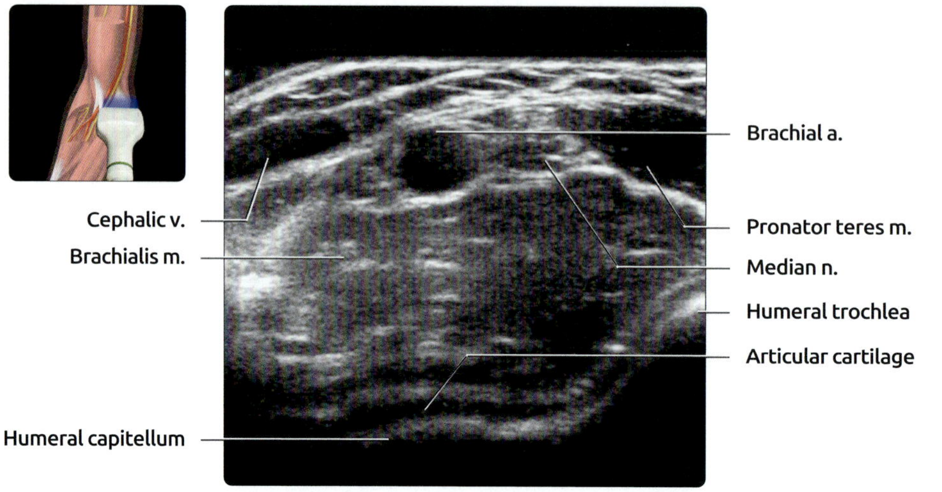

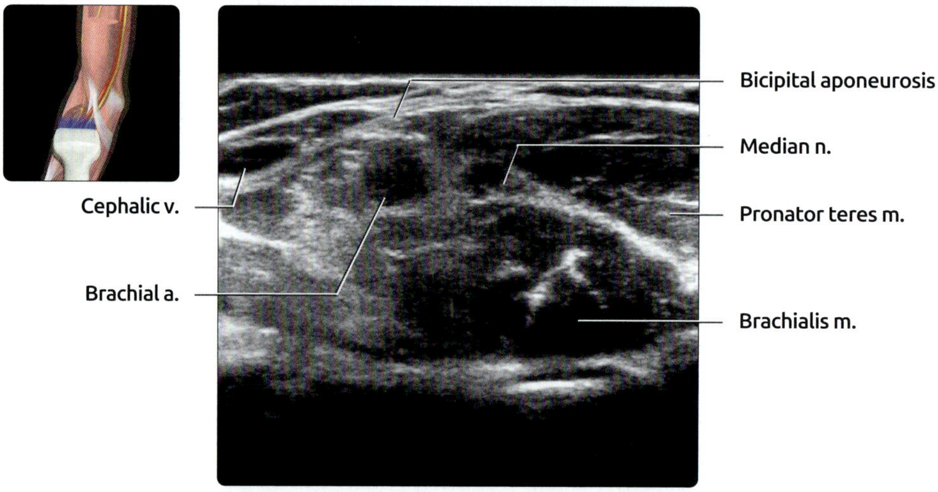

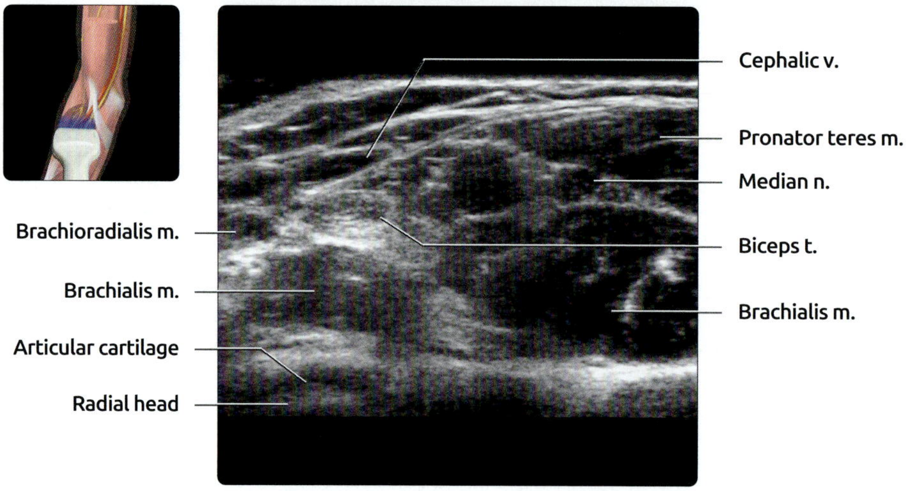

(Top) *Transverse grayscale ultrasound shows the median nerve at the distal arm. The median nerve travels deep to the bicipital aponeurosis. A fascial aponeurosis (a.k.a. lacertus fibrosis) arises from the biceps musculotendinous junction and inserts into the antebrachial fascia. The median nerve may be compressed deep to this fascia.* **(Middle)** *Transverse grayscale ultrasound shows the median nerve at the elbow region.* **(Bottom)** *Transverse grayscale ultrasound shows the median nerve at the proximal forearm. The median nerve is also prone to compression between the 2 heads of the pronator teres muscle (i.e., common flexor origin head and coronoid process head). The median nerve branches as it passes through these heads (giving rise to the anterior interosseus nerve) and may be compressed at this location (pronator teres syndrome). Pain in forearm or hand occurs with either prolonged forearm pronation or recurrent pronation and supination of forearm.*

TRANSVERSE US

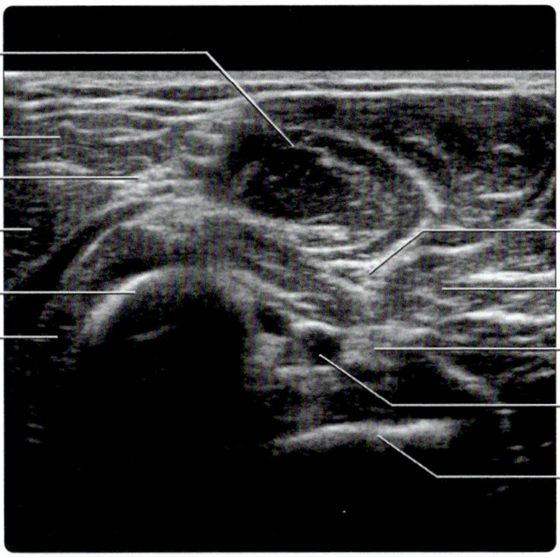

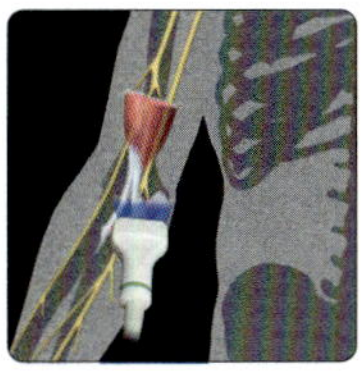

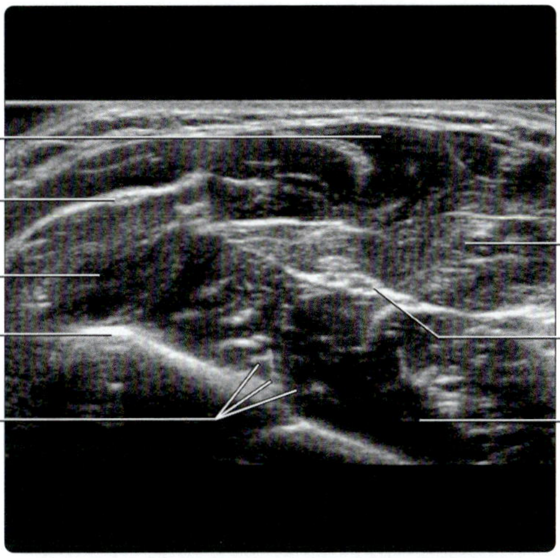

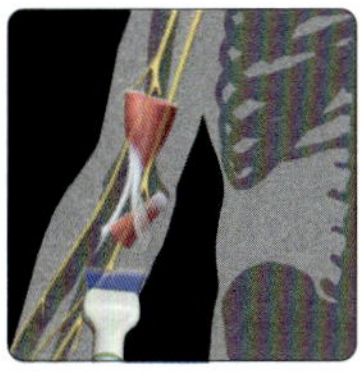

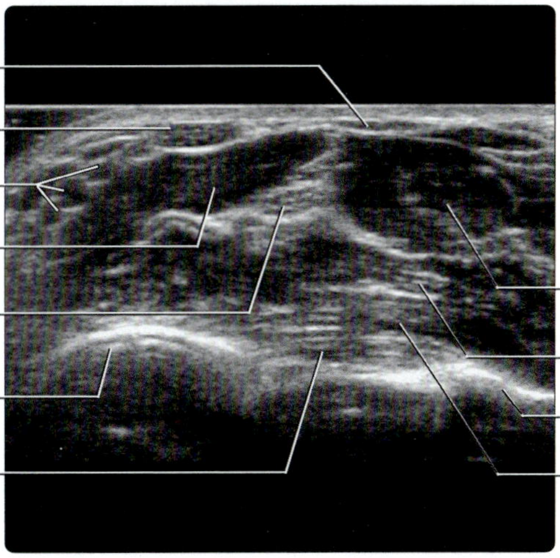

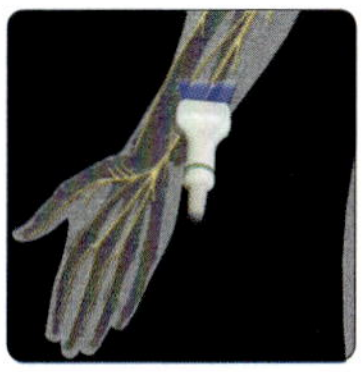

(Top) *Transverse grayscale ultrasound shows the median nerve at the proximal forearm. The anterior interosseus nerve supplies all the muscles of the anterior compartment of the forearm except the ulnar 1/2 of the flexor digitorum profundus muscle. It may be injured with fracture of the radial shaft.* **(Middle)** *Transverse grayscale ultrasound shows the median nerve at the midforearm. The median nerve is clearly identified in the middistal forearm as it travels between the flexor digitorum superficialis and profundus muscles.* **(Bottom)** *Transverse grayscale ultrasound shows the median nerve at the distal forearm.*

TRANSVERSE US: WRIST

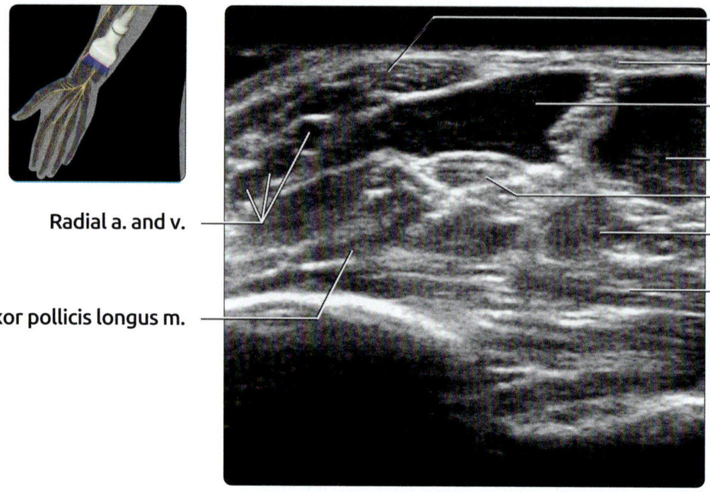

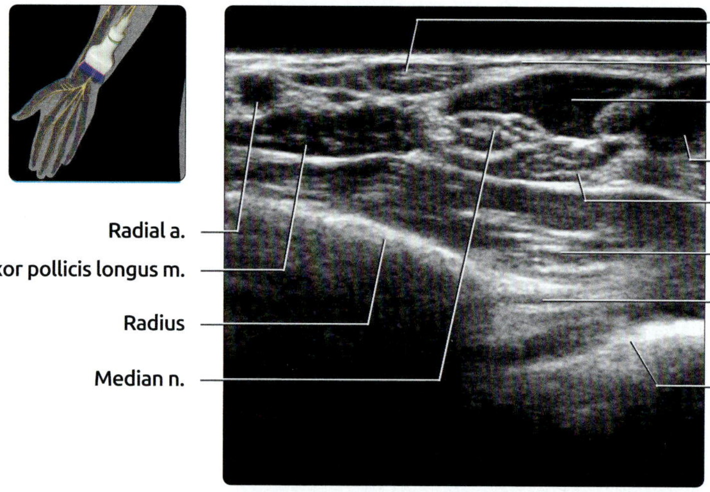

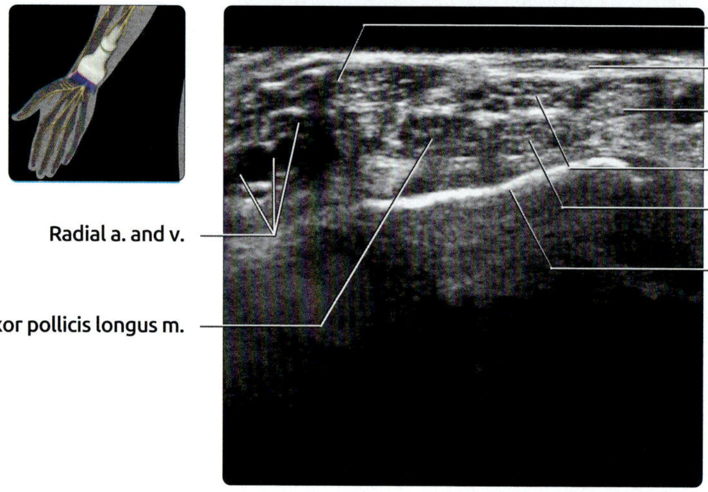

(Top) *Transverse grayscale ultrasound shows the median nerve in the distal forearm.* **(Middle)** *Transverse grayscale ultrasound shows the median nerve just proximal to the wrist. The median nerve gives off the palmar cutaneous branch in the distal forearm. This branch innervates the skin on the palm of the hand.* **(Bottom)** *Transverse grayscale ultrasound shows the median nerve just proximal to the carpal tunnel; the median nerve ascends from a deeper position (between the flexor digitorum profundus and superficialis) to lie just deep to palmaris longus tendon. Enlargement of the median nerve in patients with carpal tunnel syndrome may present clinically as a mass-like lesion in this area. In suspected carpal tunnel syndrome, the caliber of the nerve should be measured at this location.*

TRANSVERSE US, CARPAL TUNNEL

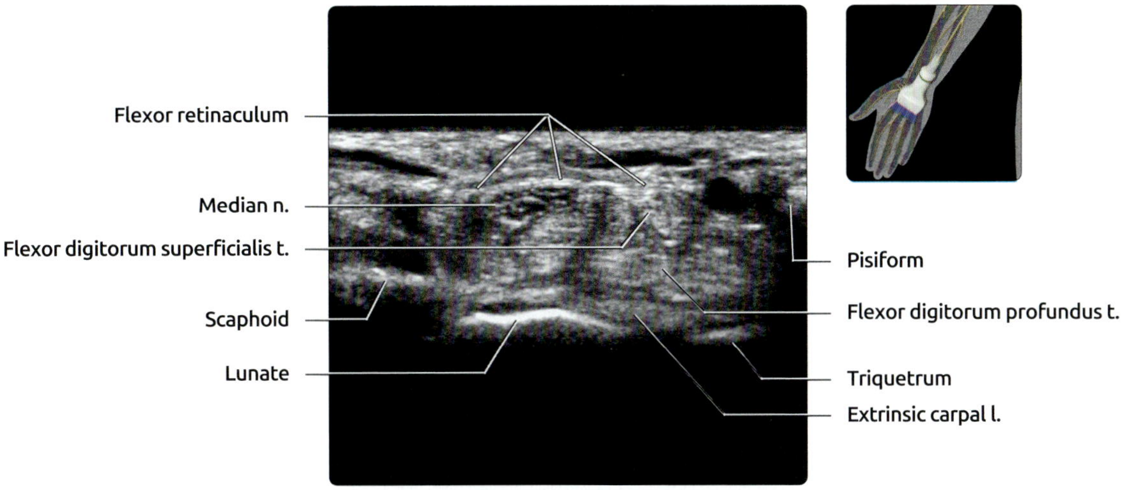

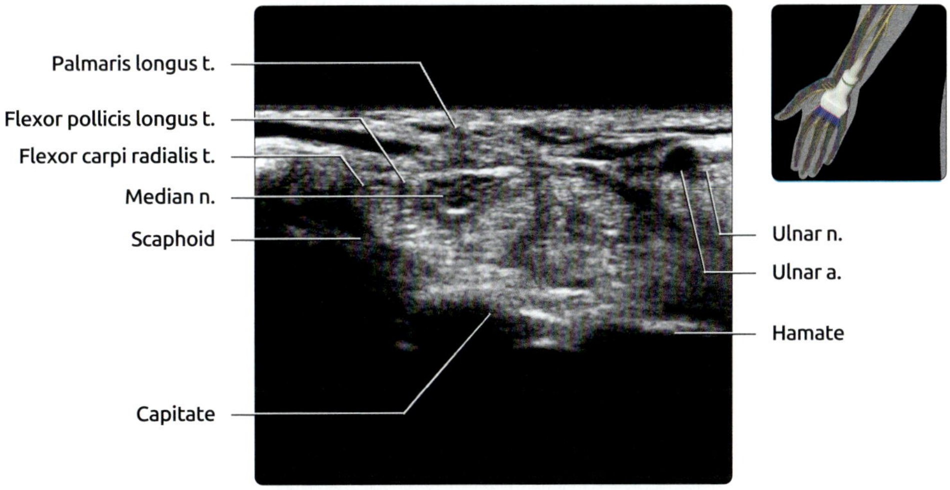

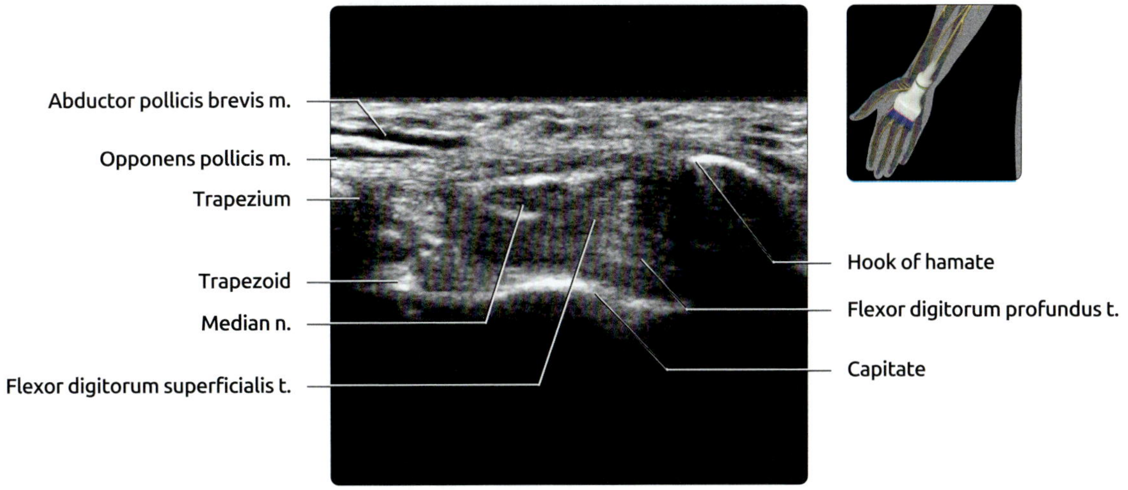

(Top) *Transverse grayscale ultrasound shows the median nerve at the carpal tunnel inlet. The best way to identify the carpal tunnel is to identify the leading and trailing edges of the flexor retinaculum. The median nerve dips deep to the retinaculum. The cross-sectional area should be measured at this location in suspected carpal tunnel syndrome.* **(Middle)** *Transverse grayscale ultrasound shows the median nerve at the midcarpal tunnel. The median nerve lies just deep to the retinaculum. It is variable in shape and changes with flexor tendon movement.* **(Bottom)** *Transverse grayscale ultrasound shows the median nerve at the carpal tunnel outlet. The cross-sectional area of the median nerve should also be measured at this location in patients with suspected carpal tunnel syndrome. This is the narrowest part of the carpal tunnel.*

TRANSVERSE AND LONGITUDINAL US

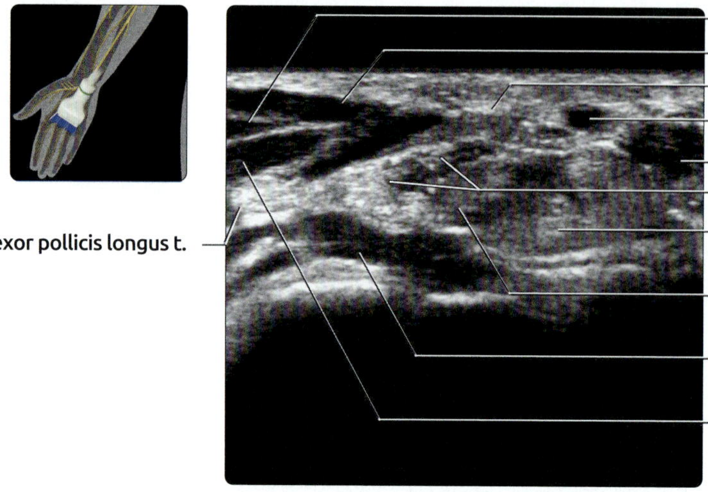

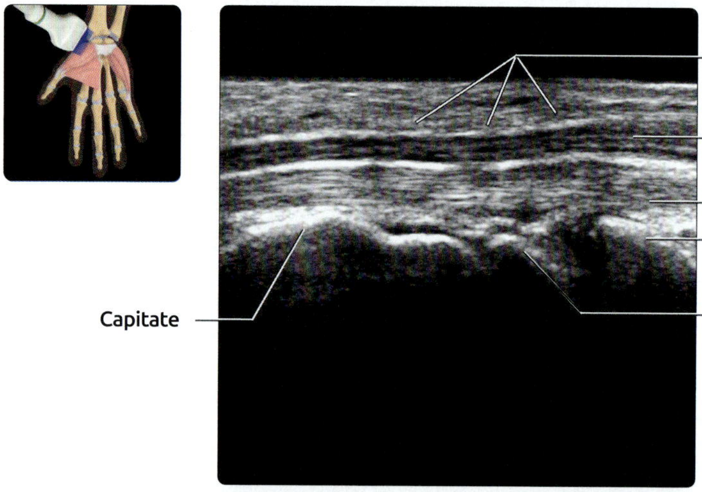

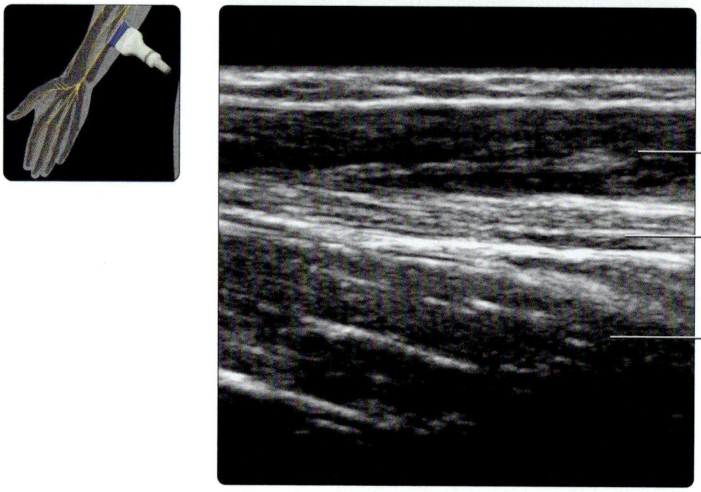

(Top) *Transverse grayscale ultrasound shows the median nerve just beyond the carpal tunnel outlet. The median nerve fairly consistently divides into its digital branches just beyond the tunnel outlet. This makes it difficult to get consistent results for nerve caliber at this location.* **(Middle)** *Longitudinal grayscale ultrasound shows the median nerve in the carpal tunnel.* **(Bottom)** *Longitudinal grayscale ultrasound shows the median nerve in the distal forearm.*

PALMAR NERVES

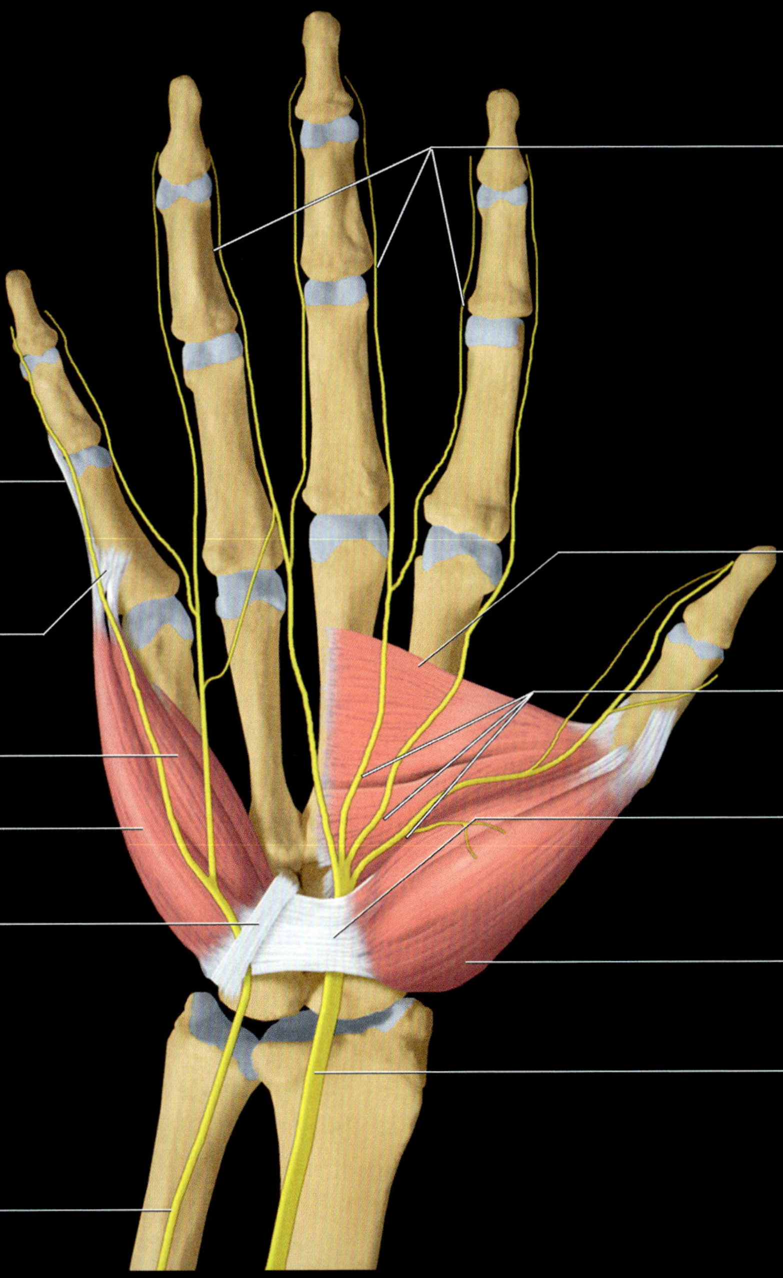

The median nerve travels deep to the flexor retinaculum (within the carpal tunnel). The ulnar nerve travels superficial to the flexor retinaculum and deep to the pisohamate ligament (within the Guyon canal). Both are prone to compression syndromes within these fibroosseous tunnels. Motor supply of the median nerve is to the thenar muscles (via the recurrent branch) and the 1st and 2nd lumbricals. The ulnar nerve motor supply is to all intrinsic muscles of the hand not supplied by the median nerve. Muscle atrophy should prompt an evaluation of the supplying nerve.

GROSS ANATOMY

Arm: Ulnar Nerve

- Arises from medial cord of brachial plexus (C8, T1)
- Located posteromedially, deep to triceps muscles in distal arm
- Cubital tunnel: Fibroosseous tunnel formed by medial epicondyle, olecranon, & cubital retinaculum (Osborne fascia)
  - Cubital tunnel syndrome
    - Pain, weakness, & numbness in 4th & 5th fingers due to compression of ulnar nerve in cubital tunnel
- May sublux anterior to medial epicondyle in 15% of normal elbows during flexion
- Anconeus epitrochlearis muscle runs on same course at cubital retinaculum, i.e., lies between cubital retinaculum & ulnar nerve
  - From medial cortex of olecranon to medial epicondyle
  - Present in 11% of normal subjects, bilateral in 25%
  - Can contribute to cubital tunnel syndrome
  - When anconeus epitrochlearis muscle is not present, it may be replaced by band of tissue called epitrochleoanconeus ligament

Forearm: Ulnar Nerve & Branches

- Passes behind medial epicondyle, enters forearm between humeral & ulnar heads of flexor carpi ulnaris (FCU)
- Courses distally between FCU & flexor digitorum profundus (FDP)
  - Supplies FCU, medial 1/2 of FDP
- Palmar cutaneous branch
  - Arises midforearm & supplies skin on ulnar 1/2 of palm
- Dorsal cutaneous branch
  - Arises distal forearm & supplies skin on ulnar side of dorsum hand
- Distally, becomes superficial & passes into wrist superficial to flexor retinaculum

Hand: Ulnar Nerve & Branches

- In Guyon canal, ulnar nerve lies radial to pisiform & at ulnar & dorsal aspect of ulnar artery
- Enters Guyon canal; small, triangular-shaped fibroosseous tunnel
  - Floor = flexor retinaculum; ulnar wall = pisiform & hook of hamate; roof = fascia continuous with flexor retinaculum (superficial palmar ligament)
  - Contains ulnar artery & ulnar nerve
  - Nerve lies between ulnar artery & pisiform or hook of hamate
  - Divides at distal tunnel into superficial sensory branch & deep motor branch
    - Superficial branch; sensory to ulnar side of palm, little finger, & ulnar 1/2 of ring finger
    - Deep branch lying medial to hook of hamate where prone to injury; motor to hypothenar muscles, all interossei, 3rd & 4th lumbricals, & adductor pollicis

ANATOMY IMAGING ISSUES

Imaging Recommendations

- Knowledge of nerve location & anatomic relationships is essential for sonographic identification
- Examination of nerves in transverse plane is preferable to longitudinal plane as it allows nerves to be followed continuously throughout limb

Imaging Approaches

- Ultrasound
  - Use of high-frequency 12- to 17-MHz transducer
  - Nerve echogenicity
    - Transverse scan: Irregularly dispersed hypoechoic dots, due to nerve fascicles separated by hyperechoic epineurium
    - Longitudinal scan: Parallel hypoechoic tracts of uniform caliber
    - Gain control & focus settings should be optimized for adequate visualization

Imaging Pitfalls

- Smaller nerves & nerve branches (1-2 mm) are difficult to identify, & their location may only be inferred by adjacent vessels
- Anisotropy artifacts may affect scanning, especially in short axis
  - Tilting transducer during scanning helps to achieve best view
- Fixed structural causes of nerve compression are more easy to identify than functional structural compression
- Ulnar nerve normally swells to mild degree in cubital tunnel though not as much as in cubital tunnel syndrome

CLINICAL IMPLICATIONS

Clinical Importance

- Direct sonographic visualization of nerves can reveal focal abnormality or injury, differentiate from extraneural tumor, & show extent of lesion & relationship to adjacent structures & vessels
- Ulnar nerve prone to compression in cubital tunnel, between humeral & ulnar origins of FCU, & in Guyon canal
  - Cubital tunnel syndrome = compression of nerve in cubital tunnel
    - Mostly idiopathic without structural cause being found
    - Nerve swelling in cubital tunnel
    - Measure nerve caliber proximal to, within, & distal to cubital tunnel
    - Compare with opposite side
  - Guyon canal compression is often structural
    - Due to ganglion cyst at Guyon canal from pisotriquetral joint, fracture hook of hamate, or nerve sheath tumor
    - Other causes are occupational neuritis, ulnar artery aneurysm, anomalous muscle, lipoma, laceration, or giant cell tumor of tendon sheath
- Dislocation of ulnar nerve over medial epicondyle can be observed on ultrasound during elbow flexion
  - ± dislocation of medial triceps muscle belly
  - Repeated dislocation of ulnar nerve may lead to ulnar neuritis, functional impairment, & symptoms of cubital tunnel syndrome

ULNAR NERVE

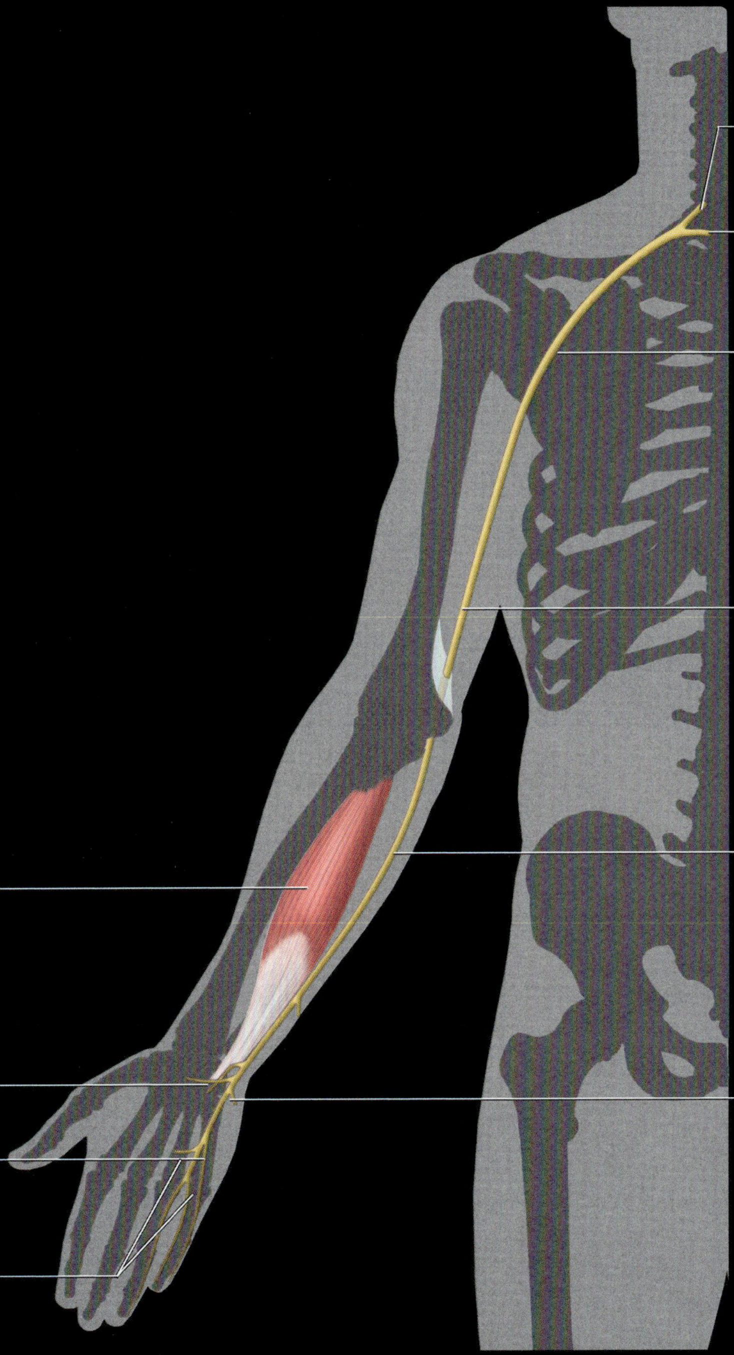

Graphic shows the course of the ulnar nerve. The ulnar nerve arises from the medial cord of the brachial plexus (C8, T1). It is located posteromedially, deep to the triceps muscles in the distal arm, where it passes posterior to medial epicondyle in the cubital tunnel (cubital syndrome due to compression of ulnar nerve in cubital tunnel). The ulnar nerve may sublux anterior to the medial epicondyle in ~ 15% of normal people, usually during flexion. The ulnar nerve passes behind the medial epicondyle, entering the forearm between the humeral and ulnar heads of flexor carpi ulnaris. The ulnar nerve then courses distally between flexor carpi ulnaris and the flexor digitorum profundus muscles, supplying the flexor carpi ulnaris muscle and medial 1/2 of the flexor digitorum profundus muscle.

CUBITAL TUNNEL

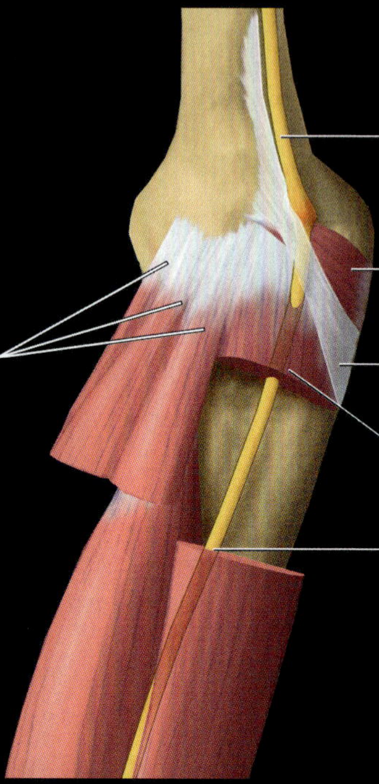

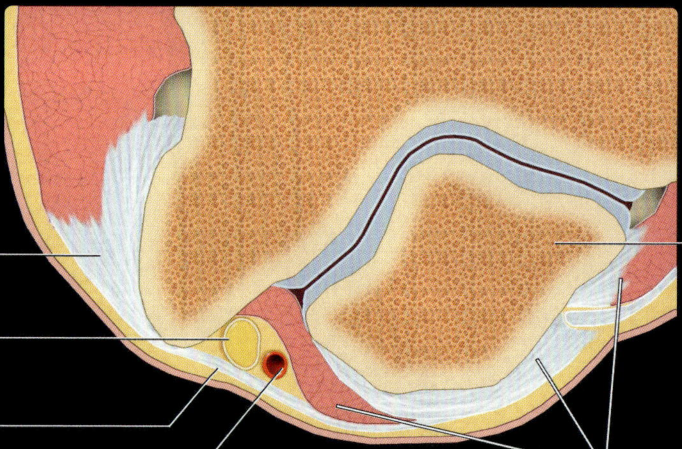

(Top) *Medial graphic shows the cubital tunnel. The anconeus epitrochlearis is an inconstant accessory muscle that lies superficial to the ulnar nerve and may compress the nerve against the medial epicondyle.* **(Bottom)** *Axial graphic shows the cubital tunnel. The ulnar nerve may be compressed within the cubital tunnel (cubital tunnel syndrome) by a mass, synovitis, posttraumatic osseous deformity, or aneurysm of the recurrent ulnar artery. The ulnar nerve may sublux out of the cubital tunnel aided by the adjacent medial head of the triceps. Recurrent subluxation, which can be found in ~ 15% of asymptomatic subjects, can lead to ulnar neuritis, which can produce similar symptoms to ulnar nerve compression.*

TRANSVERSE US, ARM

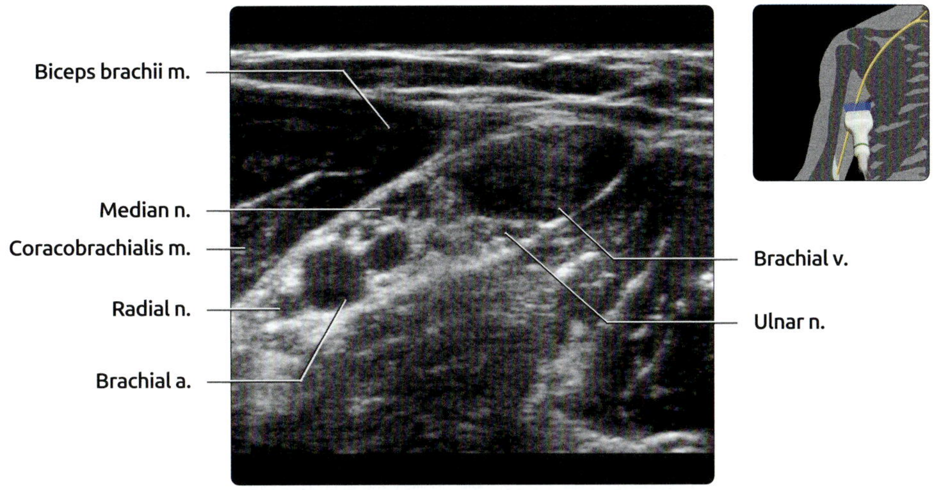

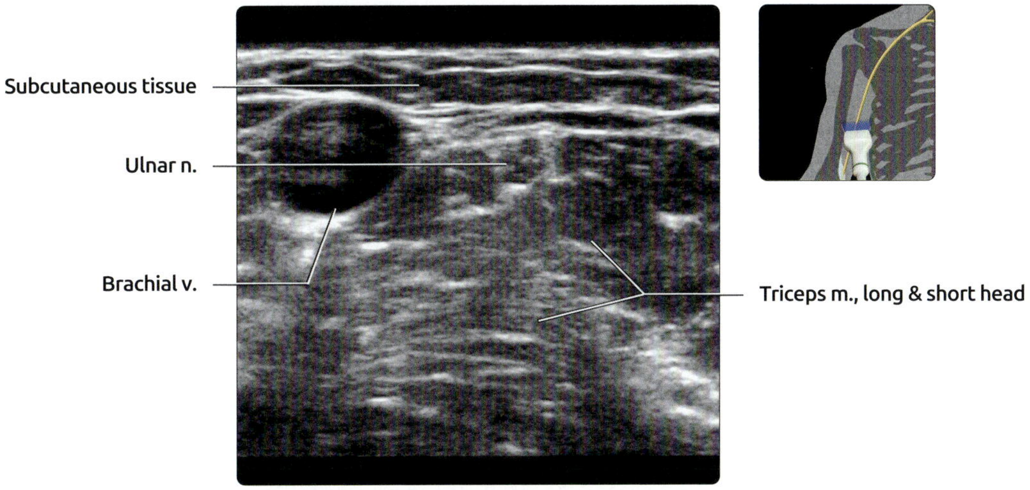

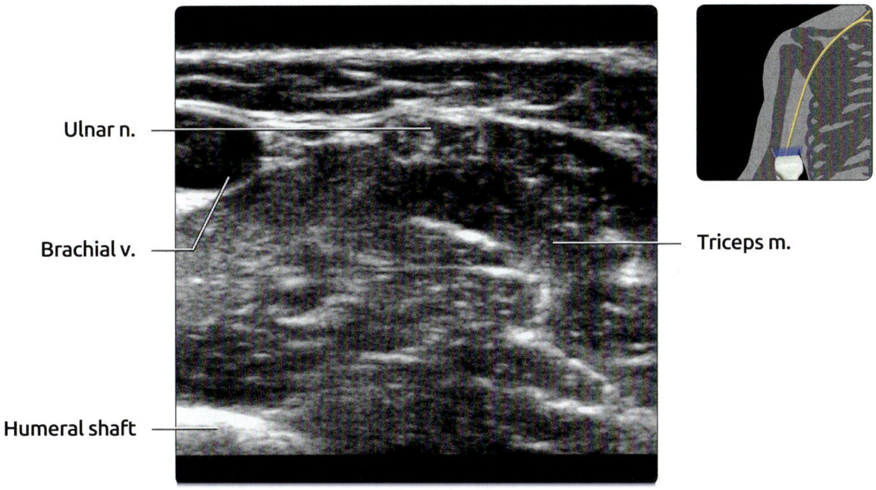

(Top) *Transverse grayscale ultrasound shows the ulnar nerve at the proximal arm. The ulnar nerve arises from the medial cord of the brachial plexus and lies just medial to the brachial vein.* **(Middle)** *Transverse grayscale ultrasound shows the ulnar nerve at the midarm. The ulnar nerve pierces the medial intermuscular septum in the midarm where it may be potentially compressed. The ulnar nerve gives off no branch in the arm.* **(Bottom)** *Transverse grayscale ultrasound shows the ulnar nerve in the distal arm. The ulnar nerve runs distally on the surface of the triceps muscle to reach the area between the olecranon and the medial epicondyle, i.e., the cubital tunnel.*

TRANSVERSE US, CUBITAL TUNNEL

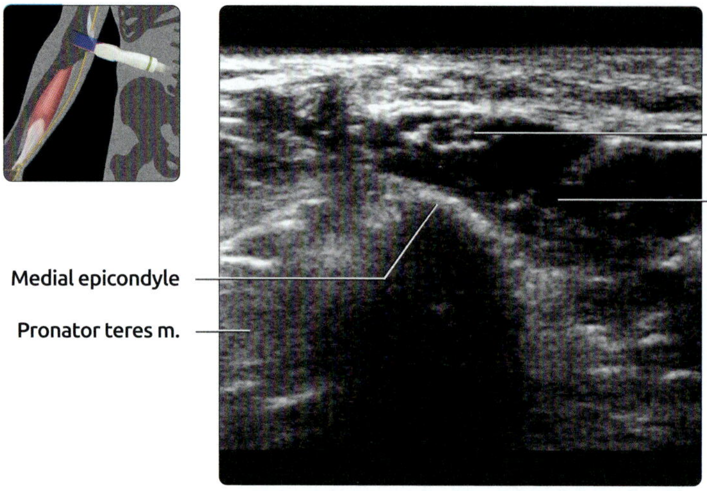

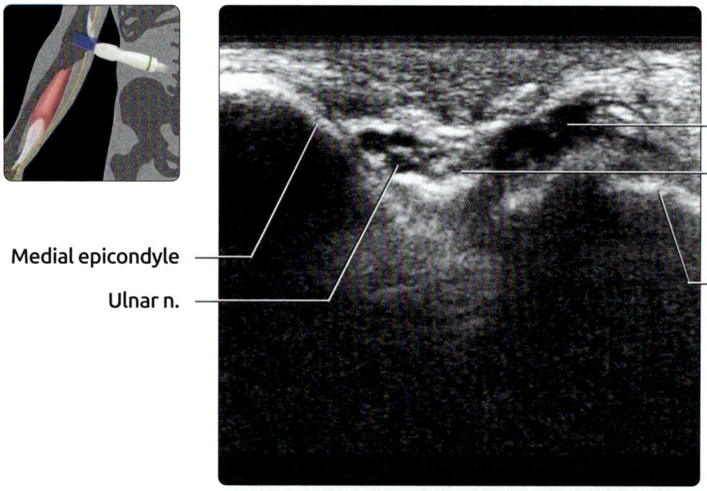

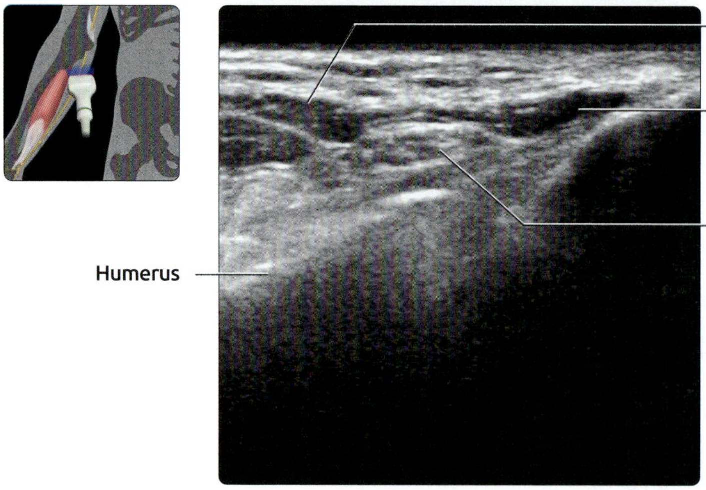

(Top) *Transverse grayscale ultrasound shows the ulnar nerve just proximal to the cubital tunnel. The cubital tunnel is the most common site of ulnar neuropathy. Neuropathy may take the form of compression from the overlying retinaculum or fibrous bands, bony impingement and tumors, subluxation of the ulnar nerve over the medial epicondyle, or direct injury.* **(Middle)** *Transverse grayscale ultrasound shows the ulnar nerve at the cubital tunnel. The cubital tunnel retinaculum (formed from an extension of investing fascia of the flexor carpi ulnaris and ligament of Osborne) forms the roof of the cubital tunnel. The retinaculum is 4 mm wide and difficult to depict on ultrasound. The floor and wall of cubital tunnel are formed by containing bones, elbow capsule, and the medial collateral ligament. The ulnar nerve normally enlarges slightly within the cubital tunnel.* **(Bottom)** *Transverse grayscale ultrasound shows the ulnar nerve just beyond the cubital tunnel.*

TRANSVERSE US, FOREARM

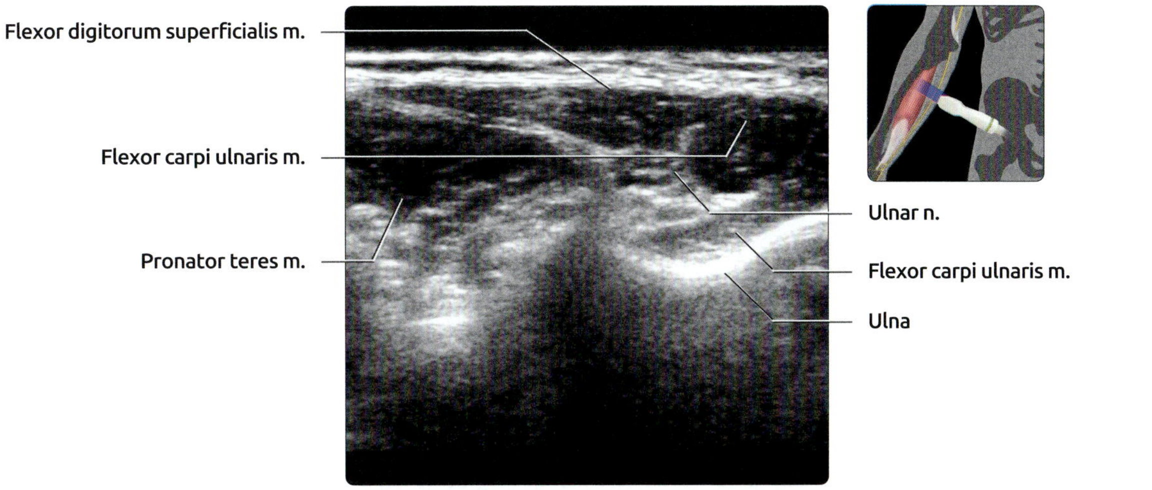

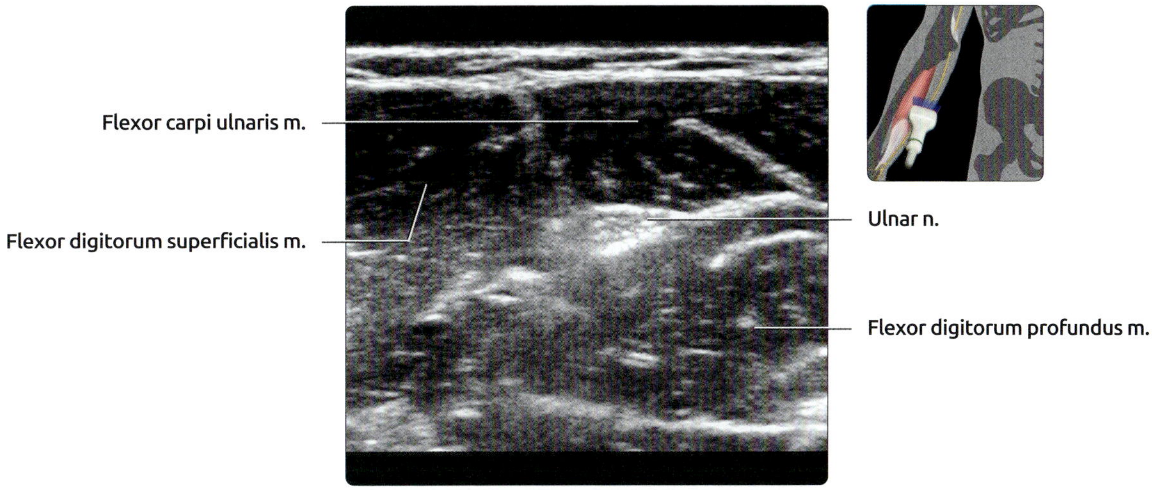

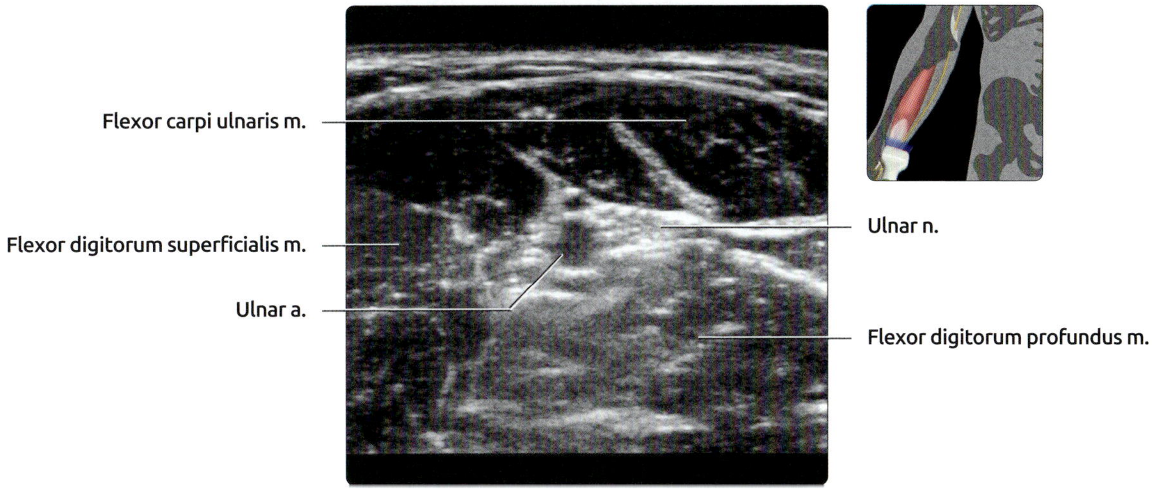

(Top) *Transverse grayscale ultrasound shows the ulnar nerve just beyond the cubital tunnel, where the ulnar nerve passes between the 2 heads (humeral and ulnar) of the flexor carpi ulnaris tendon. This is a potential site of compression.* **(Middle)** *Transverse grayscale ultrasound shows the ulnar nerve in the midforearm. The ulnar nerve pierces the intermuscular septum, dividing the flexor and extensor compartments of the forearm to descend in the flexor compartment between flexor carpi ulnaris and flexor digitorum profundus muscles.* **(Bottom)** *Transverse grayscale ultrasound shows the ulnar nerve in the distal forearm. The ulnar nerve is joined on its radial side by the ulnar artery in the distal forearm, and the 2 structures travel to the wrist, deep to the flexor carpi ulnaris muscle.*

TRANSVERSE AND LONGITUDINAL US, WRIST AND FOREARM

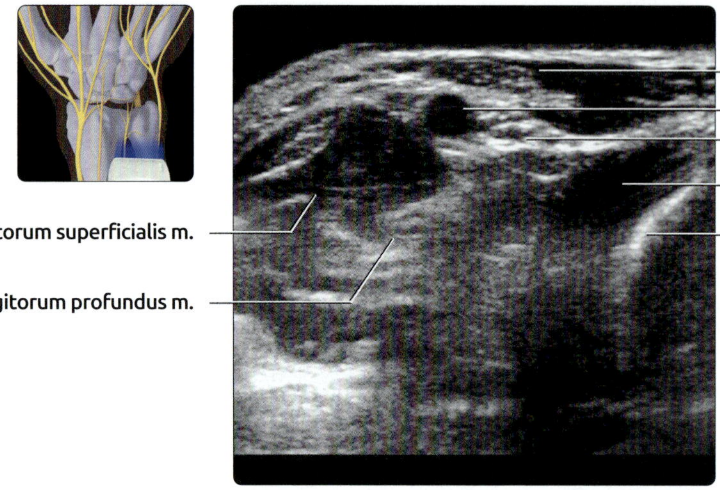

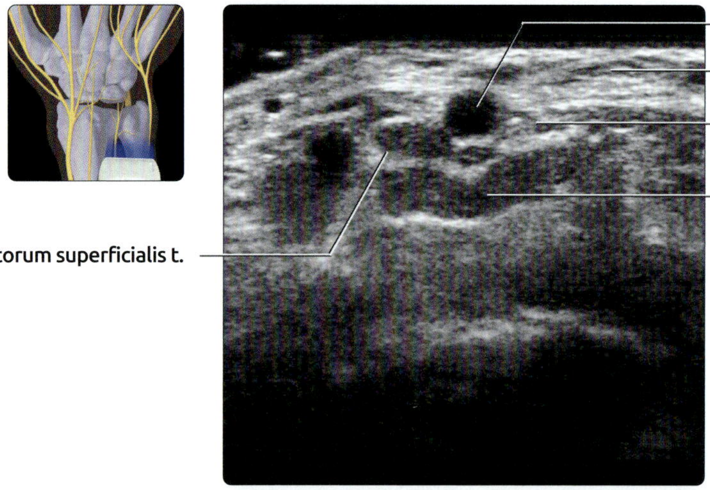

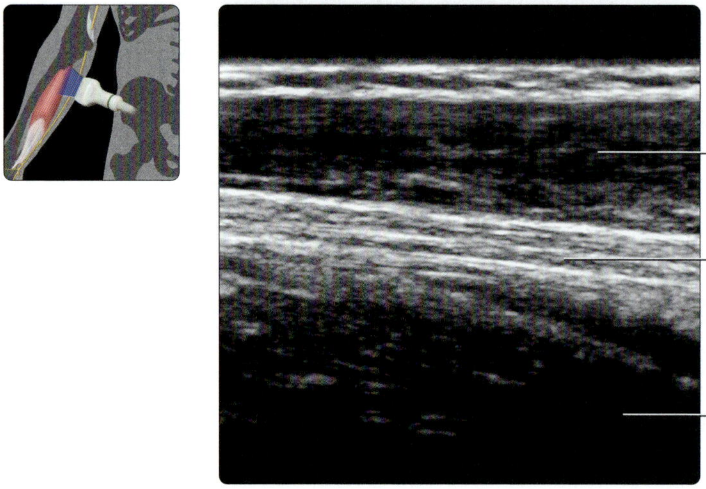

(Top) *Transverse grayscale ultrasound shows the ulnar nerve at the distal forearm. The ulnar nerve can be identified by 1st visualizing the ulnar artery and then identifying the nerve just ulnar to the artery.* **(Middle)** *Transverse grayscale ultrasound shows the ulnar nerve just proximal to the wrist. At this location, the ulnar nerve is prone to injury from dog bite, penetrating injury, and nonaccidental self-injury.* **(Bottom)** *Longitudinal grayscale ultrasound shows the ulnar nerve at the distal 1/3 of the forearm. Note the fibrillar pattern of the nerve with a mixed hypoechoic and hyperechoic pattern due to hypoechoic nerve fascicles separated by hyperechoic epineurium.*

TRANSVERSE US, GUYON CANAL

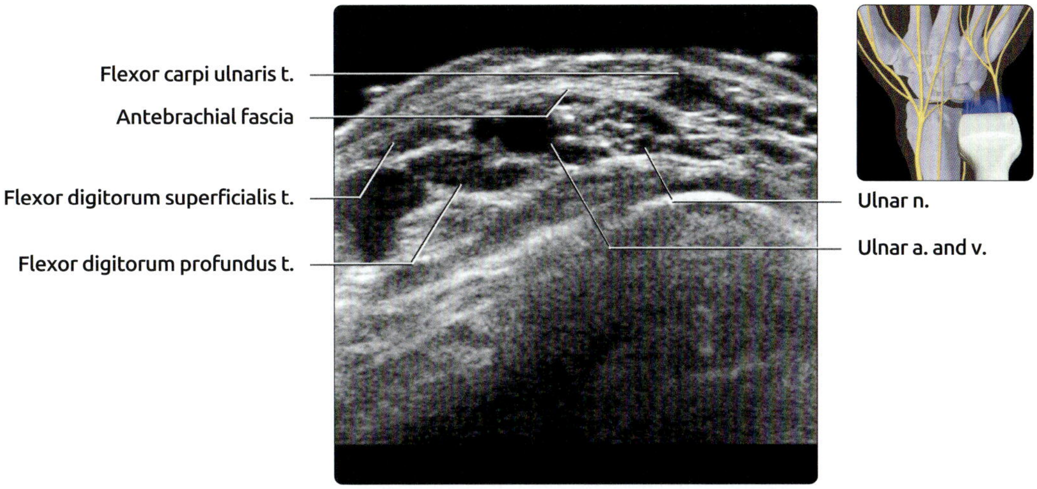

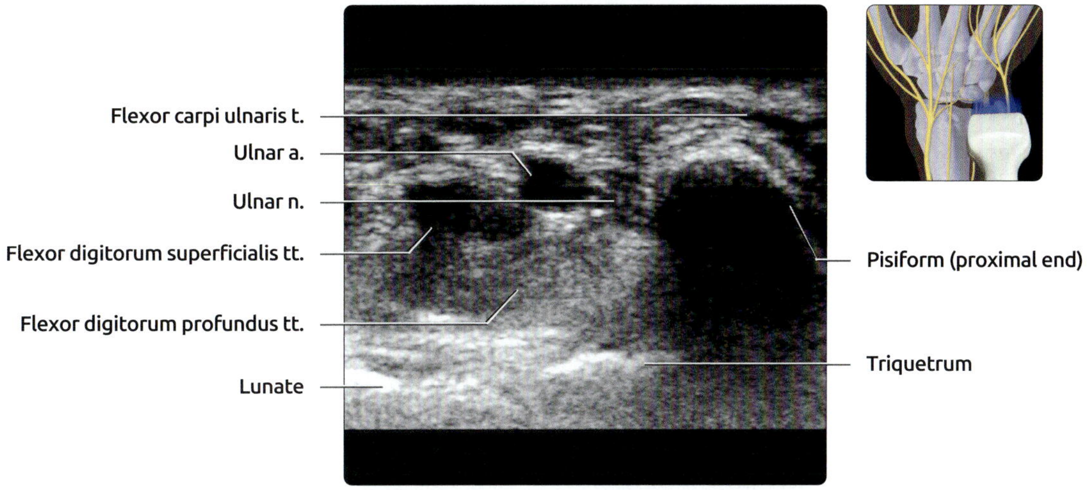

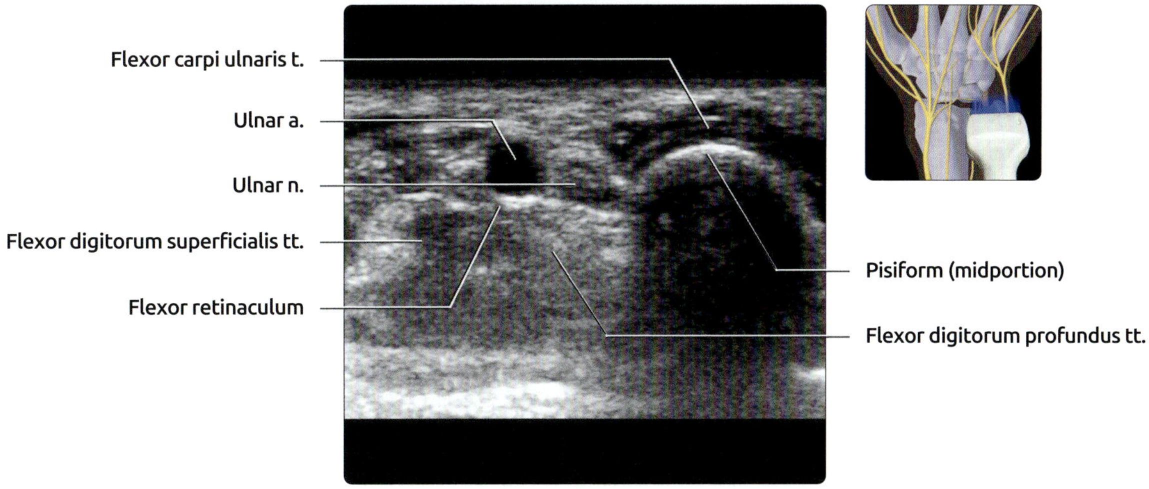

(Top) *Transverse grayscale ultrasound shows the ulnar nerve just proximal to the Guyon canal.* **(Middle)** *Transverse grayscale ultrasound shows the ulnar nerve at the proximal end of the Guyon canal. Guyon canal is the 2nd most common site of ulnar neuropathy.* **(Bottom)** *Transverse grayscale ultrasound shows the ulnar nerve at the proximal end of the Guyon canal. Ganglia arising from the pisiform-triquetral articulation are the most common cause of ulnar nerve compression in the Guyon canal.*

TRANSVERSE US, DISTAL TO GUYON CANAL

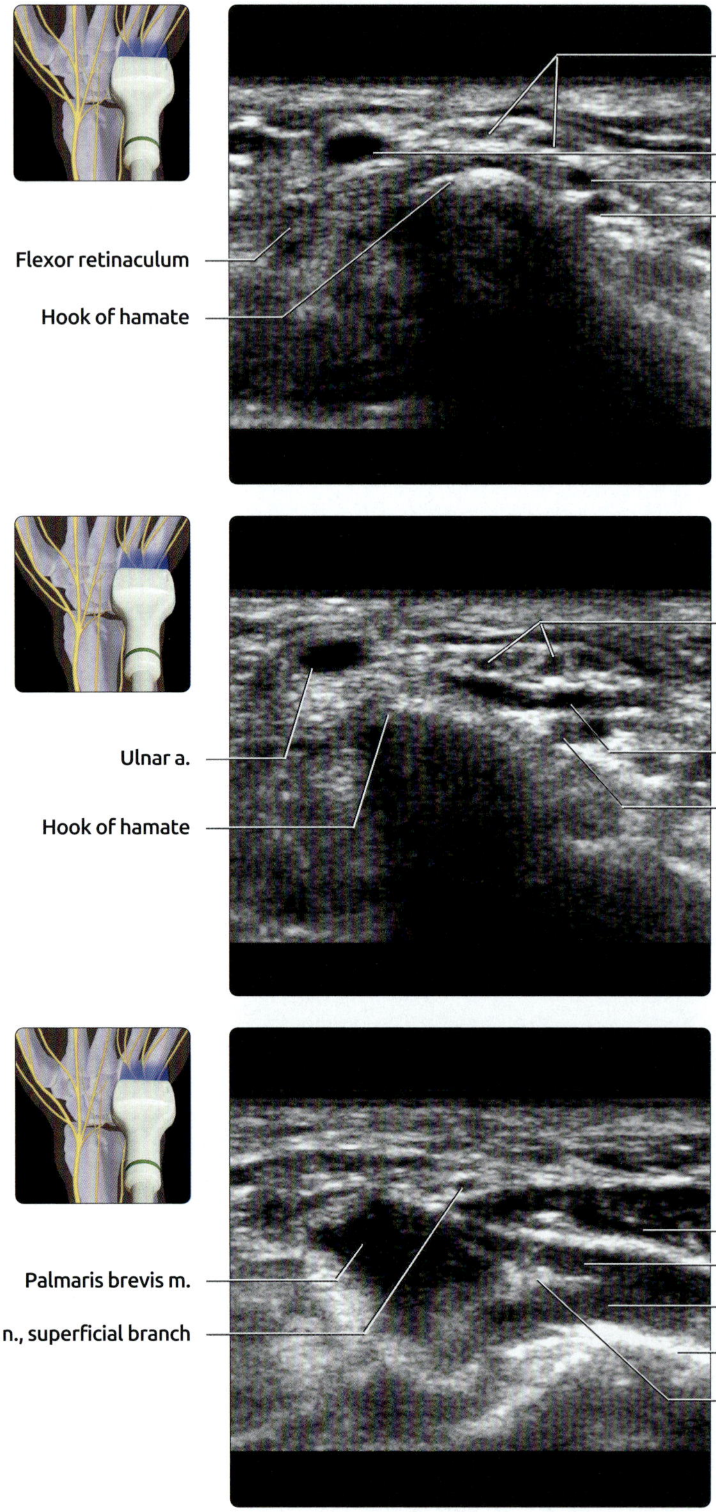

(Top) *Transverse grayscale ultrasound shows the ulnar nerve at the distal end of Guyon canal, where the ulnar nerve divides into superficial and deep branches. These pass close to the top of (superficial branch) and side of (deep branch) the hook of hamate.* **(Middle)** *Transverse grayscale ultrasound shows the ulnar nerve at the distal end of the hook of hamate level. Hook of hamate fractures (common in golfers and baseball players) and repeated blunt trauma ("handle bar" palsy) may injure the ulnar nerve as it passes by the hook of hamate. The ulnar artery can be injured at this location as well (hypothenar hammer syndrome).* **(Bottom)** *Transverse grayscale ultrasound shows the ulnar nerve at the base of the hypothenar muscles. The superficial branch of the ulnar nerve stays superficial to supply the skin, while the deep branch stays deep to supply the hypothenar muscles and the 3rd and 4th lumbrical muscles.*

PROXIMAL AND DISTAL ULNAR NERVE

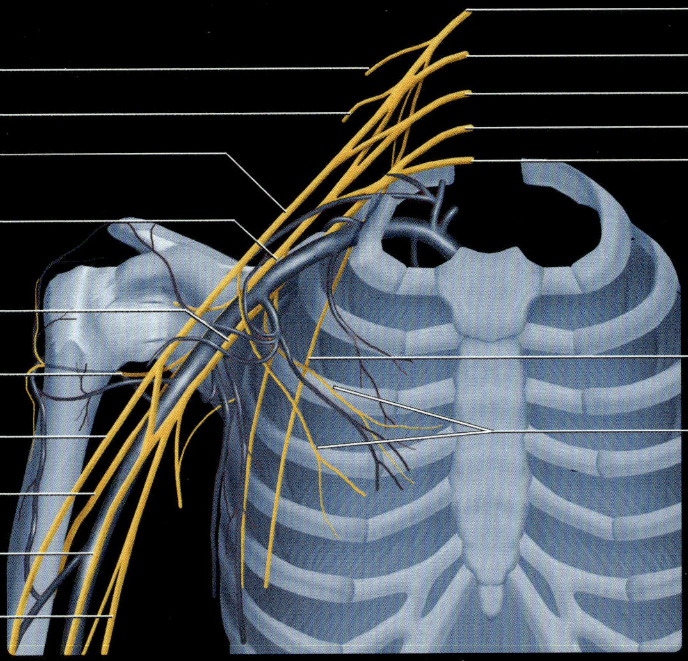

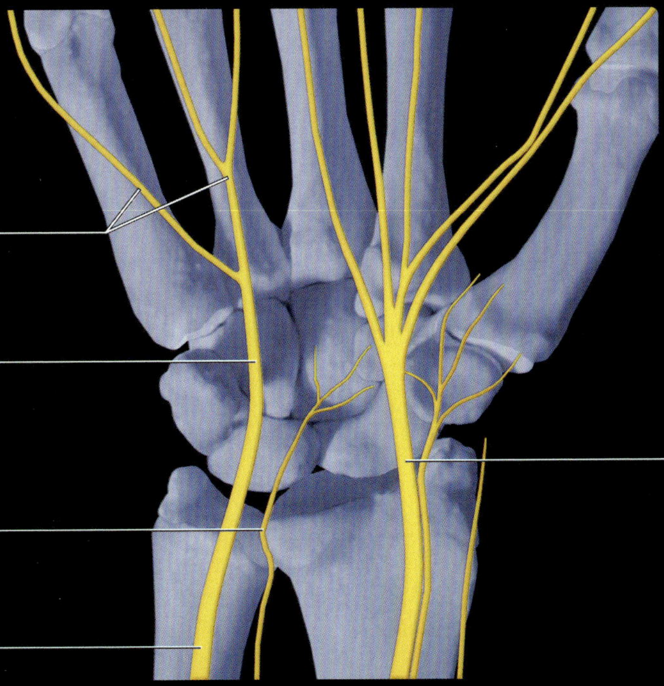

(Top) *Anterior graphic shows the ulnar nerve arising from the medial cord of the brachial plexus.* **(Bottom)** *Ulnar nerve supplies the ulnar wrist and hand. In addition to the branches shown, a dorsal branch of the ulnar nerve passes between the flexor carpi ulnaris and ulna about 5 cm proximal to the wrist to supply the dorsal ulnar wrist and the little finger. Deep and superficial branches supply motor and sensory branches to the hypothenar eminence.*

SECTION 7

Lower Extremity

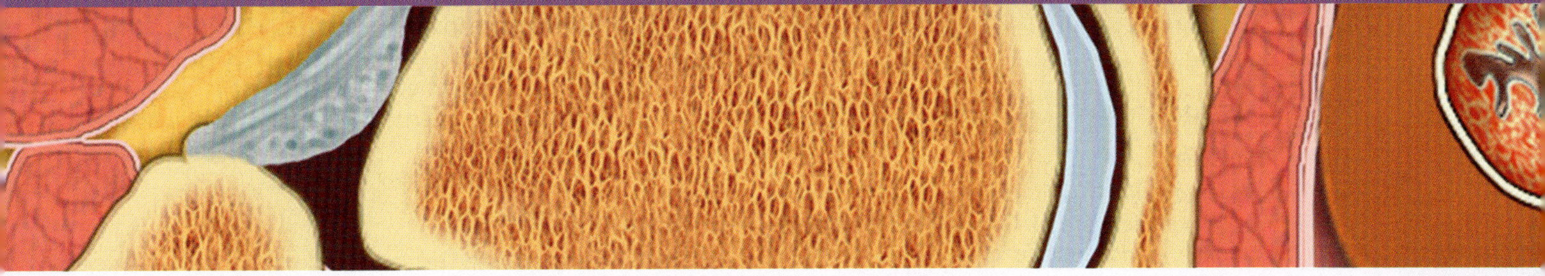

TERMINOLOGY

Abbreviations

- Origin (O)
- Insertion (I)
- Nerve supply (N)
- Function (F)

IMAGING ANATOMY

Gluteal Muscles

- Components: Gluteus maximus, gluteus medius, gluteus minimus, tensor fascia lata, piriformis, obturator internus and externus, superior and inferior gemellus, quadratus femoris
- Common Fs: Hip abduction, external rotation
- **Inferior gemellus**
  - O: Ischial tuberosity
  - I: Piriformis fossa
  - N: Nerve to quadratus femoris muscle
  - F: Hip external rotation, weak abduction
- **Superior gemellus**
  - O: Ischial spine
  - I: Piriformis fossa
  - N: Nerve to obturator internus muscle
  - F: Hip external rotation, weak abduction
- **Gluteus maximus**
  - O: Posterior gluteal line (ilium), posterior sacrum and coccyx, sacrotuberous ligament
  - I: Iliotibial tract, gluteal tuberosity
  - N: Inferior gluteal
  - F: Hip extension, abduction, and external rotation
- **Gluteus medius**
  - O: Between anterior and posterior gluteal lines (ilium)
  - I: Lateral and posterosuperior facets of greater trochanter
  - N: Superior gluteal
  - F: Hip abduction and internal rotation
- **Gluteus minimus**
  - O: External ilium between anterior and inferior gluteal lines
  - I: Anterior facet of greater trochanter
  - N: Superior gluteal
  - F: Hip abduction and internal rotation
- **Obturator internus**
  - O: Internal surface of obturator foramen and membrane
  - I: Piriformis fossa (joins with gemelli tendons)
  - N: L5, S1, S2
  - F: Hip external rotation, weak abduction
- **Obturator externus**
  - O: Edge of obturator foramen and obturator membrane
  - I: Posterior aspect of femur in trochanteric fossa
  - N: Obturator nerve
  - F: Lateral rotation of hip
- **Piriformis**
  - O: Anterior sacrum, sacrotuberous ligament
  - I: Greater trochanter (may fuse with obturator internus and gemellus muscles)
  - N: S1, S2
  - F: Hip external rotation, assists abduction
- **Tensor fascia lata**
  - O: External lip anterior iliac crest, external anterior superior iliac spine
  - I: Iliotibial tract
  - N: Superior gluteal
  - F: Hip flexion, abduction, and weak internal rotation
- **Quadratus femoris**
  - O: Lateral ischial tuberosity
  - I: Quadrate line, intertrochanteric crest of femur
  - N: L4, L5, S1
  - F: Strong hip external rotation

Rotator Cuff of Hip

- Gluteus medius and minimus tendons, trochanteric bursa, subgluteus medius and minimus bursae

Bursa

- Trochanteric
  - Greater trochanter and gluteus maximus muscle
- Ischiogluteal
  - Ischial tuberosity and gluteus maximus muscle
- Subgluteus medius
  - Greater trochanter and gluteus medius
- Subgluteus minimus
  - Greater trochanter and gluteus minimus muscle
- Gluteofemoral
  - Iliotibial tract and vastus lateralis muscle
- Obturator internus
  - Muscle and ischium
- Obturator externus
  - Synovial protrusion beneath inferior border

Piriformis Fossa

- Between posterior femoral neck and posterior medial surface of greater trochanter
- Insertion site for piriformis, superior and inferior gemelli, obturator internus

Ischial Tuberosity

- Covered by gluteus maximus muscle when hip extended, uncovered in hip flexion
- Origin of hamstrings and adductor magnus muscles

Sacroiliac Joint

- Anterior synovial joint
- Posterior syndesmosis: Tuberosities of sacrum and ilium
- F: Primarily weight transfer from axial to appendicular skeleton
  - Limited gliding and rotation
- Widening anteriorly can occur without instability
- Posterior widening indicates instability
- **Sacroiliac ligaments**
  - **Interosseous sacroiliac**: Within syndesmotic joint
    - Between tuberosities of ilium and sacrum
    - Oriented superiorly and laterally
    - Weight transferred to sacrum; sacrum displaced inferiorly; force transmitted to ligaments; ligaments pull ilii inward and compress sacrum and interlocking surfaces of sacroiliac joints
    - Extremely strong
  - **Anterior sacroiliac**: Weak ligaments
    - Mainly anterior joint capsule

- **Posterior sacroiliac**: Posterior fibers interosseous ligament
  - **Short dorsal sacroiliac**: Horizontal; lateral sacral crest S1 and S2 to posterior iliac crest
  - **Long dorsal sacroiliac**: Vertical oblique; posterior superior iliac spine to lateral sacral crest S3, S4

Posterior Sacroiliac Ligaments of Pelvis

- Primary stabilizers of pelvis
  - Little contribution from anterior structures
- Resist rotational forces and vertical shear
- **Sacrotuberous**: Superior and inferior posterior iliac spines, sacrum, and coccyx to ischial tuberosity
- **Sacrospinous**: Lateral sacrum and coccyx to ischial spine
- **Iliolumbar**: Tip of L5 transverse process to iliac crest
- **Sacroiliac ligaments**: Interosseous, long and short dorsal

Greater Sciatic Foramen

- Greater sciatic notch converted to foramen by sacrospinous ligament
- Contents: Nerves, vessels, piriformis muscle
- Passageway from pelvis to thigh
  - Structures identified as exiting pelvis above or below piriformis muscle
    - Above: Superior gluteal artery, vein, and nerve
    - Below: Pudendal nerve; internal pudendal artery and vein; nerve to obturator internus; sciatic nerve; inferior gluteal artery, vein, and nerve; posterior cutaneous nerve of thigh; nerve to quadratus femoris muscle
  - Relationship of structures
    - Most medial: Pudendal nerve
    - Most lateral: Sciatic nerve
    - Nerve to quadratus femoris muscle deep to sciatic nerve
- **Inferior gluteal artery**
  - Larger terminal branch of anterior division internal iliac artery
  - Passes between S1 and S2 or S2 and S3
  - Exits pelvis inferior to piriformis muscle
  - Posteromedial to sciatic nerve
  - Prenatally continuous with popliteal artery; remnant is artery to sciatic nerve
  - Supplies gluteus maximus, obturator internus, quadratus femoris, superior hamstrings muscles
  - **Inferior gluteal vein** travels with artery, drains into internal iliac vein
- **Inferior gluteal nerve**
  - Branch of sacral plexus
  - Exits pelvis inferior to piriformis muscle
  - Posterior to sciatic nerve
- **Superior gluteal artery**
  - Continuation of posterior division of internal iliac artery
  - Passes between lumbosacral trunk and S1
  - Exits pelvis superior to piriformis muscle
  - Superficial branch supplies gluteus maximus muscle
  - Deep branch supplies gluteus medius and minimus, tensor fascia lata muscles
  - **Superior gluteal vein** travels with artery, drains into internal iliac vein
- **Superior gluteal nerve**
  - Branch of sacral plexus
  - Exits pelvis superior to piriformis muscle
- **Sciatic nerve**
  - Exits pelvis inferior to piriformis muscle
  - Most lateral structure
- **Pudendal nerve and internal pudendal vessels**
  - Exit inferior to piriformis muscle
  - Nerve most medial structure
- **Nerve to obturator internus muscle**
  - Exits pelvis inferior to piriformis muscle

Lesser Sciatic Foramen

- Lesser sciatic notch converted to foramen by intersection (crossing) with sacrotuberous and sacrospinous ligaments
- Contents
  - **Pudendal nerve and internal pudendal vessels**
    - Pass over sacrospinous ligament
    - Exit buttock, enter perineum
  - **Obturator internus muscle**

ANATOMY IMAGING ISSUES

Imaging Recommendations

- US
  - Subcutaneous fat in gluteal region can be thick and mistaken for muscle
  - Look for investing fascia and fiber orientation to identify muscle layer
  - Anisotropy gives gluteal muscles varying echogenicity due to different orientation of fibers
    - Allows clear distinction between muscles
    - Lesions may be better demonstrated/detected by scans along and across muscle fibers to make lesion more conspicuous
  - Iliac crest, greater trochanter, and ischial tuberosity are important bony landmarks for orientation
  - Use of panoramic (extended field of view) images will help referring clinicians to understand location of lesion
    - Panoramic images should be performed in standard anatomical planes to help interpretation
    - Landmarks, especially osseous (e.g., iliac crest, sacrum, ischial tuberosity), also help interpretation of panoramic image
    - Panoramic images do not provide accurate measurements
    - Wide lesions are best assessed using dual screen/multiple contiguous images for summated linear measurement
- MR
  - Diagnose masses, anatomic anomalies, other causes of nerve compression
  - Preferred modality for sacroiliitis
- CT
  - Mainly used for assessing bone and joint (sacroiliac) abnormalities

Imaging Pitfalls

- Lower frequency transducers (5 MHz or below) should be used initially in examination to exclude deep-seated lesions

GLUTEAL MUSCLES

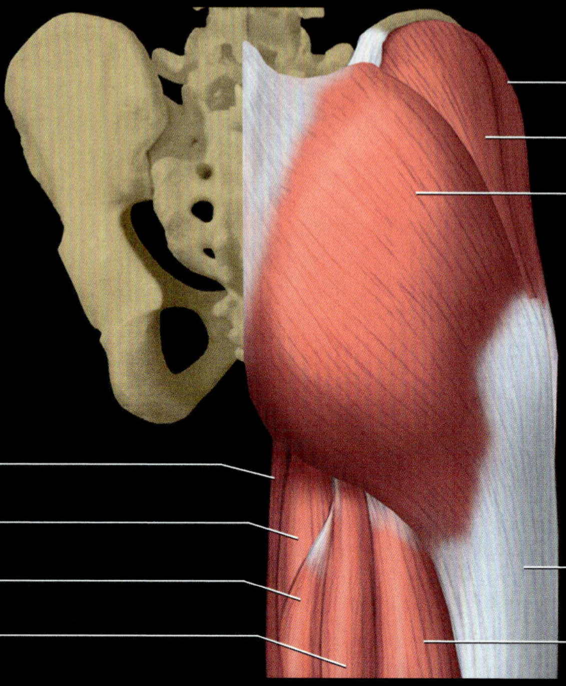

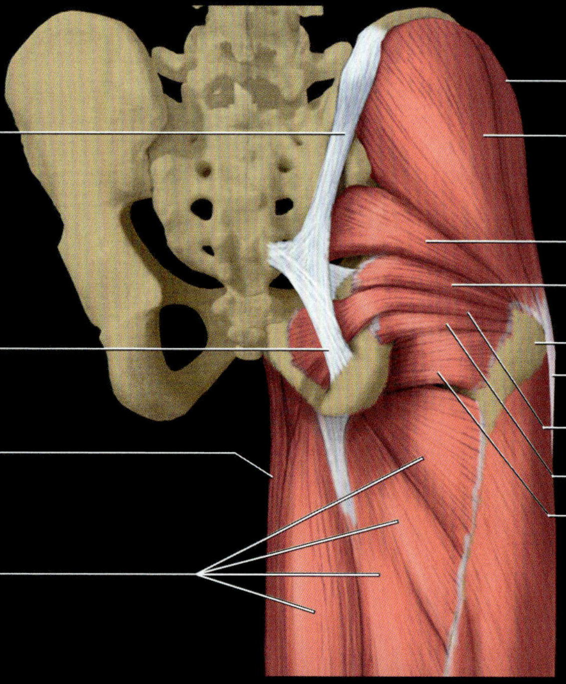

(Top) *The gluteus muscles from anterior to posterior and deep to superficial are the gluteus minimus, medius, and maximus. These muscles form the bulk of the gluteal region. The gluteus maximus, in particular, covers most of the deep gluteal structures.* **(Bottom)** *After removing the gluteus maximus, the deep muscles are visible: Piriformis, superior gemellus, obturator internus, inferior gemellus, and quadratus femoris. These all function as external rotators of the thigh. Note the piriformis muscle, which exits via the greater sciatic foramen, and the obturator internus muscle via the lesser sciatic foramen.*

SCIATIC NERVE AND GLUTEAL REGION OVERVIEW

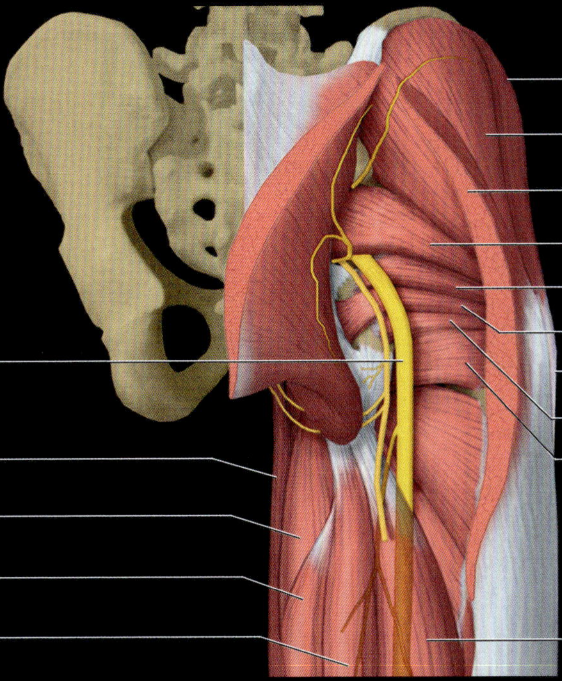

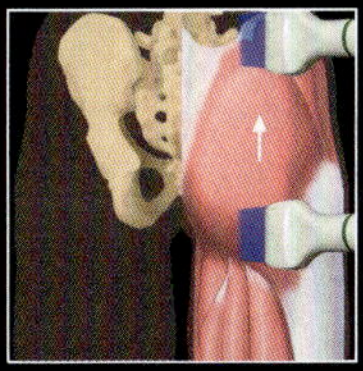

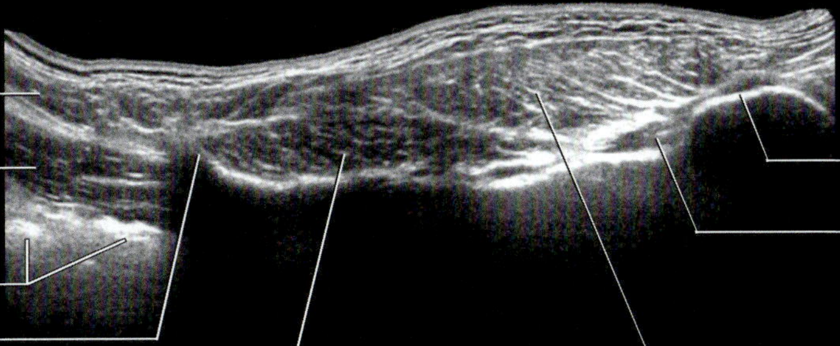

(Top) *The deep external rotators of the hip share an intimate relationship with the sciatic nerve, which emerges between the piriformis and superior gemellus muscles and then between the gluteus maximus and the rest of the deep external rotators. More distally, the sciatic nerve is covered by the biceps femoris muscle.* **(Bottom)** *Sagittal panoramic view shows the medial aspect of the gluteal region. The iliac crest separates the back muscles from the glutei. At this level, the gluteus maximus muscle forms the main muscle bulk. Inferiorly, the gluteus maximus runs across the surface of the ischial tuberosity in this extended (hip joint) position. Coverage is lost (gluteus maximus moves superiorly) when the hip joint is flexed and the ischial tuberosity is exposed.*

GLUTEAL REGION OVERVIEW

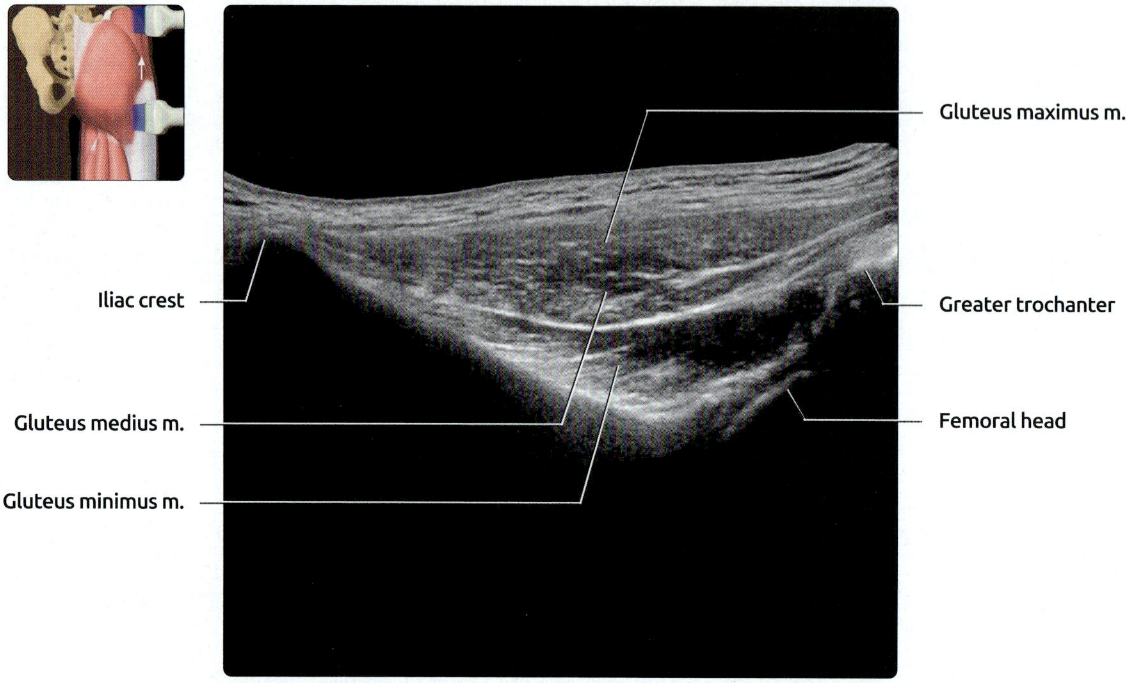

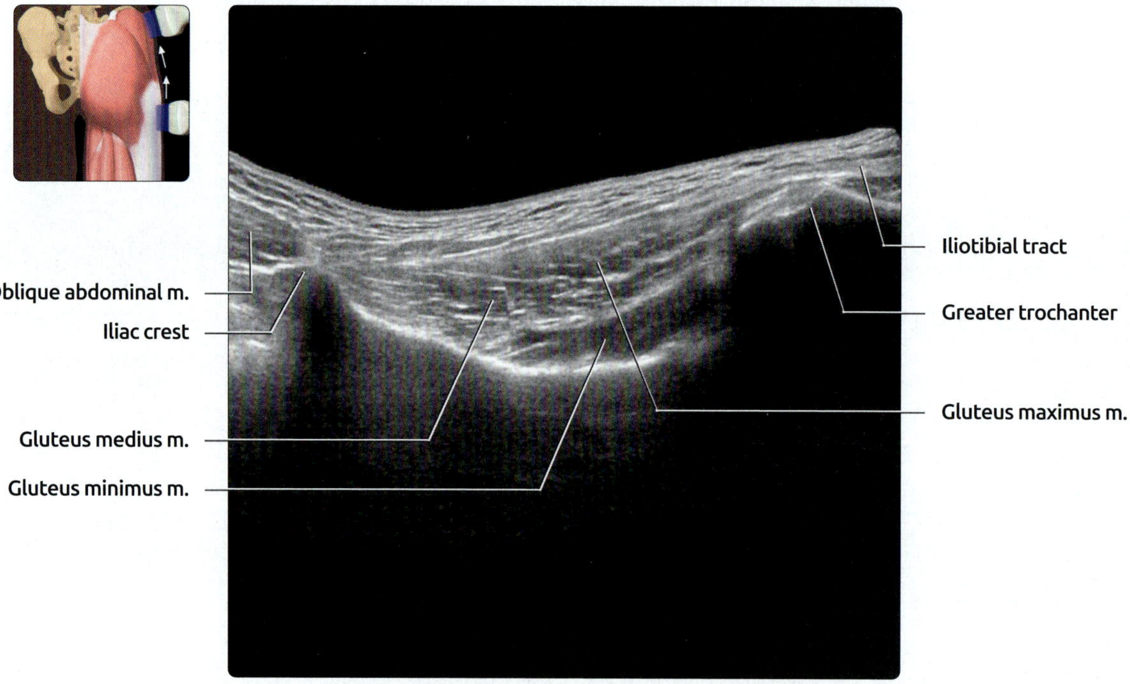

(Top) *Oblique sagittal panoramic view shows the lateral 1/2 of the gluteal region. The close relationship between the gluteus minimus muscle and the posterior aspect of the hip joint can be appreciated here. In this region, all 3 gluteal muscles are present and can be distinguished by their different echogenicity. The gluteus maximus muscle fibers are running ~ horizontally along the scan plane and, being most superficial, produce the highest echogenicity. The gluteus medius is running obliquely (lateral inferiorly) and thus shows short echogenic fibers. The gluteus minimus runs obliquely and almost horizontally toward the anterior facet of the greater trochanter. It is the deepest gluteal muscle and usually appears hypoechoic.* **(Bottom)** *Coronal panoramic view shows the gluteus medius and minimus muscles and their insertion on the greater trochanter. The iliotibial tract runs on the surface of the greater trochanter.*

SAGITTAL GLUTEAL REGION

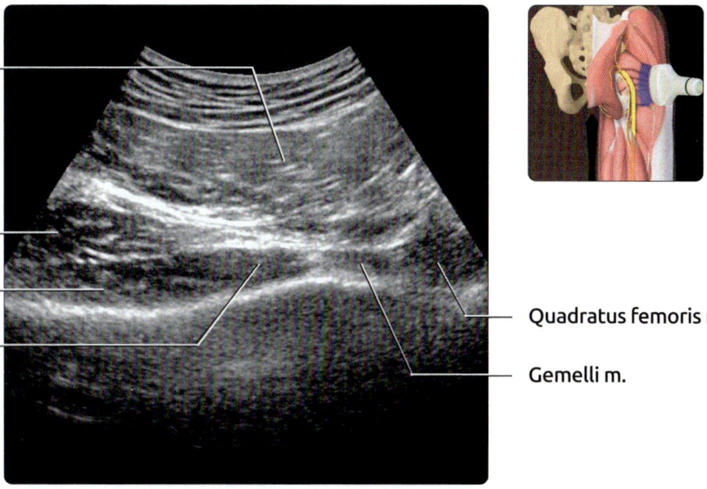

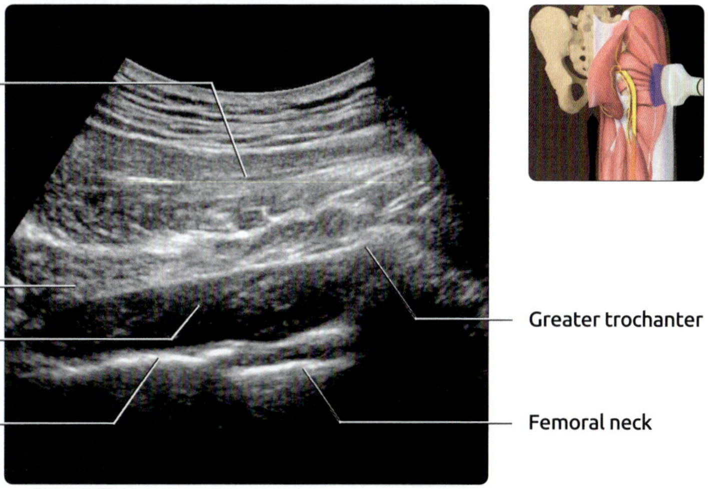

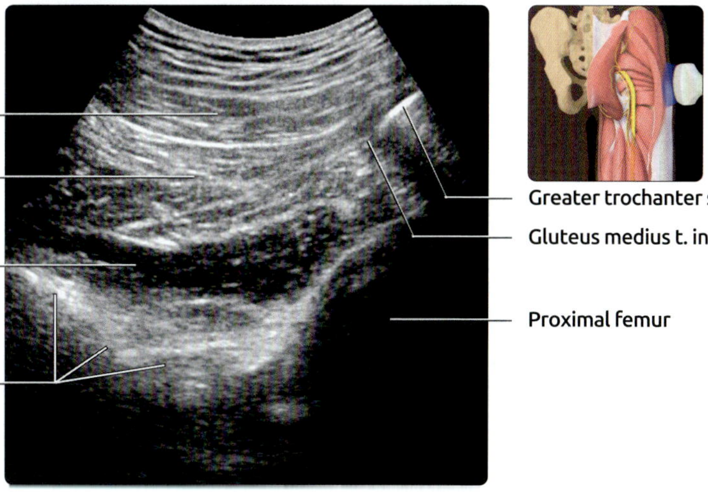

(Top) *Oblique sagittal scan in the lateral 1/2 of the gluteal region shows the 3 gluteal muscles and their relationship with the deeper, small external rotators: Piriformis, superior and inferior gemellus, and quadratus femoris muscles.* **(Middle)** *Further laterally, the femoral neck can be seen under the gluteus minimus. Again, the different echogenicity and echotexture of the 3 gluteal muscles can be appreciated.* **(Bottom)** *Further laterally, the gluteus medius tendon fibers insert onto the lateral and posterosuperior facets of the greater trochanter. The gluteus minimus tendon runs anteriorly under the gluteus medius to insert on the anterior facet of greater trochanter.*

TRANSVERSE GLUTEAL REGION

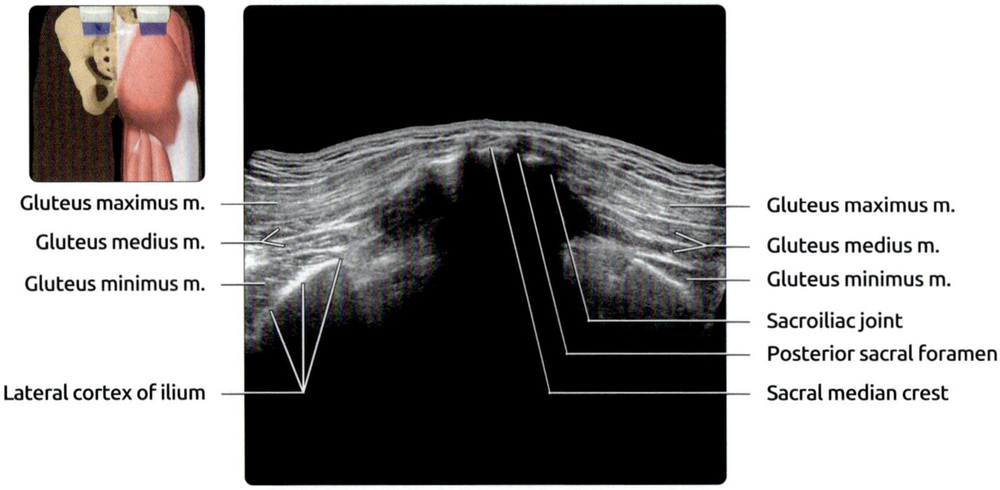

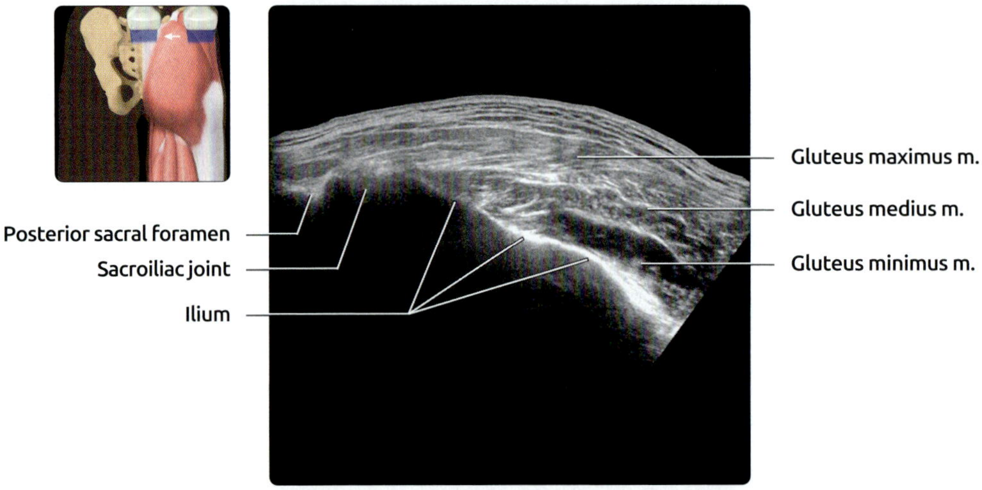

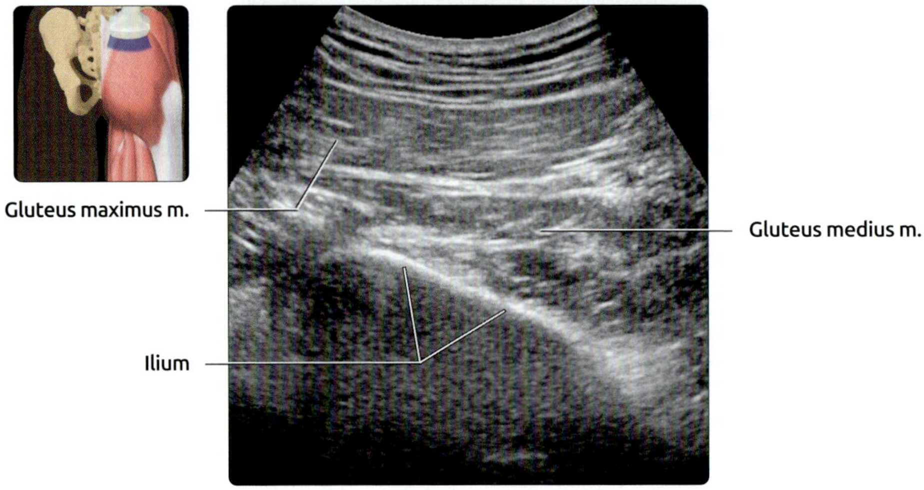

(Top) *Transverse panoramic scan shows the dorsal surface of the sacrum. This is the medial origin of the gluteus maximus muscle. Deep to this muscle, the gluteus medius and gluteus minimus muscles can be seen originating from the iliac surface.* **(Middle)** *Transverse panoramic view shows the right gluteal region. The 3 gluteal muscles show their signature difference in echogenicity and echotexture even in the horizontal plane.* **(Bottom)** *Transverse focused view of the medial gluteal region shows the fiber orientation of the different muscles. Those of the gluteus maximus are longer as they run roughly horizontally compared to those of the gluteus medius. This orderly fiber pattern is helpful in detecting intra- and intermuscular lesions in the gluteal region (orderly fiber alignment tends to be disrupted in disease).*

ISCHIAL REGION

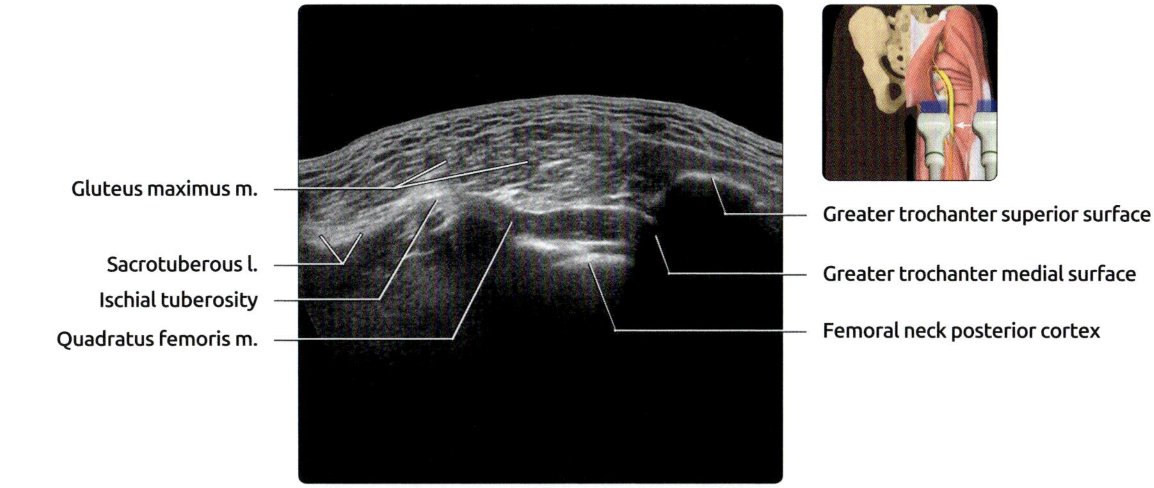

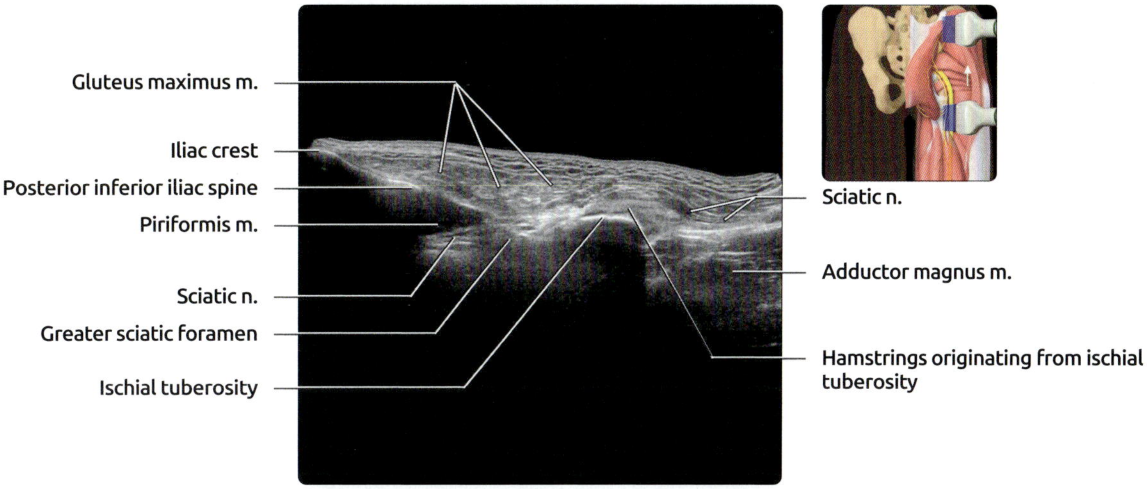

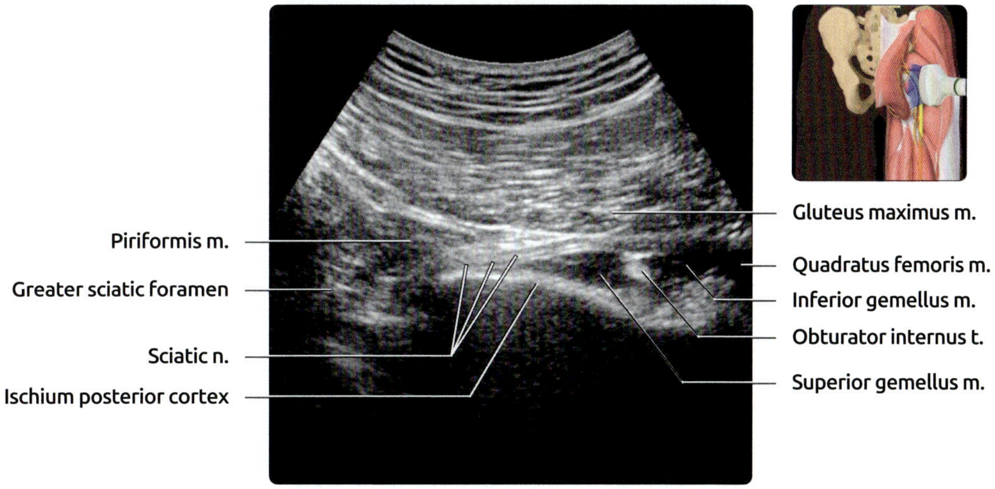

(Top) *Transverse panoramic view shows the ischial region. The gluteus maximus forms most of the bulk at this lower gluteal level. Deep to this, small external rotators of the hip can be seen (in this scan, the quadratus femoris muscle).* **(Middle)** *Sagittal panoramic scan is seen from the lower iliac bone to below the ischial tuberosity. The greater sciatic foramen is demonstrated in this scan. The piriformis and sciatic nerve exit the pelvis via this foramen and run inferolaterally, the former to insert on the greater tuberosity and the latter inferiorly to supply the leg.* **(Bottom)** *Sagittal view shows the greater sciatic foramen using the ischial tuberosity as a landmark. The sciatic nerve and its relationship with the small external rotators of the hip is shown here. The sciatic nerve emerges between the piriformis and superior gemellus muscles after exiting the pelvis through the greater sciatic foramen.*

TRANSVERSE ISCHIAL REGION

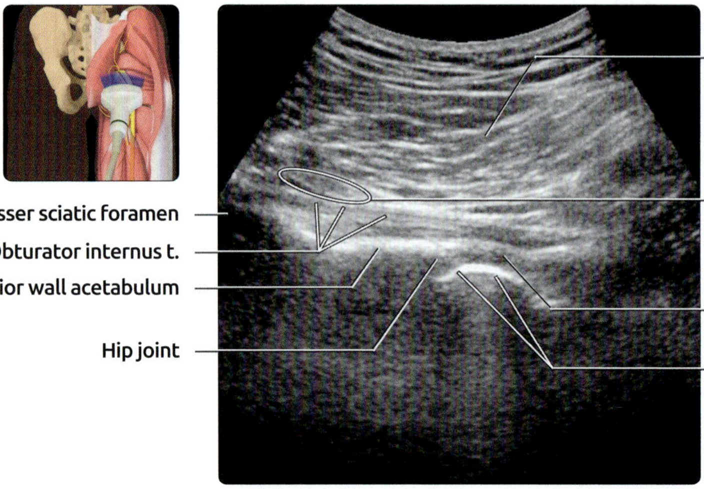

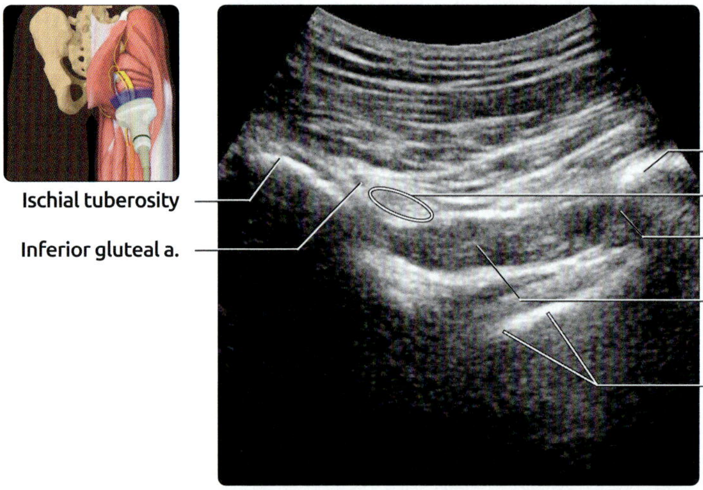

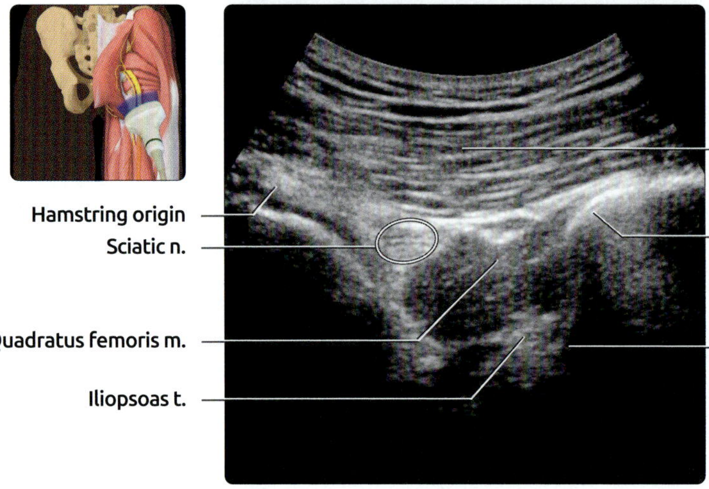

(Top) *Series of 3 transverse scans show the inferomedial aspect of the gluteal region. This image shows the obturator internus tendon exiting the lesser sciatic foramen, running laterally toward the lower aspect of the hip joint. The sciatic nerve runs superficial to this tendon.* **(Middle)** *Slightly inferiorly, this image shows the inferior gemellus muscle arising from the ischial tuberosity, running laterally over the neck of the femur and inserting at the piriformis fossa on the medial surface of the greater trochanter. The sciatic nerve runs superficial to this muscle.* **(Bottom)** *Further inferiorly, this image shows the quadratus femoris muscle arising from the lateral border of the ischial tuberosity and running laterally to insert on the quadrate tubercle of the femur. The sciatic nerve also runs superficial to this tendon.*

ISCHIAL TUBEROSITY

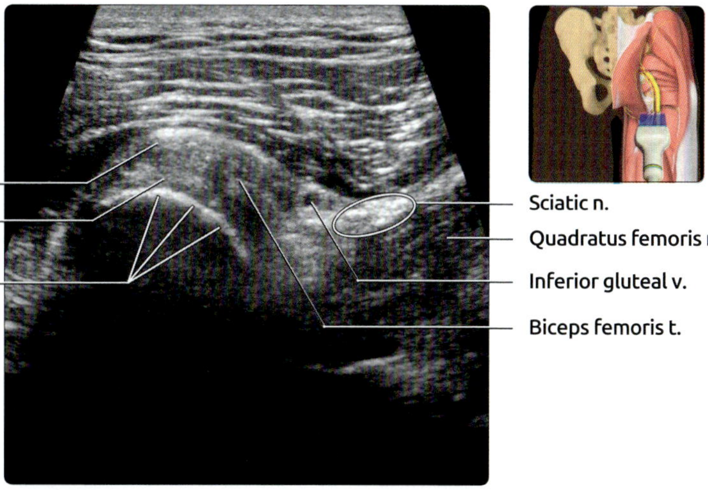

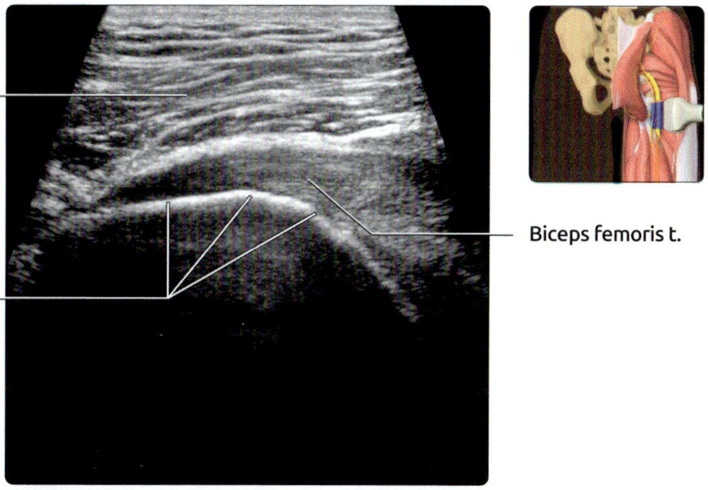

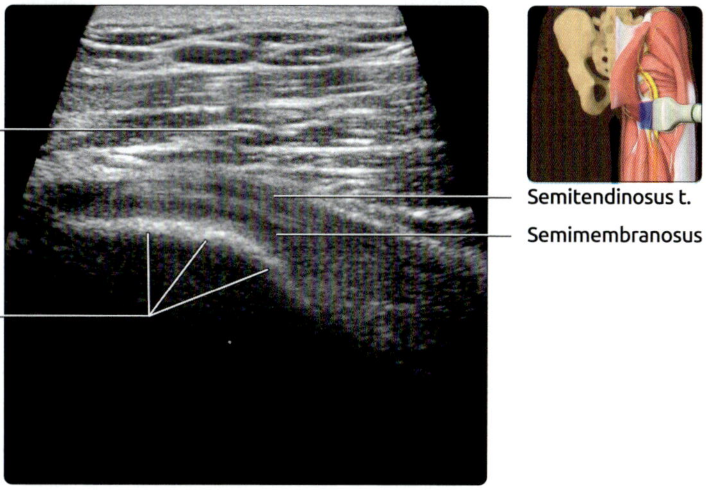

(Top) *Transverse scan shows the ischial tuberosity. The biceps femoris tendon originates lateral to the semimembranosus and semitendinosus tendons. The sciatic nerve and inferior gluteal vessels run just lateral to the ischial tuberosity. Inflammation in this region (e.g., ischiogluteal bursitis) may cause irritation of the sciatic nerve. More importantly, aspiration or injection to treat ischiogluteal bursitis should be performed with image guidance to avoid damaging the sciatic nerve.* **(Middle)** *Sagittal scan of the lateral aspect of the ischial tuberosity shows the origin of the biceps femoris tendon.* **(Bottom)** *Sagittal scan more medially on the ischial tuberosity shows the insertion of the semitendinosus and semimembranosus tendons. The semitendinosus muscle originates and runs for most of its course on the dorsal surface of the semimembranosus tendon.*

TERMINOLOGY

Abbreviations

- Origin (O), insertion (I)

Definitions

- Junctional region between thigh and trunk

IMAGING ANATOMY

Overview

- Incorporates lower abdominal wall, inguinal canal, femoral triangle, and femoral adductor muscles

Osseous Anatomy

- Pubic bone
  - **Pubic tubercle**: Small protuberance, lateral border pubic crest
    - Attachment: Inguinal ligament
  - **Superior and inferior rami**: Extend from pubis
  - **Pecten**: Ridge, posterior aspect superior pubis
    - O of pectineus muscle
    - I of conjoint tendon (internal oblique and transverse abdominis)
  - **Pubic crest**: Superior surface, anterior aspect pubic body
    - O of rectus abdominis muscle
    - I of transversus abdominis and external oblique muscles
  - **Pubic symphysis**: Cartilaginous joint between pubic bodies
  - **Superior pubic ligament**: Laterally extends to pubic tubercles

Anterior Abdominal Wall

- Abdominal wall muscles
  - **Rectus abdominis**: Paired midline muscles
    - O: Superior pubic ramus, pubic crest
    - I: Xiphoid process, costal cartilages 5-7
  - **Linea alba**: Aponeurotic junction of rectus femoris, transverse abdominis, and internal and external oblique muscles
  - **External oblique**: Most superficial
    - O: Ribs 5-12
    - I: Pubic crest, anterior iliac crest, linea alba
    - Lower border of aponeurosis contributes to inguinal ligament
  - **Internal oblique**: Between external oblique, transversus abdominis
    - O: Lateral inguinal ligament, iliac crest, thoracolumbar fascia
    - I: Pecten (conjoined tendon), pubic crest, inferior aspect of ribs 10-12, linea alba
    - O anterior to deep inguinal ring
    - I lateral to rectus abdominis muscle, posterior and medial to superficial inguinal ring
    - Arches over inguinal canal, forming roof
  - **Transversus abdominis**: Deepest
    - O: Iliac crest, posterior aspect of lateral inguinal ligament, thoracolumbar fascia
    - I: Pubic crest, pecten (conjoined tendon), linea alba
    - Remains posterior to inguinal canal

Inguinal Ligament

- Thickening inferior border of external oblique aponeurosis
- Attachments: Anterior superior iliac spine and pubic tubercle
- Separates lower extremity from pelvis
- Fascia lata attaches to inferior border
- **Subinguinal space**: Deep to inguinal ligament
  - Passageway for femoral vessels and nerve, iliopsoas muscle into femoral triangle
  - External iliac vessels become femoral vessels upon entering this space

Inguinal Canal

- Entrance: **Deep inguinal ring**
  - Located midinguinal ligament
  - Opening of evaginated transversalis fascia through which spermatic cord/round ligament pass
- Exit: **Superficial inguinal ring**
  - Division of external oblique aponeurosis lateral to pubic tubercle
- Contents: Ilioinguinal nerve; male: Spermatic cord; female: Round ligament; associated vessels
  - Covered by evaginated transversalis fascia

Adductor Musculature

- **Adductor longus muscle**: Thin tendon arises from medial superior pubic ramus
  - Overlies O of gracilis, adductor brevis, and magnus muscles
- **Gracilis muscle**: O from anterior aspect of symphysis pubis and medial aspect of inferior pubic ramus
  - O medial to adductor brevis muscle, deep to adductor longus muscle

Groin Nodes

- 2 groups of groin lymph nodes
- Superficial inguinal nodes
  - Lie in subcutaneous tissues below inguinal ligament and alongside great saphenous vein
  - Subdivided into 3 groups of lymph nodes
    - Inferior group are located distal to saphenous opening and receive drainage from lower limb
    - Superolateral group are located lateral to saphenous opening and receive drainage from lateral buttock and lower anterior abdominal wall regions
    - Superomedial group are located medial to saphenous opening and receive drainage from perineum and external genitalia
- Deep inguinal nodes
  - 3-5 nodes located in femoral triangle just medial to femoral vein
  - Cloquet node is most proximal of deep inguinal lymph nodes located just below inguinal ligament

ANATOMY IMAGING ISSUES

Imaging Recommendations

- Usual imaging pathway for assessing groin pathology is radiography followed by ultrasound
- If this does not provide necessary information, MR is next best investigation for most groin pathology

GROIN

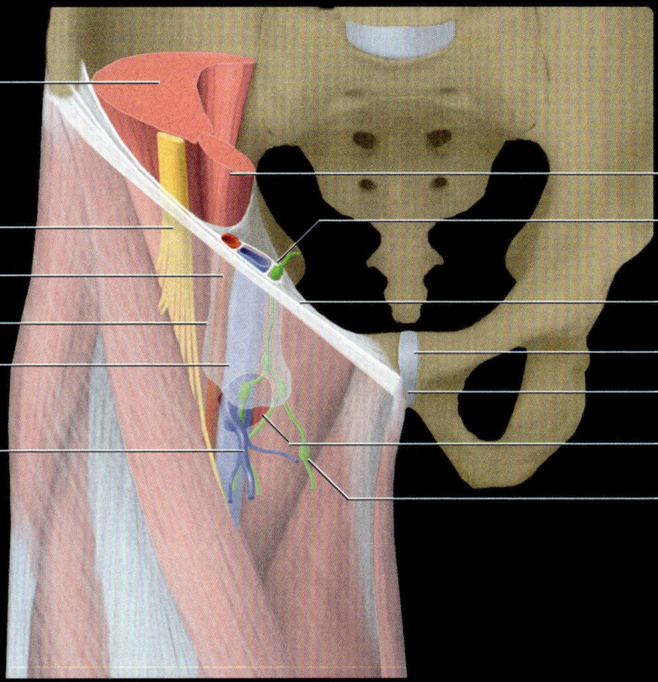

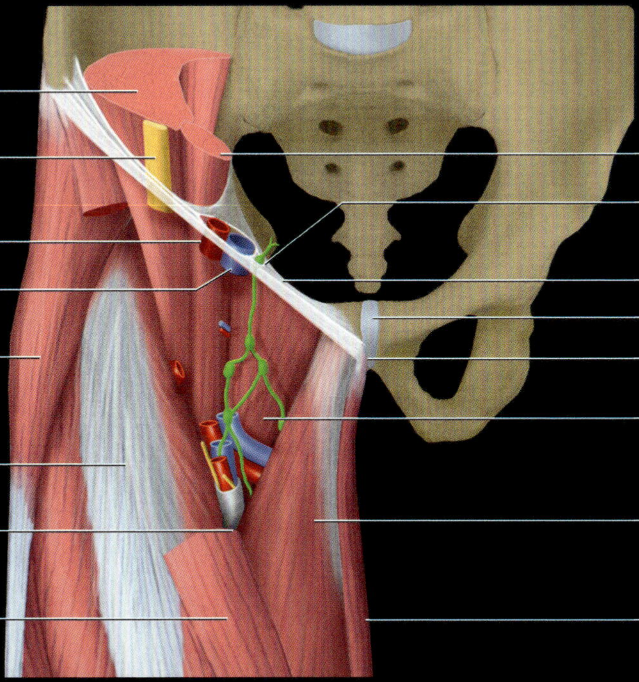

(Top) *The boundaries of the femoral triangle can be remembered by the mnemonic SAIL for sartorius, adductor longus, and inguinal ligament. The contents of the femoral triangle from lateral to medial (remembered by the mnemonic NAVeL) are the femoral nerve, femoral artery and vein, and lymphatics. The nerve lies superficial to the iliopsoas muscle. The fascia lata encases the structures of the thigh. The femoral sheath is the fascial covering over the proximal vessels. At the cut away proximal boundary, note the septa dividing the sheath into compartments. The femoral canal is the medial compartment.* **(Bottom)** *Femoral triangle is shown after removal of the fascia lata, sartorius muscle, and the vessels. The apex of the triangle is at the crossing of the sartorius and adductor longus muscles. The pectineus, adductor longus, and iliopsoas muscles form the floor of the triangle.*

GROIN MUSCLES

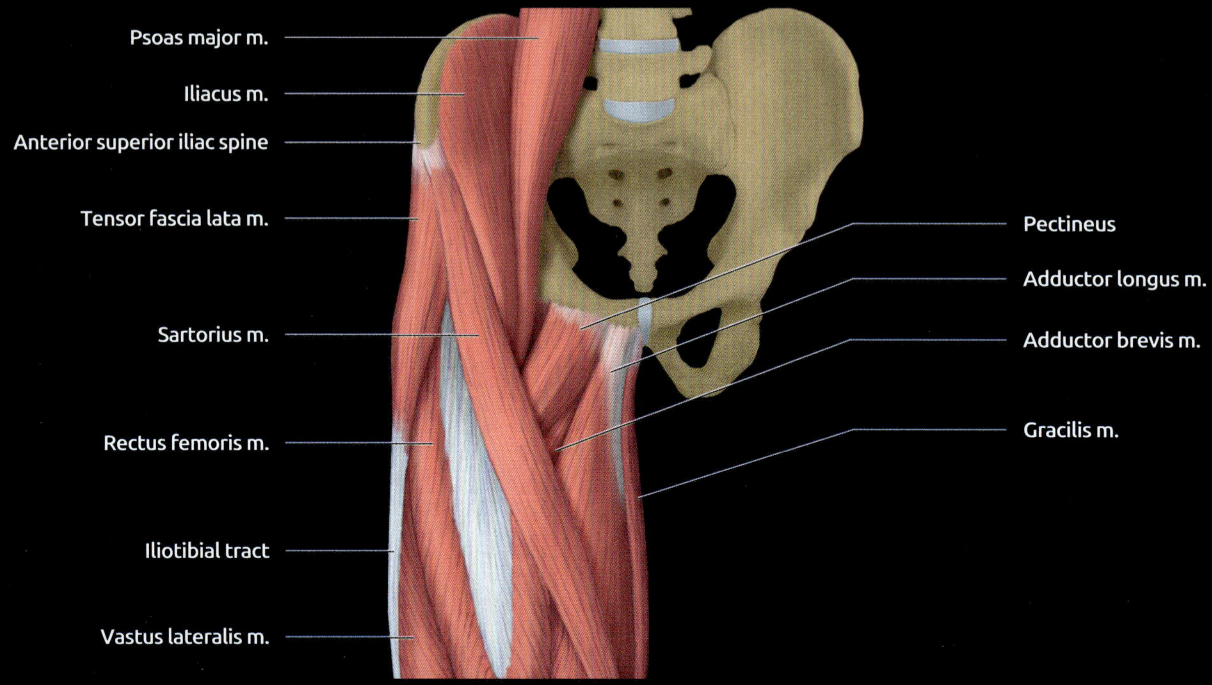

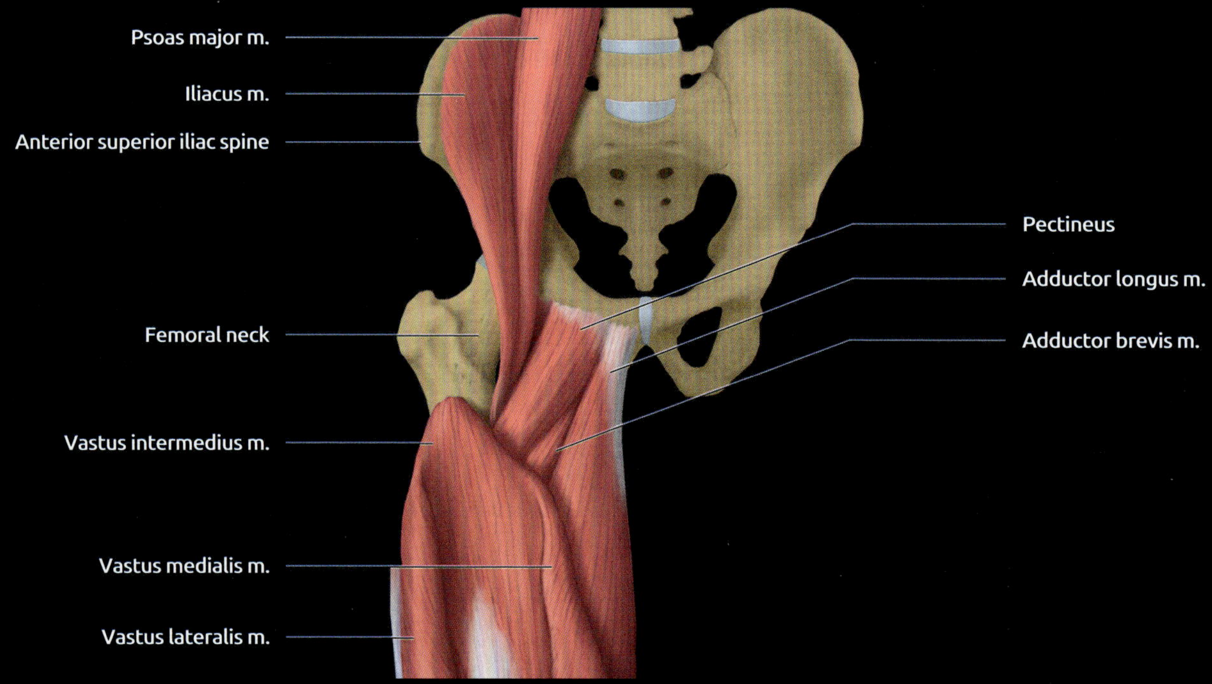

(Top) *Superficial muscles of the groin are shown. The most lateral muscle is the sartorius; its oblique course is easily appreciated. The psoas major and iliacus muscle exit from the pelvis toward the lesser trochanter. On the medial side, the most medial muscle is the gracilis. The adductor brevis muscle is deep to the adductor longus and pectineus muscles.* **(Bottom)** *After removing the superficial layer of muscles, the floor of the femoral triangle is better appreciated. The iliopsoas muscle dives deeply to its trochanteric insertion. The adductors fan out laterally from their pubic origin and constitute the major medial component at a midthigh level. The adductor magnus is in a deeper and more posterior position.*

GROIN OVERVIEW

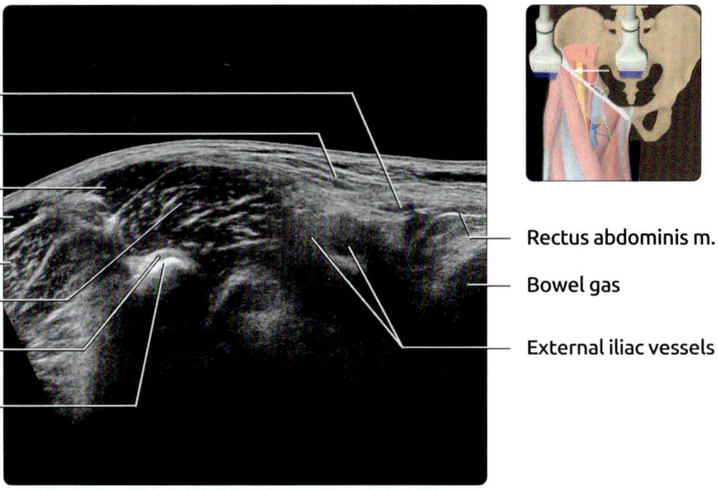

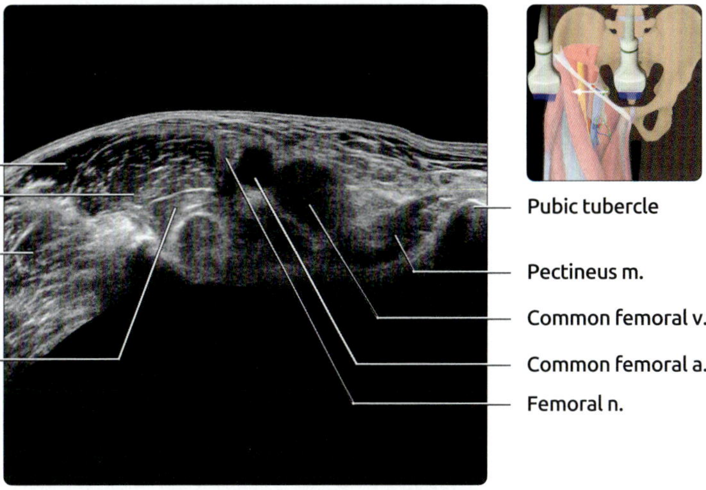

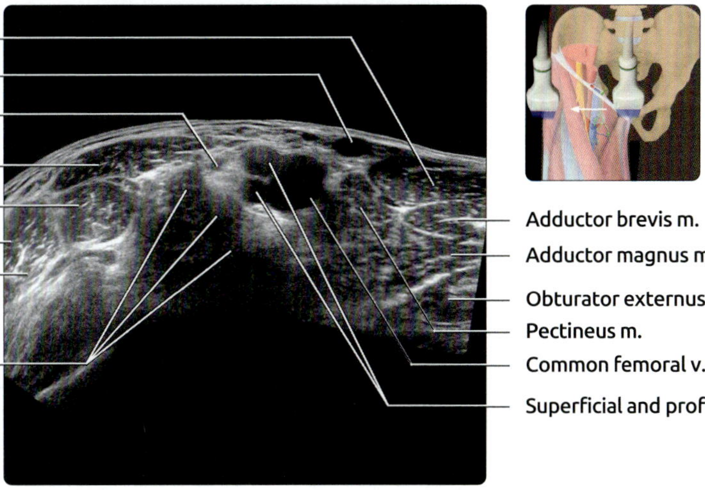

(Top) *Series of transverse panoramic scans show the groin. At this higher level, the lateral components of the groin include the sartorius and iliopsoas muscles and at the medial boundary, the rectus abdominis muscle. At this level, where most structures are above the inguinal ligament, there is bowel (containing gas, which produces shadowing artifact). At this level, vessels are named external iliac vessels.* **(Middle)** *At a lower level, through the pubic bone, the vessels have moved superficially after passing under the inguinal ligament and now become the common femoral vessels. Bowel is not normally present at this level; instead, the pectineus muscle and superior pubic ramus form the floor.* **(Bottom)** *Further inferiorly, the adductor muscles become the major medial component. Femoral vessels divide into superficial and profunda branches.*

MIDLINE GROIN

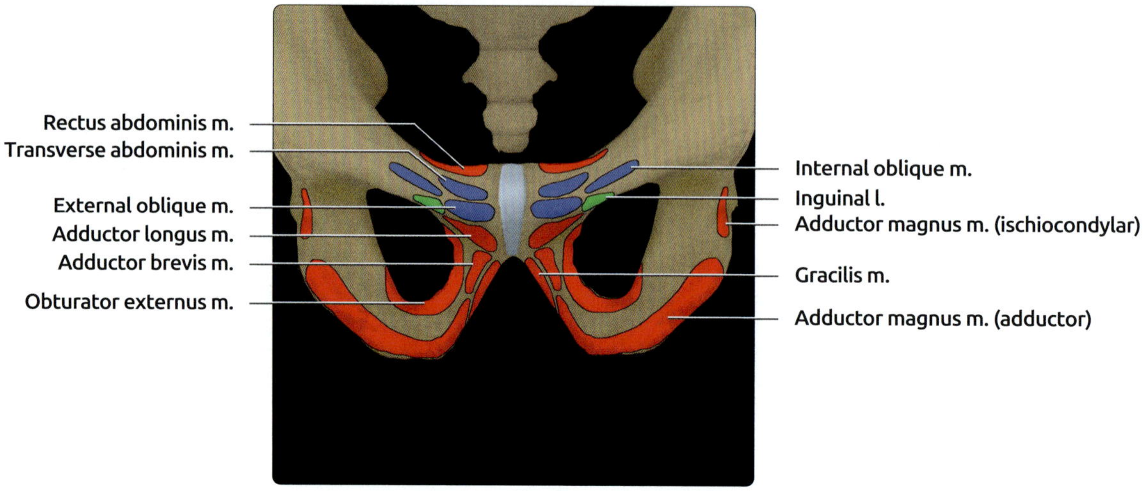

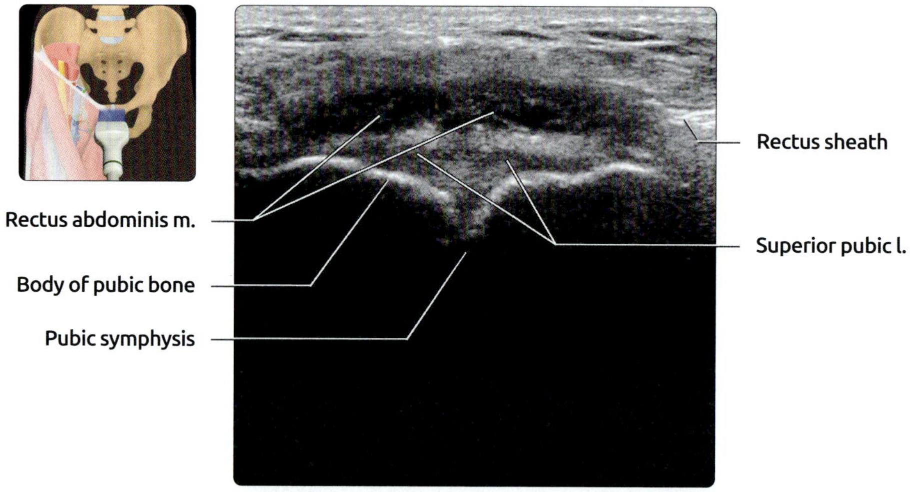

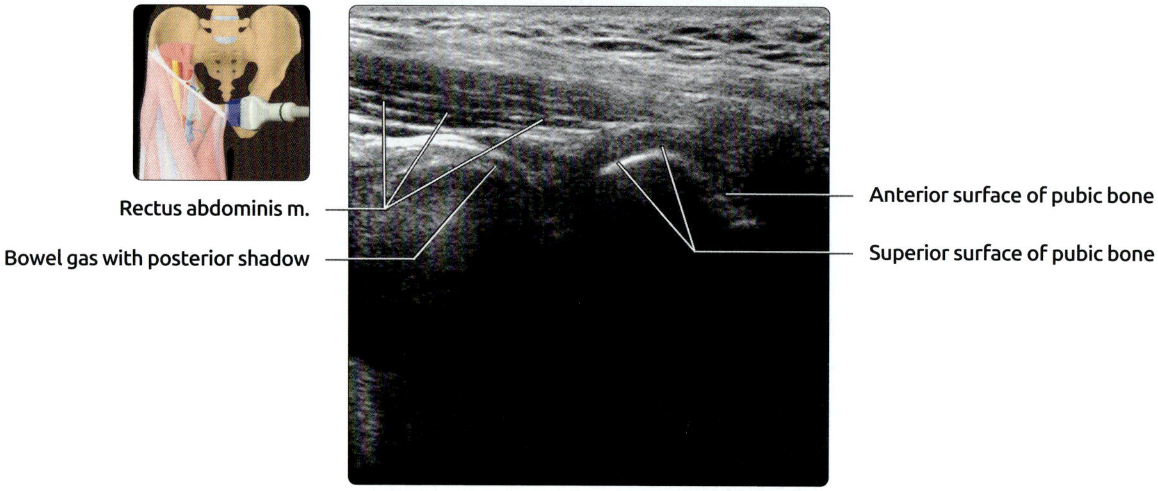

(Top) *Anterior view shows the anterior pelvis and the associated muscle attachments. A number of muscles take origin or insert onto the anterior aspect of the pelvis. These muscles aid in movement and stabilization of the trunk as well as movement and stabilization of the leg and form the medial components of the groin.* **(Middle)** *Transverse scan at the superior aspect shows pubic symphysis. The superior pubic ligament runs horizontally across the symphysis, deep to the rectus abdominis muscle origin.* **(Bottom)** *Sagittal image shows the lower rectus abdominis originating from the superior surface of the pubic bone (pubic crest). Deep to the rectus abdominis muscle is the peritoneal cavity and bowel. Bowel gas produces posterior shadowing artifact, which obscures deeper structures.*

MEDIAL GROIN

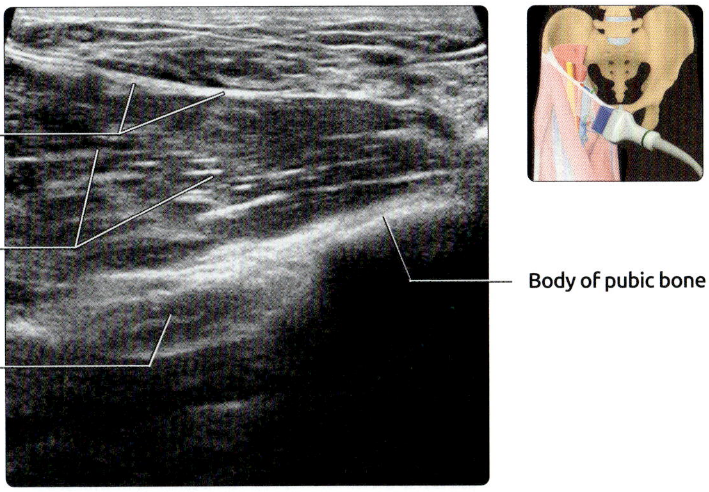

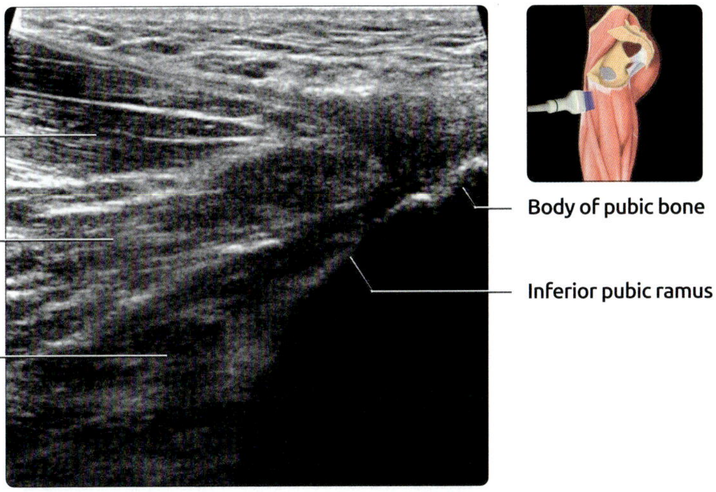

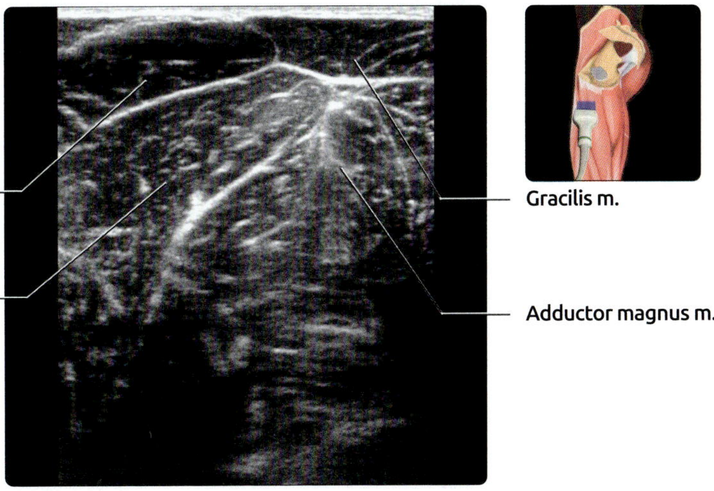

(Top) *Series of (oblique) transverse scans through the adductor muscles is shown. Superiorly, the pectineus muscle originates from the pectineal line (which is a ridge on the superior ramus of the pubic bone) and runs laterally to insert on the pectineal line of the proximal femur. The pectineal line of the proximal femur lies between the lesser trochanter and the linea aspera. The pectineus muscle runs lateral to the adductor longus muscle and superficial to the adductor brevis muscle.* **(Middle)** *Slightly inferiorly and scanning more obliquely, the adductor longus, brevis, and magnus can be seen originating from the pubic body and inferior ramus. Scanning this region requires thigh abduction and balancing the transducer head on the bony landmark. From here, individual muscles can be traced inferolaterally. Note that the scout image is a graphic in this sagittal plane.* **(Bottom)** *Oblique transverse scan of adductors using a medial approach is shown. Adductor longus and gracilis muscles form the superficial layer; adductor brevis and magnus muscles run deeper. All muscles run inferolaterally to their femoral insertions.*

CENTRAL GROIN, FEMORAL TRIANGLE

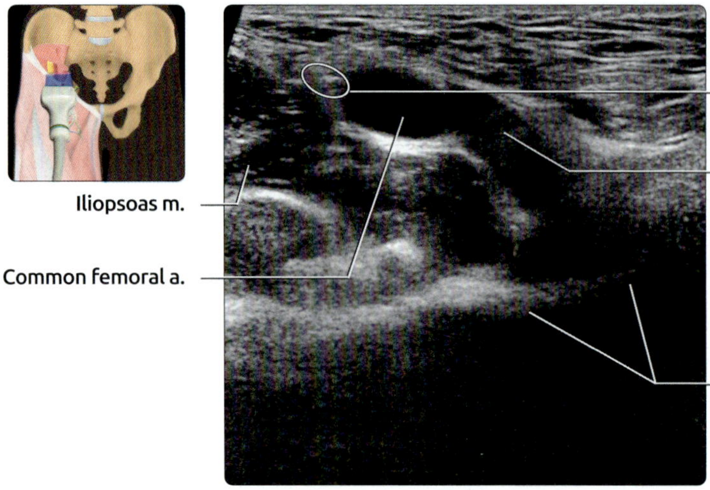

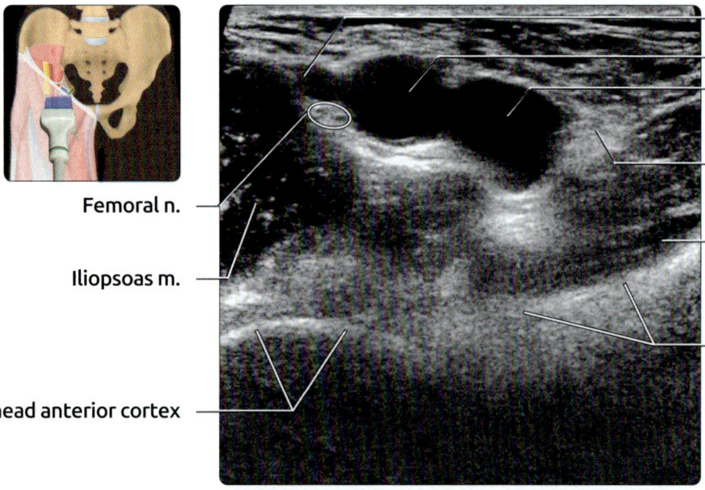

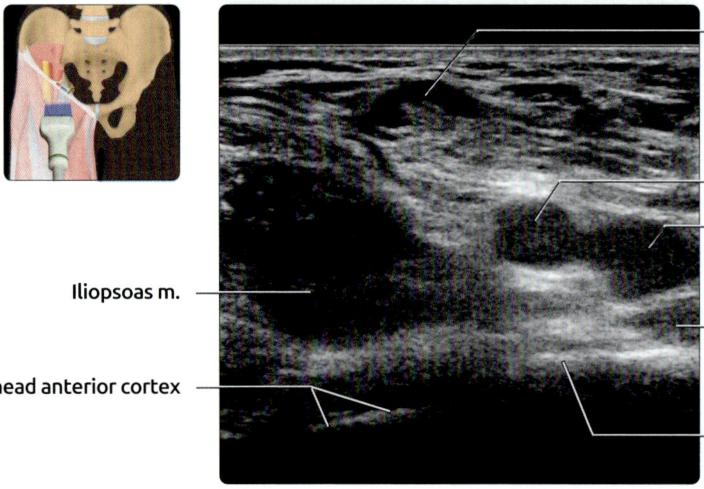

(Top) *Series of transverse scans through the femoral triangle is shown. Above the inguinal ligament, the femoral nerve, artery, and vein run medial to the iliopsoas tendon. The pubic bone acts as a hard support for the iliopsoas and pectineus muscular floor.* **(Middle)** *Slightly inferiorly, below the inguinal ligament, the femoral triangle commences. The mnemonic for the contents is NAVeL, standing for femoral nerve, common femoral artery, common femoral vein, and lymph nodes (within femoral canal). The pectineus muscle forms the floor of femoral triangle. This floor is supported by the pubic bone and hip joint.* **(Bottom)** *In addition to lymph nodes in the femoral canal, lymph nodes are frequently seen superficially (the superficial inguinal chain) in the subcutaneous tissue.*

CENTRAL GROIN, FEMORAL TRIANGLE

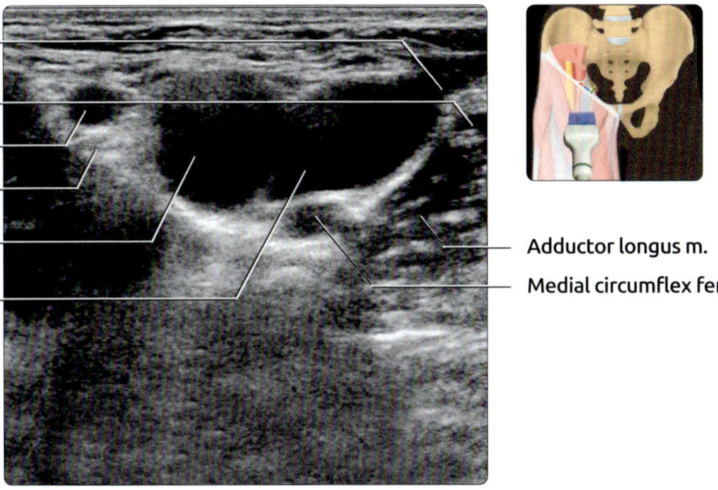

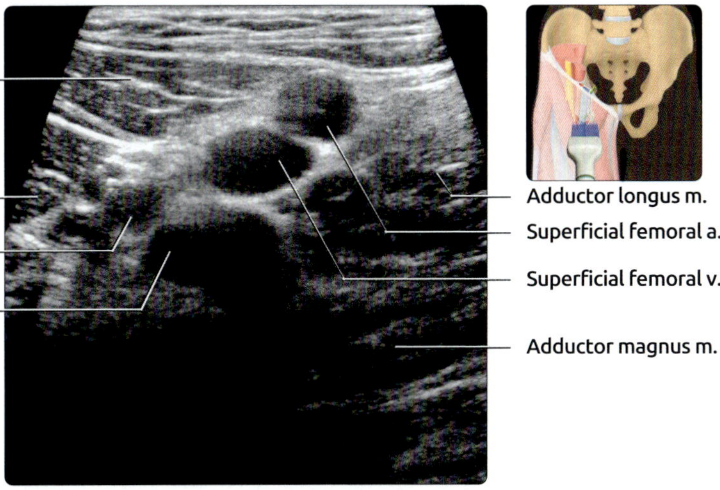

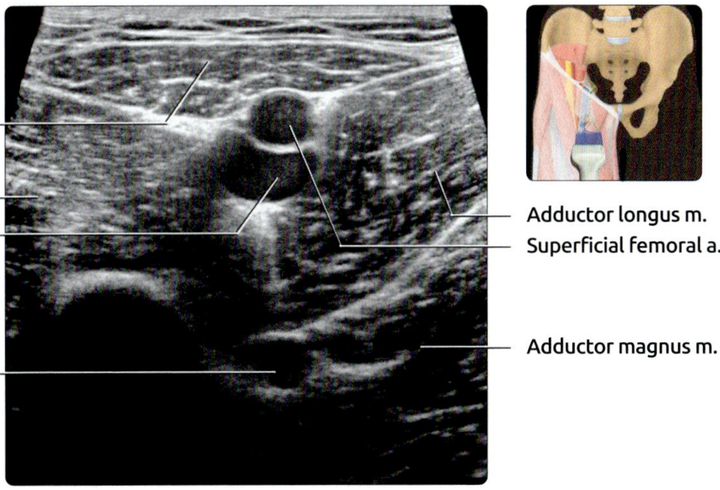

(Top) *Further inferiorly and still within the femoral triangle, the great saphenous vein perforates the anterior wall of the femoral sheath to enter the femoral vein. Also, the floor changes from the pectineus to adductor longus muscle with no bony support.* **(Middle)** *Transverse scan is shown at the apex of the femoral triangle, defined as where the sartorius muscle becomes the roof of the femoral triangle. Just before exiting the femoral triangle, the femoral vessels divide into superficial and profunda branches.* **(Bottom)** *Transverse scan shows the adductor canal. Distal to the femoral triangle, the superficial femoral vessels continue within a muscular conduit, the adductor canal. The groin muscles (the adductor longus and magnus muscles) form the medial wall and floor of this canal.*

LATERAL GROIN, ANTERIOR SUPERIOR ILIAC SPINE

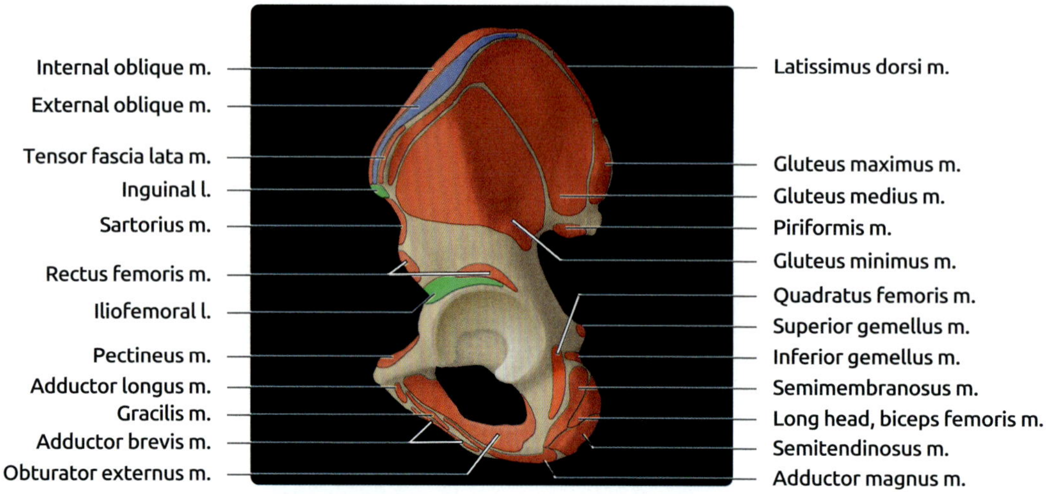

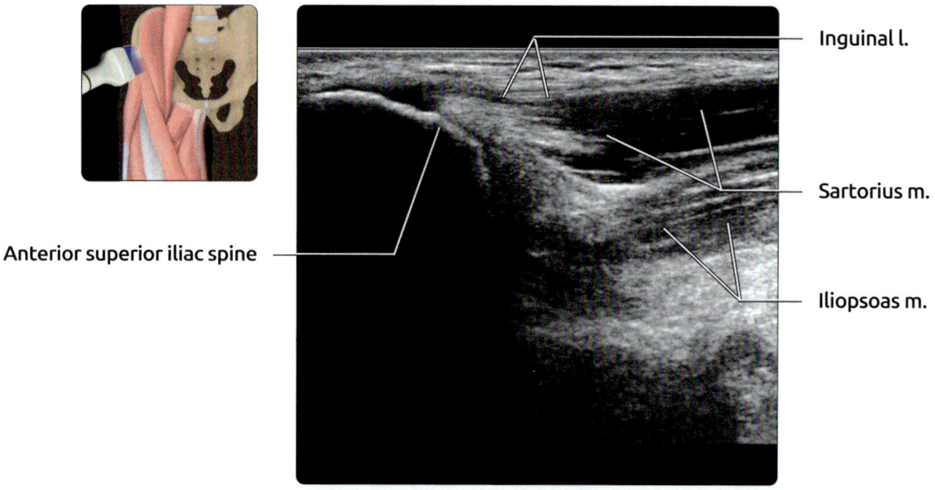

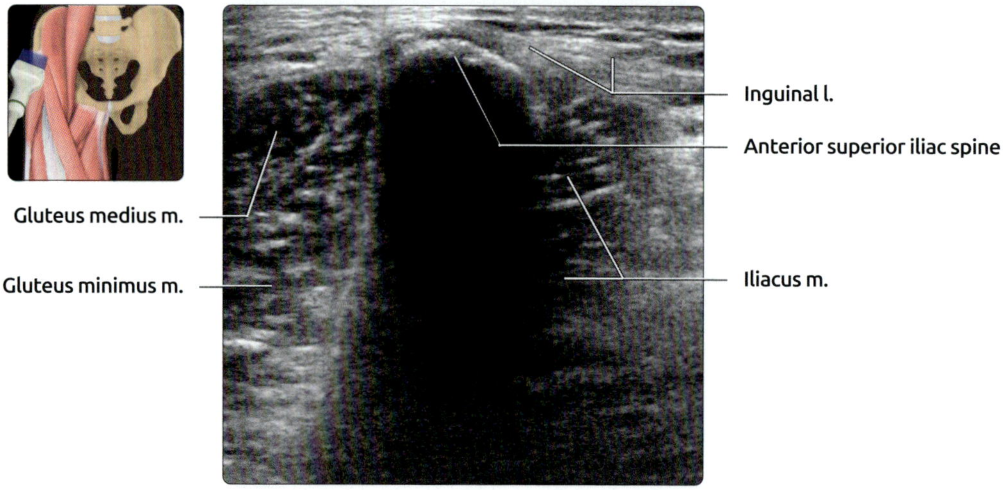

(Top) *Muscle and ligament attachments to the external surface of the pelvis are shown. Note the relatively small origin of the tendon of the adductor longus muscle. The adductor brevis muscle is just deep to the longus muscle. The origin of the gracilis muscle is lateral to the adductor brevis muscle. The adductor magnus muscle has a broad origin with its posterior fibers in close proximity to the hamstring tendon origins.* **(Middle)** *Oblique sagittal scan of the anterior superior iliac spine shows the origin of the sartorius muscle (longest muscle in the body). This slender muscle runs inferomedially over the iliopsoas muscle, covers the adductor canal, and inserts at the proximal tibia as part of the pes anserinus.* **(Bottom)** *Coronal scan of the anterior superior iliac spine shows division of muscle groups by this bone. Gluteal muscles lie lateral and hip flexors medial. The inguinal ligament originates here and runs inferomedially to the pubic tubercle.*

ACROSS GROIN, INGUINAL LIGAMENT

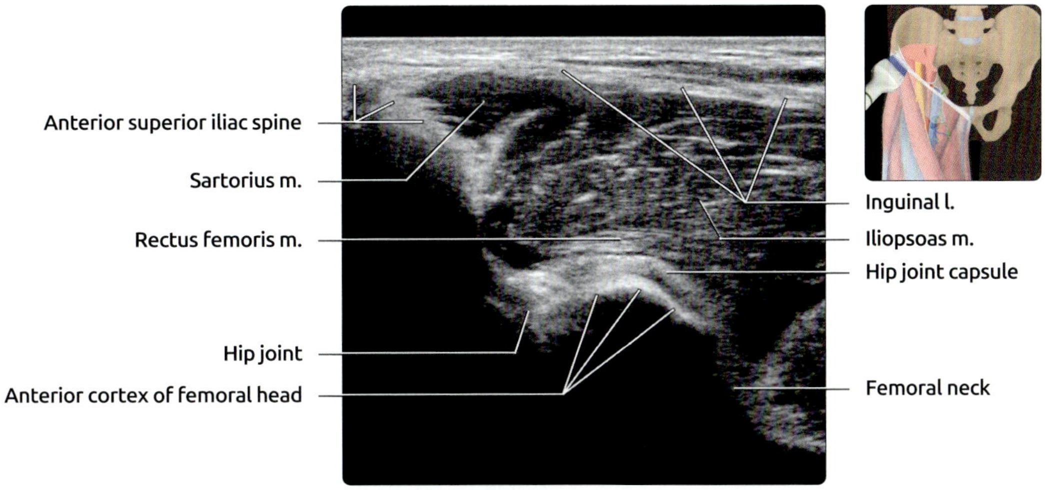

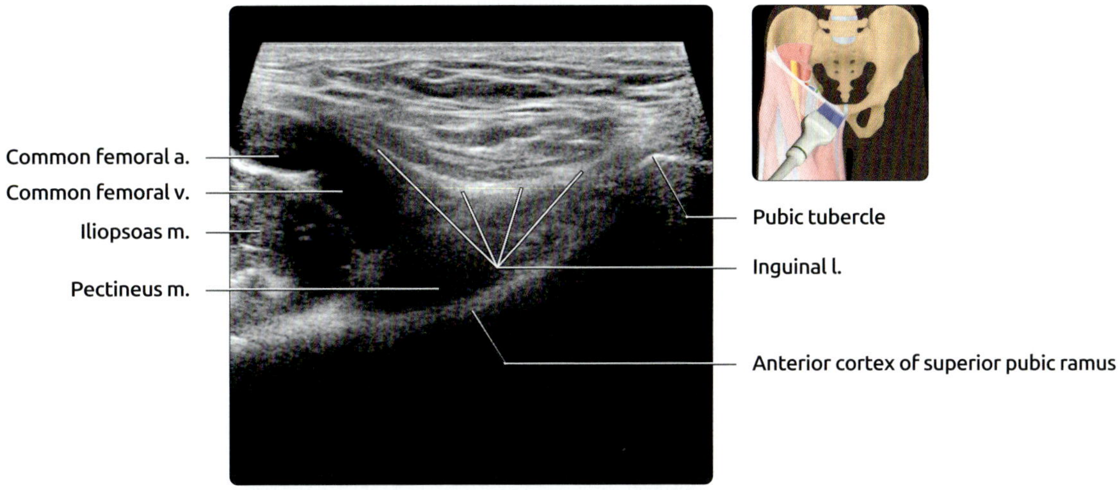

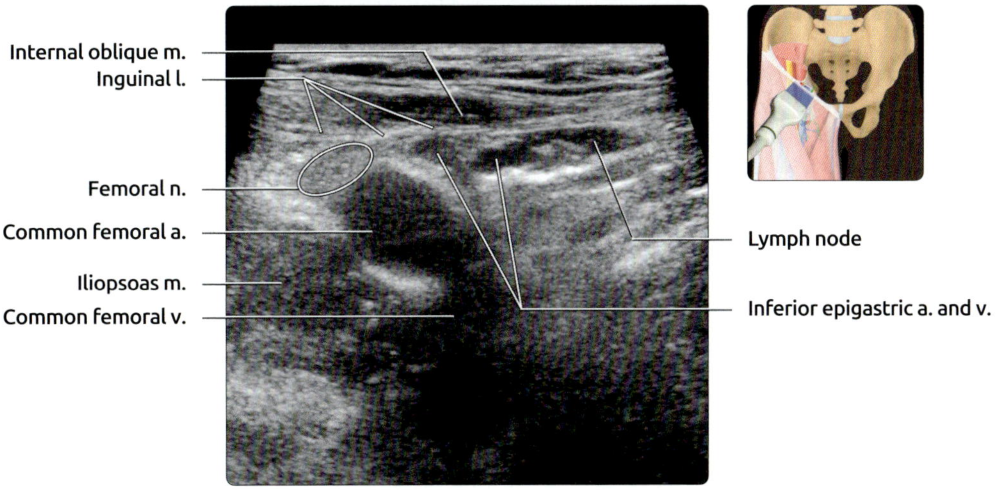

(Top) *Oblique transverse scan of the upper groin shows the inguinal ligament attachment on the anterior superior iliac spine with the sartorius muscle originating immediately inferior. The inguinal ligament runs inferomedially and covers the iliopsoas muscle.* **(Middle)** *Oblique transverse scan along the medial aspect of the inguinal ligament is shown. Medially, the inguinal ligament forms the roof of the femoral canal (covering the femoral vessels and nerve). It then runs over the pectineus muscle to insert on the pubic tubercle.* **(Bottom)** *Superficially focused oblique transverse scan over the upper femoral triangle shows the inguinal ligament forming its roof. Inferior epigastric vessels branch out from the femoral vessels and run superomedially. Clinically, the inferior epigastric vessels define the lateral boundary of direct inguinal hernias. Inguinal hernias that emerge lateral to the inferior epigastric vessels are considered indirect.*

IMAGING ANATOMY

Overview

- Ball and socket joint with wide range of motion 2nd only to glenohumeral joint

Acetabulum

- Formed by pubis, ilium, ischium
- Oriented anterior, inferior, lateral
- Covers > 50% femoral head
- **Anterior and posterior rims**: Osseous margins of acetabulum
- **Medial wall**: Quadrilateral plate ilium

Femoral Head

- 2/3 of sphere
- Covered by articular cartilage
  - Cartilage thickest superiorly
  - Cartilage thins at head/neck junction

Labrum

- Fibrocartilage lip deepens acetabular rim
- Joins transverse ligament at margins of acetabular notch
- Thickest posteriorly and superiorly
- Widest anteriorly and superiorly
- Vascular supply: Branches of obturator, superior, and inferior gluteal arteries
  - Mainly capsular surface
- Shape
  - Triangular in 66-94%
  - Variants: Rounded/blunt or absent
  - Absent labra: Constellation absent anterior labrum and small remnant superiorly
    - In 10-14% asymptomatic individuals
- Function
  - Protect cartilage: Distributes forces by maintaining synovial fluid layer between articular surfaces
  - Prevents lateral translation femoral head
- Labral tear
  - Frequently has associated paralabral ganglion/cyst
  - Easier to detect in presence of joint effusion/arthrographic contrast
  - More common anterosuperiorly and anteriorly

Joint Capsule

- 2 layers: Internal synovial and external fibrous (not usually separable on imaging)
  - External layer forms capsular ligaments
- Attachments
  - Acetabulum
    - Base of labrum anteriorly and posteriorly
    - Several millimeters above labrum superiorly
  - Femur
    - Anterior: Intertrochanteric line
    - Posterior: Proximal to intertrochanteric crest
    - Anterior attachment more lateral than posterior attachment
  - **Perilabral recess**: Between labrum and capsule
    - Smaller anteriorly and posteriorly
    - Larger superiorly

Vascular Supply

- Branches of medial and lateral circumflex femoral, deep division of superior gluteal, inferior gluteal arteries, artery of ligamentum teres (branch of obturator artery)

Innervation

- Branches from nerve to rectus femoris, nerve to quadratus femoris, anterior division obturator nerve, accessory obturator nerve, superior gluteal nerve

Major Adjacent Structures

- **Greater trochanter**
  - Insertion
    - Gluteus minimus and medius muscles
    - Gluteus minimus to anterior facet
    - Gluteus medius to lateral and posterosuperior facets
  - Separated from overlying iliotibial tract by thin trochanteric bursa (normally barely perceptible on imaging)
- **Lesser trochanter**
  - Insertion
    - Iliopsoas muscle
- **Anterior inferior iliac spine**
  - Origin
    - Rectus femoris muscle
- **Iliopsoas muscle**
  - Run inferomedially across anterior surface of hip joint
- **Femoral vessels**
  - Separated from hip joint by iliopsoas muscle

Neonatal Hip

- High success rate in treatment of developmental dysplasia demands early imaging diagnosis (with ultrasound)
- Established measurements of bony and cartilaginous components for comparison with normal reference
- Main measurements reflect depth of acetabulum and acetabular coverage of femoral head

ANATOMY IMAGING ISSUES

Imaging Recommendations

- Radiographs for assessing osseous alignment
- Ultrasound for joint effusion and surrounding soft tissue
  - Also used for real-time guided procedures: Hip joint or abscess aspiration/biopsy; intraarticular contrast injection
  - Ideal in neonates for suspected developmental hip dysplasia
- MR arthrography preferred to assess intraarticular structures

Imaging Sweet Spots

- Small joint effusions are best detected on ultrasound by hypoechoic thickening/layer over anterior surface of femoral neck
  - Best site for aspiration/injection
- Paralabral ganglion cysts should alert physician to presence of labral tear

Imaging Pitfalls

- Synovial proliferation usually shows no evidence of vascularity on color Doppler scanning and is therefore difficult to distinguish from effusion

AXIAL & SAGITTAL HIP SECTIONS

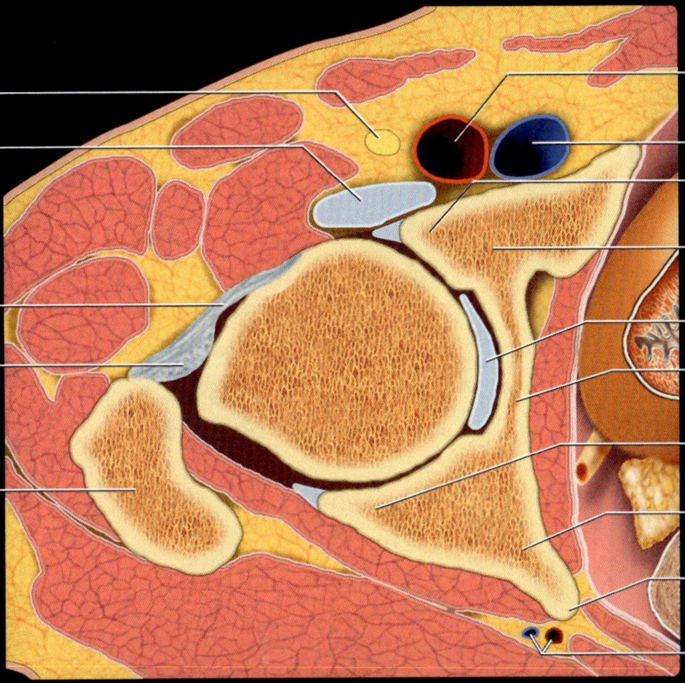

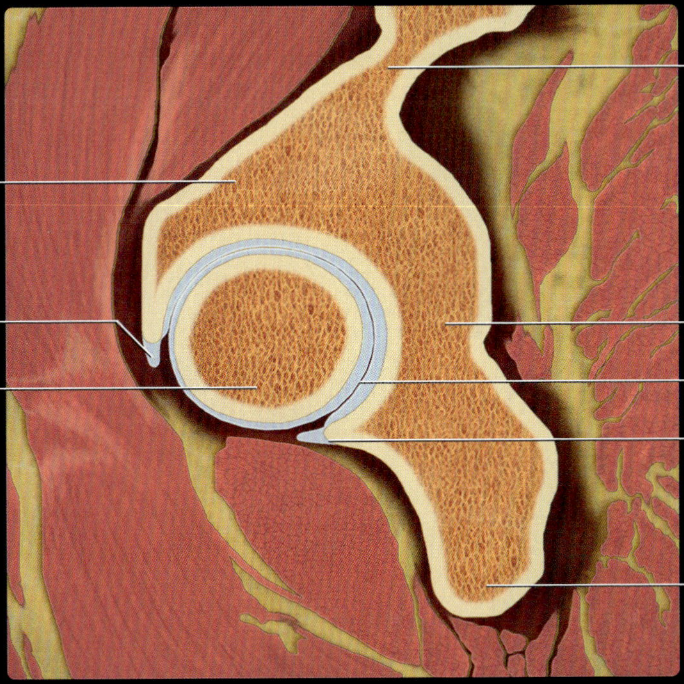

(Top) *Axial graphic representation shows the hip joint. The iliopsoas bursa is closely approximated to the joint anteriorly. The ligamentum teres is seen in profile as a flattened structure. The 2 layers of the external capsule, the more superficial longitudinal fibers and deep circular fibers, are visible.* **(Bottom)** *Sagittal representation of the hip joint nicely demonstrates the inverted Y of the acetabulum. The ilium is the stem, and the limbs are the anterior and posterior acetabular columns.*

CORONAL HIP SECTION & LIGAMENTS

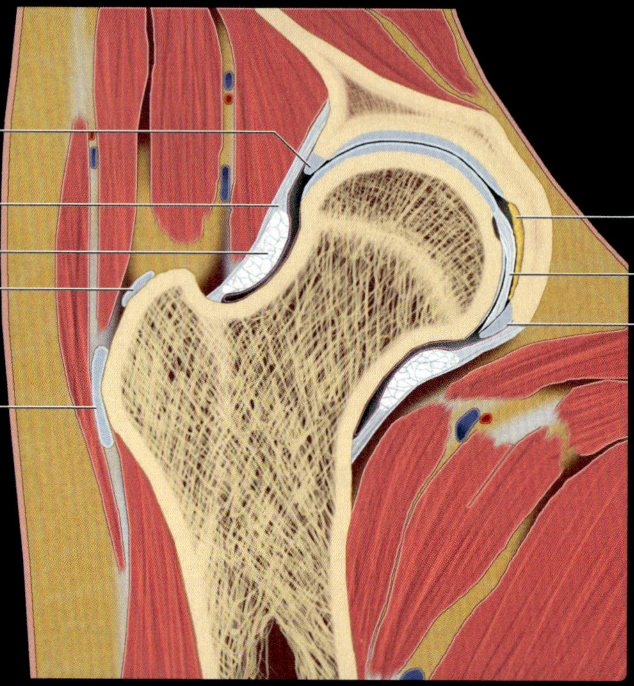

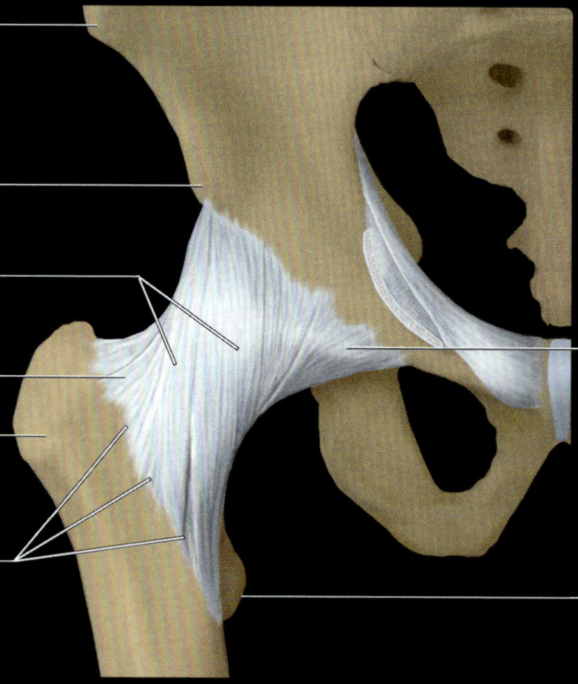

(Top) *This coronal image demonstrates many of the important structures of the hip joint. The longitudinally oriented fibers of the ligaments of the external joint capsule and the deeper circularly oriented fibers of the zona orbicularis are visible. Note the long axis of the ligamentum teres and its insertion onto the transverse ligament. The pulvinar fills the acetabular fossa.* **(Bottom)** *Anterior ligaments are shown. The longitudinal spiral of the iliofemoral and pubofemoral ligaments is well seen with attachments to the anterior inferior iliac spine and pubic aspect of the obturator foramen, respectively. The ischiofemoral ligament wraps over the superior aspect of the femoral neck to attach to the anterior femoral neck.*

HIP OVERVIEW

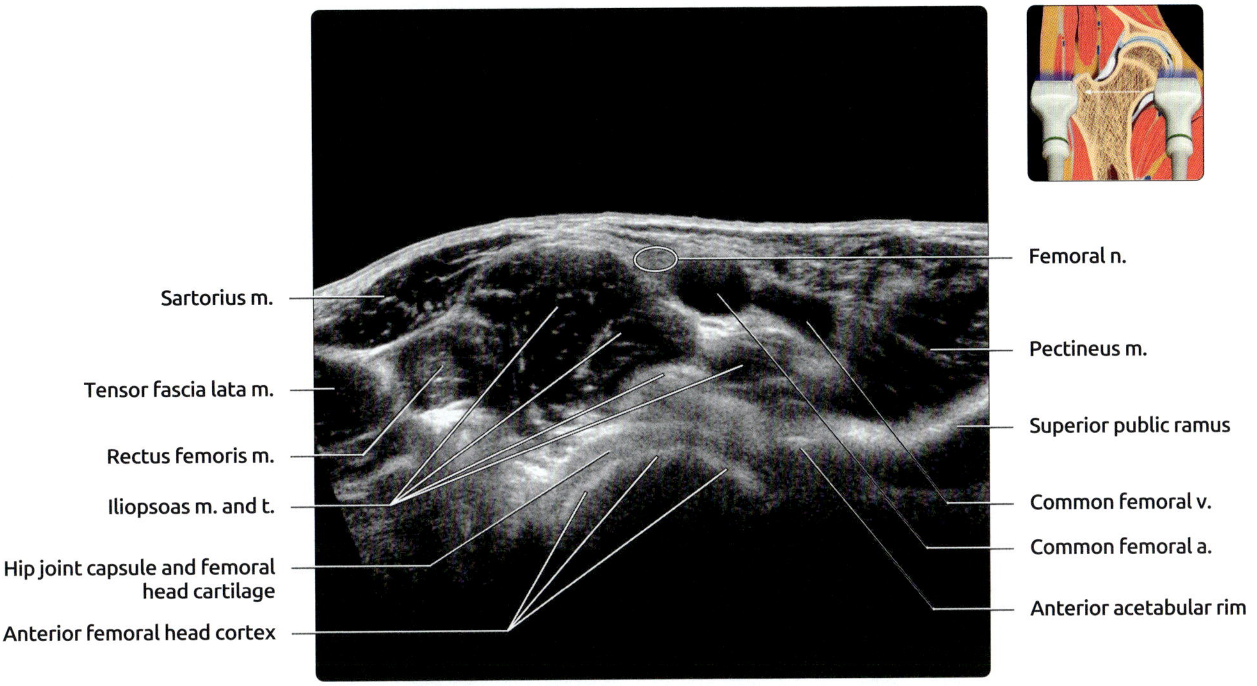

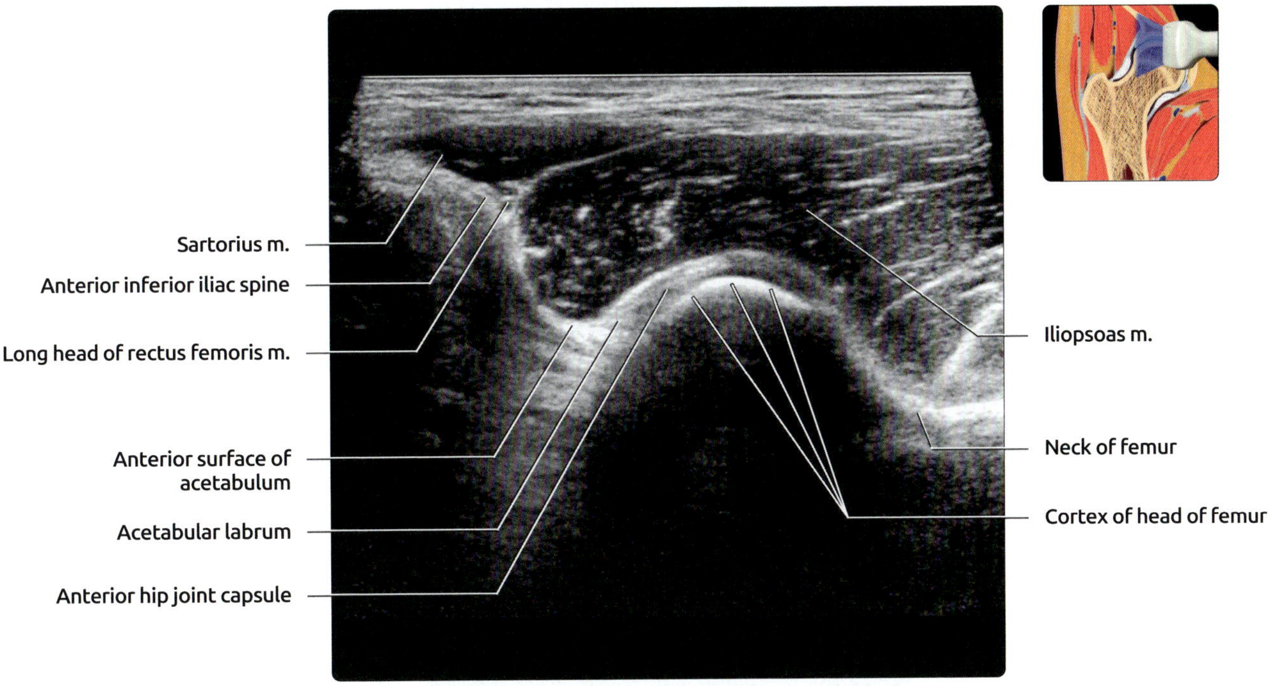

(Top) *Transverse panoramic scan shows the femoral head. Medially and superiorly, the femoral head is covered by bony acetabulum and not visible with ultrasound. Laterally, the hip joint lies deep to the iliopsoas muscle and femoral triangle. Further laterally, the hip joint is associated with the gluteal muscles.* **(Bottom)** *Oblique sagittal scan shows hip joint and anterior inferior iliac spine. The acetabular labrum, which is thickest and widest anteriorly, can be seen between the acetabular rim and joint capsule. Small joint effusions are 1st detected not over the femoral head, but over the femoral neck as a hypoechoic layer. The femoral neck's relatively safe access makes it a suitable site for guided aspiration/injection.*

ACETABULUM

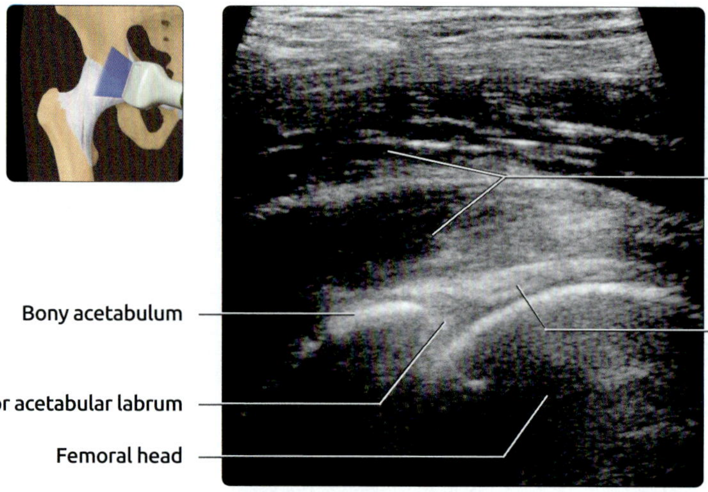

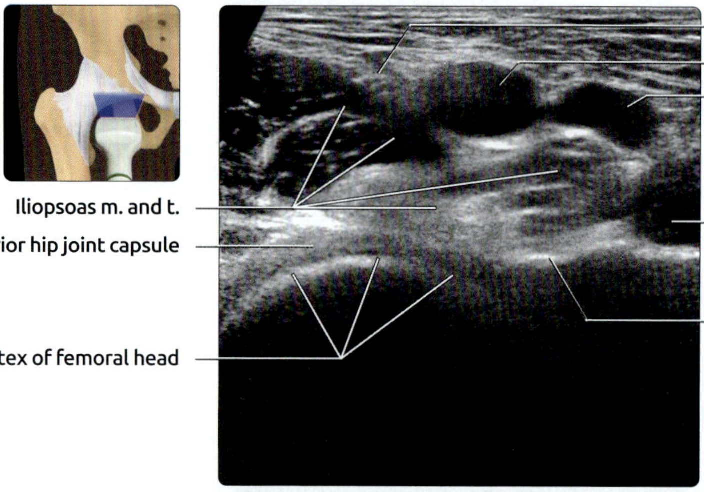

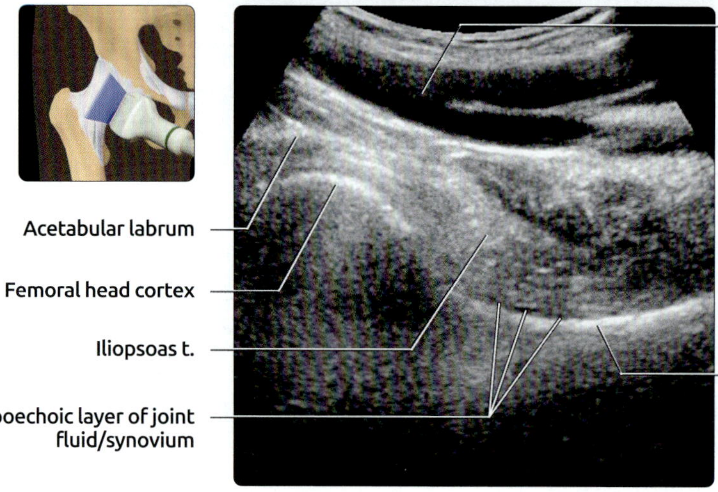

(Top) *Oblique sagittal scan along the axis of the femoral neck shows the triangular cross section of anterior acetabular labrum. Joint effusion/contrast tends to allow better delineation of its contour and lesions.* **(Middle)** *Transverse scan is shown at the level of the hip joint. Anterior femoral cortex/cartilage, hip joint capsule, and acetabular rim are well visualized. The contents of the femoral triangle are separated from the hip joint by the iliopsoas muscle and tendon, which explains why paralabral cysts may tract along the iliopsoas muscle.* **(Bottom)** *Effusions are 1st detected not over the femoral head but over the femoral neck as an increase in the normally minimal hypoechoic layer of joint fluid and synovium. This is also the safest site for joint injections (intraarticular contrast or medication).*

ILIOPSOAS INSERTION

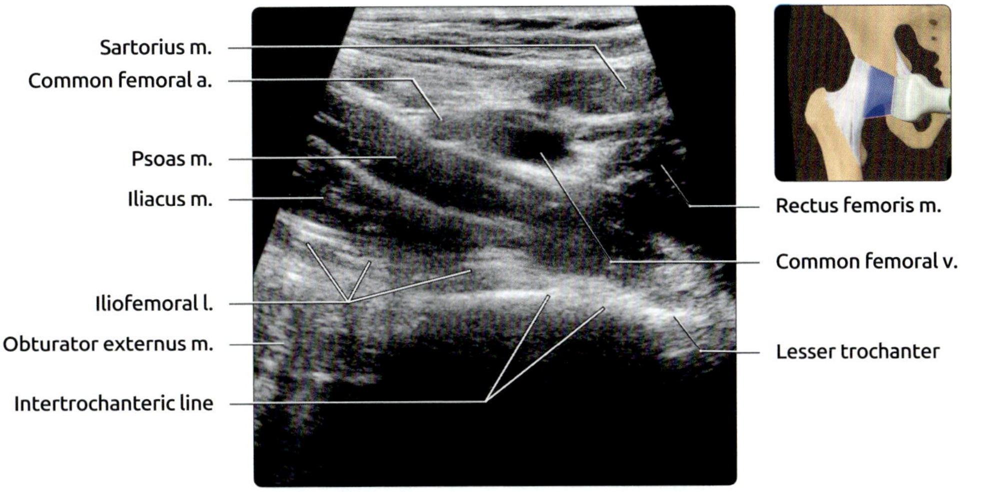

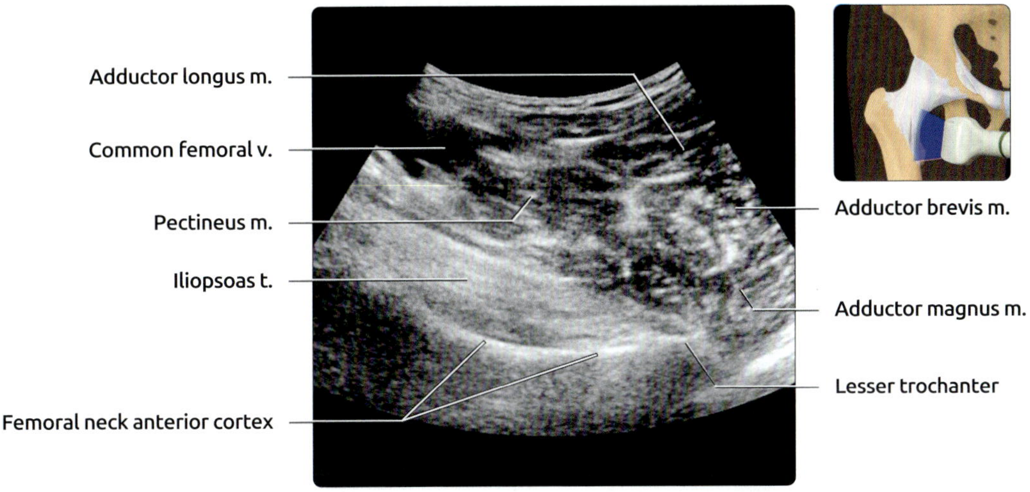

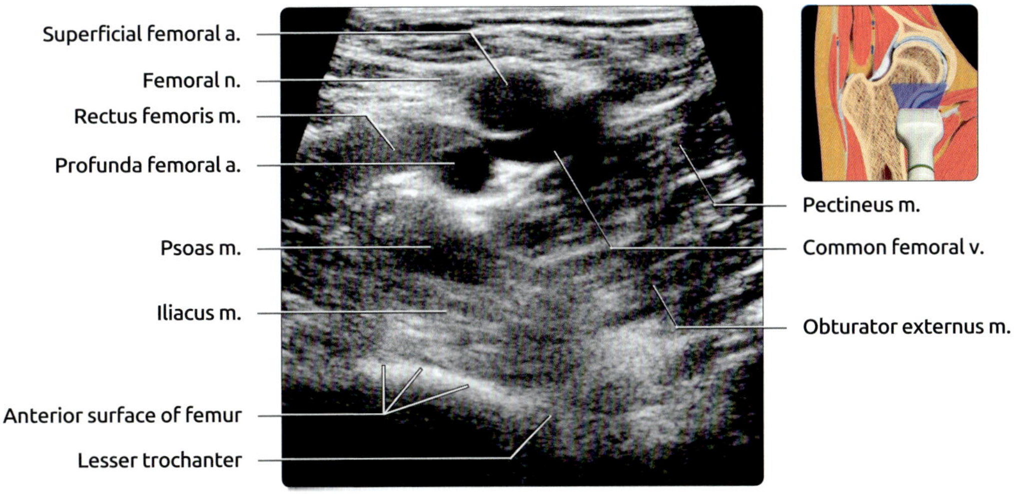

(Top) *Oblique sagittal scan along the course of the iliopsoas muscle is shown. Resolution is reduced due to depth of structures, which may be partially improved by externally rotating the hip to bring deep structures closer to the surface.* **(Middle)** *Oblique sagittal scan is shown slightly more distal. Insertion of iliopsoas tendon is at the lesser trochanter, which may be avulsed in injury. Notice the close relationship between these structures and the adductor compartment.* **(Bottom)** *Transverse scan shows distal iliopsoas tendon just before insertion. A functional division is present between the hip flexor (iliopsoas) and hip adductors. This is also the apex level of the femoral triangle where superficial and profunda vessels bifurcate.*

ANTERIOR INFERIOR ILIAC SPINE

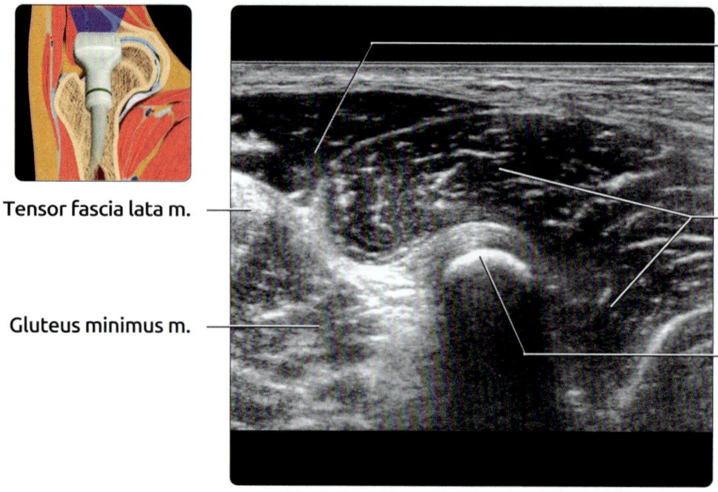

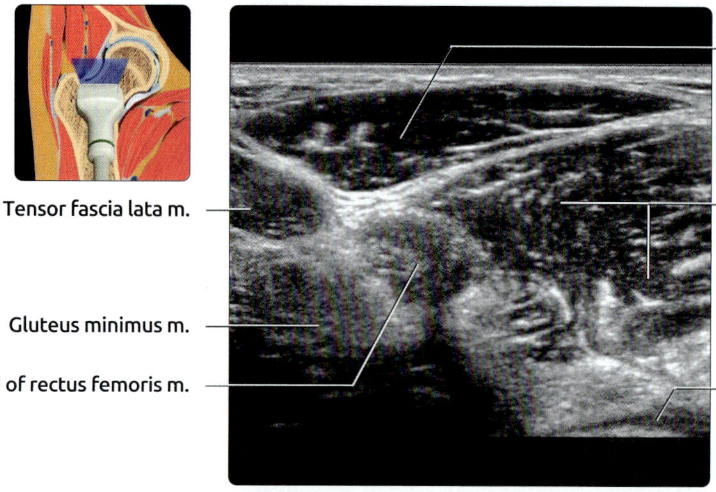

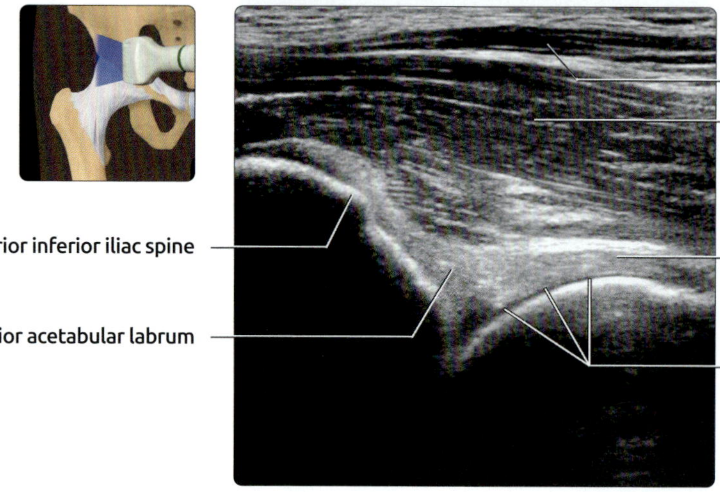

(Top) *Transverse scan through the anterior inferior iliac spine is shown. The rectus femoris muscle, together with the sartorius muscle (which originates from anterior superior iliac spine) and the iliopsoas muscle (which originates from iliac fossa and lumbar transverse processes), cross the hip joint and form the hip flexors.* **(Middle)** *Transverse scan more inferior shows the long head of the rectus femoris muscle running deep to the sartorius muscle and lateral to the iliopsoas muscle. Note that the rectus muscle is the only quadriceps muscle that crosses the hip joint and thus acts as both hip flexor and knee extensor.* **(Bottom)** *Oblique scan along the sartorius muscle at hip joint level shows it to be roughly parallel with the iliopsoas muscle. The rectus femoris lies in a more lateral plane.*

GREATER TROCHANTER

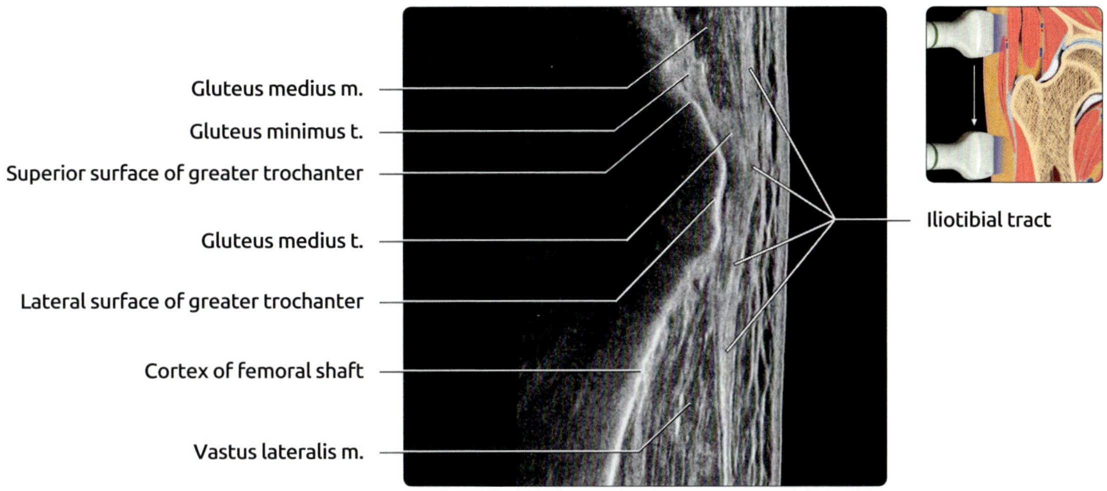

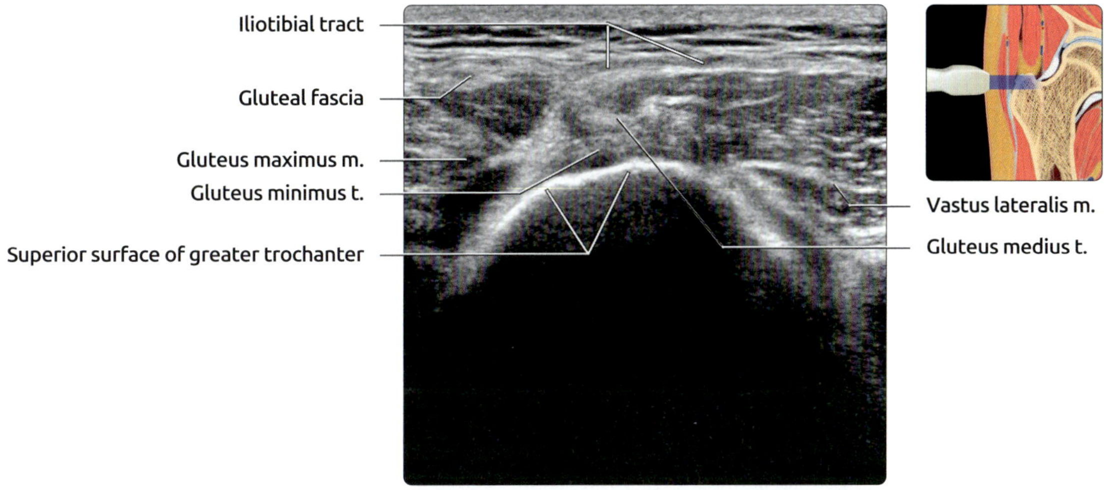

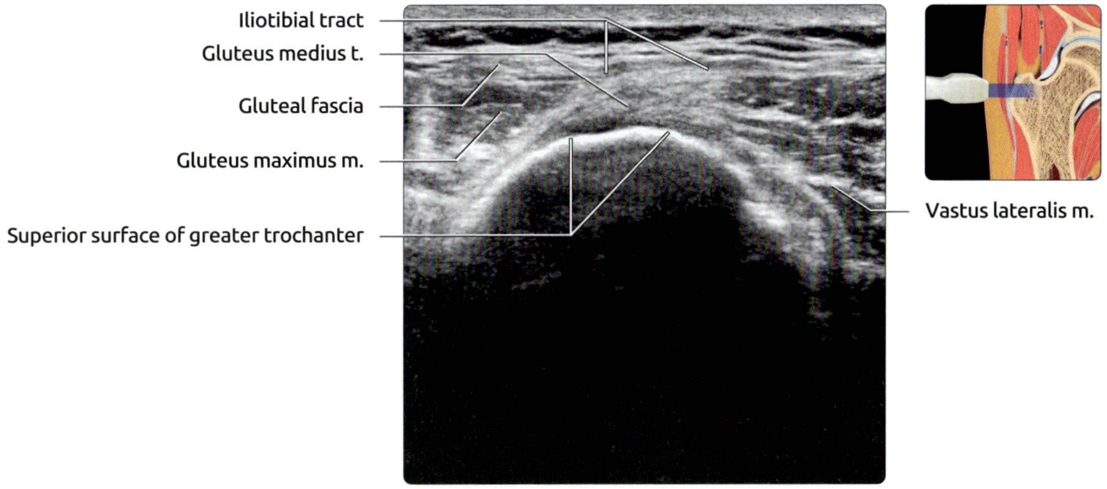

(Top) *Coronal panoramic scan shows the greater trochanter. The gluteus minimus and medius tendons insert on the anterior, lateral, and superoposterior facets of the greater tuberosity. The iliotibial tract originates from the anterior aspect of the iliac crest (as the tensor fascia lata muscle) and runs inferiorly, passing superficial to the lateral surface of the greater trochanter.* **(Middle)** *Transverse scan through the anterior facet of the greater trochanter shows insertion of the gluteus minimus tendon. The gluteus medius tendon and iliotibial tract pass superficial to this.* **(Bottom)** *Transverse scan of the lateral facet of the greater trochanter shows the insertion of the gluteus medius tendon. The iliotibial tract is closest to the trochanter at this point and may be irritated, resulting in the formation of bursitis (trochanteric bursitis).*

PEDIATRIC HIP

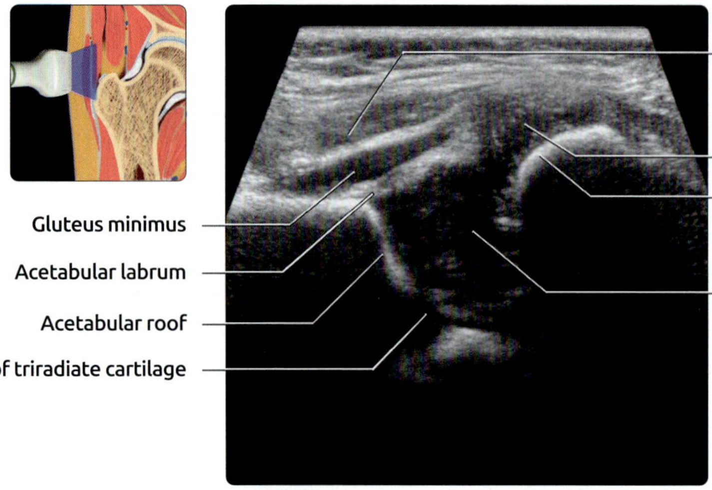

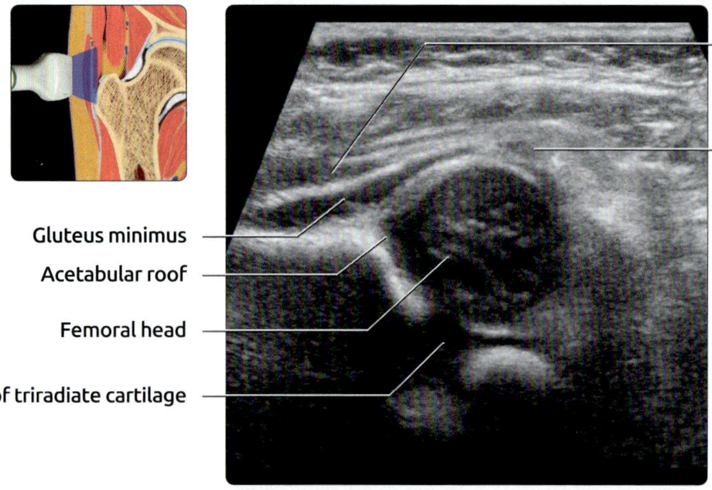

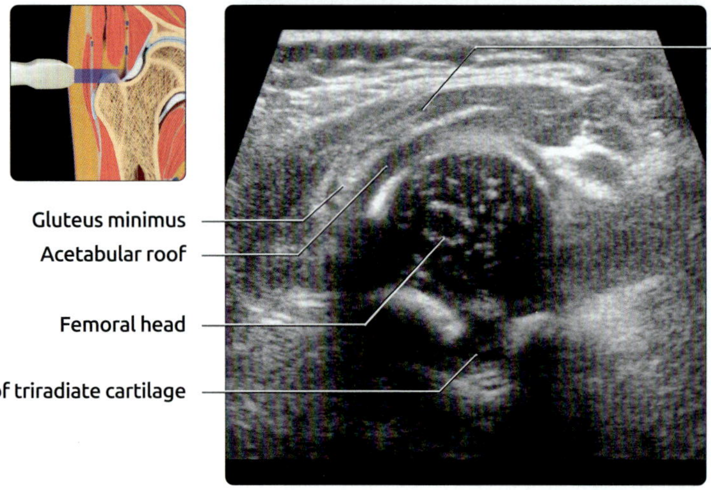

(Top) *Coronal scan shows the hip of a 1-month-old neonate with the hip in extension. The femoral head and the greater trochanter are purely cartilaginous. This allows ultrasound transmission and thus visualization of the acetabular fossa.* **(Middle)** *Coronal scan shows the hip of a 1-month-old neonate with the hip in flexion, which produces an axial section of the femoral head. The femoral head should appear round in this section. Both the femoral head and the greater trochanter are purely cartilaginous, allowing ultrasound transmission and thus better visualization of the depth of the acetabular fossa.* **(Bottom)** *Transverse scan of the femoral head with hip in extension also shows the central position of a normal femoral head on the triradiate cartilage (golf ball on tee appearance).*

PEDIATRIC HIP

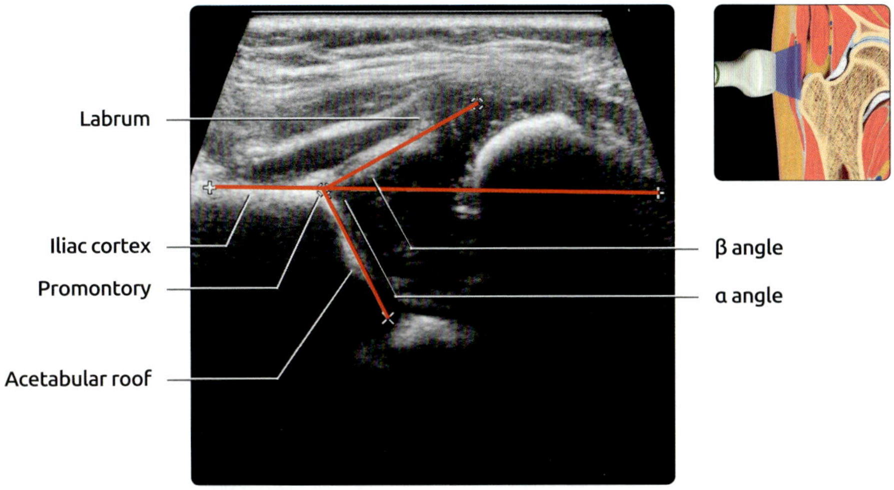

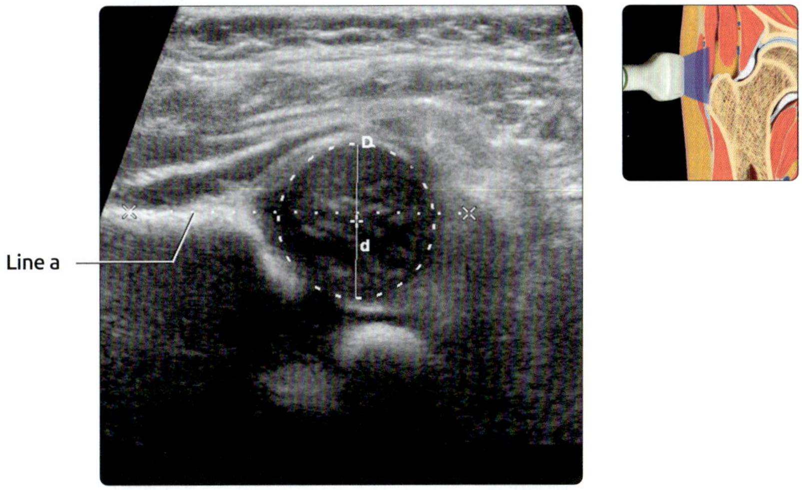

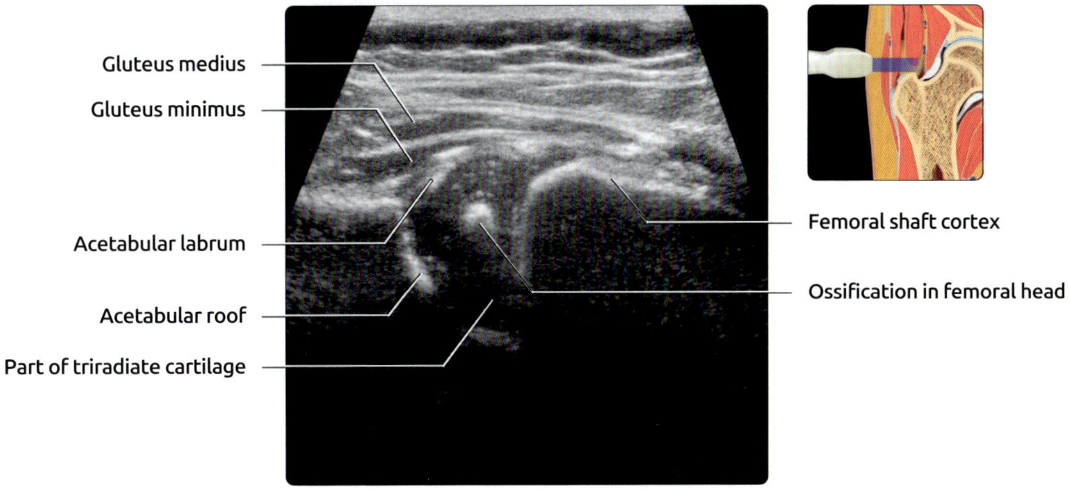

(Top) *Three lines are drawn on the standard coronal image of the pediatric hip: (A) Along the straight contour of the iliac cortex, (B) from promontory along the acetabular roof to describe the α angle, and (C) from promontory to the tip of the labrum for β angle. Normal ranges are α angle > 60° and β angle < 55°. Measurements outside of these ranges are suggestive of developmental hip dysplasia.* **(Middle)** *The proportion of the femoral head contained within the acetabulum is calculated by dividing the depth (d) of femoral head below line a by the diameter of the femoral head (D). In normal subjects, > 50% of the femoral head should be below line a.* **(Bottom)** *Coronal scan shows the hip of a 4-month-old child with the hip in extension. The center of the femoral head has begun to ossify, showing up as an echogenic focus with posterior acoustic shadowing.*

TERMINOLOGY

Abbreviations

- Function (F), insertion (I), origin (O)

IMAGING ANATOMY

Compartment Anatomy

- **Compartmental anatomy is different from functional groupings**
- Thigh divided into anterior, medial, and posterior compartments
  - **Anterior compartment**: Iliotibial tract, tensor fascia lata muscle, quadriceps muscles, sartorius muscle
  - **Medial compartment**: Gracilis muscle, adductor muscles
  - **Posterior compartment**: Hamstring muscles, short head of biceps femoris muscle, sciatic nerve
- Additional muscles at junction pelvis/thigh
  - Pectineus, iliopsoas, obturator externus, lateral femoral muscles
- Extensions from fascia lata divide compartments
  - Medial intermuscular septum: Anterior/medial
  - Lateral intermuscular septum: Anterior/lateral
  - Thin fascia separates medial, posterior compartments
- Clinical note: Compartment anatomy critical to tumor staging and biopsy planning
  - Cross-compartment extension of tumor, contamination by biopsy requires change from limb salvage to amputation

Medial Femoral Muscles

- Anterior division: Adductor brevis and longus, gracilis muscles
- Posterior division: Adductor portion adductor magnus muscle
- Common function: Hip adduction; assist hip flexion, internal rotation (except obturator externus muscle)
- **Adductor brevis**
  - O: Inferior pubic ramus
  - I: Inferior 2/3 pectineal line, superior 1/2 medial lip of linea aspera
- **Adductor longus**
  - O: Pubic body inferior to crest
  - I: Medial lip of linea aspera
- **Adductor magnus**: 2 separate muscles with different innervations and functions
  - Adductor portion
    - O: Ischiopubic ramus
    - I: Gluteal tuberosity, medial lip linea aspera, medial supracondylar line
  - Adductor hiatus between 2 divisions of muscle in distal thigh
- **Gracilis**
  - O: Inferior pubic ramus, symphysis pubis
  - I: Medial proximal tibia (pes anserine)
  - F: Also assists knee flexion
- **Obturator externus**
  - O: External margins of obturator foramen membrane
  - I: Piriformis fossa
  - F: Hip external rotation only
- **Pectineus**
  - O: Superior pubic ramus, pecten
  - I: Pectineal line femur
  - Femoral nerve ± accessory obturator nerve
  - Unclear pre- or postaxial muscle

Anterior Femoral Muscles

- Common function: Knee extension (except sartorius)
- **Rectus femoris**
  - O: Straight head = anterior inferior iliac spine; reflected head = groove above acetabulum
  - I: Superior patella, tibial tuberosity
  - F: Also hip flexion
  - Crosses 2 joints
- **Sartorius**: Tailor's muscle
  - O: Anterior superior iliac spine, notch below
  - I: Proximal medial tibia (pes anserine)
  - F: Hip flexion, abduction, external rotation; knee flexion
  - Crosses 2 joints
  - Longest muscle in body
  - Separate fascial covering
- **Vastus lateralis**
  - O: Superior intertrochanteric line femur, anterior and inferior greater trochanter, gluteal tuberosity, lateral lip of linea aspera, lateral intermuscular septum
  - I: Lateral tibial condyle (lateral patellar retinaculum) superolateral patella (quadriceps tendon)
  - Largest quadriceps muscle
- **Vastus medialis**
  - O: Entire medial lip of linea aspera, inferior intertrochanteric line, medial intermuscular septum
  - I: Tendon of rectus femoris muscle, superomedial patella (quadriceps tendon), medial condyle tibia (medial patellar retinaculum)
- **Vastus intermedius**
  - O: Anterior and lateral femoral shaft, inferior lateral lip of linea aspera, lateral intermuscular septum
  - I: Blends along deep aspect rectus femoris, vastus medialis, vastus lateralis muscles
- **Quadriceps femoris**: Rectus femoris, vastus lateralis, vastus medialis, vastus intermedius muscles
  - Common tendon of insertion onto superior, lateral, medial patella

Posterior Femoral Muscles

- Common functions: Hip extension, knee flexion
- **Biceps femoris**
  - Long head O: Ischial tuberosity (inferior, medial)
    - Common tendon with semitendinosus muscle
  - Short head O: Lateral lip of linea aspera femur, lateral supracondylar line, lateral intermuscular septum
    - Postaxial muscle
    - Not part of hamstring muscles
  - I: Fibular head, lateral condyle tibia
  - F: Also external rotation flexed knee
- **Semimembranosus**
  - O: Ischial tuberosity (superior, lateral)
  - I: Posterior medial condyle tibia, popliteal fascia
    - Some fibers extend to form oblique popliteal ligament
  - F: Also internal rotation flexed knee
  - Membranous in upper thigh

- **Semitendinosus**
  - O: Ischial tuberosity (inferior, medial)
    - Common tendon of long head biceps femoris muscle
  - I: Medial proximal tibia (pes anserine)
  - F: Internal rotation flexed knee
  - Entirely tendinous in distal thigh
- **Ischiocondylar portion of adductor magnus**
  - O: Ischial tuberosity
  - I: Adductor tubercle
  - Medial-most aspect adductor magnus muscle
- **Hamstrings**: Long head biceps femoris, semimembranous, semitendinosus, ischiocondylar portion adductor magnus muscles
  - Does not include short head biceps femoris muscle
- **Pes anserine**
  - Common aponeurosis for insertion gracilis, semitendinosus, sartorius tendons
  - Pes anserine bursa between tendons and tibia

Lateral Femoral (Gluteal) Muscles

- **Iliotibial tract/band**: Lateral thickening fascia lata
  - O: Tubercle iliac crest
  - I: Lateral condyle tibia
  - Insertion site tensor fascia lata muscle, portion of gluteus maximus muscle

Hip Flexors

- **Iliopsoas** I: Lesser trochanter
  - Iliacus
    - O: Iliac crest, iliac fossa, sacral ala, sacroiliac joint capsule
    - Femoral nerve
  - Psoas major
    - O: Lateral vertebral body and intervertebral discs T12-L5, all lumbar transverse processes
    - L1, L2, L3
  - Psoas minor
    - O: Lateral vertebral body T12, L1, and T12-L1 intervertebral disc
    - L1, L2
- Sartorius and pectineus

Femoral Triangle

- Anterior wall: Inguinal ligament
- Posterior wall: Adductor longus and pectineus muscles (medial), iliopsoas muscle (lateral)
- Medial border: Adductor longus muscle
- Lateral border: Sartorius muscle
- Apex: Crossing of adductor longus and sartorius muscles
- Contents: Femoral nerve and branches, femoral vessels, lymph node (node of Cloquet), femoral sheath
  - Structures lateral to medial at entrance
    - Mneumonic is NAVeL: Nerve, artery, vein, lymphatics
- Femoral artery/vein relationships
  - Entrance: Artery lateral
  - Apex: Artery anterior
- Femoral nerve branches within triangle
  - Saphenous nerve and nerve to vastus medialis are only branches to exit triangle
- **Femoral sheath**: Transversalis fascia covers vessels proximally
  - 3 compartments
    - Lateral: Artery
    - Middle: Vein
    - Medial: Lymph node (femoral canal)
- **Femoral canal**: Medial compartment femoral sheath
  - Anterior border: Inguinal ligament
  - Posterior border: Pubic bone
  - Medial border: Lacunar ligament
  - Lateral border: Femoral vein
  - Entrance: **Femoral ring**
    - Anterior border: Medial inguinal ligament
    - Posterior border: Superior pubic ramus
    - Medial border: Lacunar ligament
    - Lateral border: Septum between femoral canal and femoral vein
    - Open to peritoneal cavity
  - Contents: Lymphatic vessels and nodes (**node of Cloquet**), fat, connective tissue
    - Groin nodes divided into superficial and deep groups
    - Superficial nodes are situated around superficial veins
    - Deep nodes lie medial to femoral vein
    - Node of Cloquet in most medial of deep group; located medial to femoral vein at femoral triangle
  - Clinical note: **Femoral hernia**
    - Lateral and inferior to pubic tubercle
    - Travels femoral ring to femoral canal to saphenous opening to subcutaneous tissue

Adductor (Subsartorial or Hunter) Canal

- Fascial passageway for vessels midthigh
  - Boundaries are fascial surfaces of adjacent muscles
- Anteromedial border: Sartorius muscle
- Anterolateral border: Vastus medialis muscle
- Posterior border: Adductor longus and magnus muscles
- Entrance: Apex femoral triangle
- Exit: **Adductor hiatus**
  - Gap adductor magnus muscle between adductor portion and ischiocondylar portion of distal thigh
- Vessel passageway from thigh to popliteal fossa
- Contents: Femoral artery and vein, saphenous nerve
  - Nerve initially anterior to artery then medial
  - Artery anterior to vein
  - Descending geniculate artery arises in canal

ANATOMY IMAGING ISSUES

Imaging Recommendations

- US: Highest combined spatial and contrast resolution of different imaging modalities
  - Excellent for detecting and characterizing small lesions or fine anatomy, e.g., small nerve sheath tumors, internal vascularity of lesions, etc.
  - Real-time interrogation allows evaluation of muscle and tendon movement during dynamic maneuvers, e.g., for snapping hips, nerve impingement, etc.
  - Limited by depth of penetration
- CT: Best for evaluating osseous integrity
  - Soft tissue assessment inferior to US and MR
- MR: Highest contrast resolution especially when using fluid-sensitive or contrast-enhanced sequences
  - Most sensitive modality to detect tissue edema

ANTERIOR THIGH

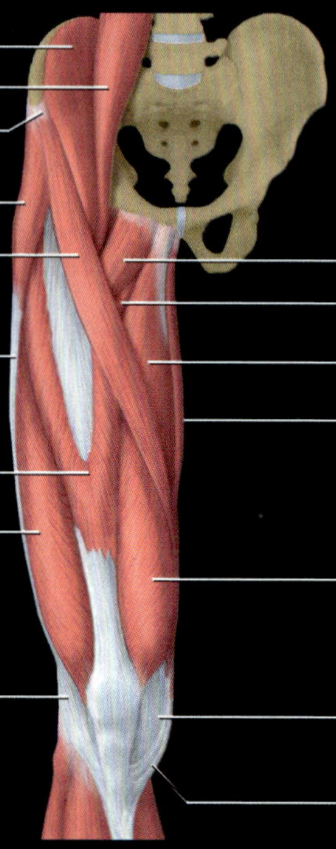

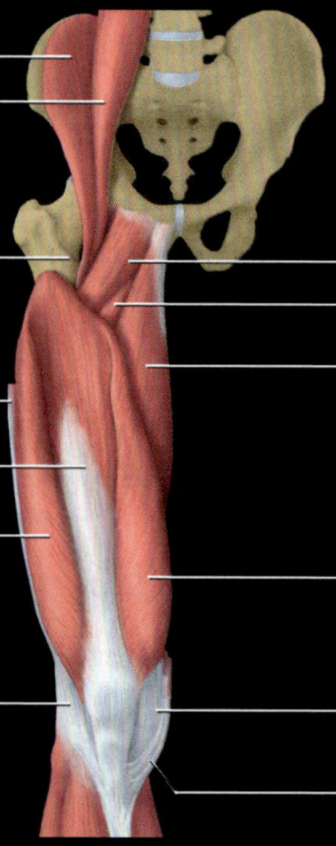

(Top) *Superficial muscles of the anterior thigh are shown. The oblique course of the sartorius muscle is easily appreciated. The adductor brevis muscle is deep to the adductor longus and pectineus muscles. The most medial muscle is the gracilis. The vastus lateralis and medialis muscles continue to their insertions as the lateral and medial patellar retinacula, respectively.* **(Bottom)** *Deep muscles of the anterior thigh are shown. The vastus intermedius muscle is deep to the rectus femoris muscle. The vastus intermedius muscle tendon blends with the rectus femoris tendon along its deep surface. The iliopsoas muscle dives deep after crossing over the pelvic brim. With removal of the sartorius muscle, a little more of the adductor brevis muscle is visible. The pes anserine tendons, the gracilis, sartorius, and semitendinosus have a conjoined insertion into the proximal medial tibia.*

OVERVIEW, ANTERIOR THIGH

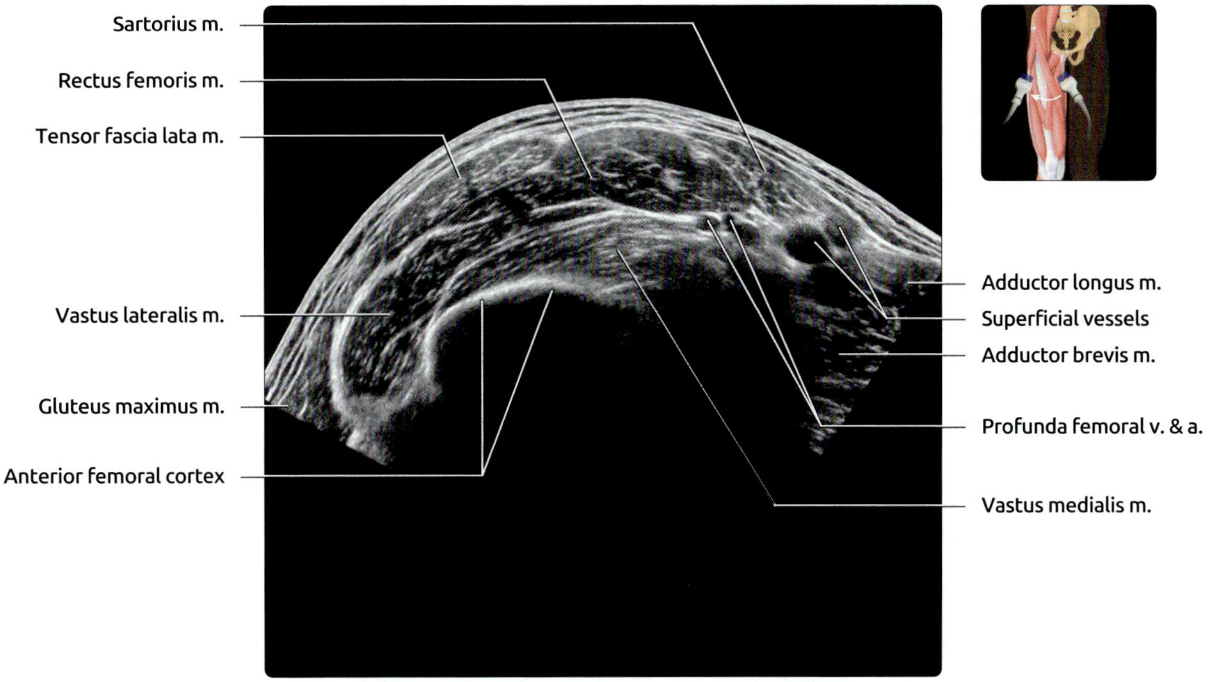

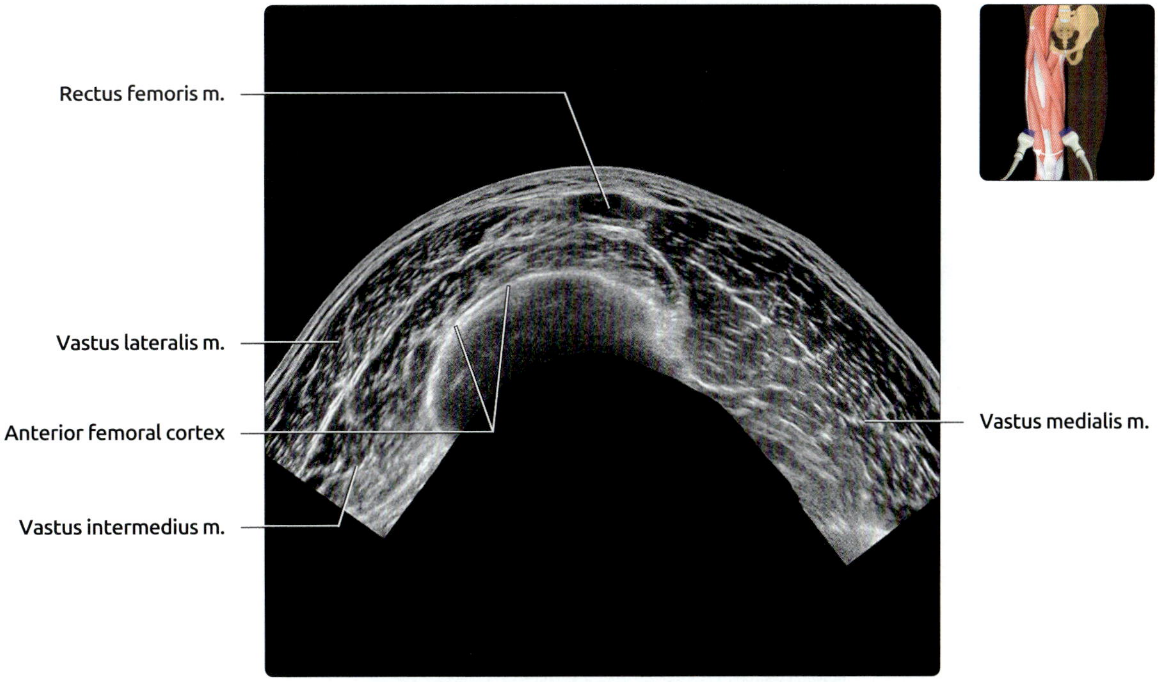

(Top) *Transverse panoramic view shows the anterior compartment of the upper thigh. The anterior compartment is composed of the quadriceps muscle, tensor fascia lata/iliotibial tract, and sartorius muscle. Medially, it is bordered by the femoral neurovascular bundle and adductor muscles. Laterally, it is bordered by the gluteal and hamstring muscles.* **(Bottom)** *In the lower thigh, the rectus femoris is much reduced in size and will continue inferiorly as the quadriceps tendon. The rest of the quadriceps continue as bulky muscles down to the suprapatellar level. They contribute directly to the quadriceps femoris tendon. The major neurovascular bundle has moved to the posterior compartment, and the only anterior compartment vessels are branches of the profunda femoris artery and vein.*

ANTERIOR COMPARTMENT, CENTRAL

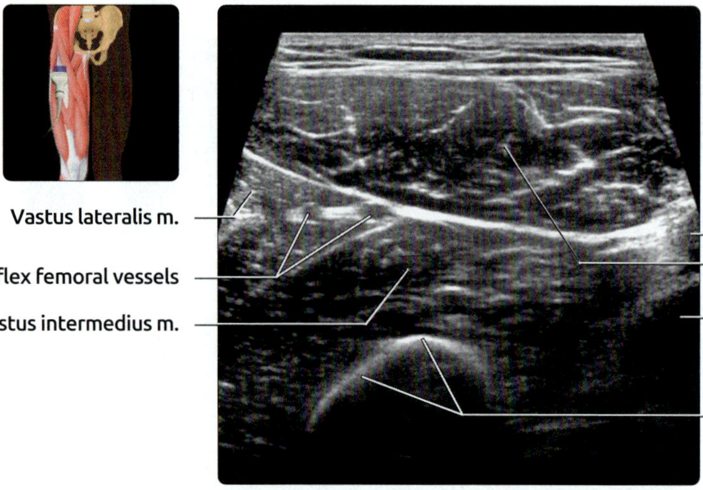

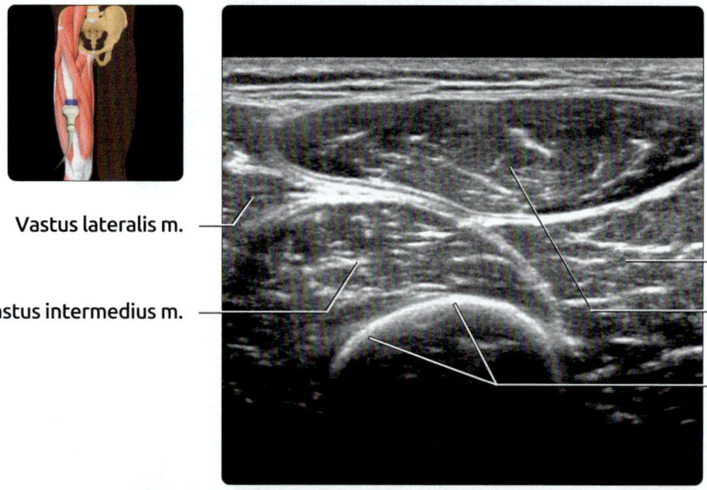

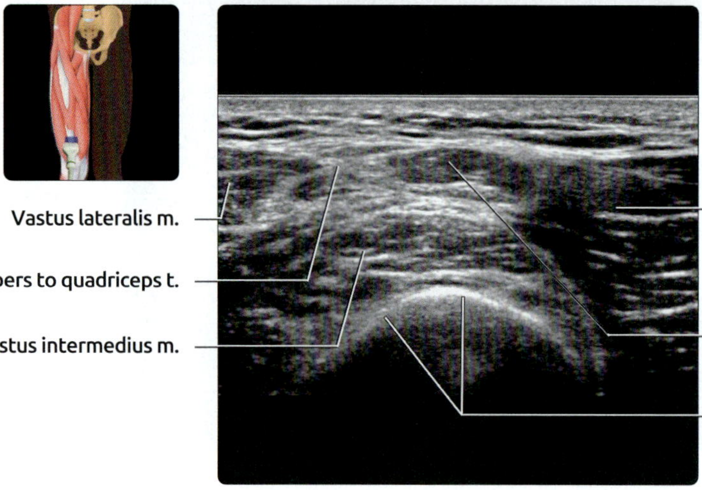

(Top) *Series of 3 transverse scans shows the central part of the anterior compartment. At the proximal level, the dominant (in terms of bulk) quadriceps muscle is the rectus femoris muscle. It is easily identified by its central position; it is lateral to the femoral neurovascular bundle (the femoral triangle). The lateral circumflex artery is important because together with the medial circumflex, its posterior counterpart, they supply the head of the femur.* **(Middle)** *Transverse scan is shown through the midthigh. The vastus medialis and intermedius have originated, and at this level, all 4 quadriceps muscles can be seen.* **(Bottom)** *Transverse scan shows the distal thigh above the level of the suprapatellar pouch. The rectus femoris has become a central tendon, which receives contributions from the vasti muscles.*

ANTERIOR COMPARTMENT, LATERAL

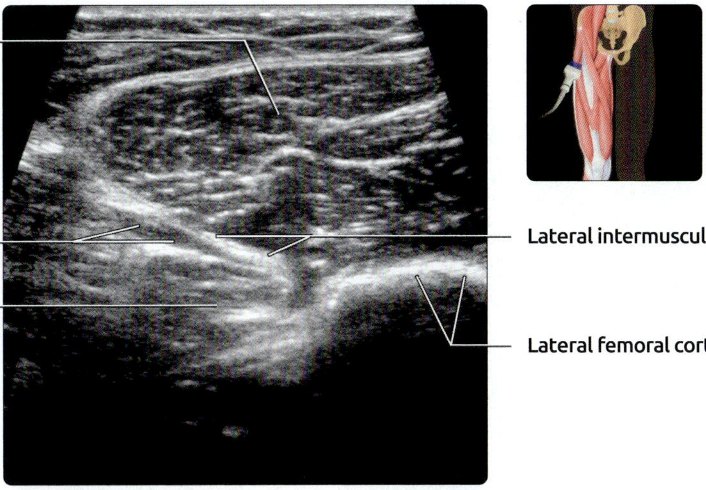

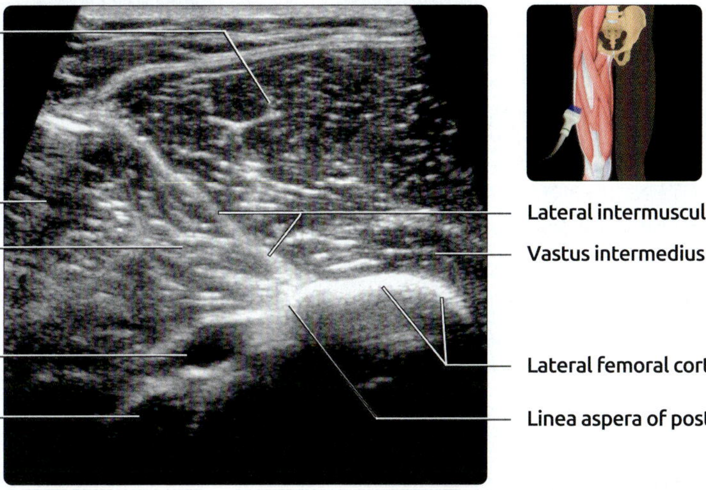

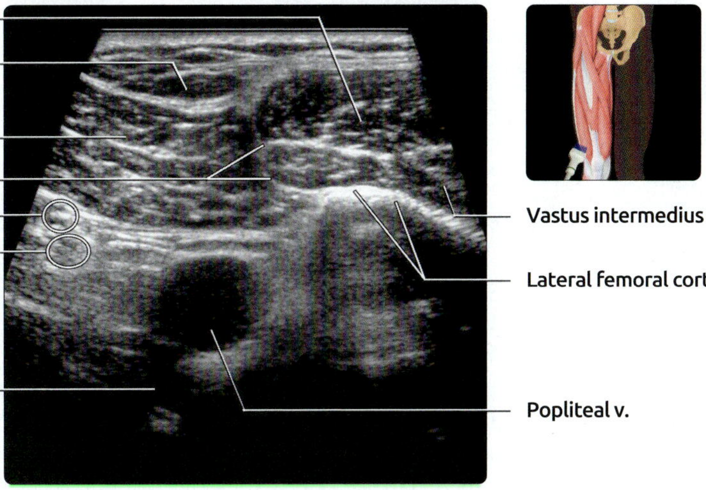

(Top) *Series of transverse scans through the lateral aspect of the thigh is shown. Proximally, the vastus lateralis muscle is dominant and borders with the gluteus maximus muscle (which inserts onto the posterior surface of the femur). The lateral intermuscular septum separates anterior from posterior compartments laterally and also acts as conduit for vessels.* **(Middle)** *Further inferiorly, the vastus lateralis borders the biceps femoris. The linea aspera is a distinct protrusion from the posterior femoral shaft where many muscles originate and insert, including the lateral intermuscular septum. On ultrasound, it is consistently seen and therefore serves as a useful bony landmark.* **(Bottom)** *At the superior popliteal level, the vastus lateralis muscle becomes less bulky (the reverse occurs with the vastus intermedius muscle), as does the bordering long head of the biceps muscle. The sciatic nerve divides into tibial and common peroneal branches.*

MEDIAL AND POSTERIOR THIGH

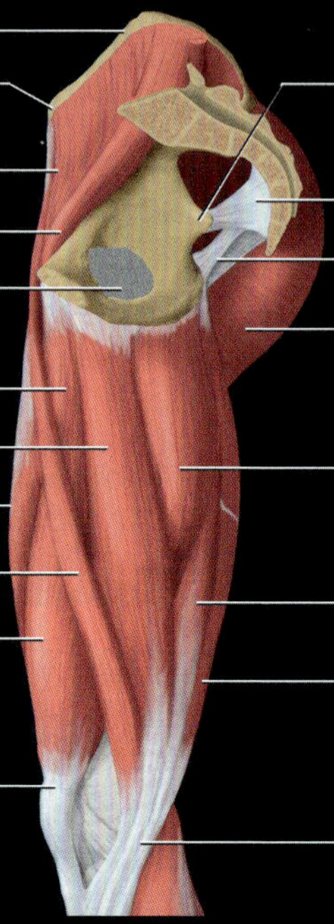

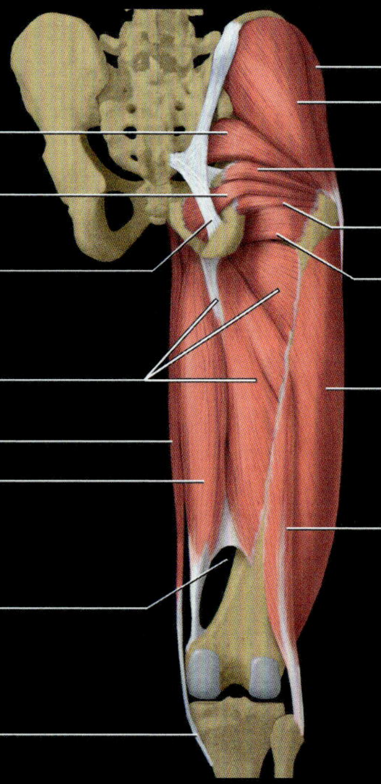

(Top) *Medial muscles of the thigh are shown. The gracilis muscle has a thin profile when viewed from the front; however, when viewed from the side, it is quite broad. The semimembranosus muscle runs along the deep surface of the semitendinous muscle and inserts onto the tibia posterior to the pes anserine tendons. Its insertion is hidden on this image. The iliopsoas muscle courses over the pelvic brim on its course to the lesser trochanter.* **(Bottom)** *Deep muscles of the posterior thigh are shown. With removal of the hamstring muscles, the expansive adductor magnus muscle is visible. The separation of its 2 heads in the distal thigh forms the adductor hiatus.*

OVERVIEW, MEDIAL THIGH

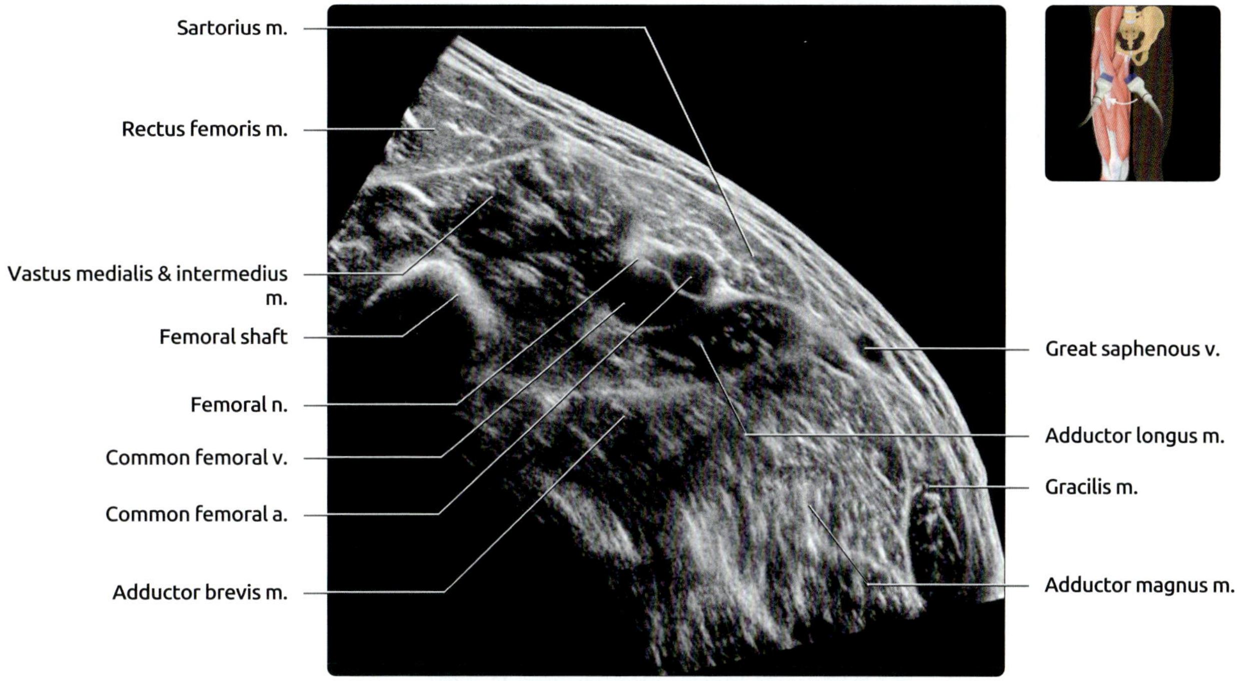

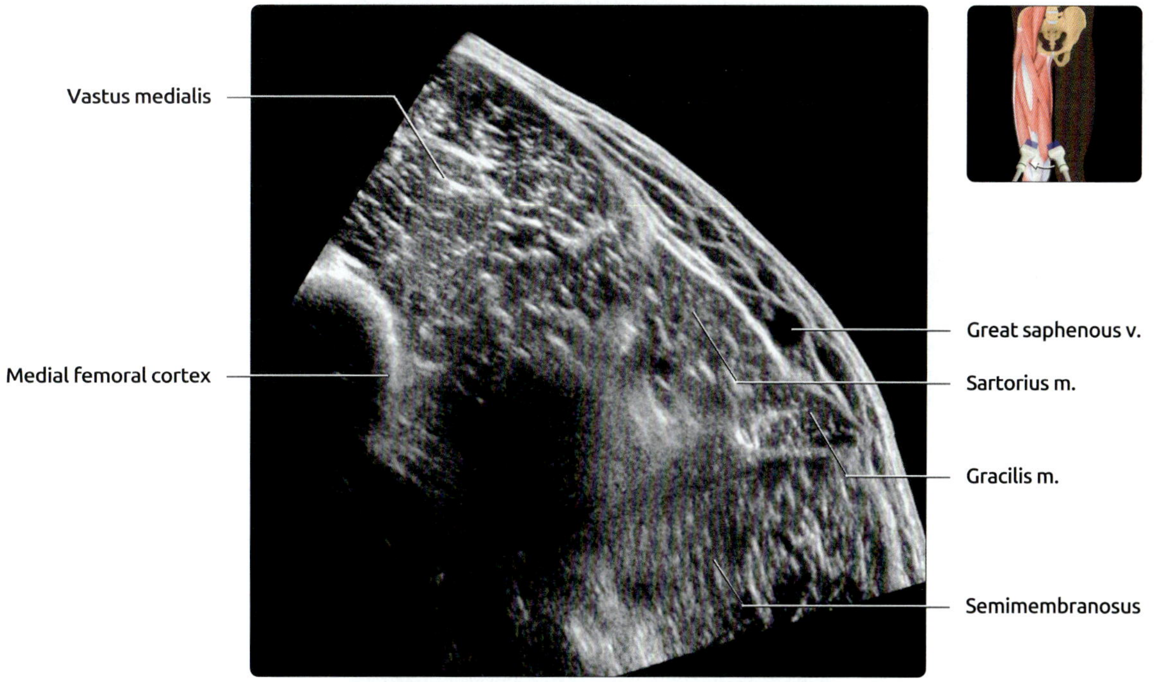

(Top) *Transverse panoramic view shows the medial compartment at the upper thigh level. At this level, all 3 adductor muscles can be seen, together with the gracilis muscle. This compartment is bordered anteriorly by the femoral triangle, which continues inferiorly as the adductor canal. Posteriorly, the medial compartment is bordered by the semimembranosus muscle.* **(Bottom)** *Transverse panoramic scan shows the medial aspect of the lower thigh. At this level, the adductor muscles have inserted superiorly; therefore, the medial compartment is much smaller and composed mainly of the gracilis muscle. The adductor hiatus, which is above this level, provides an entrance for the femoral vessels into the popliteal fossa.*

MEDIAL THIGH, ADDUCTOR CANAL

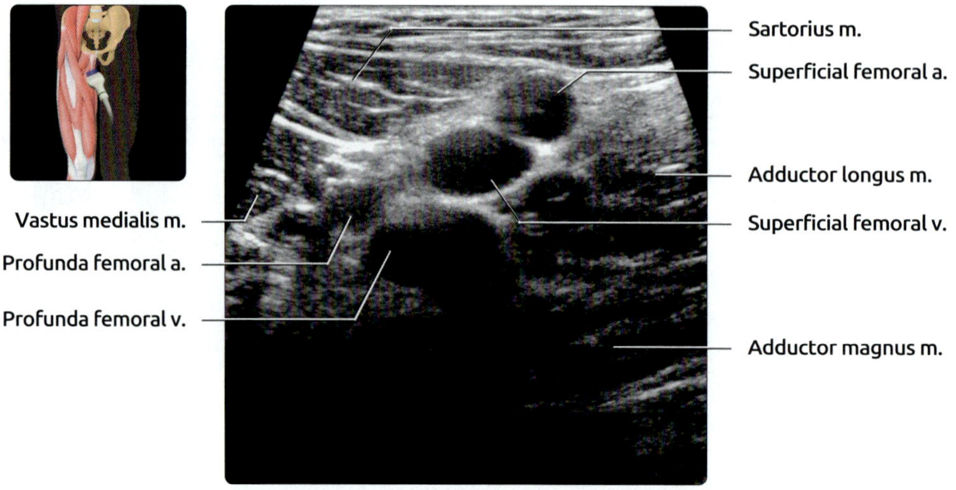

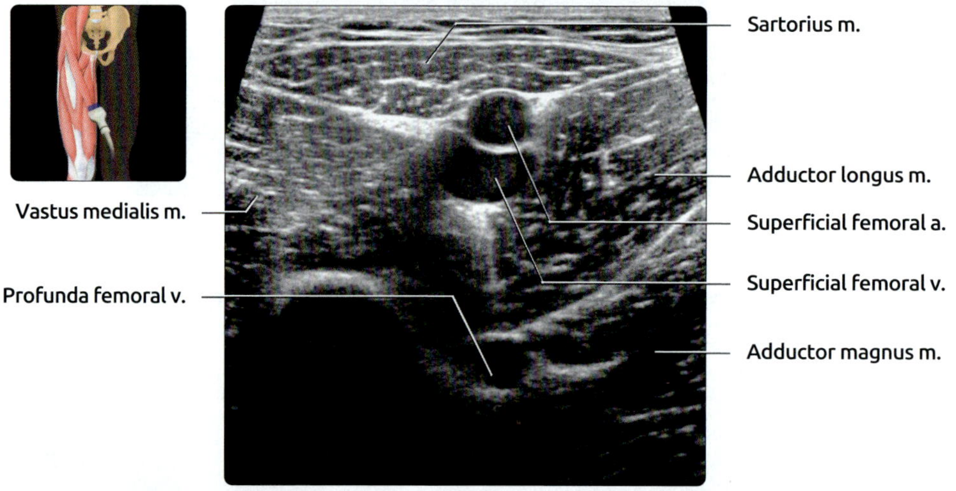

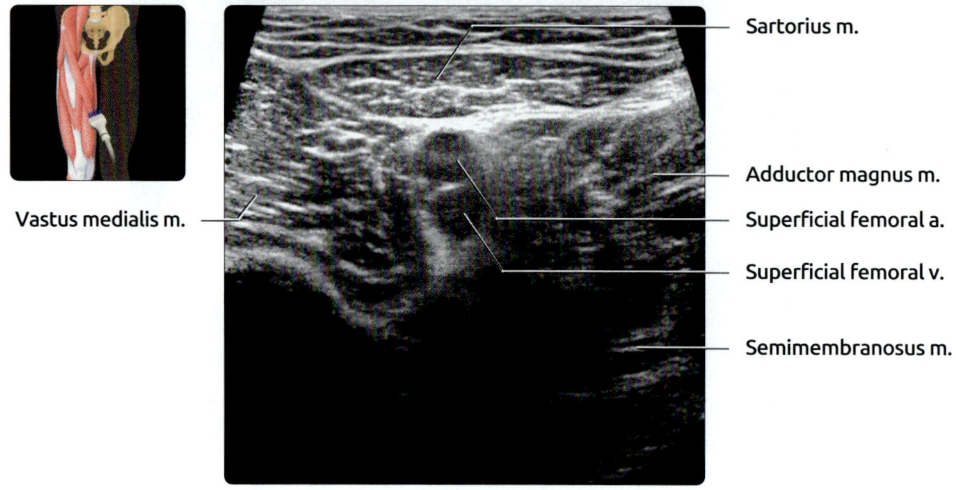

(Top) *Transverse scan shows the anterior part of the medial thigh. Above the adductor canal, the apex of the femoral triangle contains bifurcation of the common femoral vessels. The profunda will run deep to supply the thigh muscles and femur. The superficial femoral vessels continue into the adductor canal.* **(Middle)** *Transverse scan more inferiorly shows that the superficial femoral vessels are now under the cover of the sartorius muscle, indicating that they are in the adductor canal.* **(Bottom)** *Transverse scan shows the medial aspect of the midthigh. The common adductor canal continues medially and more deeply. At this level, the medial wall of the adductor canal changes from the adductor longus muscle to the adductor part of the adductor magnus muscle.*

MEDIAL THIGH

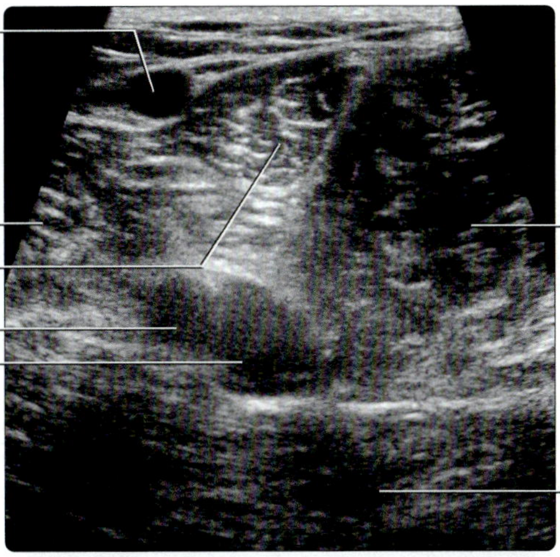

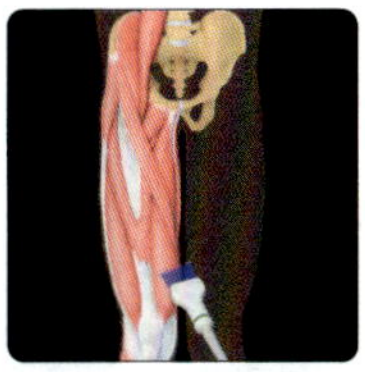

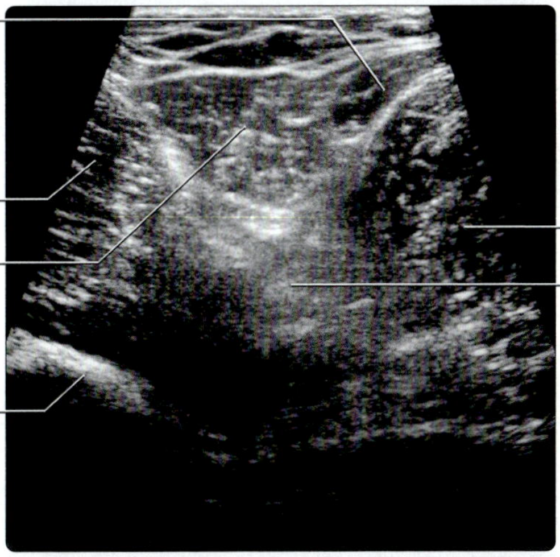

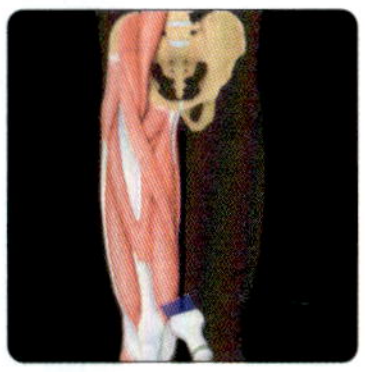

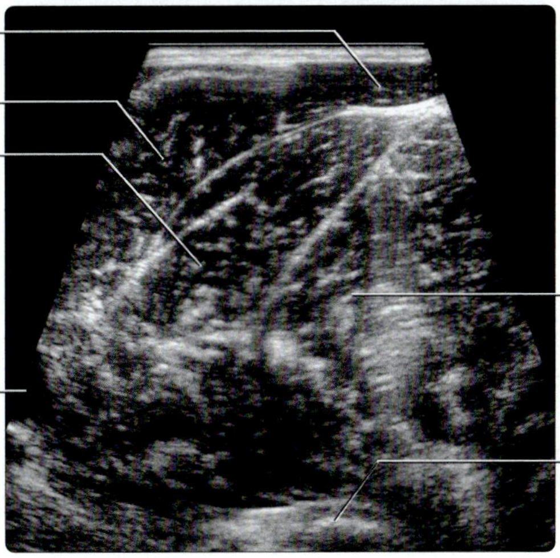

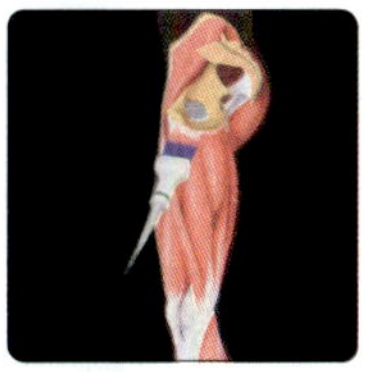

(Top) *Transverse scan using an anteromedial approach is shown. At this level, the superficial femoral vessels enter the popliteal fossa via the adductor hiatus, which is a gap in the adductor magnus muscle, between the adductor and ischiocondylar parts before their insertions (on the medial supracondylar line and adductor tubercle, respectively).* **(Middle)** *More inferiorly, only the gracilis muscle and the ischiocondylar part of the adductor magnus (acting like a hamstring) are left in the medial compartment. This space is encroached on by the vastus medialis muscle anteriorly and semimembranosus muscle posteriorly.* **(Bottom)** *Using a medial approach (from the inner thigh), the deeper muscles can be demonstrated. The adductor longus and brevis muscle plus the gracilis muscle form an anterior division, while the bulky adductor magnus muscle forms the posterior division.*

OVERVIEW, POSTERIOR

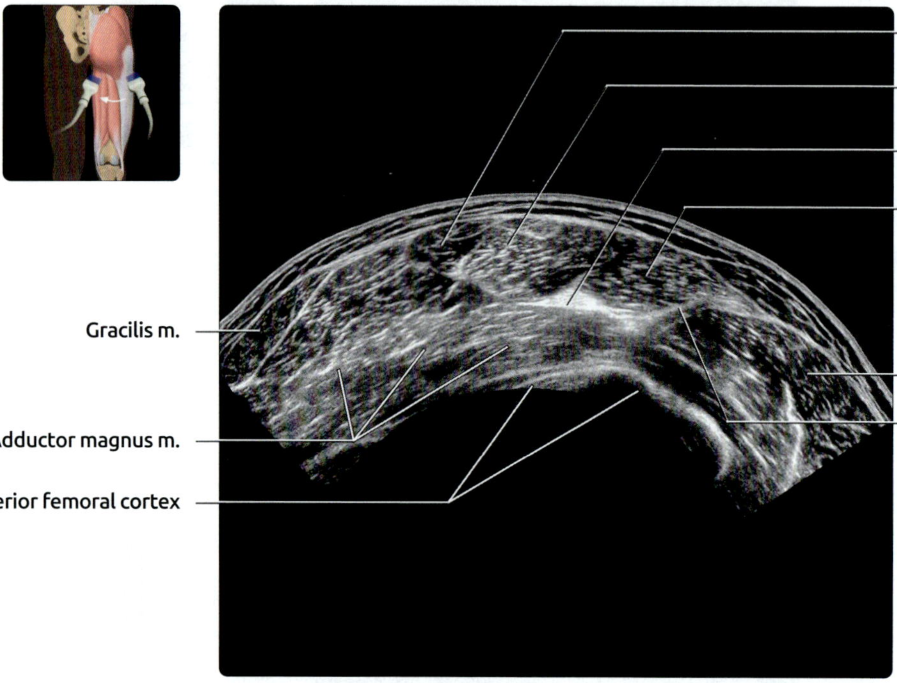

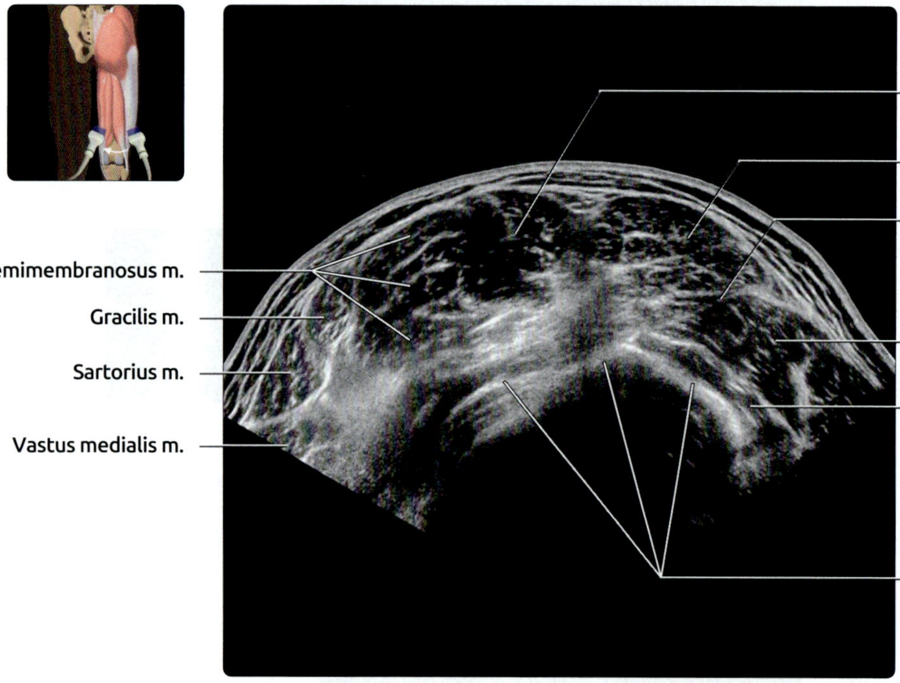

(Top) *Transverse panoramic scan shows the posterior compartment at the upper thigh level. At this level, the hamstring muscles form a small proportion of the posterior muscle bulk. The main contributions come from adductors (the medial compartment) and vastus lateralis (the anterior compartment). The sciatic nerve runs under the cover of the biceps femoris muscle.* **(Bottom)** *Transverse panoramic scan shows the posterior aspect of the distal thigh. At this more inferior level, the hamstrings form a much larger proportion of the posterior muscle bulk. In particular, the semimembranosus becomes the dominant muscle on the posteromedial side, as most of the adductors have inserted onto the femur above this level.*

POSTERIOR THIGH

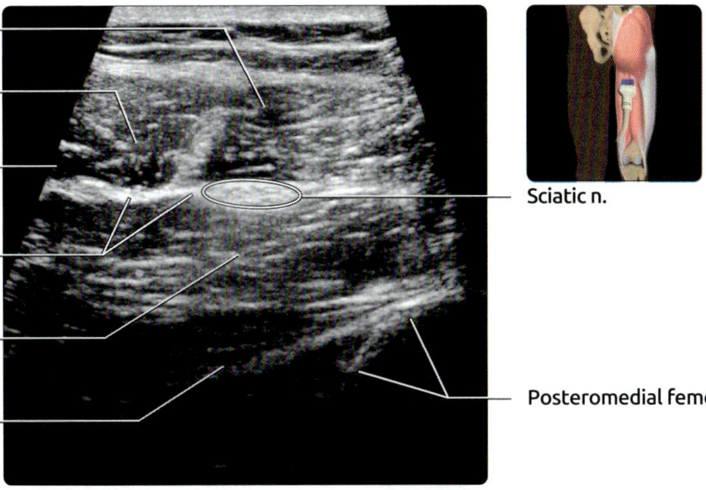

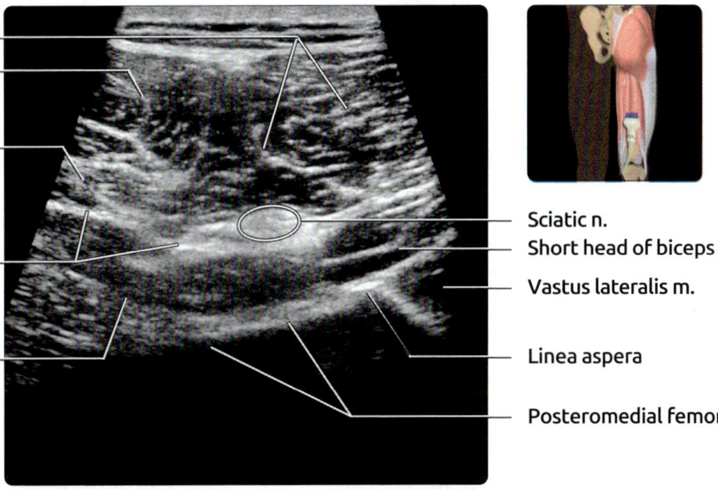

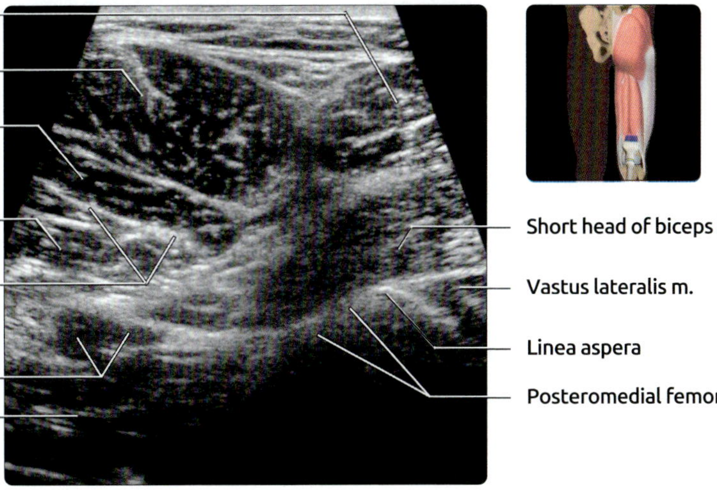

(Top) *Series of 3 transverse scans shows the central portion of the posterior thigh. After their origin from the ischial tuberosity, the semitendinosus and semimembranosus muscles run on the medial side, and the biceps femoris muscle runs on the lateral side. The sciatic nerve is under the cover of the biceps femoris for most of its way to the popliteal fossa.* **(Middle)** *At the midthigh level, the hamstrings are bulkier and the adductors smaller. The 2 groups are divided by the posteromedial intermuscular septum.* **(Bottom)** *Further inferiorly, the popliteal vessels enter via the adductor hiatus into the popliteal fossa. The superior aspect of the popliteal fossa is lined by the semimembranosus and semitendinosus muscle medially and the biceps femoris laterally.*

IMAGING ANATOMY

Femoral Vessels

- Enter thigh deep to inguinal ligament, midpoint between anterior superior iliac spine and symphysis pubis
  - Change from external iliac vessels to common femoral vessels
- Upper thigh: Vessels within femoral triangle
  - Enter: Artery lateral to vein
  - Exit: Artery anterior
- Midthigh: Vessels within adductor canal
  - Enter: Artery anterior to vein
  - Exit: Artery anterior
- Distal thigh: Exit adductor canal via adductor hiatus, enter popliteal fossa
- **Common femoral arterybranches**
  - **Superficial epigastric, superficial circumflex iliac, superficial external pudendal** arise anteriorly
  - **Deep external pudendal** arises medially
    - May branch from medial circumflex femoral
  - Divides into superficial and deep branches
  - **Superficial femoral artery**
    - Branch: **Descending genicular**
  - **Deep femoral (profunda femoris)**
    - Arises laterally in femoral triangle
    - Dives between pectineus and adductor longus muscles
    - Medial to femur, deep to adductor longus muscle
    - Branches in femoral triangle: **Medial circumflex femoral** (main supply femoral head and neck), lateral circumflex femoral, muscular branches
    - Branches in adductor canal: 3 perforating branches, descending genicular
    - Terminal branch: 4th perforating artery
- **Femoral vein**: Travels with artery
  - Tributaries: Lateral circumflex femoral vein, deep femoral vein and its associated perforating veins, descending genicular, lateral circumflex femoral, medial circumflex femoral, deep external pudendal, greater saphenous veins
  - **Greater saphenous vein**
    - Longest vein in body
    - Toes to saphenous opening (fascia lata)
    - Tributaries: Accessory saphenous, superficial epigastric, superficial circumflex femoral, superficial external pudendal veins

Femoral Nerve

- Nerve contributions from L2, L3, L4, L5
- Largest branch of lumbar plexus
- Exits plexus at lower psoas muscle
- Travels in groove between psoas and iliacus muscles
- Exits pelvis beneath inguinal ligament, lateral to femoral vessels, enters femoral triangle
- Multiple branches in femoral triangle
  - Muscular branches: To pectineus, sartorius, rectus femoris, vastus lateralis, vastus medialis, vastus intermedius muscles
  - Cutaneous nerves: Anterior femoral cutaneous, saphenous
    - Saphenous nerve exits triangle, enters adductor canal
  - Articular branches hip and knee

Sciatic Nerve

- Nerve contributions from: L4, L5, S1, S2, S3 levels
- Largest nerve in body
- 2 nerves in 1 sheath
  - Tibial nerve (medial) and common peroneal nerve (lateral)
  - Separate in popliteal fossa
- Course
  - Exits pelvis inferior to piriformis muscle
    - Anomalous relationship in 10%
  - Deep to gluteus maximus muscle, more inferiorly deep to biceps femoris muscle
    - Crosses over superior gemellus, obturator internus, inferior gemellus, quadratus femoris, adductor magnus muscles
    - Midway between ischial tuberosity and greater trochanter in transverse plane
  - Runs deep to long head of biceps femoris muscle after exiting gluteal region
- Branches arising in thigh: Articular to hip, nerves to hamstring muscles
- Supply
  - In thigh: Biceps femoris, semitendinosus, semimembranosus, ischiocondylar portion of adductor magnus muscle
- **Tibial nerve**: Larger division of sciatic nerve
  - Supplies posterior femoral muscles except short head biceps femoris muscle
- **Common peroneal nerve**
  - Oblique lateral course with biceps femoris muscle
  - Supplies short head biceps femoris muscle

ANATOMY IMAGING ISSUES

Imaging Recommendations

- Ultrasound is unique in that it assesses both morphology and functional response of vessels
  - Complete examination should include interrogation with color Doppler
  - In cases with stenosis, reflux, or reverse flow, spectral Doppler helpful to demonstrate waveform and document flow velocity
  - Dynamic maneuvers, such as transducer compression, breath holding, Valsalva, and calf augmentation are helpful to assess patency of veins on color and spectral Doppler
  - Intravenous contrast is not necessary
- High spatial resolution makes ultrasound suitable for assessing superficial nerves
  - Disruption of parallel continuity of nerve bundles suggests pathology
  - Transducer angulation may need to be changed to avoid anisotropy

THIGH ARTERIES

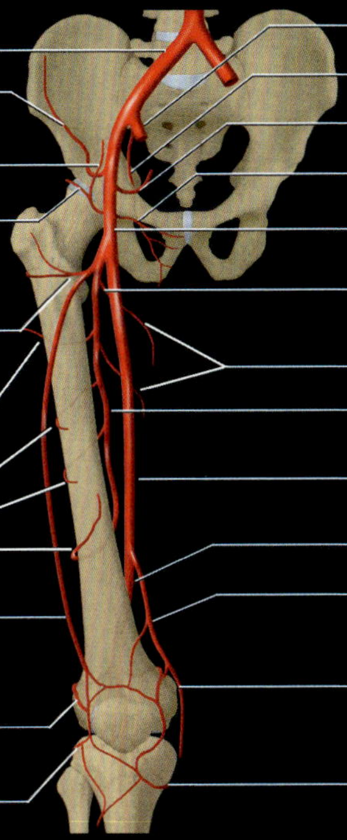

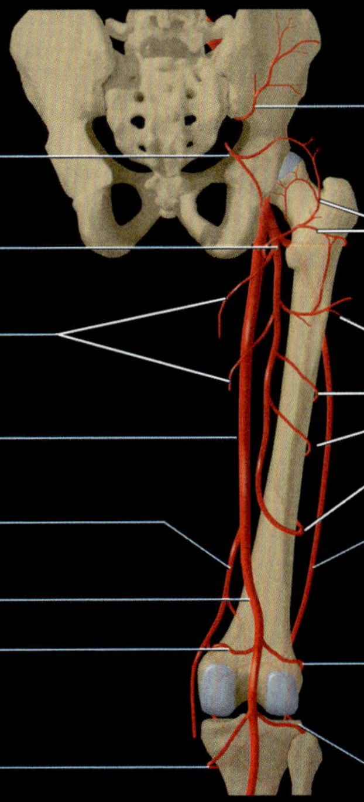

(Top) *Anterior view shows the external iliac, femoral, and popliteal arteries and their branches.* **(Bottom)** *Posterior view shows the femoral and popliteal arteries and the superior and inferior gluteal arteries of the greater sciatic notch. All of the main arteries and veins can be readily evaluated with ultrasound. CT angiography or MR angiography are the best investigations to provide an imaging roadmap of the main arteries in the lower limb, and these investigations have replaced standard angiography in this respect. Standard DSA is still the best investigation to demonstrate the smaller arteries of the lower limb (such as those around of the feet or around the hip).*

THIGH VEINS

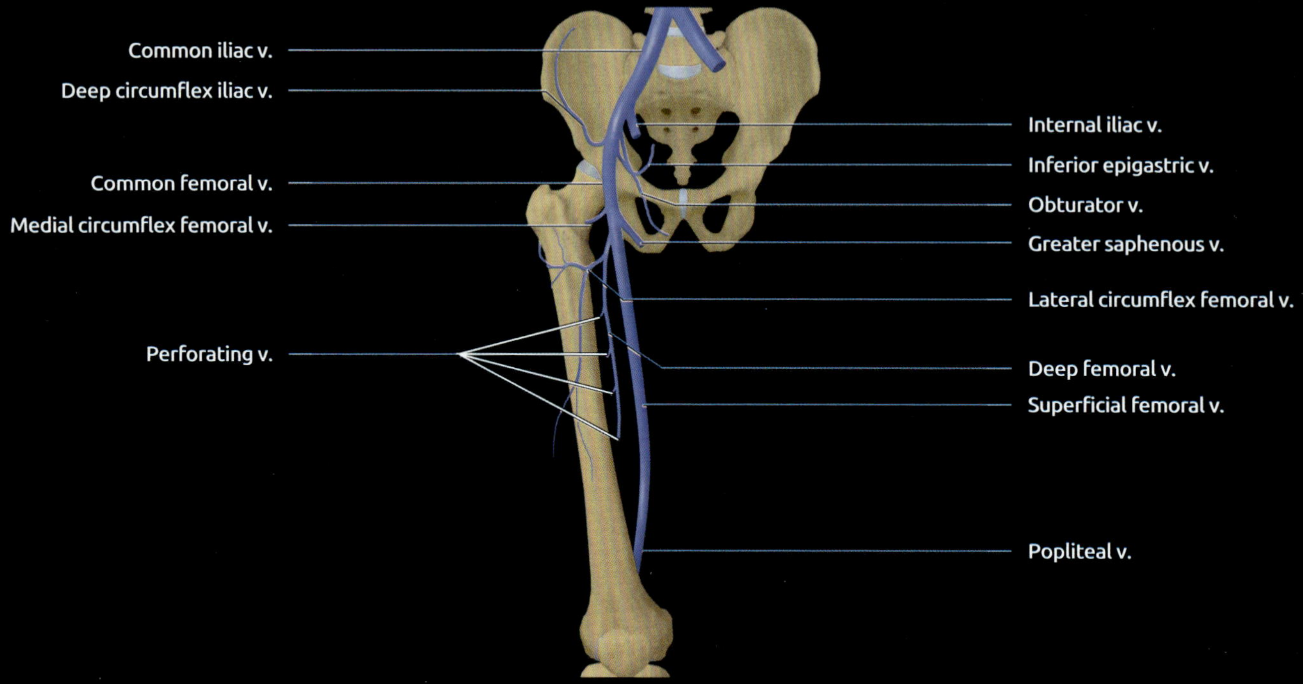

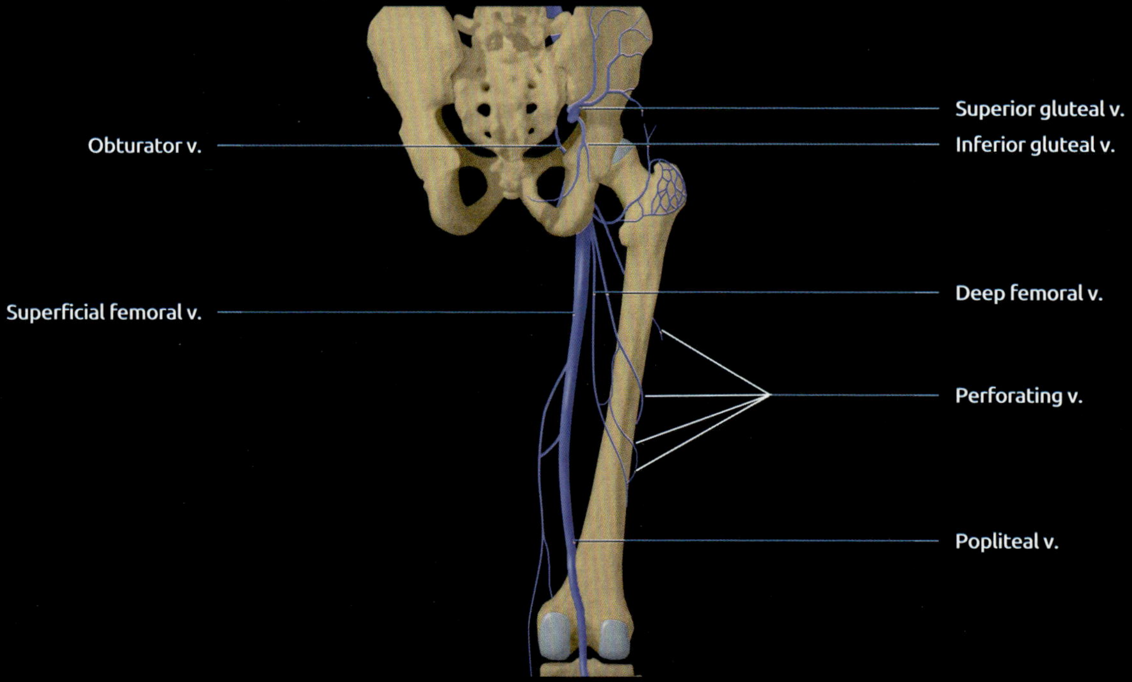

(Top) *Anterior view shows the veins of the thigh. The veins typically follow the arterial tree. The main venous drainage is the superficial femoral vein, which is actually a deep venous structure. Note the entrance of the greater saphenous vein into the femoral vein; the saphenous venous system has no accompanying artery.* **(Bottom)** *Posterior view shows the veins of the thigh and buttocks. The veins generally follow the arterial tree, although more variability is seen within the venous system.*

THIGH NERVES

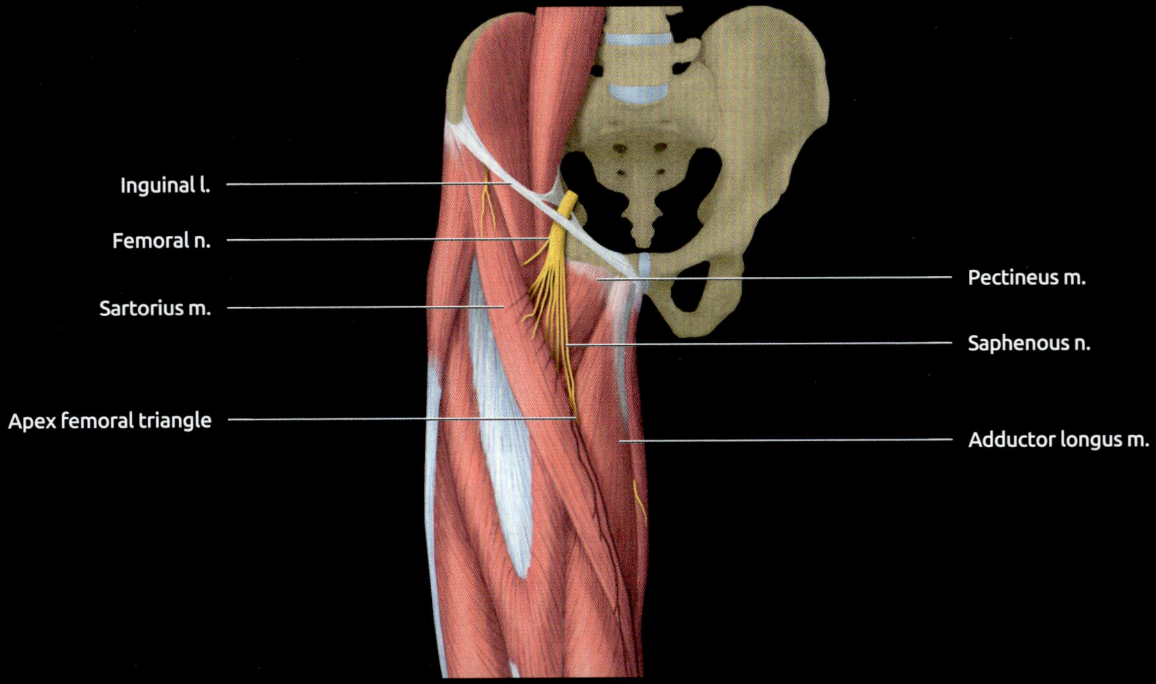

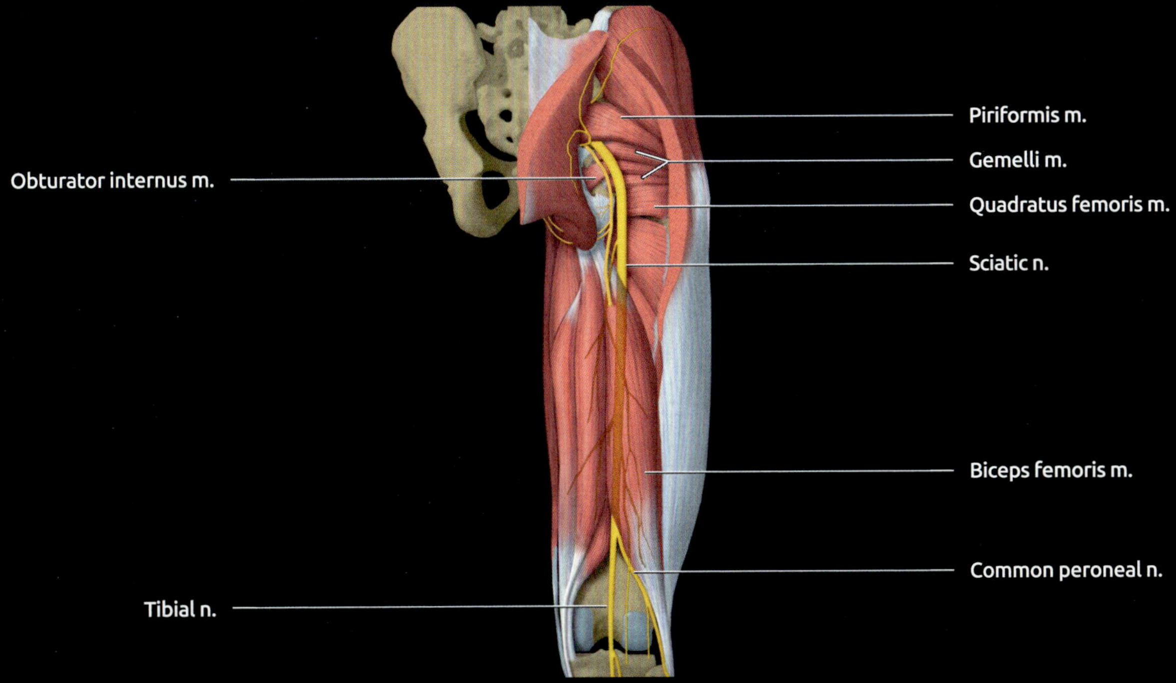

(Top) *Femoral nerve is shown. After emerging from the groove between the psoas and iliacus muscles, the nerve passes under the inguinal ligament to enter the femoral triangle. It immediately branches into muscular, cutaneous, and articular nerves. The only nerve to pass through the femoral triangle into the adductor canal is the saphenous nerve.* **(Bottom)** *The sciatic nerve enters the lower extremity by passing under the inferior border of the piriformis muscle. The sciatic nerve passes posterior to the external rotator muscles and then courses deep to the biceps femoris muscle. It separates into the tibial and common peroneal nerves in the distal thigh. The tibial nerve bisects the popliteal fossa. The common peroneal nerve follows the biceps femoris muscle around the fibular head.*

LONGITUDINAL THIGH VESSELS

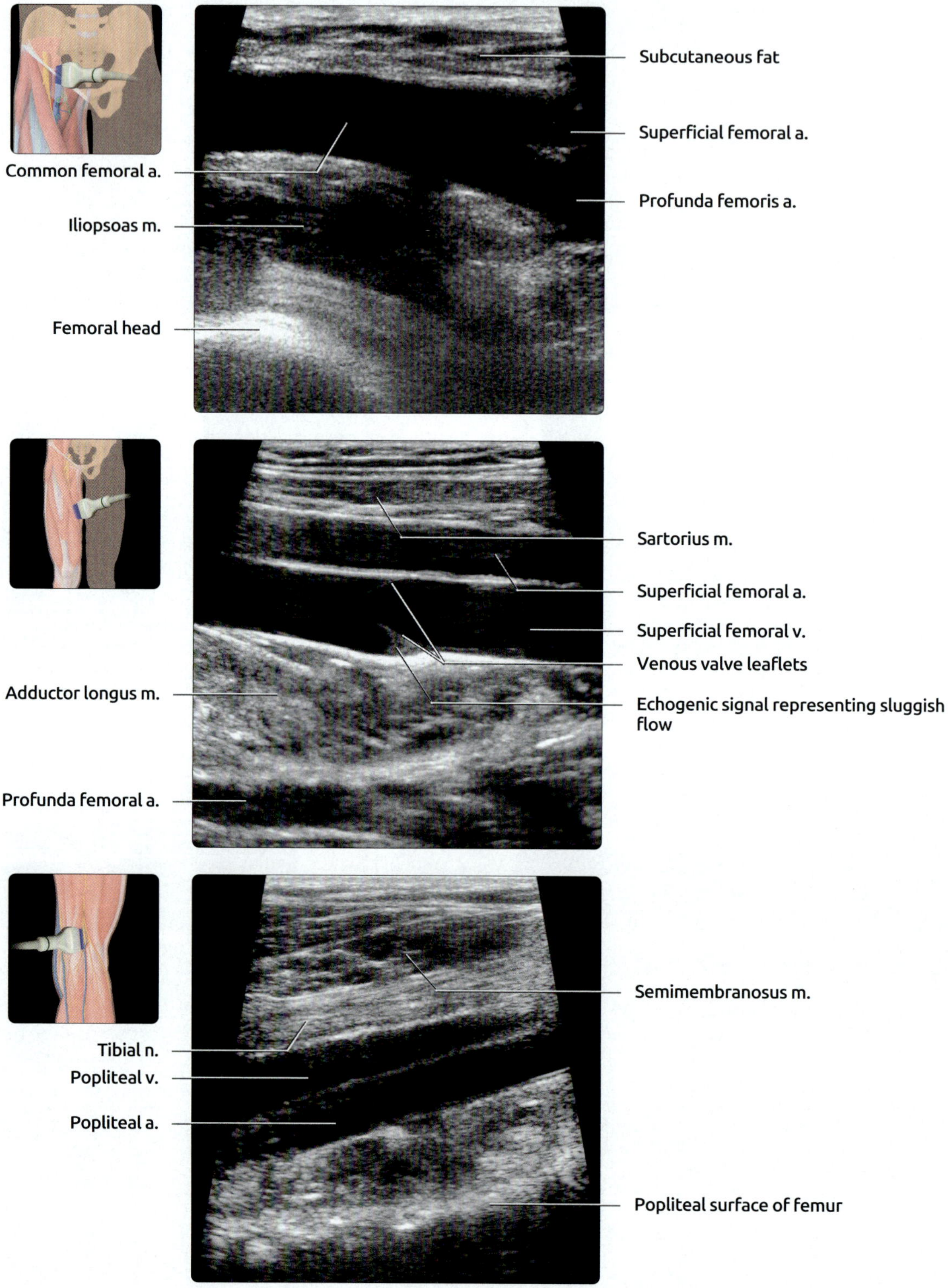

(Top) *Sagittal scan shows the common femoral artery and its bifurcation into the superficial femoral artery and the profunda femoral artery. The superficial femoral artery continues into the femoral triangle and adductor canal before ending at the adductor hiatus to become the popliteal artery. The profunda femoris artery supplies the muscles of the thigh.* **(Middle)** *Sagittal scan shows the upper adductor canal. The superficial femoral vessels are covered by the sartorius muscle. The high spatial resolution of ultrasound allows venous valves and sluggish flow to be visualized (the latter as increased echogenicity).* **(Bottom)** *Sagittal scan shows the popliteal vessels using a posterior approach. Note the reversed relationship between the vessels: The vein now lies more superficially than the artery.*

TRANSVERSE THIGH VESSELS

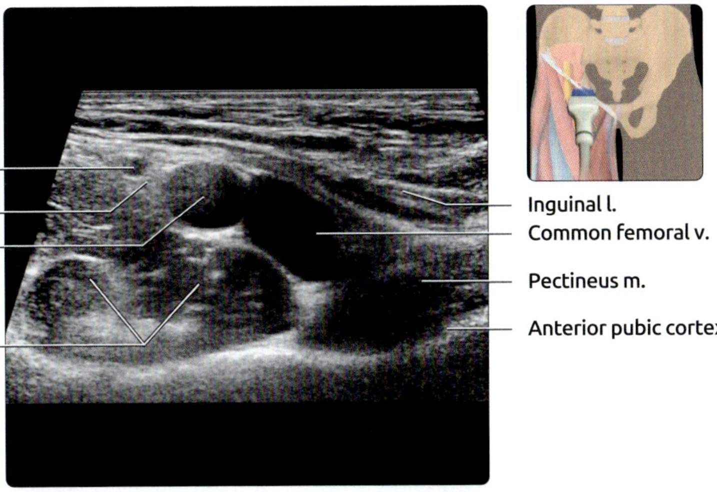

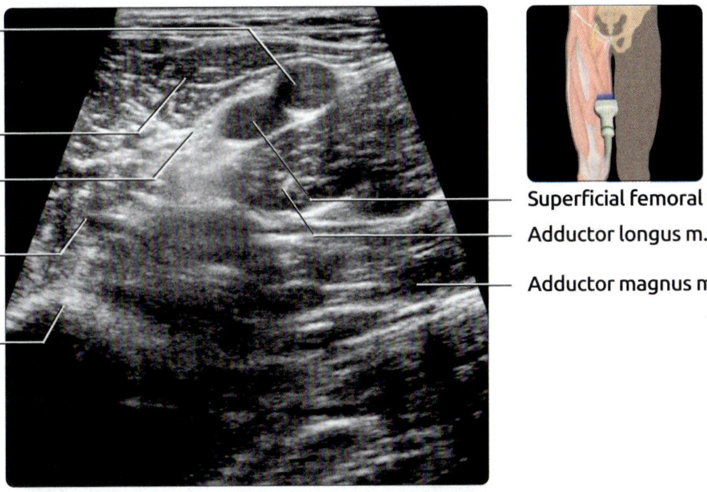

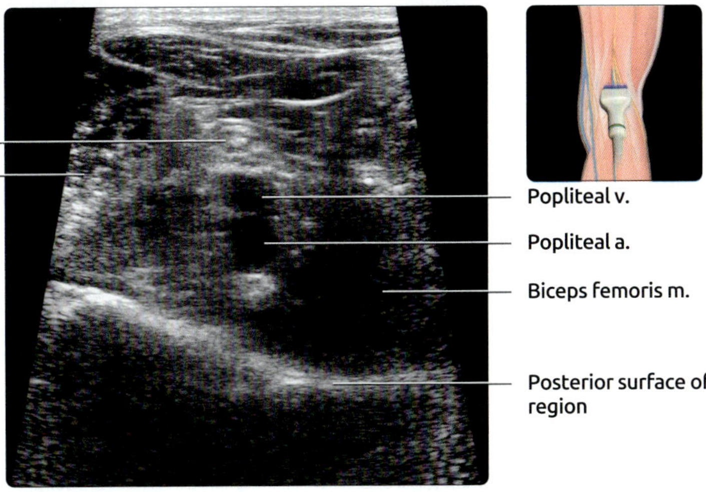

(Top) *Transverse scan shows the common femoral vessels emerging from under the inguinal ligament. The femoral nerve lies lateral to the common femoral artery. The iliopsoas muscle and pectineus muscle form the floor.* **(Middle)** *Transverse scan shows the lower aspect of the femoral triangle. The sartorius begins to cover the femoral vessels. The vastus medialis and the adductor longus muscles form the side walls of the femoral triangle. More inferiorly, the common femoral vessels bifurcate into the superficial and profunda branches.* **(Bottom)** *Transverse scan shows the posterior thigh at the upper level of the popliteal fossa. After passing through the adductor hiatus, the superficial femoral vessels enter the popliteal fossa. The hamstring muscles form the side walls of this fossa.*

COLOR DOPPLER THIGH ARTERY

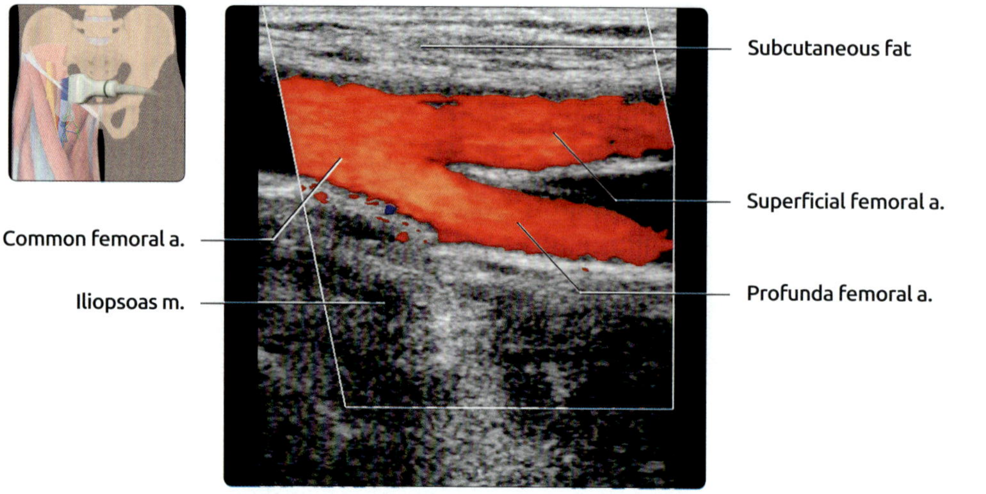

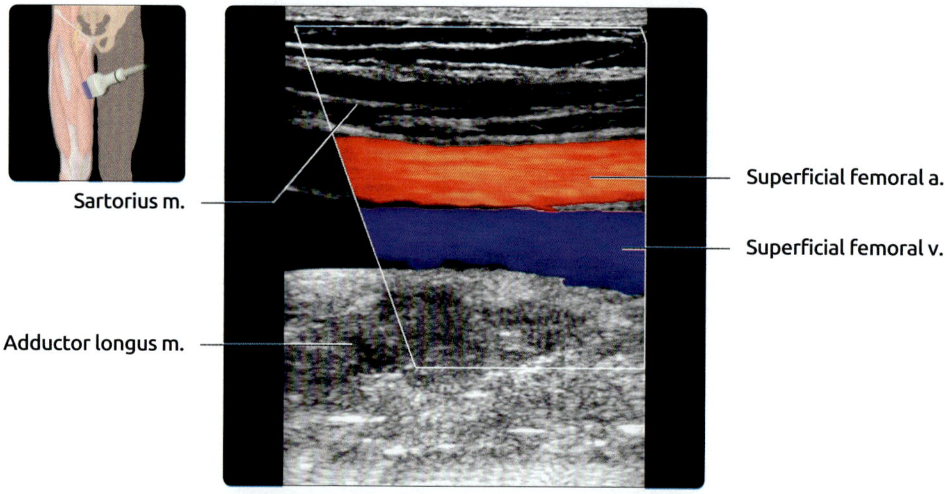

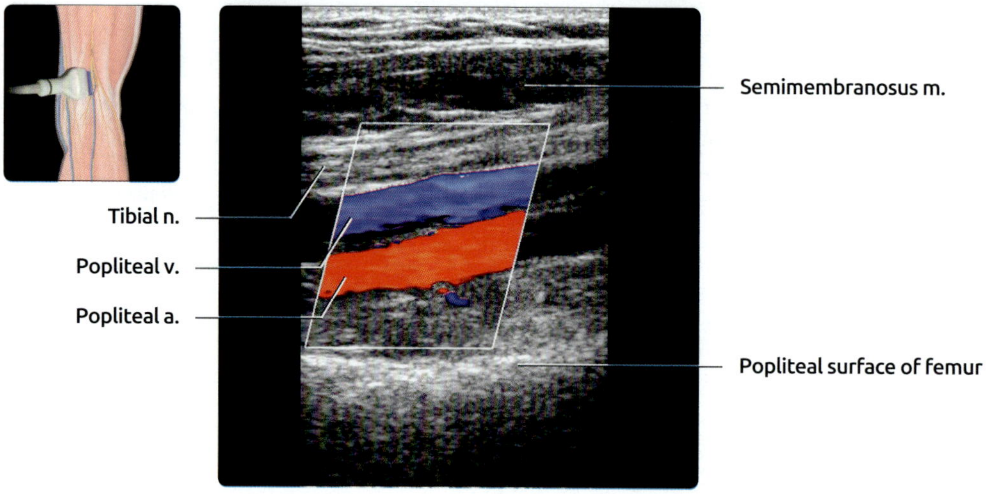

(Top) *Sagittal color Doppler scan shows the common femoral bifurcation. Color, representing flow, should fill the entire lumen with appropriate scan settings (as seen in this image).* **(Middle)** *Sagittal color Doppler scan shows the superficial femoral vessels in the upper adductor canal. The color of flow is direction dependent (the Doppler effect) and here demonstrates the opposite direction of flow between the artery (red, away from the trunk) and vein (blue, toward the trunk). Again, the color should fill the entire lumen (red superficial to blue).* **(Bottom)** *Sagittal color Doppler scan shows the popliteal vessels using a posterior approach. The vein now lies superficial to the artery (blue superficial to red).*

SPECTRAL DOPPLER THIGH ARTERIES

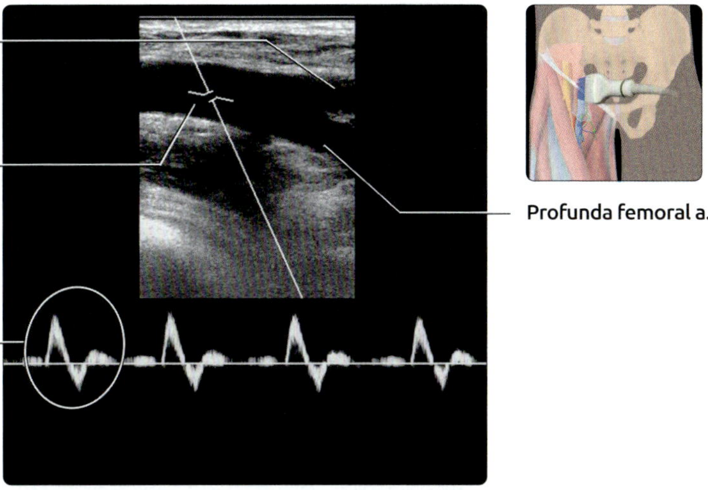

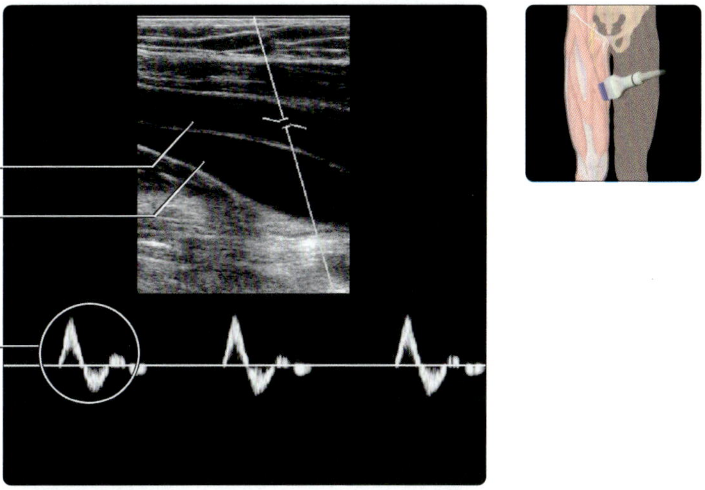

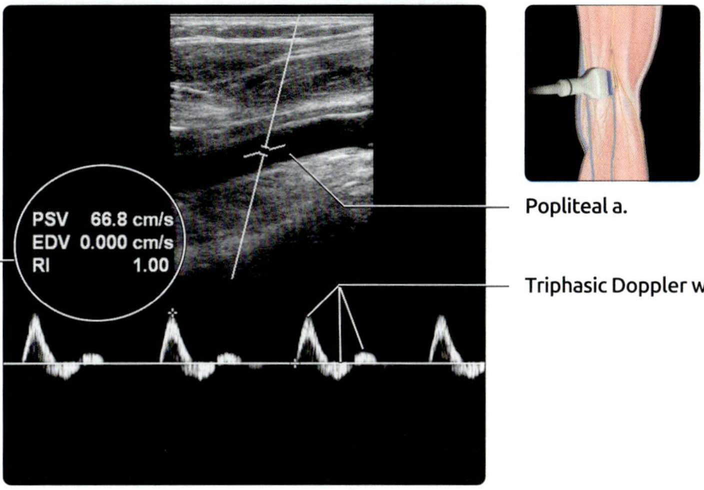

(Top) *Spectral Doppler trace at the common femoral artery is shown. The typical waveform for an artery should be triphasic: Forward sharp upstroke of forward flow (systolic), then reverse flow (early diastolic), and finally slow forward flow (late diastolic), as seen in this trace.* **(Middle)** *Spectral Doppler trace performed at the superficial femoral artery shows a similar waveform and slight decrease in the peak systolic flow. In general, as one moves further away from the heart, the peak velocity decreases.* **(Bottom)** *The typical arterial waveform is again seen in a spectral Doppler trace of the popliteal artery.*

SAPHENOFEMORAL JUNCTION

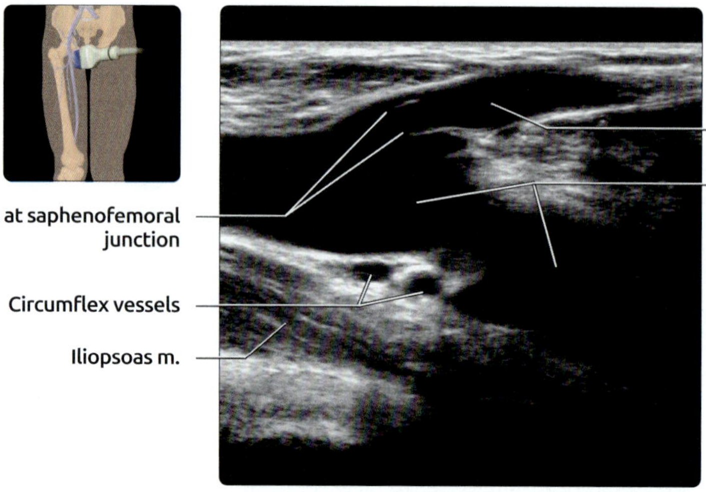

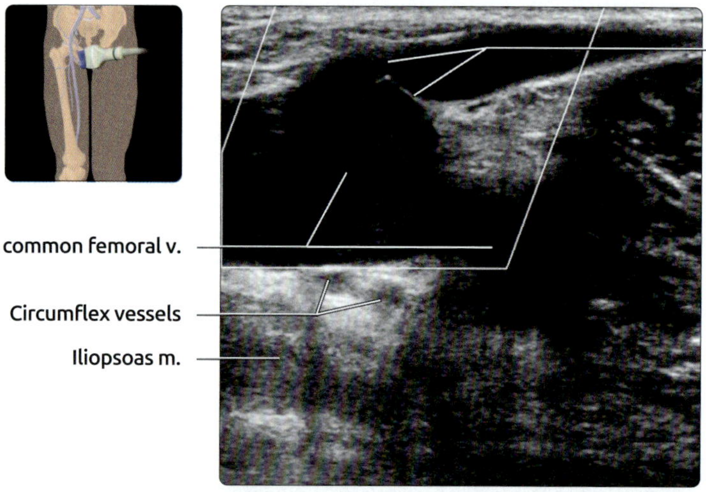

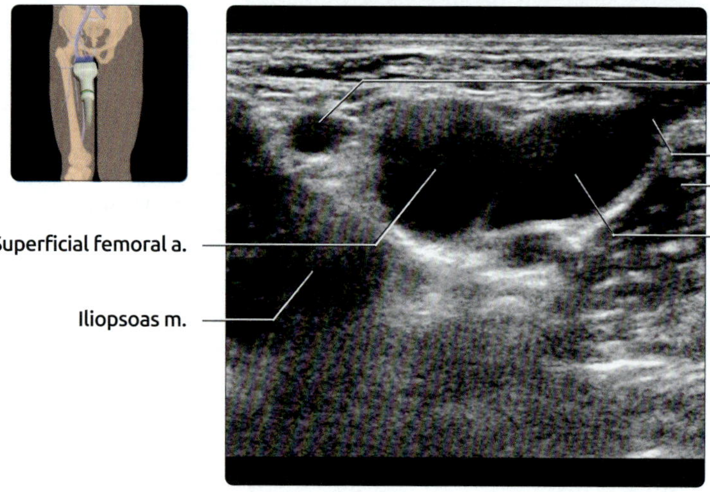

(Top) *Sagittal scan at the confluence of the great saphenous vein with the common femoral vein is shown. On this image, the valve at the saphenofemoral junction is open to allow blood from the great saphenous vein to drain into the common femoral vein.* **(Middle)** *Sagittal color Doppler scan shows the saphenofemoral junction during a Valsalva maneuver. The increased abdominal pressure is transmitted in the common femoral vein. This closes the valve at the saphenofemoral junction to prevent reflux into the great saphenous vein and also distends the common femoral vein (seen here as an increase in caliber). The lack of color flow across the valve indicates that the valve is competent.* **(Bottom)** *Transverse scan shows the saphenofemoral junction. The great saphenous vein penetrates the femoral sheath to enter the common femoral vein.*

DOPPLER SAPHENOUS VEIN

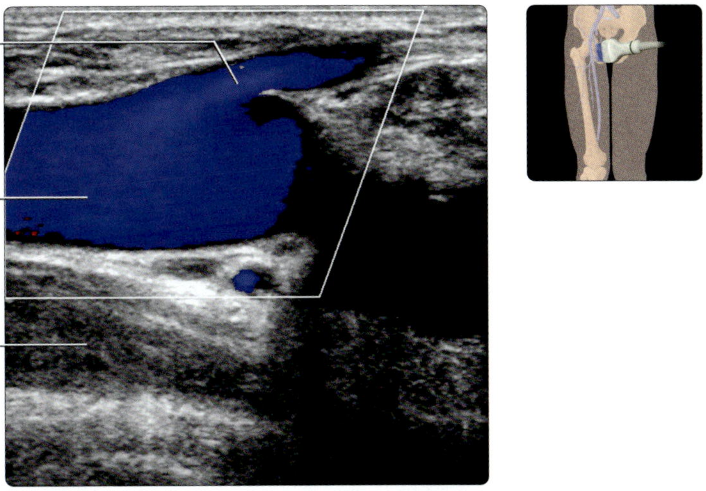

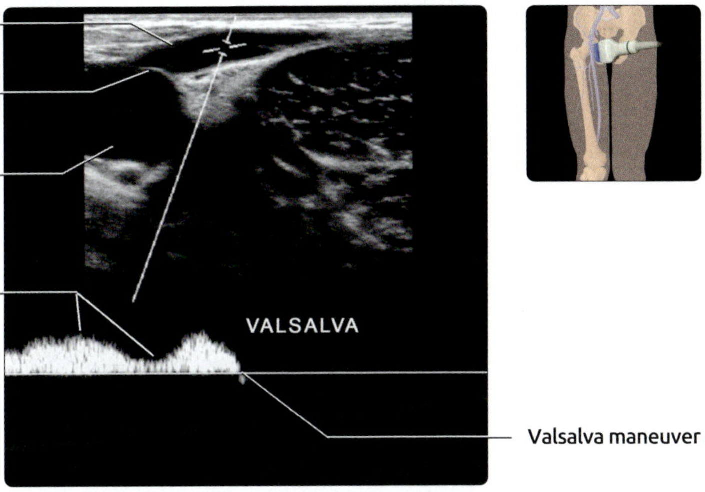

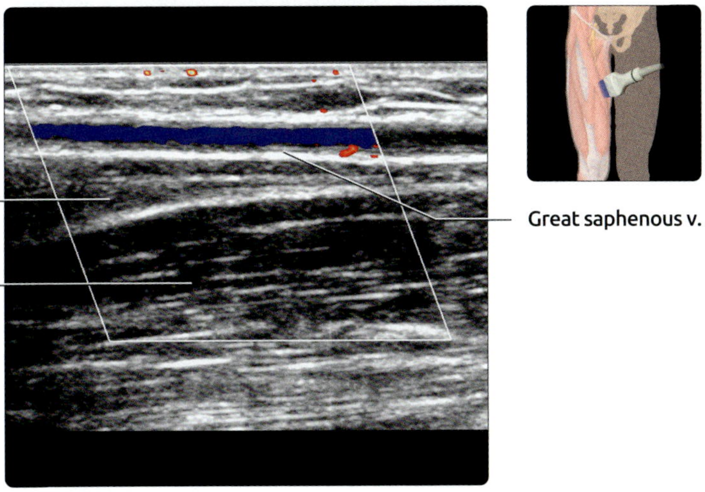

(Top) *Sagittal color Doppler scan shows the saphenofemoral junction. The color flow in both the great saphenous vein and the common femoral vein is toward the trunk.* **(Middle)** *Spectral Doppler trace of the great saphenous vein before and during a Valsalva maneuver is shown. The normal venous Doppler spectrum is biphasic due to respiratory variations. Applying a Valsalva maneuver, the valve closes abruptly at the saphenofemoral junction and stops flow in the great saphenous vein, which is seen as a zero-flow velocity on the trace.* **(Bottom)** *Oblique coronal color Doppler scan of the great saphenous vein in the medial thigh shows blood flow toward the trunk (correct direction).*

THIGH VEIN

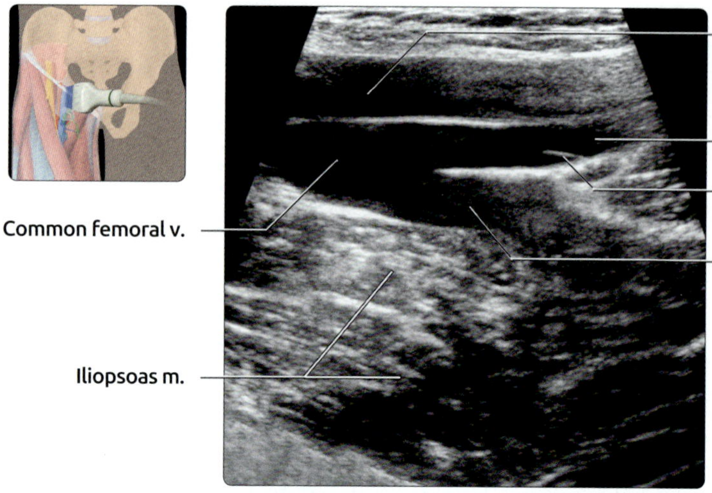

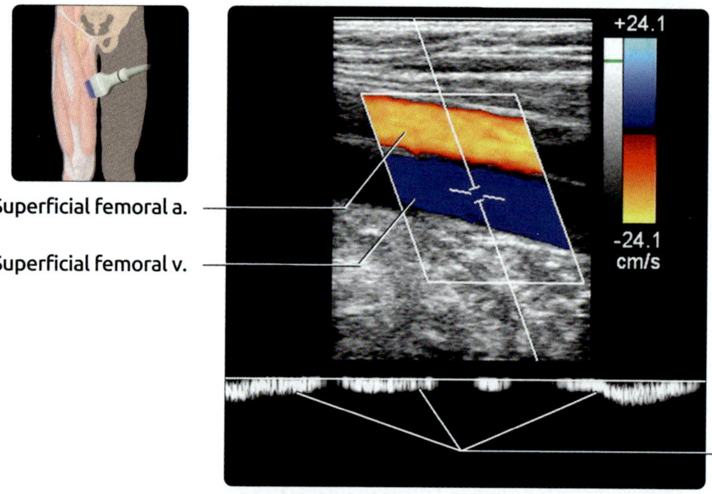

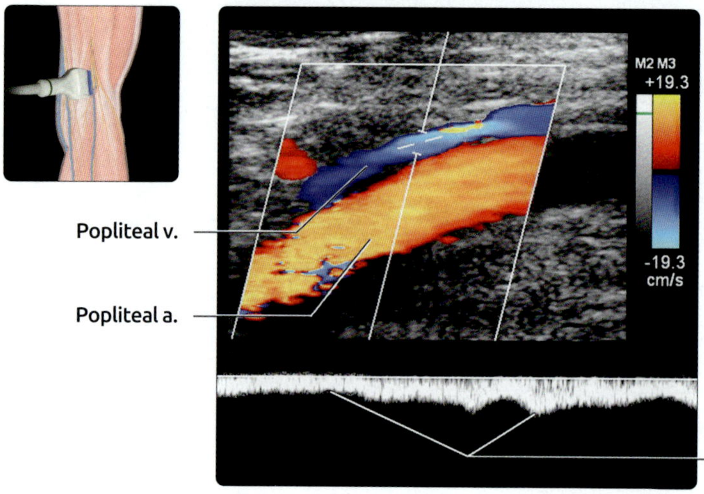

(Top) *Sagittal scan shows the confluence of the common femoral vein. The common femoral vein is formed by its main tributaries: The superficial femoral vein and the profunda femoral vein (both accompanying the course of the arteries of the same name).* **(Middle)** *Spectral Doppler trace shows the superficial femoral vein. The normal venous Doppler waveform is biphasic and varies with the phase of respiration. The flow velocity is also much slower than that in the artery.* **(Bottom)** *Spectral Doppler trace of the popliteal vein shows a similar venous waveform with respiratory phasic changes.*

FEMORAL VEIN: COMPRESSION

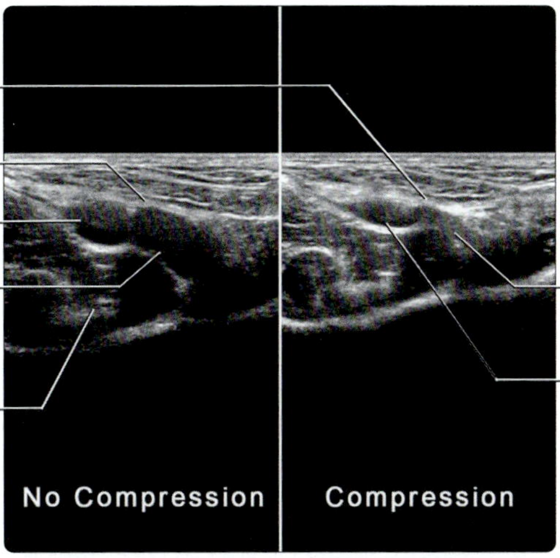

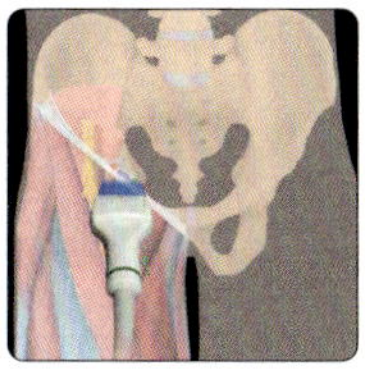

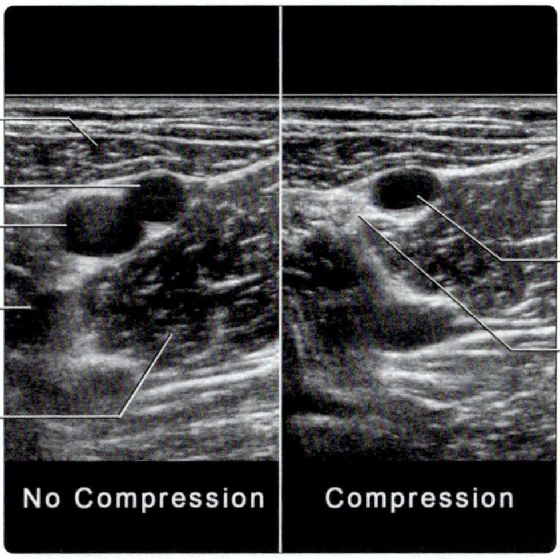

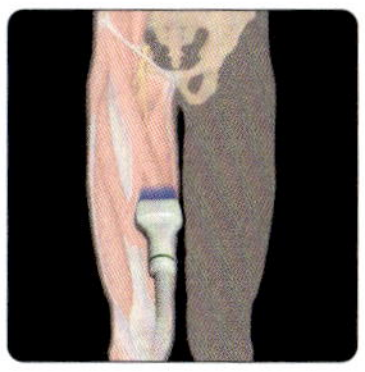

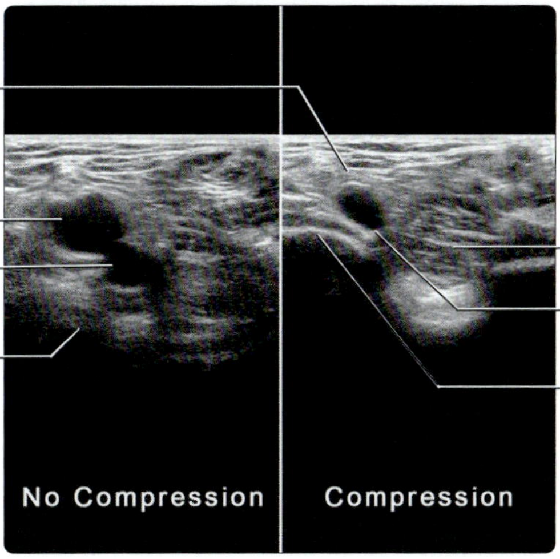

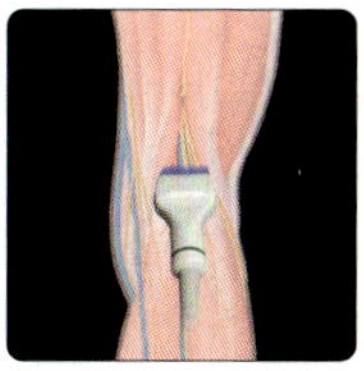

(Top) *Without and with compression split screen transverse scan shows the common femoral vessels. The left 1/2 of the image is obtained by applying normal gentle transducer pressure; the right 1/2 is scanned with firm compression. Patent veins are normally compressible with firm transducer pressure, with the exception demonstrated here due to the inguinal ligament. This is not a sign of venous thrombosis.* **(Middle)** *Without and with compression split screen transverse scan in the adductor canal shows the normal compressible superficial femoral vein. With firm pressure, the superficial artery is distorted, but the lumen is not obliterated.* **(Bottom)** *Without and with compression split screen transverse scan of the popliteal vessels shows the same response: Obliterated venous lumen and distorted arterial lumen.*

SCIATIC NERVE

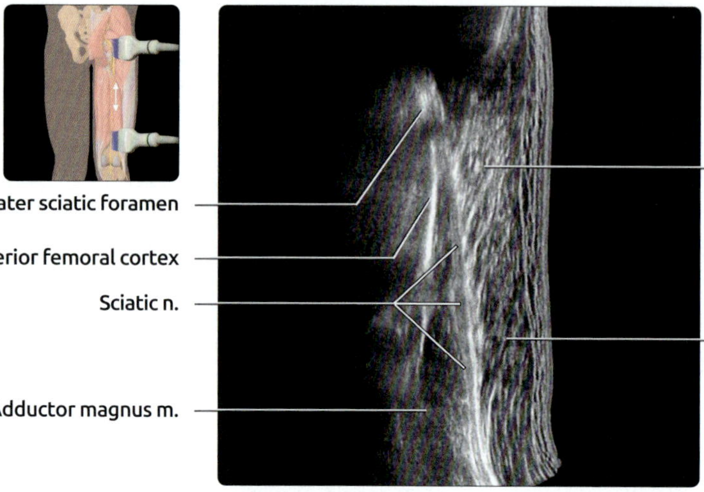

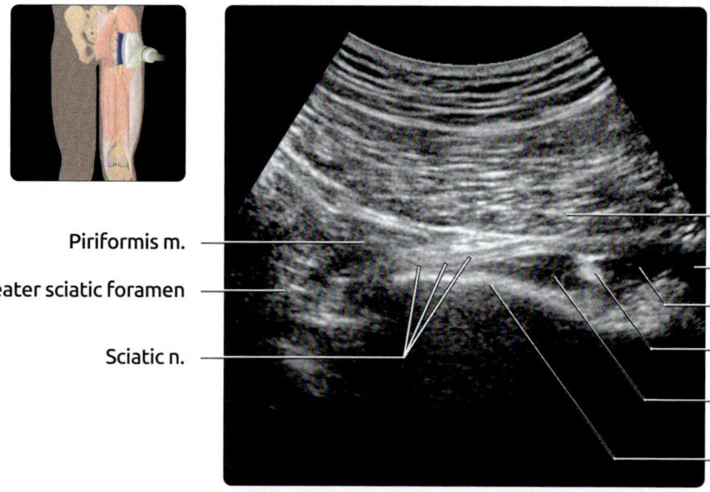

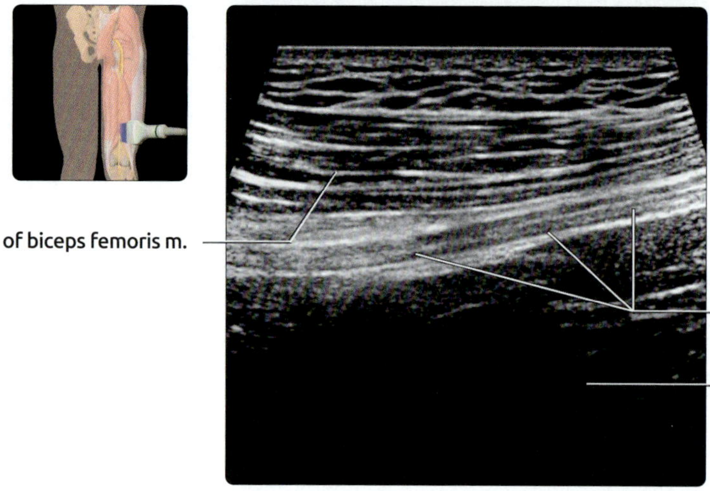

(Top) *Longitudinal panoramic view of the sciatic nerve shows its course after it has exited the greater sciatic foramen. It runs distally deep to the gluteus maximus and then deep to the biceps femoris to the popliteal fossa.* **(Middle)** *Sagittal scan shows the greater sciatic foramen using the ischial tuberosity as a landmark. The sciatic nerve and its relationship with the small external rotators of the hip is shown here. The sciatic nerve emerges between the piriformis and superior gemellus muscles after passing through the greater sciatic foramen.* **(Bottom)** *This longitudinal view of the sciatic nerve demonstrates the parallel and uniform hypoechoic tracts (representing nerve bundles). Any disruption of the uniformity or parallel course should be considered abnormal.*

TRANSVERSE SCIATIC NERVE

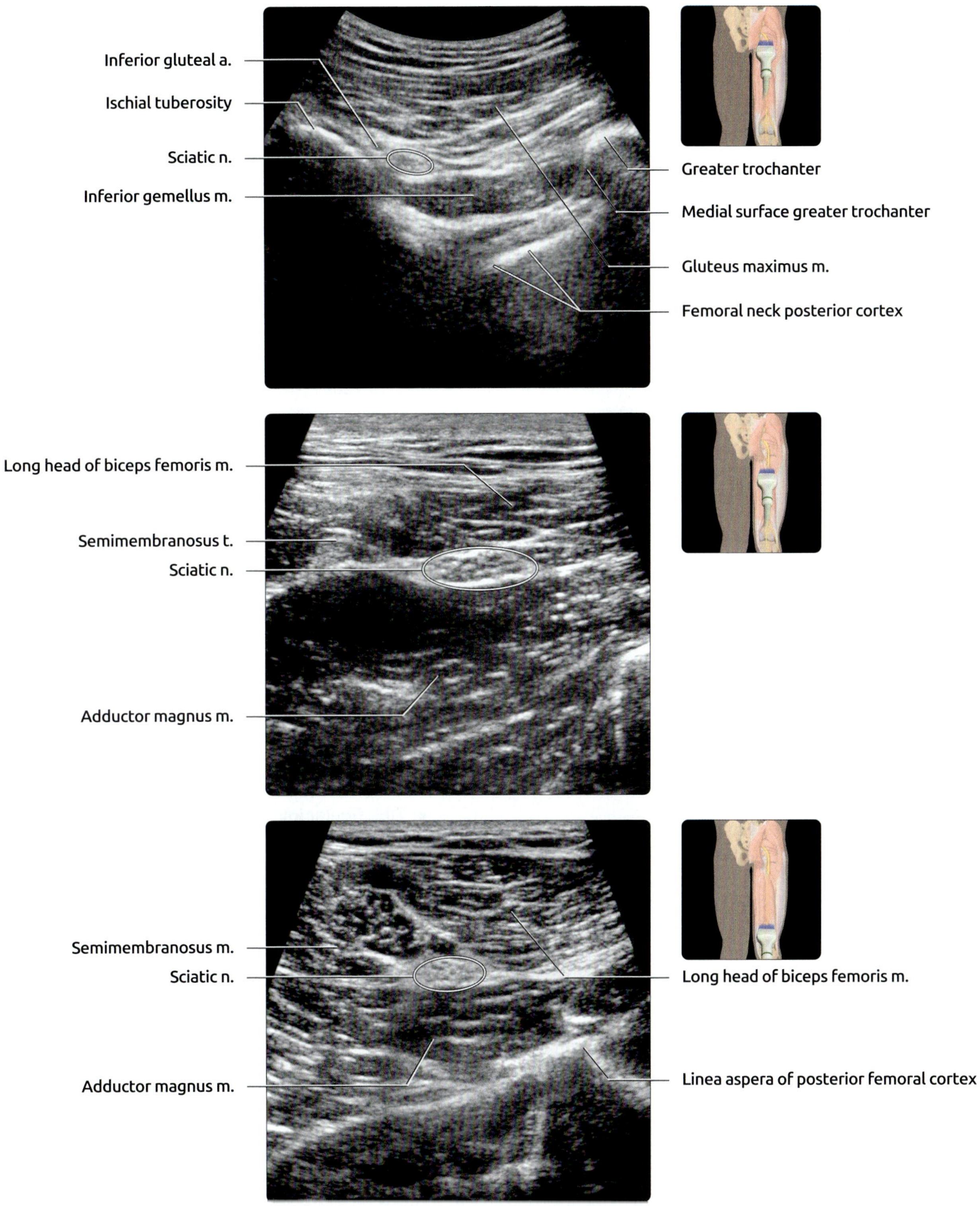

(Top) *Transverse scans at 3 different levels show the sciatic nerve. The sciatic nerve, after it has exited the greater sciatic foramen, passes superficial to the small external rotators of the hip joint (obturator internus, superior and inferior gemelli, and quadratus femoris). The inferior gluteal artery lies medial to the nerve.* **(Middle)** *At the upper thigh level, the sciatic nerve is covered by the long head of the biceps femoris muscle and is relatively superficial. The high spatial resolution of ultrasound allows individual nerve bundles to be identified as black dots in a bright matrix of epineurium.* **(Bottom)** *The same relationship with the long head of biceps femoris muscle continues inferiorly down the thigh until the sciatic nerve divides into the peroneal nerves.*

TERMINOLOGY

Abbreviations

- Origin (O), insertion (I), function (F)

GROSS ANATOMY

Overview

- Largest and most complex joint
  - Hinge joint throughout its greatest range of motion
  - Bones do not interlock; stability is maintained by ligaments, tendons, capsule, and menisci
- **Motion** of knee and relationship of osseous structures
  - In full flexion
    - Posterior surfaces of femoral condyles articulate with posterior tibial condyles
    - Lateral facet of patella is in contact with lateral femoral condyle
    - Supporting ligaments are not taut, and rotation of leg is allowed
  - During motion of extension
    - Patella slides upward on femur, engaging with femoral trochlea
    - Femoral condyles roll forward on tibial condyles and menisci
    - Lateral femoral condyle is shorter anteroposteriorly than medial and reaches full extension earlier
    - Medial tibial condyle continues to slide after lateral stops, rotates slightly laterally on femur and medial meniscus ("screwing it home")
    - This tightens cruciate ligaments turning knee into rigid pattern
  - Initiating flexion from fully extended knee
    - Requires slight medial rotation of tibia, produced by popliteus muscle
    - "Unlocks" joint, allowing remainder of motion to take place
- **Muscles acting on knee joint**
  - **Extensors**: 4 parts of quadriceps femoris
    - I: Patella
    - F: Acts together for hip flexion and knee extension
  - Rectus femoris
    - O: Anterior inferior iliac spine
  - Vastus medialis
    - O: Medial shaft of femur
  - Vastus intermedius
    - O: Anterior shaft of femur
  - Vastus lateralis
    - O: Lateral shaft of femur
  - Biceps femoris
    - O: Ischial tuberosity
    - I: Fibular head and tibia
    - F: Crosses both hip and knee joints, extending hip and flexing knee
  - **Flexors**
  - Popliteus
    - O: Arises as tendon from popliteal groove at lateral femoral condyle
    - I: Posterior surface tibia
    - F: Flexes knee and medially rotates tibia at beginning of flexion
  - Sartorius
    - O: Anterior superior iliac spine
    - I: Anteromedial tibia
    - F: Flexes both hip and knee joints, rotating thigh laterally
  - Gracilis
    - O: Pubis
    - I: Anteromedial tibia
    - F: Adducts thigh, flexes knee, and rotates flexed leg medially
  - Semitendinosus
    - O: Ischial tuberosity
    - I: Anteromedial tibia
    - F: Extends hip, flexes knee, medially rotates flexed leg
  - Semimembranosus
    - O: Ischial tuberosity
    - I: Posterior medial condyle tibia
    - F: Extends hip, flexes knee, medially rotates flexed knee

IMAGING ANATOMY

Osseous Anatomy

- **Distal femur**
  - Distal femoral metaphysis flares into medial and lateral epicondyles
  - Medial femoral condyle larger than lateral
  - Anteriorly, trochlear groove accommodates patella and is generally V-shaped
- **Proximal tibia**
  - Gerdy tubercle anterolateral just distal to joint
  - Tibial tubercle (apophysis) anterior and slightly lateral, several cm distal to joint
- **Proximal fibula**
  - Posterolateral relative to tibia
  - Fibular styloid process
- **Tibiofibular joint**
  - True synovial joint; subject to any arthritic process
  - Connects to knee joint in 20%
- **Patella**
  - Triangular sesamoid
  - Wider at base superiorly than at apex inferiorly
  - Nonarticular outer surface may develop prominent enthesopathy where quadriceps tendon insertion blends into origin of inferior patellar tendon
  - Bipartite (multipartite) patella: Always upper outer quadrant; osseous fragments may not appear to "match," but cartilage is continuous over defect

Articular Capsule

- Highly complex, noncontiguous structure
- Contributions from multiple muscles, tendons, and ligaments

Extensor Mechanism

- **Quadriceps**
  - Composed of rectus femoris: Vastus lateralis, medialis, and intermedius muscles
  - Quadriceps tendon inserts on anterior surface of patella
  - Patellar tendon is mainly composed of rectus femoris fibers that course over patella

- Patellar tendon runs from inferior pole of patellar to tibial tuberosity
- Fibers of vastus lateralis and medialis contribute to lateral and medial retinacula, respectively

Fat Pads

- Suprapatellar: Anterior and posterior to suprapatellar bursa
- Infrapatellar (Hoffa fat pad): Inferior to patella and between patellar ligament and anterior knee joint
- Posterior: Around joint margin extending into intercondylar region

Medial Supporting Structures

- Superficial (layer 1)
  - Pes anserinus: Anteromedial tibial insertion
    - Sartorius embedded in crural fascia
    - Gracilis immediately deep to sartorius
    - Semitendinosus immediately deep to gracilis
- Middle (layer 2)
  - Superficial medial collateral ligament (longitudinal and oblique components)
    - Origin: Medial epicondyle; runs slightly anteromedially to insert on tibia 5 cm distal to joint line
    - Anteriorly, longitudinal component fascia blends with layer 1
    - Posteriorly, oblique component blends with layer 3 as posterior oblique ligament
- Deep (layer 3)
  - Capsular layers [sometimes termed deep fibers of medial collateral ligament (MCL)] at midportion of knee
    - Meniscofemoral ligament: From outer superior aspect of body of meniscus to either superficial MCL or femur
    - Meniscotibial (coronary) ligament: From outer inferior aspect of body of meniscus to tibia just distal to joint line
  - More posteriorly, superficial MCL blends with capsular layers MCL
  - Posterior oblique ligament arises from superficial MCL
    - Blends with posteromedial meniscus
    - Receives fibers from semimembranosus tendon
    - Envelops posterior aspect of femoral condyle, termed oblique popliteal ligament

Lateral Supporting Structures

- Superficial (layer 1)
  - Iliotibial band anteriorly, inserting on Gerdy tubercle
  - Superficial portion biceps femoris posterolaterally, inserting on fibular styloid
- Middle (layer 2)
  - Quadriceps retinaculum anteriorly
  - Posteriorly, 2 ligamentous thickenings, which originate from lateral patella
    - Proximal: Terminates at lateral intermuscular septum on femur
    - Distal: Terminates at femoral insertion of posterolateral capsule and lateral head of gastrocnemius
- Deep (layer 3): Several thickenings in lateral part of capsule function as discrete structures
  - Lateral (fibular) collateral ligament originates from lateral femoral epicondyle, extends posterolaterally to insert on lateral fibular head
  - Arcuate ligament originates from styloid process of fibular head, interdigitates with popliteus, and inserts into posterior capsule near oblique popliteal ligament
  - Several other small and inconstant structures are located posterolaterally, which are difficult to differentiate by imaging

Internal Structures

- **Menisci**
  - Cushion, lubricate, and stabilize knee
  - Fibrocartilage
  - Only peripheral portion visualized by ultrasound
  - Medial attached to capsule throughout extent of knee
  - Lateral attached to capsule at anterior horn and far posteriorly, but by fascicles to popliteus at body and posterior horn
- **Cruciate ligaments**
  - Major stabilizing structures to anteroposterior motion
- Cruciate ligaments and central portions of menisci are not well visualized by ultrasound

Synovial Recesses of Knee Joint

- Suprapatellar, medial patellar, lateral patellar
- Increase in synovial fluid 1st accumulates in either medial or lateral recess and then suprapatellar recess
- Small amount of fluid in knee joint is normal but should not get more than small amount
- Synovial proliferation can also be best assessed at medial and lateral recesses
- Baker cyst is distension of gastrocnemius-semimembraneous bursa
- Slit-like opening occasionally connects knee joint to Baker cyst
- Synovial fluid can be expressed from knee joint into Baker cyst if this interconnection is present

ANATOMY IMAGING ISSUES

Imaging Approaches

- Ultrasound
  - Good for assessing extracapsular structures and their abnormalities but unable to satisfactorily interrogate internal structures of knee
  - Given size and complexity of this joint, ultrasound examination needs to be compartmentalized to area of complaint
    - e.g., posterior examination for popliteal mass, lateral examination for suspected iliotibial band syndrome, etc.
  - In acutely inflamed joint, ultrasound is most effective as screening tool for effusion, allowing image-guided aspiration concurrently
  - Ultrasound is ideal for assessing extraarticular abnormalities, e.g., tendinosis, ligament disruptions, bursitis, masses, etc.
  - For structures containing fibers, e.g., ligaments and tendons, longitudinal scans tend to be more helpful both in confirming course of structure and assessing integrity
    - For this reason, knee should be flexed during examination of anterior knee
    - Flexion tenses most tendinous and ligamentous structures and accentuates straight course of its fibers

EXTENSOR MECHANISM

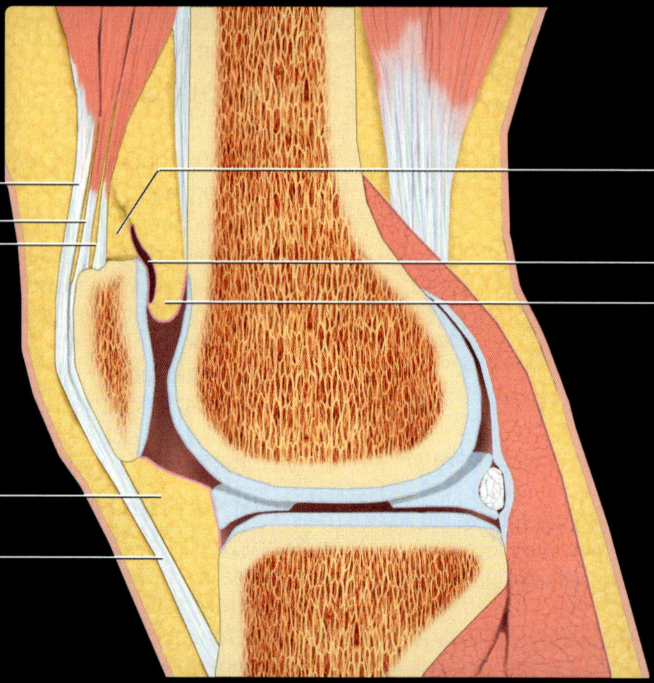

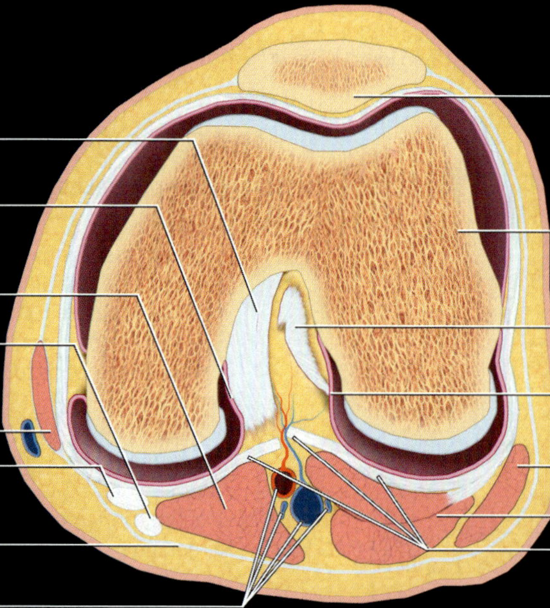

(Top) *Graphic shows trilaminar configuration of the quadriceps tendon. The superficial portion is the rectus femoris, the middle portion is the aponeurosis of the vastus medialis and lateralis, and the deep portion is the vastus intermedius tendon.* **(Bottom)** *Graphic shows the relationships of the posterior capsule and overlying structures. The most superficial layer is the crural fascia, which envelops the sartorius and otherwise confines all of the structures. The posterior capsule (white) is continuous with the synovium (pink thin line) both posteromedially and posterolaterally. Superficial to the capsule are the plantaris and gastrocnemius muscles, hamstring muscle and tendons, and the popliteal vessels.*

EXTENSOR MECHANISM

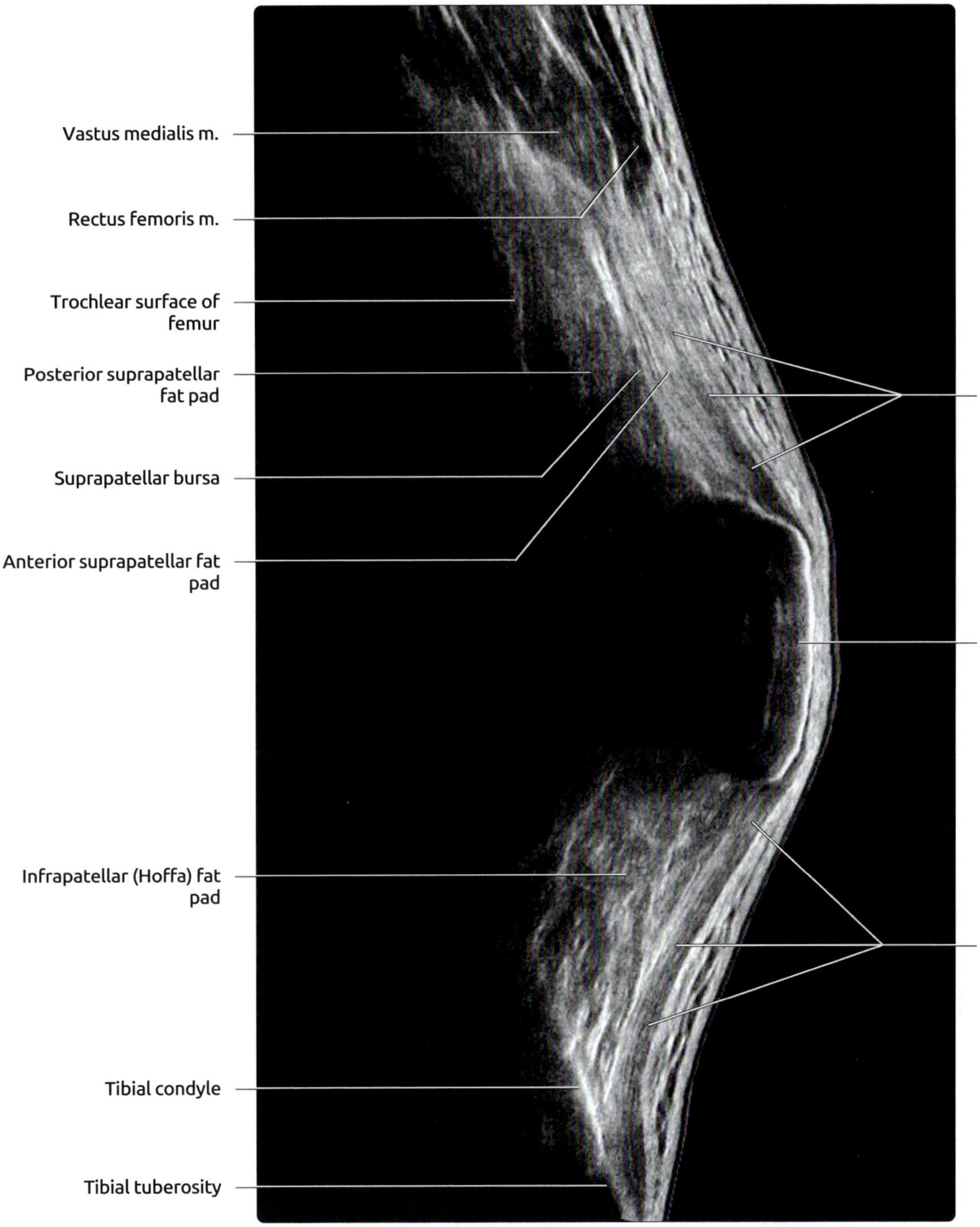

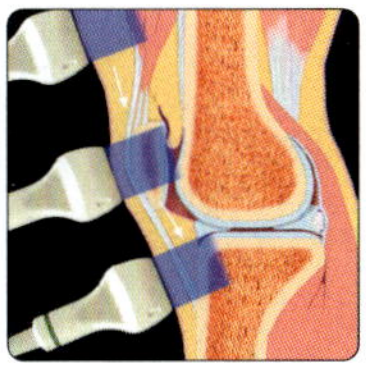

Sagittal panoramic scan of the anterior knee shows the main components of the extensor mechanism: Quadriceps muscle and tendon, patella, and patellar tendon. The scan performed here is with the knee in extension. More focused views of individual components of the anterior knee can be performed with the knee in flexion (with a pillow to support the popliteal region). Flexion increases the tension of the extensor mechanism, thus straightening the tendons. This will help highlight abnormalities in terms of tendon alignment and continuity. However, the amount of suprapatellar fluid may be decreased (due to displacement) in flexion, and assessment of the suprapatellar region in extension may be more sensitive in this respect.

EXTENSOR MECHANISM

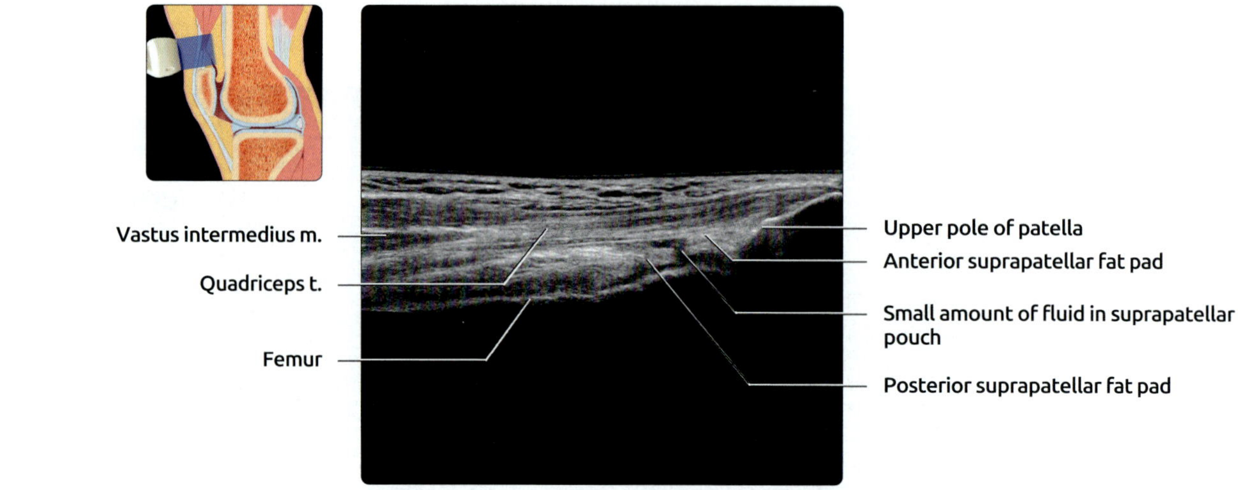

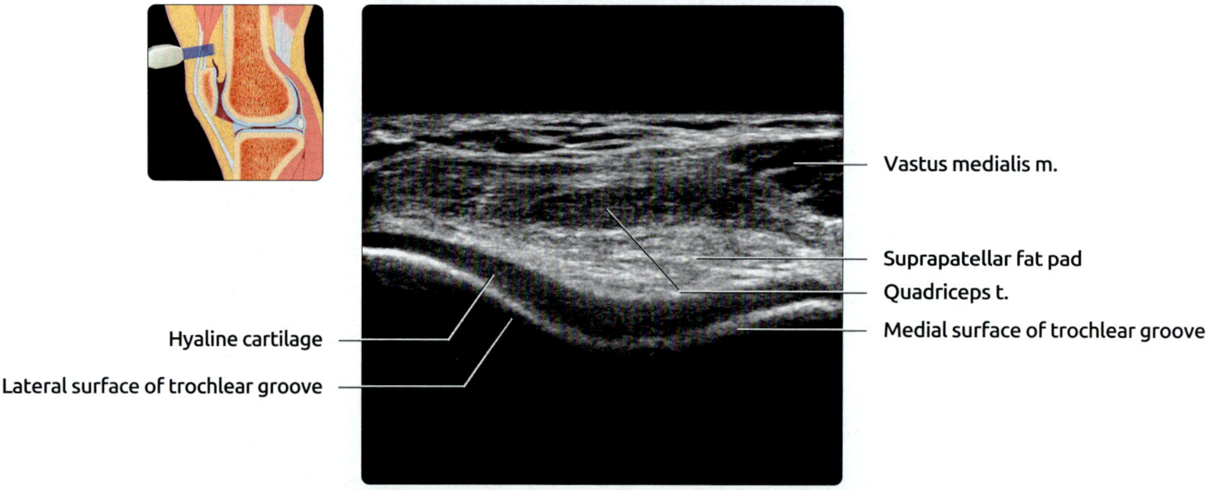

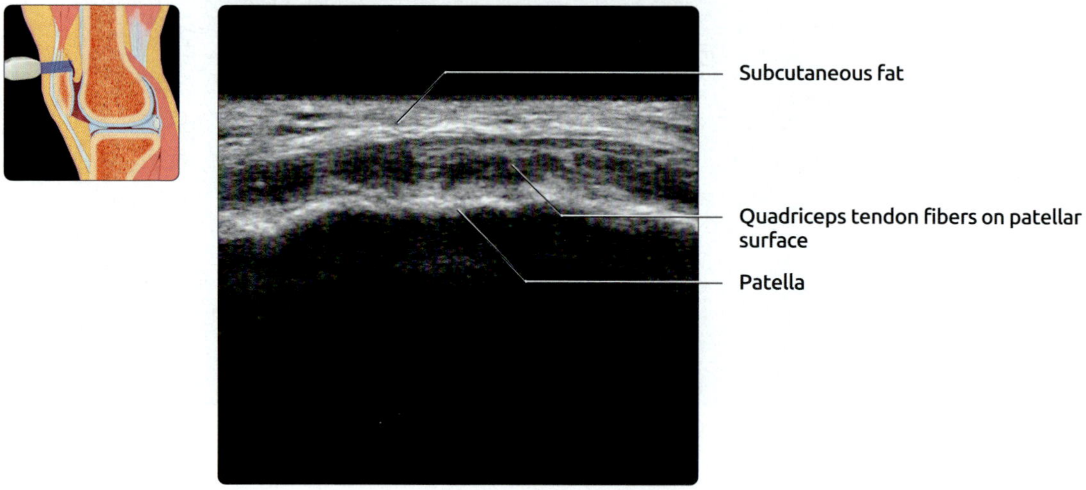

(Top) *Sagittal scan shows the quadriceps tendon. The quadriceps tendon is the direct continuation of the rectus femoris with contributions medially and laterally from the vastus medialis and the vastus lateralis, respectively, and posteriorly from vastus intermedius. The tendon inserts on the superoanterior surface of the patella. In this scan, the suprapatellar pouch can be seen. It may normally contain a small amount of fluid (several millimeters thick).* **(Middle)** *Transverse scan shows quadriceps tendon with the knee flexed. The central position of the quadriceps tendon can be appreciated here, receiving contributions from the vastus muscles. The articular cartilage of the trochlear groove can also be seen with the patella out of the way during flexion.* **(Bottom)** *Transverse scan of the anterior surface of the patella shows tendon fibers inserting or running on the patellar surface.*

EXTENSOR MECHANISM

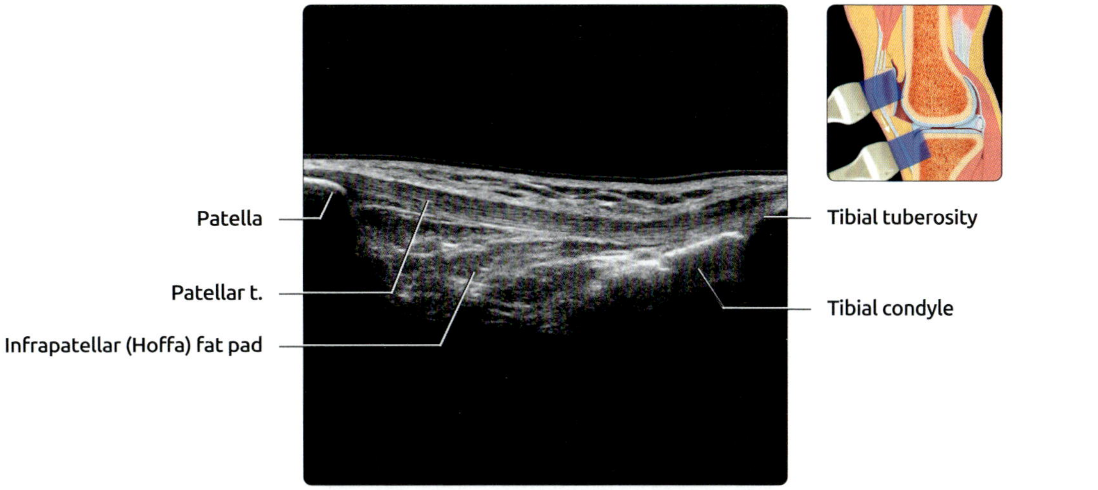

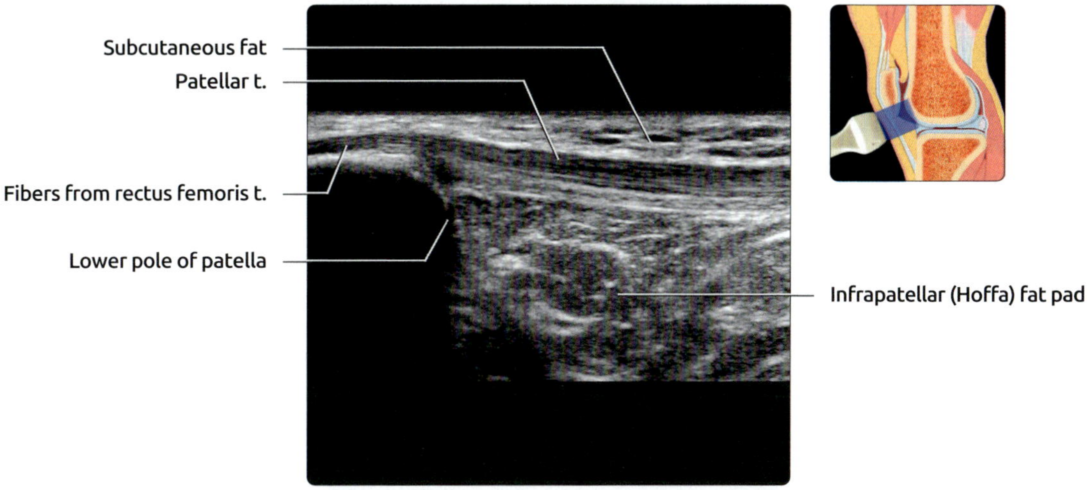

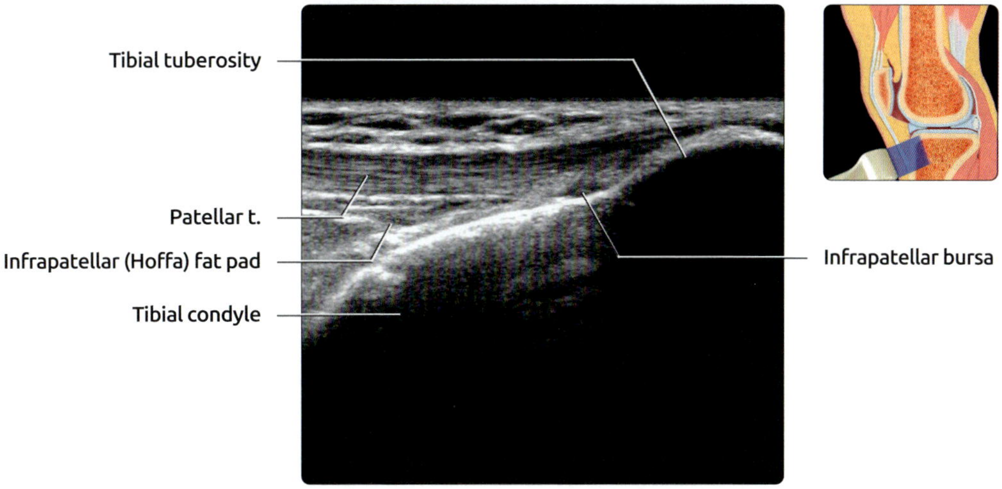

(Top) *Sagittal panoramic view shows the patellar tendon. Most of the fibers of the tendon are continuations of quadriceps tendon fibers.* **(Middle)** *Focused sagittal view of the apex of the patella shows quadriceps fibers running on the anterior surface of the patella, continuing distally as the patellar tendon. For this reason, many consider the patella a sesamoid bone. Also seen here are fibers originating from the apex of the patella itself.* **(Bottom)** *Focused sagittal scan shows the tibial tuberosity. The distal fibers of the patellar tendon are separated from the anterior surface of the tibial condyle by the infrapatellar bursa. A small amount of fluid (2-3 mm thick) may be normally present in this bursa.*

EXTENSOR MECHANISM

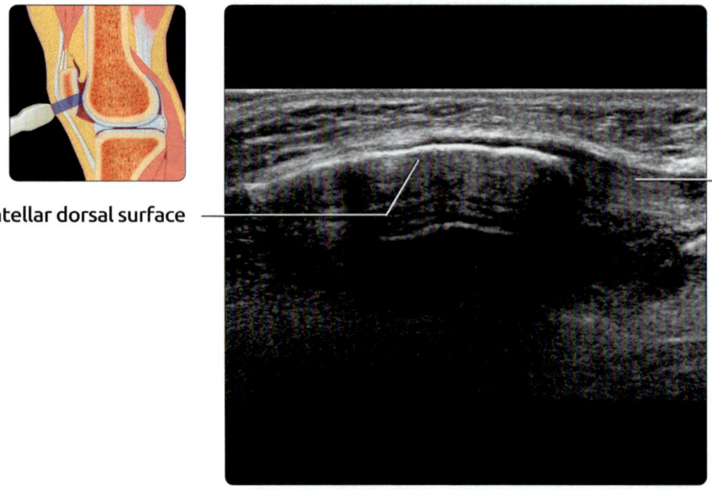

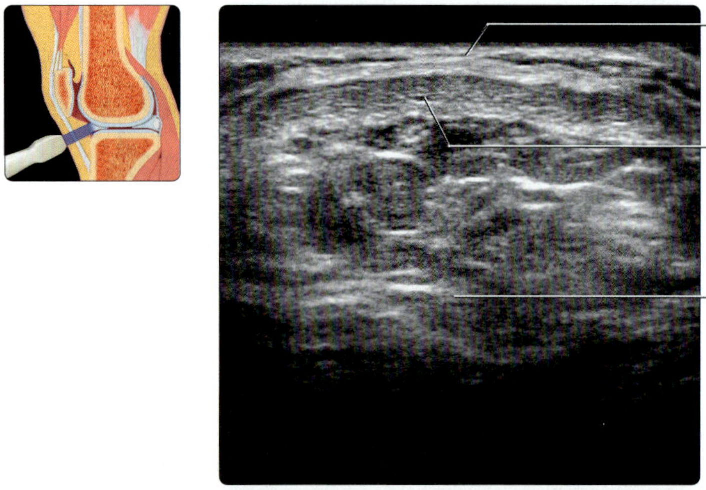

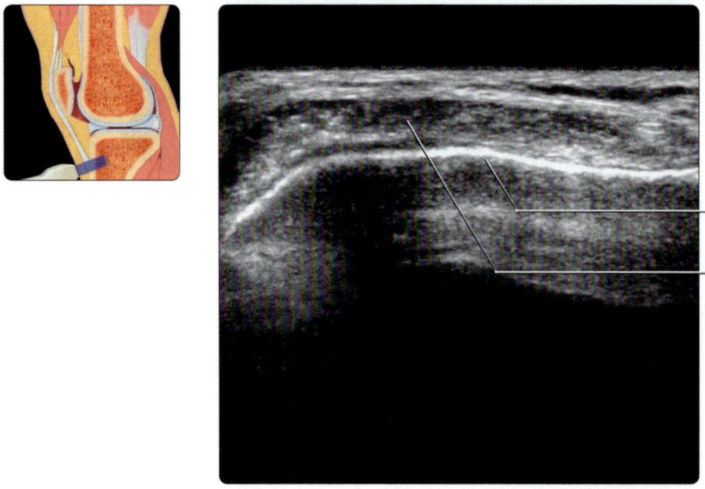

(Top) *Series of 3 transverse scans through the patellar tendon is shown. Proximally, fibers from the quadriceps tendon continue over the anterior patellar surface to contribute to the patellar tendon. On both sides, fibers from the medial and lateral patellar retinaculum also insert onto the patella and blend with the patellar tendon fibers.* **(Middle)** *At the midsubstance of the patellar tendon, the tendon assumes a flattened appearance. The fibers within the tendon can be seen as tiny hyperechoic dots. A disruption to the uniformity of these dots or a change in thickness (focal or general) of the patellar tendon suggests tendinosis.* **(Bottom)** *More inferiorly, a transverse scan shows the patellar tendon inserting onto the tibial tuberosity. The patellar tendon paratenon is seen as thin echogenic line on the anterior surface and side walls of the tendon. It becomes thickened in patellar paratenonitis. This is an infrequently encountered condition.*

EXTENSOR MECHANISM

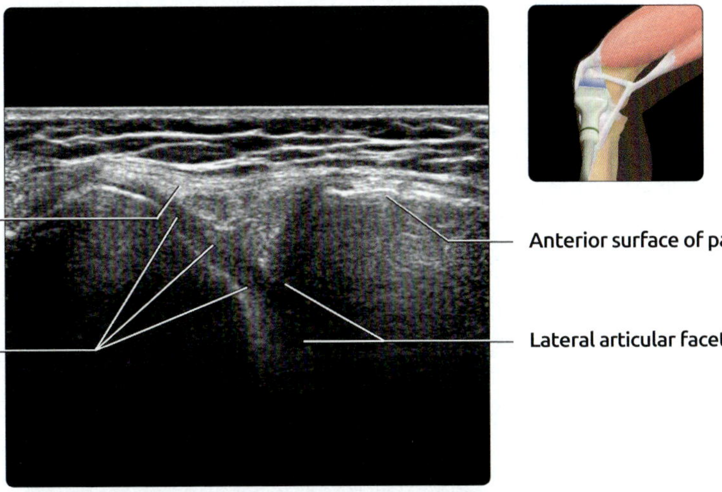

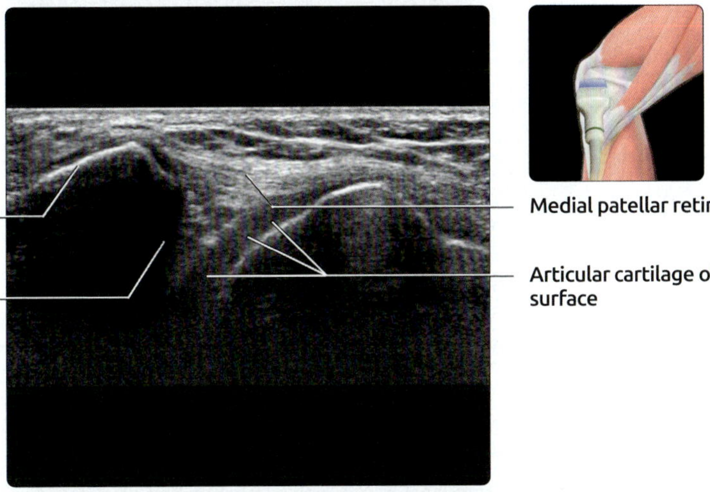

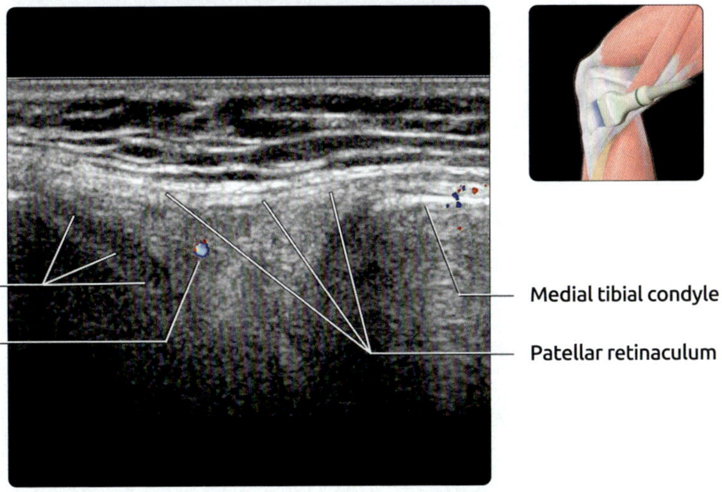

(Top) *Transverse scan at the lateral margin of the patella shows the articulation between the lateral patella and the lateral surface of the trochlear groove. This articular plane is steeper (closer to sagittal) than the medial side for mechanical advantage. As a result of the obliquity of the course of the patellar tendon with respect to the tibial axis, the patella is prone to lateral dislocation or subluxation. The steeper lateral articular plane helps prevents this.* **(Middle)** *Transverse scan at the medial margin of the patella shows the articulation between the patella and the medial surface of the trochlear groove showing the patellar retinaculum as a thick structure.* **(Bottom)** *Sagittal color Doppler scan of the medial infrapatellar region shows the patellar retinaculum inserting onto the tibia. At this level, the retinaculum has fanned out, becoming much thinner.*

MEDIAL KNEE

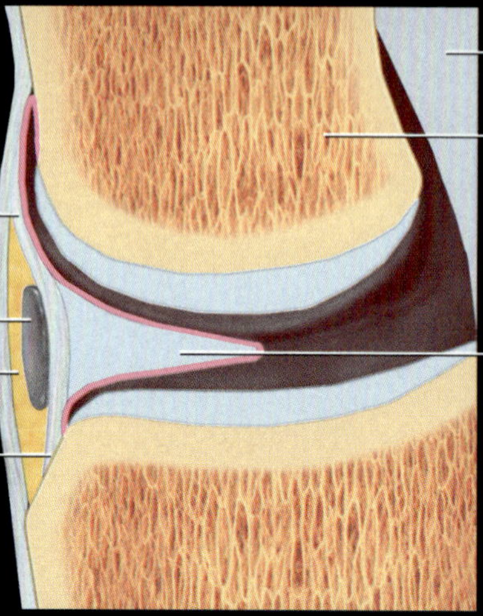

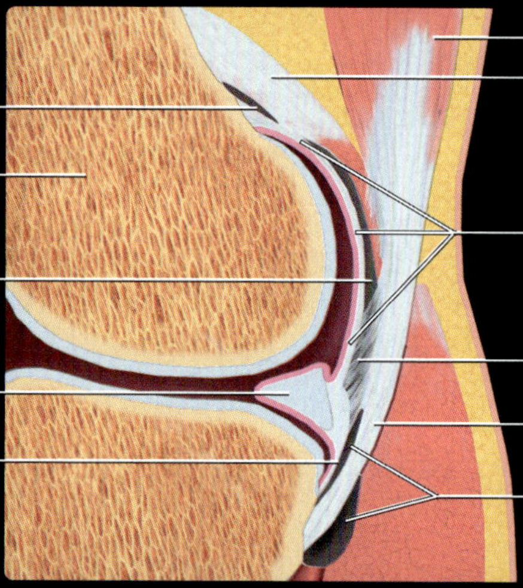

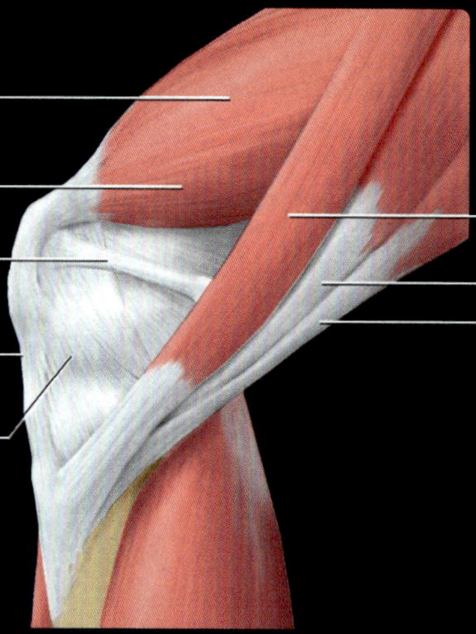

(Top) *Coronal graphic at midjoint shows superficial fibers extending from adductor tubercle to ~ 5 cm below the joint line. Deep fibers are shorter. The meniscotibial (coronary) ligament extends from the meniscus to the tibia adjacent to the joint line & the meniscofemoral ligament extends from the meniscus to the femur or superficial medial collateral ligament (MCL).* **(Middle)** *Sagittal graphic shows the far medial portion of posterior capsule. Superiorly, the capsule attaches to the cortex of the posterior femoral condyle and fuses with fibers of the medial gastrocnemius.* **(Bottom)** *Graphic shows the anteromedial attachment of pes anserinus tendons. The sartorius is the most anterior and superficial. The gracilis inserts directly adjacent and deep to the sartorius, and the semitendinosus inserts directly posterior and inferior to the gracilis. These tendons and their crural fascia comprise the superficial layer of the medial capsuloligamentous complex.*

MEDIAL KNEE

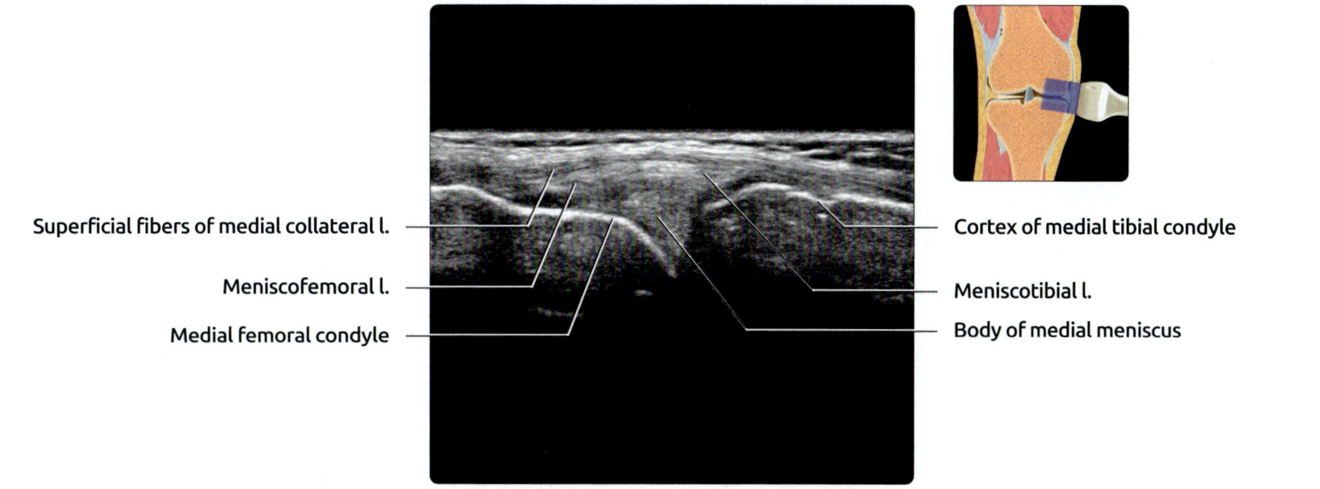

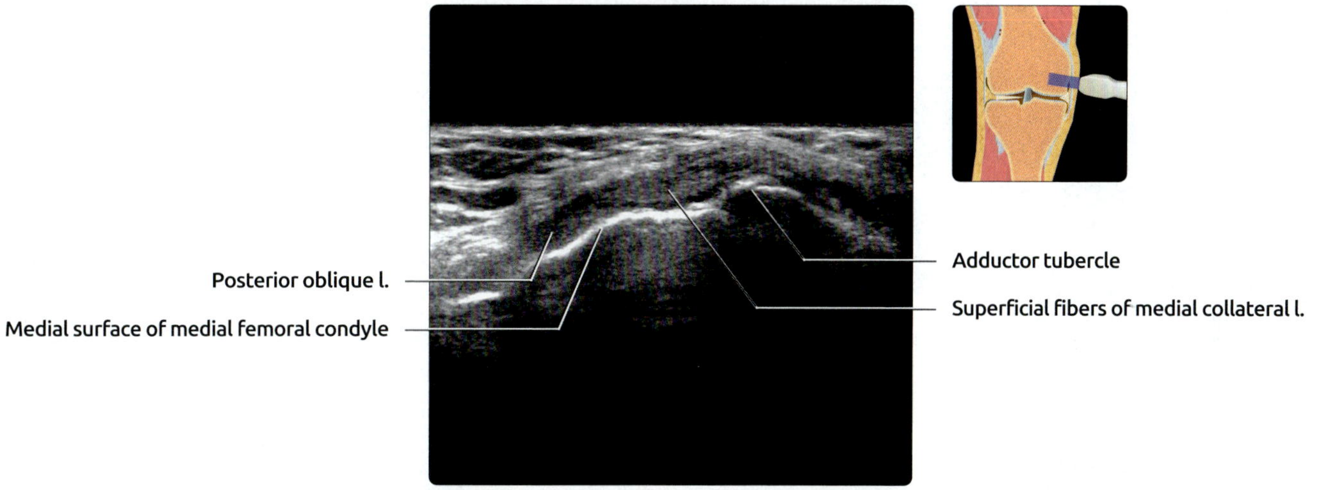

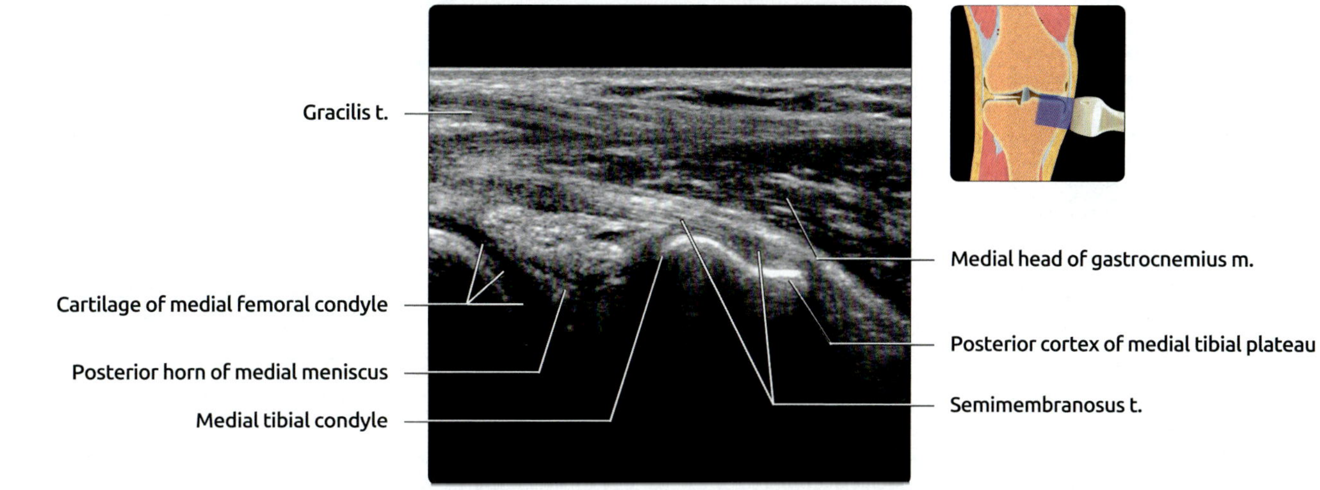

(Top) *Coronal scan shows the medial joint line. The medial collateral ligament is composed of 2 layers: Superficial and deep. The superficial fibers run from the medial femoral condyle to the tibial condyle 5 cm below the joint line. The deep layer consists of the meniscofemoral (femoral condyle to body of meniscus) and meniscotibial (body of meniscus to tibia just distal to joint line).* **(Middle)** *Transverse scan shows the medial condylar region. The superficial fibers of the medial collateral ligament arise just distal to the adductor tubercle. The popliteal oblique ligament arises adjacent to this and runs posterior to the posterior joint capsule.* **(Bottom)** *Sagittal scan shows the posteromedial joint line. The semimembranosus tendon inserts onto the posterior surface of the medial tibial condyle, just below the joint margin. Only the outer part of the meniscus is visible.*

LATERAL KNEE

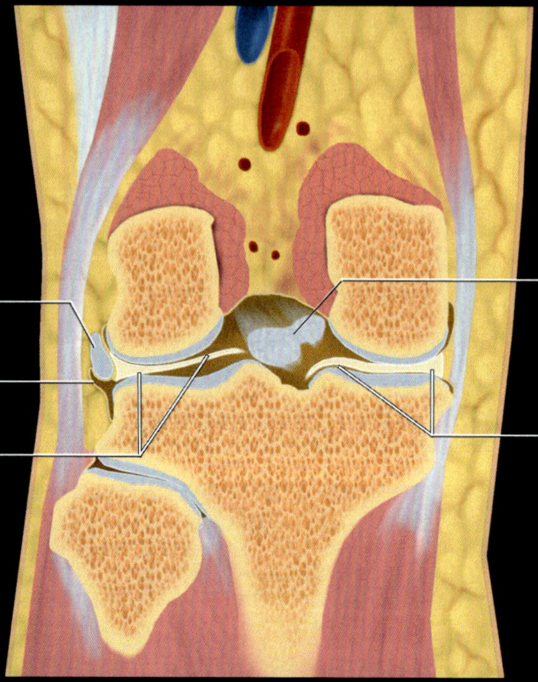

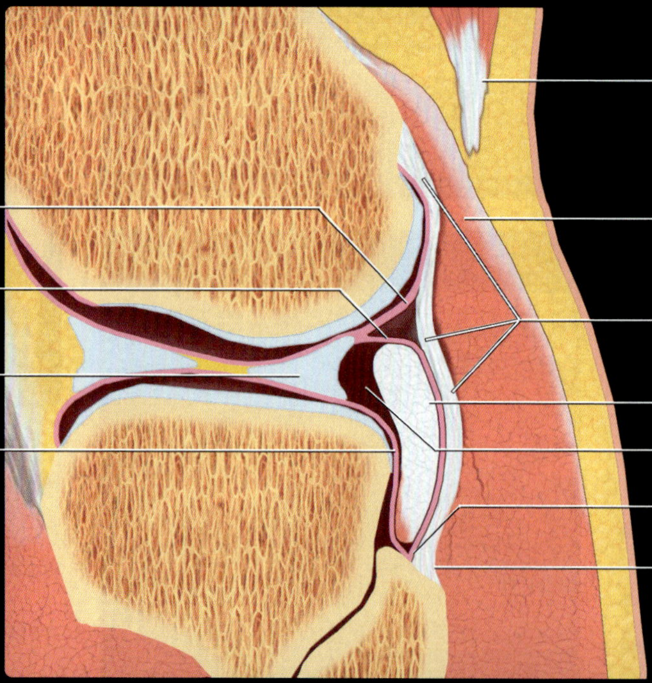

(Top) *Coronal graphic shows the roots of the posterior horns of both the medial and lateral menisci. The popliteus tendon courses through the hiatus adjacent to the body/posterior horn of the lateral meniscus.* **(Bottom)** *Sagittal graphic of lateral part of posterior capsule shows that it is intimately attached to the lateral gastrocnemius muscle. The popliteus tendon is intraarticular but extrasynovial and is attached to the capsule. Fibers from the posterolateral corner and arcuate ligament originating from the fibular head contribute to the lateral portion of the posterior capsule.*

LATERAL KNEE

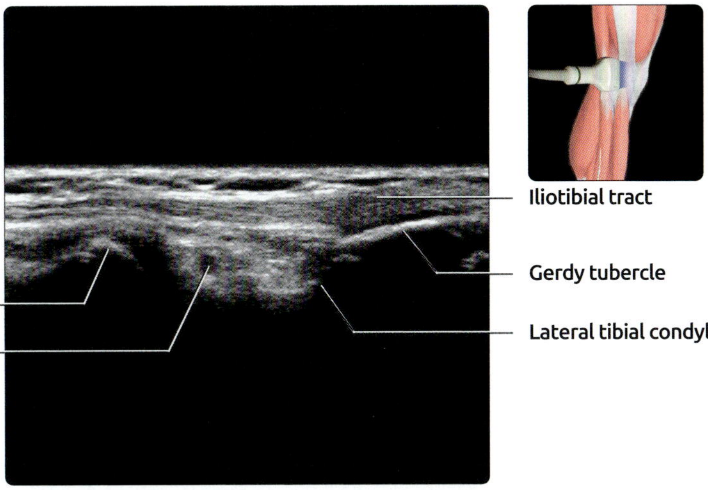

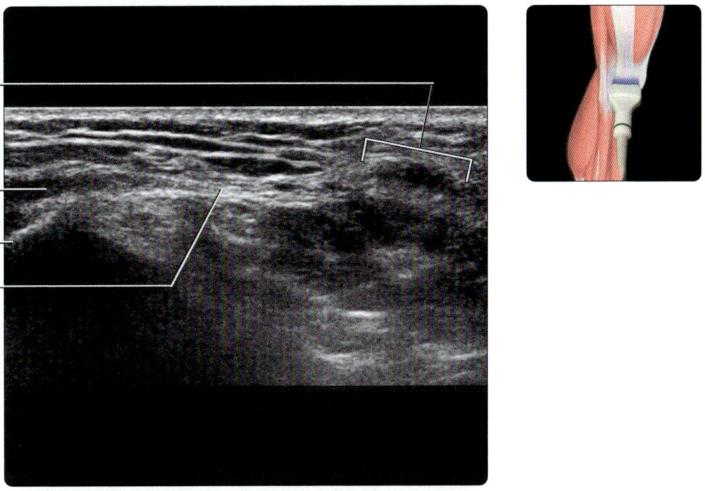

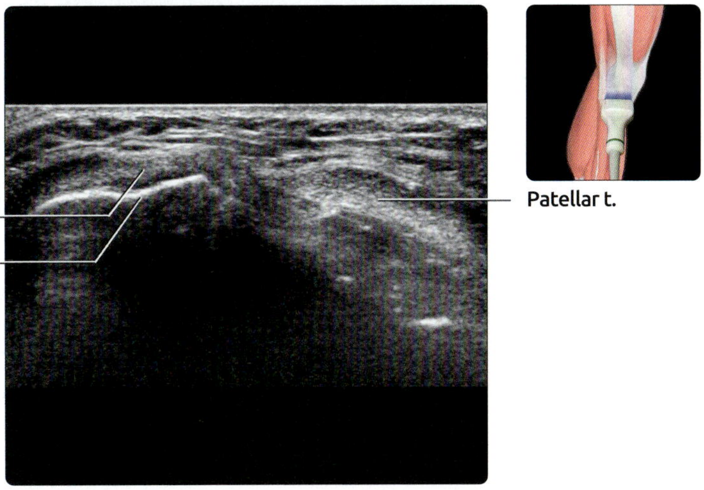

(Top) *Oblique sagittal scan shows the lateral aspect of the joint line using Gerdy tubercle as a landmark. The iliotibial tract is shown here, passing over the edge of the lateral femoral condyle [where it is subject to impingement (iliotibial tract syndrome)] before inserting on the Gerdy tubercle on the anterior surface of the lateral tibial condyle.* **(Middle)** *Transverse scan of the anterolateral knee at the level of the femoral condyle shows that the iliotibial tract and the patellar tendon are connected by fibers of the lateral patellar retinaculum.* **(Bottom)** *More inferiorly, the 2 tendons have moved closer to each other. This transverse scan shows insertion of the iliotibial tract onto the Gerdy tubercle. The patellar tendon is also approaching its tibial insertion.*

LATERAL KNEE

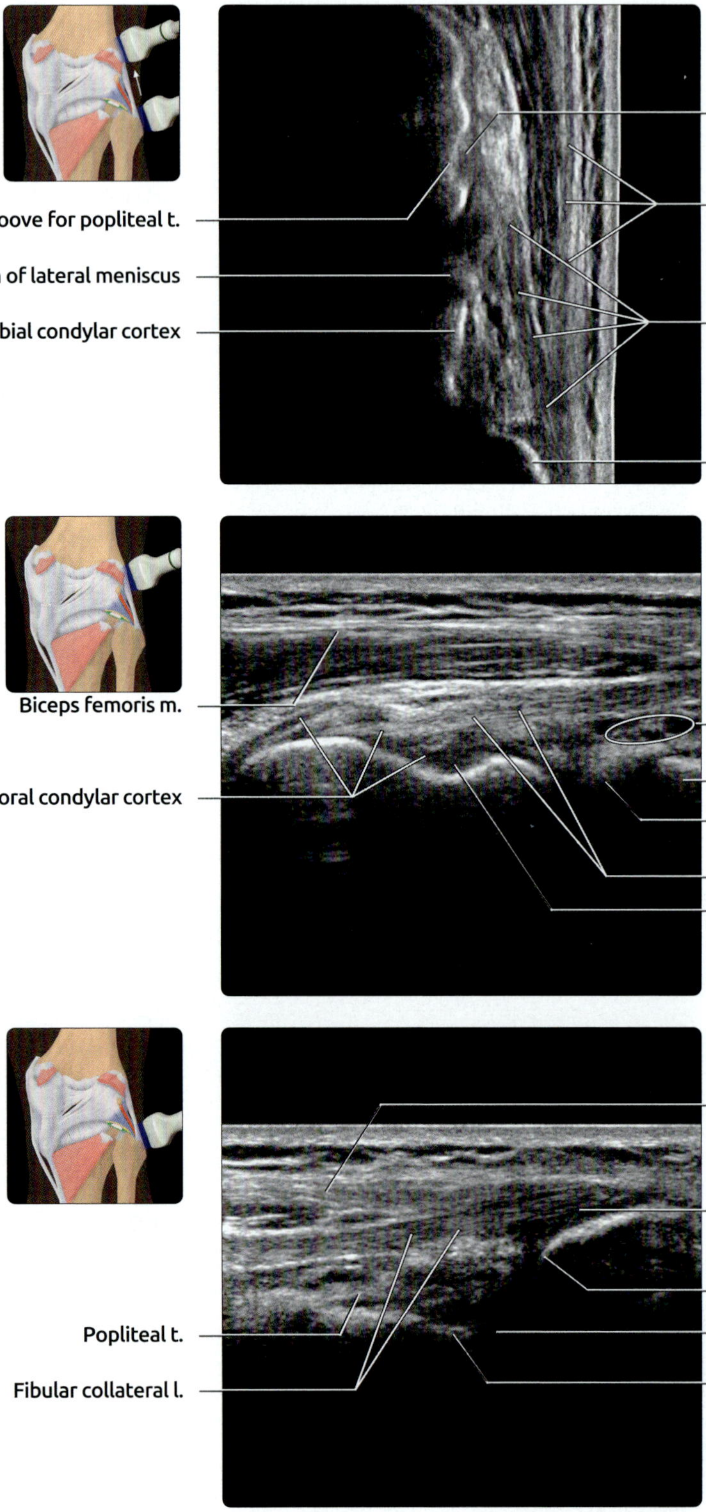

(Top) *Coronal panoramic scan shows the lateral collateral complex. The biceps femoris muscle is the dominant muscle on this side of the knee. It runs from the ischial tuberosity laterally and inferiorly to insert onto the lateral aspect of the head of the fibula.* **(Middle)** *Coronal scan of the lateral femoral condyle shows the insertion of the fibular collateral ligament. Just under this insertion is the groove for the popliteal tendon. The fibular collateral ligament is cord-like compared to the superficial component of the lateral collateral ligament, which is a flatter strap.* **(Bottom)** *The fibular collateral ligament inserts onto the lateral wall of the head of the fibula. The tendon of the biceps femoris inserts at and around the apex of the fibular head.*

POSTERIOR KNEE

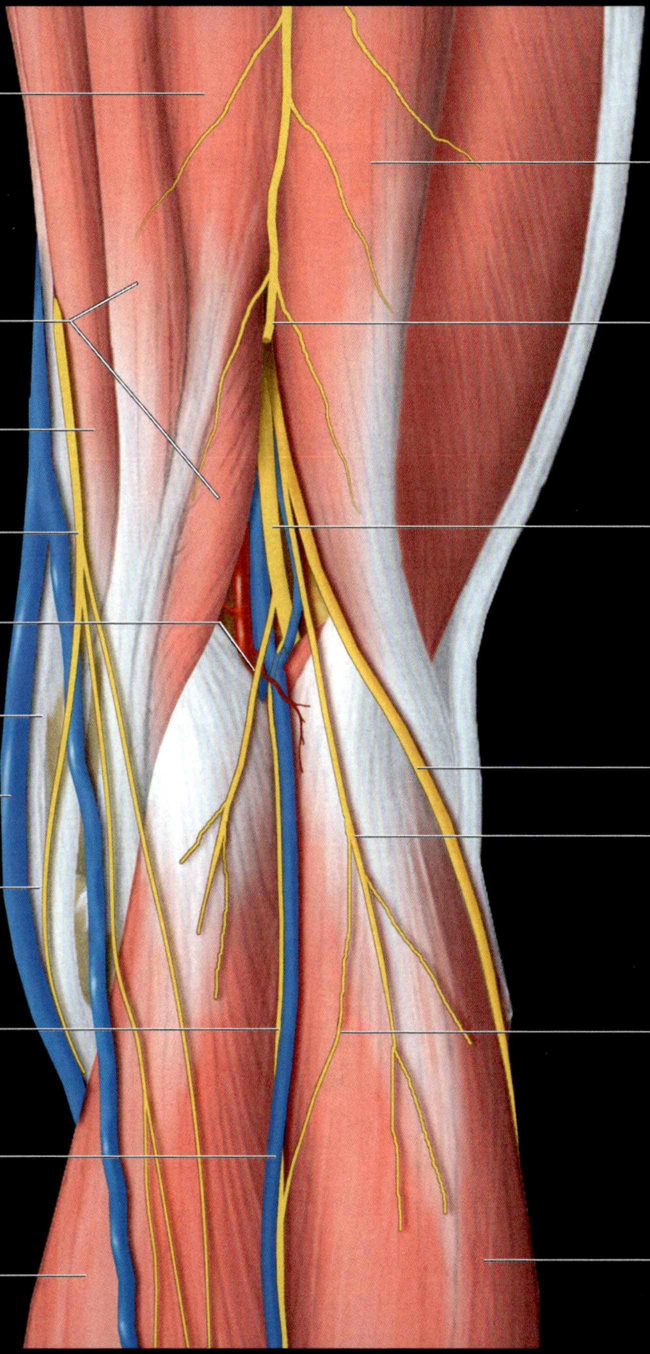

Graphic shows the superficial posterior muscles and nerves. The popliteal fossa is a diamond-shaped fossa bordered superiorly by the hamstring muscles and inferiorly by the head of the gastrocnemius muscle. Superiorly, the medial margin is represented by the semimembranosus and semitendinosus muscles, and laterally by the biceps femoris muscle. Inferiorly, the medial and lateral borders are the medial and lateral heads of the gastrocnemius muscle. This fossa contains the popliteal vessels, the sciatic nerve, and its branches. Note that the common peroneal nerve is part of the deeper nerve system, but it travels more superficially, following the posterior biceps femoris muscle and tendon until it wraps around the fibular neck.

POSTERIOR KNEE

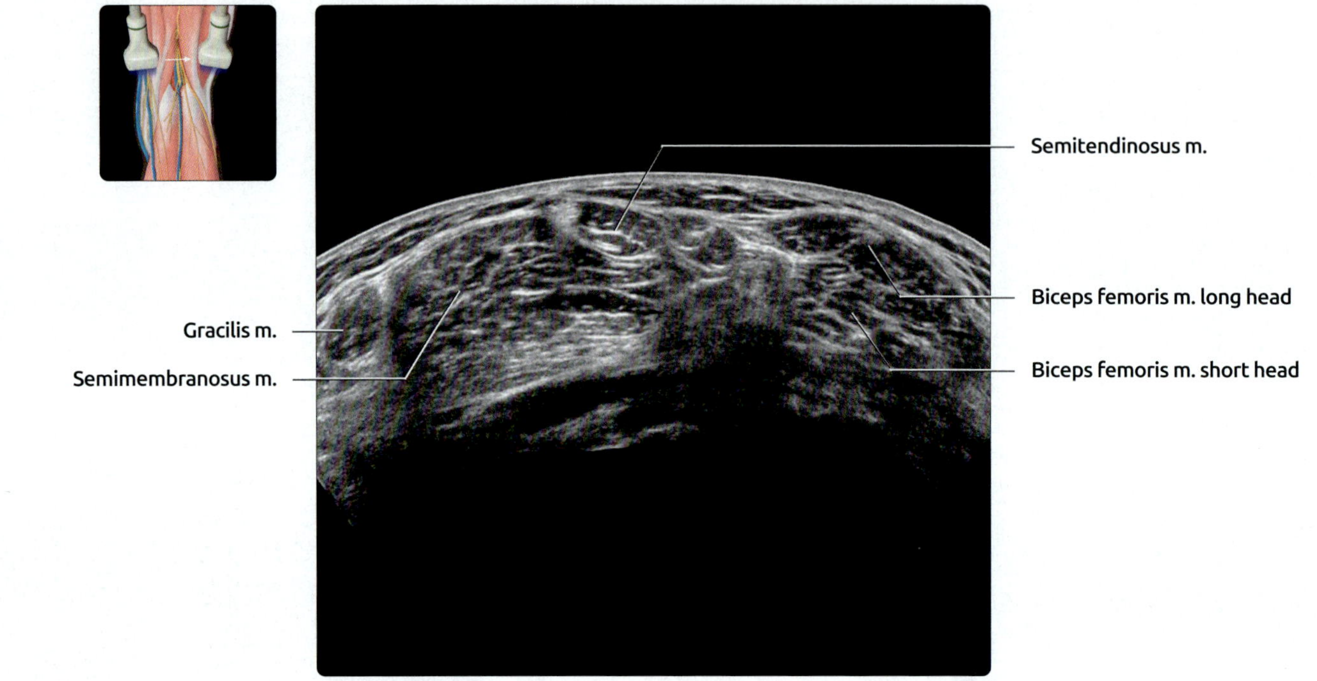

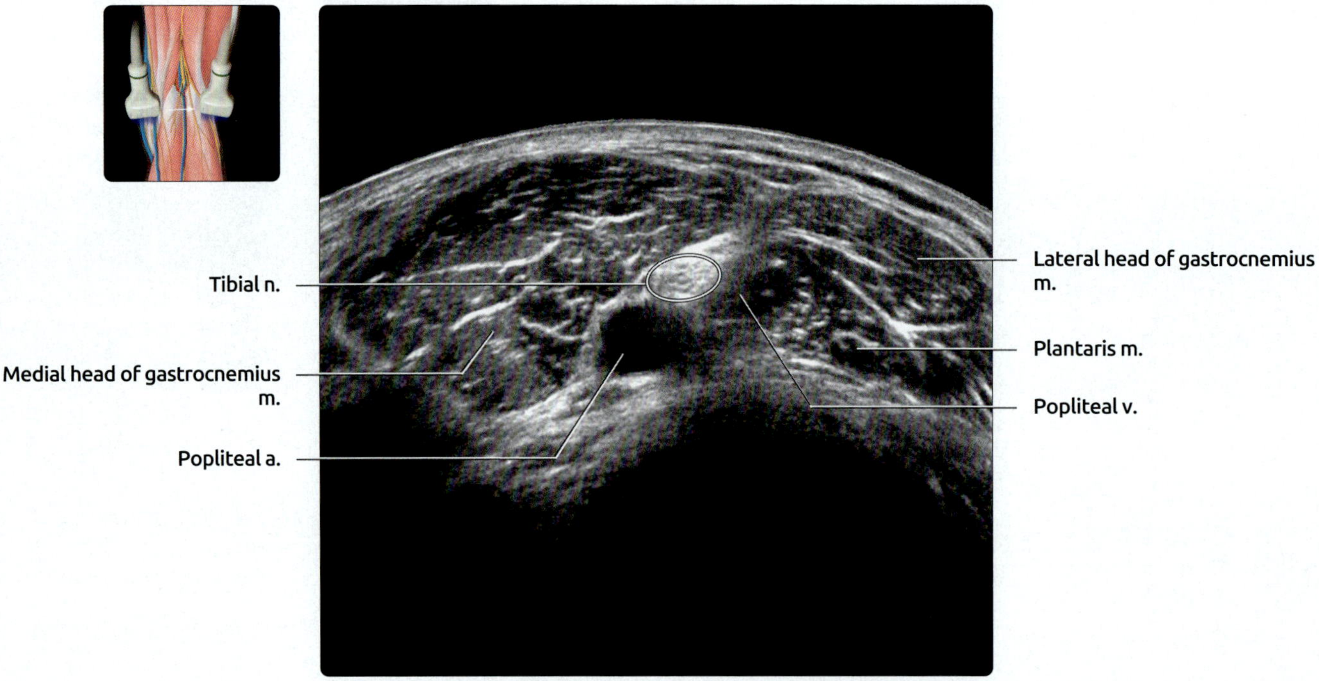

(Top) *Panoramic transverse scan of the posterior knee shows the upper popliteal fossa. At this level, the hamstring muscles form the superior boundaries (medial and lateral) of the popliteal fossa. The semimembranosus and semitendinosus muscles form the medial margin, while the biceps femoris muscle forms the lateral margin.* **(Bottom)** *Panoramic view of the posterior knee shows the lower aspect of the popliteal fossa. At this level, the boundaries of the popliteal fossa are the heads of the gastrocnemius muscle, the lateral head of gastrocnemius being the lateral margin and the medial head the medial margin. The popliteal vessels are a continuation of the superficial femoral vessels after they have emerged through the adductor hiatus.*

POSTERIOR KNEE

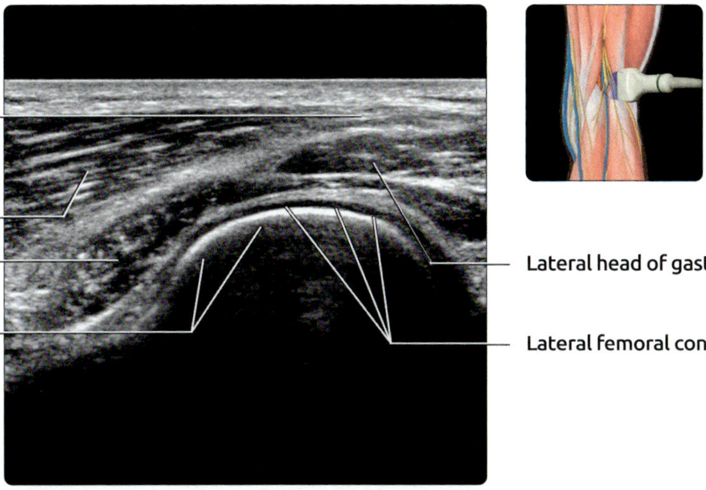

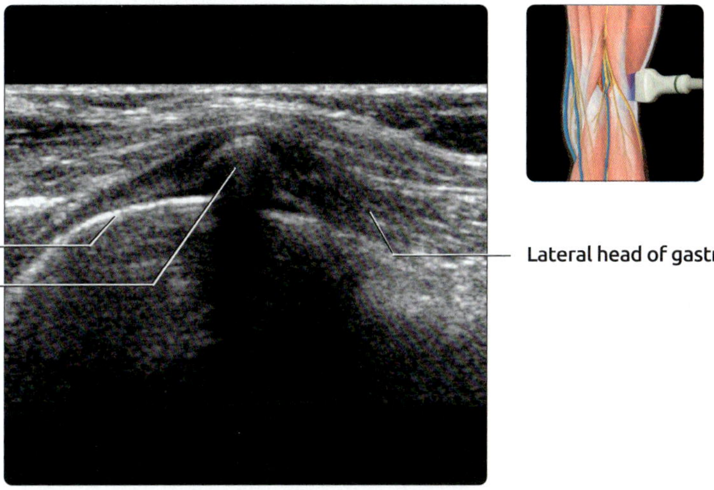

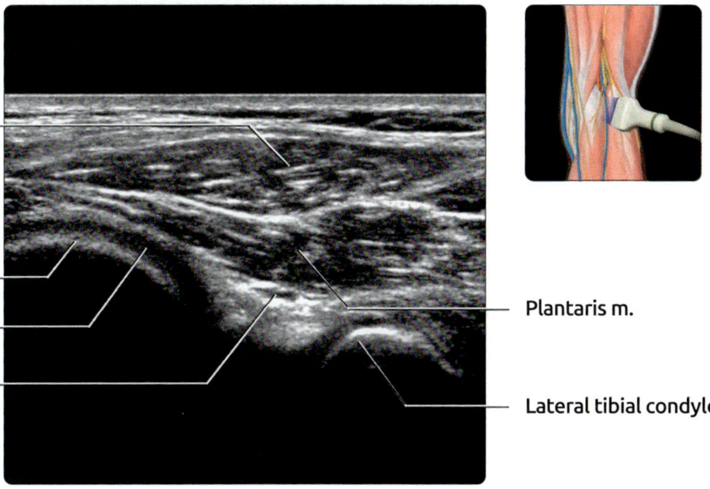

(Top) *Sagittal scan shows the lateral femoral condyle. The plantaris muscle originates slightly more superior and medial to the lateral head of the gastrocnemius. The plantaris runs inferomedially across the thigh to lie on the medial side of the Achilles tendon. More superficially, the bicipital tendon runs laterally toward its fibular head insertion.* **(Middle)** *A more lateral sagittal view shows the sesamoid bone within the lateral head of the gastrocnemius muscle, the fabella.* **(Bottom)** *A more inferior oblique scan shows both the plantaris and lateral heads of the gastrocnemius muscle running over the posterior surface of the knee joint. Further distally, the plantaris muscle becomes a thin tendon and is difficult to discern until it reaches the medial side of the Achilles tendon in the distal calf.*

POSTERIOR KNEE

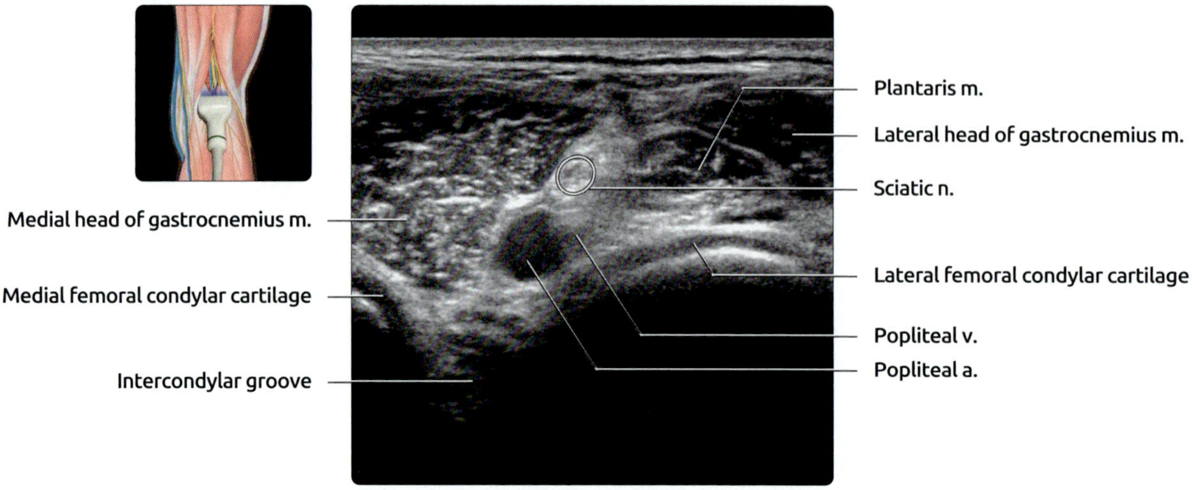

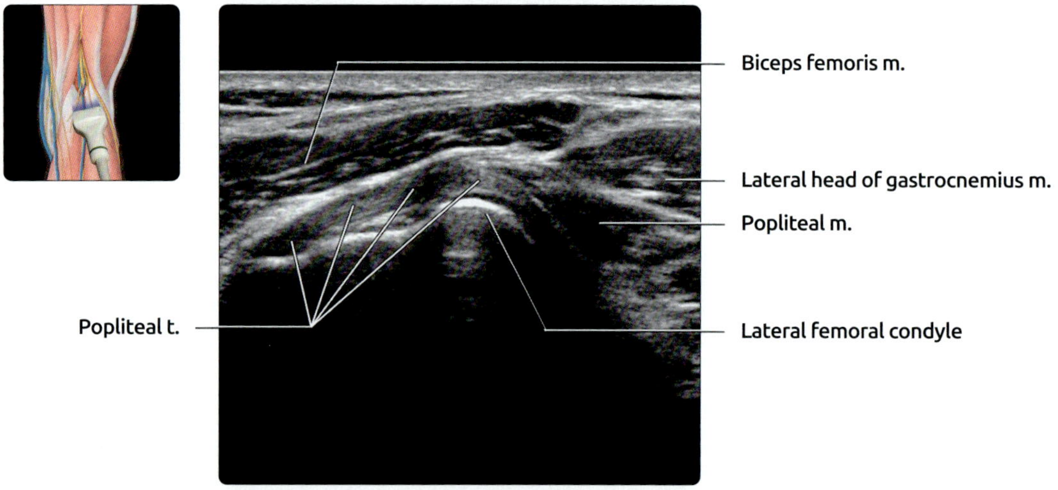

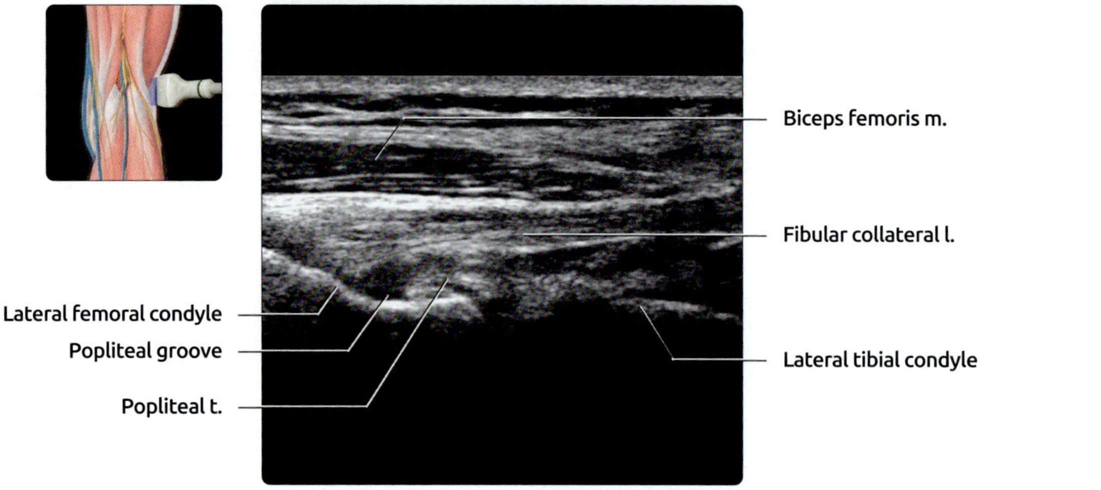

(Top) *Transverse scan shows the posterolateral aspect of the knee joint at the level of the inferior popliteal fossa. The sciatic nerve is the largest nerve in the body. In cross section, nerve fibrils are seen as hypoechoic dots separated by hyperechoic perineurium.* **(Middle)** *Scanning more laterally and inferiorly, the popliteal muscle and its tendon can be demonstrated. The popliteal tendon originates from the proximal posterior tibial surface and runs superolaterally under the lateral head of the gastrocnemius muscle to insert onto the lateral cortex of the lateral femoral condyle.* **(Bottom)** *Coronal scan shows the popliteal groove on the lateral femoral condyle and the popliteal tendon entering this groove as it heads toward its insertion.*

POSTERIOR KNEE

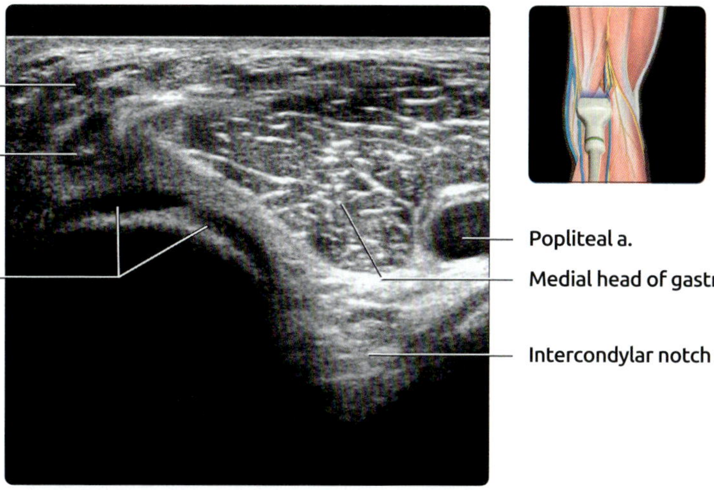

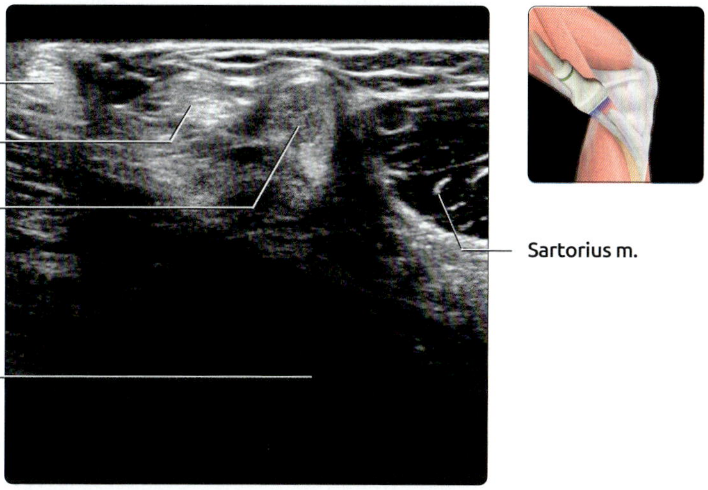

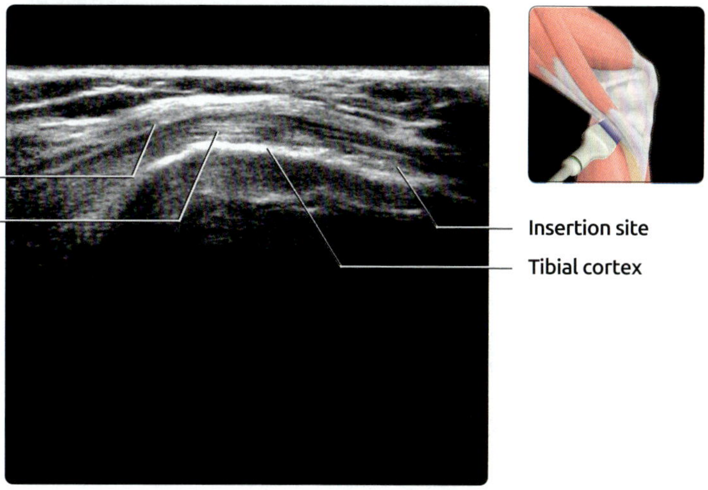

(Top) *Transverse scan shows the posteromedial knee at the level of the femoral condyles. The semimembranosus tendon can be seen running medial to the medial head of the gastrocnemius muscle and deep to the semitendinosus tendon. The semimembranosus tendon will continue inferiorly to its posterior tibial insertion, whereas the semitendinosus will be joined by others to insert on the anteromedial surface of the tibia.* **(Middle)** *More superiorly, transverse scan of the posteromedial aspect of the supracondylar region shows the medial tendons and muscles. Three of these, the sartorius, gracilis, and semitendinosus, will continue together inferiorly and insert adjacent to each other as the pes anserine on the anteromedial aspect of the proximal tibial shaft.* **(Bottom)** *Oblique transverse scan along the course of the pes anserine tendons shows their insertion on the anteromedial tibial cortex.*

GROSS ANATOMY

Osseous Anatomy

- **Tibia**
  - Proximal tibiofibular joint
    - Head of fibula and lateral condyle tibia joined by synovial-lined fibrous capsule
    - May communicate with knee joint (10%)
    - Posterolaterally located
    - Synovial joint, so can be affected by any synovial inflammatory process
  - Anterolateral tibia: Origin of anterior muscles of leg
  - Anterior border (shin): Sharp ridge running from tibial tuberosity proximally to anterior margin of medial malleolus
  - Medial tibial surface
    - Wide and flat
    - Proximally, covered by pes anserinus
    - Remainder is subcutaneous
  - Medial border of tibia: Saphenous nerve and great saphenous vein run along it
  - Posterior tibia: Origin of deep posterior muscles of leg
  - Lateral border of tibia: Ridge for attachment of interosseous membrane
  - Medial malleolus: 2 colliculi, anterior longer than posterior
  - Distal tibiofibular joint
    - Fibula articulates with tibia at fibular notch; joined by interosseous ligament
    - Strengthened by anterior and posterior tibiofibular ligaments
    - Posterolaterally located
- **Fibula**
  - Anterior fibula
    - Origin of lateral muscles of leg
  - Medial fibula
    - Origin of deep posterior muscles of leg
  - Posterolateral fibula
    - Origin of posterior muscles of leg
  - Lateral malleolus: 1 cm longer than medial malleolus

Interosseous Membrane

- Stretches across interval between tibia and fibula
- Greatly extends surface for origin of muscles
- Strong, oblique fibers run downward and laterally from tibia to fibula
- In upper part, below lateral condyle of tibia, there is opening for passage of anterior tibial vessels
- Distally, opening allows passage of perforating branch of peroneal artery
- Tibialis posterior and flexor hallucis longus take partial origin from back of membrane
- Tibialis anterior, long extensors of toes, and peroneus tertius take partial origin from front of membrane

Muscles of Leg

- Compartments separated by deep fascia, which give partial origin to several muscles
- **Posterior compartment: Superficial muscles**
  - **Gastrocnemius**
    - Origin: Medial from posterior femoral metaphysis; lateral from posterior edge of lateral epicondyle
    - Heads separated from posterior capsule by bursa
    - 2 heads unite to form main bulk of muscle
    - Join in thin aponeurotic tendon near midleg
    - Joins soleus aponeurosis to form Achilles tendon; concave in cross section; musculotendinous junction is 5 cm above calcaneal insertion
    - Nerve supply: Tibial nerve
    - Action: Plantar flexor of ankle and flexor of knee
  - **Plantaris**
    - Origin: Superior and medial to lateral head of gastrocnemius origin, as well as from oblique popliteal ligament
    - Continues deep to lateral head of gastrocnemius
    - Myotendinous junction at level of origin of soleus (muscle is 5-10 cm long)
    - Tendon then lies between medial head gastrocnemius and soleus
    - Follows medial side of Achilles to insert either anteromedially on Achilles or on calcaneus
    - Plantaris absent in 7-10% of general population
    - Nerve supply: Tibial nerve
    - Action: Acts with gastrocnemius
  - **Soleus**
    - Origin: Extensive, from back of fibular head and upper 1/3 of posterior surface of shaft of fibula, from soleal line and middle 1/3 of medial border of tibia, and from tendinous arch joining these across popliteal vessels
    - Flat, thick, powerful muscle that ends in strong tendon
    - Joins with tendon of gastrocnemius to form Achilles tendon
    - Nerve supply: Tibial nerve
    - Action: Stabilizes ankle in standing, plantarflexes ankle
    - Accessory soleus: Rare variant, arises from anterior surface of soleus or from fibula and soleal line of tibia; inserts into Achilles or onto calcaneus anteromedially to Achilles; presents as mass
- **Posterior compartment: Deep muscles**
  - **Popliteus**
    - Origin: Tendon from popliteal groove of lateral femoral condyle
    - Passes through popliteal hiatus posteriorly and medially, pierces posterior capsule of knee
    - Muscle fibers directed medially and downward to insert on posterior surface of tibia above soleal line
    - Nerve supply: Tibial nerve
    - Action: Flexes knee and medially rotates tibia with respect to femur at onset of flexion (unlocking extension "screwing home" mechanism)
  - **Tibialis posterior**
    - Origin: Interosseous membrane and adjoining parts of posterior surfaces of tibia and fibula
    - Superior end bifid; anterior tibial vessels pass forward between 2 attachments
    - Distally it inclines medially, under flexor digitorum longus
    - Grooves and curves around medial malleolus
    - Nerve supply: Tibial nerve
    - Action: Plantarflexes and inverts foot

- **Flexor digitorum longus**
  - Origin: Posterior surface of tibia, below popliteus, and medial to vertical ridge
  - Crosses superficial to distal part of tibialis posterior
  - Tendon grooves lower end of tibia lateral to that of tibialis posterior, passes around medial malleolus to foot
  - Nerve supply: Tibial nerve
  - Action: Flexes interphalangeal and metatarsophalangeal joints of lateral 4 toes; plantarflexes and inverts foot
- **Flexor hallucis longus**
  - Origin: Posterior surface of fibula, below origin of soleus
  - Passes medially, descends down posterior to midtibia
  - Associated with os trigonum posterior to talus
  - Tendon occupies deep groove on posterior surface of talus, passes around medial malleolus, under sustentaculum tali, to great toe
  - Nerve supply: Tibial nerves
  - Action: Flexes interphalangeal and metatarsophalangeal joints of great toe; plantarflexes foot

- **Lateral compartment**
  - Peroneals separated from extensors by anterior intermuscular septum and from posterior muscles by posterior septum
  - **Peroneus longus**
    - Origin: Upper 2/3 lateral surface of fibula and intermuscular septa and adjacent muscular fascia
    - Becomes tendinous few cm above lateral malleolus
    - Curves forward behind lateral malleolus, posterior to peroneus brevis
    - Nerve supply: Superficial peroneal
    - Action: Everts foot and secondarily plantarflexes foot
  - **Peroneus brevis**
    - Origin: Lower 2/3 lateral surface of fibula and intermuscular septa and adjacent muscular fascia
    - Muscle is medial to peroneus longus at origin but overlaps peroneus longus in middle 1/3
    - Tendon curves forward behind lateral malleolus, in front of peroneus longus tendon
    - Nerve supply: Superficial peroneal
    - Action: Everts foot and secondarily plantarflexes foot
  - Synovial sheath for peroneals begins 5 cm above tip of lateral malleolus and envelops both tendons; divides into 2 sheaths at level of calcaneus
  - **Peroneus quartus**
    - Accessory muscle with prevalence of 10%
    - Originates from distal leg, frequently from peroneal muscles, with variable insertion sites at foot
    - At level of malleolus, located medial or posterior to both peroneal tendons
  - **Peroneus digiti minimi**
    - Accessory with prevalence of 15-36%
    - Extends from peroneus brevis muscle around medial malleolus to foot
    - Tiny tendinous slip
- **Anterior compartment**
  - **Tibialis anterior**
    - Origin: Upper 1/2 of lateral surface of tibia and interosseous membrane
    - Tendon originates in distal 1/3; passes deep to retinaculum
    - Nerve supply: Deep peroneal and recurrent genicular
    - Action: Dorsiflexor and invertor of foot
  - **Extensor digitorum longus**
    - Origin: From upper 3/4 of anterior surface fibula
    - Descends behind extensor retinacula to ankle
    - Nerve supply: Deep peroneal
    - Action: Extends interphalangeal and metatarsophalangeal joints of lateral 4 toes, dorsiflexes foot
  - **Peroneus tertius**
    - Small, not always present
    - Origin: Continuous with extensor digitorum longus, arising from distal 1/4 of anterior surface of fibula and interosseous membrane
    - Inserts into dorsal surface at base of 5th metatarsal
    - Nerve supply: Deep peroneal
    - Action: Dorsiflexes ankle and everts foot
  - **Extensor hallucis**
    - Thin muscle hidden between tibialis anterior and extensor digitorum longus
    - Origin: Middle 1/2 of anterior surface of fibula and interosseous membrane
    - Tendon passes deep to retinacula to great toe
    - Nerve supply: Deep peroneal
    - Action: Extends phalanges of great toe and dorsiflexes foot

ANATOMY IMAGING ISSUES

Imaging Recommendations

- US
  - Owing to its smaller size, compared to thigh (less depth to penetrate), leg muscles and vessels are well demonstrated by ultrasound throughout their course
  - Relationship of different structures in posterior compartment tends to change along their courses
    - If in doubt, start from ankle level and trace structures proximally

Imaging Sweet Spots

- Deep structures of posterior compartment, such as peroneal vessels and tibialis posterior muscle, may be difficult to visualize in very muscular patients
  - Structures can be interrogated by scanning anteriorly through anterior compartment and gap between tibia and fibula
- One of most common reasons to image calf is for suspected muscle tear
  - Most common site of muscle tear is at distal myofascial junction of medial belly of gastrocnemius
  - Look carefully for preservation of arrowhead appearance of myofascial junction at this location
  - With plantaris tendon tear, this junction is intact, and there is more extensive hemorrhage deep to medial belly of gastrocnemius

ANTEROLATERAL LEG

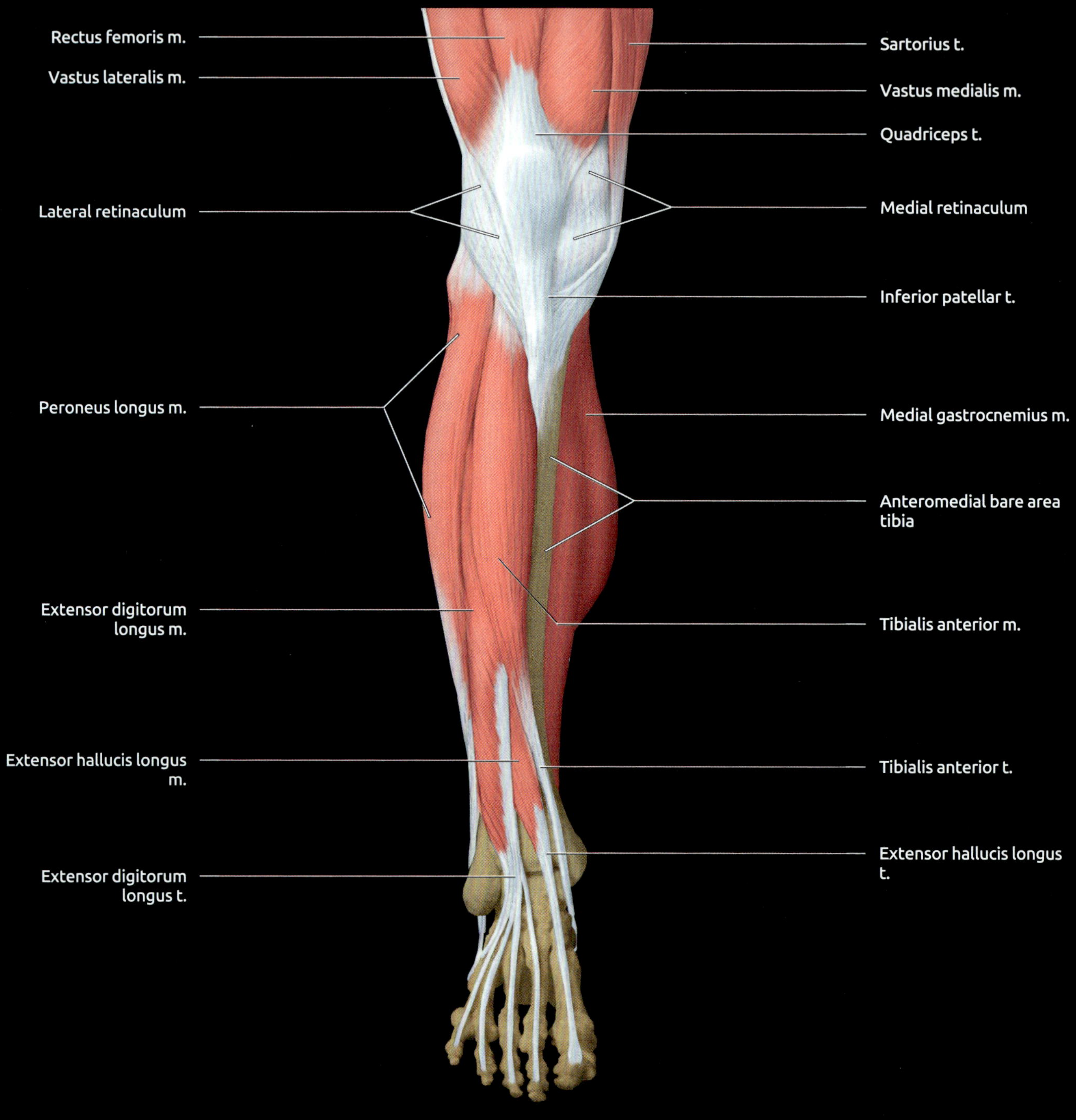

Graphic shows the anterior leg muscles. Note that nearly the entire anteromedial tibia is bare of muscle. Thus, it is poorly vascularized, resulting in slow fracture healing. The tibialis anterior has an extensive origin from both tibia and interosseous membrane and is the most substantial muscle of the anterior compartment. The extensor digitorum longus originates from the fibula. The extensor hallucis longus originates between the 2, from the fibula and the interosseous membrane. These 3 muscles retain the same orientation as they become tendinous anterior to the ankle. The mnemonic "Tom, Harry, & Dick" applies to the tendon order, from medial to lateral (tibialis anterior, extensor hallucis longus, extensor digitorum).

PANORAMIC ANTEROLATERAL LEG

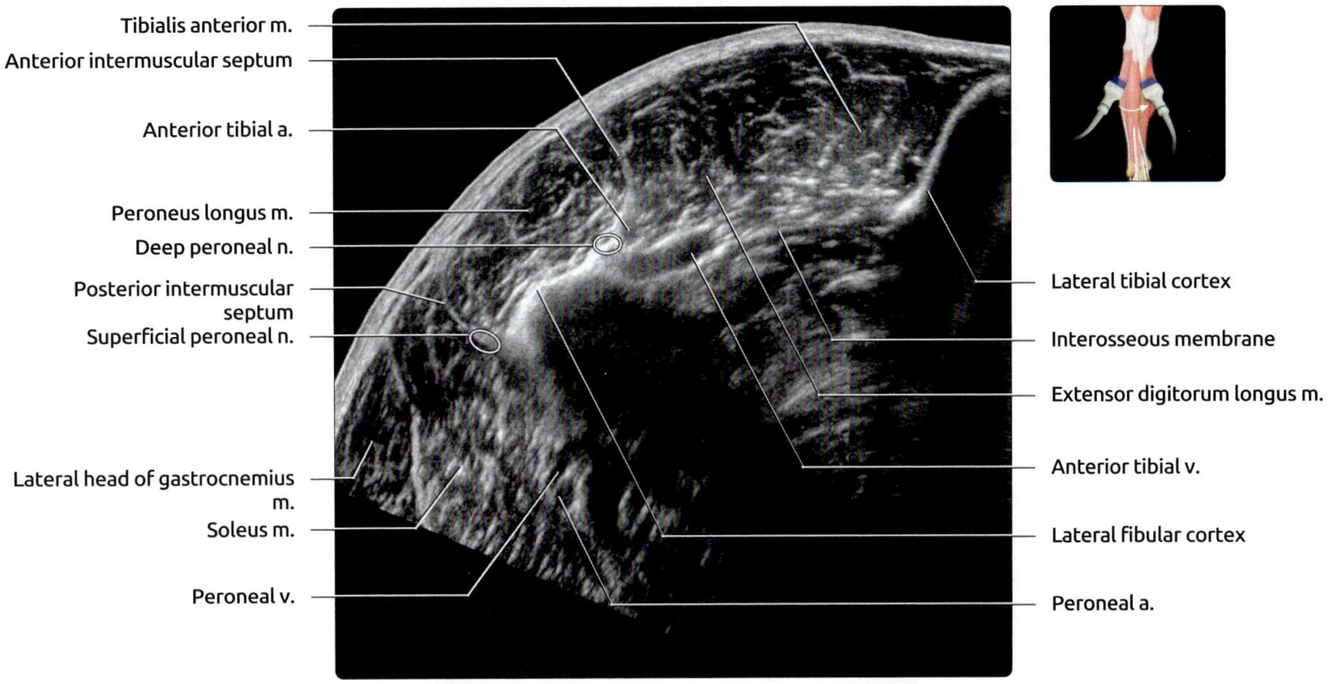

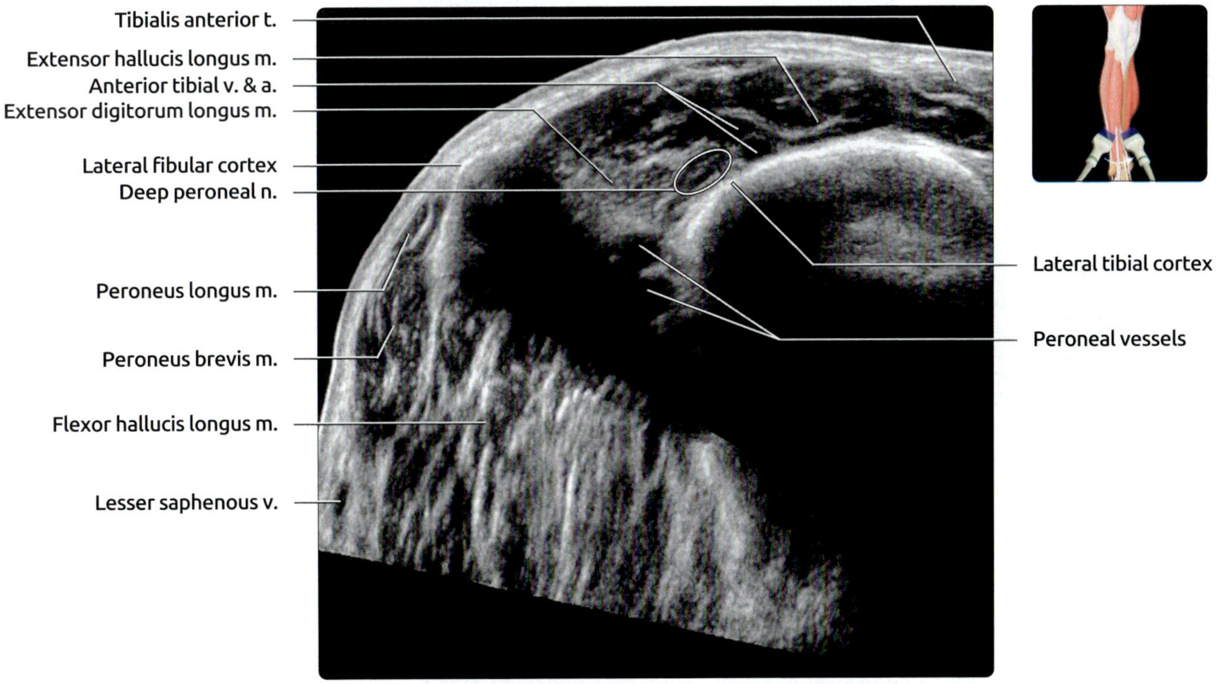

(Top) *Transverse panoramic view shows the anterior and lateral compartments of the upper leg. The crural compartments are divided by the tibia and the 2 intermuscular septa (anterior and posterior) into the anterior (extensor), lateral (peroneal), and posterior (flexor) compartments. The intermuscular septum separates the anterior from the posterior compartments.* **(Bottom)** *Transverse panoramic view shows the anterior and lateral compartments of the inferior leg. All muscles are smaller in caliber compared to those seen superiorly. An interosseous gap is still present between the tibia and fibula, but ultrasound interrogation through this gap is no longer possible and the interosseous membrane cannot be demonstrated.*

ANTERIOR COMPARTMENT

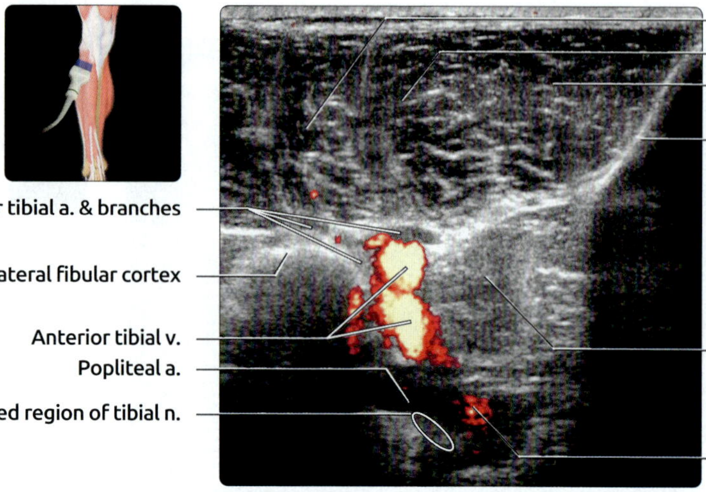

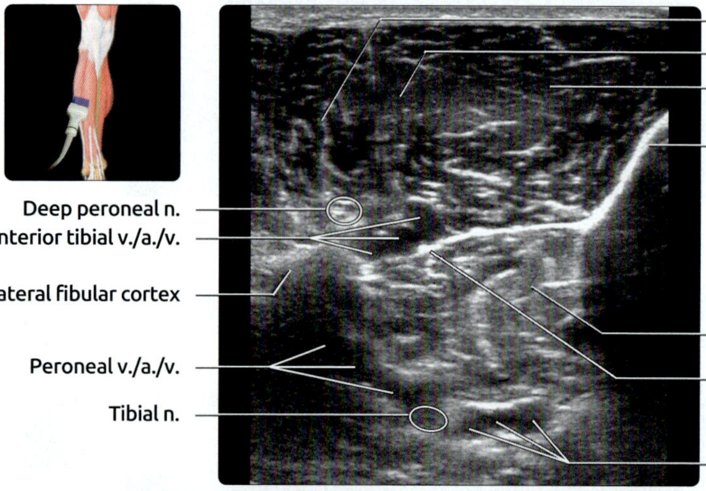

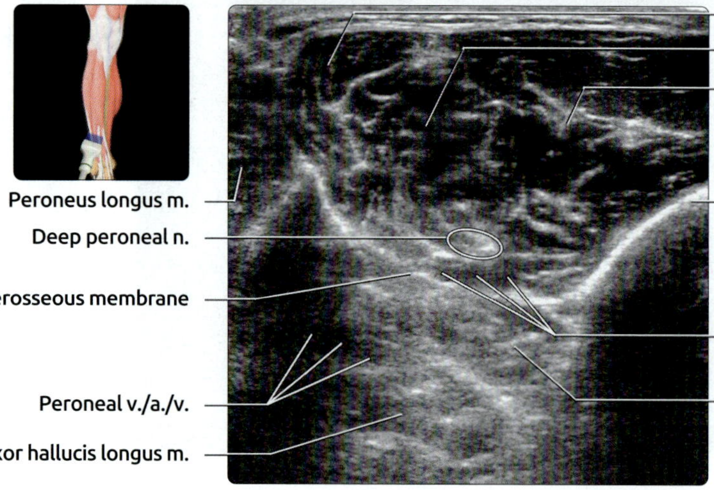

(Top) *Transverse color Doppler scan shows the anterior compartment at the upper leg level slightly below the proximal tibiofibular joint. At this level (above the superior margin of the interosseous membrane), the anterior tibial artery and vein (branches of the popliteal artery and vein) run anteriorly to supply the anterior compartment.* **(Middle)** *Transverse scan shows the anterior compartment at a slightly lower level. The tibialis anterior, extensor hallucis longus, and extensor digitorum longus muscles form the muscular component of this compartment. The deep peroneal nerve has penetrated laterally to supply this compartment. The anterior tibial vessels run on the surface of the interosseous membrane.* **(Bottom)** *Transverse scan more inferiorly shows a decrease in size of the extensor muscles and gradual migration of nerve and vessels more medially and superficially.*

ANTERIOR COMPARTMENT

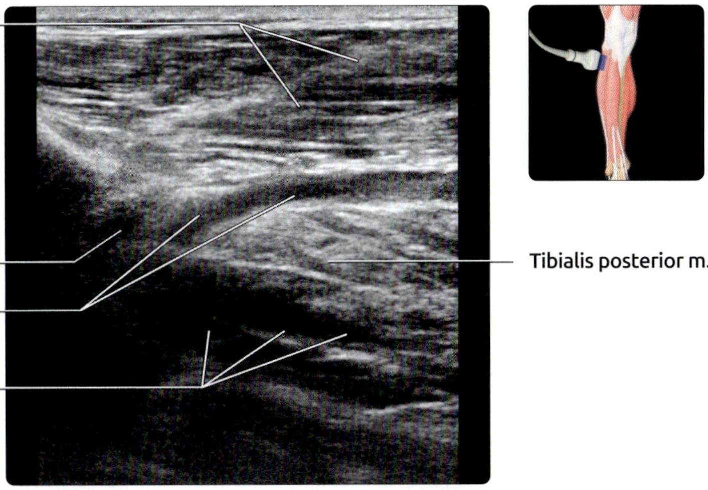

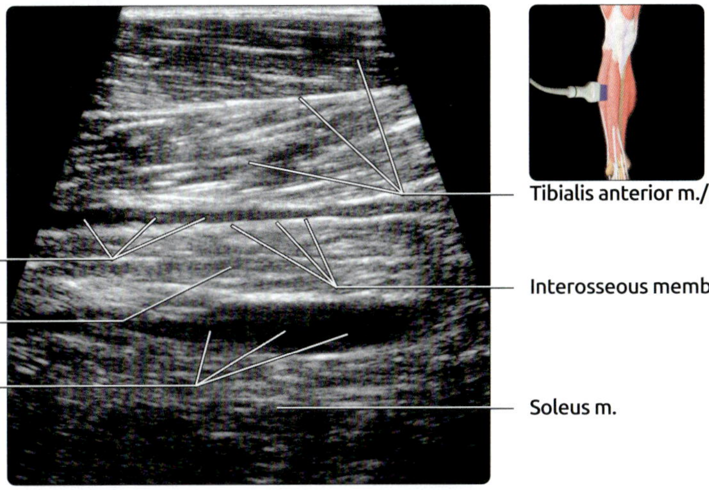

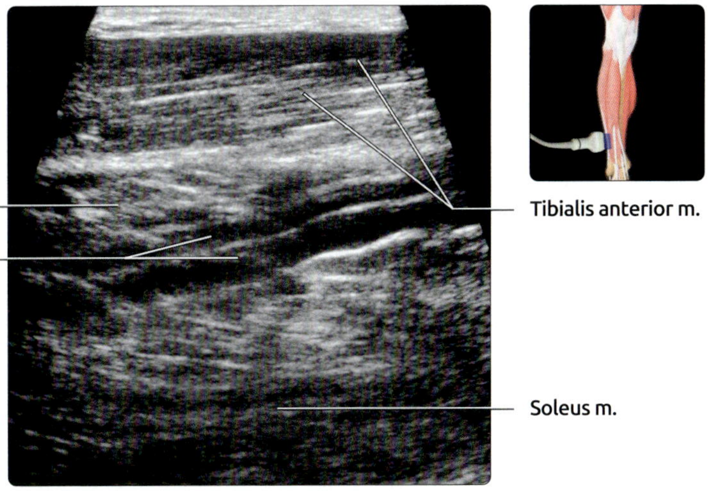

(Top) *Sagittal scan shows the anterior compartment at the upper leg level. The anterior tibial vessels pass above the superior margin of the interosseous membrane to enter the anterior compartment and supply the extensor muscles.* **(Middle)** *Sagittal scan of the anterior and posterior compartments at the midleg level shows that the interosseous gap between the tibia and fibula can be used as a window to interrogate the deep posterior structures.* **(Bottom)** *Sagittal scan shows the anterior compartment at a lower level. Within the posterior compartment, the peroneal vessels run anteriorly and just below the inferior margin of the interosseous membrane.*

LATERAL COMPARTMENT

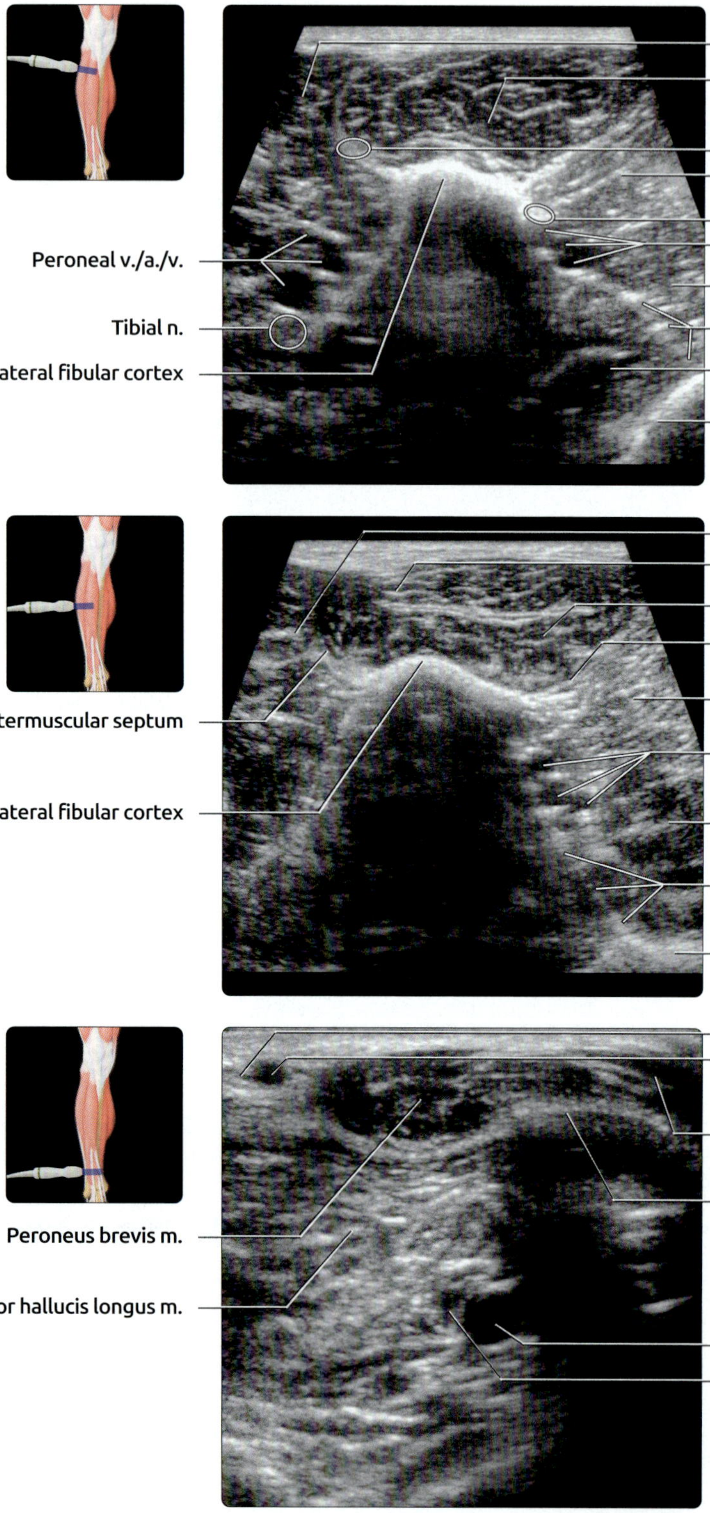

(Top) *Transverse scan shows the lateral compartment at a high level. At this level, only the peroneus longus muscle has originated and is the only muscle occupying this compartment. The superficial peroneal nerve lies on the posterior surface of the peroneus longus muscle. The deep peroneal nerve has already entered the anterior compartment after curving around the fibular neck (above this scan level).* **(Middle)** *Transverse scan at the midleg level shows that the peroneus brevis muscle has now originated and resides anterior and deep to the peroneus longus muscle. More inferiorly, the peroneus longus muscle will become smaller and turn into a tendon while the peroneus brevis will still have muscle fibers down to the level of the ankle joint.* **(Bottom)** *Transverse scan shows the level of the lateral malleolus. The peroneal tendons will run posterior to the lateral malleolus and curve around its tip to run toward the lateral foot.*

LATERAL COMPARTMENT

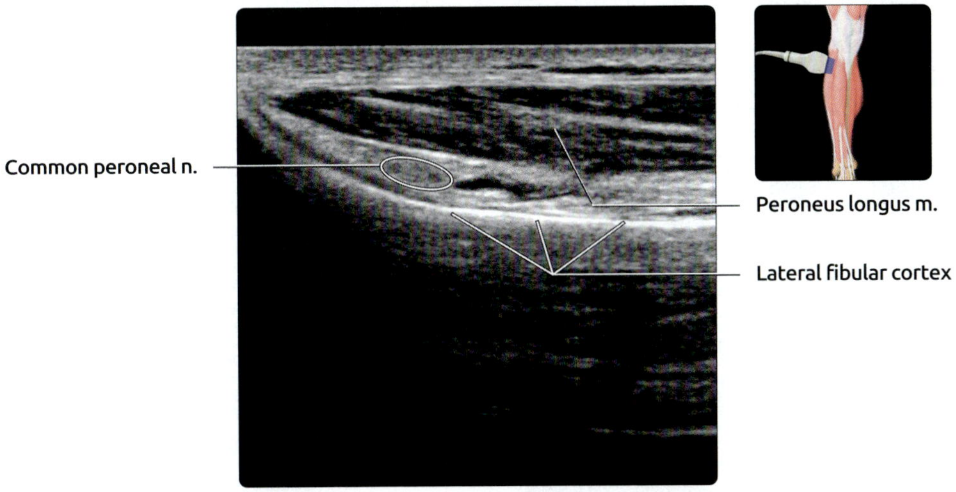

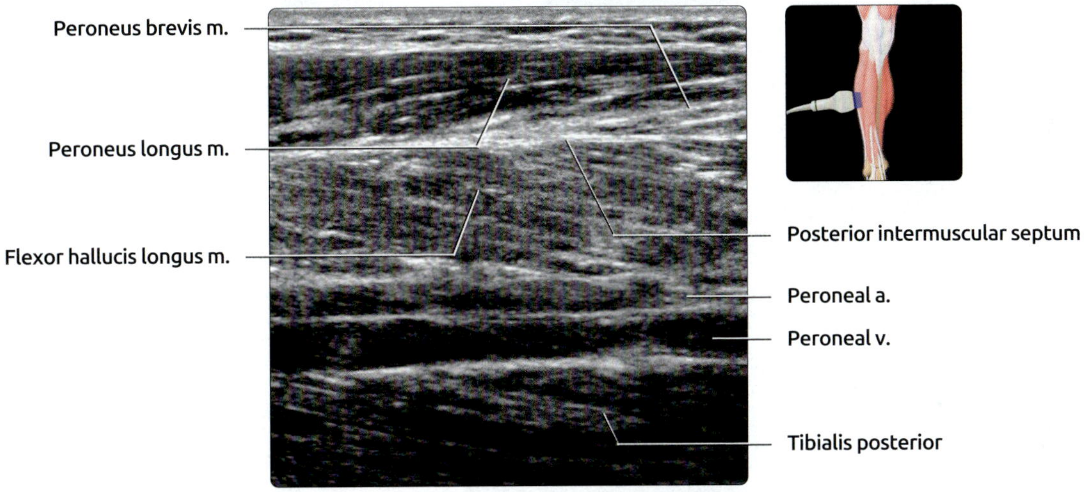

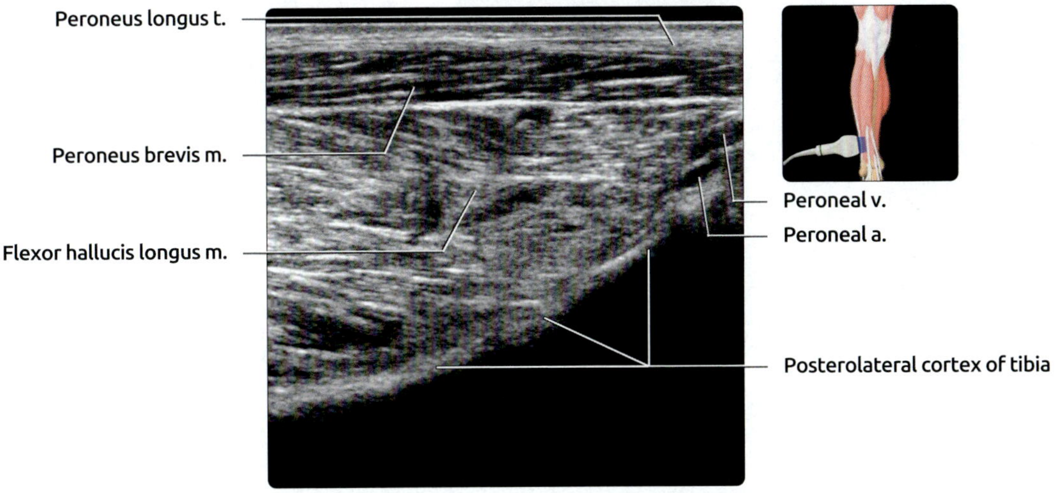

(Top) *Coronal scan shows the lateral compartment at the level of the fibular neck. The common peroneal nerve runs together with the biceps femoris tendon, then curves anteriorly around the neck of fibula to enter the lateral and then the anterior compartments.* **(Middle)** *Coronal scan of the lateral compartment at the midleg level shows the transition between the peroneus longus and peroneus brevis muscles. These are separated from the posterior compartment by the posterior intermuscular septum.* **(Bottom)** *Coronal scan further inferiorly shows that the peroneus longus muscle is now only a tendon, compared to the muscular peroneus brevis. Similarly, in the posterior compartment, the flexor hallucis longus muscle is the only muscle at this level, as the others have also become tendons.*

LATERAL LEG

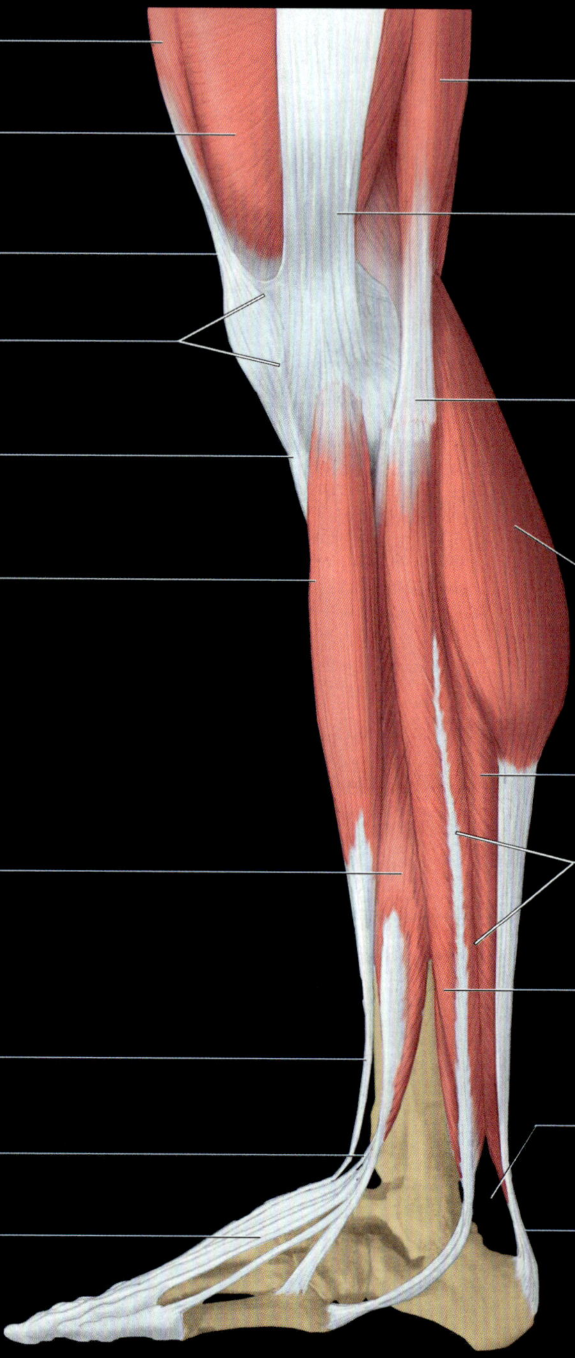

Graphic of the lateral leg shows the anterior compartment (extensors), lateral compartment (peroneals), and superficial muscles of the posterior compartment. Of the ankle tendons, tendinosis most commonly affects the Achilles tendon, the posterior tibialis tendon, and the peroneal tendons. The tibialis anterior tendon is also affected, though less commonly. As the peroneus brevis lies closest to the bone, it is more prone to trauma particularly in the retromalleolar region. Typically, this tear consists of a longitudinal split in which the overlying peroneus longus tendon pushes into the peroneus brevis tendon, splitting it into 2 sections. One may then see 3 rather than 2 tendons behind the lateral malleolus. Another reason to see 3 tendons behind the lateral malleolus is when there is a quartus tendon present.

POSTERIOR LEG

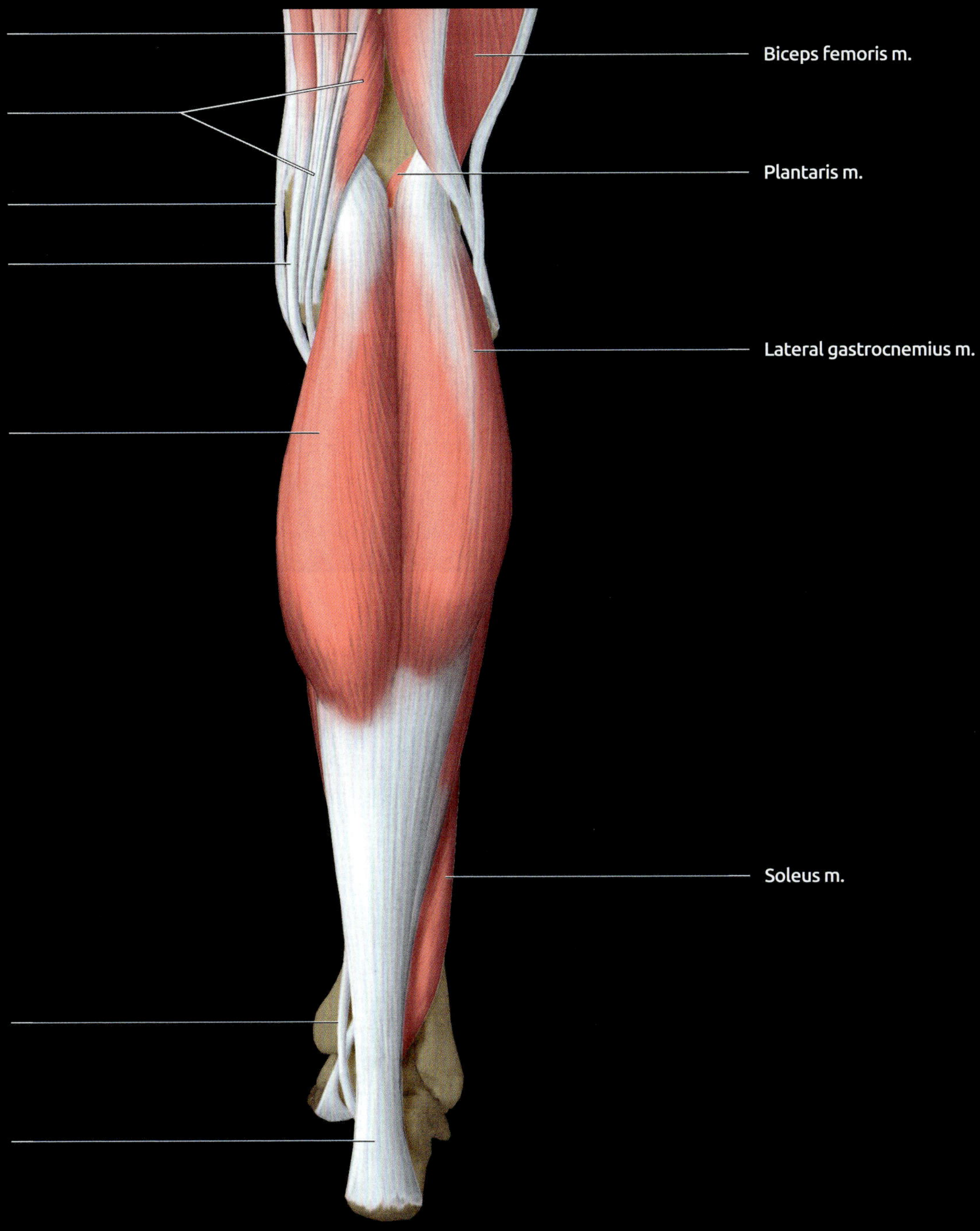

Posterior view shows superficial muscles and tendons of the leg. The gastrocnemius muscles are bulky in the proximal 1/2 of the leg and taper to an aponeurosis, which blends with the soleus more distally to become the Achilles tendon. The plantaris muscle is superficial only at its origin from the lateral femoral metaphysis, just medial to the origin of the lateral head of gastrocnemius. The muscle extends only a few centimeters before becoming tendinous, extending distally between the soleus and medial gastrocnemius. Eventually, the plantaris tendon merges with the medial side of the Achilles or inserts on the calcaneus, medial to the Achilles insertion.

PANORAMIC POSTERIOR LEG

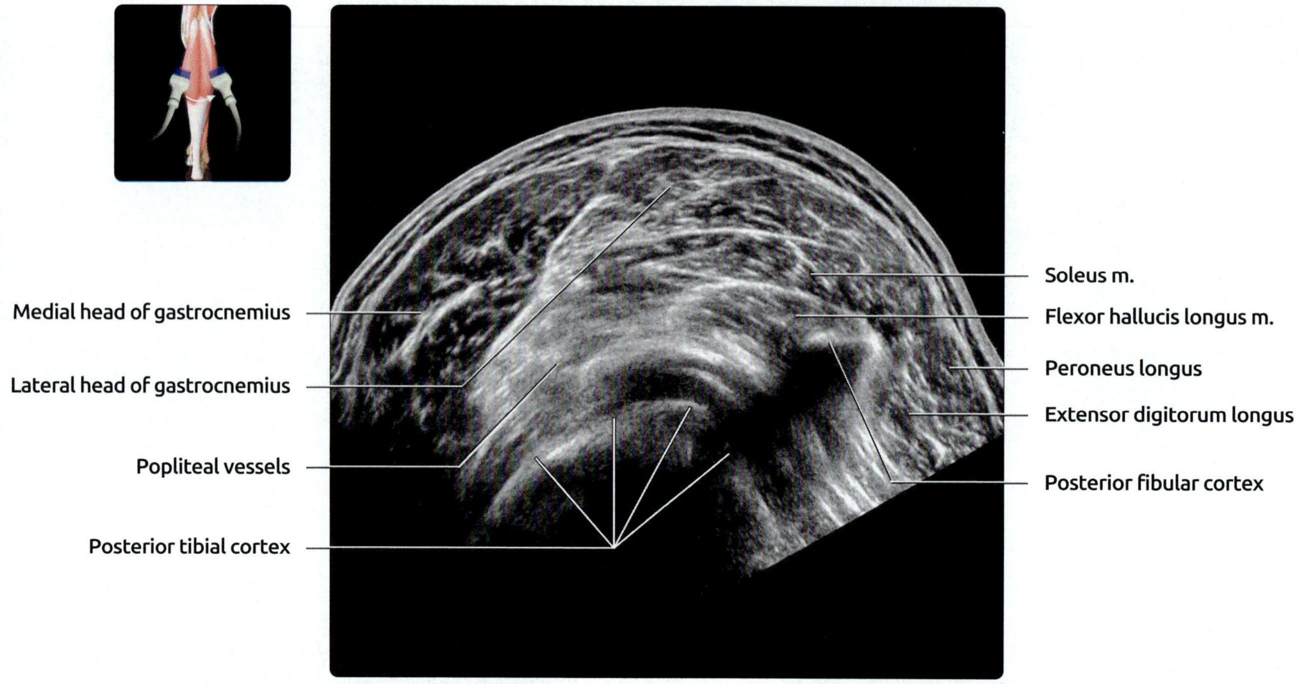

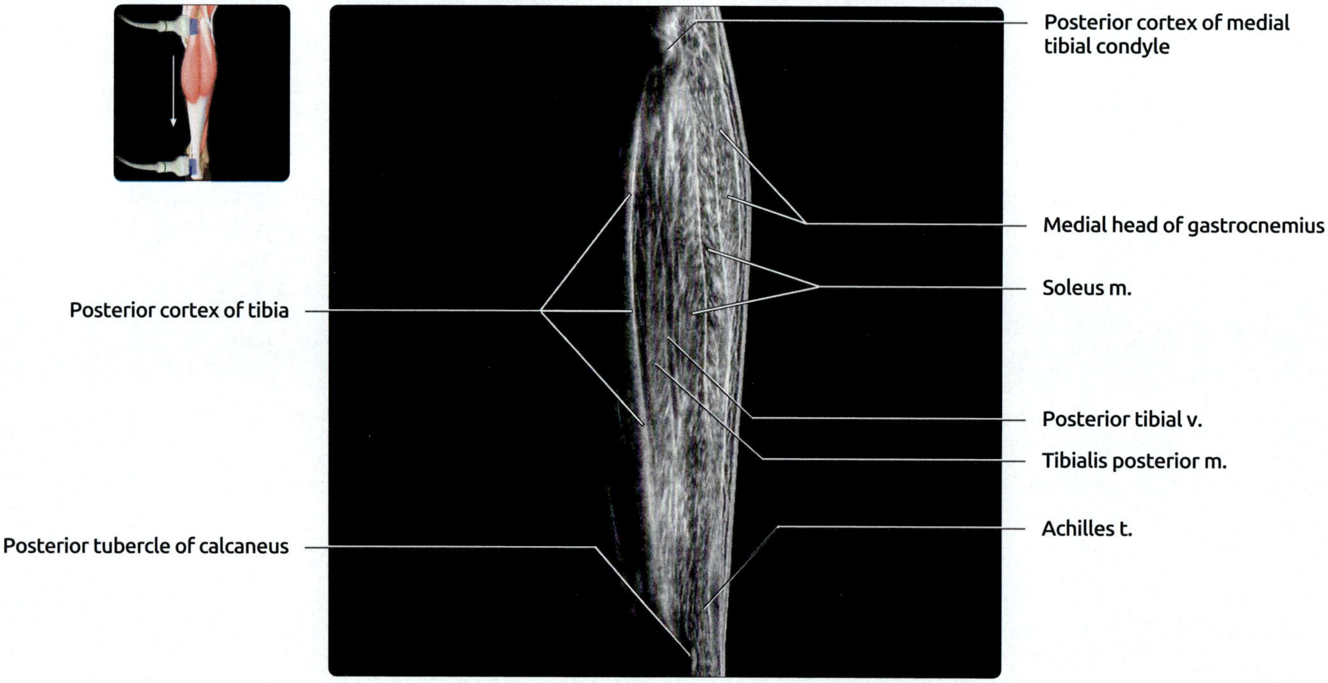

(Top) *Transverse panoramic scan shows the posterior compartment at the level of the midleg. The medial head of the gastrocnemius muscle is larger than the lateral head throughout its course. From this midcalf level downward, the soleus muscle will assume relatively more bulk as the gastrocnemius turns tendinous.* **(Bottom)** *Sagittal panoramic scan shows the posterior compartment. The gastrocnemius is muscular down to ~ the level of the midcalf. Below this level, it forms the Achilles tendon, receiving contributions from the soleus muscle. For the deep layer, the tibialis posterior muscle lies on the surface of the tibial shaft and interosseous membrane.*

TRANSVERSE POSTERIOR LEG

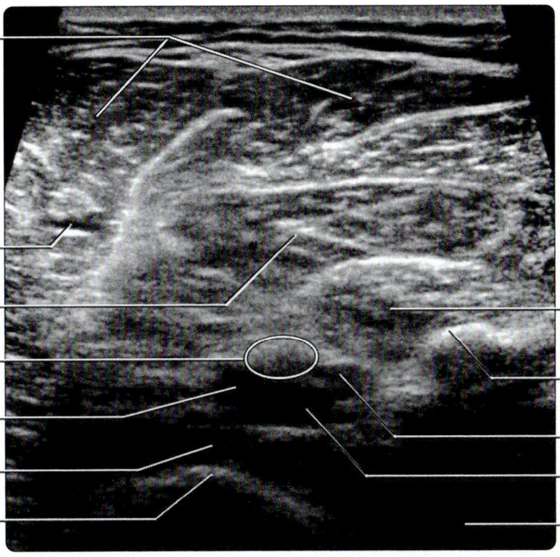

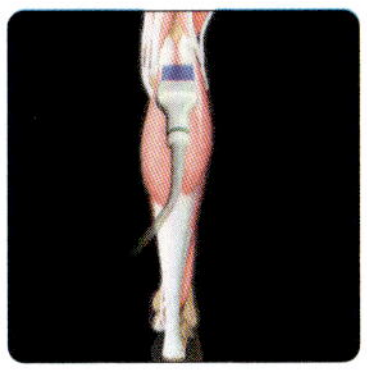

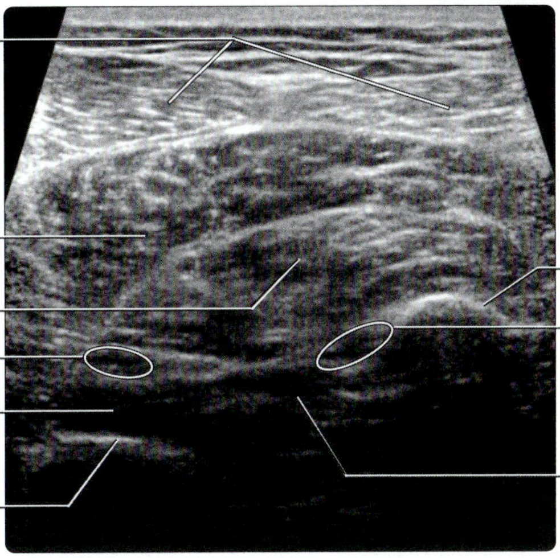

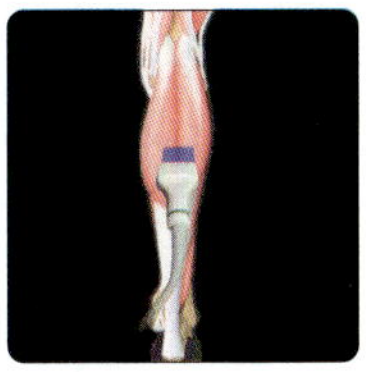

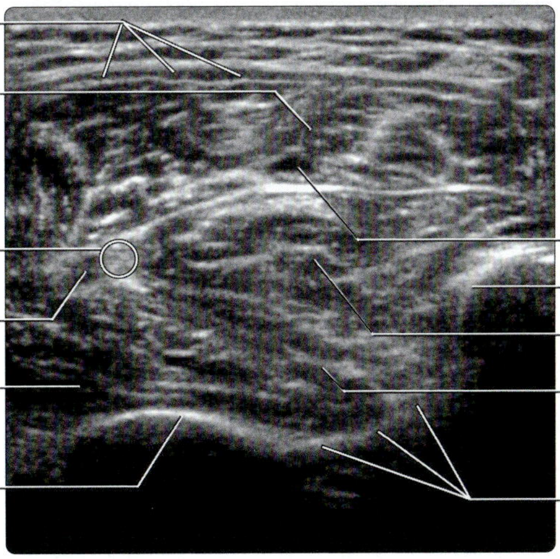

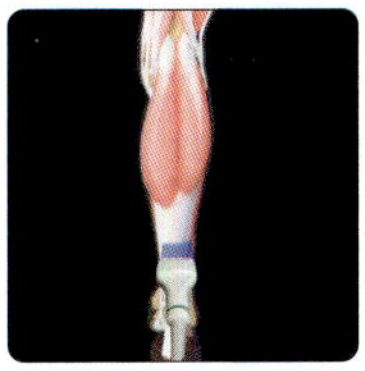

(Top) *Transverse scan is seen through the posterior compartment at the upper-calf level. For this compartment, the superficial layer consists of the gastrocnemius, plantaris, and soleus muscles. The deep layer consists of the popliteus, tibialis posterior, flexor digitorum longus, and flexor hallucis longus muscles.* **(Middle)** *Transverse scan shows the midcalf level. The tibial vessels have bifurcated into their peroneal and posterior tibial branches, which delineates a separation between the tibialis posterior and flexor digitorum longus from the rest of the calf muscles. Given the thickness of the gastrocnemius and soleus muscle, these deep muscles may be better seen using the anterior compartment as a window.* **(Bottom)** *Further inferiorly, the gastrocnemius has become a thin aponeurosis and together with the tendon from the soleus muscle will form the Achilles tendon.*

GROSS ANATOMY

Arteries

- **Popliteal artery**
  - Begins as continuation of superficial femoral artery after it has passed through adductor hiatus (inferior end of adductor canal)
  - Runs through fat of popliteal fossa
  - Relationships
    - Deep to artery: Femoral shaft, knee joint capsule, popliteus fascia
    - Superficial to artery: Semimembranosus and gastrocnemius muscles
    - **Relation to popliteal vein is variable**
      - Artery lies just deep to vein in 40%
      - Deep and medial to vein in 43%
      - Deep and lateral to vein in 9%
      - Just medial or lateral to vein in 8%
  - Gives off 4 named geniculate branches around knee
    - Superior medial and lateral geniculate arteries
    - Inferior medial and lateral geniculate arteries
  - Ends at distal border of popliteus muscle and divides into 2 branches
    - Anterior tibial artery
    - Posterior tibial artery (may be referred to as tibioperoneal trunk until branch point of peroneal artery)
- **Anterior tibial artery**
  - Smaller of 2 terminal branches of popliteal artery
  - Origin in back of leg, at distal border of popliteus muscle
  - Passes through upper part of interosseous membrane
  - Straight course down front of leg to become **dorsalis pedis artery**
  - Runs on anterior surface of interosseous membrane, deep to extensor muscles
  - Muscular branches along length
  - Malleolar branches ramify over malleoli
    - Lateral branch anastomoses with perforating branch of peroneal artery
- **Posterior tibial artery**
  - Larger of 2 terminal branches of popliteal artery
  - Main blood supply to foot
  - Passes downward and slightly medially along with tibial nerve
  - Terminates in space between medial malleolus and calcaneus
  - Within calf, artery runs just deep to transverse intermuscular septum
  - Divides into lateral and medial plantar arteries in tarsal tunnel behind medial malleolus
  - Branches
    - Circumflex fibular (may arise from anterior tibial),
      - Runs laterally around neck of fibula
    - Nutrient artery to tibia
    - Muscular branches
- **Peroneal artery**
  - Largest branch of posterior tibial artery
  - Runs obliquely downward and laterally beneath soleus muscle to fibula
  - Descends deep to flexor hallucis longus

Veins

- **Deep leg veins**
  - Paired venae comitantes follow arteries of same name
    - Anterior, posterior, and peroneal veins
  - **Popliteal vein**
    - Runs superficial to artery
    - Usually begins at inferior border of popliteus muscle from anterior and posterior tibial veins
    - Crosses from medial to lateral side of popliteal artery as it runs in popliteal fossa
    - Ends at adductor hiatus by becoming superficial femoral vein
    - Also receives small saphenous vein in popliteal fossa
- **Superficial leg veins**
  - **Great saphenous vein**
    - Longest vein in human body
    - Begins at medial border of foot and ascends anterior to medial malleolus
    - Passes obliquely upward and backward across medial surface of distal 1/3 of tibia
    - Passes vertically upward along medial border of tibia and over posterior border medial femoral epicondyle
    - Courses anteriorly along thigh
    - Passes through saphenous opening in fascia lata to join common femoral vein
  - **Small saphenous vein**
    - Extends behind lateral malleolus, ascends lateral to Achilles tendon
    - At midline of calf, in lower popliteal region, pierces popliteal fascia, and terminates in popliteal vein

ANATOMY IMAGING ISSUES

Imaging Recommendations

- Examination of lower limb vessels requires use of **morphological and functional techniques**
  - Morphology: Documentation of areas of stenosis or occlusion, sites of venous valves, aberrant branches, aneurysm formation, etc.
  - Functional: Combination of color and spectral Doppler examination
- **Arteries**
  - Require use of spectral Doppler scanning to show their phasicity; normal wave form is triphasic
    - Sharp forward flow upstroke (systolic phase)
    - Small reverse flow (early diastolic phase)
    - Final smaller forward flow (late diastolic phase)
- **Veins**
  - Require use of dynamic maneuvers
    - **Compression**: Normal vein should compress with transducer pressure (thrombus will not compress)
    - **Augmentation**: To increase venous return (to demonstrate venous flow and patency) by manually squeezing calf or by gently moving toes
      - Frequently required to demonstrate flow in deep calf veins (which normally show no color Doppler flow)
    - **Valsalva**: To increase abdominal pressure and accentuate reverse flow and incompetent venous valves

LEG ARTERIES, POSTERIOR

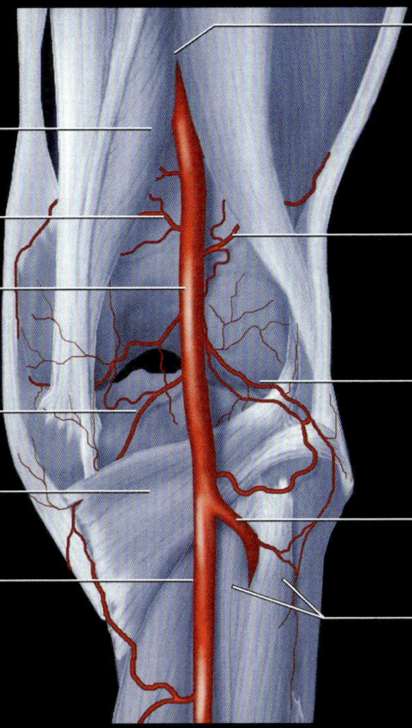

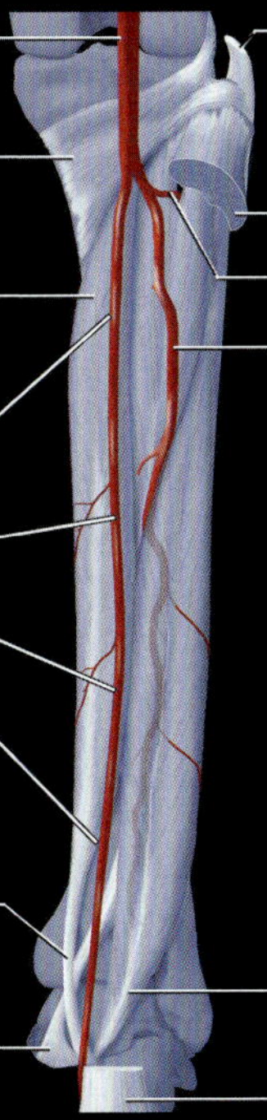

(Top) *Graphic shows the the popliteal artery, from the hiatus in adductor magnus proximally to the lower border of popliteus distally. At this point, the popliteal artery bifurcates into anterior and posterior tibial arteries. The anterior tibial is the smaller branch that extends through a slit in tibialis posterior muscle and on through the interosseous membrane to descend along the interosseous membrane down the anterior compartment. There are 4 named geniculate branches, a superior and inferior both medially and laterally.* **(Bottom)** *The posterior tibial artery passes downward and slightly medially adjacent to the tibial nerve to end in the space between the medial malleolus and calcaneus. The largest branch of the posterior tibial artery is the peroneal artery, which runs obliquely downward and laterally beneath the soleus muscle to the fibula.*

LEG ARTERIES, ANTERIOR AND ANKLE

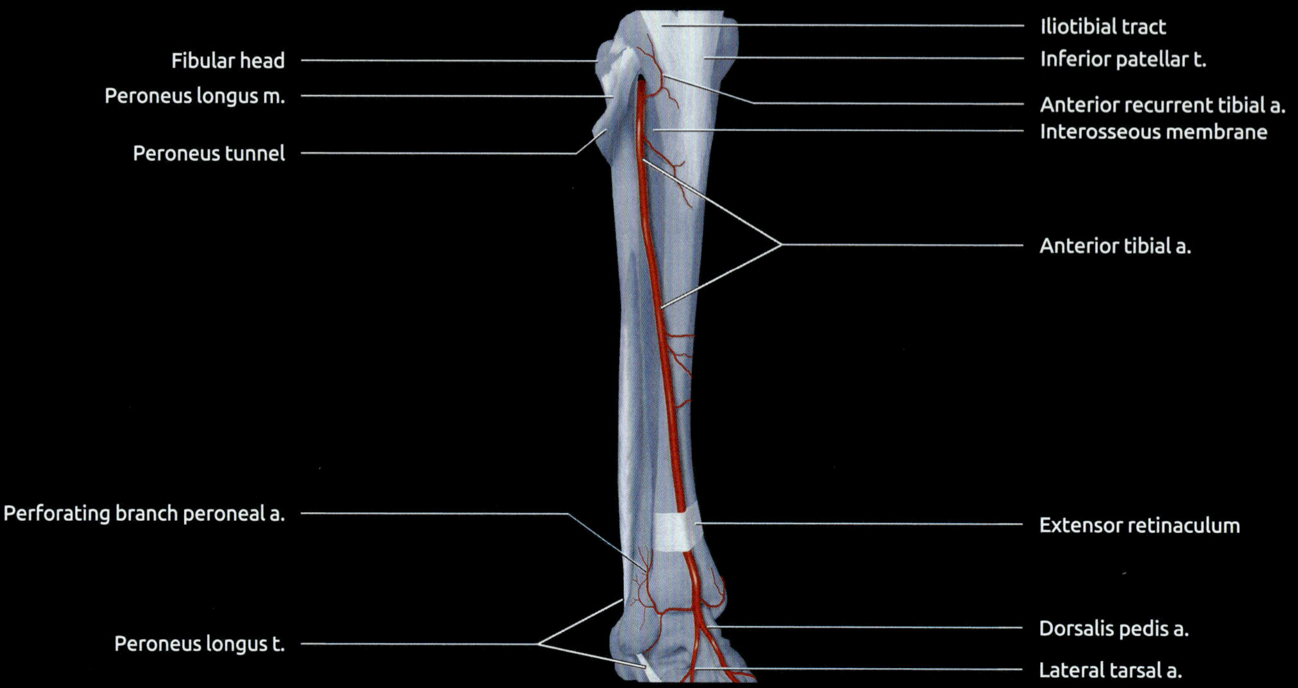

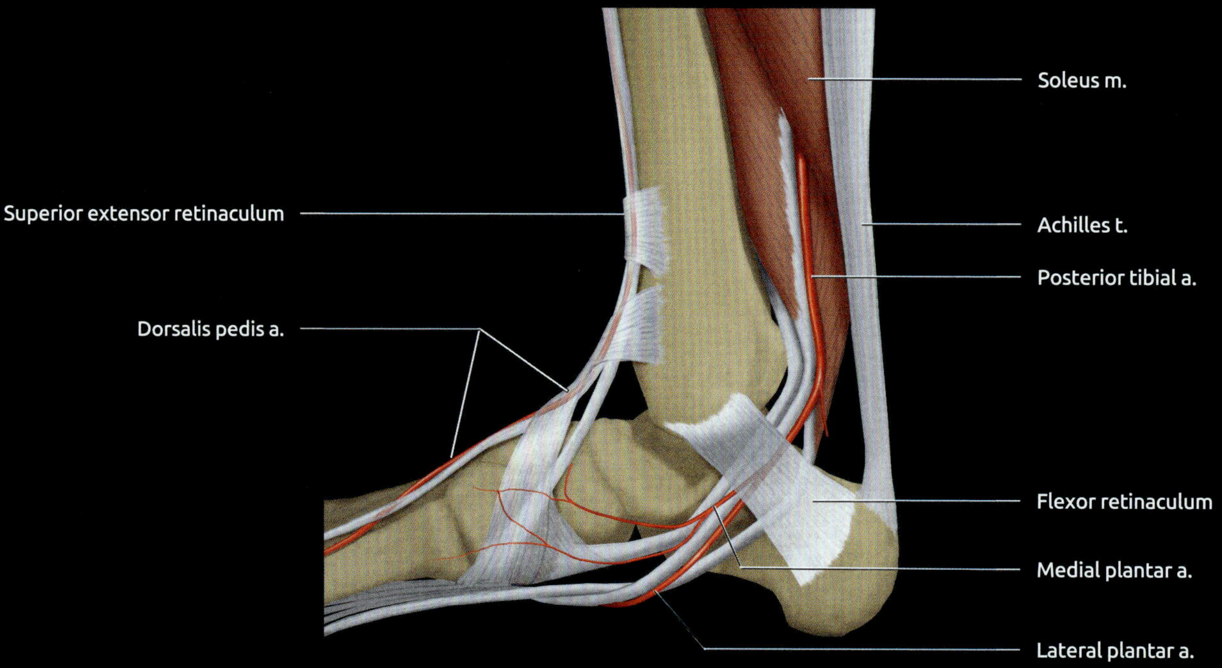

(Top) *Graphic of the anterior leg shows the anterior tibial artery perforating the interosseous septum proximally and descending along this membrane down the front of leg to terminate as the dorsalis pedis. Distally, one sees a perforating branch of the peroneal artery, which in a variant situation may provide the major blood supply to the dorsum of the foot.* **(Bottom)** *Graphic of the medial aspect of the ankle shows the distal portion of both the posterior and anterior tibial arteries. The posterior tibial artery passes between the medial malleolus and calcaneus, and divides into the lateral and medial plantar arteries. The dorsal pedis artery is a direct continuation of the anterior tibial artery and supplies the dorsum of the foot.*

LEG VEINS

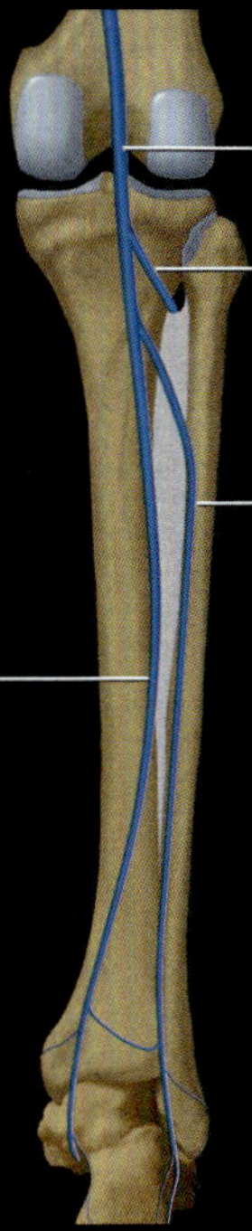

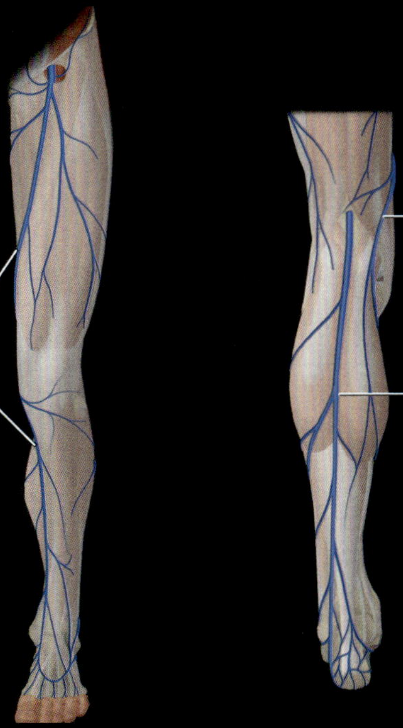

(Top) *This graphic shows the deep veins of the leg, which follow their accompanying artery of the same name and are often paired.*
(Bottom) *This graphic shows the superficial leg veins. The great saphenous vein is the longest vein in the human body. It begins in the dorsomedial area of the foot. It passes obliquely upward over the posterior medial femoral epicondyle, then anteriorly across the thigh to finally join with the common femoral vein. The small saphenous vein drains from the lateral malleolus, up the posterior calf to join with the popliteal vein in the popliteal fossa.*

POPLITEAL VESSELS

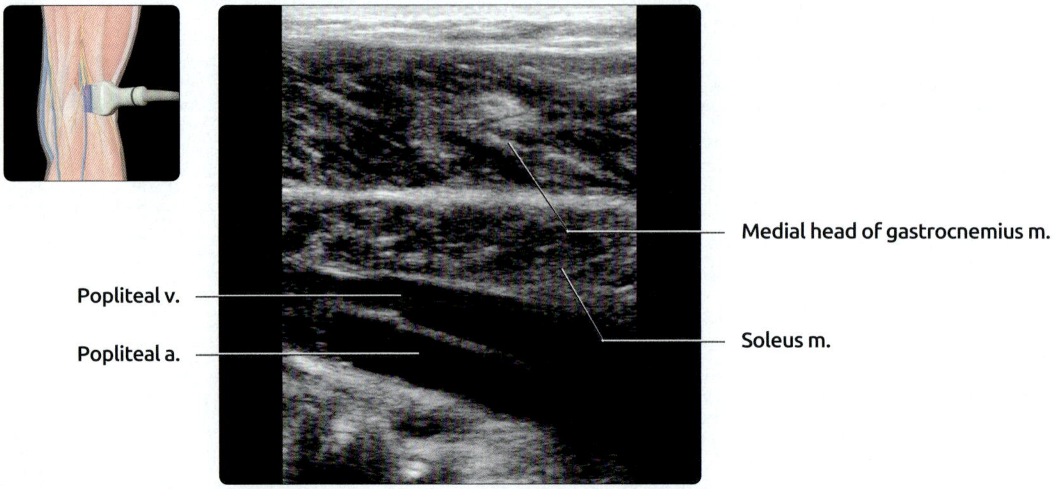

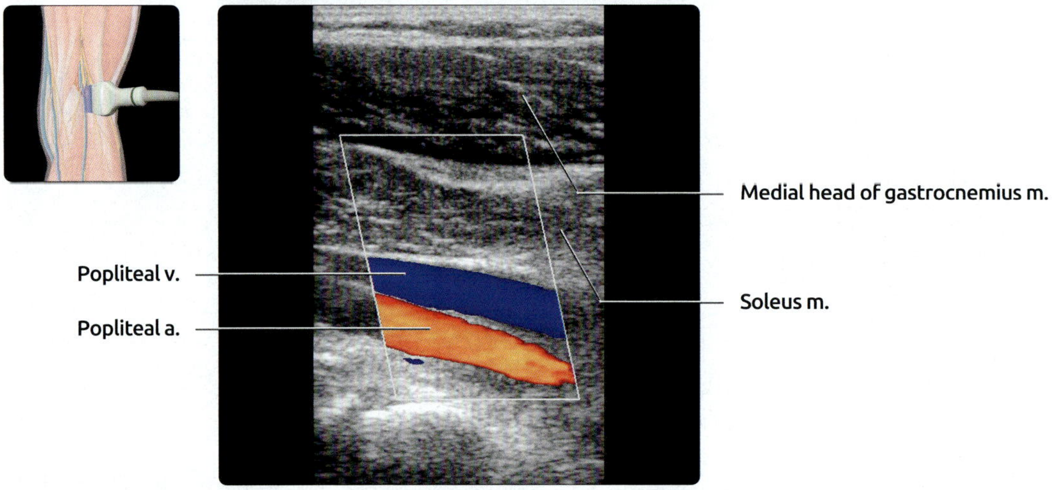

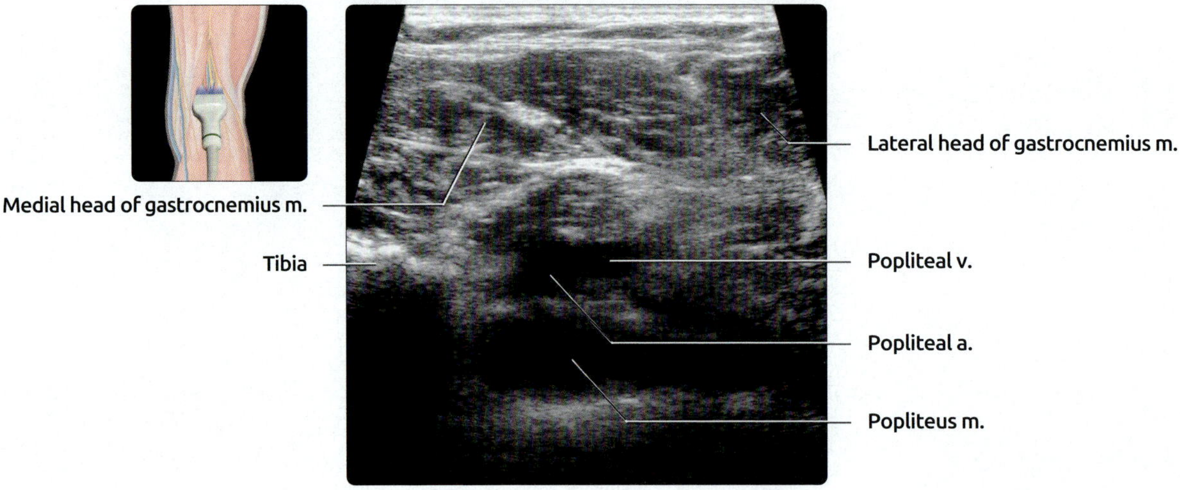

(Top) *Sagittal scan shows the lower part of the popliteal vessels. The popliteal vessels are continuations of the superficial femoral vessels after they have exited the adductor canal through the adductor hiatus. The popliteal vein lies superficial to the popliteal artery, a reverse of the relationship between the superficial femoral vein and artery. The popliteal vessels pass inferolaterally through the fat of the popliteal fossa.* **(Middle)** *Sagittal color Doppler scan of the popliteal vessels shows complete color filling of the lumina (excluding luminal thrombus) and normal flow direction: Toward the trunk (blue) for the vein and away from the trunk (red) for the artery.* **(Bottom)** *Transverse scan shows the lower part of the popliteal vessels. The popliteal vein runs superficial to the popliteal artery and usually bifurcates at the inferior margin of the popliteus muscle.*

POPLITEAL VESSEL BIFURCATION

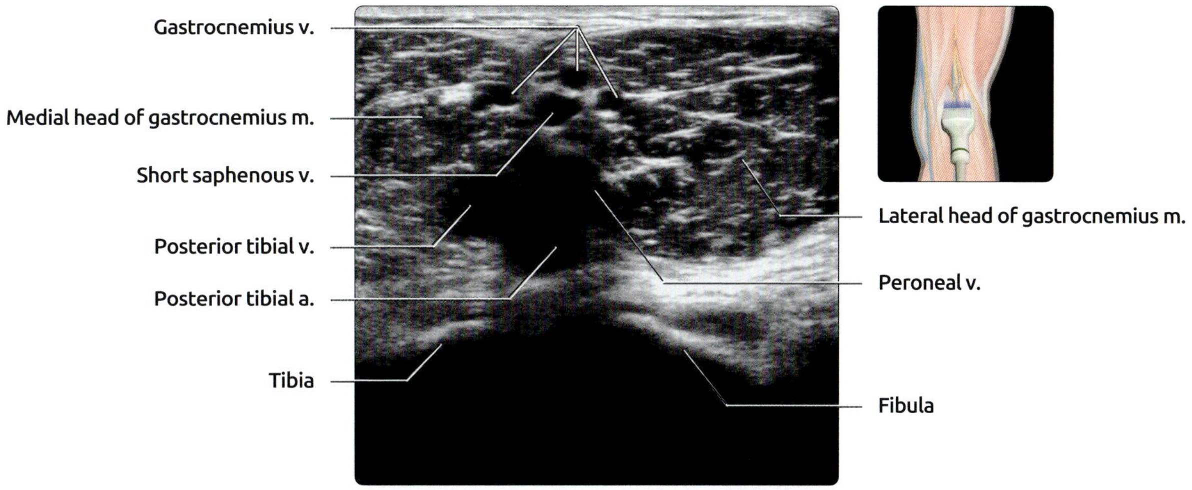

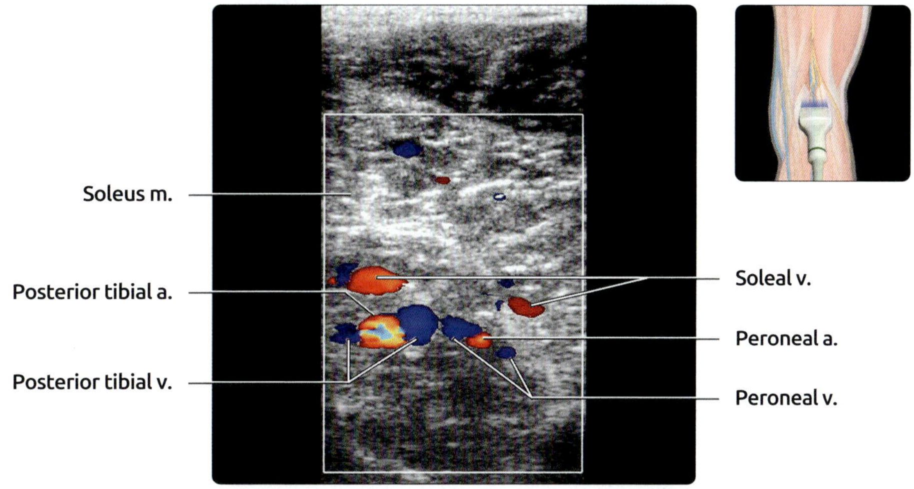

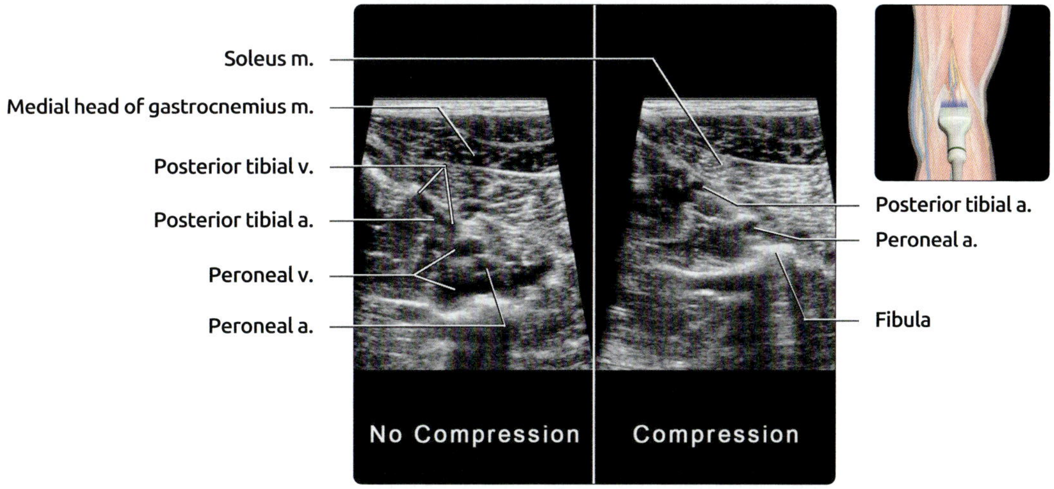

(Top) *Transverse scan below the bifurcation of the popliteal artery is shown. The popliteal artery bifurcates into the anterior and posterior tibial arteries. The posterior tibial artery is the larger of the 2 terminal branches of the popliteal artery and passes inferiorly for a short distance before giving out its largest branch, the peroneal artery.* **(Middle)** *Transverse color Doppler scan shows the posterior compartment vessels of the proximal calf. Color flow highlights the paired venae comitantes nature of the posterior tibial and peroneal veins, lying on both sides of the corresponding artery.* **(Bottom)** *Transverse scan at the midproximal calf scanning from the medial side is shown. The lumina of the peroneal and tibial veins can be compressed by firm transducer pressure (right 1/2 of image), indicating the lack of thrombus.*

TRANSVERSE POSTERIOR TIBIAL VESSELS

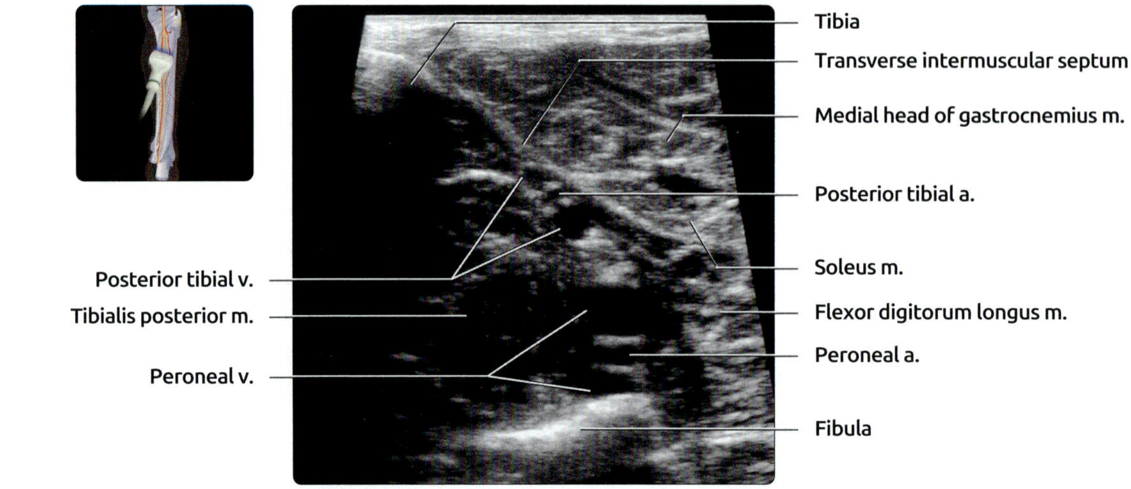

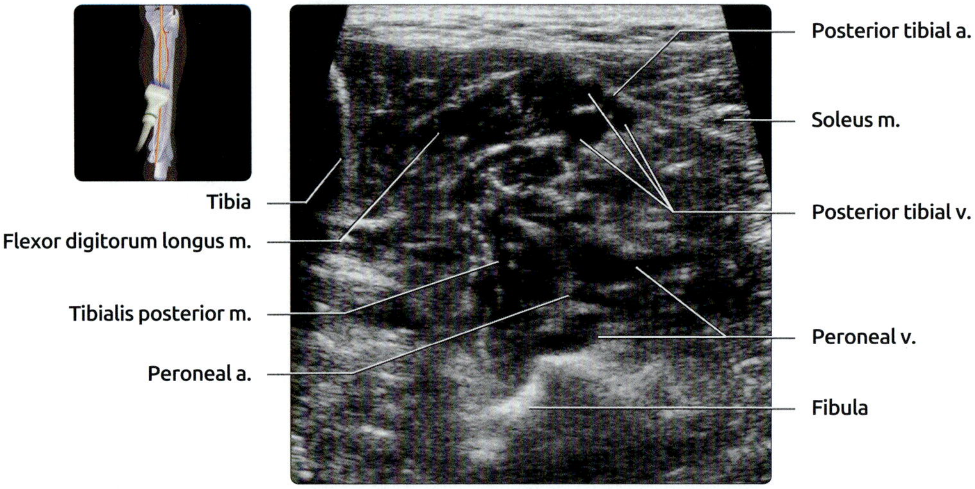

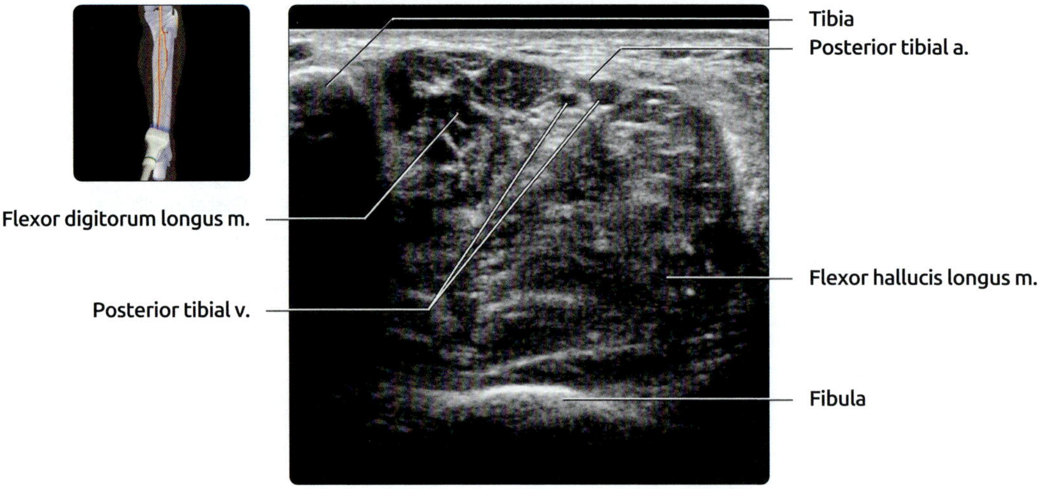

(Top) *Three transverse scans show the posterior tibial vessels using a medial scanning approach. Superiorly, the posterior tibial vessels give off their main branches, the common peroneal vessels. The posterior tibial artery lies just deep to the transverse intermuscular septum while the peroneal arteries are deep to the flexor digitorum muscle.* **(Middle)** *At the midcalf level, the posterior tibial vessels no longer line up side by side (3 in a line). Anatomical variations occur, as seen here with 3 posterior tibial veins accompanying 1 artery. The peroneal vessels continue their deeper course, running on the surface of the tibialis posterior, which separates them from the anterior leg compartment.* **(Bottom)** *Further inferiorly, the posterior tibial vessels emerge superficially as they prepare to enter the tarsal tunnel behind the medial malleolus.*

LONGITUDINAL POSTERIOR TIBIAL VESSELS

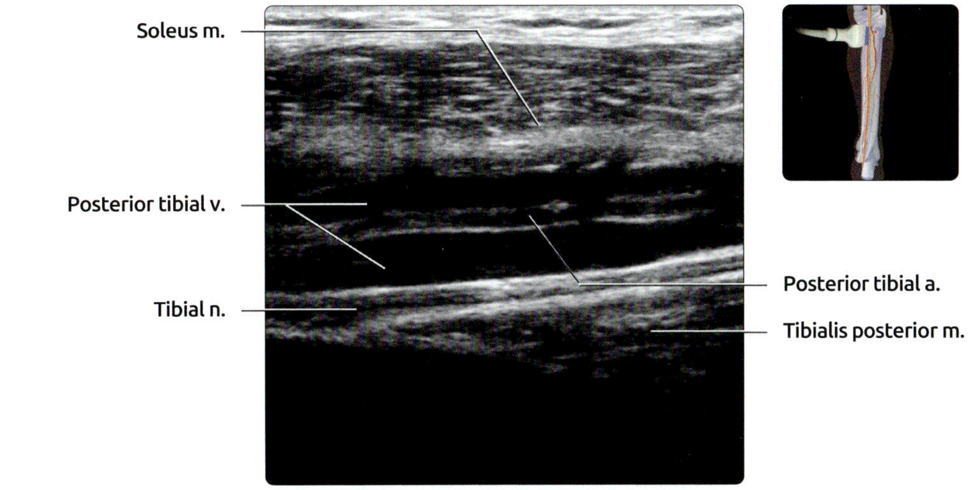

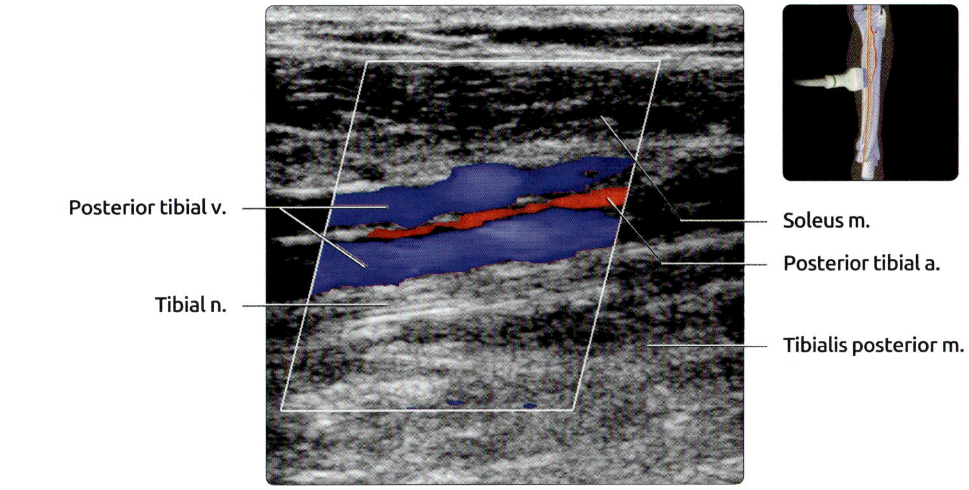

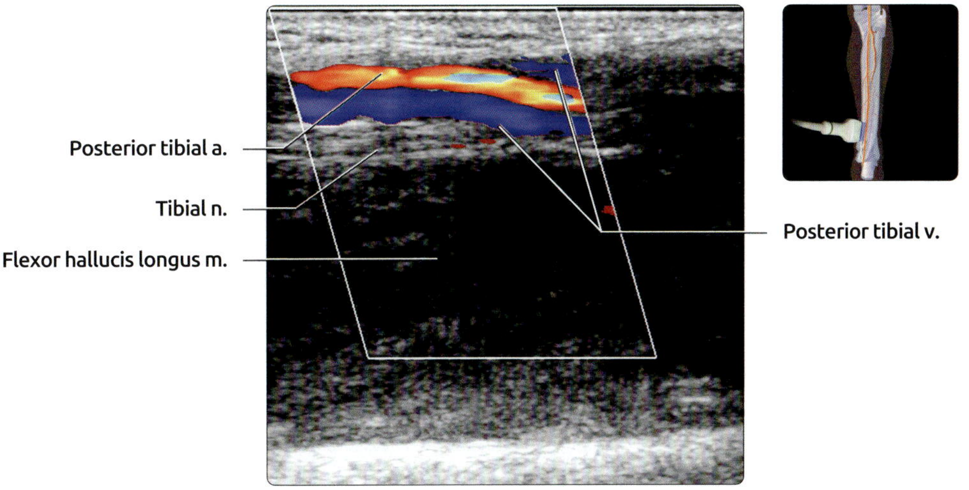

(Top) *Oblique coronal scan shows the posterior tibial vessels using a medial approach. The posterior tibial artery is accompanied by paired venae comitantes (posterior tibial veins) and the tibial nerve. They run deep to the transverse intermuscular septum of the leg.* **(Middle)** *Color Doppler scan shows the posterior tibial vessels. Notice the marked difference in caliber between the artery and the veins, all of which show normal, complete color filling and normal direction of flow.* **(Bottom)** *Further inferiorly, oblique coronal scan of the lower calf shows the distal part of the posterior tibial vessels running superficial to the flexor hallucis muscle and getting ready to curve under the medial malleolus. The posterior tibial vessels and tibial nerve run in the tarsal tunnel together with the tendon flexors of the foot.*

TRANSVERSE PERONEAL VESSELS

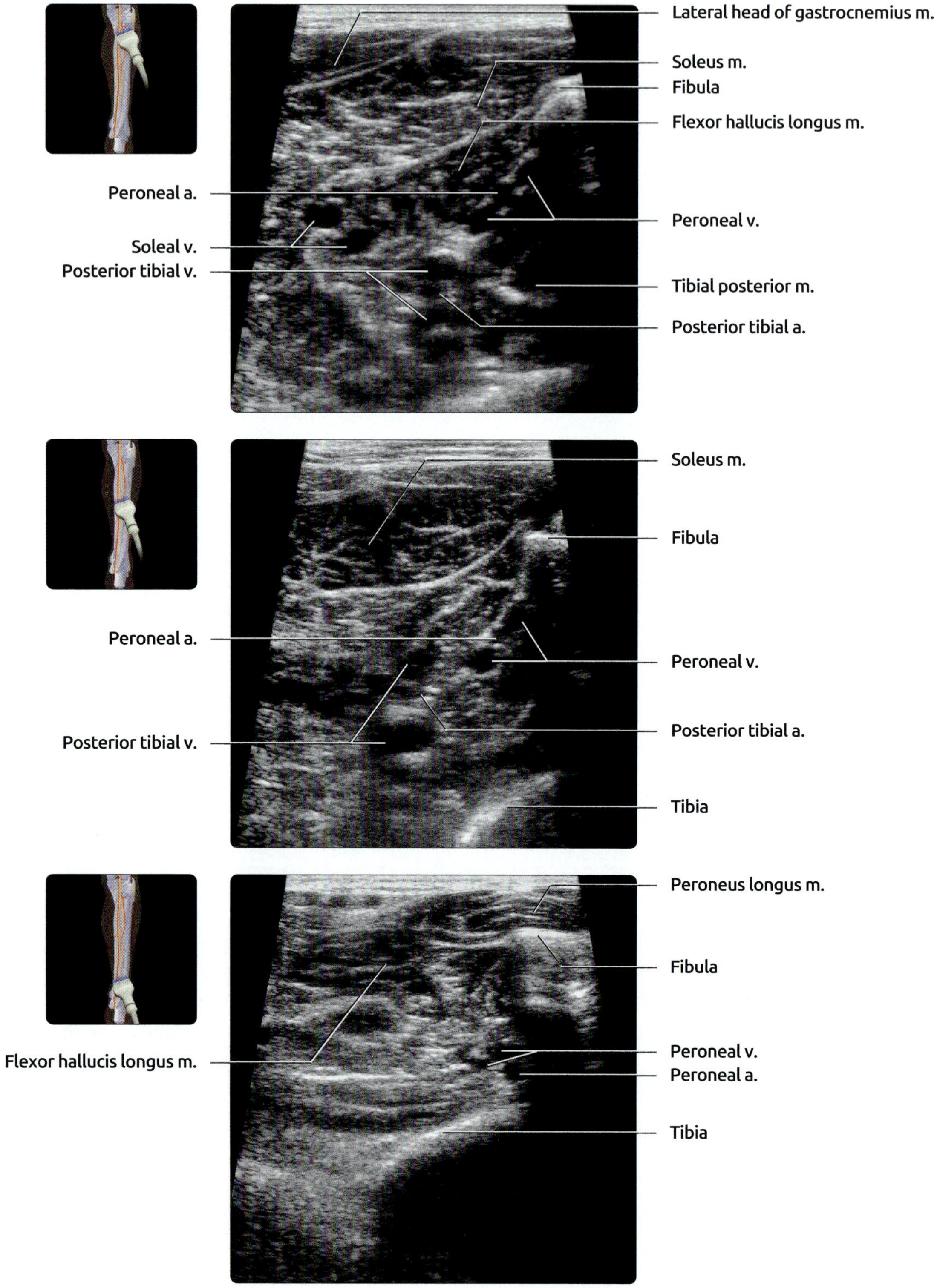

(Top) *Three transverse scans of the calf using a lateral scanning approach are shown. The peroneal artery is a branch of the posterior tibial artery. It runs obliquely toward the fibula and then runs along the medial surface of the fibula distally, sandwiched between the flexor hallucis longus muscle and tibialis posterior muscle.* **(Middle)** *More distally, the relationship between the peroneal vessels and the fibula and flexor muscles remains the same. Nutrient vessels branch out to supply the fibula and the adjacent muscles.* **(Bottom)** *Most distally, the peroneal artery runs behind the tibiofibular syndesmosis before dividing into lateral calcaneal branches, which run over the lateral and posterior surfaces of the calcaneum.*

LONGITUDINAL PERONEAL VESSELS

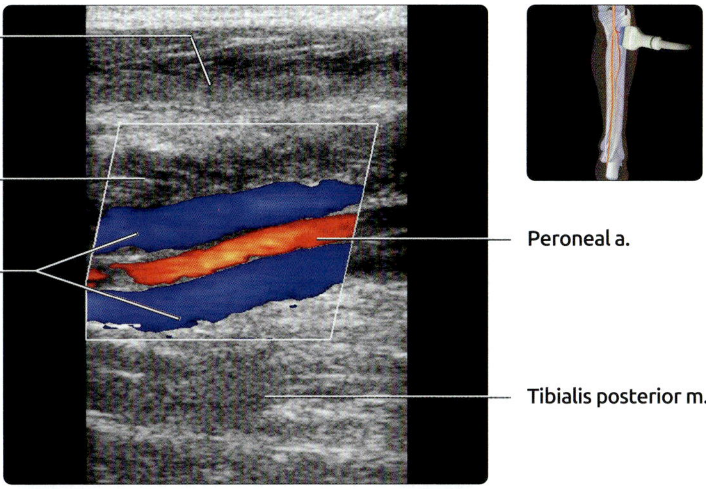

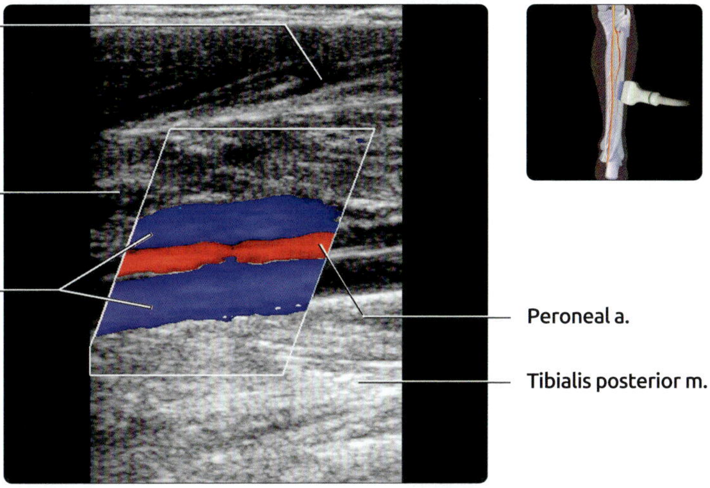

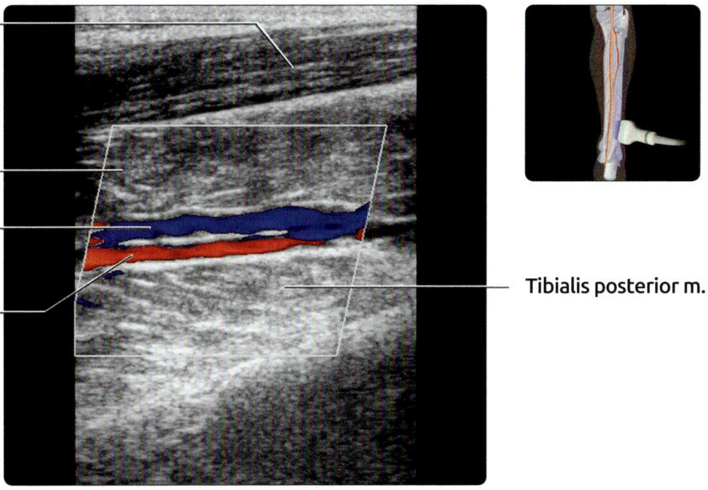

(Top) *Oblique coronal scan shows the peroneal vessels using a lateral approach. The peroneal artery, like the posterior tibial, is also accompanied by venae comitantes (peroneal veins). The difference is that there is no peroneal nerve to accompany the vessels. The peroneal nerve runs in the lateral leg compartment after curving around the neck of the fibula.* **(Middle)** *More inferiorly, oblique coronal scan shows the peroneus vessels lie deep to the flexor hallucis longus and soleus muscles while running superficial to the tibialis posterior muscle.* **(Bottom)** *Further inferiorly, the peroneal vessels are much smaller in caliber after giving out many branches. They pass posterior to the tibiofibular syndesmosis.*

TRANSVERSE ANTERIOR TIBIAL VESSELS

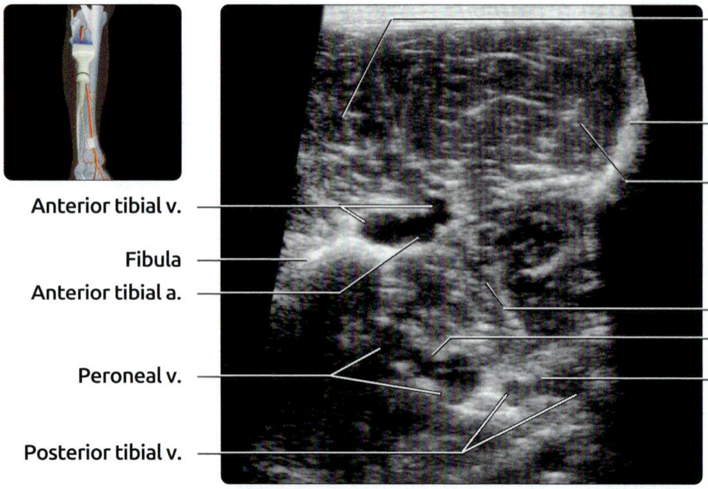

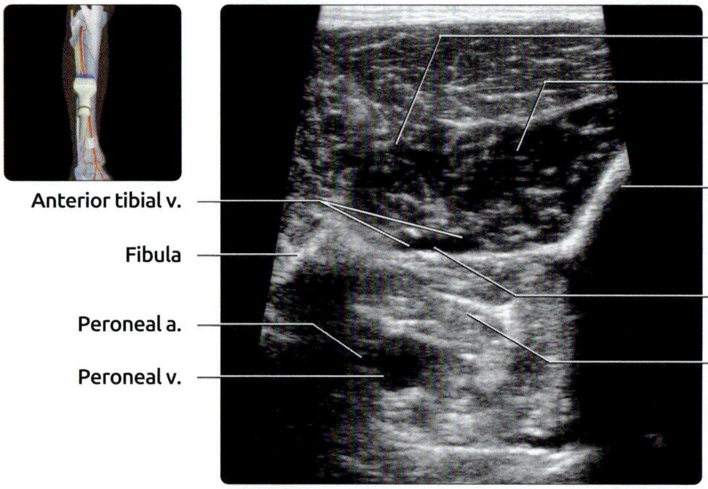

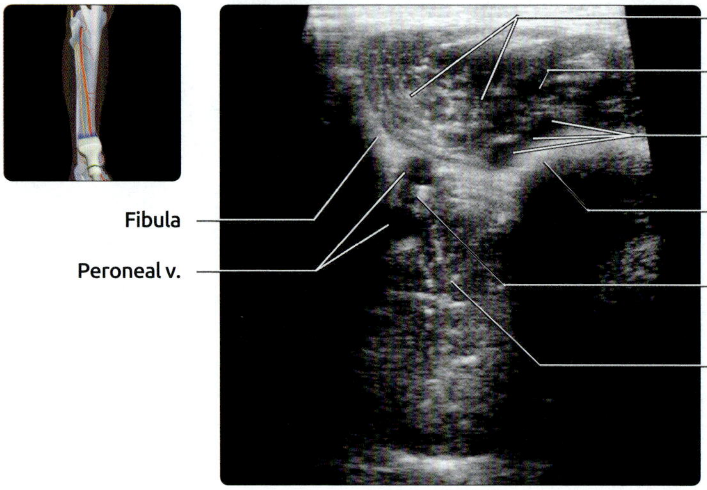

(Top) *Three transverse scans show the anterior tibial vessels using an anterior scanning approach. The anterior tibial vessels are the smaller terminal branches of the popliteal vessels (smaller compared to the posterior tibial vessels at the bifurcation). After its origin, the anterior tibial vessels run anterior through the interosseous membrane and then pass inferiorly on the anterior surface of this membrane.* **(Middle)** *At midleg level, the anterior tibial vessels continue to run in a deep position on the surface of the interosseous membrane. Through the window between the tibia and fibula, the peroneal vessels and tibialis posterior muscle are seen.* **(Bottom)** *In the most distal part of the leg, the anterior tibial vessels move medially to lie on the surface of the tibia. At this level, the peroneal artery branches run anteriorly to penetrate the interosseous membrane.*

LONGITUDINAL ANTERIOR TIBIAL VESSELS

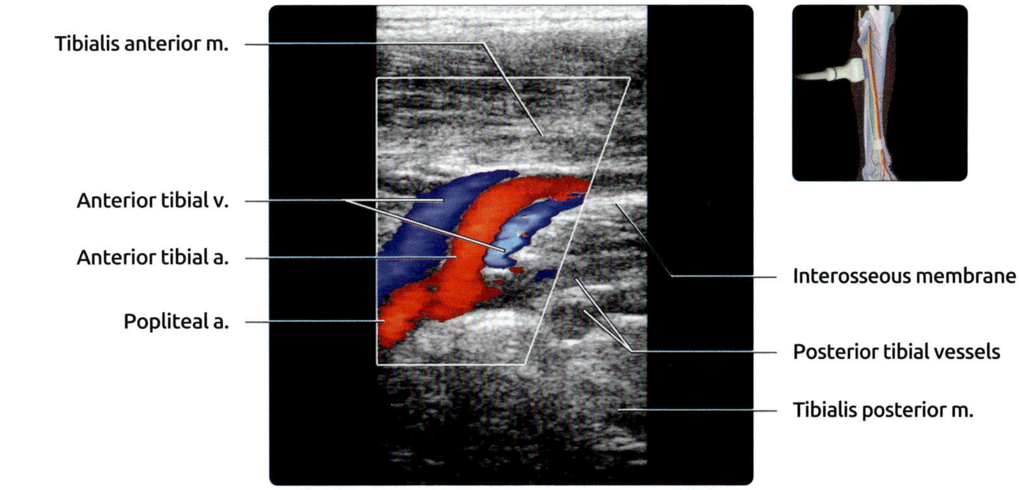

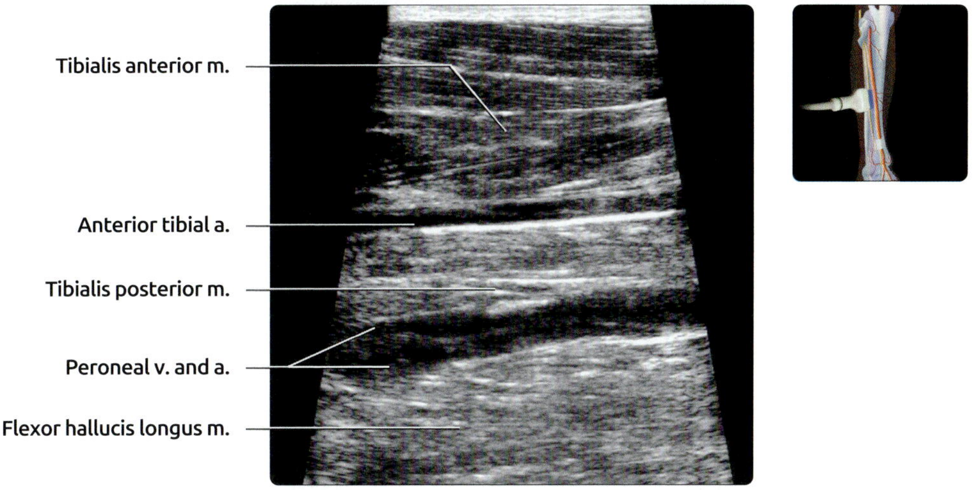

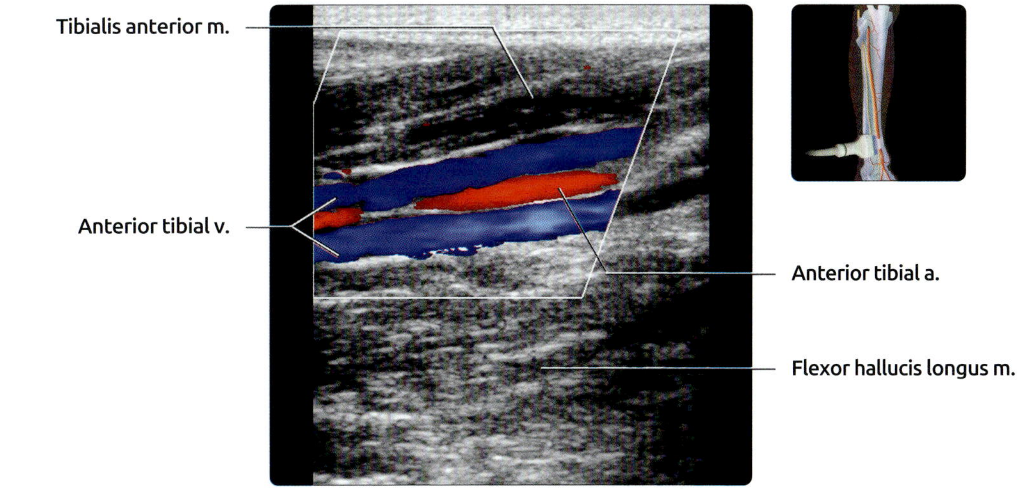

(Top) *Oblique sagittal color Doppler scan shows the anterior tibial vessels as they bifurcate from the popliteal vessels. The anterior tibial vessels run anteriorly to penetrate the interosseous membrane into the anterior compartment.* **(Middle)** *Sagittal scan shows the anterior tibial artery in the midleg level. The anterior tibial artery runs deep within the anterior compartment and on the surface of the interosseous membrane, being covered by the tibialis anterior muscle.* **(Bottom)** *Sagittal color Doppler scan shows the lower aspect of the anterior tibial vessels. The vessels begin to run more superficially, and in the foot, the anterior tibial artery continues as the dorsalis pedis artery.*

SPECTRAL DOPPLER OF LEG ARTERIES

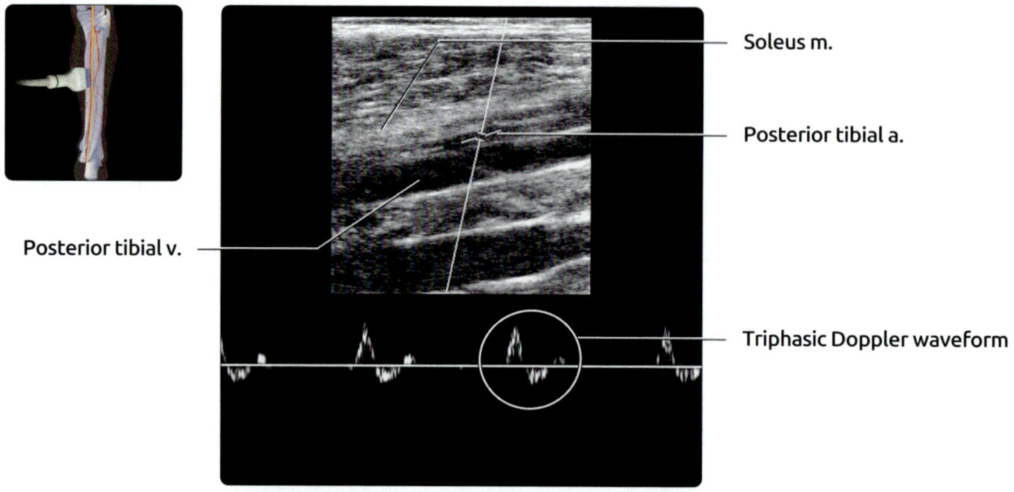

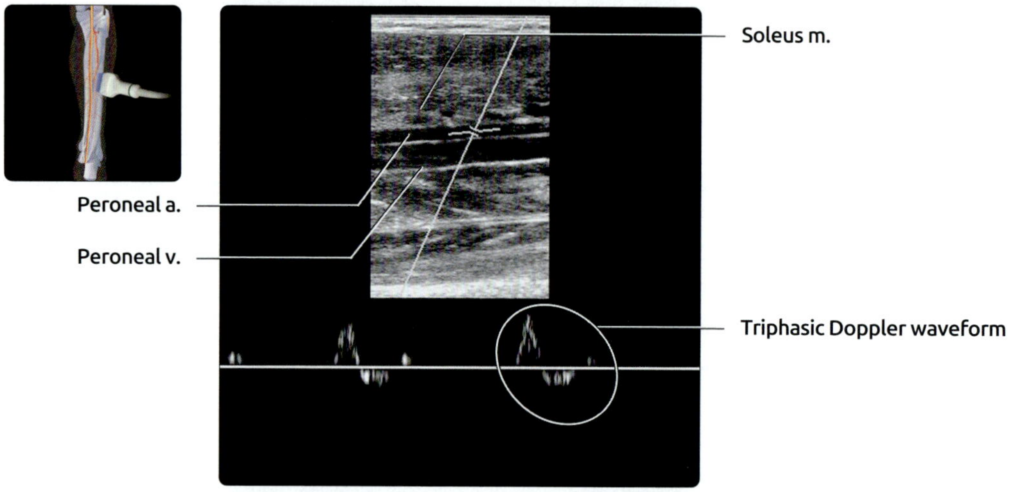

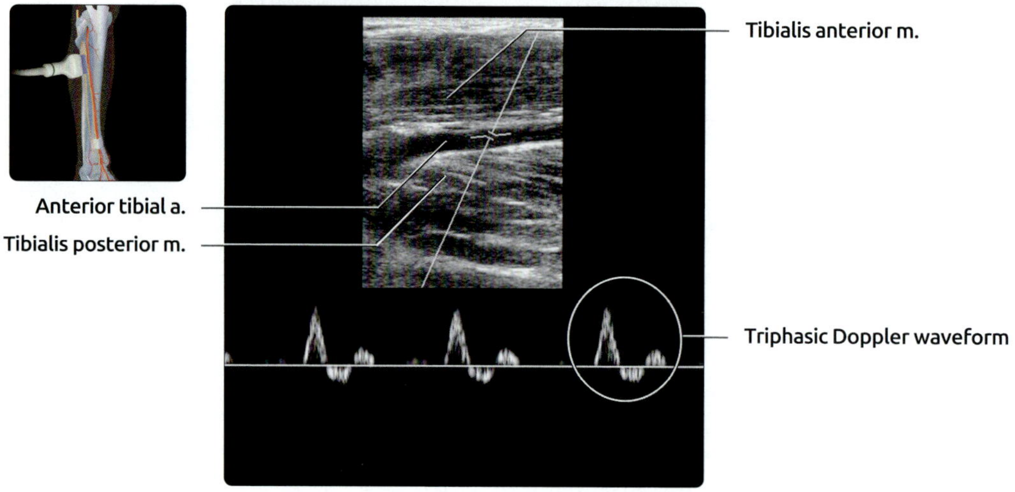

(Top) *Spectral Doppler scan of the posterior tibial artery demonstrates the typical triphasic pattern seen in normal arteries. Loss of the reverse flow in the diastolic phase is termed a biphasic pattern and is suggestive of stenosis proximal to the point of interrogation. With proximal stenosis, the peak flow velocity is also decreased.* **(Middle)** *Similarly, the peroneal artery shows a triphasic waveform on spectral Doppler.* **(Bottom)** *Oblique coronal spectral Doppler scan of the anterior tibial artery in the upper calf level also shows a normal triphasic pattern. Abnormal increase in reverse flow (negative flow in early diastolic phase) suggests increased distal resistance from stenosis.*

SPECTRAL DOPPLER OF LEG VEINS

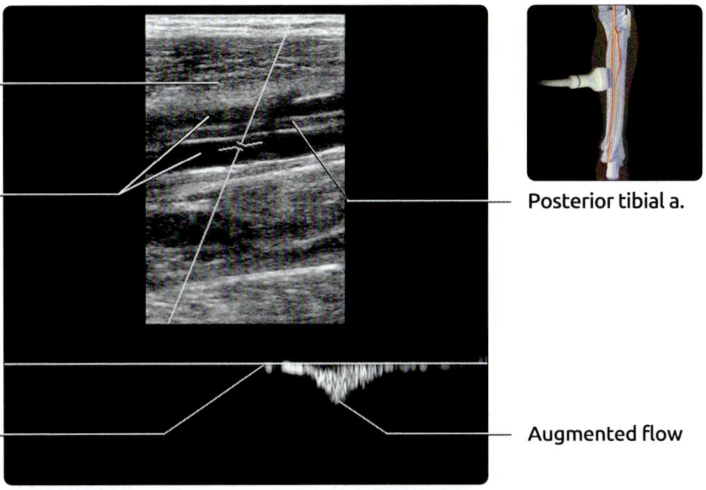

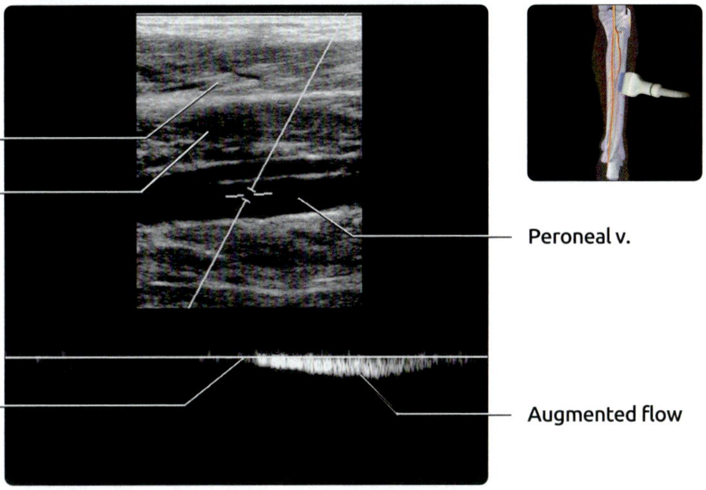

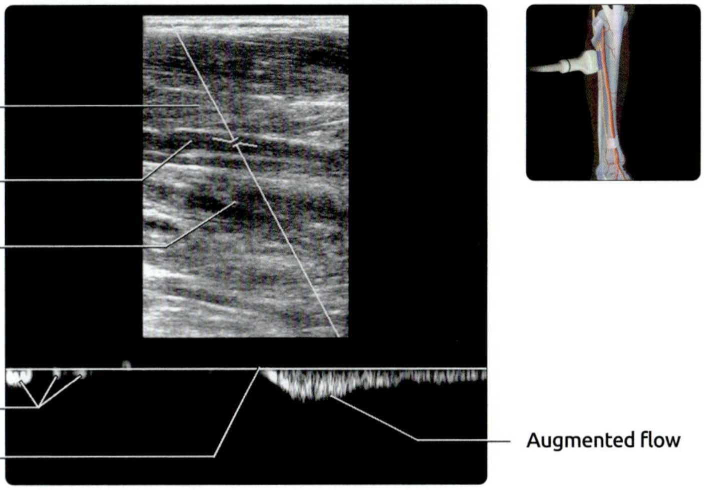

(Top) *Oblique coronal spectral Doppler scan of the posterior tibial vein shows absence of spontaneous flow at rest (left 1/2 of the spectral trace). After augmentation (squeezing calf muscles distal to transducer), a surge of blood flow toward the trunk (negative values) can be demonstrated and indicates luminal patency.* **(Middle)** *Oblique coronal spectral Doppler scan of the peroneal vein before and after augmentation shows a similar sudden surge of blood flow.* **(Bottom)** *Oblique coronal spectral Doppler scan of the anterior tibial vein demonstrates slow spontaneous flow at rest. After augmentation, a sustained surge of flow is demonstrated.*

SOLEAL VEINS

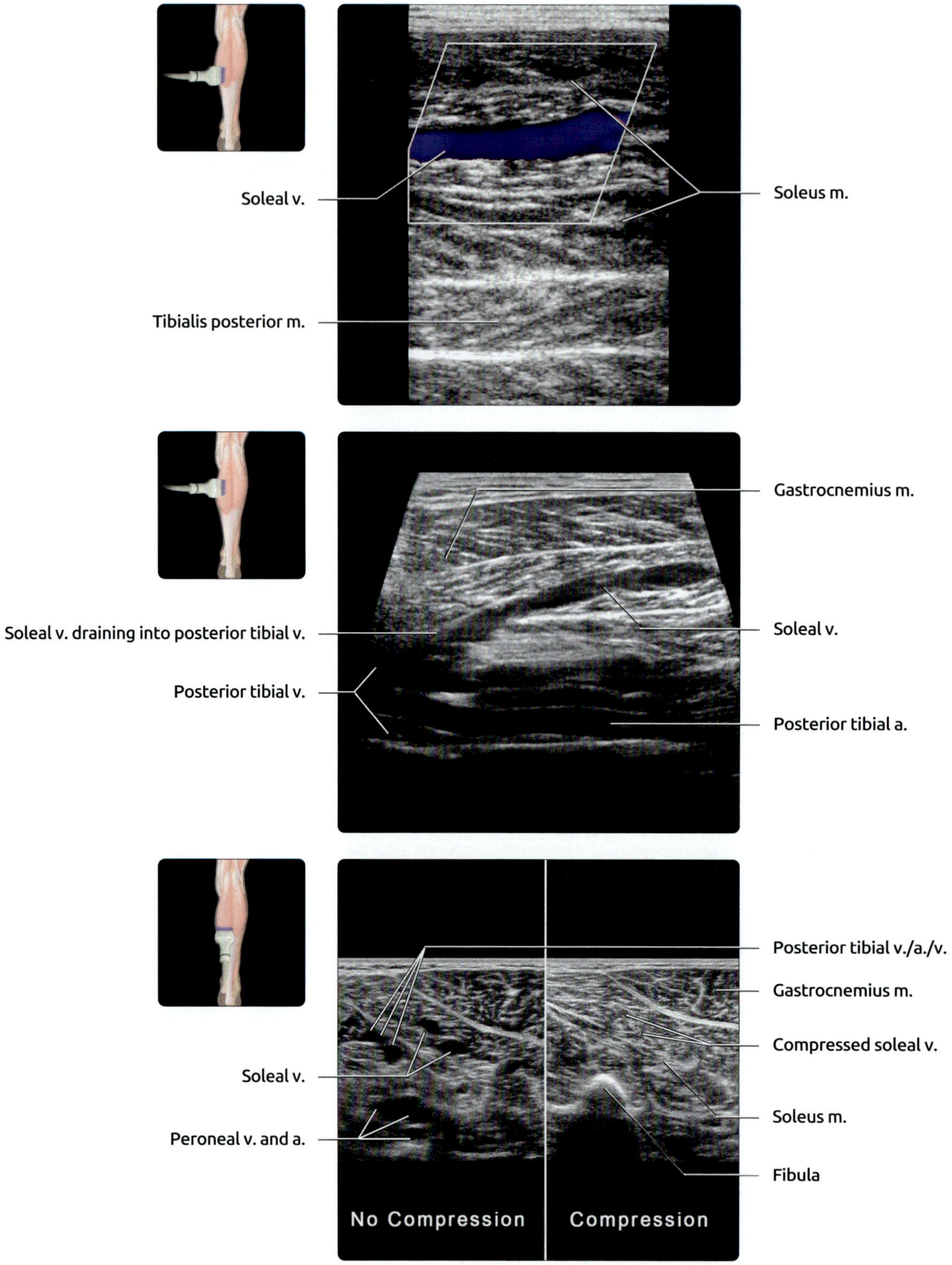

(Top) *Oblique sagittal color Doppler scan shows the posterior compartment in the midcalf level. The soleal vein shows complete color filling of its lumen indicating the absence of thrombus. Soleal veins are the most common veins to develop thrombosis and should be assessed when scanning for deep venous thrombosis.* **(Middle)** *Oblique sagittal scan of the posterior compartment shows the soleal vein running proximally and joining the posterior tibial vein. This pathway allows a soleal vein thrombus to extend into the posterior tibial vein and propagate proximally.* **(Bottom)** *Dual transverse images show the posterior compartment using a medial approach, before (left image) and during (right image) firm transducer compression. The soleal veins are completely compressed by transducer pressure, indicating an absence of thrombus.*

GASTROCNEMIUS VEINS

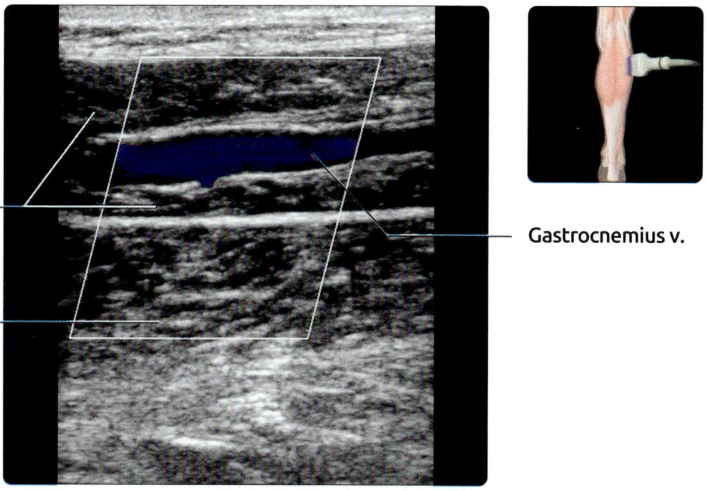

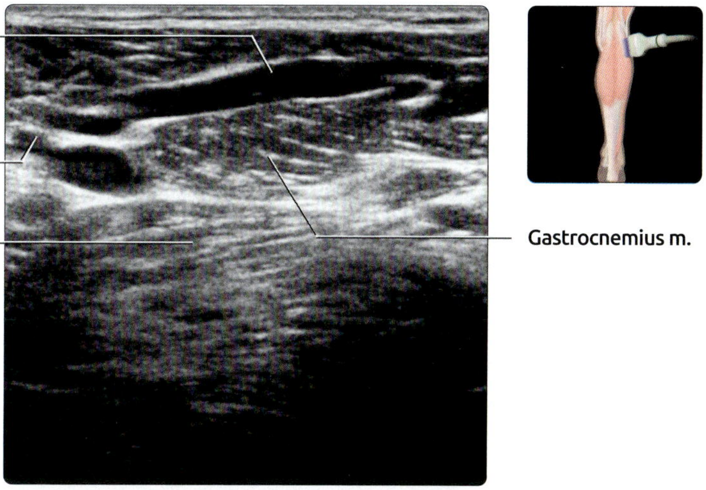

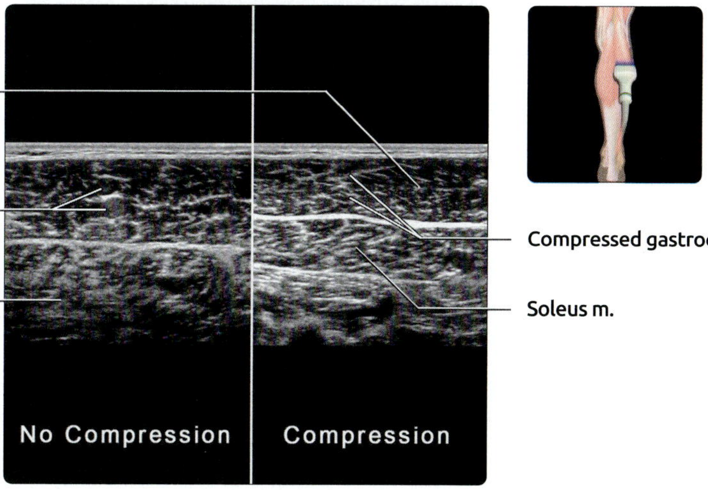

(Top) *Oblique sagittal color Doppler scan shows the posterior leg compartment. The gastrocnemius vein shows complete color filling, indicating absence of thrombus. The gastrocnemius veins are the 2nd most common site for venous thrombosis.* **(Middle)** *Oblique sagittal scan of the posterior compartment of the upper calf shows the gastrocnemius vein joining the popliteal vein, allowing a gastrocnemius vein thrombus to extend proximally to the popliteal vein.* **(Bottom)** *Dual transverse images show gastrocnemius veins before (left image) and during (right image) compression. The gastrocnemius veins are completely compressed during compression, indicating an absence of thrombus. Compression is regarded as the most reliable technique to rule out venous thrombosis.*

SMALL SAPHENOUS VEIN

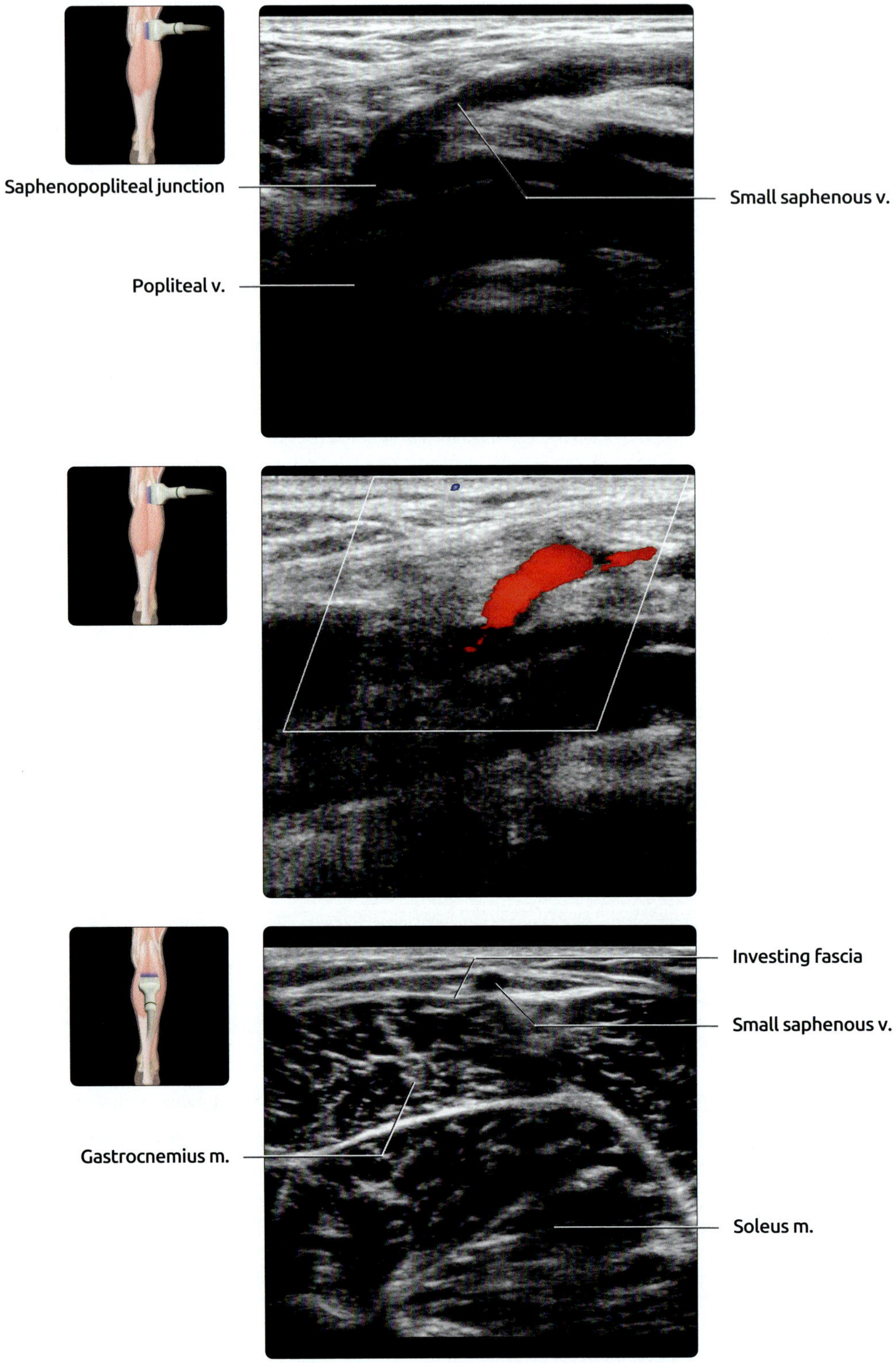

(Top) *Sagittal scan shows central popliteal fossa. The small saphenous vein running from the lateral surface of the calf turns medially and penetrates the popliteal fascia to drain into the popliteal vein. Varicose veins arise as a result of reflux and stasis due to incompetent valves. Varicose veins on the lateral side of the calf suggest an incompetent small saphenous system.* **(Middle)** *Sagittal color Doppler scan shows the central popliteal fossa with augmentation. Flow toward the popliteal vein (red) is demonstrated at the saphenopopliteal junction. Reverse flow into the small saphenous vein would suggest valvular incompetence.* **(Bottom)** *Transverse scan shows the posterior surface of the upper calf. The small saphenous vein is sandwiched between 2 layers of investing fascia. Minimal probe pressure is required to avoid inadvertent compression of the small saphenous vein.*

GREAT SAPHENOUS VEINS

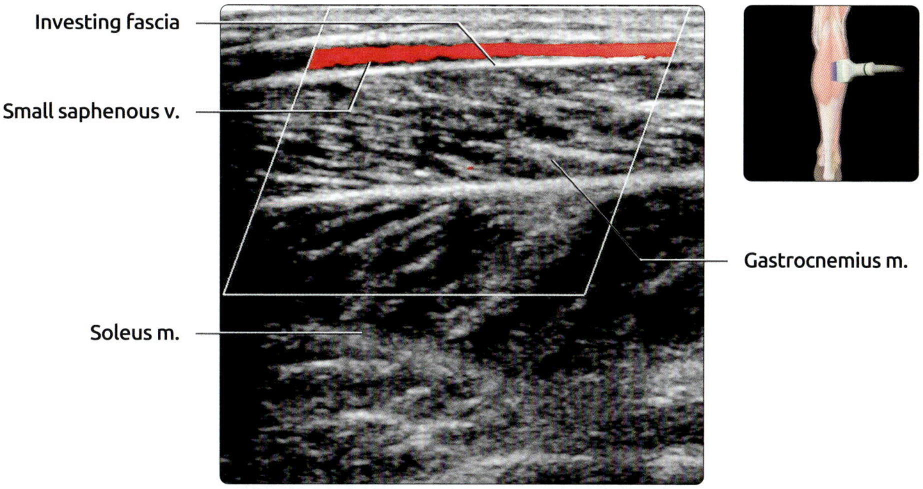

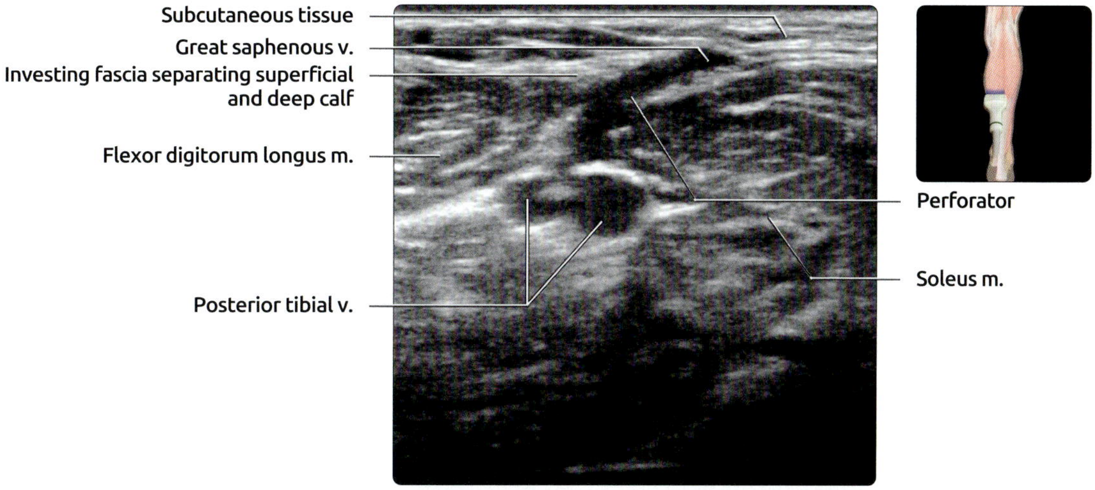

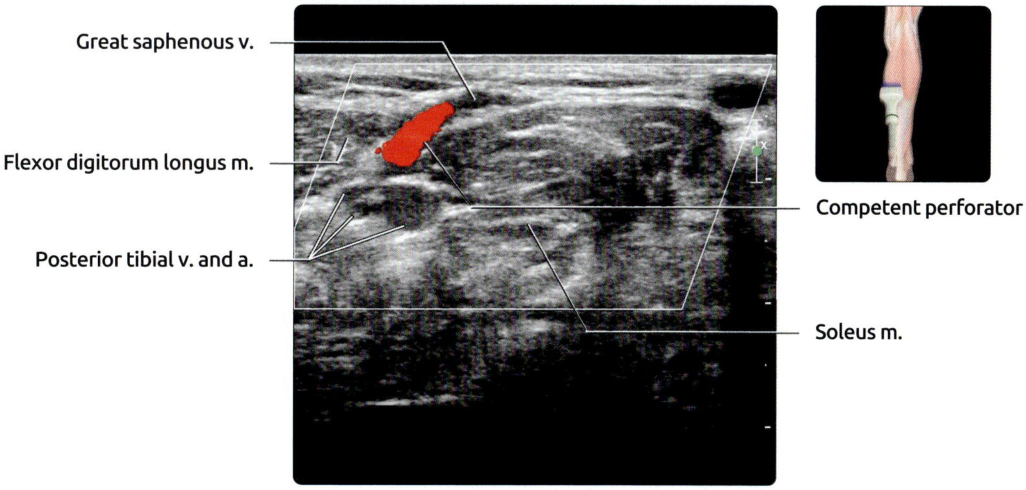

(Top) *Sagittal color Doppler scan shows upper midcalf with augmentation. The complete filling of the venous lumen of the small saphenous vein is demonstrated, indicating the absence of a superficial vein thrombus. If flow toward the trunk is demonstrated, valvular incompetence can be excluded.* **(Middle)** *Transverse scan shows the medial aspect of the lower 1/3 of the calf. A perforator arising from the great saphenous vein is seen penetrating through the investing fascia to join the posterior tibial vein.* **(Bottom)** *Transverse color Doppler scan shows the medial aspect of the lower 1/3 of the calf. With augmentation, color filling of the perforator lumen is demonstrated in the perforator between the great saphenous vein and the posterior tibial vein, indicating patency of vessel. If flow toward the posterior tibial vein can be documented on Doppler study, incompetence can be excluded.*

GROSS ANATOMY

Sciatic Nerve

- Actually 2 nerves (common peroneal and tibial) bound together

Common Peroneal Nerve

- Smaller of 2 terminal divisions of sciatic nerve
- Arises from dorsal divisions of sacral plexus (L4, L5, S1, S2)
- Separates from tibial nerve midthigh, at superior angle of popliteal fossa, runs downward laterally along medial border of biceps femoris muscle and tendon
- Leaves popliteal fossa by crossing plantaris and lateral head of gastrocnemius, passes posterior and superficial to head of fibula
- Curves anteriorly over posterior aspect of head of fibula
- Winds around lateral surface of fibular neck
  - Location at fibular head/neck puts peroneal nerve at risk in several clinical situations
    - Fibular neck fracture may result in foot drop
    - Realignment during correction of chronic knee valgus deformity may stretch nerve
- Ends between lateral side of neck of fibula and peroneus longus by dividing into 3 terminal branches
  - **Recurrent articular branch**
    - To proximal tibiofibular joint
    - Occasionally acts as conduit for intraneural transmission of ganglion arising from joint
  - **Deep peroneal nerve**
    - Arises on lateral side of neck of fibula, under peroneus longus
    - Pierces anterior intermuscular septum and extensor digitorum longus to enter anterior compartment
    - Extends down to ankle between tibialis anterior and long extensors
    - Near ankle, crossed by extensor hallucis and passes to ankle midway between malleoli
    - Muscular branches to anterior compartment and articular branch to ankle joint
    - Deep peroneal nerve crosses over to lateral side of dorsalis pedis artery at ankle joint
    - Medial and lateral terminal branches to dorsum of foot
    - Medial is main terminal branch and travels distally just medial to dorsalis pedis artery
  - **Superficial peroneal nerve**
    - Descends in substance of peroneus longus until it reaches peroneus brevis
    - Passes obliquely over anterior border of brevis and descends in groove between peroneus brevis and extensor digitorum longus
    - In distal 1/3 of leg, pierces deep fascia and divides into medial and lateral sensory branches to foot

Tibial Nerve

- Arises from ventral surface of sacral plexus
- Runs roughly along central axis of popliteal fossa
- Descends under fascial septum, which separates deep and superficial posterior muscle compartments
- In upper 2/3, lies on fascia of tibialis posterior and on flexor digitorum longus
- In lower 1/3, located midway between Achilles tendon and medial border of tibia
- Crosses posterior surfaces of tibia and ankle joint
- Posterior tibial vessels run with it, crossing in front of it from lateral to medial side
- At ankle, under flexor retinaculum, divides into lateral and medial plantar nerves
  - 1st branch of lateral plantar nerve is inferior calcaneal nerve (Baxter nerve)
    - Since this nerve supplies calcaneal periosteum, neuritis of this nerve can simulate plantar fasciitis
  - Medial plantar nerve can become compressed between navicular tuberosity and abductor hallucis muscle belly (jogger's foot)

Saphenous Nerve

- Longest branch of femoral nerve, arising 2 cm below inguinal ligament and descending via adductor canal
- Passes posterior to sartorius, descends posteromedial to knee where it pierces deep fascia
- In leg, accompanies great saphenous vein

Sural Nerve

- Purely sensory nerve that arises in popliteal fossa from tibial nerve
- Descends between 2 heads of gastrocnemius
- Pierces deep fascia midway between knee and ankle
- Courses superficial and posterior to peroneal tendons
- Accompanies small saphenous vein to lateral border of foot

ANATOMY IMAGING ISSUES

Imaging Approaches

- Ultrasound
  - Use high-frequency transducers to assess nerves
  - Neural signature on ultrasound
    - Short axis: Uniformly dispersed hypoechoic dots (nerve fascicles) separated by hyperechoic epineurium
    - Long axis: Parallel hypoechoic tracts of uniform caliber
    - Overall gain control and focus settings may have to be optimized for ultrasound to demonstrate nerve well
    - Anisotropy artifact may affect scanning, especially in short axis; tilting transducer during scanning helps achieve best view
    - Distortion of this uniform appearance suggests pathology
  - Trace medium-sized nerves by following their course as they branch from their parent
  - Small (1- to 2-mm) nerves are difficult to identify and location may only be inferred by adjacent vessels
- MR
  - Also suffers from inability to detect small nerves
  - Use of superficial small field of view coils may improve resolution

GRAPHICS, ANTERIOR AND POSTERIOR LEG NERVES

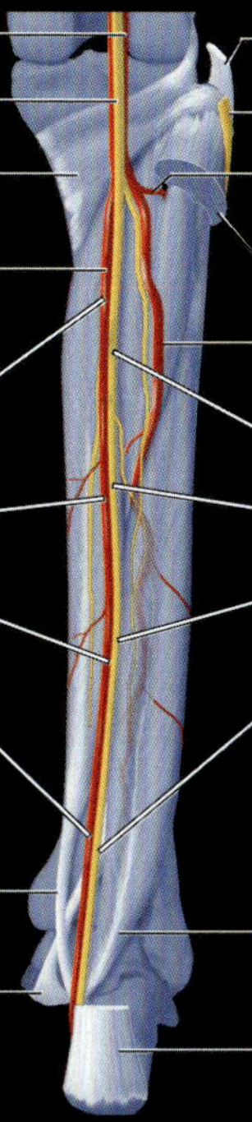

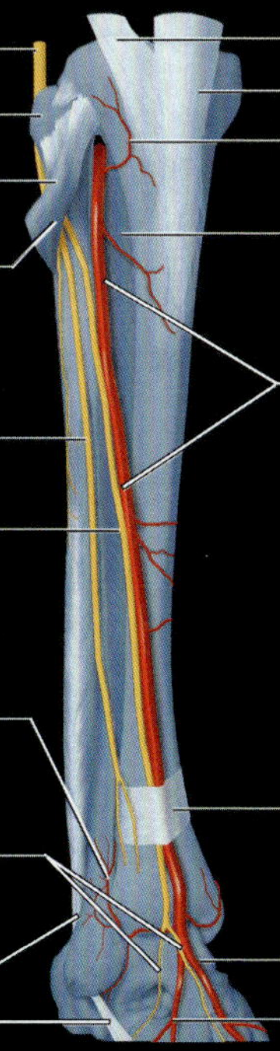

(Top) *Graphic shows posterior leg nerves and arteries. The sciatic nerve divides into common peroneal and tibial branches. The common peroneal nerve then follows the biceps femoris around the lateral fibular neck. The tibial nerve descends between the deep and superficial posterior leg muscles, paralleling the posterior tibial artery.* **(Bottom)** *Graphic of the anterior leg shows the common peroneal nerve extending through the peroneal tunnel (between the peroneus longus tendon and the fibular neck), branching into deep and superficial components. Both have muscular branches along their length; the deep peroneal nerve parallels the anterior tibial artery and terminates in the medial and lateral branches to the dorsum of the foot and ankle, respectively.*

TIBIAL NERVE

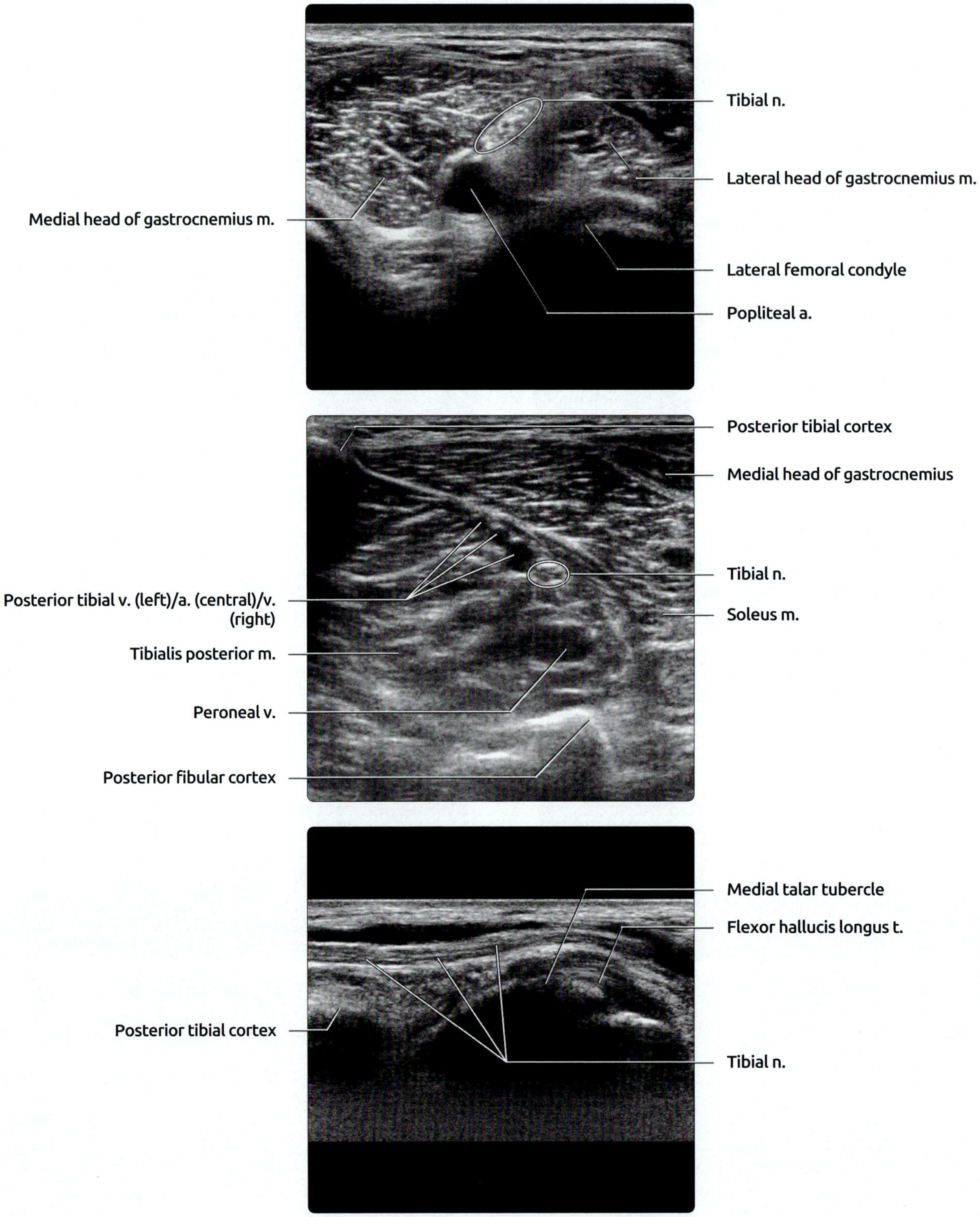

(Top) *Transverse scan shows the lower aspect of the popliteal fossa. The tibial nerve demonstrates the typical neural signature of small black dots within a bright matrix. It lies superficial to the popliteal artery.* **(Middle)** *Further inferiorly, at the level of the midcalf, the tibial nerve is deep to the soleus muscle and lies lateral to the posterior tibial vessels. Notice that even at this depth the neural signature is still visible.* **(Bottom)** *Sagittal scan shows the posterior aspect of the medial ankle. The longitudinal neural signature of the tibial nerve is well shown here with parallel hypoechoic tracts, which are roughly uniform in size. Contrast this to the more densely packed appearance of tendons.*

PERONEAL NERVE

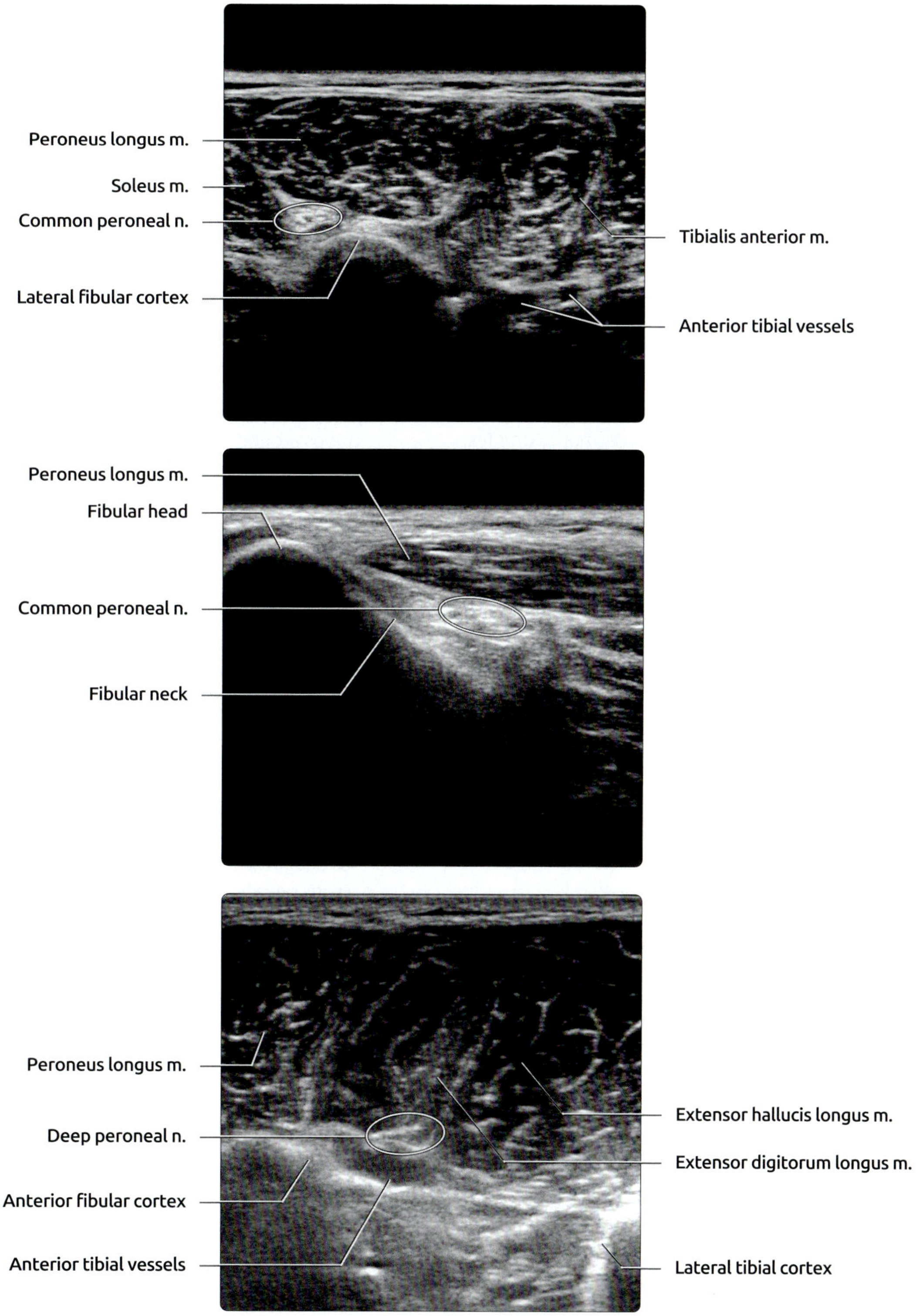

(Top) *Transverse scan of the head of fibula shows the common peroneal nerve running along the posteromedial aspect of the peroneus longus muscle. Superior to this, the nerve is covered by the biceps femoris muscle and tendon.* **(Middle)** *Coronal scan of the fibular neck shows the common peroneal nerve passing deep to the peroneus longus muscle. The nerve will then enter the anterior compartment.* **(Bottom)** *Transverse scan shows the anterior compartment at the midleg level. The deep peroneal nerve runs anterior to the anterior tibial vessels deep in this compartment.*

GROSS ANATOMY

Osseous Anatomy

- **Ankle (talocrural) joint**
  - Tibia, fibula, and talus form synovial joint
  - Supported by lateral and medial collateral ligaments
  - Mainly uniaxial hinge joint, dorsiflexion, plantar flexion, also dynamic shift of axis of rotation during dorsiflexion and plantar flexion
- **Distal tibiofibular joint**
  - Fibrous joint
  - Supported by syndesmotic ligaments
  - Synovial recess from ankle joint extends into joint
  - May have articular cartilage far distally
  - Minimal stretch ("give") during dorsiflexion: Allows increase in malleolar gap and slight fibular lateral rotation

Ligaments

- 3 sets bind ankle: Distal tibiofibular syndesmotic complex, lateral collateral ligaments, and deltoid ligaments
- **4 tibiofibular syndesmotic ligaments**
  - **Anterior and posterior tibiofibular ligaments**
    - Extend obliquely between anterior and posterior tibial and fibular tubercles, respectively
    - Distal anterior tibiofibular ligament: Bassett ligament
  - **Inferior transverse ligament**: Distal part of posterior tibiofibular ligament
  - **Interosseous ligament**: Distal thickening of syndesmotic membrane
- **3 lateral collateral ligaments**
  - **Anterior talofibular ligament**
    - Originates 1 cm proximal to lateral malleolar tip; inserts on talar neck
    - Stabilizes talus against anterior displacement, internal rotation, and inversion
    - Weakest, 1st to tear
  - **Calcaneofibular ligament**
    - Originates from lateral malleolar tip; inserts on calcaneal trochlear eminence
    - Deep to peroneal tendons
    - Lateral restraint of subtalar joint, often tears with anterior talofibular ligament
  - **Posterior talofibular ligament**
    - Extends from lateral malleolar fossa to lateral talar tubercle
    - Strongest, rarely tears
- **Deltoid ligament (medial collateral ligament)**
  - Fan-shaped, originates from anterior, apex, and posterior medial malleolus; inserts on talus, sustentaculum tali, spring ligament, and navicular
  - Deep: Posterior and anterior tibiotalar bands
  - Superficial: Tibiocalcaneal, tibiospring, tibionavicular, and posterior tibiotalar (variable) bands
- **Spring ligament (plantar calcaneonavicular ligament)**
  - Binds calcaneus to navicular, 3 components
    - Superomedial = **origin**: Sustentaculum tali; **insertion**: Superomedial navicular, tibiospring band of deltoid
    - Medioplantar oblique = **origin**: Calcaneal coronoid fossa; **insertion**: Plantar navicular
    - Inferoplantar longitudinal = **origin**: Coronoid fossa; **insertion**: Navicular beak

Retinacula

- Focal thickening of deep fascia
- Prevents bowstringing, binds tendons down
- **Superior extensor retinaculum**
  - Few cm above ankle joint
  - Attaches to anterior fibula laterally, tibia medially
  - Proximally continues with fascia cruris
  - Distally attaches to inferior extensor retinaculum
  - Binds down anterior compartment muscles
- **Inferior extensor retinaculum**
  - At ankle joint, Y-shaped, stem laterally, proximal and distal bands medially
  - Stem attaches laterally to upper calcaneus
    - Loops around extensor tendons
    - Roots extend into sinus tarsi
  - Proximal medial band has deep and superficial layers, loops around extensor hallucis longus tendon and occasionally tibialis anterior tendon
  - Distal medial band superficial to extensor hallucis longus and tibialis anterior tendons
    - Attaches to plantar aponeurosis
  - Dorsalis pedis vessels, deep peroneal nerve: Deep to all layers of inferior extensor retinaculum
- **Flexor retinaculum**
  - Attaches to medial malleolus
  - Proximally continuous with deep fascia of leg
  - Distally continuous with plantar aponeurosis
  - Abductor hallucis partly attached to it
  - Binds deep flexor tendons to tibial and calcaneal grooves
  - Lateral border of tarsal tunnel
- **Superior peroneal retinaculum**
  - Origin: Lateral malleolus, insertions vary, most commonly to deep fascia of leg and calcaneus
  - Binds peroneal tendons into retrofibular groove
- **Inferior peroneal retinaculum**
  - Continuous with inferior extensor retinaculum
  - Inserts on lateral calcaneus, peroneal tubercle (trochlea)
  - Binds peroneus brevis, peroneus longus tendons to calcaneus

Tendons

- **Anterior (extensor) compartment**
  - **Tibialis anterior tendon**
    - Most medial and largest tendon in anterior compartment
    - Inserts on medial cuneiform, base of 1st metatarsal
    - Dorsiflexes ankle, inverts foot, tightens plantar aponeurosis
    - Supports medial longitudinal arch during walking
  - **Extensor hallucis longus tendon**
    - Inserts on dorsal base of 1st distal phalanx
    - Extends 1st phalanges, dorsiflexes foot
  - **Extensor digitorum longus tendon**
    - Divides into 4 slips on dorsum of foot
    - Slips receive tendinous contributions from extensor digitorum brevis, lumbricals, and interosseous muscles

  - Each slip divides into 3: Central slip inserts on dorsal base of middle phalanx and 2 collateral slips, which reunite and insert on bases of 2nd-5th distal phalanges
  - Dorsiflexes ankle, extends toes, tightens plantar aponeurosis
- **Peroneus tertius tendon**
  - Typically part of extensor digitorum longus tendon
  - Inserts on dorsal base of 5th metatarsal

- **Lateral compartment**
  - **Peroneus longus tendon**
    - Posterolateral to peroneus brevis tendon in retrofibular groove, deep to superior peroneal retinaculum
    - Proximally has common tendon sheath with peroneus brevis
    - Separates tendon sheath at sole of foot
    - Descends inferior to peroneal tubercle, deep to inferior peroneal retinaculum
    - Curves in cuboid groove deep to long plantar ligament
    - Inserts on plantar base of 1st metatarsal, medial cuneiform
    - Plantarflexes ankle, everts foot, supports longitudinal and transverse arches during walking
    - Os peroneum is in ~ 20% of individuals
  - **Peroneus brevis tendon**
    - Anteromedial to peroneus longus tendon in retrofibular groove, deep to superior peroneal retinaculum
    - Descends superior to peroneal tubercle of calcaneus, deep to inferior peroneal retinaculum
    - Inserts into base of 5th metatarsal
    - Everts foot, limits foot inversion
- **Superficial posterior compartment**
  - **Achilles tendon**
    - Largest and strongest tendon in body
    - Commonest tendon to rupture
    - Conjoint tendon of medial and lateral gastrocnemius and soleus muscles
    - ~ 15 cm long
    - Lacks tendon sheath partially enclosed by paratenon
    - Inserts into posterior calcaneal tuberosity
    - Crescent-shaped insertion with medial and lateral projections
    - Retrocalcaneal bursa between distal tendon and calcaneal tuberosity
    - Main plantar flexor of ankle, foot
    - Midsection of Achilles tendon is supplied by peroneal artery
    - Proximal and distal portions supplied by posterior tibial artery
- **Plantaris tendon**
  - Vestigial, slender tendon, medial to Achilles tendon
  - Inserts on or medial to Achilles tendon
- **Deep posterior (flexor) compartment**
  - **Tibialis posterior tendon**
    - Crosses flexor digitorum longus tendon above ankle joint to become most posteromedial tendon
    - Shares tibial groove with flexor digitorum longus tendon
    - Inserts on navicular tuberosity, cuneiforms, sustentaculum tali, bases of 2nd-4th metatarsals
    - Main invertor of foot, aids in plantar flexion
    - Supports medial longitudinal arch
  - **Flexor digitorum longus tendon**
    - Lateral to tibialis posterior tendon in tibial groove
    - Crosses flexor hallucis longus tendon at knot of Henry (where there is also exchange of tendon fibers at this crossing of 2 tendons)
    - Divides into 4 slips, which give origin to lumbricals
    - Slips pass through openings in corresponding tendons of flexor digitorum brevis
    - Slips insert on bases of 2nd-5th distal phalanges
    - Flexes distal phalanges, assists in plantar flexion of ankle
    - When foot on ground: Maintains pads of toes on ground
    - When foot off ground: Plantar flexes 2nd-5th phalanges, aids in maintaining longitudinal arches
  - **Flexor hallucis longus tendon**
    - Passes 3 fibroosseous tunnels: (1) Between medial and lateral talar tubercles, (2) under sustentaculum tali, and (3) between 1st medial and lateral sesamoids
    - Crosses and sends slip to flexor digitorum longus at knot of Henry
    - Inserts on base of 1st distal phalanx
    - When foot on ground: Maintains pad of 1st toe on ground
    - When foot off ground: Plantar flexes 1st phalanges, aids in maintaining medial longitudinal arch
    - Weak plantar flexor of ankle
    - Innervated by tibial nerve

ANATOMY IMAGING ISSUES

Imaging Recommendations

- **Ultrasound**
  - Use ultrasound to assess tendon and ligamentous abnormalities aside from masses around ankle joint
  - Ligaments and tendons need to be assessed in both long and short axis
  - Since most structures run obliquely and tendons make acute turns around joint, clear understanding of their relationships and associated bony landmarks is helpful
- **CT**
  - Best for bony anatomy
  - Superior to MR for detecting small avulsion fragments
- **MR of ankle and hindfoot**
  - Axials optimal for ankle tendons, ligaments
  - Coronals useful for bones, cartilage, ankle, and sinus tarsi ligaments
  - Sagittals optimal for Achilles tendon, bones, cartilage, sinus tarsi ligaments

ANTERIOR ANKLE

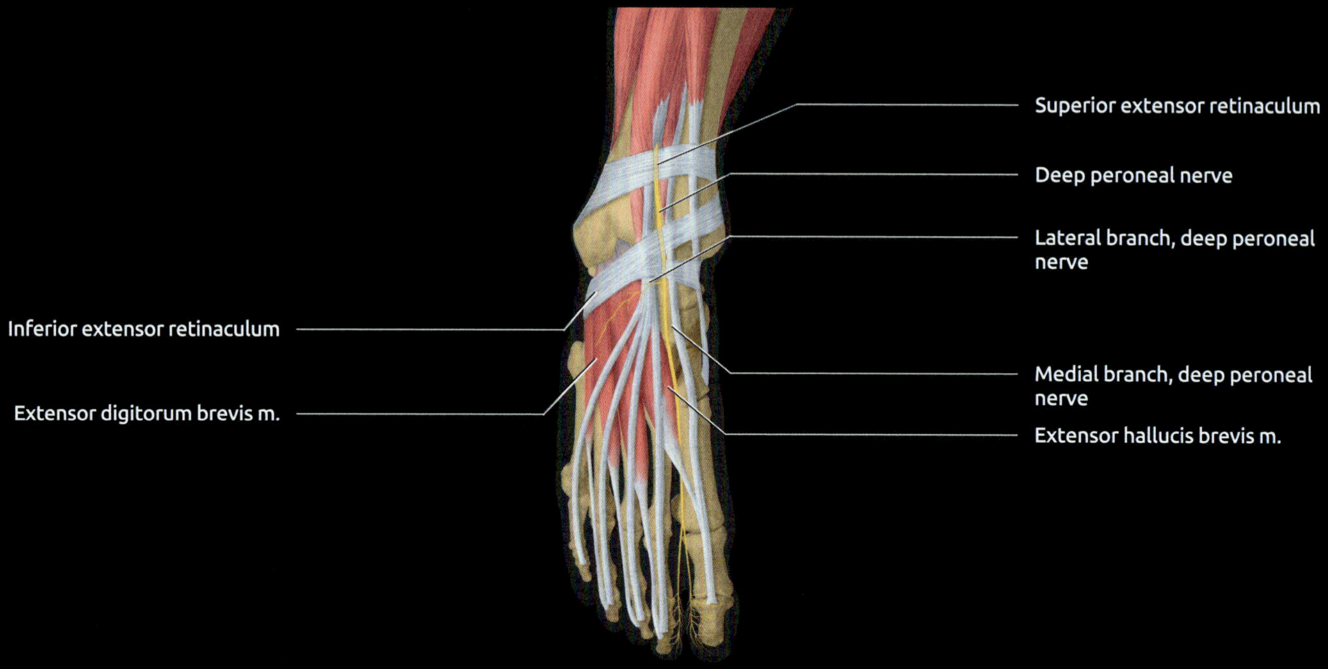

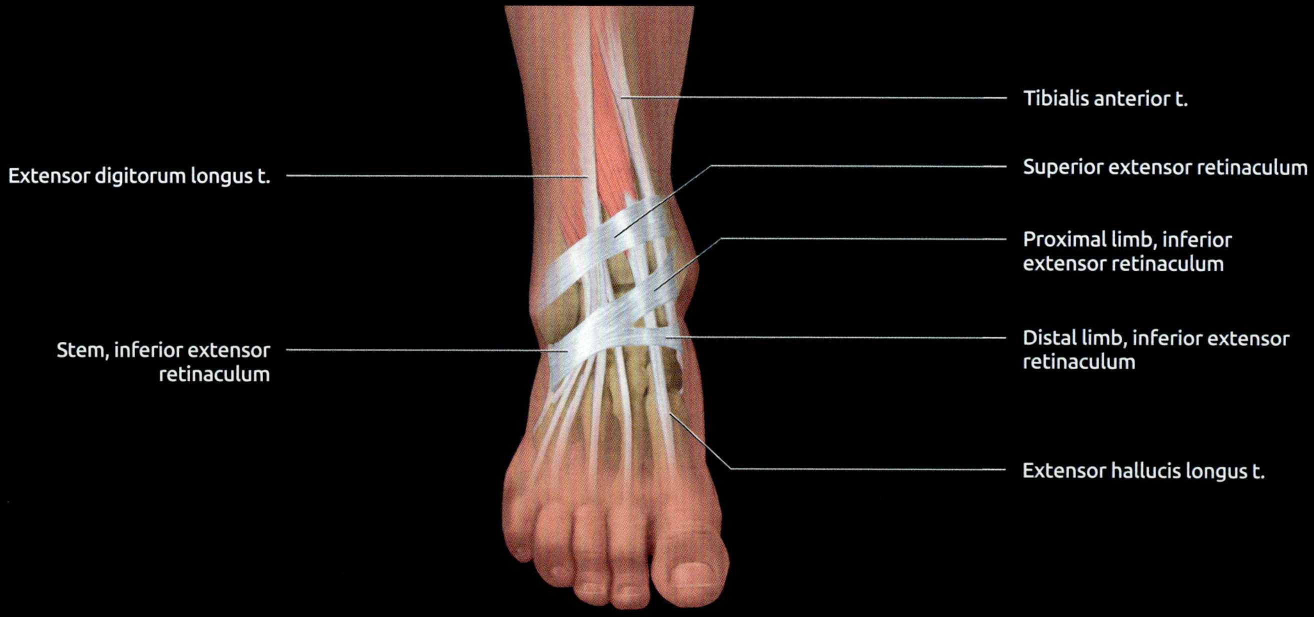

(Top) *The deep peroneal nerve travels deep to the extensor retinacula in the anterior tarsal tunnel and gives a lateral motor branch to the extensor digitorum brevis muscle. The medial branch continues dorsal to the talonavicular joint, middle cuneiform, and in between 1st and 2nd metatarsal to provide mostly sensory, but some motor, supply to the 1st web space.* **(Bottom)** *The peroneus tertius and tibialis anterior tendon insert proximally on the base of the 5th metatarsal and on the medial cuneiform and 1st metatarsal, respectively. The rest of the extensor tendons continue distally toward their insertion sites on the digits. The extensor tendons are held in place by the superior and inferior extensor retinacula.*

ANTERIOR ANKLE

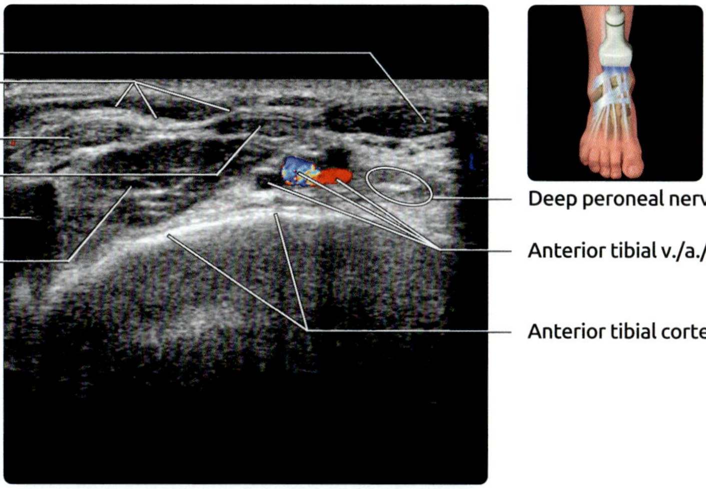

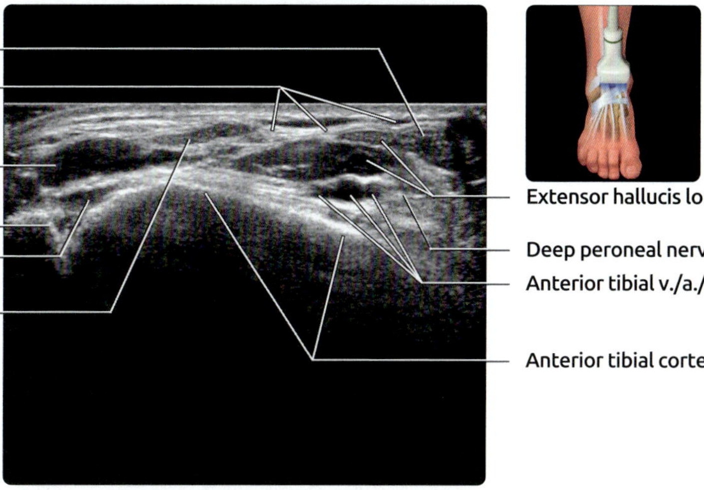

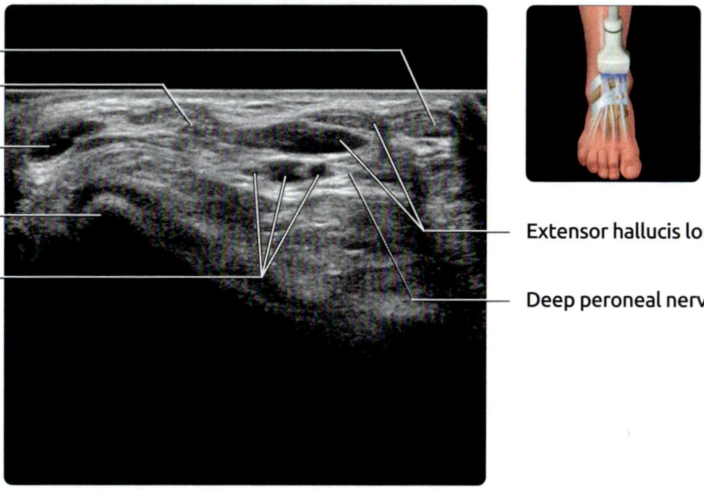

(Top) *Anterior transverse color Doppler scan above level of the ankle joint is shown. The basic relationship of the main structures of the anterior ankle hold true here and down through the ankle joint. From medial to lateral, the structures are: Tibialis anterior, extensor hallucis longus, nerve and vessels, and extensor digitorum longus (mnemonic: Tom, Harry, and Dick).* **(Middle)** *Transverse scan of the ankle joint at the level of the anterior talofibular ligament is shown. The anterior talofibular ligament, like most ankle ligaments, runs an oblique course and therefore may appear irregular or interrupted on a straight axial scan. This can be corrected by scanning obliquely along the long axis of the ligament.* **(Bottom)** *Transverse scan more inferiorly shows the asymmetrical soft tissue thickness anterior to the talus (thicker medially) due to different orientation axes of talus and foot.*

ANTERIOR ANKLE

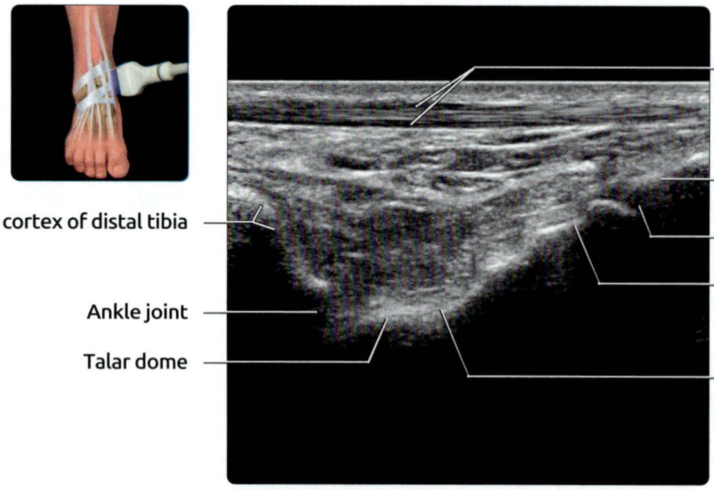

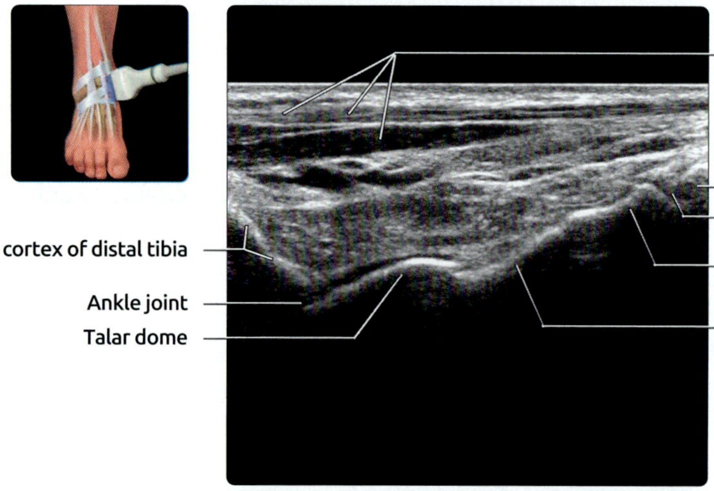

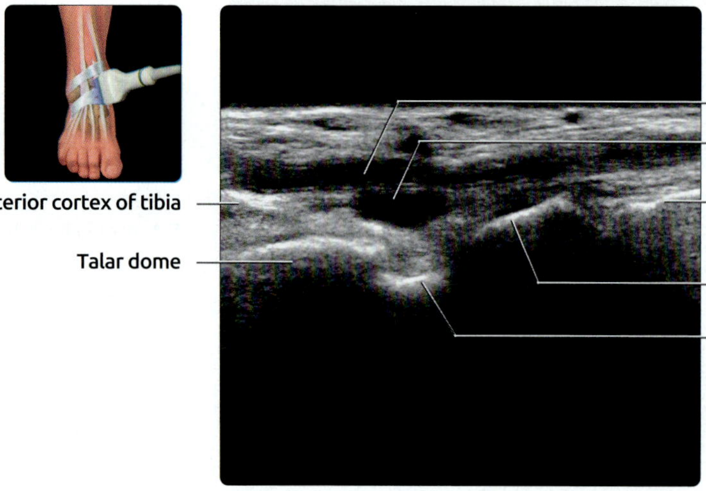

(Top) *Longitudinal scan of the lateral aspect of the ankle shows the course of the tibialis anterior tendon. A thin hypoechoic synovial layer can be seen enveloping the tendon.* **(Middle)** *Slightly more laterally, sagittal scan shows the extensor hallucis longus tendon and muscle, the latter distinguishing it from the tibialis anterior tendon.* **(Bottom)** *Further laterally, the dorsalis pedis artery and the accompanying vein(s) come into view. At this point, the subcutaneous tissue is thinnest, a result of the transverse arch of the midfoot with the apex roughly coinciding with the site of this artery. This creates a shallow bony backing for palpating the artery.*

MEDIAL ANKLE

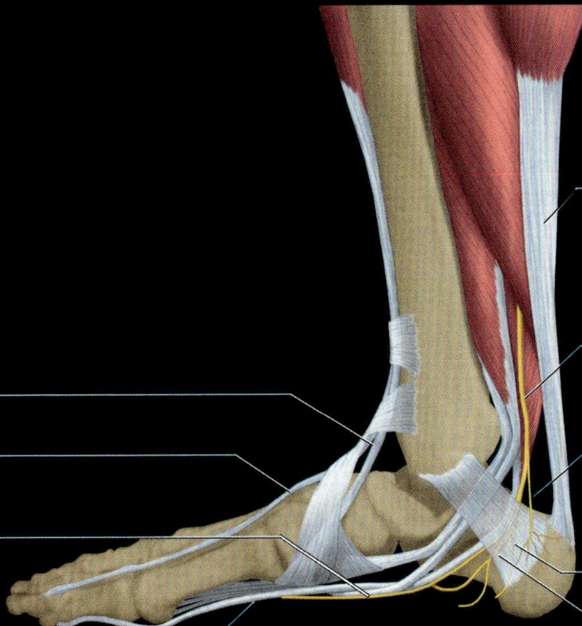

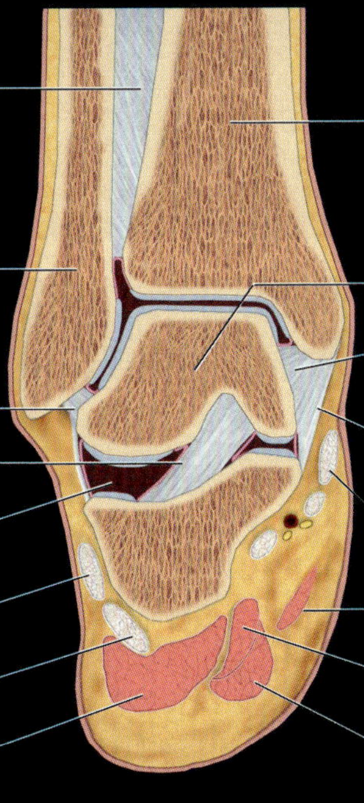

(Top) *Tarsal tunnel is seen accommodating the posterior tibial neurovascular bundle. The medial calcaneal nerve supplies the skin of the medial heel. The medial plantar nerve supplies the flexor hallucis brevis, abductor hallucis, flexor digitorum brevis, and 1st lumbrical. It carries sensation from the medial 2/3 of the plantar foot. The lateral plantar nerve supplies all other plantar muscles, including 2nd-4th lumbricals and all interossei, plus sensation from lateral 1/3 of mid and forefoot.* **(Bottom)** *Illustration of coronal section through the tibia, fibula, talus, and calcaneum shows the ankle ligaments. On the medial side, the medial malleolus is connected via the deep and superficial deltoid ligaments to the medial tarsal bones. The tibialis posterior tendon passes over the deltoid ligament as it curves below the medial malleolus. The talocalcaneal interosseous ligament, which has at least 4 discrete components, maintains apposition of the talus and calcaneus during hindfoot eversion and inversion.*

MEDIAL ANKLE

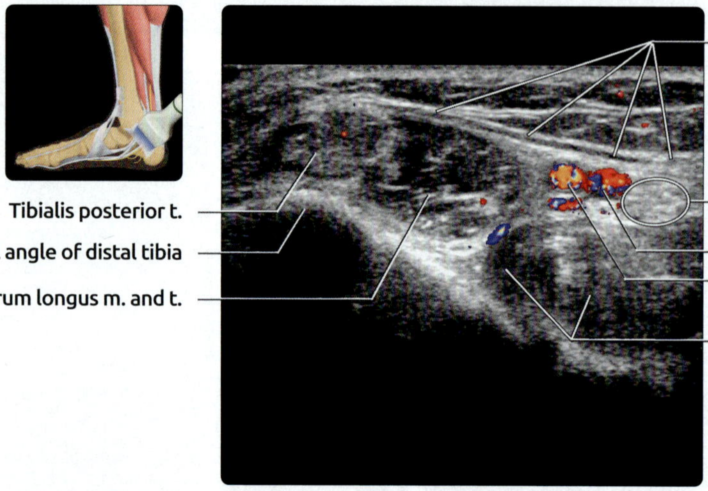

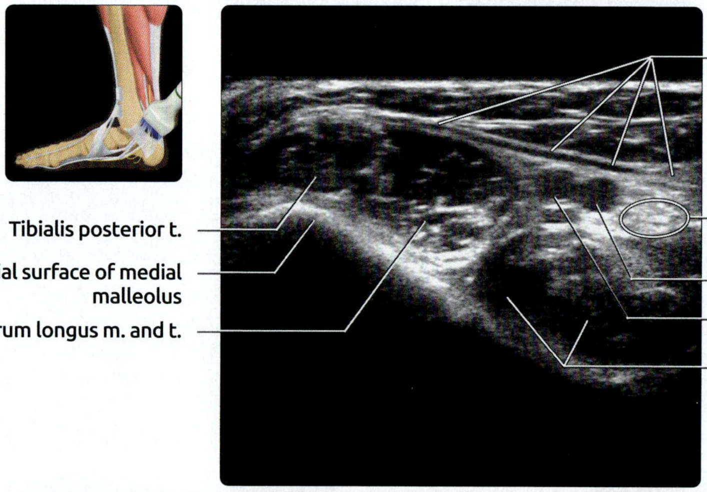

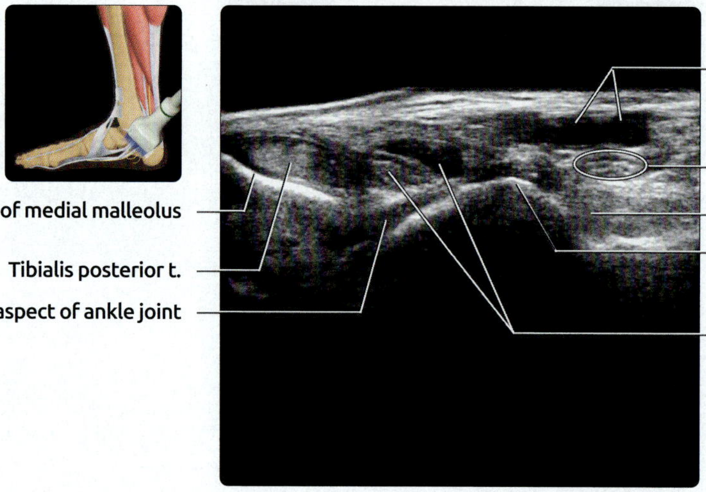

(Top) *Transverse color Doppler scan shows the posterior aspect of the distal tibia just above the medial malleolus. This area is held down by the flexor retinaculum and termed the tarsal tunnel. Within this tunnel are tibialis posterior, flexor digitorum longus, vessels and nerve, and flexor hallucis longus (mnemonic: Tom, Dick, and Harry).* **(Middle)** *Corresponding grayscale transverse scan just above the medial malleolus is shown. Image is less grainy without simultaneous color Doppler interrogation, and the nerve fibers (hypoechoic dots) give the typical neural signature.* **(Bottom)** *At the level of the talus, the tibial nerve moves anteriorly to run deep to the posterior tibial vessels and superficial to the flexor hallucis longus tendon. The flexor hallucis longus tendon runs in a groove on the posterior surface of the talus.*

MEDIAL ANKLE

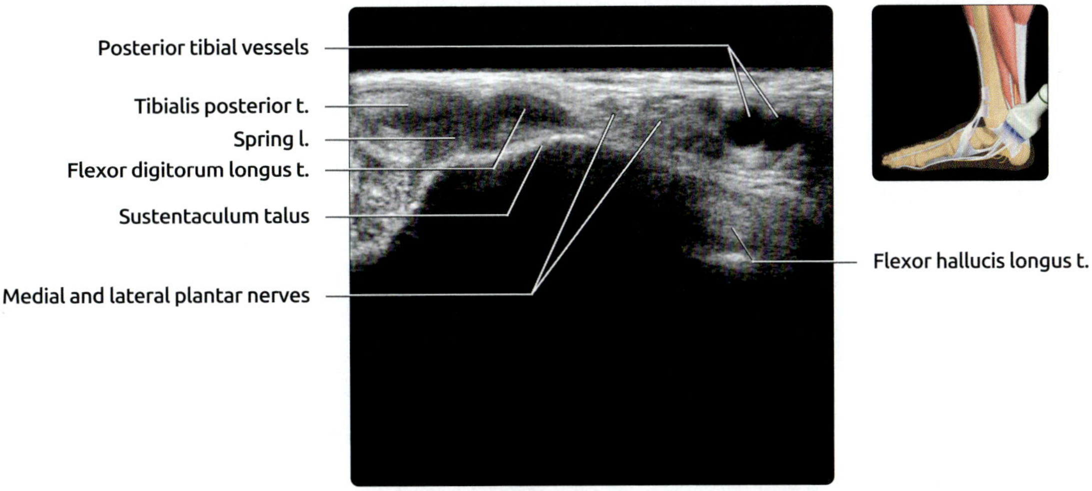

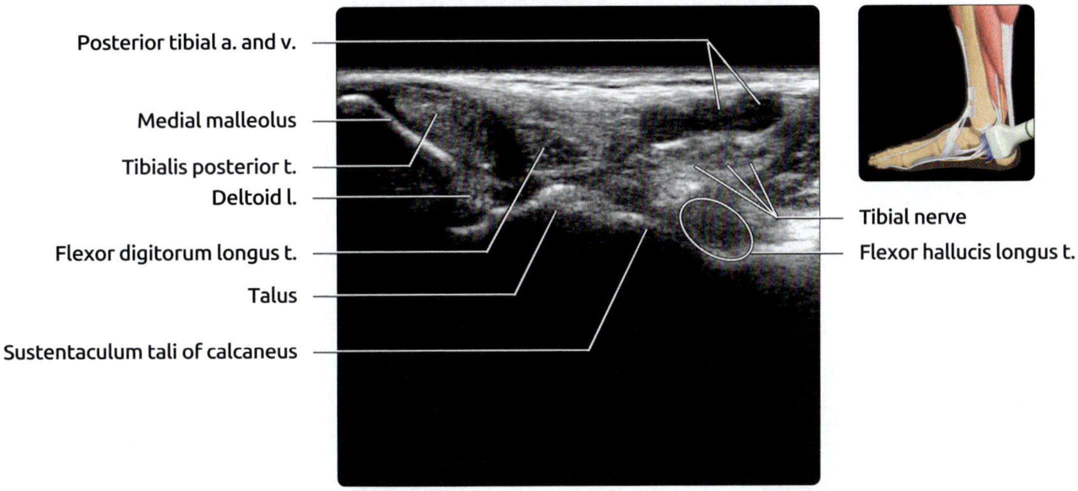

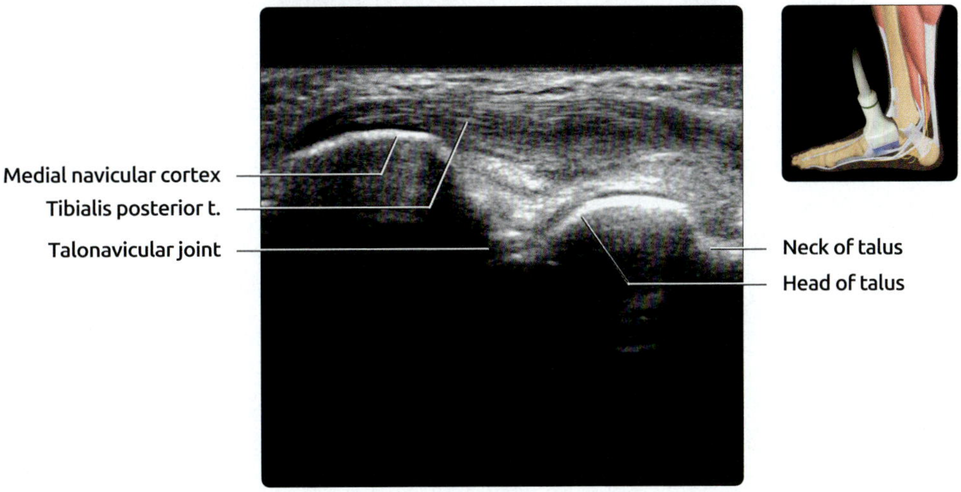

(Top) *More inferiorly, oblique transverse scan through the sustentaculum tali of the calcaneus shows that the mnemonic of Tom, Dick, and Harry still holds (tibialis posterior anteriorly and flexor hallucis longus posteriorly). The flexor hallucis longus tendon now runs in a groove behind the sustentaculum tali.* **(Middle)** *Coronal scan shows the tarsal tunnel. All structures within the tunnel curve around the medial malleolus to run forward to the midfoot. The tibialis posterior tendon runs on the surface of the deltoid ligament, flexor digitorum longus on the talus, and flexor hallucis longus on the calcaneus under the sustentaculum tali.* **(Bottom)** *Transverse image shows the long axis of the tibialis posterior tendon inserting onto the medial pole of the navicular. This insertion is sometimes accompanied by an accessory navicular bone, which may take on different shapes.*

MEDIAL AND POSTERIOR LIGAMENTS

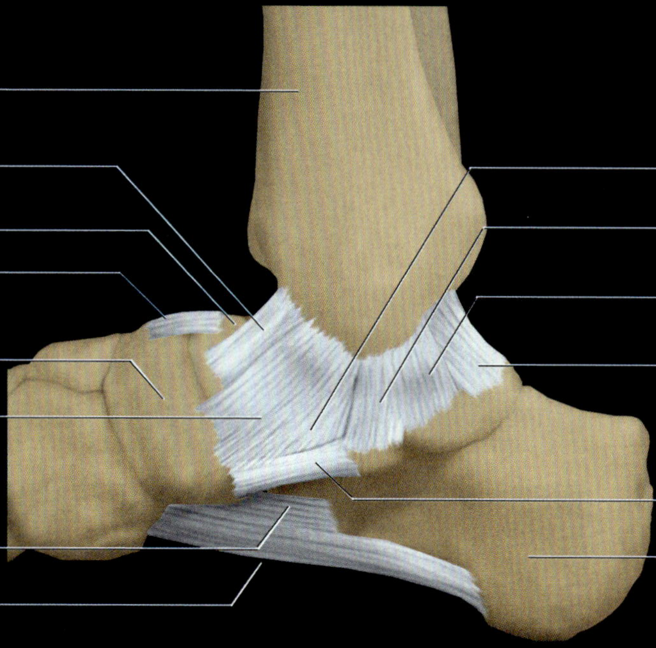

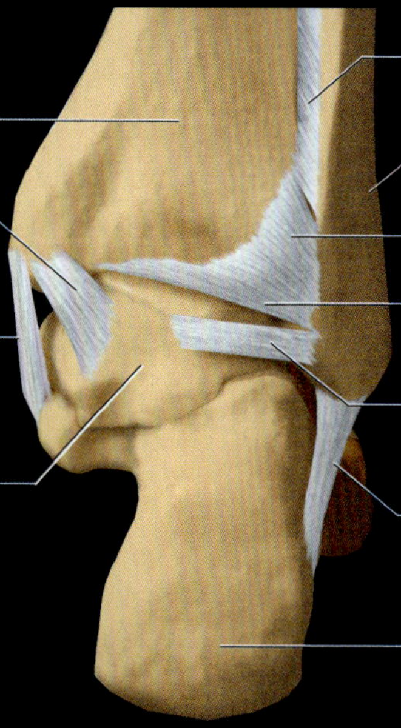

(Top) *Medial view shows the ankle. The deltoid ligament is the major supporter of the ankle. It has many variable components, but a commonly accepted division includes the deep anterior and posterior tibiotalar and superficial anterior and posterior tibiotalar, tibiocalcaneal, and tibiospring and tibionavicular bands. Note that the superficial deltoid is band-like, and distinction between its various components relies on the sites of insertion.* **(Bottom)** *Posterior view shows the ankle. The tibiofibular ligaments are obliquely oriented and their fibular origin is above the fibular fossa. The inferior transverse ligament, which is the inferior aspect of the posterior tibiofibular ligament, extends distal to the tibial posterior surface.*

DELTOID LIGAMENT

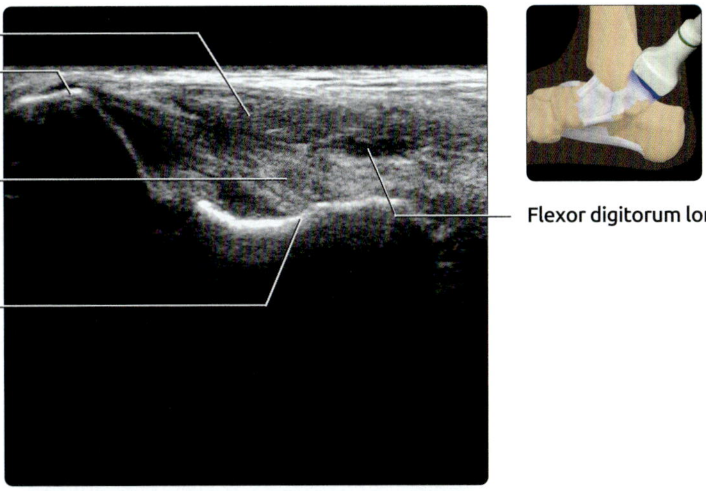

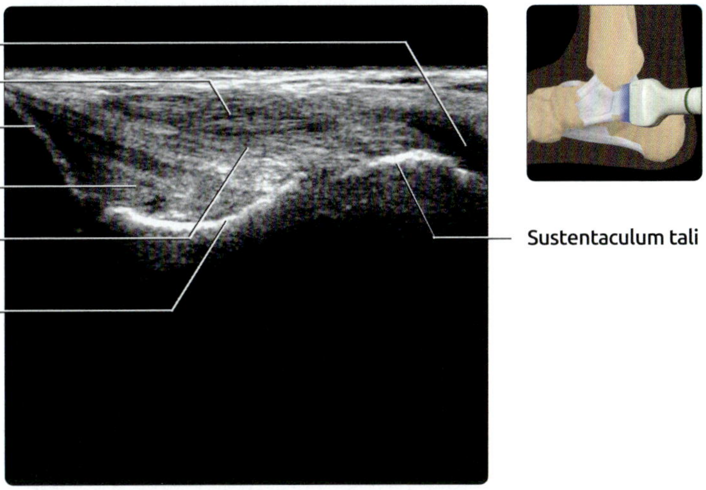

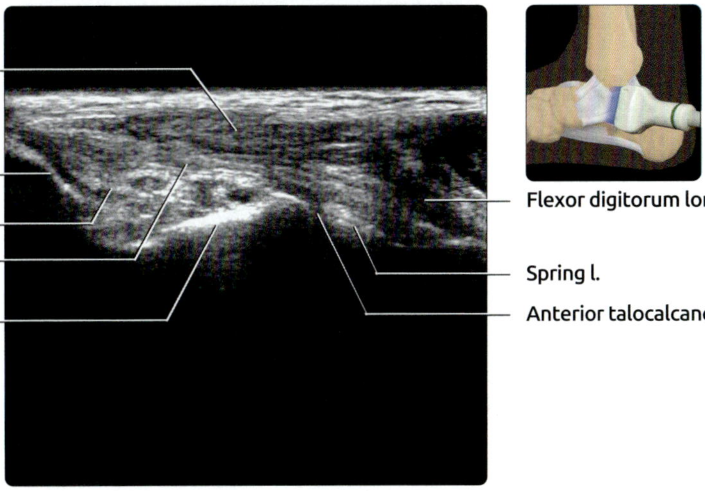

(Top) *Coronal image is seen through the posterior (posterior tibiotalar) portion of deltoid ligament, which runs from the medial malleolus to the talus. The tibialis posterior and flexor digitorum longus tendons run superficial to this ligament.* **(Middle)** *Coronal image through a more anterior portion of the deltoid ligament is shown. The superficial layer (tibiocalcaneal) runs from the malleolus to the sustentaculum tali, the deep component (posterior tibiotalar) to the talus.* **(Bottom)** *Coronal image is seen through the midportion of the deltoid ligament. The superficial layer (tibiospring) runs from the malleolus to the spring ligament, the deep component to the talus.*

DELTOID LIGAMENT

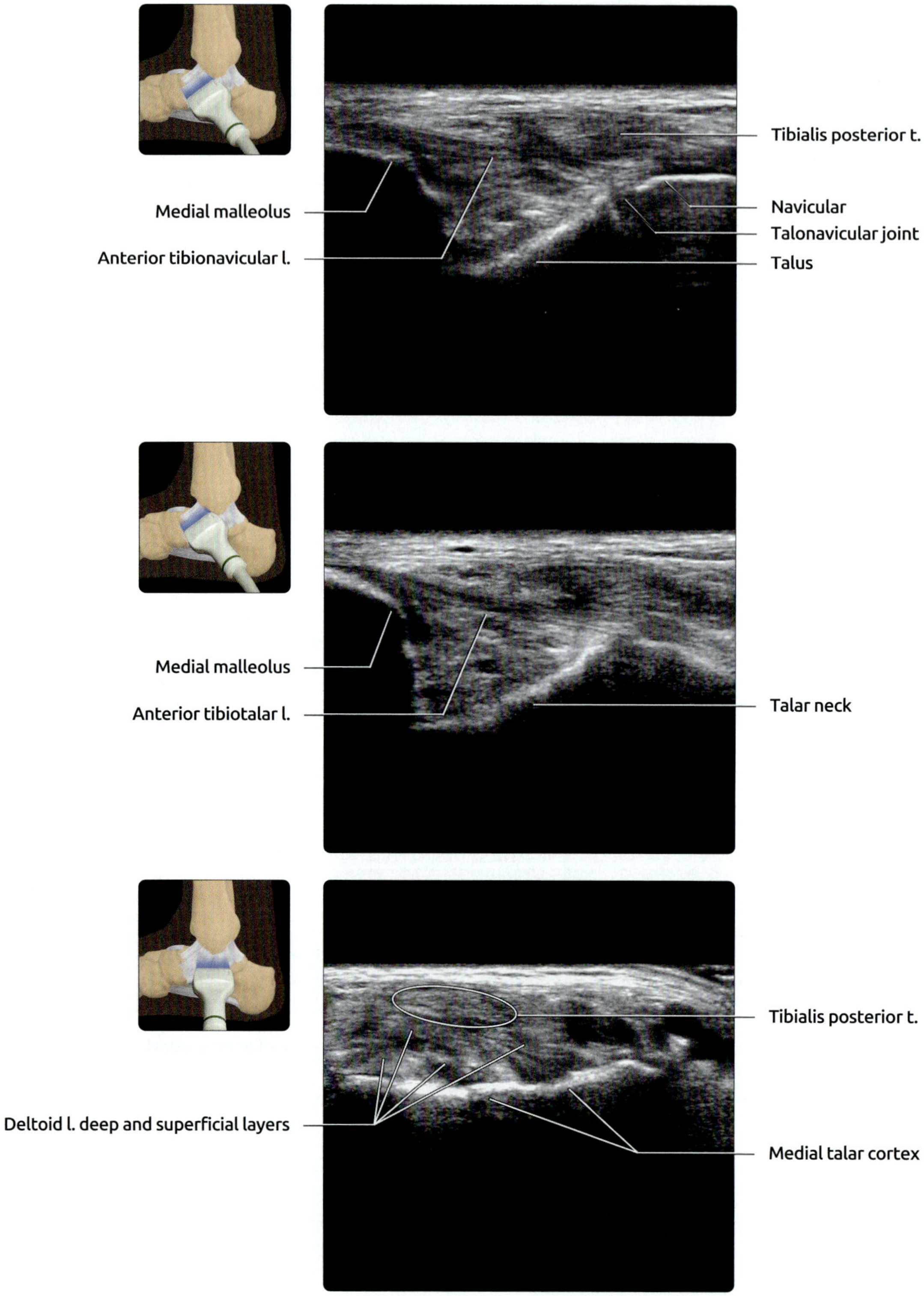

(Top) *Coronal image is seen through the anterior portion of deltoid ligament; a superficial layer (talonavicular) runs from the malleolus to the navicular.* **(Middle)** *Oblique sagittal image is seen through the most anterior portion of the deltoid ligament; fibers (anterior tibiotalar) run from the malleolus to the talar neck. From front to back, these superficial and deep components of the deltoid create a strong restraint against ankle eversion.* **(Bottom)** *Transverse scan through the medial wall of the talus shows deep and superficial components of the deltoid ligament. The tibialis posterior tendon runs superficial to this ligament.*

LATERAL ANKLE

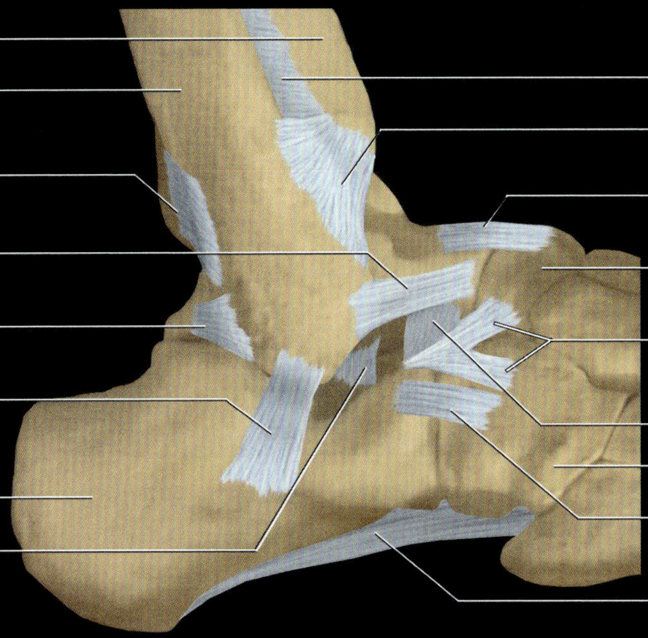

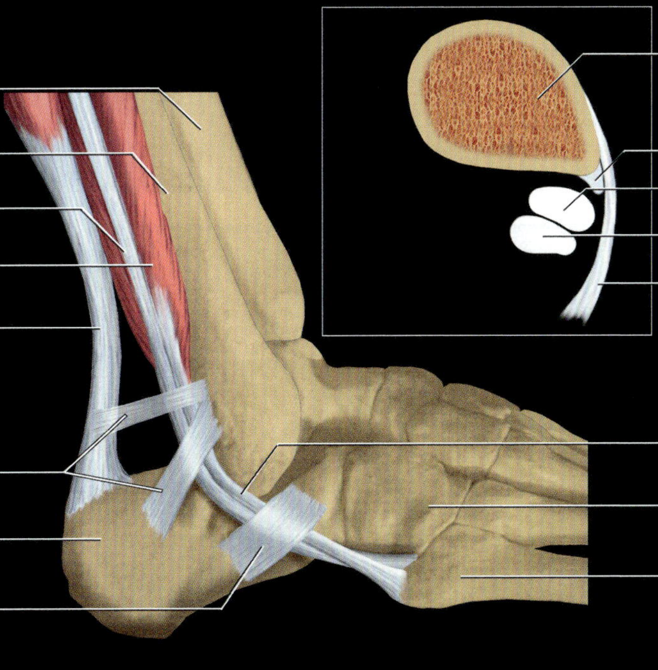

(Top) *Lateral view shows the ankle. Two groups of lateral ligaments support the ankle: (1) The syndesmotic ligaments, including the anterior tibiofibular, posterior tibiofibular, and syndesmotic ligaments, and (2) the lateral collateral ligaments, including the anterior talofibular, posterior talofibular, and calcaneofibular ligaments. Other ligaments that bind the lateral hindfoot include the talocalcaneal interosseous ligament and cervical ligament in the sinus tarsi. Note the bifurcate ligament, which extends from the calcaneus to the navicular and cuboid.* **(Bottom)** *Peroneal retinacula is shown. The superior peroneal retinaculum holds the peroneal tendons in the retrofibular groove. The retinaculum has variable insertions but typically inserts on deep fascia of the leg and on the calcaneus. The inferior peroneal retinaculum holds the peroneal tendons against the calcaneus.*

LATERAL ANKLE

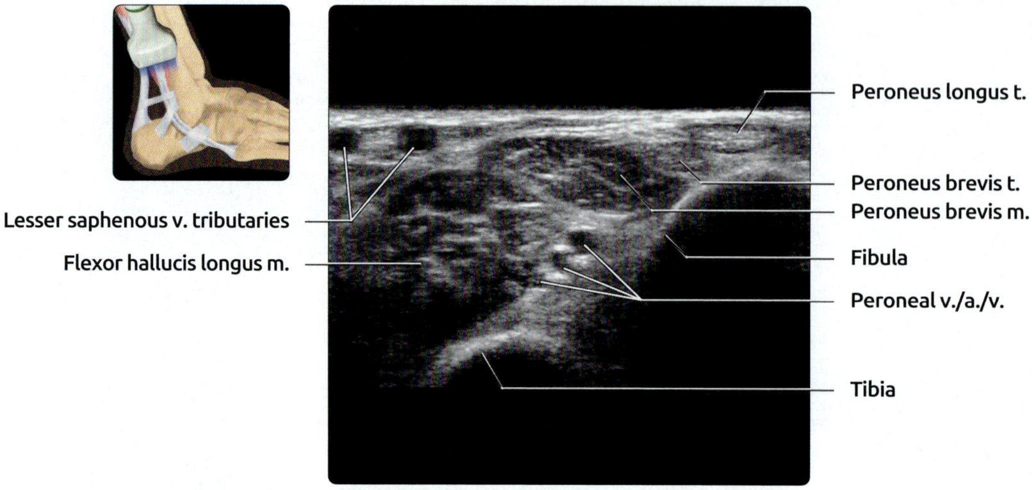

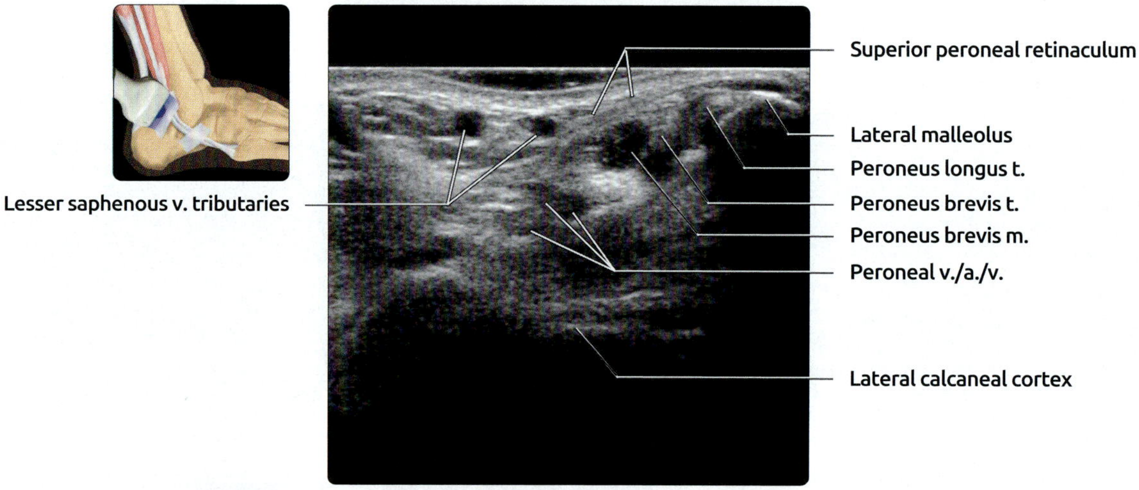

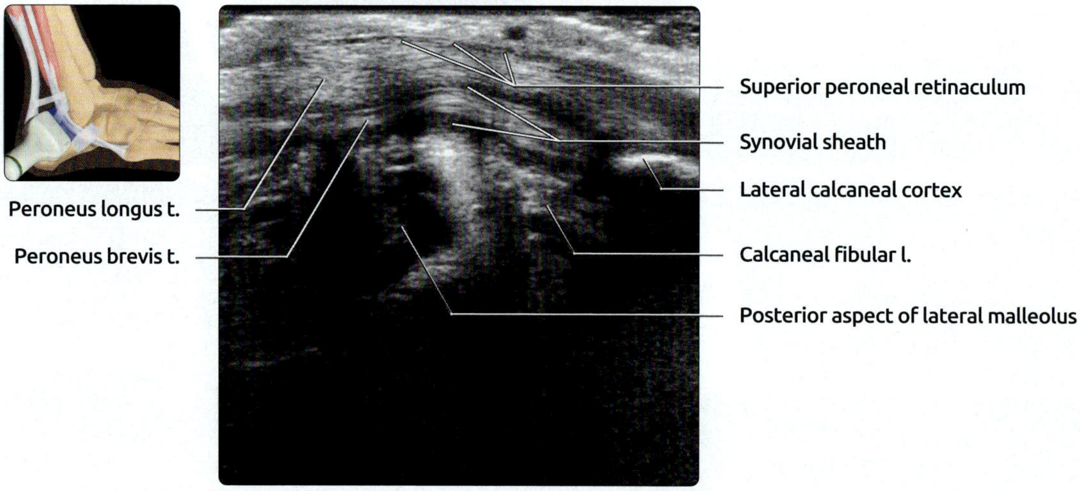

(Top) *Transverse scan of the posterior aspect of the distal fibula shows the distal peroneal muscles. The peroneus longus tendon runs anterior to the peroneus brevis. The latter is easily identified by its muscle component, which continues to just above the tip of the lateral malleolus.* **(Middle)** *Transverse scan is shown at the level just above the tip of the lateral malleolus. The peroneal tendons are now held down by the superior peroneal retinaculum, which prevents tendon subluxation (bowstringing). There is still a residual component of the peroneus brevis muscle.* **(Bottom)** *Oblique coronal scan of the posterior aspect of the lateral malleolus shows the longitudinal course of the peroneal tendons. Both tendons run inferior onto the lateral surface of the calcaneus, passing over the calcaneal fibular ligament on the way.*

LATERAL ANKLE

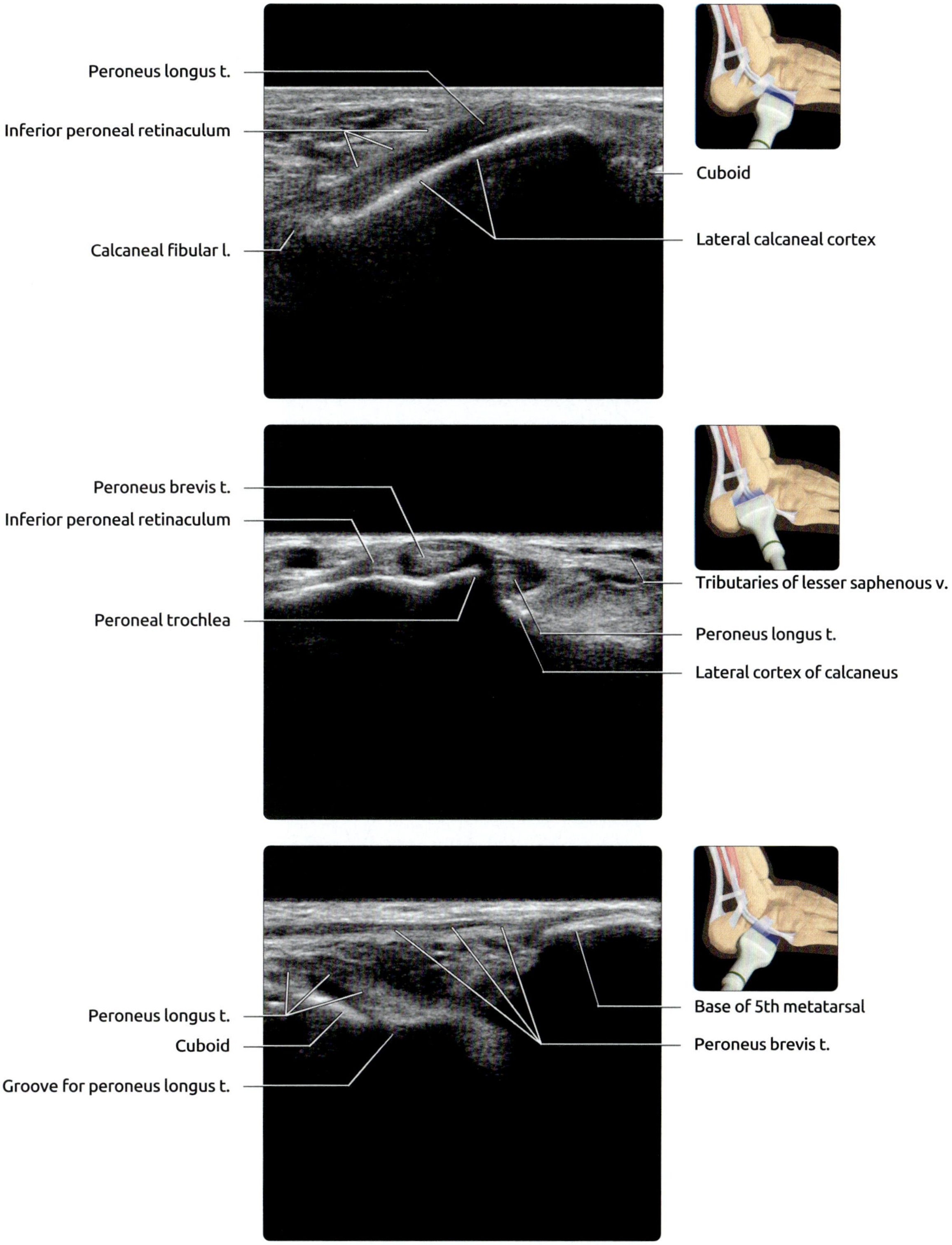

(Top) *Oblique transverse scan shows the lateral aspect of the calcaneus along the course of the peroneus longus tendon. Close to the anterior end of the calcaneus, the peroneus longus turns medially and inferiorly to enter the plantar region.* **(Middle)** *Coronal scan of the lateral surface of the calcaneus shows the peroneal tendons held down by the inferior peroneal retinaculum but partially separated by peroneal tubercle, a protuberance on the lateral calcaneal cortex. The peroneus brevis tendon lies more superior than the peroneus longus tendon at this level.* **(Bottom)** *Transverse scan of the lateral aspect of the foot shows insertion of the peroneus brevis tendon onto the base of the 5th metatarsal bone. Also shown here is the peroneus longus tendon turning inferiorly on the surface of the cuboid into a groove solely for this tendon to enter the plantar aspect of the foot.*

FIBULAR LIGAMENTS

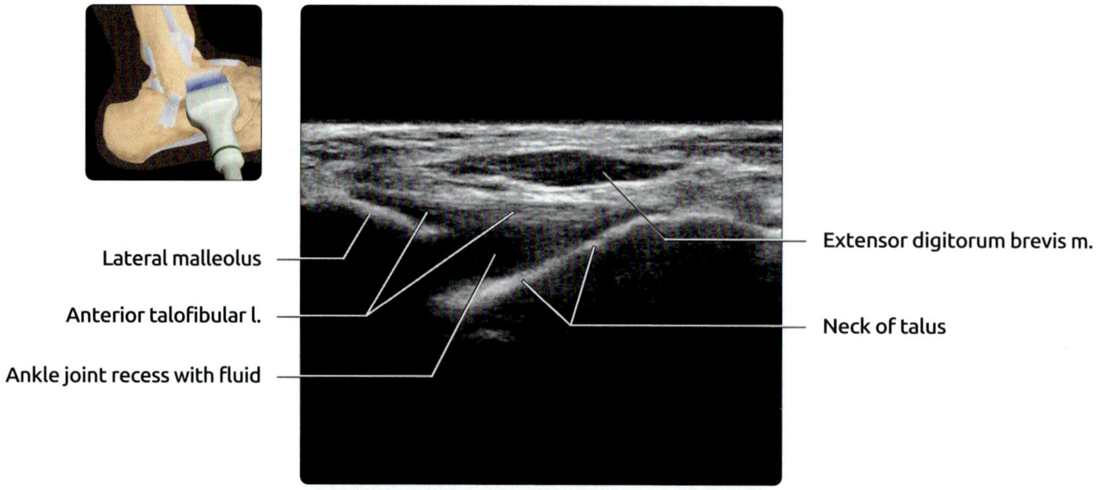

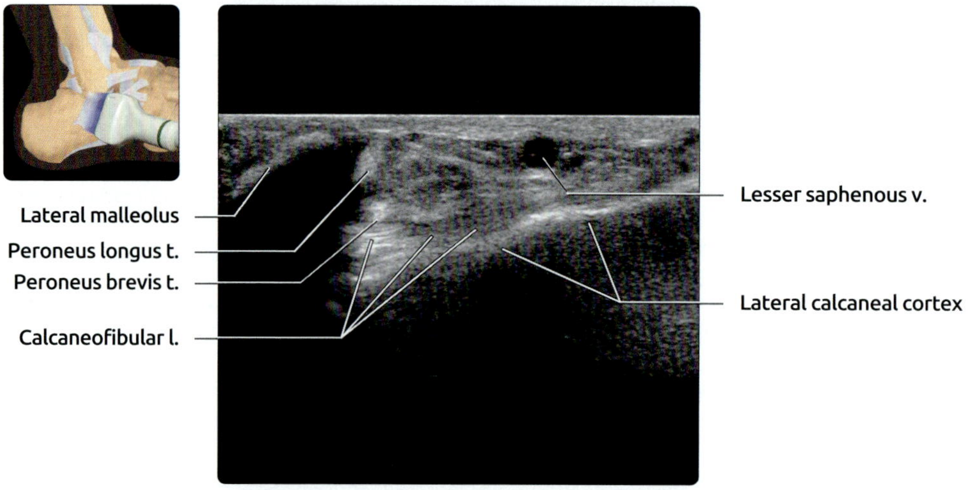

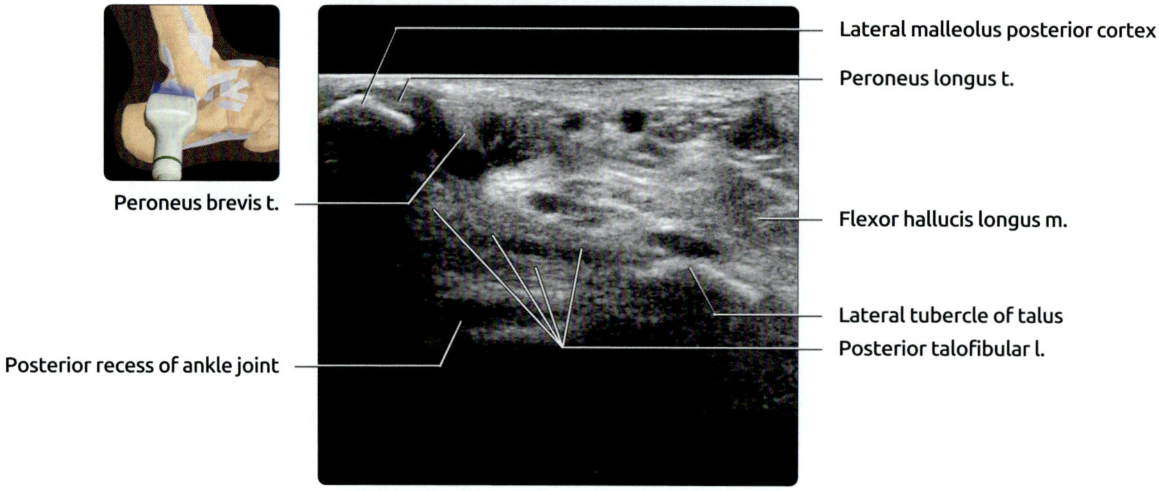

(Top) *Oblique transverse scan shows the lateral ankle along the course of the anterior talofibular ligament. An anterolateral ankle joint recess exists deep to this ligament, making it a good area to look for small effusions. The lateral fibers of the extensor digitorum brevis muscle run on the surface of this ligament.* **(Middle)** *Oblique coronal scan along the long axis of the calcaneofibular ligament is shown. The peroneal tendons can be seen curving under the lateral malleolus and running partly on the surface of this ligament.* **(Bottom)** *Transverse scan of the posterior aspect of the lateral malleolus shows the posterior talofibular ligament. Deep to this ligament is the posterior recess of the ankle joint.*

POSTERIOR ANKLE PANORAMA

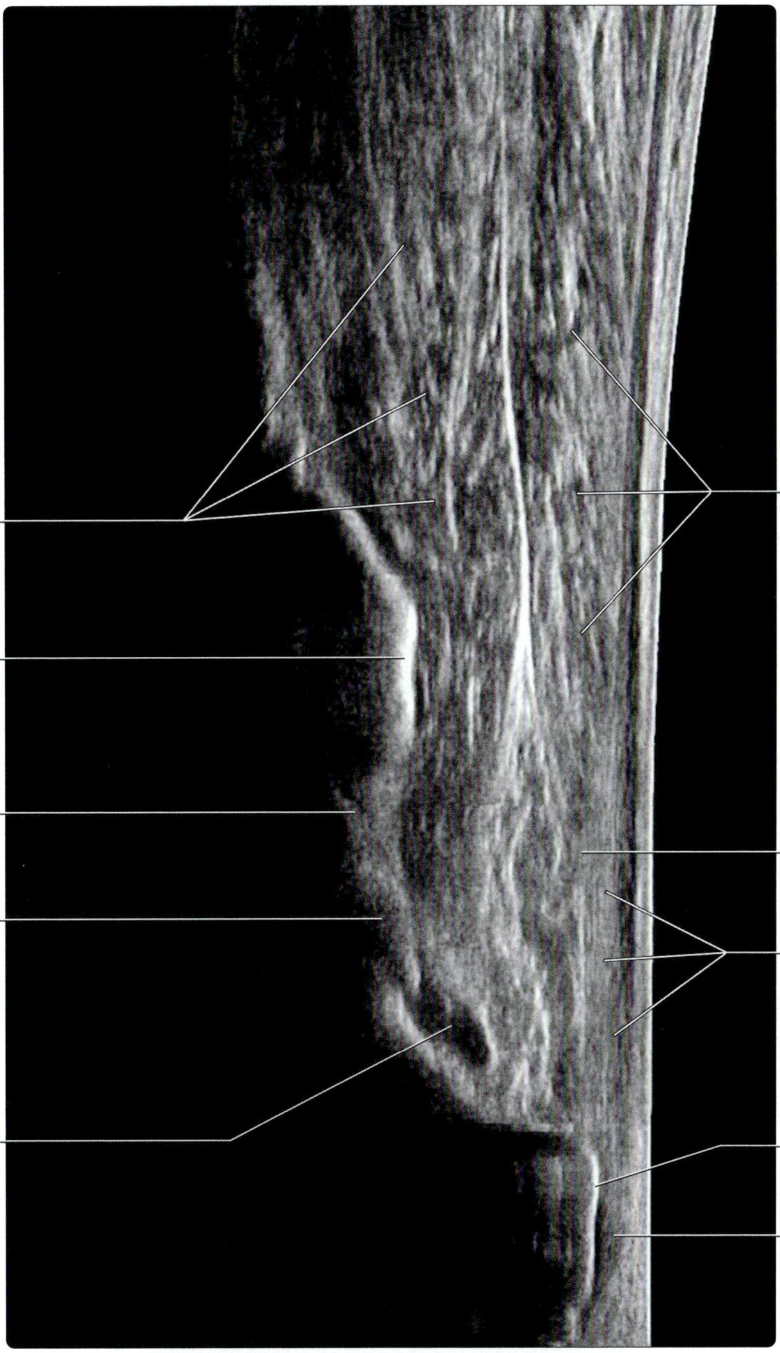

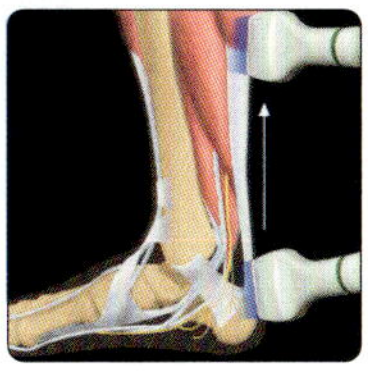

Sagittal panoramic view shows the posterior ankle. The dominant structure in the posterior ankle is the Achilles tendon, the largest tendon in the body. This runs craniocaudally just under the skin and is well demonstrated by ultrasound. The ankle should be in dorsiflexion to tighten this tendon during examination so that the organization of its fibers and continuity can be assessed. More superiorly, its contributing muscles, the soleus and gastrocnemius, can be traced. Deep to the Achilles tendon is the flexor hallucis longus muscle. An accessory soleus muscle may be seen inferomedial to the Achilles tendon. It occurs in 30% of the normal population and may be symptomatic in athletes with low-grade pain and stiffness after exercise.

ACHILLES TENDON, LONGITUDINAL

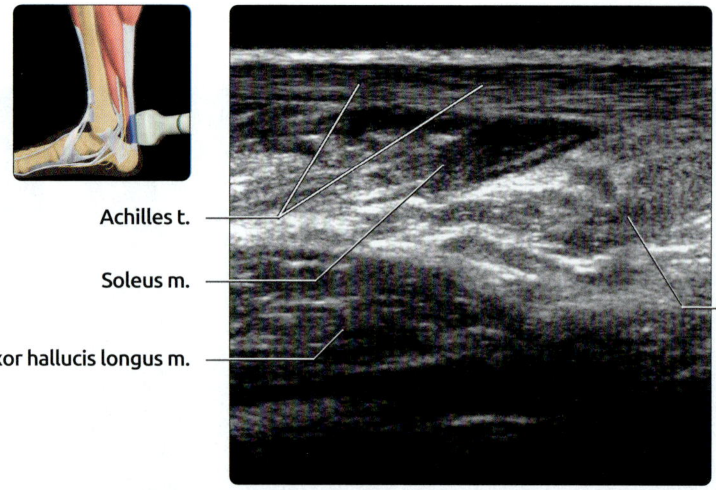

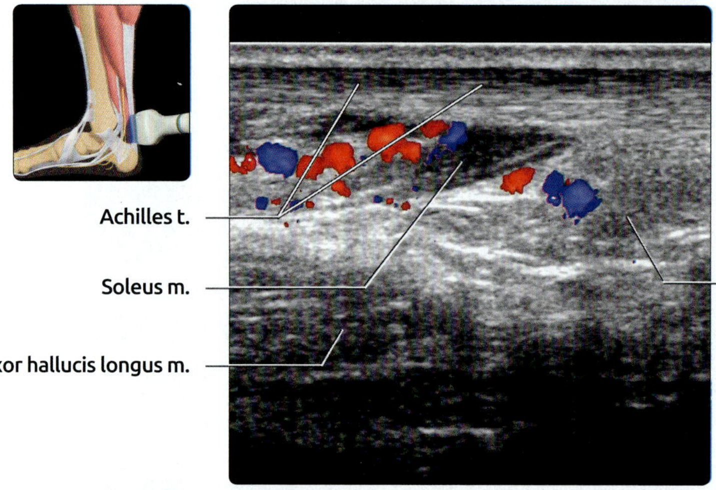

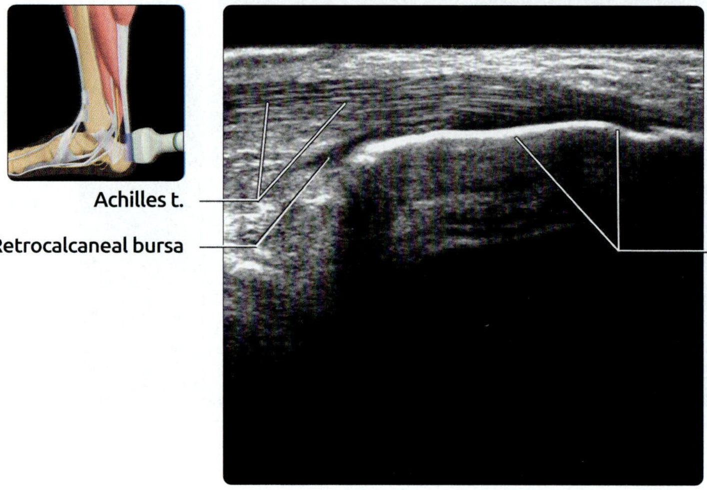

(Top) *Sagittal scan of the lower posterior leg shows the musculotendinous junction between the soleus muscle and the Achilles tendon. These structures join at an angle with a crisp border. A disruption of this sharp transition (usually due to hemorrhage and rupture of inserting muscle fibers) is suggestive of injury to the musculotendinous junction.* **(Middle)** *Color Doppler scan of the same region shows the rich blood supply to this region and the cause for extensive bruising in tendon damage at this site.* **(Bottom)** *Sagittal scan more distally shows the insertion of Achilles tendon onto the middle facet of the posterior calcaneal surface. The retrocalcaneal bursa separates the Achilles tendon from the superior facet of the posterior calcaneal surface. This bursa usually contains a tiny amount of fluid detectable on ultrasound.*

ACHILLES TENDON, AXIAL

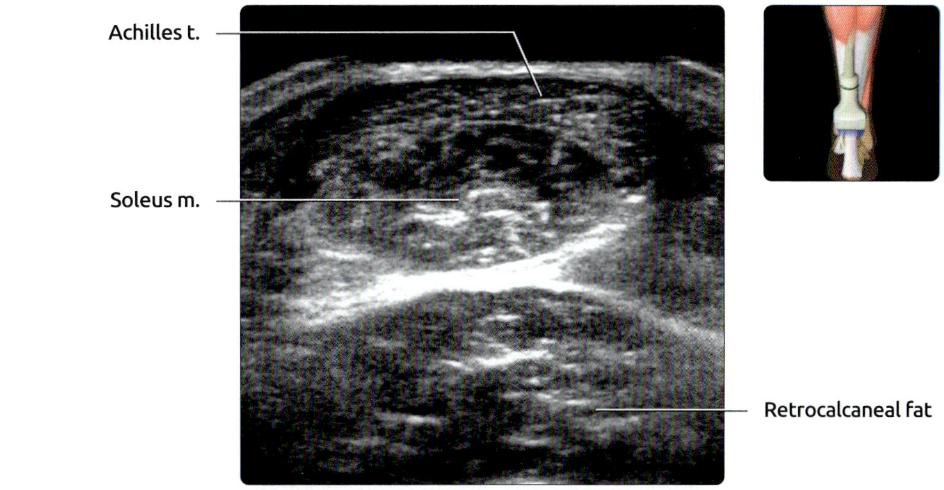

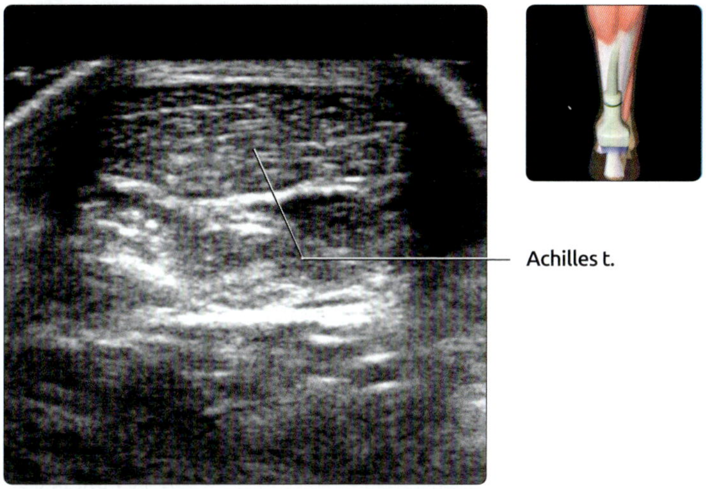

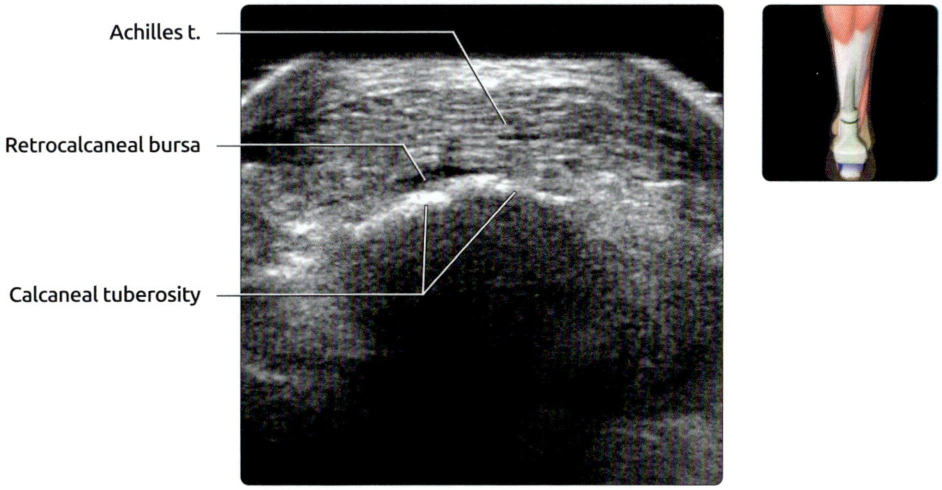

(Top) *Series of transverse scans through the Achilles tendon is shown. At a superior level, the soleus muscle can be seen deep to the Achilles tendon. The Achilles tendon begins superiorly as a continuation of the gastrocnemius aponeurosis, which then joins with the soleus tendon to form the Achilles tendon. At the level of this scan, the tendon is crescent-shaped to accommodate the soleus muscle.* **(Middle)** *Transverse scan shows the Achilles tendon at its midportion. The tendon now assumes an oval shape. The fibrils within the tendon produce a coarsely dotted appearance. Focal hypoechogenicity or change in shape of this tendon should suggest tear/rupture or tendinosis.* **(Bottom)** *At a lower level before its insertion onto the calcaneal tuberosity, the Achilles tendon becomes flattened. A small amount of fluid is frequently seen in the retrocalcaneal bursa deep to the tendon.*

TERMINOLOGY

Definitions

- **3 major divisions**
  - **Hindfoot**: Calcaneus and talus
  - **Midfoot**: Navicular, cuneiforms, and cuboid
  - **Forefoot**: Metatarsals and phalanges
- **2 columns**
  - **Medial column**: Talus, navicular, cuneiforms 1-3, digits 1-3
  - **Lateral column**: Calcaneus, cuboid, digits 4 and 5
- Some authors use 2 columns in hind and midfoot as above but divide forefoot into 3 columns
  - Medial column: 1st toe
  - Middle column: 2nd-4th toes
  - Lateral column: 5th toe

IMAGING ANATOMY

Overview

- Alignment of foot can only be assessed on weight-bearing radiographs

Arches of Foot

- Foot is arched from posterior to anterior and from medial to lateral
- **Transverse arch of foot**
  - Cuneiform bones form keystone of arch due to triangular shape
  - Major supporting structures of transverse arch
    - Spring ligament
    - Lisfranc ligament and intermetatarsal ligaments
    - Intertarsal ligaments
- **Longitudinal arch of foot**
  - From posterior process calcaneus to metatarsal heads
  - Medial side is higher than lateral
  - Apex of arch is at navicular and cuneiforms
  - Metatarsals slant downward from apex of arch to metatarsophalangeal (MTP) joint
    - This is called inclination angle
    - Inclination angle decreases from 20° at 1st to 5° at 5th metatarsal
  - Major supporting structures of longitudinal arch
    - Plantar fascia
    - Long and short plantar ligaments
    - Spring ligament
    - Posterior tibial tendon, peroneus longus tendon
- **Radiographic assessment of normal longitudinal arch**
  - Evaluate talometatarsal alignment
    - Normal: Axis of talus continues along axis of 1st metatarsal
    - Pes planus: Axis of talus points below axis of 1st metacarpal, due to flattened arch
    - Pes cavus: Axis of talus points above axis of 1st metacarpal, due to exaggerated arch

Distribution of Weight Bearing

- 50% of weight is borne on subtalar joint and calcaneus
- Remainder is transmitted via arch anteriorly to MTP joints, greatest weight on 1st toe

Bony Anatomy

- **Cuboid bone**
  - Roughly cuboidal shape
  - 1 ossification center: Ossifies between 9th fetal month and 6 months of age
  - Articulates with calcaneus, navicular, 3rd cuneiform, and 4th and 5th metatarsals, rarely head of talus
    - Dorsal ligaments (calcaneocuboid, cubonavicular, cuneocuboid, cubometatarsal) strengthen each of these articulations
    - Short and long plantar ligaments attach to plantar surface
  - Sulcus at lateral margin, under which passes peroneus longus tendon
  - 5th metatarsal base extends beyond lateral margin
- **Navicular bone**
  - Curved shape, concave proximally and convex distally
  - 1 ossification center: Ossifies in 3rd year of life
  - Articulates with talus, cuboid, cuneiforms
    - Dorsal ligaments strengthen each of these articulations
    - Single facet proximally for articulation with head of talus
    - 3 facets distally for cuneiform articulations
    - 1 facet laterally for articulation with cuboid
    - Connected to anterior process of calcaneus by bifurcate ligament
    - Connected to sustentaculum tali by spring ligament
  - Large median eminence for attachment of posterior tibial tendon is located more plantar than main body of navicular
- **Cuneiform bones**
  - Wedge-shaped, with base of wedge at dorsal surface of 2nd and 3rd cuneiforms, dorsomedial surface 1st cuneiform
  - In combination, form arch
  - 1st cuneiform (medial cuneiform)
    - Articulates with navicular, 2nd cuneiform, 1st metatarsal
    - 1 or 2 ossifications centers: Ossify in 2nd year of life
  - 2nd cuneiform (middle or intermediate cuneiform)
    - Articulates with navicular, 1st and 3rd cuneiforms, 2nd metatarsal
    - Smallest of cuneiforms
    - 1 ossification center: Ossifies in 3rd year of life
  - 3rd cuneiform (lateral cuneiform)
    - Articulates with navicular, 2nd cuneiform, cuboid, 3rd metatarsal
- **Metatarsal bones**
  - 2 ossifications centers: Shaft ossifies in 9th prenatal week, epiphysis in 3rd-4th years of life
    - 1st metatarsal has epiphysis at proximal end, others at distal end
  - 2nd-5th metatarsals have articulations at bases with adjacent metatarsals
  - 1st metatarsal
    - Largest of metatarsals
    - Articulates with 1st cuneiform, 1st proximal phalanx, sesamoids of metatarsal head
    - Variable articulation with 2nd metatarsal base

- 2nd-3rd metatarsals
  - Articulate with respective cuneiforms and proximal phalanges
  - 2nd metatarsal base recessed relative to 1st and usually 3rd
- 4th-5th metatarsals
  - Articulate with cuboid and respective proximal phalanges
  - Styloid process of 5th metatarsal extends lateral to cuboid
- **Phalanges**
  - 1st toe is biphalangeal, other toes are triphalangeal
  - 5th toe sometimes has failure of segmentation of middle and distal phalanges
  - 2 ossification centers: Shaft ossifies 9th-15th prenatal weeks, epiphysis 2nd-8th years

Musculature

- **Plantar muscles**: 4 muscle layers, numbered from superficial to deep
  - 1st layer: Abductor hallucis, flexor digitorum brevis, abductor digiti minimi
  - 2nd layer: Quadratus plantae (flexor accessorius), flexor digitorum and hallucis longus tendons, lumbricals
  - 3rd layer: Flexor hallucis brevis, adductor hallucis, flexor digiti minimi brevis, tibialis posterior tendon
  - 4th layer: Plantar interossei (3), dorsal interossei (4)
  - Peroneus longus tendon courses across all layers, from superficial plantar laterally to deep plantar medially
- **Dorsal muscles**: 2 muscle layers
  - Superficial layer: Tibialis anterior tendon, extensor hallucis longus tendon, extensor digitorum longus tendon, peroneus tertius tendon
  - Deep layer: Extensor hallucis brevis, extensor digitorum brevis
  - In forefoot, long and short extensors run side by side in single layer

Compartments

- **Plantar compartments: 4 plantar compartments are located** deep to plantar fascia
  - Divided by fascial layers
  - Lateral and medial intermuscular septa determine major compartment divisions
  - **Medial plantar compartment**
    - Contains abductor hallucis, flexor hallucis longus, and flexor hallucis brevis tendons
  - **Central plantar compartment**
    - Superficial subcompartment: Contains flexor digitorum brevis, distal portion of flexor digitorum longus tendons
    - Intermediary subcompartment: Contains proximal plantar portion of flexor digitorum longus tendon, quadratus plantae, lumbricals
    - Deep subcompartment: Limited to forefoot, contains adductor hallucis
  - **Lateral plantar compartment**
    - Contains abductor and flexor digiti minimi
  - **Interosseous compartment**
    - Contains plantar and dorsal interosseous muscles
- **Dorsal compartment**
  - Superficial layer: Extrinsic extensor tendons
  - Deep layer: Intrinsic extensor muscles

Major Ligaments

- **Plantar fascia (aponeurosis): 3 portions** extend from tuberosity of calcaneus to transverse metatarsal ligaments of toes
  - **Medial band**: Thin structure superficial to abductor hallucis muscle
  - **Central band**: Thick, strong structure superficial to flexor digitorum brevis
    - Divides into separate bands to each toe; these are linked by transverse bands
    - Distally sends septa superficially into subcutaneous fat and deep to MTP joints
  - **Lateral band**: Thin structure superficial to abductor digiti minimi
  - Medial and lateral bands sometimes terminate at level of midmetatarsals
- Long plantar ligament
  - Originates at calcaneal tuberosity, inserts at cuboid and bases of 2nd-4th metatarsals
  - Forms retinaculum for peroneus longus tendon as it courses medially on plantar aspect of foot
- Short plantar (plantar calcaneocuboid) ligament
  - Deep to long ligament, inserts more proximally on cuboid
- Plantar calcaneocuboid (spring) ligament
  - Originates on sustentaculum tali, inserts on plantar aspect navicular
- Bifurcate ligament: Originates on anterior process of calcaneus dorsally, inserts on navicular and cuboid
- Lisfranc ligament: Originates at 1st cuneiform, inserts at base of 2nd metatarsal
- Intermetatarsal ligaments: Dorsal and plantar ligaments between 2nd-5th metatarsal bases
- Transverse metatarsal ligaments: Superficial and deep ligaments between metatarsal heads

Nerves

- **Tibial nerve** divides into medial and lateral plantar branches at level of tarsal tunnel
  - **Medial plantar nerve**
    - Between 1st and 2nd muscle layers, accompanies medial plantar artery
    - Motor branches: Abductor hallucis, flexor digitorum and hallucis brevis, 1st lumbrical
    - Plantar digital nerves to 1st-3rd toes, medial aspect of 4th toe
  - **Lateral plantar nerve**: Has deep and superficial divisions
    - Motor branches: Flexor digiti minimi brevis, lumbricals, interossei, adductor hallucis
    - Superficial lateral plantar nerve: Between 1st and 2nd muscle layers
    - Plantar digital nerves to 5th toe, lateral aspect of 4th toe
    - Deep lateral plantar nerve: Between 3rd and 4th muscle layers; accompanies lateral plantar artery
- **Deep peroneal nerve**: Extends along dorsum of foot, between tibialis anterior and extensor hallucis longus
  - Motor branch: Extensor digitorum brevis

- **Superficial peroneal nerve**: Divides into medial and lateral branches at dorsum of foot
  - Sensory branches to dorsal foot
- **Sural nerve**: Lateral, superficial branch of tibial nerve
  - Extends along lateral margin of foot
  - Sensory branches to lateral foot

Arteries

- Posterior tibial artery divides into medial and lateral plantar arteries at level of tarsal tunnel
  - Plantar arteries accompany medial and deep lateral plantar nerves
- Peroneal artery accompanies superficial peroneal nerve down anterolateral aspect of ankle
  - May join or replace posterior tibial artery
- Anterior tibial artery continues into foot as dorsalis pedis artery, deep to extensor retinaculum
  - Divides into multiple branches in midfoot, forming arcade

Bursae

- Extensor digitorum brevis: Between muscle and 2nd cuneiform and metatarsal bases
- Extensor hallucis longus: Between tendon and 1st cuneiform and metatarsal bases
- Abductor digiti minimi: Between muscle and tuberosity of 5th metatarsal
- MTP joints: Dorsally between metatarsal heads and medial to 1st metatarsal head

ANATOMY IMAGING ISSUES

Imaging Recommendations

- Radiographs
  - Weight bearing, when possible
  - Standard views: Anteroposterior (dorsoplantar), lateral, oblique
- US
  - Offers highest spatial resolution for muscle and tendons
  - Particularly useful given limited depth of most soft tissue structures in foot
  - As result of transverse arch of foot, interrogation should be performed separately for medial and lateral sides for best visualization
  - Due to complex soft tissue anatomy, panoramic (extended field of view) images provide better depiction of location and extent of lesion and course of different structures
  - Dorsal foot structures tend to be very superficial and thin (tendons) or compressible (vessels)
    - To better visualize these structures, minimal compression and thick layer of coupling gel may help
- CT
  - Multidetector 1-mm images with sagittal and coronal reformations
  - 3D reconstructions often useful for analyzing relationship between bones and bone fragments
- MR
  - Better images obtained when field of view is limited to area of concern, not entire foot
  - Demonstration of bone marrow edema aside from soft tissue lesions

Imaging Sweet Spots

- Tendon movement can be easily demonstrated with US by moving the bone onto which it inserts
  - Helps with confirming tendon location and also with establishing tendon integrity
- Tarsal ligaments are usually seen as hypoechoic thickenings on cortices and between bones
  - Orientation of transducer along long axis makes ligament stand out and easier to identify

Imaging Pitfalls

- Alignment can only be reliably assessed on weight-bearing radiographs

CLINICAL IMPLICATIONS

Foot Motion

- **Supination**: Elevation of medial arch of foot
  - Combination of inversion and adduction
- **Pronation**: Depression of medial arch of foot
  - Combination of eversion and abduction
- Complex motions at multiple joints
  - Chopart (calcaneocuboid and talonavicular) joint
    - These 2 joints move together on oblique axis to produce compound motions
    - Pronation-abduction-extension to supination-adduction-flexion
  - Tarsometatarsal joints
    - Dorsiflexion and plantar flexion
    - 2nd and 3rd tarsometatarsal joints relatively immobile
    - Slight abduction of 1st tarsometatarsal joint
  - MTP joints
    - Dorsiflexion and plantar flexion
    - Abduction and adduction at 1st MTP joint

Alignment

- Normal weight bearing and gait depend on normal foot alignment
- Evaluated initially with anteroposterior and lateral weight-bearing radiographs

Malalignment

- Forefoot adductus: Medial angulation of metatarsals from axis of hindfoot
- Forefoot varus: Inversion of metatarsals resulting in shift of weight bearing to 5th metatarsal from 1st metatarsal
- Metatarsus primus varus: Medial deviation of 1st metatarsal axis relative to 2nd
- Hallux valgus (hallux abductus): Lateral deviation of 1st proximal phalanx relative to axis of 1st metatarsal
  - Valgus refers to angular deformity in vertical plane, where apex points medially
  - Abductus refers to angular deformity in horizontal plane, where apex points medially
  - Hallux valgus is misnomer, but it remains term commonly used for hallux abductus

DORSAL FOOT

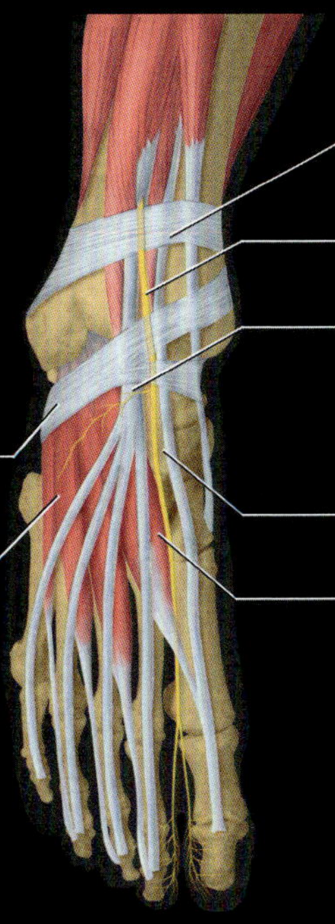

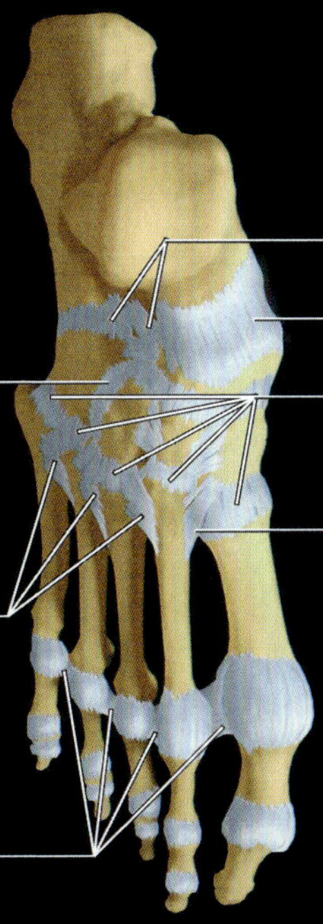

(Top) *The deep peroneal nerve travels deep to the extensor retinacula, in the anterior tarsal tunnel, and gives a lateral motor branch to the extensor digitorum brevis muscle. The medial branch continues dorsal to the talonavicular joint, middle cuneiform, and in between the 1st and 2nd metatarsals to provide mostly sensory, but some motor, supply to the 1st web space.* **(Bottom)** *Graphic of the dorsal ligaments shows dense ligaments between the tarsal bones and between tarsals and metatarsals. Ligaments are generally named for the bones they bridge. Exceptions are the bifurcate ligament, which extends from the anterior process of the calcaneus to the cuboid and the navicular, and the dorsal Lisfranc ligament, from the 1st cuneiform to the 2nd metatarsal.*

DORSAL FOOT

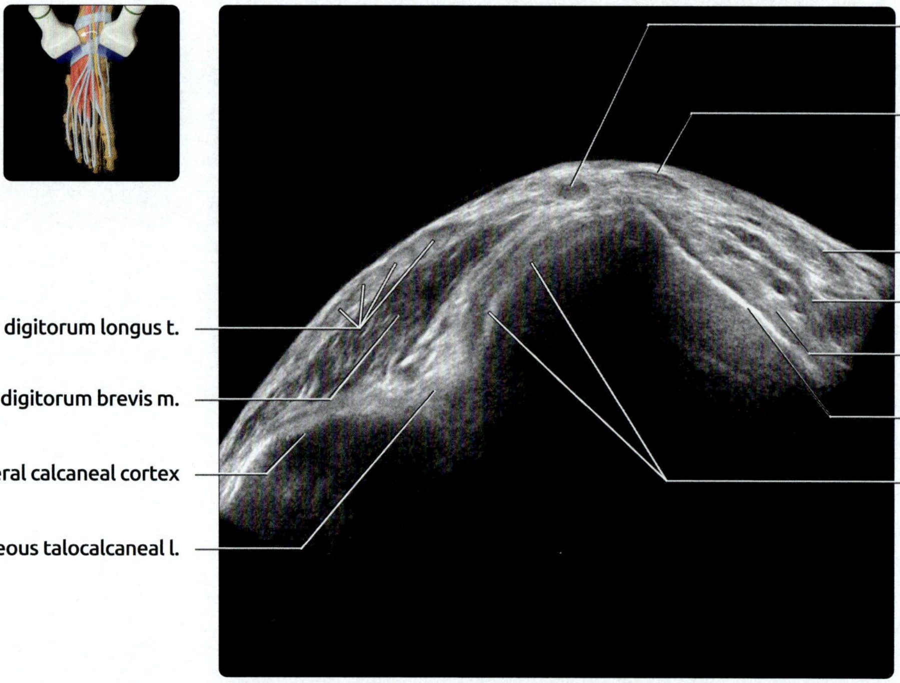

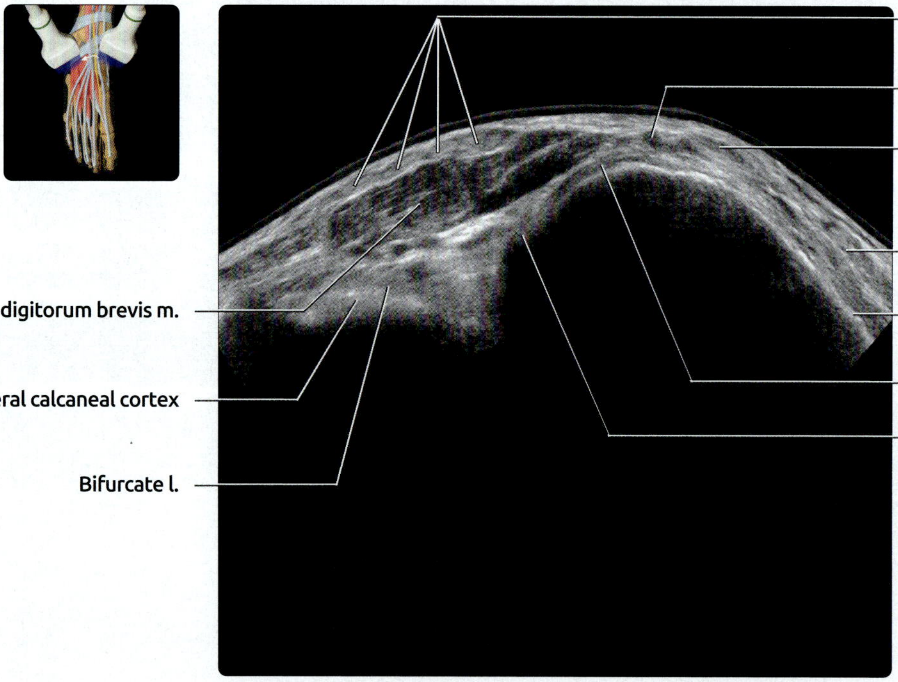

(Top) *Coronal panoramic US shows the dorsal foot at the talocalcaneus level. The foot can be divided into 2 columns: The medial column (talus, navicular, and cuneiforms) and lateral column (calcaneus and cuboid) form the anteromedial and anterolateral surfaces of the transverse midfoot arch. The dorsalis pedis artery runs at the apex of this arch.* **(Bottom)** *Coronal panoramic US shows the dorsal foot through the navicular-calcaneal level. The transverse arch of the dorsal foot creates 2 surfaces: Anteromedial and anterolateral, over which extensor tendons and small muscles lie. More focal examination of these structures requires scanning along each surface separately. Panoramic scans give an overview of anatomy. However, high-resolution imaging of the site of interest provides more details, and both should be performed as complementary techniques.*

DORSAL FOOT

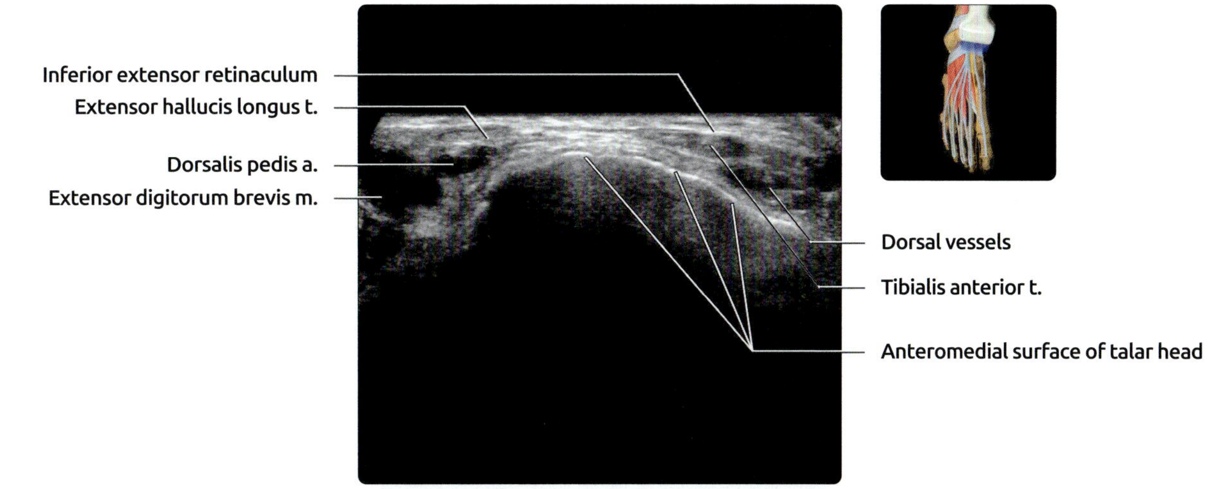

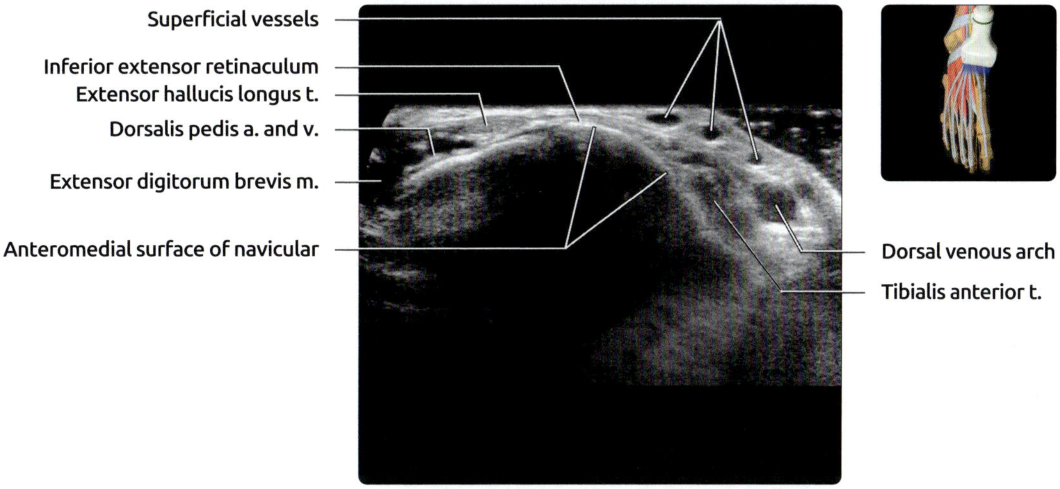

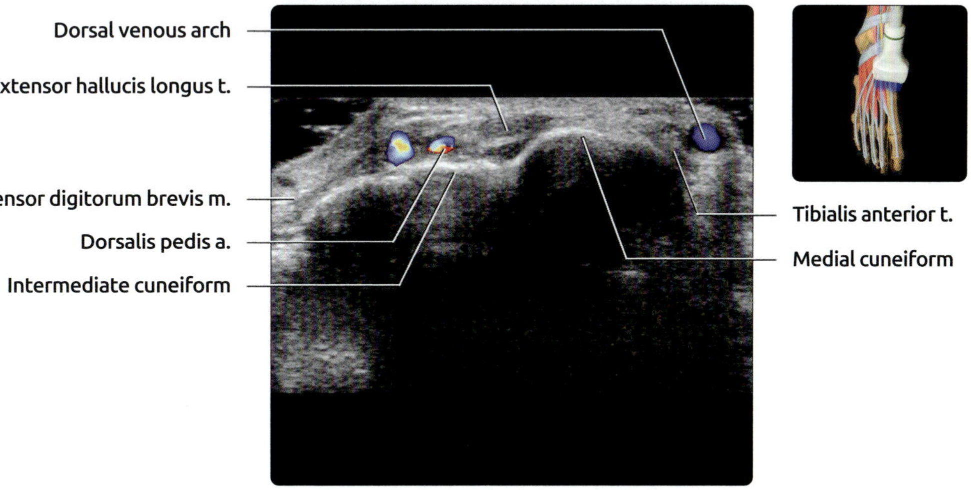

(Top) *Series of 3 coronal scans shows the dorsal medial surface of the foot. At this most proximal level, the talus is the underlying bone. The extensor hallucis can be seen partially covering the dorsalis pedis artery. More medially, the tibialis anterior tendon can be found.* **(Middle)** *At a midtarsal level, this coronal US shows the anteromedial surface of the navicular. The tibialis anterior tendon begins to turn acutely medially to curve around to the medial edge of the foot, eventually inserting on the medial cuneiform. As seen here, the use of light pressure and copious amounts of gel will allow visualization of small superficial veins, which are numerous on the dorsum of the foot.* **(Bottom)** *Coronal color Doppler US shows the level of the cuneiforms. The intimate relationship between the dorsalis pedis artery and the extensor hallucis longus tendon continues.*

DORSAL FOOT

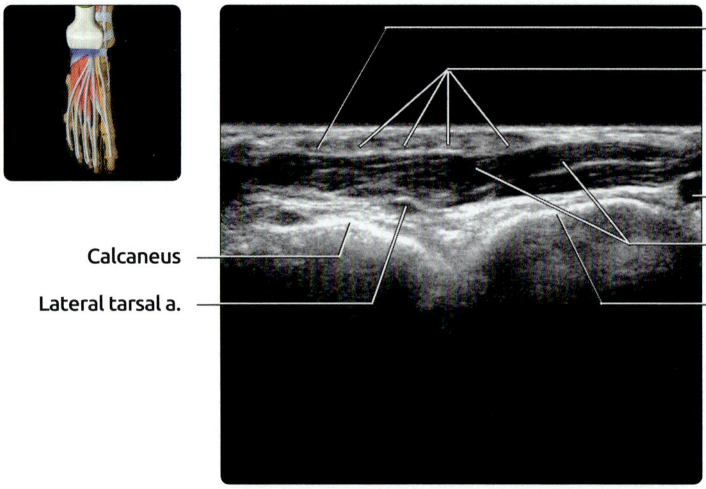

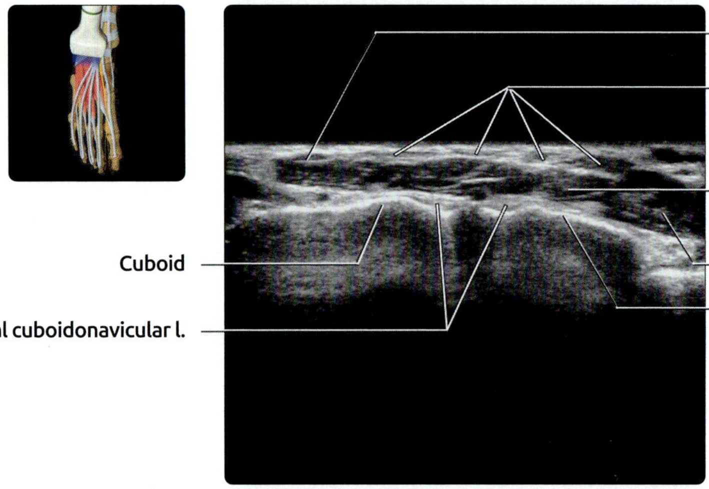

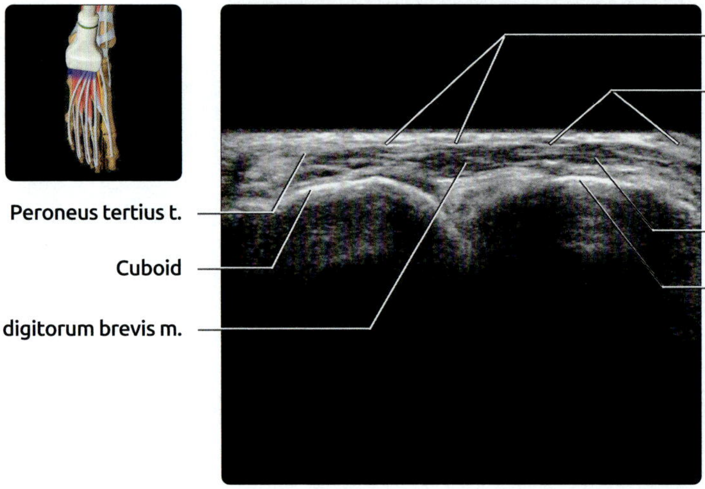

(Top) *Series of 3 coronal scans is shown over the dorsal lateral surface of the tarsus. At this proximal level, the lateral surface of the calcaneus and the talus are the underlying bones. The extensor digitorum longus tendons are close to each other and share a common synovial sheath. The extensor digitorum brevis muscle lies deep to the longus tendons. The lateral tarsal artery runs deeper and is a branch of the dorsalis pedis artery.* **(Middle)** *At the level of the midtarsus, the extensor digitorum longus tendons for individual toes begin to spread out and also flatten (the latter making them just discernible on ultrasound). At this level, the extensor hallucis brevis muscle can also be seen.* **(Bottom)** *At a more distal level, the extensor digitorum brevis muscle and the extensor hallucis brevis muscle are much thinner and will turn into tendons distally.*

DORSAL FOOT

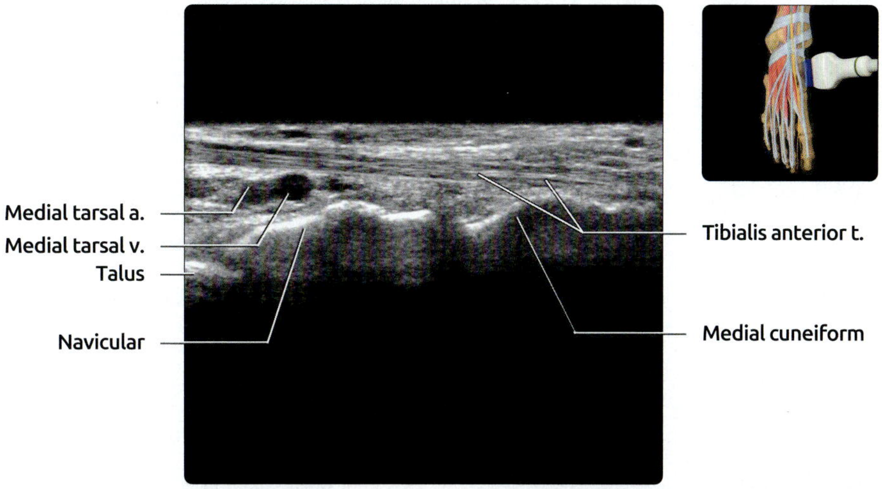

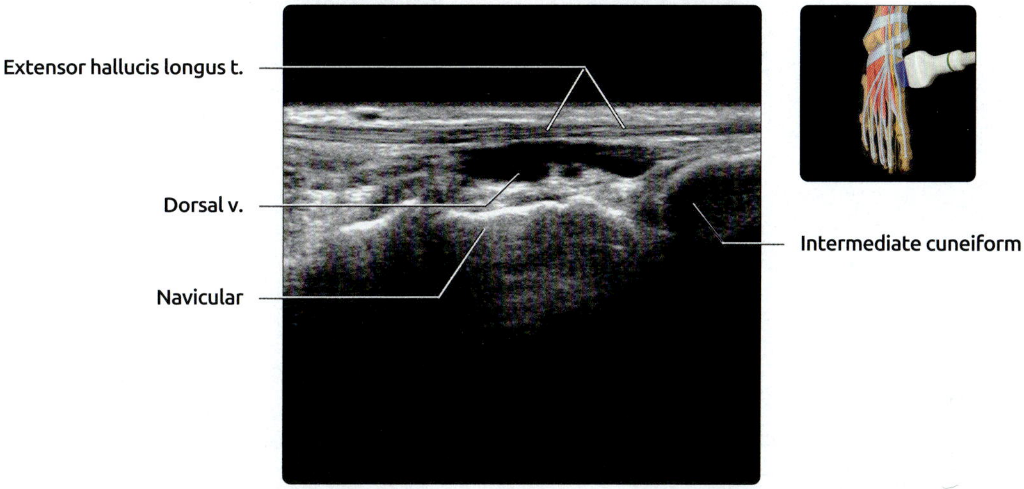

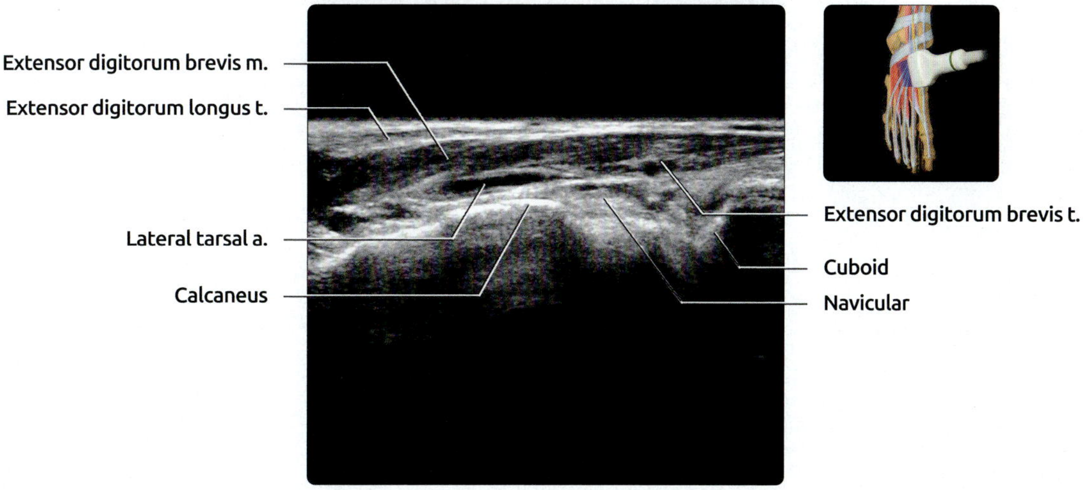

(Top) *Series of sagittal scans is shown through the tarsus. Medially, the tibialis anterior is seen running over the medial tarsal vessels (branch of dorsalis pedis) to insert on the medial and inferior surfaces of the base of the medial cuneiform.* **(Middle)** *Slightly laterally, the extensor hallucis longus tendon runs toward the big toe, passing over the intermediate cuneiform bone.* **(Bottom)** *More laterally, the extensor digitorum longus tendon and the extensor digitorum brevis muscle come into view. They lie on the surface of the calcaneus with the lateral tarsal artery coursing between the muscle and bone. The gradual tapering of the extensor digitorum brevis muscle distally can be appreciated here.*

PLANTAR MUSCLES, SUPERFICIAL LAYERS

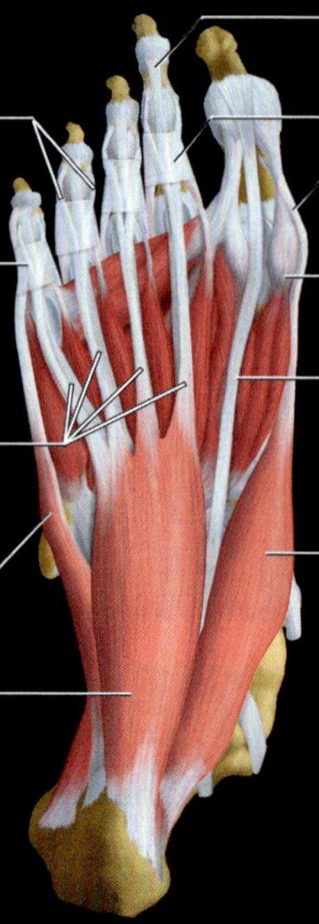

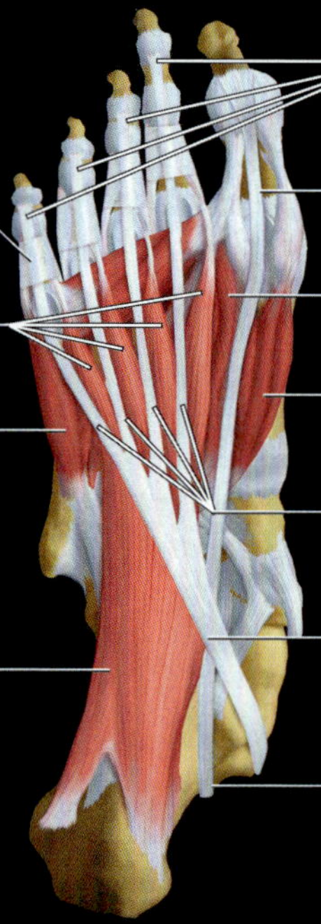

(Top) *Graphic shows the superficial layer of the plantar foot muscles: Abductor digiti minimi, flexor digitorum brevis, and abductor hallucis. The deep muscles are partially visible deep to the superficial layer. The flexor digitorum brevis tendons split in each digit (4th digit labeled), attaching at lateral aspects of the middle phalangeal bases. The flexor tendon sheaths hold the flexor mechanism in close proximity to the phalanges.* **(Bottom)** *Graphic shows the 2nd layer of plantar muscles. The interesting arrangement of interactions between flexors is well seen: Quadratus plantae inserting on flexor digitorum longus tendon and lumbricals arising from individual flexor digitorum longus tendon slips. In the toes, the flexor tendons are contained within fibrous tendon sheaths.*

PLANTAR FOOT MUSCLES, DEEP LAYERS

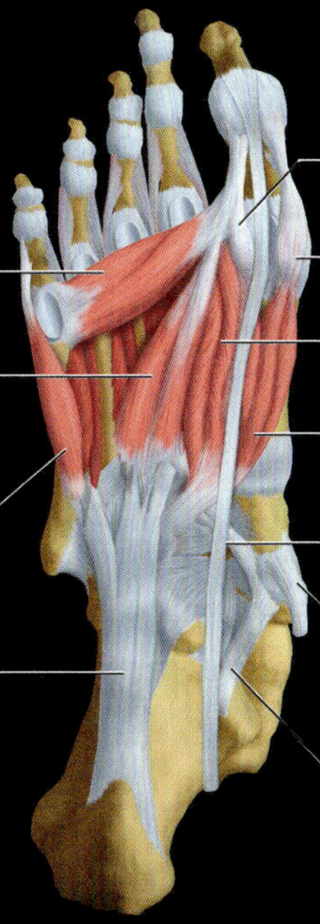

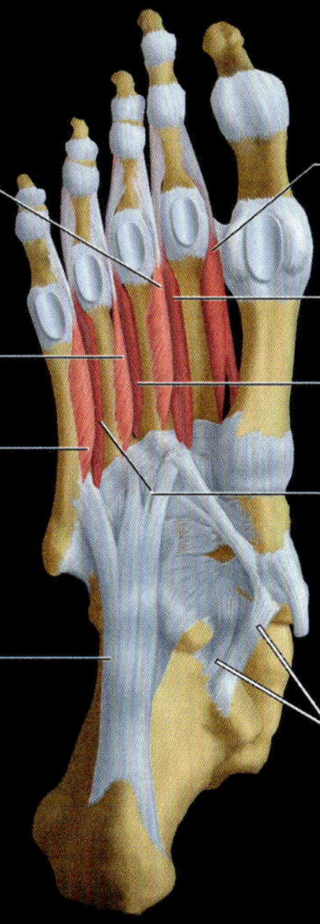

(Top) *Graphic shows the 3rd layer of plantar muscles. This layer contains the flexor hallucis brevis and flexor digiti minimi as well as the adductor hallucis muscle. The oblique head of the adductor hallucis is thick and broad, whereas the transverse head is thin.* **(Bottom)** *Graphic shows the 4th layer of plantar muscles, the interosseous muscles. There are 4 dorsal interossei. They have bipennate origins from 2 adjacent metatarsals. There are 3 plantar interossei. The 1st plantar interosseous muscle originates from the medial side of the 3rd metatarsal and inserts on the medial side of the 3rd proximal phalanx. The 2nd and 3rd plantar interossei originate on the medial sides of the 4th and 5th metatarsals, respectively, and insert on the medial side of the corresponding proximal phalanx.*

PLANTAR APONEUROSIS

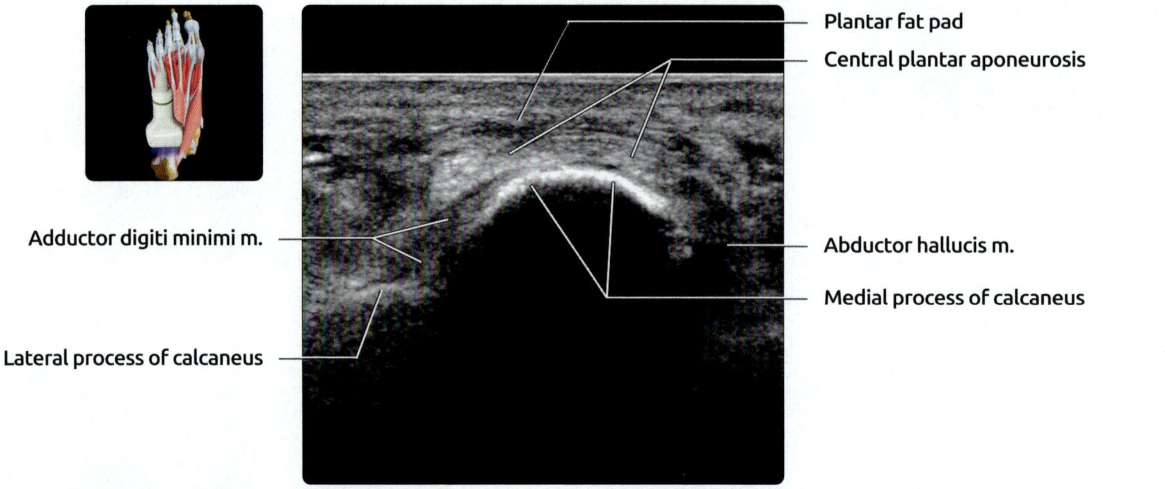

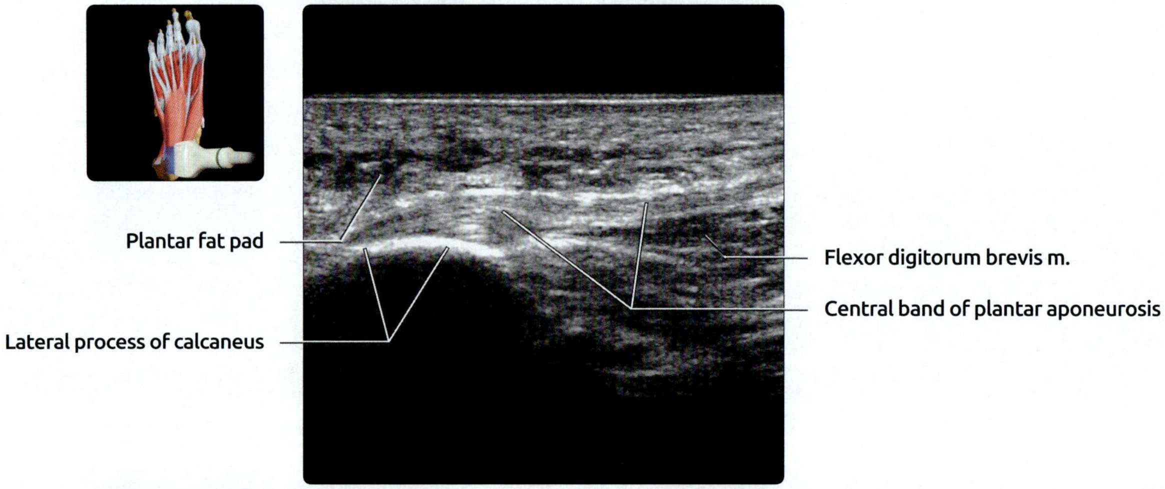

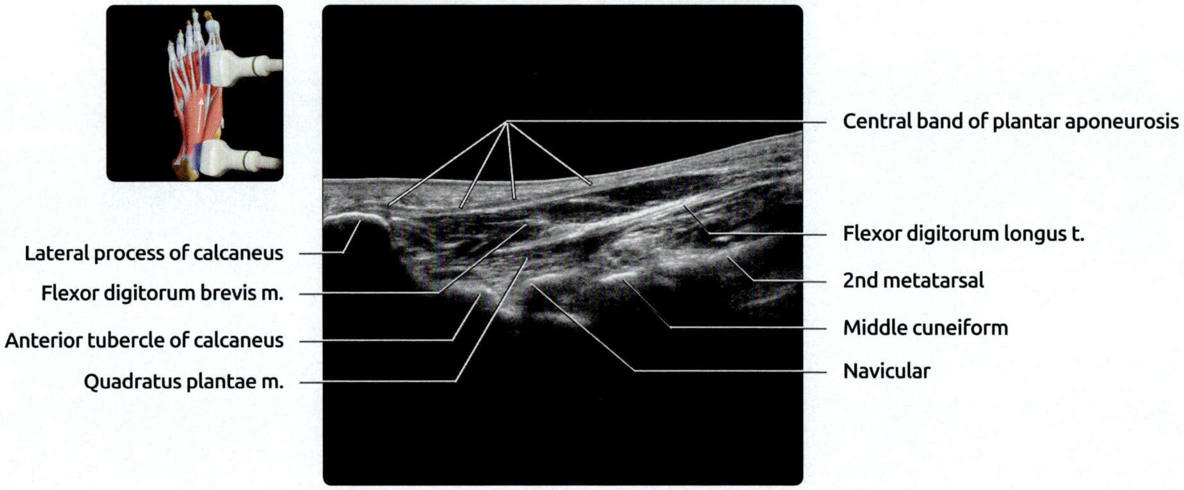

(Top) *Transverse US of the calcaneus shows the origin of the plantar aponeurosis. The plantar aponeurosis fascia consists of 3 parts: Strong, thick central band over the flexor digitorum brevis muscle; lateral band over the abductor digiti muscle; and thin medial band over the abductor hallucis muscle.* **(Middle)** *Sagittal US shows the origin of the central band of the plantar aponeurosis. This arises from the medial process of the calcaneus and runs anteriorly deep to the plantar subcutaneous fat.* **(Bottom)** *Sagittal panoramic US shows the central band of plantar aponeurosis. The central is the thickest and most easily identified component of the plantar aponeurosis. Deep to this are the flexor digitorum brevis and quadratus plantae muscles, which form the bulk of the central tarsal plantar compartment.*

SAGITTAL FOOT OVERVIEW

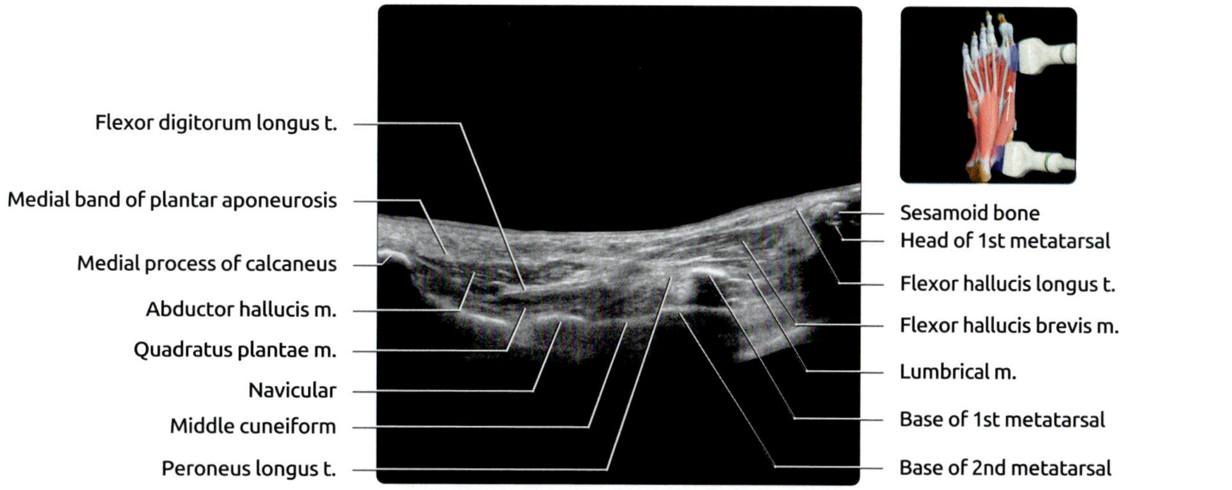

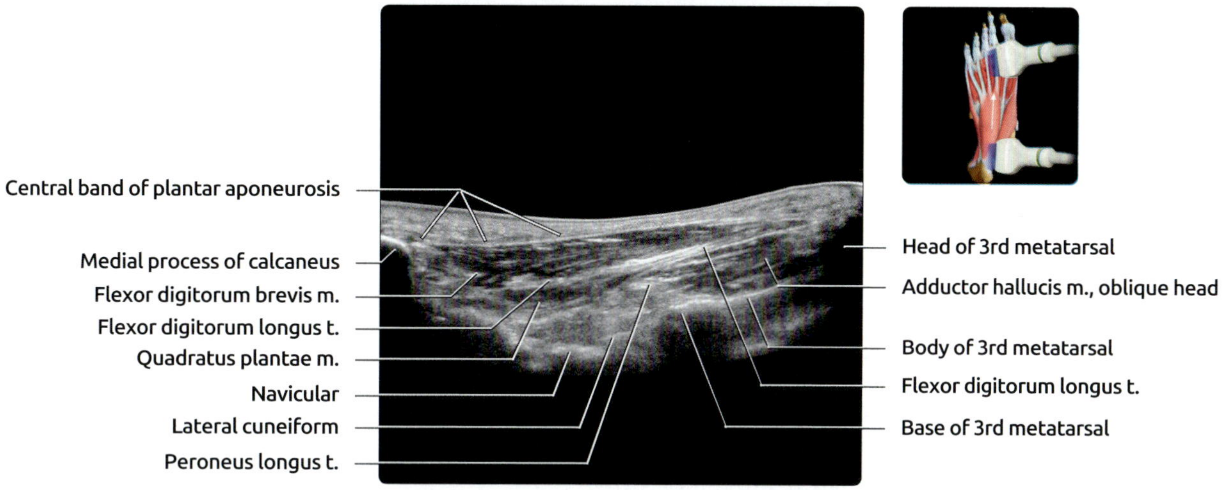

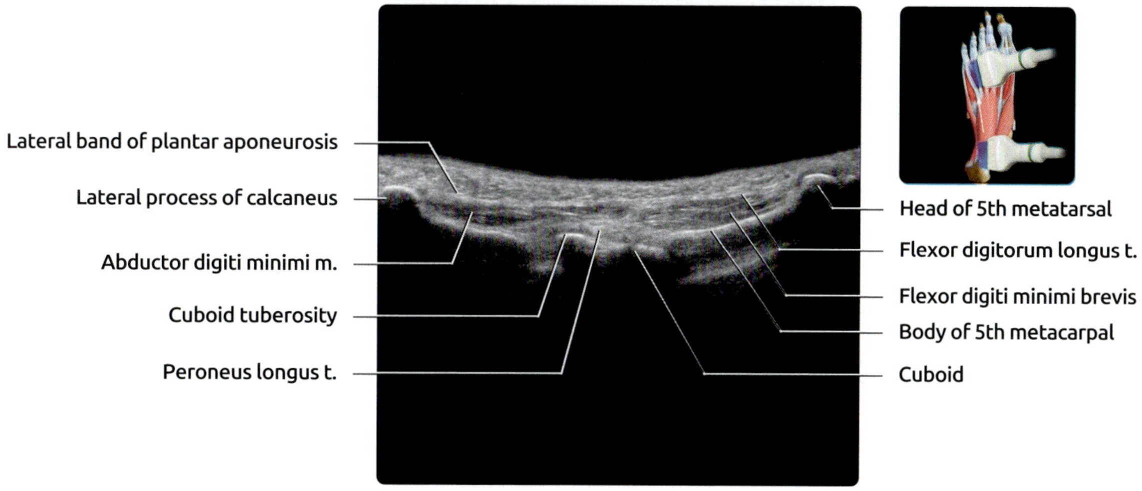

(Top) *Series of 3 panoramic sagittal scans is shown through the plantar aspect of the foot. This scan shows the content of the medial compartment: The abductor hallucis muscle, flexor hallucis longus tendon, and flexor hallucis brevis muscle, together with smaller structures.* **(Middle)** *In the central plantar compartment, the contents can be divided into layers. Superficially, there is the flexor digitorum brevis muscle and the flexor digitorum longus tendon. Deep to this layer are the quadratus plantae muscle, oblique head of the adductor hallucis muscle, and the lumbricals.* **(Bottom)** *A scan through the lateral compartment shows the abductor digiti minimi muscle and the flexor digiti minimi muscle.*

MEDIAL TARSUS

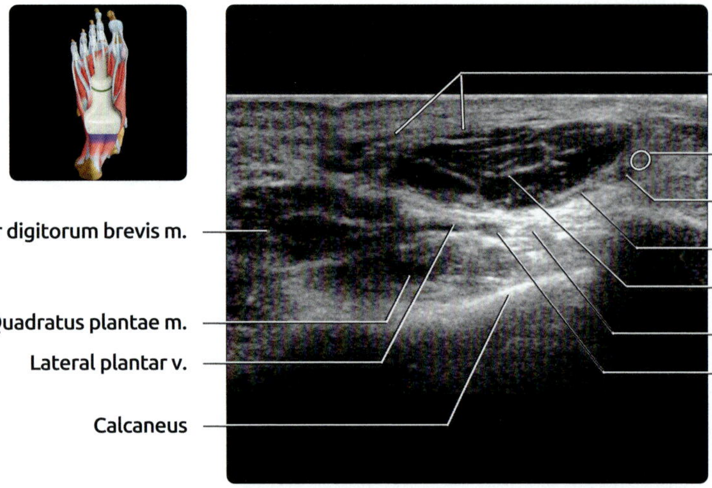

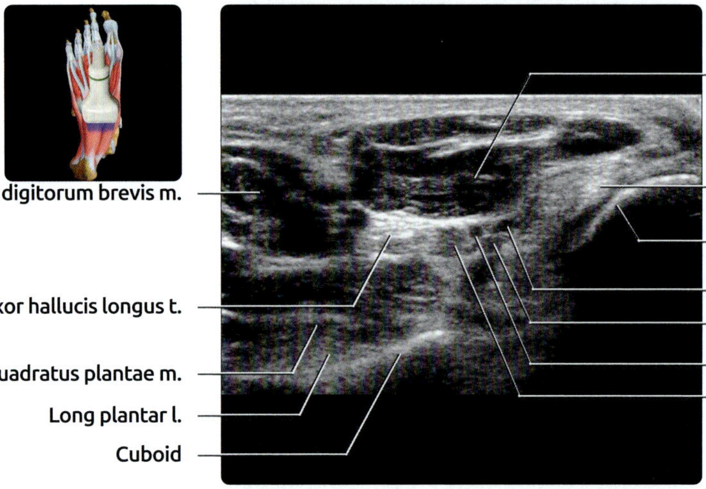

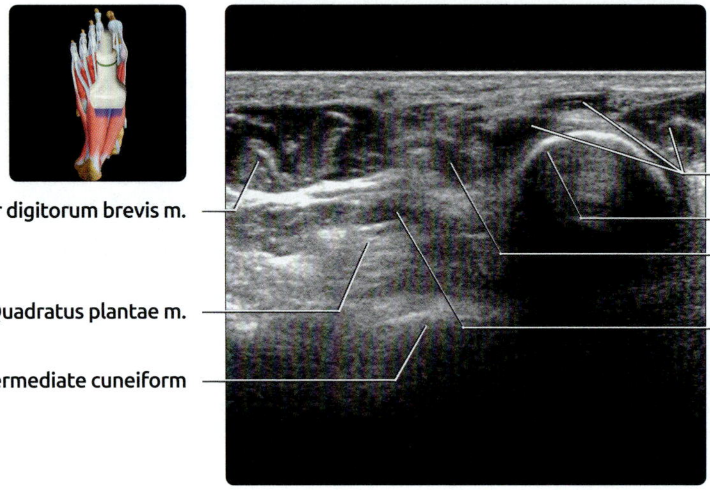

(Top) *Series of 3 coronal scans is shown through the medial 1/2 of the plantar aspect of the foot. At this proximal level, the 2 main muscles (flexor digitorum brevis and quadratus plantae) of the central compartment can be seen lying superficial to the calcaneus. The medial compartment muscle (abductor hallucis) can also be seen medially.* **(Middle)** *At the level of the midtarsus, the knot of Henry can be seen. This is where tendons of the flexor digitorum longus and flexor hallucis longus tendons cross and exchange fibers. The medial plantar neurovascular bundle lies in close proximity to this.* **(Bottom)** *More distally, the flexor hallucis longus and digitorum longus tendons separate and head toward their respective insertions.*

LATERAL TARSUS

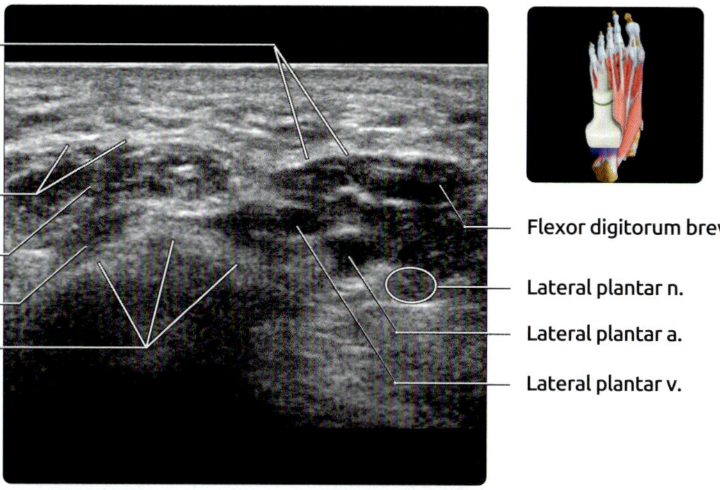

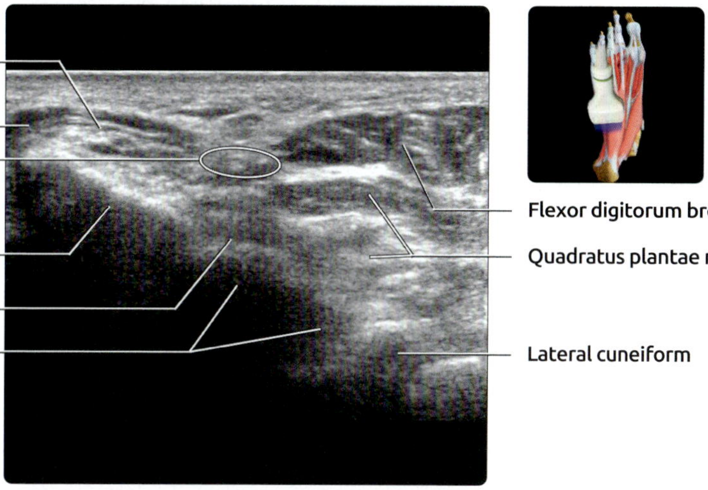

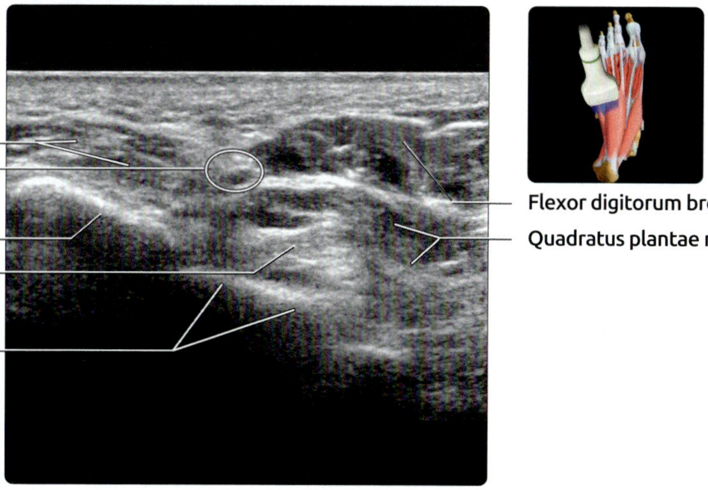

(Top) *Series of 3 coronal scans is shown through the lateral 1/2 of the plantar aspect of the foot. The lateral compartment (abductor digiti minimi) and 1/2 of the central compartment can be seen here, separated by the lateral plantar neurovascular bundle. The calcaneus forms the bony support of all structures at this level.* **(Middle)** *At the level of the midtarsus, the separation between the lateral and the central compartment by the neurovascular bundle continues. Within the lateral compartment, the abductor digiti minimi muscle moves laterally to give way to the other component of the lateral compartment, the flexor digiti minimi brevis muscle.* **(Bottom)** *Most distally, the abductor digiti minimi muscle runs on the lateral border, and only the flexor digiti minimi brevis muscle is seen on the plantar aspect of the foot.*

IMAGING ANATOMY

Arteries

- **Dorsalis pedis artery**
  - Dorsal artery of foot
  - Direct continuation of anterior tibial artery, changing name at ankle joint
  - Course and relations
    - Crosses superficial to talocrural articular capsule, talus, navicular, and intermediate cuneiform bones
    - Passes deep to inferior extensor retinaculum and distal to extensor hallucis brevis
    - Lies lateral to extensor hallucis longus tendon
    - Passes to proximal end of intermetatarsal space and curves deeply to enter sole between heads of 1st dorsal interosseous muscle
    - Forms part of plantar arch
  - Main branches
    - Medial and lateral tarsal arteries: Arise at level of navicular bone
    - Lateral tarsal artery runs under extensor digitorum brevis
    - 2 or 3 medial tarsal arteries run on medial surface of tarsus
  - Arcuate artery: Arises at medial cuneiform level and runs laterally over metatarsal bases, deep to extensor tendons; gives off branches to metatarsals and toes
  - 1st dorsal metatarsal artery: Arises just before dorsalis pedis dives deep into sole
- **Posterior tibial artery**
  - Continuation of calf vessel by same name, branch of popliteal artery
  - Runs in tarsal tunnel behind and below medial malleolus accompanied by tibial nerve
  - Divides into terminal branches (medial and plantar arteries) in tarsal tunnel
  - Medial plantar artery is smaller
    - Runs in medial aspect of foot medial to medial plantar nerve
    - Then deep to abductor hallucis and flexor digitorum brevis muscles
  - Lateral plantar artery is much larger than medial
    - Runs in lateral aspect of foot lateral to lateral plantar nerve
    - Then deep to abductor hallucis muscle and distally between flexor digitorum brevis and quadratus plantae muscles
    - Distally runs between flexor digitorum brevis and abductor digiti minimi, then between base of 1st and 2nd metatarsals, and finally joins dorsalis pedis to form plantar arch
- **Plantar arch**
  - Arterial arch at level of metatarsals
  - Formed by dorsalis pedis and lateral plantar arteries
  - Gives off branches to metatarsals and toes
- **Peroneal artery**
  - Accompanies superficial peroneal nerve down anterolateral aspect of ankle and sends off perforating branch that joins communicating branch from anterior tibial artery
  - May join or replace posterior tibial artery

Veins

- Superficial veins
  - Great saphenous vein
    - Arises from medial part of dorsal venous arch of foot
    - Runs subcutaneously over muscles and tendons to pass anterior to medial malleolus
    - Passes superiorly on medial aspect of calf and thigh to join common femoral vein at femoral triangle
  - Small saphenous vein
    - Arises from lateral part of dorsal venous arch
    - Runs subcutaneously over muscles and tendons to pass posterior to lateral malleolus
    - Passes superiorly on lateral and posterior side of calf to join popliteal vein in popliteal fossa
- Deep veins
  - Plantar venous arch: Receives venous drainage from dorsal and plantar aspects of toes and metatarsals
  - Medial and lateral plantar veins: Arise from plantar venous arch to accompany same named arteries
  - Communication exists with great and small saphenous veins
  - Medial and lateral plantar veins form posterior tibial veins in tarsal tunnel and run proximally to calf, accompanying posterior tibial artery

ANATOMY IMAGING ISSUES

Imaging Recommendations

- US
  - Provides greatest amount of information on vasculature of foot of different imaging modalities available
  - Morphology and functional status of vessels can be assessed
  - Color Doppler scans allow assessment for vessel occlusion and thrombosis
  - Spectral Doppler scans allow quantitative and semiquantitative documentation of flow velocity, resistance, waveform, etc.
  - Augmentation of foot veins usually only requires small degrees of flexion/extension of toes to produce enough venous flow for detection (unlike need to squeeze muscles in calf)

Imaging Pitfalls

- Superficial veins are easily compressed even with lightest transducer pressure
  - Thick layer of coupling gel helps act as spacer between transducer and skin to minimize pressure and allows for better visualization of vessels

DORSAL AND PLANTAR ARTERIES AND NERVES

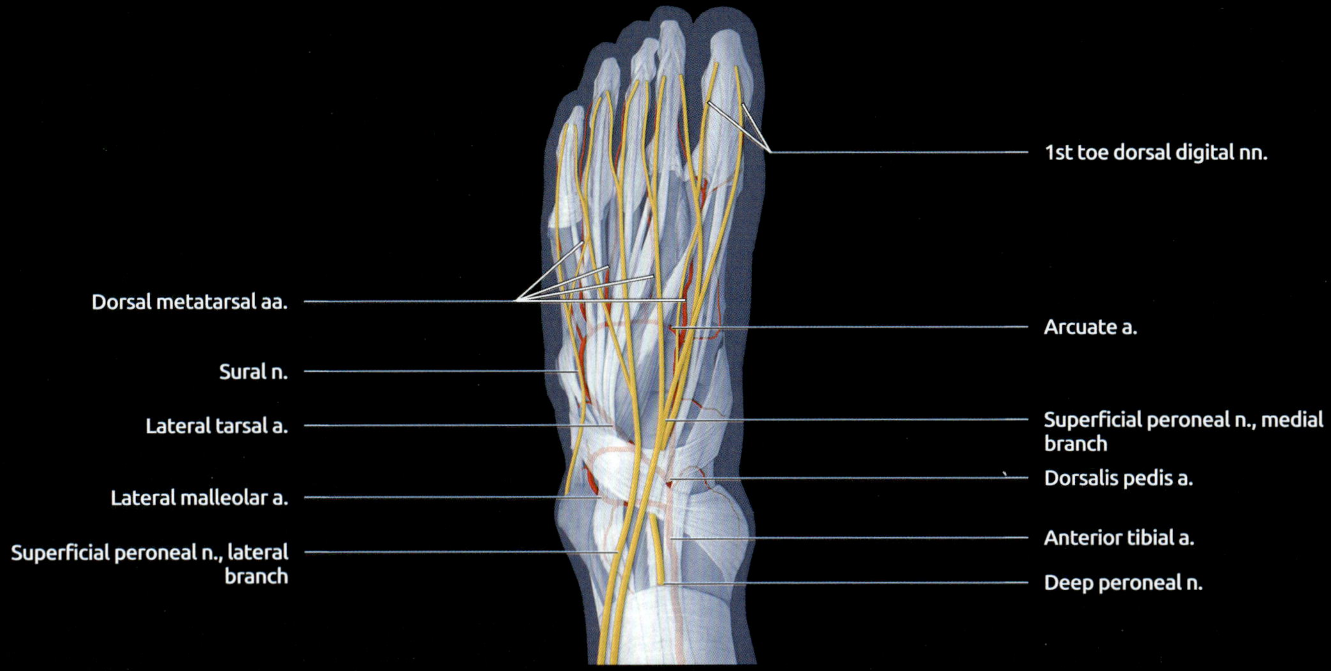

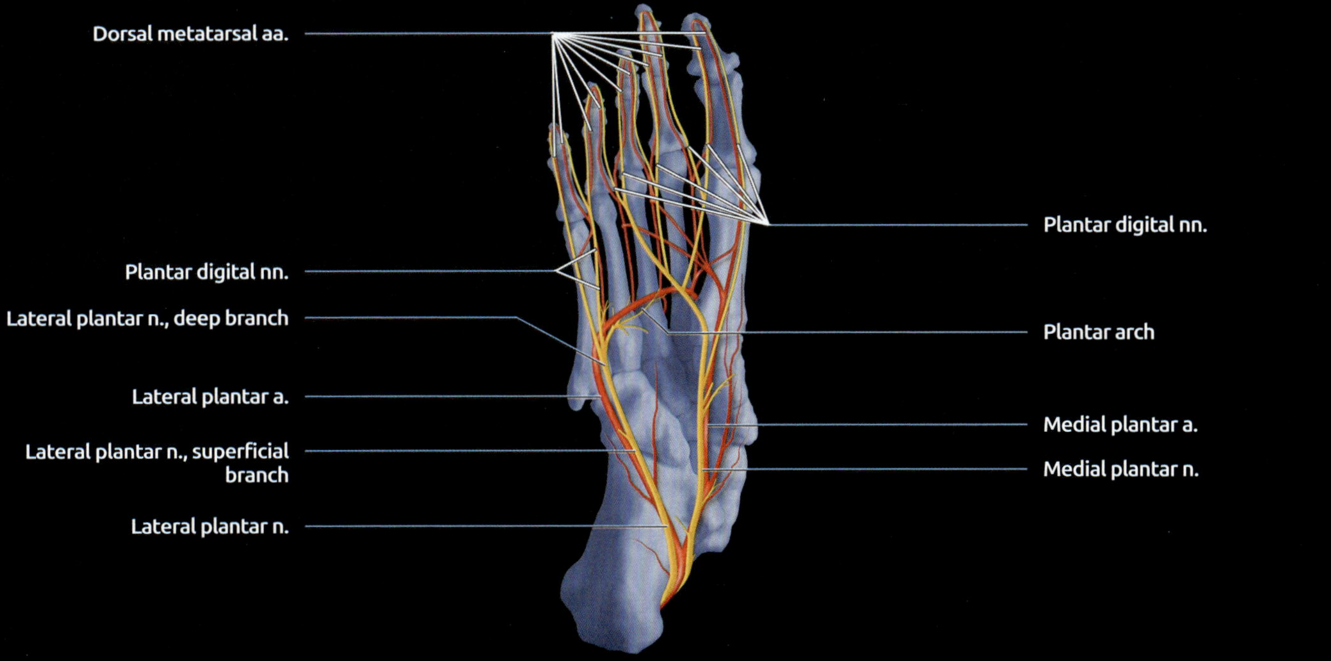

(Top) *Graphic shows arteries and nerves at the dorsal aspect of the foot. The deep peroneal nerve accompanies the anterior tibial artery. The superficial peroneal nerve divides into medial and lateral branches in distal 1/3 of leg. Anterior tibial artery is called dorsalis pedis in foot. It terminates in arcuate artery, which communicates with plantar arterial arch, and in digital vessels.* **(Bottom)** *Graphic shows arteries and nerve at the plantar aspect of the foot. The posterior tibial nerve and artery divide into medial and lateral plantar branches. These divide further into digital branches to the medial and lateral aspects of each toe. The plantar arch sends vessels dorsally as well as to the plantar aspect of the toes.*

DORSALIS PEDIS ARTERY

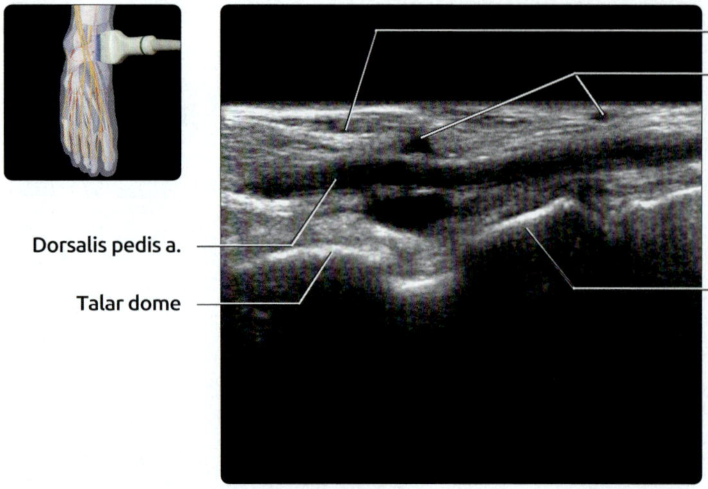

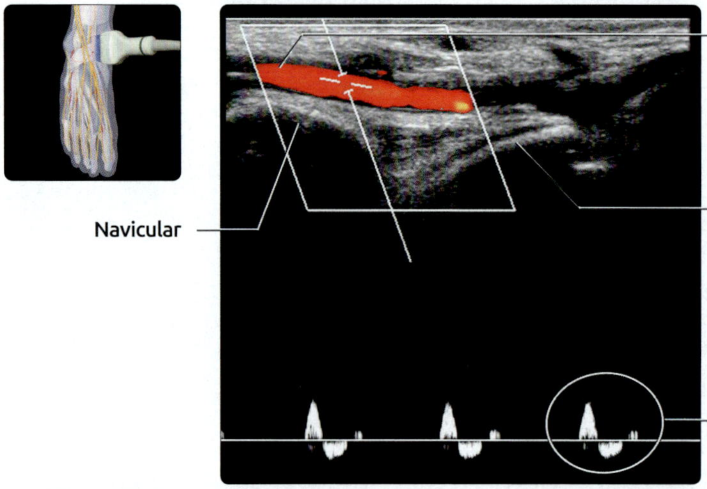

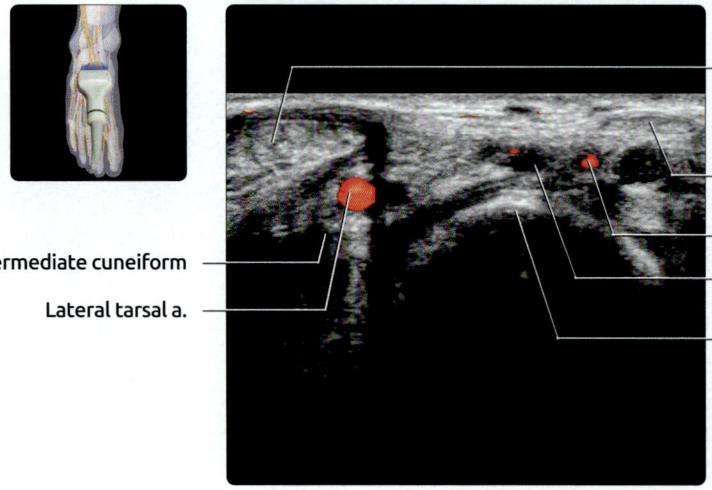

(Top) *Sagittal scan shows the dorsum of the foot. The dorsalis pedis artery runs at the anterior aspect of the transverse arch formed by the tarsal bones. It is a continuation of the anterior tibial artery and runs lateral to the extensor hallucis longus tendon.* **(Middle)** *Sagittal spectral Doppler scan of the dorsalis pedis artery shows the typical triphasic artery flow.* **(Bottom)** *Coronal color Doppler scan shows the dorsum of the foot at the level of the cuneiform. The dorsalis pedis artery gives rise to medial and lateral tarsal arteries at the level of the navicular. The dorsalis pedis artery remains lateral to the extensor hallucis longus tendon even at this distal level.*

METATARSAL AND DIGITAL ARTERIES

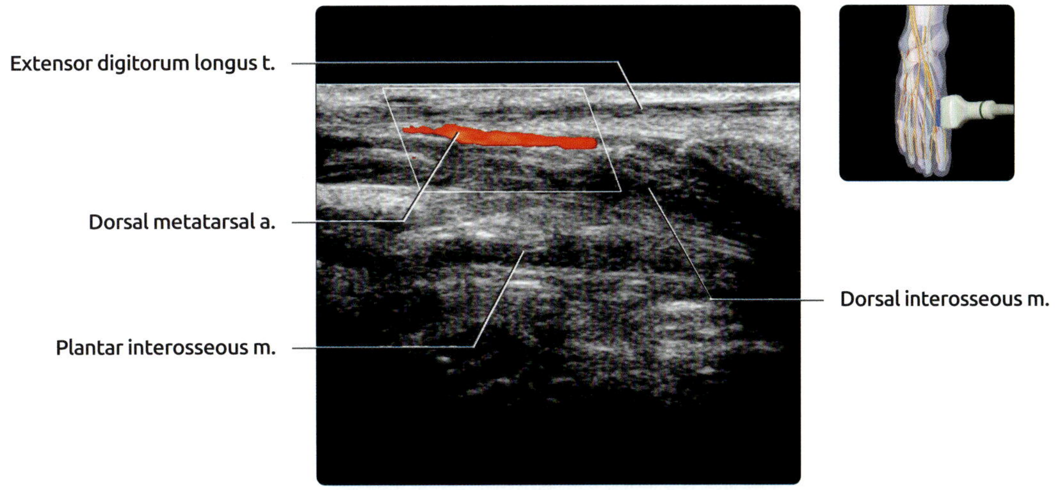

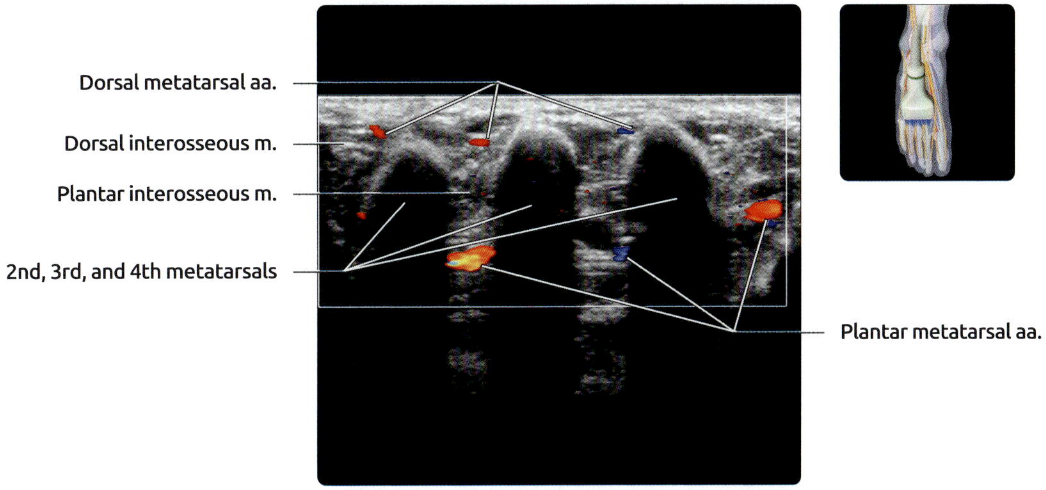

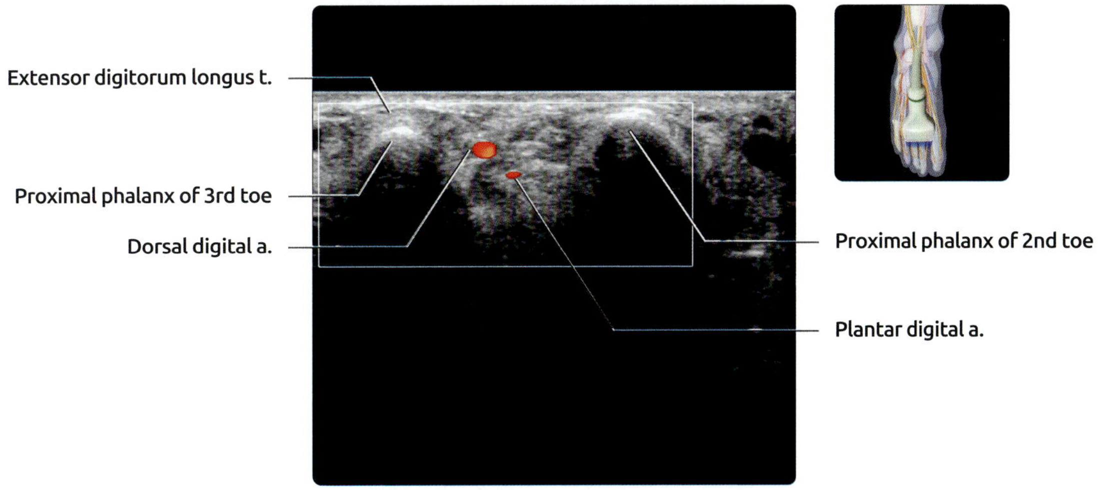

(Top) *Sagittal color Doppler shows the 2nd interosseous space. The arcuate artery gives rise to dorsal metatarsal arteries that run in the interosseous spaces to supply the metatarsals and toes. The 2 layers of interosseous muscles can be appreciated here.* **(Middle)** *Transverse color Doppler scan shows the metatarsal bodies. Both the dorsal and plantar interosseous muscles can be seen between the bones. The dorsal and plantar metatarsal arteries are also demonstrable by the use of color Doppler. The plantar metatarsal arteries arise from the plantar arch.* **(Bottom)** *Coronal color Doppler scan shows the 2nd web space. The dorsal and plantar metatarsal arteries continue distally as the digital arteries, which will divide further to run on the sides of the phalanges.*

PLANTAR ARTERIES

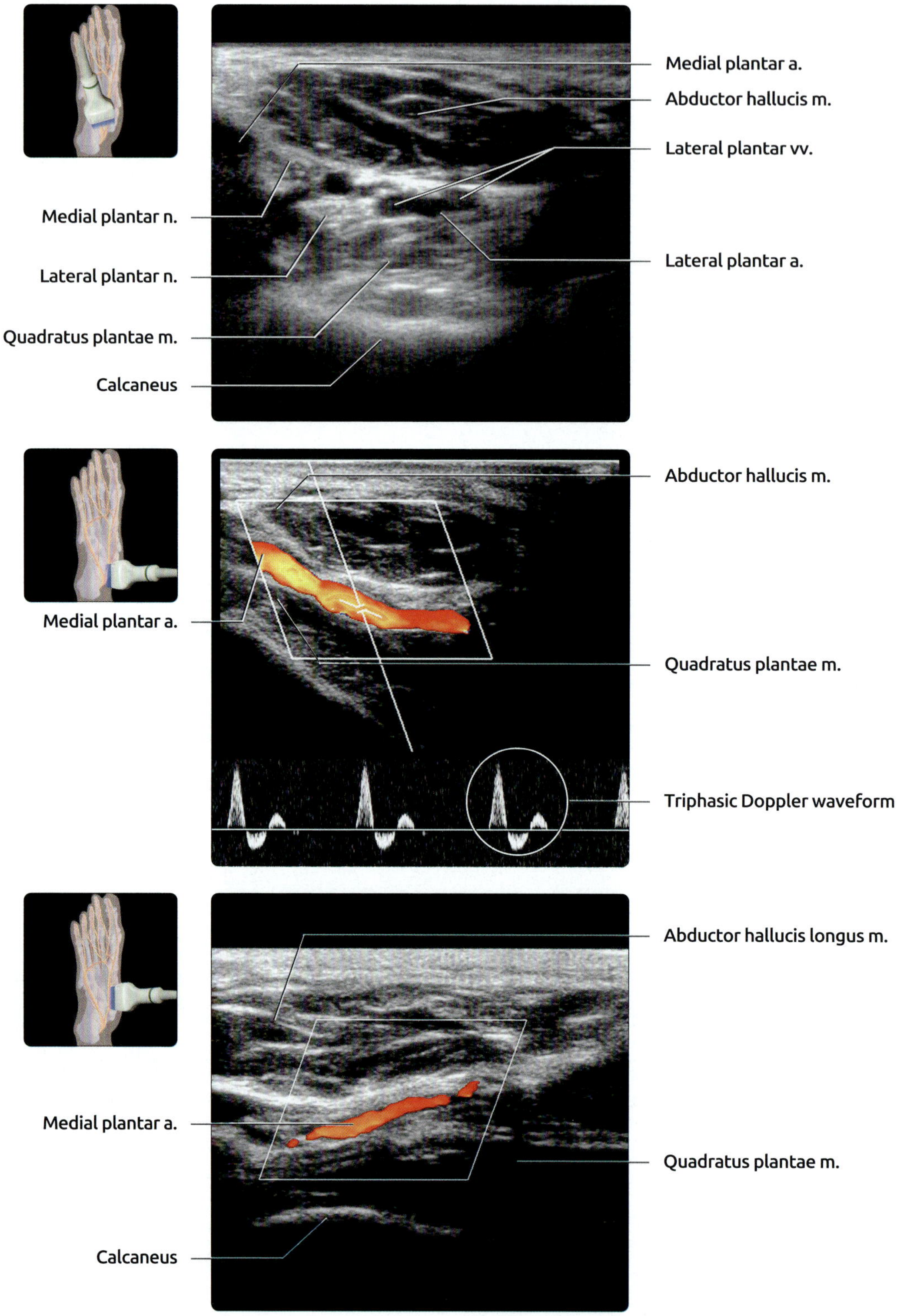

(Top) *Coronal scan shows the medial plantar compartment. The tibial nerve has divided into the medial and lateral plantar nerves, and, similarly, the posterior tibial vessels have divided into medial and lateral plantar vessels. Note that this is a scan of the left foot and the lateral side is on the right of the image.* **(Middle)** *Spectral Doppler scan of the medial plantar artery shows a typical triphasic arterial waveform.* **(Bottom)** *Oblique sagittal color Doppler scan of the medial compartment shows the medial plantar artery running between the abductor hallucis muscle and the quadratus plantae muscle.*

PLANTAR ARTERIES

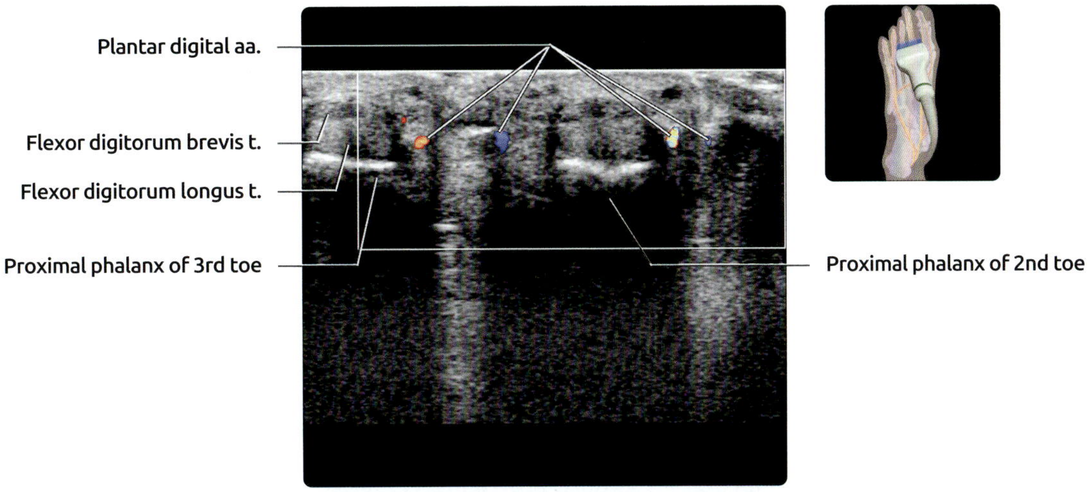

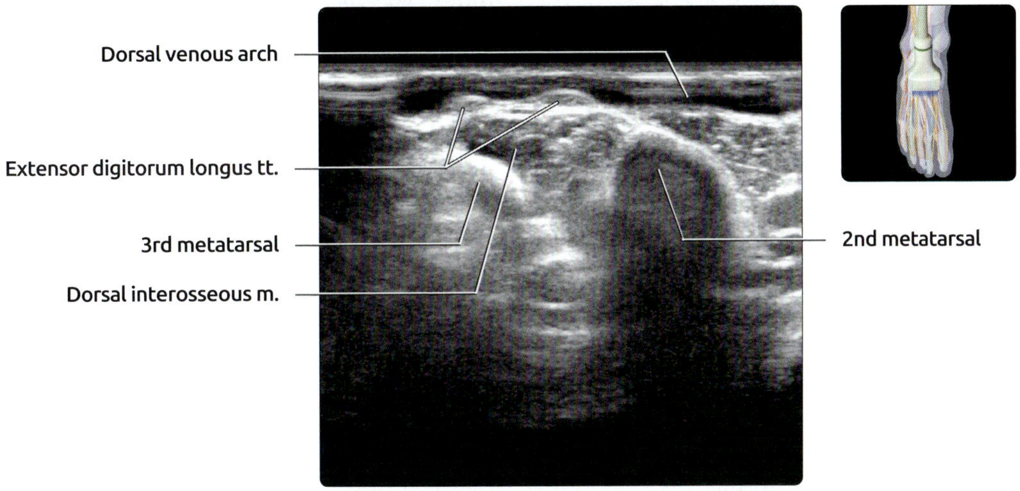

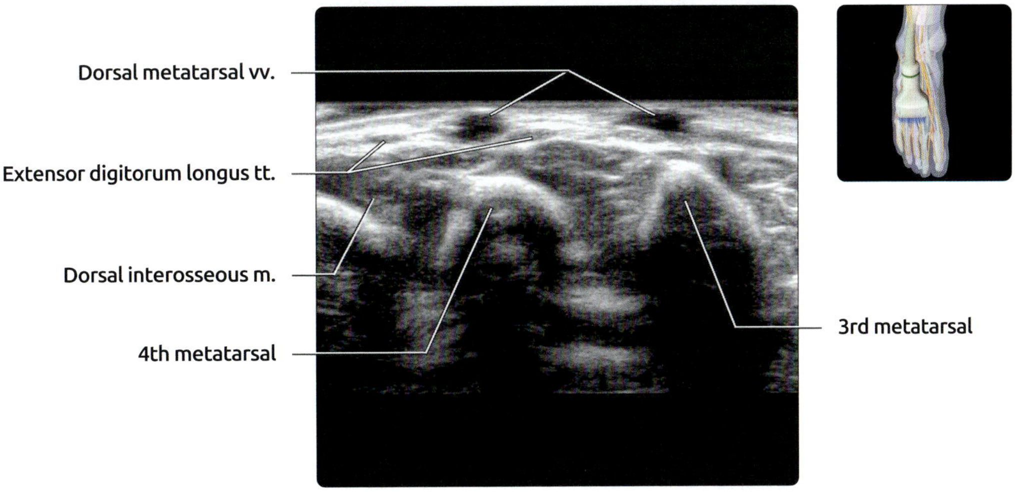

(Top) *Coronal color Doppler scan shows the plantar aspect of the toes. The digital arteries can be seen on both sides of the phalanges as they travel more distally.* **(Middle)** *Coronal scan shows the dorsum of the foot at the metatarsal level. Minimal pressure was used during this scan to prevent compression of the superficial (subcutaneous) veins. Here, the transversely running dorsal venous arch is demonstrated. The extensor tendons and the interossei run deep to this arch.* **(Bottom)** *More distally, a coronal scan of the dorsum of the foot shows the tributaries of the dorsal venous arch, the dorsal metatarsal veins.*

IMAGING ANATOMY

Overview

- In normal weight-bearing stance, all metatarsal heads are at same level and all are weight bearing
  - Metatarsophalangeal (MTP) joints are slightly extended in standing position
- Each MTP joint is separate synovial cavity

Metatarsal Bones

- 1st metatarsal
  - Largest metatarsal
  - Articulates with 1st cuneiform, 1st proximal phalanx, sesamoids
  - Medially receives attachments from tibialis anterior tendon, plantar side receives peroneus longus tendon
  - Origin of medial head of 1st dorsal interosseous muscle on its lateral surface
- 2nd-4th metatarsals
  - Gradually smaller and shorter moving laterally
  - Receive interosseous muscle on medial and lateral sides
- 5th metatarsal
  - Insertion of peroneus tertius and brevis

Phalanges

- Distal phalanx is insertion site for tendons from flexor and extensor digitorum longus
- Middle phalanx is insertion site for tendons from flexor and extensor digitorum brevis
- 2nd-4th proximal phalanges are attachments of interosseous muscles on both sides and lumbricals on medial side

1st Metatarsophalangeal Joint

- Dorsiflexion of toe important in push-off phase of gait
- Metatarsal head has 2 concave facets at plantar surface, 1 for each sesamoid, separated by ridge (crista)
- Distal articular surface of metatarsal head may be flat, rounded, or have central prominence
- Base of proximal phalanx has concave contour
- **Sesamoids**
  - Either sesamoid may be unipartite or bipartite
  - Medial sesamoid in medial head flexor hallucis brevis and abductor hallucis
  - Lateral sesamoid in lateral head flexor hallucis brevis and adductor hallucis and deep metatarsal ligament
  - Medial and lateral sesamoids joined by intersesamoid ligament
    - Intersesamoid ligament is floor of canal in which flexor hallucis longus tendon runs
  - Both are embedded in plantar plate of joint
  - **Sesamophalangeal apparatus**
    - Sesamoids fixed in position relative to 1st proximal phalanx, move relative to 1st metatarsal
    - Therefore, displaced laterally in hallux valgus
- **Plantar plate**
  - Fibrocartilaginous plantar capsular thickening extending from metatarsal neck to base of proximal phalanx
  - Incorporates sesamoids

Lateral Metatarsophalangeal Joints

- Convex articular surface of metatarsal head articulates with concave articular surface of proximal phalangeal base
- Plantar aspect of metatarsal head has rounded contour
- Dorsal aspect of metatarsal head is smaller than plantar aspect
  - Has concave or notched contour along medial and lateral margins
- Sesamoids variably present, most commonly at 5th toe
- **Phalangeal apparatus** is combination of plantar plate and proximal phalanx
- **Plantar plate**
  - Fibrocartilaginous plantar capsular thickening extending from metatarsal neck to base of proximal phalanx
  - Attached to deep transverse metatarsal ligament, plantar fascia and flexor tendon sheath, medial and lateral collateral ligaments
  - Instability may mimic Morton neuroma

ANATOMY IMAGING ISSUES

Imaging Recommendations

- Ultrasound
  - Good for assessing tendon integrity and joint alignment
    - Tendon integrity recognized by morphology (continuity of tendon and densely packed appearance of tendon fibers) and by behavior during passive movement
    - Joints can also be assessed dynamically during flexion and extension
  - Limited use in assessing for global bony alignment

CLINICAL IMPLICATIONS

Clinical Importance

- Instability of MTP joints results in pain and deformity of forefoot

Stability of 1st Metatarsophalangeal Joint

- Collateral ligaments
- Flexor and extensor hallucis brevis muscles
- Flexor and extensor hallucis longus have smaller contribution to stability

Stability of Lateral Metatarsophalangeal Joints

- Collateral ligaments
- Plantar plate
  - Rupture of plantar plate results in dorsal subluxation of MTP joint and hammer toe deformity

Short 1st Metatarsal (Morton Foot)

- Normal variant but increases stress on 2nd metatarsal
- Predisposes to osteonecrosis of 2nd metatarsal head (Freiberg infraction)

1ST METATARSAL HEAD SECTION AND TOES

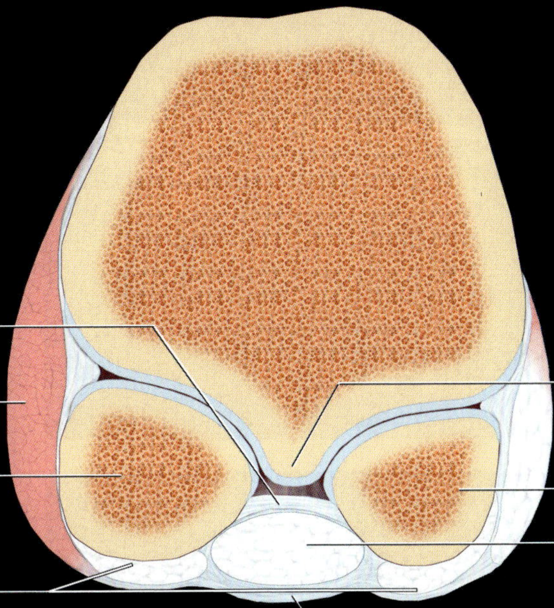

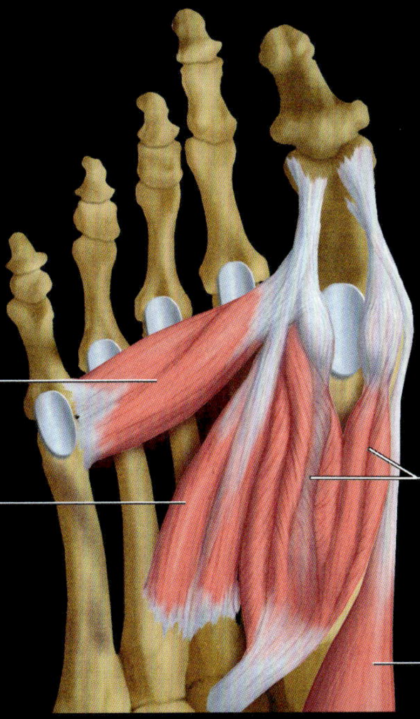

(Top) *Coronal (perpendicular to long axis of metatarsal bone) graphic shows sesamoid bone located on the plantar surface of the 1st metatarsal head, separated by the median crest of the metatarsal bone. The sesamoids are united by the intersesamoid ligament. The intersesamoid ligament forms the floor of the groove between the sesamoids; the flexor hallucis longus tendon runs along this groove.* **(Bottom)** *Axial graphic shows muscle attachments to sesamoids of the 1st metatarsophalangeal joint. Abductor hallucis and medial head of flexor hallucis brevis attach to medial sesamoid. Lateral head of flexor hallucis brevis and adductor hallucis attach to the lateral sesamoid.*

DORSUM METATARSAL

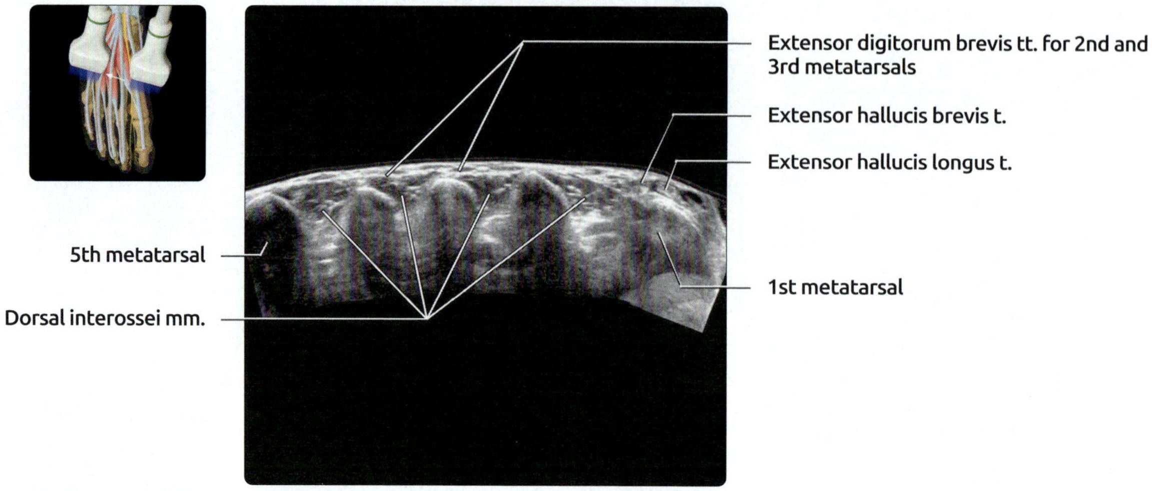

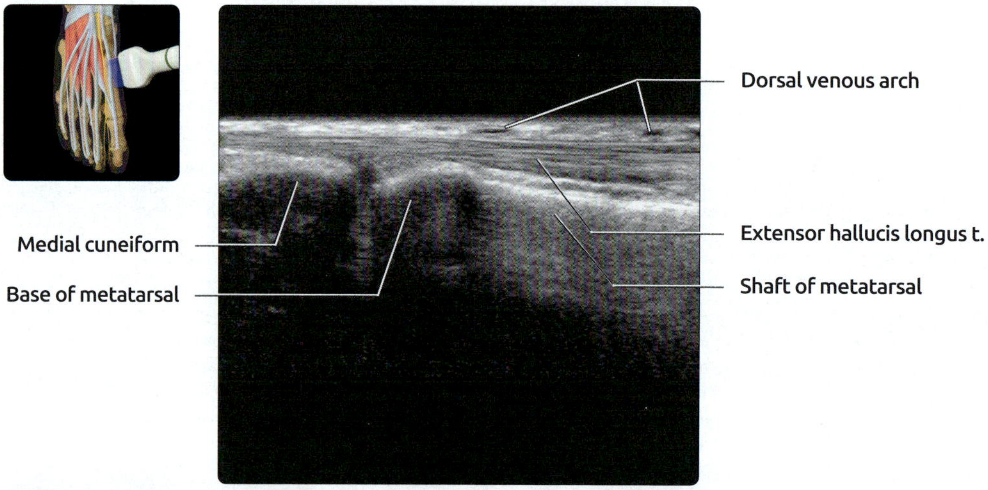

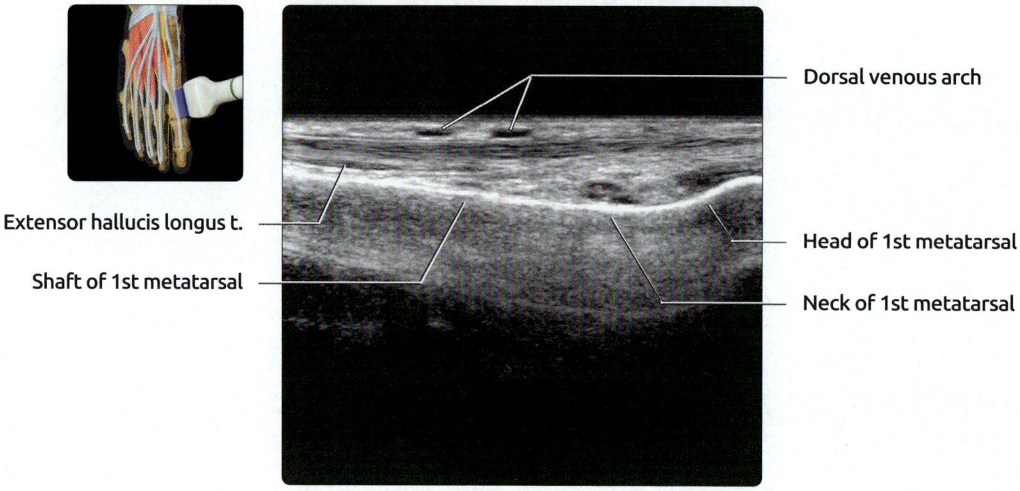

(Top) *Coronal panoramic scan shows the dorsal surface of the bodies of the metatarsals. The dorsal interosseous muscles lie between the bodies of the metatarsals. Their action is to abduct the toes. The plantar interosseous muscles act to adduct the toes. The extensor digitorum longus and extensor hallucis longus tendons receive assistance from their brevis tendons, which are deeper and approach from a lateral aspect.* **(Middle)** *Sagittal scan shows the dorsum of the big toe at the level of the tarsometatarsal joint. The extensor hallucis longus tendon is in close proximity to the bones around the joint. Note the densely packed and parallel echogenic tendon fibers.* **(Bottom)** *Sagittal scan shows the head of the 1st metatarsal. The extensor hallucis longus tendon is separated from the neck and most of the body of the metatarsal by fat.*

PLANTAR METATARSAL

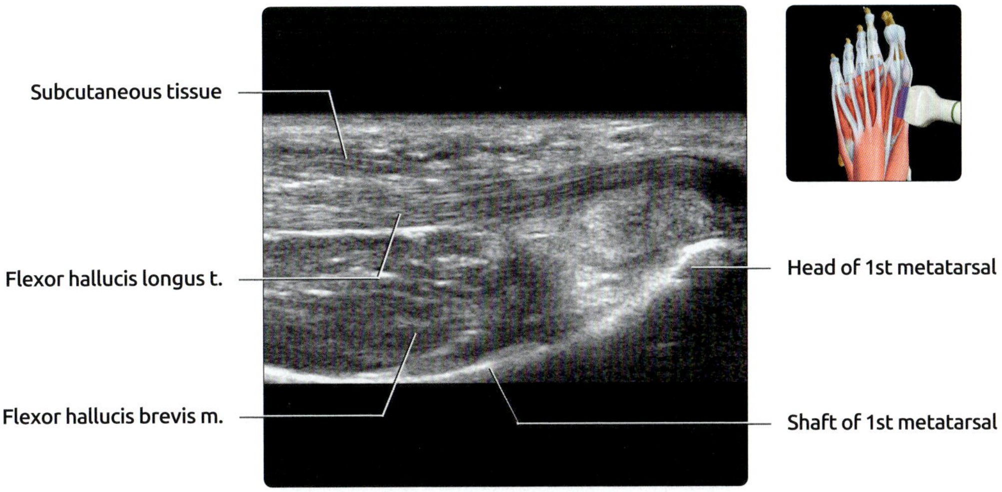

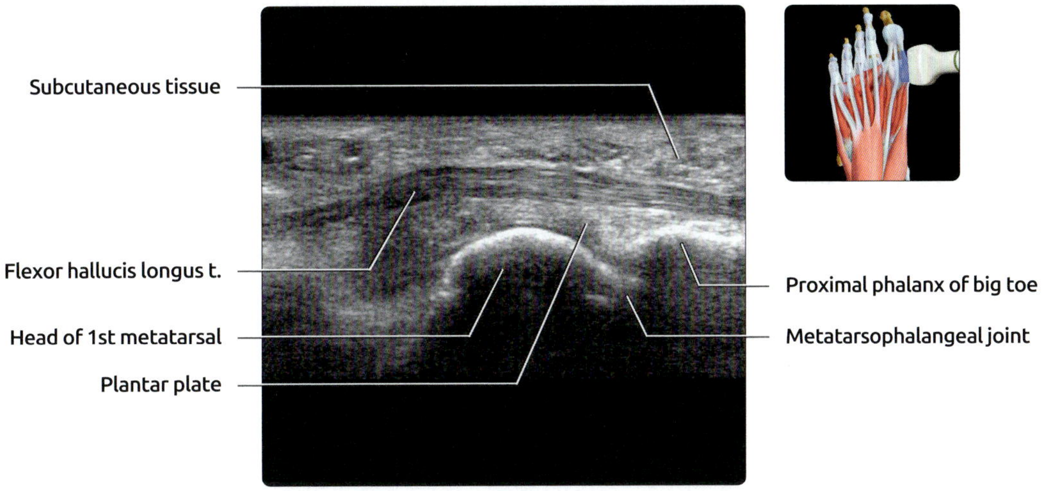

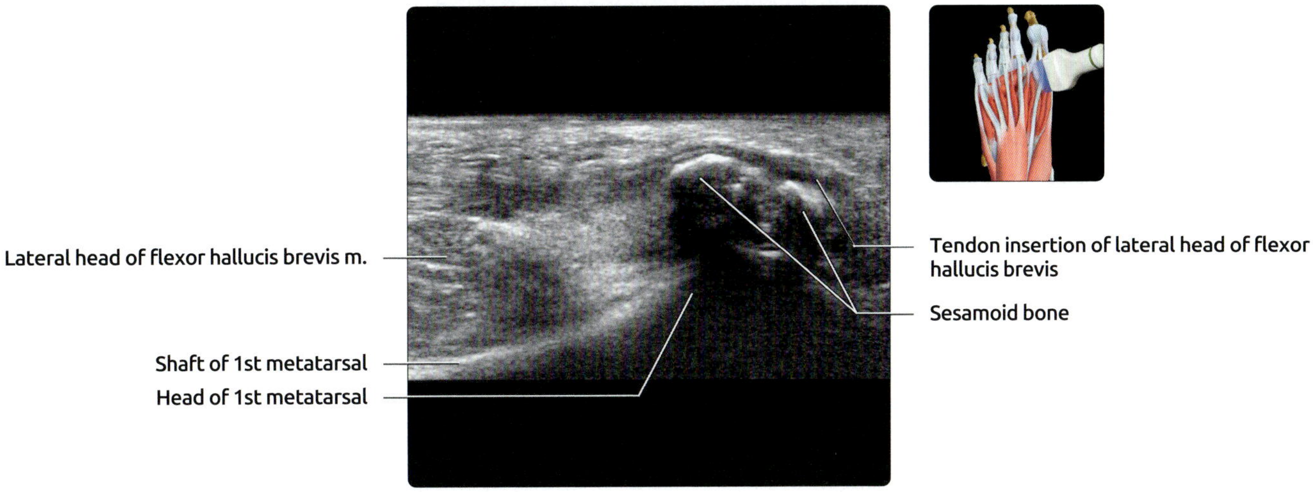

(Top) *Sagittal scan shows the plantar aspect of the head of the 1st metatarsal. The flexor hallucis longus muscle originates in the calf, while the brevis muscle arises from the cuboid and lateral cuneiform bones and tendon of the tibialis posterior.* **(Middle)** *Sagittal scan shows the center of the 1st metatarsophalangeal joint. The flexor hallucis longus tendon passes over the plantar plate of this joint to reach its insertion at the distal phalanx.* **(Bottom)** *Oblique sagittal scan shows the 1st metatarsophalangeal joint just lateral to the center. The flexor hallucis brevis tendons insert onto both sides of the base of the proximal phalanx via 2 tendons. Each tendon contains a sesamoid bone. Sesamoid bones may be bipartite as seen here, especially for the medial sesamoid.*

TRANSVERSE METATARSALS

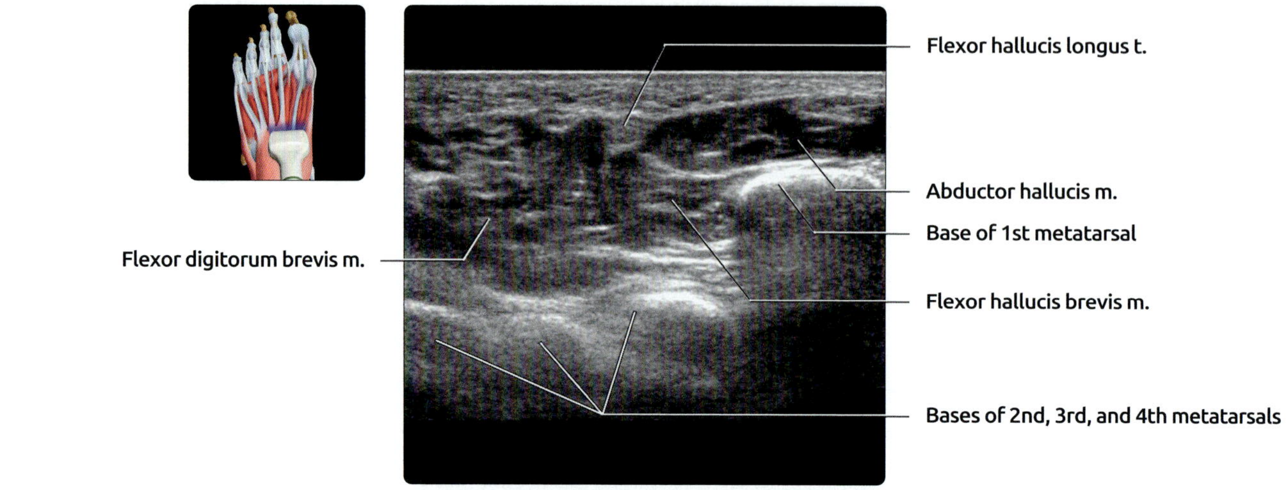

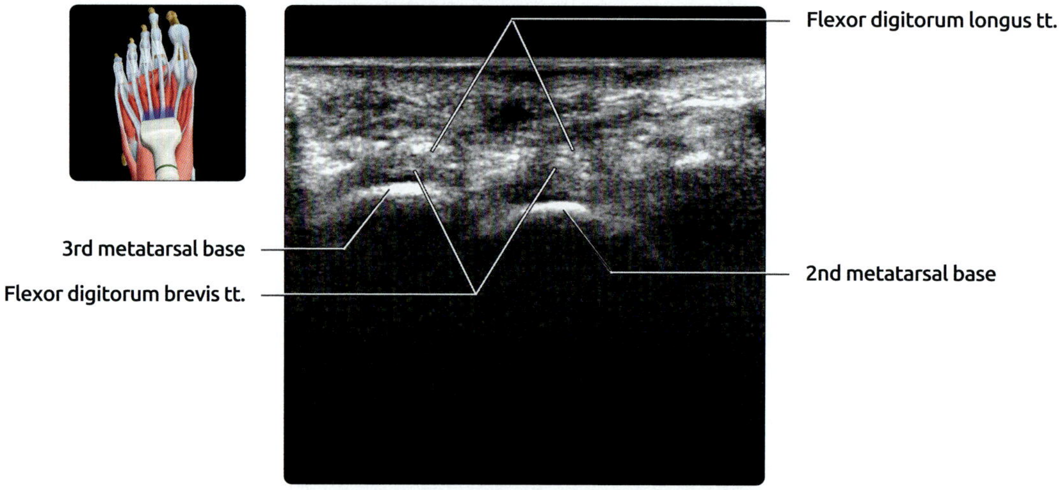

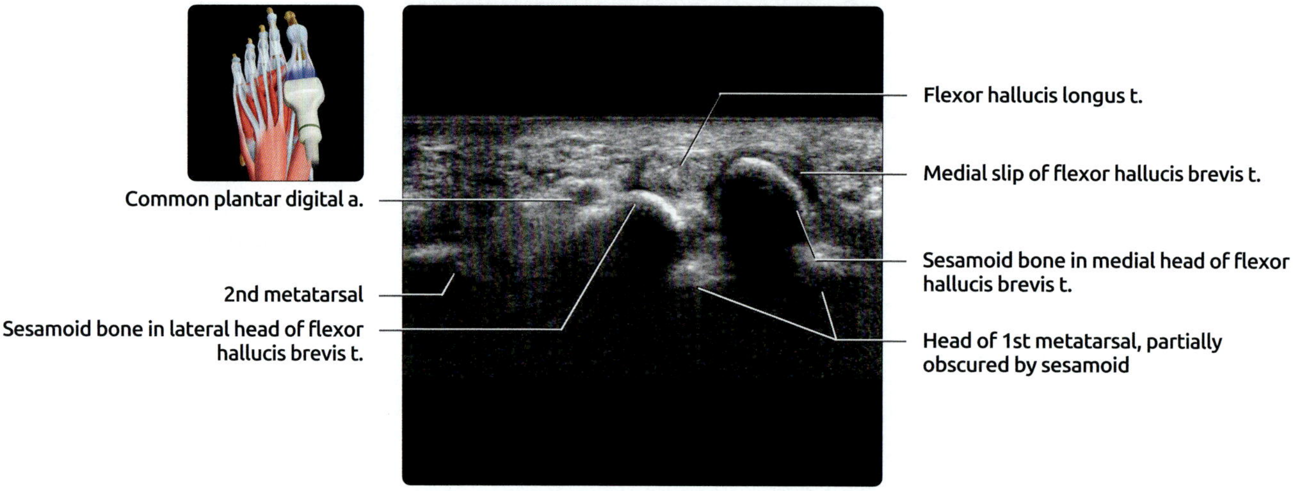

(Top) *Transverse scan shows the plantar aspect of the bases of the medial metatarsals. The abductor hallucis muscle is superficial to the 1st and 2nd metatarsals. The base of the 1st metatarsal is uniquely large compared to the rest of the metatarsals and is easily identifiable. The flexor hallucis longus tendon runs superficial to the brevis muscle.* **(Middle)** *Transverse scan shows the plantar surface of the base of the 2nd and 3rd metatarsals. The base of the metatarsals form an arch with plantar concavity. The flexor digitorum longus tendons run centrally on the plantar surfaces of the metatarsals.* **(Bottom)** *Transverse scan shows the plantar surface of the 1st metatarsal. The flexor hallucis longus runs between the 2 sesamoid bones on the plantar aspect of the 1st metatarsal. The sesamoids adhere to the medial and lateral heads of the flexor hallucis brevis tendons.*

TOE, PLANTAR AND DORSUM

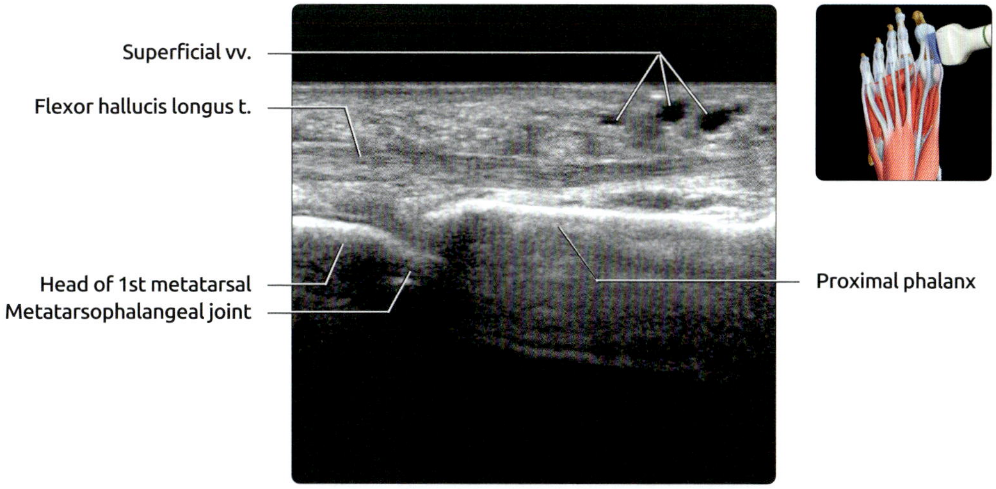

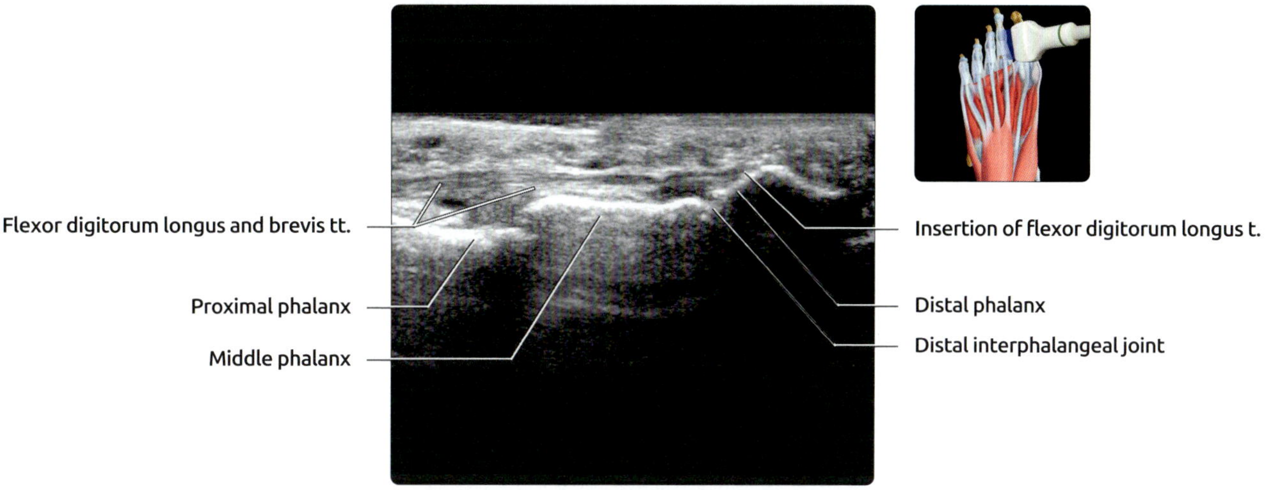

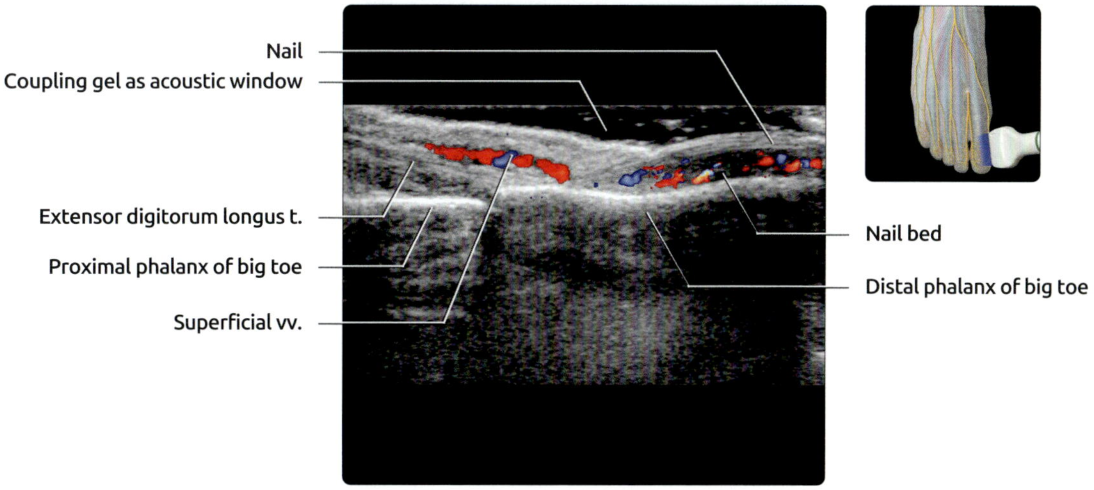

(Top) *Sagittal scan shows the plantar surface of the 1st tarsometatarsal joint. The flexor hallucis tendon runs close to the plantar surface of the phalanges and inserts at the base of the distal phalanx.* **(Middle)** *Sagittal scan shows the plantar surface of the 2nd toe. The flexor digitorum longus tendon inserts centrally at the base of the distal phalanx. The flexor digitorum brevis inserts on the sides of the base of the middle phalanx. Passive flexion of only the distal interphalangeal joint allows isolated evaluation of the tendon's movement (and integrity) of the flexor digitorum longus tendon.* **(Bottom)** *Sagittal color Doppler scan shows the dorsal surface of the distal phalanx of the big toe. The extensor digitorum longus tendon inserts on the dorsal surface of the base of the distal phalanx. The nail bed has a rich vascular supply as demonstrated by the color flow.*

SECTION 8

Obstetrics and Developmental Anatomy

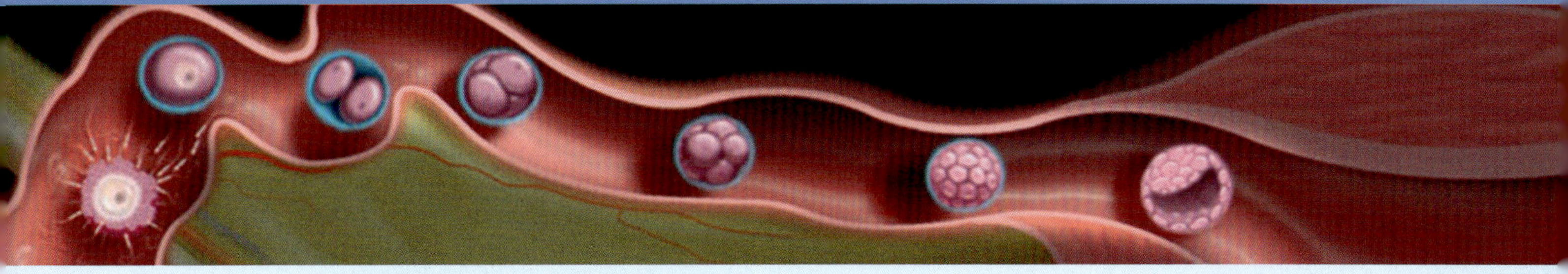

TERMINOLOGY

Definitions

- 1st trimester covers time from 1st day of last menstrual period to end of 13th postmenstrual week

EMBRYOLOGY

Embryologic Events

- 1st trimester includes
  - Ovulation
  - Fertilization
  - Cleavage
  - Implantation
  - Embryonic development
  - Organogenesis
  - Placental development
  - Umbilical cord development

Ovulation

- Primordial follicles → 5-12 primary follicles per cycle
- All but 1 degenerate, leaving single dominant follicle
- Pituitary gonadotrophin surge → ovulation → oocyte extruded onto ovarian surface
- Oocyte surrounded by tough zona pellucida as well as layers of cumulus cells
- Fimbria sweep oocyte into fallopian tube
- Remaining "empty" follicle becomes corpus luteum, producing estrogen and progesterone

Fertilization

- Occurs in fallopian tube
- Oocyte can be fertilized for ~ 24 hours
- Sperm penetrates oocyte, cell membranes fuse → zygote
- Spermatozoon and oocyte nuclei become male and female pronuclei
- Nuclear membranes disappear, chromosomes replicate in preparation for zygote cleavage

Cleavage

- Zygote → 2 cells → 4 cells → 8 cells → morula → blastocyst
- Several cell divisions result in smaller parts called blastomeres
- At 8-cell stage, compaction occurs with some cells → inner cell mass or embryoblast, some cells → peripheral trophoblast
  - Inner cell mass/embryoblast = embryonic pole of blastocyst
- 16-32 blastomeres = morula
- Morula absorbs fluid → central cavity called blastocoele within blastocyst

Implantation

- Blastocyst "hatches" from zona pellucida
- "Naked" blastocyst then interacts directly with maternal endometrium
- Trophoblast cells give rise to membranes and placenta, not embryo proper
  - Trophoblast cells at embryonic pole → syncytiotrophoblast, which burrows into endometrial lining
  - Remaining trophoblast cells become cytotrophoblast
- Maternal endometrial cells differentiate into decidual cells in response to
  - Progesterone secreted by corpus luteum
  - β human chorionic gonadotrophin produced by syncytiotrophoblast

Embryonic Development

- Bilaminar embryonic disc forms when embryoblast splits into epiblast and hypoblast
- Hypoblast = primitive endoderm
  - Hypoblast cells migrate around cavity of blastocyst to create primary yolk sac
  - Hypoblast + primary yolk sac give rise to extraembryonic mesoderm (loosely associated cells filling blastocyst cavity around primary yolk sac)
  - 2nd wave of migrating hypoblast cells create secondary yolk sac, which displaces primary yolk sac
  - Extraembryonic mesoderm splits into 2 layers, creating chorionic cavity (extraembryonic coelom)
  - Chorionic cavity separates embryo/amnion/yolk sac from chorion (outer wall of blastocyst)
- Epiblast contributes to embryo and gives rise to amnion
  - Fluid collects between epiblast and overlying trophoblast → cavity
  - Layer of epiblast differentiates into amniotic membrane separating new cavity from cytotrophoblast
- Trilaminar disc
  - Develops by process of gastrulation, which moves cells to new locations with resulting induction
  - 3 primary germ layers = ectoderm, mesoderm, endoderm
  - Body axes also determined by gastrulation
- Disc elongates and folds → series of tubular structures → major organ systems
- Ectoderm → neural plate → neural tube + neural crest cells
  - Neural tube → brain and spinal cord
  - Neural crest cells migrate from neural tube → many differing structures and cell types
- Mesoderm
  - Head mesoderm → muscles of face, jaw, and throat
  - Notochordal process
  - Cardiogenic mesoderm
  - Somites → most of axial skeleton
  - Intermediate mesoderm → genitourinary system
  - Lateral plate mesoderm → abdominal wall and gut walls
- Endoderm
  - Foregut, midgut, hindgut (oropharyngeal membrane → mouth)

Organogenesis

- **Central nervous system**
  - Arises from neural folds → neural tube + neural crest
    - Cranial/rostral 2/3 of neural tube → brain
    - Caudal 1/3 of neural tube → spinal cord, nerves
    - Neural crest → peripheral nerves, autonomic nervous system
- **Cardiovascular system**
  - Arises from cardiac tube → heart and great vessels
  - Cardiogenic precursors form 1° heart field at cranial end of embryo

- Lateral endocardial tubes brought together by embryonic folding → primitive heart tube
- Looping, remodeling, septation of primitive heart tube → definitive 4-chamber heart
- Conotruncus = primitive outflow tract that splits → ventricular outflow tracts

- **Respiratory system**
  - Foregut → respiratory diverticulum → 1° bronchial buds → 3 right + 2 left 2° bronchial buds → terminal bronchioles → respiratory bronchioles → primitive alveoli
- **Gastrointestinal system**
  - Early embryonic folding → endodermal tube → foregut, midgut, hindgut
  - Foregut (blind ending at oropharyngeal membrane) → esophagus, stomach, proximal duodenum
    - Liver, gallbladder, cystic duct, and pancreas arise from duodenal diverticula
  - Midgut (initially open to yolk sac) → distal duodenum to proximal 2/3 transverse colon
    - Future ileum elongates rapidly → 1° intestinal loop, which herniates into base of umbilical cord rotating 90°
    - During retraction into peritoneal cavity, additional 180° rotation secures normal bowl orientation with cecum right, duodenojejunal flexure left
  - Hindgut (blind ending at cloacal membrane) → distal 1/3 transverse colon to rectum
    - Terminal expansion of primitive hindgut tube → cloaca
    - Urorectal septum divides cloaca into urogenital sinus + dorsal anorectal canal
- **Genitourinary system**
  - Intermediate mesoderm → pronephros, mesonephros, metanephros
    - Mesonephros → rudimentary kidneys connected to cloaca by mesonephric ducts
    - Mesonephric ducts → ureteral bud → collecting system
    - Ureteral bud connection to metanephric blastema → induction of nephron formation
  - Bladder arises from cloaca and allantois
  - Bladder separated from rectum by urogenital sinus
- **Musculoskeletal system**
  - Upper and lower extremities develop from individual limb buds

Placental Development

- Chorionic sac initially covered in villi, atrophy of those adjacent to uterine cavity → chorion laeve
- In villi adjacent to implantation site, burrowing syncytiotrophoblast develops trophoblastic lacunae
  - Adjacent maternal capillaries expand → maternal sinusoids, anastomose with trophoblastic lacunae
  - Budding/proliferation of cytotrophoblast into syncytiotrophoblast and maternal lacunae → mature tertiary villi
  - Tertiary villi contain fully differentiated blood vessels for gas exchange in chorion frondosum
- Chorion frondosum + decidua basalis = placenta

Umbilical Cord Development

- Embryonic disc lies between amnion and yolk sac
- Embryo initially connected to chorion by connecting stalk, which arises from extraembryonic mesoderm
  - Allantois (endodermal hindgut diverticulum) arises as outpouching of yolk sac
  - Allantois and allantoic vessels extend into connecting stalk (become umbilical vessels)
- Embryonic growth and folding result in blind-ended foregut and hindgut tubes with midgut open to yolk sac
  - As body wall forms by lateral folding and midgut becomes tubular, yolk sac is "pinched off"
  - Narrow elongated neck of yolk sac = vitelline duct, which connects yolk sac to closing midgut tube
- As embryo enlarges and folds, amniotic cavity expands to encompass embryo completely except at umbilical ring
  - Connecting stalk, allantois, vitelline duct become incorporated as umbilical cord
  - Amnion continues to enlarge and forms tubular covering over incorporated cord elements → dense epithelial covering
- Progressive cord elongation and coiling occur with embryonic/fetal growth and movement

ANATOMY IMAGING ISSUES

Questions

- Developmental milestones (in weeks post last menstrual period)
  - Gestational sac (intradecidual sac sign) usually visible by 4.0-4.5 weeks
  - Yolk sac usually visible by 5.0-5.5 weeks
  - Distinct embryo with cardiac activity usually visible by 6.0-6.5 weeks
- Developmental milestones based on mean sac diameter (MSD)
  - Embryo should be visible if MSD > 25 mm endovaginal (EV) scan
- Embryo of > 7 mm in length must have cardiac activity
  - If embryo seen within visible amnion, cardiac activity should be present (expanded amnion sign)
- Gestational age assessment most accurate in 1st trimester
  - Biological variations take effect after 13 weeks
- Determination of chorionicity best done in multiple pregnancies
  - Most important factor in prognosis
- Is there evidence of increased risk for aneuploidy
  - 11- to 13-week scan can be used to adjust a priori risk of aneuploidy, determine need for invasive testing
- Is anatomy normal
  - Organogenesis is complete by end of 13th week
  - Use EV sonography for best resolution
- 1st trimester is time of complex cell multiplication and differentiation
  - Great potential for error if normal processes are not clearly understood

OVULATION AND FERTILIZATION

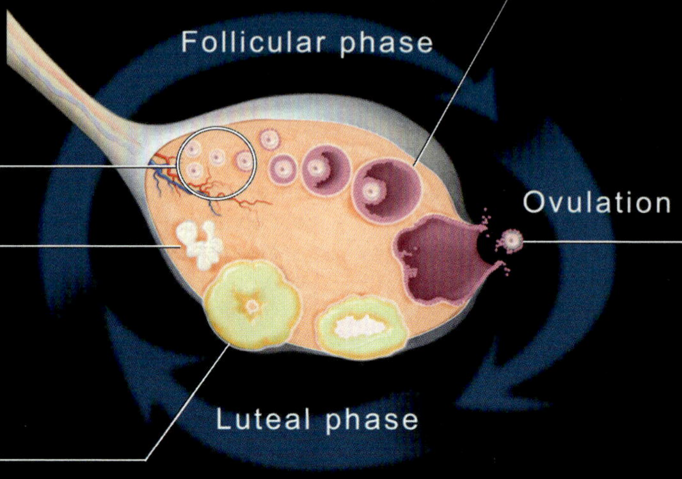

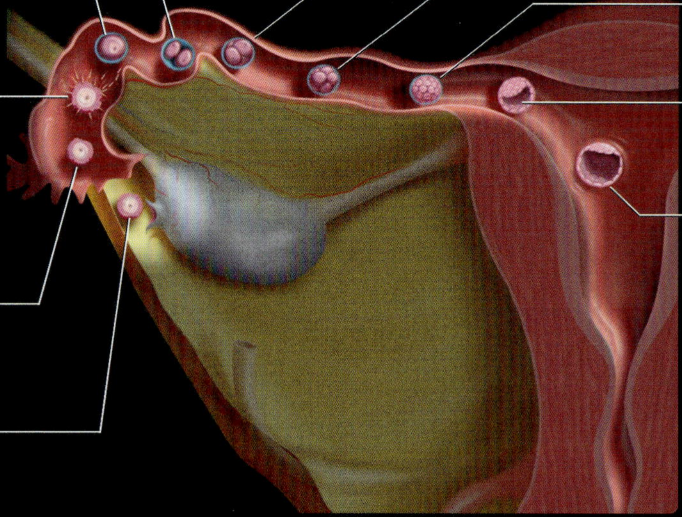

(Top) *During the follicular phase of the menstrual cycle, several follicles begin to develop; 1 becomes dominant, and eventually, a mature oocyte is extruded from the ovarian surface at the time of ovulation. The remaining follicle becomes the corpus luteum, which produces progesterone and helps to maintain the early pregnancy until the placenta is formed. If fertilization does not occur, the corpus luteum degenerates into a corpus albicans.* **(Bottom)** *The oocyte is swept into the fallopian tube where it is fertilized. The fertilized ovum divides repeatedly during passage along the tube so that by the time it reaches the endometrial cavity, a blastocyst has formed. The blastocyst "hatches" from the zona pellucida and implants into the maternal endometrium.*

CLEAVAGE AND IMPLANTATION

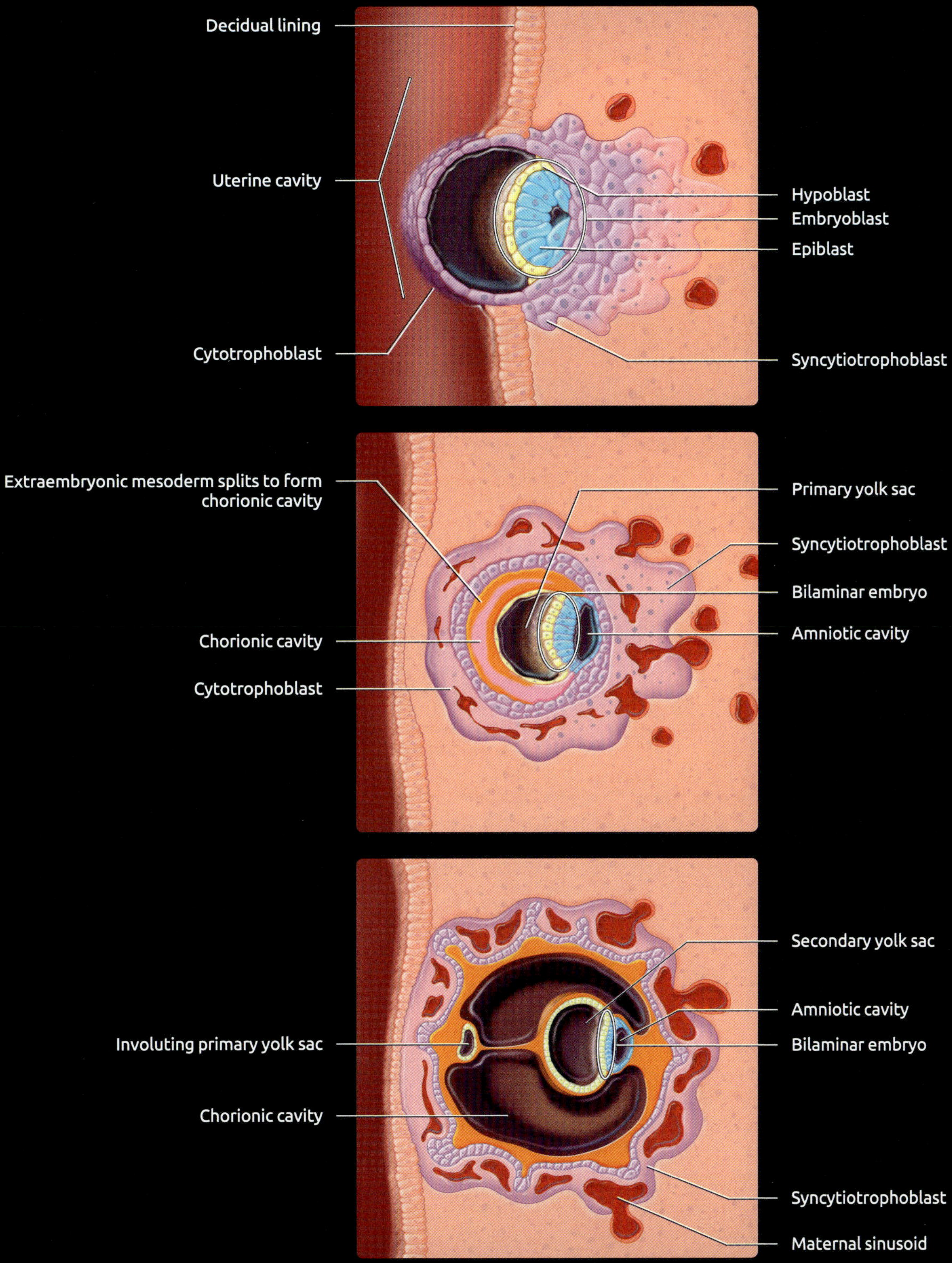

(Top) *While the dividing zygote is still in the fallopian tube (8-cell stage), cells differentiate into embryoblast and trophoblast. Syncytiotrophoblast interacts with the endometrium to form the placenta; the remainder is the cytotrophoblast. Embryoblast cells will give rise to the embryo, amnion, and yolk sac.* **(Middle)** *The embryoblast splits into 2 layers: Epiblast and hypoblast. The hypoblast gives rise to the primary and secondary yolk sacs and extraembryonic mesoderm. The latter splits, forming the chorionic cavity. The epiblast gives rise to the embryo and the amnion.* **(Bottom)** *As the primary yolk sac involutes, the secondary yolk sac develops. It is the secondary yolk sac that is visible sonographically; however, by convention, it is usually referred to as simply the yolk sac on ultrasound images. The chorionic cavity enlarges. The embryo is still a bilaminar disc.*

INTRADECIDUAL SAC SIGN

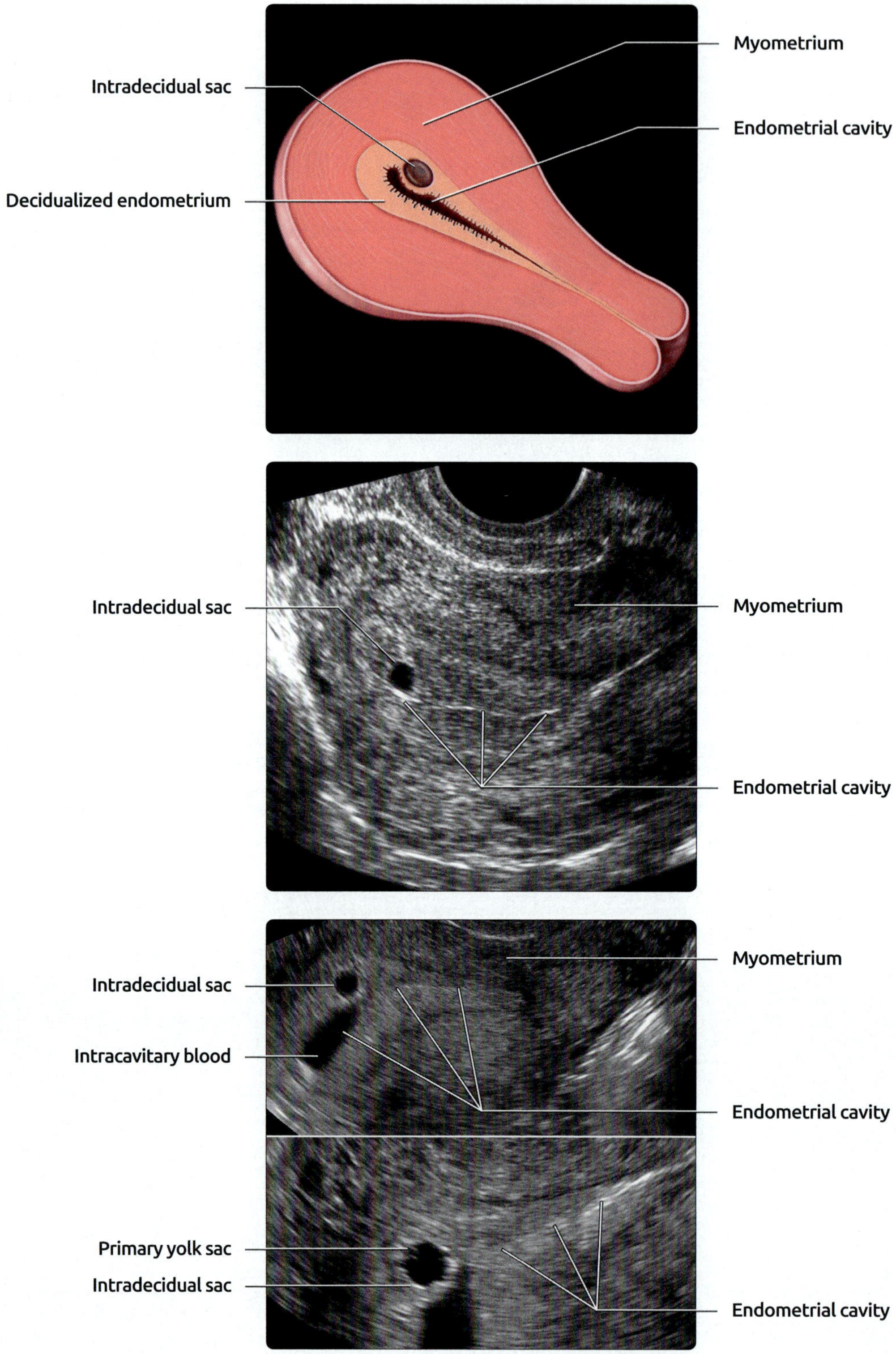

(Top) *The graphic illustrates the earliest sonographic manifestation of the embryological development. The gestational sac has burrowed into the decidualized endometrium, creating an asymmetrically placed echogenic ring with a lucent center. This was initially described as the intradecidual sac sign (IDSS). It is not always visible in early pregnancy, and it is subject to considerable interobserver variability.* **(Middle)** *The intradecidual gestational sac is an echogenic ring eccentric to the line created by apposition of the endometrial surfaces. Currently, recommended terms for such an observation are intrauterine sac-like structure or probable intrauterine pregnancy.* **(Bottom)** *In this example of IDSS, bleeding has resulted in accumulation of blood in the endometrial cavity. Again, note the eccentric location of the 4-mm diameter sac. In the lower image, a tiny circular structure within the gestational sac is likely the primary yolk sac, which can be seen with high-resolution modern transducers.*

DOUBLE DECIDUAL SAC SIGN

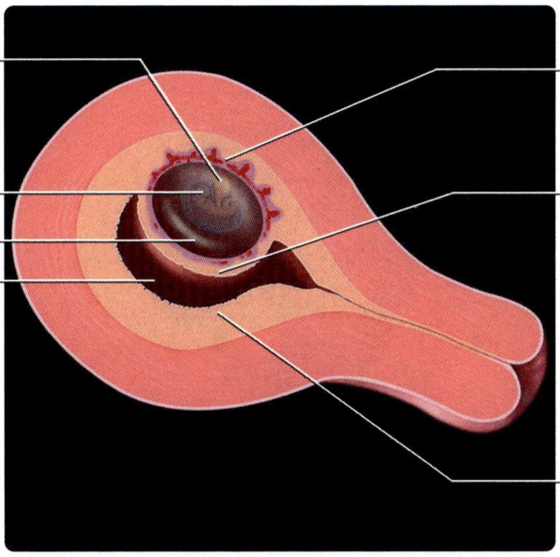

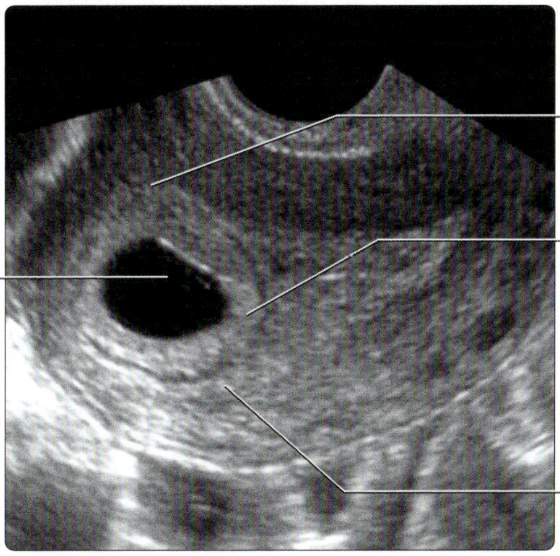

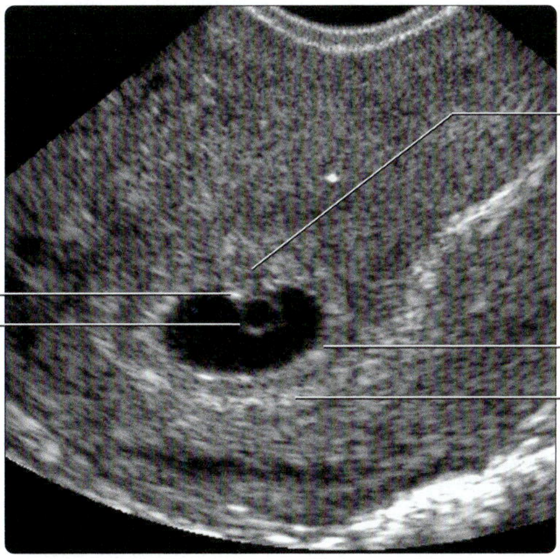

(Top) *This graphic illustrates the double decidual sac sign (DDSS). This is seen when the enlarging gestational sac protrudes from the site of implantation and starts to expand into the uterine cavity, exerting mass effect on the opposite uterine wall. The decidua covering the expanding sac is decidua capsularis, that which is being pushed ahead of the expanding sac is the decidua parietalis. The decidua basalis is where the sac is adherent to the uterine wall and marks the site where the placenta will develop. Internal structures may be seen on transvaginal scans.* **(Middle)** *The decidual layers are easily seen on this transvaginal scan. The concentric rings created by the decidua capsularis and parietalis create the DDSS. This finding is characterized as a probable intrauterine pregnancy.* **(Bottom)** *In this example, the embryo and yolk sac are visible in addition to the DDSS; this indicates a definite intrauterine pregnancy. It would also be a pregnancy of uncertain viability if there were no cardiac activity in an embryo < 7 mm in length.*

EMBRYONIC DEVELOPMENT: 6 WEEKS

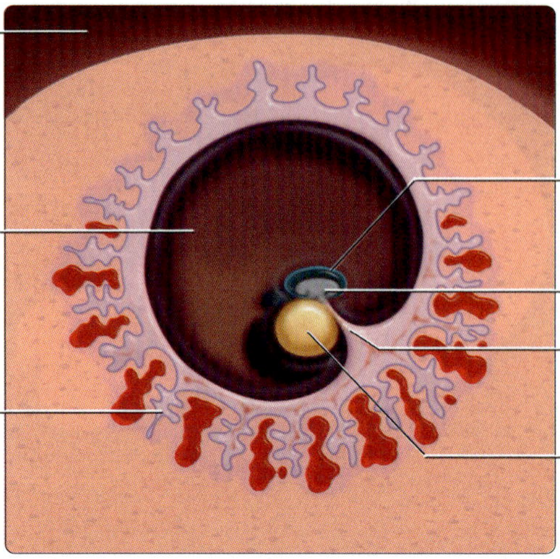

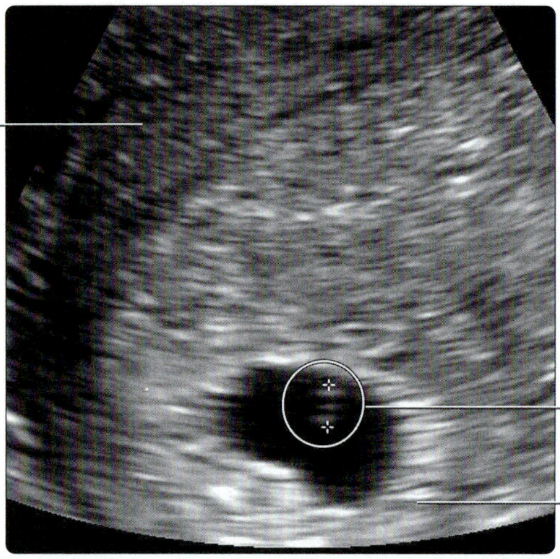

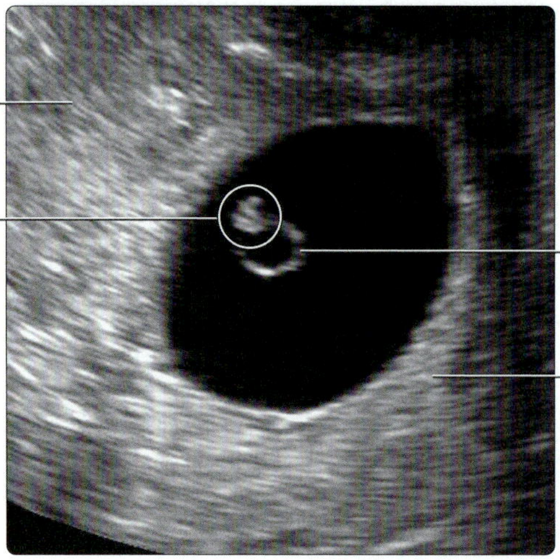

(Top) *The graphic illustrates normal early development. The embryo is intimately associated with the yolk sac so that the amnion and yolk sac appear as a double bleb with the embryo sandwiched between them. The embryo is within the amniotic sac; both the embryo and yolk sac are inside the chorionic sac.* **(Middle)** *Vaginal ultrasound at 5 weeks, 5 days from last menstrual period shows a 2-mm embryo. There are 3 linear echoes that resemble an Oreo cookie. The process of gastrulation results in cellular movement with creation of the 3 primary germ layers: The endoderm, mesoderm, and ectoderm. Despite the tiny size of this embryo, cardiac activity was visible in real time.* **(Bottom)** *Vaginal ultrasound shows the embryo immediately adjacent to the yolk sac. The amnion is not yet visible. At this gestational age, the abdominal wall is still open, and the midgut is in continuity with the yolk sac. After the abdominal wall closes, the "discarded" yolk sac is compressed between the expanding amnion and the chorion.*

EMBRYONIC DEVELOPMENT: 7-8 WEEKS

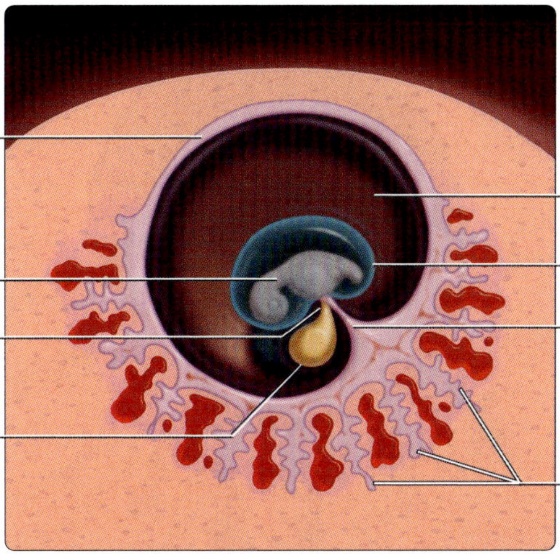

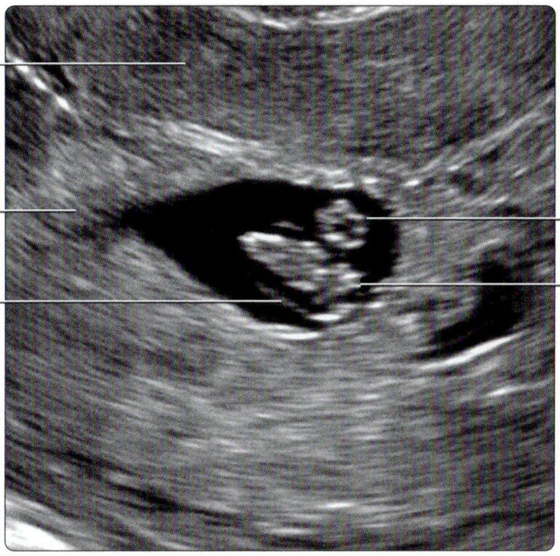

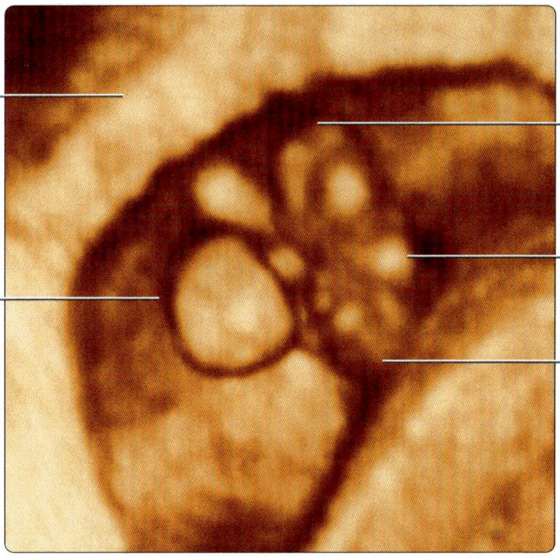

(Top) *Curvature and folding of the embryo result in closure of the abdominal wall and pinching off of the yolk sac. The elongated neck forms the vitelline duct. Eventually, the yolk sac separates from the embryo, dropping into the chorionic cavity. At the same time, it becomes clear which end of the embryo is which, and limb buds start to form. The chorion adjacent to the uterine cavity is now completely smooth. Chorionic villi in the developing placenta become more complex in structure.* **(Middle)** *Vaginal ultrasound at 7 weeks, 4 days from last menstrual period shows the yolk sac is separate from the embryo. It lies outside the amnion, which is now expanded enough to be just visible as it surrounds the embryo. Remember that the yolk sac will always be outside the amnion; the embryo lies inside the amniotic sac.* **(Bottom)** *3D surface-rendered ultrasound shows the yolk sac separate from the embryo, which now has an elongated configuration with defined cranial or crown and pelvic or rump ends. The crown rump-length is measured to confirm menstrual dating.*

EMBRYONIC DEVELOPMENT: 8-9 WEEKS

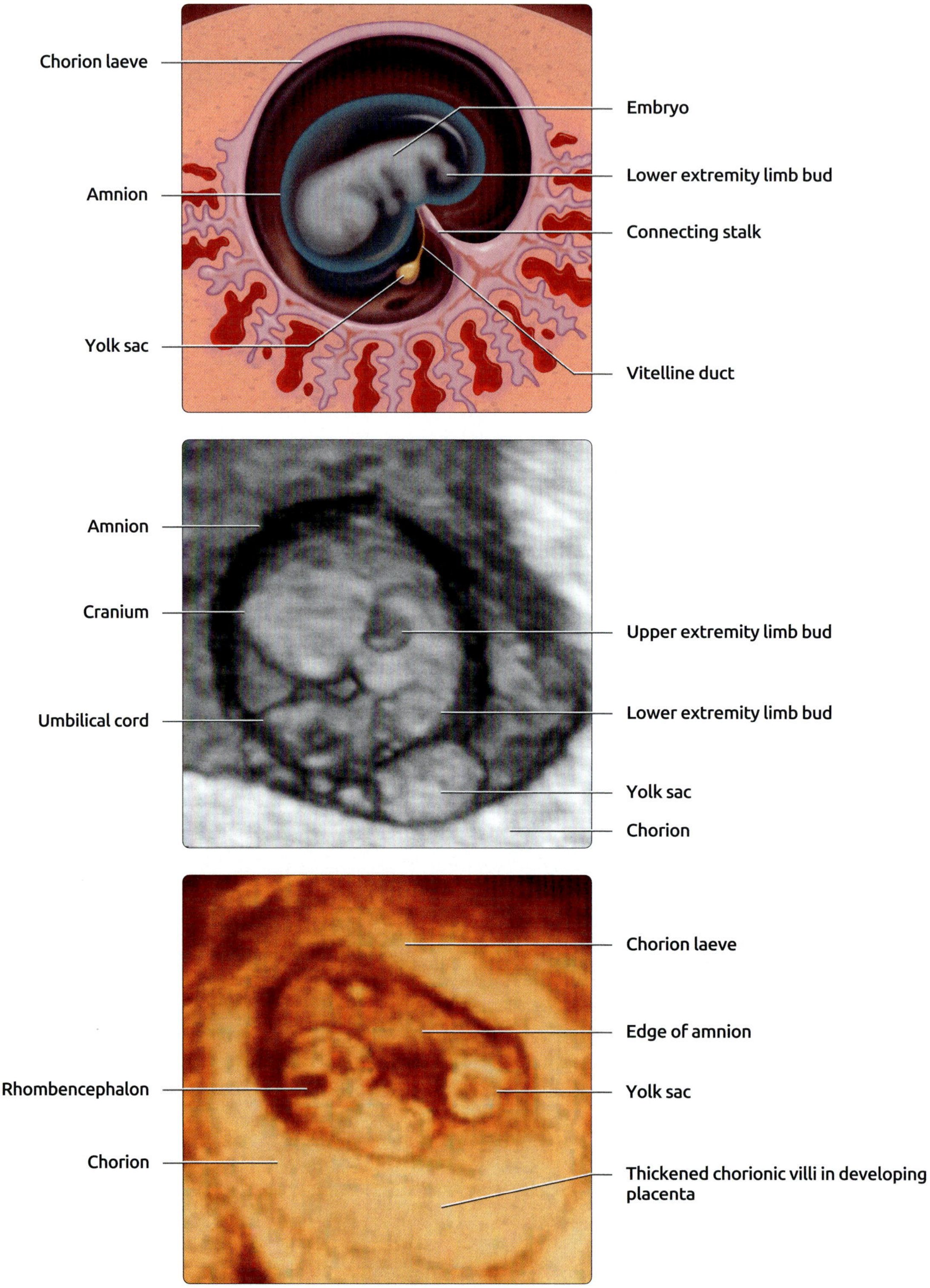

(Top) *The graphic illustrates continued embryonic development; the limb buds are evident, the head has grown dramatically, and the embryo is assuming a recognizable human form. The umbilical cord forms as a result of fusion of the vitelline duct, allantois, and connecting stalk. Once formed, it elongates rapidly until the embryo is suspended within the enlarging amniotic sac. Cord elongation allows for free mobility of the developing fetus.* **(Middle)** *3D surface-rendered ultrasound of an 8-week embryo shows the short limb buds and the relatively thick umbilical cord. The crown end is assuming a more recognizable head shape, and the embryo is curling into the typical fetal position.* **(Bottom)** *Another 3D ultrasound in the sagittal plane shows the rhombencephalic vesicle in the cranial part of the embryo. This fluid-filled space is a precursor of the 4th ventricle and should not be confused with a pathological intracranial cyst.*

EMBRYONIC DEVELOPMENT: 10-13 WEEKS

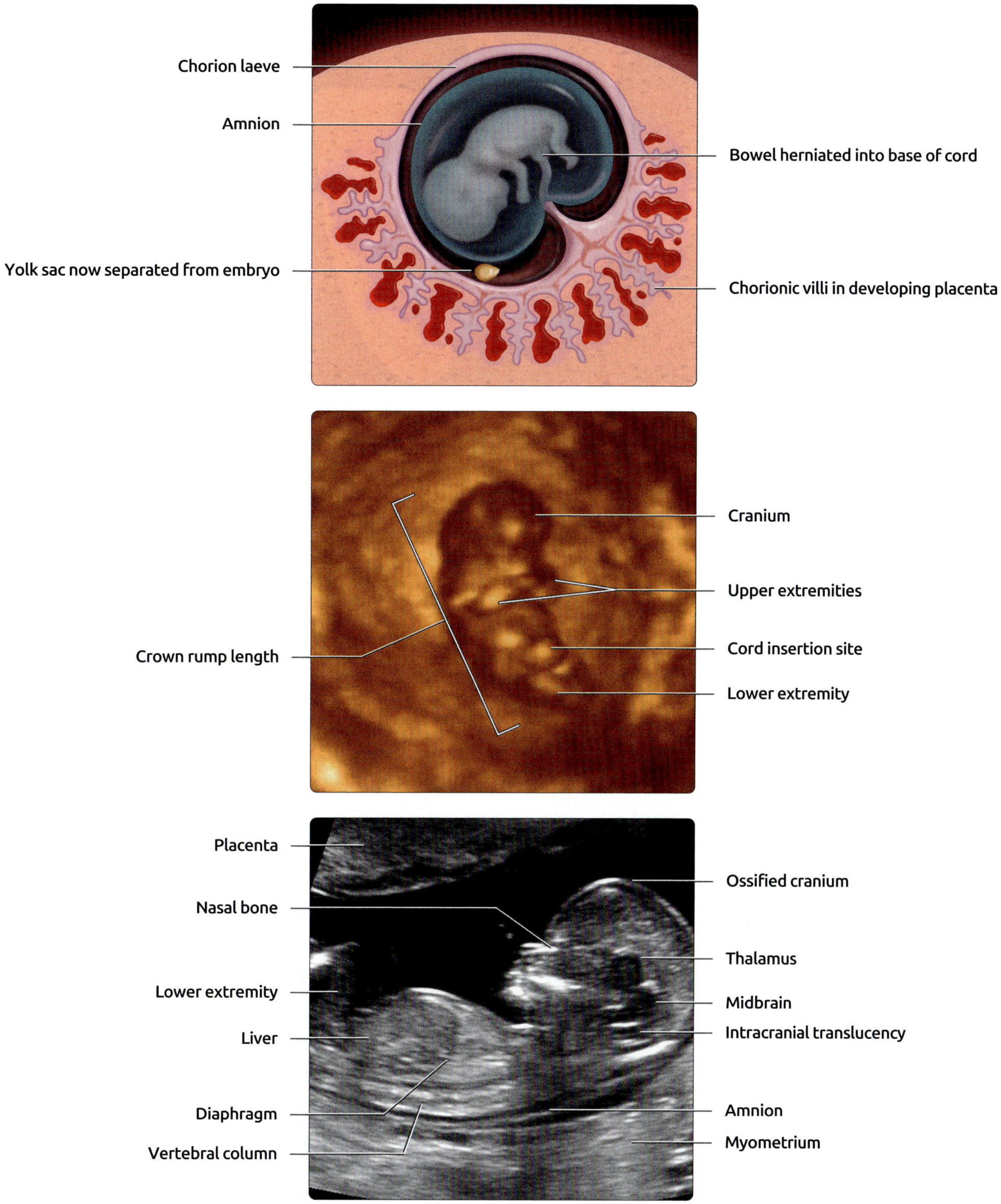

(Top) *Toward the end of the 1st trimester, the amnion fills the chorionic cavity. The membranes do not "fuse" until 14-16 weeks. The placenta continues to grow, and the chorionic villi develop an increasingly complex branching pattern.* **(Middle)** *3D surface-rendered ultrasound at 10 weeks shows increasingly recognizable fetal anatomy with a well-developed cranium and extremities. The abdominal wall cord insertion site is quite broad due to the physiologic herniation of bowel into its base. This occurs as the peritoneal cavity is too small to accommodate the rapidly growing bowel at this gestational age.* **(Bottom)** *Sagittal transabdominal ultrasound at 12 weeks, 6 days from last menstrual period shows recognizable anatomic features. Organogenesis is completed by the end of the 13th week. Growth and maturation of the various organ systems occurs during the remainder of gestation.*

ABDOMINAL WALL AND GI TRACT

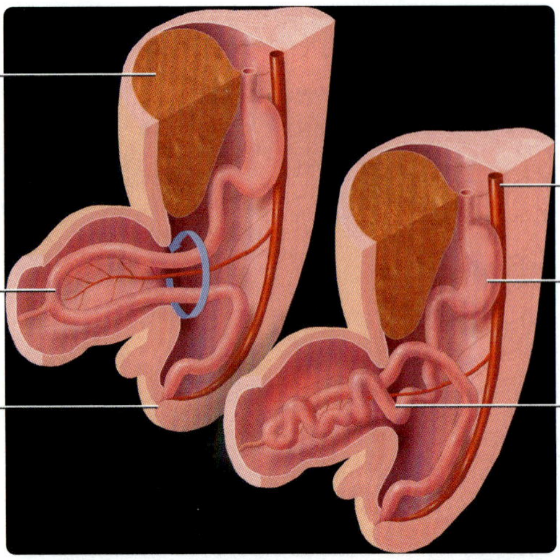

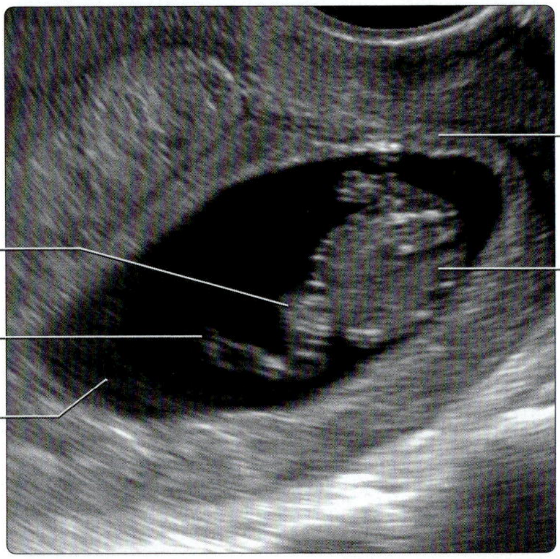

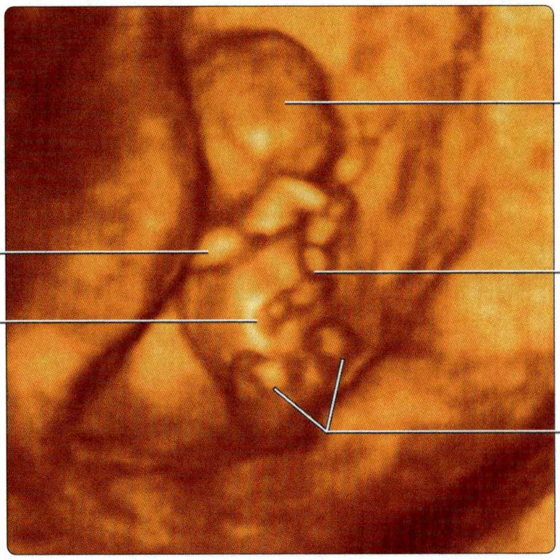

(Top) *The graphic illustrates herniation of bowel into the base of the cord in the 1st trimester. This happens as the gut tube elongates before there is adequate room to accommodate it in the peritoneal cavity. The gut undergoes a 90° counterclockwise rotation as it herniates and an additional 180° rotation as it is retracted into the peritoneal cavity. Only the gut herniates; liver is never normally seen at the base of the cord.* **(Middle)** *Vaginal ultrasound at 10 weeks, 3 days shows echogenic bowel at the base of the umbilical cord. This is normal at this gestational age and should not be misinterpreted as a bowel-containing omphalocele.* **(Bottom)** *3D surface-rendered ultrasound of a 12-week fetus shows a normal abdominal wall contour. The cord insertion is normal, and there is no residual bowel herniation. Three of the extremities are seen, the cranial contour is normal, and the cord is already coiled.*

UMBILICAL CORD DEVELOPMENT

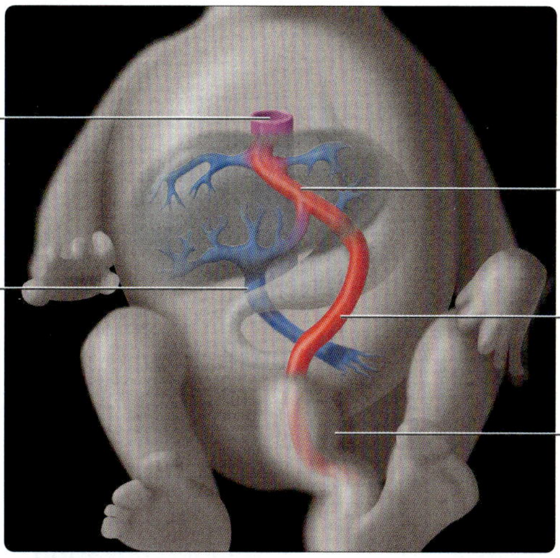

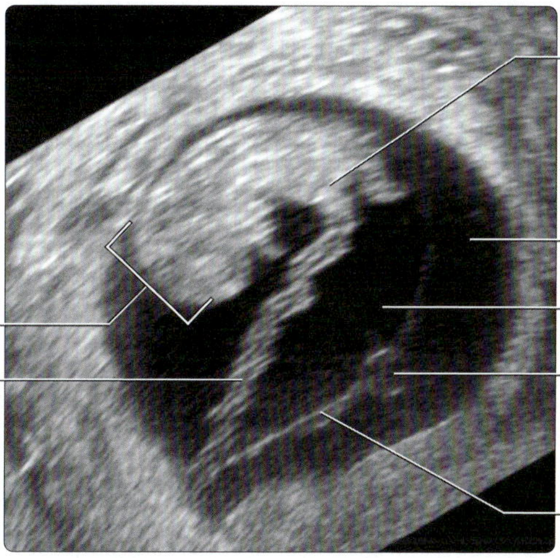

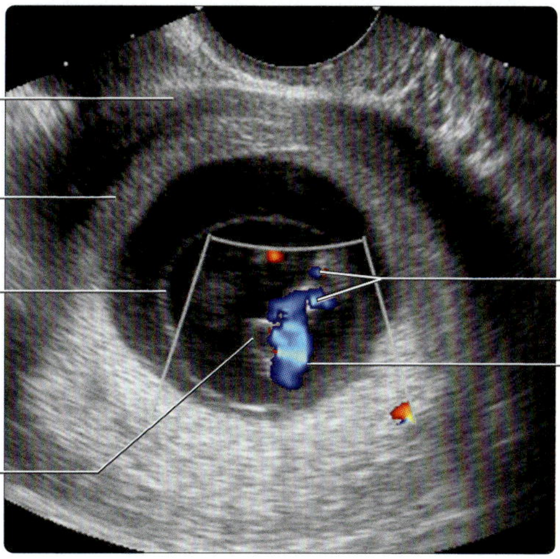

(Top) *The graphic illustrates fetal circulation in which oxygenated blood from the placenta returns to the fetus via the umbilical vein. The umbilical vein courses through the left lobe of the liver to the left portal vein, across the ductus venosus into the inferior vena cava. The umbilical cord also contains 2 arteries that arise from the internal iliac arteries.* **(Middle)** *At 10 weeks, there is some residual herniation of bowel into the base of the cord. The embryo is freely suspended within the amniotic sac by the cord, which already shows evidence of coiling. The yolk sac will be obliterated as the amnion apposes to the chorion.* **(Bottom)** *Color Doppler ultrasound shows a small umbilical cord cyst as an avascular area adjacent to the cord vessels. As a 1st-trimester finding, this is usually of no significance; the cysts form at 8-9 weeks and usually resolve by ~ 12 weeks.*

TERMINOLOGY

Definitions

- Rostral: Cranial (i.e., toward head end of embryo)
- Caudal: Tail (i.e., toward sacral end of embryo)

MAJOR EMBRYOLOGIC EVENTS

Neurulation

- Ectodermal cells give rise to dorsal, midline neural plate
  - Neural folds develop then fuse → neural tube + neural crest
- **Neural tube** → brain, spinal cord
- **Neural crest** → peripheral nerves, autonomic nervous system
- **Errors in neurulation**
  - Anencephaly
  - Cephalocele
  - Myelomeningocele
  - Chiari 2: Abnormal neurulation of hindbrain

Neuronal Proliferation

- Begins in rhombencephalon; neuroepithelial proliferation → neurons, glial cells, ependymal cells
  - Neurons "born" in ventricular zone (around central lumen) are called young neurons
    - Migrate peripherally to form mantle zone (i.e., gray matter precursor)
    - Axons extending peripherally to mantle zone establish marginal zone (i.e., white matter precursors)
    - White matter is outside, gray matter is inside
  - Glioblast cells → astrocytes, oligodendrocytes
    - Provide metabolic/structural support to neurons
  - Ependymal cells line ventricles/spinal canal
    - Produce cerebrospinal fluid
- **Errors in neuronal proliferation**
  - Holoprosencephaly
  - Agenesis of corpus callosum
  - Pituitary maldevelopment
  - Dandy-Walker malformation
  - Rhombencephalosynapsis

Histogenesis

- Process of proliferation, migration, differentiation → development of mature cerebral cortex
- Unique difference between **cerebral hemisphere neuroepithelium** and that of other parts of neural tube
  - Made up of several layers (6 in dominant neocortex)
  - Inside-out arrangement of gray/white matter
    - Unlike rest of CNS, cerebral hemispheres have white matter inside and gray matter outside
    - Mechanism poorly understood
- Specialized neurogenesis in cerebellum → gray matter of cerebellar cortex/deep cerebellar nuclei
- **Errors in histogenesis**
  - Abnormal histogenesis of periaqueductal gray matter is 1 cause of aqueduct stenosis

Neuronal Migration

- Peak activity at 11-15 weeks
  - Majority of neurons in correct location by 24 weeks
  - Migration continues up to 35 weeks
- Newly proliferated cells migrate along predetermined pathways of radial glial fibers → organized cortical layering
  - Inside-out pattern of 6 layers with newest arrivals "outside" those that migrated earlier
  - Process governed by multiple genes, circulating factors
- **Errors in migration**
  - Microcephaly
  - Megalencephaly
  - Heterotopia
  - Cortical dysplasia
  - Lissencephaly: Arrested neuronal migration
  - Phakomatoses

Myelination

- Can be detected as early as 20 weeks
- Occurs in orderly, predictable manner
  - Caudal → cranial, deep → superficial, posterior → anterior
- Continues into adulthood

Operculization

- Development of insular cortex and infolding of sylvian fissures
- During weeks 11-28
- **Errors in operculization**
  - Defective speech and language processing

Gyral and Sulcal Development

- Occurs earlier in vivo than can be detected by current imaging modalities
  - 4- to 6-week time lag before structures become visible by imaging
- Gestational age by which sulcus/fissure should always be seen
  - Callosal: US 14 weeks, MR 22 weeks
  - Sylvian: US 18 weeks, MR 24 weeks
  - Parietooccipital: US 18 weeks, MR 22-23 weeks
  - Calcarine: US 18 weeks, MR 22-23 weeks
- Continues through end of 35th week

CEREBRUM, CEREBELLUM, AND VENTRICLES

Cerebral Hemispheres

- **Notochord development**
  - Bilaminar germ disc evolves into trilaminar germ disc with ectoderm, mesoderm, endoderm
  - Mesoderm forms midline, hollow, central tube: Notochordal process
  - Notochordal process evolves into solid notochord
  - Notochord + mesoderm induce formation of neural plate
  - Neural plate grows in length and width until day 21 when neurulation begins
- Formation of **neural tube** (primary neurulation)
  - Neural plate folds elevate, forming trough (neural groove) between them
  - Neural folds fuse → neural tube
    - Neural crest cells (derived from neuroectoderm) split from neural tube as it fuses
  - Neural tube temporarily open at both ends; openings are called neuropores
    - Rostral 2/3 of neural tube → brain
    - Caudal 1/3 of neural tube → spinal cord, nerves

- Bidirectional closure begins at occipitocervical level
  - Rostral/cranial neuropore closes at day 24
  - Caudal/sacral neuropore closes at day 25
- **Primary vesicles** form by middle of 4th week
  - **Prosencephalon** (forebrain)
  - **Mesencephalon** (midbrain)
  - **Rhombencephalon** (hindbrain)
- **Secondary vesicles** form during 5th week
  - Prosencephalon → anterior telencephalon + posterior diencephalon
    - **Diencephalon** → hypothalamus, thalamus, posterior pituitary, eyes
    - **Telencephalon** → cerebral hemispheres (by sagittal cleavage), basal ganglia
  - Rhombencephalon → anterior metencephalon + posterior myelencephalon
    - **Metencephalon** → pons + cerebellum
    - **Myelencephalon** → medulla oblongata
- Tube elongates at same times as vesicles form; flexures develop at specific locations
  - **Midbrain (mesencephalic) flexure**
  - **Cervical flexure** at junction of brainstem with spinal cord
  - **Pontine flexure** develops between midbrain and cervical flexures
- **Cerebral hemispheres** formed by 11th week
  - Arise as lateral outpouchings of telencephalon
  - Grow rapidly to cover diencephalon, mesencephalon
- Cerebral hemispheres connected by **lamina terminalis** ("zip" of cranial neuropore closure)
  - Thickening of lamina terminalis at rostral end → lamina reuniens + massa commissuralis
    - Lamina reuniens → anterior commissure
    - Massa commissuralis → corpus callosum, hippocampal commissure
    - Hippocampal commissure merges with splenium of corpus callosum
  - **Corpus callosum** should be complete by 20 weeks
    - Composed of 4 parts; from front to back, these are **rostrum, genu, body, splenium**

Cerebellum

- Thickening of alar plate of rhombencephalon → rhombic lips
- Rhombic lips → cerebellar hemispheres by intense neuronal proliferation
- Rhombic lips fuse → cerebellar commissures in roof of 4th ventricle
- **Cerebellar hemispheric fusion starts cranially**, forming flocculi of hemispheres and nodulus of vermis in 9th week
  - **Proliferation and fusion continue caudally** until complete by 15th week
- Flocculonodular fissure separates flocculi of hemispheres and nodulus of vermis

Ventricles

- Cavities within brain vesicles → ventricles during weeks 4-12
  - **Lateral ventricles** develop as diverticula from telencephalic primitive ventricle
  - **3rd ventricle** develops from cavity of diencephalon
  - **4th ventricle** develops from cavity of rhombencephalon
- Foramina of Monro connect lateral ventricle to 3rd ventricle
- **Aqueduct of Sylvius** connects 3rd-4th ventricles
  - Develops from cavity of mesencephalon
- Vessels from diencephalon and myelencephalon invade ventricular walls → **choroid plexus**
- **4th ventricle roof**
  - Highly complex area
  - Ridge of developing choroid plexus divides roof → anterior and posterior membranous areas
    - Anterior superior part incorporated into choroid plexus
    - Posterior inferior persists, midline cavitation → foramen of Magendie

IMAGING ISSUES

Protocol Advice

- US
  - Use highest resolution transducer possible
  - 3D volume acquisition allows manipulation of dataset to "create" true orthogonal image planes
  - Use color Doppler to evaluate course of marker vessels
- MR
  - Use fast sequences with single slice rather than volume acquisition
  - Diffusion tensor imaging can be performed for tractography
  - T2WI for global anatomy
  - T1WI for blood, fat, myelination

Imaging Pitfalls

- Normal structures mistaken for pathology
  - Yolk sac confused with cephalocele
  - Rhombencephalic vesicle confused with posterior fossa cyst
  - Corner of ventricular atrium mistaken for choroid plexus cyst
  - Fornices mistaken for cavum septi pellucidi
- Subtle lesions that may be missed if normal development not understood
  - Absent cavum septi pellucidi
  - Heterotopia
  - Lissencephaly
  - Cortical dysplasia
- Posterior fossa
  - Rotation of vermis may be mistaken for vermian dysgenesis
  - Rhombencephalosynapsis may be confused with cerebellar hypoplasia

1ST TRIMESTER EMBRYO

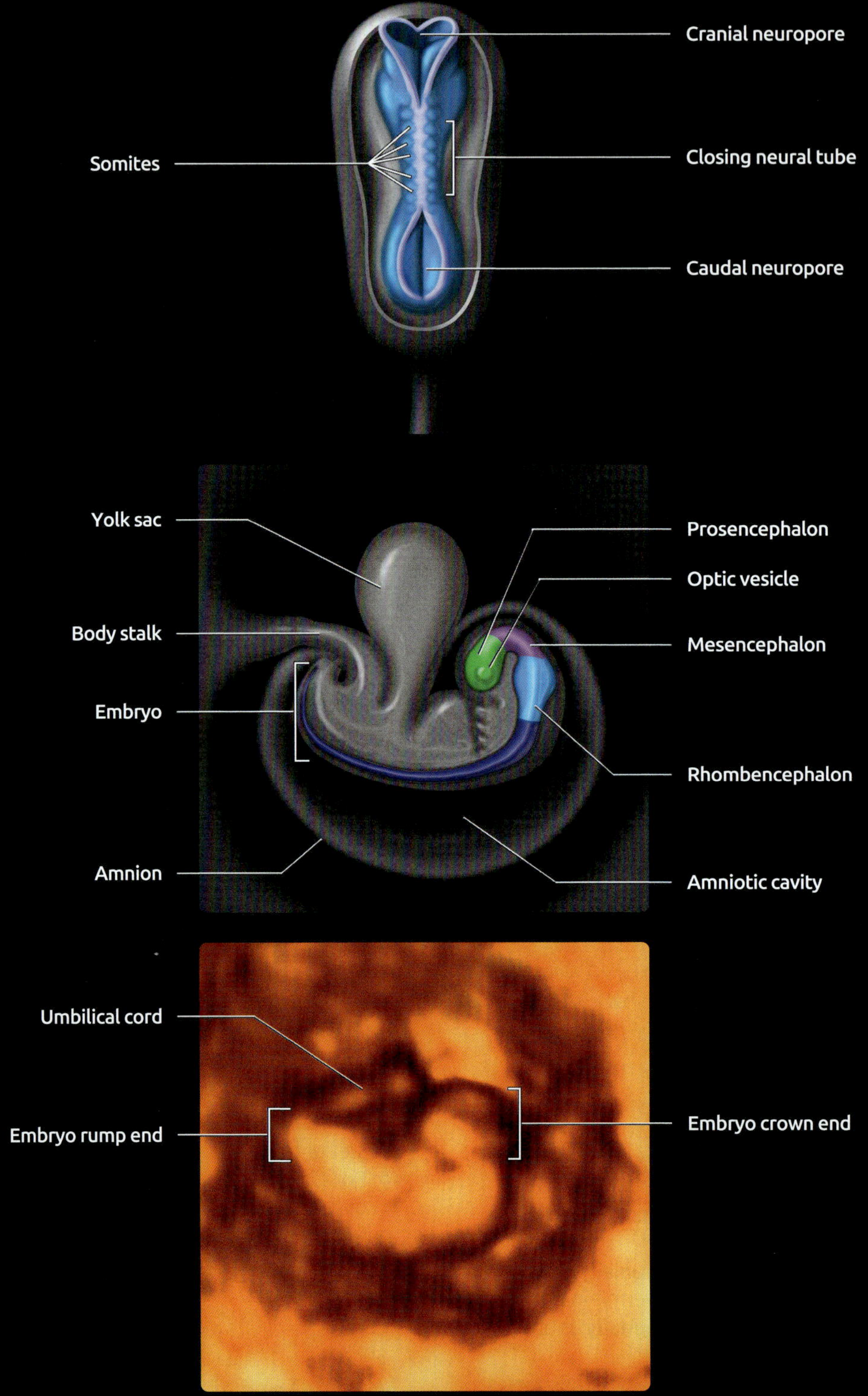

(Top) *The neural tube closes in a bidirectional manner. The cranial neuropore closes at day 24, while the caudal neuropore closes at day 25.* **(Middle)** *A series of vesicles develop at the same time the head end of the embryo enlarges and the flat embryonic disc becomes curved in profile and tubular in cross section. These are the precursors to the adult brain. The prosencephalon (green) gives rise to the forebrain, the mesencephalon (purple) to the midbrain, and the rhombencephalon (light blue) to the hindbrain.* **(Bottom)** *Surface-rendered 3D US of a 7-week embryo shows the external contour with a recognizable head end. The abdominal wall has closed, the yolk sac has detached, and the umbilical cord has formed. The torso is relatively small, the limb buds have not yet developed, but neural tube has already developed the precursors to the forebrain, midbrain, and hindbrain.*

1ST TRIMESTER EMBRYO

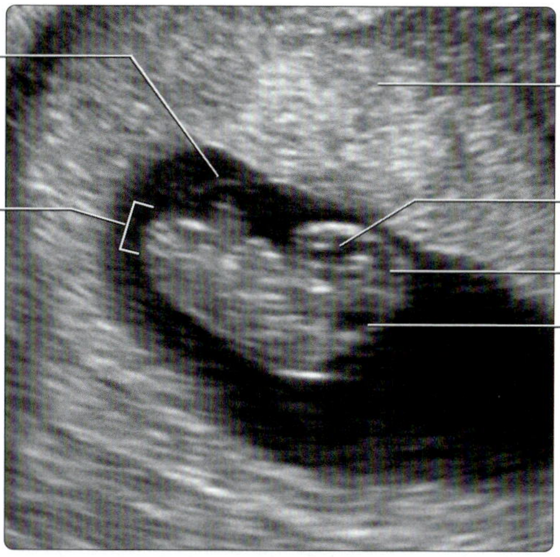

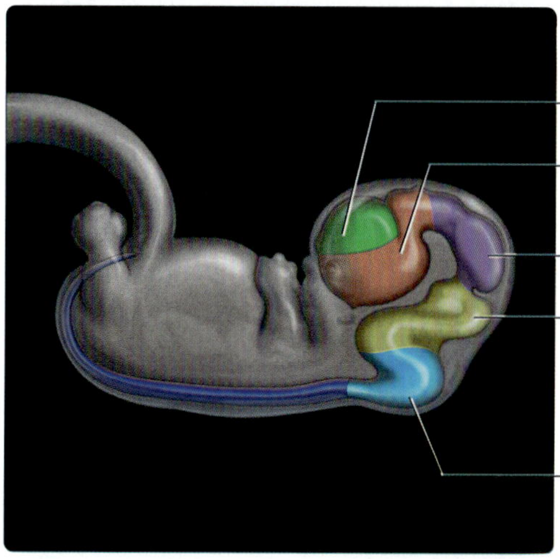

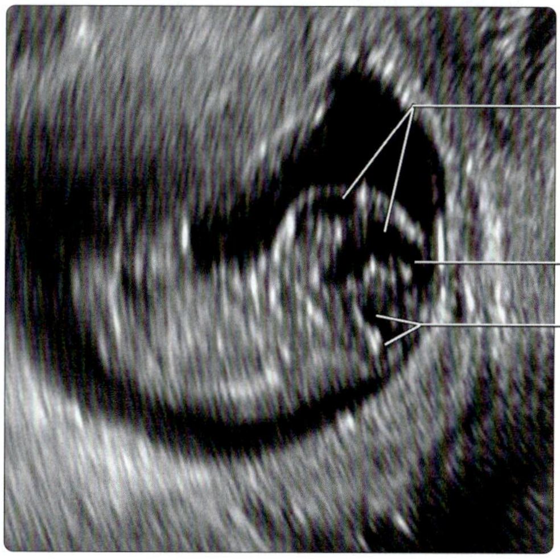

(Top) *This is a vaginal US of an 8-week, 2-day embryo in which the primitive brain vesicles are visible. This is a normal finding with high-resolution modern equipment and should not be mistaken for holoprosencephaly or other brain malformation.* **(Middle)** *With further embryonic growth, the prosencephalon gives rise to secondary vesicles known as the telencephalon and diencephalon. The mesencephalon elongates, while the rhombencephalon gives rise to the secondary vesicles, metencephalon, and myelencephalon. At this point, several flexures develop in the neural tube so that it adapts to the contour of the developing cranium.* **(Bottom)** *Another midline sagittal scan of a slightly older embryo, this time 9 weeks, 6 days, shows further neural tube growth with development of multiple flexures in the lengthening tube so that it adapts to the contour of the developing cranium. Again, it is important to recognize that this is normal for gestational age.*

1ST TRIMESTER FETUS

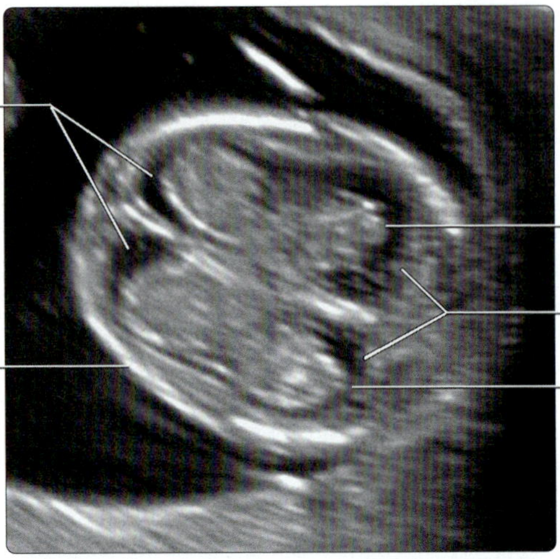

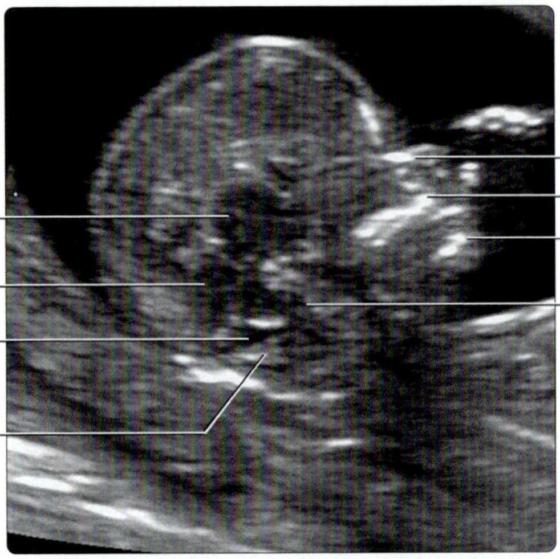

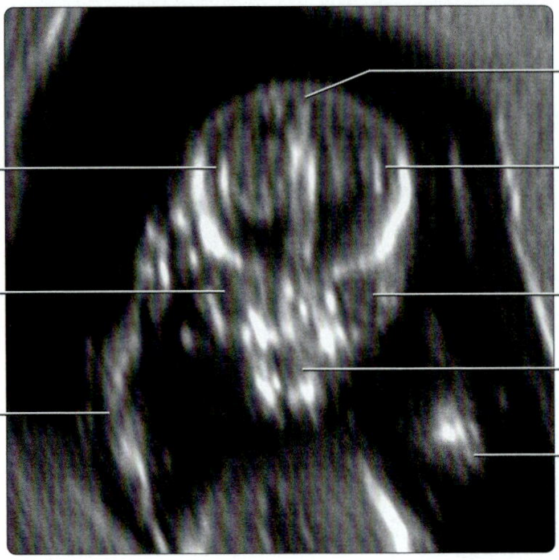

(Top) *At 12 weeks, the choroid plexus fills most of the ventricular cavity, and the brain parenchyma is thin and smooth. The echogenicity and shape of the choroid plexus on an axial US image gives rise to the butterfly sign in which the choroid forms the butterfly wings. This confirms 2 cerebral hemispheres.* **(Middle)** *Sagittal abdominal US at 13 weeks shows a normal thalamus, midbrain, and brainstem. The intracranial translucency (the future 4th ventricle) is seen between the brainstem and the choroid plexus of the 4th ventricle. Intracranial translucency assessment may be used for early detection of open neural tube defects.* **(Bottom)** *Transabdominal US at 13 weeks, 1 day shows clear separation of the cerebral hemispheres by the falx cerebri. The ocular globes are visible within the bony orbits, and the oral cavity is visible between the bright echoes of the maxilla and mandible.*

2ND TRIMESTER FETUS

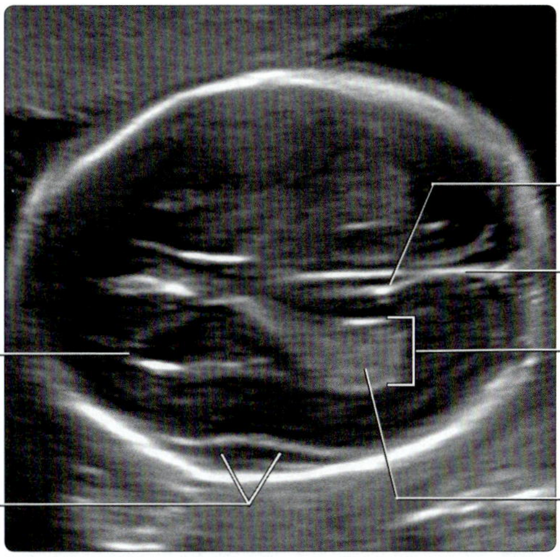

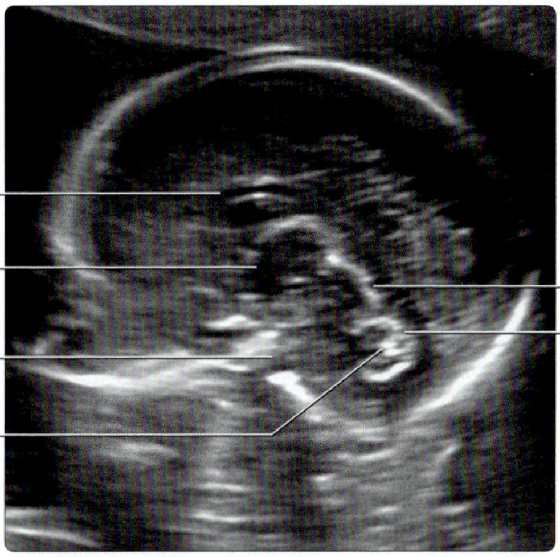

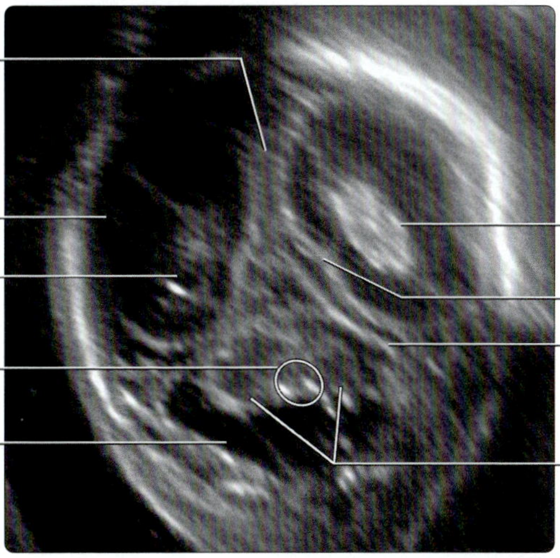

(Top) *At 18 weeks, the cerebral mantle is quite smooth with just the earliest "thumb print" indentation for the sylvian fissure and a tiny indentation at the site of the developing parietooccipital fissure. The latter marks the anatomic location at which the width of the lateral ventricle should be measured.* **(Middle)** *Sagittal transabdominal US at 20 weeks shows normal midline structures very well. Although not a standard image in a 2nd-trimester scan, this is as easy to obtain as a profile view of the face.* **(Bottom)** *Coronal image in a 22-week fetus shows another nonstandard, but easily obtainable, view of the fetal brain. Turning the transducer 90° from the standard axial images provides another way to assess symmetry of the hemispheres, ventricles, and the cortical mantle. The cisterna magna cannot be measured in this plane as it is artificially deepened by the extension into the foramen magnum.*

CAVUM SEPTI PELLUCIDI

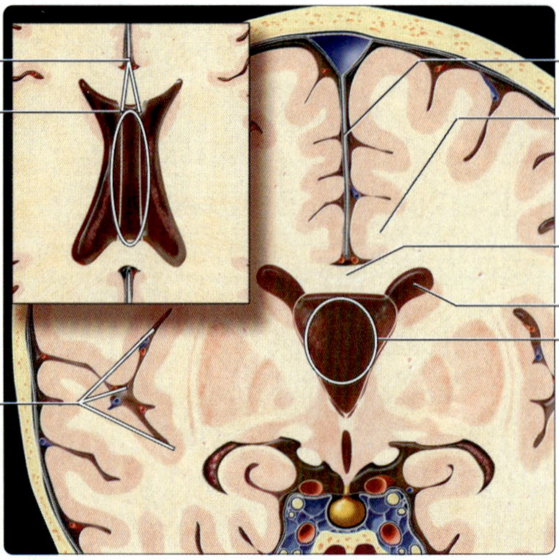

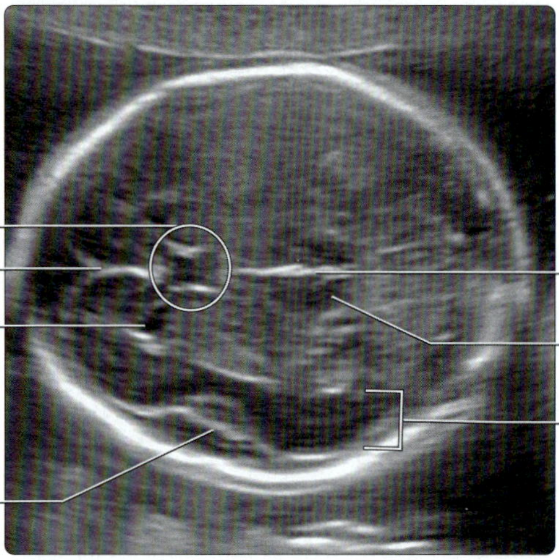

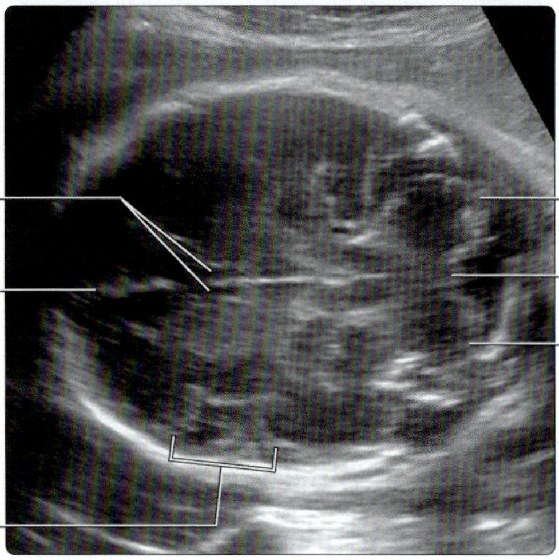

(Top) *Coronal graphic (with axial insert) illustrates the cavum septi pellucidi, which is a marker for normal midline brain development.* **(Middle)** *The cavum should be demonstrated on all routine obstetric scans between 18 and ~ 37 weeks. It is a box-shaped, anechoic space in the midline between the frontal horns of the lateral ventricles. It can also be seen on coronal views obtained transabdominally or transvaginally.* **(Bottom)** *A pitfall for the unwary is to confuse the normal fornices with the cavum. The fornices may be normal in a fetus with an absent cavum. They are seen just inferior (toward the skull base) to the normal location of the cavum. They form a series of parallel, black and white lines around the intact midline. The cavum is a box-shaped structure that interrupts the midline echo.*

SYLVIAN FISSURE

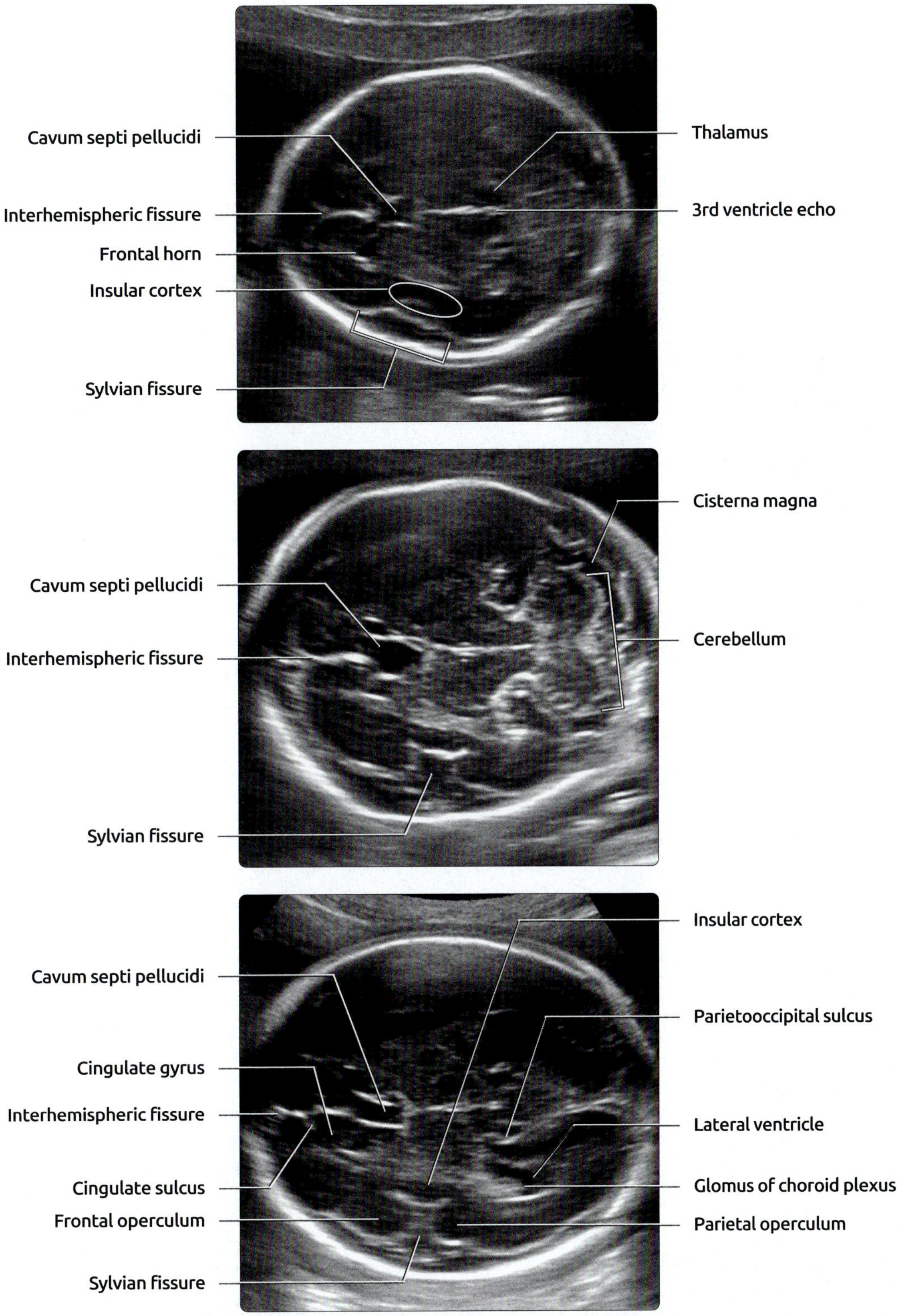

(Top) *The sylvian fissure is one of the easiest cortical indentations to see in the fetus. In this 19-week fetus, the sylvian fissure is seen as a shallow groove on the surface of the brain; it creates obtuse angles with the insular cortex. The cortical mantle is quite smooth at this gestational age. As the brain grows, the sylvian fissure deepens and the cortical mantle becomes convoluted.* **(Middle)** *As the brain grows and the surface becomes more convoluted, the sylvian fissure deepens and becomes square in profile, like an open box. This is a 25-week, 3-day gestation.* **(Bottom)** *In this fetus, at 28-weeks gestation, the process of operculization results in the sylvian fissure "closing." The floor of the box is the insular cortex, and the lid of the box is composed of the frontal and temporoparietal opercula. The sylvian fissure separates the parietal lobe superiorly from the temporal lobe inferiorly. The surface of the brain is becoming more convoluted with visible parietooccipital sulcus, cingulate sulcus, and shallow surface gyri.*

TELENCEPHALON: CEREBRAL HEMISPHERES

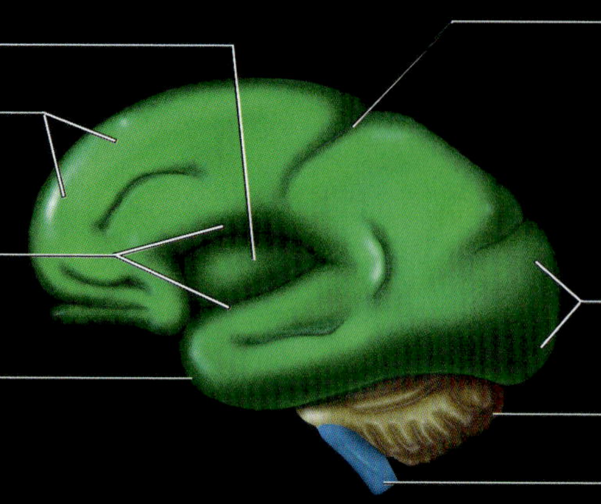

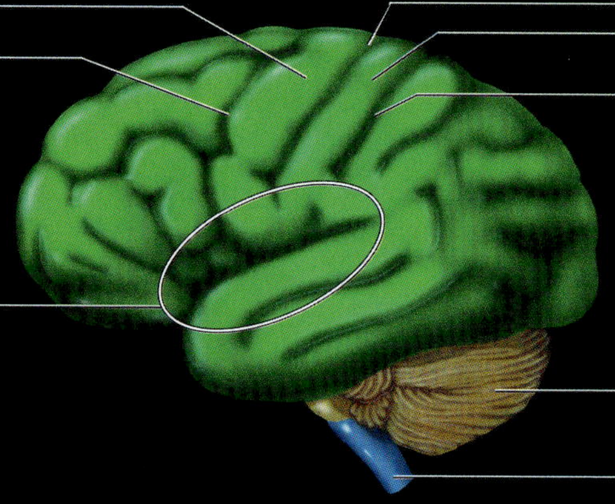

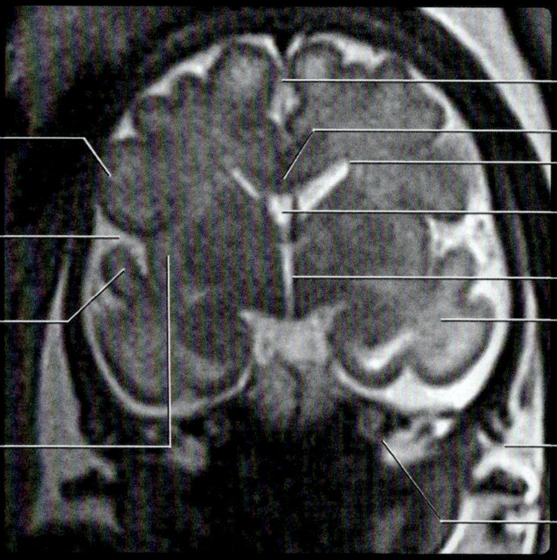

(Top) *With advancing gestational age, the forebrain enlarges dramatically in comparison to the mid- and hindbrain. The telencephalon and diencephalon arise from the prosencephalon; between them, they give rise to most of the supratentorial brain. This graphic illustrates the relative proportions of the brain arising from the prosencephalon (green), metencephalon (yellow), and myelencephalon (light blue). The mesencephalic, midbrain structures are not visible.* **(Middle)** *With advancing gestational age, multiple secondary and tertiary gyri develop and the number and complexity of the cerebellar folia increases.* **(Bottom)** *Coronal T2WI in the 3rd trimester shows the appearance of a mature brain with well-developed surface sulci and gyri and operculized sylvian fissures.*

CORPUS CALLOSUM, CINGULATE GYRUS

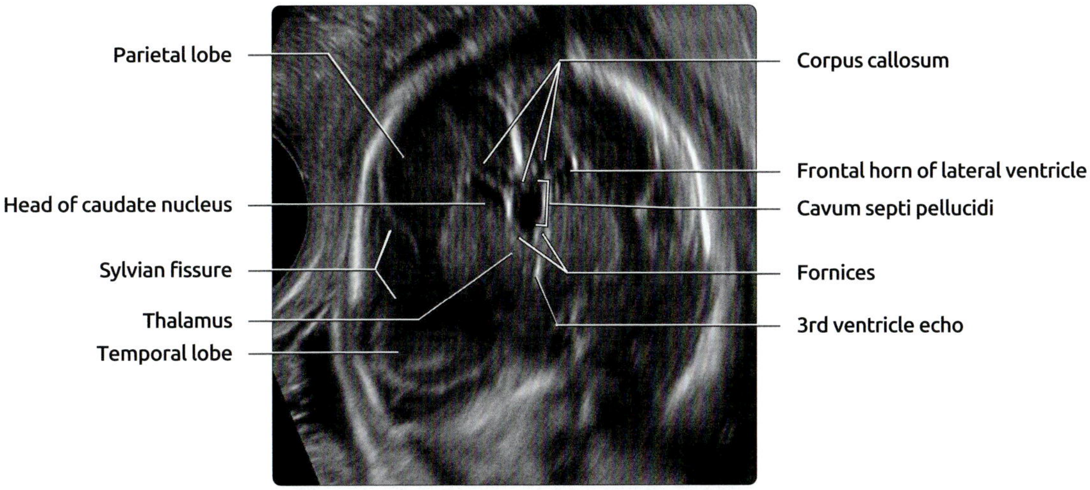

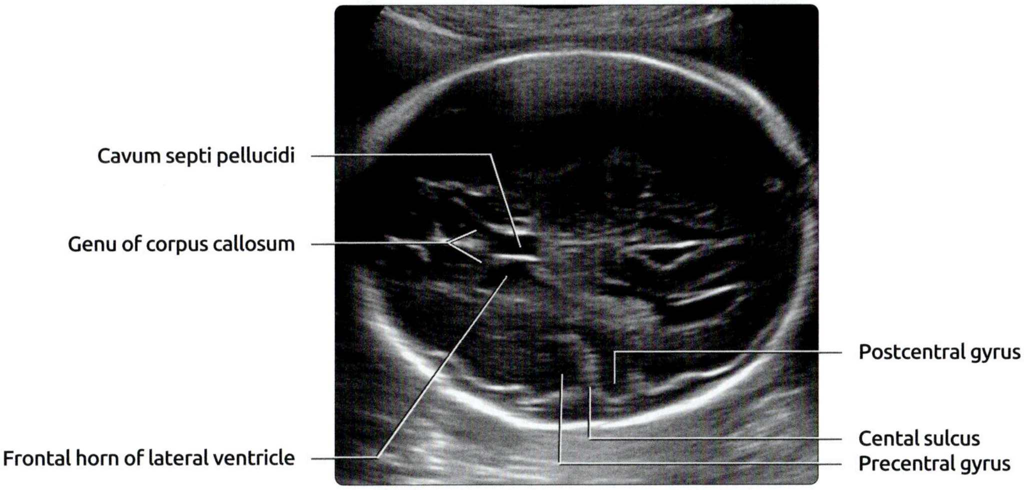

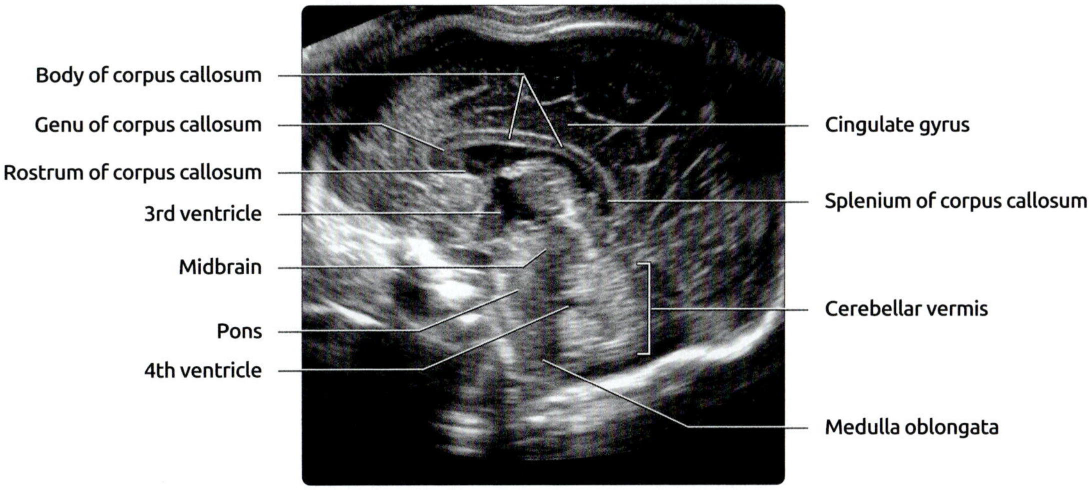

(Top) *Vaginal US allows for exquisite anatomical assessment of the fetal brain. At 21 weeks, the corpus callosum is visible, forming the roof of the cavum septi pellucidi, which sits between the frontal horns of the lateral ventricles, above the 3rd ventricle. The cingulate gyrus has not yet developed.* **(Middle)** *Even though the corpus callosum is best seen on direct sagittal view, the genu can be identified on the standard axial image obtained at the level of the cavum septi pellucidi. The genu of the corpus callosum forms the curved part of the "anchor complex" in the normal anterior brain. The stem of the anchor is created by the interhemispheric fissure. This fetus is at 28-weeks of gestation.* **(Bottom)** *If the fetus is in cephalic presentation, transvaginal US in the 3rd trimester produces exquisite images of normal brain anatomy. In this image, the cingulate gyrus is seen running parallel to the body of the corpus callosum, which is seen in its entirety.*

CORPUS CALLOSUM, CINGULATE GYRUS

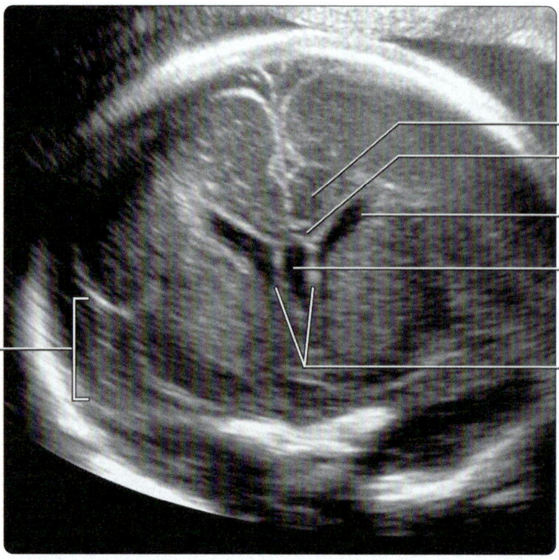

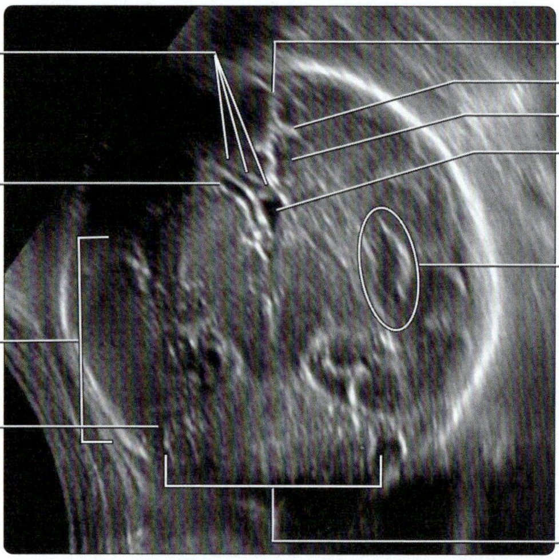

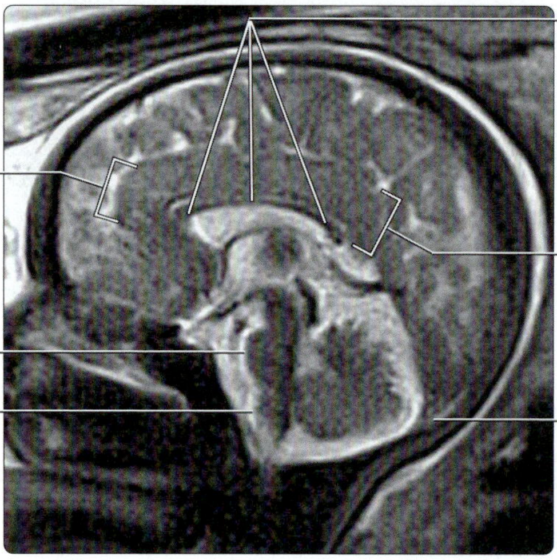

(Top) *A complete neuroanatomic survey requires multiplanar imaging. In this coronal image at 24-weeks gestation, the brain surface is still relatively smooth, but the corpus callosum and cingulate gyrus are well seen.* **(Middle)** *Later in the 3rd trimester (this is a 31-week fetus), the ventricles appear relatively small compared to scans at 20 and 24 weeks. The corpus callosum is thicker and easier to see. The cingulate gyrus and sulcus are well developed. Note that operculization of the sylvian fissure is also complete in this fetus.* **(Bottom)** *MR was performed to evaluate focal dilation of the occipital horn and confirmed occipital lobe cortical dysplasia. This midline sagittal MR shows the normal curved corpus callosum with the cingulate gyrus wrapped around it. In agenesis of the corpus callosum, the cingulate gyrus does not form, and the medial surface gyri are oriented in a sunburst pattern called stenogyria.*

PARIETOOCCIPITAL SULCUS

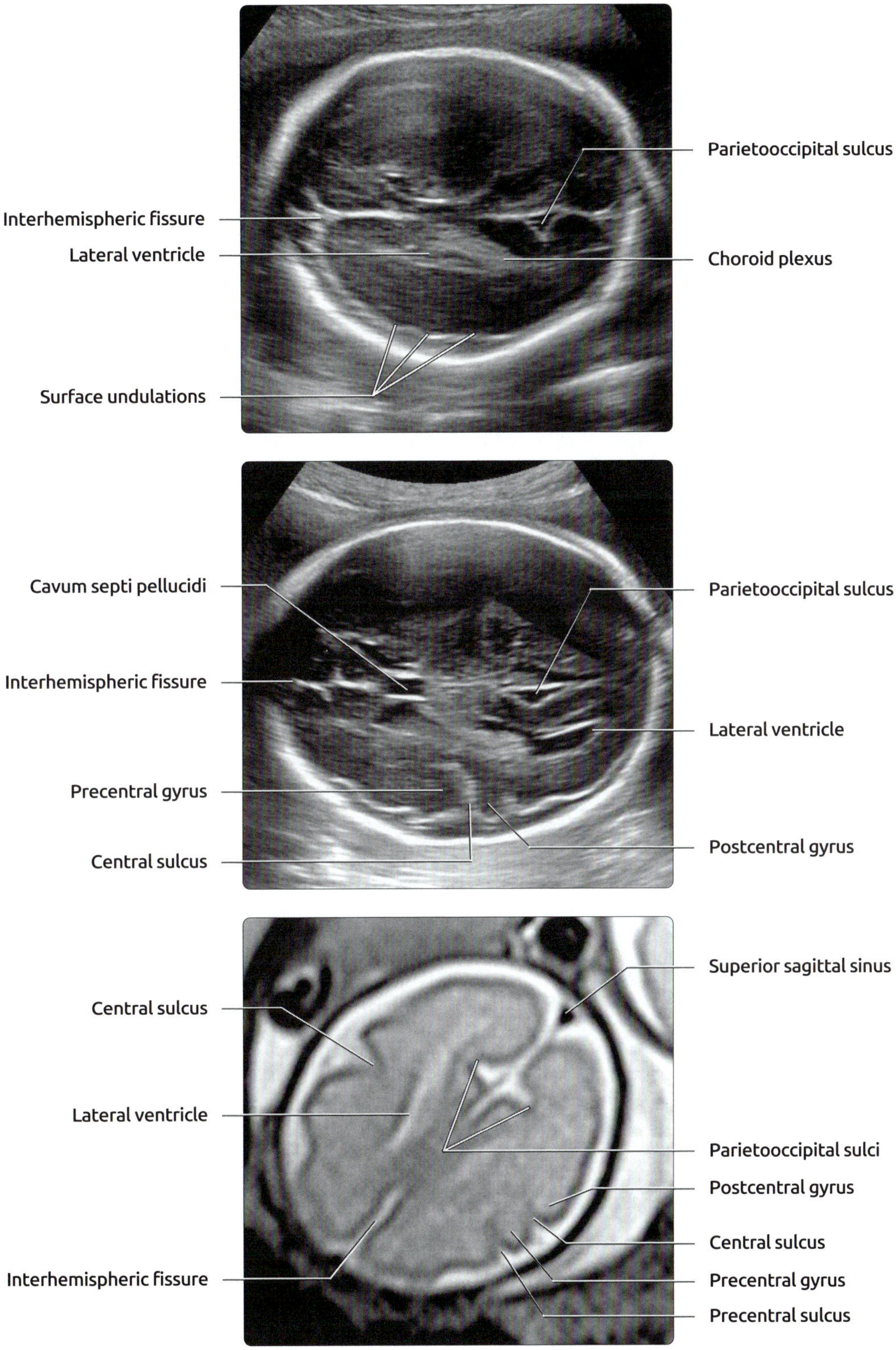

(Top) *High axial US at 25-weeks, 3-days gestational age shows a well-developed parietooccipital sulcus. The surface of the brain is starting to develop some undulations as the convexity sulci begin to form.* **(Middle)** *By 28 weeks, the parietooccipital sulcus is well established, the surface of the brain has more pronounced undulations, and there are now some named convexity sulci and gyri visible.* **(Bottom)** *At 30 weeks, the parietooccipital sulcus is clearly visible on MR. The convexity sulci are more established with clear visibility of the central sulcus and adjacent gyri.*

CALCARINE SULCUS

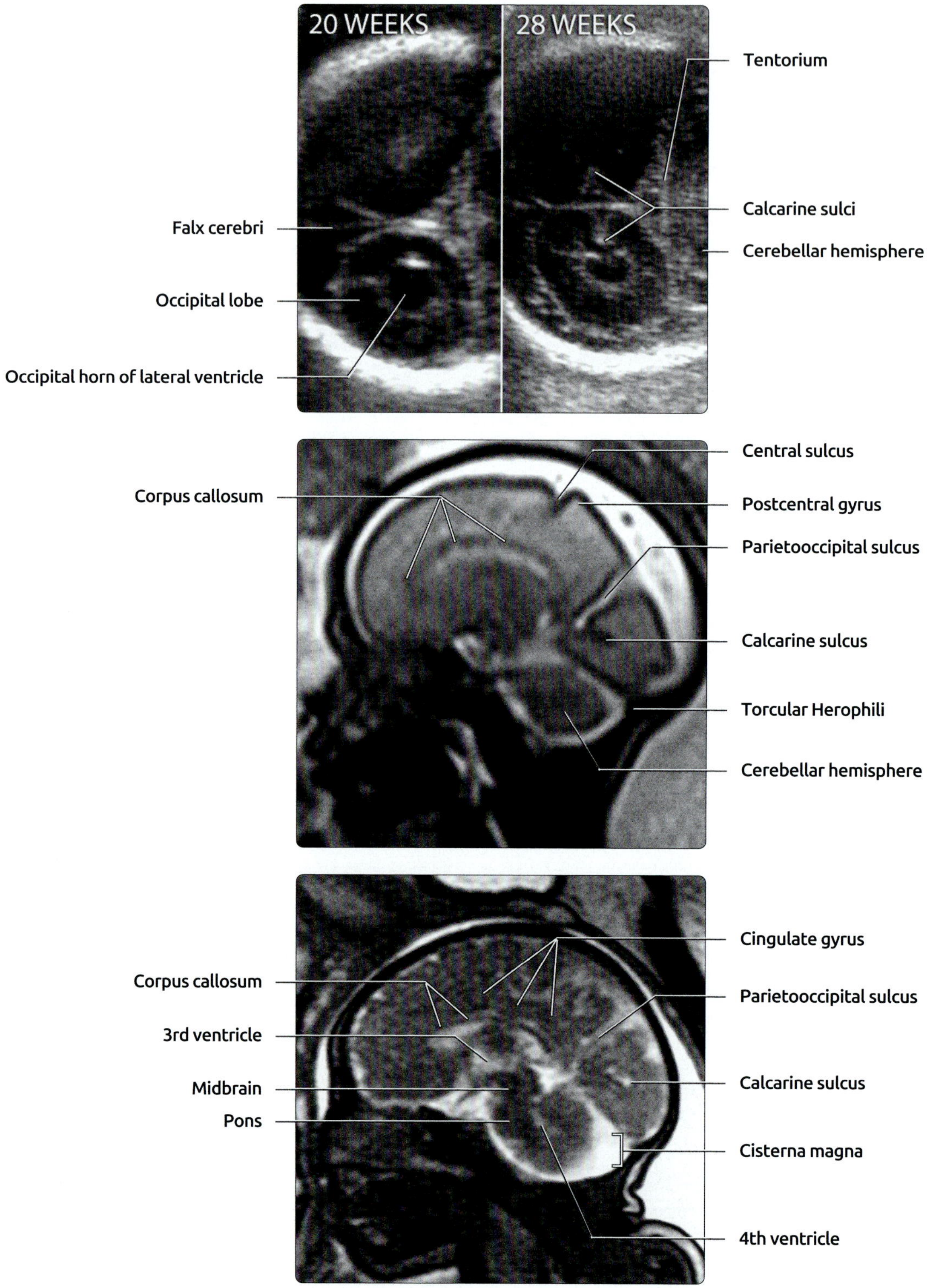

(Top) *The calcarine sulcus develops on the medial surface of the occipital lobe branching from the parietooccipital sulcus. It is best seen on the coronal plane since the US beam is then perpendicular to the plane of the sulcus. In this composite image, note how smooth the medial occipital cortex is at 20 weeks. By 28 weeks, the calcarine sulcus is easily visible in the same fetus.* **(Middle)** *Sagittal MR nicely shows the orientation of the calcarine sulcus as a branch of the parietooccipital sulcus.* **(Bottom)** *At 37 weeks, the convexity sulci are well established. Note the relative decrease in the cerebrospinal fluid volume over the surface of the brain. This is normal, as is the relative decrease in size of the ventricular system compared to the size of the brain.*

CONVEXITY SULCI AND GYRI

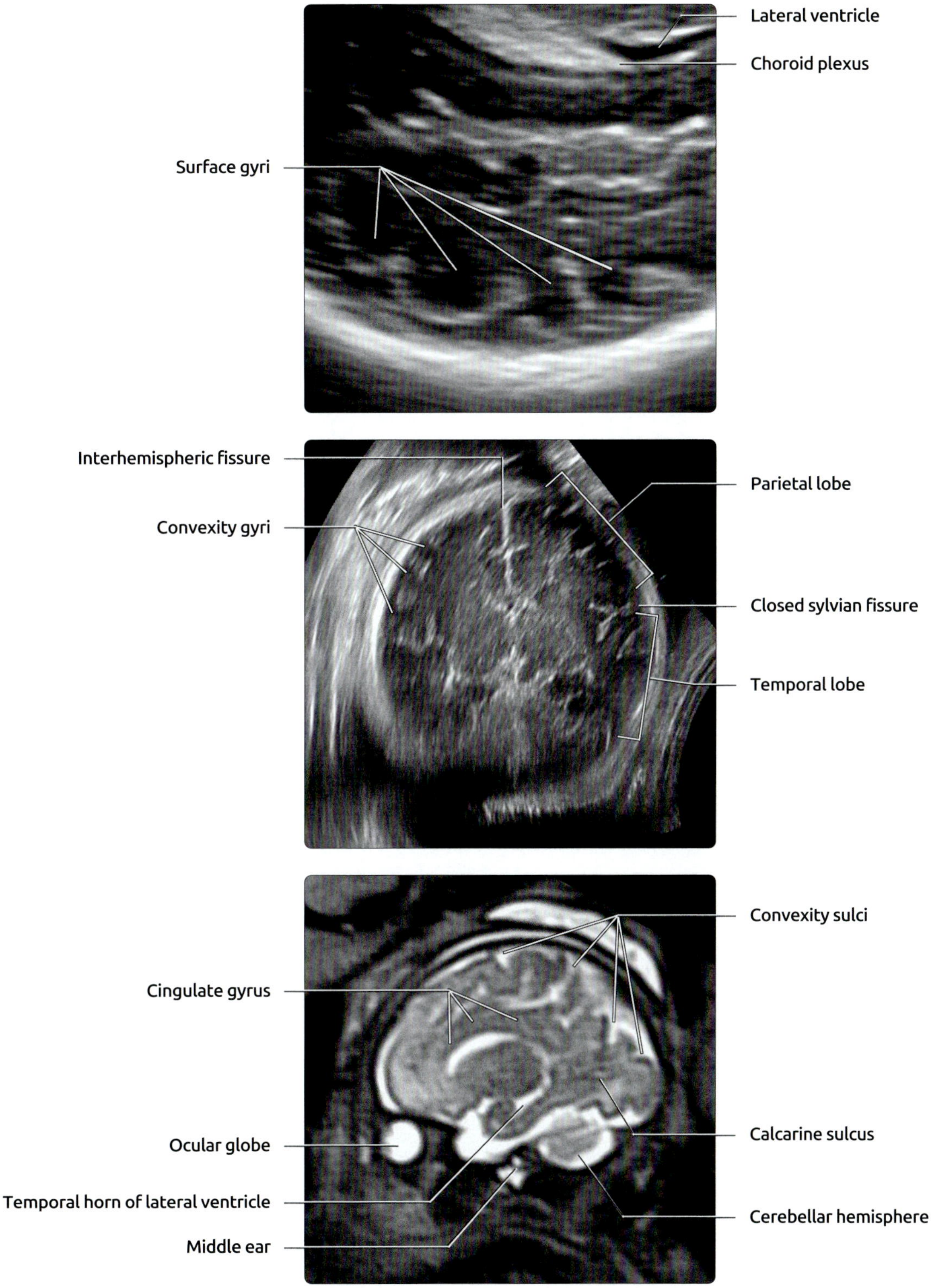

(Top) *"Zoomed" view of an axial US in the 3rd trimester shows numerous convexity sulci and gyri in the far field. There is usually loss of detail in the near field due to reverberation of the beam at the ossified skull vault. It is important to examine the brain from multiple directions as well as in multiple planes if there is any concern for structural malformation.* **(Middle)** *Coronal US in the 3rd trimester in a fetus with osteogenesis imperfecta type III shows how an underossified skull vault allows for beautiful images of the brain. This is described as the brain too-well-seen sign.* **(Bottom)** *Parasagittal MR late in the 3rd trimester shows well-developed sulci and gyri over the convexity of the brain.*

POSTERIOR FOSSA

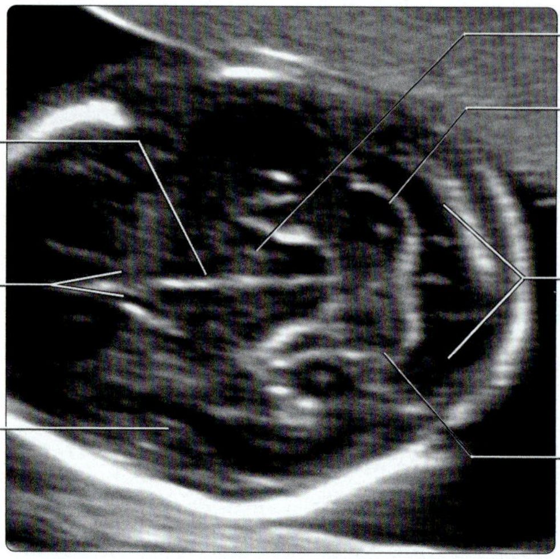

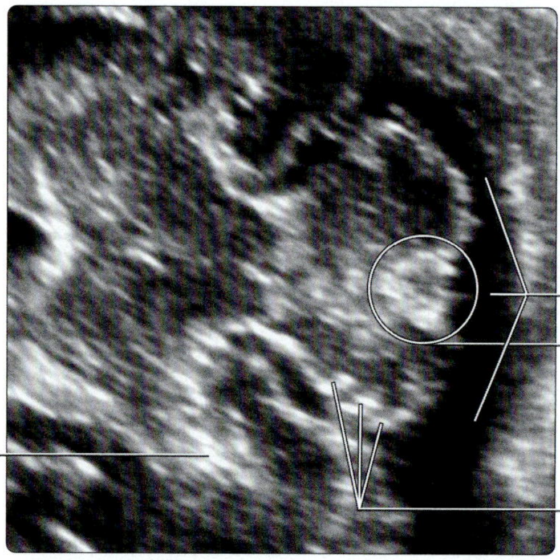

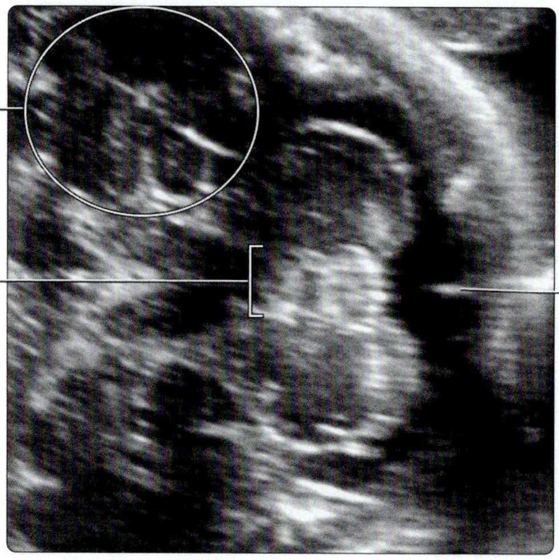

(Top) *The posterior fossa structures are evaluated on an axial oblique view. For measurement of the nuchal fold and cisterna magna depth, the cavum septi pellucidi is used as a landmark to confirm the appropriate obliquity. This image is slightly underangled, so it includes the fornices. At 18 weeks, the hemispheres are round with relatively simple architecture. The vermis is visible between the hemispheres but does not differ markedly in echogenicity from them.* **(Middle)** *With advancing gestational age, the echogenicity of the vermis increases in relation to that of the hemispheres. The cerebellar folia become visible as bright, echogenic lines around the margin of the hemispheres.* **(Bottom)** *In the late 3rd trimester, the vermis is plainly seen. Note the increasing complexity of the temporal lobe gyri. The cisterna magna is stable in size throughout gestation. It should always measure < 10 mm from the posterior surface of the vermis to the inner table of the occipital bone.*

POSTERIOR FOSSA

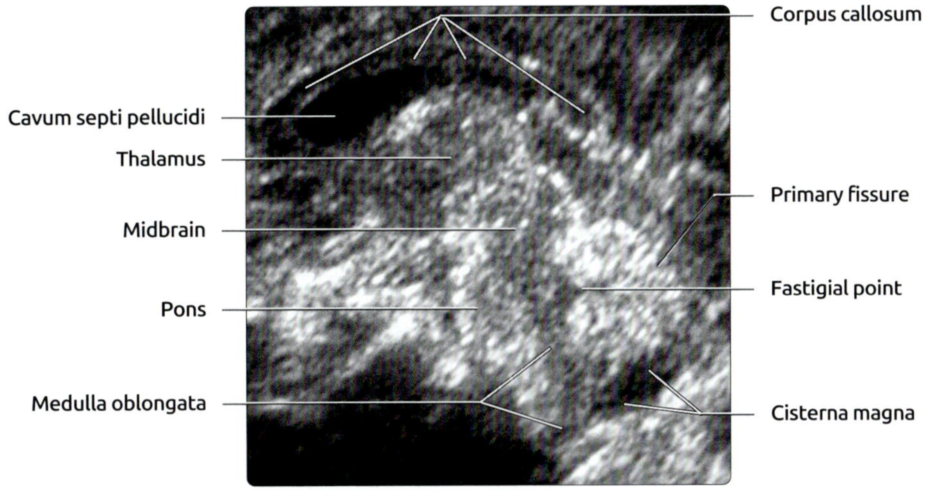

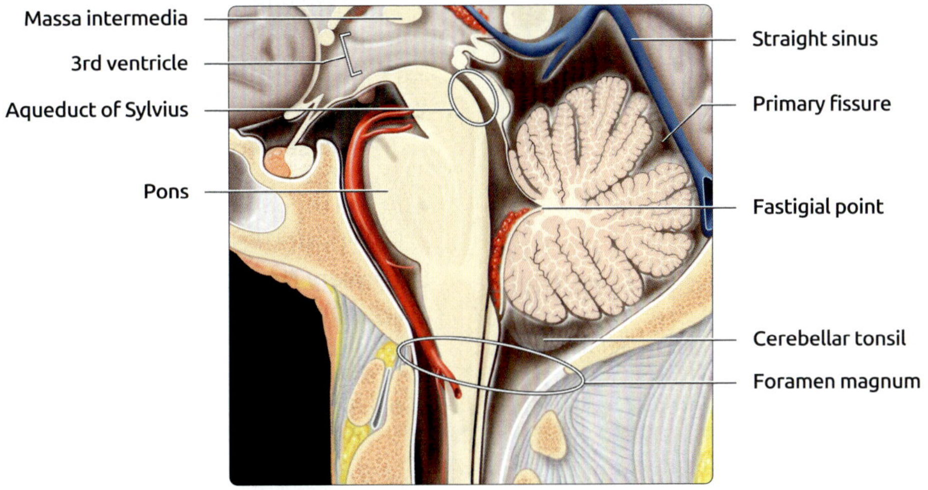

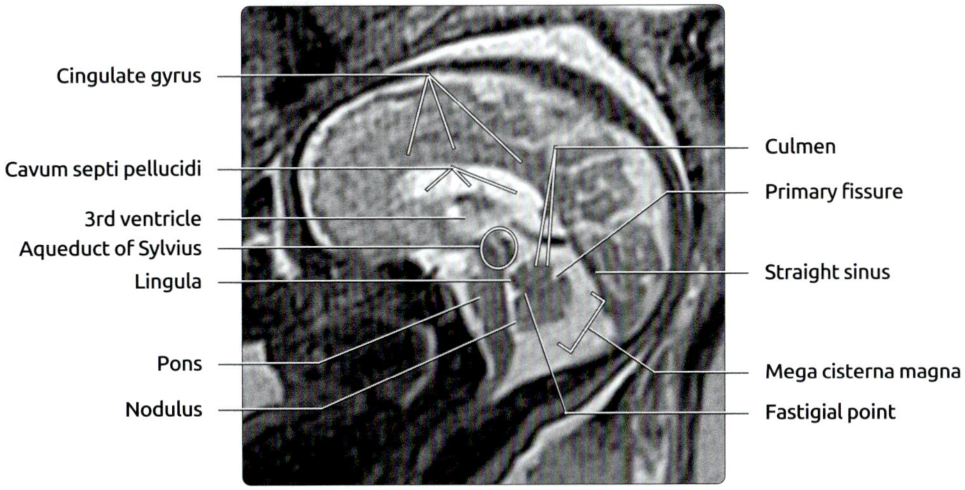

(Top) *This is a transabdominal scan of a 3rd-trimester fetus in breech presentation. Use of the metopic suture allows acquisition of a very nice sagittal image with superb detail of the posterior fossa structures.* **(Middle)** *This sagittal graphic illustrates normal anatomy and landmarks in the cerebellar vermis. The primary fissure divides the vermis into an anterior lobe (lingula, central, and culmen lobules) and a posterior lobe (declive, folium, tuber, pyramis, and uvula). The nodulus is referred to as the flocculonodular lobe.* **(Bottom)** *MR in a fetus with Gorlin syndrome (done at 37 weeks to exclude an oral cavity mass) shows normal vermian anatomy and location with a mega cisterna magna. Note the complexity of the convexity sulci as well as those on the medial surface of the brain at this gestational age.*

SPINAL CORD DEVELOPMENT

Early Embryologic Events

- 3rd week: Bilaminar germ disc evolves into trilaminar germ disc
- **Trilaminar germ disc**
  - **Ectoderm**: Part of amniotic cavity
  - **Mesoderm: Forms midline hollow central tube (notochordal process)**
    - Extends along long axis of embryonic disc
  - **Endoderm**: Part of yolk sac cavity
- Day 18: Notochord and remainder of intraembryonic mesoderm induce development of neural plate
  - Neural plate grows in length and width until day 21, when neurulation begins
  - Neural plate gives rise to most of central nervous system
- Day 21: **Hollow tube (notochordal process) evolves into solid cord (notochord)**

Neurulation

- **Primary**: Formation of cephalic spine to level of conus
- **Secondary**: Formation of spine caudal to level of conus

Primary Neurulation

- Occurs between days 18-28
  - **Formation of neural tube**
    - **Neural plate** folds and elevates, forming trough **(neural groove)**
    - Fusion of resultant neural folds
    - Before complete fusion, neuroectoderm cells give rise to neural crest cells
    - **Neural crest cells** will later migrate to various parts of body and contribute to variety of tissues
      - Autonomic nervous system, adrenal medulla, tissues of head and neck
  - Neural tube open at both ends temporarily
    - Communicates freely with amniotic fluid
    - Cranial and caudal openings of neural tube are called **neuropores**
- Simultaneously, somites paramedian to notochord differentiate to form sclerotome cells
  - Precursors to vertebral column
- Day 22-23 (4 weeks): **Neural tube closure** begins at occipitocervical level
  - Closure extends bidirectionally
  - **Neural canal**: Hollow center of neural tube later becomes
    - Ventricular system of brain
    - Central canal of spinal cord
- Day 24: Complete **cranial neuropore** closure is achieved
  - Cranial end of neural tube will become brain
- Day 26: **Caudal neuropore** closure
  - Caudal end of neural tube will become spinal cord
- Closed neural tube is required for normal development of neural arch

Dysjunction

- Final phase of primary neurulation
- Neural tube separates from overlying ectoderm
- **Premature dysjunction**
  - Perineural mesenchyme can access neural groove and ependymal lining → differentiates into fat
  - Prevents complete neural tube closure
  - May result in lipomatous malformation spectrum
    - **Intradural lipoma**
    - **Lipomyelomeningocele**: Subcutaneous fatty mass contiguous with neural placode/lipoma through posterior dysraphism
      - Lipomyelomeningocele accounts for 20-56% of occult spinal dysraphism
    - Tethered cord may result
      - Lipomatous malformation prevents normal ascent of cord as vertebral column elongates
- **Nondysjunction**
  - Failure of dysjunction to occur
  - Ectodermal-neuroectodermal tract forms → prevents mesenchymal migration
  - Results in focal or widespread spinal dysraphism and open neural tube defects
    - **Myelomeningocele**: Open neural tube defect with meningocele and neural elements
    - **Myelocystocele**: Enlarged central spinal cord canal (inner cyst) protrudes through osseous dysraphic defect into dilated subarachnoid space (outer cyst)
    - **Dorsal dermal sinus**: Midline/paramedian stratified squamous epithelial-lined sinus tract
- At end of primary neurulation, spine is formed from cephalic end of embryo through level of conus

Secondary Neurulation

- Formation of spine caudal to level of conus
- Days 28-48: Caudal neural tube forms via process referred to as secondary neurulation or canalization
  - Below or distal to posterior neuropore, undifferentiated cells form **primitive streak or caudal cell mass**
    - Once caudal neuropore closes, neural tissue is laid down as neural cord
  - Rostral neural tube extends into caudal eminence
  - **Caudal cell mass** forms vacuoles that fuse to form distal neural tube
    - Day 48: Transient ventriculus terminalis appears in future conus
    - These cells eventually form **conus medullaris, cauda equina, and filum terminale**
- Terminal cord undergoes retrogressive differentiation
  - Occurs over ensuing gestational period and into early postnatal period
- **Secondary neurulation is less precise and leads to wide range of malformations**
  - **Tethered cord**
    - Most common lesion in caudal cell mass dysplasia spectrum
    - Low-lying cord with thickened filum terminale
  - **Caudal regression**
    - **Type 1**: Foreshortened terminal vertebral column with high conus termination; severe associated anomalies
    - **Type 2**: Low-lying tethered cord with milder associated anomalies
    - Associated anomalies: Renal hypoplasia, pulmonary hypoplasia, anorectal malformations

- Associated spinal anomalies: Open dysraphism, segmentation and fusion anomalies, split cord malformations
  - **Terminal myelocystocele**
    - Hydromyelic cord terminating in skin-covered myelocystocele
    - Anorectal and visceral anomalies result in high morbidity
  - **Anterior sacral meningocele**
    - Large meningocele traversing enlarged sacral foramen produces presacral cystic mass
  - **Sacrococcygeal teratoma**
    - Primitive streak incompletely regresses and leaves caudal remnant
    - Occurs due to residual totipotent cell rests (Hensen node) → 3 cell layers with varying proportions of mature and immature elements
- **Spinal cord ascent**
  - At 12-weeks gestational age, spinal cord extends entire length of developing spinal column
  - **Vertebral column and dura** elongate disproportionately compared to spinal cord
    - **Conus medullaris ascends** relative to vertebrae; filum terminale elongates
  - **Nerve roots** exit at levels of their respective foramen
  - Nerve roots then grow longer to accommodate this relative ascension of spinal cord
    - Forms **cauda equina** (sheath of nerve roots inferior to conus)
- Conus should be at or above L3-L4 after 18-weeks gestation
  - This process extends into postnatal period
    - By 2 months of age, conus is located at adult level
      - Final position is near L1-L2 interspace
    - Conus below L2 after 1st month of life, in term infant, is probably abnormal
      - Needs evaluation for tethered cord

VERTEBRAL BODY DEVELOPMENT

Cartilaginous Stage

- Week 4: **Notochord** induces surrounding paraxial mesoderm derived from primitive streak
  - Forms paired somite blocks: Myotomes and sclerotomes
  - **Myotomes** form paraspinous muscles and skin covering
  - **Sclerotomes** divide into medial and lateral formations
    - **Form vertebral body, intervertebral disc, meninges, spinal ligaments (medial), and posterior spinal elements (lateral)**
    - Migrate from somites and surround adjacent neural tube and notochord
    - Ventral portion of sclerotome surrounds notochord and forms rudiment of vertebral body
    - Dorsal portion of sclerotome surrounds neural tube, forms precursors to neural arch, and condenses to produce spinous processes
    - Notochord will degenerate and involute where it is surrounded by vertebral body
    - **Notochordal remnant** between vertebra expands to form **nucleus pulposus** of intervertebral discs
  - **Failure of notochord induction** leads to incomplete division of neural plate → **neurenteric cyst or diastematomyelia**
- Week 6: Cartilaginous stage of vertebral development occurs when chondrification centers appear
  - **Chondrification centers** appear in each mesenchymal vertebral body
  - 2 centers in each mesenchymal vertebral body fuse at end of embryonic period
    - Forms cartilaginous **centrum**
  - At same time, centers of **vertebral arches fuse** with each other and with centrum of vertebral body
  - Spinous and transverse processes develop from extensions of chondrification centers of vertebral arch

Ossification Stage

- Vertebral ossification begins during embryonic period and ends by age 25
- Vertebral body centrum forms from fusion of ventral and dorsal primary ossification center
- At end of embryonic period, there are **3 primary ossification centers** for each vertebra, including
  - **Centrum**
  - **Each 1/2 of vertebral arch**
- Week 8: Ossification is visible
  - **Ossification begins in lower thoracic and upper lumbar regions**
  - Ossification progresses both cranial and caudal
- Week 13: 3 ossification centers present in vertebrae C1-L3
- At birth, each vertebra consists of 3 bony parts connected by cartilage

Anomalies of Vertebral Formation and Segmentation

- Abnormal vertebrae may replace normal vertebrae or be supernumerary
- **Failure of vertebral formation** (total or partial)
  - Degree and location of vertebral formation failure predicts morphology
    - Unilateral chondral center defect and failure of ossification → **hemivertebrae**
- **Failure of vertebral segmentation**
  - **Block vertebrae** with posterior elements fusion
- More severe segmentation and fusion defect → increased incidence of concurrent malformations
  - Neuraxis anomalies: Tethered cord, abnormal alignment (kyphosis, scoliosis), dysraphism
  - Visceral organ anomalies

FAILURE OF NEURAL TUBE CLOSURE

Clinical Implications

- Failure of any part of neural tube to close disrupts development of nervous system and disrupts induction of overlying vertebral arches
- Timing of event is such that other systems are also impacted
  - Look for associated visceral and anorectal malformations
- Look for consequences of abnormal innervation
  - Abnormal lower extremity positioning

NEURAL TUBE EMBRYOLOGY

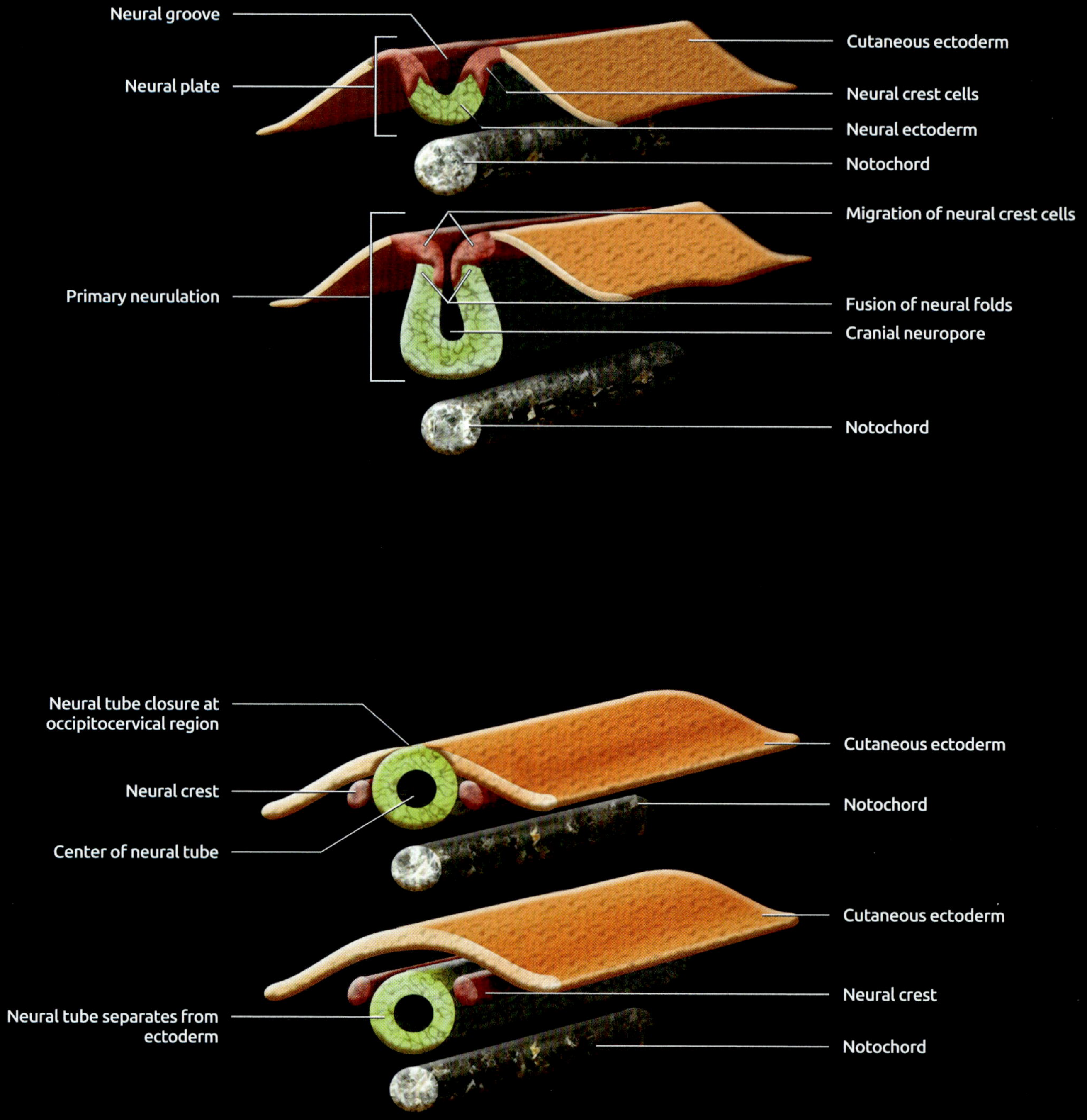

(Top) *On day 18, the notochord and intraembryonic mesoderm induce development of the neural plate. The neural plate will grow in length and width until day 21, when primary neurulation begins. The neural plate folds and resultant neural folds fuse.* **(Bottom)** *Neural tube closure begins at 4 weeks at the occipitocervical region. The hollow center of the neural tube will become the central canal of the spinal cord and ventricular system of the brain. During primary neurulation, the neural tube separates from the overlying ectoderm in a process called dysjunction. Early dysjunction results in perineural mesenchyme access to the neural groove, which differentiates into fat (intradural lipoma); it may also prevent closure of the neural tube (lipomyelomeningocele). If dysjunction fails to occur, a spectrum of open neural defects results.*

NEURAL TUBE EMBRYOLOGY

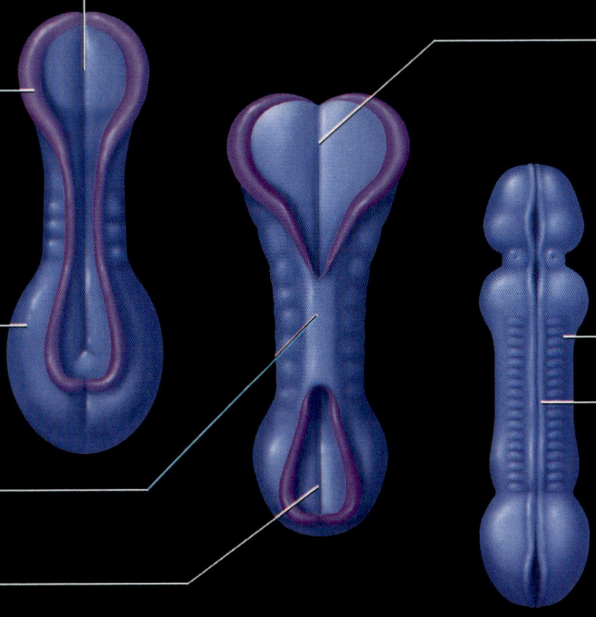

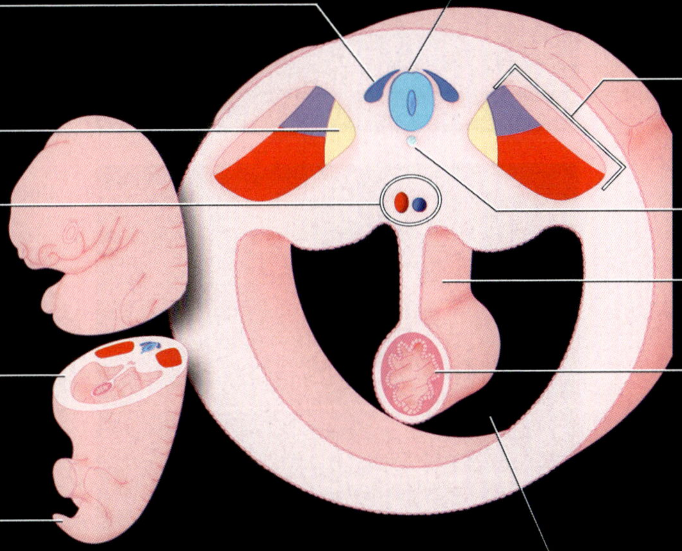

(Top) *Graphic of neural tube closure as viewed from above is shown. At either end, there is an opening, the cranial and caudal neuropore, which are open to the amniotic fluid at this stage. The cranial neuropore closes by day 24 and will become the brain. The caudal end closes by day 26.* **(Bottom)** *Cross section of an embryo shows the neural tube, which has formed dorsal to the notochord. The neural tube will form the spinal cord, while the notochord largely degenerates with remnants contributing to the intervertebral discs. Neural crest cells migrate throughout the body and give rise to diverse tissues, including ganglia of the autonomic nervous system, adrenal medulla, and tissues of the head and neck. Somites form from the mesoderm and form multiple tissues. The medial somite is the sclerotome, which will form the vertebral column. Secondary neurulation begins at he caudal eminence and forms the conus medullaris, cauda equina, and filum terminale of the spinal cord.*

FETAL SPINE

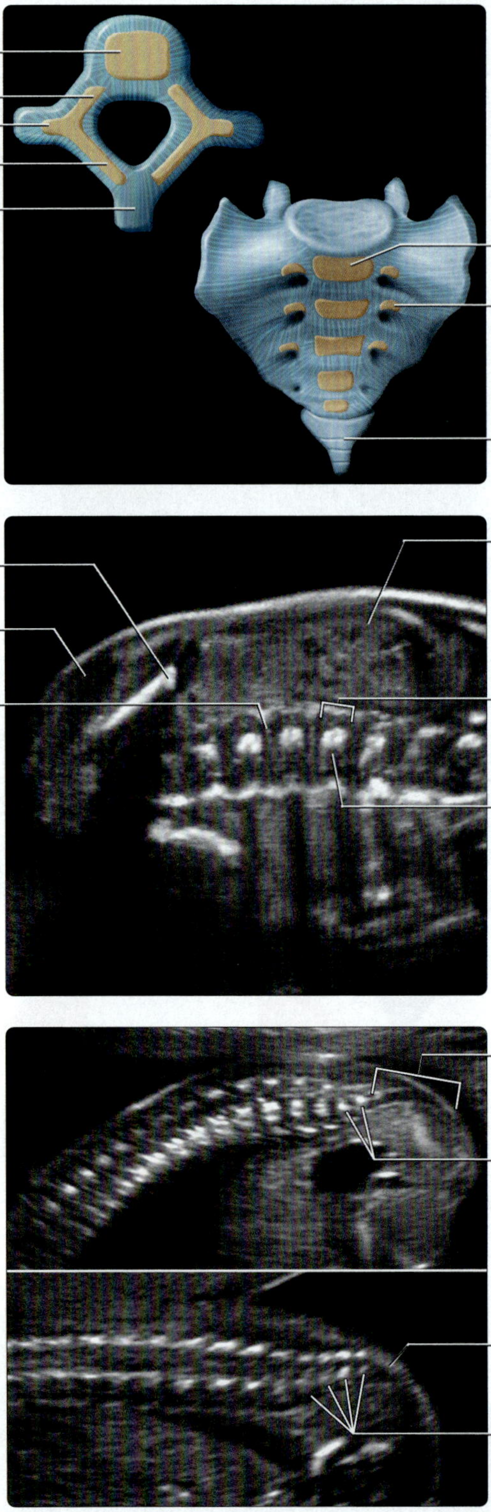

(Top) *Axial graphic shows the normal ossification centers within the developing vertebrae. The vertebral body and neural arch primary ossification centers (beige) are forming within the cartilaginous (blue) vertebral axis. Coronal graphic illustrates the normal appearance of the sacral ossification centers and cartilage. The sacrum and coccyx are the last portions of the vertebral column to ossify.* **(Middle)** *Coronal ultrasound of the lumbar spine at 22 weeks shows the ossification center (centrum) within the vertebral body; the remainder of the vertebral body is cartilaginous.* **(Bottom)** *Sagittal transabdominal ultrasound at 19.5 weeks (top) shows normal ossification and alignment of the lumbar spine. The coccyx and sacrum are unossified, which is an expected finding. At 22 weeks (bottom), there has been more complete ossification. Ossification should be complete by 25 weeks.*

FETAL SPINE

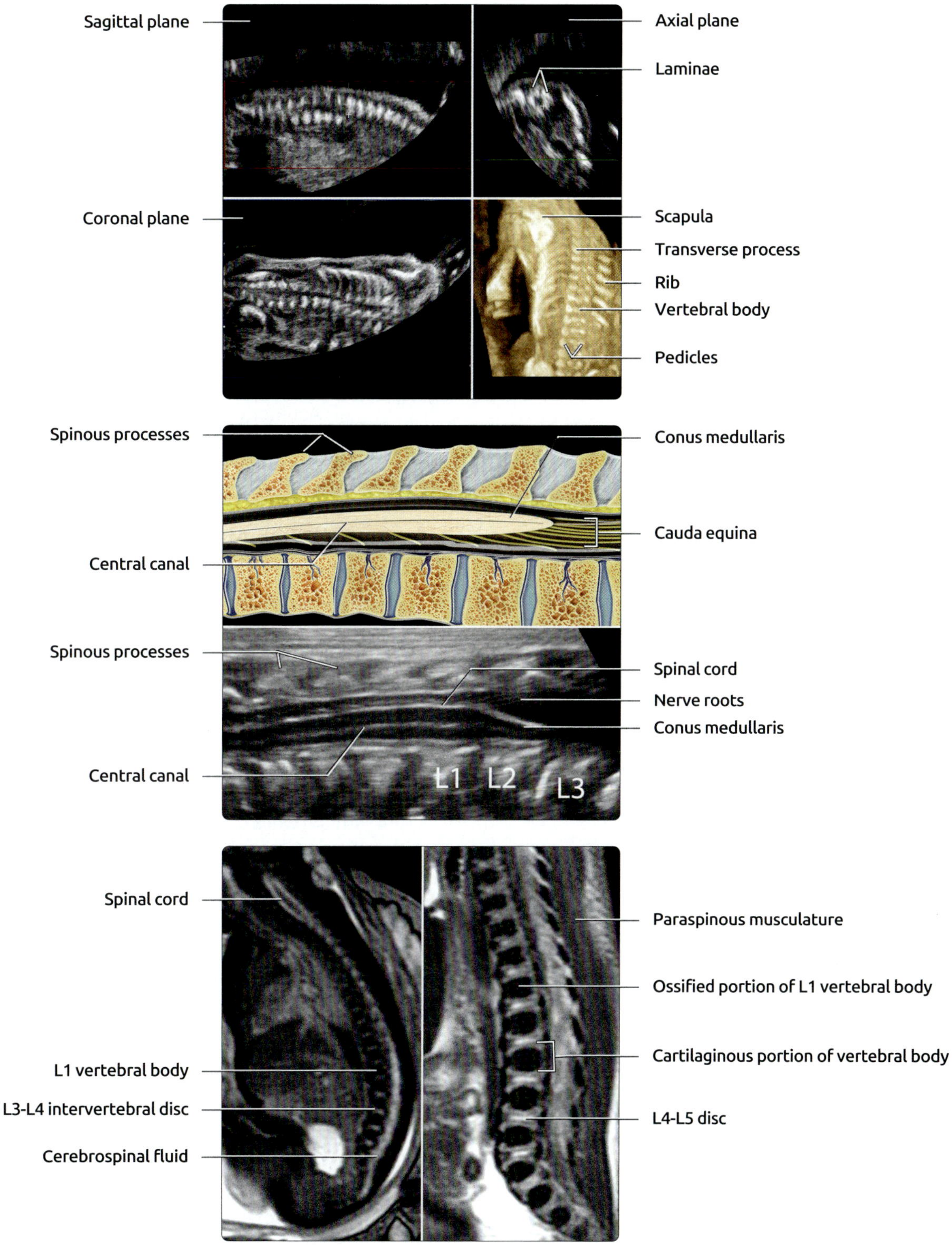

(Top) *3D ultrasound is ideal for evaluating the spine as all 3 planes can be evaluated with 1 acquisition. A 3D bone algorithm renders a skeletal view.* **(Middle)** *Sagittal graphic shows the normal appearance of the lumbar spinal column, conus medullaris, cauda equina, and central spinal canal. A correlative sagittal ultrasound of a fetal spine shows the normal hypoechoic appearance of the cord with a hyperechoic central canal. The lumbar portion of the spinal cord widens slightly compared to the thoracic portion. The normal cord ascends during gestation and should be at or above L3-L4 after 18 weeks and L1-L2 by 2 months of age.* **(Bottom)** *Sagittal T2 fetal MR (left) and newborn MR (right) show the neural axis. The ossified portion of the vertebral bodies is hypointense with hyperintense intervertebral discs. The spinal cord is surrounded by high-signal CSF. MR is an excellent tool for evaluating spinal dysraphism or cord abnormalities suspected on ultrasound.*

GENERAL CONCEPTS

Branchial Arches

- Form during 4th and 5th weeks of embryonic development
- 4 branchial arches appear as bars of mesenchymal tissue
- Branchial arches are separated by clefts
  - Branchial grooves
- Branchial arch and groove composition
  - External ectoderm
  - Internal endoderm
  - Central mesoderm
  - Migratory neural crest cells

Swellings Form on Branchial Arches

- Prominences
- Placodes
  - Migrate and fuse to form face
- Failed migration and fusion leads to common facial anomalies

Lymphatics

- Initial separate paired lymphatics
- Fuse with venous system
- Drain head, neck, upper limbs
- Failure to form or fuse leads to lymphatic disorders

NOSE, LIPS, AND PALATE

Frontonasal Prominence

- Anterior cranial bulge of tissue
- Contains forebrain

Nasal Placodes

- Develop on frontonasal prominence (FNP)
  - 5th week of embryonic life
- Bilateral, oval-shaped thickenings
- Eventually evaginate
  - Form nasal pits

Medial + Lateral Nasal Prominences

- Develop on FNP
  - 6th week of embryonic life
- Mesenchyme proliferation of nasal margins
  - Horseshoe-shaped elevations
- Deepening of nasal pit forms nasal sacs
  - Nasal sacs grow dorsal and superior
  - Initial separation between oral and nasal cavity
  - Primitive choanae forms posterior to primary palate
    - Rupture of oronasal membrane
- Medial nasal prominences merge
  - Fusion of midline medial prominence
  - Form intermaxillary segment
    - Becomes philtrum of lip

Maxillary Prominences

- 5th-8th week of embryonic life
- Start as paired swellings lateral to primitive mouth
  - Enlarge and grow rapidly toward midline
- Fuse with lateral nasal prominences
  - Lateral margins of philtrum
    - Just below nostrils

Palate

- 6th-12th week of embryonic life
- Forms from 2 primordia
- **Primary palate**
  - Innermost part of intermaxillary segment
    - From medial nasal prominence
  - Wedge-shaped segment
  - Eventual small section of adult hard palate
    - Anterior maxilla to incisor foramen
    - Includes incisor teeth
- **Secondary palate**
  - Primordia of most of hard palate and soft palate
  - Develops from maxillary prominences
    - Lateral palatine shelves
  - Palatine shelves grow toward midline and superiorly
    - Over developing tongue
  - Lateral palate shelves fuse
    - Medially with each other
    - Anteriorly with primary palate
    - Superiorly with nasal septum
  - Neural crest cells concurrently ossify palate
    - Posterior portion is without bone (soft palate)

MANDIBLE AND EARS

Mandible

- 4th-8th week of embryonic life
- Jaw is 1st part of face to form
- Paired mandibular prominences
  - Caudal boundary of primitive mouth
  - Fuse medially by end of 4th week
- Part of Meckel cartilage migrates
  - Forms incus and malleus of middle ear

Ears

- 4th-8th week of embryonic life
- Inner ear arises from hindbrain
- Middle ear arises from 1st pharyngeal pouch
- External ear from 1st branchial groove
  - Inferior and dorsal to mandibular prominence
    - Early ears are located in upper part of future neck
    - Migrate lateral and superior as mandible develops
  - Auricle from 6 swellings (hillocks)

EYES

Lens Placode

- Forms on FNP during 3rd week
- Induced by optic vesicles
  - From forebrain
  - Becomes lens vesicle and final lens of eye
- Forms optic cups
  - Large at first, then invaginate

Orbits

- From mesenchyme that encircles optic vesicle
  - Neural crest cells
- Walls of orbit from 7 skull bones
  - Superiorly: Frontal bone
  - Inferiorly: Maxilla, zygomatic
  - Medially: Frontal, lacrimal, maxilla

- Lateral: Zygomatic, frontal

LYMPHATICS

Lymph Sacs

- Begin to develop at end of 5th week
  - 2 weeks after cardiovascular system
- Develop alongside vessels
- Lymph sacs form from fusion/dilatation of adjacent mesenchymal spaces
- **6 primary lymph sacs**
  - **Paired jugular lymph sacs**
    - Subclavian and internal jugular vein junction
    - Drain head, neck, thorax, upper extremities
  - **Cisterna chyli**
    - Lymph sac below diaphragm
    - Along posterior abdominal wall
  - **Retroperitoneal (mesenteric) lymph sac**
    - Root of mesentery
    - Posterior abdominal wall, anterior to cisterna chyli
  - **Paired iliac lymph sacs**
    - Junction of iliac and posterior cardinal veins
    - Drain abdominal wall, pelvis, lower extremity
    - Join cisterna chyli
- Lymph sacs eventually become groups of lymph nodes
  - Exception is superior cisternal chyli
- Lymphatic vessels grow out from lymph sacs and make connections with venous system

Thoracic Duct

- 2 channels connect jugular sacs with cisterna chyli
  - Right and left thoracic ducts
- Anastomosis and attrition occurs between paired ducts
- Final thoracic duct anatomy
  - Superior part from left duct
  - Central part from anastomosis
  - Caudal part from right duct
- Variations of thoracic duct anatomy common

EMBRYOLOGY OF COMMON ANOMALIES

Cleft Lip and Palate

- **Isolated cleft lip**
  - Involves lip ± primary palate
    - Incisive foramen is boundary of primary and secondary palate
    - Secondary palate intact
  - Maxillary prominence fails to unite with nasal prominence
    - Results in persistent labial groove
  - Rare cases
    - Median isolated cleft lip
    - Bilateral isolated cleft lip
- **Cleft palate ± cleft lip**
  - Failure of lateral palatine processes fusion
    - Nonunion with each other
    - Nonunion with nasal septum
    - Most often involves lip and primary and secondary palate
  - Isolated cleft palate (intact lip and primary palate)
    - Posterior to incisive foramen
- **Rare facial clefts**
  - Median cleft of mandible
  - Lateral or transverse facial cleft (from mouth toward ear)
  - Oblique facial cleft (upper lip to medial margin of orbit)

Eye Anomalies

- Hypertelorism and hypotelorism
  - Optic migration follows forebrain migration
    - Holoprosencephaly: Hypotelorism, cyclopia
  - Associated with craniofacial dysostosis
    - Hypertelorism
- Absent or small eye/orbit
  - Failure of optic vesicle or lens placode to form

Hypognathia

- Insufficient 1st branchial arch
  - From poor neural crest cell migration
- Syndromes
  - Pierre Robin syndrome
    - Hypoplasia of mandible
    - Cleft palate + ear anomalies
  - Treacher Collins syndrome
    - Mandibulofacial dysostosis
    - Eye and ear anomalies

Ear Anomalies

- Low-set ears
  - Ear migration follows mandible development
    - Small chin associated with low-set ears
- Abnormal hillock development
  - Auricular appendages (tags)
  - Ear duplication
  - Anotia (absent ear), microtia (small ear)

Nose and Mouth Anomalies

- Congenital microstomia (small mouth)
  - Excessive merging of mesenchymal masses
- Absent nose
  - Paired nasal placodes do not form
- Single nostril
  - Only 1 nasal placode forms
- Bifid nose
  - Medial nasal prominences do not merge completely

Lymphangioma

- Dilated primitive lymphatic channels
  - Diffuse congenital lymphedema
  - Focal cystic mass
- Cystic hygroma
  - Failed jugular sac leads to venous connection
  - Primary fluid collection in dorsal and lateral neck
    - Multiseptated fluid
  - Associated with hydrops fetalis and aneuploidy
    - Turner syndrome most common
    - Trisomy 21 is 2nd most common
- Body lymphangioma
  - Sites
    - Axillary (most common)
    - Intraperitoneal, retroperitoneal
    - Extremities
  - Often large, infiltrative cystic mass

EMBRYOLOGY OF FACE AND PALATE

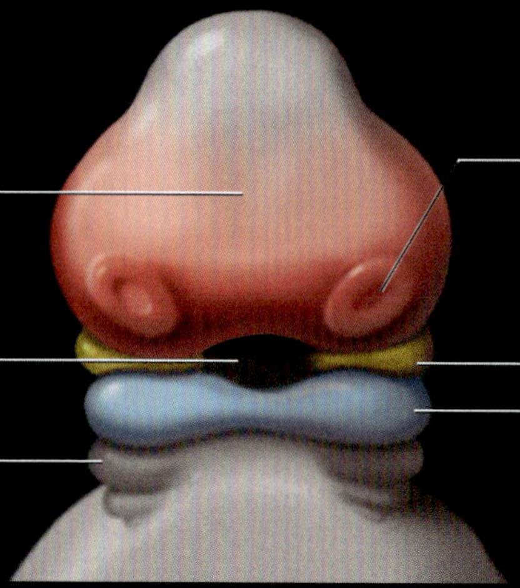

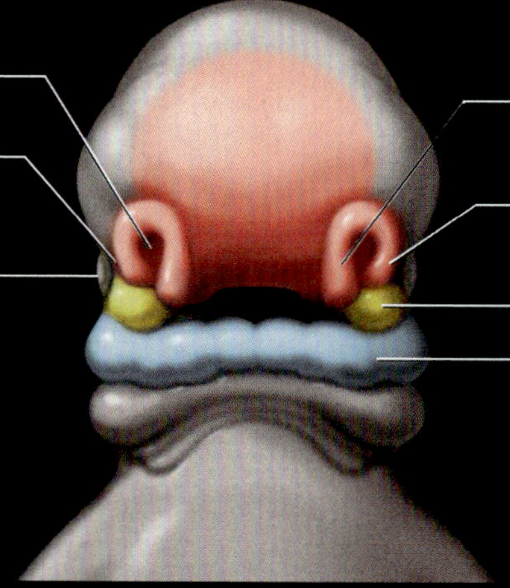

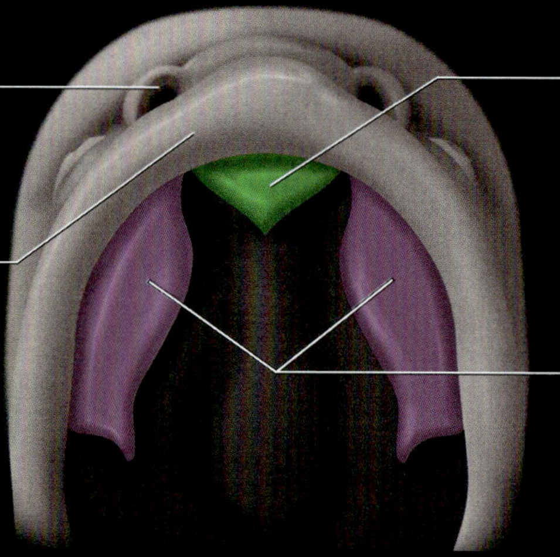

(Top) *Graphic shows a coronal view of a 5-week embryo. The face forms from 5 primordia that appear in the 4th week (frontonasal prominence, 2 maxillary prominences, and 2 mandibular prominences). By the 5th week, the mandibular prominences have fused. Nasal pits form on a pair of ectodermal thickenings, the nasal plates.* **(Middle)** *Graphic shows a coronal view of a 6-week embryo. Invagination of the nasal pits has occurred. Medial nasal processes will fuse to form an intermaxillary process and, subsequently, the upper lip philtrum. In addition, the maxillary prominences will fuse with the intermaxillary process to form an intact upper lip.* **(Bottom)** *Graphic shows an axial view of the palate at 7-8 weeks. The primary palate arises dorsally from the intermaxillary process, and the secondary palate originates from the maxillary prominence. Complete fusion occurs by the 10th week.*

EMBRYOLOGY OF FACE AND EAR

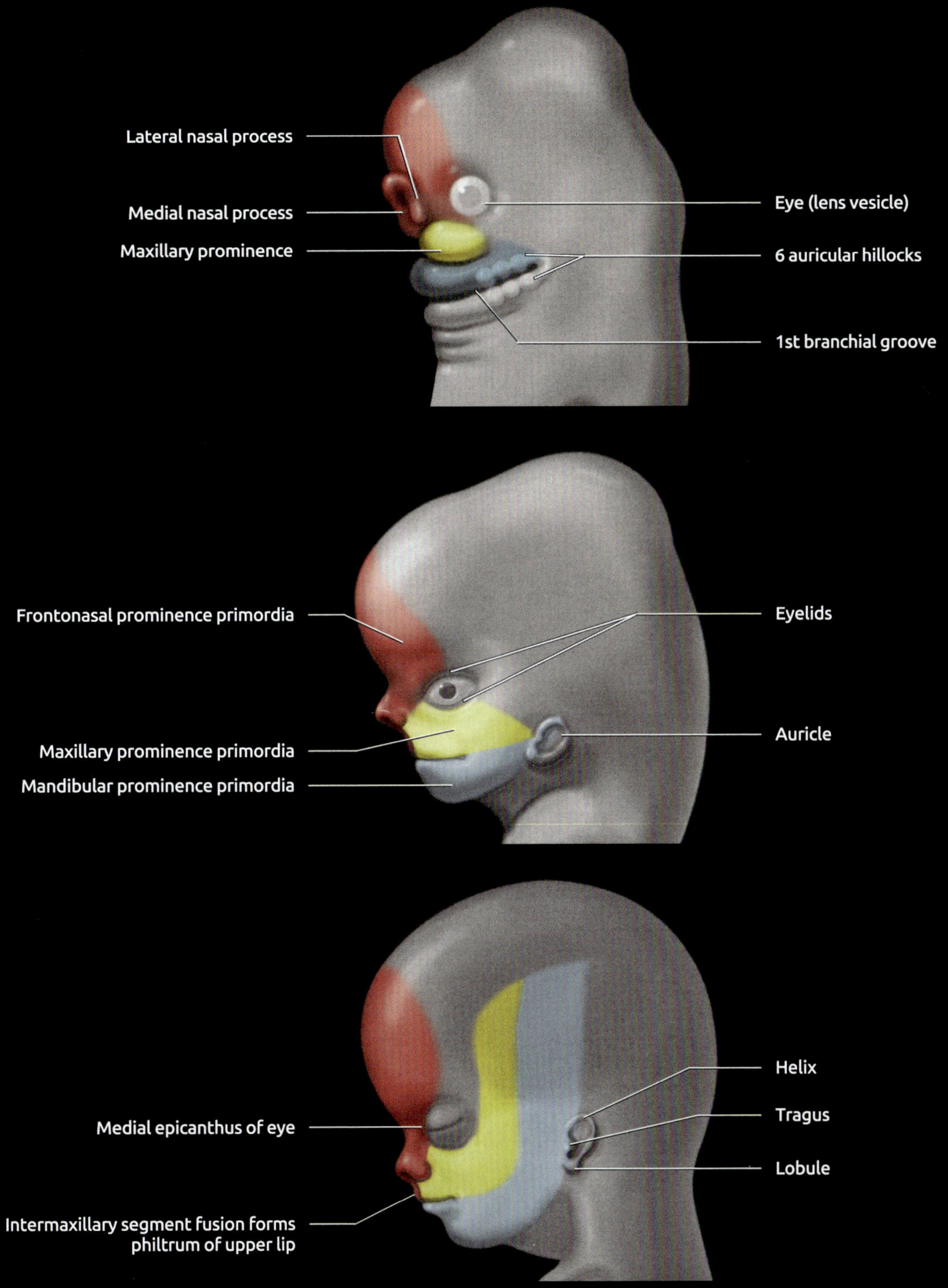

(Top) *Graphic of a 5-week embryo profile shows the lateral and medial nasal processes, not yet fused with the maxillary prominence. Arising from the 1st and 2nd pharyngeal arches, the auricular hillocks of the external ear flank the 1st branchial groove.* **(Middle)** *Graphic of a 10-week embryo profile shows the development of eyelids and the external ear. The ear position is medial and low at this time. As the mandible grows, the ear migrates superiorly.* **(Bottom)** *Graphic of a 14-week fetus profile shows that the philtrum of the lip has formed from fusion between the paired medial nasal processes. The philtrum and maxillary prominences have also fused. The ear is now at its final location with the top of the helix at the same level as the medial epicanthus of the eye.*

EYES

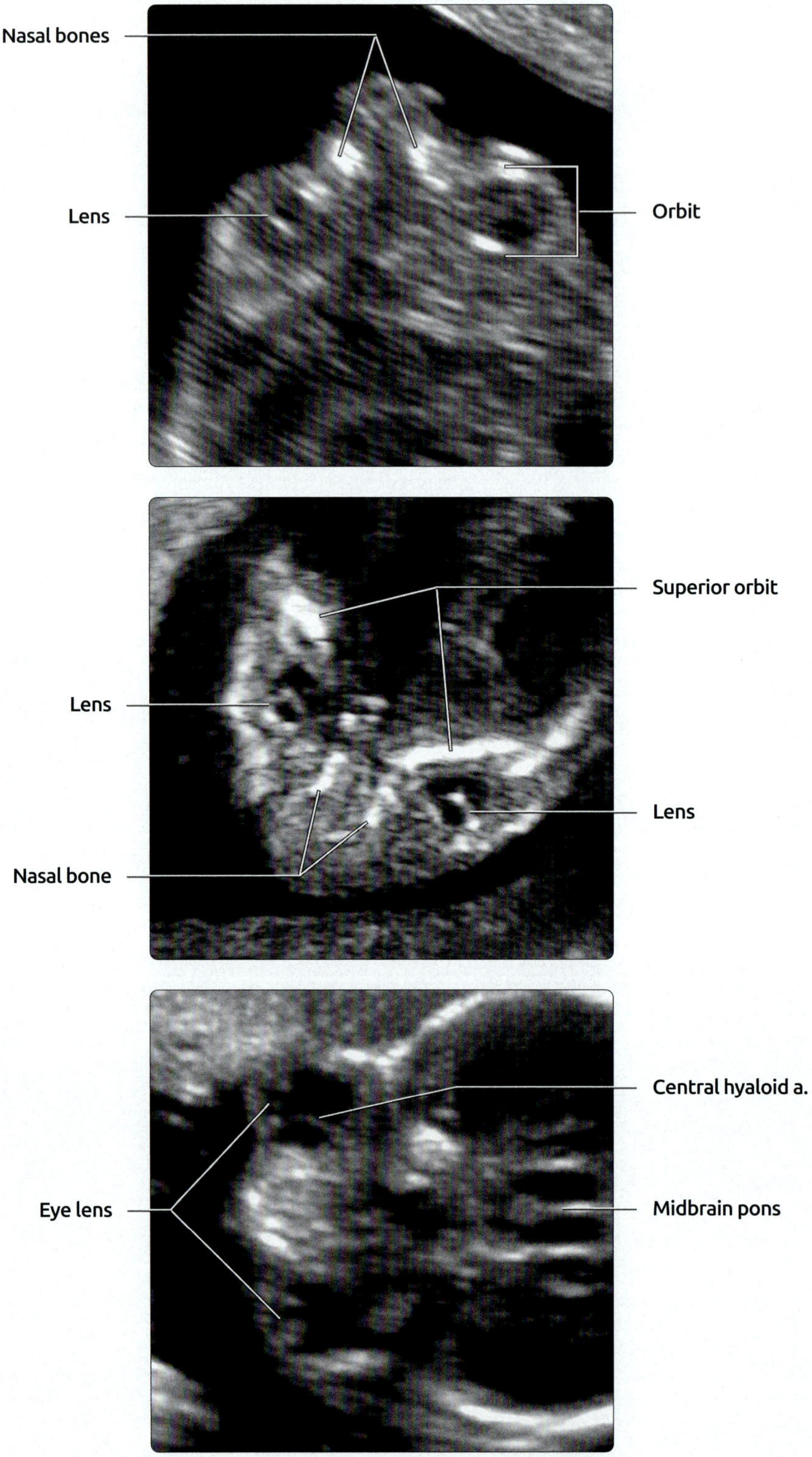

(Top) *Transvaginal axial ultrasound through the orbits of a 12-week fetus shows paired nasal bones and normal-sized orbits. The lens of the eye can be seen even at this early gestational age.* **(Middle)** *Coronal transvaginal ultrasound of a 13-week fetus shows the frontal bones and nasal bones contributing to the superior and medial borders of the bony orbit. The eyes and lens are once again seen very well.* **(Bottom)** *Axial ultrasound through the eyes in an early 2nd-trimester fetus shows the normal central hyaloid artery, which is located within the hyaloid canal. This artery supplies nutrients to the developing lens and is a normal finding at this time, usually regressing during the 3rd trimester.*

EYES

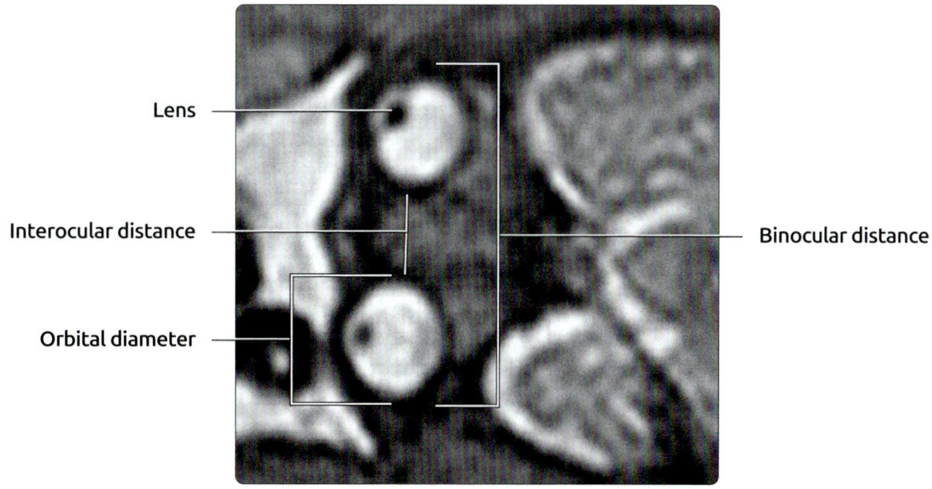

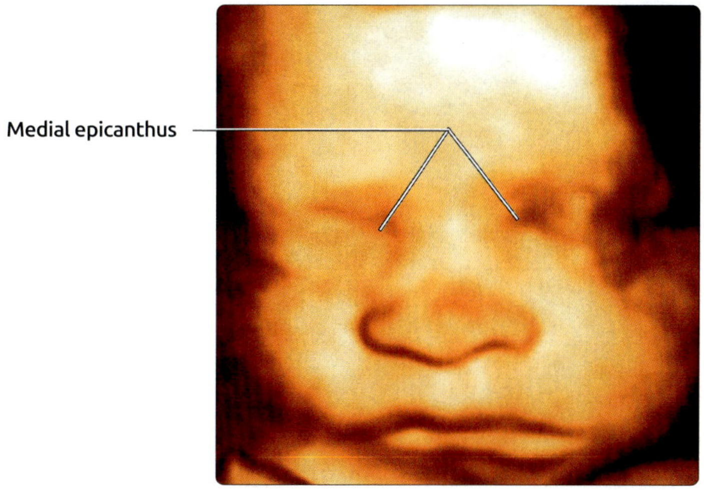

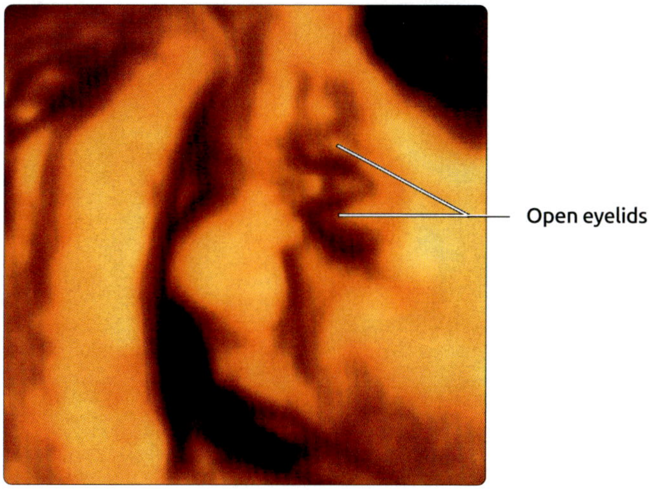

(Top) *Axial T2WI MR of a late 2nd-trimester fetus shows the orbits. MR or ultrasound can be used to measure the globe diameter, interocular distance, and binocular distance. The lens of the eye is low signal on MR.* **(Middle)** *3D ultrasound of a 3rd-trimester fetal face shows the eyes, nose, and lips. The interocular distance and the medial epicanthus of the eyes are seen well.* **(Bottom)** *3D ultrasound of a fetal profile shows open eyes. In the 3rd trimester, it is common to see the eyes open and close. In addition, globe movement is also commonly seen with real-time imaging.*

NOSE

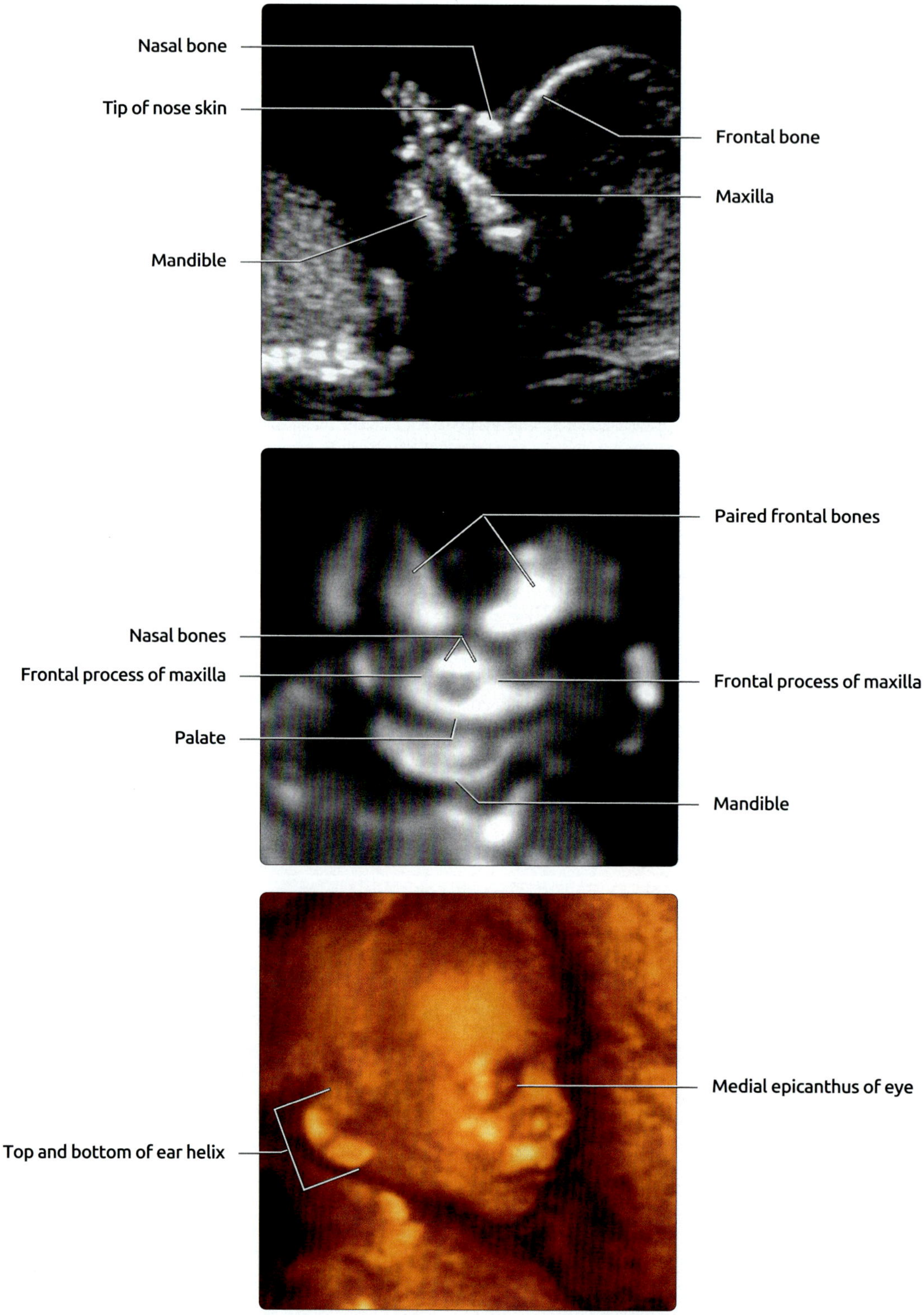

(Top) *Sagittal ultrasound of a 12-week fetus shows a normal nasal bone. The echogenic nasal bone is as bright as the frontal bone, and it is seen separately from the nasal skin.* **(Middle)** *3D ultrasound with skeletal reconstruction of a 13-week fetus shows the retronasal triangle view comprised of the paired nasal bones superiorly, the frontal process of the maxilla laterally, and the inferior primary palate.* **(Bottom)** *3D ultrasound profile view that includes the nose, eye, and ear shows normal relationships. The top of the helix should be at the same height as the medial epicanthus of the eye.*

LIPS AND PALATE

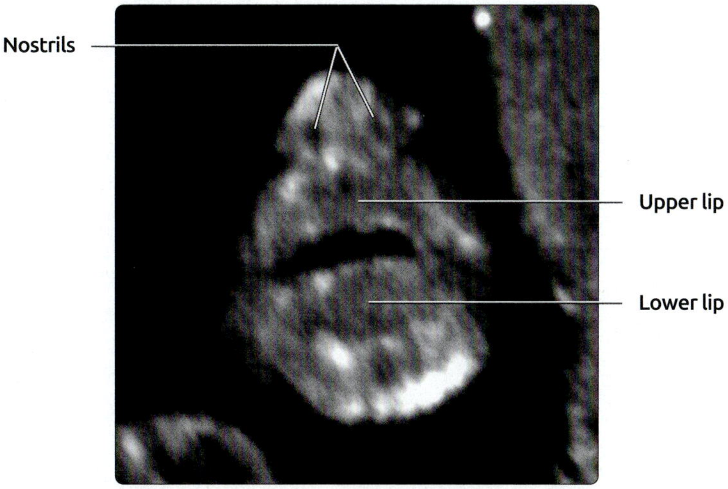

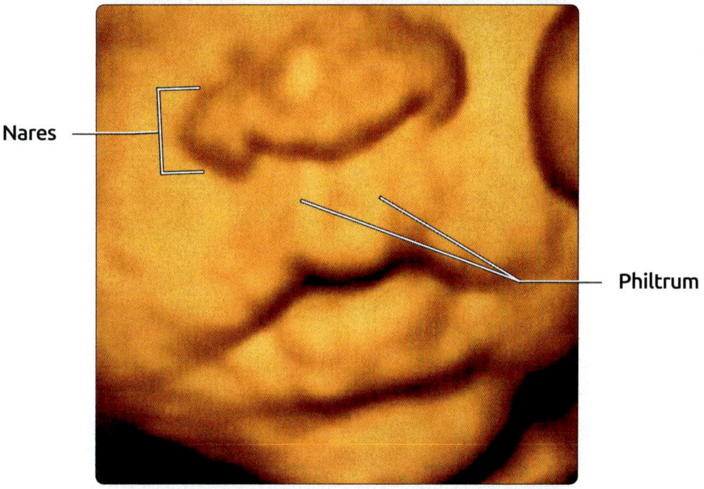

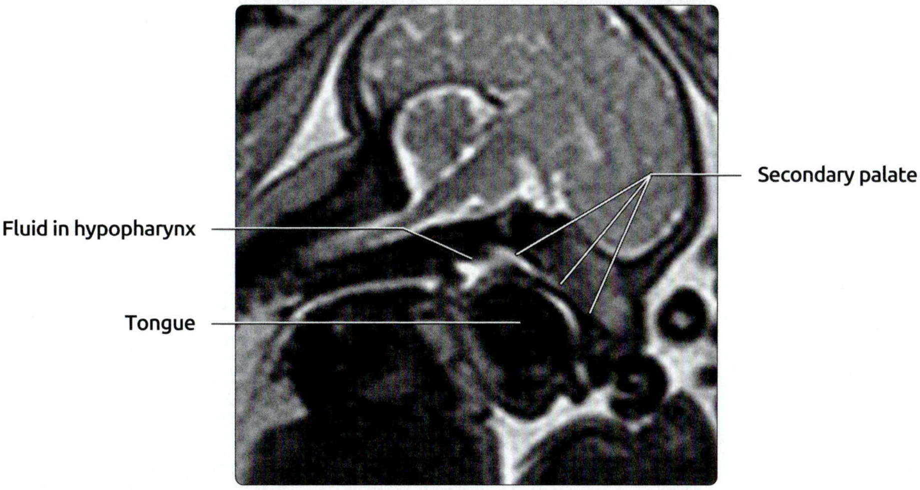

(Top) *Coronal ultrasound through the nose and lips shows the nostrils and intact upper lip. This view is considered standard for anatomy scans.* **(Middle)** *3D ultrasound with soft tissue reconstruction shows the normal rounded nares of the nose and the intact philtrum of the upper lip.* **(Bottom)** *T2WI MR of a 30-week fetus shows an intact secondary palate. A sliver of high-signal fluid in the mouth, superior to the tongue, provides excellent contrast, allowing for visualization of the palate. The fluid-filled hypopharynx is also seen extending down to the upper trachea.*

SKULL AND FACIAL BONES

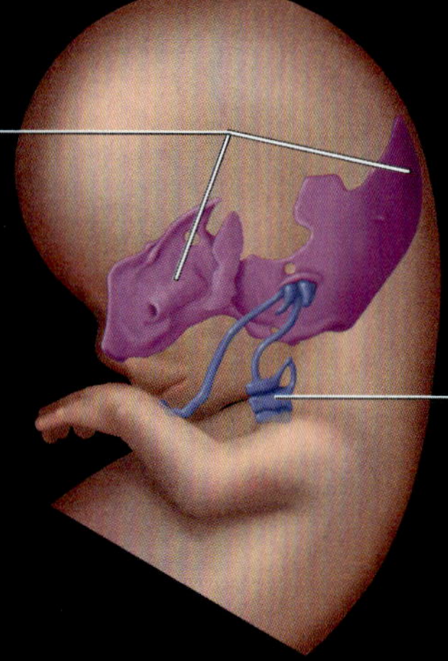

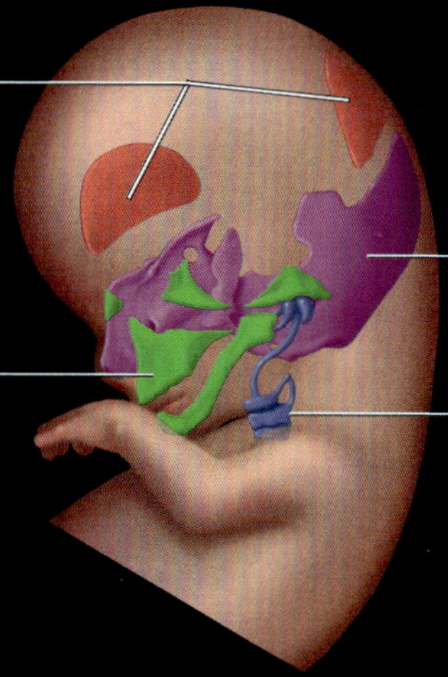

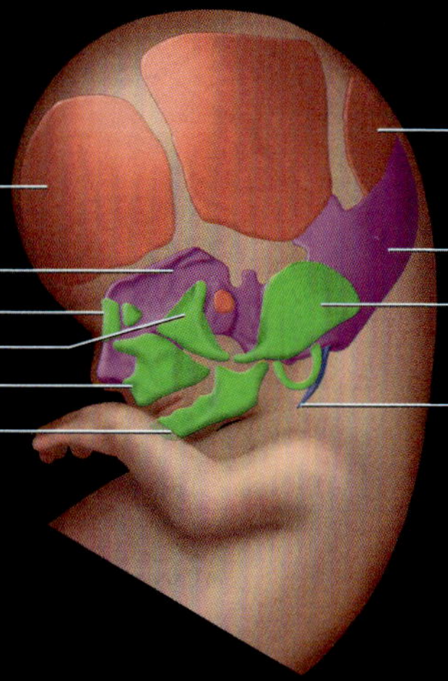

(Top) *Graphic shows the cartilaginous skeleton of the embryonic head. Purple represents the developing chondrocranium; blue represents the developing viscerocranium of the pharyngeal arches. The chondrocranium arises from the notochord and is the forerunner of the skull base. The viscerocranium, derived from the pharyngeal arches, gives rise to the facial bones.* **(Middle)** *The chondrocranium and portions of pharyngeal arches ossify as the membranous neurocranium (protective case around the brain) and the membranous viscerocranium develop.* **(Bottom)** *The final development of cranial bones from chondrocranium, pharyngeal arches, membranous viscerocranium, and membranous neurocranium is shown.*

SKULL AND SUTURES

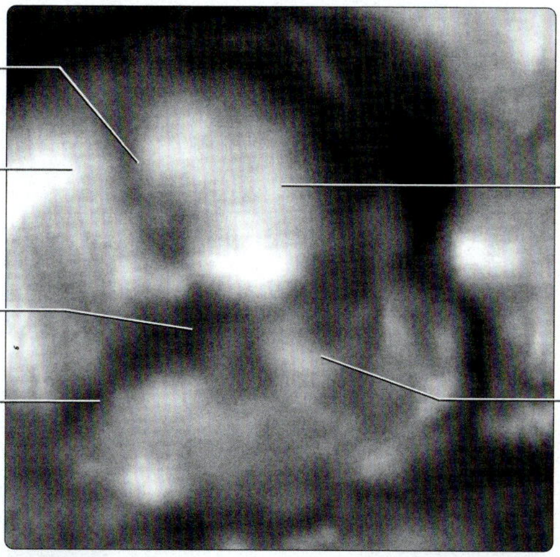

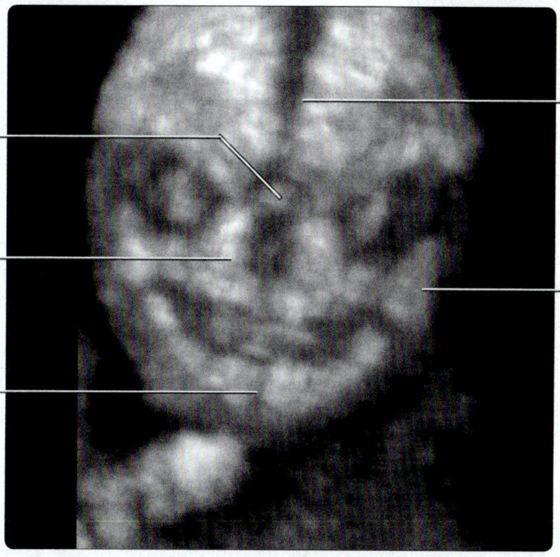

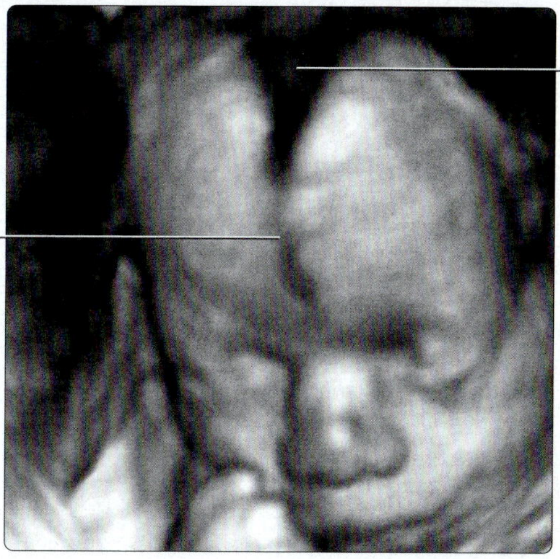

(Top) *3D ultrasound of an 18-week fetus profile shows normal skull and suture anatomy.* **(Middle)** *Coronal 3D ultrasound with skeletal reconstruction of a 20-week fetus shows normal skull bones and sutures.* **(Bottom)** *3D ultrasound of an early 3rd-trimester fetus shows the anterior fontanelle and metopic suture.*

LYMPHATIC SYSTEM EMBRYOLOGY

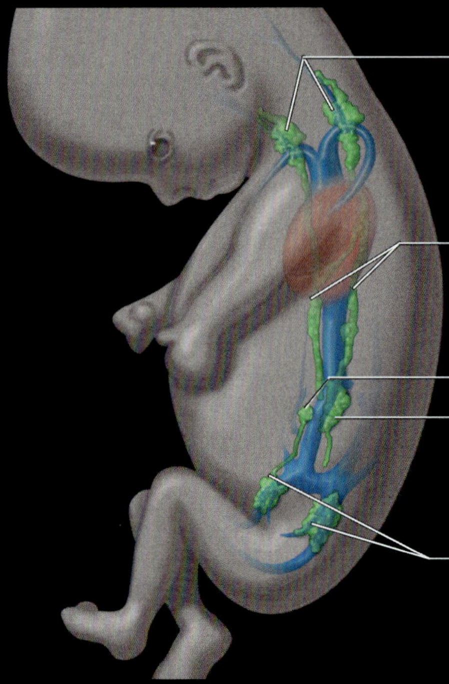

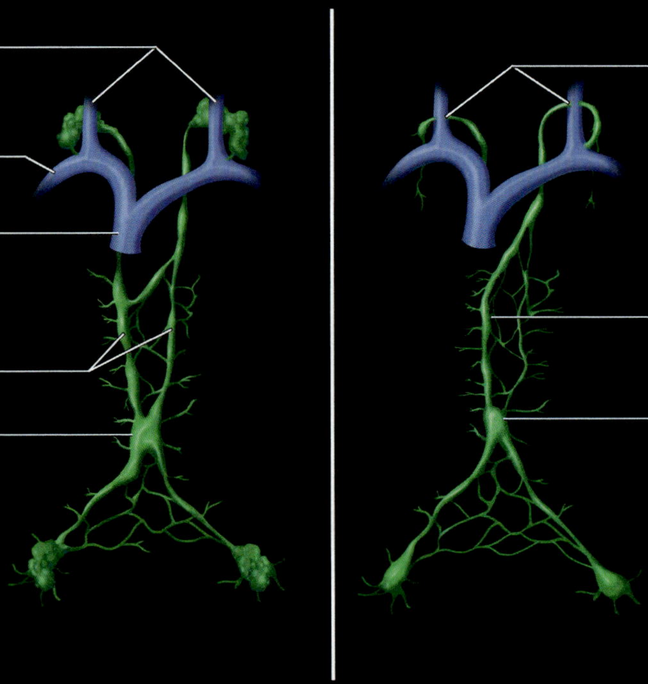

(Top) *Graphic of a 6- to 7-week embryo shows lymph sacs that collect lymph fluid before venous connections are established. Lower extremity and body lymph fluid drain into external and internal iliac veins.* **(Bottom)** *Graphic of the lymphatic system at 7 weeks (left) and 17 weeks (right) shows that, originally, there are paired thoracic ducts. In most embryos, the caudal left and cranial right ducts atrophy, resulting in a thoracic duct that crosses the midline. Upper body lymphatic drainage occurs mostly at the venous connection near the jugular-subclavian vein junction.*

CYSTIC HYGROMA

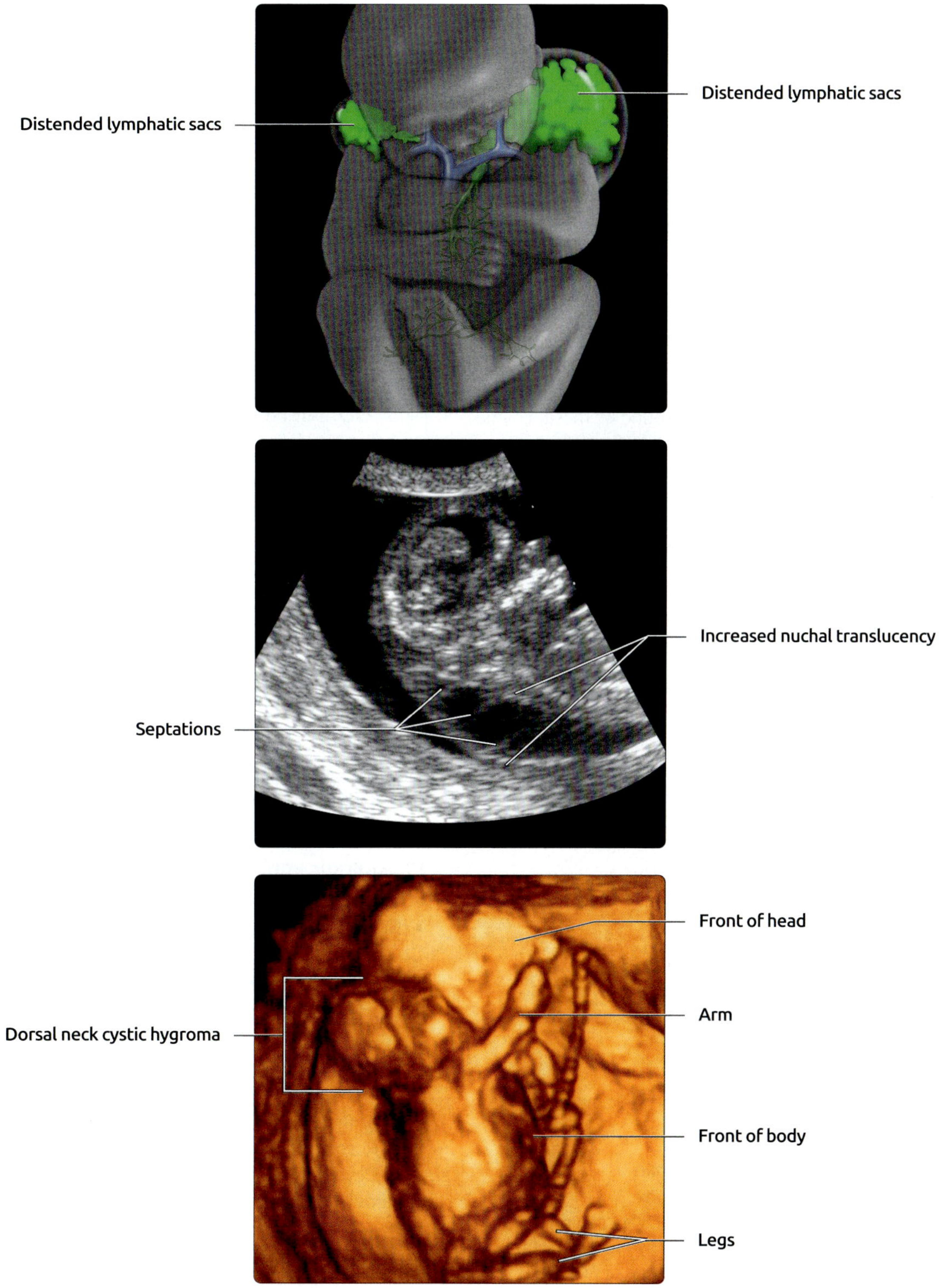

(Top) *Graphic of cystic hygroma shows massively distended jugular lymph sacs secondary to failure of establishment of venous connections at the jugular drainage site.* **(Middle)** *Sagittal ultrasound of a 12-week fetus shows cystic hygroma. The nuchal translucency is markedly increased and contains septations.* **(Bottom)** *3D ultrasound of a 13-week fetus shows posterior neck cystic hygroma. This fetus was found to have Turner syndrome.*

GENERAL CONCEPTS

Overview of Lung Development

- Larynx and trachea
  - Origin of primitive larynx from laryngotracheal groove
  - Separation of primitive trachea from foregut and developing esophagus
- Bronchi
  - Tracheal bud branches into 2 primitive bronchial buds
  - Bronchial buds are precursors of bilateral main bronchi
- Lungs
  - Sequential branching of primitive bronchi
  - Formation of distinct pulmonary lobes
  - Formation of distinct pulmonary segments
- Distal airways and lung parenchyma
  - Interaction of endodermal and mesodermal elements allow normal lung development
  - Continued branching of primitive airways
  - Progressive vascularization of surrounding mesenchyme
  - Development of alveolar-capillary interface
  - Postnatal airway development and maturation
- Pleural development
  - Developing lungs protrude into coelomic cavity
  - Separation of pleural and pericardial cavities
  - Pleural investment of lungs within bilateral hemithoraces
- 5 developmental stages
  - Embryonic stage
  - Pseudoglandular stage
  - Canalicular stage
  - Saccular stage
  - Alveolar stage

EMBRYONIC STAGE (26 DAYS TO 6 WEEKS)

Laryngotracheal Groove

- Develops 26-28 days after fertilization
- Arises caudal to primitive pharynx and 4th pair of pharyngeal pouches
- Longitudinal growth

Respiratory Diverticulum or Lung Bud

- Develops 4 weeks after fertilization
- Pouch-like outgrowth from caudal aspect of laryngotracheal groove
- Caudal growth
- Invested in mesodermal-derived splanchnic mesenchyme

Tracheal Bud

- Globular enlargement of distal lung bud
- Caudal growth from primordial pharynx
- Proximal communication with foregut through primordial laryngeal inlet

Tracheoesophageal Septum

- Longitudinal tracheoesophageal folds form on either side of developing tracheal bud
- Medial growth of bilateral tracheoesophageal folds
- Formation of tracheoesophageal septum from midline fusion of tracheoesophageal folds
- Separation of primitive trachea from developing esophagus

Primary Bronchial Buds and Branches

- Branching of primitive tracheal bud into right and left branches (5th week after fertilization)
  - Right bronchial bud: Larger and vertically oriented
  - Left bronchial bud: Smaller and horizontally oriented
- Branching of primary bronchial buds into 2 primitive lobar bronchi
  - Right superior lobar bronchus → right upper lobe bronchus
  - Right inferior lobar bronchus → primitive bronchus intermedius
    - Right middle lobe bronchus
    - Right lower lobe bronchus
  - Left superior lobar bronchus → left upper lobe bronchus
  - Left inferior lobar bronchus → left lower lobe bronchus
- Branching of primitive lobar bronchi into primitive segmental bronchi

PSEUDOGLANDULAR STAGE (6-16 WEEKS)

Important Events

- Formation of all major airway elements
  - Bronchial development complete to level of terminal bronchioles
- All bronchopulmonary segments formed by 7 weeks after fertilization

Microscopic Morphology

- Gland-like appearance of lung
- Formation of tracheobronchial cartilages, mucus glands, and cilia by 13 weeks after fertilization
- Primitive airways lined by endodermal-derived columnar epithelium
- Primitive airways surrounded by mesodermal-derived mesenchymal tissue
- Absent alveolar-capillary interface

Physiologic Implications

- Respiration not possible
- No possibility of extrauterine survival

CANALICULAR STAGE (16-28 WEEKS)

Important Events

- Continued vascularization of lung
- Continued development of primitive airways
  - Terminal bronchioles give rise to 2 or more respiratory bronchioles
  - Respiratory bronchioles give rise to 3-6 alveolar ducts
  - Development of small number of terminal saccules
- Lamellar inclusions within type 2 pneumocytes in terminal saccules with potential for surfactant production

Microscopic Morphology

- Continued enlargement of primitive airway lumina
- Continued thinning of airway epithelium
- Epithelial differentiation into type 1 and type 2 cells
- Airways separated by reduced but significant mesenchymal tissue

Physiologic Implications

- Limited surfactant production

- Respiration is possible in late canalicular stage
- Possibility of neonatal survival with intensive care and appropriate life support

SACCULAR STAGE (28-36 WEEKS)

Important Events

- Development of increasing numbers of terminal saccules
- Establishment of primitive alveolar-capillary interface
- Increased potential for surfactant production

Microscopic Morphology

- Continued airway differentiation
- Continued thinning of airway epithelium
- Some capillaries abut and bulge into developing alveoli
- Terminal sacs begin to approach morphology of adult alveoli

Physiologic Implications

- Respiration with adequate gas exchange is possible
- Survival of premature neonates with appropriate life support

ALVEOLAR STAGE (36 WEEKS TO 8 YEARS)

Important Events

- Continued development of distal airways with primordial alveoli forming along respiratory bronchioles and terminal saccules
- Development of thin alveolar-capillary membrane
- Postnatal lung development
  - 24 million terminal sacs and alveoli present at birth compared to 300 million in adult lungs
  - 5x increase of alveolar numbers within 1st year of life
  - Formation of 95% of adult alveoli by 8 years of age

Microscopic Morphology

- Continued thinning of epithelial lining of terminal sacs
- Formation of primordial alveoli
- Adjacent capillaries bulge into terminal saccules

Physiologic Implications

- Presence of nearly mature alveolar-capillary interface
- Adequate surfactant production
- Respiration possible without external support

OTHER DEVELOPMENTAL REQUIREMENTS

Volume Requirements

- Adequate intrathoracic volume required for normal pulmonary development
- Intrathoracic masses (especially diaphragmatic hernia) and chest wall abnormalities (e.g., skeletal dysplasias) restrict space for lung growth

Amniotic Fluid Requirements

- Amniotic fluid and fetal breathing required for normal lung development
- Oligohydramnios has severe adverse effect on lung development
  - Fetal compression causes decreased space for lung growth
  - Restriction of breathing movements with efflux of lung fluid into amniotic space

Circulatory Requirements

- Circulation affects pulmonary development
  - Pulmonary arterial development along developing airways
  - Vasculogenesis within primitive mesenchyme to form capillary network
  - Pulmonary vein and lymphatic development along segmental boundaries
- Right-sided obstructive lesions decrease pulmonary blood flow with subsequent poor lung development

Diaphragm Development

- Complex embryonic origin with 4 embryologic structures
  - Septum transversum: Forms most of central tendon
  - Pleuroperitoneal membranes: Bulk of diaphragm muscle, innervated by phrenic nerve
  - Paraxial mesoderm of body wall: Outer rim of diaphragmatic muscle
  - Esophageal mesenchyme: Condenses to form diaphragmatic crura

Thymus

- Arises from 3rd pharyngeal pouch
- Between 4th-7th week, thymic primordia migrate to chest
  - Posterior to sternum in superior mediastinum
  - Fuse to form bilobed gland
- Once formed, lymphocytes infiltrate gland
- Large and highly active in perinatal period
  - Continues to grow until puberty and then involutes in adulthood
- May be confused with chest mass
  - Look for thy-box to confirm it is thymus
    - Thymus is flanked by internal mammary arteries, branches of subclavian arteries
    - Color Doppler of superior mediastinum, at level of 3-vessel view, shows thymus between internal mammary arteries creating box appearance

NEONATAL LUNG

1st Breath

- Diaphragmatic contraction
- Pulmonary vascular changes
  - Fluid-filled lungs result in high-resistance pulmonary circulation
    - Small portion of cardiac output goes to lungs prior to birth
  - Pulmonary expansion with 1st breath
    - Vasodilatation
    - Increased pulmonary blood flow
    - Clearance of fetal lung fluid via lymphatics and capillaries
- Role of surfactant
  - Alveolar expansion results in surfactant discharge by type 2 pneumocytes
  - Decreased surface tension of remaining intraalveolar fluid
  - Increased surfactant activity with decreased surface area
  - Prevention of alveolar collapse during expiration

LUNG DEVELOPMENT

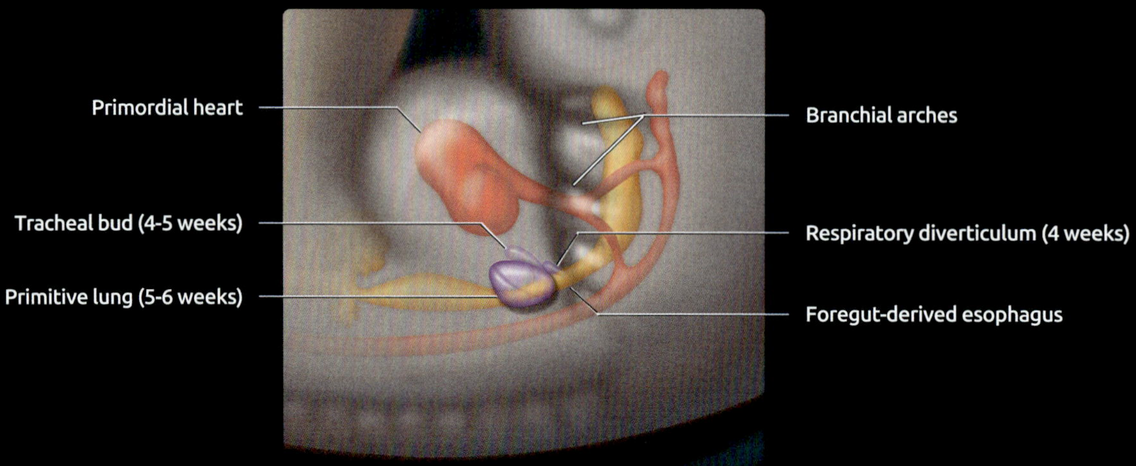

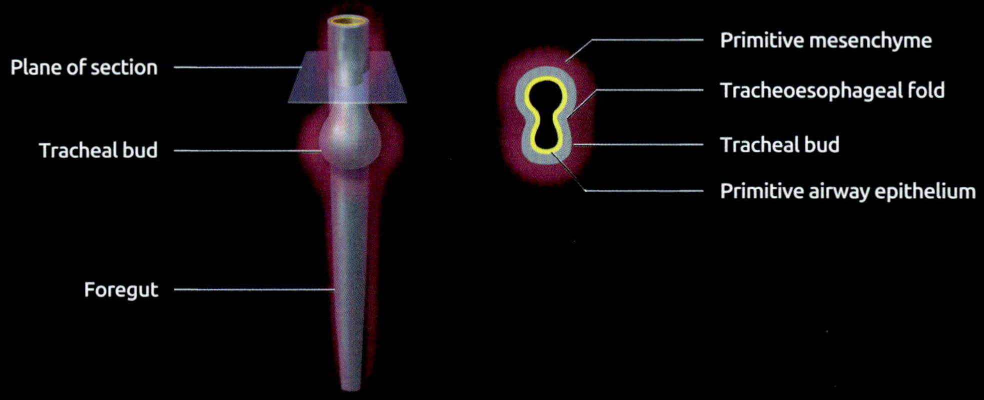

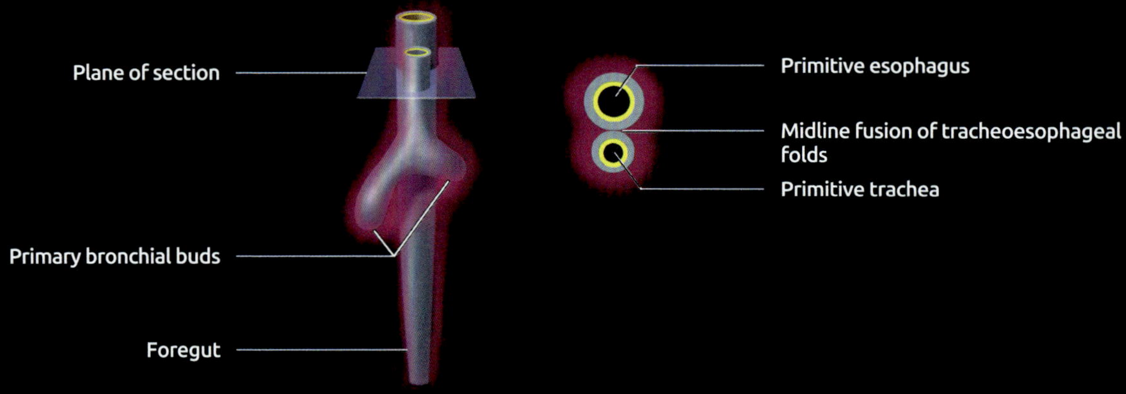

(Top) *Graphic shows the development of the primitive lung. The respiratory diverticulum arises from the laryngotracheal groove near the primordial esophagus caudal to the 4th pharyngeal pouches. The sequential evolution of the respiratory diverticulum to the tracheal bud and the primitive lung is shown. Note the close relationship of the developing tracheobronchial tree and lungs to the primitive esophagus.* **(Middle)** *Graphic shows the tracheal bud, a ventral outpouching of the foregut surrounded by mesodermal-derived mesenchyme and lined by endodermal-derived epithelium. The axial plane of section (right) shows communication between the tracheal bud and the foregut. The formation of bilateral longitudinal tracheoesophageal folds is shown.* **(Bottom)** *Graphic shows vertical development of the primary bronchial buds. The tracheoesophageal folds fuse in the midline to separate the trachea from the esophagus.*

LUNG DEVELOPMENT

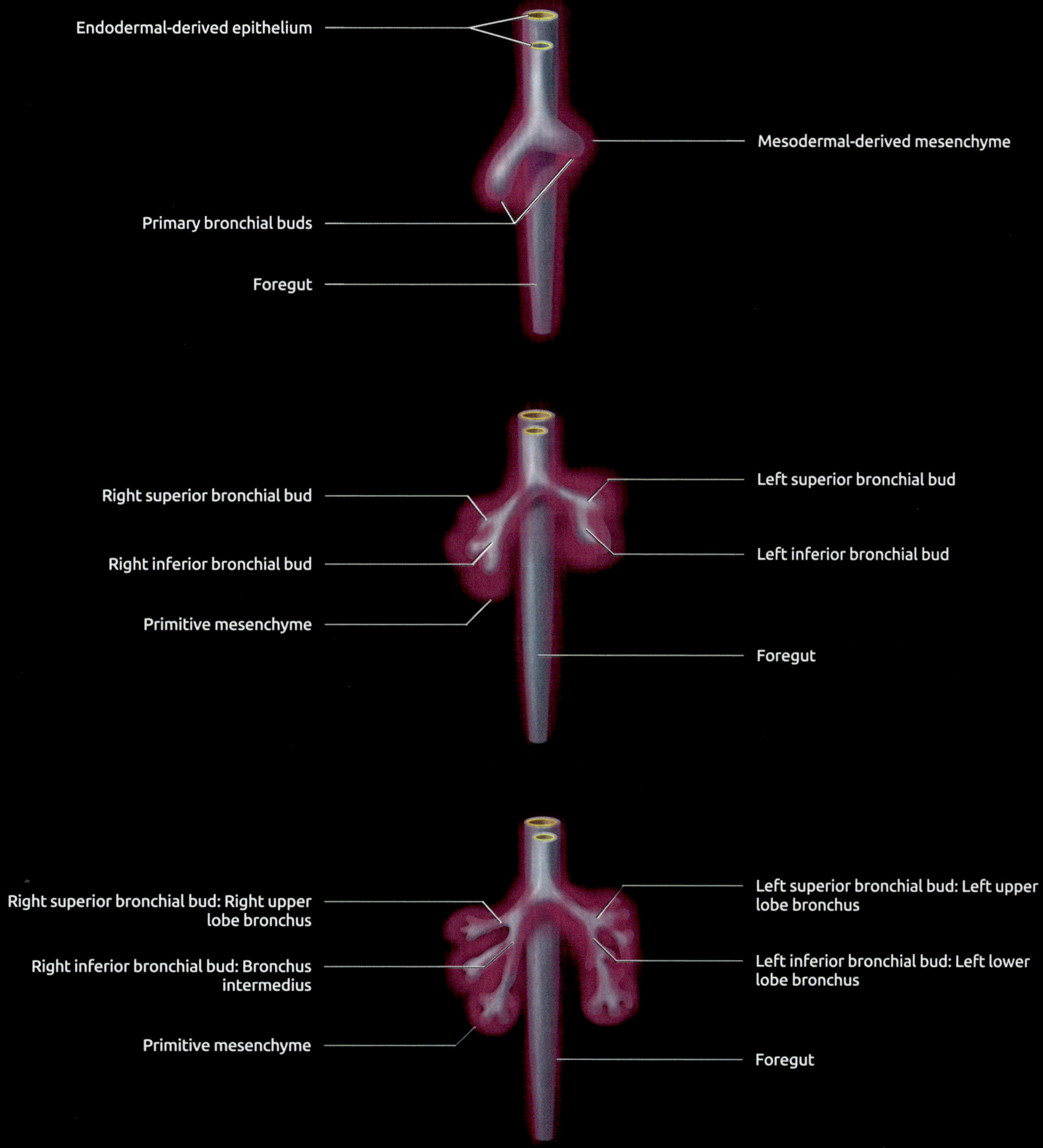

(Top) *Graphic shows the development and morphology of the bronchial buds as they invaginate into the primitive mesenchyme. The bronchial buds are the precursors of the main bronchi. Note that the right bronchial bud is vertically oriented, and the left follows a more horizontal course.* **(Middle)** *Graphic shows the tracheobronchial tree at 28 days of gestation as the right and left bronchial buds begin to divide. The developing tracheobronchial tree is surrounded by primitive mesenchyme.* **(Bottom)** *Graphic shows the developing tracheobronchial tree at 42 days of gestation with continued elongation and branching of the bronchial buds to form rudimentary lobar bronchi. Further growth and branching of the distal primitive airway forms rudimentary segmental bronchi. The rudimentary bronchus intermedius gives rise to primitive right middle and right lower lobe bronchi.*

LUNG DEVELOPMENT

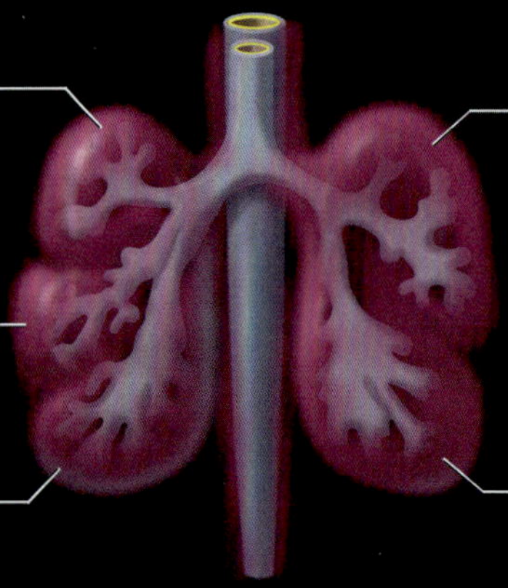

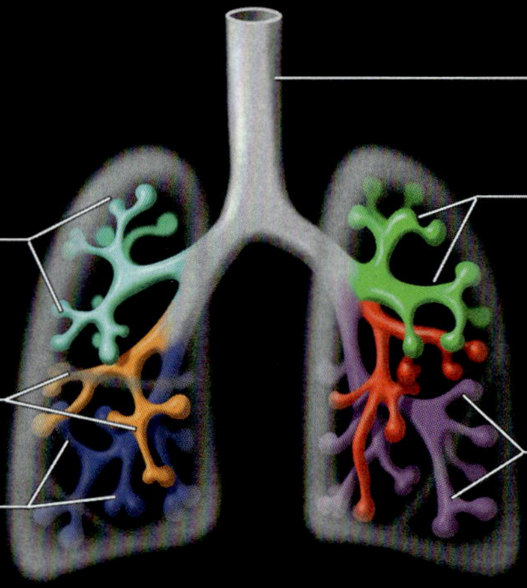

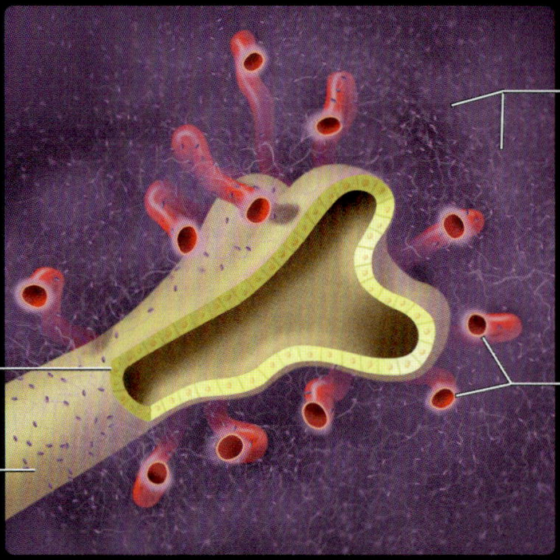

(Top) *Graphic shows tracheobronchial development at 56 days of gestation with continued branching of the primitive airways and rudimentary lung lobes.* **(Middle)** *Graphic shows tracheobronchial development at ~ 10 weeks of gestation. Note the airway differentiation into the rudimentary lobar bronchial branches (shown in different colors) and segmental bronchial branches. Note that the green and red bronchial branches represent different portions of the primitive left upper lobe. Recognizable lung lobes are present. The interaction between the primitive tracheobronchial tree and the surrounding primitive mesenchyme induces the development of lung parenchyma.* **(Bottom)** *Graphic shows the primitive airway in the pseudoglandular stage of development (6-16 weeks). The airways are blind-ending tubules. There is no alveolar-capillary interface as connective tissue separates the thick-walled primitive airway from the pulmonary capillaries. Respiration is not possible.*

LUNG DEVELOPMENT

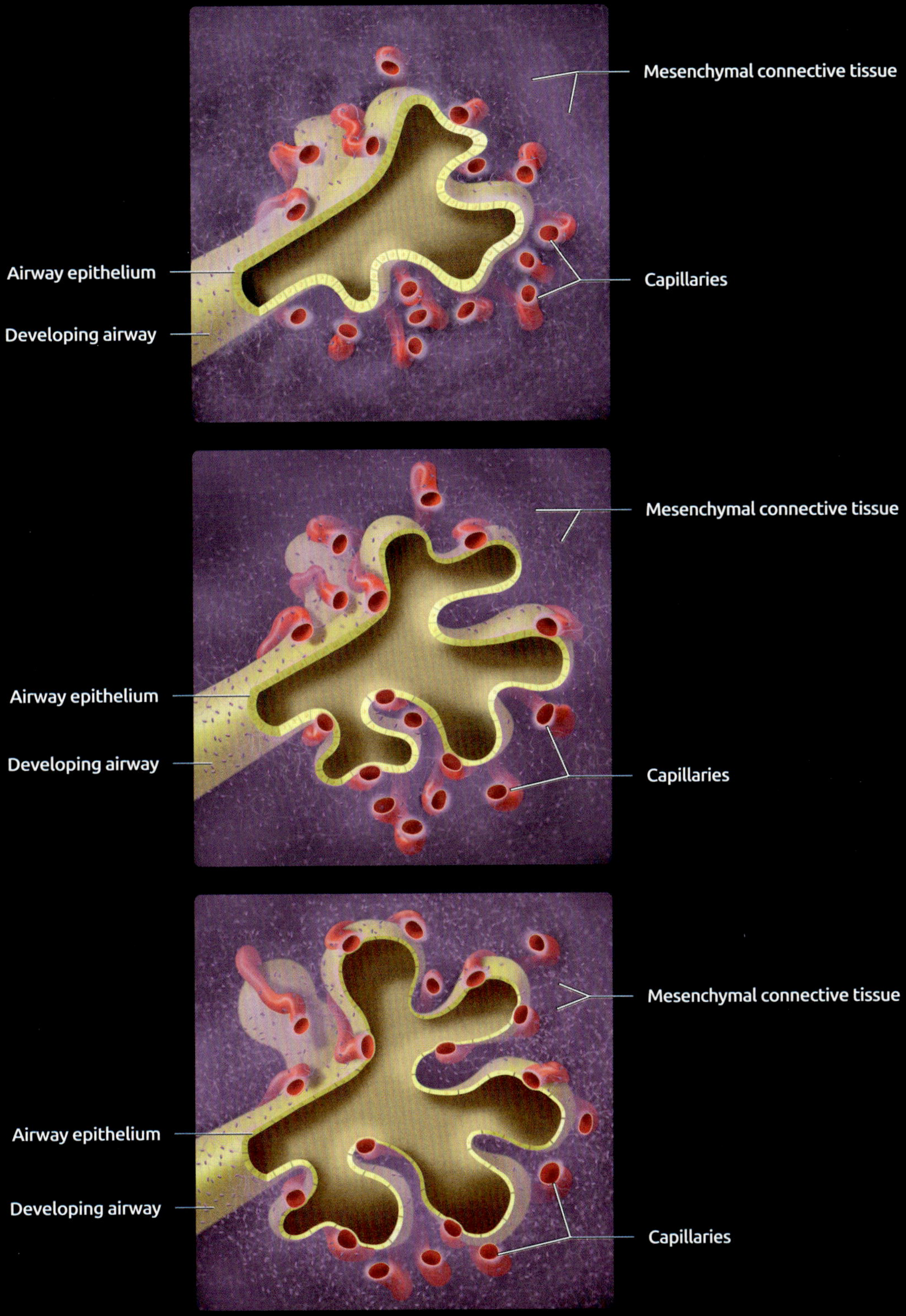

(Top) *Graphic of the canalicular stage of development (16-28 weeks) shows the airway lumen has enlarged, and the epithelium has thinned. There is an increased number of vessels within the primitive mesenchyme, some of which abut the airway wall. Respiration is possible at the end of this stage of lung development, but these infants require intensive care for survival.* **(Middle)** *Graphic of the saccular stage of lung development (28-36 weeks) shows the airway lumen continuing to enlarge. More numerous capillaries abut the wall, and some bulge into the lumen. Respiration is possible, and many infants born at this stage of pulmonary development survive with proper medical management and support.* **(Bottom)** *Graphic of the alveolar stage of pulmonary development (36 weeks to 8 years) shows continued airway enlargement with less surrounding connective tissue. The airway epithelium is thin, and many capillaries bulge into the airway lumen establishing mature alveolar-capillary interfaces. Airway development continues after birth and into childhood.*

FETAL LUNGS

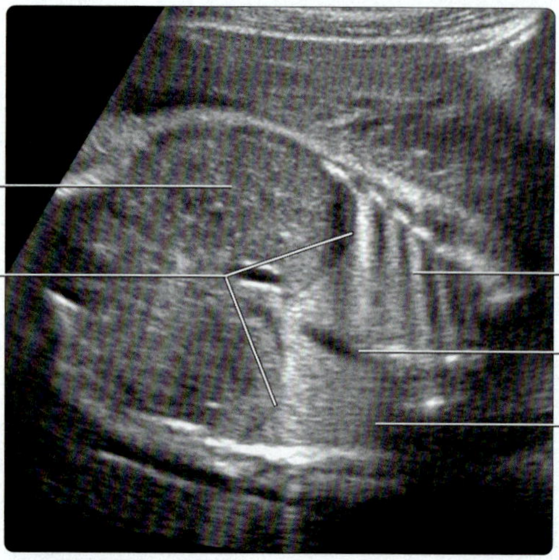

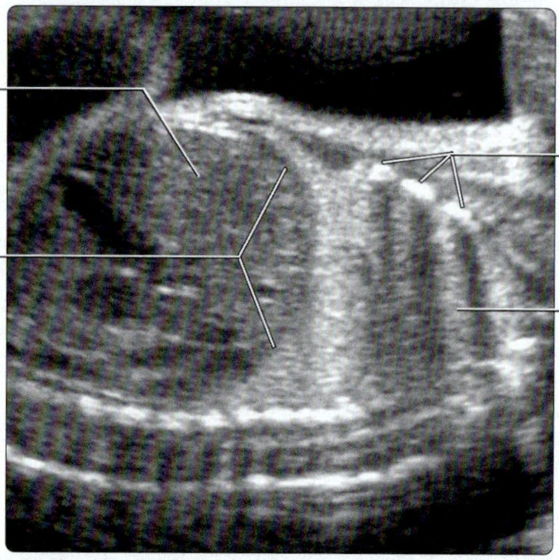

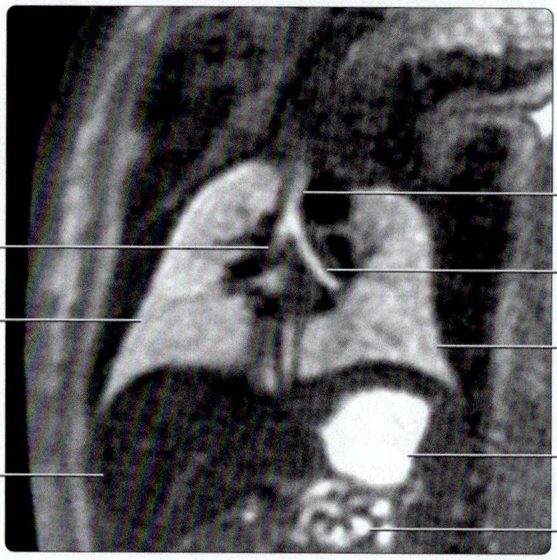

(Top) *Coronal ultrasound of a 3rd trimester fetus shows the lungs are homogeneous in echotexture and hyperechoic when compared to the liver. In the 1st trimester, the lungs and liver are similar in echogenicity, but the lungs become more hyperechoic as alveolar development progresses, creating more acoustic interfaces.* **(Middle)** *Sagittal ultrasound of the fetal chest shows the diaphragm, which is a thin, hypoechoic, muscular band. The diaphragm is best evaluated in this projection to ensure it has been seen in its entirety.* **(Bottom)** *Coronal T2WI MR of the chest shows the relatively high signal intensity lungs compared to the low signal intensity liver. Fluid-filled structures, including the trachea, bronchi, stomach, and small bowel, are all very high signal and easily distinguished on T2-weighted sequences.*

FETAL THYMUS

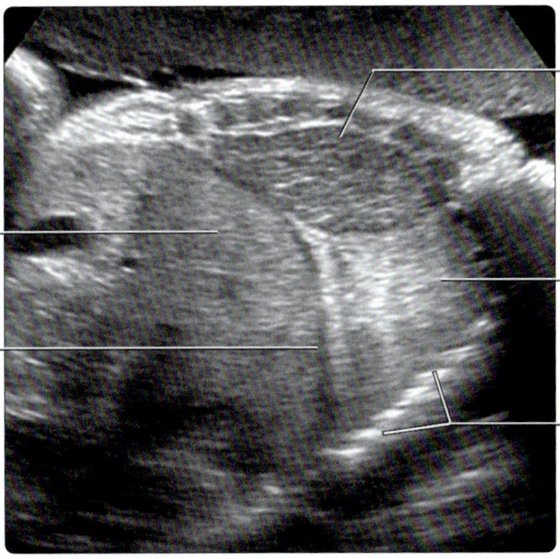

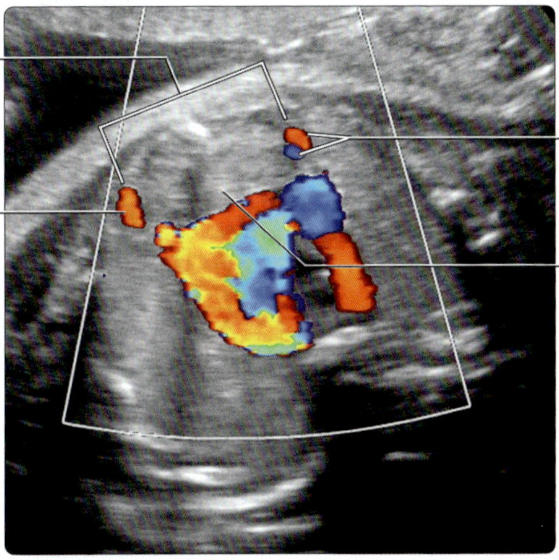

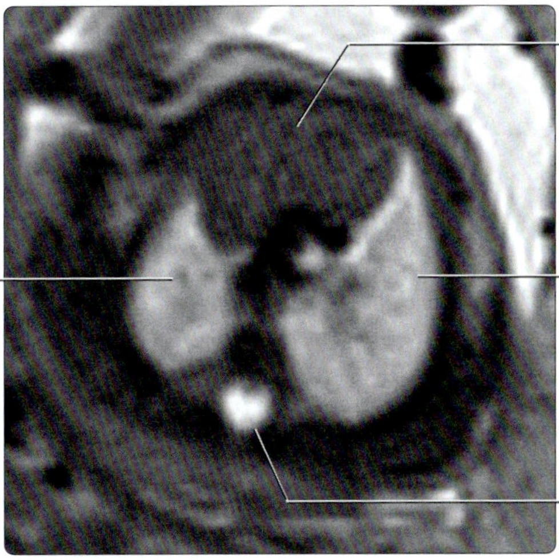

(Top) *Sagittal ultrasound of the fetal chest in the 3rd trimester shows the fetal thymus. It is hypoechoic when compared to the lungs and has a slightly reticular appearance. The thymus is often quite prominent in a fetus and should not be confused with a lung or mediastinal mass.* **(Middle)** *Axial color Doppler of the superior mediastinum at the top of the outflow tracts shows the thy-box. The internal mammary arteries, branches of the subclavian arteries, flank the thymus, creating a box appearance. Thymic measurements can be taken if there is a suspicion for hypoplasia, as in 22q11 deletion syndrome.* **(Bottom)** *Axial T2WI MR at ~ the same level shows the thymus is hypointense when compared to the lungs.*

EMBRYOLOGY OVERVIEW

Heart Fields

- Heart formed from 2 fields; primary and secondary heart fields
- Primary heart field (cardiac crescent)
  - Patterned at 16-18 days and is dependent on normal signals that establish laterality
  - Cardiac progenitor cells from **splanchnic mesoderm**
  - Angiogenic cell clusters create **endocardial tubes**
  - Body folding brings tubes ventral and midline
  - Tubes fuse in midline to create **primitive heart tube**
- Secondary heart field
  - Contributes to cardiac development at 22-28 days
  - Cardiac progenitor cells in **pharyngeal mesoderm** help develop arterial/venous poles of heart tube
    - Outflow tract, right ventricle (RV), atria
    - Contribute to myocardium, smooth muscle
  - Signaling critical in differentiation, elongation, and looping

Primitive Heart Tube

- Composed of endothelial cells surrounded by myoepicardial mantle, cardiac jelly
- From caudal → cranial (or venous → arterial) pole
  - Sinus venosus, primitive atria, primitive ventricle, bulbus cordis, truncus arteriosus
    - Aortic arch vessels sprout from arterial pole
    - Truncus arteriosus divides into ascending aorta, pulmonary artery (PA)
- Heart tube begins **contracting** around 4th week
- Effective circulation in 5th week

Looping

- Differential gene expression controlling sidedness detected as early as cardiogenic field stage
- **Differential growth** causes heart tube to fold into U shape
  - Creates typical heart shape by 28 days
- Venous pole remains anchored dorsally
- Atria migrate cephalad
- Arterial end **bends rightward** and ventral

Septation

- Occurs from 30- to 40-days gestation
- Atria
  - **Septum primum** grows down from above to meet endocardial cushions
    - Fusion with endocardial cushions closes **ostium primum**
    - Apoptosis in center of septum primum creates **ostium secundum**
  - **Septum secundum** grows to right of septum primum
  - Septum secundum fenestrates to create foramen ovale
- Atrioventricular canal
  - Initially directed toward primitive ventricle [developing left ventricle (LV)]
  - By 5th week, bulboventricular flange divides ventricles equally
  - Anterior, posterior, and lateral **endocardial cushions** grow inward
    - Cells derived from endocardium or neural crest
  - Endocardial cushion fusion creates mitral valves (MVs)/tricuspid valves (TVs), separates atria from ventricles
- Ventricles
  - Separated by muscular septum; grows as ventricles grow downward
  - Membranous septum extends from inferior endocardial cushions
  - Outflow septum extends from outflow tract cushions in bulbus cordis
- Great arteries
  - Swellings in conus and truncus create **endocardial ridges**
  - Endocardial ridges grow to separate great arteries in 7th week
  - Spiraling of great arteries results from
    - Inferior/superior orientation of truncal ridges
    - Left/right orientation of conal ridges
    - Secondary heart field directs neural crest cells through signaling pathways

ARTERIAL EMBRYOLOGY

Aortic Arches

- Paired dorsal aortae develop in mesenchyme on either side of notochord
  - Notochord is primitive axis of developing body
- Heart tube rotates into chest as cranial end of embryo bends
- Dorsal aortae follow in loop → 1st aortic arch
- Series of other arches develop and regress → adult arterial anatomy
  - 6 paired arches are blood supply for **pharyngeal arches**, connect to ipsilateral dorsal aorta
    - Right dorsal aorta, 1st, 2nd, and 5th arches regress
    - 3rd arches → internal carotid arteries
  - Left 4th arch → aortic arch
  - Right 4th arch → right subclavian artery
  - Left 6th arch → ductus arteriosus, left PA
  - Right 6th arch → right PA
- Paired dorsal aortae fuse from 4th thoracic to 4th lumbar vertebrae → single midline aorta

Conotruncus

- Conotruncus constitutes outflow tract of primitive heart tube
- Endocardial ridge growth separates truncus arteriosus into ascending aorta and PA
  - Same process creates **aortic and pulmonary valves**
- Aortic valve typically rightward of and posterior to pulmonary valve
- Initially 2 subarterial coni
  - Subpulmonary conus persists
  - Subaortic conus resorbs
    - Fibrous continuity between aortic and MVs
    - Aorta "anchored" into LV
      - Gooseneck deformity in atrioventricular canal due to lack of septal-aortic continuity

Pulmonary Artery

- Anterior and to left of aortic root

- In fetus, main PA (MPA) trifurcates into ductus arteriosus and right and left PAs
- In adults, MPA bifurcates into right and left branches as it leaves pericardium
  - Atrophic ductus arteriosus becomes ligamentum arteriosum

VENOUS EMBRYOLOGY

Sinus Venosus

- Has right and left horns
- Each horn receives **vitelline, umbilical, and cardinal** veins
- With differential growth, sinus entrance shifts rightward into developing right atrium (RA)

Systemic Vein Development

- Right anterior cardinal vein → superior vena cava (SVC)
- Left anterior cardinal vein involutes, left sinus venosus horn → coronary sinus
  - Persistent left anterior cardinal vein results in left SVC
- Vitelline veins drain yolk sac, develop into hepatic/portal venous system
- Left umbilical vein (UV) drains through **ductus venosus** to return **oxygenated** placental blood to fetal heart
- Right UV involutes
- UV also enters liver via portal sinus
  - Obliterated UV becomes ligamentum teres

CARDIAC ANATOMY

Segmental Nomenclature

- Each segment of heart is identified based on morphology
- Cardiac anatomic description is based on segments, relative locations, how connected

Normal Chamber/Vessel Morphology

- **RA** identified by limbic bands of fossa ovalis, large pyramidal appendage with pectinate muscles that extend outside appendage, and crista terminalis
  - Fetal marker for RA is systemic venous connection
- **Left atrium (LA)**: Identified by finger-like appendage, pectinate muscles confined to appendage, no crista terminalis
  - Fetal marker for LA is pulmonary venous connection
- **RV** is heavily **trabeculated** with **moderator band**
  - TV belongs to RV
    - More apically placed than MV, has septal and free wall attachments
  - Fetal marker for RV is moderator band
- **LV** is **smooth walled** with fine apical trabeculations
  - MV belongs to LV, only has free wall attachment
  - Fetal marker for LV is smooth internal contour/absence of moderator band
- **PA** should come off RV, branches early into ductus arteriosus, left and right branches
- **Aorta** should come off LV, head and neck vessels come off apex of arch
  - Continues to descending aorta, which connects with ductus arteriosus right after isthmus

CIRCULATION

Fetoplacental

- Arteries move blood **away** from heart
  - Aorta takes blood to brain and body
  - **Umbilical arteries** (internal iliac branches) carry deoxygenated blood away from heart to placenta
    - Become medial umbilical ligaments in adults
  - MPA takes blood from RV to body via ductus arteriosus
    - Only small volume circulates through lungs in fetal life (no gas exchange)
- Veins take blood to heart
  - **UV** brings oxygenated blood back to heart, from placenta
    - Oxygenated blood preferentially crosses foramen ovale to left heart to be directed to brain
  - Inferior vena cava (IVC) brings deoxygenated blood from body to RA
  - SVC brings deoxygenated blood from head to RA
- Distribution of cardiac output is very different in fetus compared to adult
  - **Ductus arteriosus** connects MPA to descending aorta
    - Allows oxygenated blood in RV to bypass lungs and be directed to body
      - Becomes **ligamentum arteriosum** in adults
  - Fetal combined cardiac output (CCO) is 55% RV, 45% LV
    - ~ 40% CCO → body via **ductus arteriosus**
    - ~ 15% CCO → lungs via branch PAs
    - ~ 30% CCO → brain via ascending aorta
    - ~ 10% CCO → body via aortic arch/isthmus and descending aorta
    - ~ 3% CCO → heart via coronaries arteries
- Streaming optimizes delivery of oxygenated blood to head
  - Oxygenated UV blood enters RA via ductus venosus and IVC
    - Jet streams preferentially across foramen ovale to LA, LV
    - Blood with highest oxygen content perfuses brain, heart
  - Deoxygenated systemic venous blood drains to RA via SVC, IVC
    - Jet streams preferentially into RV

Neonatal

- Onset of breathing increased oxygen in lungs, PAs dilate in response
  - **Decreased pulmonary arterial resistance**
- Low-resistance arterial connection to placenta removed when cord clamped
  - **Increased systemic arterial resistance**
- End result is decreased flow across ductus arteriosus, increased flow in MPA
  - Ductus arteriosus closes
- Pulmonary veins bring oxygenated blood from lungs to LA
- Increased LA pressure closes foramen ovale flap

PRIMITIVE HEART TUBE

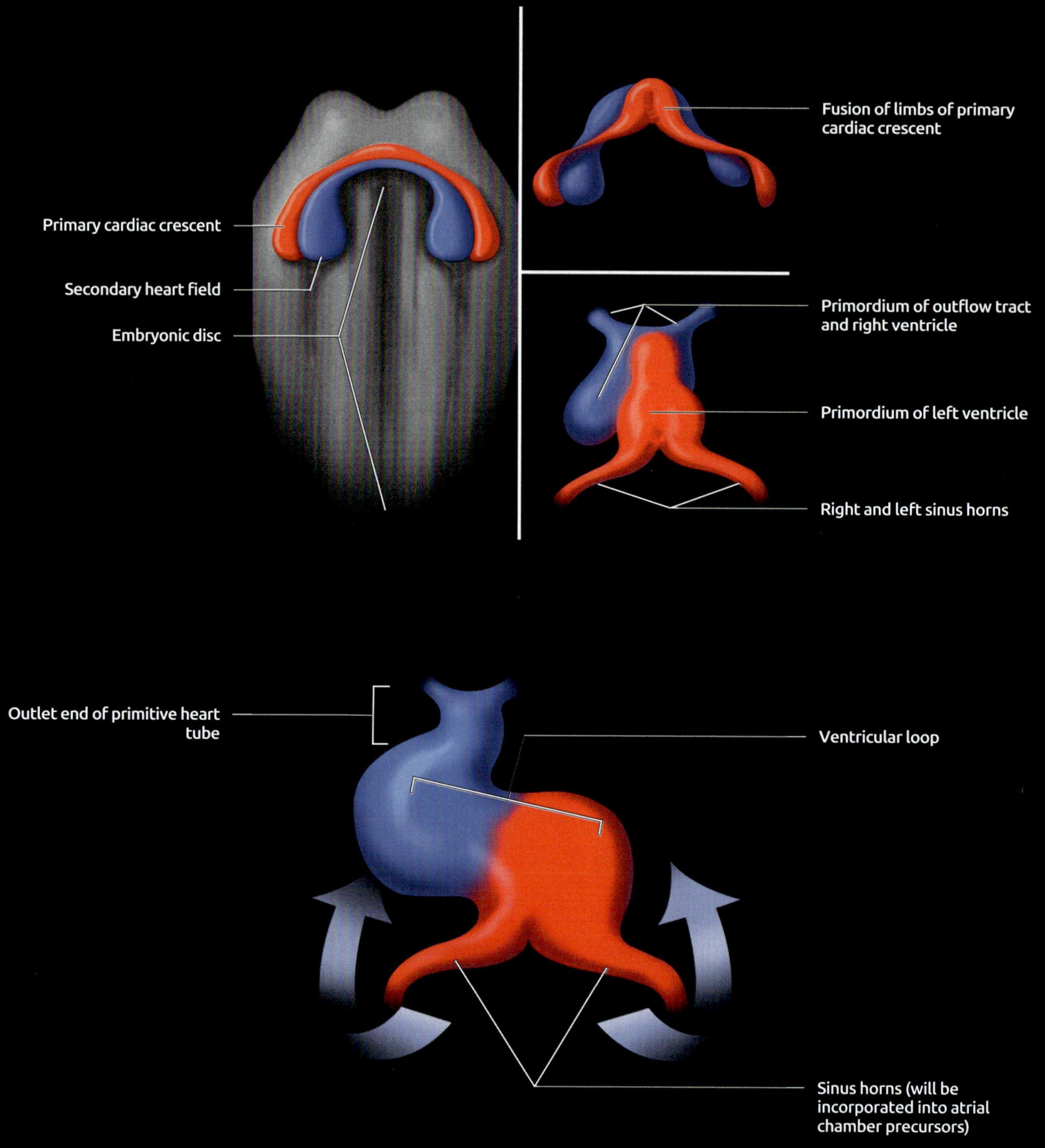

(Top) *Cells destined to form the heart derive from mesoderm, forming the primary cardiac crescent at the cranial border of the embryo. The secondary heart field lying contiguous with, but medial to, the primary cardiac crescent populates the outflow tract and primordium of the right ventricle. As the embryo elongates and folds, the limbs of the crescent come together in the midline and fuse, creating the heart tube (which moves into the thorax).* **(Bottom)** *As the straight heart tube elongates, it rotates and folds on itself due to differential growth in a standard D loop, moving the ventricles to the right and left sides. The proximal, venous pole remains anchored dorsal and will become part of the atria, while the arterial pole bends rightward and ventral and will become the outflow tract.*

4-CHAMBER HEART

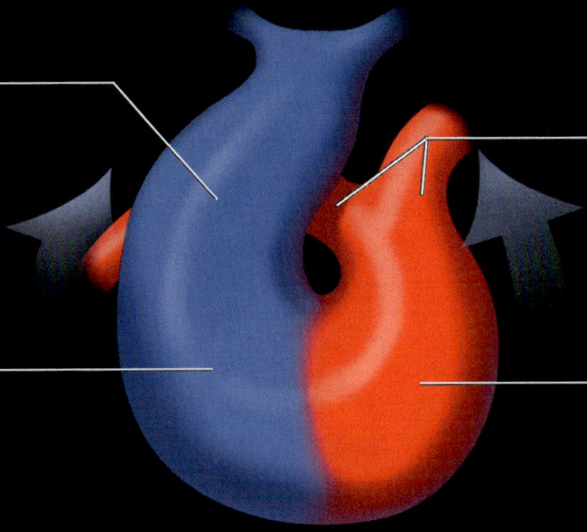

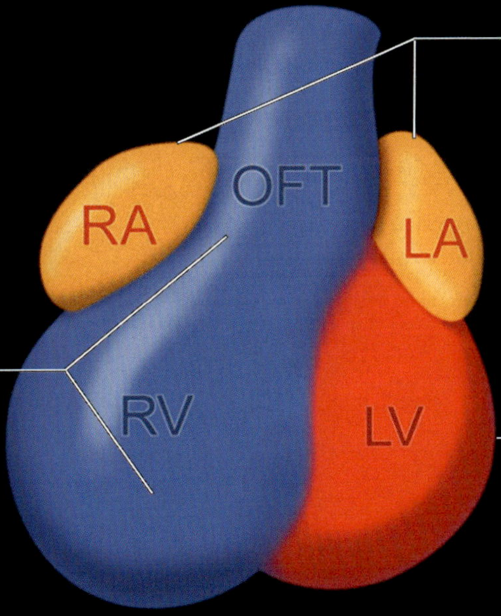

(Top) *Cellular growth leads to both the looping and development of the definitive cardiac chambers. The right and left ventricles are shown as well as the eventual right atrium and great vessels.* **(Bottom)** *Graphic shows a schematic of the early 4-chamber heart. The left ventricle is derived from the primary heart field (red), the right ventricle, and outflows from the secondary heart field (blue). The tertiary field (orange) contributes to formation of the atria and provides cellular components to the ventricles.*

VENTRICLES

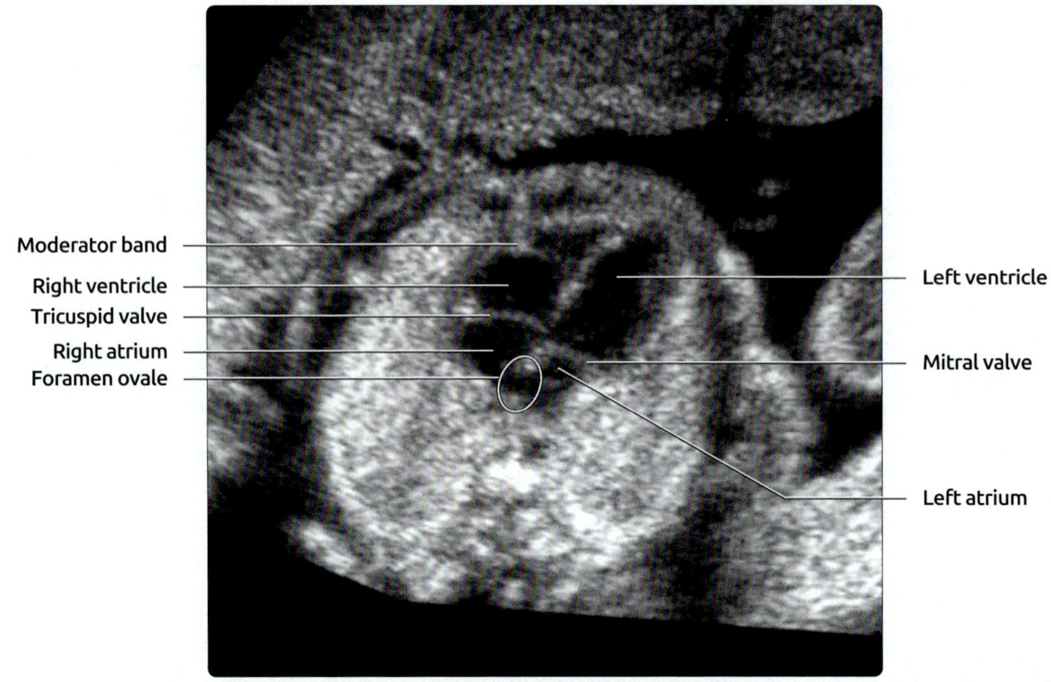

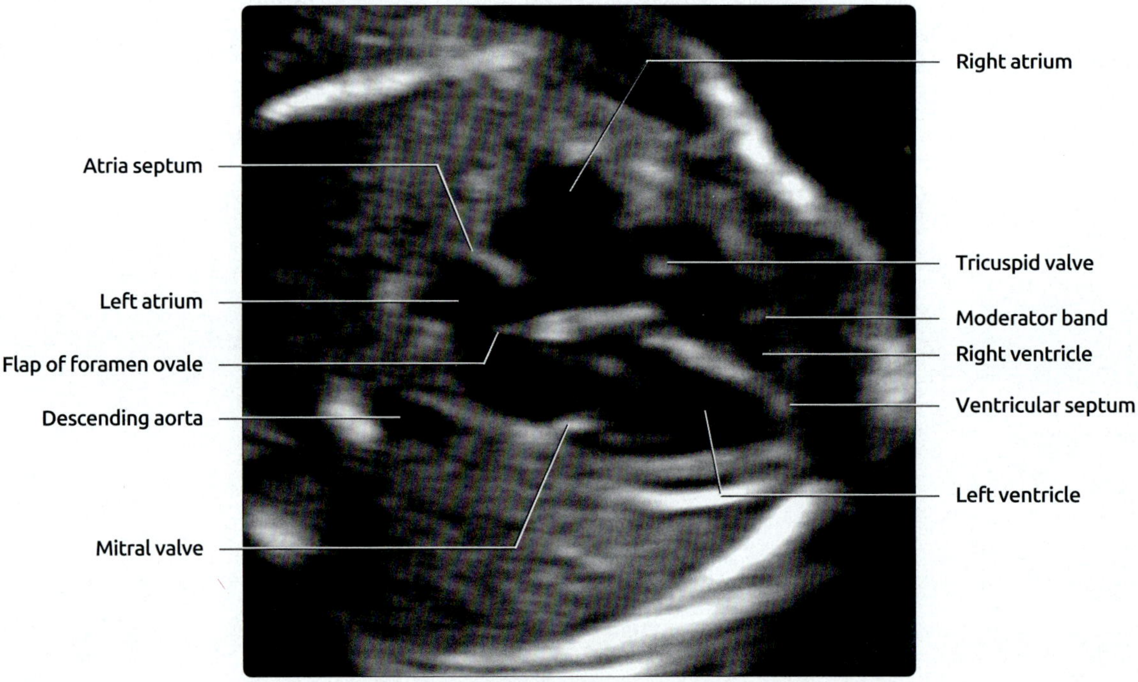

(Top) *Four-chamber view in the 2nd trimester provides excellent anatomic detail, given that the heart is about the size of a dime and is beating at 130-160 beats per minute. Note the moderator band in the trabeculated right ventricle; this can be used to identify the morphologic right ventricle, which should always be the anterior ventricle.* **(Bottom)** *This is a more zoomed-in 4-chamber view using the ultrasound machine's cardiac setting, which creates more contrast to bring out the fine details of the cardiac structures. Note that the flap of the foramen ovale is in the left atrium, which signifies right-to-left flow with the oxygenated stream of blood from the umbilical vein and ductus venosus crossing to the left to provide oxygenated blood to the brain.*

VENTRICULOARTERIAL CONNECTIONS

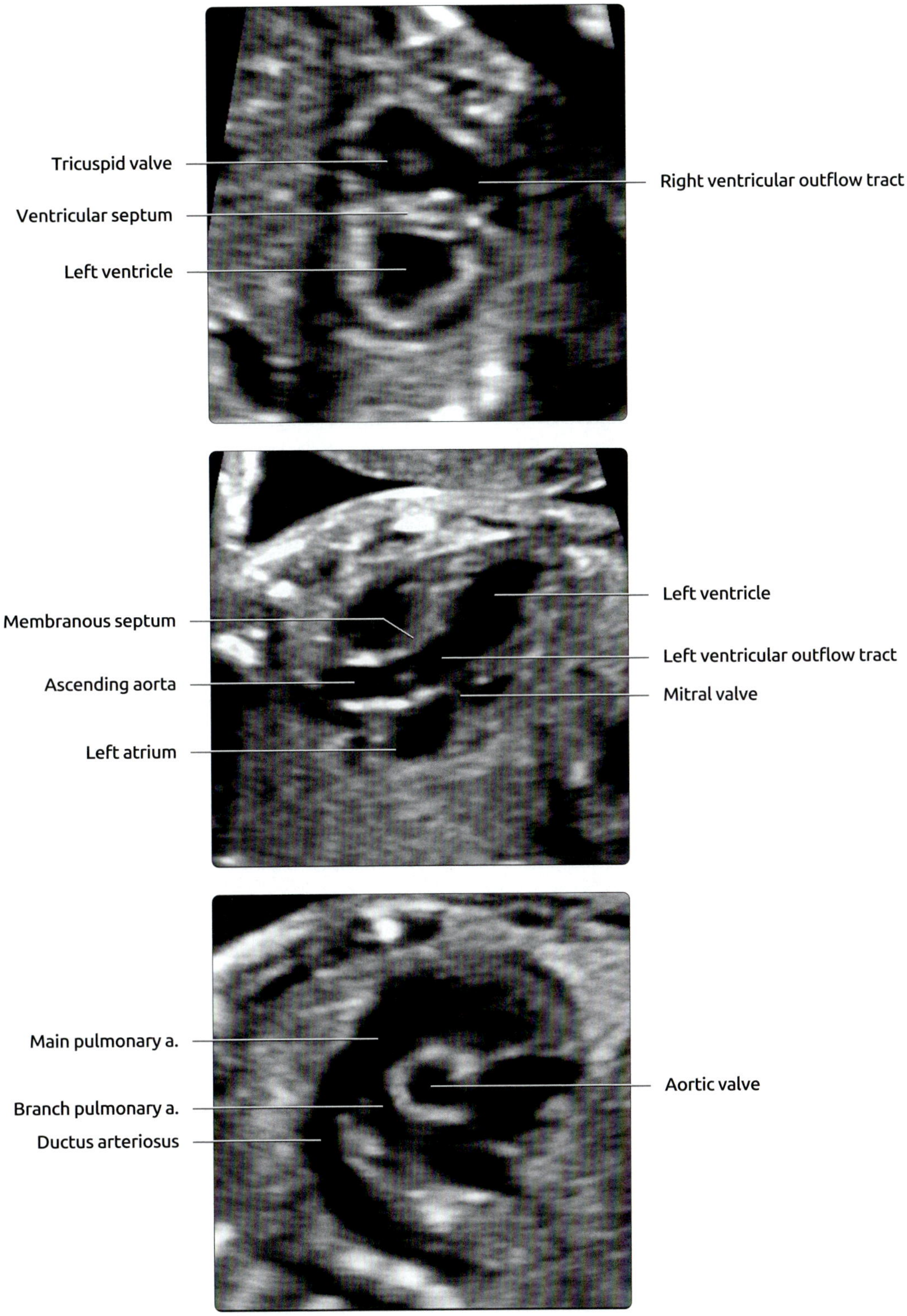

(Top) *A short-axis view through the ventricles is not a standard view in OB ultrasound but is an important component of fetal echocardiography as it allows side-by side assessment of ventricular size and function in real time. The tricuspid valve is seen in cross section in the center of the right ventricular cavity.* **(Middle)** *The left ventricular outflow tract view is the best view to evaluate the membranous ventricular septum. A defect in this area may simply be an isolated perimembranous ventricular septal defect but may also be found in association with right ventricular outflow tract or conotruncal lesions, such as tetralogy of Fallot or double outlet right ventricle.* **(Bottom)** *This is the standard right ventricular outflow tract view, which is a short-axis view at the level of the aortic valve. It allows one to lay out the main pulmonary artery and the ductus arteriosus as it runs posteriorly, toward the spine, to join the descending aorta.*

VENOATRIAL CONNECTIONS

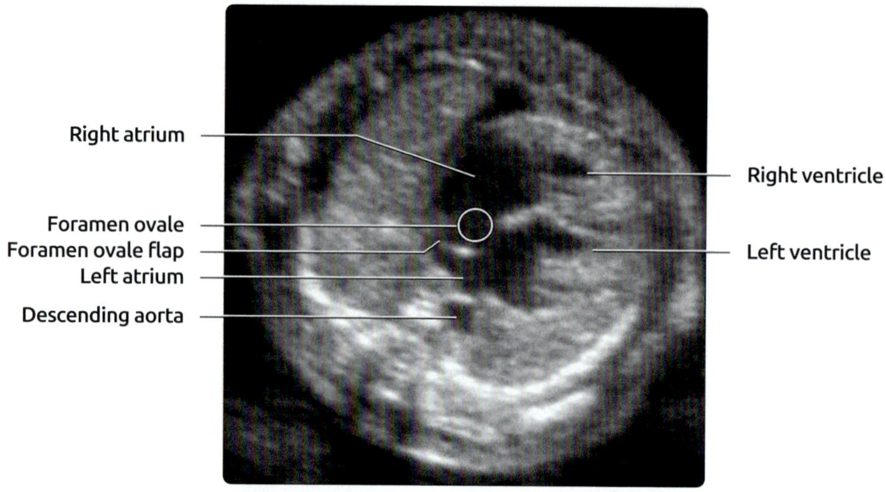

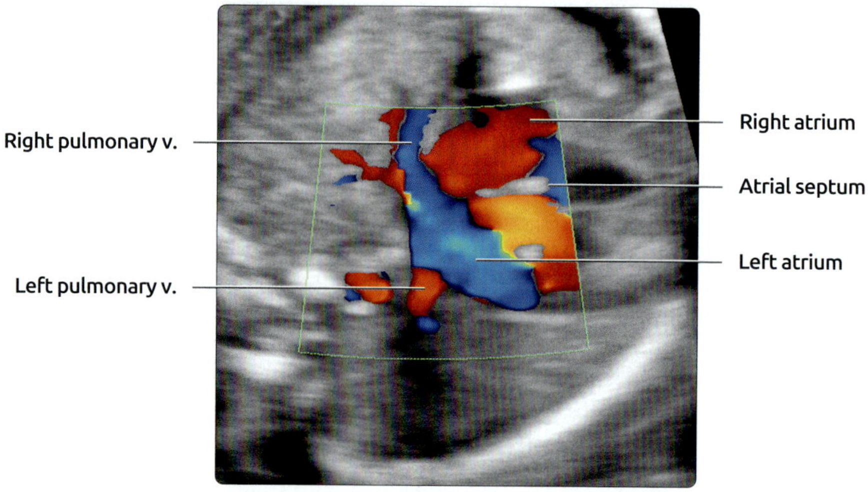

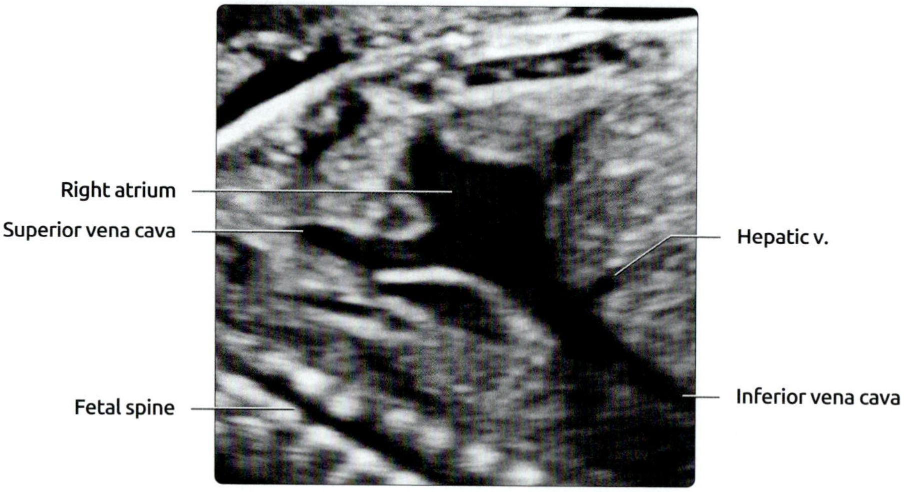

(Top) *Abnormal 4-chamber view in a monochorionic twin with ischemic cardiomyopathy following the demise of the co-twin illustrates that the atria and ventricles may be symmetric even when abnormal. The atria are enlarged due to atrioventricular valve regurgitation. The regurgitation is secondary to myocardial ischemia and impaired ventricular contraction.* **(Middle)** *Image shows a right-sided pulmonary vein that drains across the midline to the left atrium as well as a left-sided pulmonary vein draining into the left atrium. Note the right vein is blue, so the probe is anterior and to the right of the fetus.* **(Bottom)** *This is the bicaval view demonstrating systemic venous return to the right atrium. Note the hepatic vein confluence with the inferior vena cava (IVC). In azygous continuation of the IVC, the hepatic veins may drain directly to the right atrium, but they are small and should not be mistaken for the IVC. In azygous continuation of the IVC, the enlarged azygous vein is seen as 2nd vessels posterior to the heart on the 4-chamber view.*

ANATOMIC VARIANT: LEFT SUPERIOR VENA CAVA

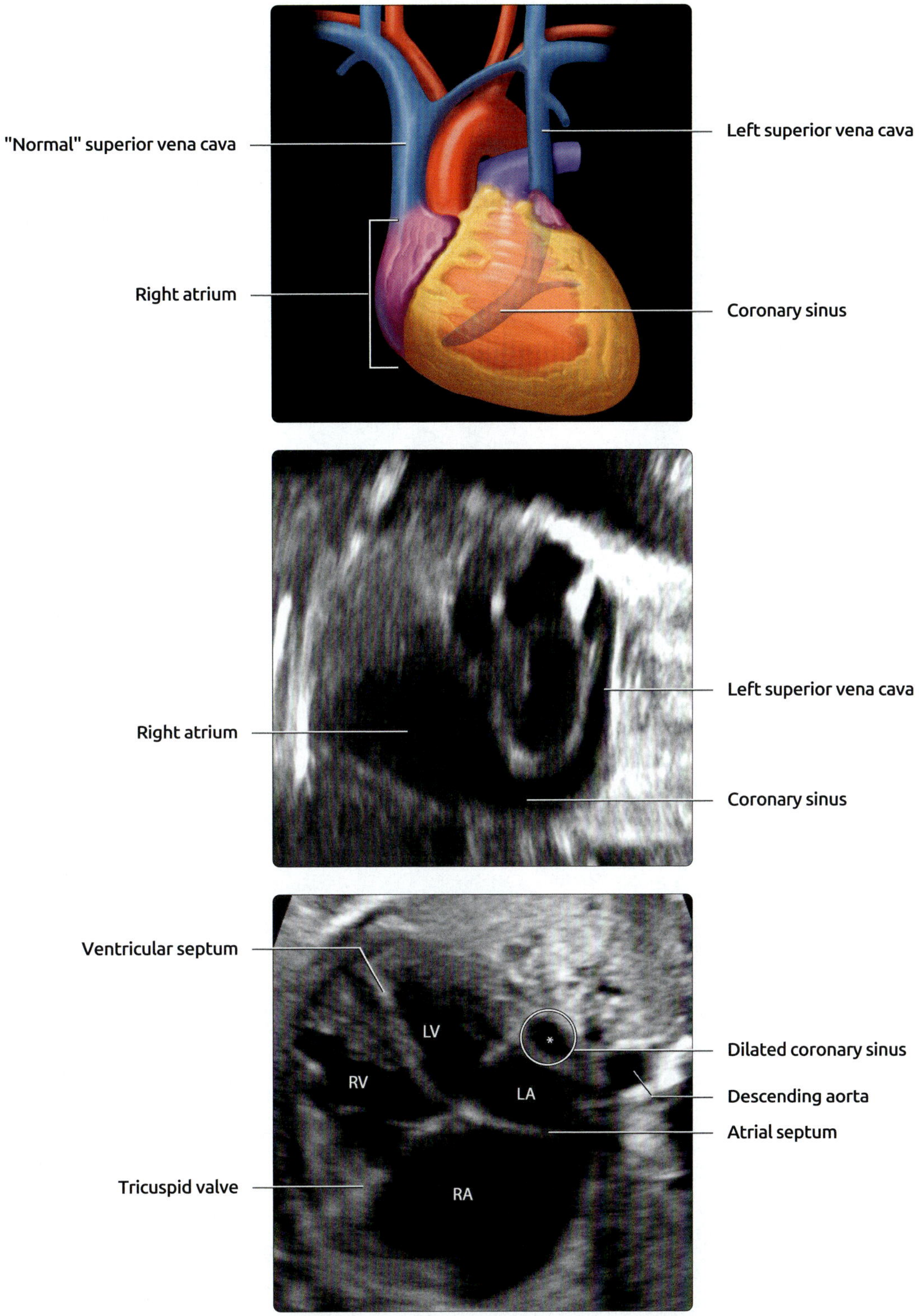

(Top) *Graphic illustrates the anatomic variant of a persistent left superior vena cava (SVC). This occurs when the left anterior cardinal vein fails to involute. The left SVC drains into the right atrium via the coronary sinus, which is enlarged due to the increased volume of blood entering it. Enlargement of the coronary sinus is visible on the 4-chamber heart view.* **(Middle)** *Sagittal view of the left SVC draining to a dilated coronary sinus and into the right atrium is shown. Be careful to check for anomalous pulmonary venous return to the coronary sinus, another important cause of dilation of this structure.* **(Bottom)** *Four-chamber view with a dilated coronary sinus (*) is shown. This finding is suspicious for a persistent left SVC.*

FETAL CIRCULATION

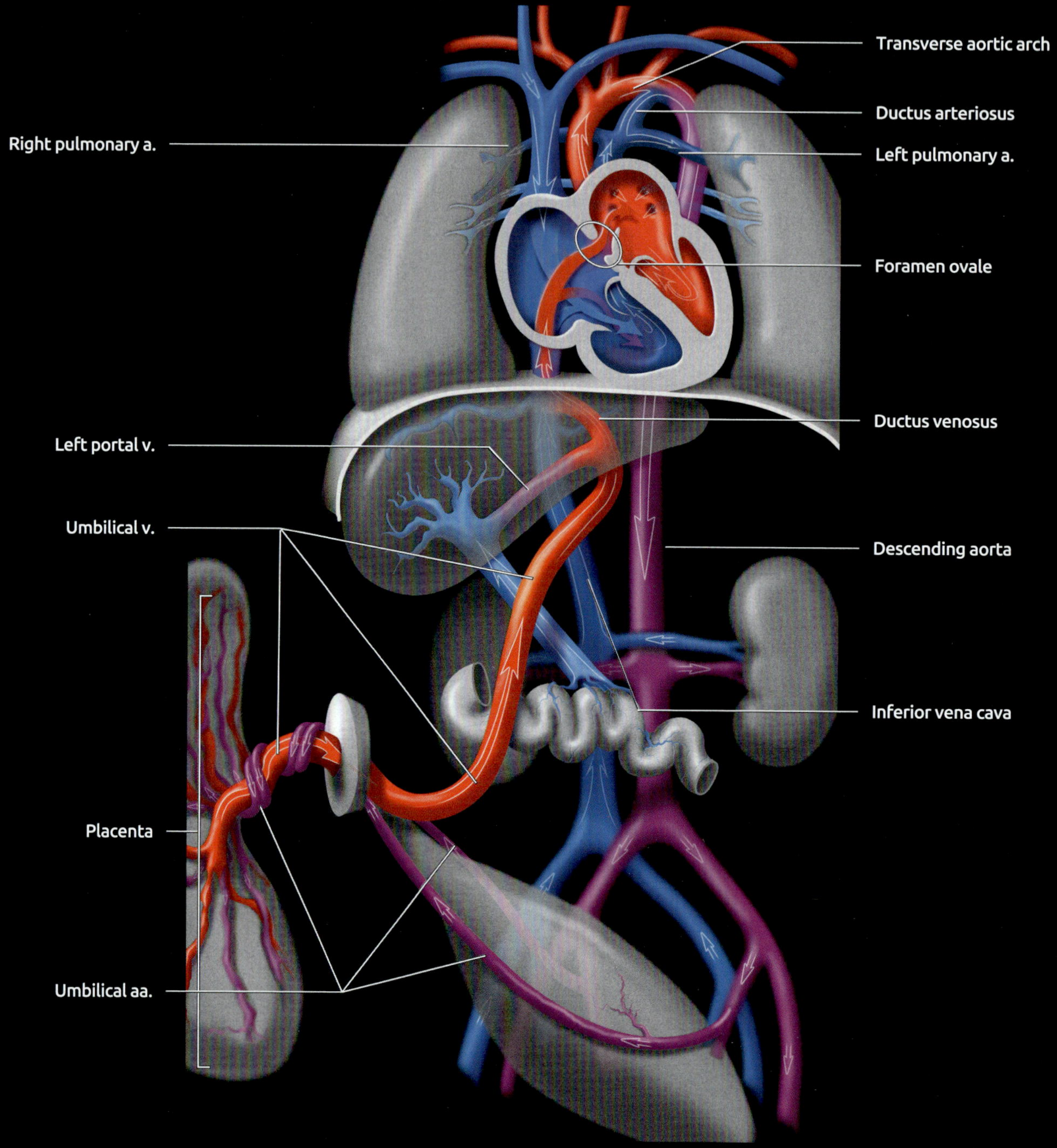

In the fetus, blood is oxygenated by the placenta and returns to the heart via the umbilical vein. This highly oxygenated blood (red) shunts through the ductus venosus and streams across the foramen ovale to the left side of the heart, supplying the head. Deoxygenated blood (blue) returns to the right atrium via the SVC and IVC. This blood preferentially flows to the right ventricle, which pumps a small amount to the pulmonary arteries but the majority is shunted across the ductus arteriosus to the descending aorta.

POSTNATAL CIRCULATION

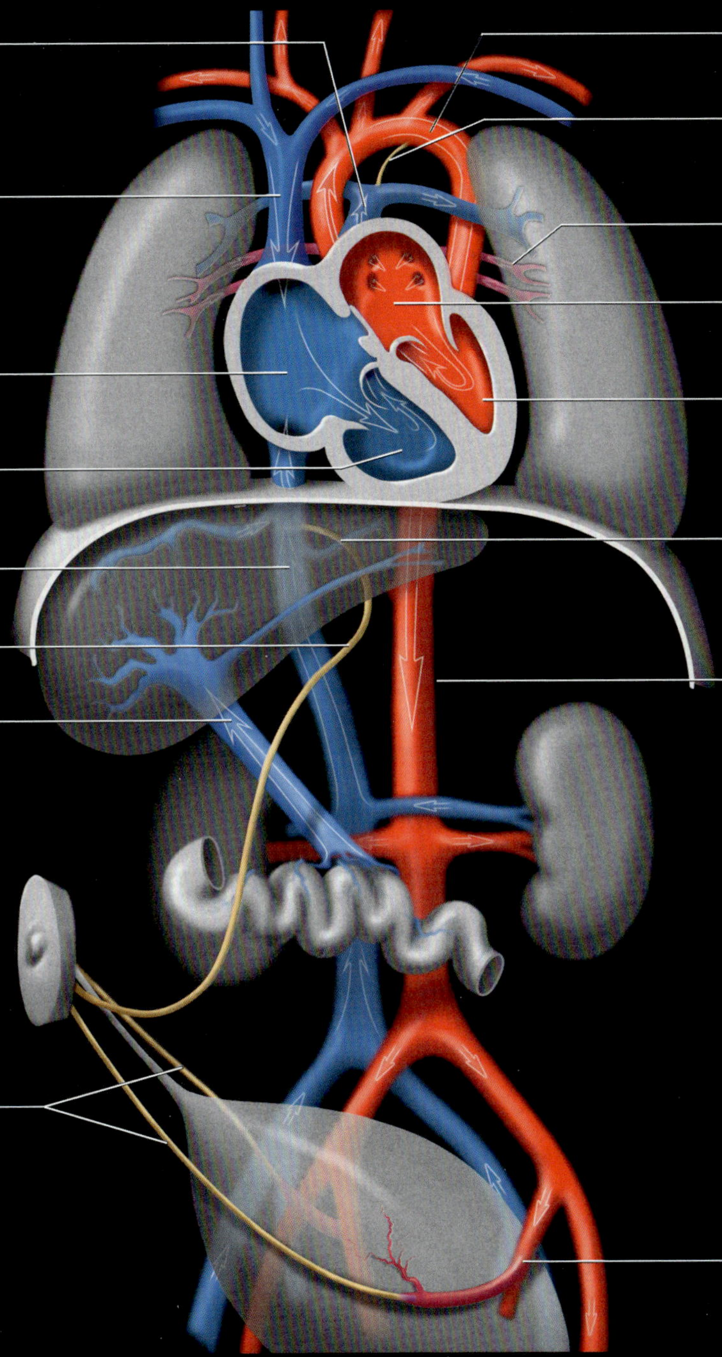

In adult circulation, blood is oxygenated in the lungs and returns to the left heart via the pulmonary veins. The left heart pumps to the body via the aorta and its branches. Deoxygenated blood returns via the SVC and IVC to the right side of the heart, which pumps deoxygenated blood to the lungs for gas exchange. The umbilical arteries become the medial umbilical ligaments, the umbilical vein becomes the ligamentum teres, and the ductus arteriosus becomes the ligamentum arteriosum.

EARLY EMBRYOLOGIC EVENTS

2nd Week: 8-14 Days Post Conception

- **Rule of 2s**
  - Embryoblast splits into 2 layers: Epiblast and hypoblast
  - Trophoblast gives rise to 2 tissues: Cytotrophoblast and syncytiotrophoblast
  - Blastocyst cavity is remodeled twice: Primary and secondary **yolk sac**
  - 2 novel cavities appear: Amnion and chorion
  - Extraembryonic mesoderm splits into 2 layers lining chorionic cavity: **Somatic** and **splanchnic mesoderm**

3rd Week: 15-21 Days Post Conception

- Bilaminar disk → trilaminar disk (gastrulation begins)
- **Gastrulation**: Process of forming 3 primary germ layers
  - **Ectoderm** → epidermis, retina, central and peripheral nervous systems, other
  - **Mesoderm** → smooth muscle, connective tissue, vessels, most of cardiovascular system, blood cells, bone marrow, skeleton, striated muscles, reproductive and excretory organs
  - **Endoderm** → epithelial linings of respiratory passages and GI tract, including glands opening into GI tract and glandular cells of liver and pancreas
- Endoderm and ectoderm separated by mesoderm in all areas of gut with 2 exceptions
  - Areas of mesodermal deficiency: Oropharyngeal membrane (future oropharyngeal cavity) and cloacal membrane (future area of urethra and anus)
- **Allantois** appears as diverticulum from caudal yolk sac on day 16 and extends into connecting stalk
  - Involved in development of bladder in humans and early blood formation
  - Becomes urachus as bladder enlarges
  - Blood vessels of allantois become umbilical arteries and veins

4th Week: 22-28 Days Post Conception

- Rapid growth results in folding of embryo
  - 4 folds: Cranial, caudal, 2 lateral
- Folding causes yolk stalk to narrow, bringing it in close proximity to body stalk
- **Umbilical ring** surrounds both yolk and body stalks
- Allantois contained within body stalk
- Lateral edges of trilaminar disk fold ventrally to form **body wall**, moving toward yolk stalk and body stalk

5th Week: 29-35 Days Post Conception

- Viscera develop in mesentery of caudal part of foregut
- Fetal stomach is suspended by 2 mesogastria, and rotation of stomach begins
  - **Dorsal mesogastrium**: Site for developing spleen, body-tail of pancreas
  - **Ventral mesogastrium**: Site for developing liver, bile ducts, head of pancreas
    - Dorsal part of ventral mesogastrium becomes lesser omentum
    - Lesser omentum includes gastrohepatic ligament and hepatoduodenal ligament
- **Foregut**: Caudal to liver bud forms esophagus, stomach, and proximal duodenum
- **Midgut**: From liver bud to 2/3 of transverse colon; opens into yolk sac
  - Gut tube begins to lengthen → primary intestinal loop (attached to yolk sac via yolk/vitelline duct)
  - Axis of loop is superior mesenteric artery (SMA)
  - Yolk duct and connecting stalk begin to merge
- **Hindgut**: Gives rise to distal 1/3 of transverse colon, descending colon, sigmoid, rectum, and upper anal canal
  - Endoderm of hindgut also forms internal lining of bladder and urethra
  - Caudal end of hindgut terminates in endodermally lined **cloaca**
    - Cloaca includes base of allantois
  - Shelf of mesodermal tissue, **urorectal septum**, sits between hindgut and base of allantois

6th-7th Weeks: 36-49 Days Post Conception

- Merging of yolk stalk and body stalk to form **umbilical cord** is complete
  - Yolk stalk atrophies: Failure to regress completely → **Meckel diverticulum**, blind outpouching from distal ileum
- **Physiologic herniation**
  - Length of midgut increases, volume of gut is greater than body can accommodate → herniation into base of umbilical cord
  - Rotates 90° counterclockwise around axis of SMA (as viewed from front of embryo)
  - Folds grow out from urorectal septum to partition cloacal membrane and divide cloaca into rectum and urogenital sinus
    - Cloacal membrane is divided into anal membrane and urogenital membrane

8th Week: 50-58 Days Post Conception

- Fusion of urorectal septum, lateral mesodermal folds, and cloacal membrane to form perineal body, partition between GI and urogenital systems
- Cloacal membrane ruptures by beginning of 8th week, creating anal opening for hindgut and ventral opening for urogenital sinus

9th Week

- Abdominal cavity has enlarged sufficiently to accommodate intestines, which begin to migrate back into abdomen

10th Week

- Upon return of intestines into abdominal cavity, rotation proceeds additional 180° for total of 270°

ABDOMINAL VESSELS

Arteries

- Major fetal arteries course anteriorly through dorsal mesenteries from aorta to supply gut and intramesenteric viscera
- Omphalomesenteric arteries supply yolk sac, then gradually fuse to form arteries in dorsal mesentery of gut
  - **Celiac** artery supplies foregut, **superior mesenteric** artery supplies midgut, **inferior mesenteric** artery supplies hindgut

Veins

- **Umbilical vein**
  - Carries oxygenated blood from placenta to fetus (major source of blood flow to liver)
  - Enters liver through ventral part of ventral mesentery (**falciform ligament** in adults)
  - Obliterated umbilical vein → **ligamentum teres**
- **Vitelline veins**
  - Paired vessels that carry blood from yolk sac to fetus in 1st few weeks of gestation
  - Give rise to venous plexus within liver
    - Precursor to hepatic and portal veins and sinusoids
  - Proximal extrahepatic veins → **portal venous system**
    - Carries blood (and nutrients) from gut to liver
  - Proximal vitelline veins → **hepatic vein precursors**
    - Carry blood from liver to heart via inferior vena cava (IVC)
- **Ductus venosus**
  - Derived from left umbilical vein (after right vein has atrophied)
  - Bypasses liver to carry umbilical vein blood to IVC and heart
  - In neonate, atrophies to become **ligamentum venosum**
- **Portal sinus**
  - In fetus, diverts some oxygenated blood from umbilical vein to liver parenchyma

ABDOMINAL ORGANS

Abdominal Viscera

- Alimentary tube
  - Foregut (esophagus, stomach, duodenum)
  - Midgut (small intestine, colon up to splenic flexure)
  - Hindgut (descending and sigmoid colon, rectum)
- Intramesenteric viscera develop from diverticula of ventral or dorsal foregut
- Supporting mesentery

Small and Large Intestine

- **Duodenum**
  - In fetus, is intraperitoneal, has mesoduodenum
    - Ventral pancreas also lies in mesoduodenum
  - Becomes retroperitoneal organ when ascending mesocolon fuses to posterior abdominal wall, "trapping" duodenum and pancreas in retroperitoneum
- **Small intestine**
  - Develops within dorsal mesentery, which elongates and persists into adulthood as small bowel mesentery
- **Large intestine (colon)**
  - Develops as straight tube within dorsal mesentery
- **Ascending** and **descending colon** usually lose their mesentery and become retroperitoneal structures in adult
  - Common variant: Ascending colon that is mobile due to persistent mesocolon (predisposes to twist and obstruct, "cecal volvulus")

Liver

- Arises from ventral bud of foregut
- Rapid growth is main factor in distortion of peritoneal spaces and mesentery
- Rotates counterclockwise and attaches to right side of diaphragm at bare area
- Rotation of liver results in **right peritoneal space** extending leftward, posterior to stomach
  - Becomes **lesser sac (omental bursa)**

Spleen

- Develops within dorsal mesogastrium, which elongates to form **gastrosplenic ligament**
  - Carries **short gastric vessels** and forms left anterior wall of **lesser sac (omental bursa)**
  - Elongated caudal parts of gastrosplenic ligament hang down (drape-like) from stomach
    - Forms **greater omentum** and **gastrocolic ligament**
  - Greater omentum and gastrocolic ligament carry **gastroomental vessels**

Pancreas

- Develops within dorsal part of **dorsal mesentery**
  - Leaves only short **splenorenal ligament**
    - Carries **splenic vessels** and tail of pancreas
    - Forms left posterior wall of **lesser sac**
- Pancreas becomes retroperitoneal organ

Peritoneal Spaces

- **Ventral mesentery** resorbs to allow communication between right and left peritoneal cavity in adults
- Variations in complex rotation, fusion, and growth of mesenteric viscera result in common variations in peritoneal and retroperitoneal spaces in adults

SELECTED DEFECTS IN DEVELOPMENT

Omphalocele

- Lack of continuous folding results in failure of closure of umbilical ring vs. failure of resolution of physiologic herniation of gut
- Pathogenetic mechanism may differ depending upon which organs are prolapsed

Gastroschisis

- Likely involves more than 1 pathogenetic mechanism
- Failure of right lateral abdominal fold, resulting in body stalk and yolk stalk not merging to form umbilical cord vs. focal weakness at juncture of right lateral fold and umbilical cord, permitting intestines to be extruded

Pentalogy of Cantrell

- Abnormality of mesodermal development of cephalic fold at ~ 14-18 days post conception

18-DAY EMBRYO

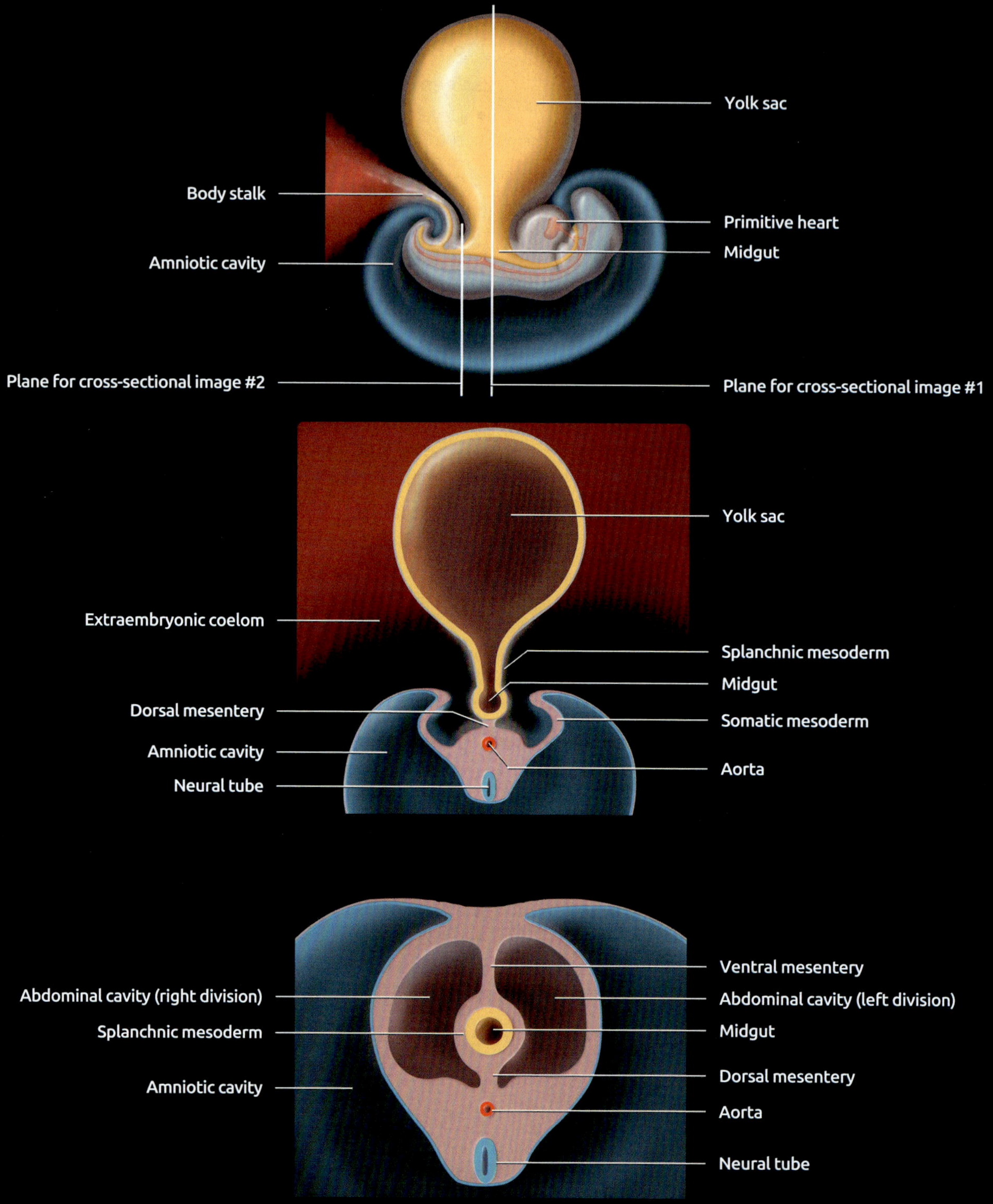

(Top) *Lateral illustration shows an 18-day embryo. The roof of the yolk sac becomes incorporated in the form of a tube as part of the primitive gut. The cranial end of the tube becomes the foregut and the caudal end becomes the hindgut.* **(Middle)** *Cross-sectional illustration along plane #1 (indicated on the lateral image) shows that the midgut has a wide communication with the yolk sac at this phase.* **(Bottom)** *Cross-sectional illustration more distally along plane #2 (indicated on the lateral image) shows that the gut is suspended by the ventral and dorsal mesenteries.*

4-WEEK EMBRYO

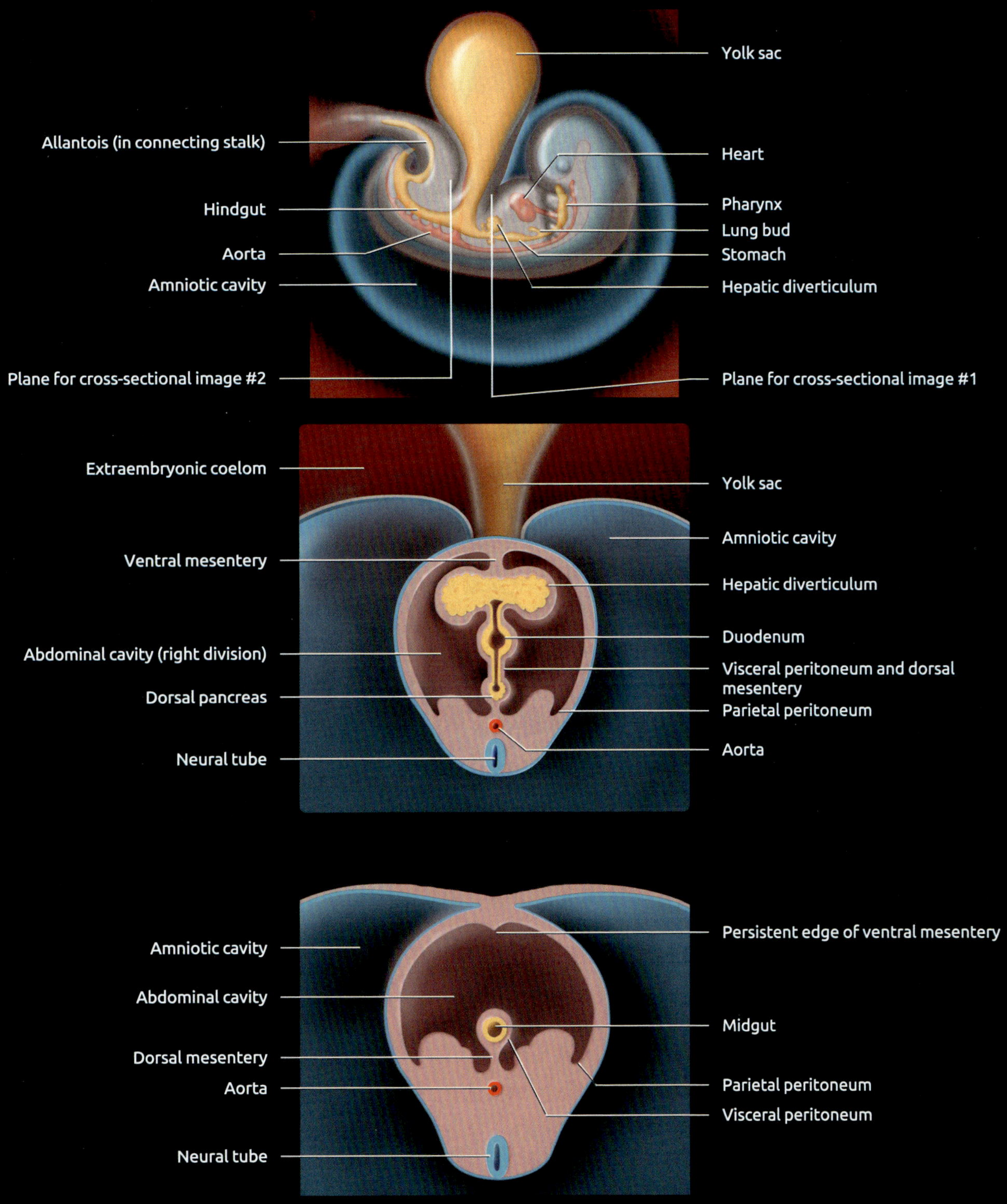

(Top) *Lateral illustration shows a 4-week embryo. The pharynx and lung bud arise from the foregut, along with the stomach. The allantois connects the body stalk to the hindgut. The yolk sac communicates broadly with the primitive gut. Errors in development include communication between the foregut branches, such as a tracheoesophageal fistula.* **(Middle)** *Cross-sectional illustration along plane #1 (indicated on the lateral image) shows that the liver arises from a ventral bud of the foregut, while the pancreas arises from the dorsal mesentery.* **(Bottom)** *Cross-sectional illustration along plane #2 (indicated on the lateral image) shows that the ventral mesentery begins to disintegrate to allow communication between right and left sides of the abdominal cavity.*

BOWEL ROTATION

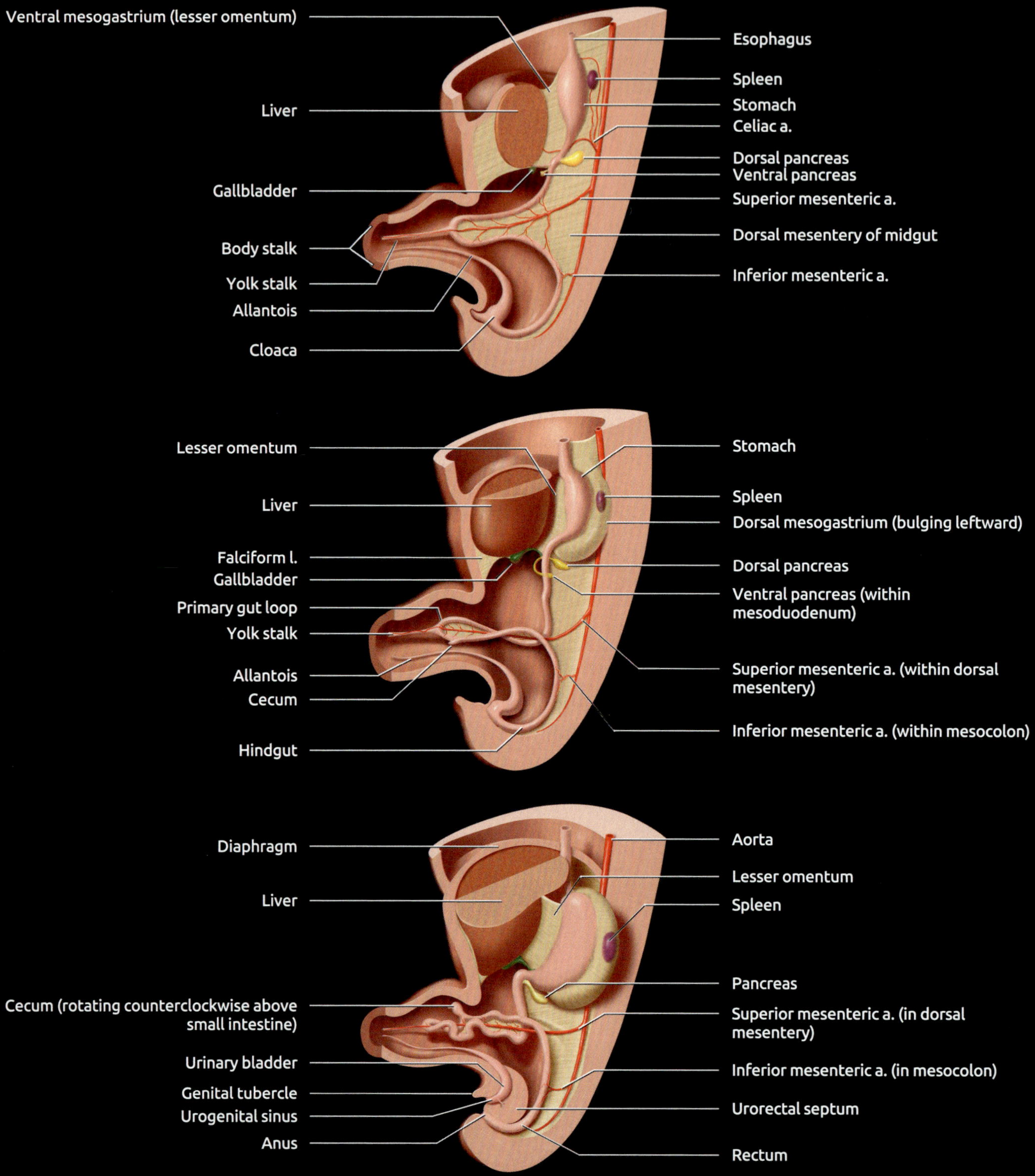

(Top) *The primary gut begins to elongate along with its dorsal mesentery. The hepatic diverticulum gives rise to the biliary tree and ventral pancreas. The arterial supply to the gut is already defined: Celiac artery (foregut), superior mesenteric artery (midgut), and inferior mesenteric artery (hindgut).* **(Middle)** *The liver expands within the ventral mesentery, which later disintegrates, leaving only the falciform ligament and the lesser omentum. The primary gut elongates and herniates out into the umbilical cord.* **(Bottom)** *The liver continues its rapid enlargement. Only the caudal part of the ventral mesentery remains (falciform ligament). The dorsal mesogastrium elongates, forming the left and caudal portions of the lesser sac. The gut continues to elongate and rotates counterclockwise (as viewed from the front) around the superior mesenteric artery within the dorsal mesentery. The urogenital sinus has separated from the rectum and anus. Common developmental errors include midgut malrotation, omphalocele, and imperforate anus.*

BOWEL ROTATION

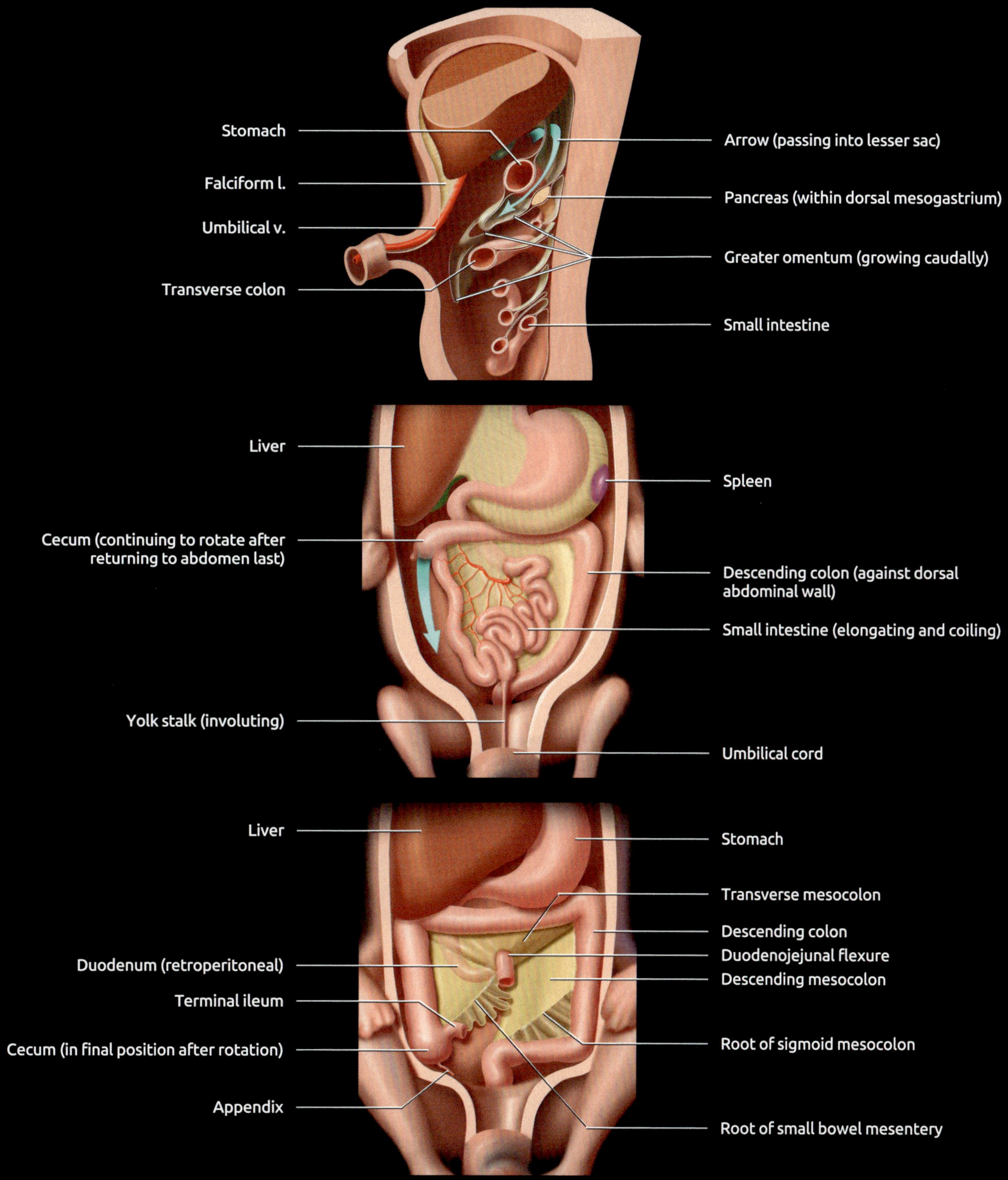

(Top) *The umbilical vein enters the liver along the caudal (free) edge of the falciform ligament. The leaves of the greater omentum elongate to the left and caudally, expanding the lesser sac and covering the transverse colon and small intestine.* **(Middle)** *At 10 weeks post conception (12 weeks by last menstrual period), the small intestine has returned to the abdomen. The yolk stalk, which connected the yolk sac to the primitive gut, is disintegrating. The cecum is the last part to return and continues to rotate in a counterclockwise direction until reaching the right lower quadrant. Errors in development include persistence of a part of the yolk stalk (Meckel diverticulum) and errors of bowel rotation.* **(Bottom)** *By 4- to 5-months gestation, the ascending and descending colon are fixed in a retroperitoneal location by fusion of their mesocolons to the posterior abdominal wall. The small bowel and transverse and sigmoid colon remain intraperitoneal, suspended by their respective mesenteries.*

MOUTH AND ANUS FORMATION

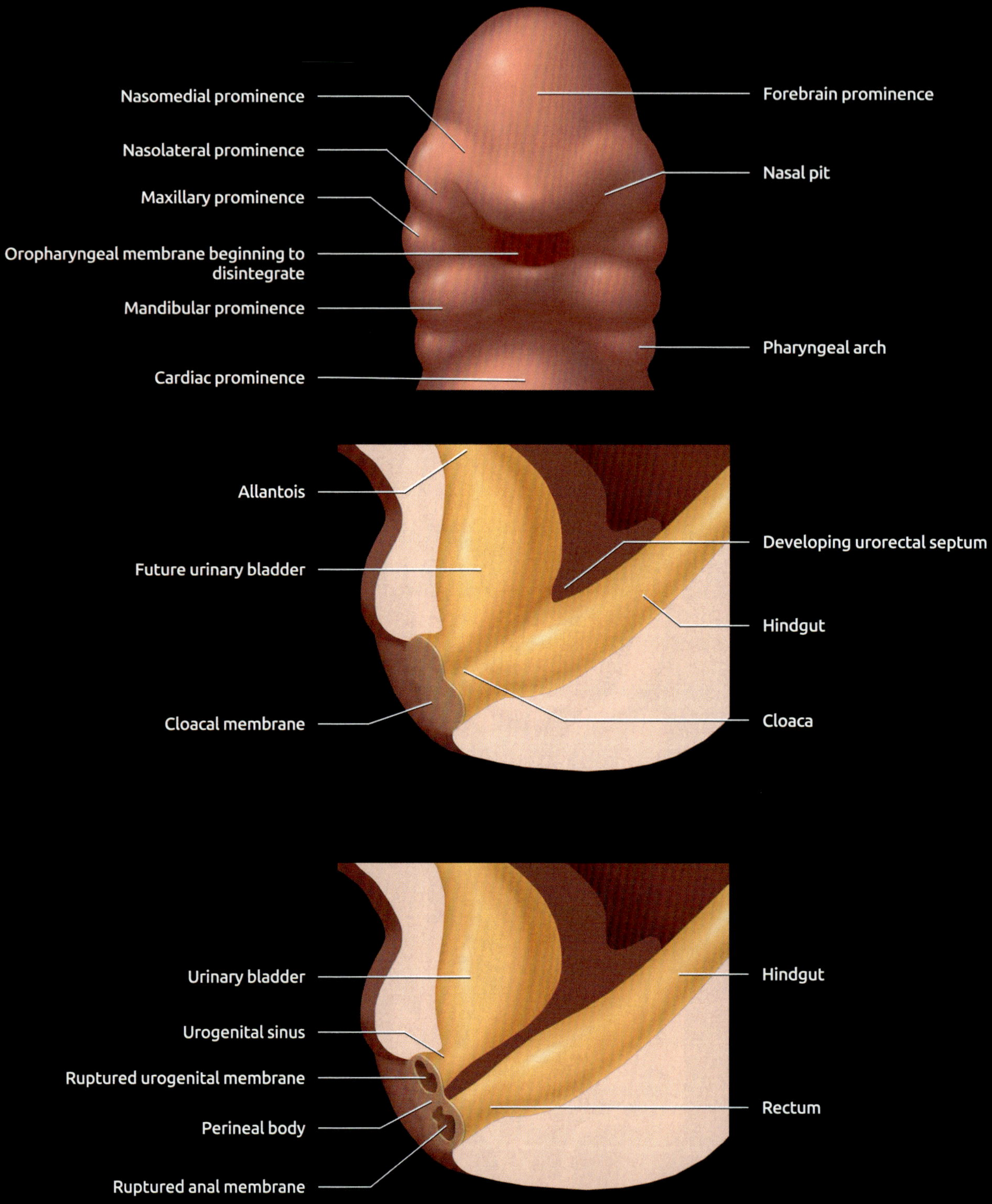

(Top) *The endoderm and ectoderm are separated by mesoderm in all areas of the gut except for the oropharyngeal and cloacal membranes. These are areas of mesodermal deficiency, which upon disintegrating became the oropharyngeal cavity and area of the urethra and anus, respectively. This graphic shows the beginning of disintegration of the oropharyngeal membrane in the 4th week. The stomodeum, or primitive mouth, is the result.* **(Middle)** *During the 6th to 7th weeks, a mesodermal shelf called the urorectal septum begins to grow between the hindgut and allantois, toward the cloacal membrane. It fuses with the cloacal membrane, dividing it into the anal and urogenital membranes, and forms the perineal body.* **(Bottom)** *By the end of the 8th week, the anal membrane ruptures, allowing access from the hindgut to the exterior of the body. The urethra forms within the urogenital sinus.*

LIVER, PORTAL, AND HEPATIC VEINS

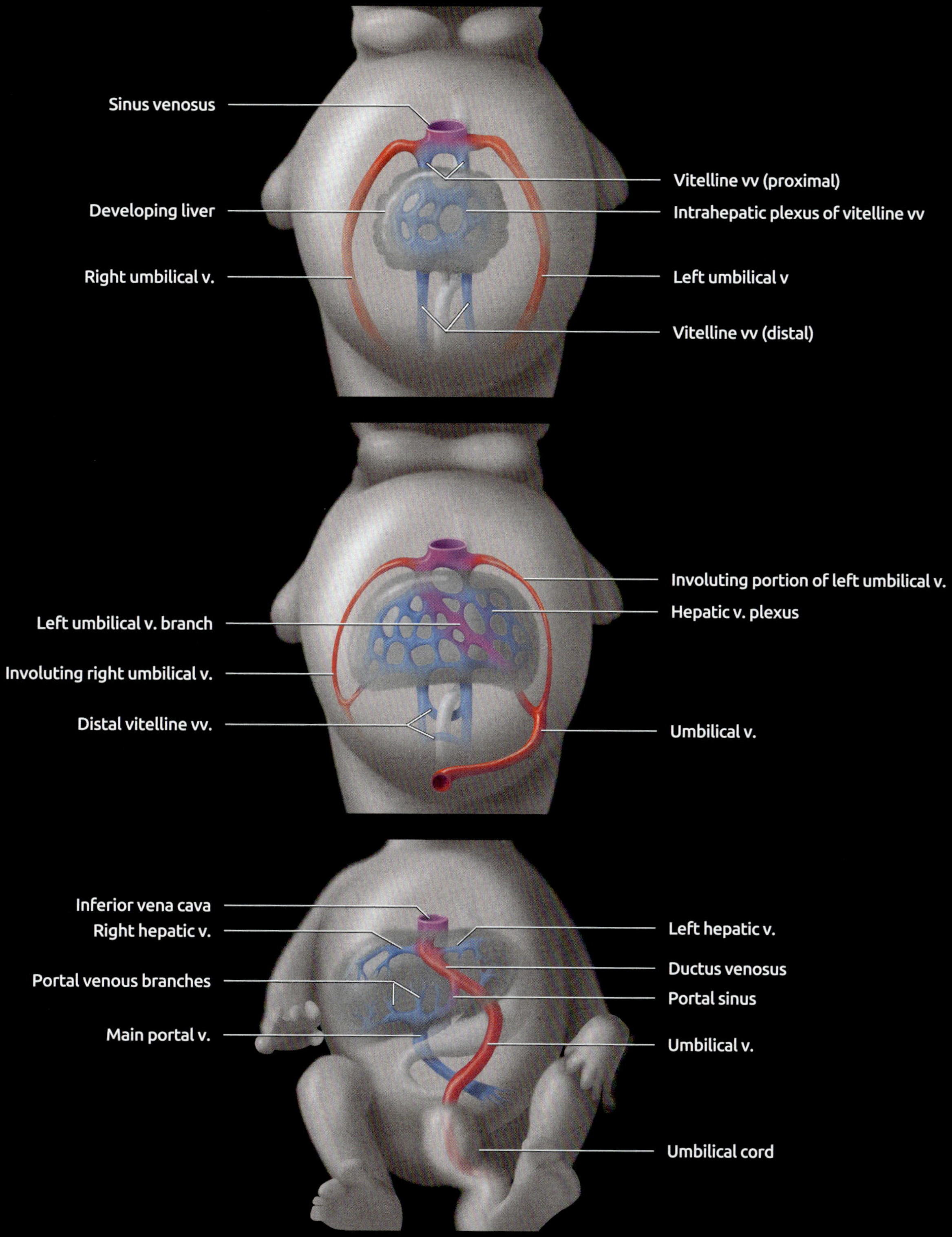

(Top) *The vitelline veins return blood from the yolk sac and branch within the liver to form the hepatic sinusoids and venous system. They unite again to form the proximal vitelline veins, which join with the (initially) paired umbilical veins to enter the sinus venosus of the heart.* **(Middle)** *All of the right and much of the left umbilical veins atrophy. The left umbilical vein sends a large branch to the liver, which anastomoses with the plexus derived from the vitelline veins. The extrahepatic (distal) vitelline veins form the precursors to the portal venous system.* **(Bottom)** *The portal sinus diverts some oxygenated blood to the liver. The proximal parts of the vitelline veins have become the hepatic veins, returning blood from the liver to the heart. The distal parts have developed into the portal venous system, returning blood from the gut to the liver sinusoids.*

FETAL ABDOMEN MR IMAGING

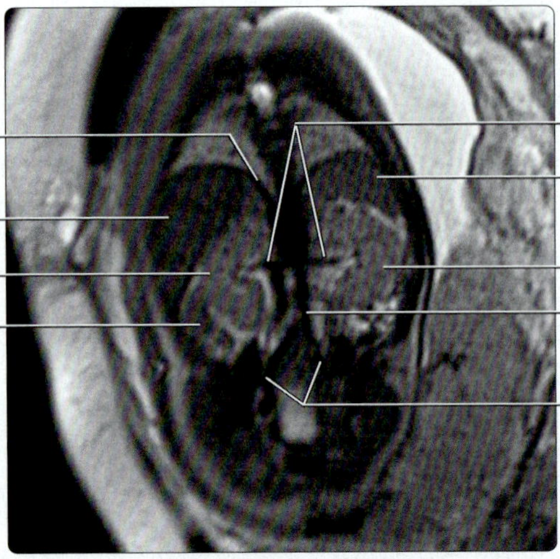

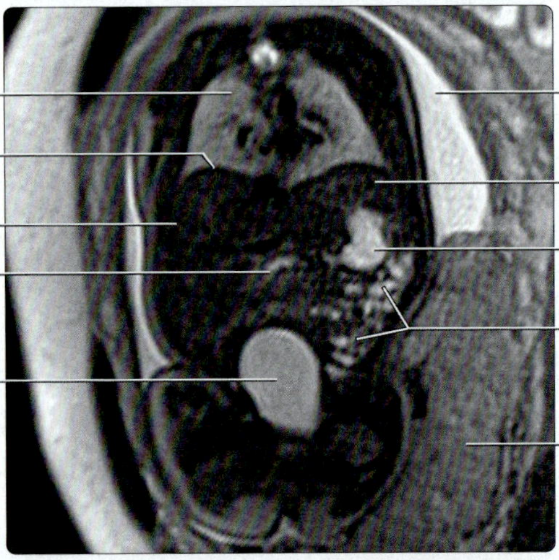

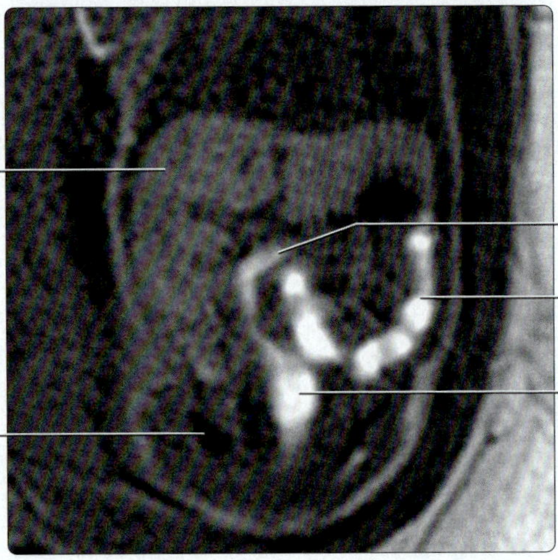

(Top) *Coronal T2WI MR through the level of the kidneys shows a portion of the ascending colon, which has a low to intermediate signal intensity, blending into the surrounding soft tissues. Both the liver & spleen are very low in signal intensity & are easily identified.* **(Middle)** *More anteriorly, the liver crosses the midline & is the main contributor to the abdominal circumference measurement. The lungs are higher in signal intensity. Fluid-filled structures, including the bladder, stomach, & small bowel, are very high signal intensity. Because of its excellent contrast resolution, MR is ideally suited for evaluating the contents of a diaphragmatic hernia.* **(Bottom)** *While T2WI MR is the primary sequence in evaluating most fetal anatomy, T1WI is particularly helpful for evaluating the colon as meconium is high in signal intensity & is obvious on this sequence. Also note the higher signal of the liver parenchyma on T1WI compared to T2WI. In addition to evaluating the colon, T1WI is valuable for evaluating for blood products, which are often high in in signal intensity.*

FETAL ABDOMEN ULTRASOUND

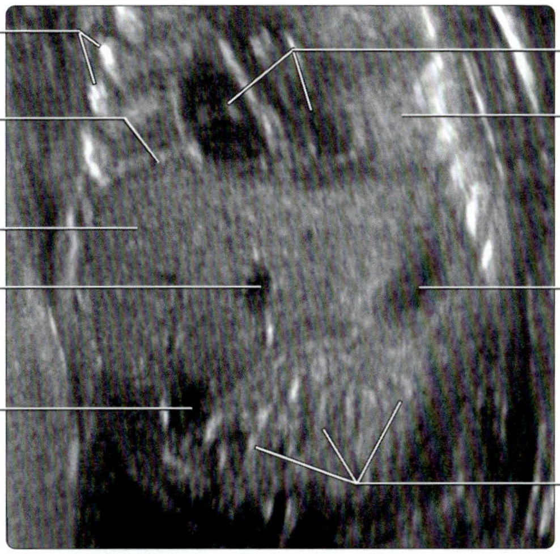

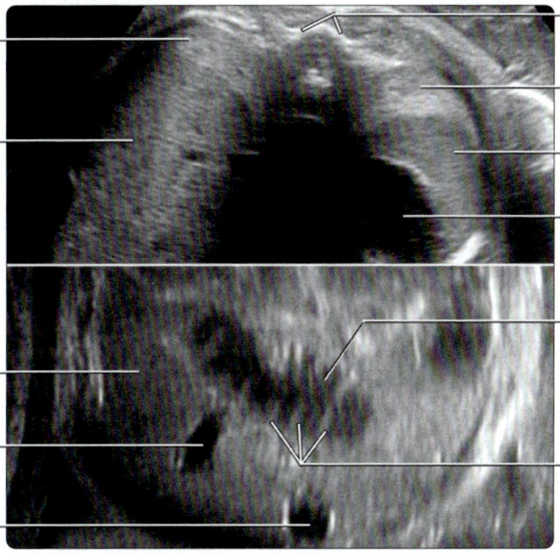

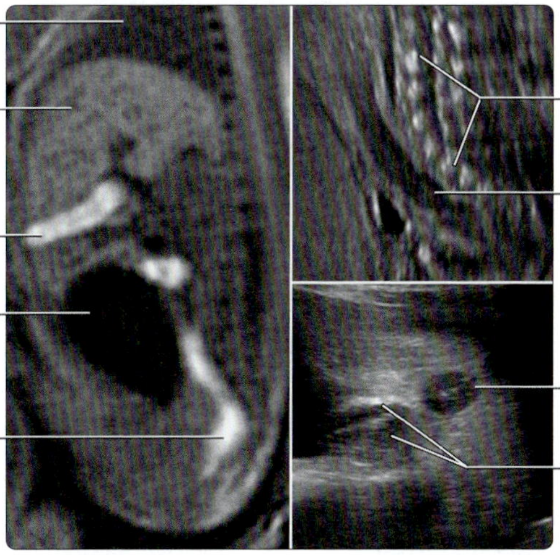

(Top) *Coronal ultrasound in a 20-week fetus shows the anatomy of the upper abdomen.* **(Middle)** *Subtle differences in echogenicity allow the identification of the various organs, as shown in this composite image of a 3rd trimester fetus. The upper oblique axial image shows the lungs are mildly hyperechoic compared to the liver and spleen. Changes in echotexture often provide clues regarding any pathologic process in a given organ. The lower image is an oblique coronal US of the transverse colon. US can show detailed anatomy, such as the colonic haustra.* **(Bottom)** *Anorectal malformations are a complex group of anomalies that require a complete evaluation of the colon down to the anus. Sagittal T1WI MR (left) shows normal rectal tapering. Following the colon on ultrasound (top right) is somewhat more difficult, but the anal dimple image (bottom right) of the perineum should be performed in any case of suspected anorectal malformation. It has a classic doughnut or target appearance with a thick, hypoechoic sphincter and hyperechoic mucosa.*

DEVELOPMENT OF URINARY TRACT

Kidney Formation

- Stages of development similar to those found in more primitive animals (e.g., invertebrates, amphibians)
  - Reflects evolutionary history
- **3 successive nephric structures** of increasing advanced design: **Pronephroi, mesonephroi, metanephroi** [pronephros, mesonephros, metanephros (singular form)]
  - **Structures develop and regress in craniocaudal sequence**
- Nonfunctional, primitive **nephrotomes** develop in cervical region
  - Vestige of **pronephroi**, which form primitive kidney in lower vertebrates
  - Regress by 4th week and are replaced by mesonephroi
- **Mesonephroi** are elongated, functional primitive kidneys, which develop from upper thoracic to 3rd lumbar level
- **Mesonephric (wolffian) ducts** 1st appear at 24 days dorsolateral to mesonephroi in thoracic region
  - Grow caudally and fuse with ventrolateral wall of bladder
  - Mesonephric ducts connect to, and drain urine from, mesonephric tubules
- After 10 weeks, mesonephric tubules cease to function and regress
  - Mesonephric ducts also regress in females but persist in males to form part of genital tract
- **Ureteric bud** (also called metanephric diverticulum) sprouts from distal mesonephric duct
  - Induces sacral mesoderm (**metanephric blastema**) to develop into **metanephroi**, definitive kidneys
  - Ureteric bud and metanephric blastema exert reciprocal inductive effects on each other
    - Ureteric bud induces metanephric blastema to form nephrons
    - In turn, metanephric blastema induces ureteric bud to bifurcate into developing calyces
- Anomalies resulting in failure of ureteric bud to interact appropriately with metanephric blastema
  - **Renal agenesis**
    - Failure of ureteric bud to come into contact with metanephric blastema
  - **Multicystic dysplastic kidney** (proposed mechanisms)
    - Ureteric bud does not appropriately signal metanephros, leading to abnormal collecting duct development with loss of nephrons, stromal expansion, and cyst formation
    - Very early ureter obstruction leads to dysplasia (metanephric tissue does not form nephrons)

Renal Ascent

- Initially, kidneys (metanephroi) lie close together low in pelvis with renal hila facing anteriorly
- Mechanism for "ascent" to final retroperitoneal flank position is not completely understood, but caudal embryonic growth is likely major contributing factor
- Blood supply changes as kidneys successively "recruit" arterial blood supply from iliac arteries and aorta
  - New, more superior arterial branches form as inferior branches involute
- With ascent, renal pelves rotate medially ~ 90°
- Ascent complete by 9 weeks when kidneys come into contact with adrenal glands
- Anomalies related to abnormal "ascent"
  - **Renal ectopia**
    - Kidney usually low in position with abnormal rotation
  - **Crossed fused ectopia and other fusion abnormalities**
    - Fusion of metanephroi prior to ascent leads to various appearances
  - **Horseshoe kidney**
    - Inferior poles of 2 metanephroi fuse
    - Become "stuck" under inferior mesenteric artery
  - **Accessory renal arteries**
    - Persistence of normally transient renal arteries during ascent

Bladder

- **Cloaca** (Latin for sewer) is common chamber with early communication between urinary, gastrointestinal, and reproductive tracts
  - Divided by **urorectal septum** into **urogenital sinus** anteriorly and **anorectum** posteriorly
    - Both structures open through perineum upon rupture of cloacal membrane
- **Urogenital sinus has 3 major components**
  - **Allantois** is cephalad portion
    - Extends from bladder to connecting stalk of yolk sac
    - Intraabdominal portion involutes to become **urachus**; in adult, this is **median umbilical ligament**
  - Middle vesicular portion becomes **urinary bladder**
  - Caudal portion forms **lower vagina** in females and **penile urethra** in males
- Distal portions of mesonephric ducts and attached ureteric ducts become incorporated into posterior bladder
  - During this process, **ureters are incorporated superiorly** in trigone
  - Orifices of **mesonephric ducts move inferomedially** and enter prostatic urethra, forming ejaculatory ducts

DEVELOPMENT OF ADRENAL GLANDS

Cortex and Medulla

- Cortex and medulla derive from 2 different tissues
  - **Cortex**
    - Forms from mesoderm
    - 3 zones: **Glomerulosa, fasciculata, and reticularis**
    - Only glomerulosa and fasciculata present at birth
    - Reticularis not recognizable until 3rd year of life
  - **Medulla**
    - Forms from neural crest cells derived from sympathetic ganglion
- **Fetal adrenal glands 10-20x larger** than adult adrenal glands relative to body size
  - May be mistaken for fetal kidneys, especially early in pregnancy
    - May potentially miss renal agenesis before oligohydramnios has developed
  - Large size is from adrenal cortex
    - Rapidly become smaller as cortex regresses in 1st year of life

DEVELOPMENT OF MALE GENITAL TRACT

Mesonephric (Wolffian) Ducts

- Persist in males to form part of genital tract
  - Epididymis
  - Vas deferens
  - Seminal vesicles
  - Ejaculatory ducts

Testes

- Form from genital ridges, which extend from T6-S2 in embryo
- **Composed of 3 cell lines**, which form primitive sex cord
  - Germ cells
  - Sertoli cells
  - Leydig cells
- **Germ cells**
  - Form in wall of yolk sac and migrate along hindgut to genital ridges
  - Form spermatogenic cells in mature testes
- **Sertoli cells**
  - **Secrete müllerian inhibiting factor**
    - Causes paramesonephric (müllerian) ducts to regress
  - In adult, form supporting network for developing spermatozoa
    - Form tight junctions (blood-testis barrier)
- **Leydig cells**
  - **Principal source of testosterone production**
  - Lie within interstitium
  - Cause differentiation of mesonephric (wolffian) ducts into male genital tract

Scrotum

- Derived from **labioscrotal folds**
  - Folds swell under influence of testosterone to form twin scrotal sacs
    - Point of fusion is **median raphe**
      - Extends from anus, along perineum to ventral surface of penis
  - **Processus vaginalis**, sock-like evagination of peritoneum, elongates through abdominal wall into twin sacs
    - Forms anterior to developing testes
    - Aids in descent of testes, along with **gubernaculum** (ligamentous cord extending from testis to labioscrotal fold)

Testicular Descent

- Between 7th-12th week of gestation, testes descend into pelvis
  - Testes are retroperitoneal throughout their descent and are intimately associated with posterior wall of processus vaginalis
  - Remain near internal inguinal ring until 7th month, when they begin descent through inguinal canal into twin scrotal sacs
- **Component layers of spermatic cord and scrotum** formed during descent through abdominal wall
  - Transversalis fascia → internal spermatic fascia
  - Internal oblique muscle → cremasteric muscle and fascia
  - External oblique muscle → external spermatic fascia
  - Dartos muscle and fascia embedded in loose areolar tissue below skin
  - Processus vaginalis closes and forms tunica vaginalis
- **Cryptorchidism** results from incomplete descent

Prostate

- In 10th week, endodermal evaginations bud from prostatic portion of urethra and develop into prostatic cords
- Under increasing levels of testosterone, these cords develop into glandular acini
- Rest of gland forms from surrounding mesenchyme, which differentiates into stroma and smooth muscle

DEVELOPMENT OF FEMALE GENITAL TRACT

Ovaries

- Male and female gonadal development identical through 7th week
- In **absence of testis determining factor** from Y chromosome, ovaries will develop
- Primitive sex cords degenerate and mesothelium of genital ridges form secondary sex cords
- **Secondary sex cords** invest primordial germ cells to form follicle cells
  - **Germ cells undergo 1st meiotic division** but further development arrested until puberty
- Like testis, ovaries also descend with help from gubernaculum

Uterus

- Formed from paired **paramesonephric (müllerian) ducts**
- Paramesonephric ducts form lateral to mesonephric ducts
  - Join with urogenital sinus medial to mesonephric ducts
  - In absence of Y chromosome, paramesonephric ducts will continue to develop and form uterus
- These paired paramesonephric ducts fuse in midline
  - Fusion forms **uterovaginal canal**, becoming uterus and upper vagina
  - Unfused portions remain as **fallopian tubes**
- **Lower vagina formed from urogenital sinus**
- Failure of müllerian duct development &/or fusion leads to spectrum of **congenital uterine anomalies**
  - Class I: Agenesis or hypoplasia
  - Class II: Unicornuate uterus
    - Single uterine horn, may have accessory rudimentary horn
  - Class III: Uterus didelphys
    - 2 separate, noncommunicating horns
  - Class IV: Bicornuate uterus
    - Concave or heart-shaped external uterine contour
  - Class V: Septate uterus
    - Normal external contour
- Müllerian duct anomalies and renal anomalies are commonly associated
  - If fetal renal anomaly is seen, consider performing postnatal pelvic ultrasound
  - Uterus is often well seen in immediate neonatal period

KIDNEY DEVELOPMENT

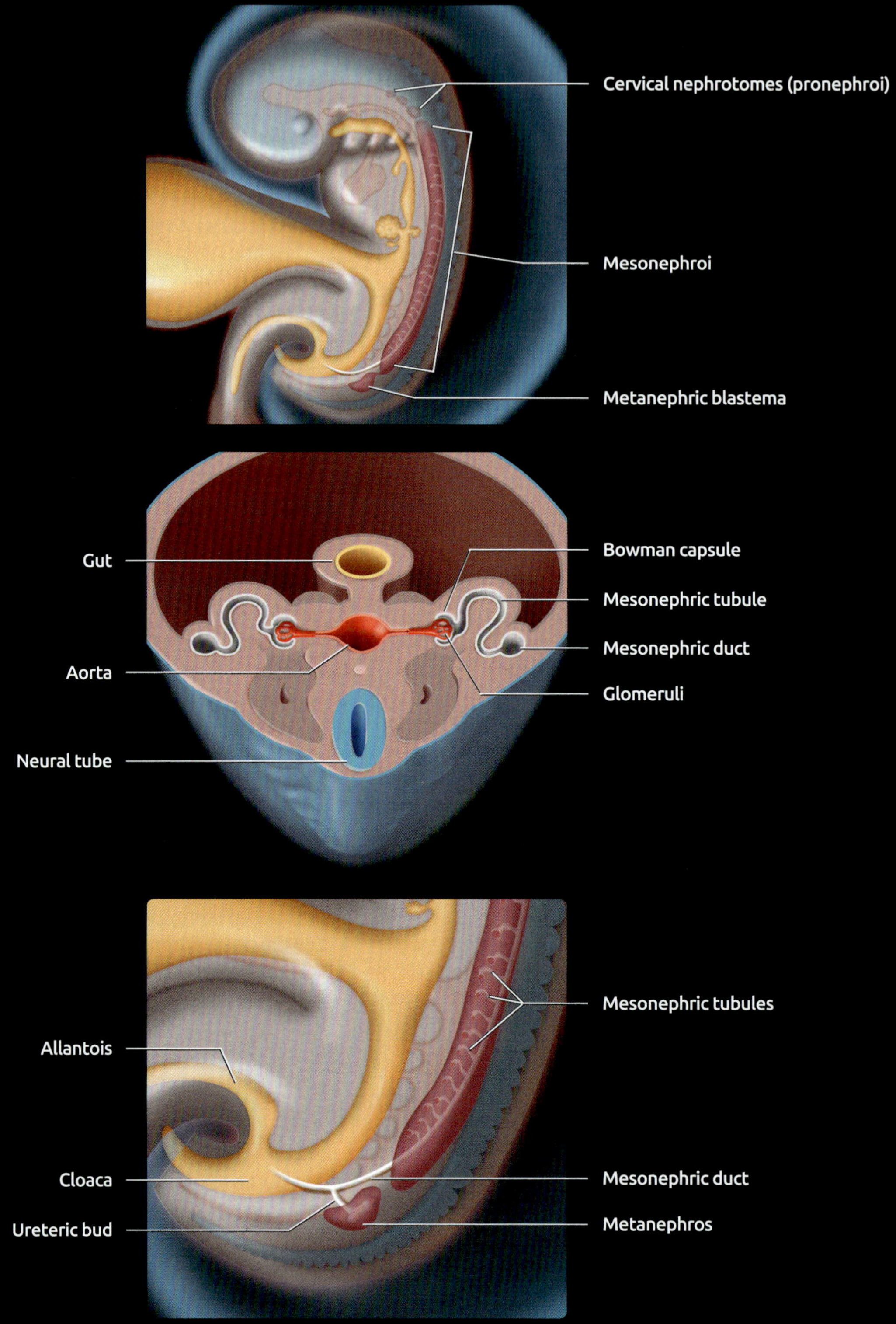

(Top) *Three sets of nephric structures develop in human embryos. They form and regress in a craniocaudal progression. The pronephroi are transitory and nonfunctional. The mesonephroi also degenerate, although the distal mesonephric duct will persist in males and form part of the genital tract.* **(Middle)** *Cross section of the embryo shows the mesonephroi. By the 4th fetal week, the mesonephric tubules and duct are formed. Branched vessels from the aorta reach the blind ends of the tubules to form glomeruli. Although these achieve an excretory function in the human embryo, they degenerate as the metanephroi form.* **(Bottom)** *The 3rd, and definitive, kidney is formed when the ureteric bud induces the metanephric blastema to form the metanephros.*

BLADDER DEVELOPMENT

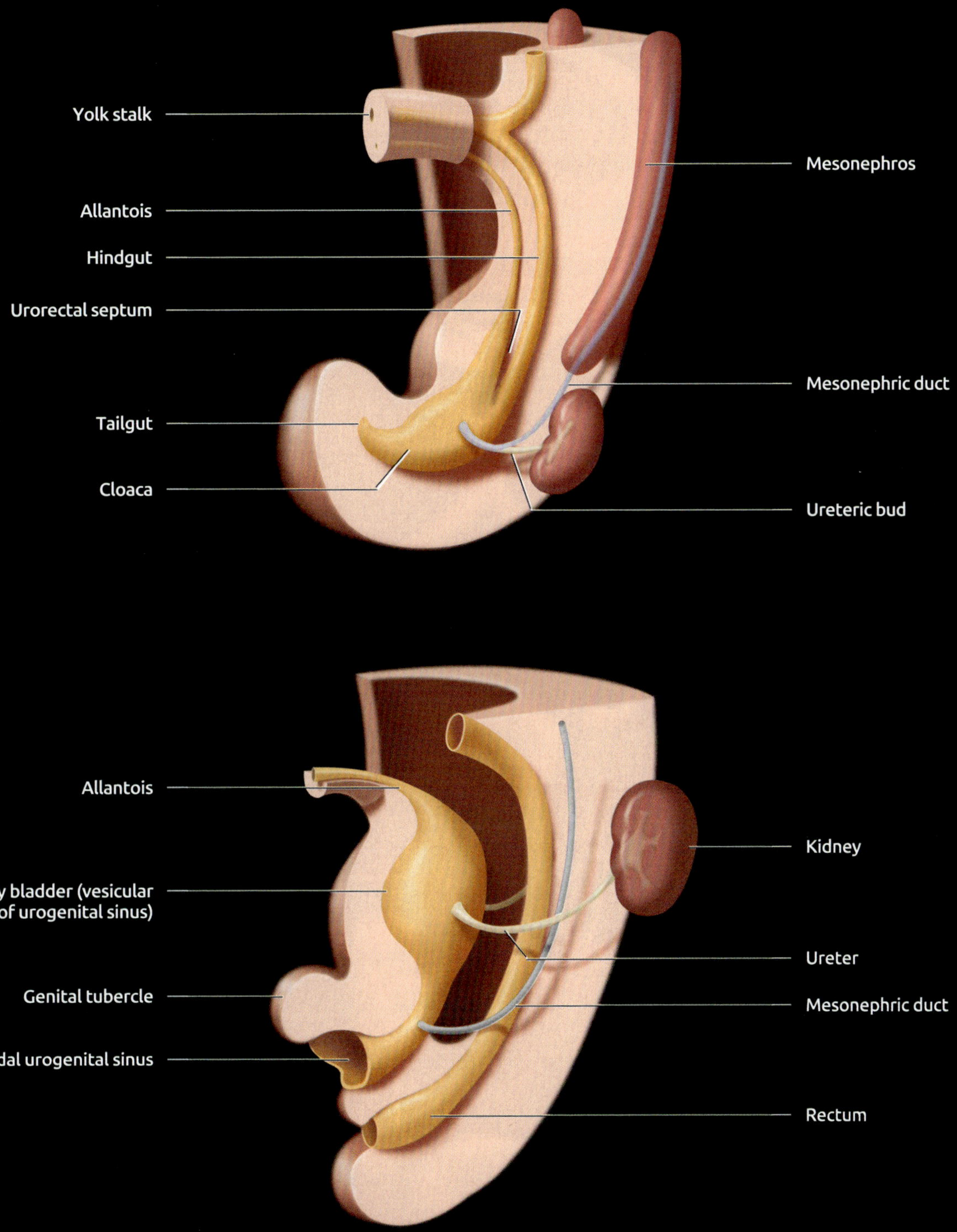

(Top) *The cloaca is a common chamber with early communication between the urinary, gastrointestinal, and reproductive tracts. Between 4-6 weeks, the urorectal septum will separate the urogenital sinus anteriorly from the rectum posteriorly.* **(Bottom)** *The largest vesicular portion of the urogenital sinus forms the bladder. It is in direct continuity with the allantois superiorly. The allantois will eventually involute to form the urachus. The caudal portion of the urogenital sinus will form the lower portion of the vagina in females and the penile urethra in males. The ureteric buds are incorporated into the posterior bladder, forming the upper portion of the trigone, while the mesonephric ducts migrate inferomedially and will ultimately insert into the prostatic urethra as the ejaculatory ducts.*

RENAL ASCENT

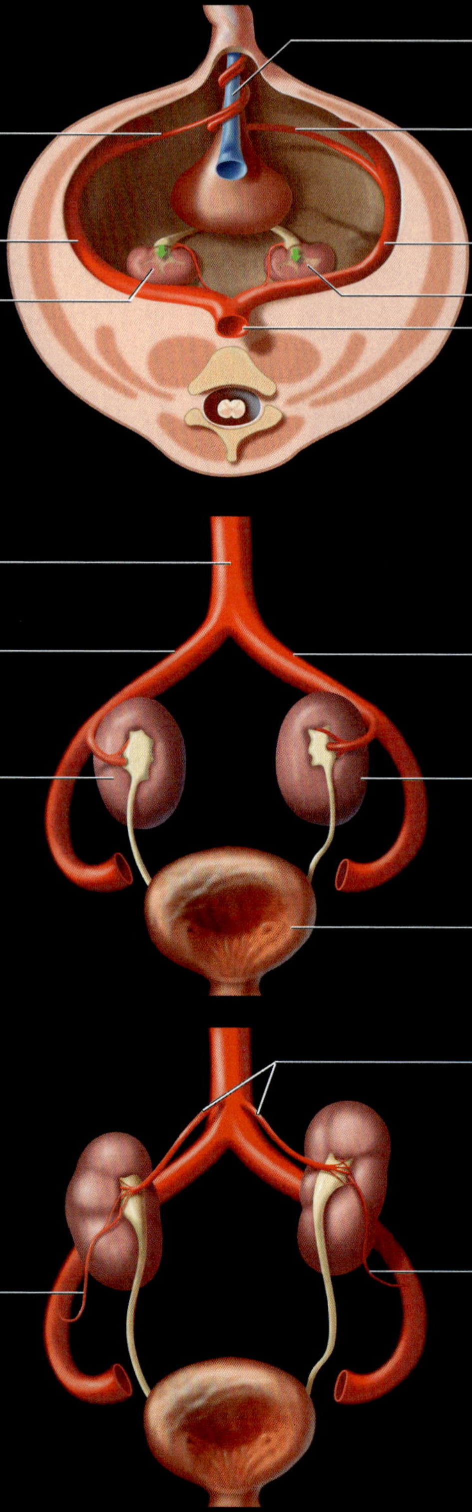

(Top) *Graphic looking down from above shows the kidneys low in the fetal pelvis. The definitive kidneys form from specialized sacral mesoderm called the metanephric blastema. Please note that while the umbilical vein carries oxygenated blood, for this set of graphics, veins will be depicted in blue and arteries in red.* **(Middle)** *The kidneys are located close together and may even be in direct continuity, leading to various anomalies of fusion. The kidneys face forward with the renal pelves directed anteriorly. When initially formed, the arterial supply is from the iliac arteries.* **(Bottom)** *With ascent, the more inferior renal arteries regress, while more superior ones are recruited.*

RENAL ASCENT

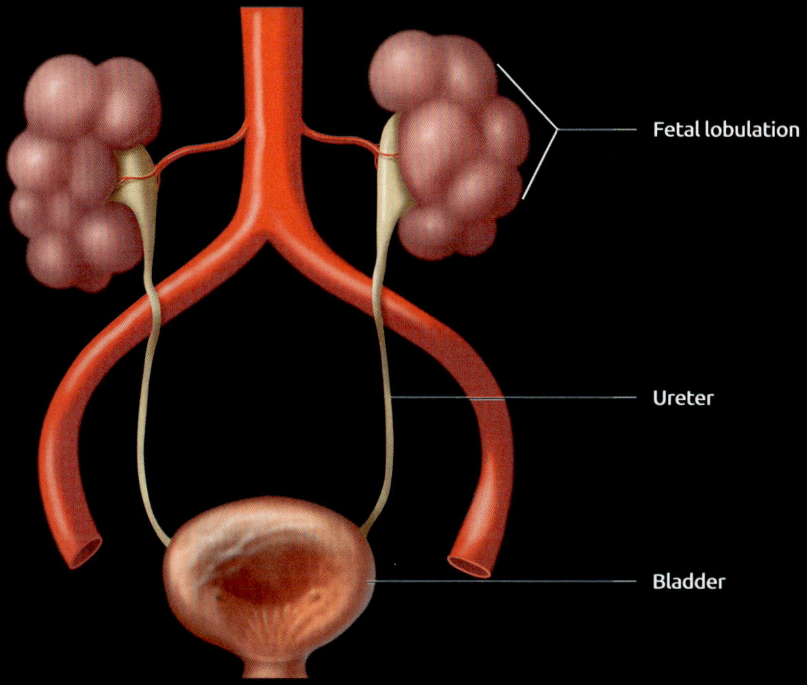

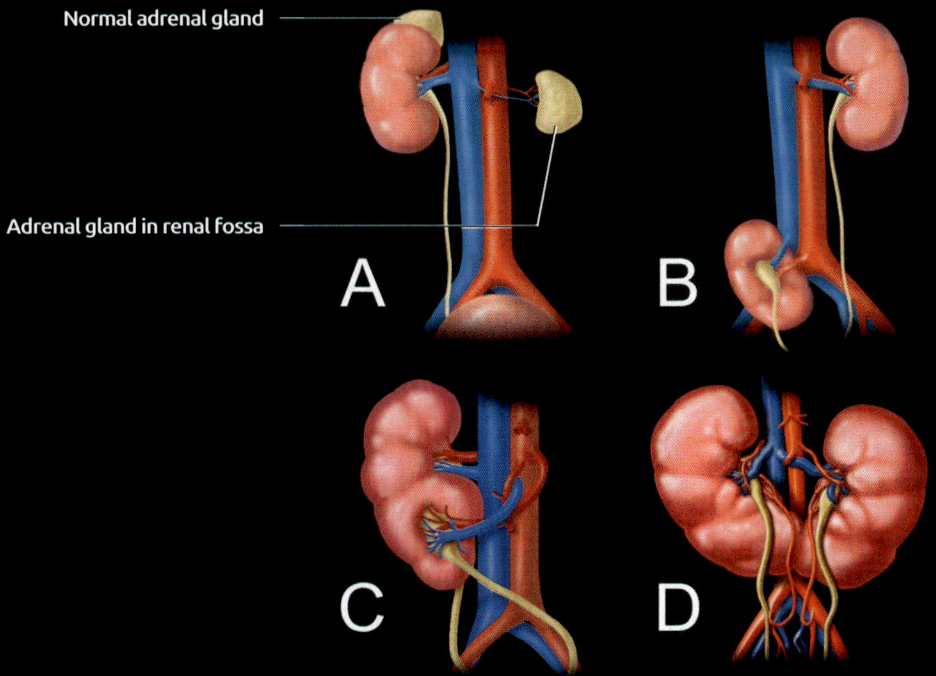

(Top) *Normally, each kidney will be left with a single main renal artery. Accessory renal arteries, a common anatomic variant, are from failure of involution of transient arteries during ascent. The fetal kidneys have a distinct lobular contour (fetal lobulation), reflecting the developmental process between the ureteric bud forming the calyces and the metanephric blastema forming the nephrons.* **(Bottom)** *Aberration in the development and ascent of the kidneys causes an array of anomalies. Renal developmental variants include unilateral renal agenesis (A), pelvic kidney (B), crossed fused renal ectopia (C), and horseshoe kidney (D). Errors of formation, fusion, and ascent lead to these anomalies.*

FEMALE GENITAL TRACT

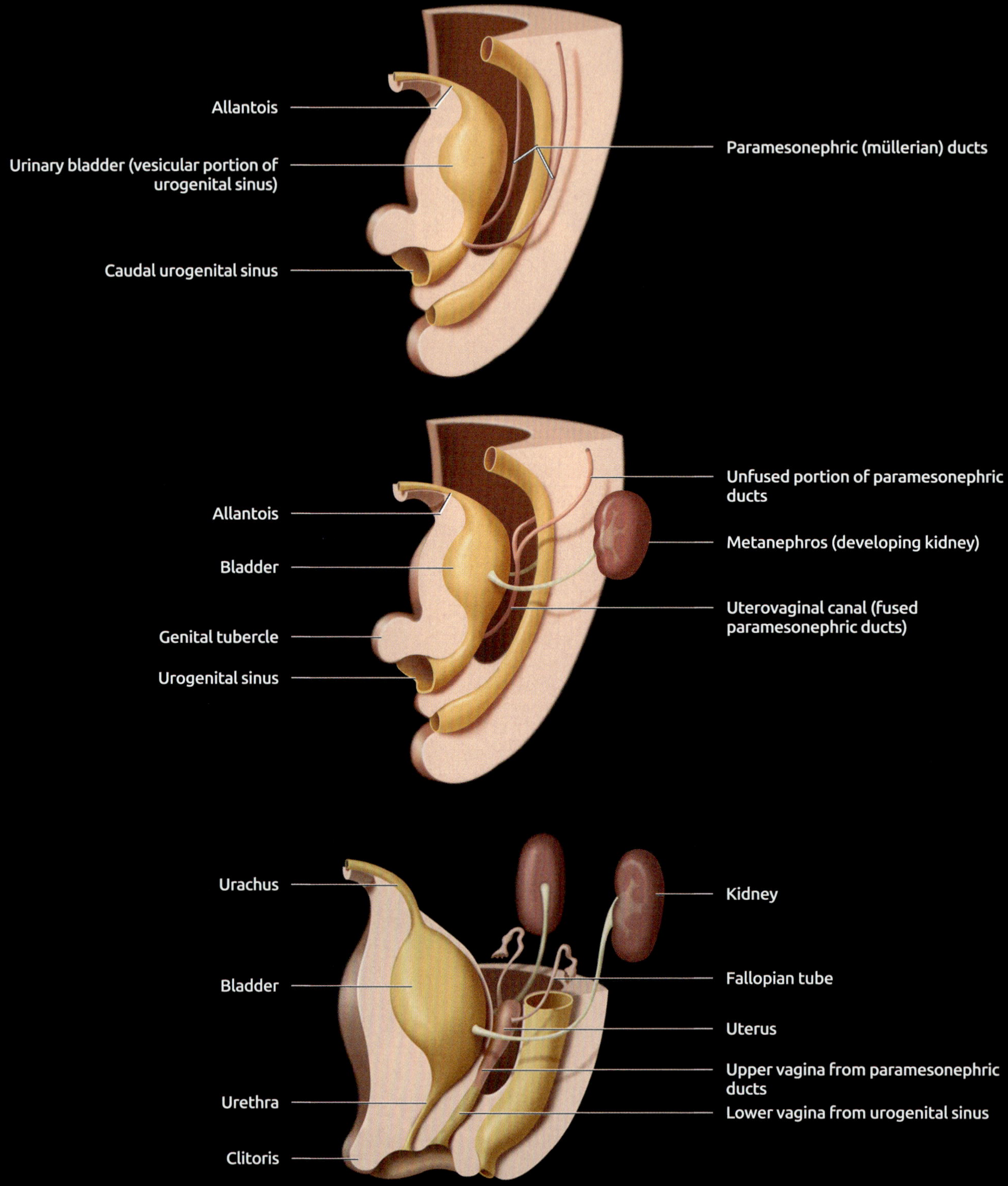

(Top) *The fallopian tubes, uterus, and upper vagina form from the paired paramesonephric (müllerian) ducts, which develop on either side of the midline lateral to the mesonephric ducts (the mesonephric ducts regress in a female fetus).* **(Middle)** *The paramesonephric ducts must meet in the midline and fuse to form the uterus and upper portion of the vagina (uterovaginal canal). The unfused portions will form the fallopian tubes. The development of the kidney (metanephros) is closely related to uterine development, and coexistent renal and müllerian duct anomalies are common.* **(Bottom)** *The distal portion of the vagina (shown in yellow) is formed from the caudal urogenital sinus, which splits to form the urethra anteriorly and the vagina posteriorly. The allantois involutes to form the urachus.*

MALE GENITAL TRACT

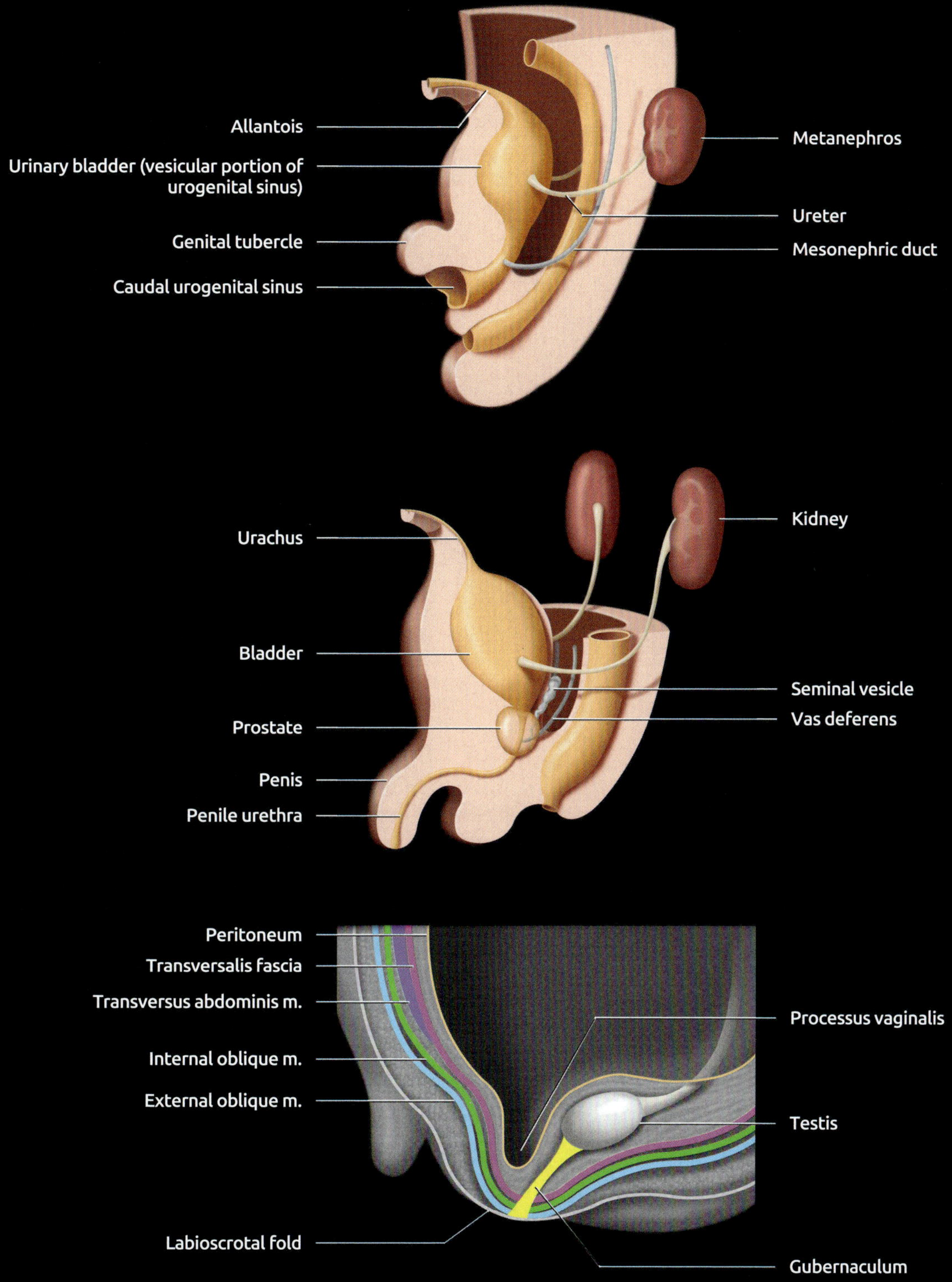

(Top) *The mesospheric ducts (only 1 shown) persist in a male and will form the epididymides, vas deferens, seminal vesicles, and ejaculatory ducts.* **(Middle)** *The caudal urogenital sinus forms the penile urethra. The allantois involutes and forms the urachus. The prostate forms from endodermal evaginations of the prostatic portion of the urethra and surrounding mesenchyme.* **(Bottom)** *The processus vaginalis is a sock-like evagination of the peritoneum, which elongates caudally through the abdominal wall. It forms just anterior to the developing testes and, along with the gubernaculum (a ligamentous cord extending from the testis to the labioscrotal fold), aids in their descent. As the processus vaginalis evaginates, it becomes ensheathed by fascial extensions of the abdominal wall, which ultimately form the layers of the scrotum and spermatic cord.*

URINARY TRACT, US

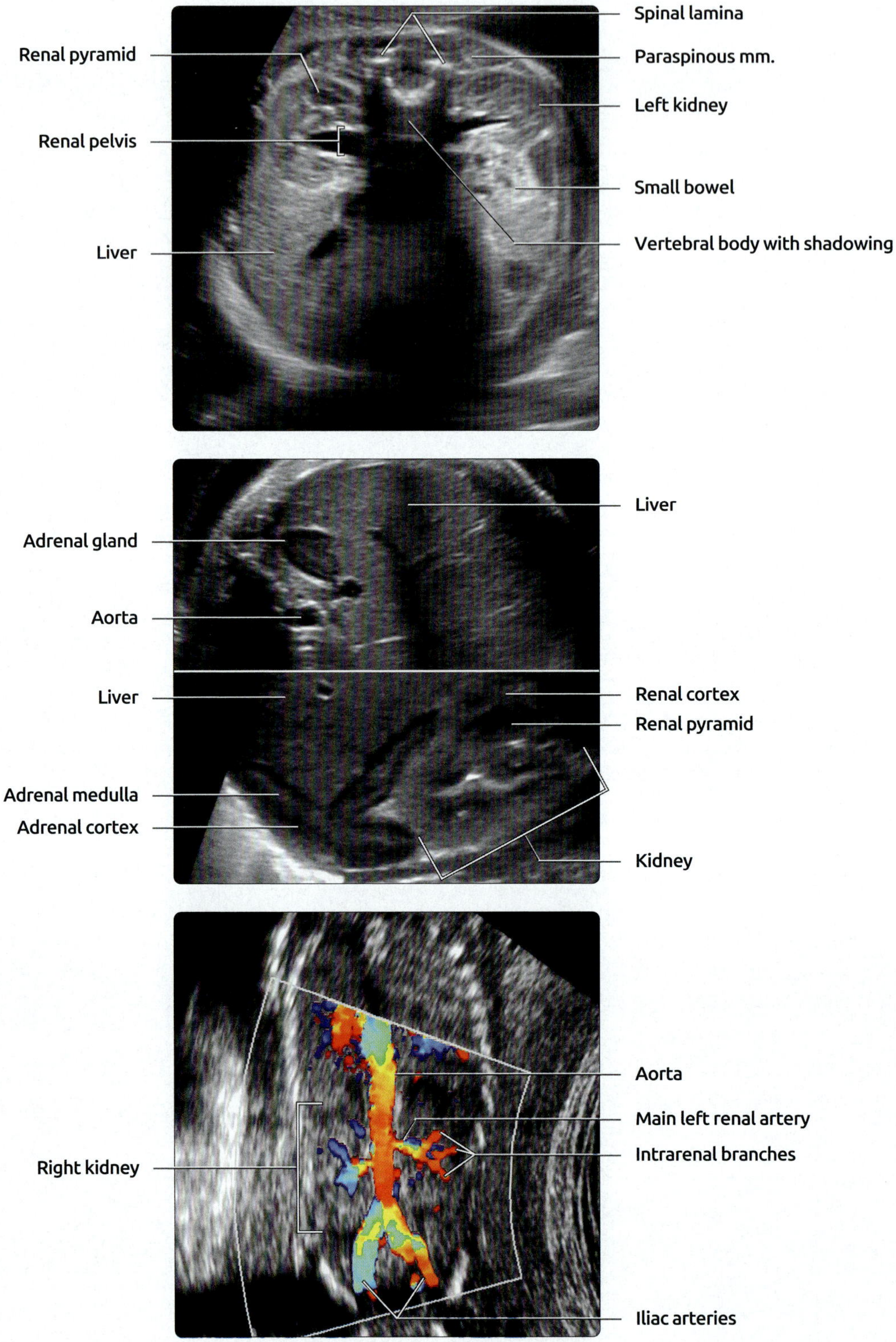

(Top) *This axial ultrasound is through the renal fossa with the spine at the 12 o'clock position. This is the best position to measure the renal pelvis diameter.* **(Middle)** *This is a composite image of an axial ultrasound of the right fetal adrenal gland (top) and sagittal ultrasound of the adrenal gland of a newborn (bottom). The fetal adrenal gland is 10-20x larger than the adult adrenal gland relative to body size. It is often quite prominent on fetal imaging and is recognized by its ice cream sandwich appearance with a hyperechoic medulla and hypoechoic cortex. It typically is folded and angular in shape with a Y or tricorn hat configuration.* **(Bottom)** *Color Doppler ultrasound shows normal renal vasculature.*

URINARY TRACT, MR

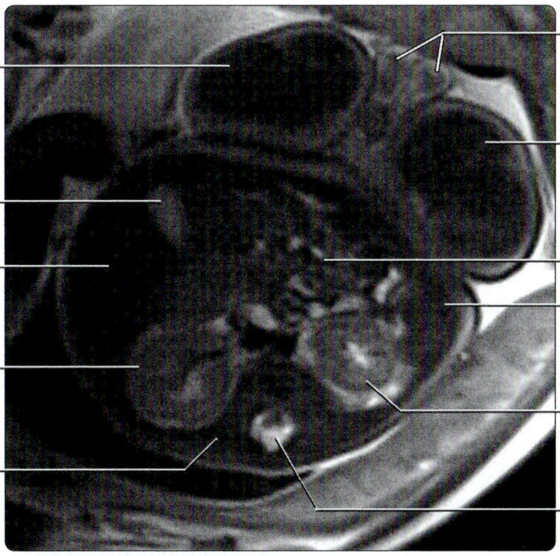

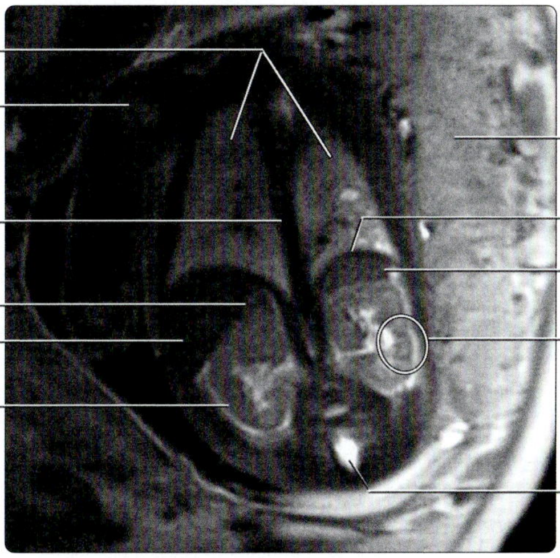

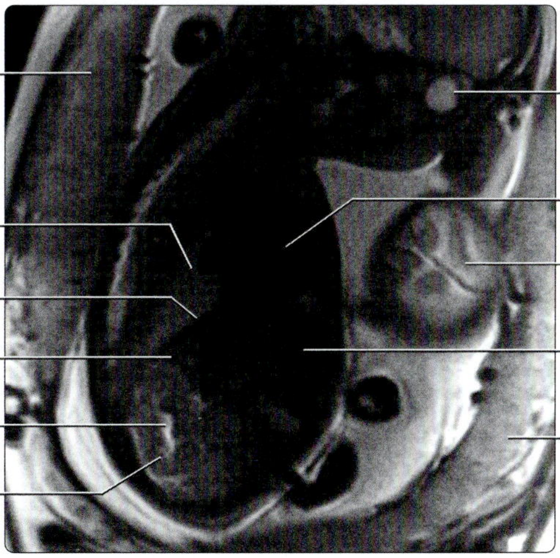

(Top) *Axial T2 MR through the abdomen, at the level of the kidneys, in a 30-week fetus shows normal anatomy. The kidneys are well seen and are higher in signal than the liver, spleen, or muscles.* **(Middle)** *Coronal T2 MR through the renal fossa shows the fetal adrenals, which are lower in signal compared to the kidneys. Normal corticomedullary differentiation is particularly well seen on the left kidney; also note the slightly "bumpy" contour (fetal lobulations).* **(Bottom)** *This image is a sagittal MR of twin A of a dichorionic twin gestation (note 2 placentas). The renal pelvis contains urine and is high signal intensity on T2-weighted imaging. Note the folded, tricorn shape of the adrenal gland.*

INDEX

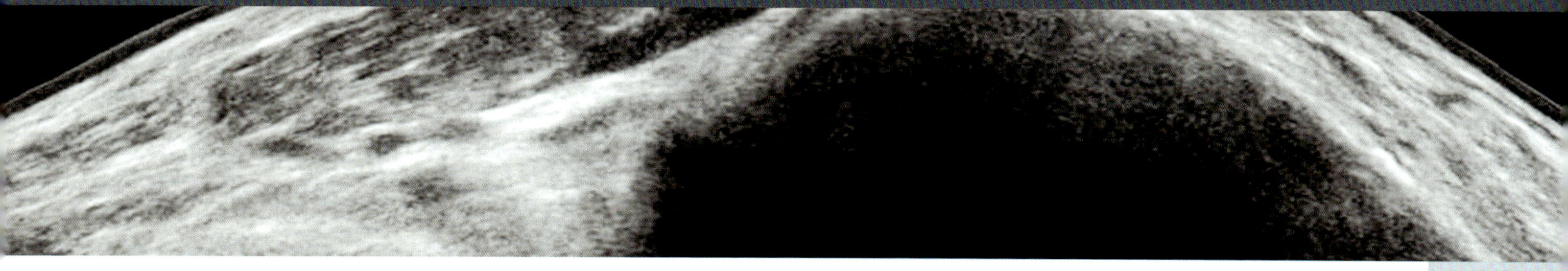

A

INDEX

INDEX

B

C

INDEX

INDEX

INDEX

D

INDEX

E

F

G

H

I

J

K

L

INDEX

M

INDEX

N

O

P

Q

R

S

T

U

V

W

X

Y

Z